MW01259974

Design of
Water Resource
Recovery Facilities

About WEF

The Water Environment Federation (WEF) is a not-for-profit technical and educational organization of 33,000 individual members and 75 affiliated Member Associations representing water quality professionals around the world. Since 1928, WEF and its members have protected public health and the environment. As a global water sector leader, our mission is to connect water professionals; enrich the expertise of water professionals; increase the awareness of the impact and value of water; and provide a platform for water sector innovation. To learn more, visit www.wef.org.

About ASCE/EWRI

Founded in 1852, the American Society of Civil Engineers (ASCE) represents more than 150,000 members of the civil engineering profession worldwide and is the oldest national engineering society of the United States. Created in 1999, the Environmental and Water Resources Institute (EWRI) is an institute of ASCE. EWRI services are designed to complement ASCE's traditional civil engineering base and to attract new categories of members (non-civil engineer allied professionals) who seek to enhance their professional and technical development.

For information on membership, publications, and conferences, contact

ASCE/EWRI
1801 Alexander Bell Drive
Reston, VA 20191-4382
(703) 295-6000
http://www.asce.org

Design of Water Resource Recovery Facilities

WEF Manual of Practice No. 8
ASCE Manuals and Reports on Engineering Practice No. 76

Prepared by the Design of Water Resource Recovery Facilities Task Force of the Water Environment Federation and the American Society of Civil Engineers/Environmental and Water Resources Institute

Sixth Edition

WEF Press

Water Environment Federation   Alexandria, Virginia
American Society of Civil Engineers/Environmental and Water Resources Institute   Reston, Virginia

New York Chicago San Francisco
Athens London Madrid
Mexico City Milan New Delhi
Singapore Sydney Toronto

Library of Congress Control Number: 2017949542

McGraw-Hill books are available at special quality discounts to use as premiums and sales promotions or for use in corporate training programs. To contact a representative, please visit the Contact Us page at www.mhprofessional.com.

Design of Water Resource Recovery Facilities, Sixth Edition

Copyright © 2018 by the Water Environment Federation and the American Society of Civil Engineers/Environmental and Water Resources Institute. All rights reserved. Except as permitted under the United States Copyright Act of 1976, no part of this publication may be reproduced or distributed in any form or by any means, or stored in a database or retrieval system, without the prior written permission of WEF and ASCE/EWRI. Permission to copy must be obtained from both WEF and ASCE/EWRI.

1 2 3 4 5 6 7 8 9   LWI   21 20 19 18 17

ISBN 978-1-260-03118-8
MHID 1-260-03118-7

Water Environment Research, WEF, and *WEFTEC* are registered trademarks of the Water Environment Federation. *American Society of Civil Engineers, ASCE, Environmental and Water Resources Institute*, and *EWRI* are registered trademarks of the American Society of Civil Engineers.

This book is printed on acid-free paper.

IMPORTANT NOTICE

The material presented in this publication has been prepared in accordance with generally recognized engineering principles and practices and is for general information only. This information should not be used without first securing competent advice with respect to its suitability for any general or specific application.

The contents of this publication are not intended to be a standard of the Water Environment Federation (WEF) or the American Society of Civil Engineers (ASCE)/Environmental and Water Resources Institute (EWRI) and are not intended for use as a reference in purchase specifications, contracts, regulations, statutes, or any other legal document.

No reference made in this publication to any specific method, product, process, or service constitutes or implies an endorsement, recommendation, or warranty thereof by WEF or ASCE/EWRI.

WEF and ASCE/EWRI make no representation or warranty of any kind, whether expressed or implied, concerning the accuracy, product, or process discussed in this publication and assumes no liability.

Anyone using this information assumes all liability arising from such use, including but not limited to infringement of any patent or patents.

Prepared by **Design of Water Resource Recovery Facilities Task Force** of the **Water Environment Federation** and the **American Society of Civil Engineers/Environmental and Water Resources Institute**

Terry L. Krause, P.E., BCEE, WEF Fellow, *Chair*

Jeanette Brown, P.E., DEE, D. WRE; Mark E. Lang, P.E., BCEE; and Kendra D. Sveum, P.E., *Volume Leaders*

Malarmagal Ahilan
Rafid Alkhaddar (ASCE Blue Ribbon
   Review Panel)
Naomi Eva Anderson
Rich Atoulikian
Eric Auerbach
Jeovanni Ayala-Lugo, P.E.
Amber Batson, P.E.
Paul Bizier, P.E., F.ASCE, D.WRE
Lucas Botero, P.E., BCEE, ENV SP
Akram Botrous, Ph.D., P.E., BCEE
Lisa Boudeman
Keith Bourgeous, Ph.D., P.E.
Gregory A. Bowden
John R. Bratby, Ph.D., P.E.
Kari Fitzmorris Brisolara, ScD,
   MSPH, QEP
John P. Brito, P.E.
Lewis Bryant, P.E.
Marie S. Burbano, Ph.D., P.E., BCEE
Misti Burkman
Chris Bye
Andres F. Onate Calderon
Onder Caliskaner, Ph.D., P.E.
Jennifer Callahan
Leonard W. Casson, Ph.D., P.E., BCEE
Stan Chilson
S. Rao Chitikela, Ph.D., P.E., P.Eng, BCEE
   (ASCE Blue Ribbon Review Panel)
Timothy A. Constantine
Bruce L. Cooley, P.E.
Emma Cooney
Chris D. Cox, Ph.D., P.E. (ASCE Blue
   Ribbon Review Panel)
Glen T. Daigger, Ph.D., P.E., BCEE, NAE
   (ASCE Blue Ribbon Review Panel)
Chris deBarbadillo, P.E.
Adam Dellinger
Timur Deniz, Ph.D., P.E., BCEE
Carlos Diaz

Bruce DiFrancisco, P.E.
Ludwig Dinkloh
Alexandra Doody, P.E.
Leon Downing
Bertrand Dussert
Na-Asia Ellis
Adam Evans, P.E.
Richard Finger
William Flores
Kristin Frederickson
Daniel Freedman, P.E.
Val S. Frenkel, Ph.D., P.E., D. WRE
John Friel
Edward W. Fritz
Rebecca Gauff
Hany Gerges, Ph.D., P.E., P. Eng
Matthew Goss, P.E., CEM, CEA, CDSM,
   LEED® AP (BD+C)
Linda Gowman
Samantha Graybill, P.E.
Bently Green
William Dana Green, P.E.
Jim Groman
Rashi Gupta, P.E.
Drew Hansen
Vaughan Harshman, P.E.
Jeff Hauser
Michael Hines, P.E.
Angela Hintz, P.E.
Anthonie Hogendoorn, MSc.
Greg Homoki, P.E.
William Hotz
Brian Huang
Christopher Hunniford, P.E.
Gary L. Hunter, P.E., BCEE, ENV SP
Jinsheng (Jin) Huo, Ph.D., P.E., BCEE
   (*Chair*, ASCE Blue Ribbon Review Panel)
Joseph A. Husband
Samuel S. Jeyanayagam, Ph.D., P.E., BCEE
Jose Jimenez

Andrew Jones
Jim Joyce
John C. Kabouris, Ph.D., P.E.
Dimitri Katehis
Morgan Knighton
Kyle Kubista, P.E.
Satej Kulkarni
Louis Lefebvre
Wayne Lem
Kevin S. Leung, Ph.D., P.E., BCEE
F. Michael Lewis
Peter Loomis, P.E.
Dusti F. Lowndes
Lee A. Lundberg, P.E.
Jose Christiano Machado Jr., Ph.D., P.E.
Laura Marcolini, P.E.
Samir Mathur, P.E., BCEE
William C. McConnell
Lauren McDaniel, MPH
Charles M. McGinley, P.E.
Michael A. McGinley, P.E.
Anna Mehrotra, Ph.D., P.E.
Henryk Melcer
Baoxia Mi (ASCE Blue Ribbon Review
    Panel)
Mark W. Miller, Ph.D.
Elizabeth Miner
Richard O. Mines, Jr., Ph.D., P.E. (ASCE
    Blue Ribbon Review Panel)
Indra N. Mitra, Ph.D., P.E., MBA, BCEE
Manny Moncholi
Ray P. Montoya, P.E.
Steve Mustard, P.E., CAP, GICSP
Garrison W. Myer
Vincent Nazareth
Maureen D. Neville, P.E.
Robert A. (Randy) Nixon
Ing. Daniel Nolasco
Helena Ochoa
David W. Oerke, P.E.
Tim Page-Bottorff, CSP, CET
Ana J. Pena-Tijerina, Ph.D., P.E.
Heather M. Phillips, P.E.
Ashley Pifer, Ph.D., P.E.
Marcel Pomerleau
Ray Porter
Coenraad Pretorius, P.E.
Pusker Regmi, Ph.D., P.E.

Leiv Rieger
Joel C. Rife, P.E.
Adam Rogensues, P.E.
A. Robert Rubin, Ph.D.
Andrew Salveson, P.E.
Domenico Santoro
Patricia A. Scanlan
Kimberly Schlauch
Harold E. Schmidt, Jr., P.E., BCEE
Kenneth Schnaars, P.E.
Megan Yoo Schneider
Sandra Schuler
Matt Seib, Ph.D.
Douglas Sherman, P.E.
Toshio Shimada, Ph.D., P.E.
Jim E. Smith, Jr, D.Sc, MASCE, BCEEM
Eric Spargimino, P.E., LEED AP
Eric T. Staunton, Ph.D.
Jennifer L. Strehler, P.E., MBA, BCEE,
    ENV-SP
Timothy H. Sullivan, P.E.
Steven Swanback
Jay L. Swift, P.E.
Alex Szerwinski, P.E.
Alex Tabb
Berrin Tansel, Ph.D., P.E., BCEE, F.ASCE,
    F.EWRI (ASCE Blue Ribbon Review
    Panel)
Anthony Tartaglione, P.E., BCEE
George Tchobanoglous (ASCE Blue
Ribbon Review Panel)
Matt Tebow, P.E.
Rachelle Tippetts
David Tomowich
K. Richard Tsang, Ph.D., P.E., BCEE
Jason Turgeon
David Ubert
David Valero
Don Vandertulip, P.E., BCEE
Ales Volcansek, P.E.
Tanush Wadhawan, Ph.D.
Trevor Wagenmaker, P.E.
Kristen Waksman
Steve Waters, P.E., P. Eng
David G. Weissbrodt, Asst. Prof., Ph.D.,
    M.Sc., Dipl.-Ing.
Jianfeng Wen
Curt Wendt, P.E., CAP

Claes Westring
Andrea Turriciano White, P.E.
Jason J. Williams, P.E.
Matthew J. Williams, P.E.
Hannah T. Wilner, P.E., PMP
Melissa K. Woo, P.E.

Paul Wood
Wade Wood, P.E.
Thomas Worley-Morse, Ph.D.
Usama Zaher, Ph.D., P.E.
Tian C. Zhang, Ph.D., P.E., BCEE, F.ASCE
  (ASCE Blue Ribbon Review Panel)

Under the Direction of the **Municipal Subcommittee** of the **Technical Practice Committee**

2018

Water Environment Federation
601 Wythe Street
Alexandria, VA 22314-1994 USA
http://www.wef.org

American Society of Civil
  Engineers/Environmental and
  Water Resources Institute
1801 Alexander Bell Drive
Reston, VA 20191-4400
http://www.asce.org

Manuals of Practice of the Water Environment Federation

The WEF Technical Practice Committee (formerly the Committee on Sewage and Industrial Wastes Practice of the Federation of Sewage and Industrial Wastes Associations) was created by the Federation Board of Control on October 11, 1941. The primary function of the Committee is to originate and produce, through appropriate subcommittees, special publications dealing with technical aspects of the broad interests of the Federation. These publications are intended to provide background information through a review of technical practices and detailed procedures that research and experience have shown to be functional and practical.

Water Environment Federation
Technical Practice Committee
Control Group

Eric Rothstein, C.P.A., *Chair*
D. Medina, *Vice-Chair*
Jeanette Brown, P.E., BCEE, D. WRE, F.WEF, *Past Chair*

H. Azam
G. Baldwin
Katherine (Kati) Y. Bell, Ph.D., P.E., BCEE
C.-C. Chang
J. Davis
C. deBarbadillo, P.E.
S. Fitzgerald
T. Gellner
S. Gluck
Michael Hines, P.E.
J. Loudon
C. Maher
S. Metzler
F. Pasquel
C. Peot
R. Pope
R. Porter
L. Pugh
J. Reeves
S. Schwartz
A. Schwerman
Andrew R. Shaw, P.E.
A. Tangirala
R. Tsuchihashi
N. Wheatley

Manuals and Reports on Engineering Practice

(As developed by the ASCE Technical Procedures Committee, July 1930, and revised March 1935, February 1962, and April 1982)

A manual or report in this series consists of an orderly presentation of facts on a particular subject, supplemented by an analysis of limitations and applications of these facts. It contains information useful to the average engineer in his or her everyday work, rather than findings that may be useful only occasionally or rarely. It is not in any sense a "standard," however; nor is it so elementary or so conclusive as to provide a "rule of thumb" for nonengineers.

Furthermore, material in this series, in distinction from a paper (which expresses only one person's observations or opinions), is the work of a committee or group selected to assemble and express information on a specific topic. As often as practicable, the committee is under the direction of one or more of the Technical Divisions and Councils, and the product evolved has been subjected to review by the Executive Committee of the Division or Council. As a step in the process of this review, proposed manuscripts are often brought before the members of the Technical Divisions and Councils for comment, which may serve as the basis for improvement. When published, each work shows the names of the committees by which it was compiled and indicates clearly the several processes through which it has passed in review, in order that its merit may be definitely understood.

In February 1962 (and revised in April 1982) the Board of Direction voted to establish a series entitled "Manuals and Reports on Engineering Practice," to include the Manuals published and authorized to date, future Manuals of Professional Practice, and Reports on Engineering Practice.

All such Manual or Report material of the Society would have been refereed in a manner approved by the Board Committee on Publications and would be bound, with applicable discussion, in books similar to past Manuals. Numbering would be consecutive and would be a continuation of present Manual numbers. In some cases of reports of joint committees, bypassing of Journal publications may be authorized.

Contents

Preface

This manual, updated from the 5th edition, continues its goal to be one of the principal references of contemporary practice for the design of municipal water resource recovery facilities (WRRFs). The manual was written for design professionals familiar with wastewater treatment concepts, the design process, and the regulatory basis of water pollution control. It is not intended to be a primer for the inexperienced or the generalist. The manual is intended to reflect current facility design practices of wastewater engineering professionals, augmented by performance information from operating facilities. The design approaches and practices presented in the manual reflect the experiences of more than 175 authors and reviewers.

This manual consists of 25 chapters, with each chapter focusing on a particular subject or treatment objective. The successful design of a municipal WRRF is based on consideration of each unit process and the upstream and downstream effects of that unit's place and performance in the overall scheme of the treatment works. Chapters 1 to 8 generally cover design concepts and principles that apply to the overall WRRF. Chapters 9 to 17 discuss liquid-train-treatment operations or processes. Finally, Chapters 18 to 25 address the management of solids generated during wastewater treatment. Since the publication of the 5th edition of this manual, key technical advances in wastewater treatment have included the following:

- Trend toward resource recovery and hence the revised name for the manual;
- Increasing goals toward energy neutrality and driving net zero;
- Use and application of modeling wastewater treatment processes for the basis of design and evaluations of alternatives, which resulted in incorporation of a new chapter summarizing Manual of Practice No. 31, *Wastewater Treatment Process Modeling*;
- Further advances with membrane bioreactors applications;
- Advancements within integrated fixed-film/activated sludge (IFAS) systems and moving-bed biological-reactors systems;
- Sidestream nutrient removal to reduce the loading on the main nutrient-removal process; and
- Advances in biosolids handling, including energy recovery and thermal hydrolysis.

In response to these advancements, this edition includes some significant changes from the 5th edition. As with prior editions, technologies that are no longer considered current industry practice have been deleted or their content minimized.

Additionally, the focus of the manual has been sharpened. Like earlier editions, this manual presents current design guidelines and practices of wastewater engineering professionals. Design examples also are provided, in some instances, to show how the guidelines and practice can be applied. However, information on process fundamentals is covered to a lesser extent than in the previous edition. Readers are referred to other publications for information on those topics.

Authors' and reviewers' efforts were supported by the following organizations:

AECOM, Piscataway, New Jersey; Buffalo, New York

American Water, Voorhees, New Jersey

Arcadis U.S., Inc., Highlands Ranch, Colorado; Buffalo, New York; White Plains, New York

Arvos Schmidtsche Schack LLC, Wexford, Pennsylvania

Automation Federation, Raleigh, North Carolina

Barge, Waggoner, Sumner and Cannon, Nashville, Tennessee

Bedrock Enterprises, Inc., Baden, Pennsylvania

Black & Veatch, Coral Springs, Florida; Indianapolis, Indiana; Overland Park, Kansas; Kansas City, Missouri; St. Louis, Missouri; Memphis, Tennessee

Brown and Caldwell, Maitland, Florida; Orlando, Florida; Charlotte, North Carolina; Nashville, Tennessee; Alexandria, Virginia; Seattle, Washington

Carollo Engineers, Costa Mesa, California; Walnut Creek, California; Littleton, Colorado; Tampa, Florida; Dallas, Texas

CDM Smith, Carlsbad, California; Irvine, California; Los Angeles, California; Denver, Colorado; Bogota, Colombia; Maitland, Florida; Miami, Florida; Orlando, Florida; Boston, Massachusetts; Manchester, New Hampshire; Albany, New York; Raleigh, North Carolina; Providence, Rhode Island; Houston, Texas; Austin, Texas; Dallas, Texas; Fairfax, Virginia; Leesburg, Virginia; Bellevue, Washington

CH2M, Tampa, Florida; Chicago, Illinois; Albuquerque, New Mexico; Herndon, Virginia; Toronto, Ontario, Canada

Corrosion Probe, Inc., Centerbrook, Connecticut

DC Water, Washington, D.C.

Donohue & Associates, Inc, Chicago, Illinois

Dynamita, Toronto, Canada

Dynamita S.A.R.L., Nyons, France

EnviroSim Associates Ltd., Hamilton, Ontario, Canada

Evoqua Water Technologies LLC, Bradenton, Florida

Garver, Dallas, Texas; Frisco, Texas

Gray and Osborne, Seattle, Washington

GREELEY and HANSEN, Chicago, Illinois; San Francisco, California

Hazen and Sawyer, Raleigh, North Carolina

HDR Engineering, Inc., Walnut Creek, California; Calverton, Maryland; Cleveland, Ohio; Nashville, Tennessee

Hubbell, Roth & Clark, Inc., Detroit, Michigan

inCTRL Solutions Inc., Oakville, Ontario, Canada

Intera, Richland, Washington

Johnson County Wastewater, Olathe, Kansas

Kennedy/Jenks Consultants, San Francisco, California

Kimley-Horn and Associates, Inc., Mesa, Arizona; Ocala, Florida; Tampa, Florida; West Palm Beach, Florida; Ft. Worth, Texas

Laura Marcolini & Associates, Inc., Cumberland, Rhode Island

Louisiana State University, Baton Rouge, Louisiana

Madison Metropolitan Sewerage District, Madison, Wisconsin

Manhattan College, Bronx, New York

Material Matters, Elizabethtown, Pennsylvania

National Automation, Inc., Spring, Texas

NOLASCO y Asociados. S. A., Buenos Aires, Argentina

North Carolina State University, Raleigh, North Carolina

SafeStart, Belleville, Ontario, Canada

Short Elliott Hendrickson Inc., St. Paul, Minnesota

Smith and Loveless, Inc., Lenexa, Kansas

Southeast Environmental Engineering, LLC, Knoxville, Tennessee

St. Croix Sensory, Inc., Stillwater, Minnesota

Stantec Consulting Services, Rocklin, California; Denver, Colorado; Tampa, Florida; Portland, Oregon

Tesco Controls, Inc., Sacramento, California

Total Safety Compliance, Mesa, Arizona

University of Pittsburgh, Pittsburgh, Pennsylvania

URS Corporation, Buffalo, New York

U.S. Environmental Protection Agency, Boston, Massachusetts

V&A Consulting Engineers, Houston, Texas

Vandertulip WateReusEngineers, San Antonio, Texas

Varec Biogas, Stafford, Texas

Veolia North America, Chicago, Illinois

Washington State Department of Ecology, Bellevue, Washington

WesTech Engineering, Salt Lake City, Utah

Xylem Inc., White Plains, New York

CHAPTER 1

Introduction

Marie S. Burbano, Ph.D., P.E., BCEE;
Andres F. Onate Calderon; and Manny Moncholi

1.0 Background

1.1 Overview

This manual is the *Design of Municipal Water Resource Recovery Facilities*, MOP 8, 6th edition, included as WEF Manual of Practice 8 and ASCE Manuals and Reports on Engineering Practice, No. 76. This 6th edition continues its goal to be one of the principal references of contemporary practice for the design of water resource recovery facilities (WRRFs). The 6th edition is an update to the 5th edition, published in 2010. This manual was written for design professionals familiar with WRRFs, the design process, treatment technologies, and the regulatory basis of water pollution control. It is not intended to be a primer for the inexperienced or the generalist.

The term WRRF is an evolution in the industry to consider wastewater as a resource with the potential to obtain valuable products (i.e., phosphorus and biogas) using treatment technologies. WRRFs as described in this MOP are used to treat municipal wastewater, which is intended to refer to those wastes treated by publicly owned WRRFs, as opposed to sanitary waste, which refers primarily to toilet wastes, and domestic wastewater, which largely encompasses household wastes without a commercial or

1

institutional component. In addition to commercial and institutional wastes, municipal wastewater often contains significant flows from manufacturing and other industrial sources. In this manual, industrial and institutional wastes are discussed only to the extent that they affect the design of municipal WRRFs.

1.2   Evolution of Municipal Water Resource Recovery Facilities

The need for community wastewater collection and treatment systems globally has evolved over a period of more than 200 years, initially being driven by the need to reduce human disease; then to eliminate gross water pollution effects, allowing native marine organisms to return to normal growth patterns and allowing full human recreational use; and, finally, to redefine wastewater as a resource with valuable products to be extracted through treatment.

In this section, a brief overview is provided of the evolution of WRRFs, which focuses on the future of municipal wastewater treatment in the United States. While this discussion reflects U.S. trends, these trends have global applications in varying degrees, and examples from other parts of the world are also cited.

According to the Clean Watersheds Needs Survey 2012, by the U.S. Environmental Protection Agency (Washington, D.C.) (U.S. EPA, 2016), 238.2 million people are serviced by publicly owned treatment works (POTWs) in the United States. Of those serviced by POTWs, 127.7 million people are served by advanced wastewater treatment, 90.4 million people are served by secondary treatment, and 4.1 million are served by less-than-secondary treatment. Additionally, there are 2281 nondischarging facilities serving 16.0 million people in the United States.

According to global records available from AQUASTAT, 298.5 km^3 of municipal wastewater are produced annually in the world, taking only into account the shares of those countries with 87.5% of the global population. Of the wastewater that is accounted for, 54.6% (163 km^3) receives some kind of treatment. From those 163 km^3 receiving treatment, about 10% of the wastewater is reused for some purpose (agricultural, industrial, etc.). For the most part, water reuse is based on water scarcity, where 90% of the countries reusing wastewater have less than the average amount of renewable water resources per capita. The general trends for growth in expenditures for global water infrastructure show the largest percent increases in Asia, with more than a 12% increase by 2019 (Global Water Intelligence, 2015). Latin America and Africa are each expected to see about a 5% to 6% increase in investment, with the remaining areas of the globe at 1% to 4% increase.

Water conservation has become more common in water-limited areas. Water conservation needs result in beneficial recycling of treated wastewater for cooling, irrigation, agriculture, drinking water, and certain classes of industrial use. As water becomes scarce, intentional recycling of wastewater into drinking water supplies is becoming more prevalent. Increased efforts to control the discharge of toxins to the nation's waterways will continue in the future. Additionally, advanced wastewater treatment practices become more prevalent.

While more WRRFs have secondary or advanced treatment processes than ever, the need for these processes is also higher than ever. Drivers of advanced treatment include the desire to recover resources from the wastewater and potential human health or habitat concerns especially with contaminants of emerging concern (CECs). As the industry shifts to recovering resources, including water, energy, and nutrients, from wastewater, treatment technology and WRRFs are advancing to meet these needs. In addition, water quality instruments are able to measure constituents at lower concentrations, thereby detecting CECs, such as pharmaceuticals and personal care products (PPCPs). The industry is still determining the human health and ecological effects of these CECs, and negative effects may result in more stringent effluent requirements.

As climate-change issues increase, municipalities will consider sustainability and carbon footprint as important criteria in evaluating alternative technologies. Changes in funding available to municipalities and wastewater management philosophy often drive the evolution and improvement of certain technologies. Along with water reuse, decentralized wastewater treatment and wet-weather flow management will play a role in future technology development and innovations (Burian et al., 2000).

In the 6 years since the publication of the 5th edition of this manual, key technical advances in wastewater treatment have included the following:

- Advances with membrane bioreactors applications;
- Advancements within integrated fixed-film/activated sludge (IFAS) systems and moving-bed biological-reactors systems;
- Biotrickling filtration for odor control;
- Increased use of ballasted flocculation;
- Sidestream nutrient removal to reduce the loading on the main nutrient-removal process;
- Use and application of wireless instrumentation;
- Use and application of modeling wastewater treatment processes for the basis of design and evaluations of alternatives;
- Advances in biosolids handling, including effective thermal hydrolysis, and improvements in sludge thickening and dewatering technologies;
- Increasing goals toward energy neutrality and driving net zero; and
- Trend toward resource recovery.

As referenced above, a needs survey is conducted by the U.S. EPA every 4 years and provides a means for assessing the present status and future direction of the nation's water pollution control efforts for WRRFs. Table 1.1 monetarily summarizes how the needs have changed from 2000 to 2012, when the latest survey was taken.

According to the latest survey (Washington, D.C.) (U.S. EPA, 2016), there still is $52.4 billion in the need for further secondary wastewater treatment (Category I), and $49.6 billion in the need for WRRFs to attain a level of treatment more stringent than secondary treatment (i.e., advanced WRRFs). Figure 1.1 summarizes the nation's needs by category in 2012 dollars, with wastewater treatment systems being the most needed.

Needs Category	2000	2004	2008	2012	2008 to 2012 Change	
					$B	%
Secondary wastewater treatment	55.0	59.8	68.0	52.4	−15.6	−23
Advanced wastewater treatment	30.5	32.9	51.4	49.6	−1.8	−3.5
Total needs	85.5	92.7	119.4	102	−17.4	−14.6

TABLE 1.1 Comparison of Treatment Needs for the 2000 to 2012 Clean Watershed Needs Survey (CWNS) (U.S. EPA, 2016) (January 2012 dollars, in billions)

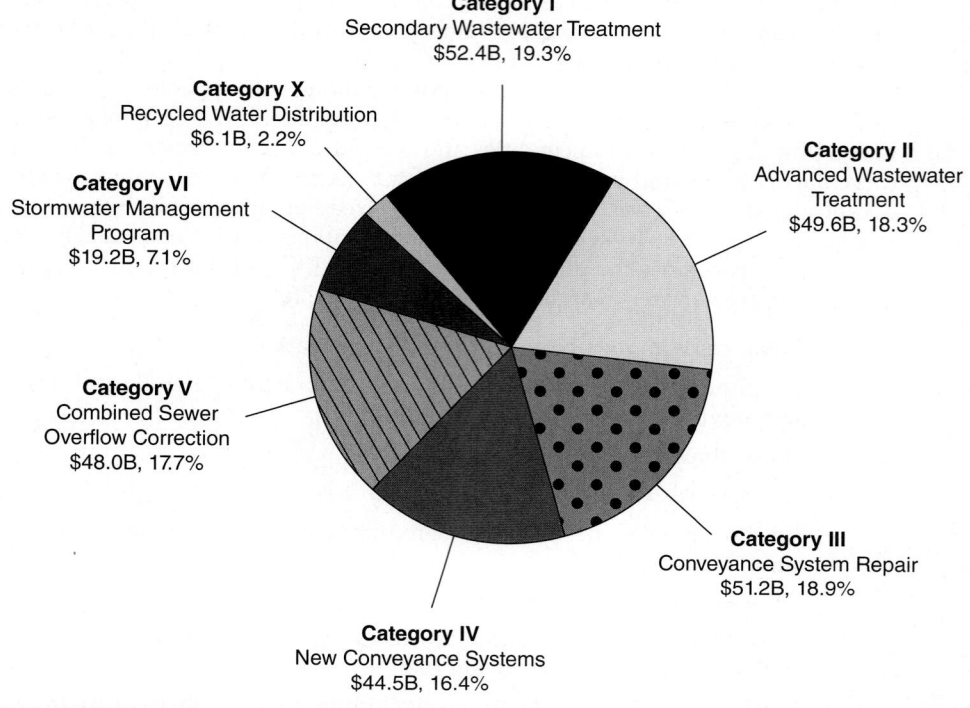

Category I
Secondary Wastewater Treatment
$52.4B, 19.3%

Category X
Recycled Water Distribution
$6.1B, 2.2%

Category VI
Stormwater Management
Program
$19.2B, 7.1%

Category II
Advanced Wastewater
Treatment
$49.6B, 18.3%

Category V
Combined Sewer
Overflow Correction
$48.0B, 17.7%

Category III
Conveyance System Repair
$51.2B, 18.9%

Category IV
New Conveyance Systems
$44.5B, 16.4%

Figure 1.1   Total documented wastewater needs (U.S. EPA, 2016).

Within wastewater treatment, the U.S. EPA projects that, by 2032, 263.6 million people will receive secondary or advanced wastewater treatment. Figure 1.2 shows the level of wastewater treatment the population in the United States has received since 1940 and how it has grown since 2012. The projected year shows 2032, if all wastewater needs are met.

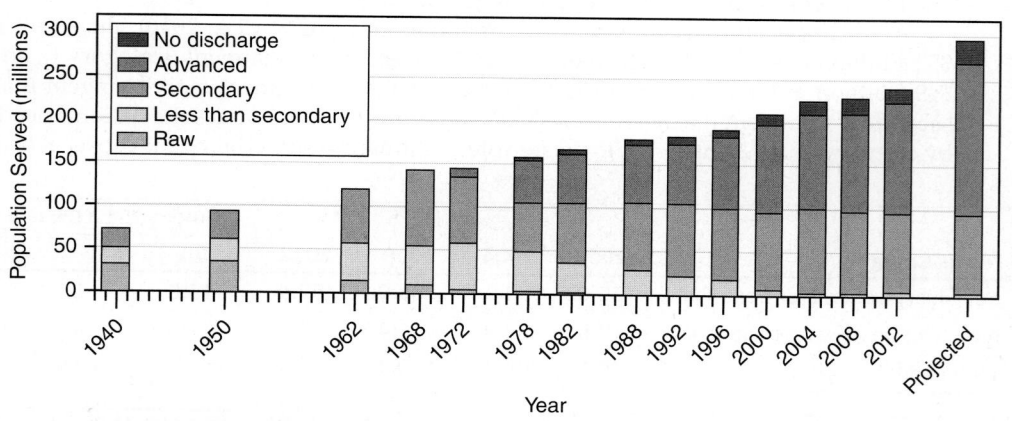

Figure 1.2   Nationwide populations served by POTWs from 1940 to 2012 (U.S. EPA, 2016).

2.0 The Designer's Role

2.1 Role

Ideally, designers translate regulatory requirements, public goals, financial constraints, and technology into WRRFs that operate reliably, economically, and unobtrusively to meet discharge standards.

Because of the continuing evolution of environmental regulations at the local, state, and federal levels, some practices and design criteria presented in this manual may be inconsistent with requirements of the regulatory agencies. Therefore, it is incumbent on design engineers to check with appropriate regulating agencies for the most current requirements.

The practice of designing municipal WRRFs—a mixture of art and science—has few hard-and-fast rules that will result in successful designs. Recognizing that experience is critical to developing an appropriate design, both regulators and owners should apply common sense and remain open to change and innovation. Arriving at the most economical and functional design for an integrated facility is critical to the development of a WRRF. In addition, the design must meet all regulatory guidelines and standards. Successful design also takes into account operability, maintainability, and safety.

Published standards for specific design and loading criteria are referenced throughout this manual. Caution should be exercised when using such standards and references for specific design or loading criteria. The original time and conditions on which the guidance was developed should be considered. Some standards may remain applicable universally, and some may not. For example, the Recommended Standards for Wastewater Facilities (Great Lakes–Upper Mississippi River Board of State and Provincial Public Health and Environment Managers, 2014), commonly called the "Ten States Standards," has served as a useful reference for regulators and designers. These guidelines were first developed in 1951 and are revised at 5- to 10-year intervals, in an attempt to maintain their applicability and reflect changes in technology. Although guidelines such as these continue to be a commonly used reference, they should not be viewed as containing absolute design values that cannot be varied. Rather, these standards offer parameters that have proved successful in typical municipal facilities with a wide range of operator capability.

2.2 Water Resource Recovery Facility Advances

Designers aware of past experiences, but amenable to change, can meet future challenges of new standards and goals, as defined by the public's changing perception of need and national priorities. Water pollution abatement, resulting in unbalanced atmospheric deterioration, energy consumption, and land degradation, is not a viable solution. Additionally, there currently is a push for utility owners to consider sustainability and carbon footprint size when evaluating project alternatives. To control the cost of treatment and address environmental effects, the designer should explore opportunities in the service area and its collection system. In designing a municipal WRRF, the designer walks a careful line between providing facilities that can respond to uncertainties of the future and excessive overdesign, the latter of which may result in the misuse of public monies for superfluous facilities and result in operational challenges. Good designs and technology selection also provide flexibility to allow modifications and additions to meet future, more stringent treatment requirements. When in doubt, innovative application of proven technology may serve the designers and their client better than less proven technology.

Three other trends are affecting the role of the wastewater design professional. One is the increased pressure on WRRFs to be good neighbors, which necessitates more emphasis on odor, noise, and visual effects. There is also a trend for aiming at zero discharge and smaller facilities in critical locations (i.e., environmentally sensitive areas). The third is a growing trend toward alternate delivery methods, including design/build (D/B), construction manager at risk (CMAR), concessions, and public-private partnerships (P3s). Whereas D/Bs and CMARs generally are financed, operated, and maintained by the public utility, concessions and P3s are often referred to as forms of privatization as the public utility generally turns over the responsibilities of operating and maintaining the facility or system to a private entity. This trend, along with overall budget concerns, can place increased pressure on municipally operated facilities to maintain competitive user-charge structures by controlling capital and operation and maintenance costs.

Additionally, a segment of environmental consideration paradigm shifts that have had a profound impact on the design of WRRFs is climate change, sustainability, and resiliency. Changing weather and precipitation patterns, rising sea levels, increasing storm events, unprecedented rainfall, hurricane and tropical storm flooding, and severe drought and fire are driving WRRFs to focus on resiliency. As a result, designers of WRRFs are addressing environmental changes by strengthening what is already in place and weaving resilience into new plans and designs. This could include increasing the height of facilities to address sea level rise, providing technologies that minimize emissions, and hardening existing facilities to tolerate stronger storm systems.

3.0 Scope and Organization of Manual

3.1 Modifications in the Latest Edition

The 6th edition of this key Water Environment Federation® (Alexandria, Virginia) (WEF) manual is intended to reflect current facility design practices of wastewater engineering professionals, augmented by performance information from operating facilities. The design approaches and practices presented in the manual reflect the experiences of 182 authors and reviewers from around the world.

This edition also includes some significant changes from the 5th edition. As with prior editions, technologies that are no longer considered current industry practice have been deleted, such as vacuum filters for sludge dewatering. While not intended to be all-inclusive, the following list describes some of the other pertinent processes and newer processes or concepts:

- Concept of sustainability;
- Energy management;
- Odor control and air emissions;
- Chemically assisted/ballast flocculation clarification;
- IFAS processes;

- Enhanced nutrient-control systems;
- Sidestream treatment;
- Process design and disinfection practices to minimize generation of total trihalomethanes and other organics monitored for potable water quality; and
- Approaches to minimizing biosolids production.

Additionally, the focus of the manual has been sharpened. Like earlier editions, this manual presents current design guidelines and practices of municipal wastewater engineering professionals. Design examples also are provided, in some instances, to show how the guidelines and practice can be applied. However, information on process fundamentals, case histories, operations, and other related topics is covered to a lesser extent than in the previous edition. Readers are referred to other publications for information on those topics.

3.2   Organization

This manual consists of 25 chapters, with each chapter focusing on a particular subject or treatment objective. The successful design of a municipal WRRF is based on consideration of each unit process and the upstream and downstream effects of that unit's place and performance in the overall scheme of the treatment works. Chapters 1 to 8 generally cover design concepts and principles that apply to the overall WRRF. Chapters 9 to 17 discuss liquid-train-treatment operations or processes. Chapters 18 to 25 deal with the management of solids generated during wastewater treatment.

Following is a brief overview of some of the major topics covered under each chapter of the manual:

- Chapter 1 presents the purpose and scope of the manual. It also presents a brief discussion of the needs for municipal wastewater treatment and has been updated for the Clean Watershed Needs Survey 2012: Report to Congress (U.S. EPA).
- Chapter 2 focuses on the overall design considerations, site selection, and configuration for a municipal WRRF and the principles of integrated facility design.
- Chapter 3 covers the site selection process for new or relocated WRRFs. The process includes the multitude of environmental, technical, and institutional variables to consider and their cost implications. The chapter then describes how area requirements are determined, candidate site identification, and the selection process. Once a site is selected, the chapter then describes how WWRF arrangement and process layout should progress. The chapter is more streamlined from the prior chapter, reorganizes some of the prior subsections, and includes a new subchapter on stakeholders and public involvement.
- Chapter 4 covers process modeling. This is an abbreviated version of MOP 31, *Wastewater Treatment Process Modeling*, 2nd edition, which was published in 2013.
- Chapter 5 covers facility hydraulics and pumping, including hydraulic considerations, unit process and other hydraulic elements, wastewater pumping, and hydraulic modeling.

- Chapter 6 brings together the topic of managing odor and air emissions, including with odor-control systems, odor regulations and community effects, odor measurement, assessing odors and air emissions, odor-dispersion modeling, odor containment and ventilation, and odor and air-emissions control.

- Chapter 7 focuses on a facility's support systems, such as general reliability criteria; electrical systems; instrumentation and control systems; heating, ventilating, and air-conditioning systems; chemical systems; and other support systems.

- Chapter 8 is a complete revamp from the 5th edition, and covers the materials of construction and corrosion control. It includes exposure conditions, forms of corrosion, design considerations, materials selection, protective coatings, and cathodic protection.

- Chapter 9 covers preliminary treatment operations, including screening, coarse solids reduction, grit removal, grease removal, septage handling, and flow equalization.

- Chapter 10 discusses primary treatment, including sedimentation, primary sludge collection and removal, floatable solids management, and various forms of advanced primary treatment, such as chemically enhanced primary treatment, plates, and enhanced high-rate clarification. It also includes the use of emerging primary treatment technologies such as filtration of primary effluent, filtration in lieu of primary clarification, microscreens, and dissolved air flotation.

- Chapter 11 deals with biofilm reactors and covers understanding and applying various biofilm design models, designing biofilm reactors for various process configurations, and evolving novel biofilm reactors, including an update about the use of fixed film processes for deammonification (anammox) in sidestream as well as mainstream application. The use of anammox-based moving bed biofilm reactor for nitrogen polishing in mainstream wastewater treatment is also included. Furthermore, this chapter has been expanded to include the biological nutrient removal (BNR) granular sludge technology.

- Chapter 12 covers all of the suspended-growth biological treatment systems, including those associated with nutrient removal. Topics include the fundamentals of biological treatment, process configurations and types, process design for carbon oxidation and nitrification, biological nutrient removal, anaerobic treatment, membrane bioreactors, wet-weather considerations, oxygen-transfer systems, secondary clarification, granular sludge, and mainstream deammonification.

- Chapter 13 deals with integrated biological treatment where integrated refers to combinations of attached and suspended growth systems in different configurations including integrated fixed-film activated sludge systems. It covers topics such as an introduction to and overview of these integrated-systems configurations, design, and the future of these systems as a technology.

- Chapter 14 presents physical and chemical processes for advanced wastewater treatment. This chapter's topics include process selection considerations, secondary effluent filtration, activated carbon adsorption, chemical treatment, membrane processes, air stripping and breakpoint chlorination for ammonia removal, and effluent reoxygenation.

- Chapter 15 covers evolving sidestream treatment approaches, including an introduction to sidestream treatment, with a focus on sidestream nitrogen and phosphorus removal and nutrient recovery design considerations. Integration of nutrient removal and recovery is presented in the form of a design example.

- Chapter 16 focuses on design and management of natural systems for wastewater treatment, including their history, conventional and alternative soil-absorption systems, pond systems, the varieties of land treatment systems, and constructed wetland treatment systems.

- Chapter 17 covers the topic of wastewater disinfection, including a discussion of technologies and regulatory considerations, impacts of upstream processes on disinfection design considerations, design considerations for chlorination, dechlorination, UV, peracetic acid (PAA), ozone, and other disinfection methods.

- Chapter 18 is an introduction to solids management. Topics covered in this chapter include residuals, applicable regulations, solids quantities and characteristics, and pretreatment methods.

- Chapter 19 discusses biosolids storage and transport approaches, including liquid biosolids, thickened biosolids, and dewatered biosolids.

- Chapter 20 covers the topic of chemical conditioning. The chapter includes theory and design of chemical conditioning systems for use in thickening or dewatering. This chapter includes discussion on chemicals used for conditioning, the feed equipment, and methods for dosage optimization.

- Chapter 21 covers the various thickening approaches, including a general overview, gravity thickening, dissolved air flotation thickening, membrane thickening, disk volute thickening, centrifugal thickening, gravity belt thickening, rotary drum thickening, and a comparison of the thickening methods and automation considerations.

- Chapter 22 presents the dewatering operations, including effect of sludge quality and the source of material to be dewatered, centrifugal dewatering, belt-filter-press dewatering, recessed plate and frame presses, screw presses, drying beds, and other dewatering systems, sidestreams, chemical conditioning, and utility requirements for each of these technologies.

- Chapter 23 discusses solids digestion and stabilization techniques. Topics include anaerobic digestion, biogas, thermal hydrolysis, pretreatment, aerobic digestion, composting, and alkaline stabilization.

- Chapter 24 discusses thermal processing, including thermal conditioning and dewatering, thermal drying, thermal oxidation, gasification and pyrolysis, emissions control, and air pollution control technology.

- Chapter 25 covers the use and disposal of biosolids, including land application, landfilling, dedicated land disposal, land reclamation and nonagricultural uses, ash use and disposal, and distribution and marketing.

The separate publication *Sustainability and Energy Management for Water Resource Recovery Facilities* (WEF et al., 2018) is considered a companion to this design manual. *Occupational Health and Safety for Water Resource Recovery Facility Design* (WEF, 2017) and *Glossary of Water Resource Recovery Facility Design Terms* (WEF, 2017) are also considered companions to this manual and are available for download at www.wef.org.

4.0   References

Burian, S. J.; Durrans, S.; Rocky, S. (2000) Urban Wastewater Management in the United States: Past, Present, and Future. *J. Urban Technol.*, **7** (3), 33–62.

Food and Agriculture Organization of the United Nations. AQUASTAT Main Database. http://www.fao.org/nr/water/aquastat/main/index.stm (accessed September 2016).

Global Water Intelligence (2015) Analysts' Report: The Global Water Market in 2016, Capital Expenditure Trends in Municipal and Industrial Water to 2019.

Great Lakes–Upper Mississippi River Board of State and Provincial Public Health and Environment Managers (2014) *Recommended Standards for Wastewater Facilities*; Health Education Services: Albany, New York.

U.S. Environmental Protection Agency (2016) *Clean Watershed Needs Survey 2012: Report to Congress*, EPA-830-R-15005; U.S. Environmental Protection Agency, Office of Water: Washington, D.C.

Water Environment Federation; American Society of Civil Engineers; Environmental and Water Resources Institute (2018) *Sustainability and Energy Management for Water Resource Recovery Facilities*; WEF Manual of Practice No. 38/ASCE Manuals and Reports on Engineering Practice No. 137; Water Environment Federation: Alexandria, Virginia.

5.0   Suggested Readings

Metcalf and Eddy Inc./AECOM (2013) *Wastewater Engineering: Treatment and Resource Recovery*, 5th ed.; McGraw-Hill: New York.

Viessman, W., Jr.; Hammer, M. J.; Perez, E. M.; Chadik, P. A. (2008) *Water Supply and Pollution Control*. 8th ed.; Pearson Education, Inc., Pearson Prentice Hall: Upper Saddle River, New Jersey.

CHAPTER **2**

Principles of
Integrated Design

Maureen D. Neville, P.E.; Misti Burkman; Alexandra Doody, P.E.;
William Dana Green, P.E.; William Hotz; Anna Mehrotra, Ph.D., P.E.;
and Don Vandertulip, P.E., BCEE

1.0 Introduction

The goal of integrated water resource recovery facility (WRRF) design is to produce a facility that achieves its defined performance goals and interfaces well with the local surroundings and other municipal systems of which the WRRF is a part. An integrated design process encourages all project stakeholders to communicate and collaborate from the planning stage through construction and startup. Integrated design places a high priority on the planning phase because upfront critical thinking and analysis with the entire project team engaged can result in fewer conflicts during final design; minimize problems during construction, operations, and maintenance; and cultivate a better long-term relationship and create more opportunities for resource recovery collaboration with the community.

This chapter provides an overview of the planning stage of integrated design, which sets the goals for the WRRF, identifies critical success factors, establishes the project team, and defines design criteria for the project. This chapter also discusses individual components of integrated WRRF design including definition of influent flows and loadings; identification of the range of conditions the WRRF should be designed to accommodate; definition of project implementation sequences and procurement mechanisms relevant to the specific technical, financial, and institutional capabilities of the implementing entity; adoption of comprehensive life-cycle cost (LCC) methodologies; and completion of evaluations using either multiple-criteria decision-making analysis or special investigations. A successful WRRF design reliably meets all performance goals, is responsive to operating and maintenance needs, and accommodates a full range of operating and environmental conditions with sufficient redundancy and flexibility.

2.0   Project Planning Considerations

WRRFs represent a significant investment by a community, and detailed planning is needed in advance of the WRRF design effort. Planning is generally completed through the development of a facilities plan to forecast influent flows, loadings, and community growth; to define other design criteria for new facilities; to incorporate current and potential future regulatory requirements and other performance goals; and to identify the recommended facilities necessary to achieve those goals. Collaboration with stakeholders early in the planning process helps to define the goals of the WRRF and to assure the facility is designed, constructed, and operated as an integrated part of the community. This section describes general considerations for planning a WRRF design project.

2.1   Performance Goals

There are universal goals common to all WRRFs, as well as goals that are community and site specific. The overarching goal common to all WRRFs is to protect human health and the environment. Some beneficial uses that often apply for the water environment include domestic and industrial water supply; recreational uses such as swimming, boating, and fishing; wildlife and aquatic life; agriculture and aquaculture; navigation; and transportation. Environmental regulations and permits define the WRRF water quality and treatment objectives, depending on whether the WRRF will be discharging to receiving waters or will be conveying treated effluent for water reuse. The permits for WRRFs discharging to receiving waters are written to protect the intended beneficial uses of those receiving waters. In some areas, there are restrictions on new discharge flows to impaired water bodies or outright bans on discharge to promote water reuse and groundwater replenishment, as is the case with Florida legislation passed in 2008 banning six ocean outfalls. WRRFs that land apply effluent (such as for irrigation or to groundwater infiltration basins) have quality requirements that protect human health. Section 3 provides an overview of common regulatory requirements and water quality objectives for WRRFs.

A variety of other community-specific performance goals may also influence WRRF design. Many areas have policies regarding work occurring within or near wetland, waterway, floodplain, other environmentally sensitive areas, and areas that may be contaminated with hazardous waste. Additional examples of locally adopted policies that influence WRRF design include but are not limited to:

- Limitations in coverage of the service area;
- Constraints concerning the type and/or level of service provided;
- Intergovernmental agreements for shared treatment capacity;
- Targeted economic development plans;
- Adopted growth rates;
- Industrial pretreatment programs;
- Biosolids management programs;
- Energy management policy and plans;
- Sustainability plans or goals;
- Water reuse programs;

- Land use requirements, such as property line setbacks, height restrictions, and noise and odor restrictions; and

- Building, fire, and electrical codes.

Many communities have adopted standards for sustainability and environmental stewardship (such as targets for incorporating sustainability principles from the Envision® Sustainable Infrastructure Rating System) or goals for energy efficiency and on-site energy production (including Leadership in Energy and Environmental Design [LEED]-compliant designs). Using established rating frameworks during project planning can help define sustainability performance goals, which can minimize operating cost by maximizing energy production and recovering value through nutrient recovery and water reuse.

Some regions have also adopted goals for flood protection at WRRFs to make infrastructure more resilient to the effects of climate change. For example, the New England Interstate Water Pollution Control Commission (NEIWPCC) recently published revisions to the TR-16 *Guides for the Design of Wastewater Treatment Works* to state that critical equipment should be protected against damage up to a water surface elevation that is 3 ft above the 100-year flood elevation (NEIWPCC, 2016).

2.2 Integrated Design

Successful WRRF facilities require sound design, responsible administration, and operator commitment. Shortcomings in any of these functions can result in process upsets and failure to meet performance goals. Employing an integrated design approach encourages the owner, operation and maintenance (O&M) personnel, and designers to communicate and collaborate from the start of WRRF planning and design through construction and startup. The goal of integrated WRRF design is to produce a facility that achieves the defined performance goals and interfaces well with the local surroundings and other municipal systems of which the WRRF is a part. Including community stakeholders in WRRF planning and design is also becoming more prevalent, not only because WRRFs must consider odor, noise, visual, and traffic impacts to neighbors, but also because more municipalities are focusing on recovery of resources (water, energy, and nutrients) from wastewater.

2.2.1 Integrated Design Process

Collaboration and frequent communication between the owner, O&M personnel, and designers is at the heart of integrated WRRF design. Integrated design also demands attention to detail and coordination as the multidiscipline design team progresses through the stages of the design from alternatives evaluations through the production of contract documents. The following subsections provide an overview of the roles of key project participants, followed by a discussion of best practices for engaging community members in the WRRF planning and design process. Descriptions of the other components of the integrated design process are included in the remaining sections of this chapter.

2.2.1.1 Project Participants   All projects begin with the identification of a need by the owner, regulator, public, or legal counsel. Projects then typically proceed through five phases—facilities planning, design, construction, startup, and operation. The roles of each participant vary with the project phases and delivery method. Table 2.1 summarizes key roles of the most common participants in a municipal WRRF design project, starting with the owner.

Participant	Roles
Owner	Pays for the design, modifications, and construction. Responsible for regulatory compliance. For owners that are public entities, final project approvals of construction contract may be required from a board of directors or city council.
Program Manager	Hired by owner to oversee project activities (commonly used when multiple projects occur simultaneously). Responsible for communication among all participants and that local requirements and performance goals are achieved.
Design Professionals	Hired by owner to develop facilities plan and design facility. For design-bid-build projects, design professionals also may serve as owner's representative during construction and startup. For alternative project delivery methods, this role may differ (some design firms may also operate the facility).
Operation & Maintenance Personnel	Provide input at various project milestones, starting with alternative development and evaluations during facilities planning through various design submittals to provide input on use of specific equipment and layout.
Contractor	Typically selected through a public bidding process. Role varies depending on the procurement approach, but the contractor is always responsible for construction of the facilities in accordance with contract documents. May also have design and/or operational responsibilities.
Regulator	Approves planning and design documents and performs reviews during construction; reviews grant and revolving fund applications, payment requests, and project closeout documentation. Involved in outfall siting, effluent dispersion, and residual disposal planning. Involvement continues after project completion with receipt/review of discharge compliance reports, operator training, and periodic inspection of the operating facilities.
Legal Counsel	Required for projects under court order or consent agreement. Legal assistance continues until all terms of the orders are complied with and the owner is released from the terms of the order or consent agreement. Status reports are filed throughout planning, design, construction, and startup phases.
Financier	Required at the time of bonding. Many public owners use short-term construction borrowing, with final bonds acquired at the end of construction. Larger projects may benefit from financial advice during planning or early in design. Public/private partnerships are also becoming more prevalent.
Process/ Equipment Vendors	Consulted for advice on functionality of their equipment, to verify that equipment will work as designed by the design professional, and that they can bid on the specifications. For specialized applications, equipment may be selected or procured early during design phase so that the key facility components are customized specifically for the particular equipment.
Community	Involvement is beneficial during all phases, particularly for owners concerned with meeting good neighbor, sustainability, or resource recovery performance goals. Community input can be facilitated in many ways, including advisory committees, public meetings, or social media.
State and Local Agencies	Involvement by any of the following might be necessary depending on project: • Local building inspectors to determine building codes/interpretations • Fire department for rulings on chemical storage, routing of alarms, etc. • Planning or zoning officials for site-review, zoning rules, and setbacks • Conservation committees or others responsible for reviewing wetlands • Local floodplain administrator for delineation of floodplain and permits • Local utilities, to discuss electrical and water supplies, drainage, etc. • Agencies that protect use of sites with historical/archaeological significance

TABLE 2.1 Project Participant Roles in Integrated WRRF Design

The owner could be a public entity (such as a municipal government or government-authorized regional district), or private (such as a commercial land development firm). All project participants are responding to the needs of the owner with the exception of the community and the regulators. Owners may choose to hire a program manager when multiple projects are occurring simultaneously and additional oversight and coordination among projects may be warranted. O&M personnel are usually employees of the owner, but alternatively could be hired through an operations contract. Integrated WRRF design encourages O&M personnel to participate in the goal-setting process during facilities planning and to review the design as it develops. This approach ensures the completed design considers operability and ease of maintenance (which are critical aspects that often lack a designer's full attention), and it ensures the operators feel that they are running "their" facility rather than one that was just turned over to them by a designer—this sense of ownership is important to the successful performance of the WRRF.

The owner may also have other staff with designated roles in the organization, such as an owner's project manager, energy manager, sustainability manager, or environmental programs manager. Including these staff in WRRF planning and design can help to achieve energy or sustainability performance goals.

Since WRRFs by their nature are complicated and multidisciplinary, the project team will include designers from a range of disciplines, including process/mechanical and civil engineers, structural engineers and architects, building mechanical and plumbing engineers, electrical engineers, site design and landscape architects, and instrumentation and controls specialists. Depending on the level of automation desired, a supervisory control and data acquisition (SCADA) system integrator may also be needed. Fluid communication between team members is crucial to ensure the design documents are thorough, and communication between the design team and the owner and O&M personnel is important to ensure the design meets the owner's requirements. Chapter 7 discusses the support systems designed by other disciplines.

2.2.1.2 Community Engagement

The level of engagement with community stakeholders during WRRF planning and design varies depending on the project. At a minimum, proactive engagement with the immediate neighbors is recommended because they are often affected by high traffic volumes, limited parking, as well as dirt, dust, and noise related to construction of the facility. These issues can be even more problematic if construction must be done outside of normal working hours or during normal periods of peak traffic. Community input is particularly valuable for topics such as facility siting and aesthetics, siting final effluent discharge locations, defining necessary level of odor mitigation, and residuals management planning. Addressing these topics during the planning process allows the design team to address concerns in a cost-effective manner. Section 11.2 provides examples of design measures to mitigate negative impacts on the community.

Community input for these "good-neighbor" considerations can be solicited using a neighborhood advisory committee that the owner and designers consult for input. Federal and state requirements largely will dictate effluent quality targets, but good-neighbor considerations are critical in finding an effluent discharge site and configuration that wins public support, or at least public acceptance. Advocacy groups also can be stakeholders in a project and can have a similar role as neighborhood advisory committees. Public meetings should have two objectives: (1) to explain the specific steps that will be taken and (2) to understand public concerns. To be effective, neighborhood meetings are not solitary events in advance of a public hearing, but regular events during the design and as construction progresses. These meetings provide an opportunity to identify and respond to issues early and effectively. Utilities have successfully used interactive websites and social media platforms to share information, design features, construction photographs, and contractor progress. These web-based tools have allowed utilities to maintain ongoing communication with the local neighborhood affected by a project.

2.2.2   Integrated Design for Resource Recovery

Municipal WRRFs are just one piece of a larger system that includes inputs and outputs of water, nutrients, and energy—all of which are finite resources. Figure 2.1 highlights many examples of how municipal WRRFs can integrate their operations with other community functions to recover resources from wastewater. Resource recovery not only improves the overall sustainability of a community but can also help reduce WRRF operating costs. When paired with energy conservation best practices, energy recovery can result in a WRRF becoming self-sufficient or even at times exporting power back to the grid.

Many WRRFs provide reclaimed water and nutrients for various uses throughout the community, ranging in purposes from irrigation water for parks or golf courses to direct potable reuse; a detailed discussion of water reuse is provided in Section 3.3. Anaerobic co-digestion of source-separated organic materials (such as food wastes, industrial organic wastes, or airport deicing fluid) with biosolids at a WRRF allows a municipality to produce renewable energy from a waste stream that would otherwise be landfilled. South Korea banned the practice of landfilling or dumping food waste down the drain, which has driven innovation to find alternative outlets for food waste materials, including through anaerobic co-digestion.

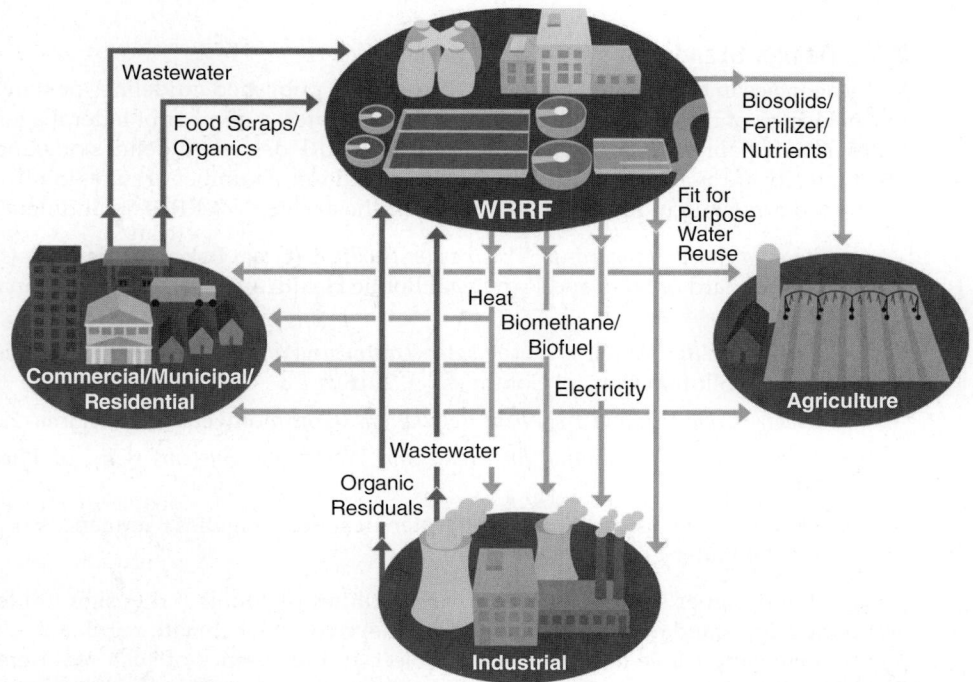

FIGURE 2.1   Opportunities for integrated resource recovery at municipal WRRFs. (Published courtesy of CDM Smith, Inc.)

In addition to anaerobic digestion, energy is being recovered in other ways, including but not limited to:

- Producing power from excess hydraulic head in effluent pipelines using turbines;
- Recovering heat directly from raw sewage or treated effluent to use as a heat source for heating and cooling buildings;
- Recovering heat from biosolids incinerators or biosolids dryers to produce electricity;
- Producing high-quality biogas for use in compressed natural gas vehicle fleets or pipeline quality biomethane.

Other WRRFs are recovering nutrients from biosolids or other high-nutrient side-streams to produce marketable fertilizer or soil amendment products for the agricultural community.

As WRRFs seek more ways to recover water, energy, and nutrients, engagement with a wider swath of the community has become more important. As part of facilities planning, some WRRF owners conduct visioning workshops with local entities, including representatives from the solid waste department, local industries, etc. These workshops help identify opportunities for collaboration to achieve resource recovery for mutual benefit: a means to provide nutrient-rich irrigation water to a nearby golf course; an opportunity to take a waste by-product from a local industry to use as a supplemental carbon source for biological nutrient removal (BNR) or as an organic carbon source for co-digestion; or a means to provide reclaimed water for a nearby waste-to-energy facility. Such opportunities are only realized through dialogue with community stakeholders.

2.3   Design Standards

Many agencies in the United States and abroad have published guidelines or standards for the design of municipal WRRFs. In the United States, a number of federal agencies, states, municipalities, and cities have published WRRF design guidelines or standards that typically are available on their websites. In addition, a number of states jointly have developed regional guidelines or standards for the design of WRRFs, including:

- *Recommended Standards for Wastewater Facilities* (Great Lakes–Upper Mississippi River Board of State and Provincial Public Health and Environment Managers, 2014);
- *Guides for the Design of Wastewater Treatment Works* (New England Interstate Water Pollution Control Commission, 2016);
- *Sewage Collection and Treatment Regulations* (Commonwealth of Virginia, 2008);
- *Design Standards Manual for Water and Wastewater Systems* (City of Phoenix, 2017); and
- Design and Construction of Wastewater Treatment Facilities (German Association for the Water Environment, 1995).

It is the designer's responsibility during facilities planning and design to identify and use design standards appropriately to achieve compliance with regulatory-driven requirements applicable for a particular project. In the absence of such requirements, the designer can still use these resources as a source of guidance, while also considering site-specific conditions.

2.4   Planning Horizon

The needs of a WRRF during the initial years of operation should be properly balanced with future facility needs. In most cases, basing design only on the objectives of any one part of this time span will compromise another part. Experience has shown that the design should primarily accommodate the projected conditions of a given design year, with allowances for proper operation ("turndown") when loading conditions may be significantly less than design year. Most WRRFs are designed for 20 years of service; however, the designer should consider if expansion or rehabilitation to handle additional loads is reasonably anticipated beyond the design year. Achieving the proper balance between the design period and beyond can be a challenge. In many cases, disregarding the future beyond the design year has resulted in abandonment of the original facility at great cost to the community. In other cases, an overly intensive focus on an uncertain future beyond the design period has resulted in facilities with operations, maintenance, or performance shortcomings during the design period. Table 2.2 (D'Antoni and Bahl, 1990) summarizes considerations involved in future planning beyond the design period. Because the reliability of loading projections declines as the time span of the projection increases, a facility process or layout commitment to an uncertain future deserves careful scrutiny if it would significantly compromise system operations during the first years after startup.

2.5   Land Requirements

After having a clear understanding of design parameters and goals for the WRRF project (including regulatory requirements and water quality objectives that are discussed in Section 3), the designer can begin evaluation of the proposed site or sites to build the WRRF. The choice of sites is often narrowed because of past investments in WRRFs, adjoining land uses, and the relationship of the collection system to existing WRRFs. Land area requirements are discussed in detail in Chapter 3. If multiple sites are evaluated, potential effects of the WRRF on the community should be considered. Site topography, local atmospheric conditions, and proximity to neighbors can all play an important role in how much odor and noise must be considered in the facility design.

2.6   Cost and Funding

A planning-level cost estimate typically is developed as part of a facilities planning effort and often is based on general facility footprints and conceptual process layouts. A detailed discussion of cost estimating is included in Section 7. Although this estimate is updated as the design progresses, it is critical to be aware of the initial cost estimate and how it was used in the development of a community's wastewater financial plan.

WRRF projects represent large capital expenses for a community and typically are funded through a mix of funding sources. User charges or rate revenues typically fund the portion of the WRRF upgrade benefiting existing customers. Funds required for capital improvements compete with other rate-funded demands, including O&M of the wastewater utility and debt service. Connection or impact fees are another revenue source used to fund WRRF improvements. Impact fees are often developed based on adopted capital improvement plans and are available to fund only that portion of project costs attributable to growth. Rate revenue and impact fees are common revenue sources to retire municipal general obligation and revenue bonds that provide long-term financing of major capital improvements.

Topics	Considerations
Design documentation	Standards and criteria for existing facilities, designed improvements, and future expansion planning (including equipment sizing; ultimate capacity layouts; and hydraulic, structural, and electrical details) should be documented as appropriate at the beginning of a project and properly archived by the owner and designer at its end. This documentation should include the hydraulic profile; process and instrumentation diagram(s); processing schematic(s); mass balances; design wastewater characteristics (including infiltration and inflow); performance objectives; peaking factors; service population; and significant sources of industrial, commercial, institutional, or specialty wastes.
Facility layout	
Ultimate facility capacity	Definition of the ultimate facility capacity is the first step in designing for the future and can be estimated from service area boundaries, population projections, land-use plans, and the available site (or needed site) for various planning periods.
Future standards	Estimates of future compliance requirements for the facility's emissions (effluent, residuals, and atmospheric) should be used to allot space for future treatment needs and other process modifications as future standards evolve.
Adjacent land use	Planned uses of adjacent land represent an external variable to be used in formulating the site needs for the future.
Site planning considerations	The first step in developing the site plans for the current design should be based on the ultimate facility capacity and needs for the future, including: • Fundamental layouts for the ultimate facility capacity showing operating and processing areas, site access, roadways, buffer zones, and drainage; • The ultimate capacity hydraulic profile to identify future hydraulic constraints and preserve and allocate hydraulic head to accommodate future expansion and distribution (pipelines and channels that are common to future expansion configurations must be carefully sized to avoid low-velocity conditions and solids deposition problems in the early low-flow years of operation); and • An ultimate electrical load analysis to form the basis for the primary electrical distribution plan.
Design details	
Isolation and diversion during construction	The design of any component of the WRRF should anticipate repairs and phased expansion while the existing facility is in full operation. Assurance of operating ease during these conditions can be incorporated to the current design through the use of isolation and diversion features, such as gates, valves, blind flanges, and stop log keys.
Structures	Often, certain structures are more appropriately designed for their ultimate capacity. Control rooms and operating centers should be sized to minimize expansion disruptions. Other buildings and facilities may be designed to accommodate alternate future uses or facilitate future expansion. Common structural details used to accommodate future expansion include knockout panels, common wall designs, wall expansions, access allowances, rebar dowel splices, and key ways.
Equipment sizing	Unit process equipment sizing reflects collective consideration of capacity with units out of service, economics of scale in construction and operation, current design criteria, and phased and ultimate capacity planning. The attractiveness of increased operational flexibility through multiple, small units should be balanced against likely savings derived through fewer, larger units. The designer should compare the future replacement of smaller units by larger capacity equipment with the progressive expansion of the current equipment; in both cases, the design should account for allocation of future spatial needs.
Electrical and instrumentation	An often overlooked consideration is expansion planning for the electrical and instrumentation systems. Space capacity should be provided in the electrical duct bank, switchgear, and control system components, to minimize future installation costs and operating inconvenience.

TABLE 2.2 Considerations Associated with Designing for the Future (D'Antoni and Bahl, 1990)

Miscellaneous revenues for funding a WRRF project include permit fees and fines, revenue from industrial surcharge fees, septage acceptance fees, intergovernmental service agreements, and public/private partnerships. These can be significant sources of revenue for the construction of new facilities. State and federal block grants, state revolving loan funds, and rural community assistance programs also may be available to complement local funding sources. A more detailed discussion of rates and financing can be found in the Water Environment Federation® (WEF) Manual of Practice No. 27, *Financing and Charges for Wastewater Systems* (WEF, 2004b).

2.7 Asset Management

Asset management serves as a tool to improve regulatory compliance; lower O&M costs; assess criticality, capacity, and LCCs of key facilities; and support environmental sustainability and stewardship. The designer should be aware of the owner's asset management programs so that the design can support established or required asset hierarchy, equipment numbering systems, data tags, and related information. If an asset management program is not in place, the owner may benefit from developing such a system as part of the design scope effort.

Asset management programs are discussed in detail in *Implementing Asset Management: A Practical Guide* (AMWA et al., 2007). In general, these programs include development of inventory systems, condition assessments, and computerized maintenance management systems (CMMSs) and electronic file management systems (EFIMS). The designer can use these systems to update equipment inventories and condition assessments and to add new equipment to the existing CMMS. The EFIMS is a readily archived database of design data, manufacturer's installation and O&M information, construction photographs, and related information, which is available to O&M staff once the project is completed.

Environmental management systems (EMSs) may be coordinated with an asset management program to improve overall environmental performance and service delivery. The International Organization for Standardization (ISO) has developed several standards for EMSs, including ISO 14001 and ISO 50001 that focus on sustainability and energy. In addition, the National Biosolids Partnership (Alexandria, Virginia) provides EMS training opportunities and a certification program for wastewater utilities that is focused on biosolids management. More information concerning EMSs can be found in Chapter 18 and *Sustainability and Energy Management for Water Resource Recovery Facility Design* (WEF, 2017).

3.0 Regulatory Requirements and Water Quality Objectives

Water quality and treatment objectives are criteria that must be established at the outset of a WRRF design. Both the current and potential future regulatory requirements must be understood so that the design can meet current water quality objectives while also being flexible for future requirements. Regulations enforcing air quality standards and management of hazardous substances can also be important.

The following subsections provide a general overview of key environmental regulations in the United States and internationally that often affect municipal WRRF design, including water quality objectives for discharge to receiving waters, the more stringent guidelines that apply for reclaimed water applications, air quality requirements, and regulation of hazardous wastes.

3.1   Overview of United States Wastewater Discharge Standards

The Federal Water Pollution Control Amendments of 1972 (Public Law 92-500) updated regulations that aim to improve water quality for human contact and recreation and eliminate pollution and introduction of toxic substances into the waterways of the United States. This legislation and updates are commonly referred to as the Clean Water Act (CWA), and amendments include the CWA of 1977 (Public Law 95-217), the CWA of 1987 (Public Law 100-4), and the CWA of 2002 (Public Law 107-303). The CWA of 1972 established technology-based effluent limits based on water quality standards developed by the various states and established a discharge permit system for point sources, exempting most nonpoint sources. The CWA of 1987 extended the discharge permit requirements to industrial dischargers and to municipal separate storm sewer systems.

3.1.1   National Pollutant Discharge Elimination System

Discharge of treated wastewater to receiving waters of the United States and management of biosolids from municipal WRRFs are primarily regulated through National Pollutant Discharge Elimination System (NPDES) permits, which are generally issued at 5-year intervals. The U.S. EPA oversees the NPDES program, although the permitting process has been delegated to many states for management and control. U.S. EPA regional offices administer and issue permits in states that have not accepted delegation. As of August 2016, 46 of the 50 states and one territory had received delegation to operate the wastewater component of their program under section 402, and only 8 states (Arizona, Michigan, Ohio, Oklahoma, South Dakota, Texas, Utah, and Wisconsin) were delegated responsibility for biosolids management under section 505.

Compliance standards are established on a national or regional perspective as the minimum quality of a facility discharge—both liquid stream and biosolids—that is acceptable under the permitting program. Water quality standards typically contain three components: designated use of a water body, water quality criteria for the type of water body, and an anti-degradation provision. If these compliance standards are not adequate to achieve the water quality standards of a receiving water body, more intensive limitations may be applied. Where technology-based effluent limits are not adequate to meet the uses established for a receiving water body, water-quality-based effluent limits (WQBELs) can be applied. Implementation of the WQBELs can be in several steps. First, a total maximum daily load (TMDL) can be established, which identifies the maximum loading (often expressed in units of mass per day) of one or more pollutants that can be discharged to a receiving water body within a defined area. This is typically developed based on water quality models and field testing. The next step is to allocate the allowable discharge mass to the upstream point source and nonpoint-source contributors to the water body. This process is often referred to as waste load allocation (WLA). In establishing the WQBELs, TMDL, and WLA, anti-degradation criteria are often applied to protect existing water quality from being degraded and to improve an existing water body when a higher beneficial use has been established than the current use of the water body. The result of these processes is a required effluent quality that is incorporated to a state or NPDES permit.

The NPDES effluent discharge standards are intended to protect and preserve beneficial uses of the receiving water body based on water quality criteria, technology-based limits, or both. National minimums, termed secondary treatment equivalency, for municipal wastewater dischargers are defined in Table 2.3. Note that secondary treatment regulations (40 CFR 403) include some exceptions and allow states to establish

Parameter	30-Day Average	7-Day Average
Conventional secondary treatment processes		
5-day BOD,[a] the most stringent of		
Effluent, mg/L	30	45
Percent removal[b]	85	—
5-day CBOD,[a] the most stringent of		
Effluent, mg/L	25	40
Percent removal[b]	85	—
Suspended solids, the most stringent of		
Effluent, mg/L	30	45
Percent removal[b]	85	—
pH, units	Within the range of 6.0 to 9.0 at all times	
Whole-effluent toxicity	Site specific	
Fecal coliform bacteria, MPN/100 mL[c]	200	400
Trickling filters and stabilization ponds (equivalent of secondary treatment)		
5-day BOD,[a] the most stringent of		
Effluent, mg/L	45	65
Percent removal[b]	65	—
5-day CBOD,[a] the most stringent of		
Effluent, mg/L	40	60
Percent removal[b]	65	—
Suspended solids,[d] the most stringent of		
Effluent, mg/L	45	65
Percent removal[b]	65	—
pH, whole effluent toxicity and fecal coliform bacteria remain unchanged		

[a] Chemical oxygen demand (COD) or total organic carbon (TOC) may be substituted for the 5-day BOD when a long-term BOD:COD or BOD:TOC correlation has been demonstrated.

[b] Percent removal requirements may be waived on a case-by-case basis for combined sewer service areas and for separate sewer areas not subject to excessive inflow and infiltration where the base flow plus infiltration is 450 L/capita/d (≤120 gpd/capita) and the base flow plus infiltration and inflow is 1041 L/capita/d (≤275 gpd/capita).

[c] Not defined in federal secondary treatment equivalency regulations, but permits typically include cited levels, often only on a seasonal basis.

[d] The state may adjust the suspended solids limits for ponds subject to U.S. EPA approval.

TABLE 2.3 Minimum National Performance Standards for POTWs (Secondary Treatment Equivalency) (40 CFR 133) (U.S. EPA, 2016f)

more stringent effluent quality requirements. Nitrogen and/or phosphorus limits are increasingly common in areas with receiving waters impaired due to eutrophication (i.e., low dissolved oxygen, excessive algal growth, etc.). NPDES effluent limitations can also be influenced by mixing zone rules that influence potential analysis of toxics. Discharges of contaminants of emerging concern (CECs), disinfection by-products

(DBPs), personal care products, and pharmaceuticals are examples of wastewater constituents that are currently unregulated but are receiving increasing attention from the scientific community and regulatory agencies.

Some states have regulations requiring daily limits. Daily limits can be statistically governing if not correctly applied, particularly in facilities serving combined systems and practicing wet-weather blending of primary and secondary effluents. The daily statistical equivalent of 30 mg/L monthly and 45 mg/L weekly is 75 to 90 mg/L. Permits containing daily limits for biochemical oxygen demand (BOD) and total suspended solids (TSS) typically use 50 mg/L.

Designers should exercise caution when applying permit limits to design criteria and anticipate pending changes in criteria when developing the site and hydraulic plan for a facility. Monthly and weekly permit limits are worst-case limits. The design criteria should be based on producing a worst-period effluent quality that complies with the permit values. This may require designing for an average annual effluent quality that is less than the critical period permitted values. Various statistical studies (Hovey et al., 1979) have indicated that secondary facilities must produce effluents with annual average BOD and TSS concentrations of 14 to 17 mg/L to ensure compliance with the 30-mg/L monthly and 45-mg/L weekly requirements.

3.1.2   Industrial Pretreatment Program

Section 403 of the CWA established an industrial pretreatment program for industrial dischargers to municipal WRRFs. General provisions (402.2) were intended to prevent the pass-through of untreated industrial waste into the nation's waterways. Section 403.5 includes criteria for prohibited discharges to municipal WRRFs, including those that are explosive, corrosive, obstructive, excessively variable, or excessively hot. Categorical discharges, defined in section 403.6—Categorical Standards, were identified in Title 40 of the *Code of Federal Regulations* (CFR) (U.S. EPA, 2016d) for 56 specific categories, applicable to various industries. The U.S. EPA estimates that as many as 45 000 facilities are regulated by these Categorical Standards. In addition to the Categorical Standards, industries that discharge to a municipal wastewater collection system must meet the municipal, state, or U.S. EPA Pretreatment Standards. Over 1600 municipal WRRFs are included in regulated pretreatment programs under section 307(b) of the CWA. Although the Categorical Standards are established, the program relies heavily on each municipality to identify its own discharge priorities and to propose solutions to account for site-specific factors. Designers are encouraged to identify the industries discharging to the WRRF during the design phase.

3.1.3   Nutrient Trading

The concept of nutrient trading as a water quality tool began in the 1980s. Nutrient trading is a structured mechanism to provide treatment options and more cost-effective construction options to all pollutant contributors in a specific stream segment or receiving water body. Design consultants may be able to reduce the structural components of a facility if nutrient trading effectively shifts loading to alternate locations. The U.S. EPA has developed a water quality trading toolkit for permit writers available online at https://www.epa.gov/npdes/water-quality-trading (U.S. EPA, 2009b).

The Water Environment Research Foundation (Alexandria, Virginia) (WERF) (http://www.werf.org) participated in funding five trading programs, and reports were prepared on the five projects under WERF project number 97-IRM-5. The WERF merged with the WateReuse Research Foundation in 2016 to form the Water Environment & Reuse Foundation (WE&RF) (http://werf.org).

3.2 Overview of International Wastewater Discharge Standards

The World Health Organization (Geneva, Switzerland) (WHO) provides guidelines for drinking water, recreational water, wastewater for use in agriculture and aquaculture, and ship and aircraft sanitation. The WHO does not provide guidelines for wastewater effluent quality, but it does support various public health programs in support of both centralized and decentralized treatment systems. Guidelines are provided for bacteriological quality of graywater in various agricultural applications. The high cost of centralized systems in the eastern Mediterranean region is regarded by the WHO as the primary constraint in expanding wastewater service, resulting in greater emphasis on decentralized systems (Bakir, 2000). The WHO Guidelines for Drinking-Water Quality were most recently published in 2011 (WHO, 2011) and are supported by separate background texts, which are available through the WHO and online at http://www.who .int/water_sanitation_health/publications/2011/dwq_chapters/en/.

European Union (EU) member countries established Council Directives in 1980 and 1994, which were updated with Council Directives 98/83/EC on November 3, 1998 (Council of the European Communities, 1998) for water intended for human consumption. The EU developed legislation regarding chemical pollution of waters with Council Directive 76/464/EEC, which was amended and supported by several other directives between 1982 and 1990. Council Directive 91/271/EEC (Council of the European Communities, 1991) established standards of wastewater treatment with tertiary and secondary treatment standards based on population equivalents and location relative to the coast. The directive defines secondary treatment as "treatment of urban wastewater by a process generally involving biological treatment with a secondary settlement." For sensitive areas, which are defined as "natural freshwater lakes, other freshwater bodies, estuaries and coastal waters which are found to be eutrophic or which in the near future may become eutrophic if protective action is not taken," removal of phosphorus and nitrogen must be implemented to achieve the required effluent quality. Population equivalents are defined by the EU as "the organic biodegradable load having a five-day biochemical oxygen demand (BOD_5) of 60 g of oxygen per day."

This directive was amended in 1998 (Directive 1998/15/EC) to include refined administrative reporting and monitoring requirements. These directives were expanded by Directive 2000/60/EC, which is referred to as the Water Framework Directive (WFD), to include management of inland surface and groundwater and protective measures for the aquatic environment. The EU wastewater standards most recently were updated by Directive (397 final) presented by the commission on July 17, 2006 (Council of the European Communities, 2006). The proposed directive includes treatment requirements for 41 chemical substances, including heavy metals. The EU directives and current legislation can be accessed online at http://eur-lex.europa.eu/legal-content/EN/TXT/?uri=CELEX%3A52006DC0398.

3.3 Water Reuse

Increasing water demands in urban areas and more stringent effluent quality requirements have led to increased interest and reliance on reuse of reclaimed water (i.e., treated wastewater) worldwide. As the world's population continues to grow and the water usable for domestic purposes represents only approximately 2% of the earth's water, the available supply must be recycled faster than typically achieved in nature. Historically, most water reuse has been for agricultural and landscape irrigation. However, commercial, industrial, and potable reuses are becoming more prevalent as a result of changing climatic conditions and urban growth water demands.

3.3.1 United States Reuse Regulations and Guidelines

The U.S. EPA does not regulate the use or quality of reclaimed water. Each individual state can choose to allow reclaimed water for a variety of purposes and develop regulations appropriate to its regional needs. Many states began regulating reclaimed water similar to their state regulations for land application of wastewater, which generally include quality and setback limits intended for wastewater treatment. In 1918 California was the first state to develop regulations for reclaimed water for agricultural use. Additional regulations were added over time, as the uses for reclaimed water diversified, effluent quality improved in response to wastewater discharge regulations, and demand for alternative water sources added value to the potential supply.

In the late 1970s, the U.S. EPA initiated development of a set of guidelines for states and industries for the use of reclaimed water. These guidelines were updated in 1992, 2004, and 2012. The *Guidelines for Water Reuse* can be obtained online from the U.S. EPA at http://nepis.epa.gov/Adobe/PDF/P100FS7K.pdf (U.S. EPA, 2012). The U.S. EPA 2012 guidelines identify the history of reclaimed water in the United States, provide U.S. and international case studies of exemplary projects, identify regulations for various categories of reclaimed water use in those states with regulations and guidelines, and provide guidance on appropriate levels of treatment for each intended use of reclaimed water. They also include a tabulation of guidelines and regulations then current in each state. Appendix C in the guidelines provides the regulatory agency and regulation title with web link to the regulation or guidance in each state. The guidelines also include process flow schematics for active indirect and direct potable reuse projects.

3.3.2 United States Reclaimed Water Quality Criteria

Water quality criteria differ by state and by intended use of the water. Designers should confirm current regulations before the start of any design. Typically, the greater the perceived human exposure and health risk from the application of reclaimed water, the higher the water quality standard tends to be. For some uses with minimal human contact, lower quality water may be allowed. Where the application may affect sensitive species or areas, such as in rehydration of a wetland in a nutrient-limited area, higher discharge water quality than required by the entity's wastewater discharge permit may be appropriate.

Irrigation and agricultural applications typically allow lower quality reclaimed water where there is limited human exposure. Where golfers or athletes are likely to come in contact with turf irrigated with reclaimed water or where edible crops are irrigated and would be in direct contact with the reclaimed water, higher standards are typically applied. For example, irrigation of a root crop such as potatoes (agricultural reuse [U.S. EPA, 2012]) warrants higher water quality compared with irrigation of sod farms or commercial nurseries. Minimum water quality standards in the United States exceed those set by the WHO (WHO, 2006) and are typically 5 to 20 mg/L BOD and 5 to 20 mg/L TSS, with a limit of 2.2 to 23 coliform colonies/100 mL set in California.

The U.S. Department of Agriculture/National Institute of Food and Agriculture (USDA/NIFA) has made funding for water reuse and research on crop application one of its key priorities (U.S. EPA, 2012).

Reclaimed water for recreational use can have high-quality standards for unrestricted recreational use, as regulated in California, Texas, Nevada, and Washington, with lower BOD values, lower bacteriological limits, and restrictions on full-body contact in constructed impoundments. Arizona and Hawaii also have regulations for restricted recreational reuse. Bacteriological quality for unrestricted recreational use ranges from 2.2 to 20 fecal coliform colonies/100 mL, while bacteriological quality for restricted use can rise to 200 fecal coliform colonies/100 mL (U.S. EPA, 2012).

Reclaimed water for environmental reuse can be used for stream augmentation, that is, creation of a flowing water feature where one does not currently exist, or as is more common today, application to rehydrate a wetland area or create a new wetland area. Several wetland applications combine wetland rehydration, water quality improvement and detention through the wetland system, and groundwater recharge to augment an underlying aquifer that is used for public water supply. Florida and Washington specify treatment, BOD, TSS, and total phosphorus limits. Florida also specifies an ammonia limit and Washington specifies bacteriological standards. Florida also has specified more stringent water quality standards where rehydration of a wetland area containing endangered species could adversely affect the species if nutrients were not strictly limited. Additional discussion of wetlands and small community systems is included in Chapter 16, Natural Systems.

Reclaimed water can also be used for most industrial non-potable water applications. Some of the higher value applications include use as cooling water at power facilities, for industrial and commercial building cooling towers, and for central heating, ventilation, and air conditioning (HVAC) or chill water systems. Reclaimed water also can be used for dust control, concrete mixtures, facility washdown, and raw water feed for process water. In some cases water is fed to reverse osmosis treatment for high-quality water in the microchip manufacturing process or as the carrier water in water-based paints used to paint new vehicles. Reclaimed process water is increasingly being used in the food and beverage industry, and the International Life Sciences Institute Research Foundation is developing guidelines for water recovery and reduction of the industry water footprint (U.S. EPA, 2012).

Groundwater recharge refers to the application of reclaimed water to areas that allow infiltration through soil strata to ultimately recharge an aquifer. This process takes advantage of soil aquifer treatment mechanisms, extended travel distances to sources used for potable purposes, and is often used where non-potable use is ultimately made of the receiving aquifer. California addresses regulation of these systems on a case-by-case basis and has new groundwater recharge regulations that are expected to be released for public comment in the first quarter of 2017. Florida and Washington have water quality standards for groundwater recharge; Washington applies its most stringent standard of Class A water to this use.

Augmentation of potable water supplies requires the highest levels of treatment and typically involves an extended public information period, pilot testing, and a study phase to implement a project. An example of the current state of the art of a large indirect potable reuse project is the Orange County (California) Groundwater Replenishment Project at 366 ML/d (100 mgd), with planning in process to expand production to 475 ML/d (130 mgd). Orange County Water District (OCWD) receives highly treated secondary wastewater from the Orange County Sanitation District and provides additional treatment to produce water exceeding current potable water criteria. Key process elements in the purification process are microfiltration and reverse osmosis membrane treatment, followed by peroxide and UV irradiation. The process includes two-stage membrane treatment (microfiltration as pretreatment for reverse osmosis), hydrogen peroxide and UV irradiation as an advanced oxidation process (AOP), and stabilization before discharge to several spreading basins or injection wells. Two-stage membrane treatment significantly removes small particles, bacteria, and viruses. The combination of reverse osmosis and an AOP provide multiple barriers for disinfection. UV supports destruction of various organics in the wastewater typically characterized as CECs and provides a high level of bacteriological disinfection, while the use of hydrogen peroxide provides a multiple disinfection barrier and is more effective against some bacteria that may be associated with small particles in the product water. When UV and hydrogen peroxide are used together, the photocatalytic oxidation of hydrogen peroxide forms a hydroxyl radical. This process achieves excellent organic destruction and disinfection. The product water can be delivered to three systems for integration with the groundwater—barrier wells for injection to control saltwater intrusion, primary spreading basins, and discharge to the Santa Ana River, which is managed to increase infiltration to the aquifer.

The Colorado River Municipal Water District in Big Spring, Texas, is currently the only direct potable reuse facility in the United States and uses the same treatment process described for the OCWD. This facility is permitted by the Texas Commission on Environmental Quality (TCEQ). A research study supported by the TCEQ and the Water Environment and Reuse Foundation (WE&RF) has conducted an intensive water quality monitoring study concluding that the water produced exceeds all U.S. and Texas drinking water standards (Steinle-Darling, 2016). Chapter 17, Disinfection, has additional discussions of advanced disinfection techniques.

The manual *Using Reclaimed Water to Augment Potable Water Resources* (WEF and AWWA, 2008) contains information regarding indirect potable reuse, including treatment technologies; the complex health and regulatory issues; the barriers, backup options, and flexibility needed to maintain system reliability; and the need to address public perception through appropriate outreach and clear communication. The WateReuse Research Foundation (now merged with WERF to form WE&RF) began intense research on potable reuse practices in 2012 raising over $6 million for 34 researcsh projects.

3.3.3   International Reuse Regulations and Guidelines

The WHO recommendations for the use of reclaimed water rely on implementation of low-cost/low-technology systems, such as stabilization ponds with long detention times. The long detention periods facilitate the settling and removal of helminth eggs and protozoan cysts, which tend to pose the highest risk to the public from exposure to reclaimed water. The historical WHO standards originally issued in 1989 were based on agricultural irrigation and limit helminth eggs to 1 egg/L and fecal coliform bacteria to 1000 fecal coliforms/100 mL for unrestricted irrigation. For public lawn irrigation, a more restrictive 200 fecal coliforms/100 mL is applied. Compliance relies on effluent testing for fecal coliforms, and the standard for helminth removal is intended as a design guideline. In 2006, the WHO updated its guidelines for the safe use of wastewater, excreta, and graywater (WHO, 2006). These guidelines expanded the bacteriological parameters, in terms of type of irrigation, crop, and exposure, and included recommended logarithmic reductions to achieve recommended health-based disability-adjusted life year targets.

There are currently no specific water reuse regulations that apply to the EU member countries. Discussions are in process, as some EU representatives recognize the relationship of using reclaimed water in agriculture. Individual countries have different levels of guidelines or regulations. On June 10, 2016, the EU Water Directors endorsed *Guidelines on Integrating Water Reuse into Water Planning and Management in the Context of the WFD* (EU, 2016). These guidelines were issued under the Common Implementation Strategy (CIS) of the WFD. The guidelines do not recommend any particular chemical, microbiological, or physical standard. Legally binding standards for water reuse have been developed by several EU member states. Some non-EU countries and international organizations have also developed specific standards that EU member countries recommend. Most of the standards that have been developed at the member state level derive from the WHO (EUWI-MED, 2007) and U.S. EPA guidelines (U.S. EPA, 2012). Two tables in the EU 2016 guidelines provide the specific water reuse standards references. Table 1 lists standards for six countries (Cyprus, France, Greece, Italy, Portugal, and Spain), and Table 2 identifies international standards referenced by EU member states (WHO, 2006; International Standards Organization Standard 16075, 2015; EPHC et al., 2006; U.S. EPA, 2012).

EU wastewater directives include water conservation and sustainability criteria. These criteria encourage the use of treated wastewater for agriculture and aquaculture because water reuse reduces the pollutant load to the receiving stream and extends limited local water supplies. The secondary and tertiary quality goals provide higher quality water than has been available in most situations. Additional information on EU water reuse regulations can be found at http://ec.europa.eu/environment/water/reuse.htm.

Australia has been developing guidelines for water recycling that are being issued in phases. Multiple documents have been developed as part of the Australian National Water Quality Management Strategy, beginning in 1994, with Policies and Principles. Phase 1, National Guidelines for Water Recycling: Managing Health and Environmental Risk (EPHC et al., 2006), established national program requirements; identified human and environmental health risk factors; defined sources; and required reclaimed water and water quality analyses, operator and user awareness training, and community involvement recommendations. Phase 2 consists of three documents. Managed Aquifer Recharge was issued in July 2009 (EPHC et al., 2009a). Two additional documents, Augmentation of Drinking Water Supplies (EPHC et al., 2008) and Stormwater Harvesting and Reuse (EPHC et al., 2009b), have been issued. EPA Victoria also released *Code of Practice Onsite Wastewater Management*, Pub. No. 891.3 (EPA Victoria, 2013) defining decentralized treatment practices. The documents include guidelines for maximum concentrations of chemicals in secondary effluent, including pharmaceuticals.

3.3.4 Contaminants of Emerging Concern in Reclaimed Water

The term "microconstituents" was adopted by WEF in 2007 (WEF, 2007b) to describe a large number of elements and compounds that are being detected in water. CEC is the term more commonly used to include a broad group of individual chemicals and classes of compounds present at trace concentrations that can include groups of compounds categorized by the end use; environmental and human health effect; or type of compound (U.S. EPA, 2012).

CECs are ubiquitous in modern life, and all living organisms have experienced some exposure. Improvements in analytical technology now allow for detection of many of these CECs at sub-nanogram-per-liter concentrations. Because many of the CECs are naturally occurring, they likely have been present in the environment forever; today, they simply can be quantified. Current treatments known to reduce the concentration of most CECs include processes with a long sludge age, membrane treatment (WEF, 2007a), and AOPs (WRRF, 2013). Designers should consider these processes if considering reduction of CECs.

3.3.5 Critical Control Point Inclusion

State agencies and utilities are adopting quality control standards for non-potable and potable reuse projects that incorporate a hazard analysis and critical control points (HACCP) design approach that had its origin in the food sector. HACCP has been adapted and is used in many countries for microbial and chemical controls in water treatment systems (WRRF, 2014), although its use is still limited in the United States. The WateReuse Research Foundation has defined the critical control point (CCP) as "a point in advanced water treatment where control can be applied to an individual unit process to reduce, prevent, or eliminate process failure and where monitoring is conducted to confirm that the control point is functioning correctly. The goal is to reduce the risk from pathogen and chemical constituents." CCPs are locations in a WRRF that have a direct impact on the quality of finished water as it relates to public health. CCPs must have both a monitoring component, which allows the operator to verify that the process is operating as intended, and a control component, which allows the operator to adjust the functioning of the unit process to improve its performance. Because some constituents are difficult to monitor directly (e.g., pathogens) and others lack accepted standard analytical methods (e.g., many chemicals), indirect measures will be needed to ensure that the WRRF and advanced water treatment facility are performing properly and that the advanced treated water will meet appropriate water quality requirements.

For each CCP there will be some type of critical limit that the process step must meet (contact time for chlorine disinfection or nitrate concentration). Set limits must be science or regulatory based, not arbitrary. Examples of CCP parameters are chlorine residual, turbidity, pressure decay rate during integrity testing, transmembrane pressure, electrical conductivity, total organic carbon (TOC), UV transmittance, UV train power, calculated UV dose, and pH. Example control point locations could include secondary clarifier effluent; MF and RO feed; chlorine contact basin inlet and outlet; and UV channel inlet and outlet. Use of CCP can lead to optimal performance rather than simply compliance. Section 13 contains citations for recent research reports by the WRRF (now WE&RF) developed to support potable reuse.

3.4   Reclaimed Water Storage and Distribution

The amount of storage required in any reclaimed water system will depend on the demands placed on the system by the customers and the supply capability from the reclamation facility. The diurnal flow pattern at the WRRF should be considered when establishing delivery rate commitments to customers. System storage and pumping rates from the reclamation facility can be minimized if agreements with customers allow for flat-rate delivery over a 24-hour period, with on-site storage by each customer. Large irrigation customers, such as golf courses, may have a water feature that can accept flow at a reduced inflow rate, while providing a significant storage volume to allow high-rate nighttime watering of the course. Where customer storage is not possible, reclaimed water distribution system hydraulic modeling can be used to simulate the system response to potential customer demands, allowing for optimal sizing of storage, pipe, and pumping systems.

Design of the reclaimed water distribution system should use similar design considerations as those used in design of a potable water system. Distribution system designers should consult AWWA M24; the 4th edition is soon to be released while the 3rd edition is still current (AWWA, 2009). If reclaimed water is being substituted for an existing potable water supply, the customer may expect the same delivery pressure and reliability as the potable system. It is good practice, where possible, to deliver the reclaimed water at a slightly lower pressure to minimize problems with cross-connection. Reliability should be considered and discussed with potential customers. Irrigation supplies may not require the same reliability as industrial reclaimed water for cooling towers that support large public buildings or business needs. Storage, pumping, and distribution main size will be significantly influenced if fire protection is included. Fire protection has traditionally been provided by potable water systems; however, conversion to feed from reclaimed water systems would liberate potable supplies for higher priority uses in the community. AWWA has released ANSI/AWWA G-481-14 *Reclaimed Water Program Operation and Management* (AWWA, 2014) and has G-485 *Potable Reuse Operation and Management* in draft with a 2017 expected publication (AWWA, 2017).

3.5 Air Quality Regulations

In the United States the Clean Air Act (CAA) establishes National Ambient Air Quality Standards (NAAQSs) for six pollutants known as criteria pollutants: carbon monoxide, particulate, lead, nitrogen dioxide, ozone, and sulfur oxides. The states establish U.S. EPA-approved State Implementation Plans, which address emission standards for stationary and mobile sources of criteria pollutants. Permits for major sources of air pollution are addressed through the Title V Operating Permit program, which is implemented by the states. The Title V program affects municipal WRRFs that operate incinerators and dryers, engines, and boilers. Specific emission reductions may be required through CAA programs, such as New Source Review (NSR)/Prevention of Significant Deterioration (PSD), New Source Performance Standards (NSPSs), and National Emission Standards for Hazardous Air Pollutants (NESHAP). Some states and local air districts have also established more stringent requirements than the federal Title V program.

Toxic air pollutants, also known as hazardous air pollutants (HAPs), also are regulated under the CAA. The U.S. EPA is working with state, local, and tribal governments to reduce air toxics releases of 187 pollutants to the environment. Examples of toxic air pollutants include dioxin, benzene, toluene, and metals, such as cadmium, mercury, chromium, and lead compounds. Examples of potential HAP sources at WRRFs include headworks, clarifiers, and the aerated zones of bioreactors. From a practical perspective, the three criteria for determining if a Title V permit is needed are (1) 190 ML/d (50 mgd) approximate flow, (2) exceeding a 5-ppm concentration of volatile organic HAPs, and (3) exceeding an industrial contribution of 30% of the facility flow. A facility meeting two of these three criteria typically must commit to federally enforceable limits to maintain emissions below the major source level, modify the process, install control equipment for emissions, or achieve equivalent reduction through pretreatment.

Chapters 6 and 24 include discussions of how air quality standards can affect WRRF design with respect to odor minimization and thermal processing of residuals, respectively.

3.6 Risk Management Plans

Under the authority of section 112(r) of the CAA, the Chemical Accident Prevention Provisions require facilities that produce, handle, process, distribute, or store certain chemicals above a threshold limit to develop a risk management program, prepare a risk management plan (RMP), and submit the RMP to the U.S. EPA. A risk management program includes prevention of release, process safety management, and emergency response. The U.S. EPA has issued Appendix F, which is Supplemental Risk Management Program Guidance specifically for municipal WRRFs (accessible at https://www.epa.gov/sites/production/files/2013-11/documents/appendix-f1.pdf).

Chemicals requiring RMPs above a certain threshold quantity that are commonly used at municipal WRRFs include chlorine gas, aqueous ammonia, and methane. For methane, the 4540-kg (10 000-lb) RMP threshold applies to the total weight of the flammable mixture of digester gases, not just the weight of methane or flammables in the mixture. However, if a municipal WRRF uses methane (or a methane mixture) as fuel or sells it as fuel (as a retail facility), the amount of methane that is used or sold as fuel is not covered under 40 CFR Part 68 (U.S. EPA, 2016a). For aqueous ammonia, the threshold applies only to the weight of ammonia in the mixture. General guidance on risk management for chemical accident prevention is updated periodically (U.S. EPA, 2009a). The designer should consider process modifications or features that limit the quantity and/or use of chemicals that would exceed the threshold amount. Sound design also accounts for chemical release prevention and control technologies.

3.7 Regulation of Hazardous Substances

Hazardous substances (termed priority pollutants) are subject to several different kinds of regulations, some based on considerations of water quality and some based on the technologies available for pollution control in different industrial sectors. The following is a brief overview of the many regulations of hazardous substances that affect municipal WRRFs in the United States.

3.7.1 Resource Conservation and Recovery Act

The Resource Conservation and Recovery Act (RCRA) regulates both hazardous and nonhazardous solid wastes. Disposal of nonhazardous wastes is allowed in underground storage tanks and municipal solid waste landfills. Hazardous substances are regulated from the point of generation, through transport, storage and treatment, and disposal. In regulatory terms, an RCRA hazardous waste is a waste that appears on one of the four hazardous wastes lists (F-list, K-list, P-list, or U-list) or exhibits at least one of four characteristics—ignitability, corrosivity, reactivity, or toxicity. The RCRA exempts domestic wastewater and municipal WRRFs from its jurisdiction through the domestic wastewater exclusion, unless the municipal WRRF uses underground storage tanks or a municipal landfill for solids disposal. The domestic wastewater exclusion applies to industrial wastes, as long as they mix with domestic wastewater before entering the municipal WRRF headworks. If this does not occur, and hazardous wastes are found to be either consciously or inadvertently received by rail, truck, vessel, or pipeline, the RCRA's permit-by-rule procedures apply. To use the permit by rule, the facility must have and be in compliance with a CWA NPDES permit. The facility can only accept wastes that meet all applicable pretreatment requirements that would be applicable to the waste as if it had been discharged to the municipal WRRF through a sewer system. In addition, the facility must obtain a U.S. EPA identification number and use the hazardous waste manifest system. The owner or operator must keep a written operating record at the facility and submit biennial reports to the governing agency (U.S. EPA or the delegated state) of each even-numbered year covering hazardous waste treatment activities during the previous calendar year.

3.7.2 Comprehensive Environmental Response, Compensation, and Liability Act

The Comprehensive Environmental Response, Compensation, and Liability Act (CER-CLA) primarily affects municipal WRRFs in the United States that are considering accepting wastewaters from Superfund cleanup sites. In these cases municipal WRRFs will be required to ensure the Superfund discharger is following all requirements of the national pretreatment program. Liabilities of the CERCLA for a municipal WRRF can also be related to wastewater exfiltration and overflow in the collection system, storage leaks, effluent and solids disposal, and other aspects of a municipal WRRF operation. If prior municipal WRRF practices, whether conscious or not, pose a threat to public health, welfare, or the environment, the municipal WRRF is liable. Liability may be shared with the permitting authority only if a municipal WRRF's NPDES permit identifies all release points, conditions under which they operate, expectation of contravention of any discharge standard, and substances likely to be received by the municipal WRRF that are found in its residual solids and effluent. Protection is not afforded to those who fail to make full disclosure. Protection is not afforded to those who allow an industry to repeatedly violate pretreatment obligations.

3.7.3 Emergency Planning and Community Right-to-Know Act

The Emergency Planning and Community Right-to-Know Act (EPCRA) establishes requirements for federal, state, and local governments, Indian tribes, and industry regarding emergency planning and "community right-to-know" reporting on hazardous and toxic chemicals. The community right-to-know provisions help increase the public's knowledge and access to information on chemicals at individual facilities, their uses, and releases into the environment. Reportable quantities for accidental releases of extremely hazardous substances (EHSs) and hazardous substances are summarized in Appendices A and B of Title 40 CFR Part 355 (U.S. EPA, 2016c). Some EHSs are also classified as CERCLA hazardous substances, defined in sections 101 and 102 of CERCLA as any elements, compounds, mixtures, solutions, and substances that, when released into the environment, may present substantial danger to public health, public welfare, or the environment. These chemicals are specifically defined to mean any substance published in Title 40 CFR Part 302, Table 302.4 (U.S. EPA, 2016b). Under sections 311 and 312, Community Right-to-Know Requirements, facilities manufacturing, processing, or storing designated hazardous chemicals must make safety data sheets (SDS, formerly known as material safety data sheets, or MSDS), describing the properties and health effects of these chemicals, available to state and local officials and local fire departments. Facilities also must report inventories of all on-site chemicals for which SDSs exist to state and local officials and local fire departments. Designers are encouraged to review the lists of hazardous chemicals regulated under the EPCRA.

3.7.4 Toxic Substances Control Act

The Toxic Substances Control Act (TSCA) regulates the manufacture, use, and disposal of toxic substances. As part of its enabling legislation, the U.S. EPA is authorized to control the risks from more than 65 000 existing chemical substances and from the use of new chemicals. The TSCA primarily regulates industries within Standard Industrial Classification Codes 20-39 (manufacturing). Typical types of industry and sectors covered under the TSCA include companies that are engaged in chemical production and importation, petroleum refining, paper production, and microelectronics manufacturing. Municipal WRRFs are regulated under the TSCA if they accept wastes contaminated with polychlorinated biphenyls (PCBs) or certain other toxic chemicals exceeding a threshold concentration. Sources of PCBs can vary and may include contaminated sediments or illegally disposed PCBs. If a WRRF experiences a spill or leak of the same substances, the TSCA specifies stringent standards for cleanup and disposal.

3.8   Summary of Regulatory Requirements

The sections above provide a broad overview of environmental regulations that may affect municipal WRRF design and operation both in the United States and internationally. Figure 2.2 illustrates common municipal WRRF activities and sources of pollutants that are potentially governed by U.S. EPA regulations (U.S. EPA, 1989). This figure is provided to indicate the range of various regulations that municipal WRRFs may need to consider but is not intended to be all-inclusive. The reader is encouraged to independently validate the information presented herein because regulations vary by location and are subject to change.

4.0   Municipal Wastewater Flow and Loading Projections

Establishing projected wastewater flows and constituent mass loadings is an important initial step in WRRF design. Specifically, the projected *variation* in flows and loadings is critical for facility design and operation. This variation is captured with the use of peaking factors, defined as a peak flowrate (or constituent mass loading) divided by the annual average flowrate (or constituent mass loading). Peaking factors can be derived to correspond to peak hour, maximum day, maximum month, or other conditions. When using process modeling, this variation can also be captured with dynamic influent data sets, including those with diurnal variation.

This section discusses sources of wastewater flow and constituents, how to estimate flowrates and loadings if no data exist, how to analyze any existing flow and loading data and identify any errors in the data, and how to select design flow and loading peaking factor values.

4.1   Wastewater Sources

Municipal wastewater primarily comprises the following, each of which is discussed below:

- Domestic wastewater;
- Industrial wastewater;
- Infiltration/inflow from the collection system; and
- Stormwater (in combined collection systems).

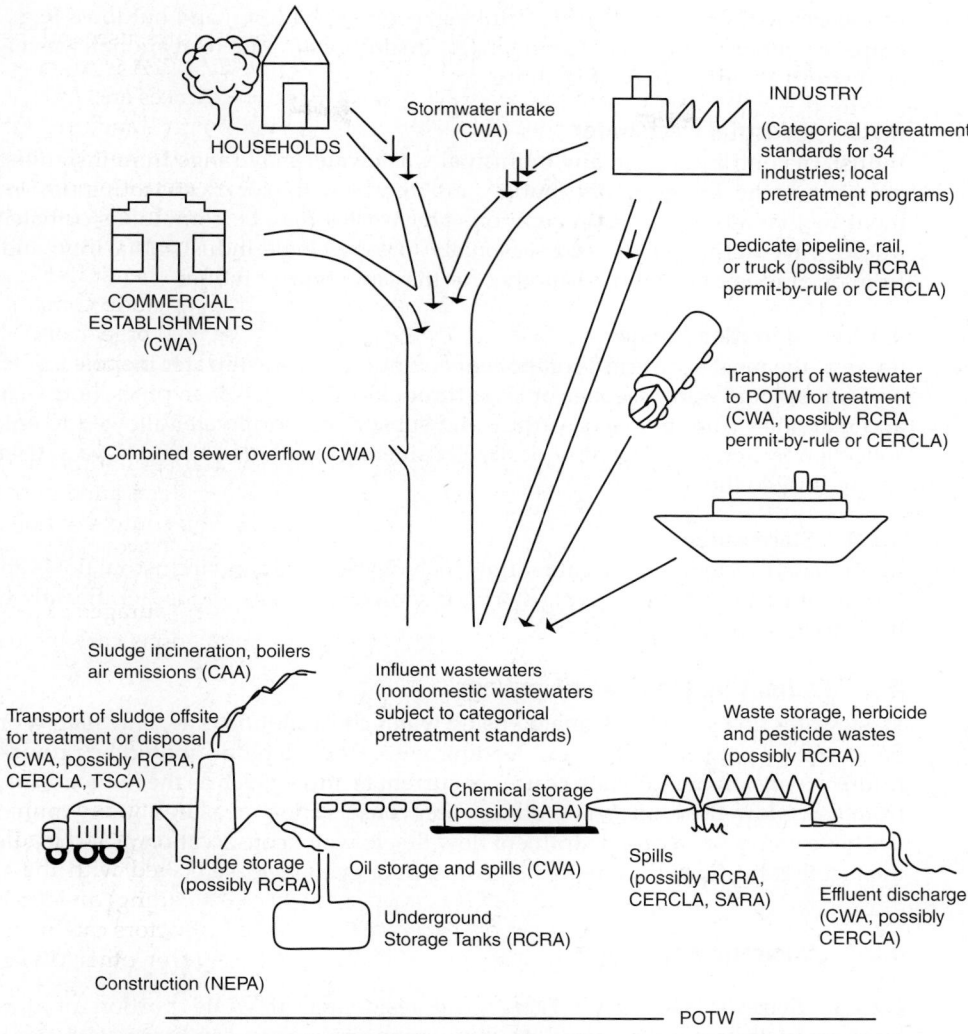

FIGURE 2.2 Activities and sources of pollutants potentially subject to U.S. EPA regulations (U.S. EPA, 1989).

Other sources may include septic tank waste (septage) generated in the surrounding, unsewered areas from septic tank cleaning contractors, and solids from sewer cleanings. Landfill leachate, water treatment residuals, and, in some instances, contaminated groundwater, possibly with low concentrations of hazardous materials, can also be discharged to municipal WRRFs. All of these sources should be accounted for to the extent possible during the development of design flows and loadings.

4.1.1 Domestic Wastewater
Domestic wastewater includes residential areas, commercial districts, institutions, and recreational facilities. Residential wastewater derives from water used in private

residences for both indoors (e.g., drinking, cooking, bathing) and outdoors (e.g., landscape irrigation) purposes. Commercial, institutional, and recreational wastewaters come from a wide variety of facilities.

4.1.2   Industrial Wastewater

Industrial contributions in any municipal wastewater may range from insignificant to many times the domestic contribution, and may be within or exceed industrial pretreatment regulations. Industrial operations and wastes may be continuous or batch produced; vary daily, weekly, and seasonally for any single industrial facility; and vary from one industrial facility to another for the same type of industry.

4.1.3   Infiltration/Inflow

Some of the most significant components of municipal wastewater include infiltration, or unintentional water seepage or leaks through collection system pipes, house laterals, and manholes, and inflow, or surface and subsurface stormwater allowed to enter the collection system. These are typically considered together as I/I and vary from one community to the next.

4.1.4   Stormwater

In combined systems, dry-weather flow includes domestic and industrial flows and I/I. Wet-weather flow occurs during storm or snowmelt events and is significantly higher than dry-weather flow.

4.2   Estimating Flows and Loadings

Wastewater flows can be estimated using water consumption data, if available, or with per capita or per source flow and loading rates. Consideration should be provided for future expansion or reduction of uses in current facilities, such as the reduction of flows from using low-flow toilets and other water conservation measures. Many municipalities have witnessed a drop in influent flow due to water conservation measures, despite the fact that loadings have remained constant or even increased.

4.2.1   Domestic Wastewater

4.2.1.1   Domestic Wastewater Flows   The residential flowrate portion of domestic wastewater is commonly based on population projections combined with per capita wastewater flow values. The population projections should take into account nonresident workers, non-permanent residents and seasonal changes in populations (e.g., heavy tourist or commercial areas). Although the seasonal population may be present in a service area for only a portion of the year, it may have a significant effect on the wastewater flow treated by the facility. Similar consideration should be included for large schools or universities, for full-class relative to no-class months, and for large populations of transients who work and visit the area during the day but maintain their permanent residence elsewhere. Demographic projections developed for wastewater planning purposes should reflect other applicable planning estimates, such as those found in relevant zoning and master plans.

Flow per capita estimates for residential wastewater can be obtained from several available references if site-specific per capita estimates are not available. One of these references, which considers increased water conservation, is shown in Table 2.4

Household Size, No. of Persons	Flowrate, gal/capita/d		Flowrate, L/capita/d	
	With Current Level of Conservation	With Extensive Conservation	With Current Level of Conservation	With Extensive Conservation
1	103	74	390	280
2	77	54	290	205
3	58	48	257	180
4	63	44	240	168
5	61	42	230	160
6	59	41	223	155
7	58	40	218	151
8	57	39	215	149

TABLE 2.4 Typical Wastewater Flowrates from Urban Residential Sources in the United States (from Metcalf & Eddy/AECOM, 2014, with permission)

(Metcalf & Eddy/AECOM, 2014). Alternatively, some U.S. state regulatory agencies use the *Recommended Standards for Wastewater Facilities* (Great Lakes–Upper Mississippi River Board of State and Provincial Public Health and Environment Managers, 2014), which recommends 380 L/capita/d (100 gal/capita/d) for use as an average design flow. In general, about 60% to 90% of water consumption reaches the sewer system, with the lower percentage applicable in semiarid regions.

Flows from commercial sources are generally included within the allowance for domestic sources. This consideration becomes less appropriate for smaller service areas, however, where commercial operations, such as laundromats, car washes, and sports events, may substantially affect the wastewater's character. If a commercial or institutional facility has many employees, for example, an estimate should be made for each contributing employee developed based on daily activities at the business location. Facilities that provide cafeterias, showers, and multiple hand-washing requirements (i.e., food services) use more water per employee than those not including these water uses. Seasonal fluctuations in these flows must also be considered.

Although it is preferable to obtain site-specific flow information, Table 2.5 provides representative commercial flows that can be used in lieu of actual data. Some typical flowrates from institutional facilities, essentially domestic in characteristic, are shown in Table 2.6. Flowrates for these wastes also vary by region, climate, and type of facility. The actual flow records from the institutions are the best source of flow data for design purposes. Note that it is possible for a single institutional contributor to dominate WRRF design flows (or loadings).

4.2.1.2 Domestic Wastewater Loadings Table 2.7 delineates the typical major pollutant composition of untreated domestic wastewater. Of the constituents listed, design loadings are typically developed for BOD, TSS, nitrogen, and phosphorus. New WRRFs should be designed for a domestic load contribution of at least 0.077 kg/capita/d (0.17 lb/capita/d) BOD, 0.09 kg/capita/d (0.20 lb/capita/d) TSS, and 0.016 kg/capita/d (0.036 lb/capita/d) TKN (total Kjeldahl nitrogen) if nitrification is

Source	Unit Typical	Flowrate, gal/unit/d		Flowrate, L/unit/d	
		Range	Typical	Range	Typical
Airport	Passenger	2.4–3.8	3	9–14	11
Apartment	Person	32–45	38	120–170	145
Automobile service	Vehicle served	6–11	8	23–42	30
Station	Employee	7–11	10	26–42	38
Bar/cocktail lounge	Seat	8–15	11	30–57	43
	Employee	8–12	10	30–45	37
Boarding house	Person	20–45	30	76–170	115
Conference center	Person	5–8	6	20–30	24
Department store	Toilet room	280–450	300	1000–1700	1100
	Employee	6–11	8	23–42	30
Hotel	Guest	52–56	53	200–215	200
	Employee	6–11	8	23–42	30
Industrial building (sanitary waste only)	Employee	12–26	15	45–98	60
Laundry (self-service)	Machine	320–413	338	1210–1560	1280
	Load	36–41	38	136–155	145
Mobile home park	Unit	100–113	105	380–430	400
Motel (with kitchen)	Guest	36–60	38	135–230	145
Motel (without kitchen)	Guest	32–53	34	120–200	130
Office	Employee	6–12	10	23–45	38
Public lavatory	User	2.4–3.8	3	9–14	12
Restaurant:					
Conventional	Customer	6–8	6	23–30	24
With bar/cocktail lounge	Customer	6–9	7	23–34	26
Shopping center	Employee	6–10	8	23–38	30
	Parking space	0.8–2.3	1.5	3–9	6
Theater (Indoor)	Seat	1.6–3	2.3	6–11	9

TABLE 2.5 Typical Wastewater Flowrates from Commercial Sources in the United States (from Metcalf & Eddy/AECOM, 2014, with permission)

required, unless available information justifies other design criteria (Great Lakes–Upper Mississippi River Board of State and Provincial Public Health and Environment Managers, 2014). If garbage grinders are used in the service area, the design domestic loads should be increased to 0.091 kg/capita/d (0.20 lb/capita/d) BOD, 0.11 kg/capita/d (0.25 lb/capita/d) TSS, and 0.021 kg/capita/d (0.046 lb/capita/d) TKN. The quantity of waste discharged by individuals on a dry-weight basis are shown in Table 2.8 (Metcalf & Eddy/AECOM, 2014).

Source	Unit	Flowrate, gal/unit/d		Flowrate, L/unit/d	
		Range	Typical	Range	Typical
Assembly hall	Guest	1.6–3	2.3	6–11	9
Church	Seat	1.6–3	2.3	6–11	9
Hospital	Bed	128–240	150	480–900	570
	Employee	4–11	7.5	15–42	30
Institutions other than hospitals	Bed	60–94	75	230–360	285
	Employee	4–11	7.5	15–42	28
Prison	Inmate	60–110	90	240–430	240
	Employee	4–11	7.5	15–42	28
School, day:					
With cafeteria, gym, and showers	Student	12–23	19	45–90	70
With cafeteria only	Student	8–15	11	30–60	42
School, boarding	Student	32–60	38	120–230	140

TABLE 2.6 Typical Wastewater Flowrates from Institutional Sources in United States (from Metcalf & Eddy/AECOM, 2014, with permission)

Water-saving efforts and water reuse will continue to increase the constituent concentrations in municipal WRRF influent. Depending on the conservation efforts planned, it might be preferable to analyze the influent for specific concentrations, rather than rely on historical data.

4.2.2 Industrial Wastewater
The actual flow and constituent loading records from industries are the best source of data for design purposes. If data do not exist, and an industrial load is or may become significant, specific sampling programs and interviews can be used to establish the effect of present operations and anticipated changes. Standard industrial flow allowances used by some U.S. cities are presented in Gravity Sanitary Sewer Design and Construction (ASCE, 2007). Daily, weekly, holiday, and seasonal variations of industrial releases should be expected, unless information to the contrary exists.

Bench-scale or pilot-plant evaluations may be necessary to develop or ensure the use of appropriate design criteria when industrial wastes are dominant. This is especially the case if the industrial constituents are thought to have an adverse impact on treatment processes, including membranes. If membranes are being used, the industrial waste should be tested on specific membranes before design selection.

4.2.3 Infiltration/Inflow
An allowable infiltration or exfiltration rate for new pipe construction is 9 L/d per meter diameter per meter length (L/d/m/m) (100 gpd/in diameter/mile) (Section 33.94, Great Lakes–Upper Mississippi River Board of State and Provincial Public Health and Environment Managers, 2014). Infiltration values can be 10 times higher in older sewers that have not undergone replacement or rehabilitation but can also be lower for newly constructed systems. In the absence of site-specific information, I/I can be estimated as

Contaminants	Unit	Concentration[a]		
		Low Strength	Medium Strength	High Strength
Solids, total (TS)	mg/L	537	806	1612
Dissolved, total (TDS)	mg/L	374	560	1121
Fixed	mg/L	224	336	672
Volatile	mg/L	159	225	449
Suspended solids, total (TSS)	mg/L	130	195	389
Fixed	mg/L	29	43	86
Volatile	mg/L	101	152	304
Settleable solids	mL/L	8	12	23
Biochemical oxygen demand, 5–d, 20°C (BOD$_5$, 20°C)	mg/L	133	200	400
Total organic carbon (TOC)	mg/L	109	164	328
Chemical oxygen demand (COD)	mg/L	339	508	1016
Nitrogen (total as N)	mg/L	23	35	69
Organic	mg/L	10	14	29
Free ammonia	mg/L	14	20	41
Nitrites	mg/L	0	0	0
Nitrates	mg/L	0	0	0
Phosphorus (total as P)	mg/L	3.7	5.6	11.0
Organic	mg/L	2.1	3.2	6.3
Inorganic	mg/L	1.6	2.4	4.7
Potassium	mg/L	11	16	32
Chlorides[b]	mg/L	39	59	118
Sulfate[b]	mg/L	24	36	72
Oil and grease	mg/L	51	76	153
Volative organic compounds (VOCs)	mg/L	<100	100–400	>400
Total coliform	No./100 mL	10^6–10^8	10^7–10^9	10^7–10^{10}
Fecal coliform	No./100 mL	10^3–10^5	10^4–10^6	10^5–10^8
Cryptosporidum oocysts	No./100 mL	10^{-1}–10^1	10^{-1}–10^2	10^{-1}–10^3
Giardia lamblia cysts	No./100 mL	10^{-1}–10^1	10^{-1}–10^2	10^{-1}–10^3

Note: $mg/L = g/m^3$.

[a] Low strength is based on an approximate wastewater flow rate of 570 L/capita/d (150 gal/capita/d). Medium strength is based on an approximate wastewater flow rate of 380 L/capita/d (100 gal/capita/d). High strength is based on an approximate wastewater flow rate of 190 L/capita/d (50 gal/capita/d).

[b] Values should be increased by amount of constituent present in domestic water supply.

TABLE 2.7   Typical Composition of Untreated Domestic Wastewater (from Metcalf & Eddy/AECOM, 2014, with permission)

Constituent	Value, lb/capita/d			Value, g/capita/d		
	Range	Typical without Ground-Up Kitchen Waste	Typical with Ground-Up Kitchen Waste	Range	Typical without Ground-Up Kitchen Waste	Typical with Ground-Up Kitchen Waste
BOD$_5$	0.11–0.26	0.15	0.20	50–120	70	93
COD	0.30–0.65	0.40	0.50	110–295	180	230
TSS	0.13–0.33	0.15	0.19	60–150	70	87
NH$_3$ as N	0.011–0.026	0.017	0.017	5–12	7.6	7.9
Organic N as N	0.009–0.022	0.012	0.013	4–10	5.4	6.0
TKN[b] as N	0.020–0.040	0.029	0.031	9–18	13	13.9
Organic P as P	0.002–0.007	0.0026	0.0029	0.39–1.8	1.2	1.3
Inorganic P as P	0.001–0.006	0.0020	0.0020	0.50–2.7	0.90	0.90
Total P as P	0.003–0.010	0.0046	0.0048	1.5–4.5	2.1	2.2
Potassium, K	0.009–0.0015	0.013	0.014	4–7	6.0	6.2
Oil and grease	0.022–0.077	0.062	0.070	10–35	28.0	32

TABLE 2.8 Quantity of Waste Discharged by Individuals on a Dry-Weight Basis (from Metcalf & Eddy/AECOM, 2014, with permission)

a function of the area served by the collection system: 0.2 to 28 m³/ha/d (Metcalf & Eddy/AECOM, 2014).

Inflow can be very high in communities with older or combined sewer systems. Although combined-sewer service may represent only a small fraction of the influent service area, inflow derived from the combined sewer service area often will dominate the design and operation of the WRRF. Precipitation-induced inflow may reflect low-buffered, often acidic rainwater and the additional pollutants derived from rooftops, roadways, and land use of the service area. Inflow can be immediate or delayed; immediate inflow refers to rain entering the sewer system directly during or immediately after the rainfall event. Delayed inflow refers to the runoff associated with the melting of an accumulated snow cover.

4.2.4 Stormwater

A designer faces special issues when a facility serves a combined-sewer service area, because oversized combined sewers and interceptors serve as traps for sediment and settleable solids. Increased quantities of influent screenings, grit, and suspended solids received at the WRRF during or following a storm reflect the extent of past accumulations in the sewer and pollutants introduced with the stormwater. This also occurs in older sanitary sewer systems that receive high inflow when a rainstorm occurs after an extended dry period.

In some combined-sewer systems, special consideration is needed for regulators, or overflow structures, which direct combined wastewater flow in excess of sewer or facility capacities to a receiving stream. These discharge locations often are referred to as combined sewer overflows. These systems can result in unintended reverse-flow conditions in cases where the receiving water elevation varies with the tidal pulse or high receiving stream elevations. Malfunctioning tide gates or backflow check gates can allow seawater to enter the collection system during both dry- and wet-weather conditions. The transient or endemic receipt of seawater at the WRRF may dictate special material selections to minimize maintenance and may impose inhibitory stresses on some unit processes (i.e., sodium and sulfide toxicity in anaerobic digestion).

4.2.5   Other Sources

Septage is expected to have higher levels of organics, grease, scum, and grit relative to raw domestic wastewater. Septage characteristics vary widely from one load to the next; data for local septage should be used for design whenever possible. In the absence of available data, septage can be assumed to contain 15 000 mg/L TSS, 10 000 mg/L volatile suspended solids, 7000 mg/L BOD, 700 mg/L TKN, and 250 mg/L total phosphorus, with a pH of 6 (Great Lakes–Upper Mississippi River Board of State and Provincial Public Health and Environment Managers, 2014).

Sewer cleanings are expected to exhibit highly variable characteristics of organically enriched grit. Sewer cleaning also can include high quantities of grease, rags, trash, and other debris. Management and treatment of the grease from sewer or wetwell cleaning need special consideration in the design of the treatment system and its components.

Landfill leachate characteristics can be observed in the form of varying soluble organic compounds and reflect the character and age of the material placed in the landfill and the amount of water that infiltrates the landfill from ground and surface sources. Leachate can contain various concentrations of heavy metals, volatile and semi-volatile organics, and color, nitrogen, phosphorus, and many other industrial chemicals. The BOD and volatile suspended solids also may be quite variable, depending on the age and condition of the landfill. Flows and concentrations vary, depending on rainfall events and integrity of the soil cover in the landfill.

Waste solids from a water treatment facility can be expected to exhibit the characteristics of TSS in the raw water supply before water treatment and any solid (i.e., powdered activated carbon) or solid-forming material (i.e., alum addition and the resultant hydroxide precipitate) added during the course of treatment. The soluble pollutant phase of these waste solids reflects the organics removed from the raw water supply and the time of storage at the WRRF. Aluminum or iron salts added during water treatment may enhance phosphorus removal at the WRRF, although not as much as chemicals specifically added for these removals. Smaller facilities may experience some problems related to flow and solids surging, unless water facility discharges—particularly filter backwashing wastes—are hydraulically equalized.

Typically, groundwater contains scaling compounds from total dissolved solids and is highly buffered. Conversely, surface waters often are slightly mineralized and contain little or no buffer. Softening, demineralization, or both may be practiced with or without accompanying changes in background alkalinity. Simple, raw water coagulation and clarification with aluminum or iron salts will add anions and deplete alkalinity. Soft, unstabilized waters will aggressively solubilize metal from the water system and customer-distribution piping. Copper, for example, may adversely affect biosolids quality and disposal schemes, or iron may adversely affect enhanced biological phosphorus removal. Copper also may affect effluent concentrations or require design for removal, as some NPDES discharge standards for copper are more stringent than drinking water supply requirements. Design of membrane treatment systems will require a thorough evaluation of specific dissolved solids in the water supply for selection of the most effective membrane.

Particular consideration should be given to in-facility recycle streams, such as digester supernatant, decant from sludge holding tanks, process drains from sludge tanks, belt press wash water, filtrate streams, centrate streams, and other recycle streams, especially if biological nutrient removal is required. This is discussed further in Section 5.

4.2.6 Community Water Supply Characteristics

The nonconsumptive portion of water used in a WRRF service area constitutes most of the wastewater routinely received at a facility. This component of wastewater reflects the character of the raw water supply, water treatment processes, and history of beneficial water use and/or reuse. Consideration of the water supply characteristics is necessary for biological nutrient removal treatment and treatment for potential reuse.

The magnitude of the available buffer (alkalinity) is important when designing one or more of the following processes: nitrification, metal salt or lime addition for phosphorus removal, pH adjustment, membrane treatment systems, and closed-system oxidation.

The scaling nature of both the water supply and the wastewater can impair equipment if processes include boilers, steam, cooling, or water seals. Chloride, sulfate, silica, sodium, and other inorganics pass through WRRFs and can affect some effluent-disposal and reuse strategies. High chlorides also influence material selections for elevated-temperature-processing schemes. During anaerobic conditions, high sulfates can result in concrete corrosion, odors, and toxic air.

Many other characteristics in the community water supply must be considered if membrane treatment is selected. These include silica, silicone (from polymers), calcium and manganese, and other parameters, depending on the type of membrane being considered. It is best to verify influent characteristics with the equipment supplier before completing the design.

4.2.7 Variability

In addition to identifying the average flows and loadings, it is important to capture the expected variability. This is typically done with peaking factors, as described in Section 4.4. It is important to note that the flowrates and concentrations will vary hourly in municipal systems, with typical values as shown in Figure 2.3 (Metcalf & Eddy/ AECOM, 2014). Generally, the smaller the system, the more variable the waste flowrates and concentrations, while larger systems may have little diurnal variations.

4.3 Analysis of Flow and Loading Data

Rather than utilize published per capita data and other rules of thumb to estimate wastewater flows and loads, it is more common for a designer to use historical data from the facility to formulate the required design flowrates and loadings. It is important to recognize that there are numerous opportunities for the introduction of sampling and analytical error into the historical data record. To assess the potential for errors associated with facility records, the designer should consider the:

- Location, method, and frequency of sample collection, especially relative to the point of introduction of in-facility recycle streams or septage (if applicable);
- Methods of flow measurement or estimation;
- Instrument installation and calibration frequency;
- Level of accuracy of the laboratory's analytical methods and procedures; and
- The basis for any calculation of any process variables.

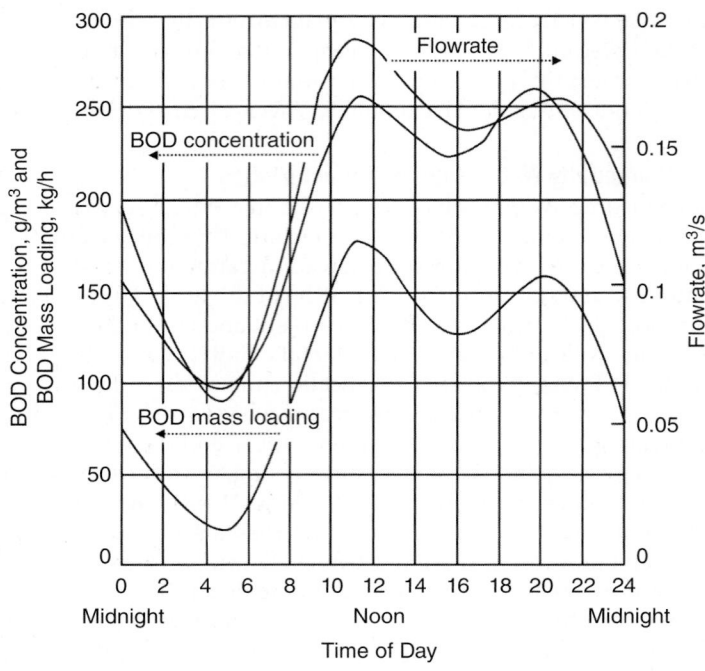

Figure 2.3   Typical hourly variations in flow and strength of domestic wastewater (from Metcalf & Eddy, Inc./AECOM *Wastewater Engineering: Treatment and Resource Recovery*, 5th ed. Copyright © 2013, The McGraw-Hill Companies, New York, NY, with permission).

In addition, specific calculations can be performed to check for data anomalies. Each of these considerations is described briefly below.

4.3.1   Location, Method, and Frequency of Sample Collection

Common locations for sample collection are shown in Table 2.9, along with considerations specific to each location. Samples can be taken as grab, interval, or composite samples, and the suitability of each sample type depends on the specific parameter being analyzed. Refer to *Operation of Nutrient Removal Facilities* (WEF, 2013a) for information about sampling location, method, and frequency.

- A **grab sample** is a single sample taken over as short a time period as possible (usually instantaneously) and as such provides a "snapshot" representation of a parameter at a specific place and time. The sample is often analyzed immediately after collection. Grab samples are best for parameters that are time sensitive.

- **Interval samples** are a series of grab samples taken at various times throughout the day. The samples are taken at a preset time interval, transferred to an individual bottle, and analyzed. Interval samples are best for parameters that change over time.

- **Composite samples** are a series of grab samples that are combined in a single container to provide concentrations over a flow- or time-based quantity. Flow-paced composite samples are taken at a preset flow volume interval. Time-paced composite samples are taken at preset time intervals. Both types of

Waste Stream	Considerations
Liquid processing train	
Raw influent	Difficult to obtain representative sample due to variability in influent.
	It should be confirmed whether samples include in-facility recycles (if any).
	Grit and floatable quantities typically not included in sample.
	For automatic samplers, clogging of sampling hose yields low suspended solids and BOD and overflow of sampling bottle yields high suspended solids.
	Results from grab samples are suspect because around-the-clock diligence is difficult to enforce.
	Validity of solids data can be checked with overall inert solids balance.
	Suspended solids error may be introduced by failure to fully filter or fully dry sample.
	BOD results may be understated with high suspended solids because classic BOD test does not keep this material in suspension.
	Inaccurate measurements of influent BOD may be encountered because of sulfides.
	Industry standard COD:BOD, BOD:TKN, and BOD:total phosphorus ratios should be used to check data and identify outliers.
Primary effluent	Generally provides a good sample, as variability associated with high suspended solids has been eliminated by the unit process.
	Inaccurate measurements of BOD may be encountered because of sulfides and nitrifying recycles.
	Data should be validated with TSS mass balance around primary clarifiers.
Aeration system	Good measure of inert suspended solids.
	Metal salt addition for phosphorus removal will result in a false high measure of volatility caused by the loss of the hydroxide water.
	Dissolved oxygen probes should be maintained and calibrated in accordance with manufacturer's recommendations to produce reliable ratings.
	Dissolved oxygen readings should be taken in a well-mixed area to prevent measurements skewed by dead zones.
	Dissolved oxygen level may not be reported at point of greatest exertion and time of greatest stress.
Return & waste	Waste activated sludge solids concentrations are one of the most critical parameters for facility operations.
	Grab samples are generally used but these reflect the hydraulic conditions at the time of sludge and so are not representative of daily average values.

TABLE 2.9   Considerations for Interpretation of Sampling Results

Waste Stream	Considerations
Secondary effluent	Unless cBOD tests are performed, BOD determinations are subject to false results caused by nitrification in the BOD bottle (likely if NH_4—N > 1.0 mg/L with NO_3—N > 3.0 mg/L, BOD_5 > suspended solids if suspended solids > 10 mg/L; this suggests nitrification in BOD bottle, spring and fall changes yielding transient false positives, or non-nitrifying facility design partially nitrifying in earlier years or in summer months).
	Grab-sampling programs may miss elevated effluent suspended solids and diurnal stresses.
	Samples for BOD analysis with residual chlorine need to dechlorinated before analysis. Samples with low microorganism concentrations may need to be seeded.
Solids-processing train	
General	Grab samples are typically in use.
	Sampling results often are higher than the real average because startup and shutdown conditions are not quantified, and sampler may be interested in best rather than representative results (especially with product solids).
	Discontinuous operations likely to have more high-side bias than continuous operations.
	When the overflow solids-storage tanks are not filled, samples of gravity thickener overflow may not be representative.
	Scales, if calibrated, represent excellent point of mass measurement.
Stabilization process	Excellent point of inert solids measurement and long-term equilibrium with applied suspended solids characteristics where inert solids have to be equal to applied inert solids.
	Supernatant suspended solids recycles often are underestimated.

TABLE 2.9 Considerations for Interpretation of Sampling Results (*Continued*)

composite samples collect the sample and transfer it into a common container where the constituents are combined to generate an averaged sample for the sampling period. Time-composite samples are best for parameters that are consistent over time and uniform (i.e., secondary clarifier/tertiary filter effluent). Flow-composite samples are best for characterizing parameters that are highly variable over the day (i.e., raw wastewater influent and periods of facility upset) and more adequately characterize loads.

Consideration should be given to the location where in-facility recycle streams enter the main process stream. Measurements from facilities with flow equalization also should be verified to determine if corresponding flow and sample measurements are upstream and/or downstream of flow equalization.

4.3.2 Methods of Flow Measurement or Estimation

Measurement error can be introduced in Venturi or magnetic flow meters that have changes in pipe size or abrupt changes in pipe direction near the device. The manufacturer should be consulted on the minimum upstream and downstream lengths of straight-run pipes for these units. Flow measurements can be considered accurate if the installed flow meter is within 10% of a calibrated secondary flow meter. Meters used for pacing chemical feed systems should have accuracy discrepancies that are less than 10%.

4.3.3   Instrument Installation and Calibration Frequency

Instrument calibration and location affect the accuracy of data. Flow-measuring devices and electronic sensors must be cleaned and calibrated on a regular basis. Maintenance requirements vary among instruments and should be coordinated with specific manufacturers' instructions. The physical location of flow-measuring and sampling devices needs to be evaluated while reviewing facility operating data. Sampling locations should be in well-mixed areas to provide a representative sample. Sampling lines should be routinely maintained and cleaned. Failure to clean sampling lines may yield erroneous results.

4.3.4   Level of Accuracy of the Laboratory's Analytical Methods and Procedures

Laboratory methods should be reviewed to establish that analytical procedures adhere to approved, best-practices protocols. Where required, samples must be preserved and/or homogenized to provide accurate results. *Standard Methods for the Examination of Water and Wastewater* (APHA et al., 2012) provides guidance on quality assurance and quality control procedures and sections on method development and evaluation, expression of results, and sample preservation. Data validity should include statistical screening of outliers, validation that the correct parameters are being tested, and verification that results are input with the proper date in operating records (date of sample collection, not the date of sample analysis).

Laboratory procedures and reporting practices should be verified. It should be confirmed that ammonia-nitrogen is not being measured in lieu of total TKN, for example, or that nitrate concentrations are reported as NO_3^-—N and phosphate is reported as PO_4^{3-}—P. It should also be verified whether conventional BOD or carbonaceous BOD (cBOD) tests are being performed. The cBOD test is the same as the test for conventional BOD, except that a nitrification inhibitor is added to prevent ammonia oxidation in the BOD bottle. The inhibited cBOD test, compared with the conventional (uninhibited) BOD test, frequently results in a 20% to 40% decrease in oxygen consumption in raw wastewater and primary effluents (Albertson, 1995), although there is disagreement as to the cause of the difference (Albertson et al., 2007; Young et al., 1995, 2005). Inhibited cBOD should not be used as the basis for process design, because it could result in underestimated waste strength and resulting undersized facilities. The cBOD test is preferred for secondary and higher quality effluents (APHA et al., 2012).

4.3.5   Basis for Process Variables

Application of process control data without understanding the basis for the data can lead to erroneous conclusions. This is especially true for solids retention time (SRT), which is recorded by most nitrifying and BNR facilities. The designer should determine whether the SRT calculated by the facility includes the solids in the suspended-growth tanks plus clarifiers, only the solids in the suspended-growth tanks, or only the solids in the aerobic portions of the suspended-growth tanks, and whether the SRT is an aerobic value or an overall system value (equal to mean cell residence time), especially in nitrifying systems.

The designer should preferably determine historic SRT from actual facility data (mixed liquor suspended solids (MLSS), effluent TSS, and waste activated sludge) rather than accepting facility reported values and should be very specific in reporting the basis of the design value.

4.3.6 Methods to Check Validity of Data

The methods to validate facility data include:

- Use of constituent ratios for raw wastewater and primary effluent;
- Determination of net yield from facility operating data; and
- Mass balances around unit processes.

Typically, constituent ratios for raw domestic wastewaters fall within the following ranges (EnviroSim Associates Ltd., 2016):

- BOD:TSS = 0.82 to 1.43;
- Chemical oxygen demand (COD):BOD = 1.80 to 2.20;
- Soluble BOD:BOD = 0.20 to 0.40;
- BOD:TKN = 4.2 to 7.1; and
- BOD:total phosphorus = 20 to 50.

Values not falling within typical ranges may indicate a sampling or analysis error or a significant nondomestic constituent. Calculation of the facility net yield also can provide a method to check the validity of facility waste activated sludge data or aerobic SRT.

Mass balances for conservative elements or pollutants allows the designer to gain a rapid understanding of the significant elements, in terms of recycles, solids captures, and validity of the sampling and measurement program. Inert solids (or total phosphorus) measurements and balances quickly can provide an overall assessment of the validity of a facility's monitoring program and determine whether the performance of a solids destruction process is correctly defined. Percent accountable mass-balance (out divided by in) closures of 100% plus or minus 10% are considered excellent; closures less than 80% and greater than 120% reveal suspect results from one or more processing points. Process modeling can be used for mass balance development, as described in Section 5.3 and in Chapter 4.

4.4 Developing Flowrate and Loading Peaking Factors

Sound design practice anticipates the range of conditions that the facility or process can reasonably be expected to encounter during the design period. The range of conditions for a facility varies from a reasonably certain minimum in the first year of operation to the maximum anticipated in the last year of the design period in a service area with anticipated growth. The reverse applies for a service area with an anticipated decline in wastewater production.

Typically, the flowrate and loading peaking factors for a given facility design project are extracted from that facility's historical record using any number of suitable methods. For example, the maximum daily flow might be calculated as the 98th percentile value of all daily flow data, or it might be taken as the absolute maximum daily flow value in a flow record that has been statistically scrubbed of its outliers.

In an effort to elucidate the differences associated with different peaking factor calculation methods, as well as to provide guidance for cases in which no historical data exist, flow and loading data from 60 U.S. WRRFs were analyzed to determine peaking factor relationships as a function of annual average flowrates. Actual average flowrates—not facility design flowrates—were used. Facilities from 20 states were represented in the analysis—34 from the northeast, 15 from the southeast, and 11 from western states, with flows ranging from 1 to 470 ML/d (0.26 to 124 mgd). Approximately two-thirds of the facilities had average daily flows less than 40 ML/d (10 mgd). Facilities serving combined collection systems were considered independently from facilities serving separate systems. This information was collated specifically for this manual of practice and has not been published elsewhere.

Frequency distributions were performed using facility flow and load data for a 3-year period for most facilities (ranged from 1 to 4 years). Parameters considered included flow, BOD, TSS, ammonia-nitrogen, TKN, and total phosphorus loads. Mass loads were calculated from flow and 24-hour composite concentrations. Data sets that were sampled less than 20% of the time (less than approximately 1.5 times per week, as a result of either infrequent sampling or occasional lapses in regular sample collection and measurement) were omitted from the analysis. Mass and flow peaking factors were determined for maximum daily, maximum monthly, and minimum daily conditions, as the ratio of each condition to the average condition.

To determine maximum day and maximum month values for flow and loadings (BOD, TSS, TKN), five methods were used and compared:

- Method 1: Daily data were filtered to exclude outliers, defined as values less than the 5th percentile or greater than the 95th percentile. This corresponds to 2 standard deviations, assuming a normal distribution of data. The maximum value of the remaining data set was used as the maximum day condition. For the maximum month peaking factor, a 30-day rolling average was calculated, and, although rare, values that included data gaps that were greater than 2 weeks were excluded. The maximum 30-day rolling average value of this conditioned data set was used as the maximum month condition.

- Method 2: Similar to Method 1, daily data were filtered to exclude outliers, but these were defined as values less than the 0.3rd percentile or greater than the 99.7th percentile. This corresponds to 3 standard deviations, assuming a normal distribution of data. Maximum day and maximum month conditions were then defined as in Method 1.

- Method 3: All daily data were included. The maximum day condition was set equal to the 95th percentile value of all daily data, while the maximum month condition was set equal to the 95th percentile value of the 30-day rolling average values.

- Method 4: All daily data were included. The maximum day condition was set equal to the 98th percentile value of all daily data, while the maximum month condition was set equal to the 98th percentile value of the 30-day rolling average values.

- Method 5: All daily data were included. The maximum day condition was set equal to the maximum value of all daily data, while the maximum month condition was set equal to the maximum value of the 30-day rolling average values.

Average conditions were taken to be the mean of the data sets. Minimum day conditions were taken to be the 5th percentile of all daily values.

Table 2.10 presents a summary of the analysis with average peaking factors (combined systems) or equations for trends as a function of average flow (separate systems). For separate systems, the general trend is a diminishing deviation from average with increasing flow (maximum peaking factors decrease, minimum peaking factors increase), while the peaking factors for combined systems were independent of flow. These trends are illustrated for separate systems flow in Figure 2.4 and for BOD in Figure 2.5, with similar trends being observed for other mass loadings. Note that only Method 3 maximum month and maximum day values are shown in Figures 2.4 and 2.5 for simplicity.

Table 2.10 can be used to set design values for maximum and minimum conditions in the absence of reliable data specific to a facility's service area, or to compare peaking factor calculations using different methods. Note that the peaking factor for a given parameter calculated using one method can be very different than the peaking factor calculated for that parameter using a different method, especially for maximum day peaking factors. For example, the average of all maximum day TSS separate system peaking factors was 3.86 when calculated with Method 5 vs. 1.64 when calculated with Method 1 (81% relative percent difference). If available, 3 full years of facility data should be used to determine peaking factors for the specific facility under consideration. Section 5 presents how the peak conditions defined in this section are applied to WRRF unit process design.

As presented above, the practice in the United States is to use BOD for determining oxygen-demanding organic load strengths in design, facility operations, and discharge permit compliance. U.S. designers and operators should be encouraged to use COD for determining oxygen-demanding mass (Ekama et al., 1984; Park et al., 1997; Mara and Horan, 2003; Melcer et al., 2004; Metcalf & Eddy/AECOM, 2014). At a minimum, routine COD sampling of raw wastewater and primary effluent should be conducted. It is possible for BOD testing to be completely eliminated, as 40 CFR 133.104 (U.S. EPA, 2016e) allows COD to be used for discharge permit compliance, if a long-term BOD:COD correlation—which would be site and sample specific—can be demonstrated.

4.5   Definition of Facility Capacity

The discussion in Section 4 considers development of design flows and loads to establish the basis around which a WRRF is designed. Sometimes it is important to consider the converse, namely, the flows and loadings an existing WRRF can treat, or its *capacity*.

The hydraulic capacity of a WRRF typically is presented in terms of its annual average daily design flow. In some instances, the capacity is presented in terms of its peak hourly design flow, maximum month flow, or other flow condition, but this should be so noted. The average daily design flow is the average flow over the 12-month/the year, when the facility is at design limits. This value is determined as the total flow for the year divided by 365 days, or more commonly as the average of the average monthly flows for 12 consecutive

	Flow		BOD		TSS		TKN	NH$_3$–N	Phosphorus
Separate collection systems									
Number of valid data sets	40		33		33		6	15	6
	Equation (Q in ML/d)[a]	Average Value[b]	Equation (Q in ML/d)[a]	Average Value[b]	Equation (Q in ML/d)[a]	Average Value[b]	Average Value[b]	Average Value[b]	Average Value[b]
Minimum daily	0.023 ln(Q) + 0.67	0.73	0.027 ln(Q) + 0.517	0.60	0.031 ln(Q) + 0.460	0.55	0.74	0.70	0.73
Maximum monthly									
Method 1[c]	−0.030 ln(Q) + 1.38	1.30	−0.046 ln(Q) + 1.44	1.30	−0.045 ln(Q) + 1.44	1.30	1.15	1.21	1.12
Method 2[c]	−0.047 ln(Q) + 1.52	1.43	−0.053 ln(Q) + 1.59	1.44	−0.050 ln(Q) + 1.66	1.51	1.25	1.29	1.20
Method 3[c]	−0.029 ln(Q) + 1.40	1.32	−0.044 ln(Q) + 1.45	1.32	−0.053 ln(Q) + 1.54	1.38	1.17	1.23	1.13
Method 4[c]	−0.053 ln(Q) + 1.50	1.40	−0.068 ln(Q) + 1.61	1.41	−0.063 ln(Q) + 1.67	1.48	1.29	1.29	1.21
Method 5[c]	−0.051 ln(Q) + 1.58	1.45	−0.041 ln(Q) + 1.57	1.45	−0.034 ln(Q) + 1.64	1.53	1.36	1.30	1.27
Maximum daily									
Method 1[c]	−0.024 ln(Q) + 1.43	1.40	−0.046 ln(Q) + 1.66	1.52	−0.051 ln(Q) + 1.79	1.64	1.27	1.32	1.25
Method 2[c]	−0.027 ln(Q) + 2.00	1.98	−0.033 ln(Q) + 2.19	2.09	0.0133 ln(Q) + 2.68	2.72	1.69	1.79	1.90
Method 3[c]	−0.023 ln(Q) + 1.47	1.40	−0.047 ln(Q) + 1.67	1.53	−0.085 ln(Q) + 1.93	1.68	1.27	1.32	1.26
Method 4[c]	−0.033 ln(Q) + 1.62	1.57	−0.056 ln(Q) + 1.94	1.77	−0.078 ln(Q) + 2.30	2.06	1.40	1.46	1.45
Method 5[c]	−0.031 ln(Q) + 3.42	2.60	0.058 ln(Q) + 2.69	2.86	0.415 ln(Q) + 2.62	3.86	3.57	2.09	3.65

TABLE 2.10 Design Peaking Factor Summary

Combined collection systems

	Flow	BOD	TSS	TKN	NH₃–N	Phosphorus
Number of valid data sets	20	19	18	2	11	3
	Average Value[d]	Average Value[d]	Average Value[d]	Average Value[d]	Average Value[d]	Average Value[d]
Minimum daily	0.68	0.60	0.53	0.67	0.66	0.73
Maximum monthly						
Method 1[c]	1.32	1.26	1.31	1.24	1.21	1.17
Method 2[c]	1.58	1.46	1.69	1.33	1.29	1.24
Method 3[c]	1.39	1.31	1.43	1.34	1.22	1.17
Method 4[c]	1.53	1.40	1.61	1.40	1.27	1.20
Method 5[c]	1.64	1.51	1.81	1.48	1.31	1.28
Maximum daily						
Method 1[c]	1.62	1.60	1.87	1.40	1.39	1.35
Method 2[c]	2.72	2.57	3.98	2.32	1.93	1.95
Method 3[c]	1.62	1.62	1.88	1.40	1.35	1.36
Method 4[c]	2.01	1.95	2.50	1.79	1.51	1.54
Method 5[c]	3.58	3.62	6.22	3.07	2.87	2.20

[a] Equation from plot of flow vs. peaking factor, as illustrated in Figures 2.4 and 2.5. Equations were not developed for TKN, ammonia-nitrogen or phosphorus due to relatively small number of valid data sets with those parameters, or for data sets from combined systems due to the weaker relations hips between peaking factors and flow.

[b] Average from all valid data sets for separate systems.

[c] Methods for calculating peaking factors are:
Method 1: Exclude daily values > 2 standard deviations from mean. Maximum value of remaining daily data = maximum day value. Maximum value of 30-day rolling average = maximum monthly.
Method 2: Exclude daily values > 3 standard deviations from mean. Maximum value of remaining daily data = maximum day value. Maximum value of 30-day rolling average = maximum monthly.
Method 3: Include all daily values. Maximum day value = 95th percentile of all daily values. Maximum month value = 95th percentile of 30-day rolling average of daily values.
Method 4: Include all daily values. Maximum day value = 98th percentile of all daily values. Maximum month value = 98th percentile of 30-day rolling a verage of daily values.
Method 5: Include all daily values. Maximum day value = maximum of daily values. Maximum month value = maximum value of 30-day rolling average values.

[d] Average from all valid data sets for combined systems.

TABLE 2.10 Design Peaking Factor Summary (*Continued*)

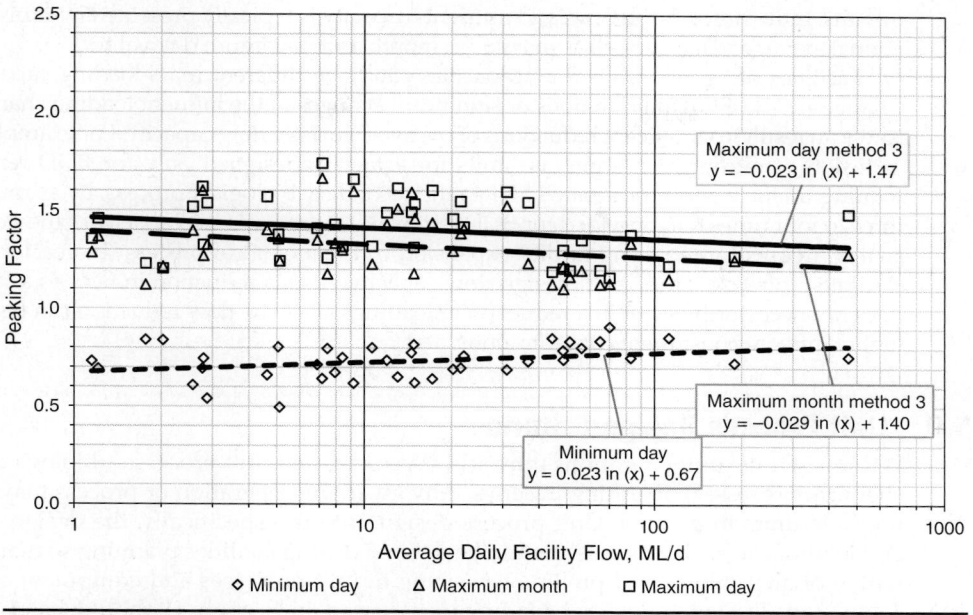

FIGURE 2.4 Flow peaking factors for separate systems.

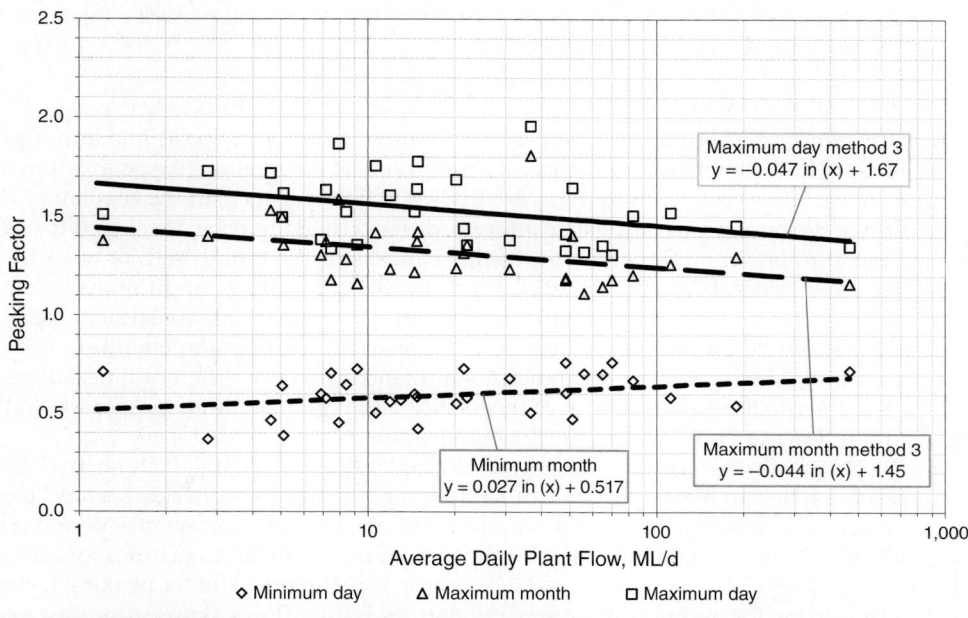

FIGURE 2.5 Biochemical oxygen demand peaking factors for separate systems.

months. Influent solids, organic, and nutrients capacity is typically presented in terms of an annual average value, as well as maximum month and maximum day values.

Facilities of equal design flow capacities can have different mass-loading capacities. Changes in the effluent limitations or significant changes in the influent loading characteristics can result in significant reductions of the average daily flow capacity. For example, the inclusion of nitrogen or ammonia limits for a facility designed only for BOD removal requires an increase in the aerated SRT in the secondary biological process, by as much as three to four times that provided for BOD removal. This reduces the capacity of the existing facility, frequently requiring facility expansion of the biological process (bioreactors and clarifiers) to recover the initial design flow capacity. Also, a reduction in flow to a facility does not necessarily result in a reduction in loadings since the flow reduction can result in higher influent constituent concentrations.

5.0   Unit Process Design Criteria

WRRFs involve unit operations (physical processes) and unit processes (biological and chemical processes) operating in series. Any given unit operation or process may have multiple units in parallel. Unit process design criteria—specifically, the design flows and loadings to each process—should be defined during facilities planning so that evaluation of alternatives and preliminary sizing of unit processes and equipment can be completed. Design flows and loadings (discussed in Section 4) should be used in conjunction with the design standards (discussed in Section 2.3) and design tools such as mass balances and process modeling to develop unit process criteria. General unit process sizing considerations are presented below. Mass balances and process modeling are also described briefly, as are two other important unit process design considerations: redundancy and turndown.

5.1   Unit Process Sizing

Unit process design should be based on required performance at maximum process loading conditions for compliance, at a minimum, at the interval (e.g., maximum month or maximum week) included in a facility's NPDES permit. With the regulatory definition of the minimum compliance interval, design criteria may be developed to establish the controlling condition(s) for the facility's most stressed month and/or week (or more stringent intervals, if appropriate). Typically, the controlling condition reflects one or more of the following constraints: maximum flow, maximum constituent mass load, most stringent effluent limitation, or most restrictive processing condition (e.g., temperature). These constraints can occur simultaneously or at different times, depending on the particular circumstances of the facility, although maximum mass loads and maximum flows are not concurrent in separate systems.

Based on the considerations concerning influent peaking factors, two influent peaks—a design hydraulic peak and a design process peak—provide a sound basis for design. The design process peak should match the facility's compliance period interval for its effluent limitations (typically, the maximum month and maximum week) under the most restrictive conditions for the control of pollutants. Higher peaking factors are appropriate for aeration requirements and hydraulic flows (maximum day load for aeration and peak hour flow for hydraulics).

Because many liquid-processing facilities are sized, in part, through hydraulic criteria that apply to some maximum condition, strategic management of flow peaks in the facility, in the collection system, or in both, offers the benefits of processing stability and cost savings. Facility design that couples peak flow mitigation opportunities found in the collection system with those of the treatment works design can proactively control and reactively respond to facility influent.

Table 2.11 provides general guidance for the range of conditions that a unit process design should accommodate. Loading variability should be considered in terms of facility influent conditions, attenuation of these conditions through wet-weather management, and individual unit process conditions with differing variability characteristics due to load modification by preceding unit process and recycle loads. The loading to any particular unit operation or process is dependent on the raw wastewater characteristics, performance of all preceding operations and processes, and characteristics of in-facility recycled sidestreams, which are a function of solids handling and disposal. Designers should consider the effects of varying process performance of upstream WRRFs and varying influent loads when sizing specific unit processes.

5.2 Mass Balances

Mass balances prepared during planning and conceptual design yield guidance concerning design quantities and major differences between treatment alternatives. More detailed mass balances prepared during preliminary design serve as the initial reference project documents to ensure a commonality of project understanding, consistent use of major design criteria and loadings, and a standard frame of reference and logic for the project team.

One of the most important applications of a mass balance is to define the design criteria for solids handling processes. Variability in solids production needs to be considered because it has a major impact on process performance. Most variability in solids production (net yield) from facility to facility is a result of differences in influent characteristics (inert suspended solids or the fraction of unbiodegradable particulate organic matter), chemical use (phosphorus removal), and biological yield. The success of the entire facility and all of its unit processes depends on the ability to remove solids from the facility. The mass balance also provides the basis for the control logic of the process and instrumentation diagrams.

Although process modeling, discussed in Chapter 4, is the preferred means for generating steady-state mass balances, this section details the steps to follow for mass balance preparation.

1. *Identify parameters to be included in balance*—The mass balance should include each of the major pollutants that has an NPDES limit or is significant for process control. At a minimum, flow, BOD (or COD), and TSS should be included and, if applicable to the facility, nitrogen and phosphorus. It also may be desirable to balance the inert solids, especially if process modeling will be used for developing the mass balance. Balances should be performed for design average, maximum month and maximum day conditions, although designers should remember that peak mass loads and peak flows rarely occur concurrently and that balances conducted with peak flows and masses will result in overly conservative design assumptions with limited physical significance.

2. *Establish the mass balance boundaries*—Although many facilities have limited sidestream facility data, it is still preferable to perform facility-wide mass balances that include all mainstream and sidestream processes. A whole-facility mass balance typically provides the best representation of the overall recycle

Element	Guidelines
Hydraulics	Pipelines and channels must be carefully sized to avoid low-velocity conditions and associated solids deposition. Typically encountered problem areas are conduits common to both existing facility and planned expansion configurations, headworks, and primary clarifiers. As troublesome deposits may be encountered even with careful design, means should be provided to facilitate cleaning. Pipelines and channels must also be sized large enough to maintain reasonable velocities and headloss at peak hour flows. Chapter 5 discusses guidelines for calculating hydraulic profiles.
Biological reactor sizing	Design for the controlling maximum month of the design life with the maximum SRT.
Biological reactor oxygen (air)	Design for the maximum daily demands in the project life and confirm equipment can turndown to meet minimum demands. Sensitivity to mixing supply and mixing needs is required, and mixing often is limiting.
Biological nutrient removal systems	Check system performance for the minimum month (weakest) wastewater quality characteristics with the maximum and minimum recycle effects and provide appropriate safeguards (backup addition of supplemental carbon source and/or metal salt) to ensure compliance with effluent standards. The success of enhanced biological phosphorus removal is influenced strongly by the solids-processing train. Resolubilization of a fraction of the removed phosphorus is likely in solids storage and under solids stabilization.
Influent alkalinity	Design for minimum monthly or weekly value (typically high rainfall month) with maximum recycle.
Seasonal nitrification temperature	Design for average temperature of the minimum month or the monthly load/temperature conditions that result in the largest mass required under aeration. Remember that nitrification systems are SRT-limited, not nitrogen-load-limited.
Unit equipment sizing (particularly pumps)	Design equipment to meet both maximum and minimum processing needs where appropriate; for example, size pumps to handle peak hour or peak instantaneous flows depending on downstream processes and design pumping station wet wells and pumps for reasonable operation during the minimum month to week condition, and design minimum return solids capacity (and, if appropriate, waste solids capacity) from minimum monthly flows (and loads) in design life with reasonably anticipated minimum MLSS concentrations and settled solids concentrations.
Waste solids	Design for minimum SRT conditions under at least the maximum monthly loading (may change with higher or lower storage) of the design life, with anticipated average or lower solids concentration under anticipated operating schedule. Provide appropriate storage with minimum SRT and anticipated average or lower solids concentration.
Solids handling unit processes	Design for the maximum day or maximum week conditions, taking into consideration the expected operational schedule for the solids equipment (for example, 24/7 vs. 5 days per week.

TABLE 2.11   Design Guidelines for the Range of Conditions

characteristics, including COD, BOD, TSS, nitrogen, and phosphorus. Available facility data, such as solids capture and cake solids, can be used for mass balance development.

3. *Prepare individual process balances*—To prepare a balance around an individual process, use appropriate process performance data to calculate the incoming flow or mass, the outgoing flow or mass, any parameter removal or destruction (e.g., solids destruction in anaerobic digestion, ammonia removal in nitrification, solids capture in gravity thickening), and any parameter generation (e.g., inorganic solids from chemical addition, nitrate removal from denitrification) to provide a balance across the process for each parameter under consideration. Check that the rate of accumulation for a given parameter, if any, is equal to the inflow minus the outflow plus generation minus destruction or capture. Parameters and suggested values required for these balances for liquid treatment and solids handling processes can be obtained in unit-process-specific chapters in this manual. However, specific values for such parameters should be generated from historical facility data whenever possible. The use of historical data is discussed in detail in Section 4.3.

4. *Couple mass balances from individual processes*—Using the process flow diagram, couple individual process mass balances by using the balanced effluent from upstream processes used as the influent to the next downstream process. Peaking factors for sludge flows and loads, and thus resultant sidestreams, should account for the actual expected hours of operation. For example, solids-related balances for facilities processing solids on a 5-day-per-week basis must be appropriately adjusted.

5. *Iterate*—Because the outflows from one unit process will impact the mass balance of any downstream unit process, it is necessary to perform iterative calculations until all recycle and conversion conditions balance (achieve equilibrium) within reasonable limits. Metcalf & Eddy (2014, pp. 1626–1636) provides an example of a detailed, iterative hand-calculated mass balance (flow, BOD, and TSS). As mentioned above, because of the iterative nature of mass balances, they are best prepared using calibrated process models (WEF, 2013b).

5.3   Process Modeling

Process modeling is the preferred means for developing unit process sizing and unit operation mass balances. Process modeling can be used in concert with peaking factor analyses to assess whether the process design will perform as needed under the range of expected conditions. Simulation programs, such as BioWin (EnviroSim Associates Ltd., Flamborough, Ontario, Canada) or GPS-X (Hydromantis Inc., Hamilton, Ontario, Canada), are available for process model development. These programs enable designers to consider, evaluate, and refine biological process configurations. Any process model should incorporate the whole WRRF to the extent possible, including sidestream treatment operations, so that the interconnection between liquid treatment processes (including both physical/chemical and biological units) and solids handling are captured in the simulations. Modeling the whole facility will capture the impact of processes that unavoidably solubilize pollutants and create special design and operating issues that vary with technologies and operating strategies used at the facility. The designer can use process modeling to evaluate different processes, especially when the facility's design objective includes the control of reintroduced pollutants. Chapter 4 includes details on process modeling.

5.4 Redundancy Requirements

Redundancy affects the overall reliability of the WRRF and cost. WRRFs should have redundant machinery and equipment and process tankage to allow maintenance work or unplanned equipment failure. Designers should determine carefully the number of treatment units to be provided for each unit process. Multiple units are required for all critical components of the WRRF. Using multiple pieces of equipment of the same size and model and by the same manufacturer facilitates maintenance and reduces spare parts inventory. Typical practice for pumps and mechanical equipment, including solids processing and odor control equipment, is to provide one standby unit in addition to those required to handle peak design flow or load. Standby units are not required for process tankage, such as clarifiers, aeration tanks, fixed-film reactors, or disinfection basins. However, hydraulic design should allow peak hourly flows, including associated sidestream flows, to be passed through the facility with the largest or longest flow path of each process unit removed from service. Redundant solids treatment trains should be considered when large quantities of solids—compared with the average daily production volume (i.e., with a digestion process)—are handled to allow for anticipated failures and differing feed-solids characteristics in addition to maintenance. This would prevent solids handling from becoming a limiting condition or bottleneck to the entire system. Applicable state and other local criteria or guidelines should be consulted and applied.

5.5 Turndown Requirements

Turndown is a big factor in facilities that have significant disparities between startup and design year flows and loadings. Typical redundancy of process units, as defined in Section 5.4, generally provides the capability for efficient operation during the initial periods of reduced loading, but minimum design conditions must be checked. The use of variable delivery capability within the range of predicted operation for pumps, chemical feed systems, blowers, and other equipment should be considered for facilities with large disparities to help accommodate a wide range of loading conditions. Examples of addressing this issue include designing four bioreactors instead of three and installing two smaller and two larger blowers instead of three large units.

6.0 Project Execution

Sections 2 through 5 discussed planning considerations for a WRRF project, including the development of design criteria. This section describes how the planning phase ties into the overall implementation of the project. It also presents potential procurement mechanisms for implementing final design and construction and defines deliverables produced for the construction of a new or modified WRRF. Determining the project procurement method is an important milestone since it impacts the execution of the project, including the project schedule.

6.1 Sequence

6.1.1 Chronological Design Evolution
Design practices for municipal WRRFs in the United States evolved from the rulemaking and guidance documents developed by U.S. EPA during the Construction Grants Program. Design practice is not static and has changed significantly since the end of the U.S. EPA grants program into a multidimensional process with distinct design and project delivery models, which localities have regulated to enhance project performance and optimize costs. Design practice has continued to evolve to consider holistic enhanced performance as presented in the Institute for Sustainable Infrastructure's Envision Certification.

Various organizations use different terminology and definitions for design practices; however, traditional design typically consists of three phases, comprising five sequential activities that are completed in series—facilities planning (process criteria, concept, and schematic design), design development (preliminary design), and final design, culminating in the preparation of contract documents for bidding and construction. Components and products of the first four design activities (process criteria through preliminary design) are summarized in Table 2.12. Alternative project delivery models such as design-build alter the typical design sequence and allow some elements of the design to be scheduled in parallel.

Facilities planning involves defining the problem and condition of existing facilities, developing and evaluating alternatives, performing an environmental assessment of alternatives, and selecting a recommended plan. Table 2.13 presents an outline for a typical facilities planning report. The typical outline is for a comprehensive, new wastewater project, including siting and effluent discharge analyses. The outline can be modified for specific projects, eliminating items not in the project or adding items not included in the outline.

General design criteria should be firmly established during the early stages of the facilities planning design process. These criteria are discussed in Sections 2 through 5 and include the planning period for the facilities (typically 20 years), flows and loads, discharge requirements, datum planes, tide levels and flood protection, standby electrical power for essential facilities, equipment and tankage redundancy requirements, methods and timeframes for unit isolation and dewatering, and general means of flow distribution and interprocess conveyance. Alternatives analyses and LCC evaluations, as described in Sections 8 and 9, should also be completed during facilities planning to determine the recommended plan.

The design development phase includes equipment and unit process sizing, layout, and configuration, structural and architectural design of facilities, environmental controls for facilities, electrical equipment sizing and instrumentation, and control process diagram development. Design development also can include special investigations such as pilot studies, as discussed in Section 10.

Final design includes contractual document preparation for the bidding and construction of the WRRF. Contractual documents generally consist of the bid proposal, general conditions of the contract, construction contract, project specifications, drawings, and addenda, but they will differ depending on the project

Design Element	Needed	Tasks Performed	Information Produced
Design Phase: Products:	**Process Criteria** **a. Schematic flow diagram** **b. Critical and desired spatial relationships between process units**		
Process	Populations and loadings Industrial waste survey NPDES permit	Preliminary design criteria plugged into unit sizes and compared to availability of area on-site Tentative process selection Client conference	Preliminary design criteria Process flow diagram Recommended alternatives
Design Phase: Products:	**Concept** **a. Approximate number, type, and size of process units** **b. General arrangement of process units and buildings on-site**		
Process	Flood elevations Hydraulic conditions Recommended alternatives	Refine unit process concepts Prepare cost studies for equipment selection Establish programs for nonprocess areas Establish facility staffing and nonprocess activities (i.e., maintenance) Define civil issues	Initial site layout Hydraulic profile Flood protection, wetland mitigation, and related issues
Architectural	Schematic flow diagram Approximate number, type, and size of process units Critical and desired relationships between process unit Client attitude and desires, appearance of project, community relations, Site: land and patterns; condition of site and adjacent area; topography, soils, and climate; boundaries and construction in area; status, terrain, and alternates	Site visit and photos Meet with client Site analysis Study function zoning alternatives of site and building Select functional zoning plan for building and site	Function zoning of site and buildings Facility elevation and rendering or perspective
Structural	Soil data Site and alternates	Evaluation of arrangement of units from structural and soil considerations Recommendations in layout for structural and construction improvements Preliminary boring program	Definition of foundation conditions
Instrumentation	Flow process Client and engineer attitude on control	Establish control and monitoring concept	Control concept

TABLE 2.12   Summary of Facility Design Procedures

Design Element	Needed	Tasks Performed	Information Produced
Design Phase:	**Schematic**		
Product:	**a. Layout of site and building(s)**		
Process	Facility hydraulic profile Establish equipment relationships Confirm planning of nonprocess areas	Layout relationship as free-bodies turn layouts over to architect for alternate studies (both site and buildings)	Layouts of hydraulic structures, rough grading, and major yard piping Preliminary layout fulfilling criteria established in "Information Needed" Hazardous areas designated Preliminary power requirements for unit processes
Architectural	Approximate staff size Type of maintenance program Type of laboratory program Type of administrative program Refinement of type, size, and relationships of process units Functional zoning plan of site and buildings Preliminary estimate of room requirements from electrical, HVAC, and instrumentation	Review functional zoning plan with client Study alternate schematic layouts and review with engineer Consult all functional groups	Layout of site and building floor plans to scale Building sections, floor and roof levels Preliminary structural grid
Structural	Ongoing schematic studies	Consultation a. Structural patterns b. Uplift c. Foundation systems	Assist in layout of site and buildings General structural recommendations
HVAC	Approximate volume of building zones	Initial HVAC room size based on similar projects and review with architect Establish system concept Review of building design	Preliminary major requirements
Electrical	Approximate size (design capacity) and location of facility	Preliminary discussions with electric utility concerning availability of a. Availability of power b. Availability of two independent power sources c. Methods of operation d. Rules and regulations Prepare estimates for architecture on space requirements	Architectural space requirements for a. Electric room b. Generation unit c. Outside substation

TABLE 2.12   Summary of Facility Design Procedures (*Continued*)

Design Element	Needed	Tasks Performed	Information Produced
Instrumentation	Description of proposed design a. Types of process equipment b. Chemicals to be used Work outside of this contract (such as pumping stations and metering stations)	Layout of flow schematic Scope of controls Written descriptions of operation	Process flow schematic with instrumentation Preliminary description of operation Computer criteria and location of control rooms
Design Phase:	**Preliminary**		
Product:	**a. Drawings of site and buildings with columns, doors, windows, walls, process equipment, electrical equipment, HVAC equipment, and control equipment**		
Process	Critical review by all design team members Refined layouts with special emphasis on problem areas Approved schematic	Confirm size of major equipment Equipment and pipe layouts Yard, process and drainage piping	Dimensioned layouts of equipment pipes and drainage Outline specifications for equipment
Architectural	Preliminary design criteria from structural HVAC, plumbing, electrical, and instrumentation Shop storage and staffing requirements Update information on process units	Design building using structural, HVAC, plumbing, electrical, Instrumentation, laboratory, shop/storage, and staffing information Select materials	Outline specifications
Structural	Layout of all buildings and outside tanks Location and estimated weights	Approximate beam, column, wall sizes, slab, thickness, loads down to foundation Establish structural grid	Drawings indicating column grid lines and rough slab, beam column, and wall structures, for all major units on the site
HVAC	Names of rooms Names of process equipment Cubic volumes of spaces Special equipment requiring steam and/or ventilation Areas requiring air conditioning Areas requiring dehumidification Odor control requirements	Calculations for supply and exhaust quantities Calculations for ventilation required to dissipate Calculations for dehumidifier Calculations for other HVAC equipment HVAC layouts coordinated with other functions Calculations for odor control system	Approximate sizes and location of HVAC Equipment Type of heat source Single line duct layouts Size and location of dehumidifiers Size and location of other Equipment

TABLE 2.12 Summary of Facility Design Procedures (*Continued*)

Design Element	Needed	Tasks Performed	Information Produced
Plumbing	Plumbing fixture locations Domestic water supply Storm and sanitary terminal locations Structural provisions for process equipment drainage a. Sump pits b. Troughs c. Floor drains Fire protection requirements based on code-building classification Preliminary design layout of laboratory and other special areas, such as kitchens or ornamental pools All information pertaining to local gas utility company and fuel for any service	Discussions pertaining to basic information requirements noted in column 1 Consultation with structural and architectural and sketches as required	Column 1 items to be answered Schematics and layouts as required
Electrical	Approximate size and location of: a. Process units and center equipment b. Instrumentation equipment c. Buildings d. HVAC equipment Instrumentation equipment Buildings	Estimate the size and type of electrical equipment needed Location and arrangement of electrical equipment	Dimensioned layout of electrical equipment and motor control
Instrumentation	Updated process control instrumentation Changes in process flow schematic Chemical requirements Computer decision Facility staffing Description of operation required	Preliminary design of process control panels Review flow schematic with process design Preliminary sizing of chemical feed facilities Coordination for architectural, structural, and electrical Panels (number of control rooms)	Process control panels cabinet size Process flow schematic with instrumentation and data logging computer Preliminary instrument tabulation Design schematics and layouts as

TABLE 2.12   Summary of Facility Design Procedures (*Continued*)

Section Number	Title	Contents
1.0	Introduction	Purpose of study and statement of need Scope Facilities plan organization
2.0	Project Background	Facility history Previous studies Related projects
3.0	Existing Facilities	Wastewater collection system and pumping stations Description of existing treatment system Evaluation of existing operations and equipment Condition and remaining life
4.0	Site Descriptions and Basic Planning Criteria	Site descriptions General design criteria Guidelines for cost evaluation
5.0	Wastewater Flows and Loads	Service population Current flows and loads Future flow estimates Conventional pollutant loads (TSS, BOD, and nutrients) Nonconventional pollutant loads Mass balance
6.0	WRRF Performance	Discharge permit requirements Design parameters for effluent quality and variability
7.0	Development and Screening of Alternatives	Identification of facility deficiencies and treatment needs Development of alternatives Screening of alternatives Recommendations for detailed evaluations
8.0	Detailed Evaluation of Wastewater Alternatives	Description of alternatives Unit processes Hydraulic design criteria Evaluation of alternatives
9.0	Residuals Handling and Treatment	Residuals quantities and quality Description of alternatives Unit processes Evaluation of alternatives
10.0	Residuals Disposal	Disposal options Siting analysis Evaluation of options
11.0	Effluent Discharge	Summary of existing conditions Development of discharge alternatives Model results/comparison to water quality criteria Evaluation of alternatives Monitoring recommendations

TABLE 2.13 Outline of Water Resource Recovery Facilities Planning, Executive Summary

Section Number	Title	Contents
12.0	Air Emission Quality Screening and Impacts	Emission estimates and source parameters Air quality modeling Odor and volatile organic compound control evaluation Control technology assessment Emission control strategy costs
13.0	Selection of Recommended Plan	Summary of alternatives and evaluations Environmental issues Public input Selection of site Selection of treatment processes Selection of residuals handling and treatment processes Selection of residuals disposal Selection of effluent discharge point
14.0	Description of Recommended Plan	Descriptions of recommended plan and components Summary of design criteria Site layout Hydraulic profile Instrumentation and controls Estimated construction and operation costs Environmental impacts Implementation schedule Permitting requirements
15.0	Financial Impacts Analysis	Current and future expenses Grant and State Revolving Fund contributions Summary of future annual costs Bonding and analyses and recommendations Estimated impacts to user charges Conclusions and recommendations
16.0	Public Participation	Citizens advisory committee Public meetings/hearings Responsiveness summary

TABLE 2.13   Outline of Water Resource Recovery Facilities Planning, Executive Summary (*Continued*)

delivery method selected. Alternative delivery methods and contract documents are discussed in more detail in Section 6.2. All decisions relating to design sizing, materials of construction, and equipment selection types should be finalized before proceeding to the final design.

6.1.2 Value Engineering

Value engineering is the application of the scientific method to study the values of systems. There are many approaches and methodologies to the practice of value engineering. A thorough description of the practice is best left to certified value specialists and cost engineers. Certified value engineering specialists are educated and trained by an organization called SAVE International (Dayton, Ohio). This international organization promotes the advancement of the value methodology, which includes value engineering, value analysis, and value management. The SAVE Value Standard provides guidance on the practice of value engineering. More information may be found on the SAVE International (2007) website (http://www.value-eng.org/pdf_docs/monographs/vmstd.pdf).

The main objective of value engineering is to minimize construction and LCCs without sacrificing the functionality of the project. Value engineering should be considered at the beginning of the design process and, in some cases, may be dictated at certain construction cost levels by local regulations. In the United States, projects receiving state grant or State Revolving Fund Program assistance may require value engineering if construction costs of the project are anticipated to exceed US $10 million.

6.1.3 Implementation Times

Even with timely project progress and approval, elapsed time from the onset of facilities planning through the completion of the facility's first year of operation is typically 6 to 8 years. Representative time requirements for project implementation are presented in Table 2.14. These should be considered minimum time requirements for a new facility or major facility expansion project having no significant facility siting,

Activity	Duration, Months
Facilities planning	8 to 12
Regulatory approval	2 to 3
Preliminary design	5 to 6
Value engineering	1 to 2
Final design	7 to 10
Regulatory approval	2 to 3
Bidding	2 to 3
Contract award	1 to 2
Construction	30 to 38
Start-up	2 to 5
First-year operations	12
Total	72 to 96

TABLE 2.14   Representative Implementation Times for Water Resource Recovery Project

environmental, or permitting considerations. Durations indicated in Table 2.14 assume the following:

- The project receives a Finding of No Significant Impact following submittal of the facilities plan and environmental assessment (failure to receive such a finding and the resultant requirement for environmental impact studies can add many months and sometimes several years to the project implementation schedule);
- The same engineer is retained for planning, design, and construction management, eliminating procurement times required for engineer selection;
- Engineering contracts for design and construction management are negotiated during the regulatory review periods at the end of facilities planning and design;
- Required permits are acquired within the design and bidding time frame; and
- There are no lengthy bid protests or rebidding activities.

Large and/or complex projects and projects undertaken in urban areas can require significantly longer time frames for implementation than those indicated in Table 2.14. In addition, projects with new outfalls or discharge points can extend indicated times by years. Projects utilizing alternative delivery methods such as design-build can have a shorter total duration since some activities are completed concurrently, but the duration of individual activities are similar.

6.2 Procurement

6.2.1 Project Delivery Alternatives

The arrangement of contractual obligations between a facility owner and consultants, contractors, and equipment vendors is referred to as the project delivery model. A number of alternative approaches for project delivery exist, and many new or innovative models have been developed since the early 1990s. Some of the more commonly used delivery models include design-bid-build, design-build, engineer-procure-construct, and construction management at risk. Delivery methods other than the traditional design-bid-build approach were developed to reduce the overall construction schedule and provide the benefits of an integrated design and construction team. The Design-Build Institute of America (Washington, D.C.) (DBIA) and the Construction Management Association of America (McLean, Virginia) (CMAA) are professional organizations dedicated to the advancement of alternative project delivery, including various design-build and construction management models. The DBIA's Manual of Practice (2009) can be referred to for additional information on design-build procurement, and the CMAA's *Capstone: The History of Construction Management Practice and Procedures* (CMAA, 2003) may be consulted for more information. The local/state regulations on procurement should be researched before selecting an appropriate alternative delivery method, as some states limit when an alternative project delivery method may be used.

6.2.2 Contract Documents

Contract documents form the legal description of work to be performed and the basis of performance for design and construction of new facilities. In most cases, contract documents consist of contract provisions (i.e., general terms and conditions), specifications, and drawings.

6.2.2.1 Contract Provisions   Contract provisions define the legal relationship between the owner and the contractor. The provisions indicate the cost of the work to be performed, schedule for performance, and other terms and conditions that direct or control the work to be performed. Some of the more common contract provisions include the following:

- Basis of award of contract. Description of how the determination of award will be calculated. The basis of award of the contract needs to be carefully defined and not subject to interpretation in the bidding documents. For example, if points are awarded to an equipment supplier during a request for qualifications phase in a procurement, it should be clearly stated whether the points awarded carry over through the request for proposal stage in an evaluated bid. Typically, award of contracts is based on the construction cost. If all items covered cannot be evaluated equally on a construction-cost basis, an evaluated LCC is most equitable. Evaluated life-cycle bids should realistically reflect the cost of all consumable products (including replacement parts). Bonus or penalty factors may be used to emphasize features that are important to the owner. The evaluation methodology should be discussed with vendors before preparing final specifications to ensure the methodology fairness.

- Payment provisions. These provisions establish the method of invoicing and payment for the work performed.

- Allocation of risks provisions. These provisions delineate who is responsible for specific risks associated with the project and allocates the risks and liabilities to the parties of the contract.

- Changed conditions. Provisions that establish the methodology for investigation, resolution, and compensation when the contractor encounters site conditions that are different from those represented in the drawings and specifications.

- Force majeure. When work on the project is delayed beyond the control of the owner or contractor, a force majeure condition occurs (such events include natural disasters, strikes, military actions, etc.). These provisions establish the rights of both the owner and the contractor in the event of such an action.

- Schedule. The owner has the right to expect a project to be completed within the time frame established in the contract. Some provisions include incentives or penalties, respectively, for meeting or exceeding the scheduled completion date. Incentives are offered to expedite the project so that it is completed early or on schedule. Liquidated damages provide compensation to the owner if the project is not completed on schedule. Delay provisions protect both the owner and the contractor in the event that circumstances beyond the contractor's control result in schedule slippage.

- Substantial completion. These provisions establish the basis for which the project (or portions of the project) is deemed to be complete and the owner takes control of the project. Work may not be completely finished, but the project is functional for its intended use. Typically, the substantial completion date is linked to the incentive/penalty clauses.

Other critical aspects of contract provisions protect the owner against contract violation of guarantee and completion time stipulations. These provisions may be for liquidated damages, consequential damages, or both types. In general, the total should not exceed the amount of the contract.

Liquidated, or delay, damages are executed for failure to complete the project during the time period specified in the contract. These damages may include any identifiable loss to the owner because of delays. Liquidated damages may cover considerations such as owner inspection costs, delay expenses from other dependent contracts, additional electric power or chemical costs experienced while awaiting operation of the new process, and penalties for failure to meet regulatory standards or permit constraints.

Consequential, or performance, damages are losses resulting from the failure of the process to work as specified. They should accurately reflect the cost of operating with or correcting the process deficiency. Performance damages should only be triggered as a last resort after all other remedies to correct the deficiencies have been exhausted. Performance damages may include the cost of installing a compliant system, the cost of added technology to bring the original system into compliance, and additional costs for power and chemical consumption. Although performance damages may be large, they assure the owner that only responsible vendors will be inclined to bid for the job. Contractors should have explicit contract language defining and limiting consequential damages.

The Engineers Joint Contract Documents Committee issues Standard General Conditions of the Construction Contract (2007), which is widely accepted for use during construction projects in the United States.

6.2.2.2 Specifications   Two general types of specifications are used to procure goods and services for construction of a project under any procurement option—the prescriptive and the performance specification. A prescriptive specification sets forth explicit criteria governing the processes, or services that are to be provided. A prescriptive specification's explicit nature complicates its preparation, but offers the owner maximum assurance of protecting the quality of the installation and easing bid comparison. In addition, it typically delineates acceptable manufacturers and suppliers and provides for the owner's consideration of "or equal" products.

A performance specification defines the input conditions and the desired objective. Vendors favor this type of procurement because it allows them greater latitude in the use of their products. This latitude can diminish the owner's control of the quality of the installation. However, these risks may be mitigated by prequalification. The prequalification procedure entails providing an opportunity, before the bidding process, for equipment manufacturers to submit qualifying information as a basis for the engineer's determination of whether the product conforms to the specifications and can thus be considered for bidding by construction contractors. In some cases, procurement of the equipment and contracting for its installation constitute separate stages, with equipment procurement preceding completion of final detailed engineering.

6.2.2.3 Drawings

6.2.2.3.1 Traditional Computer-Aided Design   Traditionally computer-aided design (CAD) drawings are developed throughout the design process to depict and represent the project in two dimensions. The design drawings are developed independently and are integrated manually between each design discipline to create a complete design.

6.2.2.3.2 Three-Dimensional Models   The use of three-dimensional models reduces the potential for conflicts between structural elements, mechanical equipment, and piping, as each layer is developed in real space, so that the designer, owner, and contractor have electronic documents for coordinating the various elements of the project. Bidding and construction drawings are developed from sections and plan views through the model.

6.2.2.3.3 Virtual Design and Construction   The use of integrated approaches like virtual design and construction (VDC) and three-dimensional modeling is becoming more standard in the industry. VDC is an overarching approach to fully describe the project, which includes visualization models and other task integration. These models provide links to bills of materials as well as construction cost estimates, and they also export to other design software, including full allocation life-cycle assessments (LCAs), which allows for evaluation of process and building performance and operation costs to be estimated during the design of a project. Drawings are integrated to other information technology to allow for more efficient use of the visualization models to predict the performance of the process, building, or structure. A project execution plan should be established at the beginning of a project to determine the desired information and uses for the models.

7.0 Cost Estimating

7.1 Introduction

Estimating construction costs is an essential component during the planning and design of a WRRF. During the development of estimated project costs, it is important to understand the documents that will be used to develop the estimate as well as how the estimate will be used, for what purpose, and by whom. This section presents different types of cost estimates, the resources available to the designers developing cost estimates, and the application of markups. It also discusses the cost-estimating process as it relates to alternative delivery projects.

7.2 Types of Estimates

The Association for the Advancement of Cost Engineering (Morgantown, West Virginia) (AACE) has developed definitions to help categorize cost-estimate types (AACE, 1997). Their Cost Estimate Classification System considers the most significant characteristics of a cost estimate, including degree of project definition, end usage of the estimate, estimating methodology, expected accuracy range, and the effort and time needed to prepare the estimate. Based on this characterization, the AACE has established five cost-estimate classes, which are defined in Table 2.15. A Class 5 estimate is based on the lowest level of project definition, and a Class 1 estimate is closest to full project definition.

End usage classification of a cost estimate can also be used to define estimate classification. However, the terminology used to define the different classifications (i.e., order-of-magnitude, budget, or definitive) tends to depend on the identity of the stakeholder (i.e., owner agency, designer, or contractor) and on the intended use for

| Estimate Class | Primary Characteristic | | Secondary Characteristic | | |
	Level of Project Definition Expressed as % of Complete Definition	End Usage Typical Purpose of Estimate	Methodology Typical Estimating Method	Expected Accuracy Range Typical +/−Range Relative to Best Index of 1[a]	Preparation Effort Typical Degree of Effort Relative to Least Cost Index of 1[b]
Class 5	0% to 2%	Screening or feasibility	Stochastic or judgment	4 to 20	1
Class 4	1% to 15%	Concept study or feasibility	Primarily stochastic	3 to 12	2 to 4
Class 3	10% to 40%	Budget, authorization, or control	Mixed, but primarily stochastic	2 to 6	3 to 10
Class 2	30% to 70%	Control or bid/tender	Primarily deterministic	1 to 3	5 to 20
Class 1	50% to 100%	Check estimate or bid/tender	Deterministic	1	10 to 100

[a] If the range index value of "1" represents +10/−5%, then an index value of 10 represents +100/−50%.

[b] If the cost index value of "1" represents 0.005% of project costs, then an index value of 100 represents 0.5%.

Table 2.15 AACE Cost Estimate Classification Matrix (AACE Recommended Practice 18R-97, 2016; reprinted with the permission of AACE International, 1265 Suncrest Towne Centre Drive, Morgantown, WV 26505 USA. Phone: 304-296-8444. Fax: 304-291-5728. Internet: http://web.aacei.org. E-mail: info@aacei.org. Copyright © 2016 by AACE International; all rights reserved)

the estimate (i.e., capital planning or project funding, engineer's estimate, and bid development).

7.3   Direct and Indirect Costs

Construction cost estimates are composed entirely of direct capital costs. These costs include:

- Land and site development costs;
- Costs of utilities, transportation, and other services to the site;
- Relocation costs;
- Materials, equipment, and labor costs, including those needed for temporary facilities and systems required to maintain existing functionality of the facility; and
- All construction costs, including contractor's overhead, profit, mobilization, permits, bonds and insurance, and construction contingencies.

Indirect capital costs consist of engineering, permitting, and legal services during construction and any other associated costs. Additional project contingencies, beyond the construction contingency listed above as a direct capital cost, should be added depending on potential for significant changes in scope or schedule, bidding climate, risk tolerance, or other larger project factors.

The sum of direct capital costs and indirect capital costs equals the total capital cost for a project. It is important to summarize direct capital costs and indirect capital costs separately and include contingencies for each category. Combining direct and indirect costs and contingencies could leave an unclear impression of the true individual estimates of construction and engineering costs.

7.4   Levels of Accuracy

The accuracy range of an estimate will depend on several factors related to the quality and depth of the input and the actual estimating process used. Other factors, such as state of the technology being considered and the quality of reference cost-estimating data, also play an important role in defining accuracy range. Generally, estimate accuracy correlates with estimate classification and thus with the level of project definition, though this can be significantly affected by extended periods between the time of the development of the estimate and the expected schedule of construction. Table 2.15 includes expected accuracy ranges as a function of the estimate class as defined by the AACE (AACE, 1997).

7.5   Quantification

The level of quantification needed for a specific estimate varies with the completeness of the design effort. For example, a Class 5 or 4 estimate, which relies on a conceptual and/or schematic design, will primarily utilize volumetric data, such as facility capacity, pipeline flowrates, and tank volume requirements. A Class 1 estimate performed near the completion of contract documents will require detailed quantification of materials for all facilities, with vendor quotes and unit cost development that are specific to the area where the project is being constructed.

The construction industry typically refers to material quantities in ways that standardize the quantification process. As the estimate evolves from project definition through the contract document preparation phases of design, the quantities will change. The quantities of construction materials used in a construction cost estimate should be organized in an orderly manner. Standard forms make it easy to organize quantities on the forms by facility and then by specification section. This organization will allow quantities to be checked by project team members not directly engaged in the preparation of the cost estimate (i.e., design team).

7.6 Cost Resources

Pricing input to construction cost estimates is derived from various sources, including cost data books and standards or one of a number of computer programs. It is important to know how the costs for a particular guide or program are structured and how they should be applied to the estimate being produced. Utilizing pricing input data specific to the geographic area where construction will take place will account for market-specific considerations such as labor and supply cost differences.

Other cost resources include the following:

- Written and telephone quotations from manufacturers and vendors;
- Quotes from contractors and subcontractors;
- Estimates from similar completed projects and bid tabulations;
- Cost curves;
- Cost/capacity ratio formulas;
- Construction indexes; and
- Minimum wage rates.

7.7 Application of Markups and Contingencies

The format of an engineer's construction cost estimate is similar to that of a general contractor's estimate for bidding. After the direct costs of material, labor, equipment, and subcontractors are subtotaled (including allowances to cover items that can be identified but not quantified at that particular level of design development), markups in the form of overhead, profit, mobilization, bonds and insurance, and construction contingencies are applied to arrive at the total bid price. Typical values of these markups are as follows:

- Overhead: 5% to 10%;
- Profit: 5% to 10%;
- Mobilization: 3% to 10%;
- Bonds and insurance: 1.5% to 2%;
- Escalation for schedule (the time period between development of the estimate and construction): based on a local inflation rate compounded by the duration; and
- Construction Contingency: 0 to 30%.

Markups are applied to the subtotal in a compounding manner. For example, the overhead markup is applied to the direct construction cost subtotal to obtain a subtotal that includes overhead. The profit markup is then applied to the subtotal that includes overhead to obtain the subtotal with overhead and profit. The compounding continues, so that construction contingency is applied to the subtotal that includes overhead, profit, mobilization, and bond and insurance. Indirect costs such as engineering and implementation and project contingency are then added to the total bid price to calculate the total project cost.

Contingency is a true capital cost that accounts for future design definition and development as well as events that experience has shown will likely occur, including small changes in scope, variation in bidding climate, and other factors. The quantification of risks associated with these factors should be considered when developing the contingency. A contingency is different from an allowance: it varies with the level of an estimate. The greater the engineering detail provided, the lower the contingency. All estimates must have a contingency, including estimates prepared near or at completion of contract document preparation.

7.8   Estimate Documentation

Complete support documentation provides defensible construction cost estimates to facility owners while helping to minimize the liability risk to the designer. Table 2.16 displays recommended minimum levels of documentation to be provided for various levels of estimates.

It is recommended that the following information also be included in the estimate for delivery to an owner:

1. *Purpose of estimate*—The facility owner, location of the project, type of facility or project, and classification and/or level of estimate should be listed.

2. *Scope of work*—Information about capacities, quantities, plans and specifications used, equipment lists or data sheets used, duration and sequencing of unique sections of the work, and site restrictions or constraints should be listed.

3. *Assumptions*—If scope data are incomplete or unavailable at the time the estimate is prepared, a list of assumptions explaining the basis of the estimate must be documented. These assumptions could include structural concrete thickness, proximity of imported fill to the site, presence of groundwater, need for sheeting, and scope that is excluded from costing (e.g., the expectation that no buried hazardous materials or contamination will be encountered.)

4. *Cost resources*—All costs used in an estimate should be referenced to the appropriate resource. Costs loaded from an estimating database should include cost resource numbers from the database. All costs not provided from a database should include support documentation. Documentation consists of vendor, supplier, or subcontractor quotes, unit cost development calculations, and productivity assumptions. Costs provided from vendors, suppliers, or subcontractors should include the name of the vendor, person supplying the information, and the date the cost was supplied. Documentation of prevailing labor rates for the region where the project will be constructed should also be included.

5. *Markups*—Indication of which markups were used for overhead, profit, mobilization, bond and insurance, escalation, and contingency should be included. Reasons why particular markups were used for each estimate should be documented. These markups may vary with the estimator, type of estimate, and market conditions at the time and location of the project.

Documentation Required	Classes 5 and 4	Class 3	Classes 1 and 2
Cost curves used to support costs for facilities, general site data, or processes	X		
Information from similar projects used as a basis for estimating the cost	X		
Engineering News Record Construction Cost Index (ENR CCI) data used to index cost curves and project costs to current values	X		
Sketches or drawings used to quantify specific facilities or processes	X	X	
Outline specification data used as a basis for cost, productivity, or scope assumptions		X	
Equipment cut sheets used as the basis for vendor equipment quotes		X	X
85% to 95% complete drawings used to quantify specific facilities, processes, or site features			X
85% to 95% specifications used as a basis for cost, productivity, or scope assumptions			X
Well-organized lists of material quantities associated with each facility that can be quantified (quantities organized by specification section and facility for easy reference between the estimate and support documentation)	X	X	X
Vendor or subcontractor quotes for equipment or services quoted for the estimate	X	X	X
ENR CCI value assigned to the estimate	X	X	X
Cost development information (crew cost and productivity assumptions)	X	X	X
Other cost development information (references to applicable cost guides used to support individual unit costs, applicable labor rates, etc.)	X	X	X
Computer spreadsheets or estimating software output that summarize the detailed cost and summary information for the estimate	X	X	X

TABLE 2.16 Recommended Levels of Cost Estimate Documentation (as cited in an earlier version of AACE Recommended Practice 18R-97, 2016; reprinted with the permission of AACE International, 1265 Suncrest Towne Centre Drive, Morgantown, WV 26505 USA. Phone: 304-296-8444. Fax: 304-291-5728. Internet: http://web.aacei.org. E-mail: info@aacei.org. Copyright © 2016 by AACE International; all rights reserved)

6. *Schedule*—The construction schedule and key delivery milestones on which the estimate is based should be described. Certain construction activities, such as general conditions and cost escalation, depend on the schedule.

7. *Allowances*—Allowances cover known scope activities that cannot be quantified or are too small to spend the time to quantify. Indicate allowances where appropriate and explain why they are being used at the level indicated in the estimate.

8. *Estimate accuracy*—Each estimate should be classified as using a system such as the AACE system referenced in this section and by describing its end usage (i.e., conceptual design, design development phase, contract document preparation).

9. *Constructability*—Any discrepancies or constructability issues noted during preparation of the estimate should be listed. How these problems were addressed in the estimate should be indicated.

10. *Qualifying language*—Each estimate should include a statement of qualification. This statement is intended to limit liability for providing construction cost estimates.

11. *Construction cost index*—Each estimate should be referenced to a construction cost index, such as that established by *Engineering News-Record* (New York, New York) (ENR) (http://www.enr.com).

7.9 Alternative Delivery Cost Development

Alternative approaches for project delivery are becoming more common for the design and construction of WRRFs, as noted in Section 6.2.1. Alternative project delivery methods can require special considerations in the development of the cost estimate. The principles discussed above will apply regardless of the delivery method; however, their application may vary compared to traditional design-bid-build delivery. Experience and judgment must be used when applying the principles of cost estimating to suit the specific conditions of a project.

The design-build model illustrates some of the special cost-estimating considerations associated with alternative delivery. Elements of design development may occur differently with a single, integrated design and construction team. For example, procurement of long lead items may occur much earlier in the project. This will allow definitive costs for this equipment to be known earlier, though connection of services to the equipment may still require design development. Depending on the scale of the equipment pricing, early knowledge of accurate pricing may result in reduction of construction contingency or elimination of escalation of costs due to schedule.

8.0 Guidelines for Life-Cycle Cost Evaluations

Private businesses and public utilities have the fiduciary duty to make prudent financial decisions regarding capital improvements and overall asset management. When evaluating the economics of capital projects, utilities are faced with analyzing many economic variables to make decisions. Economic evaluations are fairly simple when the only variable used to make decisions is capital cost. However, in many situations the cost of operating and maintaining an asset is far greater than the capital cost, and thus, it is a generally accepted industry practice to analyze the costs of an asset over its estimated lifetime. LCC evaluations are much more robust than evaluations based only on capital cost, and they allow owners to make comprehensive financial decisions consistent with advanced asset management principles.

LCC evaluations are useful in situations when there is a need to compare the economics of alternatives with capital costs, O&M costs, and varying expected life of an asset. LCC is used most commonly for evaluating the relative cost-related difference of alternatives, which is typically most useful in facilities planning and preliminary design stages. For wastewater utilities, recurring costs typically are limited to O&M costs. LCC is valuable because it allows the inclusion of O&M cost components, such as power, fuel, labor, chemicals, and asset repair or replacement. LCC evaluations also quantify the consumption of natural resources associated with the operation of the asset, such as power, natural gas, and chemicals. Thus, environmental effects can be evaluated more accurately in conjunction with the economic analysis; these multiple-criteria decision analyses (MCDAs) are discussed in more detail in Section 9 of this chapter.

8.1   Procedures for Present Value

This section presents guidelines for cost-effectiveness comparisons of WRRF project alternatives and general procedures for determining LCC. These guidelines and procedures are based on the cost-effectiveness guidelines presented in Appendix A to 40 CFR Part 35, Subpart E (U.S. EPA, 2016g).

LCC comparisons for WRRF projects can be made using present-value or equivalent uniform annual value methodologies. The present-value method is a powerful tool because future expenditures are transformed into equivalent costs in today's dollars. Present-value costs provide a method to evaluate capital costs and the annual O&M costs of alternatives on an equivalent basis. All future capital and operating costs during the planning period are converted to an equivalent value during the base year. The present worth of an alternative is the amount of money in today's dollars that must be available during the base year, with a discount rate, to pay all anticipated capital and operating costs associated with the alternative through the end of the planning period. Alternatives with the lowest present worth are the most cost-effective over the life of the project. Common parameters used in all cost-effectiveness comparisons include design lives, discount rate, equipment and structure life expectancies, and the base year for analysis. Theories of present-worth analysis, equations used to calculate present-worth costs, and tables providing discount-rate factors for the equations can be found in any engineering economics text.

A present-worth analysis should be conducted using a base year that represents the time of the present-worth analysis or the initial year of operation of the facilities under construction. Capital costs should be referenced to an ENR construction cost index. Current costs typically are used for the analysis. Inflation rates are applied for detailed financial analyses undertaken to identify the financial impacts of a selected alternative, but including inflation rates is not necessary for comparing alternatives during facilities planning.

8.2   Discount Rate

When a future amount of money is converted to its equivalent present value, the magnitude of the present amount is always less than the magnitude of the cash flow from which it was calculated. This is because for any interest rate greater than zero, all future value of the money is less than it is at present. For this reason, present-worth calculations are often referred to as discounted cash-flow methods. Regulations for the United States State Revolving Fund Program (40 CFR 35.2030[b][3], U.S. EPA, 2016g) require that cost-effectiveness analyses performed in facilities planning are based on the discount rate established by the federal government. A discount rate for each federal fiscal year is established for water resources projects in the Federal

Register. Owners also can choose to perform the cost-effectiveness analyses using a discount rate that reflects the owner's long-term, actual cost of money adjusted for inflation.

8.3   Salvage Value

Salvage value is the expected market value at the end of the useful life of the asset, which may be positive, zero, or negative. The salvage value may be negative if there are costs associated with decommissioning the asset, or the decommissioning costs are greater than the market value at the end of the useful life of the asset. Current depreciation methods approved for tax purposes typically assume a salvage value of zero, even though the actual salvage value may be positive. Note that this may force payment of extra income taxes when an asset is sold with a net realized value greater than the current book value (Blank and Tarquin, 2015).

8.4   Life Expectancies

Assets with expected useful lives shorter than the planning period have a cost associated with the replacement of the item during the planning period. Items with expected useful lives greater than the planning period have a salvage value at the end of the planning period. This can be accounted for by creating a cash flow of the costs and credits of specific items over a given time period and then performing a net present value analysis. Typically life expectancies are taken to be 15 to 20 years for all equipment and 50 years for buildings, structures, and buried pipelines. Metal structures are given a shorter life expectancy than concrete structures. Certain jurisdictions may require specified expected useful lives in LCC analyses.

8.5   Capital Costs

Capital costs typically include land acquisition costs; estimated capital construction costs, including equipment, pipelines, buildings, and structures; engineering and project administration costs; and contingencies. Equipment costs should be separately identified from other costs because of the different life expectancies used in the present-worth analysis. Methods for developing capital costs are discussed in Section 7.

8.6   Annual Operation and Maintenance Costs

Annual O&M costs are those costs paid each year to keep the facilities in a good operating condition and to preserve the useful lives of structures and equipment. Annual O&M costs include wages, salaries, and benefits; maintenance repair and replacement; energy consumption; and chemicals.

Typically, O&M costs used in analyses are for the initial year of operation. These costs can be increased over the planning period if substantial increases are expected. O&M costs typically are based on costs at the time of the analysis, without future inflation. An average cost for wages, salaries, and benefits should be obtained from the owner and applied to the expected number of facility staff. Maintenance costs, which include lubrication oils, replacement parts, and other maintenance items, are often estimated to be 1% of the equipment capital costs. More detailed analysis of maintenance costs may be warranted if such information is available. Estimated maintenance costs should be increased beyond the roughly estimated 1% of the capital cost for any equipment expected to have abnormally high maintenance requirements.

Electrical costs should be based on the average power draw and the total cost for electricity (consumption plus demand). Chemical costs should be based on estimated average use and current unit costs. Solids disposal costs are often a significant portion of the O&M costs and should be carefully established.

8.7   Interest during Construction

U.S. EPA's cost-effectiveness guidelines present the following equation for estimating interest during construction:

$$I = 0.5PCi \qquad (2.1)$$

where I = interest during construction (%),
   P = construction period (years),
   C = total project cost, and
   i = discount rate (%).

8.8   Land Costs

U.S. EPA's cost-effectiveness guidelines require that land values be escalated 3% annually to the end of the planning period. Land values also should be salvaged at the end of the planning period. The present worth for land acquisition is essentially the cost of interest through the planning period. The present worth of land costs is usually negligible in the final cost-effectiveness comparison and can be neglected. Costs should not be included if every alternative is using the same land (common factors excluded). By contrast, land cost should be included for alternatives associated with treatment by land application where the owner purchases the land. In cases where land is to be leased from another owner instead of purchased, the lease payments are included in the O&M costs or as a separate recurring annual cost of the alternative.

8.9   Sunk Costs

Cost-effectiveness analyses can include only future expenditures. Any costs already incurred for an alternative must be considered as a sunk cost and excluded from the analysis. Alternatives benefiting from sunk costs will have reduced future costs—capital, O&M, or both.

9.0   Evaluations Utilizing Multiple-Criteria Decision Analysis

Selection of alternatives for implementation at municipal WRRFs relies on the consideration of economic and noneconomic criteria. However, it is often difficult to consider appropriately the factors that are not readily quantifiable (i.e., robustness, flexibility, and operability) relative to the much easier-to-define capital and O&M costs. Further, the overall objective in a municipal WRRF project is made up of many specific objectives that often conflict with one another. One example of this is construction costs vs. environmental and social impacts: minimizing the latter could result in significant cost increases for the project.

MCDA is a methodological tool that facilitates decision making by including different kinds of criteria (i.e., economic, environmental, technical, and social) when comparing alternatives. To apply MCDA for municipal WRRF projects, it is necessary to define the alternatives to be considered, objectives that the final project must meet, criteria that will be used to measure the degree of satisfaction of the objectives by the different stakeholders, and some measure of the relative significance of the different criteria. This formal approach helps to structure the evaluation process, enables balanced consideration of various "what-if" scenarios, yields consistent and defensible decision making, and results in a consensus-based selection of solutions that meet the owner's multiple objectives.

The Triple Bottom Line (TBL) analysis is another example of a "scorecard" method for evaluating WRRF project performance in a broad (economic, social, and environmental) context to maximize project value. While TBL centers on reporting of economic, social, and environmental gains and losses, MCDA provides a framework to organize information about the trade-offs and study trade-offs. The details of MCDA are provided below; additional information on TBL can be found in Grigg (2008) and Wilcox et al. (2016).

9.1   Objectives and Applicable Criteria

A principal objective of a WRRF project is to meet anticipated treatment goals to comply with regulatory requirements, which primarily protect public and environmental health. However, there are other desirable goals, such as minimizing costs, providing process reliability and flexibility, facilitating operations, maximizing safety and security aspects, and minimizing social effects during both construction and future operation. These goals will be different for each project and can be classified generally into four main groups of objectives—economic, environmental, technical, and social. Some of them can be easily quantified, while others require a qualitative approach to perform a comparison among alternatives.

9.1.1   Economic Criteria

When comparing design alternatives from the economic point of view, it is important to include criteria related to both construction and operation costs (including personnel, energy, chemicals, and maintenance). In particular, energy requirements (e.g., aeration, pumping, heating, and mixing) must be considered with chemical requirements (e.g., metal salts for phosphorus precipitation, external carbon source to enhance denitrification efficiency, or chlorine for disinfection) and costs related to the collection and disposal of sludge. Any potential benefit from energy recovery (i.e., by means of methane production in the anaerobic digesters) should be included in the analysis. Any cost-related criterion chosen to compare alternatives, once quantified, must be normalized by expressing it as a percentage of the total budget.

9.1.2   Environmental Criteria

Implicit in municipal WRRF projects is an objective related to the protection of environmental health, and the treated effluent quality generally ameliorates the quality of the receiving water body. However, there can be negative environmental effects associated with energy consumption, chemical reagents used, treatment residuals management, and atmospheric emissions of a project. Some of these negative effects can be offset by nutrient recovery, water reuse, and energy generation.

A LCA is a useful tool for assessing the overall balance of environmental effects of a WRRF project. An LCA can be defined as a compilation and evaluation of inputs, outputs, and the environmental effect of a system through its life cycle—from the production of raw materials to the disposal of waste generated. More information on LCAs can be found in Chapter 4.

9.1.3 Technical Criteria

Technical aspects often are considered only when designing the units, equipment, and control strategies in detail. However, including technical criteria during conceptual design should ensure a reliable, flexible, robust, easy-to-operate, safe, and secure facility. Safety considerations could be considered a subset of technical criteria. Safety concerns the potential for falls, confined space entry, exposed equipment or moving parts, and chemical transport, storage, and addition for facility operators. Safety also should consider the special precautions required to reduce the level of risk to the outside community, such as chemical deliveries and truck traffic. There have been some numerical approaches based on simulation results concerning the evaluation of the technical criteria (i.e., Comas et al., 2008; Copp, 2002; Flores et al., 2007; Vanrolleghem and Gillot, 2002). When this is not possible, a qualitative comparative ranking between alternatives could be enough, making the related uncertainty and subjectivity aspects explicit.

9.1.4 Social Criteria

Social criteria are becoming increasingly important when designing WRRFs. This category of criteria includes aspects related not only to facility personnel and external workers but also to the relative effects of the facility on the neighbors and outside community. Noise, visual aesthetics (relative visual effect of the facilities—both near and distant views—from the perspective of blending in with the surroundings and having pleasing architecture and landscaping), community involvement, and odor related to air emissions are generally the distinguishing features for respective alternatives. As mentioned above, most of these criteria cannot be quantified easily, which requires a qualitative comparison between alternatives.

9.2 Evaluation Methodology

There are several evaluation methodologies for decision analysis that involve multiple objectives. Figure 2.6 shows a schematic representation of a general MCDA evaluation methodology for WRRF design (Flores et al., 2005). After a preliminary step to collect and analyze all available information, the next step is the definition of design objectives and evaluation criteria used to measure the degree of satisfaction of the objectives. Initial weight factors are assigned to determine the relative importance of the criteria. In the following step, there are a number of tasks related to the decision procedure: identification of the issue to be resolved, generation of the potential alternatives, selection of a subset of criteria defined for this specific issue, and evaluation of the proposed alternatives. This evaluation is approached as a multi-criterion method and comprises quantification; normalization of the evaluation criteria; and a weighted sum, where each alternative under evaluation obtains a score that is calculated by adding the product of each normalized criterion multiplied by its corresponding weight. The alternatives are ranked according to the score obtained. The alternative with the highest score is the one with the highest degree of satisfaction of the objectives considered, and the one recommended for implementation. The same methodology is applied iteratively to deal with each new issue that arises, until the conceptual design of the WRRF project is completed.

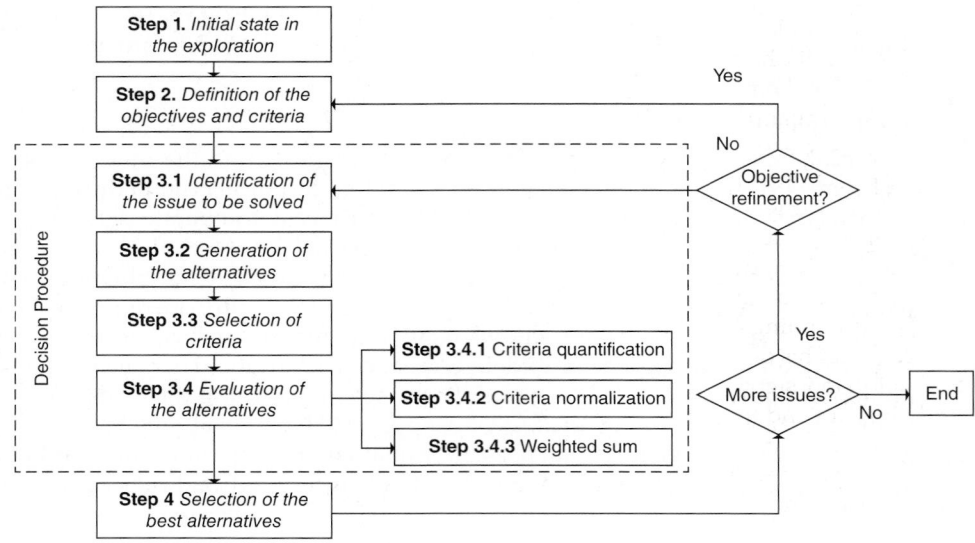

FIGURE 2.6   Schematic representation of a general MCDA evaluation methodology (Reprinted with permission from Flores, et al. (2005) "Activated Sludge Configuration During the Conceptual Design of Activated Sludge Plants Using Multicriteria Analysis," Ind. Eng. Chem. Res., 44(10), 3556-3566. Copyright © 2005 American Chemical Society).

Once the MCDA has been applied, further analysis should be carried out to investigate whether preliminary conclusions are robust or if they are sensitive to changes in the basic assumptions considered for the alternative. This sensitivity analysis is the objective examination of the effect on the output of a model as a result of changes in input parameters of the model. These changes may be made to investigate the significance of missing information, to explore the effect of a decision maker's uncertainty about his or her values and priorities, or to offer a different perspective on the problem.

In WRRF design, the context in which decisions are made greatly influences the selection of the alternative. This context is defined by the owner, design team, and all other key stakeholders, according to the weighting factor assigned to each criterion. Giving more or less weight to a determined criterion will clearly restrict some of the alternatives generated during the decision procedure. A suitable sensitivity analysis will determine which, if any, of the input parameters have a critical influence on the overall evaluation—that is, where a small change in a criterion weight can affect the overall preference order. A sensitivity analysis also provides the designer with other useful information, such as under which conditions each of the alternatives becomes the preferred one, which alternative is the best choice for the widest range of situations, and identification of each alternative's strong and weak points. The focus of MCDA and sensitivity analysis is on supporting decision making and not necessarily on defining the "right" answer.

10.0   Evaluations Using Special Investigations

During the planning phase of a municipal WRRF design, the designer and the owner may find that special investigations may be warranted. In some cases, this may involve studies using bench-scale, pilot plant, or prototype equipment. Common objectives for a bench-scale or pilot-scale study include testing process alternatives to meet stringent effluent discharge limits or solids stabilization requirements; exploring the biodegradability or chemical treatability of unconventional wastes or industrial wastes; evaluating and quantifying operational and maintenance requirements; evaluating unknown process-specific design parameters; and confirming size or layout requirements for a given unit process.

The designer may use the results of pilot studies and incorporate them into the process design for the facility. Although such investigations can provide valuable insights, they cannot guarantee subsequent process success, particularly if they are improperly structured or interpreted. Caution should be exercised if investigative findings refute understandings gained from long experience. In addition, an investigator should be aware of his or her own tendency to underrate problems, overrate the universal applicability and significance of the observations, and operate the pilot-plant unit with more diligence and tighter controls than likely to be realized with a real system. The following subsections provide guidance for developing investigative study protocols and for interpreting the results from those studies. At the end of this section, the treatability of hazardous substances is briefly discussed.

10.1   Study Scale and Duration

When developing the design and protocol for a pilot study, the investigator must first identify the study objectives. Once the objectives are known, the pilot study scale can be evaluated to determine the size that would provide the most useful results while balanced against cost and time constraints for the study. Ideally, a pilot-plant investigation would be conducted at a scale that can reveal practical design and operating problems with equipment and hydraulics. In many cases, the mixing conditions found in pilot work are far greater than those found in real systems. This is especially important in the examination of results from pilot anaerobic digesters, where performance may be enhanced by a brief period of vigorous, complete mixing. Investigators should also consider that physical factors, such as bacterial growth along walls and conduits in pilot-scale and bench studies, can affect treatability results, introducing errors.

The required duration of an investigative study also depends on the objectives. Studies intended to explore process reliability are best conducted over longer time frames so that they are given time to reflect acclimation and equilibrium to one or more variables, such as seasonal stresses, material considerations, yearly variations, diurnal patterns, or the effects of upstream system and auxiliary system upsets. If such conditions cannot be expected to occur naturally within the time and budget available, consider artificially simulating those stressed conditions. In addition, investigators should recognize that equipment reliability and maintenance needs are rarely, if ever, adequately evaluated in any pilot system unless the pilot can be operated for an extended period of time (such as a period greater than one year).

10.2   Interpretation of Study Results

Pilot performance should be evaluated under conditions that fairly represent the norm for the facility. Studies that use batch, constant-flow, and constant influent can provide insight into treatment mechanisms, but they provide limited use for assessing the ability of a system to meet performance reliability standards. If variable wastewater characteristics, loading stress, or both cannot correctly simulate actual conditions under varying influent loading conditions that the facility will experience, then average (or median) results should be adjusted to account for the reliability of the design. Many facilities are permitted for monthly average concentrations. As guidance, effluent quality for soluble constituents at maximum month conditions can be approximately 1.5 to 2.5 times greater than median observations. Thus, average effluent quality can be 40% to 60% of the maximum month values, and this must be factored into process design calculations.

The investigator also should verify that reactor flow regime is similar for the pilot- and full-scale application, as the reactor configuration (plug-flow versus completely mixed) will affect the removal of soluble substrate and the solid-liquid separation characteristics of activated sludge mixed liquors. Also important to consider is that solid-liquid separation, oxygen transfer, and mixing considerations are likely to be significantly different in a full-scale system. Environmental factors affecting oxygen transfer, such as temperature, elevation, and salinity, must be considered.

10.2.1   Considerations for Treatability Studies

When evaluating percent removal from a pilot-scale study, the investigator should consider that percent removals depend on the applied pollutant concentration. Influent and effluent pollutant levels are influenced by their soluble and suspended phases. Most treatment processes perform to some limiting residual concentration rather than at a controlling percentage. Effluent pollutants in the soluble phase reflect influent levels, the nature of the upstream reactor(s), and the performance of the upstream separator(s). If any of the three depart from the conditions that resulted in the original percent removal characterization, the percent removal may change.

Most constants are not actually constant or independent of the conditions from which they were derived. Some are linked to another, so that changing one may call for changing the other to ensure that the overall performance of the system is correctly described (such as the cell yield and endogenous decay constants used to describe net oxygen requirements and solids production from an aerobic biological stabilization process). Biokinetic constants can be expected to vary as a function of soluble and particulate loads, recycle, and the biomass's overall age and environmental conditions.

The performance of solids separation equipment and location of recycle sidestreams should be accounted for in treatability studies. Pilot-scale studies may have thickening/dewatering applications that are much more robust than those found at the full-scale facility. This will affect sidestream loadings returned upstream of the biological process. Solids processing systems that solubilize pollutants may impose transient or continuous liquid-processing stresses contributing to instability and non-compliance. Ideally, treatability studies should include consideration of these effects and, if not, be appropriately qualified to avoid overly optimistic sizing and performance assumptions.

10.2.2   Considerations for Solids Processing Studies

Solids separation, concentration, processing, and stabilization techniques can be especially difficult to investigate at the pilot scale. The results can be particularly difficult to apply in cases where physical and processing considerations are time dependent, such as in facilities with filtration backwashing and activated carbon regeneration. Investigations of thickening and dewatering require determining performance for a range of applied solids concentrations to assess the significance of this variable. The interpretation of an investigative run of limited duration should be tempered by the realization that feed-solids characteristics are highly variable and are influenced by:

- Changes in the liquid-processing train;
- Changes caused by additives, such as metal salts for phosphorus removal;
- Recycle effects imposed by the downstream solids-processing train on the unit processes ahead of those being investigated; or
- Seasonal variations in facility loading.

Frequently, excessive solids recycling associated with full and overflowing storage tanks, gravity thickeners, and poor secondary digester supernatant quality occurs unknowingly and affects process performance by increasing loads.

Investigators should also consider that the achievable degree of solids stabilization depends on the feed characteristics, including the nature and mix of raw and secondary solids, and the degree of stabilization of secondary solids.

10.3   Mechanical Design Considerations

If designs are developed from pilot testing, it is imperative to consider whether mechanical equipment used in pilot testing is representative of the full-scale condition. Pilot-scale studies generally are more frequently monitored and controlled, often on a minute-by-minute basis to optimize performance, whereas full-scale installations will operate with less human interface, thus potentially resulting in decreased performance. Furthermore, the pilot-scale equipment used will be tailored for that specific application. In full-scale installations, there is a chance that the installed equipment may not be ideal for that application. The following sections include some common mechanical design considerations that should be evaluated for the full-scale design that often are not captured during a pilot study.

10.3.1   Effects of Ragging and Stringy Materials

Influent screening in the full-scale installation should be similar to that used during the pilot testing. This is especially critical with membrane bioreactor facilities, integrated fixed-film activated sludge, and moving-bed bioreactor pilots. Inefficient mechanical screening in full-scale installations can lead to rag and string buildup on downstream equipment, which may not occur during pilot studies, either as a result of the type of screening or duration of the pilot testing. Rag formation in downstream equipment, such as mixers, pumps, and membranes, can be detrimental to performance and is a common maintenance headache. Although primary consideration should be given to providing efficient screening, in some cases fine screening is not possible and as such impellers designed for high-rag situations should be considered. Inadequate screening and resultant debris will increase the fouling propensity for membrane applications.

10.3.2   Effects of Poor Grit Removal

Because of the time limitations on pilot-scale studies, the effects of poor grit removal generally will not be captured. In full-scale applications, poor grit removal can effectively decrease the process tank volume over the course of several years. Grit will settle in poorly mixed zones in bioreactors. In some cases, grit buildup could be severe enough to affect treatment performance, especially in facilities operating at or near design capacity. The design and layout of floor-mounted diffuser configurations, including diffuser density and header spacing, should consider the effects of grit settlement.

10.3.3 Pumping

Demonstrations from pilot-scale studies that use pumping may not be representative of full-scale installation of the mechanical equipment. Items to consider include pump suction intake design and abrasiveness of the fluid being pumped. Rat-holing can occur when pumping sludge from primary and secondary clarifiers and gravity thickeners, such that water short-circuits from the top of the tank to the pump intake through the sludge layer, so that the intended sludge is not pumped. In most instances, rat-holing is caused by withdrawing solids at high rates, but it also can be attributed to the design of the sludge collection equipment and the hopper. Rat-holing can have serious effects on solids concentrations, particularly for gravity thickening, that the design should attempt to avoid. Fluid abrasiveness also may change between pilot- and full-scale studies, and thus the materials of construction for pumping systems should consider these differences.

10.4   Assessing the Treatability of Hazardous Substances

Hazardous heavy metals are encountered more routinely at higher concentrations in wastewater residuals than are hazardous organics. As conservative substances, metals merely accumulate in facility wastes and effluent. Hazardous organics, however, can act as either conservative or nonconservative compounds, depending on the compound and processes in the facility. Aeration systems and weirs may strip organics if facility influent contains hazardous materials.

Hazardous substance determinations frequently reach the limits of analytical capability. Therefore, special care should be taken to guard against the reporting of false positives because of the laboratory or sampling procedure. For municipal WRRFs that have metal limits approaching analytical detection limits, clean techniques should be used in accordance with U.S. EPA Method 1669 (U.S. EPA, 1996) to limit contamination during sampling, transport, and analysis.

A compound's chemical formula, molecular weight, solubility, Henry's law constant, organic carbon, and octanol-water partition coefficients serve to allow some prediction of the compound's relative strippability, adsorbability, and biodegradability. In general, the lower the molecular weight and the simpler the compound structure are, the higher the compound's biodegradability will be. Biodegradability also declines with the extent of halogenation, which is the ratio of the weight of chlorine to that of the total compound. High solubilities tend to support high biodegradability and low stripping and adsorption. Experiences in the field have led to the following conclusions (U.S. EPA, 1986):

- Octanol-water partition coefficient, $\log K_{ow}$. If this coefficient is greater than 3.5, the organic substance is highly adsorptive, meaning it readily partitions with the wastewater solids). If the coefficient is less than 3.5, the substance is more likely to be removed by stripping or biodegradation (some have suggested a coefficient of 2.0 as a better approximation of the strippability threshold).

- Henry's law constant. If the constant is greater than 0.024 L·atm/mol, the compound is easily stripped from solution, with strippability also increasing with a lower affinity for adsorption and a higher extent of halogenation.

- Volatile organic compounds. Most of the mass of these is not accounted for in residual solids and effluents. Stripping and biodegradation are likely removal mechanisms; adsorption is small and may not be measurable.

- Base-neutral compounds. These compounds have highly varied removal mechanisms, with the more biodegradable compounds not partitioning to the solids. Volatilization and stripping are not likely to be significant removal mechanisms.

- Acid extractables. The removal mechanism is dominated by biodegradation; many of these compounds are potentially formed during chlorination.

- Pesticides and PCBs. These compounds are adsorbed strongly, with little tendency to degrade in an anaerobic environment, yet have been shown to degrade aerobically.

- Metals. Metals are largely concentrated in the residual solids.

Removal of hazardous substances by municipal WRRFs depends not only on the form of the substances but also on the levels received in the influent wastewater and the type of treatment processes present. Treatment processes provide opportunities for volatilization (large surface areas), stripping (aeration and mixing), recycling (closed pure-oxygen systems recycle stripped volatile organic compounds, allowing for greater biodegradation), and biodegradation and chemical uses (chlorine, ozone, and organic polymers). As may be expected, removal efficiencies of municipal WRRF hazardous substances are highly variable. Further, it is almost impossible to adequately predict performance of a given facility without a detailed understanding of site-specific conditions.

Designers and WRRF operators should realize that most toxic materials, including metals and pesticides, have the ability to partition strongly to the solid phase and thus be in the residual solids. In cases where toxic wastes are in the influent, the residual solids are often more dangerous than the facility effluent.

11.0   Final Design Considerations

Integrated design, as discussed in Section 2, is a holistic approach to design, implementation, and operation of a WRRF. It begins in the planning stages, when project goals are defined with collaboration between project participants, and it continues through design development by the design team. This section presents examples of measures that can be included in final design to meet the O&M and community goals defined at the outset of the project. It also presents considerations for maintaining facility operations during construction, which can be challenging when designing upgrades to a WRRF.

11.1   Operation and Maintenance

Operability and maintainability are critical aspects of design and can be improved through collaboration with O&M staff—ideally with those who are familiar with the WRRF being designed. Direction, review, and feedback should be sought throughout the design process. In terms of operability, the designer's challenge is twofold: (1) providing for a work place that is safe, convenient, and pleasant; and (2) developing a process design and control scheme that tolerates variable loading and environmental conditions, while maintaining compliance with effluent requirements.

The appropriate level of process complexity, automation, and flexibility varies both with the size of the facility and with the required performance. The design team should meet with O&M staff to determine their preferred operating strategies and procedures relative to unit process drains, bypasses, and redundancy. The designer should also consider the level of operations staff sophistication when selecting and designing equipment and controls, with the goal of providing an operator-friendly facility that simplifies operation to the greatest degree possible.

Flexibility of operation improves the reliability of a WRRF. There are various ways of enhancing the operational flexibility, including providing sufficient land, improving accessibility, balancing flow and bypassing, interconnecting pipes and multiple treatment units, accommodating flow variations, and providing isolation valves and automated instrumentation and controls. Each of these aspects should be considered when arranging various treatment units.

The following accessibility concerns need to be considered during design to facilitate O&M of equipment:

- Desired means of access to various process components and buildings, including providing adequate area for parking, pedestrian access, deliveries of equipment, chemicals, and other materials;
- Equipment access and replacement/repair of large equipment;
- Travel patterns for each type of vehicle anticipated, including emergency vehicles.

A thorough, high-quality O&M manual should also be provided to the owner. The O&M manual is key to operational success and conveys the design intent to the operations staff. It should clearly define the tools available for an operator to achieve the effluent quality goals as well as the operating strategies to rely on. The manual should be updated over time as process and equipment changes are made or upgrades occur. Designers also can promote operator training and work with administrators to ensure adequate support for the needs of the facility.

11.2 Community Effects

Included below are strategies to mitigate negative impacts on the community and to encourage the integration of the WRRF into the community:

- Sensory—Avoid open-air turbulent mixing (and stripping) of raw and partially stabilized wastewater and minimize open-air exposure of solids processing recycles and residues before stabilization; select processing concepts that avoid generation of odors if odor-free alternatives are available (WEF, 2004a); map and consider prevailing wind directions during component sizing and when locating vents, blowers, structure penetrations, and tanks.
- Noise—Consider noise of large equipment when selecting and specifying equipment; use acoustical and architectural techniques to mitigate sound

impacts; consider building orientation, including loading zones, vents, and other building penetrations; and use barriers or other sound-wave attenuation measures within buildings, surrounding structures, and facility grounds.

- Visual—Enhance the visual appeal of structures and grounds with sensitive landscaping and architectural designs complementary to the surrounding materials; use landscaping, earthen berms, and structural shields to screen auxiliary equipment; plan the facility's buffer zone to minimize maintenance and mitigate visual effects.

- Traffic—Anticipate possible conflicts between public expectations and the needs of the WRRF at each stage and take steps to reduce these conflicts as much as practical. For example, consider establishing service schedules that do not conflict with rush hours and the presence of school children on the streets, and establish specific truck routes to the facility that avoid residential streets and neighborhood commercial centers. Require dust control and traffic control plans for each construction activity.

- Accessibility—Locate the administration building so that it is readily visible when entering the site and consider including public meeting rooms and a provision for public education. Consider a lesser level of security at the entry point to the administration building using gates that remain open during regular business hours. Comply with accessibility requirements for persons with mobility, sight, or other impairments classified in the Americans with Disabilities Act (ADA) (http://www.ada.gov). Make unsecured or unused areas available to the public for use, such as for passive or active recreation. Include signage identifying the facility and directional signs within the site consistent with sign styles within the vicinity.

- Design coordination—Schedule predevelopment meetings with building, fire, engineering, and planning staff to clarify local requirements, establish lines of communication, and identify requirements related to regulation of construction impacts and construction access routes. If required, include a public hearing related to final permitting before the policymaking body of the community. The public hearing provides those affected by the project another opportunity to voice any concerns or recommendations they have before a land-use decision is made. Mitigation measures that have not already been addressed at the time of the hearing may be added as conditions of approval for the land-use action.

11.3   Maintenance of Facility Operation during Construction

The designer should ensure that the WRRF can be operated and maintained without significant difficulties during the construction period. Compliance with permitted limits during construction is a requirement in most instances. Potential deviations from permitted limits during the construction period should be negotiated with the permitting authority before finalizing the design documents. Achieving this objective sometimes requires a rigorous analysis, special design accommodations, and in some cases the installation of temporary facilities.

The designer should map out and plan construction access and, if necessary, specify timing of construction vehicles to minimize disturbance to facility operations and abutting use traffic patterns. Interim access points and service areas for facility components may be identified to maintain continued function, while modifying existing process structures with little or no downtime. Well-planned phasing of construction components may provide opportunities for early startup of some portions of the new work.

Identifying constraints on the contractor's work and carefully planning the suggested construction sequence ensures that new facilities can be built without undue interruption of treatment efficiency. Including these steps during design also assists in minimizing change orders. The design team should consult with the owner and operations staff to identify construction constraints, such as when continued operation of a specific process unit is required, and determine a specific scenario for the sequence of events that should occur during construction. For a project utilizing a design-bid-build project delivery method, this detailed process description and sequencing should be included in the contract documents, with allowance for the contractor to propose variations. For a project using an alternative delivery method such as design-build, the contractor should be involved in determining the sequencing and whether or not temporary facilities are required. In terms of construction constraints, project specifications should identify the number of process units that can be taken out of service at any one time, allowable electrical power shutdown periods, and temporary diversion schemes. The specifications should state whether the contractor or the owner is responsible for draining and cleaning process tankage and piping.

12.0 References

Albertson, O. E. (1995) Is CBOD$_5$ Test Viable for Raw and Settled Sewage? *ASCE J. Environ. Eng. Div.*, **121** (7), 515–520.

Albertson, O. E.; Young, J. C.; Clesceri, L. S.; Kamhawy, S. M. (2007) Discussion Of: Changes in the Biochemical Oxygen Demand Procedure in the 21st Edition of *Standard Methods for the Examination of Water and Wastewater*, James C. Young, Lenore S. Clesceri, Sabry M. Kamhawy, **77**, 404–410 (2005). *Water Environ. Res.*, **79**, 453–456.

American Public Health Association; American Water Works Association; Water Environment Federation (2012) *Standard Methods for the Examination of Water and Wastewater*, 22nd ed.; American Public Health Association: Washington, D.C.

American Society of Civil Engineers; Water Environment Federation (2007) *Gravity Sanitary Sewer Design and Construction*, 2nd ed., ASCE Manuals and Reports on Engineering Practice No. 60, WEF Manual of Practice No. FD-5; American Society of Civil Engineers: Reston, Virginia.

ANSI/AWWA G-481-14 (2014) *Reclaimed Water Program Operation and Management*, May 1, 2014.

ANSI/AWWA G-485-XX (2017 Draft) *Potable Reuse Operation and Management*, AWWA, 2017 Draft.

Association for the Advancement of Cost Engineering (1997) *Cost Estimate Classification System*, AACE International Recommended Practice No. 18R-97; AACE International: Morgantown, West Virginia.

Association of Metropolitan Water Agencies; National Association of Clean Water Agencies; Water Environment Federation (2007) *Implementing Asset Management: A Practical Guide*; National Association of Clean Water Agencies: Washington, D.C.

American Water Works Association (2009) *Planning for the Distribution of Reclaimed Water*, 3rd ed., AWWA Manual of Water Supply Practices, M24; American Water Works Association: Denver, Colorado.

Bakir, H. (2000) *Sanitation and Wastewater Management for Small Communities in EMR Countries: Challenges and Strategies for Accelerated Development within the Water Resources Constraints*, Technical Note on Environmental Health; World Health Organization: Geneva, Switzerland.

Blank, T. B.; Tarquin, A. J. (2015) *Engineering Economy*, 7th ed. McGraw Hill: New York.

City of Phoenix (2017) *Design Standards Manual for Water and Wastewater Systems*; City of Phoenix, Water Services Department: Phoenix, Arizona, https://www.phoenix.gov/waterservices/publications/design-manuals/systems (accessed May 2017).

Comas, J.; Rodriguez-Roda, I.; Gernaey, K. V.; Rosen, C.; Jeppsson, U. (2008) Risk Assessment Modelling of Microbiology-Related Solids Separation Problems in Activated Sludge Systems. *Environ. Mod. Soft.*, **23**, 1250–1261.

Commonwealth of Virginia (2008) *Sewage Collection and Treatment Regulations*, 9VAC 25-790; Commonwealth of Virginia, State Water Control Board.

Construction Management Association of America (2003) *Capstone: The History of Construction Management Practice and Procedures*, Course Study Guide; Construction Management Association of America: McLean, Virginia.

Copp, J. B. (2002) *The COST Simulation Benchmark: Description and Simulator Manual*; Office for Official Publications of the European Community: Luxembourg.

Council of the European Communities (1991) Council Directive 91/271/EEC Concerning Urban Waste Water Treatment, May 21, 1991; Council of the European Communities: Brussels, Belgium.

Council of the European Communities (1998) Council Directive 98/83/EC of 3 November 1998 on the Quality of Water Intended for Human Consumption; Official Journal L 330, 0032–0054, Council of the European Communities: Brussels, Belgium.

Council of the European Communities (2006) Proposal for a Directive of the European Parliament and of the Council on Environmental Quality Standards in the Field of Water Policy and Amending Directive 2000/60/EC, July 17, 2006; Council of the European Communities: Brussels, Belgium.

D'Antoni, J. M.; Bahl, V. (1990) Designs for the Future. Abstract submitted to the 63rd Annual Water Environment Federation Technical Exposition and Conference, Washington, D.C., Oct. 711; Water Environment Federation: Alexandria, Virginia.

Design-Build Institute of America (2009) *Contract Incentives and Design-Build Acquisition*, Manual of Practice; Design-Build Institute of America: Washington, D.C., http://www.dbia.org/pubs/manualofpractice/ (accessed November 2008).

http://www.dbia.org/resource-center/Pages/Best-Practices.aspx (accessed August 2016).

Ekama, G. A.; Marais, G. v. R.; Siebritz, I. P.; Pitman, A. R.; Keay, G. F. P.; Buchan, L.; Gerger, A.; Smollen, M. (1984) *Theory, Design and Operation of Nutrient Removal Activated Sludge Processes*; Water Research Commission: Pretoria, South Africa.

Engineers Joint Contract Documents Committee (2007) Standard General Conditions of the Construction Contract. Engineers Joint Contract Documents Committee, American Council of Engineering Companies: Washington, D.C., http://www.ejcdc.org (accessed November 2008).

Environmental Protection and Heritage Council; National Health and Medical Research Council; Natural Resource Management Ministerial Council (2009a) Australian Guidelines for Water Recycling: Managed Aquifer Recharge, National Water Quality Management Strategy, Document No. 24, July 2009, Canberra, Australia.

Environmental Protection and Heritage Council; National Health and Medical Research Council; Natural Resource Management Ministerial Council (2009b) Australian Guidelines for Water Recycling: Stormwater Harvesting and Reuse, National Water Quality Management Strategy, Document No. 23, July 2009, Canberra, Australia

Environmental Protection and Heritage Council; National Health and Medical Research Council; Natural Resource Management Ministerial Council (2008) Australian Guidelines for Water Recycling: Augmentation of Drinking Water Supplies, Document No. 22, May 2008, Canberra, Australia.

Environmental Protection and Heritage Council; Natural Resource Management Ministerial Council; Australian Health Ministers' Conference (2006) National Guidelines for Water Recycling: Managing Health and Environmental Risk, Phase 1; Environmental Protection and Heritage Council: Adelaide, Australia.

EnviroSim Associates Ltd. (2016) *BioWin Process Simulator*; EnviroSim Associates Ltd: Flamborough, Ontario, Canada.

EPA Victoria, Australia (2013) *Code of Practice Onsite Wastewater Management*, Pub. No. 891.3, February 2013.

European Union (2016) Common Implementation Strategy for the Water Framework Directive and the Floods Directive, *Guidelines on Integrating Water Reuse into Water Planning and Management in the Context of the WFD*, European Union, July 2016.

EUWI-MED (2007) Joint Mediterranean EUWI/WFD Process, Mediterranean Wastewater Reuse Report, Mediterranean Wastewater Reuse Working Group (MED WWR WG), November 2007 (http://ec.europa.eu/environment/water/water-urbanwaste/info/pdf/final_report.pdf).

Flores, X.; Bonmati, A.; Poch, M.; Rodríguez-Roda, I.; Bañares-Alcántara, R. (2005) Selection of the Activated Sludge Configuration during the Conceptual Design of Activated Sludge Plants Using Multicriteria Analysis. *Ind. Eng. Chem. Res.*, **44** (10), 3556–3566.

Flores, X.; Rodríguez-Roda, I.; Poch, M.; Jiménez, L.; Bañares-Alcántara, R. (2007) Systematic Procedure to Handle Critical Decisions during the Conceptual Design of Activated Sludge Plants. *Ind. Eng. Chem. Res.*, **46** (17), 5600–5613.

German Association for the Water Environment (1995) *Design and Construction of Wastewater Treatment Facilities*; ATV-A 106E; DWA (German Association for Water, Wastewater, and Waste Management): Bonn, Germany.

Great Lakes–Upper Mississippi River Board of State and Provincial Public Health and Environment Managers (2014) *Recommended Standards for Wastewater Facilities*; Health Education Services: Albany, New York. http://10statesstandards.com/wastewater-standards.pdf (accessed August 2016).

Grigg, N. S. (2008) *Total Water Management: Leadership Practices for a Sustainable Future*; American Water Works Association: Denver, Colorado.

Hovey, W. H.; Tchobanoglous, M.; Schroeder, E. D. (1979) Activated Sludge Effluent Quality Distribution. *ASCE J. Environ. Eng. Div.*, **105**, 819–828.

Mara, D.; Horan, N. J. (2003) *Handbook of Water and Wastewater Microbiology*; Academic Press: London, United Kingdom.

Melcer, H.; et al. (2004) *Methods for Wastewater Characterization in Activated Sludge Modeling*; Project No. 99-WWF-3; Water Environment Research Foundation: Alexandria, Virginia.

Metcalf & Eddy, Inc./AECOM (2014) *Wastewater Engineering: Treatment and Resource Recovery*, 5th ed., Tchobanoglous, G., Stensel, H. D., Tsuchihashi, T., Burton, F. L. (Eds.); McGraw-Hill Education: New York.

New England Interstate Water Pollution Control Commission (2016) *Guides for the Design of Wastewater Treatment Works*, TR-16; New England Interstate Water Pollution Control Commission: Lowell, Massachusetts, http://www.neiwpcc.org/tr16guides.asp (accessed August 2016).

Park, J. K.; Wang, J.; Novotny, G. (1997) *Wastewater Characterization for Evaluation of Biological Phosphorus Removal*, Research Report 174; Wisconsin Department of Natural Resources: Madison, Wisconsin.

SAVE International (2007) *Value Standard and Body of Knowledge*; SAVE International: Dayton, Ohio, http://www.value-eng.org (accessed August 2016).

Steinle-Darling, E., Ph.D., P.E. (2016) *DPR Can (and Does!) Improve Water Quality: Full Results from Big Spring*, Water Reuse in Texas, July 15, 2016

U.S. Environmental Protection Agency (2016a) Chemical Accident Protection Provisions. *Code of Federal Regulations*, Part 68, Title 40; U.S. Environmental Protection Agency: Washington, D.C.

U.S. Environmental Protection Agency (2016b) Designation, Reportable Quantities, and Notification. *Code of Federal Regulations*, Part 302, Title 40; U.S. Environmental Protection Agency: Washington, D.C.

U.S. Environmental Protection Agency (2016c) Emergency Planning and Notification. *Code of Federal Regulations*, Part 355, Title 40; U.S. Environmental Protection Agency: Washington, D.C.

U.S. Environmental Protection Agency (U.S. EPA, 2016d) General Pretreatment Regulations for Existing and New Sources of Pollution. *Code of Federal Regulations*, Part 403, Title 40; U.S. Environmental Protection Agency: Washington, D.C.

U.S. Environmental Protection Agency (U.S. EPA, 2016e) Sampling and Testing Procedures. *Code of Federal Regulations*, Part 133.104, Title 40; U.S. Environmental Protection Agency: Washington, D.C.

U.S. Environmental Protection Agency (2016f) Secondary Treatment Regulation. *Code of Federal Regulations*, Part 133, Title 40; U.S. Environmental Protection Agency: Washington, D.C.

U.S. Environmental Protection Agency (2016g) State and Local Assistance. *Code of Federal Regulations*, Part 133, Title 40; U.S. Environmental Protection Agency: Washington, D.C.

U.S. Environmental Protection Agency (2012) *2012 Guidelines for Water Reuse*, EPA/600/R-12/618; U.S. Environmental Protection Agency: Washington, D.C., http://nepis.epa.gov/Adobe/PDF/P100FS7K.pdf (accessed August 2016).

U.S. Environmental Protection Agency (2009a) *General Guidance on Risk Management Programs for Chemical Accident Prevention* (40 CFR 68), EPA-555/B-04-001; U.S. Environmental Protection Agency, Office of Solid Waste and Emergency Response: Washington, D.C.

U.S. Environmental Protection Agency (2009b) *Water Quality Trading Toolkit for Permit Writers*; U.S. Environmental Protection Agency: Washington, D.C., https://www.epa.gov/npdes/water-quality-trading-toolkit-permit-writers (accessed August 2016).

U.S. Environmental Protection Agency (2004) *NPDES Compliance Inspection Manual*, EPA-EPA 305-X-04-001; U.S. EPA Office of Enforcement and Compliance Assurance: Washington, D.C.

U.S. Environmental Protection Agency (1996) *Method 1669: Sampling Ambient Water for Trace Metals at EPA Water Quality Criteria Levels*, EPA-821/R-95034; U.S. Environmental Protection Agency, Office of Water: Washington, D.C.

U.S. Environmental Protection Agency (1989) *Overview of Selected EPA Regulations and Guidance Affecting POTW Management*, EPA-430/09-89-008; U.S. Environmental Protection Agency, Office of Water: Washington, D.C.

U.S. Environmental Protection Agency (1986) *Report to Congress on the Discharge of Hazardous Wastes to Publicly Owned Treatment Works (The Domestic Sewage Study)*, EPA-530/SW-86-004; U.S. Environmental Protection Agency, Office of Water: Washington, D.C.

Vanrolleghem, P.; Gillot, S. (2002) Robustness and Economic Measures as Control Benchmark Performance Criteria. *Water Sci. Technol.*, **45** (4/5), 117–126.

Water Design-Build Council (2016) *Water and Wastewater Design-Build Handbook*, 4th ed. Water Design-Build Council: Edgewater, Maryland.

Water Environment Federation (2017) *Sustainability and Energy Management for Water Resource Recovery Facility Design*; Water Environment Federation: Alexandria, Virginia.

Water Environment Federation (2013a) *Operation of Nutrient Removal Facilities*, Manual of Practice No. 37; Water Environment Federation: Alexandria, Virginia.

Water Environment Federation (WEF, 2013b) *Wastewater Treatment Process Modeling*, 2nd ed., Manual of Practice No. 31; Water Environment Federation: Alexandria, Virginia.

Water Environment Federation (2007a) *Effects of Wastewater Treatment on Microconstituents*, Technical Practice Update, Microconstituents Community of Practice, Technical Practice Committee; Water Environment Federation: Alexandria, Virginia.

Water Environment Federation (2007b) *Sources of Microconstituents and Endocrine-Disrupting Compounds, Technical Practice Update*, Microconstituents Community of Practice, Technical Practice Committee; Water Environment Federation: Alexandria, Virginia.

Water Environment Federation (2004a) *Control of Odors and Emissions from Wastewater Treatment Plants*, Manual of Practice No. 25; Water Environment Federation: Alexandria, Virginia.

Water Environment Federation (2004b) *Financing and Charges for Wastewater Systems*, Manual of Practice No. 27; Water Environment Federation: Alexandria, Virginia.

Water Environment Federation; American Water Works Association (2008) *Using Reclaimed Water to Augment Potable Water Resources*, 2nd ed., Special Publication; Water Environment Federation: Alexandria, Virginia.

WateReuse Research Foundation (2013) *Potable Reuse: State of the Science Report and Equivalency Criteria for Treatment Trains*, WRRF 11-02-02, Alexandria, Virginia.

WateReuse Research Foundation (2014) *Utilization of Hazard Analysis and Critical Control Points Approach for Evaluating Integrity of Treatment Barriers for Reuse, 2014*; WRRF 09-03; WateReuse Research Foundation: Alexandria, Virginia.

Wilcox, J.; Nasiri, F.; Bell, S.; Rahaman, M. S. (2016) *Urban Water Reuse: A Triple Bottom Line Assessment Framework and Review. Sustainable Cities and Society*, **27**, 448-456.

World Health Organization (2011) *WHO Guidelines for Drinking-Water Quality*, 4th ed.; World Health Organization: Geneva, Switzerland.

World Health Organization (2006) *WHO Guidelines for the Safe Use of Wastewater, Excreta and Greywater*, Volume 1; World Health Organization: Geneva, Switzerland.

Young, J. C.; Clesceri, L. S.; Kamhawy, S. M. (2005) Changes in the Biochemical Oxygen Demand Procedure for the 21st Edition of *Standard Methods for the Examination of Water and Wastewater. Water Environ. Res.*, **77** (4), 404–410.

Young, J. C.; Riley, K. A.; Baumann, E. R. (1995) Effect of Trichloromethyl Pyridine on Carbonaceous Biochemical Oxygen Demand in Wastewater. In *Proceedings of the 70th Annual Water Environment Federation Technical Exposition and Conference*; Chicago, Illinois, Oct. 18–22; Water Environment Federation: Alexandria, Virginia.

13.0   Suggested Readings

Baird, R. B.; Smith, R.-K. (2002) *Third Century of Biochemical Oxygen Demand*; Water Environment Federation: Alexandria, Virginia.

WateReuse Research Foundation (2014) *Application of Risk Reduction Principles to Direct Potable Reuse, 2014*; WRRF 11-10; WateReuse Research Foundation: Alexandria, Virginia.

WateReuse Research Foundation (2015) *Equivalency of Advanced Treatment Trains for Potable Reuse, 2015*; WRRF 11-02; WateReuse Research Foundation: Alexandria, Virginia.

WateReuse Research Foundation (2015) Guidelines for Engineered Storage for Direct Potable Reuse, 2015; WRRF 12-06; WateReuse Research Foundation, Alexandria, Virginia.

WateReuse Research Foundation (2016) *Critical Control Point Assessment to Quantify Robustness and Reliability of Multiple Treatment Barriers of DPR Scheme, 2016*; WRRF 13-03; WateReuse Research Foundation: Alexandria, Virginia.

WateReuse Research Foundation; American Water Works Association; Water Environment Federation (2015) *Framework for Direct Potable Reuse, 2015*; WRRF 14-20; WateReuse Research Foundation: Alexandria, Virginia.

CHAPTER **3**

Site Selection and Facility Layout

Timothy H. Sullivan, P.E., and Drew Hansen

1.0   Introduction

Selecting a site for a new water resource recovery facility (WRRF), while not a common endeavor, has become more complicated in recent years as public awareness has increased and regulatory requirements governing siting have become more stringent. Selecting a site for a WRRF typically involves evaluating multiple variables affecting multiple sites to varying degrees, ultimately to justify the selection of a preferred site. The process also involves input from the public and regulatory agencies throughout the process. Because all project circumstances are different and there is no single defined approach to conducting a site selection study, these guidelines can be used to evaluate and select sites for new WRRFs, satellite campus for expanded WRRFs, or perhaps relocating an existing WRRF, or possibly for combine sewer overflow (CSO) facilities or even potable water treatment facilities.

The overall purpose of this chapter is to provide a framework for conducting site selection studies involving WRRFs as well as general facilities layout on a given site. In some instances, site selection efforts can affect the success of the entire project; well-planned and well-documented projects could avoid lengthy studies of site alternatives and/or lawsuits. At a minimum, such studies or lawsuits could increase project costs and create unnecessary delays and, in the extreme, halt the project. Therefore, having a defensible, logical, and consistent methodology to evaluate site alternatives and recommend the preferred site and process layout is critical to the success of any WRRF project. This chapter provides the tools to develop and implement a site selection and layout methodology to fit a particular project.

2.0   The Site Selection Process

2.1   General

There is no specific methodology for evaluating and selecting the most appropriate WRRF site. Although these general guidelines may apply to many site selection projects, it is important to develop a specific methodology to suit your needs. The project manager and owner must have a basic understanding of the variables affecting site selection, and work together to identify the methodology of evaluating these variables that works for their particular project.

2.2   Variables That Influence Site Selection

There are a myriad of variables that will influence the site selection process, but they all fall into one of the following categories:

- Environmental;
- Technical;
- Institutional; and
- Cost.

The relative importance of some of these variables may not become evident until the site selection study is underway. For instance, the need for a buffer zone for odor and/or noise control may not be critical when comparing multiple rural sites, but becomes critical when evaluating urban sites against rural sites. Consistency is critical to the process, particularly when it comes to public perception. Changing the quantitative and qualitative assessment once sites are identified may be perceived as altering the process to favor a political entity or influential owner(s). Thus, identifying key variables and developing qualitative and quantitative assessment frameworks early provides a consistent defensible evaluation process and avoids the possible perception of bias.

2.3   Steps in the Site Selection Process

The process generally proceeds as follows:

1. Develop a Site Selection Work Plan (which identifies public involvement, site variables, relative importance, numerical importance, etc.);
2. Identify siting variables and develop evaluation criteria;
3. Determine the WRRF technical components and minimum area requirements;
4. Identify available candidate sites;
5. Perform preliminary site assessments (and site eliminations);
6. Perform details assessment (including process layout for each site, cost assessment, etc.); and
7. Site selection and public notification.

Parallel steps to the above process involve public outreach and participation. As presented herein, the public must be involved in each step along the way. In addition to public involvement, there may also be legal, financial, and public relations work groups also proceeding parallel to this process. While we believe that legal, financial, and public relations work groups are important, the process for these is beyond the scope of this chapter. A summary list of the above steps is presented in Table 3.1.

3.0   Developing and Implementing a Site Selection Work Plan

The first step in the site selection process is to develop a Site Selection Work Plan, which includes a consistent, technically sound methodology to site selection. The work plan should cover the following:

- Evaluation Criteria:
  — Environmental,
  — Technical,
  — Institutional, and
  — Costs;
- Area Requirements;
- Selection Process:
  — Candidate Site Identification,
  — Qualitative Assessment (and site elimination),
  — Quantitative Assessment (and site elimination), and
  — Costs Comparisons and Final Selection; and
- Public Participation.

Technical Steps	Nontechnical Steps
Develop Site Selection Work Plan	Develop public involvement plan (include public participation, education, relations, and involvement)
Determine variables and evaluation criteria	Inform and discuss process with stakeholders

Determine area requirements
    Determine future flow and loadings
    Determine area requirements from local WRRFs
    Determine WRRF area requirement by process and ancillary area requirements
    Determine preliminary cost estimates
	Identify potential sites to stakeholders

Identify available sites within the minimum and maximum area requirements Assess environmental, technical, institutional, and cost conditions for each site	Identify elimination criteria and sites

Eliminate sites using Phase 1 qualitative criteria
    Eliminate fatally flawed or regulatory constrained sites
    Eliminate severe technically constrained sites
	Identify elimination criteria and sites

Eliminate sites using Phase 2 quantitative evaluate criteria
 	
    Quantify costs to correct/remediate environment, technical, and institutional issues
    Eliminate cost-prohibitive sites	

Recommend site(s) to owner	Hold town hall meeting for selected site and address technical merits of public opposition

TABLE 3.1   Site Selection Steps

The work plan should cover in detail the steps identified above and any processes to achieve these steps. For instance, it may be straightforward to identify the siting variables, but difficult to reach consensus on the evaluation criteria. As a result, workshops should be planned to assess and reach consensus on the ranking criteria.

Wastewater treatment facilities are almost universally perceived by the public to be unacceptable neighbors. Public opposition may be strong; however, with early involvement of the public in the planning process and a sincere desire to listen and mitigate their concerns, much of the opposition can be minimized. Therefore, public involvement, or participation, is one of the most important elements in selecting and evaluating alternative sites. (Planners, engineers, and owners are cautioned that regardless of the transparency and technically defensible methodology, expect that the final site will always have opposition.) This is discussed in more detail later in Section 4.4.5, Stakeholder and Public Involvement.

4.0   Developing Evaluation Criteria

4.1   General

The four categories of site selection criteria—environmental, technical, institutional, and cost considerations—are detailed below.

Environmental criteria are issues that would likely be examined in an environmental impact assessment for the project, including natural and manmade resources. They also may include regulatory constraints or aspects of the project that would require environmental permits. Adverse environmental impacts can render a site not acceptable for WRRF use.

Technical criteria are constraints or objectives of project design and implementation. They include specific engineering requirements and site-specific conditions that may affect the technical feasibility of a project (and thus costs) on a particular site. Adverse technical impacts are generally overcome by more complex designs and higher costs.

Institutional criteria are a "none-of-the-above" category for issues important to, or under the control of, other stakeholders. Site acquisition, conflict with public policy goals, and permits are some examples of this category of criteria.

Cost considerations include the purchase price for the site, costs to mitigate unfavorable site conditions, capital costs for technologies or construction practices that are required for the sites, and 20-year (or more) operating costs of project implementation on a site.

It is important to note that when comparing multiple sites, only the differences between sites are important. If sites have the same issue, that particular issue does not affect the selection process.

4.2   Environmental Considerations in Site Selection

4.2.1   General Considerations

The general rule when considering any kind of site development is to (1) first avoid environmental impacts by staying clear of the desirable or protected site feature(s); (2) minimize any impacts that cannot be avoided; and (3) mitigate for unavoidable impacts. This three-step sequence is applied during site development planning, but should be factored earlier into the site selection process.

The site selection process should result in the elimination of sites that have major environmental constraints. However, even the best sites often have sensitive features that must be factored into the design process. For example, a small wetland may not necessarily preclude selection of a particular site, but may significantly affect the layout of facilities on that site. The identification of site environmental features, such as wetlands, water bodies, floodplains, designated habitats, and historic/archaeological features, is critical in determining buffer zones, setbacks, and thus available area for each site.

4.2.2   Site Land Use

The current and previous uses of a particular site are important to its development potential. For instance, an undeveloped ("greenfield") site will be desirable, because the site preparation costs and contamination potential likely are lower, with fewer potential infrastructure conflicts than a developed industrial ("brownfield") site. However, it is possible that only brownfield sites are available. Furthermore, regulatory authorities may favor redevelopment over new development, and regulatory and/or financial incentives for redevelopment may be available.

The project manager should obtain all available information on current and former site use, including former locations of buildings, structures, rail lines, and underground tanks, and likely contaminants (often through the appropriate state regulatory agency that oversees removal of contaminated soils and groundwater). Possible sources of information include old aerial photos, soils maps, and land-use maps. Later in the site selection process, if a question remains on possible contamination, it is appropriate to perform a Phase 1 Environmental Site Assessment (ESA), to identify the presence of contaminated soil; buried, leaking fuel tanks; polychlorinated biphenyls (PCBs), buried herbicide and pesticide containers; and other dumping activities. When considering the demolition of existing structures, the presence of asbestos insulation and cement, flooring, and roofing materials should be assumed or identified. Phase 1 ESAs are typically inexpensive and can be performed quickly by experienced individuals before land purchase.

4.2.3 Surrounding Land Use
A determination of the effects of a WRRF on the surrounding land in the area should consider zoning regulations (including odor regulations), effects on adjacent property values, and compatibility with activities on neighboring properties. Constructing a facility in an industrial neighborhood rather than a residential area (particularly an affluent area) is generally more acceptable and less expensive. Wastewater treatment facilities located near airports may require U.S. Federal Aviation Administration (Washington, D.C.) approval, because airplane glide slopes may control the height of some structures. Lagoons and ponds attract birds, which also affect air-traffic operations.

If the selected site is surrounded by residences or zoned residential for future development, measures must be taken to ensure that the facility is a good neighbor during both construction and operation. Such measures include minimizing noise, odors, aerosols, air particulates, chemical hazards, insects, intrusive lighting, and traffic effects, and incorporating aesthetically pleasing architectural and landscaping features. However, all of these affect costs, which should be planned.

4.2.4 Natural Resources
In general, sites within specially designated natural areas should be avoided. The development of areas designated as wild, scenic, or recreational, under the Wild and Scenic Rivers Act (1968), or habitats of rare, threatened, or endangered species may be prohibited or, at a minimum, very difficult. Similarly, shorelines often are reserved for public use; this is especially important in urban areas, where a shortage of open shoreline access exists.

4.2.5 Historical and Cultural Significance
Federal, state, and local entities that maintain catalogs of significant archaeological and historical areas and resources should be contacted during the site selection phase to determine whether the site has historical or cultural significance. If there is reason to believe that the site might have historical or cultural significance, a preliminary site investigation should be performed by qualified archaeologists or historians.

The investigation will identify the need for preconstruction removal or preservation, the need for an on-site archaeologist or historian during construction, and mitigation measures, to ensure that sensitive features are properly preserved or documented before removal. If such resources are present, procedures prescribed by the Advisory Council on Historical Preservation (36 CFR Part 800) (U.S. EPA, 2009) and state historic preservation officials may be required to limit any adverse effects.

4.2.6 Setbacks and Buffer Zones
Setbacks (the distance between a building and the property line) are established by the building department having jurisdiction over the site. Buffer zones are recommended to reduce odor, noise, and light intrusion to the surrounding community. The amount of isolation and buffer area needed between facility processes and sensitive features and between facility processes and other property owners needs to be identified and included in the area requirements.

4.2.7 Air Quality
WRRFs can be sources of odor, chemical emissions, particulates, and aerosols, all of which must be controlled and permitted. However, air quality (like seismic activity) is not often a differentiator; generally, all the sites will require the same air quality measures equally. See MOP 8, Chapter 7 for discussion of odor regulations and community effects.

4.2.8 Noise Impacts
Noise management is an important consideration that can influence the site selection process. Communities may have noise ordinances that need to be incorporated into the comparative process. When comparing possible facility layouts, the difference in sound pressure can be quantified with the corresponding differences in distance. The equation below can be used to approximate variation in noise with distance, and to help establish buffer zones.

$$P_2 - P_1 = 10 \times \log\left(\frac{R_2}{R_1}\right)^2$$

where P_2 = sound pressure at Location 2
P_1 = sound pressure at Location 1
R_2 = distance from source to Location 2
R_1 = distance from source to Location 1

A survey to determine ambient levels of noise is recommended at any proposed WRRF site. A 3-dBA increase above ambient levels of noise by the WRRF has little or no effect on surroundings; noise level increases of 3 to 15 dBA have a moderate effect; and noise level increases of more than 15 dBA severely affect the surroundings (NCEES, 2005).

4.2.9 Hazardous Chemicals
There is always a potential for chemical spills and leaks. However, the wastewater industry has an exemplary record in chemical handling, primarily because of the training that operating personnel receive and strict adherence to procedures. Despite this record, sites that are far removed from sensitive receptors, such as schools, hospitals, daycare centers, and convalescent homes, are preferable. If there are sensitive receptors

near a proposed site, the use of chemicals and chemical safety must be addressed during the design process, leading to higher costs. These additional costs should be factored into the site evaluation process.

4.3   Technical Considerations in Site Selection

4.3.1   Proximity
An ideal site should be close to the raw wastewater collection system, treated effluent disposal point (or reuse application points), and biosolids disposal locations. It may be obvious that a site close to the existing collection system or the discharge point is preferred over a remote site. However, a remote site may be closer to the biosolids disposal location, thereby lowering biosolids hauling costs, so the proximity issue is not always clear and a more detailed analysis may be necessary.

4.3.2   Discharge Permit Conditions
When multiple sites discharge to the same water body, the site selection process is unaffected by permit conditions. However, one potential site might be close to a receiving water body that requires tertiary treatment, while an alternative site might discharge to a different receiving water body that only requires secondary treatment. Under this scenario, the area requirements and costs for additional tertiary treatment, or conveyance back to the other receiving water body, must be factored into the selection process.

4.3.3   Elevation and Topography
Low-lying sites near the point of discharge facilitate the flow of wastewater from the service area by gravity and minimize the number of pumping stations in the collection system. However, such a site also may require flood protection. Constructing dikes, process tanks, buildings, entrances, etc., above the expected high water levels provide flood protection, but these methods can be costly and may minimize the advantages of selecting a low-lying site. When contemplating the use of a site in a floodplain, a designer should contact the local floodplain management and local building authorities to identify the design flood event (100-year or 500-year floodplain) and any potential restrictions on property development.

A relatively flat site will better facilitate construction activities than a steeply sloped sight. A site with significant topographic change may provide challenges for gravity flow, significant earth cuts and fill, vehicular accessibility, and proximity between buildings and processes, all adding costs that need to be taken into consideration.

4.3.4   Geology, Hydrogeology, and Soils
Site geology, hydrogeology, and soil types significantly affect construction costs and are therefore important considerations in site selection. Unfortunately, adverse geology such as shallow rock formations and/or high groundwater conditions may not be readily apparent until soil borings are taken much later during the design process. Engineers can often get a qualitative understanding of soil conditions by working with a geotechnical consultant, or by reviewing well logs where available. If adverse geologic conditions are suspected, additional costs should be planned.

4.3.5   Seismic Activity
Seismic activity is generally not a differentiator and will not influence site selection when all potential sites are within the same seismic zone. However, if sites are within

different seismic zones, additional design and construction costs to accommodate the seismic activity need to be factored into the evaluation. Generally, the building department for each community will confirm the seismic zone that must be accommodated.

4.3.6 Transportation and Site Access

WRRFs should be located close to all-weather roads for delivery of equipment and chemicals and off-site disposal of grit, screenings, and solids. Access roads that have reasonable gradients and curves of adequate radius allow the movement of large vehicles and heavy equipment. Service from a railroad spur offers the facility the option of accepting deliveries by rail, thus lowering the unit cost of bulk deliveries (i.e., chlorine and other chemicals). Costs for nonexisting or inadequate roads or bridges need to be included in the cost analysis. Access to public transportation supports worker satisfaction and reduces parking needs.

4.3.7 Utility Services

A WRRF should have potable water, redundant electrical power, and natural gas service, all of which can be expensive. (Telephone and Internet service are also important but easily supplied.) If a site has strong potential but lacks nearby utility service, the design engineer should meet with representatives of utility companies to discuss utility sizing, extensions, and potential costs. Planning for utility extensions and capacity increases may take several years; and this process should be started early when planning and for site selection of a WRRF.

4.4 Institutional Considerations in Site Selection

Institutional criteria include issues that are important to, or under the control of, other stakeholders. Site acquisition, conflict with public policy goals, and permits are some examples of this category of criteria. State, regional, and local policies and interests also will play a role in shaping the criteria.

4.4.1 Sustainability

As described in greater detail in *Sustainability and Energy Management for Water Resource Recovery Facility Design* (WEF, 2017), consideration of sustainability goals in WRRF design has become important in recent years. This same focus on sustainability goals applies during the evaluation of alternative sites for WRRFs. Some states, and even regional and local agencies, have adopted sustainability policies and requirements that are relevant to site selection, building materials, and construction, including the requirement for Leadership in Energy and Environmental Design (LEED) certification, in some cases. For example, in Massachusetts, all state agencies are tasked with meeting "sustainable development principles," which apply not only to state projects but also to projects that require state approvals and/or funding. Some of the principles that may be pertinent during WRRF site selection (and facility layout) include the following:

- Concentrated development (i.e., discourage sprawl) and mix uses (residential, conservation, and commercial). This includes reuse of previously developed sites (i.e., brownfields);
- Advance equity (i.e., promote social, economic, and environmental justice);
- Protect land and ecosystems;
- Use natural resources wisely;

- Promote clean energy; and
- Plan regionally.

4.4.2   Land Acquisition and Ownership

The number and type of property owners of a proposed site (or adjacent sites) can significantly affect the degree of difficulty in acquiring the site. Ideally, one site has only one landowner who is a willing seller. As the numbers of sites, owners, and resistance to selling increase, so will the time and difficulty involved in acquiring the property.

Land acquisition also applies to off-site infrastructure (i.e., easements for collection mains). Ideally, pipelines can be installed within existing easements and rights-of-way; however, if extensive new easements are required, this consideration also should be factored into the site selection process.

4.4.3   Environmental Justice

Environmental justice also referred to as environmental or social equity represents the confluence of social and environmental movements and deals with the inequitable environmental burden borne by minority groups. The importance of environmental justice in the site selection of wastewater facilities has increased significantly in recent years. Today, the site selection of a WRRF is likely subject to federal, state, or regional environmental justice procedures, requiring documentation of the population characteristics of the host community, in terms of minorities and other less-represented groups, such as seniors, and evaluation of the effects of the proposed project on the well-being of those groups. Thus, evaluation and ultimate selection of a WRRF site should include evaluation of the relevant environmental justice factors.

4.4.4   Permit Requirements

The regulatory process(es) driving the site selection can influence site selection. If there are specific regulations that pertain to site selection of the project, the criteria should reflect these regulatory requirements. For example, if there are wetlands on a site, the site selection decision may be dependent, in part, on obtaining a U.S. Army Corps of Engineers (Washington, D.C.) permit to allow wetland filling. Local zoning ordinances are another example of an approval requirement that could heavily influence a site selection decision. It is imperative that all of the permits be identified during the planning process and a determination be made as to their relevance to a site selection decision.

The regulatory process(es) driving the site selection will strongly influence the choice of site selection criteria. If there are specific site selection regulations that pertain to the project, the criteria should reflect the regulatory requirements. For example, some states have, as part of their solid waste, hazardous waste, or groundwater protection regulations, specific criteria for evaluating sites for waste-handling facilities. Appropriate regulations should be reviewed when the list of criteria is compiled.

Local policies and interests also will play a role in shaping the criteria. One example of a local interest criterion is "equitable distribution of regional resources." This criterion may or may not be important, depending on the service area of the proposed facility and the interest of the involved communities in making sure that public facilities are equally spread among the communities they serve. Input on these local criteria is best obtained from local officials and citizens.

4.4.5 Stakeholder and Public Involvement

In the context of WRRF siting, the term "stakeholder" generally refers to the group(s) or individual(s) that will play a significant role in the siting process.

The project owner and/or representative plays a major role in determining the course of the project, ensuring that it remains on a predetermined course, schedule, and budget. Other stakeholders may include regulatory agencies, local officials, and members of the general public. Regulatory agencies must be convinced that the project meets regulatory requirement, whereas local officials and groups, as well as members of the public, will have interests in the project as they may be directly affected by the proposed facility.

All of these stakeholders have specific roles and appropriate opportunities for input regarding site selection; therefore, the stakeholders and roles should be determined early in the siting process. The remainder of this discussion will focus specifically on the role of the public, whose role is typically less predictable than the prescribed roles of traditional stakeholders.

The public can play a crucial role in formulating, as well as implementing, the siting approach. This is especially true for controversial projects. The project manager should develop a public participation plan that describes the form and level of public involvement. Hiring a public relations expert to help guide the process is highly recommended, particularly for large and/or complex projects. When executing public participation programs, "more," "earlier," and "full-disclosure" are key to successful public involvement, particularly in the case of controversial siting studies, when project success may depend, in large part, on public approval.

The depth and complexity of a given project will affect the level of involvement granted to the public. Straightforward projects may only include a public meeting to present the justification and results of a siting study for those community members interested in the project. For more complex projects, public contribution may include a more direct role in the siting process. In this case, the public may provide input regarding siting criteria, evaluation methods, and results. Such input generally comes from citizens' groups that are specifically formed for the purpose of providing public opinion regarding the project.

There are four aspects to consider in developing a public participation program: public education, relations, information, and involvement. The purpose of public education is to create a more informed public body. Methods of promoting public education include regular neighborhood meetings, press releases, project flyers, eRooms, and websites. Public relations refer to any type of direct interaction with the public, which can help create a positive image for the proposed project or organization. Informing the public about the site selection process (and the ramifications of such) builds awareness and generates interest. Public information does not typically involve or foster feedback. Finally, when public involvement is exercised, the public will be part of, or otherwise involved, in the decision-making process. A successful and thorough public participation program will encompass all of these forms of public engagement (WEF, 2002).

Several benefits result from actively involving and investigating the public's opinions and concerns regarding site selection of a WWRF, including building support, helping communities understand the value of the procedure, diminishing opposition, and building a more positive image. This results in a more informed, understood, and acknowledged individual (WEF, 2002). Involving the public will also lead to a more streamlined method of communication, which will enhance future interactions with the surrounding communities. Building rapport with affected persons is key to successfully involving the public and their concerns regarding the site selection process.

Public outreach is now incorporating different methods and types of communication due to advances in technology such as social media outlets. Online information streams and social media outlets, such as Facebook and Twitter, foster information delivery and uptake for the affected communities and the municipality. Diversifying the methods and types of public outreach will aid in educating and understanding how communities are affected by increasing and diversifying the audience, likelihood of information delivery, speed of questions and responses, as well as information updates to the site selection and facility development. It can be difficult to cater to a large body of people; however, incorporating social media into the public participation program can increase the attitude toward the process by providing an easier, and potentially more acceptable, method of involvement. In addition, social media sites also promote communication between end users and other people directly affected by the proposed facility location, leading to a stronger development of ideas, a deeper connection within the community, and a more robust network of information. These methods of public outreach should by no means replace conventional methods of communication and should be used to supplement already-existing and effective methods.

The key to any public participation process is organization and proper implementation of a well-conceived plan. There must be a plan developed at the outset of the project, which establishes what the public concerns and interests are anticipated to be and how, when, and where they will be addressed. With public acceptance, WRRF siting generally is much smoother than when the public is not satisfied with the siting process, results, or their opportunities to provide input. Therefore, it is in the owner's and project team's best interests to foster public communication throughout the siting process.

4.5 Cost Considerations in Site Selection

4.5.1 General

Many of the cost considerations have previously been identified, and this section helps to consolidate the costs issues. Costs of the actual WRRFs are likely the same regardless of the site, unless alternative discharge water bodies are available and thus alternative permit and WRRFs are required, or if topography requires an intermediate pumping station. Where costs are the same, they do not affect site selection. However, it is likely that additional treatment and features may be required at certain sites to be good neighbors.

In some cases, compensatory/incentive measures may be needed for increasing the regulatory and public acceptability of projects. Compensatory or incentive measures are sometimes required to gain political and public acceptance of a project, and to make a whole project and its effects more acceptable to abutters and communities. Compensatory incentives generally are not legally required, unless they are included in a contract developed between the project proponent and the host community specifically for the proposed project (i.e., a memorandum of agreement).

4.5.2 Capital Costs

Capital cost considerations for sites can include the cost difference for the following:

- Land purchase;
- Building and infrastructure demolition or relocation;
- Brownfield/site/building remediation;

- Wastewater conveyance (to the new site or discharge point);
- Mitigation of wetlands, endangered species, archaeological or historical concerns;
- Additional effluent treatment (for differing National Pollutant Discharge Elimination System [NPDES] permit limitations);
- Geotechnical and/or topography hardship;
- Electrical power, water, and gas utilities services;
- Access road and/or bridge;
- Flood protection;
- Good neighbor measures:
  — Aesthetically pleasing architectural features, and additional landscaping, and
  — Additional odor control, noise abatement measures.

The costs associated with most of these issues will be very difficult to predict early on. Where estimates can be reasonably predicted, they should be developed. However, where the planning budget is constrained or where not enough information is known, contingency factors can be applied to the overall cost estimate for each site.

4.5.3   Operations and Maintenance Cost Considerations
Operations and Maintenance (O&M) staff are likely to be identical for each site, and do not affect the site selection process. However, O&M costs for differing site conditions can include the following:

- Conveyance of wastewater (pumping to the site or discharge point);
- Land disposal cost difference (principally hauling costs); and
- O&M costs for additional treatment or mitigation.

As with capital costs, O&M costs may be very difficult to estimate, and therefore, contingency factors can be applied to the overall cost estimate for each site.

5.0   Determining Water Resource Recovery Facility Site Area Requirements

5.1   Identifying Process and Project Components
It is difficult to derive an average area requirement for WRRFs, because there are so many influencing variables, including odor control, effluent quality, biosolids handling, renewable energy systems, and many other factors. Project managers are encouraged to be conservative during this process, or to use practical and conservative assumptions to bound the site area requirements into minimum and maximum needs. To start this process, the following are identified as the major factors affecting area requirements:

- Flow and constituent concentrations;
- Effluent NPDES permit requirements;
- Solids treatment processes; and
- Ancillary facilities.

Projected flow needs to be established to determine liquid stream area and volume requirements, and concentration and thus loading is required for sizing the solids treatment processes. (These are presented in Chapter 2.) Effluent NPDES permit conditions will affect treatment and thus affect area requirements, depending upon the receiving water body. Act 451, Part 41 permit requirement will include items such as redundancy, power, floodplain protection, etc., which will affect area requirements. Redundancy must be factored into both liquid and solids treatment schemes. An alternative power source or emergency generators are required.

Solids treatment area requirements are greatly affected by the ultimate treatment and disposal process selected. Thickening, dewatering, and landfilling will require lower area requirements, whereas digestion and/or land application (and liquid sludge storage) will require more land. Project managers are cautioned to be conservative in assuming solids treatment processes, and be mindful of the concept of *WRRF of the Future*, which in part is energy neutral, recovering energy via digestion and natural gas production.

Ancillary facilities include administration, maintenance, laboratory buildings, and parking areas for each. The need for each of these facilities is likely, but potable water and clean water laboratory facilities might be combined. Likewise, maintenance facilities might be downsized in anticipation of more off-site/privatives maintenance operations, or possibly combined with an adjacent Department of Public Services yard. Area requirement will also include flow equalization, roads and greenbelts for yard piping, splitter boxes, septage or vactor receiving station, substations and generators, and other utilities, space for buffer zones, setbacks, environmental controls, and spill containment areas. Additional area requirements may include off-site sewer maintenance and/or industrial pretreatment program equipment and staff, an adjacent DPW yard, recycle program drop-off, composting facility, and the like. Project managers will need to communicate with the owner on these potential area needs.

Once the approximate processes and facilities are selected, area requirements for each can be approximated. A simple approach is to determine areas of local WRRFs similar in size and treatment and plot flow versus area. Google Earth or Google Earth Pro can be used to assist if the actual area is unknown. Figure 3.1 can be used to approximate the area requirements based on WRRF capacity. A more exacting approach is to size the various circles and boxes for each process area as described below.

5.2   Liquid Stream Area Requirements

The traditional processes to size include headwork (bar racks, influent pump station, fine screenings, and grit facilities), primary clarifiers, aeration tanks, secondary clarifiers, tertiary filters (if required), and disinfection. Hydraulic Institutes ANSI/HI 9.8 can be used to determine geometry and pump placement to determine overall pump station size. Grit and screening facilities can be sized using Chapter 9. Preliminary sedimentation overflow rates can be selected from Chapter 10 herein or from Ten States Standards for primary and secondary clarifier sizing. Biological treatment can be sized using Chapters 11 to 13. Design filtration rates can vary greatly (80 to 120 L/m²/min [2 to 3 gpm/sq. ft] to 1200 to 1600 L/m²/min [30 to 40 gpm/sq. ft] per Chapter 14) and it would be wise to use a conservative design rate to size a tertiary filtration facility. And finally, disinfection facilities can be sized using Chapter 17. Redundancy needs to be factored into the area needs for each equipment and process.

Considerable area needs to be added to the footprint of any building for process equipment, means of equipment removal, pipe galleries, electrical substations and MCC equipment, control room, hallways, lavatories, etc. Yard piping does not need to

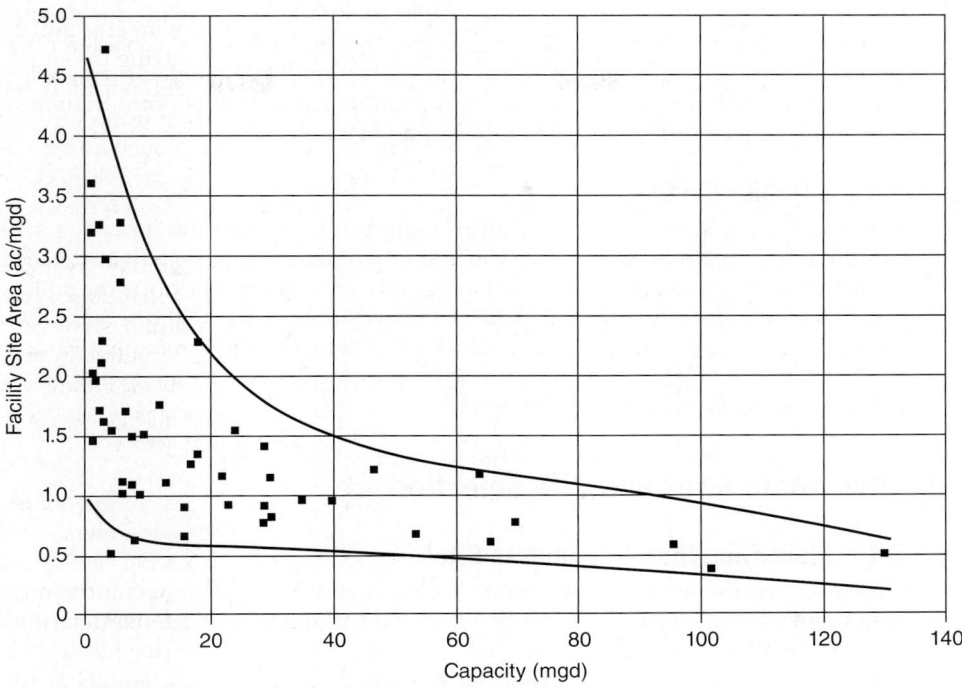

FIGURE 3.1   Facility site area versus facility capacity (mgd × 3785 = m³/d; ac × 0.4047 = ha).

be sized, but adequate room for pipe and splitter boxes need to be accounted for. An additional lift station may be necessary, depending upon site topography.

5.3   Solids Process Area Requirements

As noted above, the type of solids treatment processes have a larger impact on area requirements than liquid treatment processes, and project managers are cautioned to use conservative estimates so that when a site is selected, there is room for alternative process types. Gravity thickeners require more area than mechanical thickeners, although the difference is not significant. Similarly, centrifuges take up less space than other equipment, but again the difference is negligible. More importantly, any preliminary equipment sizing should have ample room in the building envelope for maintenance access and removal. As previously noted, digestion should strongly be considered for energy recovery, and requires significant area. A loading facility with ample truck access and trailer storage should be added to the area requirements. Solar sludge drying beds, including sludge feed conveyors, require very large areas, and could be assumed for a maximum area estimate. Chapters 19, 21, 22, and 23 can be used for preliminary design and area requirements for solids treatment processes.

5.4   Water Resource Recovery Facility Minimum and Maximum Area Requirements

Once the above assumptions and area requirements are made for each process and facility, tentative WRRF layouts can be made using the process and building circular, square, or rectangular shapes. Again, ample room should be made for yard piping, utilities, roads, parking, landscaping, buffer zones, and other considerations. The shapes should be organized into an overall square and a rectangle or irregular shape, for the minimum and maximum site area requirements. At this point, minimum and maximum acres for the planned WRRF should be known.

5.5 Satellite Campus

If area needs are large, it may be difficult to find suitable sites and it may be necessary to split up the treatment processes. The majority of the area needs will be for the liquid stream and associated buildings, and a separate area can be sized for the solids treatment facilities. Unfortunately, the total area needs (and costs) will be higher. One possible upside to a satellite campus could be that a rural solids treatment campus will likely have less odor complaints and will be located closer to the solids application point, lowering biosolids hauling costs.

6.0 Candidate Sites and Site Selection

6.1 Identifying Initial Candidate Sites

Once area requirements are determined, initial candidate sites meeting the minimum area requirements need to be identified. Initial candidate sites can be determined by any combination of the following:

- Identification of sites previously considered for similar uses or intensity of development;
- Active public solicitation for available land; and
- Using criteria to map "exclusionary areas" where the facility should not be located and identify the remaining areas as possible areas for consideration.

Typically, Geographic Information System (GIS) models and attributes are used to determine candidate areas and, sometimes, specific sites. Real estate professionals can also be used to help identify candidate sites and costs.

The criteria used to develop the list of candidates are principally the available area, but can also include proximity to existing facilities or service areas, and site availability. If information is readily available and if appropriate, features such as wetlands, floodplains, steep slopes, groundwater recharge areas, poor soils, regulatory setbacks, and dedicated public land can be mapped as "exclusionary areas," which should be eliminated from further consideration in site selection investigations. Once exclusionary areas are mapped, the candidate sites would be selected from the remaining areas.

6.2 Site Evaluation Phases

Generally, a qualitative and quantitative approach is used, typically accomplished in multiple phases, each resulting in the elimination of sites that do not meet the specified criteria defined for that particular phase. However, the number of phases is dependent on the number of sites and the complexity of the project. The goal is to systematically narrow the field, so that the least feasible sites are eliminated early in the process, leaving only the most feasible sites for detailed investigations. For the purposes of this discussion, a two-phased evaluation process is assumed.

Regardless of the phase, it is recommended that site visits be conducted before information on alternative sites is provided to the client or to the public. This is particularly important when current GIS information is unavailable, resulting in data that could be outdated and/or inaccurate. A quick "windshield survey" of surrounding land use, for example, could help verify whether the location of sensitive receptors, as interpreted from aerial photos, is accurate.

6.2.1   Phase 1 Evaluation Criteria

Phase 1 evaluation criteria generally are of two types: critical constraints (including "fatal flaws" and mandatory regulatory constraints) and technical deficiencies. Fatal flaws are issues that have a high probability of prohibiting development of the project at a specific site. Mandatory regulatory constraints are nonnegotiable conditions that the facility is required to satisfy, or protected areas that would be off-limits to the facility. Technical deficiencies are site obstacles that require significant engineering and design solutions conditions for the project to function properly.

Examples of Phase 1 critical constraints include the following:

- Land use compatibility (examples include avoiding developed residential areas and federal, state, and local parks);
- State or federally designated wetlands and/or other officially protected areas;
- Historic sites listed, pending, or recommended for inclusion in the National Register of Historic Places (state- and locally listed historic and archaeological sites also could be included here);
- State-catalogued endangered species habitat;
- State- or federally designated inactive hazardous waste disposal sites or corrective action sites or other sites known to be contaminated with hazardous wastes (depending on the severity of the contamination, this could be simply a cost issue. In addition, some of these sites designated as "brownfields" may be preferred areas for WRRF development); and
- Groundwater recharge areas.

Sites with known environmental conditions or irregular shapes need to be reevaluated to see if they meet the minimum area requirements. Wetlands, floodplains, setbacks, buffer zones, and extreme topography areas should be eliminated from each site as undevelopable, the available site area recalculated, and the site eliminated if the minimum area is not met.

There are two important points to make regarding the above criteria. First, some of the information required for decision making may not be available during Phase 1, particularly when there are a large number of sites. For example, wetlands boundaries may be inaccurate if based only on existing large-scale maps. Second, some of the criteria may apply only to certain project components. The above criteria are presented only as examples of possible Phase 1 criteria. In some cases, it may be more appropriate to apply them in a later screening phase.

Constraints and objectives mapping (often using GIS) is a particularly powerful method for conducting the Phase 1 studies. Buffer areas, mandatory setback distances, protected areas, and incompatible land uses can be identified and mapped on the candidate sites or in the study area. These constraints, or exclusionary areas, can be overlaid with sites or areas satisfying technical objectives. Sites or areas not overlapping exclusionary areas are then carried forward into the Phase 2 assessment.

6.2.2   Phase 2

At this stage, it generally is assumed that the facility could be constructed at any of the sites under consideration. It remains only to determine which site is best from the perspective of environmental, technical, institutional, and cost criteria.

It is critically important that this Phase 2 selection process is quantitative (wherever possible), transparent, and the gauging criteria for nonquantitative criteria are established prior to the evaluation process. It will be easiest to base all decisions on costs or a cost-effective analysis, so the differing environmental, technical, and institutional issues for each site should be translated to costs wherever possible.

For environmental criteria, adverse site land use conditions can be quantified by estimating the cost to remediate that site to a comparison site or an assumed standard. Similarly, costs to mitigate noise, odors, chemical hazards, etc. and aesthetically pleasing architectural and landscaping features need to be estimated. Costs to mitigate technical issues can also be estimated. Conveyance (proximity), additional treatment (permit conditions), earthwork or intermediate pump station (elevation and topography), piling or groundwater pumping (geology, hydrogeology and soils), and electrical substations (utility services) can all be estimated, or factors applied to the overall costs for a given site. In general, all the cost factors identified under Section 4.5 should be quantified wherever possible.

Some environmental and institutional criteria will be harder to quantify, and might be used to select a site where costs are very close. Any nonquantifiable criteria can be defined in terms of levels of impact, so that sites can be measured uniformly against the criterion. For example, what does "compatibility with adjacent land use" mean? More importantly, how will it be measured and quantified? Many of the definitions are subjective; therefore, it is critical that they be defined at the start of the site selection process.

Generally, the key is to determine what constitutes an acceptable versus unacceptable impact (often segregated into three qualitative categories of significant, moderate, and insignificant impacts). If there is a numerical basis that can be used to define what constitutes an acceptable versus unacceptable impact, then it should be used. If there is no numerical or even regulatory basis (i.e., "compatibility with adjacent land use"), it is important to define logical, easily applied measures that will make sense to reviewers. For example, "compatibility with adjacent land use" might be measured by the percentage of residential land use within a certain distance of the site, zoning of the same area, and plans for future use of that area. Stakeholders, particularly the public, can assist greatly with these assessments.

6.3   Evaluating the Sites and Site Selection

At this point, costs for quantified criteria for each site are presumably established. Once the project team (including stakeholders) has selected the nonquantified criteria and defined them, the criteria must be applied objectively to each site. Inevitably, each site will have its own merits and faults in each of the criteria categories, ideally each resulting in a quantifiable cost-based assessment and a nonquantifiable ranking or level of impact.

The final step in the process is to reach a site selection decision that is consistent with the established site selection criteria. The project manager can either recommend the final selection to the owner and stakeholders, or present a few alternatives with the advantages and disadvantages of each. In all but the most straightforward projects, it is generally preferable for the project manager to present the "finalists" to the owner, thereby allowing the owner to make the final choice. One approach is to make a recommendation regarding the best site in each category of criteria (i.e., "best" technical site and lowest cost site) and then allow the owner to make the final selection. Once the site selection decision is made, agreements establishing site acquisition and development procedures (and often compensation) should be finalized with the host communities, as appropriate.

7.0 Facility Arrangement and Layout

7.1 General

The section assumes that the process treatment decisions have been made, WRRFs sized, and a site selected, and that decisions on how to arrange the facilities are now required. Arrangement decisions need to be based on personnel and vehicular logistics as well as the physical connections between facilities. Facilities should be arranged to minimize the length of connecting and recycle piping, and facilities should be arranged to minimize staff personnel and vehicular travel around the site. In this manner, the arrangement of treatment processes on a facility site can affect the total life-cycle cost.

7.2 General Facility Arrangements

7.2.1 General Considerations

In general, liquid treatment processes can be arranged separately from solids treatment process. Likewise, the ancillary facilities (administration, laboratory, and maintenance buildings) can be arranged around the liquid and solids treatment facilities.

The facility arrangement should start with isolating any areas of the property deemed less suitable for development, such as wetlands, potential brownfield areas, etc. Buffer zones, setbacks, 100-year floodplains, and other similar areas should also be identified as unavailable. Solids treatment facilities should be designated to areas less prone to odor complaints. From there, the facility arrangement can start with the location of the influent sewers and the ultimate discharge location.

The liquid stream facilities layout should progress sequentially from the influent sewer and lift station to the disinfection facility and outfall. This layout can follow a linear progression, or can progress out the back, particularly when the influent sewer and effluent discharge locations are nearby. Locating raw wastewater pumping stations near the point at which sanitary sewers enter the facility site minimizes the cost of building additional lengths of deep gravity sewers. Where possible, a designer should group together similar unit processes to facilitate operation, minimize piping, and allow for expansion.

It will be helpful for the designer to a preliminary hydraulic profile to establish the profiles of principal structures. Arranging treatment processes to follow site contours helps maximize the use of natural topography and reduces both pumping requirements and excavation costs for new structures. Flow splitting is another aspect that needs due consideration.

7.2.2   Perimeter versus Internal Facilities

Many facilities should be located on the facility perimeter, within the WRRF, or for some facilities, cases can be made for both.

Administration building and laboratories (if not within the administration building), should be located on the site perimeter, to facilitate visitors, sample deliveries, and the like. Facilities unlikely to cause odors, including tertiary treatment and/or UV disinfection facilities, can also be located on the perimeter. As previously stated, solids treatment facilities, particularly those where odors are likely to occur, should be located within the facility or to an area where odor complaints are less likely to occur. A significant exception is the biosolids disposal loading area; truck traffic should be kept to a minimum within the facility, so biosolids loading facilities should be located near the perimeter or near a major roadway. Consideration of a separate access road for biosolids haulers should be considered. Chemical feed facilities have competing interests; they should be located near the application point, but minimizing internal truck traffic should be considered. A maintenance facility should be centrally located within the WRRF since that will promote efficient logistics of personnel and parts. However, access for spare parts delivery needs to be considered. Chemical bulk storage facilities and chlorine storage areas should be located along the main service road to make deliveries more convenient. A separate set of ingress and egress gates and/or loop road for chemical, parts deliveries, and solids hauling should be considered.

In locating chemical storages and dosing facilities, rules, regulations, safety, compatibility of different chemicals, and O&M aspects should be considered. For example, there should be a minimum space between the areas where the public has access and where chlorine containers are stored. Incompatible chemicals should not be stored close to each other. A lime silo should be located as close as possible to the dosing point to minimize the O&M problems resulting from blockages.

To reduce the sound level at the facility boundary, a designer should consider enclosing blowers, compressors, large pumps, centrifuges, and other equipment that operate at high speeds in buildings of appropriate, sound-attenuating construction. If possible, facilities that generate noise should be located as far away as possible from potential receptors and should erect sound walls, berms, and heavy landscaping in the surrounding area to minimize WRRF noise.

7.3   Administration, Staff, and Support Facilities

7.3.1   General Considerations

It is necessary to provide adequate support facilities to ensure efficient O&M of a WRRF and accommodate the staff who operate and maintain the facility. The extent of provision of support facilities depends on various factors, including size and complexity of the treatment process, location, philosophy (how much is done by staff vs. outside contractors), and owner preference. Some WRRFs may require a high level of support provisions, including control rooms, administration offices, lunchrooms, training/conference rooms, locker rooms, maintenance workshops, storage rooms, laboratories, toilets for staff and visitors, reception areas, and visitor education facilities. Other facilities may want to contract out all but the routine maintenance and laboratory work. In the age of "doing more with less," combining WRRFs with DPW yards, potable WRRFs, or other municipal facilities should be considered and discussed with owners.

7.3.2 Administration Offices

A small WRRF may need only a workstation for the facility operator, which could be in the same room where the computer terminal for controlling the facility is kept. A large facility may have administrative office complexes, including office areas, lobby and reception areas, and training and conference facilities.

Where administrative offices are provided and the public has access to them, they should be located near the front entrance of the facility, so that visitors may find them easily. A private office with meeting space provides a facility manager with the privacy needed to handle personnel issues, space to hold staff meetings, and easy access to facility operating records, personnel files, cost records, and O&M manuals.

If a facility has distinct organizational groups or departments, each manager may need a private office. Grouping the individual offices of a management team promotes communication among the various groups. In larger facilities, an assembly room large enough to accommodate facility personnel meetings, on-site training, visitors, and public meetings is desirable.

7.3.3 Maintenance and Storage Facilities

Maintenance and storage room facilities are dependent on the WRRF, and how much of the work is expected to be contracted out. The size and type (complexity) of the WRRF generally determines the size of the maintenance staff, the services to be performed in-house, and the tools to be provided. As previously stated, maintenance facilities are best located near the center of a facility or in the most equipment-intensive area of the facility. Facilities with a high degree of instrumentation may require a separate instrumentation shop and maintenance staff. A designer should size the area that will house maintenance equipment manuals and shop drawings in addition to equipment records. In addition, facilities for personnel to review documentation and for computerization of maintenance records, inventory, and maintenance scheduling should be provided.

The storage area for the facility's spare part and maintenance supply inventory is best located near the shop area and should be large enough to accommodate an array of shelves, bins, and drawers. Paints, lubricants, pesticides, herbicides, and similar toxic, flammable, and hazardous materials should be stored in an isolated, secure area with adequate ventilation. To facilitate deliveries, the storage area, which typically is secured, should be located adjacent to the roadway of the main facility. The building should be fully accessible to maintenance and delivery trucks and should include ramps for the passage of tractors, hand carts, and other mobile equipment.

7.3.4 Laboratory Facilities

A laboratory is necessary to analyze samples for the purpose of operational control and regulatory monitoring and reporting. The size of the facility, type of treatment provided, need to analyze samples from off-site facilities (such as CSO facilities or industrial pretreatment program samples), and extent of sample analyses to be performed on-site determine the size and layout of the laboratory (Great Lakes–Upper Mississippi River Board of State and Provincial Public Health and Environmental Managers, 2014). In instances where various components have been designed or sized based on design flow, it may be beneficial to size the lab based on the same methodology. The text *Laboratory Planning for Water and Wastewater Analysis* provides insight, figures, and direction for sizing laboratories in wastewater analysis applications (Clark, 1988).

For many small facilities, it may be cost-effective to contract all but the simplest analyses to outside laboratories. It may be economical to provide a small laboratory in the facility, for conducting routine operational control tests, and a large laboratory in a convenient location outside the facility, for conducting complex tests for the facility and tests for outsiders. During the design stage, a designer should identify the type and anticipated frequency of tests and the analytical equipment required to conduct them. Future operating and monitoring requirements should be evaluated and factored into the layout.

7.3.5   Staff Facilities

Staff facilities must comply with the requirements of the Occupational Safety and Health Administration (Washington, D.C.) and other federal, state, and local legislations and national and local standards, guidelines, and codes of practice. The designer of such facilities should give due consideration to the fact that those facilities could be used by permanent or temporarily disabled persons, male or female.

Restrooms should be provided for all O&M personnel throughout the facility, especially in areas where employees are stationed and in locker rooms.

An essential part of ensuring a safe and efficient facility is providing a training room. The room should be designed for demonstrations and presentations that will use audiovisual aids, such as DVD players, televisions, overhead projectors, video monitors, whiteboards, and easels. An emergency shelter also should be designated for employees to use during hurricanes, tornadoes, earthquakes, and other natural disasters. Large facilities may be well served by having a separate first-aid room.

In the layout of these facilities and the grounds, it is important to conform to the Americans with Disabilities Act (http://www.ada.gov) and other regulations concerning access to the disabled in publicly accessible areas. In planning and designing of these facilities, it also is important to keep in mind that these facilities may be used by both males and females. It is important to refer to local codes for additional requirements.

7.4   Other Layout Considerations

7.4.1   Roadways and Walkways

Roadways must provide access for all O&M needs (including access to double and overhead doors, access hatches, and roof- or yard-mounted H&V equipment), chemical and parts deliveries, vactor and septage receiving stations, and residuals loading areas. Pavement widths, adequate curve radii, and grades should be planned accordingly. Main roadways of 6 m (20 ft) in width and service roads of 4.9 m (16 ft) in width have been adequate at most facilities (J. M. Montgomery Consulting Engineers, 1985). However, 3.6 m (12 ft) is the minimum width recommended for one-way use.

Pavement slopes of greater than 1.5% minimize the potential for ponding and freezing of water on roadways and delivery areas. Maximum slopes should be limited to 7% for general travel and loading areas, although slopes of up to 12% may be possible for short distances, such as ramps and areas with extremely steep terrain. Slopes beyond 7% can be difficult for larger and heavier vehicles to climb and may cause safety and maintenance problems in cold weather areas subject to snow and ice. If possible, sight distances and curvature for access roads should be provided to permit safe vehicle operation at speeds of up to 56 km/h (35 mph). Sufficient visibility should be provided around tanks, buildings, and other structures. A minimum 8-m (25-ft) clear sight triangle should be maintained at all intersections and delivery areas.

It is a good practice to provide looped roads to eliminate or minimize vehicle reversal. It is also advantageous to have two entrances—one for the staff and facility visitors, and another for deliveries, septage haulers, and the removal of residuals. Emergency access should always be provided for ambulance and fire trucks. The local fire department may have specific requirements for equipment vehicle turnarounds and width and location of fire lanes.

Paved sidewalks at least 1.2 m (4 ft) wide should be provided in all areas of the facility particularly for sampling stations, buildings, tanks, and other areas requiring frequent monitoring. Lighting should also be provided, particularly for outdoor equipment.

Facilities that extend over a large area and are located in harsh-weather environments may be well served by using tunnels. Tunnels connect major process areas, serve as pipe galleries, contain electrical conduits and instrumentation cables, and perhaps allow the passage of small vehicles, such as golf carts, used for maintenance and sample collection. When used, tunnels and pipe galleries should be designed with large hatches, skylights or removable top slab sections (to facilitate the installation of piping and equipment), lighting, heating and ventilation, and provided with additional exits for emergency use.

7.4.2   Security
Access to the site must be controlled. A perimeter fence and lockable gates, minimum 2.5 m high (8 ft), should be provided. In some areas, barbed wire will be required. Public entries may be controlled via gates that remain open during business hours, while access to the remainder of the facility may require on-demand access. Security gate systems may be remotely controlled using card access or phone systems. In larger facilities, closed-circuit television may be used to control facility access and maintain security. However, in designing security measures, access to the site by outside emergency response teams should be considered. Refer to joint Water Environment Federation® (Alexandria, Virginia) (WEF)/American Society of Civil Engineers (Reston, Virginia)/American Water Works Association (Denver, Colorado) security guidelines for additional information.

7.4.3   Site Drainage
Greenbelts, roadways, and parking areas must be designed for positive drainage. Where possible, grassed swales and infiltration should be encouraged to reduce hard piping that may conflict with other facility yard piping. Ponding of stormwater should be avoided because of the potential to attract undesirable species (i.e., geese and mosquitoes) and additional maintenance requirements.

Stormwater from developed areas of WRRF sites generally cannot be discharged to receiving waters without a permit. The EPA states that a municipal WRRF with a design flow of one million gallons per day or more is considered a category 9 industrial regulated activity and requires an NPDES permit, 40 CFR 122.26(b)(14)(i)-(xi). Therefore, a designer should consult federal, state, and local requirements concerning stormwater handling.

7.4.4   Septage and Vactor Receiving Stations
The delivery of septage and vactor solids can create significant traffic, and a separate gate and access route should be provided. The unloading area should be located close to the headworks area, so wastes can drain by gravity from the holding tank to the influent raw wastewater pumping station or be pumped to the screening, metering,

and degritting facilities. The unloading area should be designed to contain and control odors and provide wash-down facilities. A separate loading and weigh station also typically is provided.

7.4.5   Visitors and Vehicle Storage and Parking

Parking should be provided for all personnel, disabled employees, and visitors. Visitor parking should be marked and placed close to the administration building. Large facilities or facilities anticipating visits from civic groups or schools should provide space for bus parking with adequate turnaround space or looped drive. Where in-facility tours are provided, the agency should provide each visitor with a protective hard hat, protective glasses, and ear protection, as appropriate. Space should be provided to store this equipment in the administration building. Additional information on safety considerations concerning design issues can be found in the WEF Manual of Practice No. 1, *Safety, Health, and Security in Wastewater Systems* (WEF, 2012).

Employee parking should be located as close as possible to the area of the facility where personnel begin and end their shift. For most facility staff, this area is the locker room.

7.4.6   Architecture and Landscaping

Pleasing architecture and attractive landscaping greatly improve a WRRF's image and provide a pleasant atmosphere for the staff. This is particularly important if the facility is located in a scenic area or in a residential neighborhood. In such cases, the additional costs of special architectural treatment and landscaping are justified. In residential areas, the buildings and site should be designed to integrate to the surrounding neighborhood. In older industrial areas, a new, aesthetically pleasing WRRF can form the nucleus for the revitalization of the surrounding area.

7.5   Layout Examples

Two layout examples are presented for discussion. Figure 3.2 presents a linear layout where the influent enters the facility at one end, the effluent at the other end, and the process flow linearly between the two. The site has a multiple-sites road providing ample room for truck access to each process area. The administration building, laboratory, and maintenance facilities are all located in one large building, which is efficient, but can complicate access when multiple vehicles are all accessing the same location. There are ample buffer zones around the perimeter and room for expansion. There is only one access road to the site, which is efficient and less expensive but the single access point can get congested and also requires a temporary access road when new pavement or pavement repairs are needed. Solids processing facilities are all located in one area, which is preferred.

Figure 3.3 in contrast presents a nonlinear WRRF on a triangular-shaped site. Like many old WRRFs, this site housed a primary WRRF initially, and grew with multiple expansions as population growth and permit limitations required it.

The flow follows a route that was dictated by site limitations. (Flow enters and leaves the site at the original locations, but progresses around the site as various process expansions were added.) The multiple-perimeter and site access roads provide good site access and supports isolated solids disposal truck traffic. The administration and

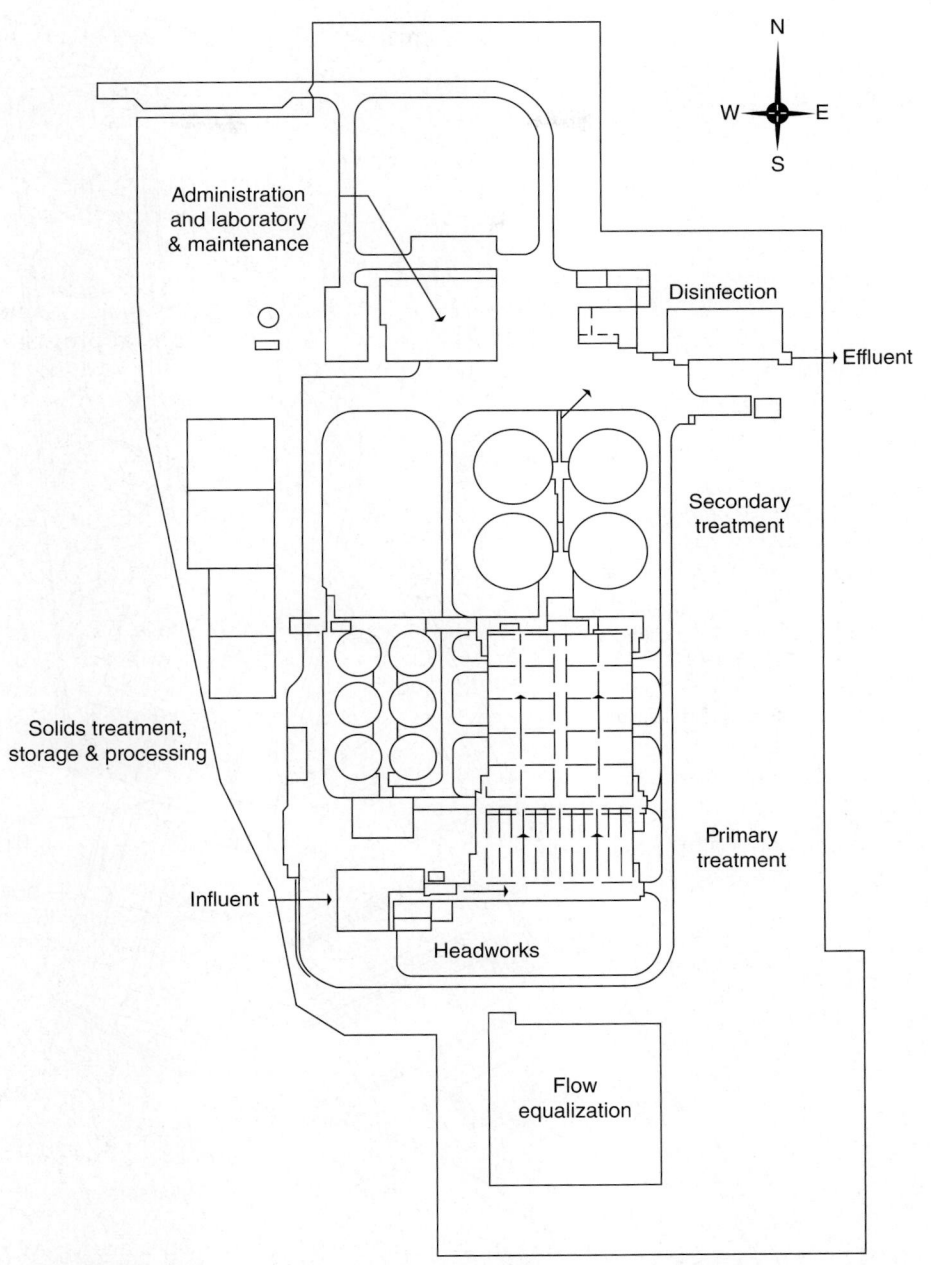

N
W ✦ E
S

Administration
and laboratory
& maintenance

Disinfection

→ Effluent

Secondary
treatment

Solids treatment,
storage & processing

Primary
treatment

Influent →

Headworks

Flow
equalization

FIGURE 3.2   Linear WRRF layout.

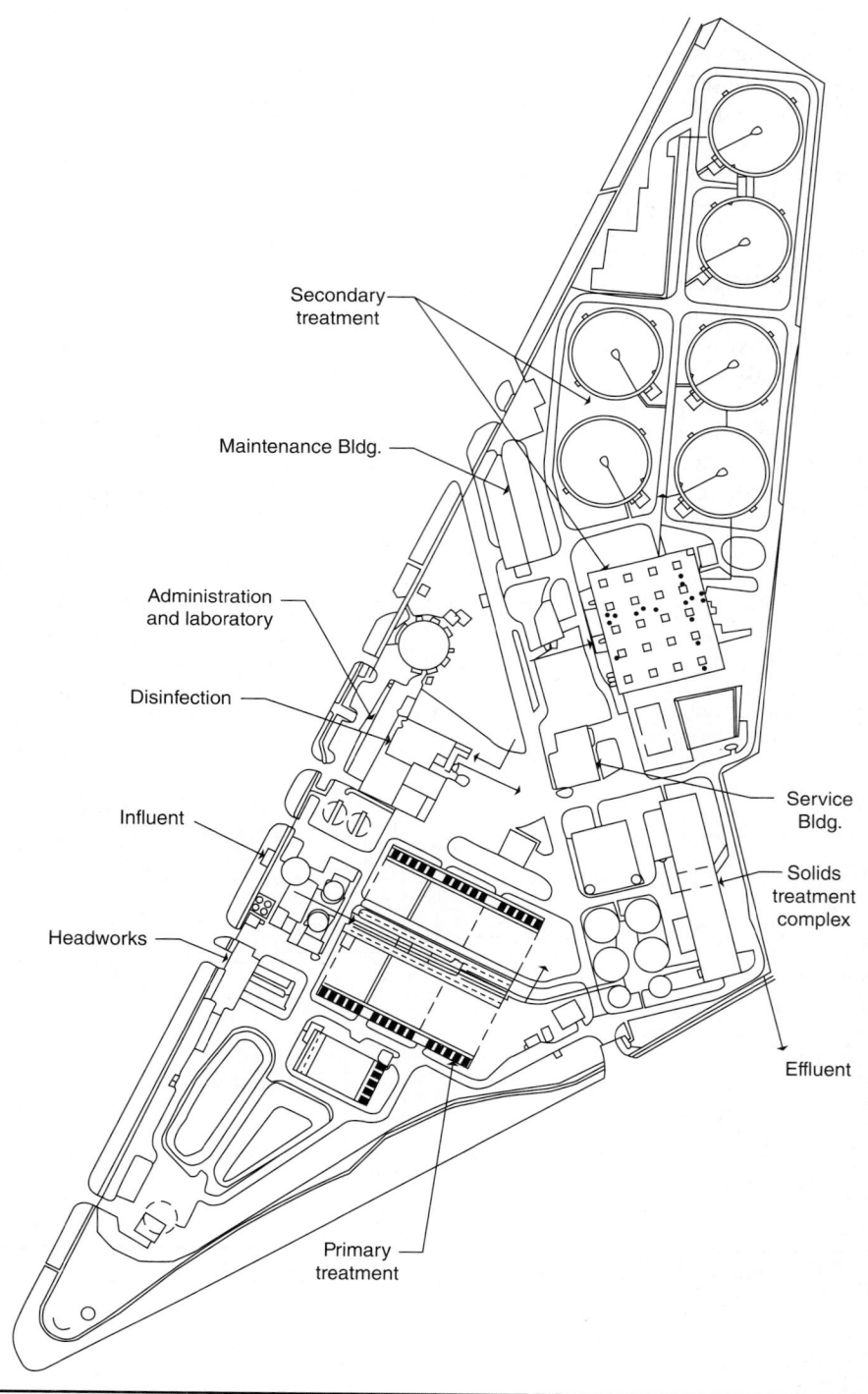

Secondary treatment

Maintenance Bldg.

Administration and laboratory

Disinfection

Influent

Headworks

Service Bldg.

Solids treatment complex

Effluent

Primary treatment

FIGURE 3.3 Nonlinear and irregular site.

laboratory buildings are combined and located on the perimeter with a parking lot, facilitating visitors. The maintenance building is separate and also on the perimeter, which supports deliveries, but requires a longer travel for maintenance staff. There is little or no buffer zones or setbacks, and several of the buildings are on the edge of the right of way. A service building supporting operations and maintenance staff, locker room and shower, and parking lot are centrally located, which is ideal. Solids processing facilities are all located in one area, which is preferred.

8.0 References

Clark, D. W. (1988) *Laboratory Planning for Water and Wastewater Analysis*; The New Mexico Water Resources Research Institute, 72.

Great Lakes–Upper Mississippi River Board of State and Provincial Public Health and Environmental Managers (2014) *Recommended Standards for Wastewater Facilities*; Great Lakes–Upper Mississippi River Board of State and Provincial Public Health and Environmental Managers: Albany, New York.

J. M. Montgomery Consulting Engineers (1985) *Water Treatment Principles & Design*; John Wiley & Sons: New York, 469.

National Council of Examiners for Engineering and Surveying (2005) *Fundamentals of Engineering Supplied-Reference Handbook*, 7th ed.; National Council of Examiners for Engineering and Surveying: Clemson, South Carolina, 150.

U.S. Environmental Protection Agency (2009) *Protection of Historic Properties*. Code of Federal Regulations, Part 800, Title 36.

Water Environment Federation (2017) *Sustainability and Energy Management for Water Resource Recovery Facility Design*; Water Environment Federation: Alexandria, Virginia.

Water Environment Federation (2012) *Safety, Health, and Security in Wastewater Systems*, 6th ed., Manual of Practice No. 1; Water Environment Federation: Alexandria, Virginia.

Water Environment Federation (2002) *Survival Guide: Public Communications for Water Professionals*; Water Environment Federation: Alexandria, Virginia.

CHAPTER 4

Modeling for Design and Operation of Biological Water Resource Recovery Processes

Jose Jimenez; Chris Bye; Ing. Daniel Nolasco; Leiv Rieger; and Tanush Wadhawan, Ph.D.

Modeling is important to the planning, design, operation, and optimization of modern water resource recovery facilities (WRRFs). A wastewater process model consists of a number of mathematical equations that describe reactions and reaction rates of biological, chemical, and physical phenomena of various unit processes. This chapter contains an overview of the fundamental concepts, protocols, and terminology that are commonly used in process modeling. The purpose of the overview is to give design engineers a broad understanding of the basics of process modeling steps, protocols available, and appreciation of the general principles involved during modeling for design and operation of water resource recovery processes. The reader who wants to understand these fundamentals in more depth is directed to reference texts that contain more details, such as the WEF Manual of Practice 31 (WEF, 2013), and the International Water Association (London, United Kingdom) (IWA) Scientific and Technical Report 21 (Rieger et al., 2012). The reader who wants to understand the fundamentals of different models is directed to references such as Henze et al. (2000, 2008) for activated sludge models; Batstone et al. (2002) for anaerobic digestion modeling; and Wanner et al. (2006) for biofilm modeling.

1.0   Applications of Process Models for the Design and Operation of Water Resource Recovery Facilities

The use of process models offers design engineers better insight into treatment challenges and requirements so that process units can be designed more efficiently. In general, there are two general categories of WRRF treatment simulation—steady-state and dynamic.

1.1   Steady-State Conditions

Steady-state spreadsheet models are often prepared by individual designers to determine tank volumes based on a set of input values and system requirements. In general, these modeling tools may predict results that can be used for planning purposes but not for detailed design since they can only provide results for one specific data point; when, in reality, WRRFs are never operated under steady-state conditions due to the nature of variable conditions that they are exposed to.

Historically, it was acceptable to use steady-state models for designing WRRFs, as large safety factors accounted for unknowns. These tools were developed when the cost of construction was lower and often federal grants were available to subsidize local utility costs. In the current economic environment, designing WRRFs with large safety factors because of limited information is not a solution. Safety factors are often justified as the application of engineering judgment, when in fact the safety factor incorporates judgments about both design criteria as well as the levels of unknowns. Some of those unknowns are created by the tools that are employed in design and could be removed by using more advanced tools and information gathering methods.

Two common uses of steady-state scenarios are:

- Overall mass balancing. Steady-state simulations are very efficient at identifying potential data sampling or analytical errors. The model always maintains mass balances, so steady-state simulation results enable errors from laboratory or monitored data to be identified.

- Long-term performance checks. Steady state can be used to estimate facility performance under various loading conditions, for example, under current or future design load conditions.

Steady-state simulations normally should NOT be used for sizing equipment that operates under dynamic conditions (such as blowers) or equipment for which design and reliable operation is dependent on peak and minimum flow or loads.

Time periods shorter than a month may not be represented well in steady state. A facility's behavior is a function of the operating conditions and the sludge retention time. Even if the operating conditions of a facility are held constant, it could take upward of a month (or three to four sludge retention times) for a typical facility to reach equilibrium after a change. This defines how much data are required.

1.2 Dynamic Conditions

The development of models of dynamic conditions was made possible by the advent of the digital computer and software to solve simulations involving simultaneous differential equations. Storer and Gaudy (1969) discussed computational analysis of transient responses to shock loading of heterogeneous populations of bacteria. Smith and Eilers (1970), at the Federal Water Pollution Control Administration office in Cincinnati, Ohio, prepared one of the first reports on the simulation of the activated sludge process.

Dynamic modeling offers the process engineer a method to make effluent predictions, size process units, and address alternatives that potentially have major cost savings or performance benefits. In effect, dynamic modeling provides a better understanding of the problem to improve engineering judgment. Dynamic models can be used to project future design conditions that cannot be tested otherwise. In some cases, there are systems that can only be fairly evaluated by dynamic modeling, such as the performance of a WRRF under peak wet-weather flow conditions.

Dynamic simulations consider the changing conditions that facilities are normally subjected to, such as variation in influent flow or loading, temperature, and operational conditions. The "raw" unprocessed simulation results consist of detailed information on the dynamic behavior of the facility. Some typical uses for dynamic simulations are summarized below. The true power of structured dynamic models lies in dynamic simulation, as these calculations cannot be repeated easily by hand or even in spreadsheets.

The following statements on dynamic simulations have been developed by the IWA Task Group on "Good Modelling Practice—Guidelines for Use of Activated Sludge Models" (GMP Task Group) involving the WEF subcommittee MEGA (Modeling Expert Group of the Americas):

1. A common dynamic simulation is the diurnal run, which contains one day (24 hours) of "typical dry weather" influent and operational data. The 24-hour period can be repeated as many times as necessary. Diurnal runs have to be used to describe cyclic processes. For example, if there is significant variation (a high peaking factor) in the influent loading throughout the day, or if the facility is inherently non-steady state, a steady-state run is not feasible or may not provide useful results, so dynamic runs are required.

   The results of dynamic simulations can be further processed to calculate averages, simple or more advanced statistics (min/max, distribution, etc.), and these can be tabulated against different running conditions to provide an easier overview for reporting. Calculating diurnal peak and minima conditions in a dynamic simulation are frequently used to determine equipment limits such as blower maximum capacity and turndown requirements.

2. Long-term dynamic simulations are frequently performed to investigate weekly, monthly, or even seasonal effects. Typical tasks for this type of simulation involve:
   a. The investigation of operational differences due to weekday and weekend loading conditions and operational control changes (e.g., no sludge wasting at weekends, or sludge liquors returned during certain shifts).
   b. A monthly dynamic run can be compared to the steady-state monthly average conditions to investigate the impact of facility dynamics on the results.
   c. The investigation of seasonal variation of the process. The length of the simulation in this case could be several months, a year, or longer.
   d. It is important to highlight the importance of complete data sets, especially when modeling long-term conditions. For example, if a year-long simulation is being modeled, but operators only collect samples three times per week, the results may not contain certain events or a specific behavior of the facility.

3. Dynamic simulation provides a unique opportunity for investigating and planning for optimal management of short-term events. These can include:
   a. Wet-weather periods or specific storms, for example, bypass planning or preparation for, and proper timing of, step-feed operation;
   b. Taking equipment or reactor volume offline for maintenance or construction activities; and
   c. Planning for equipment (e.g., pump, blower) failure.

4. Equipment modeling, controller design, and control parameter tuning are special cases of dynamic simulation. Both the facility model and relevant parts of the facility control system need to be implemented in the dynamic model. This may also be accompanied by more details such as equipment curves (i.e., blowers and pumps), sensor models, noise filtering, and actuator response curves (Amaral et al., 2016; Schraa et al., 2016).

Dynamic simulations add another dimension (time) to the analysis and therefore require more details for their setup. Influent generators can be used to generate diurnal profiles based on few diurnal data or on correlations to typical patterns or facility sizes (Langergraber et al., 2008). Dynamic runs are slower to implement and take longer to run. In accordance with the principle of "keep it as simple as possible to answer a specific question," dynamic simulations should be used only when a process question cannot be answered by steady-state simulations, or where a dynamic simulation may give a more detailed understanding of the process. For dynamic process schemes (e.g., SBRs) or controlled processes, dynamic simulation is mandatory. Results for highly dynamic processes like enhanced biological phosphorus removal (EBPR) can differ when simulated to steady state or dynamically.

2.0   Development of Commercial Simulators

With the development of more powerful personal computers, commercial simulators such as GPS-X, a product of Hydromantis, Inc. (Patry and Takács, 1990, 1994), and BioWin™ produced by Envirosim Associates Ltd. (Barker and Dold, 1997), SIMBA

(Alex et al., 1994), a product of the "Institut f. Automation und Kommunikation" (ifak), and SUMO, a product of Dynamita, which provide a platform for full-scale facility simulation, software is commonly used by consulting engineers and researchers in the wastewater field but also more and more directly at the facility level for operation support. The steps to build a model have been streamlined and facilitated by graphical user interfaces. A user first produces a process diagram of the modeled WRRF as a graphical function block diagram, establishes physical parameters and operating parameters, runs the model, and analyzes the results. These easy-to-use simulators enable users to optimize the wastewater process on their own personal computers by modeling different scenarios. However, understanding the theory behind the models is essential to properly setting up a model and interpreting the model results (Shaw et al., 2007).

3.0   Overview of Available Modeling and Simulation Protocols

Although mathematical modeling and simulation of activated sludge systems is used widely in our industry, the quality of simulation studies vary strongly depending on the project objectives, resources spent, and expertise available. Simulation protocols aim at standardizing the process of modeling and simulation and can improve the quality of the results and may reduce the required effort. In addition to a direct positive impact on the simulation study, there are several additional benefits like improved data quality for operation and design. Some of the main benefits from using simulation protocols are listed below:

- As a standardization process, protocols lead to better comparable, reproducible, and transferable results; comparison of simulation projects is indeed facilitated by use of standard procedures;

- Protocols give guidance to clearly define requirements and limitations of the obtained model and what can be reached at the beginning of the project and therefore help prevent misconceptions;

- Checking the quality of a project against standard procedures should lead to improved quality assurance/quality control (Shaw et al., 2011); and

- Inexperienced modelers and stakeholders are guided throughout the project.

In 2004, the IWA decided to synthesize the available experience into an internationally recognized industry standard and subsequently formed a task group on "Good Modelling Practice—Guidelines for Use of Activated Sludge Models" (GMP Task Group) involving the WEF subcommittee MEGA. The GMP Task Group published a scientific and technical report that outlines a *Unified Protocol* and describes a *Good Modeling Practice* with activated sludge systems (Rieger et al., 2012). The WEF adopted the steps of the GMP Unified Protocol in its Manual of Practice No. 31 (WEF, 2013), which provides additional examples and focuses on details of the GMP Unified Protocol steps.

3.1   Available Process Modeling Simulation Protocols

Several groups working on wastewater treatment have proposed activated sludge simulation protocols. The following section highlights the main features of selected protocols to identify their specific strengths. More information can be found in the GMP guidelines (Rieger et al., 2012).

STOWA (Hulsbeek et al., 2002; Roeleveld and van Loosdrecht, 2002): STOWA's main emphasis was to help end users to model their nitrogen removal facilities with the ASM1 model in a systematic and standardized way. An essential part of this protocol was the development of an easy-to-use wastewater characterization procedure. As part of the development, user groups were set up and the outcome was the result of an extensive consensus-building process. The STOWA guideline can be seen as an international standard due to its ease of use and wide application. Unfortunately, only a summary is available in English language in the form of two journal publications (Hulsbeek et al., 2002; Roeleveld and van Loosdrecht, 2002).

The **WERF** (Water Environment Research Foundation) guidelines (Melcer et al., 2003) are based on experiences from a large market (mainly North America) with authors from consulting companies, software developers, and universities. Targeted users are municipalities and consulting engineering companies including junior and intermediate modelers. The development consisted of research on wastewater characterization methods and a consensus-building process involving a large international reviewer group. The final report of 575 pages includes an extensive overview of knowledge, experience, and data and became a standard reference for wastewater characterization and simulation procedures.

BIOMATH at Ghent University (Belgium) proposed a generic calibration procedure (Vanrolleghem et al., 2003), using state-of-the-art parameter estimation methods for stepwise calibration/validation of models with a focus on the biokinetic model and sections on settling, hydraulics, and aeration. The protocol requires a high level of experimental results and takes advantage of systems analysis tools. The protocol summarizes the work of the BIOMATH research group and is mainly dedicated to experienced modelers and has been applied in academic research projects to increase process understanding or during development of new models.

The **HSG** protocol (Langergraber et al., 2004) is a generic procedure to guide modelers through all steps of a modeling project. The HSG protocol gathers the experience of specialized researchers from German-speaking countries. The focus is on a standardized structure for modeling projects. An objective-oriented approach is encouraged, but deviations from the full procedures need to be explained and documented. The importance of data quality is highlighted. The HSG protocol targets modelers from consulting firms, water boards, and municipalities. An eight-page journal publication is publicly available.

Regional protocols often focus on specific issues and constraints and may not allow generalization. **Company protocols** (e.g., Frank, 2006) are often proprietary and not easily accessible. The focus of both types of protocols is typically on practical use. **Software manuals** are focused on explaining the use of the respective software but often provide a relevant source of information on how to apply models. Some software companies provide additional support to stakeholders in their modeling work. **Published case studies** (e.g., Meijer et al., 2002; Third et al., 2007) can be used as another source of guidance but are often too specific to be used as general guidance on activated sludge modeling.

From a comparison of all of these protocols, it can be stated that the agreements outnumber the differences and where differences are evident they are mostly in the level of detail and the foci. This can be related to the background of the authors (e.g., researchers, consulting engineers, round tables, and the field of their expertise: process engineering or water management) and the targeted users. Differences may also be linked to the fact that the objectives of model use are quite different, that is, mainly for design/redesign purposes in North America, and for optimization or controller studies that require more dynamic simulations in Europe (Hauduc et al., 2009).

3.2 The Good Modeling Practices Unified Protocol

The Good Modeling Practices (GMP) guidelines consist of the GMP Unified Protocol plus examples, additional information, and case studies. The GMP Unified Protocol combines the key aspects of the protocols discussed in the preceding sections and includes the dimension of stakeholders and identifies sub-steps where other stakeholders should be involved in the decision process. A focus of the Unified Protocol is quality assurance and quality control, and therefore, every step includes certain measures on documentation and stakeholder involvement. In general, the goal is that another modeler would be able to repeat the simulations and come to similar results. A "living document" is suggested to track the work throughout the protocol steps and any change in model parameters away from the defaults need to be documented and justified in writing.

To help modelers in estimating the required effort, the GMP guidelines provide an application matrix linking typical objectives to specific requirements and efforts. The matrix consists of 12 typical modeling objectives for domestic WRRFs and 2 additional ones for industrial facilities. MOP 31 follows the steps introduced in the GMP Unified Protocol and provides additional examples and case studies.

The proposed protocol is illustrated in Figure 4.1. It comprises five main steps, which have to be reviewed and agreed on with the stakeholders before the next step is carried out (decision boxes in black).

- Step 1: Project Definition
- Step 2: Data Collection and Reconciliation
- Step 3: Facility Model Setup
- Step 4: Calibration and Validation
- Step 5: Simulation and Result Interpretation

4.0 Guidelines for Use of Activated Sludge Models Using the Good Modeling Practices Unified Protocol

Modeling of WRRF systems is used widely for research, facility design, optimization, training, and model-based process control. The quality of simulation studies can vary strongly depending on the project objectives, resources spent, and expertise available. Consideration has to be given to the model accuracy and the amount of time required for carrying out a simulation study to produce the required accuracy. A great variety of approaches and insufficient documentation make quality assessment and comparability of simulation results difficult or almost impossible. A general framework for the application of process models is needed in order to overcome these obstacles as described below.

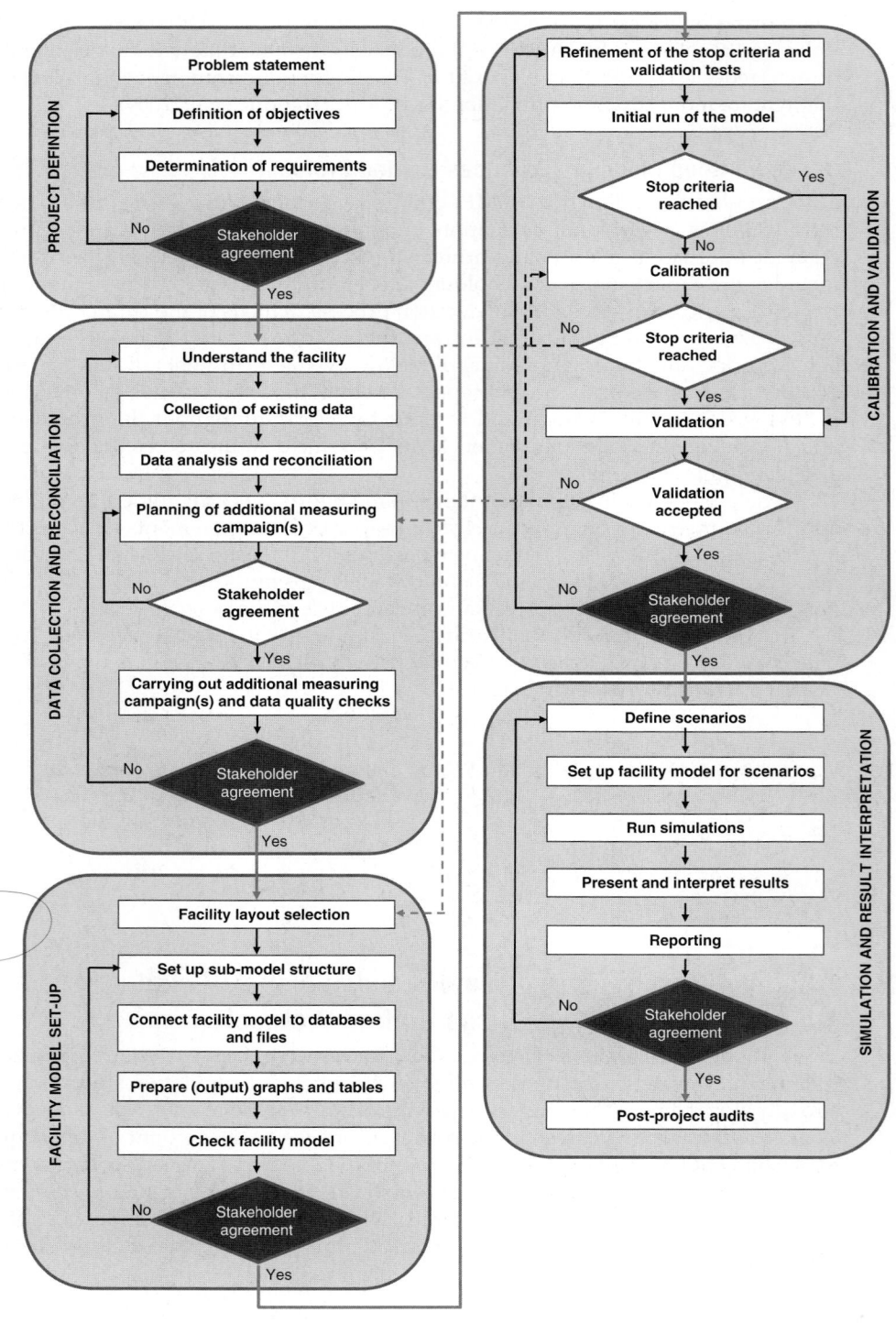

Figure 4.1 The GMP Unified Protocol.

4.1   Step 1—Project Definition

The first task in any project is to clearly define the purpose, set objectives, define requirements, and achieve an agreement with all stakeholders. In simulation projects, the focus is on the model use. Each model must have a clearly defined purpose. It is important that all stakeholders are involved in this process. Objectives are typically identified during the proposal phase of a project and further defined during the first phase of project execution. Expectations, with respect to budget, data collection needs, and level of simulation effort required, should be established when the modeling objective is defined. The objectives will be used as the basis for model selection, data collection, calibration steps, evaluation of alternatives, and conclusions.

This section summarizes different purposes based on project phases and model use. For example, most projects begin with the planning phase and end with optimization, and the models used for the different tasks typically vary in the boundaries set and the level of detail. For interested readers, MOP 31 presents a detailed section on the topic.

Defined by the project purpose and the model use, boundaries are set for the model to be used. Model boundaries determine which unit processes are included and where simplifications are acceptable. For a WRRF, the boundaries are typically the WRRF's "fence" (i.e., not including the sewer system); however, following the principle to keep the model as simple as possible to do the job, the boundaries reflect the project purpose (e.g., only include individual unit processes such as secondary treatment).

4.1.1   Planning

Planning models are typically used in the early stages of a project. The model purpose is screening options during planning of the capital improvement program, conceptual studies, and proposals. Models for planning purposes typically have limited data available, and assumptions are made when needed. Model results are used to determine the most suitable option.

A special application of planning models is in scheduling facility upgrades (Phillips et al., 2010). Models are used to develop stepwise upgrading schedules so that the required treatment performance can be maintained during the construction.

4.1.2   Design

Models are often used in the design of facility upgrades or completely new facilities (also known as greenfield designs). Typically, during WRRF upgrade projects, there are data available for the existing facility, such as influent loads and concentrations, return stream flows and loads, and biological process kinetics that can be used to establish the process model for the current configuration. However, special attention is needed if unit process changes are to be implemented as part of the upgrade. Any safety factors being applied to the design must be recognized, and decisions must be made about how the simulation results and sensitivity analysis should interface with these safety factors. Decisions on safety factors should be made during the conceptual or predesign level. Decisions made during design have significant implications on facility operation, including, but not limited to, cost, operational flexibility, and effluent quality.

In the case of greenfield projects, WRRF process models are widely used at the conceptual and predesign level. At the end of the predesign phase, most important decisions on process layouts and technologies should have been made and agreed on by the parties involved. The detailed design of new facilities starts once a decision has been made on a process layout and technologies used. Data may not exist because the facility is not yet in operation; thus, typical wastewater characteristics and flow predictions based on population and industries connected are used or data from a neighboring facility are applied, such as influent TSS, BOD, COD, TKN loads, and concentrations; the data may be used with caution.

4.1.3   Operations

Using a model to operate an actual facility possibly requires the highest level of modeling accuracy and involves using simulations on a regular basis to assess and modify facility operations. For such an application, periodic measurements and readjustments of wastewater characteristics may be valuable. This could help improve model accuracy in different seasons, for example, or as influent loads change. It should be noted that in most instances it is not recommended to make adjustments to kinetic and stoichiometric parameters.

Often process models are used by engineers to guide operational decisions or to look at trends only, where an uncalibrated process model may be used. These models are useful to demonstrate, for example, the cause/effect of equalizing or not equalizing centrate or the relative impact of operating conditions on biomass inventories.

One of the most beneficial applications of models is for communication as models can capture the knowledge and experience of different experts such as process, mechanical, and control engineers and automation and instrumentation experts. The facility model helps focusing discussions and allows a direct feedback from complex systems.

Another valuable use of models is to test different control strategies, before applying them to full-scale conditions. For this situation, a calibrated model can be used to assess different control strategies and to give the control engineer a better understanding of system dynamics. An example of this is presented by Fairey et al. (2006) and Rieger et al. (2014), who tested novel control systems using process models before implementing these in full-scale trials.

Wastewater process models can also be used for existing facilities to optimize operations for chemical dosages, power requirements, aeration requirements, reactor configurations, mixed liquor suspended solids (MLSS) concentrations, return activated sludge (RAS), and mixed liquor recycle flowrates, clarifier loads, and solids retention time (SRT). Models can be used to dynamically optimize process performance and quantify the costs for different operation strategies.

Use of models to develop probability distributions of treatment levels is becoming increasingly important with the low-level nutrient values being imposed by regulatory agencies. In this situation, models can be used to determine the ability of an existing facility to meet a lower nutrient standard by, for example, adjusting aeration or using swing zones differently. This may require rerating facility capacity, but could minimize capital investments. If exact effluent concentrations need to be predicted, the highest data and accuracy requirements apply.

4.1.4   Training

Process models are a great tool to train engineers and WRRF operators to better understand the dynamic behavior and process interactions of their facility. The models help communicate knowledge about the behavior of the system, especially highly interrelated and complex systems. A whole-facility model can be developed for a specific facility to show interactions between processes, single models can be developed for individual processes on a site, or generic "fictional" models can be used in a class-room setting to provide training on general principles. For example, models can be used to illustrate effects of increasing or decreasing MLSS inventory on sludge blanket height, waste activated sludge (WAS) rate, and aeration.

4.1.5 Research

Models help researchers to understand new and developing technologies. For example, models can be used to interpret and understand pilot test data; this information can then be used to design a full-scale system. An example of this is given by Phillips et al. (2007), who used process models to investigate pilot- and full-scale performance of an integrated fixed-film activated sludge system and then used that knowledge to design other facilities. Models are a key part of discussions on how to develop and design emerging and new processes.

4.2 Step 2—Data Collection and Reconciliation

Simulating WRRF behavior requires information in several general categories, namely, physical, operational, influent data characterization, and process model kinetic and stoichiometric parameters (Figure 4.2). Once the data have been gathered in the spread-sheet format, an extensive exercise is required to screen out poor data, and provide data reconciliation through performing mass balances.

Data collection refers to existing (process related and historical) and missing (measuring campaign setup) data. Since data collection and reconciliation is the most time-consuming step (30% to 60% of the total effort, Hauduc et al., 2009), data quality validation should play an essential role in the planning phase (design the required data quality, Rieger et al., 2010). Data required to perform the data validation step should be an essential part of the additional measurements. Understanding of process conditions based on collected data is essential to confirm the adequacy of the defined objectives

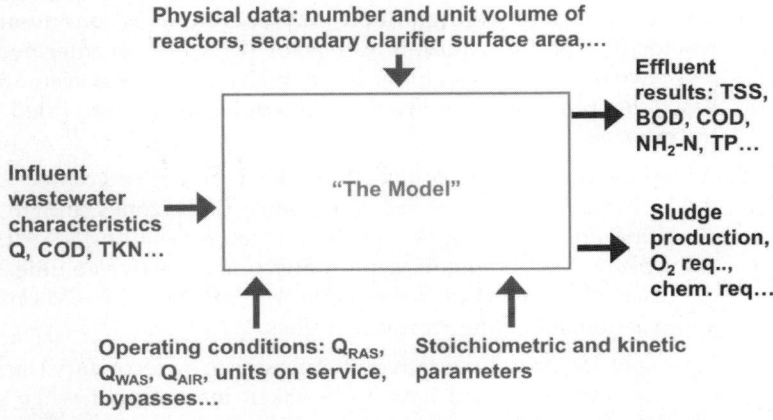

FIGURE 4.2 Summary of important data requirements for simulating a WRRF.

and of collected data. Data validation is based on data series analysis, outlier detection, mass balances, and other checks (e.g., typical component ratios). From these data, the facility model structure (processes included, flows, boundaries) is defined. The data collection step finishes with an agreement between modeler and stakeholder on the used data and the modeled processes.

Over the years, mathematical models have become more robust. From the Activated Sludge Model No. 1 (ASM1) with 13 state variables simulating an activated sludge reactor, modeling knowledge has evolved toward far more comprehensive models with four to five times more state variables. Combinations of models in powerful computer programs (simulators) can mimic the behavior of a large array of processes in a whole facility. This improvement in mathematical models, combined with powerful simulation programs, is allowing wastewater treatment engineers to simulate numerous unit operations (treatment processes) in full facilities. As models become more comprehensive (and realistic), the need for stoichiometric and kinetic parameter adjustment (calibration) has been considerably reduced. Yet, the importance of having the right physical, influent, and operating conditions remains a key aspect to obtain realistic results.

Understanding the physical, influent, and operating conditions to be entered into a simulator can be a complex task. The amount of resources (time and expenditures) devoted to this task must be a direct function of the level of calibration desired.

4.2.1 Facility Physical Data

This relates to the size of units in the treatment system: for example, aeration tank dimensions (volume, length, width, and depth), clarifier area and depths, and so on. If the objectives of simulation include assessing aeration system performance, then information such as diffuser dimensions, diffuser floor coverage, and blower configuration/ operating mode may also be required.

Physical facility data may be obtained from facility drawings (if the facility is existing) or from initial estimates based on traditional design guidelines (if the facility is being designed as a greenfield).

In the case of existing facilities, additional work will be needed to assess the actual physical configuration. For example:

1. The aeration tank configuration may operate as a continuous stirred-tank reactor (CSTR), or a plug-flow reactor (PFR), or an intermediate situation (e.g., two or more CSTRs in series). In many cases, the decision will need to be based on practical experience, estimations using simplified equations, or tracer tests.

2. Anaerobic digesters generally do not have an active volume identical to the total physical volume of the tank, since dead zones and hydraulic short-circuiting are quite common in these reactors resulting in grit accumulation among other hydrodynamic aspects affecting the active volume. The volume of the reactor to be entered in the model will need to be based on rule-of-thumb appreciation, modeling, or tracer studies.

3. Flow splitting prior to certain unit processes (e.g., secondary clarifiers) may not be even and, if so, will have to be taken into account when setting up the simulator.

These cases and many more need to be considered when setting up the physical facility model.

4.2.2   Facility Operating Data

This includes data on operational aspects such as RAS flowrates, WAS flowrates, mixed liquor recycle flows, and dissolved oxygen (DO) concentrations in aerated zones, airflow rates, return stream information (e.g., flows, concentrations of any monitored parameters), etc. This information will be collected from recent facility historical records for the purpose of model calibration. Some verification of the information (i.e., WAS flowrates) may be required and this could be part of the supplemental sampling program.

When calibrating a model to an existing or new facility, once the data have been gathered, an extensive exercise is required to screen out poor data, and provide data reconciliation through performing mass balances.

In existing facilities, flow meter calibration will need to be evaluated. If the existing flow meters are not periodically calibrated, additional analysis will be needed from the modeler to estimate the range of possible error.

Oxygen transfer efficiency is a key operational parameter that is not always readily available to the modeler. Since aeration consumes a considerable amount of energy in an activated sludge facility, oxygen transfer testing may be needed if the project justifies it. In nonexistent facilities (e.g., for greenfield designs), a sensitivity analysis around the important design parameters may be needed.

4.2.3   Influent Wastewater Characteristics

The influent flow and concentrations are essential data. This refers to raw influent wastewater (a) flowrate data, (b) concentrations of a range of variables (COD, BOD, VSS, TSS, TKN, NH_x-N, TP, PO_4-P, alkalinity, pH, etc.), and (c) temperature. It may be possible that the influent data for the WRRF to be modeled does not have all of the variables necessary to define the loading for modeling purposes. However, these can be estimated from the results of the supplemental sampling by determining appropriate relationships between the historical data and the collected data (i.e., COD/BOD ratios, volatile content of influent solids, etc.).

Influent chemical oxygen demand (COD), nitrogen (N), and phosphorus (P) fractions have to be determined according to the model used. This section focuses on the measurement of specific characteristics required for developing and calibrating process models.

One of the most important outcomes of this phase of the project is the appropriate characterization of the wastewater for modeling. A WERF project on *Methods of Wastewater Characterization in Activated Sludge Modeling* (Melcer et al., 2003) defines wastewater characterization protocols for the influent parameters commonly used in process simulators for the design, analysis, and optimization of wastewater treatment processes. The reader is referred to that publication for in-depth discussion on this topic.

Required model inputs (state variables) are typically calculated as fixed percentages (fractions) based on averages of the measured components. However, this is an assumption, and model inputs often vary over the course of a day or week or with weather conditions. Special care should be taken when intermittent industrial loads are discharged or with seasonal load variations because these atypical conditions may require specific investigations to properly characterize the influent loads.

MOP 31 provides a list of methods that can be used to characterize wastewater streams in terms of COD fractions. Experience has shown that the proposed methods may lead to different fractions altogether (Fall et al., 2011; Gillot and Choubert, 2010). This also explains why the values obtained through measurements are often modified in the subsequent calibration step.

No generally accepted method has been established yet and, therefore, the following steps are recommended (nomenclature according to Corominas et al., 2010):

- The readily biodegradable fraction (S_B) should be obtained either by respirometry or by physicochemical methods that include a flocculation step to ensure that the colloidal matter becomes part of the slowly biodegradable fraction.

- The unbiodegradable fraction (S_U) is obtained by COD analysis of a filtered (0.45-μm) effluent sample ($COD_{EFF,f\,0.45}$).

- The total biodegradable COD fraction of an influent ($S_B + XC_B$) is critical to properly simulate the oxygen demand of the process and its total removal efficiency. To obtain good values of the total biodegradable COD fraction ($COD_{TOT,B}$), long-term biochemical oxygen demand (BOD) measurements or other types of respirometry can be used.

The choice between either a physicochemical or a respirometry method is often based on available equipment and experience. In general, one can say that the former method is easier to carry out. However, it does not measure the biologically relevant property of the wastewater that is used in activated sludge models. Rather, it fractionates COD on the basis of size and not on the basis of rate of biodegradation, which may be important (e.g., for denitrification).

Nitrogen and phosphorus species such as ammonium-ammonia (NH_x-N), nitrite-nitrate (NO_x-N), phosphate (PO_4-P), total phosphorus, total nitrogen, total Kjeldahl nitrogen (TKN), and total soluble fractions are typically obtained through standard analysis. Organic nitrogen and phosphorus fractions are calculated by difference. More information on this topic can be obtained in an early extensive review on characterization methods by Petersen et al. (2003), an IWA scientific and technical report by Rieger et al. (2012), and a recent critical review of wastewater characterization methods by Choubert et al. (2013).

An important variable in practice is the concentration of total suspended solids (X_{TSS} in model notation). It consists of a volatile part (volatile suspended solids, or VSS) and an inorganic part (inorganic suspended solids; ISS = TSS − VSS). However, the approach for which X_{TSS} is introduced as a state variable in an activated sludge model is not completely described in model publications and needs to be carefully set up. Only then can it become a useful variable that can be linked to TSS measurements.

4.2.4   Experimental Methods for Kinetic and/or Stoichiometric Model Parameters

Different conditions and factors affect the degree and rate (speed) at which the compounds and contaminants of concern are removed by microbial populations in biological WRRFs. Certainly, the facility configuration and operational conditions play a major role in the prevalence of specific microbial populations and their activities, but factors as diverse and broad as wastewater characteristics and environmental and climate conditions have a strong influence as well. Eventually, in any biological wastewater treatment system, there will be a need to assess, define, and understand the facility performance with regard to the removal of certain contaminants and the response of the sludge to inhibitory or toxic compounds of interest. Moreover, from a modeling perspective, it is also of interest to assess and determine the stoichiometry and kinetic rates of the conversion processes performed by specific microbial populations.

The biological models contain many *stoichiometric and kinetic parameters* such as organism yields and maximum specific growth rates. Typically, the stoichiometric and kinetic parameters do not change appreciably for different systems *treating municipal wastewaters*. The uniformity of parameters likely reflects uniformity in the composition of municipal wastewaters, and a resultant similarity in the diversity of the microorganism populations in different systems.

The following sources of information can be used to obtain values for the kinetic and stoichiometric parameters of activated sludge models (Petersen et al., 2003):

- Default parameter values from literature. It is important to note that default parameter values are often not the originally published values (called original parameter values), but were derived from a consensus-building process in which the profession has agreed that these values are a good starting point for a modeling study (Hauduc et al., 2011).
- Full-scale facility data:
  - Average or dynamic data from grab or time/flow proportional samples,
  - Conventional mass balances of the full-scale data
  - Online data, and
  - Measurements in reactors to characterize process dynamics.

Parameter values are obtained through fitting the model until simulation results agree sufficiently with the data (i.e., a calibrated value is obtained).

Bioassays, that is, different kinds of laboratory-scale experiments with wastewater and activated sludge from the full-scale facility under study. There are a number of experimental setups that have been created that allow the direct calculation of the parameter value from the data. Such value is called a measured parameter value. In other methods, a (simplified) model is fitted to the data and one or more parameter values are estimated. In those instances, calibrated parameter values are obtained as opposed to measured ones.

Default kinetic parameters included in water resource and recovery process models generally can describe the performance of most systems treating typical domestic wastewater. However, in some instances, wastewater characteristics and the resulting bacterial community in the biological treatment process are so unusual compared to typical domestic wastewater that the system performance cannot be explained using typical kinetic parameters. Examples of such an instance include a WRRF receiving significant contributions of industrial wastewater or the use of external carbon sources with unusual degradation kinetics (e.g., methanol and glycerol). In these instances, it is beneficial to conduct kinetic studies to better understand the reason for deviations from model predictions and to adjust the relevant model parameters. In some cases (e.g., methanol addition), it may be required to select a more appropriate model or adapt the model structure. Kinetic studies can reveal possible causes for unusually slow (or fast) nitrification or denitrification rates.

Care should be taken when transferring model parameters obtained from laboratory-scale experiments with activated sludge to the full-scale installation (Gernaey et al., 2004). The execution of batch activity tests can be rather useful to (i) study the biodegradability of a given wastewater stream (municipal or industrial), (ii) determine the stoichiometric and kinetic parameters involved in the conversion of a specific compound, (iii) study the potential interactions (e.g., symbiosis and competition) between microbial populations, and (iv) assess the potential inhibitory or toxic effects of certain wastewaters, compounds, or substances. van Loosdrecht et al. (2016) provide an extensive discussion on direct methods focusing on specific parameters that can directly be evaluated from the measured data.

4.3   Step 3—Facility Model Setup

This section helps to identify and provides solutions to some of the major challenges faced by a modeler when setting up facility models. Common considerations are discussed below.

4.3.1   Managing Complexity and Runtime

The model complexity and runtime directly depends on the modeling objective. Modelers typically must make tradeoffs between increasing model complexity to create more representative models and minimizing computational runtime. For steady-state modeling, increasing the computational runtime may be a nonfactor. However, for dynamic simulations, a significant increase may not be compatible with project deadlines and objectives. Therefore, modelers typically must manage complexity and runtime. This aspect can be dealt with by merging parallel trains (if performing identical) and disabling unwanted process reactions. Some models provide the ability to enable or disable pH, DO, or precipitation reactions.

4.3.2   Modeling Input Streams

It is important to consider and include all the necessary input streams such as raw wastewater, chemicals, sidestreams, and wash water. Raw wastewater is the most important aspect of modeling that defines the sludge production and water quality, and without good data available on wastewater characterization, the modeling effort itself can be considered compromised. A majority of time setting up the correct wastewater characteristics results in a calibrated model without the need for changing any of the biokinetic parameters. Here are a few important aspects to consider.

1. Identify how many different streams enter the facility separately. If the wastewaters in these streams have significantly different characteristics, then they should be included in the model separately. Special emphasis should be given to streams receiving industrial discharge.

2. Evaluate if there is significant biological activity in the primary or secondary clarifiers, this can lead to complications in modeling and calibration.

3. If chemical addition is performed at the facility for phosphorus precipitation, denitrification, struvite, and/or odor control, then it is critical to include such streams in the model as they impact sludge production and can have an impact on effluent quality. Developing a facility-wide model that includes both liquid and solid streams is recommended.

4.3.3   Physical Configuration

It is important to accurately set up a facility by going through the piping and instrument diagrams (P&ID), process flow diagrams, and talking to facility engineers and operators. The following physical information might be needed for setting up the facility model:

1. Number of treatment units in parallel and series;
2. Reactor volume and depth;
3. Elevation, and number of blowers and diffusers (for aeration model);
4. Depth and feed point for some clarifiers; and
5. Media surface area and fill ratio for biofilm processes.

4.3.4   Selection of Sub-Models

1. Biokinetic models. There are a number of biokinetic models to choose from and it should be based on the objective of the project. Using complex models and enabling features such as pH calculations and precipitation will always increase runtime. Here are a few of the questions to consider that can help modelers to determine the complexity of the models they should use:
   a. What biokinetic reactions are needed to model and meet the objectives of the project?
   b. Are default biokinetic parameters available and established? How much effort is required to develop site-specific parameter sets?
   c. Where are these biokinetic reactions taking place (e.g., in a digester, which will require a biokinetic model with methanogenic microorganisms)?
2. Clarifier models. Modelers can choose between ideal clarifier models with fixed effluent TSS and one-dimensional clarifier models that will account for storage. Some clarifier models provide the option to calculate biochemical reactions.
3. Aeration models. Modelers can choose between a reactor where DO is specified and a reactor where airflow is input. For the latter option, it is important to have knowledge about the aeration equipment and oxygen transfer coefficient (K_{la}).

4.3.5   Operational Parameters

Within bounds, a model should account for all the operational conditions the facility staff may set. For each sub-model, modelers will need to specify a number of operational units in service, operational set points, and method of control. For complete information on the operational parameters and set points, see GMP Unified Protocol (Rieger et al., 2012).

4.4   Step 4—Calibration and Validation

Model calibration can be described as an iterative adjustment of the model parameters until simulation results match an observed set of data. Calibration in this sense does not include any specific additional measurements or modifications of the facility model. If issues are detected with respect to data quality, availability, or facility model setup, the modeler must return to steps 2 (Data Collection and Reconciliation) and 3 (Facility Model Setup). If the model is "calibrated" based on erroneous data, the predictive power of the model is compromised. Therefore, going back to previous steps is preferred over modifying (e.g., biokinetic) parameters. After the calibration is completed, a model validation step is then performed using a number of tests also defined during the Project Definition step. An acceptable validation should ensure the use of the facility model with the level of confidence required to meet the modeling objectives.

4.4.1   Calibration

Calibration typically is regarded as the process in which model parameters are adjusted until model predictions match selected sets of data linked to the performance of the actual facility. The *objective is to minimize the error* between the data sets and model predictions. It is important to remember that the objective is not to achieve a perfect fit, since the model is a simplified representation of the facility and ignores some of the inputs and processes occurring in the real world (deemed to be less important by the creators of the model). A perfect fit can be achieved with a simple set of polynomials of *N* degree where *N* is the number of data points to be matched by the model. However, the predictive power of such a model is quite limited. Overfitting, while it might reduce the total error for one particular data set, invariably will reduce the model's predictive power and increase model error for other data sets.

When evaluating the match of the model against data, it is crucial to observe all the important variables. It is preferable to fit to most of the measured variables reasonably, rather than fit perfectly to one selected (however important) component concentration and poorly to others.

In addition to minimizing model errors, there are two other objectives of the calibration procedure that often are ignored:

- The calibration should be used to *establish the field of validity* (i.e., the "design space") of the model. The authors of the model frequently include general guidelines regarding the circumstances under which the model can be used. Depending on the range of data it was calibrated to and validated on, the model might not be reliable for a particular application that is too far from the calibration conditions. For example, it is risky to develop process engineering conclusions from the behavior of a model during storm flow conditions if calibration was performed only on average or steady-state data.

- The fit achieved during calibration (and validation) can indicate the *expected accuracy* of the model under the specific circumstances. Different model variables will have different errors associated with them, but the most important performance parameters must be within a certain accuracy if the model is to be trusted for process work. In steady-state situations (assuming good quality data), these variables should be matched within 5% to 20% overall, while during dynamic runs 10% to 40% temporary deviations are not unusual even from well-established models. Secondary variables can vary even more in typical cases, without compromising the model's ability to predict general process performance truthfully.

At the beginning of any modeling project, an agreement should be reached among the involved parties regarding the target variables in the project and the stop criteria; both of these are described in detail in the sections below.

4.4.2 Validation

After the process model is calibrated, a validation step is suggested using a data set, which is independent of calibration data, but still reflects the targeted model application. Uncertainty analysis (based on data, model structure, model parameters, and other sources of uncertainty) (Belia et al., 2009) may be performed to assess the domain of validity of the model and its accuracy. Ultimately, stakeholders have to agree on the model accuracy that is reached. These situations have to be specified and the selected validation data sets should resemble the conditions under which the model is to be used for predictions. Validation tests may include:

- Engineering checks, such as the comparison of model results with similar real facility data over extended operational periods and
- Validation runs performed with data sets for the defined specific/critical conditions.

In practice, data are difficult and expensive to collect and to reconcile, which means that often the available data are dedicated to the calibration phase. In addition, all situations typically cannot be covered (i.e., it is impossible to collect data for all possible conditions). These practical concerns mean that model validity is rarely proven and is only assessed for specific situations where validation is critical to the project.

4.4.3 Target Variables

The target variables are identified based on the specific objective of the project. Three typical target variables are solids production, effluent quality, and aeration (discussed below). It is also important to identify acceptable variance in the model prediction compared to the data collected.

4.4.4 Solids Production

A steady-state period must be identified for calibrating the solids production; if the facility is experiencing a lot of dynamics due to construction or other factors, the dynamics of it should be kept in mind during calibration. Comparison between long-term average solids production predicted by the model to the data may be needed as a proof of calibration. Influent wastewater and stoichiometry of the wastewater has the most impact on solids yield, and the yield parameters must never be changed unless all other efforts to reconcile model prediction to the data have failed. A major portion of time should be spent on identifying and removing any outliers from the data set.

4.4.5 Effluent Quality

Calibration to reproduce the effluent quality should be first attempted without changing any of the biokinetic parameters. There are many factors that affect effluent quality, for instance:

a. Effluent TSS and BOD concentrations depend on the secondary clarifier performance.

b. Biological phosphorus removal is affected by influent characterization (concentration of readily biodegradable COD and volatile fatty acids present) and primary clarifier performance. High nitrate recycle from the anoxic zone or DO in the RAS deteriorates the biological phosphorus removal.

c. Denitrification is affected by the amount of carbon available, internal recycle rate, anoxic volume, and other factors. High DO in the aerobic zone and high internal recycle rates can carry over DO into the anoxic zones decreasing denitrification performance.

d. Nitrification and effluent ammonia concentration is affected by aerobic sludge retention time of the system, DO, alkalinity, and reactor configuration (plug flow versus a single CSTR), and other factors.

It is proposed to first set up the facility model in the required detail and to adapt operational settings before without changing any biokinetic parameters. The need to change biokinetic parameters often points to erroneous data or wrong transport models (mixing characteristics).

In the case of atypical conditions (e.g., significant industrial influent), it may be necessary to recalibrate the biokinetic model and values may have to be assigned to specific parameters. It is suggested to carry out sensitivity analyses to identify influential parameters (parameters that have a significant impact on the results). Engineering judgment should be used to select a few parameters to change (typically the ones that are uncertain and have rather wide typical ranges; Hauduc et al., 2011). The parameter estimation is usually carried out manually because of the over-parameterized nature of activated sludge models. However, automatic algorithms to estimate model parameters may assist in speeding up the process. Important is to define upfront on how the calibration should be carried out. Good practice is to set up a table to define and track all parameter changes.

The resulting parameter set should be validated using a data set that is independent from the calibration data but still reflects the targeted model application. Uncertainty analysis (based on data, model structure, model parameters, and other sources of uncertainty; Belia et al., 2009) may be performed, in order to assess the domain of validity of the model and its accuracy. Ultimately, the stakeholders have to agree on the reached model accuracy.

4.4.6 Aeration
Airflow rates and DO concentrations predicted by the model should be calibrated to the observed values in the aeration basins. Liquid-side mass transfer coefficient and standard oxygen transfer efficiency for process water are two very crucial parameters that are calibrated to reproduce the DO profile in the basin and match the airflow rate. The biggest challenge during aeration calibration is lack of available data; however, this calibration effort cannot be ignored as over- or undersizing of aeration equipment can lead to performance failure. Most of the facilities might not monitor airflow rates and only monitor DO concentration once or twice a day. It is very important to interview the operators and make reasonable assumptions to approximate the airflow rates. Knowledge about the valve positioning and how often the valve position is changed can be used for a rough estimate using engineering judgment.

4.4.7 Stop Criteria
Stop criteria are typically quantitative values that help identify when the modeling effort should stop. It is crucial to identify the stop criteria early on in a project. A few examples from Rieger et al. (2012) are listed in Table 4.1.

Modeling Task	Averaging Period	Target Variable	Acceptable Error Range
Develop a site-specific training for operator training	Monthly	MLSS	10%
		WAS mass load	5%
		Effluent TSS	5.0 mg/L
		NH_3-N	1.0 mg/L
		NO_x-N	1.0 mg/L
		PO_4-P	1.0 mg/L
		Airflow rates	10%
		Dissolved oxygen	0.5 mg/L

TABLE 4.1   Common Typical Target Variables and Stop Criteria for Model Calibration (Adapted from Rieger et al., 2012)

4.5 Step 5—Simulation and Result Interpretation

Simulation and Result Interpretation covers how the model can be used once it has been designed, assembled, and calibrated. There is a strong link between this step and the first step Project Definition. In Project Definition, the modeler and other stakeholders set out the scope and expectations of the model, and in Simulation and Result Interpretation, the model is used to fulfill the scope requirements, meet the majority of the expectations, and provide justification if expectations cannot be met. All important aspects of the modeling, including the results and interpretation, have to be documented in a report.

It is important to keep in mind that models are not reality. Even the best calibrated model has a level of uncertainty associated with many aspects. In general, there are three categories of uncertainty:

- Quantifiable uncertainty—parameters that are known to be uncertain and the degree of uncertainty can be reduced. An example of this is uncertainty related to model structure. The model structure may have inherent assumptions that are only approximations of reality, such as the assumption that the autotrophic half-saturation constant for oxygen is a constant. These could potentially be made more accurate through further study and model development, provided the model allows for variable values.

- Irreducible uncertainty—uncertainty that cannot be reduced by any degree of study. An example of this is uncertainty related to future flows and loads. No amount of study would be able to quantify future flows and loads beyond a certain point.

- Total ignorance—parameters that the modeler does not know are uncertain.

These items could be either reducible or not reducible. Any parameter might fall in this category, and the only method of reducing this is through learning and experience.

All these sources of uncertainty are present within models and contribute to each model's variation from reality. Project team members (stakeholders) need to be aware of this potential during projects so that appropriate measures can be taken to address uncertainty in the modeling results.

However, this does not mean that simulation results cannot be used. As discussed, simulations give valuable insight to how WRRFs function, although the exact accuracy of the output must be viewed in light of the uncertainties involved.

Reliability of modeling results is often directly dependent on the degree of effort expended by the modeler in reducing uncertainty. Once the model/simulator is selected, calibration and validation are the most common methods of reducing uncertainty of modeling results. The amount of calibration and validation done on a model should be selected according to use of the model and the budget available.

In general, time-variant data add additional uncertainty to simulation results. As raw wastewater is inherently variable, this uncertainty is present in all modeling results that examine short-term effects. Predictions of future performance also are more uncertain because they are extended further into the future. It is up to the modeler to determine how best to interpret modeling results.

Presentation of modeling results can take many different routes. In general, the degree of accuracy that is presented in modeling results should reflect the level of calibration. For example, a facility option comparison with little calibration might only report the percentage difference between the options. If a highly calibrated model was developed looking only at a short time into the future, it may be possible to report actual results with more significant digits in some of the results. Model results and limitations can be presented as follows:

- Mass balance data presentation—detailed mass balance data can be summarized and presented by a number of methods. In general, composite variables, such as BOD, COD, TSS, TKN, and total phosphorus, should be presented in data summaries. Detailed mass balance data typically are only suitable for presentation in appendices.

- Model/simulator setup data—similar to mass balance data, only significant parameters should be presented in the main body of a project report or presentation. Detailed model parameters should be included in an appendix so that future users can reproduce the modeling results if necessary. Commercial simulators typically have reporting features that can produce such detailed information automatically.

- Model/simulator assumptions—a critical component of documenting a model includes assumptions that were used in the model/simulator. It is important to model results for the user to document the degree of conservatism used in determining model parameters.

- Identification of physical modifications necessary—the modeler should include discussion of the need to discuss other modifications that would go along with the modeling effort.

- Graphic display of dynamic models results is an effective presentation method—care should be taken to not present too much information in a single graph. Comparing different time series data sets together is an effective method of displaying cause and effect.

5.0 Application of Whole-Facility Model

5.1 Structured Approach for Model Application—Case Example

The City of Galt Wastewater Treatment Plant is owned by the Region of Waterloo (ROW) and operated by the Ontario Clean Water Agency (OCWA). The facility provides tertiary treatment with activated sludge incorporating nitrification, chemical phosphorus removal, and sludge processing. Liquid treatment consists of mechanical screening, grit removal, primary clarifiers, biological reactors incorporating fine-pore diffused aeration, secondary clarifiers, single-media automatic backwash effluent filters, and ultraviolet disinfection with sodium hypochlorite backup. The solids treatment train for handling primary sludge and WAS currently includes two primary digesters, one sludge holding tank, and a digester gas handling/utilization system. When the project commenced, the facility was operated in a co-thickening mode, where WAS generated in the biological reactors was directed to the primary clarifiers (this practice has been stopped after completion of upgrades). In addition, primary sludge and WAS were processed by one primary digester followed by one secondary digester. The existing facility has a Stage 1 nominal rated capacity of 56 800 m^3/d at the time of this writing.

Some upgrades to the solids treatment train have been completed in the last several years. Primary digestion facility upgrades include new raw sludge pumps, a new primary digester roof equipped with roof-mounted vertical-flow draft tubes with shaft-mounted propeller mixers, and a sludge heating/recirculation system. A new biosolids management facility includes rotary drum thickeners for WAS thickening (which ended the previous practice of co-thickening), thickened WAS pumps, a digested sludge holding tank, centrifuges for digested sludge dewatering, and a centrate/filtrate holding tank. More recently, the secondary digester was converted to a primary digester with the two digesters currently operating in series.

5.1.1 Project Definition

A calibrated model-based scenario analysis was performed to investigate:

1. The Galt facility performance after completion of the proposed upgrades;

2. The feasibility of the Galt facility dewatering digested biosolids from other WRRFs following the recommendations of the appropriate Class EA study completed in 2008; and

3. The feasibility of diverting part or all of the wastewater flow from the Industrial Road Service Area (IRSA) from the Preston Wastewater Treatment Plant to the Galt facility following the recommendations of the Wastewater Treatment Master Plan completed in 2007.

Various combinations of these scenarios were investigated for current and estimated future Galt facility raw influent flows.

5.1.2 Process Model Calibration

A model was configured for the Galt facility and calibrated to historical data for a one-year period. Figure 4.3 shows the pre-upgrade Galt facility represented in BioWin.

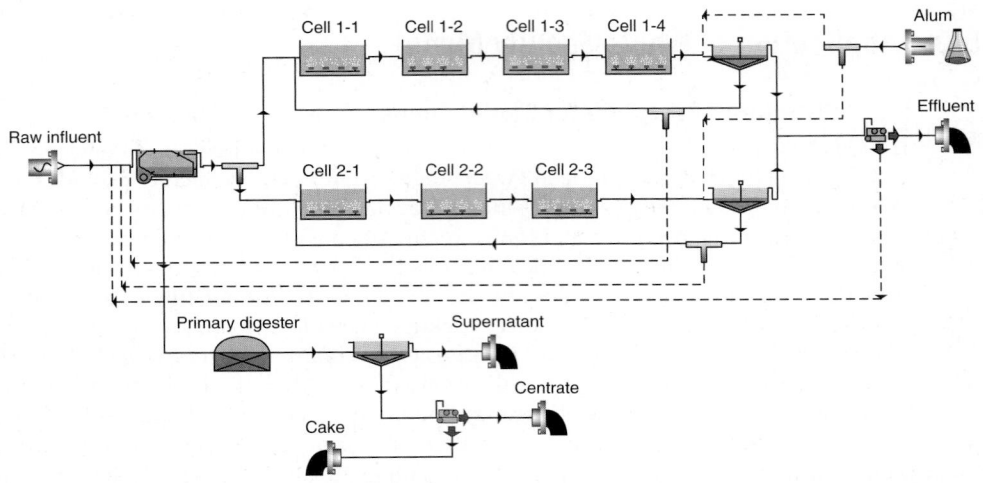

Figure 4.3   Overall model layout for the Galt facility.

Before commencing the model calibration, the facility data for the one-year calibration period were reviewed and assembled in one master spreadsheet for the entire period to facilitate transfer into the process simulator. The calculation of ratios between some of these influent characteristics and comparison to typical values is of particular importance at this stage. Table 4.2 summarizes the statistics calculated for the raw influent data. The following points were observed:

- The average TSS/BOD ratio was higher than the typical range. Cross-checks with other ratios indicated that this ratio is high because the average TSS concentration is high given the organic strength of the wastewater. There were no COD or VSS data in the historical data set to serve as a cross-check; however, the supplemental sampling data indicated that the reason for the high solids is a higher-than-typical particulate unbiodegradable COD portion in the influent.

Parameter	Average	Typical Range
Flow (m³/d)	36 941	
BOD (mg/L)	199	
TSS (mg/L)	249	
TKN (mg/L)	37.2	
TP (mg/L)	4.8	
TSS/BOD	1.39	0.70–1.22
TKN/BOD	0.19	0.14–0.24
TP/BOD	0.02	0.02–0.05

Table 4.2   Statistics Calculated for Raw Influent Parameters and Selected Ratios for the Modeling Period

Constituent Fraction	Value
Readily biodegradable (including Acetate) [gCOD/g of total COD]	0.19
Non-colloidal slowly biodegradable [gCOD/g of slowly degradable COD]	0.85
Unbiodegradable soluble [gCOD/g of total COD]	0.05
Unbiodegradable particulate [gCOD/g of total COD]	0.22
Ammonia [gNH$_3$-N/gTKN]	0.43
Soluble unbiodegradable TKN [gN/gTKN]	0.09
Phosphate [gPO$_4$-P/gTP]	0.50
Particulate biodegradable and unbiodegradable COD:VSS ratio [gCOD/gVSS]	1.7

TABLE 4.3   Wastewater Fractions and Ratios Derived from Supplemental Sampling for Use in Model Calibration

- The average TKN/BOD ratio was within the typical range.
- The average TP/BOD ratio was within the typical range.

As a result of the facility data review, additional data requirements were identified and a supplemental sampling plan was formulated and executed over a 2-week period. Samples were collected for a number of streams, including raw influent, primary effluent, secondary effluent, mixed liquor, RAS, and primary sludge. Wastewater characteristics derived from this exercise are listed in Table 4.3. Many of these values fall within ranges typically observed at North American WRRFs, with the exception of the following:

- The fraction of influent TKN in the form of ammonia (F_{SNH}) is lower than typical.
- The fraction of influent TKN that is unbiodegradable and soluble (F_{SNI}) is higher than typical. It is suspected that this may be a transient phenomenon, possibly due to periodic discharges from textile industries in the area, as there was no evidence of excess unbiodegradable soluble nitrogen on each day it was tested for. Note that testing for F_{SNI} was not in the original sampling plan because an effluent TN limit is not anticipated at this facility. However, ongoing data evaluation during the initial stages of the sampling campaign indicated high influent organic nitrogen content. Therefore, the effluent soluble organic nitrogen concentration was checked to ascertain the nature of this material. The unbiodegradable soluble nitrogen in the influent is likely the cause of the lower-than-typical F_{SNH}. Other checks (e.g., the influent TKN/COD ratio) indicate that the influent total nitrogen content is high relative to the organic strength.
- The fraction of the influent COD that is unbiodegradable particulate (F_{XI}) is on the high end of the typical range (Melcer et al., 2003). This parameter cannot be measured directly; rather, it must be inferred from a combination of other ratios including COD:BOD, COD:VSS, and TSS:BOD. Analysis of the supplemental sampling data indicated a higher-than-typical F_{XI} fraction, and this was corroborated by examination of the historical data set. The implication of using a higher-than-typical F_{XI} in the calibrated model for scenario analysis is that the model will tend to be conservative in terms of sludge production predictions.

The model was calibrated to a variety of facility parameters, including MLSS levels, as shown in Figure 4.4.

The aeration aspect of the model also was calibrated to continuous DO data that were collected as part of the sampling campaign (e.g., Figure 4.5) so that the model could provide insight into potential blower capacity limitations in the scenarios analyzed.

5.1.3   Constraints for Galt Modeling Scenarios

The calibrated model was modified to account for the completed (at that time) upgrades to the facility solids handling train. Modifications included WAS thickening (i.e., the cessation of co-thickening of WAS in the primary settling tanks [PSTs]), incorporation of a digested sludge holding tank, addition of centrifuges, and a centrate holding tank. Incorporating these changes resulted in additional loading to the facility principally in the form of returned centrate. Additional inputs to the model to represent diverted industrial flow from the IRSA and biosolids from other Region facilities were included. The modified model flowsheet used for scenario analysis is shown in Figure 4.6.

Important equipment and operational constraints incorporated into the Galt scenario modeling included information about the facility nitrifier maximum specific growth rate, the operating SRT required to maintain nitrification at winter temperatures, air supply capacity, RAS pumping limits, thickening factors achievable in the solids-handling-train unit processes, and the length of solids-handling-equipment shift runs. Clearly identifying any constraints and/or assumptions built into the modeling scenario analysis is very important for any modeling project. Key constraints are summarized in Table 4.4.

Wastewater characteristics obtained from the supplemental sampling exercise conducted in support of the Galt facility model calibration were assumed to hold for future flow conditions. Concentrations for various bulk parameters (e.g., COD, TSS) were assumed to remain at values representative of those observed over the calibration period; that is, overall mass loading on the facility is assumed to increase as a result of increasing raw influent flow.

A daily raw influent flow pattern is desirable for conducting dynamic simulation scenario analysis. Printouts of hourly flow records over 24-hour periods were obtained—these were not readily available in electronic form, which would have allowed for a more rigorous analysis of historical flow data. However, a typical flow day when there was no significant storm or snowmelt influx to the collection system was selected and normalized to establish the pattern shown in Figure 4.7.

Projected future flows obtained from the Region are listed in Table 4.5. These are plotted versus time in Figure 4.8; there generally is a linear increase in facility raw influent flows with time.

To provide a daily raw influent flow pattern for a given scenario year, the normalized daily flow pattern from Figure 4.9 was multiplied by the average flow value for that year from Table 4.5. Raw influent concentrations were assumed to remain constant throughout the day, which means that the raw influent daily mass loading pattern for wastewater constituents such as BOD tracks the daily flow pattern.

The Region commissioned a sampling campaign of the IRSA wastewater to obtain more detailed information on the concentration of various constituents. Given the large food processing component of this wastewater, it was quite high in organic strength and suspended solids. However, it was not strong in terms of nitrogen or phosphorus. Daily and overall average concentrations are shown in Table 4.6. Based on this information, it was possible to derive the wastewater characteristic fractions shown in Table 4.7.

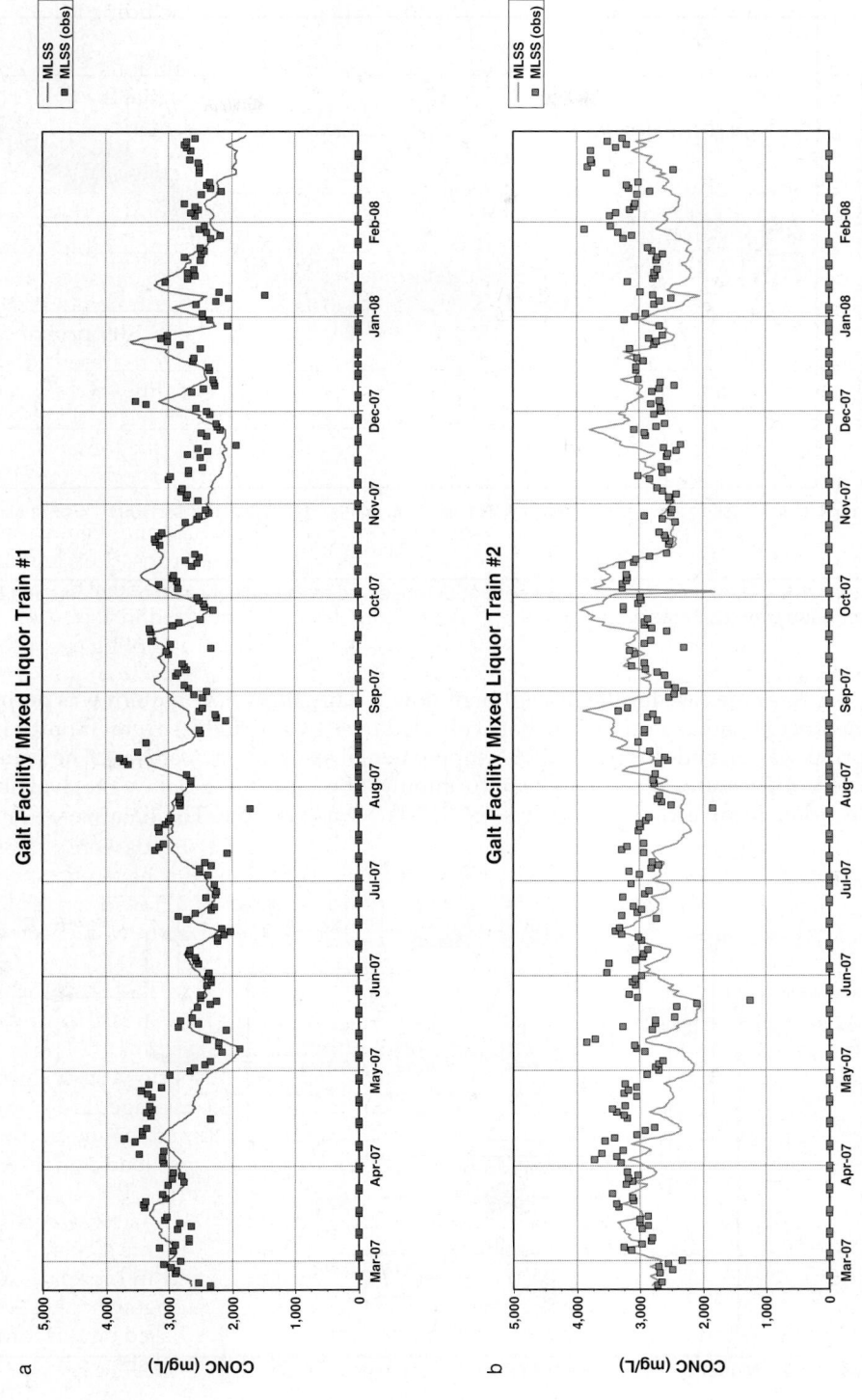

Figure 4.4 Predicted (line) versus observed (points) train #1 (a) and train #2 (b) aeration tank MLSS.

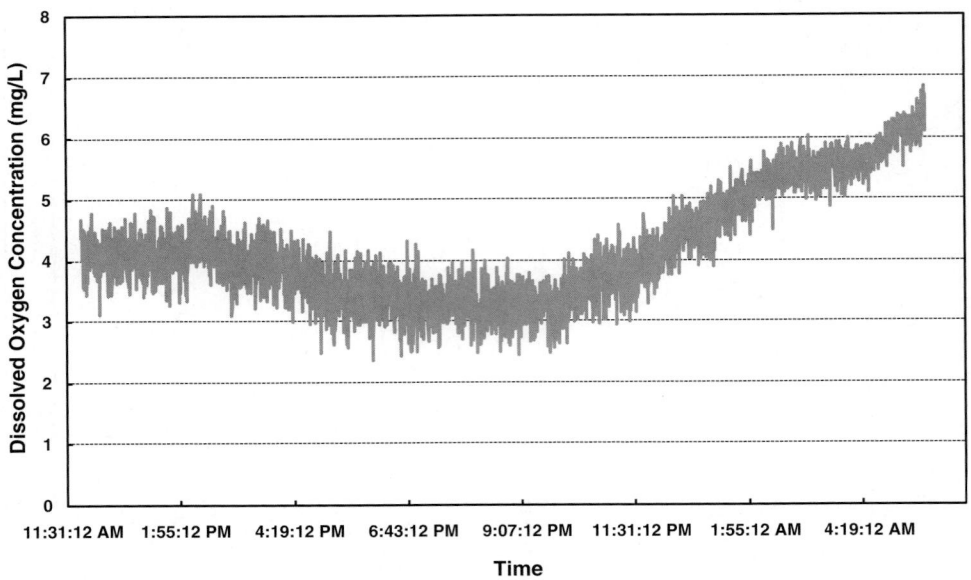

FIGURE 4.5   Measured dissolved concentration observed at the influent end of the aeration tank first pass over 18 hours.

As was the case for the raw influent flow, a daily IRSA flow pattern was desirable. The Region had extensive flow data collected over two periods (several months in the spring season, and during the IRSA supplemental sampling in October). The data consisted of flowrates recorded every 10 minutes. Because these data were available in electronic form, a rigorous analysis of the data was possible. The data were scanned,

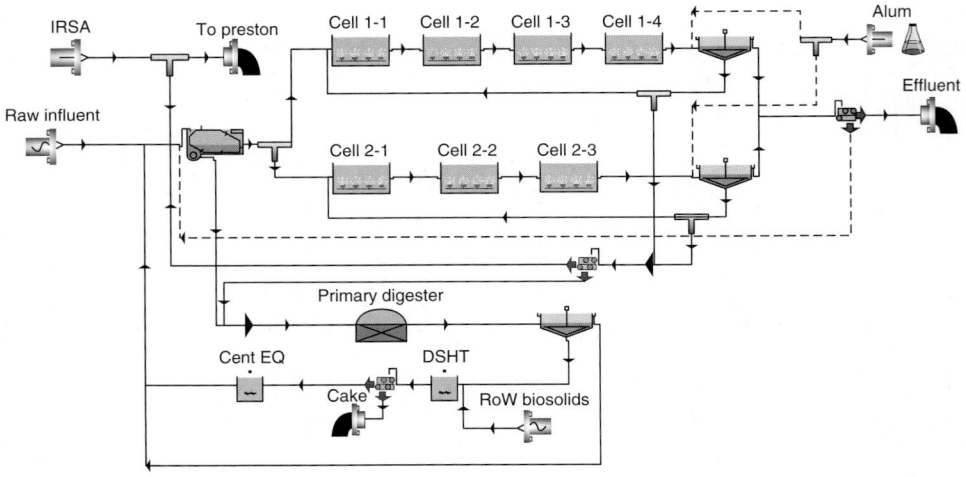

FIGURE 4.6   Model process flowsheet used for scenario analysis.

Assumption	Comments
Nitrifier growth rate of 0.85 d^{-1}	The models default value for this parameter is 0.90 d^{-1}, and this value was used in the model calibration exercise with good results. Commonly measured values in North America range from 0.8 to 1.0 d^{-1} (Melcer et al., 2003). A value of 0.85 d^{-1} was selected for the scenario analysis because that was the value measured by at a nearby Region facility receiving very similar wastewater, and it introduces a degree of conservatism to the analysis.
Operating SRT of 7.5 days	At the time of the study, the facility was operated at an average SRT of about 9 days. Based on the nitrifier growth rate discussed above and the model calibration, it appeared that the facility can achieve nitrification at winter conditions at a lower SRT. The lower SRT will help to keep MLSS levels reasonable at higher future loadings.
PST solids capture	A solids capture rate of 65% was used for the model calibration. This capture rate was assumed to prevail for future flow scenarios.
Aeration tank dissolved oxygen (DO) levels of 2 mg/L achievable	As documented in the model calibration phase, with two of four installed blowers running at 70% of capacity, the DO at the front end of the first pass of the aeration tank was at or above 4 mg/L for most of a typical day. It appeared as though the aeration equipment at the Galt facility should be able to achieve aeration tank DO levels of 2 mg/L at the future flows and loads under consideration.
Return activated sludge (RAS)	The facility is operated with the RAS rates for the final settling tanks set to approximately 100% of the average flow treated in each train. This practice has been maintained for future flow scenarios.
Thickening factors*	The concentration of certain facility solids streams is determined as a result of the degree of thickening achieved in various unit processes (e.g., primary sludge, WAS) and these have been set in the analysis according to design guidelines or actual data observed at the Galt facility as follows: • Primary sludge was assumed to achieve a concentration of approximately 3.3%, based on measurements conducted during the supplemental sampling campaign. • Thickened WAS was assumed to achieve a concentration of approximately 7%, based on typical design guidelines (Metcalf & Eddy, 2003). • Dewatered sludge was assumed to achieve a concentration of approximately 20%, based on typical design guidelines (Metcalf & Eddy, 2003).
Temperature	All scenarios were conducted at a temperature of 12°C.

*Because the model obeys a mass balance, the concentration of thickened streams determines their "flows." In the case of the primary sludge and thickened WAS streams, the flow is important in that it determines the primary digester hydraulic retention time (HRT).

TABLE 4.4 Key Assumptions Underlying Model-Based Scenario Analysis

Year	Projected Flow (m³/d)
2021	45 900
2031	51 100
2041	54 400

TABLE 4.5   Projected Raw Wastewater Flows to the Galt Facility

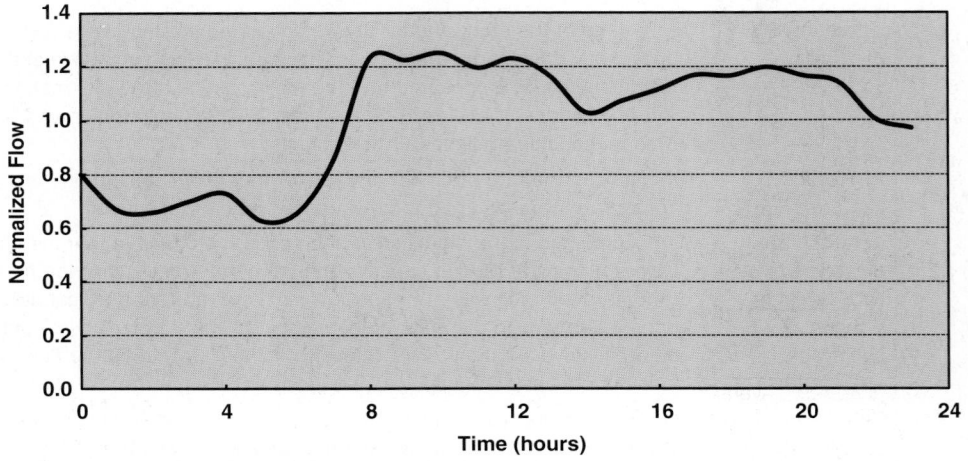

FIGURE 4.7   Normalized daily flow pattern used as a basis for scenario analysis (peak:mean ratio—1.25; min:mean ratio—0.63).

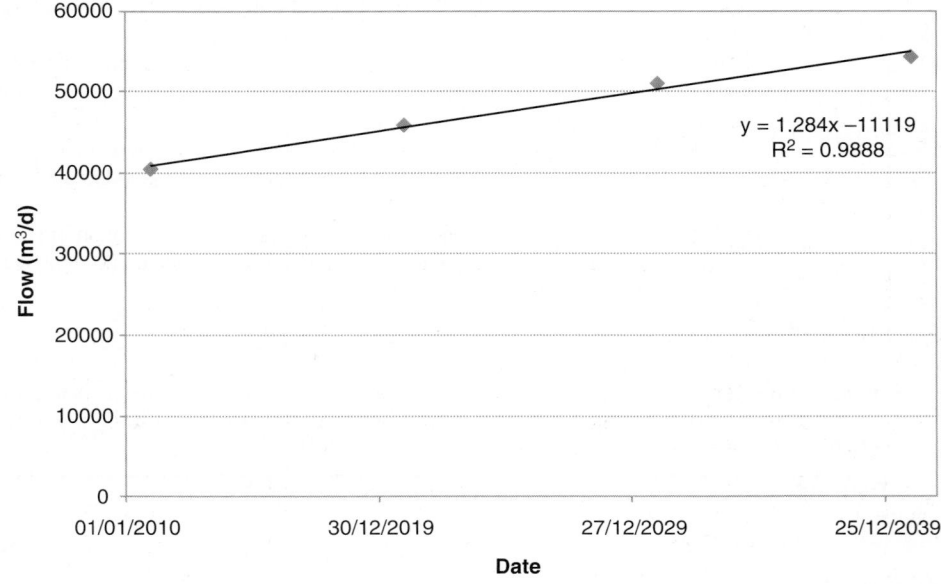

FIGURE 4.8   Raw influent flow projections for the Galt facility.

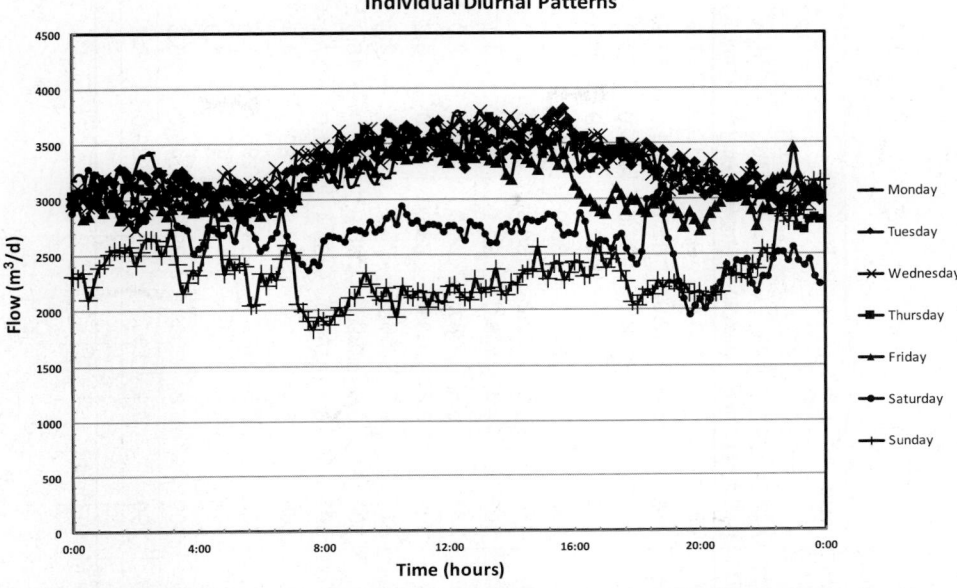

FIGURE 4.9   Average daily IRSA flow patterns based on data collected by the Region of Waterloo.

and 24-hour flow patterns for each day of the week contained in the data set were extracted and averaged to generate the flow pattern for each day of the week, as shown in Figure 4.9.

The following aspects of the average daily IRSA flow patterns are worth noting:

- The Monday–Friday patterns are quite consistent (although the Friday pattern appears to show a marked drop in flow around 4 p.m.).

- There is a clear difference in peak flows between week and weekend days.

The data were further simplified by averaging all of the weekdays and weekend days to result in typical average weekday and weekend-day flow patterns, as shown in Figure 4.10.

To be conservative, the average weekday flow pattern was selected for use in the model-based scenario analysis. The normalized pattern is shown in detail in Figure 4.11.

Projected future IRSA flows are listed in Table 4.8. Inspection of Figure 4.12 shows that there generally is a linear increase in IRSA flows with time.

To provide a daily IRSA flow pattern for a given scenario year, the normalized daily flow pattern from Figure 4.11 was multiplied by the average flow value for that year from Table 4.8. IRSA influent concentrations were assumed to remain constant throughout the day, which means that the IRSA daily mass loading pattern for wastewater constituents such as BOD tracks the daily flow pattern.

Industrial Road Influent

Date	Day	TSS (mg/L)	VSS (mg/L)	COD Tot. (mg/L)	COD 1.5μ gf (mg/L)	BOD (mg/L)	cBOD (mg/L)	TP Tot. (mg/L)	ALK (mg CaCO$_3$/L)	pH	Ca (mg/L)	Mg (mg/L)	TKN Tot. (mg/L)	NH$_3$-N (mg/L)
2008-10-15	0	1410		3130	1150	1240	1090	2.9	456	6.4	130.0	33.8	119.0	17.3
2008-10-16	1	1230		4080	1150	1640	1520	6.0	471	6.5	182.0	32.1	111.0	17.2
2008-10-18	2	884		1180		1320	941	13.8	437	6.5	168.0	37.5	145.0	25.3
2008-10-19	3	566		1350	390	1170	1100	5.2	351	6.5	162.0	29.4	68.9	4.5
2008-10-20	4	1360	960	3230	2250	1540		6.8	290	5.9	248.0	39.9	111.0	3.6
2008-10-23	5	480	480	2100	1060	1380	1310	4.9	434	6.6	172.0	39.3	113.0	17.0
2008-10-24	6	402	440	1950	1110	1330	1430	4.0	365	6.4	175.0	32.1	132.0	14.1
2008-10-25	7	400	315	900	250	408	406	18.5	338	6.8	168.0	34.0	39.0	3.0
2008-10-26	8	690	617	1560	560	788	760	3.8	329	6.8	169.0	35.3	40.0	0.2
2008-10-27	9	1700	1610	3630	1270	1740	1640	6.7	417	6.6	256.0	35.1	58.9	2.8
2008-10-28	10	1600	1100	3500	1230	1390	1320	6.2	509	6.6	189.0	35.9	92.9	16.3
2008-10-29	11	620	460	1450	170	977	1020	3.7	497	6.7	160.0	35.4	121.0	17.1
2008-10-30	12	460	353	1350	1210	894	875	5.2	388	6.4		28.6	89.0	14.4
Average		908	704	2262	983	1217	1118	6.7	406	6.5	182	34.5	95.4	11.8
Std Dev		484	434	1094	573	370	350	4.4	69	0.2	36	3.4	34.3	7.9
Max		1700	1610	4080	2250	1740	1640	18.5	509	6.8	256	39.9	145.0	25.3
Min		400	315	900	170	408	406	2.9	290	5.9	130	28.6	39.0	0.2

TABLE 4.6 Data Collected during Region of Waterloo Sampling Campaign and Used as Model Inputs

Constituent Fraction	Value
Readily biodegradable (including acetate) [gCOD/g of total COD]	0.25
Noncolloidal slowly biodegradable [gCOD/g of slowly degradable COD]	0.70
Unbiodegradable soluble [gCOD/g of total COD]	0.00
Unbiodegradable particulate [gCOD/g of total COD]	0.13
Ammonia [gNH$_3$-N/gTKN]	0.12

TABLE 4.7 Wastewater Fractions Derived from IRSA Supplemental Sampling

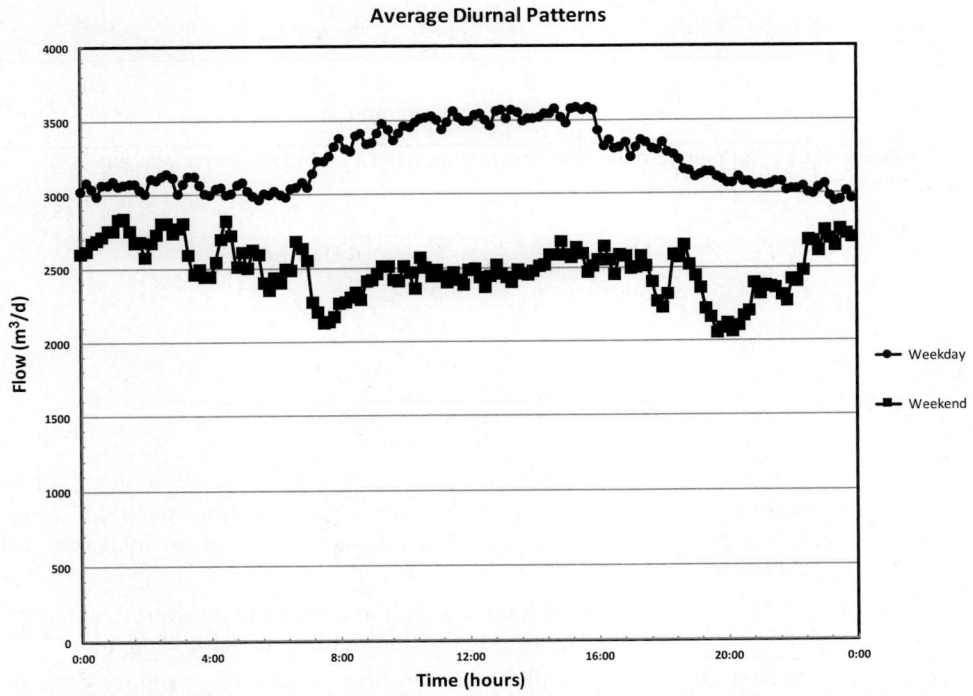

FIGURE 4.10 Average week and weekend daily IRSA flow patterns based on data collected by the Region of Waterloo.

5.1.4   Key Performance Indicators for Galt Modeling Scenarios

Key performance indicators (KPIs) to be monitored from the scenario modeling output included:

- Effluent ammonia concentration: the facility currently is required to meet an annual average limit of 2 mg/L.

- Secondary clarifier solids loading rate (SLR): Ontario Ministry of Environment guidelines currently suggest that for a nitrifying facility with 100% RAS recycle, the SLR at average flow should not exceed 120 kg/m^2/d. However, other WRRFs in Ontario and North America have achieved higher SLRs while

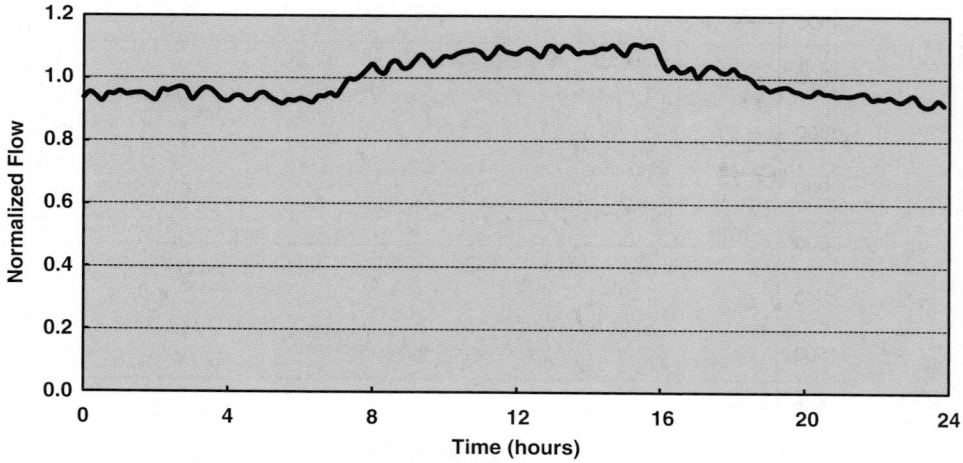

FIGURE 4.11 Normalized daily IRSA flow pattern used as a basis for scenario analysis (peak:mean ratio—1.11, min:mean ratio—0.91).

Year	Projected Flow (m³/d)
2021	3400
2031	3650
2041	3900

TABLE 4.8 Projected IRSA Wastewater Flows to the Galt Facility

maintaining stable operation. For the Galt scenarios, two SLR rates were considered; the "typical" value of 120 kg/m²/d and an increased value of 170 kg/m²/d.

- Aeration tank MLSS: MLSS was monitored as a KPI because it, along with the facility and RAS flows, impacts the final settling tank SLRs.
- Primary digester hydraulic retention time (HRT): The primary digester HRT was monitored as a KPI because of loading changes as a result of IRSA diversion, increased influent loading, and termination of the co-thickening.

5.1.5 Galt Modeling Scenarios

A number of scenarios for investigation were developed. These included:

- The impact of the proposed upgrades to the solids handling system at current raw influent flow/load conditions. For example, would separate WAS thickening help to increase the HRT of the existing primary digester?
- The feasibility of the Galt facility dewatering digested biosolids from other ROW WRRFs. For example, would the additional centrate from this processing pose a problem to the facility in terms of meeting its effluent ammonia requirement? Of particular interest was whether the Galt facility

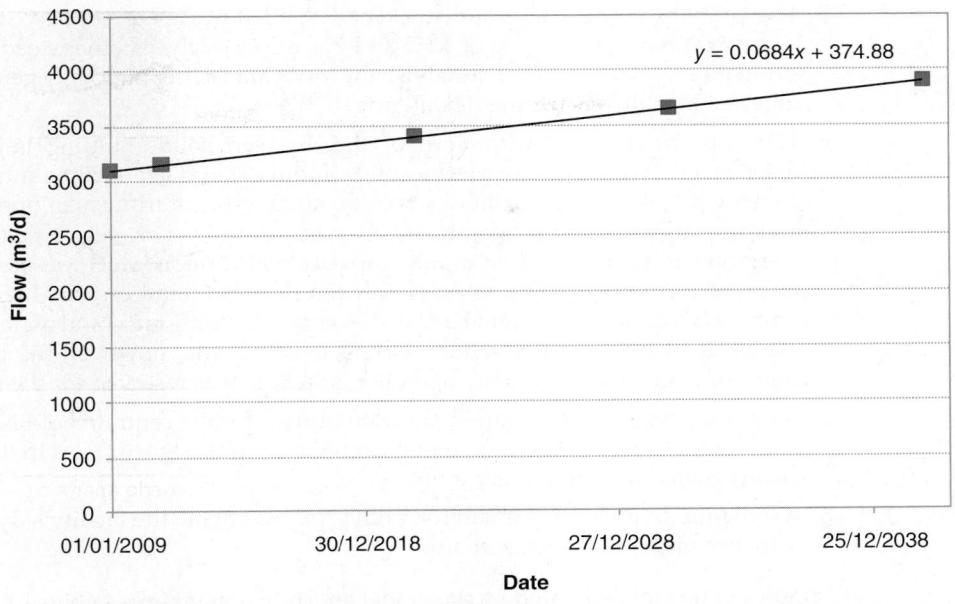

$y = 0.0684x + 374.88$

FIGURE 4.12   IRSA flow projections.

could process biosolids from other regional facilities during major capital upgrades.

- The feasibility of diverting part or all of the wastewater flow from the IRSA to the Galt facility. This would reduce the load at the Preston facility and restore capacity at that facility.

All of these scenarios were run at current and future daily average flow conditions (for the years 2011, 2021, 2031, and 2041).

5.1.6   Run Galt Modeling Scenarios

Once the modeling scenarios of interest were established, the modeling runs were performed and the KPIs of interest were monitored. The following stepwise algorithm was used for the scenario runs:

1. Flow patterns for the various model inputs (i.e., base Galt facility raw influent, external ROW biosolids, IRSA flow) were input.

2. A portion of the IRSA flow was directed to the Galt facility. Most scenarios investigated 100% of the IRSA flow being directed to Galt.

3. The appropriate RAS flow was set according to the average base raw influent flow.

4. A steady-state simulation is performed to estimate the WAS flow necessary to achieve a 7.5-day SRT. To reduce the complexity of analysis, wasting was assumed to be conducted continuously on a 24-hour-per-day, 7-day-per-week basis.

5. The primary sludge flow and thickened WAS flows were adjusted to give concentrations in the region of 3.3% and 7%, respectively. Another steady-state simulation was performed to establish the total Galt facility internally generated digested biosolids requiring dewatering.

6. The appropriate outflow pattern of the digested solids holding tank was determined, according to the biosolids inputs (i.e., Galt facility internally generated and external facility biosolids), number of centrifuges in operation, and the operation shift length. In all cases examined, a single centrifuge operating at its maximum hydraulic capacity for a 14-hour shift was assumed. It is recognized that, in some cases, this can result in unnecessarily high peak centrate flows (which are equalized in the centrate holding tank) and nonoptimal use of available digested solids holding tank volume; however, the goal of suggesting an optimized daily operation schedule was reserved for the future.

7. An appropriate outflow pattern was determined for the centrate holding tank, in order to achieve a degree of equalization of centrate return flows to the Galt facility main liquid treatment train.

8. A dynamic simulation was run for 4 days, to investigate the facility's dynamic response under the given scenario.

Examples of output generated by the model are shown in Figures 4.13 to 4.15. Some important findings from the scenario runs included:

- Upgrading the Galt facility to include WAS thickening (rather than co-thickening) results in less flow to the primary anaerobic digester. The lower flow in turn results in a longer digester HRT in comparison to current conditions; this should, along with other digester infrastructure upgrades, provide improved performance.

- At 2021 conditions, the Galt facility liquid train appears to be capable of treating the base raw influent flow, all of the IRSA flow, but only process biosolids volume from one additional medium-sized Region WRRF.

- At 2031 conditions, the final clarifier SLR begins to become critical if all of the IRSA flow is diverted to the Galt facility. This conclusion highlights the importance of perhaps addressing this facet of facility operation via secondary clarifier stress testing, upgrades, or expansion.

- Under the future flow scenarios, the model-based analysis indicates that primary digester HRT may well become a limiting factor due to the generation of additional primary and secondary sludge from the increased facility loading. Based in part on this finding, a decision was made to retrofit a secondary digester to a primary digester; this upgrade has recently been completed.

6.0 Conclusion

Linking several regional WRRFs in a model-based analysis allows for the examination of potential interactions and provides owners of these facilities with information that (1) can be used to facilitate planning on a regional basis, (2) helps decision makers maximize the capacity of existing WRRF infrastructure, (3) allows for sharing (where

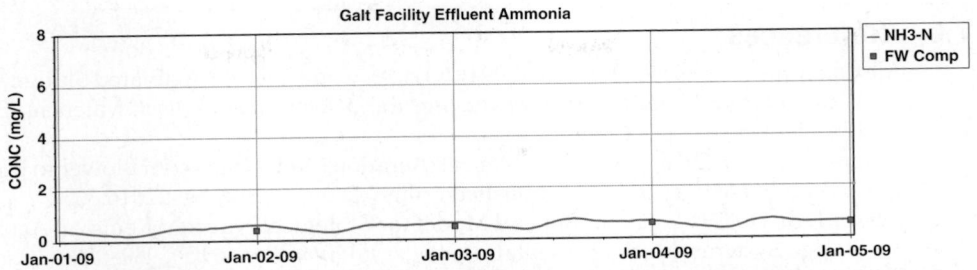

FIGURE 4.13 Predicted dynamic effluent ammonia response for upgraded facility (dashed line shows 2 mgN/L effluent permit limit) at 2021 raw influent and IRSA flows.

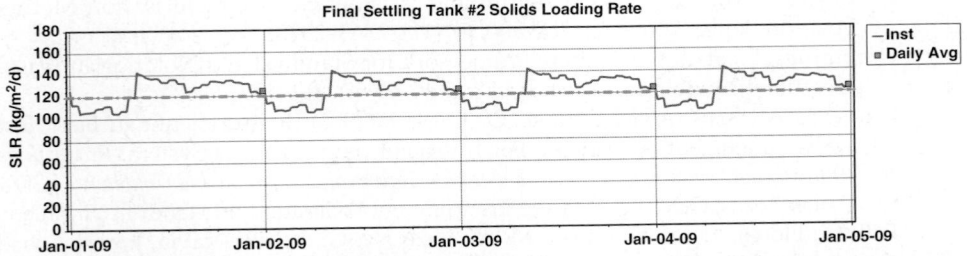

FIGURE 4.14 Predicted dynamic final settling tank solids loading rate response for upgraded facility (dashed lines show 120 kg/m²/d design guideline limit and 170 kg/m²/d suggested upper limit) at 2021 raw influent and IRSA flows.

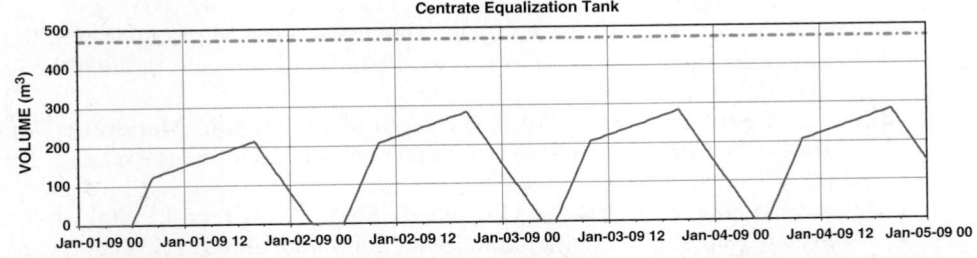

FIGURE 4.15 Biosolids facility centrate holding tank daily volume fluctuation (dashed line shows tank maximum capacity) at 2021 raw influent and IRSA flows.

possible) of excess WRRF capacity, (4) achieves maximum benefit from new infrastructure, (4) provides a basis for more detailed work related to individual unit process optimization (e.g., collection of additional data around individual unit processes, stress testing, etc.), and (5) aids in the development of rational facility upgrade strategies to accommodate future growth. This case study showed the results of such an exercise with the Galt facility serving as an example as to how decisions made at one facility could potentially affect future plans for several other regional WRRFs.

7.0 References

Alex, J.; Jumar, U.; Tschepetzki, R. (1994) A Fuzzy Controller for Activated Sludge Waste Water Plants; In *IFAC Artificial Intelligence in Real Time Control,* Valencia, Spain, Elsevier Science, pp. 61–66.

Amaral, A., et al. (2016) Towards Advanced Aeration Modelling: From Blower to Bubbles to Bulk. *Water Sci. Technol.,* published online, DOI: 10.2166/wst.2016.365.

Barker P. S.; Dold P. L. (1997) General Model for Biological Nutrient Removal Activated Sludge Systems: Model Presentation. *Water Environ. Res.,* **69**(5), 969–984.

Batstone D. J., et al. (2002) Anaerobic Digestion Model No. 1. Scientific and Technical Report No. 13, IWA Publishing, London, UK.

Belia, E., et al. (2009) Wastewater Treatment Modeling: Dealing with Uncertainties. *Water Sci. Technol.,* **60** (8), 1929–1941.

Choubert, J.-M., et al. (2013) Rethinking Wastewater Characterisation Methods—A Position Paper. *Water Sci. Technol.,* **67** (11), 2363–2373.

Corominas, L., et al. (2010) New Framework for Standardized Notation in Wastewater Treatment Modelling. *Water Sci. Technol.,* **61** (4), 841–857.

Fairey, A. W.; Shaw, A.; McConnell, O.; Cook, J. (2006) Development and Field Testing of a New Intelligent Sequencing Batch Reactor (iSBR) Control System. *Proceedings of the 79th Annual Water Environment Federation Technical Exposition and Conference* [CD-ROM], Dallas, Texas, Oct. 21– 25; Water Environment Federation: Alexandria, Virginia.

Fall, C.; Flores, N. A.; Espinoza, M. A.; Vazquez, G.; Loaiza-Návia, J.; van Loosdrecht, M. C.; Hooijmans, C. M. (2011) Divergence between Respirometry and Physicochemical Methods in the Fractionation of the Chemical Oxygen Demand in Municipal Wastewater. *Water Environ. Res.,* **83** (2), 162–172.

Frank, K. (2006) The Application and Evaluation of a Practical Stepwise Approach to Activated Sludge Modeling. *Proceedings PENNTEC 2006,* Annual Technical Conference and Exhibition; State College, Pennsylvania, July 2006.

Gernaey K. V.; van Loosdrecht M. C. M.; Henze M.; Lind M.; Jørgensen S. B. (2004) Activated Sludge Wastewater Treatment Plant Modelling and Simulation: State of the Art. *Environ. Mod. Soft.,* **19**, 763–783.

Gillot, S.; Choubert, J. M. (2010) Biodegradable Organic Matter in Domestic Wastewaters: Comparison of Selected Fractionation Techniques. *Water Sci. Technol.,* **62** (3), 630–639.

Hauduc, H.; Gillot, S.; Rieger, L.; Ohtsuki, T.; Shaw, A.; Takacs, I.; Winkler, S. (2009) Activated Sludge Modeling in Practice an International Survey. *Water Sci. Technol.,* **60** (8), 1943–1951.

Hauduc, H., et al. (2011) Activated Sludge Modeling: Development and Potential Use of a Practical Applications Database. *Water Sci. Technol.,* **63** (10), 2164–2182.

Henze, M.; Gujer, W.; Mino, T.; van Loosdrecht, M. C. M. (2000) Activated Sludge Models ASM1, ASM2, ASM2d, and ASM3, Scientific and Technical Report No. 9, IWA Publishing, London, UK.

Henze, M.; van Loosdrecht M. C. M.; Ekama, G. A.; Brdjanovic, D. (2008) *Biological Wastewater Treatment—Principles, Modeling and Design*; IWA Publishing: London, UK (ISBN 1843391880).

Hulsbeek, J. J. W.; Kruit, J.; Roeleveld, P. J.; van Loosdrecht, M. C. M. (2002) A Practical Protocol for Dynamic Modeling of Activated Sludge Systems. *Water Sci. Technol.,* **45** (6), 127–136.

Langergraber, G., et al. (2004) A Guideline for Simulation Studies of Wastewater Treatment Plants. *Water Sci. Technol.*, **50** (7), 131–138.

Langergraber, G., et al. (2008) Generation of Diurnal Variation for Dynamic Simulation. *Water Sci. Technol.*, **59** (9), 1483–1486.

Meijer, S. C. F.; van der Spoel, H.; Susanti, S.; Heijnen, J. J.; van Loosdrecht M. C. M. (2002) Error Diagnostics and Data Reconciliation for Activated Sludge Modeling Using Mass Balances. *Water Sci. Technol.*, **45** (6), 145–156.

Melcer, H., et al. (2003) Methods for Wastewater Characterisation in Activated Sludge Modeling, Project 99-WWF-3; Water Environment Research Foundation (WERF), Alexandria, Virginia, USA.

Patry, G. G.; Takács, I. (1990) Modular/multi-purpose Modelling System for the Simulation and Control of Wastewater Treatment Plants: An Innovative Approach; In *Advances in Water Pollution Control*; Briggs, R., Ed.; 385–392, Pergamon Press: Oxford, United Kingdom.

Patry, G. G.; Takács, I. (1994) GPS-X A Wastewater Treatment Plant Simulator. *Proceeding MathMod '94 Conference*; Vienna, Austria, Feb. 24; Vol. 3, pp. 456–459.

Petersen, B.; Gernaey K.; Henze, M.; Vanrolleghem, P. A. (2003) Calibration of Activated Sludge Models: A Critical Review of Experimental Designs. In *Biotechnology for the Environment: Wastewater Treatment and Modeling, Waste Gas Handling*. Agathos, S. N.; Reineke, W., Eds.; Kluwer Academic Publishers: Dordrecht, The Netherlands. pp. 101–186.

Phillips, H. M.; Shaw, A.; Sabherwal, B.; Harward, M.; Lauro, T.; Rutt, K. (2007) Modeling Fixed Film Processes: Practical Considerations and Current Limitations from a Consulting Viewpoint. *Proceedings of the 80th Annual Water Environment Federation Technical Exposition and Conference* [CD-ROM]; San Diego, California, Oct. 13–17; Water Environment Federation: Alexandria, Virginia.

Phillips, H. M.; Rogowski, S.; Anderson, W. (2010) Long-Term Operations Modelling: When Does Validation Become Recalibration? *Proceedings WWTmod2010*; Mont-Sainte-Anne, QC, Canada.

Rieger, L.; Takács, I.; Villez, K.; Siegrist, H.; Lessard, P.; Vanrolleghem, P. A.; Comeau, Y. (2010) Data Reconciliation for WWTP Simulation Studies—Planning for High Quality Data and Typical Sources of Errors. *Water Environ. Res.*, **82** (5), 426–433.

Rieger L.; Gillot, S.; Langergraber, G.; Ohtsuki, T.; Shaw, A.; Takács, I.; Winkler, S. (2012) Guidelines for Using Activated Sludge Models, IWA Scientific and Technical Report, IWA Publishing, London, UK.

Rieger, L.; Jones, R.; Dold, P.; Bott, C. (2014) Ammonia-Based Feedforward and Feedback Aeration Control in Activated Sludge Process. *Water Environ. Res.*, **86** (1), 426–433.

Roeleveld P. J.; van Loosdrecht M. C. M. (2002) Experience with Guidelines for Wastewater Characterisation in The Netherlands. *Water Sci. Technol.*, **45** (6), 63–73.

Schraa, O.; Rieger, L.; Alex, J. (2016) Development of a Model for Activated Sludge Aeration Systems: Linking Air Supply, Distribution, and Demand. *Water Sci. Technol.*, published online, doi:10.2166/wst.2016.481.

Shaw, A.; Phillips, H. M.; Sabherwal, B.; deBarbadillo, C. (2007) Succeeding at Simulation. *Water Environ. Technol.*, April, 54–58.

Shaw, A.; Rieger, L.; Takács, I.; Winkler, S.; Ohtsuki, T.; Langergraber, G.; Gillot, S. (2011) Realizing the Benefits of Good Process Modeling. *Proceedings WEFTEC 2011*, 84th Annual Technical Exhibition and Conference; Los Angeles, California, USA, October 1519, 2011.

Smith, R.; Eilers, R. G. (1970) *Simulation of the Time-Dependent Performance of the Activated Sludge Process Using Digital Computer*; U.S. Department of the Interior, FWPCA: Cincinnati, Ohio.

Storer, F. F.; Gaudy, A. F. Jr. (1969) Computational Analysis of Transient Response to Quantitative Shock Loadings of Heterogeneous Populations in Continuous Culture. *Environ. Sci. Technol.*, **3**, 143–149.

Third, K. A.; Shaw, A. R.; Ng, L. (2007) Application of the Good Modeling Practice Unified Protocol to a Plant Wide Process Model for Beenyup WWTP Design Upgrade. *Proceedings 10th IWA Specialised Conference on "Design, Operation and Economics of Large Wastewater Treatment Plants"*; Vienna, Austria, September 913, 2007, pp. 251–258.

van Loosdrecht, M. C. M.; Nielsen, P. H.; Lopez-Vazquez, C. M.; Brdjanovic, D. (2016) *Experimental Methods in Wastewater Treatment*; IWA Publishing: London, UK.

Vanrolleghem, P. A.; Insel, G.; Petersen, B.; Sin, G.; De Pauw, D.; Nopens, I.; Weijers, S.; Gernaey, K. (2003) A Comprehensive Model Calibration Procedure for Activated Sludge Models. *Proceedings WEFTEC 2003*, 76th Annual Technical Exhibition and Conference; Los Angeles, California, USA, October 11–15, 2003.

Wanner, O.; Eberl, H.; Morgenroth, E.; Noguera, D.; Picioreanu, C.; Rittman, B.; Loosdrecht, M. V. (2006) Mathematical Modeling of Biofilms. IWA Task Group on Biofilm Modeling, Scientific and Technical Report 18; IWA Publishing: London, United Kingdom.

Water Environment Federation (2013) *Wastewater Treatment Process Modeling*, 2nd ed.; Manual of Practice No. 31; Water Environment Federation: Alexandria, Virginia.

CHAPTER 5

Facility Hydraulics and Pumping

Bently Green, P.E., and John P. Brito, P.E.

1.0 Introduction

After a treatment concept has been selected and a preliminary site layout has been determined, the next step is to determine the hydraulic profile (water surface profile) or hydraulic grade line (HGL) for the water resource recovery facility (WRRF) and the unit processes therein. The objective is to ensure there is adequate head available to allow the wastewater to flow from one unit process to another and establish the appropriate control weir elevations and water surface elevations within each unit process to ensure that adequate freeboard is provided. This may include making conduits and channels large enough to meet future expansion beyond the capacity required during the design year. Sufficient hydraulic head should be provided to permit good distribution of the facility flow to all treatment processes over the range of expected flow conditions, without being excessive. During calculation of the hydraulic profile, the economics of building deeper facility structures should be considered as an alternative to pumping. Pumping stations result in higher operation and maintenance (O&M) costs and reduced reliability. Depending on the process(es) selected, intermediate pumping may be required. This is common upstream of biotowers, trickling filters, and tertiary processes.

This chapter reviews the procedures to develop the hydraulic profile. It is assumed that the designer has an understanding of basic fluid mechanics and hydraulics. There are numerous texts available to assist in this area. Several are included in the References section at the end of this chapter. Additional information on pumps and pumping hydraulics is presented in *Design of Wastewater and Stormwater Pumping Stations* (WEF, 1993), and *Pumping Station Design* (Jones, 2008).

2.0 Hydraulic Considerations

2.1 Hydraulic Profile

The hydraulic profile is based on sound hydraulic principles that determine the water level required at each treatment process for wastewater to flow through the facility. The resulting water surface profile elevations typically are presented graphically on a drawing sheet in the WRRF construction drawings. A hydraulic profile typically is prepared for the main liquid flow path extending from the facility inlet sewer to the receiving water, but one can also be prepared for ancillary flow trains, such as solids treatment and disposal facilities. The latter are particularly useful when there is gravity flow between process units. The hydraulic profile is determined for the peak flow the facility will experience (i.e., the peak wet-weather flow). However, the profile also may be determined for average and initial minimum flowrates. The hydraulic profile should present water surface elevations; hydraulic control devices, such as control valves and weirs; and critical elevations of process structures, channels, and pipelines, and the top and bottom of structures. The profile also may include ground surface elevations, pipeline sizes, and other special features that will enhance understanding of the system (see Figures 5.1 to 5.3).

2.2 Overview of Calculation Procedure

If designing modifications to an existing facility, the first step in developing the hydraulic profile is to obtain all of the as-built drawings for the facility—particularly, the civil, process, mechanical, and structural drawings. Key elevations, such as the top of structures, weirs, and outlet pipe inverts, should be surveyed to verify actual vs. plan elevations. Structure settlement over time and differences in survey datum (as well as potential changes to weir adjustments over time) are common in older WRRFs, which require reconciliation before starting on the calculations. For new facility construction, detailed drawings are not available initially, so preliminary sketches of process tank configurations, including rough dimensions and water level control concepts, and a preliminary site layout showing connecting conduits will need to be developed to provide a good starting point.

Hydraulic calculations begin at a control point—which is typically the most downstream location where the water surface elevation is known. This typically is where the facility discharges to a receiving body of water and specific regulatory criteria (such as the 100-year flood elevation) often determine the design elevation starting point. The calculation then proceeds upstream through all of the unit processes to the headworks, accounting for the headlosses of the various components and hydraulic elements. Where there are multiple tanks, the hydraulic profile is calculated along the longest flow path. Although it is commonly assumed that flows to parallel tanks will split equally, this assumption must be validated by actual calculations. Adjustments to the hydraulic design may be needed.

If the available hydraulic head at the facility inlet is not sufficient to meet the required head based on the hydraulic profile, revisions to the sizes and elevations of the hydraulic structures will be required. If available (or reasonable) revisions cannot produce the desired results, pumping will need to be considered. Pumping can be located at the facility inlet and/or elsewhere along the flow path through the WRRF. A cost-effectiveness analysis should be performed to determine the appropriate pumping location. From an O&M and reliability perspective, pumping primary or secondary effluent is preferred to pumping raw or screened wastewater.

The hydraulic profile and associated headlosses can be calculated manually using the equations presented in this chapter and those found in hydraulics and fluid mechanics texts, such as Mays (1999), Benefeld et al. (1984), Bergendahl (2008), Boulos and Nicklow (2005), Davis and Sorensen (1969), King and Brater (1963), and others.

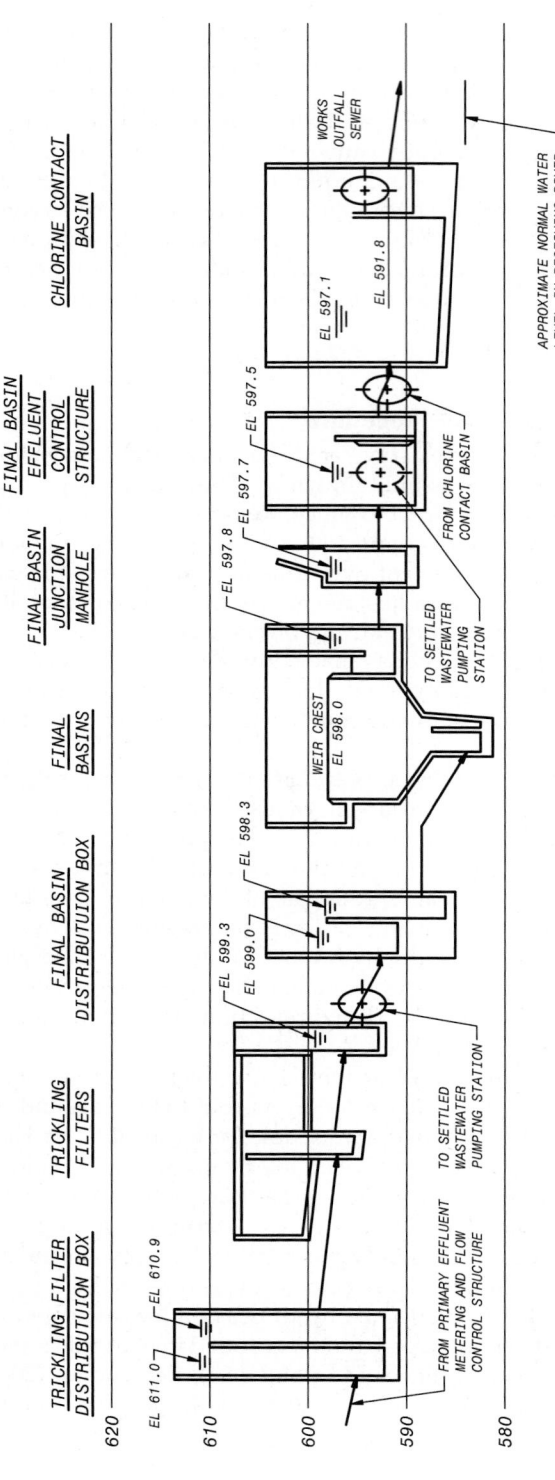

Figure 5.1 Typical hydraulic profile for influent pumping and primary treatment. Water surface elevations in feet above mean sea level represent flow of 160 000 m³/d (42 mgd).

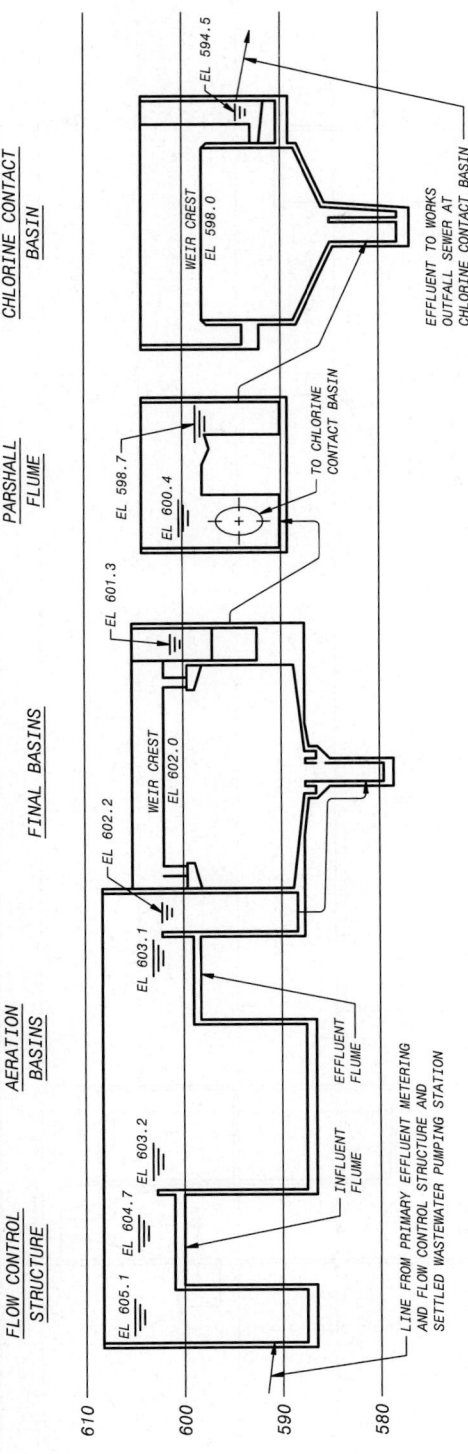

FLOW CONTROL STRUCTURE

AERATION BASINS

FINAL BASINS

PARSHALL FLUME

CHLORINE CONTACT BASIN

EL 605.1   EL 604.7   EL 603.2

EL 603.1

EL 602.2

WEIR CREST EL 602.0

EL 601.3

EL 600.4

EL 598.7

WEIR CREST EL 598.0

EL 594.5

INFLUENT FLUME

EFFLUENT FLUME

TO CHLORINE CONTACT BASIN

EFFLUENT TO WORKS OUTFALL SEWER AT CHLORINE CONTACT BASIN

LINE FROM PRIMARY EFFLUENT METERING AND FLOW CONTROL STRUCTURE AND SETTLED WASTEWATER PUMPING STATION

610

600

590

580

Figure 5.2 Typical hydraulic profile for an activated sludge facility. Water surface elevations in feet above mean sea level represent flow of 160 000 m³/d (42 mgd).

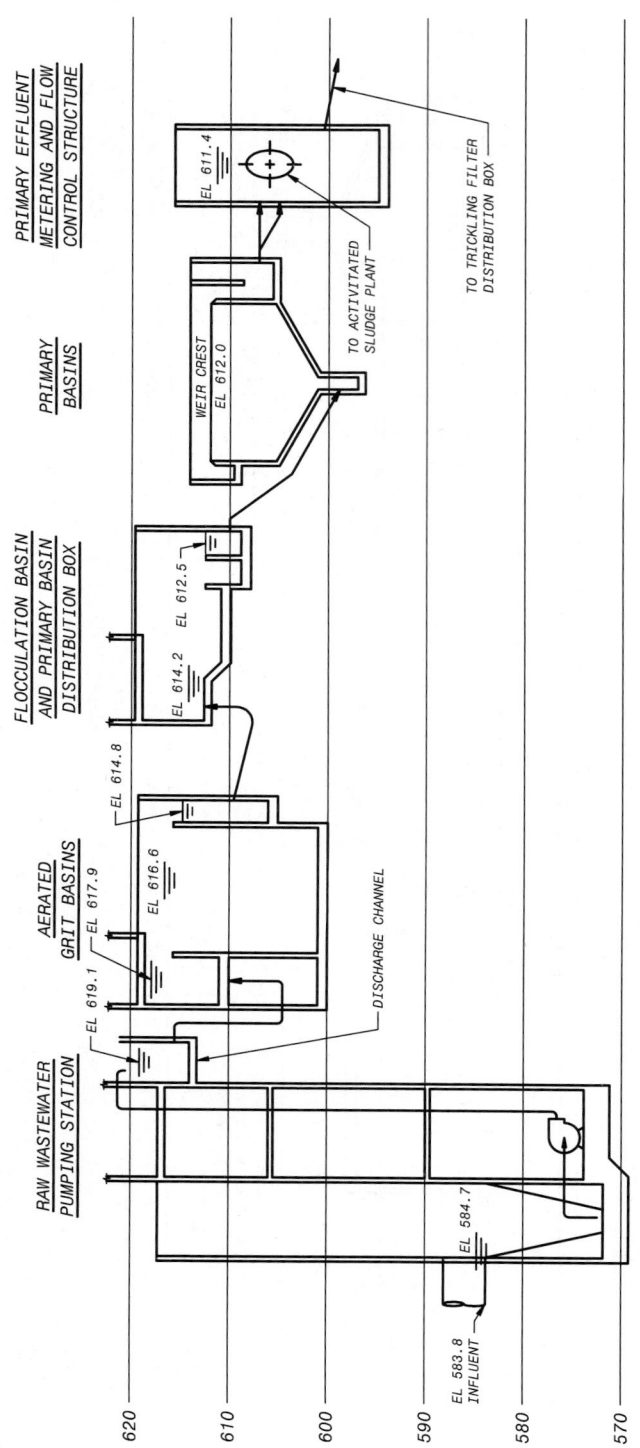

RAW WASTEWATER
PUMPING STATION

AERATED
GRIT BASINS

FLOCCULATION BASIN
AND PRIMARY BASIN
DISTRIBUTION BOX

PRIMARY
BASINS

PRIMARY EFFLUENT
METERING AND FLOW
CONTROL STRUCTURE

EL 619.1

EL 617.9

EL 616.6

EL 614.8

EL 614.2

EL 612.5

WEIR CREST
EL 612.0

EL 611.4

DISCHARGE CHANNEL

TO ACTIVITATED
SLUDGE PLANT

TO TRICKLING FILTER
DISTRIBUTION BOX

EL 584.7

EL 583.8
INFLUENT

620

610

600

590

580

570

FIGURE 5.3 Typical hydraulic profile for a trickling filter facility. Water surface elevations represent flow of 85 000 m³/d (22 mgd).

170

2.3 Initial Water Surface Elevations and Hydraulic Controls

If the receiving water body is a river or stream, the controlling elevation is typically the required flood elevation (i.e., the 100-year flood level), as calculated by accepted hydrological methods or as obtained from a governmental entities such as the U.S. Army Corps of Engineers (Washington, D.C.) or the Federal Emergency Management Agency (Washington, D.C.). The level of flood protection is established by the regulatory agency that governs the design and operation of wastewater facility. Climate change may impact controlling elevations in the future, particularly in areas with encroaching water levels or where wet-weather event frequency is significantly different from historical standards. For a storage basin or pond, the controlling water level is the maximum water level in the pond. For a larger body of water, wind forces and seiches should be considered in addition to the highest water surface elevation. For discharge to an ocean or river subject to tidal action, the controlling water level is the high tide elevation based on the selected design storm. For ocean discharge, the designer also must consider the higher specific gravity of seawater (1.025) compared with wastewater (1.00). For every 12 m (40 ft) of depth below the seawater surface, an additional 0.3 m (1 ft) of head must be provided to overcome the heavier weight of seawater. For example, an ocean outfall discharging at a depth of 60 m (200 ft) below the ocean surface would require 1.5 m (5 ft) of head (over and above the headlosses typically encountered in the outfall pipe and diffuser orifices) to overcome the density differential.

2.4 Flowrates

Typically, the peak wet-weather flowrate is used for hydraulic design to establish maximum water levels and identify minimum freeboard levels in tanks and channels. The minimum flowrate is used to identify minimum velocities, maximum heights of free-fall at weirs and channel outlets, and minimum submergences on equipment. This will identify locations where solids deposition may occur (low velocities) and where odors could be a problem (large freefall distances of raw or primary effluent).

The unit processes should accommodate the peak wet-weather flow (typically peak hourly, but may in some cases be considered maximum month), unless this flow would cause a hydraulic washout of the WRRF. In this situation, the designer should consider the use of equalization or storage basins to minimize any negative effect on the treatment process. In secondary WRRFs, the return activated sludge (RAS), trickling filter or biotower recycle flow, and other types of process recycle flows (which will vary in proportion to the influent flow depending on the process requirements of the facility) must be added to the peak flow.

2.5 Hydraulic Effect of Out-of-Service Processes

Process design typically includes unit process redundancy, which means that each of the unit processes, depending on the mechanical equipment, can have one or more units out of service. The hydraulic calculations must consider the effects of taking parallel unit processes out of service, as this will change and affect the water surface elevation, and potentially exceed the capacity of the flow conduit, weir, or overtop structural walls, etc.

2.6 Unit Process Liquid Levels and Freeboard

Each unit process should be designed hydraulically to prevent liquid from overtopping the walls of structures under all conditions. The top of structure elevation is set so that freeboard is maintained above the high-water elevation. Depending on the expected surface disturbance and relative frequency of the condition, freeboard can range from an extreme low of 150 mm (6 in.) to 1 m (3.3 ft) or more. The former may be acceptable in settling tanks, where the water surface is quiescent; the latter is appropriate in aeration tanks, where there is air bulking of the liquid (with potential foaming) and around flow conduit bends and confluences (where surface waves and splashing can occur). The typical minimum freeboard under maximum water level conditions is approximately 0.3 m (12 in.). Regulatory agencies may have established minimum freeboard requirements. Additional freeboard (or windscreens) may be necessary for locations where high winds would cause surface waves and inhibit scum removal, or where seismic conditions dictate. As a rule, there should be no submergence of control weirs under peak flow conditions. However, some submergence often is permitted when the peak flow condition is judged to be an infrequent occurrence (i.e., extreme peak wet weather), and extraordinary measures would be required to correct the condition. However, in any case, the effect on flow distribution and process performance should be considered when submergence occurs.

Many different design philosophies exist concerning the freefall allowance at weirs. A designer with a conservative approach may design to allow 80 to 150 mm (3 to 6 in.) of freefall between the weir elevation and the receiving weir trough water surface elevation. Another designer may establish a profile with no freeboard at the weirs or even allow the weirs to flood at peak flow. These philosophies are established based on the available head in the facility and the cost to pump the liquid. If sufficient head exists, without having to pump or construct excessively deep or high structures, a drop of at least 50 mm (2 in.) below the top of the weir (or the bottom of the "V" on a v-notch weir) is recommended. It should be noted, however, that excessive freefall in untreated or primary effluent will release hydrogen sulfide, which is odorous, potentially dangerous, and corrosive. Excessive freefall may also cause floc breakup at mixed-liquor splitting structures. On the other hand, additional freefall is sometimes necessary to accommodate future flows.

2.7 Facility Headloss Guidelines

The total headloss through a WRRF depends on the type of treatment processes and whether the facility includes tertiary treatment. Typical headloss guidelines for preliminary planning layout are presented in Table 5.1.

Although facilities may be designed to function well hydraulically outside typical norms, the total heads commonly found for secondary WRRFs range from 4.3 to 5.5 m (14 to 18 ft). This range of total heads applies to WRRFs for flow measurement, pretreatment, and disinfection.

In a 2008 WEF survey of WRRFs in the United States (sent to the WEF's municipal members for the development of this manual), the median headloss through a primary/secondary activated sludge (or equivalent) facility was 3.9 m (12.9 ft); in facilities with tertiary chemical precipitation or filtration, the median headloss was

Unit Process	Typical Headloss, mm (ft)
Bar screens[a]	300 to 600 (1 to 2)
Vortex or aerated grit removal	300 to 750 (1 to 2.5)
Activated sludge reactor	450 to 900 (1.5 to 3)
Primary, secondary, or tertiary sedimentation	600 to 1200 (2 to 4)
Disinfection contact tank	450 to 900 (1.5 to 3)
Tertiary filtration[b]	1000 to 3000 (3 to 10)
Conduits between unit processes, including entrance and exit losses	150 to 1050 (0.5 to 3.5)
Flow distribution boxes	300 to 750 (1 to 2.5)

[a] Allows for partial clogging.

[b] Depends on whether low head type or conventional filter is used.

TABLE 5.1   Typical Headloss through Unit Processes for Planning

5.8 m (19 ft). There was significant variability; the standard deviations were 2.9 m (9.7 ft) and 4.4 m (14.4 ft), respectively. Facilities with trickling filters or biotowers had higher total headloss, compared to traditional activated sludge facilities.

2.8   Conduit Sizing and Velocity Guidelines

Generally, a minimum velocity of 0.6 to 0.76 m/s (2 to 2.5 ft/s) at the design average flow for raw wastewater is required to prevent solids deposition in channels and pipelines. At minimum flows, velocities of 0.3 to 0.45 m/s (1 to 1.5 ft/s) are needed to transport organic matter. Achieving a particle resuspension velocity of 0.9 to 1.1 m/s (3 to 3.5 ft/s) on a daily basis can be considered in situations where it is not feasible to attain minimum flow velocity of 0.3 to 0.46 m/s (1 to 1.5 ft/s) during low-flow periods. In some cases, the minimum velocity cannot be maintained because of specific process requirements, and considerations should be given to provide access for flushing and solids removal. The velocity in conduits carrying degritted wastewater typically is at least 0.45 m/s (1.5 ft/s); the velocity in conduits conveying primary effluent typically is at least 0.3 m/s (1 ft/s). If possible, the velocity should be approximately 0.6 to 0.9 m/s (2 to 3 ft/s). Typically, conduits between process units are sized on the basis of 1 to 2 m/s (3 to 6 ft/s). Higher velocities are acceptable to minimize conduit size and cost, but headlosses can become significant.

Maximum velocities typically are not a consideration at WRRFs, because conduit and channel slopes are flat, and headlosses are minimized. Velocities rarely are high enough to be a concern. For pumping systems, the maximum velocity in suction piping should not exceed 2.4 m/s (8 ft/s), and keeping velocities below 1.5 to 2 m/s (5 to 6.5 ft/s) is more typical. The maximum recommended velocity in force mains and pump discharge piping is 3.7 m/s (12 ft/s) (Jones, 2008). However, to minimize the effects of water hammer and excessive headloss, velocities in discharge piping should be kept below 2 m/s (6.5 ft/s).

Distribution channels typically are designed with low velocities (0.3 m/s or less [1 ft/s]) to minimize headloss and ensure good flow distribution. Where low velocities result in solids deposition, aeration can be provided to keep the channel mixed. Non-aerated (pump) mixing could be considered if the channel is immediately upstream of an anoxic zone.

3.0   Hydraulic Elements

The following sections describe the principal equations used to calculate the hydraulic profile. This is only intended to be a brief summary. The designer should consult the references cited previously for additional information and equations.

3.1   Basic Equations

3.1.1   Bernoulli Equation

The Bernoulli equation is one of the fundamental equations of fluid flow and is founded in the conservation of energy. The classic Bernoulli equation has been rewritten, with the values of z being the invert of the conduit instead of the water surface, and $P_1/g = y_1$ and $P_2/g = y_2$ in eq 5.1:

$$Z_1 + y_1 + V_1^2/2g + H_L = Z_2 + y_2 + V_2^2/2g \qquad (5.1)$$

where Location "1" = downstream;

Location "2" = upstream;

Z_1, Z_2 = invert elevation of the channel bottom downstream and upstream, respectively, m (ft);

y_1, y_2 = depth of water in the channel or height of the piezometric surface above the invert of the conduit downstream and upstream, respectively, m (ft);

V_1, V_2 = velocity of flow in the channel downstream and upstream, respectively, m/s (ft/s);

H_L = headloss between 1 and 2, m (ft); and

g = acceleration caused by gravity, 9.8 m/s² (32.2 ft/s²).

3.1.2   Volumetric Flowrate

The relationship between flowrate, velocity, and the cross-sectional area of flow can be expressed as follows:

$$Q = AV \qquad (5.2)$$

where V = velocity of flow in the channel, m/s (ft/s);

Q = flow in the channel, m³/s (cu ft/s); and

A = cross-sectional area of flow, m² (sq ft).

The equation for circular pipes flowing full can be expressed as follows:

$$Q = VD^2/C \tag{5.3}$$

where Q = flow in the pipe, m^3/s (gal/min);
V = velocity of flow in the pipe, m/s (ft/s);
D = pipe diameter, m (in.); and
C = constant = 1.2732 for SI units (0.4085 for U.S. customary units).

3.2 Hydraulic Losses

The hydraulic losses in WRRFs consist of the following:

- Headloss through orifices (perforated baffle walls and gates);
- Head over weirs; and
- Friction and minor losses in channels and conduits.

These can be calculated using the equations found in hydraulic texts—several of which were mentioned at the start of this chapter.

3.2.1 Orifice Loss

Headlosses through a submerged-control gate or an inlet gate to a treatment process, openings in baffle walls, submerged launders, perforated pipes, and filter underdrains typically are calculated using the orifice equation:

$$H = (Q/CA)^2/2g \tag{5.4}$$

where H = headloss through the gate or orifice opening, m (ft);
Q = flow, m^3/s (cu ft/s);
C = gate or orifice coefficient, dimensionless;
A = area of gate or port opening, m^2 (sq ft); and
g = acceleration caused by gravity, 9.8 m/s^2 (32.2 ft/s^2).

The headloss is the difference in the water level between the upstream and down-stream sides of the gate or orifice. Values of the gate, or orifice, coefficient for calculating headloss through ports under varying conditions can be found in Davis and Sorensen (1969), King and Brater (1963), and others. The value for C can vary, but typically is 0.6 for a sharp-edged opening and 0.9 for a rounded opening; a C value of 0.6 typically is used to be conservative.

The orifice equation, as defined in eq 5.4, is applicable only when the gate or opening is completely submerged. Note that when the gate or opening is submerged on the upstream side and the gate discharges free on the downstream side, H is measured vertically from the centerline of the opening to upstream water surface. If the opening is not fully submerged, as frequently occurs at tank inlets, the water surface upstream should be calculated as a submerged weir. See King and Brater (1963) and the discussion to follow.

3.2.2 Weir Loss

Although weirs sometimes are used for flow measurement in WRRF, they more commonly serve as control devices to maintain a required water level in a unit process. Weirs are classified in accordance with the shape of the notch—rectangular, v-notch, trapezoidal, and proportional. Trapezoidal weirs are sometimes called Cipoletti weirs. Proportional weirs, sometimes called Sutro weirs, have the discharge head varying linearly with flow; they are used in grit channels to maintain constant velocity with depth. The upper edge of the weir plate (in the case of a rectangular weir) and the bottom of the notch in a v-notch or trapezoidal weir is the crest of the weir. The depth of the water over the crest is the head or H measured some distance upstream of the weir (see Figure 5.4a). This distance can be as much as 4 to 5 times H upstream (Ackers et al., 1978).

Rectangular weirs are classified as either sharp-crested or broad-crested, as shown in Figures 5.4a and 5.4b. Weirs also can be submerged or unsubmerged. Weirs typically are designed to be unsubmerged (i.e., not affected by downstream conditions).

V-notch weir angles range from 22.5° to 120°, with the 90° v-notch being the most common. Under free-flow (unsubmerged) conditions, as shown in Figure 5.5a, the head over a sharp-crested v-notch weir can be calculated using the following equation:

$$H = [Q/(C \tan \varphi/2)]^{0.4} \tag{5.5}$$

where Q = flow, m³/s (cu ft/s);
  φ = angle of the notch, degrees;
  H = head over the crest, m (ft); and
  C = weir coefficient (1.38 for SI units [2.5 for U.S. customary units] for a 90° weir).

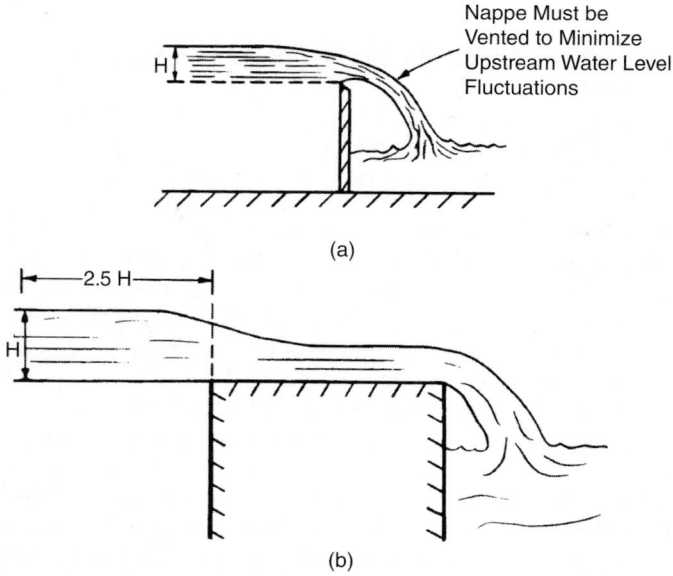

(a)

(b)

FIGURE 5.4 Weir sections: (a) sharp-crested weir and (b) broad-crested weir.

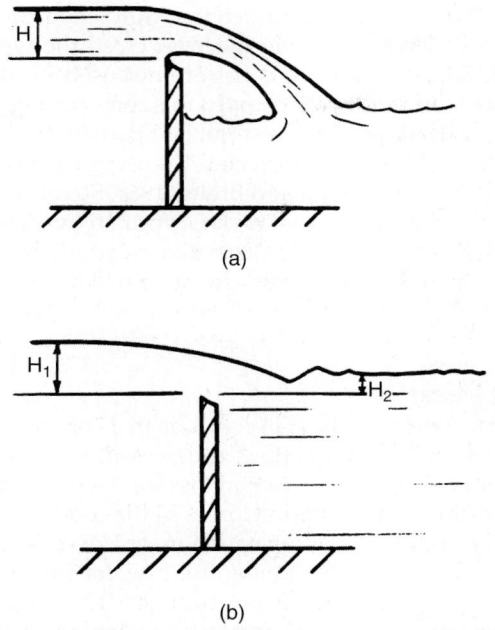

(a)

(b)

Figure 5.5 Weir flow conditions: (a) free flow and (b) submerged flow.

Head over a sharp-crested rectangular weir under free-flow conditions can be calculated using the following equation:

$$H = [Q/(C_w L)]^{0.67} \tag{5.6}$$

where Q = flow, m³/s (cu ft/s);
   L = length of weir, m (ft);
   H = head over the crest, m (ft); and
   C_w = coefficient that accounts for the approach velocity, in terms of the ratio of the weir plate depth to the head over the crest. The values range widely, depending on the approach velocity. The most commonly used value, 1.82 for SI units (3.3 for U.S. customary units), accurately represents deep-water condition upstream, such as that found in basins, with an extremely low approach velocity toward a sharp-crested rectangular weir.

Hydraulic texts often describe a rectangular weir with "end contractions." End contractions occur when the weir plate covers only a portion of the end of a channel. An adjustment is made to reduce the weir crest length, L, by $0.2H$, to account for the end contractions. This has the effect of increasing H slightly. In practice, this often is ignored, because the values of H typically are small compared with L.

Equations 5.5 and 5.6 apply under free-flow conditions. When the downstream water surface elevation rises above the weir crest, the weir becomes submerged, as shown in Figure 5.5b. Submerged weirs are not used for flow measurement, because the weir equations do not directly apply to submerged conditions and thus produce sizable errors. The head over the crest can be calculated from eqs 5.5 and 5.6; however, the discharge flow, Q, must be corrected by using curves that have been developed experimentally (Chin, 2012; King and Brater, 1963; Street et al., 1996). The basis for relationships in the referenced texts is work performed by Villemonte (1947). It should be noted that Chin (2012) includes an alternative relationship by Abu-Seida and Quraishi (1976). The head upstream of a submerged weir will be greater than for a free discharge condition.

3.2.3   Conduit Losses

Two types of flow conditions exist in conduits that connect unit processes—free surface flow, referred to as open-channel flow, and pressure flow. Headlosses in conduits connecting unit processes consist of friction loss (in the conduit) and minor losses (in conduit bends and fittings, etc.). Friction losses in the conduit can be determined using the Darcy-Weisbach equation, Manning equation, or Hazen-Williams equation. The Darcy-Weisbach equation is applicable to any fluid (water, air, chemical, etc.) and is considered the most accurate, but it is more complex in its use, as it requires an iterative solution to determine the friction factor. The Manning equation, commonly used in open channels, is applicable to closed conduits also; the Hazen-Williams equation typically is used only for full-flow conduits.

These equations and the friction characteristics for the various conduit materials can be found in hydraulics and fluid mechanics texts (e.g., Benefield et al., 1984; King and Brater, 1963; ASCE and WEF, 2007).

Darcy-Weisbach equation:

$$h = f(L/D)(V^2/2g) \tag{5.7a}$$

where h = headloss, m (ft);
f = friction factor, which is based the conduit material and Reynold's number;
L = conduit length, m (ft);
D = conduit diameter, m (ft);
V = velocity of flow, m/s (ft/s); and
g = acceleration caused by gravity, 9.8 m/s² (32.2 ft/s²).

The value for f is determined from the "Moody diagram," found in the above-referenced texts, and depends on the relative roughness of the conduit wall, e/D, and Reynold's number, where e = roughness of the conduit wall, m (ft).

$$R_N = VD/\upsilon \tag{5.7b}$$

R_N = Reynold's number (dimensionless) and
υ = kinematic viscosity of the fluid, m²/s (sq ft/s).

Values of e can be found in the above-referenced texts. For noncircular conduits, the equivalent D can be determined as follows:

$$D \text{ equivalent} = 4R \qquad (5.7c)$$

where R = hydraulic radius = area of flow/wetted perimeter.

Manning equation:

$$Q = (k/n) \, AR^{0.67}S^{0.5} \qquad (5.8)$$

where Q = flowrate, m^3/s (cu ft/s);
$\quad k = 1$ (1.486 for U.S. customary units);
$\quad A$ = area of flow, m^2 (sq ft);
$\quad R$ = hydraulic radius = A/wetted perimeter, m (ft);
$\quad S$ = slope of the energy grade line, calculated as the vertical drop, m (ft) divided by the horizontal distance, m (ft), and sometimes called the friction slope S_F, m/m (ft/ft) or friction loss, m/m (ft/ft); and
$\quad n$ = Manning's roughness factor, which can be found in the referenced texts.

Hazen-Williams equation:

$$Q = KCAR^{0.63}S^{0.54} \qquad (5.9)$$

where $K = 0.849$ (1.318 for U.S. customary units); and
$\quad C$ = Hazen-Williams roughness coefficient. Note that "C" can vary widely depending on the type of pipe material; and a value of "100" is often a default for concrete or ductile iron pipe. However, the value depends on the age and condition of the pipe. Also, a "C" value as high as 140 to 150 is common for new smooth-wall PVC pipes.

Other parameters are as defined above.

Minor losses are those headlosses that occur as a result of valves, bends, and other types of fittings. Minor losses can be accommodated in the hydraulic calculations using either the "K-factor" method or the "equivalent length" method. In the former, the headloss is a function of the velocity head, as shown in eq 5.7a.

$$H = KV^2/2g \qquad (5.10)$$

where H = headloss resulting from the fitting, etc., m (ft);
$\quad K$ = headloss factor, dimensionless;
$\quad V$ = velocity in the conduit, m/s (ft/s); and
$\quad g$ = acceleration caused by gravity, 9.8 m/s^2 (32.2 ft/s^2).

The equivalent length method converts the headloss to an equivalent length of pipe of the same diameter as the fitting or valve. Values for K and equivalent pipe length can be found in hydraulics texts cited above. The K-value method is most common in practice.

In addition to fitting losses, headloss occurs where fluids enter a pipe and exit the pipe, (i.e., entrance and exit losses). These are determined using eq 5.7a, with the entrance K ranging from 0.04 to 1.0, depending on how "rounded" the entrance is; it is commonly taken as 0.5. Exit loss, K, typically is 1.0.

The term minor loss is a misnomer when used for WRRF hydraulic assessments, as these losses typically exceed the friction losses in WRRF conduits based on the fact that conduits are relatively short and have a number of fittings.

Open-channel, or free surface, flow can be either steady or unsteady flow. In steady flow, the flowrate is constant with time and distance along the channel. This is the typical case. Unsteady flow (flow changes with distance or time) occurs in weir troughs and collection and distribution channels and will be discussed later in this chapter. In addition, steady flow can be uniform or nonuniform. In nonuniform flow, the depth of flow or the width of the conduit, or both, varies along the length of the channel. This is relatively common in WRRFs.

The calculation of the water surface profile in open channels uses the Manning equation (eq 5.8) in combination with the Bernoulli equation, previously presented as eq 5.1. In eq 5.1, H_L is as follows:

$$H_L = S_{Fave} \times L \tag{5.11}$$

where H_L = headloss between downstream and upstream locations, m (ft);
S_{F1}, S_{F2} = friction slope in the channel downstream and upstream, respectively, m/m (ft/ft);
$S_{Fave} = (S_{F1} + S_{F2})/2$ = average friction slope between downstream and upstream, m/m, (ft/ft); and
L = distance between downstream and upstream sections, m (ft).

The parameters S_{F1} and S_{F2} typically are determined using the Manning equation (eq 5.8) based on the depth of flow downstream and upstream, respectively. For open-channel flow in WRRFs, typically the downstream values are known and the upstream values are unknown and are to be determined. This is an iterative process in nonuniform flow, as the value for S_{F1} depends on the upstream depth. If there are any minor losses, they should be included. Minor losses can be determined using a K-value approach similar to pipes.

When beginning to calculate the water depth in an open channel, it is very important to calculate the critical depth. The critical depth becomes important at flow constrictions. The critical depth, D_c, for rectangular channels is determined using eq 5.12.

$$D_c = (Q^2/b^2 \times g)^{0.33} \tag{5.12}$$

where D_c = critical depth, m (ft);
Q = flowrate, m³/s (cu ft/s);
b = channel width, m (ft); and
g = acceleration caused by gravity, 9.8 m/s² (32.2 ft/s²).

Critical depth for nonrectangular channels can be determined using methods in King and Brater (1963) and others. If the depth of water downstream is greater than the critical depth, the water depth in the channel will be the downstream depth. If the depth of water downstream is less than the critical depth, a constriction is occurring, and the downstream depth at the constriction will be the critical depth.

Additional information on open-channel flow calculation can be found in hydraulics textbooks.

3.3   Hydraulic Elements

The previous sections presented the fundamental hydraulic equations typically used in the calculation of the hydraulic profile. The following paragraphs discuss specific hydraulic elements and features in a WRRF that deserve special attention.

3.3.1   Flow Distribution

To ensure proper treatment, it is essential to achieve equal flow distribution to each of the basins that make up a unit process. This requires splitting the total facility flow proportionally to the capacity of each basin or reactor. This can be accomplished through a wide variety of means.

Distribution boxes, channels, and header pipes are used for this purpose. Although modulating ports, weirs, gates, or valves can provide equal distribution, they require control systems that share the disadvantages inherent to any electrical and mechanical system (i.e., failure and high maintenance). Wherever possible, static devices, such as fixed weirs, gates, and cut-throat flumes, are preferred by facility staff for the distribution of flow rather than constantly modulating devices.

3.3.2   Splitting Structures

The use of symmetry alone will not always ensure equal flow distribution, because small differences in headloss in the flow paths to the process units will result in large differences in flow. A lack of symmetry may be caused by slight construction differences or structure settlement. Because of the long weirs associated with circular clarifiers, even slight differential settling of the clarifier structure will impair flow distribution and make a large difference in flow entering the tank or clarifier.

Weirs provide good flow distribution without the need for mechanical equipment, such as rate-controlling valves, and are less subject to clogging. As a result, they generally provide the best technical solution for flow distribution or flow splitting. Where large flows are to be split, for example, greater than 2 to 3 m³/s (50 to 75 mgd), fixed weir splitting structures can become large and complex. Often a number of weir troughs, similar to those found in rectangular clarifiers, are used to achieve sufficient weir length.

A flow-splitting structure containing weirs is constructed at a location where the flows are to be split. Figure 5.6a illustrates this type of layout for two clarifiers. Figure 5.6b illustrates an example of flow control structures for facilities with asymmetrical layouts.

In the design of the splitting structure, the water surface upstream of the splitting weirs should be as quiescent as possible. One way to achieve this is to construct the influent pipe to enter vertically in the bottom. The velocity should be low; otherwise, an impact baffle should be considered. The "rise rate" or vertical velocity in the chamber upstream of the splitting weirs should be 0.3 m/s (1 ft/s) or less, if possible. If heavy

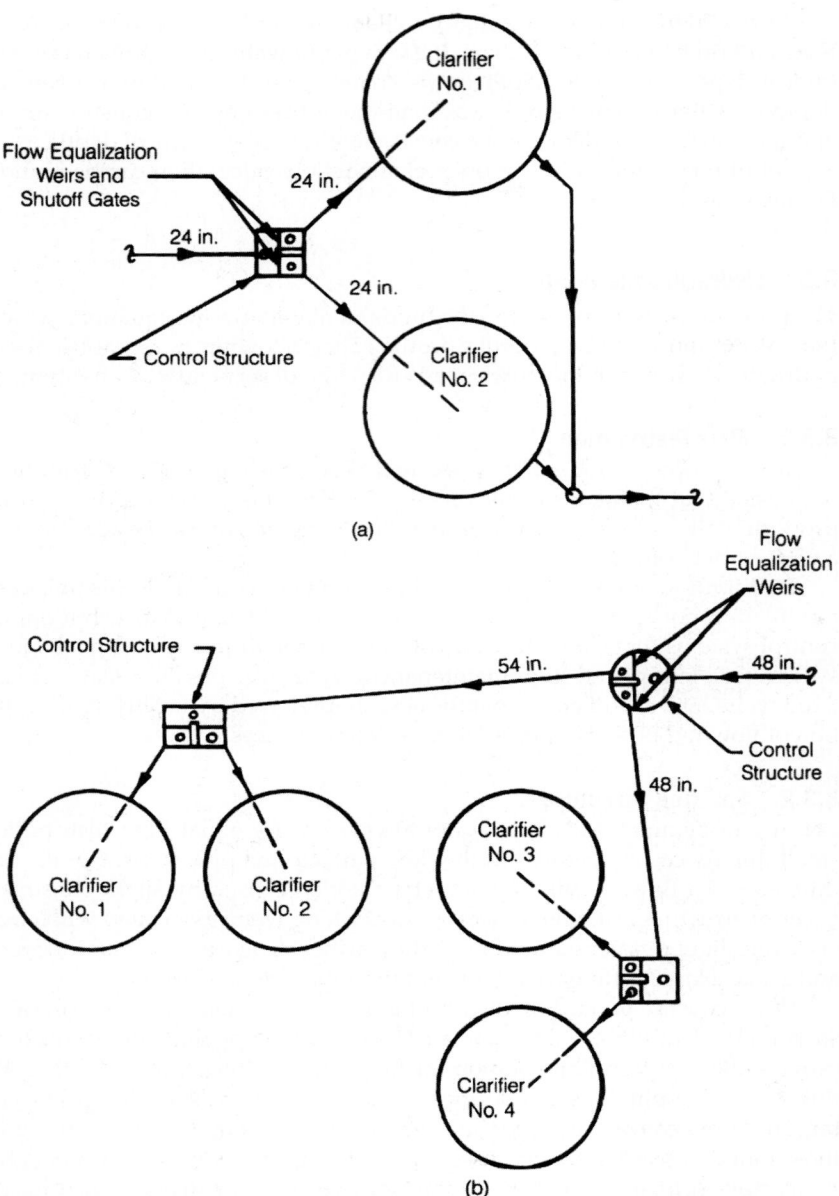

FIGURE 5.6 Typical design examples of equal flow distribution among clarifiers: (a) symmetrical layout and (b) asymmetrical layout (in. × 25.4 = mm).

solids are being transported in the liquid (i.e., raw wastewater or raw or digested sludge), higher velocities may be required to prevent deposition, but this will create more turbulence, and flow splits may not be optimum.

The top of the conduits leaving the splitting structure must be placed well below the water surface in the downstream chamber of the splitting box, to ensure submergence at all times and avoid vortexing, which will entrain air. This air will accumulate in the pipe and cause violent back flushes or be carried into the downstream process unit and cause surging.

Sizing the splitting weirs is based on eq 5.6, assuming rectangular weirs are used. If there is a desire to have an equal split, all weir lengths must be the same; if the downstream process units all do not have identical capacities, then the weir lengths are proportionately adjusted. Ideally, there should be a freefall of approximately 75 to 150 mm (3 to 6 in.) between the weir crest and the downstream water surface elevation.

For large or complex flow distribution structures, the use of three-dimensional computational fluid dynamics (CFD) modeling can be used to optimize the design.

3.3.3   Distribution Manifolds and Channels

An inlet pipe header or distribution channel also can be used to distribute flow to a unit process. However, in such cases, the inlet ports or gates to each basin should be designed with adequate headloss to ensure good distribution. As with flow distribution, equal distribution of solids to the treatment processes also should be maintained. Unless provided for in the design, the equal distribution of solids to the treatment units may not occur coincidentally with the equal distribution of flow. This is especially common where flow is distributed by use of channels. Where channels are used, such as upstream of grit basins, the wastewater flow should be mixed well, to ensure that the solids distribute evenly with the flow. For such purposes, channel aeration is often used.

Ports or gates in the distribution channel or pipe header are sometimes used to equally distribute flow from or to unit processes. To split flow equally, the headloss through inlet openings must be much larger than the total headloss in the distribution channel or header pipe. Equation 5.13 (Camp and Graber, 1968; Fair et al., 1968) presents the hydraulic relationship between the friction loss in the header pipe or distributing channel and the outlet gates or ports. The equation assumes that the outflow occurs through numerous equally spaced ports or gates.

$$h_p = \Delta h / (1 - m^2) \qquad\qquad (5.13)$$

where h_p = headloss through port, m (ft);

Δh = hydraulic grade differential along the distribution channel or pipe, m (ft); and

m = ratio of the lowest flow and the highest flow through the respective ports.

Equation 5.13 is used to determine the headloss that must be induced by the inlet gates or ports to ensure uniform distribution to all basins from a common inlet header or distribution channel. For example, to keep the flow to each basin within 5% of the other basins ($m = 0.95$), the headloss through the gate into the basin should be approximately 10 times the water level difference over the entire length of the distribution channel or pipe header. To determine the actual headloss through a gate or port, the orifice equation (eq 5.4) is applicable.

When the outlet flow from a pressurized header approximates a uniform continuous outflow along the length of the header, the headloss along the header can be estimated by calculating the headloss as if the inlet flow to the header were to be conveyed along the entire length of the header and dividing the result by 3. This estimation procedure is valid for m (ratio of the lowest flow and the highest flow through the respective ports) greater than 0.9 in eq 5.13.

As an alternative to using gated ports from a distribution channel into a basin, weirs or cut-throat flumes can be used.

Ideally, the header or distributing channel must be made large enough so that the headloss from one end to the other is minimal. This can be accomplished by using a low velocity—less than 0.3 m/s (1 ft/s); for very long channels, velocities less than 0.15 m/s (0.5 ft/s) may be appropriate.

The distribution channel formula, described in detail in *Open-Channel Hydraulics* (Chow, 1959), is a differential equation for flow with decreasing discharge. This formula can be used to analyze the hydraulic characteristics of a channel distribution header. However, most designers prefer to use a computer program to solve this equation. Benefield et al. (1984) illustrated a slightly different but equivalent method. Montgomery (1985) presented a similar method to Benefield et al. (1984).

A close approximation is to assume that the distributing channel is an open channel with constant flow (i.e., the inlet or maximum flow). Then, calculate the headloss using eqs 5.1, 5.8, and 5.11 as though the entire flow is traveling the entire length of the distributing channel. The actual headloss will be approximately one-third of the headloss calculated.

As a note of caution, if the headloss in the distributing channel is large, as sometimes occurs when an existing small facility is expanded, this will dictate that a large orifice loss be created to ensure a good flow split. This could result in excessive entrance velocities into the basin or tank and create process problems (i.e., short-circuiting and jetting flow). Adequate baffling will be needed, or the distribution channel will need to be enlarged.

Distribution channels may trap scum, unless the discharge to the reactors or clarifiers is free and open. Installing a downward opening scum gate at the end of the distribution channel will facilitate removal of the scum and minimize maintenance. The use of weirs or cut-throat flumes will minimize the scum accumulation (a cut-throat flume is simply a Parshall flume, except that the parallel throat section is removed, and the bottom is at the same elevation as the channel invert through the length of the flume. It consists only of a tapered inlet approach channel and a tapered outlet discharge channel. Flume fabricators make this as a standard product. They are not recommended for flow measurement, however).

3.4 Launders and Troughs

Effluent launders typically are found in settling tanks and filters. Launders can be of the freefall type, where the flow from the unit process passes over a weir and falls into the trough, or submerged, where the flow from the settling tank is taken off below the water surface.

3.4.1 Freefall Launders and Troughs

These launders, or channels, collect flow along their length as the flow moves downstream toward the outlet end. Conveying flow along the launders requires head to overcome the friction loss along the channel and the exchange of momentum as the water falls perpendicular to the flow stream and accelerates along the direction of flow. The side overflow formula, described in detail in *Open-Channel Hydraulics* (Chow, 1959), is a differential equation for flow with increasing discharge. Most designers prefer to use a computer program to solve this equation. Because of the low average value of the Reynolds number along the entire length of the launder, use of a higher-than-normal friction factor often is recommended (i.e., 50% higher). Because of the high degree of mixing caused by the falling flow and the corresponding distortion of the velocity profile, the velocity head alpha value, as in $\alpha V^2/2g$, can be higher than unity, up to approximately 1.3.

Thomas, in a discussion of Thomas Camp's paper on lateral spillway channels (Thomas, 1940), approximated the water surface, assuming it to be parabolic. The equations by Chow (1959) above provide the water depth at locations along the length of the launder. For design purposes, this is not necessary; in practice, the depth at the upstream end of the launder or trough typically is all that is needed to make sure there is adequate freefall from the weir notch to the water surface.

Thomas developed some simplified equations based on whether the outlet of the launder is free or submerged (Benefield et al., 1984). A free discharge occurs if the water level downstream of the end of the launder is less than the sum of the invert of the launder or trough plus critical depth. Critical depth is determined using eq 5.12 for a rectangular trough. If this condition is not true, the trough is submerged. If the discharge from the trough or launder is "free," then the maximum water level in the trough at the upstream end is as follows:

$$y_{up} = 1.73 \times D_c \qquad (5.14)$$

where y_{up} = maximum water depth in trough, m (ft); and
D_c = critical depth determined from eq 5.12 if rectangular, m (ft).

In the above equation, it is assumed that the trough or launder is level (typical), and friction loss is ignored (reasonable, as the troughs typically are short). If the trough or launder is not rectangular, or the discharge is not free, the designer should consult the reference texts.

3.4.2 Submerged Launders
Submerged launders often are used in primary sedimentation tanks in lieu of freefall launders to minimize odors. Primary effluent falling into a conventional launder will strip hydrogen sulfide and other volatile, odorous compounds. Using a submerged launder eliminates the freefall and associated odors; however, they are a bit more complex. With submerged launders, the water level in the settling tank is controlled by a level-control valve—typically a butterfly valve or butterfly gate and a level control loop sensing the water level in the settling tank. Controlling the water level is important to ensure optimum skimming of floatables.

At facilities with large peak flows, a combination of submerged orifices and v-notches may be appropriate. When peak flow occurs, the tank level rises to overflow the weirs, thus avoiding the hydraulic restrictions through the orifices. The orifice equation, eq 5.4, can be used to determine the headloss entering the submerged launder; and, depending on the configuration, there could be a fitting loss determined using eq 5.10.

3.5 Valves and Flow- and Pressure-Control Valves

Headloss through valves of any type can be determined using K-values and eq 5.10 or using the control valve equations presented below (eqs 5.15 and 5.16); the latter are more accurate. Care must be taken in the sizing of flow- and pressure-control valves. They must be sized so that the headloss through the valve is significant and the control valve operates within an acceptable opening range—not too close to its seat under minimum flow (maximum pressure) or too wide open under maximum flow (minimum pressure). Control valve sizing and headloss calculations can be determined using the flow coefficient K_v in SI units and C_v in U.S. customary units.

In SI units:

$$K_v = Q \times [1/\Delta p]^{0.5} \tag{5.15}$$

where K_v = valve coefficient;
Q = flowrate, m³/h; and
Δp = pressure drop through the valve, bar.

In U.S. customary units:

$$C_v = Q \times [1/\Delta p]^{0.5} \tag{5.16}$$

where C_v = valve coefficient;
Q = flowrate, gal/min; and
Δp = pressure drop through the valve, lb/sq in.

Also, to convert C_v to K_v,

$$K_v = C_v/1.16 \tag{5.17}$$

Values for K_v or C_v can be obtained from the valve manufacturer.

The percent "open" can be determined by calculating the K_v or C_v and comparing it with charts prepared by the valve manufacturers.

3.6 Baffles

Baffles in unit process tanks can (1) be over–under, (2) be around the end, (3) contain openings or perforations across all or a portion of the baffle, or (4) be a combination.

For over–under baffles, the headloss for "over" baffle flow can be modeled or calculated using the submerged weir equations (King and Brater, 1963); the headloss for "under" baffle flow can be calculated using the orifice equation (eq 5.4), where the area is the total area of flow under the baffle.

For around-the-end baffles, the flow takes a 180° bend. This can be calculated as minor loss using eq 5.10. The value for K varies from 2 to 3.5 (American Water Works Association, 2012). Values of 3.2 and 3.3 have been used successfully in practice in the design of baffled channel flocculators (Reichenberger, 1984). In perforated baffles, the perforations can be considered orifices, and eq 5.4 can be used.

Where over and under baffles are used in aeration basins, flow over the top is needed to avoid trapping scum. This typically is 25 mm (1 in.) or so. This occurs simultaneously with the flow under the baffle or through submerged openings in the baffle. Headlosses over the baffle and under (or through) the baffle are equal and are calculated by simultaneously solving equations for a submerged weir and a submerged port or gate.

3.7   Junctions and Confluences

Where the flow in channels and pipes join together, turbulence will be created, and there will be some loss of energy and consequently loss of head. There are a number of procedures that can be used to determine the headloss resulting from the converging flows. The simplest, though not necessarily the most accurate, is to estimate the loss using the K-value equation (eq 5.10) with an estimate of K. Benefield et al. (1984) suggested a K of 1.8 for a 90° turn in a tee and 0.6 for the straight run of a tee. Montgomery (1985) provided charts to determine the headloss through junctions. A more accurate approach uses a pressure and momentum methodology, such as that developed by the City of Los Angeles, California (1968).

$$\Delta y \times (A_1 + A_2)/2 = M_2 - M_1 - M_3 \times \cos \theta \tag{5.18}$$

(See Figures 5.7a and 5.7b.)

where Δy = change in water level in junction, m (ft);
  $\Delta y = z + D_1 - D_2$, m (ft);
  z = change in invert elevation in confluence, m (ft);

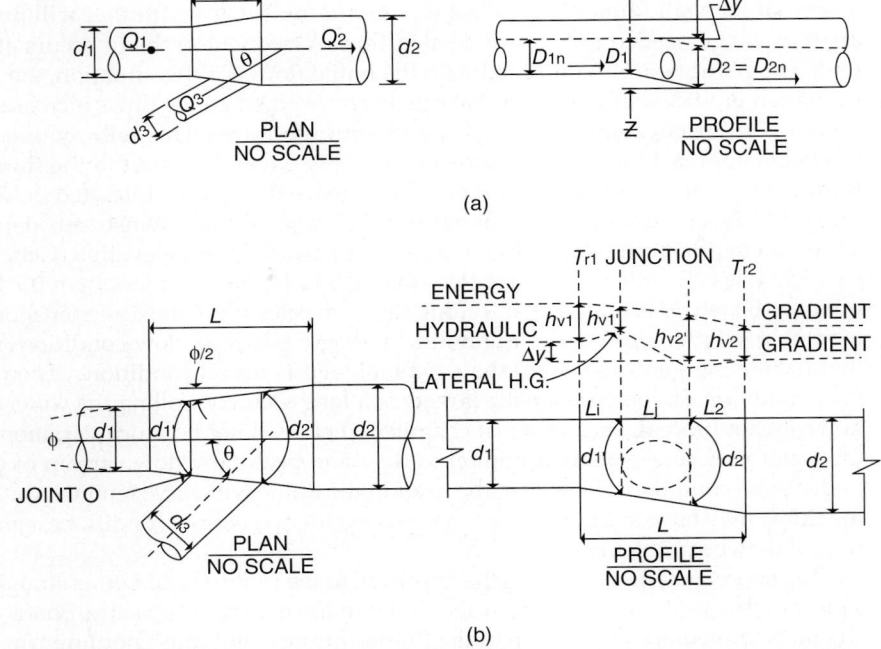

(a)

(b)

FIGURE 5.7   Confluence analysis using pressure and momentum method: (a) confluence analysis for open-channel (free surface) flow and (b) confluence analysis for pressure conduit flow (City of Los Angeles Bureau of Engineering, 1968). Image courtesy of City of Los Angeles Bureau of Engineering.

D_1, D_2 = depth of flow or distance to HGL upstream and downstream respectively, m(ft);

A_1, A_2, A_3 = area of flow, sq m (sq ft);

$M_1 = Q_1^2/A_1 \times g$, m³ (cu ft);

$M_2 = Q_2^2/A_2 \times g$, m³ (cu ft);

$M_3 = Q_3^2/A_3 \times g$, m³ (cu ft);

θ = confluence angle, degrees; and

g = acceleration caused by gravity, 9.8 m/s² (32.2 ft/s²).

For open-channel (free surface flow) conditions, eq 5.18 requires an iterative solution.

3.8   Aerated Channels

The Manning roughness coefficient used for formed concrete channels ranges from 0.013 to 0.015. The friction loss in aerated distribution channels is greater than that in unaerated channels. The suggested value of Manning's n for aerated channels with a velocity of 0.4 m/s (1.3 ft/s) is 0.035 and, for a velocity of 0.3 m/s (0.9 ft/s), is 0.0425. These values were based on experiments by Townsend in the mid-1930s (Townsend, 1935).

3.9   Flumes and Meters

Although there are several types of flumes that can be used for open-channel flow measurement in WRRFs, the Parshall flume is, by far, the most common. Parshall flume head calculation equations can be found in a number of publications (e.g., Benefield et al., 1984; U.S. Department of the Interior Bureau of Reclamation, 2001). Key to the design of Parshall flumes is to select the minimum throat width that will provide the capacity range needed for the anticipated flow. A frequent problem occurs at facilities with a large future flow compared with the initial flow. For this situation, the manufacturers can provide a throat insert that can be removed when the flows increase. Another option is to install flumes in parallel and sum the flows. This will require a careful upstream approach design to ensure a reasonably equal flow split to the flumes. Most flumes have a limit on the amount of submergence that can be tolerated before metering accuracy deteriorates (submergence is the ratio of the downstream depth to the upstream depth, both relative to the flume's upstream invert elevation. Generally, this should be less than 0.6 for throats less than 0.3 m [12 in.] and less than 0.7 for larger throats) (Benefield et al., 1984). In setting the flume elevation, the designer should check the flume submergence under minimum, average, and peak flow conditions to ensure that the submergence is within the acceptable limits for all conditions. The minimum flow condition can be problematic if there is a long weir controlling the water level in a downstream process, such as a grit chamber. There will not be much variation between the water surface elevation at minimum flow and peak flow downstream of the flume under these conditions. However, the head in the flume will vary significantly. Although the flume may have satisfactory submergence under peak flow conditions, at minimum flow, it may be problematic.

To ensure metering accuracy, the approach to the flume should be as straight as possible, and the velocity distribution should be uniform as it approaches. Sharp curves in channels immediately upstream of the flume entrance will cause nonuniform flow distribution and poor flow measuring.

Hydraulic considerations in the selection of full pipe meters, such as magnetic flow meters and sonic meters, relate to the avoidance of upstream and downstream disturbances. Upstream disturbances are more of a problem. These meters require a uniform velocity distribution across the entire cross-section in the approach piping. Valves and fittings immediately upstream of the meter are to be avoided. Typically the straight run, free of valves and fittings, upstream is 10 diameters; there are 5 diameters of straight pipe downstream. The designer always should consult with the manufacturer relative to installation. It also is important that magnetic and sonic flow meters flow full all of the time. The velocity through the flow meter should be maintained above 0.3 m/s (1 ft/s) over the entire range of flows.

3.10   Outfalls

The headloss calculation uses the same conduit friction loss equations described earlier in this chapter. Achieving uniform distribution through each outfall orifice is necessary. The same basic principles as those discussed previously in this chapter for manifolds and orifices apply. If the outfall discharges to the ocean or to brackish waters, the density of the receiving water must be considered.

Additional considerations for the proper hydraulic design of outfalls include the following: seawater purging for deep-water tunneled diffusers with long riser pipes; air in the outfall, as it affects flow capacity, buoyancy, and related structural failure, and surging and related water hammer; and aesthetics near the discharge area of the outfall. Detailed discussions on the design of outfalls are found in publications such as *Marine Outfall Systems: Planning, Design and Construction* (Grace, 1978) and *Wastewater Management for Coastal Cities: The Ocean Disposal Option* (Gunnerson and French, 1996). Design methodology for the diffuser length, depth, orientation, configuration, plume dispersion, and so on can be found elsewhere.

3.11   Telescoping Valves

Telescoping valves sometimes are used to maintain water levels in sumps and tanks or to decant supernatant. Telescoping valves will act as circular weirs, using eq 5.6, with L equal to the telescoping valve perimeter (i.e., $\pi \times$ diameter of the telescoping valve tube, for head less than 30% or so of the diameter). For heads that exceed 50% of the diameter of the telescoping valve tube, the orifice equation is applicable (eq 5.4). Both equations should be checked, and the largest value of head should be used. The value for H, in both cases, is measured from the rim of the telescoping valve tube. In addition to this loss, the friction and minor losses from the inlet of the telescoping valve to the pipe outlet will need to be added.

4.0   Unit Process Hydraulics

The hydraulic calculations for each unit process require the use of the equations discussed in the Hydraulic Elements section of this chapter and others found in hydraulic handbooks and textbooks. For example, calculation of the difference in the water surface elevation between the effluent manhole and the clarifier shown in Figure 5.8 requires the use of the following equations:

- Basic pipeline headloss calculation involving friction loss (eqs 5.7a–5.7c, 5.8, 5.9, and 5.11), minor losses, and entrance and exit losses in the pipe between the manhole and the clarifier (eq 5.10);

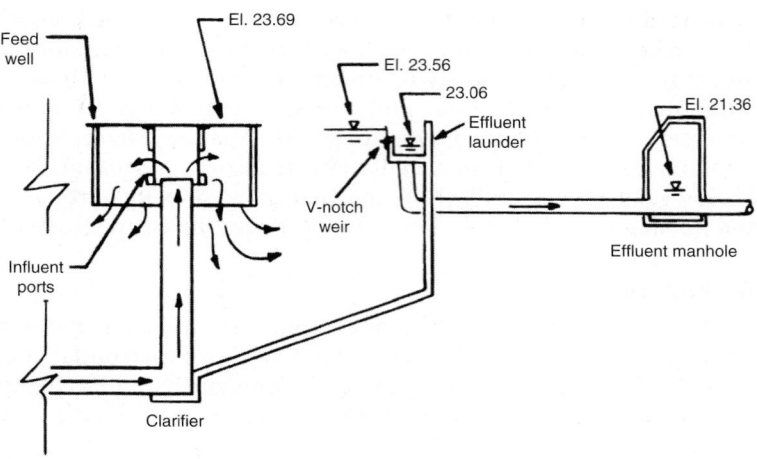

FIGURE 5.8   Unit process hydraulic for a typical clarifier.

- Weir trough equation for calculating the highest water elevation in the clarifier effluent launder (eq 5.14); and

- V-notch weir equation for calculating the head over the launder weir crest (eq 5.5).

The hydraulic designer must understand how the liquid flows through each unit process and the water depths that are required for the process. Within each unit process, devices are used to distribute flow, maintain a certain water depth, and control the flow. Typical devices include shutoff gates, weir gates, valves, ports, weirs, baffles, orifices, launders, and underdrains. Each of these devices imposes a headloss on the system and must be considered in the hydraulic calculations. Each unit process, its respective flow devices, and interconnecting piping should be analyzed carefully. The following subsections present some of the major points of consideration for many of the unit processes typically found in WRRFs. Detailed discussions of each treatment process and information related to the process design and operation are presented in subsequent chapters in this manual.

4.1   Screening

Wastewater flowing between the bars of the screen creates headloss. The headloss depends on the type of screen (coarse bar rack, bar screen, or fine screen). For manufactured screens, such as comminutors, grinders, and rotary drum screens, the headloss can be obtained from the manufacturers. However, the designer is cautioned that this headloss typically is the "clean screen" headloss. Headloss increases as the screen becomes partially clogged by debris (however, as the screen is cleaned by mechanical rakes or other means, headloss decreases. The cleaning typically is actuated based on the time interval or water level difference across the screen). Allowances should be made for a partially clogged screen, as cleaning typically is intermittent. An arbitrary allowance of 150 mm (6 in.) or more above the "clean screen headloss" typically is added.

The headloss through bar racks and manually and mechanically cleaned bar screens can be determined using the peak flow and the following equation (Metcalf & Eddy, 2013):

$$h = k(V^2 - v^2)/2g \qquad (5.19)$$

where h = headloss through the screen, m (ft);
V = velocity through bar screen (i.e., between the bars), m/s (ft/s);
v = velocity upstream of bar screen or approach velocity, m/s (ft/s);
g = acceleration caused by gravity, 9.8 m/s^2 (32.2 ft/s^2); and
k = headloss coefficient (typically 1/0.7).

The velocity through the bar screen opening can be determined by the following equation:

$$V = Q/A_o \qquad (5.20)$$

where V = velocity through bar screen (i.e., between the bars), m/s (ft/s);
Q = flowrate, m^3/s (cu ft/s); and
A_o = area of the bar screen opening m^2 (sq ft).

$$A_o = [W_{clr}/(W_b + W_{clr})] \times W \times (D_{DS}) \qquad (5.21)$$

where A_o = area of the bar screen opening, m^2 (sq ft);
W_b = width (thickness) of the bars, m (ft)
W_{clr} = clear spacing between the bars, m (ft)
W = total screen channel width, m (ft)
D_{DS} = downstream water depth, m (ft)

$$v = Q/(W_{US} \times D_{US}), \text{ m (ft)} \qquad (5.22)$$

where v = velocity upstream of bar screen or approach velocity, m/s (ft/s);
W_{US} = width of the channel upstream of the bar screen, m (ft); and
D_{US} = depth of flow upstream, m (ft).

In the above equation, D_{US} typically is unknown and must be solved for iteratively using the Bernoulli equation (eq 5.1) and knowing that y_1 is the same as D_{US}. To account for clogging of the screen, the value of A_o above can be reduced by a percentage. A 50% clogged screen frequently is used. This will increase the velocity through the screens and increase the headloss.

The designer may want to consider installing a passive overflow weir to accommodate extreme peak flows or clogged or malfunctioning screen equipment.

The headloss for fine screens uses the orifice equation and is presented as follows (Metcalf & Eddy, 2013):

$$h = (1/2g) \times (Q/CA)^2 \qquad (5.23)$$

where h = headloss through the screen, m (ft);
   g = acceleration caused by gravity, 9.8 m/s² (32.2 ft/s²);
   Q = flowrate, m³/s (cu ft/s);
   A = area of the openings submerged in the wastewater, m² (sq ft); and
   C = coefficient of discharge for the screen (typically 0.6).

In eq 5.23, the headloss is the clear water, clean screen headloss. As with bar screens, an allowance should be made for partial clogging. The screen manufacturer should be consulted.

4.2   Grit Removal

There are three generic types of grit removal systems—aerated grit chamber, vortex grit chamber, and a constant-velocity grit channel. Regardless of the type of grit removal system used, the equal distribution of both flow and grit to the individual units is important. Although the flow may be equally distributed, the equal distribution of grit may not occur. This is especially true when inlet distribution channels are used. Flow separation points (i.e., those found at sharp corners) along the inlet distribution channels can cause dead zones, where grit drops out. Once settled, the grit eventually tends to migrate to the closest basin or basins. Other than eliminating the possibility of dead zones, which is sometimes a difficult task, the inlet distribution channel can be aerated to keep the flow well mixed and the grit in suspension. The use of air to keep the grit suspended and well mixed may require odor-control facilities to treat the off-gases.

4.2.1   Aerated Grit Chamber

In an aerated grit chamber, the inflow direction and the outflow direction typically are perpendicular to each other, to achieve spiral rotation through the grit chamber. Typically, there is an overflow weir at the end of the chamber to control the water level, and there may be baffles installed in the chamber to improve performance, all of which will need to be accounted for in the hydraulics.

4.2.2   Vortex Grit Removal

There are a number of proprietary vortex-type grit removal systems. Design data must be obtained from the supplier and must be carefully followed. Of particular importance is the approach configuration and velocity. The discharge must be a free discharge at all times.

4.2.3   Constant Velocity Channels

Constant velocity channels use a proportional or Sutro weir to maintain a constant velocity in a rectangular grit channel (it also is possible to use a rectangular slot or a Parshall flume in conjunction with a parabolic-shaped grit channel to maintain a constant velocity). These systems have significant headloss, as a result of the proportional weir, which must be freely discharging under all flow conditions. See Reynolds and Richards (1995) for details.

4.3   Flow Equalization

Flow equalization can be either inline or offline. With inline flow equalization, all of the flow enters the flow equalization basin, and a constant outflow rate is maintained.

With offline flow equalization, only that portion of the flow above a given flowrate (typically the average flow) is diverted into the flow equalization basin. The accumulated flow then is released during low-flow periods, to bring the total flow to average flow for the day. Process design for flow equalization and the need for adequate mixing are discussed in Chapter 9.

The inline flow equalization is the easiest to control. Typically, the flow is pumped out using flow-controlled variable-speed pumps or is pumped in and flows out by gravity using a flow-control valve and flow meter. If the latter is used, careful selection of the flow-control valve is needed to prevent clogging, even if screened or primary treated wastewater is to be equalized.

For offline flow equalization, flow control gates or variable-speed pumps can be used. If a constant elevation side weir is used, achieving a controlled flowrate over the side weir is difficult and is not recommended. Variable-speed pumps are a better choice.

4.4 Primary Sedimentation

Headloss through circular primary sedimentation basins consists of the headloss through the influent pipeline and inlet column, headloss through the orifices at the end of the inlet column pipe in the stilling well, headloss through any flow distribution orifices within the stilling well, and head over the effluent weir. If it is a rim-feed clarifier, the total headloss includes the headloss through the inlet trough orifices, headloss in the distribution trough, and head over the effluent weir.

For a rectangular basin, the headloss includes the inlet gate, energy-dissipating tee or other system, and head over the effluent weir. For the headloss through the inlet-energy-dissipating tee, a K-value equal to 3 has been determined experimentally by the City of Los Angeles, California (Betz, 1981).

4.5 Aeration Basins

As with any unit process, equal distribution of flow to the individual basins and within each unit process is important. As stated in previous sections of this chapter, headloss is needed to ensure the equal distribution of flow. This can be generated by inlet gates (orifice loss), inlet weirs, or cut-throat flumes. The design and hydraulics have been discussed previously in this chapter. To keep solids in suspension and well mixed, the mixed-liquor channels should be aerated. However, aeration of the inlet channels is not an option in cases where the treatment process would be affected (i.e., upstream of anoxic or anaerobic basins). Channels upstream of anoxic or anaerobic zones may be mixed using propeller mixers or jet nozzles.

The mode of operation for the aeration basin should be addressed during calculation of the hydraulic profile. Aeration basins can be operated as completely mixed flow, plug flow, or stepped flow (feed) and its variations. The designer calculating the water-surface profile should anticipate that the mode of operation selected during design could change because of future unplanned operational requirements. The hydraulic analysis should be performed along the path that provides the greatest total headloss. A tank that is out of service for maintenance also should be considered. The headloss for flow through the aeration tank is small and is typically not calculated. If centrifugal aeration blowers are used, water level fluctuations between peak flow and initial minimum flowrates in the aeration basin should be kept to a minimum, so that blower performance is not affected.

The flow used in the hydraulic analysis of inlet gates, aeration basin end weir(s), and internal baffles must include the RAS and internal mixed liquor recycle (often used in biological nutrient removal facilities), as appropriate.

4.6   Biotowers and Trickling Filters

For trickling filters to operate properly, a sufficient water surface elevation difference (hydraulic head) must exist to overcome the fixed-nozzle orifice loss and rotate the distributors. The head requirement at the inlet to the distributing arm typically is approximately 2 m (6.5 ft). The designer should consult with the manufacturer for actual headloss requirements. In the analysis, recycle flowrates must be included in the total flow.

4.7   Secondary Sedimentation

Headloss calculations for secondary sedimentation basins are similar to that for primary sedimentation basins described above. The flowrate into the secondary clarifiers must include the RAS for activated sludge facilities. For trickling filter and biotower facilities, the recycled flows typically are diverted away upstream of the secondary sedimentation tanks; thus, they do not need to be included in the flow entering the secondary sedimentation tanks. However, this is not always the case.

For some activated sludge secondary clarifiers, the RAS is drawn off via a pipe or series of pipes (organ pipe clarifiers) situated within the inlet column. The headlosses in this system should be analyzed to ensure that the maximum anticipated RAS can be withdrawn and flow, by gravity, to the RAS pumps.

4.8   Disinfection Systems

Because solids settling is not as much of a concern, the basins are designed to operate at low velocities, and corresponding headlosses are low. Scum control, if desired, can be provided by a tilting pipe skimmer or similar system.

4.8.1   Chlorination

Minimizing short-circuiting in the design of chlorine contact chambers is paramount. Chorine contact basins should be designed as plug-flow reactors, with large length-to-depth or -width ratios (i.e., 40:1 or more). Internal perforated baffles should be installed just downstream from the inlet and at each "turn" in the tank. To ensure uniform flow distribution, a baffle should be installed upstream of the effluent weir (see Metcalf & Eddy [2013] for additional design information). The principal headlosses in the chlorine contact tank occur as a result of the internal baffles (described above) and the end weir.

4.8.2   Ozonation

Ozone contactors typically are designed with internal over/underbaffling, with the water level controlled by an end weir. Over baffles can be calculated as submerged weirs, and the underbaffles can be calculated as orifices. Ozone contactors frequently are analyzed using CFD modeling.

4.8.3   UV Irradiation

The water level in open-channel UV systems must be maintained as constant as possible to ensure proper disinfection. This typically is done with a constant level gate or similar device provided by the equipment supplier. It is important that this level control device have a free, unsubmerged discharge to operate properly. Also key to effective

UV disinfection is uniform velocity distribution approaching the UV lamps. Some states, such as California, require velocity profiling in conformance with the National Water Research Institute (Fountain Valley, California) (NWRI, 2012). This may require additional baffling at the inlet. Headloss data in UV systems can be obtained from the UV suppliers.

4.9   Post-Aeration

Cascade aeration, or having the effluent cascade down a series of steps, often is used for post-aeration, if there is sufficient head available. A discussion on the hydraulic design of these systems is presented elsewhere (Metcalf & Eddy, 2013). The steps are analyzed as broad-crested weirs, using eq 5.6, with $C = 3.086$.

4.10   Batch Reactors

An important hydraulic consideration for sequencing batch reactors is the rate of "unloading" or decanting. This rate is relatively high—often many times greater than the average daily flow entering the facility. This "unloading rate" will affect all downstream processes that are controlled by hydraulic loading (i.e., filtration and disinfection). Flow equalization can be provided to minimize this effect. The system supplier should be consulted. The unloading rate can be determined easily as the decant volume of the reactor divided by the decanting time.

4.11   Moving-Bed Bioreactors and Membrane Bioreactor Systems

Screens typically are used in moving-bed bioreactors to keep the fixed-film modules in the reactor. A headloss of 50 mm (2 in.) typically is allowed for each screen in the system. This allows for some clogging. Normal operating headloss is closer to 6 mm (0.5 in.) (Sen, 2008).

For membrane bioreactors (MBRs), the principal headloss occurs as the transmembrane pressure drops. This is overcome with pumping. The designer should consult the various MBR suppliers for specific headloss information. The internal recycle rates from the membrane area to the reactor inlet are significant; however, this typically is accounted for in the vendor's design.

4.12   Tertiary Processes

Common tertiary processes include mixing and flocculation and filtration. Hydraulic considerations for tertiary sedimentation are similar to that for primary and secondary sedimentation above.

4.12.1   Mixing and Flocculation

Flocculation and mixing basins typically contain perforated baffles. The headloss can be determined using the orifice equation (eq 5.4), as described previously for baffles.

4.12.2   Filtration

Gravity filters have hydraulic requirements, some of which are proprietary and peculiar to their design. For effluent gravity filters, sufficient water surface elevation must exist over the filter media, to convey the liquid through the media and the underdrains. Therefore, the manufacturers of these and other treatment units should be consulted to gather information on the water depth, velocity, and headloss requirements of the respective equipment.

In a typical filter, the hydraulics are analyzed for operation both in the filtering and the backwash modes. In the filtering mode, the headloss includes inlet losses (weirs or orifice gates, headloss through the media itself, headloss through the underdrain system, and losses through the flow metering, flow control system, and filter effluent piping).

For mono-media, dual, or multimedia filters, headloss through the media can be determined using a number of equations. The Rose equation is frequently used, although there are others, such as Karman-Kozeny, Fair-Hatch, and Hazen (Metcalf & Eddy, 2013). The Rose equation and similar equations provide the "clean filter headloss." To account for clogging of the media through operation, an arbitrary operating headloss of 1.8 to 2.7 m (6 to 9 ft) is added to the clean filter headloss. The headloss through the underdrain system depends on the type of underdrain used. Data on headloss can be obtained from the manufacturer. Headloss through the gravel support media is negligible.

The headloss in the backwash mode consists of the piping losses conveying the backwash water to the filter, control valves that control the rate of backwash water applied, headloss through the underdrain, headloss through the expanded (fluidized) bed, and headloss over the filter washwater launders. The headloss through the expanded media can be determined based on equations in Metcalf & Eddy (2013). Underdrain headloss can be obtained from the manufacturer of the system, and the headloss over the launder can be determined using the weir equation (eq 5.6). Trough design is identical to that of a clarifier launder and is described above.

Filters are subject to blinding or fouling during process upsets, so provisions for a high-water, emergency overflow may be considered to prevent damage to equipment, or overtopping of the tank walls.

4.13   Chemical Feed Systems

Chemical metering systems typically use positive displacement pumps. Some types of pumps cause a pulsing (i.e., diaphragm and solenoid pumps). These pumps generate higher velocities than average during the pulse cycle. This particularly affects the headloss in the suction piping, which, in turn, affects the net positive suction head available. Headloss calculations must consider the viscosity of the chemical. The Darcy-Weisbach equation (eq 5.7a) must be used for these calculations. Additional information is available from pump and chemical suppliers.

5.0   Pumping

The design of WRRFs confronts the engineer with a wide range of pumping applications, including raw wastewaters; treated wastewaters; mixtures of domestic and industrial wastes; raw sludges, thickened sludges, biosolids, and grit; scum containing a mixture of grease, floating solids, and trash; return and waste activated sludges (WASs); chemical solutions; flushing water; spray water and pump seal water; tank drainage; and sump pump water. Pumping of sludges and biosolids is presented in Chapter 19.

Some typical wastewater treatment applications for these pumps are described in Table 5.2. References, such as *Pump Application Engineering* (Hicks, 1971); *Hydraulic Institute Engineering Data Book* (Hydraulic Institute, 1979); *Hydraulic Institute Standards for Centrifugal, Rotary, and Reciprocating Pumps* (Hydraulic Institute, 1983); *Pumping Station Design* (Jones, 2008); *Centrifugal Pumps; Selection, Operation, and Maintenance* (Karassik and Carter, 1960); *Pump Handbook* (Karassik et al., 1986); *Pump Selection:*

Major Classification	Pump Type	Pump Description	Major Pumping Applications
Kinetic	Centrifugal (volute)	– Separately coupled – Close coupled – Submersible – Axial split – Radial split	– Raw wastewater – RAS and WAS (non-clog) – Settled primary and thickened sludge – Secondary or tertiary effluent
	Peripheral (torque-flow)	– Separately coupled – Close coupled – Radial flow – Recessed impeller	– Raw wastewater – Scum – RAS and WAS – Dilute sludge – Dilute digested sludge
	Vertical (turbine)	– Lineshaft – Submersible – Horizontally mounted axial flow – Vertical turbine solids handling (VTSH*)	– Screened wastewater (VTSH) – Primary effluent (VTSH) –Secondary or tertiary effluent
Positive displacement	Reciprocating	– Plunger	– Scum
		– Piston	– Primary, secondary, and settled sludges – Digested sludge – Thickened sludge – Chemical solutions
		– Diaphragm	– Scum – RAS and WAS – Digested sludge – Thickened sludge – Chemical solutions
	Rotary	– Lobe	– Raw wastewater
		– Screw	– Digested sludge (rotary-lobe) – Thickened sludge – Chemical solutions
		– Progressive cavity	– Scum – Sludge (when pumping primary sludge, grinder typically precedes pump) – Digested sludge – Thickened sludge
	Screw	– Spiral screw	– Raw wastewater – Settled primary and secondary sludges – Thickened sludge
	Pneumatic	– Airlift	– RAS and WAS
		– Ejector	– Raw wastewater at small installations – Scum

*VTSH is a proprietary patented design (Fairbanks Morse, Kansas City, Kansas).

TABLE 5.2 Pump Classification and Applications in Wastewater Industry

A Consulting Engineer's Manual (Walker, 1972); and others, will help in matching the demands of the design application with the characteristics of a particular pump. Metcalf & Eddy (2013) provide data on the properties of water for various temperatures.

5.1 System Curve, Pump Curve, and Pump Operation

When selecting pumps for a specific application, the design engineer must match the pumps performance with the head capacity curve for the system, taking into account the viscosity of the fluid and the range of expected operating flows.

The specific requirements of each pumping unit are determined by first calculating and plotting the system head curve. The system head curve is generated by plotting the sum of the static lift and friction headloss in the system at various (assumed) discharge flows. The static lift represents the elevation differential between the low water level in the wet well and the high point of the discharge force main or the high water level at the discharge, whichever is higher. The friction headloss is the sum of friction headlosses through the suction pipe, suction fittings, discharge pipe, discharge fittings, and force main at various flowrates. The next step is to plot the pump performance curve, or pump curve as it is commonly called, on the same graph. This is unique to the pump and available from the pump manufacturers. It relates pump capacity and discharge head (or pressure); often, efficiency, net positive suction head (NPSH), and power are plotted also.

Jones (2008) has many examples of typical system head curves, indicating portions of the total dynamic head that comprise static head and friction headloss. It is also important to note when plotting a system curve that there are the differences in the head capacity curves of one pump, two pumps, and three pumps in parallel operation. Head capacity curves for multiple pumps operating in parallel are obtained by adding the capacity of each pump at each of several given heads and plotting the results for each pump grouping.

System head and pump capacity curves can be combined to solve a number of complex pumping problems. In the case of two remote pumping stations (with differing pump hydraulic characteristics) discharging to a common force main, the system head curve would be a compound curve with a segment for each of the following conditions:

- Pump A on, pump B off;
- Pump A off, pump B on; and
- Pump A on, pump B on.

When both pumps operate, the capacity contributed by each pump will vary, depending on static head, friction losses, and individual pump head capacity curves, and will find pressure equilibrium. Therefore, pump selection for such a system is an iterative process (see Figure 5.9).

Head capacity curves for centrifugal pumps operated in a series are obtained by adding the operating head for each pump at a given capacity. An example of a series pumping arrangement would be where extreme static head conditions exist that exceeds the operating range of pumps that an owner might typically use. Series pumping increases the probability of a mechanical failure since all pumps operating in a series arrangement must be functioning properly in order to convey the desired pumping capacity. As such, series pumping arrangements should be avoided if the same objective can readily be achieved with a single reliable pump. Pump manufacturers should be consulted before designing series pump applications, to ensure proper pump selection, thereby avoiding potential cavitation or motor overload conditions.

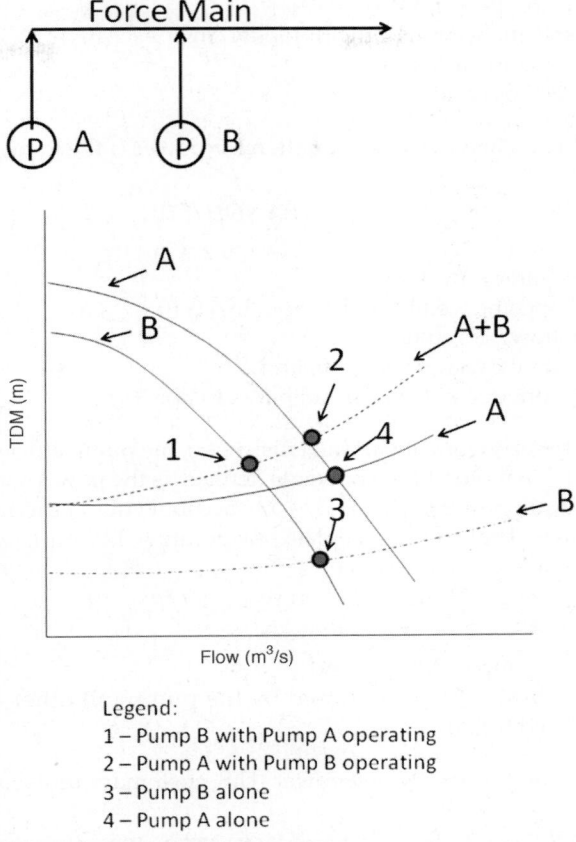

FIGURE 5.9   Multiple pumps into common force main.

A family of curves represents variable-speed pump head and capacity. Each individual curve corresponds to a discrete operating speed. The performance curves for different operating speeds can be obtained from the supplier or determined using the pump affinity laws, which relate speed, flow, head, and power. Superimposing the system head curve on the family of head capacity curves identifies specific operating points for the system.

Where variable- and constant-speed pumps discharge to a common header, a complete hydraulic analysis for all operating speeds is required to prevent recirculation cavitation. Jones (2008) has a very good discussion on pump cavitation.

5.1.1   Power Requirements

The power output necessary by a pump is determined by the energy needed by the contacting fluid, as shown in eqs 5.24a and 5.24b.

$$P = \gamma Q H \qquad\qquad (5.24a)$$

where P = water power, kW;
   γ = specific weight of the fluid, kN/m³;
   Q = flowrate, m³/s; and
   H = total dynamic head, m.

In U.S. customary units, the equation for power is the following:

$$P = \gamma QH/550 \qquad (5.24b)$$

where P = water power, hp;
   γ = specific weight of the fluid, lb/cu ft;
   Q = flowrate, cu ft/s;
   H = total dynamic head, ft; and
   550 = conversion factor from lb·ft/s to hp.

Because pumps work at varying efficiencies, the pump efficiency must be factored in before selecting a motor. Brake power is defined as the power input required by a pump (the motor output power), which takes into account volumetric, mechanical, and hydraulic energy losses. The motor nameplate power rating, kW (hp), is its output power.

$$\text{bkW} = \gamma QH/E_p \qquad (5.25a)$$

where E_p = pump efficiency; and
   bkW = brake power required by the pump (all other values are defined for eq 5.20a).

The equation for brake horsepower (U.S. customary units) is given below.

$$\text{bhp} = gQH/(550 \times E_p) \qquad (5.25b)$$

where E_p = pump efficiency; and
   bhp = brake power required by the pump (all other values are defined for eq 5.24b).

Because the brake power requirements vary with flow, head, and efficiency, changes in the pump discharge head will affect the brake power requirements. The pump brake power requirements should be checked at various points along the pump performance curve to ensure that the nameplate rating of the motor is equal to or greater than the maximum brake power required at all points.

To determine the total electrical power input to the motor for electrical power cost studies, brake power required should be divided by the motor efficiency.

5.1.2   Effects of Viscosity

The liquid handled by a pump affects its operating head and capacity, the required power input to the pump, and the construction materials. Pump type and selection must consider fluid viscosity. Although the effects of viscosity on pump performance have been tested, it is difficult to predict accurately the difference in a pump's performance when conveying a high- or low-viscosity liquid and its performance when conveying cold water. High-viscosity fluids require a higher head for a given flowrate, resulting in greater power requirements. The Hydraulic Institute (Parsippany, New Jersey) provides a procedure for analyzing viscosity effects for centrifugal pumps.

For rotary and reciprocating pumps, the pump speed should be decreased for viscous liquids. Because pump capacity depends on the pump speed, the pump capacity will decrease with increases in viscosity. Pump manufacturers should be consulted for the effect of viscosity on the pumps.

To overcome some of the effects of viscosity in sludge or biosolids pumping, polymers are sometimes added to the flow. Polymers reduce liquid viscosity by decreasing the cohesive forces between particles within the fluid. On the other hand, some types of polymers used in sludge or biosolids dewatering actually increase the headloss during pumping. In those cases, switching polymer types may reduce the pumping headloss. Chapter 19 discusses sludge pumping of wastewater in detail.

5.1.3   Variable-Speed Pumping

Variable-speed pumping applications have increased significantly over time, primarily due to their becoming more reliable and cost effective while simultaneously providing greater operating range capability versus a constant-speed pump, thus reducing the overall number of pumps required for larger and more complex pump stations. Even so, careful consideration should be given to the use of variable-speed applications if a constant-speed pumping arrangement is sufficient for the purpose and objective of the pump station, as it does add cost and complexity to the station. Jones (2008) provides an excellent description of the advantages and disadvantages of variable-speed pumping.

Affinity laws are used to assess pump performance under variable-speed conditions. These laws are used to predict centrifugal pump performance at different speeds and with different impeller diameters (WEF, 1993). For changes in speed, with the impeller diameter remaining constant, the affinity law becomes

$$Q_2/Q_1 = N_2/N_1 \qquad\qquad (5.26a)$$

$$H_2/H_1 = N_2^2/N_1^2 \qquad\qquad (5.26b)$$

$$P_2/P_1 = N_2^3/N_1^3 \qquad\qquad (5.26c)$$

where Q_1 = pump flowrate, L/s (gpm) at speed 1;
Q_2 = pump flowrate, L/s (gpm) at speed 2;
N_1 = pump rotative speed, r/min (rpm);
N_2 = pump rotative speed, r/min (rpm);
H_1 = pump head, m (ft) at speed 1;
H_2 = pump head, m (ft) at speed 2;
P_1 = pump power required, kW (hp) at speed 1, and
P_2 = pump power required, kW (hp) at speed 2

5.2   Pump Types and Applications

There are two major pump classifications defined by the Hydraulic Institute—kinetic energy pumps and positive displacement pumps. Table 5.2 lists varying types in each of these classifications, including the major applications these different types of pumps are used for in WRRFs.

5.2.1 Parallel versus Series Operation

Wastewater pumping applications can be in a series or parallel operation (or both). In general, pumps operate in series in order to overcome head conditions that a single pump is not designed to overcome. Essentially, two similar pumps in a series operation (i.e., one pump's discharge is conveyed into the suction of the second pump) will double their head capacity (operating in tandem). Similarly, for a parallel operation (i.e., two pumps discharging into a common header), the flow capacity of the combined operation of the pumps will double for any given pump flowrate. However, the system head curve (i.e., the sum of static headloss and friction headloss for any given flowrate) will determine the actual combined flowrate for the parallel operation. As such, where friction losses are greater than zero, the addition of a second pump will not double the hydraulic capacity of the pump operation, but will instead be defined by the characteristics of the system head curve (Cook, 1991).

5.3 Station Configuration

Table 5.3 presents a brief summary of the various types of pumping station configurations used at WRRFs with helpful commentary on each configuration. Additional information on pump intake design standards can be found elsewhere (Jones, 2008).

5.4 Wet-Well Types and Sizing

No single method for sizing wet wells applies to all design situations. Proper wet-well sizing considers three critical factors—detention time, pump cycle time, and turbulence at the pump intake.

As good practice, wet-well detention times generally should not exceed 30 minutes for average flow to minimize generation of unpleasant odors. In colder climates, longer detention times may be acceptable. Where such detention time limitations would be impractical, odor mitigation should be accounted for in wet-well design. Odor mitigation practice can include gas-tight covers, chemical feed systems (binding H_2S in solution), biofilters, or off-gas scrubbing systems.

Pump cycle time refers to the elapsed time between successive motor starts (i.e., the time to fill the wet well plus the time to empty it). Excessive motor wear and shortened service life result from cycle times less than the manufacturer's recommendation. Minimum cycle times range from approximately 5 minutes for 4-kW (5-hp) motors to over 30 minutes for 150-kW (200-hp) motors. The motor manufacturer should be consulted for minimum cycle time recommendations or special motor designs.

Minimum wet-well volumes can be determined from the following equation:

$$V = CQ/4 \tag{5.27}$$

where V = wet-well volume, m³ (gal);
   C = cycle time, minutes; and
   Q = pump capacity, m³/min (gal/min).

For constant-speed pumps, the minimum cycle time results if the influent flow equals 50% of the rated pump capacity. This limitation often determines wet-well

Configuration	Comments	Application
Wet pit/dry pit	– Typical for large pumping capacities – Larger footprint, surface structure recommended for electrical equipment – Ventilation and lighting required for dry pit – Subject to flooding from broken pipes; needs sump pump – Seal water system generally required – Crane/hoist for maintenance	– All pumping systems

Submersible	– Easily constructed, least costly
	– Less obtrusive, small footprint
	– Frequently used for smaller capacity stations
	– Maintenance more difficult
	– Requires crane access to remove pumps
	  – Seal water not required
	  – Moisture monitoring for motors
		– Influent, RAS/WAS
		– Intermediate and recirculation
		– Scum
		– Effluent

Submersible pumps in wet pit/dry pit	– Similar to wet pit/dry pit, except: – Seal water not required – Pumps and motors safe from flooding	– All pumping systems
Self-priming	– Pumps mounted at ground surface or shallow depth – Suction lift limited to approximately 6 m (20 ft) – Easy to maintain – Pumps may have reduced efficiency	– Influent, RAS/WAS – Intermediate and recirculation – Effluent
Self-cleaning, trench wet well	– Successful in larger lift stations – Easier to maintain than conventional design – Use in combination with wet pit/dry pit installations – Smaller footprint than conventional – Wet-well construction more complex – See Jones (2006) for details	– Influent
Vertical pumps over wet well	– Vertical turbine solids handling pumps for screened wastewater or primary effluent – Cost-effective – Small footprint – Requires crane access for maintenance of pumps	– Influent, RAS/WAS – Intermediate and recirculation – Effluent
Archimedes screw	– Low head application; limited to approximately 10 m (30 to 35 ft) lift – High efficiency – Essentially non-clog – Variable speed pumping, even though at constant rotational speed – Turbulence in pumping and discharge can release odors – Open screws should be covered for odor control – Enclosed screw interior difficult to protect from corrosion – Bearing maintenance requires heavy equipment	– Influent, RAS/WAS – Intermediate and recirculation – Effluent

TABLE 5.3   Summary of Wastewater Pumping Station Configurations at WRRFs

volumes for both single- and multiple-pump installations. For multiple-pump installations, alternating the lead pump after each pumping cycle effectively doubles the cycle time and reduces wet-well volumes accordingly. Wet-well volumes also can be optimized with strategic pump "on" and "off" settings. In multiple constant-speed installations, the required wet-well capacity represents the sum of the wet-well capacities required for the individual pumps. Such an allowance will prevent cycling when lag or standby pump units enter service.

Some designers use detention capacity in the influent sewer line to minimize wet-well volume by setting the pump "off" level above the invert elevation of the influent sewer. This practice may be acceptable in large installations, where influent wastewater velocities are sufficient to minimize solids deposition in the influent sewer. Where solids deposition issues are of significant concern, self-cleaning wet wells such as a trench or confined-type sumps may be considered. Jones (2008) offers a detailed review of these (and other) types of intakes for consideration.

Wet wells for variable-speed pumping systems can be significantly smaller than for comparably sized constant-speed stations. In determining the volume required for variable-speed pump applications, two factors need to be considered: (1) the capacity of the pump at minimum speed, and (2) pump operation at constant speed. The latter would occur if the variable-speed drive control failed and the pump would be "forced" to run in a constant-speed mode. If sufficient standby capacity is available, this may not be of concern. The capacity at minimum speed will vary with each installation and can be determined from the system head curve analysis presented previously. Equation 5.27 can be used, but the Q-value is the pumping capacity at minimum speed. If the pump has the potential to operate at constant speed for long periods of time (condition 2 above), consideration should be given to increasing the wet-well volume available to that of a constant-speed pump. This is determined from eq 5.27 using Q equal to the maximum pumping capacity.

In multiple variable-speed installations, the wet-well capacity required is the sum of the wet-well capacities required for the individual pumps. Such an allowance, similar to multiple constant-speed installations, will prevent cycling when lag or standby pump units enter service.

When determining the pump operating levels in the wet well, the design engineer needs to consider the NPSH requirements of the pump. Wet-well designs should allow adequate submergence and clearance between pump intakes to prevent turbulent currents and vortexes that could otherwise reduce pump efficiency or capacity. These requirements may dictate longer detention times than those necessary to meet pump cycling requirements. The pump manufacturer's recommendations and *Hydraulic Institute Standards for Centrifugal, Rotary, and Reciprocating Pumps* (Hydraulic Institute, 1983) should be consulted for sizing sumps and configuring wet wells. Adherence to these suggestions, based on testing by several pump manufacturers, will help ensure proper pumping station design and avoid costly future modifications. For large complex wet-well systems and very large pumps, where established design practice would be impractical, hydraulic (physical) modeling may be necessary to ensure acceptable wet-well performance (see Jones, 2008).

5.5   Pump Construction

5.5.1   Materials
Pumps operating in wastewater require materials that withstand adverse operating conditions. Universal standards for internal coatings and impeller materials may not take

precedence over those found regionally to perform acceptably well. Many of the more effective pump coatings are costly and are required to be installed by vendor-approved contractors or the coating vendors themselves. Long-lasting pump impeller materials are costly, with appreciable extended delivery time requirements, likely leading to a need for equipment prepurchase. Typical impeller material ranges from cast iron to CA15 stainless-steel 410 10BHN, which has a hardness and durability many times greater than cast iron, as outlined by the Hydraulic Institute standard ANSI/HI 9.1-9.5-2000 Pumps—General Guidelines. This also is accompanied by a proportionate cost increase over cast iron. Cast-iron impeller materials are short-lived in many wastewater applications.

Because a critical factor in the reduced life of electrical equipment is heat, selection of the type of insulation used in the pump motor depends on the operating temperature that a motor will experience. National Electrical Manufacturers Association (Rosslyn, Virginia) insulation ratings assume a motor is operating within its rated ambient temperature. Class H insulation offers 20 000 hours of life at 180°C, whereas a motor operating at the same temperature with Class A insulation will have an estimated life of only 300 hours.

Other factors to consider include enclosure type and the service factor. Some wastewater applications may require enclosures that are totally enclosed fan-cooled, that is, are minimally exposed to corrosive, but not explosive, conditions. Some wastewater applications may require totally enclosed explosion-proof enclosures. Service factors indicate the percentage of additional horsepower available without damage. A 1.20 service factor indicates that a motor can deliver 20% more horsepower without damage.

5.5.2 Seals

5.5.2.1 Mechanical Seals Centrifugal pumps often are furnished with mechanical seals to minimize leakage around the pump shaft. Mechanical seals are recommended for pumps that operate under a high suction head. Contact between the seal surfaces is maintained, in most cases, by a spring load. Mechanical seals generally are water lubricated, although other lubricating fluids may be used with specially designed seals. Clean water is needed for shaft seals. Seal water pressure generally should equal 110% of the maximum pump discharge pressure or shutoff head. Facility service water, treated effluent, or potable water can be used for seal water. If potable water is used, the supply source should be protected by an air gap arrangement to prevent backflow of the pumped liquid. If treated effluent or facility service water is used, it must be free from gritty material, which could foul the seal or score the pump shaft. Generally, seal water is supplied only when the pump operates. A solenoid valve is installed in the seal water supply line and interlocked with the pump starter. Seal water may have flow indicators (rotameters) and, with the pump cooling water pressure signal, may be transmitted to the facility supervisory control and data acquisition (SCADA) system. Drains need to be provided at all locations that seal water is used.

5.5.2.2 Packing Packing may be used to minimize leakage around the pump shaft where it penetrates the volute casing. Packing is becoming outdated and is not favored by operators. Packing is available in a wide variety of materials for specific applications. For wastewater pumps, packing is installed in the seal cage and held in place with a packing gland. Packing must be continuously cooled, properly adjusted, and lubricated while the pump is in operation. In wastewater and sludge applications, the lubricating fluid is seal water, as described above.

5.6   Pump Control Systems

To determine the proper type of control for any application, a set of parameters, including pressure, water level, and flow, must be established. Then, a control system can be selected, which will allow the pumps to produce the desired effects. However, considerations of the efficiency and/or power factor should not supersede the primary purpose of the pump control system. Nevertheless, the ever-increasing costs of power force a greater significance to power factor correction and pump efficiency. Most facilities without power factor correction are subject to appreciable financial penalties from their electricity provider.

Different processes or different pumping systems within a given process will require varying degrees of control of the primary hydraulic parameters—pressure, fluid level, and flow. The simplest system that will reliably provide effective results generally will be the most satisfactory.

No rigid rules govern the weights assigned to any of the considerations for determining the type of control most suitable for any given application. Ultimately, variables such as capital and operating costs, efficiency, power factor, reliability, operational effects, and ease of operation must be weighed against one another and the system best suited for the application chosen. Such selections are not always obvious and require thoughtful consideration of pumping effects.

The overall efficiency of a variable-speed system may exceed that of an on–off system, despite control losses. With the former system, the pump may operate against a lower average friction head, saving pump power to offset the power lost in the variable-speed control.

Constant-speed pumps will cause hydraulic surges when influent flow is less than pump capacity. Because these surges can adversely affect some biological treatment processes, variable-speed pump control may be necessary in those cases.

The selected control system must be compatible with the training and experience of the operators, or satisfactory operation will seldom be achieved.

5.6.1   Pump Actuation

5.6.1.1   Manual Control   Manual control systems generally consist of push-button stations or selector switches that energize or de-energize the pump motor starter. Manual control also can be achieved through a SCADA system. Push-button stations (sometimes called three-wire control) are electrically interlocked, so that the units have to be restarted manually after a power outage, while a selector switch (sometimes called two-wire control) remains in the "on" position and restarts automatically. Some systems require automatic restart on power outage. Manual control is essential on all systems for maintenance.

5.6.1.2   Automatic On–Off or Speed Control   Automatic control systems commonly are based on time, pressure, flow, or liquid level. Each of these is briefly described below.

5.6.1.2.1   Time   Pumps, started at regular intervals, operate for a preset length of time. Time-controlled systems are often used for sludge pumping, because sludge pumps are commonly oversized in small facilities to ensure adequate transport velocities.

5.6.1.2.2   Pressure   Pressure drop, generally sensed by a standard pressure switch in a hydropneumatic tank, is used to start the pumps in facility service water systems. Pressure also may be used to shut down positive displacement pumps to prevent damage.

5.6.1.2.3 Flow Pumps can be turned on when the required flow exceeds a certain limit or turned off when the required flow drops below a limit. Facility influent flow variations also may be used to start up or shut down return sludge or chemical feed pumps or to vary their speed.

5.6.1.2.4 Liquid Level Liquid level signals govern most of the automatically controlled constant- and variable-speed systems. Pumps are turned on or sped up as wet-well levels rise and are turned off or slowed down as wet-well levels fall. This method often controls influent and effluent pumps, sump pumps, and certain in- facility transfer pumps.

5.6.1.3 Starting In addition to power factor correction, reduced voltage starters may be necessary. Starting motors with a contactor "across the line," with no assistance other than a capacitor, likely will result in power surges with undesirable results. Some type of reduced voltage starter may be necessary to reduce line power voltage sags when a large inductive load is connected. This may be mandated by the local utility.

Reduced voltage starters are either electromechanical or solid-state. Electromechanical starters are auto-transformer, part-winding, and primary-resistor. These starters are reliable, and some accommodate a specially wound motor. Solid-state reduced-voltage starters typically use a bypass contactor to divert current, so that components do not handle the full load for the duration of the motor run time.

5.6.2 Variable-Speed Operation

Variable-speed drives, the general class of equipment used to drive wastewater pumps at varying operational speeds, typically are controlled by a signal received from a liquid level, pressure, or flow measurement system, as described in the previous section, or a manual adjustable speed control. This in turn controls the pump flow and discharge header.

5.7 Pump-Monitoring Guidelines

There is a need to store and monitor pump operating data, such as run status, amps, volts, power factor, kilowatts, and run time. For large motors, temperature and vibration data should be stored. All this can be done through the SCADA system. These data can be valuable when assessing pump operational problems and scheduling routine maintenance.

5.8 Specification Requirements for Testing and Acceptance

To ensure that pumping equipment meets specifications, testing is recommended. The extent of the testing depends on the size of the pump and how critical its operation is in the treatment system. Tests can be conducted in the factory and the field. Field testing is recommended always, regardless of the pump application.

5.8.1 Factory Inspections

Pump castings should be inspected at the factory before coatings and after the final finish. Pumps and their respective controls plus control programming should be tested at the factory before shipping. In many cases, units will need to be shipped from one factory to another for unit testing. Shipping and unit testing can be an additional cost, but is highly recommended if a mechanical drive is used (i.e., diesel or gas engine). It is very important to verify rotation, as this cannot be fixed easily later.

5.8.2   Certified Tests

Certified tests are done in a controlled environment with specific hydraulic, switchgear, and mechanical installation. Pumps are tested with specific methodology (Hydraulic Institute 2.6.5.5, 4a). Test results include pump efficiency, brake horsepower, and NPSH requirements. For large pumps, test piping configuration should match the actual design conditions.

5.8.3   Factory Witnessed Tests

These tests are performed in the factory, witnessed by the owner or owner's representative. This requires scheduling and is more costly than a certified test. Typically, the owner pays for travel, lodging, and so on. Often, the design engineer or representative is present also. These tests are typically performed on larger pumps.

5.8.4   Field Testing and Acceptance

These tests are conducted under field conditions with whatever limitations are present. Limitations may include poor metering, entrained air, vortexing, and incomplete power measurement. Field tests should be performed in accordance with Hydraulic Institute test procedures, sections 1.6, 2.6, and 11.6. A comparison between field test results and certification testing should provide an owner with guidance for acceptance of the pumping equipment.

6.0   Design Examples

6.1   Design Example—Partial Hydraulic Profile in SI Units

The hydraulic profile for the primary and secondary processes for a WRRF should be based on the plan schematic shown in Figure 5.10 and the profile shown in Figure 5.11. The invert elevation for the primary effluent channel and the elevation of the notches for the primary clarifier effluent weir should be set.

6.1.1   Input Parameters

- Design peak flow = 630 m^3/h = 0.175 m^3/s.
- RAS = 50% of the influent wastewater flow and is returned to the junction box downstream of the primary clarifiers.
- Primary and secondary clarifiers are circular, with 90° v-notches, 230 mm center to center; weir elevation shown is at the bottom of the notch.
- Primary clarifier effluent channel is rectangular in cross-section, 460 mm wide; the invert is horizontal (slope = zero).
- Each aeration basin has an effluent weir, which is rectangular and has a 1.8-m crest length.
- The mixed-liquor splitter box has two rectangular weirs, each 1.8-m crest length; there is one for each of the secondary clarifiers.

6.1.2   Assumptions

- All weirs are sharp crested; rectangular weir coefficient = 1.82; v-notch = 1.38.
- Gate and orifice coefficient = 0.60; sharp, not rounded.

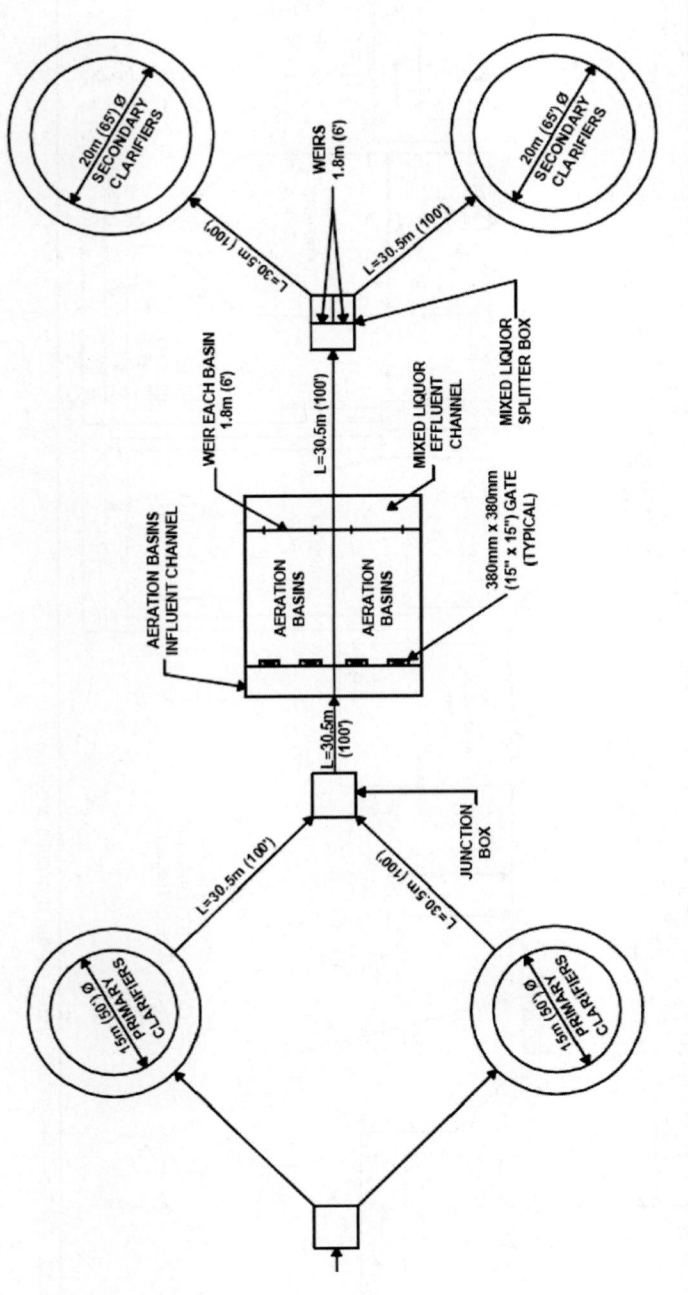

Figure 5.10 Plan—design example—partial hydraulic profile.

209

210

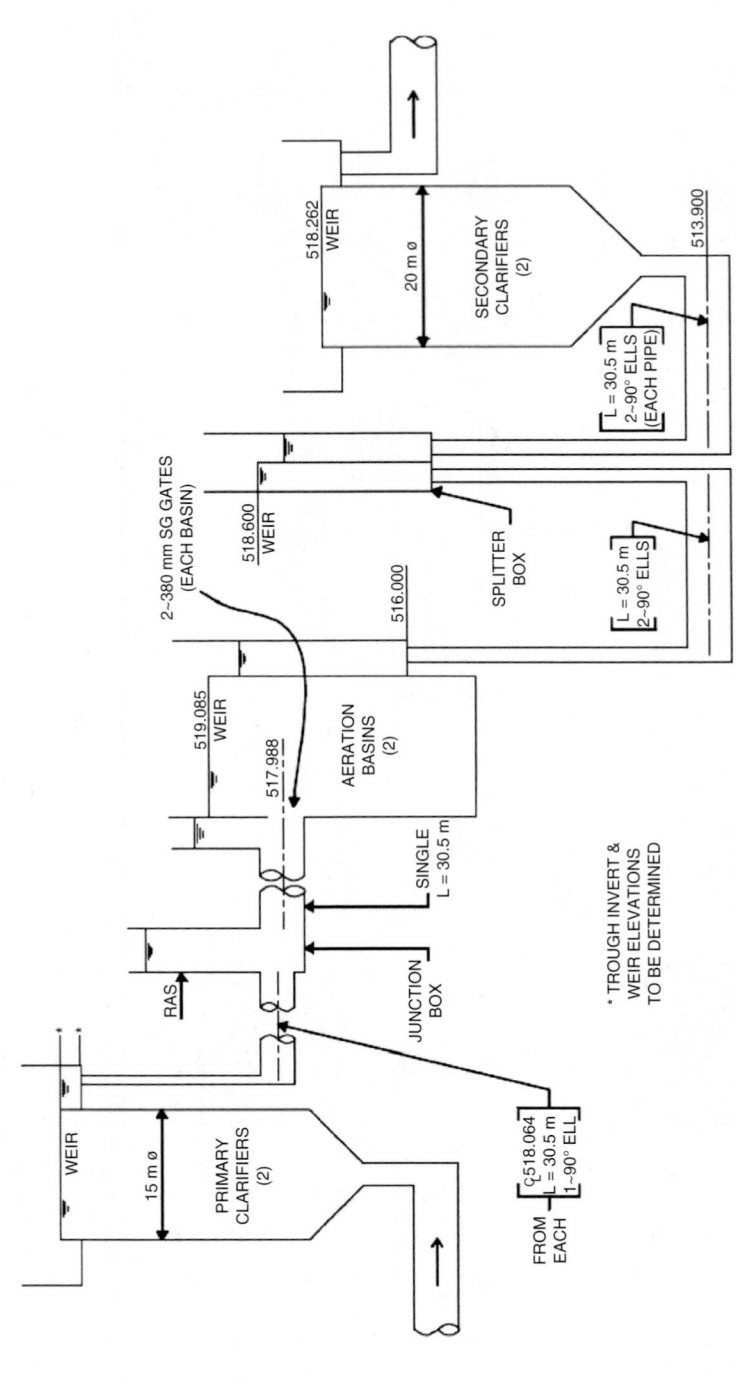

Figure 5.11 Profile—design example—partial hydraulic profile.

- Ignore end contractions at weirs.
- Manning's $n = 0.015$ for channels and pipes.
- Entrance loss coefficient, $K = 0.5$; exit loss coefficient, $K = 1.0$; 90° elbow loss, $K = 0.4$.
- Ignore losses in the aeration basin influent channel and mixed-liquor effluent channel (this is done for expediency here. These losses are likely to be very small).
- All process units are in operation (i.e., normal operation) (hydraulic profile also should be calculated, assuming one process unit is out of service, to check freeboard, etc.).

6.1.3   Calculations

Design flowrates:

Peak wastewater flow	0.175 m³/s
RAS rate (maximum)	0.088 m³/s
Peak wastewater flow + RAS	0.263 m³/s

Water surface elevation (WS El) in the secondary clarifier:

Weir crest elevation = 518.262
Number of clarifiers = 2
Flow per clarifier = 0.175 m³/s/2 = 0.088 m³/s
Perimeter of weir = $\pi \times$ diameter = $\pi \times 20$ m = 62.8 m
Number of 90° v-notches = 62.8 m/0.23 m center to center = 273
Flow per weir = (0.088 m³/s)/273 = 0.00032 m³/s
Head over the weir = $(Q/1.38)^{1/2.5}$ = (0.00032 m³/s/1.38)$^{0.4}$ = 0.035 m (eq 5.5)
WS El in secondary clarifier = 518.262 + 0.035 = 518.297

WS El in discharge side of mixed-liquor splitter box:

Select pipe size:
    Number of clarifiers = 2
    Flowrate per clarifier = (0.263 m³/s)/2 = 0.132 m³/s
Design for 0.75 m/s:
    $A = Q/V$ = (0.132 m³/s)/0.75 m/s = 0.176 m², use 460-mm diameter, A = 0.167 m²
    $V = Q/A$ = (0.132 m³/s)/0.167 m² = 0.79 m/s
    $V^2/2g$ = (0.79 m/s)²/(2 × 9.8 m/s²) = 0.032 m
Manning equation: $Q = (1/n)\ AR^{0.67}S^{0.5}$; let $K = (1/n)\ AR^{0.67}$, then $S = (Q/K)^2$
    R = diameter/4 for pipes flowing full; R = (460 mm)/4 = 115 mm = 0.115 m
    K = (1/0.015) × (0.167 m²) × (0.115m)$^{0.67}$ = 2.614
    $S = [(0.132)/2.614]^2$ = 0.0025 m/m
Pipe length = 30.5 m; friction loss = 30.5 m × 0.0025 m/m = 0.076 m
Minor losses, $H = KV^2/2g$
    Entrance loss, $K = 0.5$; 90° bend loss $K = 0.4$; exit loss $K = 1.0$
    Total K = 0.5 + 2(0.4) + 1.0 = 2.3
    Minor losses = 2.3 × 0.032 m = 0.074 m

WS El in discharge side of mixed-liquor splitter box = 518.297 m + 0.076 m
+ 0.074 m = 518.447

WS El upstream of mixed-liquor splitter box weirs:

Design total flowrate = 0.263 m³/s
Number of weirs = 2; one for each secondary clarifier
Flow per weir = (0.263 m³/s)/2 = 0.132 m³/s
Weir length = 1.8 m; rectangular weir plate
Weir coefficient = 1.82
Head over the weir, $H = (Q/C \times L)^{0.67} = [(0.132 \text{ m}^3/\text{s})/(1.82 \times 1.8)]^{0.67} = 0.116$ m
Weir crest El = 518.600
WS El upstream of mixed-liquor splitter box weirs = 518.600 + 0.116 m = 518.716

WS El in aeration basin mixed-liquor effluent channel:

Design total flowrate = 0.263 m³/s
Select pipe size:
    Design for 0.75 to 0.9 m/s
    $A = Q/V = (0.263 \text{ m}^3/\text{s})/0.75 \text{ m/s} = 0.351$ m², diameter = 0.68 m
    $A = Q/V = (0.263 \text{ m}^3/\text{s})/0.9 \text{ m/s} = 0.292$ m², diameter = 0.60 m
    Use 600-mm diameter, $A = 0.292$ m²
$V = Q/A = (0.263 \text{ m}^3/\text{s})/0.292 \text{ m}^2 = 0.90$ m/s
$V^2/2g = (0.90 \text{ m/s})^2/(2 \; 3 \; 9.8 \text{ m/s}^2) = 0.041$ m
Manning equation. $Q = (1/n) \; AR^{0.67}S^{0.5}$; let $K = (1/n) \; AR^{0.67}$, then $S = (Q/K)^2$
    R = diameter/4 for pipes flowing full; R = (600 mm)/4 = 150 mm = 0.150 m
    $K = (1/0.015) \times (0.292 \text{ m}^2) \times (0.15)^{0.67} = 5.46$
    $S = [(0.263)/5.46]^2 = 0.0023$ m/m
Pipe length = 30.5 m; friction loss = 30.5 m × 0.0023 m/m = 0.070 m
Minor losses, $H = KV^2/2g$
    Entrance loss, K = 0.5; 90° bend loss K = 0.4; exit loss K = 1.0
    Total K = 0.5 + 2(0.4) + 1.0 = 2.3
    Minor losses = 2.3 × 0.041 m = 0.094 m
    WS El in aeration basin mixed-liquor effluent channel = 518.716 + 0.070 m + 0.094 m = 518.880

WS El in aeration basin:

Design total flowrate = 0.263 m³/s
Number of weirs = 2; one for each aeration basin
Flow per weir = (0.263 m³/s)/2 = 0.132 m³/s
Weir length = 1.8 m; rectangular weir plate
Weir coefficient = 1.82
Head over the weir, $H = (Q/C \times L)^{0.67} = [(0.132 \text{ m}^3/\text{s})/(1.82 \times 1.8)]^{0.67} = 0.116$ m

Weir crest El = 519.085

WS El in aeration basin = 519.085 + 0.116 m = 519.201

WS El in aeration basin influent channel:

Design total flowrate = 0.263 m^3/s

Number of gates = 4; 2 for each aeration basin

Flow per gate = (0.263 m^3/s)/4 = 0.066 m^3/s

Gate width = 380 mm

Gate height = 380 mm

Gate area, each = (0.38 m) × (0.38 m) = 0.144 m^2

Gate orifice coefficient = 0.60

Gate headloss, $H = (1/2g) \times (Q/C \times A)^2 = [1/(2 \times 9.8)] \times [0.066/(0.6 \times 0.114)]^2 = 0.030$ m

Check velocity through the gate $V = Q/A = 0.066$ m^3/s/0.144 m^2 = 0.46 m/s OK

Headloss = 0.03 m; should give good flow distribution, as headloss in the influent channel itself likely will be very small. Note that a weir or cut-throat flume could be used in lieu of the gate.

WS El in aeration basin influent channel = 519.201 + 0.030 m = 519.231

WS El in junction box:

Design total flowrate = 0.263 m^3/s; includes RAS at this point

Use 600-mm-diameter pipe, as flow is same as from aeration basin

$V = Q/A = (0.263$ m^3/s$)/0.292$ m$^2 = 0.90$ m/s

$V^2/2g = (0.90$ m/s$)^2/(2 \times 9.8$ m/s$^2) = 0.041$ m

Manning equation. $Q = (1/n) AR^{0.67}S^{0.5}$; let $K = (1/n) AR^{0.67}$, then $S = (Q/K)^2$

  R = diameter/4 for pipes flowing full; R = (600 mm)/4 = 150 mm = 0.15 m

  $K = (1/0.015) \times (0.292$ m$^2) \times (0.15$ m$)^{0.67} = 5.46$

  $S = [(0.263)/5.46]^2 = 0.0023$ m/m

Pipe length = 30.5 m; friction loss = 30.5 m × 0.0023 m/m = 0.070 m

Minor losses, $H = KV^2/2g$

  Entrance loss, $K = 0.5$; exit loss $K = 1.0$

  Total $K = 0.5 + 1.0 = 1.5$

  Minor losses = 1.5 × 0.041 m = 0.062 m

WS El in junction box = 519.231 + 0.070 m + 0.062 m = 519.363

WS El at primary clarifier effluent trough outlet:

Design total flowrate = 0.175 m^3/s

Number of primary clarifiers = 2

Flow to each primary clarifier = (0.175 m^3/s)/2 = 0.088 m^3/s

Select pipe size:

  Design for 0.9 m/s

  $A = Q/V = (0.088$ m^3/s$)/0.9$ m/s = 0.098 m^2, diameter = 350 mm

Use 300-mm diameter, $A = 0.071$ m² (note that 350 mm could be used; designer preference)

$V = Q/A = (0.088$ m³/s$)/0.071$ m² $= 1.23$ m/s

$V^2/2g = (1.23$ m/s$)^2/(2 \times 9.8$ m/s²$) = 0.077$ m

Manning equation. $Q = (1/n) AR^{0.67}S^{0.5}$; let $K = (1/n) AR^{0.67}$, then $S = (Q/K)^2$

$\quad R = $ diameter$/4$ for pipes flowing full; $R = (300)/4 = 75$ mm

$\quad K = (1/0.015) \times (0.071$ m²$) \times (0.075$ m$)^{0.67} = 0.83$

$\quad S = [(0.088)/0.83]^2 = 0.011$ m/m

Pipe length $= 30.5$ m; friction loss $= 30.5$ m $\times 0.011$ m/m $= 0.336$ m

Minor losses, $H = KV^2/2g$

$\quad$ Entrance loss, $K = 0.5$; 90° bend loss $K = 0.4$; exit loss $K = 1.0$

For a junction box, assume total energy loss on entrance—a conservative assumption. As an alternative, the confluence pressure and momentum equation could have been used (eq 5.18). Also, a slightly larger pipe diameter could have been used to reduce the headloss.

$$\text{Total } K = 0.5 + 1(0.4) + 1.0 = 1.9$$
$$\text{Minor losses} = 1.9 \times 0.077 \text{ m} = 0.146 \text{ m}$$

WS El at primary clarifier effluent trough outlet $= 519.363 + 0.336$ m $+ 0.146$ m $= 519.845$

Maximum WS El in primary clarifier effluent channel:

Design total flowrate $= 0.175$ m³/s

Number of primary clarifiers $= 2$

Flow to each primary clarifier $= (0.175$ m³/s$)/2 = 0.088$ m³/s

Flow splits two ways in each effluent channel in circular clarifiers $= 0.088$ m³/s$/2 = 0.044$ m³/s

Primary clarifier channel width $= 460$ mm

Critical depth, $D_c = [Q^2/(B^2 \times g)]^{0.33} = \{[(0.044)^2/(0.46)^2]/9.8\}^{0.33} = 0.10$ m

Minimum invert of primary clarifier effluent channel for free flow $= 519.845 + 0.10$ m $= 519.945$

Set weir trough invert elevation at 519.97 (allows for a small freefall at end of channel)

Maximum water depth in primary clarifier effluent channel $= 1.73 \times D_c$ (because it is free flow)

Maximum water depth in primary clarifier effluent channel $= 1.73 \times 0.10$ m $= 0.173$ m

Maximum WS El in primary clarifier effluent channel $= 519.970 + 0.173$ m $= 520.143$

Set primary clarifier effluent weir crest at 520.200 (allows for a small drop)

WS El in primary clarifiers:

Weir crest elevation $= 520.200$

Total design flow $= 0.175$ m³/s

Number of clarifiers = 2

Flow per clarifier = 0.175 m³/s/2 = 0.088 m³/s

Perimeter of weir = $\pi \times$ diameter = $\pi \times 15$ m = 47.1 m

Number of 90° v-notches = 47.1/0.23 center to center = 205

Flow per weir = (0.088 m³/s)/205 = 0.00043 m³/s

Head over the weir = $(Q/1.38)^{1/2.5}$ = (0.00043 m³/s/1.38)$^{0.4}$ = 0.04 m

WS El in primary clarifier = 520.20 + 0.04 = 520.24

See Figure 5.12 for the completed hydraulic profile with calculated elevations.

6.2 Design Example—Partial Hydraulic Profile in United States Customary Units

The hydraulic profile for the primary and secondary processes for a WRRF should be based on the plan schematic shown previously in Figure 5.10 and the profile shown in Figure 5.13. The invert elevation for the primary effluent channel and the elevation of the notches for the primary clarifier effluent weir should be set.

6.2.1 Input Parameters

- Design peak flow = 4 mgd.
- RAS = 50% of the influent wastewater flow and is returned to the junction box downstream of the primary clarifiers.
- Primary and secondary clarifiers are circular with 90° v-notches, 9 in. center to center; weir elevation shown is at the bottom of the notch.
- Primary clarifier effluent channel is rectangular in cross-section, 18 in. wide; the invert is horizontal (slope = zero).
- Each aeration basin has an effluent weir, which is rectangular and 6 ft in crest length.
- The mixed-liquor splitter box has two rectangular weirs, each 6 ft in crest length—one for each of the secondary clarifiers.

6.2.2 Assumptions

- All weirs are sharp-crested; rectangular weir coefficient = 3.3.
- Gate and orifice coefficient = 0.60; sharp, not rounded.
- Ignore end contractions at weirs.
- Manning's n = 0.015 for channels and pipes.
- Entrance loss coefficient, K = 0.5; exit loss coefficient, K = 1.0; 90° elbow loss, K = 0.4.
- Ignore losses in the aeration basin influent channel and mixed-liquor effluent channel (this is done for expediency here. These losses are likely to be very small).
- All process units are in operation (i.e., normal operation) (hydraulic profile also should be calculated, assuming one process unit is out of service, to check freeboard, etc.).

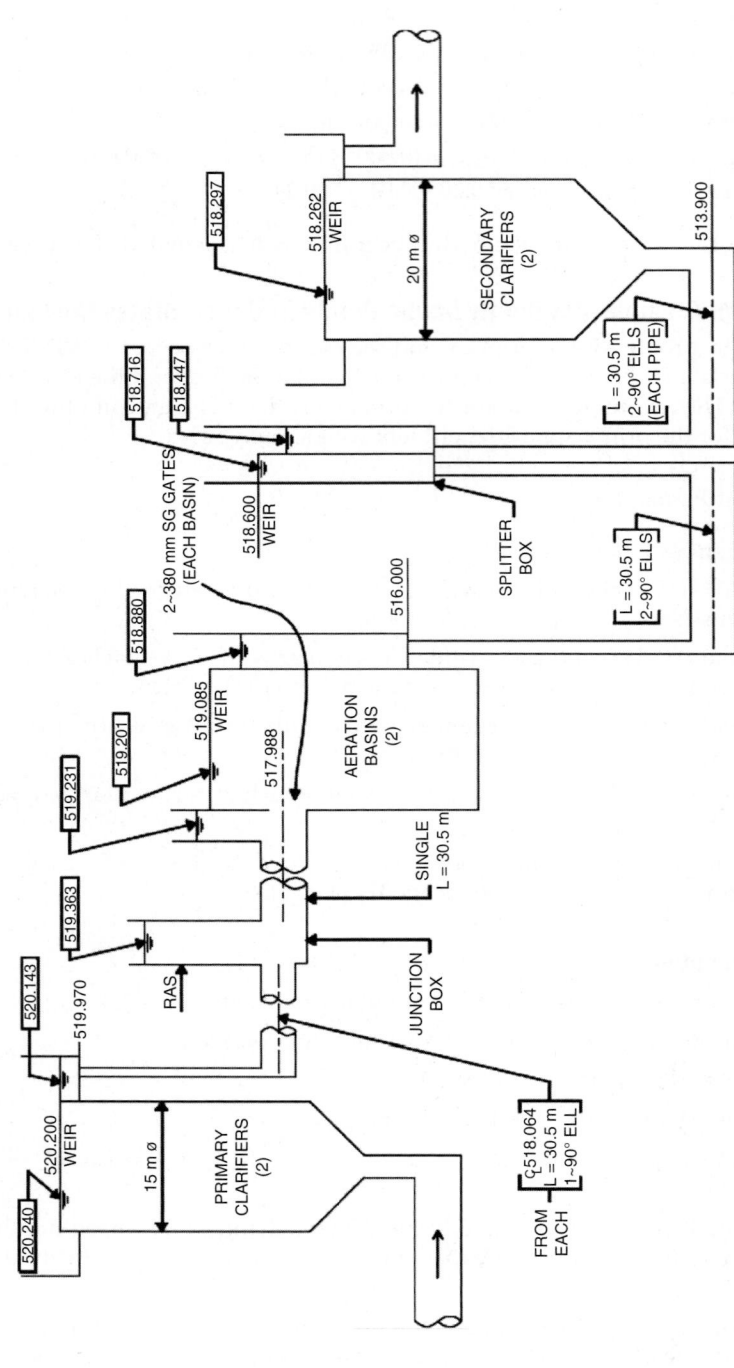

FIGURE 5.12 Profile—design example—partial hydraulic profile with calculated elevations.

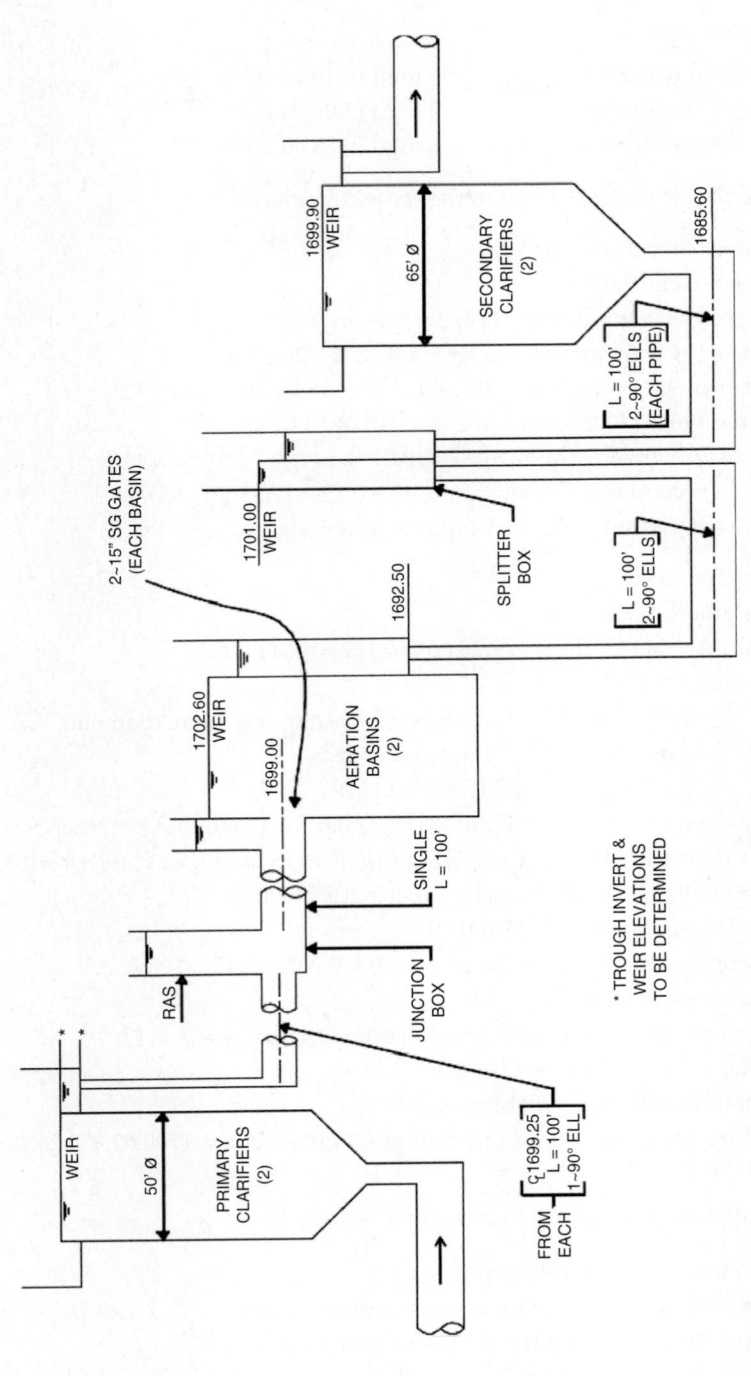

Figure 5.13 Profile—design example—partial hydraulic profile.

6.2.3 Calculations

Design flowrates:

Peak wastewater flow	4 mgd (6.19 cu ft/s)
RAS rate (maximum)	2 mgd (3.09 cu ft/s)
Peak wastewater flow + RAS	6 mgd (9.28 cu ft/s)

Water surface elevation (WS El) in the secondary clarifier:

Weir crest elevation = 1699.90
Number of clarifiers = 2
Flow per clarifier = 6.19 cu ft/s/2 = 3.09 cu ft/s
Perimeter of weir = $\pi \times$ diameter = $\pi \times 65$ ft = 204.2 ft
Number of 90° v-notches = 204.2/0.75 ft center to center = 272
Flow per weir = (3.09 cu ft/s)/272 = 0.0114 cu ft/s
Head over the weir = $(Q/2.5)^{1/2.5}$ = $(0.0114 \text{ cu ft/s}/2.5)^{0.4}$ = 0.12 ft
WS El in secondary clarifier = 1699.90 + 0.12 = 1700.02

WS El in discharge side of mixed-liquor splitter box:

Select pipe size:
Number of clarifiers = 2
Flowrate per clarifier = (9.28 cu ft/s)/2 = 4.64 cu ft/s
Design for 2.5 ft/s
$A = Q/V$ = (4.64 cu ft/s)/2.5 ft/s = 1.85 sq ft, use 18-in. diameter, A = 1.76 sq ft
$V = Q/A$ = (4.64 cu ft/s)/1.76 sq ft = 2.63 ft/s
$V^2/2g$ = (2.63 ft/s)^2/(2 × 32.2 ft/s^2) = 0.11 ft
Manning equation. $Q = (1.49/n) AR^{0.67}S^{0.5}$; let $K = (1.49/n) AR^{0.67}$, then $S = (Q/K)^2$
R = diameter/4 for pipes flowing full; R = (18 in. × 1 ft/12 in.)/4 = 0.375 ft
K = (1.49/0.015) × (1.76 sq ft) × (0.375 ft)$^{0.67}$ = 90.96
S = [(4.64)/90.96]^2 = 0.0026 ft/ft
Pipe length = 100 ft; friction loss = 100 ft × 0.0026 ft/ft = 0.26 ft
Minor losses, $H = KV^2/2g$
Entrance loss, K = 0.5; 90° bend loss K = 0.4; exit loss K = 1.0
Total K = 0.5 + 2(0.4) + 1.0 = 2.3
Minor losses = 2.3 × 0.11 ft = 0.25 ft
WS El in discharge side of mixed-liquor splitter box = 1700.02 + 0.26 ft + 0.25 ft = 1700.52

WS El upstream of mixed-liquor splitter box weirs:

Design total flowrate = 9.28 cu ft/s
Number of weirs = 2; one for each secondary clarifier
Flow per weir = (9.28 cu ft/s)/2 = 4.64 cu ft/s
Weir length = 6 ft; rectangular weir plate

Weir coefficient = 3.33

Head over the weir, $H = (Q/C \times L)^{0.67} = [(4.64 \text{ cu ft/s})/(3.33 \times 6)]^{0.67} = 0.38 \text{ ft}$

Weir crest El = 1701.00

WS El upstream of mixed-liquor splitter box weirs = 1701.00 + 0.38 ft = 1701.38

WS El in aeration basin mixed-liquor effluent channel:

Design total flowrate = 9.28 cu ft/s

Select pipe size:

Design for 2.5 to 3 ft/s

$A = Q/V = (9.28 \text{ cu ft/s})/2.5 \text{ ft/s} = 3.71 \text{ sq ft}$, diameter = 2.2 ft

$A = Q/V = (9.28 \text{ cu ft/s})/3 \text{ ft/s} = 3.09 \text{ sq ft}$, diameter = 2.0 ft

Use 24-in. diameter, $A = 3.14$ sq ft

$V = Q/A = (9.28 \text{ cu ft/s})/3.14 \text{ sq ft} = 2.96 \text{ ft/s}$

$V^2/2g = (2.96 \text{ ft/s})^2/(2 \times 32.2 \text{ ft/s}^2) = 0.14 \text{ ft}$

Manning equation. $Q = (1.49/n) AR^{0.67}S^{0.5}$; let $K = (1.49/n) AR^{0.67}$, then $S = (Q/K)^2$

$R = \text{diameter}/4$ for pipes flowing full; $R = (24 \text{ in.} \times 1 \text{ ft}/12 \text{ in.})/4 = 0.50 \text{ ft}$

$K = (1.49/0.015) \times (3.14 \text{ sq ft}) \times (0.50 \text{ ft})^{0.67} = 195.92$

$S = [(9.28)/195.92]^2 = 0.0022 \text{ ft/ft}$

Pipe length = 100 ft; friction loss = 100 ft × 0.0022 ft/ft = 0.22 ft

Minor losses, $H = KV^2/2g$

Entrance loss, $K = 0.5$; 90° bend loss $K = 0.4$; exit loss $K = 1.0$

Total $K = 0.5 + 2(0.4) + 1.0 = 2.3$

Minor losses = 2.3 × 0.14 ft = 0.32 ft

WS El in aeration basin mixed-liquor effluent channel = 1701.38 + 0.22 ft + 0.32 ft = 1701.92

WS El in aeration basin:

Design total flowrate = 9.28 cu ft/s

Number of weirs = 2; one for each aeration basin

Flow per weir = (9.28 cu ft/s)/2 = 4.64 cu ft/s

Weir length = 6 ft; rectangular weir plate

Weir coefficient = 3.33

Head over the weir, $H = (Q/C \times L)^{0.67}$; $= [(4.64 \text{ cu ft/s})/(3.33 \times 6)]^{0.67} = 0.38 \text{ ft}$

Weir crest El = 1702.60

WS El in aeration basin = 1702.60 + 0.38 ft = 1702.98

WS El in aeration basin influent channel:

Design total flowrate = 9.28 cu ft/s

Number of gates = 4; 2 for each aeration basin

Flow per gate = (9.28 cu ft/s)/4 = 2.32 cu ft/s

Gate width = 1.25 ft

Gate height = 1.25 ft

Gate area, each = (1.25 ft) × (1.25 ft) = 1.56 sq ft

Gate orifice coefficient = 0.60

Gate headloss, $H = (1/2g) \times (Q/C \times A)^2 = [1/(2 \times 32.2)] \times [2.32/(0.6 \times 1.56)]^2 = 0.10$ ft

Check velocity through the gate $V = Q/A = 2.32$ cu ft/s/1.56 sq ft = 1.5 ft/s OK

Headloss = 0.10 ft should give good flow distribution, as headloss in the influent channel itself likely will be very small. Note that a weir or cut-throat flume could be used in lieu of the gate.

WS El in aeration basin influent channel = 1702.98 + 0.10 = 1703.08

WS El in junction box:

Design total flowrate = 9.28 cu ft/s; includes RAS at this point

Use 24-in. diameter pipe, as flow is same as from aeration basin

$V = Q/A = (9.28$ cu ft/s)/3.14 sq ft = 2.96 ft/s

$V^2/2g = (2.96$ ft/s)$^2/(2 \times 32.2$ ft/s$^2) = 0.14$ ft

Manning equation. $Q = (1.49/n)\ AR^{0.67}S^{0.5}$; let $K = (1.49/n)\ AR^{0.67}$, then $S = (Q/K)^2$

   R = diameter/4 for pipes flowing full; $R = (24$ in. × 1 ft/12 in.)/4 = 0.50 ft

   $K = (1.49/0.015) \times (3.14$ sq ft) × (0.50 ft)$^{0.67} = 195.92$

   $S = [(9.28)/195.92]^2 = 0.0022$ ft/ft

Pipe length = 100 ft; friction loss = 100 ft × 0.0022 ft/ft = 0.22 ft

Minor losses, $H = KV^2/2g$

   Entrance loss, $K = 0.5$; exit loss $K = 1.0$

   Total $K = 0.5 + 1.0 = 1.5$

   Minor losses = 1.5 × 0.14 ft = 0.20 ft

WS El in junction box = 1703.08 + 0.22 + 0.20 ft = 1703.50

WS El at primary clarifier effluent channel outlet:

Design total flowrate = 6.19 cu ft/s

Number of primary clarifiers = 2

Flow to each primary clarifier = (6.19 cu ft/s)/2 = 3.09 cu ft/s

Select pipe size:

   Design for 3 ft/s

   $A = Q/V = (3.09$ cu ft/s)/3 ft/s = 1.03 sq ft, diameter = 1.1 ft

   Use 12-in. diameter, $A = 0.79$ sq ft

$V = Q/A = (3.09$ cu ft/s)/0.79 sq ft = 3.94 ft/s

$V^2/2g = (3.94$ ft/s)$^2/(2 \times 32.2$ ft/s$^2) = 0.24$ ft

Manning equation. $Q = (1.49/n)\ AR^{0.67}S^{0.5}$; let $K = (1.49/n)\ AR^{0.67}$, then $S = (Q/K)^2$

   R = diameter/4 for pipes flowing full; $R = (12$ in. × 1 ft/12 in.)/4 = 0.25 ft

   $K = (1.49/0.015) \times (0.79$ sq ft) × (0.25 ft)$^{0.67} = 30.85$

   $S = [(9.28)/30.85]^2 = 0.0101$ ft/ft

Pipe length = 100 ft; friction loss = 100 ft × 0.0101 ft/ft = 1.01 ft

Minor losses, $H = KV^2/2g$

Entrance loss, $K = 0.5$; 90° bend loss $K = 0.4$; exit loss $K = 1.0$

For a junction box, assume total energy loss on entrance—a conservative assumption. As an alternative, the confluence pressure and momentum equation could have been used (eq 5.19). Also, a slightly larger pipe diameter could have been used to reduce the headloss.

$$\text{Total } K = 0.5 + 1(0.4) + 1.0 = 1.9$$

$$\text{Minor losses} = 1.9 \times 0.24 \text{ ft} = 0.46 \text{ ft}$$

WS El at primary clarifier effluent channel outlet = $1703.50 + 1.01$ ft + 0.46 ft = 1704.97

Maximum WS El in primary clarifier effluent channel:

Design total flowrate = 6.19 cu ft/s

Number of primary clarifiers = 2

Flow to each primary clarifier = (6.19 cu ft/s)/2 = 3.09 cu ft/s

Flow splits two ways in each effluent channel in circular clarifiers = 3.09 cu ft/s/2 = 1.55 cu ft/s

Primary clarifier channel width = 1.5 ft

Critical depth, $D_c = [Q^2/(B^2 \times g)]^{0.33} = \{[(1.55)^2/(1.5)^2]/32.2\}^{0.33} = 0.32$ ft

Minimum invert of primary clarifier effluent channel for free flow = $1704.97 + 0.32$ ft = 1705.29

Set weir trough invert elevation at 1705.40 (provides a small drop to ensure free discharge)

Maximum water depth in primary clarifier effluent channel = $1.73 \times D_c$ (because it is free flow)

Maximum water depth in primary clarifier effluent channel = 1.73×0.32 ft = 0.55 ft

Maximum WS El in primary clarifier effluent channel = $1705.40 + 0.55 = 1705.95$

Set primary clarifier effluent weir crest (notch elevation) at $1705.85 + 0.25 = 1706.20$ (allows for 0.25-ft drop)

WS El in primary clarifiers:

Weir crest elevation = 1706.20

Total design flow = 6.19 cu ft/s

Number of clarifiers = 2

Flow per clarifier = 6.19 cu ft/s/2 = 3.09 cu ft/s

Perimeter of weir = $\pi \times$ diameter = $\pi \times 50$ ft = 157.08 ft

Number of 90° v-notches = 157.08/0.75 ft center to center = 209

Flow per weir = (3.09 cu ft/s)/209 = 0.0148 cu ft/s

Head over the weir, $H = (Q/2.5)^{1/2.5} = (0.0148 \text{ cu ft/s}/2.5)^{0.4} = 0.13$ ft

WS El in primary clarifier = $1706.20 + 0.13 = 1706.33$

See Figure 5.14 for the completed hydraulic profile with calculated elevations.

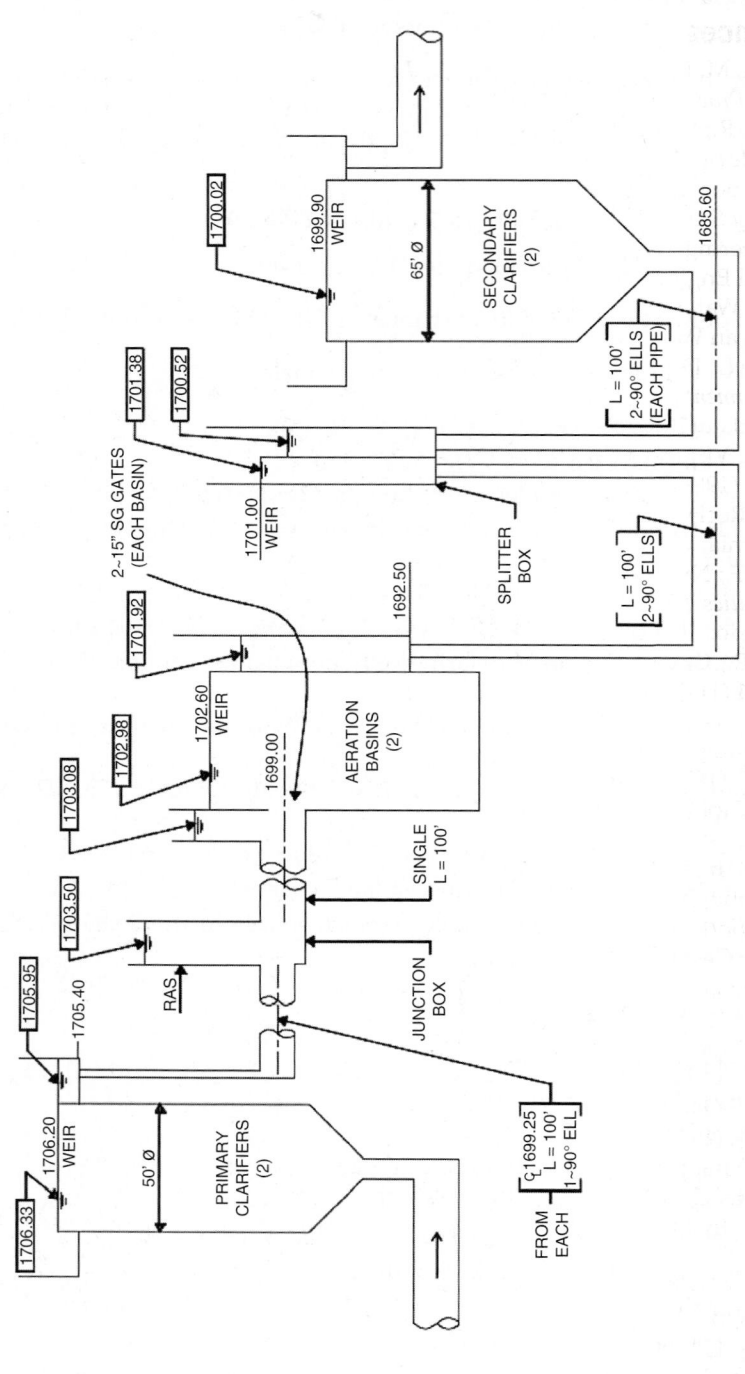

FIGURE 5.14 Profile—design example—partial hydraulic profile with calculated elevations.

7.0 References

Abu-Seida, M. M.; Quraishi, A. A. (1976) A Flow Equation for Submerged Rectangular Weirs. *Proc. Inst. Civil Eng.*, **61** (2), 685–696.

Ackers, W. R.; White, W. R.; Perkins, J. A.; Harrison, A. J. M. (1978) *Weirs and Flumes for Flow Measurement*; John Wiley & Sons: Chichester, West Sussex, United Kingdom.

American Society of Civil Engineers; Water Environment Federation (2007) *Gravity Sanitary Sewer Design and Construction*, 2nd edition; ASCE Manuals and Reports on Engineering Practice No. 60, WEF Manual of Practice No. FD-5; American Society of Civil Engineers: Reston, Virginia.

American Water Works Association (2012) *Water Treatment Plant Design*, 5th edition; American Water Works Association: Denver, Colorado.

Benefield, L. D.; Judkins Jr., J. F.; Parr, D. A. (1984) *Treatment Plant Hydraulics for Environmental Engineers*; Prentice-Hall: Englewood Cliffs, New Jersey.

Bergendahl, J. (2008) *Treatment System Hydraulics*; American Society of Civil Engineers: Reston, Virginia.

Betz, J. M. (1981) Unpublished Notes of a Head Loss Test at the Los Angeles-Glendale Water Reclamation Plant, Aug 26. Los Angeles-Glendale Water Reclamation Plant: California.

Boulos, P. F.; Nicklow, J. W. (2005) *Comprehensive Water and Wastewater Treatment Plant Hydraulics Handbook for Engineers and Operators*; MWH Soft Press: Broomfield, Colorado.

Camp, T. R.; Graber, S. D. (1968) Dispersion Conduits. *Am. Soc. Civ. Eng J. Sanit. Eng. Div.*, **94** (1), 31–39.

Chin, D. A. (2012) *Water-Resources Engineering*, 3rd ed.; Pearson-Prentice Hall: Upper Saddle River, New Jersey.

Chow, V. T. (1959) *Open-Channel Hydraulics*; McGraw-Hill: New York.

Cook, R., Tupelo, Mississippi (1991) Personal communication.

City of Los Angeles Bureau of Engineering (1968) *Hydraulic Analysis of Junctions*; City of Los Angeles Bureau of Engineering: California, http://eng.lacity.org/techdocs/sewer-ma/haj.pdf (accessed August 16, 2016).

Davis, C.; Sorensen, K. (1969) *Handbook of Applied Hydraulics*; McGraw-Hill: New York.

Fair, G. M.; Geyer, J. C.; Okun, D. A. (1968) *Water and Wastewater Engineering, Vol. 2: Water Purification and Wastewater Treatment and Disposal*; John Wiley & Sons: New York.

Grace, R. A. (1978) *Marine Outfall Systems: Planning, Design and Construction*; Prentice-Hall: Upper Saddle River, New Jersey.

Gunnerson, C. G.; French, J. A. (1996) *Wastewater Management for Coastal Cities: The Ocean Disposal Option*; Springer-Verlag: New York.

Hicks, T. A. (1971) *Pump Application Engineering*; McGraw-Hill: New York.

Hydraulic Institute (1979) *Hydraulic Institute Engineering Data Book*; The Hydraulic Institute: Cleveland, Ohio.

Hydraulic Institute (1983) *Hydraulic Institute Standards for Centrifugal, Rotary, and Reciprocating Pumps*, 14th ed.; Hydraulic Institute: Parsippany, New York.

Jones, G. M. (Ed.) (2008) *Pumping Station Design*, 3rd ed.; Butterworth-Heinemann: Burlington, Massachusetts.

Karassik, I. J.; Carter, R. (1960) *Centrifugal Pumps; Selection, Operation, and Maintenance*; McGraw-Hill: New York.

Karassik, I. J.; Krutzsch, W. C.; Messina, J. P. (1986) *Pump Handbook*, 2nd ed.; McGraw-Hill: New York.

King, H. W.; Brater, E. F. (1963) *Handbook of Hydraulics*; McGraw-Hill: New York.

Mays, L.W. (1999) *Hydraulic Design Handbook*; McGraw-Hill: New York.

Metcalf & Eddy, Inc./AECOM (2013) *Wastewater Engineering: Treatment and Resource Recovery*, 5th ed., Tchobanoglous, G., Stensel, H. D., Tsuchihashi, R., Burton, F. (Eds.); McGraw-Hill: New York.

Montgomery, J. M. (1985) *Water Treatment Principles and Design*; Wiley-Interscience: New York.

National Water Research Institute (2012) *Ultraviolet Disinfection Guidelines for Drinking Water and Water Reuse*, 3rd ed.; National Water Research Institute: Fountain Valley, California.

Reichenberger, J. C. (1984) Unpublished Notes for the Design of Three Valleys Municipal Water District, Miramar Treatment Plant, Claremont, California.

Reynolds, T. D.; Richards, P. (1995) *Unit Operations and Processes in Environmental Engineering*, 2nd ed.; PWS Publishing Company: Boston, Massachusetts.

Sen, D., Aquaregen, Mountain View, California (2008) Personal communication.

Street, R. L.; Waters, G. Z.; Vennard, J. K. (1996) *Elementary Fluid Mechanics*, 7th ed.; John Wiley & Sons: New York.

Thomas, H. A. (1940) Discussion of T. R. Camp's Paper, Lateral Spillway Channels. *Trans. Am. Soc. Civil Eng.*, **105**, 627.

Townsend, D. W. (1935) Loss of Head in Activated Sludge Aeration Channels. Transactions. *Am. Soc. Civ. Eng.*, **100**, 518.

U.S. Department of the Interior Bureau of Reclamation (2001) *Water Measurement Manual*, Chapter 8, Superintendent of Documents; U.S. Government Printing Office: Washington, D.C. (accessed August 16, 2016).

Villemonte, J. R. (1947) Submerged Weir Discharge Studies. *Eng. News-Rec.*, Dec. 866–869.

Walker, R. (1972) *Pump Selection: A Consulting Engineer's Manual*; Ann Arbor Science Publishers Inc.: Ann Arbor, Michigan.

Water Environment Federation (1993) *Design of Wastewater and Stormwater Pumping Stations, Manual of Practice No. FD-4*; Water Environment Federation: Alexandria, Virginia.

8.0 Suggested Readings

Miller, D. S. (1990) *Internal Flow Systems*, 2nd ed.; Gulf Publishing Company: Houston, Texas.

Qasim, S. R. (1999) *Wastewater Treatment Plants: Planning, Design, and Operation*, 2nd ed.; CRC Press: Boca Raton, Florida.

Vesilind, P. A. (2003) *Wastewater Treatment Plant Design*; WEF and IWA Publishing: London, United Kingdom.

Water Pollution Control Federation (1989) Technology and Design Deficiencies at Publicly Owned Treatment Works. *Water Environ. Technol.*, **1** (4), 515.

CHAPTER 6

Odor and Air Emissions Management

Charles M. McGinley, P.E., and Michael A. McGinley, P.E.

1.0 Design Basis for Managing Air Emissions

This chapter is based substantially on the information contained in the Water Environment Federation®'s Manual of Practice 25, *Control of Odors and Emissions from Wastewater Treatment Plants* (WEF, 2004). For further information, this should be the first point of reference. This chapter focuses on the characterization, assessment, management, capture, and treatment of odors and air emissions from the various processes at a water resource recovery facility (WRRF). The design engineer needs to consider methods of managing emissions from the various wastewater process units. These emission-management strategies are described in chapters specific to the individual process unit.

Before an emissions-management strategy can be planned and implemented, the design engineer must define each processes' potential and actual emissions by defining the root causes of the emissions, measuring air flowrate (AFR) of air or exhaust-gas stream, the loading rate for the pollutant(s) or odorant(s) of concern, and the performance criteria required.

1.1 Air Flowrate

A component of odor management is emission-control technology, which needs to be sized to treat a specific AFR or gas stream. The primary driver in determining the AFR is the ventilation requirements for the process unit or area from which the air is being collected. Ventilation needs are governed by many factors, including worker health, safety and comfort, fire and explosion prevention, and corrosion protection. To meet these requirements, the designer needs a thorough understanding of the process unit or area to be ventilated, including the temperature, pressure, moisture content, and gas composition of the airflow stream. To treat the exhaust gas from a boiler, engine, dryer, or thermal treatment system, the designer may require a mass-energy balance of the system to obtain the necessary airflow information.

1.2 Pollutant and Odorant Loading

Air- and odor-control systems are designed to treat a particular contaminant or group of contaminants. Knowing the maximum pollutant loading affects system sizing. The average pollutant loading also affects the life-cycle costs. The presence of compounds in the gas stream, other than the target pollutant, can enhance or hinder the unit's control efficiency or increase operations costs. For example, carbon dioxide (CO_2) increases the consumption of caustic chemical use in a packed-bed, wet-scrubber system.

Odor-control systems requiring effective treatment of more than one odorant may necessitate the use of multi-staged systems. Thus, it is necessary to understand the complexity of odorous air streams from both an olfactometric-character and a chemical-composition perspective.

1.3 Emission-Management Objective

The emission-management objectives for criteria pollutants are established through the air-quality-permitting process. In some cases, a dispersion-modeling analysis is required to demonstrate maintenance of ambient-air-quality standards.

Odor-control-technology performance criteria are established to minimize odor complaints and nuisance conditions. Community input may be needed to establish ambient-odor thresholds that serve as a basis for developing a management strategy for controlling odors.

1.4 Odor Parameters

A person's sense of smell—the olfactory sense—is the ability to detect the presence of odorant chemicals in the ambient air. A person may be able to detect odorants at extremely low concentrations, serving as an early warning or simply a marker for the presence of air emissions from a WRRF. The sense of smell has the ability to trigger emotions and memories. Odorant concentrations at detection can be an annoyance or cause a physiological response, which may lead to a complaint.

When wastewater treatment air emissions affect air quality and cause citizen complaints, investigation of those emissions may require specific odorants be measured and the odorous air be evaluated using standardized methods. Odor can be quantified and qualified using objective, scientific methods. Odor terminology is linked to standard methods. Four objective parameters of odor are concentration, intensity, persistence, and character descriptors. These are discussed in the following sections.

1.4.1 Odor Concentration

Odor concentrations, as detection and recognition thresholds, are determined using an instrument called an olfactometer. The laboratory olfactometer simulates the dilution of odor in the ambient air. Odor concentration, as detection threshold, is an estimate of the number of dilutions needed to make the actual odor emission non-detectable. Recognition threshold represents the number of dilutions needed to make the odor sample faintly recognizable. A large value for odor concentration represents a strong odor. A small value for odor concentration represents a weak odor. The standard methods used in the USA for evaluating air samples for odor concentrations are ASTM E679 and EN 13725.

1.4.2 Odor Intensity

Perceived odor intensity is the relative strength of the odor above the recognition threshold (suprathreshold). The Standard Practice for Suprathreshold Odor Intensity Measurement (ASTM E544) presents two methods for referencing intensity of ambient odors—the dynamic-scale method (procedure A) and the static-scale method (procedure B) (ASTM, 1999).

The odor intensity evaluation result are expressed in parts per million on a volumetric or molar basis (mol/1 000 000 mol [ppm_v]) for a standard reference compound. The common reference compound in air is *n*-butanol (1 ppm_v *n*-butanol equals 3.03 mg/m³ *n*-butanol). A larger value of *n*-butanol means a stronger odor. A small value of *n*-butanol means a weaker odor intensity.

1.4.3 Odor Persistence

Odor persistence is a term used to describe the rate at which an odor's perceived intensity decreases as the odor is diluted (i.e., in the atmosphere downwind from the odor source). Figure 6.1 illustrates how odor intensity decreases as the odor is diluted. Odor intensities decrease with dilution at different rates for different odors. Odor intensity is related to the odor concentration by the power law (Steven's law). Odor persistence derived from the dose-response relationship of odor concentration (dose) and odor intensity (response).

1.4.4 Odor Character and Odor Descriptors

Odor character is a nominal (categorical) scale of odor measurement. Odors can be characterized using referencing vocabulary for taste, sensation, and odor descriptors.

Numerous standard odor descriptor lists are available to use as a referencing vocabulary by odor assessors (panelists). In 1986, the International Association on Water Pollution Research and Control proposed the eight major odor descriptor categories: vegetable, fruity, floral, medicinal, chemical, fishy, offensive, and earthy (IAWPRC,

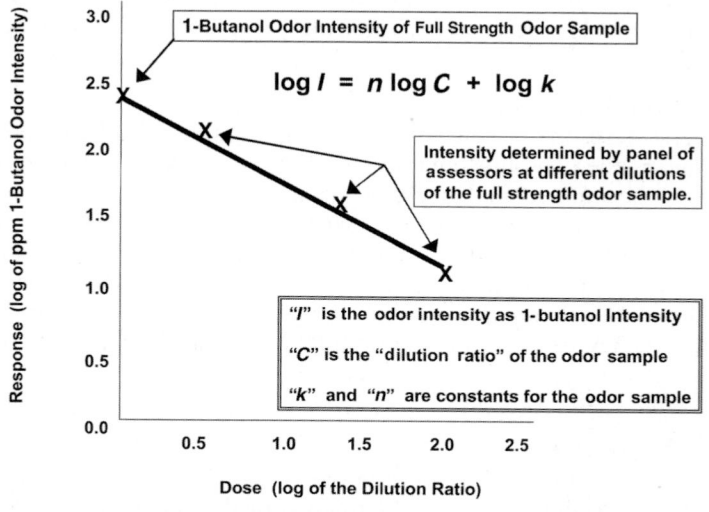

FIGURE 6.1 Relationship between odor dilution and intensity.

1986). Each of the eight major categories can have specific descriptors, which can be presented in training using exemplars. For example, the major category "vegetable" consists of a vocabulary of words that are illustrated with real-life items (exemplars), such as celery, cucumber, garlic, onion, and tomato.

These odor parameters are objective, because they are measured using scientific techniques, reference scales, and standard categories that do not depend on personal opinion.

Additionally, subjective odor parameters are:

- Hedonic tone—pleasantness versus unpleasantness;
- Annoyance—interference with comfortable enjoyment of life and property;
- Objectionable—causes avoidance or physiological effects in a person; and
- Strength—word scales, such as "faint" to "strong."

These odor parameters are subjective, because they are reported by individuals relying on their own interpretation of word scales and on personal opinions, feelings, beliefs, memories, experiences, and prejudices.

2.0   Odor Regulation and Community Effects

Air-quality regulations include established national ambient-air-quality standards, which clearly define the pollutant, primary or secondary concentration limit, corresponding averaging period, and basis for determining compliance. Emission limits for combustion sources are established through the permit-review process.

When setting performance limits for odor-control systems, preexisting ambient-air limits or performance standards are not established uniformly. Efforts to come up with a quantitative odor standard typically start with discussions of which odor levels are detectable to the affected population. Detectability or threshold refers to the minimum concentration of an odorant that produces an olfactory response or sensation. This threshold typically is determined by an odor panel consisting of a specified number of people, and the numerical result typically is expressed as occurring when approximately 50% of the panel correctly detects the odor (i.e., the geometric mean of the panel responses).

2.1   Odor Regulations and Policies

Wastewater has the potential to cause odor-nuisance complaints in the surrounding community. These odor complaints may arise from a community because of unintended emissions from sewer systems, WRRFs, or solids-processing and disposal operations.

Which levels of odors create nuisance conditions? If the goal is to avoid an odor-nuisance situation, efforts typically are made to define what constitutes an odor annoyance. The types of human responses evaluated depend on the particular sensory property measured. These properties include odor intensity, detectability, character, and hedonic tone (pleasantness/unpleasantness). The combined effect of these properties is related to the annoyance that may be caused by an odor.

2.1.1 State and Local Responsibility

The U.S. Environmental Protection Agency (Washington, D.C.) (U.S. EPA) determined that, because odors are not caused by a single pollutant, it is difficult to associate any specific health or welfare effect to a given odor concentration. Accordingly, the U.S. EPA decided to leave the establishment of odor regulations to state and local governments, particularly because the U.S. EPA is responsible for regulating public welfare and not public nuisance. A U.S. EPA report (Wahl, 1980) confirms this approach by stating that federal regulatory involvement in odor control does not seem warranted and that local and state odor-control procedures generally seem to be adequate and are probably more cost-effective than a uniform national regulatory program under the Clean Air Act (1990) (CAA).

In 1995, a survey of odor-control regulations (Leonardos, 1997) was conducted by sending questionnaires to all 50 state air-pollution-control agencies. The findings of the survey are summarized in Table 6.1. Subsequent surveys have reported no significant changes in states regulating odors; however, cities have occasionally added odor ordinances to their jurisdictions. Pope (2000) reported that as many as 40 of the 50 states apply a nuisance law. Odor nuisance is based in common law and generally states that odor is an air pollutant and is a nuisance if it "unreasonably interferes with the enjoyment of life or property." Nuisance regulations may include a statement relative to the frequency of odor effects, specifying the number of complaints received from independent households of a defined time period. Some states and local regulatory authorities have established ambient-air/odor standards or complaint-based regulations. Some of these alternative approaches to regulating odor are summarized below.

2.1.1.1 Olfactometry-Based Regulatory Approaches

Odor emissions causing nuisance complaints often are complex mixtures of numerous odorous compounds. Except for hydrogen sulfide, analytical monitoring of individual chemical compounds present in odors is often impractical. As a result, odor-sensory methods are commonly used to measure such odors rather than instrumental methods. These odor-sensory methods depend on the olfactory response of individuals who serve on panels.

There is increasing interest in basing regulatory standards on the recognition or annoyance threshold and/or dose-response studies rather than solely the detection threshold. The recognition threshold differs by compound, but is believed to be approximately three to five times the detection thresholds for most compounds. Examples of dilution-to-threshold (D/T) limits used are presented in Table 6.2. Some factors to be considered when comparing different D/T numeric odor standards are summarized in Table 6.3.

Odor nuisance/malodor/air-pollution-type regulations	25 states
No specific odor regulation	14 states
Field olfactometer or D/T	6 states
Compound-specific limits (not including total reduced sulfur)	2 states
No response to survey	3 states

TABLE 6.1 Summary of Status of Odor Regulations in the United States as of 1995 (Adapted from Leonardos, 1997)

Location	Off-Site Standard or Guideline	Averaging Times
New South Wales (Australia)	2 OU/m^3 (urban point source)	1 second (99.5% compliance)
	7 OU/m^3 (rural point source)	1 second (99.5% compliance)
	15 OU/m^3 (area source)	1 hour (99.5% compliance)
Queensland (Australia)	10 OU/m^3	1 hour (99.5% compliance)
Western Australia	2 OU/m^3	1 hour (99.9% compliance)
Tasmania (Australia)	7 OU/m^3 (poultry guideline)	1 hour (99.9% compliance)
	1 OU/m^3	3 minutes (99.9% compliance)
Ontario (Canada)	1 OU/m^3	10 minutes
Denmark	0.6 to 1.2 OU/m^3	1 minute
Siu Ho Wan WRRF (Hong Kong)	5 OU	5 seconds
The Netherlands	1 OU/m^3 (new sources)	99.5% compliance
Derby WRRF (UK)	5 OU/m^3	98% compliance
Newbiggin-by-the-Sea (UK)	5 OU/m^3	98% compliance
New Zealand	2 OU/m^3	1 hour (99.5% compliance)
Allegheny County WRRF, Pennsylvania (U.S.)	4 D/T (design goal)	2 minutes
BAAQDa (U.S.)	5 D/T	Applied after at least 10 complaints within a 90-day period
Colorado (U.S.)	7 D/T (scentometer)	
Connecticut (U.S.)	7 D/T	
New Jersey (U.S.)	5 D/T[b]	5 minutes or less
North Dakota (U.S.)	2 D/T (scentometer)	
Oregon (U.S.)	1 to 2 D/T	15 minutes
Oakland, California (U.S.)	50 D/T	3 minutes
San Diego WRRF, California (U.S.)	5 D/T	5 minutes (99.5% compliance)
Seattle WRRF, Washington (U.S.)	5 D/T	5 minutes

[a]Bay Area Air Quality District (San Francisco area, California).

[b]For biosolids/sludge-handling and WRRFs.

TABLE 6.2   Examples of D/T Limits Used (Reproduced from T. D. Mahin 2001 Comparison of different approaches used to regulate odours around the world *Water Science & Technology* **44**(9) 87–102, with permission from the copyright holders, IWA Publishing)

2.1.1.2   Approaches Based on Individual Odorants   Although hydrogen sulfide is considered the most prevalent odorous compound present in wastewater, it should not be presumed, in every case, that an odor problem is caused exclusively by hydrogen sulfide. Wastewater odors typically are sulfur or nitrogen compounds, organic acids, aldehydes, or ketones (Gostelow et al., 2001). The odors most associated with WRRFs are hydrogen sulfide and reduced-sulfur organic compounds (mercaptans, dimethyl disulfide, and dimethyl sulfide).

What model is specified for odor-dispersion analysis?	AERMOD, ASMS version 3, CALPUFF, SCREEN3, etc.
What averaging time is specified for the odor-dispersion model?	60 minutes, 15 minutes, 10 minutes, 5 minutes, 3 minutes, etc.
What laboratory AFR is specified as part of olfactometry sample analysis?	3 L/m, 20 L/m, etc.
Does the standard have to be met all of the time as predicted by the model?	100%, 99.50%, 99%, 98%, etc.
What method is used for odor sample collection?	Flux chamber, wind tunnel, etc.

TABLE 6.3   Factors to Consider When Comparing Different D/T Odor Standards

Table 6.4 lists some of the odorous compounds found in wastewater and their odor detection and recognition thresholds. Most of the odorous substances are gaseous under normal atmospheric conditions or at least have a significant volatility. The volatility is shown in the table as parts per million (ppm) and is equal to the vapor pressure. The molecular weights of these substances typically range from 30 to 150. Typically, the lower the molecular weight of a compound, the higher its vapor pressure and potential for emission to the atmosphere. Substances of high molecular weight generally are less volatile and, thus, typically have less effect for causing odor complaints.

Data in the literature on odor threshold concentrations for any particular compound may differ significantly in many cases, by tenfold or more. This is particularly true of earlier values because of inadequate equipment or methods, small panels, or large stepwise changes in odorant concentration. The existence of many different odorous compounds associated with WRRFs creates problems when using individual compounds as the basis for assessing odors. In addition, detection and odor-annoyance thresholds cited in literature and regulations vary widely for compounds such as hydrogen sulfide. Examples of ambient odorous compounds standard approaches are summarized in Table 6.5.

2.1.1.3   Odor-Intensity Approaches   Odor intensity is a measure of the strength of the odor sensation and is related to the odorant concentration, which is a different category of measurement. The intensity of an odor is perceived directly without any knowledge of the odorant concentration or degree of air dilution of the odorous sample needed to eliminate the odor.

A category scale consists of a series of numbers that refers to verbal descriptors. For example, the series numbers 0, 1, 2, 3, 4, and 5 could correspond to no odor, barely perceptible, slight, moderate, strong, and very strong. Category scales are simple to use but have certain issues that must be considered. First, the numbers are not necessarily proportional to the perceived intensities of the odors. Second, people interpret a given category scale differently; this necessitates specialized training of the panelists. Third, odor sensitivity varies considerably from person to person. Despite these issues, several municipalities use seven- or nine-point intensity scales to monitor downwind odor emissions from WRRFs.

The *n*-butanol odor-intensity reference scale E544-88 from the American Society for Testing and Materials (Conshohocken, Pennsylvania) (ASTM, 1988) can be used

Compound Name	Formula	Molecular Weight	Volatility at 25°C, ppm (v/v)	Detection Threshold, ppm (v/v)	Recognition Threshold, ppm (v/v)	Odor Description
Acetaldehyde	CH_3CHO	44	Gas	0.067	0.21	Pungent, fruity
Allyl mercaptan	CH_2:$CHCH_2SH$	74		0.0001	0.0015	Disagreeable, garlic
Ammonia	NH_3	17	Gas	17	37	Pungent, irritating
Amyl mercaptan	$CH_3(CH_2)_4SH$	104		0.0003	—	Unpleasant, putrid
Benzyl mercaptan	$C_6H_5CH_2SH$	124		0.0002	0.0026	Unpleasant, strong
n-Butyl amine	$CH_3(CH_2)NH_2$	73	93 000	0.080	1.8	Sour, ammonia
Chlorine	Cl_2	71	Gas	0.080	0.31	Pungent, suffocating
Dibutyl amine	$(C_4H_9)_2NH$	129	8000	0.016	—	Fishy
Diisopropyl amine	$(C_3H_7)_2NH$	101		0.13	0.38	Fishy
Dimethyl amine	$(CH_3)_2NH$	45	Gas	0.34	—	Putrid, fishy
Dimethyl sulfide	$(CH_3)_2S$	62	830 000	0.001	0.001	Decayed cabbage
Diphenyl sulfide	$(C_6H_5)_2S$	186	100	0.0001	0.0021	Unpleasant
Ethyl amine	$C_2H_5NH_2$	45	Gas	0.27	1.7	Ammonia-like
Ethyl mercaptan	C_2H_5SH	62	710 000	0.0003	0.001	Decayed cabbage
Hydrogen sulfide	H_2S	34	Gas	0.0005	0.0047	Rotten eggs
Indole	$C_6H_4(CH)_2NH$	117	360	0.0001	—	Fecal, nauseating
Methyl amine	CH_3NH_2	31	Gas	4.7	—	Putrid, fishy
Methyl mercaptan	CH_3SH	48	Gas	0.0005	0.0010	Rotten cabbage

TABLE 6.4 Odorous Compounds in Wastewater* (American Industrial Hygiene Association, 1989; Sullivan, 1969)

Compound Name	Formula	Molecular Weight	Volatility at 25°C, ppm (v/v)	Detection Threshold, ppm (v/v)	Recognition Threshold, ppm (v/v)	Odor Description
Ozone	O_3	48	Gas	0.5	—	Pungent, irritating
Phenyl mercaptan	C_6H_5SH	110	2000	0.0003	0.0015	Putrid, garlic
Propyl mercaptan	C_3H_7SH	76	220 000	0.0005	0.020	Unpleasant
Pyridine	C_5H_5N	79	27 000	0.66	0.74	Pungent, irritating
Skatole	C_9H_9N	131	200	0.001	0.050	Fecal, nauseating
Sulfur dioxide	SO_2	64	Gas	2.7	4.4	Pungent, irritating
Thiocresol	$CH_3C_6H_4SH$	124		0.0001	—	Skunky, irritating
Trimethyl amine	$(CH_3)_3N$	59	Gas	0.0004	—	Pungent, fishy

*Different sources report a range of values for odor-detection or recognition thresholds, particularly for compounds such as hydrogen sulfide or ammonia. The range of values used for hydrogen sulfide generally is 0.47 to 9 ppb. Odor thresholds mentioned in this manual may vary and are not intended to be definitive because of differences in odor-testing methodologies when measuring odor thresholds. The odor thresholds are included to give the reader a sense of what odor levels are potentially an issue, relative to the particular chapter of this manual in which they are included.

TABLE 6.4 Odorous Compounds in Wastewater* (American Industrial Hygiene Association, 1989; Sullivan, 1969) (*Continued*)

Location	Compound	Standard or Guideline
Victoria (Australia)	Hydrogen sulfide Methyl mercaptan Ammonia Chlorine	0.1 ppb$_v$[a] (0.14 µg/m^3) 0.42 ppb$_v$ (0.84 µg/m^3) 830 ppb$_v$ (600 µg/m^3) 33 ppb$_v$ (100 µg/m^3)
Alberta (Canada)	Hydrogen sulfide	10 ppb$_v$ (1 hour) ambient-air-quality guideline
	Ammonia	2000 ppb$_v$ (1 hour) ambient-air-quality guideline
Quebec (Canada)	Hydrogen sulfide	10 ppb$_v$ (1 hour average)
WHO (Europe)	Hydrogen sulfide	0.13 to 1.3 ppb$_v$ (0.2 to 2 µg/m^3) 30-minute average (guideline)
Japan	Hydrogen sulfide	20 to 200 ppb$_v$ (standards depend on location)
	Dimethyl disulfide	9 to 100 ppb$_v$
	Methyl mercaptan	2 to 10 ppb$_v$
	Butyric acid	1 to 6 ppb$_v$
	Ammonia	1000 to 5000 ppb$_v$
New Zealand	Hydrogen sulfide	7 µg/m^3 (5 ppbv) 30-minute average but proposed to change to 1-hour average
California (USA)	Hydrogen sulfide	30 ppb$_v$ (1-hour average) ambient-air standard. At least one air-pollution-control district in California defines an odor nuisance as being a Jerome reading of 5 ppb H2S at the property line
Connecticut (USA)	Hydrogen sulfide	6.3 µg/m^3
	Methyl mercaptan	2.2 µg/m^3
Minnesota (USA)	Hydrogen sulfide	30 ppb$_v$ (30-minute average)[b]
		50 ppb$_v$ (30-minute average)[c]
Nebraska (USA)	Total reduced sulfur	100 ppb (30-minute average)
New Mexico (USA)	Hydrogen sulfide	10 ppb$_v$ (1-hour average) or 30 ppb$_v$ (30-minute average) or 100 ppb$_v$ (30-minute average) depending on air-quality region
New York State (USA)	Hydrogen sulfide	10 ppb$_v$ (14 µg/m^3) 1-hour average
New York City	Hydrogen sulfide	1 ppb$_v$
North Dakota (USA)	Hydrogen sulfide	50 ppb$_v$ (instantaneous, two readings 15 minutes apart)
Pennsylvania (USA)	Hydrogen sulfide	100 ppb$_v$ (1-hour average)
		5 ppb$_v$ (24-hour average)
Texas (USA)	Hydrogen sulfide	80 ppb$_v$ (30-minute average)—residential/commercial 120 ppb$_v$—industrial or vacant or range lands

[a]Parts per billion by volume.
[b]Not to be exceeded more than 2 days in a 5-day period.
[c]Not to be exceeded more than two times per year.

TABLE 6.5 Examples of Regulatory Agencies with Ambient Standards for Odor-Causing Compounds[a] (Reproduced from T. D. Mahin 2001 Comparison of different approaches used to regulate odours around the world *Water Science & Technology* **44**(9) 87–102, with permission from the copyright holders, IWA Publishing)

to measure odor intensity in ambient air. Panelists match the perceived intensity of an ambient odor by comparing it with the *n*-butanol reference scale. Panelists' judgment should be reinforced by referring to *n*-butanol standards between odor-intensity measurements of the ambient air. An example of an *n*-butanol odor-intensity scale is provided in Table 6.6.

2.1.1.4 Control-Technology Approaches to Odor Best-available control technologies (BACTs) involve approaches that require the "best" odor treatment controls for new or upgraded facilities (economics are factored in sometimes). Some regulatory agencies rely more on requiring that the best odor-control technologies be used to control odors at water resource recovery or industrial facilities of a certain size than on some of the other approaches for regulating odor.

n-Butanol Reference Scale Value	*n*-Butanol Concentration in Air (ppm)	Identification	Odor-Intensity Description
n = 1	15	Very faint	Odorant present in the air that is *barely perceptible* and may not be detected if not specifically inhaling to detect an odor
n = 2	30	Very light	Odorant present in the air that activates the sense of smell, but the characteristics *may not be distinguishable*
n = 3	60	Light	Odorant present in the air that activates the sense of smell and is *distinguishable* and definite, but not necessarily *objectionable* in short durations (recognition threshold)
n = 4	120	Light-to-moderate	Odorant present in the air that activates the sense of smell, is *distinguishable* and definite, and is at times objectionable
n = 5	250	Moderate	Odorant present in the air that easily activates the sense of smell, is *very distinct* and clearly distinguishable, and may tend to be *objectionable* and/or *irritating*
n = 6	500	Moderate-to-strong	Odorant present in the air that easily activates the sense of smell, is *very distinct* and tends to be objectionable, and is at times perceived *pungent* enough to cause a person to avoid it completely
n = 7	1000	Strong	Odorant present in the air that would be objectionable and cause a person to attempt to avoid it completely
n = 8	2000	Very strong	Odorant present in the air that is so strong it is overpowering and intolerable for any length of time

TABLE 6.6 *n*-Butanol Odor-Intensity Reference Scale (ASTM E544-88; ASTM, 1988) (Reproduced, with permission from *E544-88 Standard Practices for Referencing Suprathreshold Odor Intensity*, Copyright ASTM International, 100 Barr Harbor Drive, West Conshohocken, PA 19428)

2.1.2   Ambient-Odor Limits

One particular type of ambient-odor standard stipulates that no odor be detected in specified off-site areas or beyond the property line. This type of standard is more prevalent in municipal regulations than in state regulations. The activation of the sense of smell does not necessarily cause annoyance or unpleasantness to an individual. Ambient-air-odor limits also can be established with one standard for all types of land use, or they can have multiple levels, distinguishing between residential, commercial, industrial, and non-zoned areas.

Odor also can be measured and quantified directly in the ambient air using one of two standard practices by trained inspectors. The first method uses a standard odor-intensity referencing scale (OIRS) made up of the standard odorant, 1-butanol, to quantify odor intensity. The second method uses a field olfactometer, which dynamically dilutes the ambient air with carbon-filtered air in distinct D/Ts.

2.1.2.1   Ambient-Odor Intensity   Odor intensity of the ambient air can be measured objectively using an OIRS (ASTM E544-99), as in the Standard Practice for Suprathreshold Odor Intensity Measurement (ASTM, 1999). Odor-intensity referencing compares the odor in the ambient air to the odor intensity of a series of concentrations of a reference odorant. A common reference odorant is 1-butanol. Sec-butanol is an alternative to 1-butanol for a standard referencing odorant. The air-pollution inspector, facility operator, or community odor monitor observes the odor in the ambient air and compares it with the OIRS.

The OIRS serves as a standard practice to quantify the odor intensity of the ambient air objectively. To allow comparison of results from different data sources and to maintain a reproducible method, the equivalent butanol concentration is reported, or the number on the OIRS is reported with the scale range and starting point. Figure 6.2 presents four OIRS options.

12-Point Scale	8- and 10-Point Scale	5-Point Scale
1 < 10 >	1 < 12 >	
2 < 20 >	2 < 24 >	1 < 25 >
3 < 40 >	3 < 48 >	
4 < 80 >	4 < 96 >	2 < 75 >
5 < 160 >	5 < 194 >	3 < 225 >
6 < 320 >	6 < 388 >	
7 < 640 >	7 < 775 >	4 < 675 >
8 < 1280 >	8 <1550 >	5 < 2025 >
9 < 2560 >	9 < 3100 >	
10 < 5120 >	10 <6200 >	
11 < 10 240 >		
12 < 20 480 >		

Note: <XXX> is parts per million 1-butanol equivalent odor intensity.

FIGURE 6.2   Odor-intensity referencing scales.

2.1.2.2 Ambient-Odor Concentration   In 1958, the U.S. Public Health Service (Washington, D.C.) sponsored the development of an instrument and procedure for field olfactometry (ambient-odor-strength measurement). The first field olfactometer, called a scentometer, was manufactured by the Barnebey-Cheney Company (Columbus, Ohio) (1974).

A field olfactometer creates a series of dilutions by mixing the odorous ambient air with odor-free (carbon-filtered) air. The U.S. Public Health Service method defined the dilution factor as the D/T. The D/T is a measure of the number of dilutions needed to make the odorous ambient air non-detectable.

The method of producing D/Ts with a field olfactometer consists of mixing two volumes of carbon-filtered air with specific volumes of odorous ambient air. The method of calculating D/T for a field olfactometer is the following:

$$\text{Dilution factor} = \frac{\text{Volume of carbon filtered air}}{\text{Volume of odorous air}} = D/T \qquad (6.1)$$

The field olfactometer instrument, the D/T terminology, and the method of calculating D/T are referenced in a number of existing state and local agencies' odor regulations and permits. Therefore, a field olfactometer instrument, in the hands of trained air-pollution investigators, is a proven method for quantifying ambient-odor strength.

2.2   Criteria and Hazardous Air-Pollutant Regulations

Air-emission sources from a WRRF must be reviewed before construction of the facility or completion of planned modifications. The Clean Air Act requires that a facility undergo a two-step air-permit-review process and receive approval from the reviewing agency (U.S. EPA Regional Office or delegated state or local regulating authority) before construction can begin. The attainment status of the facility site, local meteorology or terrain features, and state regulatory requirements will affect the specific emission limits that may apply to a facility. Once the facility is completed, a performance test is conducted to demonstrate that the preconstruction limits have been met. An operating-permit application then needs to be submitted and approval issued for the facility to continue operating. Careful consideration to the air-permitting requirements for an air-emission source must be made at the early planning and design phases of a project.

2.2.1   Project Planning: Pre-Permitting

The nature of the regulatory review and emission limitations depends on the type of facility, magnitude of the potential emissions, and attainment status for the area where the facility is located. If the facility is located in an area that is not in attainment of the National Ambient Air Quality Standards (NAAQS), stringent air-pollution-control requirements and emission limits will apply (CAA, Title I, Part D, Plan Requirements for Nonattainment Areas, as amended). A facility located in an area meeting the NAAQS still may be subject to a comprehensive permit review (CAA, Title I, Part C, Prevention of Significant Deterioration of Air Quality, as amended), but the resulting emission limits may be less stringent than a facility located in a non-attainment area.

To anticipate the nature of the air-permitting review and the applicable emission limitations, air permitting should begin in the project-planning stage, when detailed information about the proposed facility still is being developed. Information that can be obtained at this planning stage includes the following:

- Attainment status of criteria pollutants for the proposed facilities locations;
- Classification of the facility as to its status as a major or minor air-emission source, considering ownership of the adjacent facilities and industrial classification; and
- Potential emission rates of the new facility or net change in emissions from an existing facility.

2.2.1.1 Attainment Status   Air-quality standards in the United States are mandated by the Clean Air Act and its amendments. The U.S. EPA Office of Air Quality Planning and Standards (Research Triangle Park, North Carolina) has set NAAQS for six principal pollutants, called criteria pollutants. The six pollutants, defined in Title 40, Part 50, of the U.S. Code of Federal Regulations (CFR), are the following:

- Carbon monoxide (CO);
- Sulfur dioxide (SO_2);
- Nitrogen dioxide (NO_2);
- Ozone (O_3);
- Various categories of particulate matter, including particulate matter less than 10 mm in size (PM-10) and particulate matter smaller than 2.5 mm (PM-2.5); and
- Lead (Pb).

The U.S. EPA has identified the following two types of standards for these pollutants:

1. Primary ambient-air-quality standards, which define levels of air quality necessary to protect public health with an adequate margin of safety; and
2. Secondary standards, which define levels needed to protect the public welfare from any known or anticipated adverse effects of a pollutant.

Such standards are subject to revision. Additional primary and secondary standards may be promulgated as the U.S. EPA deems necessary to protect public health and welfare.

Geographic areas in which the NAAQS for all criteria pollutants are met are called attainment areas; areas in which one or more standards are violated are called non-attainment areas. A non-attainment area must develop and implement a plan to meet and maintain Clean Air Act standards. When a non-attaining region again meets the standard, the area can be re-designated as a maintenance area. A maintenance area is a geographic region re-designated by the U.S. EPA from non-attainment to attainment, as a result of monitored attainment of the standard and U.S. EPA approval of a plan to maintain air-quality standards for at least a 10-year period. This determination is made on a pollutant-specific basis; for example, an area can be in non-attainment for ozone and in attainment for other criteria pollutants. Because emissions of nitrogen oxides (NO_X) and volatile organic compounds (VOCs) can lead to the formation of ozone, regions designated as being in non-attainment for ozone also have more restrictive limits for NO_X and VOCs. For example, an area may be classified as a non-attainment area for ozone, resulting in lower emission threshold limits for NO_X and VOC, but be classified as attainment for carbon monoxide, sulfur dioxide, and particulate matter.

The official listing of attainment status designations is in 40 CFR 81, Subpart C—Section 107, Attainment Status Designations. This subpart of the federal regulations lists attainment areas by state and air-quality-control region. These regulations can be accessed through the online resources. The U.S. EPA also operates a web page called the Green Book, which lists non-attainment areas for criteria pollutants (https://www3 .epa.gov/airquality/greenbook/). This website offers a variety of ways (by state, county, or pollutant) to search area attainment designations.

2.2.1.2   Facility Classification   A facility's classification is determined by three factors—common owner or operator, adjacent facilities, and same industrial classification. For a municipal facility, the operation of a WRRF may be under the direction of a public works director, who also may oversee other municipal facilities. A regional authority may have responsibilities only for the wastewater treatment operations, even if the WRRF serves people in multiple communities. An operator of a privately operated facility would be responsible only for the operations at that facility. It is possible for the solids-handling facility to be privatized, even if the rest of the WRRF was run by a municipal or regional authority. In this case, the owner/operator of the solids-processing facility would have responsibility only for emissions associated with that process.

If a regional authority has responsibility for more than one WRRF and associated solids processing, the WRRFs would be considered separate facilities, as long as they are not adjacent to each other. A public roadway passing between the two facilities is not sufficient to treat the facilities as separate operations. For example, a regional authority may have several regional WRRFs that are separated geographically. The operation of the WRRFs would be considered separate facilities. If, however, the solids-processing facility for all the regional facilities was located adjacent to one of the WRRFs, then that WRRF and the solids-processing facility would be considered one facility, as they are under common ownership and are adjacent to one another.

The Clean Air Act Amendments of 1990 also groups facilities by major industrial classification. Sanitary services are part of the U.S. Department of Labor (Washington, D.C.) (U.S. DOL) Standard Industrial Classification (SIC) Major Group 4900 (the U.S. DOL main page for accessing the SIC is http://www.osha.gov/pls/imis/sic_manual .html). The subgroup (industrial group 4950) includes collection systems and refuse systems. Thus, a municipality or regional authority that operates a WRRF and a solid waste management facility on adjacent properties must consider both operations as part of the same facility.

2.2.1.3   Potential Emissions   Potential emissions are the emissions that would occur on an annual basis if the facility was operated at its design capacity continuously. Potential emissions can be reduced if there is a physical limitation that constrains the process from operating at its design capacity on a continuous basis, or a federally enforceable operational limitation was adopted, which limits operation (e.g., the total quantity of solids to be processed on an annual basis). If the potential emissions from a new source are greater than the emission thresholds for federal permit review, as presented in Table 6.7, then the facility is a new major emission source. A facility with emissions less

Pollutant	Pollutant Attainment Status	Major Source Threshold[a]	Major Modification Threshold[b]
CO	CO attainment	100	100
	CO serious non-attainment	50	50
SO$_2$	SO$_2$ attainment or non-attainment	100	40
PM-10	PM-10 attainment	100	15
	PM-10 non-attainment	70	15
PM-2.5	PM-2.5 attainment or non-attainment	100	15
NO$_2$	NO$_2$ or ozone attainment	100	40
	Ozone attainment or marginal or moderate non-attainment or ozone transport region	100	25
	Serious ozone non-attainment	50	25
	Severe ozone non-attainment	25	25
	Extreme ozone non-attainment	10	Any increase in actual emissions
VOC	Ozone attainment	100	40
	Serious ozone non-attainment or ozone-transport region	50	25
	Severe ozone non-attainment	25	25
	Extreme ozone non-attainment	10	Any increase in actual emissions
Lead	Lead attainment or non-attainment	100	0.6

[a]40 CFR 51.165 (a) (1) (iv) Major Stationary Source.
[b]40 CFR 51.165 (a) (1) (v) Major Modification.

TABLE 6.7   Emission Threshold Levels for Major Sources and Major Modifications (Tons per Year)

than major emission source thresholds is considered a minor source and is subject to state permit-review requirements.

If the proposed action is a modification to an existing facility, then emissions from the existing facility need to be quantified to determine whether the existing source is a major or minor air-emission source and whether the proposed change will result in a net increase or decrease in air emissions. If the net change in emissions is greater than the emissions thresholds for a modification, then the proposed action is a major modification. The process of determining which emissions credits apply when calculating a net change can be rather involved and may require emission-offset credits. Consultation with the governing regulatory authority may be needed to be sure that changes in net emissions are being calculated appropriately.

In the planning phase, the details associated with the proposed action may not be fully developed. Yet, estimating potential emissions is a key to determining future regulatory review. For example, a preliminary estimate of emissions for a multiple-hearth or fluid-bed incinerator can be made using emission factors. Emission rates based on generic emission factors should not be the sole method for determining emission limitations. Source-specific emission testing and vendor performance guarantees are preferred methods for setting emission limits.

State and local regulatory agencies also have established emission thresholds that determine whether a facility is subject to regulatory review. These thresholds are lower than the major source and major modification thresholds defined in Table 6.7. Facilities should consult with the appropriate regulatory authority to determine what the emission thresholds are and what information is needed to meet the state and local permit-review requirements.

Emission factors that can be used to estimate emission rates from a proposed air-emission source or to identify applicable emission limits can be found in the U.S. EPA's Technology Transfer Network (http://www.epa.gov/ttn/chief/ap42/index.html). The Reasonably Available Control Technology/Best Available Control Technology (BACT)/Lowest Achievable Emission Rate (LAER) Clearinghouse (RBLC) (U.S. EPA) is a database that contains emission limitations from across the United States (http://cfpub.epa.gov/rblc/htm/bl02.cfm). The database can be searched for emission limitations that have been adopted for various source groups.

2.2.2   Project Implementation: Permit to Construct

Emission limits for specific source categories are established under the new source performance standards (NSPS) (CAA Sec. 111, Standards of Performance for New Stationary Sources) and National Emission Standards for Hazardous Air Pollutant (NESHAP) regulations (CAA Sec. 112). These limits are minimum requirements. The prevention of significant deterioration (PSD) provisions (CAA Sec. 165, Preconstruction Requirements) can result in more stringent limits, as a result of the BACT review process. The new source-review (CAA Sec. 173, Permit Requirements) process includes an emission-control evaluation that results in the lowest achievable emission rate (LAER).

To begin the air-permit-review process, a sufficient amount of information is needed to define the process requirements, so that emission rates can be calculated and control strategies can be evaluated. However, the permit process is an iterative process between the applicant and the reviewing authority where emission limits and performance criteria are evaluated. Thus, it is best to begin the permit process before final process design decisions have been made. In a traditional design-bid-build construction process, the 30% design point is a good time to prepare and submit the air-permit application. Enough technical information has been prepared to define the process, and it is early enough in the design process to make modifications to process equipment or air-pollution-control devices.

2.2.2.1   Permit Application Requirements   The type of review and threshold levels are based on the attainment status of the region in which the facility is located. Review requirements depend on the total facility annual potential emission rate or the annual potential emission rate for the proposed modification. A typical permit application would contain the following elements, although the format and presentation would vary by the reviewing agency:

- Permit forms—each agency has a standard set of forms that permit applicants are required to use. Many agencies have the forms available electronically as downloadable documents that can be obtained from the agency website or interactive online forms that feed directly into a database.

- Process description—a detailed process description that describes the operation of the facility in sufficient detail to support and confirm emission calculations and control strategies. Often, process flow diagrams and process equipment data sheets are provided.

- Emission estimates—the basis for each pollutant emission rate should be presented with supporting information, such as emission test results from similar units, vendor performance guarantees, or mass-balance calculations.

- Control-technology assessment—this may be a determination of the BACT or LAER, depending on the attainment status of the region.

- Emission limitations—an assessment to be sure that statutory emission limitations are achieved. Both federal and state emission limits should be identified.

- Air-quality compliance—a dispersion modeling analysis may be necessary to determine whether the NAAQS or the PSD increments are exceeded.

- Special issues—state and local agencies may request additional demonstrations to show that hazardous air pollutants (HAPs), noise, and odor are within acceptable limits.

Permit-review times will vary, depending on the complexity of the permit application. To ensure that the permit is reviewed in the shortest amount of time, it is best to meet with the regulatory agency in advance, understand what information the agency needs to complete its review, and provide a permit application that is as complete as possible. A typical permit timeline is as follows:

- Determination of permit completeness—2 to 4 weeks;
- Technical review—4 to 12 weeks;
- Response to technical comments—2 to 8 weeks;
- Public comment period—4 to 8 weeks; and
- Issuance of draft permit conditions—2 to 4 weeks.

The above timeline does not include time spent preparing the initial permit application.

2.2.2.2 Federal Regulatory Requirements The federal regulatory permit-review requirements are defined in 40 CFR 51 and 52. The U.S. EPA has delegated review authority to many state and local regulatory agencies. Thus, the state and local agencies can review the application and assess compliance with the federal, state, and local permitting requirements. For state and local agencies where the U.S. EPA has retained review authority, the regional U.S. EPA office will serve as the reviewing authority for compliance with the federal regulations. The state or local agencies still will review and comment on state-specific requirements.

2.2.2.3 Non-Attainment New Source Review New source review is conducted for facilities located in areas where pollutant concentrations are greater than the NAAQS. The degree to which the standards are not being met increases the requirements to provide emission reductions and emission offsets. Emission-control strategies must demonstrate the LAER.

2.2.2.4   Prevention of Significant Deterioration   For pollutants that are in attainment with the NAAQS, a PSD review (40 CFR 52.21) is required. This review seeks to maintain ambient-air concentrations below the NAAQS by limiting air-quality effects to incremental limits above the baseline concentrations. Emission-control strategies must demonstrate that they represent BACT considering environmental, energy, and economic effects. A demonstration of potential effects to protect wildlife areas and national parks also may be required.

2.2.2.5   New Source Performance Standards   The NSPS are a set of emission standards for designated pollutants from new, modified, or reconstructed stationary source categories. The NSPS regulations are defined in 40 CFR 60.

2.2.2.6   National Emission Standards for Hazardous Air Pollutants   The NESHAP is a set of emission standards for listed HAPs emitted from specific new and existing sources. The NESHAP regulations are defined in 40 CFR 61.

In addition to the criteria pollutants regulated by the NAAQS, there is another set of federally regulated air pollutants, known as HAPs. HAPs are a set of 188 chemicals specifically regulated by the U.S. EPA that are known or believed to cause human health effects in excess of levels specified by the agency.

A source that emits more that 9 Mg/yr (10 ton/yr) of an individual HAP or more than 23 Mg/yr (25 ton/yr) of multiple HAPs is considered a significant source of HAPs. Significant sources of HAPs may be subject to the maximum-achievable control-technology regulations (40 CFR 63).

2.2.2.7   State and Local Regulatory Requirements   Even if an air-emission source is relatively small, state preconstruction permitting requirements may apply. Although the review requirements may not be as stringent, the same emission limits may apply. Most states require a BACT analysis to determine the appropriate air-pollution-control equipment and emission limits.

Many states also have special regulations governing toxic-air-pollutant emissions. Some state rules define a control-technology assessment for toxic air emissions (T-BACT). Other states require a dispersion modeling assessment to show that toxic-air-pollutant emissions are in compliance with allowable ambient levels. State air toxics programs also may establish emission limits of some HAPs.

Additional regulations also may apply, with respect to noise and odors. Special compliance demonstrations or preconstruction assessments of potential noise and odor effects may be required. Applicable standards may be developed on a case-by-case basis and include communication with the public and involvement of other interested parties. Visual plume emissions are regulated by opacity limits, as defined in many state regulations.

2.2.3   Construct: Commence Construction

To accelerate the time it takes to bring a project from concept to operation and save money, a design-build approach has been proposed as an alternative to the traditional design-bid-build approach. Although there is nothing in the Clean Air Act that would preclude a design-build approach, some cautions are warranted. The design-build approach seeks to streamline the design and procurement process, so that the proposed facility can be constructed sooner. Construction cannot commence until the air permit has been issued. To commence construction means that no person may construct a new source or alter an existing source; certain site-preparation work may begin, but no permanent facilities may be constructed.

The restriction on construction also may extend to binding agreements or obligations that make commitments to process equipment or construction services that cannot be canceled or modified without substantial loss to the owner or operator. The air-permit process may result in more stringent emission limitations, which could affect process-equipment or control-technology selection. Thus, both the facility owner/operator and the provider of design-build services must share the risks associated with delays in receiving an air permit or facility design changes, as a result of increased emissions.

2.2.4 Facility Operation: Permit to Operate

Once the proposed facility is constructed, the process of performance testing begins. The pollutant emissions to be tested are defined in the permit to construct. Testing procedures are defined in 40 CFR 60, Appendix A or methods agreed to by the reviewing authority. A testing protocol is developed that describes how the testing will be conducted and the emissions reported. If the measured emission rates are greater than emission limits established in the permit to construct, then immediate mitigation measures are implemented. Measures may include modifications to air-pollution-control equipment or re-permitting of the facility.

2.2.4.1 Federal Title V Operating Permit Program Shortly after the facility begins operation, the operating permit program ensures continued compliance and reporting of actual emissions from the major facilities. The operating permit program (Title V of the CAA) consolidates the emission limits established for the entire facility and defines a means to monitor compliance with the limits.

Once the facility is constructed, a performance test may be required by the preconstruction permit, which seeks to demonstrate that the emission limits established in the permit have been met by the operating facility. Satisfactory completion of the emission testing and submission of an operating-permit application may be needed for continued operation of the facility.

If the facility is a major source, a Title V operating permit (40 CFR 70, State Operating Permit Programs) may be needed. The operating permit identifies all air-emission sources present at the facility, summarizes the emission limitations and special conditions that have been established by the preconstruction permit, and outlines a process by which continued compliance with emission limitations can be demonstrated.

2.2.4.2 State Operating-Permit Program The state operating-permit program applies to major emissions sources that have accepted federally enforceable operating limits to restrict annual emission to levels less than the major source thresholds or large minor emissions sources, with potential emissions that are less than the major source threshold, but greater than the state operating-permit program. The state operating-permit program is similar to the federal operating-permit program, except that compliance with the program is administered through the state.

2.2.4.3 Accidental Release Prevention Program The Chemical Accident Prevention Program requirements are defined in 40 CFR 68. Facilities that use or store more than the threshold quantities must prepare and implement a risk management plan (RMP). This requirement may apply to the disinfection process of the WRRF.

The general duty clause, like much of the RMP regulation, is performance based, and the method for compliance, for the most part, is to be determined by the source. Section 112(r) of the Clean Air Act Amendments of 1990 states that owners that use extremely hazardous substances "must adhere, at a minimum, to industry standards and practices (as well as local, state, and federal laws and regulations) in order to be in compliance with the General Duty Clause." Accordingly, all potentially hazardous substances need to be stored in a building, have separate filling areas and piping and full vessel containment, and be separated in physical distance per applicable codes and standards, to ensure that no mixing can occur if a vessel loses its entire stored chemical. In addition, the fill pipes, tanks, and loading areas must be clearly marked.

2.2.4.4   Wastewater Residuals Management (CFR, Part 503)   Performance standards for the treatment and disposal of WRRF residuals required by the Clean Water Act Amendments of 1987 are contained in 40 CFR 503, also referred to as the Part 503 rules (promulgated 58 FR 9387, Feb. 19, 1993). Subpart B contains the requirements for the placement of biosolids on land-application sites. It defines the various classes of biosolids and their suitability for land application. Subpart E defines the requirements for residuals fired in an incinerator.

2.2.4.5   Air-Quality Compliance   The compliance-assurance-monitoring (CAM) requirements are defined in 40 CFR 64. The CAM rules apply to a pollutant-specific emissions unit at a major source that is required to obtain a Title V permit. The rule applies to an emissions unit subject to an emission standard or limitation, uses a control device to achieve compliance, and exceeds the uncontrolled emission criteria.

Permitting requirements under the Clean Air Act have evolved over the past 30 years. As the Clean Air Act is amended, new interpretations of existing regulations are made, and additional requirements are added. Thus, it is necessary for an air-emission source to review carefully the air-quality-permitting requirements as they apply to each location. Satisfactory completion of air-emissions testing and preparation of operating-permit applications may be needed for continued operation. Periodic monitoring, recordkeeping, and reporting may be a continued condition of facility operation.

2.3   Communicating with the Public

Unlike most parts of a WRRF, operations air emissions from the facility can have a direct, immediate, and most importantly, noticeable effect on the surrounding community. Odorous emissions cannot be completely eliminated. Odor generation is an inherent part of treating wastewater and managing wastewater residuals. Odor is an undesirable by-product that is sent directly to the community surrounding the WRRF. Odors from WRRFs often are perceived by the general public as, at least, a nuisance and, at worst, unhealthy. In addition, foul odors affect quality of life and potentially can be seen as affecting property values. For all these reasons, the community surrounding a facility is both the ultimate regulator and the arbiter of the success or failure of any odor-control program. Thus, a vital part of any odor-control program is working with the public and keeping them informed about which odors are being emitted and what measures are being taken to prevent the odors' release.

Element	Actions
Staff	Train staff on understanding the public concerns and how to respond to them
	Put a face on the facility, by designating individuals to communicate with the public
Educate the public	Invite the public to the facility
	Provide a newsletter
	Inform and involve elected officials
	Invite press to tour the facility
Provide for public feedback	Establish a 24-hour-per-day odor hotline
	Establish community odor-advisory committee
	Attend neighborhood or association meetings
Be credible	Keep to promised time frames
	Follow through on promised actions

TABLE 6.8   Communications Approaches

The following are the essential elements to a successful public outreach portion of a facility odor abatement plan:

- Understanding by management and facility staff that the facility is causing a problem and that it must be addressed;
- Commitment by management to address the problem;
- Honest and empathetic communication with neighbors about the problems and realistic expectations about what is to be done and how fast it can be done; and
- Delivering on what was promised and letting the public know about progress.

Because the public is the ultimate judge of a facility's odor-control program, it is essential that it be informed not only about the odor-control system but also about what a WRRF is all about. Table 6.8 outlines some of the steps to effectively communicating with the public.

The details of the communication plan will be dependent on the needs and involvement of the surrounding community. However, it is important that the facility take the lead and not wait for the public to come to it. Greater detail on this subject and the items listed in Table 6.8 can be found in the WEF Manual of Practice 25, *Control of Odors and Emissions from Wastewater Treatment Plants* (WEF, 2004).

2.4   Public Health versus Public Nuisance

The odors at municipal WRRFs are predominantly caused by nitrogen and sulfur compounds. The U.S. EPA lists all compounds it considers to be HAPs, and each state has its own HAP list, or it has adopted the federal HAP list. In addition, the Occupational Safety and Health Administration (Washington, D.C.) (OSHA) and the National

Institute for Occupational Health and Safety (Centers for Disease Control and Prevention, Atlanta, Georgia) (NIOSH) publish both short- and long-term exposure limits for any compound that may be injurious to individuals. Some of the compounds on these lists can be present at municipal WRRFs. However, these limits pertain to mainly to worker exposure and, as such, probably have little significance for concerns for public exposure.

However, it is important to recognize that odors can act as a powerful stimulus. Retailers use scent to increase impulse purchases. In nursing homes, scent is used in dining areas to increase appetite for some of the patients. This ability of odors to influence mood and behavior can be shaped by perceptions. Dalton (1999) reported that many health-related effects of exposure to odorants are mediated not by the direct effect of odors, but by cognitive association of odors and health. In this study (Dalton, 1999), the author found that individuals given a harmful bias reported significantly more health symptoms upon exposure to an odorant than those receiving the same odorant with no harmful effects indicated. Thus, the author concluded that prejudiced odor perceptions and reactions underscore the incredible ambiguity of odor sensation and suggested that similar non-sensory factors play a large role in people's everyday reactions to ambient odors. When odors are persistent, they can result in potential health effects.

Odor perception has been shown to affect mood, tension, stress, depression, anger, and fatigue. These conditions potentially could lead to physiological and biochemical changes with subsequent health effects (Bolla-Wilson et al., 1988; Shusterman et al., 1988; Shusterman, 1992).

In addition to specific compounds, other possible emissions that can affect human health are pathogens and bioaerosols. As with specific compounds, the workers at any facility are the most exposed and thus the most likely to be affected. However, highly debilitated individuals or immune-compromised individuals near facilities can be at greater risk than workers.

3.0 Odor Sampling and Measurement

Quantifying odorous emissions and understanding their constituent compounds is an essential part of any odor management strategy and odor-control system design. By quantifying the odorous emissions, the required degree of treatment can be determined using atmospheric dispersion modeling. Understanding the constituent compounds allows the designer to select the appropriate control technology. To accomplish these goals, samples of the emission must be collected and analyzed. There are standardized scientific methods for collecting these samples, and the method selected depends on the source of the emission and the desired data to be collected. It is vital that the correct sampling method is used, because the method used to collect the sample can have a significant effect on the results of the sampling. The designer must have a thorough understanding of how data were collected to be able to correctly interpret the results of the sampling effort. The intent of this section is not to provide a field-sampling manual, but rather to provide the designer with a framework for understanding the data that have been collected to be used in the design of an odor-control system.

3.1   Field-Sampling Approaches

The sources of emission generally are broken into the following categories for sampling procedures:

- Point sources—the emission is concentrated and released to the atmosphere through a relatively small opening. These include such sources as stacks, ductwork, and vents.

- Aerated area sources—emissions are forced through a relatively large surface by positive aeration. These sources include items such as aeration basins, biofilters, and positively aerated composting piles.

- Unaerated area sources—these emissions come from relatively large surface areas where the only driving force for the emission may be wind, natural convection, or temperature differences between the surface and the surrounding atmosphere.

- Background—samples of the background atmosphere can be taken directly. The designer is cautioned that, although such samples may be of value for periodic checks on operations or for regulatory compliance, they have little or no value in the design process because of the large number of uncontrolled variables.

Table 6.9 outlines the most common collection methods used for the various source types described above. These methods represent a range of standards, including art of sampling that has evolved over time, written standards, and combinations of these two. The designer must be aware of the applicable sampling and analysis standard for the data being collected. For example, U.S. EPA and NIOSH methods for sampling and analyzing specific compounds and groups of compounds dictate specific collection methods.

The medium used to hold and transport the collected sample to the laboratory depends on what is being sampled for. Table 6.10 summarizes the media used for various sample types.

In addition to the type of emission source and temperature and moisture content of the emission, the collection method also may be dictated by the end use of the data. Often, when sampling for regulatory compliance for specific compounds or groups of compounds, the sampling and analysis method may be specified by the regulating agency. For example, if ammonia (NH_3) release must be quantified for regulatory compliance, the NIOSH method is likely to be required by the regulating agency. However, in some cases, a less expensive approximate sampling procedure may be acceptable. For ammonia sampling, used to determine approximate concentrations for appropriate technology selection, the use of colorimetric tubes may be sufficient.

The data collected from a sample have value only if the conditions under which the sample was collected are known and the relation of those conditions to normal operations is fully understood. In addition to collecting the sample, the following are some of the additional pieces of information required to make sampling data meaningful:

- Airflow from the source;

- Loading conditions, such as wastewater flow or process operating parameters, from the source;

Source Type	Procedure	Additional Requirements
Point	• A vacuum is created to draw the sample from the source into the collection media (see Figures 6.3 and 6.4)	• A moisture or droplet trap may be required to prevent condensation from forming in the collection media • For large-diameter stacks, a traverse must be made to ensure a representative sample • A non-reactive (zero air) cooling gas may be required for high-temperature emissions
Aerated area	• A passive chimney placed in several locations with the point-source procedure used to collect the sample from the chimney (see Figure 6.5) • An alternative uses a chimney attached to a flux chamber with the point-source procedure then used to collect the sample from the chimney	• This is best done in calm conditions; wind causes a drawing effect in the chimney, which may skew results • For this alternative, a tracer gas is injected to the air stream at a measured rate. The sample is considered valid if the appropriate amount of tracer gas can be accounted for in the sample
Unaerated area	• Isolation flux chamber with sweep gas and the point-source procedure used to collect the sample from the flux chamber (see Figure 6.6)	• Sufficient time must be allowed to stabilize the air inside the flux chamber by allowing enough time for one complete air change for the volume inside the chamber with sweep gas. The sweep-gas-loading rate is limited to 5.0 L/min
Background	• Odor concentration can be measured directly using field olfactometry equipment • Specific compounds or families of compounds can be measured with colormetric tubes or specially designed sensing equipment	• Can be useful within a confined room, but, in open space, the measurements have little design value

TABLE 6.9 Sampling Methods

• Temperatures;
• Ability for the sampling source to be diluted by changes in ventilation rates or sources of makeup air; and
• Daily and seasonal variation to the process for the source being sampled and the conditions in effect at the time of sampling.

For specific information on sampling methods and field practices, the designer is referred to the specific U.S. EPA or NIOSH sampling methods that may be required and to the WEF Manual of Practice 25, *Control of Odors and Emissions from Wastewater Treatment Plants* (WEF, 2004).

Data Sought	Sample-Collection Media
Odor concentration from an odor panel Total reduced-sulfur compounds	10- to 12-L Tedlar bag with polypropylene fittings 1.0-L Tedlar bag with polypropylene fittings
Ammonia concentration (NIOSH method 6016)	Silica gel tube
Ammonia concentration (unspecified method)	Colormetric tube
Specofic VOC compounds (U.S. EPA method TO-14 and TO-15)	Evacuated Summa canister
Total non-methane VOCs (U.S. EPA method TO-25)	Evacuated Summa canister
Electronic sensors for various compounds or groups of compounds	These may be used directly from the source or may require sample collection in a Tedlar bag, with the probe inserted in the bag to take a reading

TABLE 6.10 Sample Media

Odor is measurable using standardized scientific methods. Point, area, and volume emission sources can be sampled and the samples sent to a laboratory for testing of odor parameters (i.e., odor concentration, intensity, persistence, and descriptors). Odor also can be measured and quantified directly in the surrounding air, at the property line, and in the community using standard intensity and field olfactometry practices.

3.2 Field-Sampling Methods

Odorous air samples can be collected from point emission sources (e.g., stacks or vents) and from surface-area emission sources (e.g., liquid or solid surfaces). Air samples for laboratory odor testing are typically collected in 10-L Tedlar gas-sample bags for transport to the odor-testing laboratory.

Odor sampling often is part of an odor study, odor-control-system performance test, or routine performance test at the WRRF. The purpose of odor sampling often is to compare odors from various processes at the facility or to determine if the odor-control system is performing according to specifications.

In addition to collecting samples for odor-parameter testing, the sampling plan may require that a companion sample (i.e., duplicate) be collected in a Tedlar gas-sample bag or a stainless steel silicate-lined or unlined canister for chemical compound analysis (e.g., reduced-sulfur compound gas analysis or VOC analysis). The protocol also may require testing for specific chemical odorants in the air with portable instruments (e.g., hydrogen sulfide analyzer).

3.2.1 Sampling Exhaust Stacks and Vents

An air sample from the digester sludge tank exhaust would be taken from a point source discharging from a short stack above the exhaust fan. An air sample from the gravity-belt-thickener exhaust would be taken from what was a point source before it was ducted to the gravity-belt-thickener biofilter. In addition to the air sample, the field technician also should measure velocity, pressure, and temperature of the air streams from both exhaust fans. The field technician prepares the sample tubing, 10-L Tedlar sample bag, vacuum case, and pump. The vacuum causes the sample bag to fill with odorous air from the exhaust stack. Figure 6.3 illustrates the sampling apparatus.

If the exhaust air is saturated with moisture or if the exhaust air dew point is above ambient air temperatures, additional sampling procedures must be incorporated. A moisture trap in the sampling line, before the vacuum case, is needed to collect droplets of moisture that may condense in the sampling line. Further, the sample bag may need to be prefilled with dry zero air or high-purity nitrogen to prevent warm, moist exhaust air from condensing in the sample bag. A dynamic dilution sampling probe may be needed for certain sample collection situations. A dynamic dilution sampling probe (Figure 6.4) is a device that simultaneously collects and mixes the sample from the exhaust source with a diluting gas, such as zero air. Sampling specialists must be consulted in these cases for specialized equipment.

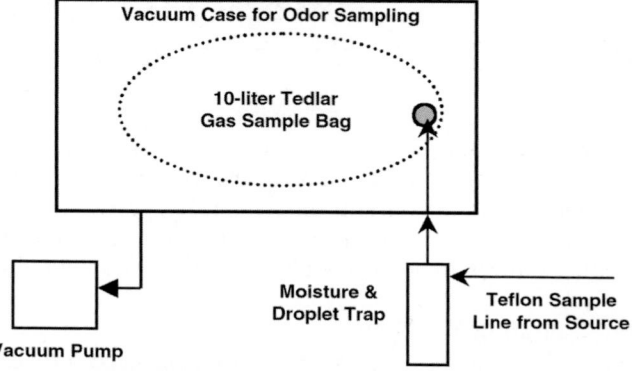

Figure 6.3   Vacuum case for odor sampling.

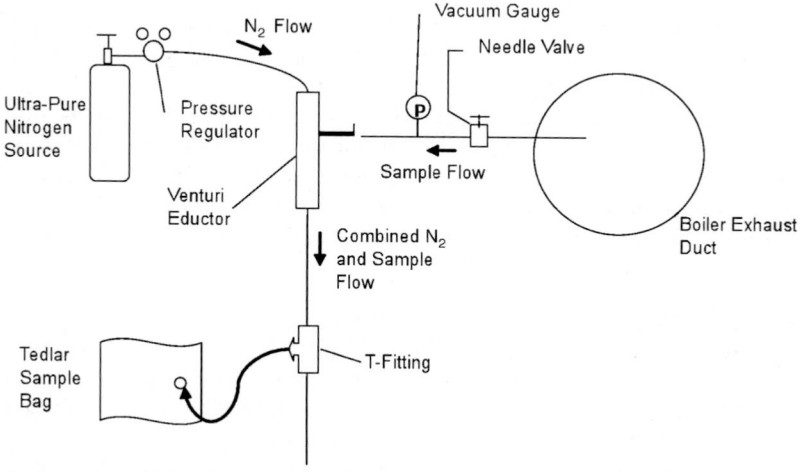

Figure 6.4   Dynamic dilution sampling probe.

The odor of exhaust air that contains oxidizing chemicals, such as ozone or chlorine, may change with time. Extra sampling precautions or procedures may be needed in these cases, and analytical laboratories must be consulted.

3.2.2   Sampling Surfaces

An odorous air sample can be collected from surfaces, sometimes called area sources. Wind speed and direction, air temperature and relative humidity, and solar radiation all affect the odorous emission rate from a quiescent surface (e.g., influent channel of primary clarifier). Aerated surfaces also are affected by the aeration blower flowrate in a diffused air process or the surface of a biofilter. Emission rates for aerated area sources (e.g., aeration basins or biofilters) are calculated by multiplying the odor concentration (i.e., pseudo-dimension of odor units per cubic meter) by the blower or exhaust fan flowrate (cubic meters per second).

A tall passive chimney or simulated stack is an apparatus used to collect aerated-surface-emission samples. Figure 6.5 illustrates the sampling method to isolate an aerated surface. An air sample (from the gravity-belt-thickener biofilter) would be taken from the surface of the biofilter that has an upward flow of exhaust air. The tall passive-chimney sampler minimizes the effects of crossflow winds at the time of sample collection. A vacuum case is used to collect the whole-air sample of exhaust air from the biofilter surface using the same bag-filling procedure described for the point-source sample collection.

An air sample (from the influent channel to the primary clarifiers) would be taken using a flux chamber floating on the surface of the influent channel. The flux chamber (also called surface-emission-isolation chamber) was originally developed in the 1970s to quantify emissions of inorganic gases from soils. Figure 6.6 illustrates the method to collect whole-air samples from quiescent liquid or solid surfaces. The flux chamber uses a flotation collar to float the chamber on a liquid surface. A clean, odor-free carrier gas (e.g., dry zero air or high-purity nitrogen) is metered into the flux chamber at a known flowrate (e.g., 5 L/min). This flow is known as the sweep air for the flux chamber. After an equilibration period of 3 to 4 residence times, a sample is withdrawn from the flux chamber at a flowrate less than the sweep AFR (e.g., 2 L/min). Similar to sampling a point source, a vacuum case and Tedlar sample bag are used to collect the sample from a flux chamber.

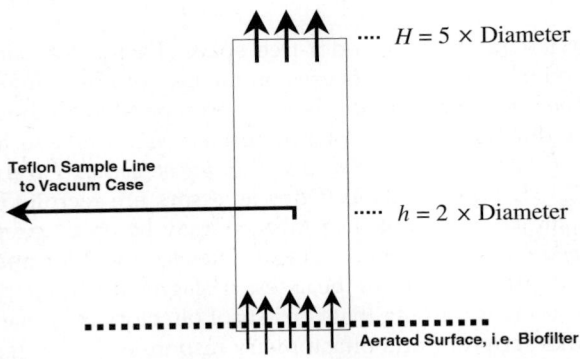

FIGURE 6.5   Tall passive chimney sampler.

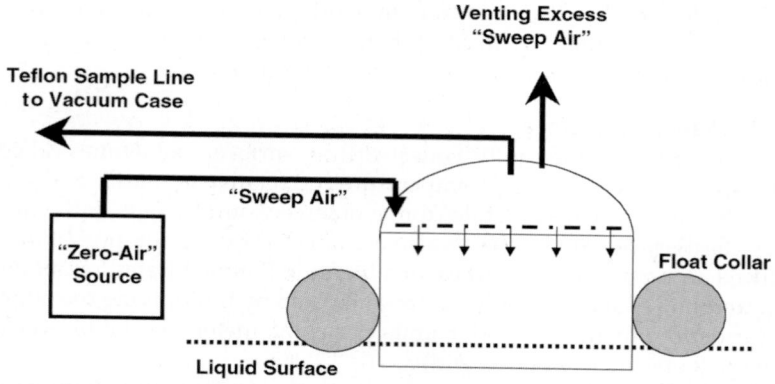

Figure 6.6   Flux chamber sampler.

The odorous emission rate for an area source is calculated by multiplying the odor concentration (odor units per cubic meter) by a sweep AFR (cubic meters per square meters per second) of the flux chamber used to collect surface-emission odor samples.

3.3   Olfactometry Standards

Odor evaluation of air samples is conducted under controlled laboratory conditions following the EN 13725 and ASTM E679 standard practices using trained panelists, like taste testers. In the early years of odor testing in laboratories, the ASTM D1391 syringe dilution technique measured odors in the laboratory from samples collected at the odor source (ASTM, 1978). In 1979, ASTM D1391 was replaced by ASTM E679, *Standard Practice for Determination of Odor and Taste Thresholds by a Forced-Choice Ascending Concentration Series Method of Limits*. In 2003, the European Union approved EN 13725:2003, *Air Quality—Determination of Odour Concentration by Dynamic Olfactometry* (EN 13725, 2003). The current edition of ASTM E679 was revised and approved in 2004, as ASTM E679-04 (ASTM, 2004), incorporating EN 13725:2003 in Appendix III (ASTM, 2004). Odor laboratories also achieve ISO 17025 accreditation to ASTM and CEN standard methods and practices.

An odor laboratory is an odor-free space. Each odor panelist (assessor), when working on odor evaluations, focuses on the task of observing the odor sample when presented. The waiting area of the assessors is separated, as much as possible, from the testing area. Odor panels consist of individuals (assessors) who are selected and trained following *Guidelines for the Selection and Training of Sensory Panel Members* (ASTM, 1981) and EN 13725 (EN 13725, 2003). Odor assessors are recruited from the community. People who smoke, use smokeless tobacco, may be or are pregnant, or have chronic allergies or asthma are excluded as candidates for the odor panel. There are standing odor-panel rules that are part of the assessor's agreement to participate in odor testing. The assessor receives training that consists of olfactory awareness, sniffing techniques, standardized descriptors, and olfactometry responses.

3.4 Analyzing for Specific Odorants

The odor observed from wastewater conveyance systems and WRRFs consists of a variety of odorants and other chemical compounds. As part of an investigation and study of odors, specific odorants and odorant chemical families can be sampled and analyzed. For example, hydrogen sulfide is a common odorant released from wastewater. Hydrogen sulfide is one of the chemical compounds of the reduced sulfur gas family of compounds. Approximately 20 reduced sulfur compounds can be identified using a gas chromatograph (GC) fitted with a chemiluminescence detector following the ASTM D5504-01 analytical procedure (ASTM, 2001). Chapter 3 of MOP 25 (WEF, 2004) presents a number of sampling and analytical methods for identifying additional odorants and odorant chemical families.

3.5 Air- and Odor-Sampling Plans

Issues of sampling and analyzing odorous criteria or hazardous air compounds generally should follow the U.S. EPA Office of Air Quality Planning and Standards, which has developed a standardized approach for the planning and quality control of air-sampling programs (see http://www.epa.gov/ttn/amtic/airtox.html for examples). The document, entitled Quality Assurance Project Plan (QAPP), has 16 sections covering all aspects of sampling, analysis, and data management.

The QAPP should contain a description of the intended project. It should include maps, site descriptions, concerns, and anticipated project outcome. Organizational charts and team contact lists are used to communicate several logistical concerns between project team members, including site-provided equipment and services, shipping instructions and chains of custody, and equipment lists. The purpose of a quality assurance/quality control (QA/QC) program is to produce data of known quality that satisfy project objectives. The QA/QC program must do the following:

- Provide a mechanism for ongoing control and evaluation of measurement data quality; and
- Provide an estimate of data quality, in terms of accuracy, precision, completeness, representativeness, and comparability for use in data interpretation.

Data validation is accomplished by comparing measured data with quality-assurance objectives to determine whether performance problems occurred. There are five analytical levels that address various data uses and the methods required to achieve the desired level of quality, as follows:

- Screening;
- Field analysis
- Engineering;
- Conformational; and
- Nonstandard.

Formal approaches for assessing precision, accuracy, completeness, degree of representativeness, and comparability should be described in the sampling plan.

3.6   Sampling Procedures

Sample methods are broken down into stack testing, area-source testing, and ambient monitoring. In addition, process data and liquid characteristics can be important to quantify potentially toxic air emissions. When a sampling program is designed, it is important to use test methods that are in compliance with appropriate regulatory jurisdictional standards. Because sampling is expensive, it is recommended that sampling and analysis programs be coupled with fate modeling, which can reduce the sampling effort.

Sample possession during all testing efforts must be traceable from the time of collection until the results are verified and reported. Sample-custody procedures provide documentation of all information related to sample collection and handling to achieve this objective. Records of all field activities should be kept in a bound log book, including observations, problems, field measurements, ambient conditions, and other pertinent information.

The data reduction, validation, and reporting procedures ensure that complete documentation is maintained throughout the program, transcription and data reduction errors are minimized, data quality is reviewed and documented, and reported results are properly qualified and in a conventional format.

Brief descriptions of calibration procedures and analytical methodology for the analysis of air samples that are collected during testing should be described. Laboratory methods include those published by ASTM, NIOSH, and others.

Individual sulfur compounds can be analyzed using a gas chromatographer with a flame photometric detector or a gas chromatographer with a sulfur chemiluminescence detector. There also is technology that uses a gas chromatographer with an electrochemical sensor. The U.S. EPA has developed ambient test methods TO-14, TO-14a, and TO-15 for gas-phase organic-compound analysis from a whole-air sample. U.S. EPA method TO-15 is widely available from many laboratories and can analyze reliably most organics of concern to below the ppb_v level. In SI units, ppb_v is expressed as the quantity moles per billion moles (mol/1 000 000 000 mol).

Amines (and carbon–nitrogen ring compounds) have no standard analytical methods. They can be analyzed on a GC–mass spectrometer (MS) that has a special cold column designed for amine analysis. However, getting the compounds to the GC–MS can be a problem. High-pressure liquid chromatography is the only universal method that can analyze these compounds at detection limits near the odor threshold. Organic acids also do not have standard analytical methods that can achieve meaningful detection limits. Siloxanes are analyzed by method TO-15, which can be used to analyze these compounds with special preparations. In addition, impingers using a methanol solution also can be used for siloxane analysis.

Wastewater liquid-phase parameter analysis methods almost always are taken from *Standard Methods for the Examination of Water and Wastewater* (APHA et al., 2017). For potentially toxic organics, U.S. EPA method 624 or 8240 is used.

All field analytical measurement data are reduced according to the QAPP and protocols in applicable standard operating procedures that describe field measurements. Information used in the calculations is recorded in sufficient detail to enable reconstruction of the final result at a later date.

Health and safety are important issues for the sampling team. This is site specific. For WRRFs, it is common to have a safety and security plan. This plan typically is attached to the QAPP as an appendix.

4.0   Assessing Odor and Air Emissions—Searching for "Root Cause" and Odor Management Stategies

Compounds that may be emitted from wastewater systems include odorants, VOCs, HAPs, and products of combustion. Accurate characterization of air-pollutant emissions is important, because emission levels may be used to determine regulatory applicability and subsequent compliance with regulatory requirements, such as the need for emission controls and construction and operating air permits. Proper characterization of air emissions also is important for selecting odor- or air-emission-control technologies.

4.1   Odor Emissions from Wastewater Systems

Fresh, aerobic wastewater contains many odorous compounds, such as indole, skatole, organic acids, esters, alcohols, and aldehydes. Odors can be released from virtually all phases of wastewater collection, treatment, and disposal, and, just as the character of the wastewater changes as it progresses through the treatment process, so do the air emissions from that wastewater. While, in general, emissions from almost all processes in the treatment process are considered objectionable, some are much worse than others, and only some have the ability to affect the community beyond the boundaries of the facility. The following three elements are required to create an odorous emission:

- Odorous compounds in the source wastewater;
- Exposed surface area, from which the compounds can be emitted; and
- A driving force that causes the compound to pass from the wastewater to the air.

Odors from WRRFs are a complex combination of a wide variety of compounds; however, for the designer, there are certain compounds and groups of compounds that contribute significantly to wastewater odors and also significantly affect the selection of the control technology. These include the following:

- Hydrogen sulfide;
- Organic sulfide compounds; and
- Ammonia and nitrogen compounds.

Hydrogen sulfide is generated from the biological reduction of sulfate (SO_4^{-2}) or thiosulfate under anaerobic conditions. Hydrogen sulfide most commonly is a concern in the headworks and early stages of the treatment process or during the anaerobic digestion of solids. Hydrogen sulfide is a colorless, potentially toxic gas with a characteristic rotten-egg odor. Death can result from exposure to hydrogen sulfide concentrations of 300 ppm_v in air. Table 6.11 summarizes the key characteristics of hydrogen sulfide gas. Hydrogen sulfide gas is moderately soluble in water and dissociates to other forms of sulfide, depending on the pH.

$$H_2S = HS^{-2} + H^- \qquad (6.2)$$

$$HS^- = S^{-2} + H^+ \qquad (6.3)$$

Molecular weight	34.08
Vapor pressure, −0.4°C	10 atm
25.5°C	20 atm
Specific gravity (vs. air)	1.19
Odor detection threshold	Approximately 1 ppb$_v$
Odor recognition threshold	Approximately 5 ppb$_v$
Odor character	Rotten eggs
Typical 8-hour weighted average exposure limit	10 ppm$_v$[b]
Imminent life threat	300 ppm$_v$

[a]Different sources report a range of values for odor-detection or recognition thresholds for hydrogen sulfide. The range of odor thresholds reported for hydrogen sulfide varies from 0.47 to 9 ppb. Odor thresholds mentioned in this manual may vary and are not intended to be definitive because of the differences in the testing methodologies used when measuring odor thresholds.
[b]Varies by state occupational safety agency.

TABLE 6.11   Characteristics of Hydrogen Sulfide[a]

The pH of wastewater has an important role in determining the amount of molecular hydrogen sulfide gas available to be released to the sewer atmosphere. Figure 6.7 shows this relationship. At pH 6.0, over 90% of dissolved sulfide is present as dissolved gas. At pH 8.0, less than 10% is available as gas for release from wastewater. Therefore, a decrease of one pH unit in wastewater can significantly increase the

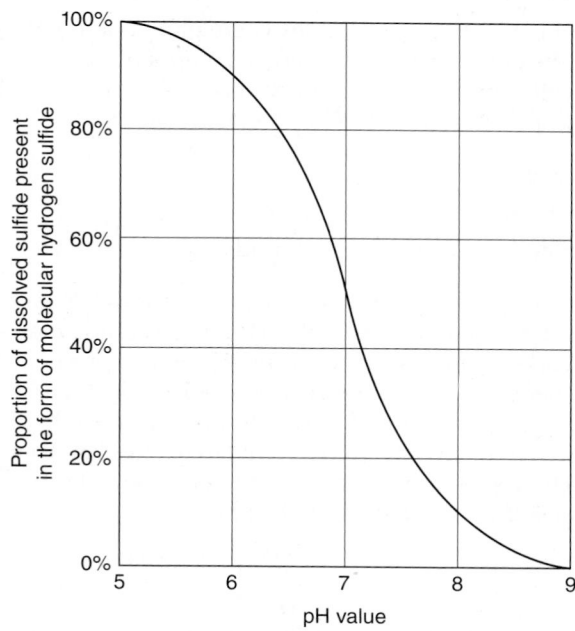

FIGURE 6.7   Relative concentrations of dissolved molecular hydrogen sulfide.

release of hydrogen sulfide gas, potentially causing odor and corrosion problems. Another major factor affecting hydrogen sulfide release is turbulence.

Corrosion is a significant problem resulting from the presence of hydrogen sulfide. Hydrogen sulfide directly attacks metals such as iron, steel, and copper. More importantly, hydrogen sulfide is oxidized biologically to sulfuric acid (H_2SO_4) by *Thiobacillus* bacteria in the presence of moisture (see Chapter 8).

4.1.1   Organic-Sulfur Compounds

Some organic-sulfur compounds, such as mercaptans, are found throughout the treatment process, while others, such as dimethyl sulfide and dimethyl disulfide, are found in solids-processing operations, such as composting. Wastewater and residuals subjected to anaerobic conditions often contain reduced sulfur compounds other than hydrogen sulfide, which can contribute to the characteristic odor. In liquid-conveyance and -treatment processes, organic sulfur compounds are likely to be of lesser importance than hydrogen sulfide, as they are present at relatively low concentrations. Removing hydrogen sulfide from these emissions often can result in significant odor reduction. However, in solids-handling processes, such as thickening, storage, and dewatering, the organic-sulfur compounds become much more prominent, often making up the dominant odorant. Removing only the hydrogen sulfide from these emissions may result in only marginal odor reduction. Some of these compounds are less soluble than hydrogen sulfide and can be difficult to treat, often requiring multiple stages to achieve adequate odor reduction. Table 6.12 lists many of these compounds that contribute to odors from wastewater and its residuals.

Substance	Formula	Characteristic Odor	Odor Threshold (ppb$_v$)	Molecular Weight
Allyl mercaptan	$CH_2\!=\!C\!-\!CH_2\!-\!SH$	Strong garlic, coffee	0.05	74.15
Amyl mercaptan	$CH_3\!-\!(CH_2)_3\!-\!CH_2\!-\!SH$	Unpleasant, putrid	0.3	104.22
Benzyl mercaptan	$C_6H_5CH_2\!-\!SH$	Unpleasant, strong	0.19	124.21
Crotyl mercaptan	$CH_3\!-\!CH\!=\!CH\!-\!CH_2\!-\!SH$	Skunk-like	0.029	90.19
Dimethyl sulfide	$CH_3\!-\!S\!-\!CH_3$	Decayed vegetables	0.1	62.13
Dimethyl disulfide	$CH_3\!-\!S\!-\!S\!-\!CH_3$	Decayed vegetables	1	94.20
Ethyl mercaptan	$CH_3CH_2\!-\!SH$	Decayed cabbage	0.19	62.10
Hydrogen sulfide	H_2S	Rotten eggs	0.47	34.10
Methyl mercaptan	CH_3SH	Decayed cabbage	1.1	48.10
Propyl mercaptan	$CH_3\!-\!CH_2\!-\!CH_2\!-\!SH$	Unpleasant	0.075	76.16
Tert-butyl mercaptan	$(CH_3)_3C\!-\!SH$	Skunk-like, unpleasant	0.08	90.10
Thiocresol	$CH_3\!-\!C_6H_4\!-\!SH$	Skunk-like, rancid	0.062	124.21
Thiophenol	C_6H_5SH	Putrid, garlic	0.062	110.18

*Different sources report a range of values for odor-detection or recognition thresholds, particularly for compounds such as hydrogen sulfide or ammonia. The range of values used for hydrogen sulfide typically varies from 0.47 to 9 ppb$_v$. Odor thresholds mentioned in this manual may vary and are not intended to be definitive because of the differences in odor-testing methodologies when measuring odor thresholds.

TABLE 6.12   Odorous Sulfur Compounds in WRRF Emissions (WPCF, 1979)*

4.1.2   Ammonia and Nitrogen Compounds

Wastewater and its residuals contain various forms of nitrogen. Much of it is present as ammonia or organic nitrogen. Ammonia typically appears in the dewatering processes and in the solids created from dewatering. This is especially true in digested solids. Ammonia is found in wastewater at concentrations of 12 to 50 mg/L (Metcalf & Eddy, 1972). However, the small quantity of ammonia in wastewater offgas at neutral pH contributes little to odor emissions, because the odors typically are dominated by sulfur compounds. Therefore, it rarely is necessary to provide an ammonia removal step in treating offgas from liquid wastewater treatment processes, unless lime or other alkaline material is used in the process to elevate the pH.

When wastewater solids are treated in a high-pH process, such as lime stabilization, or a high-temperature process, such as composting, the release of ammonia and other nitrogen-based odorants becomes a significant factor in odor emissions. Anaerobic digestion and autothermal thermophilic aerobic digestion also can cause the generation and release of ammonia and compounds, such as amines. Amines are a class of nitrogen-based odorants that often are characterized as having fishy odors. Table 6.13 lists some of the odorous nitrogen compounds and their characteristics. Because the determination of odor threshold uses human panelists and experimental methods may vary between researchers, the reported range is wide.

4.1.3   Other Wastewater Odorants

Although most odor characterization and control efforts focus on sulfur compounds and, to a lesser degree, nitrogen compounds, there are many compounds present in wastewater that contribute to its odor. Table 6.14 lists other classes of odorous compounds—acids, including volatile fatty acids, and aldehydes and ketones. Although these compounds contribute to the odor of wastewater and sludge, they seldom are the dominant odor or the target of odor-control efforts.

Substance	Formula	Characteristic Odor	Odor Threshold (ppb_v)	Molecular Weight
Ammonia	NH_3	Sharp, pungent	130 to 15 300	17.03
Methylamine	CH_3NH_2	Fishy, rotten	0.9 to 53	31.05
Ethylamine	$C_2H_5NH_2$	Ammoniacal	2400	45.08
Dimethylamine	$(CH_3)_2NH$	Fishy	23 to 80	45.08
Trimethylamine	$(CH_3)_3N$	Fishy		59.12
Pyridine	C_6H_5N	Disagreeable, irritating	3.7	79.10
Skatole	C_9H_9N	Fecal, repulsive	0.002 to 0.06	131.2
Indole	C_2H_6NH	Fecal, repulsive	1.4	117.15

*Different sources report a range of values for odor-detection or recognition thresholds, particularly for compounds such as hydrogen sulfide or ammonia. The range of values used for hydrogen sulfide typically varies from 0.47 to 9 ppb_v. Odor thresholds mentioned in this manual may vary and are not intended to be definitive because of differences in odor-testing methodologies when measuring odor thresholds.

TABLE 6.13   Odorous Nitrogen Compounds in Wastewater (IWA, 2001; WPCF, 1979)*

Substance	Formula	Odor Character	Odor Threshold (ppb$_v$)	Molecular Weight
Acids				
Acetic acid	CH_3COOH	Vinegar	0.16	60
Butyric acid	$CH_3(CH_2)_2COOH$	Rancid	0.09 to 20	74
Valeric acid	$CH_3(CH_2)_3COOH$	Sweaty	1.8 to 2630	102
Aldehydes and Ketones				
Formaldehyde	HCHO	Acrid, suffocating	370	30
Acetaldehyde	CH_3CHO	Fruity, apple	0.005 to 2	44
Butyraldehyde	$CH_3(CH_2)_2CHO$	Rancid, sweaty	4.6	72
Isobutyraldehyde	$(CH_3)_2CHCHO$	Fruity	4.7 to 7	72
Valeraldehyde	$CH_3(CH_2)_3CHO$	Fruity, apple	0.7 to 9	86
Acetone	CH_3COCH_3	Fruity, sweet	4580	58
Butanone	$CH_3(CH_2)_2COCH_3$	Green apple	270	86

*Odor-detection or recognition thresholds mentioned in this manual may vary and are not intended to be definitive because of differences in odor-testing methodologies when measuring odor thresholds.

TABLE 6.14   Other Odorous Compounds in Wastewater (IWA, 2001; Verschueren, 1983)*

4.1.4   Odorants from Industrial Sources

If a WRRF receives a high flow or load contribution from an industry, odorants present in the industrial discharge may contribute to odor emissions and may even change the character of the odors emitted at the WRRF. Depending on the industry, the odorants can vary widely, with respect to type and concentration. VOCs often have characteristic odors that can affect odor emissions from wastewater collection and treatment processes. With significant loadings of these compounds to the WRRF, these compounds sometimes can cause unique odor characteristics in the emissions from liquid treatment processes and aeration basins. For example, high loadings of wastewater from textile-dyeing processes can cause a solvent odor to be released from various unit wastewater operations. Information on specific compounds from industrial processes may be found in other references (Buonicore and Davis, 1992; Rafson, 1998).

Industries that have a high biochemical oxygen demand (BOD) discharge can contribute indirectly to downstream odor emissions. Even though the industrial waste may not be inherently odorous, a high BOD discharge can result in rapid oxygen depletion and increased sulfide generation in the receiving collection systems. High-temperature discharges can have a similar effect of increasing biological activity, reducing dissolved oxygen concentrations, and creating anaerobic conditions, leading to the increased formation of hydrogen sulfide.

4.1.5   Exposed Surface Area

In determining which odor sources may have the greatest effects on the surrounding community, it is vital to look at the available pathway for the odor to escape from the wastewater to the air. The source with the highest odor concentration may have a significantly lower effect on the surrounding community than a less odorous source with substantially greater exposed surface area. A typical example of this in a WRRF is the comparison between the headworks inlet channels and the primary clarifiers. The channels often will have one of the highest odor concentrations in the facility, but they have a small exposed surface area and thus may have little effect on off-site odors. In comparison, the primary clarifiers may have an odor concentration several times lower than the inlet channels, but may constitute the greatest contribution to off-site odor effects.

4.1.6   Driving Force

Major drivers forcing compounds from the liquid phase into the air include the following:

- Liquid turbulence;
- Air turbulence and movement over the liquid surface;
- Forced aeration (the higher the gas-to-liquid ratio, the greater the emission);
- Temperature (increased temperature causes decreased gas solubility and greater transfer rate); and
- pH (low pH favors hydrogen sulfide release, high pH favors ammonia release).

Although there are theoretical equations, such as Henry's law, to predict the mass transfer of odorous compounds from water to air, in practice, concentrations of odorants, such as hydrogen sulfide, in the headspace of sewers or enclosed tanks rarely approach the equilibrium values predicted by these equations.

Table 6.15 reviews emissions from the common areas of the wastewater liquid and solids processing.

4.2   Air Emissions from Combustion Sources

A number of processes are commonly used at WRRFs that combust fuels. For these combustion sources, odors may be less significant than pollutants, such as the following:

- Nitrogen oxides;
- Carbon monoxide;
- Hydrocarbons, including nonmethane hydrocarbon;
- Sulfur oxides;
- Particulate matter; and
- Mercury.

Nitrogen oxides, primarily nitric oxide (NO) and nitrogen dioxide (NO_2), are formed by either or both of two mechanisms—thermal or fuel NO_x. Thermal NO_x is formed by reactions between nitrogen and oxygen in the air used for combustion. Fuel NO_x results from the combustion of fuels, such as biosolids, or heavy oils that contain organic nitrogen.

Carbon monoxide results from the incomplete combustion of carbonaceous fuels, such as natural gas and digester gas. Sulfur oxides are formed when sulfur compounds found in fuels are oxidized in the combustion process. Digester gas typically contains sulfides in the form of hydrogen sulfide. Digester gas sometimes is treated to remove sulfides. Fuel oil and biosolids also contain sulfur.

Area	Emissions Characteristics
Headworks	Hydrogen sulfide is predominant odor compound
Influent channels	Very high odor concentrations
Screens	High liquid turbulence is possible
Grit separation	Aerated grit removal has high emission potential
Primary settling	Hydrogen sulfide is predominant odor compound
Splitter boxes	Large exposed surface areas for tanks
Settling tanks	Turbulance at weirs and splitter boxes result in high odor concentrations
Discharge weirs	Large exposed surface of settling tanks often results in the tanks being the most significant emission source at a WRRF
	Aerated grit removal has high emission potential
Biological treatment	Little or no hydrogen sulfide, except for anoxic zones
Suspended growth	Low concentrations of mercaptans and other organic sulfur compounds cause odors
Activated sludge aeration basins	Low odor concentration
	High turbulence and large surface area
	Typically have small effect on off-site emission, with the exception of anoxic zones
	Aeration mechanism affects emission by changing gas-to-liquid ratio
	Trickling filters have significant exposed surface
	Passively aerated trickling filters are subject to rapid changes in emissions from rapid ambient-air-temperature changes
	RAS channels or the reaeration of RAS as it enters the secondary treatment basin
Secondary settling	Very low concentrations of odorous compounds
Splitter boxes	Large surface exposed surface areas for tanks
Settling tanks	Turbulence at weirs and splitter boxes results in high odor concentrations
Discharge weirs	Generally has no effect on off-site odors
Tertiary filtration	Very low concentrations of odorous compounds
	Generally has no effect on off-site odors
	Odor potential from backwash of filters
Disinfection	Very low concentrations of odorous compounds
	Generally has no effect on off-site odors
	High turbulence possible from discharge to receiving waters
Sludge storage tanks	High odor-compound concentrations
	Reappearance of hydrogen sulfide
	Moderate exposed surface area

TABLE 6.15 Treatment-Processes—Emission Summary

Area	Emissions Characteristics
Dewatering and thickening	High odor-compound concentrations
	Ammonia may become significant, especially with digested sludge
	Dewatering method affects the character of the odor (shearing action in centrifuges may generate greater odor than other types of dewatering)
Alkaline stabilization	High odor-compound concentrations
	Ammonia release is very significant
	Dust generation is significant
	End product remains odorous for long periods of time
Composting	High odor-compound concentrations
	Dimethyl disulfide, and dimethyl sulfide are significant odorants
	Ammonia may be significant with low aeration or digested sludge
	Dust generation is significant
	Large exposed surface areas for actively composting piles
Digestion	System sealed, odors from leaks in covers or valves
	Emissions are combusted for energy production or flared, thus very little odor generation

TABLE 6.15 Treatment-Processes—Emission Summary (*Continued*)

Particulates and metals can originate from combusted material, such as biosolids that are incinerated. For a multiple-hearth incinerator, 15% to 30% of the ash content of the filter cake will become airborne in the incinerator exhaust. Emission-control equipment can remove most of this particulate matter. Small amounts of particulates also come from noncombustible ash that may be present in the fuel being burned. Particulate emissions from other combustion processes typically are low because of the low amount of ash present in the fuel.

Hydrocarbon emissions are caused by incomplete combustion of organics in fuels. For incineration and drying, hydrocarbons also can originate from volatile organics that may be present in the biosolids.

If digester gas cannot be used in engines or boilers, it is burned in flares. Flares are required to reduce odors, avoid an explosion or fire hazard, and destroy VOCs. Flares typically are designed to ignite the digester gas by passing it through a curtain of flame developed by a ring-type natural-gas pilot flame. The digester gas is deflected across the pilot flame by a baffle. Some flares or digester gas burners are designed to be highly efficient in destroying VOCs, by providing a large enclosed combustion chamber that increases retention time. Much of the organic emissions from flares are unburned methane—a pollutant that typically is not regulated in the United States.

Boilers combust fuels to heat water to produce either steam or hot water. This energy is used primarily for building heat and heating of digester biosolids. Boilers at WRRFs are designed to burn digester and natural gas and fuel oil. Boilers typically emit relatively low amounts of pollutants—especially those that burn natural gas or digester gas that had the sulfur removed before burning. Boilers can be designed to produce lower levels of emissions—primarily NO_x.

Internal combustion engines typically are used in a WRRF where there is excess digester gas available to burn as fuel for the engines. Digester gas contains methane and has approximately 60% of the energy value as methane (natural gas). The engines are used to generate power that is used in the facility or sold to power utilities. Some engines directly drive equipment, such as large air compressors or pumps. Internal combustion engines using natural gas or diesel fuel also are used as backup emergency generators. Many types of internal combustion engines can be significant sources of pollutants—primarily NO_x and CO.

Engines designed to burn digester gas also typically can burn natural gas or a mixture of natural and digester gases. Some engines are designed to be lean burning, meaning they operate with a high air-to-fuel ratio to reduce emissions. Lean-burning engines typically use a pre-chamber to ignite a lean-burning main combustion chamber.

The incineration process combusts or burns biosolids by using them as a fuel and combining the fuel with oxygen. The combustible elements of the biosolids are carbon, hydrogen, and sulfur, which are chemically combined in the biosolids as grease, carbohydrates, and proteins. If the fuel content of the biosolids is high enough, the biosolids can be burned without supplemental fuel or autogenously. However, supplemental fuel typically is required, and natural gas or fuel oil typically is used. The two common types of biosolids incinerators are multiple hearth and fluidized bed. The multiple-hearth incinerator is a cylindrical, refractory-lined steel shell containing a series of horizontal refractory hearths located one above the other. Combustion occurs in the drying, burning, and cooling zones. A fluidized-bed incinerator operates by setting the biosolids and a sand bed in fluid motion by passing combustion air through the fluid-bed zone in a homogenous boiling motion. Because the solids to be burned are surrounded by air, oxygen quickly reacts to combust the solids. Exhaust from the incinerators is vented to air-pollution-control devices. Typically, a wet scrubber is used to control particulates, and thermal oxidation sometimes is used to control hydrocarbon emissions.

Two major types of thermal dryers used to dry biosolids are direct and indirect heat. Direct-heat dryers heat air and use a fan to pull the hot air through the dryer, where the air contacts the sludge cake to dry it. This can result in large amounts of air containing particulate matter and other pollutants being discharged from the dryer, which may require highly efficient particulate control systems, such as wet electrostatic precipitators. Some manufacturers of direct-heat dryers recycle most of the air back to the feed end of the dryer, which decreases the amount of air discharged from the dryer and saves energy, reduces emissions, and reduces the size of pollution-control equipment. Direct-heat dryers can be designed to use natural gas, digester gas, or oil as a fuel to heat the air.

Indirect dryers apply heat to one side of a metal surface, and the surface is used to heat the sludge to evaporate water. An indirect system results in lower exhaust-air volumes and pollutant loads compared with most direct dryers.

4.3 Emission Estimation Methods—General Fate Models

One method frequently used to estimate VOC emissions from WRRFs is the use of general fate models. These models, as the name implies, predict the fate of VOCs as they travel through the treatment process. Fate mechanisms considered for VOCs include volatilization, sorption, and biodegradation. Volatilization is the loss or transfer of VOCs from the liquid to vapor phase. Biodegradation refers to the loss of VOCs through biological uptake and transformation, while sorption refers to the binding of VOCs to solids. The extent that any given compound volatilizes, sorbs, or is biodegraded depends on chemical-specific properties, environmental conditions, and WRRF operating conditions.

Examples of models available for estimating emissions include TOXCHEM (Hydromantis, Inc., Hamilton, Ontario, Canada) and the Bay Area Sewage Toxic Emissions (BASTE) models (CH2M Hill, Englewood, Colorado). Both of these models are proprietary and are available commercially. Other models, such as WATER9, are available from the U.S. EPA free of charge and can be downloaded from the U.S. EPA's website (http://www.epa.gov/ttn/chief/software/water/index.html). The models TOXCHEM 1 (Sterne, 2001), BASTE (BASTE, 1992), and WATER9 (U.S. EPA, 2002) estimate VOC emissions by considering each of the three mechanisms discussed previously. Each model estimates emissions on a process-unit basis, starting at the treatment influent location and solving sequentially for emissions estimates, unit by unit, in the downstream direction.

Because compound-specific properties affect the relative significance of competing fate mechanisms, most models have built-in chemical databases that include data required for modeling many of the VOCs routinely encountered in wastewater VOC emissions analyses. In addition, the models have the capability to allow the user to model other compounds not included in the databases. Examples of physical data potentially required for emissions modeling include compound molecular weight, gas- and liquid-phase diffusivities, Henry's law coefficients, biodegradation rates, and octanol-water partition coefficients.

Influent conditions also are required by each model and may include influent VOC concentration, wastewater temperature and flowrate, and solids concentration.

Emissions are estimated for each process unit and, as such, are dependent on process-unit-specific information, such as surface area, liquid depth, and whether the surface is quiescent (e.g., aerated, weir drop height, and mixed-liquor suspended-solids concentration).

Once all required compound-specific, environmental, and process-specific data are entered, emissions are solved using a generalized steady-state mass-balance approach that solves the following general equation:

$$\text{Liquid in} + \text{Liquid out} + \text{Volatilized} + \text{Biodegraded} + \text{Sorbed} + \text{Solids out} = 0 \qquad (6.4)$$

where Liquid in = mass rate of compound entering process unit in liquid phase (g/s),
Liquid out = mass rate of compound exiting process unit in liquid phase (g/s),
Volatilized = mass rate of compound volatilization (g/s),
Biodegraded = biodegradation rate (g/s),
Sorbed = rate of adsorption to sludge and subsequent loss (g/s), and
Solids out = loss of compound in liquid phase of sludge stream (g/s).

The general mass balance is applied to each process unit and simplified as required. For example, for preliminary treatment process units, biodegradation is assumed to be negligible, resulting in the biodegraded term being set to zero. While the degree of sorption is considered to be equilibrium based, the extent of volatilization is mass transfer driven and therefore requires estimation of mass-transfer coefficients. Each model uses algorithms for estimating mass-transfer coefficients.

Although the level of effort required to set up and execute the model may be intensive initially, one advantage to the modeling approach is the ability to perform sensitivity analyses or assess future conditions. For example, it is possible to perform an analysis on a specific parameter to determine the sensitivity of emissions on that parameter (i.e., wastewater temperature or flowrate). If, under future conditions, the level of flow is expected to change, the effects on emissions could be evaluated easily by changing the flowrate conditions within the model.

4.4 Odor-Control Strategies

Regardless of the emissions of concern, the general process for establishing a strategy to manage them involves the following basic steps:

- Determine the emission goals for the facility;
- Quantify the emissions from all sources;
- Determine the effect of the emission individually and collectively on the surrounding community;
- Develop control alternatives;
- Verify the effectiveness of the control measures; and
- Communicate with the public.

Each of these is examined in greater depth in the following sections.

4.4.1 Determine Emission-Reduction Goals

The primary forces driving emission goals are regulations, the need to protect worker and neighbor health, the need to prevent corrosion of facility infrastructure, and the need to prevent malodors from reaching the surrounding community.

Regarding specific compounds, regulatory drivers include the Federal Clean Air Act, which regulates primary pollutants. These rules generally apply to large facilities with thermal or energy production operations that can be affected. Many individual states also have primary pollutant regulations that may be more stringent than the Federal Clean Air Act requirements. Other states, such as California, have strict regulations on VOCs, in addition to the primary pollutants; concentrations of certain potentially toxic compounds are regulated by OSHA and NIOSH. These relate more to concentrations within process areas and generally affect ventilation requirements.

Odors can be regulated by local or state laws. These regulations vary from setting limits on odor concentrations at property lines or the nearest receptors to the more common prohibition on creating a nuisance. Ultimately, the neighbors of a facility are the regulators, and, if they are complaining, the facility is out of compliance.

In the absence of specific regulatory concentration targets, odor-concentration goals must be established. In practical terms, there is no such thing as odor free. Odors of all types, from all different types of sources, are ubiquitous. When determining a target odor concentration either at the facility property line or at the nearest receptors, the following two factors must be considered:

- The allowable odor concentration and
- The minimum duration of time that the threshold concentration persists.

Regulations from other places can be used to determine a maximum allowable nuisance threshold. Alternatively, the properties of the odorous emissions themselves also can be used.

Using Steven's law below and dose-response data from on-site odor sampling, a nuisance threshold odor concentration can be determined by setting the intensity to 3.5 (as an example).

$$I = aC^b \qquad\qquad (6.5)$$

where I = intensity,
$\quad C$ = odor concentration in D/T, and
a and b = 5 dose-response constants determined by regression analysis of the intensity ratings of an odor panel to odor samples.

Sensitivity to odors varies with individuals. In addition, odors can trigger emotional responses. Neighbors who have been sensitized to odor by past exposures may have a lower tolerance; thus, a more stringent threshold should be considered.

Once a threshold-odor concentration is established, a time of that concentration to persist must be determined. To be a nuisance, an odor effect must trigger a response or change in behavior. As with the threshold concentration, in the absence of regulatory guidance, a minimum time frame must be established.

4.4.2   Quantify Emissions from Sources

Whether dealing with primary pollutants or odors, it is vital to have an inventory of all emission sources and their contribution to overall facility emissions and effects to the surrounding community. To do this, the sources must be identified and samples taken at each source. To be an emission source, any process or location must have the following three elements:

1. Compounds of concern must be present (odorous or primary pollutants).
2. Surface area, from which the compounds can escape into the atmosphere.
3. A driving force, such as forced ventilation, temperature differences between the ambient air and the wastewater or solid surface, or exposure to wind. Any of these will move compounds from the source into the atmosphere.

Once sources are identified, air samples must be collected directly from these sources, so that emission rates can be determined and the source modeled. If primary pollutants or VOCs are a concern, these must be tested for directly. For dealing with odor, the following sample types should be collected, at a minimum:

- Odor (whole air sample, typically 10-L sample bag);
- Total reduced-sulfur compounds;
- Ammonia for solids processes;
- Continuous hydrogen sulfide monitoring at the facility inlet; and
- Exhaust rate from ventilated sources.

Air samples can be analyzed to determine the odor concentration, intensity, character, and hedonic tone. Of these, the character and the hedonic tone are of limited value. The character can be useful to see if recurring patterns of description are found. The concentration, with the source exhaust rate or flux rate, is used to calculate the odor-emission rate for the source. The intensity can be used to determine the nuisance-odor-threshold concentration.

Just as the character of the wastewater changes as it moves through the treatment process, so do the odor-causing compounds that are being emitted. Hydrogen sulfide is a significant odor-causing compound at the early stages of the treatment process; it is seldom observed at the end of the facility. Through the solids-handling processes, ammonia and dimethyl disulfide can provide a significant contribution to odor, but they do not have significant roles in other parts of the facility. It is vital to have an understanding of the emission rate for some of these compounds to be able to select and properly size an odor-control technology.

WRRFs undergo cyclic loading of various odorous compounds. Odor emissions at the initial treatment stages of the treatment process are significantly affected by the fluctuating concentrations of compounds and, in particular, hydrogen sulfide. Therefore, it is a good idea to monitor the hydrogen sulfide emission at the headworks or inlet channels continuously for at least several days. This can be accomplished using a data logging sensor, for example, OdaLog®. If possible, this should be done in advance of taking other samples, so that these samples are collected at times of maximum odor emission.

For sources with forced ventilation, measuring the airflow at the exhaust point is important for determining the emission rate for that source. Alternatively, the pressure drop across the exhaust fan, coupled with the fan curve, can be used to determine the airflow.

There are two main methods for collecting samples from sources—the lung pump for forced ventilated sources and the isolation flux chamber for unventilated sources, such as the water surfaces of channels and clarifiers. The collected air samples are analyzed by a laboratory odor panel and a separate laboratory for the compound constituents listed above. For odor-concentration analyses, different laboratories may use different presentation rates to panelists. The presentation rate used has a significant effect on the odor-concentration results. Presentation rates used in the test include 0.5, 3.0, and 20.0 L/min. It is important to be sure to know and specify the panel presentation rate. If data from past odor studies are to be used or compared with the new data, it is vital that the presentation rates are known and, if possible, are the same.

All of the above discussion pertains to samples collected at the emission source. It is possible to collect samples from the surrounding area; however, these data are of little value. The effects of an emission on the surrounding area are determined not only by the emission itself, but also by constantly changing meteorological conditions. Taking samples from the surrounding area provides only a snapshot; it is difficult to take enough samples under the right conditions to have meaningful data. Therefore, it is the greatest value to take samples at the source and use computer models to predict the effects of the emissions.

4.4.3 Determine Odor Dispersion into the Surrounding Community

Having inventoried the emission sources and determined the maximum emission in the cyclic loading of the WRRF, it is time to determine the dispersion of these emissions into the neighboring community. For this, computer modeling is used. While models are vital to determining the dispersion of emissions into the surrounding community, it is important to remember that they are estimates and approximations. The models indicate trends and must be viewed that way. With this in mind, there are some practices that should be followed:

- Sources should be modeled individually and collectively;
- Some representative individual receptor locations should be selected, and the number of times the nuisance threshold is met and exceeded should be examined; and
- When possible, use multiple years of meteorological data to avoid skewing results.

Considering sources individually provides a relative ranking of sources regarding their generation of nuisance effects. The source with the greatest emission generation may not cause the greatest odor response. Odorous emissions that are quickly dispersed may cause fewer responses by the neighbors. The velocity and height of an exhaust significantly affect this dispersion and thus the odor concentration to the surrounding area. For example, a highly odorous headworks building that is exhausted through roof ventilators at high velocity may emit more overall odor than a set of primary clarifiers that are low to the ground. However, it is possible that the clarifiers would cause a greater odor concentration in the surrounding community, because the odor is emitted over a large area, low to the ground, at a low velocity.

All sources must be considered collectively. Although a group of sources may not create a nuisance condition individually, when some of them are grouped together, nuisance-level odor may be created. Observing individual receptor location is useful. If the sum total of odor at a location is greater when all sources are modeled together rather than when they are modeled individually, there are odor levels being created from groups of sources. It may be necessary to model different groupings of sources to determine which combinations lead to nuisance odor.

4.4.4　Develop Control Alternatives

The elements following are the elements of emission control:

- Prevention;
- Capture;
- Control; and
- Exhaust.

Emission prevention can take the form of operational changes, such as eliminating turbulence that releases odor at inlet channels and pipe outfalls. New-facility-design processes should be examined for their contribution to odor, in addition to their contribution to the quality of the wastewater effluent. For example, primary clarifiers are a significant odor source in most WRRFs. Under some conditions, it can be possible to reduce or eliminate them from the treatment process.

Other prevention measures include chemical dosing of the wastewater, preventing odorous compounds from forming either chemically or biologically. Chemical dosing is most common in collection systems; however, it has been used successfully to target areas within a WRRF. Care must be taken in selecting the chemical to ensure that there is sufficient mixing and contact time for the chemical to be effective and that the chemical does not adversely affect processes, such as biological phosphorous removal at the facility.

Before an emission can be treated, it must be captured. In designing an air-collection system, several items must be considered, in addition to simply capturing and moving odors. The primary concern in these systems must be how they affect the workers. Common mistakes include dragging odorous emissions past the workers; not considering air-movement conditions when doors are left open; and providing insufficient air movement, so that workers are forced to leave doors open. Whenever possible, trapping and capturing emissions at the source with hoods or covers is desirable. However, caution must be taken to allow worker access and for the access to be easily operated, or the covers and access doors will be left off or open. Air collection must always be designed with consideration to how the area being ventilated will be used and with consideration to corrosion.

Once captured and collected, the air may need to be treated before it is exhausted. In some cases, simply creating sufficient dispersion by exhausting the air at the proper height with sufficient velocity will prevent off-site odor effects. However, when treatment is required, it is important to select the best technology for the emission being treated. As wastewater progresses through the treatment process, its character changes—so do the emissions from that wastewater and the solids removed from it.

4.4.5 Assess Effectiveness of Control Alternatives

The best way to verify effectiveness after the measures have been implemented is to sample the new emission. Some states will require sampling to verify that emission targets established in the initial modeling and design have been met. As with the initial sampling, there is little value in sampling the area surrounding the facility. This is important if air permits require ongoing verification.

5.0 Dispersion Modeling of Odors and Air Emissions

Two gases released from the same source, at different times, under apparently identical meteorological conditions, will result in two different measured concentrations at the same location downwind. These differences are caused by the stochastic (random) nature of turbulence and diffusion in the atmosphere. For this reason, exact concentration predictions are not possible, and all predictions should be regarded as estimates of the actual values.

When air pollutants, including odorous compounds, are emitted to the atmosphere, they are carried away from the emission source by wind transport and diluted by mixing with the ambient air. Topography, meteorological conditions, and source characteristics all influence pollutant concentrations at ground-level locations downwind.

When actually performing an odor-dispersion calculation or modeling for a WRRF, this chapter or any textbook approach will be insufficient in cases of unusual atmospheric or source conditions or irregular topography. Dispersion modeling, under these conditions, requires a high degree of expertise and may involve the use of other predictive or diagnostic aids. This chapter is intended to provide guidance to the design engineer regarding what information is required by the dispersion modeler to conduct the modeling analysis and provide some information regarding what information the modeler may be able to provide that will assist the engineer in designing an odor- or air-emissions-control system.

5.1   Source Characteristics

Source characteristics are the physical dimensions or parameters that define the emission point. While most discussions of dispersion modeling focus on releases from point sources (i.e., process stack or vents), many odor sources at WRRFs are unconfined releases from open basins or structures. The following discusses the characterization of some types of WRRF releases, such as area or volume sources.

5.1.1   Point Sources

The process stack or vent parameters required for dispersion modeling include stack height, inside diameter of the stack outlet, temperature of exhaust gas, and velocity of exhaust gas as it passes through the stack outlet. The stack-base elevation is evident if the process unit has a free-standing stack. If the stack exhausts through the roof of a building, the stack-base elevation is also the building-base elevation.

5.1.1.1   Stack or Vent Design   Stacks or vents from process areas discharge momentum, heat, and pollutants. The height of the plume for a process stack or vent is governed by the temperature difference between exhaust gases and atmospheric temperature (buoyancy rise) and the difference between the velocity of the exhaust-gas stream and horizontal wind speed (momentum rise). The resulting plume rise increases the dispersion of pollutants before they reach the ground. During the design of the facility, the physical parameters of the emission source should be designed to enhance or at least not hinder plume rise and dispersion. Most air-emission sources at WRRFs are neutrally buoyant (exhaust temperatures near ambient air temperatures). Thus, the exhaust velocity or momentum is important in the release of pollutants from a stack or vent.

Strategies for improving plume rise should be included in the design of all exhaust stacks or vents. Exhaust gases from several collection points should be combined where it is safe and practical to do so. By combining flow from various process areas, an increase in a contaminant concentration from one process area is diluted before being released to the environment. The increased exhaust-gas flowrate provides greater plume rise and enhances dispersion, minimizing downwind effects. Process stacks and vents should be vertically directed and uncapped. Rain caps and U-shaped vents deflect the exhaust jet, reduce plume rise, and increase rooftop concentrations. Figure 6.8 illustrates stack designs that provide adequate rain protection without adversely affecting plume dispersion (ASHRAE, 2005).

Stack-tip downwash occurs when low-momentum exhausts are pulled downward by lower pressure immediately downwind of the stack, reducing plume rise and increasing effects at downwind locations. To minimize the effects of stack-tip downwash, the exhaust-gas velocity should be 1.5 times the design horizontal wind speed. The exhaust-gas velocity should be maintained above 10 m/s (2000 fpm) to provide adequate plume rise and dispersion. A stack velocity of 12 m/s (2500 fpm) typically prevents condensed moisture from draining down the stack and keeps rain from entering. An exit nozzle should be used to attain the exhaust velocity needed to produce plume rise and avoid downwash (ASHRAE, 2005).

5.1.1.2   Building Cavity and Wake Effects   Changes in the flow of ambient air around nearby buildings and structures can influence greatly the dispersion from process stacks and vents by lowering plume rise. The recirculation zone that forms as wind passes over a building can cause exhaust gases to be drawn to air intakes and increased

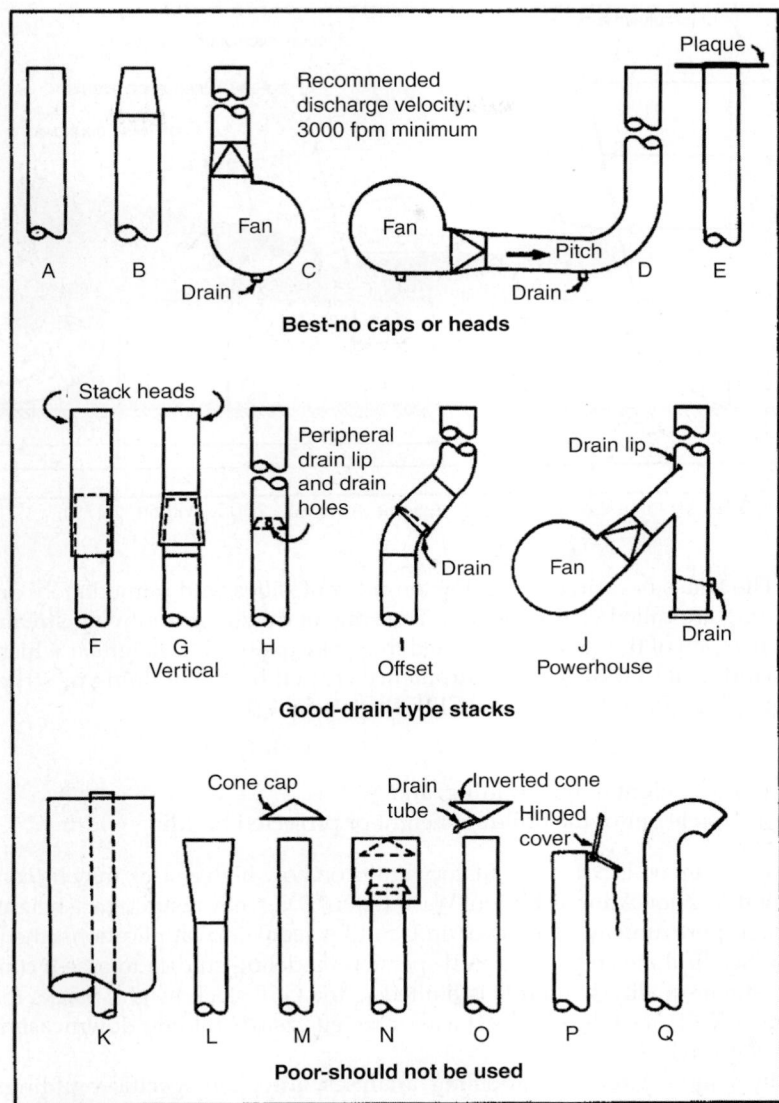

FIGURE 6.8 Stack designs providing vertical discharge and rain protection (ASHRAE, 2005) (fpm × 5.080 = mm/s).

ground-level concentrations at downwind locations. Buildings that have even moderately complex shapes can generate flow patterns that are too complicated to be treated in simple mathematical models. Most air-quality models reduce these complex structures to simple rectangular structures. To predict flow patterns from a moderately complex structure or several nearby structures requires a wind tunnel. This is depicted in Figure 6.9 (ASHRAE, 2005; Wilson, 1979).

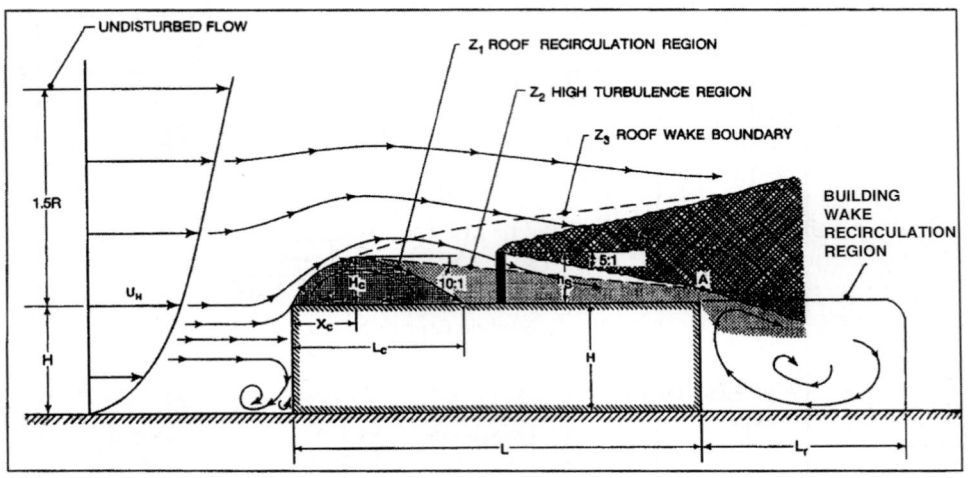

FIGURE 6.9   Building cavity and wake regions (ASHRAE, 2005; Wilson, 1979).

The zones of recirculating flow on the roof, sides, and immediately downwind of a building are called cavity regions. The zone of downward-moving streamlines further downstream of the structure is called the wake region. The height at which the horizontal wind is not influenced by turbulence created by the building or structure is called the good-engineering-practice (GEP) stack height (H_{GEP}).

$$H_{GEP} = H_B + 1.5 \times H_L \tag{6.6}$$

where H_B = height of the building, and
H_L = lesser of the building height or projected building width.

There are no U.S. EPA regulatory limits on how high a stack may be constructed. The Federal Aviation Administration (Washington, D.C.) may restrict stack heights in the vicinity of airport runways. However, in U.S. EPA regulatory applications, the portion of the stack height that can be used in dispersion-modeling studies to assess compliance with ambient-air-quality standards is limited to the GEP stack height. Stacks that are shorter than the GEP stack height must assess the effects of building downwash in dispersion-modeling analyses.

In standard refined modeling analyses, direction-specific building heights and projected widths are required as an input to the modeling analysis. The U.S. EPA Building Profile Input Program can be used to evaluate nearby structures and assign the building dimensions that would have a controlling effect on downwash effects.

5.1.2   Area Sources
Area sources are used to model low-level releases with no plume rise, such as open tanks and basins. Dispersion models have improved in their ability to adjust to the variable shape of most area sources. For large or irregular area sources, the source can be subdivided to allow for greater flexibility. The required input information for an area source includes release height, length of sides (east-west and north-south sides), orientation angle, and initial vertical dimension of area source plume.

Passive area sources are those surfaces that do not have a quantifiable vapor flux through their surfaces. Typical sources include clarifier surfaces, equalization tanks, or landfills. Mass-transfer relationships are driven by Henry's law constants and surface-boundary-layer phenomenon.

Active area sources have a quantifiable vapor flux, often driven by a fan or blower. Typical sources include aeration basins, biofilters, and aerated channels. Active surface releases may have an initial vertical dimension of an area-source plume. The velocity of air passing through the surface of the tank, basin, or channel and the density difference between moist air being released and that of ambient air drive the initial dispersion. However, the procedures to estimate this initial dispersion parameter are not defined clearly.

5.1.3 Volume Sources

Volume sources are used when the pollutant release is unconfined or has initial dispersion characteristics that are not well characterized by the standard point or area-type releases. In standard regulatory models, a virtual point-source algorithm simulates the volume source where the initial vertical and horizontal dispersion coefficients are defined. The required input information for a volume source includes release height (center of volume), initial lateral dimension of the volume-source plume, and initial vertical dimension of the volume-source plume. Several common examples are discussed below.

The building exhaust may be discharged through mushroom-type roof vents that exhaust downward toward the roof or side louvers that discharge exhaust air horizontally through the side of the building. In either case, exhaust air may be entrained in airflow around the building and initially mixed in proportion to the cross-sectional area of the building. In such cases, effects at downwind locations may be better represented by assuming that pollutants are dispersed downwind from a volume source influenced by the cavity and wake effects of that same structure.

Similarly, open doors or truck ways can be a source of odorous emissions. Odorous air from these sources is initially mixed with airflow around the structure. The effects at downwind locations may be represented by assuming that pollutants are dispersed downwind from a volume source, with initial lateral and vertical dimensions proportional to the cross-sectional area of the building. Emissions may be estimated by the odor concentration in the building times the area of the opening and the assumed velocity across the opening.

5.2 Odor-Emission Rates

Emissions are characterized as mass per unit time for point and volume sources and mass flux rates (mass per unit time per unit area) for area sources. They can be measured directly, by recording the concentration of the contaminant or odorant and multiplying by the AFR. Emission rates also may be derived from fate models, mass-balance calculations, and emission factors. Emissions from passive surfaces can be measured directly using a flux chamber, estimated using mass-emission models, or predicted through reverse modeling (back-calculating source-emission strength from the measured ambient-air concentration of the substance). Emissions from active surfaces can be measured directly using a flux chamber, recording concentration measurements near the tank surface, estimated using mass-emission models, or predicted through reverse modeling.

Emissions from volume source are expressed as mass per unit time and can be measured directly by recording the concentration of the odors and multiplying by the AFR, or estimated using mass-emission models for those process units within the structure.

5.2.1 Example Problem

A packed-bed wet scrubber was installed on a headworks facility to reduce odor effects in the neighboring community. The design AFR for the scrubber system was 4.72 m³/s (10 000 cfm). A bag sample was taken of the scrubber exhaust, and the odor-panel evaluation was conducted. The reported odor concentration was 75 (dilution ratio, i.e., odor units per cubic meter). Calculate the odor-emission rate for use in an odor-dispersion model. Define any special modeling issues.

The odor threshold is a dilution ratio. It can be illustrated as the final volume (V) of the sample divided by the initial volume.

$$\text{Dilution ratio (odor concentration)} = (V_{sample} + V_{dilution\ air})/V_{sample} \qquad (6.7)$$

While it is dimensionless, the units can be expressed as odor units per cubic meter (OU/m³). Using metric units is necessary to be consistent with the units in the dispersion model.

Multiply the odor threshold concentration by the AFR to obtain the odor-emission rate (OER). The OER will be expressed in units of odor units per second.

$$OER = OU/s\ AFR \qquad (6.8)$$

This represents the total volume of air needed to dilute odors from the scrubber stack to the odor threshold.

The odor rate may now be substituted for the mass-emission rate in the dispersion model. A simplified form of the dispersion-modeling equation is

$$X = OER \times DF \qquad (6.9)$$

where X = predicted impact (dimensionless), and
   DF = dispersion factor (s/m³).

The dispersion model may have a default emission-conversion factor to convert grams in the emission-rate term to micrograms in the predicted-concentration term. This default factor of 1×10^6 must be changed to 1.0 in order for the model to report odor units per cubic meter.

The predicted concentration term (X) is the ratio of the total volume of air needed to dilute odors from the scrubber stack to the odor threshold at a distance downwind of the plume. If the predicted concentration is 1.0, the volume of the dispersing plume is sufficient to dilute odor from the scrubber stack to the odor threshold.

5.2.2 Cautionary Note

Odor effects from different sources are not necessarily additive. Odor sources with different characteristics may be perceived differently by the receptor.

5.3 Dispersion Models

Dispersion models define the relationship between the emission source and the downwind receptor. As with any mathematical approach, dispersion models are used to predict future conditions (new source) or simulate a wide variety of conditions that would be too expensive or impossible to define using traditional measurement methods. Dispersion models can be used during the design phase of a project (when 20% to 30% of the design is complete) to assess alternative air-pollution-control strategies. Dispersion models also are used to support air-quality-impact assessments or air-quality-permit applications, to demonstrate compliance with air-quality criteria or odor-threshold levels. Such compliance demonstrations are performed when the design is 70% to 80% complete. These models are useful tools for evaluating potential effects from proposed new or modified sources, demonstrating compliance with ambient-air-quality criteria, or assessing possible causes of particular odor events at existing facilities.

Several computerized dispersion models are available from the U.S. EPA's Support Center for Regulatory Air Models (SCRAM) website (http://www.epa.gov/scram001/) and software vendors. The main reasons for using dispersion models from the U.S. EPA website are their common availability and their ability to handle multiple sources simultaneously, including point (stack), volume (fugitive), and area (aeration basin and landfill) sources. They can handle both hot exhausts and cool gases with plume buoyancy and momentum algorithms. The models listed may be used without a formal demonstration of applicability. Many of the models have been subjected to a performance evaluation using comparisons with observed air-quality data. The models have been used in toxic-air-pollutant demonstrations and odor-impact assessments.

5.3.1   Regulatory Models

Guidance for performing an air-quality analysis to demonstrate compliance for those pollutants with established ambient-air-quality standards is provided in the Guideline on Air Quality Models promulgated in 40 CFR 52, Appendix W. The U.S. EPA is primarily responsible for ensuring that the ambient-air-quality standards are protected. This authority is delegated to state and regional regulatory agencies. When conducting modeling analyses for a pollutant with an established air-quality standard, it often is necessary for a modeling protocol to be prepared, outlining the specific procedures to be followed.

Assessing the effect of toxic air pollutants in the ambient air largely has been the responsibility of state and regional regulatory agencies. The procedures vary from agency to agency. In those cases where dispersion modeling is required, the models and procedures appropriate for ambient-air-quality modeling are followed. It is recommended that a meeting with the reviewing agency be held to discuss procedures to be followed before conducting the dispersion-modeling analysis.

Compliance with odor-nuisance criteria typically is the responsibility of regional or local regulatory authorities. Odor studies may be performed as part of an odor-impact assessment or an odor-control-facility design. Such studies are used as a planning tool and may not be reviewed by a regulatory agency. However, if a regulatory agency initiates a compliance action, then odor-modeling procedures may need to be discussed before conducting the analysis.

Increasing reliance has been placed on concentration estimates from models as the primary basis for regulatory decisions concerning source permits and emission-control requirements. In many situations, such as review of a proposed source, no practical alternative exists.

5.3.2 Modeling Procedures

The general approach to an air-quality analysis or odor-impact assessment is summarized below. It follows guidance provided in the Dispersion Modeling Checklist (U.S. EPA, 1977). A modeling protocol document that summarizes key elements of the analyses should be part of a dispersion-modeling assessment:

1. Define the air-quality standard, air toxic limit, or odor-threshold level to be achieved. For pollutants with established NAAQS, the pollutant to be modeled, the period over which predicted concentrations are to be averaged, and the frequency of exceedance are defined.

2. Select the meteorological data that are most appropriate for the type of analyses being performed. Screening meteorological conditions may be entered by the user or are defined internally in the screening model. Screening meteorological data are useful when examining a limited number of sources. When evaluating effects from a larger facility where multiple emission sources are defined, a more refined model may be necessary. For refined modeling studies, 1 to 5 years of hourly meteorological data, representative of the proposed project site, may be required.

3. Obtain topographic data from the U.S. Geological Survey (Reston, Virginia). This information is available as topographic maps or electronically from digital elevation models. The need for topographic data in a dispersion model depends on the type of emission sources and surrounding topography.

4. Characterize surrounding land use. The types of land use surrounding the facility can have a significant effect on how plumes disperse and the magnitude of the predicted effect.

5. Determine the applicable dispersion model. Once the applicable standard has been defined, the type of meteorological data to be used is identified, and surrounding land use is characterized, the selection of an appropriate model should be one of the following types:
   • Screening models use limited or user-defined meteorological data and predict effects from a limited number of emission sources;
   • Refined models use hourly meteorological data and may incorporate local terrain features, and can model multiple sources and source types; and
   • Specialized models are used only under specific conditions that are consistent with their design, such as toxic releases.

6. Define the pollutant- or odor-emission rate. Emission rates are entered for each source to be modeled. Emissions may be expressed on a compound-specific basis or as odor levels.

7. Define physical-source parameters. Emission sources may be defined as point, area, or volume sources. Exhaust from a point source can be greatly influenced by nearby buildings. Building profile information may be required as input to the dispersion model.

8. Prepare modeling inputs and options. Dispersion models include control parameters, which define how the dispersion calculations will be performed. For analyses that will be used to demonstrate compliance with ambient-air-quality standards, the U.S. EPA specifies that the regulatory default options be used. In most other applications, these options are recommended also.

9. Compare model output with the corresponding standards or limits. It is necessary to understand how data will be compared with the appropriate standard before the modeling analyses are performed. The model can create a variety of output formats that cannot be obtained without rerunning the model.

5.4   Presentation of Results

In most cases, a table comparing predicted concentrations with corresponding threshold levels is sufficient. However, the output also may be presented graphically. One of the more common graphical formats is the plot of maximum concentrations. The model can prepare a plot file that contains the receptor coordinates and highest predicted concentration for that receptor. This table can be imported to a plotting package. The results are concentration isopleths, which can be plotted over a base map.

Another common output format is the frequency-of-exceedance plot. The model will create a maxi-file, which contains every occurrence that exceeds a threshold value. This can be converted to a frequency-plot file by counting the number of exceedances that occur at each receptor. This method is useful, particularly in presenting odor results where nuisance concerns are important. Where compliance is determined on a percentile basis, frequency can be related easily to percentiles by comparing the number of exceedances with the total number of predictions.

5.4.1   Averaging Periods

U.S. EPA regulatory models do not make an adjustment to the dispersion coefficients when modeling criteria pollutants (those which have NAAQS). While the modeling guidance states that the shortest time period predicted using site-representative National Weather Service (Silver Spring, Maryland) data is 1 hour, this is a conservative interpretation of the model results designed to be protective of air-quality standards. Longer averaging periods (i.e., 3-, 8-, 12-, and 24-hour averages) are arithmetic averages of these 1-hour concentrations.

For odor-dispersion modeling, prediction of odor effects for averaging periods of less than 1 hour may be needed. If the predicted concentrations need to be representative of a 10- to 15-minute averaging period, the dispersion-modeling results using dispersion coefficients could be used without modification. However, to provide an additional measure of conservatism in the assessment of odor effects, dispersion-model results can be interpreted as 1-hour averages and scaled to the averaging period of interest.

Dispersion coefficients derived from turbulence theories are not based on sampling data or a specific averaging period. The validation studies performed by the U.S. EPA used ambient-air concentrations that are averaged over a 1-hour period. Puff models are capable of shorter averaging times, if the meteorological data are available in averaging periods of less than 1 hour.

In selecting an averaging period for an odor-impact assessment, consideration must be given to the limits of such an analysis. Odor complaints are a function of odor intensity (concentration), duration (averaging period), frequency (number of odor events), and hedonic tone (offensiveness). While a person might be able to detect an odor in a breath or two (1 to 3 seconds), even an extremely offensive odor is unlikely to provoke a response, unless it occurs for a long-enough period or at a frequency to change one's actions (close a window or go inside). Odor effects that persist for a period of 3 to 5 minutes are likely to result in an odor complaint. However, it is diffi-cult to respond and verify complaints (as may be required to file an official complaint), if the peak-odor concentration is diminished after 5 minutes. Quantifying odor emis-sions at the source or effects at the facility property boundary may require a sampling period of 10 to 15 minutes, depending on the size of the sample collected and speed of the sampler pump. Even an odor that would be characterized as pleasant can result in an odor complaint, if it occurs frequently and persists for an extended period of time. Thus, an odor-impact assessment must consider many factors—not just the averaging period—in establishing an odor criterion.

5.4.2   Peak-to-Mean Scaling Factors

Cramer (1959) proposed a 1/5 power law based on observations of concentration fluctuations near the plume centerline. Because sources of emissions from WRRFs are neutrally buoyant and relatively low-level (near ground level) releases, maxi-mum effects are likely to occur at or near the plume centerline. Compliance with ambient-odor criteria also requires that the highest predicted concentrations are below the established limit; thus, peak concentrations at or near the plume center-line would be the basis for demonstrating compliance. The power law relationship is defined as follows:

$$C_1 = C_0 \times (t_0/t_1)^p \qquad (6.10)$$

where C_1 = concentration at desired averaging period,
$\quad\;\; C_0$ = initial (1-hour average) concentration,
$\quad\;\; t_0$ = initial (60-minute) averaging period,
$\quad\;\; t_1$ = desired averaging period (minutes), and
$\quad\;\; p$ = power law exponent (0.2).

Several authors, including Wang and Skipka (1993), have proposed methods for deriving stability-dependent peak-to-mean ratios. Stability-dependent peak-to-mean ratios make intuitive sense when one considers the concentration fluctuations in a looping plume under unstable conditions or a meandering plume under stable conditions.

This approach is widely used by odor-modeling practitioners as a means of apply-ing additional conservatism into the modeling analysis to address many of the inherent limitations of odor-impact assessments. Such limitations include limited sampling data to quantify the odor-emission source, sensitivity of the observer to emitted odors, and difference in sampling times vs. odor-response times.

There are several common errors in applying this methodology. The most common error is to apply the largest peak-to-mean factor for unstable conditions to predicted odor concentrations that most likely occurred under stable conditions. While providing a conservative overestimate of actual peak-odor concentrations, this can lead to over-predictions and unnecessary control strategies.

The next most common error is to define a peak concentration that is contrary to the odor-control strategy intended. The response time for an observer to detect an odor, recognize the odor, find it offensive, make a complaint, and have that complaint vali-dated as a nuisance condition can vary considerably. The peaking factors should be calculated and adjusted for the averaging period intended.

6.0   Emissions Containment and Ventilation

Technology to treat nuisance odors and VOCs that are released from municipal and industrial WRRFs is available. However, if the capture, containment, and ventilation systems are not able to keep the fugitive odors from escaping, or proper operations and maintenance (O&M) does not preserve the integrity of the system over time, then the money and the effort spent on the control devices have been wasted. Therefore, a critical element of any odor-treatment system is the effective design of containment and ventilation systems. These systems dictate the size of the odor-control system and also prevent the emissions from escaping before the control devices can treat them. This section provides information on how to contain the emissions at the source, minimize the volume of air to be treated, and capture and transport the emissions to the selected treatment device(s). These items are achieved through the following:

- Odor containment, by enclosing wet wells, pits, open channels, diversion structures, tank surfaces, solids storage, or process equipment;
- Odor capture and transport, by providing negative pressure ventilation within the enclosure;
- Odor treatment in appropriate devices; and
- Treated-air discharge with adequate dispersion to enhance dilution and minimize visual effects on the surrounding area.

6.1   Odor Containment

There are a full range of available odor-cover and containment-systems alternatives. The selection of the appropriate cover and containment system depends on several factors:

- Area climate—heat gain or freezing issues;
- Worker safety—routine access or confined-space issues;
- Ease of construction;
- Operability and maintainability—access to equipment and ease of sample collection;
- Aesthetics—visual exposure to neighbors;
- Effectiveness;
- Durability—correct choice of materials and coatings; and
- Cost of operation—reducing airflow reduces treatment costs.

With these important factors in mind, there are several different varieties of covers or enclosures available.

6.1.1   Flat Covers

Flat covers have become more popular as the cover of choice, because their installation minimizes air space between the cover and water surface. Minimizing that air space is critical to the ultimate sizing of the transfer ductwork, blower capacity, and odor-treatment technology. The smaller the odorous headspace, the smaller the volume of air

FIGURE 6.10 Flat-channel cover—Roger Road WWTP, Pima County, Arizona.

that needs to be removed and treated. This translates to a smaller capital-cost investment to construct the odor-control system and reduces the annual operating cost.

Flat covers have been designed commonly for channels, wet wells, scum pits, tanks (square, rectangular, and circular), distribution boxes, and diversion structures. They also can be effective in areas where only partial coverage of a source is required, such as primary clarifier weirs. Examples of flat-channel covers are presented in Figure 6.10, and tank covers are presented in Figure 6.11. Depending on the application, it may be necessary for a particular flat-cover installation to require internal or external structural

FIGURE 6.11 Flat tank cover—Joint Water Pollution Control Plant, Los Angeles County Sanitation District, California.

FIGURE 6.12 Flat circular tank cover with external truss—Hamilton Township Water Pollution Control Plant (WPCP), Hamilton Township, New Jersey.

support. External support is preferred in most situations, despite the sometimes nonaesthetic appearance. Internal supports are continuously exposed to corrosive gases, which could affect the covers' structural integrity and reduce their longevity. Examples of flat covers with external support for circular and rectangular tanks are illustrated in Figures 6.12 and 6.13, respectively.

When using flat covers, the designer is cautioned to provide sampling and observation access for operators in addition to removal features for access and safe entry where appropriate. Facility personnel have operated WRRFs for decades by using their senses to judge performance and influent characteristics. Restricting the operators' visual and

FIGURE 6.13 Flat rectangular tank cover with external support—91st Avenue WWTP, Phoenix, Arizona.

FIGURE 6.14   Barrel arch covers—Metro WWTP, St. Paul, Minnesota.

olfactory reference hinders their ability to effectively anticipate and react to changing wastewater conditions.

6.1.2   Barrel-Arch Covers
Barrel-arch covers have a semicircular cross section and are appropriate for relatively narrow tanks or channels and are not used for extensive wide-open surface areas. Figure 6.14 presents an example of a barrel-arch system. Items affecting the height of the arch include desired clearance, structural needs, aesthetics, and visibility of the water surface. View windows and inspection/access hatches, both at the ends and along the length of the cover, provide the operator with an extended and uninterrupted view of the water surface. However, the designer must consider the available light and fog conditions under the cover when considering view windows. When large access areas are required, barrel arches can be removed by sections or can be equipped with a rollback feature anywhere along their length. An example of a barrel arch, with the features and flexibility listed above, is shown in Figure 6.15.

6.1.3   Pitched Covers
Pitched covers are similar to barrel arches, in that they are appropriate for long distributed areas with generally narrow width. The visual effect of their exterior profile is similar also. The pitched cover slopes up evenly on both sides to a peak, like the roof on a house. One potential disadvantage of the pitched cover is that it may require, depending on the width of the opening it spans, an internal truss system to support the panel covers. This truss restricts interior visibility and eliminates the ability to roll back the covers. The covers have to be removed by section to allow access within the enclosed area. Therefore, pitched

Figure 6.15 Barrel arch covers—Coney Island WPCP, New York.

covers are most applicable to areas requiring limited visibility and minimal worker access. The application of pitched covers at WRRFs is limited.

6.1.4 Building an Extension or Enclosure

When frequent worker access to an area is required, some form of building is used. This can take the form of an extension of an adjacent structure or a stand-alone building, such as a penthouse or dome. Figure 6.16 illustrates a penthouse enclosure; Figure 6.17 is an example of a geodesic dome. Because these types of enclosures are used when frequent worker access is needed, the designer must adhere strictly to health and safety requirements. Even with high ventilation rates, the area may be considered a confined space with the associated entry requirements each time the enclosure is entered. Requirements are likely to include training courses, filling out forms, and air monitoring (both before entry and during presence in the enclosure).

The designer is cautioned that simply providing high ventilation rates does not eliminate potential safety hazards. The proper placement of ventilation-air-supply and odorous-air-exhaust registers is critical to ensuring that fresh air being supplied is directed at the workers and that dead zones of stagnant air are not created. The general rule is to design the ventilation system to withdraw odors at the point at which they are released. For example, in buildings or enclosures, the source of odor release typically is at the wastewater or solids surface. To keep the workers in a fresh air environment as they stand above the wastewater or solids surface, the ventilation-air-supply registers should be located at the enclosure ceiling over work or high-maintenance areas, while exhausting odorous air at or near the wastewater or solids surface. This way, the odors

FIGURE 6.16   Penthouse enclosure—North River WPCP, New York.

FIGURE 6.17   Geodesic dome enclosure—Fort Dix WWTP, Fort Dix, New Jersey.

are not being dragged across the worker's face, but rather the worker always is benefiting from the fresh air supply as it flows from the ceiling, across the body, and to the source, where it is removed by the exhaust register.

At most WRRFs where hydrogen sulfide and other reduced-sulfur compounds are key nuisance-odor elements, this ventilation scheme of exhausting the odors at the source before they reach worker spaces is advantageous. Hydrogen sulfide gas and other reduced-sulfur compounds are heavier than air and tend to accumulate in low-lying areas.

6.1.4.1 Dome Enclosures Although buildings and enclosures can be placed over any shape basin or unit-process odor source, the dome generally is used to cover circular basins. Circular basins at WRRFs are commonly sedimentation tanks, gravity thickener tanks, solids storage tanks, dissolved air flotation units, or trickling filters. Like the barrel arch, the dome structure has an open interior volume (no trusses are necessary for support) and is variable in its height of rise; the greater the height of rise, the greater the volume of odorous air needing treatment—particularly when occupied.

A standard feature for a dome covering a circular clarifier or sedimentation basin includes a single door, provided to gain access to the dome interior. This means that the dome is a worker-space area, and, as such, the full volume of air within the dome should conform to worker safety standards. The following options are available to reduce the volume of air needed to ventilate the open interior air space:

- Providing a fresh air supply directly over the areas where worker spaces are located when staff members are present.
- Avoiding the need for the worker to enter the confined environment of the dome. Independent accessways are available in a geodesic dome structure. These accessways are separate and independent from the environment of the internal dome. Figure 6.18 depicts a walkway open to the atmosphere, and Figure 6.19 shows an enclosed geodesic dome walkway, separate from the internal odorous air of the dome.

Figure 6.18 Geodesic dome enclosure with walkway open to atmosphere—23rd Avenue WWTP, Phoenix, Arizona.

Figure 6.19   Geodesic dome enclosure with enclosed walkway—Hamilton Township WPCP, Hamilton Township, New Jersey.

A summary of advantages and disadvantages of the walk-in cover alternative and the independent accessway cover alternative is presented in Table 6.16.

6.1.4.2   Equipment Enclosures   Previous discussions have highlighted how odors can be contained at wastewater and solids sources, such as channels, wet wells, pits, tanks, and diversion structures. The following sections address how odor released by the following wastewater and solids equipment is contained:

- Bar screens;
- Belt-filter presses (BFPs) (dewatering);
- Belt thickeners (BTs) (thickening); and
- Centrifuges (thickening and dewatering).

6.1.4.3   Bar Screens   The challenge posed by containing odors from bar screens is that they are mechanically cleaned, so the bar-rack structure, which captures the unwanted material, extends from the water surface below grade to well above (3 m or more) the concrete access floor. Based on this operation, an opening exists through the concrete access floor, through which the bar-rack passes.

Typically, odor associated with fresh screenings material removed from influent wastewater and dragged up the bar-rack does not cause off-site odors. When allowed

Walk-in cover alternative:	
Advantages	**Disadvantages**
Full access	Require high air-exchange rates for worker access
Full visibility	High volume of air to treat
– Wastewater surface	Considered confined space
– Mechanical equipment	Larger odor-control system
– Weir overflows	High external visibility
– Surface collectors	If confined space, need entry permit
Ease of O&M	Need to assess air environment before entering
Limited difference vs. uncovered	Higher cost
Easy-to-collect samples	
Independent accessway cover alternative:	
Advantages	**Disadvantages**
Worker performs tasks in pleasant environment	Limited visibility
	Limited accessibility
Minimizes air to treat	More difficult to perform O&M
Smaller odor-control system	Need to remove cover panel to access internals
Worker unable to enter confined space	
Low profile	
Motor drives are outside cover	
– Easy to maintain	

TABLE 6.16   Comparison of Process-Unit-Cover Alternatives (Walk-in Versus Independent Accessway)

to accumulate in a container, the putrescible material will become odorous. However, the opening in the concrete access floor for the bar-rack allows wastewater odors, which have been traveling with the wastewater in the sewer and facility influent channel, to escape to the atmosphere (in locations where the bar screen is out in the open) or to the room air (in locations where a building surrounds the bar screen). The key to controlling these higher off-site potential odors associated with the headspace of the wastewater channel is to prevent them from escaping to the atmosphere. The following three alternative approaches provide guidance for controlling odors from bar screens and have been proven effective in specific field situations:

1. Create a zone of negative air influence around the bar-rack opening. This approach assumes that the wastewater channel upstream and downstream of the bar-screen opening is covered, with the opening at the bar screen allowing odors to escape. Exhaust ventilation pickup points are positioned in covered channels upstream and downstream of the bar screen. The design volume of air for each position is determined by the cross-sectional area of the air space, multiplied by a face velocity that will secure that a negative air pressure is maintained across the opening.

The channel opening typically is relatively large, with a correspondingly large ventilation flow requirement. Modified plastic strip curtains are used to reduce the opening to sustain the negative air pressure influence and face velocity and reduce ventilation flow. Placed as close to the bar-rack as possible, without interfering with normal operation at both upstream and downstream channel locations, the polyvinyl chloride (PVC) strip curtains are tied to the underside of the channel cover or hung off a support bar. The lengths of the PVC strips are designed to barely reach the water surface at the high-water elevation of the channel.

The PVC strips will have a cutout at the point of overlap ranging from 1.3 to 2.5 cm (0.5 to 1 in.) in width. The ventilation is calculated by determining the cutout open area and applying a face velocity ranging from 2.5 to 5.0 m/s (500 to 1000 fpm). The lower face velocity range is applied to bar-screen channels that are indoors, while the higher range is applied to outdoor or exposed installations, where atmospheric wind conditions are involved. The zone of influence created reduces the release of wastewater-related odors at the bar screen and acts similarly to an odor hood.

2. Place a cover over the bar screen from the point at which it penetrates the concrete access floor to the top of its rise. This approach assumes that the wastewater channel upstream and downstream of the bar-screen opening is covered. There is minimal clearance between the working bar screen and the cover itself.

   These covers typically are painted metal and are provided by the bar-screen manufacturer or can be custom built. These covers or enclosures provide access to critical bar-screen mechanical areas, so that staff can perform routine O&M and observe their operation. Some O&M procedures require that panels be removed for staff access. It is imperative that the panels be replaced afterward; otherwise, the integrity of the odor containment will be lost and odors will escape.

3. Place the bar screen within a building and provide full worker-space ventilation. No special covers are required; however, the volume of air needed to capture and contain the odors is significantly greater than the first two approaches. In this approach, the ventilation layout should be designed to pick up the odors at the source, rather than drag them through the worker space of the building. Accordingly, supply registers provide fresh air at a high location, and exhaust registers are located near the floor/water surface level.

6.1.4.4 Belt-Filter Presses and Belt Thickeners   Belt-filter presses and belt thickeners (BFP-BT) both process wastewater sludge and biosolids, which can be odorous. Assuming that odor containment is necessary, the following three approaches have been applied successfully:

1. Pickup hoods have been designed to capture odors released by their operation. These hoods are suspended from the ceiling above the BFP-BT and create a negative air influence primarily to the top side of the unit. The reader is referred to the American Society of Heating, Refrigeration and Air Conditioning Engineers, Inc. (Atlanta, Georgia) (ASHRAE) *HVAC Applications Handbook—* Chapter 30: Industrial Local Exhaust Systems to Assess the Hood Ventilation Rate (ASHRAE, 2003).

Although hoods can be effective in capturing odors released from the gravity section of the BFP-BT, they are not always successful in capturing odors released from the dewatering rollers and/or the sump. The nature of the solids plays an important role with regard to odor potential. For example, fresh waste activated sludge is less odorous than stored or anaerobically digested biosolids.

The sump odor can be managed by hard-piping the gravity and dewatering section filtrate directly to the sump drain and avoiding the turbulent sump splashing that readily strips odors.

2. Some manufacturers offer a painted metal cover for their BFPs-BTs. This cover is more like an enclosure that fits like a glove and surrounds the unit, restricting operator access and visibility. In some cases, instrumentation replaces the need for operator visibility. Similar to the bar-screen enclosure, the BFP-BT enclosure has access panels that must be removed to perform certain O&M procedures. If the panels are not replaced, the ventilation and the containment are compromised. The ventilation requirements are provided by the manufacturer and will vary with the size and make of the unit. Custom enclosures also are available, with the design left to the specifications of the owner or engineer.

3. The most common approach used to contain BFP-BT odors is to place them in a building or an enclosure in the building and ventilate the whole building or enclosed space. The volume of air needed to accomplish this containment is significant, as BFPs-BTs commonly are located in large rooms with high ceilings. Worker-space ventilation requirement apply (generally 12 air changes per hour [AC/h] for the whole room volume, ASHRAE).

To reduce the ventilation volume, some owners have placed permanent walls (i.e., block walls), plexiglass walls, or more temporary walls (plastic strip curtains) around the BFPs or belt thickeners. By isolating the units in this way, the intent is that only the volume within the BFP-BT walled area needs to be contained and odor controlled. Containment is more effective with a more permanent or structured wall arrangement (block or plexiglass) when compared with the loose-fitting temporary wall style (plastic strip curtain). The plastic strip curtain is hung from the ceiling or an intermediate frame and often is movable (like a shower curtain) to open the area when the unit is out of service. The same worker-space ventilation requirement principles apply; however, the volume of the contained area is reduced considerably.

6.1.4.5 Centrifuges Centrifuges are ideal process units for dewatering or thickening sludge and biosolids when odor control is required. Centrifuges essentially are closed devices, with the exception of the filtrate and solids drop chutes. Minimal ventilation rates are necessary for containing odors. Typically, flows of 70 to 120 L/s (150 to 250 cfm) per drop chute are all that is required for containment. Taps for these pickup points can be located in the vertical leg of the drop chutes, relatively close to the centrifuge. To avoid draining filtrate or solids to the ventilation duct, the connection is positioned to pull upward from the vertical leg. The designer is cautioned to follow the centrate drainage path to avoid fugitive emissions at manholes or other discharge points.

6.2 Materials Used for Odor Containment

Selecting the appropriate materials of construction for the cover is a key concern for the designer. In general, the materials of construction should be selected to provide durability, ease of maintenance, performance, corrosion resistance, and low cost.

Site-specific conditions are important when evaluating the appropriate cover material of construction, particularly regarding the potential corrosive nature of the atmosphere being contained. Certain cover types may limit the choices of materials to be applied. In addition, the material to be selected also depends on the anticipated length of service. The proper selection of the materials of construction should be made to fit the expectations and needs of the service for which they are intended.

6.3   Access

To understand access needs and accommodate operator concerns, the designer is encouraged strongly to involve the operator and maintenance staff early in the conceptual cover design process. This way, the cover will include the appropriate appurtenances (i.e., accessways) and features that best suits the needs of those who will use it every day—the operators. The design intent is to make the cover system more operator friendly, so that routine servicing and O&M have a better chance of being done.

In general, cover accessways can be broken down into the following two broad categories:

1. Small openings that are large enough to provide operator access, but small enough to restrict personnel entrance to the covered airspace. These cover accessways would allow for the following:
   — Visual observation of wastewater and sludge surfaces; corrosion to walls, cover and metal supports; aeration patterns; spray headers and nozzles; and surface foam and scum accumulation;
   — Wastewater or sludge sampling;
   — Cleaning weirs;
   — Measuring sludge blanket depths; and
   — O&M procedures.
2. Large openings that allow for personnel access to the process unit to:
   — Conduct in-tank mechanism removal;
   — Clean an out-of-service unit; and
   — Perform a tank overhaul.

The type of cover accessway depends on the materials of construction of the cover. Cover vendors typically offer a wide range of accessways that the owner can choose from. Generally, cover accessways need to be sealed to the best extent possible, to minimize the fugitive release of odors from around the seams. Traditional gaskets made of reinforced rubber or flexible material (e.g., Hypalon [DuPont Performance Elastomers L.L.C., Wilmington, Delaware] and ethylene propylene diene monomer) currently are used to provide a good seal at the external seams.

It is better that the cover accessway has a hinged access door, particularly in an area where the accessway is used frequently by the operations staff to execute their duties. The chances of closing the cover are much higher if it is attached than if it is loose. The designer is cautioned that the hatches or doors must be light and easily opened or supplied with lift-assisting devices, or they will be left open.

6.4   Ventilation Rates

Negative air pressure must be maintained in the air space under the cover, building, or enclosure, to contain and prevent the release of nuisance odors to the atmosphere.

The negative air pressure in the enclosed air space prevents fugitive releases by creating an inward flow of air through any cracks, vents, or designed openings.

A fan is designed to create this negative air pressure in the air space by venting the odorous air at a rate that is greater than the supply of air to the same air space cavity. The volume of air vented from the space over time is referred to as the ventilation rate. The ventilation rate will vary based on site-specific information, such as the type of containment, presence or absence of workers, and whether air being diffused to the wastewater process unit is being contained.

The ventilation rate required to contain odors, reduce corrosion, provide for a safe and comfortable worker environment, and minimize the volume of air to be treated can be selected based on experience with designing similar systems. Designers typically rely on building codes, which include mechanical, fire protection, and electrical codes, and other industry standards as the basis of their ventilation designs. Unfortunately, ventilation rates and methodologies seldom are listed in state and local building codes for municipal WRRF-related spaces. Certain outside groups, such as the National Fire Protection Association (Quincy, Massachusetts) (NFPA) and Factory Mutual (Johnston, Rhode Island), provide information on recommended practice for ventilating specific areas. For example, the NFPA has developed a ventilation standard entitled *Standard 820—Standard for Fire Protection in Wastewater Treatment and Collection Facilities* (NFPA, 2016). This standard has been written in codifiable language; however, to date, it has not been adopted formally by any state code. The standard links the fire protection safety of a structure to its ventilation rate. However, the standard does not intend to assist the ventilation designer regarding the proper ventilation rate for proper worker safety, comfort, or odor-control containment (for more information about NFPA 820 as it relates to fire protection safety and electrical code classification, refer to Chapter 7).

Before developing ventilation rates for covered unit process areas, it is important to understand their accessibility to staff. Ventilation rates are different for the two categories of non-worker- and worker-accessible spaces. A higher ventilation rate is applied to covered unit process areas where staff will be present performing their required duties. As a general rule (ASHRAE), the following ventilation rates have been applied:

- Non-worker-accessible spaces—4 to 6 AC/h, and
- Worker-accessible spaces—12 AC/h.

However, today, multiple approaches are evaluated to assess independent ventilation rate needs. The limiting or higher ventilation rate determined by these various approaches then is applied generally to the covered-unit-process area.

6.4.1 Non-Worker-Accessible Space

Ventilation for non-worker-accessible spaces is used to purge the enclosure and limit corrosion to the covering system, controls, and the concrete or metal structure itself.

The appropriate ventilation flowrate in cubic meters per second (m^3/s) (cubic feet per minute [cfm]) is determined by evaluating as many as three approaches. The first approach is determined by estimating the enclosed volume, assuming the average low water or sludge level, applying an air-change rate (AC/h) to that volume, and then calculating the ventilation rate. As a general rule for non-worker-accessible areas, the minimum air-change rate should be 3 to 4 AC/h, to maintain negative pressure, and the common range should be 3 to 6 AC/h.

When comparing the first ventilation flowrate approach of applying an air-change rate to the next two approaches, one should consider at least calculating the minimum air-change rate of 3 to 4 AC/h for the space in question, to assess which rate should be considered the lower end, where protecting the internal infrastructure is an important concern.

The second approach, which is significantly more difficult to calculate, is the calculation of flowrate across all openings in the cover system, at a given pressure differential. An estimation of the cracks and seams between cover plates, valve and gate stem openings, hatches, control ports, inspection ports, and so on, must be inventoried to develop an overall opening size (area). This cover porosity (in square meters or square feet), multiplied by a specific velocity (meters per second [feet per minute]) through the opening area, will determine the ventilation system AFR, at a specific differential pressure. For example, a velocity of 6.4 m/s (1265 fpm) is required to develop a 0.02 488-kPa (0.1-in.) water gauge pressure differential between the covered air space and atmosphere. The assumption made here is that the pressure differential is equivalent to the velocity pressure. The NFPA 820 (2016) and other design guideline books state the 0.024 88-kPa (0.1-in.) water gauge differential as a design standard for odor-control-containment systems.

The third approach takes into account the ventilation required to reduce high concentrations of particular compounds. High concentrations of particular compounds, such as hydrogen sulfide, can have multiple effects, including the following:

- Corrosion to concrete and metal surfaces;
- Inability of the odor-control technology to achieve the desired exhaust concentrations; and
- Limiting access for routine sampling, observation, and maintenance operations.

The ventilation rate increase is based on site-specific information. When assigning the design ventilation rate for non-worker-accessible spaces, the values for each of the three approaches presented are evaluated and compared. Typically, to be conservative, the highest ventilation rate of the three is selected as the design ventilation rate.

Additional methods are used to determine the design ventilation rate for non-worker-accessible spaces. For example, some vendors of flat-cover systems recommend the application of a standard flowrate per unit surface area ($m^3/s/m^2$ or cfm/sq ft). Some vendors also are willing to guarantee containment when using their flat covers and applying their standard flowrate per unit surface area. These standard values vary considerably between vendors and generally do not take site conditions into consideration. It is highly recommended that, when considering using standard flowrates per unit surface area, the owner and the consultant work closely with the vendor, compare the calculated ventilation rates with the first three approaches presented above, and seek performance guarantees from the vendor.

The above ventilation flowrate discussion for non-worker-accessible spaces assumes that the liquid or solid surface beneath the cover is non-aerated. Higher ventilation flowrates are needed to account for air diffused through the liquid-solid and released at the surface. The easiest and proven approach to estimate the adjusted ventilation rate is to first calculate the ventilation flowrate as if no diffused air was present, following the procedures discussed in previous paragraphs. Next, the flowrate of air being injected or diffused to the liquid or solids matrix should be determined.

If the AFR varies, the maximum flowrate should be used, to be conservative. Then, this injected or diffused AFR should be added to the non-aerated calculated ventilation rate. The sum of the two flowrates represents the design ventilation rate for the covered and contained airspace.

The ventilation system layout for nonworker cover or enclosure areas should include exhaust points, but no forced air supply. Rather than a forced air supply, a passive makeup air device (i.e., an intake box, louver, or gooseneck pipe) is recommended and positioned strategically away from the exhaust to achieve the desired sweep direction of airflow motion. To avoid the release of odors through the makeup air device, it should be counterbalanced to close, in the event that positive pressure builds up in the enclosed headspace.

In some instances, the makeup air device will not be needed, as other sources of supply air exist. For example, when covering facility influent channels, air in the headspace of the sewer commonly provides makeup air. For other facility locations, such as loose-fitting covers, openings (that are not addressed) around bar screens, slide gates, and valve structures, makeup air will be supplied through these existing cracks and openings, and makeup air devices are unnecessary.

Special consideration should be given to ventilating the influent channel of the WRRF, based on physical characteristics of the influent sewer. For those facilities that do not have a common headspace between the influent sewer and facility influent channel, ventilation determinations should follow the approaches previously presented.

For those facilities where the influent sewer has a common headspace with the influent channel, additional ventilation volume is necessary to account for the positive pressure of the sewer air space volume. To estimate the additional ventilation volume, a review of sewer hydraulics is necessary to determine the critical flow regime that will create the worst-case positive pressure conditions.

6.4.2 Worker-Accessible Space

Worker-occupied buildings and structures typically are ventilated at higher air-change rates, in accordance with NFPA 820 guidelines (2016). Supply air typically is filtered, tempered to 12.8°C (55°F), and supplied mechanically to areas generally occupied by workers. The exhaust side of the ventilation system should draw air from spaces where hydrogen sulfide concentrations are greater or air movement is limited. Another more aggressive form of exhaust ventilation is to cover the odorous process and exhaust from within this enclosure or space (i.e., covers located over influent channels within a headworks building or installation of a hood over a grit dumpster). The common theme is to contain the odors and capture them at their source.

A summary of the ventilation-rate-evaluation approaches for both worker- and non-worker-accessible spaces is illustrated in Figure 6.20.

The tank-ventilation rates are dependent on the effectiveness of the cover or enclosure system design to be tight (i.e., without leaks). Accordingly, the final tank-ventilation rate selection also is dependent on the type of cover or enclosure system, because some are more leak resistant.

The main objective in designing the ventilation system layout for an occupied space is to provide fresh makeup air in worker-space areas in such a way that the worker is bathed in fresh air. Designing for this objective gives the worker a more comfortable environment to work in, minimizes worker-safety-related concerns, and improves worker attitude and, thereby, productivity.

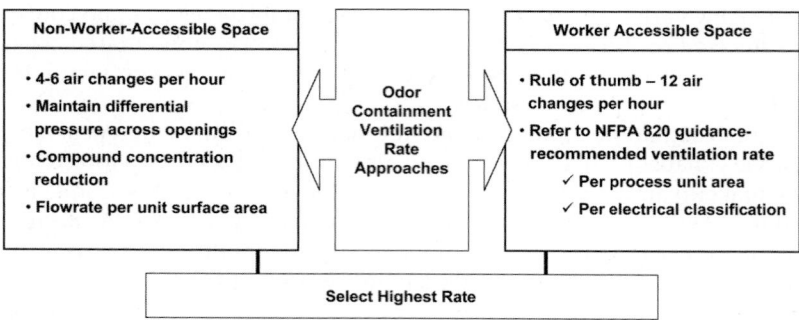

FIGURE 6.20 Summary of the ventilation rate evaluation approaches for worker- and non-worker-accessible spaces.

For worker-space areas, forced (fan) air supply and (fan) exhaust are recommended, with the exhaust AFR selected to exceed the supply AFR by no more than 10%. The higher exhaust AFR provides for negative air pressure in the building or enclosure. The NFPA 820 (2016) stipulates forced air supply and exhaust for most worker-space applications. This approach provides for better control of the worker environment and of negative air pressure in the enclosed space. Passive air supply (i.e., open louvers) and forced air exhaust provide unidirectional airflow and account for stagnant air zones for those areas of the enclosure not between the passive louvers and active exhaust registers. Odors and corrosive compounds accumulate in these stagnant air zones and create a poor working environment for staff and possible damage to the enclosure infrastructure.

The ventilation system layout for worker-space areas is site specific. However, there are a few design criteria rules of thumb to be considered, including the following:

- Position ventilation system exhaust points at or near the source of odors. Avoid fouling worker-space areas.

- Understand the odorous components of the controlled air stream. For example, hydrogen sulfide is heavier than air. To efficiently capture the hydrogen sulfide odor, the exhaust points should be low—near the floor level or water surface.

- The makeup air supply is located high—near the roof or ceiling of the enclosure, while the odorous exhaust is low—at the floor level, where the source of odors typically is found.

- This positioning also supports the concept of keeping the workers immersed in a flow of fresh air.

- Avoid positioning supply and exhaust registers at equal elevations and on the same enclosure wall, as this can support short-circuiting of the ventilation system. Short-circuiting will occur even though the correct ventilation rates are used.

- Forced air supply and exhaust are recommended (NFPA 820, 2016, stipulates forced air supply and exhaust for most applications) for worker-space areas. Forced air supply and exhaust provide the best ventilation coverage of enclosures. Passive louvered air intakes for supply air, for example, are too unidirectional and allow for short-circuiting and dead zones (areas with no apparent ventilation or air motion), where odors and harmful compounds can accumulate.

- Equip the makeup air supply and exhaust registers with volume dampers to control and balance the rate of airflow. Volume dampers are useful when field-flow-balancing the ventilation system.
- The makeup air supply should be somewhat less (10%) than exhaust to create negative air pressure within the enclosure.

These criteria will help in the design of an effective ventilation system that provides for a comfortable working environment also.

6.4.3 Confined Spaces

Odor-control structures may create circumstances that increase hazards associated with entering confined spaces. Employers have a legal and moral obligation to identify confined spaces in their facilities and effectively communicate hazards that may be present in those spaces to employees and contractors who may enter them. In states with approved OSHA plans or state regulations, employers are required to develop, maintain, and implement an effective confined-space-entry program. The designer will find it prudent to provide hazardous-gas-detection equipment and alarms for all potentially dangerous enclosed areas, even if not considered a confined space.

6.5   Ductwork and Fans

The ability to transport the captured and contained odors from the source to the odor-control equipment rests on the design of a duct network and fan. The fan provides the driving force to create the movement of odorous air through the ductwork. It is just as important to design the ductwork to minimize leaks as it is to reduce leaks from covers and enclosures. Each element in the odor-control system (i.e., covers, containment, conveying, and treating) should be considered as a link in the odor-control chain; the chain will be only as strong as the weakest link. As a result, one should insist that attention to detail be provided for the design of each element.

6.5.1   Ductwork

6.5.1.1   Shapes   Duct shapes used for odor control typically are round or rectangular. In some cases, the selection of duct shape depends on the material of construction. For example, aluminum duct construction is mostly rectangular. In other cases, tight-fitting installations may dictate the shape, as a result of limited space to route the duct.

The round shape is more efficient; that is, for the same cross-sectional area, the round shape has a smaller circumference than a rectangular duct has perimeter. As a result, less material is needed. For FRP—the material with the highest corrosion resistance to odors encountered in WRRFs—it is easier to make a round duct or pipe using filament-wound techniques that provide a high degree of quality control and minimize the need for hand layup.

6.5.1.2   Sizing Criteria   Duct-sizing criteria are based on the odorous AFR and pressure maintained within the network. A good design range for the odor-control duct is 10 to 15 m/s (2000 to 3000 fpm), with duct-pressure drop losses of 0.025 to 0.062 kPa (0.1 to 0.25 in.) water gauge per 30 m (100 lin ft) of ductwork. Velocities can exceed

15 m/s (3000 fpm); however, the designer must be prepared to enhance the design of the duct and fan to account for the following:

- Increase in pressure losses throughout the duct network;
- Modifications or redesign of the fan to transport the same volume of air at greater static-pressure requirements;
- Potential to generate noise at higher velocities; and
- Potential need for reinforcing the duct (increasing the thickness) to accommodate the increased pressure.

Pressure-drop calculations should be performed systemwide, in accordance with the ASHRAE *HVAC Applications Handbook* (ASHRAE, 2003), or Sheet Metal and Air Conditioning Contractors' National Association, Inc. (Chantilly, Virginia) (SMACNA) *HVAC Systems Duct Design Manual* (SMACNA, 1990) standards.

Accommodations must be made within the duct design to account for thermal expansion and contraction. How severe this process will be depends on extremes of temperature experienced at the site and whether ductwork is inside or outside. Expansion joints are used to address thermal expansion and contraction.

Ductwork must be sloped to a drain located at low points of the system. Expansion joints should be located at high points in the duct network, because they tend to inhibit the flow of water in the duct. The air collected at the odor source typically is moist. As temperatures decrease along the ductwork, condensation occurs, and water accumulates in the ductwork. Stainless steel and FRP, commonly used in outdoor situations, behave somewhat differently. Stainless steel is an excellent conductor of temperature and condenses the moisture out of the air stream during the cooler months. Fiberglass-reinforced plastic is more insulated and less of a conductor. Nevertheless, condensation will occur, but to a lesser degree.

Ductwork drains should include a water seal trap, which allows the condensate to drain and does not allow ambient air into and/or odorous air out of the system. The water trap should be sized based on the operating duct system pressure, to avoid forcing the water out of the trap. Local weather conditions should be considered to determine whether heat tracing is required to keep the drain line and trap from freezing in cold months.

Placing the ductwork upstream of the fan provides enhanced odor containment. The fan in the downstream position creates a vacuum (negative air pressure) along the length of the ductwork. Accordingly, any leaks within the ductwork will pull air in through the leak, thereby preventing any fugitive release of odors.

6.5.2 Fans

Fans represent the heart of the odor-control system, by creating negative air pressure at the source, providing containment, delivering the odors to the control technology, and discharging the treated air to the atmosphere. The most common odor-control fans are base-mounted centrifugal fans with backward-inclined wheels. The backward-inclined wheels ensure that the motor will not overload under varied system-pressure occurrences. The base-mounting arrangement allows for ease of maintenance on the bearings, belt, and motor. All fan housings should be drained to limit the buildup of condensed liquids. Other fan arrangements include base-mounted radial-bladed fans and in-line centrifugal fans.

Because of the corrosive nature of the odorous air stream, resistant materials are required, particularly when the fan is positioned upstream of the odor-control technology. The more common material for odor-control fans has been FRP, with a graphite-impregnated wheel. This configuration is needed to ground any static charge buildup experienced during operation. Stainless steel wheels have become more popular, especially for applications in warm climates, where FRP may be affected by heat. They also are used when particulate matter in the air stream can degrade and damage the FRP, or for high pressure and the resulting high fan revolutions-per-minute applications.

Fans also will be prone to condensation inside the fan housing. Most fans are equipped with a drain on the housing to let this water out. Unless trapped sufficiently against the pressure, the housing drain can be a source of local odors. If the housing is not drained, it can ruin the fan. Fans also will experience corrosion if they are conveying hydrogen sulfide gas. All fans have a shaft and bearing that cannot be enclosed totally or protected against corrosion. For centrifugal fans, a shaft of 316L stainless steel will be able to withstand the effects of corrosion. The use of other metals or stainless steel alloys typically yields less favorable results.

Positioning the fan within the ductwork network and odor-control technology is at the discretion of the designer. However, some advantages to the push or pull positioning are dictated by site constraints and the equipment being used.

Transporting odorous air in ductwork benefits from positioning the fan at the end of the system. Negative pressure is maintained throughout the length of the ductwork, and odors are precluded from escaping leaks, cracks, or small openings.

Certain odor-control technologies force air through the device. For example, fans associated with activated-carbon and certain biofilters almost always are found pushing through individual media beds to achieve treatment.

Wet-scrubbing odor-control technology can be operated as effectively in either a push or a pull arrangement. Although the fan materials of construction are selected to resist the corrosive nature of the odorous air stream, placing the fan downstream of wet scrubbing in a pull-through or induced-draft arrangement would eliminate the corrosive effect of the untreated odorous air stream on the fan. Downstream positioning also would prevent any odor from escaping the fan housing, duct connections, and drive shaft, which sometimes occurs with an upstream fan arrangement in front of the wet scrubber. However, the downstream wet-scrubber fan must be able to withstand the corrosive nature of the chemicals used in the scrubbing solution. For example, sodium hypochlorite (chlorine) is commonly used, with and without sodium hydroxide (caustic), in WRRF wet scrubbers.

Upstream fan positioning could minimize ductwork on single- and multiple-stage wet-scrubbing systems. Pushing odorous air through a single- or multiple-stage system allows for treated exhaust to be discharged out of the top of the unit. Pulling the air through an induced-draft arrangement requires ductwork from the top of the last unit in series be directed to a fan located somewhere below, increasing the length of ductwork.

The designer must weigh the advantages and disadvantages of either approach before selecting the appropriate fan arrangement.

6.5.2.1 Fan Balancing The airflow within the odor-ductwork system should be balanced to ensure that the design amount of air is being drawn from its intended locations. Testing, adjusting, and balancing (TAB) services are the way to confirm that design airflows are in compliance. These TAB services are commonly contracted to an independent licensed balancing firm.

The TAB services should be completed before the installation is accepted by the owner. Discrepancies must be addressed and fixed immediately, or at least before the system is accepted. The TAB services also should be used—after the system has been operating, to ascertain if design airflow levels have been maintained; and any time modifications are made to the ductwork, exhaust pickup points, or damper settings.

Provisions should be made upstream of each balancing damper to drill a hole large enough to insert a velocity probe that would be required to test the airflow within the duct. A cap should be provided to fill the hole after the probe is removed.

6.5.2.2   Dampers   The use of dampers within any ducted ventilation system typically is required. The dampers commonly are used for balancing the airflow quantities, isolating equipment from the system, or, in some cases, both. Dampers on the supply air systems should follow the guidelines of ASHRAE or SMACNA.

The following discussion focuses on dampers for foul-air-exhaust systems, starting from the exhaust point and traveling to the fan and control device. Dampers should be located on each duct connection or at each intake register. These dampers allow each point source to be balanced independently. In exhaust-register installation, a high-quality volume damper should be installed in lieu of the volume damper that may be purchased as an option with the register, as typically it is not of sufficient quality to act as a reliable and repeatable control device. Also, the branch ductwork should have its own balancing damper to balance the branches independently from the trunk duct. Permanent test ports should be installed downstream of the dampers, to allow for airflow measurements to be taken.

Each fan should have an isolation damper upstream and downstream of it. Inlet-vane dampers may be added upstream of the fan, as they act as a highly efficient means of system-flow control. Finally, all carbon vessels should have isolation dampers on both the inlet and outlet duct connections. These are important to allow the carbon adsorber vessel to be closed off to oxygen flow whenever the unit is idle. This helps to prevent bed fires.

Dampers can be manufactured using the same materials as the ductwork systems (i.e., FRP, stainless steel, aluminum, and plastic). Most commonly installed dampers are butterfly type for round-duct installations and airfoil-opposed blade type for rectangular installations.

7.0   Odor and Air Emissions Control

Controlling odors and air emissions can be achieved by preventing the formation or release of the compounds or capturing and treating the air stream from the process unit. Preventing the formation and release of odors and air emissions through the improved design and operation of the process units is discussed in the chapter for the particular unit. Under some circumstances, odors can be prevented through the treatment of dissolved sulfides in the liquid phase. For treatment of odors and air emissions in the gas phase, several technologies are discussed.

7.1   Liquid-Phase Treatment

There are many different types of control measures that can be applied to treat sulfide and other odorous compounds in the liquid phase before they can be emitted. Control methods may involve the addition of air or oxygen to reduce sulfide formation.

Another approach is the use of chemicals to halt sulfide production or react with sulfide in the liquid phase.

The following sections discuss the various chemical treatments for odor control in detail, including their chemistry and process description. For each of the liquid-phase control methods, major advantages and disadvantages are discussed, with typical dosage rates. Chemical prices vary by region and change over time, so cost information is not provided.

7.1.1 Air/Oxygen Injection

Most odor production in wastewater can be prevented if a dissolved oxygen concentration of at least 0.5 to 1.0 mg/L is maintained (U.S. EPA, 1985). Oxygen can directly oxidize odor-causing compounds, or is available for aerobic bacteria to carry out this function through metabolic processes and prevent sulfide-compound generation by preventing anaerobic conditions.

Air is a readily available source of oxygen that has been used to control hydrogen sulfide odor. If odorous compounds are present in wastewater at the point of air injection, odors will diffuse to undissolved air and escape to the atmosphere, sometimes causing increased odor. Depending on the method of air injection, turbulence also can release odors, making the problem worse.

The addition of pure oxygen gas has a major advantage over air when added to wastewater, because it is five times more soluble in water. This means that a smaller volume of gas is required to achieve the same oxygen transfer, and the possibility of gas-pocket formation is lessened. Reductions in BOD also have been reported following installation of oxygen injection equipment. The oxygen can be generated on-site for requirements greater than 900 kg/d (2000 lb/d) oxygen (O_2), or purchased commercially and delivered by truck for lesser quantities.

The equipment required for on-site storage and application includes a specialty steel, double-walled containment-pressure vessel for liquid oxygen storage, evaporator-pressurizer, control valve, pressure regulator, flow meter (rotameter), piping, and injectors (Figure 6.21). This equipment can be purchased or leased from oxygen suppliers. Liquid oxygen requires a sophisticated containment vessel to maintain the 1730-kPa gauge (250-psig) pressure required to keep it in the liquid state. When the controls call for oxygen delivery, a control valve on the tank opens and releases liquid oxygen to the evaporator. The evaporator is a pressurized, radiator-like device that raises the temperature of liquid oxygen, causing it to vaporize. When the oxygen vaporizes, it creates its own pressure; thus, compressors are not required. Another control valve and regulator at the discharge end of the evaporator meters the vaporized oxygen to the point of application.

7.1.2 Chemical Oxidation

Chemical oxidants chemically attack the odor-causing compounds and destroy them through oxidation-reduction reactions. Although some of the chemicals in this category may contain oxygen as part of their molecular structure, their primary action is to directly react with the odorous compound in the dissolved form rather than release oxygen for use by bacteria. Chlorine, hypochlorite, hydrogen peroxide, potassium permanganate, and ozone are examples of chemical oxidants.

7.1.2.1 Chlorine Compounds Chlorine may be a relatively inexpensive, powerful oxidant, and the equipment required for its use may be inexpensive and widely available.

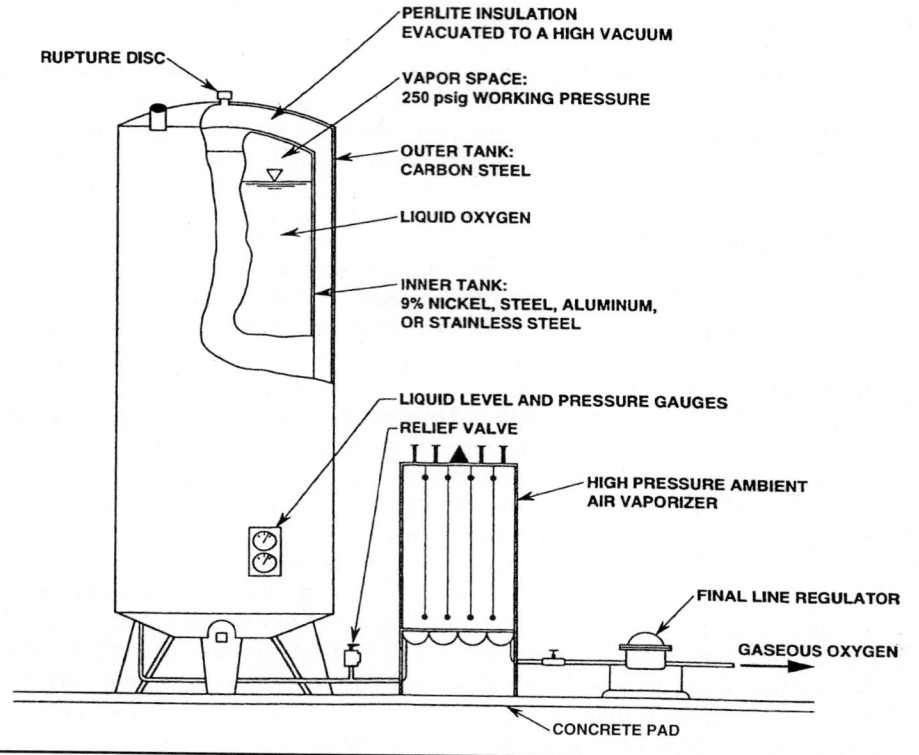

FIGURE 6.21 Liquid-oxygen injection system.

Commercially available solutions of sodium hypochlorite or calcium hypochlorite are the most common forms. The reactive component of chlorine in water is the hypochlorite ion, regardless of whether chlorine gas or hypochlorite solution is used. Wastewater pH may be affected slightly by the addition of chlorine solutions. The dissolution of chlorine gas creates an acidic product, while hypochlorite solutions are basic.

Chlorine reacts with many compounds found in raw municipal wastewater, including hydrogen sulfide. The reactions between chlorine and sulfide in municipal wastewater are as follows:

$$HS^- + 4Cl_2 + 4H_2O \rightarrow SO_4^- + 9H^+ + 8Cl^- \qquad \text{(Acidic pH)} \qquad (6.11)$$

$$HS^- + Cl_2 \rightarrow S + H^+ + 2Cl^- \qquad \text{(Basic pH)} \qquad (6.12)$$

Equation 6.11 requires 8.9 parts, by weight, of chlorine to oxidize each part of sulfide, while eq 6.12 requires only 2.2 parts of chlorine per part of sulfide. The reactivity of chlorine is a disadvantage, because it indiscriminately oxidizes any reduced compound in wastewater. These competing side reactions require overfeeding to ensure sulfide oxidation. Actual practice has shown that, depending on the pH and other wastewater characteristics, between 5 and 15 parts, by weight, of chlorine are required for each part of sulfide (U.S. EPA, 1985).

Chlorine also can act as a bactericide, because it is a strong disinfectant. Depending on the point of application and the dose, it can kill or inactivate many bacteria that cause odors. However, because chlorine is nonselective, it also will kill organisms that are beneficial to wastewater treatment processes. Therefore, care should be used if chlorine is added for odor control at the headworks of a WRRF.

When chlorine reacts with certain organic components in water or wastewater, chlorinated organics are formed. Examples of some of these compounds are chloroform, methyl chlorides, and chlorophenols. These compounds, and many others formed by the reaction of chlorine with wastewater, may be potentially toxic, carcinogenic, and impart their own objectionable odors. The potential for an increase in chlorinated VOCs should be considered before using chlorine in wastewater-odor-control applications (WEF, 2004).

7.1.2.2 Hydrogen Peroxide Hydrogen peroxide is a commonly used oxidant that chemically oxidizes hydrogen sulfide to elemental sulfur or sulfate, depending on the pH of wastewater. Hydrogen peroxide reacts with hydrogen sulfide according to the following equations:

$$H_2S + H_2O_2 \rightarrow S + 2H_2O \text{ (pH} < 8.5) \tag{6.13}$$

$$S^{2-} + 4H_2O_2 \rightarrow SO_4^- + 4H_2O \text{ (pH} > 8.5) \tag{6.14}$$

Most wastewater applications essentially are at a neutral pH, so the theoretical dosage requirement is one part of peroxide per part of sulfide. However, like other oxidant chemicals, peroxide reacts with organic material in the wastewater, so higher dosages typically are required. For some applications, successful treatment has been reported at a peroxide-to-sulfide ratio of 2:1, but other applications have required a ratio as high as 4:1 (Van Durme and Berkenpas, 1989). With proper mixing, peroxide may be fast acting, which makes it useful for addition upstream of problem locations. However, it also is consumed quickly, so the consumption rate should be taken into account when targeting a problem location.

The type of hydrogen-peroxide-feed equipment required depends on the quantity applied. For small applications requiring less than 0.08 m³/d (20 gpd), drum deliveries provide relatively simple operation and require only a metering pump, piping, valves, and an injector. Bulk storage is most economical for applications requiring more than 0.08 m³/d. Peroxide can be purchased in drum or bulk concentrations of 35% and 50%, by weight. Depending on the concentration of peroxide, storage tank materials vary, from stainless steel and high-purity aluminum at high concentrations to polyethylene at lower concentrations. Piping typically can be made of PVCs; however, nonreactive materials, such as polytetrafluoroethene and stainless steel, commonly are used for wetted pump parts. A typical hydrogen peroxide delivery system is illustrated in Figure 6.22.

7.1.2.3 Potassium Permanganate Potassium permanganate ($KMnO_4$) is a strong chemical oxidizing agent that reacts with hydrogen sulfide according to the following equations:

$$3H_2S + 2KMnO_4 \rightarrow 3S + 2H_2O + 2KOH + 2MnO_2 \text{ (Acidic pH)} \tag{6.15}$$

$$3H_2S + 8KMnO_4 \rightarrow 3K_2SO_4 + 2H_2O + 2KOH + 8MnO_2 \text{ (Basic pH)} \tag{6.16}$$

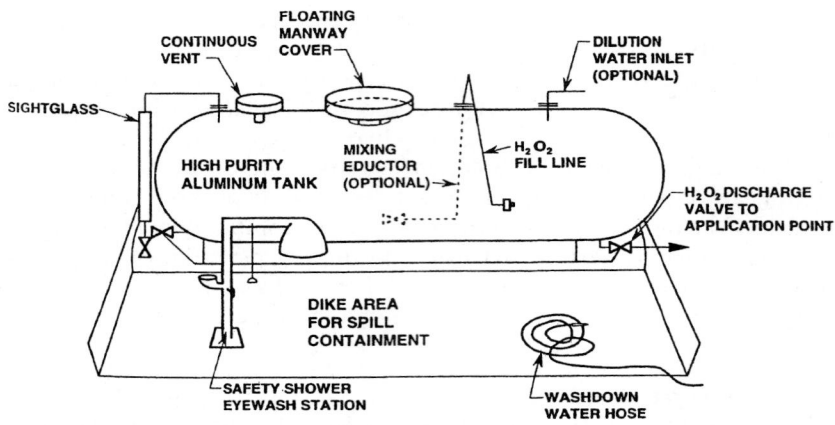

FIGURE 6.22   Bulk hydrogen peroxide storage and delivery system.

In actual practice, several reactions, ranging between these two, may take place to produce elemental sulfur, sulfate, thionates, dithionates, and manganese sulfide, depending on the local wastewater chemistry. For this reason, field studies have indicated that between 6 and 7 parts of potassium permanganate are required for each part of sulfide to be oxidized (U.S. EPA, 1985).

Potassium permanganate is available in dry crystal, granule, or pellet form and must be mixed with water to approximately a 3% to 4% solution before use. When kept dry and cool, potassium permanganate is relatively stable. However, when contaminated with organics or acids, it can become unstable and decompose, causing potentially hazardous conditions. Commercially available liquid permanganates with concentrations as high as 20% have recently been introduced to the wastewater industry. These solutions are easy to handle and apply. A typical liquid permanganate injection system consists of only a high-density polyethylene tank, metering pump(s), control panel, valves, and piping.

The equipment required for a typical potassium permanganate application can range from a simple feeder and dissolver setup to an automatic batching system. For handling large quantities of potassium permanganate, line-feed equipment has been developed, which requires limited operator involvement and is enclosed completely for minimal chemical exposure. Potassium permanganate typically has been used on an intermittent basis, because the cost of the chemical makes it prohibitively expensive to treat large flows continuously.

Potassium permanganate reactions produce manganese dioxide (MnO_2) as a by-product. Manganese dioxide is a fluffy, brown floc that is practically nonreactive and settles as chemical solids in the WRRF and will increase solids production slightly. Manganese is one of the regulated heavy metals for the beneficial reuse of biosolids. Permanganate typically is not added in large-enough amounts to be a problem on its own, but it may be a concern for facilities with biosolids that are already near their manganese limits because of other sources.

7.1.2.4 Ozone   Ozone is an extremely powerful oxidant that can oxidize hydrogen sulfide to elemental sulfur. It also is an effective disinfectant when bacteria levels are low. Although ozone reacts with practically everything in wastewater, including dissolved sulfide, its principal usage has been to treat odorous gas streams. Ozone is unstable and must be generated on-site. It also is potentially toxic to humans at concentrations of 1 ppm$_v$ or greater in air.

7.1.3 Nitrate Addition

Nitrate addition controls dissolved sulfide by two different reaction mechanisms or modes—prevention and removal. In the prevention mode, nitrate is added to fresh wastewater to be used as a substitute source of oxygen, or, more specifically, an electron acceptor. The facultative and obligate anaerobic bacteria, which are responsible for odor and sulfide generation, use dissolved oxygen, nitrate, and sulfate as oxygen sources, in that order of preference. Typically, dissolved oxygen in wastewater is depleted rapidly, and there is little nitrate present. Sulfate is typically abundant in raw municipal wastewater, so the bacteria reduce sulfate to sulfide, which causes odor and leads to corrosion problems. When nitrate is added to wastewater, the bacteria use it as their electron acceptor instead of sulfate. This results in the production of nitrogen gas and other nitrogenous compounds rather than sulfide.

In the removal reaction, nitrate can be added to septic wastewater to remove dissolved hydrogen sulfide from wastewater by a biochemical process, which converts the sulfide to sulfate. The nitrate supplies oxygen to the bacteria (likely *Thiobacillus denitrificans*) present in wastewater to metabolize hydrogen sulfide and other reduced-sulfur compounds. The removal reaction is a biochemical process, so sulfide reduction is not instantaneous, and a reaction time of 1 to 2 hours may be required for optimal effectiveness. Depending on hydraulic residence times, this could limit its effectiveness when applied on-site at a WRRF. The removal mechanism requires one-third the amount of nitrate as the prevention mechanism.

Nitrate can be used to remove preexisting dissolved sulfide. The product typically is an aqueous solution of calcium nitrate containing 0.42 kg/L (3.5 lb/gal) of nitrate-oxygen. The solution can be tailored to different applications and may contain sodium nitrate, which is more compatible with certain industrial wastewaters.

7.1.3.1 Nitrate-Reaction Mechanisms

Nitrate compounds have been obtained as dry chemicals, but the dry material requires mixing with water and settling impurities before use. The dry material is hygroscopic and tends to cement together during storage, which has caused handling problems. In untreated wastewater, a carbon source (BOD) is consumed via an anoxic denitrification reaction. Assuming a single carbon source (methanol) for simplicity, the reaction is as follows:

$$SO_4^- + 4CH_3OH \rightarrow S^- + 4H_2O + 2CH_4 + 2CO_2 \qquad (6.17)$$

[Untreated reaction]

In this process, 0.25 parts of sulfate are used, and 0.25 parts of sulfide are produced for every mole of carbon consumed.

Nitrate prevents the formation of sulfide by acting as a preferential oxygen source for the bacteria in the slime layer over sulfate. Again assuming a single carbon source for simplicity, the reaction is as follows:

$$6NO_3^- + 5CH_3OH \rightarrow 5CO_2 + 3N_2 + 7H_2O + 6OH^- \qquad (6.18)$$

[Prevention reaction]

In this process, 1.2 parts of nitrate are used, and 0 parts of sulfide are produced for every mole of carbon consumed. The nitrate requirement per unit of sulfide prevented is expressed as follows:

$$(1.2 \text{ parts } NO_3^-/\text{mol C})/(0.25 \text{ parts } S^-/\text{mol C}) = 4.8 \text{ parts nitrate per mole sulfide}$$

On a mass basis, 9.3 kg NO_3 (or 7.2 kg NO_3-O) are needed per kilogram of sulfide prevented. For a solution containing 0.42 kg/L (3.5 lb/gal) of nitrate-oxygen, this translates to 17.2 L/kg sulfide prevented (2.1gal/lb sulfide prevented). This represents the minimal stoichiometric requirements for prevention given the simplest form of organic matter. Actual dosage rates tend to be higher.

The removal mechanism uses naturally occurring bacteria to biochemically oxidize dissolved sulfide in the presence of nitrate. The dissolved sulfide may be generated upstream of the nitrate application point or may enter the line downstream, through a lateral branch. Nitrate addition causes the biochemical oxidation of sulfide according to the following reaction:

$$8NO_3^- + 5H_2S \rightarrow 5CO_2 + 5SO_4^- + 4N_2 + 4H_2O + 2H^- \qquad (6.19)$$

[Removal reaction]

This reaction occurs in the bulk flow and in the outer zone of the slime layer. Nitrate is not added in sufficient quantities to fully saturate the slime layer, so sulfide production continues in the inner zone. The sulfide is removed when it reaches the outer zone or the bulk flow. In this reaction, 1.6 parts nitrate are used for every mole of sulfide removed. On a mass basis, this requires 2.4 kg NO_3-O per kilogram of sulfide removed. For a solution of 0.42 kg/L (3.5 lb/gal) nitrate-oxygen, this requires 5.7 L/kg sulfide removed (0.7 gal/lb sulfide removed) (Hunniford, 1990).

7.1.3.2 Equipment Requirements Nitrate has the advantage of being one of the safest of all the sulfide-control chemicals to handle. Generally, nitrate-salt solutions are considered to be nonhazardous substances and are not included on federal U.S. EPA or state Comprehensive Environmental Response, Compensation, and Liability Act (CERCLA) lists. Standard nitrate solutions also are exempt from U.S. Department of Transportation (Washington, D.C.) placard requirements. A typical nitrate injection station consists of a high-density, cross-linked polyethylene tank, metering pump(s), control panel, valves, and PVC piping. Numerous configurations and materials are available for nonhazardous chemical storage systems, including horizontal, low-profile, or upright-vertical tanks. The final selection depends on local aesthetic and public-relations requirements.

7.1.4 Iron Salts

Iron salts are fast acting and often are applied just upstream of a WRRF, to remove sulfide before the headworks facilities. The iron precipitate settles rapidly in a quiescent basin. The iron precipitate adds to the overall solids production at the WRRF, with the volume dependent on the amount of sulfide treated. Even in systems with high sulfide concentrations, the added solids typically are less than 5% of the overall solids.

In addition to sulfide, iron salts also react with phosphate in wastewater and precipitate it as iron phosphate. This increases the chemical demand above the stoichiometric requirements for sulfide alone. Extensive experience in the field has established that 3.5 kg Fe/kg sulfide is an optimal dosage rate for most applications (Van Durme and Berkenpas, 1989; Wong et al., 1992). The solubility of ferrous sulfide at typical wastewater pH values only allows control of hydrogen sulfide to between 0.05 and 0.1 mg/L. Even if excess ferrous salt is added, the dissolved sulfide concentration will not be lowered below this level. In most cases, this level of treatment is satisfactory to prevent odors and corrosion. However, in areas of turbulence, hydrogen sulfide release still may be a problem. In areas of localized pH depression, such as anaerobic waste streams in which the pH drops below 6.5, ferrous sulfide partially dissociates and may release sulfide to the wastewater. Also, ferrous salts have been known to adversely affect UV disinfection equipment at WRRFs. This should be taken into consideration if ferrous salts are being considered for odor control at WRRFs where UV disinfection is performed.

Iron and other metals can combine chemically with dissolved sulfide to form relatively insoluble precipitates. The iron salt precipitates are in the form of black or reddish-brown floc, which readily settles with other solids at the WRRF. Iron salts are in widespread use in conveyance systems throughout the United States. Other metal salts, such as chromium, copper, and zinc, also react with sulfide, but these are regulated heavy metals. High heavy-metal concentrations make biosolids hazardous and disposal difficult and costly.

Both ferrous and ferric metal salts can react with dissolved sulfide. The following four types of iron salt solutions are commercially available: ferrous sulfate, ferrous chloride, ferric sulfate, and ferric chloride. The addition of the sulfate-based salts has been questioned, because sulfate can be reduced to sulfide. However, this typically is not a concern in municipal wastewater systems, because sulfate typically is present in excess, and sulfide generation is not increased significantly. Sulfate sources for wastewater include drinking water, industrial discharges, and hydrolysis of sulfur-containing organic wastes. Most commercially available iron salts are high-quality products with minimal contaminants, but purchasers should request analytical data from the supplier.

7.1.4.1 Iron Salt Reactions The reaction of ferrous (Fe^{++}) salts with dissolved sulfide is the same for both ferrous solutions, because the anionic carrier ion (sulfate or chloride) does not enter into the reaction. A variety of iron sulfide complexes may be formed, but the simplest reaction is as follows:

$$Fe^{++} + HS^- \rightarrow FeS + H^+ \qquad (6.20)$$

For this reaction, the stoichiometric amount of iron required is 1.6 kg Fe^{++}/kg sulfide.

Ferric (Fe^{+++}) salts react with dissolved sulfide in much the same way as ferrous salts, according to the following reaction:

$$2Fe^{+++} + 3HS^- \rightarrow Fe_2S_3 + 3H^+ \qquad (6.21)$$

In practice, a number of different ferric sulfide species are formed, but the result is a black precipitate of sulfide. The stoichiometric amount of iron required is 1.1 kg Fe^{+++}/kg sulfide.

7.1.4.2 Iron Solutions Ferrous sulfate ($FeSO_4$) is produced in the mining industry as a co-product with titanium dioxide. Iron and titanium are found in the same ore, and iron is removed selectively by dissolving the ore in sulfuric acid and precipitating it as ferrous sulfate heptahydrate crystals ($FeSO_4 + 7H_2O$). It also may be produced during steel pickling with sulfuric acid (the spent acid is called pickle liquor). The light-green solutions vary in strength from 8% to 16%, with 3% to 6% iron as Fe by weight and 0.06 to 0.08 kg Fe/L (0.5 to 0.7 lb Fe/gal). Solutions produced from crystals can have a pH of 2.0 or greater, if free acid is controlled. A pH of 2.0 or greater classifies the material as noncorrosive, according to the Resource Conservation and Recovery Act (RCRA); however, spills greater than 454 kg (1000 lb) are reportable. Solutions derived from pickle liquor can contain higher free acid with pH less than 1. These solutions are classified as characteristically hazardous by the RCRA, and additional controls against spills and contact with humans are required (U.S. EPA, 1990). In such a case, reportable spill quantities decrease to 0.38 m^3 (100 gal). Ferrous sulfate also is available in various dry forms (including ferrous sulfate monohydrate), which require mixing with water before use. Dry ferrous sulfate has not been used commonly for sulfide control in collection systems. The liquid solutions can be stored in FRP, high-density cross-linked polyethylene, or rubber-lined steel tanks. Solutions with pH values of 2.0 or less require precautions against spillage, as shown in Figure 6.23. The freezing point is 222°C (28°F), so insulation may be required in some locations. All fittings, piping, and valves must be of the same materials. The solution attacks many metals (e.g., brass and copper), so wetted parts should be of 316L stainless steel or resistant plastic.

Ferric sulfate [$Fe_2(SO_4)_3$] is produced by reacting natural iron-rich ores with sulfuric acid or oxidizing-spent pickle liquor. Ferric sulfate commonly is used in both water and wastewater as a coagulant and settling aid. The orange-brown liquid typically is a 45% to 50% solution with approximately 10% iron as Fe. The specific gravity is 1.44, so there is approximately 0.14 kg Fe/L (1.2 lb Fe/gal). The pH is less than 1.0, so the material is classified as a corrosion hazard, according to the RCRA. The freezing point of the ferric sulfate is 222°C (28°F), so insulation typically is not required. Otherwise, the storage and feed requirements are similar to those for ferrous sulfate.

Ferrous chloride ($FeCl_2$) can be produced by pickling steel or processing natural ores with hydrochloric acid. Ferrous chloride has a high solubility with solution

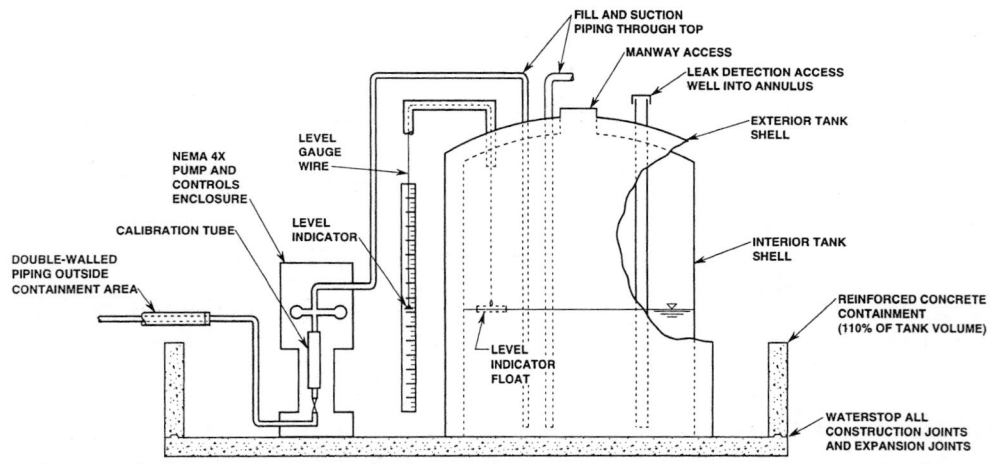

FIGURE 6.23   Typical hazardous chemical delivery system.

strengths of 18% to 28%. A solution of 23% contains approximately 10% iron as Fe by weight. The specific gravity is 1.23, so there is 0.12 kg Fe/L (1.0 lb Fe/gal). The light-green-colored solutions typically have a pH of less than 1.0, so they are classified as a characteristically hazardous material. Ferrous chloride is in widespread use for odor-control applications, because it often is the least costly iron-salt chemical. The solutions can be stored in FRP or rubber-lined steel tanks. The tanks should have spill containment. The freezing point is 220°C (24°F), so insulation may be required in some regions. Suppliers can reduce solution strength in winter to avoid crystallization of the product during cold temperatures. Ferrous chloride is corrosive and attacks most metals rapidly. All wetted parts should be of Hastelloy C, titanium, or tantalum. Aluminum, brass, and stainless steel are attacked readily and should never be used in contact with full-strength ferric chloride ($FeCl_3$). There may be concern at wet-well applications having submerged pumps or pumps with stainless steel impellers and other wetted parts. If the chemical is not sufficiently diluted by the bulk flow, prolonged contact with the stainless steel can cause pitting and etching and may require increased pump maintenance. To ensure that the chemical is adequately diluted under low-flow conditions, it is a good design practice to use a flow-paced feed control or timers.

Ferric chloride is used in the copper-engraving industry, and it commonly is applied as a coagulant in water treatment facilities, so high-purity products typically are available. Ferric chloride can be produced as a co-product with titanium dioxide from natural ores containing iron and titanium oxides. A high-quality product can be manufactured by reacting chlorine gas with iron, ferrous sulfate, or ferrous chloride. Another process involves the controlled reaction of spent steel pickling liquor, hydrochloric acid, chlorine, and scrap iron. The purity of the final product varies with the manufacturing process; purchasers should obtain information on contaminants from the supplier. Liquid ferric chloride is an orange-brown solution that is acidic and corrosive to metals. Solutions range from 28% to 47% $FeCl_3$ by weight, with the standard being 40%, which is 13.8% total iron as Fe. The specific gravity of the standard solution is 1.43, so there is 0.2 kg Fe/L solution (1.6 lb Fe/gal). The freezing point of the standard solution is 212°C (210°F), so insulation typically is not required. Storage and feed requirements are similar to those for ferrous chloride.

A blend of 1 part ferrous (Fe^{++}) to 2 parts ferric (Fe^{+++}) chloride was reported to provide optimum results in laboratory studies. For a short time, a blended product was marketed commercially. However, experience in the field showed no benefits over single solutions of ferrous and ferric chlorides, so the product was discontinued.

7.1.5 Adjustment of pH
There are two different approaches to controlling hydrogen sulfide by adjusting the pH of wastewater. One approach involves continuous pH adjustment to hold hydrogen sulfide in solution. The other approach involves using intermittent slug doses of caustic to inactivate the slime layer and minimize sulfide generation.

The objective of continuous pH adjustment is to prevent the release of hydrogen sulfide gas, by shifting the sulfide equilibrium to bisulfide and dissolved sulfide species, which remain in solution. At pH 7, approximately 50% of all sulfide exists as hydrogen sulfide, and 50% exists as the dissolved bisulfide ion. At a pH of 8, only 10% exists as hydrogen sulfide and 90% as bisulfide. A change in pH of 0.5 units can have a significant effect on the amount of hydrogen sulfide available to be emitted to the atmosphere.

Periodic slug dosing with caustic can remove all sulfide forms effectively. It is not added to shift the equilibrium, but to inactivate or kill the biological slime layer that reduces sulfate to sulfide. Exposure to high pH levels will destroy the slime layer and cause it to slough. The slime layer will begin to reform immediately, but it may take days or weeks to reach full sulfide production again. The time required for the slime layer to grow back after slug dosing is a function of pH, temperature, and time of contact.

Caustic slugging near WRRFs can adversely affect pH-sensitive treatment processes and effluent discharge limits because of locally elevated pH levels. If equalization facilities are available, they can be used to store the high-pH wastewater for slow release to the facility. If these facilities are not available, acid can be used for neutralization, but this adds to the expense. Some facilities have reported that pH levels as high as 8.5 have not adversely affected the secondary treatment process, possibly as a result of carbon dioxide buffering in the activated sludge process.

7.2   Biological Treatment

This chapter provides an overview of available biologically based odor-control systems. Additional design details are provided in other manuals of practice on this topic. The technologies used to control odors associated with wastewater conveyance and treatment historically have used chemical and physical methods. Chemical scrubbers and activated-carbon adsorbers are proven methods for the removal of odors from air streams using chemical adsorption, absorption, and chemical reactions. However, depending on the conditions, these odor-treatment processes can require a significant operational budget. Wastewater odors can be treated using purely physical and chemical methods, but bacteria and other microorganisms can accomplish this task much more efficiently, with fewer operational controls. This translates directly to cost savings for utilities and other entities responsible for the control of wastewater-related odors.

Biological odor-treatment technologies also afford other advantages compared with conventional chemical and physical treatment. Most chemicals used to control wastewater odors are strong oxidants or have pH ranges that make them dangerous or even hazardous to handle, transport, and store. The nature of biological processes precludes the use of dangerous chemicals, because the same chemicals that can adversely affect humans may have the same effect on microorganisms.

7.2.1   Biological Odor-Treatment Biochemistry

All biological odor-control technologies discussed in this chapter operate through the following three basic biochemical processes—autotrophic, heterotrophic, and biological-uptake processes.

7.2.1.1   Autotrophic Biological Processes   The term autotrophic means self-feeding. The bacteria that fall under this category are sometimes referred to as chemoautotrophic, because they use inorganic chemicals as an energy source and get their carbon for cell growth from carbon dioxide. Microbiologists use the term autotrophic to identify this general type of bacteria, which use inorganic compounds for their energy and carbon sources. In contrast, heterotrophic organisms must use organic compounds for both their energy and carbon needs.

Autotrophic bacteria typically have a thin cell wall to allow free movement of chemicals in and out of the cell. Because autotrophic bacteria do not get as much energy from the conversion of inorganic chemicals as heterotrophic bacteria get from organic compounds, they must convert more chemicals at a faster rate. Therefore, autotrophic bacteria can assimilate quickly and convert large quantities of inorganic compounds. This latter ability leads to rapid conversion of hydrogen sulfide gas in sewers to sulfuric acid and destruction of the wastewater infrastructure. In biological odor-treatment systems, these bacteria provide for the removal of hydrogen sulfide.

7.2.1.2 Heterotrophic Biological Processes   The term heterotrophic (sometimes referred to as chemoheterotrophic) is used to denote another general class of microorganisms, which consume organic compounds (i.e., compounds that contain carbon atoms). Heterotrophic organisms must obtain both their energy and their carbon for cell growth from the food they consume, whereas, in autotrophs, the carbon and energy are obtained from different sources. Heterotrophic microorganisms cannot use carbon dioxide as a carbon source for cell synthesis and must break the carbon-carbon bonds to assimilate carbon.

There are many more types of heterotrophic organisms than autotrophic organisms, because there are many more compounds in the environment that contain carbon than do not. Because of their ability to degrade a wide variety of organic compounds, heterotrophs are used in biological odor-control processes to remove organic compounds and VOCs.

7.2.1.3 Biological Uptake Processes   The third process whereby bacteria can consume and remove odor compounds can be classified as direct biological uptake. Biological uptake is different from food consumption, carbon source, and energy production, in that the odor compound is not the primary food or carbon source, but is used as a secondary component in the process of cell function, respiration, and new-cell synthesis. An example of uptake is nutrient consumption by bacteria. The odor compound is used either directly as a nutrient or processed to other compounds necessary for cell function. Uptake processes are responsible for hydrogen sulfide gas removal in aeration basins and in a process called return activated sludge (RAS) recycle. This is a primary mechanism when using aeration-basin mixed liquor for odor control. In the process of being used in reactions inside the cell wall, the odor compound is converted to a non-odorous compound or used as a building block for more cell materials. Heterotrophs are primarily responsible for this process.

All three of these biological processes (autotrophic degradation, heterotrophic degradation, and biological uptake) are used in various biological-control technologies. The primary design basis of all vapor-phase biological odor-control technologies is to build a house that is ideally suited for the specific type(s) of organism(s) that one desires to grow, and provide suitable food, oxygen, carbon dioxide, temperature, nutrients, and moisture under optimum living conditions.

In some instances, such as biotowers, specific pH regimes are created to promote intentionally the growth of one or the other of the basic bacteria types. For example, a low-pH biotower would promote hydrogen-sulfide-consuming autotrophic bacteria, while a more neutral pH would be desirable for broad-spectrum odor removal, including other organic-based odorous compounds.

7.2.2 Biofiltration Systems

Biofilters have become increasingly popular in recent years, because they are an extremely reliable and economical method of odor control, if properly designed and installed.

All biofilters consist of the following common elements:

- Air ducting and fan system;
- Air plenum;
- Underdrain piping system;
- Media-support system;
- Media; and
- Irrigation and humidification system.

The three basic biofilter configurations are the following:

- Custom in-ground biofilters;
- Custom in-vessel biofilters (both above and below ground); and
- Prefabricated in-vessel biofilters (both above and below ground).

For illustration, a typical custom in-ground biofilter installation consists of a fan connected to a network of perforated pipe, which sends the odorous air to a plenum and then through the media bed (Figure 6.24).

The following section discusses the basic components of common biofilters in more detail and provides basic design and material-selection guidelines.

7.2.2.1 Air-Ducting and Fan System   All biofilters must receive odorous air from the source to the biofilter. The movement of air through a duct and fan system is governed by the basic principles of mechanical engineering design. However, there also are biotechnology-induced, process-related issues, which must be considered in the design and selection of the system. One of the most important factors in the design of a biofilter-air-distribution system is corrosion control.

The ductwork upstream of a biofilter should be of sufficient size and material to withstand the stresses imposed by the application. The materials and construction

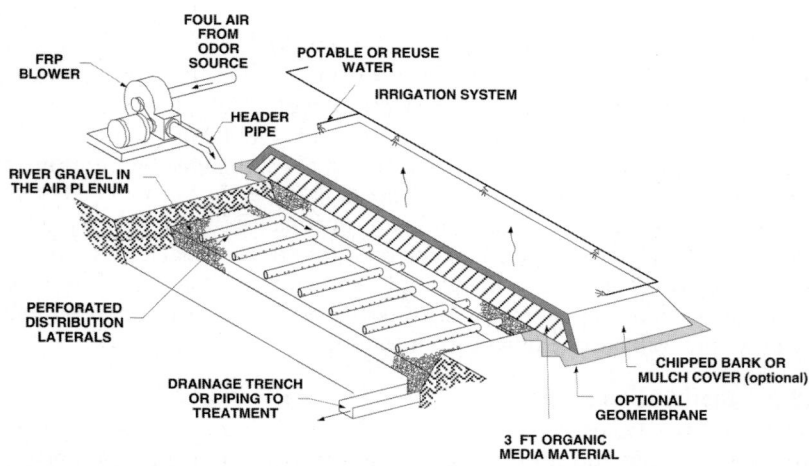

FIGURE 6.24   Typical in-ground biofilter.

methods used in the biofilter ductwork should consider that the duct may be exposed to low-pH, acid-attack conditions. Fittings, valves, dampers, and other connectors used in the construction of the ductwork also should be designed with corrosion protection in mind. Typically, the source air has a high relative humidity, and condensation occurs on the interior walls of the duct. The condensation collects and flows to the low point in the system. If the condensed water is not removed, it will fill the duct, increase headlosses, and eventually stop all airflow.

It often is an advantage to have biofilters arranged in several cells or separate units when replacing media. One cell or unit can be taken offline for media replacement without losing total odor-control capability. The use of multiple cells or units necessitates the use of isolation dampers on the separate duct headers feeding the cells or units.

The most commonly used pressure fan for biofilters is the backward-inclined impeller fiberglass centrifugal pressure fan. This type of fan can produce airflows and pressures most commonly observed in biofiltration systems. Biofilters, whether custom or prefabricated, operate over a general range of pressures between 0.12 and 3.0 kPa (0.5 and 12 in.) of water. Some soil-bed filters operate at pressures as high as 5.0 kPa (20 in.) of water. This range of pressures is needed to overcome common duct and media headlosses. Duct headlosses typically are fixed based on the airflows, diameters, lengths, dampers, valves, and other fixtures used. Media headlosses through a typical 1-m- (3-ft-) deep organic media bed can range from 0.12 kPa (0.5 in.) of water at startup to 2.5 kPa (10.0 in.) or more near the end of its useful life.

7.2.2.2 Air Plenum   Air plenums used for biofilters typically are rock-media-filled or open plenums. Media plenums contain some sort of media that surrounds the air-distribution ducting and supports the organic media, whereas open plenums do not contain media and use a specialized media-support system. Figure 6.25 shows a section of a typical earth-berm custom biofilter with a river-gravel rock plenum; Figure 6.26 illustrates a typical open-plenum biofilter arrangement applied to a custom biofilter design. The media are supported directly on a modular plastic-grid system. Uniform-length plastic support legs insert to the high-density polyethylene (HDPE) support grids to form the open plenum below, while supporting the organic media above. Air is forced upward through slots in the support grids, where it is treated in the media.

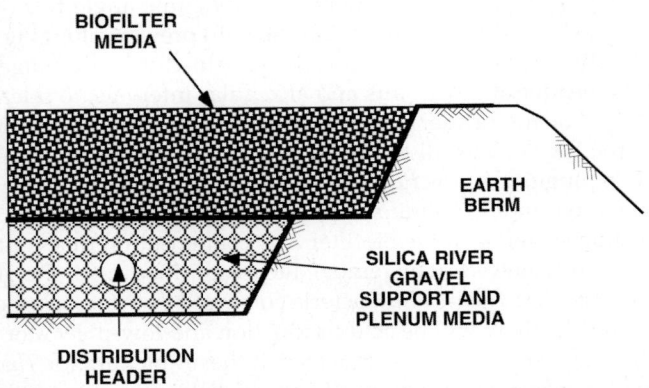

FIGURE 6.25 Typical silica-gravel-media support plenum.

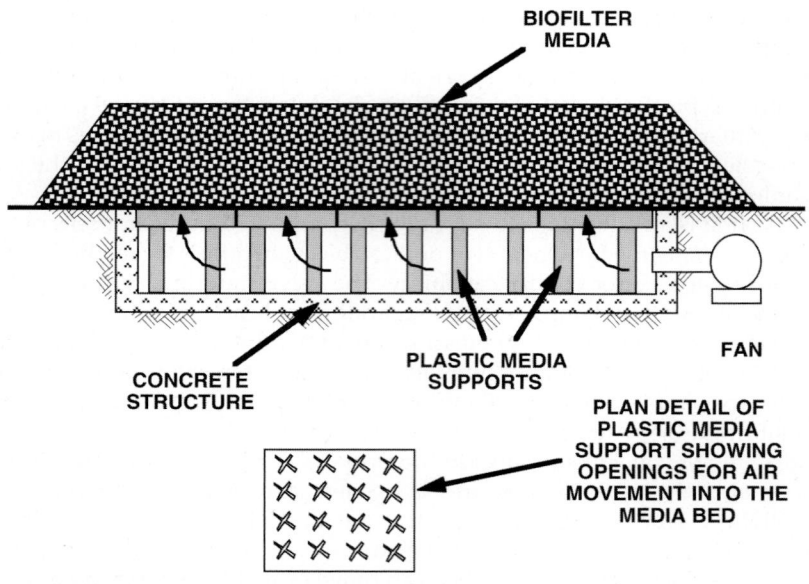

FIGURE 6.26   Typical open-plenum-biofilter arrangement.

In the case of media plenums, a perforated PVC pipe is surrounded by acid-resistant granular material. Acid production on the granular plenum media and in the biofilter media above requires that all materials in the plenum be acid resistant, including the air-distribution headers. For rock-media air plenums, it is important that the greatest headloss in the entire air-collection-and-distribution system is the headloss through perforated holes in the air-distribution piping. This ensures even distribution of air through holes in the distribution header and even application of odors to the surface of the biofilter media.

The airflow volume and the headloss design of the system dictate the diameter and spacing of the air-distribution holes in the perforated ducting. Typically, holes are drilled in plastic distribution headers at a 45-degree angle from vertical. The drilled holes are placed in a downward position to help prevent plugging from the media.

Air-distribution headers also must be provided with drainage holes drilled in the invert of the header at low points and at regular intervals, to release condensed water from the pipe, or the header must be intentionally sloped to drain. If drainage is not provided, the header can fill to the level of the air holes, causing biogrowth, fouling, and possible plugging of the holes.

Air plenums under biofilters may collect considerable water from irrigation and humidification systems on the biofilter and condensation that contributes water to the air plenum. This water must be removed to avoid flooding of the plenum and failure of the biofilter. Because *Thiobacillus* bacteria colonize all surfaces, if the air stream contains hydrogen sulfide, there will be acid production and low-pH water in the plenum.

Rock media provide more surface area in the air plenum for *Thiobacillus* colonization than open plenums and hence more acid production in the plenum than open plenums. In the case of open plenums, only the walls, floor, and underside of the support system are available for acid production, but acid generation will continue in the media above and will drip down into the plenum. In either case, the biofilter plenum must be both acid attack resistant and waterproof to perform its function. Therefore, biofilters must be provided with a positive means of draining water from the air plenum.

7.2.2.3 Underdrain Piping System Excess irrigation and condensed water must be collected and piped away for disposal in a receiving sewer or suitable treatment process. An impermeable plastic liner in the air plenum facilitates collection of the water. In the case of rock-media plenums, the floor typically is sloped and covered with a 1500-m (60 miles) or thicker HDPE liner similar to landfill-leachate-collection liners. The water then is collected in drain pipes and gravity-drained from the biofilter.

In the case of open-plenum biofilters, drainage is simple. Because the biofilter plenum is under pressure, the drain can become a route of air release if not properly trapped to withstand the pressure in the biofilter. Typically, a P-trap, with dimensions suitable to provide positive control of the maximum pressures anticipated in the plenum, is sufficient to prevent this problem. Drains also collect fines and sediment over the years and the occasional stray medium particle. For this reason, drain systems on biofilters should be equipped with cleanouts on all dead ends and long piping runs. This allows the drain piping to be jetted or flushed to remove sediment and unwanted materials, should this condition occur.

7.2.2.4 Media-Support System The media-support system provides structural support for the biological media above and typically is an integral component of the air plenum. As mentioned earlier, the two basic types of media-support systems are (1) media-based or rock-media-support systems and (2) open-plenum-support grids.

Gravel-support media or other granular material must be acid resistant to prevent damage caused by conversion of hydrogen sulfide gas to sulfuric acid. The most commonly used natural granular material is a pure silica mineral, such as common river gravel. Rounded river gravel is preferred because fractured or angular stone media have a tendency to puncture or cut the plastic liners under the rock media or plastic air-distribution ducts.

Granular media used in a plenum should be tested to ensure that they are impervious to the effects of low pH resulting from sulfuric acid formed by the biological process. Samples of proposed media should be soaked overnight in 10% sulfuric acid and the results evaluated. Many times, limestone, dolomite, and other carbonate-based rocks are found in river gravel samples, and these materials should be prohibited from use in biofilters. Carbonate rocks react with the sulfuric acid to produce calcium and magnesium sulfate and residual sludge that can foul and plug a biofilter plenum.

Rock-media-support systems offer advantages and disadvantages over open-plenum, grid-supported systems. The rock media can provide significant additional surface area for bacterial colonization and odor removal. When the foul air stream contains high concentrations of hydrogen sulfide gas, a significant amount of it can be removed in the rock-media zone. In one biofilter in Texas, hydrogen sulfide gas removal across the rock media alone was found to be 50%. The removal of hydrogen sulfide gas (and acid) in the rock media reduces acidification in the organic media above, which can extend media life and make the organic media more effective for organic odor removal.

7.2.2.5 Acid and Neutral Zones When a foul air stream being treated in a biofilter contains both hydrogen sulfide and organic odor compounds, the bacteria in the media stratify according to type. Because autotrophic organisms convert their food (hydrogen sulfide gas) faster than heterotrophic organisms, they dominate the biological system in a biofilter. If hydrogen sulfide gas is present in the air stream, the autotrophic *Thiobacillus* species outcompete their heterotrophic counterparts and produce sulfuric acid as a by-product. The acid lowers the pH of the media below the tolerance point of heterotrophic bacteria, so that no heterotrophs can grow. Only after the hydrogen sulfide gas has been removed and the *Thiobacillus* species are no longer present in the media can the VOC-consuming heterotrophic organisms survive and perform their work. The pH in the lower reaches of a biofilter treating 20 ppm_v H_2S gas can reach 0.5. Most heterotrophic bacteria cannot survive on surfaces below pH 6.0 and typically prefer a pH of 7.0 or higher. Therefore, there is a point in the media, moving upward from the bottom, where the pH changes from acidic to neutral. No significant VOC or organic-compound removal occurs below this point. This point identifies the line of demarcation between the acid and neutral zones. Hydrogen sulfide gas and ammonia are removed in the acid zone, and organic compounds and VOCs are removed in the neutral zone. It also is important to note that the zone will move upward and downward proportionally with the inlet concentration of hydrogen sulfide gas.

7.2.2.6   Biofilter Media   Above the media-support system is the workhorse of the biofilter—the media. There are many different types of media that can be used successfully in biofilters, but they all have their advantages and disadvantages. The following discussion presents significant issues associated with the use of each major type of media. The discussion of soil and other inorganic media is included here for continuity.

Extensive research has been performed to determine an optimum biofilter media. Several media have been used successfully, including compost, rice hulls, peat, soil, sand, wood chips, and various mixtures of these materials. While some mixtures may show slightly better performance in certain applications, local conditions and the specific odor compounds to be removed are the primary factors to consider when selecting biofilter media. Local media availability and economy may influence the final mixture composition. Several biofilter companies also provide custom biofilter media; however, no single medium has been shown to provide optimal treatment characteristics under all operating conditions. Cost and local availability also are primary media-selection criteria.

Media-contact time is an important factor in the overall effectiveness of a biofilter. Media-contact time commonly is expressed in terms of empty-bed contact time (EBCT), to avoid media-porosity factors. The EBCT is a function of media depth and foul-air-loading rate. The air-loading rate of a biofilter commonly is expressed in cubic meters per square meter (m^3/m^2) or cubic feet of air per minute per square foot (cfm/sq ft). The EBCT commonly is expressed in seconds of contact time.

Autotrophic organisms in biofilters (primarily *Thiobacillus*, if hydrogen sulfide gas is present) are rapid oxidizers and typically require between 15 and 30 seconds of EBCT in an organic media to accomplish their metabolic processes. This rapid oxidation occurs partly because of the relatively high solubility of hydrogen sulfide gas and partly because of the rapid metabolic rate of *Thiobacillus* bacteria. Because organic compounds are larger and more difficult to degrade, heterotrophic metabolism of VOCs is slower. Organic-compound removal in an organic-media biofilter can require from 30 to 60 seconds of contact time, depending on the solubility of the particular compounds present.

Organic compounds that are extremely difficult to degrade can require up to 75 seconds of contact time in an organic media before complete removal occurs. The

EBCT is one of the most valuable design parameters for a biofilter, because odor removal occurs only while the air is in contact with the media. Establishing design parameters based on the EBCT can be an effective starting point for the design of a biofiltration system. Once the volumes of the autotrophic and heterotrophic zones have been established through EBCT analyses, other design factors can be calculated.

7.2.2.7   Soil Media   Soil and inorganic biofilter media, such as sand and crushed gravel, perform the same function as organic media, by providing a growing surface for bacteria. They also must provide all of the same environmental and nutritional requirements for both autotrophic and heterotrophic bacterial growth. Inorganic media typically do not contain soluble nutrients available for biological uptake, so nutrients must be added to the irrigation water for inorganic media. Nutrients, such as nitrogen, phosphorous, potassium, iron, and other micronutrients, typically are added to the irrigation water for use by attached bacteria.

Soil media typically are finer grained than organic media and therefore have smaller pores through which the air must distribute. Narrow pathways through the media restrict the passage of air distribution and create higher headlosses. Some sand-based-media biofilters operate at 5.0 kPa (20 in.) of water column. The higher headlosses typically associated with sand media require a lower loading rate to avoid excessive pressures. Sand-based-media biofilters are loaded between approximately 0.3 and 0.85 $m^3/m^2/min$ (1 and 2.5 cfm/sq ft). One of the most attractive features of soil-bed filters is their simplicity of design and construction. Many soil-bed biofilters do not have plenums, and the distribution ducting simply is buried in the media. When the media are ready for replacement, the air-distribution ductwork is dug up and hauled away with the media. With a life of 10 years or possibly more, replacement of the relatively cheap air-distribution ductwork is not a major cost item.

Inorganic media must resist acid conditions caused by the oxidation of hydrogen sulfide gas. Silica sand, granite, basalt, and other pure silica minerals typically are used because of their resistance to strong mineral acids, such as sulfuric acid. Acid testing should be performed on any proposed mineral biofilter media, by soaking a washed sample overnight in a 10% solution (by weight) of sulfuric acid and observing any discoloration of the liquid or damage to the grains.

Inorganic biofilter materials are harder than organic biofilter materials and will withstand long-term media deformation better under moist and wet conditions. For this reason, soil, sand, and other inorganic media do not require replacement because of compaction, the greatest enemy of organic biofilter media.

7.2.2.8   Organic Media   Biofilters also use organic media for microorganism growth. Various studies have been performed to identify the odor-removal capacity of organic-based-media formulations. Laboratory studies using bench-scale columns often have yielded unsatisfactory results when applied full scale. There are numerous cases of a particular media showing promise during pilot studies, yet failing when applied in the field. The best test for a media is successful full-scale application under various operating conditions.

Many different formulations of organic media have been used—some more successfully than others. Table 6.17 contains a list of the most commonly used materials for biofilter media, with a rating when used in full-scale applications.

Material	Overall Rating	Comments and Observations
Yard waster compost	Excellent	Should be thoroughly composted.
		Fines smaller than 1.27 cm (0.5 in.) should be screened out.
Wood chips	Good	Best when used as a component and not a stand-alone medium. Should be aged or composted and never green.
Sludge compost	Poor	Too fine, plugging, high headloss.
Peat	Poor	Too absorbent and tends to compact.
Rice hulls	Poor	Can be used as an amendment, but has little structure when wet and tends to compact.
Tree bark	Good to excellent	Resist compaction, commonly used to top a biofilter to prevent weed growth and moisture loss.

TABLE 6.17   Typical Biofilter Organic-Media Materials

Regardless of the source, organic media must satisfy the following basic requirements of all biofilter media:

- Surface area and porosity;
- Structural support;
- Moisture retention; and
- Nutrient availability.

All media must provide sufficient surface area upon which the microorganisms can grow and perform their intended function. Organic-media biofilters exhibit an inverse relationship between surface area and air transmittance (porosity). As the media become finer, the transmittance of air becomes less because of headlosses through the smaller pores. Experience has shown that larger, more porous materials provide ample surface area, while providing excellent air movement and odor removal.

Organic media also must provide structural support for the bed and hold the pores open for optimum air movement. This is perhaps the greatest single challenge for an organic media and should be a primary selection criterion. All organic media consist primarily of cellulose and related woody materials. Cellulose fibers are held together by glue called lignin, which can be softened and dissolved over time in water. Biological activity and the production of acid in the media also aid in lignin destruction. As the lignin in the woody materials breaks down, the fibers collapse, and the media structure settles. This reduces porosity, air transmittance, and treatment effectiveness. Compaction and settlement is the greatest enemy of organic media and the single greatest factor for media failure and replacement.

7.2.2.9 Irrigation and Humidification   Irrigation and moisture control is critical in a biofilter. Too much water fills the pores with water, can cause anaerobic zones to develop, and increases the unit weight of the media. This reduces the surface area available to microorganisms and compacts the media under the weight of gravity. It has been observed that organic media exposed to heavy rains suffer more from compaction than covered organic media of the same type. The difference has been attributed to increased unit weight because of uncontrolled water addition during rain events. Therefore, irrigation and humidification of biofilters is critical to avoid overwatering, but still provide the moisture necessary for biological activity.

Biofilter media must be maintained in a moist condition to remove odors. If the inlet air is significantly below 100% relative humidity, it will tend to evaporate water from the media and dry it out. The key to understanding moisture in a biofilter is understanding the relationship between air temperature and moisture content. As air warms, it can hold more water in the vapor state (relative humidity), and, conversely, as air cools, it cannot hold as much water and will condense water (dew point). Depending on conditions when the air enters the biofilter, it will either pick up (evaporate) or deposit (condense) water. Either evaporation or condensation occurs in a biofilter and, many times, both occur simultaneously.

Biofilters typically are irrigated and humidified in the following three ways:

1. Humidification of the inlet air stream;
2. Deep irrigation of the media with hoses; and
3. Surface irrigation with sprinklers.

All three of these may be needed at times. Depending on the local relative humidity and the way foul air is collected, inlet air can be less than 100% relative humidity. If this air stream enters a biofilter and the air is not cooled below the dew point, evaporation will occur. Depending on the volume and location of deep-bed irrigation, there may not be sufficient moisture to satisfy the water demand of the air stream, and evaporation will occur. An open-bed plenum is more sensitive to fluctuations in inlet relative humidity than a rock-bed plenum, because the rock is wetted by irrigation water from above, providing a source of moisture for the air. Rock plenums provide moisture buffering of the air, which is not available in open plenums. Therefore, rock plenums act like a low-efficiency humidification system for the air stream.

To reduce the condensation effect of humidification chambers, some biofilter systems have used a low-volume, in-duct spray system immediately upstream of the plenum. Pressure-atomizing nozzles (275 800 Pa [40 psi]) operating at 4 to 20 L/h (1 to 5 gph) have been installed directly in the headers leading to biofilter plenums, to aid in humidification of the inlet air stream. The wide-angle, hollow-cone spray nozzle is attached to the end of a 316L stainless steel pipe and inserted through an adapter coupling directly to the center of the duct airflow. The spray is commonly positioned to spray toward the fan and opposite the direction of airflow, to provide more air-contact time. Figure 6.27 illustrates one typical in-duct humidification system. The lower volume of water used by the spray systems minimizes the cooling effect of the air stream and provides particulate water droplets that carry farther into the biofilter. Any excess water carried in the headers drains out to the plenum and is removed with the drainage water. These systems can be prone to plugging and may require water softeners.

Another type of air-humidification system also is a form of an in-bed irrigation system. Water is supplied through irrigation hoses to the bottom of the organic media just above the plenum. In the case of rock plenums, water flows down over the rock, keeping it wet to act as a humidification system for the air. In the case of open plenums, the irrigation is applied to the media at a depth of 150 mm (6 in.) above the plenum. The water then saturates the lower portion of the medium, which acts as a water reservoir.

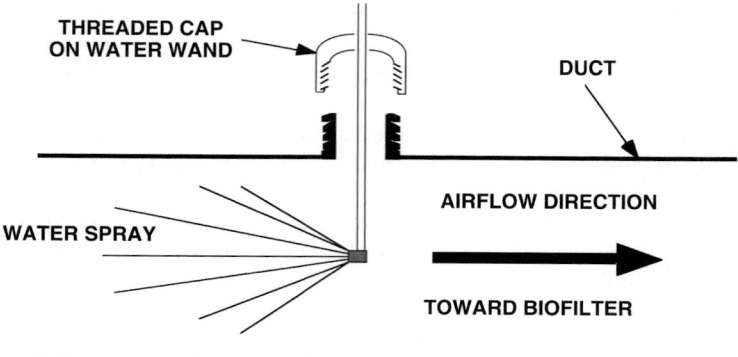

FIGURE **6.27** Common in-duct humidification spray system.

The air picks up the necessary humidity, as needed, when passing through this zone. The saturated condition in the lower zone of the organic media causes premature compaction and replacement.

The last type of irrigation available for biofilters is surface irrigation. This irrigation scheme is no more complicated than a typical residential sprinkler system and aims to provide coverage of the biofilter surface, with either a fixed or pop-up-type spray head. Surface watering is not a substitute for deep-bed irrigation or humidification. It is detrimental to the biofilter to try to irrigate totally from the surface because of saturation conditions, headloss increases, and premature compaction of the media.

Surface irrigation is used commonly only in the summer or when drying conditions are at a maximum. Surface irrigation controls media drying and cracking during dry periods. Media cracking can lead to foul air short-circuiting, and odor release, and must be controlled. Media cracking typically is a problem only for finer-grained media and is rarely experienced with larger compost mixtures.

7.2.2.10 Nutrient Control Microorganisms in a biofilter require nutrients for optimum growth and reproduction. The required nutrients are primarily nitrogen and phosphorous, with smaller amounts of sulfur, iron, manganese, magnesium, calcium, and other trace metals also being required. Typically, treated wastewater effluent contains most of the nutrients and trace minerals needed to keep a biofilter population in excellent health.

In the case of inorganic biofilter media, such as sand or pea gravel, the medium offers no nutrients to bacteria. In soil-bed biofilters, a nutrient solution must be added periodically (every 1 to 2 weeks), to provide the necessary nutrients for growth and cell function. Organic biofilter media typically will provide most (if not all) of the nutrients required for normal growth. Some studies have shown that biofilter performance can be enhanced following the addition of soluble nutrients to the irrigation water in an organic media biofilter. However, it typically is not necessary to add nutrients to an organic media.

7.2.2.11 Media Life Organic media in biofilters will need replacement occasionally because of compaction. The length of time that a biofilter can operate on a single charge of media depends on the initial mix of the media, climatic conditions to which it is exposed, production of acid in the media, and amount and type of irrigation provided.

With simple organic-based-media systems under fairly high loading, replacement may be required every 1 to 3 years. Media need to be replaced when they break down to the point that sufficient air can no longer be moved through the bed. Pressure or headloss monitoring and periodic airflow measurements will indicate when this occurs.

7.2.2.12 Media Instrumentation and Monitoring   As a minimum, biofilters should be equipped with pressure gauges on each distribution header. Monitoring the air pressure going to each cell of a biofilter and tracking the pressure readings over time will indicate when a medium has reached the end of its useful life. Media can be considered to be in need of replacement when the minimum design airflow cannot be achieved at the maximum design pressure of the fan.

Other common instrumentation for biofilters includes flow meters on each main distribution header downstream of the damper, to allow for easy balancing. Thermal mass-flow meters are commonly provided for this service, although much simpler manometers also can provide accurate data. Sample ports with threaded caps to facilitate the use of hand-held anemometers also should be provided on each main header.

It is common practice to monitor fan performance and send signals or alarms back to an operator's control panel. Local thermal-overload protection also should be provided for the fan motor.

7.2.2.13 Weed Control   Biofilters provide an excellent growing medium for seedlings and other green facilities. Wind and birds bring seeds to the biofilter, and the seeds quickly germinate to produce growing facilities. Vegetation control in operating biofilters is somewhat of an aesthetic issue, because many biofilters have been overgrown with grass, and some have been used to plant flowers and other decorative vegetation. The most detrimental effect of vegetation growth in a biofilter is short-circuiting of foul air to the surface, caused by the taproots of larger facilities. In large shrub and tree species, the seedling produces a long taproot and sends it deep within the soil to find a long-term source of water. If the taproot passes deep into the organic media, it will form a channel for the short-circuiting of air.

If it is desired to maintain a biofilter without weed or vegetation growth, hand weeding can be performed, if steps are taken to prevent excessive compaction of the media caused by foot traffic. Typically, laying some flat lumber or plywood on the media for standing is sufficient to prevent premature compaction because of foot traffic when pulling weeds. Otherwise, chemical treatments, such as systemic herbicides, can be applied on a regular schedule without damage to the microflora in the biofilter.

7.2.3 Modular/Prefabricated Biofilters

There are several prefabricated biofiltration units that can be purchased for quick installation. While all biofiltration package units use the basic principles of biofiltration, most have one or more unique features that make them attractive. Package biofiltration units are available in a variety of construction materials, media types, operational modes, and sizes. Typically, package biofiltration units are more cost-effective for air volumes up to approximately 150 m^3/min (5000 cfm), depending on specific design parameters, construction methods, and operating conditions.

7.2.4 Bioscrubbers and Biotrickling Filters

Bioscrubbers and biotrickling filters are a relatively new entry to the odor-control market in the United States, after finding general acceptance in Europe. As with any new and emerging technology, the terminology of biological scrubbing has not been widely accepted. For purposes in this chapter, the following terminology is used.

7.2.4.1 Bioscrubber A bioscrubber is a biological reactor for the removal of odors from air streams. The majority of the biomass is suspended in the recirculated mixed-liquor solution, although there may be attached growth on the packing. This type of bioscrubber often is called a suspended-growth bioscrubber. The packing in the vessel is used primarily for mass transfer and not as an attachment medium for biomass. In this configuration, the process is better suited for removal of organic odor compounds by heterotrophic organisms.

7.2.4.2 Biotrickling Filter A biotrickling filter is a biological reactor for the removal of odors from air streams. The biomass is living primarily as attached growth on the media and not in the recirculated water. Biotrickling filters commonly do not recirculate water, but operate on a once-through irrigation water scheme. In this configuration, the process is better suited for removal of hydrogen sulfide gas by autotrophic organisms.

Figure 6.28 illustrates a suspended-growth bioscrubber with a secondary oxidation reactor. A bioscrubber functions much like a conventional chemical packed-bed scrubber, except that the chemical solution has been replaced by a bioactive solution. The bioactive solution is distributed over the top of a plastic packing media in a vessel, while the odorous air is forced upward.

In suspended-growth bioscrubbers, the odor compounds are absorbed or dissolved into the bioactive solution, where the microorganisms can oxidize them partially. The compounds are then further oxidized in a separate oxidation reactor. Proper growth of a large heterotrophic biomass may require the addition of a supplemental carbon food source. In some instances, activated-sludge mixed liquor is used as the bioactive

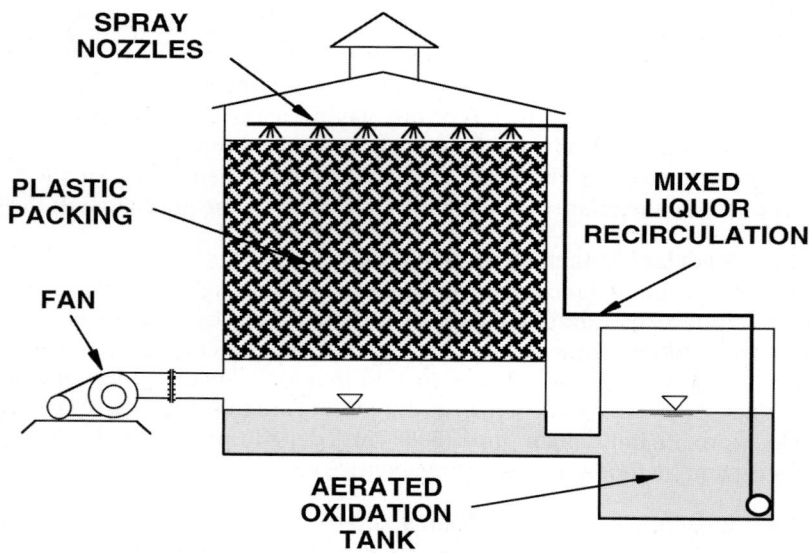

FIGURE 6.28 Typical suspended-growth bioscrubber schematic.

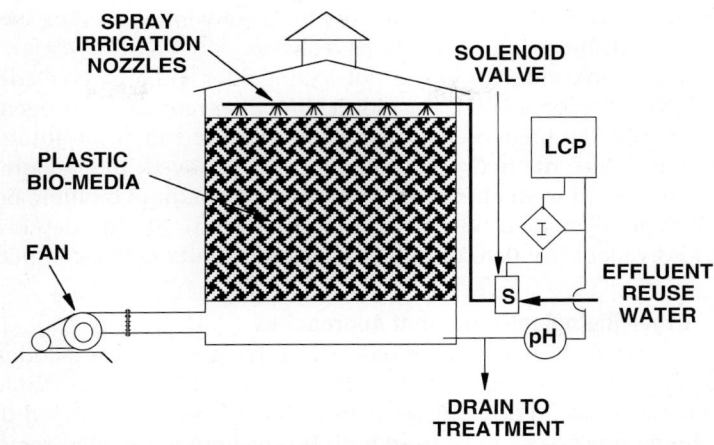

FIGURE 6.29 Typical prefabricated biotrickling filter (biotower).

solution and is recirculated from the aeration basin at a WRRF to the scrubber, and back again, so that absorbed odor compounds can be oxidized further in the aeration basin. Other systems use a portion of the RAS as the bioactive solution, with continuous blow-down to the aeration basin. Suspended-growth bioscrubbers retain an acclimated bio-mass within the scrubber sump (or external vessel), where supplemental aeration allows the time for biological oxidation. Currently, these types of bioscrubbers are not in common use, and only research applications are in service.

Biotrickling filters can be either prefabricated units, as shown in Figure 6.29, or custom designed and constructed. Constructed biotrickling filters commonly use lava rock as a medium, although some vessel-type biotrickling filters also use lava rock as a medium. Lava rock was first used as a biotrickling filter medium in Germany, where it was used successfully for many years. Improvements in biotrickling filter media design have made lava rock less common. Lava-rock biotrickling filters typically provide 85% to 98% removal of hydrogen sulfide, whereas the latest synthetic-media biotrickling filters consistently produce 99% removal. Lava rock also is much heavier than synthetic media and requires a more robust vessel. Lava rock provides less available surface area per unit volume than the latest synthetic media. Many pores in the lava rock fill with water, making them less available for oxidation of hydrogen sulfide gas.

7.2.4.3 Biotrickling Filters and Biofilters in Combination   In the United States, the majority of biotrickling filters are being used on air streams containing hydrogen sulfide gas and low-molecular-weight, sulfur-bearing organic compounds. A large quantity of sulfuric acid, which would be produced from the biological oxidation of high concentrations of hydrogen sulfide gas (25 ppm$_v$), typically is not desirable in an organic-medium biofilter. The acid lowers the pH of the media, which prevents colonization by heterotrophic organisms. It is the ability to grow heterotrophs and cost-effectively remove organic compounds that is perhaps the greatest benefit of the organic-medium biofilter. However, acid production from hydrogen sulfide prevents heterotrophic use of a portion of the media. Therefore, if a technology selectively removes hydrogen sulfide gas, it could be used as a pretreatment process for a conventional organic-medium biofilter, and the best attributes of each could be realized.

When used in combination, the biotrickling filter and organic-medium biofilter both are used to their full potential. The biotrickling filter removes hydrogen sulfide gas most effectively by providing an environment specifically suited for autotrophic organisms (*Thiobacillus*). Once the hydrogen sulfide gas is removed, the remaining organic compounds are treated most effectively in the organic-medium biofilter. Because no hydrogen sulfide remains (and no acid is being produced), the full depth and EBCT of the biofilter is available for heterotrophic colonization and organic-compound oxidation.

7.2.5 Other Biological-Treatment Approaches

Typical activated sludge aeration basins also have been used as odor-treatment technologies. The action is similar to a biofilter or bioscrubber, except that foul air is brought to the biomass instead of biomass to the foul air. It should be noted that not only activated sludge processes can be used with this technology, similar removal of odors can be achieved when aerobic digesters and solids-contact chambers are used.

A typical activated sludge mixed liquor is a combination of autotrophic and heterotrophic organisms with the ability to degrade and oxidize many different compounds. If odorous compounds are soluble or slightly soluble in water and can be degraded biologically, then aeration-basin disposal should be capable of removing them. All of the inorganic-odor compounds and most of the organic-odor compounds experienced in municipal odor control are soluble or slightly soluble in a water matrix at atmospheric pressure. The size and pressure (depth) of a diffused-air bubble in a mixed-liquor environment also greatly influences compound solubility and the direction of mass transfer across the air-water interface. In general, the solubility of an organic compound decreases with increasing molecular weight and increases with pressure. Therefore, compounds that are only slightly soluble at atmospheric pressure and temperature can be dissolved readily at depth in an aeration basin (Bowker et al., 1995).

There are several degradation mechanisms available for odor compounds in a typical activated sludge, mixed-liquor environment. Under certain circumstances, autotrophic organisms can oxidize the inorganic compounds quickly because of their high solubility and rapid metabolic processes. In addition, the reduced inorganic compounds (H_2S, HS^-, and NH_4^+) can be taken directly in the cell to be used as building blocks for proteins, cell maintenance, and cell growth. Typically, in the absence of the reduced form of sulfur and nitrogen, cells must expend energy to reduce sulfate and nitrogen compounds to a form suitable for protein synthesis. When the reduced forms of these compounds are available in the local environment, there appears to be a mechanism for the rapid and direct movement of these reduced nutrients into the cell wall. This form of direct cellular uptake appears to be rapid.

Organic compounds that are soluble or moderately soluble in water also will be degraded to a large extent. Once dissolved, these organic compounds are regarded by heterotrophic organisms in the mixed liquor as BOD and are consumed similarly. Compounds that are insoluble or slightly soluble may not be removed in this process. The air from basins treating odors has been noted to have a slightly different character than conventional activated sludge processes. No research could be found on the specific odor compounds that may be causing this odor-character change; however, they are likely the larger-molecular-weight, less soluble organic compounds. Heterotrophic organisms also are primarily responsible for direct cellular uptake of reduced inorganic compounds.

Problems also have been experienced with inadequate filtration of the inlet foul air stream before introduction to the main facility blowers. Depending on the point of collection, foul air streams can contain a significant load of particulate matter, dusts, mists, and greasy aerosols. These materials will cause damage to blower internals and coat header walls and plug diffusers. Organic mists and aerosols will form sludge deposits on the impellers or lobes of aeration blowers because of the rapid pressure change and high temperatures. Eventually, these deposits will cause imbalance of the impellers or lobes with loss of efficiency and potential failure of the blower.

The type of diffuser used also can influence the effectiveness of this process. As mentioned earlier, the size of the bubble influences the speed and effectiveness of the solubility of odor compounds to the mixed liquor. The process involved in dissolving gases into a water matrix can be described by Henry's law, which states that a gas in direct contact with water and the dissolved form of the gas in water tend to seek equilibrium. The equilibrium is dependent on the ratio of the free gas above the water and the mole fraction of the dissolved gas in solution. Essential for the rapid establishment of equilibrium between gas and water is sufficient surface area, over which Henry's law can operate. Smaller air bubbles have a much higher surface-area-to-volume ratio than larger air bubbles, which aids smaller bubbles in more rapid and complete solution of the contained gases.

Therefore, a small bubble is preferred. However, some fine-bubble diffusers have been reported to have the most problems with plugging, which may occur as part of this process. Medium- and coarse-bubble diffusers do not appear to have a plugging problem with this process; however, a larger bubble provides less surface area and therefore lowers mass-transfer rates.

The volume of foul air that can be treated technically is limited to the capacity of the aeration blowers; however, practical guidelines should be followed before sizing such a system. The source of foul air and the oxygen content of foul air should be evaluated to ensure no negative oxygen-transfer effects.

7.3 Chemical and Physical Treatment

7.3.1 Gas-Absorption Scrubbers

The basic principle of packed-bed wet scrubbers is absorption of the odor contaminant, such as hydrogen sulfide, in the gas stream into the recirculating scrubbing liquid. Gas absorption is the mass transfer of the gaseous contaminant from air to the scrubbing liquid, with subsequent reactions to a stable compound. Absorption occurs on the surface of the water, and packing is used in the tower to break up and create large liquid surfaces (i.e., maximum air-to-liquid surface area). Packing material is selected to provide adequate surface area with minimum pressure drop. Wet scrubbers can be vertical countercurrent or horizontal crossflow, with packed-bed liquid distribution at a 90-degree angle to the airflow direction. Odor scrubbers also can be once-through misting towers, in which atomizing nozzles create a fine mist spray, which is dispersed in a vessel contacting the odorous air stream.

7.3.1.1 Packed-Bed Wet Scrubbers Basic design criteria for packed-bed wet scrubbers include the selection of velocity through the scrubber vessel and the recirculation

Parameter	Value
Empty-bed gas velocity	1524 to 2540 mm/s (300 to 500 fpm)
Packing depth	183 to 366 cm (6 to 12 ft)
Scrubbant recirculation rate	4068 to 5424 g/m²·s (3000 to 4000 lb/sq ft/hr)
Gas loading rate	2441 to 3051 g/m²·s (1800 to 2250 lb/sq ft/hr)
Makeup water flow	0.0133 to 0.133 L/min·m³/min (0.1 to 1.0 gpm/1000 cfm)
Headloss through vessel	0.15 to 0.75 kPa (0.6 to 3.0 in. H_2O)
Control parameters potential	−pH (9.0 to 10.5) and oxidation-reduction (600 to 800 mV)

TABLE 6.18   Typical Design Criteria for Packed-Tower Scrubbers Treating Hydrogen Sulfide

flowrate. These and other parameters shown in Table 6.18 are selected to prevent excessive headloss (high velocity) or inadequate absorption because of low turbulence.

Scrubber velocity dictates the vessel cross-sectional area, and the mass-transfer rate dictates the packing depth. Figure 6.30 shows a typical vertical countercurrent packed-bed scrubber and its components. Horizontal or crossflow scrubbers also are used often and can be combined for gas absorption and particulate removal. Crossflow scrubbers have a vertical liquid flow, while the gas passes through the packing horizontally. Crossflow scrubbers have lower pressure drops and lower recirculation flows, and they can be used with multiple packed beds, so that more than one type of scrubbing liquor can be used. These horizontal scrubbers offer the advantage of a lower profile where headspace or elevations are a concern. Access for maintenance also is an advantage. The disadvantage of crossflow scrubbers is potentially lower odor-removal efficiency, because the bottom of the packing does not come in contact with fresh scrubbant water, but does receive the same high-strength inlet odorous air as the top of the packing.

Odorous compounds to be treated in scrubbers can include hydrogen sulfide; organic sulfides, such as methyl mercaptan, dimethyl sulfide, and dimethyl disulfide; ammonia; and, to a lesser extent, VOCs, such as ketones, organic acids, aldehydes, and alcohols. Chemical oxidation of VOCs is not efficient in packed-bed scrubbers. Carbon adsorption and thermal oxidation are the preferred processes for VOC control.

Packing depth is calculated from a determination of the number of height of transfer units and is based on Henry's law, which is applicable to gas-liquid equilibrium for dilute concentrations. Characteristics of inlet gas, including odorant concentrations, flowrate, humidity, and temperature, are required to determine packing height. Table 6.19 provides mass-transfer data from two packing manufacturers.

7.3.1.2 Chemistry   Typically, the chemical reaction in scrubbers involves the conversion of an odorous compound to a nonodorous salt. Reactions typically are acid-base or oxidation-reduction. Reactions typically are quick and ongoing, provided there is sufficient chemical available.

Caustic and hypochlorite are used for most odor-control scrubbers treating hydrogen sulfide. Conversely, ammonia and trimethylamines are absorbed with an acid

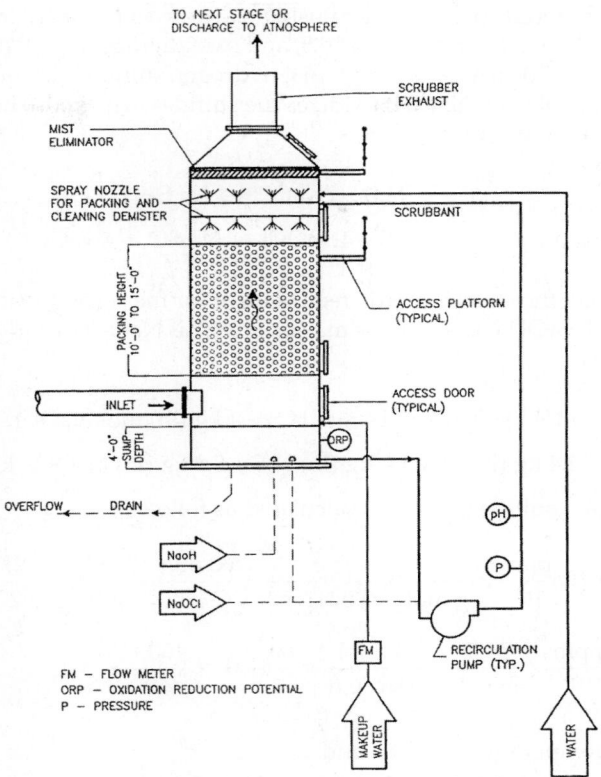

FIGURE 6.30 Typical packed-tower-scrubber schematic diagram (ft × 0.3048 = m).

Packing Manufacturer	Absorption System	G, kg/m²/h (lb/h/sq ft)	L, kg/m²/h (lb/h/sq ft)	Temperature, °C (°F)	Height of Transfer Units (HTU), kPa (in.)
1	H₂S-	6000	6500	20	3.25 (13.0)
	NaOH	(1230)	(1330)	(68)	[1-in. packing]
					4.85 (19.4)
					[2-in. packing]
					5.5 (22.0)
					[3.5-in. packing]
2	H₂S-	13 200	19 500	22	18.1 (4.0 in.)
	NaOH	(2700)	(4000)	(72)	

TABLE 6.19 Typical Mass-Transfer Performance Data for Hydrogen-Sulfide Removal

recirculation liquid (pH = 1.5 to 5) using sulfuric acid. For hydrogen sulfide removal, pH is controlled to a range of 9 to 10.5, and oxidation-reduction potential (ORP) should be greater than 700 mV. At the high pH, hydrogen sulfide is absorbed into solution, and sodium hypochlorite (NaOCl) oxidizes the sulfides captured. The equations for hydrogen sulfide removal are as follows:

$$H_2S + 2NaOH \leftrightarrow Na_2S + 2H_2O \tag{6.22}$$

$$Na_2S + 4NaOCl \leftrightarrow Na_2SO_4 + 4NaCl \tag{6.23}$$

Therefore, the stoichiometric requirement for the second reaction (eq 6.23) is as follows (actual NaOH requirements may be lower, as NaOCl includes a significant amount of alkalinity):

2 kg (lb) mol NaOH/kg (lb) mol H₂S + 4 kg (lb) mol NaOCl/kg (lb) mol H₂S or

2.4 kg (lb) NaOH/kg (lb) H₂S + 8.8 kg (lb) NaOCl/kg (lb) H₂S

Hydrogen sulfide loading is calculated as follows:

$$Q\,(m^3/min) \times \frac{ppm}{10^6} \times \frac{34g}{g\,mol} \times \frac{g}{0.0241\,m^3} \times \frac{60\,min}{h} \times \frac{kg}{1000\,g} \times \frac{273+21}{273+T} = g/h$$

$$Q\,(cfm) \times \frac{ppm}{10^6} \times \frac{34\,lb}{lb\,mol} \times \frac{lb\,mol}{386\,cu\,ft} \times \frac{60\,min}{h} \times \frac{460+70}{460+T} = lb/h \tag{6.24}$$

where T = temperature, °C (°F), and
   Q = airflow, m³/min (cfm).

The molar volume at 76 cm Hg and 21°C is 0.0241 m³/g mol (1 atm, m³/min and 70°F is 386 cu ft/lb mol).
   From the previous example, hydrogen sulfide loading is 5.1 lb H₂S/h. Chemical dosages would be as follows:

2.4 kg (lb) NaOH/kg (lb) H₂S × 2.3 kg H₂S/h (5.1 lb H₂S/hr)
   = 5.52 kg NaOH/h (12.24 lb NaOH/hr) \tag{6.25}

8.8 kg (lb) NaOCl/kg (lb) H₂S × 2.3 kg H₂S/h (5.1 lb H₂S/hr)
   = 20.4 kg NaOCl/h (44.88 lb NaOCl/hr) \tag{6.26}

Using 50% caustic, which is 1.52 kg/L (12.7 lb/gal), and 12.5% hypochlorite, which is 1.27 kg/L (10.6 lb/gal), chemical consumption required would be as follows:

NaOH:

$$\frac{5.52\,kg/h}{1.52\,kg/L \times 0.5} = 7.26\,L/h\text{ of 50\% caustic}$$

$$\frac{12.24\,lb/hr}{12.72\,lb/gal \times 0.5} = 1.90\,gal/hr\text{ of 50\% caustic} \tag{6.27}$$

NaOCl:

$$\frac{20.4 \text{ kg/h}}{1.27 \text{ kg/L} \times 0.125} = 128.5 \text{ L/h of } 12.5\% \text{ hypochlorite}$$

$$\frac{44.88 \text{ lb/hr}}{10.6 \text{ lb/gal} \times 0.125} = 33.9 \text{ gal/hr of } 12.5\% \text{ hypochlorite} \qquad (6.28)$$

Chemical-metering pumps should be sized to handle average and peak loadings. However, some reactions take time, such as the reaction of organic sulfides scrubbed by chlorine solutions. There also are interferences from carbon dioxide, which will react with sodium hydroxide. Carbon dioxide is generated in biological processes, such as fixed-film bioreactors, and air from these unit processes and slow-moving sewers often are treated in scrubbers. Carbon dioxide can consume large amounts of caustic, because it forms carbonic acid in solution. At a pH of 9, approximately 1 mol NaOH is consumed per mole CO_2, and, at a pH of 10.5, 1.6 mol NaOH are consumed per mole CO_2. There is significant additional consumption of caustic when operating a scrubber above pH 11 caused by increased carbon dioxide uptake.

With a high carbon dioxide content, it may be more cost-effective to operate at a neutral-pH (6.5 to 7.0) chlorine solution than a caustic solution. The neutral-pH chlorine solution contains 50% hypochlorous acid, which is a strong oxidant. However, this may require more packing depth and longer scrubber sump detention time.

Some research has shown that chemical oxidation of dimethyl disulfide in a misting-tower chemical scrubbing system is most effective with sodium hypochlorite neutralized to pH 6.5. Ammonia consumes hypochlorite and must be removed in an earlier stage. However, scrubbers also can treat organic sulfides in an NaOH/NaOCl solution. Equations for ammonia removal with neutral or acid solution are as follows:

$$2 \text{ NH}_3 + \text{H}_2\text{SO}_4 \rightarrow (\text{NH}_4)_2\text{SO}_4 \qquad (6.29)$$

$$2 \text{ kg (lb) mol NH}_3/\text{kg (lb) mol H}_2\text{SO}_4 \qquad (6.30)$$

or

$$2.88 \text{ kg (lb) H}_2\text{SO}_4/\text{kg (lb) NH}_3 \qquad (6.31)$$

VOC removal is widely variable in scrubbers. In fact, scrubbers may produce chlorinated organic compounds. In normal operation, VOC removal may range from 25% to a high of 60%. The Joint Emissions Inventory Program (JEIP) (South Coast Air Quality Management District, 1993) and the Water Environment Research Foundation (Alexandria, Virginia) (WERF), *Control and Production of Toxic Air Emissions by Publicly Owned Treatment Works* (WERF, 1994), are two studies undertaken to assess efficiencies of odor scrubbers for VOC control. In the JEIP study, no significant VOC removal or production was found in packed-bed scrubbers.

7.3.1.3   Overflow Rates and Makeup Water   Typically, having sufficient overflow to maintain a salt (NaCl and Na_2SO_4) concentration of less than 5% will keep the driving force of the scrubber reaction high and prevent reactions from approaching equilibrium, which is critical to scrubber operation. Sometimes, overflow rates are driven by removal of solids through particulate removal. From eq 6.23, 4 kg (lb) mol of NaCl and

1 kg (lb) mol of Na_2SO_4 are formed per kg (lb) mol of H_2S or 11 kg (lb) of salts per kg (lb) H_2S. In this example, the water overflow rate to keep the concentration of salts below 5% will be as follows:

$$\frac{11 \text{ kg NaCl/kg } H_2S \times 2.32 \text{ kg } H_2S/h}{60 \text{ min/h} \times 0.05 \times 1 \text{ kg/L}} = 8.5 \text{ L/min}$$

$$\frac{11 \text{ lb NaCl/lb } H_2S \times 5.1 \text{ lb } H_2S/hr}{60 \text{ min/hr} \times 0.05 \times 8.34 \text{ lb/gal}} = 2.2 \text{ gpm} \tag{6.32}$$

Makeup water should be continuous and is based on evaporation rates, because a scrubber will saturate the air stream with water to approximately 100% humidity. A psychrometric chart, as shown in Figure 6.31, can be used to determine the evaporation rate. Adiabatic saturation will occur in a scrubber, as the air increases in moisture content (no net heat transfer).

At 32.22°C (90°F) and an assumed 40% relative humidity, the inlet gas will contain approximately 0.012 kg (lb) of water vapor per kg (lb) dry air (da). Following

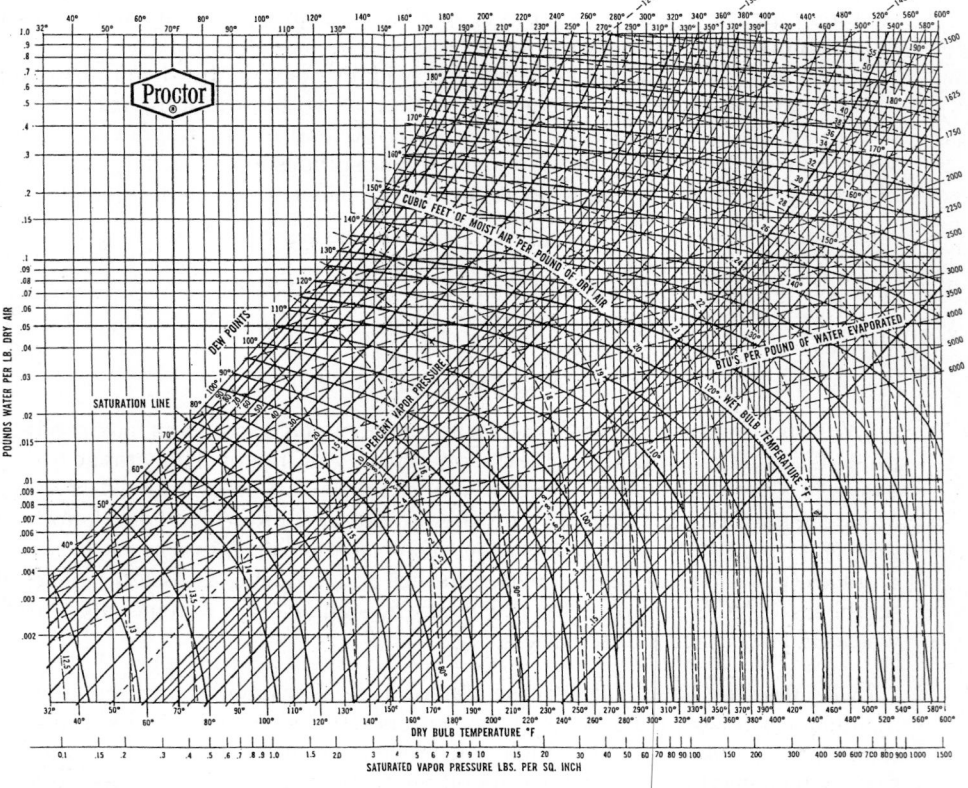

Figure 6.31 Psychrometric chart (properties of air and vapor mixtures from 0 to 316°C [32°F to 600°F]).

the adiabatic saturation line on the chart to the 100% saturation line, the saturation temperature (outlet air) is 22.78°C (73°F) and will contain 0.017 kg (lb) H_2O lb dry air (da).

Using the psychrometric chart, the gas contains approximately 0.88 m^3 air/kg da (14.2 cu ft air/lb da). Therefore, the quantity of dry air is as follows:

$$(283.4 \text{ m}^3/\text{min})/(0.88 \text{ m}^3/\text{kg da}) = 319 \text{ kg da}/\text{min}$$

$$(10\ 000 \text{ cfm})/(14.2 \text{ cu ft/lb da}) = 704 \text{ lb da}/\text{min} \tag{6.33}$$

Therefore, water evaporated is as follows:
In SI units:

$$\text{Outlet: } 319 \text{ kg da/min} \times 0.017 \text{ kg } H_2O/\text{kg da} = 5.4 \text{ kg/min} \tag{6.34a}$$

$$\text{Inlet: } 319 \text{ kg da/min} \times 0.012 \text{ kg } H_2O/\text{kg da} = 3.8 \text{ kg/min} \tag{6.35a}$$

Difference = 1.6 kg/min

$$\frac{1.6 \text{ kg/min}}{1 \text{ kg/L}} = 1.6 \text{ L/min}$$

In U.S. customary units:

$$\text{Outlet: } 704 \text{ lb da/min} \times 0.017 \text{ lb } H_2O/\text{lb da} = 12.0 \text{ lb/min} \tag{6.34b}$$

$$\text{Inlet: } 704 \text{ lb da/min} \times 0.012 \text{ lb } H_2O/\text{lb da} = 8.4 \text{ lb/min} \tag{6.35b}$$

Difference = 3.6 lb/min

$$\frac{3.6 \text{ kg/min}}{8.34 \text{ kg/L}} = 0.43 \text{ L/min}$$

Therefore, makeup water is 10.1 L/min (2.6 gpm) (assume average of 11.4 L/min [3 gpm]). A rotameter sized for 0 to 40 L/min (0 to 10 gpm) is sufficient for makeup water adjustments on the scrubber. This makeup water rate is the minimum required to avoid an approach to equilibrium with the scrubber solution.

The outlet AFR would be based on the density of the saturated air at the outlet. Using the psychrometric chart, the airflow would be as follows:

$$\frac{319 \text{ kg da}}{\text{min}} \times \frac{0.868 \text{ m}^3}{\text{kg}} = 277 \text{ m}^3/\text{min} \tag{6.36}$$

$$\frac{704 \text{ lb da}}{\text{min}} \times \frac{13.9 \text{ cu ft}}{\text{lb da}} = 9786 \text{ cfm} \tag{6.37}$$

7.3.1.4 Mist Eliminators Mist eliminators, also called demisters, are an integral part of every wet-scrubbing system and serve to remove the liquid droplets from the exit gas stream. Effective mist eliminators remove 99% to +99.9% of liquid droplets and prevent rain from coming out of the scrubbers. They also help to control chlorine odor. The spray nozzles for recirculation flow located above the packing create a mist as the water splashes on the surface. There are three basic types of mist eliminators:

1. Chevron blades (baffles);
2. Mesh type; and
3. Packing material (typically 2.5- to 6.4-cm [1.0- to 2.5-in.] balls at 25.4- to 30.48-cm [10- to 12-in.-] depth).

For all three types of systems, mist removal is by inertial impaction and centrifugal force (Schifftner and Hesketh, 1996). Figure 6.32 depicts the three types of mist eliminators.

Installed in a vertical mode, chevron blades are well suited for particulate removal and higher velocities. Chevron blades are less susceptible to plugging from sticky material. Design features of a Munters T-272 (NALCO Chemical Company, USA) are shown in Table 6.20.

Mesh-type systems have higher droplet removal efficiency than chevron blades, but are susceptible to plugging (i.e., 100% removal efficiency of 20 to 40 mm in size or 90% efficiency of droplets 3 to 5 mm in diameter). Typically, 10.2- to 15.2-cm- (4.0- to 6.0-in.-) thick mesh is applied, and velocities are 76.2 to 152.4 m/min (250 to 500 fpm). The pressure drop is affected by both vapor and liquid flowrates. At higher liquid loading rates, lower velocities are required to prevent flooding. Table 6.21 depicts a 15.2-cm- (6.0-in.-) thick mesh-pad pressure drop for a wetted and drained pad (Koch 4210—Koch Industries, USA—and ACS Style 8P—ACS Industries, Houston, Texas).

7.3.1.5 Packing Packing is applicable for mid-range velocities and can be cleaned with sprays at 14.7 to 58.7 m^3/m^2·d (0.25 to 1.0 gpm/sq ft) from the bottom up. Packing requires a support grid on the bottom and bed limiters on top and should be located approximately one-half tower diameter above the nozzles. Pressure drops at velocities of 152.4 m/min (500 fpm) are approximately 0.82 kPa/m (1 in./ft) for 2.54-cm (1-in.) balls, 0.53 kPa/m (0.65 in./ft) for 3.18-cm (1.25-in.) balls, and 0.29 kPa/m (0.34 in./ft) for 6.35-cm (2.50-in.) packing balls. Packing can be cleaned easily or removed and replaced and is a good substitute for mesh pads at velocities greater than 152.4 m/min (500 fpm), where high droplet-removal efficiency is required and particulates are a concern.

7.3.1.6 Misting-Scrubber Systems Misting scrubbers use liquid scrubbing solutions in a once-through process. The solution is sprayed at the top of the scrubber chamber and is discharged to the drain at the bottom. Because of the once-through mode, there is no accumulation of contaminants or salts. The scrubbing mist is produced by atomizing the scrubbing solution with air pressure to form a fine droplet, in the range 10 to 20 mm. The small droplets provide a large surface area for the odorous compounds to contact. As shown in Figure 6.33, gas-liquid flows within these scrubbers are concurrent, but can be vertical or horizontal. Vessel detention time ranges from 5 to 30 seconds and is dependent on the odorous compounds to be removed.

Atomization nozzles are the key element for mist scrubber performance. Small droplets equate to more surface area and smaller interstitial space between droplets for a given volume of scrubbing liquid (Hentz and Balchunas, 2000). Therefore, in theory,

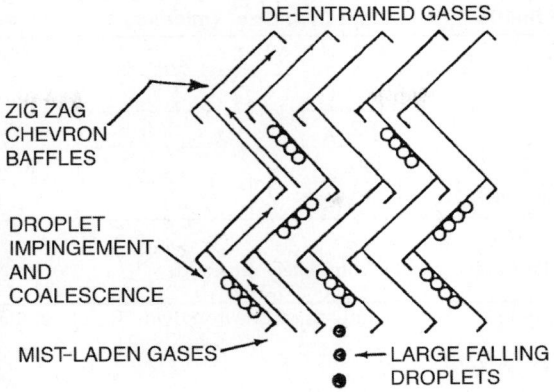

DE-ENTRAINED GASES

ZIG ZAG
CHEVRON
BAFFLES

DROPLET
IMPINGEMENT
AND
COALESCENCE

MIST-LADEN GASES →   ←LARGE FALLING
DROPLETS

CHEVRON IMPINGEMENT PRINCIPLE

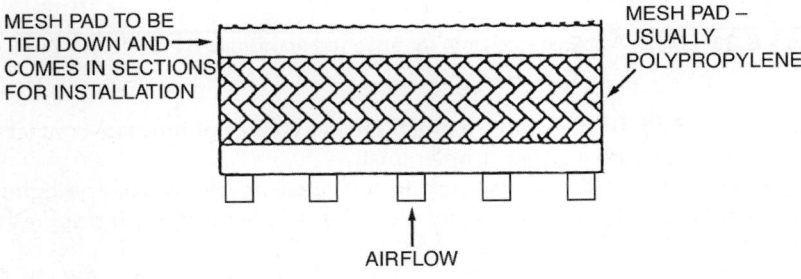

MESH PAD TO BE
TIED DOWN AND→
COMES IN SECTIONS
FOR INSTALLATION

MESH PAD –
USUALLY
POLYPROPYLENE

AIRFLOW

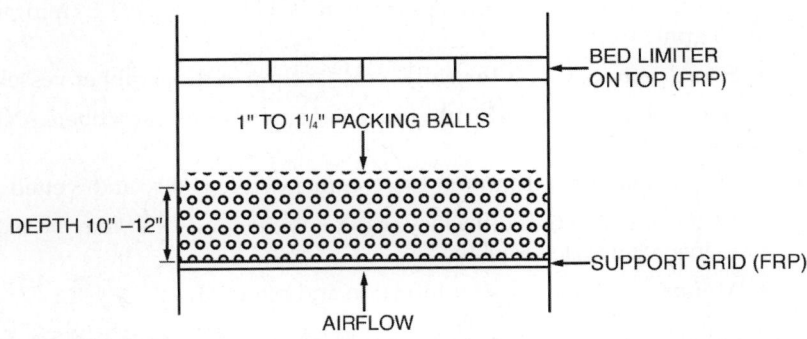

BED LIMITER
ON TOP (FRP)

1" TO 1¼" PACKING BALLS

DEPTH 10" –12"

SUPPORT GRID (FRP)

AIRFLOW

FIGURE 6.32   Typical mist eliminators in packed-bed scrubbers (in. × 2.54 = cm).

Velocity, mm/s (fpm)	Limit Drop Size* (microns)	Pressure Drop, kPa (in.)
1016 (200)	125	0.0025 (0.01)
3048 (600)	72	0.0325 (0.13)
4064 (800)	62	0.0575 (0.23)
5080 (1000)	50	0.0900 (0.36)
6096 (1200)	50	0.1300 (0.52)

*Diameter of smallest droplet that is completely removed.

TABLE **6.20**   Example of Design Features of Chevron-Blade-Type Mist Eliminator*

Velocity mm/s (fpm) Pressure Drop kPa	Koch 4210 Unit, Pressure Drop kPa (in.)	ACS Style 8P Unit, (in.)
1524 (300)	0.0375 (0.15)	0.040 (0.16)
2032 (400)	0.0800 (0.32)	0.070 (0.28)
2540 (500)	0.1050 (0.42)	0.1175 (0.47)

*Note: There are other mesh-pad manufacturers, but this information was available to the author and does not connote endorsement.

TABLE **6.21**   Mesh-Pad Pressure Drop for a Wetted and Drained Pad (15.24 cm [6 in.] Thick)*

numerous small droplets have a much greater chance of intimate contact with odorant molecules than with a larger droplet mist.

Nozzles are the key to misting and can cause problems from plugging, which may require frequent cleaning. Considerations for nozzle design and troubleshooting are as follows:

- Droplet size. Evaluate droplet size (typically approximately 20 mm), surface area, and liquid flow.
- Dilution water. Typically approximately 0.0133 to 0.133 L/min/m³/min (1 to 10 gpm/10 000 cfm).
- Spray pattern. Cover the entire cross section of the scrubber vessel.
- Mist elimination. Difficult to achieve with mist scrubbers. Needs special consideration.
- Nozzle materials. Titanium is recommended for body and wetted parts.
- Water quality. High-efficiency water softener and filter may be required. Need to inspect nozzles weekly.
- Mounting. Use lancets for insertion and removal.

There have been many fine-mist scrubbers installed in the United States; however, as an odor-control technology, there are few new applications in the wastewater-treatment-odor-control industry, and many systems have been replaced.

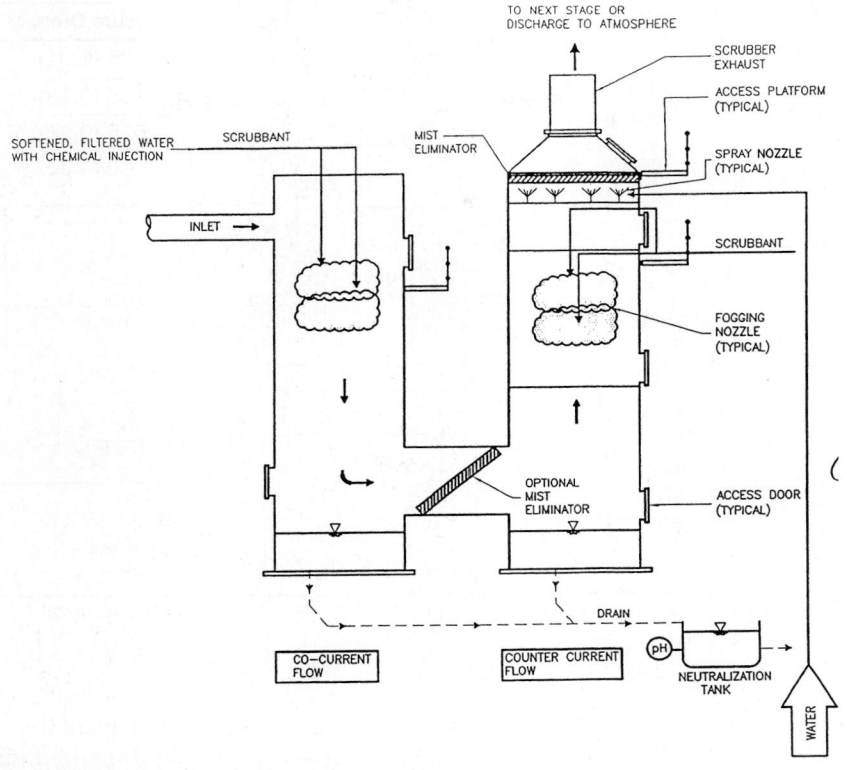

FIGURE 6.33 Typical atomized-mist-scrubber schematic diagram.

7.3.1.7 Catalytic Oxidation A proprietary scrubber, LO-CAT (Gas Technology Products, Merichem Company, Houston, Texas), uses a recirculating chelated iron catalyst that accelerates the oxidation reaction between hydrogen sulfide and oxygen to sulfur. The catalyst is regenerated, and the sulfur is removed as a slurry. The catalyst is selective for hydrogen sulfide. The advantages of this process are that it uses oxygen for oxidation, the catalyst is not consumed, and it can achieve high hydrogen-sulfide-removal efficiency (>99.9%). In large systems, the sulfur is either filtered or centrifuged to a cake. The units can be used to treat anaerobic digester gas, high in hydrogen sulfide, before use as a fuel in boilers or engines to prevent corrosion. In these cases, the catalyst is reactivated with air in a separate vessel. The catalyst could be poisoned by the presence of some organics and metals.

7.3.1.8 Multiple-Stage Scrubbers Low-profile scrubbers are package units and fit in smaller areas. The scrubbers with contact chambers can be operated in a manner that provides effective removal of ammonia, hydrogen sulfide, and organic sulfides. A typical schematic diagram is shown in Figure 6.34. Stage 1 is upflow; stage 2 is downflow; and stage 3 is upflow to the exhaust stack. In this example, stage 1 receives overflow from stages 2 and 3 sump and fresh caustic (from pH control) to treat 70% to 80% of incoming hydrogen sulfide. Sodium hydroxide and sodium hypochlorite are added to

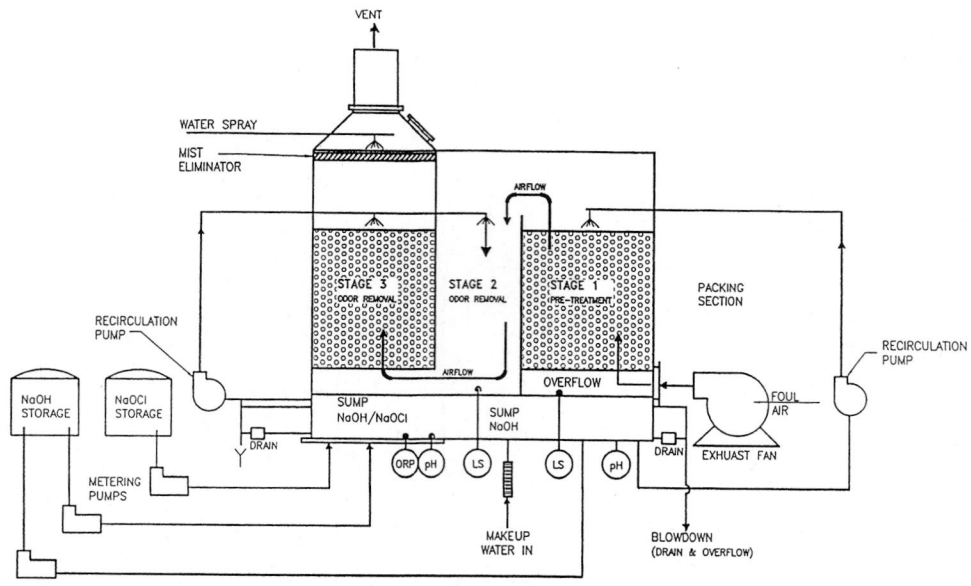

FIGURE 6.34   Typical low-profile three-stage scrubber for hydrogen sulfide removal.

stages 2 and 3, with pH and ORP control provided. One advantage of this approach is to lower sodium hypochlorite costs, because less-expensive sodium hydroxide is used to remove 70% to 80% of the hydrogen sulfide. Makeup water also can be added to each stage with separate overflows for maximum flexibility. There is approximately 1.5 m (5 ft) of packing in each stage.

For ammonia removal, stage 1 is maintained acidic with sulfuric acid addition and no overflow from stages 2 and 3. In these scrubbers, consideration should be given to using a mist eliminator following the first stage, to prevent carryover of the high- or low-pH mist to the second stage.

Advantages of this system include the following: it is low in height, it is a complete skid-mounted package unit, different chemistry can be applied based on inlet odorous compounds, and the system is rectangular (for better use of space). Another type of multiple-stage, odor-control scrubber is the L-shaped system, with the outlet of a horizontal scrubber directly coupled to the inlet of a vertical countercurrent packed-bed scrubber. This arrangement can minimize a footprint and provide flexibility in treatment with several scrubbing solutions.

7.3.2   Dry-Adsorption Systems

Granular activated carbon (GAC) has been used to control odors for over 100 years. Carbon removes odors primarily using a naturally occurring phenomenon called adsorption, in which molecules in the air stream are trapped by either an external or an internal surface of a solid. The phenomenon is similar to iron fillings being held by a magnet (Calgon Carbon Corporation, 1993).

Carbon adsorption commonly is used to treat air with low levels (less than 5 ppm_v) of hydrogen sulfide at WRRFs, but it also is used in many other applications, including removal of VOCs and other reduced-sulfur compounds. Carbon adsorption is not used for control of airborne dust or microorganisms.

In a typical carbon-adsorption system, an exhaust fan forces foul air through a 0.91-m- (3-ft-) deep bed of carbon at a velocity between 15.2 and 22.9 m/min (50 and 75 fpm), resulting in a mean bed residence time of approximately 2.4 to 3.6 seconds, at a pressure drop between approximately 1.5 and 3.0 kPa (6 and 12 in.). Treated air is then discharged directly to the atmosphere. Carbon adsorption frequently is used as a second-stage air-treatment system following packed-bed scrubbers to polish the air and remove chlorine residual before discharge; however, all water from the scrubber must be removed to prevent problems with adsorption.

The raw materials used to produce GAC may be any organic material with a high carbon content, including coal, wood, peat, and coconut shells. The GAC typically is produced by grinding the raw material, adding a suitable binder to give it hardness, recompacting, and crushing it to the correct size. Then, the carbon-based material is converted to activated carbon by thermal decomposition using a high-temperature gas, which creates a complex pore structure.

Activated carbon has an incredibly large surface area per unit volume and a network of submicroscopic pores where adsorption takes place. The walls of the pores provide the surface layer of molecules that are essential for adsorption (Deithorn and Mazzoni, 1986). Typically, 1 kg carbon has more than 90 ha of surface area, which enables the carbon to act as a molecular sponge (Worrall, 1998).

The capacity of carbon to adsorb contaminants depends on the properties of the contaminants. Large, polar molecules tend to adsorb more strongly than small, nonpolar molecules. Other factors that affect adsorption capacity are relative humidity, temperature, biological growth, and particulates. Relative humidity greater than 50% and air temperatures greater than 37.8°C (100°F) may inhibit adsorptive capacities. Biological growth and particulates can reduce airflow through the bed. It should be noted that hydrogen-sulfide-removal efficiencies are not affected by high humidity in impregnated carbons. Most carbon manufacturers publish the adsorption capacity of their carbon for inorganic hydrogen sulfide, which is the most common and prevalent odorous compound in wastewater treatment applications. Adsorption capacity for individual substances can be expressed on the basis of an empirically derived Freundlich isotherm, as follows:

$$X/M = kc^{1/n} \qquad (6.38)$$

where X/M = weight of substance adsorbed per weight of adsorbent,
c = concentration of adsorbed substance,
k = intercept of log-log plot (X/M versus c), and
$1/n$ = slope of line log-log plot.

Figure 6.35 is an example of a Freundlich isotherm. There are mixtures of compounds to be removed in an odorous air stream, which will occupy pore space. Therefore, a pilot test using the actual source of air can be used more effectively to generate an isotherm for the primary constituent.

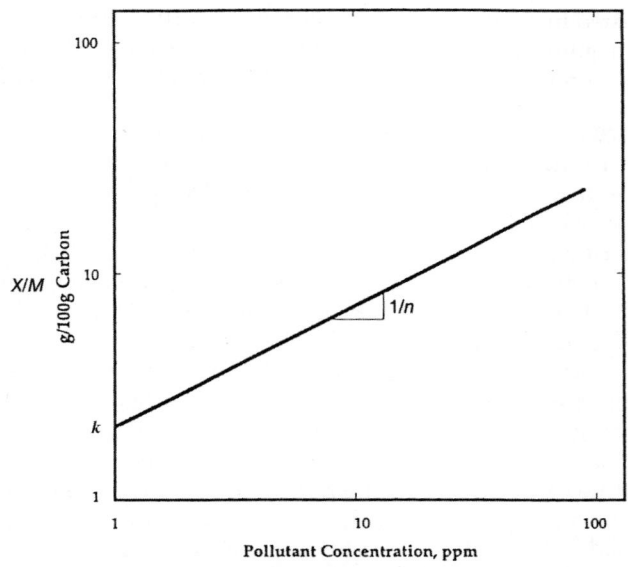

FIGURE 6.35 Example of Freundlich isotherm.

7.3.3 Types of Carbon

7.3.3.1 Impregnated Activated carbon frequently is impregnated with chemicals, such as sodium hydroxide or potassium hydroxide (KOH), to promote a chemical reaction with adsorbed acidic compounds, such as hydrogen sulfide and methyl mercaptans, enhancing the carbon's removal efficiencies and capacities. However, the impregnate reduces the carbon's adsorption capacity for other volatile and odorous organic compounds, because it takes up space on the GAC, blocking some of the adsorption pores (VanStone and Brooks, 1996). Impregnated activated carbon undergoes an exothermic reaction in the presence of oxygen, which can heat the carbon and cause smoldering or spontaneous combustion, if there is insufficient airflow to dissipate the heat. This risk is increased as a result of low ignition temperatures of impregnated carbons. Most impregnated carbons ignite somewhere between 200°C and 225°C; therefore, systems using impregnated carbon should be designed with sealed dampers, which isolate the bed if the fan shuts off. Fire-suppression systems also can be considered.

Impregnated carbons can be regenerated to restore their capacity for hydrogen sulfide removal by soaking the carbon in a sodium hydroxide or potassium hydroxide bath, to remove the sulfur that is produced when hydrogen sulfide reacts with the impregnate. This process also removes a limited amount of adsorbed organic compounds from the carbon; however, it is not possible to retrieve the full organic adsorptive capacity of the original carbon. Impregnated carbon cannot be regenerated thermally like virgin carbon, because the impregnate and adsorbed sulfur interfere with the thermal-regeneration process.

Exhausted impregnated carbon typically may be disposed of in a landfill, but transport may require hazardous waste handling. On-site regeneration by facility staff is not generally recommended because of the hazards and difficulties involved.

7.3.3.2 Virgin Carbon Virgin bituminous and coconut-activated carbons adsorb volatile and odorous organic compounds, but have a relatively low capacity to adsorb inorganic hydrogen sulfide. Therefore, virgin carbons are less common in wastewater treatment applications, where hydrogen sulfide is the dominant compound. It is effective as a final scrubber part of a multistage unit to remove VOCs. Virgin carbons have ignition temperatures of between 380°C and 425°C. Therefore, they are less likely to ignite.

New pelletized activated carbon is available with a higher hydrogen-sulfide capacity of approximately 0.3 grams per cubic centimeter (g/cc). Product data sheets of the physical properties are available from several vendors.

Unlike impregnated carbon, virgin carbon can be reactivated thermally, restoring it to near its original adsorption capacity. Facilities typically maintain a full replacement volume of carbon, which allows their system to stay online, as exhausted carbon is shipped to a reactivation site.

7.3.3.3 Catalytic Carbon Catalytic carbon is a bituminous, granular, unimpregnated activated carbon with enhanced catalytic activity. It is similar to traditional activated carbon, but its pores are finer, giving it a higher density. Theoretically, it adsorbs more hydrogen sulfide than unimpregnated activated carbon, because its catalytic sites promote a reaction between hydrogen sulfide and oxygen from the odorous air stream. More than 90% of the hydrogen sulfide reacts to form sulfates, and only a small amount goes to elemental sulfur. Most products from the hydrogen-sulfide-removal reaction on catalytic carbon are water soluble and, therefore, can be removed by washing the carbon with water. Carbon can be regenerated on-site with water, until organic loading and elemental sulfur exhausts the carbon adsorption capacity (Kazmierczak et al., 2000).

Catalytic carbon is more expensive than traditional impregnated or unimpregnated activated carbon, but it combines the benefits of both. It can be regenerated thermally to near its original adsorptive capacity, has a high ignition temperature of approximately 193.3°C to 218.3°C (380°F to 425°F) (and therefore typically does not require fire suppression), and has the VOC adsorption capacity of unimpregnated carbon, with a much higher capacity for adsorbing hydrogen sulfide.

Water washing the catalytic carbon requires more operating labor than other carbons. The wash water will be acidic and requires careful disposal. Table 6.22 provides a physical comparison of the types of carbon described.

7.3.4   Types of Activated-Carbon Applications

The most common type of adsorber—deep beds—typically consist of 0.91 m (3 ft) of carbon sized to maintain an air velocity between 15.2 and 22.9 m/min (50 and 75 fpm). Deep-bed carbon units essentially have no moving parts, making the system easy to operate and maintain.

Deep beds are manufactured commonly in single- and dual-bed designs. Dual-bed adsorbers typically are stacked one above the other, to minimize space requirements, and are used most on airflows greater than 141.6 m³/min (5000 cfm) and in applications with space restrictions, as seen in Figure 6.36, which depicts airflow through a deep-bed adsorber.

Property	Virgin	Sodium Hydroxide Impregnated	Potassium Hydroxide Impregnated	Catalytic	High Capacity
Capacity (g H_2S/cc carbon)*	0.02	0.14	0.12	0.09	0.3
Ignition temperature (°C)	380 to 425	200 to 225	200 to 225	380 to 425	380 to 425
Regeneration method for hydrogen sulfide	Thermal reactivation	NaOH	KOH	Water	One-time use
Disposal method	Reactivate	Landfill	Landfill	Reactivate	Landfill

*Based on TM-41 procedure (ASTM, 2003).

TABLE 6.22   Physical Comparison of the Four Carbon Adsorption Types

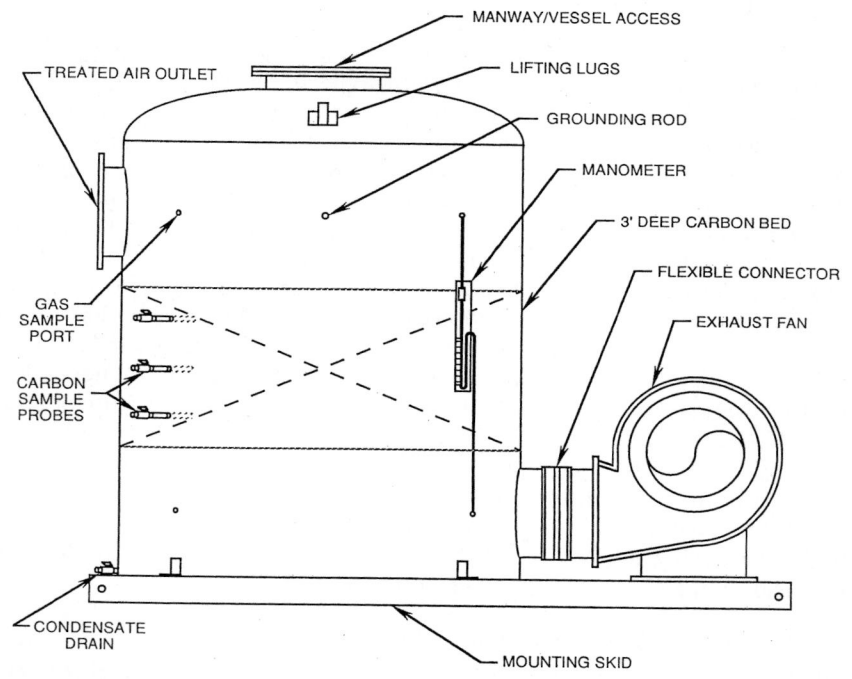

FIGURE 6.36   Typical deep-bed carbon adsorber.

Deep-bed systems are equipped with differential pressure manometers to monitor the pressure drop across the carbon and carbon-sampling probes, to allow collection of carbon samples at various depths in the bed. Systems that use impregnated carbon should be equipped with fire-suppression equipment and sealed dampers that close automatically when the exhaust fan is off.

7.3.4.1 Multiple Stage Multiple-stage carbon adsorbers typically are deep-bed adsorbers arranged in series. Each bed typically contains a different type of carbon. For example, the first stage of a multiple-stage system could be filled with impregnated carbon to remove hydrogen sulfide from an air stream, and the second stage could be filled with unimpregnated carbon for VOC removal. The primary advantage of this type of system is that carbon in each stage is removing compounds that it was designed to remove, producing more efficient use of carbon and higher odor-removal efficiencies.

7.3.4.2 Canisters Air is introduced through a central plenum, where it is uniformly distributed to several radial-flow canisters filled with catalytic carbon and arranged in parallel. The total number of canisters depends on the air volumes to be treated. Canisters are divided into chambers, allowing one compartment to be taken offline for water regeneration, while the remaining chambers remain online. When the carbon finally is spent and water regeneration no longer replenishes the carbon's capacity for oxidation-reaction to occur, the canisters are removed and replaced through side portals in the unit.

Advantages of this new technology are that it allows fully automatic water regeneration of carbon without taking the system offline, and it facilitates rapid replacement of carbon. Disadvantages are that it can be used only with catalytic carbon, the water must be filtered and purified, it requires freeze protection, and it is primarily designed for removal of hydrogen sulfide.

7.3.5 Carbon-Saturation Calculations

The adsorption capacity of activated carbon is dependent on many variables, including contaminant concentrations in air, humidity, and temperature. Most carbon manufacturers publish adsorption capacities of their carbon for particular compounds and provide assistance when estimating carbon life and replacement costs. For example, the hydrogen-sulfide-adsorption capacity of caustic impregnated carbon typically is 0.14 g/cm^3 of carbon. A calculation of the estimated life of the carbon in a typical application is shown below.

Given the following:

Air flowrate: 28.34 m^3/min (1000 cfm);
Hydrogen sulfide concentration: 10 ppm$_v$;
Volume of carbon in 1000 cfm adsorber: 1 m^3 (10^6 cm^3);
Molecular weight of H$_2$S: 34 g/g mol (lb/lb mol);
Molar volume: 24.1 cm^3/g mol (386 lb mol/cu ft) (at 21.1°C [70°F]); and
Hydrogen sulfide adsorption capacity: 0.14 g H$_2$S/cm^3 carbon.

Calculate the following:

Grams (pounds) per hour of H$_2$S (at 21.1°C [70°F]).
In SI units:

$$28.34 \text{ m}^3/\text{min} \times 10 \text{ ppm}/10^6 \times 34 \text{ g/g mol}/0.0241 \text{ cm}^3/\text{g mol} \times 60 \text{ min/h}$$
$$= 0.024 \text{ kg H}_2\text{S/h or } 24.1 \text{ g H}_2\text{S/h} \tag{6.39a}$$

In U.S. customary units:

$$1000 \text{ cfm } 10 \text{ ppm}/1\,000\,000 \times 34 \text{ lb/lb mol}/386 \text{ lb/cu ft} \times 60 \text{ min/hr}$$
$$= 0.053 \text{ lb H}_2\text{S/hr} \tag{6.39b}$$

Carbon adsorption capacity:

$$1\ 000\ 000\ \text{cm}^3 \times 0.14\ \text{g H}_2\text{S/cm}^3 = 140\ 000\ \text{g H}_2\text{S} \qquad (6.40)$$

Estimated life of carbon:

$$140\ 000\ \text{g H}_2\text{S/24.1 g H}_2\text{S/h} = 5809\ \text{h} = 242\ \text{d} \qquad (6.41)$$

(less than 1 year)

7.3.6 Other Adsorption Processes

Iron sponges are designed specifically for hydrogen sulfide removal and commonly are used in applications with hydrogen sulfide concentrations greater than 100 ppm_v, such as digester-gas cleaning for corrosion protection of combustion equipment and for reduction of sulfur dioxide emissions. An iron sponge is a vessel filled with a permeable bed of media, typically wood chips, which have been soaked in hydrated ferric oxide. Hydrogen sulfide reacts with ferric oxide, in accordance with the following equation, to form water and black, solid ferric sulfide:

$$\text{Fe}_2\text{O}_3 \cdot \text{H}_2\text{O} + 3\text{H}_2\text{S} \rightarrow \text{Fe}_2\text{S}_3 + 4\text{H}_2\text{O} \qquad (6.42)$$

An iron-sponge medium may be regenerated by flooding the vessel with water and bubbling air through the flooded bed. This process allows removal of collected sulfur in accordance with the following reaction:

$$2\text{Fe}_2\text{S}_3 + 3\text{O}_2 + 2\text{H}_2\text{O} \rightarrow 2\text{Fe}_2\text{O}_3 \cdot \text{H}_2\text{O} + 6\text{S} \qquad (6.43)$$

The media typically are replaced after three regenerations.

Iron sponges are easy to operate and cost-effective for hydrogen sulfide removal; however, attention should be given to the disposal of regeneration water and expended media. Iron sponges are not effective at removing other odorous compounds or VOCs.

Sulfatreat (Sulfatreat, St. Louis, Missouri) is a product that is used specifically for the removal of high concentrations of hydrogen sulfide similar to iron sponges. It is not effective in removing VOCs. Sulfatreat is a black, granular product that is approximately the size of pea gravel, which is placed in a vertical pressure vessel and converts hydrogen sulfide into iron pyrite (fool's gold). The system typically is located immediately downstream of a gas-liquid separator, and the air to be treated should be between 10°C and 65.6°C (50°F and 150°F) and water saturated. Single or lead-lag vessel designs are available. This is a one-time use product and is not regenerated or reactivated. Spent material typically can be disposed of at a landfill.

Impregnated activated alumina is used in deep beds like impregnated activated carbon and often is used as the adsorbent in the second stage of multiple-stage systems. It has a high capacity for hydrogen sulfide adsorption and oxidation, but also is effective at adsorbing aldehydes, sulfur dioxide, and many organic compounds. Activated alumina cannot be reactivated thermally because of impregnation, and it cannot be regenerated with water. Spent alumina typically is disposed of in a landfill. Activated alumina can be followed by a bed of virgin carbon for improved removal of VOCs. The media must be dry, because the potassium permanganate could be washed off.

7.4   Combustion Emissions Control

7.4.1   Thermal Oxidation

Thermal oxidation is a chemical process that uses oxygen or air at high temperatures to destroy odorous compounds or VOCs and often is used for pelletization or biosolids dryer operations. The process also may be referred to as combustion or incineration. Thermal oxidation works by subjecting an odorous air stream to high temperatures, in the presence of oxygen, for a sufficient amount of time, to oxidize the odorous compounds. The result of the process under ideal conditions is the oxidation of hydrocarbons to carbon dioxide and water.

Thermal oxidizers provide a broad spectrum of treatment for all types and concentrations of odorous compounds and achieve typical odor removal efficiencies in the 90% to 99% range. However, they are not applied often in the relatively dilute odorous air streams encountered in WRRFs because of their potentially high capital and operating (fuel) costs, except in the case of dryer-exhaust air treatment and in the treatment of high-strength odorous organic compounds. Thermal oxidizers are applied most efficiently on odorous air streams with high odor intensities and hydrocarbon concentrations, to keep external fuel costs low. A basic thermal oxidizer consists of the following two primary components:

1. Burners that ignite the fuel in the air stream; and

2. A chamber, which provides adequate residence time for the oxidation process.

Thermal oxidation is a relatively simple process, in theory. However, the challenge is to maintain complete combustion while keeping the process cost-effective. For complete combustion, the treated air mixture consisting of oxygen, fuel, and the odorous compounds must be subjected to the following conditions:

1. Temperature high enough to ignite the mixture;

2. Adequate residence time for the chemical reaction to occur; and

3. Turbulent mixing of the oxygen, fuel, and odorous compounds.

These three conditions are referred to as the three Ts (temperature, time, and turbulence) of combustion. The temperature, time, and turbulence determine the speed and completeness of the oxidation reactions. The temperature must be high enough to reach the ignition temperature of the compounds to be oxidized. The ignition temperatures of various fuels and compounds may be found in combustion handbooks. Most organic compounds can be destroyed at temperatures between approximately 593.3°C and 648.9°C (1100°F and 1200°F). However, most thermal oxidizers typically heat air streams to approximately 760°C and 815.6°C (1400°F to 1500°F) for 1 to 2 seconds to ensure near-complete oxidation. The compounds must be subjected to a temperature at or above their ignition temperature for a sufficient period of time to be completely oxidized. The relationship between time, temperature, and pollutant destruction is shown in Figure 6.37; the higher the temperature, the shorter the necessary residence time to achieve the same destruction percentage. Turbulence is important to provide proper mixing and a more uniform residence time in the reaction chamber. Proper mixing assists in ensuring that odorous compounds come in contact with oxygen and hot combustion products, so that the combustion process can be as complete as possible.

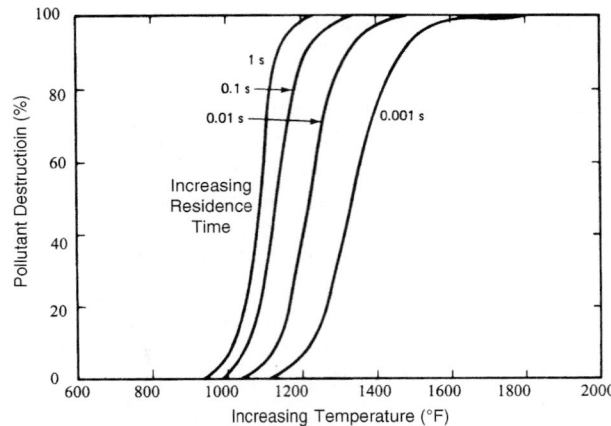

FIGURE 6.37 Effects of time and temperature on the oxidation process (Theodore and Buonicore, 1988).

The odorous compounds that are not subjected to oxygen and heat for a sufficient residence time will not be completely oxidized (i.e., short-circuiting).

The oxidation process requires sufficient concentrations of combustible reactants in the air stream to sustain the combustion process. Air streams with hydrocarbon concentrations less than the lower explosive limit (LEL), like most odorous air sources, require an external fuel source, such as fuel oil, natural gas, propane, or digester gas, to supplement the combustion process.

The four primary types of thermal oxidation processes are the following:

1. Direct combustor or flare;
2. Regenerative thermal oxidizers (RTOs);
3. Recuperative thermal oxidizers; and
4. Catalytic oxidizers.

These are discussed in detail in the sections below.

7.4.1.1 Flare   A simple flare is a device in which air and other compounds react at the burner. The combustion process must occur instantaneously, because there typically is no reaction chamber to provide residence time, and the flare can be exposed to wind. There are higher efficiency flares with reaction chambers available on the market, for example, to burn excess digester gas at WRRFs.

7.4.1.2 Regenerative Thermal Oxidizers   Regenerative thermal oxidizers are the most common form of fuel-reducing technologies available. The RTOs reduce fuel consumption by preheating the inlet air stream before routing it through the reaction chamber. Figure 6.38 is a drawing of a typical three-chamber RTO. The RTOs use multiple heat-recovery chambers filled with ceramic media to alternately capture and release heat from the combustion process. Valving in the system is sequenced, so that each

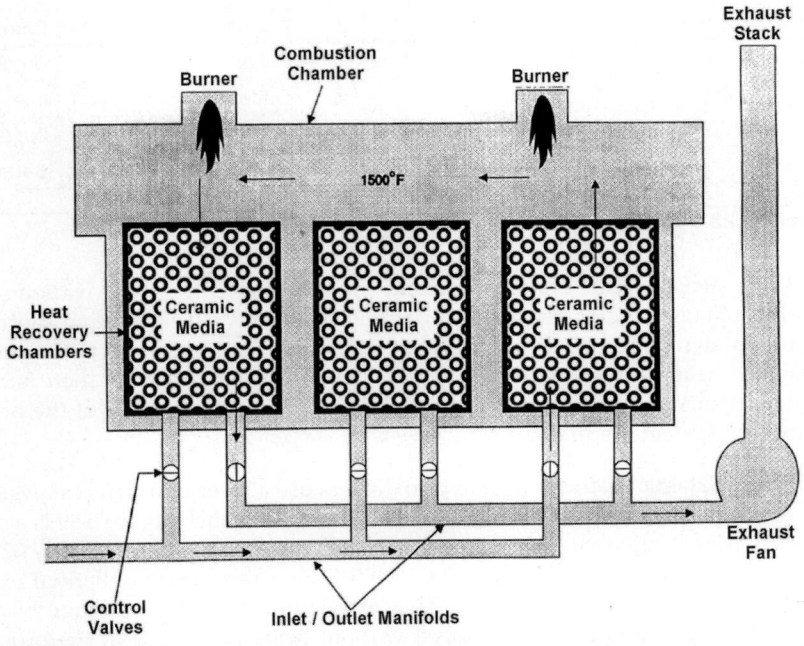

FIGURE 6.38   Regenerative thermal oxidizer (GeoEnergy) (1500°F = 815.6°C).

heat-recovery chamber captures heat from the exiting treated air in one cycle and pre-heats inlet air in the next cycle.

Three chambers are the most common for RTOs; however, for larger installations, five or even seven chambers can be constructed. Compact, modular RTOs are available in sizes as low as 0.28 m³/s (600 cfm). Two-chamber RTO systems are made by some manufacturers, which requires close coordination of the timing of opening and closing valves and of the operating cycle periods. Figure 6.38 provides a description of the following operating sequence and airflow:

- Contaminated air enters only one chamber through control valve;
- Air collects heat from media;
- Air enters oxidation chamber and is heated to 815.6°C (1500°F);
- Air leaves oxidation chamber through another chamber, which is cool;
- Hot air loses heat to media and is discharged;
- Valves switch and air enters chamber, which was previous outlet chamber (now hot);
- Clean, warm air from outlet purges third chamber into combustion chamber to treat trapped air; and
- Switching occurs when the medium is cooled (inlet cycle) to a point of practical heat exchange.

A typical cycle for a three-chamber RTO is shown in Table 6.23.

Time (seconds)	Chamber 1	Chamber 2	Chamber 3
0 to 90	Inlet	Purge	Outlet
90 to 130	Purge	Outlet	Inlet
80 to 270	Outlet	Inlet	Purge
270 to 360	Inlet	Purge	Outlet

TABLE 6.23   Typical Cycle for a Three-Chamber RTO

7.4.1.3   Recuperative Thermal Oxidizers   Recuperative thermal oxidizers are another form of fuel-reducing thermal oxidation technology. Recuperative thermal oxidizers route inlet air through the cold side of shell-and-tube or plate-type heat exchangers to transfer heat from treated to inlet air. The preheated inlet air then is routed to the reaction chamber for treatment and exhausted over the hot side of the heat-exchanger tubes to preheat the inlet air.

7.4.1.4   Catalytic Oxidizers   Catalytic oxidizers use a layer of porous catalysts to lower the temperature required for oxidation in the reaction chamber. Figure 6.39 is a schematic of a typical recuperative catalytic oxidizer. Inlet air is heated to approximately 371.1°C (700°F), compared with approximately 760°C (1400°F) in a recuperative thermal oxidizer, before passing through the catalyst beds. The catalyst facilitates the oxidation of odorous compounds on the surface of the catalyst without being consumed in the process. Catalysts

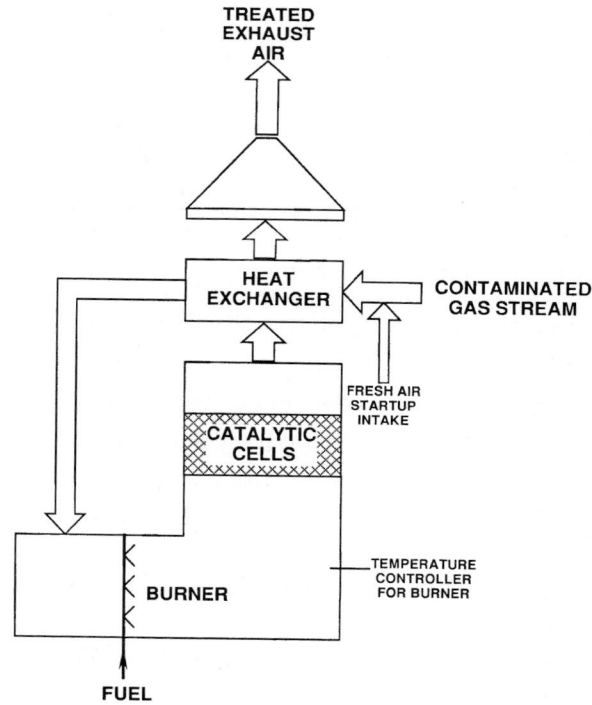

FIGURE 6.39   Diagram of recuperative catalytic oxidizer.

increase the rate of combustion reaction at lower temperatures than thermal oxidation, requiring less fuel. Catalysts typically are platinum or palladium for air toxics control. Catalysts can be poisoned by elements such as chlorine, sulfur, or particulates, and sudden increases in VOC loading can cause a temperature rise that will destroy the catalyst.

7.4.1.5 Thermal Efficiency Typically, RTOs are considered to be 90% to 95% thermally efficient, and recuperative thermal oxidizers are considered to be 65% to 70% efficient. As a rough estimate, RTOs consume 75 to 100 Btu/h/cfm of airflow, without available hydrocarbons to burn in the reaction for heat generation. Thermal efficiency is the percent of heat recovered from the heat available and is based on the oxidation-chamber exhaust temperature, as follows:

$$TE = \frac{T_{op} - T_{ex}}{T_{op} - T_{in}} \times 100 \qquad (6.44)$$

where TE = thermal efficiency (%),
  T_{op} = oxidizer or combustion chamber temperature,
  T_{ex} = oxidizer exhaust temperature, and
  T_{in} = oxidizer inlet temperature.

The following example is presented in Figure 6.40:

$$TE = \frac{1500 - 251}{1500 - 180} \times 100 = 95\% \qquad (6.45)$$

Table 6.24 refers to typical design criteria to consider. Major design considerations include the following:

- Heat capacity of air stream. The air stream must contain relatively high hydrocarbon concentrations to ensure operating efficiency. It is important to be aware of airstream LEL and VOC concentrations.

- Supplemental fuel source. A supplemental fuel source typically must be available to supplement the combustion process. Fuel-efficiency requirements and fuel costs are important.

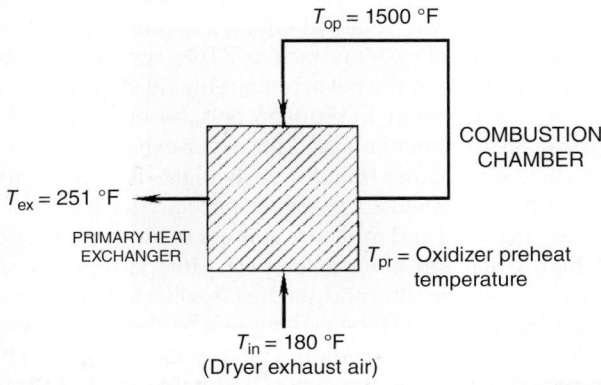

FIGURE 6.40 Example of thermal efficiency (1500°F = 816°C, 251°F = 122°C, and 180°F = 82°C).

Typical Design Criteria	
Air flowrate	Dependent on volumes to be treated
Required treatment temperature	760°C to 815.6°C (1400°F to 1500°F)
Retention time	1 to 2 seconds
Odor-removal efficiency	90% to 99%

TABLE 6.24 Typical Design Criteria for Consideration

- Particulates. Particulates in the inlet air stream must be removed before the thermal oxidizer to prevent plugging of the media, heat exchanger, and catalyst beds.
- Materials. Stainless steel inlet components should be specified because of potential corrosion in hydrogen-sulfide-laden air streams.
- Other considerations. Insurance carriers may require clean air purge for startup. The system should be accessible and expandable. Emission standards must be considered for NO_X and monitoring requirements.

7.4.2 Particulate Removal

Some odor-control systems, such as the treatment of dryer or pelletizer discharge air, require particulate removal as a first-stage treatment method and to meet emission standards. The five general classes are as follows:

1. Gravity settlers;
2. Centrifugal separators (cyclones);
3. Electrostatic precipitators;
4. Fabric filters (baghouses); and
5. Wet scrubbers, such as Venturis.

Knowledge of the particulate properties (quantity and size classification), gas stream temperature, humidity, and removal efficiencies are required for design. Protection of downstream odor-control systems such as RTOs, scrubbers, or biofilters from plugging generally is the objective of the particulate-removal step.

Cyclones are used widely in industry and, for odor-control devices, would be considered a first-stage treatment method for dryer-exhaust air. Cyclones are more efficient for larger particle sizes rather than for particulates less than 5 mm.

Baghouse filters can be used to collect particulates from hot-dryer-exhaust air with the proper filter material and moisture content of air in the proper range. Baghouses are capable of high collection efficiencies (99%) for particles as small as 0.5 mm. Air is forced through the filter or bag and, as dust is collected, the pressure drop increases and the filter must be cleaned. The bags hang in a baghouse and are cleaned intermittently online, to allow continuous operation. The bags can be cleaned by shaking or by a pulse jet of compressed air. Baghouse material suitable for hot-dryer-exhaust air could be fiberglass felt, a Teflon membrane-type bag, or polypropylene.

Baghouses typically are sized based on an air-to-cloth ratio (ACR) in terms of cubic meters per minute per square meter (cubic feet per minute per square feet) of surface area or meters per minute (feet per minute). For dryers, baghouse ACRs should be in the range 0.61 to 1.22 m/min (2 to 4 fpm). To prevent condensation and plugging of the filters, typically the exhaust-air wet-bulb (wb) temperature should be more than 10°C (50°F) from the dry-bulb (db) temperature (i.e., dryer, exhaust-air db = 98.9°C [210°F] and wb = 71.1°C [160°F]). The baghouse and possibly the duct from the dryer should be insulated to prevent cooling. Materials should be selected for corrosion protection. Consideration also should be given to using heaters to prevent condensation during startup and shutdown. Baghouses are most applicable preceding an RTO, for example, designed to prevent particulates from plugging the packing material.

Venturi scrubbers also are used often for dryer-exhaust-air treatment and can provide the additional benefit of cooling exhaust air before a scrubber or biofilter. Water is injected to the Venturi throat. High-collection efficiency is achieved by impaction of the gas stream, which fragments or atomizes the liquid into small droplets with a high density, on which the particles impact. Small liquid droplet diameters and high relative velocities are generated in a Venturi throat. The liquid (0.80 to 1.34 L/m³ [6 to 10 gal/1000 cu ft] of airflow) is distributed at right angles to the high-velocity (3658 to 7315 m/min [12 000 to 24 000 fpm]) gas flow. The higher the pressure drop through the venture, the higher the particulate removal efficiency, especially for small particulates. Venturis used for dryer-discharge-air treatment typically are operated between 2.5 and 5.0 kPa (10 and 20 in.) differential pressure. The Venturi must be followed by a separator to remove the entrained droplets. Separators also can function as condensers to cool the air, if necessary. Venturis can be built with fixed or adjustable throats or can be round or rectangular.

7.4.3 Control of Nitrogen Oxides and Carbon Monoxide

During the combustion process, nitrogen oxides are formed within the combustion gases. The following three different types of nitrogen can be identified:

- Thermal NO_x, which is formed with nitrogen and oxygen from the combustion air.

$$N_2 + O_2 \leftrightarrow 2NO \tag{6.46}$$

  The combustion temperature is the main influencing parameter (thermal NO_x is formed from 2370°F [1300°C]), which emphasizes the importance of avoiding hotspots.

- Prompt NO_x, which is formed in the flame reaction zone.

- Fuel NO_x, which is formed with nitrogen chemically bonded within the fuel. Fuel NO_x is formed at relatively low temperatures (<2370°F [1300°C]). The amount of fuel NO_x is highly influenced by the ratio of combustion air (oxygen) to fuel, and to a lesser extent by the composition of the fuel itself.

Nitrogen oxide emissions from combustion processes are found in two different forms—NO and NO_2. The majority of the NO_x emissions are nitric oxide; however, according to international standards, the emission values are expressed as mg NO_2/Nm^3, dry at reference oxygen content.

The following example demonstrates how a typical gas is analyzed and how the results are converted into one figure, identifying the NO_X content of the flue gas:

- $NO = 200$ mg/m^3;
- $NO_2 = 50$ mg/m^3;
- Flue-gas flow = 80 000 m^3/h;
- Flue-gas temperature = 392°F (200°C);
- H_2O content = 15% by volume;
- O_2 content = 6% by volume (measured wet); and
- Flue-gas pollutants reference = standard m^3 (Nm^3), dry at 11% O_2.

The quantity of NO_X for the given flue gas is determined as follows:

$$\text{Flue-gas flow} = 80\ 000 * \frac{273}{273 + 200} = 46\ 173\ Nm^3/h \qquad (6.47)$$

$$\text{Dry-flue-gas flow} = 46\ 173 \times (1\%\ \text{to}\ 15\%) = 39\ 247 \qquad (6.48)$$

$$O_2\ \text{content of the dry-flue gas} = 6\%/(1\%\ \text{to}\ 15\%) = 7.06\% \qquad (6.49)$$

$$\text{Dry-flue-gas flow at 11\%}\ O_2 = 39\ 247 \times (21–O_2)/(21–11) = 54\ 715\ Nm^3/h \qquad (6.50)$$

$$NO_X\ \text{at reference conditions} = (200 \times 46/30 + 50) \times 80\ 000/54\ 715$$

$$= 521\ \text{mg}\ NO_X\ \text{as}\ NO_2,\ \text{dry 11\%}\ O_2 \qquad (6.51)$$

There are two different approaches to reduce NO_X—primary and secondary measures. These are discussed in detail in the following sections.

7.4.3.1 Primary Measures   The primary measures reduce formation of NO_X at the source, by optimizing the design of combustion and/or the facility-operation parameters. There are a number of possible primary measures to reduce the formation of NO_X. The combustion-process equipment should be designed to ensure a complete combustion/oxidation of the flue gas. Computational fluid dynamics (CFD) modeling can be used for the design of the geometry of the combustion equipment and for the determination of the number, dimensions, and position of the secondary air nozzles. This results in effective primary NO_X reduction and co-reduction in the gas-combustion chamber.

The temperature profile within the combustion process should be kept as uniform as possible. Indeed, NO_X formation will be favored by high combustion temperatures. If hotspots occur within the combustion chamber, even if the total volume of hotspots is small compared with the total combustion chamber volume, NO_X formation within these hotspots can increase drastically the overall NO_X emission of the process.

Combustion air can be injected to different zones and levels in the combustion chamber. It is of utmost importance to be able to control the amount of air to each zone and level, depending on fuel characteristics and thermal load. Therefore, air control with frequency-converted equipment is preferred, and sufficient attention should be given to the software controlling the air supply of the process.

It is worthwhile to mention the specific mixing function that secondary and tertiary air plays in combustion processes. Secondary and tertiary air are added just downstream of the initial combustion. They are injected with relatively high velocities to enhance mixing of the flue gas, ensure complete burnout, and avoid the formation of hotspots.

The amount of excess air should be minimized to prevent the formation of fuel NO_X, while still providing enough air to avoid hotspots. Therefore, flue-gas recycling often is applied. The recycled flue gas is taken after the heat-recovery equipment and guided back to the combustion process. There, it cools down the combustion temperature; its mechanical effect is used to enhance the flue-gas mixing; and, at the same time, the oxygen content in the flue gas is not increased, because the composition of the recycled flue gas is identical to the bulk flue-gas composition.

7.4.3.2 Secondary Measures Secondary measures will remove NO_X formed during combustion by thermal reduction. The two main types of secondary measures are selective non-catalytic reduction (SNCR) and selective catalytic reduction (SCR).

The SNCR of NO_X is a process in which NO_X is bonded chemically by injecting aqueous ammonia or urea to the combustion chamber at high temperatures. The chemical reactions are represented in a simplified form in Table 6.25.

The reagents ammonia and urea have a defined temperature window, in which the reduction of NO_X is maximized.

The removal efficiencies achievable by an SNCR process vary from 50% to 70%, with an assumed NO_X content upstream of the SNCR of 400 to 800 mg/Nm^3. By applying flue-gas recycling, a removable efficiency of 80% is possible. Table 6.26 lists the theoretical and likely ratios of reagent to the amount of NO_X that is removed.

Temperature windows have a significant effect on removal efficiency. If the temperature is too high, the reagent will be oxidized, and NO_X emissions will be increased. If the temperature is too low, longer reaction times will be needed, and there will be

Reagents	Chemical Reaction	Temperature Window
Urea	$2NO + NH_2CONH_2 + \frac{1}{2} O_2 \rightarrow 2N_2 + CO_2 + 2H_2O$	900°C to 1100°C
Ammonia	$4NO + 4NH_3 + O_2 \rightarrow 4N_2 + 6H_2O$ $2NO_2 + 4NH_3 + O_2 \rightarrow 3N_2 + 6H_2O$	890°C to 1000°C

TABLE 6.25 SNCR Chemical Reactions for NO_X Control

	Theoretical	Practical
Urea	0.5	0.8 to 1.0
Ammonia	1.0	1.6 to 2.0

TABLE 6.26 Reaction Rate Ratios for NO_X-Control Reagents

reagent slippage. This will result in ammonia emissions and deposition of salts in the boiler. The salts contaminate the boiler fly ash and flue-gas residue.

The following are the main components of an SNCR:

- Reagent storage (tanks and circulation pumps);
- Reagent dosing system with control unit; and
- Reagent injection nozzles.

The type of the storage area is dependent on the selected reagent. Ammonia typically is supplied as a 25% ammonium hydroxide (NH_4OH) solution. The storage area must be designed with care and provided with all necessary precautions related to the handling of an ammonia solution. The storage and unloading area must be enclosed. All drains from the area must lead to treatment. Ammonia detectors must be provided with a sprinkling system for truck unloading.

Urea can be purchased as a solution (typically 640% urea) or as a powder. For a solution, the storage area is comparable with the storage area for an ammonia solution, with the exception that the precautions related to the handling of ammonia are not applicable. For powdered urea, a solution preparation unit must be installed. This consists of a powder storage silo, a silo-extraction and powder-dosing unit, a solution preparation tank with controlled water supply, a mixer, and circulation pumps. The system also must have heat tracing to maintain the temperature of the solution above 50°F (10°C). Below this, the urea solution will solidify. Of the two reagents, ammonia typically is substantially less expensive than urea.

The reagent is dosed to the boiler/combustion chamber using circulating pumps and flow-control valves. The circulation pumps provide a consistent pressure upstream of the control valves. The control valves are installed in pipe branches connecting the circulation-ring mains with different levels of injection nozzles.

The pressure in the ring mains must be kept high enough to provide efficient reagent distribution from the nozzles. The flow-control algorithm must take the following parameters into account:

- Flue-gas flow, which is used to estimate the total amount of NO_x in the flue gas.

- Flue gas temperature, to determine which level of nozzles is to be used. It is important to remember that the efficiency of the reduction reaction depends on the reaction temperature.

- NO_x content of the flue gas downstream of the SNCR system, to determine and control NO_x emissions.

More advanced systems also would consider the NO_x content upstream of the SNCR system, ammonia slippage, and oxygen content of the flue gas.

The injection nozzle design will influence significantly the mixing of the reagents with the flue gas at the right temperature. As noted above, this is vital in achieving the expected NO_x removal, while minimizing reagent consumption and ammonia slippage.

First, the injection levels must be determined. This is done by simulating the flue-gas temperature in the combustion chamber for varying thermal loads and fuel compositions.

Depending on the expected variations in thermal load and flue-gas composition, one, two, or even three levels of injection nozzles are provided.

Next, the penetration depth and width of the reagent in the combustion chamber is studied. The following three factors must be considered:

- Combustion-chamber geometry at the injection levels;
- Type of driving agent (water, steam, or compressed air) that will be used to ensure penetration of the reagent in the combustion chamber; and
- Type of nozzle that will be used.

It generally is preferable to inject the reagents in a narrowing section of the combustion chamber. At these narrow sections, the flow is turbulent, thus enhancing mixing. In addition, the penetration distance is reduced.

Water, compressed air, and steam are the three types of driving agents commonly used in two-phase injection nozzles. Water is inexpensive, and the penetration depth is prolonged as a result of its specific weight. In addition, the water droplets must be evaporated before the reagent is released within the flue gas. The main disadvantage of water is that injecting it to the combustion chamber disturbs the temperature profile and reduces the energy of the flue gas.

Compressed air is expensive to produce and has a lower mechanical impact per unit volume than water. However, it has virtually no effect on the temperature profile and energy content of the flue gas. For these reasons, it commonly is used in smaller boiler units.

Steam combines the advantages of water and compressed air. As with water, its mechanical impact per unit volume is higher, and it is less expensive to produce than compressed air. However, as with compressed air, it does not influence significantly the temperature profile of the flue gas.

Two-phase injection nozzles commonly are used in SNCR systems. In two-phase nozzles, the reagent and the driving agent are fed separately and combined within the nozzle and exhausted together. The size and geometry of the nozzle depends on the required flow, available pressure, required penetration depth, and spraying angle. The nozzles are to be built into the boiler walls; thus, boiler tubes must be rerouted around the nozzles. Figure 6.41 shows boiler tubes bent around a nozzle, and Figure 6.42 shows an example of nozzle spray angles.

The SCR of NO_X differs from SNCR, in that a catalyst is used. The catalyst allows the NO_X removal to take place at much lower temperatures than required in SNCR. In addition, ammonia is the only reagent used. However, the presence of the catalyst allows the removal reaction to take place in the temperature range 160°C to 450°C (320°F to 840°F). The catalyst used typically is a zeolite consisting primarily of a mixture of titanium dioxide (TiO_2) and vanadium oxide (V_2O_5). The simplified reaction equations are as follows:

$$4NO + 4NH_3 + O_2 \rightarrow (TiO_2/V_2O_5) \rightarrow 4N_2 + 6H_2O \qquad (6.52)$$

$$2NO_2 + 4NH_3 + O_2 \rightarrow (TiO_2/V_2O_5) \rightarrow 3N_2 + 6H_2O \qquad (6.53)$$

Apart from the NO_X-reduction reactions, other chemical reactions can occur. The most important are the oxidation of dioxins and furans in the temperature range 150°C

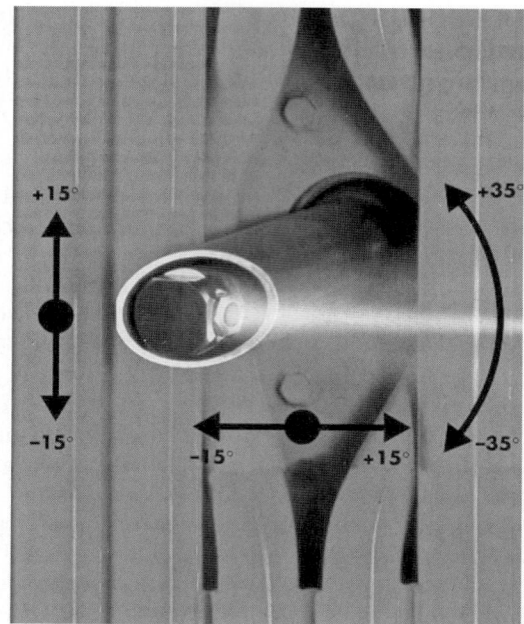

FIGURE 6.41   Boiler tubes bent to install the nozzles.

FIGURE 6.42   Nozzle penetration depth and spraying angle.

to 350°C (300°F to 660°F), oxidation of SO_2 to SO_3, oxidation of CO to CO_2, and the formation of ammonia salts. These reactions equations are as follows:

$$C_{12}H_nCl_{8-n}O_2 + (9 + 0.5n)\, O_2 \rightarrow (TiO_2/V_2O_5) \rightarrow (n\text{-}4)H_2O + 12CO_2 + (8\text{-}n)HCl \qquad (6.54)$$

$$C_{12}H_nCl_{8-n}O + (9.5 + 0.5n)\, O_2 \rightarrow (TiO_2/V_2O_5) \rightarrow (n\text{-}4)H_2O + 12CO_2 + (8\text{-}n)HCl \qquad (6.55)$$

$$SO_2 + 0.5O_2 \rightarrow (TiO_2/V_2O_5) \rightarrow SO_3 \qquad (6.56)$$

$$CO + 0.5O_2 \rightarrow (TiO_2/V_2O_5) \rightarrow CO_2 \qquad (6.57)$$

$$NH_3 + SO_3 + H_2O \rightarrow (TiO_2/V_2O_5) \rightarrow NH_4HSO_4 \qquad (6.58)$$

$$2NH_3 + SO_3 + H_2O \rightarrow (TiO_2/V_2O_5) \rightarrow (NH_4)_2SO_4 \qquad (6.59)$$

The catalyst is sensitive to poisoning by dust or the depositing of ammonia salts. Therefore, an SCR reactor typically is placed at the end of a flue-gas cleaning system for fuels where emissions of particulate matter or sulfur dioxide occur. The ammonia salts can be removed from the catalyst by heating the catalyst to a temperature higher than 280°C to 300°C (535°F to 570°F).

The NO_X-removal efficiencies achievable by an SCR process easily can be 80% to 90% and even higher, with an assumed NO_X content upstream of the SCR of 400 to 800 mg/Nm. Both the theoretical and practical ratio of NO_X to reagent for a SCR process approaches 1.0.

The following are the main components of a SCR system:

- Ammonia storage tanks and circulation pumps for the reagent solution;
- Catalytic reactor;
- Dosing system with control unit and injection nozzles; and
- Auxiliary equipment (i.e., flue-gas reheating system).

The storage tanks and associated equipment are the same as described above for an ammonia solution in an SNCR system.

The catalyst is a solid; thus, a separate reactor vessel housing the solid catalyst is used. The solid catalyst also acts as a packing and provides surface area, on which the contact of the ammonia and NO_X takes place. The solid catalyst can take one of two forms—either a pellet or a honeycomb. Each has its own configuration.

In a pellet reactor, the flue gas comes into contact with the catalyst by entering through a permeable wall, behind which catalyst particles are located. The flue gas is forced through a volume of catalyst pellets before leaving the reactor. Pellet reactors are excellent for use in low-dust applications, typically behind gas turbines, engines, or at the tail end of flue-gas cleaning systems (provided that sufficient backup systems are in place, to avoid uncontrolled dust release to the catalytic reactor). Figure 6.43 shows a typical pellet reactor.

In a honeycomb reactor, different elements are placed in a module, and different modules form a catalyst layer in the reactor. The flue gas passes through the honeycomb elements in a series of parallel channels. The size of the channels is a function of the

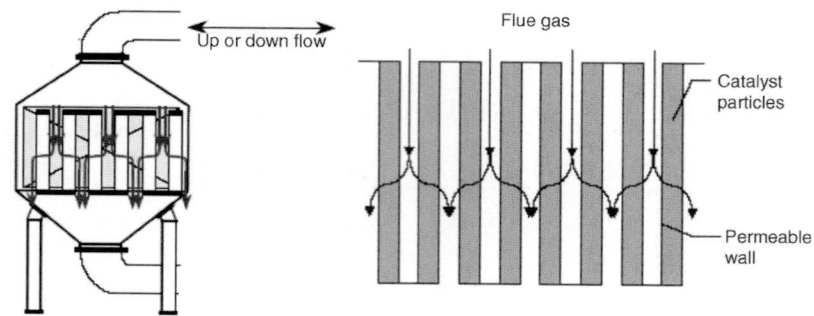

FIGURE 6.43   Pellet reactor and lateral-flow module (CRI Catalyst Company, Houston, Texas).

flowrate, allowable pressure drop, dust content, and NO_x levels. The velocity is kept low to ensure laminar flow of the flue gas through the reactor.

Honeycomb reactors are used both in high- and low-dust configurations. Typical high-dust applications include SCR systems, integrated in the boiler of coal-fired power facilities, in gas streams of metallurgical processes, or in the boiler of energy from waste facilities. Typical low-dust applications are similar to pellet-reactor applications. Figure 6.44 shows a typical honeycomb reactor and the catalyst elements.

Catalysts will lose activity over time. A proactive, analysis-based catalyst-replacement scheme is needed to reduce the overall catalyst-replacement cost, by maximizing the use of the remaining catalyst in older elements. Figure 6.45 shows a typical replacement cycle for a honeycomb reactor.

Because of the separate vessel and lower temperature operation, an SCR system generally is the final step in the flue-gas cleaning system. Therefore, it is critical to fine-tune the dosing amounts of the ammonia being dosed into the system to avoid ammonia slippage. Therefore, care must be taken to distribute the ammonia solution as evenly as

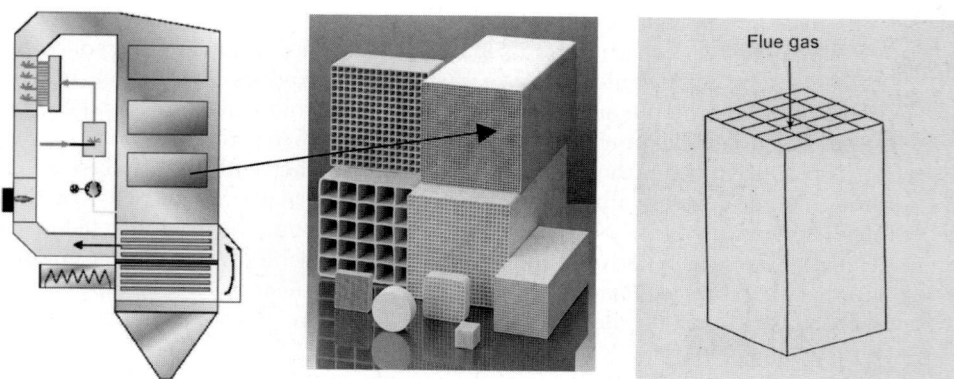

FIGURE 6.44   SCR deNO$_x$ reactor configuration and sample of typical honeycomb elements.

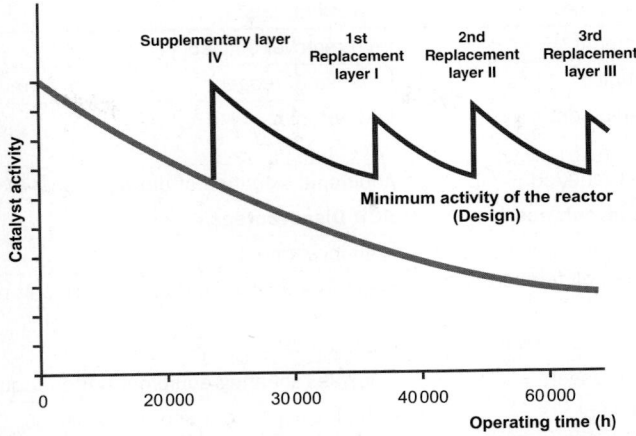

FIGURE 6.45   Typical catalyst replacement scheme for honeycomb catalyst reactors.

possible in the flue gas. The CFD analysis often is used in the design of an injection nozzle grid for the ammonia solution.

The reagent-dosing controls are similar to those described for an SNCR system, with one exception—flue gas temperature is less critical and does not need to be monitored for the system, but may be monitored for other emissions criteria.

Sulfur dioxide deactivates the catalyst; thus, it is necessary to eliminate it from the flue gas upstream of the SCR reactor. In pellet reactors, dust also must be eliminated upstream of the SCR reactor. A wet-scrubber system often is used for sulfur dioxide removal. However, a wet scrubber will reduce the flue-gas temperature to 60°C to 80°C (140°F to 175°F). The flue gas then must be reheated to 180°C to 220°C (355°F to 430°F), depending on the remaining sulfur dioxide content in the flue gas. One or a combination of the following methods can be used to reheat the flue gas:

- A low-pressure steam battery;
- A high-pressure steam battery;
- Heat exchanger; and
- An in-line burner.

The optimal combination of auxiliary equipment must be selected on a case-by-case basis. Quite often, however, a flue-gas/flue-gas-heat exchanger is used to benefit from the thermal energy of the flue gas leaving the SCR reactor. Likewise, an in-line burner often is installed, as it can be used to heat the SCR reactor to 280°C to 300°C (535°F to 570°F) and to evaporate the ammonia salts from the catalyst surface.

The necessity of a dedicated reactor and, in some cases, additional auxiliary equipment makes an SCR reactor relatively expensive if compared with an SNCR system. Table 6.27 provides a summary of the relative advantages and disadvantages of the SNCR and SCR systems.

SNCR Advantages	SCR Advantages
Low investment	High removal efficiency
Compact unit	Low use of reagent
Simple equipment and operation	Low NH_3-slip
No catalyst required	Additional oxidation of dioxin/furan and carbon monoxide
SNCR Disadvantages	**SCR Disadvantages**
Combustion process must be stable (small temperature window)	High investment
	Catalyst sensitive to dust and NH_3-salt poisoning
Lower removal efficiency	Pressure drop
High use of reagent	Flue-gas reheating equipment often required
Higher NH_3-slip	Complex equipment and operation
N_2O-formation	

TABLE 6.27   SNCR and SCR Advantages and Disadvantages

8.0   References

American Industrial Hygiene Association (1989) *Odor Thresholds for Chemicals with Established Occupational Health Standards*; American Industrial Hygiene Association: Akron, Ohio.

American Public Health Association; American Water Works Association; Water Environment Federation (2017) *Standard Methods for the Examination of Water and Wastewater*, 23rd ed.; American Public Health Association: Washington, D.C.

American Society of Heating, Refrigerating and Air-Conditioning Engineers, Inc. (2003) *HVAC Applications Handbook*; American Society of Heating, Refrigerating and Air-Conditioning Engineers, Inc.: Atlanta, Georgia.

American Society of Heating, Refrigerating and Air-Conditioning Engineers, Inc. (2005) *2005 ASHRAE Handbook FUNDAMENTALS*, Inch-Pound Edition; American Society of Heating, Refrigerating and Air-Conditioning Engineers, Inc.: Atlanta, Georgia.

American Society for Testing and Materials (1978) *Standard Test Method for Measurement of Odor in Atmospheres (Dilution Method)*, ASTM D1391; American Society for Testing and Materials: Conshohocken, Pennsylvania.

American Society for Testing and Materials (1981) *Guidelines for the Selection and Training of Sensory Panel Members—Special Technical Publication 758*; American Society for Testing and Materials: Conshohocken, Pennsylvania.

American Society for Testing and Materials (1988) *Standard Practice for Suprathreshold Intensity Measurement*, ASTM E544-88; American Society of Testing Materials: Conshohocken, Pennsylvania.

American Society for Testing and Materials (1999) *Standard Practice for Suprathreshold Odor Intensity Measurement*, ASTM E544-99; American Society for Testing and Materials: Conshohocken, Pennsylvania.

American Society for Testing and Materials (2001) *Standard Test Method for Determination of Sulfur Compounds in Natural Gas and Gaseous Fuels by Gas Chromatography and Chemiluminescence*, ASTM D5504-01; American Society for Testing and Materials: Conshohocken, Pennsylvania.

American Society for Testing and Materials (2003) *Standard Test Method for Determination of the Accelerated Hydrogen Sulfide Breakthrough Capacity of Granular and Pelletized Activated Carbon*, ASTM D6646-03; American Society for Testing and Materials: Conshohocken, Pennsylvania.

American Society for Testing and Materials (2004) *Standard Practice for Determination of Odor and Taste Threshold by a Forced-Choice Ascending Concentration Series Method of Limits*, ASTM E679-04; American Society for Testing and Materials: Conshohocken, Pennsylvania.

Barnebey-Cheney Company (1974) *Scentometer: An Instrument for Field Odor Measurement*; Barnebey-Cheney Company: Columbus, Ohio.

BASTE (1992) *Bay Area Sewage Potentially Toxics Emissions Model User's Manual*; Bay Area Air Toxics Group: Oakland, California.

Bolla-Wilson, K.; Wilson, R. J.; Bleeker, M. L. (1988) Conditioning of Physical Symptoms After Neurotoxic Exposure. *J. Occup. Med.*, **30** (9), 684–686.

Bowker, R.; King, A.; Holcomb, G. (1995) U-Tube Oxygen Dissolver Controls Odors. *Water Environ. Technol.*, 7, 20–21.

Buonicore, A. J.; Davis, W. T. (Eds.) (1992) *Air Pollution Engineering Manual*; Van Nostrand Reinhold: New York.

Calgon Carbon Corporation (1993) *Activated Carbon Principles*, TI-101-05/93; Calgon Carbon Corporation: Pittsburgh, Pennsylvania.

Cramer, H. E. (1959) Engineering Estimates of Atmospheric Dispersal Capacity. *Am. Ind. Agr. Assoc. J.*, **20**, 183–189.

Dalton, P. (1999) Cognitive Influences on Health Symptoms from Acute Chemical Exposure. *Health Phys.*, **18**, 579–590.

Deithorn, R. T.; Mazzoni, A. F. (1986) Activated Carbon—What It Is, How It Works. *Water Technol.*, **9** (8), 26–29.

EN 13725 (2003) BS *EN 13725:2003 Air Quality—Determination of Odor Concentration by Dynamic Olfactometry*; BSI: London, United Kingdom.

Gostelow, P.; Parsons, S. A.; Stuetz, R. M. (2001) Odor Measurements for Sewage Treatment Works. *Water Res.*, **35**, 578–597.

Hentz, L.; Balchunas, B. (2000) Chemical and Physical Processes Associated with Mass Transfer in Odor Control Scrubbers. *Proceedings of the Odor and VOC Emissions 2000 Conference*, Cincinnati, Ohio, April 16–19; Water Environment Federation®: Alexandria, Virginia.

Hunniford, D. J. (1990) Control of Odors and Hydrogen Sulfide Related Corrosion in Municipal Sewage Collection Systems Using a Biochemical Process: BIOXIDE. Presented at the 63rd Annual Water Pollution Control Federation Technical Exposition and Conference, Washington, D.C., October 7–11; Water Environment Federation®: Alexandria, Virginia.

International Association on Water Pollution Research and Control (1986) *Identification and Treatment of Tastes and Odors in Drinking Water*; American Water Works Association Research Foundation: Denver, Colorado, 102–120.

International Water Association (2001) *Odours in Wastewater Treatment: Measuring, Modeling and Control*; IWA Publishing, London.

Kazmierczak, M.; Loomis, P.; Johnson, M. (2000) Water Regenerable Catalytic Carbon Adsorption: A Case Study at the Broadwater Water Reclamation Facility. *Proceedings of the Odors and VOC Emissions 2000 Conference*, Cincinnati, Ohio, April 16–19; Water Environment Federation®: Alexandria, Virginia.

Leonardos, G. (1997) Odor Control Regulations in the USA, 1997 Update. *Proceedings of the National Workshop on Odor Measurement Standardization*, Sydney, Australia, Aug 20–22; University of New South Wales: Sydney, Australia.

Mahin, T. D. (2001) Comparison of Different Approaches Used to Regulate Odors Around the World. *Water Sci. Technol.*, **44** (9), 87–102.

Metcalf & Eddy (1972) *Wastewater Engineering: Treatment and Reuse*; McGraw-Hill: New York.

National Fire Protection Association (2016) *Standard for Fire Protection in Wastewater Treatment and Collection Facilities*; Standard 820; National Fire Protection Association: Quincy, Massachusetts.

Pope, R. (2000) Establishing Limits and Defining Odor Control Needs. *Proceedings of the Odors and VOC Emissions 2000 Conference*, Cincinnati, Ohio, April 16–19; Water Environment Federation®: Alexandria, Virginia.

Rafson, H. J. (Ed.) (1998) *Odor and VOC Control Handbook*; McGraw-Hill: New York.

Schifftner, K. C.; Hesketh, H. E. (1996) *Wet Scrubbers*, 2nd ed.; Technomic Publishing Company, Inc.: Lancaster, Pennsylvania.

Sheet Metal and Air Conditioning Contractors' National Association, Inc. (1990) *HVAC Systems Duct Design Manual*, 3rd ed.; Sheet Metal and Air Conditioning Contractors' National Association, Inc.: Chantilly, Virginia.

Shusterman, D. (1992) Critical Review: The Health Significance of Environmental Odor Pollution. *Arch. Environ. Health*, **47** (1), 88–91.

Shusterman, D.; Balmes, J.; Cone, J. (1988) Behavioral Sensitization to Irritants/Odorants after Acute Overexposure. *J. Occup. Med.*, **30**, 556–567.

South Coast Air Quality Management District (1993) SCAQMD Rule 1179, Emissions Inventory Report for JEIP Participating Agencies; Joint Emissions Inventory Program (JEIP) Report; CH2M Hill: Oakland, California.

Sterne, L., Enviromega, Ltd., Burlington, Ontario, Canada (2001) Personal communication.

Sullivan, R. J. (1969) *Preliminary Air Pollution Survey of Odorous Compounds*; National Air Pollution Control Administration: U.S.

Theodore, L.; Buonicore, A. J. (1988) *Air Pollution Control Equipment*; CRC Press: Boca Raton, Florida.

U.S. Environmental Protection Agency (1977) *Guidelines for Air Quality Maintenance Planning and Analysis, Vol. 10 (Revised): Procedures for Evaluating Air Quality Impact of New Stationary Sources*, EPA-450/4-77-001, PB274087; U.S. Environmental Protection Agency, Office of Air Quality Planning and Standards: Research Triangle Park, North Carolina.

U.S. Environmental Protection Agency (1985) *Design Manual, Odor Corrosion Control in Sanitary Sewerage Systems and Treatment Plants*, EPA-625/1-85-018; U.S. Environmental Protection Agency: Washington, D.C.

U.S. Environmental Protection Agency (1990) *Characteristics of Corrosivity*, 40 CFR 261.22 (Amendment to 1980 40 CFR 261.22); U.S. Environmental Protection Agency: Washington, D.C.

U.S. Environmental Protection Agency (2002) *Water 9—Technology Transfer Network Clearinghouse for Inventories and Emission Factors*; U.S. Environmental Protection Agency: Washington, D.C., http://www.epa.gov/ttn/chief/software/water/index.html (accessed May 1, 2009).

Van Durme, G. P.; Berkenpas, K. (1989) Comparing Sulfide Control Products. *Oper. Forum*, **6** (2), 12–19.

VanStone, G.; Brooks, D. (1996) Carbon Clean—A New Version of Activated Carbon Controls Odors. *Water Environ. Technol.*, **8**, 40–43.

Verschueren, K., Ed. (1983) *Handbook of Environmental Data on Organic Chemicals*; Van Nostrand Reinhold: New York.

Wahl, G. H. (1980) *Regulatory Options for the Control of Odors*, EPA-450/5-80-003; U.S. Environmental Protection Agency: Washington, D.C.

Wang, J.; Skipka, K. J. (1993) Dispersion Modeling of Odorous Emissions, 93-RA-114A.05. *Proceedings of the 86th Annual Meeting and Exhibition of the Air and Waste Management Association*, Denver, Colorado, June 13–18; Air and Waste Management Association: Pittsburgh, Pennsylvania.

Water Environment Federation (2004) *Control of Odors and Emissions from Wastewater Treatment Plants*, Manual of Practice No. 25; Water Environment Federation: Alexandria, Virginia.

Water Environment Research Foundation (1994) *Control and Production of Toxic Air Emissions by Publicly Owned Treatment Works*; Water Environment Research Foundation: Alexandria, Virginia.

Wilson, D. J. (1979) Flow Patterns Over Flat-Roofed Buildings and Application to Exhaust Stack Design. *ASHRAE Trans.*, **85**, 284–295.

Wong, P.; Mohleji, S. C.; Occiano, V. Y.; Schafer, P. L. (1992) Odor Characterization and Control by Ferrous Chloride Addition in Sewers. *Proceedings of the 65th Annual Water Environment Federation Technical Exposition and Conference*, New Orleans, Louisiana, Sept. 20–24; Water Environment Federation: Alexandria, Virginia.

Worrall, M. (1998) Case Study—Capturing Organic Vapors From Non-Condensable Gases Using Activated Carbon Adsorption Technology. *Chem. Processing*, **Jan.**

CHAPTER **7**

Support Systems

Hannah T. Wilner, P.E., PMP; Daniel Freedman, P.E.;
Matthew Goss, P.E., CEM, CEA, CDSM, LEED® AP (BD+C);
Steve Mustard, P.E., CAP, GICSP; Ken Schnaars, P.E.;
David Ubert; Ales Volcansek, P.E.; Curt Wendt, P.E., CAP

1.0 Introduction

This chapter presents a review of state-of-the-art and general design considerations for support systems of water resource recovery facilities (WRRFs), to enable the engineer to work effectively with other design disciplines when designing these systems. Numerous support systems are required to provide a fully functional and safe treatment facility that also is acceptable to the general public.

The selection, development, and design of individual wastewater treatment operations and processes require significant effort. However, even a well-designed treatment system requires well-planned and developed support systems. As a result, the design of support systems can rival facility process design in complexity and potential for innovation and energy savings, especially at larger facilities. A well-coordinated process and support-system design results in a treatment facility that effectively and efficiently provides the required degree of treatment and safety.

2.0 Reliability Criteria

In the early 1970s, when the emphasis was on building new WRRFs, the U.S. Environmental Protection Agency (Washington, D.C.) (U.S. EPA) developed reliability design criteria for mechanical and electrical support systems. The objective of the criteria was to ensure that new facilities maintained a high degree of effectiveness. In 1974, U.S. EPA published *Design Criteria for Mechanical, Electrical, and Fluid System and Component Reliability*, which still is used as a point of reference today. These reliability criteria apply not only to wastewater treatment processes, but also to mechanical and electrical support systems of the facility. The criteria, summarized below, define the minimum reliability standards for the following three classes of WRRFs:

- Class I facilities, which discharge to navigable waters that could be permanently or unacceptably damaged by effluent that was degraded in quality for only a few hours;

- Class II facilities, which discharge to navigable waters that would not be permanently or unacceptably damaged by short-term effluent quality degradations, but could be damaged by continued effluent quality degradation; and

- Class III facilities, which are not classified under classes I or II.

	Single 13.8-kV Source	Single 111-kV Source with Two Feeders to 13.8-kV	Two Independent 13.8 kV
Simple radial system			
Failures/year	1.12	0.3	0.1
Hours down/year	7.0	5.5	5.0
Primary selective system			
Failures/year	1.0	0.3	0.06
Hours down/year	5.0	3.5	3.1
Secondary selective system			
Failures/year	1.0	0.2	0.03
Hours down/year	2.7	1.2	0.8

TABLE 7.1 Power Source Reliability

After a facility classification is defined, specific design criteria to accomplish the desired reliability are found in the technical bulletin. Support systems that affect the criteria include electrical power systems, instrumentation and control systems, and other auxiliary systems, such as facility water, plumbing, and chemical systems. For example, reliability provisions for an electrical power source and its distribution depend on the particular class (class I, II, or III) of treatment facility. In general, two separate and independent sources of electrical power should be provided to the facility, either from two separate utility substations or from a single substation and a treatment-facility-based generator. If available from the electrical utility, at least one of the facility's power sources should be a preferred source—that is, a utility source that is one of the last to lose power from the utility grid if power-generating capacity is lost. Independent sources of power should be distributed to separate transformers at the facility, to minimize common-mode failures from affecting both sources. Power-distribution reliability features within the facility should also include redundant buses or feeders linked to backup equipment, dual-powered motor control centers (MCCs) with power-transfer schemes between two sources of power, coordinated breaker settings or fuse ratings, acceptable equipment locations, and standby-power-generator starting.

Table 7.1 summarizes the results of an electrical-power-source-reliability study. The study shows that two independent power sources provide a high level of reliability and that only under the most severe reliability restrictions would the expense for an additional source of power be justified.

3.0 Electrical Systems

The design of an electrical-power-distribution system for a WRRFs should be based on the system reliability that is required, which, in turn, defines the system-distribution arrangement to be implemented. The detailed design must conform to applicable national and local codes and standards. Generally, the design of the electrical system can be tailored around the requirements of the treatment facility using the types of power systems defined in the National Electrical Code (NEC) (National Fire Protection Association and American National Standards Institute, 2014). Types of distribution systems include the following:

- Simple radial system,
- Expanded radial system,

- Primary selective system,
- Secondary selective system,
- Secondary spot network, and
- Ring bus system.

These types of electrical systems are discussed later in this section.

In the absence of detailed local electrical code requirements, the detailed design typically should be in substantial accordance with the NEC, the National Electrical Safety Code (American National Standards Institute, 2017), and the requirements set forth in Occupational Safety and Health Administration (Washington, D.C.) (OSHA) Code of Federal Regulations (CFR). In addition, the design must comply with the operating rules of the local electrical utility. Therefore, it is important early in the design process to identify and thoroughly understand all of these requirements, because, on completion of construction, the contractor often is required to obtain certification of inspection and approval from all required authorities. To ensure safety, all equipment specified and installed by the contractor should meet the rigid testing standards certified by the Underwriters Laboratories Inc. (Camas, Washington) (UL) or similar test agencies, as deemed appropriate for the geographical area. Some states, such as the state of Washington, require that all electrical devices be listed or labeled by a nationally recognized testing laboratory (e.g., Underwriters Laboratories Inc., Intertek, Canadian Standards Association, and Conformance European). The states that have this requirement can provide a list of the acceptable testing laboratory and the types of equipment that the state recognized that the testing laboratory can list/label. To ensure that the electrical-system design will be compatible with the existing utility supply, sufficient time should be included in the planning phase of the project to meet with the local utility companies to define load capacity, system voltage, and any unique design considerations.

The objective of the final electrical-system design is to provide an adequately sized system that is safe and can be operated reliably and economically. Therefore, an accurate estimate of the electrical load and associated characteristics is required. As soon as the general nature of the design has been established, a preliminary analysis of the project typically is made, to determine the approximate size and location of major items of electrical equipment and to furnish the local utility company with the information required by its engineering department. Information generally desired by a utility includes plot plans showing the location of the facility and the various structures, the point of entry of the electrical service, total connected load for both lighting and power, maximum estimated demand, size and locations of motors and starting characteristics, and electrical-load-growth projection.

The physical layout of a facility affects the location of electrical equipment and the characteristics and configuration of a facility's power system. A small or compact facility layout lends itself to centralized distribution, transformer location, and radial low-voltage feeds to the individual loads. A larger and more expansive facility may be better-suited for medium-voltage distribution, looped feeders, and the location of electrical equipment, such as transformers, MCCs, and power-distribution centers, near the load. In most cases, a combination of the two systems is used.

The nature of electrical loads in a WRRF establishes the voltage levels required at the facility site. Convenience receptacles and small motors of less than 0.37 kW (0.5 hp) generally require a 120/240-V, single-phase, 60-Hz power supply. For lighting of both outdoor and large indoor areas, a more economical solution may be a three-phase, 60-Hz, 120/208- and 277/480-V power supply. Large motors (0.37 kW to up to 300 kW [400 hp]), which make up the bulk of loads in most WRRFs, typically require a low-voltage (typically 480 V), three-phase, 60-Hz power supply. Some large motors (greater than 168 kW [250 hp]) also can be powered using a medium-voltage (4160 V), three-phase, 60-Hz power supply. MCCs provide motor controls and power to low- and medium-voltage motors. Selection of operating voltage should be evaluated on a case-by-case basis to determine the best allocation for that facility and that particular client.

A facility's power load and characteristics are determined by preparing an equipment motor list and the nature of the equipment's use, such as continuous, standby, or intermittent. Included in the preliminary design is a "one-line" diagram of the power-distribution system, showing all major components of the electrical system, such as power-source transformers, switchgears, MCCs, various power and lighting distribution panels, and all motors. Figures 7.1 and 7.2 present typical electrical "one-line" diagrams for a medium-sized WRRF. The lighting load can be estimated from the areas of the various structures, by using appropriate watt-per-area factors obtained from the Illuminating Engineering Society of North America's (New York) Lighting Handbook (2011).

It is important to examine a power company's rate schedules when designing an electrical-power system. In addition to charges for actual power consumed, most power companies add a maximum demand charge. Some power companies also impose a penalty charge if a facility's power factor is lower than a certain minimum value. The maximum demand charge, frequently based on the highest rate of usage in the preceding 12 months, penalizes the user for large amounts of short-term power usage. Therefore, uniform power usage is economically important and generally can be considered during the electrical-design phase; however, the maximum demand typically is a result of the sizes selected for various large motors and the method of facility operation. Many facilities now use electrical-system-monitoring software programs that monitor energy usage at various points in the system and assist the facility operator in determining at what points in the electrical system various loads can be shed to reduce demand.

In addition to computing the connected load, the typical demand load, facility growth, and reliability should be considered. The total connected load is the sum of all electrical loads (expressed in kilovolt-amperes) connected to the power system. A designer should think of the demand load (in kilovolt-amperes) as the actual peak operating load of the facility. When calculating the demand load, maximum wastewater-flow conditions should be assumed. In designing for expansion, equipment and circuit ratings should be selected that are adequate for future load growth and allow space for additional electrical equipment and circuits. Reliability generally is maximized by providing alternate power supply sources and routes.

3.1 Supply and Distribution System Voltage

When power is obtained from a local electrical utility, the types of services available from the local utility limit the choice of power-supply voltage at the facility site for particular loads. As a result, the policy and main power-distribution system of the local power utility affect the design of the power system. Therefore, to properly coordinate the facility design with the power utility, it is important to meet with representatives of the power company early in the design process. The most important electrical characteristics of the distribution system of a power utility are its power-producing capacity, distribution voltage, and available short-circuit current.

3.1.1 Utility-Power-Supply Capacity and Redundancy

The power capacity of the distribution system should be equal, at least, to the load it will supply. Generally, "independent" means that each power feeder comes from a

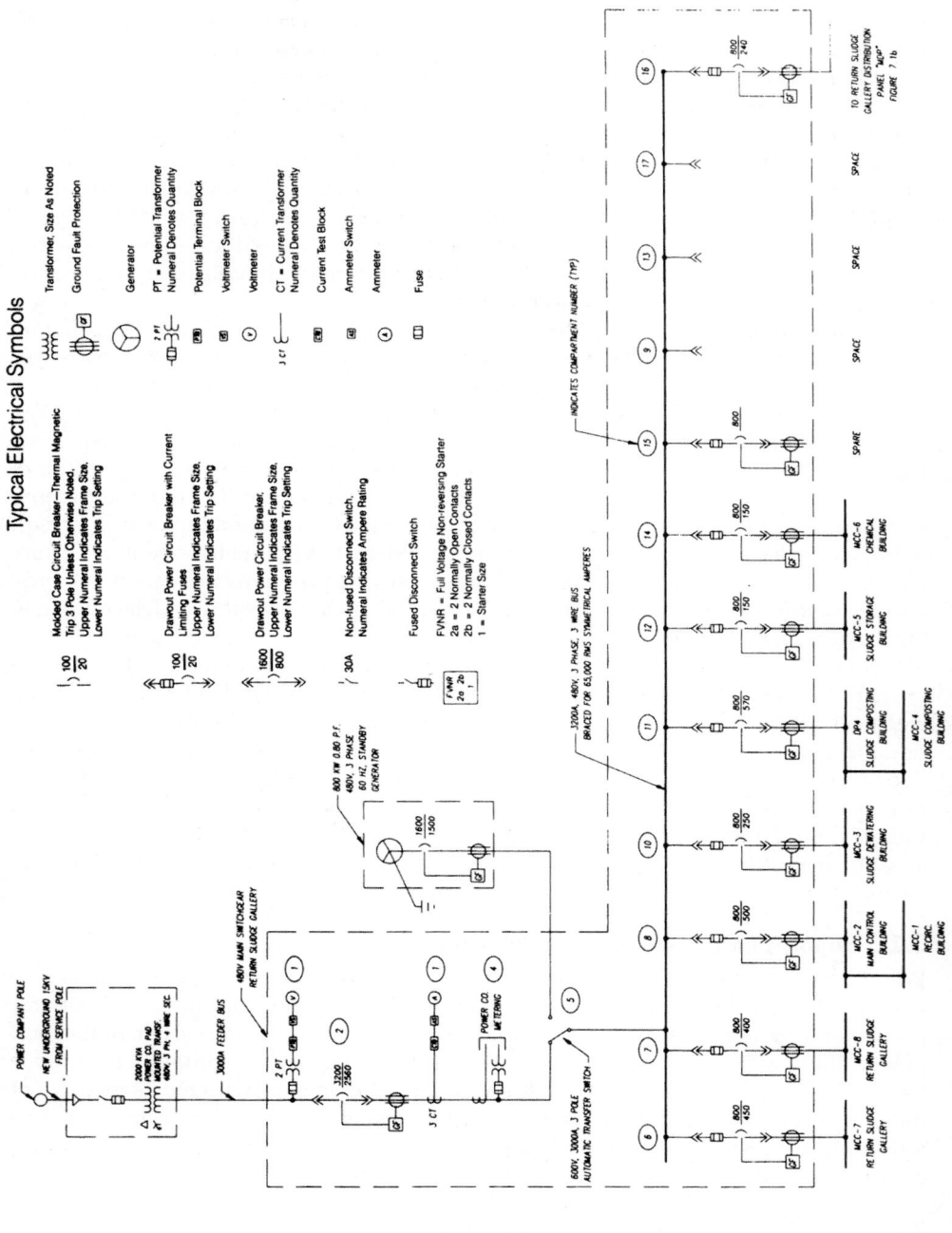

Figure 7.1 Typical electrical "one line" diagram for a medium-sized water resource recovery facility.

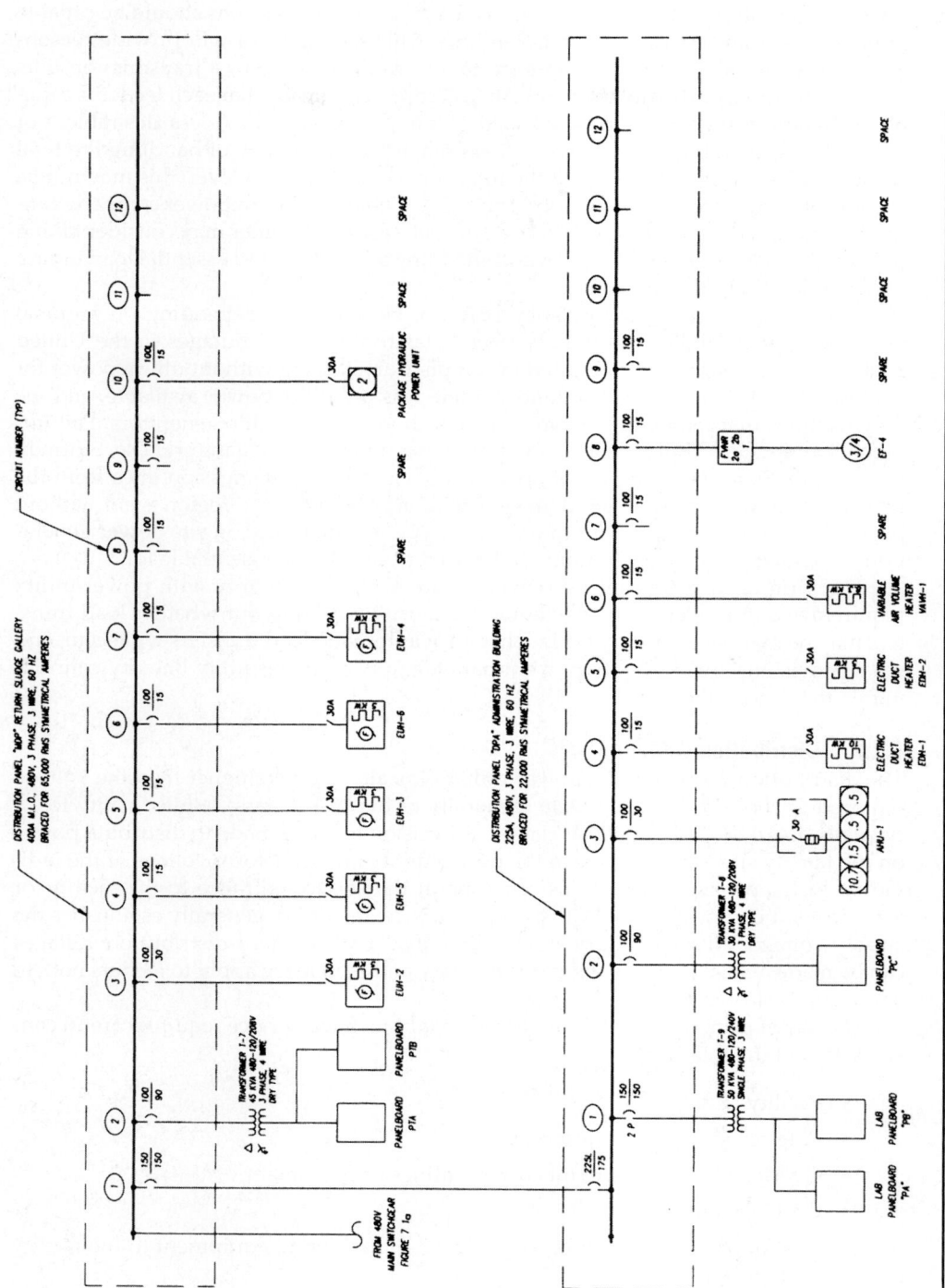

Figure 7.2 Typical electrical "one line" diagram for a medium-sized water resource recovery.

separate substation of the power company; each of these substations should be capable of being fed from separate transmission lines. This arrangement will provide reasonable protection of a complete long-term failure from a failure in a transmission line, utility substation, or distribution line. At "full capacity" means that each feeder is capable of handling the total connected load of the power system. A less desirable, but acceptable, situation is to have two feeders that are each capable of handling the total demand load and less than 50% of the total connected load. However, this may not be possible in cases where only one substation is available or where a power company cannot provide a second full-capacity feeder. In such cases, a designer may consider taking two feeders from a single substation and shedding all but the most essential load in case of power loss from the substation.

Utility-power-supply redundancy may not be adequate, depending on regional power-supply reliability. There have been catastrophic utility outages in the United States in recent years, which resulted in complete areas being without utility power for long periods of time. In these situations, there was no utility power available, and the only facilities that remained operational were those with standby generators. Failures in the Southeastern United States from hurricanes and the 1993 Inaugural Day Storm in Washington State are examples of occasions when power-transmission lines from the power plants were so severely damaged that large areas of the region were without power for extended periods of time. In these types of situations, on-site power generation or auxiliary-engine generators can be used to supply the essential load.

The form of load transfer on power failure also must comply with power-utility requirements. A power company should be consulted to find out whether load transfers may be manual or automatic, in what form automatic-load transfers will occur, and what provisions are needed to prevent paralleling two power-utility lines by connecting them to each other.

3.1.2 Distribution Voltage

The distribution voltage of the power utility typically is much higher than the voltage required for use on the facility site (generally in the 15-kV class), while facility loads typically are in the 5-kV or 600-V classes. A decision must be made to distribute power on the facility site at the voltage of the power utility and transform voltage at the individual loads, provide initial transformation at the service entrance to the facility, or apply a combination of both. With primary service, the utility generally establishes the supply voltage. With secondary service, any of several standard-distribution voltages can be made available. Here, the terms primary and secondary apply to feeders both to and from a transformer.

The use of primary metered versus secondary metered service requires careful consideration of the following:

- Facility layout,
- Rate schedules for both services,
- Overload and short-circuit current ratings of equipment,
- Cost of a substation,
- Cost of primary and secondary wiring and metering equipment to the facility for both services, and
- Facility personnel required to maintain the equipment.

The use of primary metered service generally achieves a lower electrical energy cost. With primary metered service, the utility provides and maintains feeders to the facility-owned and -maintained substation. The amount of savings may not be sufficient to compensate for the cost and inconvenience of maintaining and replacing transformers and high-voltage equipment. Long-delivery items are stocked by utility companies at all times. In addition, if the facility load is supplied by a secondary voltage of 480 V or less, the facility may not have qualified personnel to maintain high-voltage equipment and large transformers.

Meters that gauge electrical energy can be installed on utility feeders or on secondary feeders located in front of the transformers. Final determination typically is made by the utility company. Energy loss on a transformer is approximately 1% when operated at or near full load.

When the facility load is large, secondary distribution is achieved by introducing higher voltages, such as 2400 V, 4160 V, 6900 V, 7200 V, 12.47 kV, and 14.4 kV. It is common for a facility to be responsible for secondary distribution of up to 4160 V. The maximum total voltage drop between the power source and the final use for feeders and branch circuits should not exceed 5%.

3.1.3   Short-Circuit Capacity

The short-circuit capacity of a power-distribution system is the amount of power (or current) the system can supply to the point of short-circuiting. For most power-utility-distribution systems, the short-circuit capacity at the distribution line is at least 500 MVA. At points in the facility power system, capacity will be reduced by the impedance of cables and transformers between the distribution line and the short circuit and will be increased by the effect of connected motors operating as generators at the time of short-circuiting.

All equipment in a power system must have a short-circuit or interrupting current rating (KAIC rating) greater than the short-circuit current capacity at that particular point in the system. If available short-circuit capacities are excessive, they can be reduced by increasing the impedance of transformers or installing reactors in the system. A large magnitude of available short-circuit current at the incoming service will require large fault-interrupting ratings for all distribution equipment and result in a significant increase in equipment costs.

3.1.4   Selective Device Coordination and Protection

Power-distribution systems are designed to be protected against short-circuit current, overcurrent, overvoltage, undervoltage, ground fault, differential current in the system or transformer, and loss of phase. A designer should pay particular attention to adequately protecting transformers, motors, switchgears, and MCCs. Fuses and circuit-breakers protect against overcurrent, short-circuit current, or both. Other protective devices include various types of relays. However, as discussed previously, many agencies that evaluate the design of WRRFs will consider the power system in the facility as "legally required" to meet the design criteria for these facilities. The NEC requires "selective coordination" of the overcurrent devices within the power flow associated with the "legally required standby system." This is defined in NFPA 70 Article 701 and specifically paragraph 701.18—Coordination (NFPA and ANSI, 2014).

Selective device coordination isolates the problem spot within the power system, while the rest of the system operates without interruption. This is achieved by selection and selective adjustment of the overcurrent and grounding-protection devices. For proper protective-device selection, adjustment, and selective coordination, a designer should perform short-circuit calculations, load studies of the system, and a selective-device study of the components of the system. The designer may include the requirement for the study from the electrical contractor/supplier (stamped by a registered professional engineer and authorized by the local jurisdiction) in the project specifications, as the contractor may have more detailed information about the individual devices (e.g., breakers and fuses) that actually will be installed within the facility. For proper motor protection, both the motor full-load current and the starting current must be known. Large motors are additionally protected against overheating by resistive temperature detectors.

3.1.5 Arc-Flash Risk Assessment

Arc-flash risk assessment (AFRA) is a study that examines the electrical system for hazards and determines available incident energy within electrical equipment. Based on available incident energy, appropriate personal protective equipment (PPE) is determined and safe work practices are implemented. OSHA regulations (29 CFR 1910.129(d), 29 CFR 1926.28) mandate that hazards at workplace must be assessed and appropriate PPE for discovered hazards must be provided by the employer. OSHA uses NFPA 70E guidelines and methodology as a reference for electrical safety enforcement. NFPA 70E and Institute of Electrical and Electronics Engineers (IEEE) 1584-2002, provide guidance on implementing appropriate safety procedures and incident energy calculations.

Arc flash is conductance of electrical current through the ionized air, which can occur during an electrical fault. Energy released can create volatile environment surrounding the electrical equipment by discharging tremendous amount of heat and intense light in short amount of time. Results can range from minor injuries to death. The arc-flash event may create property damage and induce fines, lawsuits, etc.

AFRA study should be performed by licensed electrical engineer with experience of performing such work. While incident energy calculations can be calculated by hand, it is common practice to use computer aided modeling software. Mainstream software's use latest calculation methods; latest manufacturer component data; besides incident energy report they can produce various other reports (protective device coordination, power flow, harmonics, design verification, etc.); and can be used for producing equipment warning labels.

All switchboards, panel boards, industrial control panels, meter-socket enclosures, and MCCs that are in other than dwelling occupancies and are likely to require examination, adjustment, servicing, or maintenance while energized, are required to be evaluated for electrical safety due to arc-flash hazards. These items shall be field-marked with a warning label to inform qualified persons of potential electric-arc-flash hazards. Equipment warning labels should follow ANSI Z535.4 Safety Labels standard (Figure 7.3). AFRA study will determine the requirements for signage on the equipment, which identifies the level of incident energy available, safety distances, and the minimum PPE required to be worn by the individuals working on the energized equipment.

⚠ WARNING

Qualified Persons Only
Arc Flash and Shock Hazard
Appropriate PPE Required

REVIEW SAFE WORK PRACTICES PRIOR TO WORK

26	Inch	Arc Flash Protection Boundary		
2.2 cal/cm^2		AF Incident Energy @ Working Distance:	**18**	**Inch**

Recommended (minimum) PPE:
Arc-rated long Sleeve Shirt and arc-rated pants or arc-rated coverall or arc flash suit. Arc-rated face shield and arc-rated balaclava or arc flash suit hood, Arc-rated jacket. Hard hat, arc-rated hard hat liner, Hearing protection, Safety glasses, Heavy duty leather gloves and Leather footwear.

480 VAC	Shock Hazard		
00	Glove Class	**TESCO**	Prepared: 04/03/2017
42 Inch	Limited Approach	CONTROLS, INC. www.tescocontrols.com	Job #:T- 12345
12 Inch	Restricted Approach	PROFESSIONAL ARC FLASH SERVICES	

Location: MCC-11 MAIN

FIGURE 7.3 ARC-flash label. Courtesy of Tesco Controls, Inc.

3.2 Distribution System

Primary and secondary selective systems are the two basic configurations of dual feeders that are connected to electrical equipment, such as the switchgear or MCC. In the primary selective system, the equipment has a single bus, to which the entire load is connected. In the event of a power failure, the feeder typically supplying the power is disconnected, and the second feeder is connected to the bus to supply the load. In the secondary system, the equipment has two buses; each bus supplies 50% of the load and typically is connected to a power feeder. An open bus tie typically is provided. If either feeder fails, the faulty feeder is disconnected, and the bus tie-breaker is closed, so that the other feeder carries the entire load. In either case, the transfer may be manual or automatic. In most cases, positive means (e.g., key interlocking of switches or circuit-breakers) should be used to prevent paralleling the two feeders. To improve system reliability, redundant feeders to double-ended substations should be routed in a separate manhole and ductbank system.

In a primary selective system, a transformer primary has a switch that selects one of two feeders and supplies a single load group, such as an MCC. In a secondary selective system, the load group is designed to select one of two transformer secondaries. Both techniques can be combined to further increase reliability.

A radial-feed system is a system in which individual feeders from a central point serve individual load groups. A loop-feed system is a system in which a feeder from a central distribution point is taken to several load groups and returned to the central distribution point. The feeder then may be energized from either end. A third type of system, also referred to as a loop system (although it is actually a looped radial system), is one in which each of two feeders is taken to the primaries of several transformers to make the transformers primary selective. Figure 7.4 shows a typical dual-feed power system.

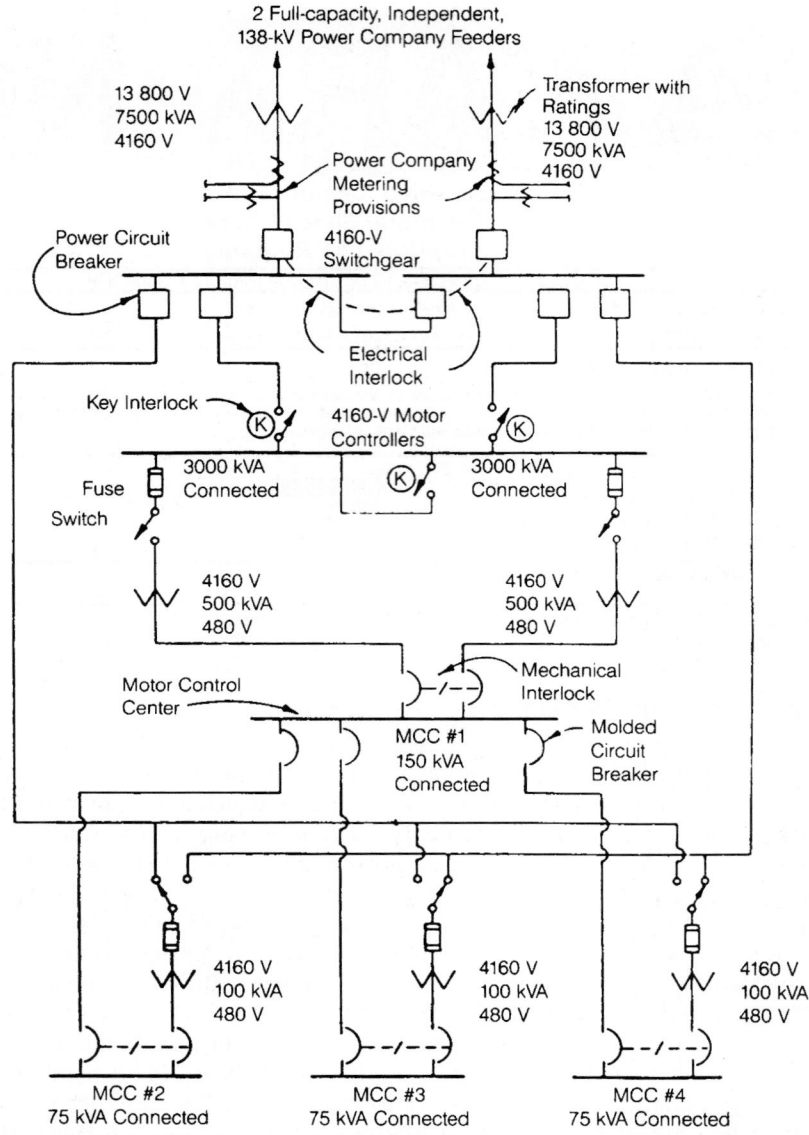

FIGURE 7.4 Example of a dual-feed power system.

3.3 Variable-Speed Drives

Variable-speed controls of electric motor loads, such as pumps and fans, allow WRRF operators to operate equipment economically, to allow the output of prime movers to match the requirements of their loads. The rising costs associated with electricity and various utility-rebate programs also are key reasons to consider the use of variable-speed drives (VSDs). In most cases, a fan or pump that varies its output by changing the drive speed is more efficient than one that varies output through throttling.

There are many types of VSDs that have been traditionally used in wastewater treatment applications. Variable-frequency drives (VFDs), direct-current drives, and eddy-current clutches (ECC) historically are the most common. However, ECCs, with their low relative energy efficiency, are not common in current wastewater projects. Permanent magnet couplings are being used more often where ECCs were used in past applications.

3.3.1 Variable-Frequency Drives

The most commonly used VFDs are those with pulse-width modulation (PWM). These VFDs use a diode on the front end with direct-current bus capacitors and a microprocessor control. The control typically is a 4- to 20-mA signal from a sensing instrument or programmable logic controller (PLC), which allows the drive to function in a closed-loop control-system operation. The drive uses a PWM output to provide power to the motor and adjust motor speed.

One consideration when applying VFDs is heating in the windings of the motor being driven. The PWM power that the VFD delivers to the motor causes additional heat in the windings, compared with an across-the-line starter. This increased heat can result in a much shorter motor life. It is important to specify "inverter duty" motors and VFD rated cable when a VFD is used, as these types of motors are designed to operate with the type of power a VFD supplies.

A typical induction motor uses fans on the end of the rotor to cool the motor. When a motor is driven at lower than full speed, the motor will have less cooling than the motor was designed for. In cases when the motor needs to be operated at less than 75% of the motor's full speed, the motor manufacturers should be consulted to see if an auxiliary cooling fan is recommended. These auxiliary fans typically are small, fractional horsepower fans that can run off the 120-VAC control circuit of the VFD and run when the motor is being operated. The designer also should use some type of imbedded winding-temperature sensor (resistance temperature detectors, temperature switches, etc.) to monitor the winding temperature and shut down the motor in the event when motor windings overheat.

VFD-rated cable is highly recommended on the load side of the VFD to minimize the impacts of the standing wave reflection and voltage spikes. Due to nature of VFD high-speed switching operation, voltage spikes can exceed 2000 V on 480 V AC systems.

Additional conditioning components may be needed as installation distance between VFD and driven equipment increases. Components such as load reactor, dV/dt filter, or active filter will have to be evaluated based on site conditions and distance.

VFD's are considered nonlinear loads and as such they induce harmonic distortion on their line side. Overall facility design must comply with IEEE 519-2014 and must consider harmonic mitigation methods. Mitigation can be achieved by specifying line reactors with 6-pulse VFDs; using 18-pulse drives with larger motors rather then 6-pulse drives; and/or active harmonic filters to mitigate a group of 6-pulse VFDs.

3.3.2 Direct-Current Drives

A commonly used type of drive on chemical feed pumps is a direct-current drive. These types of drives take the same alternating current and convert it to a direct-current bus. The drives are microprocessor-controlled and use a 4- to 20-mA DC signal for a speed reference, but use the direct current to power the motor itself. Typically, direct-current motors have a much higher low-end torque, so they are good at running at low speeds, especially when thick or viscous fluids are being pumped. Direct-current motors also have a tight speed pickup circuit, so they can be controlled tightly.

3.3.3 Permanent-Magnet Drives

A newer type of VSD that has been used successfully for more than 5 years in the municipal market is the permanent-magnet (PM) drive. This type of drive actually is a magnetic coupling between the motor shaft and the pump shaft. Permanent magnets are moved closer together, to speed the pump up, and farther apart, to slow the pump down. These types of units allow for the use of across-the-line motor starters, and the motor starts without any shaft load. After the motor is up to speed, the permanent magnets are moved closer together, based on a 4- to 20-mA DC signal to PM drive, increasing the "coupling" between the motor and the pump and, therefore, the speed of the pump shaft.

3.4 Lighting Systems

In general, lighting requirements should follow procedures recommended in the Lighting Handbook (Illuminating Engineering Society of North America, 2011). The three categories of lighting systems installed throughout WRRFs' indoor, outdoor, and emergency lighting serve various purposes. First and foremost, lighting systems are used for the safe and effective operation and maintenance (O&M) functions of the facility. In addition, adequate lighting provides a degree of security at the site at night. The lighting near moving equipment or vehicular traffic, however, should be selected to minimize glare.

3.4.1 Light-Source Considerations

The human eye responds differently to different colors of light and therefore should be considered in the selection of the light source (type of lamp). In areas where color recognition is needed, such as places with electrical or control wiring, where a multitude of colors on chemical lines is needed to ensure safety, a "whiter" light source, such as fluorescent, light emitting diode (LED) or metal halide, should be considered. Past versions of these types of lamps were lower in efficiency in the area of visual light per watt and had a much shorter lamp life than the high-pressure sodium source. However, currently, fluorescent, LED, and metal halide light sources are being used more regularly.

One reason for the use of fluorescent and metal halide light sources is that the whiter light of these sources is seen as brighter by the human eye. The yellow light on a high-pressure sodium source primarily activates the noncolor receptors in the human eye, while the whiter light will activate both the noncolor and color receptors in the eye. When both types of receptors are activated, the brain will "see" the whiter light as "brighter," even though a light meter will register the exact same light levels. This whiter light also is a better light source for use with security cameras, as the whiter light improves visual clarity and color recognition in the camera system.

With the cost of LED light fixtures decreasing, low operating cost due to exceptional efficiency and longevity, in recent years LED light fixtures are present in most of new installations.

3.4.2 Indoor-Lighting Systems

During daylight hours, properly used natural lighting minimizes the electrical lighting required indoors. Appropriately placed windows and skylights, coupled with proper surface-coating systems, will accomplish this. However, even with the best natural lighting, an electrical indoor-lighting system is required. Table 7.2 summarizes typical minimum indoor-lighting requirements.

Common indoor-lighting sources include incandescent, fluorescent, high-pressure sodium, LED, and metal halide lighting. The most efficient lighting source is LED; the least efficient source is incandescent lighting. Fluorescent and LED fixtures are commonly used in offices, areas with low ceilings, and in areas where the color of light

Typical Indoor Lighting Requirements, at the Work Plane		
Location	**Illuminance Category**	**Footcandles**
Raw material processing (cleaning, cutting, crushing, sorting, grading)		
• Coarse	C	10
• Medium	D	30
• Fine	E	50
Service spaces		
• Stairways, corridors	B	5
• Toilets and wash rooms	C	10
• Maintenance	E	50
• Motor and equipment observation	D	30
Warehousing and storage		
• Inactive	B	5
• Active: bulky items; large labels	C	10
• Active: small items; small labels	D	30
The task may be horizontal, inclined, or vertical		
IESNA—Illuminance Categories		
Designator	**Use Description**	**Value (fc)**
A	Public spaces	3
B	Simple orientation for short visits	5
C	Working spaces where simple visual tasks are performed	10
D	Performance of visual tasks of high contrast and large size	30
E	Performance of visual tasks of high contrast and small size, or visual tasks of low contrast and large size	50
F	Performance of visual tasks of low contrast and small size	100
G	Performance of visual tasks near threshold	300 to 1 000

TABLE 7.2 Typical Minimum Indoor Lighting Requirements (IESNA = Illuminating Engineering Society of North America)

is important. High-pressure-sodium, LED, or metal halide fixtures are used outdoors, in large indoor areas, and in areas with high ceilings (4.6 m [15 ft] and higher). Each of these types of lighting sources produces a different color of light.

Most LED and fluorescent systems are 120-, 240-, or 277-V single-phase systems. High-pressure-sodium and metal halide systems operate at 277 or 480 V (single-phase voltages). When using high-pressure-sodium or metal halide systems, a designer should consider the required warm-up time. As a result, the systems may not be appropriate for locations that require instantaneous lighting, such as passageways.

Lighting fixtures in areas defined as hazardous should comply with the requirements for class I, division 1, group D and, in areas where hazardous gases are contained in tanks and piping, the fixtures should comply with class 1, division 2, group D requirements. To increase reliability and minimize maintenance, all lighting fixtures in these areas should be of high quality and corrosion-resistant. The National Fire Protection Association (2014) specifies "Class" and "Division" for areas that are considered classified for WRRFs. The National Electrical Code (NFPA and ANSI, 2014) specifies the ignition temperatures of potential hazardous gases that might exist in WRRFs.

3.4.3 Outdoor-Lighting Systems

Outdoor-lighting systems generally are used for the safety of the facility staff and overall site security. Outdoor-lighting systems typically consist of pole-mounted, general-area lighting with localized, supplementary illumination at exterior process units, building entrances, and facility entrance gates. Table 7.3 summarizes typical minimum outdoor-lighting requirements. Current practice uses 480-V, single-phase, LED, high-pressure-sodium or metal halide lamps. Outdoor-lighting systems are controlled by photo cells, manual switches, lighting control system, or a combination of these. For large facilities with extensive site lighting, an economic evaluation should be performed, to determine the optimum selection of the number and height of pole-mounting luminaries needed to minimize glare and reduce spread onto adjoining properties, and to accomplish the general area lighting-system goals.

3.4.4 Emergency-Lighting Systems

Specific requirements for emergency-lighting systems typically are defined by the building-code requirements that govern construction of a facility. Emergency-lighting systems are designed to provide safe exit from a facility during a general power failure and are installed inside buildings and structures. In general, these lighting systems are self-contained, battery-operated units. Luminaries are sealed-beam incandescent or LED lamps, and the battery units are self-charging, or, in the case of more "finished" spaces, such as laboratories, control rooms, and offices, selected fixtures in the path

Location	Minimum Illumination, lx
General	5
Unoccupied site security	2
Outside entrances	50
Local task lighting	50

TABLE 7.3 Typical Minimum Outdoor Lighting Requirements

of egress from a facility are either uninterruptible power supply (UPS) (labeled for emergency-lighting services and not computer-power backup) powered or have internal battery/inverters in the selected light fixtures to provide power for these lights. Illuminated faces, either by back light, light-emitting diode, or luminescent exit lights/signs also typically are required to denote the points of exit from a structure.

3.5 Standby Power Considerations

The Recommended Practice for Emergency and Standby Power Systems for Industrial and Commercial Applications (IEEE 446-1995) (Institute of Electrical and Electronic Engineers, 1995) contains recommended engineering practices for the selection and application of emergency and standby power systems. Standby power-generating units often are used for increased reliability or for small critical loads within a facility where an affordable, additional independent power source is not available. Reciprocating engine-driven generators, fueled with gasoline (spark-ignited), diesel fuel (compression-ignited), natural gas (spark-ignited), or digester gas (spark-ignited), are used most commonly. Generators of higher capacity (typically greater than 125 kW) typically are more economical if using the compression-ignited engine. For applications requiring more than 500 kW (700 hp), gas-turbine generators sometimes are considered. Consideration should be given to providing adequate cooling and ventilation, fuel storage, starting readiness, and service availability. In addition, many areas have stringent air-quality-control ordinances that require permits to operate engines with certain types of fuel.

In large facilities with anaerobic digestion, the digester gas recovered often is used as fuel for boilers and internal combustion engines that, in turn, are used for pumping wastewater, operating blowers, or generating electricity. A survey of 40 large U.S. WRRFs (of at least 4 000-m^3/d [1-mgd] capacity) that have methane-recovery facilities showed that many of the facilities also have installed electric-generating systems that use digester gas successfully. However, many facility designers questioned whether purchasing expensive equipment was justified. The most satisfactory applications appeared to be those using gas for boiler (heating) applications and for sale to the area gas utility.

In a combined facility, cogeneration generally is defined as a system for generating electricity and producing other energy (typically hot water or steam) at the same time that is more efficient than electrical generation would be alone. A cogeneration facility typically applies waste heat recovered from a gas engine to its heating needs for processes, space heating/boiler augmentation, and nonpotable hot water.

When selecting a standby power system, a designer should consider the particular application needs. Some loads, such as computers, may not tolerate any power outage, therefore their power should be supplied from UPS system. Also, critical delays in switching to a standby utility line may affect its feasibility. In addition, the sizing of a standby power system must address the voltage dip during the starting of a large motor, because newer electronic systems and motor-starting coils are affected if the voltage decreases to less than 85% to 90% of the nominal voltage.

3.6 Other Considerations

3.6.1 Harmonics

Harmonics in an electrical system are analogous to water hammer in a hydraulic system. Nonlinear switching electrical equipment, such as VFDs, fluorescent and LED lighting, and UPSs, draws electrical current (water flow) to charge a direct-current bus

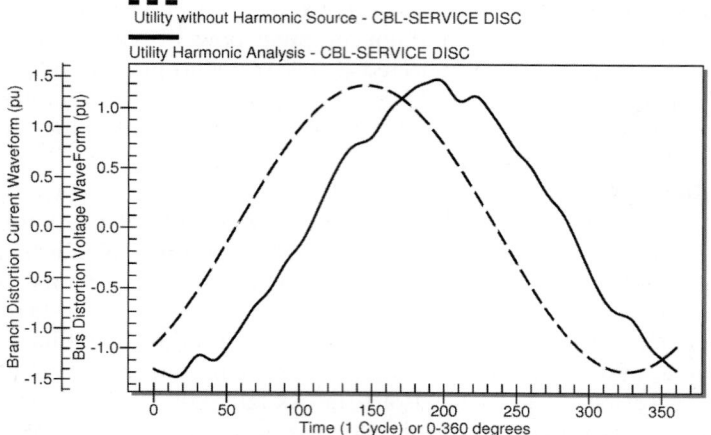

FIGURE 7.5 Harmonic distortion waveform.

at a point and then quickly stops taking current (quick shut-off of a valve). This quick shut-off causes a spike in the voltage (water pressure) at the terminals of the equipment. This spike is reflected back into the power system and also oscillates like a water hammer in a water-distribution system. Figure 7.5 shows comparison between regular and distorted sine wave.

Despite the benefits that VFDs offer in the form of energy savings, there are certain drawbacks to their use. The problem lies in the nonlinear switching mode in which VFDs (and UPS systems) obtain electrical power from an electrical system. The switching operation results in harmonic distortion on the alternating-current supply and can have a significant effect on other electrical-system equipment. Problems associated with harmonics introduced to an electrical system include the following:

- Overheating of transformers and other induction motors in the system;
- Pulsating torques in rotating equipment;
- Improper operation of computers, VFDs, and protective devices;
- Reduced power factor resulting from a distorted power factor;
- Interference with communication systems (Strope, 1994);
- Standby generator de-rating or special excitation system and generator windings; and
- Interference/failure of power factor correction capacitors in the power system.

The effect of nonlinear loads on a system depends on the total harmonic distortion (THD) of the current produced, size of the loads, and impedance of the source. The Recommended Practices and Requirements for Harmonic Control and Electrical Power Systems (IEEE 519-2014) (Institute of Electrical and Electronic Engineers, 2014) suggests that current and voltage THD limits at the point of connection should be approximately 5%. At existing facilities, pumps and fans are retrofitted for VFD control, to improve the process design and decrease energy costs. It is important to perform an evaluation of system harmonics as a result of the addition of VFDs.

Many power companies have offered incentives for users to retrofit equipment control to VFDs as part of their energy-savings-incentives programs. These programs should be investigated early in the design phase of any retrofit project. Programs of this type aid utility companies, because the use of VFDs allows the utilities to reduce their overall system demand.

3.6.2 Harmonic Mitigation Options

Many VFD and UPS manufacturers have developed units that do not create as much (or virtually none, in the case of "clean power" units) harmonic noise in power systems as traditional "6 pulse" drives. The application of harmonic filtering, by either passive filtering or active conditioning, is an option that should be considered. Passive filtering is a reasonable lower-cost alternative to active filtering and can be useful in some systems. Other systems with widely varying operations could benefit from active harmonic conditioning. The advantage of active harmonic filtering is that the systems also provide some power-factor correction, which could be a benefit to the owner to reduce/eliminate low-power-factor penalties. The application of active harmonic conditioners does need special attention when they are used with standby power generators and can affect the voltage regulation of the generator. Active harmonic-conditioner manufacturers have installation recommendations that can mitigate interference between the conditioners and generators.

3.6.3 Grounding Systems

A grounding system is designed to guard operating personnel against the possibility of injury caused by electrical shock in damp locations associated with WRRFs. At the point where the neutral is derived (the secondary of the wye-wound transformer), the system should be neutral of the low-voltage system (typically a 120- to 208-V, or 277- to 480-V, three-phase, four-wire system) and must be effectively grounded, so that a fault in any phase will trigger the protective device immediately. Electrical equipment, conductive components of equipment enclosures or mounting electrical devices, and building steel should be permanently and effectively grounded in accordance with section 250 of the National Electrical Safety Code (American National Standards Institute, 2017) and NFPA-70 Article 250—Grounding and Bonding (National Fire Protection Association and American National Standards Institute, 2014). Properly grounded equipment ensures that no voltage can reach harmful levels on the exposed conductive surface. All underground-distribution-system appurtenances (cable racks, manhole covers, and pull boxes), aerial-distribution-system equipment (lightning arresters and transformer cases) and radio communication equipment (antenna masts, antenna cables) also should be grounded.

3.6.4 Load Transfer on Power Failure

A critical factor in the design of an alternate-power-source system is the means of switching to an alternate source or to a standby power source. Both cost and complexity are functions of the transfer speed required. In addition, if the local agency dictates that the standby system is legally required, the designer must follow the requirements for legally required standby systems, as defined in the NEC (National Fire Protection Association and American National Standards Institute, 2014). If an outage of 30 minutes can be tolerated, a manual-transfer switch may be acceptable at a continuously manned facility. For an outage of several minutes, motorized-primary-system switching may be acceptable. To switch within 15 seconds, an automatic-transfer switch is required. Critical equipment that cannot tolerate a momentary outage requires a solid-state transfer switch with instantaneous switching. Special consideration should be given to the control of electrical loads that are powered from standby-generator systems. Equipment that is automatically started on generator power must have a maintained contact in the start circuit; momentary, push-button control of equipment requires restarting electrical loads on re-energization. Additionally, the use of timing circuits should be considered, to delay the start of various loads (when powered from the generator), to limit initial power, and minimize the water size requirement of the standby-generator system.

3.6.5 Hazardous Areas

Hazardous areas are classified by the National Fire Protection Association (Quincy, Massachusetts) (NFPA) (NFPA 820; National Fire Protection Association, 2016) according to the degree of risk of fire and explosions. Areas of prime concern in WRRFs are those in which methane gas is generated. Other areas of concern are those in which chlorine and sulfur dioxide are produced, and fuel is stored and handled. Corrosion is the main concern in these areas, however, and can be prevented through proper ventilation and use of corrosion inhibitors. Electrical storage batteries, which generate hydrogen, also should be considered.

The need for explosion-proof equipment depends on a number of factors, including the quantities and types of flammable liquids, gases, and dusts; type of ventilation; openness or confinement of the sources of hazard; and distance of devices from the substances. Because of the high cost of explosion-proof equipment, a designer should implement an appropriate layout and design to minimize the amount of electrical equipment and devices required within hazardous areas.

Conduit type, routing and sealing material should be considered when conduits are passing through or exiting classified areas as required by NFPA 70.

3.6.6 Seismic Protection

When electrical equipment is installed in regions prone to earthquakes, heavy electrical equipment (transformers, motors, generators, switchgears, MCCs, vendor skid packages, SCADA/UPS racks, and control panels) should be anchored to base pads or a building, as required by most national building codes (for example, International Council of Building Officials). Resilient anchorage, using either spring- or rubber-shock-vibration-isolation systems, has performed well in earthquakes; however, snubbers or stops to limit movement must be provided. In addition, conductors and connections must be sized and sagged to allow for expected movement during an earthquake, avoiding breakage of insulators and contact between adjacent conductors. This typically means wider spacing between conductors and greater sag than provided by normal non-seismic designs.

Electrical safeguards, such as the following, often are provided:

* Low-voltage and single-phase protection for motors;
* Pressure switches on the discharge piping to shut down pumps under low (ruptured pipelines) or high pressure (collapsed or blocked pipelines);

- Temperature switches to sense high-discharge piping temperature resulting from continuous water recirculation through a failed bypass-pump-control valve, a failed high-discharge-pressure switch, or a closed isolating valve; and

- Motor-winding temperature detectors to sense overloaded motor conditions or severely unbalanced electrical-power-phase and voltage conditions.

While many electrical equipment manufacturers have standard anchoring recommendations, it has become a common practice to have registered structural engineer evaluate and stamp site-specific anchoring requirements. The analysis from structural engineer will include site-specific calculations, anchoring locations, anchor type and material, and embedment depth.

3.6.7 Lightning and Surge Protection
Lightning and surge arresters (surge protection devices) limit the voltage impressed on the winding of a transformer or motor caused by lightning or switching surges. Surge arresters are connected to each phase and ground. The primary side of the main transformer generally is protected by lightning and surge arresters. Motors of 370 kW (500 hp) or greater should be protected by surge arresters and surge capacitors. This is true for motors of at least 150 kW (200 hp) that are supplied by overhead line. Surge arresters and surge capacitors should be located as close to the motor as possible. For economical reasons, lightning and surge arresters sometimes are installed in the switchgear (Type 1 SPD) and/or MCC (Type 1/2 SPD) to protect a group of motors tied to such a source.

Voltage surges often occur outside the facility, and if not quenched at the entrance point, surge can travel and amplify by the time it reaches downstream equipment. Amplification increases and can double in magnitude as distance between device requiring protection and SPD increases. It is important to implement a low-voltage surge protection device of sensitive electronic equipment such as supervisory control and data acquisition (SCADA) computers and PLC.

Lightning arrestors are needed in radio transmission lines and should be installed between radio device and outdoor mounted antenna.

3.6.8 "Clean" Power for Computer Systems
New digital-based control systems are more susceptible to power-line disturbances than older, analog-based systems. Power-line disturbances fall into one or more of the following categories:

- Power outage (short- to long-term);
- High and low voltage;
- Harmonic distortion;
- Frequency shifts;
- Short-term voltage fluctuations (surges, sags, and dips);
- Long-term voltage fluctuations (brownout); and
- Noise impulses.

A range of techniques is available to condition power-line disturbances. A designer should analyze the power requirements of each control system. One of the following power-line-conditioning techniques then can be used to clean up the electrical power:

- Isolation transformer,
- Line-voltage regulator,
- Line conditioners, and
- A UPS system.

3.6.9   Energy Savings and Design Features

In specifying electrical equipment and systems, a designer should be keenly aware of considerations that will result in energy savings and, thus, reduced operating costs for the owner. Equipment that falls into these categories includes motors and motor controls, lighting sources, power distribution transformers, and power-factor-correction capacitors. Premium energy-efficient motors should be specified for continuous-duty motors to reduce power consumption and electrical-utility charges. Modern electric motors take advantage of computer-derived analyses, high-quality materials, and improved manufacturing technology, resulting in the availability of improved efficiency motors at a slightly higher cost than standard motors. With increased operating costs associated with electrical energy, large motors should be selected and evaluated on a life-cycle cost-analysis basis, to ensure that the payback period justifies the extra costs associated with the selection of high-efficiency motors.

Lighting systems currently available are taking advantage of energy-saving lower wattage lighting fixtures, which have nearly equal illumination output of standard light fixtures and, thus, improved lumen-per-watt ratings. Along with these recent trends in lamp design has been the development of electronic ballasts, which save energy compared with the conventional magnetic-style ballasts. In addition, dimming controls can be integrated easily to the electronic-ballast circuitry. These ballasts run cooler and have fewer losses. Their only drawback is the prevalent third harmonic they introduce to the neutral-bus system, which can be accommodated with an oversized neutral.

Beginning of 2016, new high efficiency power distribution transformers (MV and LV) are required for all new installations. This direction comes from Department of Energy (DOE) efficiency regulations. In order to achieve higher efficiency, transformer size, weight, and cost has increased. Small increase in efficiency may slightly reduce overall power consumption in a single facility, but it can provide a great reduction of power consumption nationwide.

Power-factor-correction capacitors are used to improve the system-power factor, which results in less voltage drop, reduced losses, and a reduction in the total current flow. The reduction in reactive current also results in freeing up additional system capacity. Capacitors can be switched in and out of the system as loads are added and dropped offline, to ensure optimal power-factor correction (Strope, 1994).

Many facilities use the application of adding electrical-power-consumption meters for individual facility treatment processes (e.g., secondaries, aeration, and sludge handling). The addition of these meters, when tied into facility SCADA systems, can provide actual cost-of-power-consumption data to the facility operational staff, to highlight the cost of operating the facility. Many municipalities have observed an increased awareness of the facility staff of the cost of operation of the facility when they see exactly how much money is spent in large electrical-power-consumption facilities (i.e., aeration). This awareness has led staff to develop and implement power-savings measures—such as changing a thermostat setting; turning off lights in areas that are not occupied; and delaying operation of non-essential loads and certain maintenance processes for non-peak hours.

In addition, many power utilities have power-consumption rates tied to time-of-day usage. Many utilities charge more per unit (e.g., kilowatts) in times of high energy usage. In these areas, facility operations personnel sometimes can change process operations of systems that can be scheduled (e.g., sludge processing) for times of lower per-kilowatt costs, which, when done on a regular basis, can save a facility considerable monies on an annual basis.

Motor power consumption trending through SCADA in addition with specialized analytical software can give insight to health of the motors within the facility. By comparing historical data of electrical and analytical parameters (temperature, vibration), preventative maintenance can be adjusted and predictive failure analysis can be implemented.

Some municipalities and clients are pursuing the design of a facility to obtain a Leadership in Energy and Environmental Design (LEED) (U.S. Green Building Council, Washington, D.C.) (USGBC) certification. For electrical systems, the main areas of easily obtained "points" for LEED are in the area of lighting and lighting controls. Another area is the use of photo-voltaic cells to harvest power from sunlight.

4.0 Instrumentation and Control Systems

Instrumentation and control systems in WRRFs have become more important over the past several years because of their ability to provide extensive field information to the operator for control and monitoring of facility processes. There are many kinds of control and monitoring systems, and each system has strengths and weaknesses. One type of control system may be perfectly adequate for a pump station, but would be totally inadequate for a large WRRF. Therefore, proper selection of the system components is important for successful facility control-system design.

4.1 Design Standards

The design criteria for instrumentation and process control systems are based primarily on standards established by the International Society of Automation (Research Triangle Park, North Carolina) (ISA) and manufacturer guidelines. Some of the more frequently used ISA standards include:

- ISA-5.1-2009—Instrumentation Symbols and Identification
- ISA-5.2-1976 (R1992)—Binary Logic Diagrams for Process Operations
- ISA-5.3-1983—Graphic Symbols for Distributed Control/Shared Display Instrumentation, Logic, and Computer Systems
- ISA-5.4-1991—Instrument Loop Diagrams
- ISA-5.5-1985—Graphic Symbols for Process Displays
- ISA-RP12.06 series—Recommended Practice for Wiring Methods for Hazardous (Classified) Locations Instrumentation

- ANSI/ISA-18 series—Management of Alarm Systems for the Process Industries
- ISA-TR20.00.01-2007—Specification Forms for Process Measurement and Control Instruments Part 1: General Considerations
- ANSI/ISA-101.01-2015—Human Machine Interfaces for Process Automation Systems
- ANSI/ISA-62443 series—Security for Industrial Automation and Control Systems

4.2   New Construction versus Retrofit

When designing a new WRRF, standards of equipment and software systems should be established upfront, before detailed design begins. If this is not done early in the design, mismatched and incompatible equipment might result. During a retrofit or existing facility upgrade, it is good practice to analyze the current state of the facility to determine if equipment and software standards are in place and, if not, to help the owner develop standards for their system.

4.3   Project Scope Development

In developing the scope of instrumentation and control-system design, the following information should be gathered:

- Size of the facility
- Type of treatment processes
- Type of vendor supplied controls
- Amount of funds available
- New facility or modification and expansion of existing facility (retrofit)
- Current design standards and instrumentation philosophy
- Special interfaces with other control systems
- Extent of automatic controls necessary
- Ability of the owner to maintain a control system and owner equipment preferences
- Staffing requirements for the facility

After this information has been gathered, a control-system engineer determines the best approach for the control-system design and the type of instrumentation required.

4.4   Types of Diagrams

Symbols and identification codes that are used to identify instruments in WRRFs are based on ISA standards; modifications for WRRFs are detailed in Automation of Water Resource Recovery Facilities (WEF, 2013). A process and instrumentation diagram (P&ID), sometimes called a piping and instrumentation diagram, provides sufficient information for a knowledgeable person to understand the means of measurement and control of the process without detailing instrumentation. The full details of the instruments are usually documented in an instrumentation specification, data sheet, or sometimes in drawing notes.

Two types of diagrams—the process flow diagram and the P&ID—are discussed below. Sometimes a 3D/4D model is created that uses information from the P&ID and incorporates intelligent design to the background of the drawing, allowing it to be used for modeling and equipment procurement. Because ISA serves different industries with different needs, its standards allow for variations in the level of detail required by the project.

4.4.1 Process Flow Diagrams

Process flow diagrams are developed early in the design phase of a facility or during modification of the facility by the process engineer and are used as tools for the other design disciplines. Process flow diagrams do not include instrumentation symbols, except critical in-line devices like large flow meters. These diagrams provide a quick overall description of the entire facility and its many processes.

4.4.2 Process and Instrumentation Diagrams

Large projects require more detail for construction and facility operation to be performed correctly. For larger projects, P&IDs are critical in describing the entire process in detail. Included in this level of detail are all significant valves, pipe sizes, process flow streams, pipe material, process and mechanical equipment, instrumentation, control panels, basic electrical requirements, numbers and types of signals, ISA symbols, and tag designations (see Figure 7.6).

4.5 Selection and Specification of Field Instruments

The proper selection and specification of field instruments are critical to the successful operation of the WRRF. The process starts with the development of the P&ID. Detailed data sheets are then developed for all instruments. Instruments are included on all drawings defining control algorithms and monitoring requirements. Finally, special mounting details are developed, if required.

4.5.1 Considerations for Selection

Before selecting equipment, several important best practices should be considered. For outdoor field instrumentation, special attention should be given to direct sunlight on equipment and operator screens, extreme temperatures, rain, and so on. It is good practice to keep outdoor control panels and instruments out of direct sunlight, either by using sunshields or by facing equipment and control panels north, away from the direct sunlight. Sunshields are metal plates, typically used on control panels, that block direct sunlight by mounting them 2.5 to 5 cm (1 to 2 in.) off of every surface of the outside/exposed panel. These also should be used on instrumentation to prolong the accuracy and reliability of the instrument.

Certain processes may require extremely accurate data for control or monitoring. This should be considered when specifying these instruments since higher cost is generally associated with devices that provide greater accuracy. High accuracy may not always be required, as in the case of a lift station level monitoring device.

Quality should be considered when selecting equipment. A higher quality product may cost more initially; but in the long run, it may save a considerable amount of money in calibration, maintenance, and construction.

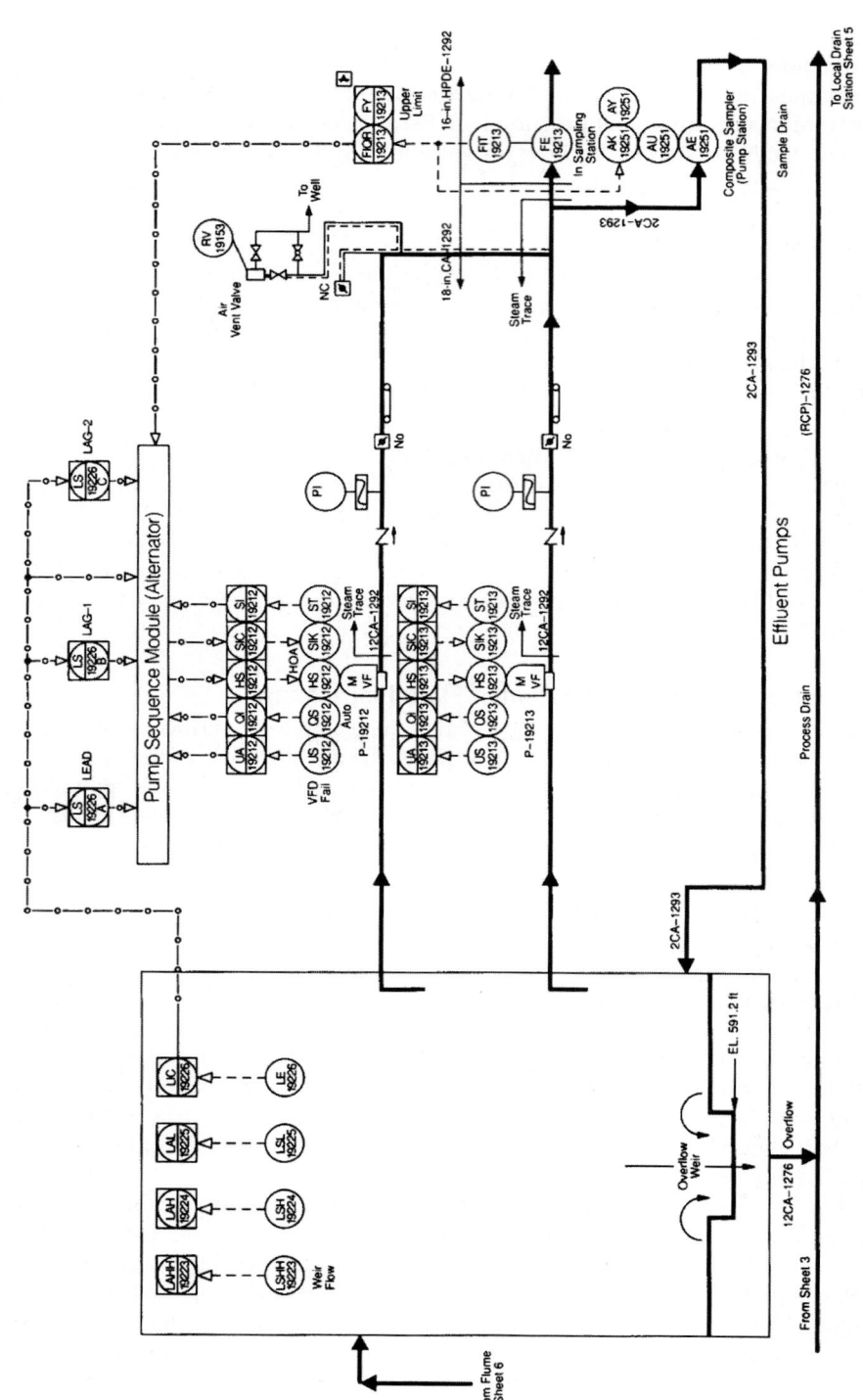

FIGURE 7.6 Typical process flow diagram.

388

4.5.2   Instrumentation Data Sheets

When designing a control system, it is important to develop a correct instrument speci-fication. A key component of an instrumentation specification is the instrumentation data sheet. Because of the complexity of present day instruments and controls, ISA has standardized both the content and form of instrumentation data sheets in older and newer formats:

- ISA-20-1981—Specification Forms for Process Measurement and Control Instruments, Primary Elements, and Control Valves
- ISA-TR20.00.01-2007—Specification Forms for Process Measurement and Control Instruments Part 1: General Considerations

Figures 7.7 and 7.8 show a typical ISA data sheet form and data sheet form explana-tions, respectively, for magnetic flow meters.

Designers either use forms directly from ISA or their own electronic versions. These standard forms can also serve as a checklist for items to be considered when specifying a given instrument. For example, a flow meter data sheet should include key informa-tion including:

- Pipe size (for in-line instruments)
- Minimum and maximum operating data (flow, pressure, temperature, and pH)
- Length between sensor and transmitter
- Pipe lengths before and after the flow sensor
- Special installation hardware required
- Environmental requirements (temperature, humidity, and corrosive or explosive atmosphere)
- Special tagging requirement
- Materials of construction for instrument and method of connection to physical system (flange)

Some specific instrument needs for the physical considerations of a flow meter are the following:

- Range of measurement
- Accuracy
- Repeatability
- Applicable vendor options
- Electrical requirements
- Spare part requirements
- Maintenance and calibration requirements
- Output signal analog or digital (linear or nonlinear)
- Required input signals
- Type of setpoint adjustment (fixed or adjustable)

MAGNETIC FLOWMETERS

Instructions for ISA Form S20.23

1. Tag number of meter only.

2. Refers to process application.

3. Show line number or identify associated vessel.

4. Give pipeline size and schedule. If reducers are used, so state.

5. Give material of pipe. If lined, plastic or otherwise non-conductive, so state.

6. Give connection type: FLANGED, DRESSER COUPLINGS, ETC.

7. Specify material of meter connections.

8. Select tube material. (Non-permeable material required if coils are outside tube).

9. Specify material of line.

10. Select electrode type: STD., BULLET NOSED, ULTRASONIC CLEANED, BURN OFF, etc.

11. Specify electrode material.

12. Describe casing: STD., SPASH PROOF, SUBMERSIBLE, SUBMERGED OPERATION, etc.

13. Give ac voltage and frequency, along with application NEMA identification of the electrical enclosure.

14. State means for grounding to fluid: GROUNDING RINGS, STRAPS, etc.

15. State power supply and enclosure class to meet area electrical requirements.

16.

17. State fluid by name or description.

18. Give maximum operating flow and units; usually same as maximum of instrument scale.

19. Give maximum operating velocity, usually in ft/s.

20. List normal and minimum flow rates.

21. List maximum and minimum fluid temperature °F.

22. List maximum and minimum fluid pressure.

23. List minimum (at lowest temp.) conductivity of fluid.

24. If a possibility of vacuum exists at meter, so state and give greatest value. (highest vacuum).

25.

26. List tag number of instrument used directly with meter.

27. Control loop function such as INDICATE, RECORD CONTROL, etc.

28. Mounting: FLUSH PANEL, SURFACE INTEGRAL WITH METER, etc.

29. Give NEMA identification of case type.

30. State cable length required between meter and instrument.

31. Span adjust: BLIND, ft/s DIAL, OTHER.

32. Give ac supply voltage and frequency.

33. If a transmitter, state analog output electrical or pneumatic range, or pulse train frequency for digital outputs, i.e., pulses per gallon.

34. List scale size and range.

35. List Scale Size and Range for indicating transmitter

36. Recorder chart drive — ELECT. HANDWIND, etc. and chart speed in time per revolution or inch per hour.

37. List chart range and number.

38. If integrator is used, state counts per hour, or value of smallest count; such as "10 GAL UNITS".

39. For control modes: (Per ANSI C85.1-1963, "Terminology for Automatic Control.") Write-in PI_f, I_f, PI_s, $PI_f D_f$, etc.

P = proportional (gain)
I = integral (auto reset)
D = derivative (rate)

Subscripts:

f = fast
s = slow
n = narrow

State output signal range, pneumatic or electronic.

40. Controller action in response to an increase in flowrate — INC. or DEC.

State auto-man. switch as NONE, SWITCH ONLY, BUMPLESS, etc.

42. Number of alarm lights in case. Give form of contacts; SPDT, SPST, etc.

43. Contact electrical load rating. Contact housing General Purpose, Class I, Group D, etc., if not in the same enclosure described in line 29.

44. Action of alarms: HIGH, LOW, DEVIATION, etc.

45. Fill in manufacturer and model numbers for meters
46. and
47. instrument after selection.

FIGURE 7.7   Typical ISA data sheet form. Taken from ISA-20-1981, Specification Forms for Process Measurement and Control Instruments, Primary Elements and Control Valves (c) 1981 by ISA—The International Society of Automation—all rights reserved. Reprinted with permission.

- Alarm switch with limit settings and amount of dead band required when returning to normal
- Communication protocols

4.5.3   Physical Drawings

It is advantageous to show field instruments on both mechanical and electrical drawings. Mechanical drawings show instrument mounting locations in the process piping. Mechanical contractors frequently install in-line instruments. For example, showing the

		MAGNETIC FLOWMETERS	SHEET ___ OF ___

The form header contains:

	MAGNETIC FLOWMETERS	SHEET ___ OF ___
	NO	BY	DATE	REVISION	SPEC. NO.	REV.
					CONTRACT	DATE
					REQ.	P.O.
					BY	CHK'D	APPR.

1	Meter Tag No.		
2	Service		
3	Location		
METERING ELEMENT			
4	CONN'S.	Line Size, Sched.	
5		Line Material	
6		Connection Type	
7		Connection Mat'ls.	
8	METER	Tube Material	
9		Liner Material	
10		Electrode Type	
11		Electrode Matl.	
12		Meter Casing	
13		Power Supply	Elect. Code
14		Grounding, Type & Matl.	
15		Enclosure Class	
16			
17	FLUID	Fluid	
18		Max. Flow, Units	
19		Max. Velocity, Units	
20		Norm. Flow	Min. Flow
21		Max. Temp.	Min. Temp.
22		Max. Press.	Min. Press.
23		Min. Fluid Conductivity	
24		Vacuum Possibility	
25			
ASSOCIATED INSTRUMENT			
26	Instrument Tag Number		
27	Function		
28	Mounting		
29	Enclosure Class		
30	Length Signal Cable		
31	Type Span Adjustment		
32	Power Supply		
33	TRANS.	Transmitter Output	
34			
35	DISPLAY	Scale Size	Range
36		Chart Drive	Speed
37		Chart Range	Chart No.
38		Integrator	
39	CONTR.	Modes	Output
40		Action	Auto-Man.
41			
42	ALARM	Contact No.	Form
43		Rating	Elec. Code
44		Action	
45	Manufacturer		
46	Meter Model Number		
47	Instrument Model Number		

Notes:

ISA FORM S20.23

FIGURE 7.8 Typical data sheet form explanations. Taken from ISA-20-1981, Specification Forms for Process Measurement and Control Instruments, Primary Elements and Control Valves (c) 1981 by ISA—The International Society of Automation—all rights reserved. Reprinted with permission.

location of the instruments in the process piping ensures that flow measuring devices have their required upstream and downstream straight run piping required to produce the required accuracy.

By using the mechanical drawings as a background, an electrical engineer can locate the instruments and place the control panels. The electrical engineer then uses the P&ID

and plan view drawings to help develop the conduit and wire schedule. Some P&IDs show different wire types included as standard symbols. The electrical engineer also develops elementary wiring diagrams to show the control interlock, permissive, and interface requirements (see Figure 7.9).

4.5.4 Instrument Mounting Details

Instrument mounting details indicate what is required for proper installation (see Figure 7.10). Mounting details show pipe sizes and materials, special fittings, and mounting fabrications required to install the instruments. A bill of materials is sometimes included. These details should include information regarding the requirements for outdoor locations, as mentioned earlier.

4.5.5 Networking Considerations

Facility-wide networks share data between computer systems and facilitate control and monitoring throughout the facility. Ethernet networks are predominantly used to

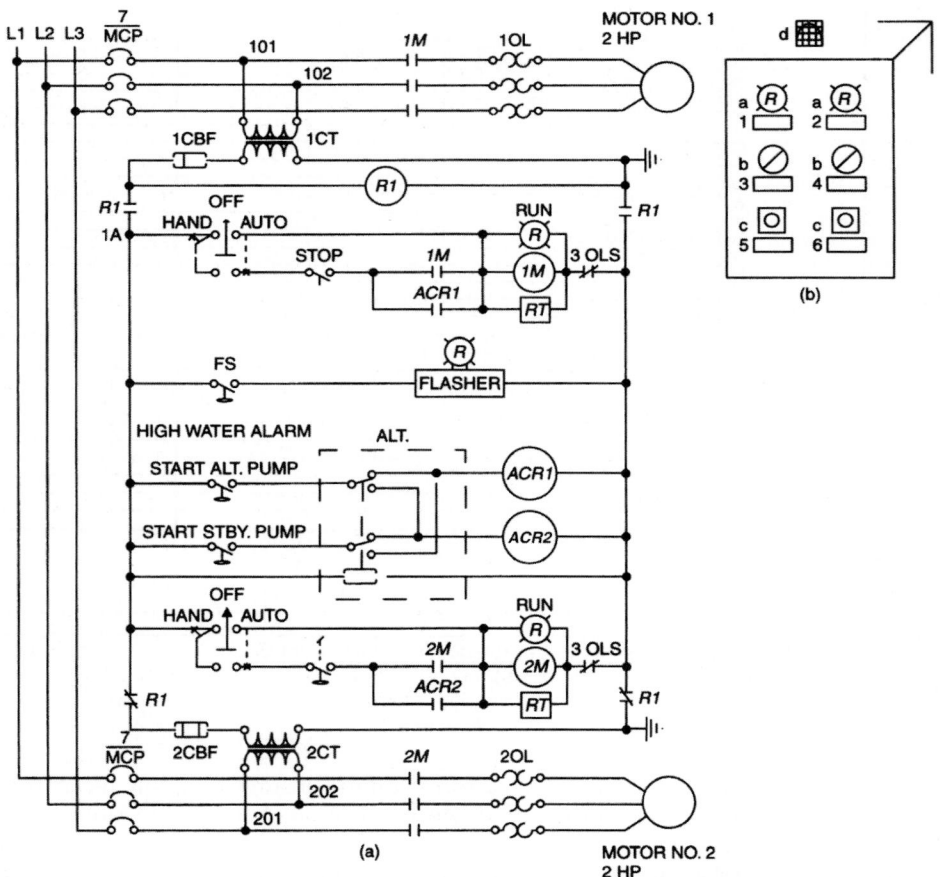

FIGURE 7.9 Typical elementary wiring diagram: (a) duplex sump pump and (b) control panel.

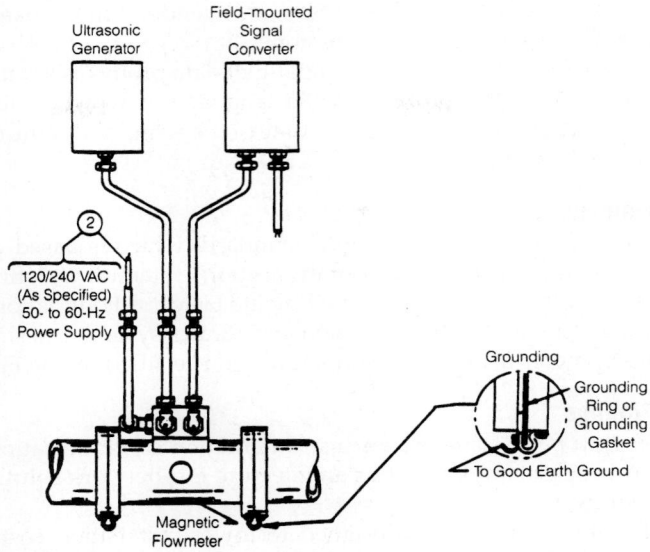

FIGURE 7.10   Typical mounting detail.

transfer data between process controllers, and across facility-wide and city-wide enterprise networks.

Modern instrumentation can provide a large amount of field data to facility operators. Using this data, operators can make better decisions and avoid major catastrophes, by detecting problems early and resolving them quickly. Networking allows control strategies to use any information within the facility. For example, chemical addition to a process in the front end of the facility can be based on information at the effluent, because all data are available at all times. This can result in cost savings of chemicals, electricity, time, and improvement of the controlled process.

4.6   Communication Networks for Instrumentation

The traditional analog 4 to 20 mA signal has been the standard way in which analog data is transmitted to control systems but is limited to one analog value for each device. Communication networks are used to reduce overall costs and obtain access to vast amounts of data that are not available using a single hardwired design, such as a 4 to 20 mA signal. This architecture simplifies design and uses less hardware when networking multiple devices to a single network.

With a communication network, each instrument can bring back a multitude of information that is stored locally at the device and then can be displayed on the human–machine interface (HMI). This can be helpful, not only for analysis of multiple points on an instrument, but also when troubleshooting issues in a large facility.

Most instruments from major manufacturers offer several communication options that are integral to the devices themselves. This technology affects the design and implementation of control systems, including the electrical wiring, the way instruments are connected, and the way control is implemented. Field instruments and other devices, such as valves, analyzers, VSDs, and vendor control panels, can be incorporated to the network for data transmission.

Many different (and usually incompatible) data protocols are used by instrumentation manufacturers. The protocol is the language in which a device communicates information. A discussion that defines some of the terms and some of the most common data protocols follows.

4.6.1   Fieldbus

Fieldbus technology is a digital signal standard, which is based on an open, vendor independent, multi-drop digital communication system. It is defined by the ISA SP50 standards committee and implemented by the Fieldbus Foundation (Austin, Texas).

When designing an instrumentation and control system, a designer should confirm that the instruments are fieldbus compatible and certified by the Fieldbus Foundation.

4.6.2   Profibus

A popular European standard for instrumentation communication is called Profibus (PI, Karlsruhe, Germany). Profibus architecture is a network solution, including hardware and software.

The Profibus system was designed to lower installation costs, while improving access to multiple data points on a single device. Currently, there are three different types, or versions, of Profibus available today—Profibus DP, Profibus PA, and Profibus FMS. Profibus DP is used in high-speed data exchange between field devices; Profibus PA is used in intrinsically safe applications; and Profibus FMS is used for communication between controllers and field devices. For a Profibus system to work, each instrument on the network must support the specified protocol.

4.6.3   DeviceNet

This device level network has become popular for use in controls and automation systems with multiple devices. DeviceNet (Rockwell Automation, Milwaukee, Wisconsin) is a common, open architecture standard. One popular use for DeviceNet is for MCCs. When not using DeviceNet, multiple wires, one for each data point, need to be installed to control and monitor equipment. It becomes costly to bring back a lot of information. Using the DeviceNet architecture can lower installation costs, improve monitoring capabilities for troubleshooting, and simplify the overall design.

4.6.4   Others

Other fieldbus and network type systems include Highway Addressable Remote Transducer (HART) (The HART Communication Foundation), which can be used over existing 4 to 20 mA loop wiring, EtherNet/IP (ODVA, Ann Arbor, Michigan), and Controller Area Network (CAN) (Robert Bosch GMBH, Stuttgart, Germany).

4.7   Control Panels

A typical WRRF has many different kinds of control panels, including vendor supplied panels for local control and custom panels for main control (including operator interface terminals).

4.7.1   Vendor-Supplied Control Panels

Vendor supplied control panels are supplied with major mechanical equipment, such as a belt-filter press, UV systems, and centrifuges. The amount and type of control these panels furnish should be consistent with the overall, facility-wide control philosophy. From a practical standpoint, enough manual control should be furnished so that vendors can prove that their mechanical equipment meets specifications independent of the facility-wide control system. It is good practice to have a means of controlling equipment by hand in case other parts of the system are out of service. This is especially true with systems that rely on networks for control data.

4.7.2 Custom Panels

Custom panels are different than vendor supplied control panels, because they generally are unique to each application and are integrated to the system to control multiple pieces of equipment or larger portions of the process. Custom panels are typically provided by systems integrators to control the input/output (I/O) in certain areas of the facility.

4.7.2.1 Local Control Panels   Typically located on or near an associated piece of equipment, local control panels (LCPs) include hand-off-auto (HOA) switches or local remote (L/R) switches and start/stop pushbuttons. Automatic or remote mode enables remote operation from the control system. Hand or local mode is generally used for maintenance. LCPs often serve as termination panels for interfacing process signals and control with a computer-based control system. The panels are designed so that each type of signal (analog or digital) is separated physically from the other to help eliminate signal interference. A 24-V direct-current (redundant) power supply is furnished for powering two wire 4 to 20 mA analog instrument loops. The panel may contain a small UPS for critical loops, which must remain powered during a power outage.

4.7.2.2 Operator Interface Terminals   Operator Interface Terminals (OITs) depict the process flow most often in graphical form. OITs vary in size and complexity, depending on the information presented. They are similar to an HMI in function but are more limited in capability. They often use a proprietary computer designed for the industrial environment and are usually mounted directly on the control panel. They usually use different operating systems than HMIs and the graphics are generally simpler. OITs can be programmed easily to display a local process or event an entire WRRF using graphical symbols. These displays can be changed quickly and easily as changes are made to the facility or process.

4.7.2.3 Motor Control Center Panels   MCCs are large control panels used to locate motor starters and local control components associated with each motor. These typically are large and can take up an entire wall. An MCC is comprised of several individual control panels or sections, often referred to as buckets. In each section or bucket, an individual motor starter is mounted to a subpanel. The MCC is used to centrally locate all motor starters, transformers, main breakers, and other high-power devices, so that large conductors do not need to run to many locations.

4.8 Supervisory Control and Data Acquisition Systems

At WRRFs, SCADA systems typically are used to control and monitor wastewater collection systems. The type of communication, or telemetry system, may vary from telephone lines, cellular, radio, satellite, or cable broadband. If a radio telemetry system is used, a radio survey should be completed before final design by a reputable firm to verify a good radio path is available.

Once field data is collected, it is sent using the telemetry system to the facility control system for use in control decisions and data recording and reporting. Custom software to interface the telemetry system with the facility control system should be avoided, because it may be difficult to get support from a vendor after the project has been completed.

4.8.1   Human–Machine Interface Systems

HMIs are used to allow the operator a graphical access point to the control system. The term HMI can be used to describe any piece of equipment that allows the operator an interface into the control system; however, this term typically is used to describe a computer system (personal computer) with a graphical interface used to control and monitor an entire facility or process.

4.8.2   Control Systems

The control system in many facilities currently is comprised of microprocessor-based equipment reading information from field devices and making control decisions for the rest of the facility based on that information. These control systems are predominantly PLC based. Careful consideration should be given when designing these systems for WRRFs, so that the end users can achieve maximum performance without excessive expenditures.

4.8.3   Independent Data Acquisition Systems

Data acquisition systems (DAS) focus primarily on data collection without control. These systems gather remote field data for modeling or analysis. These types of systems might be used to collect data, such as flow, rainfall, or pressure, on a routine basis from several lift stations in a specific area. The data can then be used to generate graphs and reports.

4.9   Facility Control Systems

This section defines two different types of digital control systems used today and their application at typical WRRFs. In addition, the control hierarchy is discussed and compared with different system architectures available.

4.9.1   Programmable Logic Controller versus Distributed Control Systems

The distributed control system (DCS) has been used predominantly in batch-type systems, with large amounts of analog data. Industries, such as steel and paper, have used these types of systems for decades. These systems are sophisticated and complex and are made up of controllers, networks, and operator interfaces (HMI devices) as one complete package. Because of its high cost, the wastewater industry has used this technology predominantly at larger facilities.

Programmable Logic Controllers (PCLs) began as standalone process controller with large number of discrete I/O and little networking. With the advancement of technology, the PLC manufacturers have improved their product offerings and capabilities, focusing a lot of attention on high speed control hardware and a multitude of communication options. Over the past several years, the gap between PLC and DCS systems has narrowed, as a result of technology and availability. Today, most WRRFs are designed using PLCs because of the simplicity and cost of this type of system.

4.9.2 Types of Process Controllers
The two basic types of process controllers are PLCs and distributed process controllers (DPCs). The PLC originally was developed to replace relay ladder logic. Because of its low cost, not only can it be used as the main controller in the WRRF, but also as a local controller for sequential control, such as backwashing a filter or controlling a UV system. In the past, PLCs have not had all of the analog control capabilities that DPCs have had. However, today's PLCs are more than capable of handling all types of control algorithms, including proportional integral derivative. Most of the more notable PLC manufacturers support a nonproprietary system architecture, which means that they support many different communication protocols, and they have made their communication protocols available for others to use.

DPCs are microprocessor based, with control logic, memory, and I/O control that are connected to other units or to a central control computer with a data or control highway. These systems are all inclusive systems, with the HMI interface integral to the overall architecture. As a result, they are predominantly proprietary, in that they do not need to interface with other products, because their hardware handles all facets of the system.

4.9.3 Building Blocks of a Facility Control System
A facility control system is composed of numerous basic building blocks, which may include the following:

- Remote terminal units (RTUs)—microcomputer-based RTUs that have the ability to transmit data, accept commands, and replace controls typically performed with relay ladder logic. These units, typically remote from the WRRF, are installed in wastewater lift stations in the collection system.
- PLCs—microprocessor-based controllers.
- Area control centers—control rooms in which an operator is on duty (on a continuous basis or for certain operating shifts only) and is responsible for the control of a certain area or group of facility processes. Equipment typically installed in this area includes PLCs, operator consoles, and certain other peripherals, such as alarms and report printers.
- Central control centers—control rooms in which operations or supervisory personnel are on duty on a continuous basis and have control or oversight responsibility for the entire facility. Equipment typically installed in a central control center includes PLCs, operator workstations with full graphic displays, alarm and report printers, and a programmer workstation for making changes to the system.
- Control network media—physical media connecting devices that provide communication among the different devices connected to control network. Included are cables, modems, processors, switches, and associated software, which provides message handling, fault detection, time synchronization, and arbitration. Physical media used for a data highway can be coaxial, fiber optic, CAT-5e, CAT-6, or other instrumentation cabling.

4.10 Process/Facility-wide Control-System Approaches
Designing a control system for WRRFs to fit the needs of the operators, facility, and budget requires much thought and expertise. Current technology is constantly moving

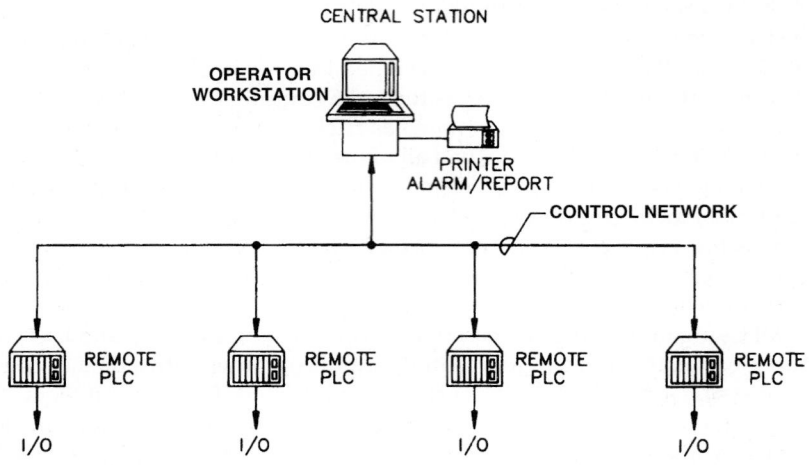

FIGURE 7.11 Small distributed control system (I/O count, 500; CPU 5 central processing unit).

ahead and therefore must be considered during the design phase, to accommodate growth and expansion in the future. Most systems are expanded easily, as long as this is incorporated to the original design. A typical block diagram for a small control system is shown in Figure 7.11. To ensure the successful migration of technology in a WRRF, owners should start with an automation plan for their entire facility before moving ahead with any major facility modifications and/or designs.

Typical facility-wide control systems provide some or all of these functions:

- Field instruments (and instrumentation network);
- PLCs (or DPCs) and remote racks;
- Control panels (with Operator Interface Terminals);
- Control network;
- Control room (and possibly area control centers);
- Operator workstations (usually personal computers running HMI software);
- Historical data server;
- Telemetry system for off-site facilities;
- City-wide network or wide area network equipment (including firewalls) for data sharing and off-site data storage requirements;
- A maintenance management system; and
- Laboratory information system.

4.11 Hazard and Operability Design

During the design phase of any project in a WRRF, consideration should be given to any potentially dangerous situations that may occur as a result of the design. Once the P&IDs are completed, or even during the initial design, a hazard and operability (HAZOP) study should be done to identify any hazards with the design. Chemical systems and electrically classified areas (i.e., explosive atmospheres) are predominantly a common source of potential danger, and proactive design approaches should be done to eliminate, as much as possible, any harm from these systems to the operators working on or around them.

4.12 Process Control Strategies

Process control strategies, or functional descriptions, are narratives that describe how the process is to be controlled. This narrative includes a listing of all I/O signals used by the strategy, any required calculations, and interfaces with other strategies, and defines HMI display requirements. The process control strategies should contain sufficient information to allow the control-system vendor to implement this strategy using its own process control language software program (i.e., ladder logic, function block, and statement list).

4.12.1 Process Control Strategy Narrative

The process control strategy narrative contains sufficient detailed information to completely describe the control requirements in a functional format. Detailed information includes items such as the following:

- Type of control—manual, local automatic, direct digital control, supervisory setpoint, or batch (sequence);
- Hierarchy of control;
- Safety and process interlocks;
- Monitoring requirements;
- Operational requirements;
- Calculations required; and
- Associated I/O signals.

4.12.2 Interaction with Other Control Strategies

Interactions with other strategies are listed in such a way that they are coordinated and that proper information and timing among the strategies are maintained.

4.12.3 Process Graphics

Graphics are often laid out in a hierarchal manner, beginning with an overview of the facility, then the process areas, and, finally, the control elements of the specific process. Typically, process graphics are generated by using a simplified P&ID to represent the process. All live data points then are added, followed by the addition of the key alarm points and any fixed text that aids the operator in controlling the process. Care must be taken to make the operator process displays that are easily understood. The high performance HMI philosophy is one example of how to accomplish that goal.

4.13 Precautions for Instrumentation and Control

The designer should provide a means for facilities to continue operating if instrumentation and control systems fail. This applies particularly to pumping systems or any system where damage could occur as a result of failure. Often, local control systems use remote instrumentation for input and control decisions. If these remote systems fail, it is good practice to have a control scheme that can compensate for the loss of controlling devices. Redundancy of equipment often is used in critical cases, but is not always cost effective. Alternative means, such as default settings and spare parts, can help when equipment fails.

4.13.1 Spare Systems

Having spare parts for important pieces of equipment is a cost effective way of providing system backup. Because of this, it is essential to provide a standard means of design criteria when selecting equipment. Standards should be set for the type of PLC or DCS controller to use as a base for all digital control systems within the facility. This allows the same I/O cards to be used throughout the entire facility, reducing the amount of spare parts required.

Additional standards for computers, timers, relays, controllers, interfaces, and so on should be set, so that spare parts inventories are reduced, and operations staff is familiar with equipment. Once these standards are established, operators should receive adequate training to support this equipment, such that any failure in the system can be corrected by facility personnel.

4.13.2 Redundancy

Providing for total redundancy of all instrumentation and control functions is impractical. Designing for damage to or failure of the central control system can be accomplished by providing a redundant processor (in case of processor or program malfunction) or separate switching at each pump motor site, to ensure that individual units can be operated manually, if necessary. This concept requires advanced emergency training for operators and well thought out sequences of operation in O&M manuals. Whenever redundancy in equipment is used, the cost of the overall system increases, including maintenance of the system. Therefore, designers should evaluate the importance of each portion of the control system, to see if equipment redundancy is warranted or if a spare part on a shelf is sufficient for backup.

In areas where earthquakes or other seismic disturbances are likely, precision equipment, such as residual analyzers, recorders, indicators, meter electronic instrumentation, switch gear, and communications systems, should be mounted rigidly, to avoid amplification of seismic accelerations. This type of equipment is a prime candidate for shake table testing qualification, because analysis generally cannot demonstrate that operating capability will survive shaking. Positive locking devices should be used to hold circuit boards in place. All mechanical switching components (e.g., relays) should be tested for their seismic response characteristics; mercury switches should be avoided. A designer also should exercise caution when using gravity or light duty, spring controlled switches. Relays often respond adequately in the energized position, but may fail in the non-energized position. Therefore, a designer should exercise caution when using friction-restrained switches and components. In addition, a designer should avoid the use of circuit board mounting on standoffs or other devices used to mount circuit boards away from the subpanel, as it may result in local resonance. Additional strengthening, such as welded supports, should also be provided.

A designer should provide communication equipment and critical instrumentation controlling equipment with a dedicated emergency power supply (possibly batteries) and a station standby power supply. Manual overrides for all automatic control systems should also be provided. Critical installations that cannot be designed to withstand seismic motion may be supported on an earthquake compatible, floor vibration isolation system designed to attenuate motion.

4.14 Control System Elements

4.14.1 Instrumentation

This equipment is vital to the data gathering and facility control operations. Instrumentation should be standardized, so that operators are familiar with equipment, allowing devices to be better maintained. This also minimizes the amount of spare parts required for the system.

4.14.2 Monitoring versus Control

Some devices are needed only for monitoring information rather than controlling equipment. Both services are vital, and the accuracy of the instrumentation should be evaluated for desired results. With the increased availability of data storage, all analog points should be monitored and recorded for future data analysis. Critical systems should use redundancy.

4.14.3 Data Systems

With the increased amounts of data being made available to operators, data systems are critical for future analysis of process systems. Also, government authorities are requiring strict compliance and data keeping for reports. Therefore, more attention is required in the historical data gathering and reporting areas of these systems and should be done during the design phase of the project, to avoid under design. Control systems are costly, but can provide a wealth of excellent information to the end user. These systems are useless if they cannot produce proper reporting information and allow the data to be accessible and usable to users of the system. It is good practice to design a WRRF control system with the end in mind.

4.15 Cybersecurity

4.15.1 Background

Cybercrime presents a major threat to the safe and reliable operation of instrumentation and control systems. Cybersecurity threats may arise from:

- Accidental or unintentional actions of an employee or contractor.
- Deliberate actions of a disgruntled employee or contractor.
- Actions from external sources such as vandals, hackers, terrorists, organized crime, or nation states.
- Environmental incidents (hurricanes, earthquakes, tornadoes, floods, fires, etc.).

Possible cybersecurity incidents include:

- Introduction of malware into control-system equipment causing a loss of control or loss of visibility of facility.
- Unauthorized access resulting in deliberate or accidental operation of facility.
- Inability to recover system after equipment failure or environmental incident.

While theft of confidential information from IT systems, such as billing systems, is a major concern, this is usually not so critical with instrumentation and control systems. However, the theft of confidential information about instrumentation and control systems, that could be used to mount an attack, is a concern.

4.15.2 Defense in Depth

The term "defense in depth" is derived from a proven military strategy successfully used for many centuries. In defense in depth, a defender uses multiple protections to provide continued resistance to attackers. In the case of military strategy this might involve using multiple fortifications or different types of protection.

In cybersecurity the defense in depth approach is similar to the military and safety related versions. It identifies all the protections that are in place to reduce the likelihood of a security event, such as malware entering a control system, causing an incident.

As illustrated in Figure 7.12, typical defense in depth elements include:

- Physical security.
- Policies and procedures.
- Network architecture.
- Antivirus and patching.
- System hardening.
- Intrusion detection and prevention.
- Incident response.

4.15.3 Cybersecurity Management Systems

A cybersecurity management system is necessary to mitigate the risk of cybersecurity incidents. A number of methodologies and frameworks exist to support the development of a cybersecurity management system. The NIST Cybersecurity Framework shown in Figure 7.13 was developed in 2013 in response to President

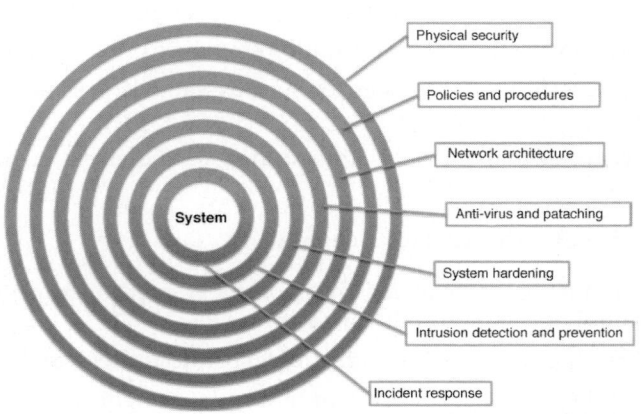

FIGURE 7.12   Defense in depth as applied to cybersecurity.

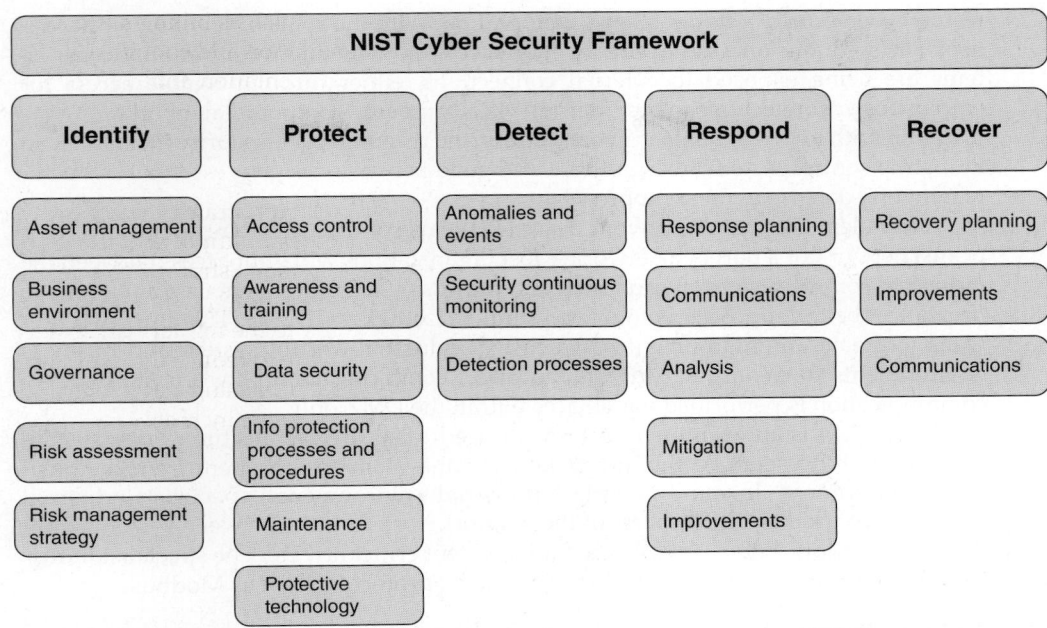

FIGURE 7.13 NIST cybersecurity framework.

Obama's Executive Order 13636. This framework identifies five high-level activities that mission critical organizations must do:

- Identify—create an inventory of all cyber assets and understand what these assets do, what hardware and software is used, how they equipment is connected, what external connections exist, etc.
- Protect—create the defense in depth protections, such as access control policies and procedures, malware protection and patching procedures, and backup and recovery strategies.
- Detect—put in place methods, automatic and manual to detect unusual activity or unauthorized access.
- Respond—establish procedures to deal with cybersecurity events.
- Recover—ensure that procedures are in place before and after cybersecurity events to ensure that it is possible to restore normal operation.

ISA/IEC62443 is the international standard for cybersecurity of industrial automation and control systems. This standard is referenced extensively throughout the NIST Cybersecurity Framework. It provides more details on how organizations should go about implementing their cybersecurity management systems.

4.15.4 Cybersecurity Design Principles

4.15.4.1 Secure Architecture Traditionally, industrial control-system components were deployed in physically secure locations, isolated from the outside world. As a result, cybersecurity was not a key consideration. With the availability of network connectivity solutions and the desire for more access to data and information, these systems are being exposed to external connections, either through remote access for employees or connections to the Internet, via a company's corporate network.

Even with improvements in cybersecurity, the risks of a cybersecurity incident in an industrial control system are significant and good network design is essential as a component of a defense in depth approach.

A key element of good network design is the use of firewalls to restrict traffic to or from a network or a part of a network. Most organizations will limit access between the industrial control-system environment and the business environment using at least one firewall. A better approach is to use a Demilitarized Zone (DMZ). This uses two firewalls to create an environment between the industrial control system and business environments to ensure that there is no direct communication between the two. All communication is performed via devices within the DMZ only.

Figure 7.14 is an example of a firewall used to secure architecture. Firewalls are not used only to separate the industrial control-system environment from the business environment. In more complex industrial environments, specialist industrial firewalls are used to isolate parts of the network, for example, between systems from different vendors, between systems with different criticality, etc. The specialist industrial firewalls are designed to support industrial protocols, such as Modbus.

4.15.4.2   Redundancy   Failure of hardware or software components can result in a major failure of an instrumentation and control system. Cybersecurity threats, such as malware or targeted denial-of-service attacks can cause a failure of components, as can inherent vulnerabilities such as hardware or software flaws.

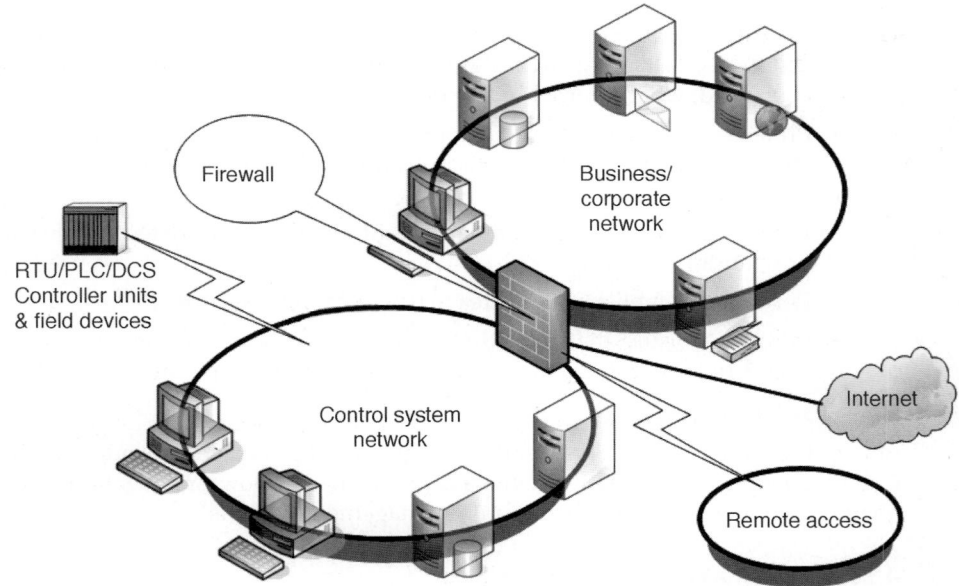

FIGURE 7.14   Secure architecture involves separating the control system network from the business network and Internet.

Where failure of a component cannot be tolerated, redundant design is required. The level of complexity in the redundancy design can vary considerably, but in general the idea is to ensure that in the event of a component failure the system can continue to operate.

4.15.4.3   Data Security   Encryption is a technique for transforming data in such a way that it becomes unreadable so that, even if someone is able to gain access to the data, they won't be able to do anything with it unless they can decrypt it. The data to be encrypted, also called plaintext or cleartext, is transformed using an encryption key, which is a value that is combined with the original data to create the encrypted data, also called ciphertext. The same encryption key is used at the receiving end to decrypt the ciphertext and obtain the original cleartext.

Data can be encrypted "at rest," that is, where it is stored and also "in transit," that is, when it is being transmitted.

In addition to helping ensure that the wrong people don't read data, encryption can also ensure that data isn't altered in transit, and verify the identity of the sender.

4.15.4.4   Remote Access   Poorly designed or controlled remote access to instrumentation and control systems can provide a major security vulnerability that can be exploited by unauthorized individuals. Authorized individuals, employees, or contractors, can also create problems that can result in system failure if not properly managed. Key objectives in providing secure remote access are as follows:

- All non-essential remote access connections to systems are removed.
- All essential connections to systems are documented and approved.
- All essential connections are secured in accordance with standards and recommended practices.
- Instrumentation and control networks are segregated from corporate and external environments using firewalls and DMZ areas.
- Remote access is limited to authorized and competent persons carrying out specific approved tasks.

4.15.5   Cyber Hygiene

The term cyber hygiene covers the day-to-day tasks that are necessary to ensure the basic elements of the defense in depth approach are in place. The key tasks are as follows:

- Access control
- Malware prevention and patching
- System hardening
- Removable media control
- Backup and recovery

4.15.5.1   Access Control   Effective access control ensures that

- All manufacturer default passwords are changed before equipment is operational.
- Passwords are kept confidential, are not shared, and are not posted for all to see.

- Passwords are changed regularly and are 'strong,' that is, they are not easily cracked.
- Any user who has changed role or has left the organization has had their access removed.

4.15.5.2 Malware Prevention and Patching   Malware is constantly changing and as a result it is difficult to prevent any possible malware attack. It is essential that

- Equipment firmware and operating system and application software is kept up to date with the latest security updates and malware detection signatures.
- The likelihood of malware being introduced into systems is minimized by carefully managing removable media access.

4.15.5.3 System Hardening   System hardening ensures that equipment has all unnecessary programs and services removed or disabled so that the likelihood of using a vulnerable program or service is reduced. The status of equipment should be carefully monitored to ensure that programs are not installed or services are not enabled unless they are necessary for the operation of the system and their use has been approved.

4.15.5.4 Removable Media Control   Removable media, such as USB drives, is a major source of malware in industrial control system environments as these are often disconnected from the Internet, the primary source for this malware. Removable devices are designed to transfer material from machine to machine and unwanted material can also be inadvertently transferred in the process. This is especially true when personal devices are used, as these may be connected to home machines with limited or no antivirus protection.

Removable media is not limited to USB drives. A CD or DVD can also contain malware. Smart phones and tablets that are charged using a USB port are also a threat as these devices contain drives that can be infected with malware.

Efforts should be made to minimize the need to use removable media by implementing alternative secure communications paths for transferring data and files. However, where removable media must be used, it is essential that:

- Access to USB ports and machine drives is restricted. Port locks can be applied which, although not impervious to a determined intrusion, can deter the casual user from inserting a device.
- Protected removable media, that is, secured when not in use and is reformatted and virus-checked before each use, is used.
- Incoming media is scanned for viruses and the contents are transferred to protected removable media before being used.
- Auto-play for USB and DVD/CD drives is disabled.

4.15.5.5 Backup and Recovery   In the event of a security incident it is essential that systems be restored to known working order as soon as possible. Backups play a key part in this process. It is essential that:

- Backups are taken at a frequency commensurate with changes in the system, so that restoration does not lose a significant amount of changes.
- Backups are logged, labeled, and stored in a readily accessible location, with copies in alternate locations in the event of a major disaster.

- Backups are tested regularly to ensure that they will work in the event they are required.

4.15.6   Detection and Prevention Systems

An Intrusion Detection System, or IDS, monitors networks or devices for malicious activity. Networks are monitored by NIDS, or Network-based Intrusion Detection Systems and devices are monitored by HIDS, or Host-based Intrusion Detection Systems.

An IDS uses signatures, similar to those used by antivirus software, to detect known attacks. Like antivirus, effective IDS protection requires regular updating of signatures. Also like antivirus, only known attacks can be detected using these signatures. However, the IDS has a normal baseline for the network or device that it can use to compare current activity against and detect new, unknown attacks. However, in this mode of operation false positives are more likely.

As illustrated in Figure 7.15, an Intrusion Prevention System, or IPS, works with an IDS to block malicious activity when it is detected. IPS must be used with extreme caution, and only in limited circumstances, in industrial control-system environments. Functionality and performance can be severely impacted if critical traffic is blocked.

Care must be taken when deploying IDS and IPS technology in an instrumentation and control environment. The IDS can generate significant additional traffic that may affect the operation of other networked devices. In addition, the operation of instrumentation and control equipment as it is not well understood by these tools and so false positives and associated inhibition of functionality can occur.

4.15.7   Standards, Guidance, and Tools

4.15.7.1   Standards and Guidance

- ISA/IEC62443 Cybersecurity of Industrial Control Systems.
- ANSI/AWWA G430-09: Security Practices for Operations and Management, (2009)

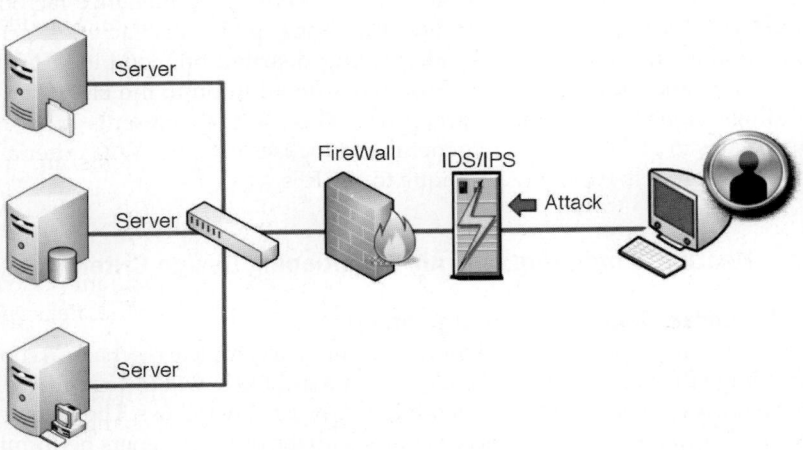

FIGURE 7.15 IDS and IPS monitoring and protecting against security attacks.

- ANSI/ASCE/EWRI 56-10 Guidelines for the physical security of water utilities (57-10 is for wastewater/stormwater)
- ASCE/AWWA/WEF Interim Voluntary Security Guidance for Wastewater/ Stormwater Utilities (2004)
- Roadmap to Secure Control Systems in the Water Sector (especially for their past control system hacks)
- NIST, SP 800-82: Guide to Industrial Control Systems (ICS) Security, June 2011
- Recommended Practice: Improving Industrial Control Systems Cybersecurity with Defense-In- Depth Strategies. Department of Homeland Security, October 2009.

4.15.7.2   Assessment Tools

- Risk Assessment and Management for Critical Asset Protection (RAMCAP) – ANSI J100
- National Association of Clean Water Agencies (NACWA) Vulnerability Self-Assessment Tool (VSAT) for water and wastewater utilities, 2010
- Cyber Security Evaluation Tool (CSET) developed per NIST 800-82

5.0   Heating, Ventilation, and Air Conditioning

Heating, ventilating, and air-conditioning (HVAC) systems for buildings and other occupied areas are important support systems in the design of WRRFs, as their primary function is to provide a comfortable and safe working environment for facility staff and operations. This section provides design team members some of the HVAC-system-design criteria and equipment commonly applied and used in WRRFs. Topics covered include specific design criteria, system-design constraints, HVAC-system descriptions, and basic design considerations for energy conservation and sustainable design.

The variety of HVAC systems incorporated to WRRFs is quite extensive. They often include mechanical systems for laboratories, vehicle-maintenance facilities, electrical rooms and buildings, administration buildings, personnel facilities, central heating (and cooling) facilities, facility-wide-heating distribution, combined heat and power generation, and other special facilities types, in addition to the HVAC systems for the buildings housing treatment processes. This section presents basic criteria and approaches to HVAC design in general and focuses on the HVAC criteria and bases of design for process-type HVAC unique to WRRFs.

5.1   Heating, Ventilating, and Air-Conditioning Design Criteria

5.1.1   Codes, Standards, and Regulations

Each project work plan must include, at the outset, a complete mechanical code search. The prevailing HVAC or mechanical codes for a project likely will include many local, county, and state codes or amendments, in addition to the list that follows. The mechanical designer needs to coordinate his or her search efforts with the other designers performing the building, fire, and life-safety code searches, to ensure that the HVAC design incorporates requirements resulting from other disciplines' codes searches. Following are the associations, agencies, societies, and codes that typically govern the design of HVAC systems.

- Air Movement and Control Association International (Arlington Heights, Illinois)
- American National Standards Institute (Washington, D.C.) (ANSI)
- Air-Conditioning and Refrigeration Institute (Arlington, Virginia)
- American Society of Heating, Refrigerating and Air Conditioning Engineers (Atlanta, Georgia) (ASHRAE)
  - 90.1 Energy Standard for Buildings Except Low-Rise Residential —Standard 90.1
  - 55.1 2013 Thermal Environmental Conditions for Human Occupancy—this standard
  - 62.1 2016 Ventilation for Acceptable Indoor Air Quality—compliance with this
  - 52.2 2012 Method of Testing General Ventilation Air-Cleaning Devices for Removal Efficiency by Particle Size
- International Energy Conservation Code
- International Mechanical Code
- International Organization for Standardization (Geneva, Switzerland).
- NEC
- National Electrical Manufacturers Association (Rosslyn, Virginia) (NEMA)
- Montreal Protocol on Substances that Deplete the Ozone Layer
- NFPA—has many codes that prescribe HVAC design features. The most significant of these are the following:
  - NFPA 820
  - NFPA 90
- OSHA
- Sheet Metal and Air-Conditioning Contractors' National Association (Chantilly, Virginia) (SMACNA)
- Underwriters Laboratories
- Great Lakes – Upper Mississippi River Board–10 States Standards for Wastewater 2014

5.1.2   Climate
Climatic data chosen for design calculations have a dramatic effect on the capital costs of HVAC systems and must be selected carefully.

5.1.3   Climatic Data Sources
Climatic data for degree days can be taken from Climatography of the United States No. 81, Supplement No. 2, 1971–2000, from the National Climatic Data Center (Asheville, North Carolina) (NCDC, 2002). Prevailing wind data are available from the NCDC in a document titled "Climatic Wind Data for the U.S." (NCDC, 1998). Design dry- and wet-bulb temperatures are available in the 2013 ASHRAE Handbook of Fundamentals, climatic data supplement.

5.1.4 Design-Temperature Frequency Levels

The HVAC designer does not typically design for the lowest winter and highest summer temperatures of record as this would impose considerable penalties on HVAC heating and cooling capital costs and reduce energy efficiencies through the resulting excess capacities. The ASHRAE data for both heating and cooling have listings for various frequency levels; 99% and 97.5% are typical, meaning that, in the latter case, 2.5% (or 219) of the winter hours would be below the design value cited. Many energy codes stipulate the frequency level to be used. Important exceptions to this frequency-level selection are cooling systems for critical electronic equipment (e.g., VFDs) that must perform during the highest expected outdoor air temperature. As a minimum, the project climate data should include the following data (and their frequency levels): heating design temperature, cooling design temperature, and cooling wet-bulb temperature.

5.1.5 Space Environmental Requirements

Space-temperature and humidity-control settings, in conjunction with ventilation quantities supplied, are prime determinates of HVAC-system capacities (and, hence, capital costs) and annual energy costs. For many space types, these criteria are stipulated by the provisions of local and national energy and mechanical codes. Nonetheless, there is some latitude in their selection, and they should be agreed on by the owner and other stakeholders.

Early (and ongoing through project's end) tabulation of environmental requirements for all HVAC-conditioned spaces is recommended and should be an integral part of the HVAC project manual. The tabulation should include, for each space description, the following:

- Heating temperatures;
- Cooling temperatures;
- Mechanical or ambient cooling choice;
- Ventilation requirements—supply, exhaust, return, and odor-control-system air;
- Hazard classification used as the design basis; and
- Installation Environment (wet, damp, corrosive, etc.).

Ventilation values for process areas should be obtained from the tables contained in the referenced NFPA 820 standard based on decisions made by the design team. For example, increased ventilation rates may allow for process area electrical declassification, allowing for the use of general purpose equipment, while decreased ventilation may result in the use and requirement of electrically classified equipment. Specific values for non-process areas will be taken from the more stringent of ASHRAE 55.1, ASHRAE 62.1, and the applicable project mechanical codes.

5.1.6 Heating, Ventilating, and Air-Conditioning Utilities

Determining the energy sources for space heating and cooling merits early investigation and resolution. The options include the following:

- Existing central heating (or cooling) facility. If there is existing central facility hot water (or chilled water), this should be the first option. The designer, as soon as is practicable, needs to make estimates of additional loads and compare them with existing excess facility thermal and distribution capacities. If central facility expansion is required, this should be determined at the outset, and the full design and cost effects should be estimated and compared with other options.

- Natural gas heating. Most WRRF process spaces require extraordinary ventilation rates compared with administrative and personnel-type spaces. Energy costs for gas heating typically are one-half that of electric heating. However, the selection and location of gas-fired heating equipment requires care be taken to meet the provisions of NFPA 820 and other codes and standards that govern systems serving hazardous areas.

- Electric heating. Small or remote spaces, with low or intermittent heating loads, may be served economically with electric heat. This is especially true for highly corrosive or explosive areas, as it is easier to specify electric-heating equipment suited to these duties.

5.1.7   Heating Loads

Design criteria for heating loads are based on guidelines established by ASHRAE. Outside-air-temperature design parameters and heat-loss calculations for structures are presented in the ASHRAE Handbook of Fundamentals (2013). Table 7.4 presents some of the commonly used values for heating-design temperatures for types of spaces typically found at WRRFs. Again, this table is for guidance, and the more stringent of the ASHRAE standards and governing mechanical codes will be used to determine the final design values.

The calculation of most process-area-heating loads is straightforward and can be performed using spreadsheet-based calculations that adhere to ASHRAE guidelines. This is especially true for areas requiring high ventilation rates. In these areas, the ventilation load often comprises 80% or more of the total heating load, the calculation of which is simply the product of the outside/inside temperature difference, the ventilation cubic feet per minute, and a constant. An exception to accepting manual-type heating-load calculations is if facility annual-energy simulation is part of the project work scope, for example, for a LEED or other compliance/certification process requiring annual-energy simulation. Many computer programs used to perform HVAC design calculations have the capability of also simulating annual energy usage.

5.1.8   Cooling Loads

Cooling-load calculations are considerably more involved and complex than those for heating. There are both sensible (temperature) and latent (humidity) cooling components. The thermodynamics of the loads themselves—for people, equipment, and solar-heat gains—are quite involved. Computer calculations of cooling loads are recommended for all but the simplest of cooling systems (e.g., small, stand-alone electrical rooms). Some HVAC equipment vendors offer HVAC-design-calculation computer programs of reasonable cost and manageable operating complexity. These programs have the added benefit of simulating energy usage with minimal additional data input. One of the better references on cooling-load calculations is the ASHRAE Handbook of Fundamentals (2013).

5.2   Design Constraints and Heating, Ventilating, and Air-Conditioning Alternative Selection Criteria

The HVAC designer first needs to satisfy the hard constraints on his or her design, such as ventilation air required by the code. Then, a consistent set of criteria should be applied to guide the development of system alternatives and their selection for

Space	Ventilation Rate	Heating Temperature Range, °C (°F)	Cooling Temperature Range, °C (°F)	Remarks
Wet well	12/6 ACH[a]	Typically no or low requirement		Less ventilation required if wet well is a confined space.
Dry well	6 ACH[b]	17 to 20 (63 to 68)	40 (104)	With proper air monitoring, some designers use return air. If raw wastewater spill can be expected, consider 12-ACH capability.
Grit-removal area	12/6 ACH[a]		40 (104)	These are a sample of the extensive listing of hazardous areas for which detailed ventilation and fire protection data are provided in NFPA 820. Refer to that standard for design parameters.
Screenings room	12/6 ACH[a]		40 (104)	
Digester-gas-control room	12/6 ACH[a]		40 (104)	
Digester-gas-compressor room	12/6 ACH[a]		40 (104)	
Enclosed grit-truck-loading area	12/6 ACH[a]	13 to 20 (55 to 68)	40 (104)	
Enclosed primary sedimentation tanks	12/6 ACH[a]	13 to 20 (55 to 68)	40 (104)	
Scum concentration tanks	12/6 ACH[a]	13 to 20 (55 to 68)	40 (104)	
Chlorine and sulfur dioxide rooms	12/6 ACH[a]	20 to 21 (68 to 70)	40 (104)	
Pump and blower rooms	6 ACH[b]	17 to 20 (63 to 68)	40 (104)	Return air may be used to conserve heating energy. Use 2-ACH outdoor-air minimum.
Filter room and dewatering areas	12/6 ACH[a]	17 to 20 (63 to 68)	40 (104)	These areas typically are not hazardous, but high odor levels often dictate 12 ACH or higher.
Garage	No requirements to 1.5 cfm/sq ft[c]	No requirement or 4.4 to 17 (40 to 62)	40 (104)	Garage ventilation dependent on activity, typical parking = 75 cfm/sq ft; maintenance with engines running = 1.0 to 1.5 cfm/sq ft.

TABLE 7.4 Suggested Design Parameters for HVAC Systems in WRRFs

Space	Ventilation Rate	Heating Temperature Range, °C (°F)	Cooling Temperature Range, °C (°F)	Remarks
Repair shops/ workshops	Varies	Varies		Prevailing codes and ASHRAE dictate ACH. Special requirements vary widely.
Electronics workshops	5 to 25	22 (72)	24 to 26 (75 to 78)	Humidity control (<60%) typically required.
Paint rooms	As required for paint booths	20 to 22 (68 to 72)		Exhaust air often needs to be filters.
Battery rooms	Approximately 1 cfm/sq ft	18 to 20 (65 to 68)	40 (104)	Varies by applicable code and especially by battery type.
Welding areas	12 ACH			12 ACH with welding, standard code ventilation when welding operations are off.
Locker rooms	Approximately 1 cfm/sq ft	20 to 22 (68 to 72)	24 to 28 (76 to 82)	Locker and restroom requirements vary widely, some are per sq feet, others per fixture.
Restrooms	Approximately 1 cfm/sq ft	20 to 22 (68 to 72)		
Offices: conference, control rooms	5 to 25 cfm/Occ	21 to 22 (70 to 72)	24 to 26 (76 to 78)	Many codes have additional lunch, and requirement prescribing a minimum percentage of outdoor air.
Laboratories	9 to 12 ACH[d]	22 (72)	24 (76)	Minimum recommended is 10 ACH; may be reduced during unoccupied periods. Humidity control a requirement with any electronic test equipment.

[a]ACH = air changes per hour. For hazardous areas, NFPA 820 requires both powered supply and exhaust, a "secure" source of fan power, monitoring of combustible gasses linked to fan-speed control and alarms. These systems may be operated at 6 ACH to save fan and heating energy when outdoor air is less than 10°C (50°F), the space is unoccupied and combustible gas monitoring is not in alarm. NFPA 820 also prescribes ventilation and control features to ensure that more hazardous areas stay at a lower pressure relative to less hazardous spaces.

[b]Some non-hazardous areas are eligible to use return air only if the strict qualifications of NFPA 820 for its use are met.

[c]NFPA 820 ventilation quantities typically are given in air changes per hour. Other governing ventilation standards and codes prescribe cubic feet per minute per square foot, cubic feet per minute per occupant (Occ), cubic feet per minute per fixture or locker, or percent of outdoor air supplied.

[d]Pressurization of various laboratory spaces (especially those handling hazardous substances), with respect to less-hazardous and non-hazardous spaces, is critical and needs addressing in both the architectural and HVAC-ventilation designs.

TABLE 7.4 Suggested Design Parameters for HVAC Systems in WRRFs (*Continued*)

incorporation to the facility design. Whether the system is serving a screenings facility or laboratory, the criteria discussed in the following sections should be considered.

5.2.1   Hazard Assessment

The design consideration that most sets WRRF HVAC design apart from other facility types is the presence of explosive gasses. The NFPA 820, referenced in the Codes, Standards, and Regulations section, sets forth strict guidelines regarding the amount of air to be supplied and exhausted from hazardous areas. It is essential for the designer to comply strictly with these requirements, to avoid liability and protect the facility occupants. Another effect of hazard assessment is that the HVAC designer must work closely with the electrical engineer, to ensure that the electrical characteristics of the HVAC equipment and associated wiring specified meet the NEC classifications required by NFPA and the NEC.

5.2.2   Corrosion Resistance

Another challenging aspect of WRRF HVAC design is the corrosive environment in which many HVAC systems operate. Chapter 8, Materials of Construction and Corrosion Control provides a detailed description of corrosion concerns in the wastewater treatment environment and provides guidance and considerations for including in the design and operation of facilities.

5.2.3   System Redundancy

The NFPA 820 requires system redundancy for ventilation systems serving hazardous areas. Inclusion of both exhaust and supply fans is stipulated by NFPA 820 in the theory that, should one or the other fail, proper air flow still will be delivered to the hazardous area. The HVAC designer should consider the need for redundancy in all spaces served. For example, there should be a redundant cooling source for VFDs that serve process-critical pumps. Central heating facilities need a standby boiler and pumping capacity. The HVAC designer weighs the potential loss of operation and the effect of partial loss of space conditioning against the increased capital costs of adding system redundancy. The designer needs to review these effects with the project manager and owner before selecting the redundancy solutions.

5.2.4   Capital, Energy, Operating, and Maintenance Costs

WRRF HVAC systems typically are designed to have a longer service life than typical commercial HVAC systems. Nonetheless, the designer needs to apply appropriate technology and attempt to meet design constraints in a cost-effective manner. Many alternatives, when developed early enough in the design process, may be considered, to reduce system costs. Questions to consider include the following:

- Can HVAC equipment be roof-mounted?
- Is it possible to consolidate multiple units?
- Can unit heaters and exhaust fans suffice in lieu of a central heating and ventilating unit with distribution ductwork?

On the other hand, with today's emphasis on energy conservation, the need to evaluate additional system-capital costs in pursuit of energy savings with a "reasonable" return on the investment is an equally important and necessary selection criterion.

Another economic factor is the cost of maintaining and operating a particular alternative. Many alternatives generate a reasonable return on their additional cost through energy savings, but their O&M costs can dilute their overall return on investment, to the point of being economically unattractive. It is contingent upon the HVAC designer to develop a full understanding of the owner's preferences, maintenance staff capabilities, and staffing levels, and use this information to assess the economics of more complex alternatives. In many instances, the simpler alternatives will be economically superior and less troublesome to the owner. The owner should be fully apprised of the alternatives' operating cost effects and tradeoffs before the HVAC system's design progresses through the design-development phase of the project.

5.2.5   Energy and Sustainable-Design Initiatives

All except a few states have enacted energy-conservation codes—most based on ASHRAE 90.1. It is contingent upon the HVAC designer to review the provisions of applicable energy codes early on in the design process. Some jurisdictions will exempt HVAC systems serving process areas, and some will not. The designer can assume that all personnel-type facilities will have to comply. Compliance is far easier to design in, than to add on when the project is going for plan or agency review.

Some jurisdictions have promulgated LEED certification, which applies to occupied buildings, for their municipal projects. The LEED certification process includes a prerequisite for minimum energy performance. Meeting this requirement, which, if not met, disqualifies the project from any certification, can be challenging and needs to commence early in the design process.

The HVAC designer needs to be proactive with the other disciplines in evaluation of energy-code compliance. A significant factor in energy-code compliance is performance of the building envelope. The designer needs to collaborate with the architect early, to determine whether the walls, windows, roofs, and other exterior closure features will comply with applicable energy codes and standards.

5.2.6   Heating, Ventilating, and Air-Conditioning Space Requirements and Discipline Coordination

Discipline coordination is one of the keys to achieve successful, easily constructed, and economical-to-operate HVAC systems. A factor here is the experience of the designer; however, there is some rigor that can be applied to the process.

- Is air-handling equipment in a location where there is sufficient wall space for (often large) outdoor intake louvers?
- Is there sufficient space around the equipment for major overhaul tasks, such as coil removal?
- Has the structural engineer provided sufficient bearing capacity for heavy equipment?

Table 7.5 lists some common coordination data that the HVAC designer provides to other disciplines and summarizes data he or she needs from them. These items are meant to be guidelines for use by all the design disciplines in assessing HVAC coordination.

5.2.7   Acoustics

WRRFs are being located more frequently in more densely inhabited areas or are one-time green-field sites now being encroached upon by residential neighborhoods. The HVAC

HVAC Design Data Required of Other Disciplines and Developed During Initial Design	HVAC Design Data Developed for Other Disciplines During Initial Design
Space data • Name/usage • Area • Volume • Equipment heat release — Maximum — Average/typical • Hazardous exhausts	Equipment rooms • Sizes • Locations
General system type and application • Central, or • Unitary	Preliminary HVAC equipment data • Weights • Vibration potentials • Sizes • Electrical inputs • Control interfaces
Environmental requirements • Ambient or mechanical cooling • Design temperatures and humidities • Redundancy needs • Acoustic criteria and constraints • Corrosion potentials	Building openings: Roofs, exterior walls (especially supply and exhaust louvers), interior walls and others • Sizes • Locations
Utilities available and costs • Electricity • Natural gas • Water	Major duct and HVAC piping alignments and sizes
Climatic data	Special exhausts and hazards Outline specifications Preliminary cost opinion

TABLE 7.5   Heating Ventilating, and Air Conditioning Preliminary Design Coordination Data Summary

designer needs to apply knowledge of fans and fan acoustics, or work with an acoustic consultant, to determine whether the supply, exhaust, odor-control, and air-conditioning equipment will comply with local noise ordinances. As with all the other design constraints discussed here, this analysis needs to commence early in the design process.

5.2.8   Future Expansion

The HVAC designer needs to consider the effects of expected future expansion on both central and unitary equipment sizing and location. The economics and other effects of providing no future expansion in the current design, providing space only, or providing some or all of the future system capacity during the current design need to be weighed.

5.2.9 Economic and Other Weighting Criteria

Ideally, the HVAC designer will document his or her system alternative selections based on economic (typically some form of life-cycle cost evaluation) or overarching project or regulatory requirements.

5.3 Heating, Ventilating, and Air-Conditioning Systems

5.3.1 Hydronic

Larger projects often use central hot water (or steam) heating. Steam has an advantage in lower heating-distribution costs, but is falling from favor, as a result of its more stringent water treatment requirements and the decreasing population of personnel experienced with its proper O&M.

Central hot water heating can justify its additional capital expenditure (compared with building-by-building unitary systems) through higher efficiency, consolidated maintenance (with a central boiler facility), ability to incorporate the heat from renewable energy sources (e.g., digester gas), and inherent safety in classified areas (compared with gas-fired equipment). Hot-water heating affords the ability to handle small-load areas by piping in a unit heater or heaters.

Care is needed in establishing the configuration of a central hot-water-distribution system. Adequate flow and head must be provided to satisfy the loads at the extremities of the system, without building in unused excess capacity, which accrues unnecessary distribution energy. Insulation of especially long distribution runs needs to be sufficient to avoid parasitic thermal losses. An alternative distribution configuration to be considered is a primary–secondary pumping system.

In a primary–secondary pumping system, there is a primary loop that is routed near the major heating loads to be served. It can be considered a "racetrack" of heating piping, to which the boilers, engine-heat recovery, process loads, and building loads are connected. The connections are in the form of two adjacent tees in the primary line. A portion of the primary flow is taken from the upstream tee, pumped to the heat source or heat load, and returned to the downstream tee. This secondary connection may contain a three-way control valve to adjust the temperature of the secondary water.

This system allows a new load (or heat source) to be tapped into the primary at any time in the future, without any need for modifying the primary pumping systems and heat sources. Variable-frequency-driven primary pumps typically are used in this application. Because the total temperature drop in the primary can be sized much higher (22°C [40°F] is not uncommon), the flow requirements can be one-half that of a direct-return system. Considerable piping, pump, and energy costs are the result.

5.3.2 Air Distribution

Two types of ventilation systems used in WRRFs are supply systems and exhaust systems. Supply systems typically are used for comfort and to replace air exhausted from the space. Exhaust systems with associated outdoor intake louvers and motorized dampers can be used for ambient cooling (in lieu of mechanical cooling) of some spaces, such as electrical rooms that do not contain sensitive electronic equipment. This is discussed in the Unitary Heating and Cooling section below.

The primary use of exhaust systems in process areas is removal of hazardous or odorous gases. The location of the supply and exhaust registers and diffusers need coordination to avoid short-circuiting of supply air to the exhaust registers. The general concept is to locate supply outlets and exhaust registers in a pattern that "sweeps" air across the entire area served.

The NFPA 820 stipulates that hazardous areas be served by both powered-supply and powered-exhaust systems. The rationale is that specified ventilation rates, essential for occupant safety and maintaining electrical classifications, will continue, even if one of the systems is out of service. The NFPA 820 also permits the reduction of supply/exhaust-air flow to 6 air changes/hour (ACH), in lieu of 12 ACH, when the outside air temperature is less than 10°C (50°F), and the space is unoccupied. This control scheme saves considerable heating energy, especially in cooler climates. The NFPA 820 also stipulates that all hazardous areas be provided with systems using 100% outdoor air; all air supplied to the space is removed completely by the exhaust system.

Hazardous areas typically are provided with 12 ACH of supply and exhaust air. For larger process areas, this leads to extraordinary air quantities, for which intake and exhaust louvers must be provided. Care is needed in the location of these louvers, to avoid short-circuiting from the exhaust louvers to the intake louvers. The referenced mechanical codes require at least 4.6 m (15 ft) of separation between exhaust and intake louvers, and, depending on the building shape, a larger separation often is warranted. Outdoor air-intake locations also need to be selected that avoid contaminated air sources, such as tank vents and truck exhausts near unloading stations.

Energy codes and standards have an effect on the design of ventilation systems, especially ductwork design and fan selection. Many energy codes penalize higher-pressure ductwork systems and fans of average efficiency. This often means that, to comply with the code, the ductwork and fans need to be physically larger. Because duct routing and fan-equipment-room space is often at a premium, these compliance effects need to be evaluated early in the design process.

5.3.3 Central Heating and Cooling

The advantages of central hot-water heating, which may offset its additional capital cost, are discussed above. One of the challenges facing the HVAC designer is the WRRF that has grown in phases, each with its own boiler system. The opportunity exists, in a facility upgrade of sufficient scope, to consolidate the boiler facility function into one location, especially if some of the existing equipment is at the end of its useful life. The heating loads served by the boilers to be replaced are connected to a primary heating loop connected to the new boilers. Such a primary loop has the added benefit of easily adding the use of the rejected heat from a combined heat and power installation anywhere in the system and allowing the use of digester gas consumed in a boiler to serve any facility-heating load.

There are few WRRFs served by central chiller facilities connected to multiple buildings. The economics of a central chiller facility versus unitary cooling and small local chillers is such that the total cooling load needs to be in the hundreds of tons for its greater efficiency to justify the additional expense. On a smaller scale, rising energy costs have increased the number of chillers that are using digester heat as the energy source for cooling. These typically are absorption chillers using heat from a digester-gas-fired boiler as the energy source to generate chilled water.

5.3.4 Unitary Heating and Cooling

By far the most common types of HVAC equipment used are unitary (stand-alone) heating and cooling units. Small, non-hazardous buildings, with defined ventilation requirements or consistent occupancy, use indirect gas-fired heating and ventilating units. Hazardous areas can use these systems for heating and ventilating, if care is taken to provide sufficient separation between the unit and the building served (typically 3 m [10 ft] from any building opening). Smaller nonhazardous process areas having only sporadic occupancy often are best served by indirect gas-fired unit heaters and rooftop (or through-the-wall) exhaust fans for summer cooling. Electric unit heaters may be considered for serving small spaces that do not have a constant ventilation load, can have significantly reduced heating temperature set-points, or both.

The last few years have seen an explosive growth in the application of VFDs, SCADA panels, and other electronic equipment susceptible to high temperature failure. Electrical equipment, such as VFDs, typically is rated for either 40 or 50°C (104 or 120°F), with 40°C being the typical selection because of reduced size and cost. It has been common practice for many years to cool electrical rooms using outdoor air. Air is supplied in sufficient quantities to produce a temperature rise of, for example, 6.7°C (12°F) above the temperature of the supply air, which, in this case, is the temperature of the outdoor air. The problem with this approach is that newer electronics, such as VFDs, will not tolerate temperature excursions above 40°C (104°F). This means that, when outdoor air rises above 33°C (92°F) for any length of time, the VFD may trip out on high temperature.

In these applications, mechanical cooling should be considered a requirement, as should the installation of redundant cooling systems. Mechanical cooling has the added benefit and also should be incorporated when the area to be cooled is located in an environment of potentially corrosive outdoor air. Locations adjacent to open tankage or with high ambient concentrations of hydrogen sulfide should not use 100% outdoor air for cooling. Hydrogen sulfide, in particular, has a destructive effect on electrical equipment, such as copper wiring, electronic circuit boards, and motor-control contacts.

The HVAC designer can coordinate with the electrical engineer to attempt to locate less temperature-sensitive equipment (e.g., transformers) in spaces that are ambiently (cooled with outdoor air), not mechanically, cooled. The designer also must review, with the electrical and process engineers, the need for cooling redundancy. Rooftop air-conditioning units or split-cooling units, with an air handler inside the electrical room and an air-cooled condenser at grade or on the roof, are appropriate unitary systems for these applications.

5.3.5 Evaporative Cooling

Many areas of the southwest United States experience summer temperatures in excess of 38°C (100°F) and require some form of cooling for operating personnel. Because of the low humidity and low coincident wet-bulb temperatures, evaporative-cooling units typically are used to provide cooling of many process and some personnel areas, in lieu of mechanical cooling. Evaporative coolers operate through one of two processes or a combination of the two—direct cooling, where water flows over a porous media located in the air stream of an air-handling unit, or indirect cooling, where water flows through a heat exchanger (e.g., a coil) located in the air stream. In the latter, the water leaving the coil is circulated through a small cooling tower to lower its temperature and increase the cooling effect.

In areas with low internal heat gain, such as a screen room, a direct evaporative cooling unit typically will suffice. In areas with high heat gain, such as a blower room, evaporative coolers that combine both indirect and direct sections into one unit often are required. Electrical room cooling can be accomplished using evaporative coolers, but the more standard practice is to use mechanical cooling to avoid the potentially high humidity that can result from the evaporative cooling process.

The HVAC designer needs to work with the plumbing and process engineers to determine the evaporative-cooler-water source. The options are to use potable water (with backflow protection) or facility service water, if it is being distributed to the building in which the cooler is located. Using service water reduces the consumption of potable water, but the service water must meet the following criteria:

- Its total dissolved solids must be below the evaporative-cooler manufacturer's recommendations, and
- It should be disinfected as it is carried over in vapor form into occupied spaces.

If these criteria are not met, potable water is the better choice.

5.3.6 Temperature and Building Automation Controls

Proper control-systems design and operation is essential to HVAC-systems design success. Planning the control strategies often is left until the later stages of design, generating cumbersome control solutions, an incomplete picture of total HVAC costs, and delayed coordination with other disciplines, such as instrumentation and electrical. The better approach is to develop HVAC-control strategies concurrently with the HVAC conceptual design.

Small remote buildings and simple systems, such as rooftop cooling, unit heaters, and exhaust fans, typically are served best with conventional electric controls. Following are options for serving the more complicated central cooling, heating, and ventilating equipment:

1. Stand-alone direct-digital-control (DDC) systems, operator-interface terminals (OITs), and an HVAC operator's console;

2. Stand-alone DDC systems, OITs, and an HVAC operator's console with discrete alarms wired to SCADA;

3. DDC interfaced via data "handshaking" through to the SCADA operator's console; and

4. SCADA implementation of HVAC monitoring and control.

Direct-digital control is the HVAC-control vendor equivalent of PLC control. The DDC modules are proprietary versions of PLCs. The DDC OITs (typically medium-sized color touch screens with input devices) are recommended at strategic locations throughout the facility, so operating personnel can troubleshoot equipment-controls operation on the spot. The options above are listed by increasing cost of installation. The second option is the one typically recommended; it provides a clear dividing line between the HVAC and instrumentation contractors, provides what most owners want regarding the segregating process and HVAC-operation monitoring, and ensures that critical HVAC and life-safety alarms from the DDC system receive 24-hour attention by the SCADA operator. The third and fourth options have merit and should be presented to the owner, as one or the other may be his or her preference.

5.3.7   Testing, Balancing, and Commissioning

These activities often lack the coordination and vigil needed to derive their full benefit. The HVAC systems continue to grow in complexity, and WRRF operating personnel often are left with insufficient knowledge to properly operate them. The LEED and other standards are promoting enhanced commissioning of systems that include more extensive engineer, contractor, and owner involvement in training, to ensure that both the design intent and operational needs of systems are made clear to all and are well-documented.

5.4   Heating, Ventilating, and Air-Conditioning Energy and Sustainable-Design Opportunities

The HVAC systems can consume 15% or more of a large WRRF's annual energy. If the facility includes a large laboratory, extensive personnel facilities, or both, the HVAC consumption percentage may be even higher. It is likely that utility costs will continue their current escalations and that future legislation will establish carbon-emission caps that will reward investments in energy-conserving systems. Many states already have instituted programs that provide grants for renewable energy systems and credits/rebates per unit of energy they produce. The HVAC designer needs to have a basic understanding of the energy-usage characteristics and energy-conservation potentials of the system designs.

5.4.1   Energy-Code Compliance

The HVAC designer's role and responsibilities for energy-code compliance is discussed in the Energy and Sustainable-Design Initiatives section. There are few, if any, jurisdictions that do not have basic energy codes that affect the HVAC design.

5.4.2   Leadership in Energy and Environmental Design and Other Sustainable-Design Effects on Heating, Ventilating, and Air Conditioning

The second "E" in LEED stands for "energy." Energy conservation, included in the LEED Standard in its Energy and Atmosphere category, is a key and fundamental component of LEED certification. The LEED credits for energy conservation are accrued in a subcategory called Energy Optimization. The LEED energy optimization currently can be applied to a wide range of WRRF building types.

Most process areas (and their HVAC systems) have unique ventilation requirements that fall outside of the LEED Energy Optimization baseline; they typically are exempt from the LEED minimum-energy-performance requirements, and a method of measuring their performance with respect to energy conservation is not yet in place. Measurement standards for process-type HVAC systems are in the works, however, and the HVAC designer needs to keep abreast of LEED and other energy-performance-standard developments. The LEED application in Energy Optimization for personnel-type buildings is well-established and applicable currently. It behooves the HVAC designer to become familiar with the energy-conservation measures for these more "standard" building types that can earn LEED credits and to learn to use the energy modeling and documentation techniques required to earn these credits.

The LEED Energy and Atmosphere category also requires "Fundamental Refrigerant Management," a prerequisite that, if not met, precludes the project from any LEED certification. The purpose of this prerequisite is to ensure that the project meets the provisions of the Montreal Protocol cited in the Codes, Standards, and Regulations section above. The LEED program also provides credits for "Enhanced Refrigerant Management" for systems that further reduce the effect of HVAC-cooling refrigerants on the environment. The use of absorption chillers discussed above is an example of a system that may be eligible for this credit.

5.4.3   Indoor-Air Quality

Another focus of LEED that is also the province of the HVAC designer is indoor-air quality. It is part of the LEED credit category called "Indoor Environmental Quality." The primary benchmarks for the associated LEED credits are ASHRAE standards 52.2 and 62.1. Additional credits are available in the HVAC design for features such as carbon monoxide monitoring or the use of carbon monoxide to actively control the amount of outdoor air introduced to the building.

5.4.4   Building Energy Modeling as Part of Heating, Ventilating, and Air-Conditioning Calculations

The case is made above that the HVAC designer needs to become proficient in the evaluation of energy-conservation measures and learning the computer techniques for modeling their energy-saving effects. Other impetuses for developing building-energy-usage models during the design process are the relentless escalation of energy prices and the probable coming of carbon caps for WRRFs. Both these forces will increase the value of detailed modeling of building-energy usage and saving potentials needed to document a facility's carbon output.

5.4.5   Renewable Energy and Energy-Recovery Systems

Renewable and recovered energy continue to gain popularity and attract local and national grants and funding. Interest in their implementation will continue to increase, because they both reduce a facility's carbon footprint and reduce energy usage and demand. There are many such opportunities at WRRFs, the most prominent of which are digester-gas-energy recovery and heat-pump-energy recovery using process water.

Natural gas prices have risen to a high enough price point to make heat-pump applications attractive. The typical WRRF has process flows, the heat from which can be recovered at relatively high coefficients of performance. Buildings remote from these process flows are candidates for more conventional ground-source heat-pump applications, where heat is derived from a piping system of considerable length buried to a depth where the earth itself provides the source of heat. In both cases, the heat pump operates like an air-conditioning unit in reverse, cooling the process flow (or the earth), and rejecting its heat to the space or system being heated. The heat-pump cycle can be reversed to function as an air-conditioning unit, cooling the system or space and rejecting the heat to the process flow or to the earth. Given an achievable coefficient of performance of 4.0 and typical natural gas and electricity rates, the heat pump can cost half, or less, in heating-energy costs than equivalent gas-fired equipment.

Many WRRFs use anaerobic digestion in their treatment process. There is an opportunity for the HVAC designer to assist the project team in the application of digester-gas-energy-recovery systems. Modification of existing thermal-recovery systems, to extend or improve process and building-heating systems, is one option. Another is the installation of prime movers or engine generators to drive pumps and blowers or generate electricity during warmer weather, when heating loads are insufficient to consume total digester-gas production.

6.0   Chemical Systems

Chemical systems are used throughout a WRRF, to support the associated wastewater treatment process or serve as the process itself. In this section, an overview of factors to be aware of before and during the design of chemical-feed systems is presented. The reasons for chemical selection (monetary; nonmonetary; and national, state, and local codes) also are presented. A discussion on the selection of application points, equipment control, and types of equipment is presented. In addition, a review of the storage (delivery), handling, feed systems, and mixing of chemicals is also discussed. Considerations for modifying existing systems then are discussed. Finally, considerations for solids management are described. The designer should also reference the specific chapter for the treatment process the chemical system is supporting. For example, Chapter 6, Odor Control and Air Emissions, should be reviewed for detailed design information regarding chemical treatments specifically for odor control.

Considerations and regulations for safety should be incorporated to the design of the facility. Chapter 2 discusses regulations regarding handling and disposing of toxic chemicals. Additionally, Material Safety Data Sheets (MSDSs) are available on the Internet or from chemical manufacturers and contain important safety, storage, and handling information. When designing chemical feed systems for a WRRF, it is important that the designer determine if any of chemical systems will require a Risk Management Plan (RMP). RMP regulations were created in response to industrial accidents. If a chemical is listed in the Code of Federal Regulations (CFR) Part 40 CFR 68.130 in the amount above the "threshold quantity" then a RMP must be prepared. The RMP program was authorized by Section 112(r) of the Clean Air Act. The Chemical Accident Prevention provisions (part 68 of Title 40 of the CFR) requires facilities that produce, handle, process, distribute, or store certain hazardous, toxic, and flammable chemicals to develop and implement a RMP, prepare a RMP, and submit that plan to the USEPA. The plans must be revised and resubmitted to the EPA every 5 years.

Procedures for choosing chemicals and chemical dosages for these applications vary and are discussed in subsequent chapters addressing specific chemical additions. Table 7.6 presents frequently used chemicals for wastewater treatment and their principal uses.

6.1   Chemical Selection

Many different chemicals are used in support of wastewater treatment (AWWA and ASCE, 2012; Metcalf and Eddy, Inc./AECOM, 2013; U.S. EPA, 1975, 1979, and 1987; WEF, 2016; White, 2010). Often, more than one chemical is suitable for a particular application. The selection of a chemical for a particular use includes many factors. These factors may depend on the size of the facility, personnel available, and maintenance requirements. Further, these can be turned into monetary and nonmonetary considerations.

6.1.1   Monetary Considerations

Cost is a significant consideration in the selection of a chemical to be used for a particular application. Capital and O&M costs need to be considered. A proper economic evaluation includes more than simply the direct cost of the chemical itself, in terms of

Chemical	Principal Use
Activated carbon	Dechlorination
	Denitrification
	Odor adsorption
	Organics removal
	Solids stabilization
Aluminum sulfate (alum)	Suspended solids removal
	Phosphorus removal
	Solids conditioning
Ammonia	Chlorine-ammonia treatment
	Anaerobic digestion
	Nutrient
	Ammonia digestion
	Nutrient
Chlorine	Ammonia removal
	Disinfection
	Grease removal
	Prechlorination
	Odor control
	Activated-sludge-bulking control
	Filter-fly control
	Solids stabilization
Ferric chloride	Phosphorus removal
	Solids conditioning
	Suspended solids removal
Ferric sulfate	Phosphorus removal
	Suspended solids removal
Ferrous sulfate	Odor control
	Phosphorus removal
	Solids conditioning
	Suspended solids removal
Hydrogen peroxide	Odor control
	Activated-sludge-bulking control
Lime	Heavy metals removal
	Suspended solids removal
	Odor control
	Phosphorus removal
	pH adjustment
	Solids conditioning
	Solids stabilization
Methanol	Denitrification

TABLE 7.6   Chemicals Frequently Used for Wastewater Treatment

Chemical	Principal Use
Ozone	Disinfection
	Odor control
	Activated-sludge-bulking control
Polymers	Suspended solids control
	Solids conditioning
Potassium permanganate	Odor control
Sodium aluminate	Suspended solids removal
	Phosphorus removal
	Sodium bisulfite
	Dechlorination
Sodium carbonate	pH adjustment
Sodium hydroxide	pH adjustment
	Odor control
Sodium hypochlorite	Disinfection
	Odor control
Sulfur dioxide	Dechlorination
Sulfuric acid	pH adjustment
	Odor control

TABLE 7.6   Chemicals Frequently Used for Wastewater Treatment (*Continued*)

quantity, concentration, and form. There may be other costs related to the effects of chemical application on the treatment facility, such as changes to the quantity and quality of solids generated and the effect of pH on downstream treatment systems. Where more than one chemical is suited for a particular application, an economic analysis should be conducted that compares the present worth or annual equivalent costs of the possible chemicals. It is important to note that potential cost savings may result if one chemical is used for multiple purposes (e.g., chlorine, which is used for disinfection, odor control, or as an aid in activated-sludge-bulking control and oxidation of ammonia and organic substances).

The required chemical dosage affects the economic analysis. Therefore, accurate estimates of chemical dosages ensure the validity of the monetary analysis. Unfortunately, practice shows that theoretical stoichiometric relationships cannot always be used accurately to predict chemical dosages. As a result, laboratory tests, such as jar tests, pilot-facility studies, or on-line studies, are recommended, to determine more accurately optimum chemical dosages before recommending the proper chemicals for a particular use.

6.1.2   NonMonetary Considerations

When selecting the optimum chemical to be used for a particular application, a designer should consider certain nonmonetary factors with economic factors. Several important nonmonetary factors are effectiveness, compatibility with other treatment processes, reliability, maintenance considerations and sustainability, and environmental effect. When designers are specifying chemical feed equipment they use a typical service life of 10 years for the equipment due to the nature of the material they are handling. Therefore, as this equipment starts to get closer to its useful service life, the maintenance needs of that equipment will also increase. As the equipment ages, the reliability and performance of that equipment will decrease resulting in higher chemical usage and costs. Service life of the equipment will have an impact on both monetary and nonmonetary aspects of operating that equipment.

The effectiveness of using a chemical for a particular application varies from facility to facility and often depends on the specific waste or operating conditions. A designer should use operating results from similar WRRFs with caution. To most accurately determine the effectiveness of a chemical, laboratory or pilot tests using the proposed chemical on an equivalent waste sample may be needed. These tests can be conducted at the treatment facility; an independent laboratory or testing can be conducted by a manufacturer to determine the optimum chemical dosage for their equipment. Conducting preliminary chemical testing on the actual facility liquid or sludge streams would also help predict appropriate chemical dosages, thereby aiding the monetary analysis.

Compatibility with other treatment processes used at the facility is an important nonmonetary factor that should be considered when selecting the appropriate chemical to use. A chemical that may be effective for one process may result in process issues for another process. For example, using aluminum sulfate (alum) for chemical phosphorus removal may affect an anaerobic digester or a dewatering process. Again, pilot- or full-scale testing may be advisable to determine the effects on other processes. This type of testing also helps in assessing the effect of introducing the chemical to the process stream at various points. It is recommended that during the project design phase, the designer doing the chemical feed systems incorporate the ability to inject chemicals at different points in order to give the operators the flexibility to make changes if their chemicals, chemical suppliers, or waste stream characteristic change. For example, a belt filter press or centrifuge should have at least two or three application points to allow for longer or shorter reaction times for the polymer to effectively interact with the feed sludge. Not all chemical feed systems will have multiple feed points, such as chlorine or sodium hypochlorite for disinfection purposes. Refer to Section 6.2 later in the chapter for further discussions regarding application points.

The reliability of the supply of a chemical also is an important factor. A chemical found to be the most economic and effective may not have a reliable and competitive supply source. This factor may negate the results of the other evaluations. It is important to include the projected quantity of chemical required in this assessment, because, although the chemical may be readily available, it may not be available at the required quantity. There are proprietary chemicals available for process addition and odor control. A designer should evaluate competitive chemicals that can produce the same result.

Finally, a designer should assess the sustainability and environmental effects associated with the use of a particular chemical. Given current concerns about the safety of effluent toxicity, chemicals selected must be demonstrated to be environmentally safe after disposal. An example is the trend requiring dechlorination of effluents where chlorine has been added to achieve disinfection. This is caused, in part, by the determination that excessive chlorine residue in effluent potentially can link with other organic chemicals to form carcinogenic substances in the receiving body of water. Additionally, effluent that has picked up a component of the chemical passed through the system could lead to out-of-bounds water quality and permit violations.

6.1.3 National, State, and Local Codes

The chemical-area design may require special requirements for fire and safety. This will involve investigating the chemical selected and collecting information, such as the MSDS and RMP. A design consultant should be aware of what is required for the facility regarding NFPA's Uniform Fire Code and any local or state codes for chemicals used.

6.2 Application Points

The optimum point for the application of a chemical to the waste stream ranges from somewhat obvious (as with chlorine) to more difficult to determine (as with chemical phosphorus removal). When selecting a location for introducing chemicals to the process stream, a designer should consider factors such as adequate mixing, effects on subsequent treatment units, and flexibility.

For any chemical addition to be effective, the chemical solution introduced to the process stream must mix adequately with the waste stream. Mixing typically is accomplished by hydraulic means, static-type mixing, mechanical mixers, or diffused aeration. Based on past problems associated with achieving mixing in real-world situations, a designer should consider performing hydraulic modeling of critical application points, to ensure that effective and desired mixing occurs. Chapter 14 of this manual contains additional information relating to rapid mixing, mixer types, fluid regimes, and design considerations.

In some cases, the point of chemical application can adversely affect downstream treatment units. For example, if the waste becomes nutrient-deficient for the biological microorganisms, phosphorus removal in advance of a biological-treatment system may decrease the efficiency of the biological system. Chemicals used to adjust pH should be applied to a point where they will not affect the secondary system.

Conversely, some chemical effects are positive. For example, the addition of alkaline chemicals (lime and caustic soda) in the phosphorus-precipitation process increases the alkalinity of the water and counteracts alkalinity destruction by nitrification in subsequent processes. Therefore, an analysis can assess potential adverse and positive effects on subsequent treatment processes and identify the mitigating measures to be incorporated to the design.

Wherever possible, flexibility should be included in the design of chemical application systems. In many cases, duplicate chemical storage and mixing tanks and the installation of several points for the introduction of a chemical to the process stream are included in a design for minor additional capital costs. Providing this type of flexibility enhances the effectiveness and may reduce the required chemical dosage rates, if minor changes in waste characteristics or other changes occur. For example, provisions to add polymers either upstream of the sludge feed pumps, or in the sludge piping prior to the belt press or directly to chemical-conditioning tanks of a belt-filter press will provide the flexibility to use different polymers that act differently when added to the solids. Because one polymer may require more reaction time than another (perhaps at a lower dose and unit cost), a design consultant should consider providing flexibility in a design as conditions warrant. Chemical lines can be small so there should be consideration to install a spare chemical feed line to the application point especially if that chemical feed line is traveling underground and if the chemical being fed to the application point has a tendency to solidify. Adding a spare chemical line needs to be determined by the designer based on the situation, cost and other factors.

Finally, a designer should ensure that adequate flushing and clean-out connections are located throughout the system and that proper safety equipment is provided in chemical-storage areas.

6.3   Equipment Selection

Selecting equipment to use at a WRRF may depend on many variables. Items that should be considered are the form of the chemical (dry, liquid, or gel), quantity to be fed, and accuracy and reliability of the equipment. The type of chemical used will affect the type of equipment supplied and the amount of redundancy required. For example, if the chemical is corrosive additional spare equipment would be good to have installed so it can be placed on line while damaged equipment is being repaired. Equipment selection often requires discussion with the owners and manufacturers, to ensure that the equipment will meet the facility requirements and clients' needs. The designer must ensure that space and proper connections (electricity, process, utility water, etc.) are provided in the design for the selected equipment and provide specifications that describe these requirements. The design consultant must ensure that there is enough clearance around the equipment to remove any internal parts and that, if required, there is proper lifting equipment. Other considerations include operation life and materials of compatibility.

6.4   Storage (Delivery), Handling, Feed Systems, and Mixing

Physical facilities associated with the storage, handling, feeding, and mixing of a chemical are dictated by the form of the chemical used, its physical and chemical characteristics, and the flow ranges of the waste stream.

The design of chemical operations involves not only the sizing of various unit operations and processes, but also necessary appurtenances. Because of the corrosive nature of many of the chemicals used, their stability, and the different forms in which they are available, special consideration should be given to the design of chemical storage, feed piping, and mixing and control systems. Selecting the proper materials that are compatible for the chemicals being used is also important. Additionally, federal government regulations for hazardous-chemical storage should be reviewed constantly for required additional security measures. The following subsections provide brief discussions of these topics.

6.4.1   Bulk Delivery and Intermediate Bulk Containers

How a chemical is stored and delivered may dictate which chemical is used. Bulk deliveries are shipped by the truck load and represent approximately 20 000 to 22 000 kg (22 to 24 tons) of dry material. For bulk deliveries of liquids, this equates to approximately 15 000 to 17 000 L (4000 to 4500 gal). Partial truck loads may be available at a premium cost. There also are intermediate bulk containers (IBCs) available from a variety of chemical producers and distributors. A typical IBC is approximately 900 kg (2000 lb) and can be as small as 500 kg (1000 lb). Liquids also can be shipped in totes, drums, and carboys. Dry chemicals are available in approximately 18- or 36-kg (40- or 80-lb) bags. The quantity of chemical storage should consider the stability of the chemical, maximum and average feed rates, availability of supply, delivery size, and costs, as discussed earlier. Bulk storage tanks must be designed so they are compatible with the chemical they are holding in the tank. These tanks may be plastic, high-density polyethylene, polypropylene, fiberglass, or steel tanks. The steel tank may need to be lined with a coating depending on the chemical in the tank.

Liquid chemical storage tanks should be slightly oversized to take a full truck plus there should be room for an additional 3 to 5 days' worth of chemical usage. For example, if a normal truck shipment is 5000 gallons the storage tank should have at least a 6000 gallon capacity so the storage tank does not have to go to a completely empty to handle a truck load. When designing liquid bulk chemical-storage facilities, it is always good to plan on allowing for extra days of storage in case the shipment is delayed for weather, truck breakdowns and other types of delays.

6.4.2 Containment and Diking

How a chemical is stored will indicate if it must be contained. Typically, a containment-type dike is constructed around a chemical-storage and feeding facility. The volume of chemical to contain generally is 110% to 125% of the largest vessel within the containment area. A designer should provide a low-point sump for collection of the spilled chemical or normal cleanup that is accessible for pumping to a drain, a waste handler, or back to the storage tank. The outdoor area may collect rain and snow, which will need to be disposed of somewhere in an acceptable manner. Floor drains are not used in certain chemical areas to prevent accidental disposal. When floor drains are located in chemical areas the design consultant must determine where that floor drain piping is connected so it does not affect other equipment such as sump pumps. Also does the drain piping need to be acid resistant? Designers must ensure that floor drains are large enough to handle the flows for wash down, spills, and other higher water conditions. If a chemical feed pump requires seal water or if a chemical high energy type mixing system is uses water in the mixing unit, then ensure that a floor drain is located next to the pump or mixing unit so water does not travel across the floor. Also when evaluating floor drains ensure that the designer is aware of the viscosity of the chemical. For example, some neat polymers have high viscosities that makes them thick so they will flow slowly and will affect floor drain systems. If the drain is too small, then the thick polymer will not flow well in the drain or it could plug the drain.

Never design a single chemical containment area for multiple chemicals because there could be a dangerous chemical reaction when certain chemicals are mixed together. Dry chemicals require some means of containment for cleanup and for protection of other areas. Designers should consult local and U.S. EPA regulations for additional spill-control and -response requirements.

6.4.3 Chemical Form

Many chemicals used in WRRFs are found in different forms (dry, liquid, gel or gas) at various stages in the chemical-handling system. In domestic wastewater treatment systems, chemicals generally are in dry or liquid form. Dry chemicals are generally converted to a solution or slurry form before introduction to the wastewater. Liquid chemicals typically are delivered to the facility in a concentrated form and may be diluted before introduction to the wastewater.

6.4.4 Gas-Chemical-Feed Systems

Chemicals can be fed in gas form. For example, chlorine is stored initially as a gas or liquid (depending on the quantity of chlorine used at a particular facility), transferred through a feeder as a dry gas, and injected as a solution to the point of application. To make the injector work, a water supply is required. Because injectors are not efficient, a large quantity of water at a high pressure is necessary. These systems may be dedicated facility water pumps or booster pumps supplied by a protected city water supply. Gaseous chemicals such as chlorine gas can also be drawn to the application point using a chemical induction unit.

6.4.5 Dry-Chemical-Feed System

A dry chemical feed system generally consists of a storage hopper, a dry-chemical feeder, a dissolving tank, and a pump- or gravity-distribution system (see Figure 7.16). Units are sized according to the volume of wastewater, treatment rate, and optimum length of time for dissolving. The feeder is sized based on the concentration of solution to be made up and the output of the dissolver to the process. Control for the feeder may be a timer for batch makeup or variable speed, depending on the degree of control necessary. Hoppers used with compressible and archable powder, such as hydrated lime, are equipped with positive hopper agitators and a dust-collection system. Dry-chemical feeders are either of the volumetric or gravimetric type. The volumetric type measures the volume of dry chemical fed; the gravimetric type weighs the amount of chemical fed.

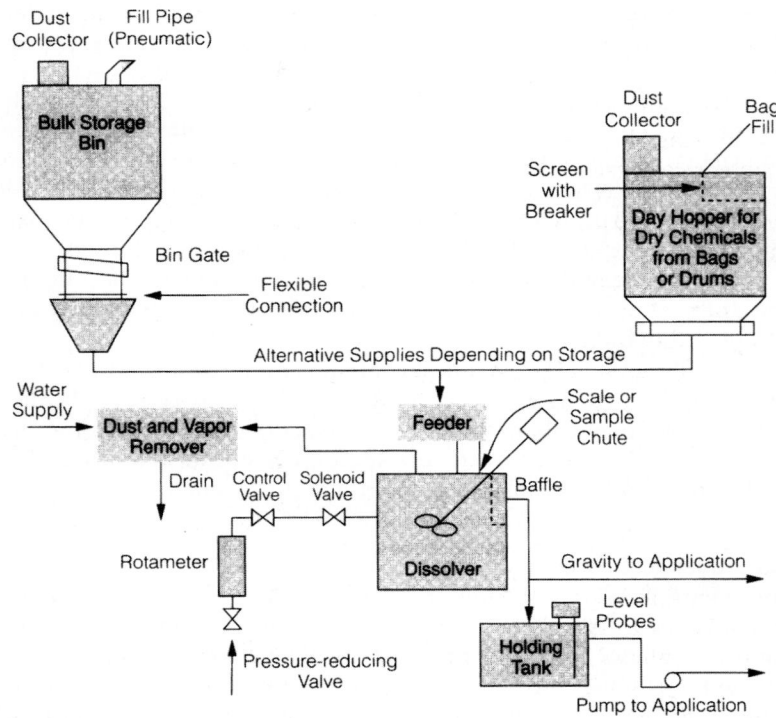

Figure 7.16   Typical dry chemical feed system.

With a dry-feed system, the dissolving operation is critical. The capacity of the dissolving tank is based on the detention time, which is directly related to the rate of chemical dissolution. When the water supply is controlled for the purpose of forming a constant-strength solution, mechanical mixers are used. After dissolving, solutions or slurries often are stored and discharged by chemical-feed pumps to the application point at metered rates.

6.4.6 Liquid-Chemical-Feed System

Liquid-chemical feed systems can include a solution storage tank, a transfer pump, a day tank for diluting the concentrated solution, and a chemical-feed pump for distribution to the application point. Typically, chemicals in WRRFs are not diluted or put in day tanks. Dilution generally is required if the chemical-feed equipment is oversized or when handling polymers, which must be diluted. Polymer feed systems also can be high mixing type systems where neat polymer is mixed with water in a mixing chamber with a high speed mixer. The polymer/water solution within the mixing chamber must be less than a 1% mixture. Typically, chemical-feed pumps draw liquid directly from the solution storage tank. For accurate metering of the chemical feed, solution-feed pumps typically are of the positive-displacement type. Chemical-feed pumps are available in a range of sizes and methods of actuation. Sizes can range from liters per day (gallons per day) to liters per minute (gallons per minute). Actuation can be either constant- or variable-speed. Variable-speed types can be electronic (solenoid-driven), silicon-controlled rectifier (direct-current-motor driven), or VFD (alternating-current-motor driven). If sodium hypochlorite is pumped or drawn to the application point the designer and contractor must be careful how the piping is installed. The designer and contractor must make sure that the chemical piping is laid in the ground so the piping is not wavy where high spots would collect gas bubbles. If gas collects in the high sections of the piping it will stop the chemical from flowing.

While chemical-handling systems may appear simple, proper design of a complete and integrated system can be as complex as that of the associated wastewater treatment process. For instance, liquid-chemical systems that are capable of freezing should be placed indoors or provided with an adequate freeze-protection design on both the tank and ancillary piping. Therefore, a design consultant should pay particular attention to the detailed design of all aspects of a chemical system.

6.4.7 Construction Materials

There are few, if any, chemicals that do not cause some type of corrosion. For this reason, construction materials are important. Most chemical-feed systems use some type of plastic material, such as PVC or CPVC, that is in contact with the chemical and/or solution. It is important to refer to chemical compatibility tables to determine the proper material that is required for the equipment, pumping internals, gaskets, piping material, and other chemical feed appurtenances.

6.4.8 Mixing: Static versus Dynamic

Getting the chemical in contact with the process is accomplished through some method of mixing. Mixing may be static or dynamic. All mixing requires using energy. Static mixing may be through the use of interferences in the pipe or process. To avoid clogging, static mixers should not be used in lines where debris may be present. Dynamic mixing is through the use of a blade on a shaft to a motor. The induced injectors also use a motor to spin an impeller at a high speed to create a vacuum and positive mixing.

6.4.9 Additional Security Measures

In response to the 9/11 attacks, the U.S. Department of Homeland Security (Washington, D.C.) directs facilities to conduct vulnerability assessments and implement counter-measures to reduce risk. The designer and owner should consult their state's department of environmental protection and be aware of necessary security measures. Specific security measures are described in more detail in the Site Security section below.

6.5 Considerations for Retrofitting, Upgrading, or Converting Existing Systems

There are many reasons a utility may need to change their existing chemical-feed systems, such as facility expansion, the economy, process changes, or risk and safety. For example, if development causes changes to influent water quality, the existing chemical processes may no longer be effective. A common example for an upgrade or retrofit is the conversion of the disinfectant chlorine to sodium hypochlorite. Many facilities making this change claim it is safer, and the chemical delivery is more reliable (some even may generate their own sodium hypochlorite) (White, 2010).

Typical design considerations for chemical systems, such as storage, handling, and mixing, as described in this chapter, also are important for changes to chemical systems. Monetary considerations also may contribute significantly to the decision of chosen technology—whether to replace or refurbish equipment. However, building on existing infrastructure poses unique challenges and considerations, as described in the Water Environment Federation® (Alexandria, Virginia) (WEF) Manual of Practice 28, Upgrading and Retrofitting Water and Wastewater Treatment Plants (WEF, 2005). The chemical-feed process often is not the only thing that requires upgrading or modification, but control-system upgrades may be needed, sometimes requiring hardware and software replacement. Another important consideration the designer must investigate is if and how the instrumentation and controls of the new system will merge with the existing system (Keskar, 2002).

6.6 Chemical Safety

Chemical safety is another important aspect to chemical feed system designs to protect the operators and maintenance staff. There are a number of different chemicals used in WRRFs and they all have safety concerns. It is important that the design consultant review the safety requirements of every chemical being considered for their facility and ensure that the proper monitoring and alarm systems are provided in the design. The design consultant must also ensure emergency showers and eyewash stations are installed throughout the chemical storage and feed areas. When designing emergency showers and eyewash stations the design consultant should install a floor drain at those locations to collect the water from the shower. Draining the water in these locations will reduce slipping accidents by facility personnel using the emergency equipment. It should be pointed out that floor drains may not be appropriate at all emergency shower and eyewash station locations, so the design consultant should consider a non-slip floor surface at those locations.

Proper ventilation systems must be provided in all chemical storage and feed areas. The design consultant must also know whether the ventilation duct work is to draw air out of the room close to the floor or higher up in the room. The design consultant must check all applicable codes to ensure that the proper air changes are met for the chemical areas.

If gas detectors are to installed in chemical storage and feed areas, the design consultant must ensure they are placed in locations where maintenance personnel can get to the equipment and perform the manufacturer's required equipment maintenance and calibration.

Warning alarms must be indicated using lights and alarm horns inside and outside the rooms and/or chemical buildings. Also the alarms must to be transmitted via the facility's SCADA to alert operators in the facility Control Room.

6.7 Considerations for Solids Management

Chemical addition to wastewater treatment processes can change solids characteristics frequently, resulting in an increase of the inert fraction of the solids, quantity of solids to be processed, and quantity of solids to be disposed. For example, lime addition for heavy-metal removal generates a chemical sludge that is difficult to dewater and dispose of properly.

Given the current challenges in disposing wastewater solids, the quantities of solids generated should be considered carefully. Because it is difficult to obtain true estimates of solids generated by chemical applications from strict chemical relationships, bench-scale testing has proven to be a valuable method to aid in these predictions. Several publications present methodologies for calculating solids quantities. These include Recommended Standards for Wastewater Facilities (Great Lakes–Upper Mississippi River Board of State and Provincial Public Health and Environment Managers, 2014), Chemical Aids Manual for Wastewater Treatment Facilities (U.S. EPA, 1979), and Process Design Manual for Suspended Solids Removal (U.S. EPA, 1975).

6.8 Instrumentation and Controls

The instrumentation and controls associated with chemical feed systems is a critical part of the success of the facility. The use of chemicals to treat wastewater must be done carefully and thoughtfully to achieve expected and positive results at the final effluent.

6.8.1 Controls

There are several ways to implement chemical dosing strategies, from simple to complex. When selecting control strategies, it is important to take into consideration chemical cost, measuring points for control, reaction times, and instruments required. Control of chemical dosing, when done correctly, can provide a significant cost savings to the annual operational cost of operation. Conversely, if a control algorithm for chemical dosing is done poorly, a significant amount of money can be wasted. In the past, it was common practice to set a chemical dosing/metering pump manually and walk away, with no automatic control associated with the pump. This is poor practice and can lead to over-feeding chemical or not providing enough chemical to accomplish positive results. Most chemical dosing should use proper control algorithms using proportional-integral-derivative (PID) control loops or a combination of the PID. These types of controls must be tuned during startup of the system to ensure proper operation of the system, and proper response of the pumps and valves associated with the installation.

6.8.2   Instrumentation

The instrumentation associated with chemical feed systems is critical to ensure proper and accurate dosing is taking place. Since many chemicals have corrosive properties, extreme care and caution should be used when selecting instruments that will come in contact with chemicals. Most instrument manufacturers can provide chemical compatibility charts providing information to assist in the selection of instruments. It is also important to consider the instruments accuracy when selecting from the different manufacturers offerings for a particular instrument. Generally speaking, a higher accuracy instrument is worth the extra cost up front to save money over the life of the instrument by providing better control response. With chemical dosing systems, often times the flow of the chemical is very small, and care should be taken to ensure the measuring instrumentation will read accurately at low flows. Also, providing additional instrumentation for safety is an effective way of mitigating issues when systems break down. For example, by adding a flow switch to a sodium hypochlorite line to ensure chemical is flowing when pumps are running, the control system can identify if there has been a break or clog in the line and shut the system down with an associated alarm. This is a good way to avoid inadvertently pumping chemicals through a broken section of pipe or damaged pump. The minor added cost of quality instrumentation in chemical feeding systems can provide monetary gain and safety, and is worth the investment.

7.0   Other Support Systems

7.1   Fire Protection

In general, the highest risk of fire and explosion is associated with wastewater collection and pumping operations and the early stages of liquid and solids processing. Specific unit processes to consider include pumping station wet wells, which handle raw wastewater or unstabilized solids, preliminary screening and grit-removal processes, primary sedimentation, anaerobic digestion, scum-collecting processes, and fuel- and chemical-handling and -storage areas.

Principal control procedures used to minimize potential fire and explosion incidents at WRRFs include risk evaluation, process and equipment controls, ventilation, construction materials, and education. These control procedures also include proper electrical classification of hazardous locations and the selection and installation of electrical equipment, motors, and devices that are suitable for these locations. Effective implementation and enforcement of the control procedures requires an adequate safety program and the cooperation of facility management and personnel and public, private, and government sectors.

Specific design criteria for fire-protection systems are based on guidelines established by applicable building codes, NFPA codes (listed in the Representative Fire-Suppression Codes section below), local codes and ordinances, and the owner's insurance carrier. To ensure compliance with codes and ordinances, the fire-protection designer must review applicable requirements with the appropriate local officials, referred to in codes as the authority having jurisdiction (AHJ), early in the design process.

7.1.1 Building Classifications

In the preliminary stages of any WRRF design, the design team (project manager, architect, and fire-protection designer) needs to specify which spaces within buildings require fire protection per the applicable building codes. Once this is determined, the design team selects the occupancy hazard of the buildings or structures, or portions thereof, and the fire-protection designer develops the appropriate fire-protection systems in accordance with the applicable fire-prevention codes. Most areas in WRRFs are classified as "industrial use groups." In addition, with the exception of administration buildings, maintenance shops or areas, repair garage areas, and control rooms, most areas are unoccupied. Local fire codes and NFPA regulations typically apply to occupied industrial, institutional, storage, and commercial buildings. Therefore, considerable judgment and local fire-code interpretations (by the AHJ) determine which buildings or portions thereof require fire protection and the type of occupancy hazard that applies. Assessment of chemical, explosive, and other hazards are multidiscipline tasks requiring project-management oversight and AHJ review.

7.1.2 Representative Fire-Suppression Codes

In addition to applicable local, county, and state fire codes, or their amendments to other codes, the following NFPA standards typically govern fire-suppression design:

- NFPA 13—Installation of Sprinkler Systems,
- NFPA 14—Installation of Standpipe Systems,
- NFPA 20—Installation of Centrifugal Fire Pumps,
- NFPA 24—Private Fire Mains and their Appurtenances, and
- NFPA 54—National Fuel Gas Code.

7.1.3 Authority Having Jurisdiction and Agency Review

Experience shows repeatedly that soliciting review of all of a project's fire- and life-safety features by the AHJ as early as possible in the design avoids costly changes later in the design. The AHJ is the final authority on fire- and life-safety-code evaluations and is empowered to require more stringent interpretations of codes or even to require exceeding code requirements. Some examples of review items appropriate for early review with the AHJ include the following:

- Applicable codes and amendments—which ones apply?
- Building occupancy and construction classifications;
- Sprinkler system requirements;
- Standpipe system needs;
- NEC hazardous classifications;
- Chemical and flammables effects on design and proposed design solutions;
- HVAC-duct-smoke-detector and fan-shutdown-design review;

- Special ventilation systems, especially those not covered by specific codes;
- Smoke control and partitioning assessments;
- Special fire-alarm-system applications;
- Fire-control-panel needs, locations, and access;
- Electrical-room suppression and alarm considerations and proposed solutions;
- Elevator and stair-system evaluations;
- Fire-hydrant-location criteria; and
- Validation of code and system-design interpretations.

7.1.4 Fire-Suppression-Systems Overview

The types of fire-protection systems most often used in WRRFs are wet-pipe-sprinkler systems, dry-pipe-sprinkler systems, pre-action-sprinkler systems, and gaseous-fire-extinguishing systems. In addition, adequate portable fire extinguishers should be located throughout the facility, for use by facility staff. Specific installation and safety requirements for these fire-protection systems typically are covered by local codes or NFPA standards. Regardless of the type of fire-suppression system(s) applied to the design, the fire-protection designer needs to collaborate with the electrical engineer, to ensure that the fire-flow and tamper-switch-alarm signaling required by NFPA standards is properly designed and specified. The designer also needs to coordinate signaling from the fire-suppression systems to any facility-wide fire detection and annunciation systems, if it also is a project requirement.

7.1.4.1 Wet-Pipe Sprinkler Systems

Wet-pipe sprinkler systems are used in applications where the temperature is above freezing. Pressurized water is discharged from a sprinkler head immediately after heat actuation. Alarm-check valves or water-flow detectors are used to activate local and remote alarms. In is important to note that, up until the last few years, most AHJs granted variances or waivers from fire-suppression systems in many process buildings. This is no longer the case. Insurers and plan reviewers, in an increasing number of jurisdictions, are adhering more strictly to code and standard interpretations.

7.1.4.2 Dry-Pipe Sprinkler Systems

Dry-pipe sprinkler systems are used in unheated areas, where the inside temperature is expected to drop below freezing. Sprinkler heads are connected to a piping system containing air under pressure. Water flow is controlled by a dry-pipe valve that will automatically open when air is released from the piping through the sprinkler heads, which open when the temperature in the area exceeds the sprinkler-temperature rating.

7.1.4.3 Pre-Action Systems

These systems commonly are used in areas where valuable equipment, such as electronics, needs to be protected from inadvertent water discharge. The pre-action system consists of automatic sprinkler heads that are connected to a piping-distribution system containing air, which may or may not be pressurized. Water flow is controlled by a pre-action valve that opens automatically, in response to signals from a fire-detection system installed in the protected area. Water flows through the distribution piping and discharges only through sprinkler heads that have been opened by the fire event. Inadvertent operation of the pre-action system is a remote possibility, because there must be two occurrences of signals from the detection system and the opening of the sprinkler heads to have water discharged to the area.

7.1.4.4 Gaseous, Foam, and Other Systems Gaseous-fire-extinguishing systems are designed for use in applications where a water discharge is undesirable or unsafe (e.g., electrical rooms). A gaseous-fire-extinguishing system uses non-halogenated products. The design is based on flooding the protected area with a gas that inhibits combustion, even in small concentrations.

The system efficacy depends on maintaining the gas-design concentration for a recommended period of time after its initial discharge. This is achieved by sizing proper flow and ensuring that all openings to the space served are properly closed upon initiating the gas discharge. A controlled opening is provided to allow for air release during agent discharge to prevent overpressurizing the protected space. The designer needs to incorporate, and coordinate with the electrical engineer, visual and aural warning signals, which allow facility personnel to evacuate the area before the agent is discharged. The NFPA standards and the AHJ will prescribe this delay timing.

For years, it has been common for AHJs to grant variances from suppression systems in electrical rooms, even if the buildings in which they were located required suppression. One rationale for this was that the discharge of water posed a greater hazard to personnel in the process of working on electrical equipment than the fire itself. These variances are granted less frequently now and likely will become rarer in the future. As a result, the fire-protection engineer and project manager should develop and present alternatives and their effects, for both wet-fire suppression and gaseous suppression, to the project owner.

7.1.5 Standpipe, Fire-Pump, and Water-Supply Systems

After the extent and flow density of the fire-suppression-system design has been determined with some accuracy and agreed upon by the AHJ, the fire-protection designer needs to initiate evaluation and testing of the water source(s) proposed for the systems. The NFPA standards prescribe detailed flow-testing methods to be performed on the proposed water sources. It behooves the design team to perform this evaluation early, to avoid having to add additional city-water connections or water-storage and fire-pumping systems later in the project. In those locations where city-water flow and pressure are insufficient for the fire demand (or unavailable), the fire-protection designer will need to collaborate with the civil sanitary engineers to develop a storage and fire-pumping and distribution system that satisfies NFPA standards and the AHJ. Many new, larger projects may require a dedicated fire loop serving the sprinkled buildings. The fire-protection engineer also needs to coordinate the building-suppression needs with the NFPA, local ordinance, and AHJ requirements for site hydrants that will be connected to the fire loop and affect the total fire-flow calculations.

Fire-suppression systems using site storage or having conventional water sources with insufficient pressure to serve the hydraulic flow and head of the sprinkler system will require fire pump(s). The NFPA standards and the AHJ determine whether there will be a need for backup power, emergency power, or engine-driven backup for the pumping systems. There also will be jockey pumps that maintain residual system pressures and fire-pump controllers, the designs of which need to be closely coordinated with the project electrical engineer.

Fire-Protection Design Data Required of Other Disciplines and Developed During Initial Design	Fire-Protection Design Data Developed for Other Disciplines During Initial Design
Space data • Hazard classifications • Special constraints, codes • Special chemical requirements	Fire alarm interface points and requirements
Water source(s) • Pressure • Flow	Water service requirements
Outside fire protection requirements	Fire-protection equipment space requirements
Emergency power alternatives	Riser and interstitial space requirements Outline specifications Preliminary cost opinion

TABLE 7.7 Fire-Protection Preliminary Design Coordination Data Summary

Standpipe systems typically will be required in building three stories or more and will add to the required fire flow rate. There are tamper switches and other alarm-signaling functions that the fire-protection designer needs to coordinate with the electrical engineer. The location of hose cabinets, hose connections, and hose and nozzle sizes and types need to be coordinated with the fire chief, the AHJ, or both.

Table 7.7 lists some common coordination data that the fire-protection designer provides to the other disciplines and summarizes data the designer needs from them. These items are meant to be guidelines for use by all the design disciplines in assessing fire-protection coordination during the early design phases.

7.2 Site Security

In the past, site-security systems consisted primarily of chain-link fencing to enclose a facility site. The purpose of fencing was to protect the facility from theft and prevent injury of the general public.

7.2.1 Vulnerability Assessment

Municipalities have been mandated to evaluate their facilities for their threat vulnerabilities by some empirical method, such as RAM-W, as developed by Sandia National Laboratories (Albuquerque, New Mexico). This method takes the municipality through an analysis of a facility's vulnerabilities and likelihood of risk and compares that risk with a cost to mitigate that risk. The municipality then can make an informed decision regarding which risks it can afford to mitigate by construction (e.g., fencing, cameras, vehicle barriers, and intrusion detection) and those that can be mitigated without capital costs (i.e., providing more security patrols).

Given these increasing concerns about the liability of a facility owner, site-security systems are becoming more important. While fencing is still the mainstay of a facility security system, electronic systems also are becoming popular, particularly at unmanned remote facilities and at facilities with reduced staffing levels during the evening and night shifts. The most common electronic systems used are intrusion alarms, which are installed at the primary structures of the facility. Intrusion alarms are connected either to the main control center of the facility, if manned during all shifts, or to other public facilities, such as the local police station.

In addition to security fencing or electronic alarms, security guards control access to many facilities. If security guards are employed at a facility, a designer should provide proper facilities to house the security personnel, such as a guardhouse at the entrance of the main facility.

7.2.2   Types of Security
The advancement of security systems have allowed for an increase in sophistication, while making the systems more cost-effective. Systems are able to communicate over many types of media. These types include

- Ethernet—wired, fiber, and wireless; and
- Traditional closed-circuit television (cable).

The Ethernet-type systems allow for multiple devices (cameras, intrusion/motion switches, and infrared motion switches) to operate on the same wires, thus reducing the amount of wired infrastructure. Also, many of these devices are offered with a power-over-Ethernet option, which eliminates the needs for an external power source and also simplifies backup power needs.

7.2.3   Camera Surveillance
Camera surveillance can be accomplished by using many types of cameras. These cameras include fixed, pan-tilt-zoom (PTZ), and infrared. Any of these types of cameras can be connected to a central television-type monitoring system. However, most systems are connected to a PC-based system with long-term digital storage and digital video recorders (DVRs). These PC/DVR systems allow for control of the PTZ camera and detection of motion in selected areas to trigger alarms, among many other features. Infrared cameras, which previously needed infrared illuminators, now are able to operate well without the need for illuminators.

7.2.4   Fence Detection
Perimeter intrusion detection also has advanced in the public sector, to allow for more sophisticated detection. Sensors can be attached on fences that can detect a person trying to climb or cut a fence. In addition, infrared or other types of area-motion-detection-type sensors can be installed to monitor large areas for a person moving in an area that is not expected to have people walking in.

7.3   Plumbing

7.3.1   General Design Criteria
Design criteria for plumbing systems frequently are established by the local codes of the municipality in which the facility is located. Many states have adopted the BOCA National Plumbing Code (Building Officials and Code Administrators International, 2006), Uniform Plumbing Code (International Association of Plumbing and Mechanical Officials, 1994), or International Plumbing Code (International Code Council, 2009) as their governing criteria. Other states have separate plumbing codes. At the project's

outset, the plumbing designer needs to perform a complete search to determine the applicable local, county, and state codes and, preferably, to confirm his or her findings with a local code-enforcement officer.

To withstand the constant use and abuse of everyday service, all equipment for a WRRF plumbing system should be of a heavy-duty industrial-grade. To minimize operating costs and conserve resources, all fixtures should be of the low-flow type. In addition, equipment should be designed to accommodate handicapped personnel, in public areas, where required.

Backflow prevention is installed at any connection between a potable-water-supply system and any other system that has the potential to contaminate the potable system. Break-tank separation is an alternative to backflow prevention and often is implemented in the design of facility water systems that use city water as the source. In such a system, potable water enters a tank through an air break to prevent the possibility of contamination. A float valve typically controls the level of the tank, and two supply pumps are recommended on the nonpotable supply, to provide redundancy. National and local plumbing codes have strict requirements for the application and use of backflow preventers (BFPs). The BFPs require regular testing, which typically produces considerable water discharge. When BFPs are located indoors, as often is the case in cooler climates, the plumbing designer needs to ensure that there is sufficient space to perform maintenance and that he or she provides a floor drain with sufficient capacity to handle the sometimes considerable flow that results from BFP testing or component failure.

Table 7.8 lists some common coordination data that the plumbing designer provides to the other disciplines and summarizes data he or she needs from them. These items are meant to be guidelines for use by all the design disciplines in assessing plumbing coordination during the early design phases.

7.3.2 Water-Supply Systems

Three types of water-supply systems typically designed for WRRFs are potable or city water, service water (nonpotable water), and effluent water. Potable water is satisfactory for drinking and is supplied to plumbing fixtures, such as sinks, water closets, urinals, showers, lavatories, emergency showers/washdown units, and water coolers. For other types of fixtures, a designer should refer to the applicable plumbing code.

Service, or nonpotable, water is supplied to heat-generation equipment, heat-transfer equipment, pump seals, and process equipment for cooling or makeup-water purposes. Cross-connections between potable and nonpotable water are not allowed, except where approved protective devices or means to prevent backflow into the potable water are installed. In some applications, effluent water is used as service water. The amount of effluent water used depends on the availability and cost of potable water, water-quality standards, and water-treatment costs. Some of the most common uses of effluent water are for irrigation systems, lawn watering, washdown and flushing systems for process equipment, chlorine dilution, cooling systems, and heat-recovery systems. The plumbing designer should recommend disinfection of facility effluent water for any use that could come in contact with operating personnel.

Rigid copper piping is recommended for each of these water-service types, except when it is to be installed in areas where the atmosphere is corrosive to copper (e.g., hydrogen sulfide). Water-heating-energy options include natural gas, electricity, and, in specialized cases, air-conditioning-heat recovery. For significant loads, such as personnel

Plumbing Design Data Required of Other Disciplines and Developed During Initial Design	Plumbing Design Data Developed for Other Disciplines During Initial Design
Utilities available and capacities • Water • Sewer • Gas	Pipe chase locations and sizes Interstitial space requirements Equipment/fixture locations and space requirements
Toilets (number required and locations)	Plumbing equipment electrical loads
Showers and lockers (number required)	(Vent locations and sizes and locations)
Roof type and drainage requirements (drains required and layout)	Potable water and sanitary flows
Equipment drains • Sizes • Locations • Flow capacities	Major piping alignments and sizes
Water outlets • Types • Locations	Control system descriptions
Laboratory services • Water • Deionized water • Gases • Drains and chemical drains • Equipment with plumbing requirements	Outline specifications
Emergency eyewashes and showers • Locations • Capacities	Preliminary cost opinion

TABLE 7.8   Plumbing Preliminary Design Coordination Data Summary

facilities, natural-gas-fired heaters are recommended, as a result of their reduced energy usage. Small, intermittent loads, such as remote washrooms, may be served nearly as economically using electric water heaters.

Emergency shower and eyewash fixtures typically are located in areas where hazardous chemicals are handled. These emergency fixtures are supplied with tempered water, either from a dedicated, local water heater or through hot-water supply and recirculation lines connected to a central water heater.

7.3.3   Sanitary Systems
Laboratory drain piping from the sink outlet to the point of dilution in the main wastewater flow should be chemical-resistant. In addition, a neutralizing tank should be provided for treating laboratory waste. Garage drains should be provided with oily water separators, to prevent oil from entering facility treatment units. Granulated chemical and vehicle storage areas, where significant amounts of sand or grit may find a way into drain piping, need special traps suited to the duty.

Floor- and equipment-drain systems located in basements should be drained by gravity to under-floor sumps, where submersible sump pumps lift the drainage to the sanitary system. Sump pumps of the non-clog-grinder type are recommended. Provisions for preventing flow backup should be considered, if tank drains are connected to the same drainage system that directly drains basements and galleries.

In general, the plumbing systems designer needs to establish and review, with the project manager, the proposed floor- and equipment-drain locations. This task should commence as early as possible, because the drain locations establish minimum floor slopes that the structural engineer applies in his design. Changes later can affect negatively the engineer's structural solutions.

Chemical rooms and storage areas should be provided with sumps and chemical-resistant pumps to discharge floor washdown water to the sewer or discharge chemical solutions from a failed tank or chemical equipment to a truck through the fill station. Current trends indicate that more AHJs are prohibiting the discharge of chemicals into the sanitary sewer system. The project team needs to determine which building features and plumbing systems will be incorporated to deal with chemical spills. A typical approach is to design physical containment that will catch and hold any likely spill for later pumping to a hazardous materials tank truck for approved disposal.

Considering the potential for piping mechanical damage and the long service life for which WRRFs are designed, the following sanitary piping materials are recommended: 38 mm (1.5-in.) and smaller, galvanized steel with malleable fittings; 50 to 152 mm (2 to 6 in.), hubless cast iron with hubless fittings, neoprene sealing, and 316 stainless-steel clamps; 203 mm (8 in.) and larger, hub and spigot cast iron with hub and spigot, rubber-gasketed joints.

7.3.4 Storm Drainage

Key issues here are coordination with the architectural design, validation of the proper code application for complying with overflow requirements, and coordination of interior piping for conflicts. Roof-drainage design requires close coordination between the architect and plumbing designer. Small flat roofs, gabled roofs, and hipped roofs may not require any interior piping or piping. Larger flat roofs require coordination of the insulation and pitch to drains to meet both the plumbing code and, potentially, energy-conservation prescriptions for roof-insulating value.

The plumbing designer needs to be sure that his or her interpretation of acceptable overflow means is met in the design. Some codes and jurisdictions allow scuppers; others prescribe separate overflow drains to the risers; and the most stringent separate overflow piping all the way to the building's storm connection. The design team also needs to review and coordinate the storm-piping routing within the building for interferences with the structure, ductwork, cranes, monorails, and other systems.

The following storm-piping materials are recommended above-grade, inside buildings: 76- to 152-mm (3- to 6-in.), hubless cast iron with hubless fittings, neoprene sealing, and 316 stainless-steel clamps; 203 mm (8-in.); and larger, hub and spigot cast iron with hub and spigot, rubber-gasketed joints. Inside buildings, below-grade, hub and spigot cast iron with hub and spigot, rubber-gasketed joints are recommended.

7.4   Fuel

Natural gas, propane gas, and fuel oil have numerous uses at WRRFs. The type of gas and its specific use vary, depending on the treatment processes used and the size and complexity of the facility. Typical fuel used at a WRRF may include the following:

- Building heating (central boiler or unit heaters);
- Hot-water heaters,
- Laboratory fixtures,
- Solids-heating systems,
- Incinerators,
- Waste-gas-burner-pilot lights,
- Emergency generators,
- Dual fuel engines, and
- Direct-drive process-equipment engines.

When available at a facility, digester gas can be used instead of other fuels for many of these fuel uses. In addition, service pressures for each of the fuel uses likely will vary and depend somewhat on specific requirements of the equipment manufacturers.

A designer should make estimates of natural-gas requirements for a WRRF early in the design process. These estimates can be used as the bases of initial discussions with a local gas utility, to determine service requirements and rates. Typically, each facility has only one connection to a local utility. However, in larger facilities, it may be more appropriate and cost-effective to have several connections, each served by an individual pressure regulator and meter.

Within the treatment-facility site, it may be necessary to develop an internal natural-gas-distribution system that then is tapped to serve individual buildings or structures. If the internal distribution system is maintained at high pressure, it may be possible to reduce the pressure at each structure, to meet the specific equipment requirements within that structure. It also may be appropriate to meter gas usage at each structure.

7.5   Compressed-Air Systems

Compressed-air systems are used throughout a WRRF. Specific applications of compressed-air systems include the following:

- Air-operating pneumatic tools,
- Odor-control systems,
- Sealed water systems (HVACs),
- Laboratory applications,
- Valve-air operators and measuring instruments, and
- Diaphragm pumps.

A clear understanding of the basic types of compressed-air systems is helpful, in all cases.

A power-driven device transforms air at some initial intake pressure (typically atmospheric) to a greater working pressure. The elements of a typical compressed-air system are the source of air (the air compressor), which is connected to a storage tank, and the receiver, in which the pressure is maintained between fixed limits. A pressure-relief valve is necessary to prevent pressure from building up beyond a safe preset limit. Intake-air filters remove dust and other particles from the air entering the compressor. A filter at the receiver discharge prevents foreign matter from causing malfunction of the regulator. The regulator, desirable in most compressed-air systems, maintains a constant pressure, regardless of the rise and fall of line pressure at the compressor.

Pneumatic systems serving instruments and close-tolerance valves should be provided with refrigerated dryers to reduce the moisture content, and coalescing filters to remove oil vapors from the air.

7.6   Communication

An owner of a WRRF often owns or leases the facility communication system from one of the many communication companies currently offering such services.

Features provided by the communication system vary, depending on the specific needs of the WWRF. Along with standard incoming, outgoing, and internal voice communications, it is possible to provide intercommunication with on- and off-site mobile vehicles, pocket personnel pagers, on-site fixed-paging systems, and data communications.

Within the treatment-facility site, current practice is to provide telephones in each major structure or building and at specific workstations within the buildings. In addition, paging speakers are located appropriately to provide full coverage of the WRRF grounds and in unmanned areas, such as service tunnels or process tankage. While telephone conduits and low-voltage electrical conduits may be routed together through common electrical-duct banks, it generally is not good practice to install telephone cable and electrical power or control wiring in the same conduit.

A designer should review specific features of the communication system with the owner of the WRRF during the design process. Once the required features of the communication system have been defined, specific proposals should be solicited from system vendors, regardless of whether the communication system is to be owned or leased.

Many communication systems (e.g., phone, intercom, and paging) are using voice-over-Internet-protocol. This type of system uses Ethernet cabling for communications between devices. It also allows for multiple devices to be connected to the same cable system, reducing the amount of wire needed for a given installation.

7.7   Location of Devices

It is important for communications devices be installed in critical areas, allowing for easy communications between facility staff. Locations also should take into consideration facility staff safety. Considerations should be made to include two-way communications devices (phone/intercom) at each critical control center/panel location, which would allow for reasonable operation while making critical facility operational viewing and control-system adjustments. Proximity and access also should be considered in tunnels, pipe galleries, and so on, which are reached easily in the event of an emergency. These communications locations also should be clearly marked (e.g., using a blue light) for easy identification.

8.0   References

American National Standards Institute (2017) National Electrical Safety Code.

American Water Works Association; American Society of Civil Engineers (2012) *Water Treatment Plant Design*, 5th ed.; McGraw-Hill: New York.

American Society of Heating, Refrigerating and Air Conditioning Engineers (2013) *Handbook of Fundamentals*, I–P ed.; American Society of Heating, Refrigerating and Air Conditioning Engineers: Atlanta, Georgia.

Building Officials and Code Administrators International (2006) *BOCA National Plumbing Code*; Building Officials and Code Administrators International, International Code Council: Washington, D.C.

Great Lakes–Upper Mississippi River Board of State and Provincial Public Health and Environment Managers (2014) *Recommended Standards for Wastewater Facilities*; Health Education Services: Albany, New York, http://www.hes.org (accessed November 2008).

Illuminating Engineering Society of North America (2011) *Lighting Handbook*, 10th ed.; Illuminating Engineering Society of North America: New York.

Institute of Electrical and Electronic Engineers (1995) *Recommended Practice for Emergency and Standby Power Systems for Industrial and Commercial Applications, IEEE Orange Book, ANSI/IEEE standard 446*; Institute of Electrical and Electronic Engineers: New York.

Institute of Electrical and Electronic Engineers (2014) *Recommended Practices and Requirements for Harmonic Control and Electrical Power Systems, ANSI/IEEE standard 519*; Institute of Electrical and Electronic Engineers: New York.

International Association of Plumbing and Mechanical Officials (1994) *The Uniform Plumbing Code*; International Association of Plumbing and Mechanical Officials: Ontario, Canada.

International Code Council (2009) *International Plumbing Code*. International Code Council: Washington, D.C.

Keskar, P. Y. (2002) *Control and Instrumentation Issues in Upgrading Existing Chemical Feed Systems in Water/Wastewater Plants*, ISA 2002 Technical Conference Paper, Chicago, Illinois; ISA: Research Triangle Park: North Carolina.

Metcalf and Eddy, Inc./AECOM (2013) *Wastewater Engineering: Treatment, Disposal and Reuse*, 5th ed.; McGraw-Hill: New York.

National Climatic Data Center (1998) *Climatic Wind Data for the U.S.*; National Climatic Data Center: Asheville, North Carolina.

National Climatic Data Center (2002) *Climatography of the United States, No. 81, Supplement No. 2, 1971–2000*; National Climatic Data Center: Asheville, North Carolina.

National Fire Protection Association (2016) *Recommended Practice for Fire Protection in Wastewater Treatment and Collection Facilities, NFPA Standard 820*; National Fire Protection Association: Quincy, Massachusetts.

National Fire Protection Association; American National Standards Institute (2014) *National Electrical Code, an American National Standard, NFPA No. 70-2014 ANSI C1-2014*; National Fire Protection Association: Quincy, Massachusetts.

National Fire Protection Association (2015) *Standard for Electrical Safety in the Workplace, NFPA Standard 70E*; National Fire Protection Association: Quincy, Massachusetts.

Strope, C. (1994) *Nonlinear Loads: The Best Defense*. Consult. Specifying Eng., Nov, 42.

U.S. Environmental Protection Agency (1974) *Design Criteria for Mechanical, Electrical, and Fluid System and Component Reliability*; U.S. Environmental Protection Agency: Washington, D.C.

U.S. Environmental Protection Agency (1975) *Process Design Manual for Suspended Solids Removal, EPA-625/1-75-003a*; U.S. Environmental Protection Agency: Washington, D.C.

U.S. Environmental Protection Agency (1979) *Chemical Aids Manual for Wastewater Treatment Facilities, EPA-430/9-79-018*; U.S. Environmental Protection Agency: Washington, D.C.

U.S. Environmental Protection Agency (1987) *Design Manual Phosphorus Removal*; EPA-625/1-87-001; U.S. Environmental Protection Agency: Washington, D.C.

Water Environment Federation (2005) *Upgrading and Retrofitting Water and Wastewater Treatment Plants, Manual of Practice No. 28*; Water Environment Federation: Alexandria, Virginia.

Water Environment Federation (2013) *Automation of Water Resource Recovery Facilities, Manual of Practice No. 21*, 4th ed.; Water Environment Federation: Alexandria, Virginia.

Water Environment Federation (2016) *Operation of Water Resource Recovery Facilities*, 7th ed., Manual of Practice No. 11; Water Environment Federation: Alexandria, Virginia.

White, G. C. (2010) *Handbook of Chlorination and Alternative Disinfectants*, 5th ed.; Wiley & Sons: New York.

9.0   Suggested Readings

American Conference of Governmental Industrial Hygienists (2004) *Industrial Ventilation, A Manual of Recommended Practice*, 25th ed.; American Conference of Governmental Industrial Hygienists: Lansing, Michigan.

American Society of Heating, Refrigerating and Air Conditioning Engineers (2004) *Energy Standard for Buildings Except for Low-Rise Residential Buildings, ASHRAE Standard 90.1*; American Society of Heating, Refrigerating and Air Conditioning Engineers: Atlanta, Georgia.

American Society of Heating, Refrigerating and Air Conditioning Engineers (2004) *Thermal Environmental Conditions for Human Occupancy, ASHRAE Standard 55-2004*; American Society of Heating, Refrigerating and Air Conditioning Engineers: Atlanta, Georgia.

American Society of Heating, Refrigerating and Air Conditioning Engineers (2004) *Ventilation for Acceptable Indoor Air Quality, ASHRAE Standard 62.1 2004*; American Society of Heating, Refrigerating and Air Conditioning Engineers: Atlanta, Georgia.

National Fire Protection Association (2009) *NFPA 90A: Standard for the Installation of Air-Conditioning and Ventilating Systems*; National Fire Protection Association: Quincy, Massachusetts.

National Fire Protection Association (2008) *Standard on Clean Agent Fire Extinguishing Systems, NFPA 2001*; National Fire Protection Association: Quincy, Massachusetts.

Sheet Metal and Air Conditioning Contractors' National Association (2006) *HVAC Duct Construction Standards—Metal and Flexible*, 3rd ed., ANSI/SMACNA 006-2006; Sheet Metal and Air Conditioning Contractors' National Association: Chantilly, Virginia.

Smeaton, R. W.; Ubert, W. H. (1998) *Switchgear and Control Handbook*, 3rd ed.; McGraw Hill: New York.

Texas Commission on Environmental Quality (2009) *Homeland Security and the TCEQ*; Texas Commission on Environmental Quality: Austin, Texas, http://www.tceq.state .tx.us/comm_exec/homelandsecurity.html (accessed May 30, 2008).

Turner, W. (2004) *Energy Management Handbook*, 5th ed.; Fairmont Press: Atlanta, Georgia.

U.S. Environmental Protection Agency (2004) *Homeland Security Strategy*; U.S. Environmental Protection Agency: Washington, D.C.

U.S. Green Building Council (2008) *LEED for Existing Buildings: Operations & Maintenance*; U.S. Green Building Council, Washington, D.C.

U.S. Green Building Council (2005) *LEED for New Construction*; U.S. Green Building Council, Washington, D.C.

CHAPTER 8

Materials of Construction and Corrosion Control

Robert A. (Randy) Nixon; Douglas Sherman, P.E.; Jim Joyce;
Christopher Hunniford, P.E; and Brian Huang

449

1.0   Common Corrosion Mechanisms

1.1   Introduction

The objectives of this Chapter are to acquaint the reader with the types of corrosion common to water resource recovery facilities (WRRFs) and with the primary means of preventing or controlling the associated corrosive damage to metal and non-metal materials of construction.

Section 1 describes the major corrosion mechanisms that cause damage to metals and non-metals in these facilities.

Section 2 explains the primary methods of corrosion control including process control, the use of alternate materials of construction, the application of barrier protection with coatings and liners, and electrochemical control of corrosion by cathodic protection.

Section 3 provides general corrosion control design guidance for all of the most common unit processes for a WRRF.

Section 4 presents corrosion control design guidelines for material selection for facility support systems. This covers Heating, Ventilating, and Air Conditioning (HVAC) systems, electrical and instrumentation systems, odor control facilities, and chemical addition storage and distribution facilities.

Controlling and preventing corrosion in WRRFs is a constant battle. The failure to control the damaging effects of corrosion can result in significant financial costs to the facility and its rate payers. There are several different corrosion-causing mechanisms common to WRRFs. Some of these corrosion mechanisms affect both metals and concrete surfaces and structures while others are intrinsic to metals corrosion only. In the following paragraphs, the corrosion damage mechanisms most commonly encountered in WRRFs and their technical explanations are discussed. A firm grasp of these corrosion damage mechanisms makes their prevention methods more understandable and identifiable to the design engineer.

Concrete is an inexpensive and very versatile material of construction, which comprises the majority of surfaces exposed to corrosion in WRRFs. In addition to its great strength and compatibility with water, it is a porous matrix of inorganic compounds with different chemistries. Most corrosive damage to concrete in WRRFs involves the chemical reaction of these hydrated cement paste matrix constituents with process chemistry and environmental exposure. Preventing these chemical reactions and slowing the penetration of aggressive corrodents, such as chlorides, into the matrix through the concrete's porous structure are fundamental to concrete corrosion prevention in water resource recovery applications.

The corrosion of metals in these facilities can best be understood when one realizes two fundamental concepts. First, metals are made from natural ores taken from the ground, and their manufacture involves adding a great deal of energy to them. Generally, that energy involves the addition of heat to bring about a fundamental change in their molecular structure. Thereafter, that energy is locked up in those metals, and each metal has the natural tendency to return to its original oxide or raw ore form. When this natural force of nature is manifested via metallic corrosion, that pent-up energy is released back as the electrochemical energy of corrosion. All metals have a particular energy level or electrochemical potential (voltage) relative to one another. When a given metal returns to its original oxide or ore form through corrosion, that potential energy, or voltage, is released causing electrical current (electrons) to flow. Differences between the electrical potential or energy levels of metals are what drive corrosion rates in both singular metal corrosion and in galvanic, or dissimilar metals, corrosion.

Second, metals can sometimes avoid corrosion by forming a natural protective barrier of oxide film or other passive film that isolates the metal from the corrosive environment or helps it resist corrosion. This tendency for a protective, passive film formation is critical to the understanding of metals corrosion. As long as the environments in which metals are exposed permit protective or passive film formation, and the stability of the film can be maintained, metals will not corrode or will corrode lightly. However, when the environment includes conditions that breakdown or destabilize the passive or protective films, the metals will actively corrode. Understanding the conditions that create, maintain, or breakdown a passive or protective film is one of the keys to making sound materials selection decisions in WRRFs.

Metals-related corrosion mechanisms are found in all municipal WRRFs and will be further identified and discussed in the following subsections of this chapter.

1.2 Corrosion due to Biogenic H₂S Related Acid Generation

Strong mineral acids, such as sulfuric acid (H_2SO_4), corrodes and destroys concrete, mortar, and most metals used in the construction of WRRFs. Under certain conditions, biological activity can generate concentrated sulfuric acid on surfaces above the waterline, which are exposed to hydrogen sulfide gas (H_2S). Typically produced in the collection system by a submerged anaerobic biofilm, hydrogen sulfide gas is converted to sulfuric acid by a second aerobic biofilm living on surfaces above the water. Figure 8.1 illustrates this process in a sewer. Just like sewers, many parts of a WRRF have these conditions and can experience this very damaging form of biological acid generation.

This dual (aerobic and anerobic) biofilm corrosion process starts with the rapid growth of a black, mucous-like biofilm on all surfaces continuously under wastewater, such as in influent pump stations, channels, grit chambers and primary clarifiers, to name a few. This biofilm is composed of anaerobic "sulfate reducing bacteria" (SRB) that consume organics

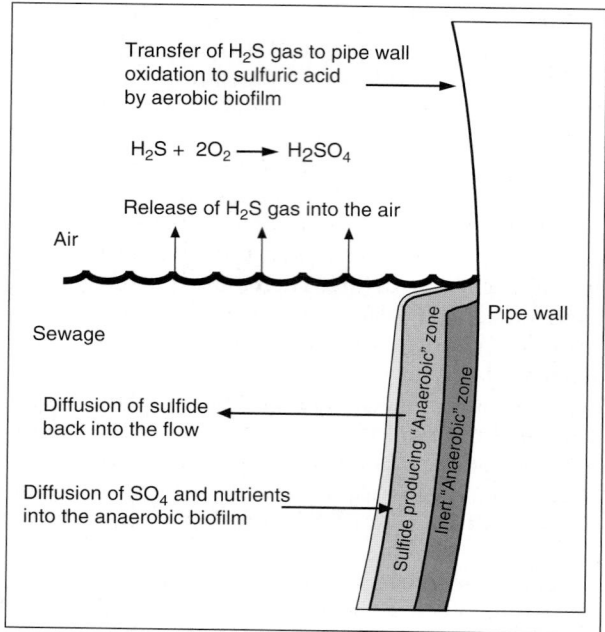

Transfer of H₂S gas to pipe wall
oxidation to sulfuric acid
by aerobic biofilm

$$H_2S + 2O_2 \longrightarrow H_2SO_4$$

Release of H₂S gas into the air

Air

Sewage

Diffusion of sulfide
back into the flow

Diffusion of SO₄ and nutrients
into the anaerobic biofilm

Pipe wall

Sulfide producing "Anaerobic" zone

Inert "Anaerobic" zone

FIGURE 8.1 Biogenic acid generation in sewers.

from the wastewater and use the sulfate ion ($SO_4^=$), a common component of wastewater, to oxidize them. The reduction of sulfate by SRB also generates alkalinity.

There is always enough $SO_4^=$ in the wastewater to generate sufficient sulfide to cause corrosion. If wastewater sulfate concentrations are high (>100 mg/l), increased hydrogen sulfide generation will result. The rate at which hydrogen sulfide is produced by the anaerobic biofilm depends upon a variety of other environmental conditions, including the concentration of organic food source (biological oxygen demand or BOD), dissolved oxygen concentration, temperature, and other factors. Because sulfide generation in wastewater is a biological process, warmer wastewater temperatures cause an increase in metabolic activity, resulting in an increase in hydrogen sulfide production.

The anaerobic biofilm releases sulfide back into the water where it immediately establishes a chemical equilibrium between four forms of sulfide: the dissolved sulfide ion ($S^=$), the dissolved bisulfide ion (HS^-), dissolved (aqueous) hydrogen sulfide gas ($H_2S_{(aq)}$), and hydrogen sulfide gas ($H_2S_{(g)}$). Only the $H_2S_{(g)}$ can exist in the air above the water as a gas, where it can be recognized by its characteristic "rotten egg" odor. It is important to note that this equilibrium is dynamic and reversible. This means that whenever $H_2S_{(g)}$ is released, dispersed, or removed by odor control the remaining dissolved forms of sulfide automatically shift to maintain the equilibrium. When this happens, $H_2S_{(aq)}$ shifts to replace the lost $H_2S_{(g)}$ and more HS^- shifts to replace the $H_2S_{(aq)}$ as shown in the equilibrium (eq. 8.1)

$$H_2S_{(g)} \leftrightarrow H_2S_{(aq)} \leftrightarrow HS^- \leftrightarrow S^= \qquad (8.1)$$

hydrogen sulfide gas	hydrogen sulfide gas (dissolved)	bisulfide ion	sulfide ion

The release of $H_2S_{(aq)}$ to the airspace as $H_2S_{(g)}$ is a function of the pH of the wastewater, the concentration of dissolved $H_2S_{(aq)}$, and turbulence of the wastewater. Figure 8.2 illustrates the relationship between $H_2S_{(aq)}$ concentration, HS⁻ concentration, and pH of the wastewater.

Figure 8.2 shows that approximately half of all the dissolved sulfide in wastewater exists in the $H_2S_{(aq)}$ form and the other half exists as HS⁻ at a neutral pH of 7.1. As the pH of the water decreases, the fraction of the releasable $H_2S_{(aq)}$ form of sulfide dominates the equilibrium and increases the release of $H_2S_{(g)}$. Turbulence of the wastewater dramatically increases the release of $H_2S_{(g)}$. When the area of the wastewater/air interface is increased by splashing (such as a spraying force main) or aeration (such as in an aerated grit chamber), the transfer of $H_2S_{(aq)}$ to $H_2S_{(g)}$ is dramatically increased. Even minor turbulence of water flowing over a weir in a primary clarifier can cause excessive hydrogen sulfide gas release and result in severe damage if dissolved sulfide is present.

Only $H_2S_{(g)}$ can be converted to sulfuric acid by the aerobic biofilm after it is released to the airspace above the water surface. This second aerobic biofilm consists primarily of Acidithiobacillus bacteria, which consume $H_2S_{(g)}$ and oxidize it to sulfuric acid according to eq. 8.2.

$$\text{Acidithiobacillus}$$
$$H_2S_{(g)} \quad + \quad 2\,O_2 \quad \leftrightarrow \quad H_2SO_4 \tag{8.2}$$

These bacteria can produce enough acid to drop the surface pH below 1 when the surface is continuously exposed to $H_2S_{(g)}$ concentrations over 25 ppm, or even at lower gas concentrations if the surface is covered and keeps the gas from escaping.

Concrete surfaces are particularly susceptible to biogenic acid corrosion. The alkaline matrix of concrete reacts quickly with the acid to produce a layer of soft corrosion product consisting primarily of calcium sulfate (gypsum). This layer of corrosion product provides some level of protection for the sound concrete below. While it is preferable to prevent the concrete from corroding in the first place, this soft corroded layer should not be intentionally washed away due to the increased rate of corrosion that will subsequently occur.

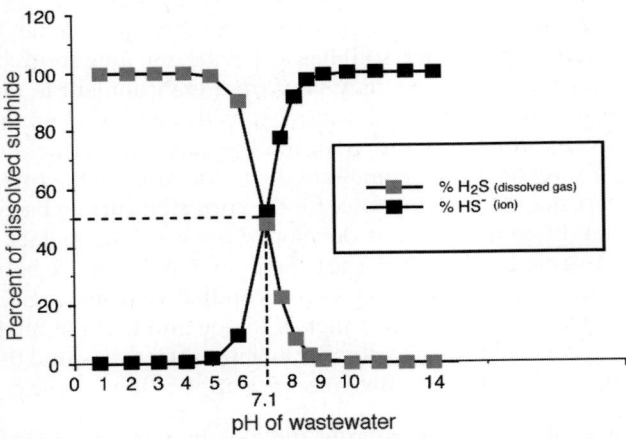

FIGURE 8.2   pH Effect on sulfide equilibrium distribution.

1.3   Electrolytic Oxygen-Driven Corrosion of Ferrous Metals

This form of metallic corrosion occurs in immersion service or cyclical wetting and drying conditions. In near-neutral solutions, like most municipal wastewater, the factors that influence the corrosion of ferrous metals in immersion service include oxygen activity, pH, temperature, flow rate, and numerous other contributors. The most important of these factors is aeration: the amount of oxygen that reaches the surface of the ferrous metal has the primary influence on the corrosion rate. After oxygen content, the relative acidity or pH of the water has the most influence on corrosion rate. At lower pH, the evolution of hydrogen tends to prevent the possibility of protective film formation so carbon steel or ductile iron will continue to corrode. In alkaline solutions, the formation of protective films is typically enhanced such that the corrosion rates are greatly reduced for carbon steels and ductile iron. This corrosion mechanism is typically identified on coated carbon steel primary and secondary clarifier rake mechanisms and in screening applications, among many other unit processes in the WRRF.

Electrolytic corrosion of ferrous metals is also very prevalent throughout WRRFs in the form of environmental exposure and weathering-related conditions. Electrical cabinets, structural steel, exposed piping, piping and equipment supports, building doors and door frames, and many other structures invariably show evidence of oxygen-driven corrosion where rain and condensation, and drying cycles result in metal damage. The rate of atmospheric exposure corrosion is primarily influenced by the extent of wetting due to humidity, condensation, and rainfall, along with the presence of atmospheric pollutants. Atmospheric corrosion is dealt with in more detail later in this chapter.

1.4   Galvanic Corrosion

Galvanic corrosion occurs in many situations at WRRFs and causes deterioration of most common metals including carbon steel, zinc, and aluminum. Galvanic corrosion occurs when two dissimilar metals with different surface electrochemical potentials are in contact with each other and are present in a common electrolyte (water). For reference, Table 8.1 presents a galvanic series for metals exposed to a certain electrolyte (in this case, seawater). The pure metals and alloys are arranged in order of their corrosion potentials relative to a reference half-cell, starting with the most electronegative (or "active") and progressing to the most electropositive (or "noble"). Note that each electrolyte (seawater, fresh water, soil) has a specific galvanic series, and the relative positions of the metals and alloys may vary from environment to environment. Galvanic series for groundwater and wastewater exposures will be close to that for seawater.

In a galvanic corrosion situation, the less noble metal (less corrosion-resistant or anodic metal) corrodes or becomes anodic to the more noble (more corrosion-resistant or cathodic) metal. The driving force for the corrosion current becomes the electrochemical potential difference that has developed between the metals. The most influential factors in galvanic corrosion rates are the potential difference between the two metals; the environmental conditions such as pH, conductivity, and chemistry of the electrolyte (water); the proximity of the two metals to one another; the relationship between the size of the exposed anodic and cathodic metal surface areas; and the polarization behavior of the metals. In water resource recovery applications, some common examples are:

1. The preferential corrosion of the zinc in galvanized steel relative to exposed and active carbon steel surfaces;

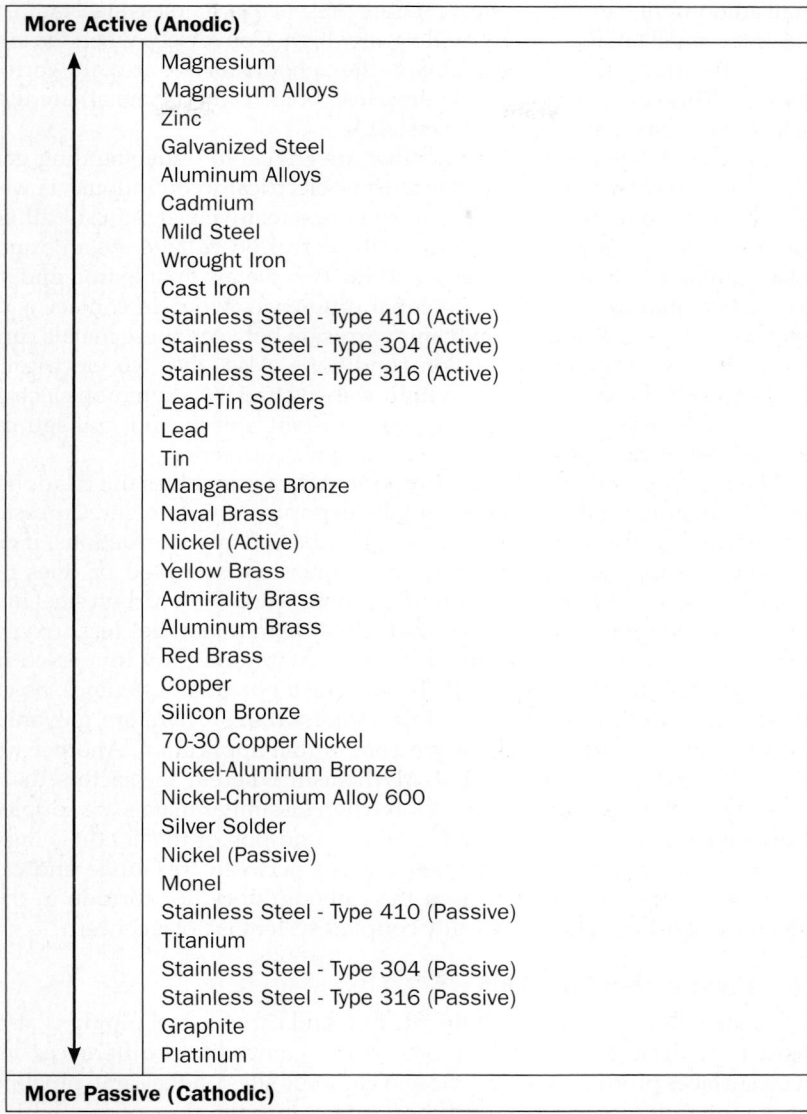

More Active (Anodic)
Magnesium
Magnesium Alloys
Zinc
Galvanized Steel
Aluminum Alloys
Cadmium
Mild Steel
Wrought Iron
Cast Iron
Stainless Steel - Type 410 (Active)
Stainless Steel - Type 304 (Active)
Stainless Steel - Type 316 (Active)
Lead-Tin Solders
Lead
Tin
Manganese Bronze
Naval Brass
Nickel (Active)
Yellow Brass
Admirality Brass
Aluminum Brass
Red Brass
Copper
Silicon Bronze
70-30 Copper Nickel
Nickel-Aluminum Bronze
Nickel-Chromium Alloy 600
Silver Solder
Nickel (Passive)
Monel
Stainless Steel - Type 410 (Passive)
Titanium
Stainless Steel - Type 304 (Passive)
Stainless Steel - Type 316 (Passive)
Graphite
Platinum
More Passive (Cathodic)

TABLE 8.1 Galvanic Series of Pure Metals and Alloys in Seawater at 25°C. Courtesy of Corrosion Probe, Inc.

2. The corrosion of immersed aluminum gates and gate frames when electrically coupled to the carbon steel reinforcing bars in concrete via anchor bolts; or

3. The active pitting corrosion of coated carbon steel relative to nearby and electrically connected stainless steel surfaces.

The anodic-to-cathodic surface area relationship is also a critical factor in the rate of galvanic corrosion. When the majority of the surface area behaves cathodically (like when small amounts of bare carbon steel pipe are present in a stainless steel tank), the corrosion rate of the small anodic surfaces will be very high. Conversely, when the anodic (corroding metal) surface area is large relative to the cathodic surface area, the corrosion rate will be lower. This can occur when stainless steel bolts (cathodic metal) are used on carbon steel pipes or pipe hangers (anodic metal).

There are a few essential factors that are critical to understanding galvanic corrosion. First, the two dissimilar metals must be electrically continuous via welds or bolted connection. Second, the two metals must have a relatively significant difference in electromotive energy, or potential, to drive the corrosion voltage. As an example, the two metals aluminum and carbon steel, and the two metals ductile iron and stainless steel have substantial electromotive potential differences when in contact with each other (refer to Table 8.1). Therefore, galvanic corrosion between these metals can be expected under most wastewater immersion conditions. Also, the two electrically connected metals must be immersed in, or continuously wetted by, a common electrolyte (liquid). If the metals, are not fully immersed, constant splash and spillage can create an immersion-like condition conducive to galvanic corrosion.

The last, and perhaps the most important, factor involves the relationship between the two dissimilar metal surfaces and the exposure environment. Corrosion is an electrochemical process in which there are cathodic and anodic reactions. If either of these reactions is suppressed or missing, then corrosion is limited or does not occur. An example is when stainless steel is used in conjunction with carbon steel in an anaerobic digester, an environment that lacks reducible chemical species (e.g., oxygen or sulfur). Even though the two metals are electrically continuous, are immersed in a common electrolyte, and have a substantial electromotive potential difference, no cathodic reactions occur on the carbon steel or stainless steel surfaces. Therefore, galvanic acceleration of carbon steel corrosion is not a problem in that application. Another way to look at this is that galvanic corrosion is a mechanism wherein the factors discussed above accelerate (rather than create) corrosion when dissimilar metals are coupled together in a corrosive electrolyte. Thus, the corroding condition of the anode is important. Even when a significant potential difference exists between the anode and cathode, if the anode (electrically uncoupled from the cathode) does not corrode in the electrolyte, galvanic corrosion in the electrically coupled system is not a concern.

1.5 Soil-Related Corrosion

Soil-related corrosion of buried ductile iron and carbon steel piping is also common in WRRFs. In these situations, microstructural composition differences in the ferrous metal surfaces produce both anodic and cathodic sites on the same pipeline. The resulting potential differences between these sites cause the flow of electrons (an electrical corrosion current) from the anodic sites to the cathodic sites. Several factors influence soil corrosion rates. The most important of these factors are the moisture content of the soil that serves as the electrolyte, the availability of oxygen, the resistivity of the soil (the inverse of conductivity), and the extent of chemical contamination in the soil. The presence of chlorides, sulfates, sulfides, and other chemical constituents can increase soil conductivity (or lower soil resistivity) and cause degradation of the passive films formed on the ferrous metal surfaces.

Soil-related corrosion of metals can also be caused by stray current effects. This generally involves cases where direct electrical current (DC) leaves one structure and jumps through the soil to another structure or pipeline. When that current again jumps or leaves the unintended electrical conduit, corrosion occurs. Stray current corrosion is very common in cities where 3rd rail electrical transit systems are used for public transportation. Stray current can occur on metal pipelines at WRRFs when those conduits cross or are adjacent to other pipelines that are protected by impressed current (IC) cathodic protection systems. Both of these exposures use DC, which is the most common source of stray current corrosion in buried pipelines.

1.6   Graphitic Corrosion of Cast and Ductile Iron

Graphitic corrosion of cast and ductile iron is another corrosion mechanism that is often encountered at WRRFs due to the extensive use of ductile iron piping and the common presence of cast iron piping in older facilities.

Graphitic corrosion involves the gradual, selective leaching of iron from cast or ductile iron pipes or structures in slightly acidic waters and soils, especially those containing sulfates and SRB. The attack usually proceeds fairly uniformly inward from the surface, leaving a porous network of mostly graphite behind. Upon initial examination, the pipe will not appear to be badly corroded. For example, a corroded cast iron pipe may have lost the majority of its iron, with no apparent change in volume or wall thickness. However, the remaining graphite network is brittle, soft, and easily scraped away. It is also worth noting that a completely graphitically corroded, buried cast iron pipe may continue to hold water under pressure until, for example, the earth around the pipe shifts.

Graphitic corrosion rates are generally low and the damage of consequence (causing of leaks) typically takes many years. Ductile iron, with its different graphite structure, is much less susceptible than cast iron to this mechanism, but is not immune. Graphitic corrosion is sometimes, incorrectly, referred to as "graphitization," which is actually a metallurgical phenomenon that occurs in carbon and low alloy steels at high temperatures.

1.7   Under-Deposit Corrosion

Localized pitting corrosion of carbon steel and ductile iron can occur in WRRF applications where bioactive sludge and organic materials collect and persist for long periods. This is very common in headworks, grit facilities, and clarifier structures. Localized deposits of these bioactive materials collect on bare metal surfaces or over breaches in protective coating systems where tubercules form from the combination of weak organic acids and the iron corrosion products. These tubercules or deposits create low-oxygen conditions at the metal surface (or in surface pits) and oxygen concentration cells form. The slow biological production of organic acids on the surface causes acidification within the pit, promoting more aggressive corrosion. This under-deposit corrosion is exacerbated if the anaerobic bacteria metabolize various sulfur species to form dilute sulfur acids, rather than strong sulfuric acid. The ever-present SRB in wastewater, which thrive in anaerobic conditions beneath biofilms and deposits, contribute to this form of microbiologically influenced corrosion (MIC). What is also important to note is that SRB-related MIC also occurs under nominally aerated environments where anaerobic microenvironments exist under biodeposits of aerobic organisms, especially in crevices built into structures. Such crevices are commonplace in coated steel clarifier rake mechanisms and in other WRRF structures. MIC is discussed further below with specific reference to stainless and carbon steels.

1.8   Erosion-Corrosion

Erosion-corrosion is the combined action of corrosion and erosion, most commonly identified in WRRFs in the presence of a moving fluid containing solid particles, which leads to rapid corrosion and accelerated loss of material. Fluid flow by itself, or most commonly flowing fluids in combination with suspended solids, can cause this form of corrosion, and carbon steel and ductile iron pipelines are most susceptible. Erosion-corrosion is most common in grit piping and in sludge-handling piping, particularly ductile iron elbows and reducing fittings where flow direction changes and velocity increases occur. This problem is well documented in the invert of return activated sludge (RAS) piping and near RAS pumps, especially in facilities with poor grit removal. The moving solids prevent the formation of any protective surface film and constantly scour the surface, increasing the rate of corrosion and loss of material.

1.9   Microbiologically Influenced Corrosion

As discussed in Section 1.7, MIC, occurs commonly in WRRFs under both anaerobic and aerobic conditions. MIC is not to be confused with corrosion due to biogenic sulfuric acid generation discussed in Section 1.2. As discussed previously, MIC typically occurs in the anaerobic phase in wastewater systems where certain conditions exist. Welds, heat-affected zones (HAZs) of welds, and crevice locations on metals are often susceptible to MIC in stagnant and low-flowing wastewater conditions. This process involves the accumulation of an underwater anaerobic biofilm that generates mild organic acids directly on the metal surface. The organic acids create the acidic conditions that initiate and propagate pitting corrosion. In WRRFs, the most common form of anaerobic MIC is associated with the ever-present SRB that accumulate on metal rake mechanisms in clarifiers and dissolved air flotation thickeners (DAFTs). Not to be confused with corrosion due to biogenic strong acid generation, this type of corrosion is also frequently observed under water in RAS and waste activated sludge (WAS) piping, where stagnant or low-flow conditions exist for extended periods of time. One caution for MIC is the need to carefully monitor the condition of new, Type 316 stainless steel primary clarifier mechanisms. Thick sludge blankets in primary clarifiers can produce MIC-favoring conditions even for 316 stainless steels, where anaerobic biofilms can cause localized pitting and crevice corrosion. This is especially true where the 316 stainless steel is roughened by grinding, scraping, or rubbing. The rough areas provide an anchor for the MIC biology and allow them to remain in thick films. Monitoring of the performance of such new metal mechanisms is prudent.

Other forms of MIC encountered in WRRFs include corrosion of stainless steels due to manganese-oxidizing bacteria and corrosion of iron and steel by iron-depositing bacteria. These mechanisms occur in biologically active natural waters. Manganese-oxidizing bacteria form complex deposits of microbial cells and other organic and inorganic debris that accelerate corrosion by changing the electrochemical behavior of stainless steels. Iron-depositing bacteria produce oxygen concentration cells that divide the metal surface into small anodic sites and large surrounding cathodic zones, which leads to localized corrosion at the anodic sites.

1.10   Localized Corrosion of Stainless Steels

Stainless steels resist corrosion differently than carbon steels and many other metals because stainless steels do not form surface protective films that are true oxide barriers that separate the metal from the environment. Stainless steels form a passive film that is dependent upon the presence of oxygen. Stainless steels resist corrosion best when they are exposed to an environment with ample oxygen present and when the surfaces are free of deposits, roughened areas, and crevices. If a portion of the metal surface is covered by coatings, biofilm buildup, gasketed connections, or other fabrication conditions that create oxygen-depleted zones, the oxygen-depleted areas become anodic relative to the well-aerated surfaces exposed to flowing conditions. These anodic areas actively corrode under such conditions if they are allowed to persist over time.

Passive film formation for any specific type of stainless steel is immediate provided oxygen is present and there are no aggressive chemicals that might disrupt or break down the passive film. It is for these reasons that the most common form of stainless steel corrosion involves localized corrosion in the form of pitting or crevice corrosion. General thinning or uniform corrosion of stainless steels is very rare and typically involves exposing the metal to a severe reducing environment where passive film formation is prevented altogether. This generally does not occur in a WRRF.

Pitting and crevice corrosion are the most common forms of stainless steel corrosion expected in municipal wastewater environments. Pitting corrosion is associated with a local discontinuity in the passive film. It can be caused by a mechanical discontinuity such as a rough weld, a covered area, or grinding damage to the metal surface. It can also be promoted by local chemical breakdown of the passive film. Chloride ion is the most common chemical agent that promotes the pitting corrosion of stainless steels. Once a pit is formed, the corrosive species accumulates and prevents passive film reformation, making the local pit environment more aggressively corrosive than the surrounding bulk environment. Hence, pitting corrosion rates can be high for stainless steels. Higher flow rates over the metal surface can reduce pitting corrosion rates because the high hydraulic shear forces lower the concentration of the corrosive species within the pit.

Crevice corrosion should basically be considered a severe form of pitting corrosion. Crevices are created by biofouling (underdeposit type corrosion as discussed in Section 1.7), gasketed flanged surfaces, or other mechanical connections or flaws in structures that create conditions where oxygen-depleted regions develop. In WRRFs, pitting and crevice corrosion mainly occur where relatively high chloride concentrations are present or where biofilm buildup can proceed under low-flow or stagnant water conditions. Chloride concentrations can become elevated in wastewater due to saltwater infiltration into collection systems in coastal or brackish water areas. This is a common issue of concern in facilities located on islands, or near coastlines, oceans or their estuaries.

The use of ferric chloride as a flocculation agent or for sulfide control in influent wastewaters can increase chloride concentrations and accelerate the chemical breakdown of surface protective passive films of both 304 and 316 stainless steels (the most commonly used grades in WRRFs). When chloride concentrations approach 300 to 400 ppm, Type 304 stainless steel suffers from pitting and crevice corrosion. Type 316 is slightly more resistant than Type 304, with its chloride pitting threshold closer to 1000 ppm under near-neutral pH wastewater conditions. Ferric chloride exposures are more severe because this chemical also lowers the pH of the environment, which makes the chloride pitting resistance of the metal even lower.

Areas in the WRRF where organic and biological deposition creates anaerobic conditions must also be monitored. This can involve underdeposit, localized corrosion of stainless steel aeration piping in aeration basins, anaerobic digesters, and DAFTs.

Areas where localized stainless steel corrosion can be problematic include:

- Crevice corrosion and pitting corrosion of primary clarifier rake mechanisms and other internal, submerged metallic components (particularly where chloride concentrations in the wastewater are high).

- Localized pitting corrosion of stainless steel exposed to moderate concentrations of ferric chloride solutions.
- Underdeposit corrosion of stainless steel at weld HAZ's and potential crevice locations where present within anaerobic sludge or stagnant wastewater.

1.11   Corrosion of Aluminum Alloys

Aluminum alloys resist corrosion by forming a surface protective barrier of aluminum oxide film that is well-adhered to its surface and, if damaged, reforms rapidly in most environments. However, aluminum alloys are also very reactive metals from a thermodynamic and electromotive standpoint. This means that aluminum will corrode actively relative to most other metals. Only beryllium and magnesium are more reactive than aluminum.

Aluminum and its alloys are also "amphoteric" meaning that they can effectively resist corrosion in waters or wet exposures with a pH range between 4.5 and 8.5. This generally covers wastewater exposure except for headspace conditions where the presence of hydrogen sulfide gas can cause biogenic formation of sulfuric acid, which can reduce the surface pH well below 4.5. This typically results in localized pitting corrosion of the alloy, or complete dissolution of the metal in severe situations. The extent of this type of pitting and the corrosion rate of the aluminum is dependent on a number of factors in wastewater headspaces. These factors include the specific metallic alloying additions, degree of atmospheric condensation, temperature, airflow (related to odor control ventilation), and varying hydrogen sulfide gas concentrations. For example, at some WRRFs, relatively new aluminum odor control covers on primary clarifiers showed early evidence of active pitting corrosion from sulfuric acid, but the corrosion rate was very low and extent of pitting over time limited. As in this case, it has been noted that in hot, arid environments there can be moisture condensation present on the underside of such aluminum covers during evenings or on a cooler day, but almost no moisture present during the heat of the day or on a warmer day. There are two possible reasons for this observation. First, the aerobic biofilm that generates the sulfuric acid requires constant moisture and average temperatures for growth and survival. During daytime or hot conditions there is no condensation on the underside of the aluminum cover, so the high temperatures and lack of water kills the aerobic biofilm. Second, it has been noted that elevated soluble aluminum concentrations (Al^{+3}) can interfere with the metabolism of the bacteria living in the aerobic biofilm, making them less productive.

Occasionally, aluminum corrosion rates are high where the metal is embedded in concrete that is constantly wetted or in cases where the aluminum is exposed to sodium hydroxide.

1.12   Corrosion of Copper and Copper Alloys

Copper is a very noble (low corrosion potential) metal that is almost totally impervious to corrosion from soils and natural waters. Protective copper oxide forms readily on the surface in aerobic environments and its noble nature means that copper is not affected by reducing acids. However, copper and brass pipe or tubing can corrode in a variety of special situations in WRRF applications. Here are some examples:

- The action of stray direct currents (DC) flowing in the ground and/or passing through the copper to ground.

- Certain conditions created by alternating current (AC) exposure—so called electrical interference.

- Thermo-galvanic effects—in certain high-temperature exposures (rare in WRRFs.)

- Where exposed to high concentrations of acids such as in H_2S-rich headspaces in the facility, where the acid-producing aerobic biofilm can exist.

- Perhaps the most observable corrosion impact is a direct sulfidation reaction (inorganic reaction with H_2S gas) to produce a black or dark blue copper sulfide (CuS) film directly on the surface. This is common in headworks facilities where H_2S gas is present and copper and brass are used for water supplies, hose bibs, and bubbler tubes. Although sulfidation is typically not a serious corrosion issue, in extremely high H_2S gas concentrations the copper sulfide surface will flake off, exposing new metal underneath and eventually resulting in a reduced metal thickness and potential failure. The copper sulfide product can also foul threads and fittings making them difficult or impossible to remove or modify should the need arise.

Direct sulfidation can also seriously impact the function of motor control centers (MCCs), wires, and electrical connections as discussed more later in this chapter. Copper components within electrical cabinets and, particularly, variable frequency drives and older motor speed control equipment corrode actively and can fail due to the presence of even low (parts-per-billion) concentrations of hydrogen sulfide gas in the local atmosphere.

Internal corrosion of copper and copper alloys, including brasses and bronzes, by clean or potable-quality waters for which copper pipe typically is used, does not typically occur unless the water flow velocity is excessive, such as in heat-exchanger tubes and pumps.

Brass is subject to dezincification by soft waters: the zinc-rich phase preferentially dissolves and creates a soft, weak, copper-rich material, similar in many ways to graphitic corrosion of grey cast iron. Copper and copper alloys are galvanically cathodic (noble) to steel, cast iron, and aluminum and slightly anodic to stainless steels. As such, these metals rarely corrode galvanically.

1.13 Atmospheric Corrosion

Atmospheric corrosion is mostly a function of sunlight exposure at ambient temperatures and the duration of wetting or wetness for the materials of construction. Sunlight affects the temperature of the material. In hot, wet climates, atmospheric corrosion rates are generally higher due to elevated surface temperatures of metals and extended times of wetness. In colder climates, corrosion rates are typically lower. However, freeze-thaw damage to concrete in temperate zones can also be an issue. Thermal cycling in arid climates has its own special effects on corrosion of structures. Hot days and cool nights with high radiant heat and evaporative cooling effects promote chloride concentration cells on the surface of metals at the water line. For instance, wet-dry cycles increase salt concentration in crevices on stainless steel and other metal surfaces causing increased rates of corrosion. Salt concentration distress of concrete and mortar is also a corrosion factor due to these cyclical thermal forces. Scaling and spalling of concrete and masonry block surfaces can occur at and just above the soil line in facilities in arid climates. These structures suffer the corrosive effects of salt concentration distress as the salts are wicked-up from the ground and into the concrete during cooling cycles. The concrete then dries out during the hot days leaving the salts behind. The salt concentration eventually exceeds the available space in the concrete capillaries, and reacts with materials there creating expansive products and deterioration (spalling) of the concrete.

This same thermal cycling causes thermal expansion and contraction of metals and concrete resulting in high surface stress on any protective coatings. This can cause disbonding of rigid, inflexible coatings from metals as well as cracking of concrete structures. These surface manifestations of thermal distress can promote further ingress of corrodents, chemicals, and water, causing higher rates of many forms of corrosion.

Hot climates also have significant impact on protective coating performance. Coatings must have good ultraviolet (UV) light exposure resistance because UV light in sunlight can destroy organic compounds. Since most coatings used at WRRFs contain organic resin compounds, the light will break down these resins, expose pigments, and cause chalking of the coatings to occur. Over time, this can be very destructive to coatings, eventually allowing corrosion of the substrate. Epoxy coatings generally have poor UV light resistance, while many polyurethane formulations perform well in UV exposure. Many coating formulations contain UV light inhibitor compounds to extend coating life in direct sunlight applications.

In marine/coastal environments, all of the factors discussed above are also important, but chloride exposure in this environment can significantly increase the severity of the corrosion. For example, the concentration effects caused by thermal cycling has even more important implications due to aggressive, chloride-driven corrosion. In addition, marine exposure conditions provide almost daily condensing environments. Therefore, the time of wetness for materials, especially metals, is much higher. This, combined with wet-dry cycling, chloride concentration effects, UV light exposure, and radiant heating, make atmospheric corrosion much higher in most coastal environments.

1.14   Other Concrete Deterioration Mechanisms

Cast-in-place concrete is the main material of construction for secondary containment areas, pump station wet wells, screen chambers, channels, and most tanks and basins in WRRFs. The predominant corrosion or deterioration mechanisms for these structures are described in the following subsections.

1.14.1   Acid Attack (Non-Biogenic)

Non-biogenic acid attack of concrete occurs in WRRFs where alum, ferric chloride, and other acids are used as coagulants, for odor control, and for pH adjustment. Alum, ferric chloride, and chemical acid storage and handling areas are where most of the non-biogenic acidic concrete degradation occurs. All three solutions have a low pH and will readily destroy the alkaline matrix of concrete.

1.14.2   Alkaline Reactions

Alkaline attack occurs in WRRFs where sodium hydroxide for process and odor control is stored and handled and where unattended leaks can occur. This is mainly a corrosion issue in the secondary containment areas.

1.15   Failure of Prestressed Concrete Cylinder Pipe and Reinforced Concrete Pipe

Prestressed Concrete Cylinder Pipe (PCCP) and Reinforced Concrete Pipe (RCP) are commonly used in WRRFs to convey influent, effluent, and intermediate process flows, and can corrode in several different ways. These pipes are almost invariably buried, which leads to the primary causes of failure. This includes breakdown of the natural passivation of the prestressed steel wires in PCCP, conventional RCP reinforcement, and/or the steel cylinders from exposure to chlorides in the soil and ground water. Failure can also include corrosion of the external cement mortar coating from acidic (low pH) soil conditions, including sulfate attack of the cement mortar coating due to high sulfate containing soils. In some cases, exposure to soft groundwater high in carbon dioxide concentration (carbonic acid) has resulted in carbonation attack of the external cement mortar coating. Point failures have also been attributed to stray current interference.

Perhaps the most damaging corrosion of these concrete pipes in WRRFs occurs on their internal surfaces as a result of air pockets and biogenic acid generation. Since these pipes are typically buried and convey flows from headworks to primaries and/or primaries to aeration, they are assumed to be flowing full of water and therefore immune to corrosion due to biogenic acid generation from an aerobic biofilm exposed to hydrogen sulfide gas. However, many facilities have suffered failures of critical, large-diameter piping due to air pockets that become trapped inside the pipes and persist long enough to eat their way through the crown and allow the soil to collapse into the pipe. When water containing dissolved sulfide (raw influent and primary effluent) are dropped into these buried pipes small bubbles are naturally entrained in the water. After the water enters the pipe these bubbles coalesce and form an air pocket in a natural high point. Although the air pocket may be only a few millimeters thick, it contains hydrogen sulfide gas and oxygen—all that is needed to start the growth of Acidithiobacillus bacteria that generate sulfuric acid. As the corrosion progresses, the material in the pocket corrodes away causing the pocket to progress upward and get larger because the pocket is continuously fed a supply of hydrogen sulfide gas and oxygen. Depending upon the concentration of hydrogen sulfide gas (which is increased in turbulent areas), this process can corrode through reinforcing steel in RCP and steel wires and cylinders in PCCP in a few years.

1.16 Chemical-Treatment-Related Corrosion

A number of chemicals are used for the treatment of wastewater and foul air (odor control) in WRRFs. Refer to Table 8.2 in this Chapter for a listing of the most commonly used chemicals, their treatment purposes, and their relative corrosivity. This table provides data for unintentional spills and accidental exposure to secondary containments and leaks, not materials provided with the systems and intended for daily use. Each chemical and its effects on concrete and metals is discussed in Section 3 of this chapter.

2.0 Major Corrosion Control Methods

Corrosion of materials involves the interaction of the material surface with, or response to, process exposure conditions that promote chemical or electrochemical reactions and result in wastage of the construction material. Corrosion of metal and non-metal materials can be prevented or controlled to low acceptable rates using one of five possible methods. They are as follows:

- **Change the process** such that the material is unaffected by the exposure conditions. This is rarely possible or economically feasible in WRRF unit processes. An exception might be to dose wastewater with ferric or ferrous chloride to reduce H_2S gas in the headspaces and therefore reduce biogenic

Chemical	Characteristics	Purpose(s)	Corrosive to Concrete	Corrosive to Metals	Protective Coatings Used
Alum	Low pH	Flocculant to enhance settling/phosphorous removal	Mildly	Yes - ferrous especially	Epoxy mostly, plastic liners for concrete
Sodium hydroxide	High pH	For raising pH	Not unless hot	Generally No	Epoxy, only for secondary containment to prevent spills.
Ferric chloride Ferric sulfate	Low pH; highly oxidizing	For phosphorous removal, as a coagulant and to reduce H_2S in influent	Yes	Yes; even most stainless steels	Epoxy and vinyl ester coatings and plastic liners on concrete
Ferrous sulfate Ferrous chloride	Low pH; not oxidant but reactive	For phosphorous removal as coagulant	Yes somewhat	Yes, not so bad for stainless steels	Epoxy mostly, liners for concrete
Lime (calcium carbonate)	High pH	For phosphorous removal and pH adjustment	No	No	No
Sodium hypochlorite	Alkaline, highly oxidizing	Chlorination/disinfection and reduces H_2S	Yes, especially rebar	Yes for ferrous and stainless steels	Epoxy mostly
Sodium bisulfite	Low pH	Dechlorination	Yes	Yes	Epoxy mostly, plastic liners on concrete
Sulfuric acid	Low pH oxidizing	pH adjustment	Yes	Yes	Epoxy and vinyl ester coatings and liners.
Hydrogen peroxide	High pH, good oxidant	H_2S removal for odor/corrosion control	No	Yes	Epoxy and vinyl ester coatings.
Organic poly; electrolytes or polymers	Neutral pH	Sludge thickening and dewatering	No	No	No

Surface preparation for all exposures listed in this Table are minimum SSPC-SP10 for Steel and SSPC-SP13 for Concrete.

Table 8.2 Chemical Addition in Wastewater Facilities: Why and How Corrosive? Courtesy of Corrosion Probe, Inc.

sulfide corrosion. Or to adjust the pH of sludge upward using chemical addition in order to reduce corrosion of ferrous metals in sludge handling systems.

- **Change the process environmental conditions** from corrosive to noncorrosive by adjusting the environment and limiting one or more of the factors that cause the corrosion. An example would be providing extra ventilation for an enclosed area containing hydrogen sulfide gas and susceptible to corrosion due to biogenic acid generation. By extracting large volumes of air from the process area the relative humidity would be reduced and the concentration of hydrogen sulfide gas would be minimized or reduced below the corrosive limit. Biogenic acid generation requires high humidity and over 2 ppm of hydrogen sulfide gas. If the process area can be continuously maintained below 2 ppm the rate of metallic or concrete corrosion will be negligible.

- **Change the materials of construction** such that the exposed material will be resistant to corrosion under the particular unit process exposure conditions. For example, 316 stainless steel performs very well for rake mechanisms in circular primary clarifiers where wastewater chloride concentrations are below 800–1000 mg/l. It is therefore an excellent choice of materials when compared to epoxy coated carbon steel or, worse still, hot dip galvanized steel. Another example would be the use of high-density polyethylene pipe or fiberglass reinforced mortar pipe for primary wastewater conveyance (gravity flow), to prevent biogenic acid corrosion. Carbon steel and unlined concrete pipe are highly susceptible to biogenic sulfuric acid corrosion in this service exposure.

- **Provide a barrier of protection** for the construction material from the process exposure conditions. This is generally achieved through the use of protective coatings or liners. This can include liquid-applied epoxy, polyurethane, polyurea, or other polymeric-based coatings or the use of manufactured thermoplastic liners such as High density polyethylene (HDPE) or polyvinyl chloride (PVC) anchored or attached sheet liners. Coatings and liners isolate the construction materials from the process exposures with materials that can withstand the corrosive environment and prevent corrosion of the construction material.

- **Use electrochemical protection methods to prevent** corrosion of metals, including metals encased in concrete. These techniques—sacrificial anode or impressed current cathodic protection methods—electrically alter the electrochemical conditions at the corroding metal surface to prevent corrosion. Cathodic protection is widely used to protect metals from corrosion in immersion exposures, such as in tanks, and in soil exposures such as buried steel or ductile iron piping.

One of these five general methods of corrosion control must be used to reduce or mitigate costly corrosion damage wherever it is identified in WRRFs. Sometimes two of these methods can be used in concert to effectively stop or prevent corrosion damage. The most common example is the use of well-adhered dielectric coatings (barrier protection) with cathodic protection (electrochemical protection) to prevent buried pipeline corrosion. Here, the coating does the lion's share of the corrosion prevention through surface isolation from the soil. However, applied coatings often have minor microscopic defects (pinholes and thin areas) that are not 100% sealed. Also, construction activity can introduce scratches and abrasions which leave small openings to the pipe surface. To provide complete corrosion control, cathodic protection provides electrochemical protection at any holes or breaches in the coating. This prevents localized corrosion, which can be aggressive at isolated breaches in the barrier protection. It also protects the integrity of the coating by preventing undercutting corrosion damage at the localized breaches in the coating film.

2.1 Process Control

There are modifications to various unit processes in a WRRF that can be used to prevent or reduce corrosion. The use of chemicals to enhance processes is common but typically related to the need to get process conditions under control. Examples of this include pH adjustment of sludge and wastewater. The pH of anaerobic digesters is critical to proper operation and production of methane, which sometimes requires the addition of chemicals to maintain a proper pH. The adjustment of pH is often necessary for many other unit processes to assure effective or efficient treatment. The addition of iron solutions and other chemicals is also commonly used to reduce dissolved sulfide in raw wastewater and process streams to control biogenic acid generation. In these cases the addition of chemicals should be carefully monitored to assure that unintended corrosion of concrete or ferrous metals does not occur.

However, there are several instances within WRRF unit processes where chemicals or pH controls cannot be feasibly or economically used to prevent corrosion. Some corrosion-preventing process modifications are presented in the following subsections.

2.1.1 Turbulence Control

As mentioned above, waters containing dissolved sulfide (which includes all waters from the influent to the aerated process—and many process return streams after that) release hydrogen sulfide gas (H_2S) to the air where it can be oxidized to sulfuric acid by aerobic bacteria in the headspace. The concentration of H_2S released from water is increased by orders of magnitude where splashing or turbulence is present. Higher concentrations of H_2S mean more sulfuric acid can be produced, leading to higher corrosion rates of concrete and steel exposed to the atmosphere.

Unit processes where high dissolved sulfide concentrations are present include influent sewers and pump station wet wells, influent channels, grit chambers, primary clarifiers, and conveyance piping to the aerated process. Once dissolved sulfide enters a continuously aerobic biological process, the sulfide is quickly consumed biologically and no longer presents a corrosion problem. Dissolved sulfide is also commonly present in biosolids processes including DAFTs, gravity thickeners (GTs), and sludge holding processes where anaerobic conditions dominate. The release of H_2S, and the subsequent corrosion, can be reduced and or averted by controlling turbulence of the water in these processes. Usually consisting of a pipe routing issue or other simple modification, reduction or elimination of turbulence can, over time, prevent significant corrosion damage.

Gravity flows should not be allowed to free-fall into pump station wet wells and pumped flows should be directed below a standing water surface. Aerated processes for grit removal should not be considered unless extraordinary corrosion control measures are implemented. Turbulence into and out of primary clarifiers should be minimized by providing gradual slopes on weir outfalls and effluent collection piping. Discharges into and out of DAFTs, sludge holding tanks, and GTs should be turbulence-free to minimize sulfuric acid corrosion of concrete and steel in the headspace.

Many unit processes require mixing of the contents to be effective. In the case of an aeration basin, the mixing is provided by the aeration system. Aeration (or gas mixing) is a common way to efficiently mix the contents of a tank or basin. When the process contains dissolved sulfide, such as an equalization basin or sludge holding/storage tank, the effect of gas mixing is to strip out high concentrations of H_2S gas, which otherwise would be oxidized to sulfuric acid. Gas mixing should be scrutinized during process design to assure that excessive corrosion will not occur.

2.1.2 Process Return Flows

Process waste flows are commonly routed back to the headworks as a convenient place to reenter the facility, dilute them and receive treatment along with the influent. However, these return flows are not essential to the primary treatment process and does not necessarily need to return to the headworks of the facility. Many solids processing and dewatering flows are anaerobic and contain high concentrations of sulfide, and in these cases there is little reason to add another significant source of dissolved sulfide to the headworks. The characteristics of process return flows should be assessed and directed to the process where treatment is best accomplished. For example, gravity thickener supernatant can be returned directly to aeration where the sulfide will be immediately bioaccumulated without release as H_2S if the supernatant is discharged below the water surface to minimize turbulence.

2.1.3 Solids Storage and Dewatering

Perhaps the most extreme, and preventable, supply of corrosive H_2S gas generated at a WRRF results from the storage, handling, and dewatering of biosolids. Primary solids and WAS are commonly blended in a storage tank to improve their dewatering characteristics. Primary solids are typically difficult to dewater unless blended and conditioned with WAS. However, primary solids are high in BOD and low in active biology, and WAS is essentially starved activated biosolids. When these two sludges are combined in the same storage tank, the WAS immediately starts to consume the BOD, but there is no dissolved oxygen present. Without oxygen the WAS organisms immediately start to reduce sulfate to dissolved sulfide, often accumulating to very high concentrations. Subsequent turbulence from mixing or dewatering processes releases the sulfide as H_2S gas, resulting in biogenic sulfuric acid generation where oxygen is present. This includes expensive belt presses, centrifuges, or other dewatering equipment, associated equipment and the building in which they are housed.

2.2 Material Selection: Natural Corrosion Resistance

There are a great number of corrosive environments within a WRRF in which the proper initial selection of construction materials can avoid corrosion damage. Some examples of this are discussed in the following subsections.

2.2.1 Aluminum

The use of aluminum slide gates in covered primary clarifiers, grit tanks, or headworks facilities often results in aggressive pitting corrosion of 6000-series aluminum alloys. Biogenic acid generation occurs in the H_2S gas-rich aerobic headspaces of those structures. The aluminum may perform well when the structures are not covered for odor control purposes; however, once covered, the aluminum surfaces can become acidic with a pH of 2.0 or lower. As previously noted, aluminum alloys are susceptible to corrosion in acidic and highly alkaline exposures. In this case, the use of Type 316L stainless steel is a much better choice of material in this aerobic headspace exposure, provided chloride ion exposure is sufficiently low.

2.2.2 Epoxy Coated Carbon Steel and Galvanized Carbon Steel

Epoxy coated carbon steel and galvanized carbon steel scum baffles and weir plates were widely used in primary and secondary clarifiers in the past. Experience has shown the galvanized steel components to be susceptible to rapid zinc corrosion below the water line. While the epoxy coated steel components performed better overall, localized cracking or failures in the coating resulted in high corrosion rates at those small anodic areas of bare carbon steel. The associated damage, combined with the high degree of difficulty for properly recoating these elements (edges, corners, etc.), made coated carbon steel a poor material choice for baffles and weirs. Today, the use of fiberglass reinforced plastic (FRP) baffles and weirs in clarifiers and headworks facilities is preferred. The FRP material—provided it is well made and any cut edges properly sealed with the appropriate resins—is inert in these environments and does not corrode. This is true for both immersion and headspace exposure conditions.

2.2.3 Concrete Pipes

Concrete pipe, including RCP, PCCP, and CCP have all been used historically for wastewater conveyance in facilities from the headworks to primary treatment and from primary clarification to aeration. Biogenic acid generation corrosion within the headspaces of these pipes has caused significant corrosion damage and many failures. High density polyethylene, heavy wall PVC, and fiberglass-reinforced mortar pipe have all been used successfully to resist biogenic acid corrosion in pipelines.

2.2.4 Concrete

As stated above, concrete is an extremely versatile material of construction that often comprises the majority of a WRRF. However, concrete is also a mixture of materials that have chemically and physically reacted in a variety of ways to produce a porous, alkaline, amorphous matrix of many different products. Therefore, concrete can react and corrode in a variety of ways, both chemically and physically. The most damaging corrosion mechanism for concrete is that caused by biogenic acid generation. The strength of concrete comes from the formation of calcium-silicate gel crystals produced from the hardening reactions of concrete. These very strong cementicious crystals are readily dissolved in sulfuric acid formed biogenetically from hydrogen sulfide. Experience has shown that the best ways to protect concrete from acidic corrosion are to prevent contact with hydrogen sulfide in a humid environment or provide surface protection. There have been many attempts to modify concrete to make it naturally acid resistant by adjusting the chemistry through admixtures and other additives. The current research shows that there are still no additives to concrete that can make it acid resistant. Any cementicious product that relies on a calcium-silicate or aluminum-silicate crystals can be destroyed by sulfuric acid regardless of additive or admixture. Liners for concrete are discussed in greater detail in Section 2.4.

2.2.5   Stainless Steel

When chloride ion concentrations in the wastewater to be treated are in the 1000 mg/l range or higher, Types 316 and 316L stainless steels are typically susceptible to localized pitting and crevice corrosion. Types 304 and 304L have even lower susceptibility to chlorides and tend to pit at 300 to 400 mg/l chloride concentrations. When the chloride concentrations are above 1000 mg/l, selecting a more resistant stainless steel alloy is important if non-metal solutions are not possible or practical. Options include the use of Type 2205 duplex stainless steel, the so-called "super duplex" stainless steels like Type 2507, or for very high chlorides, the seawater grades of stainless steel, like the 6%-molybdenum stainless steels—this includes the AL6XN and 254 SMO super austenitic stainless steels. The Pitting Resistance Equivalent Number, or PREN, is the method used by corrosion engineers to compare the relative chloride pitting resistance of various stainless steel and nickel-base alloys. Table 8.3 lists the PREN for the alloys described above.

So as you can see, proper choice of materials of construction can avert costly corrosion problems, based on the natural corrosion resistance of the materials chosen.

There are many more examples of natural corrosion resistant alternatives in these facilities, but where this is not possible, barrier protection and electrochemical protection are viable corrosion prevention options.

2.3   Protective Coatings

As stated above, coatings are used for surface barrier corrosion protection. Metallic or concrete surfaces are isolated from corrosive exposure conditions by wellbonded coatings. The term coating is generally used to describe a liquid or semisolid product that is applied to the surface of construction materials and allowed to cure into a finished product for the purpose of providing corrosion protection. Also, the term liner is used separately in this text to identify manufactured thermoplastic sheets and other types of liners that are not liquid applied. Liners are discussed later in Section 2.4.

There are a number of essential criteria to consider when selecting protective coatings for use in municipal wastewater collection systems and WRRFs. The more significant of those criteria are discussed in the following subsections.

Alloy	UNS No.	Pitting Resistance Equivalent Number*	Approx. Max. Chloride Concentration Resistance @ 95°F Neutral pH (mg/l)
304L	S30403	18	300–400
316L	S31603	23	1000
317L	S31703	28	1500
2205	S32205	34	5000
2507	S32750	38	15 000 plus
AL6XN	N08367	43	18 000 plus
254SMO	S31254	42	18 000 plus

*PREN = % Cr + (3.3 × %Mo) + (16 × %N), based on minimum composition.

TABLE 8.3   Pitting Resistance Equivalent Numbers for Stainless Steel Alloys. Courtesy of Corrosion Probe, Inc.

2.3.1   Volatile Organic Content Compliance

Concerns over personnel health and atmospheric pollution from the application of organic coatings has resulted in regulations that limit the volatile organic content (VOC) of most coatings applied in the United States. Manufacturers must formulate coating systems and coatings to meet these regulations in order to sell those products. One should check to be certain that coatings specified for a local project meet local VOC requirements.

2.3.2   Permeability and Water Resistance

Water molecules are very small and chemically active, which are two reasons why water is considered the universal solvent. To some extent, all coatings are permeable on the molecular level, allowing a certain amount of water molecules to enter a coating film. When a coating has good water resistance properties and good adhesion to a substrate, the amount of water absorbed and desorbed in and out of the coating film reaches equilibrium and the coating film retains its integrity. But, the extent to which water is absorbed or passes through a coating film (as vapor) has a substantive effect on its performance in immersion service in wastewater applications.

Based on many years of empirical data, it is widely held in the coatings industry that the best performing immersion-grade coatings for wastewater immersion and aggressive headspace conditions are those that exhibit low water absorption and low moisture vapor transfer rates. These properties are largely influenced by the relative permeability of the coating film. The extent of cross-linking achieved by a particular polymer can improve a coating's water resistance; however, the degree of cross-linking of a polymer is not the only factor affecting a coating's permeability properties. The type and quantity of fillers, driers, plasticizers, pigments, and other substances added to it can substantially alter a coating's permeability properties.

One should consult with the manufacturers of candidate coatings being considered for wastewater system projects, and request that those suppliers provide water absorption and water vapor transmission rate values for their coating products. The pertinent test standard is ASTM D1653, Standard Test Methods for Water Vapor Transmission of Organic Coating Films.

When comparing the water resistance properties of various coatings, care must be taken to ensure that the test values are all based on the same parameters. For example, water vapor transmission (WVT) values expressed as both permeance (perms) and as WVT (g/m²/24 hrs) should be compared for candidate coatings, but only when the conditions under which the tests were performed are the same. Specifically, the relative humidity and the temperature conditions must have been the same, as well as the thickness of the coatings. All of those values must be the same for the comparison of values to be meaningful and fair.

Two other test standards and practices, which are valuable for comparative purposes with coating selection in water systems, are as follows:

- ASTM D870, Standard Practice for Testing Water Resistance of Coatings Using Water Immersion.
- ASTM G9, Standard Test Method for Water Penetration into Pipeline Coatings.

Again, the test conditions and coating thicknesses for all candidate coatings should be the same when contrasting candidate products for ultimate material selection.

The permeation resistance of protection coating and liner materials is also extremely relevant, particularly for headspace exposure conditions in municipal wastewater systems. Nowhere is this more pertinent than in hydrogen sulfide gas-rich conditions in WRRF headworks, covered primary clarifiers, grit chambers, and anaerobic biosolids processes. In these applications, the coating must have high resistance to permeation by wastewater gases including H_2S, CH_4, and CO_2. These small molecules, along with H_2O, penetrate the film and can cause substrate attack or undercutting of the coating at the substrate interface. This gas permeation also can lead to direct sulfuric acid permeation of the film and acidic corrosion of the ferrous metal or concrete substrate.

In recent years, an accelerated test method for assessing the extent of gas and acid permeation in simulated wastewater headspace environments has been developed. This test method utilizes a special controlled-environment cabinet that exposes coated carbon steel coupons to a mixture of H_2S, CH_4, and CO_2 gases and a dilute sulfuric acid solution at elevated temperature. The coupons are analyzed at various time intervals using electrochemical impedance spectroscopy to measure the extent to which the coating films have been permeated.

2.3.3   Chemical Resistance

The coatings selected must provide excellent chemical resistance to the coagulants (alum, ferric chloride, etc.), the pH adjustment chemicals (such as sulfuric acid, sodium hydroxide or lime), and the disinfectants, oxidants, or dechlorination chemicals (including sodium hypochloride and sodium bisulfite) used in the WRRF, at the expected concentrations in the wastewater, at the expected temperatures. The expected water temperatures in most municipal wastewater systems in the world vary between approximately 45°F (7°C) and 110°F (43°C). These temperatures do not typically rise or fall quickly due to the large volume of wastewater being conveyed or treated.

2.3.4   Physical Exposure Resistance

Coatings must be resistant to the physical exposure conditions present in any given application. In the collection system, this can include resistance to abrasion-erosion by suspended solids such as grit and gravel in the wastewater. Abrasive conditions also occur in pump station wet wells and in recovery facility headworks and preliminary and primary treatment structures.

2.3.5   Moisture Tolerance During Application and Cure

Because most wastewater collection system and recovery facility exposures routinely involve damp, wet, or highly humid conditions, it is very important that the coating systems selected are moisture tolerant with respect to possible inhibition of cure. It is well known that many amine cured epoxy coatings are highly susceptible to "amine blush" problems. This routinely causes intercoat adhesion problems with coating systems.

Several carefully formulated blended amine cured epoxy coating systems are available in the marketplace, which are extremely resistant to moisture and amine blush reactions. These products were specifically developed for concrete and steel corrosion protection in wastewater applications.

Aromatic and aliphatic polyurethane coatings are also extremely moisture sensitive during polymerization and cure. These materials include an isocyante coreactant that is highly susceptible to reaction with the hydrogen in moisture (water), which causes cure and property development problems for polyurethane coatings. If polyurethane coatings are to be used for other performance reasons, the substrate and ambient air conditions must be carefully controlled to ensure dryness and acceptably low relative humidity.

The sources of troublesome moisture can include concrete substrates and headspace environments in wastewater systems. Most concrete tanks and structures in wastewater collection system pump stations and facilities are built, at least partially, below grade. This means condensing conditions and high moisture vapor transmission rates can create wet or highly humid circumstances for coating application work.

Concrete moisture can be measured using various test methods. Conductivity based, two-pin meters are used regularly, but really only measure moisture levels near the exposed concrete surface. Moisture vapor transmission rates are measured in accordance with ASTM 1869, a test method that relies on the weight gain incurred by a pre-weighed amount of anhydrous calcium chloride. This test method, originally developed for indoor floor coating applications, does not provide reliable data when utilized on below-grade structures exposed to the atmosphere or on concrete in wet, below-grade structures. The calcium chloride test method and the relative humidity test methods commonly used in the coatings industry are based on the application of coatings in non-immersed concrete applications. So these test methods are of limited use in WRRF applications, except for floor coating projects in buildings. The moisture levels present in below-grade concrete structures are a concern because coating cure can be detrimentally affected or there is the potential for water vapor transfer-related blistering of the coating to occur. This water-filled blister problem has very rarely been documented in wastewater tanks or structures that are normally in immersion service. Rather, these blistering problems tend to occur in non-immersed protective floor coatings where high moisture vapor transmission rates are exacerbated by large variations in humidity on either side of the coating system (constantly moist concrete on one side and relatively dry conditioned air inside buildings on the other).

Substrate dryness can also be examined quantitatively using the Plastic Sheet test in accordance with ASTM D4263. This test method continues to be useful and mostly accurate for wastewater system applications involving concrete substrates. If the source of the high moisture conditions is the air within the structure, typical ventilation and dehumidification measures should be taken.

In short, the selection of coatings that are moisture tolerant is prudent given the requisite knowledge of the expected conditions under which the coating system must be applied. If this is not possible, the conditions must be controlled and appropriate air and substrate moisture-related testing should be performed.

2.3.6 Adhesion to Substrate
For coatings to perform well in wastewater service, excellent coating adhesion must be achieved. This is, of course, largely a function of the quality of the surface preparation work performed prior to coating, specifically the provision of sufficient decontamination, achievement of adequate surface profile, and degree of cleanliness on the substrate.

Decontamination means removal of high concentrations of soluble chlorides, where likely (e.g., coastal facilities and in process areas where sodium hypochlorite is handled), and removal of wastewater residues, such as sludge or grease, from surfaces to be prepared for coating application. When coating or recoating existing concrete previously exposed to biogenic sulfide corrosion, there are some specific tests for decontamination, which are very useful. Sulfuric acid attack of concrete reduces the pH of the hydrated cement paste. Empirical experience has shown that concrete having a surface pH of 8.0 to 9.0 or higher generally indicates that reactive acid salts have been removed and no destructive concrete break down will occur below the newly applied coating. Hence, surface pH measurements are recommended to ensure good long-term coating adhesion. In addition, when concrete is exposed to sulfuric acid attack, destructive sulfate reactions tend to occur within the cement paste below the acid reaction zone. This can cause substrate degradation below the newly applied coating system. Sulfate ion concentration testing can be performed to detect whether or not threshold concentrations are present or have been exceeded in the prepared concrete substrate. If concentrations are much over 3% by weight of cement in the concrete, it can generally be assumed that more concrete needs to be removed to avert destructive sulfate reactions in the concrete beneath the new coating or liner system.

Achieving the right degree of cleanliness is also essential. Degree of cleanliness for steel substrates includes white metal blast cleaning, near-white metal blast cleaning, and other well-defined degrees of cleanliness as described and demonstrated with photo plates in SSPC VIS-1. There also are specific degree-of-cleanliness standards for power tool cleaning, ultra-high pressure water jetting, and other cleaning and preparation methods published by SSPC and NACE, which must be consulted and referenced in any well-written coating specification.

For ductile iron pipe coating work, there are specific degree of cleanliness standards for abrasive blast cleaning published by the National Association of Pipe Fabricators (NAPF), which differ substantially—for metalurgical reasons—from steel blast cleaning standards. For concrete degree of cleanliness, one should consult SSPC SP 13 for specific degree of cleanliness language.

Surface profile is the other crucial key to good coating adhesion. Surface profile requirements are mostly determined by coating film thickness and the need for mechanical anchorage and surface tension, which determine the extent of the coating's bond to the substrate. Surface profile requirements are typically prescribed by the coating manufacturer for specific coatings on specific substrates. Surface profile for metal substrates is generally specified as number of mils (thousandths of an inch) and is checked during the surface preparation work using visual profile comparators or with profile measuring methods such as pressure-sensitive tape and a micrometer. SSPC and NACE provide specific guidance for assuring the achievement of adequate surface profile on steel. For concrete, surface profile is checked in the field using visual and tactile comparative replica coupons in accordance with ICRI 310.2.

Specific test methods are also used to assure proper coating adhesion to both metal and concrete substrates.

2.3.7 Film Quality

Taking all steps necessary to avoid having pinholes or holidays in finished coating systems is important for most coating projects. But, for immersion service and for aggressive headspace or gas-phase exposures in wastewater, it is critical. Pinholes to the substrate provide paths for corrodents to reach the substrate and promote substrate corrosion and undercutting of the coating system. This is less complicated to achieve on metal substrates than on concrete substrates, although its accomplishment is crucial on both.

To avert coating film quality problems for concrete applications, it is essential that compatible (with the coatings materials) filler/surfacer materials are available and properly applied onto the concrete. These products, typically installed by spray application then trowel finishing, are used to fill "bugholes" and other voids in the concrete. If these voids, which are often larger in volume than their visible openings, are not filled, entrapped air pushes back out through the wet coating film during cure, causing pinholes to form. This air release process is called "outgassing" and it is exacerbated when concrete substrate temperatures are rising and not declining. This is why it is better to install coatings over concrete substrates when ambient and substrate temperatures are going down rather than rising.

Proper filling and surfacing of concrete substrates, even if multiple trowel-applied passes are required, is the most efficient way to prevent pinhole problems in coating systems.

2.3.8   Weathering Resistance: Gloss and Color Retention

Weathering resistance is important where applicable. This includes good resistance to ultraviolet light (resistant to chalking), moisture vapor, salt fog, and wetting and drying cycles. Coatings with good weathering resistance retain good color and gloss after long-term exposure to these weathering elements.

2.3.9   Film Build Properties

In many wastewater system structures, including clarifier mechanisms and structural steel in buildings, there are many steel shapes present that have geometrics characterized by extensive edges and corners. Coating these shapes effectively means the coatings used must provide good film build and edge hanging properties such that coating coverage is adequate. While surface preparation methods should include the breaking of sharp edges to form a radius, good edge hanging and film build properties are necessary to assure long-term corrosion protection of steel angles, channels, H-beams, bolted connections, welds, and other difficult-to-coat surfaces. Coatings that provide good film build properties are crucial to the success of corrosion control and prevention in municipal wastewater systems. Epoxy coatings are the best choices for immersion service or structural steel protection in vapor-rich headspaces. When weathering exposure is involved, epoxy-urethane systems, epoxy-fluoroethane, zinc-epoxy-polyurethane, and some acrylic coating systems have provided good performance in wastewater system applications.

2.3.10   Extended Recoat Times

When new or rehabilitation protective coating projects involve shop coating and field touch-up coating work, or where extremely large and complex structures are being coated, coatings should be selected based on optimizing their maximum recoat times. When recoat windows are exceeded, surface preparation costs go up and another potential opportunity for coating adhesion problems is created. For shop coating work where field touch-up or field topcoating is required, this can be especially important. For large, complex structures to be field coated, long recoat times are critical such that intercoat adhesion problems can be avoided. Alternating the color of each coat in the coating system is important to avoid stop-start memory loss, which can lead to intercoat failures.

Extended recoat times are especially valuable in steel pipeline coatings work (both internal and external) in which field welded joints must be field-prepared and coated, but whereas the majority of the coatings were shop-applied well ahead of the pipe installation schedule.

2.3.11 Resistance to Cathodic Disbondment

In those applications in wastewater systems where impressed current cathodic protection (ICCP) systems are used, the protective coatings selected must be resistant to cathodic disbondment. This means the coating must have sufficient dielectric strength, substrate adhesion, ionic resistance, and electron flow resistance properties, along with adequately low moisture vapor transmission characteristics, to resist the formation of blisters associated with hydrogen gas evolution under the coating. Consultation with the coating manufacturer of the coating under consideration is recommended to be certain the coating is resistant to cathodic disbondment.

This requirement applies almost exclusively to coated steel rake mechanisms in clarifiers and possibly to some immersed or buried piping in which steel substrates are protected from corrosion via the use of both protective coatings and impressed current cathodic protection systems.

2.3.12 Future Accessibility for Maintenance

All protective coating systems, even those properly applied at the outset, require ongoing maintenance to optimize their performance and cost-effectiveness. But, when future access to those coated structures is poor or not possible, the use of liquid-applied protective coatings should be avoided. This includes many applications in conveyance tunnels, trunk sewers, influent pump station wet wells to facilities, and headwork structures. Other alternatives for corrosion protection, such as selection of appropriate metal alloys or FRP, or use of thermoplastic sheet liners, should be considered where ongoing maintenance of the asset is not necessary or possible, or where replacement does not require ambient condition control. Please refer to Section 2.4 of this Chapter, which addresses protection methods that are sufficiently robust to not require ongoing maintenance.

2.3.13 Long-Term Successful Track Record of Performance

Selecting protective coatings for a given wastewater system application must also involve checking prior job references for a successful track record of performance. It is recommended that a successful performance period of at least 3 to 5 years—verified by actual reinspection—be sought. This timeframe is based on two empirical truths: (1) most coating failures occur in the first 1 to 2 years of service or sooner and (2) most wastewater system components are not reinspected routinely unless there was a requirement for a one-year or two-year warranty inspection. Such inspection requirements are a very good and sound idea.

When evaluating project references, it is best to speak directly to the wastewater agency end-user personnel and not just to the applicator or the coating product salesperson. This reference checking should include careful verification of the application contractor's experience and expertise performing similar coating projects. For example, many coating contractors have good references for airless spray applications, but may not possess or provide the requisite expertise (and equipment) to apply specialty coatings that require the use of plural-component spray equipment.

Selecting protective coatings for WRRF applications require a careful review of the principles discussed above. For some guidance with this selection process, refer to Table 8.4, which provides general guidance for generic coating system selection for WRRF projects.

2.4   Protective Liners

There are several different types of thermoplastic liner products that can be used to protect both concrete and metals from external corrosion environments in WRRFs. Most liner products are manufactured for use with concrete but a few are more suitable for metals.

2.4.1   Cast-in-Place Liners for Concrete

Most plastic liners for concrete surface applications have knobs, tees, ribs, or other types of protrusions on one side of the liner sheet, which are embedded into concrete during original construction. When the forms are removed, the smooth surface is revealed to provide a thick, manufactured, pinhole-free acid-proof barrier. Seams between liner sheets are commonly welded or sealed chemically. These types of liners can be used for either new construction or rehabilitation, although new construction is where they are found most often. Cast-in-place liners do not completely seal the concrete surface, but the embedded protrusions hold the liner very close to the surface and do not allow the external environment to reach the surface. They can therefore allow hydrostatic water and vapor pressure to escape without any chance of corrosion. Figure 8.3 illustrates some of the attachment configurations for concrete liners.

Cast-in-place liners have been used successfully for the past 70 years to protect concrete from biogenic sulfuric acid generation. The aerobic biofilm that generates the acid can live directly on the surface of the liner and the acid harmlessly runs down the liner.

2.4.2   Chemically Attached Liners

Chemically attached liners differ from cast-in-place liners in that the liner is "glued" into place with a coating-like product called "mastic." The semisolid mastic product is first applied to the dry, hard concrete surface where it anchors onto the porous concrete surface and the manufactured plastic sheet liner is then pressed into the mastic where it chemically and/or physically bonds. These chemically attached liners come in two basic forms. One type has protrusions on one side of a PVC liner which are pressed into a freshly applied epoxy mastic on the concrete. The mastic flows around the protrusions to provide the attachment mechanism when cured. Another form of chemically attached liner is a flat PVC sheet on both sides, which is pressed into a freshly applied polyurethane mastic. A pre-applied chemical additive in the PVC sheet allows chemical fusion and cross-polymerization to occur between the urethane and the PVC, resulting in a very secure bond between the two. Figure 8.4 illustrates two different forms of chemically attached liners.

Chemically attached liners can be applied over new concrete; however, their real strength may be in rehabilitation. These liners can be applied to concrete after placement and curing, or after the surface has already been corroded.

Coating System Options

Structure: Collection System	Interior or Exterior	Substrate	Recommended Surface Preparation Methods(s)	Primer	Intermediate Coat	Finish Coat or Coats	Notes
Sewer Collection Piping	Buried (exterior)	Steel	SSPC-SP10	Epoxy	Epoxy Flexible Polyurethane	Epoxy Flexible Polyurethane	Can be 2 coat self-priming
Sewer Collection Piping	Buried (exterior)	Steel	SSPC-SP2 Power Tool Cleaning	N/A	See Notes	See Notes	Tape Coating per AWWA* C214
Sewer Collection Piping	Buried (exterior)	Steel	SSPC-SP10	See Notes	See Notes	See Notes	AWWA C209 AWWA C210 AWWA C215 AWWA C216 AWWA C217
Sewer Collection Piping (same for WRRFs)	Interior or Exterior (not buried)	Steel or Ductile Iron	Steel- SSPC-SP10 Ductile Iron per NAPF 500-3	Epoxy		Acrylic Aliphatic Polyurethane for sunlight exposure and epoxy for indoor exposure	N/A
Pump Station Wet Well	Headspace	Concrete	SSPC-SP13 Clean, Sound, Decontaminated pH = 10.0 or greater	Epoxy (over filler/ surfacers)	Epoxy (over filler/ surfacers)	Epoxy (over filler/ surfacers)	Can be self-priming troweled epoxy or spray applied reinforced lining completed in one or two coats
Pump Station Wet Well	Headspace	Concrete	SSPC-SP13 Clean, Sound, Decontaminated pH = 11.0 or greater	N/A	N/A	Flexible polyurea or flexible polyurethane	Can be self-priming or with epoxy primer or can be applied in two coats for total DFT
Archimedes Screw Pump Impellers	Covered or exposed	Carbon Steel	SSPC-SP10	Epoxy	Epoxy or phenolic epoxy	Epoxy or phenolic epoxy	Blended amine cured if aggressive H_2S environment

*AWWA = American Water Works Association.

TABLE 8.4 Most Common Generic Coating Systems Used in Municipal Wastewater Systems. Courtesy of Corrosion Probe, Inc.

Structure: Collection System	Interior or Exterior	Substrate	Recommended Surface Preparation Methods(s)	Coating System Options			Notes
				Primer	Intermediate Coat	Finish Coat or Coats	
CSO Tunnels "Hot Zones" ONLY	Interior	Concrete	SSPC-SP13 and decontaminated	Epoxy	Epoxy	Epoxy	Blended amine cured typical for aggressive H2S environment
Headworks Structural Steel	Interior Headspace	Steel	SSPC-SP6	Epoxy	Epoxy	Epoxy	Blended amine cured if aggressive H_2S environment
Headworks Structural Steel	Exterior	Steel	SSPC-SP6	Epoxy	Epoxy	Polyester Polyurethane	Blended amine cured if aggressive H_2S environment
Grit Chambers Headspace	Interior	Concrete	SSPC-SP13	Appropriate filler/surfacer	Appropriate filler/surfacer	Trowel applied thick blended amine cured epoxy with blended amine epoxy gel coat	For aggressive H_2S exposure can also be handed with reinforced spray applied blended amine cured lining system.
Septage Handling Areas	Interior or Exterior of Buildings	Structural Carbon Steel	SSPC-SP6 Minimum	Epoxy	Blended amine cured epoxies	Blended amine cured epoxies	Including exterior of piping and pipe supports
Septage Storage Tanks or Flow Equalization Tanks	Interior	Concrete	SSPC-SP13	Appropriate filler/surfacer	Appropriate filler/surfacer	Epoxy Flexible Aromatic Polyurethanes	High H_2S exposures can include epoxy primers.
WRRF: Primary Clarifier	Covered Headspaces - Interior	Concrete	SSPC-SP13	N/A	Appropriate filler/surfacer	Trowel applied thick or reinforced spray applied amine cured epoxy coating.	High H_2S exposures
WRRF: Primary Clarifier	Covered Headspaces - Interior	Steel	SSPC-SP10	N/A	Appropriate filler/surfacer	Spray applied blended amine cured epoxy coating	High H_2S exposures

				Epoxy	Epoxy	Acrylic Urethane Topcoat	Weathering exposure
WRRF: Primary Clarifier	Non-covered Headspace	Steel	SSPC-SP6				
WRRF: Primary Clarifier	Non-covered Headspace	Structural Steel	SSPC-SP10	Blended amine cured epoxy	Blended amine cured epoxy	Blended amine cured epoxy coating	High H_2S exposures
Ambient Aeration Basins or Tanks	Interior Headspace	Concrete	SSPC-SP13	Epoxy primer	Flexible polyurethane lining	Flexible polyurethane lining	See notes below for filler/surfacers. Lining can be in one coat or two.
Ambient Aeration Basins or Tanks	Interior Below Water Line	Concrete	SSPC-SP13	Epoxy	Epoxy lining or flexible polyurethane lining	Epoxy lining flexible polyurethane lining	See notes below for filler surfaces to prevent aqueous carbonation.
Pure Oxygen Reactors	Interior Above Water Line	Concrete	SSPC-SP13	Epoxy primer	Flexible polyurethane or asphalt polyurethane or polyurea	Flexible polyurethane or asphalt polyurethane or polyurea	To ensure gas tightness main purpose can be done as self-priming in one or two coats.
Pure Oxygen Reactors	Interior Below Water Line	Concrete	SSPC-SP13	Epoxy primer	Flexible polyurethane or asphalt polyurethane or polyurea or reinforced epoxy-spray applied lining	Flexible polyurethane or asphalt polyurethane or polyurea or reinforced epoxy-spray applied lining	To prevent aqueous carbonation of concrete can be self-priming and applied in one or two coats.
Pure Oxygen Reactors	Interior Below Water Line	Steel or Ductile Iron	SSPC-SP10	Epoxy	N/A	Epoxy	Two coat system for corrosion protection
Secondary Clarifiers	Interior Below Waterline	Steel Clarifier Mechanisms	SSPC-SP10	Epoxy	N/A	Epoxy	N/A

TABLE 8.4 Most Common Generic Coating Systems Used in Municipal Wastewater Systems. Courtesy of Corrosion Probe, Inc. *(Continued)*

Structure: Collection System	Interior or Exterior	Substrate	Recommended Surface Preparation Methods(s)	Coating System Options			Notes
				Primer	Intermediate Coat	Finish Coat or Coats	
Secondary Clarifiers	Above Waterline	Steel Rake Mechanisms & Access Bridge	SSPC-SP6	Epoxy	Epoxy	Acrylic Polyurethane	Weathering Exposure
Trickling Filters	Interior	Concrete	SSPC-SP13	N/A	Blended amine cured epoxy systems over filler surfaces	Blended amine cured epoxy systems over filler surfaces	Can be aggressive MIC conditions can be one or two coat systems
Trickling Filters	Interior	Steel	SSPC-SP10	N/A	Blended amine cured epoxy systems over filler surfaces	Blended amine cured epoxy systems over filler surfaces	Can be aggressive MIC conditions
Tertiary Treatment RO–Brine Side Exposure	Interior	Concrete	SSPC-SP13	N/A	Epoxy systems resistant to high salt containing brines and low pH conditions	Epoxy systems resistant to high salt containing brines and low pH conditions	Can be aggressive MIC conditions can be one or two coat systems
Disinfection Chlorine Contact Tanks	Interior	Concrete	SSPC-SP13	Epoxy primer	Epoxy coatings resistant to chlorine and chlorides at near neutral pH	Epoxy coatings resistant to chlorine and chlorides at near neutral pH	Always design for higher than target chlorine residual levels can be one or two coat systems
Disinfection Chlorine Contact Tanks	Interior	Steel or Ductile Iron	SSPC-SP10	Epoxy primer	Epoxy coatings resistant to chlorine and chlorides at near neutral pH	Epoxy coatings resistant to chlorine and chlorides at near neutral pH	Always design for higher than target chlorine residual levels can be one or two coat systems

Sludge or Biosolids Exposures: and **Sludge Stabilization:** appear as section sub-headers within the table.

Dechlorination Exposure	Interior	Steel	SSPC-SP10	Epoxy primer	Epoxy coatings resistant to chlorine and chlorides at near neutral pH	Epoxy coatings resistant to chlorine and chlorides at near neutral pH	Always design for higher than target chlorine residual levels can be one or two coat systems
Dechlorination Exposure	Interior	Concrete	N/A	No coating required	No coating required	No coating required	N/A
Sludge or Biosolids Exposures:							
Gravity Thickeners	Interior	Concrete	SSPC-SP13	N/A	Blended amine cured epoxy linings	Blended amine cured epoxy linings	For High H_2S exposure conditions can be one or two coat systems
Gravity Thickeners	Interior	Steel	SSPC-SP10	N/A	Blended amine cured epoxy coatings	Blended amine cured epoxy coatings	For High H_2S exposure conditions can be one or two coat systems
Sludge Dewatering Facilities (Structural Steel)	Interior	Steel	SSPC-SP6	N/A	Blended amine cured epoxy coatings	Blended amine cured epoxy coatings	For High H_2S exposure conditions can be one or two coat systems
Sludge Stabilization:							
Anaerobic Digesters	Interior	Steel	SSPC-SP10	N/A	Blended amine cured epoxy coatings	Blended amine cured epoxy coatings	Potentially corrosive MIC conditions can be one coat or two system.
Anaerobic Digesters	Interior	Concrete	SSPC-SP13	N/A	Blended amine cured epoxy coatings or flexible polyurethane linings	Blended amine cured epoxy coatings or flexible polyurethane linings	Potentially corrosive MIC conditions typically not lined below water level.

TABLE 8.4 Most Common Generic Coating Systems Used in Municipal Wastewater Systems. Courtesy of Corrosion Probe, Inc. (*Continued*)

Structure: Collection System	Interior or Exterior	Substrate	Recommended Surface Preparation Methods(s)	Coating System Options			Notes
				Primer	Intermediate Coat	Finish Coat or Coats	
Secondary Containment Area for Alum or Sulfuric Acid or Ferric Chloride	Interior	Concrete	Abrasive Blast Clean or UHP water Blast Clean - SSPC-SP13	Epoxy	Novolac Epoxy	Novolac Epoxy	Or two coat vinyl ester system.
Secondary Containment Area for Alum or Sulfuric Acid or Ferric Chloride	Interior	Concrete	Abrasive Blast Clean or UHP water Blast Clean - SSPC-SP13	Epoxy	Vinyl Ester	Vinyl Ester	
Secondary Containment Sodium Hydroxide	Interior	Concrete	Abrasive Blast Clean or UHP water Blast Clean - SSPC-SP13	Epoxy	Epoxy	Epoxy	

Notes of Importance

1. Table 8.4 does not address exterior coatings for ductile iron pipe as it is rarely coated. When it is, the same systems recommended for steel pipe would be used except that different surface preparation requirements would apply. Refer to NAPF 500-3, Surface Preparation Standard for Ductile Iron Pipe and Fittings In Exposed Locations Receiving Special External Coatings and/or Special Internal Linings. This document was published by the National Association of Pipe Fabricators, Inc.

2. Chemical treatment exposure is addressed in Table 8.2 in this chapter.

3. Table 8.4 purposefully does not address specific surface profile requirements as those details are product and project specific.

4. Table 8.4 does not attempt to list film thickness requirements either as this is project specific or is covered by the documents referenced herein.

TABLE 8.4 Most Common Generic Coating Systems Used in Municipal Wastewater Systems. Courtesy of Corrosion Probe, Inc. (*Continued*)

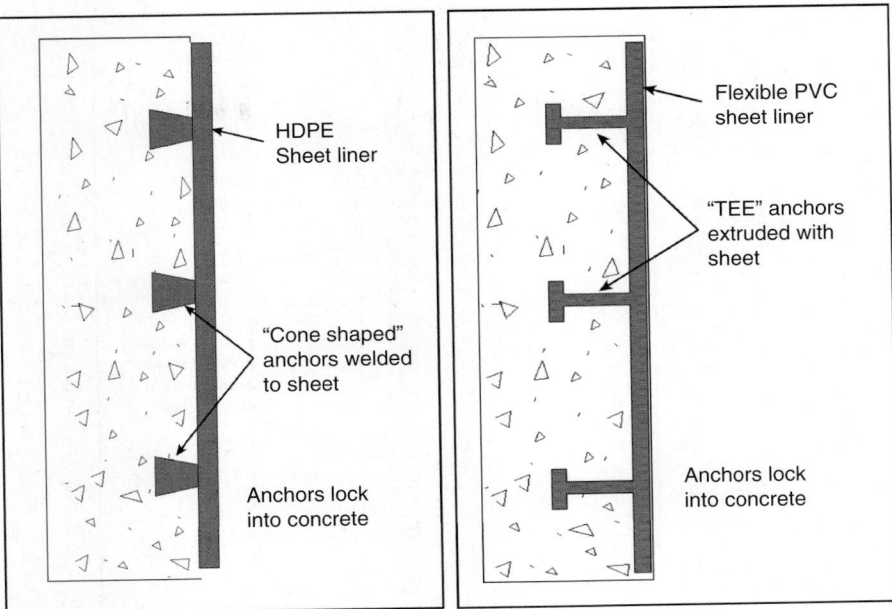

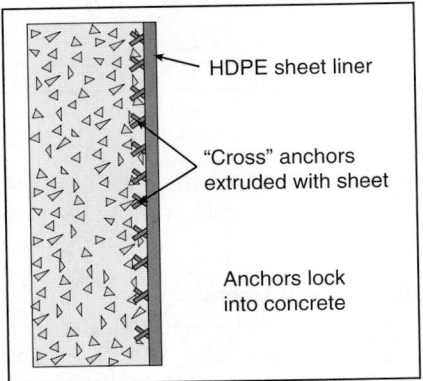

Figure 8.3 Cast-in-place liner schematics.

One of the potential drawbacks to chemically attached liners is the requirement to physically resist all hydrostatic and vapor pressure exerted against the liner. Concrete is porous and groundwater as well as water vapor, can eventually penetrate concrete from the outside to the inside. Since the chemically attached liner is "glued" to the concrete surface with an epoxy or urethane mastic, the security of the physical attachment of the epoxy or urethane mastic is critical to liner performance. This bond is not unlike the bond required for coatings, which have demonstrated an inability to withstand high hydrostatic and vapor pressure forces exerted in a deep sewer environment. Due to the reliance on a mechanical-chemical bond with the concrete surface, the preparation of the concrete surface is very important to the success of the application. In order to

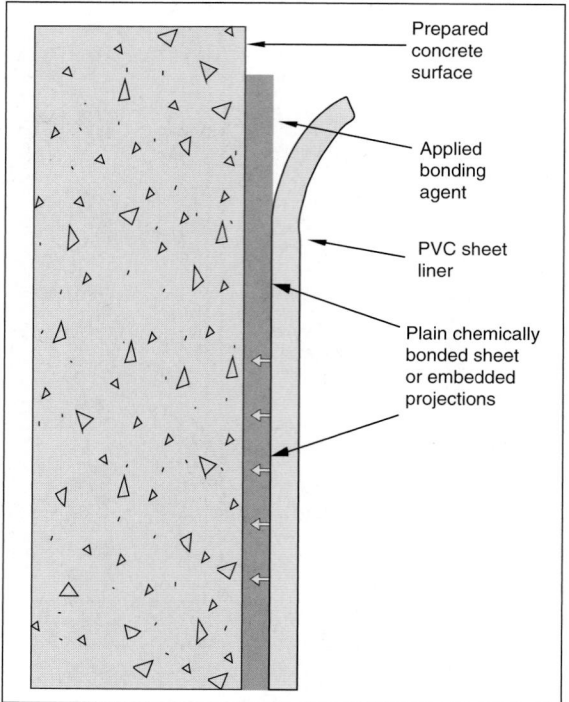

Figure 8.4   Two forms of chemically attached liners.

prevent the buildup of forces against the mastic material at this point, a low viscosity, 100% solids epoxy primer with good wetting characteristics should be applied just prior to mastic application.

2.4.3   Mechanically Attached Liners

Mechanically attached liners can be made of almost any manufactured sheet plastic material, although rigid (un-plasticized) PVC and HDPE are most common. The most significant difference is the method of attachment. Attachment of the liner to the concrete surface is mechanical, which can vary from small expansion type anchors to stainless-steel batten strips with epoxy grout anchors. The high degree of flexibility offered by mechanically attached liners makes them adaptable to a wide variety of conditions. Attachment systems can be specially designed to safely accommodate the anticipated hydraulic and hydrostatic forces. Liner thicknesses can also be easily varied to adjust to specific design needs. These liners have been used primarily for new construction, although numerous rehabilitations have been reported with equal success.

Mechanically attached liners, like the cast-in-place liners, do not resist hydrostatic and vapor forces but pass them to the outer environment. Like the cast-in-place liners, the flat inner surface of the liner is held closely against the interior of the concrete pipe but not fastened to it.

Particular care must be exercised during liner design to provide adequate structural support for the liner (both thickness and attachment) to withstand the anticipated loads with a safety factor of 2. Serious failures have resulted when these design issues and other anticipated forces upon the liner are not addressed. Designs must consider all hydraulic scenarios particularly in applications where water levels can fall or rise suddenly. The space between the concrete and the mechanically attached liner is larger than the cast-in-place liners, which allows water to flow into this space when water levels are high. If, for example, the water level in a lined tank falls rapidly, the water behind the liner may not flow out fast enough and the weight of the water can force the liner off the wall. Spacing, type, size, and material of construction of the attachment anchors is a critical factor for successful mechanically attached liner installation. Mechanically attached liners should not have spacers intentionally placed between the liner and the concrete surface. Spacers increase the volume of water behind the liner during a surcharge event and can add to the forces trying to force the liner off the wall during sudden dewatering.

2.4.4 Wax Tape

Wax tape, also called "petrolatum tape," has been used for many years in the gas pipeline industry to seal the surfaces of steel gas pipelines from the corrosive effects of soil and groundwater. Wax tape consists of a specially formulated fabric that is infused with a thick layer of sticky, waxy petroleum-based olefin compound that never hardens. It comes in rolls of various widths and is typically wound around the external surface of steel and ductile iron pipelines. Properly installed, it completely seals the surface against water, vapor and any other surface corrodents, including biogenically generated acids. Since wax tape never "cures" it is more of a liner than a coating. Also because wax tapes never cure and remain soft, sticky, and pliable their entire life, they can be easily removed for surface inspection in spots by cutting with a knife and peeling back a "window" for visual observation. Once the inspection is finished the flap of tape can be pushed back into place where it bonds with the adjacent tape without damage.

Wax tapes can be used to protect the metallic surfaces of pipelines exposed to aggressive atmospheres including hydrogen sulfide gas, corrosive soil conditions, sulfuric acid, and most chemicals at a WRRF, except hydrogen peroxide. Wax tapes perform best when applied to the pipe surface before exposure to corrosion conditions, and are recommended for ductile iron pipe risers in pump stations (biogenic acid), brass and copper pipes in headworks and pump stations (sulfidation), atmospheric corrosion, or anywhere a metal is exposed to corrosion.

2.5 Electrochemical Control

Corrosion is an electrochemical process in which metal in an oxidizing electrolyte dissolves to produce ions by an "anodic" (oxidation) reaction:

$$M \quad \rightarrow \quad M^+ \quad + \quad e^-$$

M	M⁺	e⁻	
solid metal	ion in solution	electron in metal	(8.3)

For example, iron oxidizes to ferrous ions, $Fe \rightarrow Fe^{++} + 2e^-$, or to ferric ions, $Fe \rightarrow Fe^{+++} + 3e^-$, in more oxidizing solutions. Electrons generated by oxidation reactions at anodic sites "pass through the metal" and are consumed by reduction reactions at the cathodic sites.

In acid solutions the cathodic reaction is reduction of hydrogen ions (protons) to produce hydrogen gas:

$$2H^+ \quad + \quad 2e^- \quad \rightarrow \quad H_2$$

$$\text{hydrogen} \qquad\qquad\qquad\qquad \text{gas in} \qquad (8.4)$$
$$\text{ions} \qquad\qquad\qquad\qquad\qquad \text{solution}$$

In neutral solutions the cathodic reaction involves reduction of dissolved oxygen (or other reducible species):

$$O_2 \quad + \quad 2H_2O \quad + \quad 4e^- \quad \rightarrow \quad 4OH^-$$

$$\text{dissolved} \qquad\qquad\qquad\qquad \text{hydroxyl ions} \qquad (8.5)$$
$$\text{oxygen} \qquad\qquad\qquad\qquad (\text{increase alkalinity})$$

Oxidation and reduction reactions must generate and consume equal numbers of electrons—one reaction cannot occur without the other. Anodic and cathodic microsites are not fixed, but corrosion often is concentrated where surface conditions accelerate oxidation. Corrosion reactions, therefore, convert metal to ions at anodic sites and convert solution acidity to alkalinity at cathodic sites.

As shown in Figure 8.5, the anode and cathode in a corrosion process may be on two different metals connected to form a bimetallic couple, or, as with rusting of steel, they may be two areas on the same metal surface. The electrolyte (water or soil) must conduct ions.

The electrochemical corrosion reactions involved are driven by

- Differences in the corrosion potential in galvanic (bimetallic) couples.

- The concentration of reducible species, for example, dissolved oxygen, is not uniform over the surface of a metal. Oxygen-rich areas favor reduction reactions,

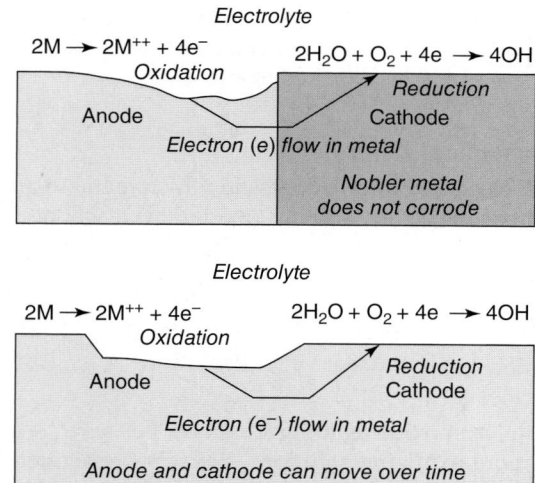

FIGURE 8.5   Corrosion cell reactions on bimetallic couple (top) and single metal (bottom).

becoming cathodic; oxygen depleted areas favor oxidation reactions, becoming anodic. Anodic and cathodic areas can move around over time.

- Metallurgical differences in the metal surface, for example, welds, may make one area more or less anodically active.

The most common electrochemical process for protecting buried or submerged metallic surfaces from corrosion is cathodic protection (CP). CP involves connecting the target metallic structure to an auxiliary anode and applying enough direct current (electrons) to satisfy the reduction reactions, making the protected surface cathodic. The oxidation reactions occur at an auxiliary anode, which is part of the CP system.

There are two basic types of CP system, the names of which indicate the method of supplying current to the electrochemical process. Galvanic or "sacrificial" anode CP systems involve one or more anodes buried or submerged and electrically connected to the target metallic structure. The anodes are of a material that is relatively more active than (anodic to) the less active (or more noble) target structure. The most common anode metals used to protect wastewater system infrastructure are magnesium and zinc. ICCP systems provide the protection current by way of an inert electrode that is immersed or buried in the electrolyte and electrically connected to the target structure, to complete the circuit that impresses the direct current (DC) on the system from an external power source. ICCP system anodes are typically high-silicon cast iron or mixed-metal oxide. The power source is typically a rectifier that converts alternating current (AC) to DC. Solar power sources can be used to supply DC in areas remote from AC power sources.

In both types of system, the applied current negatively shifts the potential of the protected metal in the contacting electrolyte such that the corroding (anodic) reactions cease and only cathodic reactions occur at the metal surface. Figure 8.6 illustrates the "freely corroding" scenario for steel, and Figure 8.7 illustrates the effect on the corrosion rate of depressing the potential with impressed or galvanic cathodic current.

Freely corroding steel, cast iron, and ductile iron in soil or water typically have a structure-to-electrolyte potential (referred to as the "native" potential), relative to a saturated copper/copper-sulfate reference electrode (CSE), of between –400 and –600 millivolts (mV). According to NACE International, the criteria for cathodic protection of steel and iron structures are

- A structure-to-electrolyte potential of –850 mV or more negative (with IR drop[1] taken into consideration), as measured with respect to a CSE; or

- A minimum of 100 mV of cathodic polarization (potential shift in the negative direction).

CP systems can be used to protect the external surfaces of buried or submerged steel, cast iron, and ductile iron pipe; the steel cylinder and reinforcing steel wires or bars in concrete cylinder pipe; the reinforcing steel in concrete structures; the external surfaces of buried portions of steel tanks; the internal, submerged surfaces of steel tanks; and the steel mechanisms in clarifiers.

---

[1] IR drop is the voltage drop associated with the soil/electrolyte. To eliminate IR drop, either the current source(s) must be interrupted or, on galvanic systems, a remote potential measurement must be taken well away from the anodes.

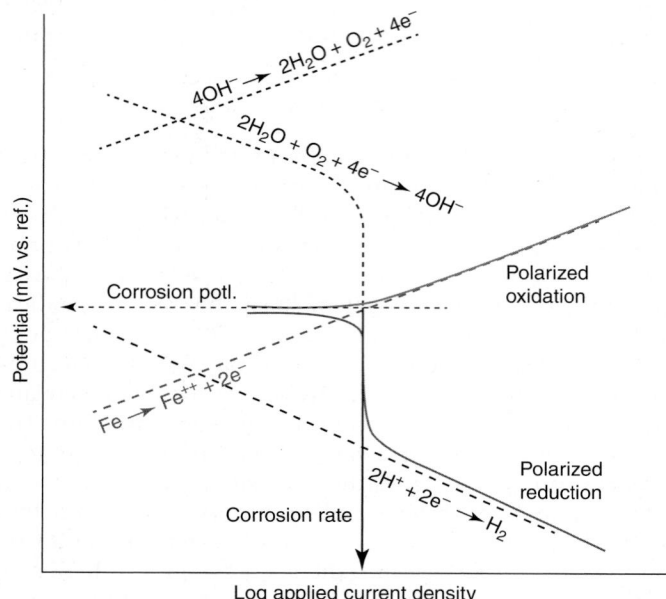

FIGURE 8.6 Electrochemical kinetics with no cathodic protection. Intersection of steel oxidation reaction line (red) and (dissolved) oxygen reduction reaction line (blue) defines the corrosion potential, at which oxidation and reduction rates are equal. This intersection also defines the steel corrosion rate (current density) at that potential. Courtesy of Corrosion Probe, Inc.

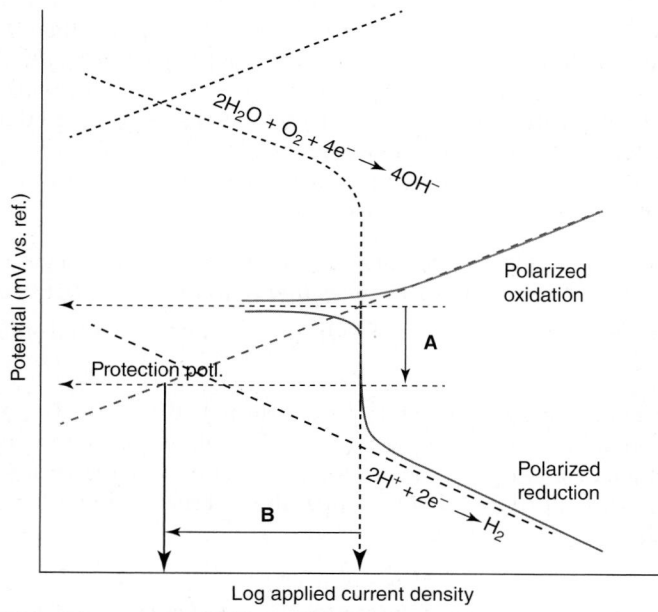

FIGURE 8.7 Electrochemical kinetics with cathodic protection. Connecting the steel to a more active metal, or to the negative terminal of a DC power supply, shifts the steel potential cathodically (negative) by A. This reduces the corrosion rate by B, which is the applied cathodic current. Courtesy of Corrosion Probe, Inc.

In the design of CP systems, several factors must be taken into account, including:

- The resistivity of the soil/water. Lower-resistivity electrolytes are generally more corrosive to steel and iron structures; however, higher-resistivity electrolytes (e.g., sandy soils) require greater driving voltages, which may require an ICCP system.

- The surface area of the target structure to be protected. The greater the bare metallic surface area of the structure, the higher the required protection current from the CP system. Very high current requirements necessitate use of ICCP systems.

- The type and effectiveness of protective coating systems on the target structure. Well-bonded, dielectric coatings reduce the bare metallic surface area to be protected. Coating systems that are not bonded to the structure can shield protection current from the underlying surfaces needing protection.

- The electrical continuity of all target structures to be protected in a single system. CP systems can protect only the metallic structures to which they are electrically connected. If, for example, there is isolation between sections of pipe, protection current will reach only the sections electrically continuous with the CP system anodes. Individual sections of cast or ductile iron pipe, steel pipe with push-on joints, or concrete cylinder pipe can be made electrically continuous by "bonding" across pipe joints with electrical cable.

- The electrical isolation of target structure from other metallic structures in the same soil/water. The lack of electrical isolation between the target structure and other metallic structures in the same electrolyte can draw current away from the target structure and prematurely deplete anodes in a galvanic anode system.

- The presence of stray electrical currents in the soil/water. Typical sources of stray current include high-voltage AC transmission power lines, direct current welding operation, electric rail systems, and ICCP systems on other pipelines or structures. Such stray currents can increase the protection current requirements of the CP system.

- The possibility of other metallic structures in the same soil/water being adversely affected by an ICCP system on the target structure. An ICCP designed to protect the target structure can cause stray current corrosion of other metallic structures in the same electrolyte, and this includes electrically isolated portions of the target structure.

- The amount of room, and accessibility of soil, for buried anodes. Galvanic anode systems can require a large number of anodes. To protect buried pipe, for example, anode beds must be located and arranged in the soil with consideration for other buried structures, the ability to access the anodes for future monitoring and replacement (soil covered by paving or structures), etc.

- Availability of power for an ICCP system. A safe, reliable source of AC power for the rectifier is needed.

- The desired design life of the CP system. The supply of protection current in galvanic systems depletes the anodes such that they will require replacement over time. The anodes in some ICCP systems may also have finite lives.

3.0 Corrosion Control Design for Unit Processes

3.1 Water Resource Recovery Facility
WRRFs, depending on their era of construction and overall purpose, consist of several varied stages of treatment. However, most WRRFs include the stages of treatment shown in the simplified flow diagram of Figure 8.8. Table 8.5 lists the recommended types of corrosion prevention design for each unit process area in a typical WRRF.

3.2 Headworks
The headworks of a WRRF generally refers to the influent pump station and screening pretreatment stages of the facility. However, headworks can be remote to the facility, such as when wastewater is collected and conveyed over long tunnel or trunk sewer distances to regional WRRFs. Headworks facilities are invariably one of the most corrosive environments in a WRRF. Turbulence of the raw wastewater strips H_2S gas out of solution. Biogenic sulfide corrosion is aggressive in the headspaces of screening chambers, wet wells, etc. Concrete needs to be protected with coatings or liners, and metal selection is important. Aluminum corrodes readily under these conditions, whereas properly selected stainless steels perform very well.

3.3 Preliminary Treatment
The headworks at a WRRF also include preliminary treatment that consists of the removal of very large suspended solids, rags and fibrous materials, abrasive grit, the equalization of flow, and septage handling.

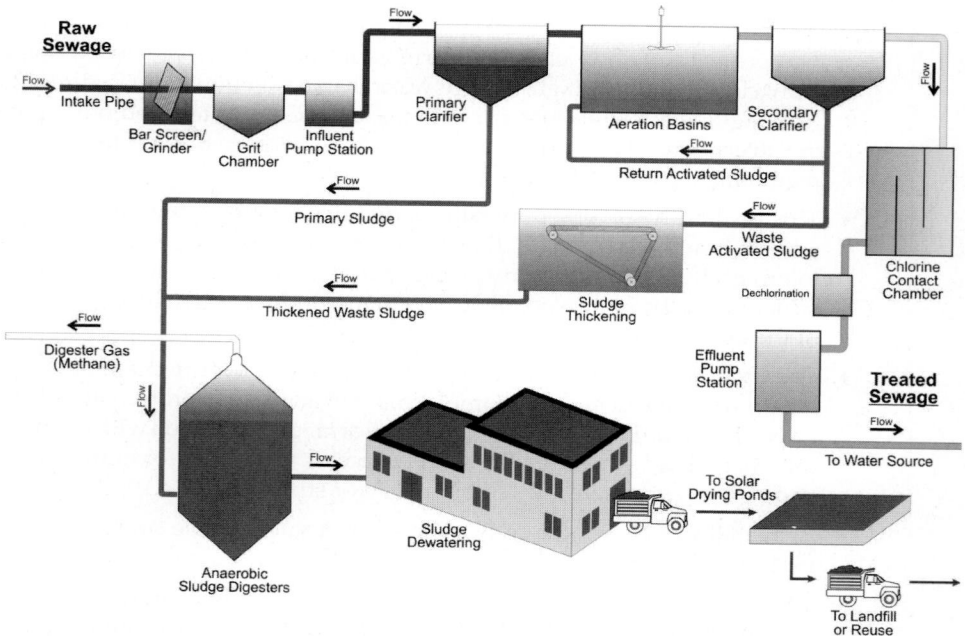

FIGURE 8.8 Simplified flow diagram of a typical water resource recovery facility.

Unit Process	Corrosion Mechanism(s)	Substrates	Recommended Corrosion Protection Methods
Influent Sewers	Biogenic sulfide headspace	Concrete	Anchored thermoplastic linings
Influent Pump Station	Biogenic sulfide headspace	Metals	Appropriately selected stainless steels or wax tape liners
Influent Pump Station	Biogenic sulfide headspace	Concrete	Anchored thermoplastic linings
Screening	Biogenic sulfide headspace	Concrete	Anchored thermoplastic linings
Screening	Immersion – electrolytic corrosion	Metals	Appropriately selected stainless steels
Grit Removal	Biogenic sulfide headspace	Concrete	Anchored thermoplastic linings
Grit Removal	Immersion – electrolytic corrosion	Metals	Appropriately selected stainless steels
Flow Equalization	Biogenic sulfide headspace	Concrete	Blended amine cured epoxy or plastic liners
Flow Equalization	Headspace or immersed	Metals	Appropriately selected stainless steels
Primary Treatment	Biogenic sulfide headspace	Concrete	Blended amine cured epoxy coating or anchored HDPE or PVC liners
Primary Treatment	Headspace	Metals	Appropriately selected stainless steels
Primary Treatment	Immersed	Metals	Blended amine cured epoxy coating
Secondary Treatment Aeration	Biogenic sulfide headspace	Concrete	Typically no biogenic sulfide corrosion
Secondary Treatment Aeration	Immersed Electrolytic Corrosion	Metals	Appropriately selected stainless steels
Secondary Treatment Aeration	Headspace	Metals	Blended amine cured epoxy coating.
Secondary Treatment Aeration	Above waterline atmospheric corrosion	Metals	Epoxy and urethane coating systems
Secondary Treatment Aeration	Immersed	Metals	Epoxy coated
Secondary Treatment Aeration	Immersed	Concrete	Uncoated – typically
Pure O_2 Reactors	Headspace	Concrete	Epoxy lined to prevent O_2 leakage
Pure O_2 Reactors	Immersed carbonation	Concrete	Epoxy lined
Trickling Filters	Headspace – Biogenic Sulfide	Concrete	Vinyl ester or epoxy coatings or plastic liners

TABLE 8.5   Corrosion Prevention/Control Methods By Wastewater Unit Process. Courtesy of Corrosion Probe, Inc.

Unit Process	Corrosion Mechanism(s)	Substrates	Recommended Corrosion Protection Methods
Trickling Filters	All concrete acidic attack	Concrete	Vinyl ester or epoxy coating or plastic liners
Bio-Nutrient Removal	Acidic corrosion	Concrete	Epoxy coatings or liners
Bio-Nutrient Removal	Acidic corrosion	Metals	Epoxy coatings
RO/Membrane Filtration	Acidic corrosion	Concrete	Epoxy coatings or liners
RO/Membrane Filtration	Acidic corrosion	Metals	Epoxy coatings
UV Disinfection	Corrosive to Concrete	Concrete and Metals	Epoxy coat concrete and ferrous metals or use properly selected stainless steels
Chlorination for Disinfection	Corrosive to Concrete	Concrete	Causes rebar corrosion. Epoxy coating
Chlorination for Disinfection	Corrosive to Ferrous Metals	Metals	Use FRP or PVC or CPVC not ferrous metals – or 6 moly stainless steel
Dechlorination with Sodium Bisulfite	Corrosive to Concrete Acid Attack	Concrete	Epoxy or vinyl ester coating
Dechlorination with Sodium Bisulfite	Metals corrosive also	Metals	Use 316 stainless steel or 2205 duplex stainless steel for piping use CPVC
Bio Solids Treatment Gravity Thickeners	Biogenic sulfide corrosion headspace	Concrete	Anchored PVC or HDPE liners.
Bio Solids Treatment Gravity Thickeners	Bio Solids Treatment Gravity Thickeners	Metals	Use proper stainless steel epoxy coated carbon steel.
DAF Units	Corrosive gases CH_4, CO_2, & H_2S	Concrete	Epoxy coated or anchored plastic liners
DAF Units	Corrosive gases CH_4, CO_2, & H_2S	Metals	Use properly selected stainless steels
Gravity Belt Sludge Thickeners or Dewatering Filters	Corrosive gases CH_4, CO_2, & H_2S	Metals	Epoxy coated ferrous metals or properly selected stainless steels
Rotary Drum Units or Centrifuges	Corrosive to metals low pH	Metals	Use properly selected stainless steels
Biosolids Stabilization			
Anaerobic Digesters	Internal	Metals	Epoxy coated carbon steel
Anaerobic Digesters	Internal	Concrete	Do not coat except to prevent leakage of gas then use epoxy or polyurethane linings
Anaerobic Digesters	Internal acidic corrosion	Metals	Use properly selected stainless steels or vinyl ester or epoxy lined ferrous metals

TABLE 8.5 Corrosion Prevention/Control Methods By Wastewater Unit Process. Courtesy of Corrosion Probe, Inc. (*Continued*)

Unit Process	Corrosion Mechanism(s)	Substrates	Recommended Corrosion Protection Methods
Support System			
HVAC (air conditioning units)	Biogenic sulfide headspace	Metals	Use 316 Stainless steel or epoxy coatings inside and out
HVAC ductwork	Biogenic sulfide headspace	Metals	Use 316 stainless steel, FRP, coated steel
Electrical	Biogenic sulfide headspace	Metals	Use 316 stainless steel, FRP, PVC, CPVC.
Odor Control Facilities	Biogenic sulfide headspace	Metals	Use 316 stainless steel, FRP, PVC, CPVC.

TABLE 8.5   Corrosion Prevention/Control Methods By Wastewater Unit Process. Courtesy of Corrosion Probe, Inc. (*Continued*)

3.3.1   Screening
The first step in preliminary treatment, is utilized to remove large solids from the wastewater. Various types of screening systems are used to remove large, hard solids from the facility's influent.

Grinders are widely used to cut, chop, and reduce the size of rags, plastic materials, and other fibrous materials.

The use of appropriately selected stainless steel equipment is the best design choice for preliminary treatment. Headspace concrete should be coated or lined to prevent biogenic sulfide corrosion.

3.3.2   Grit Removal
Grit removal is an essential step in wastewater treatment as it prevents unnecessary abrasion and erosion damage to pumps and other mechanical equipment, as well as buildup of grit in channels, pipes, and subsequent stages of treatment.

Both screening and grit removal treatment create turbulent flow conditions for wastewater. This turbulence releases dissolved wastewater gases, especially hydrogen sulfide, into the aerated headspaces above the water level, which ultimately generates sulfuric acid, as described in Section 1.2 of this chapter. The acid attacks the highly alkaline hydrated cement paste in concrete and causes it to disintegrate. It also causes aggressive corrosion of carbon steel, galvanized carbon steel, aluminum, and ductile iron metal components and equipment. Liquid-applied protective coatings are used extensively to protect metal and concrete substrates from biogenic sulfide related corrosion in screening and grit removal facilities. The actual screening and grit removal equipment is typically fabricated from stainless steels. Some flexible polyurethane liners, heavily filled and trowel-applied epoxy, and spray-applied reinforced liners have been used successfully to protect concrete in facility influent pump station wet wells, grit chambers, and screen wells. The use of anchored stainless steel or HDPE and other thermoplastic liners is more favored for the highly abrasive exposures, while the corrosive headspaces can be very effectively protected with liquid-applied coating systems.

3.3.3 Septage Handling

Septage handling involves receiving wastewater from septic tank pumpers. The septage contains rags, hair, fibrous materials, grit, and plastics, which cause treatment difficulties. Many WRRFs operate grit removal, grinding, and septage pretreatment in a separate septage receiving facility. Septage handling areas can be quite corrosive due to H_2S gas release, particularly in the aerated tank and building headspaces and in the septage unloading areas. Protective coatings are widely used to protect concrete, metal, and masonry surfaces in septage handling areas and receiving stations. This includes the headspaces for the wet wells and holding tanks and the surfaces of buildings in which septage handling processes are housed.

3.3.4 Flow Equalization

Equalization tanks are useful for both control of flow variation and waste-strength. The waste-strength control aspect involves the blending of wastewater ahead of primary treatment. This is more often done in industrial wastewater treatment than in municipal applications. If the facility influent is high in dissolved H_2S and turbulent flow greatly liberates H_2S gas into equalization tank headspaces, corrosion rates for concrete and unprotected ferrous metals and aluminum alloys can be high. Under these conditions, certain blended amine cured epoxy coating systems have performed very well for headspace corrosion protection.

3.4 Primary Treatment

The two major design types of primary settling tanks or clarifiers used are rectangular and circular. Both types are comprised of concrete structures with internal metal mechanisms designed to collect settled solids and remove the sludge material for further treatment. The rectangular clarifiers can involve single pass settling or a stacked bidirectional flow design. In either case, rotating chains and flights are used to push the settled sludge into hoppers or channels from which the primary sludge is pumped away for further treatment. Figure 8.9 shows a flow diagram for a typical stacked rectangular primary settling tank or clarifier.

Some settling tanks or clarifiers are covered, which introduces the likelihood of dramatically increased biogenic sulfide corrosion in the enclosed headspaces.

Circular settling tanks or clarifiers utilize scraper blades on turning rake mechanisms and centrifugal force to push the settled sludge into the center of the tank which leads to the suction sides of sludge pumps. Alternatively, circular clarifier design can address sludge collection with suction tubes attached to the rake mechanism. Suction tube design is tricky from a corrosion control perspective because water-side corrosion can occur on both sides of the tube. This makes coated carbon steel a poor choice for suction tubes. Figure 8.10 shows a typical circular primary settling tank design, including scraper blade options for sludge collection.

Primary treatment of wastewater, much like preliminary treatment, can involve more corrosive conditions because the sulfide and compound concentrations in the wastewater are higher than in subsequent stages of treatment. This means that greater concentrations of gaseous H_2S and other acidic wastewater gases are released into the aerated headspaces of the concrete structures. This results in more aggressive deterioration of concrete and ferrous metal substrates in primary treatment than in subsequent treatment stages.

Typical Settling Tank

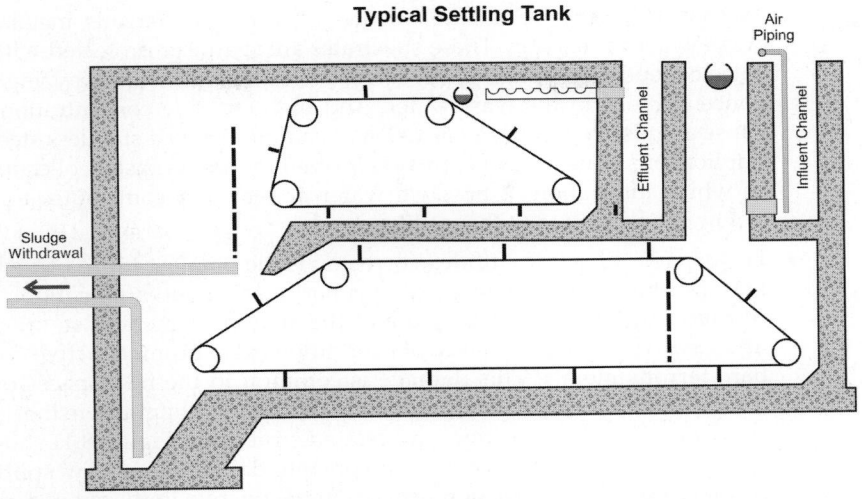

FIGURE 8.9   Profile view of a typical stacked horizontal settling tank.

The major corrosion damage experienced in primary treatment is found in the following locations in the facility:

- Piping, structural steel in buildings, and concrete floors in secondary containment areas for chemical coagulant storage and handling areas especially for ferric-chloride. This generally involves leaks and spills of the chemicals. Epoxy and vinyl ester coatings have been widely and successfully used in these applications.

FIGURE 8.10   Typical circular clarifier.

- Water-side corrosion of bare ductile iron or other ferrous metals found in wastewater immersion. These substrates are routinely protected with properly selected epoxy coatings. This corrosion can be enhanced by high chloride concentrations in the wastewater. And high chloride concentrations can also cause aggressive localized corrosion of commonly used stainless steels in these applications. This type of corrosion is often found in coastal collection systems in which infiltration of brackish water or seawater contributes to the water volume to be treated at the facility.

- Headspace "biogenic sulfide" corrosion is the greatest cause of both concrete and ferrous metals damage in primary treatment applications. Bacterial metabolism forms sulfuric acid and the alkaline cement paste in concrete is attacked. The resulting lower pH conditions also promote active corrosion of bare ferrous metals. This damage is common in the headspaces of concrete flow splitter boxes, influent channels to primary settling, and in the headspaces of covered primary clarifiers and related structures. Figure 8.11 shows a view of typical biogenic sulfide related concrete damage in a flow splitter box to primary clarifiers. Corrosion protection of concrete in these headspace zones can be successfully accomplished using anchored or adhered thermoplastic liners such as PVC or HDPE. In addition, liquid-applied coatings, such as some blended amine cured epoxies, aromatic polyurethanes, vinyl ester-based coatings, and some aliphatic polyurethanes and polyurea formulations, have proven to be effective under these headspace conditions. The more flexible coating options are commonly selected for their better existing crack bridging capacities, whereas the more rigid cured epoxy and vinyl ester liner options require specific existing crack treatment when applied over concrete substrates. Ferrous metal corrosion protection for these biogenic sulfide exposures has been very successful through the use of various amine-cured epoxy coatings.

FIGURE 8.11 Concrete attack in wastewater collection system.

3.5 Secondary Treatment

Ambient aeration tanks or basins are normally constructed from reinforced concrete with stainless steel air distribution piping and equipment. Headspace biogenic corrosion of concrete is typically not a problem in aeration tanks or in pure oxygen reactors as the reducible sulfur species should have been removed prior to this treatment stage. Cement paste losses in immersion service are not, however, uncommon in ambient air aeration tanks when the raw water in the system is low in hardness and alkalinity. The microbiological synthesis of organics in the wastewater produces CO_2 gas as a byproduct. Some of the CO_2 that remains dissolved as carbonic acid lowers the wastewater pH causing slow acidic attack of cement paste in concrete. High hardness and alkalinity buffer the formation of CO_2 as carbonic acid. A lack of carbonate species related buffering capacity in the water leads to aqueous-phase carbonation problems. This attack, which occurs below the waterline, is very gradual and is more pronounced where turbulence removes the acidic-cement reaction compounds (mainly calcium carbonate), exposing fresh cement paste to reaction. This is minimized when the aeration tanks are uncovered and most dissolved CO_2 is released into the atmosphere or where the raw water has high hardness and alkalinity.

In pure oxygen reactors, a partially pressurized headspace keeps CO_2 dissolved in the wastewater thereby reducing the wastewater pH as more of the CO_2 remains as carbonic acid. Liquid-phase carbonation damage to the concrete's cement paste can be significant in pure oxygen reactors especially where raw water alkalinity and hardness are low.

Carbonation-related concrete deterioration seldom warrants the use of protective coatings in ambient aeration tanks except where highly diffused air bubble systems are utilized. In these applications in which older ambient air nozzles were replaced with bubble diffusing systems, more severe wastewater-related carbonation of concrete has been solved using epoxy coatings. The use of protective epoxy and flexible polyurethane coatings is common for concrete substrates in pure oxygen reactors to prevent liquid-phase carbonation. In addition, it is typical to coat the headspaces of pure oxygen reactors to prevent pure oxygen gas leakage through cracks and joints in the concrete covers. Such leakage can be very dangerous if sources of ignition exist or occur on the tops of the reactors. Flexible polyurethane and epoxy liners have been utilized successfully for both immersion protection and headspace gas tightness for pure oxygen reactors.

The use of ferrous metals such as carbon steel or ductile iron is avoided in aeration or pure oxygen service due to the high corrosion rates associated with this oxygenated environment. Types 304/304L and 316/316L austenitic stainless steels are extensively used in aeration tanks and pure oxygen reactors with good success, except where chloride concentrations in the wastewater exceed the pitting resistance of those alloys. Should the chloride concentrations be close to the stainless steel alloy's threshold concentration for pitting and crevice corrosion, the pH depression caused by the retention of dissolved CO_2 in pure oxygen reactors can exacerbate localized corrosion damage. High chloride concentrations are generally found in coastal collection systems and facilities. In facilities where H_2S gas is ducted to the aeration blower intake to use the aeration tanks for odor control, only 316/316L stainless steel should be used for aeration piping. For more discussion on air piping materials, see Chapter 6.

Like primary clarifiers, rectangular and circular secondary settling tanks or clarifiers are mostly built from reinforced concrete with either coated carbon steel sludge collection mechanisms or stainless steel and plastic internal components. Biogenic headspace corrosion associated with sulfur oxidizing bacteria (SOB) should not occur in secondary clarifiers even when these structures are covered for odor control purposes. Immersed carbon steel rake mechanisms and collector flight and sludge collection equipment in secondary clarifiers are subject to the following corrosion mechanisms:

- Oxygen-driven electrolytic corrosion.
- Galvanic corrosion, if electrically continuous with more noble metals like stainless steels.
- MIC associated with anaerobic biofilm, underdeposit acidification. This is especially problematic in the sludge buildup zones.

Protective coatings, including various epoxy formulations, are employed to protect these carbon steel components in secondary treatment.

In trickling filter and rotating biological contactor applications, carbon steel and concrete substrates can be subjected to MIC and acidic attack forms of corrosion. Therefore, various amine cured epoxy and vinyl ester coating systems are widely utilized with good success to protect those substrates.

3.6 Biological Nutrient Removal

WRRFs today are regulated to ensure they meet strict chemical composition-related effluent standards. These regulations now include biological nutrient removal specifically for phosphorous and nitrogen. In general, this removal process requires the integration of either chemical treatment processes or physical unit processes with suspended growth or attached growth biological treatment. This involves either final-stage media filtration to remove phosphorous and nitrogen-bearing solids (not removed by secondary treatment), chemical addition to precipitate more phosphorous, or the use of external energy for denitrification. These nutrient removal processes involve anaerobic and aerobic treatment stages, which can be corrosive to concrete and ferrous metals requiring protection in some cases with protective coatings.

3.7 Tertiary or Advanced Wastewater Treatment

Figure 8.12 presents a flow diagram for one common type of tertiary wastewater treatment. The final stages of treatment can involve nutrient removal processes, completion of organic and inorganic solids removal, disinfection, and dechlorination of the

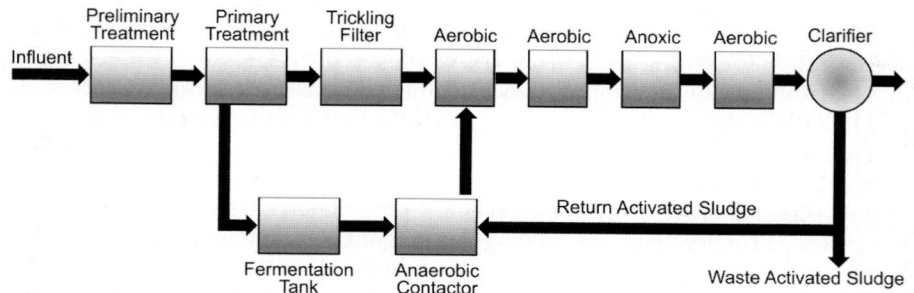

FIGURE 8.12   Example of tertiary treatment for phosphorus removal.

facility effluent. Nutrient and solids removal is accomplished through granular media filtration which may or may not be enhanced by chemical pretreatment such as through the addition of organic poly-electrolytes and/or organic or inorganic coagulants. There are several versions of media filtration processes. Most are not especially corrosive to the concrete or metals used for filter construction.

Reverse osmosis and other membrane unit processes are used in tertiary wastewater treatment for dissolved solids removal. These unit processes, which utilize salt brines (to attract wastewater through the semi-permeable membranes), can be very corrosive to the reinforcing steel in concrete and to ferrous metals and non-ferrous metals (exposed to the brines) including certain grades of stainless steel. Protective coatings are commonly used to protect concrete and metal substrates in these applications.

The two most commonly used methods of disinfection today are chlorination with sodium hypochlorite addition and irradiation with ultraviolet (UV) light. Disinfection is used to reduce the fecal coliform concentration in the treated water. Sodium hypochlorite solutions, while alkaline, are powerful oxidants and therefore can promote active corrosion of both ferrous metals and stainless steels if the residual chlorine concentrations are sufficiently high. Also, concrete chlorine contact tanks are typically lined to prevent reinforcing steel corrosion.

Dechlorination, another form of chemical addition used in water resource recovery facilities, is required in most U.S. WRRFs by law. Dechlorination is generally accomplished via the addition of sodium bisulfite solutions to the treated water to remove the remaining residual chlorine present after chlorination reactions. Dechlorination can also be achieved through the use of sulfur dioxide (SO_2), but handling SO_2 is far more dangerous than sulfite salts like sodium bisulfite. Sodium hypochlorite and sodium bisulfite can be corrosive and are generally stored in rubber-lined carbon steel, vinyl ester-lined steel, FRP, or dual laminant (FRP with fluoropolymer corrosion barrier) tanks. The piping of sodium bisulfite is typically CPVC, which is very resistant to corrosion in this application. The secondary containment areas for these chemicals can be corrosive and typically require the use of epoxy, vinyl ester, and polyurea protective coatings for concrete and metal substrate protection.

UV irradiation for disinfection is gaining popularity in new WRRF design and construction. It eliminates the need for effluent dechlorination. However, it has various operational drawbacks, as well, including energy consumption. UV disinfection is not particularly corrosive to concrete or metal substrates, but can cause aggressive degradation of certain protective coating films and thermoplastic materials, especially epoxy materials.

3.8   Chemical Treatment in the Facility

As described previously, there are a number of chemicals stored, handled, and added at various stages of the WRRF to accomplish certain objectives. These are summarized in Table 8.2 by their use and with regard to their corrosivity and protective coatings used.

3.9   Biosolids Treatment

Biosolids is the term used today when referring to the sludge collected from water resource recovery facilities. Biosolids or sludge treatment is a separate flow stream for WRRFs and can consist of several unit processes. Those separate processes are described briefly below vis-à-vis the need for and use of protective coatings for corrosion protection.

3.9.1   Sludge Conditioning

Sludge conditioning refers to the chemical or thermal treatment processes used to enhance the efficiency of dewatering and/or thickening of biosolids. Chemical conditioning involves the use of metal salts such as ferric chloride and the addition of lime to improve thickening. Organic polymers are also used to improve thickening and dewatering of sludge. The corrosivity of these chemicals has been discussed earlier in this chapter. Thermal conditioning utilizes the addition of heat to the sludge to improve its dewaterability without adding chemicals to it.

3.9.2   Sludge Thickening

Sludge thickening is accomplished in a number of different ways. Gravity thickeners, which in essence are smaller (scaled down) circular clarifiers, are commonly used for thickening primary and lime sludges. Figure 8.13 shows a schematic of a typical gravity thickener. Most gravity thickeners are concrete tanks with coated carbon steel rake mechanisms. The headspaces in gravity thickeners, typically covered for odor control purposes, can be very corrosive. Biogenic sulfide corrosion in these headspaces can cause aggressive acidic attack of concrete. Bare ferrous metals will also corrode actively in this service. As such, the use of protective coatings, including epoxies and flexible polyurethanes, is commonplace and successful if properly done.

Another common thickening method is through the use of Dissolved Air Flotation Thickening. DAF processes can release corrosive methane, H_2S, and CO_2 gases. Therefore, protective coatings are commonly used to protect concrete and ferrous metal substrates in contact with these liberated gases. DAF equipment is commonly built from stainless steel, which performs well provided chloride concentrations in the wastewater do not exceed the pitting resistance for the particular alloys used.

3.9.3   Gravity Belt Thickeners and Belt Dewatering Filters

Gravity Belt Thickeners and Belt Dewatering Filters are also used to separate solids and liquids for sludge thickening. These processes rely on the use of chemical coagulants and flocculants added upstream of the equipment. Gravity belt filter press equipment is typically constructed from stainless steel. The atmosphere above and around these filters can be corrosive and requires the use of mostly epoxy coatings to protect ferrous metal and concrete substrates in the associated buildings.

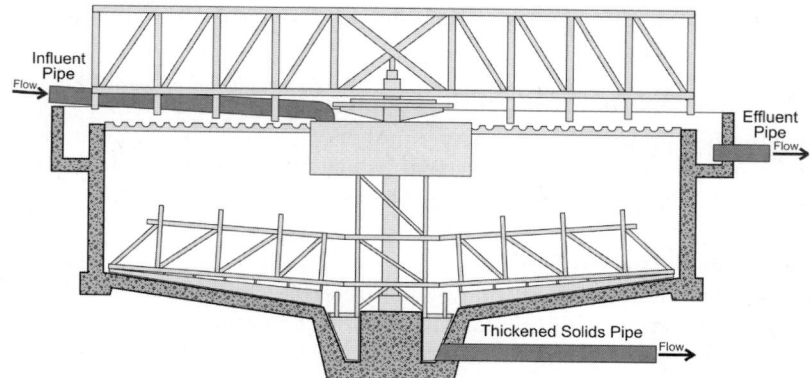

Figure 8.13   Typical gravity thickener.

3.9.4   Rotary Drum Units, Screw Presses, and Centrifuges

Rotary drum units, screw presses, and centrifuges are also commonly used for sludge dewatering and thickening in facilities today. The equipment is generally made from appropriate stainless steel alloys to resist corrosion, while equipment and structures around these unit processes are generally epoxy coated for corrosion protection.

Other types of dewatering equipment include vacuum filters, filter processes, and drying bed dewatering systems. In general, protective coatings are needed to protect the exposed ferrous metal surfaces around these unit processes. For more discussion on dewatering system, see Chapter 24.

3.9.5   Sludge or Biosolids Stabilization

Once primary sludge from various sources in the recovery facility has been thickened and dewatered sufficiently, it is common to stabilize it for safe disposal. This is most often achieved through anaerobic digestion of the sludge, another biological treatment technology. Stabilization is also accomplished with aerobic digestion, composting, alkaline addition stabilization, and through combustion processes. Corrosion and its control in digesters are discussed in the following subsections.

3.9.6   Anaerobic Digestion

While the acidic gases produced by bacterial digestion—including methane, CO_2, and H_2S—can be very corrosive, anaerobic digesters, when properly operated, do not expose metals or concrete to highly corrosive acidic conditions due to the lack of free oxygen which permits acid generation. However, when digesters are operated under cold start (no anaerobic synthesizing biology has been established) or upset conditions (free oxygen gets into the vessel and acidic conditions prevail), the corrosion of carbon steel, ductile iron, and concrete can be extremely aggressive under these conditions.

Typically, all steel substrates and all concrete substrates in the gas phase of anaerobic digesters are coated with either epoxy or flexible polyurethane coatings. This is done for potential corrosion protection and to ensure the gas tightness of the digesters. In Canada, there is a federal law that requires a coating be applied on all digester gas-phase surfaces to ensure gas tightness. In addition, the exterior of the steel covers, prone to corrosion under insulation, need to be well coated to prevent degradation over time. This is generally accomplished with epoxy coatings.

The pontoon-type steel floating covers for digesters, common in facilities built between the 1970s and 1980s, are highly susceptible to corrosion within the steel floor and roof plates due to high humidity and gas and solids leakage conditions. All these internal carbon steel surfaces, including the structural steel trusses, require good coating protection. These covers rise and are lowered through exposure to the gas formed through the digestion process.

The coatings selected for internal corrosion protection of anaerobic digesters must be selected to be resistant to extremely acidic (pH of 1.0 or less) conditions. In addition, the coatings must have very low gas and liquid permeability properties (see Section 2.3 of this chapter).

3.9.7   Aerobic Digestion

Aerobic digestion can be very corrosive and proper selection and application of protective liners and coatings is required to prevent corrosion of concrete and metal substrates.

Another form of sludge stabilization includes composting. Composting facilities can present corrosive conditions due to high humidity and acidic gas liberation from decomposition of biosolids. Accordingly, ferrous metals and other substrates must be properly coated for corrosion protection. Experience has shown repeatedly that epoxy-coated steel does well here, whereas galvanized steel does not due to the low-pH, constantly wet conditions. Zinc-rich inorganic coatings do not perform well under acidic conditions (at pH 4.5 or lower) and in constantly wetted conditions.

4.0   Corrosion Control Design for Facility Support Systems

4.1   HVAC Systems

Another challenging aspect of WRRF HVAC design is the corrosive environment in which many HVAC systems operate. Hydrogen sulfide, volatile organic compounds, and various chemicals used in facility processes require analysis for their effects on equipment. Corrosion activity is also accelerated in below-grade and tunnel areas, especially where there may be standing water. As is also the case for hazardous areas, the HVAC designer must coordinate any mechanical electrical work with the electrical engineer.

Material options for ductwork include coated steel, aluminum, stainless steel, and FRP. The minimum grade of stainless steel recommended is 316 (this recommendation for 316 stainless steel applies not only to ductwork, but to any other HVAC material selection for which stainless steel is warranted). Each material has its own corrosion-resistance advantages and considerable variations in cost. A schedule of ductwork material to be used in various areas should be developed early on in the design process and analyzed using readily available corrosion-resistance charts and tables.

Air-conditioning units and heating and ventilating units should be commercial-grade and constructed of stainless steel or steel with epoxy coatings inside and out. Air-cooled condensing units and rooftop-cooling units, like those used to cool electrical rooms, often suffer chronic failures, as a result of corrosion. Their condensing coils, especially, are aggressively attacked by hydrogen sulfide. The design should attempt to locate such equipment as far as possible from the hydrogen sulfide source. Lacking that option, the HVAC designer needs to work closely with the equipment vendor to specify special coatings or higher grade, non-copper material construction.

4.2   Electrical and Instrumentation Systems

Electrical and Instrumentation (E&I) systems mainly consist of instruments or electrical devices, electrical or electronic cables, wires, wire contacts, conduits, and containment cabinets. The major corrosion issues for this equipment include exposure to water and corrosive gases such as sulfur dioxide and hydrogen sulfide.

Most electrical devices and instruments are manufactured as totally enclosed units made from corrosion-resistant non-metals or metal alloys, for example, stainless steel. The electrical or instrumentation cables that connect those devices to their controls or power sources are coated and insulated copper or other metallic wires. The wires or cables are generally carried within plastic or metal conduits within the facility to protect them from damage and for personnel electrical safety, etc.

Conduit material selection is an important proposition when designing to prevent E&I system corrosion. Aluminum and galvanized steel conduits corrode when used in contact with H_2S gas because low pH (acidic) surface conditions develop due to biogenic acid formation. (See Section 1.1 of this chapter.) Both aluminum and zinc are amphoteric metals meaning they corrode under low pH and high pH conditions when sufficient moisture and oxygen are present. PVC, CPVC, fiberglass, and stainless steel are much better choices for conduit when H_2S gas exposure can be expected. This can typically occur in the facility headworks area, in and around grit removal, and in other primary treatment and secondary aeration areas of the facility. The fasteners used for conduit support and conduit connections should generally be stainless steel to avert corrosion problems. This is recommended throughout the WRRF.

Similarly, chlorine gas exposure can be very corrosive to galvanized steel, lower grades of stainless steel (Type 304 and 304L), and carbon steel. Therefore, the use of aluminum conduit or PVC, CPVC, or FRP conduit is highly recommended in areas where chlorine vapors are expected to be significant. This will include disinfection facilities where sodium hypochlorite is used (headspaces), stored, or handled. If the disinfection basins or structure are covered, the aggressiveness of chlorine vapors can be substantial, even for Type 316 or 316L stainless steel.

Sodium bisulfite storage and handling areas for dechlorination can also be corrosive to E&I equipment and cable conduits. This is because sulfur dioxide (SO_2) can evolve from the mixing and reactions of sodium bisulfite. SO_2 can be extremely corrosive to zinc (and galvanized steel) and aluminum. Hence, non-metals such as PVC, CPVC, or FRP are highly recommended in such areas of the facility. Stainless steel does well for fasteners and support hardware in dechlorination exposure areas.

Electrical cabinets and instrumentation cabinets that contain, house, and support switch gears, electrical transmission bases, or electronic controllers or other devices should be made from well-coated steel, as a minimum, throughout WRRFs. For corrosive areas, NEMA 4X totally sealed enclosure cabinets should be utilized. FRP electrical enclosure cabinets are a wise choice for areas exposed to hydrogen sulfide gas, chlorine gas, or sulfuric dioxide gas.

Electrical cabinets often house equipment that gives off heat. As a result, it is typical for such cabinets to be pressurized and ventilated. When such E&I equipment and cabinets will be located in corrosive environments within the facility, filtration of the air or conditioned air must be provided to prevent corrosive gases and excess moisture from coming into contact with the equipment. Much of this equipment includes copper alloys, which are susceptible to corrosion from H_2S, SO_2, and chlorine vapor exposure. Special air filtration equipment is available to prevent this contact and the resultant corrosion damage. Carbon steel, magnesium, aluminum, and various tin- and nickel-plated metals are all susceptible to corrosion driven by high moisture and corrosive gas exposure conditions.

To summarize, E&I system design to prevent corrosion should be focused on matching the exposure conditions to resistant materials as outlined above. Where this isn't possible for internal cabinet components, the following rules of thumb are suggested:

- Use well-sealed cabinets or enclosures where no internal heat is generated by the equipment.
- Be certain that cabinets are coated with weather-resistant coatings or are protected from the weather within buildings or structures. Metal cabinets are available with moisture- and UV-resistant coatings. Non-metal enclosures, especially FRP, can also be utilized.

- Where ventilation and/or pressurization of E&I enclosures is necessary, use proper air filtration systems to prevent exposure of the metal internal components to corrosive gases.

4.3 Odor Control Facilities

Odor control facilities typically contain, transport, and treat air containing corrosive gases and are exposed to some of the most extreme atmospheric conditions. Even components that are not directly exposed to the air stream may be susceptible to corrosion due to leakage or corrosive byproducts (such as low-pH blowdown water). Thus, the selection of appropriate construction materials is critical for these systems. Foul air from treatment processes may contain a combination of water vapor, hydrogen sulfide, volatile organic compounds, and carbon dioxide.

Not only is the air stream corrosive, but odor-control devices themselves can contain corrosive substances, chemicals, or both. For example, chemicals used in wet-scrubbing odor-control systems commonly are oxidants (chlorine, sodium hypochlorite, hydrogen peroxide, or potassium permanganate), high- or low-pH solutions, or both. Dry odor-control systems use abrasive and sometimes caustic oxidant-impregnated media. In general, plastics and FRP chosen to resist specific chemicals have proven to be the most successful and cost-effective materials for odor control. Carbon steel, aluminum, copper, and copper alloys are subjected to severe corrosion from chemicals and moisture conditions surrounding odor-control facilities. Stainless steel may be satisfactory for certain equipment and piping, if it is selected carefully for a compatible environment.

4.3.1 Materials Used for Odor Containment

The most common materials used at WRRFs for the containment of odors are described below; concrete, aluminum, and FRP are the most popular.

Concrete has the ability to support considerable weight, but also introduces the greatest dead loading as a cover material. Therefore, concrete covers may limit the ability of the facility maintenance staff to remove the covering system for major repairs or to provide access as needed. Furthermore, when used as a flat cover, the weight of removable concrete panels makes it more difficult to provide adequate storage for these units while removed. If space is a premium at the facility, flat concrete covers can be designed to allow for other structures to be placed on top. For example, the odor-treatment system or other plant facilities can be located on top of the flat concrete covers. Concrete also is subject to corrosion, and a protective surface/layer coating may be required to prevent corrosion. Concrete typically is the highest capital-cost investment, especially when installed as a retrofit.

Aluminum covers provide a high-tensile strength with a thin cross-sectional area in a lightweight frame. Some corrosion of this material would be anticipated, but an anodized coating should aid in corrosion prevention. In addition, aluminum covers also can have a factory-applied epoxy coating to provide enhanced corrosion protection. The lightweight nature and thin cross-sectional area of aluminum make it easier to remove and store during maintenance operations. Although typically less expensive than FRP and concrete, aluminum has a higher salvage value and theft can be a concern. Therefore, site security is an important consideration when selecting aluminum. In addition, the design of an aluminum covering system should take into account the incompatibility of aluminum with other materials, such as concrete and other metals (i.e., stainless steel). If proper separation of incompatible materials is not addressed, the aluminum can disintegrate, and the structural integrity of the system could be compromised.

Fiberglass covers provide the greatest resistance to corrosion from water resource recovery facility emissions. When installed outdoors, FRP typically requires periodic maintenance in the form of a UV inhibitor coating for durability and longevity protection, particularly when the material is exposed to direct sunlight. Fiberglass reinforced plastic also is relatively lightweight and typically can be removed by facility staff and stored during maintenance operations. The thickness of the covers will be dependent on the load-bearing requirements of installation; the greater the loading requirements, the deeper the honeycomb web structure (thickness) needs to be. Although lower in strength than concrete and aluminum, FRP unit cost falls between the two.

Stainless steel has almost exclusively been used as a cover material for flat systems and most typically for covering narrow channels. Stainless steel has good corrosion resistance and, like aluminum, provides high-tensile strength with a thin cross-sectional area. However, it is heavier than aluminum and costs more. Although heavier than aluminum and FRP, stainless-steel flat covers typically are sized to enable facility staff to remove and store them during maintenance operations. Two grades of steel are commonly used—304 and 316. Type 304 serves a wide range of applications and resists sulfuric acids at moderate temperatures and concentrations. Type 316 contains slightly more nickel than type 304 and 2% to 3% molybdenum, giving it better resistance to corrosion than type 304, especially in chloride environments that tend to cause pitting. Type 316 also resists sulfuric acid and other corrosive compounds that may be encountered at WRRFs.

Canvas-style covers have been used as an alternate means of covering channels, wet wells, tanks (square, rectangular, or circular), and basins. Typically used at smaller water resource recovery facilities to cover small (in area) tanks and channels, canvas-style covers are not as durable as concrete, FRP, or aluminum. Based on a flat-cover design, canvas covers can provide the following:

- Ease of maintenance, typically through zippered access sections;
- Containment of odors; and
- Corrosion resistance (the canvas has an outer coating that resists degradation).

Depending on the size of the opening, canvas covers may require a support structure, particularly when spanning larger openings, to prevent sagging and accumulation of rainwater and to allow for carrying snow loads. The support structure placed below the canvas cover must be capable of resisting corrosion in the headspace of the enclosed area. Zippered sections allow access to enclosed areas; however, because staff cannot walk on the canvas covers, it is not always easy to reach and open the zippered sections. The durability of canvas falls short of aluminum, FRP, concrete, and stainless steel. The cost is higher than would be expected initially (considering the need for a support structure), but will still be less than the above materials of construction.

4.3.2 Air Duct Materials of Construction

Duct materials of construction are similar to corrosion-resistant materials used in covers and enclosures and include the following:

- FRP,
- Stainless steel,

- Aluminum, and
- PVC.

For corrosive-odor environments, FRP and type 316 stainless steel are the preferred materials. Filament-wound or contact-molded vinylester FRP and type 316 stainless steel are far superior materials of construction and have become the standard for most municipal-wastewater odor-control installations. For milder odor-corrosive environments, aluminum and type 304 stainless steel are available.

Polyvinyl chloride has been used for small flow installations, but is limited to duct diameters of less than 46 cm (18 in.). There are concerns over its ability to maintain its shape over time in hostile and corrosive environments.

Below ground, the choice of materials of construction narrows. The most commonly used are the following:

- HDPE. Check with the manufacturers to be sure that the HDPE being considered is rated for below-ground use.
- FRP. Used for conditions where truck and vehicular traffic may affect the heavy stress on the buried duct (i.e., if a duct runs under a heavily traveled roadway at the facility).

4.4   Chemical Feed and Distribution Facilities

Materials of construction guidelines for chemical feed and distribution facilities are best captured in tabular form. Refer to Sections 1.16 and 3.8 of this chapter, which delineate the corrosive nature of the most commonly used chemicals in WRRFs. Table 8.6 presents guidelines for material selection for chemical feed and distribution facilities.

4.5   Buried Piping and Structures

Facility piping systems that may be affected by soil-side corrosion can include steel pipe, cast iron pipe, ductile iron pipe (DIP), RCP, and PCCP. Additionally, there are other buried steel structures and concrete with reinforcing steel.

Where soils are determined to be corrosive, sacrificial cathodic protection systems are installed to protect the exterior of DIP and steel piping that is well coated. Buried ferrous metal pipes are also commonly protected by concrete encasement or, in the case of ductile iron, by wrapping with polyethylene plastic sheeting or other forms of applied tape liners.

Concrete encasement also has the advantage of adding stability to the bedding of buried pipes. The concrete encasement also provides alkaline protection of the passive film on ferrous metal pipe and helps prevent corrosion. Often, low strength, high fly ash concrete fill is used for these dual purposes. The low strength of the material permits exposure of the pipe without pipe damage should excavation be necessary for piping modifications or repairs in the future.

Polyethylene encasement or wrapping of buried ductile iron pipe is commonly used to provide barrier protection from corrosive soils. There are cases where additional corrosion protection methods need to be used for buried ductile iron pipe, including well bonded coatings or tape wrap systems in concert with cathodic protection.

Chemical	Characteristics	Substrate	Recommended Corrosion Resistant Materials
Alum	Low pH	Concrete	Epoxy or Vinyl Ester coatings
Alum	Low pH	Carbon steel structures	Epoxy or Vinyl Ester coatings
Alum	Low pH	Piping	PVC, CPVC, FRP, or 316 Stainless Steel
Alum	Low pH	Tanks	HDPE or FRP or Stainless Steel
Alum	Low pH	Fasteners/Hardware	Stainless Steel
Sodium Hydroxide	High pH	Concrete	Epoxy Coating
Sodium Hydroxide	High pH	Carbon Steel Structures	Epoxy Coating
Sodium Hydroxide	High pH	Piping	304L Stainless Steel or Carbon Steel
Sodium Hydroxide	High pH	Tanks	304L or Carbon Steel Unlined or FRP
Ferric Chloride	Low pH Highly oxidizing High chlorides	Hardware/Fasteners	Stainless Steel
Ferric Chloride	Low pH Highly oxidizing High chlorides	Concrete	Vinyl Ester or Novolac Epoxy Coatings
Ferric Chloride	Low pH Highly oxidizing High chlorides	Carbon Steel Structures	Vinyl Ester or Novolac Epoxy Coatings use FRP framing and grating and handrails where possible.
Ferric Chloride	Low pH Highly oxidizing High chlorides	Piping	CPVC, or Appropriate FRP or Titanium
Ferric Chloride	Low pH Highly oxidizing High chlorides	Tanks	FRP
Ferric Chloride	Low pH Highly oxidizing High chlorides	Hardware/Fasteners	Titanium
Ferrous Sulfate	Low pH Reactive Reducible Hot Oxidant	Concrete	Novolac Epoxy or Vinyl Ester Coatings

TABLE 8.6   Materials Selection Guidelines for Chemical Feed & Distribution Facilities. Courtesy of Corrosion Probe, Inc.

Chemical	Characteristics	Substrate	Recommended Corrosion Resistant Materials
Ferrous Sulfate	Low pH Reactive Reducible Hot Oxidant	Carbon Steel Structures	Amine Cured Epoxy or Novolac Epoxy Coated
Ferrous Sulfate	Low pH Reactive Reducible Hot Oxidant	Piping	FRP or CPVC or appropriate stainless steel
Ferrous Sulfate	Low pH Reactive Reducible Hot Oxidant	Tanks	FRP or Stainless Steel
Ferrous Sulfate	Low pH Reactive Reducible Hot Oxidant	Hardware/Fasteners	316 Stainless Steel
Lime (Calcium Carbonate)	High pH	Concrete	No Coating Needed
Lime (Calcium Carbonate)	High pH	Carbon Steel Structures	Epoxy Coating
Lime (Calcium Carbonate)	High pH	Piping	Carbon Steel, Ductile Iron Pipe, or Stainless Steel
Lime (Calcium Carbonate)	High pH	Tanks	Unlined Carbon Steel
Lime (Calcium Carbonate)	High pH	Hardware/Fasteners	Carbon Steel or Stainless Steel
Sodium Hypochlorite	High pH Highly Oxidizing	Concrete	Protected with Increased Cover Thickness Over Rebars & Epoxy Coat or Vinyl Ester Coat
Sodium Hypochlorite	High pH Highly Oxidizing	Carbon Steel Structures	Epoxy Coat

TABLE 8.6   Materials Selection Guidelines for Chemical Feed & Distribution Facilities. Courtesy of Corrosion Probe, Inc. (*Continued*)

Chemical	Characteristics	Substrate	Recommended Corrosion Resistant Materials
Sodium Hypochlorite	High pH Highly Oxidizing	Piping	CPVC or FRP
Sodium Hypochlorite	High pH Highly Oxidizing	Tanks	FRP
Sodium Hypochlorite	High pH Highly Oxidizing	Hardware/Fasteners	Stainless Steel
Sodium Bisulfite	Low pH	Concrete	Protected with Increased Cover Thickness Over Rebars & Epoxy Coat or Vinyl Ester Coat
Sodium Bisulfite	Low pH	Carbon Steel Structures	Epoxy Coat
Sodium Bisulfite	Low pH	Piping	CPVC or FRP
Sodium Bisulfite	Low pH	Tanks	FRP or Dual Laminant
Sodium Bisulfite	Low pH	Hardware/Fasteners	Stainless Steel
Sulfuric Acid	Low pH Oxidizing	Concrete	Epoxy or Vinyl Ester Coatings
Sulfuric Acid	Low pH Oxidizing	Carbon Steel Structures	Epoxy or Vinyl Ester Coat
Sulfuric Acid	Low pH Oxidizing	Piping	304 Stainless Steel or CPVC
Sulfuric Acid	Low pH Oxidizing	Tanks	Unlined Carbon Steel (with desiccant dryer or vent) or Stainless Steel
Sulfuric Acid	Low pH Oxidizing	Hardware/Fasteners	Stainless Steel
Hydrogen Peroxide	High pH Good Oxidant	Concrete	Uncoated
Hydrogen Peroxide	High pH Good Oxidant	Carbon Steel Structures	Epoxy Coated
Hydrogen Peroxide	High pH Good Oxidant	Piping	Stainless Steel or PVC/CPVC
Hydrogen Peroxide	High pH Good Oxidant	Tanks	FRP or Stainless Steel
Hydrogen Peroxide	High pH Good Oxidant	Hardware/Fasteners	Stainless Steel
Organic Electrolytes or Polymers	Neutral pH Non-Reactive for Corrosion	Concrete	Uncoated

TABLE 8.6 Materials Selection Guidelines for Chemical Feed & Distribution Facilities. Courtesy of Corrosion Probe, Inc. (*Continued*)

Chemical	Characteristics	Substrate	Recommended Corrosion Resistant Materials
Organic Electrolytes or Polymers	Neutral pH Non-Reactive for Corrosion	Carbon Steel Structures	Epoxy Coated
Organic Electrolytes or Polymers	Neutral pH Non-Reactive for Corrosion	Piping	PVC or CPVC
Organic Electrolytes or Polymers	Neutral pH Non-Reactive for Corrosion	Tanks	HDPE, polypropylene, or FRP
Organic Electrolytes or Polymers	Neutral pH Non-Reactive for Corrosion	Hardware/Fasteners	Stainless Steel

TABLE 8.6   Materials Selection Guidelines for Chemical Feed & Distribution Facilities. Courtesy of Corrosion Probe, Inc. (*Continued*)

The use of protective coatings for PCCP and RCP pipe is very limited. Under extremely aggressive soil conditions such as low pH (<5.0) soils or seawater exposure and tidally affected soils along coastal areas, PCCP and RCP have been externally coated with epoxy, coal tar epoxy, and/or polyurethane coating systems, but this has been rare. Also, there are specific recommendations made by AWWA for enhancing the corrosion resistance of the external cement mortar coating for PCCP to improve its sulfate and chloride intrusion resistance. This includes the use of Type V Portland Cement for better sulfate resistance, lower water to cement ratios, and the use of microsilica admixtures to increase the density of the cement mortar or concrete exterior of the pipe (PCCP or RCP dependent).

5.0   Suggested Readings

Bianchetti, R. (ed.) (2000) *Peabody's Control of Pipeline Corrosion*; NACE International: Houston, Texas.

von Baeckmann. W.; Schwenk, W.; Prinz, W. (ed.) (1997) *Handbook of Cathodic Corrosion Protection*; Gulf Professional Publishing: Oxford, United Kingdom.

Morgan, J. H. (1987) *Cathodic Protection*; NACE International: Houston, Texas.

NACE International (2007) *Standard Practice: Electrical Isolation of Cathodically Protected Pipelines*; NACE SP0286-2007; NACE International: Houston, Texas.

NACE International (2011) *Standard Practice: External Corrosion Control of Underground Storage Tank Systems by Cathodic Protection*; NACE SP0285-2011; NACE International: Houston, Texas.

NACE International (2012) *Standard Test Method: Measurement Techniques Related to Criteria for Cathodic Protection of Underground Storage Tank Systems*; NACE Standard TM0101-2012: NACE International: Houston, Texas.

NACE International (2012) *Standard Test Method: Measurement Techniques Related to Criteria for Cathodic Protection on Underground or Submerged Metallic Piping Systems*; NACE Standard TM0497-2012: NACE International: Houston, Texas.

NACE International (2013) *Corrosion and Corrosion Control for Buried Cast- and Ductile-Iron Pipe*; NACE International Publication 10A292 (2013 Edition); NACE International: Houston, Texas.

NACE International (2013) *Standard Practice: Control of External Corrosion on Underground or Submerged Metallic Piping Systems*; NACE SP0169-2013; NACE International: Houston, Texas.

NACE International (2014) *Standard Practice: Cathodic Protection to Control External Corrosion of Concrete Pressure Pipelines and Mortar-Coated Steel Pipelines for Water or Waste Water Service*; NACE SP0100-2014; NACE International: Houston, Texas.

Nixon, R.; Drisko, R. (ed.) (2001) *The Fundamentals of Cleaning and Coating Concrete, Chapter 3*; The Society for Protective Coatings: Pittsburgh, Pennsylvania.

CHAPTER 9

Preliminary Treatment

Lucas Botero, P.E., BCEE, ENV SP; Joel C. Rife, P.E.;
Kendra D. Sveum, P.E.; and Alex Szerwinski, P.E.

1.0 Introduction

The purpose of preliminary treatment is to remove, reduce, or modify wastewater constituents in the raw influent that can cause operational problems with downstream processes or increased maintenance of downstream equipment. These constituents primarily consist of large solids and rags (screenings), abrasive inert material (grit), floating debris, and grease. This chapter presents descriptions of and design considerations for preliminary treatment processes. Industrial pretreatment can also be considered preliminary treatment, but it is outside the scope of this chapter.

This chapter includes separate sections addressing the handling of hauled-in septic tank waste (septage) and attenuation of high flows and pollutant loading that can disrupt the performance of downstream processes (equalization). Receiving and handling high-strength waste and fats, oils, and grease (FOG) is covered in Chapter 25.

2.0 Screening

2.1 Benefits of Screening

The main goal of screening wastewater flow is to protect downstream equipment and processes. Screening can be used to remove large objects that could damage influent pumps or block flow in raw sewage channels and piping systems, or it can remove fine objects such as human hair, which protects sensitive downstream equipment including membrane systems, fabric filters, or suspended media used in integrated fix-film activated sludge and moving bed biofilm reactor systems. The passage of rags and debris into downstream processes is one of the largest causes of equipment maintenance and failure because of jammed pump impellers, clogged sludge and scum pipelines, and imbalanced operation of rotating equipment. Floating material in downstream processes or receiving streams is an aesthetic problem and a safety hazard to operators attempting removal. Solids treatment and disposal benefits from screenings removal as this is where all screenings eventually end up, if it is not removed at the front of the facility. Several reports suggest that the effects of inadequate screening are significant and can impede proper functioning of oxygen delivery systems and mixers (Burbano, 2014) as well as result in excessive maintenance for other rotary equipment such as pumps in water resource recovery facilities (WRRF) (Borys, 2014). Removal of fine solids also benefits biosolids programs seeking commercial acceptance of Class A products. As wastewater processes continue to advance, damage associated with inert objects in wastewater becomes increasingly important. For these reasons there is a trend toward installing screens with smaller openings. As screen openings become smaller, greater amounts of organic matter are removed and it becomes important to provide screenings washer/compactors to return the organics to the wastewater flow stream. Washers/Compactors are becoming standard design practice for several reasons:

- Operator safety when handling the screenings.
- Disposal of the screenings is typically in municipal landfills and restrictions are becoming increasingly strict. U.S. Environmental Protection Agency (U.S. EPA) Method 9095B (known as the Paint Filter Liquids Test) has historically been used to regulate maximum moisture in the screenings being landfilled, requiring no free water to be present.

- Landfills in Europe require at least 45% dry solids content and less than 3% organics content for construction waste landfills and less than 5% organics content for municipal waste landfills.

- Washer/Compactors return organic matter to the liquid stream, benefitting downstream nutrient removal processes that require soluble organic matter to function properly.

- Odor from the storage of screenings is reduced because of the removal of most of the organics.

2.1.1   Impacts of Non-Dispersible Products

Non-dispersible (flushable) consumer products were introduced into the market in the last decade. These newer personal hygiene products predominately consisted of non-woven wipes that, because of their size and use, are frequently disposed of in toilets. Water utilities quickly learned that most of these products do not break up in the sewer system, leading to problems in the collection system pump stations and headworks of water resource recovery facilities (WRRFs). According to INDA (Association of the Non-Woven Fabrics Industry) and EDANA (international association serving the non-wovens and related industries) flushable products shall clear toilets and properly maintained drainage pipe systems under expected product usage conditions, be compatible with existing wastewater and compatible with existing wastewater conveyance, treatment, reuse, and disposal systems, and become unrecognizable in a reasonable period of time and be safe in the natural receiving environments. However, some WRRFs have seen an increase of these products in their screening systems, particularly those that rely on gravity collection systems.

Non-dispersible products have the potential to increase the screenable material carried to the facility due to non-dispersion in the wastewater, increase the instantaneous blinding of fine screens due to slug loads containing non-flushable material, and increase the wear and tear on screening equipment because more screenable material processing is required. They also render certain types of screens inadequate to deal with the increased load of screenable material, which causes ragging issues in processes downstream of the screens when there is an inadequate capture of screenable material, and renders screening conditioning equipment (conveyance and conditioning) inadequate due to the higher peak load of screenable material, which reduces the facility's ability to deal with slugs screening loads. Therefore, designers of headworks systems in WRRFs must take all of these factors into consideration for the proper selection and design of screening equipment that may be subjected to non-dispersible materials.

2.2   Screening Categories

Screening of wastewater can be categorized according to screen opening size as follows:

- Trash racks and bypass screens with openings greater than 36 mm (1.5 in.).
- Coarse screens with openings greater than 6 and less than or equal to 36 mm (0.25 to 1.5 in.).
- Fine screens with openings greater than 0.5 and less than or equal to 6 mm (0.25 in.).
- Microscreens with openings less than or equal to 0.5 mm (0.2 in.).

2.3   Screenings Characterization

2.3.1   Quantities

The quantity of screenings removed can vary significantly depending on the wastewater flow, wastewater characteristics, type of collection system, screen type, screen opening, screen cleaning mechanism, and effectiveness of washer/compactor equipment. Actual operating data can be used when replacing screens with equal size openings, but it is of little value if an upgrade project involves putting in screens with smaller openings.

For screen openings between 25 and 50 mm (1 and 2 in.), each 13 mm (0.5 in.) reduction of clear opening size will approximately double the volume of screenings. For screen openings smaller than 25 mm (1 in.), the volume of screenings removed increases rapidly and becomes much more a function of wastewater characteristics and efficiency of washer/compactor equipment. Removal by trash racks is more a function of wastewater characteristics than rack openings.

The quantity of screenings removed will depend on the sewer characteristics in the collection system, number and location of pumping stations, and whether or not the stations include screening. In short, gently sloping collection systems with low turbulence will contain more first-flush screenings than lengthy, steep interceptor systems, or systems with pumping stations. Slug screening loads corresponding to pump stations startup periods have been reported in some facilities (Wodrich et al., 2005). Slug loads are common in combined collection systems after first flush conditions (especially after a dry period) and in areas with significant numbers of deciduous trees at the beginning of the fall.

Another critical factor in determining screenings quantity is the type of sewer system feeding the WRRF. Previous experience indicates that combined systems produce several times the coarse screenings compared to separate systems. Peak wet-weather removal from combined systems may vary by as much as a 20:1 from average dry weather conditions on an hourly basis.

As part of the development of this manual, the Water Environment Federation (WEF) performed two surveys of utility members (2008 and 2016b), obtaining data from 358 WRRFs across the United States. Figures 9.1 and 9.2 show the normalized data from the survey and differentiates between coarse and fine screens. Based on the survey data, some general screenings generation conclusions emerged:

- The quantity of wet screenings collected proportional to influent flow seems to be higher in smaller WRRFs.

- A wide range of wet screenings is generated. Designers should study all factors affecting the possible wet screenings collection amount before making a final judgment. Adequate safety factors should be included for instantaneous peak loadings.

- Extreme variations in screening quantities were reported, from less than 0.74 to 148 L/1000 m³ (0.1 to 20 cu ft/mil gal).

- Screenings quantity generation in the United States apparently is higher than in Europe. Recent vendor studies in Europe have found an average screenings production rate of 2.4 kg/person/d compared to 4.5 kg/person/d in the United States. This is based on U.S. screenings production of 40.7 L/1000 m³ (5.5 cu ft/mil gal); 378 L/person/d (100 gal/person/d) wastewater generation rate; and

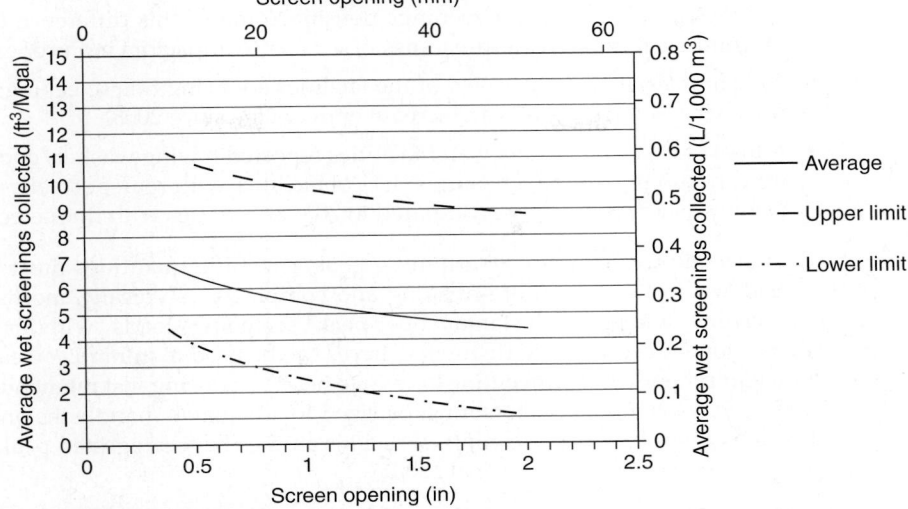

Coarse screens-wet capture rate
2008 & 2016b WEF survey data[1,2]

1. 95% confidence interval
2. Majority of data from facilities with washer/compactors and 60% volume
   reduction assumed to convert to wet

FIGURE 9.1 Screenings quantities from coarse screens.

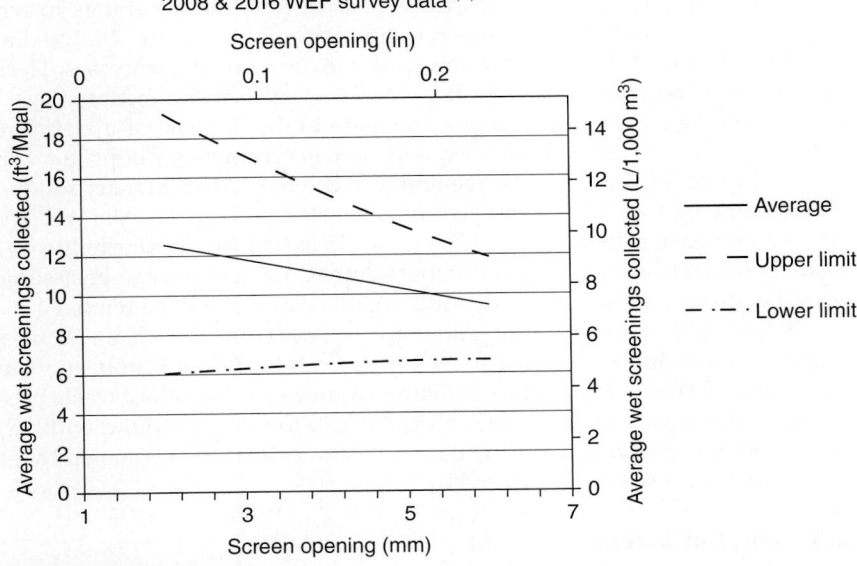

Fine screens-wet capture rate
2008 & 2016 WEF survey data[1,2,3]

1. 95% confidence interval
2. Majority of data from facilities with washer/compactors and 60% volume
   reduction assumed to convert to wet screenings
3. Capture rate assumes no coarser screen before the fine screen

FIGURE 9.2 Screenings quantities from fine screens.

800 kg/m³ (50 lb/cu ft) screenings density. Some of this difference could be attributable to more widespread use of washer/compactors in Europe.

- The survey data indicates 60% of the facilities are using fine screening; the rest use coarse screens. This is an increase of over 20% since 2008.

- A total of 60% of the wastewater facilities reported having washer/compactors for conditioning of the screenings; 40% of facilities with coarse screens reported using washer/compactors compared to 70% of facilities with fine screens.

Adequate safety factors for determining peak screening quantities must be carefully considered in the design of screening and compactors. Typically, mechanically-cleaned screens withstand instantaneous peak screening loads without special provisions. Flow-paced variable frequency drives can be used to minimize wear on the collection equipment while preventing excessive headloss during instantaneous peaks.

Screenings washer/compactors must be sized to adequately handle instantaneous peak loadings to process the screenings. Previous studies have suggested peaking factors from 4 to 6 up to 15 (Wodrich et al., 2005).

2.3.2 Physical Properties

The composition of the screenings removed will depend on the type of screen that is used. Coarse screenings consist of rags, sticks, leaves, food particles, bones, plastics, bottle caps, and rocks. These screenings are typically thought of as large debris that can immediately damage pumps and other equipment. Fine screens will remove coarse screenings along with a higher degree of rags, wipes, stringy material, and a large amount of organic matter. Fine screenings, if not removed, typically have a long-term impact on equipment and processes as it takes time for screenings to wrap around impellers or mixer blades. The effects of fine screening may not be immediately apparent but will show its benefit over time in all the downstream processes. The smaller the opening size, the more organic matter will be removed with the screenings. The microscreen opening sizes will result in an increase in the removal of material such as hair and string and the material removed will have the consistency of primary sludge.

Unwashed, un-compacted screenings can contain 10% to 20% dry solids with a bulk density ranging from 600 to 1100 kg/m³ (40 to 70 lb/ft³). In the 2008 WEF member survey, screenings density averaged 825 kg/m³ (55 lb/ft³). There is currently no distinction made between coarse, fine, and microscreenings, however, it can be assumed that the finer the screening, the more water and organic material will be removed.

Typical performance specifications for washer/compactors are 90% reduction in organic content and an increase to 50% in dry solids. Recent European vendor studies have found that 50% dry solids is difficult to achieve with standard washer/compactors. Based on the survey information, a range of 30% to 40% was found with an average of 37%. From the inlet hopper to the discharge chute, it can be estimated that the screenings volume and weight is reduced by 50% to 80%.

2.4 Types of Screening Media

Typically, there are four types of screening media used, which are bars, wedge wire, perforated plate, and mesh.

2.4.1 Bars

Historically, bars have been the most commonly used media because it is preferred for coarse screens and trash racks. Bars are available in a variety of shapes including rounded, rectangular, trapezoidal, and teardrop. Rounded bars have low capture efficiency and are only used on large opening bar racks. Trapezoidal bars have increasingly wide openings, allowing solids that pass through the narrowest opening at the front of the screen to pass through without getting trapped between the bars. Teardrop bars combine the benefits of the trapezoidal bars with better hydrodynamic flow characteristics, which minimize the headloss through the screen. Long vertical or horizontal gaps between bars can allow the passage of long, thin objects.

2.4.2 Wedge Wire

Wedge wire is a refinement of the trapezoidal bar screen used in much finer screening applications. The same narrow-to-wide opening profile is used to prevent trapping solids between the openings. The narrower openings of wedge-wire screens result in much thinner media, which is why they are called "wires" instead of "bars." Wedge-wire screens also have long, vertical gaps and are not allowed by some manufacturers of membrane bioreactor facilities because of the need to keep long, thin objects such as hair from accumulating on the membranes. A close up view of wedge-wire bars is shown in Figure 9.3.

2.4.3 Perforated Plate

Perforated-plate media (Figure 9.4) is more effective at capturing solids than bars or wedge wire when fine screening (such as hair removal) is required. The technology for perforated-plate media is constantly advancing with the size lower limit currently at 1 mm. Perforated-plate media has higher headloss because the effective open area is decreased, there are orifice losses, and the increased blinding compared to bar or wedge wire media.

2.4.4 Mesh

Mesh (Figure 9.5) is used for fine screens 1 mm and smaller because of manufacturing limitations of perforated-plate media. Mesh media is more fragile and can result in "stapling" of solids within the media, which interferes with release of captured solids

FIGURE 9.3 Wedge wire screen media (courtesy of Industrial Screen Products, Inc.).

FIGURE 9.4   Perforated plate panels (courtesy of JWC Environmental).

FIGURE 9.5   Mesh media in drum screen (courtesy of Baycor Fibre Tech, Inc.).

by the removal mechanism. To avoid clogging the mesh, high-pressure, water-jet cleaning is recommended. Openings in the mesh media can also be a source of confusion because corner-to-corner distance of the square mesh opening is slightly longer than side-to-side distance. Typically opening definition is provided for the side-to-side distance.

2.5   Screen Types

There are several screens designs and styles in the wastewater industry. The following sections summarize the various screen types based on their screening categories as listed below, however, some screens can be used in multiple categories. Table 9.1 summarizes advantages and disadvantages of various screen types.

2.5.1   Trash Racks and Bypass Screens

Trash racks are used in older facilities and in facilities receiving wastewater from combined sewer systems that can contain large objects. These are bar screens with large openings of 36 to 144 mm (1.5 to 6 in.) designed to prevent logs, timbers, stumps, bricks, and other large, heavy debris from entering treatment processes. Trash racks typically are followed by screens with smaller openings. Where space is limited, facilities sometimes have basket-type trash screens that are manually hoisted and cleaned.

Bypass screens are used for emergency screening purposes in the event the mechanically cleaned coarse or fine screen must be taken out of service. Openings range from 24 to 48 mm (1 to 2 in.). Manually cleaned trash racks and bypass screens typically are mounted on a 45° to 60° angle from the horizontal to facilitate cleaning using a rake and perforated plate drain pan. Depth should not be more than what can easily and safely be raked by operators. Mechanically cleaned trash racks are available and are mounted 75° to 80° from the horizontal. As facility size increases, it becomes unmanageable to use manual bypass screens because of the larger volume of screenings retained and the potential for channel overflows due to blinding. Trash racks are typically not provided by screening manufacturers but rather designed by a structural engineer and constructed by miscellaneous metal contractors.

2.5.2   Coarse Screens

Historically, coarse screens have been the most commonly used standalone screens in WRRFs because they provide sufficient screening without producing excessive volumes of organic materials, which eliminates the requirement for a washer/compactor. The smallest opening coarse screens can still remove organic material. For these screens, washer/compactors should be provided. The coarse screens most commonly used today are in two-stage screening processes to provide protection and reduce blinding of the fine screens.

Coarse screens are cleaned mechanically, which allows the screening media to be mounted in a more vertical position, typically 70° from the horizontal. Mechanical cleaning reduces labor cost; improves flow conditions and screening capture; reduces nuisances; and, in combined systems, better handles large quantities of storm water debris and screenings. A mechanically cleaned screen is almost always specified for new facilities of all sizes. Many types of mechanically cleaned bar screens are manufactured, including, but not limited to, chain/cable driven, single rake, multiple rake, and continuous.

Type of Screen		Advantages	Disadvantages
Coarse Screens	Chain or cable driven screens	Design in the market for many years Simple channel construction High screenings loading rate Insensitive to fat, oil, and grease (FOG)	Submerged components
	Reciprocating rake screens	No critical submerged components Widely used Allows a pivot design for servicing the unit above the channel	Low screening loading rate High overhead clearance, particularly at deep channels
	Continuous self-cleaning screen	Medium to low headroom required. Allows a pivot design for servicing the unit above the channel.	Several moving components. Components subject to wear and tear
	Arc screens	Simple design Lower capital and operational cost No drive parts under water Utilizes 100% of channel width	Limited to small to medium flow plants Not suited for deep channels
Fine Screens and Microscreens	Continuous element screens	Proven technology High screenings capture rate Allows a pivot design for servicing the unit above the channel	Numerous moving parts Some designs have components subject to wear and tear High headloss requirements High potential for grease blinding Additional motor and/or spray bars for cleaning Potential carryover of screenings
	Multiple rake screens	Widely used High screenings loading rate Low headroom requirements	Some designs have submerged components subject to wear and tear Medium headloss requirements High potential for grease blinding

TABLE 9.1   Summary of Coarse and Fine Screening Equipment

Type of Screen		Advantages	Disadvantages
Fine Screens and Microscreens (*Cont.*)	Stair screens	Low headloss requirements Allows a pivot design for servicing the unit above the channel	Stringy solids can pass through the screen Not recommended when rocks or excessive grit loads are expected Requires considerable footprint for installation compared to others
	Band screens	Low screen carryover Well suited for expanding capacity without increasing the channel size	Requires pressure water for cleaning Requires especial provisions for removing the unit from the channel Difficult to access and to maintain Large solids removal may be an issue
	Drum screens	Low screen carryover Recommended for downstream processes sensitive to screenings (MBRs, IFAS, etc.)	Require considerable footprint for installation Installation requires special hydraulic provisions High headloss required (pumping is usually required)
	Helical basket	Lowest screenings carryover potential Provides screenings washing/compaction in a single unit Recommended for downstream processes sensitive to screenings (MBRs, IFAS, etc.)	Requires significant footprint for installation Not suitable for deep channels
	Static screens	Minimal or no moving parts Well suited for smaller facilities	High headloss Operator intensive Susceptible to fast blinding
	Microscreens	Removal of very small solids	Requires special washing systems Should be preceded by larger opening screens High headloss

Notes: MBR = membrane bioreactor; IFAS = integrated fixed-film activated sludge

TABLE 9.1 Summary of Coarse and Fine Screening Equipment (*Continued*)

2.5.2.1 Chain-Driven Screens These types of screens are manufactured in several configurations: front clean/front return, front clean/rear return, and back (or through) clean/rear return. The front clean/front return type most efficiently retains captured screenings by minimizing carryover. Cables are used in place of chains on deep applications. This type of screen fell out of favor because of high maintenance requirements of submerged chains, bearings, and sprockets, but technology advances have resulted in a resurgence in popularity. Multi-rake screens (Figure 9.6) have become increasingly popular because the multiple rakes quickly clear accumulated material from the screen, allowing it to handle high screening volumes during peak flows. Damage by obstructions is prevented by mechanical or electrical torque sensing and repeated reversal of the rake movement. Currently multiple-rake screens are available for coarse and fine screening applications.

2.5.2.2 Single-Rake Screens The single reciprocating rake screen can be equipped with a back clean/back return mechanism or with a front clean/front return mechanism

FIGURE 9.6 Multi-rake bar screen (courtesy of Duperon Corporation).

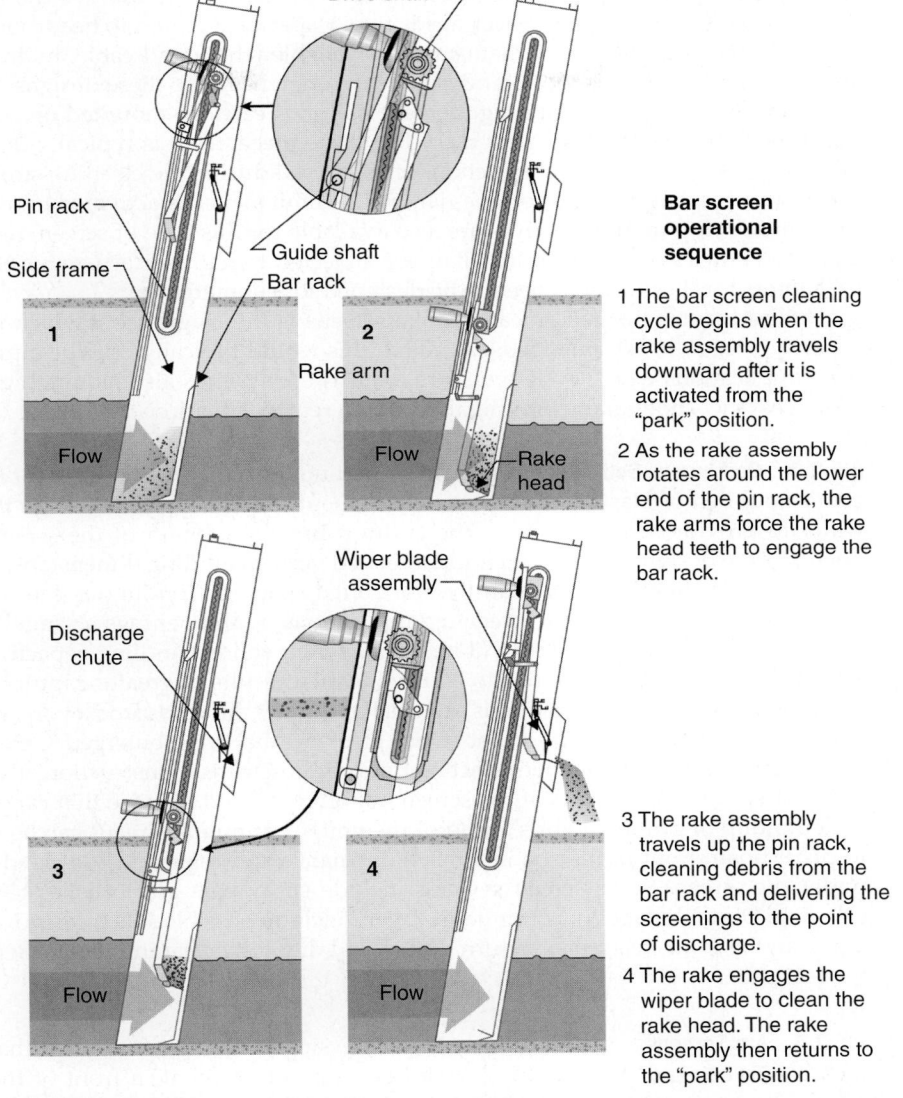

Drive shaft

Pin rack

Side frame

Guide shaft

Bar rack

1

Flow

2

Rake arm

Flow

Rake
head

Wiper blade
assembly

Discharge
chute

3

Flow

4

Flow

**Bar screen
operational
sequence**

1 The bar screen cleaning
cycle begins when the
rake assembly travels
downward after it is
activated from the
"park" position.

2 As the rake assembly
rotates around the lower
end of the pin rack, the
rake arms force the rake
head teeth to engage the
bar rack.

3 The rake assembly
travels up the pin rack,
cleaning debris from the
bar rack and delivering the
screenings to the point
of discharge.

4 The rake engages the
wiper blade to clean the
rake head. The rake
assembly then returns to
the "park" position.

FIGURE 9.7   Reciprocating rake screen (courtesy of Vulcan Industries).

that minimizes solids carryover (Figure 9.7). Although the front-clean design minimizes
carryover, back-clean is less vulnerable to jamming. Because of the limited beam
strength of the long teeth of a back-clean screen, its use is limited to larger opening
screens. The up-and-down reciprocating motion of the rake, similar to that of a person
raking a manual bar screen, minimizes the possibility of jamming. The single biggest
advantage of single-rake screens is that all moving parts are located out of the water.

Headroom requirements for reciprocating rake screens are greater than those for
other types of screens. The estimated headroom requirement can be determined by

adding the vertical depth of the screen to the discharge height above the floor plus 0.72 m (2.5 ft). The design engineer needs to pay special attention to headroom.

Although many drive mechanisms are available (chain and cable, hydraulic, and screw operated) the most popular design is the cogwheel. For these designs, the entire cleaning rake assembly, including the gear motor, is carriage-mounted on cog wheels that travel on a fixed pin or gear rack. The drive mechanism is typically designed to allow the rake to ride over obstructions encountered during the cleaning stroke. In the unlikely event that the rake becomes jammed, a limit switch is activated to turn off the drive motor. Top mounted drives are also available for this type of screen, resulting in increased efficiency. On deep installations subjected to high surface-water elevations, the motors for these screens need to be designed for submergence.

A disadvantage of reciprocating rake screens is the single rake, which limits the capacity to handle extreme loads, although this would typically only be a problem in deep applications where cycle times are long. These systems also require higher overhead clearance, potentially limiting their use in retrofit situations.

2.5.2.3 Continuous Self-Cleaning Screens

Continuous self-cleaning screens consist of a continuous belt of plastic or stainless steel elements that are pulled through the wastewater to provide screening along the entire submerged length of the screen. Screen openings are designed with both horizontal and vertical limiting dimensions; the vertical spacing is slightly larger than the horizontal spacing. Continuous screens can be used in both coarse and fine screening applications, with openings as small as 1 mm ranging up to more than 72 mm (3 in.). The greater solids-handling capacity of these screens allows smaller openings to be used, which results in greater capture of solids from the waste stream. Continuous screens have either a lower sprocket or a guide rail at the channel bottom to support the screen elements that are submerged. Careful material selection is important when selecting this type of screens. Construction of a recessed notch or stair in the channel at the screen bottom is a good practice that can help prevent buildup of grit and debris ahead of the unit. Continuous screens can be designed to pivot up and out of the channel for maintenance and removal of trapped material under the screen. Some screen systems include spray bars and brushes to improve cleaning. Disadvantages of continuous screens include possible solids carryover resulting from the front clean/back return design and difficulty cleaning screen elements on the rear side.

2.5.2.4 Arc Screens

Arc screens are similar to single-rake screens except that the bar rack is curved, and the rake mechanism has a pivoting point in front of the screen, which allows an arc motion during cleaning. These screens can be installed on the side of the channel and can be as long as necessary to provide a large surface area suited for overflow applications like combined sewer overflow (CSO) or sanitary sewer overflow (SSO). Arc screens are also used for influent screening in small WRRFs where channel depth does not exceed 3.5 m (7 ft). Arc screens can be provided with single or multi-rake mechanisms, allowing either full or partial rake rotation depending on the headroom constraints of each site. A new pin-joint design avoids the need for large headspace. An advantage of arc screens is high hydraulic capacity and simple design. Perforated plate arc screens using a swinging brush instead of a rake have recently been introduced for fine-screening applications.

2.5.3   Fine Screens

Mechanical cleaning of these screens is essential and the smaller the opening the more critical cleaning performance is for proper operation. Water sprays or brushes are typically used for cleaning these screens. Hot water provides better results than cold water for cleaning fine screens because it helps to remove grease that has adhered to the surface. Brush-cleaned screens can have a lower capture performance because over time the bristles may bend or may become wrapped in screenings. Washer/Compactors, either integral to the screen or standalone, must be used with fine screens because of the large amount of screenings and organic material removed. Fine screens typically involve mechanisms more complex than the ones found in coarse screens because of the need to remove smaller solids. Fine screens have more elaborate cleaning mechanisms compared with coarse screens, because attached organic material is more difficult to remove from smaller openings.

2.5.3.1   Continuous-Element Screens
Continuous-element screens consist of screens with an endless cleaning grid that is attached to a main drive via different configurations. Screenings are collected and conveyed to the top part of the screen and then discharged. Continuous-element screens can be provided in a variety of forms, but the most popular are the perforated plate and the belt type technologies. Perforated-plate screens (Figure 9.8) are constructed of plastic or stainless steel panels with orifices

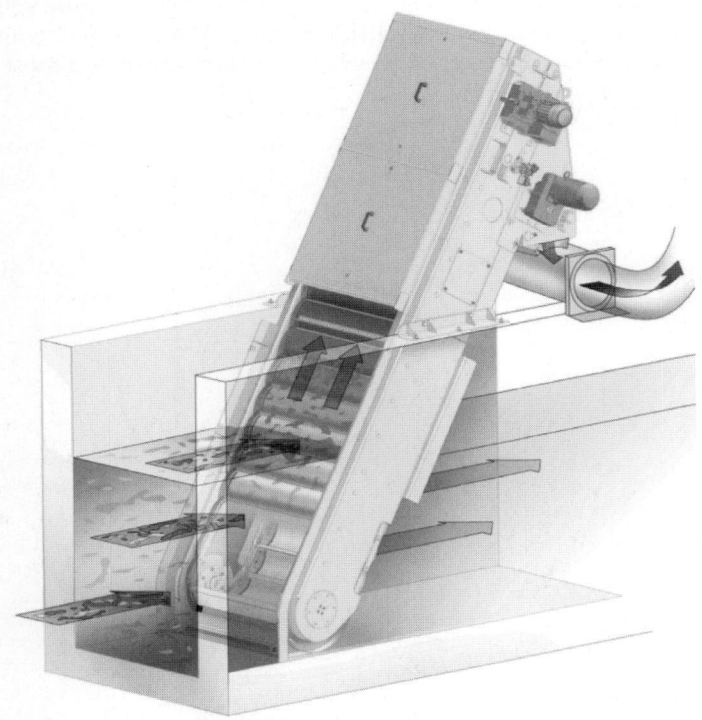

Figure 9.8   Perforated-plate screen (courtesy of Huber Technology, Inc.).

throughout the panel, provided in a stacked panel configuration with rakes at regular intervals to prevent screenings rollback. Orifices are typically round for ease of construction. A proper cleaning mechanism (water spray and/or brushes) is required to remove buildup of organic material.

Perforated-plate screens in a through-flow configuration typically require an inclination angle between 60° and 75° from horizontal. Because the screening media travels downwards on the back of the screen over the screened flow, if the cleaning mechanism is not functioning properly screenings carryover can occur with this type of screen. Continuous-element screens have one of the best screenings retentions in the industry and are often used where minimizing the screenings for downstream processes is important. Because the flow travels through the screen media twice, these screens have greater clean screen headloss than other screens and require a significant amount of flushing water.

2.5.3.2 Multiple-Rake Screens The multiple-rake screens, discussed in the coarse screen section and shown in Figure 9.6, are being used with increasing frequency in fine-screening applications. The basic construction is the same as the multi-rake coarse screens; however, the raking system for finer screens is designed to operate at a higher cleaning frequency because of the higher screenings collection. The close tolerances of the multiple rake mechanism allow this screen to be supplied for openings down to 4 mm (3/16 in.).

2.5.3.3 Stair Screens Also referred to as Step Screens®, (Figure 9.9) stair screens consist of thin, 2- or 3-mm long parallel lamellas of stainless steel with 3- to 6-mm clear openings. There are two sets of lamellas in a stair screen and most designs have one

Figure 9.9   Stair screen (courtesy of Vulcan Industries).

fixed and one moving set that rotates in and out of the screen to provide a step motion pattern that lifts the collected screenings upward until they are discharged at the top of the screen. The moving lamellas are typically connected by either chain drives or levers. The thin lamellas are vulnerable to damage by large objects, rocks, broken glass, and grit. Sometimes wedging or larger objects have been a problem with this design. Flexible lamellas at the bottom are used to prevent blockage or damage by large objects. Water flushing connections prevent accumulation of grit under the lamellas. Stair screens provide a higher open area when compared with other fine screens, because they have open slots running from the bottom to the top of the screen grid. This slot configuration allows passage of stringy solids, which can be minimized through "matted operation" in which the screen is operated with a mat of accumulated screenings created by using differential head control. The larger open area of these screens helps to minimize headloss, which makes it suitable for retrofits where this is a significant concern. Stair screens typically require a significant footprint because it is recommended that they be inclined 45° to prevent screenings rollback. To reduce footprint requirements, some designs use hooked steps, which allows the screens to be installed at a 75° inclination and allows deeper channels. Most stair screens designs can be provided with a pivoting point above the channel that allows the units to be serviced from the operations platform.

2.5.3.4 Band Screens   Band screens (Figure 9.10) are similar to perforated screens because both typically use perforated plate panels. But usually they are wider in band

Figure 9.10   Band screen (courtesy of Headworks®).

screens to accommodate the rotation of the screening plates, creating the "band" shape pattern. Lifting lips provided on each panel lift the screenings over the water and discharge them on a screenings collection trough. There are several configurations available for this technology depending on the way the wastewater flow approaches the screen, but only the centrally fed design is used in wastewater applications. Centrally fed screens are mounted 90° from horizontal with the screen panels parallel to the flow, making screen area independent of the width of the channel. Unlike other fine screens, these screens are not typically provided with pivoting elements for servicing the screen out of the channel, and require headroom or skylights if located inside buildings for removal. Headloss through these screens is typically lower than continuous element screens as the water flows into the center of the screen and out two side panels. This unique layout requires a wider concrete channel surrounding the screen to accommodate the flow pattern. Cleaning of the panels can be accomplished by brushes and spray nozzles at the top of the band which drop the screenings onto a discharge chute. The benefit of this type of screen is that the screening debris remains on the upstream side of the screen and has no chance of passing over to the downstream side.

2.5.3.5 Drum Screens   There are three types of drum screens (a) aboveground, (b) in-channel, and (c) vertical. The aboveground drum screen is mounted on legs, requires pumped influent, and can be fed internally or externally. For the internally fed screen (Figure 9.11), wastewater is fed into a distribution pan inside the rotating drum, and screenings are conveyed to the front of the screen using large, internal auger flights. Screened wastewater is then discharged out the back of the screen. The drum continuously rotates on wheels and screening media can be removed and replaced in sections. Wastewater is fed outside the rotating drum in the externally fed version (Figure 9.12) and solids are deposited to an auger below.

Gravity-flow drum screens are available. These screens are similar to band screens except they are composed of perforated plate panels joined together in a circular pattern.

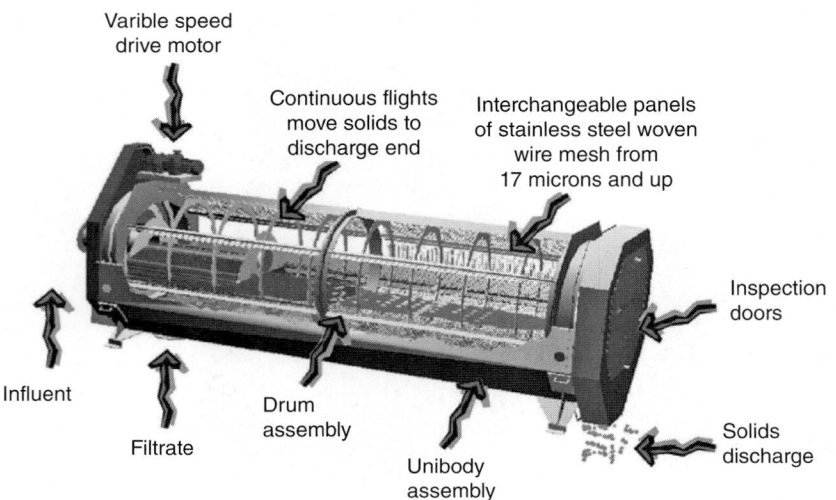

FIGURE 9.11   Internally fed drum screen (courtesy of Baycor Fibre Tech, Inc.).

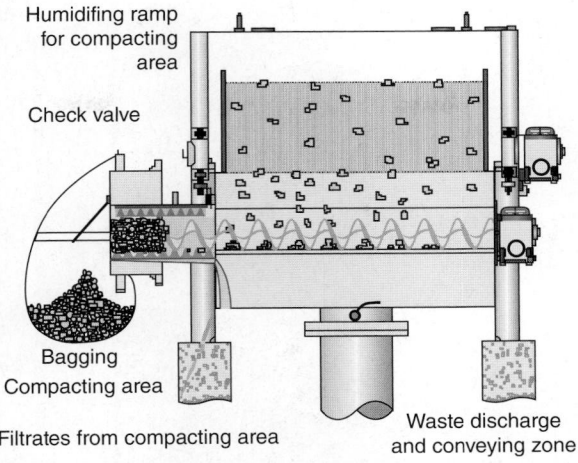

Humidifing ramp
for compacting
area

Check valve

Bagging
Compacting area

Filtrates from compacting area

Waste discharge
and conveying zone

Figure 9.12   Externally fed drum screen (courtesy of Andritz, Inc.).

Drum screens are typically fed centrally with lifting lips similar to those in band screens. The screenings are conveyed to the top of the screens where they are removed by gravity with the help of the washing system. Drum screens must be of a large enough diameter to extend above the operating floor of the channel to allow screenings discharge. They should extend into the channel to a depth sufficient to submerge enough mesh to pass the required flow at low water level.

Vertical drum-screen designs are available for installation in deep manholes. These screens typically have slotted openings to allow an auger-raking mechanism to collect the screenings from the bottom and convey them above the water.

For large flows ranging from 8 to 35 m³/s (180–800 mgd) in a channel, single-entry and double-entry drum screens have successfully been used (Figure 9.13). The drum screen rests with half of the screen in the channel as it slowly rotates around from wet side to dry side. Wastewater flows from the side into the center of the screen and then out through the screening panels. The screenings are rotated to a position above the concrete deck where they are removed by brush or spray nozzles. These massive circular screens are available in opening sizes ranging from 2 to 6 mm (1/16–1/4 in.) and designed to handle a large water differential across the screen.

2.5.3.6   Inclined-Cylindrical Screens   These screens (Figure 9.14) consist of a cylindrical screening basket, similar to the drum screens, with an internal screenings removal mechanism that is typically a helical screw. This screen is typically installed at a 30° to 45° incline. This type of screen can either be fixed with screenings removed by the internal helical screw or by an arm that rotates around the inside of the basket. The basket can consist of either perforated plates or bars. The screenings drop into an axially located hopper feeding a screw and are conveyed through an inclined washer/compactor pipe. Because of the central feed pattern, screenings carryover is low. One of the biggest advantages of this screen is that it is supplied with an integral washer/compactor. However, it requires a larger footprint and shallower influent channel compared to other screens.

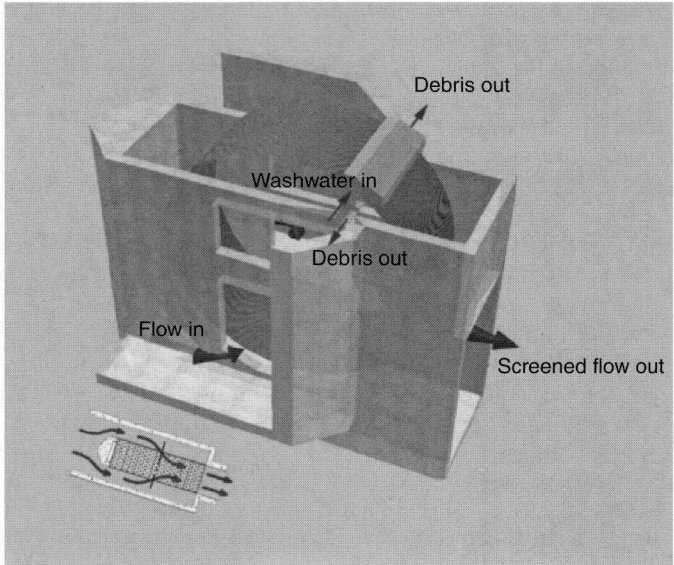

FIGURE 9.13   High flow double entry drum screen (courtesy of OVIVO).

2.5.3.7 Static Screens   Static screens (Figure 9.15) have an inclined metal sieve that screens media and allows the water to pass while retaining the solids on top. Wastewater is fed from the top of the unit, which acts as a weir, and runs down the sieve and collects the screenings. There are no moving parts associated with screenings removal

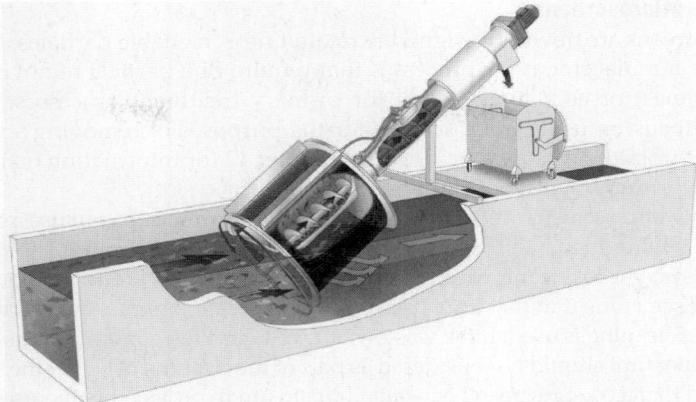

FIGURE 9.14   Inclined-cylindrical screen (courtesy of Huber Technology, Inc.).

and the solids are removed solely by gravity. Screening media is typically wedge wire with openings from 0.25 to 2 mm. Spray nozzles or brushes are provided for cleaning. Static screens can be used along channels, which provide considerable screening area. However, these types of screens can have high headloss requirements.

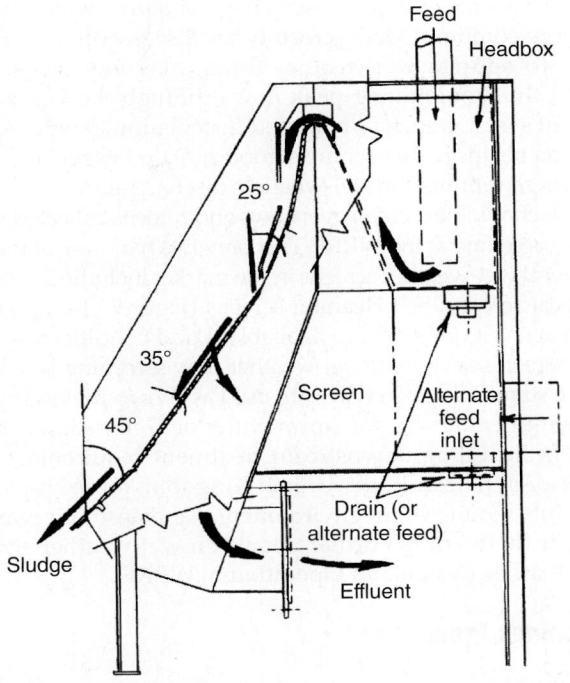

FIGURE 9.15   Static screen.

2.5.4 Microscreens

Microscreens are devices designed to retain fine, screenable particles. They are typically used as a replacement for processes that require fine particle removal (such as certain membranes) or as a replacement for primary treatment processes. This section will briefly focus on using microscreens for the purposes of removing screenable material and grit from the wastewater. Refer to Chapter 12 for information on microscreening to achieve primary treatment standards.

Some microscreens are configured similar to drum screens but use a fine mesh fabric as their screening media and are capable of removing solids from 0.1 to 6 mm on average. Special systems for cleaning these screens are typically provided and should be coordinated with the screening manufacturer based on the application required. Typically, high-pressure, water-jet cleaning is used. If the wastewater contains high concentrations of grease or scum, a hot water unit should be considered as part of the cleaning cycle for the screen. It is recommended that a coarse screen be installed upstream to protect the microscreen.

Recently, the use of a specific type of microscreen—the filter screen—has been resulted in the removal of grit particles due to the operating opening size that these units achieve. Filter screens are devices that use a rotating filter fabric as means to intercept solid particles, but can also be operated in a way to promote organic material buildup on top of the fabric to further reduce the operating opening size and increase solids capture. However, for grit removal applications, the primary mechanism for removal is fabric interception, so the selection of the fabric opening size becomes critical.

2.5.5 Stormwater/Wet Weather Flow Screens

Because of increased regulations, screening of storm water and wet-weather flows is becoming more common. Most screen types discussed have been used successfully on these flows. To address wet-weather flows, overflow weirs are provided upstream of the WRRF, thus minimizing peak flows through the facility. Separate wet-weather flow treatment systems such as ballasted flocculation, retention treatment basins (RTB), or equalization basins have become a good practice, especially where wastewater facilities have stringent limits for wet-weather discharges.

The CSO technologies screen overflows and returns collected materials to the wastewater stream for screening at the WRRF. This prevents handling at the CSO screening location. There are several wet-weather screens in the market including static, self-cleaning, horizontal, and vertical screens. Self-cleaning screens (Figure 9.16) and combination screen/grit removal systems (Figure 9.17) are available. Local conditions will determine the type of storm water screen selection. If wet-weather flow screening is not desired at the diversion point then the standard screen types are used as shown previously.

The screen-opening size for storm water or wet-weather flow screens depends on several factors including downstream treatment requirements, permit requirements, and operator preferences. In the United Kingdom, for example, fine screening at CSO discharge points requires—for environmental and aesthetic reasons—removal of solids that are bigger than 6 mm. Further details on wet-weather screening can be found in WEF's Wet Weather Design and Operation in WRRFs.

2.6 Screenings Processing

2.6.1 Conditioning

Washing and compaction of removed screenings is a critical function of the screening process, especially for fine screens. However, the amount of rock and grit removed with

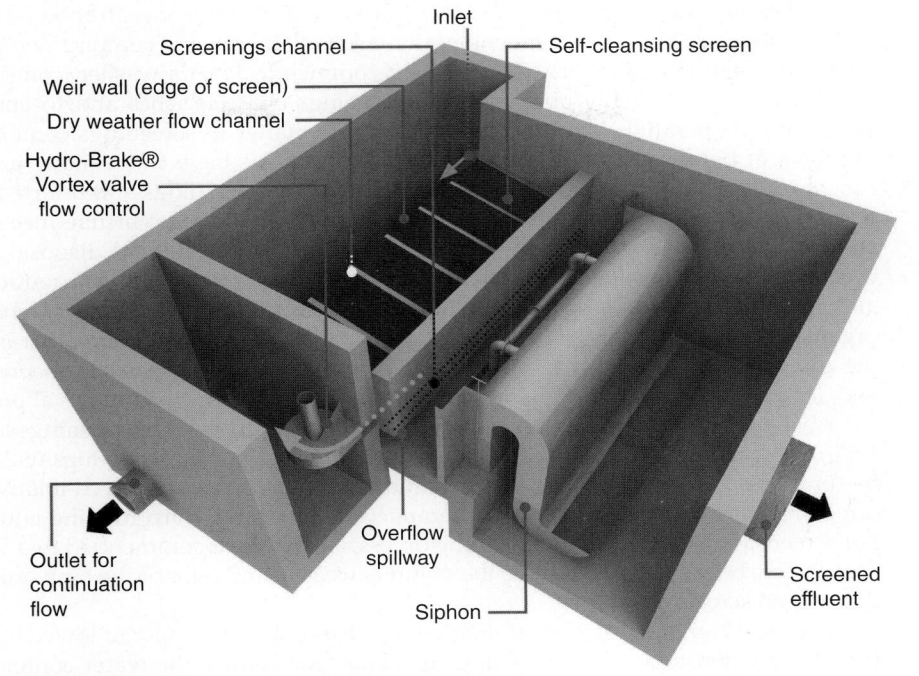

Figure 9.16 Combined sewer overflow screen (courtesy of Hydro International).

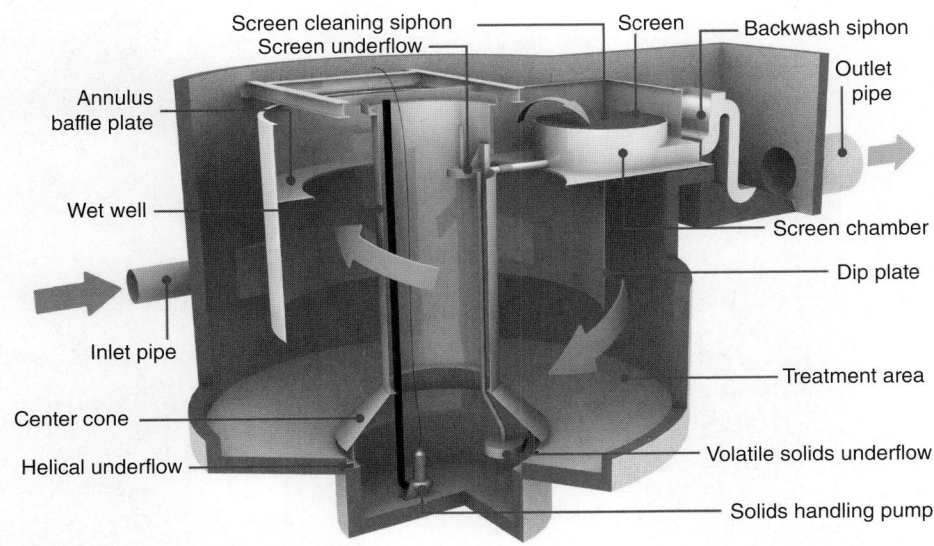

Figure 9.17 Screen/grit removal for combined sewer overflow (courtesy of Hydro International).

the screenings will need to be considered. If there is a high concentration of grit and rocks removed relative to the screenings, it may result in plugged compaction tubes and worn out rotating equipment. It is highly recommended that a washer/compactor be provided to avoid highly objectionable screenings characteristics and to ensure the safety of the operations staff, because all organic matter in screenings contains large numbers of pathogenic organisms. Washer/Compactors have two equally important functions. Washing removes organic material from the screenings and returns it to the wastewater flow. Compaction reduces the volume of screenings, which reduces costs of storage and disposal, as well as, in some cases, helping meet solid disposal requirements for screenings. Returning the organics to the wastewater stream reduces odor and handling hazards associated with the high concentrations of pathogens in the organic matter in screenings. Through solubilization, the washer/compactor increases the concentration of readily biodegradable carbon in the influent to the anaerobic and anoxic zones, providing increasingly important benefits to nutrient removal processes.

Washer/Compactors are recommended for coarse screens with openings less than 1.5 in.; however, the amount of rock and grit removed with the screenings will need to be considered. If there is a high concentration of grit and rocks removed relative to the screenings, it may result in plugged compaction tubes and worn rotating equipment. For screens with 6 mm (0.25 in.) opening or less, it is highly recommended that washer/compactors be provided to reduce the volume, weight, and odor of the high quantity of discharged screenings.

The most common form of washer/compactors is a screw auger followed by a compression friction tube (Figure 9.18). Compactors can reduce the water content of the screenings by up to 50% with a volume reduction of 60% to 85%. However, sticks and large objects in the screenings may cause mechanical breakdowns. As a good practice, controls should sense jams, automatically reverse the mechanism, and actuate an alarm

Figure 9.18   Screenings washer/compactor (courtesy of Huber Technology, Inc.).

when a motor overloads. Provisions for even distribution of the screenings in the collection area increase the efficiency of washer compactors and may be considered. Stacked auger designs are available in which the top auger conveys screenings in the opposite direction of the bottom auger, providing additional washing. Washer/Compactors can be equipped with grinders ahead of the auger to significantly reduce the quantity of screenings and provide maximum washing. In fine screening applications, separate wash tanks are frequently provided that mix and agitate the screenings in wash water. For this type of washer/compactor, screenings sluice channels provide a low-cost method of transporting screenings that uses the same water for both transport and washing. Wash water is drained through a fine sieve at the bottom of the tank and the screenings are pushed by a screw through the friction tube. This water typically has a chemical oxygen demand (COD) concentration between 1000 and 5000 mg/L. Alternatively, a hydraulically operated nozzle in the friction tube can increase pressure and improve screenings compaction. Because the wash water from washer/compactors will have solids in it, the wash water drain pipe must be designed to prevent solids accumulation, which could lead to clogging. Redundant washer/compactors should be provided for fine screen applications because of the large quantity of screenings produced. Hydraulic ram washer/compactors are not recommended because they provide little washing capability. A typical performance specification for washer/compactors is to provide a minimum of 90% removal of suspended organics.

U.S. EPA Method 9095B consists of filtering through an approximately 60-mesh paint filter to determine solids content of screenings for solids contents of at least 15% and no free water. Higher solids concentrations are analyzed using the conventional total solids test (drying at 103°–105°C).

No standard test exists for determining the organic content of washed, compacted screenings. There is little value in performing a volatile solids test because of the large amount of paper in screenings. In Europe a "quality factor" test has been used to quantify organic removal in washer systems. A discussion of this procedure is provided in Section 2.7.8 of this chapter.

2.6.2   Transport, Storage, and Disposal

The method of cleaning the screens, manual or mechanical, relates to how the screenings are removed and transported to a disposal site. When designing a manually cleaned screen, relatively shallow screening channels are needed to allow manual cleaning with a rake and a drainage plate is provided to allow drainage of the screenings before shoveling. The container used to carry the screenings to a truck or other transport may range from a wheelbarrow to a bin carried by an overhead crane or monorail. Whatever the means of conveyance, operator safety, including nonslip platforms and railings, deserves special attention during design.

In mechanically cleaned units, rakes move screenings up the screen to above-deck where they discharge to a conveyor, washer/compactor, or removable containers. Some dewatering of the screenings typically occurs as they are lifted from the wastewater. When discharging screenings directly from screen to container, sufficient clearance under the discharge chute must be available for easy placement and removal. Screens discharging to conveyors should also be provided with enough clearance to allow use of a container if the conveyor malfunctions.

Conveyors are a widely used method of mechanically transporting screenings. Belt conveyors are the most common, but screw augers are also used. If the screen discharges directly to a washer/compactor then the compaction friction tube can be extended to the screenings collection point. Friction tubes longer than 15 feet should be avoided because the screenings can become too dry, requiring excessive force to push the screenings through the tube. Augers can be designed to serve both for collection/transport and as the first part of the washer/compactor.

Belt conveyors do not require significant maintenance but, depending on odor control requirements, they may need to be covered and ventilated. Belt conveyors are typically straight, but if direction changes are needed then curved conveyors can be used, eliminating the need for multiple units. Excessive slope on a belt conveyor should be avoided to avoid roll back and prevent screenings from falling off the conveyor. Doctor blades, sometimes equipped with spray washes, are required at the end point of the conveyor or the screenings will adhere to the screen and fall into the drain pan on the return path. The conveyor or drain pan should be sloped for drainage throughout the conveyor length. Another method uses a perforated conveyor to allow water to pass beneath the conveyor and the trough drains back to the wastewater flow stream. The holes should be located to prevent dripping on the carriage assembly. Concave, ribbed belts should be used to prevent spillage and to prevent screenings from sliding back if the conveyor is steeply inclined. Because of the corrosive environment, metal components are typically stainless steel.

Screw augers are used for screenings transport, particularly when an enclosed system is required for odor control. Both shafted and shaftless augers are used but because objects are more likely to become wedged in shafted augers, shaftless augers are more common. The biggest benefit of augers is the ability to function as the first part of a washer/compactor system prior to discharge to a compaction tube. Augers can have multiple introduction points. The auger should be sloped to a drain. A well-designed spray system is required along the auger and, particularly, at the drain point. The drain point must be accessible in the event of blockage.

Another method of screenings transportation is sluice channels. Sluicing screenings is becoming more popular for a number of reasons, but is only used in conjunction with a washer or compacter. As the need for cleaner screenings increases, requiring the use of wash tubs on the washer/compactor, the sluice essentially becomes an extension of the wash tub, getting the washing process started as the screens are discharged from the screen. Sluices have no moving parts and have a low capital cost compared to conveyors. The screenings are flushed through a sloped channel using an automated flush-water connection. The flushing water serves a dual purpose of transporting the screenings and serving as wash water in washer/compactors. For this reason, sluice system design must be coordinated carefully with the supplier of the washer/compactor. The use of sluice channels can also facilitate grinding and, in some cases, conveying to a site better suited for treatment and removal. In some CSO applications, the wet-weather overflow is pumped into a sloped channel to sluice screenings into a collection pit. The main downside to sluicing is that it requires a relatively large amount of water, especially if a facility effluent water supply is not available. The sluice also requires a certain slope to convey the screenings, so the layout of the sluicing channel will need to be coordinated with the design of the screening discharge and screening conditioning.

To complete the handling of screenings, the material must be transported for disposal in accordance with local, state, and federal regulations. Landfilling is by far the most common method of disposal; however, increasingly strict European regulations for landfilling are designed to require incineration and other methods of disposal.

Onsite incineration, another method of screenings disposal, typically involves mixing the screenings with other treatment facility solids. Nevertheless, a few large facilities incinerate screenings separately because incomplete mixing of screenings with the primary sludge and waste biosolids creates unstable conditions in the incinerator. Screenings tend to clog feed mechanisms and should be ground or shredded and dewatered before being fed to the incinerator (U.S. EPA, 1979). If the screening-to-solids volume ratio is small, material should be mixed with the solids feed at a relatively constant ratio. The batch addition of screenings to an incinerator can cause uneven burning clinkers and hot spots.

Sometimes screenings hauled from the facility are mixed with municipal refuse and burned at the municipal incinerator. This is a common practice in Europe. Although screenings contain more water and have a higher organic content than municipal solid waste, the relatively small volume of screenings has negligible effect on incineration (U.S. EPA, 1979).

2.7 Design Considerations

2.7.1 Design Criteria

Design considerations for screens include opening size; screen type; equipment height and footprint; channel depth, width, and approach velocity; headloss; provisions for removing the screen for maintenance, including pivoting designs; access for futures screen replacement; discharge height to accommodate screenings conveyed or washer/compactor equipment; angle of screen; screen cover for weather protection and odor control; construction materials and coatings for overall unit; drive unit service factor, motor size, environmental rating, and enclosure; spare parts; and provision of redundant screen or bypass manual screen.

2.7.2 Equipment Selection Criteria

The following describes current preferences for wastewater screening:

- Trash racks are seldom used ahead of coarse screens or grinders, except for combined systems with significant quantities of storm water and at prisons with heavy screening quantities. A more common use of trash racks is for fine screens protection.

- Because of advances in technology, most new projects use screen openings from 3 to 12 mm (0.125–0.5 in.).

- Even finer screens (1 to 3 mm) are used for advanced processes such as membrane bioreactors (MBRs).

- Along with finer screens, washer/compactors are now standard equipment.

- In the WEF utility members survey of June 2008, the median screen opening is reported to be 9.5 mm (3/4 in.) and 53% of the facilities were operating washer/compactors.

2.7.3 Location of Screen

Design practice for the location of screening devices at the treatment facility varies. Screens can be located either upstream or downstream of influent pumps because most common raw wastewater pumps are capable of pumping screenings. However, with the recent introduction of non-dispersible products, preference has been given to screening prior to pumping. Screens are typically located upstream from the grit removal system. When fine screens are used, they could be located either upstream or downstream of grit removal, depending on the screen opening size provided and the anticipated impact on grit concentration equipment, which tends to plug with large solids. The location of fine screens sometimes requires special considerations as discussed under Section 2.7.5.

The question of whether screening equipment should be located in an enclosed structure depends on three conditions: (1) equipment design, (2) odor control needs, and (3) climate. In climates with freezing temperatures, a heated enclosure is necessary. A structure will not only protect the equipment but will ease maintenance and improve aesthetics. Regardless of whether the screening equipment will be housed, drive mechanisms of a mechanically cleaned bar screen should be enclosed. In windy areas, the screen rake and discharge chute areas need covers to prevent screenings from blowing away.

Any structure that contains a raw sewage screening device requires good ventilation to reduce accumulation of acidic and corrosive moisture and to treat odors. Ventilation rate of the enclosure is based on the ability to maintain a sufficient velocity through openings to prevent odorous gases from escaping. In areas of excessive humidity, combining good ventilation with air-drying or heating units may be warranted. Consideration should be given to venting the building air from the space under the channel covers. This will provide a more pleasant air space and will help reduce corrosion as fresh air will be pulled across the equipment as it moves to the vent under the channel. A good odor control system will be necessary to reduce the odors vented outside.

The effects of backwater, that is, upstream water level, caused by headloss through the screen are important. Many installations include an overflow weir to a bypass channel to prevent upstream surcharging if the screen becomes blinded because of power failure or mechanical problems. Screens typically precede flow-metering devices to prevent backup from affecting, measurement accuracy.

2.7.4   Hydraulic Considerations

Hydraulic considerations described herein apply to the design of a bar screen installation. For other screens and grinding devices, designers should follow manufacturer's recommendations for channel dimensions, capacity ranges, upstream and downstream submergence, power requirements, headloss information, and screening retention.

Velocity distribution in the approach channel has an important influence on screening equipment performance, regardless of screen type. A straight channel ahead of the screen ensures good velocity distribution across the screen and maximum device effectiveness. Use of anything other than a straight channel can divert too much flow to one side, resulting in accumulation of debris on that side of the screen. Hydraulic symmetry is important in the design of influent channels especially when multiple screens will be provided to prevent uneven screenings distribution. Borys (2014) evaluated the effects of uneven flow distribution prior to a screening device, which included screenings carryover due to excess velocities in localized areas of the screen and resulted in the design of a flow straightening feature in the channel (Figure 9.19).

The design engineer must ensure that the wastewater approach velocity to the screen does not fall below a self-cleaning value or exceed velocities that may force material through the screen. Lower velocity will result in more solids and grit deposited in

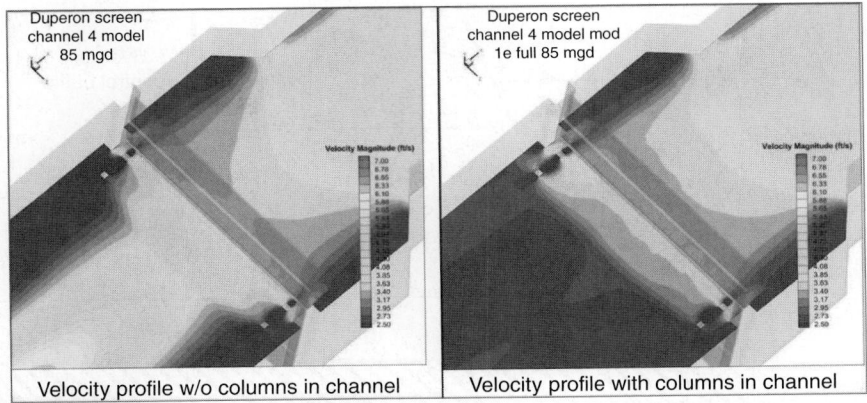

Duperon screen channel 4 model 85 mgd	Duperon screen channel 4 model mod 1e full 85 mgd
Velocity profile w/o columns in channel	Velocity profile with columns in channel

Figure 9.19   CFD modeling for proper screening design (Borys, 2014).

the channel. Ideally, the velocity in the channel ahead of the screens should exceed 0.4 m/s (1.3 ft/s) at minimum flows to avoid grit deposition. However, this is not always possible with the typical diurnal fluctuation in wastewater flows. As a reasonable compromise, the channel could be designed so that a velocity of at least 0.76 m/s (2.5 ft/s) for re-suspending solids is attained during peak-flow periods. Where significant storm water must be handled, approach velocities of approximately 0.9 m/s (3 ft/s) should be used to prevent grit deposits from compromising screen operation during storms. A method to prevent solids accumulation is the use of diffused air systems to provide solids buoyancy (Rothman and Norrköping, 2005). In addition to solving the solids issue, this method also modifies flow streamlines in the channel, improving collection of stringy solids. The flow velocity ahead of and through a bar screen affects its operation significantly. To avoid screenings pass-through, a peak flow velocity of no more than 1.2 m/s (4 ft/s) through the openings is recommended. Because the screen open area is typically around 50% of the cross-sectional area of the flow channel, one of the most challenging aspects of screen design is balancing the need for grit suspension velocity upstream of the screen with maintaining low enough velocity through the screen openings to prevent screenings pass-through. If peak-to-average flow ratios are significant, peak-flow velocities typically exceed recommended design values, causing higher than anticipated screening carryover. In these situations, hydraulic control structures, such as cut-throat flumes and partially submerged baffled walls (Figure 9.20), have been successfully used to control the water level inside the channel at peak conditions. This allows the velocities through the screen to be within the design values. Provisions must be made for the added headloss in the hydraulic profile and for the scum that could collect in the channel. Rothman and Norrköping reported that a partially submerged gate downstream of a stair screen caused excessive scum accumulation, which compromised operation (2005).

Based on the velocities required for proper screen operation, the number of screens can be evaluated to help best meet the flow requirements. Typically, designs have resulted in at least one screen rated for 100% of the peak hour flow with either a fully redundant standby screen or a manual bar rack. Recently, more designs are including multiple screens to meet the design flow with at least two screens rated for 50% of peak

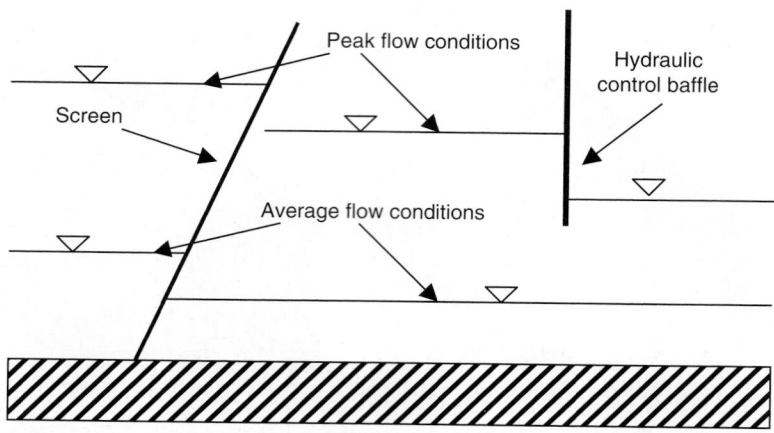

Figure 9.20 Hydraulic control structure example.

hour flow with a third channel with another 50% screen or manual bar rack. The multiple screen option will require some additional instrumentation and programming so that the SCADA system calls for the right number of screens based on influent flow, but it provides for a more efficient screening design. In installations where the water surface elevation inside the channel is less than 0.15 m (6 in.) at average flow conditions, the collection of screenings in through-flow design screens is not optimal. In these cases, incorporation of hydraulic control structures like weirs or channel constrictions downstream of the screen to raise surface water elevations at average and low flow conditions can help improve operation, but bear in mind the negative impacts of lower velocities in the channel. Some form of opening should be considered in the weir or constriction to prevent grit from settling downstream of the channel.

For facilities with multiple screen channels, automation of screen channel gates to allow for taking screen channels in and out of service according to influent flow is another method of balancing minimum channel scour velocities with maximum screen flow-through velocity.

Typical headloss allowed through a manually cleaned bar screen is 150 mm (6 in.), assuming frequent screen inspection. A nearly constant headloss for continuously cleaned mechanical screens can be maintained with a constant flow, but this headloss allowance should never be less than 150 mm (6 in.) because of intermittent cleaning cycles. Curves and tables for headloss through the screening device are available from equipment manufacturers. Fine screens can be operated at low-frequency cleaning cycles that promote development of mats in the screening elements, mostly of grease and organic material, increasing screenings capture ratio. There have been installations where screened wastewater has met primary treatment standards following this approach (Rusten and Lundar, 2006). However, it is advisable to increase cleaning frequency during wet-weather flows to minimize headloss through the screen, temporarily tolerating reduced screening capture performance.

Headloss through screens typically ranges from 50 to 635 mm (2–24 in.), depending on flow and screen-opening size. Maximum headloss can reach 1 m (3 ft) or more in large units at peak flow with fine screens or when screen cleaning frequency is inadequate. The design engineer must take steps to ensure that headloss through the screening device at maximum flows will not surcharge influent sewers enough to impair upstream services. To prevent flooding of the screening area caused by blinding of the screen because of a power failure or another problem with the screen, the design should provide for an overflow weir or gate and a parallel bypass channel allowing overflows to go around the screen. The screen manufacturer should be made aware of the maximum hydraulic differential that may occur to ensure the screen does not break under high pressure. Flow measurement in the vicinity of screening equipment can sometimes be erratic because of surges generated after cleaning cycles. This could be mitigated by arranging time readings over a minimum period (typically 10 minutes) to dampen variation in surface water elevations during surges.

Assuming continuous two-dimensional flow, the clean water headloss through a screen can be calculated using the Bernoulli equation. This equation states that the depth of flow plus the velocity head before the screen equals the depth of flow plus the velocity head after the screen plus friction losses. The actual headloss varies with the quantity and composition of the screenings allowed to accumulate between cleanings. The headloss created by a clean or partially blinded screen may be calculated based on the flow and velocity through the screen. Velocity is calculated by dividing flow by the effective area of screen openings. The effective open area is the actual open area minus a blinding factor that depends on intended mode of operation and the type of screen. Determine whether the screen frame will be recessed or will protrude into the channel area as this will affect the effective open area.

Proper sizing of screening equipment involves careful consideration between the channel width, available hydraulic grade line, the screening opening size, the maximum slot velocity through the screen at different conditions, and the minimum velocity at low flows. Of paramount importance is the estimation of the headloss through the screen, as the screen size (screening area) is correlated to this number. The recommended practice for sizing screening elements should include estimation of the surface water elevation before and after the screen, and the size of the screen shall be determined by the average slot velocity between the upstream and downstream components such that the maximum slot velocity is not exceeded. Care should be taken with conservative headloss estimations, as they will drive the size of the screen down, which could result in a screen that is smaller than what is required. Conversely, underestimation of the headloss could lead to screen overdesign or higher than predicted hydraulic grade line upstream of the screens with all its impacts to upstream processes or infrastructure (Botero, 2014).

Recent screening headloss investigations have concluded that the methods used for headloss estimation may not be very accurate, specifically with fine screens. Botero suggested that further research is needed in this area and cautioned practitioners from relying on the commonly used expressions for this purpose (2014). However, with the development of computational fluid dynamics, more accurate methods for headloss estimation in screens are now available and can be considered when sizing screening components (Botero, 2014; Borys, 2014). The expressions presented in the following subsections are the traditional headloss computation expressions.

2.7.4.1 Coarse Screens Headloss The headloss through a clean or partially blinded coarse bar screen can be represented by the following equation:

$$h = \frac{k(V^2 - v^2)}{2g} \tag{9.1}$$

where h = headloss, m (ft)

V = velocity through bar screen, m/s (ft/s)

v = velocity upstream of bar screen, m/s (ft/s)

g = gravitational acceleration, 9.81 m/s² (32.2 ft/s²)

k = friction coefficient (function of opening size and bar shape and can vary from 3.5 for a teardrop bar with 6 mm opening, to 1.4 for a trapezoidal bar 19 mm opening clean).

This equation is recommended for use in preliminary stages of the project, when the screening components have not been fully selected. For more accurate results, however, Kirschmer's equation is recommended. This equation takes into consideration the angle of the screen, which significantly affects headloss through screens because the screening area is directly proportional to the incline of the screen.

Headloss through a clean-bar screen is indicated by Kirschmer's equation:

$$H_L = b\,(w/b)^{1.33}\,h\,\sin\phi \qquad (9.2)$$

where H_L = headloss, m (ft)

b = bar shape factor (Table 9.2)

w = maximum cross-sectional width of bars facing upstream, m (ft)

b = minimum clear spacing of bars, m (ft)

h = upstream velocity head, m (ft)

ϕ = angle of bar screen with horizontal

2.7.4.2 Fine Screens Headloss   For fine screens with slotted openings, headloss can be estimated using the above methods. For fine screens with circular or rectangular openings, headloss can be estimated using the orifice headloss equation. However, similar to coarse screens, the installation angle must be considered when calculating the screen open area. A good practice to account for the angle of installation is to project the submerged screening area to plane perpendicular to the flow and use the projected area in the calculation of the headloss. Manufacturers should be consulted if a high degree of accuracy is required in the calculation because screen panels or elements may change geometry slightly to allow for the rotation of the screen components.

Bar Type	β Range (Function of Opening Size)
Sharp-edged rectangular	1.3–2.6
Rectangular with semicircular upstream face	1.83
Circular	1.79
Rectangular with semicircular upstream and downstream faces	1.67
Trapezoidal	1.3–2.5
Teardrop	0.8–1.5

TABLE 9.2   Kirshmer's Values of β

The orifice headloss equation is as follows:

$$H_L = \frac{1}{(2g)}\left(\frac{Q}{C \cdot A_e}\right)^2 \qquad (9.3)$$

where H_L = headloss, m (ft)

C = discharge coefficient (0.61 widely accepted for clean screens, however depends on the type of orifice and screen construction)

Q = flow upstream of bar screen, m³/s (ft³/s)

g = gravitational acceleration, 9.81 m/s² (32.2 ft/s²)

A_e = effective open area of submerged screen, m² (sq ft)

Headloss calculations typically include a blinding factor of up to 50% reduction of the open area for determining maximum headloss. Koch and colleagues reported that with the improvements in screening collection efficiency, lower blinding factors can be used when sizing screening systems (2014). Therefore, the selection of the screen type could have implications on the size of the screen elements and should be accounted for in the design.

2.7.5 Special Fine Screening Design Considerations

Fine screens require special design considerations because the high capture rate of these screens makes them more susceptible than coarse screens to blinding and damage from large objects. The main things to determine are if (1) fine screening requires prior coarse screening, (2) grit removal is required before fine screening, and (3) grease/scum removal is also required for proper operation.

It is good practice to include coarse screening before fine screening. There are installations with 6 mm (1.4 in.) opening bar screens that operate successfully without coarse screening pretreatment. However, for bar screens with smaller openings or perforated fine screens, coarse screening is recommended. Determining whether or not to include coarse screening or grit removal is site specific. The design engineer should evaluate operating conditions to make a final determination. Outlined below are the most important factors considered when making that determination:

- The type of sewer system. Compared with sanitary sewers, combined sewer systems carry larger objects that could be difficult to remove or even cause damage to some types of fine screens. Combined sewers also have higher first flush solids loads that could affect screening equipment.

- The type of fine screen. There are certain types of fine screens that are more likely to operate without problems without coarse screening ahead of them. For example, multiple rake screens and stair screens have proven to work properly in these conditions. In contrast, all perforated plate fine screens are more likely to require coarse screening because these screens are more prone to damage by large objects. Microscreens rely on coarse screening systems being installed upstream for proper operation.

- The wastewater influent conveyance method. If the influent to the wastewater facility is pressurized through pumps then there may be some level of coarse screening or coarse solids reduction at the pumping station(s) (either by coarse screening devices, grinding devices, or solids disintegration at the pump impeller). Certain types of pumps, like Archimedes screws, do not provide screening or coarse solids reduction capability by themselves so the designer should account for this.

- The effective size openings on the fine screens. If the fine-screening facilities require removal of solids equal to or less than 3 mm (1/8 in.), then it is a standard practice to provide coarse screening before fine screening. Additionally, grit removal before the fine screens is also a standard practice in this case.

The high capture rate of fine screens can lead to a host of new problems, if the entire screening system from channel to dumpster is not evaluated. A 2002 Water Environment Research Foundation study found that some facilities that had replaced coarse screening with fine screening experienced excessive screenings collection, overwhelming handling systems and causing backwater issues upstream. A fine screen should not be installed as a replacement for a coarse screen without first performing a hydraulic evaluation to determine if there will be any adverse effects. Therefore, the design engineer should evaluate the impacts of a fine screen during the design phase.

2.7.6 Screenings Conditioning and Handling

Design considerations for conditioning and handling of screenings vary depending on whether or not washing/compaction equipment is provided. Screenings that have not undergone washing and compaction may require some dewatering before final disposal. Special containers that allow water to be drained have been used, but typically require maintenance for proper operation, as vectors attraction and odor concerns may arise.

If screenings will undergo washing and compaction, then provisions to deliver facility effluent water is recommended as washer/compactors and the screens they serve have a high water demand. Using facility effluent water helps increase the sustainability of preliminary treatment designs. It is important to have a robust and redundant effluent water filtration system if the facility relies on a steady source of water for process purposes to prevent nozzles and smaller wye strainers from plugging.

Washing/Compaction equipment is designed to convey a maximum volume of screenings in a given amount of time; therefore, determining peak screenings capture becomes a critical factor for proper equipment selection. The design engineer should consult with the washer/compactor manufacturer to determine equipment size and safety measures. It is also important to consult facility data and the operations staff to determine if there are specific peak events that could overwhelm the system. It is a typical practice to convey reject water from washing/compaction equipment upstream of the screens.

Now that the screenings have been removed from the liquid and processed, they will need to be stored in a dumpster before they are hauled off to the landfill. Consider how the screenings will be transferred over to the company that operates the landfill and what requirements it may have on the hauled screenings. While small systems may produce enough compacted screenings to fill a small wheelbarrow, it is still back-breaking and potentially hazardous work to transfer this over to a larger dumpster. For larger systems, consider whether additional storage may be needed if the dumpster is removed from the site and hauled off to the landfill. A storage hopper above the dumpster can be used to provide a few hours of additional storage. In larger dumpsters, the processed screenings can mound up around the discharge location. A discharge chute or shuttle conveyor can be used to help spread the screenings to multiple discharge locations.

2.7.7 Automation and Instrumentation
For mechanically cleaned bar screens, the following types of controls can be used alone or in combination:

- Manual start and stop
- Automatic start and stop by timer control
- High-level switch
- Automatic duty rotation for even wear
- Flow-based controls for multiscreen operation
- Differential head-level actuated starting switch on cleaning mechanism
- Variable-speed operation with one of the following equipment items:
  — Two-speed motors,
  — Variable-frequency drive alternating current motors, or
  — Mechanical speed reduction devices.

Mechanical bar-screen installations should include three alarms: high upstream level, screen start, and screen fail. The upstream-level alarm can be an ultrasonic sensor, bubbler system, or submerged transducer. The float device may become clogged so it should not be used as a primary control device, but may be used as a high-level float switch. Care should be taken in locating a float to avoid areas of excessive turbulence or velocity in the main influent channel.

A well-designed screening system will interlock auxiliary equipment such as spray wash, conveyors, or screenings compactors with the operation of the screen. This will ensure coordination among process operations.

2.7.8 Performance Testing
Screening performance is a new concept in the wastewater screening industry. Previously, screening performance was not required by regulations. But with the advances in technology and new regulations, such as those related to CSOs, validating the performance of screens is becoming more common. An example is the maximum size particle restrictions required by advanced treatment processes. Warranty issues also may be a concern that has led to performance testing of the screens.

Reputable manufacturers test their designs before introducing them to the market; however, for the previously mentioned reasons, validating performance claims is becoming standard practice. Currently used testing parameters are discussed in the following parameters.

2.7.8.1 Screen Retention Value The screen retention value is the percentage of the total influent solids retained by a screen. The screen retention value can be computed using the following formula:

$$SRV = \frac{TSRE_{with} - TSRE_{without}}{100 - TSRE_{without}} \times 100 \tag{9.4}$$

where SRV = screens retention value, percentage

TSRE$_{with}$ = mass of total solids retained in channel and removed by screen, kg (lb)

TSRE$_{without}$ = mass of total solids retained and removed by the channel, kg (lb)

To screen for the mass of total solids in influent is established by collecting the samples on 6-mm mesh sacks for 30 minutes, and letting them dry for 20 minutes before weighing (U.K. Water Industry Research Limited 2002). Performance criteria for different fine screen technologies are presented in Table 9.3 (Cambridge et al., 2003). A recommended screens retention value of 30% or higher is recommended in the United Kingdom for installations in CSO applications and could be a good guideline value for sanitary sewer applications depending on system characteristics.

2.7.8.2 Screenings Organics Test Even though testing and quantification of organic content of screenings has not been standardized, a quality factor test is outlined below. Samples of the washed and compacted screenings are re-suspended in clean water and carefully agitated. The sample is filtered through a 100-micron mesh and the screenings are lightly dewatered by carefully pressing a sample down on top of the filter mesh with the goal of removing most of the remaining water. This procedure is repeated four times. The mass of biochemical oxygen demand (BOD) in the collected rinsing water is analyzed and divided by the dried solids mass of the remaining screenings. The lower the quality factor is, the better the performance of the tested washer/compactor. A quality factor of maximum 20 mg of BOD$_5$ per gram of dry solids is recommended as acceptable washer/compactor performance.

2.8 Screening Sustainability and Resource Recovery

With the increased focus on sustainability and resource recovery in WRRFs, design of the screening process should also consider these principles. Sustainability is enhanced by a properly designed screening and screenings washing process by reducing the pathogen and organic content of the screenings; direction of the highly soluble organic loading associated with the screenings to the liquid stream, enhancing the nutrient removal processes; and enhancement of the gas generation potential of the facility by keeping the downstream operating equipment working properly. When inadequate screening causes ragging issues, then the adverse effects on the different processes could range from increased energy demand for pumping, aeration, mixing, etc.; to inadequate material for biosolids reuse. While reuse of screenings is not commonly practiced, some opportunity for resource recovery exists with facilities equipped with incinerators that practice heat recovery.

Fine Screening Technology	SRV
Fine bar screens	30–40
Stair screens	30–40
Front entry perforated plate screens	65–75
Center flow perforated plate screens	75–85
Perforated plate drum screens	75–85

TABLE 9.3   Screens Retention Value for Various Screens

Screening Design Example

This example illustrates a few of the design dilemmas and compromises to be negotiated in a process apparently as simple as initial screening.

Assumptions:

(1) Processes downstream of the screenings facilities require removal of fine solids because they are sensitive to solids. Therefore, the screen design should prevent solids carryover through the screens at all flow conditions. A 6-mm through flow perforated plate screen opening was selected.

(2) The new plant will be built in a new area (no historical screening generation information is available). It was determined that there is no excessive fat, oil, and grease (FOG) in the influent, and flows will be pumped to the site. Assume that the pump stations feeding the plant have coarse screening capabilities. Assume a combined sewer system.

(3) A washer/compactor unit will be installed to reduce the volume of the screenings.

(4) The following items have been assumed (numerical values for each are below): screenings generation peak factor (function of the type of collection system); screening container volume safety factor; and maximum allowable blinding screen factor.

(5) The maximum number of influent channels for this facility is assumed to be four, due to area constraints.

(6) Other assumptions are displayed below.

Input parameters (numerical values for each item are below):
Q_{avg} (average flow); PF (peak factor); number of channels; number of duty channels; hydraulic elevations downstream of the screen (from hydraulic profile); screen type and opening size; and recommended screen installation angle.

Manufacturer's information required:
Percentage of screen open area and screenings volume reduction in washer/compactor.

Equations and figures used:
Orifice equation (eq 6.22); fine screen headloss (eq 9.3); average screenings quantities–fine screens (Figure 9.2).

Screenings Equipment Design

Number of total influent channels = 4
Assume 1 channel is fully redundant

Number of duty channels $= N_{channels}: = 3$

$$Q_{avg} = \frac{Q_{avgtotal}}{N_{channels}} = 0.35 \frac{m^3}{s}$$

$$PF = 3.0 = \text{peak factor}$$

$$Q_{peak} = PF \cdot Q_{avg} = 1.051 \frac{m^3}{s}$$

Channel Design

Select a channel based on the following criteria as outlined on Section 2.7.4.

$$V_{max} = 0.9 \cdot \frac{m}{s} \quad \text{Target channel velocity.}$$

$$V_{min} = 0.3 \cdot \frac{m}{s} \quad \begin{array}{l}\text{Minimum velocity to avoid deposition} \\ \text{in the channel.}\end{array}$$

$$V_{max.through.screen} = 1.2 \cdot \frac{m}{s} \quad \text{Maximum velocity to minimize solids carryover.}$$

$$V_{resuspension} = 0.76 \frac{m}{s} \quad \begin{array}{l}\text{Minimum velocity to ensure solids resuspension} \\ \text{in the channel (settled solids).}\end{array}$$

Area of Channel Required

$$A_{max} = \frac{Q_{peak}}{V_{max}} = 1.168 \text{ m}^2 \tag{1}$$

From the hydraulic profile analysis, following are the surface water elevations downstream of the screen.

$Q_{per\ channel}$	m³/s	0.35	0.45	0.55	0.65	0.75	0.77	0.88	0.92	0.95	1.05
h	m	0.65	0.68	0.71	0.74	0.76	0.77	0.80	0.81	0.82	0.85

$$h_{max} = 0.85 \cdot \text{m (at peak flow)} \qquad h_{avg} = 0.65 \cdot \text{m (at average flow)}$$

$$\text{Width}_{calc} = \frac{A_{max}}{h_{max}} = 1.374 \text{ m} \tag{2}$$

$$\text{Select Width} = 1.3 \cdot \text{m}$$

$$V_{channel.downstream.avg} = \frac{Q_{avg}}{h_{avg} \cdot \text{Width}} = 0.415 \frac{m}{s} \qquad (> 0.3 \text{ m/s, so OK}) \tag{3}$$

$$V_{channel.downstream.peak} = \frac{Q_{peak}}{h_{max} \cdot \text{Width}} = 0.951 \frac{m}{s} \qquad (\text{close to } V_{max}, \text{ so OK}) \tag{4}$$

Manufacturers information provides effective open area of a 6-mm screen at recommended angle of installation of 70° to horizontal. For this example:

$$\text{Effective}_{area.screen} = 51\%$$

Now, check the velocity through the screen openings at peak flow conditions

$$Cd = 0.61$$

Calculate the water elevation in the channel upstream at peak and average conditions

Peak conditions

$$h_{through.screen.peak.0} = \frac{1}{2g}\left[\frac{Q_{peak}}{Cd\,(h_{through.screen.peak.0} + h_{max})Width \cdot Effective_{area.screen}}\right]^2 \qquad [5]$$

By trial and error, or using numerical methods, solve for $h_{through.screen.peak.0}$.

$$h_{through.screen.peak.0} := 0.273 \text{ m}$$

$$h_{upstream.peak.0} = h_{max} + h_{through.screen.peak.0} = 1.123 \text{ m} \qquad [6]$$

$$V_{channel.upstream.peak} = \frac{Q_{peak}}{h_{upstream.peak.0} \cdot Width} = 0.72\,\frac{m}{s} \qquad [7]$$

$$V_{screen.peak.initial} = \frac{Q_{peak}}{Width \cdot (h_{upstream.peak.0}) \cdot Effective_{area.screen}} = 1.411\,\frac{m}{s} \qquad [8]$$

Average conditions

$$h_{through.screen.avg.0} = \frac{1}{2 \cdot g}\cdot\left[\frac{Q_{avg}}{Cd\cdot(h_{throughscreen.avg.0} + h_{avg})Width \cdot Effective_{area.screen}}\right]^2 \qquad [9]$$

By trial and error, or using numerical methods, solve for $h_{through.screen.avg.0}$

$$h_{through.screen.avg.0} := 0.073 \text{ m}$$

$$h_{upstream.avg.0} = h_{avg} + h_{through.screen.avg.0} = 0.723 \text{ m} \qquad [10]$$

$$v_{channel.upstream.avg.0} = \frac{Q_{avg}}{h_{upstream.avg.0} \cdot Width} = 0.373\,\frac{m}{s} \qquad [11]$$

$$V_{screen.avg.initial} = \frac{1}{2 \cdot g}\cdot\frac{Q_{avg}}{Width \cdot (h_{upstream.avg.0})Effective_{area.screen}} = 0.731\,\frac{m}{s} \qquad [12]$$

Then, repeat the same procedure as above with different channel widths. The results are outlined in the table below.

Table 1 permits one to select the minimum channel width that meets most of the V_{min}, V_{max}, V_{screen}, and $V_{resuspension}$ criteria outlined in Section 2.7.4, at peak and average flows. Typically, it is difficult to satisfy all constraining criteria with a single channel layout, so the designer must evaluate compromises and decide the best solution for each particular application. This example is no exception: widths less than 1.3 m will produce excessive velocities through the screen and widths greater than 1.3 m will have upstream velocities inadequate to resuspend solids that have settled ahead of the screen.

Perhaps the best compromise will be with a width of 1.3 m—the peak upstream velocity will be only slightly less than the criterion resuspension velocity of 0.76 m/s.

Width	m			0.9	1	1.2	1.3	1.4	1.5	1.6	1.7	1.8
Peak Conditions												
$h_{through.screen}$	m			0.434	0.382	0.303	0.273	0.246	0.224	0.204	0.187	0.172
$h_{upstream}$	m			1.284	1.232	1.153	1.123	1.097	1.074	1.054	1.037	1.022
$V_{channel.upstream}$	m/s	0.3<V<0.9		0.908	0.852	0.759	0.720	0.684	0.652	0.623	0.596	0.571
$V_{through.screen}$	m/s	V<1.2		1.781	1.671	1.488	1.411	1.341	1.278	1.221	1.168	1.120
$V_{channel.downstream}$	m/s	0.3<V<0.9		1.373	1.235	1.029	0.950	0.882	0.824	0.772	0.727	0.686
Average Conditions												
$h_{through.screen}$	m			0.131	0.111	0.083	0.073	0.065	0.058	0.051	0.046	0.041
$h_{upstream}$	m			0.781	0.761	0.733	0.723	0.714	0.707	0.701	0.696	0.692
$V_{channel.upstream}$	m/s	0.3<V<0.9		0.498	0.460	0.398	0.373	0.350	0.330	0.312	0.296	0.281
$V_{through.screen}$	m/s	V<1.2		0.977	0.901	0.780	0.731	0.686	0.647	0.612	0.580	0.551
$V_{channel.downstream}$	m/s	0.3<V<0.9		0.598	0.538	0.449	0.414	0.385	0.359	0.337	0.317	0.299

TABLE 1   Channel Design Summary

However, the V_{screen} is higher than recommended, by nearly 20%. What is the solution?

One alternative to address this problem would be to install a downstream slide gate to control the velocity (hydraulic control structure). The rest of this example pursues this option.

Calculate the required depth of water upstream of the screen to prevent solids carryover with a hydraulic control structure.

Peak flow conditions

The target velocity through the openings at peak flow = 1.2 m/s to minimize screenings carryover. Height required to get 1.2 m/s

$$V_{max.openings} = 1 \cdot 2 \cdot \frac{m}{s}$$

$$h_{req} = \frac{Q_{peak}}{Width \cdot V_{max.openings} \cdot Effective_{area.screen}} = 1.321\ m \qquad [13]$$

Calculate the headloss through the screen at peak conditions.

$$h_{through.screen.peak} = \frac{1}{2 \cdot g} \cdot \left(\frac{Q_{peak}}{Cd \cdot h_{req} \cdot Width \cdot Effective_{area.screen}} \right)^2 = 0.197\ m\ \text{clean screen} \qquad [14]$$

Hydraulic Control Structure Design

Additional headloss required to increase the depth to the height required.

$$h_{through.gate} - h_{req} - h_{max} - h_{through.screen.peak} = 0.274\ m \qquad [15]$$

$$headloss_{orifice} = \frac{1}{2 \cdot g} \cdot \left(\frac{Q}{Cd \cdot Area} \right)^2 \qquad \text{Headloss through orifice, eq 6.22} \qquad [16]$$

Q	m³/s	0.35	0.45	0.55	0.65	0.75	0.77	0.88	0.92	0.95	1.05
h	m	0.65	0.68	0.71	0.74	0.76	0.77	0.80	0.81	0.82	0.85
h_{req}	m	0.44	0.57	0.69	0.82	0.94	0.97	1.11	1.16	1.19	1.32
$h_{through.screen}$	m	0.07	0.10	0.13	0.16	0.19	0.20	0.20	0.20	0.20	0.20
h_{up}	m	0.72	0.78	0.84	0.90	0.96	0.97	1.11	1.16	1.19	1.32
$h_{throughgate}$	m							0.11	0.15	0.18	0.27
h_{op}	m							0.76	0.68	0.64	0.57
$V_{through\ screen}$	m/s	0.73	0.87	0.99	1.09	1.18	1.20	1.20	1.20	1.20	1.20

TABLE 2   Slide Gate Control Range

From the equation above, solve for $h_{opening}$ (Area $= h_{opening} \times$ width), making headloss$_{orifice} = h_{.through.gate}$

$$h_{opening} = \frac{Q_{peak}}{(h_{through.gate} \cdot 2 \cdot g)^{0.5} \cdot Cd \cdot Width} = 0.572 \text{ m}$$

Opening height required for the hydraulic control structure at peak conditions   [17]

Now, establish the slide gate control range by repeating the same procedure above with the different flow conditions. The table below summarizes the results.

Table 2 shows that the slide gate should regulate channel velocity when flows reach approximately 0.77 m³/s up to 1.05 m³/s (peak flow) since $h_{req} < h$ in the channel.

Note that the headloss calculations for flows less than 0.88 m³/s require an iterative process as illustrated in the headloss calculations shown above.

Headloss through Screen

Peak conditions

Surface water elevation
Upstream of the screen at peak

$$SWE_{peak} = h_{req} = 1.321 \text{ m} \qquad \text{clean screen}$$

Check channel velocity at peak conditions with the hydraulic control structure.

$$V_{peak.calc} = \frac{Q_{peak}}{Width \cdot SWE_{peak}} = 0.612 \frac{\text{m}}{\text{s}} \qquad \begin{array}{l} V_{peak.calc} > V_{min}, \text{ then OK} \\ \text{clean screen} \end{array} \qquad [18]$$

Assume screen blinded 40% and calculate headloss at peak conditions.

$$h_{through.screen.peak.blinded} = \frac{1}{2 \cdot g} \cdot \left[\frac{Q_{peak}}{Cd \cdot h_{req} \cdot Width \cdot (1-0.4) Effective_{area.screen}} \right]^2 = 0.548 \text{ m} \qquad [19]$$

$$SWE_{peak.blinded} = h_{max} + h_{through.gate} + h_{through.screen.peak.blinded} = 1.672 \text{ m} \qquad [20]$$

Downstream of the screen and the hydraulic control structure.

$$\text{SWE}_{\text{peak.ds}} = h_{\max} = 0.85 \text{ m}$$

Average conditions

$$h_{\text{through.screen.avg}} = \frac{1}{2 \cdot g} \cdot \left[\frac{Q_{\text{avg}}}{\text{Cd} \cdot (h_{\text{avg}} + h_{\text{through.screen.avg}}) \text{Width} \cdot \text{Effective}_{\text{area.screen}}} \right]^2 \qquad [21]$$

By trial and error, or using numerical methods, solve for $h_{\text{through.screen.avg}}$.

$$h_{\text{through.screen.avg}} = 0.073 \text{ m} \qquad \text{clean screen}$$

Surface water elevation upstream of the screen, at average conditions.

$$\text{SWE}_{\text{avg}} = h_{\text{avg}} + h_{\text{through.screen.avg}} = 0.073 \text{ m} \qquad \text{clean screen} \qquad [22]$$

Check velocity through screen at average conditions.

$$V_{\text{screen.avg}} = \frac{Q_{\text{avg}}}{\text{Width} \cdot (h_{\text{avg}} + h_{\text{through.screen.avg}}) \text{Effective}_{\text{area.screen}}} = 0.731 \frac{\text{m}}{\text{s}} \qquad [23]$$

Notes:

(1) Velocity through screen openings at peak conditions is lower than at average conditions because of the effect of the hydraulic control structure.

(2) Slide gate controls should be provided with an upstream level indicator to prevent the gate from throttling down if excessive blinding (excessive headloss) is occurring upstream of the gate. The designer also should make provisions for adequate freeboard in the channels.

(3) The slide control does reduce the $V_{\text{resuspension}}$ in the channel at flows greater than 0.77 m³/s. Section 2.7.4 outlines some control strategies to minimize solids deposition in channels.

(4) Other options to resolve the dilemma of balancing $V_{\text{resuspension}}$ against $V_{\text{through.screen}}$ might include placing grit removal channels ahead of the fine screens to minimize sedimentation that would have to be resuspended or giving the upstream channel sufficient side slope and longitudinal slope to discourage settling. The ultimate choice depends on the local situation.

Screenings Capture

If no plant data are available (even from nearby facilities), then use Figures 9.1 or 9.2 to estimate the average screening quantities and use the upper limit of the screening generation curve for estimating the screenings.

For a 6-mm opening screen:

$$\text{Screenings}_{\text{avg}} = 9 \cdot \frac{\text{ft}^3}{\text{Mgal}} = 6.732 \times 10^{-8} \frac{1}{\text{L}} \cdot \text{m}^3 \qquad [24]$$

Based on plant data or similar facilities, select an appropriate peak screening factor. For this example, assume a facility that gets significant peak screening loads during the fall due to tree leaves falling and getting carried into the sewer system.

Screenings sf = 5

$$\text{Screenings}_{.peak} = \text{Screenings}_{.avg} \cdot \text{Screenings}_{.sf} = 3.366 \times 10^{-7} \cdot \frac{1}{L} \cdot m^3 \qquad [25]$$

Washer Compactor Design

$$\text{Capacity} := \text{Screenings}_{peak} = 3.366 \times 10^{-7} \cdot \frac{1}{L} \cdot m^3$$

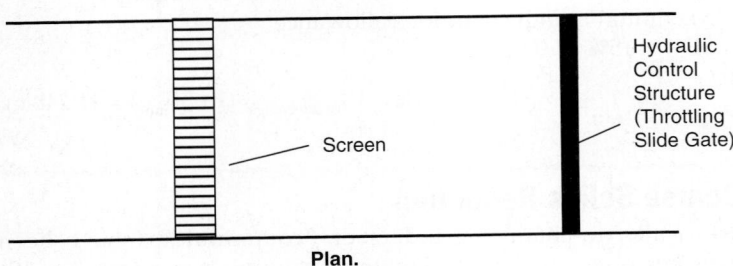

Plan.

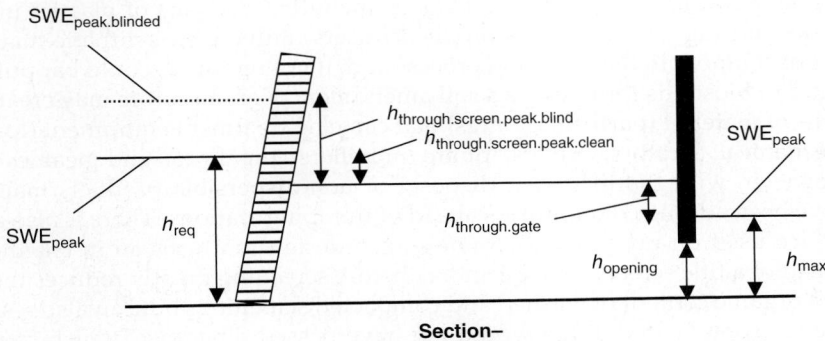

Section–
peak flow conditions.

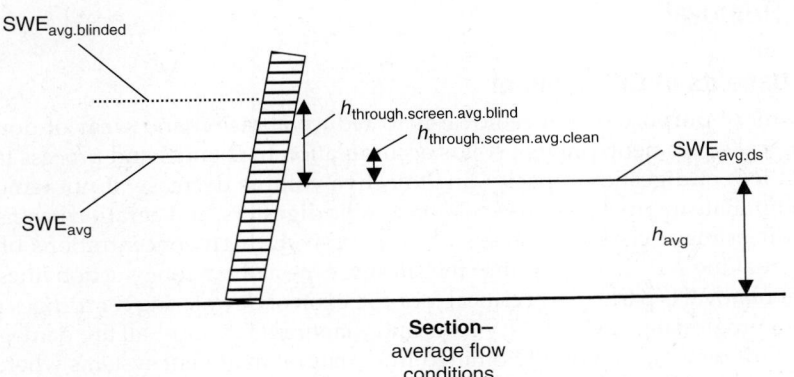

Section–
average flow
conditions.

Assume the following compactor performance (based on manufacturers' published data) $V_{reduction} = 60\%$

$$V_{wet.screenings} = Screenings_{avg} \cdot Q_{avgtotal} = 6.113 \frac{m^3}{day} \qquad [26]$$

$$V_{dry.screenings} = V_{wet.screenings} \cdot (1 - V_{reduction}) = 2.445 \frac{m^3}{day} \qquad [27]$$

Container to be sized for 4 days at average conditions
Maximum month conditions allowance
$SF_{mm} = 15\%$

$$V_{container} = 4 \cdot day \cdot V_{dry.screenings} \cdot (1 + SF_{mm}) = 11.248 \ m^3 \qquad [28]$$

3.0   Coarse Solids Reduction

Historically, in-channel grinders or "comminutors" have been used instead of screening to screen and shred material ranging in size from 6 to 19 mm (0.25–0.75 in.) without removing particles from the flow. Now comminutors are rarely used as standalone substitutes for screens. Solids from comminutors and screenings grinders have caused downstream problems, including deposits of plastics in digestion tanks and rag accumulations on air diffusers. Pulverized synthetic materials will not decompose in the digestion process and, if not removed, could bar public acceptance of biosolids for reuse as a soil amendment. Comminutors may create ropes or balls of material (particularly rags) that can clog treatment equipment (for example, mechanical aerators, mixers, pump impellers, pipelines, and heat exchangers). However, with the increased disposal of non-dispersible products many utilities have been adding comminutors ahead of pumping stations. There is also a grinding device used on raw sewage in one screen design that uses an in-channel grinder ahead of a fine screen. Using grinders before screening greatly reduces the quantity and organic content of washed and compacted screenings but can also result in passage of some material that would not have passed otherwise, which could impact the quality of biosolids produced.

4.0   Grit Removal

4.1   Benefits of Grit Removal

The primary purpose of grit removal is to reduce abrasion and wear of downstream mechanical equipment, prevent solids accumulation in channel and process tanks, and prevent loss of digestion capacity due to active volume decrease. If not removed, grit can accumulate in pipelines, channels, anaerobic digesters, and aeration basins. Settling of grit in primary clarifiers can result in excessively high concentrations of primary sludge, making it difficult to pump the sludge, especially in long suction lines.

Grit removal is particularly critical for protection of dewatering centrifuges and high-pressure progressing cavity, rotary lobe, and diaphragm pumps; all are damaged easily by grit. Grit removal is typically omitted from natural treatment systems where grit and biosolids are applied to land or allowed to accumulate on the bottoms of lagoons.

Removed grit (grit slurry) is concentrated and washed to remove lighter organic material that is captured. The degree of washing and concentration of the grit slurry is dictated by several factors, including the degree of odor control and vector attraction required during storage, and costs and regulations associated with disposal.

4.2   Grit Characterization

To allow for proper design and performance assessment of grit removal processes, sampling should be performed to quantify the amount of grit in the liquid or solids stream before and after treatment of these processes. This sampling can be performed at the following locations in the treatment process:

- Raw wastewater influent to the facility headworks
- Screened influent to the liquid stream grit removal process
- Effluent from the liquid stream grit removal process
- Influent to the grit dewatering/washing process
- Overflow from the grit dewatering/washing process
- Sampling of the washed/dewatered grit

There are three commonly used methods of grit sampling:

- Cross-channel sampling and dry sieving
- Full depth sampling and wet sieving
- Primary sludge sampling

The first two methods, cross-channel and full depth sampling are designed to sample from the grit removal process for grit in the influent and effluent. These two methods differ fundamentally in many ways, including the design of the sampling device, the hydraulic loading on the settling basin used to capture the grit, and the methods used in the sieve analysis that determines the various size fractions of the grit. For example, the cross-channel sampling method tends to indicate higher percentages of large grit than the full depth method because there are more sample points located along the channel bottom. There are advantages and disadvantages associated with both types of sampling methods. Central to the debate on which method is more appropriate is the degree that organic coating and sequestration of the grit particle affects the settling behavior of the grit particle in raw wastewater. The full depth sampling method assumes that these affects are, or can be, significant; whereas, the cross-channel method assumes that these affects are minor. For this reason, the first step should be to conduct settling velocity tests of the influent grit particles, as it is likely that these coating/sequestration effects vary from facility to facility. However, additional research is still needed within the industry to determine comparable settling velocities of different particles.

The alternative to sampling the influent and effluent from grit processes is primary sludge sampling. This method assumes that all grit settles out in the primary clarifiers, an assumption that may or may not be true for all facilities. It is used to quantify grit loads that have not been captured by an existing grit removal process, if one exists. Primary sludge sampling is combined with quantification of grit removed from the existing process to determine total quantities of grit entering a facility. To be done correctly, primary sludge sampling must involve large and frequent sample volumes, creating a severe burden on the laboratory performing the sieve analyses on the grit contained in the highly odorous and pathogenic primary sludge.

As a result of the lack of standardization in grit testing methods, the majority of grit systems are built with no grit testing performed before or after construction of the system. This may lead to poorly informed decisions being made about the effectiveness of the new grit removal systems. Without proper testing, poorly performing grit systems are accepted that increase maintenance and operating cost for downstream processes. Lack of standardization in grit sampling has created controversy around the usefulness of grit sampling data. To lead the effort in resolving these issues WEF has published *Guidelines for Grit Sampling and Characterization* that provides detailed information on the currently used grit sampling methods (WEF, 2016a).

4.2.1   Grit Quantities

Grit materials include particles of sand, gravel, other mineral matter, and minimally putrescible organics such as coffee grounds, egg shells, fruit rinds, and seeds. The quantity and characteristics of grit removed from wastewater will vary. Variables influencing grit quantity include the type of collection system (combined or separate), characteristics of the drainage area, use of household garbage grinders, condition of sewer system, sewer grades, types of industrial waste, and efficiency of grit removal. The perceived quantity of grit is affected by the efficiency of washing/dewatering equipment; therefore, upgrading washing/dewatering equipment can give the false impression that less grit is being removed because of elimination of organics and water. In most cases, grit accumulates in the collection system and is flushed into the WRRF in large quantities during high flow. Because these peak grit loading periods frequently overload poorly functioning grit systems, facility data can be of limited use when designing upgrades. In the 2008 WEF member survey, grit quantities reported were erratic. When correcting for outliers, however, quantities ranged from 3.7 to 148 L/1000 m³ (0.5 to 20 cu ft/mil gal) and averaged 37 L/1000 m³ (5.0 cu ft/mil gal).

4.2.2   Physical Properties

Grit solids characteristics will vary with solids content from 35% to 80% and volatile content from 1% to 55% (U.S. EPA, 1979). Well-washed grit should achieve a solids content of 90% with a minimum of putrescible solids. The moisture and volatile content will be influenced by the degree of washing the grit receives. The bulk density of dewatered grit will range from 1400 to 1800 kg/m³ (90 to 110 lb./ft³). In the 2008 and 2016b WEF member surveys, reported grit density ranged from 1100 to 2200 kg/m³ (70 to 140 lb/ft³) and averaged 1400 kg/m³ (90 lb/ft³). While the effects of air voids in the removed grit must be accounted for, when comparing this data to clean sand (specific gravity of 2.65 or a density of 2600 kg/m³ / 165 lb/ft³), it can be seen that removed grit characteristics differ significantly from clean sand. Because of the effects of organic coating and sequestration discussed earlier in this section, the concept of "sand equivalent size" (SES) has been developed and SES testing is a standard procedure in the full depth sampling method discussed earlier in this section. According to SES theory, grit particles in raw wastewater do not settle at the same velocity as a clean sand particle so, for example, a 200 micron grit particle in raw wastewater may settle at the same velocity as a 150 micron clean sand particle. If the SES theory is in fact true, this can result in the need to design the primary grit process to capture smaller grit particles than the target particle, if the process sizing is based on clean sand settling velocities. This is the reason for the SES test; to allow the process to be sized according to SES rather than clean sand settling velocity.

4.3   Grit Removal Processes

4.3.1   Aerated Grit Basins

In aerated grit chamber systems, air introduced along one side near the bottom causes a spiral roll velocity pattern perpendicular to flow through the tank. The heavier particles with their correspondingly higher settling velocities drop to the bottom, and the roll suspends the lighter organic particles into the tank effluent. The rolling action induced by the air diffusers is independent of flow through the tank, allowing the aerated grit chamber to operate effectively with a wide range of flows. Heavier particles that settle are moved by the spiral flow of the water across the tank bottom and then into a grit trough or hopper. Screw augers are the most frequently used method to transport grit along the bottom of the tank to the grit sump and pumps. Other methods include chain-and-flight collectors, chain-and-bucket collectors, clamshell buckets, and multiple sumps each with dedicated grit pumps. Air lifts also are used but with marginal success. The grit collection trough, depending on the chamber design, requires special design attention. A properly designed aerated grit chamber will produce relatively clean grit.

When the wastewater flows into the aerated grit chamber, particles will settle to the bottom at rates dependent on their size and specific gravity and on the velocity of roll in the tank. The rate of air diffusion and the tank shape govern the rate of roll and the size of the particle with a given specific gravity that will be removed. Diffused air serves as a velocity control method with sufficient flexibility to accommodate varying conditions. Diffused air also requires only minimal headloss through the unit. Proper design depends on an understanding of the variables affecting the airlift pumping energy and its effect on the roll pattern in the tank. Some empirical expressions include the variables influencing the roll pattern (Albrecht, 1967). The aerated grit design was evaluated by Londong (1989).

Poor performance of some aerated grit chambers has resulted in high percentages of organic matter in the removed grit, low removal efficiencies, and periodic abnormal grit loads on the collection equipment. There are several factors that can lead to poor performance in aerated grit chambers including improper tank geometry (most common); improper baffling; inconsistent air supply to diffusers, resulting in erratic roll patterns; and improper diffuser location. Based on the work of Londong, optimum aerated grit chamber design parameters are listed in Table 9.4 (1989). Following is a discussion of these and other design criteria.

- Air rates typically range from 5 to 12 L/s/m of tank length (3 to 8 cfm/ft). Rates as low as 2 L/s/m (1 cfm/ft) have been used for shallow, narrow tanks.

Criteria	Design Values	Notes
Peak hour flow detention time	3–10 minutes	Use longer times for plants with heavy grit loads and to improve fine-grit removal
Width-depth ratio	0.8–1	
Depth	3.7–5 m (12–16 ft)	
Length-width ratio	3–8:1	Longer basins provide better fine-grit removal
Bottom floor slope	30°	
Diffuser location	70% of total depth	

TABLE 9.4   Aerated Grit Basin Recommended Design Criteria

Rates higher than 8 L/s/m (5 cfm/ft) are often used for deep, wide tanks. Providing valves and flow meters is a good practice for monitoring and controlling the air flow. Aeration is typically with coarse-bubble non-clog diffusers. Separate, dedicated blowers are preferred.

- Minimum hydraulic detention time of three minutes at peak hourly flow of an aerated grit basin with proper geometry will provide a sufficient degree of coarse grit removal. Longer times should be considered if removal of fine (0.65 mesh) grit is desirable or if basins are covered and ventilated to odor control processes and to provide pre-aeration and volatile organic compound removal. It is not unusual to size aerated grit chambers for 10 to 15 minutes detention time.

- A chamber length-to-width ratio ranging from 3:1 to 8:1 is necessary, with longer tanks providing better grit removal. Square tanks have been used successfully with proper air diffuser location (perpendicular to the flow through the tank) and baffles to prevent short circuiting (Morales and Reinhart, 1984).

- A width-to-depth ratio of 0.8 to 0.9 is necessary to prevent short-circuiting down quiescent central core not affected by the roll pattern.

- Floor slope of 30° is recommended to ensure quick movement of grit into the longitudinal grit collection sump and to prevent re-entrainment of fine grit by the spiral roll. Inadequate floor slope is a common design error in aerated grit chambers.

- Incorrect width-to-depth ratio and floor slope are the most common errors in aerated grit design because historic use of swing-out knee-joint diffusers necessitated wide basins with shallow floor slopes. For this reason, use of these diffusers in not recommended.

- Tank inlet flow should enter the tank from the air header side and parallel to the spiral roll pattern to encourage the start of a roll pattern. Inlet baffles are frequently used to turn the influent flow into the spiral roll. Flow through the tank should be perpendicular to the spiral roll pattern. Tank outlet is typically perpendicular to the flow through the tank.

- When retrofitting aerated grit chambers with improper geometry, longitudinal baffles positioned approximately 1 m (3.0 ft) from the wall along the air diffusers help to decrease the effective width of the basin. Intermediate baffles can be used to control short circuiting down the quiescent central core of excessively wide basins. Baffles should provide sufficient open area at the bottom of the basin to avoid excessive velocities resulting in re-suspension of settled grit. Computational fluid dynamics modeling can be used to determine proper baffle location.

- With proper adjustment, the aerated grit tank at a level 150 mm (6 in.) below the top of the water, should produce a roll velocity of 0.6 m/s (2.0 ft/s) near the tank entrance and 0.6 m/s (1.5 ft/s) at the tank exit (WEF, 2016a).

- Screw augers are used to transport grit to a grit pocket at the end of the tank and into flooded suction recessed impeller grit pumps. Both shafted and shaftless augers are used. Shafted augers have submerged bearings requiring an extended lubrication line to the nearest point of access outside the basin; shaftless augers will require replacement of wear bars or liners. Shafted augers are used more frequently.

- Chain-and-flight collectors can be used to drag grit up an inclined slope for removal. Chain-and-bucket conveyors are used; bucket systems typically require significant maintenance.

- In lieu of auger or chain-driven collectors, top-mounted self-priming or vertical cantilever pumps, or flooded suction pumps mounted alongside the bottom of the tank can be used, but require formation of individual sumps along the length of the tank. To avoid this, top-mounted or submersible pumps are used with a traveling-bridge mechanism. Top-mounted airlifts tend to have operational problems and should be avoided.

- The simplest method of removing grit from a grit chamber is using an overhead clamshell hoist; however, this method tends to be labor-intensive and does not allow the use of grit classifier/washers and, therefore, can produce poorly dewatered grit with high organic content. For these reasons, this method tends not to be favored by operators.

4.3.1.1 Aerated Grit Removal Process Design Example

Average daily flow = 63.1 m³/min (24 mgd)
Peak hourly flow = 157.7 m³/min (60 mgd)
Procedure:

1. Size grit chamber based on six minutes detention time (peak flow) for fine grit removal.

2. Determine number of basins required.

3. Calculate basin geometry based on MOP 8 guidelines.

4. Select grit slurry treatment method.

5. Size grit pumps.

Size grit chamber 2 Volume required = 6 min 3157.7 m³/min = 946.35 m³

Determine number of basins required. Two minimum of two basins, more as facility design flow increases. Provide 4 basins for redundancy.

Calculate basin geometry:

> Volume/basin = 237 m³(62 609 gal)
> Depth of basin = 4.2 m (14 ft)
> W:D ratio = 0.8:1 (measure depth at midpoint of floor slope 2 use 30° slope, min)
> L:W ratio = 5:1
> Width of basin = 3.4 m (11 ft)
> Length of basin = 16.8 m (55 ft)

Size grit slurry pumps and grit washers:

Use 3 conical grit washers each rated for 25 lps (400 gpm) for 50 lps (800 gpm) firm capacity.

Use 4–13 lps (200 gpm) grit pumps (assuming all pumps pumping during heavy grit loading).

Provide shelf spare grit pump.

4.3.2   Vortex Grit Removal Systems

4.3.2.1   Forced Vortex   Forced vortex systems combine the hydraulic forces associated with tangential incoming flow with mechanical paddles to create and enhance a vortex pattern. A forced vortex system comprises two circular chambers. The upper chamber is where the flow enters allowing for the vortex creation that separates and settles the grit. The lower chamber is where the settled grit is stored until it is removed for further processing. Tanks are circular to encourage the vortex flow to force grit solids to the side walls where velocities are lower due to boundary layer effects. Once near the walls, the vortex and gravity continue to move the captured grit solids into the lower chamber. A straight flow pattern into the tank is critical to minimize turbulence at the inlet of the chamber. At the end of the inlet flume, a ramp allows grit that may already be on the flume bottom to slide downward along the ramp until it reaches the chamber floor where it is captured. At the center of the chamber, rotating paddles maintain a constant circulation within the chamber and help separate organics from the grit path as it approaches the lower hopper. The first step in the grit washing cycle occurs with the chamber propeller blades (or paddles). The vortex flow pattern creates a quiescent central zone in the tank to which the grit migrates while the organics stay in suspension. Grit solids are removed from the center hopper by top-mounted vacuum-primed pumps, flooded-suction pumps or (less frequently) air lifts.

Forced vortex grit systems include two basic designs: (1) tanks with flat bottoms and a plate that separates the sump from the main tank and is surrounded by a narrow slot through which grit enters the sump; and (2) tanks with a sloping bottom and a large opening to the grit hopper. Figure 9.21 shows a flat-bottom type chamber.

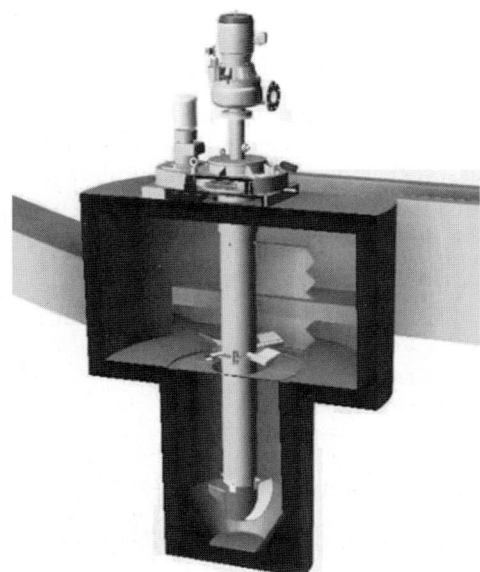

FIGURE 9.21   Forced vortex chamber, high efficiency baffle not completely shown for clarity. Top mounted pump arrangement shown. Flooded suction pump arrangement available (courtesy of Smith & Loveless, Inc.).

As the vortex directs solids toward the center, rotating paddles increase the velocity enough to lift lighter organic materials and return them to the flow passing through the grit chamber. All grit passes under the paddles for removal of organic materials before grit falls into the storage chamber. When sufficient grit has accumulated in the storage chamber, water flush lines are used to scour the sump to remove additional organics and fluidize the grit for more effective pumping. Some designs also include fluidizing vanes attached to an extended paddle shaft. The following is a listing of recommended design guidelines for mechanical vortex grit chambers.

- Ideally, flow into a vortex grit chamber should be straight and streamlined. As a good practice, the straight inlet channel length should be four to seven times the width of the inlet channel. The ideal velocity in the influent channel ranges from 0.5 to 1.1 m/s (1.6–3.5 ft/s) for the grit particle removal historically targeted in this system >105 (65 mesh or >210 microns). This ideal range should approximate flows between 40% and 80% of the peak flow. The minimum acceptable velocity for low flow is 0.15 m/s because lower velocities will not carry grit into the chamber. If velocities as low as 0.15 m/s will be experienced then provisions for flushing are necessary to move settled grit into the tank. The flushing system must avoid washing grit through the grit chamber.

- Recent designs have used baffles on the influent to the grit chamber to encourage the flow to follow the outer wall, resulting in decreased turbulence in the interior of the grit chamber at all flows. Use of these baffles have resulted in better fine grit removal with one manufacturer claiming 95% removal of particles down to 150 mesh (100 micron). The grit chamber's effluent outlet has twice the width of the influent flume, which results in a lower velocity, thereby preventing grit below the opening level from being drawn into the effluent flow.

- Sizing of proprietary vortex grit chambers is based on recommended dimensions provided by equipment manufacturers. The units, typically marketed in standard nominal sizes, are rated on a peak flow basis. Typical detention times for these units at peak design flows are short (20–30 seconds). Manufacturers should verify that the given unit dimensions have been field tested to determine performance parameters. Deviation from the recommended dimensions without the manufacturer's prior approval could void any performance guarantees.

- After selecting the vortex grit unit, additional information must be obtained from the manufacturer's drawings and design data to provide the appropriate entrance and exit channels and the concrete chamber in which to install the grit removal equipment.

4.3.2.2 Induced Vortex Induced vortex grit removal systems (Figure 9.22) are fundamentally different than mechanical vortex systems because the vortex is created by the force of the incoming flow. Induced vortex basins are, therefore, smaller in diameter and require pumped flow. Induced vortex systems can be designed to capture a higher percentage of fine grit than mechanical vortex systems but result in significantly more headloss. Headlosses of 0.5 to 0.7 m are typical for 95% removal of 105 microns 100 micron grit. The use of induced vortex system on raw wastewater has seen a rapid decline in recent years and these systems are far more commonly used in the grit dewatering processes.

FIGURE 9.22 Induced vortex grit system (courtesy of Hydro International).

4.3.2.3 Multiple-Tray Vortex The multiple-tray vortex system (Figure 9.23) is a proprietary system that has rapidly increased in popularity. It uses a flow-distribution header for distributing influent over multiple conical trays. Tangential feed establishes a vortex flow pattern in which solids settle into a boundary layer on each tray and into a center underflow collection chamber for pumped removal. The system

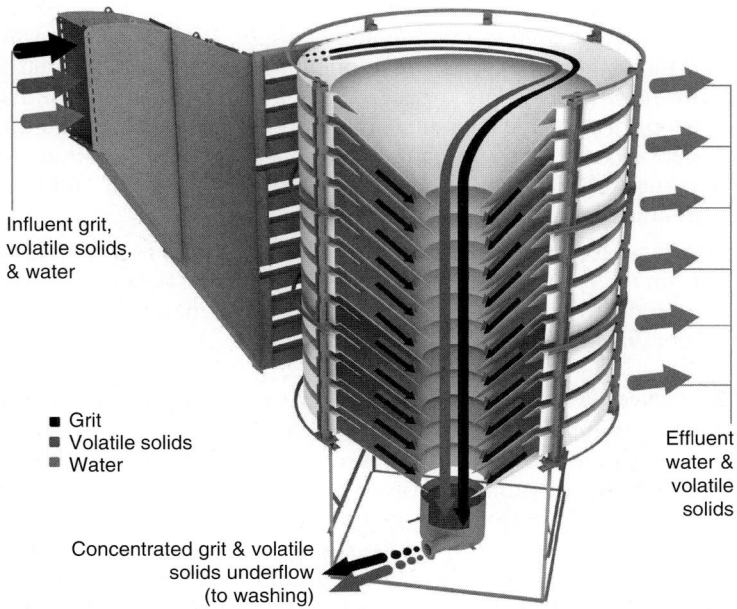

FIGURE 9.23 Multiple-tray grit separator (courtesy of Hydro International).

applies plate-settler technology to grit removal by using the stacked trays to create a large, concentrated surface area and short settling distances. For new systems, the multiple-tray system can be installed in a concrete basin with a smaller footprint than aerated systems. Multiple-tray vortex systems can be designed to remove down to 75 micron particles and multiple installations have been proven to remove 95% or 100 micron particles. For existing facilities, it can be fit into existing grit chambers or equalization basins.

4.3.3 Detritus Tanks

One of the earliest grit chambers was a constant-level, short-detention-time settling tank called a detritus tank (square tank degritter). Because these tanks settle heavy organics and grit, they require a grit-washing step to remove organic material. Some designs incorporate a grit auger and a rake that removes and classifies grit from the grit sump.

Detritus tanks are sized on an overflow rate based on particle sizes. Design considerations for tank depth include minimizing the horizontal velocity and turbulence while maintaining a short detention time (typically less than 1 minute). An additional 150 to 300 mm (6–10 in.) of depth is provided for the raking mechanism. The detritus tank relies on well-distributed flow into the settling basin. Allowances for inlet and outlet turbulence and short circuiting are necessary to determine the total area required. Thus, good design practice typically applies a safety factor of 2.0 to the calculated overflow rate as an offset for these hydraulic inefficiencies. Detritus tanks are not recommended for facilities with widely varying flows. Detritus tanks collect large amount of inorganic material that is later removed in grit concentration and classification, and they also provide proper conditions for grease removal.

4.3.4 Velocity Control Tanks

One of the earliest types of grit removal systems was the horizontal flow grit chambers, using proportional weirs or rectangular control sections (such as Parshall flumes) to vary the depth of flow and keep the velocity of the flow stream at a constant 0.3 m/s (1 ft/s). Chain and flights are used to scrape the grit either up an inclined slope for dewatering or into a hopper for removal by pump, auger, or chain-and-bucket elevator.

In designing a horizontal-flow grit chamber, the settling velocity of the target grit particle and the flow control section/depth relationship govern the length of the channel. The length of the channel must include an allowance for inlet and outlet turbulence. The cross-sectional area will be governed by the rate of flow and the number of channels. Allowances for grit storage and removal equipment are included in determining the channel depth.

Circular velocity control tanks were popular for many years. This process uses a shallow circular tank and a series of baffles to control velocity. The angle of the baffles can be changed to optimize flow control.

4.3.5 Primary Sludge Degritting

The removal of grit from primary sludge was once common practice but saw a rapid decline in use as liquid stream grit removal processes were established. Today, repeated cases of poorly performing liquid stream processes have resulted in a resurgence of primary sludge degritting. Devices used for washing and dewatering grit slurry are used to remove grit from stand-alone sludge for less cost than installing new liquid stream grit removal processes. Conventional cyclone/classifiers and induced vortex systems are used for this purpose. When degritting primary sludge, sludge solids must be pumped "thin," at concentrations of 0.5% to 1% for the process to function properly. This either requires frequent or continuous pumping of the primary sludge or an addition of dilution water. As a result of grit commingling with the primary sludge, the organic coating effect is severe and a high degree of grit washing is required to capture down to 100 micron grit in primary sludge degritting processes.

4.4   Grit Slurry Pumping

One of the most important elements of successful grit removal systems is the design of the grit slurry pumping system. Grit pumping is the most commonly used method for transporting collected grit. Successful grit pumping design carefully considers the following:

- Minimize the potential for plugging by making suction lines straight and provide flushing connections.
- Size the suction and discharge lines to prevent grit deposition with liberal flushing provisions. The minimum pipe diameter to be used is 4 inches.
- Provide fluidizing lines in the areas where the grit will be collected from the tank. Use only long radius fittings and use multiple 45° fittings.
- Use recessed impeller centrifugal pumps with suitable impeller material to withstand grit erosion. Limit the concentration of solids in the grit slurry to 1% by volume to ensure proper operation of the centrifugal pumps.
- Prevent the use of check valves on the discharge lines, and use valves with full open port configurations such as pinch or plug valves.
- Allow for redundant pumps.

4.5   Grit Slurry Processing

One of the objectives of the primary grit removal process is to minimize the organic material removed. Varying degrees of organic material, however, will always be removed with the grit. Because of effects of grease and detergent in sewage, providing sufficient detention time for removal of fine grit results in increased capture of organic material. Design of each grit removal system must, therefore, weigh the need for fine grit removal versus the higher cost of more effective grit washing equipment. Another important consideration in grit slurry processing is to avoid grit carryover and the grit washing equipment must be designed to remove the same design particle as the primary grit system associated with it. The exception to this would be grit washers that are highly effective in removing the organic coating from the grit particle and would allow the grit washer to be designed for a larger size grit particle than the primary grit removal process, as defined by the SES test.

4.5.1   Cyclone/Classifiers

Cyclone systems are used for separating grit from organics in grit slurries. Centrifugal forces that develop in the cyclone cause heavier grit and suspended solids particles to concentrate along the sides and on the bottom. Lighter solids, including scum, are removed from the center through the top of the cyclone. Cyclones operate best at constant flow and pressure. If flows depart from design flows, solids will be lost to the centrate stream.

There are two types of cyclone/classifier systems used today. The traditional cyclone/classifier causes high velocities in the cyclone chamber. Also available is a modified induced vortex technology that generates a sub-cyclonic vortex with a tangentially entering flow. Centrifugal and gravitational forces within the cylindrical unit remove grit with densities higher than water by forcing particles to the wall where they fall by gravity to the bottom; lighter organics exit with the effluent through the top. Organic materials entrapped with the grit are partially removed by scour at the bottom of the unit.

A cyclone separator concentrates the grit centrifugally, requiring a steady feed of grit slurry at an inlet pressure of 34 to 140 kPa (5–20 psi). The constant feed rate will be within the range of 800 to 1900 L/m (200–500 gpm) depending on the size of the cyclone. Cycle times for intermittent operation can vary from 5 minutes to 8 hours; peak grit loadings may require continuous operation. Frequent grit removal cycles will tend to reduce grit accumulation in the hopper and its associated compaction and plugging and will dilute the grit slurry. Removal of excessively diluted grit slurry from the hoppers causes inefficiencies, including increased energy costs for recycling reject water through the headworks.

Cyclone separator sizing is based on the cycled feed flow rate and grit slurry solids concentrations. Cyclones work best at feed concentrations of less than 1% solids. The centrifugal action created in the cyclone separators increases the solids content to an average of 5% to 15%. Approximately 90% to 95% of the feed flow rate discharges through the vortex finder at the top of the cyclone. This flow volume reduction saves transportation and storage and reduces the required classifier size.

Grit classifiers, either the inclined screw or escalator type, wash grit by separating out putrescible organics. Classifiers are sized based on settling velocity of the particles to be settled, feed flow capacity, and grit-raking capacity. For a target particle size and flow rate, the design engineer selects a minimum pool area and overflow weir length. The design engineer checks the classifier slope to ensure removal of the desired marginal particle size. Flatter slopes will remove finer grit particles.

Classifiers offered by manufacturers are inclined from 15° to 30°. In addition to slope, proper flight tip speed (r/min) and pitch (typically half or double pitch) assist in particle removal. Sectional flight construction may perform better than helicoid flights. Hardened flight edges should be used to resist the abrasive action of the grit. The screw or rake is sized to convey anticipated peak grit mass loading. An example of a typical grit cyclone and classifier is shown in Figure 9.24.

Grit slurry pumps for removing grit from chambers are sized to meet high head requirements of the cyclone separators, static head, and pipe-and-fitting friction losses. Because headloss through the cyclone is a function of flow rate and size, the manufacturer's pressure and flow rating information should be consulted. Swing-type check valve wear is common for grit pumps and rubber pinch-type check valves are recommended.

Influent piping to cyclone separators or classifiers is designed to ensure an even flow distribution in each unit. Isolation valves are necessary to allow removal of units from service for repair. Screening of the cyclone separator overflow has been found to reduce maintenance requirements by removing plastics and rags that accumulate in the system. Also, placement of cyclone separators and classifiers above and near the discharge to the disposal truck or hopper reduces the need for conveyance.

Because of high head requirements and maintenance associated with traditional cyclones, classifiers with oversized clarifier sections (Figure 9.25) are increasing

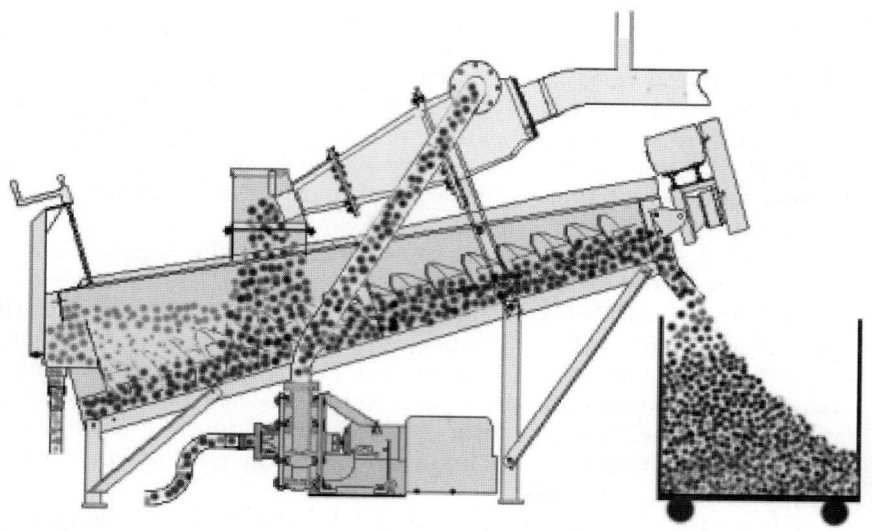

FIGURE 9.24 Cyclone/Classifier (courtesy of Wemco, Inc.).

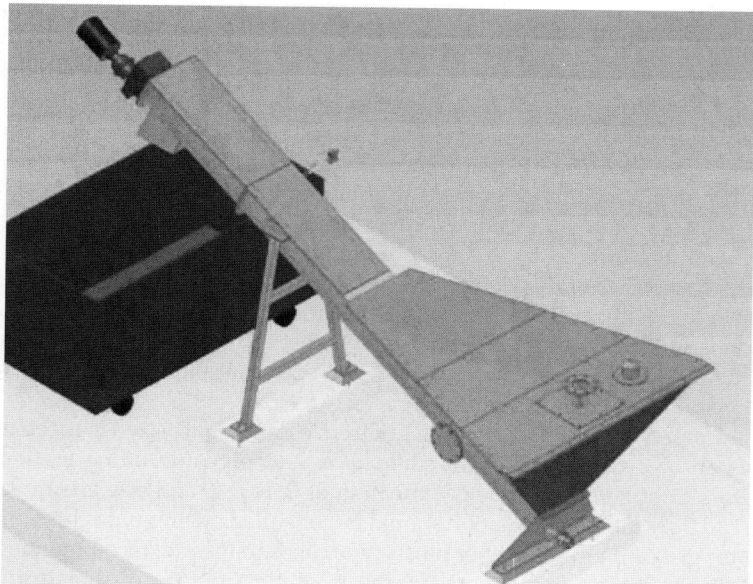

FIGURE 9.25 Grit classifier with oversized clarifier (courtesty of Westech, Inc.).

in popularity. Grit washing is accomplished using sprays on the inclined screw section. This unit has a significantly larger space requirement than the cyclone/classifier.

Traditional cyclone/classifiers are typically designed to capture down to 150 micron particles and are not designed to provide a high degree of washing with final organic content of greater than 25%. The more advanced proprietary induced vortex/escalator design is capable of capturing down to 75 micron particles and producing a final organic content of no more than 15%.

4.5.2 Conical Grit Washers
Conical grit washer technology (Figure 9.26) represents advancement in grit slurry processing resulting from strict preliminary treatment residuals disposal regulations in Europe. Stainless-steel, conical-shaped vessels are used to capture grit slurry and various systems are used to wash organics from the grit. Rotating arms within the vessel slowly mix the settled grit and a washing jet at the bottom of the unit, activated by a solenoid valve on a timer, vigorously washes it. Lighter organic material continuously overflows the unit and heavier organic material is blown off at regular intervals from a midlevel overflow. These washers create grit with low organic content. The surface flow velocity, including the wash water, should be less than 25 m/h (0.02 ft/sec) and weir overflow rate of less than 15 m²/h (0.045 sq ft/sec). Design flow to these units are typically limited to 25 L/s (400 gpm), although higher flows can be used if the units are equipped cyclone(s). Historically, target grit particle size has been 200 microns for these units, although one manufacturer has recently released a unit reportedly capable of capturing down to 100 micron particles. Washing provided by these units is superior to traditional cyclone classifiers with a final organic content of less than 5%.

4.6 Transport, Storage, and Disposal
Overhead clamshell bucket, chain-and-flight collectors, and augers can be used to remove grit from aerated grit basins, but most installations use mechanical pumping. This is because grit slurry can be pumped directly to washing and dewatering equipment.

Figure 9.26 Conical grit washer (courtesy of Huber Technology, Inc.).

Vortex or recessed impeller pumps are used to handle grit slurries. Air lifts have also been used but have a history of operational problems. It is important that air release be provided before the grit enters the cyclone/classifier or washer. Water jets or compressed air lines are used to fluidize grit that has become compacted in a hopper. Grit can be conveyed directly to trucks, dumpsters, or storage hoppers. Containers should be covered to prevent odors during storage and hauling. Conveyors frequently are used for transporting grit from handling facilities to containers. Overhead storage hoppers that discharge to truck containers avoid the need to keep a truck at the facility.

4.7 Design Considerations

4.7.1 Process Selection Criteria

The quantity and characteristics of grit and its potential adverse effects on downstream processes are important considerations in selecting a grit removal process. Other considerations include headloss and space requirements, removal efficiency, organic content, and economics.

Grit systems have traditionally been designed to remove 95% of particles larger than 200 micron (65 mesh) with a specific gravity of 2.65 (U.S. EPA, 1987). There is increasing evidence for higher percentages of grit in the influent wastewater smaller than 200 microns and complexities associated with SES theory, as explained earlier in this chapter, therefore, there is an increasing demand to design grit systems capable of removing up to 95% of 100 micron (150 mesh) grit particles to avoid adverse effects on downstream processes.

A single grit removal unit with a bypass channel will suffice for small installations (average flow 15 000 m³/d or 4 mgd) or for facilities where infrequent flows of wastewater containing grit can be tolerated in downstream processes. For facilities that are large, served by combined sewers, or with unit processes such as centrifuges that are more prone to abrasion from grit, multiple grit removal units are necessary. This allows periodic removal of units from service for cleaning, maintenance, and repair.

Similar to screen design, wastewater flow extremes must be known so that chambers can be designed to efficiently remove grit from all flows. The quantity of grit entering the treatment facility is typically greatest during peak flows when scour velocities and transport rates are highest. Grit chambers are sized to remove grit effectively at peak flows but to avoid removing excessive organic material at lesser flows.

Typically, the most used grit processes in new facilities are the vortex grit removal systems, because of compact size, mechanical simplicity, and lack of turbulence, which results in less release of odors from the process. Because of their small sumps, vortex systems are more susceptible to being overwhelmed by large quantities of grit during peak flow events and for this reason the multiple-tray vortex system manufacturer advises continuous grit pumping. Aerated grit systems, because of the use of larger tanks, are more capable of handling large, sudden quantities of grit and, when designed according to the guidelines provided in this chapter, the historically poor performance of this technology can be avoided.

4.7.2 Grit Handling

Several design standards should be included in piping layout for grit pumping:

- Length of grit pump suction piping should be minimized. Flooded suction pumps should be used whenever possible.

- Long radius horizontal and vertical bends should be minimized to reduce abrasion and plugging. Long radius bends should be used.

- Cleanouts and removable couplings should be placed at bends to readily clear blockages. Wyes can be used at bends to provide a combination gradual bend/cleanout.

- Velocity of 1 to 2 m/s (3–6 ft/s) should be maintained to keep grit and other solids moving while minimizing pipe abrasion.

- Discharge piping with nominal diameters of at least 100 mm (4 in.) should be used to avoid high scouring pressures and velocities that would cause excessive wear.

- Grit piping has traditionally been designed to be manifolded with multiple pumps and grit dewatering devices to allow operator flexibility, requiring the use of check valves and isolation valves. As a result of valve wear and frequent failure and the benefits of minimizing bends in grit piping, there is a trend to use dedicated pumps with each grit dewatering system and provide no interconnections, which allows the elimination of check and isolation valves and decreases the cost and complexity of the piping systems. If grit is transported directly to trucks or dumpsters then the container should have enough capacity to handle daily peak grit loads in the event that replacement trucks cannot arrive or leave (i.e., inclement weather). Two bays are advisable to ensure continued loading if the loaded container must remain because of a truck breakdown.

Overhead storage hoppers should be equipped to prevent grit bridging in the hopper and bypass the hopper if it fails to open. A good practice to prevent bridging is to have a minimum hopper side slope of 60°. In northern climates, storage hoppers should be located in areas that can be heated to prevent freezing. They will be odorous, so higher ventilation rates and isolation of air spaces using flexible curtains should be considered.

4.7.3   Automation and Instrumentation

Solenoid valves on grit sump flush lines are often setup to automatically fluidize the grit before pumping. Grit pumps and conveyors are operated using adjustable timers. Particularly for mechanical vortex systems, consider implementing high flow grit pump cycles to avoid overwhelming the sump when large quantities of grit are flushed into the system. Each piece of the grit washing/dewatering equipment has specific automation and instrumentation equipment. Dewatered grit conveyors can be made to automate whenever the grit washing/dewatering equipment is being used. A well-washed and dewatered grit tends to stick to belt conveyors and can be difficult to remove at the end. As a result, it can be beneficial to place the conveyor on a separate timer to allow larger piles of grit to accumulate prior to transport.

4.7.4   Performance Testing

Performance testing of grit removal systems has a history of inaccuracy and misrepresentation. Two inherent difficulties associated with grit removal testing are uneven distribution of grit in both the influent and effluent and erratic loading patterns associated with grit only being carried into the facility during high flow. Testing procedures must take into account the target grit removal performance, and the sampling procedure should be adjusted accordingly. There are several methods of testing for grit, as explained earlier in this chapter in Section 4.2 (3 gpm/sq ft). Sampling must be performed at high flows as grit is washed into the facility. Because of the "flushing" effect of high flows, sampling at normal flows and at high flows is recommended, because sampling at normal flows may not capture the bulk of the grit load entering the facility. This is particularly important for water utilities implementing water conservation programs, resulting in average sewer velocities below what is required to suspend grit. However, because sampling typically requires hiring an outside testing service, sampling at high flows can be difficult or impossible to schedule. One solution to this problem is to train the facility operations staff or other personnel capable of quickly responding to a high flow event in the grit testing procedure.

4.8   Grit Sustainability and Resource Recovery

Fully washed grit recovered from WRRFs could be reused for different purposes including pipe bedding material, road sub-base, ground cover, and as a soil stabilizer in sloped areas. In Germany, clean conditioned grit is being used as construction material when its organic contact in below 3% and the water content does not exceed 10%. Grit from WRRFs has also been combined with material from street sweeping and sewer cleaning operations and conditioned in resource recovery facilities to convert the coarse mineral material for use as aggregate in the construction industry. Proper grit removal can also be important from a sustainability standpoint, because otherwise the digestion capacity of WRRFs could be affected, which translates into less energy generation over time and keeps rotating equipment from operating at its desired efficiency point.

5.0   Grease Removal

5.1   Application and Benefits

Grease removal processes should be considered at nutrient removal treatment facilities that do not use primary clarifiers. One of the biggest disadvantages of not having primary clarifiers is that the grease-removal capability of primary clarifiers is lost. Grease in the nutrient removal process combined with long solids retention time (SRT) associated with these processes is the primary cause of foaming problems in these facilities. Long SRT facilities with a grease removal process incorporated into the headworks do not typically have foaming problems.

5.2   Grease Removal Processes

The most typically used standalone grease removal process is a side channel next to an aerated grit removal basin (Figure 9.27). The roll pattern from the aerated grit basin creates an air flotation effect and forces floating material into the side channel. The channel is separated from the aerated grit basin by a series of baffles that consist of the upper portion of the sloping side of the grit chamber. The grease is moved down the length of the channel with a series of air lances and is removed using a screw.

Two other grease removal processes include rotating pipe skimmers along the width of aerated grit basins and dissolved air flotation (DAF) processes. For the rotating

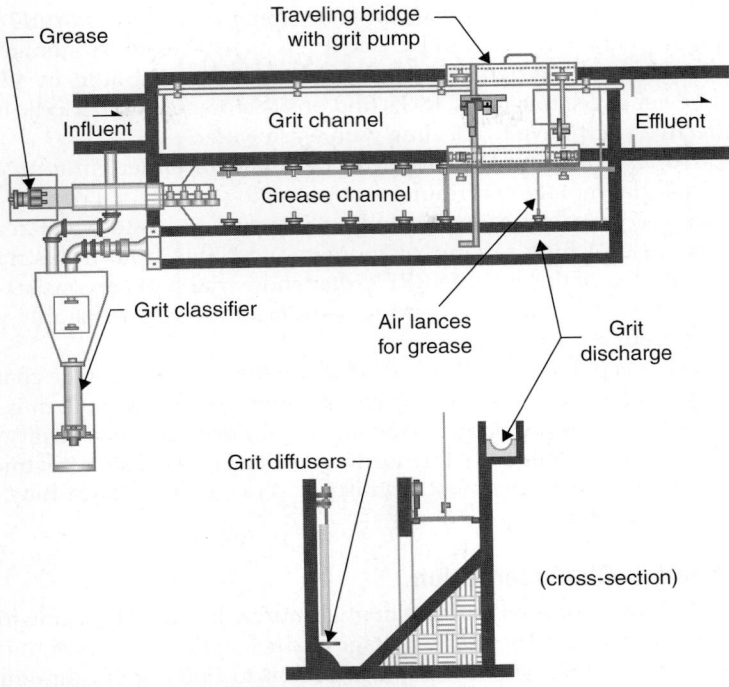

Grease

Traveling bridge with grit pump

Influent

Grit channel

Effluent

Grease channel

Grit classifier

Air lances for grease

Grit discharge

Grit diffusers

(cross-section)

FIGURE 9.27 Combination grit/grease removal process.

pipe skimmers to capture a significant amount of grease, decreased aeration must be used at the end of the grit basin. Use of standalone DAF processes for municipal raw wastewater grease removal is not typically practiced in the United States but is used in smaller facilities in Europe.

6.0 Septage Acceptance and Pretreatment

6.1 Applications

Residuals that are pumped from onsite treatment systems or holding systems are typically discharged to the nearest WRRF. This semisolid residue is referred as septage, and it can come from a variety of sources including cesspools, privies, septic tanks, grease collection programs, and holding tanks. Sources of septage include residential, commercial, and industrial activities. As a result, its composition is not uniform, requiring specific considerations for each case.

Septage acceptance and pretreatment in WRRFs is good practice for several reasons:

- Can provide an additional source of revenue.
- Can be used as a fuel source to generate energy, although its fuel value is not typically significant since it is well digested.
- Not providing septage acceptance services may invite illegal dumping.

The largest amount of septage is found going to facilities in rural areas with large populations that are not served by WRRFs, which necessitates storage or treatment of wastewater in onsite systems. Similarly, commercial contributors in isolated rural areas lacking sewer collection and WRRFs find onsite storage with periodic hauling the most cost-effective alternative for dealing with wastewater.

Industrial septage contributors can be found in either rural or urban locations. Rural industrial septage contributors may elect to store all or a portion of the wastewater onsite for various reasons, including preventing onsite water resource recovery upsets because of high organic or inorganic loadings from industrial operations or reducing the size of the onsite WRRF. Urban industrial septage consists of waste residuals from industrial wastewater pretreatment facilities used to comply with local sewer industrial pretreatment standards.

Septage acceptance in WRRFs depends on the influent septage characteristics, volume, and frequencies of deliveries. Separate septage pretreatment may not be required if a limited amount is discharged to an interceptor upstream of a facility or if a relatively small volume is discharged to the headworks of a large existing treatment facility with adequate preliminary treatment. Industrial septage acceptance may require special design considerations.

6.2   Septage Characterization

Septage is often associated with residential sources but can also come from commercial and industrial sources. Therefore, septage characteristics can vary significantly depending on several factors. However, it is common to find significant amounts of grease associated with septage, especially in commercial and residential sources.

Septage characteristics from residential sources vary widely for many reasons:

- User habits
- Tank size and design or reservoir configuration
- Pumping frequency and pretreatment programs
- Climate
- Types of in-home appliances used, such as garbage grinders or washing machines
- Local regulations
- Type of in-situ pretreatment prior to collection (holding tank, true septic tank, etc.)
- Difficulties in sampling septage
- Seasonal variations (including high real estate sales)

Characteristics of residential sources septage are presented in Table 9.5. Further discussions of residential septage characteristics are provided in other references (WEF, 1997). Typically, residential sources septage is stronger than domestic wastewater and has higher concentrations of all constituents.

Commercial source septage characteristics may vary depending on several factors including those for residential septage and other factors, like types of commercial development. A practical way to estimate commercial septage characteristics is to study historical records of similar facilities. Industrial source septage must be estimated for each user based on historical records or predicted effluent characteristics.

Constituent	Concentration [mg/L]	
	Range	**Typical**
TS	5000–100 000	40 000
SS	4000–100 000	15 000
VSS	1200–14 000	7000
BOD5	2000–30 000	6000
COD	5000–80 000	30 000
Ammonia	100–800	400
TKN	100–1600	700
Total phosphorus	50–800	250
Heavy metals*	100–1000	300

*Primarily iron, zinc, and aluminum.

TABLE 9.5   Typical Domestic Septage Characteristics (WEF, 1994)

6.2.1 Quantities

Septage quantities will vary from source to source. Most septage received at WRRFs comes from residential sources (WEF, 1994). However, commercial and industrial septage contributions may be higher than residential septage in a given locale.

6.2.1.1 Residential Sources Residential sources septage quantities are site specific. The quantity of septage delivered to a WRRF is related to the number of septic tanks in service and how often they are pumped. Intervals between septic tank pumping are associated with septic tank failure, which typically occurs every five years (WEF, 2016b).

Several factors should be considered in estimating residential sources septage quantities (WEF, 1994):

- Septic tank capacity: 1900 to 5600 L for single-family dwellings.
- Average annual per capita residential septage generation: 225 L.
- Septic tank pumping frequency: if served by a municipality then two to four years; if not then average failure rate is five years.
- Hauling vehicle capacities: small tankers of 3800 L; large tankers up to 30 000 L.

6.2.1.2 Nonresidential Sources Nonresidential septage sources include commercial and industrial. Commercial sources septage is generated from commercial activities including retail stores, motels, and restaurants (WEF, 2016b). Guidelines for estimating commercial sources septage are given in Table 9.6.

Because there are too many variables in industrial sources septage, it is not possible to generate guidelines for its quantification (WEF, 2016b). Therefore, sound engineering analysis and judgment of each situation is required to estimate the quantities of this type.

Commercial Source	Septage Generation	
Motel and guest houses	0.005 61 m³/m²·a	(6000 gal/yr·ac)
Restaurants	0.016 8–0.021 5 m³/ m²·a	(18 000–23 000 gal/yr·ac)
Office, shops, etc.	0.00093 m³/ m²·a	(1000 gal/yr·ac)
Overall average	0.00524–0.00627 m³/ m²·a	(5600–6700 gal/yr·ac)

TABLE 9.6   Typical Commercial Septage Quantities (WEF, 1998)

6.2.2 Physical Properties

Septage physical properties vary considerably and are site and source specific, so each case requires detailed analysis. Table 9.5 provides typical residential source septage characteristics. Commercial source septage has higher BOD, total suspended solids (TSS), oil and grease, and surfactant concentrations than the residential. Industrial source septage can sometimes be a combination of different types including domestic, high or low pH waste, high strength waste, high oil and grease waste, or high metal contents waste. As a result, reviewing existing sampling records is the best approach for quantifying the physical properties of commercial septage.

Septage sampling is a crucial component of a septage receiving station because of the potential effects on WRRF processes. All septage received at WRRFs should record and monitor the following components (WEF, 2016b):

- Septage identification
- Volume discharged
- Sample collection
- Sample analysis (pH, BOD, COD, TSS, total solids, ammonia, and heavy metals)
- Toxicity testing

Provisions for taking samples must be made in receiving station design to ensure a representative sample.

6.3 Design Considerations

Preliminary treatment facilities must be designed to accept septage loads from hauling vehicles. Some of these vehicles are equipped with pumping systems for loading and unloading and others discharge by gravity. Therefore, provisions for both types of vehicles should be included in the design. The following considerations are important in designing residential septage-receiving facilities (Segall and Ott, 1980; U.S. EPA, 1984):

- Septage contains hair, grit, rags, stringy material, and plastics and is highly odorous.
- Septage is difficult to feed at controlled rates unless a receiving station exists.
- Septage receiving and storage facilities with separate screening and grit removal constitute the best arrangement.
- At a minimum, 150 mm (6 in.) diameter line sizes should be provided (U.S. EPA, 1984).

Receiving facility design must account for the anticipated volume and effects of septage and odor control. If the septage is from industrial sources then careful attention must be paid to the overall design of the WRRF.

When discharging septage directly to a water resource recovery facility, equalization storage is recommended to control flow proportionately to wastewater flow. Equalization of industrial septage is typically required to prevent process upsets or to appropriately use the septage as a resource. For example, it is becoming common practice to provide separate facilities for acceptance of industrial septage with high soluble BOD or grease content and introducing this material directly to anaerobic digesters to enhance gas production. Equalization is not typically necessary where residential sources septage is discharged to an interceptor at a point far enough upstream from the facility to allow complete mixing with wastewater, provided that the total quantity represents less than 1% of the wastewater flow at that time and location. This can often be achieved by avoiding septage discharge during daily low-flow periods. Nonetheless, it is preferred to accept septage at the WRRF directly to avoid odor control issues in the collection system and properly monitor the septage characteristics as required to prevent toxic upsets.

6.3.1 Receiving Station Design

The receiving station design must account for several factors. One critical factor is the type of septage that will be accepted at the facility and how it will be unloaded from hauling vehicles. WRRFs that accept more than one type of septage may require separate receiving stations for each type of septage, which could require different pretreatment and may be incorporated at different locations or rates.

The receiving station should have a slightly sloped ramp to tilt the truck for complete drainage and to accommodate washed spillage to a central drain. The receiving station design should encourage operators of hauling vehicles to unload septage through hose connections to prevent spillage and release odors. Open channel designs where the hauling vehicles discharge over grating into open channels have also been used. This method provides for coarse screening of the septage by removing big rocks that could damage mechanical equipment downstream; however, odor control considerations may be required.

Design considerations for receiving station equipment include the following:

- Hoses and other washdown equipment should be provided. This includes steam equipment in colder climate for thawing lines and valves at a convenient location to aid cleanup by individual haulers.

- Septage unloading connections should be placed strategically to promote their widespread use.

- Quick-release discharge tubes for hose connections in the dumping station should be provided.

- For colder climates, a heater cable in the dump chamber bottom should be installed to prevent freezing.

- A discharge tube should extend below liquid level in receiving chambers to prevent release of odorous gases; the tube diameter is typically 100 mm (4 in.) (U.S. EPA, 1984).

- Positive means of measuring and recording the septage quantities should be available, like truck scales or flowmeters.

- A location for sampling incoming septage should be easily accessible.
- In locations particularly susceptible to odors, provisions for enclosing the septage receiving station may be required.
- An adequate number of unloading bays should be provided based on the maximum number of trucks that could be unloading simultaneously.
- Wash down provisions need to be in place to prevent solids deposition in open channels if used in the design.

The amount of septage entering a receiving station and the rate at which it passes through the pretreatment facility must be accurately estimated during design. Receiving facilities design requires an accurate assessment of both septage volumes and the range of daily septic flows expected. The critical limitation on a receiving station's peak flow capacity may be the number of discharge points (that is, unloading docks and hose connections). Therefore, multiple discharge points deserve consideration, if heavy traffic is expected during peak hauling periods. Similarly, the access arrangement should permit efficient queuing of several hauling vehicles in the receiving station area. Figure 9.28 illustrates a back-in bay septage station configuration.

6.3.1.1 Screening and Grit Removal   Screening and grit removal of septage is essential. As in normal water resource recovery, screening is followed by grit removal.

It has been standard to provide coarse screening with 6 mm openings for septage treatment, which will remove large solids and plastics. New process technologies requiring fine screening, however, makes it critical to understand discharge location at the facility and consider this in sizing screening facilities. There are packaged septage receiving stations available in the market, which incorporate screen and sometimes grit removal to pretreat septage. These systems have been found to operate properly in areas where low amounts of coarse solids are delivered. Conversely, in areas with high amounts of coarse solids (rags, larger solids, etc.) these package systems have proven to be less effective. Screens with forced coarse material conveyance have proven to be successful in overcoming high amounts of coarse solids.

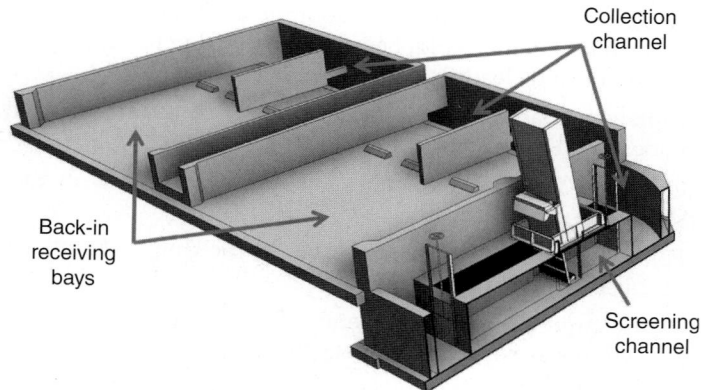

FIGURE 9.28   3D Render of septage receiving station (courtesy of Black & Veatch).

Design flow can be estimated by determining the maximum unloading rate from the hauling vehicles. Some hauling vehicles have compressors for loading and unloading, so the screening facility should be able to handle this flow. It has been suggested that it takes 10 minutes, on average, to unload septic haulers (WEF, 2016b). Other septage hauling vehicles have provisions for lifting the tank with a hydraulic jack, allowing gravity discharge of the septage. In this case, the maximum unloading rate from the tank at the beginning of the unloading operation will become the design flow.

Grit removal is also used in septage pretreatment. Grit is typically present in large quantities because of the long residence time of septage in tanks between pumpings. The concepts outlined in Section 4.0 of this chapter are applicable to the design of grit removal systems for septage.

An alternative that has been used in some facilities is to discharge the septage ahead of screening and grit removal facilities. Because of the potential high grease content of septage and its effects on screening operations, however, providing separate septage pretreatment facilities is preferred.

There are also package systems available that offer comprehensive pretreatment (screening and grit removal) in a single unit (Figure 9.29). These systems typically use a rotating drum screen with a screen conveyor for removal of the screenings, followed by an aerated grit removal system with screw conveyors for removing the grit from the aerated grit chamber. Some designs even include scum removal, which could be significant if commercial sources septage is received at the facility. This system can sometimes be accommodated inside the headworks building, minimizing the need for additional odor control facilities.

6.3.1.2 Storage and Equalization Storage and septage equalization is provided in holding basins equipped with mixers or aeration. These holding facilities control outflow of septage to downstream treatment processes to prevent hydraulic and organic shock loadings.

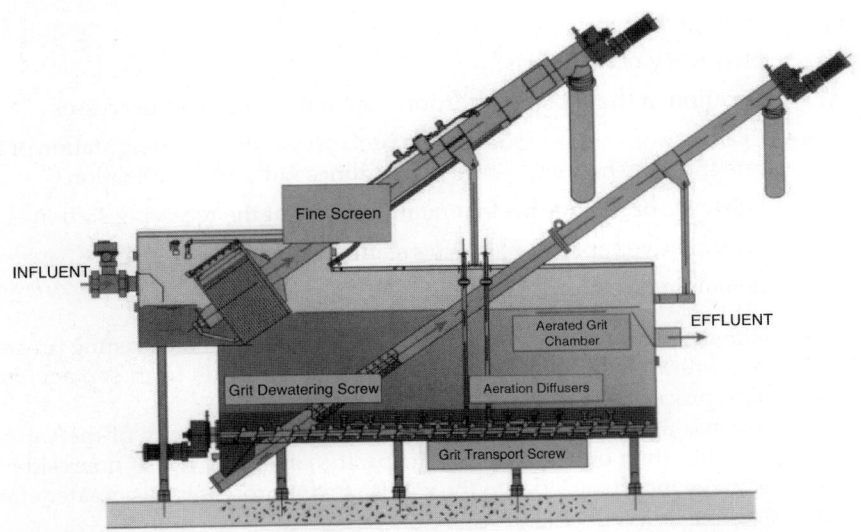

FIGURE 9.29 Septage pretreatment system (courtesy of Lakeside Equipment Corp.).

Pumping stored septage to the water resource recovery facility has been accomplished successfully with air-operated diaphragm pumps and centrifugal chopper pumps. If the septage will be added directly to a sewer or to a primary treatment unit then mechanical or diffused aeration improves treatability and prevents settling of organic solids. Aeration tends to aggravate the odor problem (because of the air stripping effect) and, therefore, requires use of enclosed tanks. In cases where outflow of the equalization tank is not required to be constant then it is considered good practice to have the septage mixed in equalization tanks to 95% of the influent septage TSS concentration within 30 minutes of starting the mixing system. This will prevent mixing equipment from operating continuously during periods of no outflow and will allow completely-mixed septage to be discharged from the tank.

The primary design criterion for a holding tank is detention time. As a rule, the holding tank storage capacity will equal at least a one-day maximum expected volume of septage. Storage of several days of peak flow may be necessary, however, depending on the sensitivity of downstream treatment processes, the expected variation among daily volumes, and the ultimate use or location where the septage will be conveyed. Design of the equalization basin requires a site-specific approach that depends on the type and magnitude of the input flow variations and the facility configuration. If other preliminary treatment functions, such as pre-aeration, are to be combined with flow equalization then the equalization basin should provide adequate detention times for these functions. Studies have shown negligible changes in the characteristics of finely screened septage after 24 hours of aeration (Condren, 1978). Detention periods of fewer than 48 hours are considered good practice, but can vary from site to site. Detailed equalization basins design and considerations will be discussed in the Flow Equalization section of this chapter.

6.3.2   Effects on Downstream Processes

Septage treatment in WRRFs can have adverse effects on downstream processes if not carefully considered. The effects of septage vary widely depending on several factors:

- Septage source
- Frequency of deliveries
- Location of the septage incorporation in the treatment processes
- Existence of septage equalization processes at the receiving station or unloading rate from the hauling vehicles in facilities without equalization
- Existence of septage pretreatment facilities at the receiving station
- Assimilative capacity of downstream processes
- Remaining capacity of existing receiving processes

To estimate the effects of septage, it is necessary to evaluate existing processes at the facility, determine the added hydraulic and organic loads from septage, and design equalization processes to attenuate negative effects.

Septage has a greater effect on the organic loading capacity of the water resource recovery facility than on the hydraulic capacity, especially if it is nonresidential. Conversely, septage incorporated into the solids stream processes has greater effects on the hydraulic capacity (WEF, 1997).

Septage incorporated into either liquid or solids processes will ultimately increase the solids generated at the facility. Screenings, grit, primary sludge, waste activated sludge, filter, and backwash sludge quantities could be increased as septage is introduced to liquid processes. Primary or biological sludge quantities that require stabilization, thickening, and dewatering could be increased if septage is introduced in liquid or solids processes. Septage incorporation in solids processes can adversely affect liquid processes by degrading sidestreams from solids that are returned to liquids.

6.3.3 Automation and Instrumentation

Automation and instrumentation of septage receiving facilities consists of flow measurement, pH, or conductivity probes that can either be alarmed or designed to automatically close the receiving station influent valve. Frequent maintenance of online probes in septage receiving lines is critical because of the potentially corrosive and abrasive nature of septage. Such probes typically must be inspected, maintained, and calibrated once per shift.

6.4 Septage Acceptance Sustainability and Resource Recovery

Septage acceptance is considered a sustainable practice from the standpoint that the organic loading is reduced at the source with anaerobic pathways, resulting in less energy requirements and lower solid yield compared to toxic systems. Additionally, less water is conveyed to the WRRF from septage operations, reducing the liquid train requirements. Conversely, septage solids have little energy value as they are well digested, so there is little biogas generation potential with them.

7.0 Equalization

Accommodating wide variations in flow rates and organic mass loadings is one of the primary challenges in the design of WRRFs. Efficiency, reliability, and control of unit process operations within the facility can be adversely affected by the cyclic nature of waste generation, resulting in possible violations of effluent standards unless treatment processes account for peak conditions. Equalization can also be used to minimize downstream processes sizing in wastewater facilities. This section will address the equalization processes for pretreatment facilities only.

Equalization is designed to achieve more constant hydraulic or organic loading of downstream treatment processes. Equalization can result in some degree of treatment. This can be beneficial if, for example, a facility is receiving high concentrations of industrial waste. Treatment in equalization can be unintended, which must be taken into account when designing the downstream process. This can be particularly important in biological nutrient removal (BNR) processes requiring maximum concentrations of soluble BOD.

7.1 Benefits

The benefits of equalization include the following:

- Flow equalization
- Waste strength equalization
- Biological treatment enhancement
- Improved secondary clarifier performance
- Improved chemical feed and process reliability
- Preconditioning of wastewater for downstream processes

- Partial primary treatment
- Toxic substances dilution
- pH dampening

Equalization usually occurs at the WRRF. However, in certain areas where the collection system has significant storage volume available such as large gravity sewers (tunnels, collectors) or sufficient storage in the pump station wetwells, equalization can also be practiced by carefully allowing storage to occur in the system (Heath, 2014).

Flow equalization basins for WRRFs are constructed to dampen wet weather and diurnal peak flow variation. An additional benefit is a reduction in the variability of concentration and mass flow of wastewater constituents by blending in the equalization basin. This more uniformly loads downstream processes with organics, nutrients, and other suspended and dissolved constituents.

Waste-strength equalization dampens the variability of the waste strength by blending wastewater in the equalization basin. For this purpose, the volume of wastewater in the equalization basin typically remains constant and flow remains variable.

Equalization will reduce the size of unit processes at a new facility or relieve overloaded unit processes at an existing facility. By providing relatively constant loading, equalization can improve the efficiency, reliability, and operability of some types of facilities. For example, activated-sludge and trickling-filter facilities that are sized for peak organic loadings could be downsized. Equalization also protects from shock loading, which adds reliability to any biological process.

The use of equalization either to reduce the size of or relieve an overloaded condition on an individual treatment process seldom would be economically justified. Nevertheless, if the cumulative effects of equalization throughout the facility are accounted for, equalization may be more cost effective than modifying existing unit processes.

7.1.1   Primary Treatment

Equalization benefits to primary treatment can be realized for facilities with influent peaking factors exceeding 2:1. For such systems, equalization benefits related to primary clarification include the following:

- Reduction in the area required for primary clarification because the units will be designed to accept the equalized flow rather than the normal peak flow.
- For an overloaded primary treatment system, a reduction of the peak flow relieves stress on the units.

Another possible benefit related to primary clarification is improved performance because of pre-aeration in the equalization basin. Studies have shown that prolonged pre-aeration will improve primary clarifier TSS removal performance up to 15% because of solids flocculation effects in the equalization basin (Roe, 1951; Seidel and Bauman, 1961). Often the flow equalization unit process is located after primary clarification to reduce operation and maintenance requirements (Ongerth, 1979).

7.1.2   Secondary Treatment

Equalization benefits secondary treatment units by dampening the flow and waste strength, which allows the process to operate at near steady-state conditions. Enhanced primary treatment efficiency resulting from equalization also reduces loads entering secondary units.

Equalization directly benefits secondary clarification by reducing peak flows that typically dictate clarifier size. Sludge settleability in secondary clarifiers can also be improved because filamentous floc formed under low dissolved oxygen conditions in aeration basins may be prevented. In existing facilities, reduced peak flows from equalization may offer an opportunity to increase the mixed-liquor suspended solids (MLSS) concentration in the aeration system and still maintain an acceptable solids loading on secondary clarifiers. An increase in MLSS concentration allows the facility to operate at the design SRT in a smaller aeration basin, thereby providing an increase in treatment capacity. This can improve the reliability of nitrification and reduce biological sludge production. Conversely, equalization may allow a reduction of the MLSS in response to lower primary effluent BOD_5, which reduces solids loadings on secondary clarifiers and improves treatment performance.

Reducing wastewater strength variations may minimize aeration requirements and associated power requirements. In addition, prolonged aeration in an equalization basin will reduce BOD approximately 10% to 20% for an inline basin equalization of raw wastewater.

7.1.3 Advanced Treatment

The reliability and efficiency of sensitive BNR systems may benefit from equalization through dampening variations of flows and mass constituents. Dampening of the mass loading to chemical coagulation and precipitation systems will improve chemical feed control and process reliability. As a result, this may reduce instrumentation complexity and cost. With biological phosphorus removal, stabilization of the BOD:phosphorus ratio—which is key to performance—may enhance reliability. A constant flow rate to filters will lead to more uniform solids loading and a higher level of performance and may allow the use of smaller filter units.

Combined BNR can be beneficial because denitrification and phosphorus removal processes perform better under equalized flows and loads conditions (Mikola et al., 2007).

7.1.4 Wet-Weather Treatment

Flow equalization has been used to reduce or eliminate the frequency of CSOs and SSOs by providing an equalization volume for all flows that exceed peak rated capacity of the WRRF. After the peak event has subsided, the equalization volume is drained to the facility at a controlled rate.

The sizing of equalization basins for this purpose depends on the design confidence interval of the storm flows. For example, certain facilities have been designed to provide flow equalization up to 95% of the storm flow occurrences per year, based on previous years' records. Other methods for determining wet-weather flow storage requirements include the Storage, Treatment, Overflow, and Runoff 1 Model (STORM) developed by the U.S. Army Corps of Engineers (1976) and a modified version of STORM. Provisions for flows exceeding the design capacity of the equalization units should be made and those provisions should depend on regulatory requirements. In locations where discharging flows above a certain threshold is allowed, equalization basins can become treatment units where primary clarification can occur and be supplemented with chlorination and dechlorination of the effluent flow.

7.2   Design Considerations

7.2.1   Peak Flow Characterization

The design methodology for equalization involves determining the necessary volume, mixing, and aeration requirements, and the control of flows leaving the basin. The first step in estimating the volume involves determining the diurnal variations of wastewater. Whenever possible, this step should be based on actual operating data. Diurnal flow patterns will vary from day to day and seasonally, depending on the nature of the community (for example, tourist area, winter residences, agricultural food processing). It is, therefore, important to select a pattern that will ensure a large enough volume for equalization, taking into account conditions such as infiltration and storm-related inflow that influences variability of flow.

7.2.2   Volume Determination

Several methods proposed for estimating equalization volume required for typically varying wastewater flows and strengths are (U.S. EPA, 1974a) as follows:

- Flow balance (mass diagram)
- Concentration balance
- Combined flow and concentration balance
- Sine wave method
- Rectangular wave method

These methods, sufficient for most WRRF applications, offer relatively simple procedures for estimating equalization volume requirements. If needed, other more sophisticated methods of equalization analysis are available (Adams and Eckenfelder, 1974; LaGrega and Keenan, 1974; DiToro, 1975; Novotny and Stein, 1976; McInnes et al., 1978; Ongerth, 1979; Roe, 1951; Seidel and Bauman, 1961; Wallace, 1968). For all sizing methods, it has been common practice to add a 15% to 20% safety factor to the calculated equalization volume.

An equalization design example is provided below.

Equalization Design Example

Input Information Required	Flow distribution over the desired period for equalization, required safety factor for the tank's volume, type of equalization required (flow, mass, etc)
Assumptions	The equalization volume determination will be based on the maximum month daily flow distribution.
Output	Required equalization volume
Notes	This example will present the equalization volume determination for flow equalization only. There are two methods presented: Numerical and Graphic.

Equalization Design Example (Numerical Procedure)

Determine the flowrate equalization volume requirements for the following data

$$\text{Average Daily Flow} = 90,800 \text{ m}^3/\text{d} = 3785 \text{ m}^3/\text{hr}$$

Flow distribution (See Table below)

Procedure

1. Since the given data has an hourly distribution, convert all flow units into volume/hr.
2. Calculate the distribution time's period. In this example the distribution provided was hourly, therefore Time period = 1 hr.
3. Calculate the difference between Qavg and the flow for each period.

Given Data			Derived Data			
Time of Day	Qavg	Flow distribution Q	Difference (Flow-Qavg)	Amount to Storage	Amount from Storage	Running Total in Storage
hr	m³/hr	m³/hr	m³/hr	m³	m³	m³
0 to 1	3785	3385	(400)	0	(400)	0
1 to 2	3785	2722	(1063)	0	(1063)	0
2 to 3	3785	2024	(1761)	0	(1761)	0
3 to 4	3785	1605	(2180)	0	(2180)	0
4 to 5	3785	1291	(2494)	0	(2494)	0
5 to 6	3785	1221	(2564)	0	(2564)	0
6 to 7	3785	1466	(2319)	0	(2319)	0
7 to 8	3785	2513	(1272)	0	(1272)	0
8 to 9	3785	4362	577	577	0	577
9 to 10	3785	5060	1275	1275	0	1852
10 to 11	3785	5235	1450	1450	0	3302
11 to 12	3785	5305	1520	1520	0	4822
12 to 13	3785	5235	1450	1450	0	6272
13 to 14	3785	4990	1205	1205	0	7477
14 to 15	3785	4746	961	961	0	8438
15 to 16	3785	4327	542	542	0	8980
16 to 17	3785	4013	228	228	0	9208
17 to 18	3785	4013	228	228	0	9436
18 to 19	3785	4048	263	263	0	9699
19 to 20	3785	4502	717	717	0	10 416
20 to 21	3785	4921	1136	1136	0	11 552
21 to 22	3785	4921	1136	1136	0	12 688
22 to 23	3785	4676	891	891	0	13 579
23 to 24	3785	4258	473	473	0	14 052

4. Calculate the amount to storage and amount from storage.

Amount to storage is zero (0) if the difference is negative, or the difference value if the difference is positive

Amount from storage is zero (0) if the difference is positive, or the difference value if the difference is positive.

5. Calculate the Running Total in Storage

Running total in storage is the cumulative addition of the amount to storage values.

Running Total in Storage = Previous Period Amount to Storage + Current Period Amount to Storage· Time Period

Note that only positive values of cumulative volume are to be tabulated. If this number falls below 0, then use 0.

For instance, for the 9 to 10 time slot:

Running Total in Storage (9–10) = 577 m³ + 1275 m³/hr·1 hr = 1852 m³

6. Determine the equalization volume by selecting the maximum cumulative volume of the period.

for this example, Vequalization = 14 052 m³

7. Include adequate safety factors

Typical safety factors for equalization basins are between 10% and 20%.

For this example, assume 15%

Vrequired = (1+15%)·14 052 m³ = 16 160 m³ – Use 16 200 m³

Equalization Design Example (Graphic Procedure)

Determine the flowrate equalization volume requirements for the following data.

Average Daily Flow = 90,800 m³/d = 3785 m³/hr

Flow distribution (See Table below)

Procedure

1. Calculate the Cumulative flow

Volume = Previous Period Stored Volume + (Outflow/Inflow) · Time Period

For instance, for the 3 to 4 time slot:

Volume = 8131 m³ + 1605 m³/hr·1 hr = 9736 m³

2. Plot the cumulative volume against time as shown on the graph below.

3. Draw a line from the origin to the end point of the cumulative flow. The slope of this line should be the average daily flow.

4. Draw a parallel line to the average flow line at the lowest point of the cumulative flow diagram. Note that if the cumulative volume also has a peak point over the average line, another tangent should be to the average flow line should be drawn.

5. Determine the equalization volume by measuring the vertical distance from the point of tangency to the average flow line, or the higher tangent line if there is one.

for this example, Vequalization = 14 000 m³

6. Include adequate safety factors

   Typical safety factors for equalization basins are between 10 to 20%.

   For this example, assume 15%

   Vrequired = $(1 + 15\%) \cdot 14\ 000$ m^3 = 16 100 m^3 – Use 16 100 m^3

7.2.3   Location within Treatment Process
Equalization basins typically are located in the facility following the grit and screening units. To minimize solids accumulation, it is placed downstream of primary clarification. Also, equalization of solids-handling sidestreams is used frequently to regulate and reduce the effect of sidestream returns on the liquid treatment system. Figures 9.30 and 9.31 show some typical equalization configurations.

7.2.4   Method of Operation
Fill-and-draw mode is the most efficient method of operating an equalization basin for flow dampening. The basin is filled during the day when peak flows are occurring, and then it is drawn down at night when the facility is receiving low flows, so it is more capable of treating excessive flow. If an equalization basin is not operated in fill-and-draw mode, it will act as a mass loading equalization basin only, assuming the basin is completely mixed.

7.2.5   Basin Configuration
Equalization basins are typically large concrete or earthen structures with sloping sides to facilitate cleaning. Earthen basins may have the side slopes lined with geo-membranes or pavement to prevent erosion. Basin floors are sloped to one side to allow collection of settled solids along one wall. It is common practice to provide more than one equalization basin to allow for flexibility. Provisions for cleaning should be included in the

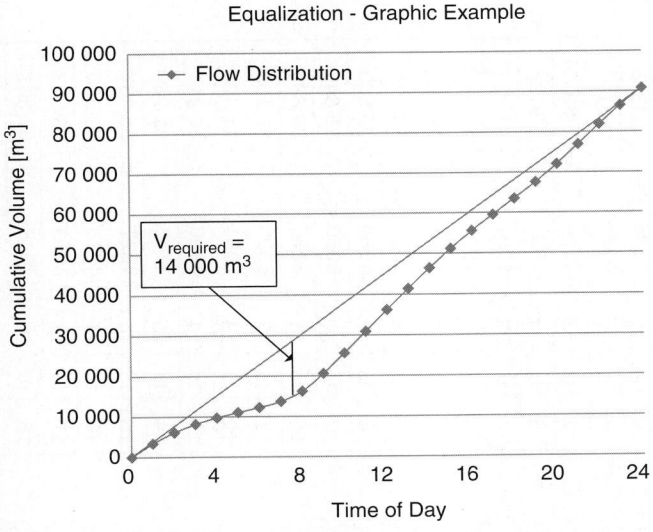

Figure 9.30

Time of Day			Max Flow Distribution	Cumulative Volume
hr			M³/hr	m³
0	to	1	3385	3385
1	to	2	2722	6107
2	to	3	2024	8131
3	to	4	1605	9736
4	to	5	1291	11 027
5	to	6	1221	12 248
6	to	7	1466	13 714
7	to	8	2513	16 227
8	to	9	4362	20 589
9	to	10	5060	25 649
10	to	11	5235	30 884
11	to	12	5305	36 189
12	to	13	5235	41 424
13	to	14	4990	46 414
14	to	15	4746	51 160
15	to	16	4327	55 487
16	to	17	4013	59 500
17	to	18	4013	63 513
18	to	19	4048	67 561
19	to	20	4502	72 063
20	to	21	4921	76 984
21	to	22	4921	81 905
22	to	23	4676	86 581
23	to	24	4258	90 839

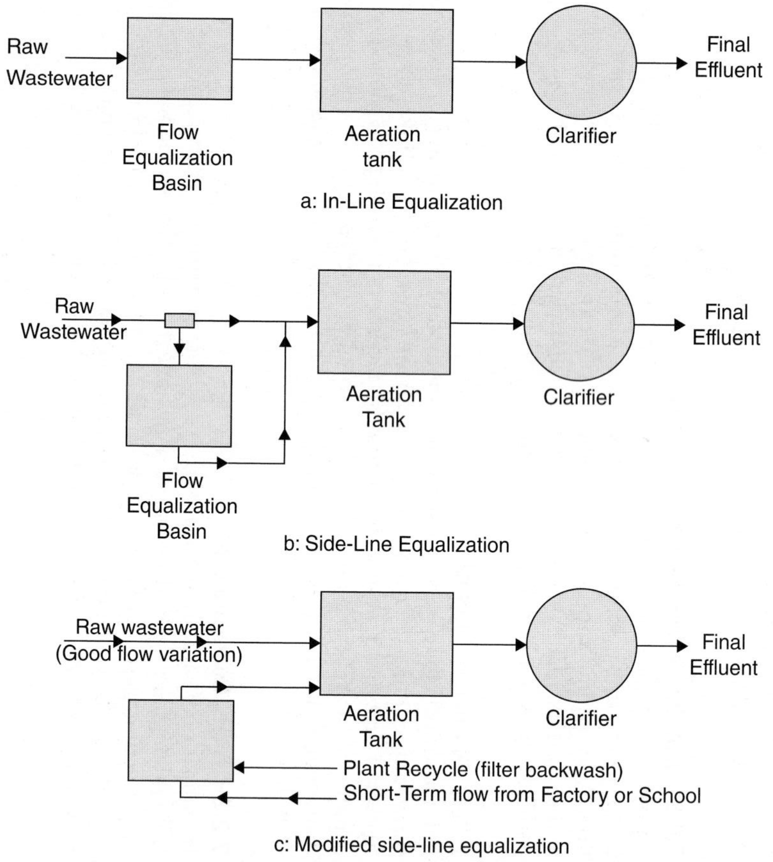

a: In-Line Equalization

b: Side-Line Equalization

c: Modified side-line equalization

FIGURE 9.31 Equalization configurations.

design. For example it is standard to provide hose connections around basins. Certain facilities have also used water cannons to assist with cleaning operations. Larger equalization basins are typically designed with roads to allow access for sludge collection equipment for cleaning. Influent structures are typically designed to allow energy dissipation to prevent basin erosion, particularly in earthen basins. Effluent structures are built out of concrete with a series of slide gates that serve as isolation or flow control acting as weirs. Equalization basins are seldom covered because they are large. Facilities providing equalization for dry-weather daily peaks and no wet-weather inflows, however, have smaller basins that could be covered with domes similar to those used for primary clarifiers.

7.2.6 Aeration and Mixing

The successful operation of equalization basins requires proper mixing and aeration. Design of mixing equipment provides for blending the contents of the tank and preventing deposition of solids in the basin. Ideally, equalization basins should be placed after screening and grit removal; however, the increased cost of facilities is generally not considered worth the relatively minor operational benefits to the equalization process. Aeration is usually provided to prevent septic conditions unless the equalization process is designed to treat only dilute wet-weather flows. Mixing requirements for blending a municipal wastewater that has a suspended solids concentration of approximately 200 mg/L range from 0.004 to 0.008 kW/m^3 (0.02 to 0.04 hp/1000 gal.) of storage (U.S. EPA, 1974a). However, mixing requirements in equalization basins typically depend on basin geometry, mode of operation, and type of mixing system used. To maintain aerobic conditions, air should be supplied at a rate of 0.156 to 0.25 $L/m^3/s$ (1.25–2.0 cfm/1000 gal) of storage.

Mechanical aerators, one method of providing both mixing and aeration, have oxygen transfer capabilities varying from 0.5 to 1.0 kg O_2/MJ (3 to 4 lb. O_2/hp × hr) in clean water under standard conditions, but the oxygen-transfer efficiency (OTE) in wastewater is lower. A reasonable OTE value for design would approximate 0.16 to 0.39 kg O_2/MJ (1.0–1.5 lb O_2/hp × hr). Minimum operating levels for floating aerators typically exceed 1.5 m (5 ft) and vary with the power and design of the unit. Low-level shutoff controls are needed to protect the unit. If the equalization basin floor is subject to erosion (earthen basins), concrete pads on the basin floor are recommended. Baffling may be necessary to ensure proper mixing, particularly with a circular tank configuration.

Power requirements to prevent deposition of solids in the basin may greatly exceed what is needed for blending and oxygen transfer. In such cases, the most economical approach might be to provide mixing equipment to keep the solids in suspension and a diffused air system to supply the air requirements. Aspirating aerators have been used in equalization basins because of the horizontal mixing component provided. Low OTE of these units, however, is a consideration.

Diffused aeration systems may use coarse- or medium-bubble diffusers. Membrane (not ceramic) fine-bubble diffusers can also be used if increased OTE is required. With variable volume systems and, therefore, variable water depths, blowers must have pressure regulation controls. For varying level conditions, positive displacement blowers are typically preferred over centrifugal blowers especially because of surge conditions. Mixing requirements are 0.5 to 0.8 L/m^3 s (30–50 cfm/1000 cu ft) of basin volume.

Equalization basins located prior to primary clarification should always be designed with the premise that some level of deposition will always occur. One method for containing solids deposition in small areas include the use of baffle walls within the equalization basins that allow settling of particles within the baffled areas and allow overflow of partially clarified wastewater to the remainder of the basin, which reduce the cleaning efforts after each use.

7.2.7 Flow Control Methods

The use of flow equalization will, in most cases, involve pumping either before or after the equalization basins. If flow is pumped to the equalization basin then the basin effluent must be controlled by a regulating device. The design engineer must recognize the acceptable flow range of the device and the range of basin operating water levels that must be throttled. For pumped flow from the equalization basin, a variable speed pumping system will be required to carefully regulate the hydraulic loadings, providing a nearly constant flow to the downstream units. Influent pumping to the equalization basin requires a pump of sufficient capacity to handle diurnal peaks.

A flow measuring device downstream of the basin must monitor the equalized flow. Instrumentation should be provided to control the pre-selected equalization rate by automatic adjustment of the basin effluent pumps or flow regulating device.

The use of waste strength equalization, typically requiring a constant volume, may need pumping to the equalization basin with a variable outflow equal to the input flow. Alternatively, if hydraulics allow, gravity discharge paced by a modulating device (valve or gate) tied to an effluent flow metering device may be used. Flexibility to account for operator inefficiency when dealing with the equalization inlet and outlet devices must be considered in the design.

7.2.8 Cleaning of Basins

Because grit removal is rarely provided ahead of equalization, grit will accumulate in the basins. Therefore, provisions for collecting these solids should be made in the design. If the primary purpose of the equalization basin is flow dampening then after the basin has been emptied following the peak flow event, primary sludge solids will be present on the basin bottom. Water cannons or strategically placed cleaning hoses, ideally supplied with facility effluent water, will allow for cleaning the basins. Other equalization basin types that do not operate in a fill/draw mode will still accumulate solids after a time and will have to be emptied for cleaning. The time between cleanings is dependent on the influent wastewater characteristics and likely will have to be established by facility operation staff based on operational experience.

Equalization basins cleaning provisions shall include water cannons, tipping buckets, siphons, and different dewatering methods (Tabor, 2014). Sloped floors have been used in certain facilities to allow solids deposited to be carried as dewatering is taking place.

Frequently, a minimum level is maintained in the equalization basin to either avoid excessive cleaning cycles or to protect the bottom of basins that are aerated with surface aerators. If required, this unused volume must be accounted for in the design.

7.2.9 Automation and Instrumentation

Flow equalization automation depends on the type of equalization to be provided and the specific facility control system of each facility. Following are some of the recommended monitoring elements required in flow equalization basins (WEF, 2016b):

- Basin liquid level
- Basin dissolved oxygen level
- Influent pH
- Mixers and/or aeration blower status
- Influent/Effluent status pumps
- Influent/Effluent flow

Additional monitoring requirements for mass equalization basins may include the following:

- COD
- TSS
- Total nitrogen
- Total phosphorus

An additional successful control strategy used in some facilities is to equalize only those flows greater than the 24-hour moving average, resulting in better control of the equalization volume.

7.3 Equalization Sustainability and Resource Recovery

Equalization operations have little resource recovery potential. However, they could provide significant sustainable features to a WRRF, as peak demand loadings and their associated larger energy requirements can be avoided when equalization is practiced.

8.0 References

Albrecht, A. E. (1967) Aerated Grit Chamber Operation and Design. *Water Sew. Works*, **114** (9), 331.

Borys, A. (2014) Influent Screening Upgrade to Address Process Impacts at EBMUD. *Proceedings of the 87th Annual Water Environment Federation Technical Exposition and Conference* [CD-ROM]; New Orleans, Louisiana, Oct. 18–22; Water Environment Federation: Alexandria, Virginia.

Burbano, M.S., *et al.* (2014) Modeling the Effect of Inadequate Screening on Nutrient Removal. *Proceedings of the 87th Annual Water Environment Federation Technical Exposition and Conference* [CD-ROM]; New Orleans, Louisiana, Oct. 18–22; Water Environment Federation: Alexandria, Virginia.

Cambridge, D.; Fullington, B.; Rom, P.; Tattersall J. (2003) Influent Fine Screening for Improved Biosolids Quality. *Proceedings of the 76th Annual Water Environment Federation Technical Exposition and Conference* [CD-ROM]; Los Angeles, California, Oct. 11–15; Water Environment Federation: Alexandria, Virginia.

Condren, A. J. (1978) *Pilot-Scale Evaluations of Septage Treatment Alternatives*; EPA-600/2-78-164; U.S. Environmental Protection Agency: Washington, D.C.

DiToro, D. M. (1975) Statistical Design of Equalization Basins. *J. Environ. Eng.*, **101**, 917.

Heath, G., *et al* (2014) Preliminary Design Of The DCDc Water Enhanced Clarification Facility. *Proceedings of the 87th Annual Water Environment Federation Technical Exposition and Conference* [CD-ROM]; New Orleans, Louisiana, Oct. 18–22; Water Environment Federation: Alexandria, Virginia.

Koch, J.; Harmon D.; Jensen T.; Moser P.; Mayhew C. (2014) With the Advancement in Screening Technology, Not all Screens are Created Equal. *Proceedings of the 87th Annual Water Environment Federation Technical Exposition and Conference* [CD-ROM]; New Orleans, Louisiana, Oct. 18–22; Water Environment Federation: Alexandria, Virginia.

LaGrega, M. D.; Keenan, J. D. (1974) Effects of Equalizing Wastewater Flows. *J. Water Pollut. Control Fed.*, **46**, 123.

Londong, J. (1989) Dimensioning of Aerated Grit Chambers. *Water Sci. Technol.*, **21**, 13.

McInnes, C. D., *et al.* (1978) Stochastic Design of Flow Equalization Basins. *J. Environ. Eng.*, **104**, 1277.

Mikola, A.; Rautiainen, J.; Kiuru H. (2007) Diurnal Flow Equalization and Prefermentation Using Primary Clarifiers in a BNR Plant. *Water Practice Technol*, **1**, (5).

Morales, L.; Reinhart, D. (1984) Full-Scale Evaluation of Aerated Grit Chambers. *J. Water Pollut. Control Fed.*, **56**, 337.

Novotny, V.; Stein, R. M. (1976) Equalization of Time Variable Waste Loads. *J. Environ. Eng.*, **102**, 613.

Ongerth, J. E. (1979) *Evaluation of Flow Equalization in Municipal Wastewater Treatment*, EPA-600/2-79-096; U.S. Environmental Protection Agency, Municipal Research Laboratory: Cincinnati, Ohio.

Process Design Techniques for Industrial Waste Treatment; Adams, C. E., Jr.; Eckenfelder, W. W., Jr., Eds.; Enviro Press: Nashville, Tennessee, 1974.

Roe, F. C. (1951) Pre-Aeration and Air Flocculation. *Sew. Works J.*, **23**, 127.

Rothman, M.; Norrköping, V. (2005) Screen Design: Practice In Europe. *Proceedings of the 78th Annual Water Environment Federation Technical Exposition and Conference* [CD-ROM]; Washington, D.C., Oct. 29–Nov. 2; Water Environment Federation: Alexandria, Virginia.

Rusten, B.; Lundar, A. (2006) How A Simple Bench-Scale Test Greatly Improved The Primary Treatment Performance Of Fine Mesh Sieves. *Proceedings of the 79th Annual Water Environment Federation Technical Exposition and Conference* [CD-ROM]; Dallas, Texas, Oct. 21–25; Water Environment Federation: Alexandria, Virginia.

Segall, B. A.; Ott, C. R. (1980) Septage Treatment. *J. Water Pollut. Control Fed.*, **52**, 2145.

Seidel, H. F.; Bauman, E. R. (1961) Effect of Preaeration on the Primary Treatment of Sewage. *J. Water Pollut. Control Fed.*, **33**, 339.

Tabor C. (2014) Wet Weather Practices: Treatment, Systems and Tools. *Proceedings of the 87th Annual Water Environment Federation Technical Exposition and Conference* [CD-ROM]; New Orleans, Louisiana, Oct. 18–22; Water Environment Federation: Alexandria, Virginia.

U.K. Water Industry Research Limited (2002) *CSO Screen Efficiency* (Proprietary Designs); U.K. Water Industry Research Limited: London, England.

U.S. Environmental Protection Agency (1974a) *Flow Equalization*. U.S. Environmental Protection Agency Technology Transfer: Washington, D.C.

U.S. Environmental Protection Agency (1979) *Process Design Manual for Sludge Treatment and Disposal*, EPA-625/1-79-011; U.S. Environmental Protection Agency: Washington, D.C.

U.S. Environmental Protection Agency (1984) *Handbook of Septage Treatment and Disposal,* EPA-625/6-84-009; U.S. Environmental Protection Agency: Washington, D.C.

U.S. Environmental Protection Agency (1987) *Preliminary Treatment Facilities Design and Operational Considerations,* EPA-430/09-87-007; U.S. Environmental Protection Agency: Washington, D.C.

Wallace, A. T. (1968) Analysis of Equalization Basins. *J. Sanit. Eng. Div., Proc. Am. Soc. Civ. Eng.,* **94,** 1161.

Water Environment Federation (1997) *Septage Handling; Manual of Practice No. 24;* Water Environment Federation: Alexandria, Virginia.

Water Environment Federation (2016a) *Guidelines for Grit Sampling and Characterization;* Water Environment Federation: Alexandria, Virginia.

Water Environment Federation (2016b) *Operation of Water Resource Recovery Facilities, 7th ed., Manual of Practice No. 11;* Water Environment Federation: Alexandria, Virginia.

Wodrich, J.; Winkler, T.; Leaf, B.; Clark, S.; Youker, B. (2005) Wet Weather Impacts On Influent Fine Screening System Design And Operation. *Proceedings of the 78th Annual Water Environment Federation Technical Exposition and Conference* [CD-ROM]; Washington, D.C., Oct. 29–Nov. 2; Water Environment Federation: Alexandria, Virginia.

9.0 Suggested Readings

Anderson, M. M., et al. (1990) Designing to Improve Grit Removal at the Point Loma Wastewater Treatment Plant. *Paper Presented at 63rd Annual Conference Water Pollution Control Federation;* San Diego, California; Water Pollution Control Federation: Washington, D.C.

Finger, R. E.; Patrick, J. (1980) Optimization of Grit Removal at a WWTP. *J. Water Pollut. Control Fed.,* **52,** 2106.

Frechen, F.; Schier, W.; Wett, M. (2006) Pre-Treatment of Municipal MBR Applications in Germany—Current Status and Treatment Efficiency. *Water Practice Technol.,* **1** (3).

Neighbor, J. B.; Cooper, T. W. (1965) Design and Operation Criteria for Aerated Grit Chambers. *Water Sew. Works,* **112** (12), 448.

Pankratz, T. (1988) *Screening Equipment Handbook for Industrial and Municipal Water and Wastewater Treatment.* Technomic Publishing Co.: Lancaster, Pennsylvania.

Sabherwal B., Kobylinski E., Keller J., Lawrence M. (2007) Novel Approach to Storm Water Management for BNR Facilities. *Proceedings of the 80th Annual Water Environment Federation Technical Exposition and Conference* [CD-ROM]; Dallas, Texas, Oct. 21–25; Water Environment Federation: Alexandria, Virginia.

U.S. Environmental Protection Agency (1974b) *Process Design Manual for Sulfide Control in Sanitary Sewerage Systems;* U.S. Environmental Protection Agency: Washington, D.C.

U.S. Environmental Protection Agency (1974c) *Process Design Manual for Upgrading Wastewater Treatment Plants;* U.S. Environmental Protection Agency Technology Transfer: Washington, D.C.

Water Environment Federation (2004) *Control of Odors and Emissions from Wastewater Treatment Plants,* Manual of Practice No. 25; Water Environment Federation: Alexandria, Virginia.

Water Pollution Control Federation (1989) Technology and Design Deficiencies at Publicly Owned Treatment Works. *Water Environ. Technol.,* **1** (4), 515.

WEMCO (1990) Hydrogritter Separator. Bull. No. 11–86, Sacramento, California.

CHAPTER 10
Primary Treatment

Akram Botrous, Ph.D., P.E., BCEE; Onder Caliskaner, Ph.D., P.E.;
Jeff Hauser; and Mark W. Miller, Ph.D.

1.0   Introduction

Primary treatment involves the separation and removal of suspended solids and floatables (fats, grease, oils, plastics, etc.) from wastewater by physical-chemical methods. Much of the solids removed are organic and comprise a significant portion of the influent chemical oxygen demand (COD) and biochemical oxygen demand (BOD). Therefore, primary treatment results in reduced total suspended solids (TSS), COD, and BOD loadings to downstream biological treatment processes. Nutrients, such as nitrogen and phosphorus, and other constituents are also partly removed to the extent they are contained in the removed solids. The TSS, COD, and BOD reductions result in smaller reactor volume requirements and lower aeration demands for biological treatment. Removal of floatables minimizes operational problems from the buildup of scum in downstream treatment processes and improves the facility's overall aesthetics by reducing visual blights and odors.

The selection and design of the primary treatment process should assess the economic effect of primary treatment on other processes throughout the water resource recovery facility (WRRF). For example, primary treatment can reduce the costs (capital and operational) for biological treatment but increase the costs for solids handling facilities that receive the solids removed by primary treatment. Primary treatment is frequently coupled with anaerobic stabilization of waste solids, which produces methane gas that can be utilized to produce valuable energy. It is important to consider the net overall cost impact of primary treatment throughout the life of the WRRF. Typically, primary treatment is most economical for larger facilities (larger than 38 ML/d [10 mgd]).

The most common form of primary treatment is quiescent (conventional) sedimentation, which includes skimming; collection; and removal of settled primary sludge, floating debris, and scum. Enhanced sedimentation methods, such as chemically enhanced primary treatment (CEPT) and high-rate clarification (HRC) can be employed to reduce the area required for primary treatment and/or to improve treatment performance. Microscreens and filters can also be used for primary treatment. Table 10.1 provides typical removal rates for TSS, COD, or BOD, phosphorus, and bacteria for some of the available primary treatment processes. The terms sedimentation and clarification are used interchangeably to describe the same process. Similarly, sedimentation tank, clarifier, and settler are synonymous.

2.0   Conventional Sedimentation

Conventional sedimentation is an effective removal method for raw wastewater suspended solids, which range from a low concentration of nearly discrete particles to a high concentration of flocculent solids. These particles tend to settle by gravity under quiescent conditions as grease and scum float to the surface. Primary clarifier design historically has been based on empirically derived design criteria. These criteria, when coupled with a theoretical understanding of the clarification process, may be used to design reliable, efficient primary clarifiers. Computational fluid dynamic (CFD) modeling can be used to refine designs.

2.1   Types of Clarifiers

Rectangular (Figure 10.1), circular (Figure 10.2), stacked (Figures 10.3 and 10.4), and plate-and-tube settlers (Figure 10.5) are four types of clarifiers. Although square tanks

Primary Treatment Process	TSS Removal (%)	COD or BOD₅ Removal (%)	Phosphorus Removal (%)	Reference
Conventional primary clarifier	25–70	25–40	5–10	Metcalf and Eddy, 2014; Steel 1979; U.S. EPA, 1987
Chemically enhanced primary treatment (CEPT)	60–90	40–70	70–90	Metcalf and Eddy, 2014; Øedegaard, 2005
High-rate clarification	30–95	35–70	70–95	Metcalf and Eddy, 2014; Stevenson et al., 2008
Primary effluent filtration[a]	60–80[b]	40–60[b]	5–10[b]	Caliskaner et al., 2015
Primary filtration	60–80[b]	40–60[b]	NA	Caliskaner et al., 2016
Rotating belt filtration/ Microscreens	30–60[b]	25–40[b]	NA	Sutton et al., 2008
Dissolved air floatation clarifier (with integrated biological treatment)	50–80	40–60	NA	Ding et al., 2015; Johnson et al., 2014

[a]Combined performance of the primary clarifier and the primary effluent filtration systems.
[b]Without chemical addition.

TABLE 10.1 Typical Primary Treatment Removal Rates

with circular sludge collection equipment ("squircles") are also seen occasionally, this configuration is prone to poor hydraulics and biological activity and should be avoided [Water Environment Federation (WEF), 2005]. Rectangular and circular clarifiers are typically used for wastewater treatment. The selection of clarifier type for a given application can be governed by facility size, local regulatory authorities, site conditions, reliability and maintenance considerations, owner preference, and economics.

2.1.1 Rectangular Clarifiers
Rectangular clarifiers range from 15 to 90 m (50–300 ft) in length, 3 to 24 m (10–80 ft) in width, and 3 to 4.9 m (10–16 ft) in depth (Metcalf and Eddy, 2014). Rectangular tanks with common-wall construction are advantageous for sites with space constraints. The sludge removal equipment in rectangular clarifiers has greater maintenance requirements than similarly sized circular tanks because drive bearings in rectangular tanks are submerged. Rectangular clarifiers are discussed in more detail later in this chapter.

2.1.2 Circular Clarifiers
Circular clarifiers range from 3 to 60 m (10–200 ft) in diameter and 3 to 4.9 m (10–16 ft) in depth (Metcalf and Eddy, 2014). Circular clarifiers can use relatively trouble-free, circular primary sludge removal equipment (drive bearings are not underwater) (WEF, 2005). Walls of circular tanks are structurally designed as tension rings, which permit thinner walls than those for rectangular tanks. As a result of such advantages, circular tanks often have a lower capital cost per unit surface area than that

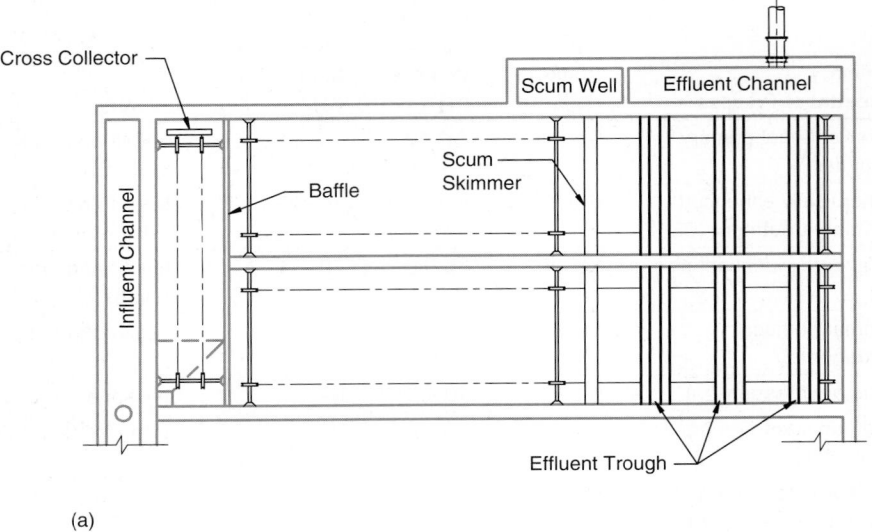

(a)

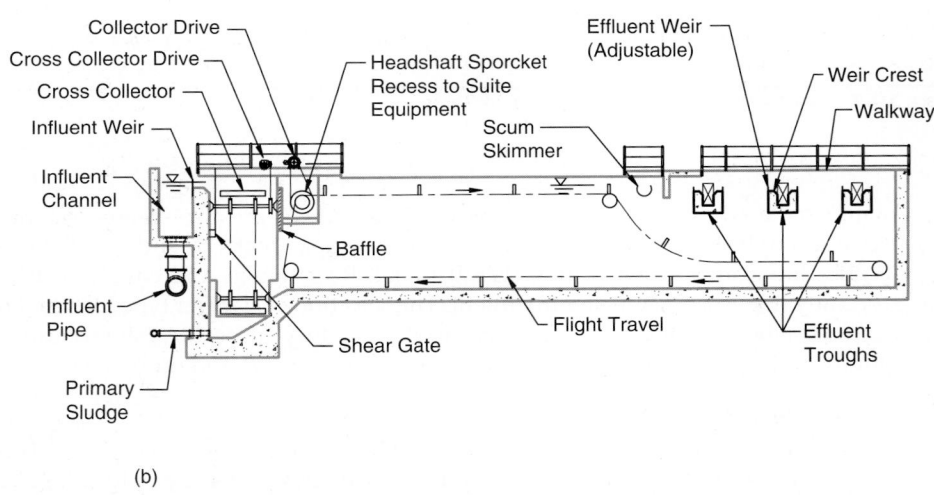

(b)

Figure 10.1 Typical rectangular primary sedimentation tank: (a) plan and (b) section (from Metcalf and Eddy, Inc., *Wastewater Engineering: Treatment, Disposal, Reuse.* 5th ed., Copyright © 2014, The McGraw-Hill Companies: New York, N.Y.).

for rectangular tanks. Circular tanks typically require more yard piping than rectangular tanks and require separate structures for flow distribution and sludge pumping. Circular clarifiers are discussed in more detail later in this chapter.

2.1.3 Stacked Sedimentation Tanks

In areas where land for treatment facilities is not available or is extremely expensive, stacked sedimentation tanks can be double-decked or even triple-decked to increase the available clarifier area without increasing the clarifier footprint. Series flow and parallel flow are two types of stacked clarifiers.

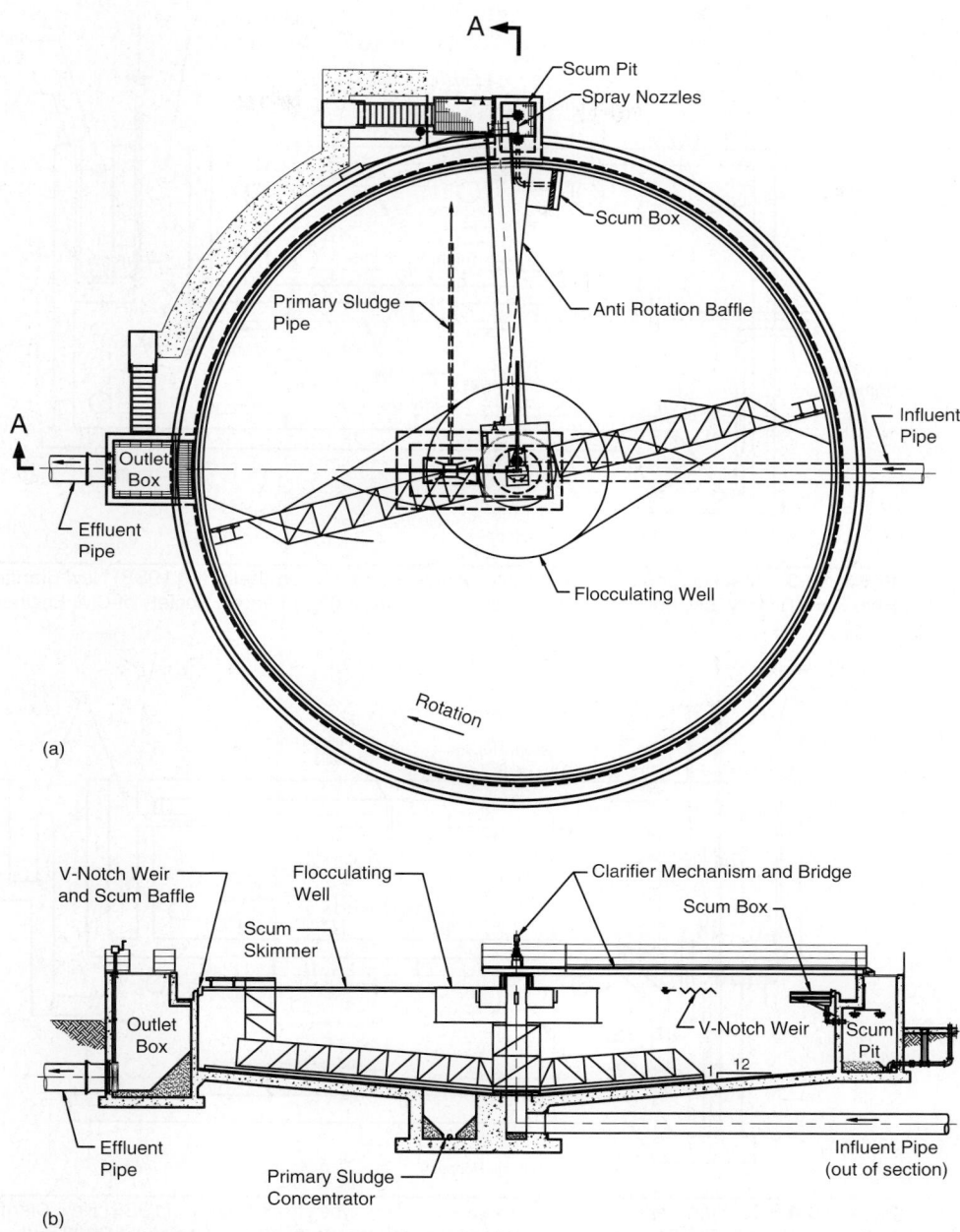

A ◄┐

Scum Pit
Spray Nozzles

Scum Box

Primary Sludge
Pipe

Anti Rotation Baffle

A
▲
└

Outlet
Box

Influent
Pipe

Effluent
Pipe

Flocculating Well

Rotation

(a)

V-Notch Weir
and Scum Baffle

Flocculating
Well

Clarifier Mechanism and Bridge

Scum Box

Scum
Skimmer

Outlet
Box

V-Notch Weir

Scum
Pit

1   12

Effluent
Pipe

Primary Sludge
Concentrator

Influent Pipe
(out of section)

(b)

FIGURE 10.2 Circular primary sedimentation tank at Merced, CA.

In the series-flow unit, wastewater enters the lower tray, flows to the opposite end, reverses direction in the upper tray, and exits the effluent channel as shown in Figure 10.3 (Kelly, 1988). Baffles straighten the flow paths and minimize turbulence at the influent point in the lower tray and at the turnaround in the upper tray.

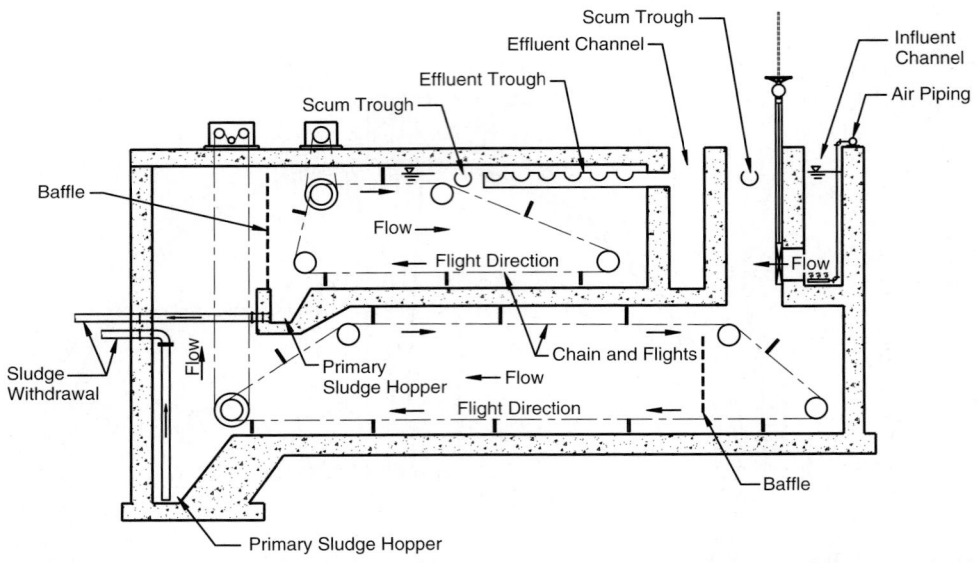

FIGURE 10.3   Stacked sedimentation tank: series-flow type (from Kelly, K. [1988] New Clarifiers Help Save History. Civ. Eng., 58, 10, with permission from the American Society of Civil Engineers).

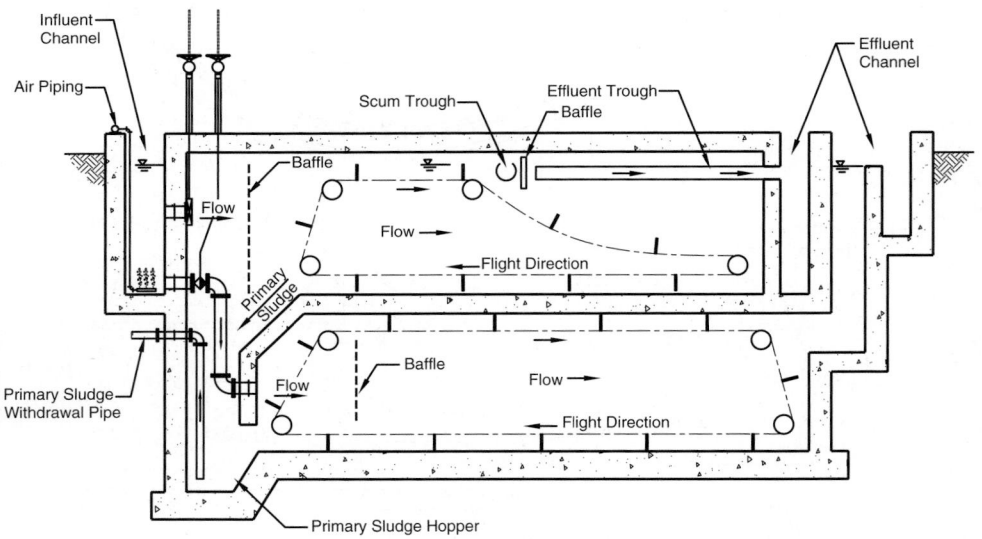

FIGURE 10.4   Stacked sedimentation tank: parallel-flow type (from Kelly, K. [1988] New Clarifiers Help Save History. Civ. Eng., 58, 10, with permission from the American Society of Civil Engineers).

In the parallel-flow unit, pipes convey wastewater from the influent channel to both the upper and lower trays as shown in Figure 10.4 (Kelly, 1988). Influent baffles in each tray straighten the flow path and minimize turbulence. The parallel-tray unit is the most common stacked configuration used for primary sedimentation.

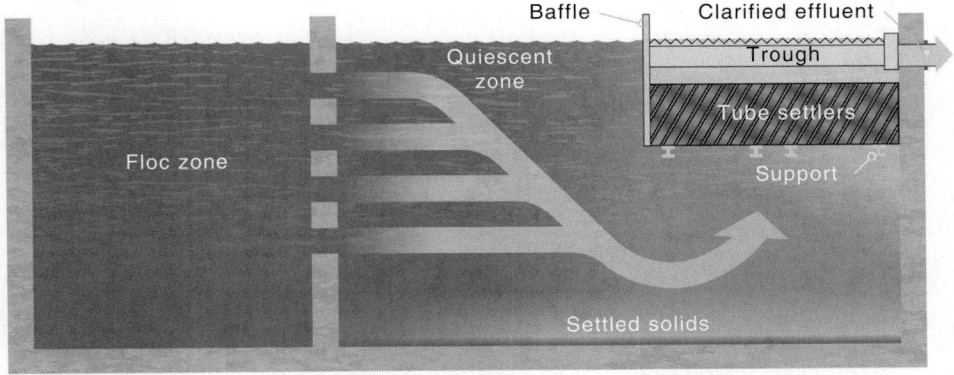

FIGURE 10.5   Typical tube settler (courtesy of Brentwood Industries Inc. Reading, PA).

Chain-and-flight collectors are used for primary sludge collection and removal from stacked tanks. Scum is removed only from the top tray.

2.1.4   Plate and Tube Clarifiers

In plate and tube clarifiers (commonly referred to as lamella or inclined plate clarifiers), bundles of parallel plates or tubes inclined at an angle of 45° to 60° to the water surface (Figure 10.5) are used to increase the effective settling surface area of the tank. The spacing between plates is typically 40 to 120 mm (1.6–4.7 in.). The increase in settling surface area is equal to the total projected horizontal area for all plates or tubes installed. Depending on the height of the plates or tubes, the angle of inclination, and the spacing between adjacent plates or tubes, the effective settling surface area can be 6 to 12 times the plan area occupied by the plates or tubes (based on 100 mm [4 in.] spacing). The performance of plate or tube settlers is similar to conventional primary clarifiers operating at the same overflow rate based on total projected horizontal area of the plates or tubes. Plate and tube settlers have been widely used in Europe, especially France, but have seen limited use in the United States. The main reason for this is the increased operational and maintenance requirements resulting from rags, debris, grease, and other solids that collect on the settlers, requiring frequent cleaning. Preliminary treatment measures such as fine screens and enhanced grease removal could minimize cleaning requirements. However, even with preliminary treatment, relatively frequent cleaning of the settlers (approximately weekly) may be required. Plate and tube clarifiers are being used more frequently for high-rate and wet-weather clarifier applications because the intermittent nature of wet-weather influent and its characteristics tend to mitigate problems with settler cleaning.

2.2   Design Considerations

Historically, primary sedimentation basin design has relied on criteria such as surface overflow rate, hydraulic detention time, depth, surface geometry, and weir loading rate. These criteria are helpful for design but are not accurate enough to permit prediction of actual sedimentation performance, which will depend on the wastewater characteristics (mainly settleability of suspended solids).

2.2.1   Surface Overflow Rate

The surface overflow rate (SOR) of the clarifier can be expressed by the following equation:

$$SOR = Q/A \qquad (10.1)$$

where SOR = surface overflow rate, m^3/m^2 d (gpd/ sq ft);
  Q = flow to clarifier, m^3/d (gpd); and
  A = surface area of clarifier, m^2 (sq ft).

Primary clarifiers historically have been designed on the basis of the SOR. Typical SOR values for conventional sedimentation range between 30 and 50 m^3/m^2 d (800 and 1200 gpd/sq ft) at average flow conditions and between 80 and 120 m^3/m^2 d (2000–3000 gpd/sq ft) with peak hourly flow (Metcalf and Eddy, 2014).

As shown in Figure 10.6, there is considerable scatter in the dataset for SOR and corresponding primary clarifier removal efficiencies, but a lower SOR will typically increase TSS, COD, and BOD removal efficiencies. As a result, some regulatory agencies have established SOR requirements for primary clarifiers (e.g., see Table 10.2).

2.2.2   Influent Wastewater Characteristics

TSS, COD, and BOD removal efficiencies during primary treatment are affected by various influent wastewater characteristics. Building on an equation first proposed by Tebbutt and Christoulas (1975), Wahlberg et al. (2003, 2005), and a Water Environment Research Foundation (WERF) (2006) study concluded that TSS, and

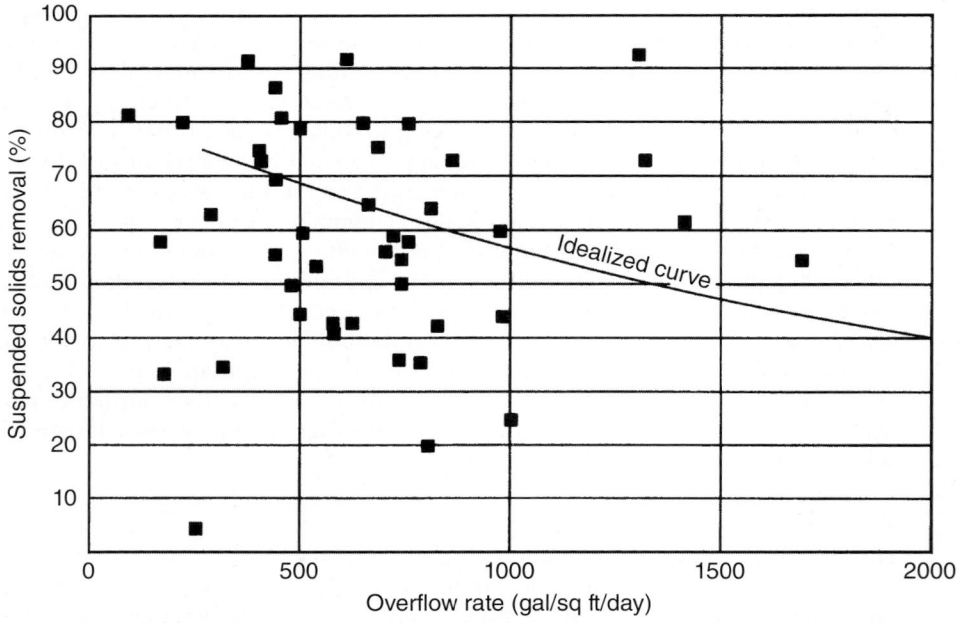

Figure 10.6   Primary clarifier suspended solids removals versus overflow rates showing idealized curve with data (gpd/ft² = 0.04075 m³/m² d).

Type of Primary Settling Tank	Surface Overflow Rates at:	
	Design Average Flow, m³/m² d (gpd/sq ft)	Design Peak Hourly Flow, m³/m² d (gpd/sq ft)
Tanks not receiving waste activated sludge	41 (1000)	61–81 (1500–2000)
Tanks receiving waste activated sludge	29 (700)	49 (1200)

TABLE 10.2 Primary Clarifier Surface Overflow Rate Requirements (Great Lakes—Upper Mississippi Board of State and Provincial Public Health and Environmental Managers, 2014). Reprinted with permission from Health Research, Inc.

COD or BOD removal efficiencies could be predicted by SOR and the following wastewater characteristics:

1. Nonsettleable TSS (TSS_{non}) concentration, which is defined as the supernatant TSS concentration after 30 minutes of flocculation and 30 minutes of settling;
2. Non-settleable COD or BOD concentration (COD_{non} or BOD_{non}), which includes soluble COD or BOD (sCOD or sBOD) and the particulate COD or BOD (pCOD or pBOD) associated with TSS_{non};
3. Primary influent TSS concentration (TSS_{PI});
4. Primary influent COD or BOD concentration (COD_{PI} or BOD_{PI}); and
5. Settling characteristics of the settleable solids or settling parameter (λ).

Removal efficiencies for TSS, COD, and BOD may be estimated with the following equations (WEF, 2005; Wahlberg et al., 2003, 2005; WERF, 2006):

TSS removal efficiency:

$$E_{TSS} = E_{TSSmax}\left(1 - e^{-\lambda/SOR}\right) \tag{10.2}$$

$$E_{TSSmax} = 1 - \left(TSS_{non}/TSS_{PI}\right) \tag{10.3}$$

Chemical oxygen demand removal efficiency:

$$E_{COD} = E_{CODmax}\left(1 - e^{-\lambda/SOR}\right) \tag{10.4}$$

$$E_{CODmax} = 1 - \left(COD_{non}/COD_{PI}\right) \tag{10.5}$$

Biochemical oxygen demand removal efficiency:

$$E_{BOD} = E_{BODmax}\left(1 - e^{-\lambda/SOR}\right) \tag{10.6}$$

$$E_{BODmax} = 1 - \left(BOD_{non}/BOD_{PI}\right) \tag{10.7}$$

where E_{TSS} = TSS removal efficiency (often reported as a percentage);
E_{TSSmax} = maximum TSS removal efficiency;
λ = settling parameter, m^3/m^2 d (gpd/sq ft);
SOR = surface overflow rate, m^3/m^2 d (gpd/sq ft);
TSS_{non} = non-settleable influent TSS concentration, mg/L;
TSS_{PI} = primary influent TSS concentration, mg/L;
E_{COD} = COD removal efficiency (often reported as a percentage);
E_{CODmax} = maximum COD removal efficiency;
E_{BOD} = BOD removal efficiency (often reported as a percentage); and
E_{BODmax} = maximum BOD removal efficiency.

The TSS, COD, and BOD of the primary effluent can be estimated with the following equations (WEF, 2005; WERF, 2006):

$$TSS_{PE} = TSS_{non} + (TSS_{PI} - TSS_{non})e^{-\lambda/SOR} \tag{10.8}$$

$$COD_{PE} = COD_{non} + (COD_{PI} - COD_{non})e^{-\lambda/SOR} \tag{10.9}$$

$$BOD_{PE} = BOD_{non} + (BOD_{PI} - BOD_{non})e^{-\lambda/SOR} \tag{10.10}$$

where TSS_{PE} = primary effluent TSS concentration, mg/L;
COD_{PE} = primary effluent COD concentration, mg/L; and
BOD_{PE} = primary effluent BOD concentration, mg/L.

Wastewater characteristics are affected by physical and biological processes occurring in the collection system, which depend on the size and slope of the collection system, flow rates, temperatures, and other factors. Where existing clarifiers can be sampled to determine the wastewater characteristics at various flows and mass loadings, the settling parameter (λ) can be estimated from historical data from a best-fit curve to eqs 10.8, 10.9, or 10.10. Although settling velocity distribution tests have been used to help quantify the settling parameter (λ), more research is required (WEF, 2005).

Figure 10.7 shows the effect of settling parameter (λ) on primary clarifier performance with increasing SOR when TSS_{PI} and TSS_{non} are 280 and 60 mg/L, respectively.

Although the preceding method represents an advancement in primary clarifier design, an in-depth review of the underlying assumptions and mathematical techniques in the WERF study is suggested before using such data alone to design a primary clarifier (Kinnear, 2004).

2.2.3 Depth

The opportunity for contact between particles and flocculation increases with depth. Hence, removal efficiency should theoretically increase with depth. In actual practice, it is uncertain whether better removals can be obtained or higher overflow rates can be applied to deeper tanks. Clarifiers must be deep enough to accommodate mechanical primary sludge removal equipment, store settled solids, prevent scour and resuspension of settled solids, and avoid washout or carryover of solids with the effluent. Shallower depths may be acceptable with continuous primary sludge removal. Typical depths range from 3 to 4.9 m (10–16 ft) (Metcalf and Eddy, 2014).

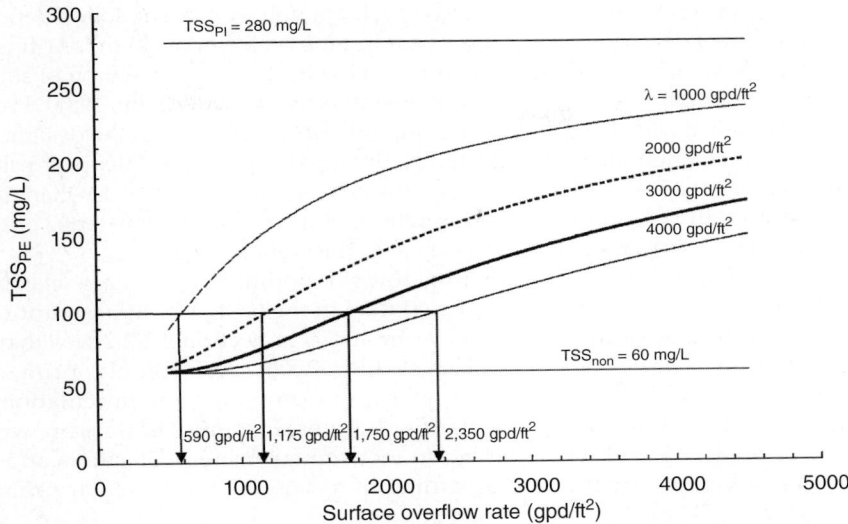

FIGURE 10.7 The effect of λ on primary clarifier performance at increasing flows for a hypothetical case in which the TSS_{PI} and TSS_{non} concentrations are 280 and 60 mg/L, respectively (WEF, 2005).

2.2.4 Hydraulic Detention Time

The hydraulic detention time typically is calculated from the required area and selected depth of the clarifier. Sufficient time for contact between solids particles is necessary for flocculation and effective clarification. As a result, some states have set limits (high and low) for detention times. Design considerations should include effects of low-flow periods to ensure that longer detention times will not cause septic conditions. Septic conditions increase potential odors, solubilization of particulate matter, and loading to down-stream processes. Typical hydraulic detention times range between 1.5 to 2.5 hours for average flow conditions with a wastewater temperature of 20°C (68°F) (Metcalf and Eddy, 2014). Detention times should be increased (SOR decreased) in cold climates to account for greater water viscosity at lower temperatures, which hinders particle set-tling. The following multiplier may be used for cold climates with wastewater tempera-tures lower than 20°C (68°F).

$$M = 1.82e^{-0.03T} \tag{10.11}$$

where M = detention time multiplier (SOR divisor), and
       T = temperature of wastewater, °C.

2.2.5 Flow Distribution and Inlet Conditions

In facilities with multiple sedimentation tanks, practice has provided for equal distribu-tion of flow and solids between the tanks. Splitter boxes and common channels some-times are used for this purpose. Automatically controlled valves or gates, together with flow measuring devices, can also be used to split the flow equally to each tank. Manually controlled valves or gates are generally not effective for flow splitting because the frac-tion of flow directed to each tank can vary substantially with changing head losses over the range of flows to be handled.

Inlet channel velocities should be high enough to prevent solids deposition. The inlet channel design typically allows a minimum velocity of 0.3 m/s (1 ft/s) at 50% of design flow. Other alternatives to minimum velocities for prevention of solids deposition are inlet channel aeration or water jet nozzles (Yee and Babb, 1985). Provisions for scum removal from the inlet channel should also be included in the design.

Sedimentation tank inlets should be designed to dissipate inlet port velocities, distribute flow and solids equally across the cross-sectional area of a rectangular tank or equally in all directions from the center-feed area of a circular tank, prevent short-circuiting, and promote flocculation before quiescent settling.

The importance of flocculation toward optimizing primary clarifier performance is now more widely recognized than in the past. A basic assumption inherent in the derivation of the performance equations in Section 2.1.2.2 was that the solids in raw wastewaters are flocculent (WERF, 2006). Accordingly, primary clarifier designs should incorporate the same care in providing for flocculation and ideal flow as is typical for secondary clarifier designs. Examples of this care would be the use of preaeration closely coupled to rectangular primary clarifiers and the use of inlet dissipation structures and baffling in rectangular and circular primary clarifiers (WERF, 2006).

Influent flow to rectangular clarifiers can be distributed by inlet weirs or by submerged ports or orifices. Practical considerations typically govern the size of openings, but principles of jet diffusion may serve as a design guide (Hamlin, 1972). The primary causes of skewed flow distributions in a tank are uneven flow distribution in the inlet channel and deflection by the baffle in the inlet zone (Yee and Babb, 1985). Locating inlet ports away from tank sides, adding partitions or baffles in the inlet zone to redirect the influent, and creating a higher headloss in the inlet ports relative to that in the inlet channel can all improve uniform flow (Yee and Babb, 1985). However, the higher headloss may break up flocs.

Inlet velocities to rectangular clarifiers typically are dissipated through some type of inlet baffle. Baffles typically are installed 0.6 to 0.9 m (2–3 ft) from the inlets. When full-width baffles are used, the top of the baffle should be 150 mm (6 in.) below the water surface to allow scum to pass and the bottom of the baffle should be 0.3 m (1 ft) below the entrance opening (Metcalf and Eddy, 2014).

Many designs of inlet baffling devices have been used with varying degrees of success. Figures 10.8 and 10.9 illustrate various inlet diffusers for rectangular sedimentation tanks. Figures 10.10 and 10.11 show the configuration of inlet diffusers and finger baffles at the Encina WRRF in Carlsbad, California. Inlet baffles with finger baffles are used at Sunnyvale, California, for influent distribution (Figure 10.12). At the Renton WRRF in Seattle, Washington, target and finger baffles are used for influent distribution (Figure 10.13).

One of the several possible combinations will describe the flow pattern in a circular clarifier. The flow pattern can be center feed with center withdrawal, center feed with peripheral withdrawal, peripheral feed with center withdrawal, or peripheral feed with peripheral withdrawal. Center feed with peripheral withdrawal is the most common type of flow pattern. Peripheral-feed configurations require horizontal pipe supports that impair flow distribution and scum collection. Center-feed configurations eliminate these disadvantages.

Center-feed circular clarifiers typically have a feed well with a diameter 15% to 25% of the tank diameter. Manufacturers' recommendations for submergence

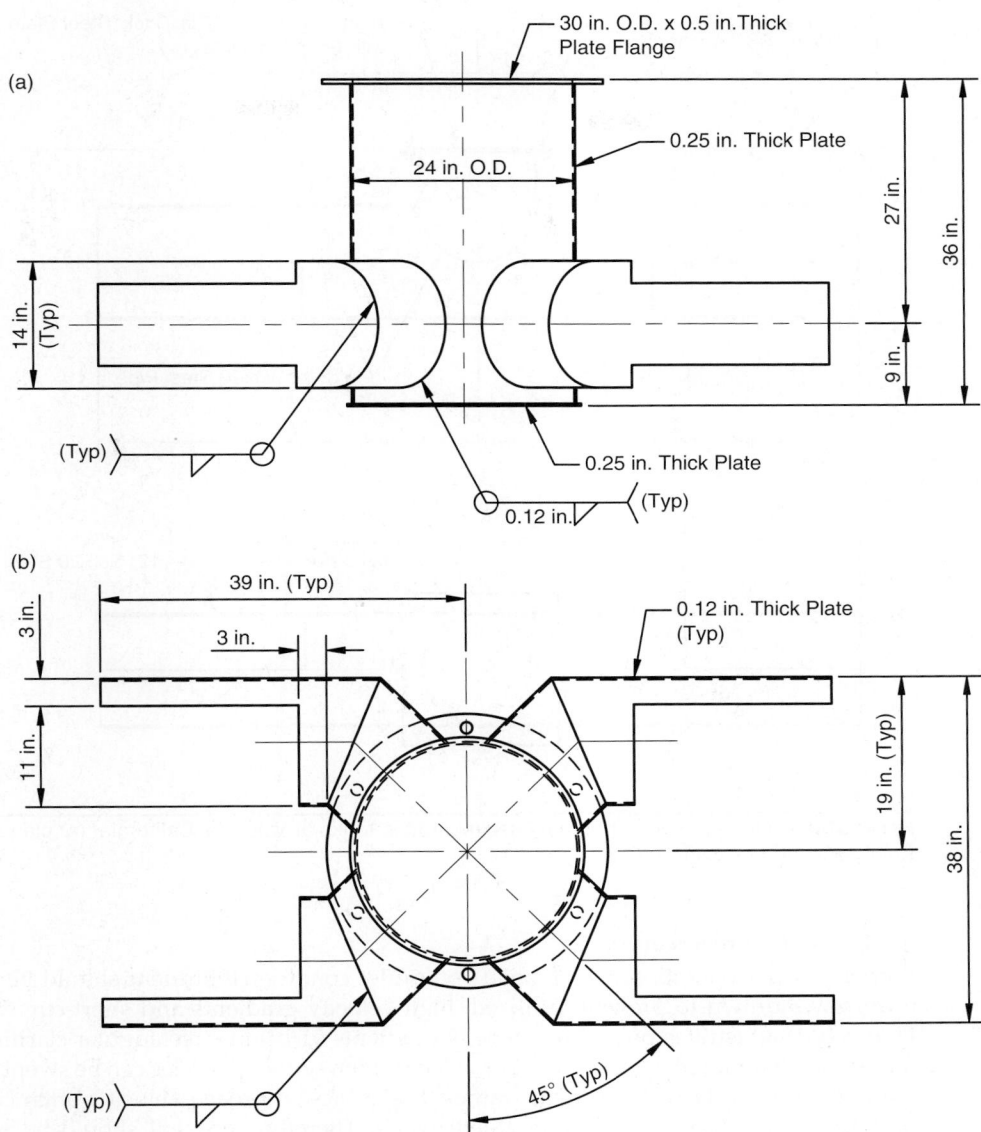

FIGURE 10.8   Inlet diffuser for primary sedimentation tanks at Long Beach, California: (a) plan and (b) elevation (in. = 25.4 mm).

vary significantly. In practice, the feed well typically has been extended at least half of the tank depth. An energy dissipating inlet should be provided to dissipate inlet pipe and port velocities and promote flocculation as the influent flow enters the feed well area. Various configurations for energy dissipating inlets have been developed by clarifier manufacturers (e.g., Figure 10.14).

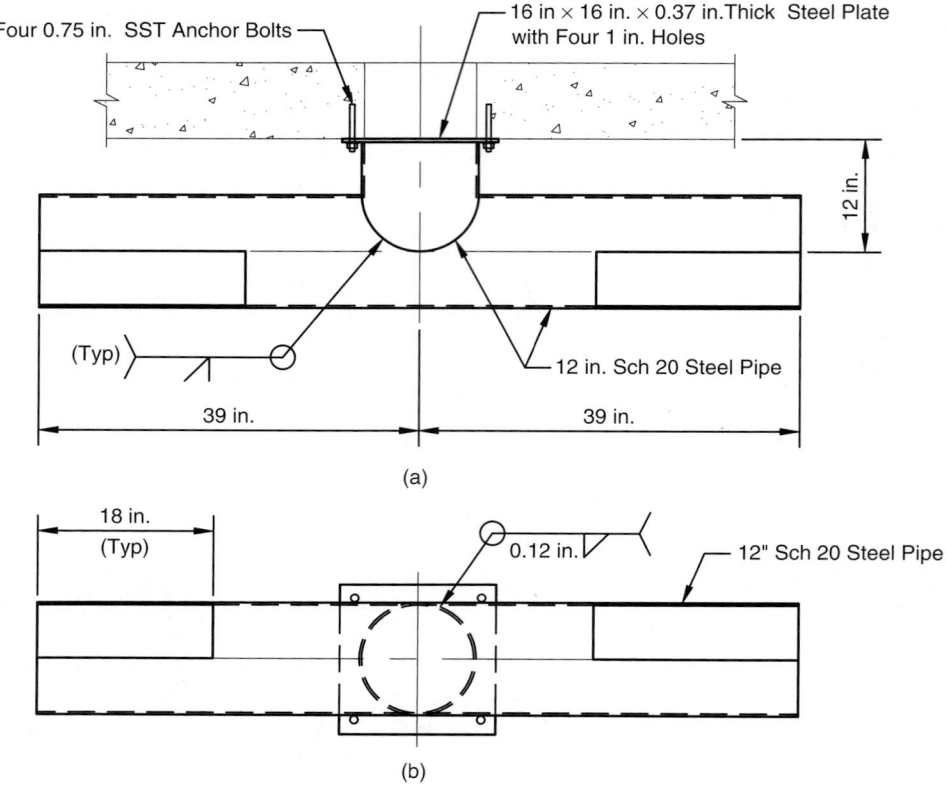

FIGURE 10.9   Inlet diffusers for primary sedimentation tanks at Valencia, California: (a) plan and (b) elevation (in. = 25.40 mm).

2.2.6   Outlet Conditions

Proper clarifier operation also depends on outlet conditions. Effluent should be uniformly withdrawn to prevent localized, high-velocity gradients and short-circuiting. Figure 10.15 illustrates prevailing velocity gradients (drift) in a rectangular clarifier. If these velocity gradients reach the scour velocity, then settled particles can be swept into the tank effluent. Density currents, rather than a high approach velocity, often cause primary sludge carryover of the effluent weirs. Therefore, effluent should be withdrawn from the tank in a manner that minimizes these currents. Typically, the effluent is withdrawn from a sedimentation tank by an overflow weir into a launder or effluent channel. The overflow weir must be leveled to control water surface elevation in the clarifier and promote uniform effluent withdrawal. Weirs may be either straight edged (Figure 10.16) or V-notched (Figure 10.17). V-notched weirs provide better lateral distribution of outlet flows than straight-edged weirs that are imperfectly leveled.

Submerged launders have also been used for effluent withdrawal (Figure 10.18). Collection pipes and launders with submerged orifices are two types of submerged launders. Orifices should be sized for uniform flow distribution. Compared with overflow weirs, submerged launders offer some advantages. Submerged launders avoid free fall of

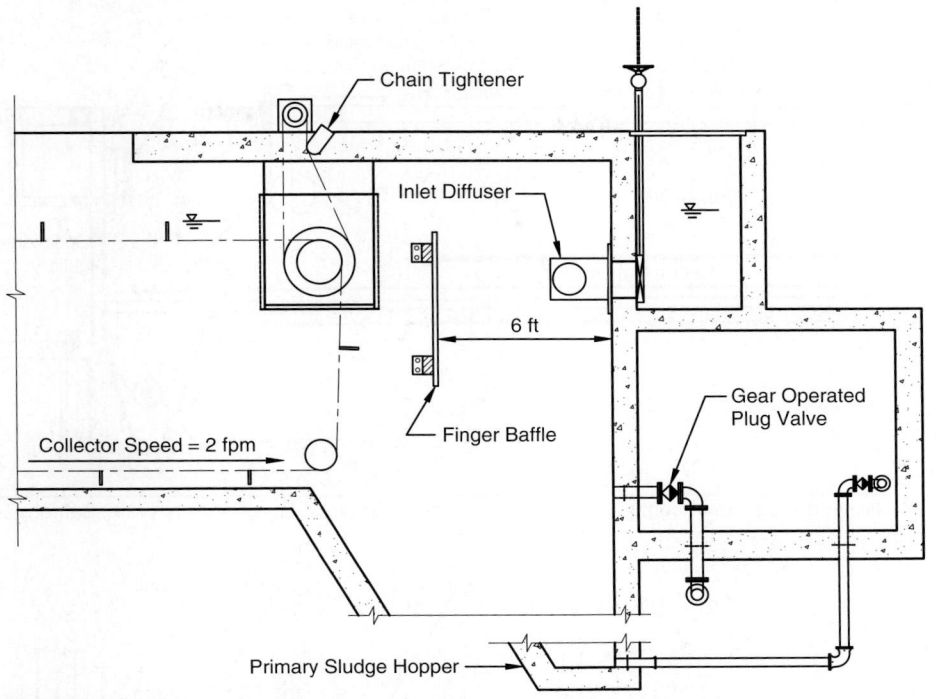

FIGURE 10.10   Inlet configuration of primary sedimentation tanks at Encina plant in Carlsbad, California (ft = 0.30485 m).

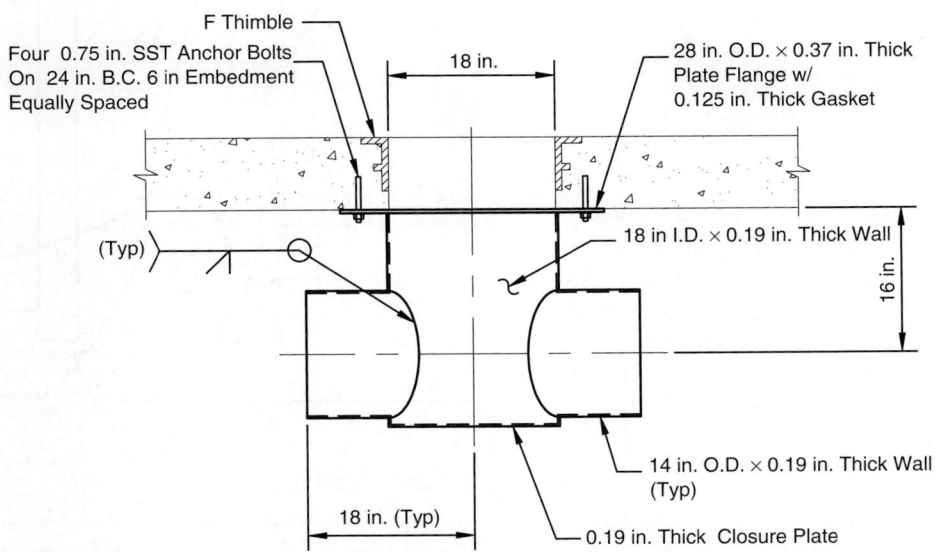

FIGURE 10.11   Inlet diffusers for primary sedimentation tanks at Encina plant in Carlsbad, California (in. = 25.405 mm).

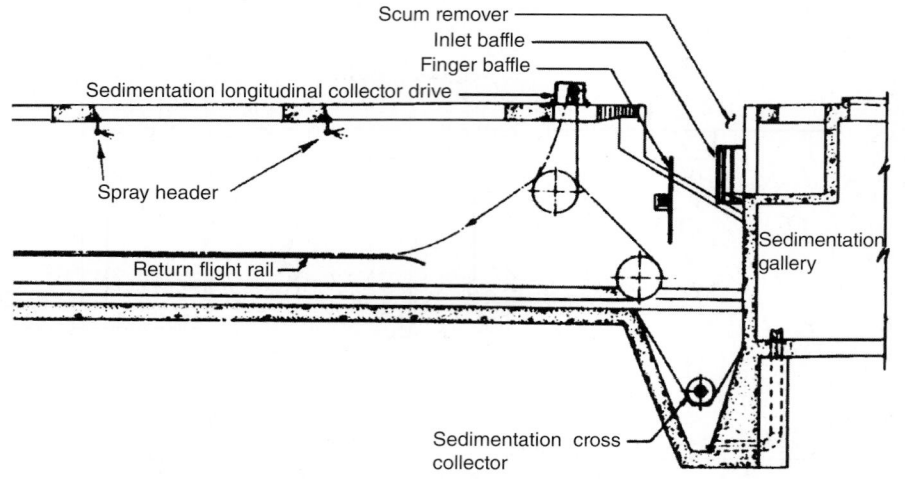

Figure 10.12   Inlet configuration of primary sedimentation tanks at Sunnyvale, California.

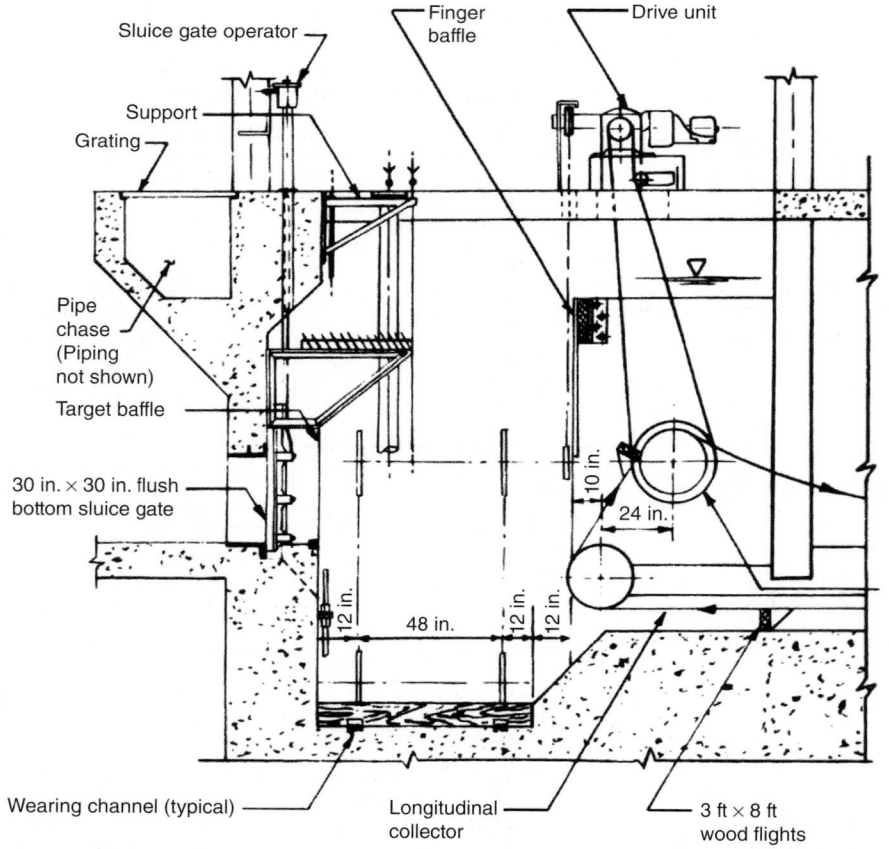

Figure 10.13   Inlet configuration of primary sedimentation tanks at Renton plant in Seattle, Washington (ft = 0.3048 m; in. = 25.405 mm).

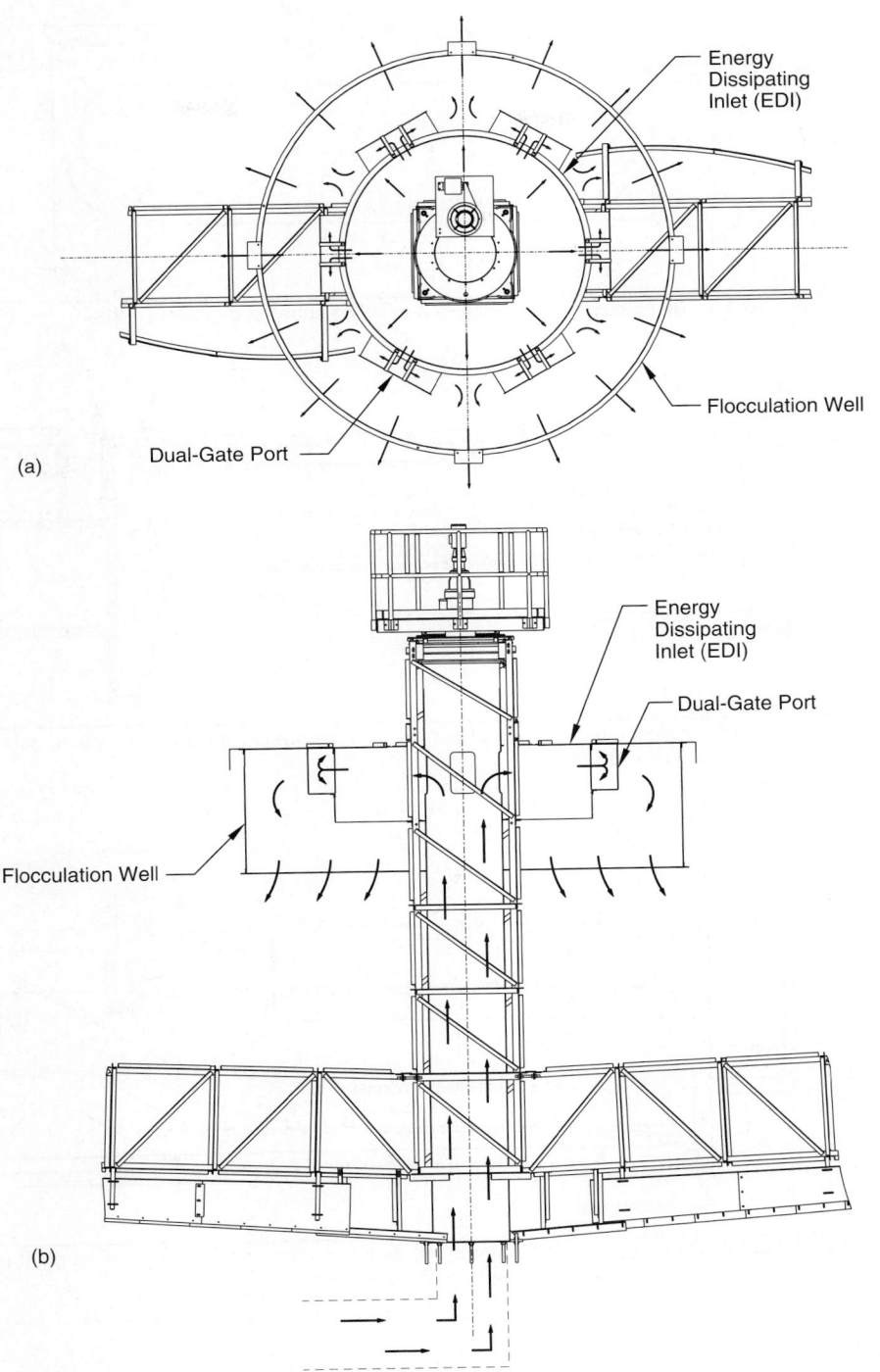

(a)

(b)

FIGURE 10.14   Energy dissipating inlet and flocculation well: (a) plan and (b) section (courtesy of WesTech Engineering, Inc.).

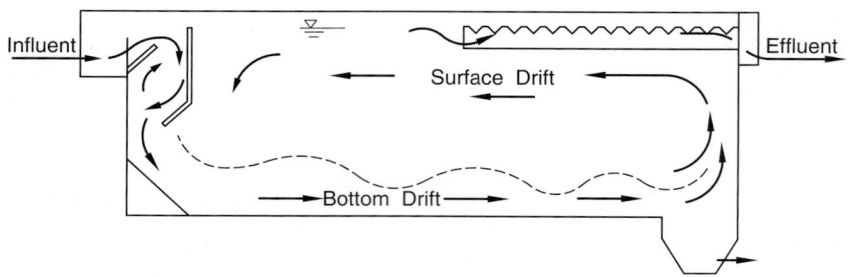

Figure 10.15 Displacement vectors in a real basin showing prevailing drifts.

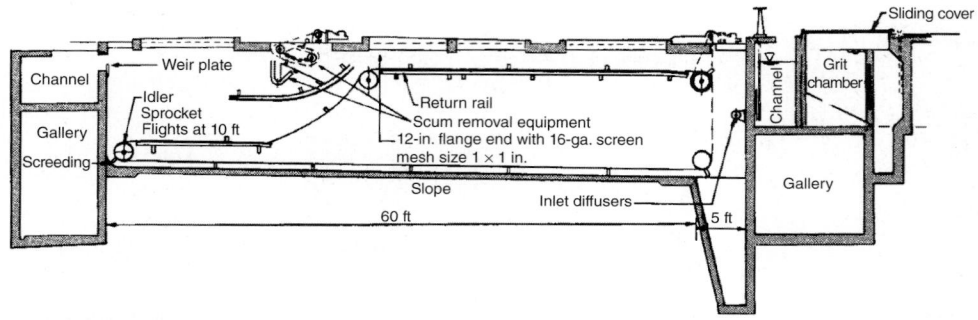

Figure 10.16 Typical cross section of primary sedimentation tanks at Valencia, California (ft = 0.30485 m; in. = 25.405 mm).

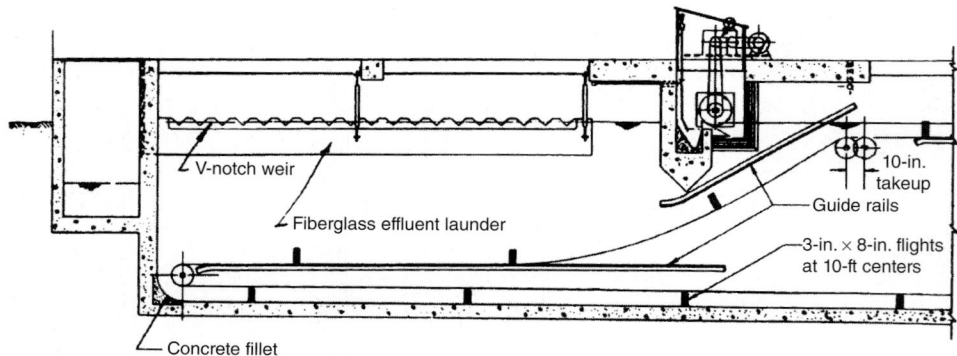

Figure 10.17 Outlet configuration of primary sedimentation tanks at the Encina plant in Carlsbad, California (ft = 0.30485 m; in. = 25.405 mm).

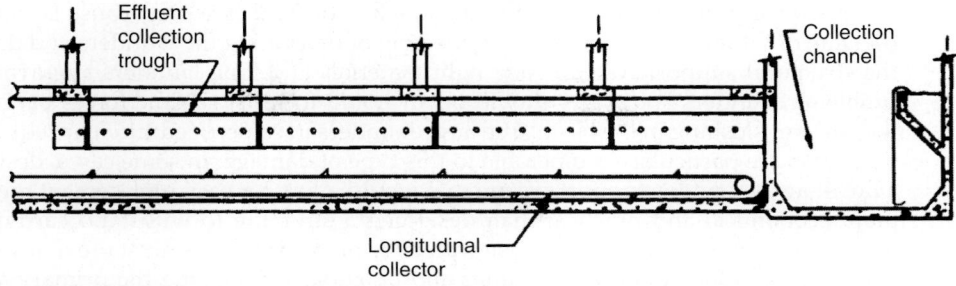

FIGURE 10.18   Outlet configuration of primary sedimentation tanks at the Renton plant in Seattle, Washington.

wastewater and release of entrained odorous gasses and allow surface skimming at the end of the tank. A disadvantage of submerged launders is that orifices sized for uniform flow distribution at average flows will not be effective at peak flows. Also, with submerged launders, downstream hydraulic controls are required to maintain desired water levels in the clarifiers. Unique launders with both submerged orifices for average flows and V-notch weirs for overflow capability are used at the Renton facility in Seattle, Washington (Uhte, 1990). This configuration allows normal flows to pass through the orifices, thus minimizing odor release while allowing peak storm flows to overflow the V-notch weirs.

Weir and launder design include long- and short-launder options. The long-launder approach assumes weir placement and length to be as important as that for secondary clarifier design. Using this approach, weirs and launders for rectangular clarifiers would be designed to cover from 33% to 50% of the basin length (AWWA and ASCE, 2012; Kawamura and Lang, 1986). Long launders control the water elevation in the sedimentation basin within a narrow range. However, they are ineffective when bottom-flowing density currents exist in the basin (Kawamura and Lang, 1986). In cold regions, long launders might not be best because fluctuating water levels with short launders would minimize ice attachment to launders and basin walls (Kawamura and Lang, 1986). The short-launder approach assumes weir length to be unimportant. A simple tank-width weir is used at the end of the tank for outlet control at Valencia, California (Figure 10.16).

Designers' opinions regarding launder spacing also differ. Some designers believe that launders should be spaced 5 to 6 m (16–20 ft) apart (AWWA and ASCE, 2012). A launder spacing of only 2.4 m (8 ft) exists at the San Jose Creek facility in Whittier, California. Launders typically are arranged transversely across the basin with chain-and-flight primary sludge collection equipment. With traveling-bridge primary sludge collection equipment, launders must be arranged longitudinally as parallel finger weirs supported on piers. Parallel finger weirs are used at Sunnyvale, California, with chain-and-flight collection equipment.

Weirs and launders for circular clarifiers typically are mounted on the peripheral wall of the tank. Experience at some facilities has demonstrated that weirs and launders should be placed at least 15% of the basin radius inboard from the periphery of the tank (AWWA and ASCE, 2012). Such placement minimizes wall flow disturbance and draws effluent from a broader area.

Launder stability is an important design consideration. Launders should be designed with provisions to relieve loadings during tank draining and prevent buoyancy uplift during tank filling.

Wave harmonics from the wind, earthquakes, or fluid flow may cause launders to oscillate or vibrate, thereby possibly deflecting or deforming the launders and damage the structural support system. New light materials and long launders aggravate this problem. Launders and weirs should be anchored to resist seismic forces because of wastewater sloshing in the basin. The large launders in center-feed, circular sedimentation tanks are particularly vulnerable to this type of damage. In some cases, designing a break-away launder support system that would allow for easy replacement might be more economical and practical than designing a structure to withstand earthquake-induced loadings from sloshing wastewater. Break-away designs must prevent sections from falling to the bottom of the tanks and potentially damaging the primary sludge collection and removal equipment.

2.2.7 Weir Loading Rate

Weir loading rates have little effect on the performance of primary sedimentation tanks, especially with side water depths exceeding 3.7 m (12 ft) (Graber, 1974). Typical weir loading rates range between 125 and 500 m³/m d (10 000 to 40 000 gpd/ft) (Metcalf and Eddy, 2014). In practice, weir loadings often do not exceed 125 m³/m d (10 000 gpd/ft) in facilities handling 3.8 ML/d (1.0 mgd) or less and 190 m³/m d (15 000 gpd/ft) in facilities handling more than 3.8 ML/d (1.0 mgd) (Merritt, 1983). Existing regulations sometimes govern weir loadings.

2.2.8 Linear Flow-Through Velocity

Currents in the settling zone may hinder settling or result in scouring of settled solids. In practice, the linear flow-through velocity (scour velocity) has been limited to 1.2 to 1.5 m/min (4 to 5 ft/min) to avoid re-suspension of settled solids (Theroux and Betz, 1959). The critical scour velocity may be calculated from the following equation (Camp, 1946):

$$V_H = \sqrt{\frac{8k(s-1)gd}{f}} \tag{10.12}$$

where V_H = critical scour velocity, m/s (ft/s);
   k = constant for type of scoured particles (unitless);
   s = specific gravity of scoured particles;
   g = gravitational acceleration, 9.81 m/s² (32.2 ft/s²);
   d = diameter of scoured particles, m (ft); and
   f = Darcy-Weisbach friction factor (unitless).

Typical k values are about 0.04 for rounded granular material and about 0.06 for non-uniform sticky and flocculent material. Typically, f values range between 0.02 and 0.03. Values for f are a function of the Reynolds number and characteristics of the settled solids surface.

2.2.9 Surface Geometry

Surface geometry has been used to control scouring of solids from high linear flow-through velocities or wind. Although the length-to-width ratio of rectangular tanks has historically been used as such a design tool, it is not considered to be reliable. Width is often controlled by the availability of primary sludge collection and removal equipment. For wider tanks, multiple sludge collection mechanisms can be used.

2.2.10 Weather Conditions

Weather conditions can affect the performance of sedimentation tanks and must be considered in their design (Wells, 1998). Wind may cause the water surface on the leeward side to be higher than on the windward side. Wind-caused turbulence may lead to unbalanced weir rates and short-circuiting (U.S. EPA, 1975). Surface skimmers should be oriented so that prevailing winds will push scum towards the collector. Design considerations for wind mitigation include orientation of tanks, installation of windbreaks or covers, increase of tank freeboard, and reduction of circular tank diameters to 37 m (120 ft) or less. Walls with a height of 1 to 1.2 m (3–4 feet) at the perimeter sometimes are used to provide wind protection for primary sedimentation tanks (Wall and Petersen, 1986).

Cold weather may also require freeze protection of surface sprays, insulation of piping, installation of underground piping at greater depths (below the freeze line), and provisions for auto draining piping that conveys intermittent flow. Because scum-collection equipment is prone to freezing, it needs adequate protection. Occasionally, these steps are insufficient in areas of severe cold and freezing; in which case, sedimentation tank covers may be required to avoid operational problems.

2.2.11 Maintenance Provisions

Two or more clarifiers will allow the process to remain in operation while a tank is out of service for maintenance or repair. The U.S. EPA Design Criteria for Reliability states that there should be a sufficient number of units of a size, such that with the largest-flow-capacity unit out of service, the remaining units shall have a design flow capacity of at least 50% of the total design flow to that unit process (U.S. EPA, 1999). Some states have increased this requirement to 75%.

Primary treatment facilities should include provisions for necessary maintenance. For example, pumps or other equipment located in buildings or vaults require access for maintenance and repair. Access measures include overhead lifting eyes, traveling-bridge cranes, access openings for use of a crane outside the structure, and adequate lighting. Sufficient clear space should be provided around pumps, meters, valves, and other equipment to accommodate maintenance and repair. This is particularly important when replacing the rotor and stator on progressing cavity pumps. Valves should be operable from floor level.

A good design will allow dewatering of clarifiers for servicing primary sludge collection equipment or removing an obstruction from an inlet baffle. Dewatering measures can include sloped bottoms for draining, permanent pumps, and piping or connections for temporary equipment. Also necessary are provisions to isolate a tank that is out of service from the remainder of the facility, which must remain in service. Primary sludge collection and removal mechanisms with their critical components remaining above the waterline merit consideration. Ample flushing ports and cleanouts are needed at critical tees, elbows, and ends.

Routine maintenance needs should be taken into account. Access to inlet distribution boxes or channels must be possible to remove scum and prevent floating material from affecting flow distribution devices. Sluice gates should be downward opening wherever possible to avoid buildup of scum on the water surface and prevent deposition of solids in the track, which impedes full closure of the gate. Corner pockets and dead ends should be eliminated when possible to minimize the potential for septic conditions; corner fillets and channeling should be used where necessary. In practice, the tops of submerged troughs, beams, and other construction features often have been sloped 1.4:1 and their bottoms have been sloped 1:1; this reduces or prevents accumulation of solids and scum

(Great Lakes-Upper Mississippi River Board of State and Provincial Public Health and Environmental Managers, 2014). Provisions should be made for cleanup after maintenance such as, for example, frequent use of hose bibs and sump pumps. Hose bibs should be provided at an approximate spacing of 15 meters (50 feet) at each tank, scum trough, sump, and pumping station.

2.2.12   Extreme Flow (Wet-Weather) Considerations

The effect and anticipated frequency and duration of extreme conditions—high and low flows, such as peak storm flows with recycle flows and tanks out of service—on sedimentation tank performance should be evaluated during design to verify that operating parameters are acceptable. The design of primary clarifiers should be flexible enough to allow successful operation during low-flow start-up conditions and high-flow conditions.

Wet-weather flows depend on the location, intensity, and duration of rainfall and characteristics of the sewer system. Therefore, wet-weather flows are more difficult to predict than dry weather flows. Substantial infiltration or inflow from sanitary sewers or the existence of combined storm and sanitary sewers might result in wet-weather flows that are several times higher than normal dry weather flows. As communities implement aggressive programs to control combined sewer overflows (CSOs) and sanitary sewer overflows (SSOs), such wet-weather flows, now routed to the treatment facility, may result in very high peak flow to average flow ratios for treatment processes. In some cases, additional or enhanced sedimentation facilities have been used to treat excessive wet weather flows at peak flow to average flow ratios greater than 5:1 (Fitzpatrick et al., 2008). Provisions for emptying and flushing tanks after storm events should be considered for facilities with additional tanks (Leffler and Harrington, 2001).

Recycle flows, such as waste activated sludge or trickling filter underflow may cause surges in flow and should be considered in sizing primary clarifiers. These surges should be avoided if possible or returned to the facility influent stream during low-flow periods. Influent pumps, typically variable speed or multiple constant speed, should be designed to provide a smooth gradual transition of flow to primary clarifiers.

2.2.13   Odor Control Considerations

Odor control is a significant design consideration for primary treatment. Odors emanate from raw wastewater with high hydrogen sulfide concentrations and anaerobic conditions. The release of hydrogen sulfide and other similarly noxious odors occurs in areas of high turbulence such as sedimentation tank inlets, effluent launders, and weirs. Turbulence increases the water surface area for gas transfer with hydrogen sulfide release to the atmosphere. Scum and settled primary sludge handling systems can also be significant sources of odors. Available odor control strategies include source control, chemical treatment, pre-aeration, and containment (covers). Chapter 6 includes additional information on the design of odor control facilities.

2.2.14   Corrosion Control Considerations

Important design considerations include corrosion control of concrete and metallic surfaces such as those of inlet and outlet structures, structural members, gratings, covers, walkways, and equipment. Most corrosion problems are associated with hydrogen sulfide gas that converts to sulfuric acid on moist surfaces. Odor containment may result in high hydrogen sulfide concentrations and humidity in contained areas.

Reducing the production or release of hydrogen sulfide gas from wastewater can minimize corrosion of primary treatment tanks, screens, and equipment. Chapter 8 contains information on corrosion control and material selection.

2.2.15 Design Considerations for Stacked Clarifiers

Typically, the design of stacked primary clarifiers is similar to conventional primary sedimentation tanks (Metcalf and Eddy, 2014). Ducoste et al. (1999) showed that CFD could be used to quantitatively evaluate the performance of different stacked clarifier designs for the 800 ML/d (211 mgd) Changi WRRF in Singapore.

The inlet and outlet design is considered a weakness of stacked tanks because wastewater flow patterns might intersect with those of the primary sludge (Lager and Locke, 1990). The lower trays of stacked clarifiers are confined spaces subject to confined space entry requirements.

Facilities with stacked clarifiers range in size from 95 to 400 ML/d (25–100 mgd), with average overflow rates between 15 and 43 m^3/m^2 d (370–1100 gpd/sq ft) and weir loading rates between 84 and 170 m^3/m d (6800–14 000 gpd/ft) (Kelly, 1988).

In Mamaroneck, New York, stacked primary clarifiers are designed for peak influent flows of 350 ML/d (92 mgd) (Kelly, 1988). The design overflow rates are 22 m^3/m^2 d (550 gpd/sq ft) at average flow and 45 m^3/m^2 d (1100 gpd/sq ft) at peak flow (Kelly, 1988).

3.0 Enhanced Sedimentation

Primary sedimentation can be enhanced by preaeration or chemical coagulation and flocculation, each of which is discussed in the following subsections.

3.1 Preaeration

3.1.1 Description

Preaeration of raw wastewater before primary clarification increases the settling parameter (λ) of the wastewater by promoting flocculation of finely divided solids into more readily settleable flocs, thereby increasing suspended solids and BOD removal efficiencies. Other benefits of preaeration include scum flotation improvement, scrubbing of volatile organic chemical (VOC) odor components, and prevention or mitigation of septicity.

3.1.2 Design Considerations

Detention times of 20 to 30 minutes are necessary for floc formation and improved TSS, COD, and BOD removals. This range exceeds the range of 10 to 15 minutes suggested for odor control. The exact quantity of air required is a function of wastewater characteristics and tank geometry. The minimum air supply is typically 0.82 m^3/m^3 (0.11 cu ft/gal).

3.2 Chemically Enhanced Primary Treatment

3.2.1 Description

Chemical coagulation and flocculation of raw wastewater promote clumping and agglomeration of finely divided solids into more readily settleable flocs, thereby increasing TSS, COD, BOD, and phosphorus removal efficiencies.

CEPT decreases TSS_{non} and increases the settling parameter (λ) and allows a higher SOR than conventional primary sedimentation. Use of CEPT minimizes the required process footprint by reducing the size of primary clarifiers and aeration basins. It can efficiently remove 60% to 90% of TSS, 40% to 70% of COD or BOD, 70% to 90% of phosphorus, and 80% to 90% of bacteria loadings (Ødegaard, 2005; Metcalf and Eddy, 2014; U.S. EPA, 1987). In comparison, conventional primary sedimentation may remove only 50% to 70% of TSS, 25% to 40% of COD or BOD, 5% to 10% of phosphorus loadings, and 50% to 60% of bacteria loading (Metcalf and Eddy, 2014; McGhee and Steel, 2007; U.S. EPA, 1987). CEPT can also capture and remove heavy metals in the influent that could adversely affect the performance of downstream biological processes and effluent quality (Johnson et al., 2008). This treatment is ideally suited for locations where ocean discharge is acceptable or with abrupt seasonal load increases (Rogalla et al., 2007). Use of CEPT is especially attractive in developing countries where construction costs are a limiting factor for the expansion of WRRFs (Harleman and Murcott 2001a, 2001b; Jordão and Figueiredo, 2005). When followed by high-rate filtration with synthetic compressible media for effluent polishing, CEPT can achieve less than 2 NTU and remove 84% of TSS and 47% of total COD (Jimenez et al., 1999). The use of CEPT for the treatment of wet-weather flows may be limited to peak-weather flow periods (Krugel et al., 2005).

Disadvantages of CEPT include an increased mass of primary sludge, production of solids that may be more difficult to thicken and dewater (particularly if lime or aluminum salts are used as coagulants), and an increase in operational cost and operator attention. Depending upon downstream processes, there may be other site-specific disadvantages to CEPT. These include the removal of too much alkalinity, which may inhibit nitrification or too much phosphorus, which may result in nutrient deficiencies in biological processes. Additional downstream process considerations are discussed in Section 8.0 of this chapter.

Removal efficiencies and coagulant doses for various CEPT facilities are listed in Tables 10.3 and 10.4. Research in Sarnia and Windsor, Ontario, indicated that with CEPT, SORs up to 98 m^3/m^2 d (2400 gpd/ sq ft) did not significantly affect effluent quality (Heinke et al., 1980). Full-scale trials in King County, Washington, demonstrated that CEPT may provide satisfactory results to support a split treatment strategy (i.e., wet-weather blending) at SORs up to 204 m^3/m^2 d (5000 gpd/ sq ft) (Krugel et al., 2005). However, a design SOR of 147 $m^3/m^2 \cdot d$ (3600 gpd/ sq ft) was selected for this facility (Melcer et al., 2014).

3.2.2 Design Considerations

3.2.2.1 Chemical Coagulants   The characteristics of wastewater including turbidity, TSS, COD or BOD, and particle weight distribution can have a significant effect on the non-settleable TSS (TSS_{non}) concentration (Narayanan et al., 2000; Neupane et al., 2006, 2008). Coagulant addition was found to lower the TSS_{non} concentration of wastewater, thereby improving overall removal efficiencies (see eqs 10.2 and 10.3) (Neupane et al., 2006, 2008).

Historically, iron salts, aluminum salts, and lime are the chemical coagulants used for wastewater treatment. Iron salts typically are the most common coagulant used for primary treatment. A potential advantage when using iron salts is a reduction in anaerobic digester gas hydrogen sulfide level and a reduction in hydrogen sulfide release at the primary clarifiers. Only a few facilities use lime as a coagulant because it is pH dependent and requires significant dosages to achieve the required pH. Lime also produces more primary sludge than metals salts and is more difficult to store, handle, and feed. To enhance

TABLE 10.3 Performance and Coagulant Dosages for Enhanced Sedimentation (Harleman and Morrissey, 1990) (with permission from the American Society of Civil Engineers)

Location	Flow (mgd)	Advanced Primary Performance						Chemical Addition		
		BOD			TSS					
		Influent (mg/L)	Effluent (mg/L)	Removed (%)	Influent (mg/L)	Effluent (mg/L)	Removed (%)	Type	Concentration (mg/L)	Duration
Point Loma								$FeCl_3$	35	Continuous
City of San Diego	191	276	119	56.9	305	60	80.3	Anionic Polymer	0.26	
Orange County								$FeCl_3$	20	8 hours
Plant No. 1[b]	60	263	162	38.4	229	81	64.6	Anionic Polymer	0.25	peak flow
Orange County								$FeCl_3$	30	12 hours
Plant No. 2[b]	184	248	134	46.0	232	71	69.4	Anionic Polymer	0.14	peak flow
JWPCP										
Los Angeles County[a]	380	365	210	42.5	475	105	77.9	Anionic Polymer	0.15	Continuous
Hyperion								$FeCl_3$	20	
City of Los Angeles[a]	370	300	145	51.7	270	45	83.3	Anionic Polymer	0.25	Continuous
Sarnia								$FeCl_3$	17	
Ontario, Canada	10	98	49	50.0	124	25	79.8	Anionic Polymer	0.3	Continuous

[a] Advanced primary treatment has since been replaced by full secondary treatment.

[b] Full secondary treatment is scheduled to replace advanced primary treatment in 2012.

Note: TSS = total suspended solids; BOD = biochemical oxygen demand.

Pollutant	In (mg/L)	Out (mg/L)	Reduction (%)
SS	233 ± 186	16.6 ± 9.6	92.9
BOD₇	167 ± 95	27.2 ± 12.7	83.7
Phosphorus	5.24 ± 2.53	0.26 ± 0.16	95.0

Note: SS = suspended solids; BOD_7 = seven-day biochemical oxygen demand.

TABLE 10.4 Results of Direct Precipitation Representing the Average of 87 Norwegian Water Resource Recovery Facilities in 1985 (Ødegaard, 1992; Reprinted from Water Science and Technology, with permission from the Copyright Holders, IWA)

sedimentation, some facilities use aluminum salts (alum). However, aluminum as a coagulant inhibits the specific methanogenic activity of bacteria needed for anaerobic digestion by approximately 50% to 70% (Noyola and Tinajero, 2005).

Coagulant selection for CEPT should be based on performance, reliability, and cost. Performance evaluation should use jar tests of the actual wastewater to determine dosages and effectiveness (Hetherington et al., 1999; Jordão and Fiogueiredo, 2005; Mills et al., 2006; Wahlberg et al., 1999). Full-scale testing should also be considered to verify bench-scale jar tests (Carter et al., 2003; Gerges et al., 2006). Coagulant dosages should be assessed under low surface-overflow rates (Peric et al., 2006). Operating experience, cost, and other relevant information drawn from other facilities should be considered during selection.

The designer of CEPT facilities should consider the effect of the process on downstream biological and solids-handling facilities. Although the additional primary sludge production is partially offset by a decrease in waste-activated-sludge production, additional solids handling equipment typically is required. At Metropolitan WRRF in St. Paul, Minnesota, a clinking problem in the multiple-hearth incinerators has been attributed to metal salt addition (WERF, 1999).

3.2.2.2 Chemical Flocculants Anionic polymers sometimes are added during the flocculation step to promote floc formation. Anionic polymers should be added as a dilute solution to ensure thorough dispersion of polymers throughout the wastewater. Evaluation of 30 different polymers at the Joint Water Pollution Control Plant (JWPCP) at Carson, California, found anionic polymers to be the most effective for CEPT (Parkhurst et al., 1976). Polymer addition (0.15 mg/L) at JWPCP increased suspended solids capture from 66% to 83%. Turbulence in the aerated headworks inlet channel is often used to provide mixing for flocculation, although this may not be optimal. Jar and full-scale testing are also recommended for selecting the flocculent and determining required dosages. Flocculent dosages should be assessed under high SORs (Peric et al., 2006). While the addition of chemical coagulants could be based on TSS load, chemical flocculants should be based on flow (Neupane et al., 2006; 2008). The use of a chemical flocculant typically helps compensate for short flocculation detention times and strengthens chemical floc, which makes the floc more resistant to mechanical disruption during treatment.

3.2.2.3 Rapid Mix During rapid mix, the first step of the coagulation process, chemical coagulants are mixed with raw wastewater. The coagulants destabilize the colloidal particles by reducing the forces (zeta potential) that keep the particles apart, which allows their agglomeration. The destabilization process occurs within seconds of coagulant addition. At the point of chemical addition, intense mixing will ensure uniform dispersion of the coagulant throughout the raw wastewater. The intensity and duration of mixing must be controlled to avoid overmixing or undermixing. Overmixing may reduce removal efficiency by breaking up existing wastewater solids and newly formed floc. Undermixing inadequately disperses the chemical, increases chemical use, and reduces removal efficiency.

The velocity gradient, G, is a measure of mixing intensity. Velocity gradients of $300 \ s^{-1}$ typically are sufficient for rapid mix, but some designers have recommended velocity gradients as high as $1000 \ s^{-1}$ (Hudson, 1981; Kawamura, 1976; Sanks, 1981). Formulae for calculating the velocity gradient for various mixer configurations are presented in other references (Camp, 1955; U.S. EPA, 1975, 1987). Rapid mix intensity had no appreciable effect on TSS_{non} concentration in one study (Neupane et al., 2006; 2008).

The optimal point for coagulant addition is as far upstream as possible from primary sedimentation tanks. If possible, several different feed points should be considered for additional flexibility. Dispersing the coagulant throughout the wastewater is essential to minimize coagulant dosage and concrete and metal corrosion (Soap and Detergent Association, 1989). To promote dispersion, multiple injection points or a chemical solution header (Figure 10.19) can be used. Flow metering devices should be installed on chemical feed lines for dosage control.

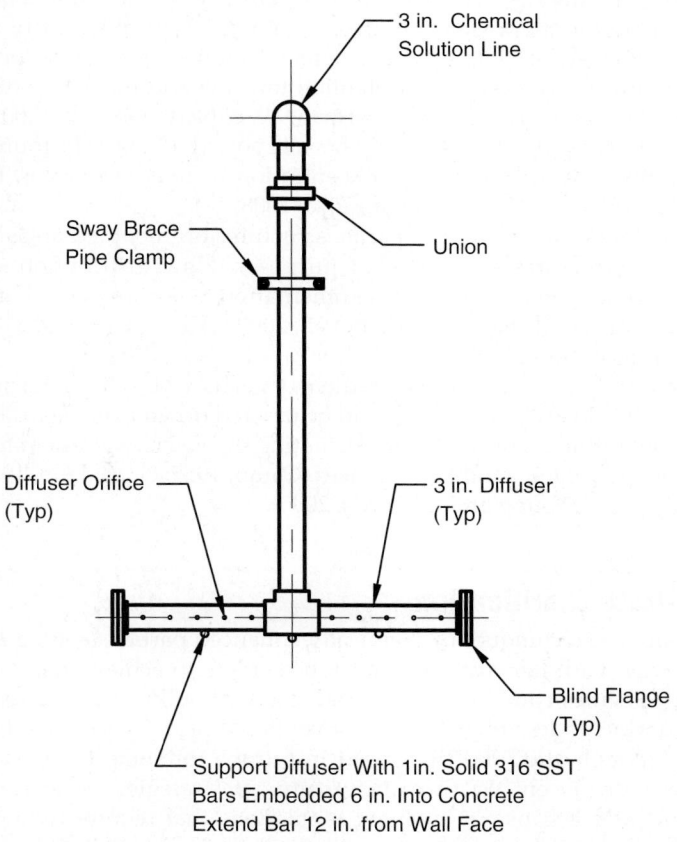

FIGURE 10.19 Chemical solution diffuser (in. = 25.405 mm).

Rapid mixing can be accomplished by mechanical mixers, inline blenders, pumps, baffled compartments, baffled pipes, or air mixers (Klute, 1985). The mixing intensity of mechanical mixers and inline blenders is independent of flow rate, but these mixers cost considerably more than others and might become clogged or entangled with debris. Air mixing eliminates the problem of debris and can offer advantages for primary sedimentation, especially if aerated channels or grit chambers already exist. Pumps, Parshall flumes, flow distribution structures, baffled compartments, or baffled pipes are methods often used to upgrade existing facilities. They offer a lower-cost but less-efficient alternative to separate mixers for new construction. These methods are less efficient than separate mixers because, unlike separate mixing, mixing intensity depends on flow rate. For wet weather applications, however, these lower-cost methods may be the most suitable because CEPT will be used under high flow conditions when hydraulic turbulence is high. Chapter 14 contains additional information on the design of rapid mix facilities.

3.2.2.4 Flocculation   During the flocculation step of the coagulation and flocculation process, destabilized particles grow and agglomerate to form large, settleable flocs. Through gentle prolonged mixing, chemical bridging or physical enmeshment of particles, or both, occur. Flocculation is slower and more dependent on time and agitation than the rapid mix step. The benefits of flocculation are dependent on the percentage of organic matter in the wastewater that is colloidal. Typically, 40% of the soluble COD is colloidal (Foess et al., 2003). Typical detention times for flocculation range between 20 and 30 minutes. Increasing the detention time beyond this range offers only marginal benefits (Andreu-Villegas and Letterman, 1976; Neupane et al., 2006; 2008). Detention times as short as five minutes have been reported. One study found that flocculation significantly reduces the TSS_{non} concentration of the wastewater, thereby improving CEPT performance (Neupane et al., 2006; 2008).

Flocculation can occur in separate structures or in baffled areas of channels, tanks, or existing structures serving other purposes. Flow distribution structures, influent wells, and inlet zones of primary sedimentation tanks are areas that promote flocculation and avoid floc breakup (Parker et al., 2000). Advantages and disadvantages of different configurations resemble those for rapid mix facilities.

Like rapid mix, the velocity gradient, G, achieved with each configuration should be checked. Velocity gradients should be tapered down from 50 to 80 s^{-1} to lower levels for optimal settling. Formulae for calculating the velocity gradient for various configurations are presented in other references (Camp, 1955; Grohmann, 1985; Moll, 1985; U.S. EPA, 1975; 1987; Young and Edwards, 2000).

4.0 High-Rate Clarification

HRC combines techniques of chemically enhanced particle settling and solids contact/recirculation with lamella plate and tube settlers to achieve rapid settling. Very high settling velocities combined with rapid flocculation kinetics can reduce required process footprints to less than 10% of conventional primary treatment (Jolis and Ahmad, 2001). Although HRC has been used in Europe for some time, its use in the United States is more recent. Wet-weather treatment strategies have been implemented at many facilities because of evolving regulations that require reductions in CSOs and SSOs and some degree of treatment for all flow received at a facility. Wet-weather facilities include storage basins that operate as flow-through clarifiers once full. After the storm event, these basins are emptied, cleaned, and readied for the next event. These basins or wet-weather clarifiers employ some combination of chemical coagulation, enhanced flocculation (ballast), and plate or tube settlers to increase allowable hydraulic loading rates. HRC units are compact, have short start-up times, and produce high-quality effluent. HRC facilities can be located at the WRRF or at remote satellite facilities that accept peak-flow diversions from the main facility; however, the operator attention required at startup may be a disadvantage for satellite facilities that typically are unstaffed. Another disadvantage of satellite facilities is the management of waste solids. Wet-weather HRC facilities in Lawrence, Kansas, and Toledo, Ohio, have been used in parallel with activated sludge-facilities to achieve secondary treatment standards while treating peak flows exceeding five times their annual average flows (Fitzpatrick et al., 2008).

4.1 Plate or Tube Settlers

Plate or tube settlers may be used to improve clarification with or without chemical addition. Plate or tube settlers increase the capacity of existing clarifiers by increasing the available settling area. Figure 10.20 gives the various settler flow patterns that have been used. Chemical coagulation with recycled sludge (dense sludge process) or floc-weighting agents (ballasted flocculation) and tubes or plates allow greater SORs. Although decreasing angle and spacing of the plate or tube will increase the projected total area, flattening the tube or plate too much or restricting the space between tubes or plates will hinder the movement of settled solids. Fine screening before plate and tube settlers allows a closer spacing without plugging, but the spacing should not be so close as to cause high velocities. Effective grit and grease removal is required before the plates (Reardon, 2005). Fischerstrom (1955) recommended a minimum spacing such that flow between plates or in tubes has a Reynolds number less than 2000, a Froude number greater than 10^{25}, and a detention time greater than three to five minutes.

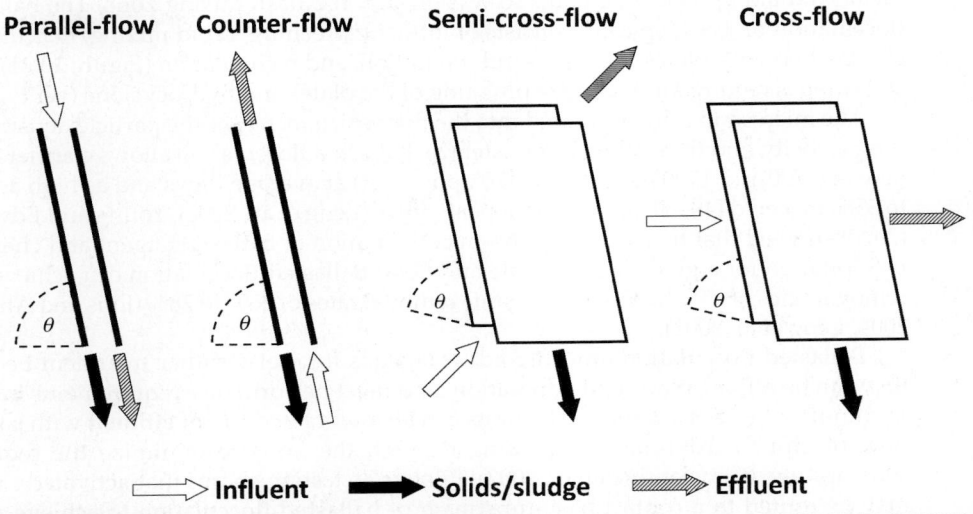

FIGURE 10.20 Plate settler flow patterns (Baur et al., 2000).

Parameter	Equation	
Lamella velocity:	$$V_l = \dfrac{Q}{A_{tp}}$$	(10.13)
Total projected area:	$$A_{tp} = n \cdot a \cdot b \cdot \cos u$$	(10.14)
Reynolds number:	$$N_{RE} = \dfrac{V \cdot R}{v}$$	(10.15)
Froude number:	$$N_{FR} = \dfrac{V^2}{Rg}$$	(10.16)

where V_l = lamella (Hazen) velocity, m/h
  Q = influent flow, m³/h
  A_{tp} = total projected surface area, m²
  n = number of inclined plates, dimensionless
  a = length of a single plate, m
  b = width of a single plate, m
  u = angle of inclination of plates from the horizontal plane, degree
  N_{RE} = Reynolds number, dimensionless
  V = velocity of the water between plates, m/s
  R = hydraulic radius = cross-sectional area/wetted area of plate, m
  v = kinematic viscosity, m²/s
  N_{FR} = Froude number, dimensionless
  g = gravitation constant, m/s²

4.2 Ballasted Flocculation

Ballasted flocculation is a generic term for an HRC that adds a ballasting agent such as microsand along with coagulants to wastewater in a flash mixing zone. The ballasted flocculation process typically consists of influent screening, rapid mixing, flocculation, clarification with plates or tubes, sand separation, and recirculation (Figure 10.21). Fine screening is required to minimize plugging of the plates and hydrocyclone (WEF, 2005).

The microsand is incorporated into the floc which increases the particle density, settling velocity, and the settling parameter (λ). Ballasted flocculation allows clarifier SORs between 700 and 1200 m³/m² d (12–20 gpm/sq ft) at average flows and as high as 2300 to 3500 m³/m² d (40–60 gpm/sq ft) at peak flows (Leng et al., 2002). Young and Edwards (2000) showed that there is an optimum combination of ballasting agent and chemical precipitation for a given settling time and SOR. Ballasted flocculation can achieve TSS removal rates of 75% to 90% and organic removal rates of 58% to 78% (Jolis and Ahmad, 2004; Leng et al, 2002).

Ballasted flocculation units are advantageous for wet-weather treatment because they can be offline, placed into operation, and meet performance requirements in 10 to 15 minutes (Leng et al., 2002). The units can be maintained full of effluent with a small flow of effluent continuously passing through the units to minimize the required start-up time (Constantine et al., 2003). Pilot-scale testing shows that activated sludge may be routed to a contact basin upstream of ballasted flocculation to achieve rapid

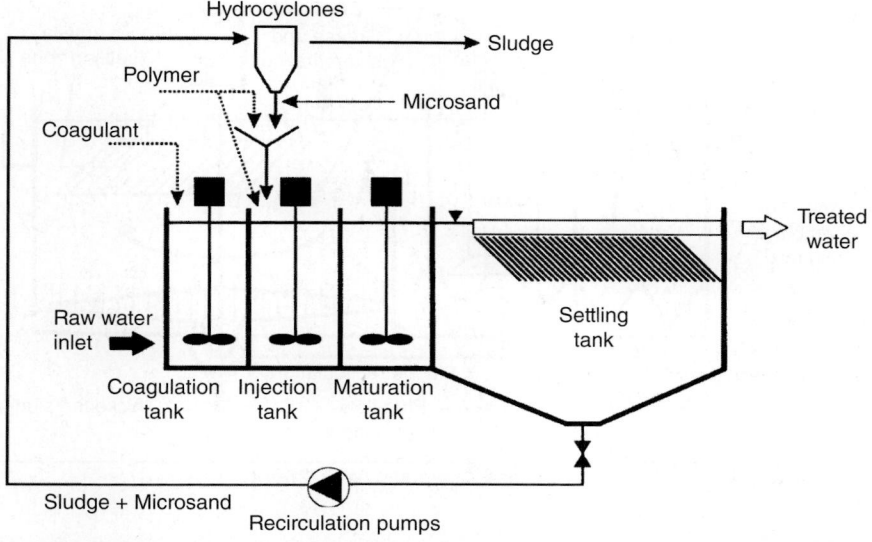

Figure 10.21 Ballasted flocculation process schematic (WEF, 2005).

uptake of soluble organic matter into the biomass (Siczka et al., 2007; Sun et al., 2008). This configuration was successfully implemented in a full-scale wet-weather treatment application (Katehis, 2011). This treatment concept is a variation of the contact-stabilization activated sludge process with ballasted flocculation acting as a high-rate secondary clarifier and the main facility's aeration basin acting as the stabilization basin. Chapter 12 provides further information regarding the contact-stabilization activated sludge process.

Special considerations for freeze protection include enclosing the units in a heated building or partially draining and refilling tanks with warmer facility effluent as required to maintain water temperature (Jolis and Ahmad, 2001).

4.3 Solids Contact/Sludge Recirculation

The solids contact/sludge recirculation or dense sludge process consists of screening, rapid mix, and flocculation followed by clarification (Figure 10.22). Sludge is thickened and recirculated to the influent end of the process where it is combined with a coagulant before the rapid mix zone. The polymer is added in the flocculation zone. The recirculation of the sludge increases the number of particles in the water, which increases particle density, settling velocity, and settling parameter (λ). Fine screening is required to minimize plugging of the tubes in the settling zone (WEF, 2005). Dense sludge units are advantageous for wet-weather treatment because they can be offline, placed into operation, and meet performance requirements in 20 to 30 minutes (WEF, 2005). Removal rates between 70% and 95% for TSS, 55% and 70% for carbonaceous BOD, and 70% and 95% for total phosphorus have been achieved at the Bay View WRRF in Toledo, Ohio, with the dense sludge process at SORs up to 2750 m^3/m^2 d (47 gpm/sq ft) (Stevenson et al., 2008).

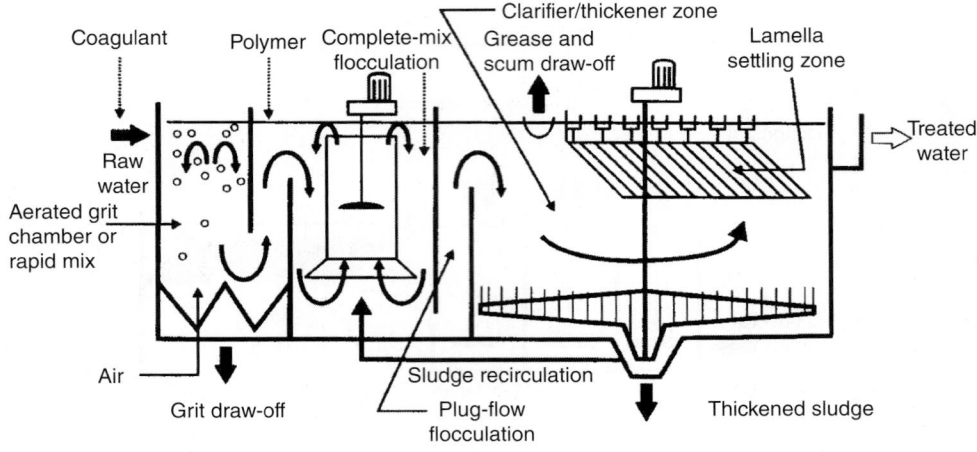

FIGURE 10.22   Dense sludge process schematic (WEF, 2005).

5.0   Emerging Primary Treatment Technologies

5.1   Introduction

Several technologies have recently emerged to provide a higher degree of primary treatment, reduced footprint, and decreased operational and maintenance requirements compared to conventional primary clarifiers. These emerging primary treatment technologies may also offer enhanced carbon diversion (i.e., increased removal of COD before biological treatment and routing of the COD to anaerobic digestion facilities) at WRRFs, resulting in additional energy and capital cost savings from decreased aeration energy requirements in secondary treatment, increased digester biogas production, and reduced sizing of secondary treatment facilities (or increased secondary treatment capacity of existing facilities). These emerging technologies can be grouped into the following general categories: (1) primary effluent filtration, (2) primary filtration, (3) rotating belt filters/microscreens, and (4) dissolved air floatation clarifier (integrated with biological treatment). Engineers need to take into account the specific design considerations when integrating these emerging technologies in existing or new WRRFs.

5.2   Primary Effluent Filtration

Testing of filtration technologies for primary effluent started in the 1980s, but the granular medium filtration technologies did not prove to be feasible for this application. Since about the year 2000, a number of new wastewater filtration technologies, such as Cloth Depth Filters (CDF) and Compressible Media Filters (CMF), have emerged as popular solids removal processes within the wastewater treatment industry. These technologies have been implemented successfully at hundreds of WRRFs as reliable tertiary filtration technologies to serve as a final treatment step before treated effluent is discharged to receiving water bodies or used as recycled water. These filtration technologies are now used to provide enhanced primary treatment because of the following advantageous attributes (i.e., compared to conventional granular medium filters):

(1) higher hydraulic and solids loading capacities, (2) lower backwash reject water (BRW) ratio, and (3) effective cleaning mechanisms for difficult wastewaters such as primary influent or effluent (Caliskaner, et al., 2015; Caliskaner, et al., 2014).

The effluent from the primary clarifier is filtered before secondary treatment. Reducing the primary clarifier effluent organic load will result in increased secondary treatment capacity (or decreased secondary treatment facilities requirements) and decreased aeration energy requirements. The filter BRW containing the solids captured from the primary effluent is diverted to the anaerobic digester after thickening. The high energy content of the captured volatile suspended solids (VSS) in filter BRW will increase biogas production in the anaerobic digester. A two-year demonstration project conducted at the Linda County WRRF (Olivehurst, CA) using both the CDF and CMF technologies proved primary effluent filtration to be reliable and efficient in reducing organic and TSS loads on the downstream process (Caliskaner et al., 2015), as compared to primary clarification alone. Primary clarifier and filtered effluent TSS values are shown in Figure 10.23. As seen in Figure 10.23, the load to the secondary treatment process was reduced and stabilized with primary effluent filtration.

Overall removal performance results for a number of constituents are presented in Table 10.5. The hydraulic performance of both filter technologies was reliable and appropriate considering the TSS loading rates were significantly higher compared to those typically seen in the filtration of secondary effluent (e.g., by a factor of 5–50). Both CMF and CDF operated at 70% to 100% of the filtration rates typically used in tertiary filtration applications. The average BRW ratio was observed to be only 5% to 10% (i.e., filtered water production rate between 90% and 95% of filter influent flow).

5.2.1   Cloth Depth Filtration Design Considerations in Primary Effluent Filtration

Although filtration mechanisms and operational principles are similar, deployment of CDF for primary treatment requires certain design and manufacturing modifications, compared to systems used in tertiary treatment applications. The main system modifications are for handling higher influent TSS loading rates and floatable material. Tertiary CDF systems design can still be used for primary effluent filtration (as in the

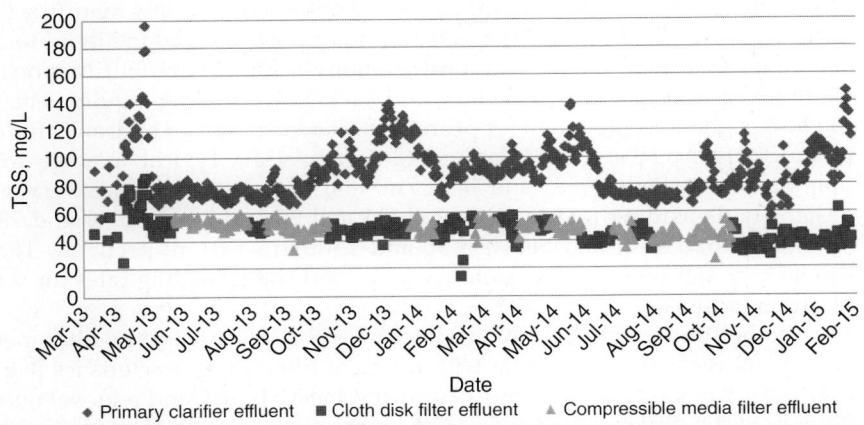

FIGURE 10.23   Primary effluent filtration TSS removal performance (Caliskaner et.al., 2015).

Constituent	Primary Effluent (mg/L)		Filter Effluent (mg/L)		Typical Removal Efficiency (%)
	Range	Average	Range	Average	
TSS	60–240	120	40–170	60	45
VSS	60–230	110	40–150	50	45
COD	260–580	390	180–400	260	30
Soluble COD	110–150	130	110–180	130	0
BOD	120–260	180	80–180	130	30
Soluble BOD	70–95	75	55–95	70	0–5
CBOD	140–190	160	100–130	120	25
Nitrate	<1	<1	<1	<1	0
Nitrite	<1	<1	<1	<1	0
Ammonia	25–35	32	25–30	30	0
Fats, Oil, Grease	ND	ND	ND	ND	ND

TABLE 10.5   Primary Effluent Filtration System Concentration Ranges and Average Removal Performances for CDF and CMF Technologies (Caliskaner et al., 2015)

two-year Linda County demonstration project discussed above), but it is recommended to utilize the modified system configuration designed for primary filtration (described in Section 5.3).

The CDF uses OptiFiber pile cloth medium on disks that are oriented vertically to reduce the required footprint. The influent wastewater enters through the influent channel, with each cloth media disk completely submerged in the tank. The filtrate is collected inside the disks and then discharged to the effluent channel. As a solids layer builds on the surface of the cloth depth medium, the tank's water level increases. When the liquid level reaches a set point of approximately 300 mm (12 in.) above the clean filter level, the CDF system automatically backwashes and removes the solids layer to restore the water level to its lower setting. Backwash shoes contact the media directly and solids are removed by the application of a vacuum from the backwash pump. The BRW ratio for primary effluent filtration was observed to be less than 10% (Caliskaner et al., 2015). An illustration of the CDF technology is presented in Figure 10.24.

Some of the important design considerations include hydraulic filtration rate, solids loading rate, cloth medium selection, and backwash handling requirements. The CDF can be designed and operated for primary effluent filtration at hydraulic filtration rates ranging between 80 and 240 L/m² min (2–6 gpm/sq ft). Typical/average design filtration rates range between 80 and 160 L/m² min (2 to 4 gpm/sq ft). Average and peak design solids loading rates are between 24 and 37 and between 49 and 73 kg/d m² (5 and 7.5 lb/d sq ft and between 10 and 15 lb/d sq ft), respectively. The designer should consider both the filtration hydraulic and solids loading rates for sizing of the filtration facilities.

Another important design consideration is the selection of the filter media. There are several types of cloth media with different filtration characteristics (e.g., nominal pore sizes ranging between 5 and 10 μm) and the hydraulic and removal performances differ between different media. Selection of the media should be made considering the design target hydraulic loading rate and TSS removal rates in conjunction with the filter

(a)

(b)

(c)

(d)

Figure 10.24   Operational cycles of cloth depth (disk) filter: (a) Filtration, (b) backwash, (c) solids removal, and (d) scum removal (Courtesy of Aqua-Aerobic Systems, Inc.).

influent characteristics. Proper handling of BRW is an essential component of a robust primary effluent filter design. Provisions to add chlorine during the backwash cycle is recommended. The BRW is thickened before being conveyed to digestion facilities. Average BRW ratios less than 10% should be targeted for average flow and TSS load conditions. For CDF and other primary effluent filtration technologies discussed below, the design criteria for hydraulic and solids loading rates should be determined considering BRW flow rates and ratios in conjunction with the target TSS removal rates.

5.2.2 Compressible Media Filtration Design Considerations in Primary Effluent Filtration

The compressible media filter (CMF) was first developed in Japan in the mid-1980s and has been installed in many facilities around the world for various applications (e.g., tertiary filtration, CSO treatment, and industrial applications). The CMF was then introduced in the United States in the 1990s with increasing applications within the last 10–15 years.

There are two prominent CMFs used in the United States. One of the CMF technologies, also known as the "Fuzzy Filter™", involves the use of mechanical compression of the filter medium. The medium has some unusual properties, including being highly porous (approximately 90%) and compressible. The density of the medium is slightly greater than that of water. Because of its low density, the filter medium is retained between two perforated plates. The filter bed consists of a large number of fuzzy balls, which are formed by shrinking synthetic fibers into a quasi-spherical shape resulting in a complex porous structure. An individual fuzzy ball is approximately 40 to 50 mm (1.5–2 in.) in diameter. The filter medium properties, such as pore size, porosity, and depth can be adjusted in response to changing influent conditions because it is compressible. Properties of the filter medium are altered by adjusting the position of the upper moveable plate.

During the filtration cycle, filter influent is introduced into the bottom of the filter into a plenum. The influent flows upward through the filter medium and is discharged from the top of the filter. On a regular operational basis, the filtration cycle is interrupted for backwashing when the terminal head loss or preset time is reached. At the inception of the backwash cycle, the influent flow is diverted to the backwash water line, and the upper plate is moved up mechanically to increase the bed volume for backwashing. Filter influent water is used in combination with an aggressive air scouring sequence to backwash the filter. During a wash cycle, the influent flow rate is reduced to the filter while an external blower supplies air in the bottom of the chamber to agitate the media. To start the next filtration cycle, the upper perforated plate is returned to its desired position to obtain the required level of medium compression. The filtration and backwash cycles for the Fuzzy Filter are illustrated in Figure 10.25.

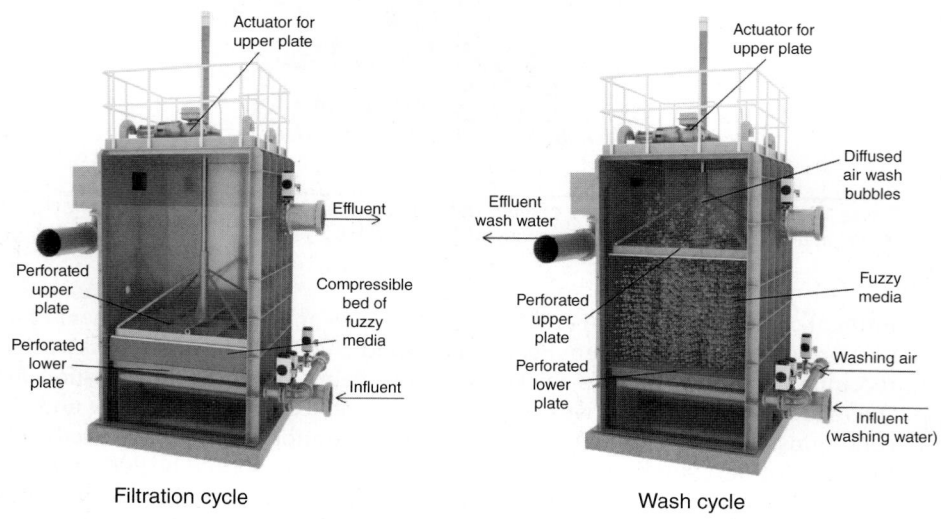

Filtration cycle   Wash cycle

FIGURE 10.25 Operational cycles of Fuzzy Filter (Courtesy of Schreiber, LLC).

Main design considerations include medium depth and compression ratio, filtration hydraulic rate, and backwash handling. Typical uncompressed medium depth is 760 to 915 mm (30–36 in.). The medium compression ratio is calculated as the compression depth (i.e., change in media depth caused by compression) divided by the uncompressed medium depth. Feasible/optimum medium compression ratios range between approximately 10% and 30%, depending on specific primary effluent characteristics (e.g., particle size distribution [PSD] and TSS). For primary effluent filtration, like tertiary treatment, the Fuzzy Filter is able to operate at high filtration rates [e.g., 400–1000 L/m² min (10–25 gpm/sq ft)] because of the low head losses resulting from the high porosity of the medium. The BRW ratio for primary effluent filtration is typically less than 10% (Caliskaner et al., 2014, Caliskaner et al., 2015).

Another prominent type of emerging CMF technology in primary treatment, known as "FlexFilter™", is configured based on down-flow filtration. FlexFilter™ uses an engineered bladder coupled with the hydraulic pressure of the influent water to laterally compress the media. Influent water fills the area behind the bladder causing the bladder to compress the media as illustrated in Figure 10.26 resulting in a conical shaped bed. Full compression is achieved as the water level reaches the inlet weir elevation. Filtration starts as influent water discharges over the influent weir and passes downward through the compressed medium. The lateral compression provides a porosity gradient with loose media at the top and compressed media at the bottom. As a result, large particles are removed at the top and fine particles are removed at the bottom. As solids are removed within the media, the influent level increases until a backwash cycle is triggered. When the influent pressure is released, the bladder relaxes forming a curved bottom allowing smooth rotation when backwashing. During backwash, a rotational fluidized bed and air scrub is used to clean the media, with air lift of the spent backwash water into troughs. Provisions to add chlorine (e.g., during backwash cycle) is recommended for all primary effluent and primary filtration technologies. The operation of FlexFilter™ is illustrated in Figure 10.26.

There are recent developments for primary effluent filtration and primary filtration applications of FlexFilter™ to remove soluble BOD in addition to particulate BOD. Biofiltration entails growing and utilizing fixed biomass (i.e., biofilm) on the filter medium

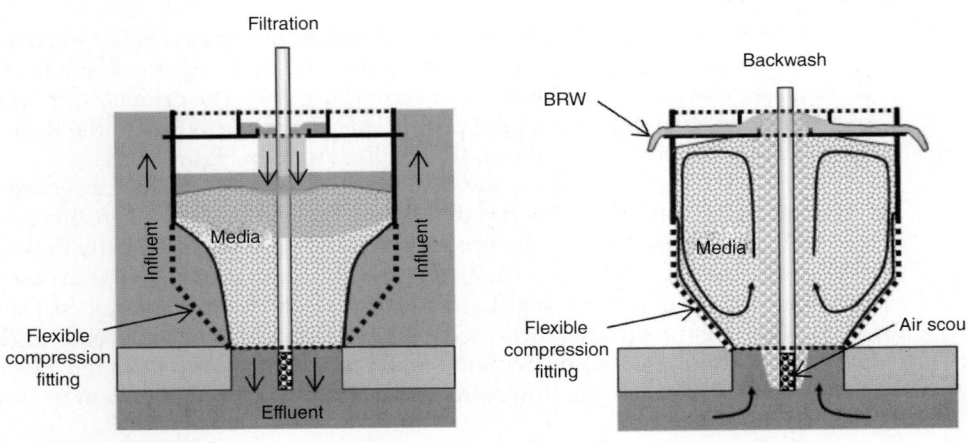

Figure 10.26 Operational cycles of FlexFilter™ (Courtesy of WesTech, Inc.).

to achieve soluble BOD removal (Caliskaner et al, 2015 and Fitzpatrick et al., 2011). Two additional important considerations for biofiltration are aeration and biomass management. Aeration is accomplished in three ways: (1) during filtration cycle—via cascade aeration occurring with the influent fall into the filter cell, (2) during backwash cycle—with the air scour used to help clean the media, and (3) during drain cycle—as more air is pulled into the media bed. Periodic chlorine addition is used to control the excess biomass attachment to the media.

Main design considerations for FlexFilter™ include medium depth, hydraulic and solids loading rates, and backwash management. Typical medium depth is 760 mm (30 in.) for TSS removal (higher medium depths may be used for primary treatment applications). Design filtration rates depend on solids and hydraulic loading and head loss development. Hydraulic loading rates up to 600 L/m² min (15 gpm/sq ft) are used for primary effluent filtration and up to 400 L/m² min (10 gpm/sq ft) for primary applications for CSO treatment. Lower filtration rates (e.g., 120–200 L/m² min; 3–5 gpm/sq ft) are more typical for biofiltration applications.

For both CMF technologies, design ranges for medium depth, medium compression ratio, and filtration rates depend on specific wastewater characteristics such as filter influent TSS loads/concentrations. The BRW is thickened before being conveyed to the digestion facilities. An average BRW ratio of less than 10% to 15% should be targeted at the average flow and TSS load conditions. The design criteria for the main filtration parameters (e.g., medium depth, compression ratio, and filtration rate) should be determined considering BRW flow rates and ratios in conjunction with the target TSS removal rates.

5.3   Primary Filtration

Filtering raw screened wastewater has recently emerged as a promising technology for primary treatment because of its significant energy and capital savings potential resulting from robust operation and high removal efficiencies (Caliskaner et al., 2016; Ma et al., 2015). Primary filtration reduces the footprint both for primary and secondary treatment facilities. Primary filtration in lieu of primary clarification can be used to increase the flow capacity of existing facilities and/or reduce the size and cost of new secondary treatment facilities by reducing the organic load to the secondary treatment system.

The filter BRW containing the solids captured from the raw wastewater is conveyed to the anaerobic digester after thickening. The high organic content of filter BRW increases biogas production in the anaerobic digester. The primary filtration performance results obtained from a pilot project conducted at a WRRF at the Rock River Water Reclamation District (Rockford, IL) are illustrated in Figure 10.27.

As illustrated in Figure 10.27, the BOD removal efficiency was observed to be approximately 50% to 55%. Observed TSS removal efficiencies were higher than 80% to 85%. The filtered TSS levels produced by the CDF were typically between 20 and 50 mg/L despite influent 24-hour composite samples that were occasionally recorded as high as 600 to 800 mg/L. In addition to capital and energy cost savings for secondary treatment and increased biogas production in anaerobic digestion resulting from high BOD and TSS removals, the stability of the filtered BOD/TSS concentrations will allow the downstream secondary treatment system to function more efficiently.

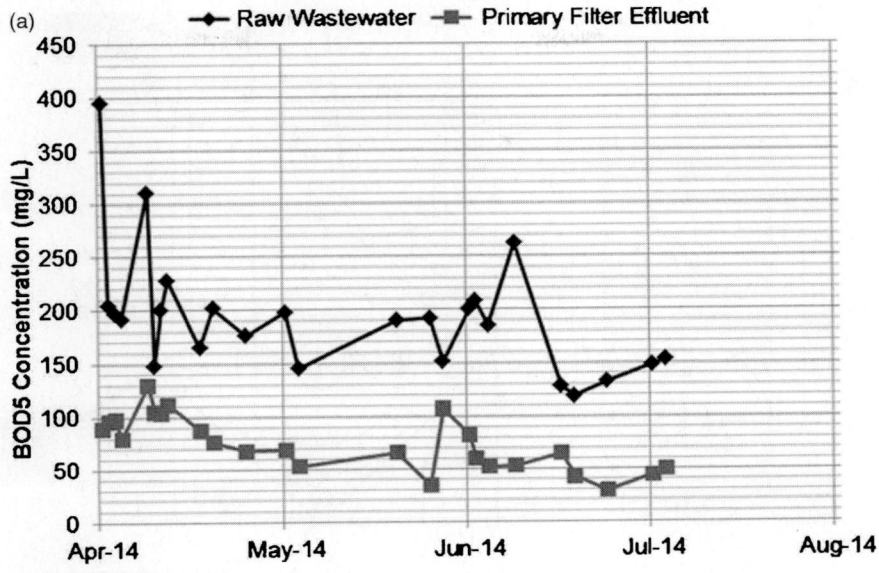

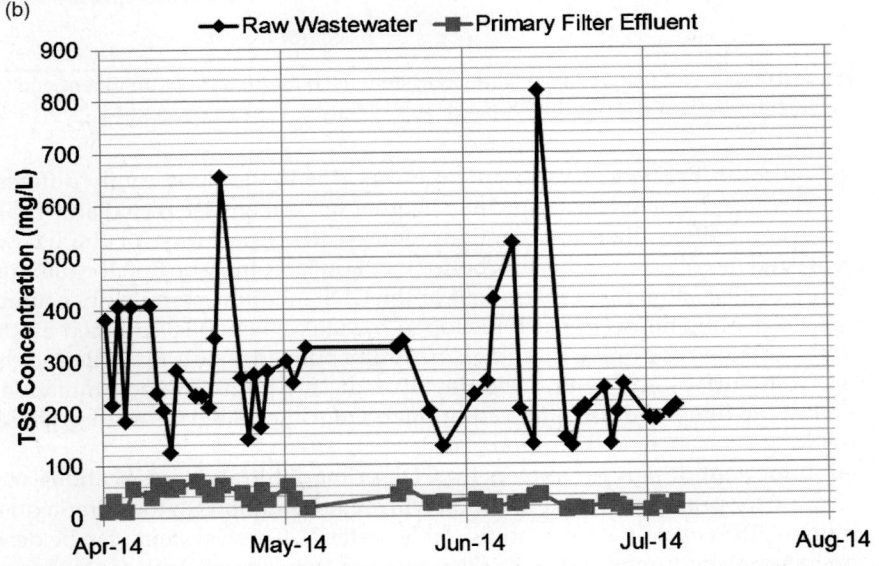

Figure 10.27   BOD and TSS reductions achieved by CDF for primary filtration: (a) BOD and (b) TSS (Caliskaner et al., 2016).

5.3.1   Design Considerations for Cloth Depth Filters in Primary Filtration

Although basic filtration principles and operational cycles are similar, CDF systems used in primary filtration applications include several design and manufacturing changes compared to CDF systems used in tertiary treatment applications, as illustrated

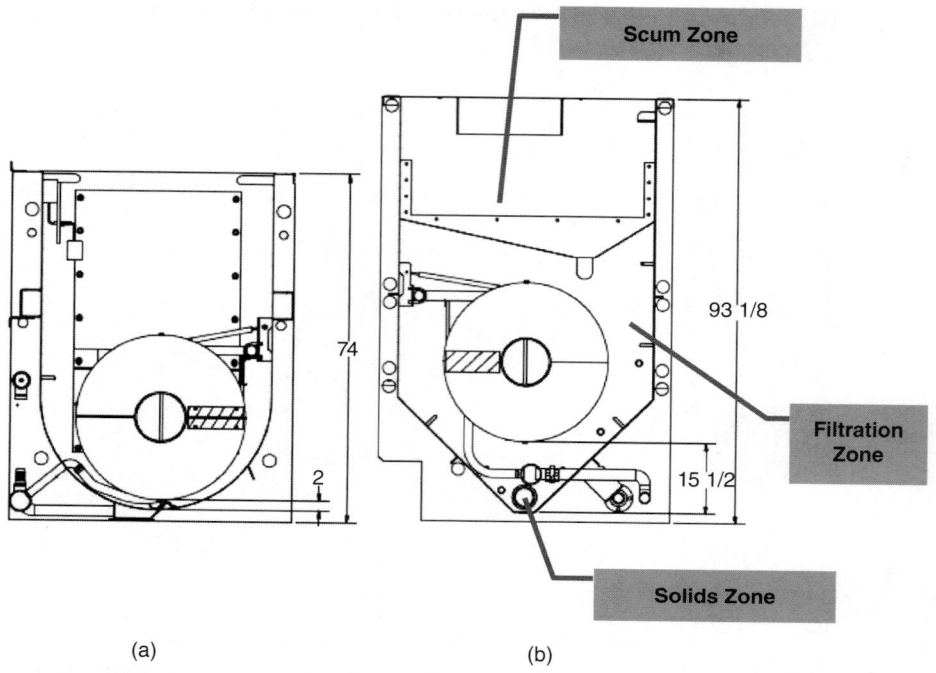

(a)   (b)

FIGURE 10.28 Cloth depth (disk) filtration system used for (a) tertiary treatment and (b) primary treatment (Courtesy of Aqua-Aerobics Systems, Inc.).

in Figure 10.28. These changes are necessary due to the significantly different influent wastewater characteristics (e.g., much higher fats, oil, grease [FOG] and TSS).

The modified filter tank configuration creates a pathway for the heavy solids to settle and be collected at the tank bottom without accumulating on the medium surface. Additionally, submerged placement of the CDF medium permits unencumbered passage of floating debris to reach the top of the tank where it is skimmed off the surface. In the middle zone of the filter tank, neutrally buoyant solids are collected on the filter medium surface and within the depth of the medium. The accumulation of solids within the filter medium offers additional removal capabilities as these solids create a working layer that aids filtration.

Important design parameters for CDF primary filtration applications are the same as those for primary effluent filtration with some variations in the criteria due to higher influent TSS concentrations and floatable material. These systems can be designed and operated at hydraulic filtration rates ranging between 60 and approximately 160 to 240 L/m² min (1.5–6 gpm/sq ft). Typical average design filtration rates are 60 to 160 L/m² min (1.5–3 gpm/sq ft). Average and peak design solids loading rates are between 24 and 37 and between 49 and 73 kg/d m² (5 and 7.5 and between 10 and 15 lb/d sq ft), respectively. Solids loading usually governs the primary effluent filtration and primary filtration designs, but the engineer should consider both the hydraulic and solids loading criteria for final design and proper sizing of the filtration system.

As in primary effluent filtration, another important design consideration is the selection of the filter medium. There are several types of cloth media with different filtration characteristics (e.g., nominal pore sizes ranging between 5 and 10 μm) and the hydraulic and TSS removal performances differ between different media. Selection of the medium should be made considering the design target hydraulic loading rates and TSS removal rates in conjunction with the filter influent characteristics.

Proper handling of BRW is an essential design component. The BRW is thickened before being conveyed to digestion facilities. For primary filtration applications, average BRW ratios less than 15% to 20% should be targeted for average flow and TSS load conditions. Final design criteria for filter hydraulic and solids loading rates should be determined considering BRW flow rates and ratios in conjunction with the target TSS removal rates.

5.4 Rotating Belt Filters/Microscreens

Rotating belt filters (RBF) and microscreens can provide treatment levels close to primary clarification in a reduced footprint. The use of RBFs and microscreens in lieu of primary clarifiers has increased since the early 2000s (particularly in Europe). Grit removal and/or coarse screening (e.g., with an opening size of 25–50 mm [1–2 in.]) are used upstream of the RBFs (especially in larger installations) to minimize operational and maintenance requirements. In the RBF, solids are removed through the use of a continuous-loop fine mesh belt screen. The solids concentration of the reject stream is typically between 3% and 8%, allowing conveyance to solids digestion or dewatering facilities without further thickening. However, RBFs are also available with integrated screw press dewatering systems to increase the solids concentration to between 20% and 40%.

Overall performance is site specific (as in all other technologies) and dependent on the influent characteristics, design, and operational conditions. Two RBF manufacturers were tested at the City of Largo WRRF. The results from both pilot tests yielded similar conclusions (Porter and Strain, 2015). The feasibility of RBFs and microscreens were evaluated (i.e., to in lieu of primary clarifiers) in another pilot project conducted for the Capital Regional District (Victoria, BC). TSS and BOD removal performances obtained for three different technologies using different mesh sizes are summarized in Table 10.6 (Sutton et al., 2008).

	Salsnes Filter*		Drum Screen with Filter Panels		Perforated Drum Filter (Roto-Sieve)	
Screen size, microns	250	350	200	250	600	1000
Effluent TSS, mg/L	102	180	109	113	185	217
Percent TSS removal	51	27	51	47	12	12
Percent BOD$_5$ removal	30	7	20	27	4	10

*Salsnes Filter provided by Salsnes sales representative. Drum screen with filter panels provided by Hydrotech. Perforated drum filter provided by Wastetech. TSS results presented are mean values derived from analysis of equal to or greater than 40 grab samples taken at a frequency of two to three times per day. BOD results are mean values based on analysis of approximately 20 percent of the samples analyzed for TSS. Results include that derived when operated with old and new 250 micron screen.

TABLE 10.6 Performance of Microscreens

5.4.1 Design Considerations

There are several RBF and microscreen configurations and types offered by different equipment manufacturers. The main differences include type and characteristics of the fine mesh, configuration and layout, cleaning mechanisms, and support structure. The fine mesh may be produced using different types of material, different thicknesses of the thread and different mesh sizes. Design engineers should evaluate the specific design and manufacturing features/differences of available RBF / Microscreen types to identify the suitable system(s).

The configurations of an RBF and microscreen system are illustrated in Figure 10.29. The belted screens move linearly and as the screen moves, it acts as a conveyor and carries captured solids out of the incoming wastewater. As wastewater enters the inlet chamber and flows through the mesh screen, the effluent is collected behind the mesh screen and discharged into the outlet chamber. Sludge that has accumulated on the belt is conveyed to the upper portion of the belt and then dropped into a hopper. High-pressure water spray and/or compressed air is used to dislodge the remaining solids off the belt. The mesh screen is also backwashed twice a day with hot water to remove any oil and grease buildup.

Similar to the primary effluent and primary filtration technologies, removal is a dynamic process in RBFs, depending on the buildup of a filter mat. A pressure transmitter

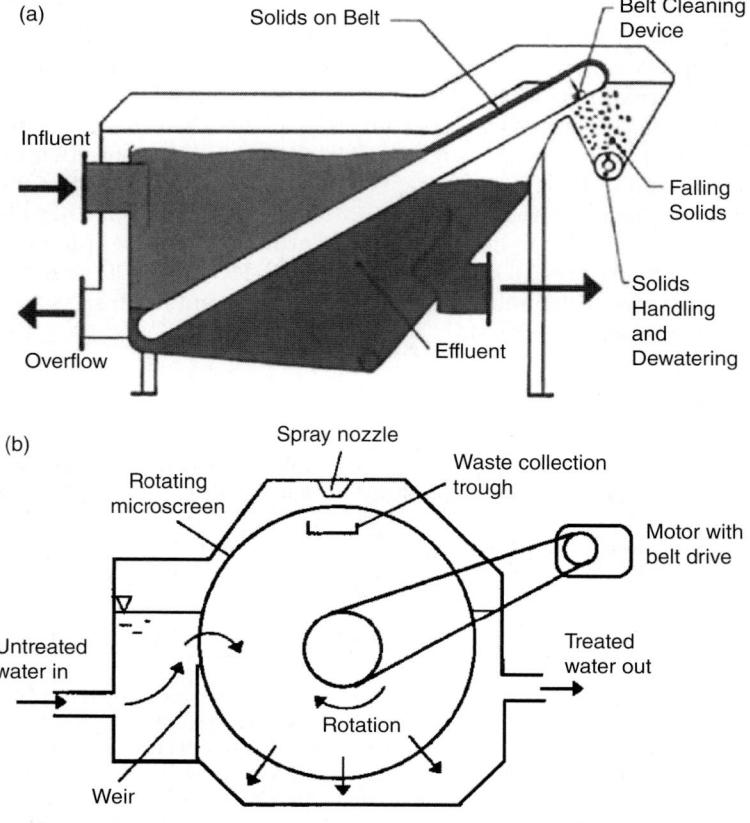

Figure 10.29 Enclosed configuration for rotating belt filters (RBF) and microscreens. (a) RBF. (b) Microscreen.

is typically used to measure the level of incoming water. The rotational speed of the belt filter is adjusted based on the water level to achieve optimum performance at variable flow rates and variable influent TSS concentrations. The belt filter is stationary if the water level is less than a preset value. As particles accumulate on the surface and build a filter mat, the water level increases triggering the rotation of the belt filter. If the water level keeps increasing while the belt filter is rotating, the speed will automatically increase. If the water level drops below a preset limit, the motor will stop until the level increases again. Operating with a proper filter mat is crucial for optimum RBF performance. After a backwash cycle (i.e., when the belt is clean) solids are removed by simple sieving mechanisms. As more solids are accumulated on the rotating belt, the filter mat is formed, enhancing removal performance over time (i.e., by removing particles smaller than the belt mesh size but larger than the effective pore size of the formed filter mat). All areas of the rotating belt filter are cleaned once during each rotation cycle.

One main difference between RBFs/microscreens and the primary effluent or primary filtration technologies discussed above is the pore size used for rejection of particles. In RBFs/microscreens, typical mesh pore sizes range between 50 and 600 μm with more typical values between 200 and 400 μm. For CDF and CMF filtration technologies effective/nominal media pore sizes are typically less than 10 μm. The PSD of the influent wastewater with respect to the mesh pore size is one of the most important considerations in RBF designs. Chemical conditioning is especially beneficial if the influent wastewater is not well suited for treatment by RBFs. Proper chemical conditioning alters the influent PSD such that a higher fraction of particles will be larger than the mesh pore size and hence easily removed by the RBFs.

Other important design considerations in addition to the mesh pore size and influent PSD include hydraulic and solids loading rates. RBFs are typically designed to operate at hydraulic loading rates ranging approximately between 2400 and 4000 L/m² min (60–100 gpm/sq ft). Design solids loading rates range between 0.3 and 0.5 kg/m² min (0.06 and 0.1 lb/sq ft min). Proper sizing of the systems requires an evaluation of the head loss as a function of hydraulic and solids loading rates, subject to the flow and load design ranges. The impact of loading rates on the TSS removal performance is summarized in Figure 10.30 based on several RBF pilot tests conducted

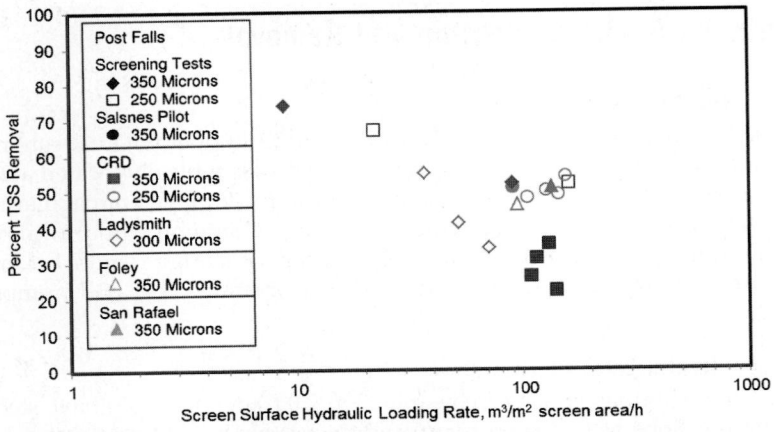

FIGURE 10.30 Effect of hydraulic loading rate on microscreen performance.

at different WRRFs (Sutton et al., 2008). Removal performance is impacted by the loading rates (i.e., both hydraulic and solids), pore size, and influent characteristics. Therefore, pilot studies are recommended before the design of full-scale installations.

5.5 Primary Dissolved Air Floatation Clarification Integrated with Biological Treatment

The primary dissolved air floatation (DAF) clarification integrated with biological treatment system combines enhanced primary treatment and sludge thickening in one unit, resulting in footprint reduction as compared to conventional technologies. This system (also known as Captivator™) is a unique carbon diversion technology that uses biosorption principles to achieve high BOD and TSS removal, as well as sludge thickening without the need for chemicals. The system blends secondary WAS with raw wastewater in a mildly aerated contact tank to promote rapid biosorption of soluble organics. The Captivator™ includes a small aerated contact tank upstream of a DAF unit to capture BOD from influent wastewater prior to the activated sludge process. The contact tank design detention time ranges between 20 and 40 minutes. WAS from the activated sludge process is returned and blended in the aerated contact tank where it acts as an absorbent (similar to contact stabilization process). The target solids concentration of the blended stream is between 400 mg/L and 600 mg/L. Following the contact tank, the blended stream flows to a DAF unit that functions as a liquid-solids separation unit, as well as a sludge thickening step (without the need for polymers). Typical sludge solids concentration is between 5% and 6%, and therefore no further thickening is necessary before being processed by anaerobic digestion. BOD and TSS removal efficiencies between 50% and 60%, and 70% and 80%, respectively were reported (Ding et al., 2015). The design hydraulic overflow rate of the DAF unit ranges between 205 and 410 m³/m² day (5000 and 10 000 gpd/sq ft). Another important design consideration is the solids loading rate with a range of 73 and 146 kg/day m² (15 to 30 lb/day sq ft).

The flow diagram for this system is shown in Figure 10.31. This technology has gone through a rapid progression in the past years with the first full-scale installation in January 2014 at the 121 000 m³/d (32 mgd) Agua Nueva Water Reclamation Facility, Arizona (Johnson et al., 2014).

6.0 Primary Sludge Collection and Removal

6.1 Description

Settled primary sludge is typically scraped into a hopper where it is removed by gravity or pumping. The hopper for rectangular tanks is typically located at the inlet end of the tank to minimize travel time of particles to the hopper. For circular tanks, the hopper is typically located in the center of the tank. Common withdrawal pipes from two or more hoppers often result in unequal primary sludge removal from the hoppers. Therefore, multiple tanks and hoppers need separate pipes and pumps or valves on each outlet.

6.1.1 Rectangular Clarifiers

The sludge hopper, which can be up to 3 m (10 ft) deep, should have steep sides with a minimum slope of 1.7:1 (vert:horiz) and the bottom should a maximum width of 0.6 m (2 ft) (Great Lakes-Upper Mississippi River Board of State and Provincial Public Health

and Environmental Managers, 2014). Sludge hopper wall surfaces should be smooth with rounded corners to avoid any solids buildup. Rectangular sedimentation tanks with longer widths often need more than one hopper to reduce hopper depth.

Chain and flight or traveling bridge sludge collectors are typically used for rectangular primary sedimentation tanks. Chain and flight collectors (Figure 10.16) consist of two loops of chains with cross scrapers (flights) attached at approximately 3-m (10-ft) intervals. Revolving flights push the settled primary sludge to the hopper at the end of the tank. Most chain and flight collectors are within a 20- to 30-foot width range (Green et al., 2007). Some installations, however, use side-by-side collectors without common walls in tanks as wide as 24 m (80 ft) or more (Metcalf and Eddy, 2014). Historically, cast iron chains and wood flights were used. Designers now select stainless steel or nonmetallic (plastic) chains and fiberglass flights almost exclusively. Typically, chain and flights for a pair of tanks move approximately 0.6 m/min (2 ft/min), driven by a single-drive unit located on the wall between the two tanks (AWWA and ASCE, 2012). Some facilities use a higher flight speed of 0.9 m/min (3 ft/min). Flights travel along the long axis of the tank and, as the upper flights move away from the primary sludge hopper, they can skim the surface, pushing floating material toward the scum removal mechanism. At the end of the tank opposite from the sludge hopper, the flights drop to the floor and drag heavy, settled material to the primary sludge hopper for removal. Single tanks can have either a single or double hopper. Multiple tanks or tanks with more than one primary sludge collection assembly (widths more than 6 m [20 ft]) typically require more than one assembly use a cross collector in a transverse trough. The cross collector, typically 1.2 m (4 ft) wide and 0.9 to 1.2 m (3 to 4 ft) deep, runs the width of one or more tanks as it conveys primary sludge to a hopper (AWWA and ASCE, 2012). Cross collectors are typically chain and flight with flights on 1.5-m (5-ft) centers, which travel along the transverse trough between 0.6 and 1.2 m/min (2 and 4 ft/min). Screw-type cross collectors, sometimes used, rotate at approximately 10 r/min (Figure 10.32) (AWWA and ASCE, 2012).

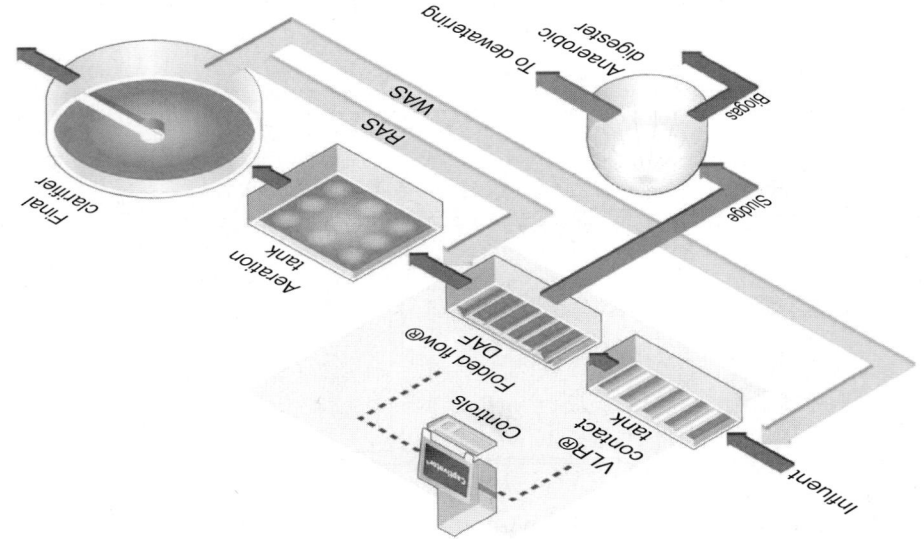

Figure 10.31 Dissolved air floatation clarifier system flow diagram.

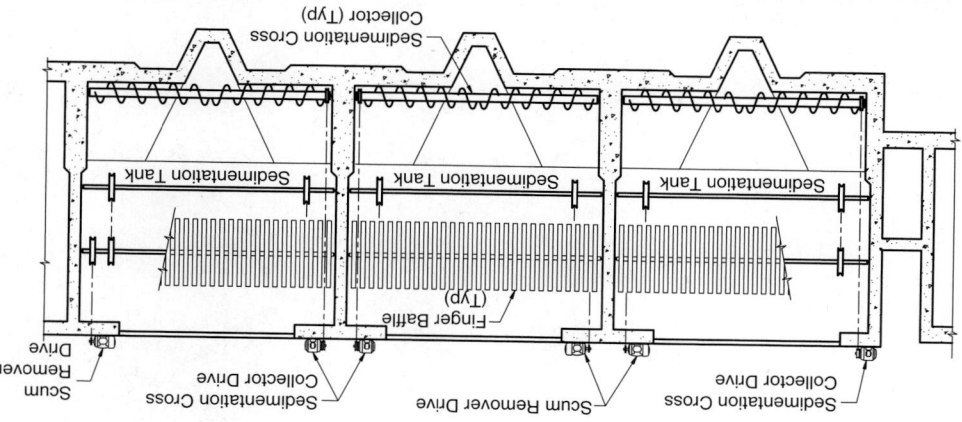

FIGURE 10.32 Typical cross collector layout for primary sedimentation tanks at Sunnyvale, California.

A traveling bridge (Figure 10.33) consists of a scraper blade mechanism mounted on a bridge or carriage that travels approximately 1.8 m/min (6 ft/min) toward the hopper on tracks or rails mounted on top of the tank. As it travels away from the hopper at approximately 3.7 m/min (12 ft/min), the mechanism, largely out of the water, acts as a skimmer, pushing floating material toward the scum removal mechanism. As it reaches the end of the tank opposite from the hopper, the mechanism drops to the floor of the tank, reverses direction, and travels toward the hopper end of the tank, pushing settled primary sludge.

Traveling-bridge collectors cannot be used if covers are required on primary sedimentation tanks. However, traveling bridges can be used if the tanks are enclosed in a building with a height that enables operation and maintenance of the equipment. Traveling-bridge collectors may span tank widths up to 30 m (100 ft). In earthquake zones, a restraining mechanism should be considered to resist derailing from seismic ground motion. In cold climates, the design of traveling-bridge collectors should provide for control of snow and ice buildup on rails or tracks. Otherwise, this buildup could derail the bridge or reduce traction between the wheel and rail.

6.1.2   Circular Clarifiers

Circular clarifiers typically have segmented rake or plow-type primary sludge collection equipment. The conventional plow-type consists of scrapers that drag the tank floor at a tip speed of approximately 1.8 to 3.7 m/min (6–12 ft/min) (AWWA and ASCE, 2012). Plows are located at an angle to the radial axis to force primary sludge towards the hopper, which is typically located at the center of the tank, as the device rotates. The center hopper is typically a vertical-sided sump where the primary sludge is removed by pumping. The rotating element of the device can be driven from either the center or the outside tank wall. Torque must be sufficient to move the densest primary sludge expected. The sludge blanket at the perimeter can become resuspended by the plow because of the high tip speed, especially with large diameter clarifiers. Several revolutions are required to move the solids to the center hopper. Albertson and Okey (1992) found that conventional plow-type scrapers typically were undersized and showed that the sludge-scraper capacity was greatly reduced by the slower speed of the plows near the center of the clarifier.

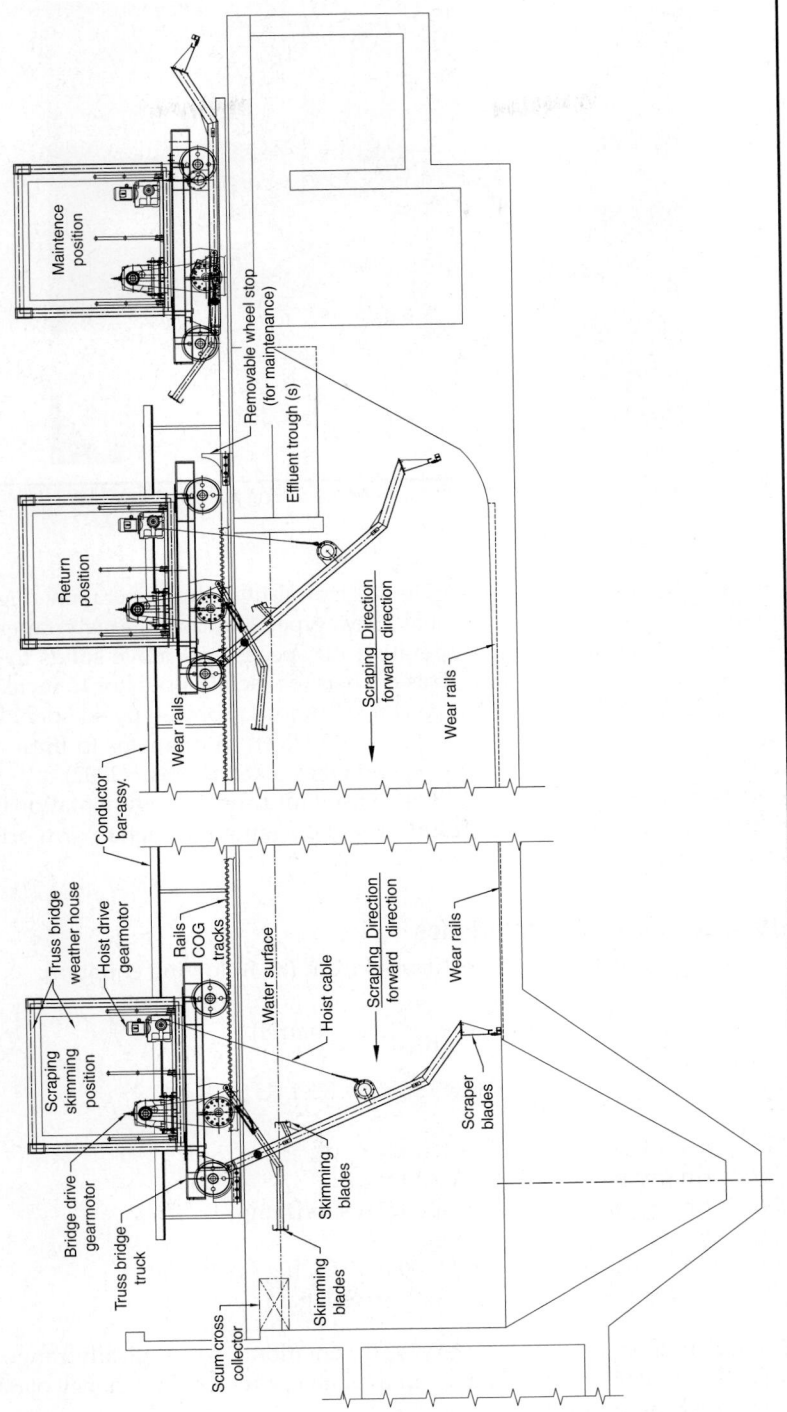

Figure 10.33 Typical section of a traveling bridge primary sludge collector (Courtesy of Ovivo).

641

FIGURE 10.34   Spiral blade clarifier (Courtesy of Westech Engineering Inc.).

More recently, deeper, tapered, segmented, spiral-type sludge scrapers (Figure 10.34) are favored over the more conventional plow type. The continuously tapered spiral-shaped scraper blades allow faster operating tip speeds and move solids to the center hopper in as little as one revolution. This enables the facility operator to increase sludge transport capacity and improve sludge concentrations. Spiral-type scrapers have been successfully employed in clarifiers up to 68 m (225 ft) in diameter to improve sludge collection/transport efficiency (Albertson and Okey, 1992; Kinnear, 2002).

The suction-type collector should not be used for primary sedimentation because of the high primary sludge density and the risk of clogging the suction arm orifices with items such as rags.

6.2   Quantities and Characteristics

Primary sludge production can be estimated from the following equation:

$$S_M = \frac{Q \times \text{TSS} \times E}{1000} \text{ (metric)} \qquad (10.17)$$

$$S_M = Q \times \text{TSS} \times E \times 8.34 \text{ (U.S.)} \qquad (10.18)$$

where S_M = mass of primary sludge, kg/d (lb/d);
$\quad\quad Q$ = primary influent flow, m³/d (mgd);
$\quad$ TSS = primary influent total suspended solids, mg/L;
$\quad\quad E$ = removal efficiency, fraction;
$\quad$ 1000 = units conversion factor ([1000 mg/L]/[kg/m³]); and
$\quad$ 8.34 = units conversion factor ([8.34 lb/mil. gal.]/[mg/L]).

TSS removal efficiencies in primary sedimentation tanks typically range between 50% and 70%. If actual removal data are unavailable, a removal efficiency of 60% can be assumed for estimating purposes.

CEPT can increase primary sludge mass by 50% to 100% (Soap and Detergent Association, 1989). The addition of approximately 20 mg/L of ferric chloride and 0.2 mg/L of polymer to the headworks before primary sedimentation increased primary sludge production by approximately 45% (Chaudhary et al., 1989). Approximately 30% of this increase results from improved suspended solids removal and the remaining 15% stems from chemical precipitation and removal of colloidal material (Chaudhary et al., 1989).

Chemical-sludge quantities can be estimated by the stoichiometric relationship between the raw wastewater constituents and coagulants. The stoichiometric quantity should be increased by approximately 35% for aluminum and iron salts to account for increased BOD, COD, and TSS removal (Mertsch, 1985; U.S. EPA, 1987).

The composition of primary sludge is variable and depends on the nature and degree of industrial development in the collection area. Table 10.7 lists typical characteristics. Some chemical sludges (particularly from lime and aluminum salts) are gelatinous, with a high water content, low suspended solids content, and high resistance to mechanical or gravitational dewatering. Feed solids composition merits careful consideration in the design of solids-handling and processing units. For further discussion of sludge characteristics, refer to Chapters 5 and 19.

6.3 Thickening

Primary sludge is thickened in primary sedimentation tanks, stabilization facilities, or separate thickening units. Primary sedimentation tanks can be operated to consistently produce a thickened solids concentration of 3% to 6% solids by allowing a 0.6 to 0.9 m (2–3 ft) blanket of solids to build up and compact the primary sludge (WEF, 2005). Higher concentrations can be achieved but often cause problems in the conveyance system (WEF, 2016).

If the facility influent is prone to septicity or other site-specific conditions that prevent sludge from settling well, then sedimentation tanks are sometimes operated with continuous withdrawal of dilute primary sludge to minimize thickening, maximize removal, and prevent anaerobic decomposition of settled solids. Anaerobic or septic conditions will result in solubilization of BOD (which may be beneficial in facilities with downstream biological phosphorus and/or nitrogen removal systems—see Sections 8.1 and 8.2), can cause rising sludge blankets resulting in the poor removal of solids, and generate odorous compounds. BOD removal efficiencies for raw wastewater with a large fraction of soluble BOD will be considerably lower than those for wastewater with a smaller fraction of soluble BOD. Solubilization and septicity are especially troublesome in hot climates (Southwest United States and Hawaii) and where collection systems have long detention times.

Soluble BOD accumulated in the sedimentation tank during attempts to thicken primary sludge at the Renton Plant in Seattle, Washington (Uhte, 1990). This was attributed to the development of septic conditions in the tank and scouring of the primary sludge blanket at peak flows, which resulted in soluble BOD release from the sludge blanket. The BOD loading to the aeration tanks was increased by approximately 20% (Uhte, 1990). Typically, primary sludge thickening should not be attempted with overflow rates greater than 100 m³/m² d (2500 gpd/sq ft) (Uhte, 1990). Such rates call for separate thickener facilities. The Water Environment Federation (2005) advises separation of thickening and clarification into separate unit processes. Additional information on sludge thickening can be found in Chapter 21.

Characteristic	Range of Values	Typical Value	Comments
pH	5–8	6	—
Volatile acids, mg/L as ascetic acid	200–2000	500	—
Heating value, kJ/kg Btu/lb	16 000–23 000 6800–10 000	—	Depends on volatile content and primary sludge composition; reported values are on a dry basis
		10 285	Primary sludge 74% volatile
		7600	Primary sludge 65% volatile
Specific gravity of individual solid particles	—	1.4	Increases with increased grit and silt
Bulk specific gravity (wet)	—	1.02	Increases with primary sludge thickness and with specific gravity of solids
		1.07	Strong wastewater from a combined system of storm and sanitary sewers
BOD_5:VSS ratio	0.5–1.1	—	—
COD:VSS ratio	1.2–1.6	—	—
Organic nitrogenV:SS ratio	0.5–1.1	—	—
Volatile content, percent by weight of dry solids	64–93	77	Value obtained with no primary sludge recycle, good gritting; 42 samples, standard deviation 5
	60–80	65	
	—	40	Low value caused by severe storm
	—	40	Low value caused by industrial waste
Cellulose, percent by weight of dry solids	8–15	10	—
Hemicellulose, percent by weight of dry solids	—	3.2	—
Lignin, percent by weight of dry solids	—	5.8	—
Grease and fat, percent by weight of dry solids	6–30	—	Ether soluble
	7–35	—	Ether extract
Protein, percent by weight of dry solids	20–30	25	—
	22–28	—	—
Nitrogen, percent by weight of dry solids	1.5–4	2.5	Expressed as N
Phosphorus, percent by weight of dry solids	0.8–2.8	1.6	Expressed as P_2O_5; divide values as P_2O_5 by 2.29 to obtain values as P
Potash, percent by weight of dry solids	0–1	0.4	Expressed as K_2O, divide K_2O by 1.20 to obtain values as K

Note: BOD = biochemical oxygen demand; COD = chemical oxygen demand; and VSS = volatile suspended solids.

TABLE 10.7   Primary Sludge Characteristics (U.S. EPA, 1979)

6.4 Transport and Handling

The primary sludge drawoff system should be designed with the capacity to allow either continuous withdrawal or intermittent withdrawal at a rate that will control the primary sludge blanket depth. If sedimentation tanks are to be operated to achieve additional primary sludge thickening, drawoff piping and pumps must be designed to handle the more concentrated sludge. Withdrawal lines should be at least 100 mm (4 in.) in diameter. As the solids concentration increases to more than 6%, the risk of plugging increases because of the greater viscosity of the thickened primary sludge and its tendency to clog the piping. For this reason, suction piping should be as straight as possible and accessible for rodding, pigging, or flushing to clear obstructions. A sight glass or solids-density meter is necessary on the suction side of the primary sludge pump to monitor the solids level. Sova et al. (2008) demonstrated that automated sludge withdrawal using a solids-density meter and controller was dependable and increased primary solids from 4.5% to 5.2%. The primary sludge line should include a sampling port and flow meter. Pumps should be positioned with the pumping element below the water surface elevation (flooded suction). When timers are used to control pump cycles, they should be capable of being set in 30-minute increments. Where separate sludge thickening is used, primary sedimentation tank sludge pumps may be set to run continuously. Chapters 5 and 19 provide a more detailed discussion of pumps.

7.0 Floatable Solids Management

7.1 Description

Removal of floating materials, or scum, is an important function of primary treatment. Fats, oil, grease, and other floating material increases the organic load to downstream treatment processes and might cause various operational troubles, including visual blights, odors, and scum buildup in downstream treatment processes.

7.2 Collection

Scum collection typically has been located at the effluent end of the rectangular primary sedimentation tank (Figure 10.16). Some facilities, however, have located scum collection on the influent end of the sedimentation tank to decrease travel distance to the collection point and ensure rapid removal of all floatables (Kemp and MacBride, 1990). Manual scum collection is used at some facilities. The primary sludge collection mechanism or a separate device may operate the automated scum removal mechanisms. The scum removal mechanism should extend the full width or radius of the sedimentation tank to prevent floating material from reaching the effluent weir and to minimize wind effects (Gelderloss et al., 2004). In circular tanks, wind velocities can exceed the outward velocity produced by a radial skimmer with an 8° to 12° angle of attack (Albertson, 2005). A skimmer that is tangential to the feed well with two converging straight sections (bent) increases the angle of attack to approximately 25° compared to a radial skimmer (Albertson, 2005). The inlet design should allow scum to enter the sedimentation tanks freely without being trapped in inlet channels or behind baffles. Slots with diverter plates should be provided on the circular feed well for this purpose. In some cases—for example, influent with high oil and grease content—a separate skimmer mechanism or water spray system is necessary for the center feed well. A hot water system is sometimes warranted for cleaning scum troughs and flushing hoppers and pipelines (Gelderloss et al., 2004). A nearly constant water surface elevation in sedimentation tanks should be maintained for proper operation of scum collection equipment.

The tilting trough (Figure 10.35), tip tubes, and sloping beach (Figure 10.36) are all used for removing scum from primary sedimentation tanks. A tripping device on the primary sludge collector activates the tilting trough which allows collected material on the surface to flow into the trough and then to a wet well. Tip tubes are partially submerged pipes open on one side and located at the water surface. An actuator and lifting screw rotate the tubes axially to allow material collected on the surface to flow into the pipes and then to a wet well. Wet well scum is pumped to another treatment process. The sloping beach is a stationary device with a collector trough. Floating material may be directed to the sloping beach by air jet sprays; water jet sprays; the primary sludge collection mechanism; or a separate, blade-type scraper. A separate, blade-type scraper moves the floating material up the beach and into a trough. Carrier water (primary effluent) flushes scum through the trough into a wet well where it is pumped to another treatment process.

Collectors for rectangular clarifiers are tilted manually with a lever, rack or pinion, worm gears, or motor-operated device. In circular clarifiers, the rotating device acts as

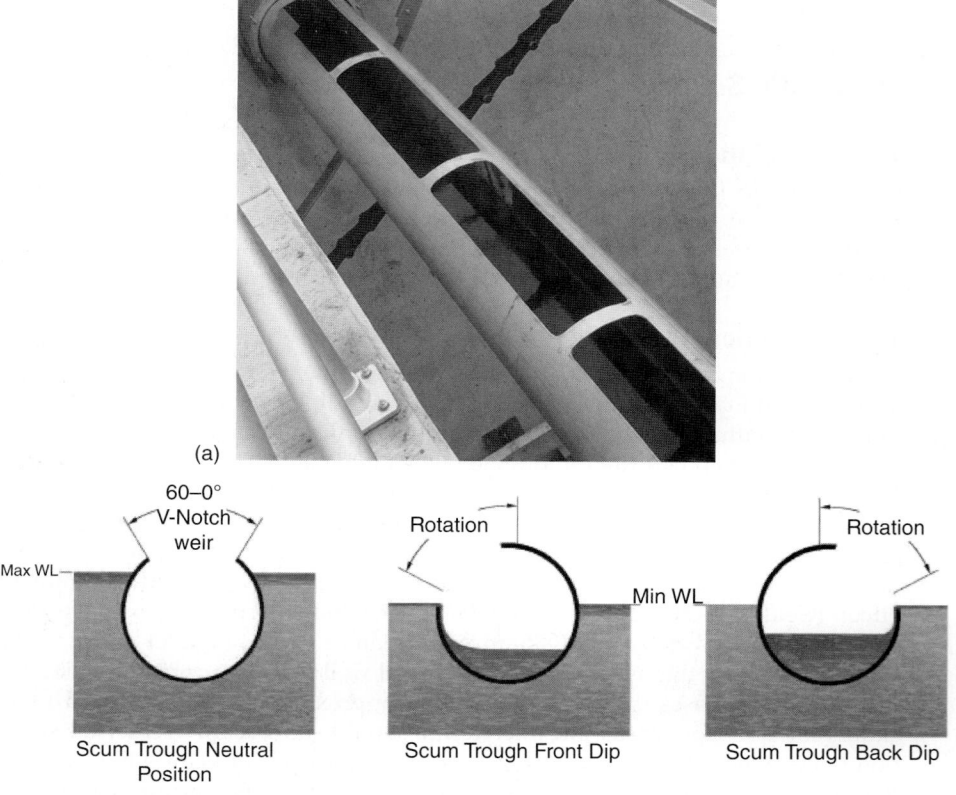

FIGURE 10.35   Tilting trough scum collector (a) picture, (b) operation positions (Courtesy of Brentwood Industries Inc. Reading, PA.).

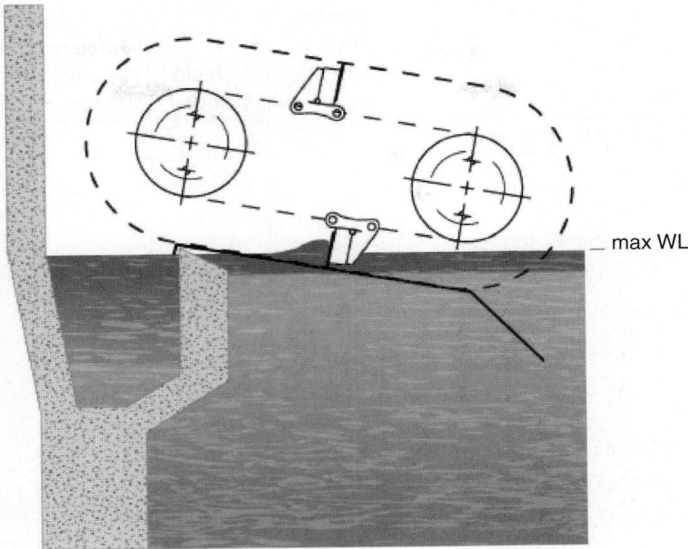

FIGURE 10.36 Sloping beach scum collector (Courtesy of Brentwood Industries, Inc. Reading, PA).

a collector for material floating on the surface, which is pushed to either a sloping beach or a tilting trough skimmer. A spring-loaded section of the rotating arm rides up the beach, wipes the floatable material into a trough, and drops back into the water on the far side. The tilting trough typically extends into the tank just short of the rotating arm. As the arm or surface collector passes, it physically tilts (rotates) the trough, allowing it to skim the surface.

7.3 Quantities and Characteristics

Scum quantities and chemical composition for 19 facilities are summarized in Table 10.8. The dry-weight scum quantities ranged between 0.1 and 19 mg/L, with a median value of approximately 5 mg/L. Quantity and chemical composition of scum are both highly variable and depend on several factors, including the degree and type of industrial development, eating establishments, commercial kitchens, and local demographics in the collection area. Chemical composition also depends on recycled sidestreams and scum removal efficiencies of upstream processes.

7.4 Transport and Handling

Progressing cavity pumps, plunger pumps, pneumatic ejectors, submersible chopper, and recessed impeller centrifugal pumps, both with and without cutting-bar attachments, have been used to pump scum. The design of scum removal equipment includes measures to keep the scum tank or hopper contents mixed during pumping to prevent scum from building up and crusting at the surface. A recirculation line from the pump discharge can be used for mixing. The scum hopper should be relatively small in size to minimize the scum holding time and the bottom of the scum tank should be sloped

Treatment Plants	Quantity (Dry Weight), mg/L	Percent Volatile	Percent Oil and Grease	Fuel Value	
				Btu/lb	kJ/kg
Northwest Berge County, New Jersey	2.3				
Minneapolis-St. Paul, Minnesota		98		5600	13 000
East Bay, Oakland California	9.8	96	91	6000	14 000
West Point, Seattle Washington	2.9				
Not Stated		89		7200	16 800
Three New York City Plants, New York	0.1–2.0		80		
Los Angeles County Sanitation Districts, California	10				
Albany Georgia	17				
Milwaukee, Wisconsin	3.1				
Centra Costa County, California	5.6				
Sacramento, California	14.4				
Passaic Valley Sewerage Commissioners, New Jersey	6				
Detroit, Michigan	3				
Wards Island, New City, New York	4.8				
Bissell Pt., St. Louis, Missouri	10.5				
Calumet, Metropolitan Water Reclamation District of Greater Chicago, Illinois	1.8	Average	Average	Average	Average
Southwest, Metropolitan Water Reclamation District of Greater Chicago, Illinois	5.3	Of all	Of all	Of all	Of all
Westside, Metropolitan Water Reclamation District of Greater Chicago, Illinois	4.8	Four is	Four is	Four is	Four is
Northside, Metropolitan Water Reclamation District of Greater Chicago, Illinois	0.5	91	73	6600	15 300
Number of plants	19	8	7	8	8
Maximum	17	98	91*	7200	16 800
Median	4.8				
Average	5.3	93	77*	6500	15 000
Minimum	0.1	89	73*	5600	13 000

*Likely saponifiable (biodegradeable) content is 60% to 70%.

TABLE 10.8   Raw Wastewater Scum Characterizations (From Primary Sedimentation Facilities) (Mulbarger et al., 1989; U.S. EPA, 1979)

(Gelderloss et al., 2004). Adequate safety provisions should be included to prevent workers from inadvertently falling into a scum tank. Large infrared heaters should be provided in cold climates to keep the scum from congealing and to improve conveyance (Gelderloss et al., 2004). Some designers use glass-lined pipe, which is kept reasonably warm (15°C [59°F] or higher) to minimize blockages (U.S. EPA, 1979). Flushing connections, pigging stations, and cleanouts should be provided where blockage could occur.

7.5   Concentration, Treatment, and Disposal

Methods that have been used for scum concentration, treatment, and disposal are listed in Table 10.9. Historically, scum has been landfilled, co-processed with wastewater treatment sludges, digested, or incinerated. The designer should make sure that the scum disposal issues are not just deferred or shifted to another process. Addition of scum to anaerobic digesters can result in increased digester biogas production; however, adequate digester mixing must be provided when scum is discharged to a digester to ensure complete digestion and to minimize scum blanket formation. Scum and sludge grinding should be considered to eliminate aesthetically offensive floating plastic and rubber articles. Scum may be concentrated by flotation and self-cleaning, rotating screens. Chemically fixed scum facilitates handling and disposal and offers possible beneficial reuse as a structural fill or interim or final landfill cover. In Boston, Massachusetts, chemical fixation of scum with lime, Portland cement, soluble silicates, and cement kiln dust yields an easily handled product with low permeability and no indicator organisms (Mulbarger et al., 1989).

8.0   Downstream Process Considerations

The selection and design of primary treatment systems should consider the objectives of the systems and the effects on downstream treatment processes (Bixio et al., 2000). Special downstream considerations are discussed below.

8.1   Generation of Volatile Fatty Acids

Primary clarifiers that precede biological nutrient removal (BNR) systems may be operated to generate volatile fatty acids (VFAs) by fermenting primary sludge to enhance biological phosphorus removal (EBPR) (Barajas et al., 2002; Christensson, 1997; Christensson et al., 1998; Skalsky and Daigger, 1995). Online pre-fermenters are known as activated primary clarifiers (APCs). In addition, fermentation of primary sludge enhances its suitability as an electron donor and increases the quantity of soluble substrate (carbon) available to support denitrification in the downstream BNR processes (Baur et al., 2002; Lee et al., 1995; Millard, 2006). A primary clarifier can be operated as a pre-fermenter by increasing the sludge blanket depth and solids retention time (SRT) to create anaerobic conditions for hydrolysis and fermentation. Fermented sludge is recirculated from the bottom hopper to the top (inlet) of the clarifier where the influent wastewater flow elutriates generated VFAs. The APCs must have sufficient depth; otherwise, increasing the operating sludge blanket will be difficult and increase the risk of sludge washout, particularly during wet-weather conditions. Sludge collection equipment must also be designed to withstand the additional torque resulting from a higher

Method	Advantages	Disadvantages
Co-processing with wastewater treatment solids and biological stabilization		
Aerobic	Partial decomposition occurs. Avoids complexity of separate handling. Widely used.	May cause grease balls to form. May cause petroleum contamination of solids affecting reuse. May degrade appearance of solids if to be reused. May cause scum buildup if remaining residue is not completely removed.
Anaerobic	Same as above for Aerobic.	Same as above for Aerobic. If digester not strongly mixed, scum layer is certain to form; some scum layer is likely with any digester; digester should be sized for this layer. Removal of scum layer is difficult except on cleaning; disposal of layer on cleaning is still difficult. Requires good decanting to avoid the introduction of significant quantities of water with scum when it is applied to the digester. Avoidance of significant water may lead to line clogging unless special precautions are taken.
Co-processing with wastewater treatment solids and incineration	Low theoretical incremental cost.	Requires good decanting. Can tax furnace if not introduced in nearly continuous manner. Special air pollution concerns with particulates and volatiles.
Individual processing without wastewater solids and incineration	Little residual ash.	Same as above Co-processing. High first and maintenance costs. Unacceptable at small plants.
Individual processing without wastewater solids and landfill	Low capital cost.	May have high operating cost. Requires good decanting to minimize volume and fluidity. Product may be unacceptable because of odors and infectious character.
Individual processing without wastewater solids and reuse as animal feed or low grade soap manufacturer	Low capital cost.	Limited application because of risks (most toxic organics tend to concentrate in grease) and superior, low-cost alternatives.

TABLE 10.9   Methods of Handling Raw Wastewater Treatment Scum (Mulbarger et al., 1989; U.S.EPA, 1979)

and thicker sludge blanket. Activated primary clarifiers are often plagued with erratic performance, are difficult to control, emit odors, and create less than optimal conditions for settling of suspended solids. Operation in a sequential cluster arrangement, where the feed is cycled between different primary clarifiers, provides better control but is limited by the number of available clarifiers. For example, if four clarifiers are available, each clarifier receives wastewater for three days and on the fourth day receives no wastewater and is allowed to continue fermentation. At the end of the fourth day the fermented sludge is pumped to the downstream BNR process and the cycle is repeated. A small amount of fermented primary sludge is retained to seed the next batch. This provides an effective SRT of four days (Latimer et al., 2007). Typically, sidestream pre-fermentation in a separate covered reactor is preferable as it is more efficient, less prone to upsets and odors, produces more VFAs, and provides better SRT control (McCue et al., 2006). For a more thorough discussion of BNR systems including EBPR and side-stream pre-fermentation, see Chapter 12.

8.2 Preservation of Available Carbon

A primary treatment system upstream of biological nutrient removal processes designed for nitrification and denitrification might be specifically designed or operated for less than optimal removals of suspended solids and BOD to preserve available carbon needed for denitrification (Tang et al., 2004; Parker et al., 2001). The designer should carefully compare the benefits and disadvantages of such changes. For a more thorough discussion of BNR systems, see Chapter 12.

8.3 Co-thickening of Primary and Secondary Solids

Waste biological sludge is sometimes discharged to the influent end of primary sedimentation tanks for co-thickening (settlement and consolidation). In several fixed-film facilities (i.e., trickling filters, biologically aerated filters [BAF], and solids contact), co-thickening of waste solids with raw wastewater has not adversely affected primary sludge settling or thickening (Kemp and MacBride, 1990). These facilities, which exist in areas with moderate climates, do not have excessive SORs, and are equipped with rapid sludge withdrawal systems to prevent increases in soluble BOD as a result of biological activity in the sludge blanket. Typical average SORs for co-thickening clarifiers range from 24 to 32 m^3/m^2 d (600–800 gpd/sq ft) with co-thickened sludge concentrations ranging from 3% to 5% (Metcalf and Eddy, 2014). Several benefits of co-thickening include enhanced primary sedimentation because of increased flocculation with secondary solids and elimination and simplification of separate secondary thickening facilities and solids-handling processes (WEF, 2005). Co-settling raw sewage with backwash from BAF appears to enhance primary treatment by 10% to 20% regardless of ferric salt addition (Rogalla et al., 2007). It should be noted that the nature of biological sludge from an activated sludge process (waste activated sludge [WAS]) is significantly different than that from fixed-film processes. Separate WAS thickening has replaced the practice of WAS co-thickening for most activated-sludge facilities.

8.4 Digester Hydrogen Sulfide Control

The primary treatment system might be designed to add ferric chloride, ferrous chloride, ferric or ferrous sulfate, or chlorine to control the content of hydrogen sulfide in digesters and digester biogas. While chlorine oxidizes reduced sulfur compounds, iron salts react with sulfide to form an insoluble iron-sulfide precipitate. This ties up the sulfide and prevents it from being released in either the primary or digestion processes. These chemicals sometimes are best added upstream where there is more turbulence for mixing. Information on odor and emission control strategies including hydrogen-sulfide control strategies for anaerobic digestion tanks can be found in Chapter 6.

8.5 Toxicity Considerations

Coagulants, such as cationic polymers, in secondary and tertiary treatment have been shown to exhibit some toxicity (Fort and Stover, 1995). The effect of using coagulants and flocculants in primary clarification is not known at this time. The addition of cationic polymers to mixed liquor or secondary effluent was found to contribute to disinfection byproducts such as N-nitrosodimethylamine (NDMA) during chlorination or chloramination (Huitric et al., 2006) The effect of adding cationic polymers to primary clarifiers on the formation of NDMA is not known at this time. When effluent discharge requirements or receiving water quality objectives include a chloride limit, polymer or ferric sulfate may be used instead of ferric chloride as the coagulant.

8.6 Sludge Disposal

Site-specific sludge disposal alternatives and the overall facility energy balance should be considered when weighing whether or not to provide primary treatment and in selecting the target TSS and COD or BOD removal efficiencies. For instance, incineration typically favors a high primary to biological solids ratio to maximize the caloric value of sludge to support autogenic combustion. Anaerobic sludge digestion also favors a higher primary to biological solids ratio, which results in higher biogas production rates.

9.0 Design Example

This section provides an example of the design of a primary clarifier. It is being designed for a WRRF with influent characteristics provided in Table 10.10. The facility needs to achieve at least 60% TSS removal and 45% BOD removal at average flow conditions.

where $TSS_{PI} = 290$ mg/L; and
  $TSS_{non} = 90$ mg/L.

	Concentration (mg/L)						Temperature (°C)	Primary influent (m³/d)
	TSS_{PI}	TSS_{non}	COD_{PI}	COD_{non}	BOD_{PI}	BOD_{non}		
Average	290	90	580	290	250	125	24	80 000*

*21 mgd; peak sanitary flow factor = 1.7; peak storm flow factor = 2.6.

Note: TSS = total suspended solids; COD = chemical oxygen demand; BOD = biochemical oxygen demand; PI = primary influent; non = nonsettleable.

TABLE 10.10 Influent Characteristics for Design Plant

Substituting these values into eq 10.3, yields the maximum TSS removal efficiency as follows:

$$E_{TSSmax} = 1 - (TSS_{non}/TSS_{PI}) = 1 - (90/290) = 0.69$$

The settling parameter (λ) was estimated as 100 m³/m² d (2500 gpd/sq ft) from facility data by plotting the TSS removal versus TSS_{PI} data and then fitting a curve with eq 10.8 and the TSS_{PI}, TSS_{non}, TSS_{PE}, and SOR data.

The new facility will be designed with an average SOR of 40 m³/m² d (1000 gpd/sq ft). This value was selected to provide the desired TSS and BOD removals, which are validated below. Using eq 10.2:

$$E_{TSS} = E_{TSSmax}(1 - e^{-\lambda/SOR}) = 0.69(1 - e^{-(100/40)}) = 0.63 \text{ (more the 60\%, acceptable)}$$

where COD_{PI} = 580 mg/L; and
COD_{non} = 290 mg/L.

Using eqs 10.5 and 10.4:

$$E_{CODmax} = 1 - (COD_{non}/COD_{PI}) = 1 - (290/580) = 0.50$$
$$E_{COD} = E_{CODmax}(1 - e^{-\lambda/SOR}) = 0.5(1 - e^{-(100/40)}) = 0.46$$

where BOD_{PI} = 250 mg/L; and
BOD_{non} = 125 mg/L.

Using eqs 10.7 and 10.6:

$$E_{BODmax} = 1 - (BOD_{non}/BOD_{PI}) = 1 - (125/250) = 0.50$$
$$E_{BOD_5} = E_{BODmax}(1 - e^{-\lambda/SOR}) = 0.50(1 - e^{-(100/40)}) = 0.46 \text{ (more the 45\%, acceptable)}$$

For the design flow with a SOR of 40 m³/m² d (1000 gpd/sq ft), the quality of the primary effluent can be estimated.

Using eqs 10.8, 10.9, and 10.10 as follows:

$$TSS_{PE} = TSS_{non} + (TSS_{PI} - TSS_{non})e^{-\lambda/SOR} = 106 \text{ mg/L}$$
$$COD_{PE} = COD_{non} + (COD_{PI} - COD_{non})e^{-\lambda/SOR} = 314 \text{ mg/L}$$
$$BOD_{PE} = BOD_{non} + (BOD_{PI} - BOD_{non})e^{-\lambda/SOR} = 135 \text{ mg/L}$$

Using eq 10.1:

where $SOR = Q/A$
SOR = 40 m³/m² d (1000 gpd/sq ft); and
Q = 80 000 m³/d (21 mgd).

The required surface area (A), for all tanks, is:

$$A = Q/SOR = (80\ 000 \text{ m}^3/\text{d})/(40 \text{ m}^3/\text{m}^2 \text{ d}) = 2000 \text{ m}^2$$

In this example, facility operations have requested that equipment is standardized for a 6-m (20-ft) rectangular tank width. A length of 70 m (230 ft) is available for the primary sedimentation tanks. Therefore, the required number of tanks is

$$N = A/(6 \text{ m} \times 70 \text{ m}) = 2000 \text{ m}^2/420 \text{ m}^2 = 4.8 \text{ tanks}$$

$$\text{Number of tanks needed} = 5 \text{ (rounded up)}$$

Using a depth of 3.65 m (12 ft), the volume of each tank can be calculated:

$V_{tank} = A_{tank} \times \text{depth} = 420 \text{ m}^2 \times 3.65 \text{ m} = 1530 \text{ m}^3$

Design flow per tank $= Q_{tank} = Q/N = (80\,000 \text{ m}^3/\text{d})/(5 \text{ tanks}) = 16\,000 \text{ m}^3/\text{d per tank}$

Hydraulic detention time $= V_{tank}/Q_{tank} = [1530 \text{ m}^3/(16\,000 \text{ m}^3/\text{d}] \times 24 \text{ hr/d} = 2.3 \text{ hrs}$ (2.5 hrs is acceptable)

Optionally, using a depth of 3 m (10 ft), the volume of each tank can be calculated:

$V_{tank} = A_{tank} \times \text{depth} = 420 \text{ m}^2 \times 3 \text{ m} = 1260 \text{ m}^3$

Hydraulic detention time $= V_{tank}/Q_{tank} = [1260 \text{ m}^3/(16\,000 \text{ m}^3/\text{d}] \times 24 \text{ hr/d} = 1.9 \text{ hrs}$

From eq 10.17:

$$S_M = \frac{Q * \text{TSS} * E}{1000} \text{ metric}$$

$$S_M = [80\,000 \text{ m}^3/\text{d} \times 290 \text{ mg/L} \times 0.63]/[(1000 \text{ mg/L})/(\text{kg/m}^3] = 14\,600 \text{ kg/d}$$

Assuming typical primary sludge solids concentration = 4% and bulk specific gravity = 1.02 (from Table 10.7), the sludge volume is calculated as follows:

$$\text{Sludge volume} = [14\,600\text{kg/d}]/[1.02(1000 \text{ kg/m}^3)(0.04)] = 358 \text{ m}^3/\text{d}$$

10.0   References

Albertson, O. E. (2005) Clarifier Scum Removal Can Be Both Simple and Efficient. *J. Environ. Eng.* **131** (2), 225–231.

Albertson, O. E.; Okey, R. W. (1992) Evaluating Scraper Designs. *Water Environ. Technol.* **4**(1), 52–58.

American Water Works Association; American Society of Civil Engineers (2012) *Water Treatment Plant Design*, 5th ed.; McGraw-Hill: New York.

Andreu-Villegas, R.; Letterman, R.D. (1976) Optimizing Flocculation Power Input. *J. Environ. Eng.*, **102**, 251.

Barajas, M. G.; Escallas, A.; Mujeriego, R. (2002) Fermentation of a Low VFA Wastewater in an Activated Primary Tank. *Water SA* **28** (1), 89–98.

Baur, R.; Bhattarai, R. P.; Benisch, M.; Neethling, J. B. (2002) Primary Sludge Fermentation—Results from Two Full-Scale Pilots at South Austin Regional (TX, USA) and Durham AWWTP (OR, USA). *Proceedings of the 77th Annual Water Environment Federation Technical Exposition and Conference* [CD-ROM]; New Orleans, Louisiana, Oct. 2–6; Water Environment Federation: Alexandria, Virginia.

Bixio, D.; van Hauwermeiren, P.; Thoeye, C. (2000) Impact of Primary Treatment Technologies on BNR: The Case Study of the STP of Ghent. *Watermatx Gent.*, Sept. 18.

Caliskaner, O.; Tchobanoglous, Downey, M. (2014) Pilot Testing of Primary Effluent Filtration for Carbon Diversion Using Compressible Medium Filter and Disk Filter. *Proceedings of the 87th Annual Water Environment Federation Technical Exposition and Conference* [CD-ROM]; New Orleans, Louisiana, Sept. 27–Oct. 1; Water Environment Federation: Alexandria, Virginia.

Caliskaner, O.; Tchobanoglous, G.; Reid, T.; Kunzman, B.; Young, R.; Ramos, N. (2015) Evaluation and Demonstration of Five Different Filtration Technologies as an Advanced Primary Treatment Method for Carbon Diversion. *Proceedings of the 88th Annual Water Environment Federation Technical Exposition and Conference* [CD-ROM]; Chicago, Illinois, Sept. 26–30; Water Environment Federation: Alexandria, Virginia.

Caliskaner, O.; Tchobanoglous, G.; Reid, Young, R.; Downey, M.; T.; Kunzman, B.; (2016) Advanced Primary Treatment via Filtration to Increase Energy Savings and Plant Capacity. *Proceedings of the 89th Annual Water Environment Federation Technical Exposition and Conference* [CD-ROM]; New Orleans, Louisiana, Sept. 24–28; Water Environment Federation: Alexandria, Virginia.

Camp, T. R. (1946) Sedimentation and the Design of Settling Tanks. *Trans. Am. Soc. Civ. Eng.*, **3** (2285), 895.

Camp, T. R. (1955) Flocculation and Flocculation Basins. *Trans. Am. Soc. Civ. Eng.*, **120**, 1.

Carter, P.; Worrell, J.; Daigger, G.; Allen, E.; Land, G. (2003). Enhanced Primary Treatment: Full-Scale Pilot Answers Many Questions. *Proceedings of the 76th Annual Water Environment Federation Technical Exposition and Conference* [CD-ROM]; Los Angeles, California, Oct. 11–15; Water Environment Federation: Alexandria, Virginia.

Chaudhary, M. R.; Shao, Y. J.; Crosse, J.; Soroushian, F. (1989) Evaluation of Chemical Addition in the Primary Plant at Los Angeles' Hyperion Treatment Plant. *Proceedings of the 62nd Water Pollution Control Federation Annual Conference*; San Francisco, California; Oct. 16–19; Water Pollution Control Federation: Washington, D.C.

Christensson, M. (1997). Enhanced Biological Phosphorus Removal: Carbon Sources, Nitrates as Electron Acceptor, and Characterisation of the Sludge Community. Ph.D. Thesis, Lund University, Lund, Scania, Sweden.

Christensson, M.; Lie, E.; Jonsson, K.; Johansson, P.; Welander, T. (1998) Increasing Substrate for Polyphosphate-Accumulating Bacteria in Municipal Wastewater through Hydrolysis and Fermentation of Sludge in Primary Clarifiers. *Water Environ. Res.*, **70** (2), 138–145.

Constantine, T. A.; Brook D.; Crawford, G.; Sacluti, F.; Black, S.; McKenna, D. (2003) The Disinfection Potential of Two Enhanced Primary Treatment Technologies Treating Wet Weather Flows at Edmonton, Gold Bar WWTP. *Proceedings of the 76th Annual Water Environment Federation Technical Exposition and Conference* [CD-ROM]; Los Angeles, California, Oct. 2–6; Water Environment Federation: Alexandria, Virginia.

Ding, H. B.; Doyle, M.; Erdogan, A.; Wikramanayake R.; Gallagher, P. (2015) Innovative use of dissolved air flotation with biosorption as primary treatment to approach energy neutrality in WWTPs. *Water Practice & Technology* Vol. 10 No. 1

Ducoste, J.; Daigger, G. T.; Smith, R. (1999) Evaluation of Stacked Secondary Clarifier Design Using Computational Fluid Dynamics. *Proceedings of the 72nd Annual Water Environment Federation Technical Exposition and Conference* [CD-ROM]; New Orleans, Louisiana, Oct. 10–13; Water Environment Federation: Alexandria, Virginia.

Fischerstrom, C. N. H. (1955) Sedimentation in Rectangular Basins. *Sanit. Eng. Div. Am. Soc. Civ. Eng.*, **81**, 1–29.

Fitzpatrick, J.; Long, M.; Wagner, D.; Middlebrough, C. (2008). Meeting Secondary Effluent Standards at Peaking Factors of Five and Higher. *Proceedings of the 81st Annual Water Environment Federation Technical Exposition and Conference* [CD-ROM]; Chicago, Illinois; Oct. 18–22; Water Environment Federation: Alexandria, Virginia.

Fitzpatrick, J.; Weaver, T., Boner, M.; Anderson, M.; O'Bryan, C.; Tarallo, S. (2011) Wet-Weather Piloting Toward the Largest Compressible Media Filter on the Planet. *Proceedings of the 84th Annual Water Environment Federation Technical Exposition and Conference* [CD-ROM]; Los Angeles, California, Oct. 15–19; Water Environment Federation: Alexandria, Virginia.

Foess, G. W.; Jarrett, P.; Williams, B. G. (2003). Economic Benefits of Chemically Assisted Primary Treatment, Step Feed, and Anoxic Zones. *Proceedings of the 76th Annual Water Environment Federation Technical Exposition and Conference* [CD-ROM]; Los Angeles, California, Oct. 11–15; Water Environment Federation: Alexandria, Virginia.

Fort, D. I.; Stover, E. L. (1995) Impact of Toxicity and Potential Interactions of Flocculants and Coagulant Aids on Whole Effluent Toxicity Testing. *Water Environ. Res.*, **67** (921), 921–925.

Gelderloss, A.; Lachlik, T.; Syed, A.; Moore, T. (2004) New Primary Scum Handling System for the City of Detroit's WWTP–An Innovative Approach. *Proceedings of the 77th Annual Water Environment Federation Technical Exposition and Conference* [CD-ROM]; New Orleans, Louisiana; Oct. 2–6; Water Environment Federation: Alexandria, Virginia.

Gerges, H.; Cortez, M.; Wei, H. P. (2006) Chemically Enhanced Primary Treatment Leads to Big Savings Oro Loma Sanitary District Experience. *Proceedings of the 79th Annual Water Environment Federation Technical Exposition and Conference* [CD-ROM]; Dallas, Texas; Oct. 21–25; Water Environment Federation: Alexandria, Virginia.

Graber, S. D. (1974) Outlet Weir Loading and Settling Tanks. *J. Water Pollut. Control Fed.*, **46**, 2355.

Great Lakes-Upper Mississippi River Board of State and Provincial Public Health and Environmental Managers (2014) *Recommended Standards for Wastewater Facilities.* Health Research, Inc., Health Education Division: Albany, New York.

Green, W. D.; Sangrey, K.; Nowak, J.; Foisy, M.; LaVergne D. (2007) From Theory to Practice: Preliminary and Primary Treatment Residuals Handling Improvements. *Proceedings of the 80th Annual Water Environmental Federation Technical Exposition and Conference* [CD-ROM]; San Diego, California; Oct. 13–17; Water Environment Federation: Alexandria, Virginia.

Grohmann, A. (1985) *Flocculation in Pipes: Design and Operation*; Verlag: New York.

Hamlin, M. J. (1972) Paper 2. Preliminary Treatment and Sedimentation. In Advances in Sewage Treatment, *Proc. Conf. Inst. Civil Eng.* London, England.

Harleman, D. R. F.; Morrissey, S. P. (1990) Chemically-Enhanced Treatment: An Alternative to Biological Secondary Treatment for Ocean Outfalls. *Proceedings of the American Society of Civil Engineers National Conference on Hydraulic Engineering*, July 30–Aug. 3; San Diego, California.

Harleman, D.; Murcott, S. (2001a) An Innovative Approach to Urban Wastewater Treatment in the Developing World. *Water 21*, June 2001, 45–48.

Harleman, D.; Murcott, S. (2001b) CEPT: Challenging the Status Quo. *Water 21*, June 2001, 57–59.

Heinke, G.W.; Tay, A. J.-H; Qazi, M. A. (1980) Effects of Chemical Addition on the Performance of Settling Tanks. *J. Water Pollut. Control Fed.*, **52**(12), 2946–2954.

Hetherington, M.; Pamson, G.; Ooten, R.J. (1999) Advanced Primary Treatment Optimization and Cost Benefit Documented at Orange County Sanitation District. *Proceedings of the 72nd Annual Water Environment Federation Annual Technical Exposition and Conference* [CD-ROM]; New Orleans, Louisiana, Oct. 9–13; Water Environment Federation: Alexandria, Virginia.

Hudson, H. E. (1981) *Water Clarification Process: Practical Design and Evaluation.* Van Nostrand Reinhold: New York.

Huitric, S., Kou J.; Creel, M.; Tang, C.; Snyder, D.; Horvath, R.; Stahl, J. (2006) Reclaimed Water Disinfection Alternatives to Avoid NDMA and THM Formation. *Proceedings of the 79th Annual Water Environment Federation Technical Exposition and Conference* [CD-ROM]; Dallas, Texas, Oct. 21–25; Water Environment Federation: Alexandria, Virginia.

Jimenez, B.; Chavez, A.; Leyva, A.; Tchobagoglous, G. (1999). Sand and Synthetic Medium Filtration of Advanced Primary Treatment Effluent for Mexico City. *Water Res.*, **34**(2), 473–481.

Johnson, P. D.; Girinathannair, P.; Ohlinger, K. N.; Ritchie, S.; Teuber, L.; Kirby, J. (2008). Enhanced Removal of Heavy Metals in Primary Treatment Using Coagulation and Flocculation, *Water Environ. Res.*, **80**(5), 472–479.

Johnson, B.; Phillips, J.; Bauer, T.; Smith, G.; Smith, G.; Sherlock, J. (2014) Startup and Performance of the World's first Large Scale Primary Dissolved Air Floatation Clarifier. *Proceedings of the 87th Annual Water Environment Federation Technical Exposition and Conference* [CD-ROM]; New Orleans, Louisiana, Sept. 27–Oct. 1; Water Environment Federation: Alexandria, Virginia.

Jolis, D.; Ahmad, M. (2001). Pilot Testing of High Rate Clarification Technology in San Francisco. *Proceedings of the 74th Annual Water Environment Federation Technical Exposition and Conference* [CD-ROM]; Atlanta, Georgia, Oct. 13–17; Water Environment Federation: Alexandria, Virginia.

Jolis, D.; Ahmad, M. (2004). Evaluation of High-Rate Clarification for Wet-Weather-Only Treatment Facilities. *Water Environ. Res.*, **76**(5), 474–480.

Jordão, E. P.; Figueiredo, I. C. (2005) Chemically Enhanced Primary Treatment – Pilot Investigation and Case Studies in Brazil. *Proceedings of the 78th Annual Water Environment Federation Technical Exposition and Conference* [CD-ROM]; Washington, D.C., Oct. 29–Nov. 2; Water Environment Federation: Alexandria, Virginia.

Katehis, D.; Sandino, J.; Daigger, G. (2011) Maximizing Wet Weather Treatment Capacity of Nutrient Removal Facilities., *Proceedings of the 84th Annual Water Environment Federation Technical Exposition and Conference* [CD-ROM]; Los Angeles, California, Oct. 15–19; Water Environment Federation: Alexandria, Virginia.

Kawamura, S. (1976) Considerations on Improving Flocculation. *J. Am. Water Works Assoc.*, **65**(6), 320.

Kawamura, S.; Lang, J. (1986) Re-Evaluation of Launders in Rectangular Sedimentation Basins. *J. Water Pollut. Control Fed.*, **58**, 12.

Kelly, K. (1988) New Clarifiers Help Save History. *Civ. Eng.*, **58**, 10.

Kemp, F. D.; MacBride, B. D. (1990) Rational Selection of Design Criteria for Winnipeg South End WPCC Primary Clarifiers. *Proceedings of the 63rd Water Pollution Control Federation Annual Conference*; Washington, D.C.; Water Environment Federation: Alexandria, Virginia.

Kinnear, D. J. (2002) Evaluating Secondary Clarifier Collector Mechanisms. *Proceedings of the 75th Annual Water Environment Federation Technical Exposition and Conference* [CD-ROM]; Chicago, Illinois, Sept. 28–Oct. 2; Water Environment Federation: Alexandria, Virginia.

Kinnear, D. J. (2004) Comparing Primary Clarifier Design and Modeling Techniques Favors Traditional Techniques Over Recent Water Environment Research Foundation Studies. *Proceedings of the 77th Annual Water Environment Federation Technical Exposition and Conference* [CD-ROM]; New Orleans, Louisiana, Oct. 2–6; Water Environment Federation: Alexandria, Virginia.

Klute, R. (1985) Rapid Mixing in Coagulation/Flocculation Processes-Design Criteria. *Chemical Water Wastewater Treatment*; Grohmann, A., Hahn, H. H., Klute, R., Eds.; Gustaf Fischer Verlag Publishers: Stuttgart Germany, New York.

Krugel, S.; Melcer, H.; Hummel, S.; Butler, R. (2005) High Rate Chemically Enhanced Primary Treatment as a Tool for Wet Weather Plant Optimization and Re-Rating. *Proceedings of the 78th Annual Water Environment Federation Technical Exposition and Conference* [CD-ROM]; Washington DC; Oct. 29–Nov. 2; Water Environment Federation: Alexandria, Virginia.

Lager, J. A.; Locke, E. R. (1990) Design Management Keeps Boston's Wastewater Program on *Track. Public Works*, **121**, 13.

Latimer, R.; Pitt, P. A.; van Niekerk, A.; Houk, T.; Deacon, S. (2007). Review of Primary Sludge Fermentation Performance in South Africa and the USA. *Proceedings of the 80th Annual Water Environment Federation Technical Exposition and Conference* [CD-ROM]; San Diego, California, Oct. 13–17; Water Environment Federation: Alexandria, Virginia.

Lee, S.; Koopman, B.; Park, S.; Cadee,K. (1995) Effect of Fermented Wastes on Denitrification in Activated Sludge. *Water Environ. Res.*, **67**(7), 1119–1122.

Leffler, M. R.; Harrington, J. (2001). SSO Elimination through Expanded Primary Treatment Capacity and Blended Effluent. *Proceedings of the 74th Annual Water Environment Federation Exposition and Conference* [CD-ROM]; Atlanta, Georgia, Oct. 13–17; Water Environment Federation: Alexandria, Virginia.

Leng, J.; Strekler, A.; Bucher, B.; Gellner, J.; Kennedy, K.; Neethling, J. B. (2002) High Rate Primary Treatment—Emerging Technologies. *Proceedings of the 75th Annual Water Environment Federation Technical Exposition and Conference* [CD-ROM]; Chicago, Illinois, Sept. 28–Oct. 2; Water Environment Federation: Alexandria, Virginia.

Ma, J.; Reid, T.; Caliskaner, O.; Binder, D.; Castillo, C. (2015) Cloth Depth Filtration of Primary Domestic Wastewater. *Proceedings of the 88th Annual Water Environment Federation Technical Exposition and Conference* [CD-ROM]; Chicago, Illinois, Sept. 26–30; Water Environment Federation: Alexandria, Virginia.

McCue, T. M.; Randall, A. A.; Eremektar, F. G. (2006) Contrasting the Benefits of Primary Clarification versus Prefermentation In Activated Sludge Biological Nutrient Removal Systems. *J. Environ. Eng.*, **132**(9), 1061–1067.

McGhee, T. J.; Steel, E. W. (2007) Water Supply and Sewerage, 6th ed.; McGraw-Hill: New York.

Melcer, H.; Carter, P.; Butler, R. C.; Moore, Z.; Caldwell, B. (2014) Bench- to Pilot- to Full-Scale Verification of CEPT Assisted Dual Use Primary Clarifier Technology at King County's Brightwater Treatment Plant, *Proceedings of the 87th Annual Water Environment Federation Technical Exposition and Conference* [CD-ROM]; New Orleans, Louisiana, Sept. 27–Oct. 1; Water Environment Federation: Alexandria, Virginia.

Merritt, F. S. (1983) *Standard Handbook for Civil Engineers*, 3rd ed.; McGraw-Hill: New York.

Mertsch, V. (1985) *Characteristics of Sludge from Wastewater Flocculation/Precipitation*. New York.

Metcalf and Eddy, Inc. (2014) *Wastewater Engineering: Treatment, Disposal, Reuse*, 5th ed.; McGraw-Hill: New York.

Millard, S. (2006) Fermentation of Primary Sludge for Volatile Fatty Acids Production. Masters Thesis, Cranfield University, Cranfield, Bedfordshire, United Kingdom.

Mills, J. A.; Reardon, R. D.; Chastain, C. E.; Cameron, J. L.; Goodman, G. V. (2006) Chemically Enhanced Primary Treatment for Large Water Reclamation Facility on a Constructed Site—Considerations for Design, Start-up, and Operation. *Proceedings of the 79th Annual Water Environment Federation Technical Exposition and Conference*; Dallas, Texas, Oct. 21–25; Water Environment Federation: Alexandria, Virginia.

Moll, H. G. (1985) *Fluid Mechanical Principles of Flocculation in Pipes*; Verlag: New York.

Mulbarger, M. C.; Trubiano, R.; Pape, L.; Gallinaro, G. (1989) Scum Management: Past Practices/New Approaches. *Proceedings of the 62nd Water Pollution Control Federation Annual Conference*, San Francisco, California, Oct. 16–19; Water Pollution Control Federation: Alexandria, Virginia.

Narayanan, B.; Karam, W.; Leveque, E. G.; Besett, T.; Baadsgarrd, M. (2000) A New Approach for Defining the Limits of Chemically Enhanced Primary Treatment. *Proceedings of the 74th Annual Water Environment Federation Technical Exposition and Conference* [CD-ROM]; Atlanta, Georgia, Oct. 13–17; Water Environment Federation: Alexandria, Virginia.

Neupane, D. R.; Riffat, R.; Murthy, S. N.; Peric, M. R. (2006) Influence of Source Characteristics, Chemicals and Flocculation on Non-Settleable Solids and Chemically Enhanced Primary Treatment. *Proceedings of the 79th Annual Water Environment Federation Technical Exposition and Conference*; Dallas, Texas, Oct. 21–25; Water Environment Federation: Alexandria, Virginia.

Neupane, D. R.; Riffat, R.; Murthy, S. N.; Peric, M. R.; Wilson, T. E. (2008) Influence of Source Characteristics, Chemicals and Flocculation on Chemically Enhanced Primary Treatment. *Water Environ. Res.*, **80** (4), 331–338.

Noyola, A.; Tinajero, A. (2005) Anaerobic Thermophilic Digestion of Sludge from Enhanced Primary Treatment of Municipal Wastewater. *Proceedings of the 78th Annual Water Environment Federation Technical Exposition and Conference* [CD-ROM]; Washington D.C., Oct. 29–Nov. 2; Water Environment Federation: Alexandria, Virginia.

Ødegaard, H. (1992) Norwegian Experiences with Chemical Treatment of Raw Wastewater. *Water Sci. Technol.*, **25** (12), 255–264.

Ødegaard, H. (1998) Optimized Particle Separation in Primary Step of Wastewater Treatment. *Water Sci. Technol.*, **37**, 43–53.

Ødegaard H. (2005) Combining CEPT and Biofilm Systems. *Proceedings of the International Water Agency Specialised Conference on BNR*; Krakow, Poland, Sept. 2005; IWA Publishing: London, England.

Parker, D.; Esquer, M.; Hetherington, M.; Malik, A.; Robison, D.; Wahlberg, E.; Wang, J. (2000) Assessment and Optimization of a Chemically Enhanced Primary Treatment System. *Proceedings of the 73rd Annual Water Environment Federation Technical Exposition and Conference on Water Quality and Wastewater Treatment* [CD-ROM]; Anaheim, California; Oct. 14–18; Water Environment Federation: Alexandria, Virginia.

Parker, D.; Barnard, J.; Daigger, G.; Tekippe, R.; Wahlberg, E. (2001) The Future of Chemically Enhanced Primary Treatment: Evolution Not Revolution. *Water 21*, June, 49–56.

Parkhurst, J. D.; Stahl, J. F.; Wuerdeman, D. J.; Yunt, F. W. (1976) Wastewater Treatment for Ocean Disposal. *Paper presented at American Society of Civil Engineers National Conference of Environmental Engineering Research, Development and Design*; Seattle, Washington; American Society of Civil Engineers: Washington, D.C., July 12–14.

Peric, M.; Riffat, R.; Murthy, S. N.; Cassel, A.; Neupane, D. R.; Wilson, T. E. (2006). Laboratory Clarifier test to Predict and Optimize Full Scale Chemically Enhanced Primary Treatment (CEPT) Process. *Proceedings of the 79th Annual Water Environment Federation Technical Exposition and Conference* [CD-ROM]; Dallas, Texas, Oct. 21–25; Water Environment Federation: Alexandria, Virginia.

Porter, J.; and Strain C. J. (2015) Alternative Technology for Primary Treatment—Pilot Testing and Process Modeling at the City of Largo WWRF, *Florida Water Resources Conference.*

Reardon, R. (2005) Clarification Concepts for Treating Peak Wet-Weather Wastewater Flows. *Fla. Water Res. J.*, January, pp. 25–30.

Rogalla, F.; Chan, T. F.; Michelet, F.; Jolly, M. (2007) Troubleshooting and Optimisation of a Large Lamella and High Rate BAF Plant for Ocean Discharge. *Proceedings of the 80th Annual Water Environment Federation Technical Exposition and Conference* [CD-ROM]; San Diego, California, Oct. 13–17; Water Environment Federation: Alexandria, Virginia.

Sanks, R. L. (1981) *Water Treatment Plant Design*; Chelsea, Michigan: Ann Arbor Science.

Siczka, J.; Sandino, J.; Onderko, R.; Sigmund, T. (2007). Biologically and Chemically Enhanced Clarification for Improved Treatment of Wet-Weather Flows. *Proceedings of the 80th Annual Water Environment Federation Technical Exhposition and Conference* [CD-ROM]; San Diego, California, Oct. 13–17; Water Environment Federation: Alexandria, Virginia.

Skalsky, D. S.; Daigger, G. (1995) Wastewater Solids Fermentation for Volatile Acid Production and Enhanced Biological Phosphorus Removal. *Water Environ. Res.*, **67** (2), 230–237.

Soap and Detergent Association (1989) *Principles and Practices of Nutrient Removal from Municipal Wastewater*; Soap and Detergent Association: New York.

Sova, R.; Wilson, R.; Crisler, C. (2008) Automated Primary and Secondary Solids Control Using Solids Density Meters. *Proceedings of the 81st Annual Water Environment Federation Technical Exposition and Conference* [CD-ROM]; Chicago, Illinois, Oct. 18–22; Water Environment Federation: Alexandria, Virginia.

Steel, W. E. (1979) Water Supply and Sewerage; McGraw-Hill: New York.

Stevenson, R.; Nitz, D.; Middlebrough, C.; Dyson, J. (2008) Operation of the Largest High Rate Clarification System for CSO Control in North America. *Proceedings of the 81st Annual Water Environment Federation Technical Exposition and Conference* [CD-ROM]; Chicago, Illinois, Oct. 18–22; Water Environment Federation: Alexandria, Virginia.

Sun, J.; Townsend, R.; Parke, S.; Dillon, J. (2008) Biologically Enhanced HRC System Solving Peak Wet Weather Flow Challenges. *Proceedings of the 81st Annual Water Environment Federation Technical Exposition and Conference* [CD-ROM]; Chicago, Illinois, Oct. 18–22; Water Environment Federation: Alexandria, Virginia.

Sutton, P.; Rusten, B.; Ghanam, A.; Dawson, R.; Kelly, H. (2008) Rotating Belt Screen: An Attractive Alternative for Primary Treatment of Municipal Wastewater. *Proceedings of the 81st Annual Water Environment Federation Technical Exposition and Conference* [CD-ROM]; Chicago, Illinois, Oct. 18–22; Water Environment Federation: Alexandria, Virginia.

Tang, C. C.; Prestia, P.; Kettle R.; Chu, D.; Mansell, B.; Kuo, J.; Horvath, R. W.; Stahl, J. F. (2004) Start-Up of a Nitrification/Denitrification Activated Sludge Process with High Ammonia Side-Stream: Challenges and Solutions. *Proceedings of the 77th Annual Water Environment Federation Technical Exposition and Conference* [CD-ROM]; New Orleans, Louisiana, Oct. 2–6 ; Water Environment Federation: Alexandria, Virginia.

Tebbutt, T. H. Y.; Christoulas, D. G. (1975) Performance Relationships for Primary Sedimentation. *Water Res.*, **9**, 347.

Theroux, R. J.; Betz, J. M. (1959) Sedimentation and Preaeration Experiments at Los Angeles. *Sew. Ind. Wastes*, **31**, 1259.

Uhte, W. R. (1990) Personal Communication.

U.S. Environmental Protection Agency (1975) *Process Design Manual for Suspended Solids Removal*, EPA-625/1-75-003a; U.S. Environmental Protection Agency: Washington, D.C.

U.S. Environmental Protection Agency (1979) *Process Design Manual for Sludge Treatment and Disposal*, EPA-625/1-79-011; U.S. Environmental Protection Agency: Washington, D.C.

U.S. Environmental Protection Agency (1987) *Design Manual for Phosphorus Removal*, EPA-625/1-87-001; U.S. Environmental Protection Agency: Cincinnati, Ohio.

U.S. Environmental Protection Agency (1999) *Design Criteria for Mechanical, Electric, and Fluid System and Component Reliability*, EPA-430-99-74-001; U.S. Environmental Protection Agency: Washington, D.C.

Wahlberg, E. J.; Wunder, D. B.; Fuchs, D. C.; Voight, C. M. (1999) Chemically Assisted Primary Treatment: A New Approach to Evaluating Enhanced Suspended Solids Removal. *Proceedings of the 72nd Annual Water Environment Federation Technical Exposition and Conference*; New Orleans, Lousiana, Oct. 9–13; Water Environment Federation: Alexandria, Virginia.

Wahlberg, E. J.; Sheridan, C.; Koho, S. (2003) Determine the Affect of Individual Wastewater Characteristics and Variances on Primary Clarifier Performance. *Proceedings of the 76th Annual Water Environment Federation Technical Exposition and Conference* [CD-ROM]; Los Angeles, California, Oct. 11–15; Water Environment Federation: Alexandria, Virginia.

Wahlberg, E. J.; Stallings, R. B.; Appleton, A. R. (2005) Primary Clarifier Design Concepts and Considerations. *Proceedings of the 78th Annual Water Environment Federation Technical Exposition and Conference* [CD-ROM]; Washington DC, Oct. 29–Nov. 2; Water Environment Federation: Alexandria, Virginia.

Wall, D. J.; Petersen, G. (1986) Model for Winter Loss in Uncovered Clarifiers. *J. Environ. Eng.*, **12**, 1.

Water Environment Federation (2005) *Clarifier Design*, 2nd ed.; Manual of Practice No. FD–8; Water Environment Federation: Alexandria, Virginia.

Water Environment Federation (2016) *Operation of Water Resource Recovery Facilities*, 7th ed., Manual of Practice No. 11; Water Environment Federation: Alexandria, Virginia.

Water Environment Research Foundation (1999) Establishing Primary Sedimentation Tank and Sedimentation Clarifier Evaluation Protocols. *Research Priorities for Debottling, Optimizing and Rerating Wastewater Treatment Plants*, Final Report, Project 99-WWF-1; Water Environment Research Foundation; Alexandria, Virginia.

Water Environment Research Foundation (2006) Determine the Effect of Individual Wastewater Characteristics and Variances on Primary Clarifier Performance. Final Report, Project 00-CTS-2; Water Environment Research Foundation; Alexandria, Virginia.

Wells, S. A.; Liberte, D. M. (1998). Winter Temperature Gradients in Circular Clarifiers. *Water Environ. Res.*, **70** (7), 1274–1279.

Yee, L. Y.; Babb, A. F. (1985) Inlet Design for Rectangular Settling Tanks by Physical Modeling. *J. Water Pollut. Control Fed.*, **57**, 12.

Young, J. C.; Edwards, F. G. (2000) Fundamentals of Ballasted Flocculation Reactions. *Proceedings of the 73rd Annual Water Environment Federation Technical Exposition and Conference* [CD-ROM]; San Diego, California, Oct. 14–18; Water Environment Federation: Alexandria, Virginia.

CHAPTER 11

Biofilm Reactor Technology and Design

Pusker Regmi, Ph.D., P.E.; Chris deBarbadillo, P.E.; and
David G. Weissbrodt, Asst. Prof., Ph.D., M.Sc., Dipl.-Ing.

1.0   Introduction: Biofilms and Biofilm Reactors in Municipal Wastewater Treatment

Biofilm reactors retain microbial cells in a biofilm attached to fixed or movable carriers. The biofilm matrix consists of microorganisms, water, and a variety of soluble and particulate components that include soluble microbial products, inert material, and extracellular polymeric substances (EPS) (Boltz et al., 2017).

Active biomass concentrations inside the biofilm are large at 10 to 60 g of volatile suspended solids (VSS)/L of biofilm compared to suspended growth reactors at 3 to 8 g VSS/L of reactor volume. Biomass in suspended growth reactors is removed from the system through sludge wastage, resulting in an average solids residence time (SRT). If the suspended biomass SRT is insufficient, the system is incapable of maintaining a substantial inventory of slow-growing or temperature-sensitive bacteria. These bacteria include autotrophic bacteria and heterotrophic bacteria, respectively. This loss of specific bacterial groups is known as washout. In biofilm reactors, microbes attached to a carrier are protected from washout as long as they do not detach from the biofilm and can grow in locations where their food supply remains abundant. Detached biofilm fragments are eliminated from the system if the suspended phase SRT is lower than the washout SRT.

If the suspended biomass SRT is less than the minimum required for a particular biomass to develop, organism will grow selectively in the biofilm. However, little research exists to describe the interaction between suspended and biofilm entrained microbes when both suspended biomass and biofilm are used to transform substrates in a single bioreactor (e.g., integrated fixed-film activated sludge bioreactors). Figure 11.1 provides a conceptual illustration of different biofilm reactor types. Biofilm reactors can be classified according to the number of phases involved—gas, liquid, solid—according to the biofilm being fixed or moving within the reactor. They also are classified according

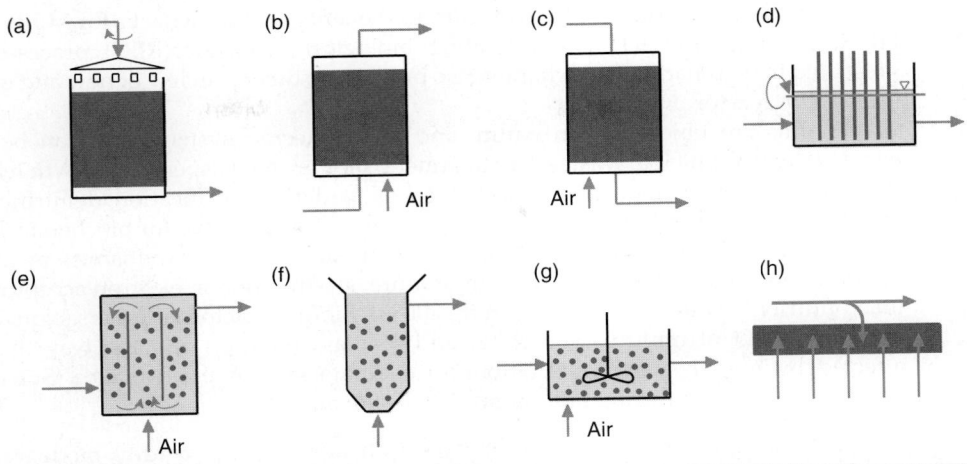

FIGURE 11.1 Types of biofilm reactors: (a) trickling filter; (b) submerged fixed-bed biofilm reactor operated as upflow or (c) downflow; (d) rotating biological contactors; (e) suspended biofilm reactor, including airlift reactor; (f) fluidized bed reactor; (g) moving bed biofilm reactor; and (h) membrane-attached biofilm reactors (Morgenroth, 2008) (with permission from IWA Publishing).

to how electron donors or acceptors are applied to seven basic types as follows (Harremoës and Wilderer, 1993; Lewandowski and Boltz, 2011):

1. Three-phase system—fixed biofilm-laden carrier material, bulk water, and air. Water trickles over the biofilm surface and air moves upward or downward in the third phase (e.g., trickling filter) (Figure 11.1a).

2. Three-phase system—fixed (or semifixed) biofilm-laden carrier material, bulk water, and air. Water flows through the biofilm reactor with gas bubbles (e.g., aerobic biologically active filters [BAFs]). Gravel is a fixed media and polystyrene beads are semifixed (Figures 11.1b and 11.1c).

3. Three-phase system—moving biofilm-laden carrier material, bulk water, and air. Water flows through the biofilm reactor. Air is introduced with gas bubbles (e.g., aerobic moving bed biofilm reactors [MBBRs]) (Figure 11.1g).

4. Two-phase system—moving biofilm-laden carrier material and bulk water. Water flows through the biofilm reactor with the electron donor and electron acceptor (e.g., denitrification fluidized bed biofilm reactor [FBBR]) (Figure 11.1g).

5. Two-phase system—fixed biofilm-laden carrier material and bulk water. Water flows through the biofilm reactor with the electron donor and electron acceptor (e.g., denitrification filter).

6. Three-phase membrane system—a microporous hollow-fiber membrane with biofilm and water on one side and gas on the other diffusing through the membrane to the biofilm (e.g., membrane biofilm reactor) (Figure 11.1h).

7. Two-phase membrane system—a proton exchange membrane separating a compartmentalized biofilm-laden anode from a compartmentalized cathode with water on both sides, but with the electron donor on one side and electron acceptor on the other (e.g., biofilm-based microbial fuel cell [MFC]).

Detailed design criteria, physical features, benefits, and drawbacks for MBBR, BAF, FBBR, trickling filter (TF), and rotating biological contactor (RBC) processes are presented in this chapter. This chapter also presents a cursory review of new and emerging biofilm reactor processes.

Biofilms are ubiquitous in nature and in engineered systems and can be used beneficially in municipal wastewater treatment. Biofilm and suspended growth reactors can meet similar treatment objectives for carbon oxidation, nitrification, denitrification, and desulphurization. The same microorganisms are responsible for biochemical reactions in both activated sludge and biofilm systems and respond in the same way to local environmental conditions (i.e., pH, temperature, electron donor, electron acceptor, and macronutrient availability) (Morgenroth, 2008a). Biofilm reactor designers should consider the effect of multiple substrates and biomass fractions and the way they are affected by mass-transport limitations to evaluate system performance. Substrates typically considered during biofilm reactor design are:

1. Soluble compounds, including electron donors (e.g., readily biodegradable chemical oxygen demand [COD], NH_4^2, NO_2^2, and H_2); electron acceptors (e.g., O_2, NO_3^2, NO_2^2, and SO_4^{22}); nutrients; and buffers (e.g., PO_4^{32}, NH_4^1, and HCO_3^2).

2. Particulate compounds, including electron donors (e.g., slowly biodegradable COD); active biomass fractions (e.g., heterotrophic bacteria and autotrophic bacteria); inert biomass; and EPS.

1.1 Biofilm Reactor Compartments

Biofilm reactors have five primary compartments: (1) influent wastewater (distribution) system; (2) containment structure; (3) carrier with biofilm; (4) effluent water collection system; and (5) an aeration system (for aerobic processes and scour) or mixing system (for anoxic processes that require bulk-liquid agitation and biofilm carrier distribution). Because design is system specific, each is discussed relative to specific biofilm reactor types in subsequent sections. Five components determine the local environment of the biofilm: (1) carrier surface (i.e., substratum); (2) biofilm (including both particulate and liquid fractions); (3) mass-transfer boundary layer (MTBL); (4) bulk liquid; and (5) gas phase (when significant). These components are illustrated in Figure 11.2.

1.2 Biofilm Processes, Structure, and Function

Two processes—mass transfer and biochemical conversion—are characteristic of all biofilm reactors and influence biofilm structure and function. Mass transport inside the biofilm is controlled by molecular diffusion. If mass transfer to the biofilm is slow com-

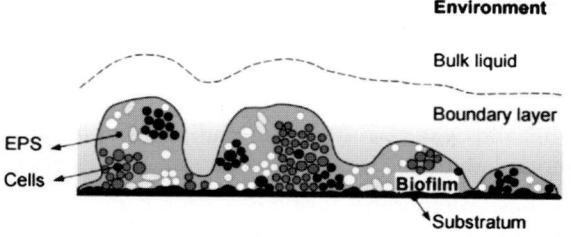

FIGURE 11.2 Schematic representation of biofilm components: (1) carrier surface (i.e., substratum) medium; (2) biofilm (particulate and liquid components); (3) mass transfer boundary layer; (4) bulk liquid; and (5) gas phase (Wanner et al., 2006; reprinted with permission from IWA Publishing).

pared to biochemical conversion, the result is strong concentration gradients for substrates within the mass-transfer boundary layer and inside the biofilm. These mass-transport limitations result in inactive zones deep inside the biofilm and have implications for the design and operation of biofilm reactors and microbial ecology. Mass transport is the primary mechanistic difference between biofilm and suspended growth reactors. Typically, full-scale operating suspended growth systems are kinetically (i.e., biomass) limited, whereas biofilm reactors are diffusion (i.e., surface-area) limited. Therefore, it is necessary to understand the interactions between mass-transport and substrate transformation processes to completely evaluate biofilm systems.

Microbial competition in a biofilm is based on local substrate concentrations and the location of different organism groups. Microbes closer to the surface have the advantage of higher local substrate concentrations. But these microbes are also the most susceptible to shear- and abrasion-induced detachment and subsequent removal from the system with the effluent stream. Microbes living near the carrier, also called the substratum, are offered some protection from detachment but are subject to reduced substrate availability.

Biofilms in wastewater treatment applications typically are mass-transfer limited. But, for high bulk-liquid concentrations or low degradation rates, the limiting substrate can fully penetrate biofilms. Even systems that typically operate in a range of conditions that generate mass-transfer-limited biofilms may be periodically subjected to conditions that result in biofilms being completely penetrated. This is caused by a noncontinuous detachment that is hydraulically triggered, induced by an organic shock load, or carrier/granule collision. Concentration gradients inside the biofilm allow for the development of different redox zones, and partial penetration can result in biofilms simultaneously having aerobic, anoxic, and anaerobic zones. The existence of aerobic and anoxic zones inside the biofilm can be beneficial, for example, in promoting processes such as simultaneous nitrification and denitrification. Once a biofilm is partially penetrated, increasing its thickness (L_F) has no treatment benefit in removing the mass-transfer-limited, depleted substrate. Therefore, biofilm reactor performance is surface-area-dependent and not dependent on the total amount of biomass in the system. In some cases, biofilm penetration can be increased by increasing bulk-phase concentrations (e.g., by increasing bulk phase O_2 concentrations in aerobic biofilm reactors). Boltz et al. (2006) and Boltz and La Motta (2007) demonstrated that the removal of organic and inorganic particles is also a function of biofilm surface area. Organics in municipal wastewater are mostly in the particulate form.

A balance between microbial growth and detachment will result in biofilms with a range of thicknesses. Biofilm thickness that exceeds the rate-limiting substrate penetration depth typically hinders system performance. Excessive biofilm thickness can have two detrimental effects on full-scale reactors that rely on either passive or dedicated biofilm thickness control mechanisms. First, it can reduce biofilm surface area. Second, it can deprive biomass near the carrier of electron donor, electron acceptor, or macronutrients. As a result, the interior biofilm is likely to become anaerobic, may produce odors and result in uncontrolled detachment of biofilm segments that are equivalent in size to its thickness, L_F. The latter is known as sloughing. Passive biofilm thickness control mechanisms are inherent to normal biofilm reactor operating conditions, including, for example, mixing resulting in continuous carrier collisions in an MBBR. Dedicated biofilm thickness control mechanisms require function-specific operating cycles such as backwashing a submerged biologically active filter or flushing a trickling filter. A majority of the biofilm thickness control mechanisms are mechanically induced and hydrodynamically mediated. Table 11.1 lists, for relative comparison, the controlled biofilm thickness range typical of the reactor types described in this chapter.

Type of Biofilm Reactor	Typical Biofilm Thickness (mm)	
	Lower Estimate	Upper Estimate
Moving bed biofilm reactor	50	500
Biologically active filters	20	300
Fluidized bed biofilm reactors	20	400
Rotating biological contactors	200	2000
Trickling filters (carbon oxidizing units)	200	2000
Membrane biofilm reactors	20	500

TABLE 11.1   Typical Biofilm Thicknesses for a Variety of Biofilm Reactors

1.3   Bulk-Liquid Hydrodynamics

Bulk-liquid hydrodynamics is an important, often overlooked, component of biofilm reactor design and simulation. A key difference between the seven previously described biofilm reactor types is mixing conditions and bulk-liquid hydrodynamics. Systems with complex bulk-liquid hydrodynamics, such as trickling filters, make mathematical model application to process design and evaluation a complicated and difficult task. Relatively simple bulk-liquid hydrodynamics, such as submerged and completely mixed systems, however, promote use of mechanistic principles during biofilm reactor design. Reactor-scale hydrodynamics are influenced by tank and biofilm carrier configuration; type; appurtenances (e.g., baffles, mixers, and aeration system); and operating mode (e.g., continuous flow, sequencing batch, and periodic backwashing) (Grady et al., 1999). Bulk-liquid hydrodynamics influence biofilms at all stages of their development (Lewandowski, 2000) and biofilm reactor design through (1) biofilm development and control and (2) biofilm surface area loading (i.e., by influencing the extent of mass-transfer resistance external to the biofilm and load distribution over the available carrier area). Modeling and design approaches typically evaluate bulk-liquid mixing, the mass-transfer boundary layer, and biofilm processes separately. This is a simplification, however, because bulk-liquid turbulence affects biofilm density and development.

Turbulent, high-shear stress environments result in planar, denser biofilms, whereas quiescent, low-shear stress environments result in rough, less-dense biofilms (van Loosdrecht et al., 1995). Boessmann et al. (2004) suggested that the planar, denser biofilms may have improved diffusivity compared with rough, less-dense biofilms. As suggested in Table 11.1, different biofilm reactor types designed to achieve identical treatment objectives may have varying characteristics because of reactor qualities. This chapter focuses on biofilm reactor scale hydrodynamics. Biofilm-scale hydrodynamics is discussed in more detail in other publications (Wanner et al., 2006).

Mass-transfer resistance external to the biofilm results in reduced flux into the biofilm. In some cases, this can be the rate-controlling process, such as in nitrification MBBR (Hem et al., 1994). The extent of external mass-transfer resistance can vary among biofilm reactor types and operating conditions and are conceptualized with the hypothetical mass-transfer boundary layer. This diffusion layer mechanistically links biofilm (micro) and bioreactor (macro) scales. Figure 11.3 illustrates the cross-sectional view of a completely mixed bulk liquid, mass-transfer boundary layer, biofilm, and carrier element (or growth substratum). Mass-transfer boundary layer thickness (L_L) is a function of bulk-liquid hydrodynamics and substrate concentration.

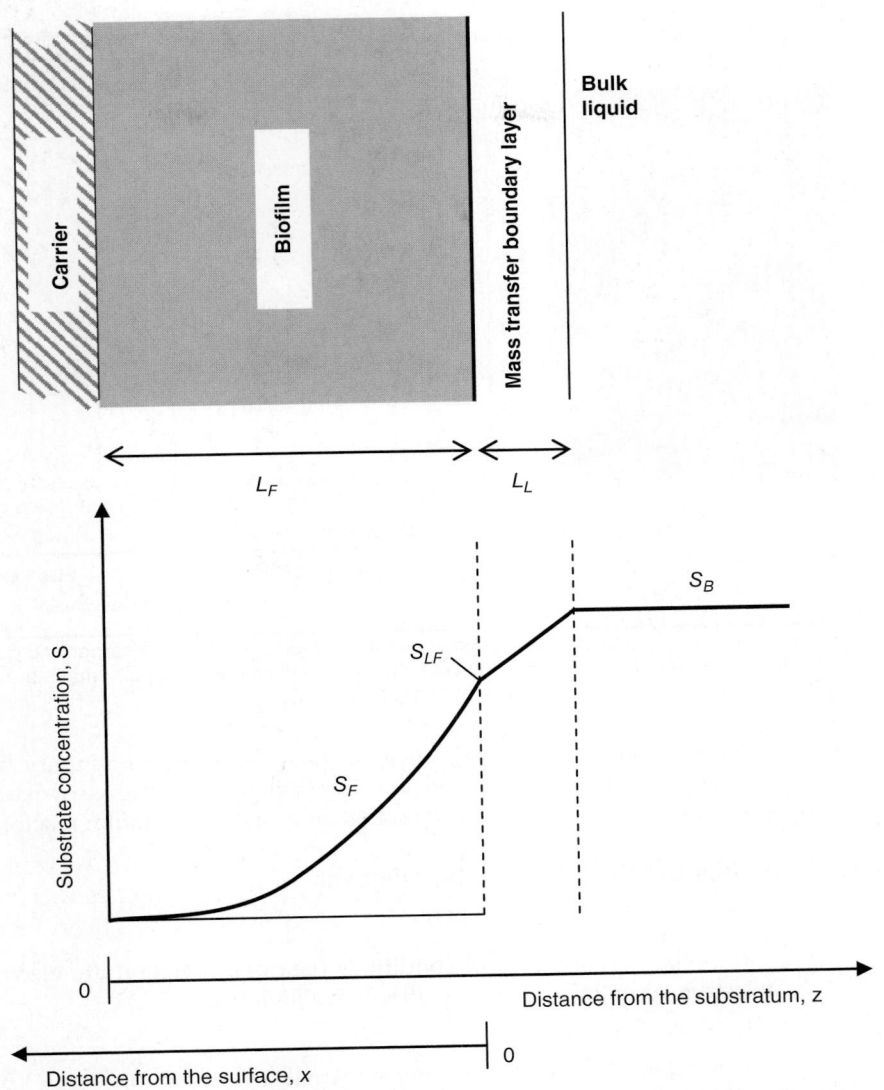

Figure 11.3 Concentration over the thickness of the biofilm (L_F) in the mass-transfer boundary layer and the bulk phase. The space coordinate can be measured either from the biofilm carrier (z, typically used for numerical simulations) or from the surface of the biofilm (x, simplifies the solution of hand calculations) (Morgenroth, 2008).

In Figure 11.4, oxygen profiles both external and internal to the biofilm are shown (Zhang and Bishop, 1994). Lines are conceptual profiles based on observed mass-transfer boundary layer thickness. Although the photograph shows that biofilms are nonplanar, porous, and heterogeneous biostructures, the simplified fundamental concepts discussed here are useful for capturing the mechanisms that govern substrate transformation in reactors. Increasing turbulence in the bulk liquids helps to reduce the

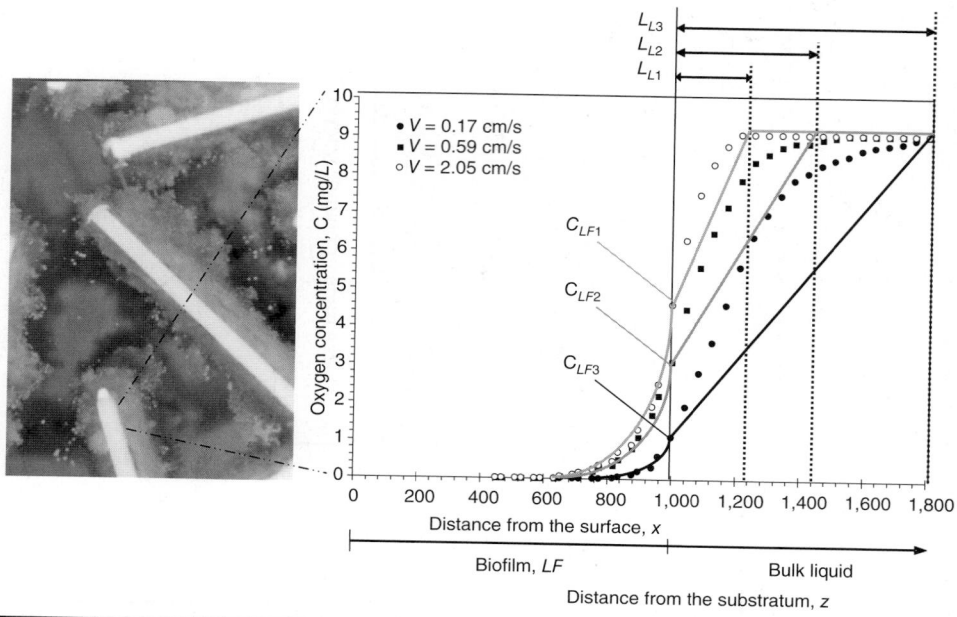

FIGURE 11.4 Photograph of a biofilm grown on a commercially available carrier and comparison of measured oxygen concentrations (circles/squares) and calculated (lines) mass-transfer boundary layer thickness (Zhang and Bishop, 1994).

effect of external mass-transfer resistance, or the mass-transfer boundary layer thickness. Even with vigorous mixing, however, there will be some degree of external mass-transfer resistance that must be considered when designing a biofilm reactor.

1.4   Biofilm Development and Detachment

Five factors affect biofilm development:

1. Bulk-phase environmental conditions (i.e., pH, temperature, electron donor, electron acceptor, and macronutrient availability);

2. Hydraulic loading rate;

3. Extent of mass-transfer resistances external to the biofilm;

4. Extent of biofilm internal mass-transfer resistances;

5. Kinetics and stoichiometry of transformation processes resulting from microbial conversion processes inside the biofilm; and

6. Detachment.

Biofilm ecology in reactors affects their activity. Cold wastewater, for example, increases solubility (thereby increasing the rate of diffusion) but retards biochemical transformation processes. Soluble substrates may, therefore, penetrate deeper into the biofilm resulting in a reactor that can support more active biomass. Figure 11.5 illustrates these effects for a nitrification MBBR with increasing active biomass concentrations during cold weather.

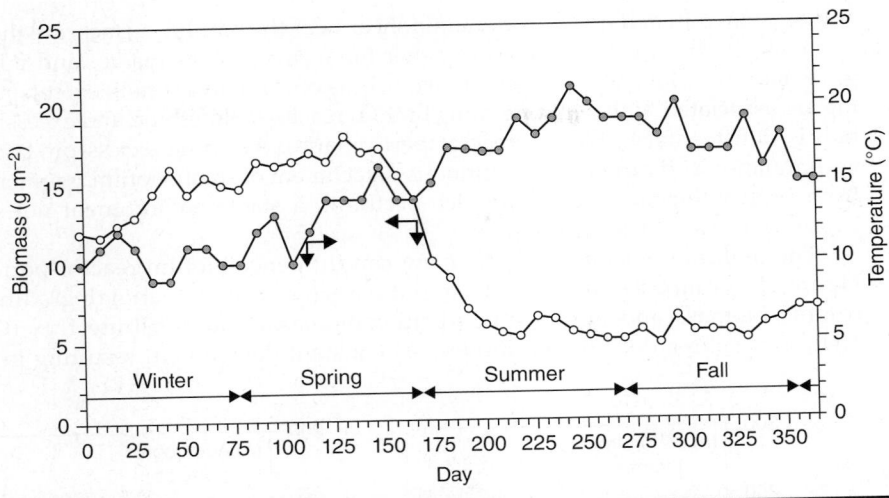

FIGURE 11.5   Seasonal biomass growth in a nitrification moving bed biofilm reactor. As the temperature increases, biofilm mass decreases (Daigger and Boltz, 2010).

Developing biofilms accumulate microbial cells, but all biofilms eventually loose particulate components. The loss of particulate components and their introduction to bulk liquid is called detachment. Bryers (1984) described four biofilm detachment processes: (1) abrasion, (2) erosion, (3) sloughing, and (4) predator grazing. Abrasion and erosion are illustrated in Figure 11.6. Abrasion, which is initiated by particle collision,

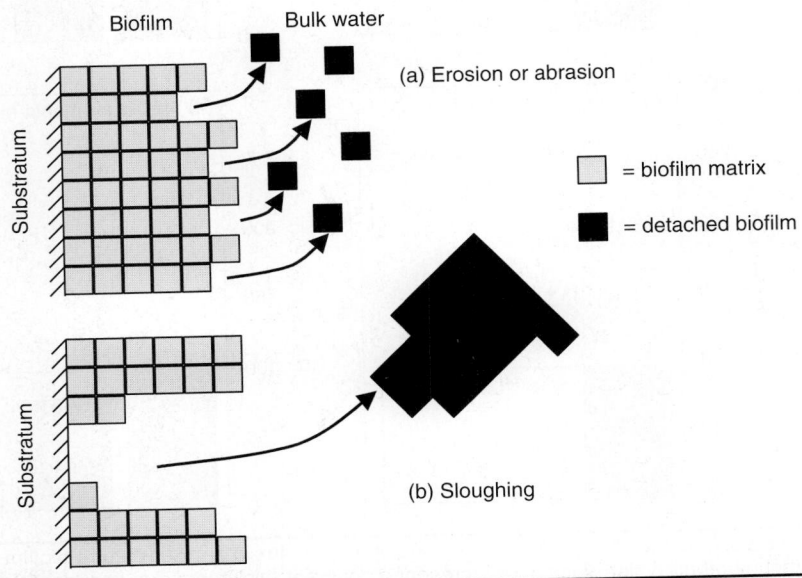

FIGURE 11.6   (a) Erosion and abrasion and (b) sloughing biofilm detachment processes (Morgenroth, 2003).

and erosion, initiated by hydrodynamic shear near the biofilm surface, are the removal of small groups of cells. Predatory higher life forms such as macro- and micro-fauna graze biofilms (Boltz et al., 2008). Abrasion, erosion, and, to a certain extent, fauna grazing are associated with well-operating biofilm reactors. Sloughing and excessive predation is detrimental to biofilm reactor performance. Avoiding excess predatory fauna accumulation and promoting continuous detachment of small biofilm fragments results from proper thickness control, which occurs in a stable environment not subject to excessive mass-transfer resistances.

The biofilm detachment mechanism can influence biofilm reactor performance. Figure 11.7 compares simulated bacterial mass (i.e., nonmethanol degrading heterotrophic biomass and autotrophic nitrifier biomass) and substrate flux (COD and NH_3-N) for three detachment modes: (1) constant detachment resulting in constant

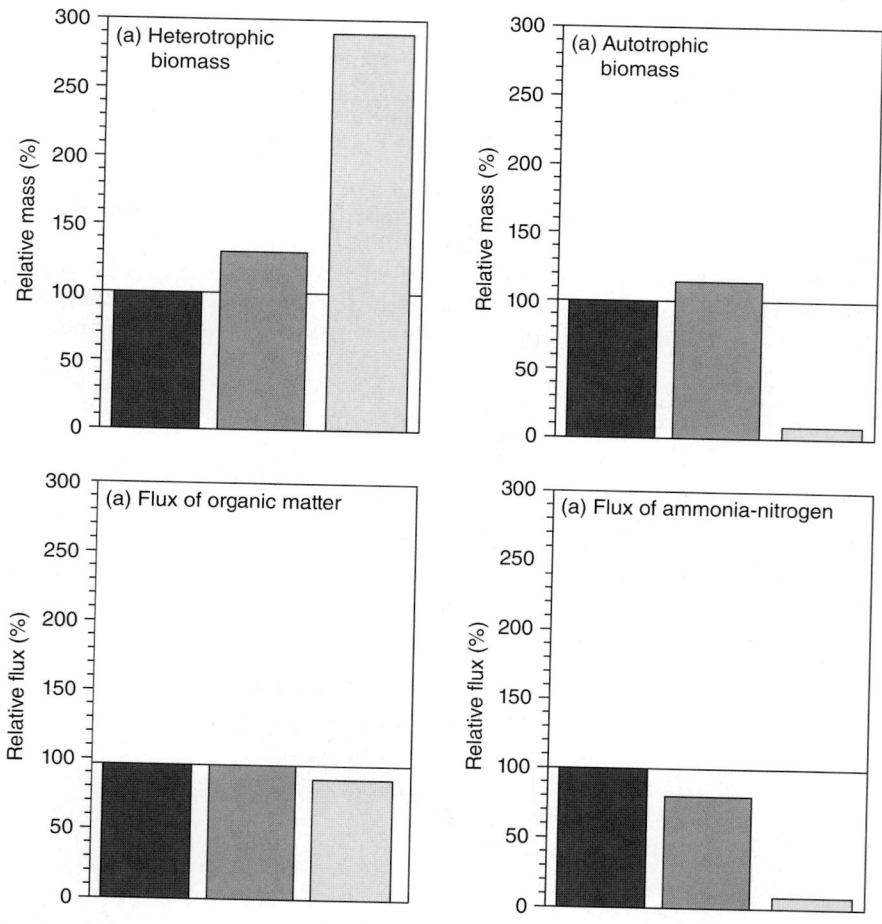

FIGURE 11.7   Simulated bacteria mass and substrate flux in a mixed-culture biofilm for different detachment intervals. Black bars represent a constant thickness; gray bars represent a one-day backwash frequency; and white bars represent a seven-day backwashing frequency (100% equals the results for constant biofilm thickness) (Morgenroth and Wilderer, 1999).

biofilm thickness, (2) daily backwashing, and (3) a 7-day backwashing interval (meant to capture biofilm sloughing) in a BAF. Analysis by Morgenroth and Wilderer (2000) suggests that an increase in the average mass of heterotrophic organisms does not produce a higher COD flux. This, in turn, suggests that the biofilm was partially penetrated and the system was diffusion rather than biomass limited. Average ammonia-nitrogen flux and autotrophic nitrifier biomass were reduced significantly after seven-day backwashing. The rapid loss of both fast- (nonmethanol degrading heterotrophic biomass) and slow-growing (autotrophic nitrifier biomass) bacterial species is advantageous to the former. Biofilm formation and detachment has a significant influence on bacterial competition for substrate in mixed-culture biofilms (Morgenroth, 2003).

2.0 Biofilm Reactor Design Approaches and Considerations

Several design approaches, biofilm reactor model, and biofilm model types typically are used in engineering practice. The primary objective of biofilm or biofilm reactor models is to predict soluble substrate flux (J) into the biofilm. This flux information can be used to obtain an estimate of the (1) overall biofilm reactor performance; (2) required active biofilm surface area; (3) electron acceptor (e.g., dissolved oxygen); (4) external electron donor (e.g., methanol or hydrogen); and (5) biosolids management requirements. This section discusses the relative benefits and limitations of general biofilm reactor design approaches. It is common practice to use more than one biofilm reactor design approach and base the final process design on a comparison of results. The design approaches and biofilm reactor models discussed here include a graphical procedure, empirical models, semiempirical models, and mechanistic mathematical models. Examples are used to demonstrate applicability of the described design approaches, facilitates the comparison of results produced by each method, and provides a basis for discussion of relative benefits and drawbacks. Because mathematical biofilm models are so widely used, this approach is discussed on more detail later in this chapter.

2.1 Simplified Biofilm Reactor Design Approaches

Improvement to a hypothetical water resource recovery facility (WRRF) was evaluated using each of the design procedures outlined above based on meeting the more stringent total nitrogen limitation of 3 mg/L on an annual average basis. Two single-stage or one two-stage MBBRs are evaluated to denitrify secondary wastewater effluent. The system must produce a NO_3-N concentration in the effluent stream of less than 1 mg/L to meet the effluent total nitrogen goal of 3 mg/L. Detailed design criteria for MBBRs are presented later in this chapter. Several assumptions were applied to each example:

- Annual average day flow rate influent to the denitrification MBBR is 34 000 (m^3/d).
- The annual average wastewater temperature is 24°C.
- The following annual average secondary (clarifier) effluent wastewater characteristics have been recorded by operation staff:
  — NO_3-N: 8 mg/L
  — dissolved oxygen = 0 mg/L

- Two, 650-m^3 empty-bed liquid volume tanks exist and can be operated in parallel or series.
- Both of the existing tanks have a 1:1 length-to-width (L:W) ratio.
- Evaluations are conducted assuming that the media has a 500 m^2/m^3 biofilm specific surface area.
- The bulk-liquid volume displaced by 50% empty-bed media fill fraction is 7.25%.
- A 100-mm headloss is permissible through the system.
- Methanol will be used as the supplemental carbon source.
- Acceptable empty-bed media fill fractions is 25% to 67%.
- Downstream filtration units are sufficiently sized to receive solids produced during denitrification.

2.1.1   Graphical Procedure

A graphical procedure can be used to determine the total hydraulic load (THL) required to decrease a substrate concentration, and by definition the biofilm surface area required to provide a desired substrate concentration remaining in the effluent stream. These items can be determined directly. The graphical procedure can be used to determine effluent substrate concentration from any series of continuous flow stirred tank (biofilm) reactors (CFSTRs) and the size required to achieve a desired substrate concentration remaining in the effluent stream.

A stepwise procedure must be used when a series of CFSTRs will be used. Antoine (1976) and Grady et al. (1999) developed the graphical procedure described here and the approach is valid for any biofilm-based CFSTR. If multiple stages are expected to have different characteristics, then the graphical method requires different flux curves to describe system response in each of the CFSTRs.

The procedure requires a graphical representation of substrate flux (J) as a function of bulk-liquid substrate concentration. This relationship between flux and bulk-liquid substrate concentration can be obtained from numerical simulations or full-scale or pilot-scale observations. In practice, this graphical procedure typically is used to extend pilot-scale observations to full-scale biofilm reactor design criteria. The process designer should recognize that the relationship between flux and bulk-liquid substrate concentration is based on the system and location. Therefore, the flux curve required to implement the graphical procedure may not be obtained from or correlate well with values reported in the literature or from different systems. As a result, the process designer should consider carefully the conditions under which the flux curve was developed before applying results. A flux curve representing mass transfer and environmental conditions characteristic of a specific system and operating mode may not be representative of different biofilm reactor types designed to meet the same treatment objectives. A flux curve generated for the same biofilm reactor type under similar operating conditions, however, may offer some direction in the absence of system-specific numerical simulation or pilot/full-scale observations.

When using the graphical procedure to evaluate pilot-plant observations, fluxes should be compared to rates in full-scale systems. Any flux that deviates significantly from those reported in published studies should be used only after careful consideration. Pilot or experimental systems may promote a greater flux than expected.

At steady-state conditions, the basis for the graphical procedure is a material balance on a biofilm-based CFSTR:

$$0 = \underbrace{Q \cdot S_{in,i}}_{\substack{\text{mass per} \\ \text{time input}}} - \underbrace{Q \cdot S_{B,i}}_{\substack{\text{mass per} \\ \text{time output}}} - \underbrace{J_{LF,i} \cdot A}_{\substack{\text{biofilm} \\ \text{transformation rate}}} - \underbrace{r_{B,i} \cdot V_B}_{\substack{\text{suspended growth} \\ \text{transformation rate}}} \qquad (11.1)$$

where Q = flow rate through the system (m³/d);
$\quad S_{in,i}$ = influent concentration of soluble substrate i (g/m³);
$\quad S_{B,i}$ = effluent, or bulk-liquid, concentration of soluble substrate i (g/m³);
$\quad J_{LF,i}$ = flux of soluble substrate i across the biofilm surface (g/m²/d);
$\quad A$ = biofilm surface area (m²);
$\quad r_{B,i}$ = rate of substrate i conversion because of suspended biomass (g/m³/d);
$\quad V_B$ = bulk-liquid volume (m³).

Assuming that transformation occurring in the bulk liquid is negligible, the "suspended growth transformation rate" (eq 11.1) can be neglected. Rearranging eq 11.1 provides the rationale for the graphical procedure:

$$J_{LF,i} = \underbrace{\frac{Q}{A} \cdot S_{in,i}}_{\text{const.}} - \underbrace{\frac{Q}{A}}_{\text{slope}} \cdot S_{B,i} \qquad (11.2)$$

The slope, or $\left(\dfrac{Q}{A}\right)$, is referred to as the operating line and represents the total hydraulic load on each stage [= 34 000 m³/d/(250 m²/m³ × 650 m³ × 0.5) = 0.4 m/d for the example conditions]. Figure 11.8 illustrates the graphical method for the denitrification MBBR example. It is assumed that the flux curves have been created based on observations in both the first and second stage of a pilot-scale denitrification MBBR treating secondary effluent similar to the example. There is a 50% empty-bed media fill fraction and supplemental methanol (MeOH) dosing based on 3:1 (g MeOH:g NO₃-N) and 1.5:1 (g MeOH:g O₂) nitrate-nitrogen and oxygen mass ratios, respectively. The graphical solution indicates that the first-stage denitrification MBBR effluent NO₃-N concentration is approximately 3.9 mg/L. The second stage effluent NO₃-N concentration is approximately 1.1 mg/L with flux rates of approximately 1.6 g/m² d and 1.1 g/m² d in the first and second stage, respectively.

2.1.2 Empirical and Semiempirical Models

Empirical models can be implemented easily either by hand or using a spreadsheet but have limited applicability because of their simplistic, "black-box" consideration of system parameters. Because environmental conditions and bioreactor configuration affect biofilm reactor performance, a system can respond differently from the description provided by an empirical model. The limited descriptive capacity of empirical models typically results from parameter values and model features based on data that was obtained from few system installations or operating conditions. Therefore, the process designer should be aware of conditions under which system-specific model parameters have been defined. Significant sources of variability in values include differences in biofilm carrier type and configuration, the extent of external mass-transfer resistances, and biofilm composition. Despite their ease of implementation, empirical

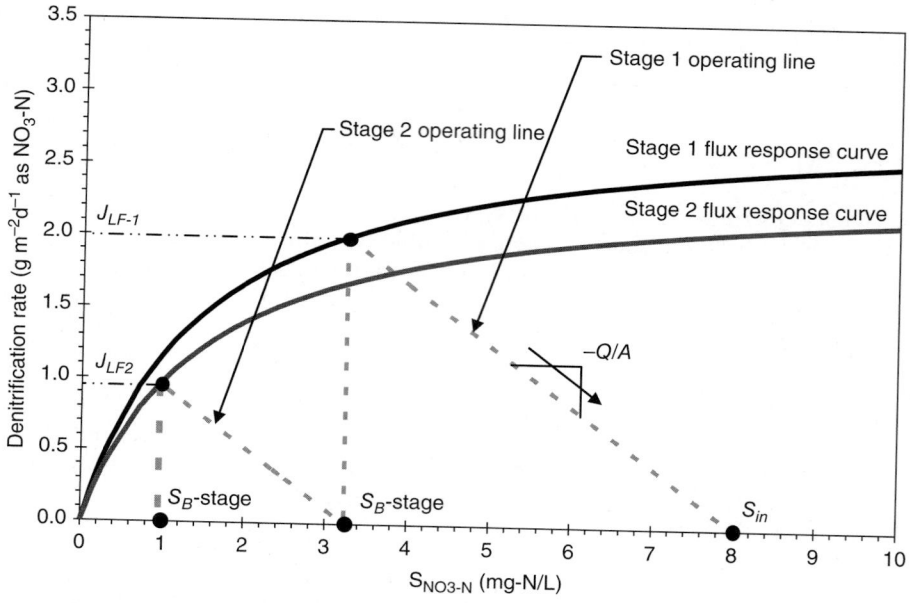

Figure 11.8 Graphical procedure for describing the response of a denitrification moving bed biofilm reactor to defined conditions, including (1) first- and second-stage operating lines and (2) flux curves based on observations at a pilot-scale denitrification moving bed biofilm reactor (Lewandowski and Boltz, 2011).

models can produce results that vary from 50% to 100% compared to actual system performance. The designer must determine if such an error is acceptable.

Coefficient values, and sometimes the empirical models, typically are created to describe system response for the removal of a specific material (e.g., five-day biochemical demand [BOD_5] removal, nitrification, and denitrification). The models can be used as an indicator of system viability for meeting treatment objectives with respect to the specific process governing transformation. Empirical models are, however, inadequate for describing complex processes such as explicit evaluation of two-step nitrification of ammonia to nitrite and then to nitrate. Therefore, empirical models have limited application in defining the conditions that either promote or deter complex processes in biological systems.

Historically, biofilm reactors have been designed using empirical criteria and models. Although this trend is changing, a majority of the design formulations presented in this chapter are empirical in nature even for newer biofilm reactor types such as the MBBR and BAF. In practice, biofilm models typically are applied to process design. Bioreactor-specific empirical models are described in the relative reactor-specific section of this chapter. Equation 11.3, however, presents a simple empirical model applicable to the denitrification MBBR example.

$$J_i = J_{i,\max,T} \cdot \left(\frac{S_{B,i,EA}}{S_{B,i,EA} + K_{i,EA}} \right) \cdot \left(\frac{S_{B,i,ED}}{S_{B,i,ED} + K_{i,ED}} \right) \tag{11.3}$$

where J_i = flux of soluble substrate i (g/m²/d);

$J_{i,\max,T}$ = reaction rate constant, for this empirical model the global maximum flux of soluble substrate i at temperature T (g/m²/d);

$S_{B,i}$ = soluble substrate i concentration remaining in the effluent stream (g/m³);

K_i = system-specific half-saturation coefficient incorporating mass-transfer resistances and other local environmental conditions (g/m³);

EA = electron acceptor;

ED = electron donor.

Temperature correction often is introduced with an Arrhenius function, which in this case is applied to the global maximum flux ($J_{i,\max,T}$):

$$\frac{k_{T_2}}{k_{T_1}} = \theta^{(T_2 - T_1)} \qquad (11.4)$$

where k = reaction-rate constant at temperature T_1 and T_2 (varies for function);

u = temperature coefficient < 1.1.

Equation 11.3 includes multiple Monod-like rate expressions. The designer should recognize that the half-saturation coefficient, K, includes system, and many times, location-specific mass-transfer resistances (Grady et al., 1999). For this reason, the values typically differ from apparent or intrinsic values reported in the literature. For illustration, eq 11.3 has model parameters that include a maximum NO₃-N flux ($J_{i,\max,T}$) of 5.2 g/m² d, methanol half-saturation coefficient ($K_{i,ED}$) of 18 mg/L as COD, and a NO₃-N half-saturation ($K_{i,EA}$) of 1.5 mg/L as N that were obtained from a nonlinear regression analysis. The methanol concentration influent to the first stage was 23 mg/L. These values were obtained from a pilot-scale denitrification MBBR and were applied to eq 11.3 to create the flux curves illustrated in Figure 11.8. Consistent with the illustrative application of the graphical procedure, supplemental carbon is assumed to have been consumed at 3:1 (g MeOH:g NO₃-N) and 1.5:1 (g MeOH:g O₂) nitrate-nitrogen and oxygen mass ratios, respectively. Equation 11.3 can be applied to calculate flux, but eq 11.1 must be rearranged, neglecting bulk-phase conversion processes, to calculate the material concentration remaining in the effluent:

$$S_{B,i} = S_{in} - \frac{J_{LF,i} \cdot A}{Q} \qquad (11.5)$$

Applying eqs 11.3 and 11.5 to the denitrification MBBR example requires an iterative procedure that can be implemented easily by hand or using an optimization tool such as the Excel™ Solver. The following equations were applied:

STAGE 1

$$J_{NO_3\text{-}N}^{STAGE\,1} = 5.2\frac{g}{d \cdot m^2} \cdot \left(\frac{S_{B,NO_3\text{-}N}^{STAGE\,1}}{S_{B,NO_3\text{-}N}^{STAGE\,1} + K_{NO^3\text{-}N}}\right) \cdot \left(\frac{S_{B,M}^{STAGE\,1}}{S_{B,M}^{STAGE\,1} + K_M}\right)$$

$$S_{B,NO_3-N}^{STAGE\,1} = 8\,\underbrace{\frac{g}{m^3}}_{S_{in}} \cdot \frac{\overbrace{J_{NO_3-N}^{STAGE\,1} \cdot \overbrace{250\,\frac{m^2}{m^3}}^{SSA} \cdot \overbrace{0.5}^{Fill} \cdot \overbrace{(650\,m^3 - 650\,m^3 \cdot 0.0725)}^{V}}^{A}}{\underbrace{34,000\,\frac{m^3}{d}}_{}}$$

$$\underbrace{}_{\frac{J \cdot A}{Q}}$$

$$S_{B,M}^{STAGE\,1} = \underbrace{23\,\frac{g}{m^3}}_{S_{in}} - \underbrace{\left\{ \left(8\,\frac{g}{m^3} - S_{B,NO_3-N}^{STAGE\,1} \right) \cdot 3\,\frac{g_M}{g_{NO_3-N}} \right\}}_{\ddot{A}S_{NO3-N} \cdot i_{MeOH}}$$

STAGE 2

$$J_{NO_3-N}^{STAGE\,2} = 5.2\,\frac{g}{d \cdot m^2} \cdot \left(\frac{S_{B,NO_3-N}^{STAGE\,2}}{S_{B,NO_3-N}^{STAGE\,2} + K_{NO^3-N}} \right) \cdot \left(\frac{S_{B,M}^{STAGE\,2}}{S_{B,M}^{STAGE\,2} + K_M} \right)$$

$$S_{B,NO_3-N}^{STAGE\,2} = S_{B,NO_3-N}^{STAGE\,1} - \frac{J_{NO_3-N}^{STAGE\,2} \cdot 250\,\frac{m^2}{m^3} \cdot 0.5 \cdot (650\,m^3 - 650\,m^3 \cdot 0.0725)}{34,000\,\frac{m^3}{d}}$$

$$S_{B,M}^{STAGE\,2} = \underbrace{S_{B,M}^{STAGE\,1}}_{S_{in}} - \underbrace{\left\{ \left(8\,\frac{g}{m^3} - S_{B,NO_3-N}^{STAGE\,2} \right) \cdot 3\,\frac{g_M}{g_{NO_3-N}} \right\}}_{\ddot{A}S_{NO3-N} \cdot i_{MeOH}}$$

After iterating, the equations converged on predetermined error criteria. The method produced, approximately, a first- and second-stage nitrate-nitrogen flux of 1.6 g/m²/d and 1.2 g/m²/d, respectively. These values are comparable with results obtained from the graphical procedure.

If sufficient data exists to allow for development of parameter values and mathematical relationships capable of describing a complete range of conditions expected when treating municipal wastewater, then empirical models can be used. The addition of model components to account for specific phenomenon encroaches on the premise of mechanistic mathematical model development. For this reason, a distinction is made between empirical and semiempirical models. Gujer and Boller (1986) and Sen and Randall (2008a; 2008c) provide an example of the latter describing nitrifying trickling filters (describing MBBRs and BAFs). The literature and biofilm reactor-specific sections of this chapter provide additional information on these semiempirical approaches. Some system manufacturers develop and use proprietary semiempirical models that are based on sufficient data collected from a variety of installations for a specific biofilm reactor type. Therefore, process designers typically cross-reference design criteria with manufacturer recommendations and seek to reconcile discrepancies.

2.2 Mathematical Biofilm Models for the Practitioner

The mass transport and biochemical transformation processes previously described are common to all biofilms; therefore, biofilms can be described by a unifying mathematical expression. Wanner et al. (2006) describe the general biofilm model. Uncertainty and complexity because of wastewater composition, differences in biofilm reactor configuration, appurtenances, operation, and bulk-liquid hydrodynamics renders the general biofilm reactor model for engineering design impractical. Therefore, a variety of simplifications, primarily in the assumed biofilm structure and spatial complexity, to overcome these factors has resulted in the several mathematical biofilm models. Biofilm models, however, are complex despite these simplifications. Furthermore, little documentation exists to aid practitioners in selection and application of biofilm models for meeting specific modeling objectives. Because of their complexity and limited application guidance, biofilm models are not widely used in engineering design, although this is changing.

2.2.1 Why Should We Use Biofilm Models as a Design Tool?

Biofilm models should be used to answer specific questions. Its quality should then be judged based on its usefulness in answering the questions. For example, the Modified Velz Equation was developed, primarily, to describe carbon-oxidation in trickling filters (Eckenfelder, 1961). The empirical model is not applicable to other biofilm reactor types nor is it in any way sufficient for describing complex processes. The following questions are relevant for the design and operation of biofilm reactors:

- What is the rate of substrate removal as a function of bulk-liquid substrate concentrations at a given location within a biofilm reactor?

- What factors are controlling substrate removal in each stage of the system? Possible factors include available biofilm surface area, total amount of biomass, or amount of specific types of microorganisms in the biofilm.

- How much substrate removal occurred throughout the system?

- How should the reactor be designed and operated to ensure that sufficient biofilm remains while avoiding flow distribution problems resulting from excessive biofilm accumulation?

- What are the mixing conditions for water in a biofilm reactor and how do they influence reactor performance? Mixing conditions directly influence flow distribution, problems associated with short-circuiting, and mass-transfer resistances from the bulk of the water to the biofilm surface.

Existing biofilm models can answer at least some of these questions.

2.2.2 Biofilm Models Used in Engineering Design

Typically, process design involves the use of mathematical biofilm models because (1) simulation models are efficiency tools that allow to evaluate quickly a variety of scenarios; and (2) empirical procedures are inadequate for providing information that is now widely considered essential to biofilm reactor design (e.g., local flux as a function of bulk-liquid substrate concentrations, biofilm composition and competition for multiple substrates, and influence of individual processes on interaction between several microbial types).

Modern biofilm reactors facilitate the use of biofilm models in engineering design. For example, MBBRs do not present the hydrodynamic or operational complexities that historically have hindered use of mathematical biofilm model-based process design. This is because they are comprised of zones that are essentially completely mixed and are continuous flowing systems, respectively. Submerged completely mixed biofilm reactors allow for the application of modern biofilm knowledge and are conducive to simulation with existing biofilm models. As a result, most of the existing whole-WRRF modeling programs have been expanded to include a submerged completely mixed biofilm reactor module that consists of a mathematical biofilm model. The process designer should understand the basis for the mathematical biofilm model, its supporting assumptions, and limitations before using these models in design.

Unfortunately, choosing a modeling approach that offers an appropriate level of complexity to meet the modeling objective can be difficult. An overview of different model approaches that are suitable for biofilm reactor design are as follows:

- One-dimensional homogeneous biofilm (single limiting substrate): This approach takes into account mass-transfer limitations and the corresponding effects on concentration profiles and substrate flux into the biofilm. It is assumed that active microbes are homogeneously distributed across the biofilm thickness. The approach is valid only if calculations are performed for the limiting substrate, which has to be determined a priori.

- One-dimensional homogeneous biofilm (multiple substrates and multiple biomass components): One key aspect of modeling biofilms is to evaluate the competition and coexistence of different bacterial groups (e.g., carbon-oxidizing heterotrophic bacteria, nitrifying autotrophic bacteria) and local process conditions (e.g., aerobic, anoxic, or anaerobic). Local process conditions can be determined by calculating biofilm penetration depths for different soluble substrates (e.g., COD, ammonia-nitrogen, oxygen, and nitrate-nitrogen). Growth of individual bacterial groups can be determined based on fluxes. To simplify calculations, it can be assumed that all bacterial groups are homogeneously distributed over the thickness of the biofilm (Rauch et al., 1999; Boltz et al., 2009a; 2009b; 2009c).

- One-dimensional heterogeneous biofilm: Different groups of bacteria are competing in a biofilm for substrate and space. One-dimensional heterogeneous biofilm models must keep track of local growth and decay of different bacterial groups and detachment to biomass distributions over the biofilm thickness (Wanner and Reichert, 1996).

Different scales of heterogeneity are relevant for biofilm reactor design. Length scale of biofilm thickness, on the order of 100 to 1000 mm, is taken into account in one-dimensional and multi-dimensional biofilm models. Substrate fluxes from these simulations can then be integrated into models describing overall reactor performance where the length-scale is on the order of 1 m. But, heterogeneities also can be observed in biofilm reactors in between these scales. For example, in some cases, patchy biofilms develop. This may occur in any under-loaded biofilm reactor such as the lower reaches of some nitrifying trickling filters. These observations confirm that some sections of the biofilm carrier are bare whereas others contain dense development. These heterogeneities between the micro and macro scale typically are not considered in biofilm models, and it is not clear to what extent they are relevant.

Wanner et al. (2006) provide a detailed description and comparison of different modeling approaches. Wanner et al. (2006) and Boltz et al. (2010) state that one-dimensional biofilm models are adequate for reactor design. Currently, most commercial software that is used for biofilm reactor design and evaluation takes into account multiple substrates and biomass fractions in either a one-dimensional heterogeneous or homogeneous biofilm. Some examples of software and references are summarized in Table 11.2.

Mathematical biofilm models are now standard design tools despite the fact that they are significantly more complex than the previously described graphical procedures, empirical and semiempirical models.

Software	Company	Biofilm Model Type and Biomass Distribution	Reference
AQUASIM™	EAWAG, Swiss Federal Institute of Aquatic Science and Technology, Dübendorf, Switzerland (www.eawag.ch/index_EN)	1-D, DY, N; heterogeneous	Wanner and Reichert (1996) (modified)
AQUIFAS™	Aquaregen, Mountain View, California (www.aquifas.com)	1-D, DY, SE and N, heterogeneous	Sen and Randall (2008a,b,c)
BioWin™	EnviroSim Associates Ltd., Flamborough, Canada (www.envirosim.com)	1-D, DY, N, heterogeneous	Wanner and Reichert (1996) (modified), Takács et al. (2007)
GPS-X™	Hydromantis Inc., Hamilton, Canada (www.hydromantis.com)	1-D, DY, N, heterogeneous	Hydromantis (2002)
Pro2D™	CH2M HILL Inc., Englewood, Colorado (www.ch2m.com/corporate)	1-D, SS, N(A), homogeneous (constant L_F)	Boltz et al. (2009a,b,c)
Simba™	ifak GmbH, Magdeburg, Germany (www.ifak-system.com)	1-D, DY, N, heterogeneous	Wanner and Reichert (1996) (modified)
STOAT™	WRc, Wiltshire, England (www.wateronline.com/storefronts/wrcgroup.html)	1-D, DY, N, heterogeneous	Wanner and Reichert (1996) (modified)
WEST™	MOST for WATER, Kortrijk, Belgium (www.mostforwater.com)	1-D, DY, N(A)[a], N[b], homogeneous[a], heterogeneous[b]	Rauch et al. (1999)[a], Wanner and Reichert (1996) (modified)[b]
SUMO	DYNAMITA (http://www.dynamita.com/)	1-D, DY, N; heterogenous	Wanner and Reichert (1996) (modified), Takács et al. (2007), Dynamita, 2016

Note:
1-D = one-dimensional
DY = dynamic
N = numerical
N(A) = numerical solution using analytical flux expressions

TABLE 11.2 Biofilm Models Used in Practice (adapted from Boltz et al., 2009b)

2.2.4 Limitations of Biofilm Models for the Practitioner

Mathematical biofilm modeling has advanced reactor design. Furthermore, the models have been instrumental in explaining the mechanisms that result in mass-transfer limitations, stratification of bacteria over the biofilm thickness (and throughout a biofilm reactor), competition between bacterial groups, and factors affecting the development of multi-dimensional heterogeneous biofilm morphologies. Adoption of biofilm models has been slow; additional research and development is needed to overcome limitations of existing models. There are several limitations of which process designers should be aware:

- Biofilm models are complex and they typically exclude important factors. Most models focus on small-scale heterogeneities inside the biofilm to better describe soluble substrate flux. Many approaches, however, fail to describe overall reactor operation (e.g., backwashing, flow distribution) as a function of system appurtenances (e.g., aeration system, mixing devices).

- Too many models are available. The choice of modeling approach depends on the specific question to be addressed. This is different from activated sludge modeling where mathematical models are directly applicable. The design engineer must be aware of the biofilm model type being used, its simplifying assumptions, and the resultant limitations. If this information is not clearly documented, then results from the model should be used with caution.

- There are significant deficiencies in approaches for calibrating biofilm models. Without reliable and transparent approaches for model calibration, the developed models will not be accepted as a robust design tool.

- Existing biofilm models adequately describe soluble substrate flux, but little basic research exists to allow for development of mathematical terms describing the fate of particulate matter.

- The mass-transfer boundary layer that links the one-dimensional biofilm model with the bioreactor compartment is well understood. Unfortunately, there is a paucity of mathematical descriptions based on a fundamental understanding of the features that influence mass-transfer resistances external to the biofilm (Boltz et al., 2011a).

3.0 Moving Bed Biofilm Reactors

The MBBR is a two (anoxic) or three (aerobic) phase system with a buoyant free-moving plastic biofilm carrier that requires energy (i.e., mechanical mixing or aeration) to ensure uniform distribution throughout the tank. These systems can be used for municipal and industrial wastewater treatment. The process includes a submerged biofilm reactor and liquid-solids separation unit. The installations include several process configurations and effluent water quality standards for carbon oxidation, nitrification, and denitrification. The MBBR process is capable of processing wastewater to meet effluent water quality standards ranging, for example, from the U.S. Environmental Protection Agency definition of secondary treatment (30 mg/L total suspended solids [TSS] and 30 mg/L BOD_5 monthly average) to more stringent nitrogen limits (advanced wastewater treatment standard total nitrogen less than 3 mg/L). According to Rusten et al. (2006), the first MBBR installed in Norway (see European Patent No. 0.575,314 and

U.S. Patent No. 5,458,779) has been routinely inspected and no plastic biofilm carrier wear had been observed after 15 years of continuous operation. Benefits of MBBR include:

- It can meet similar treatment objectives as activated sludge systems for carbon-oxidation, nitrification, and denitrification, but requires a smaller tank volume than a clarifier-coupled activated sludge system.
- Biomass retention is clarifier independent. Therefore, solids loading to the liquid-solids separation unit is reduced significantly compared to activated sludge systems.
- Because it is a continuous flow process, it does not require a special operational cycle for biofilm thickness control. Hydraulic headloss and operational complexity is minimized.
- It offers much of the same flexibility to manipulate system flowsheet (to meet a specific treatment objective) as the activated sludge process. Multiple reactors can be configured in series without the need for intermediate pumping or return activated sludge pumping (to accumulate mixed liquor).
- It can be coupled with a variety of different liquid-solids separation processes including sedimentation basins, dissolved air flotation, ballasted flocculation, and membranes.
- It is well-suited for retrofit installation into existing municipal WRRF such as those based on activated sludge (McQuarrie and Boltz, 2011).

Research and development supporting MBBR-process commercialization resulted from a political agreement among North European countries to make a substantial reduction of approximately 50% in nutrient discharge to the North Sea from 1985 to 1995 (Hem et al., 1994). Since then, many free-moving plastic biofilm carriers have been used in different MBBR configurations. In addition, a range of pollutant loading and bulk-phase external carbon sources in denitrification MBBRs and dissolved oxygen in carbon-oxidation/nitrification MBBRs concentrations have been applied, and system response evaluated (Lazarova and Manem, 1994).

3.1 General Description

An MBBR may be a single reactor or several reactors in a series. Typically, each MBBR has a length-to-width ratio (L:W) in the range of 0.5:1 to 1.5:1. Plans with a L:W greater than 1.5:1 can result in nonuniform distribution of free-moving plastic biofilm carriers. The MBBRs contain a plastic biofilm carrier volume ranging from 25% to 67% of the liquid volume. This parameter is referred to as the carrier fill. Sieves typically are installed with one MBBR wall and allow treated effluent to flow through to the next treatment step while retaining the free-moving plastic biofilm carriers. Carbon-oxidation, nitrification, or combined carbon-oxidation and nitrification MBBRs use a diffused aeration system to uniformly distribute plastic biofilm carriers and meet process oxygen requirements. Plastic biofilm carriers in denitrification MBBRs are homogenized by mechanical mixers. Each component is submerged. Plastic biofilm carriers must be removed before draining and servicing or repairing air diffusers. Figure 11.9 depicts the Williams-Monaco WRRF, Commerce City, Colorado, a two-train bioreactor that consists of four MBBRs in series.

FIGURE 11.9   Moving bed biofilm reactor at the Williams-Monaco water resource recovery facility, Colorado. This installation consists of two parallel trains each with four moving bed biofilm reactor in series.

3.1.1   Plastic Biofilm Carriers

The biofilm carriers described here typically are extruded or molded from either virgin or recycled high-density polyethylene. Table 11.3 summarizes characteristics and manufacturers of several commercially available plastic biofilm carriers. The carriers are slightly buoyant and have a specific gravity between 0.94 and 0.96 g/cm^3. Both native and biofilm-covered plastic biofilm carriers have a propensity to float in quiescent water. In operating MBBRs, they are uniformly distributed throughout the bulk of the liquid by the aeration system, liquid recirculation, or mechanical mixing. Biofilms primarily develop on the protected surface inside of the plastic biofilm carrier. For this reason, the specific surface areas of plastic biofilm carriers listed in Table 11.3 exclude areas that are not inside plastic carrier. Plastic biofilm carriers have a bulk specific surface area, net specific surface area, bulk liquid volume displacement, and net liquid volume displacement. These terms are defined as follows:

- Bulk specific surface area: biofilm area per unit volume of plastic biofilm carriers, or $\left(\dfrac{\text{m}^2 \text{ of biofilm}}{\text{m}^3 \text{ of plastic biofilm carrier}}\right)$;
- Net specific surface area: biofilm area per unit bioreactor volume, or $\left(\dfrac{\text{m}^2 \text{ of biofilm}}{\text{m}^3 \text{ of reactor volume}}\right)$;
- Bulk liquid volume displacement: liquid volume displaced per unit volume of plastic biofilm carriers, or $\left(\dfrac{\text{m}^3 \text{ of liquid displaced}}{\text{m}^3 \text{ of plastic biofilm carrier}}\right)$;
- Net liquid volume displacement: liquid volume displaced per unit bioreactor volume, or $\left(\dfrac{\text{m}^3 \text{ of liquid displaced}}{\text{m}^3 \text{ of reactor volume}}\right)$.

Manufacturer	Name	Bulk Specific Surface Area, Weight, Gravity	Nominal Carrier Dimensions (Depth; Diameter)	Carrier Photo
Veolia Inc.	AnoxKaldnes™ K1	500 m²/m³	7.2 mm; 9.1 mm	
		145 kg/m³		
		0.96–0.98		
	AnoxKaldnes™ K3	500 m²/m³	10 mm; 25 mm	
		95 kg/m³		
		0.96–0.98		
	AnoxKaldnes™ Biofilm Chip (M)	1200 m²/m³	2.2 mm; 45 mm	
		234 kg/m³		
		0.96–1.02		
	AnoxKaldnes™ Biofilm Chip (P)	900 m²/m³	3 mm; 45 mm	
		173 kg/m³		
		0.96–1.02		
Infilco Degremont Inc.	ActiveCell™ 450	450 m²/m³	15 mm; 22 mm	
		134 kg/m³		
		0.96		
	ActiveCell™ 515	515 m²/m³	15 mm; 22 mm	
		144 kg/m³		
		0.96		
Siemens Water Technologies Corp.	ABC4™	600 m²/m³	14 mm; 14 mm	
		150 kg/m³		
		0.94–0.96		
	ABC5™	660 m²/m³	12 mm; 12 mm	
		150 kg/m³		
		0.94–0.96		
Entex Technologies Inc.	BioPortz™	589 m²/m³	14 mm; 18 mm	

TABLE 11.3 Plastic Biofilm Carrier Characteristics as Reported by Manufacturer (adapted from McQuarrie and Boltz, 2011)

Bulk specific surface area, based on 100% carrier fill, is characteristic of a specific plastic biofilm carrier and is reduced proportionately. Hence, the net specific surface area is characteristic of a specific plastic biofilm carrier and carrier fill. For example, if a plastic biofilm carrier has a 500-m²/m³ bulk specific surface area, then the net specific surface area at 50% carrier fill is 250 m²/m³. Similarly, the net liquid volume displacement at 50% carrier fill is 0.0725 for a plastic biofilm carrier having a characteristic 0.15-bulk liquid volume displacement.

Figure 11.10 illustrates how biofilm thickness on plastic carriers varies depending on reactor conditions. Biofilm thickness does not become excessive because of the turbulent motion of the carriers in the reactor. Therefore, effective surface area reduction resulting from increasing biofilm thickness is not a critical factor for the design engineer. The rate of soluble substrate transformation in biofilm systems is defined in terms of mass flux (J), which has the units of g/m²/d. Therefore, it is convenient to quantify loading rate in similar terms. The net specific surface area of a plastic biofilm carrier is directly related to the calculation of MBBR pollutant loading. The volumetric load can be multiplied by the net specific surface area (a) to calculate a surface loading rate. This is expressed mathematically by eq 11.6.

$$\text{Loading Rate} = \frac{Q \cdot S_{in}}{V_B \cdot a} [=] \frac{\text{g}}{\text{m}^2 \cdot \text{d}} \qquad (11.6)$$

The plastic biofilm carriers listed in Table 11.3 maximize protected bulk specific surface area and preserve an adequate open space through which significant advective flow can pass. As previously discussed, mass-transfer resistances external to the biofilm surface are reduced if, among other factors, a sufficiently high water velocity exists in this region. Furthermore, conditions in an MBBR promote development of relatively thin

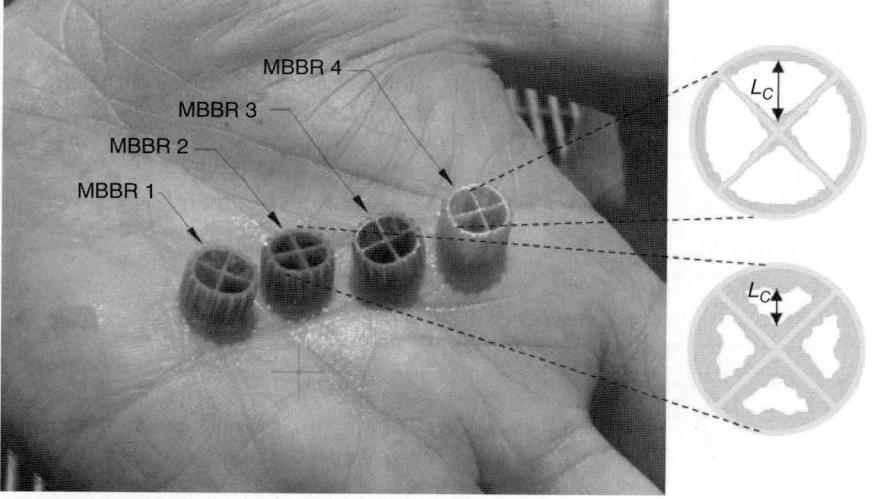

FIGURE 11.10 Photograph of biofilm carriers taken from MBBRs in series and illustrative renderings of how biofilm thickness on the carriers can vary from reactor to reactor based on electron acceptor and donor conditions (Boltz et al., 2009b).

and dense biofilms, which is characteristic of effective biofilm thickness control (see Section 11.1 for additional discussion). Larger plastic biofilm carriers allow for sieves to be constructed with larger openings. As a result, hydraulic headloss is reduced per unit sieve area. Other factors affecting MBBR plastic biofilm carrier properties include cost of manufacturing and transportation.

3.1.2   Media Retention Sieves

Plastic biofilm carriers are retained in an MBBR by horizontally configured cylindrical sieves or vertically configured flat sieves (Figure 11.11). Aerobic zones typically contain cylindrical sieves; anoxic zones contain flat-wall sieves. The cylindrical sieves extend horizontally into the upward-flowing air bubbles imparted by the diffuser grid. As a result, the air scours accumulated debris from the sieve surface. Energy imparted by the mechanical mixers is insufficient to dislodge debris accumulated on the wall sieve. Therefore, flat-sieve scour is accomplished in a denitrification MBBR with a sparging air-header. Removing the debris retained on a sieve aids in maintaining hydraulic throughput. The sieves and their supporting structural assemblies, if required, typically are constructed from stainless steel and may be wedge-wire, mesh, or perforated plates.

3.1.3   Aeration System

Low-pressure diffused air is applied to aerobic MBBRs. The airflow enters the reactor through a network of air piping and diffusers that are attached to the tank bottom. Airflow has the dual purpose of meeting process oxygen requirements and uniformly distributing plastic biofilm carriers throughout the aerobic MBBR. To promote uniform distribution of the plastic biofilm carriers, the diffuser grid layout and drop pipe arrangement provides a rolling water circulation pattern as illustrated in Figure 11.12. Fine- and coarse-bubble diffusers have been used in free-floating plastic biofilm carrier reactors (both are pictured in Figure 11.13).

Historically, process oxygen requirements and distribution of plastic biofilm carriers in MBBRs have been achieved with coarse-bubble aeration systems. Coarse-bubble diffusers typically used in MBBRs are plastic or stainless-steel pipes with circular orifices along the underside. These coarse-bubble diffusers are less affected by scaling and fouling because of the large dimension and turbulent airflow through the discharge

FIGURE 11.11   Horizontal cylindrical sieves over coarse-bubble diffuser grid (left) and vertical flat-panel wall sieves with sparging air-header (right).

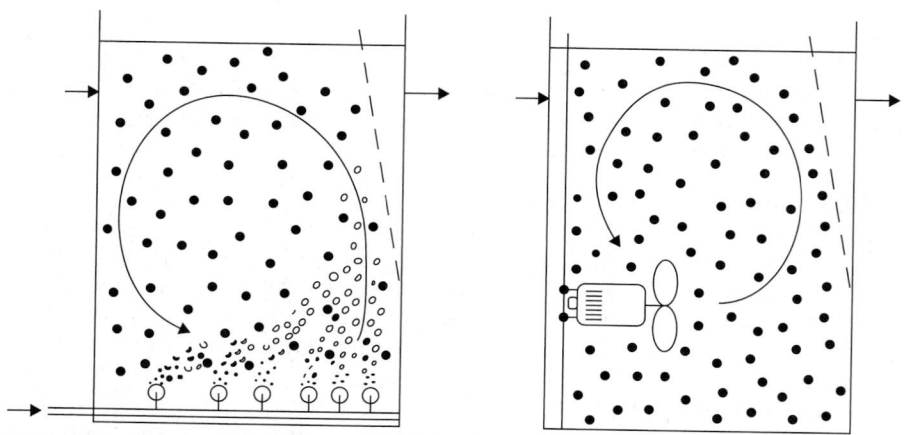

Figure 11.12   Rolling-water circulation pattern induced by a diffused aeration system (left) and mechanical mixers (right) (Ødegaard et al., 1994).

Figure 11.13   Coarse-bubble diffusers in an MBBR (left) and fine-bubble tube diffusers in an MBBR (right).

orifice (Stenstrom and Rosso, 2008). As a result, coarse-bubble diffusers require less maintenance than fine-bubble diffusers. Figure 11.13 illustrates that the coarse-bubble diffusers are designed with a structural end support that enables it to withstand the weight of plastic biofilm carriers when the MBBR is taken out of service and drained.

3.1.4   Mechanical Mixing Devices

Denitrification MBBRs use mechanical mixers to agitate the bulk of the liquid and to distribute uniformly plastic biofilm carriers. The mechanical mixers may be platform (dry motor) mounted hyperbolic or rail-mounted submersible (wet motor) units. Typically, these mixers agitate liquids with biological suspended solids concentrations up to 10 000 g/m³. State-of-the art submersible mechanical mixers typically have a maximum 120-rpm impeller speed and a minimum of three blades per impeller. These features are designed to meet process objectives and minimize potential for impeller damage resulting from abrasion induced by the plastic biofilm carriers. Figure 11.14 shows a denitrification MBBR and submersible mechanical mixers in a MBBR.

FIGURE 11.14 Anoxic MBBR (left) and basin internals of a swing-anoxic MBBR (right). The molded fiberglass impellers were among the first generation of denitrification MBBRs. Many of these impellers were damaged by the free-moving plastic biofilm carriers.

3.2 Process Flow Sheets and Bioreactor Configurations

Relevant considerations when selecting a MBBR configuration include site-specific treatment objectives, wastewater characteristics, site layout, existing basin configuration (if a retrofit), system hydraulics, existing treatment scheme (if applicable), and the potential to retrofit existing tanks. Figure 11.15 illustrates carbon oxidation, nitrification, and denitrification MBBR flow sheets. Although the process mechanical features of a MBBR are typically consistent, biofilms grown in carbon oxidation, nitrification, denitrification, and combined carbon oxidation and nitrification MBBRs are variable and depend on local environmental conditions (see Section 11.1 for additional information). For example, the MBBRs pictured in Figure 11.9 were designed to achieve carbon oxidation, nitrification, and partial denitrification with a process configuration similar to the modified Ludzack–Ettinger process. Table 11.4 demonstrates variability in the system's biofilm characteristics from the first reactor (R1) to the fourth reactor (R4).

Plastic biofilm carriers in the first denitrification MBBR (R1) have a brownish color and contain 150% more biomass than the dark biofilm in the second denitrification MBBR (R2). The first aerobic MBBR (R3) has the thickest biofilm development in the series, and the final MBBR (R4) has approximately 50% of the biomass in the first aerobic MBBR (R3).

3.2.1 Carbon Oxidation

Biodegradable soluble organic carbon is quickly consumed in an MBBR. Figure 11.16 illustrates filtered-COD flux (i.e., soluble COD), as a function of filtered-COD load for two types of plastic biofilm carriers (with different specific surface areas). Figure 11.17 illustrates for the same two plastic biofilm carriers the "obtainable" COD removal rate per unit biofilm area as a function of total-COD load. Particulate organic matter is entrapped by a biofilm; subsequently a portion of the entrapped biodegradable organic particulates is hydrolyzed and the resulting soluble organic carbon used. The remaining entrapped particulate exits the biofilm before being hydrolyzed with detached biofilm fragments. Suspended organic particles exit the MBBR in the effluent stream with the detached biofilm fragments and are subject to removal in the liquid-solids separation unit.

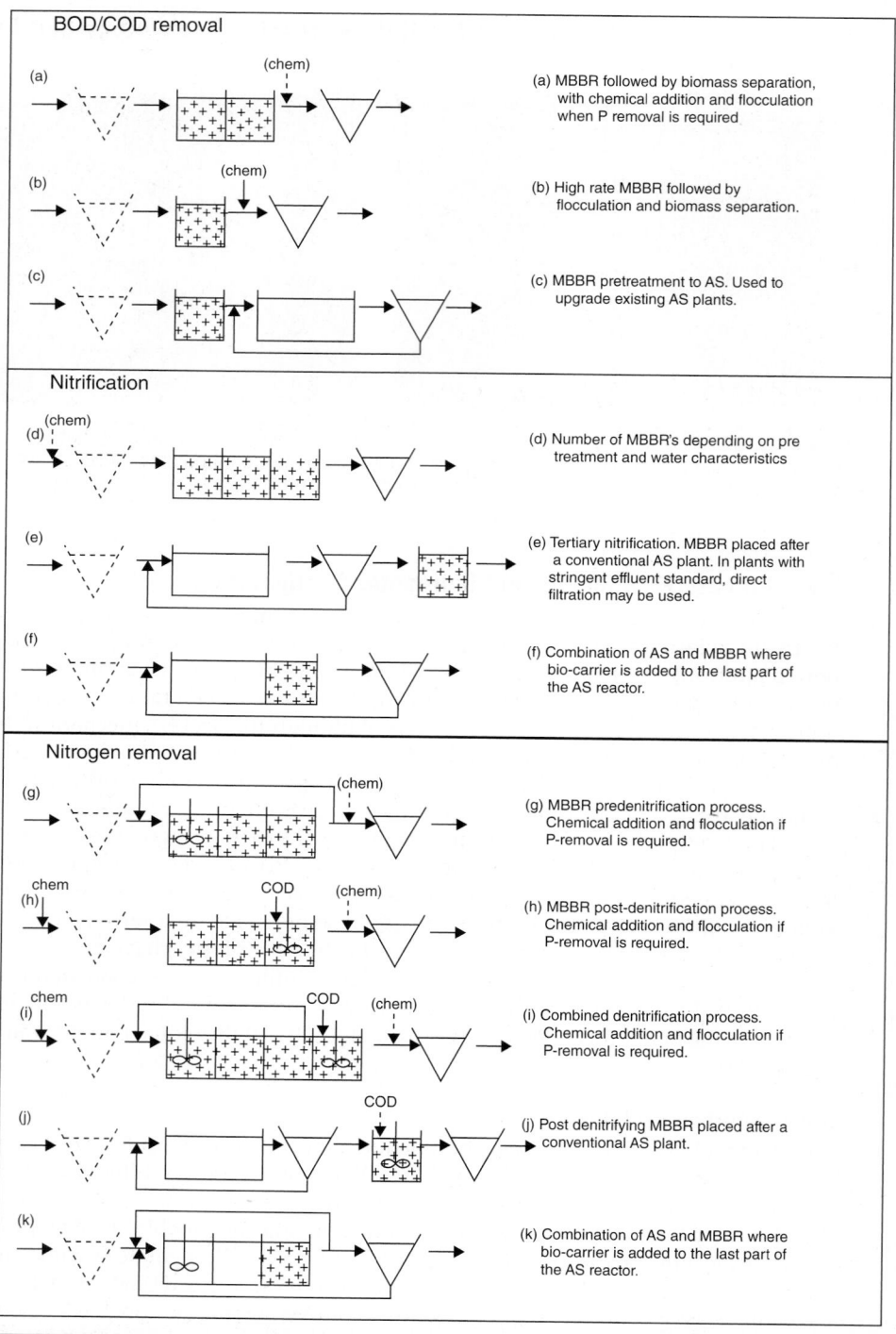

FIGURE 11.15 Typical process flowsheets for carbon oxidation, nitrification, combined carbon oxidation and nitrification, and denitrification in an MBBR treating municipal wastewater (Ødegaard, 2006). Rectangular tanks with crosses are MBBRs and rectangular tanks with mixers are anoxic (COD = chemical oxygen demand; BOD = biochemical oxygen demand; P = phosphorus; and AS = activated sludge).

Parameter/Reactor (R)	R1–Anoxic	R2–Anoxic	R3–Aerobic	R4–Aerobic
Function	Denitrification	Denitrification	Carbon oxidation	Combined carbon oxidation and high-rate nitrification
Biomass per net specific surface area	9.4 g SS/m^2	6.1 g SS/m^2	28 g SS/m^2	12.9 g SS/m^2
Carrier fill	57%	57%	60%	60%
Solids per unit tank volume	2680 g SS/m^3	1740 g SS/m^3	8400 g SS/m^3	3870 g SS/m^3
Photograph of media taken from tank				

TABLE 11.4   Example of the Variation of Biofilm Across Four Moving Bed Biofilm Reactor (MBBR) in Series (SS = Suspended Solids) (McQuarrie and Boltz, 2011)

Based on this premise, Ødegaard et al. (2000) used the "obtainable" COD removal rate, which is defined as the influent total-COD concentration less the soluble-COD concentration remaining in the effluent stream. This "obtainable" removal rate of COD at 100% biomass separation suggests that high removal efficiencies were obtained in the pilot-scale carbon-oxidizing MBBR at high organic loading and provided good

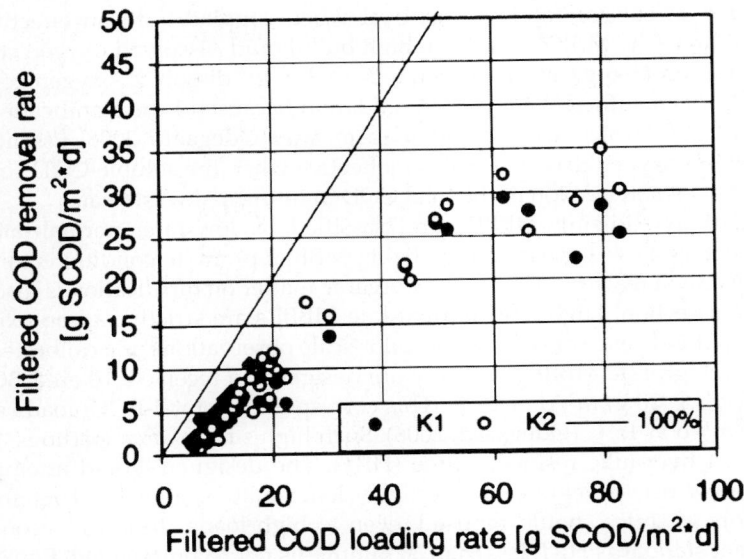

FIGURE 11.16   Filtered chemical oxygen demand (COD) flux (1.2-μm pore opening) for two types of plastic biofilm carriers with different specific surface areas as a function of filtered COD load (Ødegaard, 2000).

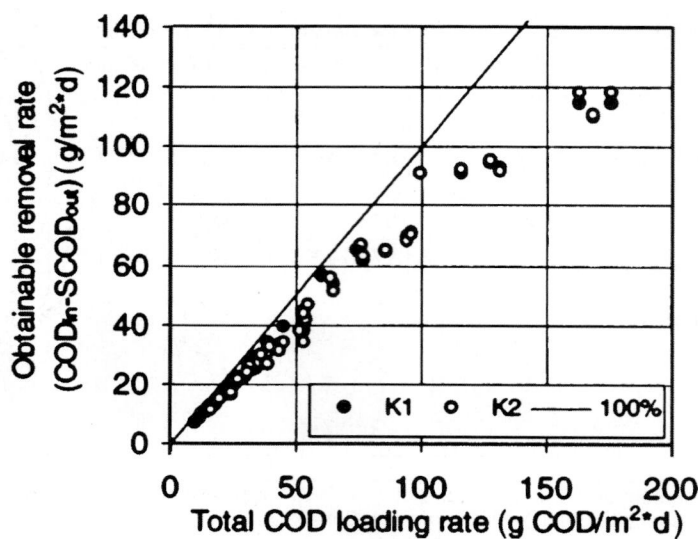

FIGURE 11.17   Total chemical oxygen demand (COD) flux for two types of plastic biofilm carriers with different specific surface areas as a function of bulk-liquid COD concentration (Ødegaard, 2000).

liquid-solids separation. For the pilot-scale MBBR evaluated, these graphs demonstrate that approximately 30 g/m²/d filtered COD flux was attainable for filtered COD loading greater than 60 g/m²/d. They also show that a total COD load up to a 60 g/m²/d resulted in substantial COD removal when coupled with an effective liquid-solids separation. The MBBRs require a high bulk-liquid dissolved oxygen concentration for nitrification (4–6 g/m³); however, a 2- to 3-g/m³ dissolved-oxygen concentration has been proven sufficient for carbon oxidation because of the significant particulate and colloidal COD fraction in municipal wastewater (Ødegaard, 2006). Essentially, additional dissolved-oxygen driving force is ineffective when the soluble-COD concentration is a relatively small fraction of the total COD in municipal wastewater.

Carbon-oxidizing MBBRs are classified as low-rate, normal-rate, or high-rate bioreactors. Low-rate carbon oxidizing MBBRs promote conditions for nitrification in downstream reactors. The effect of organic matter on nitrification is discussed in a subsequent section. High- and normal-rate MBBRs are strictly carbon-oxidizing bioreactors. In the absence of site-specific pilot-scale observations or a calibrated mathematical model, high-rate MBBRs typically are designed to receive a filtered BOD$_5$ load in the range 15 to 20 g/m²/d at 15°C. This corresponds to total-BOD$_5$ loads as high as 45 to 60 g/m²/d at 15°C (Ødegaard, 2006). Such high surface area loadings, however, result in short hydraulic residence time (HRT). The designer should attempt to achieve a plug-flow reactor regime so that higher loads may be allowed. Therefore, at least two reactors in series should be used, even at high loads. To reach secondary treatment effluent standards HRT, less than 30 minutes is not recommended. Equation 11.4 can be applied to adjust these rates and describe MBBR performance at different wastewater temperatures. Ødegaard (2006) observed biomass yield in a carbon-oxidizing MBBR approximately equal to 0.5 g suspended solids per gram of filtered COD transformed.

The settling characteristics of detached biofilm fragments suspended in the MBBR effluent stream deteriorates as BOD_5 (or COD) load increases (Ødegaard et al., 2000; Melin et al., 2004). Consequently, process performance may be limited by the solids separation unit. An increased BOD_5 (or COD) load may be imparted on the MBBR when coupled with a high-efficiency solids separation such as chemically enhanced secondary clarification, dissolved air flotation, or coupled with a solids contact reactor clarifier. A majority of the organic matter in municipal wastewater is colloidal or particulate (Levine et al., 1985; 1991). Biofilms growing in MBBRs readily consume soluble organic matter, however, like other biofilm reactors, they may be poor bioflocculating units.

Medium-rate MBBRs designed for meeting basic secondary treatment standards typically are designed for a loading of 5 to 10 g $BOD_5/m^2/d$ at 10°C, depending on choice of final separation method. Values in the higher range are used when coagulation occurs before the separation unit; values in the lower range are used without coagulation. Properly designed, chemically enhanced secondary clarification units improve water quality by reducing carbon-based oxygen demand, suspended solids, and phosphorus. Table 11.5 summarizes BOD, COD, and phosphorus removal at four WRRFs with MBBR followed by chemically enhanced secondary clarification. Each normal-rate MBBR installation was designed to receive a 7 to 10 g/m²/d total-BOD_5 loading rate at 10°C.

3.2.2 Nitrification

Nitrification in MBBRs has been extensively studied using synthetic and municipal wastewater (Hem et al., 1994; Rusten et al., 1995a; Æsøy et al., 1998). Like all biofilm reactors, the rate of ammonia-nitrogen oxidation in an MBBR is influenced by organic load, bulk-liquid dissolved oxygen concentration, bulk-liquid ammonia-nitrogen concentration, temperature, pH, and alkalinity. Ammonia-nitrogen oxidation has been achieved in the MBBR process flowsheets illustrated in Figure 11.15 (d to e). This chapter defines (tertiary) nitrification (Figure 11.15e) as the ammonia-nitrogen oxidation process in a biofilm reactor treating secondary effluent that meets the following criteria: BOD_5:TKN ≤ 1.0 and soluble BOD_5 ≤ 12 g/m³. A combined carbon oxidation and nitrification MBBR (Figure 11.15d) is defined as a unit that receives an organic load exceeding these conditions. Sufficient bulk-liquid total-BOD_5 and ammonia-nitrogen concentrations result in competition between heterotrophic and autotrophic nitrifying

WRRF	BOD₇[a] IN (g/m³)	BOD₇[a] OUT (g/m³)	COD IN (g/m³)	COD OUT (g/m³)	Total Phosphorus IN (g/m³)	Total Phosphorus OUT (g/m³)
Steinsholt[b]	398	10	833	46	7.1	0.3
Tretten[c]	361	4	—	—	7.3	0.1
Svarstad[c]	—	—	403	44	5.1	0.25
Frya[c]	181	5	—	—	8.6	0.21

Notes: BOD_7 = seven-day biochemical oxygen demand; COD chemical oxygen demand; WRRF = water resource recovery facility.
[a]BOD_5/BOD_7 = ~0.86.
[b]1996–1997.
[c]Data from 2000–2002.

TABLE 11.5 MBBR Coupled with Chemically Enhanced Secondary Clarification Performance (Ødegaard et al., 2004)

organisms growing inside a mixed-culture biofilm. When competing for dissolved oxygen in a combined carbon oxidation and nitrification MBBR, the faster growing heterotrophic organisms may overgrow the slower developing autotrophic nitrifiers at high bulk-liquid soluble BOD_5 concentrations (Wanner and Gujer, 1984). If a high bulk-liquid soluble BOD_5 concentration exists and the bulk-liquid dissolved-oxygen concentration is insufficient to penetrate the faster growing heterotrophic bacteria that have overgrown the autotrophic nitrifiers, then the slower growing (autotrophic) bacteria will washout of the biofilm. Therefore, as biofilm reactor BOD_5 loading increases, the bulk-liquid dissolved oxygen concentration must be increased to maintain a constant ammonia-nitrogen flux.

Figure 11.18 illustrates (total) ammonia-nitrogen flux in an MBBR for various BOD_5 loads and bulk-liquid dissolved-oxygen concentrations. While studying a pilot-scale combined carbon oxidation and nitrification MBBR receiving primary effluent, a (tertiary) nitrification MBBR receiving secondary effluent and maintaining a 4- to 6-g/m³ bulk-liquid dissolved-oxygen concentration in both units, Hem et al. (1994) observed:

- Total-BOD_5 load of 1 to 2 g/m²/d resulted in nitrification rates from 0.7 to 1.2 g/m²/d,
- Total-BOD_5 load of 2 to 3 g/m²/d resulted in nitrification rates from 0.3 to 0.8 g/m²/d, and
- Total-BOD_5 load greater than 5 g/m²/d resulted in virtually no nitrification.

Rusten et al. (1995a) described ammonia-nitrogen flux as a function of bulk-liquid ammonia-nitrogen concentration in an MBBR as a first-order process when bulk-liquid ammonia-nitrogen is the rate-limiting substrate. They described it as a zero-order

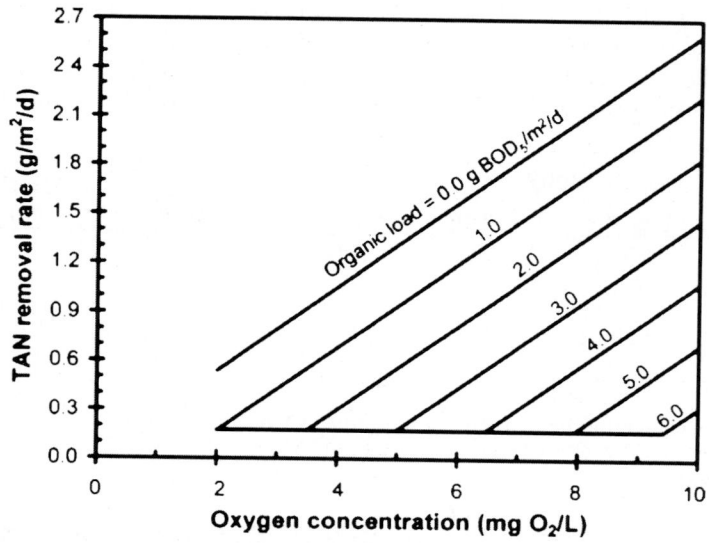

FIGURE 11.18 Effect of organic load and bulk-liquid dissolved oxygen concentration on ammonium (or ammonia-nitrogen) flux (Rusten et al., 2006).

process when bulk-liquid dissolved oxygen is the rate-limiting substrate. The researchers used eq 11.7 to describe ammonia-nitrogen flux in a MBBR.

$$J_{NH_3-N} = k \cdot (S_{B,NH_3-N})^n \tag{11.7}$$

where J_{NH_3-N} = ammonia-nitrogen flux (g/m²/d);
$\quad k$ = rate constant (m/d);
$\quad S_{B,NH_3-N}$ = bulk-liquid ammonia-nitrogen concentration (g/m³);
$\quad n$ = reaction order constant.

The reaction-order constant, n, is assigned the value 0.7 for MBBRs (Hem, 1991). However, the rate-constant value, k, varies because of its dependence on local environmental conditions such as primarily soluble BOD_5 load. Rusten et al. (1995a) reported k-values in the range of 0.4 to 0.7 m/d for pilot-scale combined carbon oxidation and nitrification MBBRs treating pre-precipitated primary effluent at 10°C. The following k-values at 10°C and respective conditions can be applied to combined carbon oxidation and nitrification MBBR design: k = 0.40 m/d with no primary clarifier; k = 0.47 m/d with primary clarification or pre-denitrification; k = 0.50 m/d with primary clarification and pre-denitrification; k = 0.53 m/d with chemically enhanced primary clarification. The rate constant of 0.6 to 0.7 m/d at 15°C can be used in tertiary nitrification MBBR applications following secondary treatment.

When applying eq 11.7 to describe nitrification in an MBBR, the ratio $\dfrac{\text{bulk} - \text{liquid dissolved oxygen concentration}}{\text{bulk} - \text{liquid ammonia} - \text{nitrogen concentration}}$, $\dfrac{S_{B,O_2}}{S_{B,NH_3-N}}$ is used to identify the transition whereby ammonia-nitrogen flux transforms from being ammonia-nitrogen limited to oxygen limited as a function of bulk-liquid ammonia-nitrogen concentration in the reactor effluent. The ratio has been assigned the value 3.2 m/d by Rusten et al. (1995b; 2006). The calculated ammonia-nitrogen flux is adjusted to reflect the influence of site-specific wastewater temperature with eq 11.4.

In practice, a designer may assume a constant transition ratio $\dfrac{S_{B,O_2}}{S_{B,NH_3-N}} = 3.2$.

However, the transition point is influenced by stoichiometric coefficient for the electron donor and acceptor and, to a lesser extent, the material diffusivity influenced by liquid temperature. The diffusivity of a material i is characterized by the aqueous-phase diffusion coefficient, $D_{aq,i}\left(\approx \dfrac{D_{F,i}}{0.8}\right)$, where $D_{F,i}$ is the diffusion coefficient of substrate i inside the biofilm (m²/d) (Stewart, 2003; Horn and Morgenroth, 2006). The temperature dependence of $D_{aq,i}$ is calculated using the following relationship:

$$\frac{D_{aq,i} \cdot \text{viscosity}}{T} = \text{constant}, \quad \text{or} \quad D_{aq,i,T} = D_{aq,i,25°C} \cdot \frac{T}{25°C} \cdot \frac{\text{viscosity}_{25°C}}{\text{viscosity}_T}$$

where T = temperature (°C), and
$\quad$ viscosity = kinematic (m²/d).

Figure 11.19 illustrates nitrification rates observed at a pilot-scale tertiary nitrification MBBR (Kaldate et al., 2008). The data is identified as being in the ammonia-nitrogen rate-limiting region or oxygen rate-limiting region. Figure 11.19 also illustrates the previously

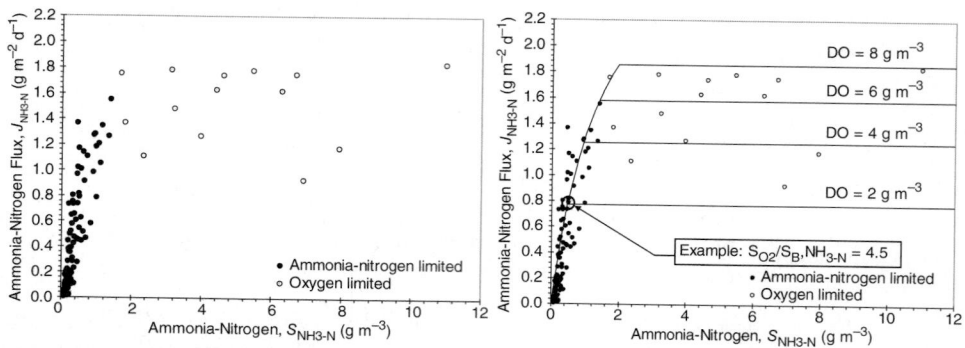

FIGURE 11.19   Observed ammonia-nitrogen concentrations in a second-stage nitrification MBBR grouped according to the rate-limiting substrate as defined by eq 11.14 (left) (Kaldate et al., 2008). Empirical nitrification MBBR model for various bulk-liquid dissolved oxygen concentrations (model: $J_{NH_3-N} = k \cdot \theta^{(T-20)} \cdot (S_{B,NH_3-N})^n$, $n = 0.7$, $\theta = 1.10$, $k = 0.7$, and $T = 18°C$). Nonlinear regression analysis performed using DataFit© v9.0.59 (Oakdale Engineering, California; www.curvefitting .com). The average bulk-liquid dissolved oxygen concentration for observations in the oxygen-limiting region is 6.8 g/m³ (Hem et al., 1994; McQuarrie and Boltz, 2011).

described empirical MBBR nitrification model. Equation 11.7 was applied to the pilot-plant data and four different oxygen rate-limiting regions corresponding to the bulk-liquid dissolved-oxygen concentrations 2, 4, 6, and 8 g/m³. To illustrate, the transition between near first- and zero-order response (horizontal line) with dissolved oxygen of 2 g/m³ the following is calculated:

$$S_{B,NH_3-N} = \frac{1}{v_{ED,EA}} \cdot \left[\frac{D_{F,O_2 T}}{D_{F,NH_3-N,T}} \right] \cdot S_{B,O_2}$$

$$= \frac{1}{4.57 \dfrac{g_{O_2}}{g_N}} \cdot \frac{\left[0.000200 \dfrac{m^2}{d} \cdot \dfrac{18°C}{25°C} \cdot \dfrac{0.077 \dfrac{m^2}{d}}{0.081 \dfrac{m^2}{d}} \right]}{\left[0.000197 \dfrac{m^2}{d} \cdot \dfrac{18°C}{25°C} \cdot \dfrac{0.077 \dfrac{m^2}{d}}{0.081 \dfrac{m^2}{d}} \right]} \cdot 2 \dfrac{g_{O_2}}{m^3}$$

$$= 0.44 \frac{g_N}{m^3}$$

Figure 11.19 helps to illustrate how operating at higher dissolved oxygen concentrations increases ammonia nitrogen flux in the oxygen limited region but provides little benefit once ammonia becomes rate-limiting at approximately 2.5 g/m³.

The design bulk-liquid dissolved oxygen concentration is not available entirely for ammonia-nitrogen oxidation in a combined carbon oxidation and nitrification MBBR. Heterotrophic-organism activity resulting from the presence of biodegradable organic matter reduces the dissolved oxygen concentration available for ammonia-nitrogen oxidation. Rusten et al. (1995a) observed that a soluble-BOD₅ load of 0.5 g/m²/d reduced the dissolved oxygen concentration available for nitrification by 0.5 g/m³.

Rusten et al. (2006) estimated that a 1.5-g/m²/d soluble-BOD₅ load reduces the dissolved oxygen concentration available for nitrification by 2.5 g/m³. Ammonia-nitrogen flux values calculated using eq 11.7 should be applied to combined carbon oxidation and nitrification MBBR design only if (1) the k-value is representative of site-specific environmental conditions, or (2) the flux has been verified by a calibrated mathematical model that considers competition in mixed culture biofilms (see Table 11.2).

Siegrist and Gujer (1987) and Rusten et al. (1995a) recommend a minimum alkalinity of 75 mg/L as $CaCO_3$ (1.5-meq/m³). Szwerinski et al. (1986) and Zhang and Bishop (1996) state that the ratio bulk-liquid bicarbonate-to-dissolved oxygen (as $mgCaCO_3$:mgO_2) should be greater than 6.25 to avoid nitrification being alkalinity-limited in a nitrifying biofilm reactor. Nordeidet et al. (1994) reported that (tertiary) nitrifying biofilm reactors may become orthophosphate (PO_4-P) limited at bulk-liquid concentrations less than approximately 0.15 g P/m³. Phosphorus limitation in tertiary biofilm reactors is discussed at greater depth in the post-denitrification MBBR section.

Given the following assumptions and treatment objectives, ammonia-nitrogen flux can be estimated and the volume of a single-stage nitrification MBBR receiving settled effluent from secondary treatment calculated.

- Wastewater temperature = 10°C;
- Targeted effluent ammonia-nitrogen concentration = 2 g/m³;
- Bulk-liquid dissolved oxygen concentration is kept constant at 6 g/m³;
- Soluble BOD₅ load is less than 0.5 g/m²/d;
- Nitrification MBBR receives 3785 m³/d of partially nitrified secondary effluent;
- A 16 g/m³ ammonia-nitrogen concentration in the influent stream;
- Plastic biofilm carrier bulk-specific surface area is 500 m²/m³ at a 50% carrier fill.

Solution:

1. Given a bulk-liquid oxygen concentration of 6 g/m³, estimate the ammonia-nitrogen concentration corresponding to the point whereby flux transitions from being oxygen-limited to being ammonia-nitrogen limited. Reduce the bulk-liquid dissolved oxygen concentration by 0.5 g/m³ to account for heterotrophic bacteria activity inside the biofilm.

$$S_{B,NH_3\text{-}N} = \frac{1}{v_{ED,EA}} \cdot \left[\frac{D_{F,O_2,T}}{D_{F,NH_3\text{-}N,T}} \right] \cdot S_{B,O_2}$$

$$= \frac{1}{4.57\frac{g_{O_2}}{g_N}} \cdot \left[\frac{0.000200\frac{m^2}{d} \cdot \frac{10°C}{25°C} \cdot \frac{0.077\frac{m^2}{d}}{0.081\frac{m^2}{d}}}{0.000197\frac{m^2}{d} \cdot \frac{10°C}{25°C} \cdot \frac{0.077\frac{m^2}{d}}{0.081\frac{m^2}{d}}} \right] \cdot \left(6\frac{g_{O_2}}{m^3} - 0.5\frac{g_{O_2}}{m^3} \right)$$

$$= 1.2\frac{g_N}{m^3}$$

2. The reactor effluent ammonia concentration 1.2 g/m³ is suitable given the design objective of less than 2 g/m³. Calculate the ammonia-nitrogen flux as a function of the above-defined (Step 1) bulk-liquid ammonia-nitrogen concentration.

$$J_{B,\text{NH}_3\text{-N}} = k \cdot \theta^{(T_2 - T_1)} \cdot (S_{B,N})^n$$

$$= 0.7 \cdot 1.1^{(10-15)} \cdot (1.2)^{0.7}$$

$$= 0.5 \frac{g}{m^2 \cdot d}$$

The zero-order ammonia-nitrogen flux is less 0.5 g/m²/d Therefore, the estimated ammonia-nitrogen flux for the given design example is 0.5 g/m²/d.

3. Rearranging eq 11.1 and calculate the biofilm area required to meet the treatment objective.

$$A = \frac{Q \cdot (S_{in,\text{NH}_3\text{-N}} - S_{B,\text{NH}_3\text{-N}})}{J_{\text{NH}_3\text{-N}}}$$

$$= \frac{3{,}785 \dfrac{m^3}{d} \left(16 \dfrac{g}{m^3} - 2 \dfrac{g}{m^3} \right)}{0.5 \dfrac{g}{m^2 \cdot d}}$$

$$= 105\,980 \text{ m}^2$$

4. The nitrification MBBR volume is calculated.

$$A = \frac{A}{a}$$

$$= \frac{105{,}980 \text{ m}^2}{500 \dfrac{m^2}{m^3} \cdot 0.5}$$

$$= 424 \text{ m}^2$$

An alternate method for sizing a nitrification MBBR when pilot-plant data is available is the graphical biofilm reactor design approach described in Section 11.2. Figure 11.20 presents ammonia-nitrogen flux curves obtained from pilot data collected from a pilot-scale system consisting of two reactors in series (Kaldate et al., 2008). Using the pilot-scale nitrification MBBR data and assuming a design dissolved oxygen of 6 g/m³, a first-stage MBBR operating line intersects the ammonia-nitrogen flux curve at 1.04 g/m²/d. For the second reactor, an ammonia nitrogen flux curve is determined assuming a dissolved oxygen of 4 g/m³, and the second-stage MBBR operating line intersects the ammonia-nitrogen flux curve at 0.36 g/m²/d. For this graphical design example, the

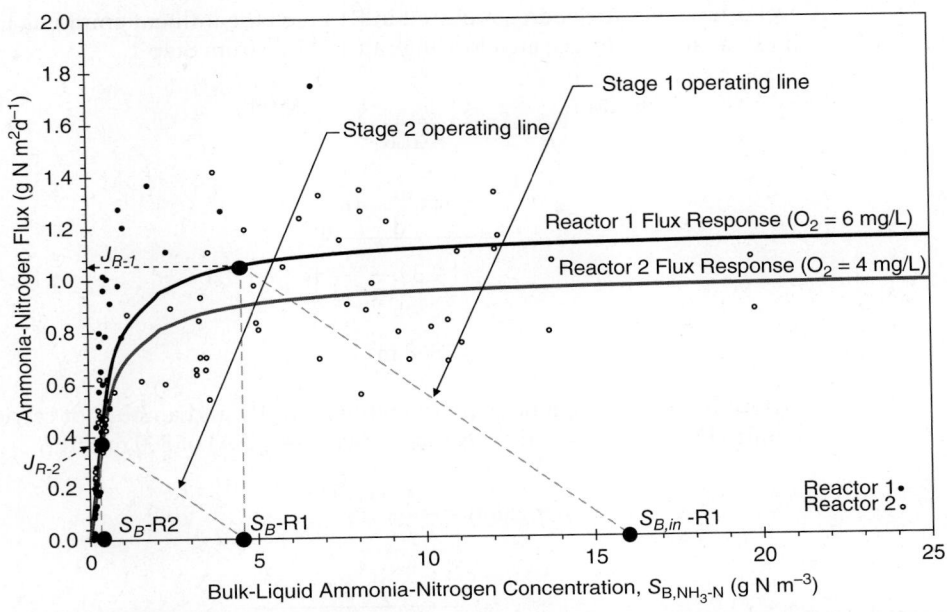

FIGURE 11.20 Graphical design procedure applied to the nitrification MBBR example. Ammonia-nitrogen flux curves based on data obtained from a pilot-scale two-stage nitrification MBBR (Kaldate et al., 2008).

reactor volume determined above for a single reactor (424 m³) is subdivided into two equal-volume reactors in series.

1. Determine the design loading rate for reactor 1 (R1) by using R1 flux from Figure 11.20. Ensure the required ammonia oxidation is achieved in R1 such that the influent ammonia concentration to reactor 2 (R2) is in the range of 4 g/m³.

$$\text{R1 design loading rate} = \frac{J_{\text{NH}_3\text{-N}}}{\left(\dfrac{S_{in,\text{NH}_3\text{-N}} - S_{B,\text{NH}_3\text{-N}}}{S_{in,\text{NH}_3\text{-N}}}\right)}$$

$$= \frac{1.04\,\dfrac{\text{g}_N}{\text{m}^2 \cdot \text{d}}}{\left(\dfrac{16\,\dfrac{\text{g}_N}{\text{m}^3} - 4\,\dfrac{\text{g}_N}{\text{m}^3}}{16\,\dfrac{\text{g}_N}{\text{m}^3}}\right)}$$

$$= 1.39\,\frac{\text{g}_N}{\text{m}^2 \cdot \text{d}}$$

2. Determine the carrier area required in R1 given the influent ammonia loading, and the design surface area loading rate (SALR) from Step 1.

$$A = \frac{Q \cdot S_{in,NH_3\text{-}N}}{SALR}$$

$$= \frac{3785\frac{m^3}{d} \cdot 16\frac{g}{m^3}}{\left(1.39\frac{g_N}{m^2 \cdot d}\right)}$$

$$= 43\ 568\ m^2$$

3. Determine the fill fraction given the volume of R1 and amount of carrier area required based on a media specific surface area (SSA) of 500 m^2/m^3.

$$\text{Fill fraction} = \frac{A}{V \cdot a}$$

$$= \frac{43\ 568\ m^2}{\left(212\ m^3 \cdot 500\frac{m^2}{m^3}\right)}$$

$$= 41\%$$

4. Determine the design loading rate, carrier area, and fill fraction for R2 using the same calculation procedure.

Reactor 2 requires 31 540 m^2 of carrier area in a 212 m^3 resulting in a 30% media fill fraction. The designer should note the difference in carrier area required with a two-reactor design vs a single reactor design.

3.2.3 Denitrification

Denitrification in an MBBR has also been extensively studied using synthetic and municipal wastewater (Rusten et al., 1995b; Rusten et al., 1996; Aspegren et al., 1998; Bill et al., 2008). Like all biofilm reactors, the rate of denitrification in an MBBR is influenced by the external carbon source, bulk-phase carbon-to-nitrogen ratio (C:N), wastewater temperature, bulk-liquid dissolved oxygen concentration, and bulk-liquid macronutrient concentration (primarily phosphorus). Nitrogen removal using the MBBR process has been achieved using the flowsheets illustrated in Figure 11.15 (g to k) for both pre-denitrification and post-denitrification.

Pre-denitrification typically is used in the activated sludge process. Pre-denitrification activated sludge configurations such as the modified Ludzack Ettinger (MLE), have been well documented (Grady et al., 1999). The pre-denitrification MBBRs are situated upstream of combined carbon-oxidation and nitrification MBBRs. The electron acceptor nitrate/nitrite-nitrogen is supplied by an internal recirculation stream that directs nitrified MBBR effluent to the pre-denitrification MBBR. The internal recirculation ratio, or $\frac{Q_R}{Q_{in}}$, is typically in the range 2 to 4, but may be as high as 6. There is a practical upper limit on the effective recirculation ratio, but this must be evaluated on a site specific basis. Additional increase in the recirculation flow rate beyond the effective limit has been found to reduce overall denitrification effectiveness (Ødegaard, 2006).

Pre-denitrification MBBR performance is primarily dependent on the availability of soluble BOD_5 in the influent wastewater stream. When ample soluble-BOD_5 concentration exists, pre-denitrification MBBRs can achieve 50% to 70% nitrogen removal. Dissolved oxygen inhibits anoxic biochemical transformation processes. Combined carbon oxidation and nitrification MBBRs operate at a relatively high bulk-liquid dissolved oxygen concentration (i.e., 3 to 6 g/m^3). Therefore, the internal recirculation stream may also have a high dissolved oxygen concentration. Aerobic reactions have an energetic advantage over denitrification, and will take precedence resulting in reduced soluble-BOD_5 for denitrification. Therefore, the presence of dissolved oxygen in the internal recirculation stream must be considered when assigning a pre-denitrification MBBR volume and assessing the availability of soluble-BOD_5 for denitrification. Dissolved oxygen is converted to its nitrate equivalence with the dissolved oxygen-to-nitrate-nitrogen mass ratio ($g-O_2$:$g-NO_3$-N) of 2.86:1. Table 11.6 lists ranges of denitrification rates that a designer could expect to observe in a pre-denitrification MBBR. Nitrate/nitrite-nitrogen transformation rates in a pre-denitrification MBBR are typically in the range 0.3 to 0.6 g NO_3-$N_{eq}/m^2/d$ (at 10°C). Variation in the observed nitrate-nitrogen transformation rates is a result of different wastewater characteristics and environmental conditions.

Post-denitrification MBBRs require the addition of a supplemental electron donor (i.e., an external carbon source), but do not require recirculation of a nitrified effluent stream to receive the electron acceptor nitrate/nitrite-nitrogen. These MBBRs are beneficial when the stream influent to pre-denitrification MBBR has insufficient soluble-BOD_5 concentration to promote the desired nitrate/nitrite-nitrogen conversion. Bill et al. (2008) demonstrated that some commercially available, readily biodegradable electron donors result in higher nitrate/nitrite-nitrogen flux than previously described for pre-denitrification MBBRs. They use readily biodegradable, low-molecular weight compounds that typically are measured as soluble-BOD_5 in raw sewage. Therefore, a post-denitrification MBBR is beneficial when a high-volumetric efficiency design is required to meet treatment objectives in a compact footprint. Post-denitrification MBBRs are capable of achieving nearly complete nitrate/nitrite-nitrogen reduction within a short hydraulic retention time (Ødegaard, 2006). The nitrate/nitrite-nitrogen load and flux selected for post-denitrification MBBR design is influenced by (1) the type of external carbon used, (2) wastewater temperature, (3) acceptable residual soluble-BOD_5 concentration, and (4) phosphorus availability.

Reference	Nitrate/Nitrite-Nitrogen Flux (g NO_3-Neq/$m^2 \cdot$ d)
Gardermoen WRRF (Rusten et al., 2007)	0.40–1.10
FREVAR WRRF (Rusen et al., 2000)	0.15–0.50
Crow Creek WRRF (McQarrie and Maxwell, 2003)	0.25–0.80
NRA WRRF	0.20–0.40

TABLE 11.6 Comparison of Pre-Denitrification MBBR Performance

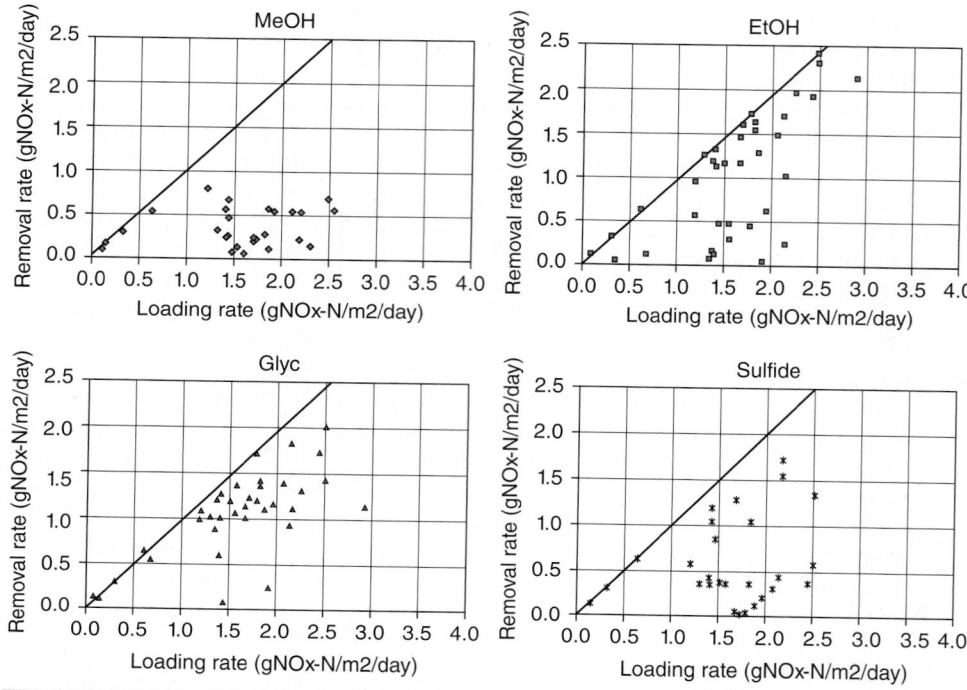

FIGURE 11.21 Nitrate/Nitrite removal rates as a function of bench-scale MBBR loading rate for electron donors methanol (MeOH), ethanol (EtOH), glycerol (Glyc), and sulfide at 20°C (Bill et al., 2008).

Nitrate/nitrite-nitrogen removal in four bench-scale MBBRs (treating synthetic wastewater) configured to simulate post-denitrification using four different supplemental carbon sources—methanol, ethanol, glycerol, and sulfide—is illustrated as a function of nitrate/nitrite-nitrogen load in Figure 11.21. Ethanol resulted in the most substantial biofilm development (see Figure 11.22), and had the highest observed nitrate/nitrite-nitrogen flux, which approached a maximum rate of 2.5 g $N/m^2/d$ at 20°C. Similarly, Aspegren et al. (1998) reported that a pilot-scale post-denitrification MBBR had a maximum denitrification rate approximately equal to 2.5 g $N/m^2/d$ and 2.0 g $N/m^2/d$ (at 16°C) when ethanol and methanol were used as the supplemental electron donors, respectively. Supplemental carbon source selection typically is dependent on cost and availability of commercial electron donors and, sometimes, the denitrification rate required at cold temperatures. Despite improved efficiency in denitrification using ethanol as the supplemental electron donor, operational costs typically result in a design optimized for performance, capital, and life-cycle cost. The beneficial reuse of waste products, such as spent aircraft deicing fluid, has been used as the external carbon source in post-denitrification MBBRs (Rusten et al., 1996). Typically, loading rates applied to post-denitrification MBBR design range from 1 to 2 g NO_3-N_{eq}/ m^2/d. Figure 11.23 illustrates effluent nitrate/nitrite-nitrogen concentrations in 24-hour-flow proportional samples collected from a pilot-scale post-denitrification MBBR. The figure illustrates that effluent nitrate/nitrite-nitrogen concentrations approach an

FIGURE 11.22 Nitrate/nitrite removal rates as a function of bench-scale moving bed biofilm reactor loading rate for electron donors methanol (MeOH), ethanol (EtOH), glycerol (Glyc), and sulfide at 20°C (Bill et al., 2008).

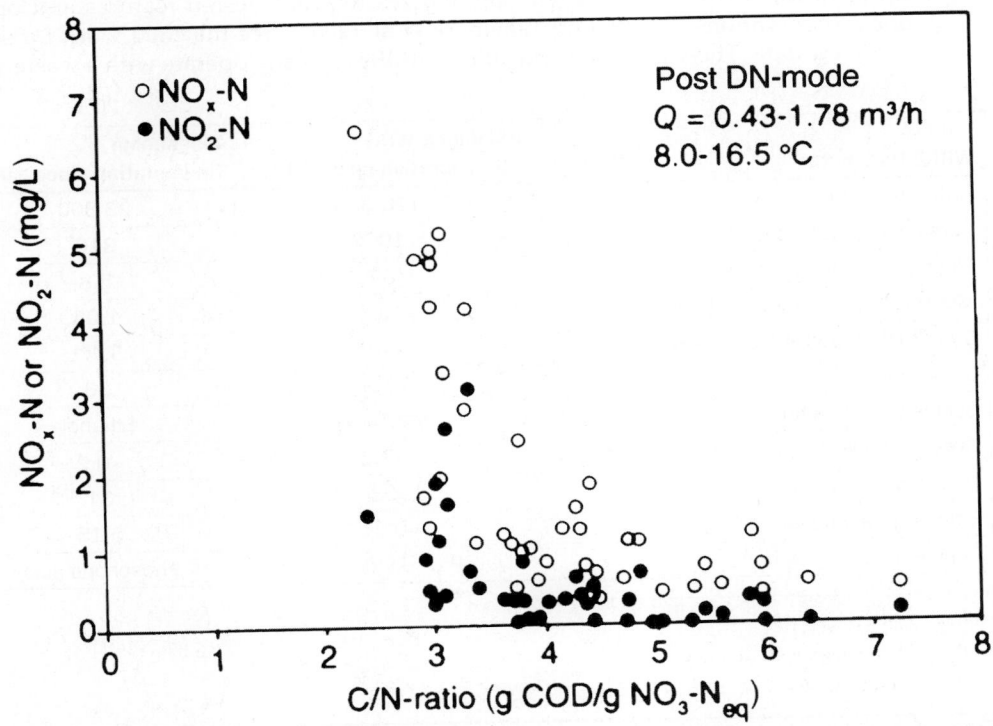

FIGURE 11.23 Post-denitrification moving bed biofilm reactor effluent nitrate/nitrite-nitrogen and nitrite-nitrogen concentrations as a function of carbon-to-nitrogen ratio (C:N) (Rusten et al., 1995b).

asymptotic minimum value when operating a post-denitrification MBBR with a carbon-to-nitrogen ratio (C:N) from 4 to 6 g COD/g NO_3-N_{eq}.

These observations are consistent with other pilot-scale post-denitrification MBBR investigations and full-scale operations (Aspegren et al., 1998; Täljemark et al., 2004). Lower C:N ratios are applicable to less stringent nitrate/nitrite-nitrogen effluent water quality standards. Higher carbon-to-nitrogen ratios may be applied to increase the rate of denitrification. Operating a post-denitrification MBBR with a C:N ratio higher than 4 to 6 g COD/g NO_3-N_{eq} increases the risk of residual soluble-COD concentration in the final anoxic tank effluent stream. The acceptable soluble-COD concentration is dependent on external carbon source cost, effect on downstream processes, and effluent water-quality standards. A postaeration zone containing media may be required to oxidize the remaining COD. Post-denitrification MBBRs typically have two equally sized anoxic zones and, sometimes, a postaeration zone.

Aspegren et al. (1998) reported an observed biomass yield resulting from nitrate-nitrogen removal in a post-denitrification MBBR using ethanol or methanol as the external carbon source in the range 0.2 to 0.3 g suspended solids per gram of COD transformed. Little information exists describing the startup period required to achieve a quasi steady-state with respect to nitrate/nitrite-nitrogen concentration remaining in the effluent stream of a post-denitrification MBBR. Rusten et al. (1995c) reported that the Lillehammer WRRF required approximately 4 to 6 weeks to obtain complete nitrate-nitrogen conversion.

The Sjölunda and Klagshamn WRRFs, Malmö, Sweden, operate full-scale post-denitrification MBBRs. Table 11.7 summarizes relevant design features and operational observations reported by Täljemark et al. (2004). See the cited work for study overview data. These post-denitrification MBBRs typically operate with a wastewater

WRRF Parameter	Sjölunda WRRF (in operation since 1997)	Klagshamn (in operation since 1999)
Flow rate (m³/d)	126 000	23 800
Nitrate-nitrogen load (kg/d)	1960	310
Effluent total nitrogen (g/m³)	6.8	5.8
Nitrate-nitrogen removal rate (g/m²/d)	1.05	1.05
CN (g COD added/g NO_3-N removed)	4.4:1	5.4:1
Carrier fill (%)	50	36
Supplemental carbon	Methanol	Ethanol
Overall sludge yield (g SS/g COD removed)	~0.2	~0.2
Mixing power (W/m³)	23	31
Effluent total phosphorus (g/m³)	0.21	0.15
Supplemental phosphorus	Phosphoric acid	Phosphoric acid

Notes:

COD = Chemical oxygen demand

SS = suspended solids

TABLE 11.7 Comparison of Sjölunda and Klagshamn Water Resource Recovery Facilities (WRRFs), Malmö, Sweden, Full-Scale Post-Denitrification Moving Bed Biofilm Reactor Design Features and Operational Observations (Täljemark et al., 2004)

temperature of 10 to 20°C. The performance of the reactors typically exceeds 90% nitrate/nitrite-nitrogen reduction when load is 0.8 to 1.2 $g/m^2/d$. Neither of these post-denitrification MBBRs are followed by postaeration reactors to oxidize any soluble-COD remaining in the effluent stream. Periodically, nitrate/nitrite-nitrogen removal at both of these post-denitrification MBBRs have been rate limited by phosphorus availability.

Where possible, combined pre- and post-denitrification MBBRs should be considered when high levels of nitrogen removal are required. This helps to optimize performance, efficiency, and operational flexibility. Most MBBRs designed for high levels of nitrogen removal use combined pre- and post-denitrification reactors (Ødegaard, 2008; Rusten and Ødegaard, 2007). The benefit of using combined pre- and post-denitrification is that removal efficiency becomes independent of the availability of biodegradable carbon source and temperature of the raw water. In addition, use of external carbon source to reach the effluent nitrogen standard is minimized.

3.2.4   Phosphorus Limitations (Focus on Denitrification)

Macronutrients, such as phosphorus, are required to complete biochemical transformation processes including nitrification and denitrification. The electron acceptor (i.e., nitrate-nitrogen or nitrite-nitrogen), external carbon source (e.g., methanol or ethanol), or macronutrients (primarily phosphorus) may be rate-limiting in a post-denitrification biofilm reactor. The soluble material orthophosphate is an indicator of phosphorus that is readily available for use in biofilm reactors. Particulate phosphorus assimilation and endogenous respiration are other phosphorus sources. There is a paucity of information describing both practical and applied aspects of simultaneous phosphorus and nitrogen reduction to low concentrations. Design engineers must be aware, however, of the potential effect of low orthophosphate concentrations in the influent to a post-denitrification MBBR. They must fully evaluate the need for additional process components required to ensure that the system is capable of meeting treatment objectives over the range of expected operating conditions.

deBarbadillo et al. (2006) report that the nitrate/nitrite-nitrogen concentration remaining in the effluent stream of a pilot-scale denitrification filter (using methanol as the external carbon source) increased when the influent orthophosphate concentration-to-influent nitrate/nitrite-nitrogen concentration ratio, $\frac{S_{in,PO_4-P}}{S_{in,NO_3-N}}$, was less than 0.02 g P/g N. These findings are illustrated in Figure 11.24.

When incorporated into a WRRF that produces effluent water with low nitrate-nitrogen and total-phosphorus concentrations, upstream unit processes may require optimization to meet phosphorus requirements in the post-denitrification biofilm reactor (Boltz et al., 2012). In some cases, it may be necessary to provide a supplemental phosphorus source (e.g., commercially available phosphoric acid). Andersson et al. (1998) reported that the ability to inject commercially available phosphoric acid into the influent stream of a full-scale post-denitrification MBBR improved the rate of denitrification. Researchers observed that an influent 1-g/m^3 orthophosphate concentration resulted in a nitrate-nitrogen removal about 70% with the effluent orthophosphate concentration of 0.1 g/m^3. Although the mechanism is not clearly understood, it typically is accepted that denitrification may proceed when the phosphorus availability is less than the stoichiometric requirement Callieri et al. (1984) suggested that bacteria may reduce their biomass yield and alter their phosphorus content when insufficient phosphorus is available for normal biochemical transformation processes. However, such conditions may result in a reduced denitrification rate or poor system response to dynamic loading conditions.

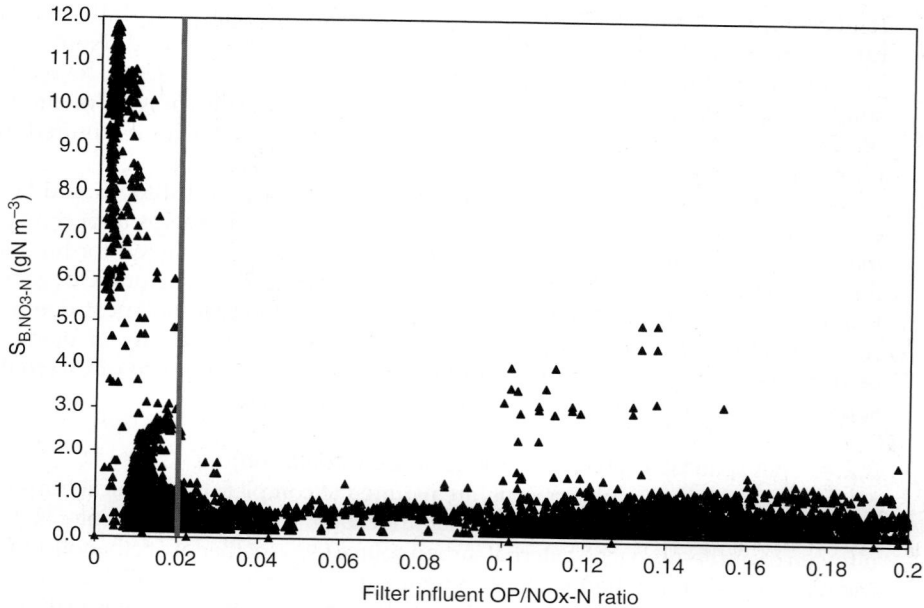

FIGURE 11.24   Nitrate-nitrogen concentration remaining in the effluent stream of a pilot-scale post-denitrification biofilm reactor (using methanol as the external carbon source) as a function of the influent orthophosphate concentration-to-influent nitrate/nitrite-nitrogen concentration

(deBarbadillo et al., 2006). The line drawn vertically from $\dfrac{S_{in,PO_4\text{-}P}}{S_{in,NO_3\text{-}N}} = 0.02$ empirically represents

the ratio below which the nitrate-nitrogen concentration remaining in the effluent stream begins to increase (Boltz et al., 2012).

3.3   Design Considerations

Successful MBBR design includes appropriate pretreatment, provisions for handling plastic carriers, well-designed aeration system and media retention sieves, properly specified mixers, and appropriate solids separation.

3.3.1   Preliminary and Primary Treatment

Biofilm reactors, including the MBBR, require proper preliminary treatment. Robust screening and grit removal is recommended to prevent sieve blinding and long-term accumulation of inert material such as rags, plastics, and sand in the tank. Once accumulated, these materials are difficult to remove. Manufacturers typically recommend no larger than 6- to 12-mm screen spacing if primary treatment is also provided. Fine-screens (3 mm) are recommended for secondary installations without primary treatment. Scum must be removed from the system because of its potential to blind the media-retention sieves. Tertiary or add-on MBBR processes receiving wastewater that received significant upstream treatment do not require additional screening. Table 11.8 provides a list of screen spacing for installations at selected full-scale WRRFs incorporating the MBBR process.

Facility	Pretreatment	Details
Lillehammer WRRF[a] (Lillehammer, Norway)	Step screens/grit removal	15-mm (coarse screens) followed by 3-mm (fine screens)
Gardemoen WRRF[a] (Oslo, Norway)	Step screen/grit removal	6 mm
Crow Creek WRRF[a] (Cheyenne, Wyoming)	Self-cleaning filter screen	10-mm × 15-mm
Yavne Municipal WRRF[c] (Yavene, Israel)	Medium screen, sedimentation lagoon, fine screen	15-mm (coarse screens) followed by 6-mm (screens)
Western WRRF[b] (Perth, Australia)	Step screen/grit removal	3-mm (fines screens)
Mao Point WRRF[a] (Wellington, New Zealand)	Step screen/grit removal	3-mm (fine screens)

[a]Facility includes primary treatment.
[b]Facility does not have primary treatment.
[c]Tertiary MBBR process.

TABLE 11.8   Preliminary Treatment at Full-Scale Moving Bed Biofilm Reactor (MBBR) Installations (adapted from McQuarrie and Boltz, 2011)

3.3.2   Plastic Biofilm Carrier Media

Plastic biofilms carriers typically are delivered to the site in sacks of known volume. The carriers are introduced to the MBBR simply by elevating the sacks and allowing the plastic biofilm carriers to fall into the tank (see Figure 11.25). A recessed impeller pump can be used to transfer the plastic carriers out of the water-filled MBBR to perform aeration system maintenance. The MBBR aeration or mixing system should remain engaged while the pump transfers carriers to temporary storage. Ideally, a dedicated basin or container must be available for interim plastic biofilm carrier storage to ensure that the original carrier fill volume is returned to the emptied tank. If the carrier fill is transferred to another MBBR for temporary storage, then the best way to restore original carrier fill is to pump the plastic carriers back into the emptied tank, drain both

FIGURE 11.25   Plastic biofilm carrier installation and 50% carrier fill in a dry MBBR before startup.

MBBRs, and measure the average bed height at several locations. The plastic carrier bed depths should be equal.

It is common for light foam to float on the water surface during MBBR startup and while mature biofilm is developing. If necessary, a defoaming agent may be used. It is important, however, to ensure that the defoaming agent is compatible with the plastic carriers by consulting with the manufacturer. Even when agitated, plastic biofilm carriers have a propensity to float when introduced to the water-filled tank, but will disappear within a few days. Carbon-oxidation likely will be observed after 2 to 15 days; ammonia-nitrogen oxidation will proceed after approximately four weeks but may take 60 to 120 days to reach a quasi steady-state for biofilm thickness, mass, and ammonia-nitrogen flux. Approximately four to six weeks may be required before nitrate/nitrite-nitrogen removal occurs because the denitrifying biofilms will not develop until sufficient nitrate-nitrogen is present.

3.3.3   Aeration System

The MBBRs are not compatible with all commercially available diffused aeration systems that typically are used for aerobic (biological) municipal wastewater treatment. The piping network and air diffusers must (1) provide air that meets process oxygen requirements; (2) have a reasonable oxygen-transfer efficiency; (3) promote rolling-water circulation pattern that uniformly distributes plastic biofilm carriers; (4) structurally withstand the weight of biofilm-covered plastic carriers (unit weight of a biofilm is approximately equal to unit weight of water) when the tank is drained; (5) be robust with infrequent maintenance requirements.

Coarse-bubble diffusers produce bubbles with a diameter of 6 to 12 mm (compared to 2-mm diameter typical of fine-bubble diffusers). These bubbles rise rapidly through the plastic biofilm carrier laden water column. Therefore, a rolling-water circulation pattern can be generated with a coarse-bubble diffuser grid that covers a majority of the tank bottom, although complete floor coverage is not recommended. Multiple drop pipes with individual valves for modulation provide added flexibility to induce a rolling pattern. Coarse-bubble diffusers typically used in MBBRs are 25-mm diameter stainless-steel pipes with 4- to 5-mm diameter orifices that are spaced approximately 50-mm apart along the underside of the diffuser pipe (see Figure 11.26, top). The air diffuser typically is anchored approximately 0.25 m above the tank bottom. The coarse-bubble diffuser orifice must be smaller than the plastic biofilm carrier to avoid air-pipe and orifice plugging. Pham et al. (2008) reported that a 2-m deep tank (with a length-to-width ratio equal to 1) produced 2.1%, 3.1%, and 2.5% clean-water oxygen transfer efficiency per meter of water submergence at a 13.6-m^3/hr airflow rate when containing 0%, 25%, and 50% carrier fills, respectively. These clean-water trials demonstrate, for the conditions tested, that the presence of plastic biofilm carriers increases coarse-bubble aeration system oxygen transfer efficiency by 20% to 40%. Clean-water oxygen transfer efficiency design values for full-scale operating MBBRs with coarse-bubble diffusers are 3.0% to 3.5% per meter of water submergence. The MBBR-specific coarse-bubble air diffusers have been designed with a 0.8-alpha (a) factor and 0.95-beta (b) factor (Johnson and Boltz, 2013).

When using fine-bubble diffusers, biofilm carriers must periodically be removed to service or replace diffusers. The ideal MBBR-diffused air system would operate at a point that optimizes both oxygen transfer efficiency and mixing capacity. Manufacturers have examined the capability of fine-bubble-diffusers to meet the air-induced plastic carrier mixing requirements and desirable maintenance characteristics typical

FIGURE 11.26   Photograph of the underside of a coarse-bubble air diffuser commonly used in MBBRs showing the structural support (top). Disc-type, fine-bubble air diffuser network configured in a T-pattern promotes the rolling water circulation pattern and uniformly distributes plastic biofilm carriers.

of coarse-bubble diffusers historically used in MBBRs. The bubble-rise velocity character-istic of fine-bubble diffusers does not create the rolling-water circulation pattern required to uniformly distribute the plastic biofilm carriers. Therefore, the diffuser-grid layout may require modification to promote a rolling-water circulation pattern (see Figure 11.27, left). The fine-bubble diffuser configuration required to promote a rolling water-circulation pattern has a negative effect on oxygen transfer efficiency. Pham et al. (2008) reported that a 2-m deep tank (with a length-to-width ratio equal to 1) produced 7.1%, 5.8%, and 4.9% clean-water oxygen transfer efficiency per meter of water submergence at

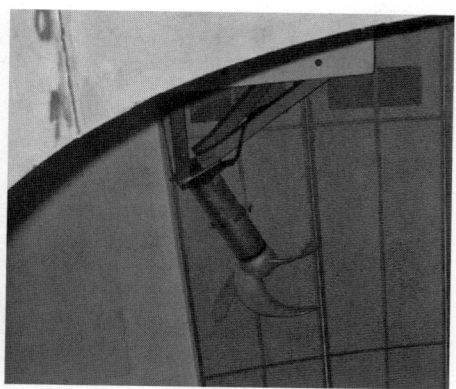

FIGURE **11.27** Mechanical mixers on rail-mounts (left) and top-mount (right) for denitrification MBBR.

a 13.6-m³/hr air-flow rate when containing 0%, 25%, and 50% carrier fills, respectively. These clean-water trials demonstrate, for conditions tested, that the presence of plastic biofilm carriers decreases fine-bubble aeration system oxygen transfer efficiency by 20% to 30%.

3.3.4 Media Retention Sieves

Sieves used to retain media, and their supporting structural assemblies if required, typically are constructed with stainless steel. The sieve may be wedge-wire or mesh with approximately 6-mm spacing, but can be procured as perforated plates with 5- to 6-mm diameter orifices. Two sieve configurations typically are used in MBBRs: horizontally configured cylindrical sieves (carbon oxidation/nitrification MBBRs) and vertically mounted wall sieves (denitrification MBBRs). Horizontal sieves are attached to the reactor wall by cast-in-place wall thimbles or by inserting wall sleeves through poured or core-drilled holes in the reactor wall. The cylindrical sieves are typically submerged 35% to 65% of the side water depth (see Figure 11.11, left). Depending on the selected sieve length, it may be necessary to add a structural sieve support assembly.

Vertically mounted flat panel sieves are attached to wall-fixed brackets. The brackets extend outward from the wall and create approximately a 0.15- to 0.3-m space between the sieve panels and the reactor wall (see Figure 11.11, right). Forward flow passes through the sieves, then through the void space between the sieve panels and the wall, and finally into the next treatment step through the concrete wall openings located at the liquid surface. A full concrete wall is used to divide MBBRs for two reasons: (1) the concrete wall provides structural support for the sieve brackets and panels; (2) the rigid segregation of MBBRs promotes a completely mixed bulk liquid (i.e., eliminates the potential for back mixing with the downstream MBBR).

The sieve area is defined based on the allowable hydraulic headloss across the MBBR wall. However, typical sieve design allows for a maximum 50- to 100-mm headloss (at the peak hydraulic flow, which is typically measured as the peak hour flow) across each sieve-containing wall. Proper sieve design is primarily related to system hydraulics. Critical design parameters include sieve loading rate, approach velocity, and the MBBR length-to-width ratio. These terms are defined as follows:

- Sieve loading rate: wastewater flow rate (including recirculation streams) applied per unit screen area, or $\left(\dfrac{\text{flow rate}}{\text{sieve area}}, \dfrac{\frac{m^3}{hr}}{m^2} \right)$.

- Approach velocity: wastewater flow rate (including recirculation streams) in an MBBR divided by the reactor cross-sectional area, or $\left(\dfrac{\text{flow rate}}{\text{side depth} \cdot \text{tank} \cdot \text{width}}, \dfrac{\frac{m^3}{hr}}{m^2} \right)$.

- Length-to-width ratio: 0.5:1 to 1.5:1; MBBR L:W > 1.5 may be possible, but also may result in plastic biofilm carrier migration. The designer must ensure that the approach velocity is below the recommended limit.

Based on the design hydraulic headloss across each MBBR sieve-containing wall, the corresponding sieve loading rate is selected by the MBBR manufacturer using empirical criteria that accounts for sieve material and the influence of plastic biofilm carrier fill. Sieve loadings are typically 50 to 60 m/hr, but values up to 85 m/hr have been applied with sufficiently low approach velocity. Once determined, the total sieve area is divided by the area of a sieve fabricated with a standard wedge-wire (or perforated plate) panel length. Typically, cylindrical sieves have a 16- to 24-inch diameter. Their length varies but is typically 12 feet. Under peak flow conditions, the approach velocity should be less than 30 to 35 m/h. Higher approach velocities will cause the media to migrate with the flow and accumulate on the sieves. The result is reduced sieve hydraulic throughput, increased hydraulic headloss, poor oxygen transfer in the zone where media accumulates, and likely an oxygen deficiency because of incomplete use of the MBBR volume. When an approach velocity greater than 30 m/hr cannot be avoided, it may be prudent to reduce the sieve loading rate (Leaf et al., 2011).

Provisions for filling and draining an MBBR train are necessary. Small wall openings with screens typically are installed near the floor of the reactor to allow for equalization of water level between reactors during fill and drain periods. The designer also must consider how the tank will be drained without removing the plastic biofilm carriers. Perhaps water can be withdrawn from a location immediately downstream of the last MBBR wall or with an underdrain system.

3.3.5 Mechanical Mixing

There are special requirements for proper mechanical mixer design for denitrification MBBRs. Early MBBR designs used molded fiberglass blades, which did not hold up well in the abrasive environment induced by the plastic biofilm carriers (see Figure 11.14). These early designs also used a paint-protected motor and gear housing, which quickly gave way to the abrasive environment ultimately, exposing the metallic surface to corrosive conditions. State-of-the art mixer placement and tank orientation has resulted from full-scale operational experience. Early designs placed the mixers near the tank bottom. As a consequence, poor mixing of the plastic biofilm carriers resulted in their accumulation at the water surface and in the tank corners.

Manufacturers have developed a mechanical mixer specially designed to withstand the abrasive MBBR environment. The mixer uses a stainless-steel backward-curve

Water Resource Recovery Facility	Operating Mode, Mixing Energy (Media Fill Fraction)
NRA WRRF[a] (Oslo, Norway)	Pre-denitrification, 10 W/m³ (54%)
	Post-denitrification, 8 W/m³ (52%)
	Post-denitrification, 5 W/m³ (14%)
Sjolunda WRRF[b] (Malmo, Sweden)	Post-denitrification, 23 W/m³ (50%)
Klagshamn WRRF[b] (Malmo, Sweden)	Post-denitrification, 31 W/m³ (50%)
South Adams County[b] (Colorado)	Pre-denitrification, 19 W/m³ (57%)
Norman M. Cole Jr., Pollution Control Plant (Virginia)	Post-denitrification, 31 W/m³ (45%)

[a]Measured consumption.
[b]Motor label.

TABLE 11.9   Observed Mixing Energy Required per Unit Moving Bed Biofilm Reactor Volume (adapted from McQuarrie and Boltz, 2011)

propeller with a round bar welded along its leading edge to avoid damage to the plastic biofilm carriers and impeller wear. The mixer has a large-diameter impeller with a fairly low rotational speed (90 rpm at 50 HZ and 105 rpm at 60 HZ). Unlike activated sludge, the plastic biofilm carriers described in this section will float in quiescent water. As a result, the mixers need to be located near the water surface but not so close as to create an air-entraining vortex because dissolved oxygen inhibits denitrification. In addition, a slight negative inclination of mixer orientation helps maintain the rolling-water circulation pattern and uniformly distribute plastic biofilm carriers (see Figure 11.27). It is especially important to use rail-mounted units to facilitate access to the mixer when maintenance is required. These specially designed mechanical mixers are typically sized to input 25 W/m³. However, the references listed in Table 11.9 provide the designer with a full-scale operating basis for estimating the mixing energy requirement per unit MBBR volume.

3.3.6 Solids Separation
The MBBR process performance is dependent of a liquid-solids separation unit. Biomass accumulation in an MBBR is, however, independent of the settler. Therefore, the MBBR process offers considerable flexibility in terms of the type of process that can be used for liquid-solids separation. Typically, the suspended solids concentration in the MBBR effluent stream is at least an order of magnitude lower than typical of activated-sludge bioreactors. As a result, a variety of different solids separation processes have been paired with MBBRs. Representative examples of solids separation following MBBRs are summarized in Table 11.10. The MBBR can be combined with compact, high-rate solids separation technologies such as dissolved air flotation and ballasted flocculation. Tertiary units may discharge directly to a filtration unit or coagulation/flocculation/sedimentation (with lamella plate settlers) basins. Circular or rectangular secondary clarifiers also may be used (especially in the case of retrofit application in which the clarifiers already exist). Figures 11.28 and 11.29 present representative long-term performance of conventional clarifier settling and dissolved air flotation following MBBRs designed and sized for nitrogen removal. These figures

MBBR Facility	Separation Technology	Design Rate (m³/m²·hr)
Yavne Municipal WRRF[b]	Rectangular clarifiers	1
South Adams WRRF[a]	Reuse existing clarifiers	1.0–1.8
Crow Creek WRRF(1)	Reuse existing clarifiers	1.1–2.2
Lillehammer WRRF(1)	Flocculation/settling	1.3–2.2
Gardemoen WRRF(1)	Flocculation/flotation	3.1–6.4
Nordre Follo WRRF[a]	Flocculation/flotation	5–7.5
Sjolunda WRRF[b]	Dissolved air flotation	—
Skreia WRRF[a]	Ballasted flocculation	45–70

[a]Multi-stage MBBRs.
[b]Tertiary MBBR.

TABLE 11.10 Solids Separation Examples at Moving Bed Biofilm Reactor (MBBR) Installations (adapted from McQuarrie and Boltz, 2011)

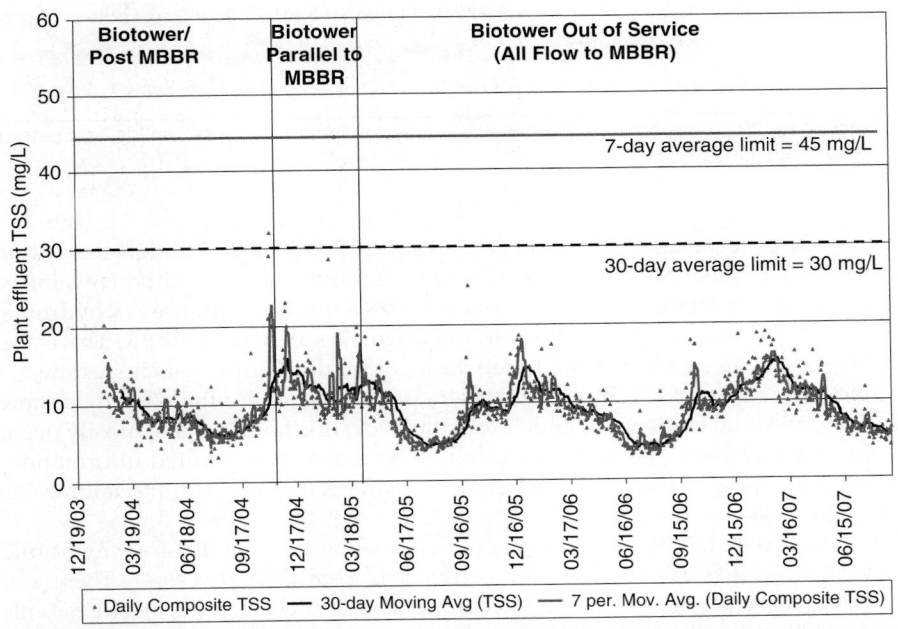

FIGURE 11.28 Settled clarifier effluent suspended solids concentration following MBBR.

showing long-term performance provide the designer with a range of expected effluent suspended solids concentrations for the two liquid-solids separation technologies.

3.3.7 Hydraulic Limitation and Loss of Media

Leaf et al. (2011) presented a review of hydraulic limitations of the IFAS and MBBR including case studies of hydraulic failures, which resulted in media loss. While it is not the intent to replicate work presented earlier, a brief summary is considered suitable here.

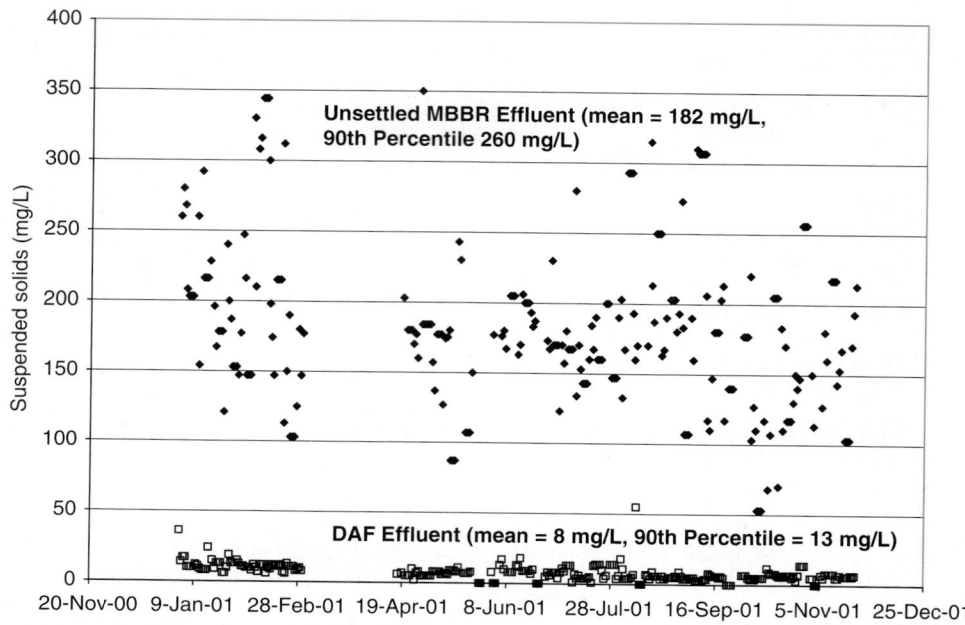

Figure 11.29 Dissolved air flotation influent and effluent suspended solids concentration following an MBBR process.

The mechanical components are designed with similar design standards for both MBBR and IFAS systems. Each of these process mechanical components influences the hydraulic throughput of MBBR and IFAS systems. The instances of hydraulic failures and media loss are greater with IFAS systems compared to MBBR. There are six IFAS systems that have been subjected to a hydraulic failure, which resulted in plastic biofilm carrier loss. These failures were public and under the scrutiny of news media. The anecdotal evidence suggests that the hydraulic failures have mostly occurred during construction. Therefore, under these circumstances, limited information exists to describe the conditions that ultimately resulted in the hydraulic failure and biofilm carrier loss.

Leaf et al. (2011) gives account of five case studies that illustrate hydraulic failures, which resulted in plastic biofilm carrier loss from IFAS processes. These case studies were primarily based on press record, and a limited technological evaluation of the respective systems. The main aim was to identify general mechanisms of hydraulic failure and biofilm carrier loss and recommend applicable design guidelines to avoid hydraulic failure and biofilm carrier loss. The case studies include IFAS installations at:

- Broomfield, Colorado, USA
- Raisio, Finland
- Groton, Connecticut, USA.
- Hooksett, New Hampshire, USA
- Mamaroneck, New York, USA

The hydraulic characteristics of moving bed reactors are typically measured by approach velocity and screen hydraulic loading rate.

3.3.7.1 Modes of Flat Screen Failure A loss of flat screen hydraulic throughput is a mode of system hydraulic failure. The absence or inefficiency of air scour device can contribute to the retention of screen accumulation. Plastic biofilm carriers or trash accumulation on the screen surface due to insufficient screen area or excessive approach velocity, which causes hydraulic throughout substantially reduced and peak hydraulic flows, may result in basin overflow and plastic biofilm carrier loss.

3.3.7.2 Modes of Cylindrical Screen Failure In the case of cylindrical screen, if not reinforced, could lead to structural failure, which causes plastic biofilm carriers to migrate to the downstream unit processes. Trash accumulation due to insufficient pretreatment, would cause hydraulic throughput to substantially reduce and peak hydraulic flows that may result in basin overflow and plastic biofilm carrier loss. Biofilm carrier accumulation on the screen surface resulting from insufficient screen area causes too high media fill fraction, improper depth of submergence, insufficient underneath aeration intensity, combined with excessive approach velocity causes reduction of hydraulic throughput and peak hydraulic flows and may result in basin overflow and plastic biofilm carrier loss.

3.3.7.3 Design Considerations to Avoid Hydraulic Limitation and Media Loss Good preliminary treatment for screening and grit removal to prevent inert material accumulation within the bioreactors, as needed. Design features to overcome hydraulic limitations inherent to the IFAS process are as follows:

- Redirecting flow perpendicular to normal basin flow scheme
- Bypassing wet weather flows around the IFAS zone
- Split internal mixed liquor flow using series of pumps
- Process control with instrumentation to overcome hydraulic limitations during wet weather conditions

3.3.7.4 Design of Retention Screen

- Typical headloss of 50 to 150 mm across each screen-containing wall at the peak hydraulic flow (McQuarrie and Boltz, 2011)
- The hydraulic loading rate is typically less than 60 m^3/m^2 hr under all flow conditions
- Cylindrical screen submergence of 35% to 65% of the reactor side water depth
- Cylindrical screen with structural support to resist forces exerted by plastic biofilm carrier.

4.0 Biologically Active Filters

Biological wastewater treatment and suspended-solids removal are carried out in BAFs under either aerobic or anoxic conditions. In a BAF, the media acts simultaneously to support the growth of biomass and as a filtration medium to retain filtered solids. Accumulated solids are removed from the BAF through backwashing. There is a direct interaction between the media characteristics and the process, because the configuration (sunken media or floating media), and flow and backwash regimes depend on media density. Media may be natural mineral, structured plastic, or random plastic.

The BAF reactor can be used for carbon oxidation or BOD removal, only, combined BOD removal and nitrification, combined nitrification and dentrification, tertiary nitrification, and tertiary denitrification. Once the raw wastewater has undergone screening, grit removal, and primary treatment, the BAF process can include full secondary treatment for a facility or can be constructed for operation in parallel to an existing secondary treatment process. Using BAF as a tertiary treatment process for nitrification and/or denitrification as an upgrade to existing secondary processes is common. A typical process flow diagram for four different BAF options is provided in Figure 11.30.

4.1   Biologically Active Filter Configurations

Historically, the acronym BAF has meant "biological aerated filters" and the term typically has been used to refer to aerated biofilters used in secondary treatment. However, the acronym BAF is being expanded herein to cover all "biologically active filters," including those that operate under anoxic conditions for denitrification, which have been referred to as denitrification filters. The BAF reactors can be characterized into groups according to their media configurations and flow regime (Debarbadillo et al., 2010):

- Downflow BAF with media heavier than water. This general category includes both the Biocarbone® reactors commercially marketed in the 1980s for secondary and tertiary treatment and packed-bed tertiary denitrification reactors such as Tetra Denite® filters. These BAFs are backwashed using an intermittent counter-current flow regime.

- Upflow BAF with media heavier than water. This includes BAF reactors for secondary and tertiary treatment that use expanded clay and other mineral media, such as the Degremont Biofor®. These BAFs are backwashed using an intermittent concurrent flow regime.

- BAF with floating media. This includes BAF with polystyrene, polypropylene, or polyethylene media, such as the Kruger Biostyr®. These BAFs are backwashed using an intermittent counter-current flow regime.

- Continuous backwashing filters. These filters operate in an upflow mode and consist of media heavier than water that continuously moves downward, countercurrent to the wastewater flow. Media is directed continuously to a center air lift where it is scoured, rinsed, and returned to the top of the media bed.

- Non-backwashing, submerged filters. These processes consist of submerged, static media and are often referred to as submerged aerated filters (SAF) although there has been recent work in applying this technology with anoxic conditions for denitrification. Solids are intended to be carried through the reactor and removed through a dedicated solids separation process.

This section provides a detailed description of each type of BAF reactor, followed by practical design considerations and guidance.

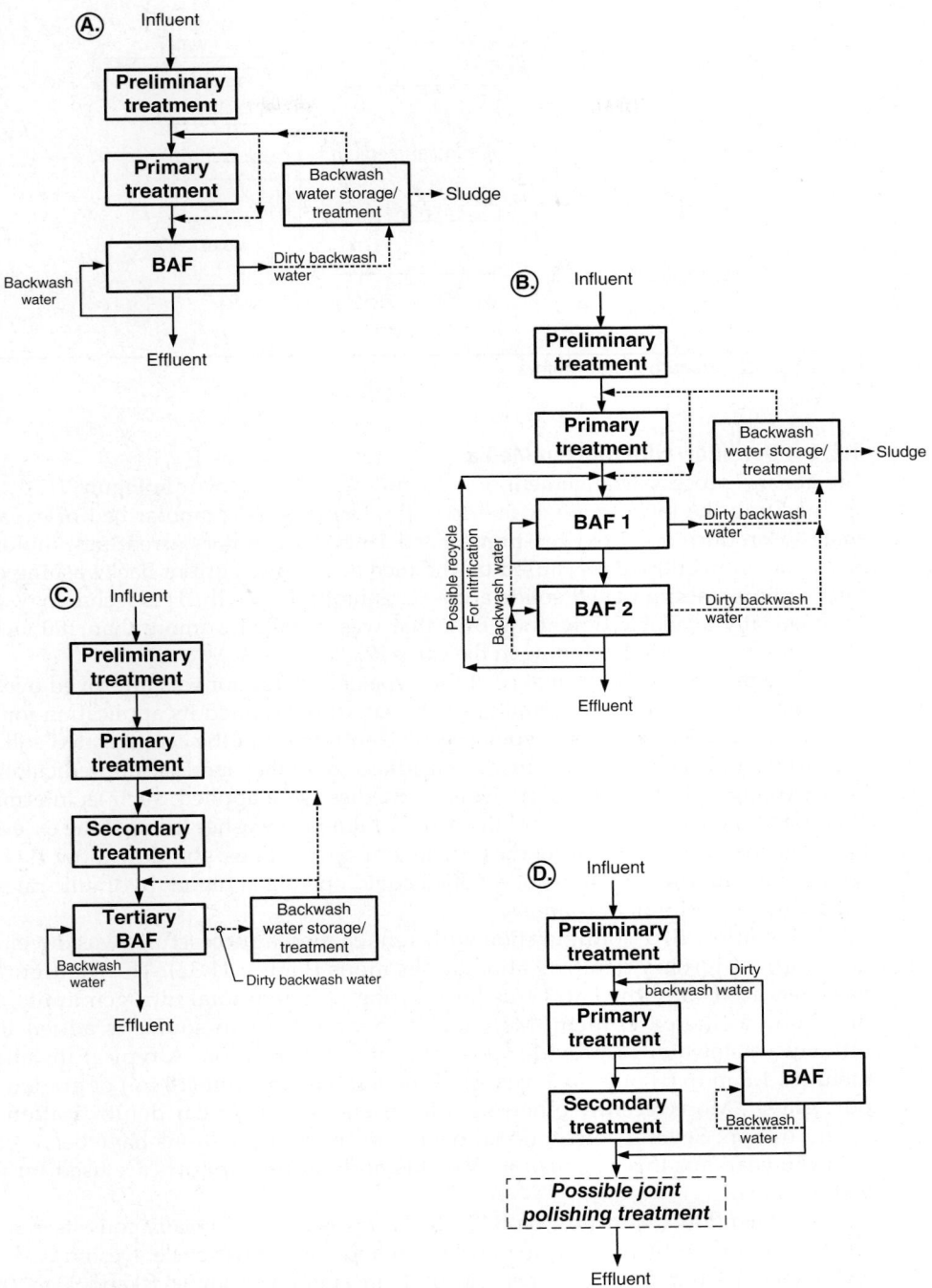

FIGURE 11.30   Typical process flow diagram for four different biologically active filter options.

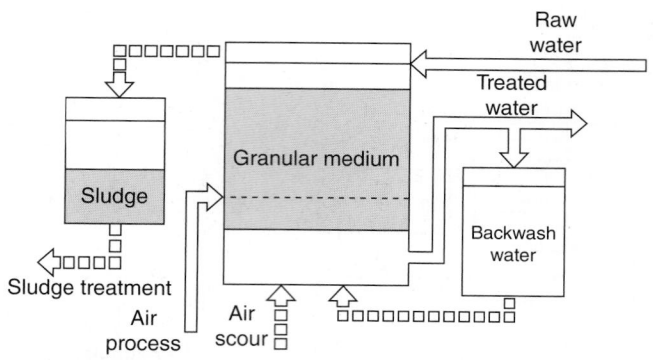

FIGURE 11.31   Biocarbone downflow.

4.1.1   Downflow with Sunken Media

The general process arrangement of a downflow BAF is shown in Figure 11.31. Air is sparged into the lower zone of the downflow submerged granular bed of expanded shale to produce good oxygen-transfer efficiency by counter-current gas-liquid flow and a circuitous flow path caused by the media. Counter-current backwashing of the filter removes accumulated solids and excess biofilm growth. This technology was a commercially available downflow BAF that was installed in more than 100 facilities throughout the world beginning in the early 1980s.

While the process performance of this type of BAF reactor was improved over previous practice, the counter-current air and water-flow limited its application for BOD removal and nitrification. Air would become entrapped in the accumulated solids on the surface and at the top of the media. Headloss could then increase unpredictably and backwashing would be necessary. Some remedies were applied, such as intermittent aeration at higher rates to expand the bed, or mini-backwashes to expel the excess solids from the surface. For secondary treatment applications, the downflow BAF was replaced by upflow configurations, which could operate at higher hydraulic rates and handle wider hydraulic variations.

A downflow BAF configuration with sunken media successfully was developed, however, for tertiary denitrification applications (Figure 11.32). This configuration has been used since the late 1970s for meeting stringent total nitrogen limits while providing a filtered effluent. Methanol or another carbon source is added to the influent wastewater to provide substrate for denitrification. A typical installation includes 1.8 m (6 ft) of 2- to 3-mm sand media over 457 mm (18 in.) of graded support gravel. More recently other manufacturers offer a similar denitrification BAF configuration. Several conventional deep-bed filter installations have been retrofit over the years for this application. Various underdrain supports are used for these installations.

In a downflow denitrification BAF, the backwash cycle typically consists of a brief air scour followed by an air-water backwash and water rinse cycle. Design backwash water and air scour flow rates are typically 15 m^3/m^2/h (6 gpm/sq ft) and 90 m^3/m^2/h (5 cfm/sq ft), respectively. The filter influent and backwash piping are similar to that of conventional filters. Backwash water usage is typically 2% to 3% of the average flow

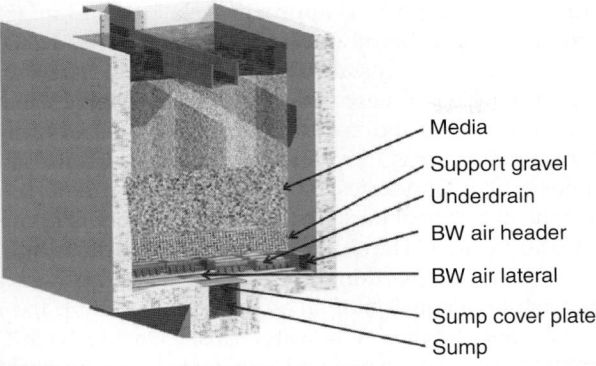

FIGURE 11.32   Downflow BAF configuration with sunken media.

being treated. Nitrogen gas accumulates within the media and is released by pumping backwash water up through the media bed for a short duration. The denitrification capacity between nitrogen release cycles typically ranges from 0.25 to 0.5 kg NO_x-N/m^2 (0.05 to 0.10 lbs NO_x-N/sq ft) (Severn Trent, 2008).

4.1.2   Upflow with Sunken Media and discrete backwash
The upflow mode of BAF operation through a sunken granular bed has been used in more than 185 installations worldwide for BOD removal, nitrification, and denitrification. Its general process arrangement is shown in Figure 11.33. Solids are trapped mostly in the lower part of the media bed during normal operation and are backwashed as required by increasing the hydraulic rate and applying scour air. As the backwash consists of concurrent scour air and backwash water, the accumulated solids travel up through the media bed before being released at the top.

    Three types of media can be used depending on the application. The media consists of expanded clay or expanded shale either in the form of spherical grains (with an effective size of 3.5 or 4.5 mm) or as angular grains (with an effective size of 2.7 mm). The media form a submerged, fixed bed in the bottom of the reactor, typically a height of 3 to 4 m (9.8 to 13.1 ft), with approximately 1 m (3.3 ft) of freeboard zone above the bed.

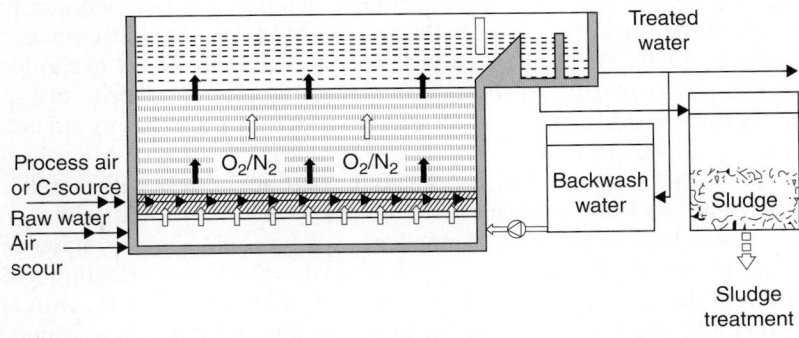

FIGURE 11.33   Biofor upflow.

The clean surface area of grains is approximately 1640 m²/m³ (500 sq ft/cu ft). Influent water is introduced to the bed through a filter plenum and nozzle air/water distribution system. The nozzles are installed in a false floor located approximately 1 m (3.3 ft) above the filter floor. The influent flow must be fine screened to prevent blockage of the nozzles. Backwash water and scour air are introduced through the same plenum/nozzle system. Process air is introduced through separate air diffusers located in the media bed above the inlet nozzles.

To reduce the quantity of backwash water and the risk of media loss, the backwash starts with a drain down. The duration of the drain down is affected by the level of solids accumulation and determines the need for more vigorous backwashing. The concurrent backwash then consists of an air scour to break up the media, followed by an air/water wash, and finally by a water rinse. During backwashing, the solids are pushed from the bottom of the bed and transferred into the freeboard zone between top media level and the discharge, and on to waste.

Backwash water quantities equivalent to several times the volume of the freeboard zone are needed to reduce solids levels to discharge limits as they are released directly from the filter. The effect of the discharge of these solids depends on the actual treatment objectives and number of BAF cells. This can be addressed by incorporating a "filter-to-waste" step at the end of the backwash cycle. Alternately, increasing the total volume of flushing water may be necessary to improve the effluent quality following backwashing (Michelet et al., 2005).

A key issue with backwash of sunken media systems is the potential for "boils" during backwashing. For even backwashing, the water flow must be well distributed across the plan area of the BAF, and, therefore, the headloss across the distribution system must be greater than the headloss through the bed. If the bed becomes blocked because of high loads or insufficient backwash, then its headloss becomes the controlling factor. The flow will short circuit through the line of least resistance. This will result in a "boil", or violent eruption of the flow through the point of least resistance. Similar short circuits and boils can also occur if the nozzles are blocked. These boils during backwashing can result in excessive media loss.

4.1.3   Upflow Biologically Active Filters with Floating Media

These processes use a floating bed of media to provide biological surface area and filtration. This process was first used in industrial filtration and drinking water denitrification (Roennefahrt, 1986). Coarse-bubble aeration diffusers were introduced at the bottom of the media to enhance the contact of air, water, and biomass (Rogalla and Bourbigot, 1990). While the Biostyr® process uses lightweight expanded polystyrene (specific gravity of 0.05), a process using recycled polypropylene with a specific gravity slightly lower than 1, the Biobead® (Brightwater F.L.I.), also has found large application in the United Kingdom.

The Veolia unit (see Figure 11.34) is partially filled with small (2- to 6-mm) polystyrene beads. Process objectives determine selection of the bead size; larger beads can be more heavily loaded and smaller beads typically achieve higher process performance. The beads, which are lighter than wastewater, form a floating bed in the upper portion of the reactor, typically a height of 3 to 4 m (9.8 to 13.1 ft), with approximately 1.5 m (4.9 ft) of free zone below the bed. The top of the bed is restrained by a ceiling fitted with filtration nozzles to evenly collect the treated wastewater. The clean surface area of spherical beads is 1000 to 1400 m²/m³.

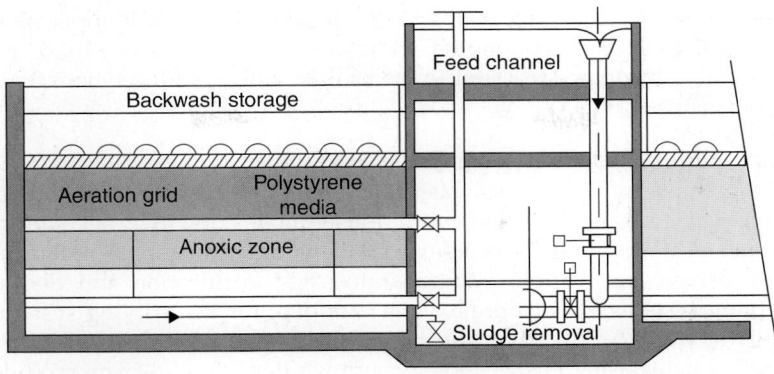

FIGURE 11.34 Biostyr filter.

In the bottom of the reactor, influent is distributed by troughs formed in the base of the cells. The troughs are covered with plates, which have gaps at intervals to allow the flow to enter the cells and backwash wastewater to be collected. There is no need for a filter underdrain because the media does not require support. Process air is distributed through diffusers located along the bottom of the reactor or within an aeration grid in the media bed; the latter being used if an anoxic zone is required for nitrogen removal. Only treated wastewater comes in contact with nozzles. Backwashing consists of counter-current air scour and backwash water flow. Solids are removed through the shortest pathway at the bottom of the reactor. Since 1993, when Toettrup et al. (1994) presented information regarding process status, the Biostyr® process has been widely used in European and the United States. Another option, Biobead®, is similar, except that the media is larger and heavier, using polypropylene or polyethylene with a density of approximately 0.95. Wastewater flow enters at the bottom of the reactor, and the flow is distributed by a grid or specially designed distribution system. For small cells, a simple system using one central feed and distribution plate arrangement is sufficient. For larger cells (greater than 5.5 m × 5.5 m (18 ft × 18 ft), a more sophisticated arrangement is required, such as horizontal slots staggered across the cells. The slot size needs to be carefully designed for even distribution, especially at low flows (Cantwell and Mosey, 1999).

To prevent the media loss, a metal grid is fixed near the top of the reactor. Process air is supplied from a grid located below or within the media bed. By placing the process air grid within the media, it is possible to achieve some solids removal at the bottom of the bed. Because of specific gravity, it is relatively easy to release the accumulated suspended solids during backwashing, requiring only a relatively low head. Typically, backwashing consists of a combination of partial drainage, air scour, and countercurrent water flush. The dirty backwash water is removed over the outlet weir or through a bottom drain. Recovery of solids retention following backwash may take some time until the bed is sufficiently packed again, during which the effluent from the reactor may be recirculated through the facility.

Upflow floating BAF also may require a certain number of mini-backwashes (typically four to eight and, in extreme cases, more than 10) to bump the filter, remove some solids, and lower the headloss to achieve a complete filtration cycle of 24 or 48 hours (time between two "normal" backwashes). The requirement for mini-backwashes plus normal backwashes can generate significant backwash wastewater. During the demonstration testing in San Diego, California, a single stage BOD removal application, the floating media BAF generated a volume of backwash wastewater between 10.3% and 13.9% of the influent flow, compared to a sunken media BAF, which produced between 7.4% and 7.9% (Newman et al., 2005).

4.1.4 Moving Bed, Continuous Backwash Filters

Moving bed, continuous backwash filters operate in an upflow mode and consist of media heavier than water that continuously moves downward, countercurrent to the wastewater flow. These filters are used widely for tertiary solids and turbidity removal but also have been applied to separate stage nitrification and denitrification. For nitrifying systems, air or pure oxygen is added; for denitrifying systems, a source of readily biodegradable carbon substrate such as methanol is added. Two commercially offered systems using this technology supply filter cells as 4.65 m² modules with center airlift assembly. The effective media depth is typically 2 m, and sand media size typically ranges from approximately 1 to 1.6 mm.

Moving bed filters backwash continuously at a low rate. A typical unit is shown in Figure 11.35. Influent wastewater enters the filter bed through radials located at the bottom of the filter, moves up through the downward-moving sand bed, and the effluent flows over a weir at the top of the filter. The media, with the accumulated solids, is drawn downwards to the bottom cone of the filter. Compressed air is introduced through an airlift tube extending to the conical bottom of the filter and rises upward with a velocity of greater than 3 m/s (10 ft/s) creating an airlift pump that lifts the sand at the bottom of the filter up the center column. The turbulent upward flow in the airlift provides scrubbing action that effectively separates solids from the media before discharging to the filter wash box.

Moving bed filter manufacturers typically set the reject weir to provide a continuous wash water flow rate of about 10 to 12 gpm per filter module. This is the equivalent to a wash water rate of approximately 10% of the forward flow at an average filter loading rate of 4.9 m/h (2 gpm/cu ft).

The backwash frequency is quantified by the bed turnover rate. If used for solids removal only, moving bed filters media turnover rates range from 305 to 460 mm/h or four to six bed turnovers per day. To maintain sufficient biomass in the filter for denitrification, the bed turnover rate must be reduced to approximately one to three turnovers per day or 100 to 250 mm/h.

4.1.5 Non-Backwashing, Open-Structure Media Filters

These processes consist of submerged, static media to support the growth of biofilm for BOD removal, nitrification, or denitrification, but solids are intended to be carried through the reactor. This type of BAF typically is referred to as a SAF. If suspended solids removal is required beyond adsorption and capture in the biofilm, then it is carried out in a separate downstream process. A diagram of a typical simple SAF system is shown in Figure 11.36. The system arrangement depends on the supplier and duty of the SAF. This section is adapted in part from Rundle (2009).

The SAF media used may be plastic (either random or structured) or mineral. The media must be open in structure to prevent blockage by accumulated solids. Structured plastic media systems must include provisions for retaining the media. Mineral media has a high specific gravity and is unlikely to be dislodged in normal use. In the United Kingdom, blast furnace slag is readily available and is used for both carbonaceous and nitrification applications.

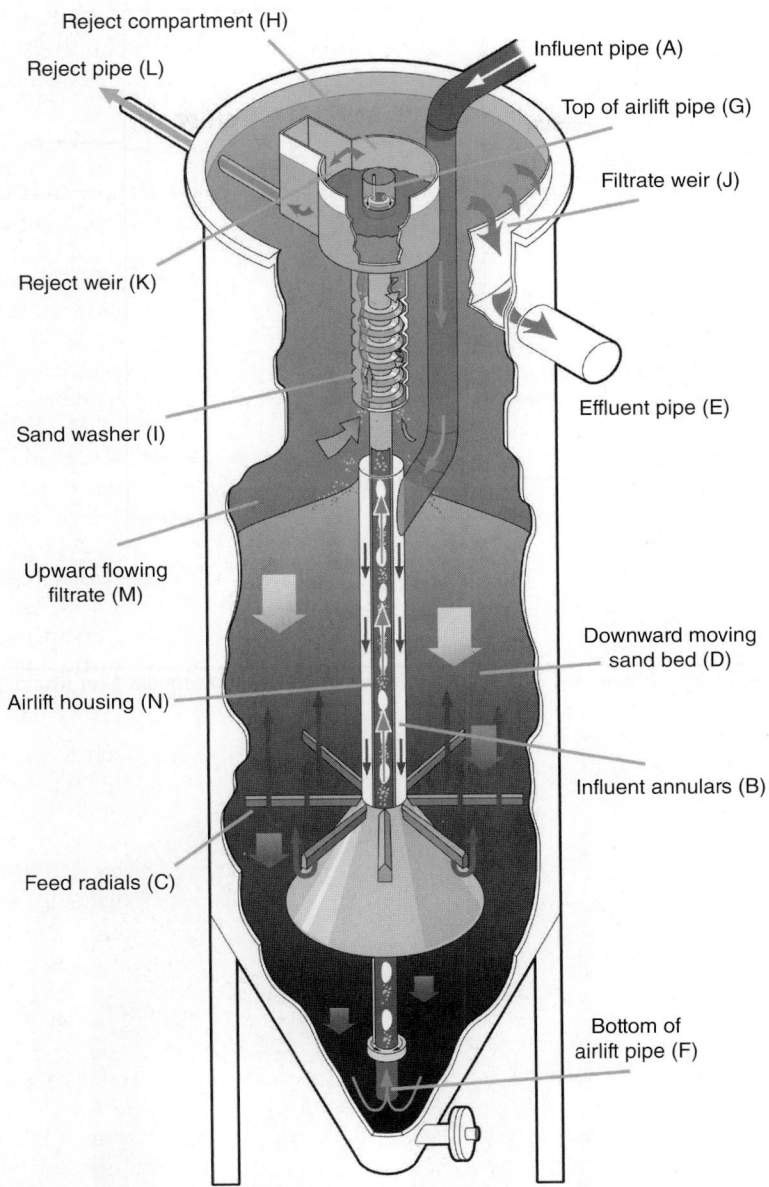

Reject compartment (H)

Reject pipe (L)

Influent pipe (A)

Top of airlift pipe (G)

Filtrate weir (J)

Reject weir (K)

Sand washer (I)

Effluent pipe (E)

Upward flowing filtrate (M)

Downward moving sand bed (D)

Airlift housing (N)

Influent annulars (B)

Feed radials (C)

Bottom of airlift pipe (F)

Figure 11.35 Schematic of moving bed denitrification filter (courtesy of Parkson Corp.).

Influent wastewater typically is introduced at the bottom of the reactor. Larger systems may have more than one cell in series or have a more sophisticated flow distribution system, similar to upflow BAF, or use the air and the headloss to distribute the fluids when used in a downflow mode (Sulzer Biopur®). Proper distribution of water and air has to be assured to scour all parts of the media and prevent anaerobisis in the outer zones (Cooper-Smith and Schofield, 2004).

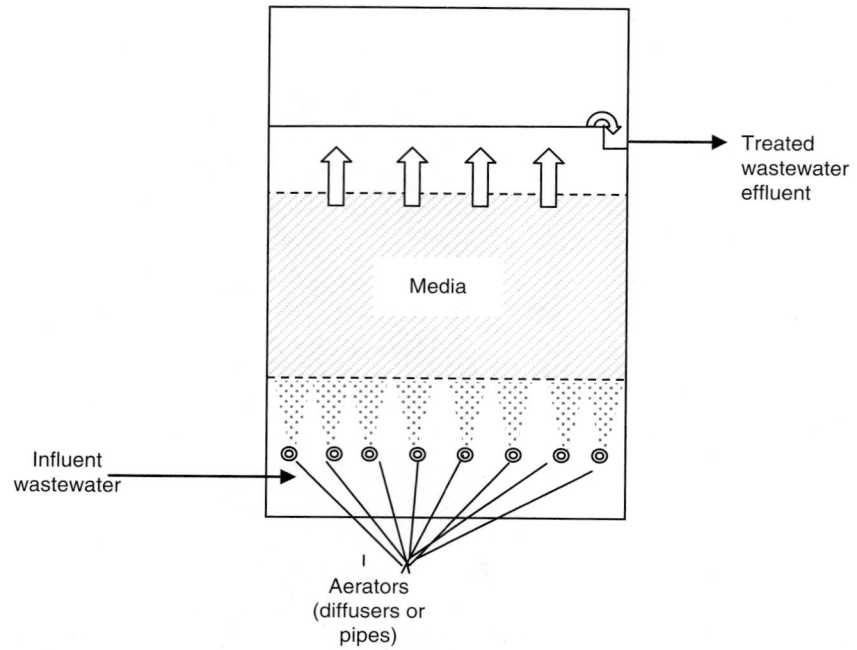

Treated
wastewater
effluent

Media

Influent
wastewater

Aerators
(diffusers or
pipes)

FIGURE 11.36   Schematic of non-backwashing, open-structure media filter (Rundle, 2009).

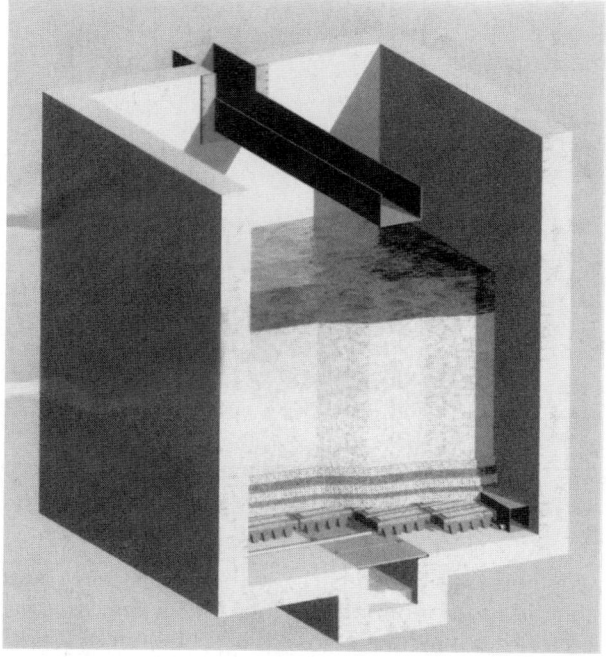

FIGURE 11.37   Mineral media upflow submerged activated filter with block under drain (courtesy of Severn Trent).

For SAFs using mineral media, the air and influent distribution systems are combined with a floor system designed to support the heavier media. This system configuration is shown in Figure 11.37. Influent wastewater enters via a central channel in the base of the reactor, which is covered by plates. The plates are covered by rows of specially designed concrete underdrain blocks that are fitted with interlocking plastic jackets. For upflow SAF, treated effluent is discharged over a weir or trough at the top of the cell. For downflow SAF, piping or channels in the bottom collect the treated effluent.

In addition to process needs, aeration is required to prevent media blockage. The air can be supplied by either a grid consisting of perforated distribution pipes or diffusers installed at the bottom of the SAF. Though improved air distribution may be achieved by using diffusers, some studies have shown that in packed beds, air bubble coalescence results in little to no oxygen transfer advantage (Hodkinson et al., 1998). In some WRRFs, aeration scouring was insufficient to keep the media clean resulting in deteriorated performance. Therefore, it is good design practice to maintain a backwash or backflush option where air and water flows can be increased to scour the media periodically. Jet aeration, in which air is injected into a moving stream of water typically via a Venturi and dispersed into fine bubbles, also has been coupled with a shallow SAF facility for small communities (Daude and Stephenson, 2004).

4.2 Media for Use in Biologically Active Filter Reactors

Several media types are available for use in BAF reactors. Media selection is integral to treatment objectives, flow and backwashing regimes, and specific process equipment manufacturer. Media typically can be categorized as mineral media and plastic media. In most cases, mineral media is denser than water and plastic media is buoyant. The media needs to resist breakdown from abrasion during backwashing and chemical degradation by constituents in municipal wastewater. Commercially available BAF systems and their media are listed in Table 11.11.

4.3 Backwashing and Air Scouring

Backwashing filters maximizes capture and run times and guarantees proper effluent quality. Proper backwashing requires filter bed expansion and rigorous scouring, followed by efficient rinsing. Poor filter cleaning will result in shortened filter runs, accumulation of solids, and deteriorating performance. Accumulation of solids and media (mudballing) produces shortcircuiting of water flow and can result in excessive media loss.

Feed characteristics and type of treatment provided by the BAF affect solids production and frequency of backwashing. For wastewaters with high suspended solids concentrations, a significant portion of solids is removed by filtration. Inert solids will be retained within the media until removed by backwashing, but biological solids may be degraded depending upon retention time. Inorganic salts of iron or aluminum, which may be added to the influent for phosphorus removal, will form precipitates within the media bed and increase backwash frequency. Solids growth for tertiary BAF systems is typically low, so backwashing is relatively infrequent (one backwash per 36–48 hours). When solids content is low, foam caused by detergent in the wastewater may be a problem because the scour aeration is concentrated in a small surface. Foam also can be an issue during process startup. Netting is recommended across the surface of the cells to keep the foam from blowing about the site.

Reactor characteristics and media type influence backwash frequency. More openly structured media capture fewer solids. This reduces backwash frequency, but effluent

Process	Supplier	Flow Regime	Media	Specific Gravity	Size (mm)	Specific Surface Area (m²/m³)
Astrasand®	Paques/ Siemens	Upflow, moving bed	Sand	>2.5	1–1.6	
Biobead®	Brightwater F.L.I.	Upflow	Polyethylene	0.95		
Biocarbone®	OTV/Veolia	Downflow	Expanded shale	1.6	2–6	
Biofor®	Degremont	Upflow	Expanded clay	1.5–1.6	2.7, 3.5 and 4.5	1400–1600
Biolest	Stereau	Upflow	Pumice/ pouzzolane	1.2		
Biopur	Sulzer/Aker Kvaerner	Downflow	Polyethylene		Structured	
Biostyr®	Kruger/Veolia	Upflow	Polystyrene	0.04–0.05	3.3–5	1000
Colox™	Severn Trent	Upflow	Sand	2.6	2–3	656
Denite®	Severn Trent	Downflow	Sand	2.6	2–3	656
Dynasand®	Parkson	Upflow, moving bed	Sand	2.6	1–1.6	
Eliminite®	FB Leopold	Downflow	Sand	2.6	2	
Submerged activated filter	Severn Trent	Up/down	Slag	2–2.5	28–40	240
			Washed gravel	2.6	19–38	

TABLE 11.11 Commercially Available Biologically Activated Filter Reactor Systems and Media (Debarbadillo et al., 2010)

wastewater may contain higher suspended solids concentrations. Fine mineral media typically have the best solids retention characteristics but tend to require more frequent backwashing.

Intense backwashing regimes have been developed to clean rapid gravity filters used in potable water treatment (Fitzpatrick, 2001). The bed typically is fluidized to allow the grains to separate and move freely and to remove as much accumulated material as possible. However, fluidization is avoided in BAFs; instead, the removal of excess biomass and accumulated solids is achieved during backwash by intense media contact and air scouring in a slightly expanded media bed.

Table 11.12 provides a comparison of typical BAF backwashing requirements and Section 4.1 described backwashing for each type of BAF configuration. Final backwashing requirements and duration typically are developed in collaboration with the BAF manufacturer. For example, the backwash sequence for an upflow sunken media BAF typically includes drain down, air scour, air and water scour (may include cycling between air only and air/water scour), a water-only rinse, and filter-to-waste when the backwashed cell initially is placed back in operation. Thus, backwash water is delivered only to the BAF cell for a portion of the total duration. Media, hydraulic and organic loading rates, and treatment objectives influence the frequency and duration of each step. It is often adjusted during facility commissioning and long-term operation.

	Backwash Water Rate, m/h (gpm/ft²)	Air Scour Rate, m/h (scfm/ft²)	Total Duration, (Min[c])	Total Backwash Water Volume Per Cell[c]	Total Backwash Wastewater Volume Per Cell[d]
Upflow, sunken media normal BW energetic BW[a]	20 (8.2) 30 (12.3)	97 (5.3) 97 (5.3)	50 25	9.2 m³/m² (225 gal/sq ft) 9.2 m³/m² (225 gal/sq ft)	12 m³/m² (293 gal/sq ft) 10 m³/m² (245 gal/sq ft)
Upflow, floating media normal BW	55 (22.5)	12 (0.65)	16	2.5 m³/m³ media[e] (18.7 gal/cu ft media)	2.5 m³/m³ media[e] (18.7 gal/cu ft media)
Mini-BW[b]	55 (22.5)	12 (0.65)	5	1.5 m³/m³ media[e] (11.2 gal/cu ft media)	1.5 m³/m³ media[e] (11.2 gal/cu ft media)
Downflow, sunken media	15 (6)	90 (5)	20–25	3.75–5 m³/m² (90–120 gal/sq ft)	3.75–5 m³/m² (90–120 gal/sq ft)
Upflow, moving bed[f]	0.5–0.6 (0.2–0.24)	Continuous through air lift	Continuous	55–67 m³/d (14 400–17 300 gpd)	55–67 m³/d (14 400–17 300 gpd)

[a]Energetic backwash once every one to two months depending on trend in "clean bed" headloss following normal backwash.

[b]Mini-backwash applied as interim measure when pollutant load exceeds design load.

[c]Backwash duration reflects total duration of the typical backwash cycle, which includes valve cycle time and pumping and nonpumping steps. The duration of each step is adjustable via programmable logic controller and supervisory control and data acquisition control systems.

[d]The total backwash wastewater volume includes drain and filter to waste steps where applicable.

[e]Backwash volume requirements for upflow floating media BAF typically are based on media volume rather than cell area because depths vary.

[f]Continuous backwash filter backwashing is based on a standard 4.65 m² cell and a typical weir setting for reject flow of approximately 2.3–2.8 m³/h/cell (10–12 gpm/cu ft/cell).

TABLE 11.12 Summary of Biologically Active Filter Backwashing Requirements (Degremont, 2008; Kruger, 2008; Severn Trent, 2004; Parkson, 2004; Debarbadillo et al., 2010)

4.4 Biologically Active Filter Process Design

Several factors influence process design for BAF systems. As discussed earlier, mass-transfer limitations into the biofilm often limit substrate removal performance, and the media-specific surface area available for biofilm attachment and substrate flux affect biofilm reactor design. Several physical conditions within BAF systems also significantly affect performance including oxygen availability and air flow velocity, filtration velocity, media packing density, and backwash efficiency. These all factors affect external mass transfer, and indirectly, penetration into the biofilm. Because of the importance of these parameters, and perhaps because of uncertainty of actual media-specific surface area, BAF performance results typically are expressed as a function of substrate volumetric loading rates rather than surface area.

Deterministic modeling of BAFs based on kinetic expressions is complicated, as biofilms are complex and highly dynamic structures. Therefore, uncertainty of prediction persists because of the many variables involved that affect the degree of soluble and particulate substrate diffusion, rate of biomass growth, biofilm density, and type and quantity of microorganisms in the biofilm. The filtering capability of BAFs makes the already difficult task of quantifying the degree of particle hydrolysis even more important in BAFs than in activated sludge systems or other biological processes with downstream solids separation. However, although continued development and calibration of biofilm models is needed, these models provide an excellent tool for evaluation and development of more tailored designs.

Parameters governing treatment capacity of BAF are as follows:

- Substrate loading (volumetric loading rates in terms of kg BOD/m³/d or kg N/m³/d), which will determine media volume. Guidelines for design loading rates have been compiled from the literature and are a function of wastewater characterization, substrate flux, temperature, and physical conditions in the biofilm reactor as discussed above. Design guidance is provided based on flow regime and typical media and backwashing practices in use for different BAF reactors.

- Filtration rate, or total volume of wastewater applied per area of media per unit time (m³/m²/d), also is used to determine filter surface area. Filtration velocity affects system headloss, solids capture, and air and water distribution within the media, diffusion, and detention time.

- Solids holding capacity, which will determine backwash frequency.

4.4.1   Secondary Treatment

This section reviews criteria for BAFs designed for carbon oxidation and suspended solids removal in secondary treatment.

Volumetric BOD loading rates vary widely in the literature for upflow BAFs designed for secondary treatment, ranging from 1.5 to 6 kg/m³/d. Table 11.13 provides COD and BOD applied loading rates and removal efficiencies for several pilot- and full-scale references.

Average and peak hydraulic loading rates for secondary treatment systems typically range from 4 to 7 m/h and 10 to 20 m/h, respectively. Because BAFs for secondary treatment typically are placed immediately downstream of primary clarification, the applied volumetric mass loading rate is almost always the limiting design parameter (see Table 11.14).

For simultaneous secondary treatment and nitrification, the carbon loading at lower temperatures needs to be limited to less than 2.5 kg BOD/m³/d (Rogalla et al., 1990). Simultaneously, a total Kjeldahl nitrogen (TKN) loading removal rate of 0.4 kg N/m³/d can be obtained.

The backwash frequency for BAFs designed for secondary treatment is related to the applied organic and TSS load, degree of particle hydrolysis taking place within the media, biomass yield, and solids retention capacity of the media. Because of the higher biomass yield of heterotrophic bacteria and higher applied TSS loadings, BAFs for secondary treatment (BOD removal) need to be backwashed at least once per day. More frequent backwashing results in less hydrolysis of particulate BOD, which in turn results in lower oxygen demand and higher backwash waste solids quantities. Phipps and Love (2001) calculated biomass observed yields in the range of 0.43 to 0.48 mg biomass as COD generated per mg substrate COD consumed. They also determined that 40% to 46% of applied particles underwent hydrolysis in a full-scale upflow BAF (with

Water Resource Recovery Facility	BAF Applied Volumetric Loading		Removal Efficiency (%)	Source and Notes
	kg BOD$_5$/m³·d	kg COD/m³·d		
Roanoke, Virginia (upflow, sunken media)	2.8		80 (2001) 80–96 (2006)	First-stage BAF performance testing data from two-stage C + N BAF system (2006) (Degremont, 2008).
Seine Centre, Colombes, France (upflow, sunken media)	2.8 (average)	5.9 (average)	86 72	2000 full-scale data
King County, Seattle, Washington (pilot, upflow sunken media)	4.4		80	Neethling et al. (2002) (average loading and removal efficiency)
Binghamton, New York (pilot, upflow sunken media)	BOD$_5$: 3.7–7.8		60–75	Average of three test phases during five-month pilot test, March 1999–July 1999

TABLE 11.13 Chemical Oxygen Demand and Biochemical Oxygen Demand Applied Loading Rate and Removal Efficiencies for Selected Full- and Pilot-Scale Biologically Activated Filter Facilities

Type of BAF	Applied Volumetric Loading (kg/m³·d) (lb/d/1000 cu ft)	Hydraulic Loading (m³/m²·h) (gpm/cu ft)	Removal Efficiency (%)
Upflow sunken or floating media, backwashing[a,b]	BOD: 1.5–6 (94–370) TSS: 0.8–3.5 (50–220)	3–16 (1.2–6.6)	BOD: 65–90 TSS: 65–90
Upflow, sunken media[c]	10		
Upflow, floating media[c]	8		
Submerged, non-backwashing[d]	BOD: 0.8–1.5 (50–94) @ 20°C	2–12 (0.8–5) @ 20°C	BOD: 85–95

Notes: Design loading rates depend on specific wastewater characteristics and level of treatment required.
[a]Degremont, 2007.
[b]Kruger, 2008.
[c]German Association for Water, Wastewater and Waste, 1997.
[d]Severn Trent, 2008.

TABLE 11.14 Typical Biologically Active Filter (BAF) Loading Rates for Secondary Treatment (Debarbadillo et al., 2010)

media heavier than water) treating conventional primary clarifier effluent for carbon removal (backwash frequency of once per day). The main limitation of any BAF remains the solids storage capacity. The volume that can be accumulated between backwashes is 2.5 to 4 kg TSS/m³/d depending on the media selection, water velocity, and water temperature (Degremont, 2007).

Backwash water typically contains 500 to 1500 mg/L suspended solids, but this varies with the type of treatment, cycle time, and water used. The BOD removal produces biomass from growth of the microorganisms, which convert degradable material into new cells, carbon dioxide, and water, similar to that of activated sludge. Sludge production is typically 0.7 to 1 kg solids per kg BOD removed. In a two-stage SAF/BAF in Aberdeen, the German ATV Standard equation to predict solids from activated sludge was successfully applied (Jolly, 2004; German Association for Water, Wastewater and Waste [ATV-DVWK], 2000).

Below is a design example for a submerged, upflow BAF system for secondary treatment without nitrification:

Determine the total volume of BAF media, total BAF reactor filtration area, and number of BAF cells required to achieve BOD_5 and TSS removal efficiencies (E_{BOD} and E_{TSS}) of at least 90% when treating domestic wastewater. Determine BAF backwash wastewater volume and solids concentration. Assume the following conditions apply for this example:

- Influent (including returns) maximum month flow rate: $Q_0 = 94\ 800\ m^3/d$;
- Influent (including returns) flow peaking factor: PF = 2.8;
- BOD_5 after primary settling: $C_{BOD5} = 220$ mg/L;
- TSS after primary settling: $C_{TSS} = 150$ mg/L;
- BAF media height: $H_M = 4$ m;
- BAF effluent used as backwash water;
- BAF backwash return flow equalized and combined with other return flows at head of facility.

1. Calculate BOD_5 and TSS load to the BAF system:

   BOD_5 load = $(Q_0)(BOD_5)/1000 = (94\ 800)(220)/1000 = 20\ 856$ kg/d;
   TSS load = $(Q_0)(TSS)/1000 = (94\ 800)(150)/1000 = 14\ 220$ kg/d (31 284 lbs/d).

2. Assume maximum volumetric applied loading rates (Table 11.14):

   $BOD_5 = 3$ kg/m³/d for 90% removal efficiency;
   TSS 5 1.6 kg/m³/d for 90% removal efficiency.

3. Calculate total BAF media volume (V_M) required:

   $V_{1BOD} = 20866/3 = 6955$ m³;
   $V_{2TSS} = 14220/1.6 = 8888$ m³ TSS load is limiting.

4. Calculate total BAF filtration area (A) required based on volumetric loading and filter depth:

   $A_{vol} = V/H_M = 8888/4 = 2222$ m² (23 909 cu ft).

5. Calculate total BAF filtration area based on maximum hydraulic loading rate of 20 m/h:

   $A_{hyd.} = (94\ 800/24)(PF)/20 = (3950)(2.8)/20 = 553$ m² , 2222 m², A_{vol} is limiting.

6. Select standard cell size, A_{cell} (144 m²), provided by BAF manufacturers.

7. Calculate number of BAF cells required assuming one backwash per cell per 24 hours:

   $$n = 2222/144 = 15.4$$
   $$N = 15.4 + 15.4/24 = 16 \text{ BAF cells}$$

Note: Depending on the initial capacity needs compared to design capacity, the designer should consider incorporating a redundant BAF unit for reliability and ease of maintenance.

8. Check BAF media retention capacity (see Table 11.13).

   Assume 2.5 kg/m³ cycle solids retention capacity:

   Total media retention capacity = (2.5)(16)(144)(4) = 23 040 kg/cycle;

   Biomass yield = Y = 0.7 2 1 kg TSS/kg BOD removed (assume Y = 1.0);

   Solids production = $(Y)(BOD_5 \text{ load})(E_{BOD})$;

   = (1.0)(20 856)(0.90) = 18 770 kg/d = 782 kg/h;

   Backwash frequency (23 040)/(782) = 29 hours.

9. Check maximum hydraulic loading rate with one cell in backwash and one cell out of service:

   $$(Q_0)(PF)/(N-2)(A_{cell}) = (94\ 800/24)(2.8)/(16-2)(144) = 5.5 \text{ m/h} < 20 \text{ m/h}$$

10. Calculate BAF backwash wastewater volume and solids concentration based on one backwash per cell per day (volume of backwash wastewater produced per media volume, Vol_{BW}, from Table 11.12):

    Assume Vol_{BW} = 3 m³/m³ media;

    Volume of backwash wastewater produced per backwash, V_{BW};

    $$V_{BW} = (Vol_{BW})(H_M)(A_{cell}) = (3)(4)(144) = 1728 \text{ m}^3.$$

    Backwash wastewater solids concentration, C_{BW};

    $$C_{BW} = (Y)(BOD_5 \text{ load})(E_{BOD})/(N)(V_{BW});$$
    $$C_{BW} = (1.0)(20\ 856)(0.9)(1/16)(1/1728) = 679 \text{ kg/m}^3 = 679 \text{ mg/L}.$$

This set of calculations represents an initial estimate of the BAF facility sizing. Development of the final design is typically an iterative process between the design engineer and the process equipment manufacturers being considered. Refinements typically are made by incorporating a combination of the manufacturer's experience and more detailed process modeling results.

4.4.2 Nitrification

Temperature, effluent requirements, fluid velocities (air and water), and loading influence nitrification capacity. As discussed in Section 2.0, biochemical transformation processes occurring in the biofilm are dependent on substrates diffusing in and

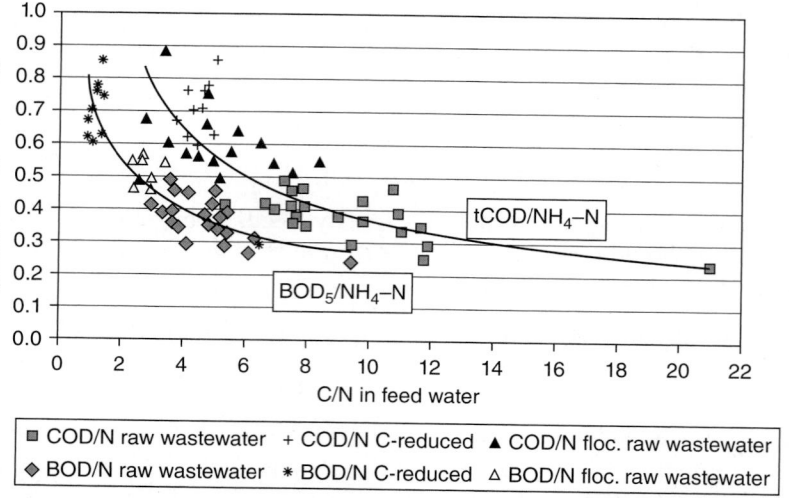

FIGURE 11.38 Nitrification rate for differently pretreated raw wastewaters as a function of C/N (temperature adjusted to 12°C; expanded clay) (Rother, 2005) (BOD = biochemical oxygen demand; COD = chemical oxygen demand).

out of the biofilm. Reaction rates or level of treatment achieved are defined by the rate limited substrate. Bulk phase ammonia, alkalinity, oxygen, and COD concentrations affect nitrification. As COD loadings increase, oxygen tends to become the rate limited substrate. Competition for oxygen intensifies, and the heterotrophic respiration at the outer layers of the biofilm will reduce availability of oxygen for nitrification in the deeper layers (Wanner and Gujer, 1985). Rogalla et al. (1990) found that nitrification tends to decrease when biodegradable COD loadings approach 4 kg/m^3/d. The influence of the C/N ratio on nitrification is illustrated in Figure 11.38 (Rother, 2005).

In BAFs, increasing fluid velocities increase external mass transfer, which leads to higher nitrification rates (Tschui et al., 1993). Under constant volumetric loading rates of 1.3 to 1.4 kg NH$_3$-N/m^3/d and 0.65 ± 0.2 kg CBOD$_5$/m^3/d, Husovitz et al. (1999) observed a 17% increase in ammonia mass removal (up to 1.26 kg NH$_3$-N/m^3/d) as the hydraulic loading rate was increased from 5.1 to 15.8 m/h. If temperature, effluent quality, and removal efficiency are not the limiting factors, then ammonia removal of 80% to 90% at ammonia loads between 2.5 and 2.9 kg/m^3/d can be achieved (Peladan et al., 1996; Peladan et al., 1997). On a full-scale demonstration cell (surface 144 m²) at 22°C, 91% removal of NH$_3$-N at loadings up to 2.3 kg/m^3/d was observed (Pujol et al., 1994). A summary of typical loading rates for nitrification applications is provided in Table 11.15.

Similar to water velocity, the increase of process air velocity improves nitrification rates as higher process air velocities increase turbulence and external mass transfer (Figure 11.39) (Tschui et al., 1993; Tschui et al., 1994).

When BAFs are operated for significant periods under reduced ammonia loading conditions, the inventory of biomass also will decrease. An example of this was shown in testing by Tschui et al. (1994) where volumetric ammonia removal rates decreased by 30%

Type of BAF	Applied Volumetric Loading (kg/m^3 · d) (lb/d/1000 cu ft)	Hydraulic Loading (m^3/m^2 · h) (gpm/cu ft)	Removal Efficiency (%)
Upflow, sunken or floating media, backwashing[a,b] following primary treatment	BOD: <1.5–3 (<94–188) TSS: <1.0–1.6 (<62–100) NH$_3$-N: <0.4–0.6 (<31–62) @ 10°C <1.0–1.6 (<62–100) @ 20°C	3–12 (1.2–5)	BOD: 70–90 TSS: 65–85 NH$_3$-N: 65–75
Upflow, sunken or floating media, backwashing[a,b] following secondary treatment	BOD: <1–2 (<62–125) TSS: <1.0–1.6 (<62–100) NH$_3$-N: <0.5–1.0 (<31–62) @ 10°C <1.0–1.6 (<62–100) @ 20°C	3–20 (1.2–8.2)	BOD: 40–75 TSS: 40–75 NH$_3$-N: 75–95
Upflow, floating media, backwashing[c] following secondary treatment	NH$_3$-N: 1.5 (94)		
Upflow, sunken media, backwashing[c] following secondary treatment	NH$_3$-N: 1.2 (75)		
Submerged, non-backwashing[d] following secondary treatment	NH$_3$-N: 0.2–0.9 (12–56) @ 20°C	2–12 (0.8–5) @ 20°C	NH$_3$-N: 85–95

Notes: The design loading rates depend on specific wastewater characteristics, upstream treatment processes and level of treatment required.

[a]Degremont, 2007.
[b]Kruger, 2008.
[c]German Association for Water, Wastewater and Waste, 1997.
[d]Severn Trent, 2008.

TABLE 11.15 Typical Biologically Active Filter (BAF) Loading Rates for Nitrification (Debarbadillo et al., 2010)

after a transition from operation under non-NH$_3$-N limiting conditions to lower volumetric NH$_3$-N loading rates. This is an important consideration for separate-stage nitrification BAF applications in which some nitrification can occur in the main secondary facility during summer months. The BAF will need to be able to treat higher ammonia loads when temperature drops and upstream nitrification decreases.

Nitrification performance also depends on long-term loading of the reactor. Excess biomass accumulates in the media that can be available when peaks are applied because they typically are associated with higher velocities or concentrations, which allow deeper penetration of the substrates into the biofilm. Figure 11.40 illustrates that maximum instantaneous removal rate can be twice the average applied load, up to the maximum capacity of the reactor.

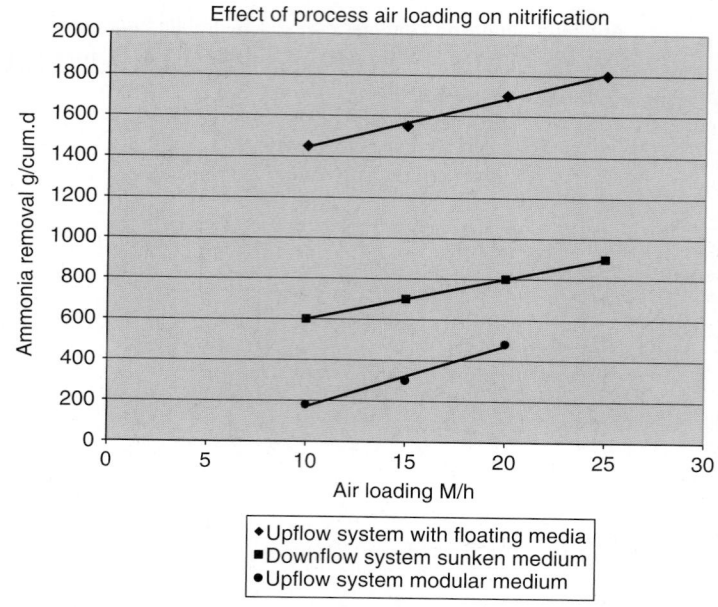

FIGURE 11.39   Effects of air velocity on nitrification (Tschui et al., 1993).

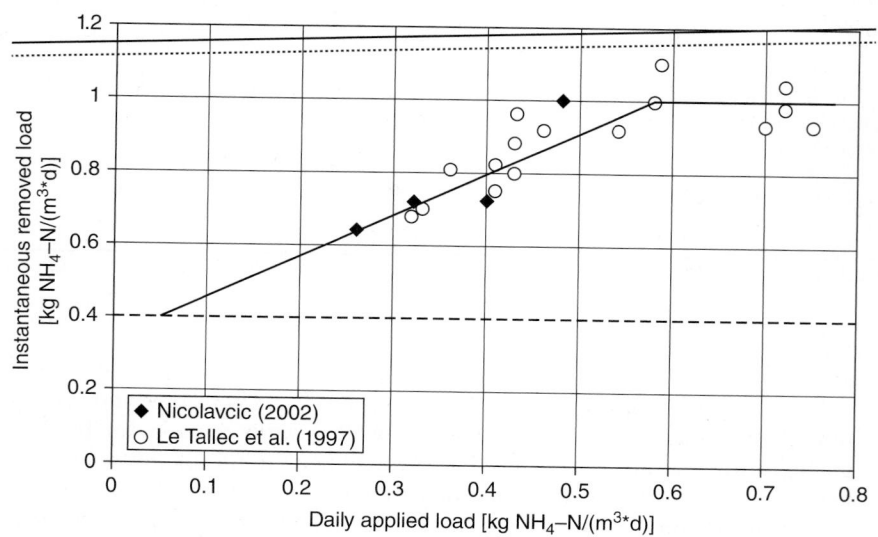

FIGURE 11.40   Maximum nitrification vs long-term average loading (Le Tallec, 1997).

Lower operating temperatures have a significant effect on nitrification. Comparing three types of tertiary nitrifying BAFs (downflow mineral media and either floating or modular upflow plastic media), the following long-term temperature dependency of nitrification was established (Tschui et al., 1994):

$$r_{V,\text{NH}_4\text{-N}(T)} = r_{V,\text{NH}_4\text{-N}(T=10°C)} \cdot e^{k_T(T-10)} \tag{11.8}$$

where $r_{V,\text{NH}_4\text{-N}(T)}$ = volumetric nitrification rate at temperature, T (°C);
$r_{V,\text{NH}_4\text{-N}(T=10°C)}$ = 5 volumetric nitrification rate at T = 10°C;
k_T = temperature coefficient (Arrhenius factor) = 0.03/°C.

These authors found that the temperature coefficient was 0.03/°C for all three types of media studied (corresponding to an Arrhenius factor of 1.04).

Nitrifying organisms have a relatively low solids yield [about 0.05 kg per kg N removed (Downing et al., 1964a; 1964b)]. Most of the sludge produced by tertiary BAFs is derived from the suspended solids removed by filtration. A portion of these solids undergo hydrolysis and heterotrophic degradation. The net sludge production is about 0.5 to 0.8 kg per kg of solids removed.

4.4.3 Combined Nitrification and Denitrification

Nitrogen removal can be accomplished by either oxidizing the ammonia in a first stage followed by reducing the nitrate in a second stage where an external carbon source is added (referred to as post-denitrification), or by recycling the nitrified effluent to a denitrification stage before nitrification (pre-denitrification). In pre-denitrification, the nitrified effluent is recycled to a separate anoxic BAF reactor located upstream of the reactor. In some upflow, floating media BAF configurations, a portion of the nitrified effluent may be recycled to an anoxic zone in the bottom of the media (U.S. Patent No. 6632365).

If sufficient carbon is available and the anoxic zone is large enough, then nitrogen removal is proportional to the recycle but with a diminishing return. Recycle typically is limited to ratios of 3 or total nitrogen removals of 75% because of the excessive hydraulic load and oxygen recycle. The minimum nitrate concentration achievable assuming complete nitrification and denitrification and ignoring nitrogen uptake because of cell synthesis is given by:

$$\text{NO}_3\text{-N}_{EFF} = \frac{\text{NH}_3\text{-N}_{INF}}{R+1} \tag{11.9}$$

where $\text{NH}_3\text{-N}_{INF}$ = ammonia concentration in influent;
$\text{NO}_3\text{-N}_{EFF}$ = nitrate concentrations in effluent respectively;
$R(5Q_R/Q_{in})$ = recirculation ratio.

Recycling treated wastewater has the advantage of increasing upflow velocity in both pre-denitrification and nitrification reactors, which increases the reaction rate.

Ryhiner et al. (1993) tested a pre-denitrification configuration using submerged structured media BAFs with a final polishing filter to ensure low nitrogen and suspended solids. Approximately 60% to 70% $\text{NO}_3\text{-N}$ removal was achieved in the pre-denitrification reactor at loading rates ranging from 0.1 to 0.6 kg $\text{NO}_3\text{-N}/\text{m}^3/\text{d}$. During a one-year study, Pujol and Tarallo (2000) achieved approximately 68% $\text{NO}_3\text{-N}$ removal (or 0.9 kg $\text{NO}_3\text{-N}/\text{m}^3/\text{d}$) with typical wastewater feed to the pre-denitrification BAF and up to 90% removal when additional substrate (methanol) was added. A range of

Type of BAF	Applied Volumetric Loading (kg/m³ · d) (lb/d/1000 cu ft)	Hydraulic Loading (m³/m² · h) (gpm/cu ft)	Removal Efficiency (%)
Upflow, sunken media[a] separate BAF stages (pre-denitrification + nitrification)	N-NO₃: 1–1.2 (62–75)	10–30 (4–12)	NO₃-N: 75–85
Upflow, floating media[b] combined anoxic/ aerated BAF stage	1–1.2 (62–75)	12–21.5 (4.9–8.8)	NO₃-N: 70 without supplemental carbon; 85% with supplemental carbon

Notes: Design and performance are dependent on wastewater characteristics, upstream treatment processes, effluent goals, and readily biodegradable carbon substrate.
[a]Degremont, 2007.
[b]Ninassi et al., 1998.

TABLE 11.16   Typical Biologically Active Filter Loading Rates (BAF) for Pre-Denitrification (Debarbadillo et al., 2010)

recycle rates (150%–350%) and corresponding filtration velocities (9.4 to 16.9 m³/m²/h) also were tested. Design guidance for pre-denitrification BAF systems is provided in Table 11.16.

4.4.4 Tertiary Denitrification

This section focuses on process design criteria for post-denitrification BAF including half-order nitrogen removal kinetics, temperature, supplemental carbon and nitrogen release cycle requirements, and hydraulic, volumetric mass, and solids loading.

Volumetric loading (and removal) rates vary widely, with citations ranging from 0.2 to 4.8 kg/m³ (15–300 lbs/1000 cu ft) (U.S. Environmental Protection Agency [U.S. EPA], 1993; WEF, 2013; Metcalf and Eddy, Inc./AECOM, 2013; Degremont, 2007). Many post-denitrification filters are preceded by an activated sludge biological nutrient removal process. Because some denitrification is achieved upstream, filter influent NO_x-N concentrations are typically less than 10 mg/L. In these cases, hydraulic considerations govern post-denitrification design and many installations are operating at mass loading rates of approximately 0.3 to 0.6 kg/m³/d (20–40 lbs/d/1000 cu ft). The range of loading rates for upflow post-denitrification BAF reactors tends to be higher than for post-denitrification sand filters because they are not designed for the same level of TSS removal and are not as limited by hydraulics. A summary of typical volumetric and hydraulic loading criteria for different types of denitrification filters is provided in Table 11.17.

Harremoës (1976) suggested that denitrification filter kinetics are dependent on diffusion of substrate into pores in the biofilm. Zero-order heterogeneous reactions in a pore that is only partially penetrated by the substrate result in half-order reaction kinetics for nitrate concentration. Half-order reaction kinetics result in the following expressions for a plug flow reactor (Harremoës, 1976; Hultman et al., 1994):

$$r_{DN} = \frac{dS}{dt} = -k_{\frac{1}{2}} \cdot \sqrt{S_{B,NO_3\text{-}N}}$$

or

$$\sqrt{S_{B,NO_3\text{-}N}} - \sqrt{S_{in,NO_3\text{-}N}} = -\frac{1}{2} \cdot k_{\frac{1}{2}} \cdot \bar{t}$$

(11.10)

Type of BAF	Applied Volumetric Loading (kg/m³ · d) (lb/d/1 000 cu ft)	Hydraulic Loading (m³/m² · h) (gpm/cu ft)	Removal Efficiency (%)
Downflow, sunken media[a]	NO₃-N: 0.3–3.2 (20–200)	4.8–8.4 (2–3.5) average	NO₃-N: 75–95
		12–18 (5–7.5) peak	
Upflow, sunken media[b]	NO₃-N: 0.8–5 (50–300)	10–35 (4 to14)	NO₃-N: 75–95
Upflow, sunken media[c]	2 (125)		
Upflow, floating media[c]	1.2–1.5 (75–94)		
Moving bed, continuous backwash[d]	NO₃-N: 0.3–2 (20–120)	4.8–5.6 (2–4) average 13.4 (6) peak	NO₃-N: 75–95

Notes: Selected loading rates are dependent on treatment objectives, upstream processes, wastewater characteristics, and carbon source.
[a]Severn Trent, 2004; U.S. Environmental Protection Agency, 1993.
[b]Degremont, 2007.
[c]German Association for Water, Wastewater and Waste, 1997.
[d]deBarbadillo et al., 2005.

TABLE 11.17   Typical Biologically Active Filters (BAF) Loading Rates for Post-Denitrification

where r_{DN} = denitrification rate per unit volume of filter (mg/L/min, or g/m³/min);
S_{B,NO_3-N} = bulk-liquid nitrate-nitrogen concentration (effluent) (mg/L);
S_{in,NO_3-N} = nitrate-nitrogen concentration in the influent stream (mg/L);
= half-order reaction coefficient per unit volume of filter [(mg/L)$^{1/2}$/min, or (g/m³)$^{1/2}$/min]; = filter empty bed retention time (min);
h = filter media height (m);
q_A = filter surface hydraulic loading rate (m³/m²/min).

Substituting $S_T = 0.87$ dissolved oxygen, mg/L + 2.47 NO₃-N, mg/L + 1.53 NO₂-N mg/L for S_B and S_{in} (McCarty et al., 1969):

$$\sqrt{S_{B,T}} - \sqrt{S_{in,T}} = -\frac{1}{2} \cdot k_{\frac{1}{2}} \cdot \frac{h}{q_A} \tag{11.11}$$

or

$$\sqrt{\frac{S_{B,T}}{S_{in,T}}} = 1 - \frac{a \cdot h}{\sqrt{S_{in,T}}} \tag{11.12}$$

where,

$$a = \frac{k}{2 \cdot q_A}$$

Profile sampling conducted at different media depths by Harremoës (1976), Hultman et al. (1994), and Janning et al. (1995) have shown the half-order kinetic model reasonably correlates to observed values. A summary of half-order kinetic constants for several post-denitrification BAF is provided in Table 11.18. Because of relatively limited information on half-order rate constants at full-scale facilities, several examples also have been included. The rate constant is calculated from influent and effluent data under the assumption that plug flow conditions were achieved as demonstrated in earlier work.

There is a significant difference in design hydraulic loading rates for upflow post-denitrification BAF and post-denitrification BAFs that serve as the final filtration step (typically downflow BAF with sunken media or moving bed continuous backwash filters). Average hydraulic loading rates for denitrification sand filters typically range

Facility	Temperature (°C)	Half-Order Constant[a] (Based on NO$_3$-N) [(mg/L)$^{1/2} \cdot$ min]	Half-Order Constant[a] (Using NO$_3$-N, NO$_2$-N, and DO) [(mg/L)$^{1/2} \cdot$ min]	Notes
Pilot scale downflow filter, 3–4 mm gravel media[b]	6–19	0.25–0.90	N/A	Based on profiles through depth of media.
Full-scale sunken media denitrifying BAF, Denmark[c]	11.7	0.09–0.15	N/A	Based on profiles through depth of media. Values transposed from half-order constants reported per unit of media surface area.
De Grote Lucht STP, The Netherlands (full scale)[d]	9–16	0.28	0.46	Based on inlet and outlet concentrations. February through May 2003
De Grote Lucht STP (full scale)[d]	19–24	0.32	0.50	Based on inlet and outlet concentrations. July through September 2003
Hagerstown, Maryland (pilot)[d]	14–15	0.23	0.38	Based on inlet and outlet concentrations. Average loading conditions, full denitrification
Hagerstown, Maryland (pilot)[d]	14–15	0.36	0.64	Based on inlet and outlet concentrations. Peak loading conditions

[a]Half-order reaction coefficients are reported per unit volume.
[b]Harremoes, 1976.
[c]Janning et al., 1995.
[d]deBarbadillo et al., 2005.

TABLE 11.18   Half-Order Kinetic Constants for Tertiary Denitrification Filters

from 4 to 9 m/h. Peak hour rates typically are limited to 18 m/h with one cell out of service for backwashing. When upflow BAF are used for post-denitrification applications, larger media size allows for higher hydraulic loading rates of 10 to 35 m/h, but solids retention capability may be reduced.

Hydraulic loading rates also affect contact time within the filter. Original denitrification filter design curves related percent NO_3-N removal to empty-bed detention time (EBDT) (Savage, 1983). Data from pilot and full-scale systems superimposed onto existing curves suggested that NO_x-N removals of 90% could be achieved at hydraulic residence time of 10 minutes at temperatures ranging from 13 to 21°C (deBarbadillo et al., 2005).

In addition to removal of solids from influent wastewater, biomass is produced within the filter. Typically, a biomass yield coefficient of 0.4 g biomass COD produced/g methanol COD consumed (approximately 0.4 g VSS/g methanol consumed) is adequate for post-denitrification systems using methanol as the carbon source. When estimating solids quantities, it is assumed that approximately 10% of the biodegradable solids removed and produced undergo hydrolysis.

Post-denitrification sand filters (downflow sunken media and moving bed) typically serve as the final solids separation step and can produce an effluent with average TSS of 5 mg/L or less. Downflow, sunken media filter manufacturers have found that as much as 9.8 kg/m² solids an reliably captured between backwashes (Severn Trent, 2004). Normal design procedures account for backwash frequencies based on 4.9 kg/m². Data from a moving bed, continuous backwashing denitrification filter pilot test in Hagerstown, Maryland, suggest that solids loading rates to the filters should be limited to 2.45 kg/m² or less on average to maintain an average effluent TSS concentration of 5 mg/L or lower (Schauer et al., 2006). This limitation is specific to denitrification mode under the conditions tested and at a media recirculation rate. Other BAF configurations, including upflow with sunken and floating media, have installations that achieve final effluent TSS as low as 5 mg/L; with larger media sizes, however, overall solids filtration is not as robust and the effluent may be comparable to high quality secondary clarifier effluent.

For tertiary denitrification systems, such as filters, BAFs, and MBBRs with a post-anoxic fixed-film zone, the supplemental carbon source is vital to system operation. Methanol feed requirements can be estimated using eq 11.13, as follows (McCarty et al., 1969):

$$S_M = 2.47 \text{ (NO}_3\text{-N removed)} + 1.53 \text{ (NO}_2\text{-N removed)}$$
$$+ 0.87 \text{ (Dissolved oxygen removed)} \qquad (11.13)$$

Equation (11.13) applies specifically to methanol. The COD requirement for denitrification varies depending on the carbon substrate used. The amount of substrate COD stabilized by 1 mg of oxygen is (Copp and Dold, 1998; Melcer et al., 2003):

$$COD/NO_3\text{-N} = 1/(1 - Y) \qquad (11.14)$$

where Y = heterotrophic yield coefficient (mg biomass COD formed per unit substrate COD used).

A value of 0.66 is typically used for aerobic respiration. For denitrification, a 2.86 conversion factor is incorporated into the equation to account for the amount of nitrate required to accept the same number of electrons. This yields the following expression for COD requirements for anoxic growth:

$$COD/NO_3\text{-}N = 2.86/(1 - Y) \qquad (11.15)$$

An estimate of supplemental carbon substrate requirements can be made by factoring the appropriate anoxic yield coefficient into the calculation. Determination of anoxic yield coefficients for different substrates has been a topic of recent research (Mokhayeri et al., 2006; Nichols et al., 2007; Cherchi et al., 2008). For example, incorporating a methanol anoxic yield coefficient of 0.38 into eq 11.15 results in a requirement of 4.6 mg COD/mg NO_3-N denitrified. Based on a COD-to-methanol ratio of 1.5, the methanol requirement is 3.07 mg methanol/mg NO_3-N denitrified.

4.4.5 Phosphorus Removal Considerations

Similar to those discussed for activated sludge systems, several methods are available to remove phosphorus:

- Pre-precipitation using metallic salts (typically iron or aluminum) in primary settling;
- Precipitation using metallic salts at the biofilter stage; and
- Biological removal.

Primary-stage precipitation is widely used because high-rate settling often is combined with BAFs, which achieves some phosphorus removal in chemically enhanced primary treatment. Multipoint dosing can be applied to reach low-effluent concentrations, but phosphorus limitation has to be prevented to allow efficient biological reactions (Odegaard, 2005). Precipitation by adding iron salts to the biofilter is possible, but the increase of mass of solids removed by the biofilter results in higher backwash frequency. When ferric chloride was used in a two-stage facility and precipitant was added in the second (nitrifying) stage, filter run times were reduced and a smaller medium was required to retain the floc (Sagberg et al., 1992).

Biological phosphorus removal has been developed for suspended growth systems, in which alternation of anaerobic and aerobic zones by recycling biomass encourages phosphorus uptake (Barnard, 1974). To achieve biological phosphorus removal with fixed biomass, the alternation can take place only in time, not in space. Therefore, a two-reactor alternating system was tested, using two pilot biofilters in series, anaerobic and aerobic, and a six-hour cycle (Gonçalves and Rogalla, 1992). The system successfully achieved biological phosphorus removal and was adapted for nitrogen removal with one cell out of five switched into anaerobic mode (Gonçalves et al., 1994a; 1994b). However, the additional expense in valving arrangements has prevented its application on full scale.

At some facilities, the need to meet stringent effluent total phosphorus limits while operating a tertiary nitrification or denitrification process is difficult. Adequate phosphorus is needed for microbial growth and insufficient phosphorus will limit the ability of tertiary BAF to achieve treatment goals. An evaluation of pilot and full-scale post-denitrification BAF performance data suggests that nitrate removal goals can be achieved in practice at orthophosphate to nitrate/nitrite-nitrogen ratios of 0.02 (deBarbadillo et al., 2006). This subject is covered in more detail in Section 3.2.4 of this chapter.

4.5 Facility Design Considerations for Biologically Activated Filter Facilities

Several issues must be considered in the design of the BAF reactors, supporting facilities, and upstream and downstream processes.

4.5.1 Preliminary and Primary Treatment

Fine screening should be implemented, if possible, at multiple locations in the facility depending on the type of BAF used. Though the BAF influent may have been screened several times, it is imperative to include a screen immediately upstream of the inlet for upflow BAFs with nozzled bottom floors. A simple bag screen or manual flat screen with less than 2.5-mm openings will protect the BAF if an automatic fine screen is provided upfront. Dedicated automatic screens are needed, however, if influent wastewater screening is poor, some inlets are not screened (e.g., septage, imported sludges), the facility is surrounded by trees, or channels are not covered.

For a facility with a smaller footprint, high-rate primary treatment often is used. This may be achieved by chemically enhanced primary treatment (CEPT), high-rate lamella settlers, or ballasted (sand or dense sludge) flocculation and settling.

4.5.2 Backwash Handling Facilities

The BAF backwash facilities and equipment typically include effluent clearwell, backwash water pumps (for concurrent systems only), air scour blowers, backwash waste equalization tank and return pumps, and all automatic valves, instruments, and controls required for automatic initiation and sequencing. Equipment and facilities must be sized adequately to handle air and water rates and volumes necessary for effective backwashing. During backwashing, the effluent flow from the BAF may stop or decrease, which must be accounted for in design and operation of any downstream treatment process, such as UV disinfection.

In multiple-stage BAF systems with varying requirements, facilities and equipment sizes are based on the largest cell to avoid separate sets of equipment for each stage. Final effluent taken directly from the effluent channel of the last stage of a multistage BAF system or via a final clearwell is used for backwash water. The effect of backwash return streams must be accounted for in design. If an interstage clearwell is used, then the ability to maintain minimum flow to the downstream BAF cells must be considered to avoid flow interruptions and effects on downstream and disinfection processes.

Provisions for mixing should be considered for large backwash waste tanks to prevent solids settlement. Backwash waste may contain some media, which can accumulate over time. The backwash waste return pumping system should be designed to avoid drawing media into pumps and rising mains. Some installations use media recovery systems including transfer pumps, settlement zones, or baskets.

Backwash waste is returned to the head of the WRRF and the solids are removed in the primary settling tanks. This improves performance of the primary settling tanks because biological aeration filter (BAF) biosolids adsorb some BOD and simplify pumping and handling of primary solids by improving rheology (Michelet et al., 2005).

Alternatively, the backwash wastestream can be treated using a dedicated solids separation system. This can be of particular benefit at larger facilities (more than approximately 100 000 m³/d [26 mgd]), if the existing primary settling tanks have solids handling limitations, or if there are multiple BAF stages. Several technologies can be used, such as a ballasted flocculation and settling system, a solids contact/sludge recirculation system, or a dissolved air flotation thickener (DAF). The DAFs have been used in this application at a number of European installations.

4.5.3   Process Aeration

Blowers or compressors supply process air that is distributed by either a grid of pipework or diffusers located at or near the bottom of the reactor. While the air flows up the reactor through the media, oxygen is dissolved in the water and diffuses into the biofilm. The passage of air bubbles also helps to maintain clear flow channels through the medium (Rundle, 2009).

All aeration studies on BAF facilities have observed higher oxygen uptake rates than typically found in the activated sludge process, which is consistent with reduced volume and hydraulic retention time. Stensel et al. (1984) measured oxygen uptake rates (OUR) from 121 to 250 mg/L/h in a 1.7-m tall reactor, which was 3.0 to 3.2 times greater than the observed rates in clean water tests performed on the same equipment. This was attributed to transfer of dissolved oxygen in water to the biomass and sparged air bubble area being in direct contact with the biofilm, allowing a second mechanism of direct transfer of oxygen from air bubbles to the biofilm. Lee and Stensel (1986) and Canziani (1988) had similar findings.

In field demonstrations, oxygen transfer test of BAFs at full-depth of 3.6 m using the off-gas method showed process water oxygen transfer efficiencies (OTE) of 1.6% to 5.8% per meter (0.5%–1.8% per foot) for floating media (average 4-mm diameter) and 3.9% to 7.9% per meter (1.2%–2.4% per foot) for mineral media (3–5 mm) at nominal design conditions (Redmon et al., 1983).

Additional studies yielded the following results:

- Rogalla and Sibony (1992) measured oxygen transfer rates of 7% to 15%.
- Pearce (1996) measured clean water oxygen transfer efficiencies of 10% to 17% in a downflow BAF pilot with 2-m depth and 3.3-mm angular media.
- Shepherd et al. (1997) measured oxygen transfer efficiencies of 7.9% to 10.3% in upflow BAF with 4-m depth of 2- to 3-mm silica sand.
- Laurence et al. (2003) reported oxygen transfer efficiencies of approximately 20% from off-gas testing of a 3-m depth upflow BAF with floating media and a 4-m depth upflow BAF with sunken media during side-by-side pilot testing in New York City.
- Stenstrom et al. (2008) measured oxygen transfer efficiencies of 5.8% to 21.1% for 3.6 m of 3- to 5-mm rock media and 13.1% to 29% for 3.6 m of 4-mm Styrofoam spheres in pilot reactors.

Process air distribution systems in BAFs include (Rundle, 2009):

1. Simple pipes with sparge holes drilled at intervals positioned in media or near the floor of the filter. Coarse bubble aeration through sparging pipes is used widely.

2. Diffusers placed on a pipe grid at the floor of the reactor to obtain even air distribution at low airflow rates, rather than to produce smaller bubbles for improved oxygen transfer efficiency. Although diffusers are more efficient for oxygen transfer than coarse-bubble sparging in open aeration basins, a comparison between coarse- and

fine-bubble aeration did not reveal any difference (Harris et al., 1996). In a comparison of coarse- and fine-bubble aeration in reactors with and without plastic random media, fine-bubble diffusers were found to be more efficient without packing and inefficient without (Hodkinson et al., 1998). The media-sheared coarse bubbles favored dispersion into smaller bubbles with a larger surface area and improved oxygen transfer. The fine bubbles, however, coalesced into larger bubbles and reduced oxygen transfer.

3. Injection of air under the plenum, frequently used to scour filters during backwash, also can be used during filtration. In this design, an air blanket is formed under the false floor in the plenum chamber; air enters the cell via holes in the specially designed combined nozzle. At low airflows, only the upper holes are used (Figure 11.41), but as air flow and pressure increase, blanket depth increases and more holes are used. This system provides efficient aeration, but requires periodic chemical cleaning to prevent biological growth from blocking the air holes, causing poor air distribution and increasing energy costs (Holmes and Dutt, 1999; Springer and Green, 2005).

Several factors complicate process air control in BAF reactors (Rundle, 2009):

1. The facilities operate primarily as plug flow systems, so that the dissolved oxygen at the top of the reactor does not represent the dissolved oxygen concentration within the media.

2. Oxygen transfer not only takes place from dissolved oxygen in water, but also occurs by direct interfacial transfer from gas to biofilm, which cannot be accounted for with a dissolved oxygen probe.

3. In systems aerated by a coarse-bubble air grid, the minimum flow to provide effective distribution of air can exceed process requirements.

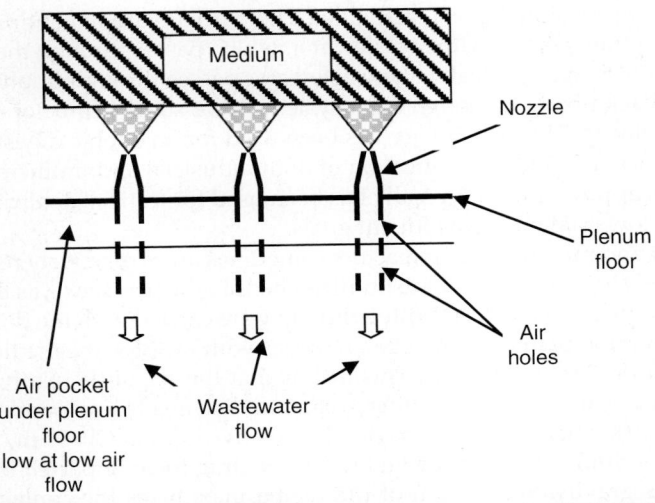

FIGURE 11.41 Operation of combined nozzles at low airflow (Rundle, 2009). Reprinted with permission from the Chartered Institution of Water and Environmental Management.

Blower selection is important for efficient facility operation. As solids accumulate in the media, filter headloss increases, which can affect the air flow. When several BAF cells receive air from a common main, backwashed cells will have the lowest headloss and will take more air flow. This balancing issue can be mitigated by providing individual blowers for each BAF cell. For larger facilities a centralized blower station with a common air main, air pipes feeding each cell are fitted with a mass flow meter (measuring velocity, pressure, and temperature). The meter is used to control a modulating valve, which balances air flow to the cells.

4.5.4 Supplemental Carbon Feed Facilities

In tertiary denitrification BAF, and in some pre-denitrification systems, an external carbon substrate (electron donor) must be dosed. Methanol typically has been used for this purpose. Increasingly, alternative carbon sources are being considered including ethanol, acetic acid, and sugar solutions. The chemical properties for the selected carbon source must be evaluated and accounted for in design. Carbon dosage control is important for tertiary denitrification systems. Overfeeding wastes chemical and increases the BOD and COD of the final effluent, which could be an issue for facilities with stringent BOD limits. Underfeeding the carbon source reduces the amount of nitrate removed, and the facility may not achieve the desired effluent nitrate or total nitrogen concentration. Several control alternatives include manual, flow-paced, feed-forward, and feed-forward and feedback with effluent concentration control.

5.0 Expanded and Fluidized Bed Biofilm Reactors

Expanded and fluidized bed biofilm reactors (EBBRs and FBBRs) are attached-growth systems with a range of applications in aerobic, anoxic, and anaerobic biological treatment. They use small media particles that are suspended in vertically flowing wastewater, so that the media becomes fluidized and the bed expands. Individual particles become suspended once the drag force of the flowing wastewater overcomes gravity and they are separated from each other. The particles are in continual relative motion but are not transported by the wastewater, which passes through the bed at a relatively fast rate (30–50 m/h). Ideally, the wastewater passes through in plug-flow mode with minimal backmixing, although most systems have some degree of recycle to maintain vertical velocity. This technology has been used for anaerobic digestion, carbon oxidation, nitrification, and denitrification of both industrial and municipal wastewaters. In municipal applications, it typically has been used for tertiary denitrification in facilities that have low total nitrogen effluent goals.

Airlift and moving bed bioreactors sometimes are erroneously referred to as "fluidized beds." However, neither are fluidized beds in the true sense, as the particles are not suspended in a vertical flow. With airlift, they are carried with the fluid flow in an internal recirculation path (Figure 11.1e); whereas with MBBRs, the media are carried in the horizontal flow (Figure 11.1g). This means that there is little relative motion between the media and wastewater in either system; unlike in a true expanded or fluidized bed (Figure 11.1f), where there is considerable relative motion (30–50 m/h).

Particle fluidization is achieved when the drag force imparted by the vertical flow overcomes gravity. Suspension of the media maximizes the contact surface between microorganisms and wastewater. It also increases treatment efficiency by improving mass transfer because there is significant relative motion between the solid (biofilm) phase and the flowing wastewater. Because media tend to be naturally occurring materials, they are relatively inexpensive. Because of the balance of forces involved in particle fluidization and bed expansion, the smallest particles are found at the top and the largest at the bottom. Therefore, the media particles should be graded to a relatively tight size range.

The degree of bed expansion determines whether a bed is deemed expanded or fluidized. The transition lies between 50% to 100% expansion over the static bed height. This discussion assumes the upper limit: beds less than double static bed height (100% expanded) are considered expanded; those more than double the static bed height (100% expanded) are fluidized. A lower degree of bed expansion is advantageous, because it requires a lower flow velocity, less energy, and increases effective biomass concentration, which reduces the footprint. In aerobic processes, however, it increases volumetric oxygen demand because of increased biomass concentration.

Starting with a static bed that partially fills the column, the vertical flow velocity is chosen to expand the bed to its initial design height, typically 50% of available height. Microorganisms attach to the media particles and grow to form a biofilm. This results in an increase in particle size but a decrease in composite particle (bioparticle) density. Thus, despite their initial differences, all media tend towards a similar specific gravity (approximately 1.1) as the biofilm thickness increases (Figure 11.42). Because of the size increase and density decrease, the bed continues to expand until it reaches full design height at which point biofilm thickness must be controlled.

Because biofilm thickness control is necessary to constrain the expanded bed within the confines of the reactor to better control the bioreactor performance. For example, biofilm thickness can be controlled to maintain optimum thickness and maximum reaction rate by choosing bioparticle concentration based on volume of support media added and degree of bed expansion.

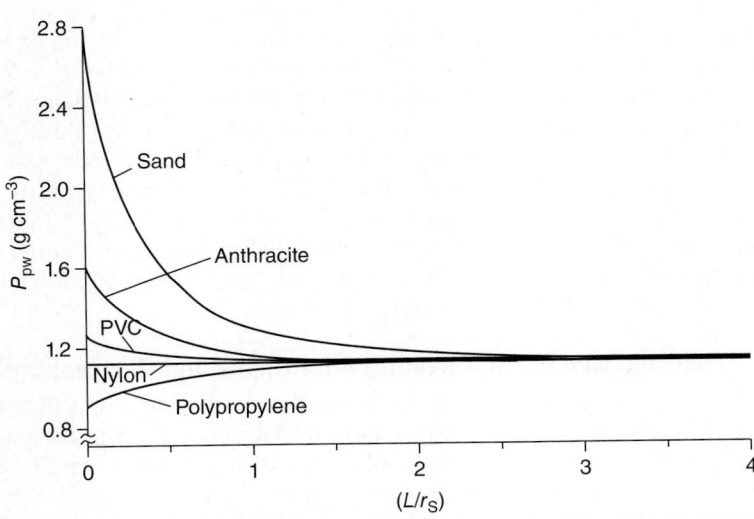

FIGURE 11.42 Overall densities of solid support particles with attached biomass (Atkinson and Black, 1981).

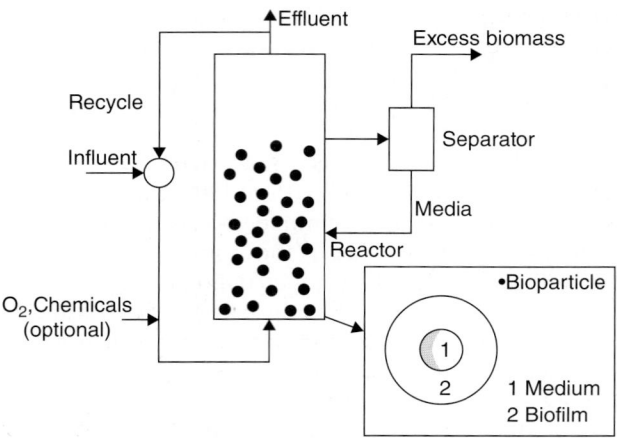

FIGURE 11.43   Process flow diagram of a fluidized bed biofilm reactor (Shieh and Keenan, 1986).

Whether this generic technology is used for aerobic, anoxic, or anaerobic processes, all are based on a similar basic design (Figure 11.43). This design consists of a column in which the particles are fluidized and the bed expanded and a recycle line that is used to maintain a fixed, vertical hydraulic flow. In this way, bed expansion is kept constant and bioparticles are retained irrespective of influent flowrate.

Anoxic and anaerobic designs are the simplest; aerobic designs require a system to supply oxygen. Aeration typically is achieved during recycle, during which influent wastewater mixes with recycled effluent from the top of the bed. If aeration is conducted within the fluidized bed, then a significant volume of gas disturbs the fluidized state by causing turbulence and increased force of interparticle collisions. This can dislodge biofilm. Nevertheless, this approach is used sometimes. The advantage of adding air to the recycle stream is that biomass is not stripped from the media by turbulence of rising gas bubbles and, therefore, the treated effluent typically has a lower concentration of suspended solids (Jeris et al., 1981; Oppelt and Smith, 1981).

Process flow enters at the bottom of the reactor and flows through a distribution system to ensure even dispersion and uniform fluidization. Silica sand (0.3–0.7 mm diameter) and granular activated carbon (GAC; 0.6 to 1.4 mm) typically are used. Other materials, however, have been used at pilot scale, such as 0.7 to 1.0 mm glassy coke (McQuarrie et al., 2007). Small carrier particles (1 mm) provide a large specific surface area for biofilm growth (up to 2400 m^2 m^{-3} when expanded 50%), which is one of the key advantages of this process technology (Figure 11.44).

5.1   Fluidized Bed Biofilm Reactor Advantages and Disadvantages

The key advantage of the FBBR configuration is the large specific surface area for biofilm growth. This area results in a high concentration of active biomass (Table 11.19), a high rate of reaction, and a small footprint. However, owing to high biomass concentration, aerobic processes can be oxygen limited. Another disadvantage can be the degree of recycle required to maintain upward velocity for bed expansion and bioparticle fluidization, which can increase pumping costs. The recycle pump, however, only has to overcome the sum of frictional resistances and density difference between the

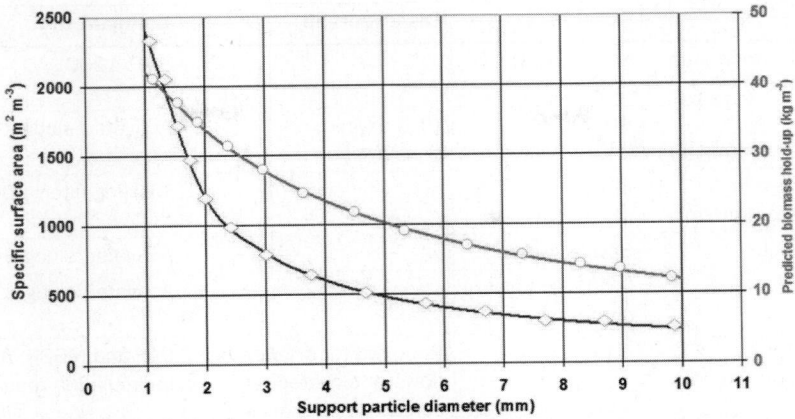

Figure 11.44   Specific surface area of biomass support material particles (à) and predicted biomass hold-up (o)—(dry weight), based on 50% bed expansion and a calculated volume for 500-μm deep biofilms of 80% water content.

Treatment	MLVSS (mg/L)
Carbon oxidation	12 000–15 000
Nitrification	8000–40 000
Denitrification	30 000–40 000

Table 11.19   Biomass Concentration in Fluidized Bed Biofilm Reactor

fluid in the recycle line (aeration gas and wastewater) and the expanded bed (wastewater and biofilm-coated media). Pressure drop across the recirculation (fluidizing) pump is significantly less than fluid height in the bioreactor. With processes that have a gas phase within the bed—either through aeration or gas evolution such as denitrification or anaerobic digestion—bioparticle loss and excess biomass in the treated effluent can require post-treatment sedimentation or filtration.

Specific surface area in an FBBR is dependent on media size (0.3–1.4 mm) and degree of bed expansion (50%–100%) but typically ranges from 1000 to 3000 m²/m³, which is higher than for other systems (Table 11.20). This large surface area allows biomass concentrations of 15 000 to 40 000 mg/L VSS to be maintained in the FBBR (Grady et al., 1999). Thus, the volumetric efficiency of an FBBR can be as much as 10 times that of an activated sludge system (Rabah and Dahab, 2004b). Because of its high biomass concentration, the FBBR can operate efficiently at short retention times (less than three minutes) for a denitrification process, even with high nitrate loading rates (greater than 70 kg $NO_3{}^2N/m^3$d) (Green et al., 1994).

Clogging and the resulting high headloss can be avoided in FBBRs compared to the use of small media in packed beds (Shieh and Keenan, 1986). Another advantage is that the fluidized bed retains a thin active biofilm around the entire particle; whereas in fixed beds, only the parts of the media that are not in contact with other particles can

Feature	Advantage	Disadvantage	Compare to
Specific surface area (m^2/m^3)	1000–3000		BAF: 1200
Biomass concentration (mg/L)	40 000	High oxygen consumption	Activated sludge: 3000
Biofilm specific surface area (m^2/m^3)	3000		Trickling filter: 300
Biomass age	Several weeks		Activated sludge: 10 days
F:M ratio	0.001 for tertiary nitrification		Activated sludge: 0.2–0.4
Wastewater recirculation		Required for dry weather flows or elevated concentrations of pollutant	BAF and MBBR: normally, no recirculation required
Backflushing	Not required		BAF requires 10% of treated effluent to be used
Aeration	Highly efficient counter-current systems can be used	Nitrogen-depleted air, oxygen-enriched air or pure oxygen normally required	Aeration in upflow BAF is inefficient co-current. MBBR highly inefficient, as 80% of aeration energy is required for media suspension

TABLE 11.20   Summary of Fluidized Bed Biofilm Reactor Advantages and Disadvantages

develop biofilm. Thus, the surface area provided in FBBRs is 10 times greater than the surface area provided in equivalent downflow fixed beds (U.S. Filter/Envirex, 1997).

The main disadvantages of the fluidized bed system recognized by U.S. EPA in 1993 were:

- Limitations on reactor size (in terms of maintaining the height-to-diameter ratio);
- Energy requirements (as a result pumping to maintain high recycle ratios);
- Difficulties in biomass control and media selection (media loss and biomass in the effluent through particles that were too small or of insufficient specific gravity and allowing biofilm to become too thick); and
- Imprecision in process control because of difficulty in monitoring biomass concentration.

In addition, liquid distributors often were costly for large systems, long startup was required for biofilm formation, and clogging in the flow distributor could prevent uniform fluidization (Rabah and Dahab, 2004a). Sutton and Mishra (1994) state that widespread use of FBBR technology has been hampered by mechanical scale-up issues, slow development of economically attractive commercial systems, and proprietary constraints. Nevertheless, the technology continues to be developed and some solutions have been introduced. For example, glassy coke media and biofilm thickness control using an internal bioparticle recycle system that, in tertiary nitrification, allows consumption of removed biofilm by protozoa and metazoa in the system (Dempsey, 2003; 2007; Dempsey et al., 2006).

5.2   Fluidized Bed Biofilm Reactor Technology Status

5.2.1   Installations
In 1999, more than 80 full-scale FBBRs had been installed in North America and Europe. Two-thirds of these were treating industrial wastewater; the rest were treating municipal. Nicollela et al. (2000) stated that the use of particulate biofilm reactors can be considered a mature technology for which good design and scale-up guidelines are available.

Widespread use of this technology, however, has stalled recently. Treatment of a variety of wastewaters has been investigated at both the laboratory and pilot-scale, which may allow expansion or upgrading of facilities to meet more stringent future standards, especially for ammonia, nitrate, carbonaceous biochemical oxygen demand (CBOD), and TSS. In particular, anaerobic FBBRs have found a market for the treatment of high-strength industrial wastewaters with COD concentration greater than 1000 mg/L because the high costs of oxygenation can be avoided and valuable biogas can be recovered. The compact nature of FBBRs may allow upgrading within existing structures because complete skid-mounted systems can be directly shipped to site, requiring only piping and electrical connections to make them fully operational. In addition, modules can be added to either augment current processes or to add new ones (Shieh and Keenan, 1986).

5.3   Process Design

5.3.1   Typical Design Parameters
Vertical flow velocity, recirculation, and flow distribution are key FBBR process design parameters discussed in this section.

5.3.1.1   Vertical Flow Velocity   The U.S. EPA recommends an upward velocity range for 0.5 mm silica sand of 36 to 60 m/h; others recommend a range of 30 to 36 m/h (U.S. EPA, 1993; Metcalf and Eddy, Inc./AECOM, 2013). Although the acceptable range of upflow velocities appears to be 30 to 60 m/h, many pilot-scale reactors have operated at 30 to 36 m/h (Shieh and Keenan, 1986). The upflow velocity is chosen according to the size and specific gravity of the support media particles (Table 11.21). Basically, vertical

Media	Size (mm)	U_{mf} (m/h)	$U_{50\%}$ (m/h)	Reference
Glassy coke	0.7–1.0	9.0	30	Dempsey et al. (in press)
Granular activated carbon	1.7	10.2	—	Coelhoso et al. (1992)
Silica sand	0.5–1.0	22.2	90	Dempsey et al. (in press)

U_{mf} = Upward velocity for minimum fluidization.
$U_{50\%}$ = Upward velocity for 50% bed expansion.

TABLE 11.21   Upward Velocity Required to Fluidize Different Media

drag force must exceed the downward force of gravity, so that particles are fluidized and the bed expands. Once this state is achieved, further increasing the flow velocity increases the degree of bed expansion. As biofilm develops, the size of the particles increases (Figure 11.42). These two changes in the physical characteristics of the particles result in continuous bed expansion until it reaches design height, at which point biofilm thickness control must be initiated.

The upward flow velocity of wastewater does not have a significant effect on biomass shearing in an FBBR because of laminar flow conditions. Grady et al. (1999) stated that superficial velocities used in FBBRs result in low Reynolds numbers (less than 10), and thus surface shear forces are small, resulting in a relatively thick biofilm. However, if there is a significant gas phase, then biomass stripping by bubbles (single, swarms, and coalesced) can be severe. Typically, however, accumulation of biofilm on the fluidized media is affected primarily by bioparticle collisions, rather than by fluid velocity past the biofilm surface (Shieh and Keenan, 1986). For this reason, the degree of bed expansion can affect the rate of biofilm sloughing and, consequently, the degree of additional biofilm control. That is, the greater the degree of bed expansion, the more space there is between bioparticles resulting in fewer interparticle collisions and less biofilm attrition.

A key factor that can affect upward flow velocity is the amount of entrained gas because of its effect on fluid density. Most facilities will have a control system for varying gas flow according to the dissolved oxygen concentration, which inevitably affects the degree of bed expansion. For systems with in-bed aeration, the higher the gas flow rate, the lower the bed expansion, owing to the decrease in fluid (gas and liquid) density. For systems with external aeration, the same effect occurs because of increased upward drag of the rising gas bubbles reducing inlet pressure to the pump. Therefore, best practice is to install a flow meter between the fluidizing (recirculation) pump and bed so that it can be controlled at the correct value (e.g., using an inverter to vary the pump speed). This approach will also minimize the energy consumed for bed expansion.

5.3.1.2 Recirculation To maintain a constant degree of bed expansion, the vertical flow velocity if fixed and differences in influent flow rates are addressed by recycling a fraction of process effluent. For tertiary nitrification, the basic design rule is to size the reactor so that it can take the full influent flow rate without recirculation during extreme wet weather, when wastewater is at its most dilute.

Although experimental reactors used for denitrification often are used without recirculation, this mode is unsuitable for WRRFs, where recycle flows of 2 to 5 times influent are used to maintain constant upflow velocity during diurnal variations and to protect against shock loading by diluting the influent. For example, the denitrification FBBR at Himmerfjarden WRRF (Figure 11.45) employs recirculation to compensate for diurnal flow variations and ensure constant upward velocity (Bosander and Westlund, 2000). In aerobic processes, recirculation also is required for re-aeration, because of low solubility of oxygen in wastewater.

5.3.1.3 Flow Distribution The influent distribution manifold is a critical design feature in the FBBR. This distribution system must be balanced to

- Disperse influent kinetic energy by achieving uniform distribution of flow across the entire reactor cross section;
- Support media and prevent it from falling through the manifold;

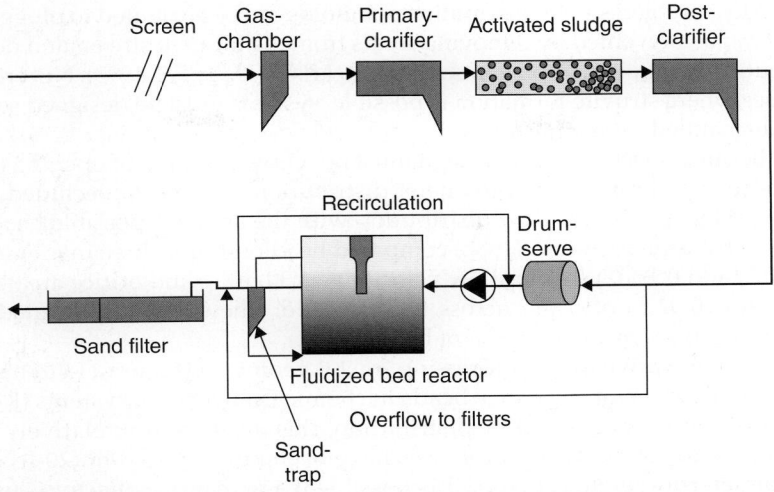

Figure 11.45   Flow schematic of the Himmerfjarden Water Resource Recovery Facility, Sweden (Bosander and Westlund, 2000).

- Avoid plugging;
- Minimize biofilm shearing because of turbulence or particle collisions, thereby promoting uniform biofilm buildup throughout the media;
- Minimize headloss.

Two important design aspects regarding distribution of the wastewater flow should be examined. First, uniform distribution of incoming wastewater across the base and perpendicular to the flow avoids spouting, caused by rising jets of wastewater rising (Rabah and Dahab, 2004a). Spouting tends to disrupt distribution of bioparticles and substantially increases their collision frequency at the base of the bed. This can prevent biofilm formation and lead to media loss through abrasion. It also increases the amount of short circuiting through the reactor. In extreme cases, a portion of the bed at the reactor bottom can remain static and largely inactive. Second, the kinetic energy associated with incoming wastewater must be dissipated to ensure that turbulence in the region is minimal.

Most pilot-scale FBBRs introduce wastewater upward into the reactor through a perforated plate. Some pilot-scale reactors include a layer of gravel above the plate to improve distribution and prevent media from clogging the plate (Jeris et al., 1981). However, clogging of the static gravel bed by materials entrained in the wastewater flow through biofilm growth or chemical precipitation or crystallization (e.g., struvite formation) is likely. In contrast, downward-facing nozzles have been used by Sigmund (1982) and Dempsey et al. (2006). At full scale, wastewater is introduced through wide-bore pipework with downward-facing nozzles. Wastewater delivered into the reactor makes a 180° turn, with much of its kinetic energy dissipated. Sigmund (1982) describes two, full-scale distribution systems and states that headloss through the manifold should be at least as much as that through the fluidized bed.

Most problems with distribution manifolds can be attributed to plugging. However, this can be prevented by removing solids from the influent stream and designing a distribution manifold that prevents media backflow (U.S. EPA, 1993). Nevertheless, in processes where struvite formation is possible, systems must be designed so that they can be dismantled for maintenance.

Because successful inlet design must not allow clogging of openings by solids, use of porous plates and small-diameter distribution systems is precluded. The designer must achieve uniform flow distribution with the smallest possible headloss through the distributor. Sigmund (1982) compared headlosses for three manifold orifice areas (2, 0.5, and 0.1%). When upflow velocity was 30 m/h, and orifice area was 2%, headloss was 0.02 m of water across the manifold. The 0.5% area resulted in 0.32 m of headloss; 0.1% resulted in 8 m of headloss.

An innovative distributor using fractal geometry (Figure 11.46) has been developed to fluidize ion exchange beads in chemical absorption systems (Kearney, 2000). In addition to excellent flow distribution, this design has a relatively low pressure drop (0.7 to 1.4 m water). Processes have been built up to 6-m (20-ft) diameter and a diameter-to-height ratio of 2:1 is possible if a matching collector is used at the top of the bed to ensure vertical flow streaming. Furthermore, lower iteration of the fractal pattern compared to that used for ion exchange beads is needed to produce sufficiently wide-bore flow channels that should not clog. Cost of this distributor is, however, related to its cross-sectional area, making it more expensive for large-scale applications.

5.3.2 Media

The media should allow colonization by a variety of microorganisms that can develop into a firmly attached biofilm. Small particles provide a large surface area, and the material should be available in a narrow size to minimize classification (stratification) upon fluidization. To minimize pumping energy, there should be sufficient difference in the specific gravity between the wastewater and the material to allow adequate bed expansion at an upward velocity that does not result in an excessive recirculation ratio. Particles must not have a specific gravity too close to that of water, however, or be so small that they become buoyant or form aggregates by attaching to each other. Also, the

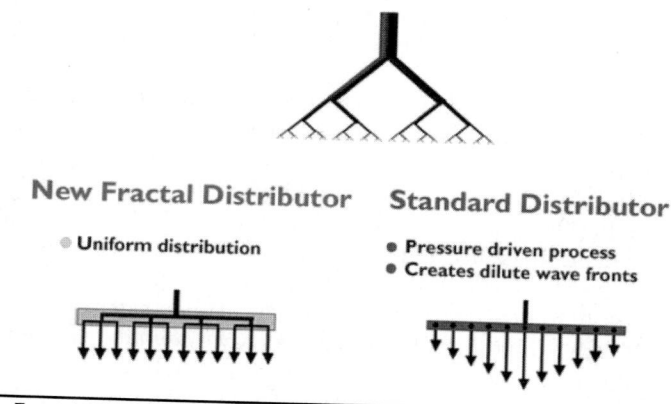

FIGURE 11.46   Fractal and conventional flow distributors.

material should be inexpensive. For these reasons, mineral particles of about 1-mm nominal diameter typically are used.

In one of the first trials of FBBR technology for municipal wastewater treatment, Oppelt and Smith (1981) reported that anthracite coal (1 mm) and cinders (0.5, 1.2, and 2.0 mm) were unsuitable for CBOD oxidation because of "frequent separation of the fluid bed, plugging of the columns and interconnecting piping and excessive media loss in the effluent." Nevertheless, some of these problems may have been caused by the scale of operation (column diameter was only 10 cm) and the method of aeration, which involved injecting oxygen at the base of the bed, which may have caused churn turbulence within the bed. However, they achieved more success when they used fine (0.5 mm) silica sand, which is inexpensive and readily available. Jeris et al. (1974), however, found that activated carbon was more suitable than silica sand.

Silica sand used in FBBRs typically has a diameter of 0.3 to 0.7 mm and GAC is 0.6 to 1.4 mm. Shieh and Keenan (1986) found that 0.3 mm GAC resulted in the highest biomass concentration. The GAC has some properties that make it more desirable than silica sand, for example lower density, macro-porosity, good adsorptive characteristics, homogeneous biofilm thickness along the reactor, and easy startup or restart. In response to variations in superficial upflow velocity, the bed height stability increases with increased media size and density (Figures 11.47 and 11.48) (Shieh and Keenan, 1986). The Y-axis in both of these figures represents bed expansion, while the X-axis represents upflow velocity. Figure 11.48 includes three curves with varying media size (t_m). Figure 11.49 includes three curves with varying media density (r_m). Although smaller media particles reduce the energy required for fluidization and provide a larger specific surface area for biofilm formation, if the media is too small (0.3–0.9 mm silica sand) and the biofilm is too thick (greater than 200 mm), bioparticles may aggregate into "golf balls" and operation may become unstable (Grady et al., 1999; Cooper, 1986).

Coelhoso et al. (1992) showed that differences occurred in biofilm thickness, mass-transfer resistance, and solids production with GAC compared to silica sand. When thick biofilms (800 mm) were allowed to develop on GAC, the reactor operated under a diffusion-controlled regime. In contrast, biofilms were much thinner when silica sand was used and the reactor functioned in a kinetic-controlled regime. Easier startup was observed with GAC, when grey biofilm formed within three days. Although many differences were observed, N removal rates were not significantly different: GAC at 5.3 to 8.6 kg NO_3^2-N/m^3/d; silica sand at 5.4 to 10.4 kg NO_3^2-N/m^3/d.

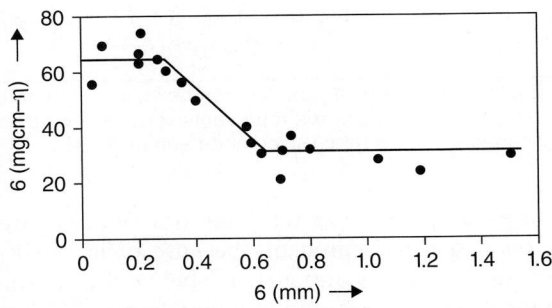

FIGURE 11.47 Biofilm dry density vs biofilm thickness (Shieh and Keenan, 1986).

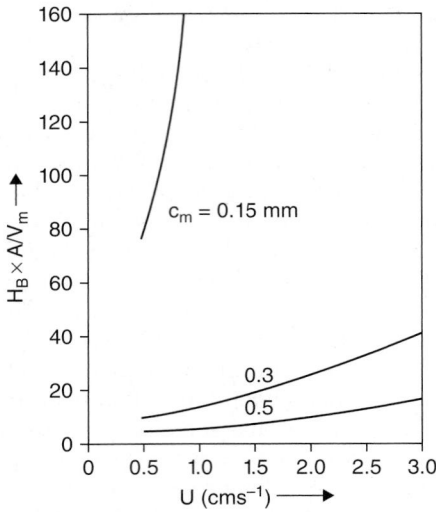

Figure 11.48 Effect of media size on fluidized bed biofilm reactor bed expansion (Shieh and Keenan, 1986).

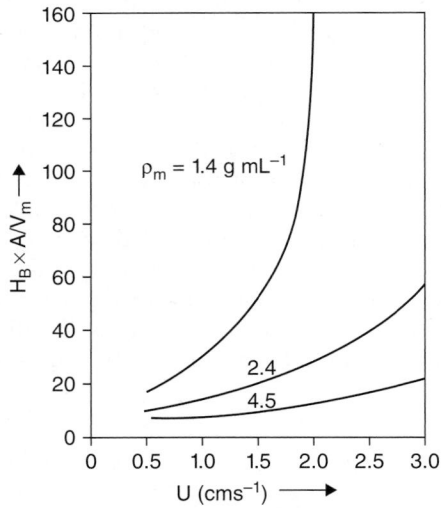

Figure 11.49 Effect of media density on fluidized bed biofilm reactor bed expansion. The *y* axis represents bed expansion and the *x* axis represents vertical flow velocity. Three curves show the effect of changing media density (ρm) (Shieh and Keenan, 1986).

The most important advantage reported for GAC compared to silica sand, was the rapid startup period. This is probably because of its adsorptive capacity, irregular shape, and porosity (microbes in pores are sheltered from shear forces). In one study, initial biofilm development occurred in a few days on GAC media, while biofilm did not develop on silica sand media for several weeks (Jeris and Owens, 1975). In a later study, initial biofilm development also was found to occur more rapidly on GAC than on silica sand. Coelhoso et al. (1992) reported that a bed of GAC showed homogeneous biofilm thickness (800 mm) throughout the reactor, because of the more similar specific gravity of biofilm and media. However, when silica sand was used, the difference in specific gravity caused stratification—particles with thicker biofilms moved to the top of the bed and thinner biofilms moved to the bottom. For this reason, GAC or media with a similar porosity and specific gravity to GAC, such as glassy coke, are typically better than silica sand.

Fundamental studies of microbial adhesion and fouling show that materials that are prone to biofilm formation have a rough or porous surface. In a comparison of silica sand (nonporous) with three types of porous, diatomaceous earth (all four media approximately 0.4- to 0.6-mm diameter), Yee et al. (1992) found that media with 6- or 30-mm pores allowed the fastest rate of methanogenic biofilm formation and organic carbon removal. Pores of this size are substantially larger than the typical size of Bacteria or Archaea (both approximately 1 mm) and provide protected zones for initial attachment. Although diatomaceous earth has not become widely used in expanded or fluidized beds, another porous material, glassy coke, manufactured from bituminous coal, has been used successfully at pilot scale for tertiary nitrification of municipal wastewater (Dempsey et al., 2006). Because coke is carbon-based and porous, it has a substantially lower specific gravity than silicon dioxide-based, nonporous sand, which means that the energy consumption for particle fluidization and bed expansion is lower (see Table 11.21). Coke is approximately 90% carbon (atomic mass of C = 12) and 10% ash; whereas silica sand is mainly silicon dioxide (molecular mass of SiO_2 5 46).

5.3.3 Biofilm Thickness Control

Continued growth of biofilm causes an increase in bioparticle volume and degree of bed expansion, resulting in the bed exceeding its design volume. Therefore, it is necessary to have a system to control biofilm thickness and prevent bioparticles leaving with the treated effluent. The expanded bed height must be controlled continuously at a fixed height or intermittently between a minimum and maximum. To maintain a constant biomass inventory, the rate of biomass wasting must balance its rate of growth. Therefore, for faster-growing systems such as carbon oxidation or heterotrophic denitrification, the rate of biofilm growth and consequent bed expansion is higher than for slower growing systems such nitrification, autotrophic denitrification, and anaerobic digestion.

Any control system should maintain the initial particle inventory. Atkinson (1981) states that "either an impeller has to be added at the top of the bed to provide increased attrition, or particles which elutriate from the bed have to be recycled through a shear field, or the flow through the bed has to be increased by recycle to further expand and agitate the bed." An alternative approach is to deliberately remove bioparticles from near the top of the bed, strip the biofilm and return the cleaned particles. A variety of systems have been developed to control biofilm thickness, using either internal control as advocated by Atkinson (1981), or external control. An example of internal control involves a rotating perforated disc at the top of the bed that strips excess biomass and allows the cleaned particles to sink back into the bed; treated effluent then carries off the stripped biofilm (Bosman and Hendricks, 1981).

Historically, a common approach involved removal of bioparticles for external control. For example, in their FBBR for CBOD oxidation of primary effluent, Jeris et al. (1981)

used a pump to transport bioparticles from the top of the bed onto a vibrating screen, which removed excess biofilm. A variation of this system was used for an anthracite-based fluidized bed to denitrify nuclear industry wastewater. In that case, particles with thick biofilms that elutriated from the bed were captured on a vibrating screen before being pumped back to the base of the bed; a sedimentation tank captured stripped biofilm (Francis and Hancher, 1981). In the Dorr-Oliver oxitron process, the fluidized sand bed height was controlled within a band of 0.6 to 0.7 m by intermittently pumping bioparticles from the upper region of the bed using a "a rubber-lined centrifugal pump operating in conjunction with a hydrocyclone and media washer" (Sutton et al., 1981).

To avoid operational problems of mechanical pumps and screens, Basin Water/Envirogen's fluidized bed reactor uses an airlift system for biofilm control (Frisch, 1998a; 1998b). A newer method is to recycle bioparticles from the top of the bed to the bottom via an eductor pump, where excess biofilm is stripped off during passage through the turbulent regions in the eductor and the base of the bed (McQuarrie et al., 2007). Advantages of this system include its simplicity, as it requires no sensor, actuator or moving parts; and the fact that, in a tertiary nitrification process, stripped biofilm is consumed by protozoa and metazoa during its passage up through the bed (Dempsey et al., 2006).

Control of the expanded bed height provides the most direct and convenient means to maintain optimum biofilm thickness. In conventional denitrification, the expanded bed height is not continuously controlled because, if it were, the separated biomass would be carried away as a dilute sludge requiring further thickening before final disposal, thus increasing the costs. To minimize these costs, the bed height is allowed to fluctuate over a range and is controlled once a day to produce a thickened sludge that can be captured.

The denitrification FBBR at Himmerfjarden WRRF in Sweden produces an average of 0.5 g VSS/g NO_3^{2-}-N (Bosander and Westlund, 2000). In earlier work, U.S. EPA (1993) reported that biomass production in denitrifying FBBRs was 0.4 to 0.8 g TSS/g NO_3^{2-}-N consumed. In the Himmerfjarden FBBR, biomass is sheared from media using a pump inside a central cone at the top of the bed, producing sludge of 2000 to 4000 mg/L VSS. In contrast, the U.S. Filter/Envirex proprietary FBBR growth control system produces wasted solids concentrations of 5000 to 15 000 mg/L TSS (U.S. Filter/Envirex, 1997). At Himmerfjarden, the excess biomass is removed at the top of the bed, where a sand trap captures any escaping particles and the cleaned silica sand is dropped back into the bed. In contrast, the Rancho, California denitrification FBBR uses periodic backwashing to control biofilm thickness (MacDonald, 1990).

5.3.4 Aeration

There are two approaches to supplying oxygen, either in the bed or in the recirculation loop. Oxygen typically is supplied using air, nitrogen-depleted air, oxygen-rich air, or pure oxygen. Using air to supply oxygen may not be the least expensive solution. Because oxygen is sparingly soluble, the high concentration of active biomass that characterizes EBBR and FBBR processes typically means that the dissolved oxygen concentration at the bed inlet achieved using air is insufficient to keep the entire bed aerobic. By increasing the oxygen content of the aeration gas, the dissolved oxygen concentration can be increased by the same proportion. For example, if oxygen saturation using air (20.9% O_2) is 10 mg/L then that using pure oxygen (100% O_2) is almost 48 mg/L.

Nitrogen can be removed from air using vacuum or pressure swing molecular sieve technology, which has been considerably improved and costs lowered recently. An alternative is to add oxygen to air from a cryogenic source, which has also become cheaper recently. Although it is also possible to raise the dissolved oxygen concentration by pressurization, pressure vessels are expensive and are not typically considered practical for wastewater treatment. Key factors affecting the rate of oxygen transfer from the gas phase to the wastewater include the surface area of gas bubbles and the relative velocity between the gas and liquid phases.

5.4   Pilot Testing

The advantage of in-bed aeration is that oxygen can dissolve from rising gas bubbles to replace dissolved oxygen consumed by the microorganisms. However, achieving an oxygen transfer rate (OTR) to at least match the oxygen consumption rate (OCR) of the microorganisms is difficult. If OTR is less than OCR, then the dissolved oxygen concentration will fall along the bed and within the biofilm until it becomes rate-limiting. Also, because the gas bubbles and wastewater are both rising, their relative velocity is lower than in a counter-current bubble contactor. In fact, Fonseca et al. (1986) cite a study that demonstrated that the OTR of a fluidized bed with 1-mm particles was 20% that of a bubble column, compared to 200% when 6-mm diameter particles were used. Therefore, this theoretical advantage of increased oxygen transfer in the main FBBR has practical limitations when small (approximately 1 mm) media are used.

Another significant disadvantage of in-bed aeration is that rising gas bubbles can transport bioparticles and entrain them in either recirculation or effluent flows. This phenomenon can be caused by either small gas bubbles attaching to bioparticles and making them buoyant or gas bubbles coalescing and forming gas slugs that transport bioparticles by bulk flow. For example, at high gassing rates (superficial liquid velocity/superficial gas velocity less than 0.4), churn-turbulence occurs, fluidization breaks down, and particles can be transported (Lee and Buckley, 1981). Furthermore, this research shows that the violent movements associated with churn-turbulence could make initial biofilm formation or its retention a problem. Thus, increased turbulence can result in biofilm stripping and increased TSS in the treated effluent.

If bioparticles exit the bed, they must be trapped to avoid being lost with the effluent or, if they enter the recirculation flow, they may either damage the fluidizing pump or be damaged during passage. Thus, to prevent transport of bioparticles by rising gas, Oppelt and Smith (1981) fitted bubble traps to the top of their multistage, pilot-scale FBBR with in-bed aeration for CBOD oxidation. An alternative approach is to generate gas bubbles that are small or sparse enough not to coalesce as they rise through the bed by using, for example, a static mixer to generate submillimeter bubbles. Thus, by careful choice of process technologies, otherwise troublesome operational issues can be overcome.

The alternative to in-bed aeration is to oxygenate in the recirculation flow by using a cone, bubble column, or static mixer (Jeris et al., 1981; Cooper and Wheeldon, 1981; Dempsey et al., 2006). There are several advantages to this approach: biofilm disruption by gas bubble turbulence is avoided; transport of bioparticles by gas slugs is prevented; and the driving force for oxygen uptake is maximized. Because these systems work with counter-current flow, the relative velocity between the two phases is maximized.

By judicious design, the downward velocity can be such that turbulent conditions prevail in the aeration device and the injected gas is broken up into tiny bubbles. This increases the gas-liquid interfacial area for mass transfer of O_2, which then increases OTR. Furthermore, it is possible to obtain up to 25% gas holdup in downflow bubble columns, which is typically accepted as ideal in static mixing systems. If pure oxygen is used, then it is possible to dissolve almost all of the gas but difficult to strip metabolically produced CO_2, which inhibits biological processes. In contrast, when nitrogen-depleted or oxygen-enriched air is used, the rising gas bubbles become depleted in oxygen, but CO_2 is extracted by the residual inert gases (N_2 and Ar) and transported out of the system.

Table 11.22 presents typical design criteria for denitrifying FBBRs that used silica sand media.

5.5 Fluidized Bed Biofilm Reactor Design Models

Because of the complex nature of mathematically modeling the biological activity of fluidized bed biofilm reactors, the reader is referred to existing literature on this topic. Shieh and Keenan (1986) provide an extensive discussion of FBBR modeling theory, whilst Grady et al. (1999) provide a more recent review of this topic.

Parameter	Unit	Value Range	Typical
Packing:			
Type		Sand	Sand
Effective size	mm	0.3–0.5	0.4
Sphericity	Unitless	0.8–0.9	0.8–0.85
Uniformity coefficient	Unitless	1.25–1.50	≤1.4
Specific gravity	Unitless	2.4–2.6	2.6
Initial depth	m	1.5–2.0	2.0
Bed expansion	%	75–150	100
Empty-bed upflow velocity	m/h	36–42	36
Hydraulic loading rate	$m^3_{effluent}/m^2_{bioreactor\ area}{\cdot}d$	400–600	500
Recirculation ratio	Unitless	2:1–5:1	2:1–5:1
NO_3^--N loading:			
13°C	kg/m³·d	2.0–4.0	3.0
20°C	kg/m³·d	3.0–6.0	5.0
Empty-bed contact time	min	10–20	15
Methanol–NO_3^--N ratio	Unitless	3.0–3.5	3.2
Specific surface area*	(m²/m³)	1000–3000	2000–3000
Biomass concentration[1]	mg/L	15 000–40 000	30 000–40 000

Source: Metcalf and Eddy (2003); U.S. EPA, 1993; Sadick et al., 1994; Sadick et al., 1996.
*Grady et al., 1999.

TABLE 11.22 Design Criteria for Denitrifying Fluidized Bed Biofilm Reactors (FBBRs)

5.6 Design Considerations

5.6.1 Nitrification

For tertiary nitrification of activated sludge settled effluent (5 to 25 mg/L NH_3-N) using 1-mm particles of glassy coke as the support medium, 1 m^3 of expanded bed is required to remove each kg NH_3-N per day. Whereas for nitrification of sludge filtrate (1000 mg/L NH_3-N) or centrate, only 0.5 m^3 expanded bed is required because this process produces so much acid that alkali is added and, therefore, it can be operated at the optimum pH (7.8 to 8.0). However, it would probably be too expensive to use pH control for tertiary nitrification because of the large volume of wastewater needing treatment, which is typically the entire works flow.

Because of the high rate of oxygen consumption by nitrifying bacteria, the maximum bed depth is typically between 5 to 6 m. All oxygen has been consumed by then, even when using nitrogen-depleted or oxygen-rich air. Under these conditions, each meter of bed depth can oxidize up to 2 mg/L NH_3-N. Thus, a 5-m deep bed can oxidize up to 10 mg/L NH_3-N per recycle (Dempsey and Boltz, 2010).

Although the stoichiometric amount of oxygen for nitrification is 4.6 kg O_2/kg N, the oxygen supply needs to exceed this (e.g., 5 kg O_2/kg N) to allow for organic matter consumed by heterotrophs. However, the actual oxygen supply will depend on the amount of organic matter (CBOD plus TSS) that needs to be removed from the influent. Nevertheless, it may not always be possible to control the degree of CBOD and TSS removal and, therefore, more or less oxygen than 5 kg/kg may be needed. Because of competition for oxygen, the hydraulic loading rate should be less than 40 m^3 secondary effluent per m^3 of expanded bed per day to achieve less than 5 mg/L NH_3-N; and less than 25 m^3/m^3/d to achieve less than 0.5 mg/L NH_3-N (Dempsey et al., 2006).

At technical-scale, it has been found that rate-limiting concentrations of dissolved oxygen (dissolved oxygen) and NH_3-N at the top of the bed were both 1 mg/L (Dempsey et al., 2005). This finding means that if the effluent concentration of NH_3-N has to be less than 1 mg/L, then the loading rate must be reduced. It also means that if the dissolved oxygen concentration falls below 1 mg/L, then the NH_3-N effluent concentration will rise. Therefore, if the process is to be used at maximum efficiency, then the bioreactor must be designed to supply oxygen fast enough to meet the metabolic requirements of nitrifying bacteria. This includes control of oxygen supply based on the dissolved oxygen concentration at the top of the bed and use of nitrogen-depleted or oxygen-enriched air and a counter-current bubble-column or a static mixer.

5.6.2 Tertiary Denitrification

Empirical design methods are presented in the U.S. EPA Nitrogen Control Manual (U.S. EPA, 1993). Onsite pilot testing is recommended to generate design criteria. The designer must first define the following design criteria:

- Influent flow rate,
- Influent and effluent nitrate concentration, and
- Minimum operating temperature.

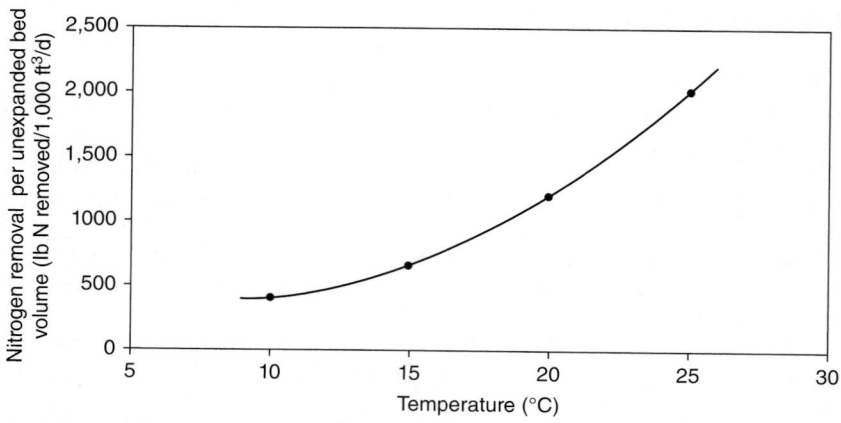

FIGURE 11.50   Fluidized bed biofilm reactor nitrogen removal rate vs temperature (U.S. Environmental Protection Agency, 1993).

Figure 11.50 shows that the nitrogen-removal rate in an FBBR is 6.4 kg NO_3^{2-}-N/m³/d (400 lb/d/1000 cu ft) at 10°C, which is the U.S. EPA (1993) design recommendation. This removal rate assumes methanol as the carbon source and sand media (0.3–0.6 mm). The U.S. EPA (1993) also gives a loading rate in terms of surface area of 0.8 to 3.4 kg NO_3^{2-}-N/1000 m²/d, with an average of 1.84 kg NO_3^{2-}-N/1000 m²/d. However, Metcalf and Eddy, Inc./AECOM (2013) recommend design loading rates in the range of 2 to 6 kg NO_3^{2-}-N/m³/d. All of these rates are based on unexpanded bed volumes. For denitrification FBBRs, loading and removal rates are calculated for unexpanded bed volumes, although using expanded bed volume would make it easier to compare to other processes and technologies, which typically are based on bioreactor volume.

The concept of hydraulic loading rate (HLR) can be confusing for FBBRs, where the diameter of the reactor and required upflow velocity dictate the flow rate (HLR) through the bed. Thus, HLR equals the influent plus recirculation flow rates. For single-pass operation (e.g., extreme wet weather) influent HLR equals bed HLR and there is no recycle. During normal operation, however, influent flow rate is less than bed HLR and, therefore, there is a degree of recycle. The recycle flow rate is variable and equal to the bed HLR minus the influent flow rate. It varies automatically, without need for electronic monitoring and control. The required nitrogen-removal rate, the expected biomass concentration, and its specific treatment rate dictate the nitrogen-loading rate for a given process. The U.S. EPA (1993) recommends a bed hydraulic loading range of 880 to 1470 m³/m²/d (36–61 m/h) and Grady et al. (1999) recommend a bed hydraulic loading rate of 576 to 864 m³/m²/d (24–36 m/h).

5.7   Design Example for Denitrification

This design example (Table 11.23) was adapted from U.S. EPA (1993). It uses a nitrate loading rate of 6.4 kg NO_3^{2-}-N/m³/d, which was derived empirically. Assuming a media

Characteristic	Influent–FBBR	Effluent Limits
Minimum monthly temperature	15°C	
Average flow (m³/d)	18 930	
Peak week flow (m³/d)	28 396	
TSS (mg/L)	15	30
CBOD (mg/L)	3	30
COD (mg/L)	33	—
TKN (mg/L)	1.8	—
$(NO_3 + NO_2)$-N (mg/L)	23.4	7
NH_3+ -N (mg/L)	0.05	2
TN (mg/L)	26.5	10

Notes: TSS = total suspended solids; CBOD = carbonaceous biochemical oxygen demand; COD = chemical oxygen demand; TKN = total Kjeldahl nitrogen; TN = total nitrogen.

TABLE 11.23 Design Criteria for Denitrifying Fluidized Bed Biofilm Reactor (FBBR) (U.S. EPA, 1993)

specific surface area of 2000 m²/m³ results in a loading rate of 3.2 kg NO_3^--N/1000 m²/d. However, other sources recommend lower loading rates. Metcalf and Eddy, Inc./ AECOM (2013) suggest 2.0 to 6.0 kg/m³/d as an appropriate range for nitrogen loading rate, depending on temperature. Pilot testing is recommended to confirm the effectiveness of the fluidized bed process with the site-specific wastewater and to determine an appropriate nitrate loading rate for use in facility design.

1. Calculate nitrate removed:

    (23.4 – 7 mg/L)(18 930 m³/d)/1000 = 311 kg NO_x-N/d

2. Calculate reactor volume:

    Assume nitrate loading rate = 6.42 kg NO_x-N/m³/d (Note that the nitrate loading rate is provided for the unexpanded media volume; see Figure 11.50 to correct for temperature effect.)

    Calculate volume of reactor:

    Volume = (311 kg NO_x-N/d)/(6.42 kg NO_x-N/m³/d) = 48.4 m³

    Assume two reactors in service and one standby; 3.65-m (12-ft) diameter reactor (area = 10.5 m² (113 sq ft)).

    Calculate bed height:

    (48.4 m³)/[(10.5 m²/reactor)(2 reactors)] = 2.3 m

    Use 3-m high bed with 1.8 m of freeboard for solids separation for a 4.9-m high reactor, based on manufacturer's standard.

    Total volume of in-service reactors = 2(10.5 m²)(3 m) = 63 m³

3. Calculate HRT at peak week flow:

    (63 m³)(1440 min/d)/28 396 m³/d = 3.2 min

4. Check total hydraulic load:

   Total hydraulic load5 influent flow/area

   $(18\,930 \text{ m}^3/\text{d})/(10.5 \text{ m}^3)/2 = 901 \text{ m}^3/\text{m}^2/\text{d}$ at average flow

   $(28\,396 \text{ m}^3/\text{d})/(10.5 \text{ m}^3)/2 = 1352 \text{ m}^3/\text{m}^2/\text{d}$ at peak week flow

5. Calculate actual nitrogen loading rate based on selected reactor:

   $(311 \text{ kg NO}_x\text{-N/d})/(63 \text{ m}^3) = 4.94 \text{ kg N/m}^3/\text{d}$

   Calculate recycle rate:

   Maintain reactor flow rate equal to peak flow rate of 20 m³/min because the peak flow rate provides adequate fluidization for the media in this example. Total hydraulic load should be between 880 and 1470 m³/m²/d.

6. Calculate methanol required:

   Calculate nitrate removed: 311 kg/d;

   Assume 3 kg methanol/kg N removed;

   Methanol = (311 kg/d) (3 kg methanol/kg N removed) = 933 kg/d;

   (933 kg/d) (1 L/0.79 kg) = 1181 L/d;

   Methanol dose = 50 L/h.

   Alternatively, methanol requirements can be computed from the following equation: Methanol = 2.47 $(NO_3\text{-N})$ + 1.53 $(NO_2\text{-N})$ + 0.87 dissolved oxygen.

   If the influent dissolved oxygen to denitrification is assumed to be 3 mg/L, then the methanol requirement (neglecting residual methanol) is:

   $(311 \text{ kg})(2.47) + (0.87)(3)(18\,930 \text{ m}^3/\text{d})/1000 = 818 \text{ kg (1800 lb)}$

7. Calculate biomass produced:

   From suspended growth systems, use 0.18 kg VSS/kg COD removed;

   COD removed = (933 kg methanol/d)(1.5 kg COD/kg methanol) = 1400 kg/d;

   VSS = (0.18 kg VSS/kg COD)(1400 kg COD/d) = 252 kg VSS produced/d.

   At 75% volatile:

   TSS produced = 252/0.75 = 336 kg TSS/d = 1.08 kg TSS/kg NO_3^{2-}-N;

   Typical biomass production is 0.4 to 0.8 kg TSS/kg NO_x-N.

   Calculate excess biomass flow rate (assume 1% solids):

   Flow = (336 kg TSS/d)(1/0.01)(1 L/kg)(1/1440 min/d) = 23.3 L/min

8. Calculate horsepower (hp) of pumps:

   Fluidization pump: total pump capacity is 19 720 L/min (peak week flow);

   With two pumps: 19 720/2 = 9860 L/min.

A typical fluidized-bed configuration requires a fluidization pump with approximately 12.2 m total dynamic head (TDH). This should be verified for the actual reactor configuration.

$$kW = [(Q)(r)(a)(\text{TDH})]/[(\text{Pump efficiency})(\text{Motor efficiency})]$$

where $\quad$ Q = flow (m³/s),
$\quad$ r = density of water,
$\quad$ a = acceleration of gravity,
$\quad$ TDH = 12.2 m (assumed), and
$\quad$ Pump efficiency = 75% (assumed).
$\quad$ (Note that hp = kw/0.746.)
$\quad$ kW = [(9860 L/min)/(1000 L/m³)/(60 s/min)](1000 kg/m³) (9.81 m/s²)
$\quad\quad$ (12.2 m)]/[0.75 × 0.9] = 29 kW
$\quad$ (hp = 29 kW/0.746 = 39 hp)

Calculate horsepower of growth control pump:
Base flow rate on biomass flowrate of 11.5 L/min for each pump.
kW = [(11.5 L/min)/(1000 L/m³)/(60 s/min))(1000 kg/m³)(9.81 m/s²)(12.2 m)]/
$\quad$ [0.75 × 0.9] = 0.03 kW
(hp = 0.03 kW/0.746 = 0.04 hp)

5.8 Performance of Fluidized Bed Biofilm Reactor Fauna

In a study of tertiary nitrification of activated sludge settled effluent using a pilot-scale EBBR, Dempsey et al. (2006) found that the process also removed up to 56% CBOD and 62% TSS from the influent. Removal of these materials was attributed to the activities of protozoa (free-living and stalked) and metazoa (rotifers, nematodes, and oligochaetes), as shown in Figure 11.51. Soluble organic matter probably was metabolized by heterotrophic bacteria, which were in turn consumed by protozoa and rotifers. Activated sludge flocs carried over from the preceding process and sloughed biofilm from the EBBR were probably consumed primarily by the worms. In this way, approximately 90% of the incoming organic matter (soluble or suspended) was mineralized to carbon-dioxide, water, and ammonia, thus increasing the actual rate of nitrification above the apparent

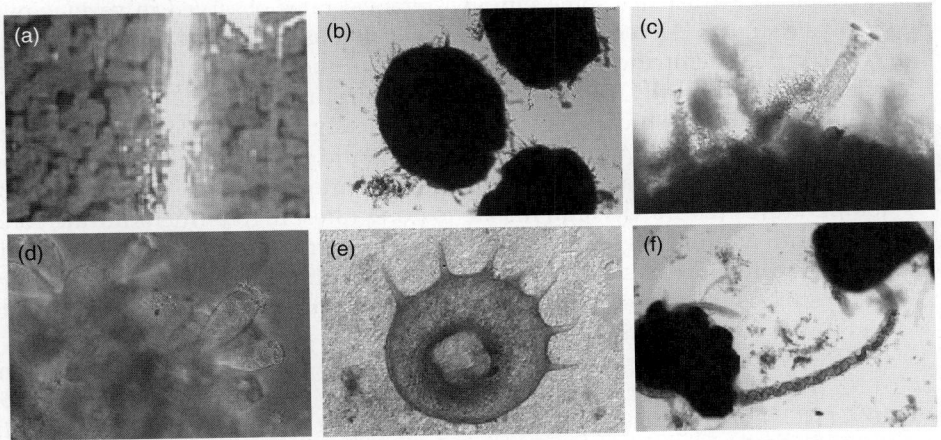

FIGURE 11.51 Particulate biofilms with associated protozoa and metazoan from expanded bed: (a) bioparticles in expanded bed; (b) bioparticles with surface attached; (c) close-up of rotifer attached to bioparticle; (d) stalked protozoa on surface of particulate biofilms; (e) testate amoeba grazing on biofilm; and (f) oligochaete worm grazing on bioparticles (Dempsey et al., 2006).

one, which is measured using the difference between inlet and outlet ammonia-nitrogen concentrations. Madoni (1994) provides an excellent explanation of these phenomena and their relation to similar processes in natural waterbodies.

Bergtold et al. (2007) reported destruction by bactivorous nematode worms; Ratsak and Verkuijen (2006), Liang et al. (2006), and Huang et al. (2007) observed it with oligochaete worms.

5.9   Process Performance

5.9.1   Nitrogen Removal Rate

Nitrogen removal rate is the single most important parameter quantifying denitrification FBBR performance. This parameter is useful because it accounts for differences in bed volume and nitrogen loading rate between various FBBR systems. The nitrogen removal rate depends on biomass concentration, biomass surface area, temperature, carbon source, bulk nitrate concentration, nutrient availability, and hydrodynamic and physical conditions.

Green et al. (1994) reported that nitrate removal efficiency always was greater than 97% with nitrate concentrations in the effluent less than 3 mg/L and nitrite concentrations less than 1 mg/L when the nitrate loading rate was 10.8 kg NO_3^--N/m^3/d and the retention time was 3 minutes. These authors also showed that the FBBR can operate efficiently at retention times shorter than 3 minutes and at nitrate loading rates higher than 15.8 kg NO_3^--N/m^3/d. In fact, when this FBBR was operated at a loading rate of 21.7 kg NO_3^--N/m^3/d and a hydraulic retention time of 1.5 minutes, it removed nearly 100% of the influent NO_3^--N. However, this system required careful control of the biofilm thickness to achieve reliable performance. The maximum nitrate removal rate achieved by the pilot-scale FBBR was 30.5 kg NO_3^--N/m^3/d. Although based on groundwater treatment, the results of this study are likely to be applicable to secondary effluent if the organic matter (CBOD) content is not too high.

Table 11.24 displays some nitrogen removal rates observed in various pilot- and full-scale studies. All removal rates are calculated using the unexpanded bed volume.

Observed denitrification rate in an FBBR is likely to be mass-transfer limited for nitrate levels typically found in nitrified municipal wastewaters (Shieh and Keenan, 1986). Some FBBR studies have cited empirically derived N removal rates as high as 15 to 30 kg NO_3^--N/m^3/d. However, it should be noted that some of these studies are not representative of full-scale operations as they operated at optimal temperatures (20–30°C), high nitrate loading rates, high effluent nitrate concentrations and nonlimiting orthophosphate concentrations ($P > 1$ mg/L). Nitrogen removal rates in full-scale facilities are substantially less than those observed in these studies due to lower temperatures, lower bulk liquid nitrate concentration, and lower nitrate loading rates.

The denitrification rate at the Rancho California WRRF was 3.5 kg NO_3^--N/m^3/d at 20°C (MacDonald, 1990). The denitrification rate in the Himmerfjarden FBBR was 1.7 kg NO_3^--N/m^3/d. The facility received an influent of 18 mg/L NO_3^--N at temperatures as low as 10°C and produced an average effluent concentration of 1.9 mg/L NO_3^--N.

5.9.2   Temperature

At temperatures between 15 and 25°C, the nitrogen removal rate doubles with each 5°C increase in temperature (Figure 11.50). Shieh and Keenan (1986) reported that the

Scale	Temperature (°C)	S_{in} (mg N/L)	N Removal Rate (g N/g VSS · d)	N Removal Rate (kg N/m³ · d)	Reference
Pilot	18–23	5–100	N/A	5.4–20.7	Jeris and Owens, 1975
Pilot	N/A	6.6–30	0.033–0.243	0.69–3.28	Hermanwicz and Cheng, 1990
Pilot	30	15–300	0.141–2.575	3.23–18.7	Hirata and Meutia, 1996
Pilot	N/A	676–1500	N/A	11.8–17.7	Chen et al., 1996
Pilot	23	1000	0.41	12	Rabah and Dahab, 2004b
Pilot	N/A	N/A	N/A	5.3–8.6	Coelhoso et al., 1992
Full	10–20	18	N/A	1.7	Bosander and Westlund, 2000
Full	20	20	0.1	3.5	MacDonald, 1990

*N/A = not available.

TABLE 11.24   Nitrogen Removal Rates Reported in Literature

optimal temperature for development of denitrifying biofilms lies between 20 and 30°C. Most pilot-scale studies have been performed at optimal temperatures. Coelhoso et al. (1992) ran experiments at 26°C and Rabah and Dahab (2004b) operated their pilot-plant at temperatures of 21 to 25°C. However, Bosander and Westlund (2000) reported consistent denitrification of wastewater at 10 to 20°C.

It is likely, from the standpoint of microbial ecology, that operation within any particular temperature range will select for microorganisms with temperature optima in that range (e.g., psychrophiles have optimum temperatures less 15°C). Therefore, pilot-plant studies under laboratory conditions may be misleading and should be conducted at WRRFs to ensure design criteria are obtained at typical process temperatures.

6.0   Rotating Biological Contactors

6.1   Introduction

RBC design criteria presented in presented in detail in WEF (2010) (see Biofilm Reactors; WEF MOP No. 35). The discussion of the RBC is limited to a brief introduction with few important design considerations in this section. As a secondary treatment process, RBC has been applied where average effluent water-quality standards are less than or equal to 30-mg/L BOD$_5$ and TSS. When the RBC is used in conjunction with effluent filtration, the process is capable of meeting more stringent effluent water-quality limits of 10-mg/L BOD$_5$ and TSS. Nitrification RBCs can produce effluent having less than 1-mg/L ammonia-nitrogen remaining in the effluent stream. The RBC employs a cylindrical, synthetic media bundle that is mounted on a horizontal shaft. Figure 11.52 illustrates the shaft-mounted media. The media is partially submerged (typically 40%) and slowly (1–1.6 rpm) rotates to expose the biofilm to substrate in the bulk of the liquid (when submerged), and to air (when not submerged). Detached biofilm fragments suspended in the RBC effluent stream are removed by solids separation units. The RBC process

FIGURE 11.52   Photograph of rotating biological contactor cylindrical synthetic media bundle mounted on a horizontal shaft (left) and rotating biological contactor covers (right).

typically is configured with several stages operating in series. Each reactor-in-series may have one or more shafts. Parallel trains are implemented to provide additional surface area for biofilm development.

Media-supporting shafts typically are rotated by mechanical-drives. Diffused air-drive systems and an array of air-entraining cups that are fixed to the periphery of the media (to capture diffused air) have been used to rotate the shafts. The RBC process has the following advantages: operational simplicity, low energy costs, and rapid recovery from shock loadings. The literature has documented several examples of RBC failure resulting from shaft, media, or media support system structural failure; poor treatment performance; accumulation of nuisance macro fauna; poor biofilm thickness control; inadequate performance of air-drive systems for shaft rotation. State-of-the-art biofilm reactors such as the MBBR and BAF can provide equivalent or improved effluent water quality with reduced susceptibility to macrofauna infestation and reduced physical footprint.

7.0   Trickling Filters

During the first five decades of use, trickling filters included soil and rock biofilm carriers and their design was scattered and empirical in nature. During the 1950s and 1960s, Dow Chemical Co. (Midland, Michigan) began early experimentation with modular plastic packing media (Bryan, 1955; Bryan and Moeller, 1960). Other studies during that time that resulted in development of accepted design protocol (Howland, 1958; Schulze, 1960; Eckenfelder, 1961, 1963; Atkinson et al., 1963; Galler and Gotaas, 1964; Germain, 1966). In the early 1970s, when U.S. EPA issued its definition of secondary treatment standards, the trickling-filter was regarded as being unable to produce water quality that consistently met published standards (Parker, 1999). This was partly because of poor secondary sedimentation tank design. In 1979 in Corvallis, Oregon, Norris and co-workers (1980, 1982) followed a rock-media trickling filter with a small aeration basin and a flocculator clarifier. The researchers demonstrated that WRRF effluent water quality could be significantly improved by bioflocculation in the solids contact basin and improved clarification. The researchers referred to the combined units as the trickling filter/solids contact process. Chapter 14 described combined suspended-growth and biofilm bioreactors. The German definition of instantaneous hydraulic application rate, or Spülkraft (SK), was described by ATV-DVWK in 1983. Albertson (1995) advanced the use of Spülkraft with a timing mechanism that adjusted rotary-distributor speed with an electric drive.

7.1   General Description

The trickling filter is a three-phase system with fixed biofilm carriers. Wastewater enters the bioreactor through a distribution system and trickles down over the biofilm surface and air moves upward or downward in the third phase. Biofilm develops on biofilm carriers. Trickling filter components typically include a distribution system, containment structure, rock or plastic biofilm carrier, underdrain and ventilation system. Figure 11.53 illustrates a modern trickling filter cross-section. Trickling filters that are treating wastewater produce TSS, which means that liquid-solids separation is required. This is achieved with either circular or rectangular secondary sedimentation basins. The secondary segment of a trickling filter process typically includes an influent pumping station, trickling filter, trickling filter recirculation pumping station, and liquid-solids separation unit process.

7.1.1   Distribution System

Primary effluent (fine-screened and degritted wastewater) is either pumped or flows by gravity to a trickling filter distribution system. The distribution system uniformly distributes wastewater over the trickling filter biofilm carriers in intermittent doses. The distributors may be hydraulically or electrically driven. Intermittent application allows for resting, or aeration, periods. Efficient influent wastewater distribution results in proper media wetting. Poor media wetting may lead to dry media pockets, ineffective treatment zones, and odor. There are two types of systems: fixed-nozzle and rotary

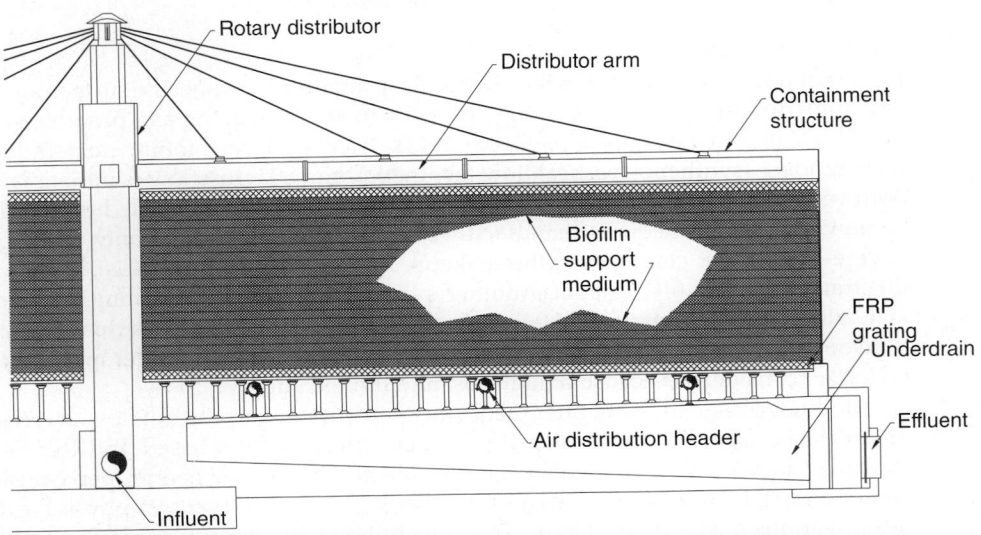

Figure 11.53   Typical trickling filter components and cross section (Daigger and Boltz, 2010).

FIGURE 11.54   Hydraulically driven rotary distributors use variable frequency drive-controlled gates that either open or close distributor orifices that adjust with varying pumped flows to maintain a constant preset rotational speed (left). On the right is an electrically driven rotary distributor (Daigger and Boltz, 2010) (courtesy of WesTech Engineering, Inc.).

distributors. Because their efficiency is poor, distribution with fixed nozzles should not be used (Harrison and Timpany, 1988).

Hydraulic rotary distributors use retardant back spray orifices to slow rotational speed, while maintaining desired pump flow rate. Figure 11.54 illustrates both a modern, hydraulically driven rotary distributor that uses gates that either opens or closes distributor orifices to adjust rotational speed and an electrically driven rotary distributor. The use of a variable frequency drive allows for more precise control of distributor-arm rotation. Electrically driven rotary distributors have motorized drive units that control distributor speed independent of the wastewater flow.

7.1.2   Biofilm Carriers

Ideal trickling filter biofilm carriers, or media, provide a high specific surface area, low cost, high durability, and high enough porosity to avoid clogging and promote ventilation (Metcalf and Eddy, Inc./AECOM, 2013). Trickling filter biofilm carriers include rock, random (synthetic), vertical flow (synthetic), and 60° crossflow (synthetic) media. Both vertical-flow and crossflow media are constructed with corrugated plastic sheets. Some vertical-flow media is manufactured with corrugated sheets only while others have e-other sheet corrugated (the makeup are smooth plastic panels). Figure 11.55 illustrates trickling filter media. Another synthetic media that is commercially available, although not typically installed, is vertically hanging plastic strips. Horizontal redwood or treated wooden slats also have been used as trickling filter media, but are no longer considered because of high cost and limited supply.

Modular plastic trickling filter media (i.e., self-supporting vertical flow or crossflow modules) is used almost exclusively for new trickling-filter-based WRRFs. Several trickling filters using rock-media exist provide good service when properly designed and operated. Figure 11.56 compares filter media; Table 11.25 presents physical properties of various media types. Figure 11.57 illustrates self-supporting crossflow modular plastic trickling filter media. Synthetic media allows for higher hydraulic loadings and

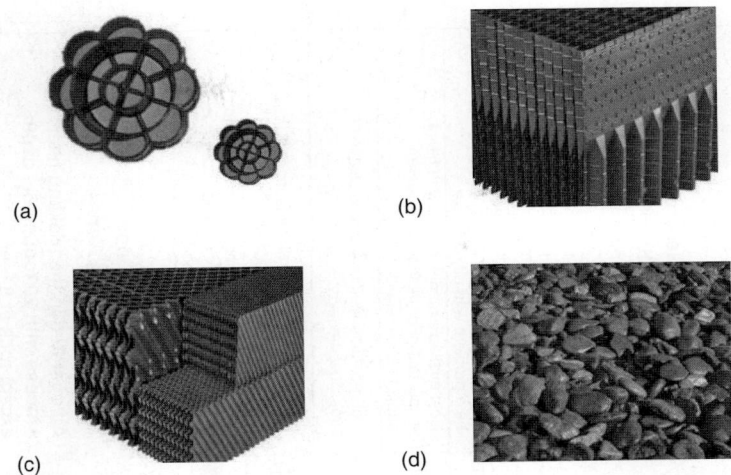

(a)

(b)

(c)

(d)

FIGURE 11.55   Trickling filter media. (a) Random, (b) vertical, (c) crossflow, and (d) rock.

enhanced oxygen transfer compared to rock-media because of the higher specific surface area and void space. Rock media has, ideally, a 50-mm diameter, but may range in size. Rounded rock trickling filter media helps mitigate issues associated with rigid, rock (slag) media. The slag-type rock contains numerous crevices that can retain water and accumulate biomass. Because of structural requirements associated with the large unit weight of the rock media, the trickling filters are shallow in comparison to synthetic media towers and are susceptible to excessive cooling. The water retained inside crevices in the slag-type rock media may then expand and sever rock fragments. This can result in the production and accumulation of fine material. The accumulation of both fine material and retained biomass is a primary contributor to rock-media trickling filter plugging (Grady et al., 1999). Typically, rock media has low specific surface area, void space, and high unit weight. Although recirculation is common, the low void ratio in rock-media trickling filters limits hydraulic application rates. Excessive hydraulic application can result in ponding, which results in limited oxygen transfer and poor bioreactor performance. Existing rock-media trickling filters may sometimes be improved by providing forced ventilation, solids contact channels, or deepened secondary clarifiers that include energy dissipating inlets (EDIs) and flocculator-type feed wells. Replacement or deepening of the trickling filter using plastic media often is required if rock media quality is poor, space is limited, and WRRF expansion is expected. A well-designed and operated rock-media trickling filter can provide high-quality effluent. Grady et al. (1999) suggest that for organic loads of less than 1 kg $BOD_5/d/m^3$, rock- and synthetic-media trickling filters are capable of equivalent performance. However, as organic load increases synthetic media will result in fewer nuisance problems and reduced plugging.

Synthetic trickling filter media has a high specific surface area and void space, and low unit weight. Because of the reduced unit weight, synthetic-media trickling filters can be constructed at depths more than three times that for a comparably sized rock-media trickling filter. Modular plastic media is manufactured with the following specific surface areas: 223 m^2/m^3 (68 sq ft/cu ft) as high-density; 138 m^2/m^3 (42 sq ft/cu ft)

Relative Difference in Trickling Filter Media (A) (B)	Effluent Quality – Low Load: Soluble Bod	Low Load: TSS	Low Load: Total Bod	Low Load: Nitrification	High Load: Soluble BOD	High Load: TSS	High Load: Total BOD	Operating characteristics: Hydraulic Residence Time	Oxygen Transfer	Wetting	Filter Flies	Biogrowth	Plugging	Qualitative Factors: Ease of Installation	Air ventilation	Media Strength	Tower Construction	Media Height	Media Surface Area	Media Void Space	Weight of Media	General Recommendations
Horizontal	●	■	●	●	●	■	●	○	○	+	○	■	⊕	○	+	○	○	○	●	⊕	○	Surface area low, not recommended for trickling filter use. Best used with sludge recycle to optimize media.
Vertical	⊕	○	○	⊕	+	○	+	■	○	○	+	+	+	⊕	●	○	○	⊕	⊕	⊕	⊕	Optimum use is at high loading to the filter especially where plugging might be a problem with other media.
60° Crossflow	+	+	+	+	⊕	+	⊕	○	○	○	⊕	○	○	⊕	○	⊕	○	⊕	⊕	⊕	⊕	Best available media for use with filters at low to moderate loading or where nitrification is required.
45° Crossflow	○	○	○	○	⊕	○	○	○	○	●	○	○	●	+	○	⊕	○	⊕	⊕	⊕	⊕	Not recommended in lieu of available 60° crossflow media.
Random	⊕	○	○	○	○	●	○	⊕	○	■	○	●	○	○	○	●	●	○	⊕	⊕	⊕	Best used with rotary distributors and at low to moderate loading.
Rock	○	○	○	■	■	○	■	+	■	⊕	■	○	■	■	■	+	■	■	■	■	■	Replacement of existing rock media, with high rate types usually not justified. Best used at low to moderate loading.
Comments	Little significant difference	Laminar flow increases bioflocculation	Little significant difference	Crossflow gave best results	Vertical & crossflow near equal	Laminar flow increases bioflocculation	Differences increase at high loading	Geometrical complexity increases time	May not control biokinetics	Poor wetting, results in nuisances	Prevalent w/poor wetting	Poor distribution in some media	Blinding or solids bridging	Significant placement differences	Horizontal gives best cross ventilation	Verification by testing encouraged	Self-supporting media simplest	Weight or media compaction problems	Biofilm thickness also important	High rate type best	Rock significantly higher	

Notes:

(A) + Best ⊕ Good ○ Average ● Poor ■ Worst

(B) **All operated in the trickling filter process with 20 foot media depth except rock at 8 feet**

Important: Costs must be considered on a site specific basis

FIGURE 11.56 Relative comparison of trickling filter media (CH2M HILL, 1984) (courtesy of CH2M).

Media Type	Material	Nominal Size, m (feet)	Bulk Density, kg/m^3 (lb/cu ft)	Specific Surface Area m^2/m^3 (sq ft/cu ft)	Void space (%)
Rock (river)		0.024–0.076 (0.08–0.25)	1442 (90)	62.3 (19)	50
Rock (slag)		0.076–0.128 (0.25–0.42)	1601 (100)	45.9 (14)	60
Corrugated plastic modules[a]					
60° Crossflow	PVC	0.61 × 0.61 × 1.22 (2 × 2 × 4)	24.0–44.9 (1.5–2.8)	100 and 223.1 (30, 48, and 68)	95
Vertical flow	PVC	0.61 × 0.61 × 1.22 (2 × 2 × 4)	24.0–44.9 (1.5–2.8)	101.7 and 131.2 (31 and 40)	95
Random pack[b]	PP	0.185 ø × 0.051 H 7.3" ø × 2" H	27.2 (1.7)	98.4 (30)	95

[a]Manufacturers of corrugated plastic modules are (formerly) BF Goodrich, American Surf-Pac, NSW, Munters, (currently) Brentwood Industries, Jaeger, and Marley (SPX Cooling).
[b]Manufactures of random media are (formerly) NSW Corp. and (currently) Jaeger.
[c]Manufacturers of plastic strips are (formerly) NSW corp. and (currently) Jaeger.

TABLE 11.25   Physical Properties of Commonly Used Trickling Filter Media (lb/cu ft × 16.02 = kg/m³ᵇ sq ft/cu ft × 3.281 = m^2/m^3)

as medium-density; 100 m^2/m^3 (30 sq ft/cu ft) as low-density. Both vertical and cross-flow media are reported to effectively remove BOD_5 and TSS (Harrison and Daigger, 1987; Aryan and Johnson, 1987). Research shows, however, that different synthetic media provide different treatment efficiencies despite being manufactured with virtually identical specific surface areas. The designer should carefully consider effects of media type and configuration on trickling filter effluent water quality.

Plastic modules with a specific surface area of 89 to 102 m^2/m^3 are well suited for carbon oxidation and combined carbon oxidation and nitrification. Parker et al. (1989) recommended a medium-density crossflow media and were against the use of high-density crossflow media in tertiary nitrification applications. This argument is supported by the pilot application data and conclusions of Gujer and Boller (1983, 1986) and Boller and Gujer (1986), which show lower nitrification rates for lower-density media. Researchers claim that lower rates occur with high-density media because of the development of dry spots below the interruption points in the media. Higher density media has more interruptions and, therefore, is wet less effectively. Using medium-density media will reduce plugging. Vertically oriented modular plastic media is suited for high-strength wastewater (perhaps industrial) or high organic loadings such as with a roughing filter. Other advantages of vertical flow include more effective biomass flushing and less complicated geometry, which enhances air movement. In some cases, crossflow media has been placed in the top layer to enhance wastewater distribution.

FIGURE 11.57   Self-supporting crossflow modular plastic trickling filter media (courtesy of Brentwood Industries, Inc., Reading, PA).

7.1.3   Containment Structure

Rock and random plastic media are not self-supporting when stacked and require structural support to contain the media within the bioreactor. Containment structures typically are precast or panel-type concrete tanks. When self-supporting media such as plastic modules is used, other materials such as wood, fiberglass, and coated steel are used as containment structures. The containment structure avoids splashing and provides media support, wind protection, and flood containment. Trickling filters are well known for the nuisance macrofauna such as filter flies and snails. A properly designed containment structure increases operator flexibility and allows control of nuisance macrofauna. It can include a variety dosing alternatives and possibly a flooding the filter.

7.1.4   Underdrain System and Ventilation

The trickling filter underdrain system is designed to meet two objectives: collect treated wastewater for conveyance to downstream unit processes and create a plenum that allows for the transfer of air throughout the media (Grady et al., 1999). Clay or concrete underdrain blocks typically are used for rock media because of the required structural support. A variety of support systems, including concrete piers and reinforced fiberglass grating, are used for other media types. Figure 11.58 illustrates field-adjustable plastic stanchions and fiber-glass-reinforced plastic grating on the concrete floor of a trickling filter containment structure. The volume created between concrete and media bottom creates the underdrain.

FIGURE 11.58 Field-adjustable plastic stanchions and fiber-glass-reinforced plastic grating on the concrete floor of a trickling filter containment structure. The volume created between concrete and media bottom creates the underdrain (courtesy of Brentwood Industries, Inc., Reading, PA).

The vertical flow of air through the media can be induced by either mechanical means (forced or fan ventilation) or natural air draft. Natural air ventilation results from a difference in ambient air temperature outside and inside the trickling filter. The temperature causes air to expand when warmed and contract when cooled. The net result is an air-density gradient throughout the trickling filter. Depending on the differential condition, an air front either rises or sinks, which results in continuous airflow through the bioreactor. Natural ventilation may become unreliable or inadequate in meeting process air requirements when neutral temperature gradients do not produce air movement. Such conditions may be daily or seasonal and can lead to odorous anaerobic biofilm conditions and poor performance. Therefore, mechanical forced air ventilation typically is included. Forced-air ventilation is accomplished by adding low-pressure fans to circulate air continuously. When using ventilation, the designer should ensure that the air is uniformly distributed to provide oxygen to all biofilm in the reactor.

7.1.5 Trickling Filter Pumping Station

The pumping station lifts primary effluent and recirculates unsettled trickling filter effluent (also known as underflow) to the influent stream. Typically, trickling filter underflow is recirculated to the distribution system to achieve the hydraulic load required for proper media wetting and biofilm thickness control. The intent of recirculating bioreactor effluent is to decouple hydraulic and organic loading. Although effluent from the secondary clarifier can be recirculated, this not common practice because it may lead to the hydraulic overloading of secondary clarifiers. Influent pumping typically is selected to allow underflow to flow by gravity to the suspended growth reactor (or solids contact basin), secondary clarifier, or other downstream process. Submersible or vertical turbine pumps are used. Weir positioning in the wet well typically allows for one pumping station.

7.1.6   Hydraulic and Pollutant Loading

Trickling filters are classified by the intended mode of pollutant degradation and loading, including carbon oxidation, combined carbon oxidation and nitrification, or nitrification. Organic loading is expressed as kg/d/m³ of filter media as BOD_5 or COD. General practice is to ignore the organic load imparted by recirculation streams, but the designer should account for the effects of recycle flow and pollutant loading (specifically ammonia-nitrogen) on treatment efficiency. The total organic load (TOL) may be calculated using:

$$TOL = \left(\frac{BOD_5 \text{ applied}}{\text{media volume}} \right) = \left(\frac{Q_{in} \cdot S_I}{V_M} \right) \cdot \frac{1 \text{ kg}}{10^3 \text{ g}} \qquad (11.16)$$

where V_M = trickling filter media volume, m³;
Q_{in} = influent to the trickling filter (typically primary effluent), m³/d; and
S_I = influent BOD concentration, g/m³.

Nitrifying trickling filter data is expressed in terms of surface-based ammonia-nitrogen loading rate. Trickling filter hydraulic loading rate is calculated without and with recirculation. The wastewater hydraulic load (WHL) excludes recirculation and can be calculated using the following equation (m³/d/m²).

$$WHL = \frac{Q_{in}}{A} \qquad (11.17)$$

where A = biofilm area in reactor (m²); and
Q_I = influent flowing from upstream unit processes.

The THL is used to gauge media wetting and biofilm thickness control, and considers trickling filter influent flowing from upstream unit processes, Q_I, and the recirculation stream, Q_R. The total hydraulic load can be expressed mathematically by eq 11.18

$$THL = \frac{Q_I + Q_R}{A} \qquad (11.18)$$

where Q_R = recirculation stream.

Current practice, and the standard for this section, is to reference hydraulic loading in units of cubic meter per day per square meter of plan trickling filter area (m³/m²/d) rather than referencing the required hydraulic application per unit biofilm growth area.

7.2   Process Flow Sheets and Bioreactor Configuration

7.2.1   Process Flow Diagrams

The trickling filter process typically consists of preliminary treatment (including screening and grit removal), primary clarification, bioreactor, secondary clarification, and disinfection unit processes. Recirculation methods influence the process flow. There are two types of recirculation. The first allows for direct recirculation to the trickling filter, and the second passes flow through a primary clarifier before entering the trickling filter. Four trickling filter process flow diagrams, including both single- and two-stage trickling filters, are shown schematically in Figure 11.59. Recirculation dilutes influent wastewater and dampens organic variability in the influent because of diurnal fluctuations. Clarifying trickling effluent will enhance performance of a subsequent trickling filter in two-stage operation. The designer must ensure that the recirculation flow required for wetting and biofilm thickness control does not exceed the limiting hydraulic loading rate for the sedimentation tank. When two trickling filters are operated in series (rather than parallel) several studies have shown that the second stage trickling

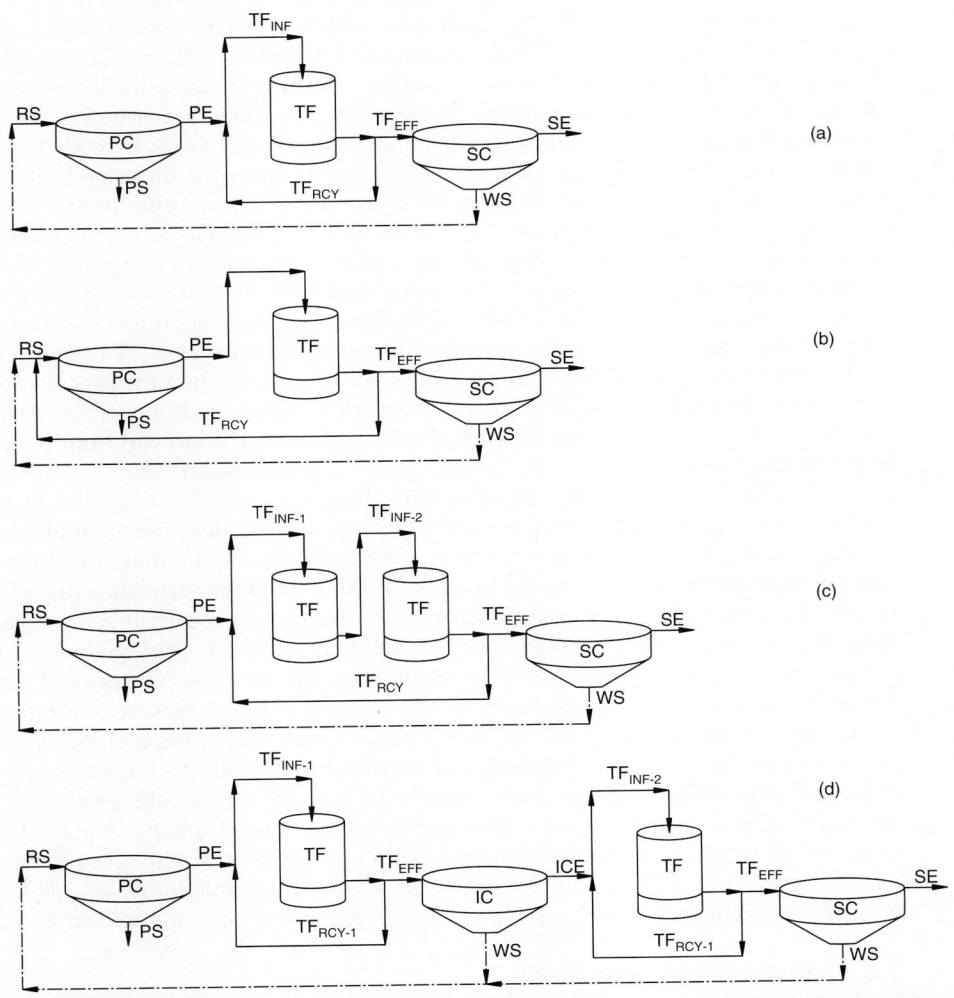

FIGURE 11.59 Typical flow diagrams for the trickling filter process: (a and b) Single-stage trickling filter process; (c) two-stage trickling filter process; (d) two-stage trickling filter process with intermediate clarification (RS = raw wastewater; PC = primary clarifier; PS = primary sludge; PE = primary effluent; TF$_{INF}$ = trickling filter influent; TF = trickling filter; TF$_{EFF}$ = trickling filter effluent; TF$_{RCY}$ = trickling filter recycle; SC = secondary clarifier; WS = waste sludge; SE = secondary effluent; IC = intermediate clarifier; and ICE = intermediate clarifier effluent) (Daigger and Boltz, 2010).

filter performance is not adversely affected by the absence of a clarifier between first and second stage units. However, there are indications that certain wastewaters containing high concentrations of soluble BOD_5 (likely from an industrial source) will result in excessive biofilm growth and subsequent excessive biomass production. These solids can adversely affect the second-stage trickling filter if not removed by intermediate settling. The design of settling tanks in two-stage trickling filter systems also is affected by the recirculation pattern. The designer should consider the most economical method of securing acceptable effluent quality.

The practice of alternating the lead trickling filter in a primary-secondary trickling system is referred to as an alternating double-filtration (ADF) system. This concept is beneficial when applied to modular plastic-media trickling filters. Gujer and Boller (1986) and Parker et al. (1989) observed patchy biofilm growth in the lower section of pilot-scale nitrifying trickling filters (NTFs). The researchers attributed the patchy growth to dry spots. Boller and Gujer (1986) advocated use of the ADF mode to enhance performance of the NTF. Aspegren et al. (1992) also observed improved nitrification using the ADF mode vs a single-stage NTF because of virtual elimination of biofilm patchiness. Use of the ADF approach with two trickling filters in series encourages full-depth biofilm development in both trickling filters. The lead trickling filter should be switched every three to seven days to ensure that both contain a steady-state biofilm developed along the entire bioreactor length. Operating trickling filters in series (rather than parallel) may result in added nitrification without requiring ADF operation.

Sludge handling also affects the trickling filter process. Each of the process flow diagrams illustrated in Figure 11.59 implies that waste biological solids are removed by cosettling the biological sludge with the primary sludge before withdrawal from the system. Many facilities exist that separately handle primary and secondary sludge. For example, primary sludge may be thickened by gravity thickeners and trickling filter humus by gravity belt thickeners. The benefits must be evaluated by the designer. In principle, however, sludge cosettling is sound practice. It does, however, require that operators consistently withdraw the solids from the process, and that designers provide equipment and means to maintain a near zero sludge blanket if necessary. A common operational issue that arises from improper maintenance of a solids inventory is "rising sludge." In any trickling filter application that results in nitrification, the produced nitrate (NO_3^-) may be further reduced to nitrogen gas (N_2) in an anoxic sludge-blanket macroenvironment. The N_2 (g) can become entrained in the sludge blanket and float clumps of biomass to the sedimentation basin surface. This biomass may float over weirs and degrade secondary effluent water quality. Improper maintenance of a primary clarifier sludge blanket is also a consideration. When combined with waste biological sludge, $sbBOD_5$ that exists in primary sludge may generate odor. The mechanism for odor control strategies are discussed later. Alternatively, $sbBOD_5$ may hydrolyze and reenter the bulk liquid as $rbBOD_5$. This can result in an increased trickling filter TOL and diminish bioreactor performance.

7.2.2   Bioreactor Classification

Trickling filters are often categorized as roughing, super-high rate, high-rate, and low-rate. Within this section, trickling filters are discussed based on mode of operation or process application: (1) roughing, (2) carbon oxidation, (3) carbon oxidation and nitrification, and (4) nitrification (Daigger and Boltz, 2010). Table 11.26 summarizes typically accepted defining criteria for each operational mode. Roughing filters receive high-hydraulic and high-organic loadings and require the use of vertical-flow media to minimize plugging. Although they may provide a high-quantity organic load removal per

Design Parameter	Roughing	Carbon Oxidizing (cBOD)	cBOD and Nitrification	Nitrifying
Media used	VF	RA, RO, XF, or VF	RA, RO, or XF	RA or XF
Wastewater source	Primary effluent	Primary effluent	Primary effluent	Secondary effluent
Hydraulic loading, $\frac{m^3}{d \cdot m^2}$ (gpm/sq ft)	52.8–178.2 (0.9–2.9)	14.7–88.0 (0.25*–1.5)	14.7–88.0 (0.25*–1.5)	35.2–88.0 (0.6–1.5)
Contaminant loading, $\frac{kg}{m^3 \cdot d}$ (lb BOD$_5$/d/1000 cu ft)	1.6–3.52 (100–220)	0.32–0.96 (20–60)	0.08–0.24 (5–15)	N/A
$\frac{g}{m^2 \cdot d}$ (lb NH$_3$-N/d/1000 sq ft)	N/A	N/A	0.2–1.0 (0.04–0.2)	0.5–2.4 (0.1–0.5)
Effluent quality, mg/L (unless noted)	50%–75% BOD$_5$ conversion	15–30 BOD$_5$ and TSS	<10 BOD$_5$ <3 NH$_3$-N	0.5–3 NH$_3$-N
Predation	No appreciable growth	Beneficial	Detrimental (nitrifying biofilm)	Detrimental
Filter flies	No appreciable growth	No appreciable growth	No appreciable growth	No appreciable growth
Depth, m (ft)	0.91–6.10 (3–20)	≤12.2 (40)	≤12.2 (40)	≤12.2 (40)

Notes: VF = vertical flow; RA = random pack; XF = crossflow; RO = rock; and TSS = total suspended solids.

*Applicable–shallow trickling filters; gpm/ft^2 = gallons per minute per square foot of trickling filter plan area; gpm/sq ft × 58.674 = m^3/m^2·d (cubic meter per day per square meter of TF plan area); lb BOD$_5$/d/1000 cu ft × 0.016 0 = kg/m^3·d (kilograms per day per cubic meter of media); and lb NH$_3$-N/d/1000 sq ft × 4.88 = g/m^2·d (grams per day per square meter of media).

TABLE 11.26 Trickling Filter Classification (Daigger and Boltz, 2010)

unit volume, their settled effluent still contains substantial BOD$_5$. Roughing filters provide approximately 50% to 75% soluble BOD$_5$ conversion and may receive total loadings of 1.5 to 3.5 kg BOD$_5$/d/m^3. Carbon-oxidizing trickling filters provide settled effluent of 15 to 30 mg/L for both BOD$_5$ and TSS, and may receive BOD$_5$ loadings of 0.7 to 1.5 kg/d/m^3. Combined carbon oxidation and nitrification trickling filters provide effluent BOD$_5$ less than 10 mg/L and NH$_3$-N less than or equal to 0.5 to 3 mg/L (after solids separation). These trickling filters may receive BOD$_5$ loadings less than 0.2 kg/d/m^3, and TKN loadings of 0.2 to 1.0 g/d/m^2. Tertiary nitrifying trickling filters provide 0.5 to 3 mg/L effluent NH$_3$-N when receiving a clarified secondary effluent and NH$_3$-N loadings of 0.5 to 2.5 g N/d/m^2.

7.2.3 Hydraulics

Recirculation, distributor operation, biofilm thickness, and macro fauna accumulation affect wetting of trickling filter media. Albertson and Eckenfelder (1984) postulated that the active biofilm surface area in a trickling filter is dependent upon biofilm thickness and media configuration, and that increased biofilm thickness reduces active surface area. The net result of excess biomass accumulation is reduced trickling-filter performance. The researchers stated that for medium-density crossflow media with 98 m^2/m^3 specific surface area, a 4-mm biofilm thickness would cause a 12% reduction of active biofilm area, assuming that the entire media has been appropriately wetted.

Poor trickling filter media wetting results in reduced effluent water quality. Crine et al. (1990) found that the wetted area-to-specific surface area ratio of trickling filters ranged from 20% to 60%. The lowest values for wetting occurred with high-density random-pack media. Many of the newer hydraulically based design formulations incorporate a term that allows for specific surface-area reduction because of distributor inefficiency in media wetting. The interrelationship of liquid residence time, dosing, and media configuration on BOD_5 removal kinetics has not been addressed, and additional research is required. Increasing the average hydraulic application rate reduces the liquid residence time, but has been proven to increase wetting efficiency.

Conventional practice is direct recirculation of the trickling-filter underflow. Recirculation systems with recycle of settled effluent and direct recycle were compared at a facility in Webster City, Iowa (Culp, 1963). No significant differences in results were obtained when trickling filters operated simultaneously on the same settled wastewater under either winter or summer conditions. Researchers have concluded that it is the amount of recirculation and not the arrangement that is the more important for optimizing performance of rock-media trickling filters. The absence of intermediate settling between the first and second trickling filter operating in series does not adversely affect performance of the second-stage unit. The recirculation ratio is typically in the range 0.5 to 4.0. Dow Chemical Co. studies, which are summarized in Figure 11.60, demonstrate that vertical-flow corrugated media require an average application rate higher than 0.5 L/m^2/s to provide maximum BOD_5 removal efficiency (Bryan, 1955; 1962; Bryan and Moeller, 1960). Shallow towers using crossflow media have used hydraulic rates of 0.39 to 1.08 m^3/m^2/h.

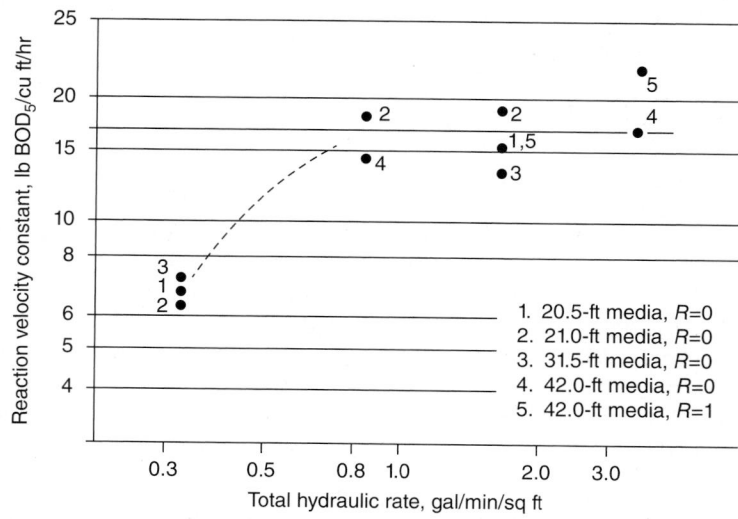

FIGURE 11.60 Effect of hydraulic application rate on five-day biochemical demand removal (Bryan, 1955, 1962; Bryan and Moeller, 1960).

Slowed distributor operation benefits trickling-filter facilities because of interrupted flow (periodicity of dosing), increased wetting efficiency (percent of media wetted), and controlled biofilm thickness. The designer should consider recirculation capabilities, the effect of reverse thrusting jets, or use of speed control on the distributor to enhance performance or improve operation. Another useful process control parameter is the dosing rate, Spülkraft. Methods for calculating dosing rate (mm/pass) are given by eqs 11.19 and 11.20.

$$DR = \frac{Q_I + Q_R}{A} \cdot \frac{1\,000 \dfrac{mm}{m}}{N_a \cdot \omega_d \cdot 1\,440 \dfrac{min}{day}} \tag{11.19}$$

or

$$DR = \cdot \frac{THL \cdot 1\,000 \dfrac{mm}{m}}{N_a \cdot \omega_d \cdot 1\,440 \dfrac{min}{day}} \tag{11.20}$$

where N_a = number of arms on the distributor, and
v_d = distributor rotational speed (rev/min).

Here, Spülkraft is in mm/pass. The typical hydraulically driven distributor in North America operates in the range of 2 to 10 mm/pass. Table 11.27 lists recommended operating and flushing dosing rates for rotary distributors. Higher dosing rates are recommended for higher organic loading rates to provide biofilm thickness control. Wastewater characteristics, temperature, or media type may influence dosing rate. It also may be beneficial to periodically use a higher flushing dosing rate for 5% to 10% of the 24-hour operating period. Work by Albertson (1989a; 1989b; 1995) and Parker et al. (1995; 1997; 1999) demonstrate that biofilm control measures enhance the trickling filter process when operating and flushing dosing rate values are used. These enhancements include improved performance, reduced odors, reduced power use for recycling,

Total Organic Load (kg/m³·d) (lb BOD₅/d/1000 cu ft)	Operating Dosing Rate (mm/pass) (in./pass)	Flushing Dosing Rate (mm/pass) (in./pass)
<0.4 (<25)	25–75 (1–3)	100 (4)
0.8 (50)	50–150 (2–6)	150 (6)
1.2 (75)	75–225 (3–9)	225 (9)
1.66 (100)	100–300 (4–12)	300 (12)
2.4 (150)	150–450 (6–18)	450 (18)
3.2 (200)	200–600 (8–24)	600 (24)

Note: Actual values are site specific and vary with media type.

TABLE 11.27 Operating and Flushing Dosing Rates for Distributors (Daigger and Boltz, 2010)

reduced nuisance organisms, and elimination of heavy sloughing cycles (Albertson 1989a; 1989b; 1995). Parker et al. (1995) described the use of both distributor speed control and variable frequency drive controlled recirculation pumps to maintain constant trickling filter hydraulic application. Pilot studies demonstrated mechanically driven distributor dosing did not improve performance of a nitrifying trickling filter. There is little research describing the effect of hydraulic transients on synthetic trickling filter media and their effect on media life.

7.3   Oxygen Requirements and Air Supply Alternatives

Trickling filters require oxygen to sustain aerobic biochemical reactions. Several researchers have demonstrated that at least some portion of roughing, carbon oxidizing, combined carbon oxidizing and nitrification, and NTFs may operate under oxygen-limited conditions (Schroeder and Tchobanoglous, 1976; Kuenen et al., 1986; Okey and Albertson, 1989b). Ventilation is essential to maintain aerobic conditions. Current design practice requires adequate sizing of underdrains and effluent channels to permit free airflow. Passive devices for ventilation include vent stacks on the periphery, extensions of underdrains through side walls, ventilating manholes, louvers on the sidewall of the tower near the underdrain, and discharge of effluent to the subsequent settling basin in an open channel or partially filled pipes. However, these methods may not be adequate if high performance is required or in the presence of low natural draft.

7.3.1   Natural Draft

One method to determine the amount of natural draft is to require 1 m^2 of ventilating area for each 3 to 4.6 m of trickling filter circumference. Another gauge is 1 to 2 m^2 of ventilation area in the underdrain area per 1000 m^3 of media. Inlet openings to rock-media underdrains have a recommended nonsubmerged combined area equal to at least 15% of the trickling filter surface area. Drains, channels, and pipes should be sized to prevent submergence of greater than 50% of the cross-sectional area under design hydraulic loading. Forced ventilation should be provided for covered trickling filters. Benzie et al. (1963) studied 17 rock-media-based WRRFs in Michigan and demonstrated that airflow was stagnant when ambient and wastewater temperatures were equal. Researchers also concluded that, during winter months, recirculation has a cooling effect on natural draft.

Schroeder and Tchobanoglous (1976) proposed the following equation to determine the natural draft in synthetic-media trickling filters.

$$\Delta P = 353 \cdot D \cdot \left(\frac{1}{T_C} - \frac{1}{T_H} \right) \tag{11.21}$$

where ΔP = pressure head resulting from differential temperature, mm H$_2$O;

T_C = colder temperatures (ambient air or air inside trickling filter), °K;

T_H = hotter temperatures (ambient air or air inside trickling filter), °K;

$T_m = \dfrac{T_C - T_H}{\ln\left(\dfrac{T_C}{T_H} \right)}$ = average temperature inside the trickling filter, °K; and

D = media depth, m.

The log-mean pore temperature, T_m, can be replaced by the hotter temperature, T_h, to obtain a less conservative estimate of the average pore air temperature. Some trickling filters do not have adequate oxygen daily for at least part of the year. Air stagnation results in odors and performance variability. Furthermore, little data exists to provide guidance on defining the amount of natural airflow through rock- or synthetic-media trickling filters. The velocity of air is low except when there is a large difference between air and wastewater temperature or if the trickling filters are deep (6–12 m). In the case of natural draft, if the wastewater is colder than ambient air, then the air will flow down. Alternatively, if the ambient air is colder than the wastewater, then the airflow will be up. Because a constant temperature differential does not occur naturally, power ventilation by mechanical means is recommended.

7.3.2 Forced Ventilation

Most new and improved trickling filter-based WRRFs use low-pressure fans to force airflow. This practice, known as power ventilation, offers a wintertime benefit of limiting cold airflow and minimizing cooling of wastewater. Powered ventilation and enclosed trickling filters also can destroy odorous compounds in influent wastewater and prevent excessive ventilation during winter or during periods of high air–water temperature differentials. Trickling filters may be ventilated in either an upward or downward airflow pattern. Dow Chemical presented a calculation to determine differential pressure as a function of airflow rate ($m^3/m^2/h$) through VFM for natural draft and mechanical fans. Differential air pressure for natural draft typically is insufficient during some portion of the day. Whether the trickling filter is 1.5 m (5 ft) or 6 m (20 ft) deep, the driving force is low. With temperature differentials between ambient air and water of less than 62.8°C (5°F), airflow in the trickling filter can stagnate. Humidity differences also drive airflow, but ambient air values vary widely and are unpredictable. Downflow may reduce or eliminate trickling filter odors if incorporated with good flushing hydraulics.

Recommended minimum airflow for design of power ventilation has been developed. However, additional research is required to determine airflow rates necessary to maximize kinetics. Influent BOD_5 and effluent $rbBOD_5$ are used to determine the required airflow rate. The oxygen-transfer rate used to determine airflow is 5% for carbon oxidizing trickling filters and 2.5% for both combined carbon oxidation and nitrification and NTFs. The higher air rate used for NTFs is to ensure that all areas of the trickling filter have airflow. Losses through vertical corrugated media can be described by using the following equation:

$$\Delta P = R \cdot \left(\frac{v^2}{2 \cdot g} \right) \tag{11.22}$$

where v^2 = superficial air velocity, m/s;
　　g = acceleration because of gravity, 9.81 m/s²;
　　R = tower resistance, velocity heads lost per unit tower depth, kPa/m; and
　　ΔP = total headlosses, kPa.

The term, R, is the total sum of individual headlosses through the trickling filter. Assuming adequate inlet and underdrain openings, the main loss through the trickling filter will be the packing loss (R_p) of the filter media. Dow Chemical proposed that R can be determined using eq 11.23 for medium density (89 m²/m³) VFM (Metcalf and Eddy, Inc./AECOM, 2013):

$$R_p = 10.33 \cdot D \cdot e^{(1.36 \times 10^{-5}) \left(\frac{L}{A} \right)}$$

(11.23)

where R_p = packing headloss in terms of velocity heads;
L = liquid loading, kg/h;
A = trickling filter plan (slab) area, m²; and
D = trickling filter media depth, m.

To estimate total headloss through the trickling filter, the value of R_p determined by eq 11.21 should be multiplied by a factor of 1.3 to 1.5 to account for minor headlosses such as the inlet and underdrain. Because there are few data for other media types, the following multipliers should be used to determine R_p (Metcalf and Eddy, Inc./AECOM, 2013):

- Crossflow: 100 m²/m³ (30 ft²/ft³) = 1.3 3 R_p
- Crossflow: 138 m²/m³ (42 ft²/ft³) = 1.6 3 R_p
- Rock: 39 to 49 m²/m³ (12–15 ft²/ft³) = 2.0 3 R_p
- Random: 100 m²/m³ (30 ft²/ft³) = 1.6 3 R_p

7.4 Trickling Filter Design Models

Numerous investigators have attempted to delineate the fundamentals of the trickling filter process by developing relationships among variables that affect operation. Existing process models range from simplistic empirical formulations to complex numerical models. Analyses of operating data have established equations or curves and have led to development of various empirical formulae. Unfortunately, many models exist and there no industry standard. Trickling filter models may be classified as dissolved-organic-loading models, particulate-organic-loading models, hydraulic-loading models, and mass-transfer models. Although these formulae include many variables that affect operations, none can predict actual performance. Designers need to assess which equation best fits a particular situation, particularly when meeting discharge permit requirements. Several formulae are discussed below: National Research Council, Galler and Gotaas, Kincannon and Stover, Velz equation, Schulze equation, Germain equation, Eckenfelder formula, Institution of Water and Environmental Management (IWEM) formula, and Logan model.

7.4.1 National Research Council Formula

The National Research Council (NRC) formula (1946) resulted from an analysis of operational records from rock-media trickling filter WRRFs serving military installations. The NRC analysis is based on two principles: The amount of contact between media and organic matter removed depends on trickling filter dimensions and number of passes; the greater the effective contact, the greater the performance efficiency. However, the greater the applied load, the lower the efficiency. Therefore, the primary determinant of efficiency in a trickling filter is the combination of effective contact and applied load. Organic loading primarily influences trickling filter efficiency. Hydraulic loading modifies the efficiency; increased rate equals increased efficiency. For the 34 WRRFs selected for the study, the efficiency curve best fitting a plot of the parameter "applied load per effective contact area" (W/VF) is captured in eqs 11.24 and 11.25 for single- and second-stage trickling filters, respectively.

$$E_1 = \frac{100}{1 + 0.0085 \cdot \left(\dfrac{W_1}{V \cdot F}\right)^{0.5}} \tag{11.24}$$

and,

$$E_2 = \frac{100}{1 + \dfrac{0.0085}{1 - \dfrac{E_1}{100}} \cdot \left(\dfrac{W_2}{V \cdot F}\right)^{0.5}} \tag{11.25}$$

where E_1 = BOD$_5$ removal efficiency through first-stage and settling tank, %;
$\quad$ W_1 = BOD$_5$ loading to first or single stage, not including recycle, kg/d;
$\quad$ V = volume of stage (cross-sectional area 3 media depth), acre-feet (Note: 1 ac ft = 1 233 m^3);

$\quad$ F = number of organic material passes = $\dfrac{1 + \dfrac{Q_R}{Q}}{\left[1 + (1 - P) \cdot \left(\dfrac{Q_R}{Q}\right)\right]^2}$;

$\quad$ $\dfrac{Q_R}{Q}$ = dimensionless recirculation ratio;

$\quad$ P = a weighing factor, for military trickling filters, with rock media = 0.9;
$\quad$ E_2 = BOD$_5$ removal efficiency through second stage and settling tank, %; and
$\quad$ W_2 = BOD$_5$ loading to second stage, not including recycle, kg/d.

Equations 11.24 and 11.25 are empirical but represent the average of data for rock-media trickling filter-based WRRFs, both with and without recirculation. Because of the nature of their development, the NRC formulae include several limitations and conditions:

1. Military wastewater has a higher strength (250–400 mg/L) than average domestic wastewater.

2. Effect of temperature on performance is not considered (most of the studies were in the Midwest and South).

3. Clarifier practice when the formulae were developed favored shallow units that were hydraulically loaded higher than current practice allows, resulting in excessive BOD$_5$ and TSS losses.

4. Applicability may be limited to stronger-than-normal domestic wastewater because no factor is included to account for differing treatability rates of lower strength wastewater.

5. The formula for second-stage trickling filters is based on the existence of intermediate settling tanks following first stage. Figure 11.61 compares trickling

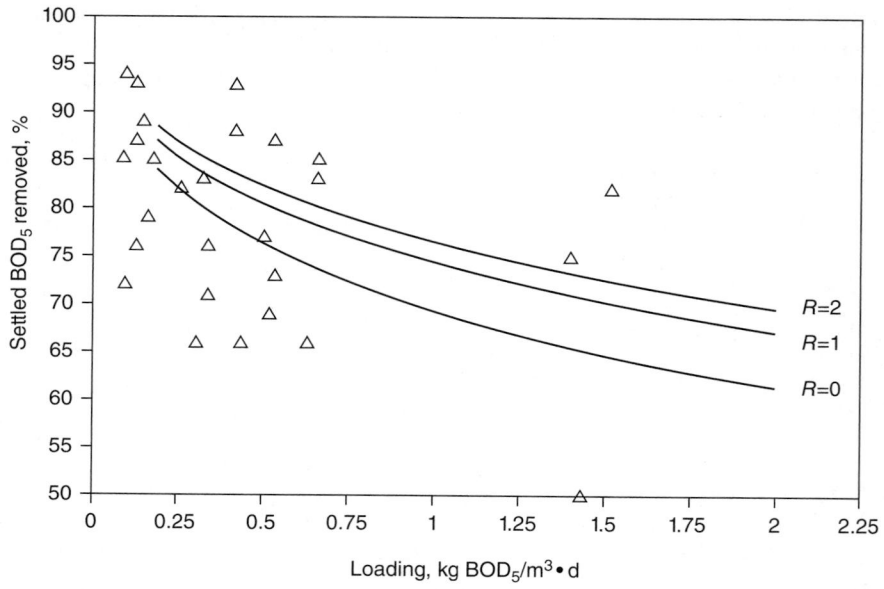

FIGURE 11.61   Comparison of trickling filter operating data with performance predicted by the National Research Council formula (kg/m³·d × 0.0624 = lb/d/cu ft) (NRC, 1946).

filter operational data for recirculation ratios of 0 to 2 with predicted values using the first- or single-stage NRC formula with a similar range of recirculation ratios (NRC, 1946). This figure shows that actual trickling filter performance may deviate substantially from predicted removals. Scattered data for loadings less than 0.3 kg/m³/d (20 lb/d/1000 cu ft) could be biased by lack of a BOD_5 test that inhibits nitrification. Inadequate flushing, poor ventilation, or an inefficient clarifier design could have contributed to poor performance data. Thus, the foregoing variables should be accounted for when designing trickling filters based on the NRC formula curves of Figure 11.62.

7.4.2   Galler and Gotaas Formula

Galler and Gotaas (1964) attempted to describe the performance of rock-media trickling filters with multiple regression analysis of data from existing WRRFs. Based on analysis of extensive data (322 observations), the following was developed:

$$S_e = \frac{K \cdot (Q \cdot S_i + Q_R \cdot S_e)^{1.19}}{(Q + Q_R)^{0.78} \cdot (1 + D)^{0.67} \cdot ra^{0.25}} \quad (11.26)$$

where K = coefficient = $\dfrac{0.464 \cdot \left(\dfrac{43\,560}{\pi}\right)0.13}{Q^{0.28} \cdot T^{0.15}}$;

Q = flow rate (ML/d);

Q_R = recirculation flow rate (ML/d);

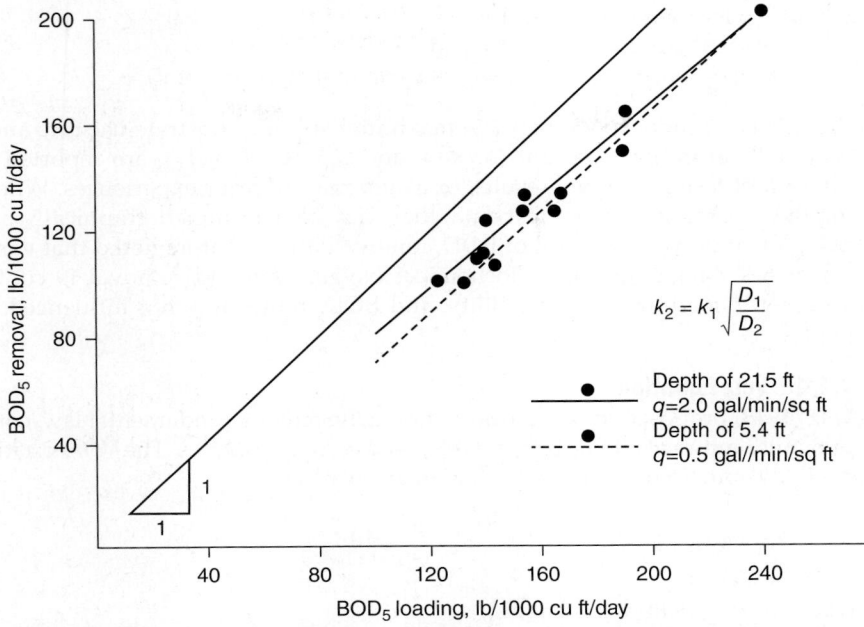

$$k_2 = k_1 \sqrt{\frac{D_1}{D_2}}$$

Depth of 21.5 ft
q=2.0 gal/min/sq ft
Depth of 5.4 ft
q=0.5 gal//min/sq ft

FIGURE 11.62 Biochemical oxygen demand removal vs loading and depth of plastic media from simultaneous loading studies at Midland, Michigan, facility (lb/d/1000 cu ft × 0.016 02 = kg/m³·d; gpm/sq ft × 0.679 = L/m²·s).

D = trickling filter depth (m);
S_e = settled trickling filter effluent as BOD_5 at 20°C (mg/L);
S_i = trickling filter influent as BOD_5 at 20°C (mg/L); and
ra = trickling filter radius (m).

The Galler and Gotaas formula recognizes recirculation, hydraulic loading, trickling filter depth, and wastewater temperature as important variables for performance. Deeper trickling filters performed better in their analysis. They indicated that recirculation improves performance but established a 4:1 ratio as a practical upper limit. A statistical analysis of experimental data, performed by Galler and Gotaas (1964), resulted in a high coefficient of multiple determination (R^2 = 0.97). Hydraulic flow rate was unimportant in determining bioreactor efficiency. The BOD_5 loading correlated most closely with performance; BOD_5 loading controlled performance.

7.4.3   Kincannon and Stover Model

Kincannon and Stover (1982) developed a mathematical model based on a relationship between the specific substrate use rate and total organic loading, which followed a Monod plot to determine required biofilm area (A_s):

$$A_s = \frac{\left(\dfrac{8.345 \cdot q \cdot S_i}{\mu_{\max} \cdot S_i} \right)}{S_i - S_e} - K_b \tag{11.27}$$

where S_i = influent BOD_5 (mg/L);

S_e = effluent BOD_5 (mg/L); and

K_b = proportionality constant of specific surface area (m²).

Biokinetic parameters, namely the maximum specific substrate use rate and Monod-type half-saturation constant (or m_{max} and K_b, respectively), are reported based on pilot-plant tests, full-scale results, or a summary of prior experiences. When extracting these parameters from test data, they may be determined graphically by plotting BOD_5 loading vs the inverse of BOD_5 removed. Investigators noted that variability in correlated data is normal. Biochemical oxygen demand removal is controlled by volumetric loading and treatability, and BOD_5 removal is not influenced by media depth.

7.4.4 Velz Equation

Velz (1948) proposed the first formulation delineating a fundamental law, compared to previous empirical attempts that were based on data analyses. The Velz equation relates the BOD_5 remaining at depth D mathematically by:

$$\frac{S_e}{S_i} = 10^{-k \cdot t} \qquad (11.28)$$

where S_i = influent BOD_5 (mg/L);

S_D = BOD_5 (mg/L) removed at depth D, meters;

t = residence time (d); and

k = Velz first-order rate constant (d⁻¹).

The formula depicted in eq 11.26 implies that k_V is constant for all hydraulic rates; however, Albertson and Davies (1984) presented evidence that k_V varies with the hydraulic rate. The Velz equation is presented because of its foundation in currently used design formulations, namely the Schulze and Eckenfelder equations.

7.4.5 Schulze Formula

Schulze (1960) postulated that the time of liquid contact with the biological mass is directly proportional to trickling-filter depth and inversely proportional to the hydraulic loading rate. This is expressed by eq 11.29.

$$t_c = \frac{c \cdot D}{THL^n} \qquad (11.29)$$

where t_c = liquid contact time (d);

c = constant (dimensionless);

D = trickling filter depth (m);

THL = hydraulic loading rate (m³/m²/d); and

n = exponent on hydraulic loading (dimensionless).

Combining the time of contact with the first-order equation for BOD_5 removal in an adaptation of the Velz theory, Schulze derived the following formula:

$$\frac{S_e}{S_i} = e^{\frac{-k_s \cdot D}{THL^n}} \qquad (11.30)$$

where S_e = soluble BOD_5 in trickling filter effluent stream (mg/L);
$\quad S_i$ = soluble BOD_5 in trickling filter influent stream (mg/L);
$\quad k_S$ = Schulze coeffecient (d^{-1} when n = 1);
$\quad D$ = trickling filter depth (m);
$\quad n$ = exponent on hydraulic loading (dimensionless); and
$\quad$ THL = hydraulic loading rate ($m^3/m^2/d$).

Equation 11.28 is similar to that proposed by Velz. However, Velz's constant, k, was not formulated to consider hydraulic load. For a given wastewater strength, the hydraulic rate is proportional to the loading rate. Thus, volumetric organic loading may still be the controlling process variable. The value of k published by Schulze (based on U.S. customary units) for a rock-media trickling filter with a 1.8-m (6-ft) depth at 20°C was 0.69 d^{-1}. The dimensionless constant characteristic of rock-media trickling filter, n, was found to be 0.67. The common temperature correction value of $u = 1.035$ could be applied to determine k_t as follows by

$$k_t = k_{20} \times 1.035^{(T-20)} \quad \text{for} \quad k_s \tag{11.31}$$

and

$$k_t = A_s \times k_S$$

where k_t = temperature-corrected coefficient value (d^{-1} when $n = 1$);
$\quad k_S$ = Schulze coefficient (d^{-1} when $n = 1$); and
$\quad A_s$ = clean surface area of the media, m^2.

7.4.6 Germain Formula
Germain (1966) applied the Schulze formulation to a synthetic-media trickling filter:

$$\frac{S_e}{S_i} = e^{\frac{-k_G \cdot D}{THL^n}} \tag{11.32}$$

where S_i = soluble BOD_5 in trickling filter influent stream (typically primary effluent excluding recirculation) (mg/L);
$\quad$ THL = hydraulic loading rate ($m^3/m^2/d$); and
$\quad k_G$ = Germain coefficient (d^{-1} when $n = 1$).

Values of k_G and n are related to media configuration, clarication efficiency, dosing cycle, and hydraulic rate; k_G is a function of wastewater characteristics, media depth, specific surface area, and media configuration. Therefore, because a high degree of interdependency exists between k_G and n, this must be considered in data comparisons. Germain (1966) reported that the value of k_G of 0.24 $(L/s)^n/m^2$ (0.088 gpm^n/ft^2) for a synthetic-media trickling filter 6.6 m (21.5 ft) deep treating domestic wastewater with a value of 0.5 for n. This vertical-flow media had a clean surface area of 89 m^2/m^3 (27 sq ft/cu ft). Correction of k_G for the high BOD_5 concentration represented by the Institution of Water and Environment Management (IWEM) model, , resulted in similar $k_G\left(\frac{150}{360}\right)^{0.5}$ predictive values from these two models for plastic media operating in the loading range of 0.2 to 1.5 $kg/m^3/d$ (12.5 to 93.6 lb/d/1000 cu ft) at 20°C.

In tests designed to determine the effects of recirculation on BOD_5 removal, Germain (1966) found no statistically significant difference. However, the relatively tall 6.6-m (21.5-ft) tower resulted in high influent application rates, ensuring adequate wetting of media. This conforms to the practice of using recirculation for shallow filters where influent hydraulic rates are low and wetting efficiency would likely suffer. Equation 11.30 is used widely for synthetic-media trickling filter analysis and design. The k_G data were developed from more than 140 pilot studies performed by Dow Chemical and many more by other media suppliers. Most of these tests used a trickling filter media depth of 6 to 7 m (20–22 ft).

7.4.7   Eckenfelder Formula

Eckenfelder (1961) and Eckenfelder and Barnhart (1963) described a modified trickling filter formula to account for trickling filter media specific surface area (a, m^2/m^3). The formula proposed for soluble BOD_5 removal can be expressed mathematically by eq 11.33.

$$\frac{S_e}{S_i} = e^{\frac{-k_s' \cdot a^{(1+b)} \cdot D}{THL^n}}$$ (11.33)

where k_s' = overall treatability coefficient based on soluble BOD_5 $[(m^3/d)^{0.5}/m^2]$;
  D = depth (m); and
  THL = hydraulic loading rate ($m^3/m^2/d$).
  b = surface area modifier for surface loss with increasing area.

With recirculation, eq 11.33 can be extended and expressed mathematically by eq 11.34.

$$\frac{S_e}{S_i} = \frac{e^{\frac{-k_s' \cdot D}{WHL^n}}}{\left(1 + \frac{Q_R}{Q}\right) - e^{\frac{-k_s' \cdot D}{WHL^n}}}$$ (11.34)

Using the Eckenfelder formula and $k_s a = a \times k_s$, eq 11.34 can be rewritten as eq 11.35, this is known as the Modified Velz Equation.

$$S_e = \frac{S_i}{\left(\frac{Q_R}{Q} + 1\right) \cdot e^{\left[\frac{k_s \cdot a \cdot D \cdot \theta^{(T-20)}}{\left[WHL\left(\frac{Q_R}{Q}+1\right)\right]^n}\right]} - \frac{Q_R}{Q}}$$ (11.35)

7.4.8   Institution of Water and Environmental Management Formula

The Institution of Water and Environment Management (IWEM) developed a formula describing the BOD_5 in trickling filters having rock, random-packed plastic media (rock and random synthetic), or modular plastic media. Equation 11.36, resulting from a multiple regression analysis, follows:

$$\frac{S_i}{S_e} = \frac{1}{1 + k_{IWEM} \cdot \theta^{(T-15)} \cdot \left(\frac{a^m}{VLR^n}\right)}$$ (11.36)

where S_i = influent BOD_5 (mg/L);
   k_{IWEM} = kinetic coefficient ($m^{m-1}\ d^{n-1}$);
      u = temperature coefficient;
      a = media specific surface area (m^2/m^3);
      m = reduction factor for surface loss with increasing area;
   VLR = volumetric hydraulic loading rate ($m^3/d/m^3$) of trickling filter media; and
      n = hydraulic rate coefficient.

Equation 11.34 has reported coefficients that account for 90% of data variability:

- k_{IWEM} = 0.0204 (rock and random); 0.40 (modular plastic).
- u =1.111 (rock and random); 1.089 (modular plastic).
- m = 1.407 (rock and random); 0.732 (modular plastic).
- n = 1.249 (rock and random); 1.396 (modular plastic).

The model was developed using data collected from tests performed on a strong domestic wastewater with primary effluent concentrations of 360-mg/L BOD_5, 240-mg/L TSS, and 52-mg/L NH_3-N. The model predicts a continuous performance curve from low- to high-rate loadings. The trickling filter depths from which the samples were collected range from 1.74 to 2.10 m, biofilm growth areas ranged from 1.0 to 5.0 m^2, and loadings were 0.3 to 16 kg/m^3/d. The IWEM model is temperature sensitive which may be caused by site-specific wastewater characteristics and data reduction procedures. The NRC equations agree with the IWEM's projection based on an influent strength of 360 mg/L BOD_5 at loadings up to 1.0 kg/m^3/d.

7.4.9   Logan Trickling Filter Model

The Logan trickling filter (LTF) model is based on characterizing modular plastic trickling filter media as a series of inclined plates covered with a thick, partially penetrated, biofilm. The rate of soluble COD removal is determined using a numerical model to solve transport equations that describe biochemical transformation rates resulting from diffusion through a thin liquid film and into the biofilm. Although the model was calibrated using a single data set for only one type of plastic trickling filter media, a variety of laboratory, pilot-plant, and full-scale trickling filter studies claim that LTF accurately predicts soluble COD removal (Logan et al., 1987a; 1987b; Bratby et al., 1999; Logan and Wagenseller, 2000). Unlike kinetic models, the LTF cannot be collapsed into a single equation. Thus, a computer program is required to use this approach. The theoretical basis of the LTF model is reviewed, and example calculations are provided below. The computer model of Logan et al. (1987a; 1987b) was developed to predict soluble BOD_5 removal in plastic media trickling filters as a function of plastic media geometry. The LTF model has not been tested or adapted for use with rock- or random-media trickling filter design.

A disadvantage of kinetic models, such as the Velz equation, is that new kinetic (k_{20}) and hydraulic (n) constants may require determination for each type of trickling filter media. The LTF model requires only that the media geometry be measured and input. Consequently, there was no need to recalibrate the model for new plastic media types. The actual, computer-code-based model (written in FORTRAN) was given the name TRIFIL2; the LTF computer program uses tabulated values for a range of conditions for specific media. Dissolved organics that compose soluble COD in wastewater are assumed to be equally distributed into a five-component molecular size. As the waste-water flows over the biofilm, the dissolved organics diffuse into the biofilm and are removed at a rate close to that predicted. Small molecules diffuse faster than larger ones and are predicted to be removed more efficiently. Soluble COD is not included in the model because it is assumed to be removed by particle bioflocculation. Temperature affects water viscosity (m), which affects fluid film thickness and thus retention time in the trickling filter media. Changes in chemical diffusion coefficients (D) with temperature (T) are adjusted by the usual assumption that is constant (Welty et al., 1976). The $\frac{D \cdot m}{T}$ model is available free at http:/www.engr.psu.edu/ce/enve/logan/bioremediation/trickling_filter/model.htm. Additional information on the model can be obtained from the original publications cited below, as well as a chapter in Logan (1999).

7.4.10 Selecting a Trickling Filter Design Model

Design engineers may use various empirical criteria and design formulations for sizing trickling filters. The NRC (eqs 11.24 and 11.25) or Galler and Gotaas (eq 11.26) formulae typically are used for rock-media trickling filter design. The Schulze equation (eqs 11.29 and 11.30), Eckenfelder equation, and IWEM equation are used for both rock- and plastic-media trickling filter design over a wide range of media specific surface areas and depths. The coefficients k and n vary, however. (The word "coefficient" is used to describe k [or K] and n because they are neither constants nor treatability factors.) Bruce and Merkens (1970, 1973) conducted simultaneous testing of two trickling filters at the same flow and BOD_5 loading but at a 4:1 ratio in the application rate because one was 7.41 m (24.3 ft) deep and the other was 2.1 m (6.9 ft) deep (Bruce and Merkens, 1970; 1973). From this study it was concluded BOD_5 efficiency would be independent of depth. Further work showed that the BOD strength also affect BOD_5 efficiency. Thus, the value of k of the hydraulically driven equations could be modified as a function of media depth (D) and influent BOD strength (L) by eq 11.35 where D_1 is 6.1 m (20 ft) and L_1 is 150 mg/L:

$$k_2 = k_1 \cdot \left(\frac{D_1}{D_2}\right)^{0.5} \left(\frac{L_1}{L_2}\right)^{0.5}$$

(11.37)

where k = Velz's constant;
     D = depth (m); and
     L = length (m).

Simultaneous tests were also conducted by the Dow Chemical using 1.65-m and 6.55-m trickling filters. The results of these studies (Figure 11.62) demonstrate that trickling filter performance is controlled by organic loading, and that k value of the deep trickling filter is exactly 50% of that for the shallow trickling filter (i.e., $[1.6/6.6]^{0.5} = 0.5$).

The variation of k with trickling filter depth is an important consideration. A k-value developed for specific depth should not be used for a different depth without proper modification. Using data from installations and simultaneous tests, Albertson and Davies (1984) showed that k could be used for any trickling filter configuration if corrected for depth. However, this research also indicated that inadequate wetting might produce lower k values than optimum.

The Eckenfelder, Germain, Schulze, and Velz equations are similar. Because the coefficients k (or K) and n must be, or have been, empirically derived, background data are influenced by a variables such as hydraulic rate, dosing mode, temperature, wastewater characterization, media configuration and depth, ventilation, and other unknown test-specific factors. The equations have proven effective in modeling specific WRRF data, but also have been proven to deviate significantly from observed results. When modifying the configuration, the value of the treatment constant, k (or K) changes, even when considering the same trickling filter media and wastewater.

The NRC, Germain, and Eckenfelder equations may be used for rock–media trickling filters, although results are highly variable. Designers may use each of these models before making a decision. Improved hydraulic application systems provide control of ponding, nuisance macrofauna, and odors. The designer may want to attempt to account for improved performance with increased dosing rates. A well-designed and operated rock-media trickling filter can provide high-quality effluent. Grady et al. (1999) suggest that for low organic loads (less than 1 kg $BOD_5/d/m^3$), rock- and synthetic-media trickling filters are capable of equivalent effectiveness at low to moderate organic loading.

Synthetic-media CTFs are designed with the Eckenfelder, Germain, or LTF models. Use of the Germain coefficient k (for A_sK) is justified because of the lack of an adequate, properly compiled database that allows for effective separation of A_s and K. Historical pilot- and full-scale data are impaired by lack of proper dosing intensity and hydraulic rate. In addition, many pilot-plants used were equipped with continuous-flow nozzles, which are known to be inefficient. The Eckenfelder equation often is used to define rbCOD removal efficiency. The beneficial effect of recirculation is reflected in this formula were derived from low application rates of standard rate rock-media trickling filters. The literature values of n were derived from continuous-flow studies; to properly compare k values, use of 0.5 for n is suggested.

Municipal wastewater k_{20} values were generated from studies by Dow Chemical on medium density, 89 m^2/m^3, vertical-flow media with hydraulic applications rates ranging from 0.176 to 0.244 $(L/s)^{0.5}/m^2$ and a trickling filter media depth of 6.55 m. This has evolved to a common design k_{20} value of 0.203 $(L/s)^{0.5}/m^2$ at 6.1 m (20 ft). This k_{20} value is used with a minimum wetting rate of 0.51 $L/m^2/s$. As trickling filter depth decreases, recycle must increase to maintain minimum wetting flow. This criterion must not be ignored when replacing rock-media with synthetic media less than 4-m (13.1-ft) deep. Hydraulic rates have been 20% to 50% of the minimum wetting rate established by Dow Chemical. Therefore, wetting efficiency should be considered when rerating or optimizing an existing facility. There is little evidence to resolve the question of wetting effectiveness as a function of mode of application; additional research is required. Drury et al. (1986) demonstrated improved process performance by simply replacing rock media with synthetic media in a shallow-bed trickling filter (approximately 1 m). Designers must recognize that performance of crossflow media trickling filters 1- to 2.4-m deep may not exceed that of rock-media trickling filter because of wetting limitations (i.e., existing recirculation pumping facilities).

7.5 Combined Carbon Oxidation and Nitrification

Combined carbon oxidizing and nitrification trickling filters may be accomplished using synthetic or rock-media. The effect of combined carbon oxidation and nitrification in rock-media trickling filters is an artifact of under-loading based on soluble COD. Parker (1998) noted that design of combined carbon oxidation and nitrification in synthetic media trickling filters is empirical.

The U.S. EPA (1991) conducted a survey of 10 combined carbon oxidation and nitrification facilities, six of which used the trickling filter/solids contact process. The manual for nitrogen control recommends BOD_5 loading (g/m²/d) to achieve both carbon oxidation and nitrification in a single trickling filter (U.S. EPA, 1993). The kinetics of combined BOD_5 removal and nitrification are complex. Lack of fundamental research supporting the combined carbon oxidation and nitrification process contributes to the empirical design procedures presented herein. Because of the facultative heterotrophic biofilms, researchers such as Biesterfeld et al. (2003) have demonstrated that recirculation sometimes results in denitrification.

The rate of nitrification in combined carbon oxidation and nitrification trickling filters will be influenced by many factors such as influent wastewater characteristics, hydraulics, ventilation, and media type. The U.S. EPA (1975) summarized full- and pilot-scale rock-media trickling filter data from Lakefield, Minnesota; Allentown, Pennsylvania; Gainesville, Florida; Corvallis, Oregon; Fitchburg, Massachusetts; Ft. Benjamin Harrison, Indiana; Johannesburg, South Africa; Salford, England. Figure 11.63 illustrates these data and shows the relationship between BOD_5 volumetric loading with nitrification efficiency.

Recommendations proposed by U.S. EPA (1975) include an organic matter loading limit of 0.16 to 0.19 kg BOD_5/m³/d required to achieve approximately 75% nitrification. Bacterial cellular synthesis of ammonia by heterotrophic bacteria for cell growth contributes to the complexity of estimating nitrification (NH_3-N) in trickling filters. Figure 11.63 indicates that recirculation typically improved nitrification, particularly for efficiencies greater than 50%. Stenquist et al. (1974), reporting on combined carbon

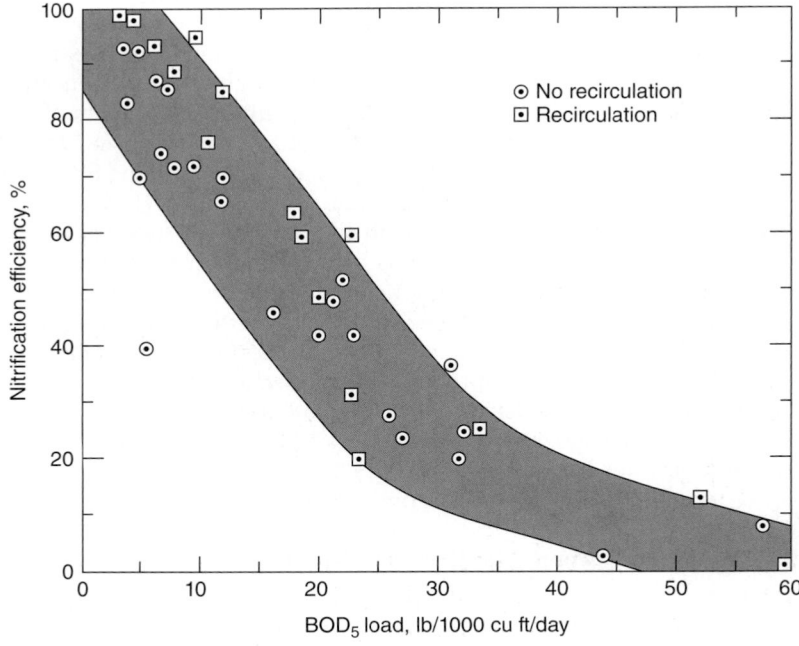

FIGURE 11.63   Effect of organic load on nitrification efficiency of rock trickling filters (lb/d/1000 cu ft × 0.016 02 = kg/m³·d).

oxidation and nitrification in both synthetic- and rock-media trickling filters, related organic loading to the level of nitrification achieved. These researchers determined that 89% NH_3-N removal occurred at an organic loading of 0.36 kg/m³/d. The nitrification capacity of trickling filter was found to be a function of surface BOD_5 loading (kg BOD_5/ m²/d of trickling filter media). Bruce and Merkens (1975) demonstrated that effluent BOD_5 and COD had to be less than 30 and 60 mg/L, respectively, to initiate nitrification and complete nitrification occurred with an effluent BOD_5 less than 15 mg/L. Harremoës (1982) concluded that soluble BOD_5 would have to be less than 20 mg/L for the initiation of nitrification and will reach maximum rates when effluent filtered or soluble BOD_5, is 4 to 8 mg/L. Figure 11.64 illustrates results from full- and pilot-scale studies at Stockton, California. This figure provides data for vertical-flow media and support for the work of Harremoës (1982).

Lin and Heck (1987) reported successful operation of a trickling filter/SC process with trickling filters designed for combined carbon oxidation and nitrification with 1.5 mg/L effluent NH_3-N at 13°C. The trickling filter design was based on 0.2 kg BOD_5/ m³/d and a TKN loading of 0.051 kg/m³/d using 98-m²/m³ crossflow media. Complete nitrification occurred with summer BOD_5 loadings up to 0.32 kg/m³/d.

Parker and Richards (1986) presented results of tests conducted at Garland, Texas, and Atlanta, Georgia. The results are presented as percent ammonia-nitrogen removal vs average BOD_5 loading (lb/d/1000 cu ft) are shown in Figure 11.65 (a) and (b). Daigger et al. (1994) presented an evaluation of three full-scale crossflow media trickling filters achieving combined carbon oxidation and nitrification as shown in Figure 11.65 (c).

Removal of organic nitrogen was reported to vary between 21% and 85% (U.S. EPA, 1975). Rock-media trickling filter studies at Gainesville and Johannesburg indicated

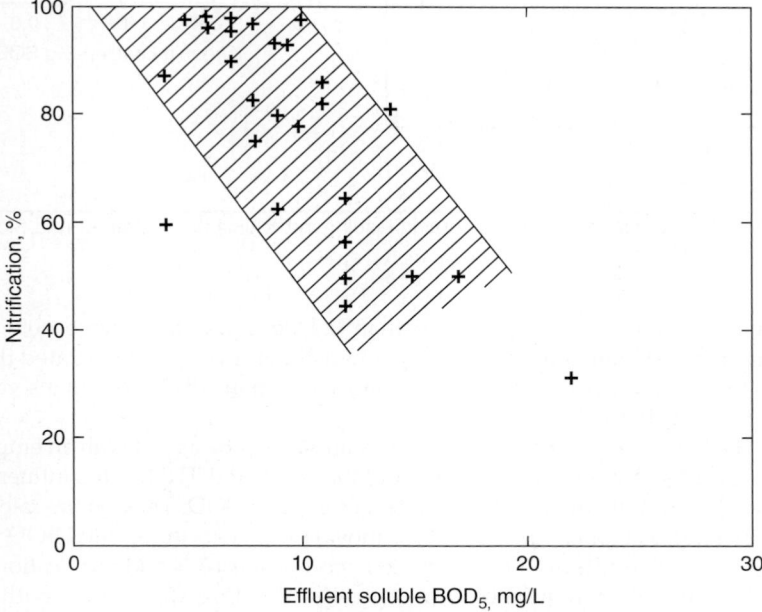

Figure 11.64 Relationship between nitrification efficiency and soluble biochemical oxygen demand in the effluent of a vertical media trickling filter at Stockton, California.

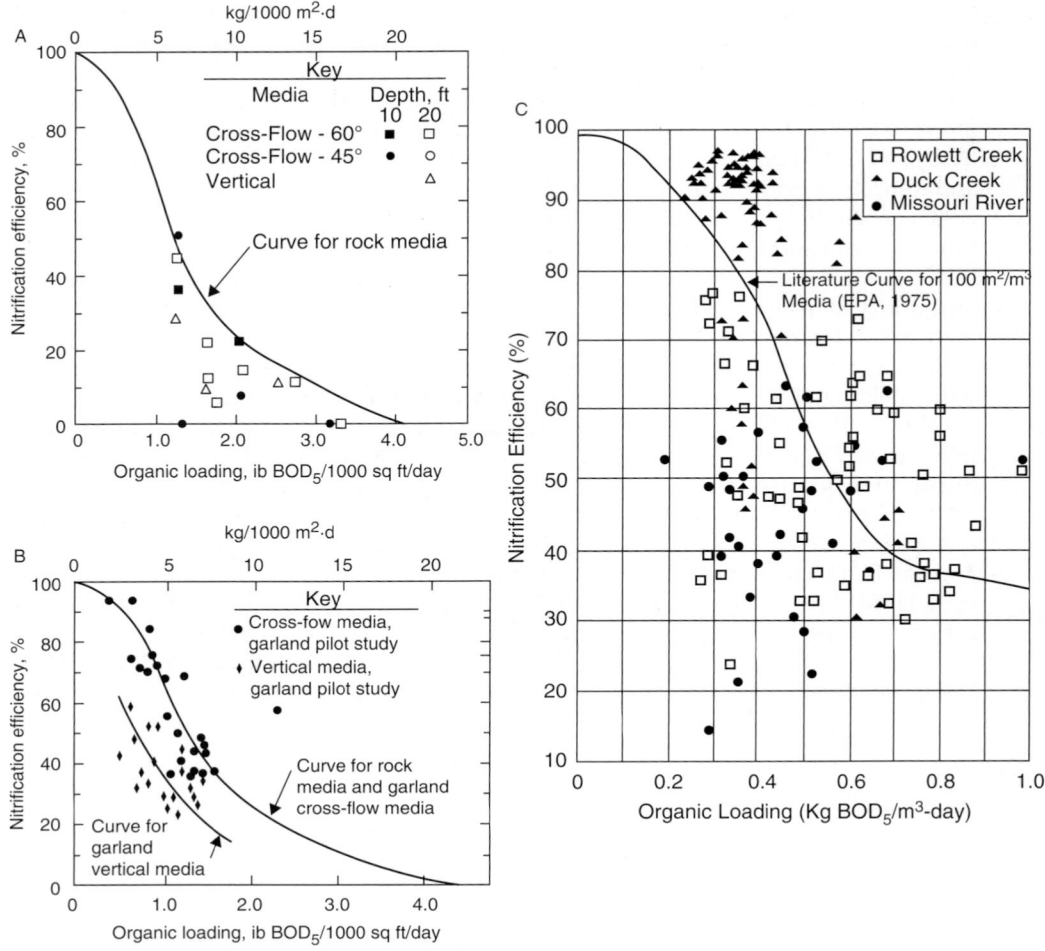

FIGURE 11.65   Nitrification efficiency vs organic loading in trickling filters (Parker and Richards, 1986; Daigger et al., 1994).

that BOD_5 loadings must be less than 0.2 kg/m³/d to remove 60% to 85% of the ammonia-nitrogen. Studies at Bloom Township, Illinois, demonstrated that nitrification was temperature dependent (Baxter and Woodman, 1973). Removals varied from 30% to 70% from 10 to 23°C.

The Wauconda, Illinois, facility, for example, was designed with an empirical approach that considered the organic loading and the BOD_5 and TKN of the influent. These procedures did not fully account for the effect of influent BOD_5:TKN on the rate of nitrification. Figure 11.66 is an illustration of TKN removal rate versus the applied BOD_5:TKN ratio. The TKN removal results from the combined effect of synthesis and nitrification.

Trickling filters require biofilm control by flooding, flushing, or both to produce the lowest ammonia-nitrogen concentrations. In the case of a combined carbon oxidizing and nitrification trickling filter, the designer must include provisions to control growth

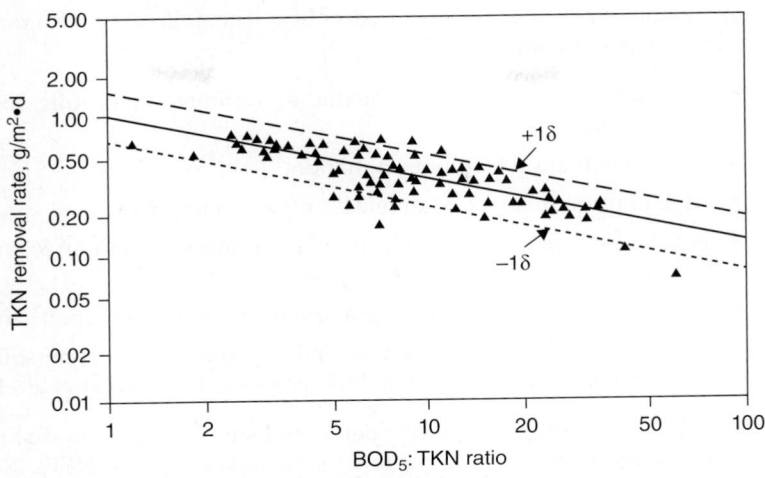

FIGURE 11.66   Study of nitrification at temperatures below 20°C for facilities in Stockton and Chino, California; Garland, Texas; and the Twin Cities Metro facility at St. Paul, Minnesota (median $\bar{y}$ = 0.460 ± 0.175; $\bar{x}$ = 11.081 and ≈ 15°C TKN_{ox} = 1.086 [BOD_5:TKN]$^{-0.44}$; g/m²·d × 0.204 8 = lb/d/1000 sq ft).

of heterotrophic biofilms and provide predator control to avoid grazing of the delicate nitrifying biofilm.

Daigger et al. (1994) found that oxidation of BOD and NH_3-N in trickling filters can be characterized by the following:

$$VOR = \left(\frac{L_{in} + 4.6 \cdot (NO_X\text{-}N)}{10^3 \, \frac{g}{kg}} \right) \cdot \left(\frac{Q}{V_m} \right) \tag{11.38}$$

where VOR = Volumetric oxidation rate, kg/m³/d;
$\quad\quad L_{in}$ = Influent BOD concentration, mg/L (or g/m³);
$\quad$ NO_x-N = amount of ammonia oxidized, mg/L (or g/m³);
$\quad\quad\; Q$ = Influent low rate, m³/d; and
$\quad\quad V_m$ = Volume of filter media, m³.

Using eq 11.38, the volumetric oxidation rate for three trickling filter facilities with modular plastic media was found to vary from 0.75 to 1.0 kg/m³/d.

7.6   Nitrifying Trickling Filters

Using a nitrifying trickling filter, or NTF, to treat secondary effluent (versus primary effluent as previously discussed) is a reliable and cost effective means to control NH_3-N. Mulbarger (1991) evaluated the literature to understand biological wastewater treatment processes. The researcher postulated that NTFs are effective in the range of effluent NH_3-N concentration greater than or equal to 2 mg/L. The NTFs are affected by oxygen availability, temperature, organic matter and NH_3-N in the influent wastewater stream, media type, and process hydraulics. The following design practices help optimize NTF performance:

- Medium-density crossflow media to optimize hydraulic distribution and oxygenation,
- Power ventilation to avoid stagnation,
- ADF to promote more complete biofilm development,
- Polished secondary effluent to avoid microbial competition for substrates in the biofilm,
- Maximum wetting efficiency to avoid formation of dry spots, and
- Storage and control of NH_3-N laden supernatant from solids processing operations to even out diurnal NH_3-N variability (Parker et al., 1997).

Low energy consumption, stability, operational simplicity, reduced sludge yield, and improved sludge settleability are a few of the advantages of NTFs. Reduced sludge yield because of nitrifying biofilms have led to construction of NTFs solids separation. Predatory macrofauna can negatively affect performance; therefore, the designer must include a way to manage solids and predator-laden water resulting from predator treatment cycles. Subsequent sections present design and operational features dedicated to macrofauna control. Current design practice may include alternating double filtration (ADF). Hawkes (1963), Gujer and Boller (1986), Parker et al. (1989), and Wik (2000) have demonstrated that alternating the sequence of NTFs in series improves system performance. The researchers observed patchy biofilm growth near the bottom of the NTFs. The ADF may allow for more complete biofilm coverage in both bioreactors. The lead NTF typically is alternated every three to seven days; therefore, current practice suggests construction of a minimum of two NTFs. Although the responsible mechanism has not been identified, the use of high NTFs with 6 to 12.2 m (20–40 ft) media depths have demonstrated good performance. Some NTFs have been constructed with depths as great as 12.8 m (42 ft) with excellent results. Recirculation should be minimized to maintain maximum NH_3-N concentration in the influent, up to 12 mg/L, and reduce pH depression of nitrification. To maximize nitrification, a depth of 6 to 12 m (20–40 ft) typically is optimal for producing a high hydraulic rate and maintaining a maximum zero-order kinetic region. Shallower trickling filters can be used in series.

Parker (1998; 1999) further illustrated that performance differences between trickling filter media types are most clear for tertiary NTFs. Table 11.28 demonstrates that on a unit-area basis, zero-order nitrification rates are greater for crossflow media than vertical-flow media. Because NTFs are compared on a unit-area basis, it is easier for the designer to evaluate site-to-site data. In each of the cases listed in the table, ammonia-nitrogen fluxes were greater for crossflow than vertical-flow media. As previously discussed, the postulated factors contributing to enhanced crossflow performance are improved oxygen transfer efficiency because of the increased number of media interruption (mixing) nodes (Gujer and Boller, 1986; Parker et al., 1989). Because of the low biofilm accumulation associated with autotrophic nitrifying biofilms, denser crossflow media typically is used in NTFs. The higher specific surface area characteristic to medium- and high-density crossflow media yields increased volumetric nitrification rates. The combined effect of higher ammonia-nitrogen flux and media density may result in volumetric uptake rates approximately three times higher than can be expected

Location	Investigator(s)	Media Type	$J_N^0 \left(\dfrac{gN}{d \cdot m^2 \text{ Bioflim}} \right)$	Temperature Range (°C)
Central Valley, Utah	Parker et al. (1989)	XF 140	2.3–3.2	11–20
Malmo, Sweden	Parker et al. (1995)	XF 140	1.6–2.8	13–20
Littleton/Englewood, Colorado	Parker et al. (1997)	XF 140	1.7–2.3	15–20
Midland, Michigan	Duddles et al. (1974)	VF 89	0.9–1.2	7–13
Lima, Ohio	Okey and Albertson (1989a)	VF 89	1.2–1.8	18–22
Bloom Township, Illinois	Baxter and Woodman (1973)	VF 89	1.1–1.2	17–20

TABLE 11.28 Nitrification Rates for Vertical (VF) and Crossflow (XF) Media (Parker, 1998, 1999)

with vertical-flow media (Parker, 1998). These observations are from pilot-scale trickling filter studies, typically with continuous hydraulic application. Recommendations related to materials are presented in this chapter because they are sound. Full-scale verification of the performance gradient has not been reported, and additional research is recommended.

7.6.1 Kinetics and Design Procedures

The trickling filter bioreactor has been described previously as a PFR with large axial dispersion that allows for no exchange of reactant with the environment outside the physical system boundaries. Because of complexities in separating bioreactor compartments, such as external diffusion and internal diffusion/reaction, during tests, most design formulations fail to separate compartments. The NTF design models are based on flux, which is consistent with state-of-the-art biofilm process modeling. Advanced NTF models based on biofilm kinetics exist and have been referenced in the section dedicated to design models and formulations. These models are based on assumptions that may limit their applicability, or require information that is not readily available. However, the models provide process insight well beyond the design formulations presented in this section. The engineer is referred to the literature for determination of applicability and procedure. Three simplistic NTF design models are presented in this section:

- Gujer and Boller model (1986),
- Modified Gujer and Boller (1986) model, and
- Albertson and Okey (1988) model.

The NTF (nitrification) kinetic regime changes from zero- to first-order from the entrance plane of the bioreactor to the exit as the NH_3-N concentration decreases. Then NH_3-N, rather than oxygen, becomes the rate-limiting substrate in the first-order regime.

Results presented by Okey and Albertson (1989b), which are illustrated in Figure 11.67, were obtained from five different NTF facilities. These data were not

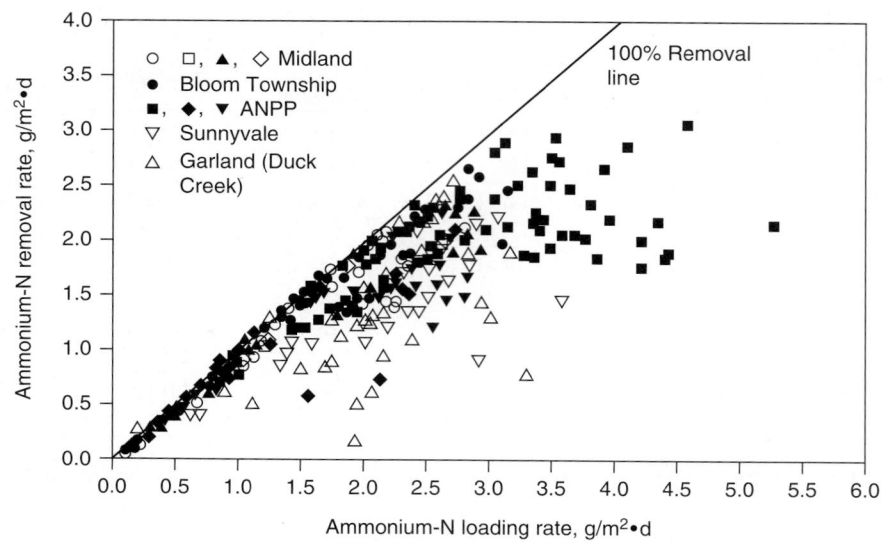

Figure 11.67   Area ammonium-nitrogen load vs the observed rate of ammonium-nitrogen removal (g/m²·d × 0.204 8 = lb/d/1000 sq ft).

corrected for temperature, and all test trickling filters relied on natural draft ventilation. Therefore, wastewater characterization, temperature, substrate availability, media type, and hydraulic application rate may have caused variability in the date. The data suggest that the NH_3-N flux will approach 100% removal for loadings less than 1.2 g/m²/d.

The typical NH_3-N profile in the upper portion of an NTF will exhibit a straight-line reduction of NH_3-N at a rate controlled by oxygen availability. The rate of removal will decrease as the rate-limiting substrate switches from oxygen to NH_3-N. Consequently, the rate of nitrification in a NTF is not constant with bioreactor depth.

The concentration of TSS in the secondary effluent has an appreciable effect on tertiary nitrifying biofilm reactors (Parker et al., 1989). The biomass suspended in the bulk liquid will compete with the biofilm for available substrate, particularly for oxygen.

Andersson et al. (1994) demonstrated that the maximum zero-order nitrification rate in a pilot-scale NTF apparently decreased from approximately 2.6 g N/d/m² to approximately 1.8 g N/d/m² when effluent TSS concentration exceeded 15 mg/L. These findings are illustrated in Figure 11.68. According to the figure, nitrification in NTFs will approach oxygen-limiting conditions when the bulk liquid TSS is less than 15 mg/L. There is little research describing the degree of benefit as TSS concentrations are reduced well below 15 mg/L.

7.6.2   Gujer and Boller Model

The Gujer and Boller NTF model was developed based on stoichiometry, Fick's law and Monod-type kinetics. Gujer and Boller (1986) presented eq 11.39 for NTF design.

$$J_N(S,T) = J_{N,\max}(T) \cdot \frac{S_{B,N}}{K_N + S_{B,N}} \cdot e^{-k \cdot z} \tag{11.39}$$

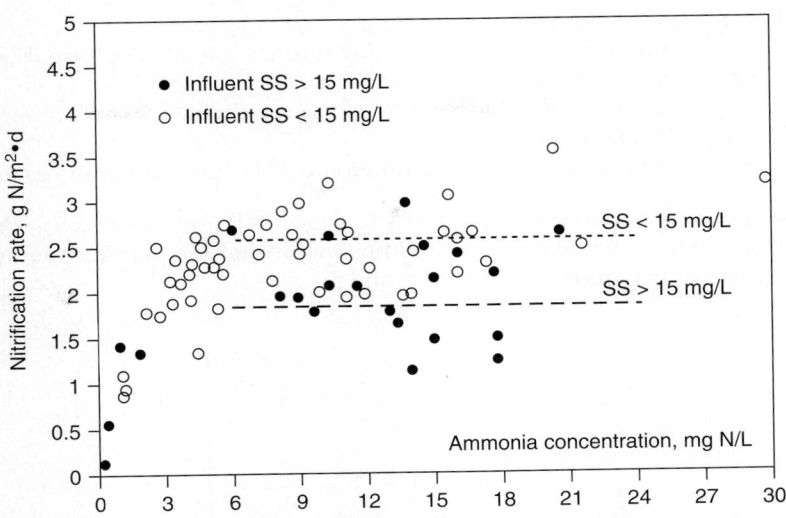

FIGURE 11.68   Impact of bulk liquid total suspended solids concentration on nitrification in a pilot-scale nitrifying trickling filter (Anderson et al., 1994; reprinted from *Water Science and Technology*, with permission from the copyright holders, IWA).

Here $J_N(S,T)$ = ammonia-nitrogen flux at $S_{B,N}$ (g N/m²/d);

   $J_{N,max}(T)$ = maximum ammonia-nitrogen flux at temperature T (g N/m²/d)

   $= \dfrac{J_{O_2,max}(T)}{4.3}$;

   $J_{O_2,max}(T)$ = maximum dissolved-oxygen flux at temperature T (g O₂/m²/d) deter-

   mined from literature or pilot testing;

   $S_{B,N}$ = bulk-liquid ammonia-nitrogen concentration (g/m³);

   K_N = half-saturation coefficient for ammonia-nitrogen (g N/m³);

   = 1.0 g N/m³; and

   T = temperature, °C.

Based on a "line-fit" relationship, $J_N(z,T) = J_N(0,T)/e^{-k/z}$, the researchers developed two solutions for the design of NTFs. The first accounts for a change in the rate of nitrification with NTF depth ($k > 0$) (eq 11.40), and the second assumes no decrease in the rate of nitrification with NTF depth ($k = 0$) (eq 11.40).

$$\frac{a \cdot J_{N,max}(T)}{k \cdot v_h} \cdot (1 - e^{-k \cdot z}) = S_{in,N} - S_{B,N} + K_N \cdot \ln\left(\frac{S_{in,N}}{S_{B,N}}\right) \tag{11.40}$$

where $k = 0$

$$\frac{z \cdot a \cdot J_{N,max}(T)}{v_h} = S_{in,N} - S_{B,N} + K_N \cdot \ln\left(\frac{S_{in,N}}{S_{B,N}}\right) \tag{11.41}$$

where a = specific surface area (m^2/m^3);

k = empirical parameter describing nitrification rate decrease, 1/m;

= 0 to 0.16, typical 0.1;

v_h = NTF hydraulic load (with or without recirculation) (m^3/d/m^2);

z = NTF depth (m); and

$S_{in,N}$ = ammonia-nitrogen concentration in NTF influent stream, g/m^3.

These equations can be solved directly to size an NTF for a desired $S_{B,N}$. When recirculation is used, an iterative solution routine that includes eq 11.42 is required because of the effect recirculation has on both v_h and $S_{in,N}$:

$$S_{N,i} = \frac{S_{0,N} + R \cdot S_{B,N}}{1+R}$$

$$R = \frac{S_{0,N} - S_{in,N}}{S_{in,N} - S_{B,N}}$$

(11.42)

where $S_{0,N}$ = ammonia-nitrogen concentration in the influent stream before being mixed with the recirculation stream.

The ammonia-nitrogen concentration in NTF influent stream, $S_{in,N}$, will be less than $S_{0,N}$ when recirculation is applied.

7.6.3   Modified Gujer and Boller Model

Parker et al. (1989) modified eq 11.39 to account for oxygen transfer efficiency variability among modular plastic media types and operating conditions. The revised expression follows.

$$J_N(z,T) = E_{O_2} \cdot \frac{J_{O_2,max}(T)}{4.3} \cdot \frac{S_{B,N}}{K_N + S_{B,N}} \cdot e^{-kz}$$

(11.43)

where E_{O_2} = dimensionless NTF media effectiveness factor.

Gujer and Boller (1986) reported, based on their experience, an E_{O_2} value in the range 0.93 to 0.96 for K_{S,O_2} = 0.2 g O$_2$/m^3 and the temperature range 5 to 25°C. Parker et al. (1989), on the other hand, observed lower E_{O_2} values (in the range 0.7–1.0) and claimed that a departure from E_{O_2} = 1.0 accounts for wetting inefficiency, biofilm grazing by predatory macro fauna, or competition between autotrophic nitrifying and heterotrophic bacteria for dissolved oxygen. The researchers recommended that medium-density crossflow media be used in NTF applications, and that E_{O_2} may range from 0.7 to 1.0. High-density crossflow media had a corresponding E_{O_2} approximately equal to 0.4 (Parker et al., 1995). According to Parker et al. (1995), is the zero-order $E_{O_2} \cdot \frac{J_{O_2,max}(T)}{4.3}$ ammonia-nitrogen flux. The maximum dissolved-oxygen flux reflects the oxygen transfer efficiency of the selected modular plastic media, which was determined by researchers using the Logan trickling filter model (Logan et al., 1987a). The coefficient K_{S,O_2} determined for the Central Valley WRRF, Utah, was between 1 and 2 mg/L (Parker et al., 1989). Additional research is required to establish values for a wide variety of operating conditions.

7.6.4 Albertson and Okey Model

The empirical design procedure proposed by Albertson and Okey (1988) can be summarized as the sum of the medium-density NTF media for zero-order and first-order regions. The design procedure includes two steps:

1. Determine trickling filter media volume based on zero-order kinetics using medium density (138-m²/m³) media and an NH₃-N flux (J_N) of 1.2 g/m²/d over a temperature range of 10 to 30°C. Below 10°C, adjust the rate using $u = 1.045$.

2. Determine trickling filter media volume based on first-order kinetics using a rate (J_N), which equals the following formulation and does not have a temperature correction between 7 and 30°C:

$$J'_N = J_N^{avg} \cdot \left(\frac{S_{N,e}}{S_{N,TRAN}}\right)^{0.75} = 1.2\frac{g}{d \cdot m^2}\left(\frac{S_{N,e}}{S_{N,TRAN}}\right)^{0.75} \tag{11.44}$$

where $S_{N,TRAN}$ = a transition NH₃-N concentration (mg/L) that can be determined from Figure 11.69.

This concentration is dependent on the degree of oxygen saturation and temperature. The designer can determine the total media volume by adding the volume required

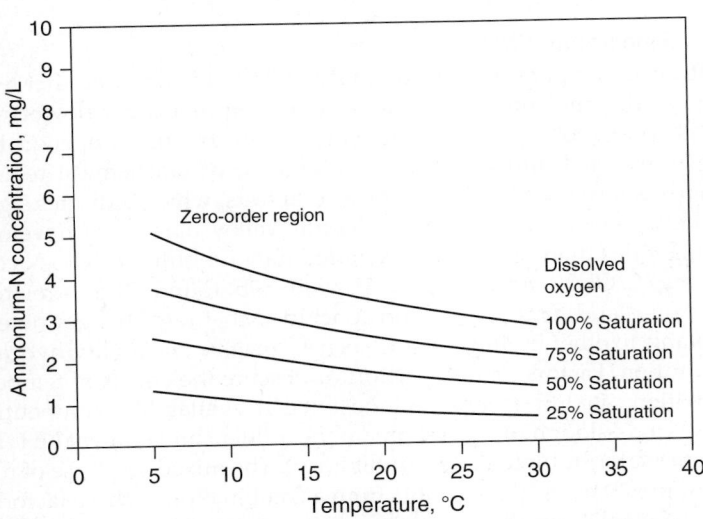

Figure 11.69 Transitional ammonium-nitrogen concentrations as functions of temperature (the transitional region below the 100% saturation line may be either zero order or first order, depending on the oxygen concentration).

for both zero- and first-order kinetic realms. The above design procedure stipulates several conditions be met:

- Ratio of BOD_5 to TKN ≤ 1.0;
- Filtered $BOD_5 \leq 12$ mg/L;
- $Q(1 + R)/A \geq 0.54$ L/m²/s (0.8 gpm/sq ft);
- Carbonaceous BOD_5 and TSS ≤ 30 mg/L for medium density (138-m²/m³) (42-sq ft/cu ft) media;
- Power ventilation; and
- Distributor control to provide instantaneous application rate, DR, of 25 to 75 mm/pass and flushing greater than or equal to 300 mm/pass.

7.6.5 Comparison of NTF Models

Wall et al. (2001) found that both the model of Gujer and Boller (1986) and Albertson and Okey (1988) provide good prediction of general NTF performance under average NH_3-N loading conditions. The designer should note that the comparison performed by Wall et al. (2001) did not include the modification to the model of Gujer and Boller (1986) proposed by Parker et al. (1989). For effluent ammonia, however, the models typically showed more significant peaks and troughs than sample data.

The Gujer and Boller model predicted peaks more exaggerated than the model of Albertson and Okey (1988). No justification was presented for the models' inability to account for peak NH_3-N loading conditions. Parker et al. (1995) demonstrated that the modified Boller and Gujer model (eq 11.22) effectively predicted NTF effluent NH_3-N loading concentrations under both average and peak conditions. Example results of the study performed by Parker et al. (1995) are illustrated in Figure 11.70.

7.6.6 Temperature Effects

The effects of temperature are variable in NTFs. It is reported that temperature affects zero-order (higher) nitrification rates more than first-order (lower) nitrification rates. Research does not explain if temperature effects are dampened because of liquid viscosity (external diffusion resistance limited) or biochemical reaction. Figure 11.71 summarizes tertiary NTF data from several tests, which indicates significant temperature effects on nitrification rates. Central Valley data were developed from higher hydraulic rates than typical and excluded data for effluent NH_3-N concentrations less than 5 mg/L. Okey and Albertson (1989a; 1989b) found little correlation between nitrification rates and temperature and concluded that rate changes noted by others were attributable to other limiting factors such as oxygen availability, hydraulics, and NH_3-N concentration. Factors that can distort or obscure the effects of temperature and cause perturbations in test results include oxygen availability, competitive heterotrophic activity, solids-sloughing cycles, predators, influent and effluent NH_3-N concentrations, and wastewater-induced effects (inhibitory). The mixed response of nitrification rate to temperature changes likely results from a combination of these factors.

7.6.7 Hydraulic Application

Optimal hydraulic requirements for promoting maximum nitrification rates are still unknown. Okey and Albertson (1989b) and Gullicks and Cleasby (1986; 1990) presented

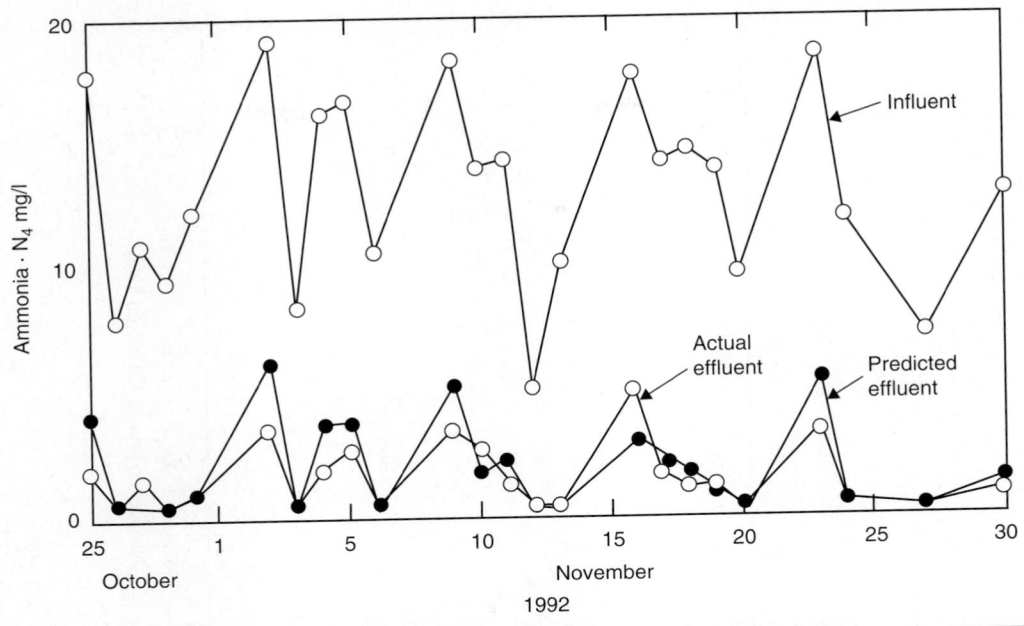

FIGURE 11.70   Actual and predicted effluent form a nitrifying trickling filter (Parker et al., 1995). Predicted effluent was calculated using the modified Gujer and Boller model.

data from studies indicating that increasing the application rate $(L/m^2/s)$ increased the rate of NH_3-N oxidation. Application rates of more than 1 $L/m^2/s$ produced the best results. Okey and Albertson (1989b) noted that hydraulic effects were complex and might be interwoven with oxygen availability. The effects of hydraulics were found to be more significant in the zero-order range and difficult to discern in the first-order range of less than 4 mg/L NH_3-N. Rearranged data taken from the Arizona nuclear pilot-plant study illustrate the effect of effluent NH_3-N concentration on the rate of nitrification and hydraulic application (Okey and Albertson, 1989a; 1989b). The nitrification rate depended on effluent NH_3-N concentration and available oxygen. If effluent NH_3-N exceeds 5 mg/L, then rates are high and are consistent with the findings of Parker et al. (1990).

7.7   Design Considerations

The following sections presents specific information related to selection and construction of reactors and equipment associated with trickling filters.

7.7.1   Distribution System

Methods of supplying wastewater to the trickling filter distributor include gravity feed, dosing siphons, and pumping. The conveyance selected depends on the hydraulic gradient available and the distributor. Distributors require piping between conveyance systems and the trickling filter distribution system. Where the trickling filter is not designed for continuous dosing, a pump or dosing tank and siphon may precede the distribution system.

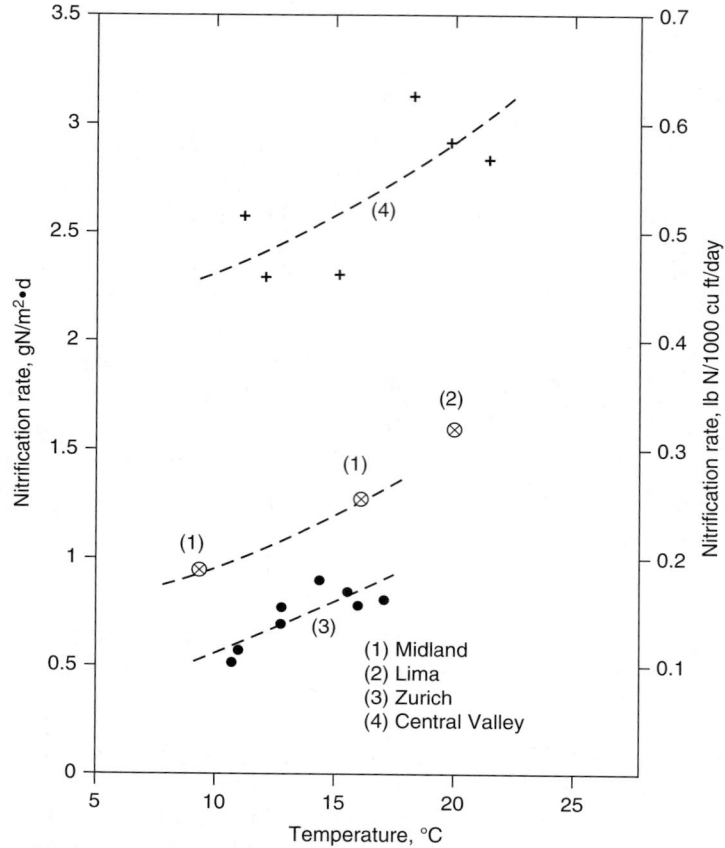

FIGURE 11.71   Effect of temperature on nitrification rate of nitrifying trickling filters.

Flow distribution is an important feature in a trickling filter system. Flow must be applied at a rate that keeps the media wetted and unclogged. Uneven application of flow and insufficient flow rates for adequate biofilm control will result in poor performance. Odors will result as solids build up and clog the trickling filters and growth of nuisance organisms will increase. Most new trickling filters are circular to accommodate a rotary distributor. Rock filters also may use rotary distributors, fixed nozzles, continuous feed, or periodic dosing with siphons or a sequenced pumping arrangement. New facilities should use rotary distributors, which are discussed in this section. If a rectangular unit is upgraded with a rotary distributor, then special provisions for wetting media outside the diameter of the rotary distributor should be made, or unwetted media should be removed because the wet-dry area will provide a breeding area for undesirable fauna such as filter flies. The need for and benefits of providing a means of controlling the instantaneous application rate, or DR, was reviewed in the discussion of hydraulics.

7.7.2 Hydraulic Propelled Distributors
The conventional rotary distributor is propelled by thrust from hydraulic discharge as pumped water contacts splash plates from the trailing side of the distributor

FIGURE 11.72   Hydraulically propelled rotary distributor (HydroDoc™). Courtesy of WesTech
Engineering, Inc.

arms as shown on Figure 11.72. Little attention has been paid to the velocity of rotation, which, in part, dictates the instantaneous dosing (mm/pass of an arm). A typical dosing rate for conventional rotary distributor systems is 2 to 10 mm/passes at
0.2 to 1.5 min/rev. The rotary distributor typically is equipped with two to six arms.
The distributed flow may be staggered for full coverage per arm. That is, each arm
may provide 50% or 100% coverage per revolution. Providing appropriate flushing
intensity is difficult with rock media operating at typical application rates of 0.2 to
0.6 m³/m²/h (0.08–0.25 gpm/sq ft). No minimum speed has been specified for a
hydraulically propelled distributor. To increase flushing, distributor speeds have been
reduced with reverse thrusting jets, as shown in Figure 11.73. Some hydraulically
driven units have stalled or stopped rotating at speeds from 4 to 20 min/rev. Unless

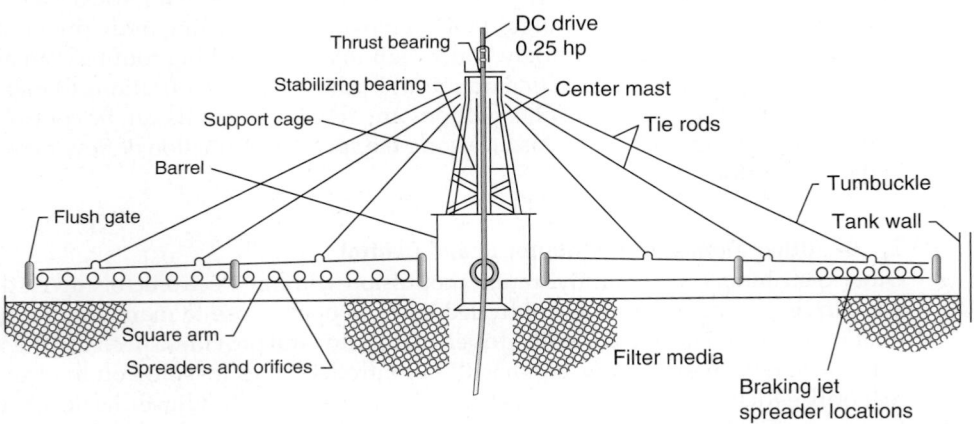

FIGURE 11.73   Typical rotary distributor with braking jets and electrical drive.

trickling filters with hydraulically propelled distributors receive nearly constant flow, however, most cannot operate at minimum speed during average to peak diurnal loading. Such an accommodation would result in the distributor stopping during low flow periods. Thus, even with reverse jets, conventional hydraulically propelled distributors often will have unpredictably limited capabilities of maintaining, and especially fluctuating, the desired DR. Either a mechanical VFD or a hydraulically propelled rotary distributor with sliding gates controlled by programmable logic control (PLC) can help avoid these limitations.

7.7.3 Electronic or Mechanically Driven Distributors

Electrically driven rotary distributors may have either a center or peripheral drive. Such an apparatus can typically be retrofitted easily at low cost. The units can be programmed to operate at varying DR as required to optimize BOD_5 removal, nitrification, or macrofauna control. The center-driven unit will be anchored to nonrotating parts of the influent structure, as shown in Figure 11.74. Where no upper steady bearing exists, support must be installed with a stationary shaft to provide a platform for the drive unit. This can be located in the mast support for the arm guy wires. Where an upper steady bearing does exist, the stationary shaft into this assembly can be extended to support the drive assembly.

A peripherally mounted electric drive can be used instead of a center drive. The traction drive can either use the inside or top of the wall. By spring loading the drive wheel, it can operate with irregularities in the wall. A rotary union is used to transmit power. The arrangement is similar to that of a traction drive clarifier. It should be emphasized that hydraulic motors with equally wide speed ranges may be used instead of electrically propelled units.

Electrically propelled units with remote variable-speed controllers and timers can operate independently of flow. This is particularly advantageous for WRRFs without recycle flow or sufficient recycle flow to minimize the rotational speed. With two units, optimum DR and biofilm control requirements can be determined. For example, the flushing would be best conducted during low-flow and loading periods, such as 1:00 A.M. to 6:00 A.M. coincidentally, this is also when clarification capacity is at a maximum. Optimum DR can be determined by simultaneously operating trickling filters at different operating DR, evaluating $rbBOD_5$ removal, and adjusting individual trickling filter distribution speed accordingly. Daily high-intensity flushing routines can also be programmed for the units to define optimum flushing DR and durations to maximize performance. Once these two DR conditions are defined, the units can be controlled to optimize operating conditions. Also, varying the speed as a function of flow may result in the best performance.

7.7.4 Other Means for Distributor Speed Control

Other distributors maintain hydraulic propulsion without electronic or motor drives. This drive uses pneumatically controlled gates to open/close to maintain the desired rotational speed. The hydraulically driven distributor still provides speed control inherent to electric drives. This system has PLC-controlled gates installed on the front and rear orifices of the outer section of each arm that proportionally adjusts the flow between the forward and reverse direction for speed control.

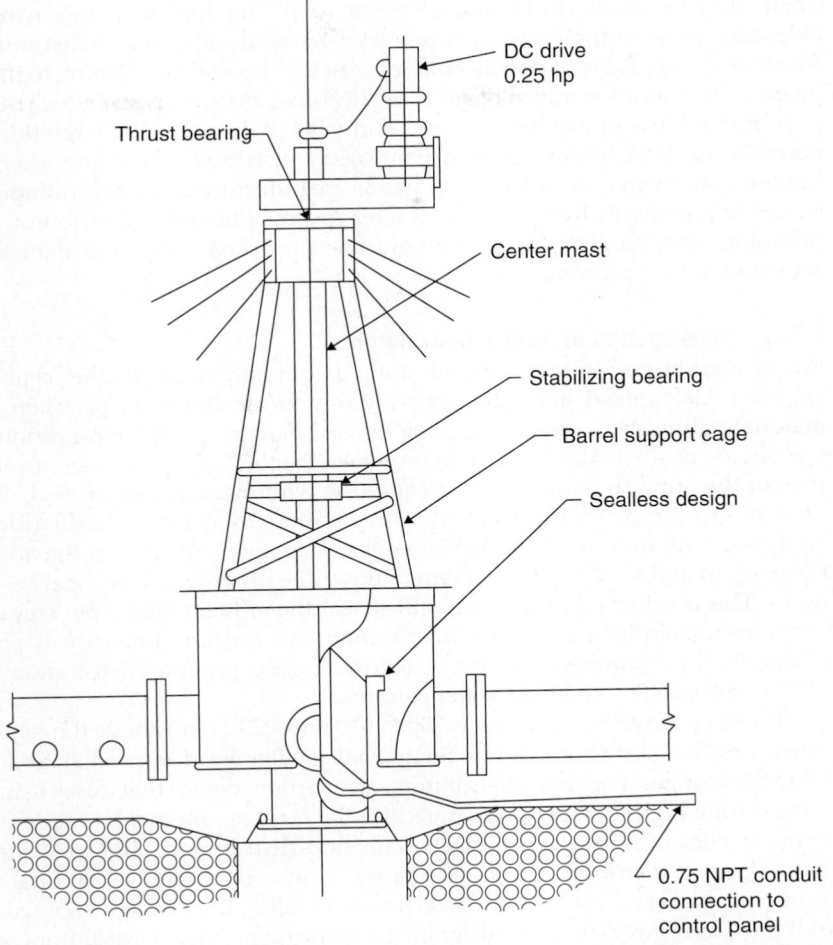

FIGURE 11.74 Electrically driven distributor with overflow (sealless) arrangement (hp × 0.745 7 = kW) (NPT = national pipe thread).

7.7.5 Trickling Filter Pumping Station or Dosing Siphon

Most trickling filters use recirculation pumps, which are typically constant-speed, low-head centrifugal units designed to operate with a total head of the trickling filter media depth, 2.0 to 3.0 m of static head, and friction losses. The VFD-controlled motors are now typical fixtures on process pumps. Both submerged or non-submerged (dry-pit) vertical pumps have been used extensively. Pump intake screens are typically unnecessary because the recirculated flow is typically free of clogging solid materials. Hydraulic computations are always necessary. Computations for minimum flow are necessary to ensure adequate head to drive hydraulically driven distributors and computations for maximum flow indicate the head required to ensure adequate discharge capacity. The net available head at the horizontal center line of the distributor's arm and other points may be calculated by deducting the following applicable losses from the available static head: entrance loss, drop in level in the dosing tank as distributor pipes are filled (only applicable to dosing siphons), friction losses in the piping to the distributor, proper allowance for minor headlosses, headloss through distributor riser and center port, friction loss in distributor arms, and velocity head of discharge through nozzles necessary to start hydraulically driven rotary distributor. Trickling filter distribution head requirements are set by a system's manufacturer. Despite pumping headloss, power requirements for the trickling filter process (including distributor, recirculation pumping, and auxiliary powered equipment) are typically less than those for the activated-sludge process.

7.7.6 Construction of Rotary Distributors

Rotary distributor arms are typically tubular, but may come in other shapes (e.g., rectangular). Galvanized mild steel and aluminum are the most common construction materials, although stainless steel may be used in more corrosive conditions. A series of nozzles are positioned in the arm to provide either 50% or 100% coverage of the unit per pass of the arm. These nozzles are equipped with manually controlled slide gates for flow and splash plates to wastewater distribution. In many cases, distributors may be equipped with four arms in a high-low flow arrangement. Two of the arms operate at flows up to and slightly above average flow. The other two arms operate during peak flows. This is achieved by interior baffles near the influent feed pipe. This arrangement provides maximum wastewater distribution and flushing intensity, if practiced. The hydraulic head required to drive a distributor and provide distribution ranges from 410 to 1000 mm (16–40 in.) of water column.

The head for minimum flow is 300 to 610 mm (12–24 in.) above the center line of the orifices on the distributor arms. Somewhat greater head is needed to accommodate wide flow ranges. For some distributors, an overflow device that doses using additional arms during high-flow periods can reduce the head requirement. Maintaining the flow to the nozzle at the minimum velocity enhances distribution. High velocities will result in inadequate distribution at higher flows. Units operating at rotational velocities of 1 rev/min can exert centrifugal force; newer trickling filter designs operating at speeds of 8 to 50 min/rev will exert insignificant centrifugal force. Distributors require a seal between the fixed influent column and the rotary section. Older designs have various types of water traps, mercury seals, or packed mechanical seals to prevent water from leaking between fixed and rotary parts. One type of seal is an overflow arrangement without a lower seal. This type creates no friction on the mechanism and requires no maintenance; the head required, however, is higher than the modern mechanical seal. The modern mechanical seal, with a double neoprene seal with a stainless steel seal ring, also requires no maintenance and needs less head than the sealless design. When older units are upgraded, improvements often include one of these arrangements.

7.7.7 Filter Media

As previously described, ideal trickling filter media provides a high specific surface area, low cost, high durability, and high enough porosity to avoid clogging and promote ventilation (Metcalf and Eddy, Inc./AECOM, 2013).

7.7.7.1 Media Selection   The design engineer must make an informed decision regarding the selection of media for specific trickling filter applications, including the construction of new facilities and retrofit of existing facilities. A common upgrade for rock-media trickling filters is replacement of existing filter media with a synthetic media. In some cases, the existing rock media may require vertical wall expansion to contain additional synthetic media. Drury et al. (1986) demonstrated that existing rock-media trickling filters can be improved simply by using existing volume and replacing rock with synthetic media. This is because of increased specific surface area, ability to increase hydraulic loading, and improved ability to control biofilm growth. Changing the media can help address problems such as severe odor generation and deterioration of media or to expand capacity using existing footprint and assets. Table 11.29 provides a list of guidelines for the best available synthetic trickling filter media for a specific application. Crossflow media typically will perform better than vertical-flow media in low- to medium-organic loading scenarios. However, if the TOL becomes substantial, then biofilm accumulation will be so great that performance in the trickling filter will be hindered by the crossflow media. Parker (1999) suggested that this efficiency change illustrates the "switchover effect" in which efficiency switches over from crossflow to vertical-flow media at high TOLs that is not observed in the other studies.

Gullicks and Cleasby (1986) have demonstrated the importance of synthetic media wetting. Crine et al. (1990) have demonstrated that lava rock and random media wetting effectiveness decreased with increasing specific surface area. The researchers found that wetting effectiveness was only 0.2 to 0.6. In some cases, media can be combined such that media efficient in hydraulic distribution are in the upper layers and media less prone to excessive biofilm accumulation compose the remainder. Upper portions of the trickling filter will receive higher organic loading and lower layers will receive little organic loading. Many reports indicate that denser media, such as random and crossflow, that are effective for flow redistribution are more prone to solids retention and fouling (Boller and Gujer, 1986; Crine et al., 1990; Gullicks and Cleasby, 1986; 1990; Onda et al., 1968; Parker et al., 1989). Applications such as treating strong wastewaters, pretreatment with fine screens, and roughing tend to produce thicker biofilms. Vertical media types are preferred for these applications. Media types may evolve as wetting and use of the media surface are better understood.

Type	Specific Surface Area (m²/m³)*	Roughing	Carbon Oxidizing	Carbon Oxidizing and Nitrification	Tertiary Nitrifying
Rock	40–60		✓	✓	
Wood	45	✓	✓	✓	
Random	85–110		✓	✓	
	130–140				✓
Vertical Flow	85–110	✓	✓	✓	
Crossflow	130–140				✓
	85–110		✓	✓	
	130–140				✓

*m²/m³ × 0.3048 = sq ft/cu ft.

Table 11.29 Applications for Trickling Filter Media

Typically, rock media are not used for new WRRFs; nonetheless, existing units may often be part of an expansion or upgrade. Performance may be enhanced by modifying the distributor speed or power ventilation or by adding solids contact or using the dual biological process as described earlier.

7.7.7.2 Filter Media Depth In North America, rock-media trickling filters typically are 1- to 2-m deep, but may be as deep as 2.4 m. This depth limitation is associated with lack of adequate ventilation produced by natural draft and an increased tendency to pond. In Europe, deeper filters are common; units in Arnheim, Netherlands, were constructed at 4.9-m deep but equipped with power ventilation. Comparative data are lacking for deep, power-ventilated rock filters and shallow, natural-draft rock-media.

Synthetic-media trickling filters typically are constructed between 5- and 8-m deep, although units up to 12.8-m deep exist. The limiting depth is because of height aesthetics, serviceability, pumping requirements, and structural design. The increased depth has no implication on biological treatment efficiency. Increasing the trickling filter depth is typically worthwhile to reduce the minimum flow required for high wetting efficiency. In taller filters that have high loadings, oxygen deficiency may occur in the uppermost layers. However, adequate ventilation and biofilm control measures can prevent problematic odors.

The effect of trickling filter media depth on bioreactor performance has been treated as a matter of controversy in previous design manuals. Several investigators suggest that volume, irrespective of depth, controls performance (Bruce and Merkens, 1970; 1973; Galler and Gotaas, 1964; Kincannon and Stover, 1982; NRC, 1946). Recent research indicates that performance is dependent on specific surface area (which may translate to bioreactor volume when considering identical media types), and not trickling filter depth. Essentially, performance is governed by substrate availability. In a combined carbon oxidation and nitrification trickling filter with a stratified depth, carbon-oxidizing facultative heterotrophic dominate biofilms near top layer of media. Multispecies biofilms, including both facultative heterotrophs and autotrophic nitrifiers, live near the center, and autotrophic nitrifying biofilms may exist in greater numbers in the lower layers. The facultative heterotrophy-dominated biofilm near the surface exists as a function of carbon-based substrate and oxygen; the autotrophic nitrifying biofilm exists as a function of ammonia-nitrogen and oxygen (in addition to the absence of carbon-based substrates). A given biofilm with a given bacterial density will have a capacity to oxidize a finite mass of carbon-based substrates. Therefore, the "layer" of trickling filter containing a carbon-oxidizing biofilm will vary as a function of the influent load. This can theoretically be achieved with a tall or shallow trickling filter. Practical limitations are based on media wetting and site constraints for shallow trickling filters, and the aforementioned constraints for tall trickling filters. Most, if not all, of the improved performance with depth noted by some investigators is likely a result of improved hydraulic distribution. The average hydraulic rate should exceed 0.5 $L/m^2/s$ to ensure maximum performance.

7.7.7.3 Structural Integrity The choice of rock media often is governed by locally available materials or cost of transportation. Field stone, gravel, broken stone, blast-furnace slag, and anthracite coal have been used. Whatever material is chosen, it should be sound, hard, clean, free of dust, and insoluble in wastewater constituents. There is some difference in opinion as to the optimum size. A common specification requirement is that 95% or more of the media pass 2600-mm^2 mesh screens and be retained on 1600-mm^2 mesh screens. The pieces typically are specified to be uniform in size, with all three dimensions as nearly equal as possible. The material should not disintegrate under service conditions. Frequently, the material is specified to be substantially sound as determined by the sodium-sulfate soundness test. Specifications for placing rock-media include: (a) when placing trickling filter media, breakage and segregation of differently sized particles must be prevented; (b) media will be screened and cleaned immediately before placing to eliminate as many fine sediments, or stone fragments, as possible; (c) media will be placed by a method that does not require heavy traffic of any type on the top of those media already placed; (d) placing media by means of belt conveyor, wheelbarrow, or bucket crane will be acceptable.

Synthetic trickling filter media, specifically the bundle type (0.61 m × 0.61 m × 1.22 m), are the most common in new WRRFs. Bundle media are manufactured from PVC and random media are manufactured from polyethylene or polypropylene. Testing procedures herein apply to bundle media, but it is equally important that the random media have sufficient strength to resist subsidence because of the combined weight of media, water, and biomass.

Consideration should be given to long-term (96 hour) and short-term (less than two hour) test and the ability of either to predict media strength over what will hopefully be a minimum 20-year life. The PVC is suitable as a structural material as long as deformation, or creep, loading is not exceeded. The material fails by deformation, which can be a slow process that persists if loading is maintained.

New trickling filter media is stronger than 10-year-old trickling filter media because PVC weakens with time. In addition, the plasticizer dissipates and the media become brittle. Because of the high initial strength-to-weight ratio, exceptionally thin media may not be adequate and may shorten useful life. Typically, this is a problem of not understanding the relationship between short-term testing results and long-term load capacity. Test temperature is important because PVC loses strength as the temperature exceeds 18 to 21°C. The load testing should be conducted at the maximum water temperature. The database temperature from media suppliers is 23 +/− 1°C (73 +/− 2°F); however, this temperature may not satisfy specific duty requirements. Mabbott (1982) introduced the short-term compression test to assess media strength and reported that the modulas of elasticity and corresponding media strength dropped drastically with increasing temperature (see Table 11.30).

When wastewater temperatures exceed 30°C (86°F), all structural testing should be conducted at the maximum operating temperature of the media. The designer must carefully consider heat buildup coinciding with trickling filter shut down and its effect on modular plastic media structural integrity. The issue is amplified when trickling filters are covered by a dome that may prevent air from easily escaping.

Temperature, T (°C)	21	40	49	60
Ratio modules of elasticity at T/T_{70} (%)	100	85	75	50
Comparable minimum test load (lb/sq ft)	500	425	375	250

TABLE **11.30**   Change in Plastic Media Strength with Temperature

Aerobic biochemical reactions typically proceed within a temperature range of 5 to 40°C, which is the upper limit for growth of mesophilic bacteria (Grady et al., 1999). These temperatures may be observed in activated sludge systems, but are not common operational temperatures for trickling filters. Pore (internal biofilm) temperature approaches equilibrium with air temperature inside the trickling filter, which may be in the range of 10 to 30°C. In addition to ambient conditions, the amount of biomass present, biomass condition, and mode of ventilation affect trickling filter internal temperatures. Harrison (2007) showed that temperature control is an important consideration during emergency shutdown or installation of filter media. When procuring media, a good design practice is to specify a service temperature that exceeds actual water temperature to provide adequate protection during unplanned conditions. A service temperature of 38 to 49°C (100–120°F) would not be unreasonable for warmer climates, high organic loading rates, or for filters where partial plugging or temperature concerns exist. An alternative may be to provide rotary sprinklers within domed trickling filters for heat dissipation during shutdown periods, emergency, or otherwise.

7.8   Design Examples
The following are examples of trickling filter sizing based on the equations and materials presented in earlier sections.

7.8.1   Example 11.1: Biofilter Design for Carbonaceous Biochemical Oxygen Demand Limitations
Determine the size of a trickling filter with plastic filter media for providing secondary effluent that will provide an average effluent soluble BOD (S_e) of 15 mg/L.

where $q = 6630$ m³/d or 76.7 l/s; $R = 0.75$;
      $Li = 135$ mg/L; $Si = 75$ mg/L or 75 g/m³;
      Specific surface area = 90 $K_{20} = 0.19$ (l/s)$^{0.5}$/m²; and
      m²/m³ (27 sq ft/cu ft);
      $D = 5.49$ m (18 ft);   Water temperature = 14°C.

1. Determine the size of the trickling filter.

   (a) Determine the k_{20} using eq 11.37:

$$k_2 = k_1 \cdot \left(\frac{D_1}{D_2}\right)^{0.5} \cdot \left(\frac{L_1}{L_2}\right)^{0.5}$$

$$k_2 = k_1 \left(\frac{6.1 \text{ m}}{5.49 \text{ m}}\right)^{0.5} \left(\frac{150 \text{ g/m}^3}{135 \text{ g/m}^3}\right)^{0.5}$$

$$k_2 = (0.19)(1.054)(1.054)$$

$$k_2 = 0.21 \text{ (l/s)}^{0.5}/\text{m}^3$$

(b) Temperature correct for k_2 using eq 11.31:

$$k_t = k_{20}/1.035^{(T-20)}$$

$$\theta_{14} = 1.035^{(14-20)}$$

$$= 0.814$$

(c) Calculate the allowable total hydraulic application rate, THL, using eq 11.35:

$$\ln\left[\frac{R+1}{S_i/S_e + R}\right] = \frac{-K_s D\, 1.035^{(T-20)}}{(\text{THL})^n}$$

$$\ln\left(\frac{0.75+1}{75/75+1}\right) = -\frac{(0.21)(5.49\text{ m})(0.814)}{(\text{THL})^{0.5}}$$

$$\ln(0.292) = -1.232$$

$$\text{THL}^{0.5} = \frac{-0.938}{-1.232} = 0.762$$

$$\text{THL} = 0.580\text{ L/m}^2 \cdot \text{s}$$

Since $R = 0.75$, then

$$\gamma = q \times R = (76.7\text{ l/s})\,0.75$$

$$= 57.5\text{ l/s}$$

$$\text{THL} = \frac{(76.7 + 57.5)}{A} = 0.580\text{ L/m}^2 \cdot \text{s}$$

then $A = 231\text{ m}^2$

2. Check to see whether the trickling filter BOD loading and size seem reasonable based on the experience-based criteria.

   (a) Calculate organic loading rate, TOL, to the trickling filter:

$$\text{TOL} = \frac{\text{BOD applied}}{\text{volume of media}}$$

$$\text{TOL} = \frac{(6,630\text{ m}^3/\text{d})(135\text{ g/m}^3)\,1\text{ kg}/10^3\text{ g}}{(231\text{ m}^2)(5.49\text{ m})}$$

$$= \frac{895\text{ kg/d}}{1268\text{ m}^3}$$

$$= 0.706\text{ kg/m}^3 \cdot \text{d}\ (44\text{ lb BOD/d} \cdot 1\,000\text{ ft}^3)$$

Compare modeling results of TOL = 0.706 kg/m³/d with experience-based values in Table 11.26. Table 11.26 indicates that effluent of secondary quality standard may be possible with a TOL as high as 0.96 kg/m³/d. Consider local conditions and other trickling filter case histories in the area where treatment is to occur. It may be that assumptions for rational equations were not correct. Reassess assumptions, consider using a different equation, conduct a pilot test, or defer to actual trickling filter experience.

7.8.2 Example 11.2: Nitrification Trickling Filter Design

Design plastic media trickling filter towers for tertiary nitrification.

where,

Flow, $m^3/d = 37\,840$ (438 L/s or 10 mgd);   Influent TKN-N = 28 mg/L (1060 kg/d);
Influent BOD, $Li = 20$ mg/L (757 kg/d);   Influent NH_4^+-$N_i = 25$ mg/L;
Influent SBOD, $Si = 8$ mg/L (303 kg/d);   Effluent TKN-$N_e = 1.5$ mg/L; and
SAA – 138 m^2/m^3;   Water temperature = 12 °C.

1. Check whether criteria for tertiary nitrification design are met.

Ratio of BOD_5 to TKN   = 20/28 kg BOD/kg N;
= 0.71 kg BOD/kg N, which is less than 1.0 kg BOD/kg N; and
Soluble BOD_5 = 8 mg/L, which is less than 12 mg/L.

2. Determine media surface area for zero-order nitrification using the first step of the design procedure with 138-m^2/m^3 media and $k_n = 1.2$ g/m^2/d. From Figure 11.69, transitional ammonium-nitrogen (N_T) concentration is approximately 3.2 mg/L at 75% saturation and 12°C.

Total Kjeldahl nitrogen   = [(438 L/s)/(1 000 L/m^3)] (86 400 s/d)
[(25 × 3.2 mg/L)/(1000 mg/g)] (1000 L/m^3)
= 825 000 g/d
Media surface area   = (825 000 g/d)/(1.2 g/m^2/d)
= 687 500 m^2

3. Determine the media surface area for first-order nitrification using the second step of the design procedure using eq 11.44.

Total Kjeldahl nitrogen   = (438/1000) (86 400) (3.2 × 1.5/1000) (1000)
= 64 330 g/d
k_n = 1.2 $(N_e/N_t)^{0.75}$
= 1.2 $(1.5/3.2)^{0.75}$
= 0.68 g/m^2/d
Media surface area   = (64 330 g/d)/(0.68 g/m/d)
= 94 600 m^2

4. Determine the total media volume required.

Total media surface area   = 687 500 + 94 600
= 782 100 m^2
Total media volume   = (782 100 m^2)/(138 m^2/m^3)
= 5670 m^3

5. Calculate maximum tower surface area based on the minimum flow rate of 0.54 L/m^2/s stipulated by the design procedure.

Maximum tower surface area   = (438 L/s)/(0.54 L/m^2/s)
= 811 m^2
Minimum depth   = 5670 m^3/811 m^2
= 6.99 m

6. Determine number and size of towers. Two biotowers operating in parallel with a total area of 775 m² (8330 sq ft) are suggested.

Diameter	$= [(4/r)\,(775)]^{0.5}$
	$= 22.2$ m
Depth	$= 5670$ m³$/775$ m²$= 7.32$ m
WHL	$= (438$ L$/$s$)/(775$ m²$)$
	$= 0.57$ L$/$m²$/$s
TOL	$= 0.13$ kg$/$m³$/$d

7.8.3 Example 11.3: Organic and Hydraulic Loading

Determine the TOL, SOL, WHL and THL at an existing trickling filter WRRF.

where,

Two, 25.9-m (85-ft) diameter plastic media trickling filters have a media depth of 4.27 m (14 ft).

$q = 15\,000$ m³$/$d; $r = 7500$ m³$/$d;

$S_{i,total} = 140$ mg$/$L; $S_{i,soluble} = 90$ mg$/$L or 90 g$/$m³; and

Specific surface area $= 98$ m²$/$m³ $S_i - 90$ mg$/$L.

(30 sq ft$/$cu ft);

1. Find the influent BOD load to the trickling filter.

   (a) Calculate influent total and soluble BOD loads in kg$/$m³$/$d.

$$\text{BOD applied} = q\,Li = (15\,000\text{ m}^3/\text{d})(140\text{ g}/\text{m}^3)\,1\text{ kg}/10^3\text{ g}$$

$$= 2100\text{ kg}/\text{d}$$

$$\text{SBOD applied} = q\,Si = (15\,000\text{ m}^3/\text{d})(90\text{ g}/\text{m}^3)\,1\text{ kg}/10^3\text{ g}$$

$$= 1350\text{ kg}/\text{d}$$

   (b) Determine trickling filter media area and volume.

$$\text{Area } A_1 = \pi\,(\text{dia})^2/4$$

$$A_1 = 3.14\,(25.9\text{ m})^2/4$$

$$A_1 = 526.6\text{ m}^2$$

$$\text{Total Area} = (\#\text{ units})(A_1) = 2(526.6\text{ m}^2)$$

$$A_T = 1053.2\text{ m}^3$$

$$\text{Total Volume } V_T = A_T\,(D)$$

$$= (1053.2\text{ m}^2)(4.27\text{ m})$$

$$= 4497\text{ m}^3$$

(c) Calculate total and soluble organic loading of the trickling filter (TOL and SOL, respectively, using eq 11.20).

$$\text{TOL} = \frac{\text{BOD applied}}{V_M} = \frac{2100 \text{ kg/d}}{4497 \text{ m}^3}$$

$$= 0.467 \text{ kg BOD/d} \cdot \text{m}^3$$

$$\text{SOL} = \frac{\text{SBOD applied}}{V_M} = \frac{1350 \text{ kg/d}}{4497 \text{ m}^3}$$

$$= 0.30 \text{ kg SBOD/d} \cdot \text{m}^3$$

2. Determine hydraulic loading rates to the trickling filter system.

(a) Calculate wastewater hydraulic loading to the trickling filter using eq 11.22.

$$\text{WHL} = \frac{q(1000 \text{ m}^3/\text{ML})}{A}$$

$$\text{WHL} = \frac{(15\,000 \text{ m}^3/\text{d})}{1053 \text{ m}^2}$$

$$= 14.2 \text{ m}^3/\text{d} \cdot \text{m}^2$$

(b) Calculate total hydraulic loading to the trickling filter using eq 11.21.

$$\text{THL} = \frac{\text{total pumped flow}}{A} = \frac{q+r}{A}$$

$$\text{THL} = \frac{(15\,000 \text{ m}^3/\text{d} + 7500 \text{ m}^3/\text{d})}{1053 \text{ m}^2}$$

$$= 21.4 \text{ m}^3/\text{d} \cdot \text{m}^2$$

(c) Calculate recirculation rate.

$$R = \frac{r}{q} = \frac{7500 \text{ m}^3/\text{d}}{15\,000 \text{ m}^3/\text{d}} = 0.5$$

(d) Determine surface loading rate to the filter using eq 11.21.

SLR = TOL/SSA = 0.467 kg/d·m³/98²/m³ = 4.76 × 10⁻³ kg/d·m²

7.8.4   Example 11.4: Biofilter Classification and Distributor Adjustment

Determine filter classification and adjustment of the distributor speed for the trickling filters in Example 11.1.

where,

The distributor is operating at 1.5 rpm ($N = 0.667$ minutes per revolution). The secondary system has been experiencing extreme sloughing events. The distributor has four distributor arms.

$q = 1.1$ ML/d or 1100 m³/d;
$S_{i,total} = 140$ mg/L
Specific surface area $= 98$ m²/m³
  (30 sq ft/cu ft);

$r = 0.55$ ML/d or 550 m³/d;
$S_{i,soluble} = 90$ mg/L or 90 g/m³; and
$S_{i,soluble} = 90$ mg/L.

1. Determining trickling filter classification by comparing actual load values vs those in Table 11.26.

   (a) Organic loading: calculated TOL of 0.467 kg/m³/d falls in the lower range for carbon oxidizing trickling filters (0.32 to 0.96 kg/m³/d), suggesting that this is a lightly loaded conventional trickling filter.

   (b) Hydraulic loading: calculated THL is 21.4 m³/m²/d, which is in the lower range of the suggested range of 13.7 to 88 m³/m²/d. To increase the hydraulic load and rate, one approach would be to increase recirculation. It might also be good to consider slowing down the distributor or establishing a periodic flushing cycle. Finally, the manufacture of the filter media or distributor should be contacted for advice.

2. Calculate the distributor dosing rate and determine if changes in pumping or distributor speed should be made.

   (a) Calculate existing dosing rate.

   $$DR = \frac{(N)(q + r)(1000 \text{ mm/m})}{A(9)(1440 \text{ min/d})}$$

   $$= \frac{(0.667 \text{ min/rev})(15\,000 \text{ m}^3/\text{d} + 7500 \text{ m}^3/\text{d})\,1000}{(1053 \text{ m}^2)(2 \text{ arms})(1440)}$$

   $$= 4.95 \text{ mm/pass}$$

   (b) Comparing the above dosing rate with recommended values in Table 11.27, the recommended rates are 50 to 150 mm/pass, which is 10 to 30 times less than desired. Consideration could be given to adding a mechanically driven distributor and/or increasing recirculation pumping.

3. A discussion with suppliers for distributors indicates that the normal rotational speed would be 10 minutes per revolution for conventional operation and 40 minutes per revolution for flushing. Determine the dosing rate if $R = 1.0$ and the manufacturers rotational speeds are maintained.

   (a) Calculate the existing dosing rate.

   $$DR_{normal} = \frac{(10 \text{ min/rev})(15\,000 \text{ m}^3/\text{d} + 15\,000 \text{ m}^3/\text{d})\,1000}{(1053 \text{ m}^2)(2 \text{ arms})(1440)}$$

   $$= 142 \text{ mm/pass}$$

An evaluation of these changes indicates that slowing down the arms or increasing recirculation could enhance flushing and possibly reduce the magnitude of sloughing events.

8.0 Emerging Biofilm Reactors

New biofilm processes are emerging on the field of water resource recovery in the United States. We present here novel designs using anaerobic and aerated membrane biofilm reactors, and various suspended-biofilm reactors involving airlift and internal circulation regimes, granular sludge particles, and anammox microbial processes.

8.1 Membrane Biofilm Reactors

Membranes are efficient to filter water, separate gases, and transfer gases to and from liquids (Aybar et al., 2014). Gaseous substrates, for example dioxygen, dihydrogen, or methane, are delivered to biofilms forming membrane outer surface (Timberlake et al., 1988; Clapp et al., 1999; Lee and Rittmann, 2000). These processes are referred to as membrane biofilm reactors (MBfR), and membrane-aerated bioreactors (MABR) when used to deliver oxygen (Brindle and Stephenson, 1996a; Lee and Rittmann, 2000; Nerenberg and Rittmann, 2004; Syron and Casey, 2008).

Besides membrane sheets (Semmens, 2005), hollow-fiber membranes with external diameters of 280 μm are used in MBfRs to provide specific surface areas as high as 5100 m^2/m^3 (Pankhania et al., 1999; Adham et al., 2004). Microporous, hydrophobic materials foster high gas transfer rates (Yang and Cussler, 1986). Small pores prevent bubbling at low gas-supply pressures (Weiss et al., 1996). Unlike membrane bioreactors (MBRs) in which membranes filter water, pores of MBfRs are filled with gas, thus unlikely to foul with solids or microbial biomass. However, pores may become wetted, what hampers gas transfer rates.

The microporous hollow fibers are collected into a gas-supplying manifold at one end and sealed at the opposite end (Figure 11.75). Pressurized gas in the fiber lumen (interior) diffuses through the dry pores into the biofilm. When used in such "dead-end" mode, the entire gas passes into the biofilm, allowing high gas-use efficiencies. The gas flux to the biofilm can be modulated by controlling the gas pressure inside the membrane (Martin et al., 2010).

MBfR biofilms are subject to "substrate counter-diffusion." One substrate (electron donor or acceptor) diffuses into the biofilm from the bulk liquid. The other is supplied by diffusion across the membrane surface without traversing a liquid boundary layer, allowing greater fluxes. The liquid diffusion layer at the biofilm-liquid interface retains the gaseous substrate inside the biofilm. Substrate-rich conditions near biofilm attachment surfaces are beneficial to certain microbial processes. Substrate counter-diffusion is much less efficient with thick biofilms that decrease substrate fluxes significantly:

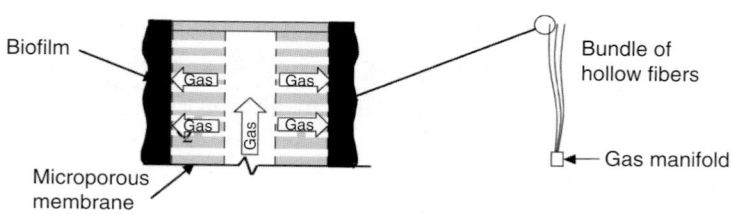

FIGURE 11.75 Section of fiber (left) and schematic of hollow-fiber membrane bundle (right).

e-donor and e-acceptor become rate-limiting on opposite sides of the biofilm, and only the middle section is metabolically active (Essila et al., 2000). Preventing biofilm accumulation is essential.

MBfRs are implemented for diverse applications and with different gases (Clapp et al., 1999; Grimberg et al., 2000; Syron and Casey, 2008). Methane-based MBfRs enable the co-metabolic reduction of trichloroethylene and trinitrophenol. Air-based MBfRs allow for nitrification and denitrification. Hydrogen-based MBfRs reduce arsenate, bromate, chromate, selenate, and trichloroethane (Chung et al., 2006a; Downing and Nerenberg 2007a; Chung et al., 2006b; Chung et al., 2006c; Chung and Rittmann, 2007).

8.1.1 Hydrogen-Based MBfRs

H_2-based MBfRs were first developed for drinking water treatment: the H_2 e-donor is used to reduce nitrate via biological autotrophic denitrification, and other oxidized contaminants (Ergas and Reuss, 2001; Lee and Rittmann, 2002; Nerenberg and Rittmann, 2004).

Advantages of H_2 over organic e-donors include (1) lack of human health toxicity, (2) use by indigenous bacteria without bioaugmentation, (3) low solubility preventing overdosing, (4) low biomass yield ($Y_{H2} = 0.4\ Y_{ethanol}$) resulting in less excess biomass, and (5) onsite H_2 generation. *Disadvantages* include (1) use of a combustible gas except if hydrogen is generated via electrochemical process, (2) aggregation of individual membranes into a single biofilm when membranes are packed at high densities, and (3) lack of experience at full scale.

Pilot testing demonstrated effective removal of nitrate and perchlorate from groundwater (Adham et al., 2004; Rittmann et al., 2004) and drinking water (Figure 11.76).

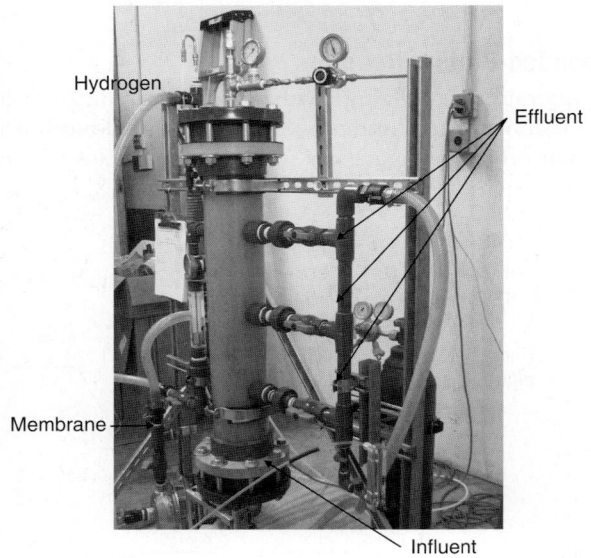

FIGURE **11.76** Small pilot membrane biofilm reactor for groundwater treatment (courtesy of APT, Inc).

Tertiary denitrification of wastewater was achieved (WaterReuse Foundation, Alexandria, Virginia).

8.1.2   Oxygen-Based MBfRs

O_2-based MBfRs provide "passive aeration" and can achieve concurrent COD-removal, nitrification, and denitrification (Timberlake et al., 1988; Suzuki et al., 1993; Brindle and Stephenson, 1996b; Brindle et al., 1998). Nitrification occurs in inner portions of the biofilm, close to the membrane filled with air or O_2. Denitrification and BOD-removal occur in outer portions with low concentration of O_2 dissolved in the bulk liquid phase (Schramm et al., 2000; Semmens et al., 2003).

Pilot testing showed that O_2-based MBfRs are effective to remove COD from high-strength brewery wastewater at surficial rates of 71 g $COD/m^2/d$ (Brindle et al., 1999). Simultaneous removal of COD and total nitrogen has been achieved using hollow fibers and membrane sheets. Ammonium was nitrified at up to 0.5 g$N/m^2/d$. However, rates decreased along membrane leakage and biofilm sloughing events (Semmens, 2005).

Hybrid (suspended and attached growth) MBfRs processes (HMBP; Figure 11.77) are efficient for BOD- and N-removal from wastewater (Downing and Nerenberg, 2007b). This process is similar to a cord-type IFAS retrofitted with hollow-fiber membranes instead of cords in the activated sludge tank. This process fosters (1) retrofitting of existing activated sludge tanks, (2) nitrification at short bulk-liquid solids retention times, (3) maximal use of influent BOD for denitrification, (4) lower energy demand since bubbled aeration is replaced by passive diffusion and water recirculation loops are avoided, and (5) nitrogen removal via nitrite (Downing and Nerenberg, 2008a; 2008b).

MBfR research targets efficient and cost-effective full-scale applications. Configurations aim for high specific surface area, good mixing, and control of biofilm accumulation.

8.2   Suspended-Biofilm Reactors

Reactor configurations involving fluidization of biofilm particles exist since the early 1980s, but are continuously rediscovered. Suspended-biofilm reactors involve either biofilms grown on carrier materials (e.g., inert particles in biofilm airlift reactors or

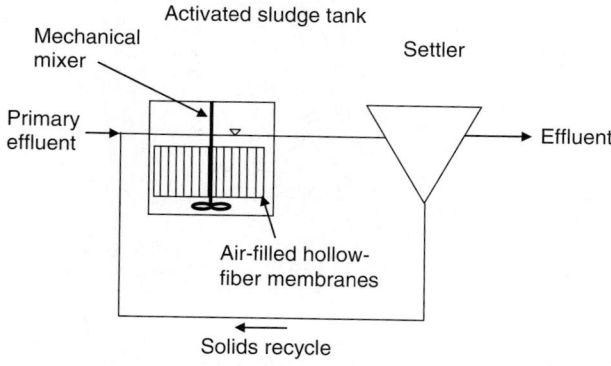

FIGURE 11.77   Hybrid membrane biofilm reactor process layout.

polymeric carriers in moving-bed bioreactors) or "self-granulated" biofilms (also called granules). The latter result from microbial assembly in compact and fast-settling bioaggregates, and are encountered under diverse redox conditions (anaerobic, anoxic, aerobic, and combinations of them).

8.2.1 Biofilm Airlift Reactors

Biofilm airlift reactors were developed to oxidize organic matter, sulfide, and ammonia aerobically (Heijnen et al., 1993). The tower process configuration is divided vertically into riser and downcomer sections (Figure 11.78). The air supplied traverses the riser from foot to top. The upward bubbling provides mixing and biofilm form around inert carrier particles in response to high upflow velocities and biomass wash-out regime installed by hydraulics (high dilution rates). The commercial CIRCOX® process is characterized by high loading capacity (4–10 kg COD/m³/d), short HRTs (0.5–4 h), fast-settling velocities of biomass (50 m/h), and concentrated biomass (15–30 g/L) (Frijters et al., 2000; Nicolella et al., 2000). Nitrification is easily achieved. A modified configuration includes an anoxic compartment for denitrification (Frijters et al., 2000) to treat nitrogen at volumetric loadings of 1 to 2 kg N/m³/d. Anaerobic biofilm airlift reactors (also called gas-lift reactors) use, for example methane, dihydrogen, or dinitrogen gases, to provide fluid circulation. These gases can be formed internally from microbial conversions occurring in the reactor (e.g., methanization), similar to what happens in up-flow anaerobic sludge blankets (UASB, see Section 8.2.2).

8.2.2 Granular Sludge Reactors: Anaerobic Digestion and Nutrient Removal

Granules are compact, particulate, and mobile biofilms that result from microbial auto-aggregation without external carrier material (Morgenroth et al. 1997; Beun et al. 1999; Tay et al. 2001; van Lier et al. 2008; Weissbrodt et al. 2013a; Johnson et al. 2015). The microbial assembly holds in a matrix of EPS (McSwain et al., 2005; Seviour et al., 2012).

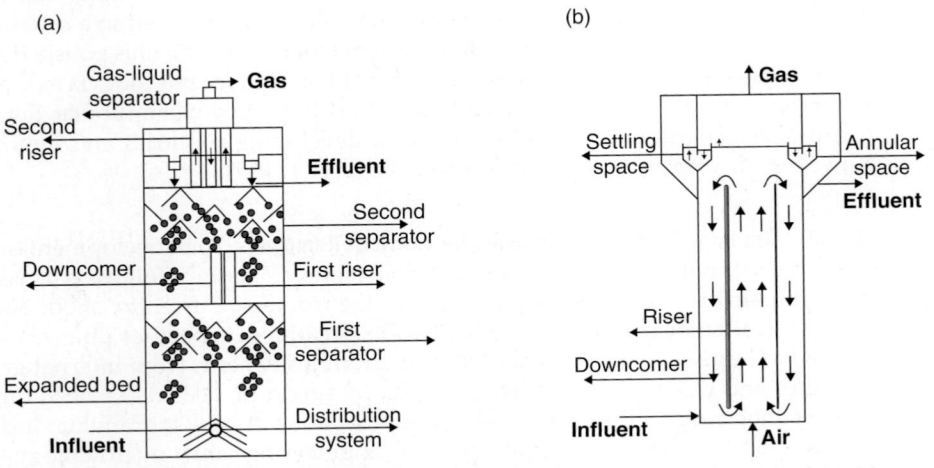

Figure 11.78 (a) Configuration for biofilm airlift reactor and (b) configurations for internal circulation reactor (after Nicolella, 2000).

Granules are defined as "aggregates of microbial origin, which do not coagulate under reduced hydrodynamic shear, and which settle significantly faster than activated sludge flocs" (de Kreuk et al., 2005b; 2007). Granules differ from flocs in metrics, material density, settling properties, and internal gradients, and behave like biofilms (Weissbrodt et al., 2014b). Granules display large diameters of <0.5 to 3 mm (Liu and Tay, 2002). Their density range from 1.05 to 1.30 kg/m³ depending on fractions of inorganics (Winkler et al., 2013): granules settle much faster than activated sludge flocs (Mancell-Egala et al., 2014). Granular sludge fosters process intensification via efficient solid-liquid separation, high biomass retention, and high volumetric treatment capacity (Morgenroth et al., 1997; van Lier et al., 2016), and facility-wide integration. It applies to ranges of redox conditions from anaerobic digestion (Lettinga, 1995; Lettinga et al., 1997; Beun et al., 1999; Bachman & Upendrakumar, 2000; Suthanthararajan et al., 2004; Batstone et al., 2005; Cakir and Stenstrom, 2007; Giraldo et al., 2007; van Lier et al., 2008; Angelidaki et al., 2011; van Lier et al., 2016) to biological nutrient removal (BNR) (de Kreuk et al., 2005a; Kishida et al., 2006; Bassin et al., 2012; Winkler et al., 2012b; Lochmatter et al., 2013; Weissbrodt et al., 2014a; Pronk et al., 2015b) and deammonification (van Dongen et al., 2001; Volcke et al., 2005; van der Star et al., 2007; Vlaeminck et al., 2008; Ni et al., 2009; Al-Omari et al., 2012; Regmi et al., 2012; de Clippeleir et al., 2013; Nifong et al., 2013; Wett et al., 2013; Lotti et al., 2014; Klaus et al., 2015; Perez et al., 2015; Han et al., 2016; Yang et al., 2016).

8.2.2.1   Granular Sludge Technologies for Anaerobic Digestion   UASB reactors were developed in the 1970s based on the propensity of slow-growing anaerobic organisms to form particulate biofilms (i.e., to "granulate") under loading and hydraulic constrains and up-flows of methane gas formed during digestion (Lettinga, 1995). The technology is well established. It evolved with expanded granular sludge bioreactors (EGSB) (Rinzema et al., http://dx.doi.org/10.1016/0032-9592 (93) 85014-7) (ref) and internal circulation (IC) reactors (Pereboom and Vereijken, 1994). IC reactors consist of two sequential UASB processes configured over the height of a tower design (Figure 11.78). The bottom compartment contains an expanded bed of granular sludge that converts organic matter to biogas in a high-rate process. The gas is collected in a separator and lifts water and sludge to the upper compartment where the effluent is polished at lower rate. The gas is separated and the sludge returned back to reactor foot via a down pipe. Anaerobic granular particle has also been found in polyethylene tubular designs for decentralized digestion of livestock waste in developing countries and remote areas (Kinyua et al., http://dx.doi.org/10.1016/j.bej.2015.11.017).

8.2.2.2   Granular Sludge Technologies for Nutrient Removal   The development of granular sludge technologies for BNR (e.g., original Nereda® design) followed on successes of biofilm airlift and anaerobic granular sludge reactors. Currently, about 30+ facilities are installed or under design world-wide. Treatment capacities range from small (5k p.e.) to large (3 mio p.e.) scales. BNR in granules benefits from internal gradients of solutes and redox (de Kreuk et al., 2005a; Morgenroth, 2008a). Niches of microorganisms metabolize nutrients across granule sections. A robust granular sludge process for BNR relies on conjunction of engineering and microbiology factors (Weissbrodt, 2012; Weissbrodt et al., 2014a). BNR is achieved in sequencing batch reactors (SBR) (de Kreuk et al., 2005a; Yilmaz et al., 2008). An integrated management of wastewater, fluid mechanics, SBR phase lengths, redox conditions, and microorganisms is needed (Zima et al., 2007; Zima, 2008; Weissbrodt et al., 2014b; Giesen et al., 2015; Pronk et al., 2015b; Derlon et al., 2016)(Verstraete et al., 2007; de Clippeleir, 2012; Vlaeminck et al., 2012; Weissbrodt, 2012).

SBR cycles comprise fill/draw, aeration, and settling phases. EBPR is obtained by alternating phases of anaerobic organic-feast and of aerobic or anoxic organic-starvation, like in conventional systems. The anaerobic selector is engineered and controlled (1) to select for polyphosphate-accumulating organisms (PAOs) or glycogen-accumulating organisms (GAOs) when EBPR is not requested, and (2) to prevent filamentous bulking (de Kreuk and van Loosdrecht, 2004; Weissbrodt et al., 2013a; 2014c). Conditions that select for slow-growing PAOs and GAOs contribute to stabilize granules. The influent is fed anaerobically in an up-flow regime through the settled bed of granules in the absence of e-acceptors (i.e., dissolved oxygen, nitrite and nitrate). COD gets hydrolyzed, fermented into volatile fatty acids (VFAs), and stored by PAOs and GAOs. The bed geometry and hydraulic regime are designed to ensure full depletion of COD prior to aeration. Mathematical modelling supports the design of feeding conditions to engineer microbial selection (Weissbrodt et al., 2014c). Flow mechanics should cope with the wastewater temperature and salinity (Bassin et al., 2011; Winkler et al., 2012a; Pronk et al., 2014). Impacts of wastewater compositions and particulate substrates need consideration in process design and operation (de Kreuk et al., 2010; Pronk et al., 2015a; Wagner et al., 2015b). Phase lengths need extension to deal with slowly biodegradable COD, like for conventional BNR processes (Weissbrodt, 2012; Weissbrodt et al., 2014b). An anaerobic mixed batch phase is an efficient measure to complement the feeding phase, to this end. During aeration, redox gradients in granules are managed by dissolved oxygen control for nitrification in the aerobic biovolume and denitrification in the anoxic biovolume of granules. This is achieved either simultaneously or alternately (Lochmatter et al., http://dx.doi.org/10.1016/j.water res.2013.07.030). Phosphorus is taken up by PAOs aerobically and/or anoxically. Aiming for a PAO-based granular sludge process is advantageous to (1) stabilize granules, (2) remove phosphorus at no extra cost and achieve more stringent quality criteria, and (3) to possibly recover phosphorus from waste granular sludge. Full BNR with granular sludge relies on SRT control (15–20 days) and purge of excess sludge rich in phosphorus (Weissbrodt et al., http://dx.doi.org/10.1016/j.watres.2013.08.043; Winkler, http://dx.doi.org/10.1016/j.watres.2011.03.024; Bassin et al., http://dx.doi.org/10.1002/bit.24457). Fully granulated processes are obtained at bench scale. Pilot and full-scale processes are mostly hybrid by comprising a substantial fraction of flocs. By analogy to IFAS (Veuillet et al., 2014), hybrid granular sludge can propel process efficiency and stability by a separate management of microbial processes in granules and flocs (Weissbrodt et al., 2014b)(Han et al., 2016). Existing flow-through activated sludge installations are retrofitted using different alternatives with, for example hydrocyclones and screens to mechanically select for granules (Wett et al., http://dx.doi.org/10.2175/1 06143015X14362865227319; Han et al., http://dx.doi.org/10.1016/j.biortech.2016.08.115). Granular sludge is also implemented for deammonification (or partial nitritation and anammox, PN/A) by slow-growing aerobic and anaerobic ammonium-oxidizing organisms (van der Star et al., http://dx.doi.org/10.1016/j.watres.2007.03.044; Perez et al., http://dx.doi.org/10.1016/j.watres.2014.08.028; Winkler et al., http://dx.doi.org/10 .1016/j.watres.2011.10.034; Vlaeminck et al., http://dx.doi.org/10.2166/wst.2008.731; de Clippeleir et al., http://dx.doi.org/10.1007/s00253-011-3222-6), using the engineering and ecological principles described under § 8.2.3 hereafter.

8.2.3   Anammox Biofilm Reactors and Moving Bed Biofilm Reactors (MBBRs)

Since discovery in the mid-1990s (Mulder et al., 1995), the deammonification process based on anaerobic ammonia oxidation (anammox) has gained popularity to treat nitrogen-rich waste streams. Deammonification is a two-step process, which utilizes the symbiotic relationship of two genera of bacteria, ammonia oxidizing bacteria (AOB) and anaerobic ammonia oxidizers (AMX). AMX are chemolithoautotrophs and members of the order Planctomycetes. They use ammonia as an electron donor and nitrite as an acceptor, producing dinitrogen gas without the need for an organic carbon source or electron donor (Strous et al., 1999b). In the nitritation step, approximately half of the ammonia is oxidized to nitrite by AOB. The remaining ammonia and the produced nitrite are then converted to dinitrogen gas during the second step by AMX. Nitrate is produced as a byproduct at approximately 12% of the influent N. The process is ideal for high-strength ammonium wastes (>200 mg N/L) and low in organic carbon (C:N ratio lower than 1), such as digester supernatant.

Successful deammonification requests providing different redox environments for AOB and AMX, and decoupling their SRTs. A variety of reactor configurations and biomass separation techniques can provide conditions to facilitate deammonification. Major alternatives involve two-step (two sludge: AMX and AOB reactions take place in separate reactors) and one-step (single sludge: all biological reactions take place in one reactor) processes.

8.2.3.1   Two-Step Deammonification Processes
A two-step process consists of an aerobic reactor for AOB reaction (i.e., nitritation), followed by a second reactor for anammox. In the SHARON®-ANAMMOX® process, the aerated SHARON® reactor is optimized for partial nitrification (i.e., inhibition of nitrite oxidation into nitrate) prior to a fully anoxic reactor for anammox. The first reactor in a two-step process is operated to achieve 55% conversion of ammonia to nitrite. The consumption of the remaining ammonia and nitrite is targeted in the subsequent anammox reactor. It is crucial to prevent nitrite oxidation by suppressing nitrite oxidizing bacteria (NOB) activity in the nitritation reactor. NOB are suppressed by DO manipulation, inhibitory chemicals, and short SRTs. However, when these measures are used in conjunction, the relative influence of one particular factor is difficult to distinguish and remains a topic of research. Retention of AMX biomass is important to sustain deammonification in the second process. Recent techniques involve wasting through hydrocyclone, fixed film media, granular sludge, and other selective retention devices. The fixed film media and granular sludge are commonly used in an upflow reactor for anammox.

8.2.3.2   One-Step Deammonification Process
The challenges of the one-step approach for deammonification is relatively different from two-step processes. Combining the steps in a one-reactor system significantly lowers the investment costs and avoids controlling two reactors. Preliminary full-scale investigations showed that one-stage autotrophic nitrogen removal emits less of the environmentally harmful nitric and nitrous oxides than the two-stage process (Kampschreur et al., 2008). Since, AMX and AOB grow in the same reactor aeration levels should be managed to activate both microbial processes. It can go via continuous or intermittent aeration. Decoupling of the SRTs of the two guilds is the major challenge for the success of the one step process.

Location	Year	Size (m³)	N load (kgN/d)	kg N/m³/d
Malmö, Sweden[a]	2011	4 × 50	200	
Växjö, Sweden[a]	2011	300	320 / 430	
James River, USA[a]	2013	393	253	0.63
South Durham, USA[a]	in construction	318	303	0.95
Holbæk, Denmark[a]	2012	600	120	
Grindsted, Denmark[a]	2013	140	100	
Hattingen, Germany[b]	2003	230	120 (180)	0.5
Dalian XiaJiaHe, China[b]	2009		2200	
Stockholm, SE[b]	2007	1400	600	0.3

[a]ANITA Mox™
[b]DeAmmon®

TABLE 11.31 Installation of MBBR Based Deammonification Systems World-wide for Reject Water Treatment (modified after Lackner et al., 2014)

8.2.3.3 Moving-Bed Biofilm Reactor MBBRs are continuous operation systems, which provide no control of biomass wasting (else than implemented as an IFAS process). In an MBBR system, carrier media provides the surface area for slow-growing organisms (e.g., AMX) to adhere. Bacteria grow as a biofilm attached to the support media. At a certain thickness, the biofilm provides substrate gradients for the organism residing at different layers. AMX, which grow slower and are inhibited by high levels of DO, are found in the deeper anoxic layers of the biofilm while the aerobic AOB are found on outer layers. Nitrite is produced by AOB and penetrates in the biofilm deeper than oxygen thus providing substrate for AMX. The support media harbor growth competition between AOB and NOB, which is a major concern in MBBR configurations (Pellicer-Nacher et al., 2010). Aeration- and DO-control are important to prevent NOB activity in MBBRs. Two major MBBR deammonification systems are available to treat reject water, the Veolia-AnoxKaldnes ANITAMox™ and DeAmmon® process (Table 11.31).

8.2.3.4 Mainstream Tertiary Nitrogen Polishing with Anammox MBBR Dilute and cold conditions at mainstream are not well-suited for the suppression of NOB, short-cut nitrogen removal, and deammonification. This is a challenge for full-scale implementation. Due to lower growth rates and activities associated with AMX, nitrogen removal rates decreases substantially at lower temperatures (Isaka et al., 2008; Vazquez-Padin et al., 2011). At bench, anammox processes have been successfully operated at temperatures ≤20°C for waste streams with low COD:N (Cema et al., 2007, Dosta et al., 2008, Isaka et al., 2008, Vazquez-Padin et al., 2011). AMX processes have been demonstrated at 20°C (Hendrickx et al., 2012). A stable operation of nitritation-anammox on long term was obtained with low concentration of ammonium in the influent in laboratory-scale SBR at 12°C (Hu et al., 2013) and RBC at 15°C (de Clippeleir et al., 2013).

Regmi et al (2015) studied nitrogen polishing of mainstream nitritation–denitritation system effluent via anammox at 25°C in a fully anoxic 0.45-m³ MBBR over 385 days. Unlike other anammox-based processes, a very fast startup of the anammox MBBR in less than 30 days was demonstrated, despite nitrite-limited feeding conditions. Nitrogen removal was very stable within a wide range of nitrogen inputs. An AMX activity of up to 1 gN/m²/d comparable to other biofilm-based systems was observed. Nitrate production is generally assumed to limit nitrogen removal by AMX. However, the anammox MBBR demonstrated ammonia, nitrite, and nitrate removal at limited COD availability. Nitrogen removal might have been aided by denitratation by denirifying heterotrophs. Alternatively, AMX have recently displayed ability to use short-chain fatty acids to reduce nitrate to nitrite. Overall, the feasibility of nitrogen polishing in an anammox MBBR is demonstrated.

9.0 References

Abwassertechnische Vereinigung (ATV) (1983) *German ATV Regulations—A135*; Grundsätze für die Bemessung von einstufigen Tropfkörpern und Scheibentauchkörpern mit Anschluwerter über 500 Einwohnergleichwerten, D–5205; St. Augustine, Germany.

Adham, S.; Gillogly, T.; Nerenberg, R.; Lehman, G.; Rittmann, B. E. (2004) *Membrane Biofilm Reactor Process for Nitrate and Perchlorate Removal*; AWWA Research Foundation: Denver, Colorado.

Æsøy, A.; Ødegaard, H.; Bentzen, G. (1998) The Effect of Sulphide and Organic Matter on the Nitrification Activity in a Biofilm Process. *Water Sci. Technol.*, **37** (1), 115–122.

Albertson, O. E. (1989a) Slow Down That Trickling Filter! *Wat. Env. Tech. (Oper. Forum.)*, **6** (1), 15–20.

Albertson, O. E. (1989b) Slow Motion Trickling Filters Gain Momentum! *Water Environ. Technol. (Oper. Forum.)*, **6** (8), 28–29.

Albertson, O. E. (1995) Excess Biofilm Control by Distributor-Speed Modulation. *J. Environ. Eng.*, **121** (4), 330–336.

Albertson, O. E.; Davies, G. (1984) Analysis of Process Factors Controlling Performance Plastic Bio–media. *Proceedings of the 57th Water Pollution Control Federation Conference*; New Orleans, Louisiana, October; Water Pollution Control Federation: Washington, D.C.

Albertson, O. E.; Eckenfelder, W. (1984) Analysis of Process Factors Affecting Plastic Media Trickling Filter Performance; *Proceedings of the Second International Conference on Fixed Film Biological Processes*; Washington, D.C.

Albertson, O. E.; Okey, R. W. (1988) *Design procedure for Tertiary Nitrification*. Prepared for American Surfpac Inc.: West Chester, Pennsylvania.

Andersson, B.; Aspegren, H.; Nyberg, U.; la Cour Jansen, J.; Ødegaard, H. (1998) Increasing the Capacity of an Extended Nutrient Removal Plant by Using Different Techniques. *Water Sci. Technol.*, **37** (9), 175–183.

Andersson, B.; Aspregren, H.; Parker, D. S.; Lutz, M. (1994) High Rate Nitrifying Trickling Filters. *Water Sci. Technol.*, **29** (10–11), 47–52.

Antoine, R. L. (1976) *Fixed Biological Surfaces—Wastewater Treatment*; CRC Press Inc.: Cleveland, Ohio.

Aryan, A. F.; Johnson, S. H. (1987) Discussion of a Comparison of Trickling Filter Media. *J. Water Pollut. Control Fed.*, **59**, 915–918.

Aspegren, H. (1992) *Nitrifying Trickling Filters, A Pilot Study of Malmö, Sweden*; Malmö Water and Sewage Works: Malmö, Sweden.

Aspegren, H.; Nyberg, U.; Andersson, B.; Gotthardsson, S.; Jansen, J. (1998) Post Denitrification in a Moving Bed Biofilm Reactor Process. *Water Sci. Technol.*, **38** (1), 31–38.

Atkinson, B.; Busch, A. W.; Dawkins, G. S. (1963) Recirculation, Reaction Kinetics and Effluent Quality in a Trickling Filter Flow Model. *J. Water Pollut. Control Fed.*, **35**, 1307–1317.

Aybar, M.; Pizarro, G.; Boltz, J. P.; Downing, L.; Nerenberg, R. (2014) Energy Efficient Wastewater Treatment Via the Air-Based, Hybrid Membrane Biofilm Reactor (Hybrid MBfR). *Water Sci. Technol.*, **69** (8), 1735–1741.

Barnard, J. L. (1974) Cut P and N without Chemicals. *Water Waste Eng.*, **11**, 41–44.

Baxter and Woodman Environmental Engineers (1973) *Nitrification in Wastewater Treatment: Report of the Pilot Study*; Prepared for the Sanitary District of Bloom Township: Illinois.

Benzie, W. J.; Larkin, H. O.; Moore, A. F. (1963) Effects of Climactic and Loading Factors on Trickling Filter Performance. *J. Water Pollut. Control Fed.*, **35** (4), 445–455.

Beun, J. J.; Hendriks, A.; van Loosdrecht, M. C. M.; Morgenroth, E.; Wilderer, P. A.; Heijnen, J. J. (1999) Aerobic Granulation in a Sequencing Batch Reactor. *Water Res.*, **33** (10), 2283–2290.

Biesterfeld, S.; Farmer, G.; Figueroa, L.; Parker, D.; Russell.; P. (2003) Quantification of Denitrification Potential in Carbonaceous Trickling Filters. *Water Res.*, **37** (16), 4011–4017.

Bill, K.; Bott, C.; Yi, P. H.; Ziobro, C.; Murthy, S. (2008) Evaluation of Alternative Electron Donors in Anoxic Moving Bed Biofilm Reactors (MBBRs) Configured for Post-Denitrification. *Proceedings of the 81st Annual Water Environment Federation Technical Exposition and Conference* [CD-ROM]; Chicago, Illinois; Oct 18–22; Water Environment Federation: Alexandria, Virginia.

Boessmann, M.; Neu, T. R.; Horn, H.; Hempel, D. C. (2004) Growth, Structure and Oxygen Penetration in Particle Supported Autotrophic Biofilms. *Water Sci. Technol.*, **149** (11–12), 371–377.

Boller, M.; Gujer, W. (1986) Nitrification in Tertiary Trickling Filters Followed by Deep Filters. *Water Res.*, **20**, 1363.

Boltz, J. P; La Motta, E. J. (2007) The Kinetics of Particulate Organic Matter Removal as a Response to Bioflocculation in Aerobic Biofilm Reactors. *Water Environ. Res.*, **79**, 725.

Boltz, J. P.; La Motta, E. J.; Madrigal, J. A. (2006) The Role of Bioflocculation on Suspended Solids and Particulate COD Removal in the Trickling Filter Process. *J. Environ. Eng.*, **132** (5), 506–513.

Boltz, J. P.; Goodwin, S. G.; Rippon, D.; Daigger, G. T. (2008) A Review of Operational Control Strategies for Snail and Other Macrofauna Infestations in Trickling Filters. *Water Pract.*, **2** (4), 1-16.

Boltz, J. P.; Johnson, B. R.; Daigger, G. T.; Sandino, J. (2009a) Modeling Integrated Fixed Film Activated Sludge (IFAS) and Moving Bed Biofilm Reactor (MBBR) Systems I: Mathematical Treatment and Model Development. *Water Environ. Res.*, **81**, 576–586.

Boltz, J. P.; Johnson, B. R.; Daigger, G. T.; Sandino, J.; Elenter, D. (2009b) Modeling Integrated Fixed Film Activated Sludge (IFAS) and Moving Bed Biofilm Reactor (MBBR) Systems II: Evaluation. *Water Environ. Res.*, **81**, 555–575.

Boltz, J. P.; Daigger, G. T.; Johnson, B. R.; Hiatt, W.; Grady, Jr., C. P. L. (2009c). Expanded Process Model Describes Biomass Distribution, Free Ammonia/Nitrous Acid Inhibition and Competition Between Ammonia Oxidizing Bacteria (AOB) and Nitrite Oxidizing Bacteria (NOB) in Submerged Biofilm and Integrated Fixed-Film Activated Sludge Bioreactors. *Proceedings of the Water Environment Federation Nutrient Removal Conference* [CD-ROM]; Washington, D.C., Jun 28–Jul 1; Water Environment Federation: Alexandria, Virginia.

Boltz, J. P.; Daigger, G. T. (2010) Uncertainty in Bulk-Liquid Hydrodynamics and Biofilm Dynamics Creates Uncertainties in Biofilm Reactor Design. *Water Sci. Technol.*, **61** (2), 307–316.

Boltz, J. P.; Morgenroth, E.; Sen, D. (2010) Mathematical Modeling of Biofilms and Biofilm Reactors for Engineering Design. *Water Sci. Technol.*, **62** (8), 1821–1836.

Boltz, J. P.; Morgenroth, E.; Brockmann, D.; Bott, C.; Gellner, W. J.; Vanrolleghem, P. A. (2011) Systematic Evaluation of Biofilm Models for Engineering Practice: Components and Critical Assumptions. *Water Sci. Technol.*, **64** (4), 930–944.

Boltz, J. P.; Morgenroth, E.; Daigger, G. T.; deBarbadillo, C.; Murthy, S.; Sørensen, K.; Stinson, B. (2012) Method to Identify Potential Phosphorus Rate-Limiting Conditions in Post-Denitrification Biofilm Reactors Designed for Simultaneous Low-Level Effluent Nitrogen and Phosphorus Concentrations. *Water Res.*, **46** (19), 6228–6238.

Boltz, J. P.; Smets, B.; Rittmann, B. E.; van Loosdrecht, M. C. M.; Morgenroth, E.; Daigger, G. T. (2017) From Biofilm Ecology to Reactors: A Focused Review. *Water Sci. Technol.*, **75** (7–8), 1753–1760.

Bosander, J.; Westlund, A. D. (2000) Operation of Full-Scale Fluidized Bed for Denitrification. *Water Sci. Technol.*, **41** (9), 115–121.

Bosman, J.; Hendricks, F. (1981) The Technologies and Economics of the Treatment of a Concentrated Industrial Effluent by Biological Denitrification Using a Fluidised-Bed Reactor. In *Biological Fluidized Bed Treatment of Water and Wastewater*, Cooper, P. F., Atkinson, B., Eds.; Ellis Horwood for Water Research Laboratory, Stevenage Laboratory: Chichester, United Kingdom, pp. 222–233.

Bratby, J. R.; Fox, B.; Parker, D. S.; Fisher, R.; Jacobs, T. (1999) Using Process Simulation Models to Rate Plant Capacity. *Proceedings of the 72nd Annual Water Environment Federation Technical Exposition and Conference* [CD-ROM]; New Orleans, Louisiana, Oct 10–13; Water Environment Federation: Alexandria, Virginia.

Brindle, K.; Stephenson, T. (1996a) The Application of Membrane Biological Reactors for the Treatment of Wastewaters. *Biotechnol. Bioeng.*, **49** (6), 601–610.

Brindle, K.; Stephenson, T. (1996b) Nitrification in a Bubbleless Oxygen Mass Transfer Membrane Bioreactor. *Water Sci. Technol.*, **34** (9), 261–267.

Brindle, K.; Stephenson, T.; Semmens, M. J. (1998) Nitrification and Oxygen Utilisation in a Membrane Aeration Bioreactor. *J. Membr. Sci.*, **144** (1–2), 197–209.

Brindle, K.; Stephenson, T.; Semmens, M. J. (1999) Pilot-Plant Treatment of a High-Strength Brewery Wastewater Using a Membrane–Aeration Bioreactor. *Water Environ. Res.*, **71** (6), 1197–1204.

Bruce, A. M.; Merkens, J. C. (1970) Recent Studies of High Rate Biological Filtration. *J. Water Pollut. Control*, **2**, 449.

Bruce, A. M.; Merkens, J. C. (1973) Further Studies of Partial Treatment of Sewage by High-Rate Biological Filtration. *J. Water Pollut. Control*, **5**, 499.

Bruce, A. M.; Merkens, J. C. (1975) Pilot Studies on the Treatment of Domestic Sewage by Two-Stage Biological Filtration—With Special Reference to Nitrification. *J. Water Pollut. Control*, **74**, 80-100.

Bryan, E. H. (1955) Molded Polystyrene Media for Trickling Filters. *Proceedings of the 10th Purdue Industrial Waste Conference*; Purdue University: West Lafayette, Indiana; pp. 164–172.

Bryan, E. H. (1962) Two-Stage Biological Treatment: Industrial Experience. *Proceedings of the 11th South Municipal Industrial Waste Conference*; University of North Carolina: Chapel Hill, North Carolina; p. 136.

Bryan, E. H.; Moeller, D. H. (1960) Aerobic Biological Oxidation Using Dowpac. *Proceedings of the Conference on Biological Waste Treatment*; Manhattan College: New York.

Bryers, J. D. (1984) Biofilm Formation and Chemostat Dynamics: Pure and Mixed Culture Conditions. *Biotech. Bioeng.*, **26**, 948–958.

Callieri, D. A. S.; Núñez, C. G.; Díaz Ricci, J. C.; Scidá, L. (1984) Batch Culture of Candida utilis in a Medium Deprived of a Phosphorus Source. *App. Microbiol. Biotech.*, **19**, 267–271.

Cantwell, A.; Mosey, F. (1999) Recent Applications and Developments of the Biobead System; *Proceedings of the BAF3 Conference*; Cranfield University: Cranfield, England.

Canziani, R. (1988) Submerged Aerated Filters IV–Aeration Characteristics. *Ingegneria Ambientale*, **17** (11/12), 627–636.

CH2M HILL (1984) A Comparison of Trickling Filter Media Internal Project Report; CH2M HILL: Denver, Colorado.

Cherchi, C.; Onnis-Hayden, A.; Gu, A. Z. (2008) Investigation of MicroCTM as an Alternative Carbon Source for Denitrification, *Proceedings of the Water Environment Federation 81st Annual Technical Exposition and Conference* [CD-ROM], Chicago, Illinois; Oct 18–22; Water Environment Federation: Alexandria, Virginia.

Chung, J.; Rittmann, B. E. (2007) Bio-reductive Dechlorination of 1,1,1-Trichloroethane and Chloroform Using a Hydrogen-Based Membrane Biofilm Reactor. *Biotechnol. Bioeng.*, **97** (1), 52–60.

Chung, J.; Li, X. H.; Rittmann, B. E. (2006a) Bioreduction of Arsenate Using a Hydrogen-Based Membrane Biofilm Reactor. *Chemosphere*, **65** (1), 24–34.

Chung, J.; Nerenberg, R.; Rittmann, B. E. (2006b) Bioreduction of Soluble Chromate Using a Hydrogen-Based Membrane Biofilm Reactor. *Water Res.*, **40** (8), 1634–1642.

Chung, J.; Nerenberg, R.; Rittmann, B. E. (2006c) Bioreduction of Selenate Using a Hydrogen-Based Membrane Biofilm Reactor. *Environ. Sci. Technol.*, **40** (5), 1664–1671.

Clapp, L. W.; Regan, J. M.; Ali, F.; Newman, J. D.; Park, J. K.; Noguera, D. R. (1999) Activity, Structure, and Stratification of Membrane-Attached Methanotrophic Biofilms Cometabolically Degrading Trichloroethylene. *Water Sci. Technol.*, **39** (7), 153–161.

Coelhoso, I.; Boaventura, R.; Rodrigues, A. (1992) Biofilm reactors—An Experimental and Modeling Study of Wastewater Denitrification in Fluidized-Bed Reactors of Activated Carbon Particles. *Biotechnol. Bioeng.*, **40** (5), 625–633.

Cooper, P. F. (1986) *The Two Fluidized Bed Reactor for Wastewater Treatment. In Process Engineering Aspects of Immobilized Cell Systems*; Webb, C., Black, G. M., Atkinson, B., Eds.; The Institution of Chemical Engineers: Rugby, United Kingdom, 179–204.

Cooper, P. F.; Wheeldon, D. H. V. (1981) Completer Treatment of Sewage in a Two-Fluidised Bed System. In *Biological Fluidized Bed Treatment of Water and Wastewater*; P. F. Cooper and B. Atkinson. Cooper, P.F., Atkinson, B., Eds; Ellis Horwood for Water Research Laboratory, Stevenage Laboratory: Chichester, United Kingdom, pp. 121–144.

Cooper-Smith, G.; Schofield, I. (2004) Submerged Aerated Filters, Coming of Age for AMP4; *Proceedings of the 2nd National CIWEM Conference*; September; Wakefield, United Kingdom; Chartered Institution of Water and Environmental Management: London, England.

Copp, J. B.; Dold, P. L. (1998) Comparing Sludge Production under Aerobic and Anoxic Conditions. *Water Sci. Technol.*, **38** (1), 285–294.

Crine, M.; Schlitz, M.; Vandevenne, L. (1990) Evaluation of the Performances of Random Plastic Media in Aerobic Trickling Filters. *Water Sci. Technol.*, **22** (1/2), 227–238.

Culp, G. L. (1963) Direct Recirculation of High-Rate Trickling Filter Effluent. *J. Water Pollut. Control Fed.*, **35** (6), 742–747.

Daigger, G. T.; Boltz, J. P. (2010) Trickling Filter and Trickling Filter-Suspended Growth Process Design and Operation: A State-of-the Art Review. *Water Environ. Res.*, **83**, 388-404.

Daigger, G. T.; Heinemann, T. A.; Land, G.; Watson, R. S. (1994) Practical Experience with Combined Carbon Oxidation and Nitrification in Plastic Media Trickling Filters. *Water Sci. Technol.*, **29** (10–11), 189–196.

Daude, D.; Stephenson T. (2004) Cost-Effective Treatment Solutions for Rural Areas; Design of a New Package Treatment Plant for Single Households. *Water Sci. Technol.*, **48** (11), 107–113.

de Kreuk, M. K.; van Loosdrecht, M. C. M. (2004) Selection of Slow-Growing Organisms as a Means for Improving Aerobic Granular Sludge Stability. *Water Sci. Technol.*, **49** (11–12), 9–17.

de Kreuk, M.; Heijnen, J. J.; van Loosdrecht, M. C. M. (2005a) Simultaneous COD, Nitrogen, and Phosphate Removal by Aerobic Granular Sludge. *Biotechnol. Bioeng.*, **90** (6), 761–769.

de Kreuk, M. K.; McSwain, B. S.; Bathe, S.; Tay, S. T. L.; Schwarzenbeck, N.; Wilderer, P. A. (2005b) Discussion Outcomes. In: *Aerobic Granular Sludge*, Bathe, S.; de Kreuk, M. K.; McSwain, B. S.; Schwarzenbeck, N. (Eds.); International Water Association: London, U.K., pp. 153-169.

de Kreuk, M. K.; van Loosdrecht, M. C. M. (2006) Formation of Aerobic Granules with Domestic Sewage. *J. Environ. Eng.*, **132** (6), 694–697.

de Kreuk M. K.; Kishida N.; van Loosdrecht M. C. M. (2007) Aerobic Granular Sludge– State of the Art. *Water Sci. Technol.*, **55** (8-9), 75-81.

de Kreuk, M. K.; Kishida, N.; Tsuneda, S.; van Loosdrecht, M. C. M. (2010) Behavior of Polymeric Substrates in an Aerobic Granular Sludge System. *Water Res.*, **44** (20), 5929-5938.

deBarbadillo, C.; Rectanus, R.; Canham, R.; Schauer, P. (2006) Tertiary Denitrification And Low Phosphorus Limits: A Practical Look At Phosphorus Limitations On Denitrification Filters. *Proceedings of the 79th Annual Water Environment Federation Technical Conference and Exposition* [CD-ROM]; Dallas, Texas, Oct 21–25; Water Environment Federation: Alexandria, Virginia.

deBarbadillo, C.; Shaw, A.; Wallis-Lage, C. (2005) Evaluation and Design of Deep-Bed Denitrification Filters: Empirical Design Parameters vs. Process Modeling. *Proceedings of the 78th Annual Water Environment Federation Technical Conference and Exposition*, Washington, D.C., Oct 12–15; Water Environment Federation: Alexandria, Virginia.

deBarbadillo, C.; Rogalla, F.; Tarallo, S.; Boltz, J. P. (2010) Factors Affecting the Design and Operation of Biologically Active Filters. *Proceedings of the WEF/IWA Biofilm Reactor Technology Conference*, Portland, Oregon.

Degremont (2007) *Water Treatment Handbook*, 7th ed.; Lavoisier SAS: France.

Dempsey, M. J. *Nitrification Process*, U.S. Patent 6,572,773, 2003.

Dempsey, M. J. *Fluid Bed Expansion and Fluidization*, U.S. Patent 7,309,433, 2007.

Dempsey, M. J.; Boltz, J. P. (2010) Tertiary Nitrification in an Expanded Bed Biofilm Reactor. *Proceedings of the WEF/IWA Biofilm Reactor Technology Conference*, Portland, Oregon.

Dempsey, M. J.; Porto, I.; Mustafa, M.; Rowan, A. K.; Brown, A.; Head, I. M. (2006) The Expanded Bed Biofilter: Combined Nitrification, Solids Destruction, and Removal of Bacteria. *Water Sci. Technol.*, **54** (8), 37–46.

Dempsey, M. J.; Lannigan, K. C.; Minall, R. J. (2005) Particulate-Biofilm, Expanded-Bed Technology for High-Rate, Low-Cost Wastewater Treatment: Nitrification. *Water Res.*, **39**(6), 965–974.

Downing, L.; Nerenberg, R. (2007a) Kinetics of Microbial Bromate Reduction in a Hydrogen-Oxidizing, Denitrifying Biofilm Reactor. *Biotechnol. Bioeng.*, **98** (3), 543–550.

Downing, L.; Nerenberg, R. (2007b) Performance and Microbial Ecology of the Hybrid Membrane Biofilm Process (HMBP) for Concurrent Nitrification and Denitrification of Wastewater. *Water Sci. Technol.*, **55** (8–9), 355–362.

Downing, L.; Nerenberg, R. (2008a) Effect of Oxygen Gradients on the Activity and Microbial Community Structure of a Nitrifying, Membrane-Aerated Biofilm. *Biotechnol. Bioeng.*, **101** (6), 1193–1204.

Downing, L.; Nerenberg, R. (2008b) Total Nitrogen Removal in a Hybrid, Membrane-Aerated Activated Sludge Process. *Water Res.*, **42** (14), 3697–3708.

Downing, A. L.; Tomlinson, T. G.; Truesdale, G. A. (1964a) The Effect of Inhibitors on Nitrification in the Activated Sludge Process. *J. Inst. Sewer Purif.*, **6**, 537.

Downing, A. L.; Painter, H. A.; Knowles, G. (1964b) Nitrification in the Activated Sludge Process, *J. Inst. Sewer Purif.*, **2**, 130.

Drury, D. D.; Carmona, J.; Delgadillo, A. (1986) Evaluation of High Density Cross Flow Media for Rehabilitating and Existing Trickling Filter. *J. Water Pollut. Control Fed.*, **58** (5) 364–366.

Eckenfelder, W.W. (1961) Trickling Filter Design and Performance. *J. San. Eng. Div., Am. Soc. Civ. Eng.*, **87**, 33–45.

Eckenfelder, W. W. (1963) Performance of a High Rate Trickling Filter Using Selected Materials. *J. Water Pollut. Control Fed.*, **35**, 1536.

Eckenfelder, W.W, and Barnhart, E. L. (1963) Performance of a High-Rate Trickling Filter Using Selected Materials. *J. Water Pollut. Control Fed.*, **35** (12), 1535–1551.

Ergas, S. J.; Reuss, A. F. (2001) Hydrogenotrophic Denitrification of Drinking Water Using a Hollow Fibre Membrane Bioreactor. *J. Water Supply Res. Technol. Aqua*, **50** (3), 161–171.

Essila, N. J.; Semmens, M. J.; Voller, V. R. (2000) Modeling Biofilms on Gas-Permeable Supports: Concentration and Activity Profiles. *J. Environ. Eng.*, **126** (3), 250–257.

Fitzpatrick, C. S. B. (2001) Factors Affecting Efficient Filter Backwashing. *Proceeding from the International Conference on Advances in Rapid Granular Filtration in Water Treatment*, Chartered Institution of Water and Environmental Management: London, England.

Francis, C. W.; Hancher, C. W. (1981) Biological Denitrification of High-Nitrate Wastes Generated in the Nuclear Industry. In *Biological Fluidized Bed Treatment of Water and Wastewater*; Cooper, P. F., Atkinson, B., Eds.; Ellis Horwood for Water Research Laboratory, Stevenage Laboratory: Chichester, United Kingdom, pp. 234–250.

Frijters, C.; Vellinga, S.; Jorna, T.; Mulder, R. (2000) Extensive Nitrogen Removal in a New Type of Airlift Reactor. *Water Sci. Technol.*, **41** (4–5), 469–476.

Frisch, S. (1998a) *Biomass Separation Apparatus and Method*, U.S. Patent 5,788,842, 1998.

Frisch, S. (1998b) *Biomass separation apparatus and method with media return*. U.S. Patent 5,750,028, 1998.

Galler, W. S.; Gotaas, H. G. (1964) Analysis of Biological Filter Variables. *J. Sanit. Eng. Div., Am. Soc. Civ. Eng.*, **90** (6), 59.

Germain, J. E. (1966) Economical Treatment of Domestic Waste by Plastic Medium Trickling Filters. *J. Water Pollut. Control Fed.*, **38**, 192.

German Association for Water, Wastewater and Waste [ATV-DVWK] (2000) Standard ATV-DVWK-A 131 E, Dimensioning of Single-stage Activated Sludge Plants, German ATV-DVWK Rules and Standards.

Gonçalves R.; Rogalla, F. (1992) Continuous Biological Phosphorus Removal in a Biofilm Reactor. *Water Sci. Technol.*, **26** (9–11), 2027–2030.

Gonçalves, R. F.; Le Grand, L.; Rogalla, F. (1994a) Biological Phosphorus Uptake in Submerged Biofilters with Nitrogen Removal. *Water Sci. Technol.*, 29(10–11), 135–143.

Gonçalves, R. F.; Nogueira, F. N.; Le Grand, L.; Rogalla, F. (1994b) Nitrogen and Biological Phosphorus Removal in Submerged Biofilters. *Water Sci. Technol.*, **30** (11), 1–12.

Grady, L. E.; Daigger, G. T.; Lim, H. (1999) *Biological Wastewater Treatment*, 2nd ed.; Marcel Dekker: New York.

Green, M.; Shnitzer, M.; Tarre, S.; Bogdan, B.; Shelef, G.; Sorden, C. J. (1994) Fluidized-Bed Reactor Operation for Groundwater Denitrification. *Water Sci. Technol.*, **29** (10–11), 509–515.

Grimberg, S. J.; Rury, M. J.; Jimenez, K. M.; Zander, A. K. (2000) Trinitrophenol Treatment in a Hollow Fiber Membrane Biofilm Reactor. *Water Sci. Technol.*, **41** (4–5), 235–238.

Gujer, W.; Boller, M. (1983) Operating Experience with Plastic Media Tertiary Trickling Filters for Nitrification. In *Design and Operation of Large Treatment Plants*, de Emde, V, Tench, H. B., Eds.; Pergamon: Oxford, United Kingdom.

Gujer, W.; Boller, M. (1986) Design of a Nitrifying Trickling Filter Based on Theoretical Concepts. *Water Res.*, **20**, 1353.

Gullicks, H. A.; Cleasby, J. L. (1986) Design of Trickling Filter Nitrification Tower. *J. Water Pollut. Control Fed.*, **58** (1), 60–67.

Gullicks, H. A.; Cleasby, J. L. (1990) Cold–Climate Nitrifying Biofilters: Design and Operation Considerations. *J. Water Pollut. Control Fed.*, **62** (1), 50–57.

Harremoës, P. (1976) The Significance of Pore Diffusion to Filter Denitrification. *J. Water Pollut. Control Fed.*, **48** (2), 377–388.

Harremoës, P. (1978) *Biofilm Kinetics in Water Pollution Microbiology*, Vol. 2; Michell, R., Ed.; Wiley and Sons: New York.

Harremoës, P. (1982) Criteria for Nitrification in Fixed Film Reactors. *Water Sci. Technol.*, **13**, 167.

Harremoës, P.; Wilderer, P. A. (1993) Fundamentals of Nutrient Removal in Biofilters. *Proceedings from the 9th Annual EWPCA-ISWA Symposium*; München, Germany, May 11–13; Abwassertechnische Vereinigung e.V.: St. Augustin, Germany.

Harris S. L.; Stephenson, T.; and Pearce, P. (1996) Aeration Investigation of Biological Aerated Filters using Off-Gas Analysis. *Water Sci. Technol.*, **34**, 307.

Harrison, J. R. (2007) Personal Communication. *Shutdown of Covered Biofilters.*

Harrison, J. R.; Daigger, G. T. (1987) A Comparison of Trickling Filter Media. *J. Water Pollut. Control Fed.*, **59**, 679.

Harrison, J. R.; Timpany, P. L. (1988) Design Considerations with the Trickling Filter Solids Contact Process. *Proceedings of the Joint Canadian Society of Civil Engineers, American Society of Civil Engineers National Conference on Environmental Engineering*; Canadian Society of Civil Engineers: Vancouver, British Columbia.

Hawkes, H. A. (1963) *The Ecology of Waste Water Treatment*; Pergamon Press: Oxford, England.

Heijnen, J. J.; van Loosdrecht, M. C. M.; Mulder, R.; Weltevrede, R.; Mulder, A. (1993) Development and Scale-Up of an Aerobic Biofilm Airlift Suspension Reactor. *Water Sci. Technol.*, 27 (5–6), 253–261.

Hem, L. (1991) *Nitrification in a Moving Bed Biofilm Process*. Unpublished Ph.D. Dissertation, The Norwegian Institute of Technology, Trondheim, Norway.

Hem, L.; Rusten, B.; Ødegaard, H. (1994) Nitrification in a Moving Bed Reactor. *Water Res.*, **28** (6), 1425–1433.

Hodkinson, B. J.; Williams J. B.; Ha, T. N. (1998) Effects of Plastic Support Media on the Diffusion of Air into a Submerged Aerated Filter, *J. Chart. Inst. Water Environ. Manage.*, **12**, 188.

Holmes, J.; Dutt, S., (1999) Coln Bridge (Huddersfield) WWTW Biopur Plant Process Design and Performance. *Proceedings of the BAF3 Conference*; Cranfield University: Cranfield, England.

Horn, H.; Morgenroth, E. (2006) Transport of Oxygen, Sodium Chloride, and Sodium Nitrate in Biofilms. *Chem. Eng. Sci.*, **61** (5), 1347–1356.

Howland, W. E. (1958) Flow Over Porous Media as in a Trickling Filter. *Proceedings of the 12th Purdue Industrial Waste Conference*; Purdue University: West Lafayette, Indiana.

Huang, X.; Liang, P.; Qian Y. (2007) Excess Sludge Reduction Induced by Tubifex tubifex in a Recycled Sludge Reactor. *J. Biotechnol.*, **127** (3), 443–451.

Hultman, B.; Jonsson, K.; Plaza, E. (1994) Combined Nitrogen and Phosphorus Removal in a Full-Scale Continuous Upflow Sand Filter. *Water Sci. Technol.*, **29** (10/11), 127–134.

Husovitz, K. J.; Gilmore, A.; Delahaye, N. G.; Love, K. R.; Little, J. C. (1999) The Influence of Upflow Liquid Velocity on Nitrification in a Biological Aerated Filter. *Proceedings of the Water Environment Federation 72nd Annual Water Environment Federation Technical Conference and Exposition* [CD-ROM]; New Orleans, Louisiana, Oct 10–13; Water Environment Federation: Alexandria, Virginia.

Janning, K. F.; Harremoes, P.; Nielsen, M. (1995) Evaluating and Modelling of the Kinetics in a Full-scale Submerged Denitrification Filter. *Water Sci. Technol.*, **32** (8), 115–123.

Jeris, J. S.; Owens, R. W. (1975) Pilot-Scale, High-Rate Biological Denitrification. *J. Water Pollut. Control Fed.*, **47** (8), 2043–2057.

Jeris, J. S.; Beer, C., et al. (1974) High-Rate Biological Denitrification Using a Granular Fluidized-Bed. *J. Water Poll. Control Fed.*, **46** (9), 2118–2128.

Jeris, J. S.; Owens, R. W., et al. (1981) Secondary Treatment of Municipal Wastewater with Fluidized Bed Technology. In *Biological Fluidized Bed Treatment of Water and Wastewater*; P. F. Cooper and B. Atkinson. Cooper, P.F., Atkinson, B., Eds; Ellis Horwood for Water Research Laboratory, Stevenage Laboratory: Chichester, United Kingdom, pp. 112–120.

Johnson, C.; Boltz, J. P. (2013) Aeration System Design in Integrated Fixed-Film Activated Sludge (IFAS) and Moving Bed Biofilm Reactors (MBBR) Using Stainless Steel Pipe Diffusers, Manifold, and Down Pipes. *Proceedings of the 86th Annual Water Environment Federation Technical Exposition and Conference* [CD-ROM]; New Orleans, Louisiana, Oct 5–9; Water Environment Federation: Alexandria, Virginia.

Jolly, M. (2004) *Aberdeen (Nigg) Wastewater Treatment Works-1st Year of Operation*. CIWEM 2nd National Conference, Wakefield.

Kaldate, A.; Holst, T.; Pattarkine, V. (2008) Moving Bed Biofilm Reactor Pilot Study for Tertiary Nitrification of HPOAS Wastewater at Harrisburg AWTF. *Proceedings of the 81st Annual Water Environment Federation Technical Exposition and Conference* [CD-ROM]; Chicago, Illinois; Oct 18–22; Water Environment Federation: Alexandria, Virginia.

Kearney, M. M. (2000) Engineered Fractals Enhance Process Applications. *Chem. Eng. Prog.*, **96** (12), 61–68.

Kincannon, D. F.; Stover, E. L. (1982) Design Methodology for Fixed-Film Reactors, RBCs and Trickling Filters. *Civ. Eng. Pract. Design*, **2**, 107.

Kuenen, J. G.; Jørgensen, B. B.; Revsbech, N. P. (1986) Oxygen Microprofiles of Trickling Filter Biofilms. *Water Res.*, **20** (12), 1589–1598.

Laurence A.; Spangel A.; Kurtz W.; Pennington R.; Koch C.; Husband, J. (2003) Full-Scale Biofilter Demonstration Testing in New York City. *Proceedings of the 76th Annual Water Environment Federation Technical Exposition and Conference* [CD-ROM], Los Angeles, California; Oct 11–13; Water Environment Federation: Alexandria, Virginia.

Lazarova, V.; Manem, J. (1994) Advances in Biofilm Aerobic Reactors Ensuring Effective Biofilm Activity Control. *Water Sci. Technol.*, **29** (10–11), 319–327.

Le Tallec, X., Zeghal, S., Vidal, A., Lesouef, A. (1997). Effect of Influent Quality Variability on Biofilter Operation. *Water Science & Technology*, Vol. 36, No. 1, pp. 111–117.

Leaf, W.; Boltz, J. P.; McQuarrie, J. P.; Menniti, A.; Daigger, G. T. (2011) Overcoming Hydraulic Limitations of the Integrated Fixed-Film Activated Sludge (IFAS) Process. *Proceedings of the 84th Annual Water Environment Federation Technical Exhibition and Conference* [CD-ROM]; Los Angeles, California, Oct. 15–19; Water Environment Federation: Alexandria, Virginia.

Lee, J. S.; Buckley, P. S. (1981) Fluid Mechanics and Aeration Characteristics of Fluidised Beds. In *Biological Fluidized Bed Treatment of Water and Wastewater*; Cooper, P.F., Atkinson, B., Eds; Ellis Horwood for Water Research Laboratory, Stevenage Laboratory: Chichester, United Kingdom, pp. 62–74.

Lee, K. M.; Stensel, H. D. (1986) Aeration and Substrate Use in a Sparged Packed-Bed Biofilm Reactor. *J. Water Pollut. Control Fed.*, **58**, 1066–1072.

Lee, K.-C.; Rittmann, B. E. (2000) A Novel Hollow-Fiber Membrane Biofilm Reactor for Autohydrogenotrophic Denitrification of Drinking Water. *Water Sci. Technol.*, **41** (4–5), 219–226.

Lee, K.-C.; Rittmann, B. E. (2002) Applying a Novel Autohydrogenotrophic Hollow-Fiber Membrane Biofilm Reactor for Denitrification of Drinking Water. *Water Res.*, **36** (8), 2040–2052.

Lettinga, G. (1995) Anaerobic Digestion and Wastewater Treatment Systems. *Antonie van Leeuwenhoek, Int. J. General Molecular Microbiol.*, **67** (1), 3-28.

Lettinga, G.; Field, J.; Van Lier, J.; Zeeman, G.; Hulshoff Pol, L. W. (1997) Advanced Anaerobic Wastewater Treatment in the Near Future. *Water Sci. Technol.*, **35** (10), 5-12.

Levine, A. D.; Tchobanoglous, G.; Asano, T. (1985) Characterization of the Size Distribution of Contaminants in Wastewater: Treatment and Reuse Implications. *J. Water Pollut. Control Fed.*, **57**, 805–816.

Levine, A. D.; Tchobanoglous, G.; Asano, T. (1991) Size Distribution of Particulate Contaminants in Wastewater and Their Effect on Treatability. *Water Res.*, **25** (8), 911–922.

Lewandowski, Z. (2000) Structure and Function of Biofilms. In *Biofilms: Recent Advances in their Study and Control*, Evans, L.V., Ed.; Harwood Academic Publishers: Australia.

Lewandowski, Z.; and Boltz J. P. (2011) Biofilms in Water and Wastewater Treatment. In *Treatise on Water Science*, Wilderer. P. (Ed.); Academic Press: Oxford, U.K., Vol. **4**, 529–570.

Liang, P.; X. Huang, et al. (2006) Excess Sludge Reduction in Activated Sludge Process through Predation of Aeolosoma hemprichi. *Biochem, Eng. J.*, **28** (2), 117–122.

Lin, C. S.; Heck, G. (1987) Design and Performance of the Trickling Filter/Solids Contact Process for Nitrification in a Cold Climate. *Proceedings of the 60th Annual Conference of the Water Pollution Control Federation*; Philadelphia, Pennsylvania; Water Pollution Control Federation: Alexandria, Virginia.

Liu, Y.; Tay, J. H. (2002) The Essential Role of Hydrodynamic Shear Force in the Formation of Biofilm and Granular Sludge. *Water Res.*, **36** (7), 1653–1665.

Logan, B. E. (1999) *Environmental Transport Processes*; John Wiley and Sons: New York.

Logan, B. E.; Wagenseller, G. A. (2000) Molecular Size Distributions of Dissolved Organic Matter in Wastewater Transformed by Treatment in a Full-Scale Trickling Filter. *Water Environ. Res.*, **72** (3), 277–281.

Logan, B. E.; Hermanowicz, S. W.; Parker, D. S. (1987a) Engineering Implications of a New Trickling Filter Model. *J. Water Pollut. Control Fed.*, **59** (12), 1017–1028.

Logan, B. E.; Hermanowicz, S. W.; Parker, D. S. (1987b) A Fundamental Model for Trickling Filter Process Design. *J. Water Pollut. Control Fed.*, **59** (12), 1029–1042.

Mabbott, J. W. (1982) Structural Engineering of Plastic Media for Wastewater Treatment by Fixed Film Reactors. *Proceedings of the First International Conference on Fixed Film Processes*; Kings Island, Ohio; U.S. Environmental Protection Agency: Washington, D.C.

MacDonald, D. V. (1990) Denitrification by Fluidized Biofilm Reactor. *Water Sci. Technol.*, **22** (1–2), 451–461.

Madoni, P. (1994) A Sludge Biotic Index (SBI) for the Evaluation of the Biological Performance of Activated-Sludge Plants Based on the Microfauna Analysis. *Water Res.*, **28** (1), 67–75.

Martin, K. J.; Downing, L. S.; Boltz, J. P.; Shrout, J. D.; Na., C.; Nerenberg, R. (2010) The Hollow-Fiber Membrane Biofilm Reactor: Current State of the Technology and Challenges in Development. *Proceedings of the WEF/IWA Biofilm Reactor Technology Conference*, Portland, Oregon.

McCarty, P. L.; Beck, L.; Amant, P. S. (1969) Biological Denitrification of Wastewaters by Addition of Organic Materials, *Proceedings of the 24th Industrial Waste Conference*, Purdue University, 1271–1285.

McQuarrie, J. P.; and Boltz, J. P. (2011) Moving Bed Biofilm Reactor Technology: Process Applications, Design, and Performance. *Water Environ. Res.*, **83**, 560-575.

McQuarrie, J.; Dempsey, M. J.; Boltz, J. P.; Johnson, B. (2007) The Expanded Bed Biofilm Reactor (EBBR)—An Innovative Biofilm Approach for Tertiary Nitrification. *Proceedings of the 80th Annual Water Environment Federation Technical Exposition and Conference* [CD-ROM]; San Diego, California, Oct 13–17; Water Environment Federation: Alexandria, Virginia.

Melcer, H.; Dold, P. L.; Jones, R. M.; Bye, C. M., Takacs, I.; Stensel, H.D.; Wilson, A.W.; Sun, P.; Bury, S. (2003) *Methods for Wastewater Characterization in Activated Sludge Modeling*; IWA Publishing: London, England; Water Environment Federation: Alexandria, VA.

Melin E.; Ødegaard, H.; Helness, H.; Kenakkala, T. (2004) High-Rate Wastewater Treatment Based Nitrification MBBRs. In *Chemical Water and Wastewater Treatment VIII*; Hahn, H., Hoffman, E., Ødegaard, H., Eds.; IWA Publishing: London, England, pp. 39–48.

Metcalf, L.; Eddy, H. P. (1916) *American Sewerage Practice*, Volume III—Disposal of Sewage; McGraw-Hill: New York.

Metcalf and Eddy, Inc./AECOM (2013) *Wastewater Engineering: Treatment and Resource Recovery*, 5th ed.; McGraw Hill: New York.

Michelet, F.; Jolly, M.; Chan, T.; Rogalla, F. (2005) Troubleshooting SAF and BAF Biofilm Reactors on Full Scale, *Proceedings of the Water Environment Federation 78th Annual Conference and Exposition*, Washington, D.C.

Mokhayeri, Y.; Nichols, A.; Murthy, S.; Riffat, R.; Dold, P.; Takacs, I. (2006) Examining the Influence of Substrates and Temperature on Maximum Specific Growth Rate of Denitrifiers, *Water Sci. Technol.*, 54 (8), 155–162.

Morgenroth, E. (2003) Detachment: An Often Overlooked Phenomenon in Biofilm Research. In *Biofilm in Wastewater Treatment*; Wuertz, S., Bishop, P., Wilderer, P., Eds.; IWA Publishing: London, England

Morgenroth, E. (2008a) Modelling Biofilm Systems. In: *Biological Wastewater Treatment—Principles, Modelling, and Design*; Henze, M.; van Loosdrecht, M. C. M., Ekama, G.; Brdjanovic, D., Eds.; IWA Publishing: London, England.

Morgenroth, E. (2008b) Biofilm Reactors. In: *Biological Wastewater Treatment—Principles, Modelling, and Design*; Henze, M.; van Loosdrecht, M. C. M., Ekama, G.; Brdjanovic, D., Eds.; IWA Publishing: London, England.

Morgenroth, E.; Sherden, T.; van Loosdrecht, M. C. M.; Heijnen, J. J.; Wilderer, P. A. (1997) Aerobic Granular Sludge in a Sequencing Batch Reactor. *Water Res.*, **31**(12), 3191–3194.

Morgenroth, E. T.; Wilderer, P. A. (2000) Influence of Detachment Mechanisms on Competition in Biofilms. *Water Res.*, **34** (2), 417–426.

Morgenroth, E.; Kommedal, R.; Harremoës, P. (2002) Processes and Modeling of Hydrolysis of Particulate Organic Matter in Aerobic Wastewater Treatment—A Review. *Water Sci. Technol.*, **45**(6) 25–40.

Mulbarger, M. C. (1991) Fundamental Secondary Treatment Insights. *Proceedings of the 64th Annual Conference of the Water Pollution Control Federation*; Toronto, Canada: Water Environment Federation: Washington, D.C.

National Research Council (1946) Sewage Treatment at Military Installations. *Sew. Works. J.*, **18**, 787.

Nerenberg, R.; Rittmann, B. E. (2004) Reduction of Oxidized Water Contaminants with a Hydrogen-Based, Hollow-Fiber Membrane Biofilm Reactor. *Water Sci. Technol.*, **49** (11–12), 223–230.

Newman, J.; Occiano, V.; Appleton, R.; Melcer, H.; Sen, S.; Parker, D.; Langworthy, A.; Wong, P. (2005) Confirming BAF Performance for Treatment of CEPT Effluent on a Space Constrained Site. *Proceedings of the 78th Annual Water Environment Federation Technical Conference and Exposition*, Washington, D.C., Oct 12–15; Water Environment Federation: Alexandria, Virginia.

Nichols, A.; Hinojosa, J.; Riffat, R.; Dold, P.; Takacs, I.; Bott, C.; Bailey, W.; Murthy, S. (2007) Maximum Methanol-Utilizer Growth Rate: Impact of Temperature on Denitrification, *Proceedings of the Water Environment Federation 80th Annual Technical Exposition and Conference* [CD-ROM], San Diego, California; Oct 13–17: Water Environment Federation: Alexandria, Virginia.

Nicolella, C.; van Loosdrecht, M. C. M.; Heijnen, J. J. (2000) Wastewater Treatment with Particulate Biofilm Reactors. *J. Biotechnol.*, **80** (1), 1–33.

Nordeidet, B.; Rusten, B.; Ødegaard, H. (1994) Phosphorus Requirements for Tertiary Nitrification in a Biofilm. *Water Sci. Technol.*, **29** (10–11), 77–82.

Norris, D. P.; Parker, D. S.; Daniels, M. L. (1980) Efficiencies of Advanced Waste Treatment Obtained with Upgrading Trickling Filters. *J. Environ. Eng.*, **50** (9), 78–81.

Norris, D. P.; Parker, D. S.; Daniels, M. L.; Owens, E. L. (1982) High Quality Trickling Filter Treatment without Tertiary Treatment. *J. Water Pollut. Control Fed.*, **54** (7), 1087–1098.

Ødegaard, H. (2006) Innovations in Wastewater Treatment: The Moving Bed; IWA Publishing: London, England.

Ødegaard, H. (2008) The Use of the Moving Bed Biofilm Reactor (MBBR) Technology for Industrial Wastewater Treatment. *Proceedings of the International Water Association Specialized Conference on Industrial Water Treatment Systems*; Amsterdam, The Netherlands, Oct 2–3; IWA Publishing: London, England.

Ødegaard, H.; Gisvold, B.; Strickland, J. (2000) The Influence of Carrier Size and Shape in the Moving Bed Biofilm Process. *Water Sci. Technol.*, **41** (4–5), 383–391.

Ødegaard, H.; Rusten, B.; Wessman, F. (2004) State of the Art in Europe the Moving Bed Biofilm Reactor (MBBR) Process. *Proceedings of the 77th Annual Water Environment Federation Technical Exposition and Conference* [CD-ROM]; New Orleans, Louisiana, Sep 16–18; Water Environment Federation: Alexandria, Virginia.

Okey, R. W.; Albertson, O. E. (1989a) Diffusion's Role in Regulating Rate and Masking Temperature Effects in Fixed-Film Nitrification. *Water Environ. Res.*, **61** (4), 500–509.

Okey, R. W.; Albertson, O. E. (1989b) Evidence of Oxygen Limiting Conditions During Tertiary Fixed-Film Nitrification. *J. Water Pollut. Control Fed.*, **61**, 510.

Onda, K.; et al. (1968) Mass Transfer Coefficients Between Gas and Liquid Phase in Packed Columns. *J. Chem. Eng. Jpn.*, **1**, 56–62.

Opatken, E. J. (1980) Rotating Biological Contactors-Second Order Kinetics. *Proceedings of the 1st National Symposium on Rotating Biological Contactor Technology*, Vol. I, EPA-600/9-80-046a; U.S. Environmental Protection Agency: Washington, D.C.

Oppelt, E. T.; Smith, J. M. (1981) United States Environmental Protection Agency Research and Current Thinking on Fluidised-Bed Biological Treatment. In *Biological Fluidized Bed Treatment of Water and Wastewater*; Cooper, P. F., Atkinson, B., Eds.; pp. 165–178.

Pankhania, M.; Brindle, K.; Stephenson, T. (1999) Membrane Aeration Bioreactors for Wastewater Treatment: Completely Mixed and Plug-Flow Operation. *Chem. Eng. J.*, **73** (2), 131–136.

Parker et al. (1990) New Trickling Filter Applications in the USA. *Water Sci. Technol.*, **22**, 215.

Parker, D. S. (1999) Trickling Filter Mythology. *J. Environ. Eng.*, **125** (7), 618–625.

Parker, D. S. (1998) Establishing Biofilm System Evaluation Protocols. WERF Workshop: Formulating a Research Program for Debottlenecking, Optimizing, and Rerating Existing Wastewater Treatment Plants. *Proceedings of the 71st Annual Water Environment Federation Technical Exposition and Conference* [CD-ROM]; Orlando, Florida, Oct 3–7; Water Environment Federation: Alexandria, Virginia.

Parker, D. S.; Jacobs, T.; Bower, E.; Stowe, D. W.; Farmer, G. (1997) Maximizing Trickling Filter Nitrification Through Biofilm Control: Research Review and Full Scale Application. *Water Sci. Technol.*, **36** (1) 255–262.

Parker, D. S.; Lutz, M.; Andersson, B.; Aspegren, H. (1995) Effect of Operating Variables on Nitrification rates in Trickling Filters. *Water Environ. Res.*, **67** (7), 1111–1118.

Parker, D. S.; Lutz, M.; Dahl, R.; Berkkopf, S. (1989) Enhancing Reaction Rates in Nitrifying Trickling Filters through Biofilm Control. *J. Water Pollut. Control Fed.*, **61** (5), 618–631.

Parker, D. S.; Richards, T. (1986) Nitrification in Trickling Filters. *J. Water Pollut. Control Fed.*, **58** (9), 896–902.

Pearce, P. A. (1996) Optimisation of Biological Aerated Filters. *Proceedings of the BAF2 Conference*; Cranfield University: Cranfield: England.

Peladan, J. G.; Lemmel, H.; Tarallo, S.; Tattersall, S.; Pujol, R. (1997) A New Generation of Upflow Biofilters With High Water Velocities. *Proceedings of the International Conference on Advanced Wastewater Treatment Processes*; Leeds, United Kingdom; Aqua Enviro Ltd.: Wakefield, United Kingdom.

Peladan, J. G.; Lemmel, H.; Pujol, R. (1996) High Nitrification Rate With Upflow Biofiltration. *Water Sci. Technol.*, **34** (1–2), 347–353.

Pereboom, J. H. F.; Vereijken, T. (1994) Methanogenic Granule Development in Full-Scale Internal Circulation Reactors. *Water Sci. Technol.*, **30** (8), 9–21.

Perez, J.; Isanta, E.; Carrera, J. (2015) Would a Two-Stage N-Removal Be a Suitable Technology to Implement at Full Scale for the Use of Anammox for Sewage Treatment? *Water Sci. Technol.*, **72** (6), 858-864.

Pham, H.; Viswanathan, S.; Kelly, R. (2008) Evaluation of Plastic Carrier Media on Oxygen Transfer Efficiency with Coarse and Fine Bubble Diffusers. *Proceedings of the 81st Annual Water Environment Federation Technical Exposition and Conference* [CD-ROM]; Chicago, Illinois; Oct 18–22; Water Environment Federation: Alexandria, Virginia.

Phipps, S. D.; Love, N. G. (2001) Quantifying Particle Hydrolysis and Observed Heterotrophic Yield for a Full-Scale Biological Aerated Filter. *Proceedings of the 74th Annual Water Environment Federation Technical Exposition and Conference* [CD-ROM]; Atlanta, Georgia, Oct 13–17; Water Environment Federation: Alexandria, Virginia.

Pujol, R.; Hamon, M.; Kandel, X.; Lemmel, H. (1994) Biofilter: Flexible, Reliable Biological Filters, *Water Sci. Technol.*, **29** (10–11), 33–38.

Pujol, R.; Tarallo, S. (2000) Total Nitrogen Removal in Two-Step Biofiltration, *Water Sci. Technol.*, **41** (4–5), 65–68.

Rabah, F. K. J.; Dahab, M. F. (2004a) Biofilm and Biomass Characteristics in High-Performance Fluidized-Bed Biofilm Reactors. *Water Res.*, **38** (19), 4262–4270.

Rabah, F. K. J.; Dahab, M. F. (2004b) Nitrate Removal Characteristics of High Performance Fluidized-Bed Biofilm Reactors. *Water Res.*, **38** (17), 3719–3728.

Ratsak, C. H.; Verkuijlen, J. (2006) Sludge Reduction by Predatory Activity of Aquatic Oligochaetes in Wastewater Treatment Plants: Science or Fiction. A Review. *Hydrobiologia*, **564** (1), 197–211.

Rauch, W.; Vanhooren, H.; Vanrolleghem, P. A. (1999) A Simplified Mixed-Culture Biofilm Model. *Water Res.*, **33** (9), 2148–2162.

Redmon, D. T.; Boyle, W. C.; Ewing, L. (1983) Oxygen Transfer Efficiency Measurements in Mixed Liquor Using Off-Gas Techniques. *J. Water. Pollut. Control Fed.*, **55**, 1338–1347.

Rittmann, B. E.; Nerenberg, R.; Stinson, B.; Katehis, D.; Leong, E.; Anderson, J. (2004) Hydrogen-Based Membrane Biofilm Reactor for Wastewater Treatment. *Water Sci. Technol.* (in press).

Roennefahrt, K. W. (1986) Nitrate Elimination with Heterotrophic Aquatic Microorganisms in Fixed-Bed Systems with Buoyant Carriers. *Aqua*, **5**, 283–285.

Rogalla, F.; Bourbigot, M.-M. (1990) New Developments in Complete Nitrogen Removal with Innovative Biological Reactors, *Water Sci. Technol.*, **22** (1–2), 273–280.

Rogalla, F.; Sibony, J. (1992) Biocarbone Aerated Filters–Ten Years After: Past, Present and Plenty of Potential. *Water Sci. Technol.*, **26** (9–11), 2043–2048.

Rogalla, F.; Ravarini, P.; DeLarminat, G.; Courtelle, J. (1990) Large Scale Biological Nitrate and Ammonia Removal. *Water Environ. J.*, **4** (4), 319–329.

Rother, E. (2005) Optimising Design and Operation of the Biofiltration Process for Municipal Wastewater Treatment, Ph.D. dissertation, Schriftenreihe WAR, Band 163, Darmstadt, ISBN 3-932518-59-4.

Rundle, H. (2009) Good Practice in Water and Environmental Management: Biological and Submerged Aerated Filters, Chartered Institution of Water and Environmental Management (CIWEM), Aqua Enviro Technology Transfer, Wakefield, U.K.

Rusten B.; Hem L.; Ødegaard, H. (1995a) Nitrification of Municipal Wastewater in Moving Bed Biofilm Reactors. *Water Environ. Res.*, **67** (1), 75–86.

Rusten B.; Hem L.; Ødegaard, H. (1995b) Nitrogen Removal from Dilute Wastewater in Cold Climate Using Moving-Bed Biofilm Reactors. *Water Environ. Res.*, **67** (2), 65–74.

Rusten, B.; Ødegaard, H. (2007) Design and Operation of Nutrient Removal Plants for Low Effluent Concentrations. *Water Pract.*, **1** (5), 1–13.

Rusten, B.; Siljudalen, J. G.; Bungun, S. (1995c) Moving Bed Biofilm Reactors for Nitrogen Removal: from Initial Pilot Testing to Start-Up of the Lillehammer WWTP. *Proceedings of the 73th Annual Water Environment Federation Technical Exposition and Conference* [CD-ROM]; Miami, Florida, Oct 14–17; Water Environment Federation: Alexandria, Virginia.

Rusten, B.; Wien, A.; Skjefstad, J. (1996) Spent Aircraft Deicing Fluid as External Carbon Source for Denitrification of Wastewater: from Waste Problem to Beneficial Use. *Proceeding of the 51st Purdue Industrial Waste Conference*; Purdue University: West Lafayette, Indiana.

Rusten, B.; Eikebrokk, B.; Ulgenes, Y.; Lygren, E. (2006) Design and Operations of the Kaldnes Moving Bed Biofilm Reactors. *Aquacult. Eng.*, **24**, 322–331.

Ryhiner, G., Sørenson, K.; Birou, B.; Gros, G. (1993) Biofilm Reactors Configuration for Advanced Nutrient Removal. *Proceedings of the 2nd International Specialized Conference on Biofilm Reactors*, Paris, France; IWAQ: United Kingdom.

Sagberg, P.; Dauthille, P.; Hamon, M. (1992) Biofilm Reactors; a Compact Solution for Upgrading of Waste Water Treatment Plants. *Water Sci. Technol.*, **26** (3–4), (733–742).

Savage, E. S. (1983) Biological Denitrification Deep Bed Filters. *Paper presented at the Filtech Conference*, Filtration Society, London, England.

Schauer, P.; Rectanus, R.; deBarbadillo, C.; Barton, D.; Gebbia, R.; Boyd, B.; McGehee, M.; (2006) Pilot Testing of Upflow Continuous Backwash Filters For Tertiary Denitrification and Phosphorus Removal. *Proceedings of the 79th Annual Water Environment Federation Technical Exposition and Conference* [CD-ROM]; Dallas, Texas, Oct 21–25; Water Environment Federation: Alexandria, Virginia.

Schramm, A.; De Beer, D.; Gieseke, A.; Amann, R. (2000) Microenvironments and Distribution of Nitrifying Bacteria in a Membrane-Bound Biofilm. *Environ. Microbiol.*, **2**(6), 680–686.

Schroeder, E. D.; Tchobanoglous, G. (1976) Mass Transfer Limitations on Trickling Filter Design. *J. Water Pollut. Control Fed.*, **48**, 772.

Schulze, K. L. (1960) Load and Efficiency of Trickling Filters. *J. Water Pollut. Control Fed.*, **32**, 245.

Semmens, M. I. (2005) *Membrane Technology: Pilot Studies of Membrane Aerated Bioreactors*; Water Environment Research Foundation: Alexandria, Virginia.

Semmens, M. J.; Dahm, K.; Shanahan, J.; Christianson, A. (2003) Cod and Nitrogen Removal by Biofilms Growing on Gas Permeable Membranes. *Water Res.*, **37** (18), 4343–4350.

Sen, D.; Randall, C. W. (2008a) Improved Computational Model (AQUIFAS) for Activated Sludge, Integrated Fixed–Film Activated Sludge, and Moving-Bed Biofilm Reactor Systems, Part I: Semi-Empirical Model Development. *Water Environ. Res.*, **80**, 439–453.

Sen, D.; Randall, C. W. (2008b) Improved Computational Model (AQUIFAS) for Activated Sludge, Integrated Fixed-Film Activated Sludge, and Moving-Bed Biofilm Reactor Systems, Part II: Multilayer Biofilm Diffusional Model. *Water Environ. Res.*, **80**, 624–632.

Sen, D.; Randall, C. W. (2008c) Improved Computational Model (AQUIFAS) for Activated Sludge, Integrated Fixed-Film Activated Sludge, and Moving-Bed Biofilm Reactor Systems, Part III: Analysis and Verification. *Water Environ. Res.*, **80**, 633–646.

Severn Trent (2004) E-mail correspondence from David Slack of Severn Trent-Tetra Process Technologies, May.

Severn Trent (2008) E-mail correspondence from Don McCarty of Severn Trent Water Purification, December.

Shepherd, D.; Young, P., Hobson, J (1997) Biological Aerated Filters and Lamella Separators: Evaluation of Current Status, WRc Report No. PT2061; Water Research Commission: Swindon, United Kingdom.

Shieh, W. K.; Keenan, J. D. (1986) Fluidized Bed Biofilm Reactor for Wastewater Treatment. *Adv. Biochem. Eng. Biotechnol.*, **33**, 133–169.

Siegrist, H.; Gujer, W. (1987) Demonstration of Mass Transfer and pH Effects in a Nitrifying Biofilm. *Water Res.*, **20**, 971.

Sigmund, T. W. (1982) Simulation of Diurnal Operation of the Fluidized Bed System for Wastewater Treatment, M.S. Thesis, University of Wisconsin, Madison, Wisconsin.

Springer, A.; Green S (2005) Colne Bridge BAFF Process Improvements, *Proceedings of Conference on The Design and Operation of Activated Sludge and Biofilm Systems*, Horan, Aqua Enviro Ltd.

Stenquist, R. J.; Parker, D. S.; Dosh, T. J. (1974) Carbon Oxidation-Nitrification in Synthetic Media Trickling Filters. *J. Water Pollut. Control Fed.*, **46** (10), 2327–2339.

Stensel, H. D.; Brenner, R.C.; Lubin, G. (1984) Aeration Energy Requirements in Sparged Fixed Film Systems. *Proceedings of the International Biological Fixed Film Conference*, Washington, D.C., July; U.S. Environmental Protection Agency: Washington, D.C.

Stenstrom, M. K.; Rosso, D. (2008) Aeration and Mixing. In *Biological Wastewater Treatment—Principles, Modelling, and Design*; Henze, M., van Loosdrecht, M. C. M., Ekama, G., Brdjanovic, D., Eds.; IWA Publishing: London, England.

Stenstrom, M. K.; Rosso, D.; Melcer, H.; Appleton, R.; Occiano, V.; Langworthy, A.; Wong, P. (2008) Oxygen Transfer in a Full-Depth Biological Aerated Filter. *Water Environ. Res.*, **80** (7), 663–671.

Stewart, P. S. (2003) Diffusion in Biofilms. Guest Commentaries. *J. Bacteriol.*, **185** (5), 1485–1491.

Strous, M.; Fuerst, J. A.; Kramer, E. H. M.; Logemann, S.; Muyzer, G.; van de Pas-Schoonen, K. T.; Webb, R.; Kuenen, J. G.; Jetten, M. S. M. (1999a) Missing Lithotroph Identified as New Planctomycete. *Nature*, **400** (6743), 446–449.

Sutton, P. M.; Mishra, P. N. (1994) Activated Carbon-Based Biological Fluidized-Beds for Contaminated Water and Wastewater Treatment: A State-of-the-art Review. *Water Sci. Technol.*, **29** (10–11), 309–317.

Sutton, P. M.; Shieh, W. K.; et al. (1981) Dorr-Olivers' Oxitron SystemTM Fluidised-Bed Water and Wastewater Treatment Process. In *Biological Fluidized Bed Treatment of Water and Wastewater*; Cooper, P. F., Atkinson, B., Eds; Ellis Horwood for Water Research Laboratory, Stevenage Laboratory: Chichester, United Kingdom, pp. 285–305.

Suzuki, Y.; Miyahara, S.; Takeishi, K. (1993) Oxygen-Supply Method Using Gas-Permeable Film for Wastewater Treatment. *Water Sci. Technol.*, **28** (7), 243–250.

Syron, E.; Casey, E. (2008) Membrane-Aerated Biofilms for High Rate Biotreatment: Performance Appraisal, Engineering Principles, and Development Requirements. *Environ. Sci. Technol.*, **42** (6), 1833–1844.

Szwerinski, H.; Arvin, E.; Harremoës, P. (1986) pH-Decrease in Nitrifying Biofilms. *Water Res.*, **20**, 971.

Taljemark, K.; Aspegren, H.; Gruvberger, N.; Hanner, N.; Nyberg, U.; Andersson, B. (2004) 10 Years of Experiences of a Nitrification MBBR Process for Post-Denitrification. *Proceedings of the 77th Annual Water Environment Federation Technical Exposition and Conference* [CD-ROM]; New Orleans, Louisiana, Sep 16–18; Water Environment Federation: Alexandria, Virginia.

Timberlake, D.; Strand, S.; Williamson, K. (1988) Combined Aerobic Heterotrophic Oxidation, Nitrification and Denitrification in a Permeable-Support Biofilm. *Water Res.*, **22** (12), 1513–1517.

Toettrup, H.; Rogalla, F.; Vidal, A.; Harremoes, P. (1994) The Treatment Trilogy of Floating Filters: From Pilot to Prototype to Plant. *Water Sci. Technol.*, **29** (10–11), 23–32.

Tschui, M.; Boller, M.; Gujer, W.; Eugster, J.; Mäder, C. (1993) Tertiary Nitrification in Aerated Biofilm Reactors. *Proceedings of the European Water Filtration Congress*, Ostend, Belgium.

Tschui, M.; Boller, M.; Gujer, W.; Eugster, C.; Mäder, C.; Stengel, C. (1994) Tertiary Nitrification in Aerated Biofilters. *Water Sci. Technol.*, **29** (10–11), 53–60.

U.S. Environmental Protection Agency (1975) *Process Design Manual for Nitrogen Control*; U.S. Environmental Protection Agency, Office of Wastewater Management: Washington, D.C.

U.S. Environmental Protection Agency (1991) *Assessment of Single-Stage Trickling Filter Nitrification, EPA-430/09-91-005*; U.S. Environmental Protection Agency, Office of Wastewater Management: Washington, D.C.

U.S. Environmental Protection Agency (1993) *Nitrogen Control Manual, EPA/625/R-93/010*; U.S. Environmental Protection Agency, Office of Wastewater Management: Washington, D.C.

U.S. Filter/Envirex. (1997) *The Fluid Bed for Denitrification of Municipal and Industrial Wastewater*. U.S. Filter/Envirex: Waukesha, Wisconsin.

van Loosdrecht, M. C. M.; Eikelboom, D.; Gjaltema, A.; Mulder, A.; Tijhuis, L.; Heijnen, J. J. (1995) Biofilm Structures. *Water Sci. Technol.*, **32** (8), 35–43.

Velz, C. J. (1948) A Basic Law for the Performance of Biological Filters. *Sew. Works J.* **20**, 607.

Wall, D.; Frodsham, D.; Robinson, D. (2001) Design of Nitrifying Trickling Filters. *Proceedings of the 74th Annual Water Environment Federation Technical Exposition and Conference* [CD-ROM]; Atlanta, Georgia, Oct 13–17; Water Environment Federation: Alexandria, Virginia.

Wanner, O.; Gujer, W. (1984) Competition in Biofilms. *Water Sci. Technol.*, **17**, 27–44.

Wanner, O.; Gujer, W. (1985) A Multispecies Biofilm Model. *Biotech. Bioeng.*, **28**, 313–328.

Wanner, O.; Reichert, P. (1996) Mathematical-Modeling of Mixed-Culture Biofilms. *Biotech. Bioeng.*, **49** (2), 172–184.

Wanner, O.; Eberl, H.; Morgenroth, E.; Noguera, D.; Picioreanu, C.; Rittmann, B.; Van Loosdrecht, M. C. M. (2006) *Mathematical Modeling of Biofilms, Scientific and Technical Report No. 18*; IWA Publishing: London, England.

Water Environment Federation (2013) *Operation of Nutrient Removal Facilities*, Manual of Practice No. 37; Water Environment Federation: Alexandria, Virginia.

Weiss, P. T.; Oakley, B. T.; Gulliver, J. S.; Semmens, M. J. (1996) Bubbleless Fiber Aerator for Surface Waters. *J. Environ. Eng.*, **122** (7), 631–639.

Welty, J. R.; Wicks, C. E.; Wilson, R. E. (1976) *Fundamentals of Momentum, Heat and Mass Transfer*, 2nd ed.; John Wiley and Sons: New York.

Wett, B. (2006) Solved Upscaling Problems for Implementing Deammonification of Rejection Water. *Water Sci. Technol.*, **53** (12), 121–128.

Wett, B.; Omari, A.; Podmirseg, S. M.; Han, M.; Akintayo, O.; Gómez Brandón, M.; Murthy, S.; Bott, C.; Hell, M.; Takács, I.; Nyhuis, G.; O'Shaughnessy, M. (2013) Going for Mainstream Deammonification from Bench to Full Scale for Maximized Resource Efficiency. *Water Sci. Technol.*, **68** (2), 283-289.

Wik T. (2000) Strategies to Improve the Efficiency of Tertiary Nitrifying Trickling Filters. *Water Sci. Technol.*, **41** (4–5) 477–485.

Yang, M.-C.; Cussler, E. L. (1986) Designing Hollow-Fiber Contactors. *Am. Inst. Chem. Eng. J.*, **32** (11), 1910–1916.

Yang, Y.; Zhang, L.; Han, X.; Zhang, S.; Li, B.; Peng, Y. (2016) Determine the Operational Boundary of a Pilot-Scale Single-Stage Partial Nitritation/Anammox System with Granular Sludge. *Water Sci. Technol.*, **73** (9), 2085–2092.

Yee, C. J.; Hsu, Y.; et al. (1992) Effects of Microcarrier Pore Characteristics on Methanogenic Fluidized-Bed Performance. *Water Res.*, **26** (8), 1119–1125.

Yilmaz, G.; Lemaire, R.; Keller, J.; Yuan, Z. (2008) Simultaneous Nitrification, Denitrification, and Phosphorus Removal from Nutrient-Rich Industrial Wastewater Using Granular Sludge. *Biotechnol. Bioeng.*, **100**(3), 529–541.

Zhang, T. C.; Bishop, P. L. (1994) Experimental Determination of the Dissolved Oxygen Boundary Layer and Mass Transfer Resistance Near the Fluid-Biofilm Interface. *Water Sci. Technol.*, **30** (11) 47–58.

Zhang, T. C.; Bishop, P. L. (1996) Evaluation of Substrate and pH Effects in a Nitrifying Biofilm. *Water Environ. Res.*, **68** (7), 1107–1115.

CHAPTER **12**

Suspended-Growth Treatment Processes

Timur Deniz, Ph.D., P.E., BCEE; John R. Bratby, Ph.D., P.E.;
Eric T. Staunton, Ph.D.; Claes Westring; and Andrea Turriciano White, P.E.

1.0 Introduction

Suspended-growth systems are biological treatment processes based on the growth and retention of a suspension of microorganisms. These microorganisms convert biodegradable, organic wastewater constituents and certain inorganic fractions into new cell mass and by-products, both of which then can be removed by settling, gaseous stripping, and other physical means. Suspended-growth systems for wastewater treatment are predominantly aerobic processes, typically referred to as activated sludge, with a variety of reactor configurations and flow patterns. Strictly anaerobic suspended-growth processes for liquid-phase treatment are also in use.

This chapter presents fundamentals of aerobic activated sludge treatment and the process configurations that can be used. It covers process design for carbon oxidation and nitrification of ammonia, process configurations for nitrogen and phosphorus removal, anaerobic processes, membrane bioreactors (MBRs), design considerations for wet-weather conditions, oxygen-transfer systems, and secondary clarification design. The chapter concludes with a comprehensive process design example that illustrates several aerobic suspended-growth systems for different applications and performance requirements.

1.1 Process Description

Figure 12.1 presents a typical flow-through suspended-growth activated sludge process.

Wastewater is combined with biomass and other solids in the reactor where mixing and aeration provided. Typically, the process operates in a continuous-flow mode, but can also be operated as a batch process. Contents of the reactor, referred to as mixed liquor, consist of wastewater; microorganisms; microbial cell debris; and inert,

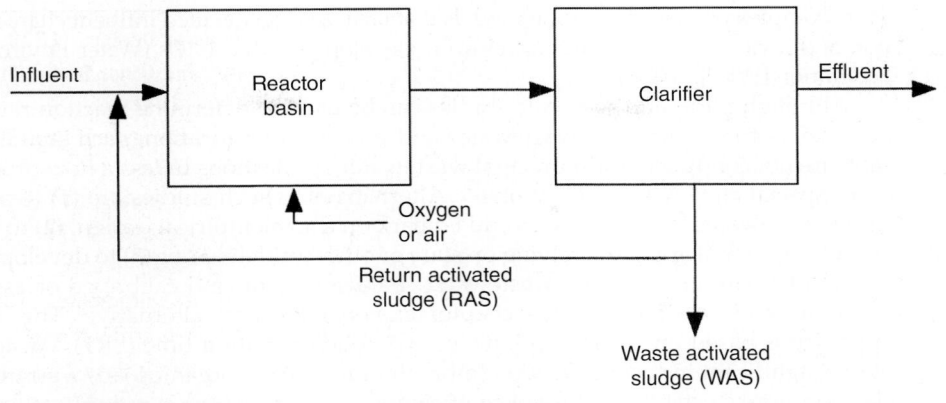

FIGURE 12.1   A typical activated sludge process.

biodegradable, and nonbiodegradable suspended and colloidal matter. The particulate fraction of the mixed liquor is termed mixed-liquor suspended solids (MLSS). Suspended-growth anaerobic processes are similar in concept but without aeration or mixing of the solids resulting in a stratified sludge blanket of microorganisms.

After sufficient time for biological reactions, mixed liquor is conveyed to a separate settling basin (clarifier) or other solids-liquid separation step that separates MLSS from treated wastewater and produces clarified effluent. Settled MLSS are then recycled as return activated sludge (RAS) to the aeration basin to maintain a concentrated microbial population for efficient degradation of influent wastewater constituents in an economically sized reactor. Microorganisms and cell debris are generated continuously, which must be removed from the system through wasting. Wasting can be continuous or periodic and typically is from the clarifier or return sludge line, although removal from the aeration basin is an alternative. Retention of MLSS and effluent clarification also can be accomplished with synthetic, microfiltration membranes. The MLSS recirculation and wasting requirements remain with this alternative.

1.2   Historic Overview

The activated sludge process received its name from and was developed based on a series of experiments conducted in Manchester, England. Widespread use of the process did not begin until the 1940s. Aeration-basin hydraulic retention time (HRT) was one of the first design parameters used; short HRTs were chosen for what was considered to be weak wastewater and long HRTs for strong wastewater. Loading criteria eventually were developed, typically relating to the mass of biochemical oxygen demand (BOD) applied per day per mass of microbial solids present in the aeration basin.

Design equations were developed based on the concepts of microbial growth kinetics and mass balances back in the 1960s and 1970s. Eckenfelder (1966), McKinney (1962), Lawrence and McCarty (1970), and Ramanathan and Gaudy (1971) developed design approaches that yield similar results (Gaudy and Kincannon, 1977). These design approaches were based on behavior and performance of microorganisms of the activated sludge process as characterized by measures such as BOD, total suspended solids (TSS), and kinetic parameters and coefficients representing a diverse biological population.

More complex process simulation models that are based on detailed influent characterization and a variety of microorganisms were developed in the 1990s (Water Environment Federation [WEF], 2013b).

Although pilot- and full-scale studies can be used to determine reaction rates and parameters for a particular wastewater and process configuration, such studies typically are not conducted for municipal wastewater applications unless a new process or other special circumstance is involved. Alternatives to such studies are (1) to assume certain wastewater characteristics and embark on a semiempirical design, (2) to use an entirely empirical approach relying on state or other guidelines, or (3) to develop a process model based on detailed wastewater characterization and calibrated or assumed constants and coefficients. This chapter explores the first alternative. This design approach is based on mass of organisms and solids retention time (SRT). Wastewater with a significant industrial fraction (more than 10% of the organic load) warrants special attention to establish values of coefficients.

Capabilities of the activated sludge process, through its ability to enrich biological populations to achieve specific objectives, have increased since the 1990s. The addition of unaerated zones to a nitrifying activated sludge process can result in effective removal of inorganic nitrogen through biological denitrification and enhanced biological phosphorus removal (EBPR). Over the last decade, the use of membranes with pore sizes of 0.1 μm allowed for higher reactor MLSS concentrations resulting in smaller reactor volumes and effluent of exceptional clarity.

In the wide variety of activated sludge process configurations and applications in use today, the fundamental biological processes at work are the same.

1.3   Activated Sludge Environment

An activated sludge process uses a suspension of diverse microorganisms to treat wastewater. The dry weight of these microorganisms is 95% or more organic in composition. Suspension of microorganisms in an activated sludge process is typically 70% to 90% organic and 10% to 30% inorganic substances because of inert materials in the wastewater. Composition of the organic fraction of biomass is approximated by the empirical formula $C_5H_7O_2NP_{0.2}$ (Grady et al., 2011). Successful facility performance depends on a microbial community that will oxidize the waste materials and form a flocculent biomass that is readily removed by gravity separation.

Heterotrophic organisms that require biodegradable organic matter for energy and new cell synthesis typically dominate the microbial population. Autotrophic bacteria that oxidize ammonia to nitrite and nitrate use inorganic materials for energy and cell synthesis. Such autotrophs typically are present in varying concentrations. A well-designed activated sludge system provides an environment that promotes growth of desired microorganisms and inhibits those that contribute to poor sludge settleability and foaming; it also can control nuisance organisms that may appear.

Most bulking microorganisms are filamentous bacteria. An excess of filaments protruding from flocs are believed to prevent biomass compaction. Some researchers contend that an ideal floc contains just the right mixture of filamentous microorganisms and floc formers, with the filaments forming the backbone of the floc (Jenkins et al., 2003; Sezgin et al., 1978).

Detailed discussions of the microbiology of biological treatment systems is presented elsewhere (Jenkins et al., 2003; Grady et al., 2011; Metcalf & Eddy, Inc./AECOM, 2014; U.S. Environmental Protection Agency [U.S. EPA], 1987).

Biological nitrogen removal is achieved through ammonia oxidation followed by denitrification, which is the use of nitrate by biomass as an electron acceptor to oxidize carbon substrate. The end result of this process is the conversion of nitrate to nitrogen gas that is released to the atmosphere.

Excess biomass that must be wasted will remove the portion of the influent phosphorus that is incorporated into the biomass. Biological phosphorus removal can be enhanced by enriching the bacterial culture, which results in organisms that can retain greater amounts of phosphorus than a typical aerobic biomass. Given the expected phosphorus content of the biomass to be wasted, removal can be estimated from a mass balance across the system.

Aeration is an essential component of activated sludge process for two purposes. Oxygen serves as terminal electron acceptor for the aerobic organic matter degradation by the heterotrophic bacteria and ammonia oxidation to nitrate by nitrifiers. Aeration also provides mixing to keep the activated sludge solids in suspension.

1.4   System Components

A basic suspended-growth system consists of several interrelated components:

- Single or multiple reactors designed as completely mixed flow, plug flow, or intermediate patterns and sized to provide adequate SRT, organic loading, or other criteria resulting in an HRT of 2 to 3 hours minimum up to 24 hours or more.
- An oxygen source and equipment to disperse atmospheric, pressurized, or oxygen-enriched air to the aeration basin at a rate sufficient to keep the system aerobic.
- A means of mixing the aeration basin to keep solids in suspension.
- A clarifier, membranes, or period of settling to separate the suspended solids from treated wastewater.
- A method of collecting and returning sludge from the clarifier or recycling concentrated solids from membrane zones back to the aeration basin. Since all the treatment and solids settling is accomplished in single reactors, sludge return is not required with a sequencing batch reactor (SBR) system.
- A means of wasting excess biomass and accumulated nonbiodegradable influent solids from the system.

2.0   Process Configurations and Types

Suspended-growth reactors, used in activated sludge and biological nutrient removal (BNR) facilities, have been designed in many different configurations. They have to be categorized by basin shape, loading rates, feeding and aeration patterns, and other features. Combinations of features from several categories give design engineers an array of choices. Basic activated sludge facilities have often been called by such reactor descriptions as:

- Completely mix;
- Plug flow;

- Sequencing batch reactor; and
- Combination (capable of being operated in more than one configuration).

For smaller facilities, low-load processes (such as oxidation ditches and SBRs) are common in part because of simplicity of operation and reliable performance. For larger facilities, conventional plug flow (some with configuration flexibility) is favored. Plug flow often is favored because completely mixed activated sludge (CMAS) reactors can promote growth of filamentous bacteria that hinder sludge compaction. This growth, however, is a site-specific issue related to dissolved-oxygen concentration and other factors such as the rate at which substrate is applied and total available biomass. Plug-flow reactors generally offer more flexibility if they need to be converted for BNR. Specifically, anaerobic and anoxic zones can be created. Compartmentalization by using baffles or walls within a tank can provide flexibility for intermediate zones that can be used for more than one reactor.

The SBRs also are used widely, especially for smaller facilities. More than several thousand SBR facilities are in operation in the world. Designs incorporating operational flexibility can remove nutrients to low levels (Young et al., 2008).

2.1   Reactor Types

Categorization of reactor types leads to the definitions of complete mix, plug flow, and SBR.

2.1.1   Complete Mix

By definition, a CMAS reactor has uniform characteristics throughout the contents of the entire reactor. In this configuration, shown in Figure 12.2, the CMAS influent wastewater is distributed rapidly throughout the basin, and operating characteristics of MLSS, respiration rate, and BOD are uniform throughout. Because the total body of basin liquid has the same quality as the basin effluent, only a low level of food is available at any time for the large mass of microorganisms present. This characteristic is cited as the primary reason why CMAS can handle surges in organic loading and toxic shocks (to a limited extent) without producing a change in effluent quality. As mentioned above, CMAS systems can promote the growth of filamentous bacteria that settle poorly. Nevertheless, many CMAS facilities produce excellent results if properly operated. Design of a selector tank before CMAS can help limit the growth of filamentous bacteria. Control of dissolved oxygen is simpler with CMAS geometry because of the approximately uniform oxygen demand throughout the reactor.

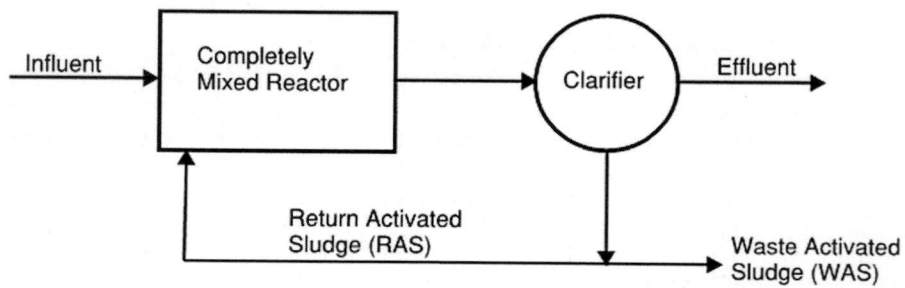

Figure 12.2   Completely mixed reactor.

The CMAS basins are typically square, round, or rectangular. Tank dimensions may be controlled by the size and mixing pattern of the aeration equipment and local site considerations. Surface turbine units were popular for providing aeration for CMAS tanks although diffused air became the preferred technology today. Factors that influence mechanical aeration mixing effectiveness include length-to-width ratio, mixing power per unit of volume, and the locations of feed points and outlet structures. Achieving complete mixing in a real-world basin is difficult but can be attempted. Any square or circular basin with a reasonable detention time and level of mixing intensity can be considered a completely mixed reactor, regardless of the type of aeration system used. The length-to-width ratio of a basin typically should be maintained at less than 3:1 to remain primarily complete mix using mechanical aeration and no baffles. Multiple mechanical aeration units in long, narrow basins—for example, length-to-width greater than 5:1—create a mixing pattern that starts to resemble plug flow. If diffused air is used, then full tank width influent feed and effluent removal weir structures typically are provided as good practice. Multiple feed points and withdrawal weirs along opposite sides of a rectangular aeration basin also could be used for this purpose. Oxidation ditches may be viewed as a completely mixed reactor even though they have some plug-flow characteristics. Some consider it to be completely mixed because the influent concentration of substrate is immediately diluted by the large mixed-liquor flow to a value nearly equal to that of the aeration-basin effluent. To accurately model such closed-loop flow, 10 or more CMAS cells in series need to be used with at-large rate of recirculation from the last to the first.

2.1.2 Plug Flow

Plug flow and basins in series are discussed together because a plug-flow reactor can be viewed as several small, completely mixed basins in series. Plug-flow basins used for municipal activated sludge facilities are 5 to 9 m (15 to 30 ft) wide and up to 120 m (400 ft) long (length-to-width ratio more than 10:1). Long basins may be constructed as single-pass tanks, side by side, or in a folded arrangement.

The ideal plug-flow configuration has a relatively high organic loading at the influent end of the basin. The food-to-microorganism (F:M) ratio is reduced over the length of the basin as organic matter in wastewater is assimilated. At the downstream end of the basin, oxygen consumption shifts increasingly toward endogenous respiration. The high organic loading at the head end of this process (i.e., high F:M) discourages most types of filamentous bacteria growth and results in better sludge settling compared to a completely mixed reactor if sufficient dissolved-oxygen concentrations are maintained. Keeping a low dissolved-oxygen level too long, however, may encourage filamentous growth. As discussed elsewhere in this chapter, polysaccharide formation may result from high loading and low dissolved-oxygen concentrations at the inlet end.

Whereas a completely mixed reactor is able to handle surges in loading, plug-flow configurations have a superior ability to avoid "bleed-through" or passage of untreated substrate during peak flows. Plug-flow reactors also have an advantage where high-effluent dissolved-oxygen concentrations are desirable. In a completely mixed reactor configuration, the entire tank contents would have to be maintained at the elevated dissolved-oxygen level to achieve that objective. Control of dissolved-oxygen concentrations in a plug-flow system can be complicated if a wide range of oxygen demands and at multiple locations are expected.

2.1.3 Sequencing Batch Reactors

The SBR process involves a fill-and-draw, completely mixed reactor in which both aeration and clarification occur in a single reactor. Settling is initiated when aeration is turned off. When settling time is up, a decanter device is used to withdraw supernatant. The sequential phases comprise a cycle with defined time intervals to achieve certain objectives. The bulk of MLSS remains in the reactor during the cycle with periodic wasting. Specific treatment phases are illustrated as a percentage of reactor volume in Figure 12.3 as an example. The phases of each cycle include:

- Fill (raw or settled wastewater fed to the reactor);
- React (aeration/mixing of the reactor contents);
- Settle (quiescent settling and separation of MLSS from the treated wastewater);
- Draw/decant (withdrawal of treated wastewater from the reactor); and
- Idle (delay period before beginning the next cycle and might include removal of waste sludge from the reactor bottom).

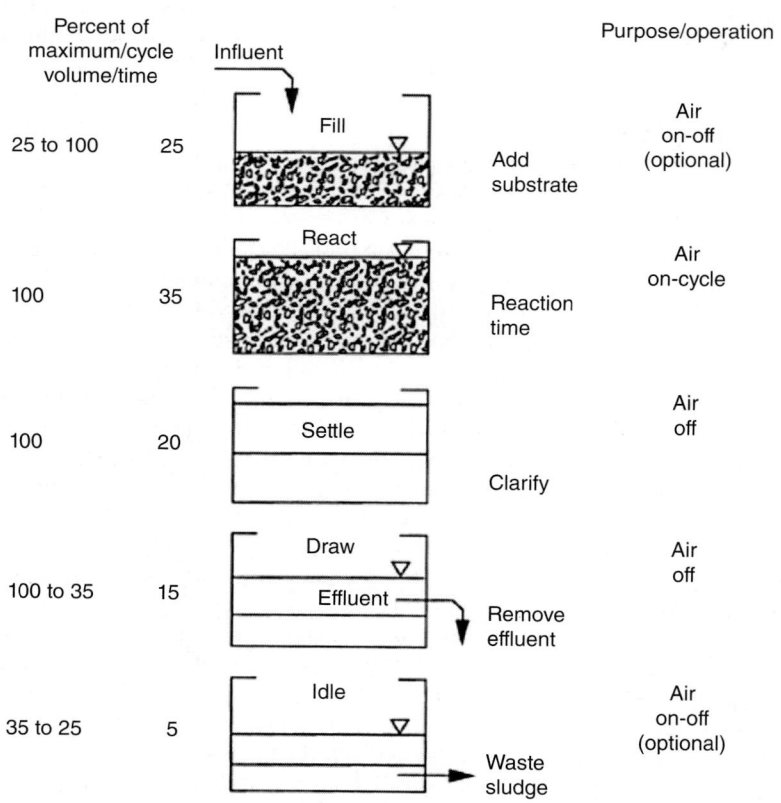

FIGURE 12.3   Typical sequencing batch reactor operation for one cycle.

The idle phase may be omitted and sludge wasted at the end of the reactor draw phase. Cycles and phases may vary with each reactor. Because of the batch nature of the process, flow equalization or multiple reactors are required to accommodate the continuous and varying inflow of wastewater to the facility.

Advantages of SBR include elimination of a secondary clarifier and RAS pumping systems, high tolerance for short-duration peak flows and shock loadings, operational flexibility, and clarification that occurs under nearly ideal quiescent conditions. Disadvantages include the potential for sludge bulking at low F:M ratios, the inability to effectively chlorinate RAS for filament control, and the need for multiple reactors for reliability, for adequate equalization, or to accommodate long-duration peak flows. Equalization of effluent decant also might be required for subsequent downstream treatment, conveyance, or discharge to small, hydraulically limited receiving waters. The intermittent cycle extended aeration system (ICEAS) was developed in Australia as a modification to the typical SBR (Goronszy, 1979). Influent feeds continuously to the reactor during all cycles as in a continuous-flow system; but withdrawal is intermittent, similar to the SBR system. Continuous influent feed addresses some of the disadvantages of SBRs cited above. The pre-react zone of the ICEAS provides a higher F:M selector to limit the growth of filaments, and also allows operation with a single reactor so that only two reactors are required for redundancy.

Another SBR concept is the patented cyclic activated sludge system (CASS). It features plug-flow initial reaction conditions and a complete-mix reactor basin. Each reactor basin is divided by baffle walls into three sections (zone 1: selector; zone 2: secondary aeration; zone 3: main aeration). For municipal applications, these sections are in the approximate proportions of 5%, 10%, and 85%. The MLSS are continuously recycled from zone 3 to the zone 1 selector to remove the readily biodegradable soluble substrate and favor the growth of the floc-forming microorganisms. The sludge return rate causes an approximate daily cycling of biomass in the main aeration zone through the selector zone. Proponents contend that the selector is self-regulating for any load condition and operates under anoxic conditions during aerobic periods and anaerobic reaction conditions during nonaerated periods. The system can be operated such that EBPR is also achieved. The completely mixed nature of the main reactor provides flow and load balancing and a tolerance to shock or toxic loading.

More than 1000 SBR-type facilities are operating in the United States. Approximately 80% of the facilities have flows of 4000 m³/d (1 mgd) or lower; 70% have flows of 1900 m³/d (0.5 mgd) or lower. Few, if any, are as large as 40 000 m³/d (10 mgd). Larger facilities that have been constructed include the Kung Ming, China (190 000 m³/d or 50 mgd); Cardiff, Wales (300 000 m³/d or 80 mgd); and Quakers Hill, Australia (57 000 m³/d or 15 mgd); and the Ringsend facility in Dublin, Ireland (490 000 m³/d or 130 mgd).

SBRs can be modified to provide carbonaceous oxidation, nitrification, and BNR. Nitrification takes place at the highest rates during the react phase and portions of the fill period when aeration is practiced. Because SBRs typically are designed and operated at long SRTs and low F:M, partial or complete nitrification is observed in nearly all facilities treating municipal wastewater. Denitrification can be achieved when aeration is reduced or stopped so that anoxic conditions form but reaction rates will be depressed if the reactor is not thoroughly mixed; mechanical mixing might be warranted. Continuous-feed intermittent withdrawal SBR systems achieve higher denitrification rates compared to a batch feed process, including in the sludge blanket during the settle and decant phases, because carbon (BOD) is supplied continuously throughout the cycle. Conditions for EBPR can be created by incorporating a phase without aeration with readily available carbon substrate and low nitrate concentrations, such as at the beginning of a cycle. The flexibility of SBR allows for upgrades from regular carbon oxidation to BNR without costly construction.

2.2 Loading Rates

Activated sludge processes also can be classified by loading or organic-feed rate. Common terms are conventional, low rate, and high rate. Table 12.1 provides a summary of general characteristics for various processes. Table 12.2 presents typical ranges for relevant design parameters.

2.2.1 Conventional

Conventional loading rates apply to plug-flow or CMAS systems with an F:M loading of approximately 0.2 to 0.5 kg BOD/d·kg mixed-liquor volatile suspended solids (MLVSS). These systems can obtain BOD removal efficiencies in the range of 85% to 95%. Conventional rate system MLSS design concentrations often range from 1500 to 3000 mg/L. Design MLSS concentrations increased considerably over the years because of improvements in the oxygen-transfer capability of aeration devices, clarifier performance, and understanding of system concepts.

   An important consideration in the design of conventional systems is that nitrification might occur, even when not desired. This often happens with low-loading conditions during summer months or high SRTs because of wasting practices. When nitrification occurs, denitrification may occur in the final clarifiers resulting in "rising sludge" problems when nitrogen gas buoys and floats the biomass floc. Approaches used to limit nitrification and unwanted denitrification include reducing the SRT and the dissolved-oxygen concentration to reduce nitrification or increasing the dissolved oxygen before clarification. In warm climates, SRT reduction to prevent nitrification might adversely affect floc formation and secondary clarifier performance.

2.2.2 Low Rate

Low-rate (also called extended aeration) facilities are characterized by the introduction of pretreated (e.g., screened and degritted) wastewater directly to an aeration basin with a long HRT, high MLSS concentration, high RAS rate, and low sludge wastage. This system, initially used in the United States for flows of approximately 4000 m³/d (1 mgd) or lower, often incorporated complete-mix reactors. During the past few decades, low-rate systems have been applied to larger sizes in the shapes of oxidation ditches and similar shapes.

   A particular advantage of using long HRTs (typically 16 to 36 hours) is that they allow the facility to operate effectively over widely varying flow and waste loadings and lower overall solids production. Stable solids are often advantageous for subsequent solids-handling processes. Secondary clarifiers must be designed to handle variations in hydraulic loadings and high MLSS concentrations associated with this process.

   One of the process goals is to maintain the biomass in a highly endogenous respiration phase. Because microorganisms are undergoing aerobic digestion in the aeration basin, more oxygen is required than for other single-stage systems. Many low-cost, low-rate facilities experience a dissolved-oxygen deficiency when waste load is high because the design may not include automation to increase aeration in proportion to load or the operations staff fails to maintain such provisions. In some cases, the long SRT and

Process Modification	Flow Model	Aeration System	BOD Removal Efficiency (%)	Remarks
Conventional	Plug flow	Diffused-air, mechanical aerators	85–95	Use for low-strength domestic wastes; process is susceptible to shock loads
Complete-mix	Continuous-flow stirred-tank reactor	Diffused-air, mechanical aerators	85–95	Use for general application; process is resistant to shock loads but is susceptible to filamentous growths
Step feed	Plug flow	Diffused-air	85–95	Use for general application for a wide range of wastes
Modified aeration	Plug flow	Diffused-air	60–75	Use for intermediate degree of treatment where cell tissue in the effluent is not objectionable
Contact stabilization	Plug flow	Diffused-air, mechanical aerators	80–90	Use for expansion of existing systems and package facilities
Extended aeration	Plug flow	Diffused-air, mechanica aerators	75–95	Use for small communities, package facilities, and where nitrified element is required; process is flexible
High-rate aeration	Continuous-flow stirred-tank reactor	Mechanical aerators	75–90	Use for general applications with turbine aerators to transfer oxygen and control floc size
Kraus process	Plug flow	Diffused-air	85–95	Use for low-nitrogen, high-strength waste
High purity oxygen	Continuous-flow stirred-tank reactors in series	Mechanical aerators (sparger turbines)	85–95	Use for general application with high-strength waste and where limited space is available at site; process is resistant to slug loads area of land is available; process is flexible
Oxidation ditch	Plug flow	Mechanical aerators (horizontal axis type)	75–95	Use for small communities or where large area of land is available; process is flexible
Sequencing batch reactor	Intermittent-flow stirred-tank reactor	Diffused-air	85–95	Use for small communities where land area is limited; process is flexible and can remove nitrogen and phosphorus
Deep shaft reactor	Plug flow	Diffused-air	85–95	Use for general application with high-strength waste; process is resistant to slug loads
Single-stage nitrification	Continuous-flow stirred-tank reactors or plug flow	Mechanical aerators, diffused-air	85–95	Use for general application for nitrogen control where inhibitory industrial waste is not present
Separate stage nitrification	Continuous-flow stirred-tank reactors or plug flow	Mechanical aerators, diffused-air	85–95	Use for upgrading existing systems, where nitrogen standards are stringent, or where inhibitory industrial waste is present and can be removed in earlier stages

TABLE 12.1 Operational Characteristics of Activated Sludge Process (from Metcalf & Eddy, Inc., 2014, with permission from the McGraw-Hill Companies)

Process Modification	θ_c, d	F:M, lb BOD_5 Applied/d/lb $MLVSS$[a]	Volumetric Loading (lb BOD_5/d/10^3 cu ft)	MLSS (mg/L)	V/Q·h	Q_r/Q
Conventional	5–15	0.2–0.4	20–40	1500–3000	4–8	0.25–0.75
Complete-mix	5–15	0.2–0.6	50–120	2500–4000	3–5	0.25–1.0
Step feed	5–15	0.2–0.4	40–60	2000–3500	3–5	0.25–0.75
Modified aeration	0.2–0.5	1.5–5.0	75–150	200–1000	1.5–3	0.05–0.25
Contact stabilization	5–15	0.2–0.6	60–75	(1000–3000)[b] (4000–10 000)[c]	(0.5–1.0)[b] (3–6)[e]	0.5–1.50
Extended aeration	20–30	0.05–0.15	10–25	3000–6000	18–36	0.5–1.50
High-rate aeration	5–10	0.4–1.5	100–1000	4000–10 000	2–4	1.0–5.0
Kraus process	5–15	0.3–0.8	40–100	2000–3000	4–8	0.5–1.0
High-purity oxygen	3–10	0.25–1.0	100–200	2000–5000	1–3	0.25–0.5
Oxidation ditch	10–30	0.05–0.30	5–30	3000–6000	8–36	0.75–1.50
Sequencing batch reactor	NA	0.05–0.30	5–15	1500–5000[e]	12–50	NA
Deep shaft reactor	NI	0.5–5.0	NI	NI	0.5–5	NI
Single-stage nitrification	8–20	0.10–0.25 (0.02–0.15)[d]	5–20	2000–3500	6–15	0.50–1.50
Separate stage nitrification	15–100	0.050–0.20 (0.04–0.15)[d]	3–9	2000–3500	3–6	0.50–2.00

[a]MLVSS = mixed liquor volatile suspended solids.
[b]Contact unit.
[c]Solids stabilization unit.
[d]Total Kjeldahl nitrogen/MLVSS.
[e]MLSS varies depending on the portion of the operating cycle.
Note: lb/10^3 cu ft × 0.016 0 = kg/m^3·d.
lb/d/lb = kg/kg·d.
NA = not applicable.
NI = no information.

TABLE 12.2 Design Parameters for Activated Sludge Processes (from Metcalf & Eddy, Inc., 2014, with permission from the McGraw-Hill Companies)

excess dissolved oxygen at night will allow some nitrification, causing a daily, but non-coincidental, nitrification/denitrification cycle.

Some common problems with long extended aeration systems include continuous loss of pinpoint floc and the tendency to lose MLSS following short-term periods of low influent loading intensity such as on weekends. Long HRTs combined with a long clarification time also can result in denitrification, leading to rising sludge in secondary clarifiers. This condition, combined with the lack of primary sedimentation to remove floatables, requires use of effective skimming devices in final clarifiers. Guo et al. (1981) suggest that the average MLSS concentration should not fall below 2000 mg/L. In cold climates, low temperatures likely will impair performance of the extended aeration process unless heat loss is controlled. Use of open surface aerators would be at a disadvantage to covered aerators or diffused air and mechanical velocity control mixers.

The patented Cannibal solids reduction process is similar to extended aeration activated sludge with the addition of physical and biological solids processing of sidestreams. Although systems have been in operation for several years, knowledge of the process is still evolving (Johnson et al., 2008; Novak et al., 2006; Roxburgh et al., 2006). Physical treatment of the MLSS in a sidestream consists of fine-screening to remove fibrous materials and periodic removal of heavy particulates by a hydrocyclone. The screenings are primarily volatile and fibrous, and represent 20% to 30% of the MLSS that might be expected in a conventional process. The main reactor is operated at a moderate SRT of 8 to 10 days, and MLSS are wasted to a second reactor that is operated as an SBR with intermittent aeration. The net result is low observed yield of biomass from the system. Labelle et al. (2015) evaluated the Cannibal process at a full-scale facility operated at an SRT of 400 days. The estimated sludge yield was 0.14 g TSS/g chemical oxygen demand (COD) removed, including a trash and grit yield of 0.06 g TSS/g COD. Under similar influent and operating conditions, conventional activated sludge (CAS) process would have a sludge yield of 0.23 g TSS/g COD, which leads to an estimated sludge reduction of 0.09 g TSS/g COD removed attributable to the slow degradation of unbiodegradable influent particulate organics. However, to achieve this additional sludge yield reduction, more electricity must be used.

2.2.3 High Rate
High rate is the term applied to an activated sludge system characterized by a short HRT and a high organic loading rate (OLR). MLSS concentrations may vary from 800 to as high as 10 000 mg/L; F:M ratios are higher than those used in conventional systems. Process integrity depends on maintaining the biomass in a relatively high rate of growth. Although high-rate systems can produce an effluent quality approaching that of a conventional system, they encourage a higher fraction of dispersed organisms than in higher SRT systems. This can result in turbid clarifier effluent. Therefore, high-rate systems must be operated with special care. For example, inadequate RAS flowrates, insufficient wasting, and high sludge flux rates make the clarifiers of these systems more sensitive to washout.

2.3 Feeding and Aeration Patterns
Changing the number and location of feed points of an activated sludge aeration basin can appreciably alter acceptable loading rates and quality of clarifier effluent.

2.3.1 Conventional

Conventional activated sludge design typically would introduce influent to the head end of a rectangular basin. The RAS could be mixed with the influent before the tank or be added separately. Keeping the RAS separate facilitates subsequent conversion to other feed patterns (e.g., step feed or contact stabilization), if such flexibility is important. If RAS is blended with the influent ahead of multiple aeration basins operating in parallel, then care must be taken to ensure that the influent is well mixed before flow splitting occurs.

2.3.2 Contact Stabilization

Contact stabilization is a modification of the activated sludge process in which the feed point is moved downstream in the aeration tank (or into a separate tank). This provides a relatively short detention time for the MLSS to be in contact with the feed stream before mixed liquor leaves the reactor for solids separation. The RAS is added to the tank inlet separately and aerated before being blended with the mainstream influent. Because the upstream end of the aeration basin contains liquid at the RAS concentration instead of the MLSS concentration, a given volume of aeration tank would contain a larger mass of mixed-liquor solids and, therefore, longer SRT. The longer SRT increases the time that the microorganisms are under aeration and allows for the metabolization of substrate that would otherwise not be removed because of the shortened HRT of the process. The reduced HRT results in less opportunity to oxidize ammonia and remove organic nitrogen. The process originally was developed to remove readily absorbed, soluble BOD, although particulate BOD is removed as well. Soluble organics absorbed and suspended organics adsorbed or enmeshed by the microorganisms in the short HRT zone subsequently are stabilized when returned to the reaeration zone. An existing aeration basin's capacity might, therefore, be readily increased by conversion to contact stabilization, if there is flexibility in feed-point location, separate RAS return, and sufficient aeration capacity and distribution.

2.3.3 Step Feed

The step-feed process, a modification of a plug-flow reactor, allows entry of influent wastewater at two or more points along the length of the aeration basin. With this arrangement, the oxygen uptake rate (OUR) becomes more uniform throughout the basin. Other operating parameters are similar to those of the conventional process. RAS typically would be added to the aeration basin in a separate conduit at the inlet end of the basin. Step-feed configurations typically have diffused aeration. An existing plug-flow reactor can be modified for step feed by simply dividing the basin into compartments and redirecting the flow so that each compartment receives wastewater input. A step-feed tank configuration is shown in Figure 12.4.

One of the main advantages of step-feed process is decreased solids loading to secondary clarifiers. Because RAS enters separately at the head of the aeration basin, mixed-liquor concentration decreases downstream as a function of the number and location of influent feed points, each of which further dilutes the mixed liquor. On the other hand, feeding influent near the head of the aeration basin increases the loading on the secondary clarifiers and, consequently, increases the RAS concentration. Shifting influent to downstream feed points lowers the solids loading rate and RAS concentration, thereby allowing higher hydraulic loading rates with less danger of clarifier solids overload.

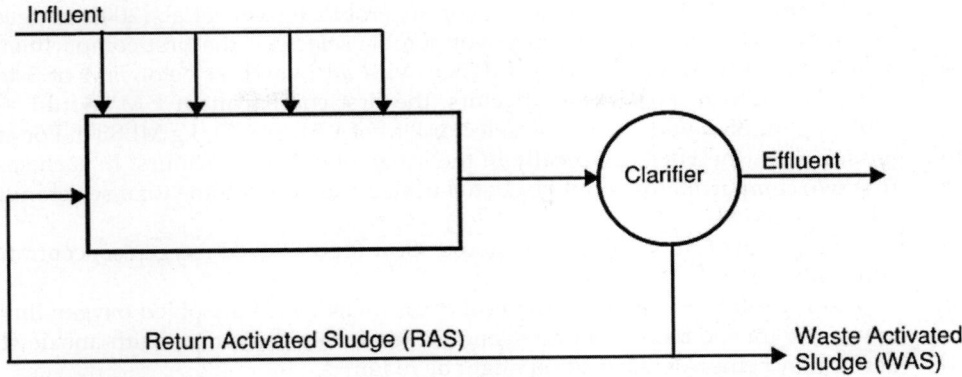

FIGURE 12.4 Step-feed process.

2.4 Selectors

A variety of microorganisms affect settleability of MLSS, and environmental conditions can be altered to favor or select one type of bacteria over others (Jenkins et al., 2003). Mixed liquors that are low in nutrients, dissolved oxygen, or F:M tend to favor growth of filamentous bacteria that have high surface area-to-volume ratios. Many of these filamentous bacteria, which hinder settling, can be placed at a growth-rate disadvantage if mixed liquor is subjected to periods of high F:M. Organisms with the greatest ability to rapidly uptake soluble substrate and store it internally for use later during low concentration conditions tend to be those that are more flocculent and settle better.

Selector basins at the head end of main aeration basin can take advantage of these dynamics. A schematic drawing showing three selectors in series is shown in Figure 12.5. Selectors may or may not be compartmentalized like this. A single basin or even the head end of a long, narrow plug-flow aeration basin may be adequate to obtain improved results. To help reduce longitudinal mixing and overcome variations in waste flow and strength, it is recommended that three or more compartments be used to take full advantage. Albertson (2007) and the Water Environment Research Foundation (WERF) (2006a) present case studies and general guidelines for selector implementation and design.

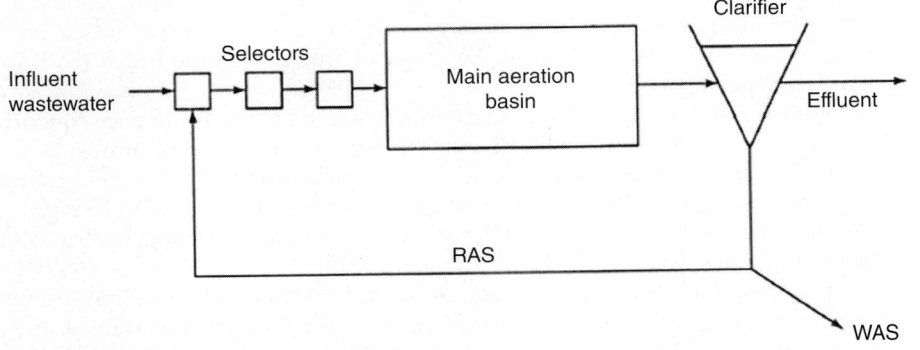

FIGURE 12.5 Selector system configuration.

Selectors may be aerobic, anoxic, or anaerobic. Jenkins et al. (2003) suggest that selectors have at least three zones. For aerobic selectors, the first compartment F:M should be 10 to 12 kg COD/kg MLSS·d with an overall selector F:M of 3 to 6 kg COD/kg MLSS·d. For anoxic selectors, the first compartment F:M should be 6 kg COD/kg MLSS·d with an overall selector F:M of 1.5 kg COD/kg MLSS·d. For anaerobic selectors, the HRT is typically in the range of 0.75 to 2.0 hours. In each case, the first two compartments should be equal in size and one-half the total selector volume when combined.

The design of aerobic selectors should allow for dissolved-oxygen concentrations of 1 to 2 mg/L.

Anoxic selectors can be mechanical or air mixed, with dissolved oxygen limited to low levels for the latter. If nitrate concentrations might interfere with anoxic selector performance, then denitrification might be required.

Operating at high F:M and low dissolved oxygen can lead to formation of polysaccharides, an intermediate product formed by microbes in an attempt to metabolize BOD. Polysaccharides are not readily biodegradable in digestion and can detrimentally affect solids dewatering by reducing throughput and dewatering cake solids concentrations. Pilot-scale or facility performance data may be used to further improve selector design and define expected performance for a given situation.

Table 12.3 summarizes advantages and disadvantages of the three types of biological selectors.

2.5   Other Variations

2.5.1   Pure Oxygen

Primary advantages claimed by manufacturers of the pure oxygen process include reduced power for dissolving oxygen into mixed liquor, improved biokinetics, ability to treat high-strength soluble wastewater, reduction in bulking problems from dissolved-oxygen-deficit stress, and, with covered reactors, off-gas emissions and odors are contained.

Pure-oxygen systems in the past were characterized by high MLSS concentrations (3000 to 8000 mg/L) and relatively short HRTs (1 to 3 hours). Many of these systems for municipal wastewater are operated at MLSS concentrations of 1000 to 3000 mg/L while maintaining relatively short HRTs. This may be due, in part, to *Nocardia* foam accumulations in the reactors of some municipal facilities operating at high MLSS concentrations. Operation of older facilities at lower MLSS might also be a result of solids loading limitations of secondary clarifiers that were sized based on overflow rate or without consideration of the effect of sludge volume index (SVI) on solids-handling capacity.

For enclosed reactors, enriched oxygen gas is fed to the headspace concurrent with the wastewater flow. Mechanical mixers entrain the enriched atmosphere into the mixed liquor, and maintaining a constant gas pressure within the tanks maintains the oxygen feed (Figure 12.6). A dissolved-oxygen concentration of 4 to 10 mg/L typically is maintained in the mixed liquor. Less than 10% of the inlet oxygen vents off from the last stage of the system.

For open-reactor systems, oxygen is injected or entrained into a liquid stream entering the reactor (U.S. EPA, 1979). Many open-reactor systems are online worldwide, but most are small and used for municipal applications, although there are many large, open-tank systems for industrial wastewater treatment. An open-tank pure-oxygen

Selector Type	Advantages	Disadvantages
Aerobic	Simple process, no additional internal recycle streams,* other than RAS	Does not reduce oxygen requirements
	Relies on basin geometry, not nitrification	Requires more complex aeration system design to meet maximum oxygen uptake rate in the initial high F:M zone
Anoxic	Tends to buffer nitrification (recovers approximately 3.5 lb alkalinity as $CaCO_3$/lb of NO_3^--N denitrified)	Cannot be used with a process that does not nitrify Uses an additional recycle stream
	Lowers oxygen demand in a nitrification process (recovers approximately 2.86 lb O_2/lb of NO_3^- reduced)	Requires care in design and operation to minimize the introduction of oxygen in the anoxic zone; poor system design can induce low-dissolved oxygen bulking
	The initial high F:M region occurs in the anoxic zone, with the high oxygen demand met by NO_3^- instead of oxygen	
Anaerobic	Simple design, no internal recycle other than RAS	Does not reduce oxygen requirements May not be compatible with long SRTs
	The simplest of selector system to operate	Requires care in design and operation to minimize the introduction of NO_3^- and oxygen in the anaerobic zone
	Can be used for biological phosphorus removal	Poor system design can induce low-DO bulking

*Recycle stream may not be needed for selector but to extend denitrification (lb/lb = kg/kg).

TABLE 12.3   Comparison of Biological Selectors (Sykes, 1993)

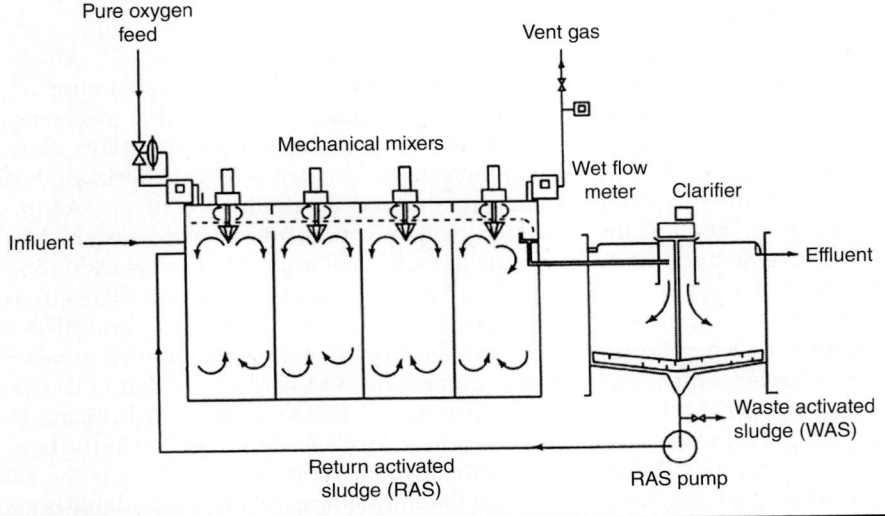

FIGURE 12.6   Closed-tank, high-purity oxygen system schematic.

system can be used in combination with surface aeration to eliminate pH depletion due to carbonic acid formation. Typically, the open tank that receives the pure oxygen is the first aerobic cell that is 25% to 40% of the total aeration volume. In systems where all cells include surface or diffused aeration to accommodate average and lower facility loading, pure oxygen can be injected to satisfy peak-loading aeration demands. Covered-tank pure-oxygen systems reduce pH values because of high partial pressure of carbon dioxide, which is not stripped out by the nitrogen gas in air leaving the liquid surface as in conventional aeration basins.

When the pH is depressed below 6.5 to 7.0, nitrification will attenuate, and the system can require a longer SRT, greater aeration tank volume, and, perhaps, additional final clarifier capacity. These effects have led to consideration of separating the carbonaceous BOD (cBOD) removal and nitrification stages when using oxygen. Some design engineers suggest that the first stage of a two-stage system receive oxygen and the second stage receive air. Another option is to open up the last cell of the train to air rather than pure oxygen.

Covered tanks using oxygen have provisions for warning of potential explosions that could result from the presence of combustible, volatile hydrocarbons in influent wastewater. A detector system is used to purge the tanks automatically with air if the volatile hydrocarbon level becomes excessive. Covered tanks capture volatile organic chemical (VOC) emissions; therefore, off-gas volume from pure-oxygen systems is approximately 1% of that leaving a typical air system.

An atmosphere of high-purity oxygen and carbon dioxide in the reactor basin requires careful selection of construction materials. Compared with air, this atmosphere is more corrosive and reactive with organic compounds such as oils and greases. Some facilities have experienced corrosion of materials used in downstream conveyance channels and secondary clarifiers. Suppliers of high-purity oxygen systems have evaluated materials suitable for safe and reliable construction.

Mechanical surface turbines keep the reactor mixed. For deep tanks, submerged turbines or surface turbines with extended shafts are used to provide additional mixing blades closer to the bottom.

2.5.2 Oxidation Ditch

In a classical oxidation ditch system, wastewater and mixed liquor are pumped around an oval pathway (racetrack) by brushes, rotors, or other mechanical aeration devices and/or pumping equipment located at one or more points along the flow circuit. Figure 12.7 shows oxidation ditches with alternative horizontal- or vertical-shaft aerators to maintain tank motion and aerate ditch contents. As mixed liquor passes the aerator, the dissolved-oxygen concentration increases sharply and then declines as the flow traverses the circuit. Oxidation ditches typically operate in an extended aeration mode with long HRTs (12 to 24 hours) and SRTs (10 to 30 days). However, oxidation ditches can be designed for shorter HRTs and SRTs in warmer climates. Depending on the relative location(s) of wastewater inlet, mixed-liquor outlet, sludge return, and aeration equipment and control, oxidation ditches also can achieve nitrification and denitrification. For BOD removal or nitrification, the influent typically enters the reactor near the aerator and the effluent exits the tank upstream of the entrance. To enhance denitrification, influent should enter the ditch where dissolved-oxygen level is lower so that influent readily biodegradable organic matter can be used for denitrification.

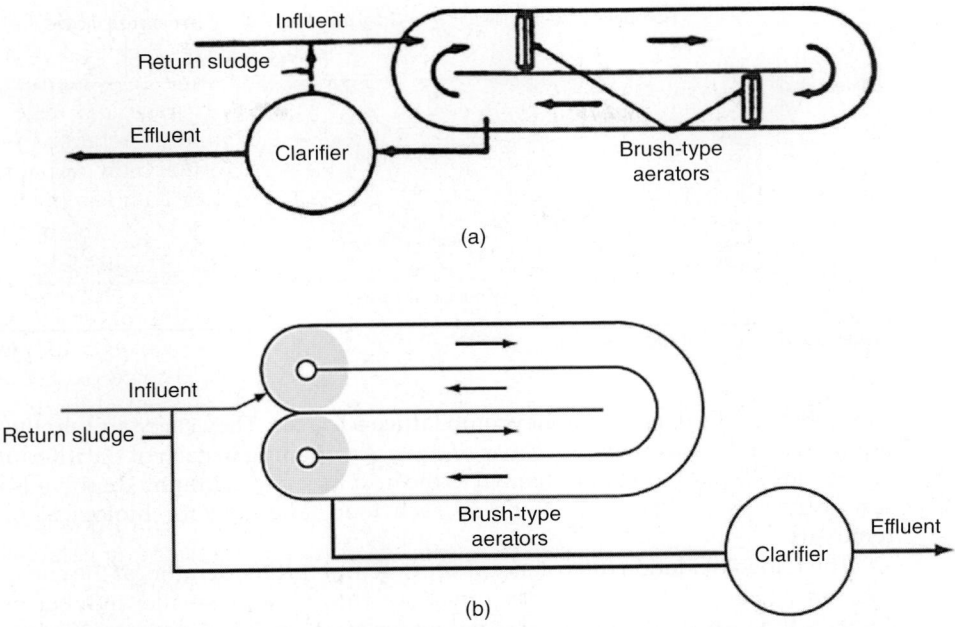

FIGURE 12.7   Oxidation ditch reactors: (a) simple loop and (b) folded loop.

Oxidation ditches have depths ranging from approximately 0.9 to 5.5 m (3 to 18 ft) and channel velocities from 0.24 to 0.37 m/s (0.8 to 1.2 ft/s). Oxygen ditch depths can be as much as 7 m (24 ft) when surface turbines and draft tubes are utilized. Ditch geometry must be compatible with aeration and mixing equipment and should be coordinated with the manufacturers. Mechanical brushes, surface turbines, and jet devices are used to aerate and move the liquid flow. Combinations of diffused aeration and submersible mixers also have been employed (Christopher and Titus, 1983). Several alternative designs of intrachannel clarifiers have been developed to provide for separation and return of MLSS to the ditch. The inability to readily modify the RAS rate and the reduced flexibility of taking a reactor or clarifier out of service independently have led to the demise of the intrachannel clarifier concept.

Since 1973, approximately 10 000 oxidation ditch facilities have been constructed in the United States. They are widely used in small- to medium-sized communities (5000 to 50 000 population) and with flows of 1900 to 19 000 m³/d (0.5 to 5.0 mgd), although some are much larger. Advantages include simple operation, reliable performance, and cost-effectiveness.

The vertical loop reactor (VLR) shown in Figure 12.8 is an aerobic activated sludge biological treatment process similar to an oxidation ditch. Wastewater in a VLR circulates in a vertical loop around a horizontal divider baffle. Proponents assert that overall oxygen-transfer efficiency (OTE) for a VLR is higher than that of an equivalent conventional oxidation ditch.

Another variation of the oxidation ditch concept is the concentric loop. There are more than 500 of these systems in the United States. In this process, mixed liquor is aerated and propelled around a series of concentric loops or a single loop by partially

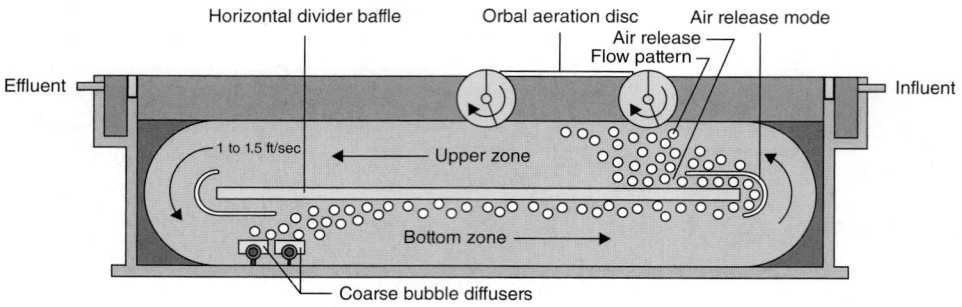

FIGURE **12.8**   Vertical loop reactor.

submerged disks that have honeycomb lattice surfaces. They carry air into the mixed liquor, which keeps moving because of drag. An opening in each of the interior walls allows flow to pass from one channel to the next in series. Different dissolved-oxygen concentrations can be maintained in each loop, allowing for biological nitrogen removal.

The folded oxidation ditch concept uses vertical turbine aerators at the end of baffle walls at which point flow reverses direction. This allows a smaller number of larger aerators than possible with brush aerators for large facilities, increases the power-to-volume ratio in the aeration zone, and operates with low dissolved-oxygen concentrations entering this zone, thereby increasing OTE.

Some designers choose to combine diffused aeration for OTE and independent mechanical mixing to control velocity in the ditch. Submerged mixers of different configurations have included jets, horizontal propellers, paddles on a vertical shaft at divider wall ends, and others. Aspirating propellers on hollow shafts with and without blowers for air supply also have been used.

A combination of oxidation ditch and SBR technologies resulted in the phased isolation ditch. They have been used principally in Denmark, although facilities have been constructed in the United States in Ocoee, Florida, and Lewisburg, North Carolina, and in Germany, Greece, China, and Australia (Tetreault et al., 1987).

2.5.3   Aerated Lagoon

Aerated lagoons are partially mixed, aerated reactors with long HRTs and without clarifiers or sludge return. Aerated, lagoon-style reactors consisting of lined earthen basins with HRTs of one or more days and with clarifiers and sludge return can be used to create activated sludge processes. Reactor shapes historically were square or rectangular, which is mixing limited, leading to development of oxidation ditches. Horizontal flow in a ditch keeps particles in suspension at lower energy input than mechanical or diffused aeration in lagoon reactors.

Another concept is to aerate and mix lagoon-style reactors with less energy by installing rows of diffused aerators near the bottom and then sequencing them on and off or suspending them from floating aeration piping. This extended aeration system was first introduced to the United States in 1986. Typically, these systems have lower capital costs and potentially lower operations and maintenance costs compared to conventional extended aeration basins with lower HRTs. The latter might be more cost-effective to construct where soil conditions are not favorable for a lined earthen basin. Design flows for facilities using aerated lagoon reactors are typically 400 to 190 000 m³/d (0.1 to 50 mgd). The sequencing of aeration also can create conditions allowing for denitrification.

2.5.4 Integrated Systems

Integrated fixed-film activated sludge (IFAS) systems incorporate inert support media into the activated sludge reactors. This allows fixed-film biomass to grow on the media and augment the microbial population of the mixed liquor. Integrated systems are presented in Chapter 13.

2.5.5 Emerging Processes

2.5.5.1 Granular Sludge Granular sludge formation has been well studied under anaerobic conditions and used in anaerobic treatment processes since the 1980s; however, the first "full-scale" facility utilizing granule formation under aerobic conditions was installed in 2005 and is patented as the Nereda® process. There are currently over 20 installations of the Nereda® process for the treatment of municipal and industrial wastewater in the world, but there has yet to be a full-scale installation in the United States. Granular sludge processes reduce, on average, capital costs by 25%, operation and maintenance costs by 25%, and facility footprint by 75% in comparison to CAS treatment processes (Giesen et al., 2015). These savings are accomplished because the granules form distinct layers: the outer layer is the aerobic layer for heterotrophic growth and nitrification, an anoxic layer for denitrification below the aerobic layer, and an anaerobic layer at the core of the granule. This enables simultaneous COD, nitrogen, and phosphorus removal in a single aerobic tank and granule formation also improves sludge settling ability (Welling et al., 2015). Sturm et al. also found that granule formation is possible under aerobic conditions treating low-strength municipal wastewater similar to typical COD concentrations of municipal wastewater in the United States (Sturm et al., 2015). See Chapter 11 for more information regarding granular sludge.

2.5.5.2 Absorption/Bio-Oxidation Process The absorption/bio-oxidation (AB) process for carbonaceous COD removal is an emerging process used for carbon redirection and energy recovery. In the A-stage, a high-rate activated sludge system is used to absorb particulate and colloidal organic matter and sludge is wasted to anaerobic digestion for increased biogas production. In the B-stage, nonabsorbed organic matter and influent ammonium is oxidized in a similar manner to CAS. Due to the emerging nature of this process, typical SRTs, loading rates, and removal rates for the A- and B-stages are not well established.

2.6 Solids Separation

All suspended-growth systems count on successful separation of the MLSS from the process effluent. Since the development of the activated sludge process nearly a century ago, clarification has been used for this purpose. Membrane filtration for MLSS retention has been developed and widely applied in recent years. In special cases, flotation and centrifugation have been used.

2.6.1   Clarifiers

There are many aspects of clarifier design that pertain to making the process effective for separating suspended solids in the activated sludge process. The technology has evolved to the point that nearly all such clarifiers are either circular or rectangular, and equipped with energy-dissipation inlets, surface skimming, and scraper or hydraulic suction sludge-removal mechanisms.

2.6.2   Membranes

Adding membrane separation technology to the activated sludge process and the resulting MBR configuration now can compete with mainstream processes. This is particularly true when filtration or further membrane treatment would be required for a high-quality effluent, or where a compact footprint is needed.

Advantages of MBR process configuration include a nearly solids-free effluent, modular configuration with small footprint, reduced downstream disinfection requirements, ability to retrofit existing reactors, and elimination of adverse sludge settling properties. Disadvantages include capital costs, increased power requirements for aeration, ongoing membrane replacement requirements, and constrained ability to accommodate peak flows.

3.0   Process Design for Carbon Oxidation and Nitrification

Eckenfelder (1966), Lawrence and McCarty (1970), McKinney (1962), and McKinney and Ooten (1969) led the way in developing a quantitative understanding of the activated sludge process. The work of Lawrence and McCarty (1970) was particularly significant in providing a more unified approach emphasizing the importance of SRT. This approach is the basis for carbon oxidation and nitrification process design and led to more sophisticated software models (WEF, 2013b).

3.1   Carbon Oxidation

Figure 12.9 illustrates a typical suspended-growth system flow diagram. Presentation of the fundamental relationships and development of design equations for carbon oxidation can be found in the references (Grady et al., 2011; Lawrence and McCarty, 1970; Metcalf & Eddy, Inc./AECOM, 2014; Ritmann and McCarty, 2001). A set of design equations for sizing systems based on completely mixed reactors is presented below.

$$\mathrm{HRT} = \frac{V}{Q} \tag{12.1}$$

$$\frac{1}{\mathrm{SRT}} = \frac{\mu_{max}S_e}{K_s + S_e} - b \tag{12.2}$$

$$X = \frac{\mathrm{SRT}}{\mathrm{HRT}} Y_{net}(S_o - S_e) \tag{12.3}$$

$$Y_{net} = \frac{Y}{1 + b\,\mathrm{SRT}} \tag{12.4}$$

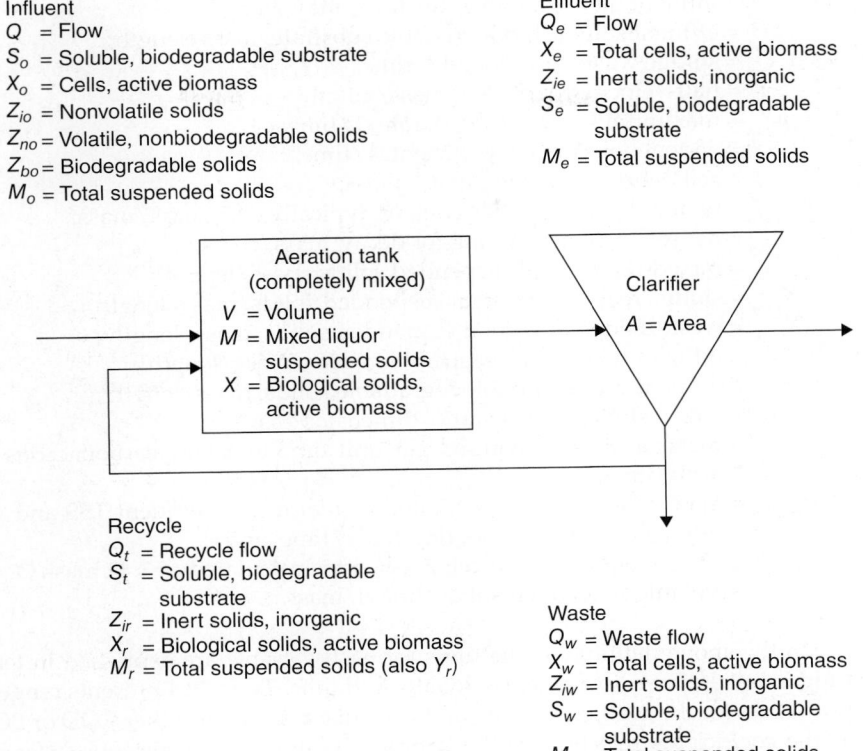

Influent
Q = Flow
S_o = Soluble, biodegradable substrate
X_o = Cells, active biomass
Z_{io} = Nonvolatile solids
Z_{no} = Volatile, nonbiodegradable solids
Z_{bo} = Biodegradable solids
M_o = Total suspended solids

Effluent
Q_e = Flow
X_e = Total cells, active biomass
Z_{ie} = Inert solids, inorganic
S_e = Soluble, biodegradable substrate
M_e = Total suspended solids

Aeration tank
(completely mixed)
V = Volume
M = Mixed liquor suspended solids
X = Biological solids, active biomass

Clarifier
A = Area

Recycle
Q_t = Recycle flow
S_t = Soluble, biodegradable substrate
Z_{ir} = Inert solids, inorganic
X_r = Biological solids, active biomass
M_r = Total suspended solids (also Y_r)

Waste
Q_w = Waste flow
X_w = Total cells, active biomass
Z_{iw} = Inert solids, inorganic
S_w = Soluble, biodegradable substrate
M_w = Total suspended solids

FIGURE 12.9 Nomenclature for activated sludge flowsheet (volatile and nonvolatile represent organic and inorganic solids, respectively).

$$X_{\text{MLSS}} = \frac{\text{SRT}}{\text{HRT}} \left\{ \frac{Y_{\text{net}}(S_o - S_e)}{f_v} + \left(f_d \frac{Y_{\text{net}}(S_o - S_e)}{f_v} \right) + Z_{\text{io}} + Z_{\text{no}} \right\} \tag{12.5}$$

$$\text{SRT} = \frac{V X_{\text{TSS}}}{M_{w,\text{TSS}}} = \frac{V X_{\text{TSS}}}{X_{w,\text{TSS}} Q_w + M_{e,\text{TSS}} Q_e} \tag{12.6}$$

$$\alpha r = \frac{Q_r}{Q} = \frac{X}{X_r - X} = \frac{X_{\text{TSS}}}{X_{r\text{TSS}} - X_{\text{TSS}}} \tag{12.7}$$

$$R_c = Q(S_o - S_e) - B(Q Y_{\text{net}}(S_o - S_e)(1 + f_d \times b \times \text{SRT}) \tag{12.8}$$

where V = aeration tank volume, length^3;
Q = wastewater inflow, length^3/time;
Q_r = sludge recycle flow, length^3/time;
X = reactor biological solids, mass/length^3;
X_r = sludge recycle flow biological solids, mass/length^3;
Y = true cell yield, mass/mass;

S_o = influent biodegradable substrate, mass/length^3;
S_e = effluent soluble biodegradable substrate, mass/length^3;
Q_w = sludge waste flow, length^3/time;
K_s = half-velocity coefficient, mass/length^3;
μ_{max} = maximum specific growth rate, 1/time;
b = endogenous decay coefficient, 1/time;
f_d = cell debris coefficient, mass/mass;
f_v = biomass volatile solids content, typically 0.85, mass/mass;
Y_{net} = net cell yield accounting for decay, mass/mass;
X_{TSS} = mixed-liquor total suspended solids, mass/length^3;
X_{rTSS} = sludge recycle flow total suspended solids, mass/length^3;
X_{wTSS} = waste sludge flow total suspended solids, mass/length^3;
Z_{io} = influent nonvolatile suspended solids, mass/length^3;
Z_{no} = influent volatile nonbiodegradable solids, mass/length^3;
αr = return sludge recycle ratio, dimensionless;
R_c = mass of oxygen required per unit time to satisfy carbonaceous oxidation, mass/time;
M_{wTSS} = mass of total solids generated or removed as effluent TSS and waste activated sludge (WAS) per day, mass/time; and
B = oxygen equivalent of cell mass, often calculated as 1.42 mass O_2/mass VSS (volatile suspended solids), mass/mass.

In the above definitions, the units for each variable are expressed in terms of the fundamental dimensions of mass, length, and time. Table 12.4 presents ranges and typical values for the coefficients. The units for substrate can be either COD or BOD as long as the coefficients selected are the appropriate units. Carbonaceous oxygen demand determined by eq 12.9 requires the use of biodegradable COD or ultimate BOD for the influent substrate concentration S_o. The reader is cautioned that the units of mass for

Coefficient	Unit	Range	Typical Values
μ_m	g VSS/g VSS·d	3.0–13.2	6.0
K_s	g bCOD/m³	5.0–40.0	20.0
Y	g VSS/g bCOD	0.30–0.50	0.40
b_n	g VSS/g VSS·d	0.06–0.20	0.12
f_d	Unitless	0.08–0.20	0.15
θ values			
μ_m	Unitless	1.03–10.08	1.07
k_d	Unitless	1.03–1.08	1.04
K_s	Unitless	1.00	1.00

*Adapted from Henze et al., 1987; Barker and Dold, 1997 (provided in the original reference); and Grady et al., 2011.

TABLE 12.4   Activated Sludge Kinetic Coefficients for Heterotrophic Bacteria at 20°C (from Metcalf & Eddy, Inc., 2014, with permission from the McGraw-Hill Companies)*

reported coefficients can be in terms of total or volatile solids and be based upon active biomass, total biomass that includes cell debris, or total mass.

More recent computer simulations often divide influent COD into multiple fractions including:

rbCOD—readily biodegradable COD;

nbCOD—nonbiodegradable COD; and

sbCOD—slowly biodegrable COD.

Each fraction is further divided into soluble, colloidal, and particulate components with each fraction having different responses to the variable conditions in the treatment process. A detailed waste fractionation study can be performed to generate influent data for modern computer models although these studies can be costly. The reader is referred to Chapter 4 for more detailed discussion of the benefits of calibrating models and the data required for the proper fractionation of the influent wastewater.

In design, it is important to account for all components of incoming wastewater that will influence solids production and oxygen demand. Influent microorganisms are assumed to be negligible relative to those in the reactor but can influence oxygen demand patterns in a system's aeration tank (Grady et al., 2011). Nonvolatile suspended solids can be estimated as the difference between influent TSS and VSS. Nonbiodegradable VSS (organic) are approximately 40% of influent organic or VSS (Dague, 1983). Nonbiodegradable volatile solids cannot be directly measured but must be estimated.

Biodegradable VSS are presumed to be rapidly adsorbed into the biomass and subsequently hydrolyzed (solubilized) and are reflected in the BOD or COD of the influent and ignored as a specific component of suspended solids. Slowly hydrolysable biodegradable volatile solids, however, can significantly affect system kinetics and mass balances.

The above equations allow for estimation of process volume, waste sludge production, RAS ratio, and oxygen required for an activated sludge system for cBOD removal. Design of a system requires determination of several items:

- Volume of the aeration basins, V;
- Quantity of sludge to be wasted, M_{wTSS};
- Total oxygen demand, R_c;
- Sludge recycle requirements; and
- Size of clarifiers.

By convention, the stated HRT of suspended-growth reactors typically excludes the RAS flow and is based only on the flow influent to the process. Alternative definitions of HRT or other parameters can be used but their basis needs to be indicated. The actual flow through the reactor, including recycle flow, often is reported as a design or operations parameter. In that case the basis for the flow should be indicated.

The SRT of eq 12.2 represents the theoretical SRT required to achieve a required effluent soluble substrate concentration. Typical designs are based on higher SRT values to provide for biological solids that have desired settling properties. For carbonaceous substrate removal by heterotrophic organisms, the theoretical SRT is used rarely.

Rather, an SRT is selected based on experience. A safety factor can also be applied to establish a value for design (Rittmann and McCarty, 2001).

SRT is the total mass of solids in a system divided by the rate at which solids are wasted. It is based on reactor volume only. Including secondary clarifier solids in the total would be a more fundamentally correct approach but would require an estimate of solids in the clarifier, which is an operational variable and beyond the designer's control. Additionally, sludge in clarifiers is often anoxic or anaerobic and including it in an SRT definition could result in undersizing the system. For systems with low SRT and short HRT, such as pure-oxygen activated sludge systems, the difference between the two methods of SRT calculation can be significant. Clear representation of the basis of the calculation is recommended in any case.

The true and net yield coefficients in the above equations characterize the production of biomass. An additional term, observed yield, Y_{obs}, is calculated from data gathered at operating activated sludge facilities and is equal to the sum of effluent ($M_{e,TSS}$) and waste solids ($M_{w,TSS}$) divided by the mass of substrate removed over the same period of time.

Oxygen requirements determined from eq 12.8 considers only carbonaceous oxygen demand. In that equation, the second term is the oxygen equivalent of the biomass plus cell debris. An activated sludge system intended to remove cBOD that has a design or operating SRT equal to or greater than the theoretical minimum SRT for nitrification will nitrify to some degree, and therefore will experience additional oxygen demand, and Y_{obs} will include nitrifier biomass. In that case, the oxygen equivalent of the nitrifier biomass and cell debris produced should be subtracted from the additional oxygen demand resulting from nitrification. Nitrification oxygen demand and biomass production are discussed in the next section.

It is important that in-facility recycle streams from solids processing be accounted for in the concentration and variability of influent. Although expected to be low in soluble BOD, return flows from dewatering of aerobically or anaerobically digested WAS could have significant, particulate BOD and nonbiodegradable solids. Consideration also should be given to different potential modes of process failure, including an underloaded facility failing to perform when subjected to design conditions without adequate time for the system to acclimate (see Chapter 4).

The equations above are for a single, completely mixed reactor. Plug-flow reactors are more efficient for first-order reactions that occur in the activated sludge process but are mathematically too complex to solve directly with a simplified model. Process modeling, as presented by the WEF (2013b) and discussed in Chapter 4, is required to assess plug-flow and other reactor configurations. The equations for cBOD removal in a complete-mix reactor might be a reasonable approximation of a plug-flow system for typical wastewaters because the slower rate of particulate substrate removal (Metcalf & Eddy, Inc./AECOM, 2014) and RAS flows result in a more uniform substrate concentration within the reactor.

3.2   Nitrification

Nitrogen contained in municipal raw wastewater occurs predominantly in organic and ammonia nitrogen forms. Typical concentrations of total nitrogen (TN) in domestic wastewaters range from 20 to 85 mg/L, with a medium strength of 40 mg/L (Metcalf & Eddy, Inc./AECOM, 2014). Approximately 40% of the total occurs as organic and 60% as ammonia. Typically, less than 1% is present as nitrate or nitrite unless influenced by industrial waste contributions. Influent nitrogen concentrations, as with other constituents, have trended upward in areas with increasing water conservation efforts or reduced infiltration and inflow. Additional information on influent quality characteristics is presented in Chapter 3. Please note that nitrogen concentrations in this chapter will be expressed in mg as N/L unless noted otherwise.

The growth of new cells will remove some of the influent nitrogen. This nitrogen removal will be approximately 12% of the net biomass generated. An additional fraction of the influent TN is nonbiodegradable or removed as particulates (1% to 2% of influent TKN, can be higher with novel solids processing such as thermal hydrolysis). As a result, approximately 80% to 85% of the influent nitrogen may be available for oxidation. For given conditions of SRT and cBOD removal, nitrogen removal by assimilation can be estimated as a percentage of the net biomass plus cell debris produced. Nitrogen assimilation depends on the ratio of BOD to nitrogen in the influent, and thus can be significant in systems treating wastewater with high concentrations of organic carbon.

Nitrification is the aerobic oxidation of ammonium to nitrate often through a nitrite intermediate. The reaction is often carried out by two distinct groups of autotrophic bacteria where nitroso- species (e.g., *Nitrosomonas*) oxidize ammonium to nitrite (ammonium oxidizing bacteria; AOB) and nitro- species (e.g., *Nitrobacter*) oxidize nitrite to nitrate (nitrite oxidizing bacteria, NOB). Recently, a member of the *Nitrospira* was observed to perform complete oxidation of ammonium to nitrate in single cell (Daims et al. 2015) though this pathway is not described extensively in this text. Some members of the Domain *Archaea* are known to oxidize ammonium aerobically (e.g., *Nitrosoarchaeum limnia*) but the stoichiometry is the same as the *Bacteria* nitrifiers. The net reaction is shown below.

$$NH_4^+ + 1.5O_2 \rightarrow NO_2^- + 2H^+ + H_2O$$

$$NO_2^- + 0.5O_2 \rightarrow NO_3^-$$

$$NH_4^+ + 2O_2 \rightarrow NO_3^- + 2H^+ + H_2O$$

Each gram of ammonia oxidized to nitrate (both expressed as N) will result in:

- 4.57 g of oxygen consumed;
- 7.14 g of alkalinity (as calcium carbonate) destroyed; and
- 0.15 g of new cells (nitrifiers) produced.

The synthesis of new nitrifying biomass results in oxygen and alkalinity consumption slightly less than that predicted by stoichiometry.

The degree of biological nitrification will depend on the mass of nitrifying organisms allowed to remain in the system. Their presence depends on relative growth rates of the autotrophic populations involved, system SRT, and other conditions such as temperature, ammonia, organic substrate, and dissolved-oxygen concentrations. For a given maximum MLSS concentration, the fraction of autotrophic biomass will be limited based on the heterotrophic biomass that is present. Denitrification, whether intended as part of the process or incidental—such as that which occurs in clarifiers or low dissolved-oxygen zones of aerobic reactors—can recover some of the consumed alkalinity and reduce consumption of oxygen due to anoxic heterotrophic growth.

Biological oxidation of ammonia to nitrate can be achieved in combined cBOD removal and nitrification (single-stage) systems or in separate nitrification (two-stage) systems.

The degree of nitrification in a combined, single-stage process depends on system SRT, provided that a population of nitrifying autotrophs can be maintained. The degree of nitrification, therefore, is governed to a large extent by design parameters (HRT and SRT for a nitrification system). Two-stage systems allow some separation of carbonaceous and nitrogenous oxidation processes. In the first stage (aeration basin with clarification and sludge recycle), most of the cBOD removal occurs and nitrification is limited. The second stage (separate aeration basin and clarifier with sludge recycle) maintains more favorable conditions for nitrification. Two-stage systems have been found to be more costly.

Presentation of the fundamental relationships and development of design equations for nitrification can be found in the references (Grady et al., 2011; Metcalf & Eddy, Inc./ AECOM, 2014; Rittman and McCarty, 2001; U.S. EPA, 1993). A set of design equations for nitrification in complete-mixed, suspended-growth systems is presented below.

$$SRT_{Nmin} = \frac{1}{\mu_N - b_N} \tag{12.9}$$

$$\mu_N = \mu_{N,max} \left(\frac{N_e}{K_N + N_o} \right) \left(\frac{DO}{K_o + DO} \right) \tag{12.10}$$

$$SRT_{design} = SRT_{N,max} (SF) \tag{12.11}$$

$$Y_{Nnet} = \frac{Y_N}{1 + b_N SRT_{design}} \tag{12.12}$$

$$M_{NTSS} = Q \left\{ \frac{Y_{Nnet(N_o - N_e)}}{f_v} + \left[f_d \times b_N \times SRT_{design} \times \frac{Y_{Nnet(N_o - N_e)}}{f_v} \right] \right\} \tag{12.13}$$

$$R_n = 4.57Q(N_o - N_e) - 2.86(N_o - N_e - NO_{3e}) - B \times f_v \times M_{NTSS} \tag{12.14}$$

where SRT_{Nmin} = minimum SRT for nitrification, time;
  μ_N = nitrifier specific growth rate, 1/time;
  $\mu_{N,max}$ = maximum nitrifier specific growth rate, 1/time;
  N_o, N_e = influent and effluent oxidizable nitrogen concentrations, respectively, mass/length^3;
  K_N, K_o = half-velocity constants for ammonia and oxygen, respectively, mass/ length^3;
  Y_N = nitrification yield coefficient, mass/mass;
  Y_{Nnet} = net nitrification yield coefficient, mass/mass;
  DO = reactor dissolved-oxygen concentration, mass/length^3;
  SRT_{design} = design SRT, time;
  SF = safety or design factor, dimensionless (SRT_{design} = SF × SRT_{Nmin});
  b_N = endogenous decay coefficient for autotrophs based on biomass in aerated zone, 1/time;
  f_d = cell debris coefficient, mass/mass;
  f_v = biomass volatile solids content, typically 0.85, mass/mass;

M_{NTSS} = mass of total autotrophic solids generated or removed as effluent TSS or wasted, per day, mass/time;

R_N = mass of oxygen required per unit time to satisfy nitrification oxygen demand, mass/time; and

NO_{3e} = effluent nitrate nitrogen, mass/length^3.

The basic approach to design of a suspended-growth nitrification process is the same as for carbon oxidation and begins with determining an appropriate design SRT. Table 12.5 presents ranges and typical values for the kinetic coefficients. The relatively slow growth rate of AOB causes nitrifying systems to be slow to recover following process upsets (e.g., low dissolved oxygen, depressed pH, and toxic inhibition) or to large changes in influent oxidizable nitrogen concentrations. A factor of safety is applied to the minimum necessary SRT for increased performance reliability based on consideration of variations in nitrogen loading, process performance requirements, and environmental factors. Equating the peak safety factor to the ratio of peak-to-average influent TN concentrations or to a combination of loading variations and an additional safety factor has been illustrated (U.S. EPA, 1993; Metcalf & Eddy, Inc./AECOM, 2014).

Equation 12.13 estimates the mass of total autotrophic solids generated or wasted. For combined carbon oxidation and nitrification, this quantity is added to carbonaceous waste mass for total mass generated.

It is important that in-facility recycle streams from solids processing be accounted for in the concentration and variability of influent nitrogen. Sidestreams are often generated from mechanical thickening and dewatering processes, which frequently occur during the day shift. This is the same diurnal period that corresponds to typically high loading to the water resource recovery facility (WRRF) and can cause a "double peak" of loading to the WRRF. By providing sidestream storage, the high load in recycle streams can be bled back to the system overnight when flows and loads are typically lower. This "load equalization" can significantly improve biological performance and stability and aid in facilities needing to meet very low discharge permits (e.g., 0.1 mg/L

Coefficient	Unit	Range	Typical Values
μ_{mn}	g VSS/g VSS·d	0.20–0.90	0.75
K_N	g NH$_4$·N/m^3	0.5–1.0	0.74
γ_N	g VSS/g NH$_4$·N	0.10–0.15	0.12
b	g VSS/g VSS·d	0.05–0.15	0.08
K_o	g/m^3	0.40–0.60	0.50
θ values			
μ_n	Unitless	1.06–1.123	1.07
K_N	Unitless	1.03–1.123	1.053
K_{dn}	Unitless	1.03–1.08	1.04

*Adapted from Henze et al., 1987; Barker and Dold, 1997 (provided in the original reference); and Grady et al., 2011.

TABLE 12.5 Activated Sludge Nitrification Kinetic Coefficients 20°C (from Metcalf & Eddy, Inc., 2014, with permission from the Mcgraw-Hill Companies)*

ammonium-N). Alternatively, the sidestream nitrogen load could be treated separately. The reader is referred to Chapter 15 for additional detail on sidestream treatment.

Consideration also should be given to potential modes of process failure that can include an underloaded facility failing to perform when subjected to design conditions without adequate time for the system to acclimate to higher nitrogen loads (WERF, 2006b).

For eq 12.11, the dissolved-oxygen concentration is considered to be a system average. It is recognized that some nitrification can occur in portions of an aeration basin where dissolved oxygen may be low (Albertson and Coughenour, 1995; Applegate et al., 1980; Smith, 1996). Conversely, the mechanism of dissolved-oxygen penetration into biological floc will limit the dissolved oxygen available to entrained autotrophic organisms, which could result in the average reactor dissolved oxygen overpredicting nitrification performance particularly when a relatively high amount of organic carbon leads to large biological flocs (Metcalf & Eddy, Inc./AECOM, 2014).

3.3   Design Considerations

Temperature, dissolved oxygen, nutrients, toxic and inhibitory wastes, pH, and the inherent variability of wastewater affect performance of activated sludge systems.

3.3.1   Temperature

Temperature will affect reaction rate, stoichiometric constants, and oxygen-transfer rates (OTRs). Most temperature corrections used in biological treatment designs follow the modified van't Hoff-Arrhenius equation:

$$K_{T_2} = K_{T_1} \theta^{T_2 - T_1} \tag{12.15}$$

where K_{T_1} = a specific kinetic, stoichiometric, or mass-transfer coefficient at temperature T_1;

K_{T_2} = a specific kinetic, stoichiometric, or mass-transfer coefficient at temperature T_2; and

θ = temperature correction factor, dimensionless.

Tables 12.4 and 12.5 include θ values for heterotrophic and autotrophic bacterial kinetics. The range of θ values for k in aerated lagoons range from 1.06 to 1.12. Note that temperature correction factors are approximate and should be reviewed for appropriateness. Also, nitrification kinetic coefficients are presented routinely as either 15°C or 20°C; the basis should be confirmed.

3.3.2   Dissolved Oxygen

In systems designed for cBOD removal, a minimum average tank dissolved-oxygen concentration of 0.5 mg/L is acceptable under peak loading conditions and 2.0 mg/L under average conditions. Using low values increases OTE but can lead to filament formation and poor settleability. In nitrifying systems, a minimum average tank dissolved-oxygen concentration of 2.0 mg/L under all conditions is reasonable.

3.3.3   Nutrients

An adequate nutrient balance is necessary to ensure an active biomass that settles well. Nutrients refer to nitrogen, phosphorus, and trace metals that are necessary for biological growth (Metcalf & Eddy, Inc./AECOM, 2014). Systems with higher SRT are expected to require fewer nutrients in the influent because nutrients released during endogenous respiration become available for growth of active biomass. Because nutrient requirements depend on SRT, they can be based on excess biomass and cell debris produced. The minimum nitrogen requirement should be 12% and the phosphorus requirement should be 2% of the excess biomass and cell debris generated. Normal domestic wastewater typically contains ample nutrients. Wastes with substantial industrial contributions might require nutrient addition.

3.3.4 Toxic and Inhibitory Wastes

The presence of certain inorganic and organic constituents can inhibit or destroy suspended-growth system microorganisms. A listing of many of these is presented elsewhere (Grady et al., 2011). Nitrification processes are particularly sensitive to toxic inhibition (U.S. EPA, 1993; Water Research Commission, 1984).

3.3.5 pH

The pH of mixed liquor should range from 6.5 to 7.5 for optimum cell growth in cBOD removal systems. Nitrifying systems are more sensitive to system pH because the rate of growth of these organisms is a function of pH over the range of 6.5 to 7.5 (U.S. EPA, 1993). Pure oxygen systems often depress pH more than air systems because the former lacks nitrogen gas flow to help strip dissolved carbon dioxide (formed in respiration) from the mixed liquor. Unless stripped out by downstream channel aeration or a similar process, at least some carbon dioxide is recirculated through clarifiers and back to reactors.

To avoid pH reduction, a residual alkalinity of at least 60 mg/L (as calcium carbonate) for either pure oxygen or conventional aeration systems should be provided. Operating at 50 mg/L level is minimal, and a value of 80 to 100 mg/L would better maintain a stable pH under varying conditions.

3.4 Design Approach

3.4.1 Influent Characteristics

Municipal loads and wastewater characteristics typically vary with season, day of the week, and hour of the day. Influent characterization is discussed in Chapter 3. Unless these variations are addressed in the design of a facility, process performance can be affected significantly. The aeration basin/secondary clarifier combination is vulnerable to high levels of such change. Excessive hydraulic peaks shift aeration basin solids inventory to the clarifiers, which may not be able to contain them. Temperature changes may adversely affect the settling of solids in the clarifier, resulting in a loss of solids to the effluent and affecting the rates of reactions. Increases in organic load may lead to deterioration of mixed-liquor settling and turbid effluent high in suspended solids. Periodically, toxic compounds in the influent may reduce significantly biological activity in the aeration basin and result in poor process performance. In some areas, long-term trends of increasing wastewater strength resulting from efforts to reduce infiltration and inflow into the collection system or from water conservation significantly can reduce nominal hydraulic capacity of a facility. Such trends of increasing influent concentrations might become apparent only through examination of data spanning many years.

Activated sludge facilities can include features to account for influent variations. Such features include flow equalization, larger clarifiers, alternative aeration basin feed patterns, and greater RAS capacities.

Process models can be used to evaluate flow and load patterns to quantify the effects of peak flows, such as solids inventory shifting, and effluent quality variations. Furthermore, higher levels of in-facility sensing and automation facilitate flow diversion strategies that allow process units to be reduced in size from those with less sophistication.

3.4.2 Volume of Aerobic Reactor Basins

Sizing of aeration basins is based on two important factors. The first is sufficient time to remove soluble and particulate substrates (and oxidize ammonia nitrogen, if required) and to allow biomass activity to return to a declining growth or endogenous level. The second is maintenance of flocculent, well-settling MLSS that can be removed effectively by gravity settling.

For municipal systems, it is recommended that process design be based on $S_e = 0$. Experience has shown its exact value to be somewhat unpredictable. It is inadvisable to reduce aeration tank volume by assuming a higher S_e value because discharge permits typically are based on total BOD or COD, which includes the contribution of effluent organic solids consisting of biomass and microbial products.

Equation 12.2 is based on the Monod kinetic relationship and is not well supported by data derived from full-scale activated sludge systems. For this reason, the relationship between S_e and SRT given by eq 12.2 typically is not used for design. When it is used, the value of SRT is predicted to be unrealistically low and is scaled upward by a safety factor (Dague, 1983; Grady et al., 2011; Lawrence and McCarty, 1970; Metcalf & Eddy, Inc./AECOM, 2014; U.S. EPA, 1993; Water Research Commission, 1984). Alternatively, information from the literature, similar facilities, or pilot-facility studies may be useful for estimating SRT and other kinetic parameters.

From a practical point of view, selection of SRT for cBOD removal systems is not based on kinetic considerations but rather on experience. Typically, design is based on providing a high-enough value of SRT for the system to yield a flocculent sludge that settles well and produces a clear effluent. Figure 12.10, which represents a nonfilamentous sludge grown on a soluble waste (glucose plus yeast extract), shows that a minimum SRT value of approximately 3 days is required (Bisogni and Lawrence, 1971). In practice, an SRT of 1 to 5 days typically is used during warm weather and up to 15 days during cold weather. Nitrification may well occur in these ranges and should be taken into account during design. Values of SRT outside this range are selected in situations where environmental or performance conditions warrant lower or higher values.

In warm climates where nitrification is not desirable, SRT values of 1 to 2 days are used. Also, long SRT values are often used in extended aeration systems where secondary goals require minimization and stabilization of the excess sludge solids generated. A later chapter discusses the challenges of meeting federal regulations (Section 503) for stabilization by providing long SRTs. Figure 12.11 presents suggested ranges of SRT for BOD removal and nitrification at various temperatures (Bisogni and Lawrence, 1971). Once a design value of SRT has been selected, eq 12.7 can be used to estimate the required aeration tank volume.

Selection of MLSS may be determined by trial and error in the design process. Optimizing the aeration tank and clarifiers design should be based on the SRT required for wastewater treatment, oxygen-transfer limitations, solids settling characteristics, and the allowable solids loading rate to the secondary clarifiers. Conventional air activated sludge system MLSS concentrations ranging from 1500 to 3000 mg/L often are used. It is possible, however, for these systems to accommodate higher concentrations. Extended aeration systems are frequently designed for up to 4000 mg/L MLSS. Pure-oxygen and membrane systems can be operated at MLSS concentrations of more than 10 000 mg/L,

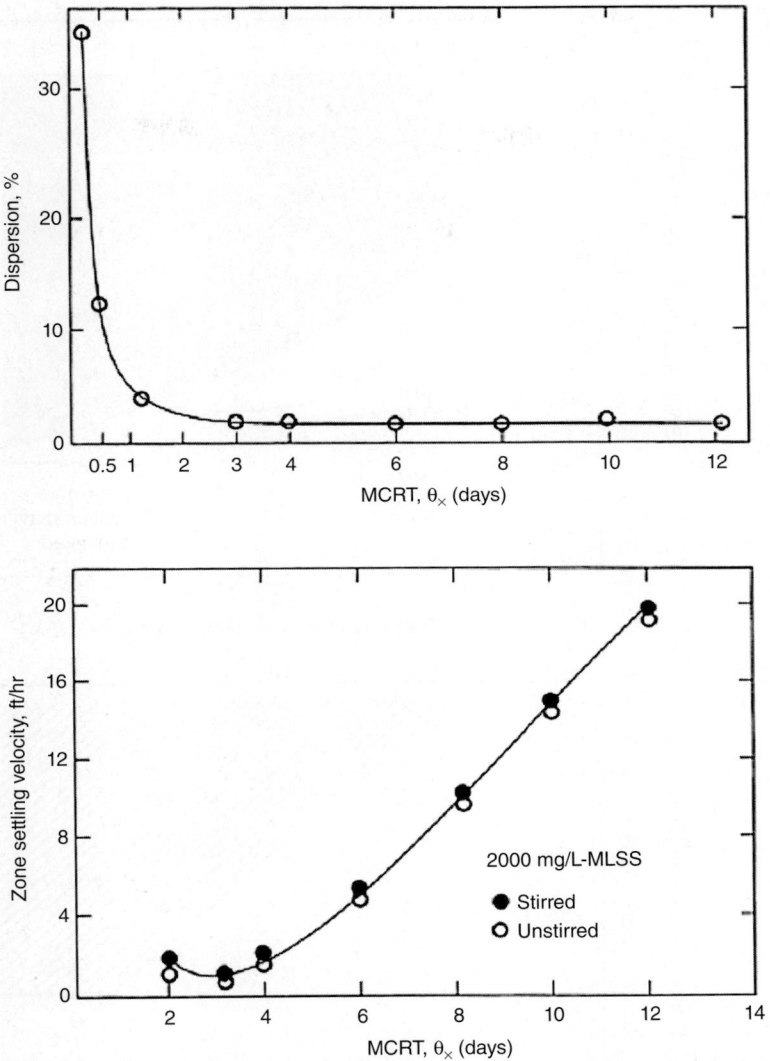

Figure 12.10 Effects of mean cell residence time (MCRT) on the amount of dispersed growth in activated sludge effluent and the settling velocity (ft/hr × 0.3604 8 = m/h) of activated sludge mixed liquor (Bisogni and Lawrence, 1971).

but foaming problems may be observed at values greater than 2000 mg/L for systems with mechanical surface aerators.

Solids settling and thickening properties often dictate final selection of MLSS concentration for systems with secondary clarification. For air activated sludge systems, design for concentrations more than approximately 5000 mg/L is seldom economical (Eckenfelder, 1967). Figures 12.12 and 12.13 show suggested values as functions of SVI and temperature. Approaches for evaluations based on settling and thickening characteristics of the MLSS are discussed later in this chapter. In pure-oxygen systems, the

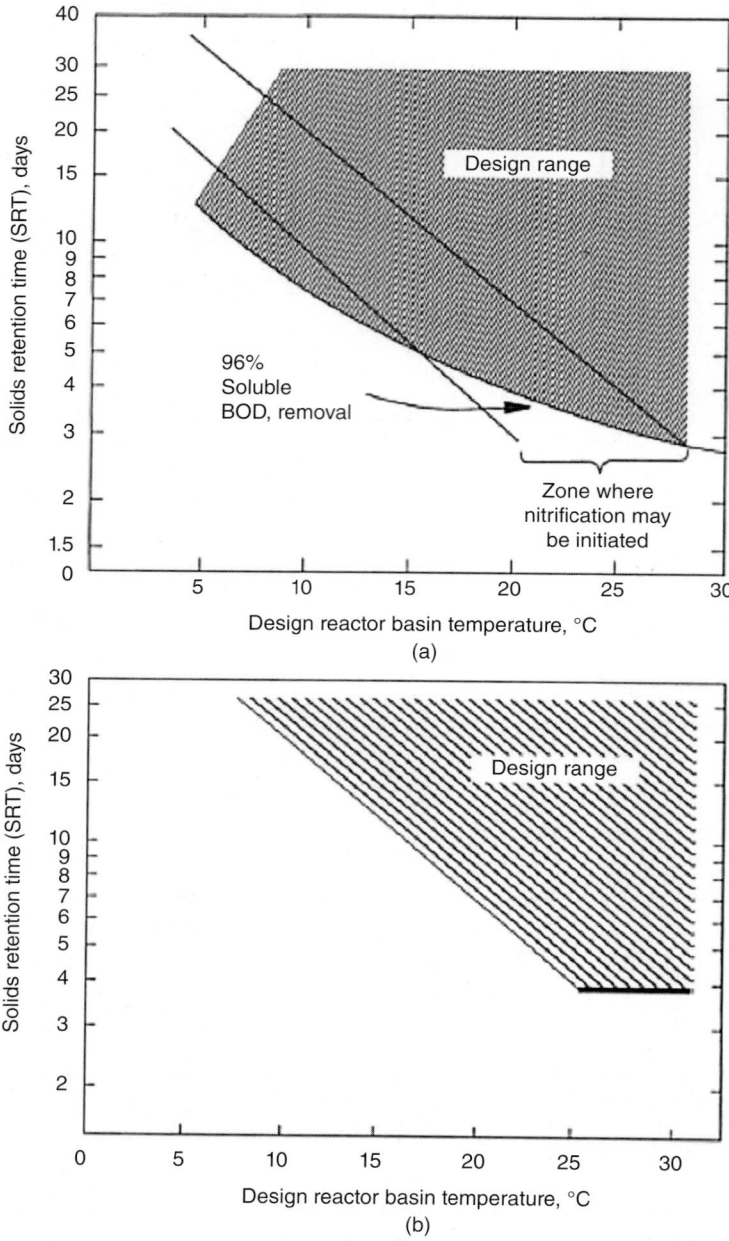

Figure 12.11   Design solids retention time for (a) carbonaceous biochemical oxygen demand removal and (b) single stage nitrification (toxicity not present, mixed-liquor suspended solids washout controlled at pH 7.5 to 9.0).

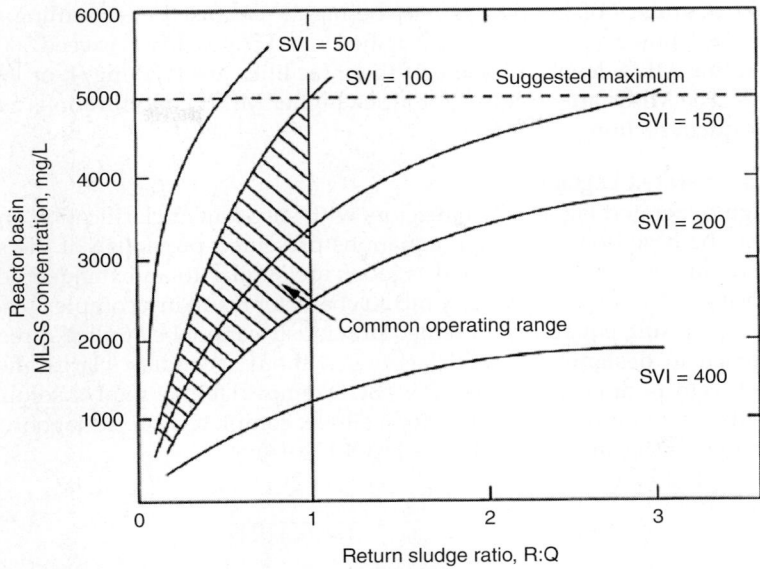

FIGURE 12.12   Design mixed-liquor suspended solids (MLSS) vs sludge volume index (SVI) and return sludge ration (high-rate sludge removal mechanism) at a reactor basin temperature of 20°C.

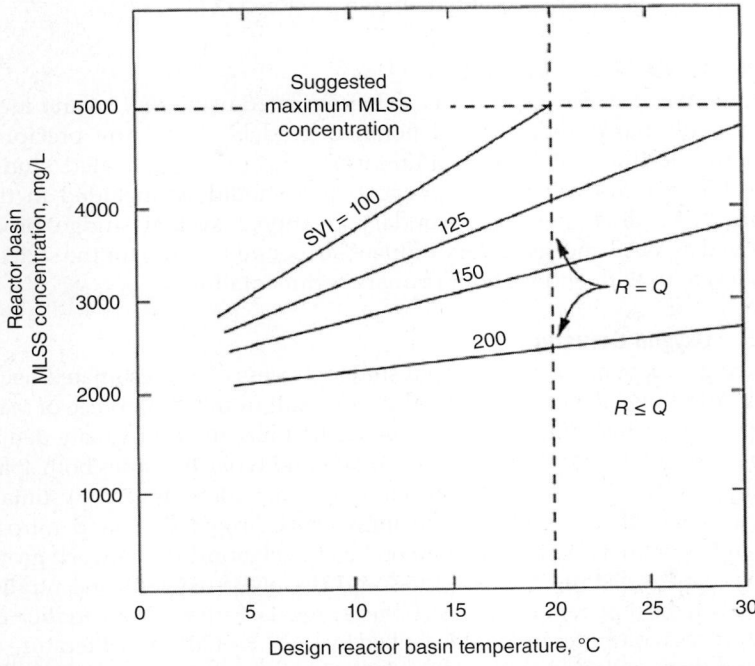

FIGURE 12.13   Suggested maximum mixed-liquor suspended solids (MLSS) design vs temperature and sludge volume index (SVI) at aerator temperature, not ambient temperature (e.g., at 20°C and SVI = 150 mL/g, MLSS should not exceed 3300 mg/L).

upper boundary of the figures may be higher because better-settling sludges can be generated; however, most facilities in the United States do not exceed 5000 mg/L MLSS. Operating MLSS levels for some of these facilities are 1500 mg/L or less in warm climates. The MBRs are designed for much higher MLSS concentrations as discussed in a subsequent section.

3.4.3 Aerated Lagoons

Design of aerated lagoon-style reactors with subsequent clarification and sludge recycle can be based on the design equations presented previously if the system is completely mixed. Although aerated lagoons are similar to an extended aeration system, deposition of solids and partially mixed character result in a complex reactor configuration. As a result, equations presented previously cannot be applied directly. A common approach to designing aerated lagoons without secondary clarification and sludge recycle is to assume that the observed BOD removal (either total or soluble BOD) can be described by first-order kinetics. For a single, completely mixed lagoon, the first-order equation is (Metcalf & Eddy, Inc./AECOM, 2014):

$$\frac{S_e}{S_o} = \frac{1}{1 + k_1 \, \text{HRT}} \tag{12.16}$$

where k_1 = observed BOD removal rate constant $(1/t)$.

Temperature affects reported values of k_1, which have ranged from 0.25 to 1.0 d^{-1} for overall BOD removal (Metcalf & Eddy, Inc./AECOM, 2014). Additional details on aerated lagoon design are presented elsewhere (Reed et al., 1998).

3.4.4 Waste Sludge Generation

The amount of sludge generated can be estimated using eq 12.6 and includes nonvolatile, volatile biodegradable, and nonbiodegradable VSS. Any precipitates that form from the addition of iron or aluminum salts to the activated sludge process for phosphorus removal or other purpose also should be included in this calculation. Figure 12.14 illustrates net secondary treatment system sludge production (to be removed as WAS and secondary effluent suspended solids) for the stated waste characteristics, both with and without primary sedimentation.

3.4.5 Oxygen Demand

The oxygen demand of an activated sludge process can be estimated using eqs 12.9 and 12.14. Additional oxygen demand also can result from the presence of readily oxidizable compounds in the influent such as sulfide that has an approximate demand of 2 mg/L oxygen per mg·L sulfide (as S). Oxygen demand typically varies both spatially and temporally in a suspended-growth system. Temporal variations can be estimated from statistical analyses of data collected for influent loadings (cBOD and nitrogenous oxygen demand). Spatial variations depend on kinetic relationships between growth rates of the biomass and substrate removal rates and dissolved-oxygen concentrations. They also depend on the flow regime and HRT of the process. Estimates of variation can be obtained from process computer models described by the WEF (2013b) or literature data. Table 12.6 presents data collected in the United Kingdom for long, narrow (plug flow) aeration tanks (L/W is greater than 20) (Boon and Chambers, 1985). Estimates of nitrogenous demand assumed that nitrification progressed uniformly along the entire tank length.

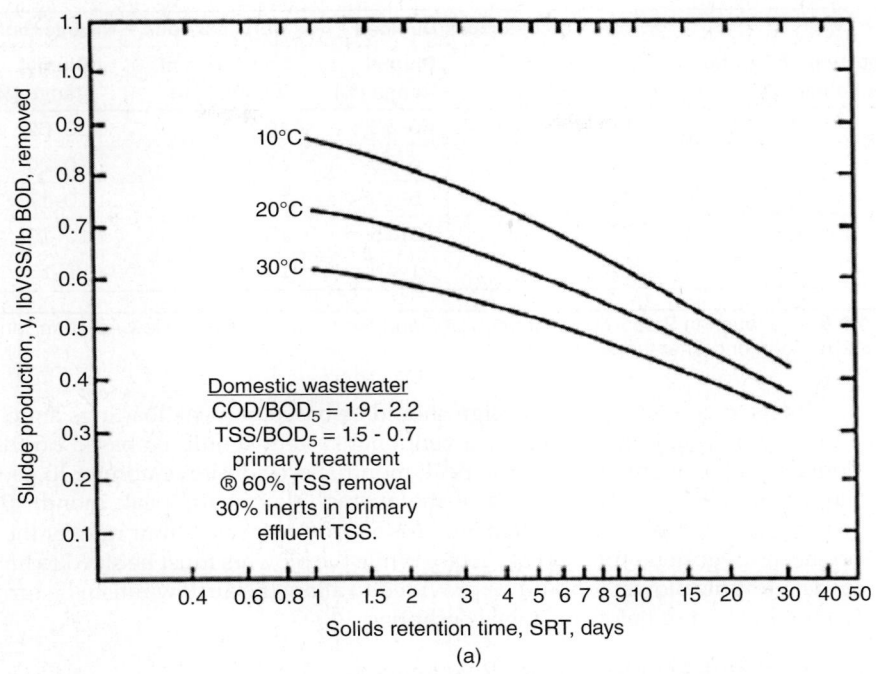

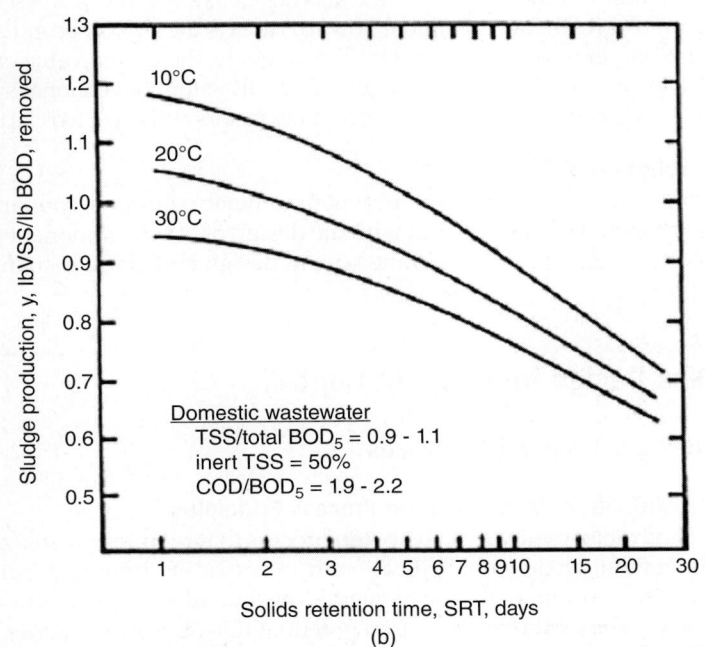

FIGURE 12.14 Net sludge production vs solids retention time and temperature (a) with primary treatment and (b) without primary treatment (lb/lb = kg/kg).

Proportion of Aeration Tank Volume (%)	Carbonaceous Demand		Carbonaceous + Nitrogenous Demand	
	Proportion of Demand (%)	Diurnal Range (%)	Proportion of Demand (%)	Diurnal Range (%)
20	60	40–85	46	33–62
20	15	5–20	17	10–20
20	10	5–15	14	10–17
20	10	5–15	13	10–16
20	5	<1–10	10	7–13

TABLE 12.6   Variation in Proportion of Oxygen Demand the Length of a Plug Flow Aeration Tank ($L/W > 20$) (Boon and Chambers, 1985)

Total oxygen demand for design should be based on peak loadings anticipated. As a minimum, requirements for a conventional system should be based on the 24-hour demand of the average day of the peak month. Some designers prefer to use the peak day or the peak 4-hour demand of the average day of the peak month. Basing the requirement on the peak day demand plus 50% of the peak 4-hour rate for the peak day has been suggested (Young et al., 2008). While the blowers must be sized to handle peak loads, multiple blowers should be provided to allow the airflow rate to be turned down to treat loads at or below average conditions.

3.4.6   Return Activated Sludge Requirements

Requirements for RAS pumping capacity can be estimated from eq 12.8 for an assumed clarifier underflow concentration in the absence of site-specific solids settling characteristics. The ratio of RAS flow to influent flow (αr) affects the size of secondary clarifiers without influencing the size of aeration tanks. As a guide, the design value of αr should range from 20% to 100% of the average facility design flow for conventional systems and up to 150% for some systems depending on peak-flow factors and expected clarifier performance.

3.4.7   Solids-Liquid Separation

Design of solids separation systems, whether membranes or secondary clarifiers, is an important function that is integral with the design of other components of a suspended-growth system. Details of membrane system design and clarifier sizing appear later in this chapter.

4.0   Process Design for Nutrient Control

4.1   Nitrogen Removal Processes

4.1.1   Nitrification/Denitrification Process Principles

Biological nitrogen removal is a two-step process that requires nitrification in an aerobic environment followed by denitrification in an anoxic environment. An anoxic environment is defined as one with low oxygen levels but adequate electron acceptor present (e.g., nitrate). The oxidation reduction potential (ORP) in anoxic zones typically ranges from −50 to 50 mV. As with all biological activity, these reactions are affected by the specific environmental conditions in the reactor including pH, wastewater temperature, dissolved-oxygen concentration, substrate type and concentration, and the presence or absence of toxic substances

Nitrification is the sequential oxidation of ammonium-nitrogen to nitrite-nitrogen and then to nitrate-nitrogen. Biological denitrification reduces nitrate-nitrogen (or nitrite-N; see Section 4.1.3) to nitrogen gas, which is sparingly soluble and can be released to the atmosphere. Denitrifying organisms are primarily facultative hetero-trophs that reduce nitrate in the absence of molecular oxygen. A relatively broad range of heterotrophic bacteria can denitrify.

Autotrophic denitrification has been reported using alternate electron donors (e.g., molecular hydrogen or elemental sulfur) (Tang et al., 2012; Sun et al., 2012). This process frequently consumes alkalinity as opposed to heterotrophic denitrification, which pro-duces alkalinity. Because heterotrophic mechanisms predominate wastewater treatment, this discussion is restricted to pathways using organic compounds as an energy source.

Nitrate can be used in lieu of ammonia as a nutrient for bacterial growth. Therefore, if ammonia levels are low in a reactor, and nitrate is present, then the system will not be nutrient limited.

Biological mechanisms and stoichiometry for biological denitrification are rela-tively well established. Basic reactions for the reduction of nitrate to nitrogen gas are as follows:

1. Reduction sequence.

$$NO_3^- \rightarrow NO_2^- \rightarrow NO \rightarrow N_2O \rightarrow N_2$$

2. Overall reduction using a methanol carbon source.

$$NO_3^- + \frac{5}{6}CH_3OH \rightarrow \frac{1}{2}N_2 + \frac{5}{6}CO_2 + \frac{7}{6}H_2O + OH^-$$

3. Overall reaction, including cell synthesis ($C_5H_7O_2N$): CH_3OH carbon source, and nitrate nitrogen source.

$$NO_3^- + 1.08\ CH_3OH + 0.24\ H_2CO_3$$
$$\rightarrow 0.056\ C_5H_7O_2N + 0.47\ N_2 + 1.68\ H_2O + HCO_3^-$$

4. Overall reaction, including synthesis: CH_3OH, carbon source, and ammonia nitrogen source.

$$NO_3^- + 2.5\ CH_3OH + 0.5\ NH_4^+ + 0.5\ H_2CO_3$$
$$\rightarrow 0.5\ C_5H_7O_2N + 0.5\ N_2 + 1.45\ H_2O + 0.5\ HCO_3^-$$

5. Overall reaction, including synthesis: municipal wastewater carbon source and ammonia nitrogen source.

$$NO_3^- + 0.345\ C_{10}H_{19}O_3N + H^+ + 0.267\ HCO_3^- + 0.267\ NH_4^+$$
$$\rightarrow 0.655\ CO_2 + 0.5\ N_2 + 0.612\ C_5H_7O_2N + 2.3\ H_2O$$

There are many texts and references for more information (Barnes and Bliss, 1983; Ekama et al., 1984; Grady et al., 2011; McCarty et al., 1969; Parker et al., 1975; Pitter and Chudoba, 1990; Sharma and Ahlert, 1977; Stensel et al., 1973; Tchobanoglous et al., 2003; U.S. EPA, 1993).

Overall denitrification results that interest the design engineer can be summarized as follows:

- Nitrate is converted to nitrogen gas in a step-wise manner; NO and N_2O are also gaseous and can be released from solution as alternate end products (Section 4.1.5).
- Oxygen equivalency is 2.856 mg O_2/mg NO_3-N reduced to N_2. For the other steps, the oxygen equivalency is:
  1. NO_3-N reduced to NO_2-N = 1.142 mg O_2/mg NO_3-N reduced to NO_2;
  2. NO_2-N reduced to N_2 = 1.713 mg O_2/mg NO_2-N reduced to N_2;
  3. NO_2-N reduced to NO-N = 0.571 mg O_2/mg NO_2-N reduced to NO-N;
  4. NO-N reduced to N_2O-N = 0.571 mg O_2/mg NO-N reduced to N_2O-N; and
  5. N_2O-N reduced to N_2 = 0.571 mg O_2/mg N_2O-N reduced to N_2.
- Alkalinity equivalency is 3.57 mg $CaCO_3$/mg NO_3-N.
- Heterotrophic biomass production is approximately 0.4 mg VSS/mg COD removed.

Because nitrification only oxidizes ammonium to nitrate and nitrite, denitrification must be used if TN removal is required. This step requires the presence of both a degradable carbon source and nitrate. This can be achieved in three ways:

1. Supplying an exogenous carbon source such as methanol or acetate to the denitrification zone or reactor;
2. Using cBOD in the wastewater as a degradable carbon source by either
   - Recycling a large amount of nitrified mixed liquor back to an anoxic reactor or
   - Diverting a portion of the raw influent or primary effluent flow to a zone containing nitrate; and
3. Using endogenous carbon present in cell mass as the degradable carbon source.

The amount of nitrification removed by secondary treatment systems is limited by the amount of refractory dissolved organic nitrogen (rDON) present in the facility influent and created in the biological treatment processes (WERF, 2008). The rDON is the organic nitrogen in soluble compounds that is not easily removed by biological treatment. Typical levels of rDON in WRRF effluent range between 0.5 and 1.2 mg N/L with most values in the 1 to 2 mg N/L (Pagilla, 2007). Figure 12.15 summarizes facility effluent DON (dissolved organic nitrogen) from several facilities. Even if the system is designed to remove all the ammonia and nitrate, significant soluble residual TN will remain under the best of circumstances.

Biological denitrification rates have been evaluated and studied by researchers, both in laboratory and full-scale operations. A wide range of rates have been reported, as shown in Figure 12.16 (Christensen and Harremöes, 1972; Parker et al., 1975). Several variables have been shown to significantly affect biological denitrification kinetics, including:

- Carbon substrate type and concentration;
- Dissolved-oxygen concentration;
- Alkalinity and pH; and
- Temperature.

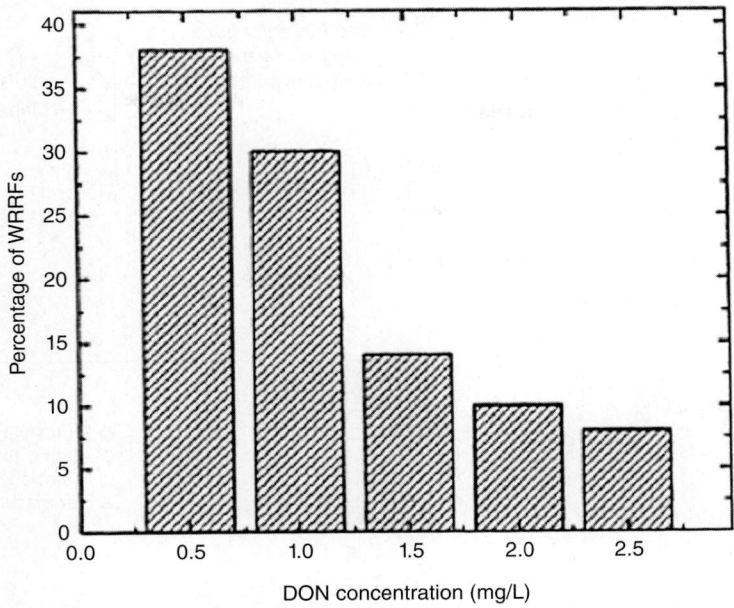

Figure 12.15 Summary of effluent dissolved organic nitrogen (DON) concentration (0.45 μm filtration) from 188 Maryland and Virginia WRRFs (Pagilla, 2007).

The most critical variables are the type and concentration of carbonaceous substrate available in the mixed liquor. Two primary substrate conditions have been identified for suspended-growth denitrification (Grau, 1982):

1. Denitrification under noncarbon-limiting conditions; and
2. Denitrification under carbon limiting conditions.

In sBNR systems, the first set of conditions typically corresponds to those found in preaeration anoxic tanks (first anoxic or preanoxic tanks); the second set corresponds to conditions in postaeration anoxic tanks (second anoxic tanks) when external carbon is not added or RAS endogenous denitrification tanks.

Of the several mathematical models for predicting denitrification rates, the most common are:

- Monod-type relationships; and
- Zero-order equations (with respect to nitrate).

The following equations (based on 20°C temperature) list kinetic expressions in common use (Grau, 1982). Table 12.7 summarizes typical values for the Monod kinetic coefficients.

1. Monod denitrification rate expression:

$$r_{v,NO} = \frac{1 - Y_H}{2.86 Y_H} \mu_{max,H} \eta_g \left(\frac{S_s}{K_s + S_s} \right) \left(\frac{S_{NO}}{K_{NO} + S_{NO}} \right) X_{b,h} \qquad (12.17)$$

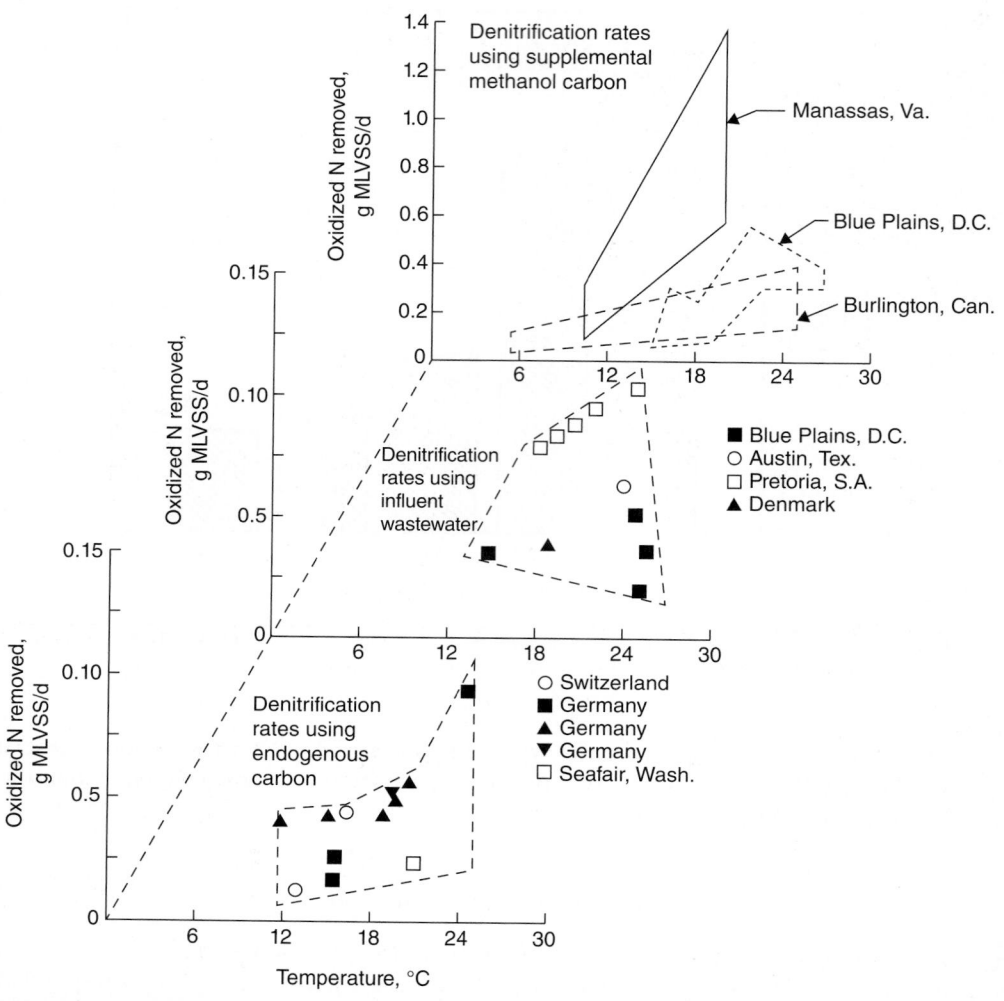

FIGURE 12.16   Specific denitrification rates on suspended-growth systems.

where $r_{V,\mathrm{NO}}$ = reaction rate per unit volume nitrate and nitrite nitrogen, mg nitrate/nitrite N/L·d;

Y_H = biomass yield coefficient, mass/mass;

$\mu_{\max,H}$ = maximum specific growth rate of heterotrophs, 1/time;

η_g = correction factor for μ_H under anoxic conditions, dimensionless;

S_s = soluble biodegradable COD substrate, mass/length³;

K_S = half-saturation coefficient organic substrate, mass/length³;

S_{NO} = soluble material concentration nitrate and nitrite nitrogen, mass/length³;

K_{NO} = half-saturation coefficient nitrate-nitrite, mass/length³; and

$X_{b,h}$ = particulate heterotrophic biomass, mass/length³.

Coefficient	Symbol	Typical Range	Suggested
Maximum specific growth rate of heterotrophs	$\mu_{max} \cdot H$	3–13	4.0–6.0
Heterotrophic biomass yield	Y_H	0.46–0.69	0.67
Half-saturation coefficient organic substrate	K_{COD}	10–180	10–20
Half-saturation coefficient nitrate-nitrite	K_{NO}	0.06–0.5	0.2–0.5
Correction factor for μ_H under anoxic conditions	η_g	0.5–1.0	0.8
Half-saturation coefficient for dissolved oxygen for heterotrophic biomass		0.10–0.28	0.1–0.2
Mass nitrogen per mass of COD in biomass	$C_{X,B}$	0.06–0.12	0.06–0.086
Decay coefficient for heterotrophic biomass	b_H		0.05

TABLE 12.7 Monod Kinetic Coefficients (Baillod et al., 1990; Henze et al., 1987)

2. Zero-order denitrification rate expression:

$$r_{v,NO} = k \tag{12.19}$$

where k = reaction rate coefficient, mass/length$^3 \cdot$time.

Selected zero-order rate coefficients, as reported in the literature, are tabulated in Table 12.8.

The correction factor (η_g) applied to the specific growth rate of heterotrophs was proposed to account for observed reductions in the growth of heterotrophs under anoxic conditions (Batchelor, 1982; Henze et al., 1987). This reduction is a composite number that accounts for (1) the part of the heterotrophic biomass that cannot use nitrate as an electron acceptor and (2) the slower growth of microorganisms in the presence of nitrate compared to oxygen.

Optimal pH for denitrification ranges from 6.5 to 8.5. The following equation has been used to model pH effects on specific growth rates. The effect of pH on denitrification rates is illustrated in Figure 12.17.

$$r_{X,NO} = r_{X,NO,max}\left(\frac{1}{1+10^{5.5-pH}+10^{pH-9}}\right) \tag{12.20}$$

Denitrification rates are influenced significantly by temperature; therefore, temperature correction factors must be selected carefully. The effect of temperature on denitrification rates has historically been modeled using different equations and is currently modeled using commercially available modeling packages (e.g., BioWin or GPS-X). The reader is referred to Chapter 4 for additional information on process modeling.

Unit process configurations for biological nitrogen removal can be simulated through use of process modeling techniques using International Water Association (IWA)-type activated sludge models (ASM). Process modeling is described in detail elsewhere (WEF, 2013b). However, there are several considerations specific to nitrogen removal modeling that are noted here.

Key design criteria for nitrogen removal common to biological treatment configurations include SRT, temperature, recycle rates, and dissolved-oxygen concentrations

k	Units*	Temperature Range, °C	Substrate	Type system	Reference
0.720	mg NO_3-N/mg VSS·d	12–24	Raw and settled wastewater	Modified Ludzack–Ettinger plug flow (k_1)	Ekama et al. (1984)
0.100 8	mg NO_3-N/mg VSS·d	12–24	Raw and settled wastewater	Modified Ludzack-Ettinger plug flow (k_2)	Ekama et al. (1984)
0.072	mg NO_3-N/mg VSS·d	—	Endogenous	Wuhrmann plug flow (k_3)	Ekama et al. (1984)
0.086	mg NO_3-N/mg TSS·d	17–25	Wastewater (first anoxic)	Bardenpho™	Barnard (1975)
0.062	mg NO_3-N/mg VSS·d	22	Primary effluent	Acclimated plug flow	Kang et al. (1990)
0.031	mg NO_3-N/mg TSS·d	17–25	Endogenous (second anoxic)	Bardenpho™	Barnard (1975)
0.68	mg NO_3-N/mg VSS·d	20	Sodium citrate	Laboratory batch	Dawson and Murphy (1972)
0.03–0.11	mg NO_3-N/mg VSS·d	15–27	Wastewater	Various	Parker et al. (1975)
0.21–0.32	mg NO_3-N/mg VSS·d	25	Methanol	Laboratory complete mix	Becarri et al. (1983)
0.12–0.60	mg NO_3-N/mg VSS·d	20	Methanol	Various	Parker et al. (1975)
0.192	mg NO_3-N/mg MLVSS·d	9	Activated sludge	Pilot facility	Mulbarger et al. (1970)
0.593	mg NO_3-N/mg MLVSS·d	22.5	Activated sludge	Pilot facility	Mulbarger et al. (1970)
0.062	mg NO_3-N/mg VSS·d	22	Methanol	Acclimated plug flow	Kang et al., (1990)
0.15	mg NO_3-N/mg VSS·d	20	Methanol	Laboratory complete mix	Sutton et al. (1975)
0.062–0.070	mg NO_3-N/mg TSS·d	—	Glucose		Paskins et al. (1978)

*MLVSS = mixed liquor volatile suspended solids; TSS = total suspended solids; and VSS = volatile suspended solids.

TABLE 12.8 Zero-Order Denitrification Coefficients

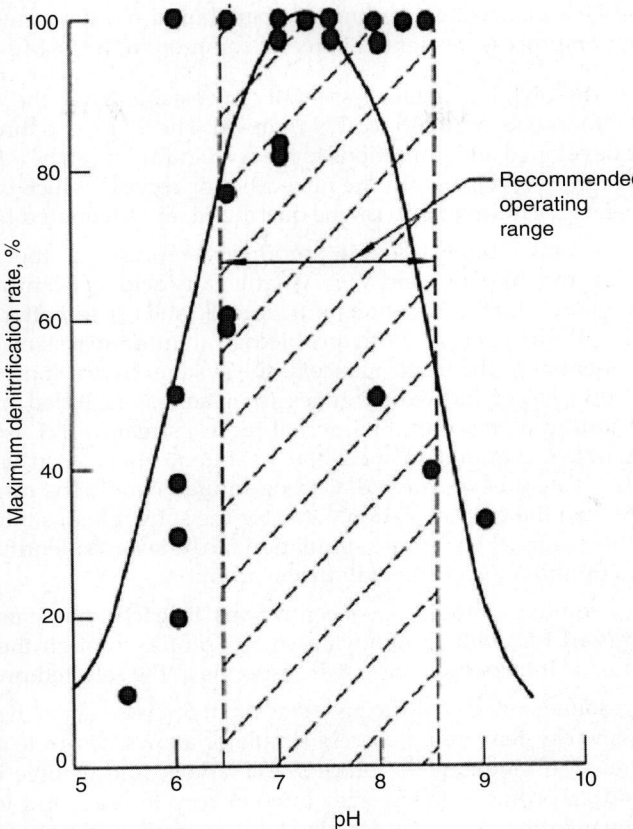

FIGURE 12.17   Effect of pH on kinetic coefficient for maximum specific rate of denitrification.

throughout the process. These criteria should be established before use of a process simulator, but can be refined throughout the modeling task as design details are optimized.

For a biological nutrient treatment system, an initial screening for typical process configurations (e.g., modified Ludzack-Ettinger or Bardenpho™) should be completed before using a process simulator. The screening process may refine the number of configurations to be modeled. The benefit of using a process simulator is that any number of different configurations can be simulated in a reasonably short period of time, but an initial screening may result in a more efficient modeling effort. The process simulator can then be used to optimize the design and help the designer select the appropriate configuration for the treatment system.

Initial sizing should be based on industry standard criteria (SRT for each reactor, mixed liquor recycle (MLR), etc.). This provides a base model for the selected configuration, again improving the overall efficiency of the modeling task. From the base model, various iterations and configuration modifications can be evaluated to come up with the optimized design.

Several key wastewater parameters that should be monitored during the design and evaluation process (presented in terms common to the ASM models):

- S_O, dissolved oxygen—essential in establishing the anoxic or aerobic environments required for TN removal. The S_O profile through the reactor can be developed and limitations to the configuration can be addressed. Issues such as high S_O levels within the mixed-liquor recycle, which can be detrimental to an anoxic environment, can be quantified and accounted for in design.

- S_S, soluble biodegradable products—represents the associated readily biodegradable COD and VFA (volatile fatty acid) concentration available. Most simulators further fractionate S_S into S_F and S_A to better simulate biological phosphorus removal. With any biological nutrient system, the availability of S_S is essential to the performance of the system. By tracking the concentration of S_S, the size of individual zones (or reactors) included in the model can be optimized along with the internal recycle streams. If S_S is not available in the quantities required to meet the treatment goals, then supplemental carbon sources may be warranted. Carbon-limited conditions can be seen when S_S is less than the half saturation value for the substrate in a particular anoxic zone. If this occurs, then process simulation can be used to identify system deficiencies and optimize supplemental carbon addition.

- S_{NH}, soluble ammonia—concentrations in each zone can be monitored and adjusted to optimize removal. An S_{NH} profile through the basin will provide valuable information on how it responds to the selected process configuration.

- S_{NO}, soluble nitrate/nitrite N—removal of S_{NO} is essential in a biological nitrogen removal system, and a process simulator allows design to quantify the level of removal throughout the process. The anoxic and aerobic reactor will then be sized accordingly, along with internal recycle rates, to meet treatment goals. Similar to the approach described for S_S, availability of S_{NO} in an anoxic zone provides valuable information on how to design the nutrient removal system. For example, if the S_{NO} levels in an anoxic zone are low (i.e., less than K_{NO} values), and there is still adequate S_S in the system, then further S_{NO} removal can be achieved by increasing the mixed-liquor recycle rates (to supply more S_{NO} to the anoxic zones). Conversely, if the anoxic zones are showing higher levels of S_{NO}, then it may be possible to reduce mixed-liquor recycle rates and save power.

- S_{ALK}, alkalinity—available alkalinity (or higher pH) is a key requirement for ammonia removal. A process simulator allows a designer to identify any alkalinity deficiencies within the process and adjust the configuration accordingly. Biological nitrogen removal replenishes system alkalinity as described previously.

- X_{TSS}, mixed liquor suspended solids (MLSS)—solids inventory within the process can be tracked, providing the designer with information on the appropriate size and associated capacity of the treatment system. Impacts of basin sizing to the X_{TSS} are easily quantified, and adjustments made to meet treatment goals.

A significant advantage of using a process simulator for the design of a treatment system is the efficiency with which optimization of the treatment configuration can be achieved. Multiple treatment scenarios and configuration alternatives can be evaluated.

Completing a sensitivity analysis on a basin configuration is a common practice to help determine the final basin layout. Ideally, one parameter is adjusted at a time allowing the designer to see the resulting impacts. The sensitivity analysis, however, can be completed for any number of variables. An example for this would be to quantify performance based on a range of mixed-liquor recycle flowrates. Most commercial process simulators can provide a dynamic simulation, which can aide in a sensitivity analysis.

4.1.2 Process Configurations

Suspended-growth processes for nitrogen removal can be grouped into three categories: single, dual, and triple sludge.

4.1.2.1 Single-Sludge Processes

4.1.2.1.1 Wuhrmann and Ludzack-Ettinger Wuhrmann (1954) proposed the single-sludge configuration for nitrogen removal shown in Figure 12.18. The Wuhrmann approach typically is referred to as post-denitrification.

Without addition of an exogenous electron donor (i.e., carbon source), the design relies on residual organic matter passing through the first stage or the endogenous respiration of biomass to provide the energy sources for denitrification. If complete nitrification (thus, complete carbon oxidation) is achieved, then endogenous respiration would provide the principal energy source. Nitrogen removals of 29% to 89% have been achieved in bench- and pilot-scale studies (Christensen and Harremöes, 1972; Christensen et al., 1977; Gundelach and Castillo, 1976; Horstkotte et al., 1974; Johnson and Schroepfer, 1964; Timmermans and Van Haute, 1982; Wuhrmann, 1954, 1964).

Variations of the Wuhrmann design have been developed to supply an exogenous electron donor to the anoxic stage. These consist of either bypassing the first stage with a portion (e.g., 15%) of the influent flow or supplying a suitable carbon supplement, such as methanol, directly to the anoxic zone.

The Ludzack-Ettinger configuration shown in Figure 12.19 reverses the sequence of anoxic and aerobic stages in the Wuhrmann design (Ludzack and Ettinger, 1962). The advantage of this design is the provision of influent BOD to the anoxic stage as an exogenous electron donor.

Barnard (1973a) proposed the modified Ludzack-Ettinger (MLE) configuration that incorporates an internal recycle (Q_{IR}) of mixed liquor from the aeration stage to the anoxic stage (Figure 12.20). This modification increases both the denitrification rate and overall nitrogen removal efficiency and provides control over the fraction of nitrate

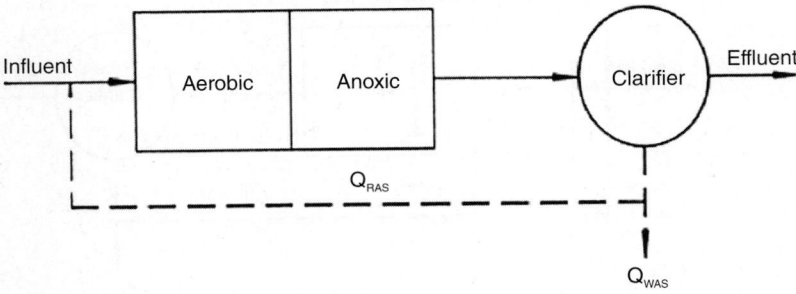

Figure 12.18 Wuhrmann process for nitrogen removal (RAS = return activated sludge and WAS = waste activated sludge).

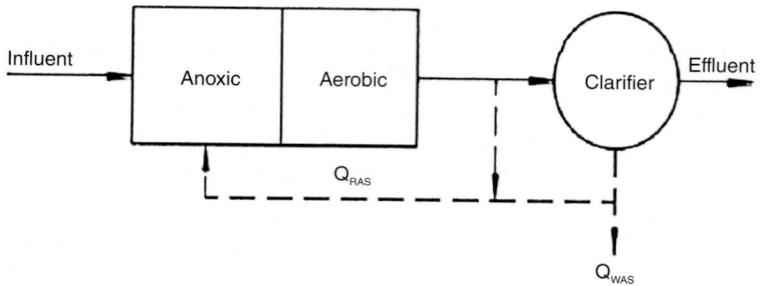

Figure 12.19   Ludzack-Ettinger process for nitrogen removal (RAS = return activated sludge; WAS = waste activated sludge).

removed through variation of the internal recycle ratio. In addition, higher denitrification rates are attained because the anoxic reactor receives a source of readily biodegradable COD. This allows smaller anoxic volumes for a given nitrate removal requirement compared to the Wuhrmann and Ludzack-Ettinger processes.

This process can be used when nitrification is occurring, and denitrification is required to recover alkalinity, lower overall oxygen demand, and provide a better settling sludge. The process effluent typically will contain between 6 and 10 mg/L of nitrate nitrogen and is the most common method of achieving nitrogen removal.

4.1.2.1.2 Four-Stage Bardenpho   The four-stage Bardenpho process consists of a series of four anoxic and aerobic zones with recycling of mixed liquor from the first aerobic zone to the first anoxic zone at a rate as high as four to six times the influent flowrate (Barnard, 1973a, 1973b, 1974, 1976, 1983; Ekama et al., 1984; Irvine et al., 1982; Kang et al., 1990). This process (Figure 12.21) is intended to achieve more complete nitrogen removal than is possible with a two- or three-stage process. Complete denitrification cannot be attained with preaeration anoxic zones because part of the aerobic-stage effluent is not recycled through the anoxic zone. The second anoxic zone provides for additional denitrification using nitrate produced in the aerobic stage as the electron acceptor and endogenous or supplemental organic carbon as the electron donor.

The second anoxic zone is capable of almost completely removing the nitrate in the aeration tank effluent, provided the size is adequate and supplemental carbon is added.

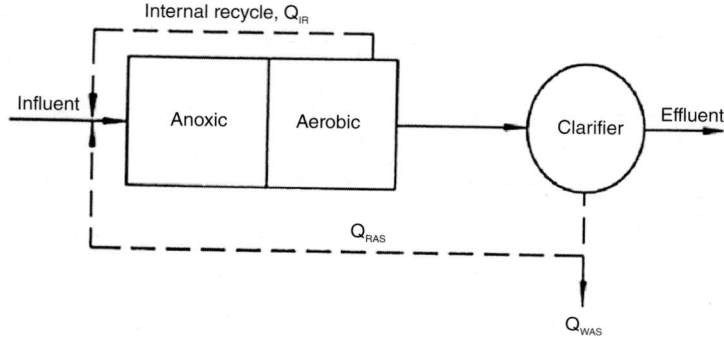

Figure 12.20   Modified Ludzack-Ettinger process for nitrogen removal (IR = internal recycle; RAS = return activated sludge; WAS = waste activated sludge).

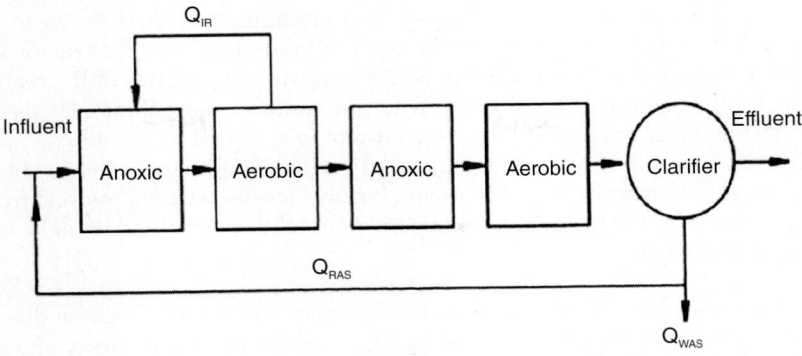

FIGURE 12.21 Four-stage Bardenpho™ process for nitrogen removal (IR = internal recycle; RAS = return activated sludge; WAS = waste activated sludge).

The final aeration stage strips residual gaseous nitrogen (N_2) from solution and minimizes phosphorus release in the final clarifier by increasing the oxygen concentration.

The ability to successfully use the Bardenpho process to achieve an effluent concentration of TN as low as 2 to 4 mg/L depends on the ratio of oxidizable nitrogen to carbon in the influent to the activated sludge process and on the use of supplement carbon addition. Ekama et al. (1984) report that the TKN:COD ratio must be less than 0.08 to obtain complete denitrification.

4.1.2.1.3 Denitrification with methane/biogas   Methane is the most reduced form of carbon available and theoretically offers greater nitrate/nitrite reduction potential on a mole basis than any other carbon source. Facilites that utilize anaerobic digestion have a source of this compound on-site, which could eliminate or minimize the need for purchase of supplemental carbon. At the time of this writing, this technology is still at lab scale with significant decline in performance after operating for 1 to 2 years (Kampman et al., 2014). At full scale, care will need to be exercised to prevent incomplete mass transfer and the release of methane, a potent greenhouse gas.

4.1.2.1.4 Step Feed   This process is essentially the same as conventional step feed where a portion of the influent to the process is fed to one or more points downstream of the head of the reactor. The difference in step-feed nitrogen removal is that each of the feed points has an anoxic zone for nitrogen removal. Step-feed nitrogen removal has been implemented at several full-scale facilities in a variety of configurations and number of feed points. Figure 12.22 shows a schematic of a three-pass step-feed system with an additional zone (similar to the Bardenpho process) (Johnson et al., 2003).

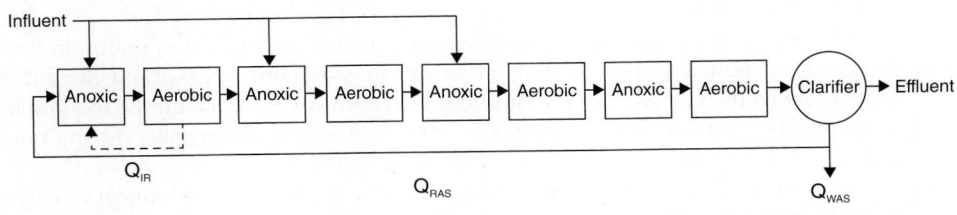

FIGURE 12.22 Three-pass step-feed nitrogen removal system with secondary anoxic zone (IR = internal recycle; RAS = return activated sludge; WAS = waste activated sludge).

The primary benefits of the step-feed configuration include, as in conventional treatment, capacity improvement for a given volume or reduced reactor volumes for a specific capacity due to the higher MLSS at upstream portions of the reactors. In addition, for nitrogen removal, the step-feed process reduces or eliminates the need to recycle nitrate back to the anoxic zones. Nitrate is supplied to the anoxic zones from the upstream aerobic reactor directly, except for the first pass. In the first pass, the only nitrate supply is from the RAS stream. For this reason, nitrified recycle from the end of the first pass to the head of the first pass is sometimes provided to make best use of the available carbon.

4.1.2.1.5 Simultaneous Nitrification and Denitrification   In this process, the dissolved-oxygen level in an aeration tank is reduced to allow heterotrophs to denitrify and autotrophs to convert ammonia to nitrate/nitrite in a single basin. Typically, dissolved-oxygen levels in these basins are less than 1 mg/L. This process has the advantage of eliminating the need for recirculation streams in the aeration basin and reduces aeration requirements due to increased OTR. It does not, however, use influent carbon as efficiently as does the MLE or Bardenpho process. Also, there is some risk of reduced sludge settleability as a result of the low oxygen levels present simultaneously with elevated levels of soluble COD. Research has shown that these conditions can result in filamentous bulking (Jenkins et al., 2003). For more information, see *Shortcut Nitrogen Removal—Nitrite Shunt and Deammonification* (WEF, 2015).

4.1.2.1.6 Oxidation Ditch   Extended aeration oxidation ditch systems are readily adaptable for carbon oxidation, nitrification, and denitrification as described above (Barnes and Bliss, 1983; Barnes et al., 1983; Stensel, 1978; Van der Geest and Witvoet, 1977). Oxidation ditches are reactors that induce significant velocity and recirculation flows in basins with a race-track or other type configurations. Horizontal rotors, slow-speed mechanical aerators or rotating disks, or draft tube aerators provide aeration and force to move the mixed liquor at one or more locations in the ditch. Also, submerged mixers and conventional diffused aeration can be combined to provide aeration and mixing power independently of each other. Dissolved-oxygen concentration will be highest at points of aeration and will subsequently decrease because of oxygen uptake by the biomass as the mixed liquor moves around the looped reactor. After sufficient travel time, zones of simultaneous nitrification and denitrification will form and may go to truly anoxic conditions upstream from aeration devices, as illustrated in Figure 12.23. Feed points typically are located in the anoxic zones to provide carbon for denitrification. The location and size of these anoxic zones will vary with time because oxygen uptake and transfer rates will vary with wastewater quality and flow. Therefore, reliance on this mechanism for denitrification requires a comprehensive control system to monitor and control dissolved oxygen throughout the basin.

The energy input for mixing and aeration must be controlled to maintain the mixed liquor in suspension. This system must afford sufficient operational flexibility with adjustable weirs, variable speed, or two-speed aerators for varying the oxygen input to match diurnal and seasonal changes in oxygen demand. Otherwise, during periods of low loading, necessary anoxic zones will not develop.

The variable levels of dissolved oxygen that are present in oxidation ditches can be used to promote simultaneous nitrification and denitrification or true anoxic zones within a single ditch. Additionally, oxidation ditches can be configured in series of

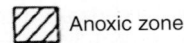

Anoxic zone

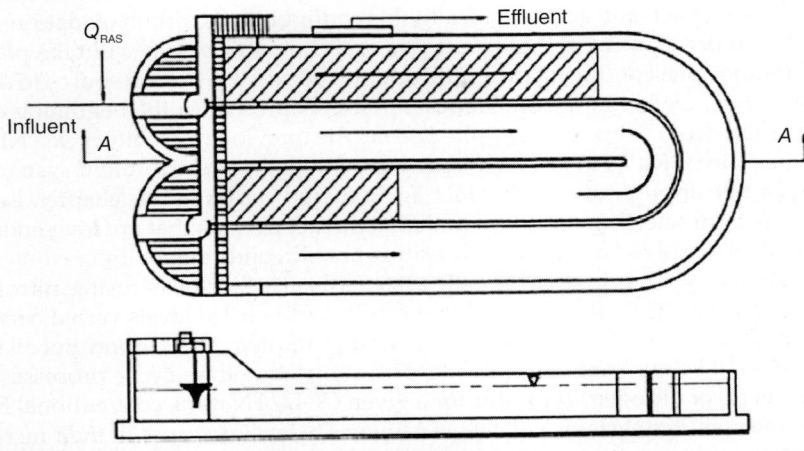

FIGURE 12.23   Oxidation ditch process for denitrification.

concentric reactors with different oxygen levels in each, or simply as an aerobic reactor as described previously in nitrogen removal processes.

In typical oxidation ditch reactors used for nitrogen removal, rates of both nitrification and denitrification will be low because of the relatively long SRTs required for nitrification, low concentration of readily biodegradable COD, and marginal dissolved-oxygen concentrations for either nitrification or denitrification. Oxidation ditch systems have the same limitations in removing nitrogen as other suspended-growth processes and can be designed to remove all but the lowest levels of nitrate/nitrite, subject to carbon availability. Large masses of mixed liquor in the system can compensate for low reaction rates. As with other nitrogen removal processes, highly variable influent flows pose a challenge to achieving a consistently low effluent nitrogen concentration. Nitrogen removals greater than 90% have been reported with oxidation ditch processes but most operate at 5 to 10 mg/L effluent nitrate levels (Rittmann and Langeland, 1985).

An anoxic zone can precede the oxidation ditch as in a typical MLE configuration using the ditch as the aerobic zone. The forward momentum of the moving mixed liquor is often sufficient that a small fraction can be diverted for the internal recycle without the use of an internal recycle pump as is required in a traditional MLE system.

4.1.2.2   Time-Cyclic Processes

4.1.2.2.1   Sequencing Batch Reactor   Biological nitrogen removal can be accomplished in SBRs by creating, in one reactor, the proper cycle of aerobic and anoxic conditions in time sequence (Abufayed and Schroeder, 1986; Alleman and Irvine, 1980; Arora et al., 1985; Irvine et al., 1983; Palis and Irvine, 1985; Silverstein and Schroeder, 1983). Control strategies for BNR take into account reaction time, tank water level, and mixed-liquor dissolved-oxygen concentrations. The SBRs appear well suited for relatively small systems with highly variable wastewater flow and strength. Similar to conventional processes, successful operation depends on efficient clarification, which is accomplished in the same reactor. For nitrogen removal, fill and react phases are subdivided into mixed fill, mixed react, and aerobic react. In this configuration, carbon oxidation and nitrification will occur in the aerobic react phase and denitrification will take place in anoxic conditions present during mixed fill and mixed react. A carbon source to support denitrification, needed in the anoxic react phase, is present in the beginning of each cycle from the feed. Nitrate is supplied from the previous aerobic cycle. Nitrification is attained in SBRs, as in any suspended-growth biological treatment system, by designing for the appropriate aerobic SRT as discussed earlier in this chapter. Denitrification results from selecting mixed fill and mixed react periods that are long enough to allow use of all dissolved oxygen, thus creating anoxic conditions.

A survey was done of 10 SBR systems in the Northeast using nitrogen removal (Young et al., 2008). It was found that facility effluent TN levels varied between 2.5 and 9.5 mg/L. None of these facilities was adding supplemental carbon, but all were operating at well below their design loads. This shows that time-cyclic processes can achieve low levels of nitrogen. Typically, for a given COD/TN ratio, conventional flow through activated sludge systems will achieve lower TN levels because of their increased ability to change the amount of nitrogen brought back to the anoxic zones through recycle systems. Time-cyclic processes are limited in nitrogen removal capacity by the amount of nitrogen present at the beginning of their anoxic periods.

4.1.2.2.2   Continuous-Feed Intermittent Decant Systems   In these systems, feed continuously enters to the reactor, to help improve overall nitrogen removal (Peters et al., 2004). These systems typically are segregated into a first stage that continuously feeds downstream SBR tanks. Some variations do continuous feed within a single tank. Sludge from the SBR tanks is recycled back to the upstream tank continuously (from the SBR that is currently under aerate mode). In this way, these systems approach the carbon usage efficiency of conventional flow-through activated sludge systems.

4.1.2.2.3   Alternating Aeration   Nitrogen removal in a single-reactor activated sludge process using intermittent aeration is feasible (Barth and Stensel, 1981; Schwinn and Hotaling, 1988). Required equipment and operational requirements include:

- Timers on blower or aerators to provide aerobic/anoxic cycling; and
- Proper tankage and solids inventory for nitrification and denitrification.

Phased isolation ditch processes alternate aeration within multiple oxidation ditch reactors to create aerobic/anoxic cycling. The feed is also cycled between reactors so that the anoxic period coincides with the introduction of carbon for denitrification. Alternating aeration has been proposed as a method to selectively inhibit NOB (Kornaros et al., 2010) and promote shortcut nitrogen removal (Section 4.1.3).

4.1.2.2.4   Dual Sludge and Integrated Systems   Separate sludge systems, by definition, house various process stages in physically separate tanks, each with their own clarifier and return-sludge systems. Because EBPR depends on exposure of a single biomass population to various environmental conditions in different process stages, multiple-sludge systems typically are best suited for nitrogen removal only.

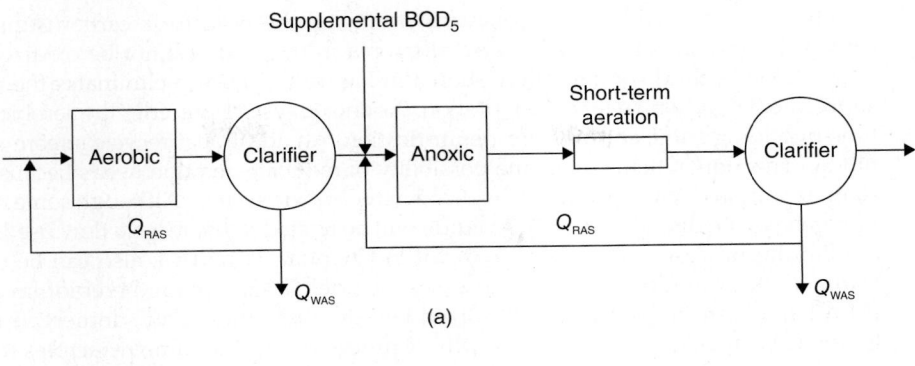

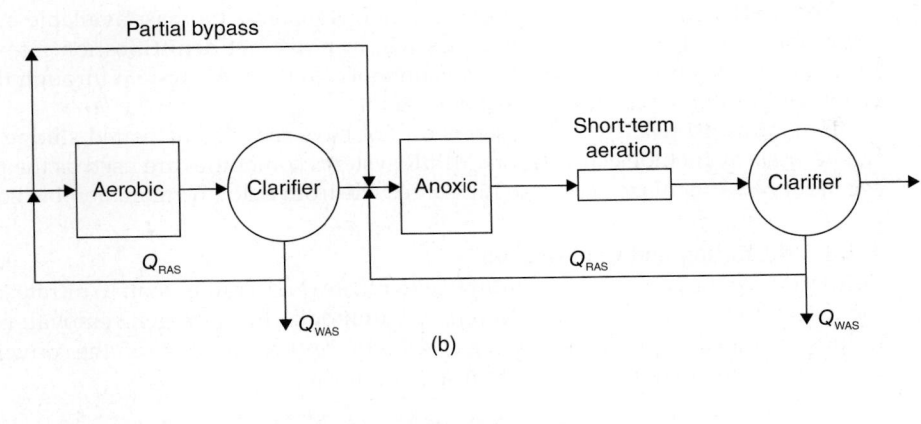

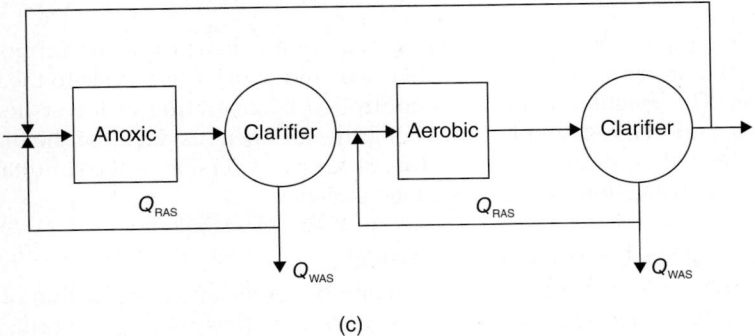

FIGURE 12.24   Dual-sludge processes for nitrogen removal (RAS = return activated sludge; WAS = waste activated sludge) (Grady et al., 2011).

Figure 12.24a-c shows dual-sludge configurations. In Figure 12.24a, the aerobic system first performs carbon oxidation and nitrification. Then, an external carbon source supplements the nitrate-laden stream before contacting the denitrifying biomass in the anoxic system. The system in Figure 12.24b uses the same configuration, except that a portion of influent wastewater fed to the second stage supplies the organic carbon to the anoxic system. Although this system eliminates the need for a carbon supplement, some TKN will pass through because ammonia in the feed will not be oxidized in the anoxic zone. A third configuration, shown in Figure 12.24c, also eliminates the need for supplemental carbon. In this configuration, the anoxic system precedes the aerobic system, thus providing sufficient BOD for denitrification. An additional recycle stream supplies nitrate. This flow scheme offers the possibility of reducing aeration needs because a substantial portion of the BOD can be oxidized in the anoxic system. Although some oxidized nitrogen will be discharged, its magnitude will be related to the recycle flow used.

Biofilm processes, discussed in detail in Chapters 11 and 13, also can be used for nitrogen removal. This group of processes includes IFAS, moving-bed biofilm reactors (MBBRs), denitrification filters, fluidized bed denitrification, and submerged rotating biological contactors. The IFAS and MBBR processes use the same principles for nitrogen removal as suspended-growth systems.

The IFAS media addition to anoxic zones will increase biomass available in a denitrification zone. The additional biomass will improve net denitrification rates within the zone. A suspended-growth SRT is maintained in the IFAS system through the recycle of RAS from the secondary clarifier system.

In contrast to IFAS, MBBR systems do not have recycled activated sludge and do not necessarily include clarification. MBBR systems sometimes are used as the primary secondary treatment process with nitrate recycle from a downstream aerobic zone.

4.1.3   Nitritation and Denitritation

Nitritation, or the conversion of ammonia to nitrite (NO_2^-) rather than to nitrate, reduces the amount of oxygen required to remove ammonia. For nitrogen removal, nitrite is reduced in anoxic zones to nitrogen gas but requires less carbon than the conversion of nitrate because of the lower oxidation state of nitrite.

$$NH_3 \rightarrow NO_2^- \rightarrow N_2$$

This process is used primarily to treat nitrogen-rich and carbon-poor liquors from the dewatering of anaerobically digested sludge prior to recycle to the main treatment trains. The reaction to nitrite is controlled by operating at lower dissolved-oxygen levels, and denitritation is done using the endogenous decay of the biomass. Because dissolved-oxygen levels are low, this process is a version of the simultaneous nitrification/denitrification process discussed earlier.

4.1.4   Anaerobic Ammonium Oxidation

Anaerobic ammonium oxidation (anammox) is the anoxic oxidation of ammonium to nitrogen gas carried out by members of the *Planctomycetes*, which utilizes nitrite as the terminal electron acceptor. The energy-yielding reaction is shown in eq 12.21.

$$NH_4^+ + NO_2^- \rightarrow N_2 + 2H_2O \tag{12.21}$$

In this process, nitrite is concurrently oxidized to nitrate to yield reducing equivalents for carbon dioxide fixation and cell growth. The net reaction (including cell synthesis) increases the stoichiometry from 1:1 ammonium:nitrite. The stoichiometry of anammox metabolism has been found to vary from 1.11 to 2.00 nitrite-N reduced per ammonium-N oxidized (Dapena-Mora et al., 2004; Staunton and Aitken, 2015; Strous et al., 1999; van de Graaf et al., 1996) and the variation is thought to be due to the effect of SRT; with a longer

SRT, the stoichiometry approaches 1:1 (Lee et al., 2013). The generally accepted reaction that includes cell synthesis is shown in eq 12.22 (Strous et al., 1998).

$$NH_4^+ + 1.32NO_2^- + 0.066HCO_3^- + 0.13H^+$$
$$\rightarrow 1.02N_2 + 0.26NO_3^- + 0.066CH_2O_{0.5}N_{0.15} + 2.03H_2O$$

(12.22)

The low energy yield (357 kJ/mol) of eq 12.21 coupled with the autotrophic nature of the bacteria involved leads to a very slow growth rate with an estimated doubling time of 11 to 14 days under warm sidestream conditions. Although not well established, the doubling time at the reduced temperature found in mainstream treatment has been estimated to be 35 days (Lotti et al., 2014). Achieving anammox in the mainstream treatment process requires a feed with the proper stoichiometry of ammonium to nitrite (alternatively, produce nitrite in situ; see Section 4.1.4), a low BOD-to-N ratio to prevent out competition of anammox bacteria by denitrifying bacteria for nitrite via shortcut denitrification (Section 4.1.3), and efficient biomass retention in order to maintain an adequate SRT for the slow growing biomass.

4.1.5 Nitritation/Anammox

The dominant form of influent nitrogen to most influent wastewater treatment facilities is ammonium-N, and a fraction of the influent N must be selective oxidized to nitrite prior to further treatment by anammox bacteria. In addition to the above requirements for anammox, NOB must be selectively inhibited. Similar control techniques for shortcut nitrogen removal can be used for nitritation/anammox systems. Nitritation and anammox can be designed to occur in the same reactor or in separate reactors.

Nitrogen removal by nitritation/anammox lends itself to a two-sludge AB-type system (see Section 2.5.5.2). In the A-stage, BOD is removed and is either oxidized or redirected to anaerobic digestion for biogas production; in the B-stage, nitrogen is removed by nitritation/anammox. It is important to note that anammox produces nitrate as a side product and a tertiary polish step may be necessary to reach very low nitrogen discharge permits. At the time of this writing, very few mainstream nitritation/anammox systems had been developed and all mainstream systems require an operational sidestream system.

4.1.6 Nitric and Nitrous Oxide Emissions

Nitrification and denitrification produce nitric oxide (NO) and nitrous oxide (N_2O) as by-products and intermediates, respectively. NO reacts in the atmosphere to form ground-level ozone (Chatfield et al., 2010), and N_2O is a potent greenhouse gas with a global warming potential approximately 300´ greater than CO_2 (IPCC, 2007).

The mechanism of by-product formation is unknown, but several factors have been associated with increased NO and N_2O production. Nitrifying systems have been shown to produce nitrous oxide when operated under low dissolved oxygen, with a high nitrite or ammonium concentration, low SRT, and low temperature, or with the presence of toxic compounds (Kampschreur et al., 2009). Release of nitrous oxide as an end product from denitrification has been linked to low concentration of dissolved oxygen, and limited availability of COD amendable to denitrification (Kampschreur et al., 2009). A recent survey of municipal wastewater treatment facilities in the United States found 0.1% to 1.8% of influent TKN emitted as nitrous oxide (Ahn et al., 2010).

4.2 Enhanced Biological Phosphorus Removal Processes

Nitrogen and phosphorus removal systems can achieve low levels of nutrients in facility effluents. Depending on the effluent quality required, treatment could be by biological means alone, by chemical means, or by a combination. This section deals only with biological removal of phosphorus. Chemical treatment for phosphorus removal is presented in Chapter 14.

Phosphorus is essential for metabolism during biological growth. Therefore, all biological processes remove phosphorus from wastewater to varying degrees. The VSS in CAS processes contain 1.5% to 2.5% phosphorus. Assuming a phosphorus content of 2% (0.02 mg P/mg VSS), and 0.5 mg of VSS produced per milligram of BOD removed, then about 1.0 mg/L of phosphorus is converted to cell mass per 100 mg/L of BOD removed. Therefore, a CAS process reduces the influent phosphorus by 1 to 2 mg/L.

Phosphorus removal in excess of metabolic requirements can be achieved by using EBPR. This process relies on the selection and proliferation of a specialized heterotrophic microbial population capable of storing soluble phosphorus in excess of their minimum growth requirements. These organisms, collectively called phosphate-accumulating organisms (PAOs), can sequester up to 0.38 mg P/mg VSS and, as a result, mixed liquor from an EBPR system can contain 0.06 to 0.15 mg P/mg VSS (Wentzel et al., 2008). The higher the mixed-liquor PAO fraction, the greater are the phosphorus content of the waste sludge and the amount of phosphorus removed.

4.2.1 Process Principles

The EBPR processes consist of anaerobic and aerobic zones and, in combined nitrogen and phosphorus removal configurations, anoxic zones. The anaerobic zone provides a competitive advantage to PAOs and is defined as not intentionally containing usable dissolved oxygen or nitrate. The ordinary heterotrophic organisms (OHO) cannot compete for the same substrate in the anaerobic zone because of the absence of an electron acceptor (oxygen or nitrate).

The PAOs do not grow in the anaerobic zone, but consume and convert readily available organic material (i.e., VFAs) to energy-rich carbon polymers called poly-β-hydroxyalkanoates (PHAs). The energy required for this reaction is generated through breakdown of stored polyphosphate (poly-P) molecules, which results in phosphorus release and an increase in the bulk liquid-soluble phosphorus concentration in the anaerobic stage. Magnesium and potassium ions are concurrently released to the anaerobic medium together with phosphate, in the molar ratio $P:Mg^{2+}:K^+ \approx 1.0:0.33:0.33$. PAOs also break down glycogen, which is stored within the cells, to provide substantial reducing power for PHA production (Erdal et al., 2004; Filipe et al., 2001b; Mino et al., 1987).

The PAOs metabolize the internally stored PHA in the anoxic and aerobic zones and use the energy to take up all of the soluble orthophosphates released in the anaerobic zone and additional phosphorus present in the influent to renew the stored polyphosphate pool. Phosphorus uptake in excess of metabolic requirement is possible because the energy released by PHA oxidation is significantly greater than the energy required for PHA storage. The PAOs also use PHAs as a carbon source for growth.

Effluent from the EBPR reactors is low in phosphorus because the removed soluble phosphorus is stored in the biomass that is removed from the system with the WAS. RAS containing stored polyphosphate is recirculated to the head of the anaerobic zone to seed the incoming flow. Some of the energy and carbon is used to restore the glycogen pool for reactions to continue when mixed liquor is recirculated to the head of the

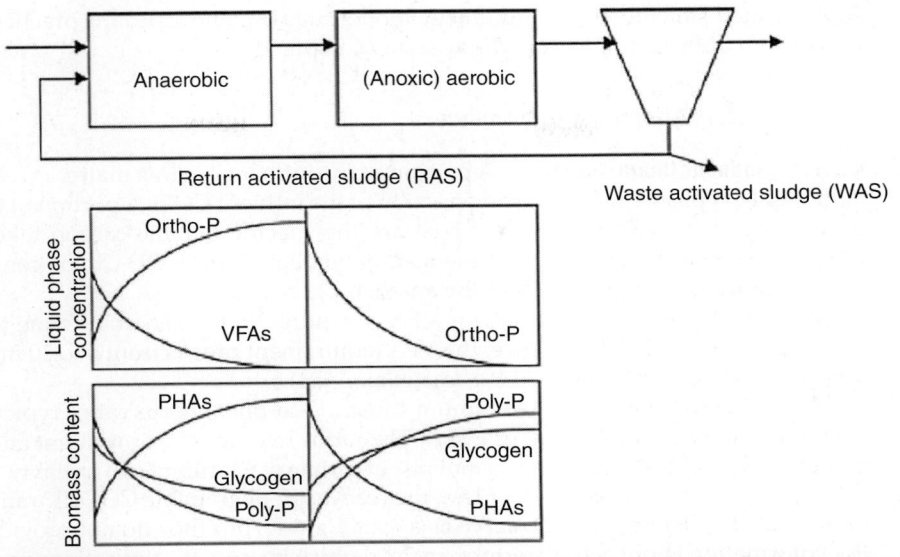

FIGURE 12.25   Typical concentration patterns occurring in enhanced biological phosphorus removal (EBPR) systems (WEF, 2013a).

anaerobic zones. The events that take place in the anaerobic and aerobic stages are illustrated in Figure 12.25.

There are two basic guidelines designers should follow when designing phosphorus, or combined nitrogen and phosphorus removal systems:

1. Minimize oxygen and nitrate/nitrite to anaerobic zones. The anaerobic environment is most beneficial to developing biological phosphorus removal populations when there are no or low levels of electron acceptors present in the anaerobic zone.

2. Minimize oxygen to anoxic zones. High dissolved oxygen or entrained air in the nitrified recycle stream or reactor feed will reduce the carbon available for denitrification.

Internal facility sidestreams can contribute a significant nutrient load to secondary treatment system influent. The recycle load depends on both the nutrient and the solids-handling system at the facility. Typically, those facilities without solids digestion systems have low levels of nutrient return. Facilities with aerobic digestion may have low levels of ammonia return but higher nitrate returns that need to be considered in design. Higher levels of phosphorus in the dewatering liquor are also common, but typically not at high levels. Anaerobic digestion facilities typically have the highest levels of ammonia and phosphorus in the return stream although nitrate is not present. All nutrient removal facilities with digestion must consider the effects of the recycle system on the design of the mainstream nutrient removal process.

Because of the inherent complexity of phosphorus, and combined nitrogen and phosphorus removal systems, the use of commercial process simulators that incorporate sophisticated stoichiometric and kinetic models has become standard practice for process design. Process models are discussed in Chapter 4.

4.2.2 Factors Affecting Performance

4.2.2.1 Influent Characteristics A portion of the influent organic matter is readily biodegradable, typically measured as a fraction of the influent COD. A portion of this readily biodegradable COD is VFAs, which are the specific organic carbon taken up by PAOs. However, the remainder of the readily biodegradable COD can be converted to VFAs by fermentation reactions in the anaerobic zone.

In general, more than 25 mg/L as VFA is required in the anaerobic zone to accomplish significant EBPR. In practice, the VFA requirement ranges from 5 to 10 mg/L VFA per mg/L phosphorus removed (Ekama et al., 1984).

Table 12.9 summarizes the minimum substrate-to-phosphorus ratios typically used to estimate the potential extent of EBPR. The ratios refer to reactor influent and should account for recycle loads and removals in primary clarifiers. To reliably produce 1.0 mg/L total phosphorus (TP) or less, the readily biodegradable COD:TP ratio should be at least 15:1. These ratios do serve as a rough guide, but they do not provide definitive information about actual performance or which process configurations provide the most efficient use of the available substrate (Bratby et al., 2012a).

The salinity of the influent wastewater can affect EBPR performance. This has significance in some applications where seawater or brackish water is used for secondary purposes such as toilet flushing, or in cases of seawater intrusion into sewerage systems, or in some industrial wastewaters. Short-term increases in salinity can be detrimental with complete inhibition exhibited at 0.2% salinity (Welles et al., 2016). However, long-term tests on acclimated cultures show that EBPR organisms successfully adapt and sustained salinities up to 5.2% can be treated. Anaerobic kinetic rates increase up to 3.5% salinity but decrease markedly as salinities increase further. Aerobic kinetic rates gradually decrease as salinities are increased from freshwater levels, and at 6.9%, the

Substrate Measure	Substrate:P[a]	Remarks
cBOD$_5$	20:1	Provides a rough/initial estimate. Based on typically available facility data.
sBOD$_5$[b]	15:1	Better indicator than cBOD$_5$.
COD	45:1	More accurate than cBOD. Not measured by all facilities.
VFA	7:1 to 10:1	More accurate than COD. Involves specialized lab analysis.
rbCOD[c]	15:1	Most accurate. Measures VFA formation potential. Accounts for VFA formation in the anaerobic zone. Specialized laboratory analysis.

[a]Minimum requirements.
[b]Soluble BOD.
[c]Readily biodegradable COD.

Table 12.9 Minimum Substrate to Phosphorus Requirements for EBPR

phosphate uptake is negligible (Welles et al., 2016). At salinities higher than approximately 5.2%, chemical phosphorus removal would likely be necessary.

4.2.2.2 Integrity of the Anaerobic Zone
The most important function of the anaerobic zone of an EBPR process is PAO selection, which is a rapid reaction if adequate, rapidly biodegradable substrate is available. In some instances, the anaerobic zone is also required to generate VFAs through fermentation. This is a slower reaction.

Although the definition of anaerobic condition is zero dissolved oxygen, in practice such conditions are established at levels of less than 0.2 mg/L. For optimal EBPR the PAOs need to pass through an anaerobic zone in which the required ORP appears to be −300 mV, or less (Barnard et al., 2016).

There are often insufficient VFAs in wastewater influent for substantial EBPR. Therefore, in such cases the anaerobic zone assumes particular importance in fermenting readily biodegradable COD. It is for this reason that the exclusion of dissolved oxygen and nitrates from the anaerobic zone is important (Ekama et al., 1986; Barnard et al., 2016).

Sources of dissolved oxygen and nitrates that threaten the integrity of the anaerobic zone are listed in Table 12.10. The presence of these two oxygen sources causes a reduction of the effective anaerobic volume. Consequently, this will decrease anaerobic contact time between the PAOs and the substrate (VFAs), which could potentially compromise phosphorus removal. In addition, the presence of nitrate and dissolved oxygen will provide competing organisms access to the substrate. For example, 1.0 mg of nitrate-N will steal readily biodegradable organics needed for the removal of 0.7 mg of phosphorus by supporting denitrification. Likewise, the presence of 1.0 mg of dissolved oxygen will deprive the substrate needed for the removal of 0.3 mg phosphorus by facilitating normal heterotrophic activity (BOD oxidation). In addition, dissolved oxygen in the anaerobic zone can trigger filamentous growth. Design engineers and operators must take steps to minimize the introduction of dissolved oxygen and nitrate to the anaerobic zone as far as possible.

4.2.2.3 Aerobic Zone Impacts
In EBPR systems, the PAOs are enriched with stored PHA in the anaerobic zone, and the mixed liquor has high levels of soluble phosphorus.

Source	Introduces
Pre-aeration	Dissolved oxygen
Influent screw pumps	Dissolved oxygen
Free-fall over weirs[a]	Dissolved oxygen
Excessive turbulence[a]	Dissolved oxygen
Aggressive mixing in the anaerobic zone	Dissolved oxygen
RAS flow	Nitrates, dissolved oxygen
Backflow from aerobic to anaerobic zone	Dissolved oxygen
Internal mixed liquor recycle[b]	Nitrates and dissolved oxygen

[a]Upstream of the anaerobic zone.
[b]In nitrogen removal systems.

TABLE 12.10 Common Sources of Dissolved Oxygen and Nitrates

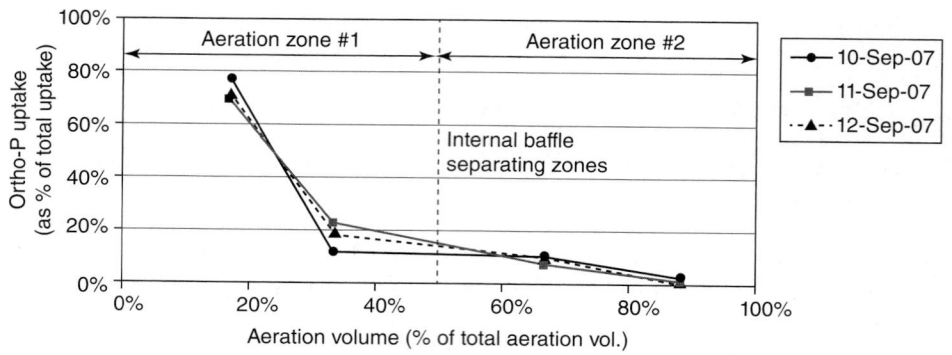

In the anoxic and aerobic zones, phosphorus is then taken up by the PAOs. In general, approximately 75% of soluble phosphorus removal occurs in the first 20% of aeration volume, as shown in Figure 12.26 (Jeyanayagam, 2007). Organism growth only occurs under aerobic conditions (Wentzel et al., 2008).

Both the dissolved-oxygen level and the aerobic SRT affect the competition between PAOs and glycogen-accumulating organisms (GAOs). Carvalheira et al. (2014) showed that PAOs possess a higher affinity for oxygen and have a kinetic advantage over GAOs at low dissolved-oxygen concentrations. Further, extensive aerobic HRTs can be detrimental to EBPR with secondary release occurring toward the end of the aerobic phase, likely used as an energy source for cell maintenance when PHA degradation is complete. This can result in lower poly-P available for VFA uptake in the anaerobic zone, resulting in more VFAs available for the GAOs.

Despite the evident advantage of lower dissolved-oxygen levels that favor PAOs over GAOs, there is often a practicable limit to the dissolved-oxygen concentration. For example, in nitrogen and phosphorus removal facilities, both the lower dissolved-oxygen limit and the aerobic SRT are dictated by the needs for nitrification. There are also sludge settleability issues that may dictate the lower dissolved-oxygen limit. In such cases, compromise in design is needed. However, the effect of dissolved-oxygen concentration is not completely clear. Jimenez et al. (2014) showed that at one facility operated as an anaerobic/aerobic (A/O) process with simultaneous nitrification-denitrification, very good biological phosphorus removal performance was obtained with significant phosphate release occurring in the anaerobic zone, and phosphorus uptake in the aerobic zone, despite the low operating dissolved-oxygen concentrations between 0.1 and 0.4 mg/L. Effluent phosphate concentrations are approximately 0.1 mg P/L, which does not support prior understanding that dissolved-oxygen levels should exceed approximately 1.5 mg/L for effective phosphorus uptake. Confirmatory laboratory tests demonstrated that phosphorus uptake did not occur with nitrite as electron acceptor. Nitrates were used as electron acceptors by denitrifying PAOs, although lower phosphorus uptake rates were obtained under anoxic conditions. The low operating dissolved oxygen and the low effluent NO_x-N concentrations (less than 1 mg/L) at this facility likely contributed to very low dissolved oxygen and NO_x-N returned to the anaerobic zone by the RAS, allowing significant fermentation in the anaerobic zone and generation of VFAs that enhanced P release.

4.2.2.4   pH   In the anaerobic zone of EBPR processes, there is evidence that GAOs take up VFAs faster than PAOs when the pH of the anaerobic zone is less than 7.25. PAOs take up VFAs faster than GAOs when the pH is greater than 7.25. In the aerobic zone, the rates of phosphorus uptake, PHA consumption, and biomass growth are highest when the pH is increased to above 7.0. At lower pH, PAO growth rates are inhibited, whereas the GAOs are largely unaffected by pH (Filipe et al., 2001a, 2001b). From these studies, it appears that adjustment of pH to at least 7.0 would be beneficial to maximize EBPR performance.

Other workers have found that adjustment of pH to near-neutral values is beneficial to EBPR, with minimal EBPR activity at pH values below approximately 5.2 (Tracy and Flammino, 1985; Chapin, 1993) as presented in Figure 12.27.

4.2.2.5   Solids and Hydraulic Retention Times   The effect of SRT on EBPR performance is complex. An increase in SRT is accompanied by (1) an increase in non-PAO (heterotrophic) activity leading to increased VFA production and enhanced phosphorus removal; (2) a reduction in the wasting rate of phosphorus-rich PAOs leading to reduced phosphorus removal; and (3) an increase in the MLVSS phosphorus content due to an increase in PAO fraction. These phenomena are attributed to the low endogenous decay rate of PAOs: 0.05 day^{-1} on a COD basis compared to 0.24 day^{-1} for aerobic heterotrophs (Wentzel et al., 2008).

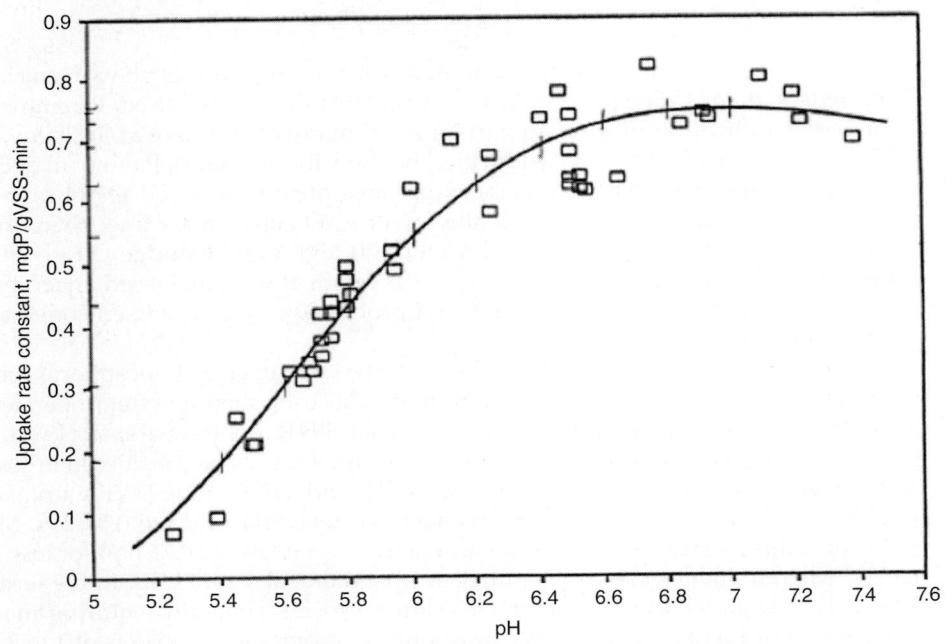

FIGURE 12.27   Effect of pH on the phosphate uptake rate constant (Tracy and Flammino, 1985).

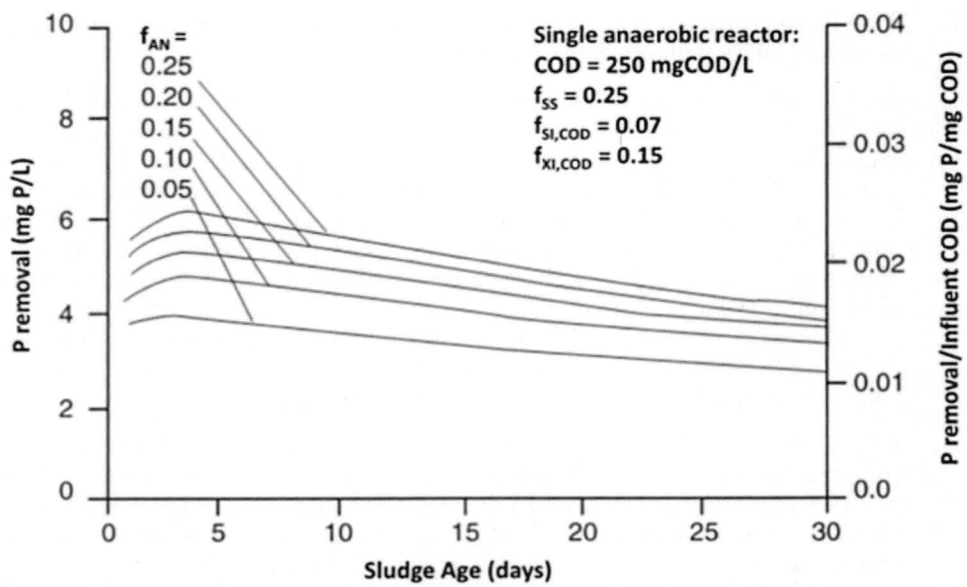

FIGURE 12.28 Predicted phosphorus removal versus sludge age for various anaerobic mass fractions (f_{AN}) for a single anaerobic reactor system treating unsettled wastewater with a total COD of 250 mg/L, with characteristics as shown. f_{ss} = influent readily biodegradable fraction of influent biodegradable COD; $f_{SI'CODi}$ = influent unbiodegradable soluble COD fraction; $f_{XI'CODi}$ = influent unbiodegradable particulate COD fraction (adapted from Wentzel et al., 2008).

Figure 12.28 exemplifies the combined effects of the various phenomena. At low SRTs (less than 3 days), an increase in SRT results in increased P removal and increased utilization of influent carbon for P removal (P removal/influent COD). However, as the SRT is increased further (beyond about 3 days), P removal decreases with increasing SRT. This behavior is because of the increased OHO (non-PAO) mass in the system at higher SRT, the lower PAO active mass and, therefore, the lower mass of P removed from the system via the wasted sludge. At given SRTs, the figure also shows increased phosphorus removal with increased anaerobic volume fractions (f_{AN}), due to enhanced VFA production from increased conversion of fermentable COD.

Figure 12.28 also illustrates that to achieve the same degree of phosphorus removal, a system with longer SRTs will require more substrate than a system operated at a shorter SRT. This was also pointed out by Stensel (1991) and Fukase et al. (1982).

Anaerobic phosphorus release and aerobic uptake must be considered in selecting the overall system and individual-zone SRT and HRT values. VFA uptake is a relatively rapid process, requiring an anaerobic zone SRT of 0.3 to 0.5 days. Most of the time, this corresponds to a nominal anaerobic zone HRT of 0.75 hour or less. However, when influent VFAs are insufficient to sustain substantial EBPR, the fermentation of biodegradable organic matter is a slower process, generally requiring anaerobic zone SRTs of 1.5 to 2 days, which corresponds to anaerobic zone HRTs of 1 to 2 hours or more. Hence, if the influent wastewater contains significant concentrations of VFAs, then a relatively short anaerobic zone SRT and HRT can be used. If, on the other hand, fermentation is required in the anaerobic zone to generate VFAs, then a longer anaerobic zone SRT and HRT should be considered. In applying these guidelines, it should be noted that, depending on the mixed-liquor biomass concentration, the required HRT would vary for different systems (see, for example, the University of Cape Town [UCT] system, later).

At SRT values greater than 4 days and at temperatures greater than 15°C, nitrification will tend to occur, and process configurations that include anoxic zones for denitrification of nitrate in the recycle flows must be used. In nitrifying systems, the RAS can be a significant source of nitrate, and this should be adequately addressed in the design of the system configuration (see discussion in the following sections).

4.2.2.6 Temperature   There have been inconsistencies in the effects of temperature on EBPR performance reported from various investigations, principally attributed to the use of different substrates, process configurations, and measurement methods (Wentzel et al., 2008). Temperature influences EBPR systems in a variety of ways, including fermentation, lysis, and nitrification.

McClintock et al. (1991) showed that, at a temperature of 10°C and an SRT of 5 days, the EBPR function would "wash out" before other heterotrophic functions. Mamais and Jenkins (1992) also showed that EBPR ceases when the SRT temperature combination is below a critical value. Erdal et al. (2002, 2003) investigated this phenomenon and showed that in EBPR systems, the main effect of SRT is on PHA and glycogen polymerization reactions. While PAOs washed out of the system, ordinary heterotrophs (non-PAOs), which do not exhibit glycogen metabolism, continued to grow in the aerobic zone down to shorter SRTs.

Early researchers reported that from 5°C to 24°C, EBPR efficiency is unchanged at lower temperatures compared to higher temperatures (Barnard et al., 1985; Daigger et al., 1987; Ekama et al., 1984; Kang et al., 1985; Sell, 1981). Mamais and Jenkins (1992) found the optimum temperature for aerobic phosphorus uptake was between 28°C and 33°C. Jones and Stephenson (1996) suggested that the optimum temperature was 30°C for anaerobic release and aerobic uptake of phosphate. Brdjanovic et al. (1997), using laboratory-scale SBRs, found the optimum temperature for anaerobic phosphorus release and acetate uptake was 20°C. For aerobic phosphorus uptake, however, a continuous increase was obtained for temperature values up to 30°C. The stoichiometry of EBPR was found to be insensitive to temperature changes.

Panswad et al. (2003) reported lower EBPR performance at higher temperatures, which may be attributed to decreases in phosphorus content and PHA storage caused by longer anaerobic contact times. Similar findings were reported by Wang and Park, (1998). Based on full-scale facility data and laboratory-scale investigation, Rabinowitz et al. (2004) reported decreased rate of EBPR at temperatures above approximately 30°C. This was attributed to reduced rates of phosphorus release and uptake. The researchers also concluded that loss of EBPR can lead to sludge bulking because the anaerobic zone does not function as a selector (no soluble COD uptake). At the microbial level, the reason for lower EBPR performance at warmer temperatures is likely related to increased GAO competition for substrates in the anaerobic zone. The colder temperatures provide selective advantage to PAOs although higher temperatures cause a population shift from PAOs to GAOs.

Improved cold-weather EBPR performance has been reported by several investigators. Helmer and Kuntz (1997) and Erdal et al. (2003) reported that, despite slowing

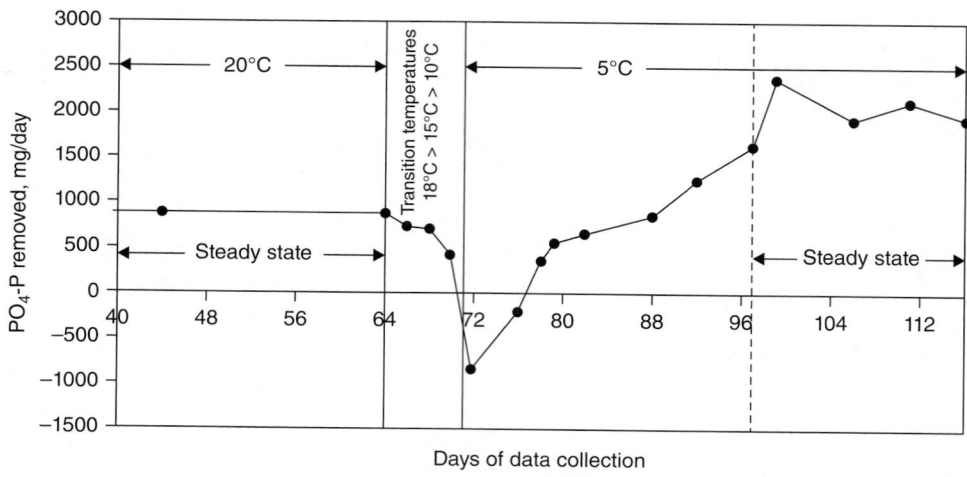

FIGURE 12.29   Effect of acclimation on cold-temperature performance of enriched enhanced biological phosphorus removal populations (Erdal et al., 2002).

reaction rates, EBPR performance can be significantly greater at 5°C compared to 20°C. Citing work reported by other investigators, Stensel (1991) attributed better cold-weather EBPR performance to a population shift to slower growing psychrophilic organisms with higher yields. The findings of Erdal et al. (2002) presented in Figure 12.29 show the importance of cold-weather acclimation and the resulting improved EBPR performance.

Brdjanovic et al. (1998b) showed the link between temperature and aerobic SRT required for EBPR. At low temperatures, the biomass fails to grow fast enough and PAOs are washed out of the system, as shown in Figure 12.30. In EBPR systems the aerobic SRT should be long enough to oxidize the PHA stored within the cells during the anaerobic phase. Therefore, the minimum aerobic SRT depends on the PHA conversion kinetics and the maximum achievable PHA storage capacity in the cells. Also superimposed on Figure 12.30 are the minimum aerobic SRTs typically assigned to BNR facilities for stable nitrification at different temperatures. From this chart it appears that in facilities that practice both nitrification and EBPR, the operating aerobic SRT will likely be greater than the minimum required for stable EBPR.

4.2.2.7   Solids Capture   Effluent TP consists of two components: soluble phosphorus and particulate phosphorus. Efficient EBPR can reduce the effluent soluble phosphorus to approximately 0.1 mg/L. Particulate phosphorus represents solids-associated phosphorus. Hence effluent total solids and the phosphorus content of the solids dictate its value. The effect of effluent solids on effluent TP is illustrated Figure 12.31 (WEF, 2013a). For example, if the effluent TSS is 10 mg/L with a VSS content of 75% and the phosphorus content of the mixed liquor is 0.06 mg/mg VSS (6%), then the effluent particulate phosphorus concentration would be 0.45 mg/L. Hence, controlling the effluent solids through design and operation of final clarifiers and effluent filters is important in achieving low effluent TP.

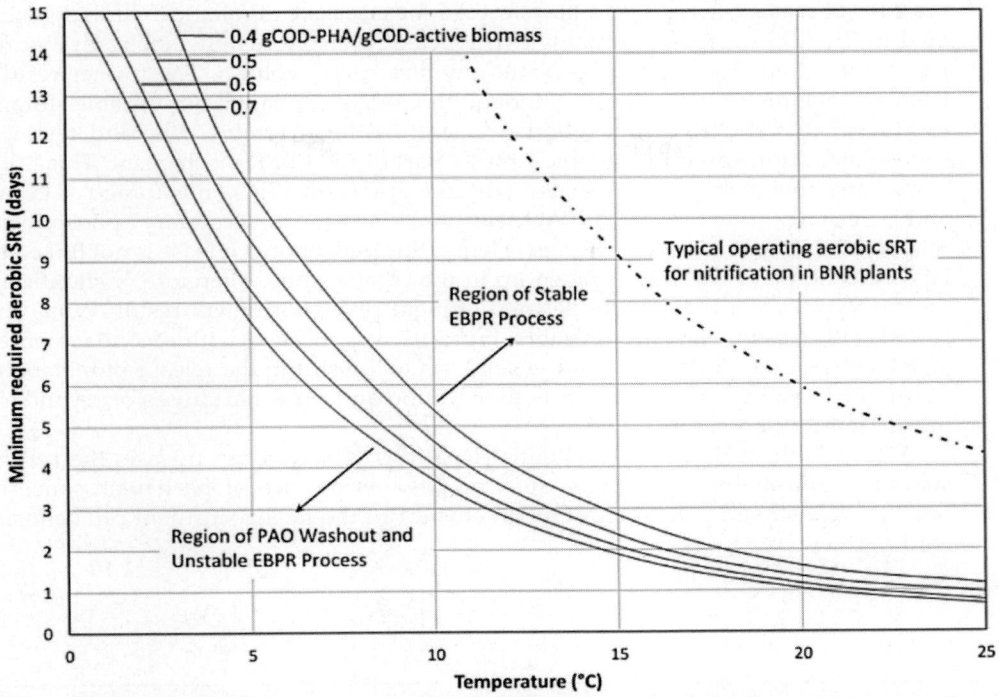

FIGURE 12.30 Minimum required aerobic SRT as function of temperature and the storage capacity of the cell (0.4 to 0.7 g PHA-COD/g COD-active biomass), based on laboratory SBR systems, showing regions (above) where stable EBPR occurs and (below) where washout and unstable EBPR occurs (adapted from Brdjanovic et al., 1998b).

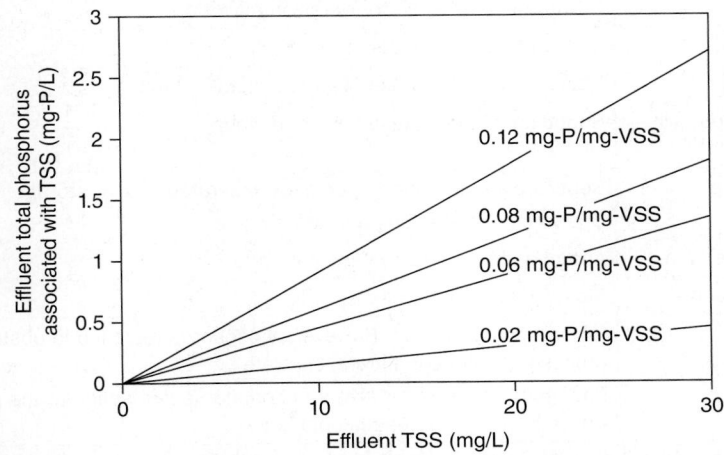

FIGURE 12.31 Contribution of effluent TSS to effluent total phosphorus for different mixed liquor solids phosphorus contents, assuming MLVSS:MLSS ratio = 0.75 (WEF, 2013a).

4.2.2.8 Secondary Release and Recycle Load Management   Biological sludge generated by the EBPR process contains two types of phosphorus: metabolically bound phosphorus and stored polyphosphate granules called volutin. The former results from normal microbial synthesis, although the polyphosphate is an unstable storage product that is depleted (phosphorus release) in the anaerobic zone and restored (phosphorus uptake) in the aerobic zone as part of the EBPR mechanism. This "primary" anaerobic release is associated with the uptake of VFAs and storage of PHAs and is desired and necessary for PAO selection. In contrast, "secondary" phosphorus release occurs without carbon storage. Hence, this phosphorus release is not linked to PAO selection and will not be taken up in the aerobic zone. Therefore, if significant secondary release occurs, then elevated effluent phosphorus will result. While the stored polyphosphate, being unstable, typically is associated with secondary phosphorus release, conditions that cause cell lysis will result in the release of metabolic phosphorus as well. Table 12.11 lists the location and potential causes of secondary release in EBPR processes.

While Table 12.11 lists all potential locations of secondary release, the return streams from sludge operations, such as dewatering, are of particular concern. Total recycle streams can amount to 20% to 30% of the facility influent phosphorus loading.

Location	Cause of Phosphorus Release
Primary clarifier	Co-settling of primary and EBPR sludges
	Poor solids capture during thickening and dewatering operations may return phosphorus rich solids to the primary clarifier where secondary release could occur
Anaerobic zone	VFA depletion due to oversized anaerobic zone
Anoxic zone	Nitrate depletion due to oversized anoxic zone
Aerobic zone	Long SRT leading to cell lysis
Final clarifier	Septic conditions caused by deep sludge blanket
Primary sludge gravity thickener	Septic conditions caused by deep sludge blanket
Sludge storage	Septic conditions due to poorly or unaerated sludge storage
	Due to cell lysis in long aerated storage
Anaerobic digestion	An anaerobic conditions and cell lysis
Aerobic digestion	Mostly due to cell lysis
Dewatering	No significant release. However, phosphorus released in upstream processing will be in filtrate/centrate
	Poor solids capture may return phosphorus rich solids to the primary clarifier where secondary release could occur

TABLE 12.11   Location and Potential Causes of Secondary Phosphorus Release

Return streams often occur intermittently in many facilities causing significant variation in nutrient loadings and short-term peak loads that could overwhelm the EBPR process. For example, if dewatering operations occur over one shift, 5 days per week, then recycle loading could potentially be 4 times the loading generated by a 24/7 operation. The complex microbial consortium in a single-sludge system has limited ability to respond quickly to influent variations by self-adjusting. The period of acclimation is directly influenced by SRT, MLSS, and the magnitude and duration of peak loads. Within limits, higher SRT and MLSS enhance microbial diversity and system robustness although extremely high and persistent loadings can overwhelm EBPR capability potentially resulting in regulatory noncompliance.

The quantity and the quality of these streams vary based on the technology used in the solids processing operations. For example, sludge thickening using belt-filter dewatering generally generates two times more recycle flow (filtrate) compared to centrifuge dewatering because of the amount of washwater used in the dewatering operation. This will affect the recycle hydraulic load although returned phosphorus mass load will remain unchanged. Equalization of the return streams should be designed to allow for the volumes and intermittent nature of the streams.

Use of anaerobic digesters is of particular concern at EBPR facilities. The recycle stream from anaerobically digested sludge dewatering can contain up to 900 to 1100 mg/L of ammonia and 100 to 800 mg/L of phosphorus. Actual recycle loads will depend on how much of the released nitrogen and phosphorus are chemically precipitated as struvite ($MgNH_4PO_4$), brushite ($CaHPO_4 \cdot 2H_2O$), and vivianite [$Fe_3(PO_4)_2 \cdot 8H_2O$]. The precipitated phosphorus is discharged with the dewatered sludge and the amount of phosphorus in the return streams is thereby reduced.

Recycled phosphorus will reduce process influent BOD:TP ratios, which could potentially convert a typically phosphorus-limited (excess substrate) EBPR system to a substrate-limited condition with a likelihood of elevated effluent phosphorus.

4.2.2.9 Key Design Considerations The complex nature of the EBPR process demands careful design considerations. Adequate flexibility should be incorporated in designs to allow facility operators to respond to changing operating conditions. Below is a summary of key design considerations for reliable EBPR performance.

- The EBPR process is sensitive to influent characteristics. A minimum of two years of facility data (preferably more) should be used for the purpose of characterizing the influent for adequate process modeling and design. Recycle loads from sludge operations can modify the process influent characteristics significantly and should be characterized as well as the influent.

  EBPR performance is sensitive to influent loads, particularly following periods of low loads (e.g., after weekends and in the mornings after nighttime lows). This phenomenon has been termed *Monday phosphate peaks*. The reason is that PHA accumulation becomes exhausted during low loads. When influent loads suddenly increase, the depleted PHA reserves prevent phosphorus uptake to the extent required to match the increased influent phosphorus loads (Brdjanovic et al., 1998a). One approach to mitigate this delayed response to the low load-high load transition is to provide influent equalization (Filipe et al., 2001c). If this is implemented, care should be taken in the design to avoid appreciable depletion of VFAs in the contents of the equalization tank.

Depending on the effluent phosphorus and the effluent ammonia requirements, influent equalization could be beneficial.

- Most EBPR systems also are required to achieve nitrification, which must be optimized first because it is the controlling process. Next, the EBPR capability can be maximized by removing process and operational bottlenecks and considering chemical addition, if required. This approach will reduce chemical use although enhancing phosphorus removal reliability.

- The anaerobic zone should be adequately sized to accommodate PAO selection, VFA production, and RAS denitrification (in nitrifying EBPR systems). When ORP levels in the anaerobic zone are not low enough, extensive HRTs in this zone can be detrimental with secondary release occurring in some cases. However, longer HRTs are beneficial when ORPs can be maintained below −300 mV (Barnard et al., 2016).

  Excessive secondary influent flows should be allowed to bypass the anaerobic zone, to the anoxic zone, to avoid diluting the anaerobic zone with higher ORP flow, particularly at times of low influent VFAs. By the same token, splitting the RAS return flow so that only part of it is returned to the anoxic zone, with the rest to the anoxic zone, increases the HRT of the anaerobic zone and reduces the mass rate of nitrate returned with the RAS (Barnard et al., 2016).

  These concepts, influenced by James L. Barnard, were implemented in the Brasilia South and North facilities in 1992 in Brazil. Figure 12.32 shows a schematic of the EBPR reactors. RAS flow is divided between the anaerobic and anoxic zones. Primary effluent flow is divided between the anaerobic, anoxic, and final aeration zones. In general, flows higher than the maximum month are diverted away from the anaerobic zone, mostly to the anoxic zone. During storm events, excessive flows are diverted to the final aeration zone for contact treatment thereby alleviating the solids loading to the secondary clarifiers. In the main, diversion of flows is passive, using rectangular (cutthroat) flumes or

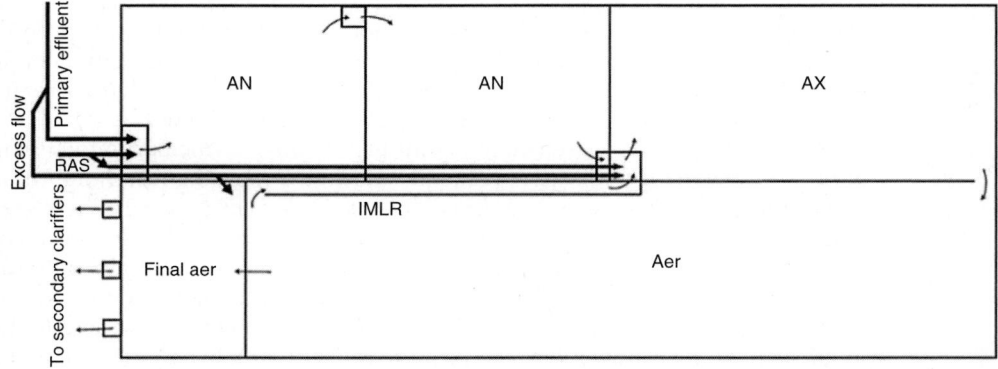

FIGURE 12.32 Example schematic of EBPR reactor incorporating split flows to maintain integrity of anaerobic zones and encourage VFA production (Bratby, 1984).

side weirs. Measuring devices at the flumes and a number of gates allow monitoring and adjustments, if required (Bratby 1984).

- Structures should be designed to achieve even flow split. Uneven flow distribution can cause operational challenges and lead to inefficiencies. For example, the improved performance of underloaded clarifiers typically cannot compensate for reduced performance of overloaded units.

- Design conditions that entrain air upstream of the reactor (e.g., unflooded screw pumps, free-fall weirs, turbulence, etc.) should be avoided.

- The different densities of process influent and return sludge requires careful design of the intimate blending/mixing of these two streams. Poor mixing will reduce contact duration between organisms and substrate. This could potentially lead to reduced VFA production in the anaerobic zone and lower EBPR efficiency.

- Strategically placed and properly designed baffles can enhance EBPR performance, while minimizing surface oxygenation.

- While primary clarifiers remove solids and increase the active biomass fraction of the MLSS, excessive BOD removal in these units can deprive EBPR of biodegradable substrate.

- The A/O swing cells can be considered if significant influent load fluctuations are anticipated.

- Access to waste sludge from the end of the aeration zone needs to be provided. The phosphorus content of the biomass would be highest at this point. In addition, this will keep the sludge fresh and prevent/minimize secondary phosphorus release. This wasting strategy will also allow tighter SRT control. However, since this will result in greater volumes of waste sludge, dissolved-air flotation (DAF) thickening of WAS would be advantageous in this case. Mixed-liquor wasting also greatly facilitates SRT control.

- Strategies to enhance settleability and minimize foaming also can be incorporated. The causes and control of filamentous growth are provided elsewhere in this chapter.

- The following conditions that favor phosphorus release should be avoided if possible:
  - Mixing and storing primary and secondary sludges;
  - Co-settling EBPR sludge in the primary clarifier;
  - Septic conditions in final clarifiers due to deep sludge blanket;
  - Anaerobic or aerobic digestion of primary and EBPR waste sludge; and
  - Unaerated storage or long aerated storage of EBPR sludge.

- Recycle streams from sludge processing operations could impose significant additional nutrient loadings, overwhelming the EBPR process. The magnitude of the problem is dependent on the type of sludge processing and handling operations. The effect of recycle streams could be minimized by:
  - Equalizing recycle flows;
  - Scheduling sludge processing/conditioning operations;
  - Treating the sidestreams with chemicals to precipitate phosphorus; and
  - Recovering phosphorus through struvite precipitation.

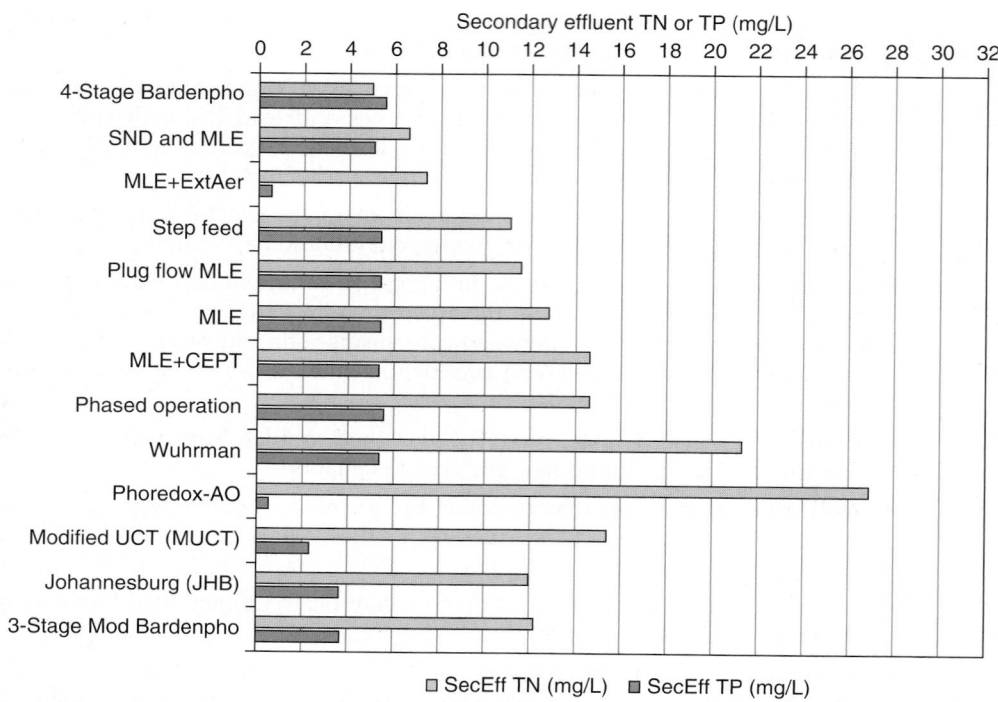

FIGURE 12.33 Summary of relative performances of alternative BNR configurations (Bratby et al., 2012a).

4.2.3   Process Configurations for Enhanced Biological Phosphorus Removal

Figure 12.33 compares the nitrogen and phosphorus removal performances of a number of configurations, based on one set of actual wastewater characteristics, generally characterized as carbon limited (Bratby et al., 2012a). The results shown were obtained using a proprietary process simulator. The secondary effluent values are specific to the wastewater and should not be taken as general. However, the *relative* performances are probably characteristic of the respective process configurations.

All of the configurations included primary clarifiers, except the (MLE+ExtAer) configuration, which modeled an MLE process without primary clarifiers. The excellent results achieved are explained by the reduction in the mass of biodegradable COD through primary clarifiers, which reduces the fermentable COD. In this case, despite the absence of an anaerobic zone in the MLE process, the retention time was sufficient to cause fermentation of readily biodegradable COD. Although the overall P removal per unit secondary influent COD is higher for settled wastewater, the overall P removal is generally higher per unit influent COD for unsettled wastewater.

The following sections describe selected EBPR process configurations. Most variations of nitrogen and phosphorus removal systems are designed to minimize the return of oxygen and nitrate/nitrite to the anaerobic zone.

4.2.3.1   Anaerobic/Oxic (Phoredox and A/O) System   The A/O process was first developed in the 1970s as the "Phoredox" system (Barnard, 1976) and later patented in the

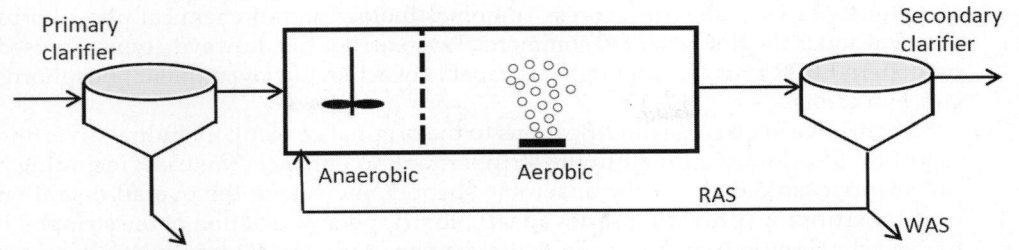

FIGURE 12.34   Schematic of Phoredox (A/O) process.

United States in the early 1980s as the A/O process. It entails a simple process configuration consisting of an anaerobic zone followed by an aerobic zone, shown schematically in Figure 12.34. Typically, the anaerobic zone HRT is between 30 and 45 minutes to select for PAOs. Longer anaerobic HRTs often are required to promote fermentation in the zone to improve performance. The A/O configuration can be used with any type aerobic reactor and over the full range of aerobic SRTs.

In general, there has been mixed success with A/O EBPR facilities due to the difficulty of avoiding nitrification and the return of nitrates to the anaerobic zone with the RAS.

4.2.3.2 PhoStrip Process   The PhoStrip process, illustrated in Figure 12.35, combines biological and chemical phosphorus removal. It diverts part of the phosphorus-rich RAS (typically 10% to 30% of influent flow) to an anaerobic stripper where sludge settles and phosphorus is released. The phosphorus-rich stripper supernatant is then precipitated with lime, while the biomass, stripped of phosphorus, returns to the

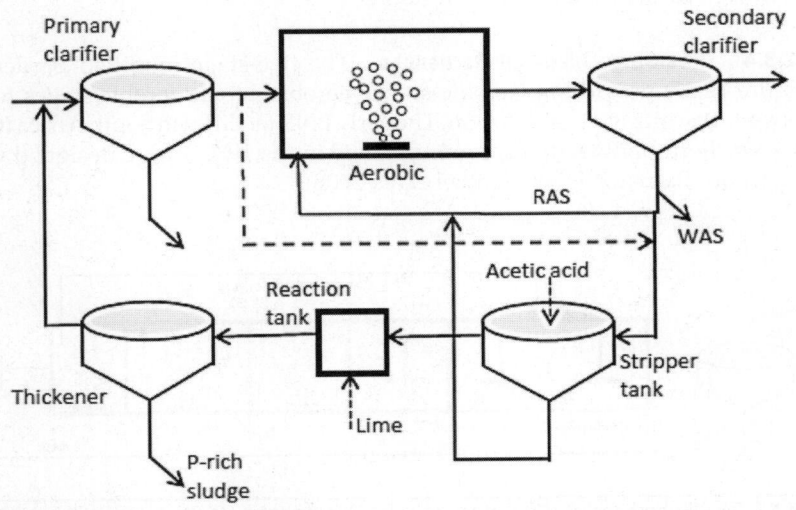

FIGURE 12.35   Schematic of PhoStrip process.

aeration tank. The PhoStrip process combines biological and chemical phosphorus removal and is the first patented commercial system. It is not, however, typically used in modern EBPR facilities although it is capable of achieving low effluent phosphorus concentrations.

There have been some modifications to the original concept, including diverting a portion of primary effluent to the stripper tank to enhance P release; including a pre-stripper tank ahead of the anaerobic stripper, increasing the overall detention time for stripping; providing series anaerobic strippers; elutriation of the stripped P by recycling settled sludge around the stripper tank; and incorporating anoxic zones for denitrification upstream of the aerobic reactors. When nitrification occurs, without a formal denitrification zone, the pre-stripper tank would accept underflow from the secondary clarifier containing nitrates. Stripper underflow typically has high concentrations of soluble BOD and would provide the carbon source for denitrification. The pre-stripper tank hydraulic detention time is approximately 2 hours. Up to 70% denitrification has been observed (Kang, 1988; Matsch and Drnevich, 1987).

4.2.3.3 Three-Stage Modified Bardenpho and Anaerobic-Anoxic-Oxic System
Figure 12.36 shows the flow schematic of a typical three-stage anaerobic-anoxic-oxic (A^2/O) process. Mixed liquor is recycled from the end of the aerobic stage to the anoxic stage for denitrification at an internal recycle rate typically ranging from 100% to 400% of the influent flow. Clarifier underflow returns to the first stage of the anaerobic reactor with the reactor feed. This process was implemented in the Bushkoppies WWTP in Johannesburg, commissioned in 1985 (Pitman et al., 1982, 1988). Table 12.12 shows operating data with typical influent sewage, and with the discharge of a yeast factory effluent to the incoming sewer. This waste likely fermented in the sewer and in the anaerobic zone to form VFAs. The results demonstrate the importance of influent wastewater characteristics. The impact of nitrates recycled back to the anaerobic zone is also apparent from the results since relative nitrate concentrations in the RAS are reflected by secondary effluent nitrate concentrations.

4.2.3.4 Five-Stage Modified Bardenpho
The five-stage modified Bardenpho process (Figure 12.37) provides anaerobic-anoxic-aerobic-anoxic-aerobic stages for removal of phosphorus, nitrogen, and carbon. The early BNR facilities in South Africa (Goudkoppies and Northern BNR facilities, commissioned in the 1970s) were designed with this configuration (Barnard, 1976; Wentzel et al., 2008).

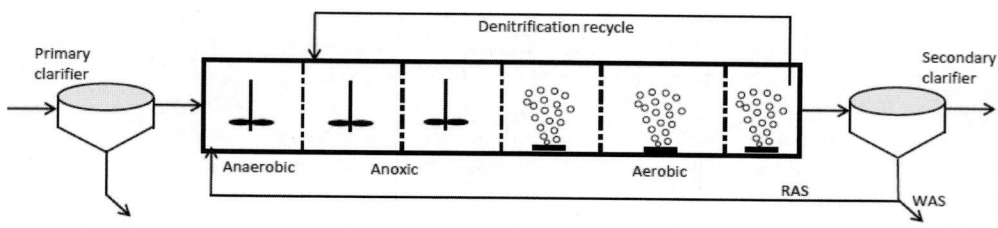

Figure 12.36   Schematic of three-stage modified Bardenpho (A^2/O) process.

Operating Condition	Location	Temp (°C)	COD (mg/L)	Total P (mg/L)	PO$_4$-P (mg/L)	Nitrate (mg N/L)	TKN (mg N/L)	Ammonia (mg N/L)	TSS (mg/L)
No primaries	Secondary influent	22	500	9.0	5.4	0	42	26	230
	Secondary effluent	23	38	2.0	1.6	12	1.5	0.4	8
No primaries+	Secondary influent	22	640	8.0	4.2	0	36	23	330
Yeast factory effluent to sewer	Secondary effluent	23	53	0.7	0.3	8.6	1.4	0.5	7

TABLE 12.12 Performance of the Bushkoppie Facility Under Different Conditions (Pitman et al., 1988)

917

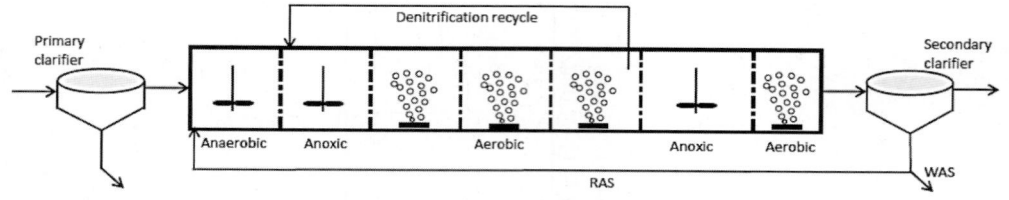

FIGURE 12.37 Schematic of five-stage modified Bardenpho process.

Table 12.13 summarizes the basic design and performance information for several five-stage facilities that have been in operation for several years. The Palmetto, Florida, WRRF, which began operation in October 1979, was the first in the United States to use this process (Burdick and Moss, 1980; Stensel, 1980). Most of these facilities have required supplemental chemical addition (metal salts and/or carbon) to meet effluent TP limits of less than 1.0 mg/L. Facilities using this process use a variety of aeration methods, tank configurations, pumping equipment, and methods of solids handling.

4.2.3.5   University of Cape Town and Virginia Initiative Plant Processes   Researchers at the University of Cape Town developed the UCT and modified UCT (MUCT) processes shown in Figures 12.38 and 12.39. In the UCT process, both the RAS and aeration tank contents are recycled to the anoxic zone, and the contents of the anoxic zone are then recycled to the anaerobic zone. This recycle sequence decreases the amount of nitrates returned to the anaerobic zone. The internal recycle can be controlled to maintain near-zero nitrates in the effluent from the anoxic reactor, thereby ensuring that little nitrate will be returned to the anaerobic reactor. The Virginia Initiative Plant (VIP) process is very similar to the UCT process (Daigger et al., 1988; Grady et al., 2011).

The anaerobic recycle (anoxic to anaerobic zone) is typically around 100% of the influent flow, and the actual anaerobic HRT is approximately twice the HRT of an A/O system. This is because in the UCT process, biomass is conveyed to the anaerobic zone via a mixed-liquor recycle rather than RAS. Therefore, to maintain the required anaerobic SRT, larger anaerobic volumes are required because of the dilution effect of the anaerobic recycle.

In the MUCT process, the anoxic zone is divided into two reactors. RAS enters the first reactor, and internal recycle from the aeration tank enters the second anoxic reactor. Internal recycle to the anaerobic zone comes from the first anoxic reactor. The MUCT process was intended to eliminate nitrate recycle to the anaerobic tank while limiting the actual HRT in the anoxic zone to one hour.

For wastewaters with low influent BOD:P ratios, the UCT processes can achieve both phosphorus removal and partial nitrogen removal up to 6 to 8 mg/L. It is reported that near-zero nitrate recycle can be maintained for TKN:COD ratios up to 0.14 (Ekama et al., 1983).

4.2.3.6   Johannesburg and Westbank Processes   The Johannesburg (JHB) process, shown in Figure 12.40, also was developed in South Africa (Nicholls et al., 1987). The distinguishing feature of this process is the RAS denitrification zone before the

TABLE 12.13 Basic Design Information for Existing Modified Bardenpho™ Systems

Parameter	Palmetto, Florida	Kelowna, British Columbia, Canada	Orange County, Florida (phase III)	Fort Myers, Florida Central	City of Cocoa, Florida	Tarpon Springs, Florida	Johannesburg, South Africa (Goudkoppies)
Design flow, m³/d	5300	22 500	28 400	41 600	17 000	15 100	150 000
Final effluent standards							
Total nitrogen, mg N/L	3	6	3	3	7.3	6.3	3
Total phosphorus, mg P/L	1	2	1	0.5	0.7	3.1	1
Aeration mode	Submerged turbine	Submerged turbines	Carrousel™—mechanical surface aerators	Carrousel™—mechanical surface aerators	Fine-bubble	Carrousel™—mechanical surface aerators	Mechanical surface aerators
Primary settling	Yes	Yes	No	No	No	No	Yes
Sludge handling	Anaerobic digestion for primary sludge	DAF[b] thicken WAS[c]—land application gravity thicken primary	Belt press dewatering—landfill	Belt press dewatering—landfill	Belt press dewatering—landfill	Belt press dewatering—landfill	Anaerobic digestion of primary sludge
Anaerobic volume, m³	228	1760	2271	3680	1440	1500	6240
First anoxic volume, m³	614	3520	4013	3500	1500	2300	14 400
Aeration volume, m³	1060	7920	12 643	18 900	6000	5800	44 100
Second anoxic volume, m³	496	1760	2196	4100	1300	2600	14 400
Second aeration volume, m³	228	2640	379	1000	260	200	8100
Total volume, m³	2626	17 600	21 501	31 200	10 500	12 300	87 240

Parameter	Palmetto, Florida	Kelowna, British Columbia, Canada	Orange County, Florida (phase III)	Fort Myers, Florida Central	City of Cocoa, Florida	Tarpon Springs, Florida	Johannesburg, South Africa (Goudkoppies)
Clarification surface area, m²	230	1960	2100	3220	2100	930	12 350
Filtration	Yes	Yes	Yes	No	Yes	Yes	
Chemical addition							
Type	Alum	Alum	Alum	Alum	Alum		
Dose	90 gpd[a]		20–40 mg/L	50 mg/L			
Internal recycle							
Type	Vertical axial flow	Vertical axial flow	Horizontal axial flow	Vertical turbine	Horizontal axial flow	Vertical turbine	Archimedes screw
Ratio	4:1	4–6:1	4–6:1	4:1	4–6:1	4:1	4.5:1 to 18:1

[a]gpd = 3.785 L/d = 0.0038 m³/d.
[b]DAF = dissolved air flotation.
[c]WAS = waste activated sludge.

TABLE 12.13 Basic Design Information for Existing Modified Bardenpho™ Systems (Continued)

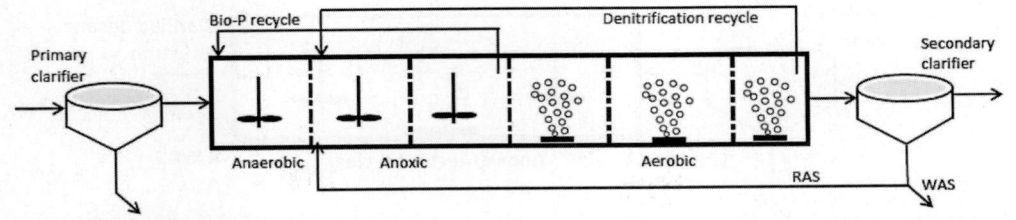

FIGURE 12.38   Schematic of UCT (VIP) process.

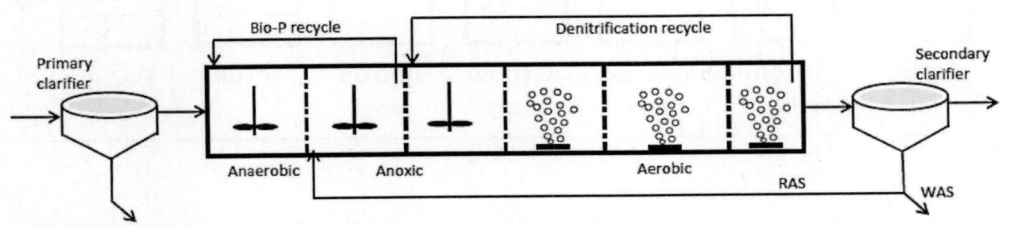

FIGURE 12.39   Schematic of modified UCT process.

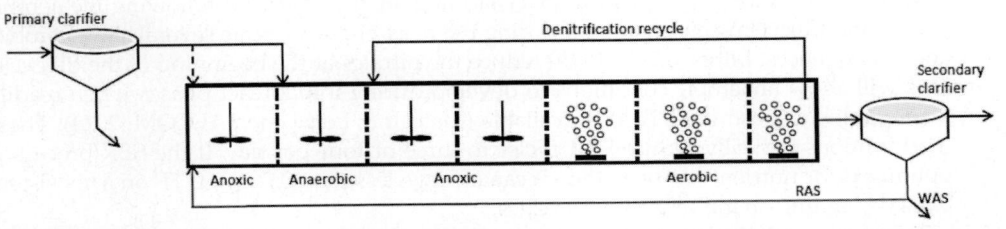

FIGURE 12.40   Schematic of Johannesburg process. Dotted line signifies the modification for the Westbank process.

anaerobic zone. The concept behind this process is that endogenous respiration within the RAS provides the carbon needed to denitrify the RAS before it enters the anaerobic zone. This process has three primary benefits when compared to other combined removal systems: (1) anaerobic zone mixed liquor is at full concentration; (2) using endogenous respiration for nitrogen removal does not require carbon from the feed, thus resulting in efficient carbon usage; and (3) denitrified mixed-liquor recycle stream is eliminated.

A modification of the JHB process resulted in the Westbank Process (Oldham and Rabinowitz, 2001; Stevens et al., 1999). This process aims to improve denitrification rate by adding either a small fraction of the process influent or supplemental carbon to the RAS denitrification zone.

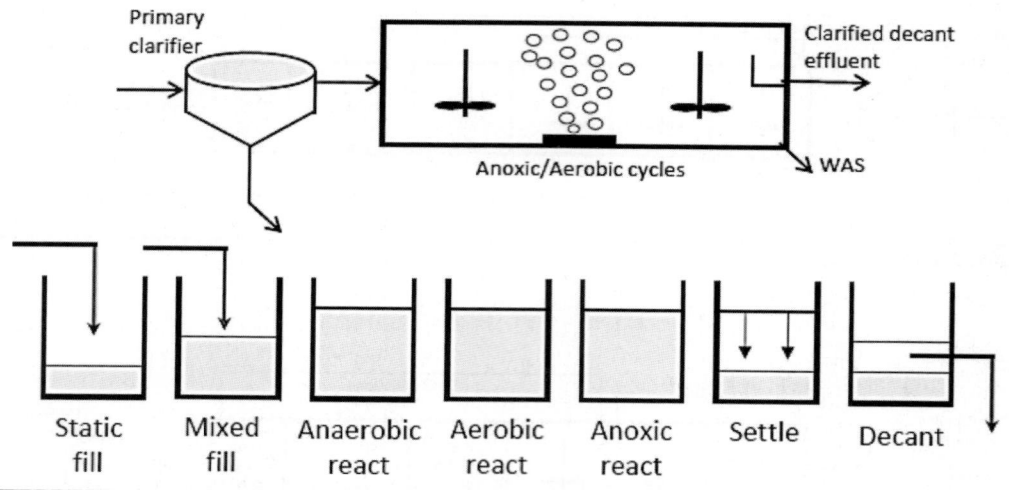

FIGURE 12.41   Schematic of sequencing batch reactor type processes.

4.2.3.7 Time-Cyclic Processes   The SBRs can be operated to achieve combined carbon and nitrogen oxidation, nitrogen removal, and phosphorus removal by controlling the sequence and duration of cycles as shown in Figure 12.41 (Ketchum and Liao, 1979). Phosphorus removal is achieved by depleting the nitrate produced during the aerobic phase. Two ways in which this can be accomplished are (1) adding an anoxic period following the aerobic period and (2) cycling air on and off during the react phase to create several short aerobic-anoxic sequences. Either method will reduce the nitrates at the beginning of the fill cycle. This will allow anaerobic conditions to develop during initial react phases when readily biodegradable substrate (VFAs) is available (Metcalf & Eddy, Inc./AECOM, 2014). These modifications typically reduce SBR cycles to three or four per day. If the SBR process is optimized for nutrient removal, then it can achieve less than 0.7 mg/L TP on a consistent basis depending on influent characteristics.

The time-cyclic, phased isolation ditch process configuration developed in Denmark uses a pair of oxidation ditches operated in an alternating mode to achieve EBPR. Phased isolation ditches and alternating aeration processes can be configured for combined nitrogen and phosphorus removal by adding an initial anaerobic zone for EBPR. While it may be possible to maintain informal but localized A/O zones within an oxidation ditch or a similar looped reactor by carefully controlling dissolved-oxygen levels, the most typically implemented EBPR configuration includes an external anaerobic basin for PAO selection followed by the oxidation ditch where phosphorus uptake occurs (WEF, 2013).

4.3   Adding External Carbon to Suspended-Growth Nutrient Removal

Carbon availability for BNR is a significant issue when facilities are tasked with meeting stringent nutrient limits. Readily biodegradable carbon sources such as VFAs are crucial for the EBPR process. Wastewater-derived VFA sources include:

- Collection system. If collection systems have long detention times and relatively warm conditions, then fermentation occurs resulting in the conversion of readily biodegradable organics to VFAs. This is one of the most common sources of VFAs.

- Anaerobic zone of the EBPR tank. Often, the anaerobic zone is sized to facilitate fermentation and enhance the VFA pool.

- Off-line sludge fermentation. It is possible to generate VFAs by fermenting primary and/or WAS solids in an off-line tank.

- Primary clarifiers. Dedicated primary clarifiers operating at long SRTs (active primaries) can generate VFAs. Primary sludge fermentation occurs within the accumulated sludge blanket releasing VFAs into the supernatant.

- Facility recycle. Supernatant from primary sludge gravity thickening is also a potential source of VFAs.

If the wastewater-derived VFA content is insufficient, a supplemental carbon source could be added. Supplemental carbon can be added to the main feed stream or as a separate stream directed to a specific zone. The external carbon source can be a separate stream imported into the treatment facility, such as methanol for denitrification and acetic acid to the anaerobic zone to sustain EBPR, or industrial waste products could be used. The carbon source could also be internally generated within the facility, such as fermentation of primary sludge, or of mixed liquor. Some external carbon types, such as methanol and ethanol, are highly inflammable. Facilities for handling these chemicals must be designed according to appropriate fire and safety codes.

Acetic acid (CH_3COOH) typically is available as 100% (glacial), 56%, or 20% solutions. Unless dilute solutions significantly less than 84% (nearing the properties of water) are used, the design of acetic acid storage facility must include freeze-protection measures. Glacial acetic acid storage would most likely require provisions for heating and, in warm climates, it may be necessary to consider an inert gas blanket or floating cover because of the low flash point. Storage tanks, piping, and appurtenances must be corrosion resistant and the facilities must meet all applicable code requirements.

While research on pure substrate has shown that acetic acid is associated with the highest phosphorus release, continued use of this carbon source can lead to the proliferation of GAOs that compete with PAOs and reduce EBPR efficiency. For this reason, a mix of acetic and propionic acid often is recommended. Additional discussion of the effects, sources, and generation of carbon are presented later.

Because of the expense of adding pure chemicals such as acetic acid, some facilities have considered industrial wastes as supplemental carbon sources. These could be used directly, typically for denitrification, or after some fermentation to produce VFAs for EBPR. These sources include manufacturing waste streams from sugar, molasses, brewery, dairy (typically whey), and acetic acid solutions from some pharmaceutical manufacturers. When using such sources, it is important to ensure they are free of contaminants and the supply is reliable. Published data provide typical characteristics as well as kinetic and stoichiometric coefficients for different carbon sources (deBarbadillo et al., 2008; Onnis-Hayden and Gu, 2008; Sigmon et al., 2014). Abu-garrah and Randall (1991) researched the effect of several organic substrates on biological phosphorus removal. Ratios of phosphorus uptake per COD used and COD used per milligram per liter of phosphorus removed are summarized in Table 12.14. This work suggests that acetic acid is the most effective chemical substrate for biological phosphorus removal enhancement.

Substrate	mg/L Phosphorus Uptake[b]/ mg/L COD Used	mg COD Used[c]/mg P Removed
Formic acid	0	Infinity
Acetic acid	0.37	16.8
Propionic acid	0.10	24.4
Butyric acid	0.12	27.5
Isobutyric acid	0.14	29.1
Valeric acid	0.15	66.1
Isovaleric acid	0.24	18.8
Municipal wastewater	0.05	102[d]

[a]An SRT of 13 days was used for all experiments.
[b]Total phosphorus uptake in aerobic zone.
[c]COD used and phosphorus removed in total system.
[d]Value obtained with highly aerobic wastewater.

TABLE 12.14 Effect of Organic Substrate on Enhanced Biological Phosphorus Removal (Abu-garrah and Randall, 1991)[a]

Adding external carbon benefits nutrient removal when there is inadequate carbon in the native process feed to serve either as an electron donor for nitrogen removal or as a source of VFAs to drive biological release, and subsequent uptake, of soluble phosphorus.

In nitrogen removal, when there is limited carbon, external carbon addition improves denitrification by serving directly as the electron donor for the reaction. In many cases, carbon can be dosed based on stoichiometric relationships. In contrast, the effects of external carbon on phosphorus removal, as discussed earlier, are not as direct. There are two pathways by which external carbon addition enhances biological phosphorus removal:

1. If the external carbon feed contains VFAs and it is added to a zone with anaerobic conditions, then external carbon will act as a carbon source for PAOs, thus improving overall biological phosphorus removal.

2. If the external carbon source does not contain VFAs or is added to an anoxic zone, then primary use is likely for denitrification, reducing nitrate/nitrite levels within the reactor and the amount that will be returned to the anaerobic zone by internal recycle streams.

In addition, if the carbon source contains fewer nutrients than are required by biological growth, as is the case for methanol, additional nutrient removal will be achieved through microbiological growth incorporating the nutrient into the biomass (assimilatory pathway).

There are two other, potentially negative effects. The first effect is increased energy usage because of increased aeration demands. In nitrogen removal, most carbon will be used up in the anoxic zones. However, it is not uncommon to have some bleed-through of carbon into the aerobic zones, thus increasing aeration demands. Endogenous respiration of additional biomass grown on the external carbon also increases aeration demands. In phosphorus removal, addition of VFAs directly increases demands. This is a result of storage of VFAs as poly-β-hydroxyalkanoates (PHAs) in the anaerobic zones and its subsequent metabolism in the aerobic zones. Thus, while VFAs are added in the anaerobic zones, their oxygen demand is expressed in the downstream aerobic zones.

The second effect is that adding carbon will increase the amount of biomass, which decreases secondary treatment capacity and puts additional load on the solids-handling system.

4.3.1 Dosage Locations
For the purposes of nutrient removal, there are three significant locations in a suspended-growth process where external carbon can be beneficially added:

4.3.1.1 Process Feed
External carbon can be added directly to the wastewater stream feeding the process. This can occur either directly before the reactor or in one of the upstream processes. However, if the goal of supplemental carbon addition is to reduce the amount of nitrate from a biological nitrogen and phosphorus removal process, then it would be most beneficial to add the external carbon directly to the anoxic zone and ensure there is adequate mixing.

Since primary sludge fermentate is typically odoriferous, adding it at the headworks or the primary clarifiers could increase odors to be treated at the facility.

4.3.1.2 Anoxic Zone Feed
Feeding external carbon directly to anoxic zones can be beneficial to both nitrogen and phosphorus removal, when the zones have inadequate carbon. By reducing the nitrate levels recycled to the anaerobic zones, overall biological phosphorus removal is improved.

The external carbon addition system needs to be designed to maximize distribution of the external carbon source within the anoxic zone and minimize any short-circuiting to the inlet of the downstream aerobic zone. These goals can be achieved by a combination of locating the external carbon discharge point near the inlet to the anoxic zone mixer or by distributing the carbon across the anoxic zone inlet, and by designing multiple anoxic zones in series, with the carbon added to the first anoxic zone. The use of "chimney baffles" is an ideal feature for carbon addition.

4.3.1.3 Anaerobic Zone Feed
If process influent does not contain adequate VFAs to support the needed biological phosphorus removal levels, then the addition of VFAs to the anaerobic zones typically will improve biological phosphorus removal. Similar to external carbon addition to anoxic zones, good mixing is important.

Most VFA streams added to anaerobic zones include either pure acetic acid or a mixture of acetic and propionic acids. An external carbon feed of pure acetic acid could eventually promote the growth of GAOs, which compete directly with PAOs and disrupt the EBPR process. A mixture of acetic and propionic acids, typically found in the fermentate from primary sludge fermenters, provides the best results for EBPR.

It is possible for too much VFA to be added to anaerobic zones (Johnson et al., 2006; Neethling et al., 2005). The GAOs will use any VFA in excess of what is needed to achieve required effluent phosphorus levels, thus improving their competitive position in the mixed liquor. Dosing of VFA to anaerobic zones should, therefore, be carefully controlled to the minimum needed to reliably achieve effluent phosphorus goals.

4.3.2   Primary Sludge Fermentation

Anaerobic fermentation of primary sludge results in conversion of organic particulate material into soluble VFAs, including acetic, propionic, and butyric acids. The amount of VFAs formed depends on influent (and primary sludge) characteristics, temperature, and the design of the fermenter, particularly the hydraulic and SRTs provided in the fermenter. The benefit of using fermentate is that, once the capital investment is made, fermentation has a low operating cost compared to purchasing an external carbon source.

A characteristic of many primary sludge fermenters is the return of considerable suspended solids with the fermentate. This represents a significant and unwanted burden for BNR activated sludge reactors and for secondary clarifiers. The high suspended solids in the fermentate returned to liquid stream processes are often due to the general preference of using gravity thickening for solids-liquid separation of the fermented primary sludge. However, settling of the fermented primary sludge is hampered by the colloidal nature of the fermented solids and the inevitable gases produced during the fermentation process.

An added benefit of controlling the fermentate TSS to low levels is that more precise control of the fermenter SRT is possible. This is particularly important when operating SRTs are lower, at around 1.5 days for example.

Another design issue is water-temperature sensitivity of the fermentation process. Colder water typically will reduce the amount of VFAs created in the fermentation process. Therefore, the operating SRT of the fermenter needs to be adjusted depending on the temperature within the fermenter.

Different primary sludge fermentation configurations are described below.

4.3.2.1   Activated Primary Sedimentation
Activated primary sedimentation uses primary clarifiers to accumulate and ferment primary sludge (Figure 12.42). Primary sludge is allowed to accumulate in the primary tanks with a residence time in the blanket of 1 to 3 days. Approximately 50% of the underflow sludge is recycled back to the inlet to the clarifiers to elutriate the VFAs, thereby passing them to the primary effluent and downstream biological processes (Pitman et al., 1988). Although no additional unit processes are required for this approach, it can result in high solids loading rates to the primary clarifiers. It also can be difficult to control the SRT and HRT of the fermenting

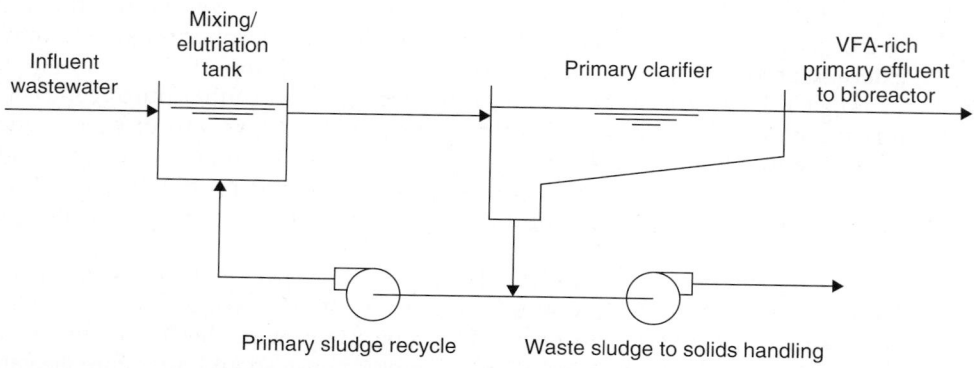

FIGURE 12.42   Activated primary sedimentation tank (Rabinowitz et al., 2011).

sludge with an increased potential for methane and sulfide formation, which reduces the net VFA production. As with all fermenters and sludge thickeners, fibrous materials can accumulate and the sludge collection mechanism and tank depth must accommodate a deep and thicker sludge blanket. Primary effluent TSS will generally deteriorate when primary clarifiers serve the dual purposes of clarification and fermentation, which would increase the solids load to secondary processes.

4.3.2.2 Static Fermenter Instead of operating primary clarifiers as fermenters, an alternative is to operate the primary clarifier normally, but concentrate the primary sludge in a gravity thickener. The sludge age within the blanket of the gravity thickener is controlled to provide fermentation of the thickened sludge (Oldham and Stevens, 1984). Primary sludge pumped to the fermenter displaces supernatant that is returned to the main process. Elutriation water also can be added. Thickened sludge is transferred for subsequent processing, and the SRT is controlled by blanket depth in relation to the primary sludge withdrawal rate. The thickener depth and sludge-removal mechanism need to accommodate the thicker and deeper sludge blanket required for the necessary SRT. A disadvantage of this system is that a significant part of the VFAs that develop within the blanket are passed to sludge processing and are not used in the BNR process. Recycling of part of the thickened sludge to the inlet of the thickener could improve capture of the VFAs.

A more efficient approach utilizing static fermenters is the unified fermentation and thickening (UFAT) process developed at the Durham Advanced Wastewater Treatment Plant (AWWTP) owned by Clean Water Services in Oregon, described by Baur et al. (2002) and shown schematically in Figure 12.43. The process consists of two thickeners in series. The first gravity thickener receives primary sludge. An appreciable sludge blanket is developed in this first gravity thickener, and it is here that the majority of fermentation takes place. The fermenter SRT is controlled by varying the solids pumping rate. Supernatant and thickened/fermented sludge are both then passed to a second gravity thickener for solid-liquid separation. The supernatant from this second thickener comprises the fermentate that is returned to the liquid stream process. The thickened fermented sludge is pumped to the anaerobic digesters for further processing.

Data reported at the Durham AWWTP for the UFAT fermentation process during pilot trials measured specific VFA generation rates that ranged from 0.05 to 0.11 kg VFA/kg VSS for SRTs of 1.6 to 5.5 days and temperatures approximately 20°C (Baur et al., 2002). The specific VFA generation rate is dependent on the temperature and the effective SRT of the fermenting solids, but from operating data at the Durham AWWTP, the average appears to be approximately 0.025 kg VFA/kg VSS. Operating data show that the suspended solids content in the fermentate returned to the BNR reactors varies from approximately 420 to 900 mg/L, with an average of 560 mg/L. This relatively wide variation in fermentate suspended solids demonstrates the difficulty of efficiently settling fermented solids. Despite avoiding methane formation during acid fermentation, some gas is evolved within solids floc particles that retard settling velocities and exacerbate high effluent suspended solids levels (Bratby et al., 2012b).

4.3.2.3 Complete-Mix Fermenter In complete-mix fermentation, sludge is fed to a completely mixed tank that overflows to the inlet of primary clarifiers, or to gravity thickeners, or to some other means of solid-liquid separation. In the case of using primary clarifiers, sludge is wasted directly from the fermenter. A disadvantage is possible overloading of primary clarifiers. In this regard the use of gravity thickeners is an advantage.

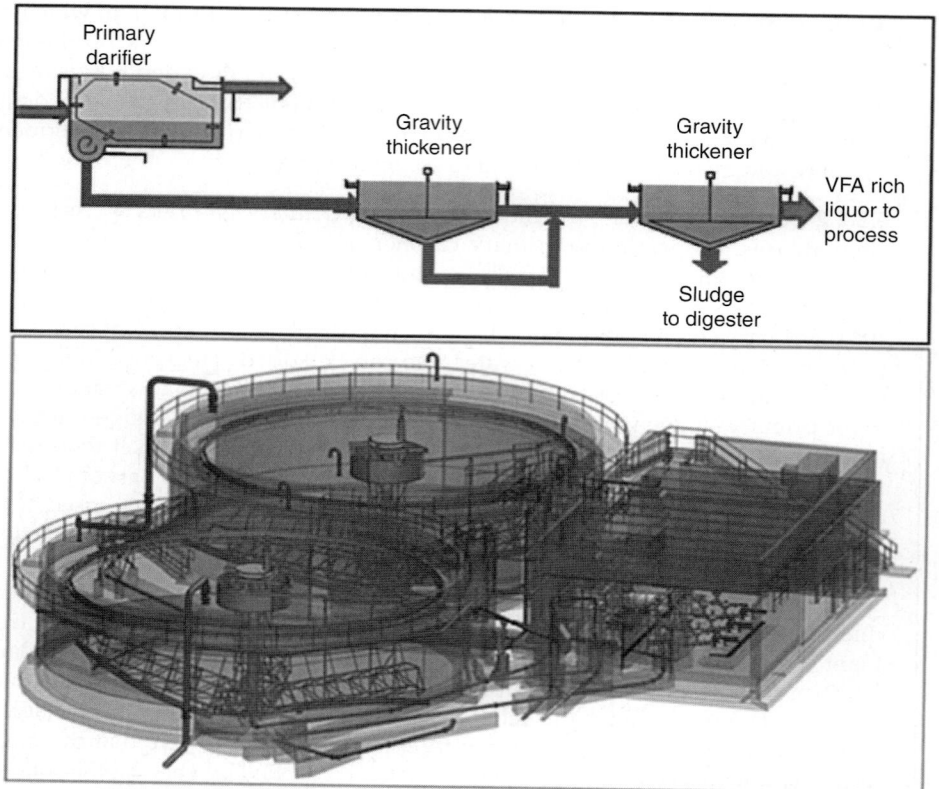

Figure 12.43 Schematic example of the UFAT static fermenter.

The first large full-scale application of using complete-mix fermenters with gravity thickeners was at the Bonnybrook WWTP in Calgary, Canada (Fries et al., 1994).

Another example of using complete-mix fermenters is at the Mason Farm WWTP owned by the Orange Water and Sewer Authority (OWASA) in Carrboro (Chapel Hill), North Carolina. The facility has operated a primary sludge fermenter for many years. Originally the fermenter was constructed within the shell of an old digester. An internal wall was constructed 6 feet from the old digester wall to form an annular space and an inner gravity thickener tank. Primary sludge was pumped from the primary clarifiers into the annular space of the fermenter. The annular space was completely mixed 20 hours per day using jet mixing pumps. This well-mixed suspension of fermenting sludge flowed from the annular space to the inner tank as fermentate was drawn out of the central gravity thickener.

An undesirable characteristic of the original fermenter design was high fermentate solids from the gravity thickener. Fermentate TSS concentrations returned to the activated sludge BNR process ranged widely from 500 mg/L to 5000 mg/L, with an average of approximately 2500 mg/L. Subsequent upgrades eliminated gravity separation of fermented solids and converted the whole volume within the fermenter to a complete-mix fermenter, using a jet mixing system. Solid-liquid separation of the fermented sludge is achieved using a gravity belt thickener (GBT). The fermenter HRT and SRT are directly based on the primary sludge feed rate, which lessens the potential for methane and sulfide production (Bratby et al., 2005).

Filtrate from the GBT is pumped to the anaerobic zones of the BNR activated sludge system, and thickened sludge is pumped to the anaerobic digesters. Although the design did not provide for separation of the GBT filtrate and washwater streams, typical TSS concentrations of the fermentate are approximately 280 mg/L. By separating the cleaner GBT filtrate from more heavily solids-laden belt washwater, TSS in the fermentate stream returned to the mainstream process could be reduced further, probably to less than 100 mg/L, thereby reducing even further the impacts of fermentate TSS on the mainstream process. A schematic of the fermenter configuration at the OWASA facility is shown in Figure 12.44a. Specific VFA generation values for the OWASA fermenter have been recorded at about 0.16 kg VFA/kg VSS (Bratby et al., 2012b).

An alternative to the OWASA configuration is to include a gravity thickener to thicken the feed to the fermenter, as shown schematically in Figure 12.44b. The advantage of this configuration is that primary sludge withdrawal can be independent of the fermenter

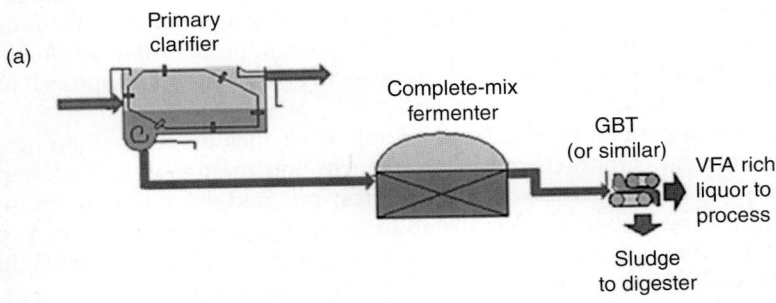

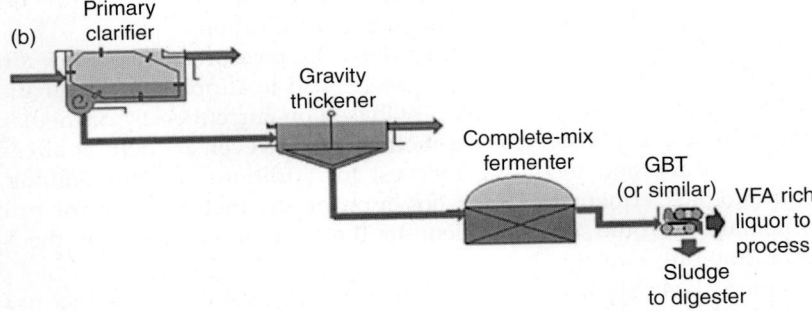

FIGURE 12.44 Complete mix fermenter concept (a) at the OWASA facility (Bratby et al., 2012b), and (b) with pre-thickening before the fermenter. In both cases clarification of the fermentate could be by gravity belt thickener (GBT) or rotating drum thickener (RDT) or similar.

operation, thereby maximizing primary clarifier performance. The HRT/SRT in the fermenter is dependent on the rate of sludge withdrawal from the gravity thickener. Therefore, the control system logic can be quite simple: depending on the temperature, the operator assigns a fermenter SRT that directly controls the gravity thickener underflow pump rate. Level detection in the fermenter then governs the rate of sludge withdrawal from the fermenter, to maintain a constant level. Methane detectors in the headspace within the fermenter tank signal the operator to adjust the SRT if required, to maintain the sludge in acid digestion mode.

4.3.2.4 Evaluation of the Feasibility of Primary Sludge Fermenters
In a given BNR facility (without altering the BNR configuration) the use of primary sludge fermentation to enhance EBPR is likely to be more cost-effective on a life-cycle cost basis than the use of coagulant chemicals to remove phosphorus, despite the higher capital costs of the fermenter (Rabinowitz and Fries, 2010).

Comparisons between various fermenter configurations, including the expected specific VFA production and the impacts of fermentate TSS on the size of facility processes are shown in Table 12.15. The underlying assumption is that facility capacity must be maintained equal to that without the return of gravity thickener and/or fermentate solids. If gravity thickeners already exist at a given facility where fermenters are to be installed, then the impacts of TSS return from fermenters may represent a lower impact than suggested by the results. The impacts of fermentate solids should be evaluated on a case-by-case basis. The results ignore differences in sludge production yields using fermentate, acetic acid, or methanol. From deBarbadillo et al. (2008) and Onnis-Hayden and Gu (2008) this appears to be a reasonable approach and the error is likely to be small.

For comparison purposes, the results also include corresponding costs for two supplemental carbon chemicals: acetic acid and methanol. A 20-year present worth analysis was used assuming a 7% interest rate and 4% inflation. From the results it appears to be cost-effective to install primary sludge fermenters when compared with the corresponding cost for acetic acid. However, this is only true when the fermenter is designed to maximize the specific VFA generation rate. It is evidently not cost-effective to install a fermenter and limit the specific VFA production by limiting the SRT in the fermenter. In this case, it is probably more cost-effective to use chemicals, even in the case of using acetic acid to enhance denitrification.

In many cases, for supplementing the carbon required to meet low nitrogen limits, methanol would be used instead of acetic acid to supplement denitrification, given the high cost of the latter. In this case, based on current costs for methanol, it would simply be less expensive to use methanol. However, if there is already sufficient reactor volume and including the cost for additional reactor volume required to accommodate fermentate TSS is not appropriate, then the cost for primary sludge fermentation would be equivalent to the cost of methanol, if the specific VFA generation rate is relatively high.

Figure 12.45 shows a compilation of results relating the solids recovery in the primary sludge fermenter process with the total unit cost of one particular fermenter system (based on the OWASA system without prior gravity thickening), in terms of dollars per pound per day of VFA generated, expressed as COD. The importance of the specific VFA generation is evident in this chart. On a cost basis, the fermenter only

TABLE 12.15 Analysis of Alternative Fermenters for a Specific Primary Sludge Quantity and Specific BNR Facility. Assumptions of Additional Reactor Volume Required to Maintain a Baseline MLSS of 3000 mg/L (Bratby et al., 2012b).

Process	Primary Sludge (lb/d TS)	(lb/d VS[d])	Assumed Primary Sludge Thickener TSS Recovery (%)	(lb/d)	Assumed Fermentate TSS Recovery (%)	(lb/d)	VFA Production (as COD) (lb VFA/lb VSS)	(lb VFA/d)	Construction Cost of Fermenter ($M)	Assumed SRT[g] Days	Capital Cost of Additional Aeration Basin Volume[h] ($/gal)	($M)	Total Unit Cost ($ per lb/d COD)	Operating Cost ($/year per lb/d COD)	20 yr Present Worth ($ per lb/d COD)	Total Present Worth Cost ($ per lb/d COD)
OWASA[a]	21 549	18 317	—	0	90.0	2155	0.160	2931	5.60	21	1.7	3.07	2960	0	0	2960
OWASA[b]	21 549	18 317	85.0	3232	90.0	323	0.160	2931	2.65	21	1.7	5.07	2635	0	0	2635
UFAT[c]	21 549	18 317	—	0	82.0	3879	0.025	458	0.43	21	1.7	5.53	13 025	0	0	13 025
UFAT	21 549	18 317	—	0	82.0	3879	0.025	458	1.98	21	1.7	5.53	16 412	0	0	16 412
Acetic acid[e]	—	—	—	—	—	—	—	—	—	—	—	—	1000	292	4380	5380
Methanol[f]	—	—	—	—	—	—	—	—	—	—	—	—	1000	58	876	1876

[a] Primary sludge without gravity thickening.
[b] Gravity thickened primary sludge.
[c] Utilizing existing gravity thickeners.
[d] Assumes VS of primary sludge = 85%.
[e] Acetic acid cost assumed $0.80/lb COD ($6.00 per gal 80% acetic at 896 000 mg COD/L).
[f] Methanol cost assumed $0.16/lb COD ($1.59 per gal methanol at 1 188 000 mg COD/L).
[g] Assumed 4-stage Bardenpho for nitrogen removal and maximum 30-day average winter SRT.
[h] Assumes that a MLSS of 3000 mg/L is maintained. Also ignores differences between sludge production from the soluble COD in fermentate, acetic acid, and methanol.

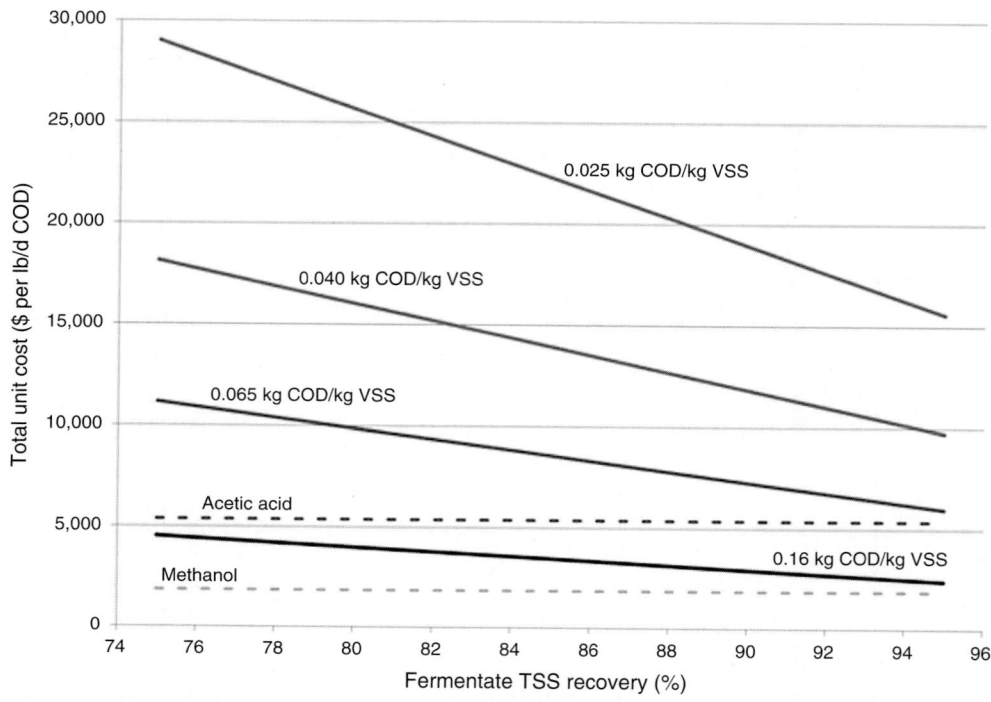

FIGURE 12.45   Impacts of solids recovery in one particular fermenter system on the unit cost of carbon generation. Acetic acid and methanol cost estimates included for comparison (Bratby et al., 2012b).

becomes competitive with supplemental carbon, particularly when compared with current methanol costs, when the specific VFA generation rate is at least 0.16 lb COD per lb feed VSS to the fermenter. At this specific VFA generation rate, the impact of fermentate TSS actually diminishes in importance.

When comparing the benefits of fermenters with supplemental carbon chemical addition, the concept of sustainability or greenhouse gas (GHG) emissions, particularly from methanol and acetic acid production, should also be considered. This would likely be a positive factor when deciding for implementation of a primary sludge fermentation system.

It is clear from Figure 12.45 that a fundamental parameter governing the feasibility of installing primary sludge fermenters is the specific VFA generation potential of the particular primary sludge. Therefore, prior to design, it is highly desirable to measure this parameter. Figure 12.46 shows a simple apparatus devised for use in most laboratories to conduct bench tests. It essentially comprises the following:

- Two-liter Erlenmeyer filtering flasks;
- Magnetic stirrers;
- Stirring bars;
- Rubber bungs and "bung-borer" to allow pipe inserts; and

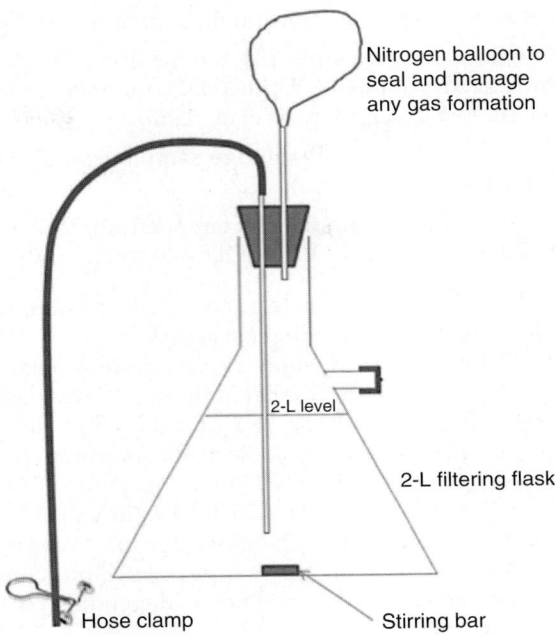

Nitrogen balloon to
seal and manage
any gas formation

2-L level

2-L filtering flask

Hose clamp   Stirring bar

FIGURE 12.46 Laboratory batch apparatus for determining specific VFA generation potential of
primary sludge.

- Balloon (loosely) filled with nitrogen gas to safely accommodate any gas
  formation and allow variations in volume during feeding and sample
  extraction.

The procedure used with this apparatus is as follows:

1. Set up three such sets of apparatus, each one at a different HRT: the first at 3 days
   HRT; the second at 4.25 days; the third at 6 days.
2. The experiment should continue uninterrupted for 3 weeks (21 days).
3. The corresponding daily feed volumes are 670, 470, and 330 mL, respectively.
4. Arrangements should be made to also feed sludge on the weekend.
5. To avoid fermentation of the primary sludge sample, a sample of primary
   sludge should be taken every day toward the end of the day, sufficient for
   feeding all three of the flasks, and left in the laboratory alongside the flasks (to
   come to the same room temperature), ready for feeding the next morning.
6. The experiments should be conducted at ambient (laboratory) temperature.
   Immediately after a sample is withdrawn from the flask, the pH and temperature
   should be measured.
7. Before adding the daily feed volume to the flask, an equal volume of fermented
   sludge should first be withdrawn from the flask for analysis. After the daily
   volume of feed, the volume within the flask would return to 2-L.

8. Analyses that should be done on the fermented sludge are:

   a. On whole sludge sample: pH; temperature; TSS; VSS.
   b. On filtered sample: COD; ffCOD; ammonia-N; and VFA, where ffCOD = floc-filtered COD (Mamais et al., 1993).

9. Before adding the primary sludge sample to each flask, a sample should be taken for the following analyses:

   a. On whole sludge sample: pH; temperature; TSS; VSS; TKN.
   b. On filtered sample: COD; ffCOD; ammonia-N; VFA.

Simplified VFA determinations (e.g., the Hach method, that can be done in most laboratories) are usually sufficient in most cases.

Figures 12.47 through 12.49 show typical results obtained with one particular primary sludge. Figure 12.47 shows that at the end of the experimental period, the pH is steady at approximately 5.1 at an SRT of 6 days and steady-state conditions were achieved. Figure 12.48 shows that volatile acid concentrations at an SRT of 6 days and at steady state reached approximately 1350 mg/L. Figure 12.49 shows that the specific VFA generation value increased from 0.20 lb VFA/lb VSS at 3 days SRT/HRT to 0.28 lb VFA/lb VSS at 6 days SRT/HRT. Therefore, for this particular sludge, the full-scale fermenter was designed to allow up to 6 days HRT but controlled to allow varying HRTs (achieved by variable sludge feed flows) depending on sludge temperatures and

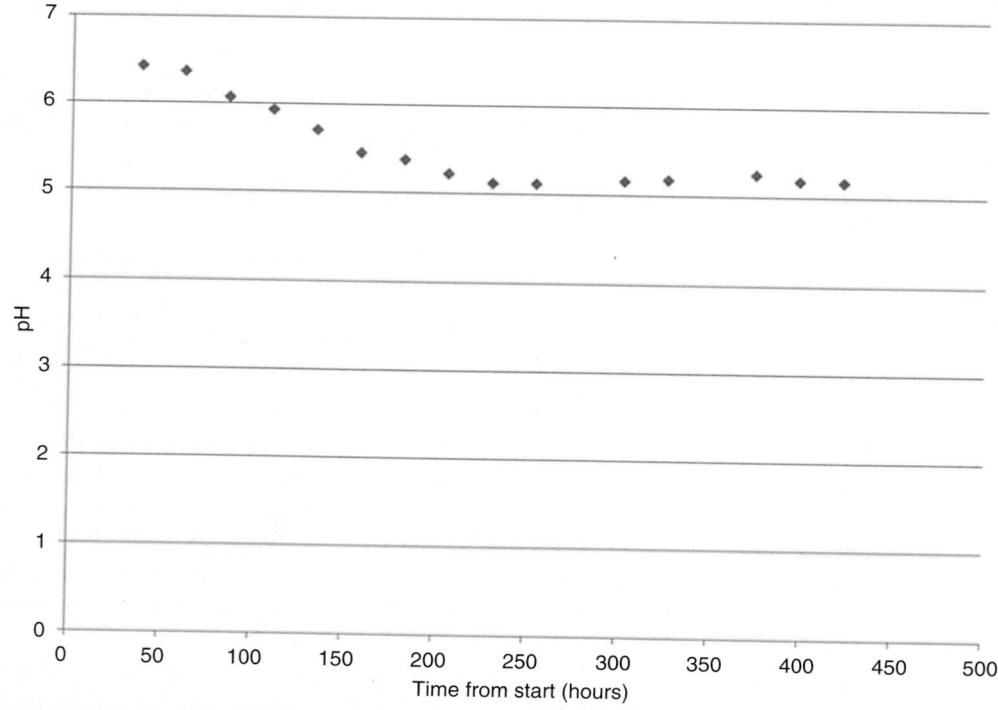

FIGURE 12.47   pH of fermenting sludge with time in laboratory apparatus.

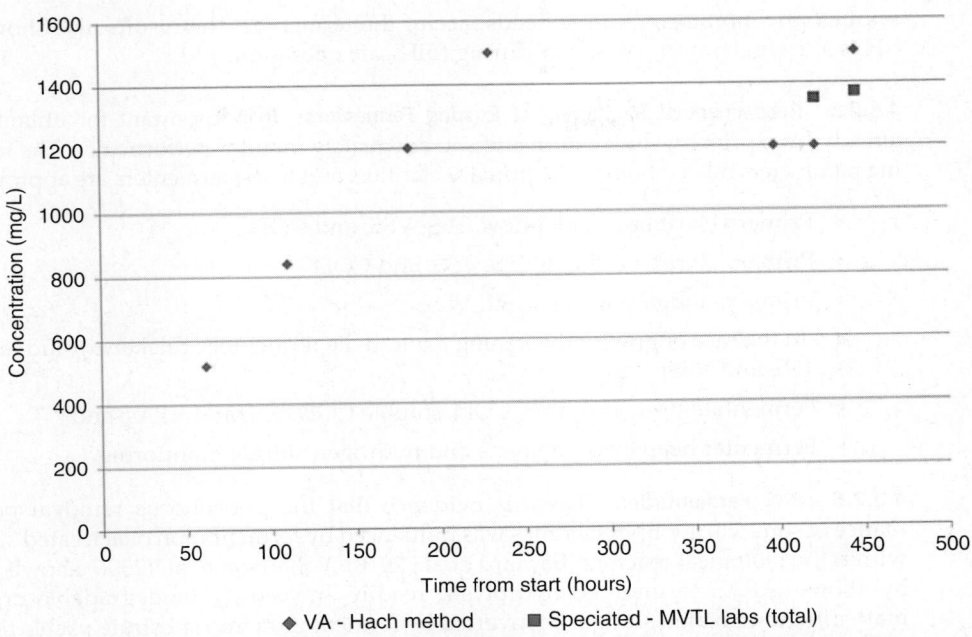

FIGURE 12.48 Fermented sludge VFAs with time in laboratory apparatus. Results show volatile acids measured in the laboratory with speciated VFAs analyzed in a specialized laboratory.

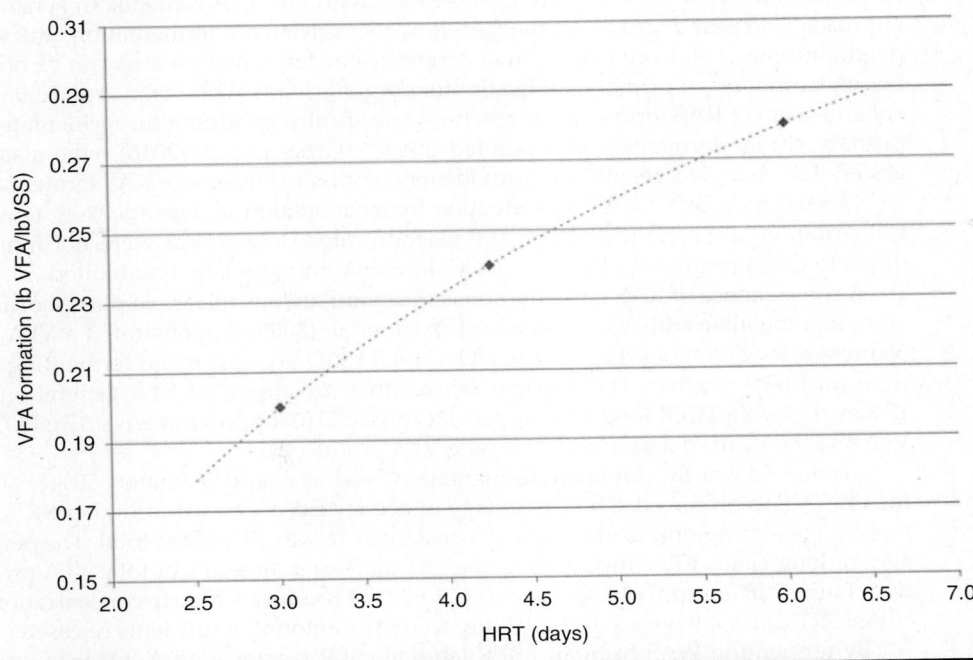

FIGURE 12.49 VFA formation potential results determined at each SRT at steady state in laboratory apparatus.

on methane monitoring in the headspace of the fermenter. The results also show that pH is a useful control parameter during full-scale operation.

4.3.2.5 Recommended Monitoring of Existing Fermenters
It is important for utilities that already have primary sludge fermenters to adequately monitor performance. The following parameters related both to the primary clarifiers and to the fermenters are appropriate:

- Primary clarifier influent flow, TSS, VSS, and COD;
- Primary clarifier effluent TSS, VSS, and COD;
- Primary sludge flow, TSS, and VSS;
- (In the case of gravity thickening prior to the fermenter): thickened sludge flow, TSS, and VSS;
- Fermentate flow, TSS, VSS, COD, soluble COD, pH, and VFAs; and
- Fermenter headspace methane and hydrogen sulfide monitoring.

4.3.2.6 RAS Fermentation
There is evidence that the phosphorus removal performance of some earlier BNR facilities was influenced by fermentation of activated sludge within the biological reactors (Barnard et al., 2016). Vollertson et al. (2006) showed that by allowing RAS to undergo hydrolysis, readily and slowly biodegradable organic matter in the activated sludge is converted by fermentation into substrate usable for the EBPR process. Some of the readily biodegradable COD produced by this process, but not stored by the organisms, was available for subsequent denitrification reactions.

In Denmark, nearly 30 WRRFs implemented fermentation of the RAS between 1996 and 2006 because of the relatively fresh sewage with low VFA contents. In general, the approach is to pass 4% to 7% of the RAS flow to a sidestream fermentation tank with a retention time of 30 to 40 hours. Process rates in the fermentation tank can be boosted by small amounts of primary sludge or other organic loads (Vollertson et al., 2006). The enhancement of RAS fermentation reactions specifically by adding the fermentate from primary sludge fermenters was pointed out by Barnard et al. (2016), who also suggested that an added benefit is to provide inoculation to encourage RAS fermentation.

The required SRT for VFA production by fermentation is dependent on temperature. Yuan et al. (2011) found that the reaction rates slowed, but were not inhibited down to temperatures of 14°C, but at 4°C, fermentation was largely inhibited.

A consequence of RAS fermentation is the simultaneous release of phosphorus and ammonia together with VFA generation. Yuan et al. (2009) demonstrated a VFA yield (expressed as COD) of 0.14 g VFA-COD/g total COD (TCOD) when fermenting WAS from an EBPR process. The nutrient release that accompanied VFA generation was 17.3 mg PO_4-P/g TCOD and 25.8 mg NH_4-N/g TCOD (or approximately 0.12 g PO_4-P/g VFA-COD produced and 0.18 g NH_4-N/g VFA produced).

Acetic acid was the dominant fermentation product at approximately 50% to 66% of the total VFAs produced. The percentage of acetic acid decreased at higher SRTs. The percentage of propionic acid was fairly constant at 16% to 18% of the total. The percentages of long-chain VFAs increased as the SRT increased. In terms of total VFA production, higher SRTs (up to 10 days at 20°C to 22°C) produced higher concentrations of VFAs. SRT did not have a significant impact on the amount of nutrients released.

By fermenting WAS from an EBPR lab-scale SBR reactor with A/O cycles, with a synthetic wastewater feed, Yuan and Oleskiewicz (2010) found a VFA production of approximately 155 mg VFA-COD/g VSS and nutrient release of approximately 25 mg PO_4-P/g VSS and 31 mg NH_4-N/g VSS, or approximately 0.16 g PO_4-P/g VFA-COD produced and 0.20 g NH_4-N/g VFA produced.

Using a metric for potential phosphorus removal by EBPR of 0.10 g P/g VFA-COD, the amount of VFA produced in the experiments of Yuan and Oleskiewicz (2010) was evidently insufficient to remove the additional load of phosphorus released. However, by applying struvite precipitation and recovery to the fermentate, the liquor after struvite precipitation was rich in VFAs, but low in nutrients. This was the approach suggested by Yuan et al. (2010, 2011).

However, Barnard et al. (2011) pointed out that mixed-liquor/RAS fermentation is still possible by appropriate design of the EBPR configuration. By inducing the anaerobic zone to function as a mixed-liquor/RAS fermenter (referred to as unmixed in-line fermenter, UMIF, by Barnard et al., 2011) and situating this zone at the front end of the liquid train, a significant portion of biodegradable particulates (including slowly biodegradable COD) is accumulated and then fermented given sufficient SRT of the biomass. This could be achieved by allowing the sludge to stratify in this zone, thereby further promoting fermentation, and lifting the biomass periodically with intermittent mixing. The UMIF approaches the function of a primary sludge fermenter. It was suggested that this approach could provide a more cost-effective and operationally effective alternative to primary sludge fermenters. If primary sludge fermentation is practiced, the addition of this fermentate to the mixed-liquor fermenting zone would enhance even further fermentation of absorbed sbCOD by inoculation (Barnard et al., 2016).

To maximize the effect of UMIF, only a portion of the RAS should be introduced to this zone. From Vollertson et al. (2006) and Barnard et al. (2011) 5% to 10% appears to be appropriate with a retention time of the solids of 40 to 50 hours. If the full RAS flow is directed toward the fermenting anaerobic zone, extensive secondary release of phosphorus could occur and VFAs released would be insufficient for EBPR (deBarbadillo and Barnard, 2014).

4.4 Other Design Considerations

4.4.1 Baffles

The purpose of baffles is to alter hydraulic characteristics within the reactor. There are several types of baffles used in reactors:

- Interzone baffles to separate unaerated and aerated zones within a reactor to prevent back-mixing between the aerated and unaerated zones;
- Intrazone baffles to create "plug-flow" characteristics in a reactor;
- Intrazone baffles in an aerated zone to prevent short-circuiting in a pass with two different diffuser densities;
- Intrazone baffle to create a "racetrack" zone;
- Mixing chimney to mix two flows before they enter a zone;
- Nitrified mixed-liquor recycle pump baffle to reduce the amount of oxygen recycled to the anoxic zone; and
- Foam trapping baffles to direct foam to waste.

Baffles can be constructed from a variety of materials suitable for the intended service in mixed liquor including wood, fiber-reinforced plastic with appropriate frames, fasteners, and anchorage and, most commonly, concrete (cast-in-place or precast). The variable flow patterns and currents within a mixed or aerated reactor can cause fatigue-related failure of flexible materials such as woven metal or synthetic fabrics.

Baffles tend not to be water-retaining walls but do need to be designed to resist differential pressure resulting from headloss. The design should provide an opening at the bottom of the baffle to prevent excessive forces when draining. The opening should allow the operator to wash the contents from one zone into another for cleaning. Low openings in overflow baffles should be sized to promote sufficient headloss to achieve the desired overflow and water surface profile at most flows.

4.4.1.1 Interzone Baffles

The interzone baffle between an anaerobic and an anoxic zone should be an overflow baffle. There should be positive headloss between the anaerobic zone and anoxic zones to prevent nitrified mixed liquor that is recycled to the front of the anoxic zone from bleeding into the anaerobic zone.

An interzone baffle to separate an unaerated zone from a subsequent aerated zone prevents aerated mixed liquor from being recycled into the unaerated zone. Failure to do this can inhibit phosphorus release or denitrification or encourage low dissolved-oxygen bulking. The bulk of the flow should pass over the top of the baffle. When the air is turned on in the aerobic zone, the water level will rise (approximately 1%). The headloss across this baffle must account for this rise to prevent backflow from the aerated zone. An example of this concept is shown in Figure 12.50.

Submerged weir creates surface head loss to induce foam transport from unaerated to aerated zone

FIGURE 12.50 Example of submerged weir to allow foam transport across zones, especially from unaerated to aerated zones. Note the slight drop in head that prevents foam trapping.

The interzone baffle between a "swing zone" and a subsequent aerated zone is similar to a baffle between an unaerated and an aerated zone.

The interzone baffle to separate an aerated zone from an unaerated zone prevents aerated mixed liquor from the aerated zone overwhelming the unaerated zone. In deep aeration tanks, this baffle may consist of two baffles forming a chimney to allow for some deoxygenation of the mixed liquor to take place.

4.4.1.2 Intrazone Baffles to Create Plug-Flow Characteristics   The intrazone baffle often is used to create plug-flow characteristics within a zone. Headloss across these baffles is minimal. Design depends on the type of mixer or type of aeration being used.

When a submerged horizontal propeller mixer is used, the flow pattern tends to be predominantly under the first baffle and then predominantly over the second baffle. The flow division at average flow should not be less than 50% to 75% of the desired path. For example, at least 75% of the flow entering the zone would pass under the upstream baffle and 75% of the flow leaving the zone would pass under the downstream baffle.

This type of arrangement often is used in zones where the biological kinetics are second order (e.g., phosphorus release) or a "selector" effect is desired. These are also common in deep aeration tanks (more than 8 m deep) to create plug-flow characteristics in an aerated zone.

Intrazone baffles to prevent aeration-induced short-circuiting are different from a baffle to create plug-flow characteristics. These baffles tend to start 0.6 m above the floor and terminate 0.6 m below the water surface. These baffles are often installed between a zone with a high diffuser density and one with a lower density. This baffle reduces longitudinal short-circuiting down an aeration pass with two different diffuser densities.

4.4.1.3 Intrazone Baffles to Create a "Racetrack" Zone   It is easier to mix a long, narrow anoxic zone by creating a racetrack anoxic zone. Staff at Thames Water in the United Kingdom developed this approach. The zone is mixed by big blade mixers that maintain the average linear velocity above 0.3 m/s. This baffle should be at least 50 mm below the water surface because foam will tend to be trapped upstream of the submerged horizontal propeller mixers.

4.4.1.4 Mixing Chimney   The purpose of a mixing chimney is either (1) to direct the flow below the water surface or (2) to blend two streams (e.g., RAS and primary effluent). A chimney can act as a preanoxic zone in a biological phosphorus configuration. The chimney also can be used to mix a volatile acids stream or a methanol feed into the mixed liquor before it is dispersed. An example of a chimney baffle is shown in Figure 12.51.

If the flow is being split between different reactors, then hydraulics should be checked to ensure that the momentum results in solids that split equally.

4.4.1.5 Nitrified Mixed-Liquor Pump Baffle   When the intake of the nitrified mixed-liquor pumps is within a fine-bubble aerated zone, the intake should be boxed in by a baffle. The height of the baffle is a couple of meters off the floor. The net flow into the baffle should be less than the rise rate of an air bubble (e.g., 0.25 to 0.30 m/s). In one case where this was not done, the dissolved oxygen in the anoxic zone was 0.5 mg/L, stimulating sludge bulking.

Figure 12.51   Example of chimney baffle (blends influent flow, RAS and IMLR recycles). Outlet at bottom. Slot allows foam to pass.

4.4.2   Mixing

The purpose of mixing is to maintain solids in suspension and to blend different incoming streams within the tank. The type and the size of mixers depend on the size and the shape of the tank that they are required to mix. When baffles are installed to separate different zones in a larger tank, the mixers should be positioned so that they do not induce localized backflow into the preceding zones. The size, speed, and format of the mixers should be such as to minimize surface turbulence and should not entrain oxygen into the mixed liquor if mixing is within unaerated zones.

4.4.2.1   Blending Streams   If two or more separate streams such as primary effluent, mixed-liquor recycle, and RAS enter the tank, the mixing energy required for effective blending of the separate streams could be considerably more than required to keep solids in suspension (Pretorius et al., 2015; Samstag and Wicklein, 2014). In this case a far superior approach is to provide a *mixing chimney*, described previously, prior to entering the full volume of the tank. Without a properly designed mixing chimney, the plumes from the separate streams could pass from one zone to the next, thereby reducing the effective volume for biological reactions. With an effective mixing chimney, the mixer in the tank would then be sized just to maintain solids in suspension.

4.4.2.2   Mechanical Mixing—Maintaining Solids in Suspension   The earliest BNR facilities had mixers sized for approximately 10 to 16 W/m³. Many of these earlier installations were found to entrain significant amounts of oxygen in unaerated zones, either through vortex formation or by excessive surface turbulence. Some more recent designs still appear to be based on relatively high mixer powers either through tradition or from some published literature.

However, many other designs reflected the realization that high mixer powers were unnecessary and, more importantly, detrimental to anaerobic or anoxic zone performance. Some of these designs reduced mixer powers to below 5 W/m³ with some as low as 3 W/m^3 when mixing chimneys are provided and the blending of separate streams is not a requirement.

The type of mixer is a factor in mixing effectiveness for a given power level. Using the criterion for adequate mixing of less than approximately 10% for the coefficient of variation of mixed-liquor solids concentrations at depths throughout the basin (where coefficient of variation, CoV = standard deviation/overall average), Samstag and Wicklein (2014) showed that field tests comparing four manufacturer mixer designs—three vertical hydrofoil mixers and one vertical hyperboloid mixer—required 2.9, 1.3, 4.1, and 4.0 W/m^3 respectively, for a CoV of 10% with mixed-liquor solids and without separate stream blending required. The most efficient mixer in these tests was a vertical hydrofoil mixer configured with three downward-pumping, relatively large-diameter curved impeller blades. The top of the blades was approximately 2 m above the floor and projected downward to approximately 1 m above the floor. Other tests with horizontal impeller mixers showed that these mixers have similar mixing efficiencies (in terms of W/m^3 required to maintain a CoV of 10%) as the vertical hyperboloid mixer and the least efficient vertical hydrofoil mixer. Pumped jet mixers (without the air in jet aeration systems) were found to be an order of magnitude less efficient, requiring from 30 to 50 W/m^3 to maintain a CoV of 10%. Introducing air to jet aeration systems improves mixing considerably.

It has been suggested that mixer manufacturers should provide information on CoV in activated sludge mixed liquors since traditional approaches of specifying mixers based on a power level (such as W/m^3) ignore relative efficiencies of different impeller and mixer designs (Pretorius et al., 2015). Specifications should ideally be in terms of CoV values less than 10%, for example, and should take into account mixer design and tank dimensions. The important issue of blending separate influent streams and the presence or not of *chimney baffles* should be taken into consideration, since mixing energy will be significantly higher if the mixer also needs to achieve blending of the streams within the tank.

4.4.2.3 Aeration—Maintaining Solids in Suspension

For coarse-bubble aeration with diffusers along one or two sides of the tank, inducing a spiral-roll pattern, a common criterion for aeration mixing is 17 to 40 Nm^3/h·m (3 to 8 scfm/ft) header length (where Nm^3 is the volume of air under standard conditions of 0°C and 1 atmosphere). An alternative criterion commonly presented in the literature is 0.9 to 1.5 Nm^3/h·m^3 (16 to 26 scfm/1000 cu ft) tank volume. An issue with this criterion is that for a given tank horizontal area, the air requirement for mixing a 6-m-deep tank, for example, would be twice the air required for a 3-m-deep tank.

One manufacturer provided an alternative basis for designing air-mixing requirements by adjusting the unit volumetric airflow based on the tank depth. In this case, by recalculating the values provided, the criterion for mixing becomes approximately 4 Nm^3/h·m^2 (0.23 scfm/sq ft) for spiral-roll mixing and for channel mixing in activated sludge systems. For channel mixing, other common criteria are in the range of 12 to 30 Nm^3/h·m (2 to 6 scfm/ft) header or channel length.

A commonly used design criterion for fine-pore diffused aeration with full-floor coverage to maintain mixed liquor in suspension is 2.2 Nm^3/h·m^2 (0.13 scfm/sq ft). This value has traditionally been applied irrespective of whether the flow entering the aeration basins is raw influent or primary effluent. However, some designers account for the removal of heavier material in primary clarifiers upstream of aeration tanks and adjust the minimum airflow required to maintain mixed-liquor solids in suspension accordingly. With primary clarifiers, the minimum airflow required is within the range 0.9 and 1.5 Nm³/h·m² (0.05 and 0.09 scfm/sq ft). On-site testing showed that at airflow rates down to 0.9 Nm³/h·m² (0.05 scfm/sq ft) after 2 weeks of testing, solids settling to the floor of the aeration basins was not detected (Yunt, 1980).

Other limitations for minimum airflow may override the above values, including the minimum airflow per diffuser to maintain air distribution throughout the grid and to prevent clogging of the diffuser, as well as the air requirements to meet process oxygen requirements.

In racetrack systems, some manufacturers use a mechanical mixer to maintain mixing and the aeration grid to provide aeration. In order to prevent the air bubbles interfering with the mixing pattern, a minimum distance between the mixer and the downstream aeration grid must be maintained. The mixer or diffuser vendor may provide this to the designer. Typical values are either the width of the channel or the water depth depending on the size of the racetrack. To maintain solids in suspension, the velocity in the channel should be 0.3 m/s (1 ft/s).

4.4.2.4 Effect of Mixing on Flocculation   Mixing systems maintain solids in suspension by imparting velocity gradients in the liquid, which prevent settlement of individual floc particles. Velocity gradients can promote flocculation of mixed-liquor floc particles or, if too large, can cause rupture of the floc particles. If floc breakup occurs to an excessive degree, then effluent quality could deteriorate due to dispersed solids that are incapable of settling during the retention times provided in secondary clarifiers.

The concept of the root mean square velocity gradient, G, describes the intensity of shear of fluids at a given viscosity (Camp and Stein, 1943). For diffused aeration systems, the G value is computed from the work done by the drag forces on bubbles as they rise to the surface. Camp (1955) derived the following expression for G in diffused-air systems:

$$G = [(Q_{avg} \times \gamma \times h)/(60 \times V \times \mu_w)]^{0.5} \qquad (12.23)$$

where G = root mean square velocity gradient (s^{-1}),
  Q_{avg} = average volumetric flowrate of air through the water depth (m³/min),
      which, assuming the average occurs at mid-depth, and
    = $Q \times 10.33/(10.33 + h/2)$ (m³/min),

where 10.33 = atmospheric pressure at sea level (m water),
      Q = standard airflow rate through the diffusers (Nm³/min),
      γ = specific weight of water (= 9789 N/m³ at 20°C),
      h = liquid depth above the diffusers (m),
      V = volume of liquid in tank (m³), and
      μ_w = absolute or dynamic viscosity of water (= 0.001 Ns/m² for water at 20°C)
       (for MLSS greater than approximately 5000 mg/L, use the viscosity of the suspension).

For a mechanical aeration system:

$$G = [P/(V \times \mu_w)]^{0.5} \text{ (s}^{-1}) \qquad (12.24)$$

where P = aerator power input (N·m/s), and
   1.0 N·m/s = 1.0 W = 0.001 kW.

For pipe flow:

$$G = 52 \times (f/D)^{0.5} \times v^{1.5} \ (\text{s}^{-1})$$ (12.25)

where f = Darcy-Weisbach friction factor (dimensionless),
   D = pipe diameter (m), and
   v = velocity (m/s).

   Das et al. (1993) investigated the degree of floc integrity at 24 activated sludge facilities employing coarse-bubble, fine-pore, and mechanical aeration systems. For both the coarse-bubble and fine-pore aeration facilities, effluent TSS started to increase as G values increased above approximately 75 s^{-1}. The coarse-bubble systems appeared to deteriorate at a faster rate than the fine-pore systems at higher G values. The mechanical aeration systems followed a similar general concept although there was a difference that depended on the distance between the point of shear at the aerator and the distance to the point of removal of mixed liquor from the aeration tank. In other words, in some cases there was the opportunity for re-flocculation depending on the distance.

   The observation that flocs will generally reform if energy levels and velocity gradients are reduced and sufficient time is provided prior to secondary clarifiers is very important and should guide designers. This also points to the benefits of tapered aeration in diffused aeration systems. Apart from economizing on overall energy by tapering aeration to match oxygen demands, the G values within the aeration basin would be gradually reduced from the inevitably higher values at the head end of the basins, to lower values at the end. This provides opportunity to develop adequate flocculation before the mixed liquor passes to the clarifiers.

   Unfortunately, many aeration tanks discharge via a weir to an aerated channel at the end of the aeration basin. The free fall over the weir can impart considerable disruptive energy that can damage flocs. Das et al. (1993) noted the percentage of dispersed particles before and after such weir free falls. The change occurring when the free fall was approximately 0.5 m was relatively insignificant. However, at 1.2-m free fall the percentage of dispersed particles increased by approximately 1.6 to 4.0 times. One mitigating factor to note is that in many designs the highest free falls would occur at lower flows, which would provide more time for re-flocculation before the clarifiers. Weirs and channels are generally designed for peak-flow conditions, and water levels could be higher and free falls after weirs lower under high-flow conditions.

   After weir transitions with appreciable free falls, it is important to ensure adequate flocculation either in conveyance piping or in aerated channels before the clarifiers. Aerated channels are often equipped with coarse air diffusers. This is not generally conducive to good flocculation. If possible it would be preferable to use fine-pore diffusers. Conveyance piping often imparts gentler shear forces compared with aerated channels and, in general, tend to promote flocculation rather than breakup of flocs.

   In this regard the design of the secondary clarifiers plays an important role. Adequate design of energy-dissipating inlets and flocculation feedwells will help to promote re-flocculation prior to settling.

4.4.2.5 Activated Sludge Foam Control   The incidence of foaming organisms can be a major activated sludge operating challenge. They manifest themselves as a problem particularly when the design of the activated sludge reactor is such that surface foam layers are trapped and foaming organisms can accumulate. In the absence of adequate design features, foam will eventually accumulate somewhere (or all over the surface) of the aeration tanks. Eventually the foam will bridge and harden and will not flow along the surface of its own accord. This becomes a burden for operations personnel as they try to remove the foam using high-pressure hoses. Often the best recourse at this stage is to remove the foam using a vacuum truck service, if access is possible. It is usually preferable to provide design features that avoid foam formation and accumulation.

Baffles that are commonly included in BNR facilities should promote the separation of distinct zones within the reactor but, at the same time, induce movement of foam through the reactor to a point where it can be wasted from the system in the secondary clarifiers, or within the reactor. The design of internal baffles was discussed earlier.

The foaming propensity of nocardioforms such as *Gordonia amarae* is mainly caused by free-floating dispersed filaments, more so than by floc-bound filaments (Narayanan et al., 2010). With subsurface withdrawal from reactors or inadequately designed baffles within the reactor that cause foam trapping, free-floating filaments tend to proliferate. Parker et al. (2014) posited that free-floating filaments are more likely to attach to the bubbles produced by fine-pore diffused aeration and accumulate at the surface. Floc-bound filaments are heavier and would require more bubble attachment, which may require a DAF environment, not the turbulent diffused-air environment of aeration tanks.

Jenkins et al. (2003) discussed a number of measures to control foam in activated sludge reactors. For nocardioforms these included aerobic selectors, anoxic selectors, anaerobic selectors, chlorination, cationic polymer addition, and surface wasting. For *Microthrix parvicella* the measures included reducing SRT, providing plug-flow conditions, avoiding intermittent aeration or low dissolved oxygen, eliminating foam trapping, and adding a polyaluminum chloride coagulant. Several of the measures, such as lowering SRT or increasing dissolved oxygen, could be contrary to other fundamental objectives, such as nitrogen and phosphorus removal. It is also common to experience foaming issues despite the best design efforts to avoid them.

Parker et al. (2014) described the concept of the classifying selector, which takes advantage of the propensity of free-floating nocardioforms, *M. parvicella* and other foam-causing filaments, to attach to bubbles. Even if visible foam is not formed, there will be a surface-rich stratum of the organisms formed at the surface of the mixed liquor. Successful designs of this concept situate a surface baffle at a strategic location to impede passage of the foam or surface layers, and continuously withdraw the foam or surface layers by means of a downward-opening weir gate that discharges to a pump sump. By this means, the bulk concentration of foam-causing organisms is maintained at low levels, thereby preventing their accumulation and nuisance foam formation.

If continuous withdrawal is practiced, this is most straightforward if DAF thickeners are provided to thicken WAS. By this means, the classifying selector can be combined with mixed-liquor wasting, thereby greatly improving SRT control. DAF thickeners are largely immune to hydraulic loading issues. A further alternative that has been successfully applied is to apply the concept described to an aerated RAS channel or a common RAS reaeration tank. This fulfills the objective while avoiding excessive flows to the thickener. SRT control could also be combined with the surface drawoff at this point, if required. It is important that the entire mixed-liquor inventory passes the classifying selector installation. Examples of suitable locations are a stretch of aerated mixed-liquor channel between the aeration tanks and the clarifiers, a stretch of aerated RAS channel, or a common RAS reaeration tank.

Necessary features of the classifying selector process are:

1. Provide a stretch of channel or aerated section with fine-pore diffused air upstream of the baffle and drawoff to promote flotation of the dispersed filaments.

2. Avoid foam trapping within the overall activated sludge system by adequate design of baffles and providing surface overflow to the clarifiers.

3. Continuous or semicontinuous pumping from the classifying selector sump to avoid foam accumulation.

Operation of the classifying selector as briefly described above has been effective at maintaining foam-causing organisms at manageable levels and avoiding their proliferation in the basins. By continuous withdrawal of surface layers where foam-causing organisms are concentrated, foam accumulations within the basins are not usually seen. However, some unusual events such as influent slugs of fats, oil, and grease, or if the classifying selector is out of service for a time, or other unusual occurrence could cause higher-than-normal accumulations. This should be taken into consideration in design, and the classifying selector station should be capable of handling increased foam accumulations and function as a foam-wasting station. The thickening process should also be designed to receive this increased load.

5.0 Anaerobic Treatment of Wastewater

5.1 Introduction

Anaerobic treatment offers many advantages over conventional aerated activated sludge systems including lower energy consumption, the potential for energy recovery, low sludge production, operational simplicity, low land area requirements, improved sludge dewatering, ability to store sludge for long periods, simple designs, and less noise from mechanical equipment. While anaerobic treatment offers many benefits, significant limitations exist including lower removal efficiencies for organics, suspended solids, and pathogens and essentially no nutrient removal with basic process configurations. As a result, anaerobic treatment requires aerobic posttreatment to meet standard effluent criteria for secondary and advanced treatment.

In anaerobic treatment, aeration is not required, thus eliminating the large energy demand needed to supply process air in aerobic processes. Because organic matter is converted to methane, the process may produce energy for the facility. Anaerobic treatment processes typically are more energy efficient for higher strength wastewater (greater than about 300 mg BOD/L), because under these conditions, the fraction of methane dissolved in the effluent becomes insignificant relative to the total methane production (Cakir and Stenstrom, 2007).

The anaerobic decomposition of organic compounds yields less energy for the microorganisms, resulting in lower biomass yields. Typically, anaerobic treatment reduces the overall biomass yield by a factor of 6 to 8 when compared to aerobic treatment (Tchobanoglous et al., 2003). The reduced biomass results in lower sludge production, which decreases handling and hauling costs, yielding a savings of approximately 10%

compared to an aerobic process (Speece, 1996). Because of low sludge production, nutrient requirements are less than for aerobic biological treatment. Anaerobic treatment can reduce the influent BOD and TSS by about 65% to 80%, yielding effluent concentrations of around 40 to 130 mg/L (Khalil et al., 2008; Noyola et al., 2006; Oliveira et al., 2006; von Sperling and Oliveira, 2008).

A survey published in 2008 identified about 3000 WRRFs (all industrial) using anaerobic treatment process (Totzke, 2008). Approximately, 60% of the thousands of industrial WRRFs utilize the upflow anaerobic sludge blanket (UASB) process making it the dominant anaerobic technology for industrial wastewater treatment (Latif et al., 2011). Due to the slow growth rate of anaerobic bacteria, anaerobic treatment of municipal wastewaters is now primarily used in warm climates where the wastewater temperature remains above 15°C.

In search of ways to provide affordable wastewater treatment, communities in India, the Middle East, and Latin America have been advancing the use of anaerobic treatment for domestic wastewater treatment. At this time, numerous full-scale facilities with capacities of up to 164 000 m^3/d (1000 mgd) are in operation in these regions using the UASB process or variations of that process (Draaijer et al., 1992; Florencio et al., 2001; Giraldo et al., 2007; Heffernan et al., 2011; Monroy et al., 2000; Sato et al., 2006; Schellingkhout and Collazos, 1992; Seghezzo, 2004; Vieira et al., 1994).

Limited use of anaerobic treatment for wastewater in municipal facilities has been documented in the United States and Canada, but there is potential for its use in the U.S. South, where winter wastewater temperatures are greater than 15°C to 20°C. Advances in the anaerobic treatment of municipal wastewater, including use of two-stage reactors, coupled anaerobic reactor/digester combinations, UASB/hybrid reactors, and the expanded granular sludge bed (EGSB) reactor likely will expand use of this technology into more temperate climates (Switzenbaum, 2007).

Anaerobic treatment processes can be categorized as suspended growth, fixed growth, and hybrid processes. Most existing full-scale municipal WRRFs with anaerobic treatment of the liquid stream use the UASB process, which is considered a hybrid process (Malina and Pohland, 1992; Sutton, 1990; Van Lier, 2008). With high-strength wastewaters, dense granules that consist of a microbial consortium characterize the UASB process. When used for municipal wastewater treatment, however, granular sludge typically does not form, and the UASB process can be considered a suspended-growth process.

5.2 Microbiology

As with anaerobic sludge stabilization processes, anaerobic processes for wastewater treatment rely on a consortium of facultative and anaerobic bacteria to degrade organic materials. In anaerobic treatment processes, a series of reactions convert organic materials in the wastewater to carbon dioxide, methane, and additional biomass. Four major groups of biological reactions comprise anaerobic decomposition: (1) hydrolysis, (2) acidogenesis, (3) acetogenesis, and (4) methanogenesis. In hydrolysis, strictly anaerobic and facultative anaerobic bacteria convert the biodegradable COD (large organic polymers including proteins, carbohydrates, and lipids) to simpler, soluble monomeric compounds like amino acids, sugars, and long-chain fatty acids. In acidogenesis, there is a further breakdown into VFAs. Following hydrolysis and acidogenesis, fermentative bacteria convert hydrolysis products to acetate, carbon dioxide, and hydrogen. In the final step, methanogens convert the acetate to methane and carbon dioxide or convert hydrogen and carbon dioxide to methane and water. More information on the details of anaerobic decomposition can be found in several references (Grady et al., 2011; Henze et al., 2008; Jördening and Winter, 2005; Pavlostathis and Giraldo-Gomez, 1991; Speece, 1996; Vaccari et al., 2006).

5.3 Process Configurations

Anaerobic treatment eliminates three of the biggest constraints on process loading that occur in aerobic processes: (1) oxygen-transfer rates; (2) solids flux limitations, and (3) high-energy inputs for aeration that hinder floc formation (Speece, 1996). Although the lack of these limiting factors in anaerobic processes enables much higher mass loading rates, different constraints are imposed by low growth rates and high half-saturation coefficients of bacteria in the microbial consortia present. The slow growth rate of anaerobic bacteria and their sensitivity to temperature places a premium on biomass retention; the high half-saturation coefficients encourage use of staged reactors to increase performance.

A wide range of reactor types exists for anaerobic treatment, and some of these are illustrated in Figure 12.52. For municipal applications, however, most installations are based on the UASB reactor. Also, around 60% of the thousands of full-scale anaerobic treatment facilities for the treatment of industrial wastewater are UASB-based reactors (Latif et al., 2011). Therefore, the remainder of this discussion will focus on the UASB process. Information on other anaerobic reactor types is available elsewhere (Chernicharo, 2007; Libhaber and Orozco-Jaramillo, 2012; Malina and Pohland, 1992; Nicolella et al., 2000; Speece, 1996; van Haandel et al., 2006).

5.4 Upflow Anaerobic Sludge Blanket

5.4.1 Description

The UASB process is an anaerobic wastewater treatment technology that incorporates two vertically stacked zones in one structure. On the bottom is an anaerobic reactor that contains the sludge blanket, and above this is a gas-liquid-solid (GLS) separator. In the GLS zone, deflection plates and collection hoods are used to capture the biogas while allowing suspended solids to settle and return to the reaction zone. One of the keys to successful application of UASBs is an efficient GLS design. Gas collected at the top of the reactor can be vented, flared, or burned for heat or power generation. Venting gas directly to the atmosphere is not recommended because methane is a potent greenhouse gas. Burning biogas requires special burners and, potentially, treatment to remove hydrogen sulfide and other contaminants contained like siloxane (Noyola et al., 2006). Gas hoods typically have triangular cross sections so that the sloped outside surfaces create a settling zone that increases with distance from the top of the digestion zone. Feed is introduced as uniformly as possible across the bottom of the sludge zone and flows vertically (Figure 12.53).

The UASB processes share the same advantages and disadvantages discussed above that are common to all anaerobic processes. The UASB reactors provide economical removal of large fractions of the influent organics; but removals are not high enough to meet secondary treatment standards; pathogens and colloidal solids are not adequately removed; and UASBs do not provide any significant nutrient removal. As a result, aerobic posttreatment must be provided for most applications. Due to high carbon removal and low nutrient removal, posttreatment for nitrogen and phosphorus removal may require chemical coagulation or addition of a supplementary carbon source depending on the nutrient to be removed and the process configuration (Ahn et al., 2006; Aiyuk et al., 2004; Foresti et al., 2006; Kalyuzhnyi et al., 2006; Li et al., 2007; Tilche et al., 1994). Extended startup periods (approximately 12 to 20 weeks) are required as a result of the slow growth of the anaerobic biomass unless the process can be seeded with an anaerobic sludge.

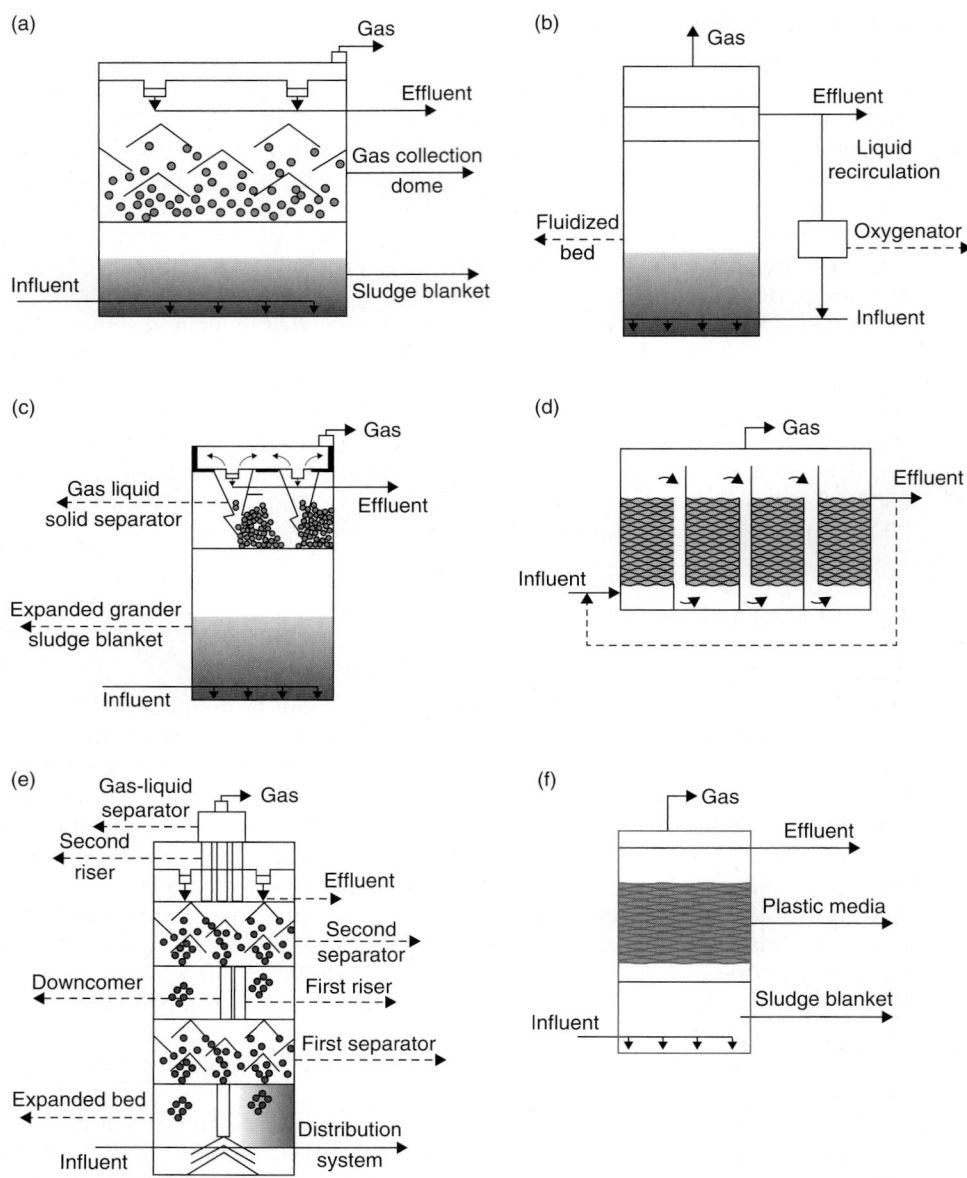

Figure 12.52 Schematic illustrations of several types of anaerobic reactor configurations: (a) upflow sludge blanket; (b) biofilm fluidized bed; (c) expanded granular sludge bed; (d) anaerobic baffled reactor; (e) internal circulation; and (f) anaerobic hybrid reactor (Nicolella, 2000).

Similarly, recovery from toxic shocks occurs slowly. Because of the increasing number of anaerobic treatment facilities around the world, seed sludge can be readily obtained.

Sludge in UASB processes treating mostly soluble wastes tends to form dense granules; however, granulated sludge does not form with dilute wastewaters containing low

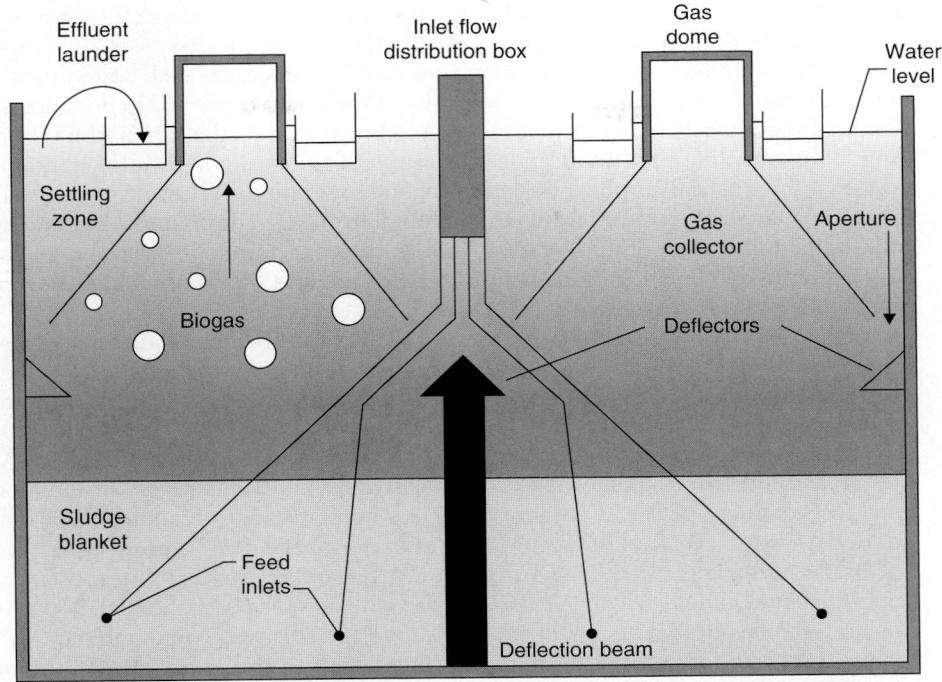

Figure 12.53   Schematic illustration of upflow anaerobic sludge blanket (UASB) reactor (van Lier, 2003).

concentrations of COD and high concentrations of TSS (Holshoff Pol et al., 2004). With flocculent (nongranulated) sludge, upflow velocities are limited to a maximum of about 1.0 m/h so that the majority of the sludge remains in the reaction zone. Gases generated in the sludge blanket and slow-settling particles of sludge flow up from the sludge zone and enter the GLS zone where the gas is captured, and the suspended solids either exit the process or are returned to the reactor. Odors are a potential problem if the biogas is allowed to escape from the gas collection system. Biomass grown in the sludge blanket remains there until wasted directly from the sludge zone or is gradually allowed to fill the sludge zone, become entrained in the liquid stream, and exit the reactor in the effluent.

Power is only needed for pumping so that when sufficient head is available to allow gravity flow, power requirements are low. Because of the relatively high concentration of biomass maintained in the reactor (30 to 40 g/L), the depth of the sludge zone (2 to 4 m), and the construction of the GLS on top of the reactor zone, land requirements are significantly lower than for most other wastewater treatment technologies. Although the relative amount of biogas generated by UASB processes treating domestic water is low due to the low concentration of COD in the influent wastewater, sufficient quantities may be available in warmer climates to generate enough energy to make the process self-sufficient (van Haandel and Lettinga, 1994).

5.4.2   Installations
Gatze, Lettinga, and coworkers at the University of Wageningen (Wageningen, the Netherlands) first developed the UASB process in the 1970s as a unique anaerobic

treatment technology. Between 1983 and 1992, UASBs were studied at laboratory scale and later at demonstration scale. Initially the UASB process was developed for full-scale use in industrial applications because the process is well suited for treating warm, soluble, high-COD wastewaters. The UASB reactors proved to be successful for high-strength wastes from industries such as breweries, distilleries, and food processing. Although there are thousands of successful full-scale, high-rate anaerobic processes in industrial applications, design and operation experience with full-scale municipal facilities remains somewhat limited. In 1989, Kanpur, India, became the first full-scale demonstration of UASB technology treating municipal wastewater, and by 2011 more than 60 installations had been reported, and this is not an extensive list of installations (Heffernan et al., 2011) (see Table 12.16). Most of these facilities are in India and Latin America. Despite the number of UASBs being installed to treat municipal wastewater, limited design and performance data are available from operating full-scale facilities to judge the effectiveness and performance of UASBs for treatment of municipal wastewater.

Selected results from several of these studies are reproduced in Table 12.17. At best, UASBs provide approximately 80% removal of BOD_5 and TSS, thus confirming the need for posttreatment of UASB effluent to meet secondary and advanced treatment standards. An extensive statistical evaluation of performance data from treatment facilities in Brazil found mean removals of BOD_5 and TSS of 72% and 67%, respectively (von Sperling and Oliviera, 2008). UASB processes with aerobic posttreatment, however, had mean removals of 88% for BOD_5 and 82% for TSS. This was comparable to the mean values reported for facilities using the activated sludge process (85% for BOD_5 and 76% for TSS). Occasionally, poor removal rates for suspended solids for UASB processes without posttreatment have been attributed to washout of sludge. Sato et al. (2006) suggested that removal efficiencies of full-scale units could be improved by proper operation and maintenance.

5.4.3 Design Considerations

Successful application of anaerobic treatment requires good mixing, contact between influent wastewater and biomass, and retention of the biomass in the reactor (van Haandel et al., 2006). The first full-scale UASB reactors for domestic wastewater were sized based on experimental results from pilot-scale studies (van Haandel and Lettinga, 1994). Even with the increasing number of full-scale installations, a thorough characterization of the wastewater is considered essential and pilot testing desirable, before designing a new UASB (Henze et al., 2008). Current full-scale installations in developing countries strive for low cost and simplicity and, typically, have only screening and grit removal for pretreatment, and aerobic or facultative lagoons for posttreatment. Pretreatment to remove fats, oils, and suspended solids should provide enhanced performance and, in many situations, is essential. More sophisticated process configurations have been proposed that maintain the basic advantages of anaerobic treatment while adding enhanced removal of dissolved organic matter, suspended solids, and nutrients. Key design considerations include reactor dimensioning, upflow velocity, GLS design, estimation of sludge and biogas production, design of the flow distribution, odor control, provisions for scum removal, and materials selection.

5.4.3.1 Reactor Sizing
As with any biological suspended-growth treatment process, expected bacterial growth rates control reactor biomass inventory, while minimum biomass settling velocities dictate the surface area for the solids separator and, for

Facility	Country	Design Flow (MLD)	Reactor Volume (m³)	Start-up, Date (year)	Comments
Bucaramanga	Colombia	31–47	3 × 3300	1990	UASB and facultative pond (Giraldo et al., 2007; Seghezzo 2004; Schellingkhout et al., 1992)
	Brazil	0.164	2 × 2286	2008	Heffernan et al., 2011
Campina Grande	Brazil	0.6	160	1989	Pedregal Township (Giraldo et al., 2007)
Sumare	Brazil	0.228	67.5	1992	Vieira et al., 1994
	Brazil	100	16 × 2705	2007	Heffernan et al., 2011
Recife	Brazil	28	810	1997	Florencio et al., 2001
Mirzapur	India	14		1994	Post treatment with FAL (Seghezzo, 2004)
	India	120	20 × 1960	2003	Heffernan et al., 2011
Kanpur	India	36		1994	Mixed tannery and domestic waste (Seghezzo, 2004)
Faridabad	India	20, 45, and 50	7000, 16 000, and 18 000	1998, 1998, and 1999	Polishing pond post treatment (Sato, et al., 2006)
Sonipat	India	30	11 000	1999	Polishing pond post treatment (Sato et al., 2006)
Gurgaon	India	30	11 000	1998	
Panipat	India	35 and 10	13 000 and 10 000	2000 and 1999	Polishing pond post treatment (Sato et al., 2006)
Yamunanagar	India	25 and 10	9000 and 3500	2000 and 2002	Polishing pond post treatment (Sato et al., 2006)
Karnal	India	40	14 000	2000	Polishing pond post treatment (Sato et al., 2006)
Ghaziabad	India	56 and 70	20 000 and 26 000	2002 and 2002	Polishing pond post treatment (Sato et al., 2006)
Noida	India	27	14 000	2000	Polishing pond post treatment (Sato et al., 2006)
Agra	India	78	10 000	2004	Polishing pond post treatment (Sato et al., 2006)
Saharanpur	India	38	28 000	2000	Polishing pond post treatment (Sato et al., 2006)
Tapeyanco–Atlamaxac Tlaxcala	Mexico	2.6	2200	1990	Post treatment by lagoons (Monroy et al., 2000)
Fideicomiso Alto Rio Blanco. Istaczoquitlan	Mexico	108	5 × 16 740	1994	Monroy et al., 2000
Middle East Facility 1	UAE	49	8 × 2645	2008	Heffernan et al., 2011

TABLE 12.16 Selected Existing, Full-Scale UASB Reactors Domestic Wastewater

Parameter	Reported Removals (%)
COD removal	56–79
BOD removal	45–81
TSS removal	45–81
Coliforms	70–90
Helminth eggs	up to 100

TABLE 12.17   Average UASB Performance Reported in Latin America and India (Khalil et al., 2008; Giraldo et al., 2007)

UASB reactors, the cross-sectional area of the reactor. For anaerobic process, the controlling growth rates are those of the slowest growing methanogens with maximum specific growth rates on the order of 0.12 day^{-1}. Because of low growth rates and difficulty in predicting the minimum growth rate for the diverse consortium of microorganisms in the sludge blanket, recommended safety factors on SRT for anaerobic processes are high, at approximately 3 to 10 (Henze et al., 2008; Speece, 1996). Figure 12.54 provides an estimate of the required SRT as a function of temperature for treating domestic wastewater in a UASB reactor. Operating experience gained from demonstration and full-scale UASBs, rather than explicit measurements of sludge settling velocities, provides the basis for current guidelines for sizing of reactors and GLS separators.

For domestic wastewater, sizing the reactor based on HRT provides a practical approach because for low-strength wastewater (COD < 1000 mg/L) the hydraulic load limits the design (Chernicharo, 2007; Henze et al., 2008). An average HRT for a single-stage UASB treating domestic wastewater is approximately 6 hours. Values reported in the literature range from 4 to 10 hours. Current design criteria for HRT for UASB reactors are provided in Table 12.18. Regardless of the method used to size the reactor, the expected SRT must still be estimated to ensure adequate design.

Use of the OLR is appropriate for high-strength domestic wastewaters because the organic load rather than the hydraulic load limits design. Care must be taken, however,

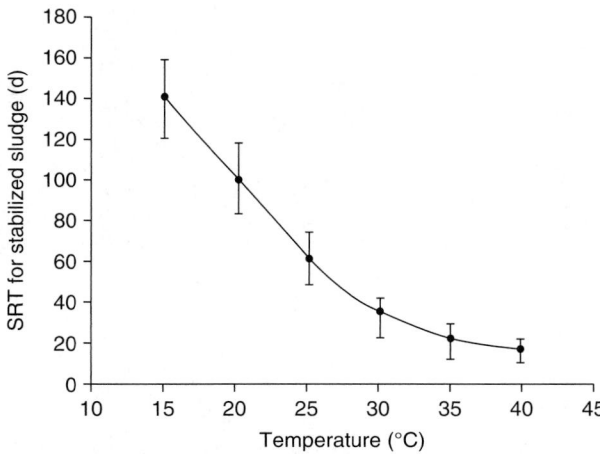

FIGURE 12.54   Required solids retention time (SRT) for domestic wastewater treatment as a function of temperature (Henze et al., 2008; reprinted with permission from IWA Publishing).

| Sewage Temperature (°C) | Hydraulic Detention Time (hour) | |
	Daily Average	Minimum (during 4 to 6 hours)
16 to 19	>10 to 14	>7 to 9
20 to 26	>6 to 9	>4 to 6
>26	>6	>4

TABLE 12.18 Recommended Hydraulic Detention Times for UASB Reactors Treating Domestic Wastewater (Lettinga and Hulshoff Pol, 1991; reprinted from *Water Science and Technology*, with permission from the Copyright Holders, IWA)

in defining and applying the OLR as the term can apply to the applied load, removed load, or the converted load (van Haandel and Lettinga, 1994). For domestic wastewater, constraints imposed by biomass settling velocities will limit the OLR to 1.5 to 3.0 kg $COD_{applied}$/ $m^3 \cdot d$. The OLRs for high-strength wastewaters with a significant amount of particulate COD are listed in Table 12.19 and are presented as an example of the OLR limits of the process.

Until mathematical models for anaerobic treatment become more advanced, prediction of effluent water quality must be done using empirical relationships between HRT and performance (Table 12.20 and Figure 12.55) (Chernicharo, 2007; Latif et al., 2011; van Haandel et al., 2006). Careful judgment must be exercised in the use of these empirical relationships, as considerable scatter exists in the limited performance data from full-scale facilities (see Figure 12.56), and data are only available for operation under tropical conditions.

For domestic wastewater, current criteria call for superficial upflow velocities to be maintained below approximately 1.0 m/h with average velocities in the range of 0.4 to 0.8 m/h. A design value of 0.75 m/h has been widely used for UASB reactors in India. UASB reactor dimensions and upflow velocities are interrelated. Typical reactor heights range from 3 to 5.0 m, with a common value of 4.5 m. Greater heights may be required, however, for wastewaters with high suspended solids concentrations (Wiegant, 2001). Settler compartments comprise 1.5 to 2.0 m of this total height.

Successful UASB operation depends on proper hydraulic distribution of the feed flow to prevent channeling of the wastewater through the sludge blanket and to avoid the formation of dead corners in the reactor. Flow must be divided proportionately to each reactor and then uniformly distributed to the numerous feed points located across the bottom of the sludge blanket. The recommended density of feed inlet points

Temperature (°C)	OLR (kg $COD_{applied}$/$m^3 \cdot d$)
15	1.5–2
20	2–3
25	3–6
30	6–9
35	9–14
40	14–18

TABLE 12.19 Permissible OLRs in Single-Step UASB Reactors in Relation to the Temperature for Wastewater with 30% to 40% COD in Suspended Solids (adapted from Henze et al., 2008).

Parameter	Empirical Equation[a]
Efficiency of COD removal	$E_{COD} = 100 \times (1 - 0.68 \times t^{-0.35})$
Efficiency of BOD removal	$E_{BOD} = 100 \times (1 - 0.70 \times t^{-0.50})$
Final effluent BOD and COD	$C_{eff} = S_0 - \dfrac{E \times S_0}{100}$
Final effluent TSS	$C_{ss} = 102 \times 2^{-0.24}$

[a]Empirical equations (wastewater temperature from 20°C to 27°C).

TABLE 12.20 Empirical Equations for Estimating UASB Reactor Performance and Effluent Water Quality (Chernicharo, 2007; reprinted with permission from IWA Publishing)

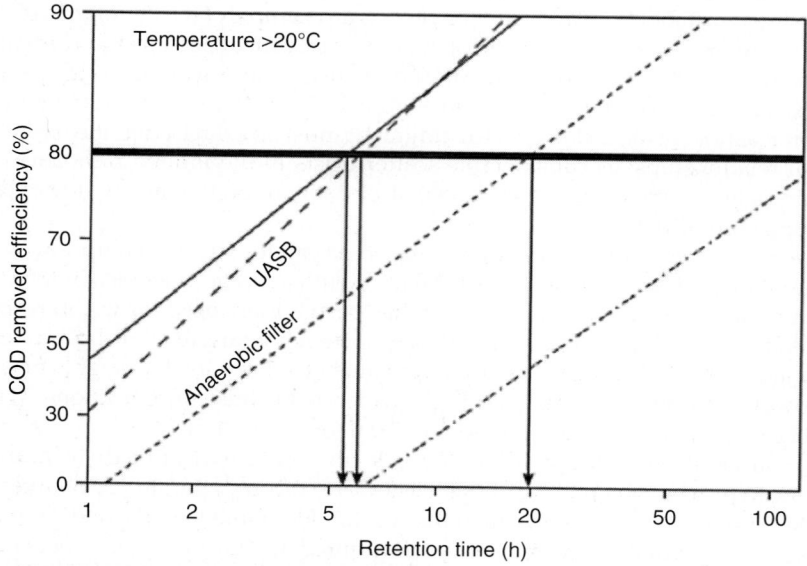

FIGURE 12.55 Emprical relationships between HRT and COD removal performance of UASB and Anaerobic Filters (van Haandel et al., 2006).

is currently approximately one for every 2.0 m². Higher densities are recommended for low influent concentrations of organics where low gas production increases the risk of channeling and short-circuiting. Table 12.21 presents guidelines on influent flow distribution. Table 12.22 presents summary guidelines for the main hydraulic criteria, and Table 12.23 provides other design criteria for UASB reactors treating domestic wastewater.

5.4.3.2 Gas-Liquid-Solid Separation As with any suspended-growth biological treatment process, retention of solids in the UASB process is critical. In UASB reactors, the need to separate gas and solids from the liquid stream complicates settler designs.

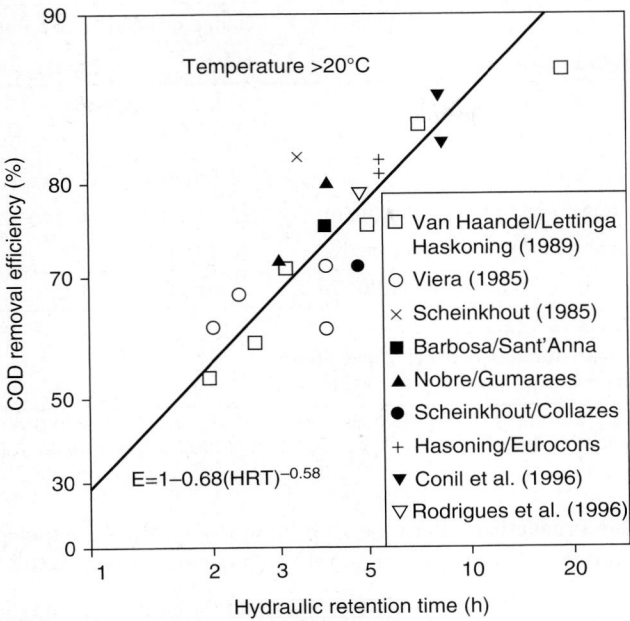

Figure 12.56 Experimental data on chemical oxygen demand (COD) removal efficiency in upflow anaerobic sludge blanket (UASB) reactors as a function of hydraulic retention time (van Haandel et al., 2006).

Suggested guidelines for the GLS are to provide a minimum slope for the settler bottom of 45° to 60°, to provide an overlap of 10 to 20 cm for the deflectors under the entrance to the settling zones, to provide a surface area for the openings between the gas collectors of 15% to 20% of the reactor surface area, and to provide a gas collector height of 1.5 to 2.0 m (van Lier, 2003).

Sludge Type	OLR (kg $COD_{applied}/m^3 \cdot d$)	Influent Area of Each Distributor (m²)
Relatively dense and flocculent (concentration 20–40 kg TSS/m³)	<1.0 to 2.0	1.0 to 2.0
	>3.0	2.0 to 5.0
Dense and flocculent (concentration > 40 kg TSS/m³)	<1.0	0.5 to 1.0
	1.0 to 2.0	1.0 to 2.0
	>3.0	2.0 to 3.0
Granular	<2.0	0.5 to 1.0
	2.0 to 4.0	0.5 to 2.0
	>4.0	>2.0

Table 12.21 Preliminary Guidelines for Flow Distributors in UASB Reactors (Lettinga and Hulshoff Pol, 1991; reprinted from *Water Science and Technology*, with permission from the Copyright Holders, IWA)

	Range of Values, as a Function of Flow		
Criterion/Parameter	**For Q_{avg}**	**For Q_{max}**	**For Q_{peak}**[a]
Hydraulic volumetric load (m³/m³·d)	<4.0	<6.0	<7.0
Hydraulic detention time (h)[b]	6 to 9	4 to 6	>3.5 to 4
Upflow velocity (m/h)	0.5 to 0.7	<0.9 to 1.1	<1.5
Velocity in apertures to the settler (m/h)	<2.0 to 2.3	<4.0 to 4.2	<5.5 to 6.0
Surface loading rate in the settler (m/h)	0.3 to 0.8	<1.2	<1.6
Hydraulic detention time in the settler (h)	1.5 to 2.0	>1.0	>0.6

[a]Flow peaks with duration between 2 and 4 hours.
[b]Sewage temperature in the range of 20°C to 26°C.

TABLE 12.22   Summary of the Main Hydraulic Criteria for the Design of the UASB Reactors Treating Domestic Sewage (Chernicharo, 2007; reprinted with permission from IWA Publishing)

5.4.3.3 Gas Production   Because COD is conserved, the expected mass of methane produced can be estimated from a COD balance around a reactor as follows:

$$COD_{T,O} = COD_{T,e} + COD_{T,s} + COD_M \cdot \Delta COD_R \qquad (12.26)$$

$$COD_M = Q(COD_{inf} - COD_{eff}) - Y_{obs}QCOD_{inf} - V_R(\Delta X_R) \qquad (12.27)$$

where COD_M = COD converted into methane (kg COD/d);
$COD_{T,O}$ = influent total COD concentration (kg COD/m³);
$COD_{T,e}$ = effluent total COD concentration (kg COD/m³);
$COD_{T,s}$ = solids COD concentration (kg COD/m³);
ΔCOD_R = change in COD inventory in the reactor (kg COD/m³);
Q = average flow (m³/d);
COD_{inf} = influent total COD concentration (kg COD/d);
COD_{eff} = effluent total COD concentration (kg COD/d);
Y_{obs} = coefficient of solids production in terms of COD = 0.11 to 0.23 kg COD$_{sludge}$/kg COD$_{applied}$;
V_R = volume of sludge blanket in reactor (m³); and
ΔX_R = change in reactor solids concentration (kg/m³).

The waste sludge term ($Y_{obs}QS_o$) only applies if solids are wasted separately. If solids are wasted in the effluent, then this term is not necessary. When evaluating operating facilities, attention must also be given to the net change in the solids inventory in the reactor. For design purposes, steady-state operation ($\Delta X_R = 0$) is assumed.

The volume of a mole of methane at reactor operating conditions of temperature and pressure can be calculated from the ideal gas law:

$$V_m = \frac{nRT}{P} \qquad (12.28)$$

where V_m = volume of one mole of gas (m³);
n = number of moles of gas;
P = pressure (atm);

Criterion/Parameter	Range of Values
Influent Distribution	
Diameter of the influent distribution tube (mm)	75 to 100
Diameter of the tube exit mouth (mm)	40 to 50
Distance between the top of the distribution tube and the water level in the settler (m)	0.20 to 0.30
Distance between the exit mouth and the bottom of the reactor (m)	0.10 to 0.15
Influence area of each distribution tube (m²)	2.0 to 3.0
Biogas Collector	
Minimum biogas release rate (m³/m²/h)	1.0
Maximum biogas release rate (m³/m²/h)	3.0 to 5.0
Methane concentration in the biogas (%)	70 to 80
Settler Compartment	
Overlap of the gas deflectors in relation to the opening of the settler	0.10 to 0.15
Minimum slope for settler walls (°)	45
Optimum slope of the settler walls (°)	50 to 60
Depth of the settler compartment (m)	1.5 to 2.0
Effluent Collector	
Submergence of the scum baffle or the perforated collection tube (m)	0.20 to 0.30
Number of triangular weirs (units/m² of the reactor)	1 to 2
Production and sampling of the sludge	
Solids production yield (kg TSS/kg $COD_{applied}$)	0.10 to 0.20
Solids production yield, in terms of COD (kg COD_{sludge}/kg $COD_{applied}$)	0.11 to 0.23
Expected solids concetration in the excess sludge (%)	2 to 5
Sludge density (kg/m³)	1020 to 1040
Diameter of the sludge discharge pipes (mm)	100 to 150
Diameter of the sludge sampling pipes (mm)	25 to 50

TABLE 12.23 Other Design Criteria for UASB Reactors Treating Domestic Sewage (Chernicharo, 2007; reprinted with permission from IWA Publishing)

R = universal gas constant;
= 8.2057 4587 × 10⁻⁵ atm·m³/mol·K; and
T = operating temperature (K).

The volume of methane produced is then calculated from the COD equivalence of methane:

$$Q_M = COD_M \frac{V_m}{K_{COD}}$$

(12.29)

where K_{COD} = COD corresponding to one mole of methane (64 g COD/mol); and
Q_M = volume of methane gas produced (m³/d).

Theoretical methane production from anaerobic treatment is 0.35 Nm^3 CH_4/kg COD removed (Nm^3 = volume at 273 K and 1 atm), although total biogas production will be about 0.5 m^3/kg COD removed assuming the biogas is 70% methane. The actual methane yield will depend on the substrate composition, sulfate concentration, water temperature (because it changes the solubility of methane), and conversion of some substrate to substances not oxidized in a COD test. Reported production rates vary from 0.06 to 0.25 m^3 CH_4/kg COD removed (Arceivala, 1995; Giraldo et al., 2007; Noyola et al., 1988).

Methane has a solubility in water of about 1 mmole/L at atmospheric pressure, which is equivalent to 64 mg/L of influent COD converted (van Haandel and Lettinga, 1994). Loss of methane in the effluent and from the reactor surface can be a significant fraction of the total methane generated from low-strength domestic wastewaters, especially at lower temperatures since methane solubility in water increases with decreasing temperature. Methane oversaturation has also been observed in anaerobic reactors from 1.34 to 6.9 times saturation values due to mass transfer limitations or losing up to 100% of the methane generated in the effluent. Thus, current research is focusing on utilizing methods to extract dissolved methane from the effluent including degassing membranes and downflow hanging sponge (DHS) reactors (Crone et al., 2016).

5.4.3.4 Sludge Production Sludge yields in anaerobic systems are directly related to the COD converted to methane, the class of organic compounds degraded, and the concentration of inert solids in the feed. Biomass yields in anaerobic processes range from 0.35 g/g COD for carbohydrates and 0.20 g/g COD for proteins down to 0.038 g/g COD for fats (Speece, 1996). Sludge yields from the anaerobic treatment of municipal wastewater will be higher due to the presence of inert solids. Reported values of the sludge yield from UASB processes treating domestic wastewater range from 0.10 to 0.20 g TSS/g COD applied, although higher values have been reported (van Haandel and Lettinga, 1994; Yu et al., 1997). Because loss of suspended solids in the effluent can be significant, the actual sludge production can be significantly less.

5.4.3.5 Alkalinity Because of the relatively high partial pressure of CO_2 in enclosed anaerobic reactors, sufficient alkalinity must be present in the wastewater to prevent depression of the pH below 6.0 to 6.5. For low-strength domestic wastewaters, however, supplemental alkalinity typically is not required (van Haandel and Lettinga, 1994). Alkalinity may vary from the bottom to the top of the sludge blanket of UASB reactors. To maintain neutral pH concentrations in the base of the reactor with wastewaters that have low alkalinity, low nitrogen, and high organic concentrations, supplemental alkalinity may be required. The alkalinity requirement can be reduced by recycling a portion of the flow from the top of the reactor to the base of the reactor (Speece, 1996; Wentzel et al., 1994). More information on the chemical equilibria in anaerobic reactors can be found in Speece (1996) and van Haandel and Lettinga (1994).

6.0   Membrane Bioreactors

6.1   Introduction

An MBR is a combination of suspended-growth activated sludge biological treatment and membrane filtration equipment performing the critical solids-liquid separation function that is traditionally accomplished using secondary clarifiers. Low-pressure membranes (either microfiltration [0.07 to 2.0 μm] or ultrafiltration [0.008 to 0.2 μm]) are typically used in MBRs (Metcalf & Eddy, Inc./AECOM, 2014).

There are two general types of membrane systems that can be used in MBRs: (1) pressure driven (in-pipe cartridge systems that are located external to the bioreactor); and (2) vacuum driven (immersed systems that are designed for installation within the bioreactor). Immersed membrane technologies using hollow-fiber or flat-sheet membranes are the most popular because they operate at lower pressures (or vacuums), can more readily accommodate variations in solids, and typically provide a lower life-cycle cost, particularly for municipal facilities. Pressure-driven systems are more prevalent in industrial systems where waste characteristics, such as high temperatures, require the use of ceramic membranes. In its simplest form, an immersed MBR can combine the functions of an activated sludge aeration system, secondary clarifiers, and tertiary filtration in a single tank. In most cases, however, membranes are immersed in a tank separate from the bioreactor.

6.2 Components and Configurations

Unlike clarifier-based activated sludge processes, MBRs use membranes to separate biological solids from the mixed liquor. The membrane pore sizes are minute—often smaller than the pore sizes of filter papers used for laboratory analysis—so the separation of solids from liquids is essentially complete, and all biological solids are retained in the process for use as return sludge or for wasting.

Although pore sizes are minute in the microfiltration or ultrafiltration range, they are not able to capture soluble organic compounds, metals, or trace contaminants such as pharmaceutical and personal care products (PPCPs), priority pollutants, or endocrine disrupting compounds (EDCs). Although the biological process of an MBR may adsorb or reduce such contaminants, the filtration mechanism is not adequate to directly filter these materials from the wastewater (Maeng et al., 2013; Ternes and Joss, 2006; WERF, 2007).

It is important to understand the distinction between membrane equipment systems and MBRs. MBRs are biological processes that use membranes for the separation of the mixed-liquor solids from the water that will be discharged. Under current practice, membrane equipment systems include membranes, frames, programmable logic controllers (PLCs), and other critical elements such as permeate pumps and turbidity instrumentation. There are several manufacturers who produce membranes and membrane equipment systems for use in MBRs. There are also several firms that represent specific manufacturers and/or offer package MBR systems that include biological process design as well as membrane equipment. Most offer a choice of purchasing the equipment only or of purchasing a package that includes equipment and process design responsibility.

6.2.1 Responsibility for Process Performance

The process performance of an MBR system is often regulated by effluent concentrations of BOD, COD, ammonia, TN, phosphorus, TSS, and turbidity. Membrane equipment can only control the concentration of the TSS and turbidity. The remaining criteria are governed by biological process design and area affected by SRT, dissolved-oxygen concentrations, recirculation rates within the process, volatile acid concentrations, and other design parameters.

6.2.2 Overview and Applications

The MBRs are used to treat both municipal and industrial wastewaters. There are numerous potential benefits of MBR systems:

- Biomass is completely retained resulting in consistently high-quality final effluent with suspended solids concentrations of less than 1 mg/L and of suitable quality to be used as feed for reverse osmosis systems.
- Compared to CAS facilities with clarifiers, the effluent quality is less dependent on the MLSS concentration and sludge properties.
- Secondary clarifiers and effluent filters can be eliminated, thereby reducing facility footprint.
- Because systems can operate at high MLSS concentrations, for a given biomass inventory, the aeration basin volume can be reduced, further reducing facility footprint.
- For a given SRT, the volume of the bioreactor is less, because of the higher MLSS concentration. Alternatively, for a given process volume, an MBR process can operate at longer SRTs than a CAS facility, reducing sludge production.
- Capital costs have fallen significantly, although reinforced concrete costs have increased. The capital cost of a new MBR is often comparable or less than that for an equivalent conventional facility having tertiary filtration using granular media or membranes (Lin et al., 2014).
- Modular nature allows for ease of expansion and flexibility in configuration, making them a popular option for facilities looking to retrofit older technology.
- System can operate within a wide range of SRTs, resulting in increased flexibility and more options for optimization.
- System is robust enough to handle elevated MLSS concentrations for short periods of time allowing for flexible solids wasting schedules.
- Processes are easily automated; operator requirements are reduced because operators are not required to closely manage sludge settleability issues.
- A physical mechanism to remove pathogens is provided.
- Low-turbidity effluent reduces downstream disinfection requirements; high transmissivity means less energy required for UV disinfection. Effluent has minimal chlorine demand so less is required to achieve target residual concentration.

The MBRs often become viable for facilities requiring high-quality effluent for reuse or for discharge to sensitive receiving waters and for facilities with significant land area restrictions (both new facilities and retrofits).

The MBRs can offer attractive treatment options for ski, golf, and other resort communities that are not connected to a municipal sewer system and have a particularly high demand for irrigation water. The MBRs provide resorts with the ability to treat wastewater on-site in a compact facility and reclaim water for non-potable reuse.

Sewer "mining" or "scalping" facilities represent another potential application of MBR technology for water reclamation. In rapidly expanding suburban areas, potential users of reclaimed water are often not located near the main WRRF, and installing distribution systems to convey the reclaimed water can be difficult or expensive. By locating remote MBR facilities near reclaimed water users, these problems can be avoided. The "satellite" or scalping facilities can extract or mine wastewater from interceptor sewers and then deliver the treated effluent directly to users.

Although the advantages of MBR systems are numerous, MBRs are not suitable for every wastewater treatment application. Some potential disadvantages of MBRs include:

- Flows above design capacity in a CAS facility typically result in incomplete treatment and may result in overflows from the secondary clarifiers. In a membrane facility, however, there is a hydraulic limitation to how much treated water the membranes can produce. If the actual flows are greater than design or membranes are fouled during a high-flow period, then flows beyond system capacity will need to be diverted to another location for treatment or contained in a holding tank. As an alternative, MBR system tanks can be designed with additional freeboard to hold volume in emergency situations.

- In light of the hydraulic limitations inherent in any membrane system, particular attention must be paid to redundancy and availability of spare parts for all system components.

- Limited peaking ability to handle typical influent peak-flow conditions. As a result, MBRs often are designed with a maximum daily or hourly flow peaking factor of 2.0 to 2.5. Any flows beyond this threshold can be equalized in an upstream holding tank, equalized within the freeboard volume of the bioreactor tanks, or additional membranes can be installed to provide the required peak capacity.

- In some cases, peak-flow capacity can be affected by polymers introduced into facility return streams or by contaminants in the raw wastewater.

- As MBRs are a relatively new technology, there is a limited amount of data available to verify long-term performance.

- Since there are limited data available, it is difficult to predict life expectancy and long-term performance. However, information provided from existing full-scale facilities and from the membrane manufacturers Kubota and GE-Zenon that it is possible to achieve membrane lifetimes of greater than 10 years (Lin et al., 2012; Santos et al., 2011).

- Operating conditions often favor the formation of foam. Newer MBR facilities are taking this into account during the design stage and include foam management options such as surface wasting to prevent accumulation.

- Proper care must be taken to optimize chemical usage for membrane cleaning to limit the effect of purchasing chemicals on operating costs.

- Facilities consume more energy than CAS facilities operating at an equivalent SRT. Major consumers of energy include air scour blowers and RAS recycle pumps.

- Membrane replacement cost affects life-cycle cost analyses.

- Although highly automated, and often remotely controlled, MBRs must be closely monitored to detect changes in flux rates and permeability before they escalate. Maintaining a proactive cleaning schedule can help avoid such situations.

- Membrane equipment systems are unique, having different configurations and shapes depending upon the manufacturer. There is, therefore, a need to prepurchase or preselect the membrane manufacturer before completion of the design phase (Santos et al., 2011).

- To achieve reliable membrane performance, a minimum SRT is required in the MBR to adsorb and synthesize wastewater constituents before their exposure to the membrane. The minimum SRT required for effective membrane performance coincides with the minimum SRT associated with nitrification. Systems are, therefore, more economically applicable to nitrifying systems.

6.2.3   Influent Quality

Just as with any other activated sludge process, the influent wastewater quality to an MBR can vary significantly with geographic location and composition (proportion of domestic/industrial wastewater). Although pretreatment to remove grit and screenable materials is critical for operation and maintenance of membrane systems, in many respects, influent quality is not as important as MLSS concentration and SRT. These parameters define the quality of the material to which the membranes will be exposed and within which the membranes will be expected to operate.

6.2.4   Effluent Quality

The effluent from an MBR process is essentially free of suspended solids and macrocolloidal material. An MBR facility also can be designed to remove nutrients from the wastewater, similar to conventional BNR processes, with minor modifications.

Typically, effluent from MBR facilities contains less than 1 mg/L TSS, less than 5 mg/L cBOD, less than 0.2 nephelometric turbidity units (NTUs), and low levels of bacteria. When membranes with pore sizes in the micro- and ultrafiltration ranges are used, MBRs alone will not remove dissolved solids, and the membranes themselves will have no effect on pH or alkalinity.

Effluent from MBR facilities can be discharged to sensitive areas, reused on public access sites, or further treated by nanofiltration or reverse osmosis. Table 12.24 summarizes the typical effluent quality produced from a municipal MBR facility that is designed to achieve nutrient removal.

6.3   Process and Equipment Design Approach

6.3.1   Biological Process Design

Biological design of MBR systems has been reported for a variety of different combinations of effluent criteria relating to ammonia, TN, and phosphorus. Design criteria are, therefore, becoming available for distinct types of treatment applications including nitrification; nitrification with chemical addition for phosphorus removal; nitrogen removal through nitrification and denitrification; nitrogen removal with chemical phosphorus removal; and combined biological nitrogen and phosphorus removal. Figures 12.57 to 12.60 provide simplified flow schematics of MBR systems, illustrating a few basic process configurations.

Parameter	Units	Values
cBOD$_5$	mg/L	<5
TSS	mg/L	<1
Ammonia	mg/L as N	<1
Total nitrogen (with pre-anoxic zone)	mg/L	<10
Total nitrogen (with pre- and post-anoxic zones)	mg/L	<3
Total phosphorus (with chemical addition)	mg/L	<0.2 (typical) <0.05 (achievable)
Total phosphorus (with Bio-P removal)	mg/L	<0.5
Turbidity	NTU	<0.2
Bacteria	Log removal	Up to 6 log (99.9999%)
Viruses	Log removal	Up to 3 log (99.9%)

TABLE 12.24   Typical Municipal MBR Effluent Quality

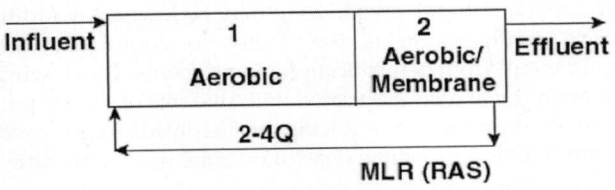

FIGURE 12.57   Nitrifying membrane bioreactor.

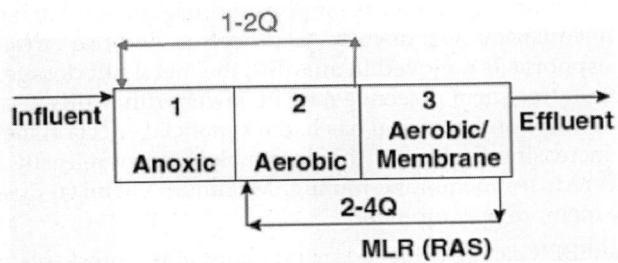

FIGURE 12.58   Membrane bioreactor with two-stage pumping for nitrogen.

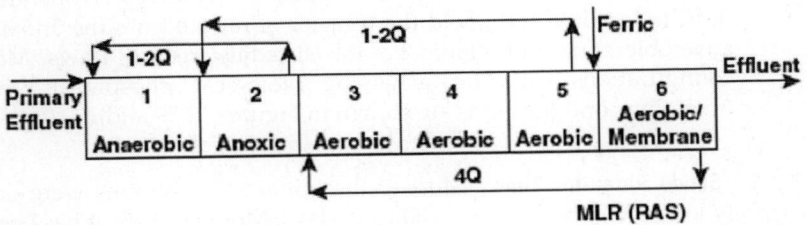

FIGURE 12.59   Traverse City membrane bioreactor design for nitrogen and phosphorus removal.

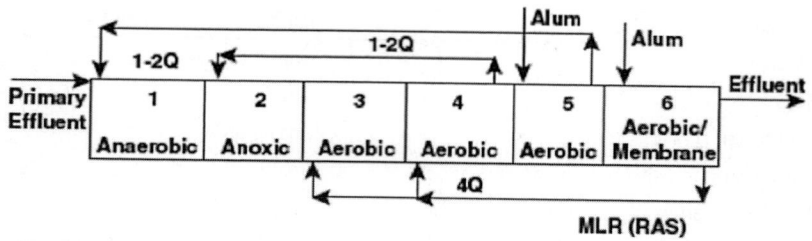

FIGURE 12.60   Loudon County, Virginia, membrane bioreactor.

In addition to the configurations illustrated above, various advanced biological processes can be combined or incorporated into an MBR design. Some of these include:

- Nitrogen removal incorporating recycle of mixed liquor to the upstream aerobic zone, combined with recycle of mixed liquor from just before the membrane zone to the anoxic zone (Figure 12.58). In this way, oxygen concentration in the stream being recycled to the anoxic zone may be lower. An additional benefit of a dual-recycle configuration is the ability to completely decouple solids recycle requirements from denitrification requirements. The downside is the higher capital and operational costs associated with two sets of recycle pumps instead of one. An alternative approach for reducing the dissolved-oxygen concentration in the recycle stream is to design a small deaeration zone upstream of the anoxic zone.

- Supplemental addition of an external carbon source such as methanol to a postanoxic zone to further enhance denitrification can be effective, particularly in facilities required to reduce TN to less than 5 mg/L.

- The addition of chemicals for phosphorus removal can be practiced with MBRs in a similar manner as for CAS processes. Because virtually all the particulate phosphorus is removed in an MBR, the metal salt dosage required to achieve a certain treatment objective may be lower with MBRs. Occasionally, metal salts for phosphorus removal has had a beneficial effect on membrane permeability, as increasing the size of the flocs makes for more easily filtered mixed liquor, and reduces membrane fouling. Maximum chemical doses may be limited by the membrane equipment.

- The EBPR can be achieved using many of the processes proven to support the growth of PAOs. In this case, the preservation of soluble organic material is even more important than for nitrogen removal, and the unintended recycle of dissolved oxygen to the anoxic zone must be avoided. The operator also would need to monitor and avoid the transfer of nitrate from the anoxic zone to the anaerobic zone and should be able to adjust recycle flows. Many alternate configurations exist for achieving biological phosphorus removal. Two operating configurations are shown in Figures 12.59 and 12.60.

6.3.1.1   Solids Retention Time   Most of the initial MBR systems were designed with extremely long SRTs, ranging from 30 to 70 days. More recently, it has been found that the optimal design SRTs for MBRs range from 20 to 50 days. This is due to the increased production of extracellular polymeric substances (EPS) and soluble microbial products (SMP) when operating at too short or too long of an SRT. Increased EPS and SMP production has been correlated to increased membrane fouling propensity. Therefore, controlling or reducing EPS and SMP production will reduce membrane fouling. The EPS and SMP production is a highly complex process and is controlled by not only operating at an optimal SRT but also by operating the activated sludge process at favorable growth conditions such as preventing extreme temperature fluctuations, shock organic loading, pH fluctuations, low dissolved-oxygen concentrations, and high salinity concentrations (Lin et al., 2014). Also, longer SRTs improve the microbial diversity of the bioreactor and have been correlated with improved biodegradation of pharmaceuticals in wastewater (Maeng et al., 2013).

To maintain membrane performance, a minimum retention of the raw wastewater before exposure to the membranes is required. This retention is required to allow influent colloidal matter to be adsorbed into flocs before reaching the membranes.

6.3.1.2 Mixed-Liquor Suspended Solids Concentration

Immersed MBR systems typically have operated with MLSS concentrations in the membrane tanks between 8000 and 12 000 mg/L, with occasional operation between 15 000 and 18 000 mg/L. Operating in this range has reduced the bioreactor volume required and minimized waste sludge handling and stabilization. High MLSS (>20 000 mg/L), however, have been shown to reduce membrane permeability because of greater SMP production and to reduce the aeration alpha factor, leading to higher aeration energy requirements (Shen et al., 2012). Current design practice is to assume the MLSS concentration to be closer to 8000 to 10 000 mg/L to ensure reasonable OTE. Operators should carefully monitor the MLSS concentration to ensure that it does not become excessively high and does not exceed the manufacturer's recommendation.

6.3.1.3 Oxygen Transfer

At MLSS concentrations higher than intended by the design process, the demand for oxygen can increase significantly, because of the higher concentration of biological activity and higher associated SRT. In some cases, demand can exceed the volumetric capacity of typical oxygenation systems. The operator may observe a decrease in the dissolved-oxygen concentration that can be maintained in the aerobic zones. The oxygen-transfer capacity of the aeration system must also be carefully understood. Further, the high MLSS concentration itself may affect transfer efficiency by reducing the alpha factor as indicated in Figure 12.61 (WERF, 2002). Immersed membranes typically are provided with shallow coarse-bubble air to agitate the membranes as a means to control fouling. This "membrane aeration" provides some oxygenation, but at low efficiency.

Typical dissolved-oxygen concentrations in the various zones of an MBR process are:

- Anaerobic: 0.0 to 0.1 mg/L;
- Anoxic: 0.0 to 0.5 mg/L;
- Aerobic: 1.0 to 3.0 mg/L; and
- Membranes: 2.0 to 6.0 mg/L.

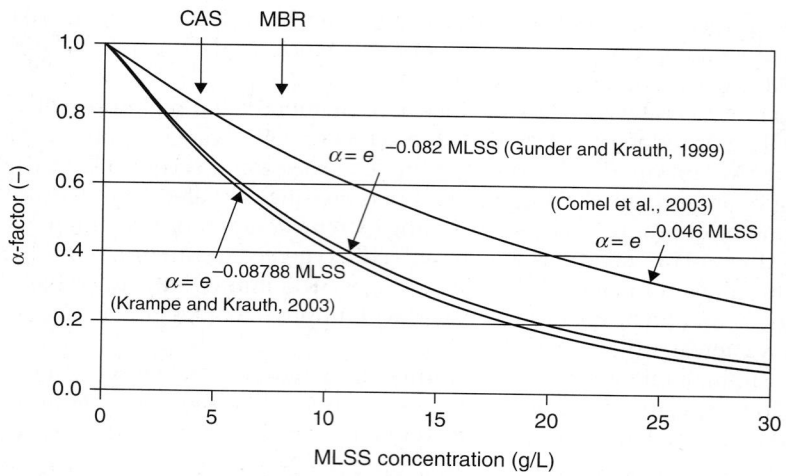

Figure 12.61 Effect of MLSS concentration on aeration alpha factor (WERF, 2002; reprinted with permission of the Water Environment Research Foundation).

6.3.2 Equipment System Design

The primary difference of MBRs compared to traditional biological suspended-growth systems is that microorganisms are more completely separated from the water using membranes, resulting in a higher quality effluent than can be achieved with secondary clarifiers and granular media filters.

For successful operation, membrane equipment systems for MBRs include various combinations of air scour, backwash, relaxation, and chemical cleaning systems to maintain performance and permeability. Membranes can be operated across a range of flows, or flux, as long as an adequate differential pressure is provided across the membranes. Immersed MBR systems have limited differential pressure (less than 10 psi), and, hence, flux rates are more restricted than with external, pressure-driven MBRs. Experience with membrane equipment suggests that average flux between 14 to 25 L/m²·h (8 to 15 gpd/sq ft) are sustainable when the MLSS are 15 000 mg/L or less. Typical membrane equipment system components may include but is not limited to membranes and support frames; permeate and backpulse pumps; PLCs, instrumentation, and controls; air scour delivery systems; and membrane cleaning delivery systems.

The two main subgroups of MBR configurations are immersed membrane systems and in-line membrane systems. Although in-line systems operate at higher flux and can require a smaller footprint than immersed MBR systems, the latter are more common in municipal applications because of significantly lower operating costs and less frequent cleaning requirements.

Some types of immersed membranes for use in MBRs include:

- Unsupported hollow-fiber membranes;
- Reinforced hollow-fiber membranes;
- Stationary flat-plate membranes; and
- Rotating flat-plate membranes.

Depending on the manufacturer, however, membranes also may differ in pore size, composition, cassette configuration, operating procedures, and maintenance requirements. Details on specific products must be obtained directly from the manufacturer. (Note: In this manual, the term cassette refers to the largest membrane assembly removable by a crane. Depending on the manufacturer, this also may be referred to as a rack or module.)

6.3.3 Equipment System Procurement

An ideal time to procure the membrane equipment system may be after the preliminary process design and facility layout is completed, and before the detailed design of facility structures or equipment systems. The completed preliminary process design should determine the SRT, process flow diagram, and bioreactor zone volumes, complete with the associated MLSS concentrations and predicted treated effluent quality. This information defines the specific conditions within which the membrane equipment system will be required to perform its solids-liquid separation function, and which are required to be included within the membrane equipment bid documents, such as SRT, temperature, MLSS concentration, and the addition of metal salts, if any. At this point, various design flows and durations are known, as are the redundancy requirements. This early procurement also allows for a competitive process and defines the responsibility of each party.

Many design aspects of an MBR facility are dependent upon the specific requirements and configurations of the selected membrane equipment. This equipment-specific information is required before the initiation of final design. This is especially true for submerged membrane systems where tank volumes and shapes and equipment piping design are unique for each manufacturer. The membrane tank dimensions, nature and design of the backpulse and chemical cleaning systems, the building layouts for the equipment, and the size and operation of the blowers that provide air to scour the membranes are all affected by the particular type of membrane equipment that is selected.

6.4 Pretreatment

6.4.1 Fine Screens

It is normal practice that fine-screening equipment with a maximum of 1- to 2-mm openings is provided to protect membranes from debris and fibrous materials. Typically, these screens are installed downstream of 6-mm screens, either at the headworks or following primary clarification. Additional measures include placing covers over the membrane tanks, or fine screening a portion of the mixed liquor as it is returned from the membrane tanks to the bioreactor.

6.4.2 Primary Clarifiers

Primary clarification is not specifically required for an MBR, although, just as for other activated sludge systems, they can reduce the total energy required for aeration and overall volume of the bioreactor. Primary clarification provides the additional benefit of settling out some of the undesirable trash and skimming off scum and floatables that would otherwise be removed by fine screening. Some membrane manufacturers will allow less stringent fine screening if primary clarification is included in the process.

6.5   Membrane Bioreactor Design

6.5.1   Mixed-Liquor Recycle Pumping

Just as in any CAS process, sludge must be recycled from the solids-liquid separation device back to the front of the biological process to redistribute biomass. In the case of MBRs, the recycle can be 200% to 400% of the facility flow, and a minimum recycle is required to flush the membrane area and to control the concentration of MLSS in the area of the membranes. If the recycle rate is too low, then the MLSS in the membrane tank will escalate rapidly, making operation unsustainable. The main objective of the high recycle rate is to redistribute the sludge inventory and to minimize membrane fouling associated with elevated MLSS concentrations.

It is important to note that the recycle of mixed liquor from the membrane area can contain high concentrations of dissolved oxygen, approximately 2 to 6 mg/L, instead of the virtual absence of dissolved oxygen in the return sludge from a clarifier. This oxygen cannot be controlled because the airflow provided by the air scour delivery system must be sufficient to provide a minimum shearing action across the surface of the membranes. If mixed liquor is recycled from the membrane tank to an anoxic zone, then the denitrification process will be less efficient. To account for the elevated dissolved oxygen in the recycle stream, anoxic zones in MBRs with single recycle streams and without dedicated deaeration zones must be larger to compensate for the reduction in denitrification efficiency. Whereas a CAS facility may have an anoxic zone that is 15% to 20% of the total bioreactor volume, an MLE-based MBR facility with a single recycle stream may have one that is 20% to 40%. Alternatively, mixed liquor can be recycled from the membrane tank to an aerobic zone, and then the aerated mixed liquor at lower dissolved-oxygen concentration can be pumped to the anoxic zone. This reduces the anoxic zone volume requirement and conserves soluble substrate for denitrification.

Mixed-liquor recirculation in an MBR system can be designed in one of two ways:

1. Pumping mixed liquor from the bioreactor to the membrane tanks and returning the mixed liquor from the membrane tanks to the bioreactor by gravity; or

2. Allowing gravity flow of mixed liquor from the bioreactor to the membrane tanks and pumping of the mixed liquor from the membrane tank to the bioreactor.

The first approach requires pumping $(R + 1)Q$, whereas the second requires pumping RQ (where R is the mixed-liquor recirculation ratio and Q is the influent flow). Different pump types can be used for mixed-liquor recirculation, although submersible or high-capacity end-suction centrifugal pumps are most common. Axial-flow pumps are well suited for this application because of the high flow and low head requirements.

Recirculation pumps should be sized to ensure:

- Sufficient flow to avoid buildup of MLSS in tanks and ensure proper solids distribution between the biological process tanks and membrane tanks; and

- Sufficient nitrate return flow to the unaerated zones at the head of the bioreactor to achieve required levels of predenitrification and the target effluent nitrate concentration.

One membrane manufacturer designs their facilities using a proprietary jet aeration system to scour the membranes. In these systems, the recirculation pumps are used to achieve the proper mixed-liquor flowrate and head through the jets at the base of the membrane modules. These systems require that the mixed liquor be pumped from the bioreactors to the membrane tanks.

In systems where the mixed-liquor recirculation is used solely to dilute solids in the membrane tank, the recirculation pump is sized for two to four times the annual average or maximum month rates of flow. If the pumps are also used as part of a denitrification system, then they may be sized as large as three to eight times the average daily flow. Some systems use recirculation pumps as part of the two-phase jet system, which combines fluid transfer with air scour energy. Total dynamic head is based on headlosses through the pump and piping systems, including the jets where applicable.

6.5.2 Mixing

The unaerated zones in the bioreactor typically are equipped with a dedicated mixer in each zone to keep the solids in suspension. Unaerated zones may include deaeration (or deoxygenation) zones, preanoxic and postanoxic zones, and anaerobic zones. Although submersible mixers are common, vertical shaft fixed-mounted mixers also can be used. The mixers ensure adequate mixing within each zone for proper contact between biomass, substrate, and electron acceptor (nitrate or oxygen) and complete volume use without short-circuiting across any zone.

6.5.3 Aeration (Bioreactors, Membrane Tanks)

Biological process aeration is provided by typical aeration systems used in CAS processes, including fine-bubble aeration, coarse-bubble aeration, and jet aeration. Fine-bubble diffusers with full floor coverage are the most common type of aeration system used in the aerobic zones of bioreactors because of their higher OTE. Tubular or disc-type membrane fine-bubble diffusers are most typically used.

Air scour is used by immersed membrane systems to create shear forces and turbulence across the surface of the membranes to keep the solids away from the membranes and to maintain optimum conditions for flow through the membranes. Typical rates of air scour are 0.2 to 0.6 Nm^3/h per square meter of membrane area (0.01 to 0.03 scfm/sq ft). Most air-scour systems operate continuously, although some systems include intermittent or varying flow. If the air scour system fails, then the transmembrane pressure (TMP) will rise quickly, possibly causing alarm conditions and the need for cleaning to restore normal operation. Due to the high airflow requirement to scour the membranes, air scouring is the most energy-intensive process in MBR treatment, which in the past caused energy requirements for MBR treatment to be three times greater than CAS treatment. However, in recent years, membrane manufacturers have developed much more efficient air scouring systems that reduce energy requirements of MBR processes so that energy requirements are only 10% to 30% greater than CAS with tertiary treatment and a finer quality effluent is produced with MBR treatment (Krzeminski et al., 2012).

6.5.4 Permeate Pumping and Gravity Permeation

The driving force for membrane filtration can be accomplished by either a pumping system or by gravity. External membrane systems can only use a pumping system because the mixed liquor must be pumped across the face of the membranes. With immersed membranes, permeation is achieved by applying slight suction to draw the clean water through the membrane. Gravity siphon systems can be used if site conditions are suitable, whereas pumped permeation systems can be used in all cases.

There are many possible configurations for permeate pumping systems and many different types of permeate pumps. The simplest configuration is a dedicated permeate pump per membrane train. Permeate pumps can be end-suction centrifugal or positive displacement (PD) rotary lobe type. Each membrane train is equipped with a permeate header that connects all the membrane cassettes within the train. When end-suction centrifugal pumps are used, some means of removing entrained air from the permeate needs to be included to prevent the pumps from losing prime. An air separator connected to a vacuum pump or a venturi system can be used to remove entrained air.

Because rotary lobe pumps can handle a higher percentage of entrained air, the permeate header typically is connected directly to the suction side of a self-priming rotary lobe pump without any air separator.

A feature of rotary lobe pumps is that they can reverse the direction of flow by reversing the direction of lobe rotation. When rotary lobe pumps are used in MBR systems, they most often serve double duty as both the permeate and backwash pumps.

The permeate pump typically is equipped with a variable frequency drive (VFD). A dedicated magnetic flow meter and a turbidi meter typically are located downstream of each permeate pump.

6.5.5 Instrumentation and Process Control Systems

Each membrane equipment manufacturer assembles its system to include a variety of monitored and controlled instrumentation and equipment. The PLCs provide several critical functions: monitoring equipment alarms and setpoints; trending of operating information such as transmembrane pressure and flow; controlling or shutting down equipment; automating control of certain operating procedures; and executing operator-initiated or event-triggered activities.

Membrane PLCs typically will be connected to the facility PLC or supervisory control and data acquisition (SCADA) system for the exchange of operating information and for the transfer of commands such as the control and sequencing of events that involve both membrane and general facility equipment. Examples of the latter include the coordination of valves, pumps, and gates for the isolation of a membrane tank for a chemical cleaning procedure.

6.6 Membrane Bioreactor System Equipment

As described earlier in this chapter, a typical facility that includes an MBR system consists of the following major unit processes:

- Preliminary treatment system (headworks);
- Biological process tankage (bioreactor);
- Biological process blowers;
- Membrane filtration system;
- Air scour blowers;
- Backpulse system (manufacturer-dependent);
- Mixed-liquor recirculation system;
- Cleaning system;
- Posttreatment; and
- Waste sludge treatment and disposal.

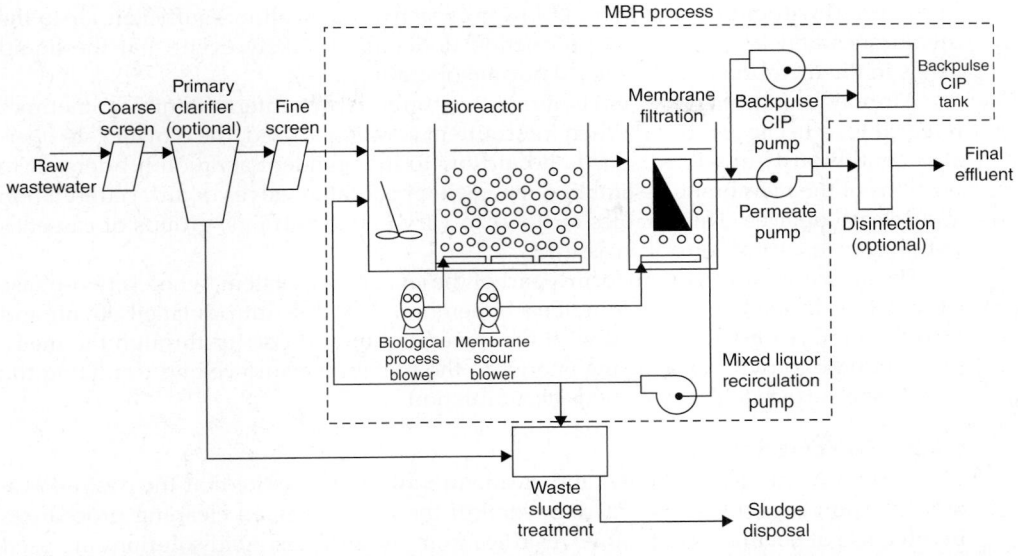

FIGURE 12.62   Process flow diagram of a typical membrane bioreactor facility (courtesy of CH2M Hill).

A schematic of the major unit processes of a complete MBR facility is shown in Figure 12.62. The equipment described in this section will be limited to the MBR process equipment, specifically the equipment used in the bioreactor and in the membrane filtration system.

6.6.1   Process Air and Air Scour Equipment Systems

6.6.1.1   Biological Process Blowers   Depending on the size of the facility, the biological process aeration blowers can be either PD or centrifugal. The process blower system is designed as a common group of blowers (duty plus online standby) that provides air to all biological process trains. All of the blowers discharge into a common air supply manifold that delivers air to the individual diffuser grids in each aerobic zone. The process aeration blowers typically are separate from air scour blowers, although the two systems may share a common standby.

6.6.1.2   Air Scour Blowers   The membrane air scour blower system typically is designed as a common group of blowers with installed standby units. The PD or centrifugal blowers are used. All blowers discharge into a common membrane air manifold that delivers air to the air header above each membrane tank (train). Each membrane cassette is connected to the air header above each membrane tank using flexible hose or rigid piping.

Membrane manufacturers dictate the design airflow rate of the air scour blowers. Once the airflow rate per membrane cassette is specified, the air scour blower is sized based on the maximum cassette spaces in the tank and the maximum possible liquid level in the membrane tank. To supply the proper airflow rate for the number of membrane cassettes initially installed, the blower airflow rates could be reduced by adjusting the VFD, inlet control valves, or inlet control vanes or by resheaving the blowers. This approach oversizes the blowers, but it provides the flexibility to add membranes (if required) without having to add blower capacity. It also allows sufficient air to the membranes and keeps them in production under the unlikely event that the liquid levels in the membrane tanks exceed normal operating level.

Air may be supplied to the membranes continuously or intermittently (sometimes referred to as cyclic aeration). When intermittent aeration is used, the blowers are operated continuously at a fixed speed and airflow to independent aeration headers, or to portions of the membrane assemblies, using either actuated valves or air-accumulation and release devices. This enables airflows to vary between trains, groups of cassettes within a train, or portions of a cassette.

The air scour blowers may form a part of the jet aeration system, where a two-phase jet system is located at the bottom of each membrane module introducing both air and mixed liquor. Air bubbles blend with the mixed liquor and rise up through the membrane bundle, providing scouring energy to the membrane surface and fluidizing the membrane surface to prevent solids accumulation.

6.6.2   Cleaning Systems

Membranes must be maintained and cleaned regularly to ensure that the desired system filtration capacity is provided. Several of the recommended cleaning procedures involve use of chemicals to remove residues from the surfaces. Acid solutions are used to clean the membranes of inert deposits and dilute chlorine solutions eliminate organic growth and fouling.

Cleaning chemicals can be injected in-line as the membranes are backwashed or membranes can be cleaned-in-place (CIP). Cleaning solution from the backpulse tank is blended with the chemical flow before being backwashed into the membranes or transferred into the tank.

Other chemical feed systems that may be found in an MBR facility include:

- Coagulants for phosphorus removal (e.g., ferric chloride, aluminum sulfate (alum), sodium aluminate); and

- Sodium hydroxide or lime for pH and alkalinity control.

6.6.3   Chemical Feed Systems

Chemical cleaning can range from a fully manual procedure to a semi- or fully automated system. The fundamental task is the same: any membrane train can undergo a chemical clean without affecting the operation of other trains (other than increased flow to maintain capacity). This is achieved by adding (manually or automatically) a chemical solution to the inside and outside of the membrane with the membrane cassettes remaining in place within the membrane tanks, known as CIP. For smaller systems, individual cassettes can be temporarily located in a dedicated cassette cleaning tank. Cleaning chemicals used, frequency, and duration of cleanings are different for various membrane systems.

Chemical membrane cleanings can be classified into one of two types: maintenance or recovery cleaning. Maintenance cleaning events occur from once per day to once per week, and each is less than 2 hours in duration. The purpose of maintenance cleaning is to increase time between recovery cleans. Maintenance cleans use lower concentrations of chemical than recovery cleans. Recovery cleaning frequency varies between membrane manufacturers and installations but typically ranges from once every 2 months to once every 6 months. Duration of recovery cleans is from 6 to 24 hours.

A fully automated cleaning system consists of a backpulse water storage tank, backpulse pumps, a flow meter, and chemical metering systems. Stored permeate from the backpulse tank is backpulsed through the membrane at a specified flowrate, and the appropriate chemical is injected directly into the permeate header using the chemical metering pumps to achieve the desired chemical concentration. In some cases, the backpulse storage tank is equipped with a heating system to allow hot-water cleaning.

Separate chemical metering systems are used for each cleaning chemical used, which includes sodium hypochlorite and citric acid. Each metering station is equipped with an appropriate chemical holding tank, a pair of chemical dosing pumps (one duty, one online standby), and a calibration column. The chemical metering stations are designed to deliver chemical for the following functions:

- Sodium hypochlorite for maintenance cleaning, for recovery cleaning, and to flush the CIP/backpulse tank to prevent contamination and biogrowth in the tank; and

- Citric acid for maintenance and recovery cleaning.

6.6.4   Backpulse Pumping and Backwashing

Backpulse systems are only included in MBRs using hollow-fiber membranes. These systems are either equipped with a separate backpulse pump or use the permeate pump for backpulsing. During backpulsing, the direction of flow is reversed, and the membranes are flushed from the inside out using permeate stored in a backpulse tank. In some systems, the backpulse pump is also used for CIP. Reversible self-priming pumps can be used for dual duty: permeation and backpulsing. Centrifugal permeate pumps also can be used for backpulsing without the need for a separate backpulse pump by changing the direction of flow through use of automatic valves and piping.

For systems that require backpulsing, a portion of the effluent water is diverted into a holding tank that is used as the reservoir for both regular and chemically enhanced backpulsing. Larger MBRs with multiple trains may not require a backpulse tank if the permeate header is a large-enough reservoir. Because only one train at a time is typically in backpulse mode, the supply of permeate water will always exceed the demand if enough trains are in operation.

6.6.5   Waste Activated Sludge Management Systems

Most MBR applications have been in smaller facilities that have aerobic sludge stabilization followed by biosolids disposal, either by liquid land application or by dewatered biosolids reuse. At larger facilities, thickening the waste sludge for anaerobic or aerobic digestion and then dewatering the biosolids is more typical. To date, little research has been conducted on thickening and/or high-rate stabilization of waste MBR sludge. The research that has been conducted has not found any significant differences in sludge thickening or stabilization characteristics. It has been found, however, that gravity thickening is not as successful because of the already-high solids concentration of the mixed liquor before thickening (WEF 2006; WERF, 2002).

A few facilities also use membranes for thickening WAS before further treatment or disposal. These membrane thickening applications typically use the same type of membranes as are used in the MBR, but operate at much lower flux (3 to 8 L/m²·h [2 to 5 gpd/sq ft]). These systems are capable of thickening WAS up to 4% to 5% total solids concentration.

6.7   Anaerobic Membrane Bioreactors

Although MBRs tend to produce a higher quality effluent compared to CAS treatment, aerobic MBR treatment has a higher overall cost for both the initial capital investment and higher operational costs (Lin et al., 2011). Recent research has focused on the feasibility of using MBR treatment of municipal wastewater under anaerobic conditions (AnMBR) to significantly reduce operational costs. Not only will this remove the need for aeration to treat the wastewater, but anaerobic treatment will also result in the production of methane enabling the potential of the methane to be utilized to offset the energy requirements for treatment. However, anaerobic treatment alone is not sufficient to meet secondary treatment standards, especially at lower temperatures (<20°C). Thus, the addition of an MBR to anaerobic reactors has been investigated to see how this improves treatment efficiency and reduces energy requirements.

From lab-scale and pilot-scale testing of AnMBRs, a high-quality treated effluent has been achieved in terms of COD, suspended solids, and pathogen removal even at lower temperatures (Ozgun et al., 2013; Shin et al., 2014; Smith et al., 2013). However, it should be noted that AnMBR treatment is not capable of removing nutrients from wastewater, and therefore, the treated effluent has the potential to be used for agricultural purposes or additional treatment might be necessary to reduce nutrient concentrations. There is also yet to be a full-scale installation of an AnMBR. In order to make full-scale AnMBR installations a reality, future research needs to focus on less-energy-intensive membrane fouling prevention techniques and methods of removing dissolved methane in the permeate at low temperatures (Smith et al., 2013). Similar to aerobic MBR treatment, many AnMBR studies have used backflushing and biogas sparging, which have been successful in preventing membrane fouling; however, these methods are highly energy intensive. Shin et al. (2014) instead used granular activated carbon (GAC) in the AnMBR reactor to clean the membranes through physical contact and movement. This method was successful in producing a high-quality effluent and significantly reduced energy requirements for fouling prevention, and further research should continue to evaluate using GAC and other less-energy-intensive fouling control methods in AnMBRs (Shin et al., 2014).

7.0   Wet-Weather Considerations

7.1   Introduction

Since the passage of the Clean Water Act in 1972, nearly all municipal facilities in the United States have implemented a minimum of secondary treatment. Regulatory attention has shifted to the capture and treatment of wet-weather overflows and bypass flows that can significantly affect receiving water quality.

Suspended-growth systems typically are not designed to treat peak wet-weather flows and loads. Consequently, such facilities are unable to provide adequate treatment when significant wet-weather conditions occur. Suspended-growth systems are particularly sensitive to peak flows because of the potential washout of biomass when secondary clarification is overloaded. The loss of biomass can result in excess effluent suspended solids, decreased treatment, and slow, post-storm recovery. Nitrification and EBPR processes are slow to replenish losses of the key microorganisms involved.

Common practice has been to provide preliminary and primary treatment for all flows, bypass excess peak wet-weather flows around the secondary treatment processes, and mix secondary effluent with the bypassed flow before disinfection and discharge. The continuation of this practice for existing or new treatment facilities is subject to regulatory approval. At some facilities, a ballasted flocculation-type process (i.e., ActiFlo®) is used to treat wet-weather flow. This process can achieve sufficient solids removal due to the dilute BOD concentration of wet-weather flow that does not result in effluent limit exceedance. Effluent from the ballasted flocculation is disinfected before disposal. Many facilities are installed in the United States, and the largest one with 950 000 m³/d (250 mgd) peak-flow treatment capacity is under construction at the Blue Plains Advanced WWTP (Washington, DC).

This section outlines wet-weather management strategies that can be used to enhance treatment and minimize overflows. Some of these methods have been successfully implemented in full-scale applications although others have a limited track record.

7.2   Flow Reduction

Wet-weather flows to suspended-growth processes can be reduced by equalization, or diversion through parallel preliminary and primary treatment processes before mixing or discharge. Combinations of treatment and storage should be investigated as part of preliminary planning for most wet-weather treatment projects to establish the potential to optimize cost and pollutant removal efficiency.

7.3   Aeration Tanks

The following approaches decrease the MLSS concentration in the secondary clarifier feed, which reduces solids loading and increases the peak flows that can be accommodated.

7.3.1   Aeration Tank Settling

Aeration tank settling (ATS) as illustrated in Figure 12.63 describes the practice of turning off the air to all or just the later parts of aeration tanks during peak flows (Nielsen et al., 2000). Without aeration, the MLSS begin to settle, and the solids concentration sent to the secondary settling tanks is reduced. By reducing the suspended solids concentration during peak-flow events, clarifier capacity is increased when it is most needed. Most of the recent literature on this subject has been published by a manufacturer who has patented a version of ATS called STAR® ATS (WEF, 2005). This system combines ATS with an internal mixed-liquor recycle stream and a high-level process control system. The recycle stream transfers mixed liquor from the last zone of the aeration tank (without air or mixing) to a preaeration anoxic zone and extends the period of time for which ATS can be effective.

In another ATS concept, process air is turned off and the RAS flow is reduced to about 20% of the influent flow. The combination of reduced mixed-liquor concentration and RAS flow increases the clarifier hydraulic capacity by 50% during storms (Reardon, 2004).

An evaluation of the effect of ATS, based on common practice in the United States using the Vesilind equation with the Daigger SVI correlation for the settling coefficients, is presented in Figure 12.64 (WEF, 2005). The figure shows the estimated increase in clarifier capacity that results from a decrease in the mixed-liquor concentration. Assuming the secondary settling tanks are clarification limited, the effect of ATS is most

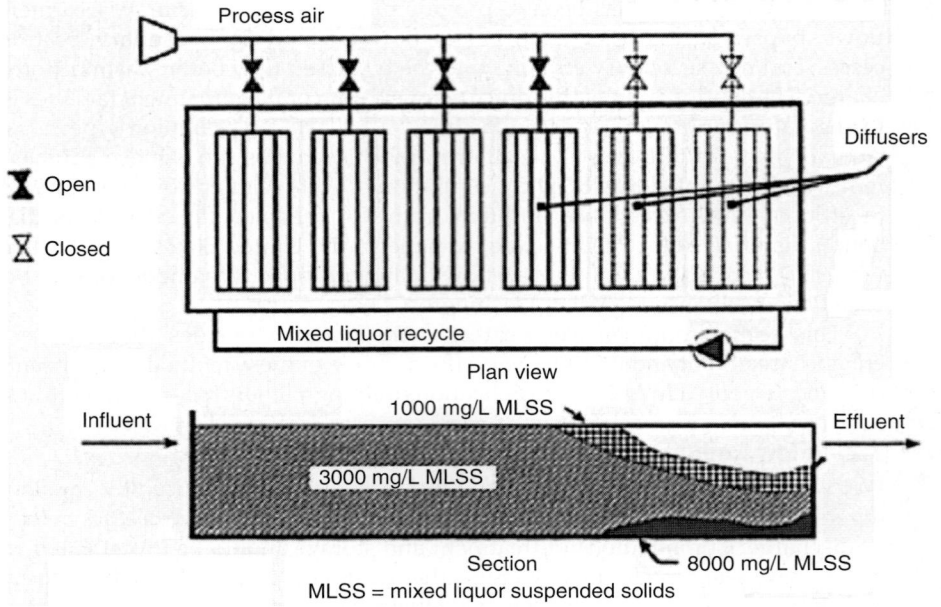

FIGURE 12.63 Aeration tank settling (Nielsen et al., 2000; reprinted from Water Science and Technology, with permission from the copyright holders, IWA).

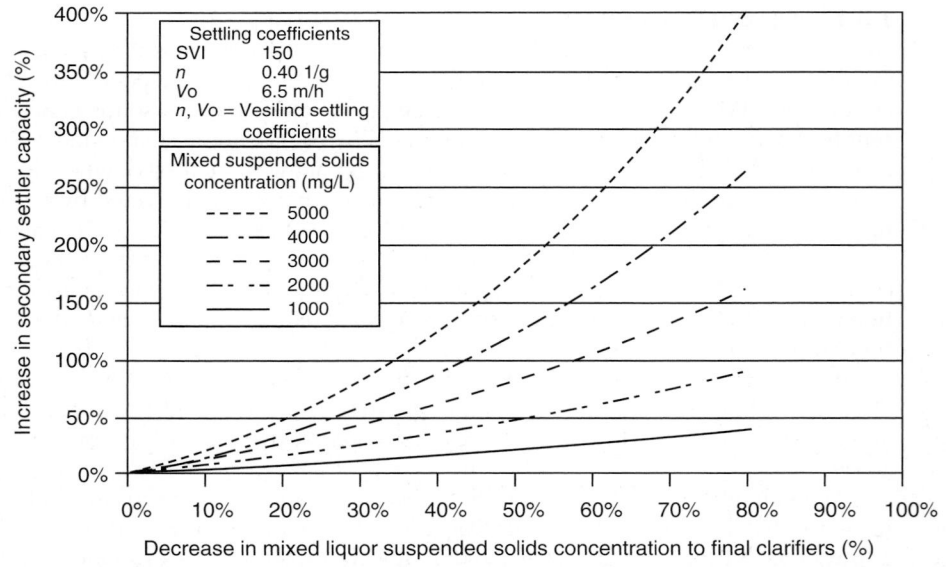

FIGURE 12.64 Aeration tank settling potential to treat peak flows (n = Vesilind coefficient calculated using the Daigger sludge volume index [SVI] correlation with an SVI of 150).

pronounced at higher mixed-liquor concentrations. For an SVI of 150 mL/g and a mixed-liquor concentration of 3000 mg/L, a 50% drop in the mixed-liquor concentration increases the clarifier capacity by more than 80%.

7.3.2 Step-Feed or Contact-Aeration Mode

Switching to a step-feed or contact-aeration mode during peak flows allows a greater mass of MLSS to be stored in the initial portions of the aeration tank and minimizes the MLSS concentration fed to the secondary clarifiers. Step-feed operation can provide a relatively high degree of treatment while accommodating higher flows. By varying the number and location of aeration tank feed points during wet-weather flow events, the suspended solids concentration in the aeration tank effluent can be reduced and the capacity of the secondary settling tanks increased significantly.

In conventional and complete mixed activated sludge processes, both the aeration tank influent and RAS are added to the beginning of the aeration tank resulting in a relatively uniform concentration of suspended solids throughout. By converting to step feed, an MLSS gradient can be created with high solids concentration at the beginning and a lower concentration at the end. This minimizes the solids loading applied to the secondary clarifiers and provides a greater solids inventory and larger SRT for a given tank volume. The step-feed configuration becomes a contact stabilization process when all the influent flow is added to a zone at the end of the aeration tank. With a contact stabilization mode of operation, a balance must be maintained between clarifier capacity increase and reduced process performance as the contact zone volume is decreased. Although process performance might suffer, the solids retained in the aeration tanks facilitate rapid process recovery after the high flows subside.

Research and full-scale implementation of step feed for control of wet-weather flows has demonstrated that secondary treatment standards can be met while switching between conventional and step-feed modes of operation (WEF, 2005). Switching to a step-feed mode can be difficult for BNR process configurations that need to retain nitrification and EBPR capabilities. The ease and cost of modifying an existing CAS process to be able to switch to a step-feed configuration during peak flows depends on the design of each facility. Care must be taken to provide adequate aeration capacity in zones not originally designed to receive influent flow. Likewise, the effect of transient solids load on the final clarifiers when switching to and from step feed must be accommodated.

7.4 Secondary Clarifiers

Secondary clarifiers often limit a facility's ability to treat wet-weather peak flows. Options to increase the secondary clarification capacity so that higher peak flows can be treated include:

- Polymer addition. One of the simplest wet-weather management strategies is to add polymer to the mixed liquor to increase solids settling velocity and reduce the SVI, thereby increasing the rated capacity of existing final clarifiers. Testing is required to determine polymer type and dose.

- Wet-weather clarifiers. Additional final clarifiers can be constructed to manage wet-weather flows. These units are placed online as needed to accommodate wet-weather peak flows. The East Bank Plant in Jefferson Parish, Louisiana, uses this strategy.

- Inclined plates and tubes. Plates or tubes installed at an angle in a clarifier will significantly increase the settling area available within a given footprint.

Inclined plates or tubes significantly increase the allowable upflow velocity in a clarifier (based on horizontal tank area) by increasing the settling area by a factor of approximately 8 to 10, thereby allowing a higher peak flow to be treated in a given tank surface area. They have been used in secondary clarifiers. Researchers in Germany have investigated their use at the end of the aeration tanks and at the entrance to the secondary settling tanks (Buer, 2002; Plass and Sekoulov, 1995). Plates or tubes in these locations reduce the MLSS concentration entering the secondary settling tanks, thereby increasing peak-flow capacity of the secondary settling tanks. Because of the potential for plugging when used for high-solids-concentration suspensions and because of algal growth, inclined plates or tubes might best be applied to separate wet-weather clarifiers that are used infrequently.

8.0   Oxygen-Transfer Systems

8.1   Introduction

Transfer rates for diffused air systems are typically reported as oxygen-transfer efficiency (OTE), expressed as a percentage or OTR, expressed in units of mass/time. Table 12.25 summarizes the general characteristics of primary types of oxygen-transfer equipment in use for 5 or more years. More detailed information is contained in the literature (U.S. EPA, 1989, 2010).

Mechanical devices are typically rated on the basis of OTR or aeration efficiency, expressed in units of mass/time/unit of power. As a secondary function, aeration devices furnish sufficient energy for mixing. Ideally, mixing energy should be sufficient to thoroughly disperse dissolved substrate and oxygen throughout a given segment of an aeration tank and keep MLSS suspended. This does not necessarily mean that both soluble and suspended materials should be uniformly mixed throughout the entire aeration tank. For example, plug-flow tanks and reactors with point-source oxygen addition do not rely on uniformity for proper operation.

The power required to satisfy oxygen demand depends on substrate and biomass concentrations, flowrate, and reactor volume. The power for mixing depends on aeration tank volume and, to a lesser degree, MLSS concentration. For certain combinations of biomass and substrate concentrations and other variables listed above, power requirements for mixing may exceed those for oxygen transfer. In systems with high biomass concentrations, oxygen demand will control power requirements; under low substrate loading, mixing may control power requirements; in plug-flow systems, mixing may control near the effluent end of aeration basin; for aerated lagoons, mixing often dictates power requirements. Additionally, some oxygen-transfer equipment has a minimum airflow rate that must be maintained for proper air distribution or equipment function. Such equipment requirements must also be considered when determining the minimum airflow requirement of the system.

8.2   Diffused Aeration

Diffused aeration, defined as the injection of a gas (air or oxygen) below a liquid surface, covers all equipment described in this section. Hybrid equipment that combines

Equipment Type	Equipment Characteristics	Processes where Used	Advantages	Disadvantages	Reported Clean Water Performance[a]	
					SOTE (%)	SAE (kg/kW·h[b])
Diffused air						
Porous diffusers	Ceramic, plastic, flexible membranes, dome, disk, panel, tube, plate configurations, total floor grids, single or dual roll, fine bubble	High-rate; conventional, extended, step, contact stabilization, activated sludge systems	High-efficiency; good operational flexibility; turndown approximately 5.1	Potential for air-or water-side clogging; typically require air filtration; high initial cost; low alpha	15–45	1.9–6.6
Nonporous diffusers	Fixed orifice; perforated pipe, sparger, slotted tube, valved orifice, static tube; coarse bubble; typically single or dual roll; some total floor grids	Same as for porous diffusers	Do not typically clog; easy maintenance; high alpha	Low oxygen transfer efficiency; high initial cost	9–13	1.3–1.9
Jets	Compressed air and pumped liquid mixed in nozzle and discharged fine bubble	Same as for porous diffusers	Good mixing properties higher SOTE than nonporous diffusers	Limited geometry; clogging of nozzles; requires blowers and pumps; primary treatment required; low SAE	15–24	2.2–3.5
Mechanical surface Radial flow, low speed (20–100 r/min)	Low output speed; large diameter turbine; floating, fixed-bridge, or platform mounted; used with gear reducer	Same as for porous diffuser	Tank design flexibility; high pumping capacity	Aerosols some icing in cold climates; initial cost higher than axial flow aerators; gear reducer may cause maintenance problems unless adequate service factor provided	—	1.5–2.1

TABLE 12.25 Characteristics of Aeration Equipment (Arora et al., 1985; Boyle, 1996; Goronszy, 1979; Groves et al., 1992; Wilford and Conlon, 1957)

Equipment Type	Equipment Characteristics	Processes where Used	Advantages	Disadvantages	Reported Clean Water Performance[a]	
					SOTE (%)	SAE (kg/kW·h[b])
Axial flow, high speed (900–1800 r/min)	High output speed; small diameter propeller, direct, motor-driven units mounted on floating structure	Aerated lagoons and reaeration	Low initial cost; may adjust to varying water level; flexible operation	Some icing in cold climates; poor maintenance accessibility; mixing capacity may be inadequate	—	1.1–1.4
Horizontal rotor	Low output speed; used with gear reducer; steel or plastic bars, plastic discs	Oxidation ditch, applied either as an aerated lagoon or as an activated sludge	Moderate initial cost; good maintenance accessibility	Subject to operational variable, which may affect efficiency; tank geometry is limited	—	1.5–2.1
Submerged turbine	Units contain a low-speed turbine and provide compressed air to diffuser rings, open pipe, or air draft; fixed-bridge application; may employ draft tube	Same as for porous diffusers, oxidation ditches	Good mixing; high capacity input per unit volume; deep tank application; operational flexibility; no icing or splash	Require both gear reducer and blower; high total power requirements; high cost	—	1.1–2.1 (Typical) 2.0–3.0 (Drain tube turbine)
Aspirating	Same as axial flow; high speed	Aerated lagoons; temporary installations	Low cost flexible operation	Same as axial flow, high speed	—	0.5–0.8

[a] Manufacturers data in clean water at standard conditions; diffused air units expressed as SOTE and SAE mechanical devices as SAE. Range of values accounts for different equipment, geometry, gas flow, power input, and other factors (SAE—wire-to-water).

[b] Wire-to-water SAE for diffused air calculated from compression relationship where ambient temperature = 30°C, submergence = 4.3, barometric pressure = 100 ha (1 atm), and blower/motor efficiency = 70%.

TABLE 12.25 Characteristics of Aeration Equipment (Arora et al., 1985; Boyle, 1996; Goronszy, 1979; Groves et al., 1992; Wilford and Conlon, 1957) (*Continued*)

gas injection with mechanical pumping or mixing equipment is also classified herein as diffused aeration equipment. These hybrid devices include jet aerators and turbine-sparger aerators.

The wastewater treatment industry has witnessed the introduction of a wide variety of air diffusion equipment. In the past, the various devices were classified as either fine-bubble or coarse-bubble diffusers. In recent years, additional diffusion equipment including medium bubble and ultra-fine bubble have been introduced to the industry. Since the demarcation between air diffusion equipment based on bubble size has become increasingly difficult to define, air diffusion systems have been categorized herein by the physical characteristics of the equipment. In the following discussion, air diffusion devices are divided into three categories: porous diffusers, nonporous diffusers, and jet aerators.

8.2.1 Porous Diffuser Systems

Porous diffusers are commonly used because of their relatively high OTE compared to other aeration technologies. An excellent reference on this subject was published by the U.S. EPA in cooperation with the American Society of Civil Engineers (ASCE) Committee on Oxygen Transfer (U.S. EPA, 1989). Much of the information presented in this section was derived from that source. Further information can be found on these specific topics in Groves et al. (1992) and U.S. EPA (1985, 1999).

Numerous materials have been used to manufacture porous diffusers. They typically can be divided into two categories: rigid materials of ceramic, or plastic, and perforated membranes of thermoplastics or elastomers.

The oldest and most common rigid porous diffuser is produced from ceramic media, including alumina, aluminum silicates, and silica. Media consist of rounded or irregular-shaped mineral particles bonded together to produce a network of interconnected passageways through which compressed air flows. As air emerges from the diffuser surface, pore size, surface tension, and airflow rate interact to produce a characteristic bubble size. Currently, the most common rigid porous diffusers are manufactured from aluminum oxide. Porous plastics are made from thermosetting polymers. The two most common are high-density polyethylene (HDPE) and styrene-acrylonitrile. Porous plastic materials are lighter in weight; are inert in composition; and, depending on the actual material, may have greater resistance to breakage. Disadvantages include the brittleness of some plastics and lack of quality control of others.

Membrane diffusers differ from rigid diffusers because they do not contain a network of interconnected passageways to diffuse gas. Instead, they utilize small, individual orifices (perforations) patterned across the membrane during fabrication to control gas passage. Perforation size, number, and pattern vary widely. Perforations are produced by slicing, punching, or drilling holes or slits in the membrane. Each hole acts as a variable aperture opening. The slit or hole size will affect bubble size (and, therefore, OTE) and backpressure. Typical slit or hole size is 1 mm, although perforated membrane panels of thin polyurethane are punched with significantly smaller slits (Boyle, 1996).

Perforated membranes are available in a variety of membrane materials, shapes, and perforation patterns. Two types of membrane materials commonly used are thermoplastics and elastomers. Typically, with elastomers, mostly ethylenepropylene dimers (EPDMs) are used for municipal applications. EPDMs contain carbon black, silica, clay, talc, oils, and various curing and processing agents. Oils make up a significant proportion of the mix and give the membrane its flexibility. Each manufacturer has its own formulation with distinctively different characteristics, including tensile strength, hardness, elongation at failure, modulus of elasticity, tear resistance, creep, compression set, and resistance to chemical attack. Thus, selection of the membrane for a particular application requires a thorough analysis of the material to ensure that it will perform effectively over the expected design life of the material. For membranes installed in industrial applications, the fouling potential, composition, and temperature of the wastewater must also be considered during membrane material selection.

Porous diffusers are available in plates, panels, domes, disks, and tubes; some are shown in Figures 12.65 to 12.67. Dome diffusers are an older technology, consisting of a ceramic material mounted on a polyvinyl chloride (PVC) saddle attached by a center bolt. They are typically 180 mm (7 in.) in diameter and have downturned edges. Disk diffusers are currently the most commonly applied porous diffuser shape in aeration tanks. Discs vary in diameter from approximately 180 to 240 mm (7.0 to 9.5 in.) for ceramic and porous plastic materials and from 180 to 500 mm (7 to 20 in.) for perforated membranes. Like the dome, the disk is typically mounted on a PVC saddle but may be fastened by either a center bolt or a peripheral clamping ring.

Most tubular diffusers have the same general shape of 500 to 610 mm (20 to 24 in.) in length with an outside diameter of 63 to 80 mm (2.5 to 3 in.). Materials used include ceramics, porous plastics, and perforated membranes.

Polyurethane membrane panels typically come in 1.2-m (4-ft) widths of variable lengths (1.2 to 1.8 m [4 to 6 ft] in 0.6-m [2-ft] increments). The base plate may be constructed of reinforced cement compound, fiber-reinforced plastic, or type 304 stainless steel. The panels are placed on the flat bottom surface of an aeration tank and fastened with anchor bolts (Figure 12.67).

With the exception of old plate designs, each porous diffuser is equipped with a flow control orifice to ensure uniform air distribution to each diffuser. Typical airflow rates for domes range from 0.014 to 0.071 m³/min (0.5 to 2.5 cfm). For disks, the range

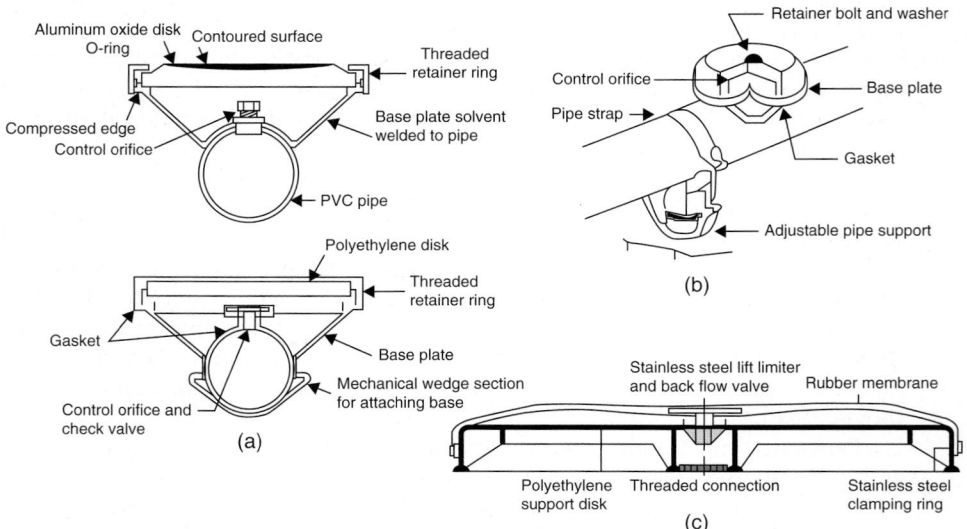

FIGURE 12.65   Selected porous diffusers: (a) disks, (b) dome, and (c) perforated membrane.

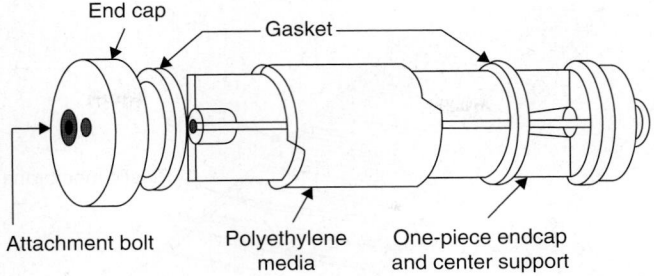

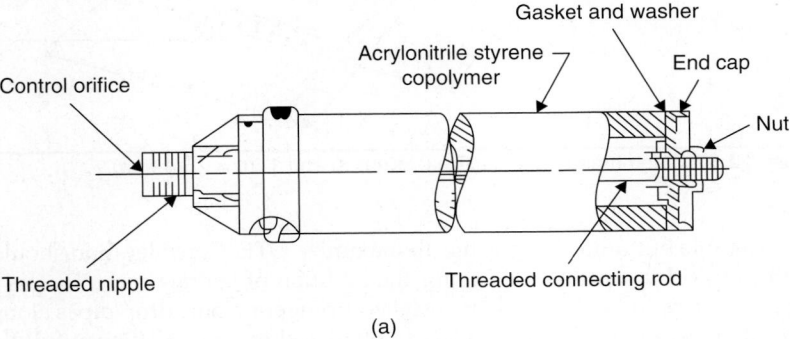

(a)

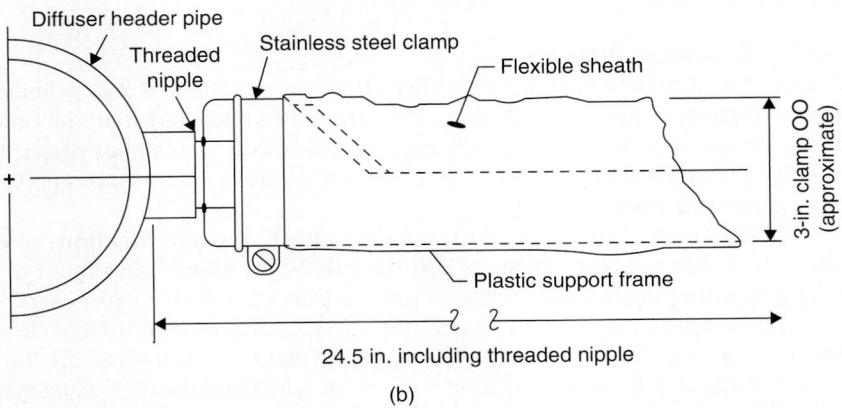

(b)

Figure 12.66 Selected porous tubes: (a) rigid plastic and (b) flexible perforated membrane.

is 0.014 to 0.08 m³/min (0.5 to 3.0 cfm) for ceramics and 0.03 to 0.6 m³/min (1 to 20 cfm) for perforated membranes, depending on their size. For tubes, the range is 0.03 to 0.14 m³/min (1 to 5 cfm).

Plates, panels, domes, and disks are typically installed in a total floor coverage configuration, but plates also have been placed along the sides of aeration tanks to generate single- or dual-roll spiral mixing patterns. Disks and domes are arranged in a grid pattern

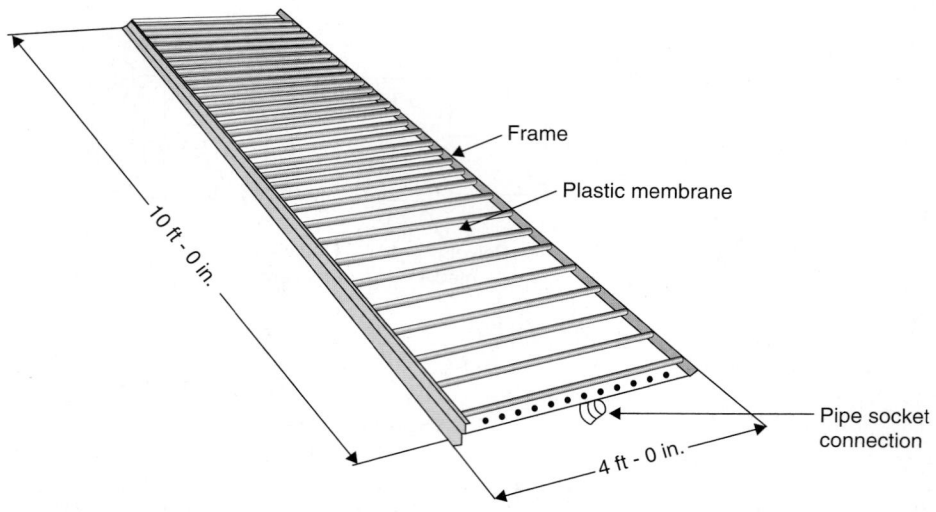

Frame

Plastic membrane

Pipe socket connection

10 ft - 0 in.

4 ft - 0 in.

Figure 12.67   Membrane panel (1 ft = 0.3048 m and 1 in. = 25.4 mm).

with variable but uniform spacings to maximize OTE. Consideration should be given to a diffuser grid layout that allows for the addition of laterals and diffusers if conditions were to change. Tube diffusers are installed from removable drop pipes along one or both sides of an aeration tank. They also can be placed in a more efficient full-floor coverage configuration. Diffuser grids and configurations can be designed to allow removal of the diffusers for cleaning without process interruption.

8.2.2   Nonporous Diffusers

Nonporous diffusers have larger orifices than porous devices and primarily produce coarse bubbles. They are available in a variety of styles and materials ranging from holes drilled in pipes to specifically engineered orifices in metal or plastic fabrications (Figure 12.68). Perforated piping, spargers, and slotted tubes are common examples of nonporous diffusers.

Valved orifice diffusers include a check valve to prevent backflow when the air is shut off. There are other types of diffusers that also allow adjustment of airflow by changing either the number or size of orifices through which air passes.

Typical system layouts for fixed and valved nonporous diffusers closely parallel those for porous diffuser systems. The most prevalent configurations are the single- and dual-roll spiral patterns using either narrow- or wideband diffuser placement. Mechanical lift-out headers with either swing joints or removable diffusers, which allow removal for cleaning without process interruption, are common. Cross-roll and full-floor coverage patterns may be used. Fixed and valved orifice diffusers are used where fouling may be a problem. These applications include aerated grit chambers, channel aeration, sludge and septage storage tank aeration, flocculation basin mixing, aerobic digestion, and industrial waste applications.

The static tube, another type of nonporous diffuser, resembles an airlift pump, except that the tube has interference baffles placed within the riser. These baffles are intended to mix the liquid and air, shear coarse bubbles, and increase contact time. With

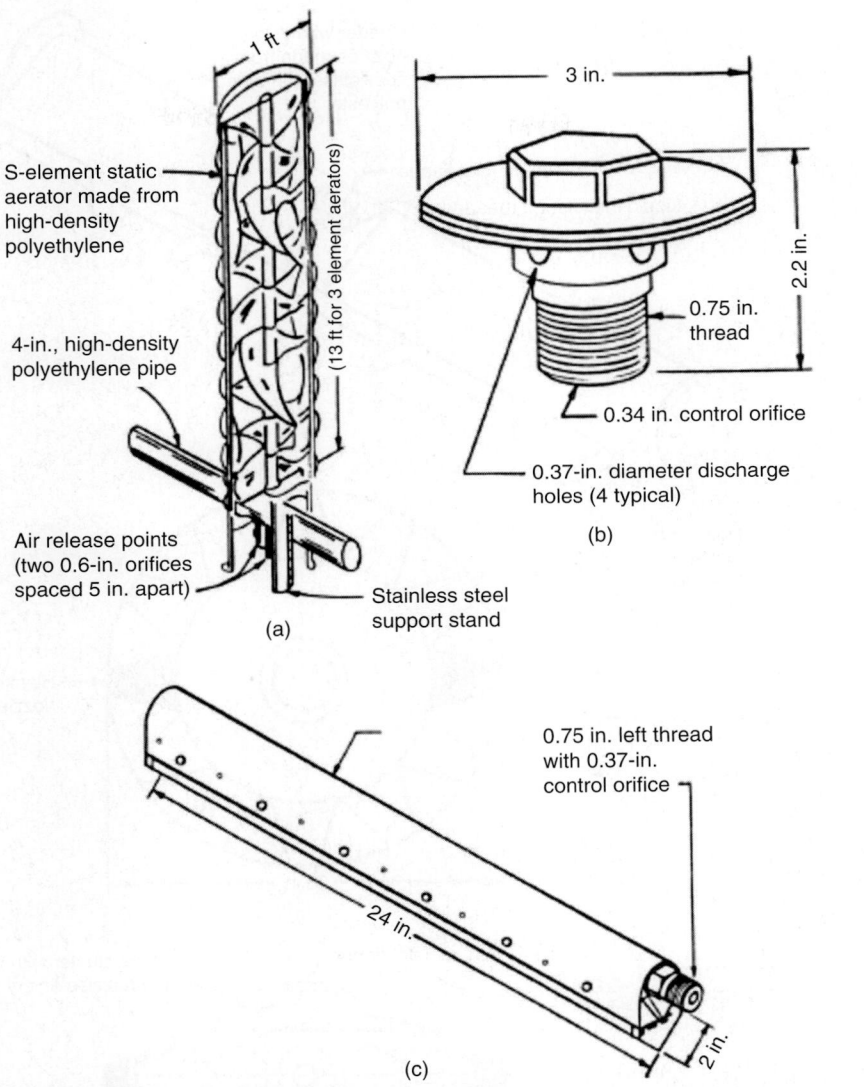

Figure 12.68   Selected nonporous diffusers: (a) static tube, (b) orifice, and (c) tube (1 in. = 25.4 mm).

this type of system, the tubes are typically 1.0 m (3 ft) in length and anchored to the basin floor in a full-floor coverage pattern.

8.2.3   Jet Aeration

Jet systems combine liquid pumping with air diffusion. The pumping system circulates mixed liquor in the aeration basin, ejecting it through a nozzle assembly. Air is typically supplied from a blower and introduced to the mixed liquor before it is discharged through the nozzles. Jets are often configured in either cluster or directional arrangements as shown in Figure 12.69. Distribution piping and nozzles are typically made of fiberglass.

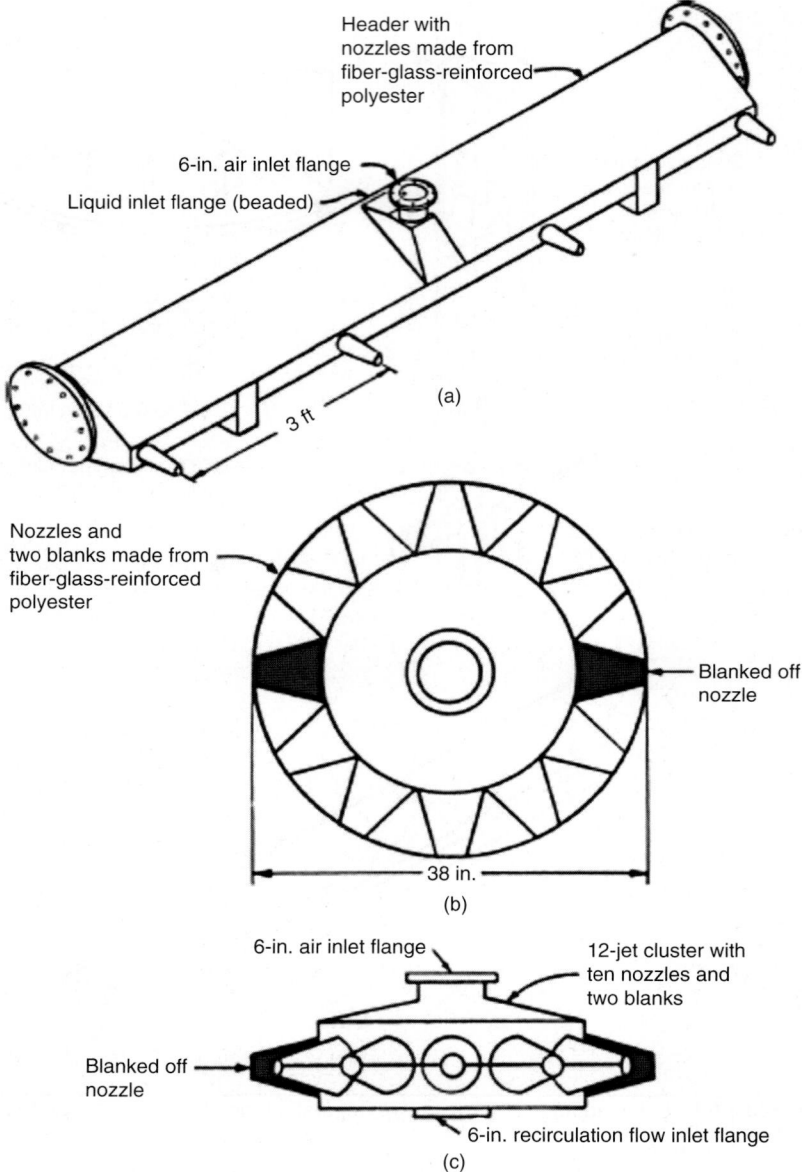

Header with
nozzles made from
fiber-glass-reinforced
polyester

6-in. air inlet flange

Liquid inlet flange (beaded)

3 ft

(a)

Nozzles and
two blanks made from
fiber-glass-reinforced
polyester

Blanked off
nozzle

38 in.

(b)

6-in. air inlet flange

12-jet cluster with
ten nozzles and
two blanks

Blanked off
nozzle

6-in. recirculation flow inlet flange

(c)

Figure 12.69 Jet diffusers: (a) bender-type, (b) plan of radial-type, and (c) elevation of radial-type (1 ft = 0.3048 m and 1 in. = 25.4 mm).

Since the recirculation pump is usually a constant-capacity device, turndown is accomplished by varying the air supply rate from the blower. A typical nozzle has a 30-mm (1-in.) throat through which air and mixed liquor pass. To overcome potential clogging problems, some systems are equipped with self-cleaning features.

8.3   Mechanical Surface Aerators

Surface aerators can be grouped into four general categories: radial flow low speed, axial flow high speed, aspirating devices, and horizontal rotors. Each is used widely and has distinct advantages and disadvantages, depending on the application.

Surface aerators are typically float, bridge, or platform mounted. Platform and bridge designs should address torque and vibration. Bridges should be designed for at least four times the maximum moment (torque and impeller side load) anticipated. The aerator manufacturer can provide the magnitude of this moment. The efficiency and power draw of platform- and bridge-mounted aerators are sensitive to changes in the depth of impeller submergence. An increase in submergence results in increased fluid pumpage at an increase in power draw and can decrease gearbox life expectancy. High-speed (radial and axial) surface aerators are most often float mounted, providing portability and low cost.

Some surface aerators are equipped with submerged draft tubes that tend to mix by bringing liquid from the bottom of the basin up through the tube and into the impeller. Draft tubes are commonly used for deeper aeration basins and oxidation ditches. All of the mixed liquor pumped through a draft tube is dispersed into the air. Without draft tubes, a portion of the pumped fluid flows beneath the liquid surface and is not aerated. Mixed liquor creates liquid momentum that tends to circulate around the aerator. Designers should recognize that mechanical aerators provide point-source oxygen input. Pumped mixed liquor flows radially outward from the aerator with decreasing velocity. Dissolved oxygen reaches its maximum near the impeller blade where surface turbulence is greatest and decreases as fluid flows back below the surface of the aeration tank toward the aerator.

As an alternative to the draft tube, an auxiliary submerged mixing impeller can be provided. This submerged impeller will increase the amount of liquid pumped from the bottom of the basin. The impeller typically has an axial-flow design to maximize pumping efficiency. Submerged impellers and draft tubes, however, increase system power requirements. The optimum location of the impeller depends on its configuration. Radial-flow impellers typically are located 0.5 to 0.7 times the impeller diameter above the tank bottom; axial-flow impellers are located at 60% to 65% of the tank depth, measured from the water surface. Water depths using unsupported shafts (no bottom bearings) span up to 9 m (30 ft). With this unsupported length, shafts can transmit high side loads that the gearbox must be designed to withstand.

The action of surface aeration devices, particularly splashing from high-speed units, can generate mist with attendant health concerns and nuisance odors. Odors can result from insufficient oxygen supplied by the aerator or influent wastewater containing sulfides or other volatiles. Mist can freeze in cold climates, coating equipment and walkways with ice. Such freezing can cause hazards to facility staff and equipment. Splashing effects can be minimized with proper geometric design of the aeration tank and use of deflector plates. In cold climates, the impact of basin heat loss induced by surface aerators on biological process should be evaluated during design.

8.3.1   Radial Flow Low Speed

Radial-flow low-speed aerators can provide a higher standard aeration efficiency (SAE) than high-speed machines and are good mixing devices. Clean water SAEs based on wire-to-water transfer range from 0.42 to 0.59 kg O_2/MJ (2.5 to 3.5 lb O_2/hp/hr). Efficiency depends on many variables, including the design of the impeller itself, tank geometry, effects of adjacent walls, input power to tank volume, impeller size and speed, number of units, location, and other factors that are less well understood.

Scale-up from shop to field applications is difficult to achieve, requiring full-scale testing for proper evaluation.

These aerators are available in several configurations. The simplest is an impeller that operates at the water surface. In another configuration, submergence of the impeller can be adjusted, typically through the use of movable weirs, to control power draw and oxygen transfer.

Low-speed aerators typically operate in the range of 20 to 100 rpm and include a gearbox to reduce the impeller speed below that of the motor. The gearbox should have a service factor rating of 2.0 or higher to ensure mechanical reliability. Without a service factor, adjacent aerators operating at different speeds or units placed close to tank walls may cause drive overloads.

The aerators are available in power increments up to 150 kW (200 hp), with either floating or fixed mounting structures. They are also available as two-speed units, with power turndown ratios of approximately 50% at the lower speed. Impellers span up to 3.7 m (12 ft) in diameter and operate at top peripheral velocities of 4.6 to 6 m/s (15 to 20 ft/s). These aerators can be provided with VFDs to provide more flexibility to meet the oxygen demand.

8.3.2   Axial Flow High Speed

High-speed aerators are typically used for stabilization lagoons where dispersed organism growth or benthic deposits exert oxygen demands. This limited application stems from concerns regarding shearing of sludge floc, which could impair settling. Icing and mist generation are also concerns.

High-speed aerators have limited depth of mixing and oxygen-transfer capabilities. Typically, high-speed units exhibit lower wire-to-water SAEs (0.20 to 0.38 kg/MJ [1.0 to 2.25 lb/hp/hr]) than low-speed devices. As with low-speed devices, performance of high-speed units is affected by basin geometry and other factors. High-speed units are available in standard motor increments up to 93 kW (125 hp) and are most often mounted on floating structures.

8.3.3   Aspirating Devices

Another aeration device is the motor-driven propeller aspirator. One such device, shown in Figure 12.70, consists of a 1.2-m (4-ft) long hollow shaft with an electric motor at one end and a propeller at the other. Propeller rotation draws air from the atmosphere through the shaft. Air velocity and propeller action create turbulence, forming small bubbles from which oxygen is dissolved. These devices can be positioned at various angles to reach different levels for aeration, mixing, or circulation. Portable units can be mounted on booms or floats in aeration tanks and oxidation ditches. An aspirator with a disk rather than a propeller disperses bubbles at a 90° angle to the shaft. Operation during cold weather has been reported to cause icing of the aspirator pipe at the air inlet end, shutting off the air supply. Manufacturers do offer solutions to problems of freezing, including heat tracing. Typical wire-to-water SAEs for these devices range from 0.13 to 0.21 kg O_2/MJ (0.75 to 1.25 lb O_2/hp/hr).

8.3.4   Horizontal Rotors

This type of unit, available in several configurations, has a horizontal impeller (rotor) (Figure 12.71). The impeller agitates the surface of the basin, transferring oxygen and concurrently moving the liquid in a horizontal direction.

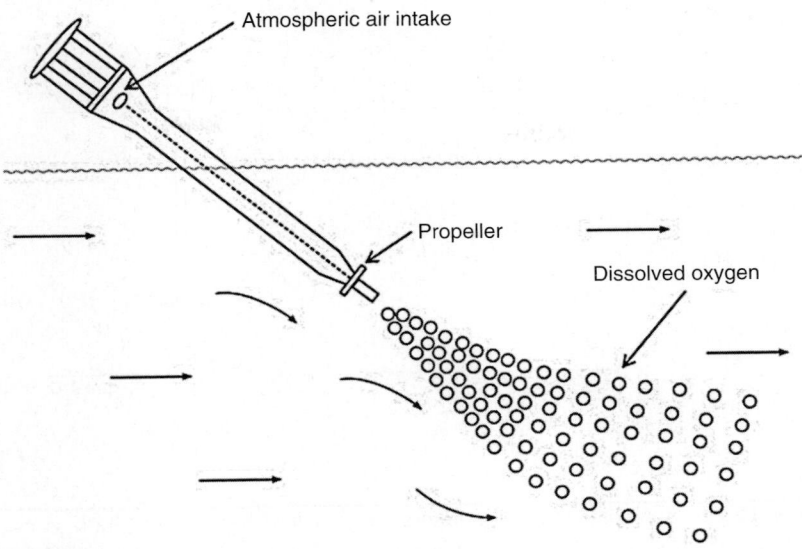

FIGURE 12.70   Aspirating device.

(a)   (b)

FIGURE 12.71   Horizontal rotor (a) with splash cover and (b) blades with splash cover removed.

Clean water wire-to-water SAEs of these rotors approximate those of low-speed surface aerators: 0.42 to 0.59 kg O_2/MJ (2.5 to 3.5 lb O_2/hp/hr). Small changes in rotor submergence do not affect transfer efficiency but will affect power draw. Lesser submergence will decrease total oxygen transfer; however, the mass of oxygen transferred per unit of power input remains approximately the same. Units are made in various sizes up to a maximum length of approximately 7.6 m (25 ft). Two rotors with one centrally located drive can aerate a 15-m (50-ft)-wide, 3.7-m (12-ft)-deep ditch. As is the case with other surface aeration devices, icing, mist, and heat loss are potential problems.

8.4   Submerged Turbine Aerators

A submerged turbine consists of a motor and gearbox drive mounted on the tank, one or more submerged impellers, and piped air from a blower to a point below the impeller.

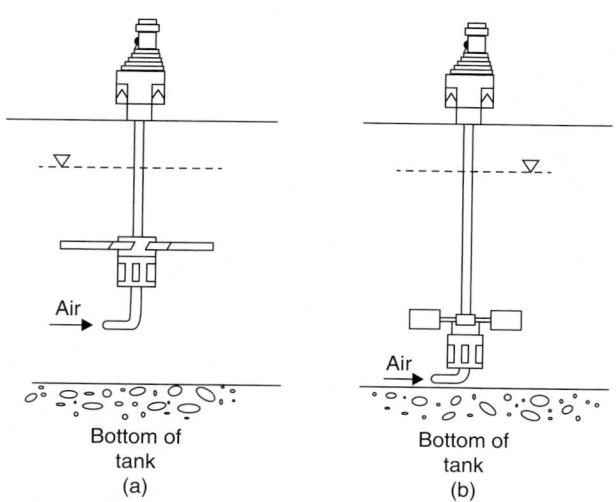

FIGURE 12.72   Submerged turbine aerators: (a) axial flow and (b) radial flow.

Impeller designs vary, but typically are either the axial- or radial-flow type (Figure 12.72). With an axial-flow impeller, pumped mixed liquor has sufficient velocity to drive released air downward and disperse it across the bottom of the tank. In the radial-flow design, air flows into the impeller, mixes with the liquid, and disperses outward, driven by the impeller blades. With either type, the operator must carefully control the airflow rate. Too much air will cause the turbine to flood, reducing the amount of oxygen transferred and the pumping capacity of the impeller, and may cause mechanical damage.

Air enters the turbine through either an open pipe or a diffuser ring. Although an axial-flow unit can transfer a higher percentage of oxygen than a radial-flow unit, a radial-flow unit can handle a higher volume of air per kilowatt. Both units can be designed to transfer oxygen at rates of up to several hundred milligrams per liter per hour in clean water. This maximum rate exceeds that typically required because even industrial applications with high-BOD wastewater seldom result in more than the OUR of 200 to 300 mg/L·h, and municipal facilities are typically designed for OURs of 30 to 100 mg/L·h.

The area of influence (tank area that is aerated) of submerged turbines is somewhat smaller than that of surface aerators. Low-speed axial-flow units achieve maximum influence. The area of influence will vary from 4 to 12 m²/kW (30 to 100 sq ft/hp), depending on reactor geometry and size. Turbine speeds are a function of impeller design and power input. Most operate in the range of 37 to 180 rpm. The gearbox service factor should be 2.0 or higher to accommodate hydraulic loads imposed by adjacent walls and other aerators.

Submerged turbines transfer varying amounts of oxygen, depending on the air-to-mixer kilowatt ratio, mixer design, and basin geometry. Clean water wire-to-water SAEs of turbines are typically reported to range from 0.30 to 0.59 kg O_2/MJ (1.75 to 3.50 lb O_2/hp/hr), including air and mixer power requirements. These SAEs are slightly lower than those of slow-speed surface aerators, but a submerged turbine offers adjustable gross oxygen input by modulating airflow rate. When used as an aerator or a mixer (as in a combined nitrification-denitrification reactor), the submerged unit may offer some advantage.

In those areas where basin cooling is a concern, submerged turbines only slightly agitate the surface with a minimal impact on basin heat loss. Use of submerged turbines in circular or square tanks requires baffling to prevent excessive rotation of tank contents. Rectangular tank baffling requirements may be less than those for circular tanks. The turbine manufacturer can provide baffling recommendations based on tank geometry.

Available submerged turbine aerators match common motor sizes up to 112 kW (150 hp). Special designs include motors up to and greater than 260 kW (350 hp). Air-flow rates vary from 0.23 to more than 8.0 m^3/min (8 to more than 300 scfm) per aerator, and OTEs range from 15% to 35% in clean water, depending on impeller configuration and depth.

Another submerged turbine aerator consists of a downward-pumping, airfoil-type impeller in a draft tube, with an air sparge ring mounted directly below the impeller (Figure 12.73). Coarse bubbles are sheared into smaller ones by flow energy from the high–pumping-rate impeller. Bubbles are forced downward through the draft tube and

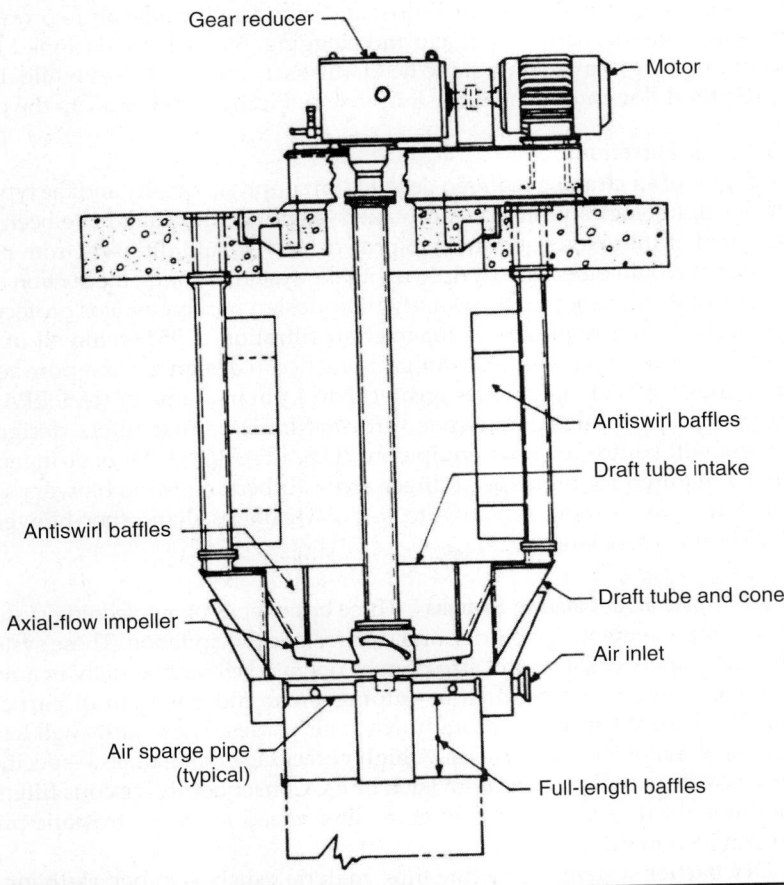

Figure 12.73 Draft tube submerged turbine aerator.

baffles before rising to the surface. A primary element of this design is the axial-flow impeller. Introduction of air disrupts the axial-flow pattern, resulting in some reduction in pumping efficiency. Such aerators, typically deck mounted above the tank, can be installed in a conventional CMAS tank or in a total barrier oxidation ditch with a J-tube. Tank depths ranging from 7.6 to 9.0 m (25 to 30 ft) are used in CMAS applications. These units are furnished with high-speed impellers (up to 180 r/min) handling sparged gas flows up to 18 m^3/min (650 scfm). Clean water wire-to-water SAEs are reported in CMAS to range from 0.56 to 0.8 kg O$_2$/MJ (3.3 to 5.0 lb O$_2$/hp/hr) (Updegraff and Boyle, 1988). In total barrier oxidation ditch configurations, however, draft tube turbines produced low clean water wire-to-water SAEs ranging from 0.24 to 0.3 kg/MJ (1.4 to 2.0 lb/hp/hr) (Boyle et al., 1989).

8.5   Air Supply System

Although selection and design of air diffusion systems typically receive the most attention, proper design of a supply system is critical to ensure that overall process goals are met. An air supply system consists of three basic components: air filters and other conditioning equipment (including diffuser cleaning systems), blowers, and air piping. Air filters remove particulates such as dust and dirt from the inlet air to protect blowers and diffusers from mechanical damage and clogging. Blowers are designed to develop sufficient pressure to overcome static head, diffuser, and line losses while delivering air at the required flowrate to the diffusion system. Piping conveys air to the diffusers.

8.5.1   Air Filtration

The degree of air cleaning required depends on supply air quality and the type of blower and diffuser technologies installed. Air intake systems traditionally have been designed with integrated air filtration equipment to protect blowers and diffusers from particulates and debris that would otherwise be drawn into the system through the suction of the blowers.

Cleaning efficiency is the primary filter design parameter. For protecting blowers, a typical minimum requirement for inlet air filtration is 95% removal of particles with diameters of 10 μm and larger. Standard practice in design for fine-pore aeration devices is to remove 90% of all particles greater than 1 μm in diameter (U.S. EPA, 1989). Manufacturers of perforated membrane diffusers indicate that filters designed to protect blowers will suffice for their equipment (U.S. EPA, 1989). More stringent air filtration may be required for high-speed direct-drive air bearing turbo blowers since they have been found to be more sensitive to fine particulates than other blower technologies (Neville and Turriciano, 2013).

8.5.1.1   Types of Air-Cleaning Systems   Three basic types of air-cleaning systems are in use: viscous impingement, dry barrier, and electrostatic precipitation. These systems, manufactured in a variety of forms and sizes, can be operated either manually or automatically.

Viscous filters remove dust by impingement and retention of particles on a labyrinth of oil-coated surfaces through which air passes. These units will handle dust and effectively remove large particles. A high percentage of small, low-specific-gravity particles, however, will pass through such units. Consequently, viscous filters work best as a preliminary device. A coarse viscous filter ahead of an electrostatic precipitator can perform well in dusty areas.

Dry barrier systems use a fine filter material, such as paper, cloth, or felt, to entrap particles. These systems typically comprise of a coarse prefilter followed by a fine filter.

The prefilter typically consists of a sheet of fiberglass cloth mounted on a frame. Fine filters are housed in racks behind the prefilter. Such replaceable systems occupy little space and offer easy maintenance.

Bag house collectors are an option for large facilities. Baghouse collectors are dry barrier units constructed as steel enclosures that house sets of cloth-stocking tubes. The tubes are precoated with filter aid before being placed in service and after each cleaning. Efficiency increases during a filter run because retained particles increase effectiveness of the filtering medium. A baghouse collector that is properly installed and maintained will protect up to recommended standards if the atmosphere is not too laden with finer particulates and smoke. Size, expense, and precoat requirements of baghouse filter systems have, however, reduced their selection for new installations.

Electrostatic precipitators impart an electric charge to particles so that they can be removed by attraction to elements of opposite polarity. These units require 30% to 50% of the area of baghouses and have relatively simple maintenance needs. This type of device will remove fine dust particles and protect up to recommended standards when operated at velocities lower than 120 m/min (400 ft/min). These devices are especially useful in areas with particulate-laden atmospheres. A combination of a prefilter, electrostatic precipitator, and final filters housed in a filter plenum is sometimes used. The precipitator helps reduce the replacement frequency of final filters.

8.5.1.2 Filter Selection Of the air-cleaning systems available, replaceable filter units are the simplest to construct and operate. Their capital costs are approximately 12% of those of electrostatic precipitators. Baghouse dust collectors are bulky and more expensive, but relatively maintenance free. Replaceable air filters are a good selection except where poor air quality would require replacement of fine elements more frequently than once per year. In such cases, electrostatic precipitators might be cost-effective. Combinations of prefilters and final filters or prefilters with electrostatic precipitators and final filters may be an advantage where air quality is poor.

8.5.1.3 Design Considerations In addition to the design recommendations of filter manufacturers, other needs require special attention. Because a WRRF must operate continuously, facilities for equipment maintenance and weather protection of the air intake structure are needed. Good louvers and an ample chamber between the louvers and filters are essential. In freezing climates, preheating the air might be necessary to prevent snow or water vapor from freezing onto the filters. A simple method of preheating relies on ducts and dampers to direct part of the airflow inside the blower building. Additional preventative measures for freezing include maintaining low surface velocities and low intake velocities at intake loads or louvers. Designers need to exercise care in locating the filter inlet to prevent drawing in excessively moist air. Weather louvers can prevent entrainment of moisture during rains, which could soak the filter medium and reduce its performance and throughput capacity, causing the suction line to collapse. Housing for air filters should consist of corrosion-resistant materials. Blower buildings and air intakes should be located away from processes and facilities that may release corrosive gases or house corrosive chemicals.

8.5.2 Blowers and Compressors

As shown in Figure 12.74, many different types of blowers can be used to supply air for aeration processes. Historically, two main types of blowers were used at WRRFs: rotary PD and centrifugal units. Within the last 10 years, 2 additional technologies have been

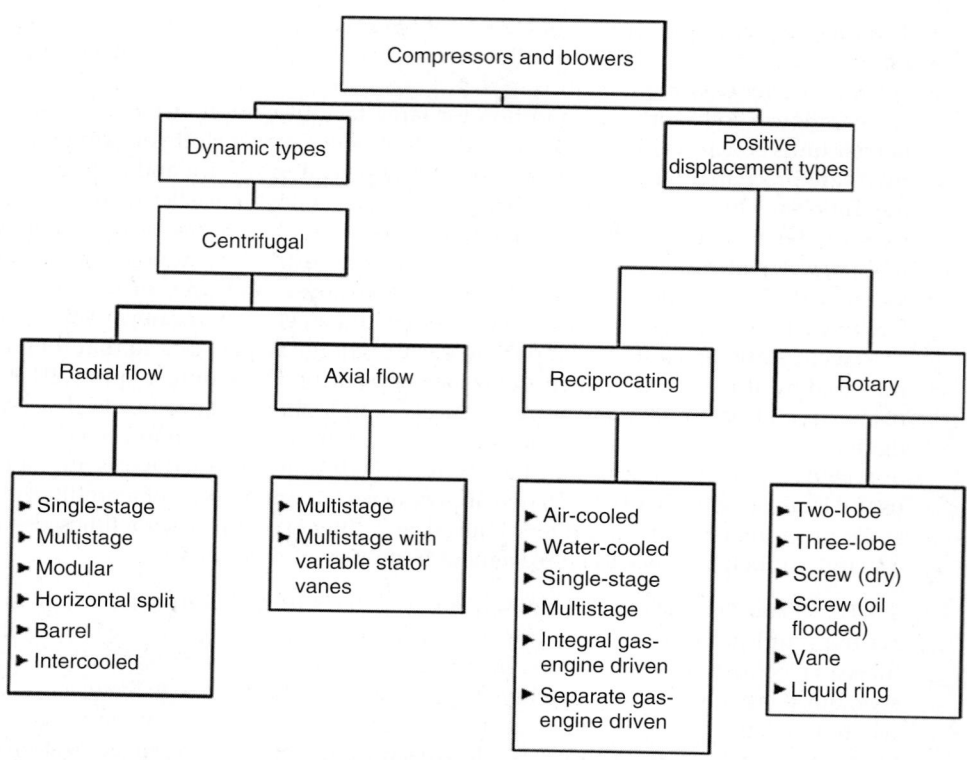

FIGURE 12.74   Compressor and blower types.

introduced to the market: high speed direct-drive single-stage centrifugal blowers (turbo blowers) and low-pressure screw compressors. Information on each of these technologies, much of which is derived from Neville and Turriciano (2013) and Aerzen (2017), is presented in this section.

The PD blowers are constant-volume machines capable of operating over a wide range of discharge pressures. Rotary lobe PD blowers consist of bi- or tri-lobed impellers mounted on parallel, counter-rotating shafts. A finite volume of air is trapped as each impeller passes the inlet. This volume of air is displaced in a positive manner from the blower inlet to the blower discharge. The PD blowers are a good selection for applications with significant variations in water level and temperature. They offer a low capital cost and high turndown, but can be less efficient and require more maintenance than other technologies. When operated at a constant speed, PD blowers discharge at a constant flowrate. Discharge airflow can be varied using a VFD to adjust motor speed in response to air demand. The PD blowers are loud and require inlet and discharge silencers. Sound-attenuating enclosures can be added to further reduce noise pollution.

Centrifugal blowers are considered variable-flow, constant-pressure machines with one or more impellers mounted on a single rotating shaft. The rotating impeller(s) draws air into the blower along the axis of the shaft, accelerating the air radially outward and discharging it perpendicular to the shaft. Centrifugal blowers are not as loud as PD blowers and can supply a wide range of airflows over a relatively narrow range of discharge pressures. Centrifugal blowers are categorized by the number of impellers and type of drive. Machines with one impeller mounted to the shaft are classified as single-stage centrifugal blowers and those with multiple impellers are classified as multistage centrifugal blowers. Multistage centrifugal blowers typically have good efficiencies at the design point, but efficiencies drop significantly as the machine turns down to the minimum airflow. Discharge airflow from multistage centrifugal blowers is controlled either by using an inlet throttling valve that restricts airflow on the suction side of the blower or by using a VFD to vary the motor speed. VFD control is seldom feasible for systems with limited variation in water level and with discharge pressures that change only slightly in response to varying airflow.

Single-stage centrifugal blowers can be geared or direct drive (gearless). Geared units include a motor, gearbox, and impeller. The gearbox enables the impeller to rotate faster than the drive and is either separate from or integral to the impeller and volute assembly. Often, single-stage integrally geared blowers use inlet guide vanes and discharge variable diffuser vanes to optimize blower efficiency and adjust blower output to meet air demand without varying speed. The guide and diffuser vanes provide dual-point control that allows the blowers to operate at a high efficiency over the full range of operating conditions without a VFD.

For single-stage direct-drive (turbo) blowers, the impeller is mounted directly to the shaft of a high-speed motor. The direct coupling of the permanent magnet motor and impeller enables turbo blowers to operate at high speeds ranging from 20 000 to more than 40 000 rpm. Turbo blowers have either air-foil bearings or magnetic bearings that do not require lubrication. A high-frequency VFD is used to vary the speed of the blowers in response to air demand. Turbo blowers offer a compact footprint, low maintenance requirement, low noise, high efficiency, and high turndown of rated capacity when compared to other blower technologies. However, turbo blowers have been found to be sensitive to environmental conditions and fine particulates and should not be installed outdoors. Additional inlet air filtration should be vetted when considering air-bearing turbo blowers. Due to the electronic components integral to the technology, electrical conditions must also be properly evaluated, regardless of bearing type, for each installation.

Low-pressure screw compressors combine rotary lobe PD blower and screw compressor technologies. Depending on discharge pressure, these machines may operate at an efficiency comparable to turbo blowers. Low-pressure screw compressors maintain many of the features of rotary lobe PD blowers, such as being tolerant of frequent starts and stops and ability to operate over a wide range of turndown in flow or pressure. VFDs are used in combination with low-pressure screw compressors to vary the discharge airflow in response to air demands. Turndown for low-pressure screw compressors is comparable to that of a PD blower and can offer greater turndown than a turbo blower in some applications.

Blower capacity can be specified as standard or actual gas flowrate. Standard conditions are defined by the American Society of Mechanical Engineers (ASME) as air at 20°C, 100-kPa (1-atm) pressure, and a relative humidity of 36%. The most practical method of defining blower capacity is to use the actual volume of air per unit of time. The range of discharge pressures, flows, motor horse powers, efficiencies, and turndown capacities differ by blower type. Typical values of these parameters can be found in the U.S. EPA (2010) for a majority of the blower technologies discussed in this section.

8.5.2.1 Turndown and Energy Efficiency   The selected blower system must be capable of supplying the volumes of air necessary to meet varying oxygen demands over the design life of the facility. To maximize operational flexibility and efficiency, blower systems should be designed to meet the maximum air requirement while still providing sufficient turndown to meet the minimum governing air requirement. In some instances, it may be necessary to install multiple blowers of smaller and/or varying capacity to efficiently or adequately meet the full range of air requirements. Therefore, in addition to site conditions, blower sizing and selection should take into account the following:

- Initial cold-weather minimum and hot-weather maximum air requirements based on existing influent loadings;

- Future minimum and maximum air requirements based on design influent loadings;

- Initial and future airflow requirements for mixing, depending on the number of diffusers and aeration tanks in use;

- Minimum airflow requirements for oxygen-transfer equipment; and

- Energy efficiency of blowers to achieve greatest efficiency over the widest range and at the most common conditions expected.

The aeration process typically accounts for 25% to 60% of the total WRRF energy use (WEF, 2009). With rising power costs and the need for sustainable designs, improving aeration system efficiency has become increasingly important. A life-cycle cost analysis (LCCA) including the equipment, operating, and maintenance costs projected over the life of the equipment can be used to compare multiple aeration system alternatives to identify the most cost-effective and/or efficient design. Guidance for developing blower LCCA can be found in Neville and Turriciano (2013). Additional guidance for efficient aeration system design can be found in the U.S. EPA (2010) and the WEF (2009).

8.5.2.2 Selection   As previously discussed, each type of blower has distinct operating characteristics. A blower must be compatible with the normal operating mode of the treatment system. Other factors such as noise, maintenance, and operator preference also are considered. If the aeration system is designed for operation with a fairly constant water depth, typical of most aeration basins, and with clean and maintained diffusers, then a centrifugal blower will be a good choice. Conversely, if the system will be operated over a wide range of depths, as in an SBR, then a PD blower or compressor might be a better selection.

Both discharge pressure and the weight of air vary with inlet temperature. For this reason, blowers are sized to provide the required airflow rate at the maximum inlet air temperature anticipated. Blower motors are sized to deliver the required airflow rate at the highest inlet temperature expected.

8.5.3   Air-Piping Materials

Primary considerations in piping material selection are strength and potential deterioration because of corrosion, thermal effects, and other environmental factors. Piping materials commonly used include carbon steel, stainless steel, and PVC. Piping materials less commonly used include ductile iron, fiberglass-reinforced plastic, and HDPE. Because blower discharge pressures are typically less than 100 kPa (15 psi), thin-walled pipe may be used with adequate provisions for protection from physical damage.

Discharge air temperature in excess of 93°C (200°F) are not uncommon as discharge temperature increases with increasing discharge pressure. Therefore, the pipe and accessories (supports, valves, gaskets, and so on) must be designed accordingly. Because thermal stresses caused by pipe heating and cooling can be significant, provisions for pipe expansion and contraction including cooling loops or expansion joints are typically needed. Blower discharge piping is often insulated to protect workers from potential burns; help attenuate noise; and minimize heat loss to the blower room. Air intake piping may also be insulated to prevent condensation (in cold climates) and to help attenuate noise. Discharge piping should be sloped away from the blower with condensate drains installed at low points occurring upstream of the diffused aeration equipment.

Because of the potential for corrosion at the interface between the atmosphere and liquid in aeration tanks, stainless steel piping material is often used. For diffuser systems, stainless steel droplegs are connected to the diffuser manifolds and headers, typically stainless steel or PVC. The choice of material depends on the structural requirements of the diffusers, diffuser connection, temperature within the pipe, and whether a diffuser-cleaning system will be provided. For diffusers with cleaning systems, the designer must check the compatibility of the piping material with the cleaning gas or liquid. The PVC is typically selected because of its inert characteristics. However, PVC temperature limits need to be considered in warm climates, industrial applications with high water temperatures, and deep tank applications. Typically, PVC is only used below the water surface.

Stainless steel is typically selected for tube diffuser systems because of the cantilever load applied by the diffusers and its corrosion resistance. However, PVC also has been used successfully. The PVC is more typically used with disk or dome diffusers because they are mounted to the top of a header; thus, weight and buoyant forces transmitted through the connection to the header are minimal.

Because basins are periodically drained and left empty, PVC piping should be protected from sunlight deterioration. Titanium dioxide and carbon black are typically used in PVC piping for UV protection. Whenever possible, drained basins should be filled with clear water to protect PVC from UV deterioration. Other design considerations include provisions to protect against the effects of pipe freezing and thawing when tanks are empty.

8.5.4 Air-Piping Design

Piping should be sized so that headloss in the supply lines and headers is small compared to that across the diffusers. Typically, if headlosses in the air piping between the last positive flow split (valve or control device) and the farthest diffuser are less than 10% of the headloss across the diffusers, then good air distribution can be maintained in the basin. Control orifices for diffusers are an important design consideration in piping design.

Basic fluid mechanics principles are used to size air-piping systems. Designers typically use standard calculation procedures such as those developed by Darcy-Weisbach. Several handbooks describe these procedures (Hoffman, 1986; Streeter and Wylie, 1979; U.S. EPA, 1989). Calculations must include corrections for temperature rise during compression, altitude (barometric pressure), and the specific weight of air at design temperature and pressure. Headloss calculations should account for both maximum summer air temperature and the temperature rise from air compression at the maximum expected airflow rate.

Losses through fittings and valves can be calculated using headloss coefficients and velocity heads. Typical coefficients can be found in texts and handbooks (Hoffman Air and Filtration Systems, 1986; Streeter and Wylie, 1979; U.S. EPA, 1989). Actual values selected should be verified by the manufacturer. The designer must also consider losses due to fouling or plugging of porous diffusers.

8.5.5 Pure-Oxygen Generation

On-site production of pure oxygen (also called high-purity oxygen) can be accomplished by cryogenic means, pressure swing adsorption (PSA), or vacuum swing absorption (VSA). The cryogenic air separation process produces liquid oxygen (LOX) by the liquefaction of air, followed by fractional distillation to separate air components, mainly nitrogen and oxygen. The PSA systems have been used in the past, but have become obsolete by the more cost-effective VSA process (Figure 12.75).

The VSA process is a minor variation of the proven PSA process. Both systems operate similarly with relatively low operating pressures (typically less than 340 kPa [50 psig]) and at or near ambient room temperatures (10°C to 27°C [50°F to 80°F]). The VSA/PSA process relies on swings in pressure to cycle zeolite molecular beds from an absorption stage (high pressure) to a desorption and regeneration stage (low pressure) and then back to an absorption stage (high pressure).

The process starts by providing ambient air through an intake filter to a blower or compressor. Pressurized air is then sent through an aftercooler and onto zeolite beds where the molecular sieve material absorbs nitrogen, carbon dioxide, water vapor, and any residual hydrocarbons that may be present. The remaining oxygen gas passes through the zeolite beds and is stored in a collection pressure vessel before being transported to the point of use.

The primary difference between VSA and PSA systems is that VSA systems operate at a lower pressure, in some cases, because of the proprietary nature of the molecular sieve. VSA systems use a vacuum blower to remove nitrogen and other gases from the

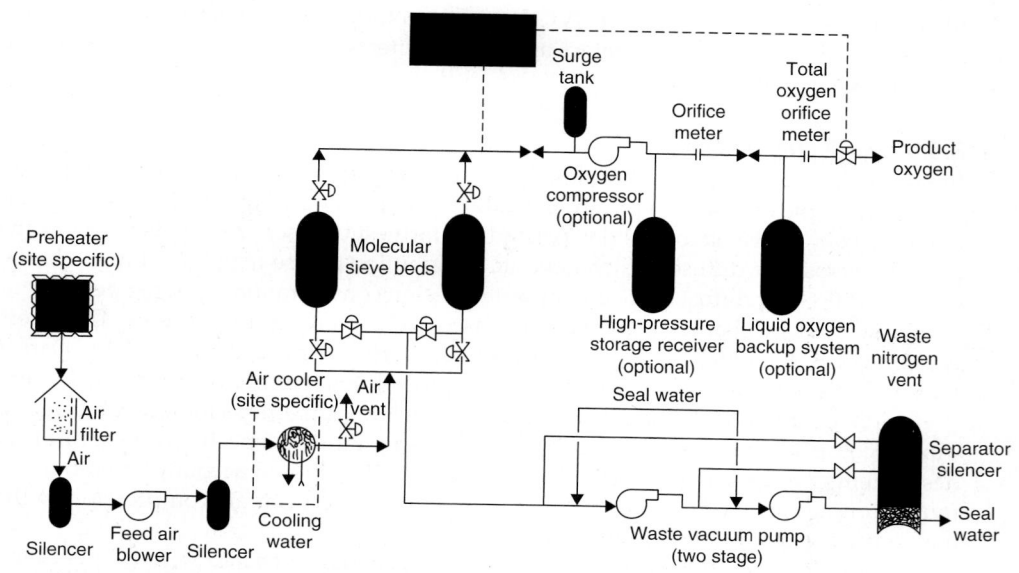

FIGURE 12.75 Typical two-bed vacuum swing adsorption process schematic.

zeolite media. This vacuum condition provides for slightly faster and more efficient purge (or "desorption") of the nitrogen-rich bed.

The VSA system typically uses backup LOX storage tanks for emergency situations or for peak periods when additional oxygen is required. On-site LOX storage tanks can be purchased or leased in sizes up to approximately 57 000- to 80 000-L (15 000- to 20 000-gal) capacity (approximately 64 to 86 Mg [70 to 95 ton] of oxygen) in either vertical or horizontal configurations. Tanker trucks typically load on-site LOX tanks, which need to be refilled periodically to account for evaporative losses. In some large industrial-urban areas, some manufacturers have underground pipe network oxygen delivery systems for their primary industrial and municipal clients.

Cryogenic oxygen generation and VSA separation are efficient. The amount of power generated (0.001 19 MJ/kg O_2 [0.15 kWh/lb O_2]) is representative of these technologies (compared to approximately 0.0015 MJ/kg O_2 [0.20 kWh/lb O_2] generated for a PSA unit). Minimum continuous output of cryogenic units is 27 000 to 36 000 kg/d (30 to 40 ton/d), whereas, for VSA units, it is 450 to 900 kg/d (0.5 to 1 ton/d). Design of either system for turndown is important. With a VSA system, turndown to 20% of full-generation capacity is possible. With a cryogenic system, turndown to only 50% of the full-load generation rate is expected. With both systems, LOX storage is provided for peak oxygen demand periods and oxygen generation equipment downtime.

Advantages of the VSA system over the cryogenic system include decreased energy use (in terms of dollars per unit volume of pure oxygen produced), simple equipment and operations, and lower maintenance requirements. VSA systems may also better protect workers because both operating pressures and temperatures are closer to normal ambient conditions. For these reason, cryogenic systems are no longer offered by pure oxygen system manufacturers.

8.5.6 Aeration System Control

Oxygen demand in aeration basins fluctuates in response to many factors including diurnal influent loading, intermittent discharge of sidestreams from dewatering activities, dilution from storm events, and changes to air, water, and other wastewater characteristics. Fluctuations in oxygen demand result in changes in dissolved oxygen concentrations over time if the air supplied is not adjusted in response to fluctuating oxygen demands. Dissolved oxygen control strategies can be very simple or quite complex depending on the system.

Manual control of dissolved oxygen requires operators to record dissolved oxygen measurements daily and make changes to the airflow and flow splits between basins and zones daily, sometimes up to a few times a day. This method of dissolved oxygen control often does not match actual dissolved oxygen demand, providing too much or too little oxygen leading to poor process performance and inefficient aeration system operation. Replacing manual dissolved oxygen control with automated dissolved oxygen control can provide considerable energy cost savings when designed and implemented correctly because it allows for the airflow to aerated zones to adjust to the changing conditions within the basins rather than over-aerating for portions of the day. It has been estimated that by adding tight control of dissolved oxygen to the aeration process, facilities can save between 10% and 30% in total energy cost (U.S. EPA, 2010).

Automated dissolved oxygen control uses real-time dissolved oxygen measurements from dissolved oxygen probes as inputs to the process controller. The process controller provides control output to the aeration system. The aeration system responds to the control output by adjusting airflow from the blower and/or the valve position on the diffuser droplegs at the basin to maintain the target dissolved oxygen setpoint. Major components of an automated dissolved oxygen control system include:

- Dissolved oxygen probes—For control, dissolved oxygen probes should be installed near the center or close to the inlet of the aerated zones. For monitoring dissolved oxygen intrusion, dissolved oxygen probes can also be installed in anoxic or anaerobic zones.

- Blower airflow control—The total airflow supplied to the aeration system is controlled by modulating the airflow rate delivered by the blowers. Blower airflow control is discussed in Section 8.5.2.1.

- Basin/zone airflow control—Total airflow is typically divided between multiple basins, aeration zones, and diffuser grids. The airflow split between basins, zones, and grids can be controlled by making manual adjustments to flow control valves or by using automated flow control valves. Implementing a most open valve control scheme with automated valves can improve efficiency by maintaining valves as open as possible and minimizing system pressure.

- Process control system—The process controllers receive feedback from the dissolved oxygen probes, compares the basin dissolved oxygen readings to dissolved oxygen setpoints, and sends signals to the air controller mechanism to make adjustments as needed.

Dissolved oxygen is typically used for aeration control since it is a good indicator of nitrification and is a reliable measurement. In recent years, alternative control parameters have been implemented in place of dissolved oxygen for aeration system control. For example, as real-time ammonia analyzers have become more reliable, there have been several WRRFs that have successfully implanted aeration-based aeration system control. Ammonia-based aeration control utilizes ammonia concentrations as the input to the process controller as more direct measurement of nitrification. Detailed information about specific dissolved oxygen control strategies, recent advancements in control strategies, and emerging technologies for using control parameters other than dissolved oxygen can be found in the U.S. EPA (2010).

8.6   Mixing Requirements

The aeration system must be able to meet the minimum process air requirement without dropping below the minimum mixing air requirement or the minimum equipment air requirement. Since the process air requirement during the low-influent load condition can be less than the mixing air requirement and/or the minimum equipment air requirement as recommended by the diffuser manufacturer, it is imperative to evaluate minimum air requirements to determine which value will govern design.

For pure-oxygen systems, supplemental mixing is needed. Likewise, in extended aeration facilities, OURs are typically low, and mixing requirements often control the rate of energy input depending on efficiency of the aerator. Additional mixing system information and design guidance can be found in Section 4.4.2.2 for mechanical mixing and Section 4.4.2.3 for mixing by aeration.

8.7   Aerator Design and Testing

Historically, many methods have been used to test and specify aeration equipment, which has led to confusion and misrepresentation of equipment performance. Furthermore, equipment suppliers, consultants, and users often use different nomenclature when they report capabilities.

The preparation of explicit equipment specifications is vital to installation of efficient, cost-effective devices. To enable a supplier to properly specify and quote aeration equipment, the prospective user or design engineer must provide accurate and detailed information about system requirements and constraints. The supplier can then provide equipment performance information based on reliable clean water test data and sound judgment based on experience from previous field applications.

Evaluation of reliable clean water oxygen-transfer test data is only one step in understanding aeration system capabilities. Measuring oxygen transfer in the field with aeration equipment operating under actual process conditions can be used to provide additional information about the capabilities of installed oxygen-transfer devices. Engineers must choose carefully among the various field test methods available and apply them with care. The next sections discuss specification and testing protocols that can be applied to measure aeration equipment performance capabilities.

8.7.1 Equipment Considerations

As the first step in proper equipment selection, the design engineer should determine aeration system field requirements. Important elements in defining these requirements are:

- Site location, elevation above sea level, and ambient high summer and low winter air temperatures;
- Aeration tank volume, process water depth, and basin configuration;
- Oxygen demand—minimum, average, and maximum plus spatial and temporal distributions;
- Minimum airflow requirements of the diffused aeration equipment;
- Mixing requirements—capability to maintain specified MLSS concentrations in suspension;
- Process water temperature—minimum, average, and maximum;
- Process water transfer characteristics—range of alpha and beta factors anticipated;
- Operating dissolved-oxygen concentrations (mg/L);
- The MLSS concentrations (mg/L)—minimum, average, and maximum;
- Desired type of system, specified efficiency (standard oxygen-transfer efficiency [SOTE], SAE), construction materials, required performance testing, and necessary quality control for installation; and
- Penalties for not meeting performance guarantees.

Equipment suppliers should give users detailed mechanical and structural requirements and performance characteristics of the equipment, including reliable clean water performance data. In most cases, clean water test data provide the primary basis for specification of aeration equipment. These data must be reported at standard conditions and supplemented with a description of the conditions under which they were derived. This information will allow the prospective user to judge usefulness of the data for the specific application. Engineers and users should insist that basic data collection and analysis conform with protocols outlined in the ASCE (2007), which is summarized in a following section.

The prospective user or design engineer must translate clean water performance results to applicable field conditions. The engineer must specify tests that are directly scalable to the field or have evidence to support scale-up from shop-to-field geometry. For example, full-floor grids typically are scalable if submergence, diffuser density, and gas flowrates are true to full-scale design. Other diffused air and mechanical aeration systems are more complex. Often, engineers will require full-scale clean water testing when scale-up is in doubt.

The importance of using the most appropriate alpha values cannot be overstated; therefore, the engineer must exercise informed judgment based on the fundamentals of aeration, equipment being considered, and past experiences. Based on experience with equipment applications, the supplier should confirm appropriateness of alpha values being considered.

The design process for aeration equipment is detailed for diffused air systems in the literature (U.S. EPA, 1989). The same process may be used for mechanical or other aeration systems up to the point of calculating standard oxygen-transfer rates (SOTRs) for appropriate basin configurations and design loads. At that point, the designer may use SAEs to estimate numbers of units and standard power requirements. Aerator spacing may be determined from the characteristics of the selected transfer device. Finally, mixing may be evaluated based on equipment placement.

As part of the design process, the engineer must be aware of the many factors that affect performance of oxygen transfer; some of these factors are summarized in Table 12.26.

8.7.2 Clean Water Testing

The ASCE (2007) describes the measurement of OTR as the mass of oxygen dissolved per unit time in a unit volume of clean water by an oxygen-transfer system operating at a given gas flowrate and power input condition. This method applies to laboratory-scale devices with water volumes of a few liters and full-scale systems and is valid for many different mixing conditions and process configurations.

The method is based on dissolved-oxygen removal from the test water volume by the addition of sodium sulfite with a cobalt catalyst, followed by transfer studies of reoxygenation to near-saturation level. Test water volume dissolved-oxygen inventory is monitored during reoxygenation by measuring concentrations at several points that best represent tank contents. These dissolved-oxygen concentrations can be measured in situ

Factor	Accounts for Effects	Source of Information
α	Process waste on $K_L a$	Field tests; experience
β	Process waste on C_{st}^*	Calculate based on total dissolved solids
θ	Mixed liquor temperature on $K_L a$	1.024 or experimental
τ	Mixed liquor temperature on C_∞^*	Calculate
Ω	Ambient pressure on C_{st}^*	Calculate
C	Process dissolved oxygen	Select
F	Diffuser fouling/deterioration	Field tests; experience

TABLE 12.26 Correction Factors and the Information Source for Transforming Clean Water Parameters to Process Conditions

or on samples pumped from the tank. The method specifies a minimum number, distribution, and range of dissolved-oxygen measurements at each determination point.

Data obtained at each sampling point are then analyzed by a simplified mass-transfer model to estimate the apparent volumetric mass-transfer coefficient (K_La) and the equilibrium spatial average dissolved-oxygen saturation concentration (C^*) (ASCE, 2007; Jiang and Stenstrom, 2012). Nonlinear regression analysis is used to fit the dissolved-oxygen profile measured at each sampling point during reoxygenation to the model's mathematical equation. In this way, estimates of K_La and C^* are obtained at each sampling point. After these estimates are adjusted to standard conditions, the system SOTR is calculated from the aeration tank volume (V) and estimates of K_La and C^* at each of the n sampling points as follows:

$$\text{SOTR} = V \frac{\sum_{i=1}^{n} K_L a_{20i}\, C^*_{\infty 20i}}{n} \qquad (12.30)$$

where $K_L a_{20}$ = sampling point value of $K_L a$ corrected to 20°C, time^{-1};
$\quad\quad C^*_{\infty 20}$ = sampling point value of steady-state dissolved-oxygen saturation concentration corrected to 20°C, mass/volume;
$\quad\quad C^*$ = equilibrium spatial average dissolved-oxygen saturation concentration, mass/volume;
$\quad\quad V$ = aeration tank volume, volume; and
$\quad\quad n$ = sampling point.

Frequently, the SAE is calculated as the SOTR divided by power input. The SOTE in percent also can be estimated for diffused air systems by:

$$\text{SOTE} = \frac{\text{SOTR}}{W_{O_2}} \times 100 \qquad (12.31)$$

where W_{O_2} = mass flow of oxygen in the gas feed stream, mass/time.

It is important to use consistent definitions during aeration testing, subsequent data analysis, and final result reporting. A consistent nomenclature has been established with more logical and understandable terminology, which will eliminate much of the difficulty in interpreting aeration literature (ASCE, 2007). Standard conditions for oxygen-transfer tests are defined as water temperature of 20°C, barometric pressure of 100 kPa (1 atm), and dissolved-oxygen concentration of zero.

8.7.3 Transformation of Clean Water Test Data to Process Water Conditions

The OTR of a particular aeration device is typically expressed as either an SOTR or an OTR$_f$. Calculations of a field transfer rate from a standard value can be performed as follows (U.S. EPA, 1989):

$$\text{OTR}_f = \alpha\, F\, \text{SOTR}\; \Theta^{T-20}\, \frac{(\tau\beta\,\Omega\, C^*_{20-c})}{C^*_{20}} \qquad (12.32)$$

where OTR$_f$ = oxygen-transfer rate estimated for the system operating under process conditions, mass/time;
$\quad\quad$ SOTR = standard oxygen-transfer rate of new diffuser, mass/time;

α = average process water $K_L a$/average clean water $K_L a$ (both with new diffusers);

β = process water C_{st}^*/clean water C_{st}^*;

C_{st}^* = tabular value of dissolved-oxygen surface saturation concentration at actual process water temperature, a barometric pressure of 100 kPa (1.0 atm), and 100% relative humidity, mass/length^3;

C = average process water volume dissolved-oxygen concentration, mass/length^3;

Θ = empirical temperature correction factor assumed to equal 1.024, unless aeration system tests show a different factor;

F = fouling factor, process water SOTR of a diffuser after a given time in service/SOTR of a new diffuser in the same process water;

C_{20}^* = steady-state value of dissolved-oxygen saturation at infinite time at 20°C and a barometric pressure of 100 kPa (1.0 atm), mass/length^3;

τ = temperature correction factor for dissolved-oxygen saturation, C_{st}^*/C_{s20}^*;

C_{s20}^* = tabular value of dissolved-oxygen surface saturation concentration at 20°C, barometric pressure of 100 kPa, and 100% relative humidity, mass/length^3;

Ω = P_b/P_s (for tanks below 6 m [20 ft] in depth); or $\Omega = (P_b + \gamma_{wt} d_e - P_{vt})/(P_s + \gamma_{wt} d_e - P_{vt})$ (for tank depths deeper than 6 m [20 ft]);

wt = weight density of water at process conditions, mass/length^3;

P_{vt} = saturated vapor pressure of water at process temperature, force/length^2;

P_b = barometric pressure under field conditions, force/length^2;

P_s = standard barometric pressure, force/length^2;

d_e = effective saturation depth, length = $1/\gamma_w [C_\infty^*/C_{st}^* (P_s - P_{vt}) P_b + P_{vt}]$; and

C_∞^* = determination point value of steady-state dissolved-oxygen concentration.

Table 12.26 serves as a guide for applying eq 12.32 and indicates the source of information for the parameters needed to estimate OTR$_f$. Values of C_{s20}^* and SOTR are obtained from the clean water test described above. The value of C should represent the desired process water dissolved-oxygen concentration averaged over the aeration basin volume. Temperature and atmospheric pressure correction factors are estimated from process design.

Review of the components of this equation reveals a parameter (F) introduced in fine-pore aeration technology analysis (U.S. EPA, 1989). This parameter attempts to account for impairment in diffuser performance caused by fouling or material deterioration. This fouling factor, a dynamic term, depends on diffuser type and wastewater characteristics. In studies of fine-pore diffusers (ceramic, porous plastic, and perforated membrane), values of F ranged from 0.5 to 1.0 (U.S. EPA, 1989). The rate at which this value changes is described by a fouling rate term (f_F). For further details on this parameter, refer to the U.S. EPA (1989). The most frequent cause of fouling of fine-pore diffusers is due to mineral scale buildup, which can be addressed by periodic in situ acid cleaning. In fine-pore aeration systems where provisions are made for periodic in situ cleaning to remove mineral scale, use of a fouling factor may not be necessary. The advantages of diffuser cleaning are discussed by Rosso and Stenstrom (2006).

Alpha is one of the most controversial and investigated parameters for oxygen transfer. Alpha has been found to be dependent on wastewater characteristics, diffuser type, airflow rate, diffuser placement, basin geometry, system operating parameters, and flow regime. In addition, its value varies spatially and temporally. With the same waste-water, alpha will typically be lowest for aeration devices generating fine bubbles and highest for coarse bubbles and surface aeration systems. Measurements of OTR_f in munic-ipal facilities indicated that F values for fine-pore devices averaged approximately 0.4, with a range of 0.1 to 0.7 (U.S. EPA, 1989). These were mean weighted values for the entire aeration basin. Individual measured values were significantly lower at the influent end of the plug-flow reactor. Diurnal variations of the mean weighted F values were represented by a maximum-to-average ratio of approximately 1.2 and a minimum-to-average ratio of approximately 0.86 (U.S. EPA, 1989). Information on alpha for mechanical aeration equip-ment is scarce and, when available, some is less reliable than for fine-pore devices. Values from 0.3 to 1.1 have been reported in the literature (Boyle et al., 1989). Additional refer-ences discuss alpha for other systems and wastewaters (Doyle and Boyle, 1985; Mueller and Boyle, 1988; Stenstrom and Gilbert, 1981; U.S. EPA, 1983, 1989).

8.8 Process Water Testing

Once aeration equipment is operating under process conditions, its performance should be compared with calculated design estimates. Process water testing provides the best and most reliable source of data on F and on the effects of system design. All methods for testing equipment during process operation are referred to as respiring system tests. The ASCE published *Standard Guidelines for In-Process Oxygen Transfer Testing* (1997).

Typically, testing methods can be categorized according to the rate of dissolved-oxygen concentration change with respect to time in a given reactor (or reactor seg-ment). Systems with a dissolved-oxygen rate of change of zero are described as being in a steady-state condition; all others are as nonsteady. If influent wastewater is diverted from a reactor for testing, these tests are referred to as batch tests. The term continuous test applies where influent wastewater flow is not diverted.

Several respiring system test methods that do not require a direct measure of OUR are broadly categorized as mass balance, off-gas, inert-tracer, and non-steady-state methods. The mass balance method requires measurement of influent and effluent liquid flows. The off-gas method is based on the mass balance of oxygen across a sys-tem and requires measurements in inlet and exit gas streams. The inert-tracer method indirectly measures OTR by determining the transfer rate of a radioactive or stable inert gas tracer. For non-steady-state methods, the reactor dissolved-oxygen level is adjusted at the beginning of the test to be either higher or lower than the steady-state dissolved-oxygen concentration. The OUR data, though not required, are often collected to ensure relatively constant operating conditions during an evaluation. Several references include a comprehensive review of available test methods for field oxygen-transfer measurements (ASCE, 1997; Doyle and Boyle, 1985; U.S. EPA, 1983, 1985). Tables 12.27 and 12.28 provide assumptions necessary to use steady- or non-steady-state tests. Selec-tion of the best method often depends on economics, degree of precision and accuracy required, process conditions, and other considerations.

8.9 Aeration System Maintenance

The principal objective in the design of the aeration system is to develop an effective system with the lowest possible cost, maintaining a balance between initial investment and long-term operations and maintenance expenditures. The manufacturers' recom-mended maintenance requirements should be followed for all aeration system

Assumptions	Test Conditions
Aeration volume DO is constant and the reactor contents are completely (uniformly) mixed	Test time is short, they are not required for non-steady-state test
Reactor influent flow is constant	Test time is short, variation in influent wastewater flow is negligible during test period; recycle sludge flow is held constant
Influent dissolved oxygen is constant	Test time is short, variations in influent wastewater flow and DO are negligible during test period; recycle sludge flow is held constant
Aeration volume oxygen uptake is constant	Test time is short; variations in influent wastewater flow and DO are negligible during test period (note that uptake rate may be difficult, if not impossible, to moderately determine for systems with high organic loadings)
Effective oxygen transfer rate is constant	Test time is short; alpha value remains constant during the test period

TABLE 12.27   Assumptions Necessary to Develop Equations for Continuous Steady and Non-Steady-State Tests (ASCE, 2007; Doyle and Boyle, 1985; U.S. EPA, 1983, 1985)

Assumptions	Test Conditions
Aeration volume DO is constant and the reactor contents are completely (uniformly) mixed	Steady state conditions have been achieved and maintained, they are required for non-steady-state test.
Recycle sludge flow is constant zero	Recycle flow rate maintained constant or discontinued
Recycle DO is constant	Steady operation of recycle system if in use during the test period (for example, sludge blanket level constant)
Aeration volume oxygen uptake rate is constant	Aeration volume biological solids and recycle sludge flow remain constant; carbonaceous and nitrogenous substrates are near zero during the test (if nitrification occurs in the test system)
Effective oxygen transfer rate is constant	Carbonaceous substrate is near zero during the test; alpha value remains constant during the test period

TABLE 12.28   Assumptions Necessary to Develop Equations for Steady- and Nonsteady-State Batch (ASCE, 2007; Doyle and Boyle, 1985; U.S. EPA, 1983, 1985)

equipment. Blower maintenance requirements vary by blower type and can range from routine filter changes to periodic lubrication, oil changes, and rotor balancing.

Maintenance and cleaning requirements for porous diffuser systems may include acid cleaning, routine membrane flexing and bumping, and the periodic draining of tanks or removal of equipment for washing, scrubbing, or membrane replacement. In situ acid cleaning of porous diffusers can significantly reduce operating costs and extend the life of diffusers (Rosso and Stenstrom, 2006). Maintenance and cleaning frequencies are dependent on a number of factors including wastewater characteristics such as fouling potential and differ by diffuser type and manufacturer. Provisions to provide the equipment necessary to complete routine maintenance should be considered during design. Such provisions may include access to high-pressure hoses, chemicals for in situ cleaning, automated diffuser flexing and bumping, and the ability to drain tanks or remove diffuser grids while tanks are still in service.

9.0   Secondary Clarification

9.1   Introduction

Gravity clarification traditionally has been used to separate MLSS from effluent in suspended-growth systems. Table 12.29 lists many of the factors that affect clarifier performance. Design considerations to address these factors that are common to all shapes

Category	Factors
Hydraulic and load factors	Wastewater (ADWF, PDWF, PWWF)*
	Surface overflow rate
	Solids loading rate
	Hydraulic retention time
	Underflow recycle ratio
External physical features	Tank configuration
	Surface area
	Depth
	Flow distribution
	Turbulence in conveyance structure
Internal physical features	Presence of flocculation zone
	Sludge-collection mechanism
	Inlet arrangement
	Weir type, length and position
	Baffling
	Hydraulic flow patterns and turbulence
	Density and connection currents
Site conditions	Wind and wave action
	Water temperature variation
Sludge characteristics	MLSS concentration
	Sludge age
	Flocculation, settling and thickening characteristics
	Type of biological process

*ADWF = average dry weather flow, PDWF = peak dry weather flow, and PWWF = peak wet weather flow.

TABLE 12.29 Factors that Affect Clarifier Performance (Ekama et al., 1997; reprinted with permission from IWA Publishing)

and sizes of clarifiers are discussed in the following paragraphs. These are followed by separate sections on rectangular and circular clarifiers. Further details and more in-depth analysis of the design of secondary clarifiers can be found in Water Environment Federation Manual of Practice No. FD-8 (WEF, 2005).

9.2   Suspension Characteristics and Settleability

9.2.1   Characteristics

Figure 12.76 illustrates the settling characteristics used to categorize suspensions. All four types of settling occur in activated sludge clarifiers. Types I and II represent the settling of individual discrete and flocculated particles that occurs in the upper reaches of the clarifier, resulting in an effluent with low suspended solids. Type III, or zone set-tling, occurs with concentrated suspensions that settle at lower velocities because water is displaced as the suspension settles. The result is clear liquid above the settling zone and settling velocities that decrease with increasing concentration. Type III settling of the sludge blanket is the most important for clarifier design and represents the behavior of the sludge blanket wherein concentration increases with depth although solids are continuously removed as RAS at the underflow concentration. In Type IV, or compres-sion settling, particles are in contact and further settling can only occur by compression. Type IV can exist in the sludge zone at the tank bottom.

Work by Coe and Clevenger (1916), Dick and Ewing (1967), Dick and Young (1972), and Yoshioka et al. (1957) advanced the solids flux approach to clarification. For a clarifier operating at a steady state, a constant flux of solids is moving downward (Figure 12.77). The total mass flux of solids is the sum of the mass flux resulting from hindered settling due to gravity and the mass flux resulting from bulk movement of the suspension. The solids flux across any arbitrary boundary resulting from hindered settling is:

$$SF_g = X_i V_i \qquad (12.33)$$

where SF_g = solids flux resulting from gravity, kg/m²·h (lb/hr/sq ft),
  X_i = solids concentration at point in question, g/m³ (lb/cu ft), and
  V_i = settling velocity of solids at concentration X, m/h (ft/hr).

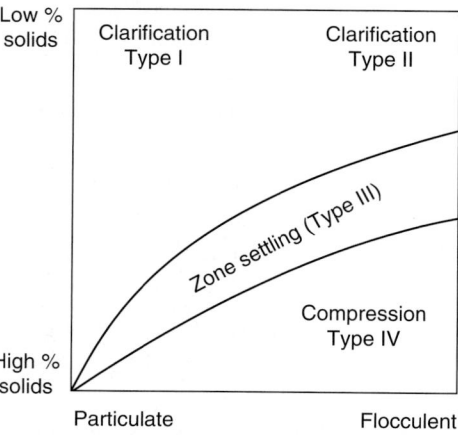

FIGURE 12.76   Relationship between solids characteristics and sedimentation processes.

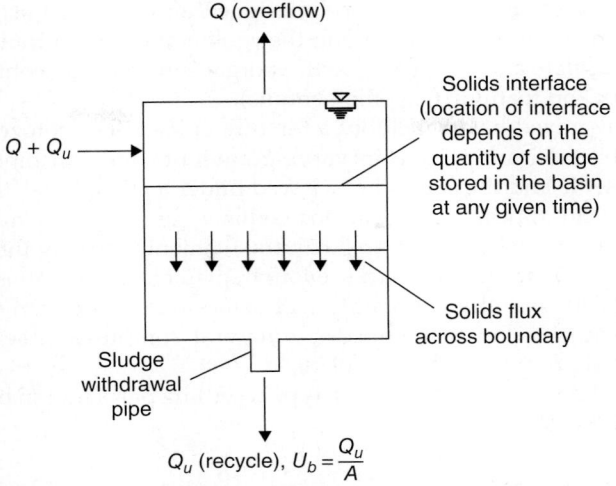

Figure 12.77 Settling basin at steady state (U_b = bulking downward velocity, m/h or ft/hr, and A = required area, m² or sq ft) (from Metcalf & Eddy, Inc./AECOM *Wastewater Engineering: Treatment and Resource Recovery*, 5th ed. Copyright © 2014, The McGraw-Hill Companies, New York, N.Y., with permission).

The solids flux resulting from underflow is:

$$SF_u = X_i U_b \tag{12.34}$$

and

$$U_b = Q_u / A \tag{12.35}$$

yielding

$$SF_u = X_i Q_u / A \tag{12.36}$$

where SF_u = solids flux resulting from underflow, kg/m²·h (lb/hr/sq ft),
$\quad U_b$ = bulk downward velocity, m/h (ft/hr),
$\quad Q_u$ = underflow flowrate, m³/h (cu ft/hr), and
$\quad A$ = required area, m² (sq ft).

The total solids flux, SF_t, in kg/m²·h (lb/hr/sq ft), is the sum of these two components:

$$SF_t = X_i V_i + X_i U_b \tag{12.37}$$

The total solids flux represents the maximum rate that solids can be continually applied to a clarifier for a given underflow rate, MLSS concentration, and characteristic settling velocity at concentration X_i. The characteristic settling velocity is, in turn, a function of the settleability of solids.

9.2.2 Factors That Affect Settleability
It is not possible to predict the settling characteristics of a particular suspension of microorganisms from day to day. Microbial makeup is the primary factor affecting

activated sludge settleability. A well-designed and operated activated sludge system provides an environment promoting the proliferation of desired microorganisms that readily flocculate and controls growth of organisms that can contribute to poor sludge settleability and foaming (e.g., filamentous).

Healthy mixed liquor includes a mixture of bacteria, protozoa, and metazoa. Filamentous bacteria are present in varying amounts and can hinder the ability of the sludge to settle and thicken. When viewed under a microscope, they are typically long and stringy in appearance. Ideal floc contains just the right mixture of filamentous microorganisms and floc formers, with the filaments forming the backbone of the floc (Figure 12.78a). If the floc lacks enough filaments, then it is likely to break up (Figure 12.78b) and effluent quality will deteriorate. If too many filaments exist, then bulking may develop that impedes zone and compression settling (Figure 12.78c) (Jenkins et al., 2003; Sezgin et al., 1978).

The settling velocity of Type I or Type II settling particles can be predicted by Stokes law (eq 12.38):

$$v = \frac{2}{9} \frac{\rho_p - \rho_f}{\mu} g R^2 \qquad (12.38)$$

where v = settling velocity, length/time;
ρ_p = density of particles, mass/length^3;
ρ_f = density of fluid, mass/length^3;
μ = viscosity of fluid, mass/time/length;
g = acceleration due to gravity, length/time^2; and
R = particle radius, length.

From Stokes' law, if a particle suspension contains particles that are too small or their density is too close to the density of the fluid, then the particles will settle at a negligible rate. Many of these solids are not removed in a typical final clarifier. They have a low tendency to flocculate or have sheared from floc particles because of excessive turbulence in the aeration basin or in the conveyance system. If their removal is required, especially to achieve very low effluent phosphorus limits, then a tertiary filtration may be required.

Suspended solids in activated sludge mixed liquor settle better in warmer temperatures. Reed and Murphy (1969) have investigated this and noted that at 0°C sludge settled 1.75 times faster than sludge at 20°C (MLSS = 2000 mg/L; Figure 12.79). The effect became less pronounced as solids concentration increased. Wilson (1996) also quantified the effect of temperature.

As mixed liquor is aerated, transported to the clarifier, and settled, the degree or status of floc formation may change. Diagnostic testing, such as proposed by Wahlberg (2001), can provide useful information about system performance and deficiencies.

9.2.3   Measures of Settleability

Two basic approaches are used in measuring sludge settleability: (1) volume of settled sludge after a given period of time and (2) settling velocity of the sludge/liquid interface during zone settling.

The SVI has long been a common measure of sludge settleability. It is the volume in milliliters occupied by 1 g of the MLSS following 30 minutes of quiescent settling.

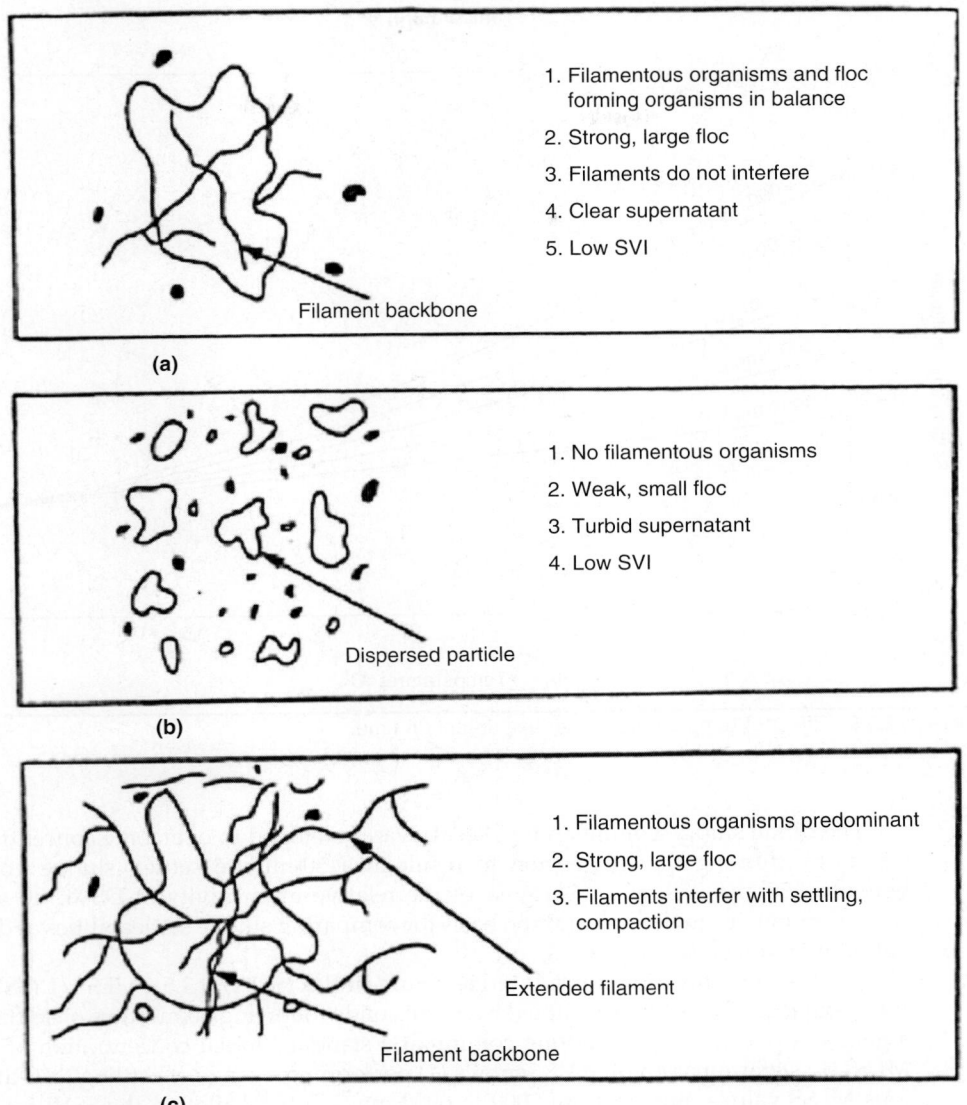

FIGURE **12.78** Effect of filamentous organisms on activated sludge structure: (a) ideal, nonbulking floc, (b) pinpoint floc, and (c) filamentous, bulking (Ekama et al., 1997; reprinted with permission from the copyright holders, IWA).

The traditional method is carried out in a 1- or 2-L settling column or graduated cylinder, but other methods produce more representative results. Standard methods specify gently stirring the sample during settling to eliminate or minimize wall effects (American Public Health Association et al., 2012).

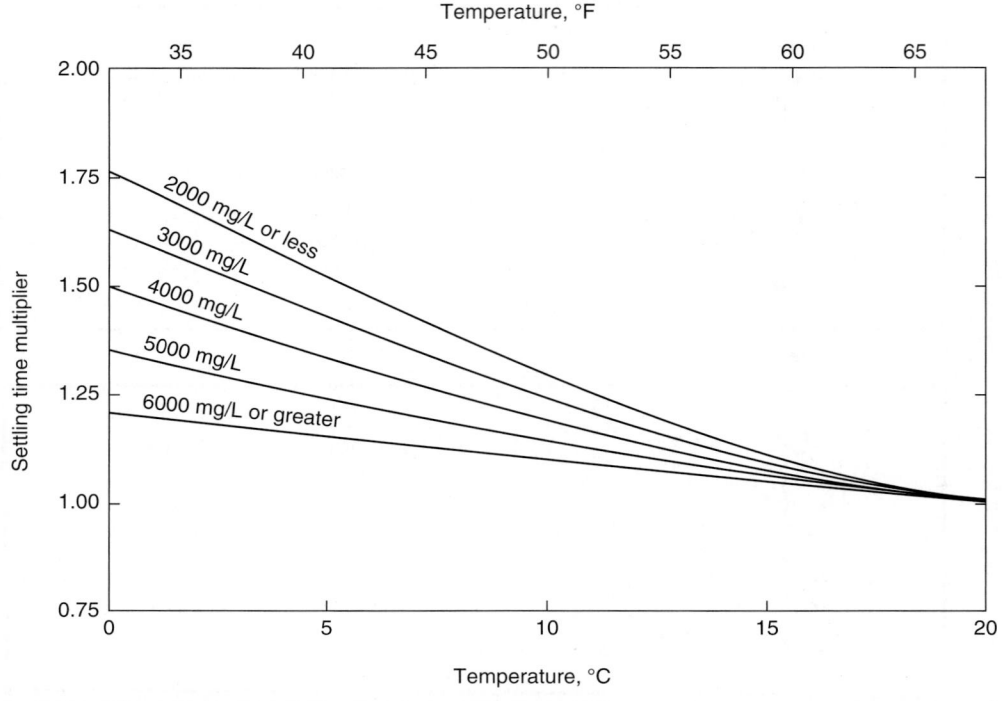

FIGURE 12.79   Effect of temperature on settling detention time.

The dilute sludge volume index (DSVI) was developed to overcome concentration effects by diluting the suspension to result in a 30-minute settled sludge volume between 150 and 250 mL/L. Because of the relative insensitivity of $DSVI_{30}$ to solids concentration, it provides a common basis for comparing sludge settleabilities at different times and facilities.

Another measure is the stirred sludge volume index (SSVI) at 3.5 g MLSS/L ($SSVI_{3.5}$). It is defined as the volume occupied by 1 g of solids following 30 minutes of settling in a gently stirred (at 1 rpm) settling column at a standard initial concentration of 3.5 g MLSS/L. Determination of $SSVI_{3.5}$ entails (1) performing a range of settling tests at various MLSS values ranging from 2000 to 6000 mg/L, (2) calculating the SSVI for each concentration, (3) developing an SSVI-concentration graph, and (4) obtaining the SSVI value at 3500 mg/L by interpolation.

9.2.4   Techniques to Improve Settleability

The settleability of activated sludge MLSS can be influenced by the loading of the biological reactor, transport, and clarifier/RAS system design.

9.2.4.1   Food-to-Microorganism Ratio Control

The F:M is an important design consideration relative to settleability. High values can lead to dispersed growth that gives preference to free-swimming bacteria and other organisms that neither settle well nor effectively incorporate into flocs. Decreasing this loading parameter results in filamentous microorganisms and endogenous respiration and decay, resulting in fragments of decomposition that do not readily settle out. At intermediate values of F:M, the MLSS may or may not settle well, depending on other variables such as nutrients, dissolved-oxygen levels, turbulence, and possible toxicity.

9.2.4.2 Dissolved-Oxygen Concentration   In a plug-flow reactor, more oxygen is required at the head end of the tank to prevent a drop in dissolved-oxygen and bulking conditions. Jenkins et al. (2003) described the interaction of aeration basin dissolved oxygen and F:M relative to bulking conditions for completely mixed reactors (Figure 12.80). The figure shows that higher loading rates can be achieved with reasonable settling characteristics if the dissolved oxygen is kept relatively high.

9.2.4.3 Selectors   Discussed earlier in this chapter, selectors help control the settleability of the activated sludge MLSS.

9.2.4.4 Process Configuration   With ample design flexibility, an operator can take positive steps to ensure good settleability. In low F:M systems, a good approach is to operate at least the first portion of the aeration tank in plug-flow configuration. This configuration creates an environment somewhat like a selector to limit the growth of low F:M bulking organisms.

Operating in a step-feed mode so that some or all of the influent flow can be added at several points along the length of the aeration tank is sometimes advantageous. Typically, influent is split equally among two to four points and return sludge is added only to the first pass of the aeration tank, as described earlier in this chapter. This type of design, for a given tank volume and F:M (or SRT), allows lower solids loading rates (SLRs) on the final clarifiers. Step feed also allows the oxygen demand to be more evenly distributed along the length of an aeration tank.

9.2.4.5 Selective Wasting and Foam Control   *Nocardia*-type organisms are notorious for accumulating at the activated sludge reactor, open channel, and final clarifier surfaces.

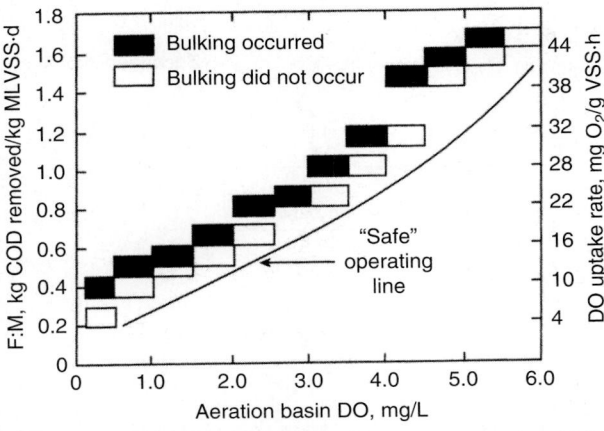

FIGURE 12.80 Bulking and nonbulking conditions in completely mixed aeration basins (COD = chemical oxygen demand; DO = dissolved oxygen; and MLVSS = mixed-liquor volatile suspended solids) (Jenkins et al., 2003; reprinted with permission from IWA Publishing).

Many facilities with sludge that is older than 5 days have this problem. Aeration basins typically are designed with overflow weir outlets that move floatables downstream to splitter box structures or aerated channels feeding the clarifiers. If not removed upstream, these nuisance filamentous organisms enter final clarifiers and rise to the surface where they must be skimmed off to prevent odor or loss of solids to the effluent. Parker et al. (2003) introduced the concept of adding mechanical skimming devices, such as spiral or chain-and-flight blades to move floatables up a beach and into a hopper located at the end of the aerated channel. Another method of foam and floatable control includes spray nozzles, which are not effective for *Nocardia*-type foam.

If the foam is passed into the clarifier, it tends to accumulate behind the baffles of the flocculation zone. It is recommended to have a means to remove the foam using automated mechanical provisions, by lowering of the baffle and causing it to overflow a bit at peak hour flow, or by periodically lowering a gate in the baffle. Spray nozzles aimed at moving the scum through small ports in the baffles may help some but often are not sufficient when used alone.

9.2.4.6   Chemical Addition   Chemical addition can enhance clarifier performance by eliminating excess filaments, changing floc size and shape, or inducing flocculation. Some bulking sludges can be controlled by RAS or sidestream chlorination. A typical design for a low (5- to 10-hour) HRT system uses 0.002 to 0.008 kg chlorine (Cl_2)/kg MLSS·d (2 to 8 lb Cl_2/d/1000 lb MLSS), with chlorine added to the RAS system. Longer HRT systems might need chlorine added to a sidestream or multiple points in the aeration tanks (Figure 12.81). Hydrogen peroxide can be substituted for chlorine in many cases. Further design and sizing details can be found elsewhere (Jenkins et al., 2003). The RAS chlorination, however, can interfere with nitrification. One full-scale study revealed that, to maintain BNR capability, the chlorine dose needs to be less than 0.001 kg Cl_2/kg MLSS·d (Ward et al., 1999). The study also reported that, following chlorine inhibition, nitrification recovered faster than EBPR after chlorine addition was stopped.

To improve flocculation, addition of cationic polymers at concentrations of less than 1 mg/L has been shown to improve mixed-liquor settleability. In rare instances, alum

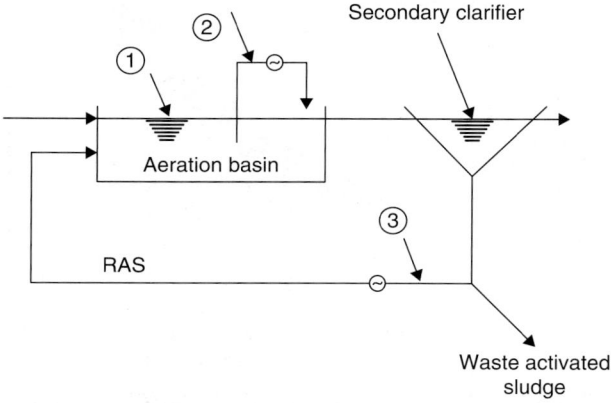

Figure 12.81   Chlorine dosing points for bulking control (Jenkins et al., 2003; reprinted with permission from IWA Publishing).

also has been used. The selection of inorganic salts, polymers, or other flocculent aids should be based on laboratory studies, including jar tests.

9.2.4.7 Ballast Addition High-density, rapidly settling materials can be added to the activated sludge process and subsequently incorporated in the biological floc and as such is present throughout the secondary treatment system (all zones of the bioreactors and the secondary clarifiers). The activated sludge suspended solids settle and thicken quickly in the clarifiers. Higher MLSS concentrations can be carried in the bioreactors as the secondary clarifiers can handle much higher SLRs due to the fast settling rates of the ballasted flocs. One such example is the "BioMag" process, which employs magnetite as the ballast. A full-scale demonstration of the BioMag process increased the MLSS from 2000 to over 5000 mg/L (McConnell et al., 2010). When measuring MLSS of the BioMag process, care must be exercised to separate the mixed liquor from the ballast, or artificially high measurements will result.

In the BioMag process, magnetite is recovered from the WAS at a rate of 92% to 95% before the WAS is sent to sludge processing to be thickened and disposed of with the rest of the facility sludge. As an inert mineral, the magnetite can be used indefinitely and does not require replacement, but supplementation of the magnetite lost in the WAS is required.

Due to the rapid settling of the ballasted floc, higher mixing energies are required to keep the floc in suspension. Even with higher mixing, facilities that have employed the BioMag process often report magnetite accumulation in bioreactor channels, especially corners and other difficult-to-mix areas.

9.2.4.8 Energy Gradient Optimization Energy gradient optimization is important to grow activated sludge floc or protect it in transit from the aeration tank to the settling zone of the clarifier. As mixed liquor leaves the aeration basin, the floc may be well formed if gentle mixing, such as that achieved with fine-bubble aeration, is used in the reactor. Use of jets, high- or low-speed mechanical aeration, or submerged turbines can tear up floc. Reformation should be achieved before mixed liquor enters the quiescent zone of the clarifiers for settling. This may be achieved by adequate detention time in aerated mixed-liquor conveyance channels.

If flow splitting over weirs is used, then fall height requires attention. Falls of as much as one meter have not destroyed some mixed liquors, but the nature of the floc may affect the result.

Within the clarifier, inlets must dissipate influent mixed-liquor energy, distribute flow evenly, reduce density short-circuiting and current effects, minimize blanket disturbances, and promote flocculation. Das et al. (1993) demonstrated that velocities in excess of 0.6 m/s (2 ft/s) would cause deflocculation. The incoming energy can be used to promote flocculation, as discussed later.

9.3 Clarifier Sizing Approaches

Providing adequate surface area and depth is critical in clarifier design. Two criteria that define the area requirements are surface overflow rates (SORs) and SLR. The SOR is the clarified effluent flow divided by effective surface area, which is defined as the dimensions to the inner tank wall and not the weir or orifice locations. The SLR is the mass loading of solids (total flow to the clarifier, including RAS, multiplied by the MLSS concentration) divided by the area.

Some regulations specify numeric limits for surface area sizing; others accept justification calculations for loading. These in turn may include solids flux analyses and other such site-specific information.

9.3.1   Surface Overflow Rate

The SOR used by design engineers, based on average dry-weather flow (ADWF) and surface area, ranges from 0.5 to 2 m/h (300 to 1000 gpd/sq ft) for activated sludge secondary clarifiers. Some facilities are known to operate without difficulty at the upper end of this range and produce a high-quality effluent. In many documented cases, diurnal or maximum pumping peak rates of 2.7 to 3.1 m/h (1600 to 1800 gpd/sq ft) do not exceed capacity. In other cases, poor clarification efficiency is encountered when the average and peak SORs are on the lower end of typical design ranges.

A survey of consulting firms resulted in preferred SORs, shown in Table 12.30. Randall et al. (1992) recommend average and maximum SORs based on the clear water zone, which is the free settling zone above the maximum height of the sludge blanket. Their recommendations, presented in Table 12.31, show peak criteria are three times the average, which may not apply in many cases.

The Ten States Standards or similar guidelines may need to be applied (Great Lakes-Upper Mississippi River Board [GLUMRB], 2014). In some cases, these capacity ratings were developed from clarifier designs in operation decades ago and do not reflect potential improvements in the design of inlet and outlet structures, depth, sludge collectors, and sludge removal that have been shown to increase allowable rates. It is projected that fully optimized clarifier designs will have 15% to 20% higher hydraulic capacity than pre-1970 clarifier designs having the same side water depth (SWD) (WEF, 2013a).

Flow	Circular Clarifiers		Rectangular Clarifiers	
	Range	Average	Range	Average
Average	0.5–1.19	0.95	0.5–1.19	0.95
	(400–700)	500	(400–700)	500
Peak	1.70–2.72	2.09	1.70–2.72	2.09
	(1000–16 000)	(1200)*	(1000–16 000)	(1200)*

*Ten of 15 firms use 2.04 m^3/m^2·h (1200 gpd/sq ft).

TABLE 12.30   Preferred Overflow Rates (m^3/m^2·h [gpd/sq ft])

Hydraulic Condition	Moderate CWZ[a] 1.83–3.05 m	Deep CWZ 3.05–4.57 m
Average SOR (m^3/m^2·h)	0.091 CWZ^2	0.278 CWZ
Maximum SOR (m^3/m^2·h)	0.182 CWZ^2	0.556 CWZ

*CWZ = clear water zone.

TABLE 12.31   Clarifier Overflow Rate Limitations (Randall et al., 1992)

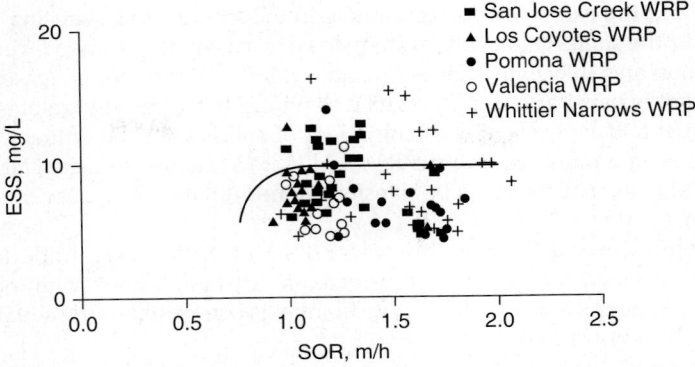

FIGURE 12.82 Typical solids concentration— depth profile assumed in flux analysis (Ekama et al., 1997; reprinted with permission from IWA Publishing).

A correlation between effluent suspended solids and SOR developed for several facilities indicates that an effluent TSS of less than 20 mg/L can be achieved at SORs ranging from 1.0 to 2.0 m/h (Figure 12.82). Such correlations can be misleading because they do not account for the effects of temperature, peaking factors, SVIs, geometrical details, RAS flowrate, and RAS concentration. Because the literature is limited, designs for specific sites should be conservative or based on experimental testing (Tekippe and Bender, 1987). Unbalanced load testing (loading multiple clarifiers at different rates to evaluate performance) at existing facilities undergoing expansion is encouraged. If such testing is not feasible, column settling investigations can be undertaken to help establish design criteria.

9.3.2 Solids Loading Rate

In establishing the maximum allowable SLR, most design engineers prefer to keep the average SLR (including full RAS capacity) in the range of 100 to 150 kg/m²·d (20 to 30 lb/d/sq ft) and peak SLR at 200 to 240 kg/m²·d (40 to 50 lb/d/sq ft). Rates greater than 240 kg/m²·d (50 lb/d/sq ft) or more have been observed in some well-operating facilities with low SVI, well-designed clarifiers, and effective solids removal. Approaches to determining the limiting SLR are presented below and include solids flux analyses and operating strategies.

Solids flux analyses are valuable in determining if refinements to simple solids loading criteria are worthwhile and helping operators run the clarification process. The assumption of flux theory is that solids are continuously removed from the clarifier as they reach the design underflow concentration and that settling characteristics of the suspension are known. A detailed mathematical analysis of the flux theory is presented in Water Environment Federation MOP FD-8 (2005).

Solids concentration-depth profiles consist of four zones: (1) clear water zone (h_1); (2) separation zone (h_2); (3) sludge storage zone (h_3); and (4) thickening and sludge-removal zone (h_4) (Ekama et al., 1997). The fundamental premise of the flux theory is that during overloaded conditions (applied solids flux greater than the limiting flux), a critical zone settling layer (sludge storage zone, h_3) develops in the sludge blanket, which limits the conveyance of solids to the bottom of the tank. Consequently, all of the solids that enter the storage zone from the separation zone are not transferred to the thickening zone below, and the excess solids accumulate in the storage zone, causing it to expand. As it expands, the solids concentration remains constant in the storage layer. The depth of the separation and thickening zones (h_2 and h_4), however, do not increase substantially. The continued expansion of the storage layer will result in the sludge blanket reaching near the effluent structure level, causing a loss of solids with the effluent. At this point, the storage layer cannot expand further, and storage capacity of the clarifier is exhausted. The solids flux that could not be transferred through the storage layer is then lost with the effluent.

When the applied solids flux is less than the critical flux (underloaded condition), all of the applied solids can be transferred to the tank bottom, eliminating the need for solids storage. As a result, the sludge blanket is composed of the separation (h_2) and the thickening (h_4) zones only.

9.3.2.1 State Point Analysis   The state point analysis (SPA) is a graphical approach derived from solids flux theory (Keinath, 1985; Keinath et al., 1977). The SPA incorporates MLSS concentration and suspension settling characteristics, surface area available for thickening, and influent and RAS flowrates into the model. It can be used to assess different design and operating conditions.

Figure 12.83 illustrates the components of the SPA. By definition, the state point is the point of intersection of the clarifier overflow rate (OFR) and underflow rate (UFR). As summarized in Table 12.32, the position of the state point and the location of the UFR line relative to the descending limb of the flux curve determine whether the clarifier is underloaded, critically loaded, or overloaded (Figures 12.84 through 12.88).

The WERF's Clarifier Research Technical Committee (CRTC) Protocol provides guidance with respect to the development and application of the SPA (WERF, 2001). Metcalf & Eddy, Inc./AECOM (2014) presents an example on the use of SPA in operation and design.

9.3.2.2 Daigger Approach   Daigger (1995) and Daigger and Roper (1985) developed a clarifier operating diagram (Figure 12.89) by plotting allowable SLR as a function of RAS solids concentration based on SPA and suspension settling velocities predicted as a function of unstirred SVI. The lines represent the limiting flux for the SVI shown.

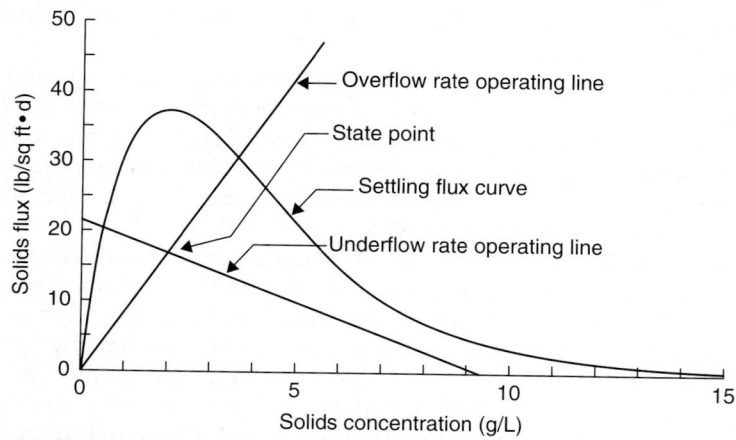

Figure 12.83   Elements of state point analysis (WERF, 2001).

Location of State Point	Location of Underflow Line	Condition of Clarifier	Potential Corrective Action
Within the flow curve (Figure 12.83)	Below the descending limb of the flux curve	Underloaded	None
Within the flux curve (Figure 12.84)	Tangential to the descending limb of the flux curve	Critically loaded	Increase RAS rate to become underloaded
Within the flux curve (Figure 12.85)	Above the descending limb of the flux curve	Overloaded	Increase RAS rate to become underloaded
On the flux curve (Figure 12.86)	Below the descending limb of the flux curve	Critically loaded	Reduce clarifier load solids to become underloaded
On the flux curve (Figure 12.87)	Above the descending limb of the flux curve	Overloaded	Increase RAS rate to become critically loaded
Outside the flux curve (Figure 12.88)		Overloaded	Reduce clarifier load solids to become underloaded

TABLE 12.32   Interpretation of the State Point Analysis

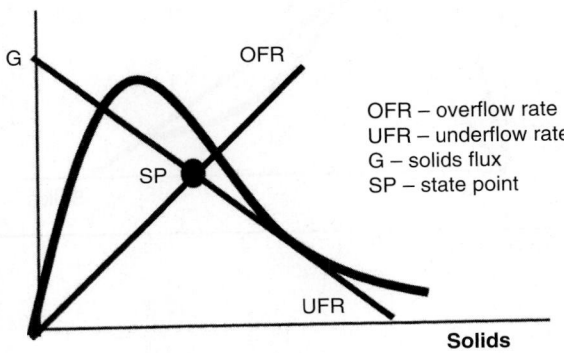

OFR – overflow rate
UFR – underflow rate
G – solids flux
SP – state point

FIGURE 12.84   Critically loaded clarifier.

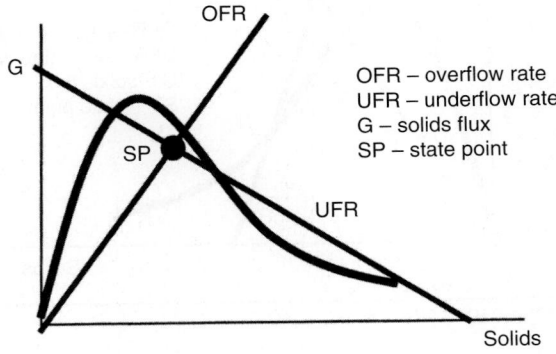

OFR – overflow rate
UFR – underflow rate
G – solids flux
SP – state point

FIGURE 12.85   Overloaded clarifier.

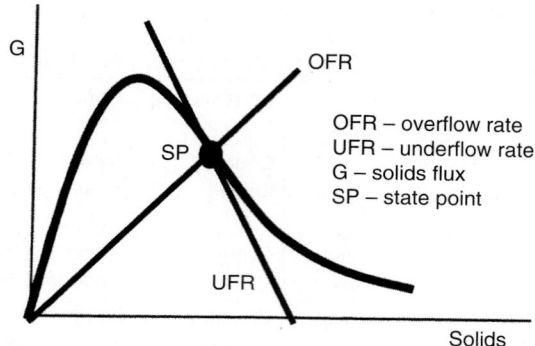

FIGURE 12.86   Critically loaded clarifier.

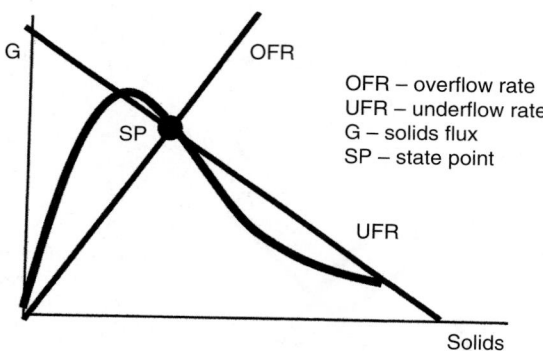

FIGURE 12.87   Overloaded clarifier.

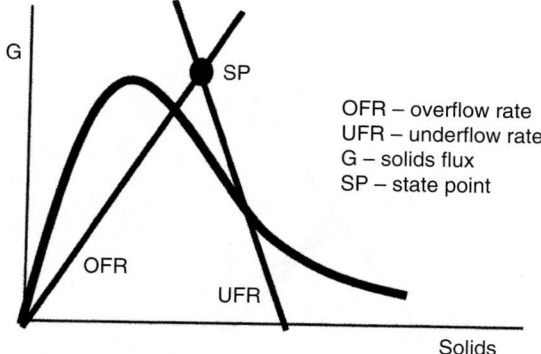

FIGURE 12.88   Overloaded clarifier.

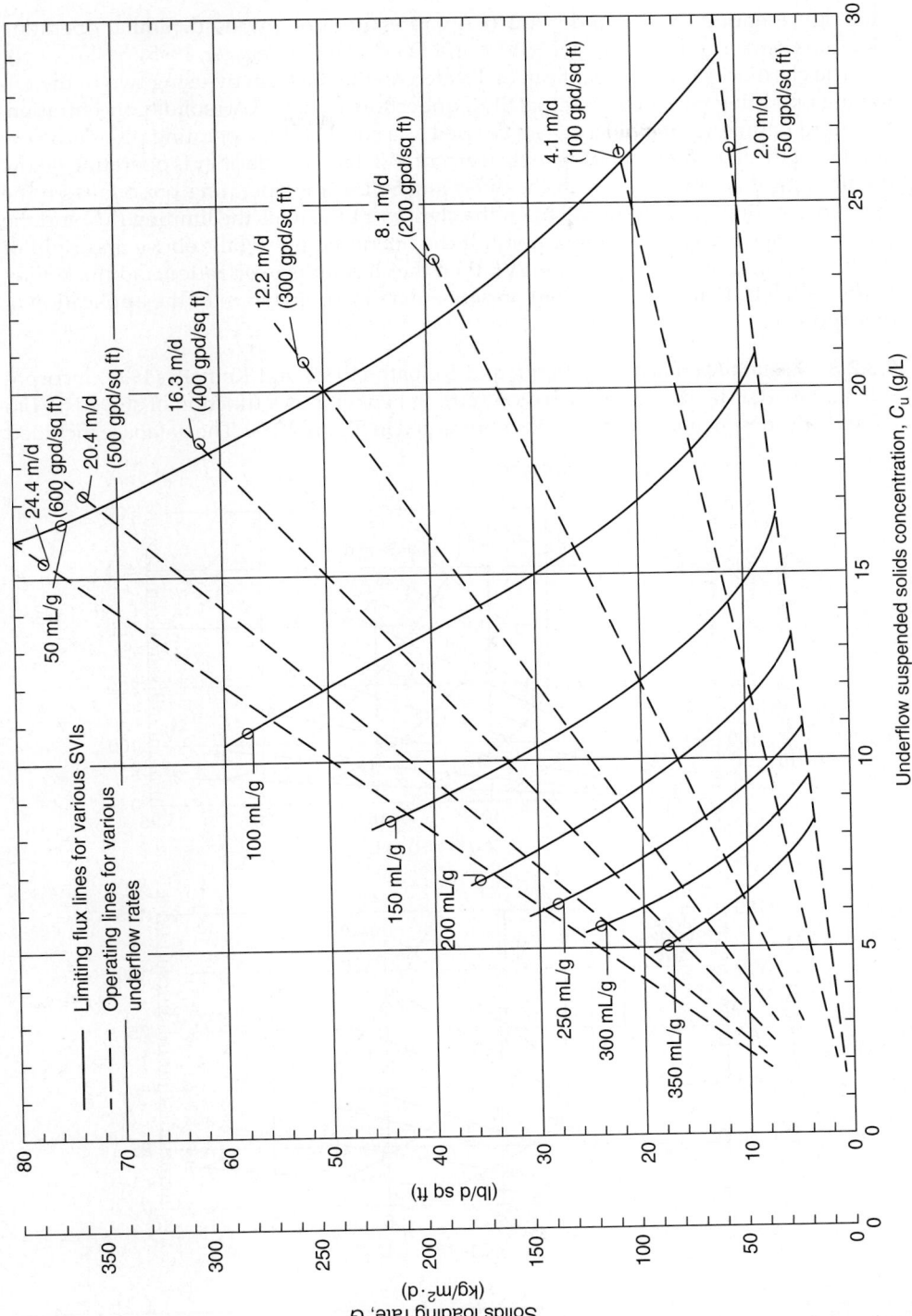

FIGURE 12.89 Daigger operating chart (Daigger, 1995).

Lines representing various underflow (RAS) rates are superimposed. Similar operating diagrams can be generated using $SSVI_{3.5}$ and DSVI values (Daigger, 1995).

The clarifier operating point can be located on the diagram by using two of the following operating parameters: actual SLR, underflow rate, or RAS solids concentration. The third parameter, if available, can be used as a check. If the operating point is below and left of the line corresponding to the current SVI, then the clarifier is operating below the limiting flux associated with the operating SVI. If the operating point falls on the line representing the current SVI, then the clarifier SLR equals the limiting flux and the clarifier is operating at its failure point. If the operating point falls above and right of the line representing the operating SVI, then the clarifier is overloaded and thickening failure is likely. Jenkins et al. (2003) present a detailed illustration of the application of this approach.

9.3.2.3 Keinath Approach   Wahlberg and Keinath (1988) and Keinath (1990) incorporated a broader database of suspension settling behavior as a function of stirred SVI to develop the design and operating chart presented in Figure 12.90. The database included

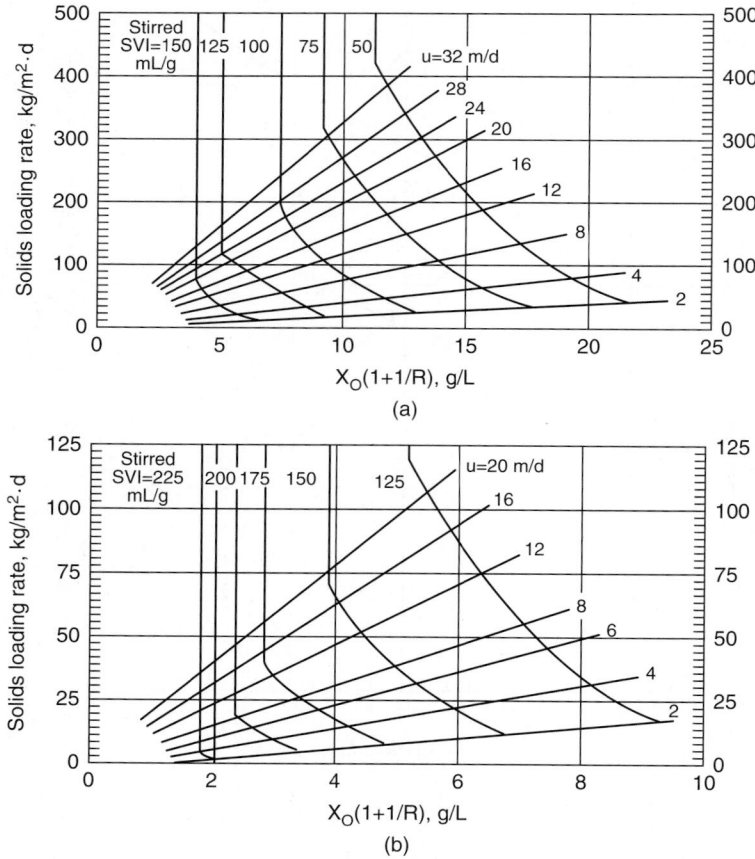

FIGURE 12.90   Keinath operating chart (Keinath, 1990).

information from 21 full-scale facilities that varied in size, geographic location, mode of operation, method of aeration, and type and amount of industrial wastewater input. None of the suspensions tested were chemically amended.

Results obtained using the Keinath operating charts differ substantially from the Daigger approach because of the differences between settling velocities correlated to stirred and unstirred SVIs, especially at high values. Daigger (1995) developed such a correlation between stirred and unstirred test data, but good correlation is neither transferable from facility to facility nor valid over a wide range of MLSS concentrations.

Keinath (1990) outlined the use of the design and operating chart (Figure 12.90) for secondary clarifiers according to the thickening criterion and evaluating various economic trade-offs to determine a cost-effective design. He also presented examples to demonstrate the effect of corrective strategies such as RAS control or conversion to step feed on ameliorating thickening overload conditions in an operating secondary clarifier. An example of this is given in Water Environment Federation MOP FD-8 (2005).

9.3.2.4 Wilson Approach   Wilson (1996) presented a simplified method of evaluating secondary clarifier performance using the settled sludge volume (SSV or V_{30}) from a 30-minute settling test. He proposed that the SSV is correlated to the initial settling velocity (ISV), which also represents the required SOR, provided that it is adjusted, where appropriate, for temperature, volatile solids content, and chemical addition. The relationships of this approach are:

$$R_{min} = SSV/(10^3 - SSV) \tag{12.39}$$

$$ISV = V_0 \times \exp(-4 \times SSV/10^3) \tag{12.40}$$

where R_{min} = minimum RAS rate (%);
   ISV = initial settling velocity (m/h);
   SSV = 30-minute settled volume (mL/L); and
   V_0 = sludge settling characteristic velocity (m/h).

Figure 12.91 presents a family of curves relating ISV (or clarifier SOR) to SSV for various values of V_0, assuming V_0 (m/h) is 0.3 to 0.5 times temperature in degrees Celsius. Wilson concluded that the model compared well with the empirically validated German Abwassertechnische Vereinigung (ATV) approach as well as the model developed by Daigger (1995).

The Wilson approach entails determining ISV, which is also the maximum surface overflow rate (SOR_{max}), using Figure 12.91 or eq 12.35; R_{min} can be derived from eq 12.34. These values are then compared with SOR and RAS rates determined from facility operating data. Finally, the clarifier safety factor (CSF) and return safety factor (RSF) are calculated as follows:

$$RSF = RAS\ rate/R_{min} \tag{12.41}$$

$$CSF = SOR_{max}/SOR \tag{12.42}$$

A CSF value of less than 1.0 indicates clarifier overload. If CSF and RSF are both greater than 1.0, then the clarifier is underloaded. If CSF is more than 1.0 and RSF is less than 1.0, then the clarifier is most likely overloaded and the operating condition should be confirmed using other methods, such as the Daigger approach.

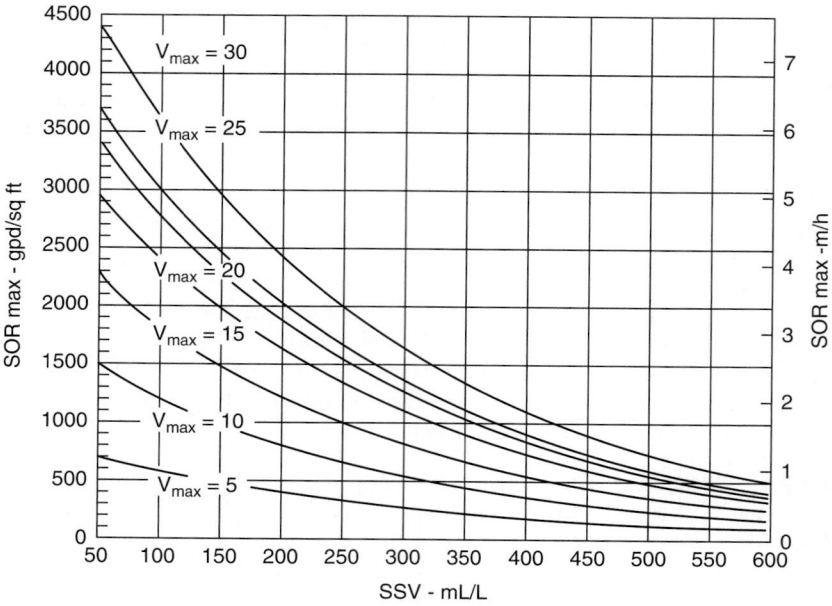

FIGURE 12.91   Wilson model (Wilson, 1996).

9.3.2.5   Ekama-Marais Approach   Ekama et al. (1997) characterized final clarifier behavior based on solids loading limited by (1) the solids flux (Criterion I) and (2) the SOR (Criterion II). These two limiting criteria are defined as functions of settleability testing data, expressed in terms of Vesilind coefficients V_o and n, overflow rate, underflow rate, and the stirred zone settling velocity (SZSV). The mathematical relationship for a given sludge is expressed graphically in the form of a design and operating chart (Figure 12.92). This defines the limiting overflow rate at various MLSS concentrations and RAS rates. Figure 12.92 illustrates that as the overflow rate increases, the recycle ratio must also increase along the Criterion I boundary up to a limiting maximum of the former. Further description of this approach with illustrations of its relationship to SPA and other methods are available (Ekama et al., 1997).

9.3.2.6   Computation Fluid Dynamics   Computational fluid dynamics (CFD) are computer models based on the Navier-Stokes equations to model the flow of viscous fluids based on the laws of conservation of mass, energy, and momentum. This approach attempts to reconcile hydrodynamics, settling, turbulence, sludge rheology, flocculation, temperature, and dynamic flow through a clarifier. Models can be based on flow in one, two, or three dimensions with increasing dimensions increasing the run-time and computational power required for a solution. Several commercially available packages are available for CFD modeling, including 2Dc, Current 2D, FLUENT, and others. Models are especially useful when flow through the clarifier is important (e.g., changing location of the influent).

9.3.3   Side Water Depth

Providing adequate tank depth is critical to good, consistent performance of any activated sludge clarifier. Selection of SWD is based on the size of the unit or the type of

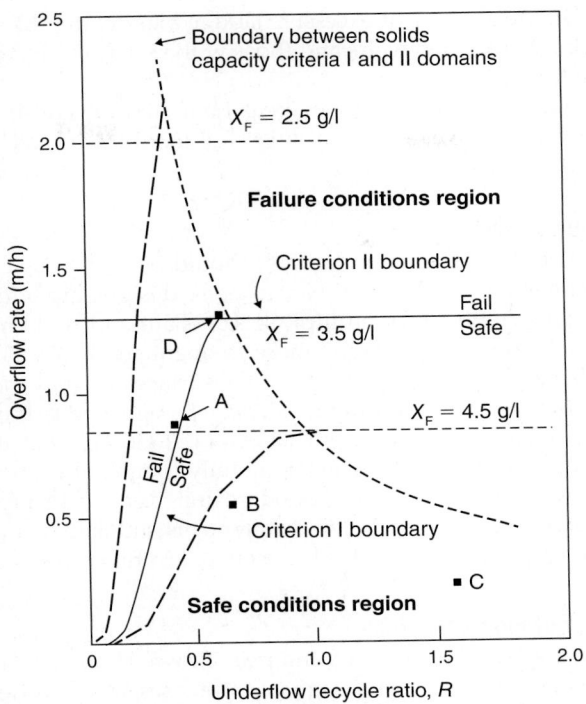

FIGURE 12.92   Design and operating chart. Graphs based on suspension with $V_0 = 5.93$ m/h and $n = 0.43$ m³/kg (Ekama et al., 1984).

biological process preceding it. The trend in design practice is to make circular clarifiers deeper than in the past. Recommended values range from 3 to 4.6 m (10 to 15 ft) as a function of diameter. The distance of the sludge blanket from the effluent weir has a direct relationship to effluent quality (Miller and Miller, 1978). For circular tanks, based on historical operating data, Parker (1983) demonstrated the positive effect of SWD on effluent quality. At similar SORs, the average concentration of effluent TSS from a settler decreased as depth increased. Variability in effluent quality also decreased with increasing depth.

In the ATV standards (1973, 1975, 1976, 1988, 1991), tank depth is calculated from four functional depths: (1) clear water zone, (2) separation zone, (3) sludge storage zone, and (4) thickening and sludge-removal zone. The SWD determined by this method is typically more than 4 m (13 ft). A 2008 telephone survey of several large consulting engineering firms and equipment suppliers specializing in U.S. WRRF design found that most large activated sludge secondary clarifiers have depths from 4 to 5 m (14 to 16 ft) diameters up to 50 m (150 ft). Optimum depth is a function of tank shape as presented later in this chapter.

9.3.4   Weir Loading

Many regulatory statutes include weir loading in a design criterion. Most design engineers believe that significantly higher weir loading rates would not impair performance and that placement and configuration have greater effects on a clarifier performance, particularly in the absence of excessive sludge blanket depths and high-flow energies near the weirs. Misaligned weirs and those with excessive algae growth can cause flow imbalance within clarifiers.

Many regulations limit maximum (peak hour) allowable weir loadings to 250 m³/m·d (20 000 gpd/ft) for small treatment facilities (less than 4000 m³/d [1 mgd]) and 375 m³/m·d (30 000 gpd/ft) for larger facilities.

9.3.5 Redundancy

All activated sludge secondary clarifiers should be able to be taken out of service for periodic maintenance and repair. Not all tank designs are equivalent in this regard. Rectangular tanks with one or multiple sets of chain-and-flight mechanisms are considered more problematic and require more frequent servicing. Circular designs, in contrast, have drives that are accessible from the surface and may include options in which the collector blades and support members can be winched to the surface.

Many regulators require some provision for redundancy that allows one or more units to be taken out of service while maintaining fully compliant treatment. In some states that allow the WRRF to have one spare, redundant unit to ensure that reclaimed water users do not have supply interruptions. Some requirements stipulate that at least 75% of the unit process design capacity remain if the largest parallel unit is taken out of service.

9.3.6 Effect of Flow Variations

Clarifier sizing is based on average and peak flows. Though such a procedure can produce an extremely conservative design in some cases, it is considered necessary because little is known regarding the mechanisms by which flow variations affect clarifier efficiency, and generalized quantitative relationships are not available. It is reasonable, however, to expect clarifier performance to reflect peak instantaneous flowrate loading, although some dampening effect is inevitable.

9.3.7 Summary of Sizing Steps

Many of the principles of design discussed above can be integrated into an approach that meets numerous objectives simultaneously. A step-by-step approach is given below.

1. Determine the operating MLSS concentration range of the facility to maintain an acceptable F:M ratio, SRT, and effluent quality under various flow and mass loading conditions. For most municipal facilities, the range is 1000 to 4000 mg/L.

2. Determine the anticipated range of the MLSS SVI (or ISV). Select a statistically high value that would seldom be exceeded by the full-scale operating facility. This maximum design value should be based on an analysis of existing records, pilot-facility data, or information from similar full-scale facilities. In 2008, the authors contacted several process experts from large firms and found that most use 90% to 95% values obtained from statistical analyses of existing facility data. Where no data exist, they use SVI values of 150 to 200 mL/g for conventional and extended aeration systems and 120 to 125 mL/g for systems with selectors and BNR facilities with anaerobic or anoxic basins. For those using the 90%, some apply a factor of safety of 20% to the resulting surface area of the tankage.

3. Provide for 20% to 100% of ADWF RAS pumping rate capacity (up to 150% for extended aeration systems or others using high MLSS concentrations). The RAS rate increases higher than 80% of ADWF may be counterproductive because of the increased hydraulic load on the clarifier.

4. Determine the maximum theoretical SLR (function of SVI) using solids flux analysis to arrive at the solids loading limit and resulting surface area. This result should be checked against the governmental regulations that may place limits on solids loading. If the regulation values are smaller than that derived by the methods discussed, then a case is sometimes made to obtain a waiver that allows the higher loading.

5. Select an overflow rate to achieve the required effluent quality that is based on influent wastewater flowrate characteristics. The overflow rate to produce a specific effluent TSS concentration has not been extensively researched but is known to vary with the geometry of the tank inlet structure and depth. This rate probably will not exceed the solids-limiting value determined in Step 4. Allowance also should be made for taking a tank out of service for maintenance or repair (see text above for criteria).

6. Select a depth to provide adequate solids clarification, thickening, and storage. Allow 0.6 to 0.9 m (2 to 3 ft) for thickening; 1 m (3 ft) or more for buffering; and 2.4 m (8 ft) for clarification. More buffering depth is needed if diurnal or influent pumping flow variations or peak-flow conditions are atypically large (e.g., greater than 2:1). Approximately 0.6 m (2 ft) of freeboard should be added to determine overall tank wall height. More detailed methods of clarifier depth requirement analysis are available in addition to the application of CFD modeling (ATV, 1973, 1976; Ekama et al., 1997).

7. Provide a reasonable weir length and place the weirs at strategic locations. Block off notches in rectangular tank launders at the outlet end of the tank and add baffles, as needed, to eliminate problems with MLSS updraft near the effluent structure.

8. Select a mechanism for sludge removal. Plows, spiral-curved blades, chain and flight sytems, or hydraulic suction systems are available depending on shape.

9. Consider other details to improve secondary clarifier performance:
   - Flocculation inlet zones, preferably separated from the rest of the tank by baffles;
   - Midlength or midradius energy-dissipation baffles (these may not be needed if a flocculating feedwell is provided);
   - Full-radius skimmers (rotating trough or beach type for small tanks), full width for rectangular tanks, or partial radius beach type for large radial-flow tanks (multiple blades, anti-rotation baffles, and spray nozzles may be added); and
   - Launder covers, algae removal mechanism, or chlorine addition equipment, if necessary, for algae control.

9.3.8 Shapes

The current consensus of design engineers is that there are no significant shape advantages between circular and rectangular clarifiers if all of the design details are well done and in accordance with modern guidelines such as those of Water Environment Federation MOP FD-8 (2005).

Design engineers consider two basic shapes viable: longitudinal and crossflow. By far, the most common is the longitudinal design (WEF, 2005).

Most design engineers and operators prefer circular clarifiers for activated sludge and specialty suspended-growth systems treating municipal wastewater. The reliability of the mechanisms for circular tanks typically is cited as the primary reason. Extensive details and text on these designs are given in a following section of this chapter and in Water Environment Federation MOP FD-8 (2005).

Except for a few square designs, square, hexagon, and octagon shapes are designed with center or peripheral feed to establish an internal radial-flow pattern. Some square tanks may be loaded on one side and effluent taken off on the opposite side, but these are rare.

For square designs with radial flow, sweeping sludge from the corner area is a problem. Corner sweep mechanisms exist, but many of them have had mechanical problems and have fallen into disfavor. In recent years, changes to eliminate corner sweeping have become common. Fillets in the corners to enable simple circular sweeps have been used in some, and new, circular inner vertical walls have been used in others. Hexagonal and octagonal tanks typically have adequate corner filleting to accommodate simple circular mechanisms.

All designs in this radial-flow noncircular category have an issue with launder shape. If they are made circular, then corner areas that are difficult to skim are created. If the weirs are placed along the straight walls, then the flow patterns are distorted and automated brushing for algae control is not possible. In view of all these considerations, tanks in this category have become highly unpopular.

9.3.9   Batch and Other Clarification

The SBR processes are favored by some engineers because they do not require separate clarifiers, making them economical. Nonetheless, proper provisions are needed to ensure a clear, high-quality supernatant when aeration and mixing are terminated. The SBR processes are discussed in an earlier section of this chapter.

Several devices for decanting are illustrated in a U.S. EPA report (James M. Montgomery Consulting Engineers, 1984). Although most commercially available decanting devices function reasonably well, some of the initial designs led to excessive TSS discharges at the onset of the decant cycle. The turbulence of aeration transferred MLSS into the decanter, and the TSS subsequently left the basin when the draw/decant cycle started. This problem has been resolved by changes in decanter design or provisions to return the initial decant to the reactor and continue the return until clarity improves to a satisfactory level. Many SBR facilities have sometimes experienced significant foam buildup; the decant system should be designed accordingly. Most successful designs keep the foam out of the effluent discharge by either incorporating baffles around the decanter or decanting from below the liquid surface. Foam subsequently remains in the reactor, or it can be removed by separate skimming devices.

Sludge recycle and ballast addition allow considerably higher hydraulic loading and excellent suspended solids removal. These advantages, however, are offset by higher costs for polymers, ballast microsand or magnetite, and recirculation energy. A relatively low level of inlet geometry sophistication has been found necessary for these designs. They are not typically used to clarify activated sludge mixed liquor but have been tested with some success for use on a temporary basis to handle wet-weather peaking flows.

9.4 Rectangular Design

Rectangular clarifiers have been used for activated sludge mixed-liquor settling for nearly a century and are found frequently in large facility applications, although they are used for all sizes. Common wall construction and space-saving footprints are attractive features. In addition, their galleries can accommodate piping and pumps. Rectangular feed channels in large facilities also can involve some degree of common wall construction.

9.4.1 Flow Patterns

Most activated sludge rectangular clarifiers have longitudinal flow patterns (Figure 12.93). A transverse-flow option has been introduced recently, but few units of this type are built and operating. An important distinction is that longitudinal tanks can have concurrent, countercurrent, or crosscurrent sludge removal. When the main liquid stream reverses itself, it is referred to as a folded flow pattern. Rectangular clarifiers placed on top of each other are referred to as stacked clarifiers; one arrangement incorporates a folded flow pattern in a vertical arrangement.

Longitudinal flow is a pattern in which the influent flow proceeds in a direction parallel to the longitudinal of the tank. The flow pattern closely resembles plug flow in theory, although sedimentation is taking place along the vertical axis. Dye studies have shown that the rectangular design does not achieve ideal plug flow because of some short-circuiting.

In the transverse flow design, the influent flow enters from a channel along the long side of a rectangular tank. Effluent weirs are placed on the opposite long side of the tank to give a conventional crossflow pattern (Figure 12.94). If the effluent weirs are located along the influent side of the tank, then this becomes a folded flow pattern. In transverse clarifier designs, sludge withdrawal is accomplished by a traveling suction mechanism, making the provision of sludge hoppers unnecessary. Or hoppers can be placed approximately 10 m (33 ft) apart along the short width of the tank where an embedded collection header with orifices can be placed.

Stacked clarifiers consist of settling tanks, located one above the other, operating in parallel, often with a common water surface. In this sense, they become modular units.

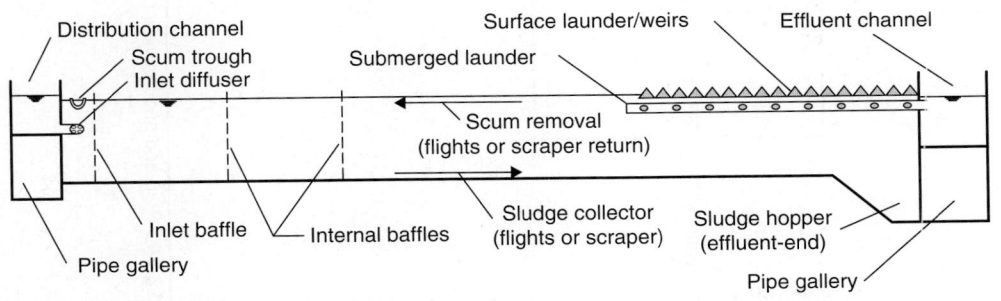

FIGURE 12.93 Rectangular clarifier design features and nomenclature (hopper locations may vary) (Ekama et al., 1997; reprinted with permission from IWA Publishing).

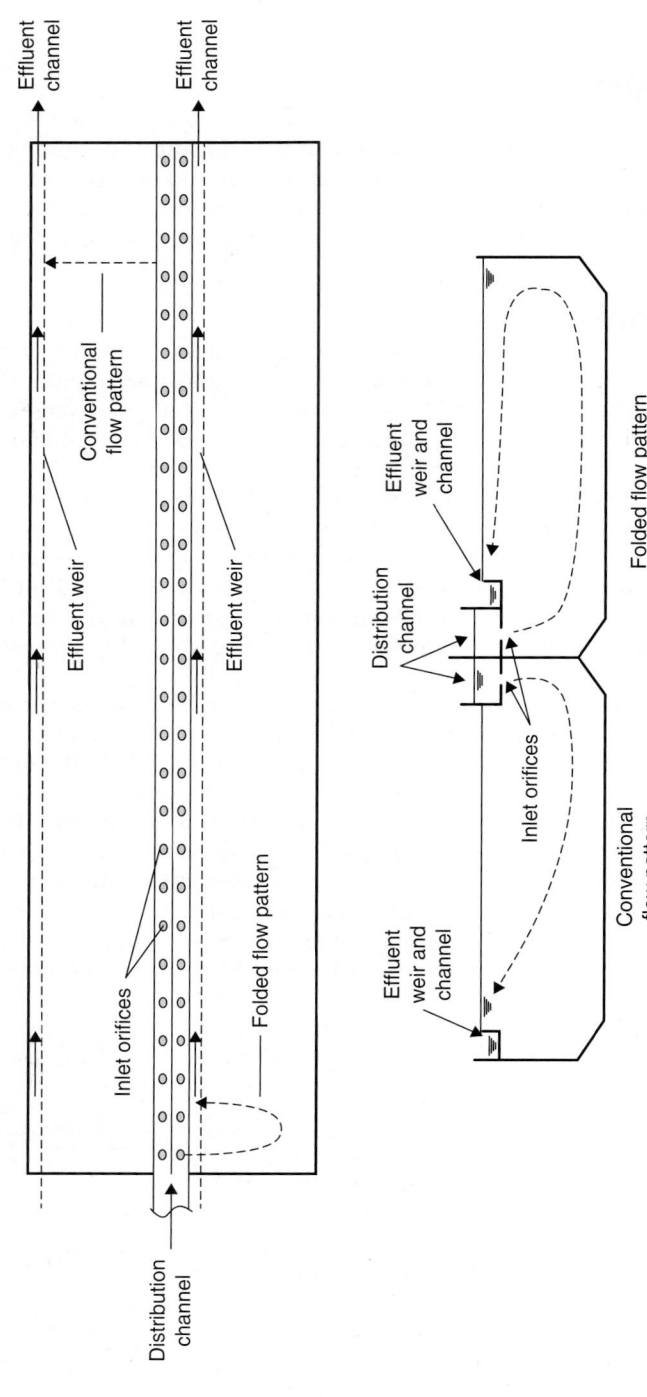

Figure 12.94 Plan and section view of transverse-flow rectangular clarifier.

Stacking increases the clarifier surface area without increasing facility footprint. They are also called tray clarifiers and can be double-decked or even triple-decked. Most stacked clarifier designs are similar to conventional rectangular clarifiers in terms of influent and effluent flow patterns and solids collection and removal. Stacked clarifiers are covered in more detail in Chapter 10 and later in this section.

In concurrent flow in a longitudinal clarifier, the clarified liquid and sludge flow proceed down the length of the tank in the same direction. Inlet baffles or diffusers are designed to distribute the flow across the width of the clarifier and dissipate the inlet energy. The bulk of the mixed-liquor solids settle to form a blanket interface. Settling action of the solids and removal of the sludge flow stream produce a density current along the bottom of the tank. The density current imparts a momentum that moves the sludge efficiently along the length of the clarifier to the hopper on the downstream end. Flights or scrapers assist the sludge movement.

The sludge hopper can be placed at the influent end so that the sludge is removed more quickly. In this case, the sludge flow reverses itself and is called a countercurrent sludge-removal flow pattern. The sludge hopper can be placed at approximately mid-tank so that the sludge does not have to travel to the end of the tank for its removal. In most rectangular secondary clarifiers, the sludge hoppers are at the opposite effluent end or midtank.

9.4.2 Dimensions

Once the area and number of tanks needed for redundancy and normal size constraints have been established, specific tank geometric details can be defined. There are acceptable minimum ratios of length to width that effectively limit the maximum size of rectangular tanks. The length-to-width ratios of longitudinal rectangular clarifiers may range from 1.5:1 to 15:1. A minimum length-to-width ratio of 3:1 was recommended to prevent short-circuiting, but it is typically greater than 5:1 (U.S. EPA, 1974a). Some references recommend that the length of the rectangular clarifier should not exceed 10 times the depth (Metcalf & Eddy, Inc./AECOM, 2014). Typical depth ranges from 3.5 m (12 ft) to 6 m (20 ft). However, this length-to-depth ratio has been exceeded with success at larger facilities. The length, width, and depth dimensions should be proportioned so that horizontal flow velocities are not excessive.

9.4.2.1 Length Rectangular clarifiers are seldom greater than about 100 m (300 ft) in length and are typically 30 to 60 m (100 to 200 ft) long. In small clarifiers, such as those used in package facilities, a minimum flow length of 3 m (10 ft) from inlet to outlet should be used to prevent short-circuiting (U.S. EPA, 1974a). There is concern for possible suspended solids carryover with the increased hydraulic flow at the weir as the clarifier length is shortened. The ultimate length of a tank is limited by stress on the collection mechanism and need to transport sludge the entire length of the tank. For long tanks and for tanks with midtank hoppers, multiple collector systems can be used.

9.4.2.2 Width For many years, the effects of deflection, buoyancy, and weight of wooden flights restricted rectangular clarifiers with a single flight system in each tank to a nominal width of 6 m (20 ft). Fiberglass composite materials have allowed single flight systems to span widths of up to 10 m (33 ft). Multiple parallel flights can be constructed in wider tanks provided with columns or partial walls to support parallel collector sprockets.

A significant disadvantage of multiple parallel flights is the larger percentage of units taken out of service to repair a single mechanism. Also, the flow patterns might not be as stable compared to long, narrow tanks. Therefore, baffles between sections of wide tanks should be considered to direct the flow longitudinally.

9.4.2.3 Depth Current practice is to provide a depth of approximately 4 to 5 m (approximately 12 to 16 ft) for activated sludge rectangular clarifiers. Differences depend on peak flows, sludge loading storage requirements, and available recycle capacity. Some have reported success with rectangular clarifiers that were only 3 m (10 ft) deep (Stahl and Chen, 1996; Wahlberg et al., 1993, 1994). Crosby (1984a, 1984b) studied the effects of sludge blankets and their maintenance and found that the top of the blanket determines the depth available for clarification. This means that relatively shallow tanks with minimal blanket levels often perform as well as deeper tanks with thicker blankets. Deeper tanks should be considered depending of the direction of sludge movement and high peak flows.

Shallow clarifiers can limit storage and thickening capability of secondary clarifiers in an activated sludge system. This, in turn, may decrease the RAS concentration and increase pumping demands. Ample depth is recommended for storage of solids and thickening during sustained peak flows and when solids loads exceed recycle capacity (Boyle, 1975).

9.4.2.4 Flow Distribution to Parallel Units Uniform flow distribution among clarifiers is critical to good performance. Once an overloaded tank starts to lose part of its blanket, improved performance of the other parallel, underloaded tanks is not sufficient.

The flows should be distributed to each tank in proportion to their respective surface areas. Weir inlets that discharge vertically into the tank should be avoided because this may exacerbate density current effects. Fixed submerged orifices do not need to be able to accommodate the expected flow range or to include storm inlet gates. Submerged inlet gates provide flexibility for a range of flows and allow for tank isolation.

Equal flow distribution to all tanks is required. Headloss across the gate or orifice should be at least 10 times the total headloss of the feed channel or pipe.

Positive flow-splitting structure flumes and flow meters coupled with automatic valves also have been used. Symmetry and effluent launder elevations should not be relied upon for flow distribution.

Mixed-liquor feed channels should be gently aerated with maximum velocities from 0.2 to 0.4 m/s (0.7 to 1.3 ft/s) to prevent floc breakup.

9.4.2.5 Inlet Geometry Inlets dissipate energy, minimize the effects of density currents, distribute flow across the tank width, and promote flocculation. For front end hopper arrangements, inlets and associated baffles should not result in sludge blanket disturbance or scouring.

The degree of floc formation by the time the mixed liquor reaches the clarifier inlet varies from facility to facility. A baffled flocculation zone at the inlet end uses some of the inlet energy to do the mixing (Barnard et al., 2007; Kalbskopf and Herter, 1984). Inlets to the flocculation zone can be quite different from those found to be most effective in tanks that do not have such internal baffling.

9.4.2.6 Flow Distribution within Clarifiers The introduction of flow to an individual tank is accomplished by multiple inlets that are situated and sized to uniformly distribute flow over the width of the clarifier. In a 6-m (20-ft)-wide tank, there are typically 3 to 4 inlet openings. Maximum horizontal spacing between inlets is 2 to 3 m (6.5 to 10 ft). Inlet baffles or diffuser elements typically are placed in the flow path of the inlet stream. Solid target baffles to deflect the flow or perforated (finger) baffles to break up any jetting action and disperse the flow have been used. Single and double rows of slotted board baffles represent another option. These features establish flow impingement to dissipate energy and promote flocculation.

Headloss though transverse perforated or slotted inlet plates should be approximately four times the kinetic energy or velocity head of the approaching flow (WEF, 2005). This often results in slotted openings of less than 5-cm (2-in.) width.

Inlet design is more complicated for transverse-flow tanks because the inlet channel extends the length of the tank; detailed hydraulic analysis is warranted.

9.4.2.7 Inlet Design There are many different inlet designs for longitudinal-flow rectangular tanks (Figures 12.95 through 12.99). Some include a flocculation zone baffled off from the quiescent portion of the tank and others do not. For the degree of floc formation observed by Kalbskopf and Herter (1984), the data in Figure 12.100 show that baffling off a portion of the head of the tank produced the best results. Data by Stahl and Chen (2006) show that excellent effluent quality is obtained at several shallow concurrent flow clarifiers equipped with only the impingement-inducing diffusers. These findings were published before the modern testing protocols outlined in the WERF report (2001); therefore, the degree of floc formation in each case was not quantified. Stahl and Chen (2006) reported on fine-bubble aeration facilities with low-velocity aerated channel transport of mixed liquor to the clarifiers. It is probable, therefore, that the floc was well formed. This suggests that if the floc entering the tank is well formed, then impingement diffusers may give excellent results. If, however, the floc is not well formed, then a separate baffled area, such as recommended by Krebs et al. (1995) and Barnard et al. (2007) (Figure 12.101), would be cost-effective.

For concurrent-flow tanks, density current problems can be minimized by positioning the inlet lower in the tank without placing it in the thickening zone, which is typically reserved for the bottom 1 m (3 ft) of the tank. Locating the inlet too low may scour the solids on the bottom and lead to resuspension. Inlet apertures should be positioned

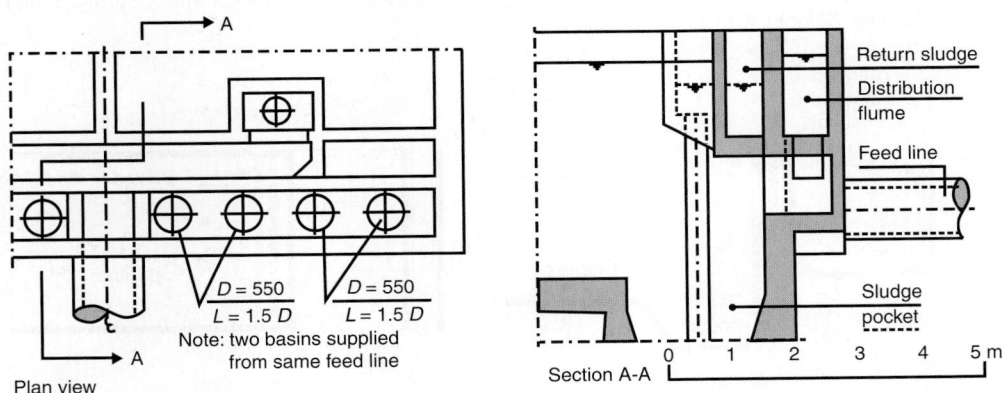

FIGURE 12.95 Inlet design of Larsen (1977) to avoid floc breakup (note that D is in millimeters).

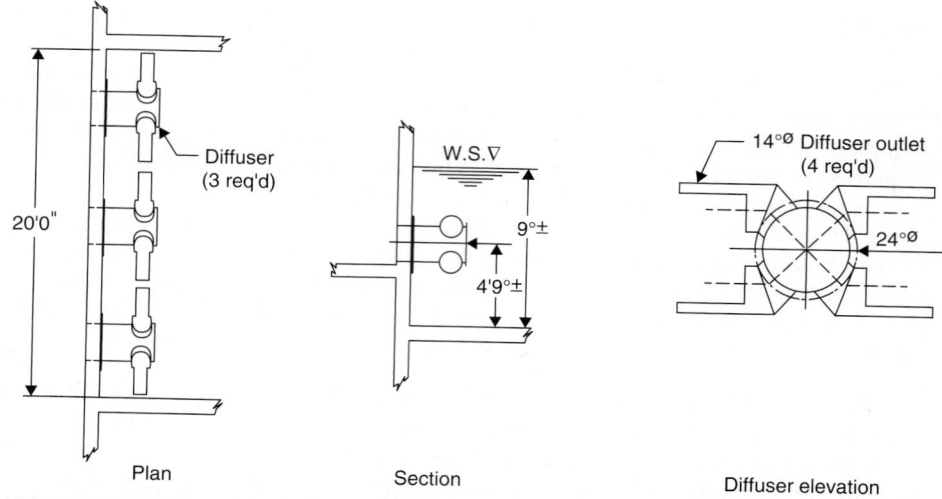

Plan   Section   Diffuser elevation

FIGURE 12.96 Secondary clarifier inlet used by the Los Angeles County Sanitation Districts (1 in. = 52.54 cm; 1 ft × 0.3048 = m).

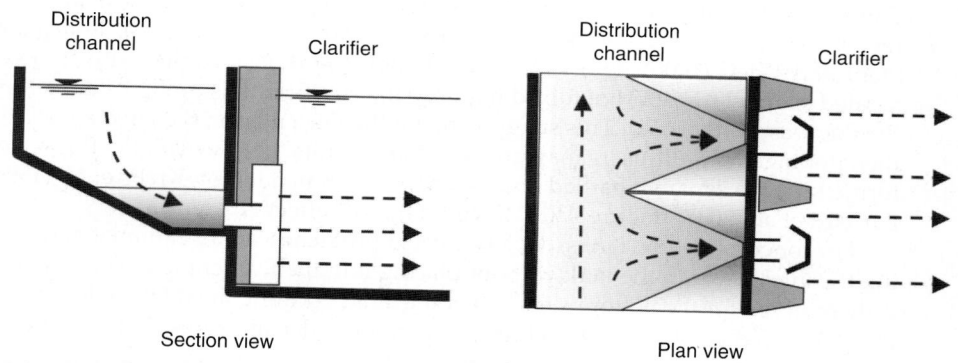

Section view   Plan view

FIGURE 12.97 Distribution channel with funnel-shaped floor (Krauth, 1993) with a Stuttgart inlet (Popel and Weidner, 1963).

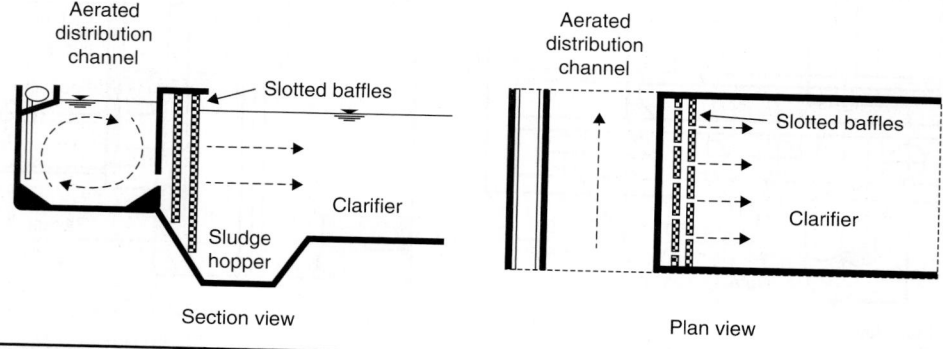

Section view   Plan view

FIGURE 12.98 Aerated distribution channel (Krauth, 1993) with two staggered slotted baffles to dissipate inlet energy.

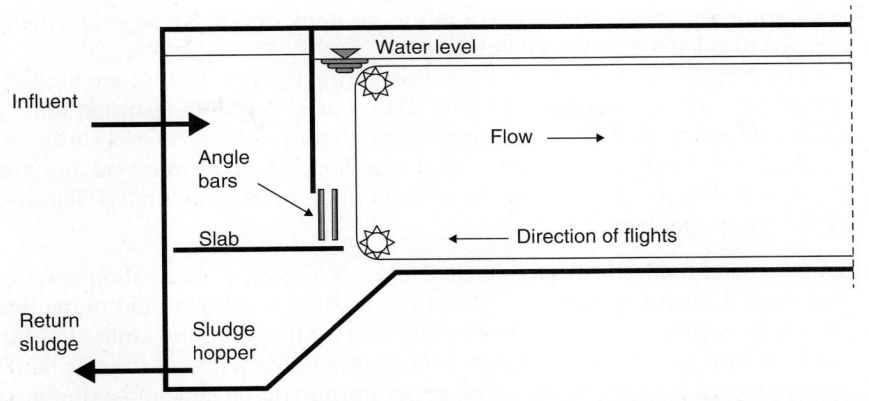

FIGURE 12.99 Aerated distribution channel with horizontal slab deflecting inlet flow energy from a sludge hopper at the inlet end (Krebs et al., 1995).

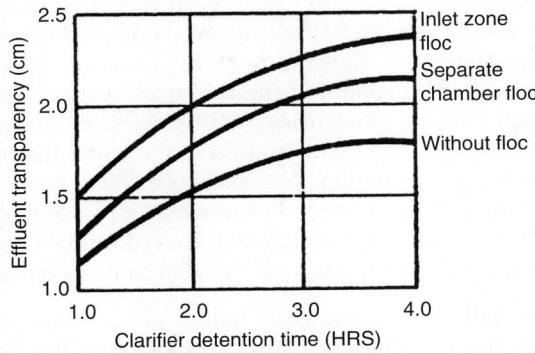

FIGURE 12.100 Improvement of effluent transparency with flocculation zone (Kalbskopf and Herter, 1984).

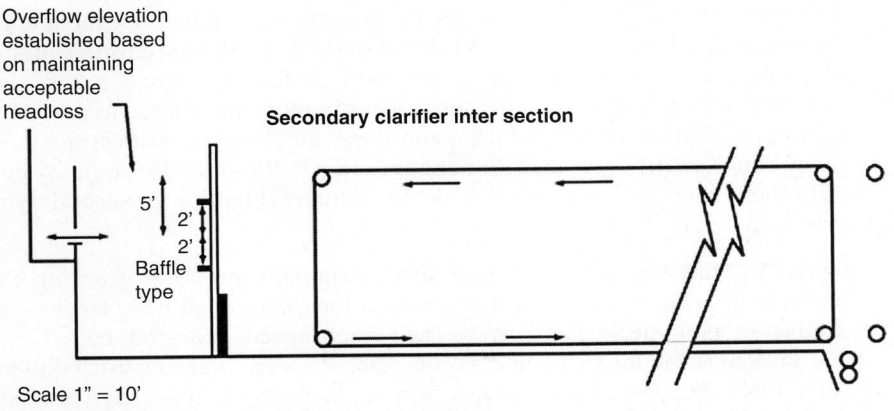

FIGURE 12.101 Inlet for new secondary clarifiers (Barnard et al., 2007).

from approximately 2-m (6.5-ft) depth to midtank depth. Krebs et al. (1995) provide a method to calculate an inlet height.

For the impingement-inducing diffusers, inlet port velocities are limited to a range of 0.075 to 0.150 m/s (0.25 to 0.5 ft/s). Das et al. (1993) demonstrated that velocities in excess of 0.6 m/s (2 ft/s) may cause deflocculation of the activated sludge solids.

For countercurrent clarifiers with sludge hoppers at the inlet end, horizontal baffles to prevent density current flow into the hopper are recommended. Figures 12.99 and 12.101 show conceptual examples.

9.4.2.8 Inlet Baffles and Flocculation Zones
If a baffled flocculation zone is designed, then inlet diffusers typically are not used. A baffle will be located immediately downstream of the inlet openings to prevent jetting of flow into the tanks. The target baffles can be simple walls, solid or perforated, spanning the width of the clarifier. Krebs et al. (1995) proposed a solid baffle in the upper portion of the tank and a double row of slotted openings in the lower portion. Barnard et al. (2007) recommended a similar design, after studying several options.

Mau (1959) showed that a single vertical row of slotted baffles was effective in distributing flow. However, a second row of vertical slotted baffles, where the boards are opposed to the slots of the first baffle, improved energy dissipation and performance by causing flow impingement. Kawamura (1981) recommended the installation of three sets of perforated baffles spanning the full cross section. Okuno and Fukada (1982) observed the best removal efficiencies from baffles with 5% open areas.

Mechanical flocculation is widely used in the water treatment industry. Most activated sludge clarifier flocculation zones have been adequately mixed by strategic direction of the incoming flow streams. Recommended G value of flocculation should be in the range of 30 to 70/s (Parker et al., 1971). The required volume of the inlet flocculation zone is calculated by a residence time of 8 to 20 minutes required for good flocculation.

9.4.2.9 Interior Baffles
Internal cross baffles are considered to enhance settling and provide a more clarified effluent. Various types of baffles have been investigated in rectangular tanks (solid, perforated, and combinations), and the effects of each can be quite different. In addition, each of these baffle types can be sized differently or possibly configured in series.

For concurrent longitudinal rectangular tanks, it is uncommon to add baffles downstream of the flocculation zone baffle (if one is provided) because the intent is to have the sludge blanket flow to the outlet end of the tank. Baffles are effective in some designs of countercurrent tanks, keeping most of the sludge in the front part of the tank and away from the outlet weir area. Mechanisms to move the sludge to the front end must contend with these baffles, which complicates the design of both systems. As a result, many countercurrent designs do not have cross baffles initially and may be retrofitted with them later. An extensive discussion of internal baffles for secondary clarifiers is provided by WEF (2005).

9.4.2.10 Stacked Clarifiers
Camp (1946) originally proposed stacking clarifiers for use in both primary and secondary clarification. Also called tray clarifiers, they can be double- or triple-decked and can be used where space is constrained.

Stacked secondary activated sludge clarifiers were first constructed in Japan in the early 1960s. Because of space constraints, rectangular clarifiers have been stacked two or three deep. Osaka City has operated stacked facilities with satisfactory performance for more than 20 years (Yuki, 1991). In the United States, stacked clarifiers were first

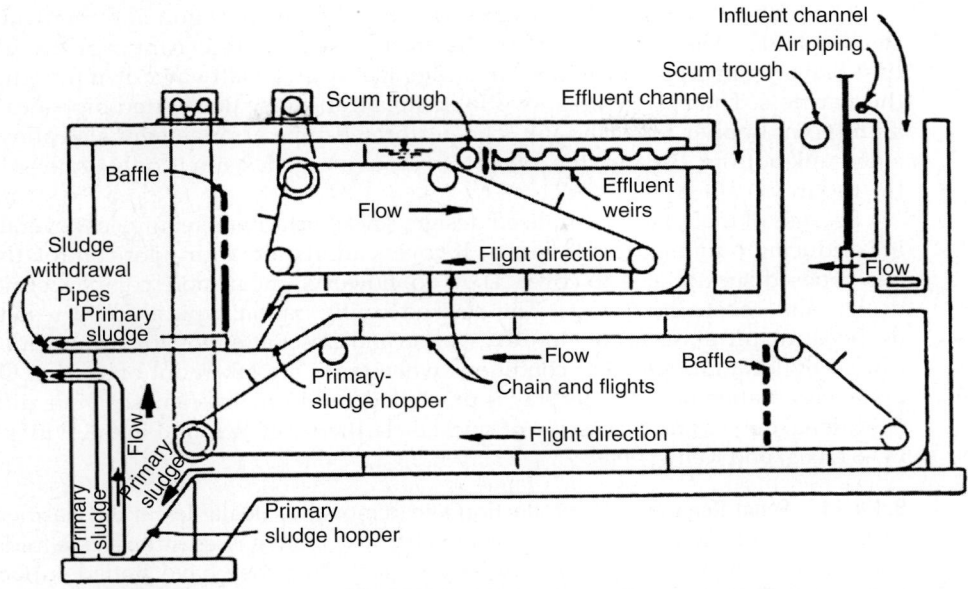

FIGURE 12.102   Stacked rectangular clarifier, series-flow type (Kelly, 1988).

constructed at the Mamaroneck, New York, treatment facility in 1993, and have been constructed in Salem, Massachusetts, at the South Essex Sewerage District; the Deer Island Treatment Facility in Boston, Massachusetts; the Ulu Pandan and Changi East facilities, both in Singapore; and the Stonecutters Wastewater Treatment Plant in Hong Kong. In theory, overflow and weir loading rates should be similar to conventional rectangular secondary clarifiers (Metcalf & Eddy, Inc./AECOM, 2014).

There are two types of flow regimes for stacked clarifiers: parallel and series. Parallel flow is the most common stacked clarifier configuration (Figure 12.102). Although Figure 12.102 shows countercurrent sludge removal, the concept would be appropriate for concurrent as well. In the less common series-flow configuration, wastewater enters the lower tray, flows to the opposite end, reverses direction in the upper tray, and exits the effluent channel (Figure 12.103).

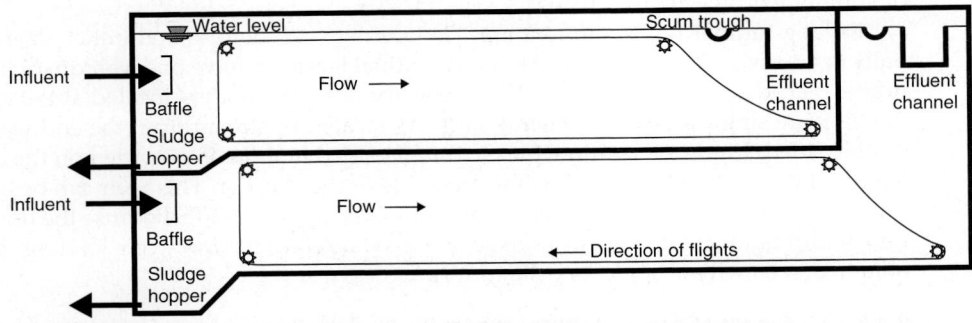

FIGURE 12.103   Stacked rectangular clarifier; parallel-flow type showing double-sided weirs at same water surface elevation.

Chain-and-flight mechanisms are used for sludge collection and removal from stacked tanks. The arrangement of the stacked secondary clarifiers at the Changi East Plant in Singapore provides for sludge withdrawal halfway down the length of the clarifiers. There, scrapers move solids to a transverse perforated pipe located in a midpoint hopper for each tank. The perforated-pipe arrangement also allows for a flat tank bottom. Because the lower tray is submerged, scum is only removed from the top tray.

Because of their more centralized design, stacked clarifiers require less overall piping, reducing pumping requirements. If covers are required for odor control, there is less exposed surface area to cover. They do, however, incur more complex structural design and construction costs. The stacked configuration typically will result in a deeper structure and require more excavation and closer attention to the buoyant effect caused by local groundwater conditions when tanks are taken out of service. Operational observation of the lower tray is precluded and its maintenance is more difficult. Additional details and discussion of stacked clarifiers can be found in the WEF's MOP FD-8 (2005) and Kelly (1988).

9.4.2.11 Scum Removal   The collection and removal of floatables in the clarifier concerns both inlet and outlet areas of the tank because foam passes through the middle of the tank unimpeded. If influent baffles are used, then they have slotted surfaces or include small downward-opening gates to allow scum to pass through.

For rectangular tanks with concurrent sludge removal, chain-and-flight collectors often are brought to the surface and move scum to the front of the tank for collection. For countercurrent designs, scum moves downstream and a collection barrier is needed upstream of the effluent launder area.

A slotted roll pipe situated across the width of the tank can be used to remove scum. It is positioned at the point where scum is concentrated by the movement of the sludge flights on the surface of the clarifier. Scum flows into the horizontally mounted pipe when the slot is rotated below water level. The pipe can be rotated manually or automatically. Other designs use a separate set of chains and flights or spiral flight arrangement to move the scum up a beach and into a trough.

Adjustable pan skimmers are also used on rectangular clarifiers. Water sprays can be used to move scum toward the pan and prevent the setup of foam.

9.4.2.12 Outlets   The most common outlet for rectangular clarifiers is the surface launder. Some designs provide submerged launders, which are also called outlet tubes or submerged pipes with orifices.

In longitudinal tanks, effluent surface launders can be oriented either longitudinally or transversely (Figure 12.104). Longitudinal launders have one weir on each side, unless placed against a side wall. Transverse (or lateral) launders located at the end of the tank are single sided, and lateral launders located upstream from the end wall are double sided. In transverse-flow tanks, a single-sided launder is provided on the entire length of the outlet wall, which is the long side of the clarifier. This weir can be on the same side as the inlet (folded-flow pattern) or on the opposite side. Because the flow per unit width is much smaller in transverse tanks, a relatively low weir loading rate is obtained even with a single, one-sided weir.

9.4.2.13 End-Wall Effect and Other Launder Design Considerations   Inlet density currents cause higher velocities in the liquid flow above the sludge layer and can cause an

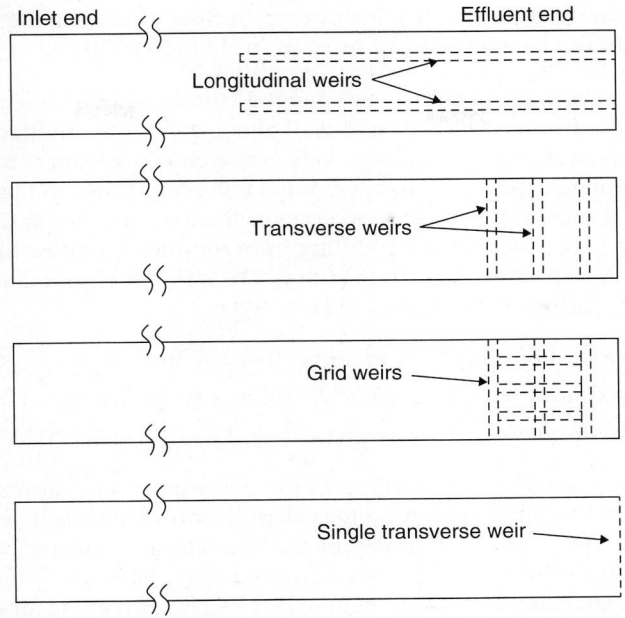

FIGURE 12.104   Plan views of typical surface weir configurations.

updraft along the end wall (Anderson, 1945). To allow resettlement of floc particles caught in this updraft, overflow is eliminated within a distance from the end wall equal to the tank depth (ATV, 1991). Alternatively, deflection baffles can be installed below the weirs to deflect the upwelling caused by the density current.

Heavy cross winds on open tanks can easily cause sloshing and surging of water over the weirs. To counteract this, fiberglass launders can either be substantially braced or covered or the weir area can be provided with more freeboard to shelter it. Concrete launders with fiberglass weir plates also can be used. Bridge-type mechanisms can pass between the launders and then be supported from the floor.

Adjustable weir plates should be used with launders so that they can be accurately leveled with the outlet weir of other clarifiers in parallel (Institute of Water Pollution Control, 1980). A v-notched weir is preferred to a straight-edged weir because they are less sensitive to slight differences in elevation and unbalanced flow caused by wind. Weir troughs should be designed so that they will not be submerged at maximum design flow and with a velocity of at least 0.3 m/s (1 ft/s) at one-half the design flow to prevent solids deposition (GLUMRB, 2014).

9.4.2.14 Weir Loading Rates   Surface launders in longitudinal rectangular clarifiers extend throughout the downstream 20% to 35% of long tanks; for short to moderately long tanks, this may increase to 50% or more. Launder spacing of 3 m (10 ft) is representative for larger tanks with widths of at least 20 ft.

For small tanks, the peak-hour weir loading rates should be limited to 250 m³/m·d (20 000 gpd/ft). This limit can be applied for the upturn zone of larger tanks. In larger tanks outside the upturn zone of an end-wall effect, the weir loading rates can be limited to 375 m³/m·d (30 000 gpd/ft). In any case, upflow velocity in the immediate vicinity of the weir should be limited to 3.7 to 7.3 m/h (12 to 24 ft/hr).

9.4.2.15 Submerged Launders Submerged launders can be oriented longitudinally or transversely. To counteract the end-wall effect, outlets are omitted from this area. Submerged launders require automatic valves or weirs downstream to control water levels. Typical arrangements of submerged outlet tubes are shown in Figure 12.105.

Analysis of several headlosses serves as the basis for design for the outlet tube system. These losses include loss resulting from confluence of flow to the orifices, through the orifice, and friction through the tubes. The ATV (1995) and Gunthert and Deininger (1995) suggest the following hydraulic criteria:

- Orifice diameter: 25 to 45 mm (1.0 to 1.75 in.);
- Maximum velocity at tube exit: 0.6 m/s (2 ft/s); and
- Maximum velocity through orifices: 0.6 to 1.0 m/s (2.0 to 3.3 ft/s).

Tubes are placed 30 to 35 cm (12 to 14 in.) below the water surface. Because the water layer above the tubes cannot be regarded as part of the clear water zone, the entire water depth of the clarifier should be increased to some degree over that of a conventional clarifier with effluent launders (Ekama et al., 1997).

Submerged launders allow scum to be concentrated at the far end of the tank. With submerged outlets, variation in water level needs to be considered when designing the scum removal systems.

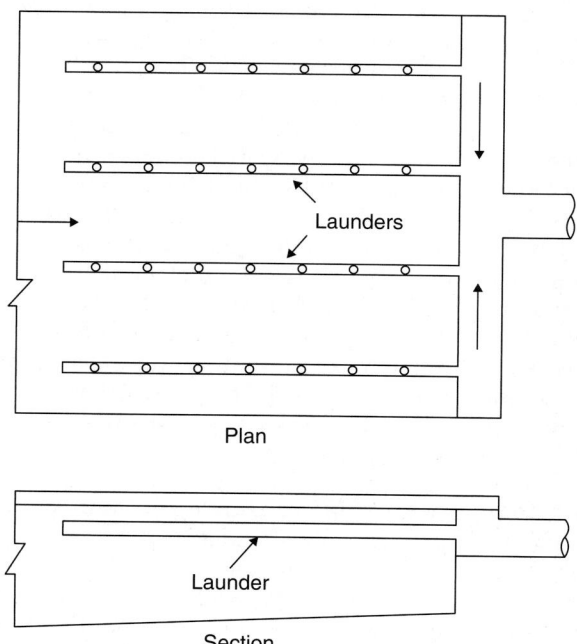

Plan

Section

Figure 12.105 Submerged launders consisting of pipes with equally spaced orifices.

9.4.2.16 Sludge Withdrawal   Most rectangular clarifiers have sloping floors and hoppers at one or more locations. Floor slope is typically 1%. Hoppers at either end of the tank are more effective with shorter runs of piping through galleries to the recycle pumps. Midtank hoppers are sometimes used when internal baffles are provided, because the gap provided by the hoppers and between the sludge collectors are a convenient location for a baffle. Midtank hoppers can also have transverse collection systems such that sludge removal is slightly different from tank to tank.

The trend for longitudinal, rectangular secondary clarifiers has been to follow the concept of Gould (Figure 12.106a), where hoppers are placed at midtank or the effluent end. These tanks typically are used in large facilities and are designed to minimize density currents and to avoid other hydraulic problems.

The typical hopper shape for rectangular clarifiers is an inverted pyramid with a rectangular opening on top. The sides are recommended to have the slope of at least 52° from horizontal to prevent solids from accumulating on the upper walls. A single rectangular tank may have two or more withdrawal hoppers, each equipped with a withdrawal pipe. Separate controls for each hopper are absolutely necessary.

The effluent end hopper design conceptually provides a more ideal solution for minimizing the breakup of the biological flocs because the sludge transport now takes place in the same direction as the bottom density current. The sludge is kept out of the relatively turbulent region of the inlet. Furthermore, the longer sludge detention time, resulting from the effluent end hopper arrangement, can enhance the flocculation and the dynamic filtration effects on the flocculent particles. Wahlberg et al. (1993) showed that rectangular tanks with effluent end sludge collection can perform exceptionally well up to SORs of 3.4 m/h (2000 gpd/sq ft).

Effluent end hoppers result in a large amount of solids transported into the effluent region, which increases the potential of solids washout. In addition, bulk horizontal

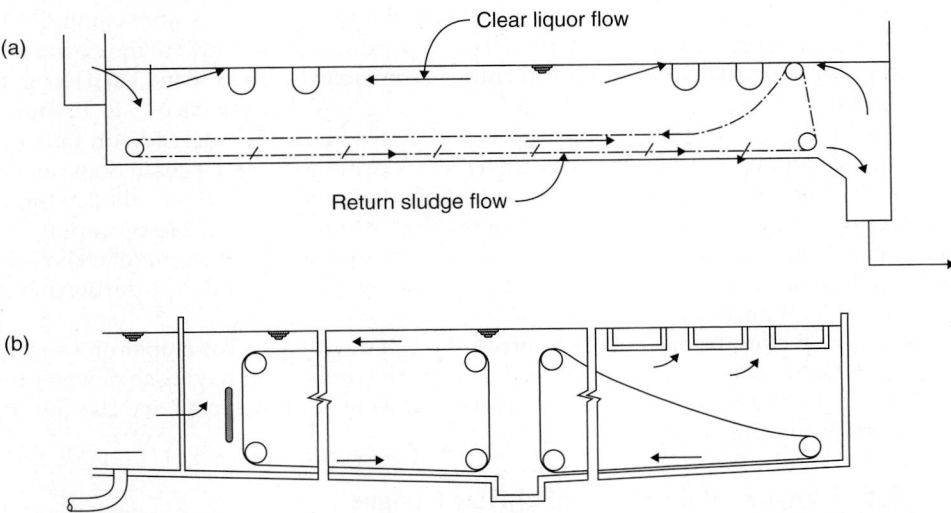

FIGURE 12.106   (a) Gould tank Type I with sludge hopper at outlet end and (b) Gould tank type II with sludge hopper at midpoint (Ekama et al., 1997; reprinted with permission from IWA Publishing).

flow through the tank consists of both the effluent and RAS; increasing RAS rates to remove additional sludge can be self-defeating.

In rectangular tanks exceeding 40 m (130 ft) in length, the sludge hopper can be situated one-half to two-thirds of the way toward the end wall. This is referred to as a Gould Tank-Type II. Two chain-and-flight mechanisms move sludge in the direction of flow in the first half of the tank and against the direction of flow in the second half (Figure 12.106b). The countercurrent flow pattern that develops on the surface (primarily by the density current on the bottom and relative to the main direction of flow) causes the effluent to travel a long and circuitous path and minimizes short-circuiting.

Two or more hoppers can be placed in the middle of the tank with the first receiving the bulk of the sludge. After the bulk of the sludge and effluent are removed, the remaining velocities are so small that the lighter sludge can be transported easily to the second hopper (Wilson and Ballotti, 1988). There are different sludge loading conditions in the first and second parts of the tank, and different flight speeds can be used.

Sludge-removal systems for rectangular tanks typically feature chain-and-flight scrapers. Traveling bridge, hydraulic suction, and reciprocating flight-collector systems have been used.

In chain-and-flight scrapers, the flights are attached to two parallel chains driven by sprockets and move slowly along the clarifier floor, scraping the settled sludge to collection hoppers. The sprocket wheels are mounted on rotating shafts. On their return path, the flights can be designed to rise above water surface elevation and transport scum to collectors. This requires use of four rotation points or sprockets. If the flights are not used to move scum on the surface, then only three rotation points are required. Sometimes, a fifth sprocket can be used to help guide and hold down the flights along the bottom of long tanks.

Historically, redwood or metal flights and steel chains have been used but stainless steel or nonmetallic flights, chains, and sprockets have become more common. Typical flights measure 5 to 6 m long (16 to 20 ft), depending on the width of the clarifier. Newer flights can be 10 m (33 ft) long and entire systems can be up to 90 m (300 ft) long. Flights generally are spaced at 3-m (10-ft) intervals and travel at speeds of 5 to 15 mm/s (1 to 3 ft/min). Plastic-wear shoes fixed to the flights allow them to slide on rails near the surface and wear strips on the bottom of the clarifier so that the chain does not bear the full weight of the flights. An adjustable rubber scraper should be attached to the bottom edge and the sides of at least some of the flights to provide complete scraping and prevent unwanted stationary sludge deposits. The total length of the flight chain is limited by the stresses exerted on the chains. The number of chains and the direction of removal depend on the hopper location.

A traveling-bridge collector can be equipped with either a scraper or a suction system. These systems were developed to solve the problem of having to dewater the tank for chain-and-flight maintenance. However, there are few secondary clarifier installations in the United States.

9.5   Circular and Other Radial-Flow Designs

Clarifiers equipped with rotating mechanisms have gained in popularity through the years because they have a reputation for being mechanically the most trouble free and reliable. Most of the tanks used in suspended-growth, secondary applications are circular in plan view and have radial-flow patterns. Square, hexagonal, and octagonal tanks are somewhat like circular in the form of the hydraulic flow regimes that are typically established, but have certain differences that limit their popularity. If fillets are used in the corners and simple collection mechanisms are used, these alternate shapes have nearly all the advantages of circular tanks. For purposes of this chapter, tanks of these shapes are considered essentially equivalent to circular tanks.

Circular tanks have the disadvantage of taking more footprints for equivalent-capacity rectangular units built with common wall construction. The former requires more inlet, outlet, and sludge piping. Square, hexagonal, and octagonal units have some common wall construction, but this advantage over circular units is offset by a requirement for thicker walls.

The circular shape requires separate structures for flow splitting ahead of the tanks and for sludge pump stations. For flow splitting, the most common and effective involves feeding the structure at a low elevation, causing flow to rise vertically and then divide by flowing over two or more weirs. Another concept is to provide overflow weirs or orifices along an aerated channel that has low horizontal velocities. On some large facilities, modulating butterfly gates with computer-controlled operators have been used successfully.

For sludge removal, it is important to have independently measured and controlled withdrawal for each clarifier. Many engineers will find that measuring hydraulic flows for this purpose is adequate. Some include solids concentration measurement that enables mass flux monitoring without sampling. Such instrumentation typically is located in a separate pump station serving multiple circular tanks.

9.5.1 Flow Patterns

Circular tanks are fed from a center inlet pipe or through ports or inlet baffles at the perimeter. Effluent is withdrawn over weirs or orifices near the surface at the tank wall or from launders supported away from the wall. Figure 12.107 presents a simplified illustration of these variations. The velocity vectors shown represent general movement when clarifiers have a thin sludge blanket at the bottom.

For center-feed clarifiers, the inlet velocities and density current combine to create a "donut roll" pattern of flow in which the vectors go radially outward across the bottom (or surface of the sludge layer), up along the wall, and radially inward at the surface. In peripheral-feed clarifiers, a reverse rolling pattern is created. If there is a tangential velocity established by the inlet design details, then the roll described above will have a spiraling pattern. Movement of the collection and skimming mechanisms can further add to the rotating and spiral velocity pattern of the liquid in the tank. Recognition of these internal velocities is important to prevent them from becoming excessive and scour solids from the sludge blanket, degrading effluent quality.

9.5.2 Diameter

Clarifier mechanisms are available more than 70 m (200 ft) in diameter although the upper limit is considered to be 50 m (150 ft). For larger diameter clarifiers, wind can create surface currents that upset the radial-flow pattern and concentrate scum downwind.

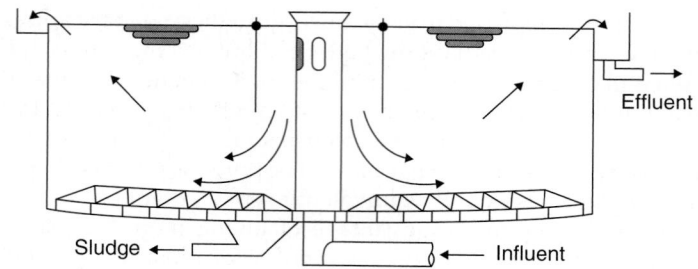

(a) Circular center-feed clarifier with a
scraper sludge removal system

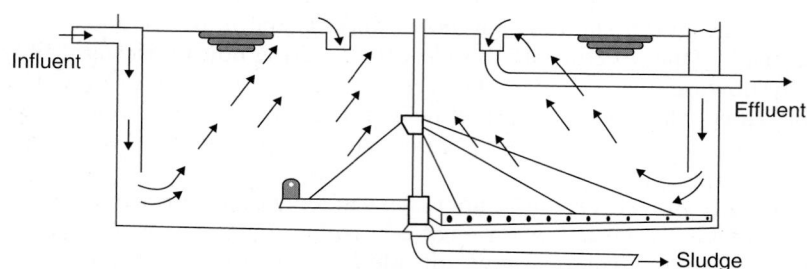

(b) Circular rim-feed, center take-off clarifier with a
hydraulic suction sludge removal system

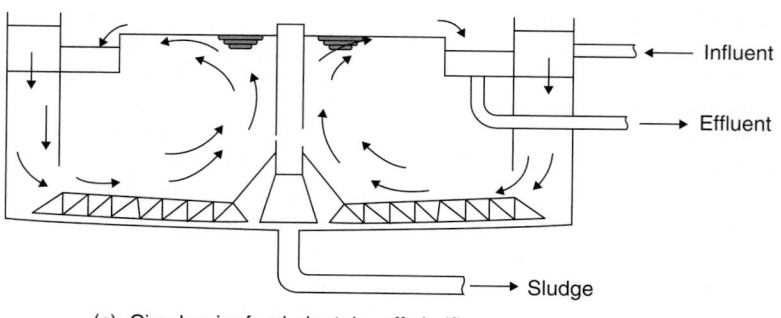

(c) Circular rim-feed, rim take-off clarifier

Figure 12.107   Typical circular clarifier configurations and flow patterns.

9.5.3 Side Water Depth

Before the early 1980s, circular secondary settling tanks often had depths of 2.4 to 3.0 m
(8 to 10 ft) as measured at the wall. Performance data showed that a shift to greater
depths resulted in lower effluent suspended solids and more resistance to upset from
hydraulic peaking. Increased RAS concentrations also occurred with increasing depth,
and guidelines were developed (Table 12.33).

In the 1984 survey, some of the largest environmental engineering consulting firms
in the United States reported using design depths of 4 to 5 m (12 to 15 ft) for activated

Tank Diameter, m (ft)	Side Water Depth, m (ft)	
	Minimum	**Suggested**
Up to 12 (40)	3 (10)	3.7 (12)
12–21 (40–70)	3.3 (11)	3.7 (12)
21–30 (70–100)	3.7 (12)	4 (13)
30–43 (100–140)	4 (13)	4.3 (14)
>43 (140)	4.3 (14)	4.6 (15)

TABLE 12.33   Minimum and Suggested Side Water Depths for Activated Sludge Clarifiers

sludge clarifiers (Tekippe, 1984). Nearly 25 years later, the authors contacted many of the same firms and found that the tabulated values are still followed although some have even recommended going a foot deeper at the largest sizes. In Europe, it is not uncommon to see new tank designs that are not this deep.

In the early 1980s, data by Parker (1983) and others quantified the value of deeper circular tanks (Figures 12.108 and 12.109). The results suggest that, to obtain effluent suspended solids of 10 mg/L, depths of over 5 m (15 ft) may be required for overflow rates above 0.85 m/h (500 gpd/sq ft). Additional data and discussion to back up the advantages of greater depth are presented in Water Environment Federation MOP FD-8 (2005).

Although Table 12.33 provides useful guidelines for design engineers in the United States, there are more sophisticated ways of determining depth requirements. One of the most sophisticated is the ATV approach that has been developed and used in

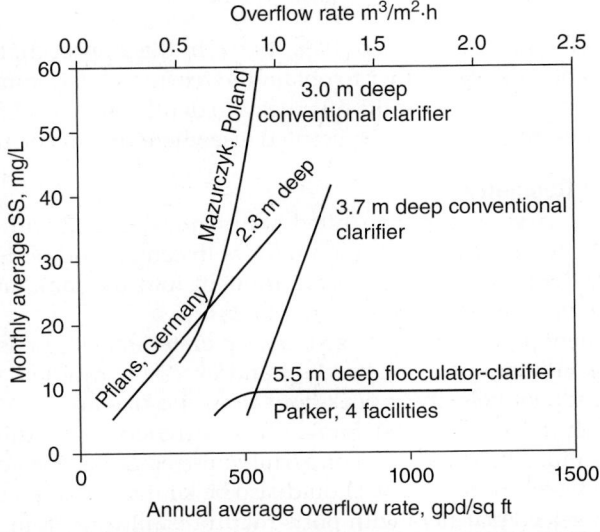

FIGURE 12.108   Performance response curves for conventional and flocculator clarifiers (gpd/sq ft × 0.0016984 = m³/m²·h) (Tekippe and Bender, 1987; Parker and Stenquist, 1986).

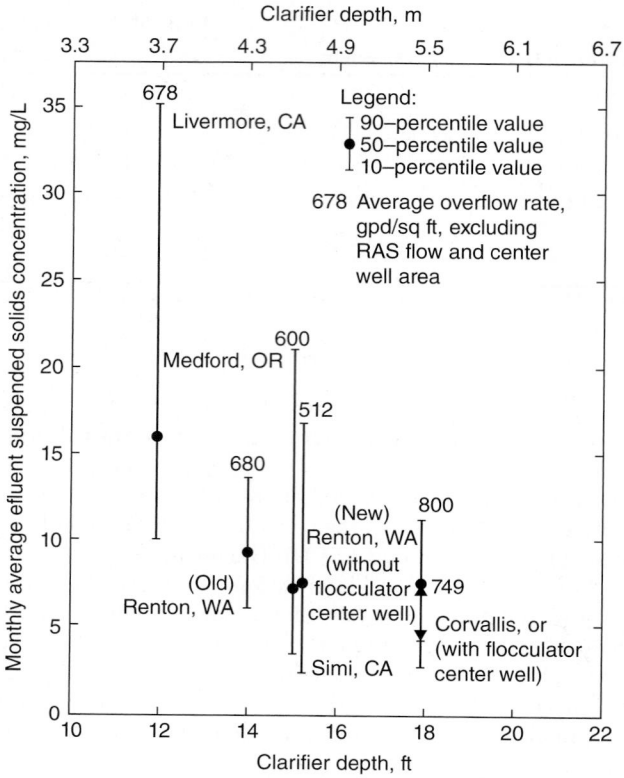

Figure 12.109   Effect of clarifier depth and flocculator center well on effluent suspended solids (gpd/sq ft × 0.0016984 = m³/m²·h) (Parker, 1983).

Germany (ATV, 1973, 1975, 1976, 1988, 1991). In this approach, four functional depths are defined and added together to obtain the recommended minimum tank depth. For larger tanks, the common results often lead to depths of 4 m (12 ft) or more. Additional details and design example are presented elsewhere (Ekama et al., 1997).

9.5.4   Inlet Geometry

Most clarifiers designed in the United States are equipped with the mechanism drive located at the top of the center column for both center- and peripheral-feed styles. For center feed, the pipe also must bring influent into the tank and transmit rotational torque from the drive into the bottom foundation.

The influent pipe should be sized to keep material in suspension but to keep velocities low enough to avoid floc breakup and excessive headloss. Many manufacturers design the influent velocity at peak hour flow and maximum RAS flow, not to exceed approximately 1.4 m/s (4 ft/s). Some other designers lower this to about half of this value to minimize floc breakup. For peripheral-feed tanks, the velocity of inflow to the distribution feed trough or skirt should also be kept to less than this value.

For center-feed clarifiers with ports that transmit flow from the feed pipe into the feedwell, port velocities should not exceed feed-pipe velocities discussed above. Most center-feed columns have four rectangular opening ports. They are often submerged, although some designs may leave the top several centimeters of the ports exposed. Instead of ports, another popular feed-pipe opening is to connect two segments of pipe with four vertical structural steel channels welded to each pipe exterior.

For peripheral-feed tanks, some designs have a raceway with multiple ports at its bottom. Others have an open raceway in which tangential dispersion of influent is achieved by introducing a directional spiral feed pattern. For those inlets with multiple ports, the port spacing and size is performed by equipment manufacturers that have computerized hydraulic models for this purpose. Most design engineers specify, for a given range, that relative flows leaving different ports do not vary by more than 5% (or such value) from the total flow divided by the number of ports.

In many center-feed inlet designs, inlet ports discharge freely into the inlet feedwell. In some, however, deflectors are constructed just downstream to break up the jetting velocities into the inlet baffled area. Likewise, for peripheral-feed tanks with multiple port bottom openings, a deflector plate typically is located immediately downstream of each port opening. This diffuses inflow and prevents jetting of flow down below influent skirt and into the settling zone.

As shown in Figure 12.110, center-feed tanks also can be fed with horizontal or vertical pipes that discharge freely at their end. Some of these pipes can also be equipped with a bell-mouth outlet that reduces the release velocity into the tank center.

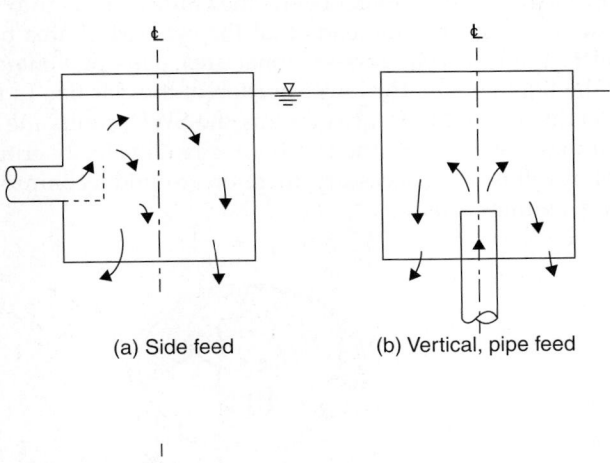

(a) Side feed   (b) Vertical, pipe feed

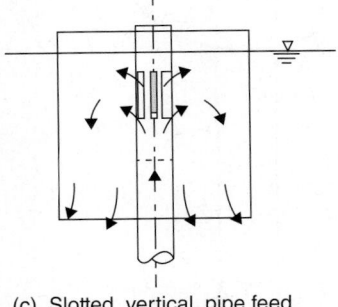

(c) Slotted, vertical, pipe feed

Figure 12.110 Various conventional center-feed inlet designs: (a) side feed, (b) vertical pipe feed, and (c) slotted, vertical pipe feed.

It is important to have a termination baffle or an upturned elbow on a horizontal feed pipe so that it does not release flow with any residual horizontal velocities. Such unbalanced velocity vectors can disturb internal flow patterns of the clarifier and affect effluent quality.

9.5.4.1 Center Feed The standard center-feed inlet design that is typically used for activated sludge is shown in Figure. 12.111. For activated sludge facilities, the feedwell diameter is often 20% to 25% of the tank diameter. Criteria for downward velocity of flow determine the diameter of the simple feedwell. Some designers and manufacturers advise that feedwell diameters should not exceed 10 to 15 m (35 to 45 ft) regardless of tank size. Likewise, downward flow velocities leaving the feedwell are often limited to about 0.7 m/min (2 or 2.5 ft/min).

The top elevation of the feedwell is generally designed to extend above the water surface at peak-hour flow. A few ports are cut into the top portion of the baffle to allow scum to move from the feedwell into the tank proper. It is common to place four such openings equidistant around the baffle.

A typical center feedwell extends down from as little as 30% to as much as 75% of the tank depth. Several manufacturers recommend that submergence be 25% to 50% of the SWD. It is also common that the center feedwell bottom edge be located about 0.3 m (1 ft) below the bottom of the center-feed pipe ports. It must be low enough so that flow jetting out of the ports does not get below the baffle and out into the settling zone.

One design concept recommends that the cylindrical area below the feedwell be about equal to the feedwell cross-sectional area. This prevents a theoretical velocity increase as the liquid enters the lower portion of the clarifier. In this case, the opening under the feedwell would be measured as the SWD minus the feedwell depth. This requirement may conflict with the clarifier feedwell velocity criteria and SWD criteria discussed above. It is often necessary, therefore, to find a compromise that meets most of the criteria simultaneously.

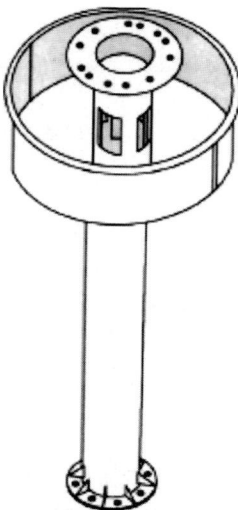

Figure 12.111 Standard center inlet design.

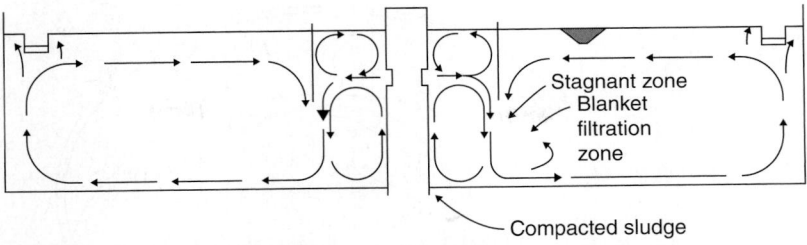

FIGURE 12.112 Typical velocity pattern of center feed tank.

In some conventional tanks, the feedwell rotates with the sludge scraper mechanism, whereas, in others, it remains stationary. The feedwell can be supported from the bridge or from the sludge collector mechanism. If it is supported by the bridge and does not rotate, then care should be taken to avoid aligning feed-pipe ports with scum port openings.

Figure 12.112 illustrates a typical velocity pattern resulting from use of the simple center-feed inlet. The incoming density current creates a "waterfall" effect. Crosby (1980) also reported that influent velocity vectors can be distorted by the sludge collector riser pipes if they pass in front of the inlet ports of the feed pipe.

McKinney (1977) recommended a flat circular baffle (Figure 12.113) to reduce the harmful effects of the cascading influent flow of activated sludge mixed liquor. The baffle is most valuable in tanks with plows and central hoppers for sludge removal. It prevents scouring of the sludge hopper and facilitates plowing of sludge radially inward as the influent flow moves in the opposite direction.

For activated sludge clarifiers, the bottom elevation of the center feedwell has a significant effect on performance. The relative level of the sludge blanket surface must then be considered in both design and operation of the clarifier. Crosby (1980) showed that better performance could be obtained with a center feedwell bottom that is either well above the sludge blanket or somewhat below the top of the sludge blanket. A shallow blanket separated from the well bottom is considered optimal for sludges that settle well, but not for sludges that settle poorly. In the latter case, it may be possible for an operator to improve performance by carrying a relatively thick blanket that provides some degree of solids filtration and settling. Operating with the bottom of the feedwell at nearly the same elevation as the top of the sludge blanket was discouraged.

In many facility designs, mixed-liquor activated sludge arriving at the clarifiers is not fully flocculated. Performance can be improved, however, by using a separate flocculation zone. Simply increasing the size of the center feedwell is one approach.

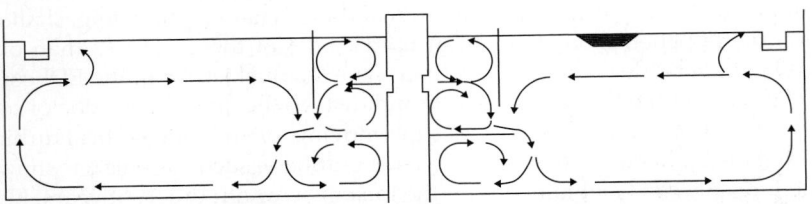

FIGURE 12.113 Circular baffle provided to reduce cascade effect in influent mixed liquor flow.

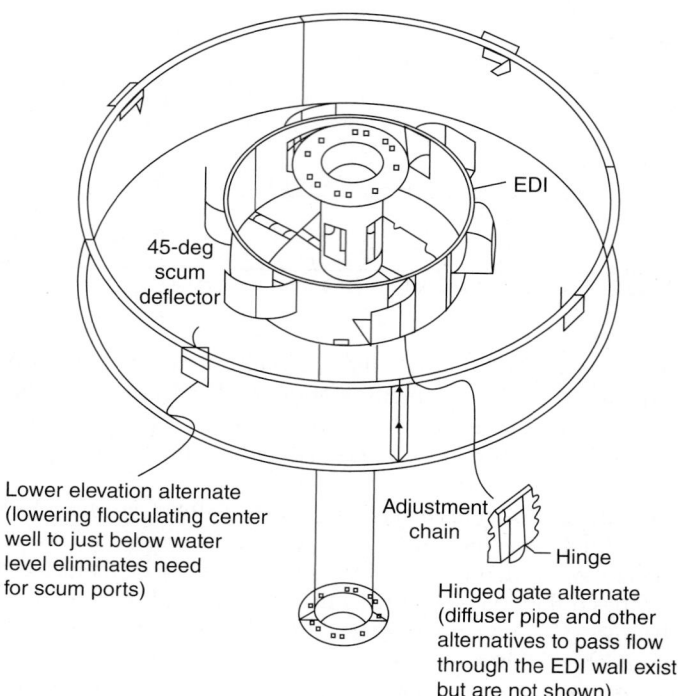

EDI

45-deg
scum
deflector

Lower elevation alternate
(lowering flocculating center
well to just below water
level eliminates need
for scum ports)

Adjustment
chain

Hinge

Hinged gate alternate
(diffuser pipe and other
alternatives to pass flow
through the EDI wall exist
but are not shown)

FIGURE 12.114   Center-column energy-dissipating inlet (EDI) and flocculation baffle.

Some have provided mechanical flocculators within this zone, whereas others have provided an energy-dissipating inlet (EDI) to distribute the flow into the flocculation zone (Figure 12.114).

Kinnear (1998), Wahlberg et al. (1994), Parker et al. (1971), and others have studied sizing of the flocculation centerwell. It has been shown that a detention time of approximately 20 minutes achieves well over 90% of the obtainable degree of floc formation. Therefore, a rule of thumb has been to size the flocculation well to obtain 20 minutes of residence time at average dry-weather flow with an additional allowance of 50% for RAS flow. A simpler approach is to set it equal to 30% to 35% of the clarifier diameter. If the well is too large, then influent can plunge and short-circuit the well.

Depth of projection into the clarifier by the flocculation well also is an important design criterion. Many design engineers have arbitrarily set this at a value equal to approximately one-half of the tank depth at the location of the baffle. With a sloped floor, this would be a little deeper than one-half of the SWD. In more recent designs supported by results from computational fluid dynamics modeling, shallower flocculation baffle penetrations have been used. Some of these are less than one-half of the SWD. If the baffle is too shallow, then some residual jets from the EDI could fall below the bottom of the flocculation baffle and disturb the quiescence zone of settling.

In some early designs, several slow-moving, pitch-blade vertical turbines were provided to obtain floc formation. Parallel operation of such systems has shown that equivalent results can be obtained with the mixers on or off. In recent years, EDIs have been used to obtain adequate mixing within the flocculation zone, and mechanical mixers are rarely, if ever, used.

Tekippe (2002) reported on numerous clarifier inlet and EDI designs. Figure 12.114 shows a popular one that uses the simple hinged-gate alternate. The designer intended to allow operators to set the adjustment chain to increase or decrease velocity entering the flocculation zone. Tightening down the chain would increase headloss, inlet velocities, and stirring. In practice, many operators simply set the hinged gate at one location (e.g., one-half or two-thirds open) and do not make further adjustments.

The diameter of the EDI often is set at approximately 10% to 13% of the tank diameter. Some design engineers use a detention time of 8 to 10 seconds. Making the EDI too large subtracts from the volume of the flocculation zone and increases downward velocities.

Data such as that shown in Figure 12.115 support use of EDIs with tangential release of flow. Stirring was shown to be the best method of forming floc and delivering low effluent turbidities. Because many operators do not adjust the hinged gates, designers have tried to improve performance by replacing the hinged gates with the curved chutes, which has become a popular alternative (Figure 12.114).

In some recent comparisons, it was discovered that the curved chutes resulted in excessive jetting into the flocculation zone. Studies by Esler (1984), Haug et al. (1999), and others at Hyperion (City of Los Angeles) demonstrated that the provision of an EDI with such chutes actually performed worse than adjacent tanks with no EDI at all. A similar side-by-side comparison for a trickling filter final clarifier was performed at Central Weber, Ogden, Utah. An EDI with curved chutes and its associated flocculation baffle did not perform as well as an old, large simple inlet well with a bottom and diffusers containing a lattice structure around its lower perimeter (Tekippe, 2002).

These findings led to modifications of the EDI to capitalize on impingement for energy dissipation. One example is the double-gated EDI. Because double gates are provided, one can be opened more than the other to create adjustable degrees of rotation within the flocculation baffle. Full-scale test results showed this design to be better than that with curved chutes (Figure 12.116).

The side-by-side, full-scale studies conducted at the Hyperion Wastewater Treatment Plant, serving Los Angeles, California, led to an innovative design involving

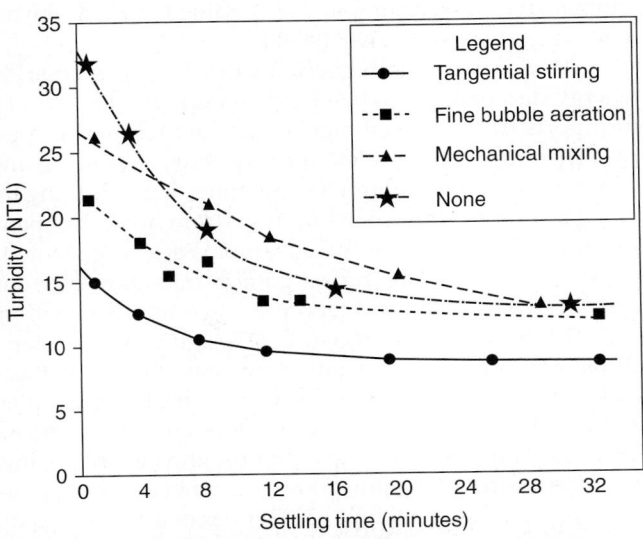

FIGURE 12.115 Reported results of different flocculation methods.

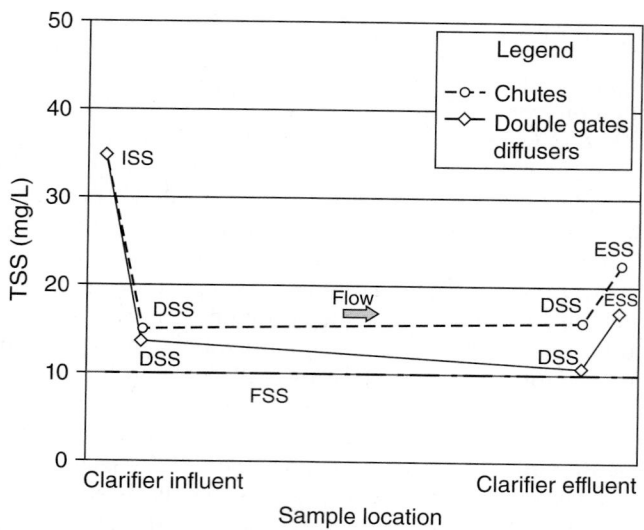

FIGURE 12.116 Diagnostic test results of different energy-dissipating inlet designs at Central Weber, Utah (overflow rate [OFR] = 1.4 m/h [825 gpd/sq ft]) (ISS = influent suspended solids; DSS = dispersed suspended solids; FSS = flocculated suspended solids; and ESS = effluent suspended solids).

multiple diffusers located around the perimeter at the bottom of the EDI (Figure 12.117) (Haug et al., 1999). In this arrangement, EDI effluent was conducted downward through eight, 0.6-m (24-in) openings that had 32 small 0.35-m (14-in.)-diameter diffuser pipes that were paired off to impinge against each other. Small openings are provided at the surface for passage of scum. This design was found to be far superior to alternatives using curved chutes and better than providing no EDI at all. All 36 clarifiers at Hyperion were converted to this design, which has become known in the industry as the "LA EDI." Additional details regarding studies leading to this design are presented in Water Environment Federation MOP FD-8 (2005).

Figure 12.118 illustrates another recently developed and marketed EDI design (flocculating energy-dissipating feedwell [FEDWA]). In this arrangement, flow enters through four ports from the feed pipe. Opposite each port is a pair of vertical baffles that form a corner. An opening is left midway between corners, and flow from adjacent corners impinges as it goes through the openings. Upon leaving this opening, it is split at 90% and again forced to impinge on flow from adjacent openings. This process is repeated one more time before the mixed liquor is discharged into the flocculation zone. Developers of the FEDWA inlet report good results; however, full-scale, side-by-side tests have not been conducted and reported for comparison with most alternatives yet.

Figure 12.119 shows another recent EDI design using the step-down energy gradient of multiple rows of concentric, offsetting slotted board baffles for impingement. It has been used with reported effectiveness at Melbourne, Australia. Figure 12.120 depicts another EDI concept in which all flow entering from the feed-pipe ports is forced to rise vertically and leave the EDI through veined openings of the surface. The pitch of the vanes creates a tangential flow pattern in the surrounding flocculation zone.

Barnard et al. (2007) and Tekippe (2002) present a discussion of several EDI designs and of a concept that releases the center-feed inlet flow vertically into a flocculation zone

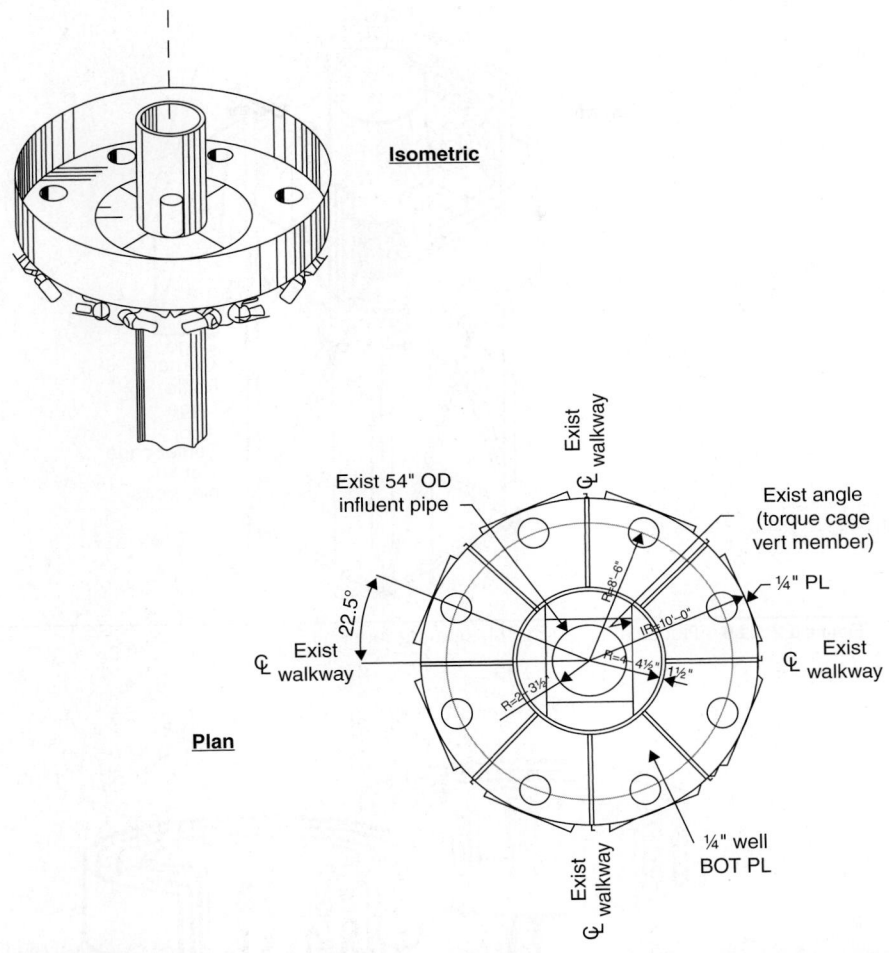

Isometric

Plan

Exist 54" OD influent pipe

Exist angle (torque cage vert member)

¼" PL

Exist ℄ walkway

Exist ℄ walkway

Exist ℄ walkway

Exist ℄ walkway

22.5°

¼" well BOT PL

Figure 12.117 Los Angeles, California, energy-dissipating inlet (EDI) patent drawing and plan view.

without an EDI. More than 100 installations of the center well shown in Figure 12.121 exist worldwide. It is referred to as the SOLE (side outlet low energy) stilling well design.

Tekippe (2002) presented data that showed a bell-mouthed, vertical-release inlet pipe that created a gentle boil at the surface performed better than a parallel, full-scale clarifier with an EDI at a large activated sludge facility in the United Kingdom. Barnard et al. (2007) likewise reported good performance of the Stickney facility of the Metropolitan Water Reclamation District of Greater Chicago, Chicago, Illinois, in which the inlet pipe is a cone-shaped concrete structure that allows for a vertical release into the flocculation zone without feed-pipe ports. A center-drive mechanism is used and its torque is transmitted through vertical bars into the concrete cone. Effluent TSS in the range of 5 to 7 mg/L has been observed at overflow rates of 2.0 to 2.2 m/h (1200 to 1300 gpd/sq ft).

To prevent odors and unsightliness (Figure 12.122), it is important to move floatables out of the flocculation zone. In early years of design, the top elevation of the flocculation

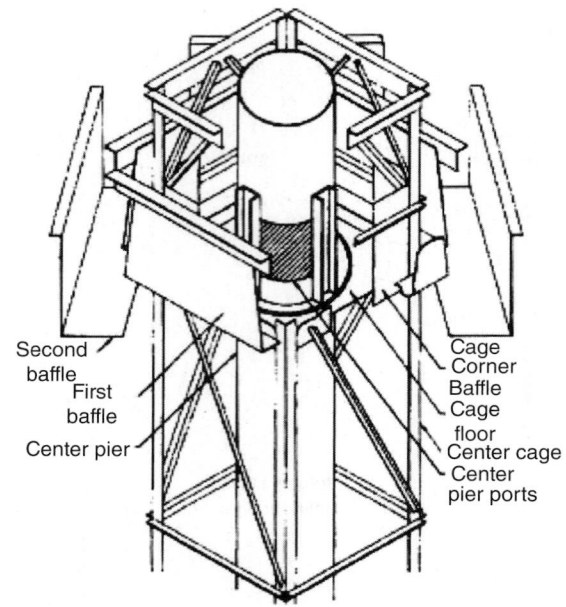

Second
baffle
First
baffle
Center pier

Cage
Corner
Baffle
Cage
floor
Center cage
Center
pier ports

FIGURE 12.118   Flocculating energy-dissipating feedwell.

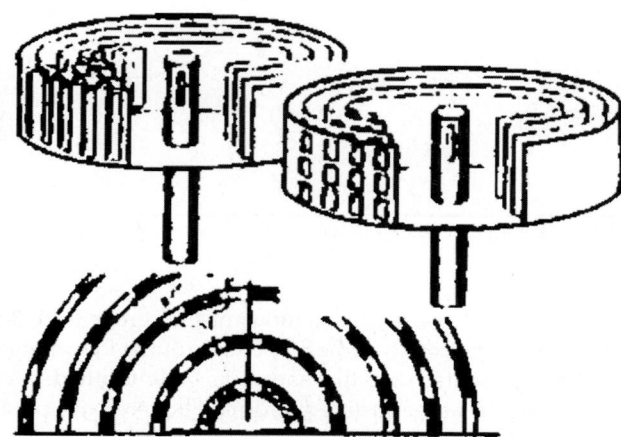

Multilayer energy dissipating inlet column
(Medic, US patent APPL No. 11/373,749)

FIGURE 12.119   Multilayer energy-dissipating inlet column (MEDIC) design.

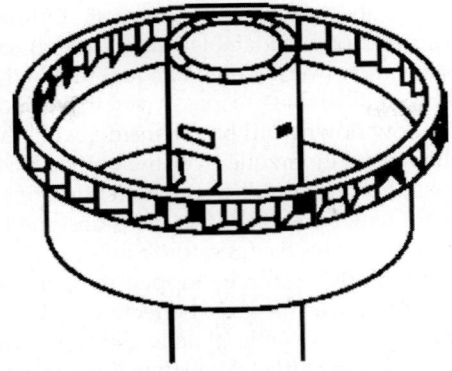

FIGURE 12.120   Energy-dissipating inlet with top-release vanes.

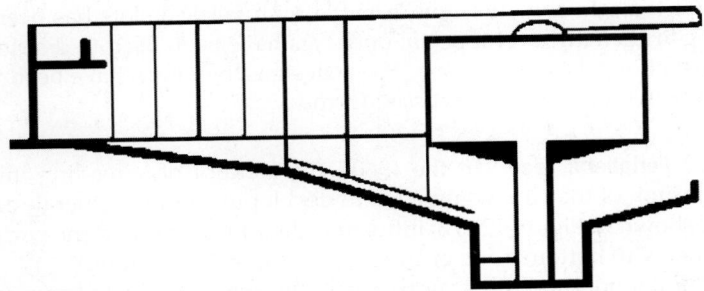

FIGURE 12.121   Side outlet low energy (SOLE) stilling well (Barnard et al., 2007).

FIGURE 12.122   A flocculation baffle that traps floatables creates odors.

baffle was set to project above the water surface at all flowrates. This design resulted in confinement of foam and other floatables even though scum ports were provided. At other sites, the top elevation was lowered to equal that of the bottom of the v-notch effluent weirs. This allowed floatables to pass over the top of the flocculation baffle, but directed most of the flow downward on the inside. At high flows, however, supernatant would flow into the flocculation zone over the baffle. This, of course, would dilute the contents of the flocculation zone and shorten detention time of the incoming flow.

To avoid this problem, some facilities have designed the flocculation baffle to be adjustable upward. This allows an operator to raise its level so that it projects above the water surface but, at high flows, can be topped to flush the floatables out into the tank proper. In some designs, it was most cost-effective to mount the flocculation baffle in a rigid position and bolt an adjustable plate at the top. Careful adjustment of this plate would allow the flocculation baffle to overflow only at the desirable peak-flow periods. Some designers also are altering spray nozzle design to move the scum through the ports more effectively or providing a scum removal mechanism inside of the flocculation baffle if it is set high enough to prevent overtopping.

Several center-feed clarifiers are designed to release flow into a zone near the bottom of a tank. In some designs, a baffle with vertical slots has been used. In others, rotating arms with several portal openings have been used to distribute the incoming flow just above the sludge zone. These designs, however, have been rarely used in the United States and are not discussed further.

9.5.4.2 Peripheral Feed

In the 1960s, the concept of spreading inlet energy over a large portion of the tank volume led to development of peripheral-feed circular clarifiers. As shown in Figure 12.123, influent is distributed around the perimeter by using of a channel with bottom ports or by creating a spiral-roll pattern.

Several model and full-scale dye tests have been conducted on peripheral-feed clarifiers (Dague, 1960). Results have indicated that peripheral-feed tanks have a higher hydraulic efficiency than center-feed models. Specifically, full-scale, activated sludge tests conducted at Sioux Falls, South Dakota, showed that in addition to better hydraulic efficiency, peripheral-feed tanks also achieved higher suspended solids removal than existing-center-feed design. The latter, however, did not employ the flocculation centerwell concept developed in more recent years.

In some designs, a headloss across the orifices of approximately 25 mm (1 in.) at average flow was used to obtain reasonably uniform distribution of flow around the perimeter of the tank. For facilities with large peaking factors, some maldistribution of flow and solids occurred. Design criteria were changed to provide more headloss (approximately 60 mm, or 2.5 in.) for better distribution at average flow. Peaks of more than 3-to-1 accommodated the higher loss. At low flows, headlosses across the orifices can be low and do not achieve good distribution. Under these conditions, however, overflow rates are low and clarifier performance may still be satisfactory. Minimum flow distribution, therefore, typically is not considered a limiting design criterion. For facilities with extreme peaking factors, a special overflow provision in the baffle wall or tank wall can be added (Figure 12.123).

For these inlet designs, the feed channel/zone is baffled off from the body of the settling liquid. As such, floatables can accumulate on the inlet zone surface and generate odors and objectionable aesthetics if not removed. Provisions for this are discussed below in the section on skimming systems.

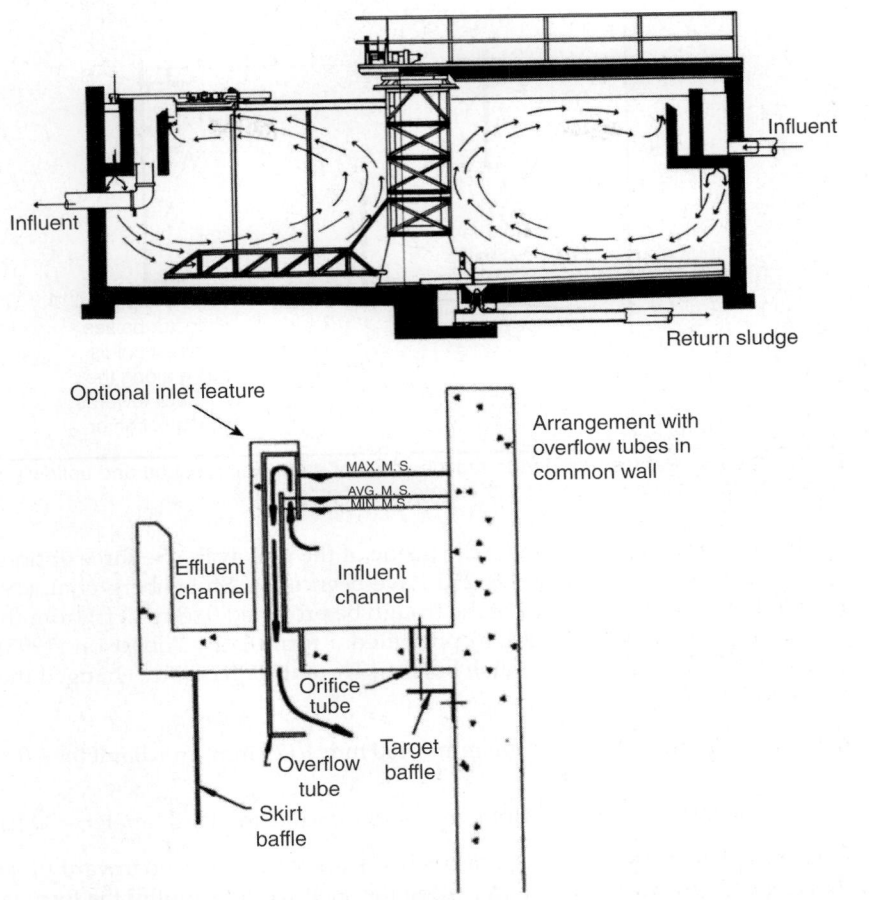

FIGURE 12.123   Peripheral feed clarifier flow pattern.

9.5.5   Interior Baffles

For many years, circular clarifiers were constructed without interior baffles, except for the inlet well. In the 1970s and early 1980s, research engineers, including Crosby (1980), McKinney (1977), and others, found that performance of activated sludge clarifiers could be improved significantly by using strategically located interior baffles. Figure 12.124 illustrates another baffle that was found effective to help confine the sludge blanket to the central portion of the clarifier. Crosby (1980) initially developed this baffle, which extends from the bottom. It has been referred to as his "midradius" baffle although its best location may not be at the midradius point.

Center-feed, activated sludge clarifiers often create an updraft of suspended solids along the outer wall. Crosby (1980) and McKinney (1977) independently arrived at a solution: constructing a perimeter baffle to deflect flow back toward the center of the tank. The conceptual design of these two options is shown in Figure 12.125a–f. Further refinements in this design are illustrated in Figure 12.125a, 12.125b, and 12.125c. For designs with the trough on the outside of the tank wall, the Crosby design shown in Figure 12.125a is most appropriate.

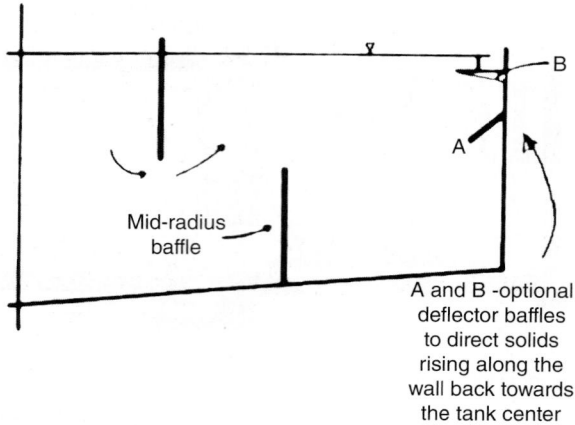

Mid-radius
baffle

A and B -optional
deflector baffles
to direct solids
rising along the
wall back towards
the tank center

FIGURE 12.124   Baffles provided to reduce effect of outer wall rebound and upflow (note that a tank typically would not have more than one such baffle).

For tanks with the trough on the inside of the tank wall, the three options shown in Figure 12.125b, 12.125c, and 12.125d have been used. Stukenberg et al. (1983) recommended that the bottom shelf of the trough be projected 0.63 m (2 ft) from the radius of the outlet weir. The WEF (2013a) presented a formula by Albertson (1995) that gave minimum dimensions for the shelf bottom. The author has since changed the equations to the following:

$$\text{Minimum shelf bottom, mm} = 460\text{ mm} + (25\text{ mm/m})(\text{diameter} - 9\text{ m})$$

$$[\text{Minimum shelf bottom, mm} = 18\text{ in.} + (0.3\text{ in./ft})(\text{diameter} - 30\text{ ft})]$$

The concept is that if the trough bottom was sufficiently wide, no inward projecting shelf (horizontal baffle) is needed. Eliminating the shelf would simplify the formwork in construction and eliminate the objectionable settlement of solids on the shelf. Some engineers have reduced the latter problem by adding the fillet option shown in Figure 12.125c. Others contend that even if the minimum shelf bottom criterion of Albertson is met, the tank will produce a lower effluent TSS value if the shelf is provided to at least the radius of the scum baffles and perhaps as far as the 0.63 m (2 ft) value of Stukenberg.

For cantilevered double launders that are constructed too close to the tank wall, updraft solids typically escape. Parker et al. (1993) developed a special slotted baffle oriented at 45° from horizontal, to deflect the updraft solids away from the wall and below the effluent trough for clarifiers of this design (Figure 12.125f). The baffle was constructed of strips of fiberglass roofing material that spanned from one support bracket to another. A spacing of approximately 35 to 50 mm (1.5 to 2 in.) was allowed to permit a small flow to rise and leave the outer weir. Most of the flow, however, was deflected to the inner weir, and the suspended solids were projected toward the tank center. At Lincoln, Nebraska, this arrangement reduced effluent TSS from 35 to 28 mg/L.

Others have attempted to minimize the updraft problem by reducing the number of notches or raise the outer weir of such a design to encourage most of the flow to leave the tank by way of the inner weir. Blocking the outer weir completely is not recommended because it then creates a dead space between the outer weir and the wall.

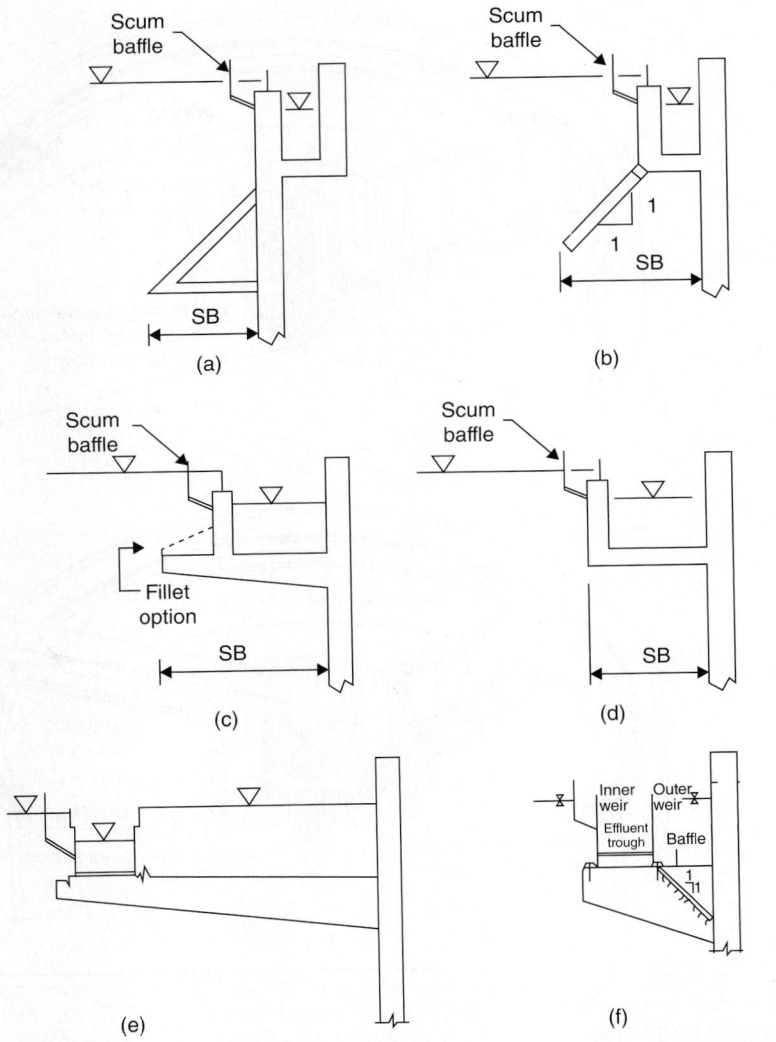

FIGURE 12.125 Alternative peripheral baffle arrangements: (a) Stamford, (b) unnamed, (c) McKinney (Lincoln), (d) interior trough, (e) cantilevered, and (f) cantilever with deflectors.

9.5.6 Scum Removal

Activated sludge clarifiers often experience scum formation because of denitrifying sludges and foams (such as *Nocardia* filamentous bacteria). In the United States, it is common practice to remove floating materials from the surfaces of secondary clarifiers. For circular tanks, a variety of skimming mechanisms have been designed and are operated with varying degrees of capacity and success. The most common system used for center-feed tanks is shown in Figure 12.126a. It features a rotating skimmer arm and wiper that travels around the outer edge of the tank next to a scum baffle. It moves the floatables onto a beach or egress ramp connected to a scum removal box. The skimmer

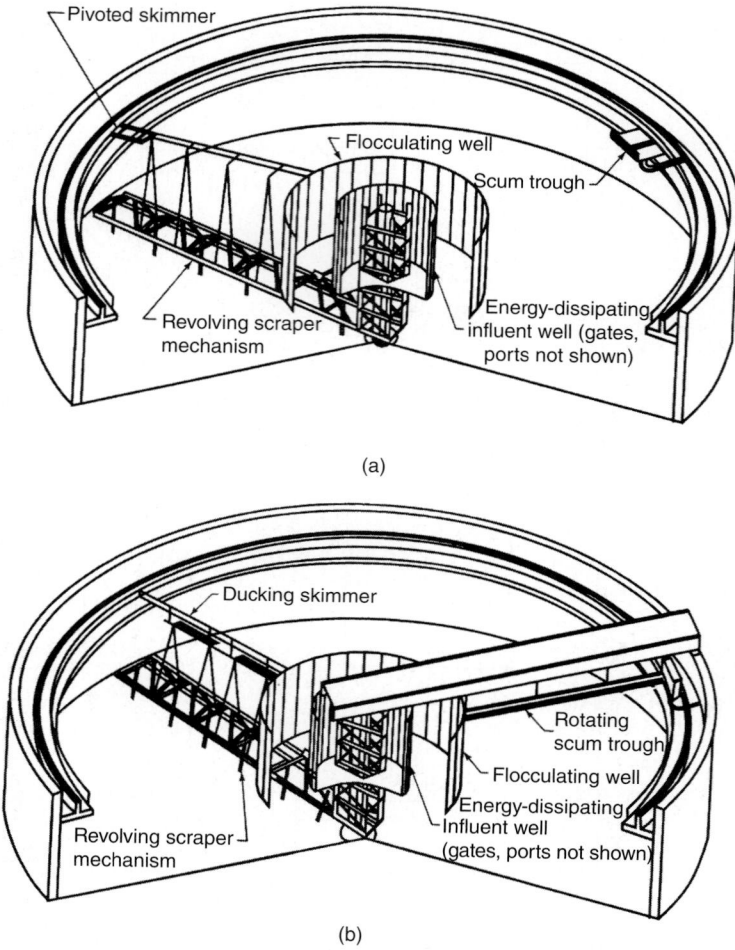

FIGURE 12.126 Alternative skimming designs for circular clarifiers: (a) revolving skimmers and fixed scum trough and (b) rotary ducking skimmers (Ekama et al., 1997; reprinted with permission from IWA Publishing).

blade is most effective if it is attached tangentially to the feed baffle, rather than perpendicular to it. The resulting pitch angle of the tangential design helps move floatables to the outer area of the tank.

Some scum boxes are equipped with an automatic flushing valve located on the center-most end of the box. The valve is mechanically actuated with each pass of the skimmer. It results in a water flush of the solids into the box hopper bottom and discharge pipe. The flush volume and duration are typically adjustable.

This scum trough often extends several meters from the scum baffle toward the center of the tank. Some designs extend this to the flocculating or center feedwell and thereby obtain full radius skimming. For shorter scum troughs, a system is provided to move the floatables toward the outer scum baffle. A fixed, flexible antirotation baffle, supported from the bridge and extended down to the surface of the tank, is used. The baffle is placed at an angle to the skimmer arm that intersects the tank water surface. The resulting scissors-like movement pushes the scum outward.

Another method of moving floatables out toward the scum baffle is the use of water surface sprays. It is a good idea to locate the fixed scum beach on the down-wind side of a tank. Yard piping arrangements may or may not make this option economically attractive.

Another skimming concept, known as the "ducking skimmer," is shown in Figure 12.126b. In this design, a skimmer board is connected to the sludge-removal mechanism through a hinged, counter-weighted assembly. It pushes the floatables toward a fixed, rotating trough that turns into position as the skimmer board approaches and trips a trigger switch. When the board reaches the rotating trough, it ducks under the trough, and its counterweights return it to the surface to continue rotation around the tank. This device has an advantage of offering full-radius scum removal. Separate flushing is not required, but some designs feature a deeper cut opening at the inner end of the rotating trough to take on more water, which moves the floatables into the collector box at the other end of the trough.

Some installations have a reported high amount of maintenance associated with the ducking skimmers. Issues have included controls, bearings, actuators and binding of the rotating trough. Subsequent designs have been made more robust.

A third type of skimmer involves use of a full-radius traveling beach that rotates with the drive cage and discharges into a central, annular well, from which the scum is pumped out. A stationary, hanging flap that has its lower edge just below the water surface bends as needed to push the scum up the beach as it travels below.

Peripheral-feed clarifiers typically remove scum from the peripheral-feed channel. One design is to feed the tank unidirectional and locate a small scraper and beach or overflow weir arrangement described above at the end of the feed channel. Scum removal is facilitated by feeding the channel fed in one direction and decreasing the cross-sectional area with distance around the tank. This can be achieved by making the channel progressively narrower or by decreasing its depth by sloping the floor upward. The latter design allows a fixed-width blade to fit the channel. If the channel becomes increasingly narrow, then a narrow fixed, flexible, or hinged skimmer blade arrangement has been used to accommodate the decreasing width (Figure 12.127). The weir gate can be carefully adjusted so that scum overflows only at peak-flow rates. In other designs, the weir gate is motorized and mechanically lowered as the skimming arm approaches.

There have been incidents in which the feed channel foam problem has become so severe that it overflowed the wall, dropping foam directly into the effluent channel. At Denver Metro in Colorado, this problem led to the conversion of 10 peripheral-feed tanks to center feed. There are, however, hundreds of peripheral-feed tanks, and most correctly designed units do not have this problem.

9.5.7 Outlets

Outlets for most circular center-feed clarifiers consist of a single perimeter v-notch weir that overflows into an effluent trough. Alternatives to this include cantilevered or suspended double weir troughs and submerged-orifice collector tubes.

For peripheral-feed designs, a singular perimeter weir is used in one concept. Another includes provisions of a square, octagonal, or circular double-sided launder suspended from the bridge or other structural support near the center of the tank.

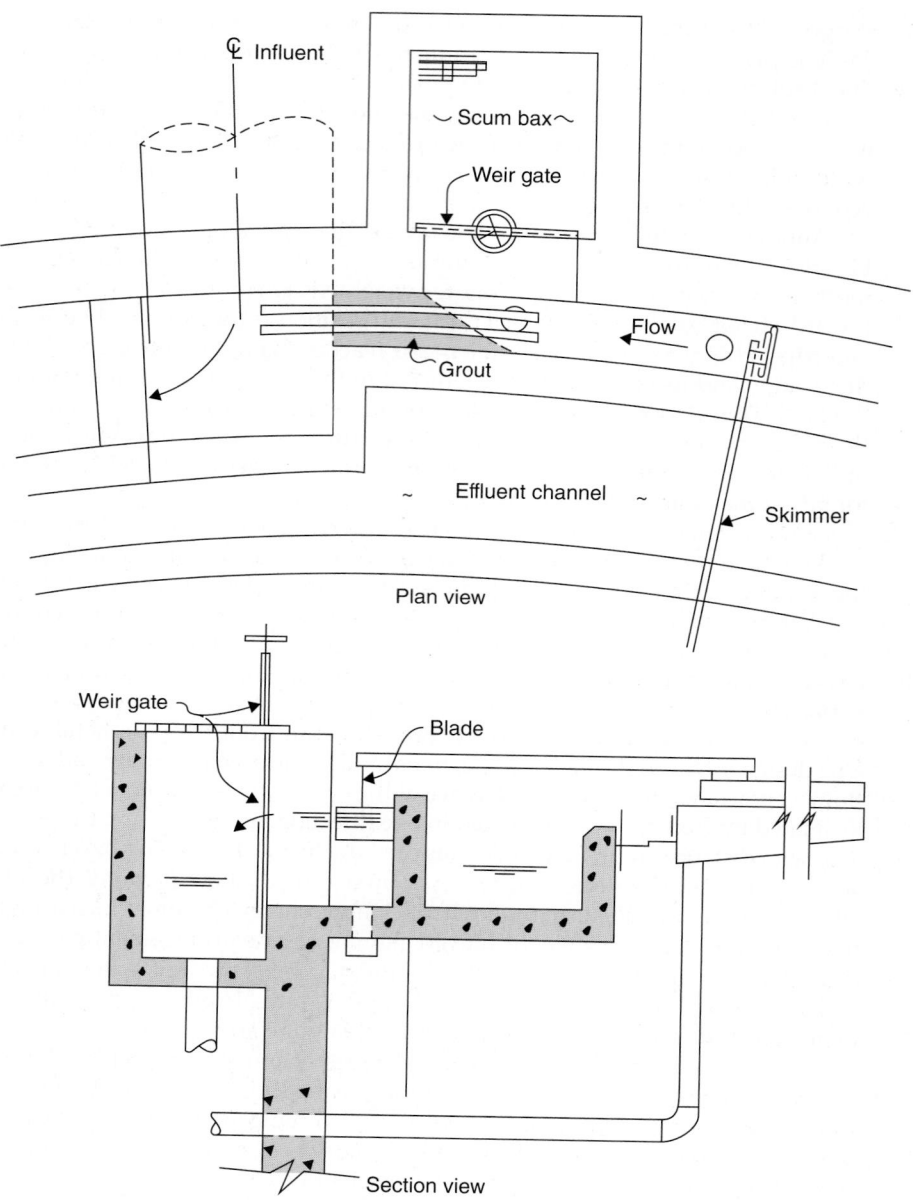

FIGURE 12.127 Plan and section of effective variable width influent channel skimming design for peripheral feed clarifiers.

In many states, regulations allow the weir loading that results from simply building a perimeter weir. In others, regulations include a weir loading limit expressed in flow-rate divided by length of weir. For example, the Ten States Standards limit weir loading to 250 m³/m²·d (20 000 gpd/ft) for facilities with average flows less than 0.04 m³/s (1 mgd) and to 375 m³/m²·d (30 000 gpd/ft) for larger facilities (GLUMRB, 2014).

9.5.7.1 Peripheral Weir   There are two common designs for peripheral weir outlets for circular tanks. In the first, a concrete trough is constructed on the inside of the tank wall. The weir plate is then bolted at the top of the inward face of the trough wall. In the second, the weir plate is bolted to the inside of the tank wall. A concrete effluent trough is then constructed outside the tank wall. This can be less costly to build if soil elevation outside the tank can be used beneficially in formwork construction.

The most common type of weir plate involves placement of 90° v-notches at 150- or 300-mm (6- or 12-in.) intervals. This design allows a balance of relatively low increases in water surface elevation when flows increase with an allowance for imperfect leveling with reasonably good distribution. In contrast, a flat weir plate is susceptible to unbalanced withdrawal if the weir is not perfectly level or if wind effects on the surface are significant.

Some designers use square notches, which result in wider level changes with flow changes and are more prone to partial blockage due to leaves, algae strings, and other debris.

9.5.7.2 Cantilevered Double or Multiple Launders   In early years, and perhaps in some areas today, regulations that limit weir loading to sufficiently low values have enticed designers to go with multiple weirs, serpentine weirs, and other ways to increase the weir length for a given diameter of tank. Requirements in recent years have been relaxed in many design guidelines. Nevertheless, this cantilever double launder concept remains. It offers the opportunity for solids moving up along to the wall to resettle before inward flow takes them to the outer weir. Anderson (1945) and others recommend that the outer weir be at least 25% of the tank radius from the wall.

9.5.7.3 Launders Suspended from the Bridge   For some small, circular tanks, the double-sided launder design discussed previously are suspended from the bridge. Necessary structural trusses are constructed to stabilize this form of outlet. This concept is used most widely with peripheral-feed clarifiers that use the spiral-influent design. Peripheral-feed tanks that use orifice feeding often have an inward-projecting trough that is constructed with a wall common to the feed trough.

9.5.7.4 Submerged Orifices   Few circular tanks have been constructed with submerged orifices for effluent removal. A typical design has a circular pipe located near the wall with evenly spaced circular orifices cut into the top. A downstream hydraulic control device is required to maintain a level within the clarifier. Advantages and disadvantages of this concept were presented previously.

9.5.8 Sludge Withdrawal

There are two basic types of sludge-removal mechanisms: plows and hydraulic suction used for activated sludge circular tanks.

For square tanks, spring- or counterweight-loaded corner-sweep plows have been used to gather sludge from the corner areas outside the fixed-sweep circular area. If tanks are sufficiently deep, filleted tank walls can be used to fill in the corners so that circular mechanisms without corner sweeps can be used. This is preferable because corner sweeps are notorious for having mechanical problems. Hexagonal and octagonal shapes typically include the same fillet concept.

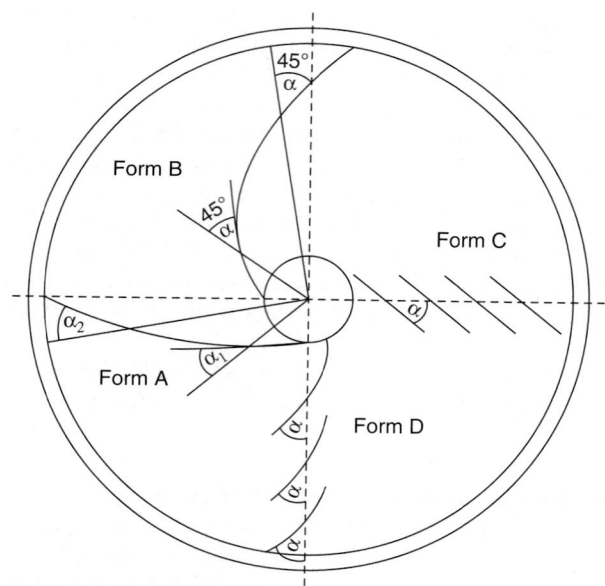

FIGURE 12.128 Scraper configuration in Germany (Gunthert, 1984). Form A is the "Nierskratzer" type, where $a_1 > a_2$. Type B is a logarithmic spiral with a constant at 45°. Types C and D are "window shade" type scrapers.

9.5.8.1 Scrapers There are several basic scraper designs used; Figure 12.128 shows four different types. The multiblade plows, using straight scraper blades, have been used most extensively in the United States. The designs using curved blades are typically referred to as spiral plows and have been used for decades in Europe.

For small tanks, two single spirals typically suffice. For larger tanks, two spirals may be added at 90° points from the others and extend only partway to the wall. Although blade angles of between 15° and 45° have been used, 30° has become popular in the United States. Suppliers report that the vast majority of new scraper mechanisms used for activated sludge are of the spiral design, as opposed to the multibladed "window shade" type.

In early years, the tip speed of spiral scrapers was approximately 3 m/min (10 ft/min). Based on several facility improvement projects, Albertson (1995) and others have recommended values as high as 10 m/min (30 ft/min). These faster speeds, as well as deepening the spiral blades closer to the center of the tank, give this system a relatively high sludge-transport and removal capacity. The higher speeds do induce some stirring of the tank contents, especially in smaller tanks, which may affect clarification. Some designs feature variable-speed drives.

9.5.8.2 Hydraulic Suction For activated sludge treatment with partial or complete nitrification, the occurrence of denitrification in clarifiers can cause solids to float and effluent quality to degrade. In the 1960s, the concept of hydraulic suction was engineered to assist in removing sludge more rapidly. Kinnear (2002) showed that this concept was more effective to maintain lower sludge blankets. Data in Figure 12.129 show that increasing sludge blanket depth results in higher effluent suspended solids in some

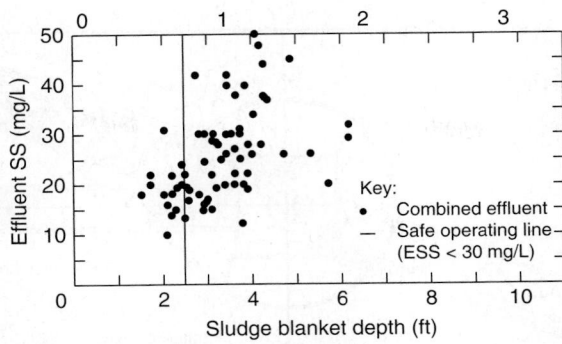

FIGURE 12.129   Effect of sludge blanket depth on effluent suspended solids (ESS) at pure oxygen activated sludge facility; sludge volume index = 51 to 166 mL/g, with an average of 86 mL/g (Ekama et al., 1997; reprinted with permission from IWA Publishing).

cases, but, regardless of the blanket depth, an adequate clear water zone is necessary for good performance.

To lift the sludge, a hydraulic head deferential is established by use of pumps or adjustable valves to move solids into the collector arms. There are two fundamentally different types of hydraulic suction removal mechanisms. The first, typically called an organ or riser pipe, has a separate collector pipe for each suction inlet orifice. The V-shaped plows direct the sludge to these pipes.

The other type has a single or double arm extending across the full radius of the tank. The arm is tubular and has several orifices. It is typically referred to as a manifold design but is also known as header, tubular, or Tow-Bro in recognition of Townsend and Brower, who developed it.

Figure 12.130 shows a typical riser-pipe clarifier design. The horizontal runs of the riser pipes are stacked vertically—an orientation that induces more tank stirring than they were horizontal. Most designers prefer the latter. Each riser pipe of this mechanism is fitted with an adjustable telescoping weir, movable sleeve, or ring arrangement that allows the operator to adjust the flow independently for each suction inlet. Advantages and disadvantages are detailed in the WEF's MOP FD-8 (WEF, 2005).

Riser-pipe designs often feature tubes that enter the collection box and pass in front of the central feed-pipe ports, which often are located just below the box. This interference can deflect the inflow and may result in some jetting of flow into the EDI or center feedwell. The relative sizes of pipe, their orientation, and proximity to these ports need to be considered in facility design and specifications.

Riser-pipe clarifiers include a mechanical seal between the center column and each return sludge well. If the seal leaks, then there is a loss of or decrease in water level differential between the tank and the well. Lower differentials can result in lower RAS rates or no sludge removal.

The manifold-type hydraulic suction mechanism contains multiple orifices along its radial length. Some clarifiers have a single tube; larger ones have two tubes opposite each other (Figure 12.131).

The orifice openings are sized and spaced in the factory to obtain a near-optimum pattern of collecting solids from the floor. The hydraulic formulas and orifice spacing

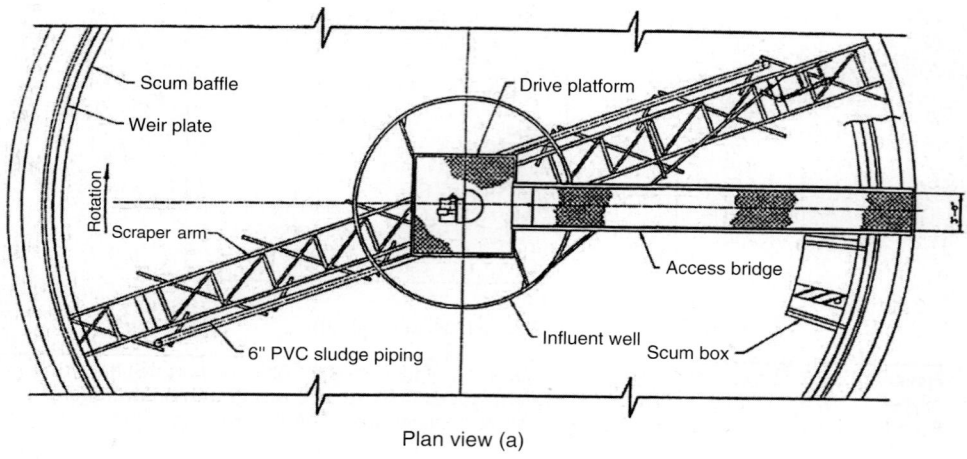

Plan view (a)

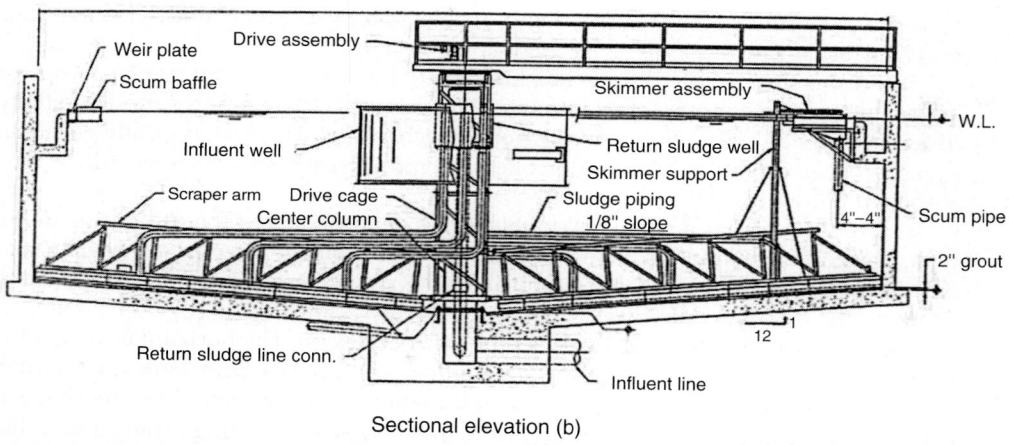

Sectional elevation (b)

Figure 12.130 Hydraulic sludge removal design with suction pipes.

criteria are beyond the scope of this text. Some designs do have adjustable openings; the tank, however, needs to be taken out of service to make the required adjustments. In view of this, most operators do not change the settings, even though adjustments are possible.

With the manifold design, plugging of any particular orifice cannot be checked without dewatering the tank. Instrumentation that compares flow to headloss can be used to determine if some plugging has occurred, and some RAS stations are designed to provide back flushing.

The manifold-type hydraulic suction device has gained in popularity compared to the rise pipe alternative. The main advantage of the manifold is that it can be coupled directly to RAS pumps or a wet well with a substantial head deferential, which allows suction of relatively dense sludges. Supporters of the design that argue lower RAS

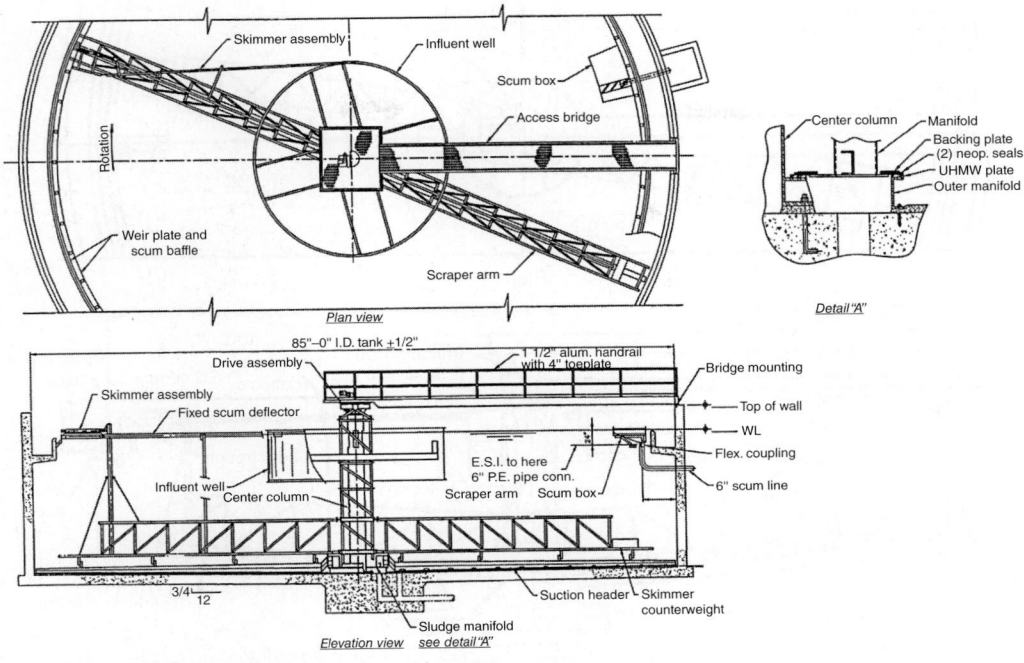

FIGURE 12.131 Hydraulic sludge removal using typical suction header, or tube, design.

flowrates are feasible. Results of a side-by-side test at a pulp and paper facility in Washington reportedly has shown that the manifold type was able to obtain 1% RAS, whereas a parallel riser pipe achieved 0.6% (Ekama et al., 1997). At that facility, WAS is removed exclusively from manifold-equipped clarifiers.

A serious design issue with the manifold device is obtaining a good seal at the bottom. This device requires two seals, one at each side of the rotating collar that moves with the suction tubes. In some early designs, the combination of silt or grit deposition and suction from the RAS pumps led to abrasion and wear of the seal. If wear is excessive or suction adequate, then relatively low TSS water can be sucked through these seals, thereby defeating the purpose of hydraulic suction. This leaking water can dilute the RAS and eventually reduce flows through the orifices. Replacing the seals requires tank dewatering. There are two or more successful modern seal designs. Figure 12.131 (detail A) shows a design with the seals located several inches above the floor to reduce the problem of grit abrasion.

9.5.8.3 Hoppers Traditionally, most U.S. activated sludge clarifiers using scraper mechanisms were equipped with trapezoidal hoppers (Figure 12.132a). Depending on tank size, these hoppers are typically 1 m (3 ft) deep and have walls with slopes of at least 50° above horizontal. Other types of hoppers have been developed to prevent "ratholing" and dilution of the RAS. One type consists of deep conical or annular sludge hoppers as shown in Figure 12.132b. The rotating mechanism has stirrups that reach into the annular hopper to prevent bridging.

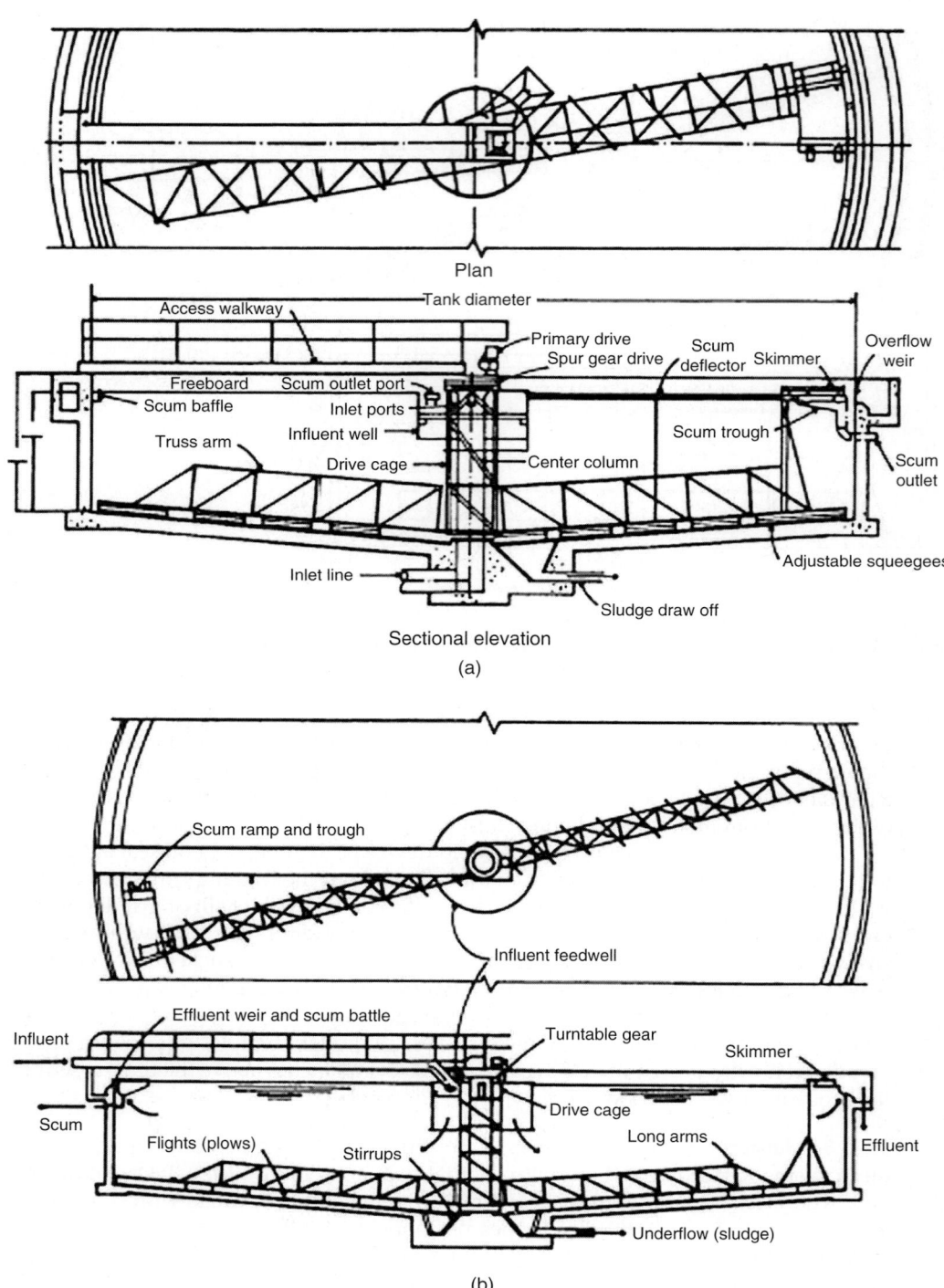

Plan

Tank diameter

Access walkway

Primary drive

Spur gear drive

Scum deflector

Skimmer

Overflow weir

Freeboard

Scum outlet port

Scum baffle

Inlet ports

Influent well

Scum trough

Truss arm

Drive cage

Center column

Scum outlet

Inlet line

Adjustable squeegees

Sludge draw off

Sectional elevation

(a)

Scum ramp and trough

Influent feedwell

Effluent weir and scum battle

Influent

Turntable gear

Skimmer

Scum

Drive cage

Flights (plows)

Stirrups

Long arms

Effluent

Underflow (sludge)

(b)

FIGURE 12.132   Circular clarifier with (a) trapezoid and (b) annular sludge hopper alternatives.

Another design concept was to make the sludge hopper longer and narrower and extend radially outward a distance of up to 25% of the tank radius. A plate with several orifices was used to withdraw the sludge more uniformly over this larger radial distance. Details of this design are given by Albertson and Okey (1992).

Even for tanks with hydraulic suction, some engineers design a separate deep trapezoidal hopper at the bottom of the activated sludge clarifiers from which to waste sludge. They believe that thicker sludge can be achieved in this way.

9.5.8.4 Collection Rings and Drums Within the last decade, sludge and drums and other variations have been developed to assist in removal of sludge plowed to the center by spirals or multiple plows. Two such devices are shown in Figure 12.133a and 12.133b. In the sludge-ring design, an annular area with multiple orifices is provided to remove sludge continuously from a full radius around the center column.

Some engineers have been concerned about plugging potential of sludge-ring orifices, resulting in development of the sludge drum. This design has only two large openings, one at the interior end point of each spiral blade. The opening is fixed relative to the blade end and the drum rotates with the mechanism. There are other minor variations of this fundamental design (Figure 12.133). In this design, only a flat "washer-shaped" top plate of the drum rotates and two inverted U-tubes lift RAS into the drum. Some of these

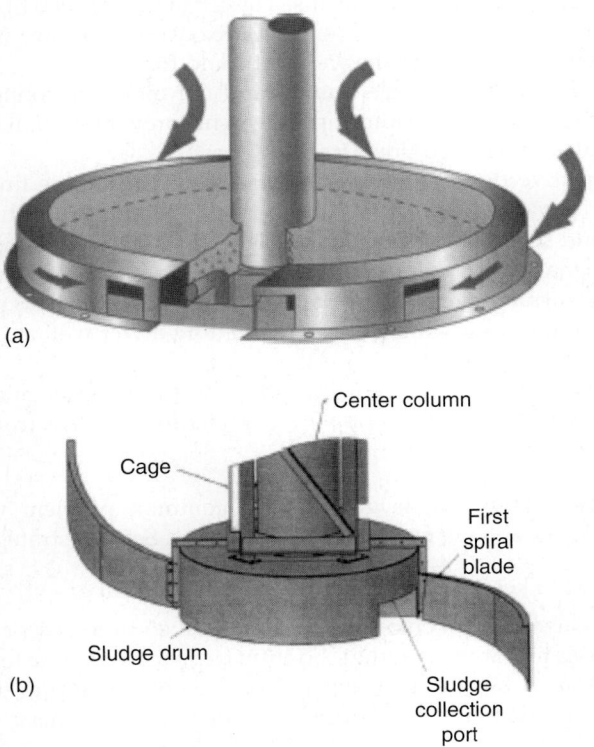

(a)

(b)

FIGURE 12.133 The (a) below top image and (b) below lower to remove solids from activated sludge final clarifiers.

devices are patented and remain relatively novel, but are reported to be effective in specific installations.

9.5.8.5 Drive Location   Most clarifiers in the United States are driven from the center column or from a fixed bridge that spans the full width of the tank. In Europe, it is common to have a drive located at the tank wall. This powers rubber tires that ride on the top of the tank wall and rotate the bridge that spans the tank diameter and is pivoted in the middle.

9.5.8.6 Floor Slopes   Most clarifiers with plow or spiral mechanisms in the United States have a constant floor slope of 1 on 12. Albertson and Okey (1992) have promoted the use of a dual-slope floor that provides a steeper slope in the center for large tanks. This steeper slope provides for greater depth and sludge compaction.

For hydraulic suction clarifiers, not much sloop is needed. Because it is not necessary to move the sludge across the floor, relatively flat floors with slope of 1% or 2% for draining are common.

9.5.9 Other Considerations

9.5.9.1 Return Activated Sludge Pumping   For activated sludge facilities, many designers elect to couple the suction side of RAS pump manifolds to the sludge-removal hoppers or hydraulic suction mechanisms. These pump stations, therefore, do not have a wet well. They do not, however, expose mixed liquor to air, where odors could be released or scum problems in the wet well could form.

It is important that a single pump—and not more—is connected to each circular clarifier. Such single, direct-piping arrangements prevent suction of dilute mixed liquor from one tank and reduced flows from another.

An alternative design provides for each clarifier sludge line to discharge into a wet well by way of a flow-control valve. Such a valve allows independent discharge of sludges and separate control of each. The RAS pumps then operate on a level control signal to maintain the desired level in the wet well. In facilities with many circular clarifiers, this arrangement offers the advantage of fewer RAS pumps. It does, however, generate the disadvantages of maintaining a wet well and its associated scum and odor problems.

Symmetry is never an acceptable principle to use to balance the withdrawal of sludge from parallel clarifier hoppers. Independent control from each hopper is an absolute necessity.

9.5.9.2 Algae Control   Algae growth is a common problem with many activated sludge clarifiers that have weirs and open troughs. Several strategies are available:

- Covers to keep the launder areas dark to prevent growth.
- Periodic release of chlorine solution (derived from gaseous chlorine) through a diffuser line at the launders. Sodium hypochlorite has a tendency to precipitate calcium carbonate and clog the orifices of a chlorine diffusion line. Lower concentration of hypochlorite solutions can reduce this problem.
- Physical control using brushes and sprays. For circular clarifiers, brushes can be attached to the rotating skimmer arms. This has been successful in many applications but does require periodic adjustment and replacement of the

brushes. Rotating water jet spray systems are also available for circular clarifiers. This can clean irregular shapes of tanks and weirs, including cantilevered, double-weir launders, and could be used in rectangular clarifiers as a fixed-grid system.

9.6   Control Strategy and Facilities Design

The performance of activated sludge clarifiers has a significant effect on effluent water quality, aeration basin MLSS concentration and performance, and efficiency of solids-handling facilities. The operator needs a control strategy and information to make correct, timely decisions to ensure efficient operation of the activated sludge clarifier system. Elements of strategy include making proactive adjustments to RAS rates, WAS rates to keep the system in balance, and blanket heights; information is needed to determine when to take processes out of service at the most opportune times. In small, conservatively designed facilities, this process can be done with relative information. In large, complex facilities with many units operating in parallel, however, it has been found most cost-effective to reduce manpower by providing increasing sophistication in instrumentation, controls, and automation. A recent survey of more than 110 wastewater facilities at 45 utilities in the United States indicates that only 10% use primary or secondary clarifier sludge blanket-level monitoring instrumentation, and approximately 5% to 10% use suspended solids concentration analyzers (Hill et al., 2001).

More comprehensive coverage of instrumentation and controls of secondary clarifiers can be found in the WEF's MOP FD-8 on clarifier design (2005) and in MOP 11 (2016).

9.6.1   Key Parameters

The key process variables that need to be monitored for efficient and cost-effective clarifier solids control are:

- Effluent water quality;
- RAS rate and concentration;
- WAS rate and concentration;
- Settleability of mixed-liquor solids;
- Quantity of solids retained in the clarifiers (a function of blanket thickness/volume and concentration);
- Quantity of solids in the aeration basin(s);
- Mechanism speed;
- Mechanism mechanical loads and torques; and
- Facility influent flows and waste loads, including fluctuations that may result in transfers of large amounts of solids from the aeration basins to the clarifiers.

Online measurement of effluent TSS or turbidity can be provided for individual or multiple clarifiers for real-time monitoring of performance

Strategies and equipment used to measure and make adjustments, where possible, to these variables are covered below. In some cases, management of certain variables merely consists of monitoring and rarely making adjustments unless warnings are given.

9.6.2   Return and Waste Activated Sludge Strategies

The amount of solids retained in the clarifiers can be monitored by frequent manual or automated measurement of sludge blanket depth and WAS/RAS solids concentration. Monitoring the concentration and volume of the WAS and MLSS retained in the aeration basin is a critical part of any control strategy. The MLSS and WAS concentration and RAS recycle rate measurements are used routinely to adjust WAS withdrawal rate and to maintain consistent steady-state performance.

Manual sludge sample collection followed by gravimetric TSS analysis in the facility laboratory is the most frequently practiced method for monitoring activated sludge MLSS, RAS, and WAS concentration fluctuations. Typically, facility staff collects one to three sludge samples throughout the day and analyzes these samples for TSS by applying standard laboratory methods and procedures or using high-speed centrifuges. Standard laboratory TSS analysis is relatively time-consuming (typically 2 to 3 hours) and, because of time constraints, typically is completed a few times per day at larger facilities and less frequently at smaller facilities. Solids determination by centrifugation of activated sludge samples takes only 15 to 20 minutes and is widely practiced in many facilities. The TSS measurement by calibrated photometric methods also is used.

For more automation, several sludge concentration analyzers are commercially available for online measurement and monitoring of MLSS and WAS concentrations. Continuous solids concentration measurements allow for tracking solids inventory fluctuations in real time and getting more accurate representation of system performance. In addition, automated solids inventory monitoring avoids human errors and reduces the time required for sampling and sample processing. On the other hand, installation and operation of online instrumentation requires additional expense, more specialized operator skills for calibration and servicing of sensors, and frequent equipment field testing to avoid potential errors caused by inaccurate readings or instrument drift. Therefore, the right level of deployment of such equipment is best defined on a site-specific basis.

Higher degrees of automated activated sludge solids inventory control systems have been reported at medium and large WRRFs (Ekster, 1998, 2000; Hinton-Lever, 2000; Samuels, 2000; Wheeler et al., 2001). An activated sludge system performance optimization study was completed at the 93 000-m^3/d (25-mgd) Burlington Skyway WWTP in Halton, Ontario, Canada (Wheeler et al., 2001). Results have proven that automation of secondary clarifier sludge blanket-level monitoring combined with close monitoring and control of activated sludge solids inventory can yield significant improvement of effluent quality at minimal additional expense.

Monitoring sludge blanket, activated sludge solids inventory, and facility influent flow variations improves understanding of clarifier performance. A sudden increase in sludge blanket depth in the secondary clarifiers at typical influent flows and loads and well-operating sludge withdrawal pumps typically indicates deterioration of sludge settleability. Nearly all facilities manually collect samples and perform tests to measure this, although automation was achieved on an experimental basis in 1990 in Japan.

Accurate influent flow measurement and monitoring are essential for efficient control of the clarification process. In many facilities, influent flowrate is used as a main activated sludge system control parameter, and RAS rate is adjusted proportionally. Primary effluent or final effluent flows might be more appropriate if flows are equalized or recycle streams are significant. As a minimum, online flow measurement is recommended for continuous monitoring of facility influent, RAS, and WAS flowrates.

9.6.2.1 Sludge Age Control Maintaining a proper SRT or sludge age is crucial to proper function of any biological process as it is one of the key variables for selecting the microbial community and function of the treatment system. Proper SRT control allows for more consistent MLSS, oxygen demand, and sludge disposal. Sludge can be wasted from the settled sludge in the clarifiers or can be wasted hydraulically from the aeration tanks.

When wasting off the RAS, the decision on how much sludge (volume) to waste depends on accurate measurement of the solids content of the RAS relative to the mixed liquor. The measurement of solids content is typically based on grab samples, which may not always be an accurate representation of conditions within the system. This method of sludge wasting can result in variable SRT

Hydraulic wasting involves wasting a set fraction of the aeration tank volume on a routine basis. The calculation for hydraulic wasting is very straightforward; the fraction wasted per day is the inverse of the target SRT. Wasting hydraulically generates a waste sludge of the same concentration as the mixed liquor in the aeration tank, which is, by definition, thinner than the settled sludge in the clarifier. The thin sludge can generate issues for downstream solids-handling equipment (e.g., hydraulically overloading thickening equipment).

9.6.3 Concentration and Density Measurement

Sludge concentration measurements are used in activated sludge treatment to give operators the information needed to optimize process performance. In the past, sludge concentration instrumentation has found limited application in full-scale facilities, mostly because of inconsistency and inaccuracy. Analyzer instrumentation problems typically were caused by the presence of air bubbles, sensor fouling, or a change in water color. The new generation of equipment has built-in provisions to mitigate these problems and can provide consistent and accurate readings. Reliable sludge blanket and concentration analyzers are commercially available and have a proven track record.

Several different measurement methods or types of equipment are used including light-emitting (optical), ultrasonic, and nuclear solids analyzers. Table 12.34 summarizes key areas of implementation of the various sludge concentration and density measurement technologies. Some of the commercially available analyzers are combined with sludge blanket level detectors, which generally amplify the benefits of automatic sludge monitoring and control. Additional details are provided in the WEF's MOP FD-8 (2005).

Device installation varies depending on the type of instrument and recommended manufacturer installation details. The best location of in-line sludge concentration measurement devices is on a vertical line with an upflow. The solids concentration measurement device must be installed where sludge is well mixed and accurately represents the actual concentration. The operation range of the instrument must match the range of the measured solids concentration. Measurement devices must be located so that they

Typical applications	Accuracy	Notes
Light-emitting (optical) analyzers		
Sludge density meters: Sludge with solids concentrations from 0.1% to 6% solids. Turbidemeters: Wastewater of turbidity between 0.01 and 10 000 NTU (TSS of 1 to 3000 mg/L).	±0.5% of full instrument span	Avoid use for • Primary sludge higher than 3% solids; • Thickened sludge; and • Wastewater with visibly apparent color
Ultrasonic analyzers		
Sludge with solids concentrations from 0.1 to 10% solids.	±0.5% of full instrument span	Avoid use for • Wastewater; and • An MLSS concentration lower than 2 000 mg/L
Nuclear analyzers		
Sludge with solids concentrations greater than 4% and lower than 15% solids.	±0.5% to 1% of full instrument span	Avoid use for • Wastewater • MLSS, RAS, and WAS; and • Low concentration primary sludge

TABLE 12.34   Areas of Application of Sludge Concentration and Density Measurement Equipment

are easy to access and maintain. The analyzer probe must be easily removable for service without shutting down process piping and disturbing the operation of the sludge pumping system.

Sludge sample lines must be large enough to prevent plugging. It is recommended to provide a flushing tap next to the instrument and a sample box, so that samples can be collected manually at the point of instrument installation for calibration purposes.

For large WRRFs, it is preferred that separate measurement devices be installed on sludge withdrawal lines from the individual clarifiers to gain a better control over the operation and performance of these units. Sludge density measurement devices must be installed coupled with sludge flow measurement devices. The displays of the sludge concentration and flowrate measurement instrumentation must be located adjacent to each other for direct observation and comparison.

In-line solids analyzers are used for measuring RAS, and WAS concentrations. When measuring MLSS concentration in aeration basins, analyzer sensors are directly immersed in the basins and mounted on holders off the walls. If a wall-mounted optical solids concentration analyzer is used, then the sensor should be immersed at least 0.04 m (1.5 in.) below the activated sludge tank water surface and should be located a minimum of 0.15 m (6 in.) away from the aeration basin wall. If the wall is bright and reflective, then distance from the sensor to aeration basin wall should be at least 0.3 m (12 in.). Installing the optical sensor too close to a wall can cause infrared light backscatter, resulting in a higher intensity signal. Optimum self-cleaning of immersed suspended solids analyzers is achieved by turning the sensor surface into the flow direction.

9.6.4 Sludge Blanket Depth Measurement

Sludge blanket depth is a key indicator of secondary clarifier performance. The depth of the sludge blanket is the distance from the clarifier surface to blanket top. The blanket thickness is the distance from the top of the sludge blanket to the bottom of the clarifier. The sludge blanket typically varies daily within certain predictable limits because of diurnal flow fluctuations. The blanket depth also may vary because of process changes induced by facility operators. Day-to-day fluctuations in a facility operated under relatively stable conditions are relatively slow and are typically limited to within 0.3 to 0.6 m (1 to 2 ft). Significant and abrupt changes in sludge blanket depth in clarifiers typically are caused either by a large increase in influent flow (transient flow conditions) or by a stoppage or malfunction of the sludge collection and/or withdrawal systems. This parameter also can be influenced by several activated sludge system performance changes, and its fluctuations over time provide critical information about overall health of the activated sludge system.

Sludge blanket depth at full-scale treatment facilities is most typically determined by manual measurements using calibrated clear plastic tube (also named core sampler or sludge judge). The key disadvantage of manual measurement is that it is a discrete sample measurement that gives only a snapshot representation of the sludge blanket level at a given time and location. Variables, such as the sampling location and time, location of the sludge collection mechanism at the time of the measurement, speed of tube descent, ambient light conditions, and subjectivity of operator readings and sampling skills, contribute to the sometimes limited benefits of manual sludge blanket measurement.

One key advantage of the manual plastic tube sampler is that it also allows for collection of a sludge sample in which TSS concentration is representative of the average solids concentration of the sludge blanket, a parameter which could be used to calculate the sludge blanket SRT and, ultimately, to determine the optimum sludge withdrawal rate. Manual plastic tube samplers are reliable, are inexpensive, require little maintenance, and can be easily replaced if damaged. In addition, one manual plastic tube sampler can be used to monitor multiple clarifiers.

Another type of manual equipment for sludge blanket depth measurement is sight glass. This type of sludge blanket finder consists of a sight glass and light source attached at the lower end of a graduated piece of aluminum pipe approximately 38 mm (1.5 in.) in diameter. The sight glass is lowered carefully into the clarifier through the zones of clear liquor and individual particles until the top of the homogenous sludge blanket is observed.

For small facilities and facilities with clarifiers where the sludge blanket does not vary significantly over time, manual sludge blanket depth measurement is generally adequate. If blanket depth is to be used for RAS rate control, manual depth determinations are not adequate during bulking events, and return of solids to the aeration basin will lag behind the influent load.

In medium and large WRRFs with more complex activated sludge and solids-handling systems, installation of instrumentation for continuous sludge blanket measurement interlocked with automated control of secondary sludge withdrawal systems warrants consideration. The benefits of automated sludge blanket depth measurements for WAS and RAS flowrate control have also been documented at a number of full-scale WRRFs (Bush, 1991; Dartez, 1996; Ekster, 1998, 2000; Hinton-Lever, 2000; Hoffman and Wexler, 1996; Rudd et al., 2001; Samuels, 2000).

Automated sludge blanket depth measurement is recommended for WRRFs with significant variations of diurnal influent water quality and quantity and associated frequent shift of sludge blanket levels. In cases where sludge blanket level fluctuations are frequent (changes of more than 0.32 m [1 ft] up and down several times per day) and clarifiers are relatively shallow (SWD of less than 3.66 m [12 ft]), use of VFD motors for the sludge withdrawal pumps is recommended. If VFD-controlled motors are used, then sludge blanket monitoring instrumentation and pump control equipment operation can be interlocked to automatically adjust the clarifier sludge pump withdrawal rate to keep the sludge blanket at an optimum, near-constant level.

Most commercially available sludge blanket-level detectors are based on ultrasonic or optical measurement of sludge concentration. These devices have provisions for compensating sensor measurement for temperature, fouling, and aging.

Ultrasonic sludge blanket level analyzers are subject to "blinding" by gas bubbles generated as a result of uncontrolled denitrification. The gas bubbles, when trapped on the surface of the sonic sensor, alter the readings. Therefore, the ultrasonic sensors must be designed with cleaning provisions. Typically, small utility water pumps are installed on the rail above the sensor or right on the sensor. These pumps typically run intermittently and wash the sensors to maintain accuracy of the instrument readings.

The use of optical sludge blanket level detectors is limited by their higher costs and relatively lower accuracy. Optical analyzers are subjected to interference by accumulation of solids on the analyzer sensor and by light reflection from nearby objects (smooth walls and sunlight reflecting tank and equipment surfaces).

Blanket level detectors must be installed in locations that do not cause interference with normal operation of the sludge collection and removal system. Typically, the stationary sludge blanket meters are installed on the catwalk or on the side rail of the clarifiers. The stationary ultrasonic sludge blanket sensors are mounted 4 to 8 cm (1.5 to 3 in.) below the liquid surface. They are equipped with skimmer guards to protect the sensors from damage.

The best location for measuring sludge blanket depth is where the actual depth is equal to the average clarifier depth. In circular clarifiers with inward-sloping floors, this point is typically one-third of the distance from the outside wall of the clarifier to the center. In rectangular clarifiers, the most appropriate location of routine sludge blanket measurement is typically at the midpoint of clarifier basin length. Because clarifier configuration, type, and size vary, the most representative location for measurement of the average sludge blanket depth should be based on a series of manual sludge blanket measurements at three to five locations along the clarifier radius or length.

Typically, sludge collection arms of circular clarifiers rotate approximately once every 15 to 30 minutes, and the sludge collection mechanism (scraper or suction header) movement disturbs the sludge blanket. If the sludge blanket is measured manually, then depth readings should be taken when the sludge collection mechanism (bridge) is perpendicular to the measuring location. Taking the sludge blanket level measurement at this location minimizes the effect of sludge collection mechanism movement.

Automated sludge blanket level analyzers typically take continuous (several times per second) interface level readings. This enables the operating staff to observe the sludge blanket behavior in real time. Blanket depth measurement instrumentation can produce an "average" sludge blanket level or interface level by averaging the sludge profile at present intervals of 15 to 60 seconds, which eliminates wide changes in the blanket level readings caused by sludge collection rake passage or temporary short-term upsets.

Individual automated sludge blanket level analyzers are recommended to be installed in all clarifier units of the WRRF rather than in only one. Comparison of sludge blanket behavior of the individual units can help identify any potential problems related to uneven flow distribution among the clarifiers, malfunction of sludge collection and withdrawal systems, or other site-specific events that cause individual clarifier units to perform differently.

Selecting the most appropriate instrumentation for the specific application is critical for reliable monitoring and control of clarifier solids concentration and sludge blanket. Most sensors perform well under ideal conditions that manufacturers use to determine their specifications for accuracy, reproducibility, and other key operational parameters. However, sensor performance in the field can be unsatisfactory and require a period of calibration and adjustment to the site-specific conditions of the application (Hill et al., 2001). Instrumentation field testing can provide information needed to select the most appropriate equipment. On-site testing, which can be costly and time-consuming, is the most reliable way to select the best monitoring system. Information from organizations that specialize in evaluation of water and WRRF monitoring equipment, such as the Instrumentation Testing Association, Henderson, Nevada, is recommended.

9.6.5 Equipment and Instrumentation

Instrumentation to monitor clarifier drive units provides protection for clarifier drive gearbox and sludge collection flights/arms. Typical monitored parameters are torque, power, and motion detection.

Torque gauges or motor power provides indirect monitoring of clarifier sludge concentration (Wilkinson, 1997). This approach, however, is relatively simplistic and inaccurate because torque gauges and power monitors are designed to provide protection of the clarifier driver mechanisms against overload rather than to indicate solids concentration.

Suppliers of clarifier drives can provide drive torque monitoring devices. High-torque and high-high torque warning, alarm, and shutoff switches typically are installed at each clarifier drive mechanism. Torque indication can be read from a scale, which is expressed as a percentage of the maximum torque load. A high-torque condition is represented by a torque load at 40% to 50%; a high-high torque is 80% to 85% of the maximum design. The high-torque condition is alarmed, and the high-high condition stops the drive.

Some equipment manufacturers offer positive torque overload protection of the clarifier drives, which allows it to produce a controlled preset maximum torque. The drive will run continuously at this torque, but, when needed, it will safely produce a higher, controlled, short-term running torque to keep the solids in the clarifier moving. When the drive with a positive torque overload protection experiences load demand above the high-high (cutoff) level, it will simply slip without overheating or overstressing. This type of drive overcomes process upset without risk of damaging the sludge collection equipment.

Similar to torque, the clarifier drive motor power (measured in watts) or current draw/amperage (measured in amperes) could be monitored to provide motor and drive overload protection.

The clarifier sludge collection mechanism motion can also be monitored. Typically, loss-of-motion switches are installed on the clarifier drives to detect when they stop moving.

10.0   Suspended-Growth Biological Treatment System Design Example

The following example shows one way of designing an activated sludge system using the information presented earlier in this chapter. The example is brief and does not address all of the analysis, concerns, and safety factors that should be included in a real full-scale facility design. The first section provides hand calculations that can be used for simple nitrification and denitrification systems. The second section uses the IWA's ASM 1 model for the same criteria (Henze et al., 2000).

An existing WRRF is nearing its design capacity and requires a new suspended-growth biological treatment train to allow it to treat future flows projected over the 20-year planning period. Based on flow and load projections and facility mass balances, the design influent loading conditions for the new activated sludge system are:

Flow

Average day	$= 37\ 850\ m^3/d$
Maximum month	$= 45\ 420\ m^3/d$
Peak hour	$= 75\ 700\ m^3/d$

Biochemical oxygen demand (BOD_5) loading

Average concentration	$= 140\ mg/L$
Average mass load	$= 5299\ kg/d$
Maximum month mass load	$= 6359\ kg/d$

Total suspended solids (TSS) loading

Average concentration	$= 90\ mg/L$
Average mass load	$= 3407\ kg/d$
Maximum month mass load	$= 4088\ kg/d$

Total Kjeldahl nitrogen (TKN) loading

Average concentration	$= 30\ mg/L$
Average mass load	$= 1135\ kg/d$
Maximum month mass load	$= 1363\ kg/d$

Total phosphorus (TP) loading

Average concentration	$= 7\ mg/L$
Average mass load	$= 265\ kg/d$
Maximum month mass load	$= 318\ kg/d$

Analysis of historic data indicates the following conditions are applicable to the system:

- Influent BOD has a BOD_{ult}:BOD_5 ratio of 1.46.
- Influent TSS is 80% VSS.

- Influent VSS is 40% nonbiodegradable under the SRT conditions anticipated.
- Design SVI is 150 mL/g.
- Process temperatures are 15°C average, 20°C maximum week, and 12°C minimum week. Historic data indicate maximum month load conditions can occur in either the warm or the cold season.
- For the diffused aeration system under consideration at appropriate diffuser submergence conditions, refer to Table 12.35.
- Over the aeration basin, αF, values of 0.3, 0.4, and 0.7 are anticipated at the first, middle, and final aeration zones, respectively. In addition, 45% of the total oxygen demand is exerted in the first zone, 35% in the second, and 20% in the final.
- A β value of 0.98, pressure of 100 kPa (1 atm), and a $C^*_{\infty, 20}$ of 10.5 mg/L are appropriate for the wastewater and site.

In the base case, the facility has the following effluent requirements, which are not anticipated to change over the planning period:

Maximum monthly BOD$_5$	= 20 mg/L
Maximum weekly BOD$_5$	= 30 mg/L
Maximum monthly TSS	= 20 mg/L
Maximum weekly TSS	= 30 mg/L
Maximum monthly NH$_4$-N	= 3 mg/L
Maximum weekly NH$_4$-N	= 5 mg/L

Airflow Per Diffuser,* m³/h Diffuser (scfm/Diffuser)	Diffuser Density, Diffuser/m² (Diffusers/100 sq ft)	SOTE (%)
0.850 (0.5)	1.94 (18)	29
	2.58 (24)	32
	3.23 (30)	34
	4.84 (45)	38
1.699 (1.0)	1.94 (18)	27
	2.58 (24)	28
	3.23 (30)	30
	4.84 (45)	32
2.55 (1.5)	1.94 (18)	26
	2.58 (24)	27
	3.23 (30)	28
	4.84 (45)	30
	1.94 (18)	26

*Based on US Standard Conditions of temperature of 23°C (68°F) and absolute pressure of 101 kPa (14.7 psia) with a resulting air density of 1.205 kg/m³ (0.075 lb/cf).

TABLE 12.35 Anticipated Aeration Diffuser Characteristics

The goal of the analysis of the base case is to provide initial estimates for aeration basin volume, aeration system capacity, and secondary clarifier size for the new suspended-growth biological treatment system.

10.1   Aeration Basin Volume

Based on review of the discharge limits, target effluent concentrations of 10 mg/L BOD_5, 15 mg/L TSS, and 2 mg/L NH_4-N will be used for design calculation purposes.

Design engineers should document the model and associated parameters used. In this case, for both heterotrophic organic removal and nitrifier conversion of ammonium to nitrate, biological activity will be modeled using processes of growth and endogenous respiration with debris production consistent with the design equations presented in the chapter. Table 12.36 provides the model parameters. This case assumes that hydrolysis of particulate organic material is not rate limiting and that influent organic material converts quickly to soluble constitutes.

It is also necessary to decide upon and document the reactor model. In this case, a single, completely mixed, suspended-growth biological reactor followed by a clarifier without biological activity, similar to the reactor configuration shown in Figure 12.9, is assumed. With these selections for the biological activity model and the reactor model, the analytical solutions presented in the chapter can be used.

Because the system is required to nitrify, the design SRT is calculated for the nitrification requirements by combining eqs 12.10 and 12.11. The cold-weather condition will be the constraining case for aeration basin volume. A design dissolved-oxygen concentration of 2.0 mg/L in the aeration basins and a minimum safety factor of 2.0 will be used.

$$\frac{1}{SRT_{min}} = (\mu_{aut})\left(\frac{S_{NH_4-N}}{K_N + S_{NH_4-N}}\right)\left(\frac{S_{O2}}{K_O + S_{O2}}\right) - b_{aut}$$

$$\frac{1}{SRT_{min}} = (0.44)\left(\frac{2}{0.66 + 2}\right)\left(\frac{2}{0.5 + 2}\right) - 0.06 = 0.205 \text{ d}^{-1}$$

$$SRT_{min} = 4.9 \text{ d}$$

$$SRT_{design} \geq SRT_{min} \times SF = 4.9 \times 2.0 = 9.8 \text{ d}$$

Use SRT_{design} = 10 d.

For preliminary sizing of components, assume the system operates at a 10-day SRT under all conditions. The cold-weather-condition effluent concentrations of organic substrate and ammonium are calculated by:

$$S_{b,e} = K_S \frac{1 + b_h SRT}{SRT \times (\mu_h - b_h) - 1} = 20 \frac{1 + (0.11)(10)}{10 \times (3.5 - 0.11) - 1}$$

$$S_{b,e} = 1.3 \text{ mg BOD}_{ult}/L$$

Effluent soluble organic substrate mass rate for the maximum month flow is

$$(45\ 420 \text{ m}^3/\text{d})(1.3 \text{ g/m}^3)/1000 = 59 \text{ kg BOD}_{ult}/\text{d}$$

Symbol	Description	Unit	Value at 20°C	Value at 12°C
μ_h	Heterotrophic biomass maximum specific growth rate	g VSS/g VSS d	6.0	3.5
K_S	Heterotrophic biomass growth rate expression half velocity coefficient	mg BOD_{ult}/L	20	20
γ_h	Heterotrophic biomass true yield	g VSS/g BOD_{ult}	0.40	0.40
b_h	Heterotrophic biomass decay coefficient	1/d	0.15	0.11
μ_{aut}	Nitrifier biomass maximum specific growth rate	g VSS/g VSS d	0.75	0.44
K_N	Nitrifier biomass growth rate expression ammonium half velocity coefficient	mgN/L	1.0	0.66
K_O	Nitrifier biomass growth rate expression dissolved oxygen half velocity coefficient	mgO_2L	0.50	0.50
γ_{aut}	Nitrifier biomass true yield	g VSS/g NH_4-N	0.15	0.15
b_{aut}	Nitrifier biomass decay coefficient	1/d	0.10	0.06
f_d	Debris (non-biodegradable) fraction coefficient	g VSS/g VSS	0.15	0.15
$i_{bm,ThOD}$	Theoretical oxygen demand of active biomass	g ThOD/g VSS	1.42	1.42
$i_{bm,N}$	Nitrogen content of active biomass	g N/g VSS	0.12	0.12
$i_{bm,P}$	Phosphorus content of active biomass	g P/g VSS	0.024	0.024
$i_{X???,ThOD}$	Theoretical oxygen demand of debris	g ThOD/g VSS	1.42	1.42
$i_{X???,N}$	Nitrogen content of debris	g N/g VSS	0.12	0.12
$i_{X???,P}$	Phosphorus content of debris	g P/g VSS	0.024	0.024
$i_{bm,FSS}$	Fixed solids associated with active biomass	g FSS/g VSS	0.176	0.176
$i_{X???,FSS}$	Fixed solids associated with debris biomass	g FSS/g VSS	0.176	0.176

TABLE 12.36 Example Problem, Parameter Values

Effluent ammonium mass rate for the maximum month flow is:

$$(45\ 420\ \text{m}^3/\text{d})(0.6\ \text{g/m}^3)/1000 = 27\ \text{kg NH}_4\text{-N/d}$$

$$S_{\text{NH4-N},e} = K_N \frac{1+b_{\text{aut}}\text{SRT}}{\text{SRT} \times (\mu_{\text{aut}} \frac{S_{O2}}{K_o+S_{O2}} - b_{\text{aut}})-1} = (0.66)\frac{1+(0.06)(10)}{10 \times (0.44 \frac{2}{0.5+2} - 0.06)-1}$$

$$S_{\text{NH4-N},e} = 0.6\ \text{mg NH}_4\text{-N/L}$$

Aeration basin sizing will be based on cold-weather, maximum month loadings. The design solids productions for the maximum month are:

Fixed solids from the influent = $(4088\ \text{kg TSS/d})(1-0.8) = 818\ \text{kg FSS/d}$

Nonbiodegradable VSS from the influent = $(4088\ \text{kg TSS/d})(0.8)(0.4) =$
1308 kg VSS/d

$$P_{x,\text{Xha}} = \frac{Y_h(\text{Mass BOD}_{\text{ult in}} - \text{Mass BOD}_{\text{ult out}})}{1+b_h\text{SRT}} = \frac{0.4(6359 \times 1.46 - 59)}{1+(0.11)(10)}$$

The production of active heterotrophic biomass is:

$$P_{x,\text{Xha}} = 1757\ \text{kg VSS/d}$$

The production of debris associated with the heterotrophic biomass is:

$$P_{x,\text{Xh},i} = \frac{f_d \times b_h \times \text{SRT} \times Y_h(\text{Mass BOD}_{\text{ult in}} - \text{Mass BOD}_{\text{ult out}})}{1+b_h\text{SRT}}$$

$$P_{x,\text{Xh},i} = \frac{0.15 \times 0.11 \times 10 \times 0.4(6359 \times 1.46 - 59)}{1+(0.11)(10)}$$

$$P_{x,\text{Xh},i} = 290\ \text{kg VSS/d}$$

The fixed solids created through and associated with heterotrophic activity are:

$$P_{x,\text{Xh,FSS}} = (i_{\text{bm,FSS}} \times P_{x,\text{Xha}}) + (i_{i,\text{FSS}} \times P_{x,\text{Xh},i})$$

$$P_{x,\text{Xh,FSS}} = (0.176 \times 1757) + (0.176 \times 290)$$

$$P_{x,\text{Xh,FSS}} = 360\ \text{kg FSS/d}$$

The mass rate of nitrogen assimilated into heterotrophic biomass components is:

N assimilated by heterotrophic biomass = $(i_{\text{bm},N} \times P_{x,\text{Xha}}) + (i_{\text{iN}} \times P_{x,\text{Xh},i})$

N assimilated = $(0.12 \times 1757) + (0.12 \times 360)$

N assimilated = 254 kg N/d

As noted above, it is assumed that all of the influent TKN is hydrolyzed to NH_4-N. The amount of ammonium used by the nitrifiers is then equal to the influent mass TKN rate minus the effluent mass rate and the nitrogen assimilated into the heterotrophic biomass:

$$NH_4\text{-}N \text{ used by nitrifiers} = 1363 - 27 - 254 = 1082 \text{ kg N/d}$$

$$P_{x,Xna} = \frac{Y_{aut}(\text{Mass } NH_4 \text{ Nutilized})}{1 + b_{aut}\text{SRT}} = \frac{0.15(1082)}{1 + (0.06)(10)}$$

The production of active nitrifier biomass is:

$$P_{x,Xna} = 101 \text{ kg VSS/d}$$

The production of debris associated with the nitrifier biomass is:

$$P_{x,Xhni} = \frac{f_d \times b_{aut} \times \text{SRT} \times Y_{aut}(\text{Mass } NH_4 \text{ Nutilized})}{1 + b_{aut}\text{SRT}}$$

$$P_{x,Xhni} = \frac{0.15 \times 0.06 \times 10 \times 0.15 \times (1082)}{1 + (0.06)(10)}$$

$$P_{x,Xn,i} = 9 \text{ kg VSS/d}$$

The fixed solids created through and associated with nitrifier activity is:

$$P_{x,Xn,FSS} = (i_{bm,FSS} \times P_{x,Xna}) + (i_{i,FSS} \times P_{x,Xn,i})$$

$$P_{x,Xn,FSS} = (0.176 \times 101) + (0.176 \times 9)$$

$$P_{x,Xn,FSS} = 19 \text{ kg FSS/d}$$

The total cold weather maximum month VSS and TSS solids productions are then:

$$P_{x,VSS} = 1308 + 1757 + 360 + 101 + 9 = 3535 \text{ kg VSS/d}$$

$$P_{x,TSS} = 3535 \text{ kg VSS/d} + 818 + 360 + 19 = 4732 \text{ kg TSS/d}$$

These results generate a 74% VSS content of the waste sludge solids and mixed liquor. The resulting total observed yield is:

$$\frac{4732 \text{ kg TSS/d}}{6359 \text{ kg BOD}_5/d} = 0.74 \text{ kg TSS/ kg BOD}_5 \text{ removed}$$

Both results should be compared to similar values from historic facility operating data to judge the validity of the assumptions. In some cases, when the situation matches the noted assumptions, Figure 12.14 may be used to calculate solids production as a function of BOD_5 removed.

A review of Figure 12.13 suggests a maximum design MLSS of 2800 mg/L for 12°C and an SVI of 150 mL/g. For this case, a value of 2500 mg/L will be considered as a lower MLSS is conservative for aeration basin sizing.

Using the definition of SRT and maximum month cold weather:

$$\text{SRT} = \frac{V \times \text{MLSS}}{P_x}$$

The required aeration basin volume can be calculated from:

$$V = \frac{(10\,\text{d})\,(4732\,\text{kg TSS/d})}{(2500\,\text{mg/L})}\left(\frac{10^6\,\text{mg}}{1\,\text{kg}}\right)\left(\frac{1\,\text{m}^3}{10^3\,\text{L}}\right) = 18\,928\,\text{m}^3\,\text{total basin volume}$$

This volume corresponds to an HRT of 11.4 hours based on average influent flow. A review of typical activated sludge design parameters listed in Table 12.2 suggests this value is somewhat greater than that provided at typical CAS systems.

10.2   Aeration Requirements

For brevity, aeration requirements for the maximum month condition under cold temperatures are considered below. In the detailed design of an actual activated sludge aeration system, peak day, maximum month, average month, and minimum month under all temperature conditions must be considered to establish maximum system requirements, verify proper turndown design, and ensure that adequate mixing occurs at minimum gas flowrates. A detailed example addressing the entire range of diffused aeration system design issues can be found elsewhere (Wilford and Conlon, 1957). In addition, publications by Johnson (1996) and Johnson and McKinney (1994) present models and techniques of comparing various types of diffusers and their arrangements to predicted performance.

A polynomial equation to predict $K_L a_T$ is presented as a function of velocity gradient, floor coverage, lateral spacing, and airflow rate per unit volume. Coefficients of the model are based on regression analyses of numerous facilities.

The mass of organic substrate oxygen demand equivalents entering the system at maximum month conditions is estimated to be:

$$(6359\,\text{kg BOD}_5/\text{d})(1.46\,\text{kg BOD}_{\text{ult}}/\text{kg BOD}_5) = 9284\,\text{kg BOD/d}$$

Using the nitrogen (-III) oxidation state as the reference, ammonia and ammonium have 0 theoretical oxygen demand (ThOD) and nitrate has a –4.57 mg ThOD/mg N equivalency. Therefore, an estimate of the nitrate production is required to accomplish an oxygen demand balance around the system.

The amount of ammonium used by the nitrifiers has previously been calculated as:

$$NH_4\text{-N used by nitrifiers} = 1082\,\text{kg N/d}$$

A small portion of this ammonium is assimilated into nitrifier biomass components. The mass rate of NH_4-N assimilated into nitrifier biomass is calculated as:

$$\text{N assimilated into nitrifier biomass} = (0.12 \times 101) + 0.12 \times 9 = 13\,\text{kg N/d}$$

The amount of nitrate produced is calculated as:

$$\text{Mass rate of nitrate production} = 1082 - 13 = 1069\,\text{kg NO}_3\text{-N/d}$$

And the ThOD leaving the system as nitrate is:

$$\text{Nitrate oxygen demand} = -4.57 \times 1069 = -4885\,\text{kg ThOD/d}$$

The mass of oxygen demand leaving the system as organic substrate was previously established to be 59 kg BOD_{ult}/d.

Oxygen demand equivalents leaving the system as heterotrophic and autotrophic biomass components are estimated as:

$$\text{Biomass ThOD} = i_{bm,\text{ThOD}} \times (P_{x,\text{Xha}} + P_{x,\text{Xna}}) + i_{Xi,\text{ThOD}} \times (P_{x,\text{Xhi}} + P_{x,\text{Xn},i})$$
$$= 1.42 \times (1757 + 101) + 1.42 \times (360 + 9) = 3162 \text{ kg ThOD/d}$$

The aeration requirements during cold-weather maximum month conditions are calculated using a steady-state oxygen demand balance around the system while recognizing that the aeration system provides an input of oxygen at OTR_f and that oxygen has a 21 g ThOD/g O_2 equivalency. As discussed below, the complete design of the aeration system requires consideration of other conditions in addition to the maximum month case.

At steady state, the oxygen demand input to the system equals the oxygen demand leaving the system resulting in the following balance equation:

$$9284 + OTR_f(-1 \text{ g OD/g O}_2) = -4885 + 59 + 3162$$
$$OTR_f = 9284 + 4885 - 59 - 3162 = 10\ 948 \text{ kg O}_2/\text{d}$$

This oxygen demand is met through aeration of the activated mixed liquor, and this value represents the estimated OTR_f in eq 12.31 for the system at the process conditions under consideration.

Using eq 12.31 and the known and assumed conditions, the OTR_f:SOTR can be calculated for each zone of the aeration basin.

At 12°C, $\tau = 1.19$, and at the site conditions, $\Omega = 1.0$. A process water dissolved-oxygen concentration of 2.0 mg/L will be used for each zone. For the first zone,

$$\frac{OTR_f}{SOTR} = \alpha F\, \Theta^{T-20} \frac{(\tau\beta\Omega C^*_{\infty 20} - C)}{C^*_{\infty 20}}$$

$$\frac{OTR_f}{SOTR} = 0.3 \times 1.024^{(12-20)} \frac{(1.19 \times 0.98 \times 1.0 \times 10.5 - 2.0)}{10.5} = 0.24$$

For the middle zone,

$$\frac{OTR_f}{SOTR} = 0.4 \times 1.024^{(12-20)} \frac{(1.19 \times 0.98 \times 1.0 \times 10.5 - 2.0)}{10.5} = 0.32$$

For the final zone,

$$\frac{OTR_f}{SOTR} = 0.7 \times 1.024^{(12-20)} \frac{(1.19 \times 0.98 \times 1.0 \times 10.5 - 2.0)}{10.5} = 0.56$$

Based on historic oxygen use patterns,

First zone, $OTR_f = 0.45$ (10 948 kg O_2/d) = 4927 kg O_2/d;

Middle zone, $OTR_f = 0.35$ (10 948 kg O_2/d) = 3832 kg O_2/d; and

Final zone, $OTR_f = 0.20$ (10 948 kg O_2/d) = 2190 kg O_2/d.

The SOTR for each zone is then calculated as:

First zone: $\qquad$ $\text{SOTR} = \dfrac{4927 \text{ kg O}_2/\text{d}}{0.24} = 20\,529 \text{ kg O}_2/\text{d}$

Middle zone: $\qquad$ $\text{SOTR} = \dfrac{3832 \text{ kg O}_2/\text{d}}{0.32} = 11\,975 \text{ kg O}_2/\text{d}$

Final zone: $\qquad$ $\text{SOTR} = \dfrac{2190 \text{ kg O}_2/\text{d}}{0.55} = 3982 \text{ kg O}_2/\text{d}$

For this condition, diffuser flux equals 1.699 standard m³/h per diffuser. A diffuser density of 4.84 diffusers/m² is selected for the first zone (SOTE of 32%); 3.23 diffusers/m² for the middle zone (SOTE of 30%); and 1.94 diffusers/m² for the final zone (SOTE of 27%). The required airflow to each zone is then calculated as follows.
First zone:

$$q_s = (20\,529 \text{ kg O}_2/\text{d}) \left(\frac{1 \text{ kg air}}{0.23 \text{ kg O}_2} \right) \left(\frac{1}{1.289\,4 \text{ kg/m}^3} \right) \left(\frac{1}{24} \right) \left(\frac{1}{0.32} \right)$$

$$= 9013 \text{ m}^3/\text{h (standard conditions)}$$

Second zone:

$$q_s = (11\,975 \text{ kg O}_2/\text{d}) \left(\frac{1 \text{ kg air}}{0.23 \text{ kg O}_2} \right) \left(\frac{1}{1.289\,4 \text{ kg/m}^3} \right) \left(\frac{1}{24} \right) \left(\frac{1}{0.30} \right)$$

$$= 5257 \text{ m}^3/\text{h (standard conditions)}$$

Final zone:

$$q_s = (3982 \text{ kg O}_2/\text{d}) \left(\frac{1 \text{ kg air}}{0.23 \text{ kg O}_2} \right) \left(\frac{1}{1.289\,4 \text{ kg/m}^3} \right) \left(\frac{1}{24} \right) \left(\frac{1}{0.27} \right)$$

$$= 1748 \text{ m}^3/\text{h (standard conditions)}$$

The number of diffusers can then be calculated as follows:
First zone:

$$\text{Diffusers} = \frac{5257 \text{ Nm}^3/\text{h}}{1.584 \text{ Nm}^3/\text{h/diffuser}} = 3319 \text{ diffusers}$$

Middle zone:

$$\text{Diffusers} = \frac{9013 \text{ m}^3/\text{h}}{1.699 \text{ m}^3/\text{h/diffuser}} = 5305 \text{ diffusers}$$

Final zone:

$$\text{Diffusers} = \frac{1748 \text{ Nm}^3/\text{h}}{1.584 \text{ Nm}^3/\text{h/diffuser}} = 1104 \text{ diffusers}$$

At a liquid depth of 6.0 m, the floor area of each zone is 1000 m².

The chosen diffuser density can then be checked as follows.
First zone:

$$\frac{5305}{1027} = 5.17 \text{ diffusers}/\text{m}^2$$

Middle zone:

$$\frac{3319}{1027} = 3.32 \text{ diffusers}/\text{m}^2$$

Final zone:

$$\frac{1104}{1027} = 1.10 \text{ diffusers}/\text{m}^2$$

For both the first and the middle zones, calculated diffuser densities are slightly higher than assumed diffuser densities. Use of calculated values will provide a conservative design because SOTE at these higher densities will be higher than initially assumed SOTEs. The calculated diffuser density at the final zone is less than the initial assumption and, therefore, an iterative approach is required to determine a suitable diffuser flux and density. Extrapolating known diffuser density and SOTE and using the methodology above, airflow equals 2364 m³/h (standard conditions) with 1391 diffusers at a density of 1.39 diffusers/m², which is suitable at maximum month conditions.

In the detailed design of the aeration system, additional conditions must also be examined to ensure adequate peak capacity and suitable turndown capabilities are provided. Peak requirements may be based on the peak day of maximum month or, in some cases, on the peak 4-hour oxygen demand. The minimum recommended diffuser flux, associated SOTE, and resultant SOTR must be compared at minimum oxygen demand conditions (minimum month in the initial years) to determine if the minimum diffuser flux allowed for proper diffuser operation will control the minimum aeration rate. In addition, the range of conditions for the initial phase of operation should be considered. These conditions likely will suggest that only a portion of the total diffusers be provided initially with blanks provided for additional diffusers over the life of the aeration basins.

The total airflow for maximum month, cold-weather conditions is:

$$\text{Total maximum month airflow} = 9013 + 5257 + 1748 = 16\ 018 \text{ standard m}^3/\text{h}$$

As noted in the previous section for full-floor grid configurations, a value of 0.6 L/m²·s is often used as a target minimum airflow for mixing purposes. Assuming a minimum recommended diffuser flux of 0.85 standard m³/h per diffuser, at minimum airflow conditions, the proposed system provides the following values:
First zone:

$$\frac{(5305 \text{ diffusers})(0.85 \text{ Nm}^2/\text{h}/\text{diffusers})\left(\dfrac{1000 \text{ L}}{1 \text{ m}^3}\right)}{(1000 \text{ m}^2)(3600 \text{ s}/\text{min})} = 1.25 \text{ L}/\text{m}^2/\text{s}$$

Middle zone:

$$\frac{(3319 \text{ diffusers})(0.85 \text{ Nm}^2/\text{h}/\text{diffusers})\left(\dfrac{1000 \text{ L}}{1 \text{ m}^3}\right)}{(1000 \text{ m}^2)(3600 \text{ h})} = 0.78 \text{ L}/\text{m}^2/\text{s}$$

Final zone:

$$\frac{(1104\,\text{diffusers})(0.85\ \text{Nm}^3/\text{h}/\text{diffuser})\left(\dfrac{1000\ \text{L}}{1\,\text{m}^3}\right)}{(1000\ \text{m}^2)(3600\ \text{s}/\text{h})} = 0.26\ \text{L}/\text{m}^2/\text{s}$$

The first and middle zones indicate that mixing should not be a controlling concern. In the final zone, mixing requirements may play a role in the minimum aeration rate.

10.3 Secondary Clarification

Figure 12.13 was reviewed to find a maximum design MLSS of 2800 mg/L for 12°C and an SVI of 150 mL/g. Sizing of the aeration basin was based on an MLSS of 2500 mg/L because lower MLSS values produce conservative aeration basin volume; higher MLSS are conservative for secondary clarifier sizing, and, therefore, a value of 2800 mg/L will be used. This shows that MLSS, aeration basin volume, and clarifier surface area are interdependent. By using an MLSS greater in clarifier sizing than in aeration basin sizing, however, the system will be able to operate at an SRT greater than the design value at design conditions. Alternative process configurations, including step feed and contact stabilization, result in spatial variations in mixed-liquor concentration along the aeration basin. This can result in lower mixed-liquor concentrations flowing to secondary clarifiers and can affect clarifier sizing.

Design SVI for this system has been established as 150 mL/g. The RAS pumping capacity is assumed to be 20% to 100% of average daily flow (8000 to 40 000 m³/d).

A maximum allowable solids loading rate can be determined iteratively from Figure 12.89 (assuming unstirred SVI is reported) and a solids balance around the conceptual junction point of RAS and influent flows. For the first iteration, an RAS concentration of 9000 mg/L is assumed. From the graph at an SVI of 150 mL/g, the allowable solids loading rate is 186 kg/m² ·d (38 lb/d/sq ft) and the operating underflow is approximately 0.849 m/h (500 gpd/sq ft). The required surface area at peak flow and 2800 mg/L MLSS can be calculated from:

$$\frac{[75\,700\ \text{m}^3/\text{d} + (0.849\ \text{m}/\text{h} \times 24\text{h}/\text{d} \times \text{SA})] \times 2800\ \text{mg}/\text{L} \times \dfrac{1000\text{L} \times 1\,\text{kg}}{1\text{m}^3 \times 10^6\ \text{mg}}}{\text{SA}}$$

$$= 186\ \text{kg}/\text{m}^2\text{d}$$

Solving for SA (clarifier surface area) = 1643 m²

The underflow is then 0.849 m³/m² ·h × 1643 m² × 24 h/d = 33 478 m³/d, which is within the pumping capacity range provided. The MLSS and RAS rates used in the calculation can be checked using a mass balance around the conceptual junction point of influent and RAS flows. The 47 mg/L is the influent nonvolatile and nonbiodegradable volatile solids.

(75 700 m³/d) (47 mg/L) + (33 478 m³/d)(9000 mg/L) = MLSS (75 700 m³/d +33 478 m³/d)

MLSS = 2762 mg/L

This finding indicates that, at design conditions, to maintain a 2800-mg/L MLSS, an RAS flow less than 30 200 m³/d (8 mgd) will be required and, therefore, the surface area calculated is appropriate and conservative.

With the clarifier surface area calculated above, the surface OFR is 1.06 m³/m²·h at average conditions and 2.13 m³/m² ·h at peak-flow conditions. These values are slightly higher than average but within the range of acceptable values. If two clarifiers were used, then this loading would equate to each being approximately 30 m (100 ft) in diameter. Allowance should be made for one tank out of service. Currently some state regulations and U.S. EPA construction grant program guidance recommend that no more than 25% of design capacity be lost with the largest unit out of service. If two tanks are used, then the surface area of each can be increased by 50% to meet this criterion and the diameter of each would increase accordingly to approximately 38 m (125 ft). This redundancy allowance would result in a total secondary clarification surface area of 2268 m². A final step in circular clarifier sizing would be to check availability of the modular size or manufacturers' preferences of the proposed sludge collection system and to provide appropriately dimensioned tanks of the next larger diameter to those calculated above.

10.4   Summary for the Base Case

The new 37 850-m³/d (10-mgd) suspended-growth biological treatment train will achieve BOD_5 and ammonia plus ammonium effluent limits with the following unit process sizing:

- Total aeration basin volume of 18 008 m³;

- A maximum month, cold-weather-condition oxygen requirement of 10 882 kg O_2/d met using a diffused aeration system with total airflow rate of 17 913 standard m³/h; and

- Secondary clarification surface area of 1483 to 2268 m², depending on the number of units provided and degree of reliability/redundancy appropriate.

10.5   Addressing Nutrients

The base case addressed design of a system with effluent BOD_5 and ammonia plus ammonium-N limits. In many cases, activated sludge systems are now also called upon to remove nitrogen and phosphorus. As discussed earlier, nutrient removal processes use configurations with multiple-staged anaerobic, anoxic, and aerobic zones. Successful design and operation of nutrient removal systems require that the complex interaction between numerous competing processes and components is understood and addressed. The complexity of biological activity and reactor behavior and the number of variables important to nutrient removal require the use of computer models for detailed solutions. Some insight, however, can be developed with simple hand calculations.

For example, the potential effect of a total effluent nitrogen limit of 10 mg N/L on the base system can be calculated. A simplistic calculation can be used to identify a target amount of denitrification to be accomplished to meet this potential total effluent nitrogen limit. As previously calculated, the effluent ammonium plus ammonia-N at the maximum month cold-weather condition was 0.6 mg N/L. Allowing for an effluent TSS of 15 mg/L at 80% VSS and 12% N of VSS results in 1.4 mg N/L of effluent particulate N. To meet the effluent TN requirement, the effluent nitrate concentration must be below the effluent TN limit (10 mg N/L) minus the sum of the effluent ammonia (0.6 mg N/L) and effluent particulate N. That is, the allowable effluent nitrate concentration is calculated as 10 mg N/L − (0.6 mg N/L + 1.4 mg N/L) = 8 mg NO_3^--N/L. For the maximum month flow, this concentration results in 363 kg N/d of nitrate leaving the system.

The amount of nitrate produced was calculated previously as 1085 kg N/d; this nitrate leaves the system through the effluent and WAS flows, which equal the influent flow. The effluent nitrate concentration at the design condition considered under the base case was:

$$(1085 \text{ kg N/d})(10^6 \text{ mg/kg})/(45\ 420 \text{ m}^3/\text{d} \times 1000 \text{ L/m}^3) = 24 \text{ mg/L}$$

Thus to meet the TN limit, the effluent nitrate must be reduced from 24 mg N/L to less than 8 mg N/L, and a mass of greater than 722 kg N/d of nitrate must be denitrified.

Using a mass balance around the aerobic zone, and the bifurcations following it, for a preanoxic configuration with complete nitrate removal in the preanoxic zone, the mass of nitrate returned to the anoxic zone is found to be equal to:

$$\text{Mass returned} = \text{mass produced} \times (Q_{RAS} + Q_{RECYLCE})/(Q_{influent} + Q_{RAS} + Q_{RECYCLE})$$

In this case, the mass desired to be returned, the mass produced, and the influent flow are known. Using Q_{RAS} of 30 200 m³/d, solving for $Q_{RECYCLE}$ results in a minimum recycle flow of 60 139 m³/d, or 132% of the maximum month influent flow.

The Refling-Stensel equation, with an adjustment for temperature, can be used to calculate a preliminary estimate of a required anoxic zone volume. Previously, design MLSS was determined to be 2500 mg/L with a VSS content of 74%, yielding the design MLVSS of 1850 mg VSS/L for the condition under consideration.

$$r_{x,NO,12oC} = \left[0.03 \times (\text{F:M}) + 0.029\right] \times 1.06^{(12-20)}$$

$$\frac{722 \text{ kg N/d}_{removed}}{V_{anoxic} \times 1.850 \text{ kg VSS/m}^3} = \left[0.03 \frac{6359 \text{ kg BOD}_5/\text{d}}{V_{anoxic} \times 1.850 \text{ kg VSS/m}^3} + 0.029\right] \times 1.06^{(12-20)}$$

$$V_{anoxic} = \frac{[722 \text{ kg N/d}_{removed} \times 1.06^{(20-12)}] - 0.03 \times 6359 \text{ kg BOD}_5/\text{d}}{0.029 \times 1.850 \text{ kg VSS/m}^3}$$

Solving for V_{anoxic} results in:

$$V_{anoxic} = 17\ 894 \text{ m}^3$$

This volume equals an HRT of 11.4 hours based on average influent flow and reflects the low denitrification rate associated use of slowly biodegradable substrate and biomass decay reported as the basis of the equation (Grady et al., 2011). This is a large volume that, if implemented as an unaerated zone before the aerated zone of the base case, would affect kinetic and mass balance calculations results for the base case. While this calculation illustrates the possible significance of a new TN limit, it should be used as an indicator of the gravity of the situation and not as the final design or planning value.

Designers also can use the heterotrophic growth rate expression to estimate potential maximum rate of denitrification at an optimum point in the preanoxic zone. Previous calculations generated 940 mg VSS/L of active heterotrophic biomass (i.e., 2500 mg/L × 1692 kg VSS/d active biomass/4732 kg TSS/d total = 894 mg VSS/L of active heterotrophic biomass). A version of eq 12.18 appropriate to the parameter unit basis can be used to estimate the optimal point rate of nitrate use by the heterotrophic growth based on three assumptions: (1) kinetic and stoichiometric parameters from the base case are used, (2) substrates (readily biodegradable organics and nitrate) are not rate limiting in an early part of the zone and dissolved oxygen is not present, and (3) a fraction of heterotrophs are using nitrate, η, of 0.5.

$$r_{yNO,12oC} = \frac{Y_h \times ibm,ThOD\text{-}1}{2.86 \times Y_h} \times \eta \times \mu_h \frac{S_b}{K_s + S_b} \frac{S_{NO_3\text{-}N}}{K_{NO_3\text{-}N} + S_{NO_3\text{-}N}} \frac{K_{o,h}}{K_{o,h} + S_o} \times X_{ha}$$

$$r_{yNO,12oC} = \frac{0.4 \times 1.42\text{-}1}{2.86 \times 0.4} \times 0.5 \times 3.5 \times (1) \times (1) \times (1) \times (940 \text{ mg/L}) = 591 \text{ mg NO}_3\text{-N/L} \cdot \text{d}$$

The negative result indicates nitrate is being used because concentration is decreasing. This value represents a potential maximum value that might be achieved. The comparable value nitrate use rate from the Refling-Stensel equation computed above is 40 mg NO_3-N/L·d. This exercise illustrates the potential for process configuration optimization that can be developed with use of more sophisticated modeling tools.

A final, simple calculation that can be used to investigate the implications of preanoxic denitrification is the possible effect on oxygen requirement and aeration system. As noted previously, when the nitrogen (-III) oxidation state is taken as the reference, nitrate has a −4.57 mg ThOD/mg N equivalency. In denitrification, neglecting any minor assimilative nitrate use, dinitrogen is produced in kind with nitrate consumption and dinitrogen has a −1.71 mg ThOD/mg N equivalency. Thus the net ThOD change with nitrate consumption in the denitrification process is −2.86 mg ThOD/mg N. The 722 kg N/d of nitrate removal considered in these calculations to achieve a 10 mg N/L total effluent nitrogen is equivalent to 2065 kg ThOD/d. In the base case without denitrification, it was determined the aeration system would have to provide 10 948 kg O_2/d for the design situation considered. If preanoxic denitrification were implemented, it might reduce the oxygen requirement and aeration system usage by approximately 20%.

As noted above, complex models are required for a thorough analysis of nutrient removal systems. The following paragraphs illustrate the principles of modeling. For this example, the ASM 1 model will be used (Henze et al., 2000). The ASM models differ in some fundamental ways from the hand calculations described above:

1. It is a death/regeneration model. This means the decay rates and fraction of particulate decay products used are lower than the conventional Monod approach.

2. It is used for single, completely stirred tank reactors (CSTRs). Thus, for long length-to-width tanks, or tanks in series, multiple ASM models must be run to characterize the performance.

3. It can be run in either steady-state or dynamic mode, depending on the needs of the project.

4. Influent parameters of the ASM 1 models are primarily COD. BOD is not used in these models. So influent BOD data must be converted to COD.

The ASM 1 matrix—called a Peterson matrix—is shown in Table 12.37. In this matrix, the rate expressions are listed in the left column, and the state variables are given across

Model Components (i): Component Units: Processes (j):	1. S_I mg COD/L	2. S_S mg COD/L	3. X_I mg COD/L	4. X_S mg COD/L	5. $X_{B,H}$ mg COD/L	6. $X_{B,A}$ mg COD/L	7. X_P mg COD/L	8. S_O mg COD/L	9. S_{NO} mg N/L	10. S_{NH} mg N/L	11. S_{ND} mg N/L	12. X_{ND} mg N/L	13. S_{ALK} meq N/L
1. Aerobic growth of heterotrophs		$-1/Y_H$			$+1$			$-(1-Y_H)/Y_H$		$-i_{XB}$			$-i_{XB}/14$
2. Anoxic growth of heterotrophs		$-1/Y_H$			$+1$				$-(1-Y_H)/2.86Y_H$	$-i_{XB}$			$v_{13,2}$
3. Aerobic growth of autotrophs						$+1$		$-(4.57-Y_H)/Y_H$	$1/Y_A$	$-i_{XB}-1/Y_A$			$v_{13,3}$
4. "Decay" of heterotrophs				$1-f_p$	-1		f_p					$i_{XB}-f_p\,i_{XP}$	
5. "Decay" of autotrophs				$1-f_p$		-1	f_p					$i_{XB}-f_p\,i_{XP}$	
6. Ammonification of soluble organic nitrogen										$+1$	-1		$+1/14$
7. "Hydrolysis" of entrapped organics		$+1$		-1									
8. "Hydrolysis" of entrapped organics nitrogen											$+1$	-1	

Stoichiometric coefficients (v_{ij}):

$v_{13,2} = (1-Y_H)/(14\times 2.86 Y_H) - i_{XB}/14$

$v_{13,3} = i_{XB}/14 - 1/7Y_A$

TABLE **12.37** Activated Sludge Model No. 1 Stoichiometry Matrix (Henze et al., 2000; reprinted with permission from IWA Publishing)

Process	Units	Rate Equation
1. Aerobic growth of heterotrophs	mg COD/L·d	$\mu_H \times \{S_S/(K_S + S_S)\} \times \{S_O/(K_{O,H} + S_O)\} \times X_{B,H}$
2. Anoxic growth of heterotrophs	mg COD/L·d	$\eta_g \times \mu_H \times \{S_S/(K_S + S_S)\} \times \{K_{O,H}/(K_{O,H} + S_O)\} \times \{S_{NO}/(K_{NO} + S_{NO})\} \times X_{B,H}$
3. Aerobic growth of autotrophs	mg COD/L·d	$\mu_A \times \{S_{NH}/(K_{NH} + S_{NH})\} \times \{S_O/(K_{O,A} + S_O)\} \times X_{B,A}$
4. "Decay" of heterotrophs	mg COD/L·d	$b_H \times X_{B,H}$
5. "Decay" of autotrophs	mg COD/L·d	$b_A \times X_{B,A}$
6. Ammonification of soluble organic nitrogen	mg N/L·d	$k_a \times S_{ND} \times X_{B,H}$
7. "Hydrolysis" of entrapped organic	mg COD/L·d	$k_h[(X_S/X_{B,H})/(K_X + (X_S/X_{B,H}))] \times [\{S_O/(K_{O,H} + S_O)\} + \eta_h \times \{K_{O,H}/(K_{O,H} + S_O)\} \times \{S_{NO}/(K_{NO} + S_{NO})\}] \times X_{B,H}$
8. "Hydrolysis" of entrapped organic nitrogen	mg N/L·d	$[X_{ND}/X_S] \times k_h \times [(X_S/X_{B,H})/(K_X + (X_S/X_{B,H}))] \times [\{S_O/(K_{O,H} + S_O)\} + \eta_h \times \{K_{O,H}/(K_{O,H} + S_O)\} \times \{S_{NO}/(K_{NO} + S_{NO})\}] \times X_{B,H}$

TABLE 12.38   Activated Sludge Model No. 1 Process Rate Expressions (Henze et al., 2000; reprinted with permission from IWA Publishing)

the top row. Table 12.38 provides the forms of the rate expressions. These rate expressions are a growth (μ_{max}) or decay (b) term multiplied by a series of Monod-type terms (switching functions) that adjust a particular rate according to the conditions present in the CSTR. Table 12.39 provides the recommended values for the kinetic and stoichiometric constants, which are the values that will be used in this example.

To get the uptake rate (mg/L·d) of a particular component, all stoichiometric coefficients in that component's column are multiplied by the rate expression in that stoichiometric coefficients row.

This matrix of equations must be solved simultaneously to get the effluent quality from that reactor. Computers and numeric solution routines are used in this solution; the math is available through commercial and academic simulators.

10.5.1   Influent Characterization and Fractionation

The BOD, TSS, and VSS data presented earlier can be used to fractionate the wastewater into COD fractions. The relevant parameters are:

Maximum month BOD$_5$ loading	= 6359 kg/d
Maximum month TSS loading	= 4088 kg/d
VSS	= 80%
Nonbiodegradable VSS	= 40% of the VSS
Raw wastewater BOD$_{ult}$/BOD$_5$ ratio	= 1.46
Chemical oxygen demand of VSS	= 1.42 g COD/g VSS
Filtrate nonbiodegradable COD (f_{si})	= 5% of the total COD
Fraction of soluble, readily biodegradable COD	= 60%

Symbol	Unit	Value at 20°C	Value at 10°C
Stoichiometric parameters			
Y_A	mg cell COD formed (mg N oxidized)$^{-1}$	0.24	0.24
Y_H	mg cell COD formed (mg COD oxidized)$^{-1}$	0.67	0.67
f_P	Dimensionless	0.08	0.08
i_{XB}	mg N (mg COD)$^{-1}$ in biomass	0.086	0.086
i_{XP}	mg N (mg COD)$^{-1}$ in products from biomass	0.06	0.06
Kinetic parameters			
μ_H	Day^{-1}	6.0	3.0
K_S	mg COD L^{-1}	20.0	20.0
$K_{O,H}$	mg O$_2$L^{-1}	0.20	0.20
K_{NO}	mg NO$_3$-N L^{-1}	0.50	0.50
b_H	Day^{-1}	0.62	0.20
η_g	Dimensionless	0.8	0.8
η_h	Dimensionless	0.4	0.4
k_h	mg slowly biodegradable COD (mg cell COD×d)$^{-1}$	3.0	1.0
K_X	mg slowly biodegradable COD (mg cell COD)$^{-1}$	0.03	0.01
μ_A	Day^{-1}	0.80	0.3
K_{NH}	mg NH$_3$-N L^{-1}	1.0	1.0
$K_{O,A}$	mg O$_2$ L^{-1}	0.4	0.4
b_A	Day^{-1}	0.05 to 0.15	0.05 to 0.15
k_a	L (mg COD×d)$^{-1}$	0.08	0.04

TABLE 12.39 Activated Sludge Model No. 1 Parameters and Typical Values at Neutral pH (Henze et al., 2000; reprinted with permission from IWA Publishing)

Based on these parameters, the VSS (on a mass basis) is made up of the following fractions:

Total VSS	= 3270 kg/d
Biodegradable VSS	= 1894 kg/d
Nonbiodegradable VSS	= 1263 kg/d

Total COD of the influent is calculated as follows:
Particulate nonbiodegradable COD:

$$X_i = \text{nonbiodegradable VSS (kg VSS)} \times \text{COD of the VSS}$$

$$X_i = 1263 \times 1.42 = 1793 \text{ kg COD/d}$$

Particulate biodegradable COD $= 1894 \times 1.42 = 2690$ kg COD/d

Total COD,

$$COD_T = \left(\frac{BOD_5 \times \dfrac{BOD_{ult}}{BOD_5}}{1 - Y_H \times f_d} + X_i \right)\left(\frac{1}{1 - f_{si}}\right)$$

$$COD_T = \left(\frac{6359 \times 1.46}{1 - 0.4 \times 0.15} + 1793 \right)\left(\frac{1}{1 - 0.05}\right)$$

The soluble COD fractions are therefore:

$$S_i = COD_T \times f_{si} = 12\,284 \times 0.05 = 614 \text{ kg COD/d}$$
$$S_s = COD_T - X_i - X_s - S_i = 12\,284 - 1793 - 2690 - 627$$
$$= 7174 \text{ kg COD/d}$$

The ASM 1 actually groups a portion of the soluble biodegradable COD into the slowly biodegradable fraction (X_s). Therefore, the final X_s and S_s fractions are:

$$X_s = 2690 + (1 - 0.6) \times 7174 = 5549 \text{ kg COD/d}$$

$$S_s = 7174 \times 0.6 = 4304 \text{ kg COD/d}$$

The fractionation of nitrogen follows a similar logic. Unlike the hand-calculation example, it is necessary to determine what the actual ammonia concentration is so that the particulate nitrogen fractions can be determined. For this example, the following parameters were used:

Maximum month TKN loading	= 1135 kg/d
Ammonia fraction of TKN	= 70%
Soluble nonbiodegradable nitrogen	= 4%
Particulate biodegradable nitrogen	= 5%
Nitrogen content of the VSS	= 12%

These parameters result in the following values:

Ammonia loading (S_{NH})	= 1135 × 0.7
Ammonia loading (S_{NH})	= 795 kg/d
Particulate biodegradable nitrogen (X_{nd})	= 1135 × 0.05
Particulate biodegradable nitrogen (X_{nd})	= 57 kg/d
Soluble biodegradable nitrogen (S_{nd})	= 200 kg/d
Particulate non-biodegrable nitrogen	= 38 kg/d
Soluble nonbiodegradable nitrogen	= 1135 × 0.04
Soluble nonbiodegradable nitrogen	= 45 kg/d

The ASM 1 model does not explicitly handle the nonbiodegradable nitrogen fractions. Soluble nonbiodegradable nitrogen simply passes through the CSTR and is added

Component		Units	Value	Units	Value
$X_{B,H}$	Heterotrophic organisms	kg/COD·d	0	g COD/m³	0.0
$X_{B,A}$	Autotrophic organisms	kg/COD·d	0	g COD/m³	0.0
X_P	Particulate products	kg/COD·d	0	g COD/m³	0.0
X_i	Inert particulates	kg/COD·d	1793	g COD/m³	39.5
X_S	Particulate organics	kg/COD·d	5664	g COD/m³	124.7
S_S	Soluble organics	kg/COD·d	4462	g COD/m³	98.2
S_{NH}	Soluble ammonia N	kg N/d	795	g N/m³	17.5
S_{NO}	Soluble nitrate/nitrite N	kg N/d	0	g N/m³	0.0
S_{ND}	Soluble organic N	kg N/d	200	g N/m³	6.2
X_{ND}	Biodegradable particulate organic N	kg N/d	57	g N/m³	1.3
S_O	Oxygen	kg O²/d	0	g O²/m³	0.0
S_{ALK}	Alkalinity	moles/d	220	moles/m³	4.8

TABLE 12.40 ASM 1 Model Influent at 45 420 m³/d Flow

to the effluent TKN. The particulate nonbiodegradable nitrogen is associated with the particulate nonbiodegradable COD and can be handled as a fraction of that material.

The ASM 1 model does not use loads in calculations, which means that this must be converted to concentrations. The maximum month flow of 45 420 m³/d results in the values shown in Table 12.40.

10.5.2 Modification of Decay Rate for Activated Sludge Models

The decay rate (*b*) and fraction of particulate products (f_D) must be changed to reflect use in a death/regeneration model. This is done with the following equations:

$$f_{D,asm} = \frac{f_D \times (1 - Y_H)}{1 - f_D \times Y_H}$$

$$b_{asm} = \frac{b}{1 - Y_H \times (1 - f_D)}$$

10.5.3 Nitrification Only

The first design example looked at nitrification in a single CSTR and resulted in an SRT of 10 days and targeted an MLSS value of 2500 mg/L. It also was assumed that the recycled activated sludge rate would be 50% of the influent flowrate.

By iteratively changing the basin volume until the target MLSS was achieved, the ASM model resulted in a basin volume of 18 600 m³ compared to the hand calculation of 18 008 m³. Table 12.41 provides the full results for the model. The ASM model predicted 0.4 mg NH_4-N/L; the hand calculation provided 0.6 mg NH_4-N/L. In addition, the ASM model predicted an actual oxygen demand (AOR) of approximately 10 400 kg/d; the hand calculation gave 10 800 kg/d. All values are well

				Reactor Concentrations						Secondary
			Feed	#1	#N/A	#N/A	#N/A	#N/A		Effluent
$X_{B,H}$	Heterotrophic organisms	g COD/m³	0.0	1077						8.6
$X_{B,A}$	Autotrophic organisms	g COD/m³	0.0	74						0.6
X_P	Particulate products	g COD/m³	0.0	510						4.1
X_I	Inert particulates	g COD/m³	39.5	954						7.6
X_S	Particulate organics	g COD/m³	126.3	15						0.1
S_S	Soluble organics	g COD/m³	100.6	3.8						3.8
S_{NH}	Soluble ammonia N	g N/m³	17.5	0.37						0.37
S_{NO}	Soluble nitrate/nitrite N	g N/m³	0.0	15						14.9
S_{ND}	Soluble organic N	g N/m³	6.2	0.4						0.4
X_{ND}	Biodegradable particulate organic N	g N/m³	1.3	0.9						0.0
S_O	Oxygen	g O₂/m³	0.0	2.0						2.0
S_{ALK}	Alkalinity	moles/m³	4.8	2.6						2.6
VSS	Volatile suspended solids	g/m³		1850						14.8
TSS	Total suspended solids	g/m³		2501						20.0
BOD_5	Biological oxygen demand	g O₂/m³								7.5
Oxygen requirements										**Total**
Oxygen uptake rate		mg/L·h		23						
Actual oxygen demand (AOR)		kg/d		10 367						10 367

TABLE 12.41   Single Tank with Nitrification

within the expected accuracy given the different kinetic and stoichiometric parameters used.

The example problem then segregates the aerobic zone into three sections to determine the staging of the aeration system. For this to be a real exercise, that means that each grid is equivalent to a single CSTR; otherwise, each grid would have the same aeration demand. Hence, the ASM model was set up with three tanks in series to better determine the oxygen split in the system. Results are shown in Table 12.42.

The oxygen split predicted by the ASM model is 57%, 26%, and 17%, compared to the design example assumption of 45%, 35%, and 20%. This shows that the ASM model is able to provide more specific information on the design of the aeration system. The staged system also results in a lower overall effluent ammonia level (0.1 mg/L versus 0.4 mg/L for the single tanks system).

10.5.4 Denitrification System Design

The example also designs an anoxic zone into the system. The goal is to achieve an effluent TN value of 8 mg/L. If the aerobic SRT is kept the same (i.e., 10 days), then the total system SRT must go up by the same proportion. The nitrified mixed-liquor return

Design of Water Resource Recovery Facilities

			Feed	Reactor Concentrations					Secondary Effluent
				#1	#2	#3	#N/A	#N/A	
$X_{B,H}$	Heterotrophic organisms	g COD/m³	0.0	1098	1089	1068			8.6
$X_{B,A}$	Autotrophic organisms	g COD/m³	0.0	74	74.7	74.9			0.6
X_P	Particulate products	g COD/m³	0.0	499	503	508			4.1
X_I	Inert particulates	g COD/m³	39.5	954	953.8	953.8			7.7
X_S	Particulate organics	g COD/m³	126.3	30	12.3	8.9			0.1
S_S	Soluble organics	g COD/m³	100.6	7.4	2.7	2.0			2.0
S_{NH}	Soluble ammonia N	g N/m³	17.5	2.65	0.3	0.1			0.10
S_{NO}	Soluble nitrate/nitrite N	g N/m³	0.0	11	13.9	15.2			15.2
S_{ND}	Soluble organic N	g N/m³	6.2	0.6	0.4	0.3			0.3
X_{ND}	Biodegradable particulate organic N	g N/m³	1.3	1.3	0.9	0.8			0.0
S_O	Oxygen	g O₂/m³	0.0	2.0	2.0	2.0			2.0
S_{ALK}	Alkalinity	moles/m³	4.8	3.2	2.9	2.8			2.8
VSS	Volatile suspended solids	g/m³		1868	1853	1839			14.8
TSS	Total suspended solids	g/m³		2524	2504	2486			20.0
BOD₅	Biological oxygen demand	g O₂/m³							6.4
Oxygen requirements									Total
Oxygen uptake rate		mg/L·h		40	18	12			
Actual oxygen demand (AOR)		kg/d		6009	2675	1790			10 475

TABLE 12.42 Three Tanks in Series with Nitrification

rate of 132% of the influent flow remains the same; by trial and error, the anoxic volume was increased until the desired 8 mg/L TN was achieved. The resultant anoxic volume was 3000 m³ compared to 17 000 m³ calculated in the design example, and MLSS was calculated to be approximately 2400 mg/L at a total SRT of 11.2 days. Table 12.43 shows the results of this simulation. For comparison purposes, the three nitrate uptake rates are shown below:

- Refling-Stensel = –40 mg NO_X-N/L·d;
- Second method = 591 mg NO_X-N/L·d; and
- ASM 1 model = –149 mg NO_X-N/L·d.

			Feed	Reactor Concentrations				#N/A	Secondary Effluent
				#1	#2	#3	#4		
$X_{B,H}$	Heterotrophic organisms	g COD/m³	0.0	961	977	977	969		6.0
$X_{B,A}$	Autotrophic organisms	g COD/m³	0.0	70	70.4	70.7	70.8		0.4
X_P	Particulate products	g COD/m³	0.0	527	529	531	534		3.3
X_i	Inert particulates	g COD/m³	39.5	954	953.8	953.8	953.8		5.9
X_S	Particulate organics	g COD/m³	126.3	58	27.8	14.0	9.3		0.1
S_S	Soluble organics	g COD/m³	100.6	15.4	5.5	3.2	2.3		2.3
S_{NH}	Soluble ammonia N	g N/m³	17.5	6.68	2.2	0.4	0.1		0.12
S_{NO}	Soluble nitrate/nitrite N	g N/m³	0.0	0.7	3.8	5.7	6.4		6.4
S_{ND}	Soluble organic N	g N/m³	6.2	0.6	0.4	0.4	0.3		0.3
X_{ND}	Biodegradable particulate organic N	g N/m³	1.3	2.0	1.4	1.0	0.8		0.0
S_O	Oxygen	g O_2/m³	0.0	0.0	2.0	2.0	2.0		2.0
S_{ALK}	Alkalinity	moles/m³	4.8	5.1	4.6	4.3	4.2		4.2
VSS	Volatile suspended solids	g/m³		1809	1801	1792	1785		11.1
TSS	Total suspended solids	g/m³		2444	2433	2422	2413		15.0
BOD_5	Biological oxygen demand	g O_2/m³							5.0
Oxygen requirements									**Total**
Oxygen uptake rate		mg/L·h			31	19	12		
Actual oxygen demand (AOR)		kg/d			4736	2915	1848		9500

TABLE 12.43   Denitrification System Simulation Results with 132% Nitrified Recycle

Table 12.43 provides the reduced overall AOR (9500 kg/d versus 10 500 kg/d) as a result of the denitrification; the AOR requirement distribution has flattened out in the three aerobic reactors to 50%, 31%, and 19% because of the soluble uptake in the anoxic zones.

In this particular system, the denitrification rate is limited by the available nitrate in the anoxic zone. The nitrate level of 0.7 mg N/L in the anoxic zone is close to the half saturation value ($K_{NO} = 0.5$) for nitrate. Therefore, additional denitrification could be obtained by increasing the nitrified recycle rate (NRCY). In Table 12.44, the NRCY was increased until the anoxic zone nitrate reached approximately 1 mg/L, resulting in an NRCY of 190% of the influent flowrate. This resulted in a drop in the effluent TN to 6.33 mg/L. The AOR also dropped to 9400 kg/d.

			Feed	#1	#2	#3	#4	#N/A	Secondary Effluent
				Reactor Concentrations					
$X_{B,H}$	Heterotrophic organisms	g COD/m³	0.0	964	976	976	970		6.0
$X_{B,A}$	Autotrophic organisms	g COD/m³	0.0	69	69.3	69.7	69.8		0.4
X_P	Particulate products	g COD/m³	0.0	528	530	531	533		3.3
X_i	Inert particulates	g COD/m³	39.5	954	953.8	953.8	953.8		5.9
X_S	Particulate organics	g COD/m³	126.3	49	25.6	14.0	9.6		0.1
S_S	Soluble organics	g COD/m³	100.6	12.6	5.2	3.2	2.4		2.4
S_{NH}	Soluble ammonia N	g N/m³	17.5	5.48	2.0	0.4	0.1		0.13
S_{NO}	Soluble nitrate/nitrite N	g N/m³	0.0	1.0	3.6	5.2	5.8		5.8
S_{ND}	Soluble organic N	g N/m³	6.2	0.6	0.4	0.4	0.4		0.4
X_{ND}	Biodegradable particulate organic N	g N/m³	1.3	1.8	1.4	1.0	0.8		0.0
S_O	Oxygen	g O₂/m³	0.0		2.0	2.0	2.0		2.0
S_{ALK}	Alkalinity	moles/m³	4.8	5.1	4.6	4.4	4.4		4.4
VSS	Volatile suspended solids	g/m³		1804	1798	1791	1785		11.1
TSS	Total suspended solids	g/m³		2438	2429	2420	2412		15.0
BOD₅	Biological oxygen demand	g O₂/m³							5.1
Oxygen requirements									**Total**
Oxygen uptake rate		mg/L·h			30	20	13		
Actual oxygen demand (AOR)		kg/d			4558	2955	1912		9424

TABLE 12.44   Denitrification System with 190% Nitrified Recycle

11.0 References

Abufayed, A. A.; Schroeder, E. D. (1986) Kinetics and Stoichiometry of SBR Denitrification with a Primary Sludge Carbon Source. *J. Water Pollut. Control Fed.*, **58**, 398–405.

Abu-gharrah, Z.; Randall, C. (1991) The Effect of Organic Compounds on Biological Phosphorus Removal. *Water Sci. Technol.*, **23**, 585.

Aerzen (2017) How to Select the Most Effective Blower Technology for Wastewater Applications. *Water Online.* Aerzen, n.d. Web. January 23, 2017.

Ahn, J. H.; Kim, S.; Park, H.; Rahm, B.; Paguilla, K.; Chandran, K. (2010) N₂O Emissions from Activated Sludge Processes, 2008–2009: Results of a National Monitoring Survey in the United States. *Environ. Sci. Technol.*, **44**, 4505–4511.

Ahn, Y. T.; Kang, S. T.; Chae, S. R.; Lee, C. Y.; Bae, B. U.; Shin, H. S. (2006) Simultaneous High-Strength Organic and Nitrogen Removal with Combined Anaerobic Upflow Bed Filter and Aerobic Membrane Bioreactor. *Desalination*, **202**, 114–121.

Aiyuk, S.; Amoako, J.; Raskin, L.; van Haandel, A.; Verstraete, W. (2004) Removal of Carbon and Nutrients from Domestic Wastewater Using a Low Investment, Integrated Treatment Concept. *Water Res.*, **38**, 3031–3042.

Albertson, O. E. (1995) Clarifier Performance Upgrade. *Water Environ. Technol.*, **7** (3), 56–59.

Albertson, O. E. (2007) The Case for Staging Biological Reactors. *Proceedings of the 80th Annual Water Environment Federation Exposition and Conference*; San Diego, California, October 10–13; Water Environment Federation: Alexandria, Virginia.

Albertson, O. E.; Coughenour, J. (1995) Aerated Anoxic Oxidation Denitrification Process. *J. Environ. Eng.*, **121**, 720.

Albertson, O. E.; Okey, R. W. (1992) Evaluating Scraper Designs. *Water Environ. Technol.*, **1**, 52.

Alleman, J. E.; Irvine, R. L. (1980) Storage-Induced Denitrification Using Sequencing Batch Rector Operation. *Water Res. (G.B.)*, **14**, 1483.

American Public Health Association; American Water Works Association; Water Environment Federation (2012) *Standard Methods for the Examination of Water and Wastewater*, 22nd ed.; American Public Health Association: Washington, D.C.

American Society of Civil Engineers (1997) *Standard Guidelines for In-Process Oxygen Transfer Testing*, ASCE-18-96; American Society of Civil Engineers: Reston, Virginia.

Anderson, N. E. (1945) Design of Final Settling Tanks for Activated Sludge. *Sew. Works J.*, **17**, 50.

Applegate, C. S.; Wilder, B.; Deshaw, J. R. (1980) Total Nitrogen Removal in a Multi-Channel Oxidation System. *J. Water Pollut. Control Fed.*, **52**, 568–577.

Arceivala, S. J. (1995) Experiences with UASB for Sewage Treatment in India. *J. Indian Assoc. Environ. Manag.*, **22**, 90–94.

Arora, M. L.; Barth, E. F.; Umphres, M. B. (1985) Technology Evaluation of Sequencing Batch Reactors. *J. Water Pollut. Control Fed.*, **57**, 867–875.

ASCE (2007) *Measurement of Oxygen Transfer in Clean Water*, ASCE/EWRI 2-06.

ATV (Abwassertechnische Vereinigung) (1991) Arbeitsblatt A 131-Bemessung von einstufigen Belebungsanlagen ab 5000 Einwohnerwerten. Abwassertechnische Vereinigung e.V.: St. Augustin, Germany.

ATV (Abwassertechnische Vereinigung) (1995) Bemessung und Gestaltung getauchter, gelochter Ablaufrohre in Nachklarbecken. *Korrespondenz Abwasser*, **42** (10), 851–852.

ATV (Abwassertechnische Vereinigung) (1973) Arbeitsbericht des ATV-Fachausschusses 2.5 Absetzerfarhen. Die Bemussung der Nächklarbecken von Belebungsanlagen. *Korrespondenz Abwasser*, **20** (8), 193–198.

ATV (Abwassertechnische Vereiningung) (1975) *Leh-und Hanbuch der Abwassertechnik*; Wilhelm Ernst & Sohn: Berlin, Germany.

ATV (Abwassertechnische Vereiningung) (1976) Erläuterungen und Ergänzungen zum Arbeitsbericht des ATV-Fachausschusses 2.5 Absetzverfarhen. Die Bemessung der Nachklärbecken von Belebungsanlangen. *Korrespondenz Abwasser*, **23** (8), 231–235.

ATV (Abwassertechnische Vereiningung) (1988) Sludge Removal Systems for Secondary Sedimentation Basins for Aeration Tanks. Working Report of the ATV Specialist Committee 2.5-Sedimentation Processes. *Korrespondenz Abwasser*, **35** (13/88), 182–193.

Baillod, C. R.; Crittenden, J. C.; Mihelcic, J. R.; Rogers, T. N.; Grady, C. P. L., Jr. (1990) *Critical Evaluation of the State of Technologies for Predicting the Transport and Fate of Toxic Compounds in Wastewater Facilities*; Final Report Project 90-1; Water Pollution Control Federation Research Foundation: Alexandria, Virginia.

Barnard, J. (1976) A Review of Biological Phosphorus Removal in the Activated Sludge Process. *Water SA*, **2**, 136.

Barnard, J. L. (1973a) Biological Denitrification. *Water Pollut. Control (G.B.)*, **72**, 6.

Barnard, J. L. (1973b) Biological Nutrient Removal without the Addition of Chemicals. *Water Res. (G.B.)*, **9** (485), 15–104.

Barnard, J. L. (1974) Cut P and N without Chemicals. *Water Wastes Eng.*, **11**, 33.

Barnard, J. L. (1975) Nutrient Removal in Biological Systems. *Water Pollut. Control (G.B.)*, **74** (2), 143.

Barnard, J. L. (1983) Background to Biological Phosphorus Removal. *Water Sci. Technol.*, **15**, 1.

Barnard, J. L.; Fitzpatrick, J. D.; White, J.; Steichen, M. T. (2007) How Important Is the EDI in Final Clarifiers? *Proceedings of the 80th Annual Water Environment Federation Technical Exposition and Conference*; San Diego, California, October 13–17; Water Environment Federation: Alexandria, Virgina.

Barnard, J.; Dunlap, P.; Steichen, M. (2016) Rethinking the Mechanisms of Biological Phosphorus Removal. *Proceedings of WEF/IWA Nutrient Removal and Recovery*, 2016, July 10–13, Denver, CO.

Barnard, J.; Houweling, D.; Steichen, M. (2011) Fermentation of Mixed Liquor for Removal and Recovery Phosphorus. *Proceedings of WEF/IWA/WERF Nutrient Recovery and Management*, 2011, January 9–12, Miami, FL.

Barnard, J.; Stevens, G.; Leslie, P. (1985) Design Strategies for Nutrient Removal Plant. *Water Sci. Technol.*, **17** (11/12), 233–242.

Barnes, D.; Bliss, P. J. (1983) *Biological Control of Nitrogen Wastewater Treatment*; E. and F.N. Spon: London.

Barnes, D.; Foster, C.; Johnstone, D. W. M. (1983) *Oxidation Ditches in Wastewater Treatment*; Pitman Publishing Inc.: Marshfield, Massachusetts.

Barth, E. F.; Stensel, H. D. (1981) International Nutrient Control Technology for Municipal Effluents. *J. Water Pollut. Control Fed.*, **53**, 1691–1701.

Batchelor, B. (1982) Kinetic Analysis of Alternative Configurations for Single-Sludge Nitrification/Denitrification. *J. Water Pollut. Control Fed.*, **54**, 1493–1504.

Baur, R.; Bhattarai, R.; Benisch, M.; Neethling, J. (2002) Primary Sludge Fermentation - Results from Two Full-Scale Pilots at South Austin Regional (TX, USA) and Durham AWWTP (OR, USA). *Proceedings of 75th Annual Water Environment Federation Technical Exposition and Conference*, Chicago, IL, September 28–October 2, Water Environment Federation: Alexandria, Virginia.

Bisogni, J. J.; Lawrence, A. W. (1971) Relationships between Biological Solids Retention Time and Settling Characteristics. *Water Res. (G.B.)*, **5**, 753.

Boon, A. G.; Chambers, B. (1985) Design Protocol for Aeration Systems: U.K. Perspective. *Proceedings of the Seminar Workshop Aeration Systems Design, Testing, Operations, and Control*, Boyle, W. C., Ed., EPA-600/9-85-005; U.S. Environmental Protection Agency, Risk Reduction Engineering Laboratory: Cincinnati, Ohio.

Boyle, W. C. (1996) Fine Pore Aeration: An Update on Its Status. In *Enhancing Design and Operation of Activated Sludge Plants*, Proceedings of the Central States Water Environment Association Seminar, Madison, Wisconsin.

Boyle, W. C.; Stenstrom, M. K.; Campbell, H. J., Jr.; Brenner, R. C. (1989) Oxygen Transfer in Clean and Process Water for Draft Tube Turbine Aerators in Total Barrier Oxidation Ditches. *J. Water Pollut. Control Fed.*, **61**, 1449–1463.

Boyle, W. H. (1975) Don't Forget Side Water Depth. *Water Wastes Eng.*, **12** (6).

Bratby, J. (1984) Brasilia Advanced Wastewater Treatment Plants First of Their Kind in South America to Remove Nutrients. *19th Conference of the Interamerican Sanitary and Environmental Engineering Association (AIDIS)*, Santiago, Chile, 11 to 16 November. (Aidis@aidis.org.br).

Bratby, J.; Fevig, S.; Jimenez, J. (2012b) The Dirty Secret of Primary Sludge Prefermenters. *Proceedings of 85th Annual Water Environment Federation Technical Exhibition and Conference*, New Orleans, LA, September 29–October 3.

Bratby, J.; Jimenez, J.; Parker, D. (2012a) The Best Bang for the Buck: Making the Most of Available Influent Carbon for Nutrient Removal. *Proceedings of 85th Annual Water Environment Federation Technical Exhibition and Conference*, New Orleans, LA, September 29–October 3.

Bratby, J.; Schuler, P.; Willis, J.; Gottschalk, W.; Redmon, D.; Parker, D. (2005) Less than 0.2 mgP/L and 3 mgN/L—Meeting These New Limits While Still Using Existing Infrastructure at OWASA's Mason Farm WWTP. *Proceedings of 78th Annual Water Environment Federation Technical Exhibition and Conference*, Washington, D.C., October 29–November 2; Water Environment Federation: Alexandria, Virginia.

Brdjanovic, D.; Slamet, A.; van Loosdrecht, M.; Hooijmans, C.; Alaerts, G.; Heijnen, J. (1998a) Impact of Excessive Aeration on Biological Phosphorus Removal from Wastewaters. *Wat. Res.*, **32** (1), 200–208.

Brdjanovic, D.; van Loodsdrecht, M.; Hooijmans, C.; Alaerts, G.; Heijnen, J. (1997) Temperature Effects on Physiology of Biological Phosphorus Removal. *J. Environ. Eng.*, **123** (2), 144–153.

Brdjanovic, D.; van Loosdrecht, M.; Hooijmans, C.; Alaerts, G.; Heijnen, J. (1998b) Minimal Aerobic Sludge Retention Time in Biological Phosphorus Removal Systems. *Biotechnol. Bioeng.*, **60** (3), 326–332.

Buer, T. (2002) Enhancement of Activated Sludge Plants by Lamellas in Aeration Tanks and Secondary Clarifiers. *International Waster Association World Water Congress*; International Water Association: London, England.

Burdick, C.; Moss, O. (1980) Operating Experience with the Bardenpho Process at Palmetto, Florida. Paper presented at Annual Joint Technical Conference of the Florida Section American Water Works Association; Florida Pollution Control Association; Florida Water Pollution Control Operators Association; Orlando, Florida.

Bush, D. (1991) Sludge Blanket Monitoring Improves Process Control, Cuts Costs. *Water Eng. Manag.*, **138** (9), 26–27.

Cakir, F. Y.; Stenstrom, M. K. (2007) Anaerobic Treatment of Low Strength Wastewater. *Proceedings of the 80th Annual Water Environment Federation Exposition and Conference* [CD-ROM], San Diego, California, October 10–13; Water Environment Federation: Alexandria, Virginia.

Camp, T. (1955) Flocculation and Flocculation Basins. Trans. *ASCE*, **120** (1), 1–16.

Camp, T. R. (1946) Sedimentation and the Design of Settling Tanks. *Trans. Am. Soc. Civ. Eng.*, **3** (2285), 895.

Camp, T.; Stein, P. (1943) Velocity Gradients and Internal Work in Fluid Motion. *J. Boston Soc. Civil Eng.*, **30** (4), 219–237.

Carvalheira, M.; Oehman, A.; Carvalho, G.; Eusebio, M.; Reis, M. (2014) The Impact of Aeration on the Competition between Polyphosphate Accumulating Organisms and Glycogen Accumulating Organisms. *Water Res.*, **66** (12), 296–307.

Chapin, R. (1993) The Inhibition of Biological Nutrient Removal by Industrial Wastewater Components and High Acetic Acid Loadings. M.S. Thesis, Virginia Polytechnic Institute and State University, Blacksburg, Virginia.

Chatfield, R. B.; Ren, X.; Brune, W.; Schwab, J. (2010) Controls on Urban Ozone Production Rate as Indicated by Formaldehyde Oxidation Rate and Nitric Oxide. *Atmos. Environ.*, **44**, 5395–5406.

Chernicharo, C. A. L. (2007) *Anaerobic Reactors*; IWA Publishing: London, England.

Christensen, M. H.; et al. (1977) Combined Sludge Denitrification of Sewage Utilizing Internal Carbon Sources. *Prog. Water Technol.*, **8**, 589.

Christensen, M. H.; Harremöes, P. (1972) Biological Denitrification in Water Treatment: A Literature Study, Report. No. 2–72; Department of Sanitary Engineering, Technical University of Denmark.

Christopher, S.; Titus, R. (1983) New Wastewater Process Cuts Plants Cost 50 Percent. *Civ. Eng.*, **53** (5), 39.

Coe, H. S.; Clevenger, G. H. (1916) Determining Thickener Unit Areas. *Am. Inst. Mining, Metal, Petrol. Eng. AIME*, **55**, 3.

Crone, B. C.; Garland, J. L.; Sorial, G. A.; Vane, L. M. (2016) Significance of Dissolved Methane in Effluents of Anaerobically Treated Low Strength Wastewater and Potential for Recovery as an Energy Product: A Review. *Water Res.*, **104**, 520–531.

Crosby, R. M. (1980) *Hydraulic Characteristics of Activated Sludge Secondary Clarifiers*, EPA Contract No. 68–03–2782; U.S. Environmental Protection Agency: Cincinnati, Ohio.

Crosby, R. M. (1984a) *Evaluation of the Hydraulic Characteristics of Activated Sludge Secondary Clarifiers*; U.S. Environmental Protection Agency, Office of Research and Development: Washington, D.C.

Crosby, R. M. (1984b) *Hydraulic Characteristics of Activated Sludge Secondary Clarifiers*, EPA-600/2-84-131; U.S. Environmental Protection Agency: Washington, D.C.

Dague, R. R. (1983) A Modified Approach to Activated Sludge Design. Paper Presented at 56th Annual Water Pollution Control Federation Conference, Atlanta, Georgia; Water Pollution Control Federation: Washington, D.C.

Dague, R. R. (1960) Hydraulic of Circular Settling Tanks Determined by Model-Prototype Comparisons. M.S. Thesis, Iowa State University, Ames, Iowa.

Daigger, G. T. (1995) Development of Refined Clarifier Operating Diagrams Using an Updated Settling Characteristics Database. *Water Environ. Res.*, **67**, 95.

Daigger, G. T.; Roper, R. E., Jr. (1985) The Relationship between SVI and Activated Sludge Settling Characteristics. *J. Water Pollut. Control Fed.*, **57**, 859–866.

Daigger, G.; Randall, C.; Waltrip, G.; Romm, E. (1987) Factors Affecting Biological Phosphorus Removal for the VIP Process, a High-Rate University of Cape Town Type Process. *Biological Phosphate Removal from Wastewaters*, Ramadori, R., Ed.; Pergamon Press: Oxford, England.

Daigger, G.; Waltrip, E.; Romm, E.; Morales, L. (1988) Enhanced Secondary Treatment Incorporating Biological Nutrient Removal. *J. Water Pollut. Control Fed.*, **60**, 1833–1842.

Daims, H.; et al. (2015) Complete Nitrification by Nitrospira Bacteria. *Nature*, 528, 504–509.

Dapena-Mora, J.; Campos, L.; Mosquera-Corral, A.; Jetten, M. S. M.; Méndez, R. (2004) Stability of the ANAMMOX Process in a Gas-Lift Reactor and a SBR. *J. Biotechnol.*, **110**, 159–170.

Dartez, J. (1996) Return Activated Sludge Control. *Water Wastes Digest*, **36** (5), 22–24.

Das, D.; Keinath, T.; Parker, D.; Wahlberg, E. (1993) Floc Breakup in Activated Sludge Plants. *Water Environ. Res.*, **65** (2), 138–145.

Dawson, R. N.; Murphy, K. L. (1972) The Temperature Dependency of Biological Denitrification. *Water Res. (G.B.)*, **6**, 71.

deBarbadillo, C.; Barnard, J. (2014) Mixed Liquor Fermentation and Other Alternatives for Providing VFA to EBPR Processes. *Proceedings of 87th Annual Water Environment Federation Technical Exhibition and Conference*, New Orleans, LA, September 27–October 1.

deBarbadillo, C.; Barnard, J.; Tarallo, S.; Steichen, M. (2008) Got Carbon? *Water Environ. Technol.*, **20** (1), 49–53.

Dick, R. I.; Ewing, B. B. (1967) Evaluation of Activated Sludge Thickening Theories. *J. Sanit. Eng. Div., Am. Soc. Civ. Eng.*, **93** (SA4), 9.

Dick, R. I.; Young, K. W. (1972) Analysis of Thickening Performance of Final Settling Tanks. *Proceedings of the 27th Industrial Waste Conference*; Purdue University: West Lafayette, Indiana.

Doyle, M.; Boyle, W. C. (1985) Translation of Clean to Dirty Water Oxygen Transfer Rates. In *Proceedings of Seminar/Workshop Aeration Systems Design, Testing, and Operations Control*, EPA-600/9-85-005; U.S. Environmental Protection Agency: Cincinnati, Ohio.

Draaijer, H.; Maas, J. A. W.; Schaapman, J. E.; Khan, A. (1992) Performance of the 5 MLD UASB Reactor for Sewage Treatment at Kanpur, India. *Water Sci. Technol.*, **25** (7), 123–133.

Eckenfelder, W. W., Jr. (1966) *Industrial Water Pollution Control*; McGraw-Hill: New York.

Eckenfelder, W. W., Jr. (1967) Comparative Biological Waste Treatment Design. *J. Sanit. Eng. Div.; Proc. Am. Soc. Civ. Eng.*, **93** (SA6), 157.

Ekama, G. A.; Barnard, J. L.; Gunthert, F. W.; Krebs, P.; McCorquodale, J. A.; Parker, D. S.; Wahlberg, E. J. (1997) Secondary Settling Tanks: Theory, Modeling, Design, and Operation, Scientific Technical Report No. 6; IWA Publishing: London, England.

Ekama, G.; Dold, P.; Marais, G.vR. (1986) Procedures for Determining Influent COD Fractions and the Maximum Specific Growth Rate of Heterotrophs in Activated Sludge Systems. *Water Sci. Technol.*, **18**, 91.

Ekama, G.; Marais, G.; Siebritz, I. (1984) Biological Excess Phosphorus Removal. In *Theory, Design and Operation of Nutrient Removal Activated Sludge Processes*; Water Research Commission: Pretoria, South Africa.

Ekama, G.; Siebritz, I; Marais, G. (1983) Considerations in the Process Design of Nutrient Removal Activated Sludge Processes. *Water Sci. Technol.*, **15**, 283–318.

Ekster, A. (1998) Automatic Waste Control, Machine Sampling, Calculations; Adjustments Ensure Better Performance of Activated Sludge Systems. *Water Environ. Technol.*, **10** (8), 21–22.

Ekster, A. (2000) Automatic Waste Control. *Water21*, **August**, 53–55.

Erdal, U.; Erdal, Z.; Randall, C. (2002) Effect of Temperature on EBPR System Performance and Bacterial Community. *Proceedings of the 75th Annual Water Environment Federation Technical Exposition and Conference*, Chicago, Illinois, September 28–October 2; Water Environment Federation: Alexandria, Virginia.

Erdal, U.; Erdal, Z.; Randall, C. (2003) The Mechanisms of EBPR Washout and Temperature Relationships. *Proceedings of the 76th Annual Water Environment Federation Technical Exhibition and Conference*, Los Angeles, California, October 11–15; Water Environment Federation: Alexandria, Virginia.

Erdal, U.; Erdal, Z.; Randall, C. (2004) Biochemistry of Enhanced Biological Phosphorus Removal and Anaerobic COD Stabilization. *Proceedings of International Water Association, Biennial Conference*, Marrakech, Morocco, September; International Water Association: London, England.

Esler, J. K. (1984) Optimizing Clarifier Performance. Paper Presented at 61st Annual Water Pollution Control Federation Conference, New Orleans, Louisiana; Water Pollution Control Federation: Washington, D.C.

Filipe, C.; Daigger, G.; Grady, C. (2001a) pH as a Key Factor in the Competition Between Glycogen-Accumulating Organisms and Phosphorus-Accumulating Organisms. *Water Environ. Res.*, **73**, 223–232.

Filipe, C.; Daigger, G.; Grady, C. (2001b) Effects of pH on the Rates of Aerobic Metabolism of Phosphate-Accumulating Organisms and Glycogen-Accumulating Organisms. *Water Environ. Res.*, **73**, 213–222.

Filipe, C.; Meinhold, J.; Jorgensen, S.; Daigger, G.; Grady, C. (2001c) Evaluation of the Potential Effects of Equalization on the Performance of Biological Phosphorus Removal Systems. *Water Environ. Res.*, **73** (3), 276–285.

Florencio, L.; Takayuki, M.; de Morais, J. C. (2001) Domestic Sewage Treatment in Full-Scale UASBB Plant at Mangueira, Recife, Pernambuco. *Water Sci. Technol.*, **44** (4), 71–77.

Foresti, E.; Zaiat, M.; Vallero, M. (2006) Anaerobic Processes as the Core Technology for Sustainable Domestic Wastewater Treatment: Consolidated Applications, New Trends, Perspectives, Challenges. *Rev. Environ. Sci. Biol. Technol.*, **5**, 3–19.

Fries, M.; Rabinowitz, B.; Dawson, R. (1994) Biological Nutrient Removal at the Calgary Bonnybrook Wastewater Treatment Plant. *Proceedings of 67th Annual Water Environment Federation Technical Exposition and Conference*, Chicago, Illinois, October 15–19.

Fukase, T.; Shibeta, M.; Mijayi, X. (1982) Studies on the Mechanism of Biological Phosphorous Removal. *Jap. J. Water Pollut. Res.*, **5**, 309.

Gaudy, A. F.; Kincannon, D. F. (1977) Comparing Design Models for Activated Sludge. *Water Sew. Works*, **123**, 66.

Giesen, A.; van Loosdrecht, M.; Robertson, S.; de Buin, B. (2015) Aerobic Granular Biomass Technology: Further Innovation, System Development and Design Optimization. *Proceedings of the 89th Annual Water Environment Federation Technical Exposition and Conference*, Chicago, Illinois, September 24–28, Water Environment Federation: Alexandria, Virginia.

Giraldo, E.; Pena, M.; Chernicharo, C.; Sandino, J.; Noyola, A. (2007) Anaerobic Sewage Treatment Technology in Latin-America: A Selection of 20 Years of Experiences. *Proceedings of the 80th Annual Water Environment Federation Exposition and Conference* [CD-ROM], San Diego, California, October 10–13; Water Environment Federation: Alexandria, Virginia.

GLUMRB-Great Lakes-Upper Mississippi River Board of State and Provincial Public Health and Environmental Managers (2014) Recommended Standards for Wastewater Facilities. Health Research Inc.: Albany, New York.

Goronszy, M. (1979) Intermittent Operation of the Extended Aeration Process for Small Systems. *J. Water Pollut. Control Fed.*, **51**, 274–287 .

Grady, C. P. L., Jr.; Daigger, G. T.; Love, N. G.; Filipe, C. D. M. (2011) *Biological Wastewater Treatment*, 3rd ed.; CRC Press: Boca Raton, Florida.

Grau, P. (1982) Recommended Notation for Use in the Description of Biological Wastewater Treatment Processes. *Water Res. (G.B.)*, **16**, 1501.

Groves, K. P.; Daigger, G. T.; Simpkin, T. J.; Redmon, D. T.; Ewing, L. (1992) Evaluation of Oxygen Transfer Efficiency and Alpha Factor on a Variety of Diffused Aeration Systems. *Water Environ. Res.*, **64**, 691–698.

Gundelach, J. M.; Castillo, J. E. (1976) Natural Stream Purification under Anaerobic Conditions. *J. Water Pollut. Control Fed.*, **48**, 1753–1758.

Gunthert, F. W.; Deininger, A. (1995) Final Settling Tanks—New Aspects and Design Approaches. *Proceedings of the 7th IAWQ Conference on Design and Operation of Large WWTPs*. Institute for Water Quality and Waste Management, Technical University of Vienna: Austria.

Guo, P. H. M.; Thirumurthi, D.; Jank, B. E. (1981) Evaluation of Extended Aeration Activated Sludge Package Plants. *J. Water Pollut. Control Fed.*, **53**, 33–42.

Haug, R. T.; Cheng, P. P. L.; Hartnett, W. J.; Tekippe, R. J.; Rad, H.; Esler, J. K. (1999) L.A.'s New Clarifier Inlet Nearly Doubles Hydraulic Capacity. *Proceedings of the 72nd Annual Water Environment Federation Technical Exposition and Conference*, New Orleans, Louisiana, October 9–13; Water Environment Federation: Alexandria, Virginia.

Heffernan, B., Van Lier, J. B., & Van der Lubbe, J. (2011). Performance Review of Large Scale Up-flow Anaerobic Sludge Blanket Sewage Treatment Plants. *Water Sci. Technol.*, **63** (1), 100–107.

Helmer, C.; Kuntz, S. (1997) Low Temperature Effects on Phosphorus Release and Uptake by Microorganisms in EBPR Plants. *Water Sci. Technol.*, **37** (4–5), 531–539.

Henze, M.; (1987) A General Model for Single-Sludge Wastewater Treatment Systems. *Water Res. (G.B.)*, **21**, 505.

Henze, M.; Grady, Jr.; C. P. L.; Gujer, W.; Marais, G. V. R.; and Matsuo, T. (1987) Model for the Single-Sludge Wastewater Treatment. *Water Res. (G.B.)*, **21**, 505.

Henze, M.; Gujer, W.; Mino, T.; Van Loosdrecht, M. (2000) *Activated Sludge Models ASM1, ASM2, ASM2d and ASM3*; IWA Publishing: London, England.

Henze, M.; van Loosdrecht, M. C. M.; Ekama, G. A.; Brdjanovic, D. (2008) *Biological Wastewater Treatment Principles, Modeling and Design*; IWA Publishing: London, England.

Hill, B.; Manross B.; Nutt, S.; Davidson, E.; Andrews, J.; Haugh, R.; Reich, R. (2001) State-of-the-Art WWTP Sensing and Control Systems. *Proceedings of the 74th Annual Water Environment Federation Technical Exposition and Conference*, Atlanta, Georgia, October 14–18; Water Environment Federation: Alexandria, Virginia.

Hinton-Lever, A. (2000) Stable Mate for STW Bosses. *Water Waste Treat.*, **43**, 9.

Hoffman Air and Filtration Systems (1986) *Compressor Engineering*, 3rd ed.; East Syracutse, New York.

Hoffman, C. G.; Wexler, H. M. (1996) A Progressive Approach to the Control of Return Sludge Pumping. *Water Eng. Manag.*, **143** (2), 30–31.

Holshoff Pol, L. W.; de Castro Lopes, S. I.; Lettinga, G.; Lens, P. N. L. (2004) Anaerobic Sludge Granulation. *Water Res.*, **38**, 1376–1389.

Horstkotte, G. A.; Niles, D. G.; Parker, D. S.; Caldwell, D. H. (1974) Full-Scale Testing of a Water Reclamation System. *J. Water Pollut. Control Fed.*, **46**, 181–197.

Institute of Water Pollution Control (1980) *Unit Processes: Primary Sedimentation*. Manual of British Practice in Water Pollution Control, ME168JH; Maidstone: Kent, England.

Intergovernmental Panel on Climate Change (2007) Contribution of Working Group III to the Fourth Assessment Report; Cambridge University Press: New York.

Irvine, R. L.; Ketchum, L. H.; Breyfogle, R.; Barth, E. (1983) Municipal Application of Sequencing Batch Treatment. *J. Water Pollut. Control Fed.*, **55**, 484–488.

Irvine, R. L.; Ketchum, L. H.; Breyfogle, R.; Barth, E. F. (1982) Summary Report— Workshop on Biological Phosphorus Removal in Municipal Wastewater Treatment. Annapolis, Maryland.

James M. Montgomery Consulting Engineers (1984) *Technology Evaluation of Sequencing Batch Reactors*; U.S. Environmental Protection Agency, Office of Research Development: Cincinnati, Ohio.

Jenkins, D.; Richard, M.; Daigger, G. (2003) *Manual on the Causes and Control of Activated Sludge Bulking, Foaming, and Other Solids Separation Problems*, 3rd ed.; CRC Press–IWA Publishing: London, UK.

Jeyanayagam, S. (2007) So, You Want to Remove Phosphorous? Part 1-Enhanced Biological Phosphorus Removal. *The Buckeye Bulletin*, December 2007.

Jiang P.; Stenstrom M. K. (2012). Oxygen Transfer Parameter Estimation: Impact of Methodology. *J. Environ. Eng.*, **138** (2), 137–142.

Jimenez, J.; Wise, G.; Burger, G.; Du, W.; Dold, P. (2014) Mainstream Nitrite-Shunt with Biological Phosphorus Removal at the City of St. Petersburg Southwest WRF. *Proceedings of 87th Annual Water Environment Federation Technical Exhibition and Conference*, New Orleans, Louisiana, September 27–October 1.

Johnson, B. R.; Daigger, G. T.; Crawford, G.V.; Wable, M.V.; Goodwin, S. (2003) Full-Scale Step-Feed Nutrient Removal Systems: A Comparison between Theory and Reality. *Proceedings of the 76th Annual Water Environment Federation Exposition and Conference*, Los Angeles, California, October 11–15; Water Environment Federation: Alexandria, Virginia.

Johnson, B. R.; Daigger, G. T.; Novak J. T. (2008). Biological Sludge Reduction Process Modeling with ASM-Based Models. *Proceedings of the 81st Water Environment Federation Technical Exposition and Conference and Exhibition*, October 18–21, Chicago, Illinois; Water Environment Research Foundation: Alexandria, Virginia.

Johnson, B.; Narayanan, B.; Baur, R.; Mengelkoch, M. (2006) High-Level Biological Phosphorus Removal Failure and Recovery. *Proceedings of 79th Annual Water Environment Federation Exposition and Conference*, Dallas, Texas, October 21–25.

Johnson, T. L. (1996) Practical Implications of Diffused Aeration System Model for Design and Operation. Paper presented at 69th Annual Meeting Central States Water Environment Association, St. Cloud, Minnesota.

Johnson, T. L.; McKinney, R. E. (1994) Modeling Performance of Full-Scale Diffused Aeration Systems as a Function of Mixing. Paper presented at American Society of Civil Engineers National Conference of Environmental Engineers, Boulder, Colorado; American Society of Civil Engineers: Reston, Virginia.

Johnson, W. K.; Schroepfer, G. J. (1964) Nitrogen Removal by Nitrification and Denitrification. *J. Water Pollut. Control Fed.*, **36**, 1015–1036.

Jones, M.; Stephenson, T. (1996) The Effect of Temperature on Enhanced Biological Phosphorus Removal. *Environ. Technol.*, **17**, 965–976.

Jördening, H.-J.; Winter, J. (2005) *Environmental Biotechnology: Concepts and Applications*; Wiley-VCH: Weinheim, Germany.

Kalyuzhnyi, S.; Gladchenko, M.; Mulder, A.; Versprille, B. (2006) New Anaerobic Process of Nitrogen Removal. *Water Sci. Technol.*, **54** (8), 163–170.

Kampman, C.; Temmink, H.; Hendrickx, T.; Zeeman, G.; Buisman, C. (2014) Enrichment of Denitrifying Methanotrophic Bacteria from Municipal Wastewater Sludge in a Membrane Bioreactor at 20°C. *J. Hazard. Mater.*, **274**, 428–435.

Kampschreur M. J., Temmink H., Kleerebezem R., Jetten M. S. M., van Loosdrecht M. C. M. (2009) Nitrous Oxide Emission during Wastewater Treatment. *Water Res.*, **43**, 4093–4103.

Kang, S. (1988) Biological Nitrification and Denitrification in the PhoStrip™ Process. *Proceedings of 61st Annual Water Pollution Control Federation Conference*, Dallas, Texas, October 3–7; Water Pollution Control Federation: Alexandria, Virginia.

Kang, S. J.; Bailey, W.; Astfalk, T.; Jenkins, D.; Aukamp, D. (1990) Biological Nutrient Removal Technologies at the Blue Plains Wastewater Treatment Plant. Paper presented at 63rd Annual Water Pollution Control Federation Conference, Washington, D.C., October 7–11; Water Pollution Control Federation: Alexandria, Virginia.

Kang, S.; Hong, S.; Tracy, K. (1985) Applied Biological Phosphorus Technology for Municipal Wastewater by the A/O Process. *Proceedings of the International Conference on Management Strategies for Phosphorus in the Environment*; Selper, Ltd.: London, England.

Kawamura, S. (1981) Hydraulic Scale Model Simulation of the Sedimentation Process. *J. Am. Water Works Assoc.*, **73** (7), 372–379.

Keinath, T. M. (1985) Operational Dynamics and Control of Secondary Clarifiers. *J. Water Pollut. Control Fed.*, **57**, 770–776.

Keinath, T. M. (1990) Diagram for Designing and Operating Secondary Clarifiers According to the Thickening Criterion. *Res. J. Water Pollut. Control Fed.*, **62**, 254–258.

Keinath, T. M.; Ryckman, M. D.; Dana, C. H.; Hofer, D. A. (1977) Activated Sludge Unified System Design and Operation. *J. Environ. Eng. Div.; Proc. Am. Soc. Civ. Eng.*, **103**, 829.

Kelly, K. (1988) New Clarifiers Help Save History. *Civ. Eng.*, **58**, 10.

Ketchum, L.; Liao, P. (1979) Tertiary Chemical Treatment for Phosphorus Reduction Using Sequencing Batch Reactors. *J. Water Pollut. Control Fed.*, **51**, 298–304.

Khalil, N.; Sinha, R.; Raghav, A. K.; Mittal, A. K. (2008) UASB Technology for Sewage Treatment in India: Experience, Economic Evaluation and Its Potential in Other Developing Countries. *Proceedings of the Twelfth International Water Technology Conference*, IWTC12 2008; Egyptian Water Technology Association: Alexandria, Egypt.

Kinnear, D. J. (2002) Evaluating Secondary Clarifier Collector Mechanisms. *Proceedings of the 75th Annual Water Environment Federation Exposition and Conference*; Chicago, Illinois, Sept. 28–Oct 2; Water Environment Federation: Alexandria, Virginia.

Kinnear, D. J.; Williams, R.; Olsen, C.; Kennedy, K.; Johnson, H.; Pollack, E.; Vitasovic, C. (1998) Theoretical and Field Comparison of Clarifier Feedwell Sizing. *Proceedings of the 71st Annual Water Environment Federation Exposition and Conference*; Dallas, Texas, October 3–7; Water Environment Federation: Alexandria, Virginia.

Kornaros, M.; Dokianakis, S. N.; Lyberatos, G. (2010) Partial Nitrification/Denitrification Can Be Attributed to the Slow Response of Nitrite Oxidizing Bacteria to Periodic Anoxic Disturbances. *Environ. Sci. Technol.*, **44**, 7245–7253.

Krauth, K. H. (1993) Abwassertechnik. Vorlesungskript. University of Stuttgart: Germany.

Krebs, P.; Vischer, D.; Gujer, W. (1995) Inlet Structure Design for Final Clarifiers. *J. Environ. Eng. Am. Soc. Civ. Eng.*, **121** (8), 558–564.

Krzeminski, P.; van der Graaf, J. H.; van Lier, J. B. (2012). Specific Energy Consumption of Membrane Bioreactor (MBR) for Sewage Treatment. *Water Sci. Technol.*, **65** (2), 380–392.

Labelle, M. A.; Dold, P. L.; Comeau, Y. (2015). Mechanisms for reduced excess sludge production in the Cannibal process. *Water Environ. Res.*, **87** (8), 687–696.

Larsen, P. (1977) On the Hydraulics of Rectangular Settling Basins; Report 1001; Department of Water Resources Engineering, University of Lund: Sweden.

Latif, M. A.; Ghufran, R.; Wahid, Z. A.; Ahmad, A. (2011). Integrated Application of Upflow Anaerobic Sludge Blanket Reactor for the Treatment of Wastewaters. *Water Res.*, **45** (16), 4683–4699.

Lawrence, A. W.; McCarty, P. L. (1970) Unified Basis for Biological Treatment Design and Operation. *J. Sanit. Eng. Div., Proc. Am. Soc. Civ. Eng.*, **96**, 757.

Lee, W.; Kwak, J.; Bae, P.; McCarty, L. (2013) The Effect of SRT on Nitrate Formation during Autotrophic Nitrogen Removal of Anaerobically Treated Wastewater. *Water Sci. Technol.* **68**, 1751–1756.

Lettinga, G.; Hulshoff Pol, W. (1991) UASB–Process Designs for Various Types of Wastewaters. *Water Sci. Technol.*, **24** (8), 87–107.

Li, X-M.; Guo, L.; Yang, Q.; Zeng, G-M.; Lio, D-X. (2007) Removal of Carbon and Nutrients from Low Strength Domestic Wastewater by Expanded Granular Bed-Zeolite Bed Filtration (EGSB-ZBF) Integrated Treatment Concept. *Proc. Biochem.*, **42**, 1173–1179.

Libhaber, M.; Orozco-Jaramillo, Á. (2012) *Sustainable Treatment and Reuse of Municipal Wastewater: For Decision Makers and Practicing Engineers*; IWA Publishing: London, U.K.

Lin, H.; Chen, J.; Wang, F.; Ding, L.; Hong, H. (2011) Feasibility Evaluation of Submerged Anaerobic Membrane Bioreactor for Municipal Secondary Wastewater Treatment. *Desalination*, **280** (1), 120–126.

Lin, H.; et al. (2012) Membrane Bioreactors for Industrial Wastewater Treatment: A Critical Review. *Crit. Rev. Environ. Sci. Technol.*, **42** (7), 677–740.

Lin, H.; et al. (2014). A Critical Review of Extracellular Polymeric Substances (EPSs) in Membrane Bioreactors: Characteristics, Roles in Membrane Fouling and Control Strategies. *J. Membr. Sci.*, **460**, 110–125.

Lotti, T.; Kleerebezem, R.; van Erp Tallman Kip, C.; Hendrickx, T. L. G.; Kruit, J.; Hoekstra, M.; van Loosdrecht M. C. M. (2014) Anammox Growth on Pretreated Municipal Wastewater. *Environ. Sci. Technol.*, **48** (14), 7874–7880.

Ludzack, F. J.; Ettinger, M. B. (1962) Controlling Operation to Minimize Activated Sludge Effluent Nitrogen. *J. Water Pollut. Control Fed.*, **34**, 920–931.

Maeng, S. K.; Choi, B. G.; Lee, K. T.; Song, K. G. (2013) Influences of Solid Retention Time, Nitrification and Microbial Activity on the Attenuation of Pharmaceuticals and Estrogens in Membrane Bioreactors. *Water Res.*, **47** (9), 3151–3162.

Malina, J., F.; Pohland, F. G. (1992) *Design of Anaerobic Processes for the Treatment of Industrial and Municipal Wastes*; Technomic Publishing: Lancaster, Pennsylvania.

Mamais, D.; Jenkins, D. (1992) The Effects of MCRT and Temperature on Enhanced Biological Phosphorus Removal. *Water Sci. Technol.*, **26** (5–6), 955–965.

Mamais, D.; Jenkins, D.; Pitt, P. (1993) A Rapid Physical-Chemical Method for the Determination of Readily Biodegradable Soluble COD in Municipal Wastewater. *Water Res.*, **27** (1), 195–197.

Matsch, L.; Drnevich, R. (1987) Biological Nutrient Removal. In *Advances in Water and Wastewater Treatment*; Ann Arbor Science: Chelsea, Michigan.

Mau, G. E. (1959) A Study of Vertical-Slotted Inlet Baffles. *Sew. Ind. Wastes*, **31**, 1349–1372.

McCarty, P. L.; Beck, L.; St. Amant, P. (1969) Biological Denitrification of Wastewaters by Addition of Organic Materials. *Proceedings of the 24th Industrial Waste Conference*; Purdue University: Lafayette, Indiana.

McClintock, S.; Randall, C.; Pattarkine, V. (1991) Effects of Temperature and Mean Cell Residence Time on Enhanced Biological Phosphorus Removal. *Proceedings of the 1991 Specialty Conference on Environmental Engineering*; Krenkel, P., Ed.; Reno, Nevada, July 10–12; American Society of Civil Engineers: Reston, Virginia; 319–324.

McConnell, W.; Moody, M.; Woodard, S. (2010) Full-Scale BioMag Demonstration at the Mystic WPCF and Establishing the Basis-of-Design for a Permanent Installation. *Proceedings of the 83rd Annual Water Environment Federation Technical Exposition and Conference* [CD-ROM]; New Orleans, Louisiana, Oct 2–6; Water Environment Federation: Alexandria, Virginia.

McKinney, R. E. (1962) Mathematics of Complete Mixing Activated Sludge. *J. Sanit. Eng. Div., Proc. Am. Soc. Civ. Eng.*, **88**, SA3.

McKinney, R. E. (1977) *Sedimentation, Tank Design, and Operation*; University of Kansas: Lawrence.

McKinney, R. E.; Ooten, R. J. (1969) Concepts of Complete Mixing Activated Sludge. *Transactions of the 19th Annual Conference on Sanitary Engineering*, Bulletin of Engineering and Architecture No. 60; University of Kansas: Lawrence, Kansas.

Metcalf & Eddy, Inc./AECOM (2014) *Wastewater Engineering Treatment and Resource Recovery*, 5th ed., Tchobanoglous, G., Stensel, H., Tsuchihashi, R., Burton, F., Abu-Orf, M., Bowden, G., Pfrang, W. (Eds.); McGraw-Hill Education: New York, New York.

Miller, M. A.; Miller, G. Q. (1978) Activated Sludge Settling in High Purity Oxygen Systems—A Full-Scale Operating Data Correlation. Paper Presented at 51st Annual Water Pollution Control Federation Conference; Anaheim, California, October 1–6; Water Pollution Control Federation: Washington, D.C.

Mino, T.; Arun, V.; Tsuzuki, Y.; Matsuo, T. (1987) Effect of Phosphorus Accumulation on Acetate Metabolism in the Biological Phosphorus Removal Process. *Proceedings of the IAWPRC Specialized Conference*; Rome, Italy, September 28–30; IWA Publishing: London, England.

Monroy, O.; Famá, G.; Meraz, M.; Montoya, L.; Macarie, H. (2000). Anaerobic Digestion for Wastewater Treatment in Mexico: State of the Technology. *Water Res.*, **34** (6), 1803–1816.

Mueller, J. A.; Boyle, W. C. (1988) Oxygen Transfer under Process Conditions. *J. Water Pollut. Control Fed.*, **60**, 332–341.

Mueller, J.; Boyle, W.; Pöpel, H. (2002) *Aeration: Principles and Practice*; CRC Press: Boca Raton, FL.

Mulbarger, M. C.; et al. (1970) Modifications of the Activated Sludge Process for Nitrification and Denitrification. Paper presented at 43rd Annual Water Pollution Control Federation Conference, Boston, Massachusetts; Water Pollution Control Federation: Washington, D.C.

Narayanan, B.; de Leon, C.; Radke, C.; Jenkins, D. (2010) The Role of Dispersed Nocardioform Filaments in Activated Sludge Foaming. *Water Environ. Res.*, **82** (6), 483–491.

Neethling, J.; Bakke, B.; Benisch, M.; Gu, A.; Stephens, H. (2005) *Factors Influencing the Reliability of Enhanced Biological Phosphorus Removal*; Water Environment Research Foundation: Alexandria, Virginia.

Neville, M.; Turriciano, A. (2013) Don't Blow It! Lessons Learned from Direct Drive Turbo Blower Designs for Energy Efficiency. *J. New Engl. Water Environ. Assoc.*, **47** (1), 46–54.

Nicholls, H.; Osborn, D.; Pitman, A. (1987) Improvements to the Stability of the Biological Phosphate Removal Process at the Johannesburg Northern Works. *Proceedings of the IAWPRC Specialized Conference*, Rome, Italy, September 28–30; IWA Publishing: London, England.

Nicolella, C.; van Loosdrecht, M. C. M.; Heijnen, J. J. (2000) Wastewater Treatment with Particulate Biofilm Reactors. *J. Biotechnol.*, **80**, 1–33.

Nielsen, M. K.; Bechmann, H.; Henze, M. (2000) Modelling and Test of Aeration Tank Settling (ATS). *Water Sci. Technol.*, **41** (9), 179–184.

Novak, J. T.; Chon, D. H.; Curtis, B. A.; Doyle, M. (2006) Reduction of Sludge Generation Using the Cannibal™ Process: Mechanisms and Performance. *Proceedings of the Water Environment Federation Residuals and Biosolids Management Specialty Conference*, Cincinnati, Ohio, March 12–14; Water Environment Research Foundation: Alexandria, Virginia.

Noyola, A.; Capdeville, B.; Roques, H. (1988) Anaerobic Treatment of Domestic Sewage with a Rotating-Stationary Fixed Film Reactor. *Water Res.*, **12**, 1585–1592.

Noyola, A.; Morgan-Sagastgume, J. M.; Lopez-Hernandez, J. E. (2006) Treatment of Biogas Produced in Anaerobic Reactors for Domestic Wastewater: Odor Control and Energy/Resource Recovery. *Rev. Environ. Sci. Biotechnol.*, **5**, 93–114.

Okuno, N.; Fukuda, H. (1982) Analysis of Existing Final Clarifiers and Design Consideration for Better Performance. Paper Presented at the 1st German–Japan Conference on Sewage Treatment and Sludge Disposal; Tsukuba Science City, Japan.

Oldham, K.; Rabinowitz, B. (2001) Development of Biological Nutrient Removal Technology in Western Canada. *Can. J. Civ. Eng.*, **28** (Suppl. 1), 92–101.

Oldham, W.; Stevens, G. (1984) Initial Operating Experiences of a Nutrient Removal Process (Modified Bardenpho) at Kelowna, British Columbia. *Can. J. Civ. Eng.*, **11**, 474–479.

Oliveira, S. M. A. C.; Parkinson, J. N.; von Sperling, M. (2006) Wastewater Treatment in Brazil: Institutional Framework, Recent Initiatives and Actual Plant Performance. *Int. J. Technol. Manag. Sustainable Dev.*, **5** (3), 241–256.

Onnis-Hayden, A.; Gu, A. (2008) Comparisons of Organic Sources for Denitrification: Biodegradability, Denitrification Rates, Kinetic Constants and Practical Implication for Their Application in WWTPs. *Proceedings of 81st Annual Water Environment Federation Technical Exhibition and Conference*, Chicago, Illinois, October 18–22.

Ozgun, H.; Dereli, R. K.; Ersahin, M. E.; Kinaci, C.; Spanjers, H.; van Lier, J. B. (2013) A Review of Anaerobic Membrane Bioreactors for Municipal Wastewater Treatment: Integration Options, Limitations and Expectations. *Sep. Purif. Technol.*, **118**, 89–104.

Pagilla, K. (2007) Organic Nitrogen in Wastewater Treatment Plant Effluents. Presentation at Water Environment Research Foundation and Chesapeake Bay Science and Technology Advisory Committee Workshop; Baltimore, Maryland, Sept. 27–28; Water Environment Research Foundation: Alexandria, Virginia.

Palis, J. C.; Irvine, R. L. (1985) Nitrogen Removal in a Low-Loaded Single Tank Sequencing Batch Reactor. *J. Water Pollut. Control Fed.*, **57**, 82–86.

Panswad, T.; Doungchai, A.; Anotai, J. (2003) Temperature Effect on Microbial Community of Enhanced Biological Phosphorus Removal Systems. *Water Res.*, **37** (2), 409–414.

Parker, D. S. (1983) Assessment of Secondary Clarification Design Concepts. *J. Water Pollut. Control Fed.*, **55**, 349–359.

Parker, D. S.; Stone, R.; Stenquist, R.; Culp, G. (1975) *Process Design Manual for Nitrogen Control*, EPA 62511-75-007; U.S. Environmental Protection Agency: Washington, D.C.

Parker, D. S.; Geary, S.; Jones, G.; McIntyre, L.; Oppenheim, S.; Pedregon, V.; Pope, R.; Richards, T.; Voigt, C.; Volpe, G.; Willis, J.; Witzgall, R. (2003) Making Classifying Selectors Work for Foam Elimination in the Activated Sludge Process. *Water Environ. Res.*, **75**, 83–91.

Parker, D. S.; Kaufman, W. J.; Jenkins, D. (1971) Physical Conditioning of Activated Sludge Floc. *J. Water Pollut. Control Fed.*, **43** (9), 1817–1833.

Parker, D. S.; Stenquist, R. J. (1986) Flocculator–Clarifier Performance. *J. Water Pollut. Control Fed.*, **58**, 214–219.

Parker, D.; Bratby, J.; Espring, D.; Hull, T.; Kelly, R.; Melcer, H.; Merlo, R.; Pope, R.; Shafer, T.; Wahlberg, E.; Witzgall, R. (2014) A Critical Review of Nuisance Foam Formation and Biological Methods for Foam Management or Elimination in Nutrient Removal Facilities. *Water Environ. Res.*, **86** (6), 483–503.

Parker, D.; Brischke, K.; Petrik, K. (1993) Optimizing Existing Treatment Systems. *Water Environ. Technol.*, **5** (2), 50–53.

Pavlostathis, S. G.; Giraldo-Gomez, E. (1991) Kinetics of Anaerobic Treatment: A Critical Review. *Crit. Rev. Environ. Control*, **21** (5/6), 411–490.

Peters, M.; Newland, M.; Seviour, T.; Broom, T; Bridle, T. (2004) Demonstration of Enhanced Nutrient Removal at Two Full-Scale SBR Plants. *Water Sci. Technol.*, **50**, 115–120.

Pitman, A.; Trim, B.; van Dalsen, L. (1988) Operating Experience with Biological Nutrient Removal at the Johannesburg Bushkoppie Works. *Water Sci. Technol.*, **20** (4-5), 51–62.

Pitman, A.; Venter, S.; Nicholls, H.; van Dalsen, L.; Soelbat, J. (1982) Operating Experience with Nutrient Removal Activated Sludge Plants in Johannesburg. *IMIESA, J. Inst. of Municipal Eng. Southern Africa*, **November**, 61–69.

Pitter, P.; Chudoba, J. (1990) *Biodegradability of Organic Substrates in the Aquatic Environment*; CRC Press: Boca Raton, Florida.

Plass, R.; Sekoulov, I. (1995) Enhancement of Biomass Concentration in Activated Sludge Systems. *Water Sci. Technol.*, **32**, 151–157.

Popel, F.; Weidner, J. (1963) Ueber einige Einflusse auf die Klarwirkung bon Absetzbecken. *GWF-Wasser/Abwasser*, **104** (28), 796–803.

Pretorius, C.; Wicklein, E.; Rauch-Williams, T.; Samstag, R.; Sigmon, C. (2015) How Oversized Mixers Became an Industry Standard. *Proceedings of 88th Annual Water Environment Federation Technical Exhibition and Conference*, Chicago, Illinois, September 24–28, Water Environment Federation: Alexandria, Virginia.

Rabinowitz, B.; Daigger, G.; Jenkins, D.; Neethling, J. (2004) The Effect of High Temperatures on BNR Process Performance. *Proceedings of 77th Annual Water Environment Federation Exposition and Conference*, New Orleans, Louisiana, October 3–6.

Rabinowitz, B.; Fries, K. (2010) Primary Sludge Fermenters in BNR Plants: Are They Cost-effective for Meeting Effluent Phosphorus Limits? *Proceedings of 83rd Annual Water Environment Federation Exposition and Conference*, New Orleans, Louisiana, October 2–6.

Rabinowitz, B.; Neethling, J.; Barnard, J.; Baur, R.; Abraham, K.; Erdal, Z.; Oldham, W. (2011) Fermenters for Biological Phosphorus Removal Carbon Augmentation. Water Environment Research Foundation (WERF) White Paper.

Ramanathan, M.; Gaudy, A. F., Jr. (1971) Steady State Model for Activated Sludge with Constant Recycle Sludge Concentration. *Biotechnol. Bioeng.*, **13**, 125.

Randall, C. W.; Barnard, J. L.; Stensel, H. D. (1992) Design and Retrofit of Wastewater Treatment Plants for Biological Nutrient Removal. In *Principles of Biological Nutrient Removal*; Technomic Publishing Co.: Lancaster, Pennsylvania.

Reardon, R. (2004) Clarification Concepts for Treating Peak Wet Weather Wastewater Flows. *Fla. Water Res. J.*, **56** (1).

Reed, S. C.; Murphy, R. S. (1969) Low Temperature Activated Sludge Settling. *J. Sanit. Eng. Div., Proc. Am. Soc. Civ. Eng.*, **95**, 747.

Reed, S. C.; Crites, R. W., and Middlebrooks, E. J. (1998) *Natural Systems for Waste Management and Treatment*, 2nd ed.; McGraw-Hill: New York, pp. 173–284.

Rittmann, B. E.; Langeland, W. E. (1985) Simultaneous Denitrification with Nitrification in Single-Channel Oxidation Ditches. *J. Water Pollut. Control Fed.*, **57**, 300–308.

Rittmann, B; McCarty, P. (2001) *Environmental Biotechnology: Principles and Applications*; McGraw-Hill: New York.

Rosso, D.; Stenstrom, M. K. (2006) Economic Implications of Fine-Pore Diffuser Aging. *Water Environ. Res.*, **78** (8), 810–815.

Roxburgh, R.; Sieger, R.; Johnson, B. R.; Rabinowitz, B.; Goodwin S.; Crawford, G. V.; Daigger, G. T. (2006) Sludge Minimization Technologies—Doing More To Get Less. *Proceedings of the 79th Annual Water Environment Federation Exposition and Conference*, Dallas, Texas, October 21–25; Water Environment Federation: Alexandria, Virginia.

Rudd, D.; Alaica, J.; Samules, C.; Geith, W. (2001) Canada's Largest Plant Uses Ultrasonic Detection and Digital Radio Technology. *Environ. Sci. Eng.*, **13** (9), 18–21.

Samstag, R.; Wicklein, E. (2014) A Protocol for Optimization of Activated Sludge Mixing. *Proceedings of 87th Annual Water Environment Federation Technical Exhibition and Conference*, New Orleans, LA, September 27–October 1.

Samuels, C. (2000) Control System Improved Activated Sludge and Sludge Thickening Process. *Environ. Sci. Eng.*, **12** (11), 23–27.

Santos, A.; Ma, W.; Judd, S. J. (2011) Membrane Bioreactors: Two Decades of Research and Implementation. *Desalination*, **273** (1), 148–154.

Sato, N.; Okubo, T.; Onadera, T.; Ohashi, A.; Harada, H. (2006) Prospects for a Self-Sustainable Sewage Treatment System: A Case Study on Full-Scale UASB System in India's Yamuna River Basin. *J. Environ. Manag.*, **80** (13), 198–207.

Schellingkhout, A.; Collazos, C. J. (1992) Full-Scale Application of the UASB Technology for Sewage Treatment. *Water Sci. Technol.*, **25** (7), 159–166.

Schwinn, D. E.; Hotaling, J. H. (1988) Modifying Existing Activated Sludge Plants to Achieve Nitrification and Denitrification. Paper Presented at 62nd Florida Water Resources Conference.

Seghezzo, L. (2004) Anaerobic Treatment of Domestic Wastewater in Subtropical Regions. Ph.D. Thesis, Wageningen University, Wageningen, the Netherlands.

Sell, R. (1981) Low Temperature Biological Phosphorus Removal. *Proceedings of 54th Annual Water Pollution Control Federation Conference*, Detroit, Michigan, October 4–9; Water Pollution Control Federation: Washington, D.C.

Sezgin, M.; Jenkins, D.; Parker, D. S. (1978) A Unified Theory of Filamentous Activated Sludge Bulking. *J. Water Pollut. Control Fed.*, **50**, 362–381.

Sharma, B.; Ahlert, R.C. (1977) Nitrification and Nitrogen Removal. *Water Res. (G.B.)*, **11**, 897.

Shen, Y. X.; Xiao, K.; Liang, P.; Sun, J. Y.; Sai, S. J.; Huang, X. (2012) Characterization of Soluble Microbial Products in 10 Large-Scale Membrane Bioreactors for Municipal Wastewater Treatment in China. *J. Membr. Sci.*, **415**, 336–345.

Shin, C.; McCarty, P. L.; Kim, J.; Bae, J. (2014). Pilot-Scale Temperate-Climate Treatment of Domestic Wastewater with a Staged Anaerobic Fluidized Membrane Bioreactor (SAF-MBR). *Biores. Technol.*, **159**, 95–103.

Sigmon, C.; Weirich, S.; Douville, C. (2014) The Best Carbon for the Job: Using the 2010 WERF Protocol to Choose an External Carbon Alternative for Enhanced Nitrate Removal. *Proceedings of 87th Annual Water Environment Federation Technical Exhibition and Conference*, New Orleans, Louisiana, September 27–October 1.

Silverstein, J.; Schroeder, E. D. (1983) Performance of SBR Activated Sludge Processes with Nitrification/Denitrification. *J. Water Pollut. Control Fed.*, **55**, 377.

Smith, A. L.; Skerlos, S. J.; Raskin, L. (2013) Psychrophilic Anaerobic Membrane Bioreactor Treatment of Domestic Wastewater. *Water Res.*, **47** (4), 1655–1665.

Smith, G. (1996) Increasing Oxygen Delivery in Anoxic Tanks to Improve Denitrification. *Proceedings of the 69th Annual Water Environment Federation Exposition and Conference* [CD-ROM], Dallas, Texas, October 5–9; Water Environment Federation: Alexandria, Virginia.

Speece, R. E. (1996) *Anaerobic Biotechnology*; Archae Press: Nashville, Tennessee.

Stahl, J. F.; Chen, C. L. (2006) Review of Chapter 8. Rectangular and Vertical Secondary Settling Tanks. *Proceedings of the 79th Annual Water Environment Federation Exposition and Conference*, Dallas, Texas, Oct. 21–25; Water Environment Federation: Alexandria, Virginia.

Staunton, E. T.; Aitken, M. D. (2015) Coupling Nitrogen Removal and Anaerobic Digestion for Energy Recovery from Swine Waste Through Nitrification/Denitrification. *Environ. Eng. Sci.*, September 2015, **32** (9), 750–760.

Stensel, H. (1980) Performance of First U.S. Full-Scale Bardenpho™ Facility. *Proceedings of U.S. EPA International Seminar on Control of Nutrients in Municipal Wastewater Effluents*, San Diego, California.

Stensel, H. (1991) Principles of Biological Phosphorus Removal. In *Phosphorus and Nitrogen Removal from Municipal Wastewater—Principles and Practice*, Sedlak, R., Ed.; Lewis Publishers: Boca Raton, Florida.

Stensel, H. D. (1978) Carrousel Activated Sludge for Biological Nitrogen Removal. In *Advances in Water and Wastewater Treatment Biological Nutrient Removal*; Ann Arbor Science: Chelsea, Michigan.

Stensel, H. D.; Loehr, R. C.; Lawrence, A. W. (1973) Biological Kinetics of Suspended Growth Denitrification. *J. Water Pollut. Control Fed.*, **45**, 249–261.

Stenstrom, M. R.; Gilbert, R. G. (1981) Effects of Alpha, Beta; Theta Factors on Specification, Design and Operation of Aeration Systems. *Water Res. (G.B.)*, **15**, 643.

Stevens, G.; Barnard, J.; Rabinowitz, B. (1999) Optimizing Nutrient Removal in Anoxic Zones. *Water Sci. Technol.*, **39** (6), 113–118.

Streeter, V. L.; Wylie, E. B. (1979) *Fluid Mechanics*, 7th ed.; McGraw-Hill: New York.

Strous, M.; Lan, C.-J.; Jetten, M. S. M. (1999) Key Physiology of Anaerobic Ammonium Oxidation. *Appl. Environ. Microbiol.*, **65**, 3248–3250.

Strous, J.; Heijnen J.; Kuenen J. G.; Jetten M. S. M. (1998) The Sequencing Batch Reactor as a Powerful Tool for the Study of Slowly Growing Anaerobic Ammonium-Oxidizing Microorganisms. *Appl. Microbiol. Biotechnol.*, **50**, 589–596.

Stukenberg, J. F.; Rodman, L. C.; Touslee, J. E. (1983) Activated Sludge Clarifier Design Improvements. *J. Water Pollut. Control Fed.*, **55**, 341–348.

Sturm, B.; Faraj, R.; Amante, T.; Wagner, B.; Kiani, F.; Waybenais, D. (2015) Strategies for Operating Aerobic Granular Sludge Reactors for Low-Strength Municipal Wastewater. *Proceedings of the 89th Annual Water Environment Federation Technical Exposition and Conference*, Chicago, Illinois, Sept. 24–28; Water Environment Federation: Alexandria, Virginia.

Sun, Y.; Nemati, M. (2012) Evaluation of Sulfur-Based Autotrophic Denitrification and Denitritation for Biological Removal of Nitrate and Nitrite from Contaminated Waters. *Bioresour. Technol.* doi:10.1016/j.biortech.2012.03.061.

Sutton, P. M. (1990) Anaerobic Treatment of High Strength Wastes: System Configurations and Selection. Paper Presented at Anaerobic Treatment of High Strength Wastes, Milwaukee, Wisconsin, December 3–4; University of Wisconsin: Milwaukee.

Sutton, P.M.; Hurvid, J.; Hoeksema, M. (1975) Low-Temperature Biological Denitrification of Wastewater. *J. Water Pollut. Control Fed.*, **47**, 122–134.

Switzenbaum, M. S. (2007) Overview of Challenges in Providing Anaerobic Wastewater Treatment. *Proceedings of the 80th Annual Water Environment Federation Exposition and Conference* [CD-ROM]; San Diego, California, October 10–13; Water Environment Federation: Alexandria, Virginia.

Sykes, J. C. (1993) Biological Selector to Minimize Sludge Bulking. *Waterwood Rev.*, **Sept/Oct.**

Tang, Y.; et al. (2012) Comparing Heterotrophic and Hydrogen-Based Autotrophic Denitrification Reactors for Effluent Water Quality and Post-treatment. *Water Sci. Technol. Water Supply*, **12**, 227.

Tchobanoglous, G.; Burton, F. L.; Stensel, H. D. (2003) *Wastewater Engineering Treatment and Reuse*, 4th ed.; McGraw-Hill: New York.

Tekippe, R. J. (2002) Secondary Settling Tank Inlet Design: Full Scale Test Results Lead to Optimization. *Proceedings of the 75th Annual Water Environment Federation Exposition and Conference*, Chicago, Illinois, September 28–October 2; Water Environment Federation: Alexandria, Virginia.

Tekippe, R. J.; Bender, J. H. (1987) Activated Sludge Clarifiers: Design Requirements and Research Priorities. *J. Water Pollut. Control Fed.*, **59**, 865–870.

Ternes, T. A.: Joss, A. (2006) *Human Pharmaceuticals, Hormones and Fragrances; The Challenge of Micropollutants in Urban Water Management*. IWA Publishing: London, England.

Tetreault, M. J.; Rusten, B.; Benedict, A. H.; Kreissl, J. F. (1987) Assessment of Phased Isolation Ditch Technology. *J. Water Pollut. Control Fed.*, **59**, 833–840.

Tilche, A.; Bortone, G.; Forner, G.; Indulti, M.; Stante, L.; Tesini, O. (1994) Combination of Anaerobic Digestion and Denitrification in a Hybrid Upflow Anaerobic Filter Integrated in a Nutrient Removal Treatment Plant. *Water Sci. Technol.*, **30** (12), 405–414.

Timmermans, P.; Van Haute, A. (1982) Removal of Nitrogen Compounds in Drinking Water and Wastewater-Comparison Between a One- and Two-Sludge System for Denitrification Using Internal Wastewater Carbon. Paper Presented at 14th International Water Supply Association Congress and Exhibition, Zurich, Switzerland.

Totzke, D. E. (2008) *Anaerobic Treatment Technology Overview*; Applied Technologies: Brookfield, Wisconsin.

Tracy, K.; Flammino, A. (1985) Kinetics of Biological Phosphorus Removal. *Proceedings of 58th Annual Water Pollution Control Federation Conference*, Kansas City, Missouri, October 6–10.

U.S. Environmental Protection Agency (1974) *Extended Aeration Sewage Treatment in Cold Climates*, EPA-660/2-74-070; U.S. Environmental Protection Agency: Washington, D.C.

U.S. Environmental Protection Agency (1979) *Full-Scale Demonstration of Open Tank Oxygen Activated Sludge Treatment*, EPA-600/2-79-012; U.S. Environmental Protection Agency: Cincinnati, Ohio.

U.S. Environmental Protection Agency (1983) *Development of Standard Procedures for Evaluating Oxygen Transfer Devices*, EPA-600/2-83-102; U.S. Environmental Protection Agency: Cincinnati, Ohio.

U.S. Environmental Protection Agency (1985) *Summary Report Fine Pore (Fine Bubble) Aeration Systems*, EPA-625/8-85-010; Water Engineering Research Laboratory, U.S. EPA, Washington, D.C.

U.S. Environmental Protection Agency (1987) *Summary Report: Causes and Control of Activated Sludge Bulking and Foaming*, EPA-625/8-87-012; U.S. Environmental Protection Agency: Cincinnati, Ohio.

U.S. Environmental Protection Agency (1989) *Design Manual: Fine Pore Aeration Systems. Center for Environmental Researcy Information, Risk Reduction Engineering Laboratory*, EPA-625/1-89-023.

U.S. Environmental Protection Agency (1993) *Process Design Manual for Nitrogen Removal*; U.S. Environmental Protection Agency: Washington, D.C.

U.S. Environmental Protection Agency (1999) *Wastewater Technology Fact Sheet: Fine Bubble Aeration*, EPA 832-F-99-065.

U.S. Environmental Protection Agency (2010) *Evaluation of Energy Conservation Measures for Wastewater Treatment Facilities*, EPA 832-R-10-005; U.S. Environmental Protection Agency: Washington, D.C.

Updegraff, K. F.; Boyle, W. C. (1988) *Technology Assessment of Draft Tube Submerged Turbine Aerators*, U.S. EPA Field Evaluation I.A. Seminars; U.S. Environmental Protection Agency: Washington, D.C.

Vaccari, D. A.; Strom, P. F.; Alleman, J. E. (2006) *Environmental Biology for Engineers and Scientists*; John Wiley & Sons, Inc.: New York.

van de Graaf, A. A.; de Bruijn, P.; Robertson, L. A.; Jetten, M. S. M.; Kuenen, J. G. (1996) Autotrophic Growth of Anaerobic Ammonium-Oxidizing Micro-organisms in a Fluidized Bed Reactor. *Microbiology*, **142**, 2187–2196.

Van der Geest, A. T.; Witvoet, W. C. (1977) Nitrification and Denitrification in Carrousel Systems. *Prog. Water Technol.*, **8**, 653.

van Haandel, A. C.; Kato, M. T.; Cavalcanti, P. F. F.; Florencio, L. (2006) Review Anaerobic Reactor Design Concepts for the Treatment of Domestic Wastewater. *Rev. Environ. Sci. Biotechnol.*, **5**, 21–38.

van Haandel, A. C.; Lettinga, G. (1994) *Anaerobic Sewage Treatment: A Practical Guide for Regions with a Hot Climate*; John Wiley and Sons: Chichester, England.

van Lier, J. B. (2003) *Design and Operation of UASB for Treatment of Domestic Wastewater*; Wageningen University and Lettinga Associates Foundation (LeAF): Wageningen, The Netherlands.

van Lier, J. B. (2008) High-Rate Anaerobic Treatment: Diversifying from End-of-the-Pipe Treatment to Resource-Oriented Conversion Techniques. *Water Sci. Technol.*, **57** (8), 1137–1148.

Vieira, S. M. M.; Carvalho, J. L.; Barigan, F. P. O.; Rech, C. M. (1994) Application of the UASB Technology for Sewage Treatment in a Small Community at Sumare, Sao Paulo State. *Water Sci. Technol.*, **30** (12), 203–210.

Vinton, R. H.; Mace, G. R. (1996) Most Open Valve Control and Cascade Control of Multiple Compressors to Improve Aeration Efficiency and Cut Costs. *Proceedings of the 79th Annual Water Environment Federation Exposition and Conference* [CD-ROM]; Dallas, Texas, October 21–25; Water Environment Federation: Alexandria, Virginia.

Vollertson, J.; Petersen, G.; Borregaard, V. (2006) Hydrolysis and Fermentation of Activated Sludge to Enhance Biological Phosphorus Removal. *Water Sci. Technol.*, **53** (12), 55–64.

von Sperling, M.; Oliveira, S. C. (2008) Comparative Performance Evaluation of Full-Scale Anaerobic and Aerobic Wastewater Treatment Processes in Brazil. *Proceedings IX Latin American Workshop and Symposium on Anaerobic Digestion*; Easter Island, Chile, October 19–23; International Water Association: London.

Wahlberg, E. J. (2001) *WERF/CRTC Protocols for Evaluating Secondary Clarifier Performance*, Project 00-CTS-1; Water Environment Research Foundation: Alexandria, Virginia.

Wahlberg, E. J.; et al. (1994) Evaluation of Activated Sludge Clarifier Performance Using the CRTC Protocol: Four Case Studies. *Proceedings of the 67th Annual Water Environment Federation Technical Exposition and Conference*; Chicago, Illinois, October 15–19; Water Environment Federation: Alexandria, Virginia.

Wahlberg, E. J.; Keinath, T. M. (1988) Development of Settling Flux Curves Using SVI. *J. Water Pollut. Control Fed.*, **60**, 2095–2100.

Wahlberg, E. J.; Stahl, J. F.; Chen, C. L.; Augustus, M. (1993) Field Application of the CRTC Protocol for Evaluating Secondary Clarifier Performance: Rectangular, Co-Current Sludge Removal Clarifier. *Proceedings of the 66th Annual Water Environment Federation Technical Exposition and Conference,* Anaheim, California, October 3–7; Water Environment Federation: Alexandria, Virginia.

Wang, J.; Park, J. (1998) Effect of Wastewater Composition on Microbial Populations in Biological Phosphorus Removal Processes. *Water Sci. Technol.,* **38** (1), 159–166.

Ward, D.; Oldham, W.; Abraham, K.; Jeyanayagam, S. S. (1999) Resolution of Capacity Constraints and Performance Variations in Biological Nutrient Removal Process. *Proceedings of the 72nd Annual Water Environment Federation Technical Exposition and Conference,* New Orleans, Louisiana, October 9–13; Water Environment Federation: Alexandria, Virginia.

Water Environment Federation (2005) *Clarifier Design,* Manual of Practice FD–8; McGraw-Hill: New York.

Water Environment Federation (2006) *Membrane Systems for Wastewater Treatment;* McGraw-Hill: New York.

Water Environment Federation (2009) *Energy Conservation in Water and Wastewater Facilities,* Manual of Practice No. 32; Water Environment Federation: Alexandria, Virginia.

Water Environment Federation (2013a) *Operation of Nutrient Removal Facilities,* Manual of Practice No. 37; Water Environment Federation: Alexandria, Virginia.

Water Environment Federation (2013b) *Wastewater Treatment Process Modeling,* 2nd ed., Manual of Practice No. 31; Water Environment Federation: Alexandria, Virginia.

Water Environment Federation (2016) *Operation of Water Resource Recovery Facilities,* 7th ed., Manual of Practice No. 11; Water Environment Federation: Alexandria, Virginia.

Water Environment Research Foundation (2001) *Membrane Bioreactors: Feasibility and Use in Water Reclamation;* Water Environment Research Foundation: Alexandria Virginia.

Water Environment Research Foundation (2002) *Membrane Technology: Feasibility of Solid/Liquid Separation in Wastewater Treatment;* Water Environment Research Foundation: Alexandria, Virginia.

Water Environment Research Foundation (2006a) *Develop and Demonstrate Fundamental Basis for Selectors to Improve Activated Sludge Settleability;* Water Environment Research Foundation: Alexandria, Virginia.

Water Environment Research Foundation (2006b) *Effects of Biomass Properties on Submerged Bioreactor (SMBR) Performance and Solids Processing;* Water Environment Research Foundation: Alexandria, Virginia.

Water Environment Research Foundation (2007) *Fate of Pharmaceuticals and Personal Care Products Through Municipal Wastewater Treatment Processes,* 03CTS22UR. Water Environment Research Foundation: Alexandria, Virginia.

Water Environment Research Foundation (2008) *Dissolved Organic Nitrogen (DON) in Biological Nutrient Removal Wastewater Treatment Processes;* Water Environment Research Foundation: Alexandria, Virginia.

Water Research Commission (1984) *Theory: Design and Operation of Nutrient Removal Activated Sludge Processes;* Water Research Commission: Pretoria, South Africa.

Welles, L.; Prabodini, M.; Lopez-Vazquez, C.; Hooijmans, C.; van Loosdrecht, M.; Brdjanovic, D. (2016) Long-term Effect of Salinity on Enhanced Biological Phosphorus Removal. *Proceedings of WEF/IWA Nutrient Removal and Recovery,* July 10–13, Denver, Colorado.

Welling, C.; Kennedy, A.; Wett, B.; Johnson, C.; Rutherford, B.; Baumler, R.; Bott, C. (2015) Improving Settleability and Enhancing Biological Phosphorus Removal through the Implementation of a Hydrocyclones. *Proceedings of the 89th Annual Water Environment Federation Technical Exposition and Conference*, Chicago, Illinois, September 24–28.

Wentzel, M. C.; Moosbrugger, R. E.; Sam–Soon, P. A. L. N. S, Ekama, G. A.; Marais, G. V. R. (1994) Tentative Guidelines for Waste Selection, Process Design, Operation and Control of Upflow Anaerobic Sludge Bed Reactors. *Water Sci. Technol.*, **30** (12), 31–42.

Wentzel, M.; Comeau, Y.; Ekama, G.; van Loosdrecht, M.; Brdjanovic, D.; Lopez-Vazquez, C. (2008) Enhanced Biological Phosphorus Removal. In *Biological Wastewater Treatment Principles, Modelling and Design*, Henze, M., van Loosdrecht, M., Ekama, G., Brdjanovic, D., (Eds.); IWA Publishing: London, U.K.

Wheeler, G. P.; Hegg, B. A.; Walsh, C. (2001) Capital Defense. *Water Environ. Technol.*, **13** (9), 126–131.

Wiegant, W. M. (2001) Experiences and Potential of Anaerobic Wastewater Treatment for Tropical Regions. *Water Sci. Technol.*, **44** (8), 107–113.

Wilford, J.; Conlon, T. P. (1957) Contact Aeration Sewage Treatment Plants in New Jersey. *Sew. Ind. Wastes*, **29**, 845–855.

Wilkinson, H. J. (1997) Continuous Online Analyzers Can Help Operators Work More Efficiently. *Oper. Forum*, **14** (7), 16–20.

Wilson, T. E. (1996) A New Approach to Interpreting Settling Data. *Proceedings of the 69th Annual Water Environment Federation Exposition and Conference*, Dallas, Texas, October 5–9; Water Environment Federation: Alexandria, Virginia.

Wilson, T. E.; Ballotti, E. F. (1988) Gould Tanks, Rectangular Clarifiers that Work. Paper Presented at 61st Annual Water Pollution Control Federation Conference, Dallas, Texas, October 3–7; Water Pollution Control Federation: Alexandria, Virginia.

Wuhrmann, K. (1954) High-Rate Activation Sludge Treatment and Its Relation to Sheam Sanitation. *Sew. Ind. Wastes*, **26**, 1–27.

Wuhrmann, K. (1964) Nitrogen Removal in Sewage Treatment Processes. *Verhandlungenden Int. Verein Limnol.*, **15**, 580.

Yoshioka, N.; Hotta, Y.; Tanaka, S.; Naito, S.; Tsugami, S. (1957) Continuous Thickening of Homogenous Flocculated Slurries. *Chem. Eng. Tokyo (Kagaku Kogaku)*, **21**, 66.

Young, T.; Crosswell, S.; Wendell, J. (2008) Comparison of Nitrogen Removal Performance in SBR Systems. *Proceedings of the 81st Annual Water Environment Federation Exposition and Conference*, Chicago, Illinois, October 18–22; Water Environment Federation: Alexandria, Virginia.

Yu, H.; Tay, J.-H.; Wilson, F. (1997) A Sustainable Municipal Wastewater Treatment Process for Tropical and Subtropical Regions in Developing Countries. *Water Sci. Technol.*, **35** (9), 191–198.

Yuan, Q.; Oleszkiewicz, J. (2010) Biomass Fermentation to Augment Biological Phosphorus Removal. *Chemosphere*, **78**, 29–34.

Yuan, Q.; Sparling, R.; Oleszkiewicz, J. (2009) Waste Activated Sludge Fermentation: Effect of Solids Retention Time and Biomass Concentration. *Water Res.*, **43**, 5180–5186.

Yuan, Q.; Sparling, R.; Oleszkiewicz, J. (2011) VFA Generation from Waste Activated Sludge: Effect of Temperature and Mixing. *Chemosphere*, **82**, 603–607.

Yuki, Y.; Takayanagi, E.; Abe, T. (1991) Design of Multi-Story Sewage Treatment Facilities. *Water Sci. Technol.*, **23** (10–12), 1733–1742.

Yunt, F. (1980) Results of Mixing Efficiency Tests with Norton Dome Aeration System at LA Glendale Treatment Plant. Los Angeles County Sanitation Districts, Whittier, California. (Cited in Mueller J., Boyle W., Pöpel H. (2002) *Aeration: Principles and Practice*; CRC Press, Boca Raton, Florida.)

12.0   Suggested Readings

Albertson, O. E. (1987) The Control of Bulking Sludge: From the Early Innovators to Current Practice. *J. Water Pollut. Control Fed.*, **59**, 172–182.

Albertson, O. E. (1993) Carbonaceous BOD Test: More Trouble Than It's Worth? *Water Environ. Technol.*, **5** (11), 16.

Albertson, O. E. (1995) Circular Secondary Sedimentation Tanks. *Proceedings of the WERF/IAWQ Secondary Clarifier Assessment Workshop*, Dallas, Texas; Water Environment Research Federation, Alexandria, Virginia.

Albertson, O. E.; Alfonso, P. (1994) Upgrading the Maxson WWTP Clarifier Performance. *Water Environ. Technol.*, **7** (3), 56.

Albertson, O. E.; Waltz, T. (1997) Optimizing Primary Clarification and Thickening. *Water Environ. Technol.*, **9** (12), 41–45.

ATV (Abwassertechnische Vereiningung) (1991) *Dimensioning of Single Stage Activated Sludge Plants Upwards from 5000 Total Inhabitants and Population Equivalents*, ATV Rules and Standards, Wastewater-Waste, UDC 628.356:628.32.-001.2(083), Issue No. 11/92.

Bain, R. E.; Johnson, W. S. (1988) Probing for Improved BNR, Automating Suspended Solids Control Helps a Plant Maintain Consistent Nutrient Removal. *Oper. Forum*, **15** (2), 23–26.

Bassett, B. D.; et al. (1994) Oxidation Ditch Conversation to VIP Process. *Proceedings of the 67th Annual Water Environment Federation Exposition and Conference* [CD-ROM]; Chicago, Illinois, October 13–17; Water Environment Federation: Alexandria, Virginia.

Benefield, L. D.; Judkins, J. F., Jr.; Parr, A. D. (1984) *Treatment Plant Hydraulics for Environmental Engineers*; Prentice Hall: Englewood Cliffs, New Jersey.

Benefield and Randall (1980) *Biological Process Design for Wastewater Treatment*; Prentice Hall: Englewood Cliffs, New Jersey.

Bidstrup, S. M.; Grady, C. P. L., Jr. (1988) SSSP Simulation of Single-Sludge Processes. *J. Water Pollut. Control Fed.*, **60**, 351–361.

Buchan, L. (1984) Microbiological Aspects. In *Theory, Design and Operation of Nutrient Removal Activated Sludge Process*; Water Research Commission: Pretoria, S. Africa.

Burke, T.; Hamilton, I. M.; Tomlinson, E. J. (1985) Treatment Process Management and Control, Instrumentation and Control of Water and Wastewater Treatment and Transport Systems. *Proceedings of the 4th International Association on Water Quality (IAWPRC) Workshop*, Houston, Texas; Denver, Colorado; IWA Publishing: London, England.

Buttz, J. (1992) Secondary Clarifier Stress Test at Laguna WWTP, Santa Rosa, California, Report. CH2M Hill: Oakland, California.

Buttz, J. (1992) Estimation of Current Laguna WWTP Capacity and Results of Plant Testing Program. Memorandum from CH2M Hill to City of Santa Rosa, California.

Camp, T. R. (1936) A Study of the Rational Design of Settling Tanks. *Sew. Works J.*, **8**, 742–758.

Camp, T. R. (1953) Studies of Sedimentation Basin Design. *Sew. Ind. Wastes*, **25**, 1–758.

Casey, J. P.; et al. (1975) *Non-Bulking Activated Sludge Process*, U.S. Patent 3,864,246.

Cashion, B. S.; Keinath, T. M. (1983) Influence of Three Factors on Clarification in the Activated Sludge Process. *J. Water Pollut. Control Fed.*, **55**, 1331–1337.

Chao, J. L.; Trussell, R. R. (1980) Hydraulic Design of Flow Distribution Channels. *J. Environ. Eng. Div., Am. Soc. Civ. Eng.*, **106** (EE2), 321–324.

Chapman, D. T. (1983) The Influence of Process Variables on Secondary Clarification. *J. Water Pollut. Control Fed.*, **55**, 1425–1434.

Chapman, D. T. (1985) Final Settler Performance During Transient Loading. *J. Water Pollut. Control Fed.*, **57**, 227–234.

Cherchi, C.; Onnis-Hayden, A.; Gu, A. Z. (2008). Investigation of MicroCTM as an Alternative Carbon Source for Denitrification. *Proceedings of the 81st Annual Water Environment Federation Exposition and Conference* [CD-ROM]; Chicago, Illinois, October 18–22; Water Environment Federation: Alexandria, Virginia.

Chudoba, J.; et al. (1973) Control of Activated Sludge Bulking II: Selection of Micro-organisms by Means of a Selector. *Water Res. (G.B.)*, **7**, 1389.

Copp, J. B.; Dold, P. L.; (1998) Comparing Sludge Production under Aerobic and Anoxic Conditions. *Water Sci. Technol.*, **38** (1), 285–294.

Crosby, R. M. (1983) *Clarifier Newsletter*; Crosby, Young, and Associates: Plano, Texas.

Dakers, J. L. (1985) Automation and Monitoring of Small Sewage Treatment Works. Instrumentation and Control of Water and Wastewater Treatment and Transport Systems. *Proceedings of the 4th International Association on Water Quality (IAWPRC) Workshop*, Houston, Texas; Denver, Colorado; IWA Publishing: London, England.

Daigger, G. T.; Roper, R. E., Jr. (1985) The Relationship Between SVI and Activated Sludge Settling Characteristics. *J. Water Pollut. Control Fed.*, **57**, 859–866.

Deeny, K. J.; et al. (1988) Evaluation of Full-Scale Activated Sludge Systems Utilizing Powdered Activated Carbon Addition with Wet Air Regeneration. Paper presented at 61st Annual Water Pollution Control Federation Conference, Dallas, Texas; Water Pollution Control Federation: Alexandria, Virginia.

Deep Shaft Technology Inc. (1988) *Manufacturer's Literature*; Simmons Group of Companies: Calgary, Alberta, Canada.

Dick, R. I.; Vesilind, P. A. (1969) The Sludge Volume Index—What Is It? *J. Water Pollut. Control. Fed.*, **41**, 1285.

Dittmar, D. (1987) Secondary Sedimentation Evaluation Operating Data Review. Tech. Memo. Municip. Metro. Seattle, Washington.

Dold, P.; Takacs, I.; Mokhayeri, Y.; Nichols, A.; Jinojosa, J.; Riffat, R.; Bailey, W.; Murthy, S.; (2007). Denitrification with Carbon Addition—Kinetic Considerations. *Proceedings of the WEF/IWA Nutrient Removal Specialty Conference* [CD-ROM], Baltimore, Maryland; Water Environment Federation: Alexandria, Virginia.

Duncan, D. (2000) Plant Automates Sludge Measurement in Clarifier. *WaterWorld*, **16** (2), 34–36.

Eikelbloom, D. H. (1977) Identification of Filamentous Organisms in Bulking Sludge. *Process Water Technol.*, **8**, 153.

Eikelbloom, D. H. (1982) Biosorption and Prevention of Bulking Sludge by Means of a High Floc Loading. In *Bulking of Activated Sludge: Preventative and Remedial Methods*, Chambers, B., Tomlinson, E. J. (Eds.); Horwood Ltd.: Chichester, England.

Eikelboom, D. H.; van Buijsen, H. J. J. (1981) *Microscopic Sludge Investigation Manual*; TNO Research Institute of Environmental Hygiene: Netherlands.

Ekama G. A.; Marais, G. V. R. (1986) Sludge Settleability and Secondary Settling Tank Design Procedures. *Water Pollut. Control*, **85** (11), 101.

Ekama, G. A.; et al. (1996) *Theory, Modelling Design and Operation of Secondary Settling Tanks*, Sci. Tech. Rep. Ser.; International Association of Water Quality: London.

Franklin, R. J. (2001) Full-Scale Experiences with Anaerobic Treatment of Industrial Wastewater. *Water Sci. Technol.*, **44** (8), 1–6.

Garrett, M. T.; Jr. (2007) How to Avoid a Solids Washout in an Activated Sludge Plant. *Proceedings 10th International Water Association Specialised Conference, Design, Operation and Economics of Large Wastewater Treatment Plants*, Vienna, Austria, Sep.

Gaudy, A. F.; Gaudy, E. T. (1980) *Microbiology for Environmental Scientists and Engineers*; McGraw-Hill: New York.

Gould, R. H. (1943) Final Settling Tanks of Novel Design. *Water Work Sew.*, **90**, 133–136.

Gould, R. H. (1950) Wards Island Plant Capacity Increased by Structural Changes. *Sew. Ind. Wastes*, **22**, 997–1003.

Gujer, W.; Henze, M. (1991) Activated Sludge Modelling and Simulation. *Water Sci. Technol.*, **23**, 1011.

Gujer, W.; Larsen, T. A. (1995) The Implementation of Biokinetics and Conservation Principles in ASIM. *Water Sci. Technol.*, **31** (2), 257.

Hale, F. D.; Garver, S. R. (1983) Viscous Bulking of Activated Sludge. Paper Presented at 61st Annual Water Pollution Control Federation Conference, Atlanta, Georgia; Water Pollution Control Federation: Washington, D.C.

Harleman, D.; Murcott, S. (2001a). An Innovative Approach to Urban Wastewater Treatment in the Developing World. *Water21*, **June**, 44–59.

Harleman D.; Murcott, S. (2001b). CEPT: Challenging the Status Quo. *Water21*, **June**, 57.

Henze, M.; Gujer, W.; Mino, T.; Matsuo, T.; Wentzel, M. C.; and Marias, G. V. R. (1995a) Activated Sludge Model No. 2., Sci. Tech. Rep. No. 3; International Association of Water Quality: London, England.

Henze, M.; Gujer, W.; Mino, T.; Matsuo, T.; Wentzel, M. C.; and Marias, G. V. R. (1995b) Wastewater and Biomass Characterization for the Activated Sludge Model No. 2: Biological Phosphorus Removal. *Water Sci. Technol.*, **31** (2), 13.

IAWPRC Task Group on Mathematical Modelling for Design and Operation of Biological Wastewater Treatment Processes (1995) Activated Sludge Model No. 2, Sci. Tech. Rep. No. 3; International Association of Water Quality: London, England.

Jimenez, J. A.; Parker, D. S.; Bratby, J. R.; Schuler, P. F.; Campanella, K. V. and Freedman, S. D. (2005) In the Absence of the Blending Policy: A Novel High Rate Biological Treatment Process. *Proceedings of the 78th Annual Water Environment Federation Exposition and Conference* [CD-ROM]; Washington, D.C., October 29–November 2; Water Environment Federation: Alexandria, Virginia.

Jenkins, D. (1988) Selectors—Bulking Control Effective Contemporary Wastewater Treatment Processes. Workshop note, Dep. Eng. Prof. Develop.; University Wisconsin, Madison, Wisconsin.

Kato, M. T.; Field, J. A.; Lettinga, G. (1997) The Anaerobic Treatment of Low Strength Wastewater in UASB and EGSB Reactors. *Water Sci. Technol.*, **36** (6–7), 375–382.

Keinath, T. M. (1976) *Design and Operational Criteria for Thickening of Biological Sludges*, NTIS No. PB-262 967; Springfield, Virginia.

Keller, J.; Schultze, M.; Wagner, D. (2001) Lawrence Kansas: Detailed Design Issues for a Ballasted Flocculation System. *Proceedings of the 74th Annual Water Environment Federation Exposition and Conference* [CD-ROM]; Atlanta, Georgia, October 13–17; Water Environment Federation: Alexandria, Virginia.

Kelly, K. F.; O'Brien, D. P.; McConnell, W.; Morris, J. (1995) Two-Tray Clarifiers Yield Positive Results. *Water Environ. Technol.*, **7** (12), 35.

Kluitenberg, E. H.; Cantrell, C. (1994) *Percent Treated Analysis of Demonstration Combined Sewer Overflow Control Facilities*, Technical Memorandum, RPO-MOD-TM17.00; Rouge River National Wet Weather Demonstration Project: Wayne County, Michigan.

Lee, S. E.; et al. (1982) The Effect of Aeration Basin Configuration on Activated Sludge Bulking at Low Organic Loadings. *Water Sci. Technol.*, **14**, 407.

Lettinga, G.; Rebac, S.; Zeeman, G. (2001) Challenge of Psychrophilic Anaerobic Wastewater Treatment. *TRENDS Biotechnol.*, **19** (9), 363–370.

Lutge, T. V. (1969) Hydraulic Control Utilizing Submerged Effluent Collectors. *J. Water Pollut. Control Fed.*, **41**, 1451–1455.

Lynggaard-Jensen, A.; Harremöes, P. (1996) Sensors in Wastewater Technology. *Water Sci. Technol.*, **33**, 1.

Manning, William T.; Jr.; Garrett, M. T., Jr.; Malina, J. F., Jr. (1999) Sludge Blanket Response to Storm Surge in an Activated Sludge Clarifier. *Water Environ. Res.*, **71**, 432–442.

Marais, G. V. R.; Ekama, G. A. (1984) *Theory, Design and Operation of Nutrient Removal Activated Sludge Processes*; Water Research Commission: South Africa.

McCarty, P. L. (1975) Stoichiometry of Biological Reactions. *Prog. Water Technol.*, **7**, 157.

McClintock, S. A.; Randall, C. W.; Pattarkine, V. M. (1993) Effects of Temperature and Mean Cell Residence Time on Biological Nutrient Removal Processes. *Water Environ. Res.*, **65**, 110–118.

McHugh, S.; O'Reilly, C.; Mahoney, T.; Colleran, E.; O'Flaherty, V. (2003) Anaerobic Granular Sludge Bioreactor Technology. *Rev. Environ. Sci. Biotechnol.*, **2**, 225–245.

McKinney, R. E. (1971) Newsletter of Environmental Pollution Control Services, Inc.: Lawrence, Kansas.

Mokhayeri, Y.; Nichols, A.; Murthy, S.; Riffat, R.; Dold, P.; Takacs, I. (2006) Examining the Influence of Substrates and Temperature on Maximum Specific Growth Rate or Denitrifiers. *Water Sci. Technol.*, **54** (8), 155–162.

Morton, R.; Henry, B.; Kriebel, S. (2000) Using Shotcrete for Repair of Concrete Structures in a High-Purity Oxygen Activated Sludge System. *Proceedings of the 73rd Annual Water Environment Federation Technical Exposition and Conference* [CD-ROM], Anaheim, California, October 14–18; Water Environment Federation: Alexandria, Virginia.

Mulbarger, M. C.; Zacharias, K. L.; Nazir, F.; Patrick, D. (1985) Activated Sludge Reactor/ Final Clarifier Linkages: Success Demands Fundamental Understanding. *J. Water Pollut. Control Fed.*, **57**, 921–928.

Namkung, E.; Rittmann, B. E. (1987) Estimating Volatile Organic Compound Emissions from Publicly Owned Treatment Works. *J. Water Pollut. Control Fed.*, **59**, 670–678.

Narayanan, B.; Hough, S. G. (2004) Taking the "Waste" Out of Waste Activated Sludge— New Process Configuration Uses Waste Activated Sludge to Treat Wastewater More Efficiently. *Proceedings of the 77th Annual Water Environment Federation Technical Exposition and Conference* [CD-ROM]; New Orleans, Louisiana, October 9–13; Water Environment Federation: Alexandria, Virginia.

Nichols, A. J.; Jinojosa, J.; Riffat, R.; Dold, P.; Takacs, I.; Bott, C.; Bailey, W.; Murthy, S. (2007). Maximum Methanol-Utilizer Growth Rate: Impact of Temperature on Denitrification. *Proceedings of the 80th Annual Water Environment Federation Exposition and Conference* [CD-ROM], San Diego, California, October 10–13; Water Environment Federation: Alexandria, Virginia.

Oliveira, S. C.; von Sperling, M. (2008) Performance and Reliability of Post-treatment Options for the Anaerobic Treatment of Domestic Wastewater. *Proceedings IX Latin American Workshop and Symposium on Anaerobic Digestion*, Easter Island, Chile; October 19–23.

Oliveira, S. C.; von Sperling, M. (2008) Reliability Analysis of Wastewater Treatment Plants. *Water Res.*, **42**, 1182–1194.

Parker, D. S.; Kaufman, W. J.; Jenkins, D. (1970) *Characteristics of Biological Flocs in Turbulent Regimes*, SERI Report No. 70-5; University of California: Berkeley, California.

Parker, D. S.; Barnard, J. ; Daigger, G. T.; Tekippe, R. J.; Wahlberg, E. J. (2001) The Future of Chemically Enhanced Primary Treatment: Evolution not Revolution. *Water21*, **June**, 49.

Parker, D. S.; Esquer, M.; Hetherington, M.; Malik, A.; Robison, D.; Wahlberg, E. J.; Wang, J. K. (2000) Assessment and Optimization of a Chemically Enhanced Primary Treatment System. *Proceedings of the 73rd Annual Water Environment Federation Technical Exposition and Conference* [CD-ROM], Anaheim, California, October 14–18; Water Environment Federation: Alexandria, Virginia.

Pettit, M.; Gary, D.; Morton, R.; Friess, P.; Caballero, R. (1997) Operation of a High-Purity Oxygen Activated Sludge Plant Employing and Anaerobic Selector and Carbon Dioxide Stripping. *Proceedings of the 70th Annual Water Environment Federation Exposition and Conference* [CD-ROM]; Chicago, Illinois, October 18–22; Water Environment Federation: Alexandria, Virginia.

Pike, E. B.; Curds, C. R. (1972) The Microbial Ecology of the Activated Sludge Process. In *Microbial Aspects of Pollution*, Sykes, G., Skinner, F. A., (Eds.); John Wiley and Sons, Inc.: New York.

Rich, L. G. (1974) *Unit Operations of Sanitary Engineering*; Clemson University: South Carolina.

Riddell, M. D. R.; Lee, J. S.; Wilson, T. E. (1983) Method for Estimating the Capacity of an Activated Sludge Plant. *J. Water Pollut. Control Fed.*, **55**, 360–368.

Roberts, P. V.; Munz, C.; Dandliker, P. (1984a) Modeling Volatile Organic Solute Removal by Surface and Bubble Aeration. *J. Water Pollut. Control Fed.*, **56**, 157–163.

Roberts, P. V.; et al. (1984b) *Volatilization of Organic Pollutants in Wastewater Treatment: Model Studies*, EPA-600/2-84-047; U.S. Environmental Protection Agency: Cincinnati, Ohio.

Rusten, B.; Hem, L.; Odegaard, H. (1995a) Nitrification of Municipal Wastewater in Moving-Bed Biofilm Reactors. *Water Environ. Res.*, **67**, 75–86.

Rusten, B.; Hem, L. J.; Odegaard, H. (1995b) Nitrogen Removal from Dilute Wastewater in Cold Climate Using Moving-Bed Biofilm Reactors. *Water Environ. Res.*, **67**, 65–74.

Rusten, B.; Kolkinn, O.; Odegaard, H. (1997) Moving-Bed Biofilm Reactors and Chemical Precipitation for High Efficiency Treatment of Wastewater from Small Communities. *Water Sci. Technol.*, **35** (6), 71.

Rusten, B.; Siljudalen, J. G.; Bungun, S. (1995) Moving-Bed Biofilm Reactors for Nitrogen Removal: From Initial Pilot Testing to Start-Up of the Lillehammer WWTP. *Proceedings of the 68th Annual Water Environment Federation Technical Exposition and Conference* [CD-ROM], Miami Beach, Florida, October 21–25; Water Environment Federation: Alexandria, Virginia.

Rusten, B.; Siljudalen, J. G.; Nordeidet, B. (1994) Upgrading to Nitrogen Removal with the KMT Moving-Bed Biofilm Process. *Water Sci. Technol.*, **29** (12), 185.

Samstag, R. W. (1988) Studies in Activated Sludge Sedimentation at Metro: Final Report, In-house Rep.; Municip. Metro. Seattle, Washington.

Seghezzo, L.; Zeeman, G.; van Lier, J.B.; Hamelers, H. V. M.; Lettinga, G. (1998) A Review: The Anaerobic Treatment of Sewage in UASB and EGSB Reactors. *Biores. Technol.*, **65**, 175–190.

Shao, Y. J.; Jenkins, D. L. (1997) Polymer Addition as a Solution to Nocardia Foaming Problems. *Water Environ. Res.*, **69**, 25–27.

Sizcka, J.; Sandino, J.; Onderko, R.; Sigmund, T. (2007) Biologically and Chemically Enhanced Clarification for Improved Treatment of Wet-Weather Flows. *Proceedings of the 80th Annual Water Environment Federation Exposition and Conference* [CD-ROM], San Diego, California, October 10–13; Water Environment Federation: Alexandria, Virginia.

Stephenson, T.; Judd, S.; Jefferson, B.; Brindle, K. (2000) *Membrane Bioreactors for Wastewater Treatment*; IWA Publishing: London, England.

Tekippe, R. J. (1986) Critical Literature Review and Research Needed on Activated Sludge Secondary Clarifiers. Report prepared for Municipal Environmental Research Laboratory; Office of Research and Development; U.S. Environmental Protection Agency: Washington, D.C.

Torpey, W. N. (1948) Practical Results of Step Aeration. *Sew. Works J.*, **20**, 781–788.

Tomlinson, E. J.; Chambers, B. (1979) Methods for Prevention of Bulking in Activated Sludge. *Water Pollut. Control*, **78**, 524.

Treybal, R. E. (1968) *Mass Transfer Operations*; McGraw-Hill: New York.

U.S. Army Corps of Engineers (1976) *HEC L7520, Generalized Computer Program, Storage, Treatment, Overflow, Runoff Model "STORM" Users Manual*; Davis, California.

U.S. Environmental Protection Agency (1977) Small Community Wastewater Treatment Facilities: Biological Treatment Systems, EPA Technol. Transfer; U.S. Environmental Protection Agency: Washington, D.C.

U.S. Environmental Protection Agency (1978) *A Comparison of Oxidation Ditch Plants to Competing Processes for Secondary and Advanced Treatment of Municipal Wastes*, EPA-600/2-78-051; U.S. Environmental Protection Agency: Washington, D.C.

U.S. Environmental Protection Agency (1979) *Inspector's Guide: To Be Used in the Evaluation of Municipal Wastewater Treatment Plants*; U.S. Environmental Protection Agency: Washington, D.C.

U.S. Environmental Protection Agency (1980) *Estimate of Effluent Limitations to be Expected from Properly Operated and Maintained Treatment Works*; U.S. Environmental Protection Agency, Office Water Program Operations: Washington, D.C.

U.S. Environmental Protection Agency (1981) *Performance of Activated Sludge Processes: Reliability, Stability, Variability*, EPA-6500/2-81-227; U.S. Environmental Protection Agency: Washington, D.C.

U.S. Environmental Protection Agency (1982) *Technology Assessment of the Deep Shaft Biological Reactor*, EPA-600/2-82-002; U.S. Environmental Protection Agency: Washington, D.C.

U.S. Environmental Protection Agency (1983a) *Design Manual for Municipal Wastewater Stabilization Ponds*, EPA-625/1-83-015; U.S. Environmental Protection Agency: Cincinnati, Ohio.

U.S. Environmental Protection Agency (1984a) *Needs Survey*; U.S. Environmental Protection Agency: Washington, D.C.

U.S. Environmental Protection Agency (1984b) *Technical Support Document for Proposed Regulations Under Section 304(d)(4) of the Clean Water Act, as Amended*, PB-85-111397; U.S. Environmental Protection Agency, Office Water Program Operations: Washington, D.C.

U.S. Environmental Protection Agency (1986a) *A Perspective on Performance Variability in Municipal Wastewater Treatment Facilities*, EPA-600/D-86–064; U.S. Environmental Protection Agency: Washington, D.C.

U.S. Environmental Protection Agency (1986b) *Summary Report, Sequencing Batch Reactors*, EPA-625/8-86-011; Technol. Transfer; U.S. Environmental Protection Agency: Washington, D.C.

U.S. Environmental Protection Agency (1987a) *Analysis of Full-Scale SBR Operation at Grundy Center, Iowa*, EPA-600/J-87-065; U.S. Environmental Protection Agency: Washington, D.C.

U.S. Environmental Protection Agency (1987b) *Post Construction Performance of Schreiber Counter-Current Aeration Facilities*, EPA-600/2-87-089; U.S. Environmental Protection Agency: Washington, D.C.

U.S. Environmental Protection Agency (1988) *Toxicity Reduction Evaluation Protocol for Municipal Wastewater Treatment Plants*, EPA-600/2-88-062; U.S. Environmental Protection Agency: Washington, D.C.

U.S. Environmental Protection Agency (1990a) *Biolac™ Technology Evaluation*, EPA-430/09-90-014; U.S. Environmental Protection Agency: Washington, D.C.

U.S. Environmental Protection Agency (1990b) *A Preliminary Assessment of High Biomass Systems*, EPA-600/9-90-036; U.S. Environmental Protection Agency: Washington, D.C.

U.S. Environmental Protection Agency (1990c) *Assessment of the Biolac™ Technology*, EPA-430/09-90-013. U.S. Environmental Protection Agency: Washington, D.C.

U.S. Environmental Protection Agency (1990d) Captor Study of Moundsville/Glendale Municipal Wastewater Treatment Works. *Proceedings U.S. EPA Municipal Wastewater Treatment Technology Forum*, EPA-430/09-90-014; U.S. Environmental Protection Agency: Washington, D.C.

U.S. Environmental Protection Agency (1992a) *Evaluation of Oxidation Ditches for Nutrient Removal*, EPA-832/R-92-003; U.S. Environmental Protection Agency: Washington, D.C.

U.S. Environmental Protection Agency (1992b) *Technical Evaluation of the Vertical Loop Reactors Process Technology*, EPA-832/R-92-007. U.S. Environmental Protection Agency: Washington, D.C.

U.S. Environmental Protection Agency (1996) *Needs Survey*; U.S. Environmental Protection Agency: Washington, D.C.

U.S. Environmental Protection Agency (1997) *1996 Clean Water Needs Survey Report to Congress*, EPA-832/R-97-003; U.S. Environmental Protection Agency, Office Water: Washington, D.C.

Van den Eynde, E.; et al. (1982) Relation Between Substrate Feeding Pattern and Development of Filamentous Bacteria in Activated Sludge. In *Bulking of Activated Sludge: Preventive and Remedial Methods*, Chambers, B.; Tomlinson, E. J., Eds.; Horwood Ltd.: Chichester, England.

Van der Roest, H. F.; Lawrence, D. P.; van Bentem, A. G. N. (2002) *Membrane Bioreactors for Municipal Wastewater Treatment, STOWA*; IWA Publishing: London, England.

van Niekerk, A. M. (1986) Competitive Growth of Flocculant and Filamentous Micro-organisms in Activated Sludge. Ph.D. thesis, University of California, Berkeley, California.

Weber, W. J., Jr. (1972) *Physiochemical Processes for Water Quality Control*; John Wiley and Sons, New York.

Wahlberg, E. J.; Merrill, D. T.; Parker, D. S. (1995) Troubleshooting Activated Sludge Secondary Clarifier Performance using Simple Diagnostic Tests. *Proceedings of the 68th Annual Water Environment Federation Technical Exposition and Conference* [CD-ROM], Miami Beach, Florida, October 21–25; Water Environment Federation: Alexandria, Virginia.

Water Environment Federation (2013) *Automation of Water Resource Recovery Facilities*, Manual of Practice No. 21; Water Environment Federation: Alexandria, Virginia.

Water Environment Federation (2015) *Shortcut Nitrogen Removal—Nitrite Shunt and Deammonification*; Water Environment Federation: Alexandria, Virginia.

Wheeler, M. L.; et al. (1984) The Use of a Selector for Bulking Control at the Hamilton, Ohio, U.S.; Water Pollution Control Facility. *Water Sci. Technol.*, **16**, 35.

White, M. J. D. (1975) Settling of Activated Sludge, Technical Report TR11; Water Research Centre: Stevenage, United Kingdom.

White, M. J. D. (1976) Design and Control of Secondary Settling Tanks. *Water Pollut. Control*, **75** (4), 459.

Wilson, T. E.; et al. (1984) Operating Experiences at Low Solids Retention Time. *Water Sci. Technol.*, **16**, 661.

CHAPTER 13

Integrated Biological Treatment

William C. McConnell; Leon Downing; Satej Kulkarni;
Laura Marcolini, P.E.; and Coenraad Pretorius, P.E.

1.0   Introduction to Integrated Biological Treatment

1.1   Introduction

Integrated biological treatment (IBT) processes are two-stage, series, dual, or coupled processes. In this manual, the term "integrated biological treatment" is used to denote (1) conventional coupling in series of two or more different reactors, at least one of which is a fixed-biofilm reactor, and (2) the integrated fixed-film activated sludge (IFAS) process. It is recognized that the IFAS process differs from the conventional coupled-in-series processes, in that the dual suspended-growth and biofilm processes occur within the same reactor.

Much of the terminology used to describe IBT processes is covered in other chapters of this manual that discuss the individual parent processes. Chapter 11, Biofilm Reactor Technology and Design, and Chapter 12, Suspended-Growth Biological Treatment, provide descriptions of common terms.

1.2   Process Objectives of Integrated Systems

Conventional IBT systems use a fixed-biofilm reactor (first stage) in series with a suspended-growth biological reactor (second stage). The fixed-biofilm reactor typically consists of a biological tower, and the suspended-growth reactor is typically an aeration basin or small contact channel. This combination results in a two-stage coupled unit process that has unique design parameters with treatment efficiency capabilities that often exceed those of the individual parent systems. The system may also consist of three stages, with the first and the last suspended-growth systems, and the second a fixed-biofilm growth system. The IFAS process is yet another configuration, developed subsequent to conventional IBT systems, that combines both fixed film and suspended growth in the same reactor tank.

Designers have used integrated processes widely to utilize the advantages that these combined systems provide. Advantages of fixed-biofilm processes include their shockload resistance, volumetric efficiency, low energy requirements, and low maintenance requirements. Advantages of suspended-growth biological treatment processes include high-quality effluent and the flexibility to operate under various treatment modes to achieve different effluent quality objectives. By combining these two processes in IBT systems, designers capitalize on the advantages of each individual parent (single-stage) process.

Typically, in conventional IBT processes, the fixed biofilm primarily removes soluble five-day biochemical oxygen demand ($sBOD_5$). The suspended-growth process—and the suspended-growth/biofilm processes in IFAS systems—can provide a variety of functions, including flocculation to capture colloidal substrate and improve clarification, removal of residual $sBOD_5$, nitrification, denitrification, and enhanced biological phosphorus removal to meet advanced wastewater treatment requirements.

For design of IBT systems, the fixed-film process and the suspended-growth process must be viewed as a combined system and developed accordingly. For example, in conventional IBT systems, removal of biodegradable organic matter drives design of the fixed-biofilm reactor. Design of the suspended-growth process must then consider the amount of biodegradable organic matter removed in the fixed-biofilm process and establishment of conditions necessary for flocculation, as well as the treatment goals (e.g., nitrification or denitrification) of the combined process.

This chapter does not repeat design guidance for individual treatment processes already described—refer to Chapter 11 for biofilm processes and to Chapter 12 for suspended-growth biological treatment. Described in this chapter are special descriptions, factors, and design considerations for IBT systems.

2.0   Overview of Integrated Biological Treatment Systems

2.1   Biofilm and Activated Sludge Systems with Segregated Return Activated Sludge

Many combinations of treatment process trains are possible, depending on the type of parent process used, loading to the treatment units, treatment goals, and the point at which biological or recycle solids are reintroduced to the main flow stream (Harrison and Daigger, 1984). Conventional integrated biological processes are divided into two groups: those that have low to moderate organic loadings applied to the fixed-film reactor and those with high (roughing) organic loadings.

The processes also are classified in terms of return of mixed liquor from the suspended-growth stage. The process schematics in Figure 13.1 illustrate the various, common alternative methods for returning secondary sludge, or reaeration, and includes terms used by designers to distinguish process modes. The figure is not a comprehensive list of IBT systems; other process configurations are also discussed in this section.

2.1.1   Trickling Filters/Biofilters

2.1.1.1   Trickling-Filter/Solids Contact   The trickling-filter/solids contact (TF/SC) process uses a trickling filter that has low to moderate organic loads followed by a small, aerated solids contact tank or channel with secondary clarifiers that polish the trickling filter effluent through flocculation and additional organics removal (Norris et al., 1982). Because most organic substrate is removed in the trickling filter, there is less filamentous microbial growth in the aerated solids contact tank or channel and, therefore, TF/SC flocs are more fragile than traditional activated sludge flocs. For that reason, flocculator clarifiers typically are used with the TF/SC process.

By combining the trickling filter with a solids-contact tank or channel, Parker and Matasci (1989) found the trickling-filter reactor size is smaller than required when used alone. The magnitude of the size reduction will depend on the specific application and effluent requirements.

Conventional TF/SC does not include return solids reaeration before the solids contact tank or channel. However, it has been observed that reaerating the return sludge before the solids contact tank can enhance bioflocculation and may allow for a smaller tank or channel to be used. When both solids reaeration and solids contact are used, the acronym TF/SCR is used.

One significant benefit of the TF/SC process is the relatively low power requirements for the suspended-growth system because of the ability of the trickling filter to remove the majority of the soluble substrate ($sBOD_5$) (Witzgall et al., WEFTEC 2013). Another benefit is the ability to upgrade existing rock trickling filters through polishing the fixed-growth effluent by using return activated sludge (RAS) as a bioflocculating agent (Krumsick et al., 1984; Matasci et al., 1988). The TF/SC process has also been shown to achieve consistent nitrification even in cold climate conditions (Parker et al., 1998).

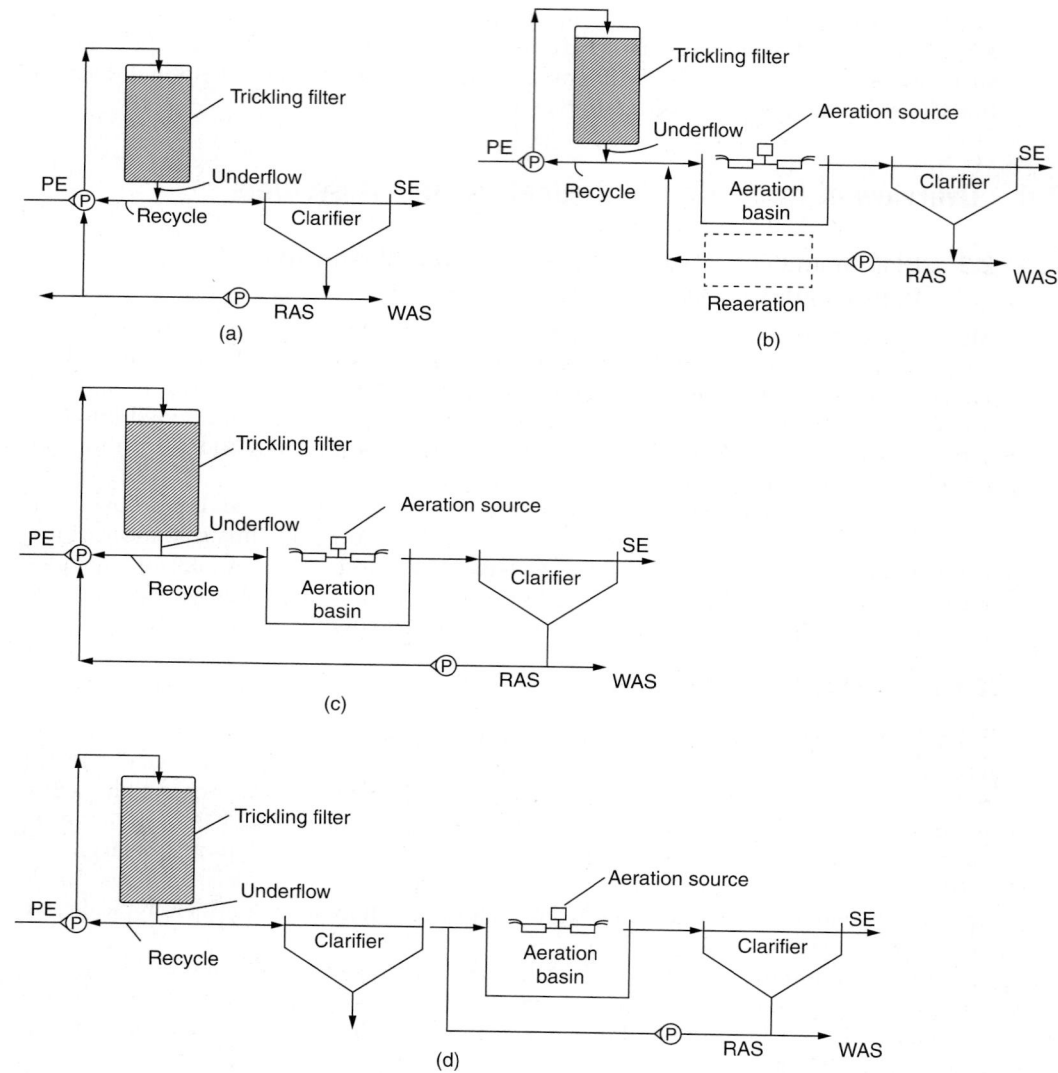

Figure 13.1 Combined process operations: (a) activated biofilter, (b) trickling-filter solids contact and roughing-filter activated sludge, (c) biofilter-activated sludge, and (d) trickling-filter activated sludge (PE = primary effluent and SE = secondary effluent).

2.1.1.2 Roughing Filter Activated Sludge A common method of upgrading existing activated sludge facilities is to install a roughing filter ahead of the activated sludge process.

As shown in Figure 13.1, both TF/SC and roughing filter activated sludge (RF/AS) processes have the same process flow schematic. However, with RF/AS, a much smaller trickling filter is used. This requires a larger suspended-growth biological reactor to provide a significant amount of oxygen for $sBOD_5$ removal and biomass stabilization.

This differs from the TF/SC process, in which the trickling filter is larger and provides almost all of the $sBOD_5$ removal, thus allowing the solids contact channel or tank to be smaller and provide enhanced solids flocculation and effluent clarity. Because some of the $sBOD_5$ is left to be metabolized in the activated sludge tank in the RF/AS process, filament growth is usually sufficient and results in a more stable floc.

Facility-specific circumstances that influence the balance between capital and operating expenses often will help determine whether it is best to use the TF/SC or RF/AS process at a particular facility. During periods of reduced loadings, RF/AS facilities can be operated as TF/SC, thus using less suspended-growth biological reactor volume and less aeration. Typical loading rate ranges for the RF/AS and TF/SC processes will be discussed in Section 3.

2.1.1.3 Trickling-Filter Activated Sludge The trickling-filter activated sludge (TF/AS) process is designed for high organic loads similar to those of RF/AS or biofilter activated sludge (BF/AS, to be discussed below). A unique feature of TF/AS, however, is that an intermediate clarification step is provided between the biofilm and suspended-growth biological reactors. The intermediate clarification removes sloughed solids from the biofilm reactor effluent before the effluent enters the suspended-growth reactor. This solids removal step reduces the effects the sloughed solids may have on the suspended-growth process dynamics. Because an intermediate clarification step is used, the TF/AS process is not a directly coupled system. The TF/AS schematic is illustrated in Figure 13.1.

The benefit of the TF/AS integrated process is that solids generated from carbonaceous biochemical oxygen demand ($cBOD_5$) removal in the trickling filter are removed before second-stage activated sludge treatment. This is often a preferred approach at facilities where ammonia removal is required, as the suspended-growth stage of the process can be designed to be dominated by nitrifying microorganisms. The intermediate clarification step can reduce required nitrifying aeration volume, which is a significant advantage for larger facilities or facilities processing higher strength wastewater (Daigger and Boltz, 2011).

The claim of significantly reduced oxygen demand or improved solids settleability from use of intermediate clarification is not universally accepted. This step is often eliminated because of the opposite claim: improved sludge settleability in the secondary clarifier when the trickling-filter sloughs directly to the suspended-growth reactor. The reduced oxygen demand afforded by intermediate clarification may not justify the capital and operating costs associated with intermediate clarification.

2.1.1.4 External Nitrification Biological Nutrient Removal Activated Sludge The external nitrification biological nutrient removal activated sludge (ENBNRAS) process is characterized by inserting a trickling filter midway through the suspended-growth system, with the aim of feeding it an influent with reduced organic load so as to encourage nitrification (Hu et al., 2000). It is used at facilities where both nitrogen and phosphorus removals are performed. In these facilities, the primary effluent (or potentially raw influent) is fed to a suspended-growth system anaerobic zone to maximize phosphorus release and the uptake of readily biodegradable substrate. The effluent from the anaerobic zone, low in readily biodegradable substrate, is fed to an intermediate clarifier with the clarifier overflow fed to the nitrifying trickling filter for nitrification. Clarifier underflow is combined with nitrified trickling-filter effluent and fed to an anoxic zone for denitrification, before polishing in an aerobic zone and secondary clarification, as shown in Figure 13.2 (Muller et al., 2006).

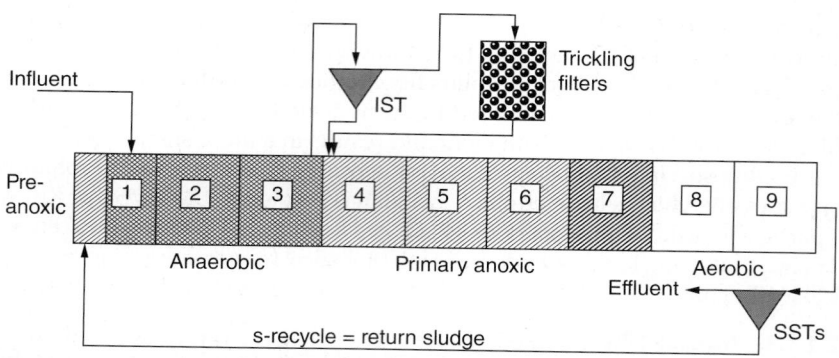

FIGURE 13.2   Process flow diagram for external nitrification biological nutrient removal activated sludge (Muller et al., 2006). Reprinted with permission from the South African Water Research Commission.

2.2   Biofilm and Activated Sludge Systems with Common Return Activated Sludge

These variations of IBT processes are characterized by the use of the RAS stream from the suspended-growth process in direct contact with the biofilm reactor/attached growth media.

2.2.1   Activated Biofilter

Figure 13.1 illustrates the activated biofilter (ABF) process schematic. The ABF uses a biofilm reactor to treat low to moderate organic loads; a final clarifier follows immediately, with no suspended-growth biological reactor in between.

Unlike conventional trickling-filter operation, return sludge is incorporated with the primary effluent and recycled over the biofilm media. Thus, suspended-growth contact is achieved as return sludge and primary effluent are mixed and passed through the biofilm reactor. Improved solids settleability usually occurs with the ABF process. A theory of why the ABF produces low sludge volume indices (SVIs) is that the high food-to-microorganism ratio ($F{:}M$) and plug flow of the trickling filter allow heterotrophic bacteria to be more competitive than filamentous bacteria, similar to what has been observed in plug-flow systems (refer to Chapter 12 for more information). Because the return sludge is incorporated with the primary effluent, high-rate plastic or redwood media must be used in the ABF biofilm reactor rather than rock, due to concerns about plugging.

Without short-term aeration following the biofilm reactor, the ABF process has been observed to perform poorly in cold climates. Therefore, to overcome these problems, the ABF process was later modified to include a relatively small aeration basin downstream of the ABF tower. This integrated configuration is known as the biofilter activated sludge (BF/AS) process (Harrison and Timpany, 1988).

2.2.2   Biofilter Activated Sludge

The BF/AS process is similar to the RF/AS, similar that the RAS is recycled to the biofilm reactor as it is in the ABF process. Also, similarly to the RF/AS, the BF/AS process is designed for high organic loads. With the BF/AS process, incorporating the RAS recycle to the biofilm reactor has sometimes reduced bulking from filamentous bacteria, especially with food-processing wastes. This may be a result of the plug-flow effect, as discussed above in connection with the ABF process. Although it usually improves solids settleability, sludge recycle to the biofilm may reduce oxygen available to the biofilm as some of it is consumed in the liquid phase.

2.2.3   Integrated Fixed-Film Activated Sludge
The IFAS process is a different configuration of IBT systems, in which the attached- and suspended-growth environments are combined into one bioreactor system, so the attached and suspended biomass are in close proximity. The IFAS is the combination of a moving-bed bioreactor (MBBR) for attached growth and a RAS system (for suspended growth). The media allow the biomass in the bioreactor to be increased beyond that possible with suspended growth alone, without increasing the solids load on the secondary clarifiers.

3.0   Design of Conventional Integrated Biological Treatment Systems

3.1   Design Considerations
Design considerations for conventional IBT processes are similar to those for the parent processes. The designer should review appropriate parent processes and special design considerations for integration.

   The design of integrated biological processes is a balance between two biological treatment reactors; neither the biofilm nor the suspended-growth process can be sized separately. Good design includes consideration of the total integrated facility, including interactions among all units, especially the secondary clarifier (Harrison and Timpany, 1988). A number of factors, described below, must be considered with combined processes.

3.1.1   Ancillary Facilities
Several key design considerations are important for ancillary unit processes before and after integrated biological processes. Key considerations for preliminary treatment, primary treatment, secondary clarification, and odors are discussed below.

3.1.1.1   Preliminary Treatment   Preliminary treatment requirements for integrated processes are similar to other biofilm and suspended-growth processes. High-performing grit removal and screening facilities help to prevent accumulation in the integrated unit process. For trickling-filter-based integrated processes, special attention should be given to grit removal and screening performance. Given the characteristics of trickling-filter reactors, grit accumulation can be detrimental to hydraulic flow as grit can quickly accumulate in underflow structures. For screening, excessive ragging can occur within the biofilm media, causing plugging and decreasing performance.

3.1.1.2   Primary Treatment   Studies have shown that both the amount and characteristics of suspended solids in primary effluent can significantly affect the performance and solids yield from integrated processes (Matasci et al., 1986; Newbry et al., 1988). The quantity of colloidal and dissolved chemical oxygen demand (COD) after primary treatment can have a significant impact on the biofilm structure and performance in the integrated process. In addition, high–primary-effluent solids can lead to clogging issues in the downstream trickling-filter reactors. The effect will vary, depending on the type of filter media used in the first-stage process. Rock-filter media has small void spaces relative to other media types. Therefore, the effects of suspended solids in primary effluent are minimized because the incoming solids are retained longer. The added solids retention provided in rock media may promote anaerobic degradation and odor generation. The solids retention may also cause plugging.

3.1.1.3   Secondary Clarification   Solids separation in secondary clarifiers with integrated systems is similar to that of suspended-growth reactors in activated sludge systems. The settling characteristics are dominated by the suspended-growth characteristics. Compared to conventional biofilm processes, integrated processes do exhibit improved settleability as the suspended-growth portion of the reactor provides increased bioflocculation. Refer to Chapter 12 for a discussion on secondary clarifier design.

3.1.1.4   Odor   Most odor problems occur at facilities that have industrial loads or inorganic constituents (nitrogen or sulfur) that tend to be odor producing. Evidence indicates that odors result from (1) waste characteristics, (2) poor hydraulics, (3) lack of ventilation, (4) high percentages of certain industrial wastes, and (5) excessive biofilm thickness often due to high organic loads. Solids recycle has been reported to increase odors at some facilities. Odors tend to be more severe with biofilm than with suspended-growth reactors. The design engineer should consider carefully the possibility of odor problems in the design.

If odor problems are occurring in the headworks or primary clarifiers, then odor control should be a significant design consideration for the biofilm reactors in integrated processes. Stated differently, biofilm reactors used in integrated processes with low to moderate loading can generate as many odors as integrated processes with heavy organic-load fixed-film reactors.

3.1.2   Integrated Process Infrastructure Considerations
Within integrated treatment unit processes, several distinct design considerations must be included for proper design of flow distribution, solids recycle lines, and the fixed media itself.

3.1.2.1   Flow Distribution   Integrated process performance can depend significantly on the type of distribution system. As with conventional trickling filters, rotary distributors are more efficient and less troublesome to operate than are fixed-nozzle distributors (Harrison and Timpany, 1988). With integrated processes, where additional solids can be present in the distributed process influent, proper feed distribution becomes even more critical. Mechanically actuated rotary distribution systems provide the flexibility to control and automate feed and flushing rates. This can provide the ability to actively manage the biofilm through regular flushing and control of the Spulkraft rate. Flushing of solids from the trickling filter in an integrated process can have less impact on effluent TSS quality as compared to conventional trickling-filter processes, as the flushed solids are incorporated into the suspended-growth matrix and settled in the final clarification step.

3.1.2.2   Solids Recycle   Sludge recycle has been used extensively in the treatment of food-processing wastes in both the ABF and BF/AS integrated process modes. Sludge recycle using horizontal redwood media has been observed to increase performance compared to plastic media. This is significant because horizontal media have a lower specific surface area than plastic media and demonstrate poorer performance if solids are not recycled over the media (Harrison and Daigger, 1987). Several facilities, however, have discontinued use of sludge recycle in RF/AS systems because of increased odors and media fouling. Thus, solids recycle is not generally practiced in integrated systems; however, design consideration to allow the operational flexibility to recycle suspended-growth solids to the upstream biofilm process should be considered during design. This can provide future operations with increased flexibility for future potential strategies. Recycle rates ranging from 15% to 100% of the RAS should be considered.

3.1.2.3 Fixed-Biofilm Media Systems that recycle solids to the fixed-biofilm system, such as the ABF or BF/AS, should be limited to applications where filter media have little tendency to plug, such as with horizontal-redwood or vertical-plastic media. Crossflow, random, and rock media, in particular, have a complex configuration that may promote solids accumulation and cause plugging when recycling solids.

Studies indicate that the performance of vertical media can meet or exceed the performance of crossflow media at high organic loads that occur with BF/AS, RF/AS, and TF/AS processes (Daigger and Harrison, 1987). At low organic loads used with the TF/SC process, crossflow media will outperform other types. Some designers have used a combination of media to minimize plugging and maximize performance.

For example, several facilities have used crossflow media as the top layers to maximize distribution and vertical media at depths greater than 1.2 m (4 ft) to minimize plugging. Compaction of filter media, media collapse, or plugging problems have occurred with nearly all types, emphasizing the need for care in selection, design, and installation (Daigger and Harrison, 1987; Richards and Reinhart, 1986).

Special care should be taken in media selection for fixed-biofilm reactors that operate under high organic loading. At high hydraulic or organic loading rates, a thick biofilm may develop and typical module bearing strength may be exceeded. There is evidence that media with complex geometry (crossflow and random) may accumulate more biomass than less complex-shaped media. Parallel studies of filter media, literature searches, and case history related to filter media selection with integrated processes should be part of the design of integrated processes.

Most integrated process biofilm reactors are designed with high-rate media, which are often supplied with a walking surface of high-strength media for the top layer. The top or walking surface of the media should be specified to protect against hydraulic shear and UV-light deterioration. Many designers also specify that grating be placed on the media surface from the access ladder or walkway to and often around the center column to allow safe maintenance of the distributor.

3.1.3 Solids Considerations

3.1.3.1 Solids Production Solids production from an integrated process is dependent in part on the type of media used in the biofilm unit process and the operation of the suspended-growth process. If the design involves a retrofit of an existing rock media trickling filter, then the amount of biologically produced solids depends on the organic and hydraulic loading applied. For rock media, the solids produced (observed yield) typically range from 0.4 to 0.7 kg total suspended solids (TSS)/kg BOD_5 (kg produced solids/kg primary effluent BOD_5). With use of plastic or redwood media, the amount of solids digestion and effective solids retention time (SRT) is reduced, which results in higher overall system yield. Solids production with synthetic media often ranges from 0.8 to 1.0 kg TSS/kg BOD_5, including when an integrated process is used.

3.1.3.2   Sloughing   Natural biological sloughing that occurs from biofilm reactors can be managed by varying the hydraulic wetting rate. Surveys indicate that integrated processes with small suspended-growth reactors (ABF or TF/SC) tend to be more susceptible to variations in mixed-liquor volatile suspended solids (MLVSS) concentration, which can result from sloughing, than integrated processes with large suspended-growth reactors (RF/AS or BF/AS). Some operators have found, however, that they can accommodate even extreme sloughing events with comparatively little effect on effluent quality (Boller and Gujer, 1986). This is especially true when compared to conventional biofilm reactors. The designer should refer to Chapter 11 for guidance on hydraulic load design to minimize drastic variations in sloughing from biofilm reactors.

3.1.3.3   Snails   As with conventional trickling-filter applications, snails can accumulate in the biofilm reactor of an integrated process. However, integrated processes differ from conventional trickling-filter applications in that snails can settle in transfer structures or aeration channels where mixing energies are low. At the Ryder Street Wastewater Treatment Plant in Vallejo, California, the first part of the aeration basin was effectively converted into an aerated grit basin, with a grit cyclone and classifier employed to achieve snail shell removal (Tekippe et al., 2004). As an alternative, some designers include sloped (approximately 3%) biofilm reactor floors that drain into a low-velocity, aerated flow-transfer box before the suspended-growth reactor to encourage the settling and removal of snails (Slezak et al., 1998). A submersible non-clog pump can pump the snail slurry to the grit removal facilities. Overall, the same snail management considerations for conventional biofilm reactors should be included in the design of an integrated treatment process.

3.1.4   Final Effluent Quality

Effluent with less than 20 mg/L BOD_5 typically is achieved with good integrated process design; 10 mg/L BOD_5 has been achieved without advanced treatment at some facilities.

For the majority of integrated biological processes, the amount of particulate BOD_5 in the treated effluent will control effluent quality. Therefore, a key factor in obtaining good effluent quality is the use of modern clarifier design as noted above. The particulate BOD_5 removal in an integrated treatment process will typically be higher than a conventional biofilm process due to the bioflocculation that occurs upstream of solids separation.

When effluent quality issues are associated with nitrogen or phosphorus removal, design decisions become more complex because biological nutrient removal (BNR) requires significant readily biodegradable carbon. Integrated processes are excellent at preferentially removing the readily biodegradable carbon that also drives BNR. Therefore, design analysis of integrated processes must include evaluating the trade-offs in chemical addition (methanol for denitrification, metal salts for phosphorus removal) versus biological removal without chemical addition.

3.1.5 Site Considerations

Integrated processes often require slightly less land than other biological treatment processes for two reasons: (1) ability to construct tall, fixed-film reactors with heights of 4.9 to 9.8 m (16 to 32 ft) and (2) use of slightly higher loadings of both the biofilm and suspended-growth systems.

3.1.6 Energy

Historically, designers considered that biological treatment with biofilm reactors required 25% to 50% of the energy of comparable suspended-growth systems. With the emergence of fine-bubble diffused air and good suspended-growth design, however, design evaluations often show the energy gap between biofilm and suspended-growth reactors has narrowed. A full energy balance that considers pumping, aeration, and mixing energy for alternative technologies should be completed for each unique facility as site-specific conditions will have a large impact on the overall energy balance.

With integrated processes, energy savings are somewhat dependent on the flexibility of equipment operation, such as turndown, when units are operated at less than full design load (Harrison et al., 1984). Typically, high power rates will tend to favor integrated processes that have large biofilm reactors (ABF or TF/SC processes).

3.1.7 Existing Treatment Units

Potential cost savings from using existing equipment often dictate the type of integrated process selected. If existing aeration basins are already available, then economics often favor the use of integrated processes with large suspended-growth reactors such as BF/AS, RF/AS, or TF/AS. If an existing trickling-filter facility is being upgraded, however, then economics often favor the use of process modes with small, suspended-growth reactors such as the TF/SC integrated process.

3.2 Process Design

Numerous mathematical models are available for predicting the performance of the fixed-film reactor operating as the sole method of biological treatment (refer to Chapter 11). Commercially available process simulation models also include integrated process technologies in the suite of options. Common design practice for sizing reactors for integrated processes is often based on either pilot- or full-scale information.

Table 13.1 presents design criteria typically used by engineers for sizing integrated processes. In all instances, rational process design equations should be used to check the design because translating data from one situation to another may not always be successful.

The inability to control interdependent variables makes it impossible to precisely predict performance of integrated processes (Albertson and Eckenfelder, 1984). Design engineers often choose both conservative design loads and variable process modes for integrated processes that produce the desired effluent quality and solids characteristics under a wide range of operating loads.

3.2.1 Design Criteria

The fixed-biofilm reactor in the RF/AS, BF/AS, or TF/AS process removes from 30% to 50% of the incoming organic load measured as $cBOD_5$ and approximately 40% to 80% of the incoming soluble organic load as $sBOD_5$. With the TF/SC and ABF processes, an even higher degree of $sBOD_5$ removal is achieved in the trickling-filter stage, close to

System	Appropriate Design Criteria	
	Range	Commonly Used
Activated biofilter		
Media type	High rate	High rate
Five-day biochemical oxygen demand loading, kg/m^3·d (lb/d/1000 cu ft)	0.2–1.2 (12–75)	0.5 (30)
Hydraulic loading, L/m^2·s (gpm/sq ft)	0.5–3.4 (0.8–5.0)	1.4 (2.0)
Filter mixed liquor suspended solids, mg/L	1500–3000	2000
Trickling filter/solids contact		
Media type	Rock or high rate	High rate
Five-day biochemical oxygen demand loading, kg/m^3·d (lb/d/1000 cu ft)	0.3–1.6 (20–100)	0.8 (50)
Hydraulic loading, L/m^3·s (gpm/sq ft)	0.1–1.4 (0.1–2.0)	0.7 (1.0)
Channel mixed-liquor suspended solids, mg/L	1500–3000	2000
Hydraulic retention time, hours	0.2–2.0	0.75
Solids retention time, days	0–2	0.1
Return activated sludge, mg/L	6000–12 000	8000
Diffused air, L/m^3·s (scfm/1000 gal)	250–500 (2000–4000)	370 (3000)
Mechanical, W/m^3 (hp/mil·gal)	12–26 (60–130)	20 (100)
Food/mass, mg/kg·s (lb/d/lb)	2.3–23 (0.2–2.0)	11.6 (1)
Roughing filter, trickling filter, or biofilter with activated sludge		
Media type	High rate	High rate
Five-day biochemical oxygen demand loading, kg/m^3·d (lb/d/1000 cu ft)	1.2–4.8 (75–300)	2.4 (150)
Hydraulic loading, L/m^2·s (gpm/sq ft)	0.5–3.4 (0.8–5.0)	0.7 (1.0)
Basin mixed liquor suspended solids, mg/L	1500–4000	2500
Hydraulic retention time, hours	0.5–4.0	2.0
Solids retention time, days	1.0–7.0	3.0
Food/mass, mg/kg·s (lb/d/lb)	5.8–13.9 (0.5–1.2)	10.4 (0.9)
Total available oxygen, g/kg (lb/lb)	600–1200 (0.6–1.2)	900 (0.9)
Oxygen typically used, g/kg (lb/lb)	300–900 (0.3–0.9)	600 (0.6)
Conventional activated sludge		
Solids retention time, days	5–15	10
F/M foot to mass, mg/kg·d	0.3–0.5	0.4
Hydraulic retention time, hours	4.0–8.0	7.2 min (@12°C)

TABLE 13.1   General Design Criteria for Fixed Film and Activated Sludge Systems with Segregated Return Activated Sludge

full removal. The ABF process has been unsuccessful, however, in consistently achieving good effluent quality (less than 30 mg/L BOD and TSS) as organics loads approach 1.0 to 1.6 kg BOD/m^3·d (60 to 100 lb BOD/d/1000 cu ft).

The suspended-growth reactors in the non-nitrifying RF/AS, BF/AS, or TF/AS integrated processes range from 15% to 50% the size that would be required if a first-stage fixed-biofilm reactor is used. With the TF/SC and BF/AS processes, the suspended-growth reactor size is typically 5% to 20% of the basin volume that would be required if no fixed-biofilm stage is used.

With the ABF process, suspended-growth contact is achieved through hydraulic retention as combined recycle solids and primary effluent pass through the filter media.

3.2.2 Fixed-Biofilm Reactor Stage

Biological towers, trickling filters, or roughing filters are the most common fixed-film reactors used in IBT systems. If actual pilot-plant information is unavailable, then standard design practice is to size the first-stage fixed-biofilm reactor based on BOD$_5$ loading and theoretical removal for the particular process mode. As listed in Table 13.1, loadings to integrated processes such as ABF or TF/SC can range from 0.2 to 1.20 kg/m^3·d (10 to 75 lb BOD$_5$/d/1000 cu ft) but typically average around 0.5 kg/m^3·d (30 lb BOD$_5$/d/1000 cu ft). Integrated systems with fixed-biofilm reactors for higher loads such as RF/AS, TF/AS, or BF/AS typically are designed for first-stage loadings of 1.20 to 5 kg/m^3·d (75 to 300 lb BOD$_5$/d/1000 cu ft) and average design loading of 2.4 kg/m^3·d (150 lb BOD$_5$/d/1000 cu ft).

The fixed-biofilm stage can be designed with the use of computer models, like the Logan trickling-filter (LTF) model, or with empirical equations, like the modified Velz and Germain equations. Refer to Chapter 11 for trickling-filter design equations and Chapter 4 for process modeling information. Alternatively, commercially available process simulation models, such as BioWin, GPS-X, or others, can be used. Over the years these models have been expanded to improve the ability to model attached growth systems.

Sizing the fixed-biofilm reactor based on BOD$_5$ removal is not a trivial task because predicting BOD$_5$ removal is difficult. Biomass in the trickling filter removes both particulate and soluble BOD$_5$. Conventional empirical design models typically do not distinguish between particulate BOD$_5$ and soluble BOD$_5$ removal rates or the fixed-biofilm effluent BOD$_5$, but consider merely the fixed-biofilm influent BOD$_5$ and the total process effluent BOD$_5$.

Particulate BOD$_5$ that is not degraded in the trickling filter will most likely be degraded in the suspended-growth reactor, with a corresponding oxygen demand. Thus, to determine the oxygen required for the suspended-growth process, the amount of particulate BOD$_5$ degraded in the fixed biofilm is critical. Removal of particulate BOD$_5$ over a range of loadings can be determined by pilot studies. More particulate BOD$_5$ is expected to be degraded as the BOD$_5$ loading to the fixed biofilm is decreased.

Although sizing fixed-biofilm reactors in IBT systems based on organic loading and removal is the most common method, it is still important to use the hydraulic loading rate to check and verify design. Because higher organic loads tend to be used with integrated processes rather than with single-stage fixed-biofilm reactors, higher hydraulic loads of 0.7 to 1.4 L/m/s (1 to 2 gpm/sq ft) are more common. Many designers consider it good practice to control sloughing through operation at high hydraulic loading for short periods on a weekly basis. This and other strategies for film control, such as distributor speed control, are described in Chapter 11.

After the fixed-biofilm reactor is sized, another design criterion for highly loaded fixed-biofilm reactors is proper air ventilation. A study of case histories indicates that with sufficient ventilation the fixed-film effluent will have high concentrations of dissolved oxygen even at loadings of 3.2 to 4.8 kg $BOD_5/m^3 \cdot d$ (200 to 300 lb $BOD_5/d/1000$ cu ft) for high-rate media.

One media manufacturer recommends $0.85 \ m^2$ of vent area per $100 \ m^3$ of filter media (2.6 sq ft/1000 cu ft) for air ventilation. A review of operating facilities indicates that natural ventilation can become limiting where less than 0.0004 to $0.0006 \ m^2$ of vent area per kilogram of primary effluent BOD_5 per day (2 to 3 sq ft/d/1000 lb BOD) is provided.

Most biological towers used for integrated processes have natural ventilation. Natural ventilation in colder climates typically includes provisions to close louvers or vents during winter months. Likewise, if both natural and forced ventilation capabilities are provided, vents must have provisions for closure. Where forced-air ventilation is required, manufacturers' recommendations vary.

One trickling-filter manufacturer recommends a minimum of $0.06 \ m^3/min \cdot kg \ BOD$ (1.0 cfm/lb BOD_5). Another manufacturer recommends a minimum of $0.16 \ m^3/min \cdot kg$ BOD (2.6 cfm/lb BOD_5) for towers loaded at 3.2 to 4.8 kg $BOD_5/d \cdot m^3$ (200 to 300 lb $BOD_5/d/1000$ cu ft). The trickling-filter section of Chapter 11 provides a procedure to calculate air requirements based on the BOD_5 load removed in the biological tower. For forced ventilation, regardless of the method used to calculate airflow, air scrubbing, covers, or both might be required where extreme odor problems exist or where a fixed-film system is constructed in an area sensitive to odors. An assessment of odor potential and provisions for control are necessary with any biological tower design.

3.2.3 Suspended-Growth Biological Reactor Stage

Once the fixed-biofilm reactor size is selected, the suspended-growth reactor is sized to ensure that the desired effluent quality is maintained. The most common method for sizing suspended-growth reactors in integrated processes is based on SRT, the same approach used with suspended-growth systems, as discussed in Chapter 12, and covered later in this chapter. Suspended-growth reactors may also be evaluated based on $F{:}M$ or hydraulic retention time (HRT).

An older approach involves calculations based on $sBOD_5$ removal. The modified Velz equation has been used for predicting effluent $sBOD_5$ concentrations from biofilm reactors to size the second-stage suspended-growth process (Albertson and Eckenfelder, 1984). When applied to a wide range of design conditions, however, these empirical methods have resulted in up to 100% error in sizing (Matasci et al., 1989). This last approach is used in tandem with the SRT approach to check $sBOD_5$ breakthrough in the aerated solids contact tank of the TF/SC process. In practice, this approach is not widely used because of the wide range of coefficients that are difficult to define.

If the characteristics of the fixed-biofilm effluent can be readily defined, then conventional suspended-growth mass balance and biomass growth relationships can be used to size the suspended-growth reactor. Unfortunately, this cannot be done easily. First, the fixed-biofilm design is focused on substrate removal rather than biomass production. Thus, the trickling-filter effluent substrate concentration is well defined, but the trickling-filter active heterotrophic biomass concentration is not. Second, the trickling-filter effluent suspended solids are composed of many constituents including biomass, cell debris, inerts, and unmetabolized substrate. Precise prediction of the concentrations of these constituents is needed for the mass balance and biomass growth equations, and this is not straightforward. Third, both aerobic and anaerobic metabolisms can occur in the biofilm of the trickling filter, and thus the biomass yields under each condition vary significantly. Finally, accumulated biomass sloughs at a nonsteady rate, which can significantly affect mixed-liquor suspended solids (MLSS) concentrations. Therefore, the volume of the suspended-growth reactor is sized using system net yield rather than influent substrate.

Designers of integrated processes must include or consider proper turndown capability, step feed, and various points of sludge recycle. With integrated processes, turndown capability is a key to providing good effluent quality while the facility is not fully loaded. For example, several integrated process designs operate with one of several basins out of service during initial operation, often because biological towers outperform expectations, especially during warm temperatures or when organic loading is more readily biodegradable than originally assumed.

Options for process optimization include (1) keeping large suspended-growth reactors online rather than using small contact channels; in at least one case, operators found that the larger basin could be operated with a lower dissolved oxygen concentration and less aeration power than the smaller contact channel requiring higher concentrations; (2) returning sludge midway along a plug-flow, second-stage, suspended-growth reactor so that excessive SRT does not result in problems with nitrification or denitrification; (3) using the integrated process only when stringent effluent standards apply. When less-stringent effluent standards apply, the suspended-growth reactor is taken offline or used as a transfer structure with no sludge recycle. At this type of facility, high winter flows are accommodated by using the trickling-filter process alone; the integrated process is used during critical (low) flow conditions.

Flexibility is a key in the design of the suspended-growth reactor. Many facilities with integrated processes are designed to operate in various modes. Often, only marginal additional capital expense is required to provide piping for operating in both BF/AS and RF/AS modes or TF/SC and TF/SCR. Some designers prefer the use of fine-bubble air diffusion in the suspended-growth reactor to promote bioflocculation, prevent the shearing of biological floc, and minimize aeration cost.

3.2.4 Solids Retention Time

Use of the SRT approach for sizing suspended-growth reactors following fixed-biofilm systems has become the prevalent method for design (Harrison and Timpany, 1988). Table 13.1 correlates the total organic loading (TOL) to SRT. An SRT in the contact channel for TF/SC may range from 0.2 to 2 days. For a combined BF/AS facility, an SRT of 1 to 7 days may be more appropriate. Using high organic loading on the trickling filter will require increased SRTs in the suspended-growth portion of the combined process.

Typically, the SRT calculation is similar to that presented in Chapter 12. It is the mass of solids under aeration divided by the rate at which solids are wasted from the system each day, either intentionally or unintentionally.

To estimate SRT_{SG}, a steady-state analysis can be used, where the rate at which suspended solids wasted from the process must equal the waste solids production rate. The waste solids production rate is determined by the biological reactions occurring in both the fixed biofilm and the suspended-growth biological reactor. Consequently, the waste solids production calculations must consider the pollutant loadings placed on the entire system, not just the organic matter contained in the trickling-filter effluent.

Biomass grown in the trickling filter will slough off and pass into the suspended-growth biological reactor. Such growth must be accounted for in the total process waste solids production calculation.

It is common to estimate solids production using a net process yield Y_n, as described in Chapter 12. Even though soluble substrate may remain in the trickling-filter effluent, it will typically be removed in the suspended-growth bioreactor. Consequently, Y_n can be used in the following equation to calculate solids wasting rate:

$$P_x = Q \times Y_n \times BOD_5 \tag{13.1}$$

where P_x = mass active solids wasted (kg/d),

$\quad Q$ = volumetric flow rate (ML/d),

$\quad Y_n$ = solids yield (kg TSS/kg BOD_5 removed), and

$\quad BOD_5$ = total influent BOD_5 concentration (mg/L).

The solids yield, Y_a, for IBT systems is lower than for conventional activated sludge (CAS) systems. However, because the retention of biomass in a trickling filter increases as TOL is decreased, the solids yield value typically is influenced more by the TOL on the fixed biofilm than by the SRT of the suspended-growth biological reactor. Solids yield decreases as TOL is decreased. Solids yield values are approximately 0.7 to 0.9 kg TSS/kg BOD_5 removed. The SRT can then be calculated using the above solids wastage rate with the mass of MLSS and the suspended-growth reactor volume:

$$SRT_{SG} = X_{SG} \times V_R / P_x \tag{13.2}$$

where X_{SG} = net active solids in the suspended-growth reactor (mg/L),

$\quad V_R$ = reactor volume (ML), and

$\quad P_x$ = mass of active solids wasted (kg/d).

The active biomass in the fixed-film reactor may be neglected in the calculations. The calculated SRT is an "effective" SRT in which residence time in the suspended-growth reactor may be reduced to below 0.5 days for TF/SC processes; a value of 2 to 3 days may be sufficient with highly loaded filters such as RF/AS, TF/AS, or BF/AS. In the case of the ABF, which does not have a suspended-growth reactor, the SRT is for the fixed-biofilm portion only and since the liquid volume in the TF cannot be accurately determined, the SRT cannot be accurately calculated.

The values in Table 13.1 for TF/AS are significantly less than the values of 5 to 15 days found in the literature for CAS. Consequently, significant reduction in suspended-growth reactor size is possible. However, the type of filter media used in the fixed-biofilm reactor can significantly affect the solids yield and result in the need to increase the SRT_{SG} or the size of the suspended-growth reactor.

3.2.5 Food-to-Microorganism Ratio

The $F:M$ is calculated based on the amount of BOD_5 (or cBOD) in the primary effluent (neglecting removal through the fixed-biofilm reactor). The solids inventory, M, is expressed in terms of MLVSS associated with the amount of mixed liquor in the suspended-growth reactor only. In comparing $F:M$ of suspended-growth reactors (see Table 13.1) for the RF/AS, BF/AS, and TF/AS processes, it is observed the values are two to three times greater than those typically used for CAS. The $F:M$

calculations typically are not used for sizing the suspended-growth reactor of the TF/SC process.

3.2.6 Hydraulic Retention Time

Use of HRT has diminished in importance for sizing suspended-growth reactors with both CAS and integrated processes. As a check, for treating domestic wastewater, HRTs of 4 to 8 hours are commonly used with CAS. Corresponding HRTs with integrated processes for heavily loaded systems (BF/AS, TF/AS, or RF/AS) typically range from 2 to 4 hours (Table 13.1).

Integrated processes with fixed-film reactors receiving light organic loads (TF/SC) are sized for less than 1 hour HRT (Parker and Matasci, 1989). For the ABF process, however, HRT is not applicable because there is no specific suspended-growth reactor present.

3.2.7 Mixing and Aeration Requirements

Aeration requirements are determined by oxygen demand similar to suspended-growth systems, based on either mass balance calculations or process model simulations as discussed in Chapter 12. Typical designs for CAS may be based on oxygen uptake rates (OURs) averaging 30 to 50 mg O_2/L·h, with peak days from 60 to 80 mg O_2/L·h. Suspended-growth reactors following highly loaded fixed-film reactors (RF/AS, BF/AS, or TF/AS) may have OURs approximately 50% that of CAS. Suspended-growth reactors following lightly loaded fixed-film reactors, such as those in the TF/SC mode, are typically designed for OURs of 10 to 20 mg O_2/L·h (Harrison and Timpany, 1988).

A survey of 12 integrated process facilities included a rational basis for calculating the amount of oxygen required in the suspended-growth reactor at various fixed-biofilm organic loads preceding the second-stage process (Harrison et al., 1984). The use of rational equations or nomographs may be useful in predicting long-term, average oxygen requirements under various loadings. However, good design practice ensures that adequate oxygen is available to supply the needs under a wide range of uncontrollable events such as low flows, sloughing, temperature variation, or other interrelated environmental variables, and during startup. Process models can be used to check oxygen demands under a variety of different conditions.

Mixing requirements should consider flocculation as discussed further below.

3.2.8 Solids Production and Settleability

Solids yield or production from IBT systems often is dominated by biomass sloughing and storage occurrences within the fixed-biofilm reactor. The characteristics and amount of inert solids in the primary effluent will also be a factor in determining actual yields from the total integrated process.

Figure 13.3 shows various solids yields under several suspended-growth SRTs and at low and high organic loads to the first-stage, fixed-growth reactor (Newbry et al., 1988). The upper band in Figure 13.3 occurred when effluent suspended solids from the primary clarifier were approximately 50% higher than for the lower band.

Figure 13.3 also demonstrates that nonbiodegradable solids in the primary effluent may affect net solids yield from an integrated process. It further illustrates that significant reductions in solids yield may not occur simply by increasing the SRT beyond 1 to 2 days. For example, one study indicated that, for integrated processes with an SRT greater than 1 day, little measurable benefit resulted from increasing the suspended-growth reactor size to reduce sludge yield (Newbry et al., 1988).

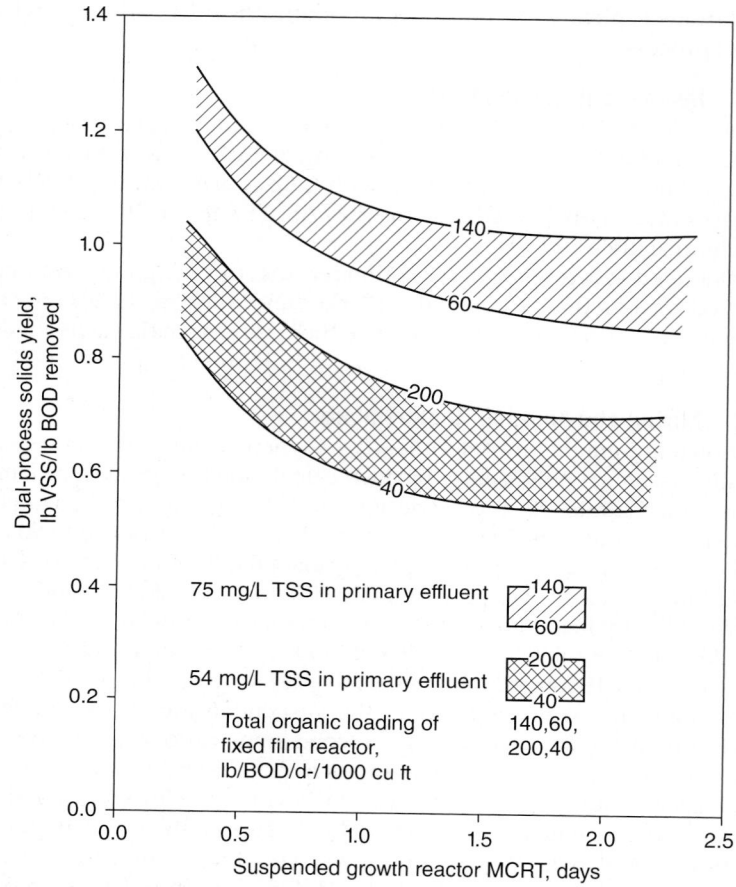

FIGURE 13.3   Combined process solids yield as a function of suspended-growth reactor mean cell residence time (MCRT) and total organic loading of fixed-film reactor with crossflow media. Solids yield varies with wastewater characteristics (lb BOD/d/100 cu ft, on the fixed-film reactor).

A comparison of 43 integrated WRRFs indicated that average solids yields were 0.7 kg TSS/kg primary effluent BOD, but values ranged from 0.15 to 4 kg TSS/kg BOD (Harrison et al., 1984). It was concluded that sloughing and varying performances from primary clarifiers were a significant cause of variability of produced solids. In addition, case studies indicate that solids from integrated processes that use rock-filter media in the first-stage trickling filter may yield 35% to 50% of those for systems with high-rate filter media (Harrison et al., 1984).

Reviewing the data from five integrated systems, Gharagozian et al. (2011) noted that at high organic loading to the trickling filter (greater than 1.6 kg $BOD_5/m^3 \cdot d$ or 100 lb $BOD_5/d/1000$ cu ft), the system solids yield decreased. The authors concluded that this was likely due to anaerobic conditions developing in the trickling filter under high loads, and mention that the exact load at which anaerobic conditions develop would depend on several factors, including the type of media used, effluent recirculation rates, whether the trickling filter is covered, and ventilation rates.

Many design engineers consider integrated processes to produce better settling sludge than conventional suspended-growth systems. Although integrated processes are relatively resistant to shock loads, process variability and sludge-bulking problems may occur with integrated processes similar to suspended-growth processes. This is particularly true for integrated processes that allow significant quantities of sBOD$_5$ to enter the suspended-growth reactor. If, on the other hand, the process is designed where the sBOD$_5$ loading to the suspended-growth reactor is consistently low, then a consistently good settling sludge will be produced and higher clarifier solids loading rates can be used (Parker et al., 1993).

In a work by Motta et al. (2003), an air-induced minimum velocity gradient of 20/s was achieved in the solids-contact tank using a fine-bubble aeration system, which consistently produces low SVI and good effluent quality. A velocity gradient of 20/s is equivalent to an air flux of 0.15 Nm^3/m$^2 \cdot$h (0.008 scfm/sq ft).

The process designer should carefully evaluate sludge-settling characteristics when sizing secondary clarifiers. The concentration of RAS in an integrated process will be similar to that of a suspended-growth process. RAS concentrations ranging from 0.5% to 1.5% and averaging approximately 0.8% TSS typically will occur with integrated processes. After allowing for the expected sludge settleability, secondary clarifiers can be designed using the same approach as for suspended-growth systems, discussed in Chapter 12.

3.2.9   Biochemical Oxygen Demand and Total Suspended Solids Removal

One study comparing various organic loads ranging from 0.74 to 3 kg BOD/m$^3 \cdot$d (46 to 200 lb BOD/d/1000 cu ft) in the fixed-film reactor showed that sBOD$_5$ in treated effluent decreased and effluent quality improved with increased SRTs in second-stage suspended-growth reactors (Newbry et al., 1988). Effluent sBOD$_5$ concentration, however, was determined to be independent of organic loading to the first-stage fixed-film reactor. Varying the organic load to the fixed-film reactor did not change the concentration of sBOD$_5$ in the effluent of the suspended-growth step, as illustrated in Figure 13.4.

Researchers have shown that increasing organic loads to the fixed-biofilm reactor results in increased amounts of nonsettleable suspended solids as illustrated in Figure 13.5 (Newbry et al., 1988). With integrated processes, the second-stage suspended-growth treatment reduces the amount of effluent nonsettleable suspended solids by increasing total system SRT. A sharp reduction in effluent suspended solids results from providing a minimum of a 1-day SRT.

3.2.10   Nitrogen Removal

Except for facility upgrades and the ENBNRAS process described earlier, integrated processes typically have not been cost-effective for achieving nitrification, compared to a single-stage, activated sludge process. This is because the suspended-growth portion of integrated processes typically has lower nitrification rates at decreased temperature than a single-stage suspended-growth process. Nitrification, then, typically drives the reactor volume, which is a function of SRT. In the ENBNRAS configuration, nitrification is achieved in the trickling filter receiving low organic loads.

Nitrification in integrated processes is a common adaptation and can be achieved through use of conventional, lightly loaded IBT systems. Sizing IBT systems for nitrification is similar to sizing for BOD$_5$ removal. Following are two design approaches for nitrification with IBT systems.

In the first design approach, the SRT calculation may include only the second-stage suspended-growth stage if it is separated from the fixed-biofilm stage by intermediate

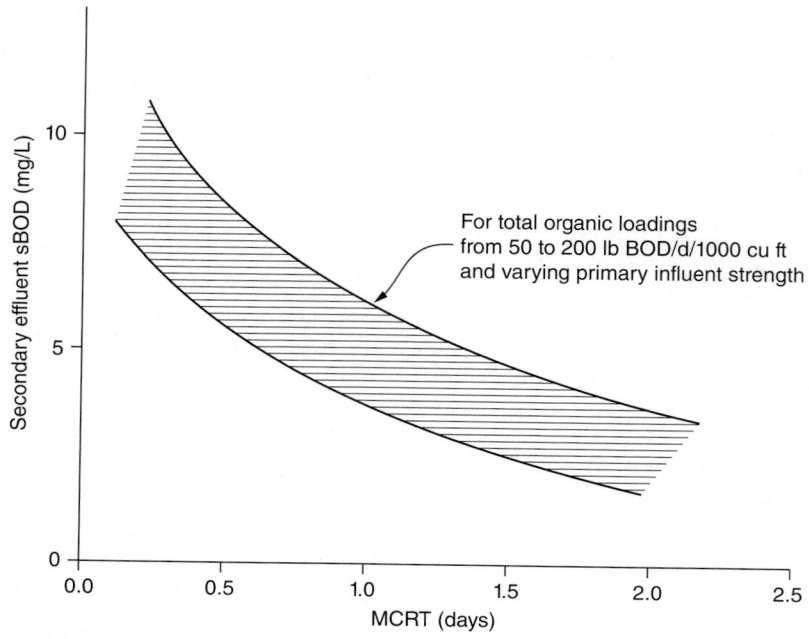

FIGURE 13.4   Effluent soluble biochemical oxygen demand (sBOD) of combined processes with crossflow filter media for various mean cell residence times (MCRT) and total organic loadings. The sBOD will vary with wastewater characteristics.

clarification, as is the case with the TF/AS process. The upstream trickling filter and the intermediate clarifier are used to remove the bulk of the organic matter from the wastewater. The downstream activated sludge system is then designed to nitrify.

The intermediate clarifier removes the bulk of the biomass produced in the trickling filter, thereby allowing a significant reduction in aeration basin size compared to when an intermediate clarifier is not provided. Without intermediate clarification, the aeration basin, which is sized based on SRT, would be approximately the same volume with or without the fixed-biofilm reactor.

This approach is often preferred because it provides a more positive means of determining the actual SRT of the second-stage suspended-growth system. The TF/AS process avoids burdening the second-stage system with heterotrophic biomass from the first-stage system that would otherwise compete with nitrifiers for the available oxygen.

Without intermediate clarification, nitrifiers that grow on the fixed-biofilm reactor may slough off, and seed the downstream suspended-growth reactor. This will allow a reduction in the required suspended-growth reactor SRT. Daigger et al. (1993) developed a procedure for accounting for nitrification in moderately loaded fixed biofilms and the seeding of nitrifiers into the suspended-growth reactor that allows for a reduction in SRT requirements (Daigger et al., 1995; Parker and Richards, 1994). Daigger et al. (1993) showed that significant ammonia removal can be obtained in TF/SC systems operated at nitrification safety factors less than 1.0 (nitrification safety factor = SRT/ minimum SRT for nitrification), if nitrification occurs in the trickling filter. Nitrification would not have occurred in the suspended-growth units during these conditions if

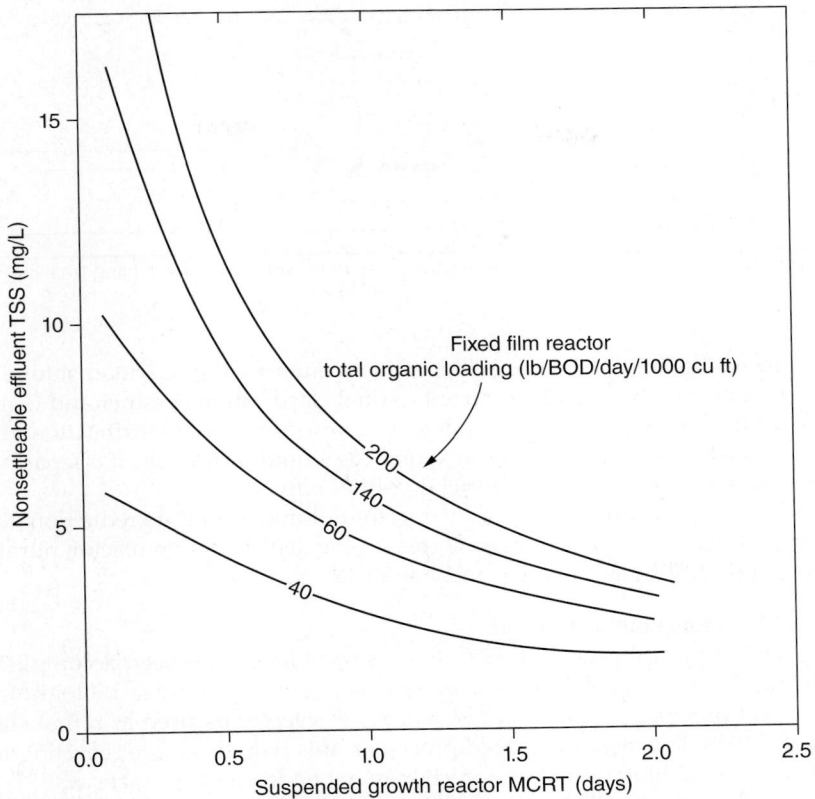

Figure 13.5 Nonsettleable effluent total suspended solids (TSS) of dual process with crossflow filter media for various mean cell residence times (MCRT) and total organic loadings. Nonsettleable TSS are measured in the supernatant of the settleable solids test. Secondary effluent TSS will exceed the nonsettleable TSS.

nitrifiers had not been seeded into the units by the upstream nitrifying trickling filter. This confirms the seeding effect of the TF/SC process.

The second design approach involves achieving nitrification with use of a separate solids reaeration tank or channel, as is sometimes accomplished with the TF/SC process. Solids reaeration typically provides two to four times more SRT per unit of aeration volume than does contact through a conventional aeration basin or channel, due to the higher solids concentration. However, there needs to be ammonia in the feed to the reaeration tank at sufficient levels to support nitrifier growth, which implies some ammonia in the secondary effluent.

Balmer et al. (1998) presented an exception to the two conventional nitrification design approaches. This alternative approach involves a tertiary nitrifying trickling filter that returns 100% of its flow to the anoxic/aerobic suspended-growth reactor for denitrification (Figure 13.6).

Flamming et al. (WEFTEC 2011) also presented an exception where in place of installing solids reaeration basins, IFAS media were installed in existing SC tanks to increase the system SRT and augment nitrification.

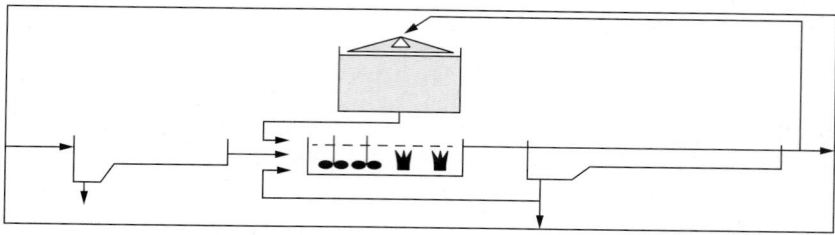

FIGURE **13.6**   Integrated biological treatment system with tertiary nitrifying trickling filter.

Instead of all the trickling-filter effluent entering the activated sludge bioreactor, a portion of the flow may be directed to final clarification (Vestner and Gunthert, 2001). Using the existing activated sludge system to provide denitrification eliminates the need to add an additional carbon source (as would be the case if a separate denitrification process were used for the trickling-filter effluent).

The denitrification rate and the reactor volume for nitrate reduction depend on the available carbon source, oxygen ingress of the denitrification reactor, nitrate concentration, and MLVSS, as discussed in Chapter 12.

3.2.11   Phosphorus Removal

Removing phosphorus with IBT systems has historically been accomplished through chemical rather than biological means. This is because most biological phosphorus removal processes require that an anaerobic selector be used as a first stage (Morgan et al., 1999). For most integrated processes, this would require recycling mixed liquor over the fixed-biofilm reactor, which may not be feasible or practical.

Studies of phosphorus removal with integrated processes indicate that chemical addition of aluminum or ferric chloride minimally affects BOD_5 removal in the biological system. Thus, many operators prefer the chemical phosphorus removal approach because of its more stable and predictable performance.

Some design techniques, however, have produced effluent with low phosphorus concentration by biological means only. An example of this is the use of fermentation to produce volatile fatty acids (VFAs) and to encourage the growth of phosphorus-removing microorganisms. This method minimizes or eliminates the need to chemically remove phosphorus with integrated processes (Kalb et al., 1990).

3.3   Specific Process Design by Integrated Biological Treatment System

3.3.1   Trickling-Filter Solids Contact

The TF/SC process is defined by the following criteria (Parker and Bratby, 2001):

- The majority of the $sBOD_5$ removal occurs in the trickling filter.
- Return sludge solids are mixed with the trickling-filter effluent (underflow) rather than the primary effluent.
- The primary function of the aerated solids-contact tank is to increase solids capture and reduce particulate BOD_5 through flocculation.
- The SRT of the solids contact tank is less than 2 days, if no nitrification is required.

- The solids-contact tank hydraulic retention time is one hour or less based on total flow, including recycle.
- The solids-contact tank is not designed to nitrify, although nitrification may occur in the trickling filter.

When the TF/SC process is designed and operated as a trickling filter with a solids contact channel or tank it is called Mode I operation (Parker and Matasci, 1989). When both solids reaeration and solids contact are used, the acronym TF/SCR is used (also called Mode III operation) (Parker et al., 1994; Slezak et al., 1998).

Figure 13.1a shows Mode I and Figure 13.1b shows Mode III operation schematics. When solids contact is eliminated and solids reaeration is the only type of suspended-growth process used, the acronym TF/SR is used and describes the Mode II operation. This design and operation mode is not commonly used and is not shown in Figure 13.1.

Mode I typically is used when the TOL on the trickling filters is so high that a significant portion of sBOD$_5$ remains to be removed in the aerated solids-contact tank. Therefore, the majority of the solids inventory must remain in the aerated solids-contact tank for removal of the residual sBOD$_5$. Because the aerated solids-contact tank is larger in Mode I, both the bioflocculation and flocculation objectives can be simultaneously achieved, without the need for return sludge reaeration (Parker and Bratby, 2001).

3.3.2 Roughing Filter Activated Sludge

The RF/AS process is attractive for treating higher strength wastewaters, such as from industrial sources, because of the relatively low energy consumption per unit of BOD$_5$ removed on the trickling filter and the enhanced settling characteristics of the MLSS. The fixed-biofilm reactor in the RF/AS process consumes approximately 40% to 70% of the incoming sBOD$_5$. Because the roughing filter acts as a biological selector in removing sBOD$_5$, good SVI values result.

In the RF/AS process, the roughing filter is typically 15% to 30% of the size required if treatment had been accomplished through the use of the trickling filter alone. The roughing-filter loading, however, is about four times greater than that used for the TF/SC process. Because the loading rates are greater, the RF/AS process also has longer suspended-growth SRTs (up to 85% longer) and higher MLSS concentrations (Table 13.1). The hydraulic retention time in the aeration basin is typically 35% to 70% of that which is required with the use of the activated sludge process alone.

3.3.3 Trickling Filter Activated Sludge

The trickling-filter process (with dedicated clarification) and the activated sludge process operated in series is what this manual has defined as the TF/AS process. Table 13.1 displays common process parameters for the TF/AS process.

When TF/AS is used for nitrification, the clarification step can reduce the required nitrifying aeration volume 60% to 80%. This is a significant advantage for larger facilities and/or facilities processing higher strength wastewater.

3.3.4 External Nitrification Biological Nutrient Removal Activated Sludge

To determine if the ENBNRAS process is an attractive option, a process model, such as those used for fully suspended-growth system BNR activated sludge systems, can be used. Design criteria for nitrifying trickling filters including BOD removal, as discussed in Chapter 11, can be used to confirm expected performance. Should the ENBNRAS process prove competitive, pilot testing is recommended, due to the limited number of full-scale applications.

Muller et al. (2006) tested the ENBNRAS process at the Daspoort STP in Pretoria, South Africa. Two rock trickling filters in series were used and reported nitrification rates of 1.63 and 0.95 g N/m² media·d, respectively, treating effluent with an average temperature of almost 22°C. Because the effluent ammonia-nitrogen concentration averaged 0.58 mg/L, the nitrification rate in the second trickling filter may well have been substrate limited.

3.3.5   Design Example

An existing rock trickling filter must meet $cBOD_5$ limits under its current National Pollutant Discharge Elimination System permit conditions. The parameters below can be used to size a solids-contact tank for the trickling-filter effluent (TF/SC process) that will achieve an $sBOD_5$ of 5 mg/L.

Input Parameters
    Flow = 70 L/s (1.6 mgd),
    Water temperature = 12°C,
    Trickling-filter TOL = 0.48 kg BOD/m³·d (30 lb BOD_5/d/1000 cu ft),
    Final effluent BOD_5 = 30 mg/L,
    TF effluent $sBOD_5$ = 15 mg/L,
    TSS = 35 mg/L,
    VSS = 28 mg/L, and
    Primary effluent total BOD_5 = 140 mg/L.

Assumptions
    Q_{RAS} = 0.25Q, (therefore, R = [1.6 + 0.4]/1.6 = 1.25)

Calculations

1. Determine the $sBOD_5$ in the solids contact influent channel with RAS.

   Solids contact influent S_0 = [(1.0)(15 mg/L) + 0.25 (5 mg/L)]/(1.0 + 0.25) = 13 mg/L

2. Calculate the solids contact HRT using the following criteria:

   K_{20} = 3.0 × 10²⁵ L/g·min

   Θ = 1.035 (at 12°C)

   MLVSS (X_v) = 2000 mg/L

   $\ln (S_e/S_O) = [-K_{20}\Theta^{T-20}X_V] \times HRT$

$$HRT = \frac{\ln\left(\dfrac{S_e}{S_O}\right)}{-(K_{20}\Theta^{T-20}X_v)}$$

$$= \frac{\ln\left(\dfrac{5.0}{13}\right)}{-(3.0\times10^{-5})(1.035^{12-20})(2000)}$$

   HRT = 21 min

   HRT at $Q + Q_{RAS}$ = (21)(1.25) = 26 min

3. Compare the calculated HRT with empirical values in Table 13.1. Use the HRT calculated in Step 2 to calculate contact tank volume.

4. Calculate the contact tank volume.

Volume = (25 min)(60 s/min)(0.070 m³/s) = 105 m³ (3708 cu ft)

5. Determine net solids production per day and the SRT of the solids-contact tank. At the low loading, the rock filter net yield is estimated at 0.50 kg VSS/kg BOD₅ removed.

Net solids production = (0.5 kg/kg)(0.140 kg/m³)(0.07 m³/s)(86 400 s/d) = 423 kg VSS/d (931 lb/d)

$$SRT = \frac{(105\ m^3)(2.0\,kg\ VSS/m^3)}{423\,kg/d}$$

SRT = 0.5 days

Output Summary

HRT at $Q + Q_{RAS}$ = (20)(1.25) = 25 min

Solids-contact tank volume = 105 m³ (3708 cu ft)

SRT = 0.5 days

4.0 Design of Integrated Fixed-Film Activated Sludge Systems

4.1 Introduction

Integrated fixed-film activated sludge (IFAS) facilities are those that incorporate a fixed-film process component (a biofilm growth) within activated sludge basins. The IFAS facilities operate with a level of MLSS concentration similar to CAS systems with a sludge age that improves sludge settleability and provides carbonaceous removal, nitrification, and/or BNR. Figure 13.7 depicts typical IFAS configurations showing process influent, RAS, and internal nitrate recycle streams for BOD removal, nitrification, and BNR configurations (total nitrogen [TN] > 5 mg/L and total phosphorus [TP] > 1 mg/L, depending on influent characteristics, basin volumes, and temperatures). Figure 13.7 does not depict enhanced nutrient removal with post-anoxic zones and external carbon, which can meet TN < 5 mg/L.

The IFAS facilities may incorporate three types of biofilm process media or carriers within the activated sludge basin(s): fixed-bed media, free-floating plastic media that move within the bed, and sponge-type carriers that move within the bed. Fixed-bed media are held in frames (Figure 13.8a). Free-floating media are held within the basins or zones (anoxic or aerobic) using screen or sieve assemblies (Figure 13.8b). Sponge-type biofilm carriers are typically allowed to move longitudinally within the aerobic zone and are recycled or pumped back to the front of the aerobic basins using the process' media recycle system (Figure 13.8c). The typical range of MLSS incorporated in IFAS processes is from 1500 to 3500 mg/L, and the upper limit is determined by the secondary clarifier capacity. Therefore, when

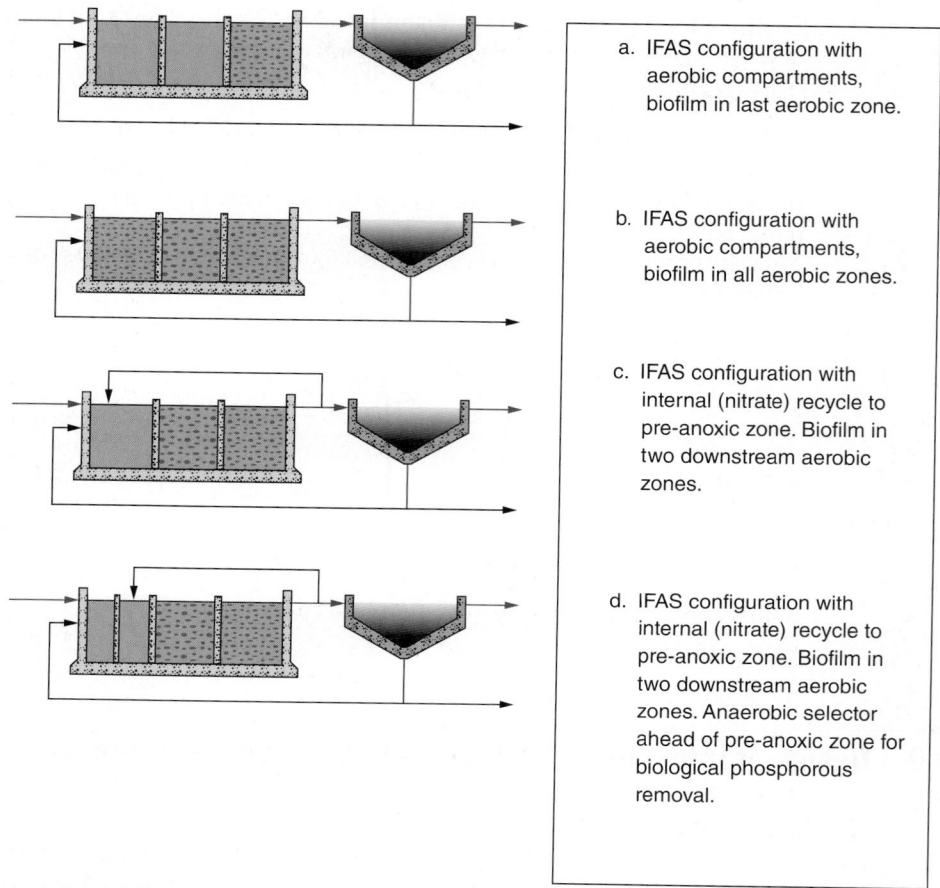

a. IFAS configuration with aerobic compartments, biofilm in last aerobic zone.

b. IFAS configuration with aerobic compartments, biofilm in all aerobic zones.

c. IFAS configuration with internal (nitrate) recycle to pre-anoxic zone. Biofilm in two downstream aerobic zones.

d. IFAS configuration with internal (nitrate) recycle to pre-anoxic zone. Biofilm in two downstream aerobic zones. Anaerobic selector ahead of pre-anoxic zone for biological phosphorous removal.

FIGURE 13.7   IFAS process configurations showing aerobic (BOD removal and nitrification) and biological nutrient removal. Enhanced nutrient removal configurations (post-anoxic basins with external carbon dosing) not shown (courtesy of Veolia Water Technologies).

designing IFAS facilities, designers must decide whether to construct new secondary clarifier infrastructure or to limit the amount of work the MLSS is able to provide, balancing reliance between suspended and fixed-film (biofilm) growth. Additionally, the designer must consider SRTs that will allow for adequate sludge-settling characteristics.

The principle of the IFAS system is to enhance BOD removal and nitrification above the removal that could have been achieved using MLSS in a suspended-growth-only reactor of the same volume. The interaction and exchange between the biofilm and the mixed liquor increases the design complexity relative to activated sludge and MBBRs (Chapters 12 and 11, respectively). Facility modeling is typically used to verify process options during facility planning and/or in permitting with state agencies. The designer should ensure that the accuracy of the model has been published or verified in full-scale systems. In addition to the effluent quality, the model should be able to predict the

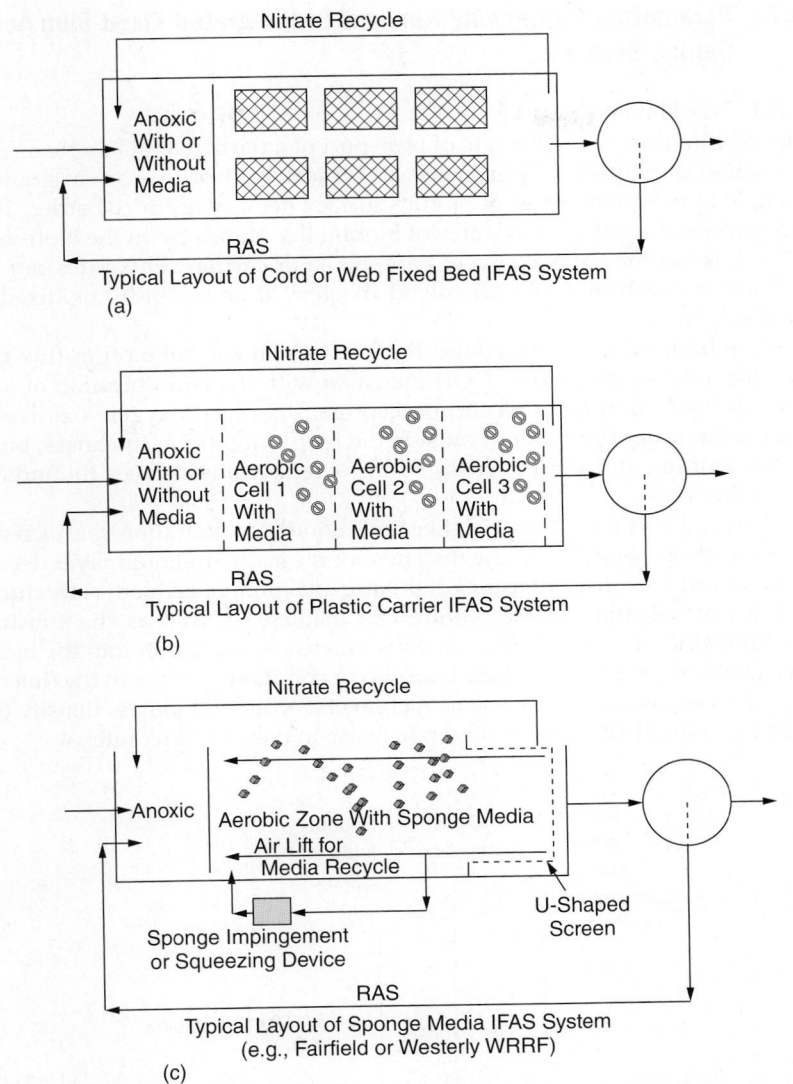

Figure 13.8   (a) Typical layout of cord or web fixed-bed medial integrated fixed-film activated sludge system. (b) Typical layout of plastic carrier media integrated fixed-film activated sludge system. (c) Typical layout of sponge media integrated fixed-film activated sludge system.

MLSS and MLVSS, substrate profiles, sludge production, oxygen demand, and airflow requirements; the model also must be fairly accurate in predicting the quantity of biofilm biomass.

Biofilm models discussed in Chapter 4 of this manual provide accurate prediction of substrate flux rates (of COD, NH_4-N, NO_x-N, sludge VSS, inert material) that influence substrate removal.

4.2   Parameters Influencing Removal in Integrated Fixed-Film Activated Sludge Systems

4.2.1   Biofilm Flux Rates

The biofilm flux rate is the rate of transport of a particular substrate or electron acceptor across the liquid-biofilm interface. Typical units of flux are in grams of substrate removed per square meter of biofilm surface per day ($g/m^2 \cdot d$) or $kg/1000\ m^2 \cdot d$. The biofilm surface is the surface area of biofilm that develops on the biofilm media or carriers. It is not the surface area of bare media or carriers. Flux rates can be in terms of parameters such as COD, dissolved oxygen, ammonium-N, oxidized-N, VSS, and inert solids.

It is important to understand the factors that control biofilm flux rates. The flux rate for a substrate such as COD increases with the concentration of substrate COD and the concentration of electron acceptors (dissolved oxygen, oxidized-N forms) in the mixed liquor (bulk liquid outside the biofilm). Biofilm thickness, biomass density in the biofilm, and the thickness of the stagnant liquid layer (boundary layer) also affect flux rate.

Figure 13.9 shows that a higher bulk liquid concentration can increase concentrations inside the biofilm. As the thickness of the stagnant liquid layer decreases, as may be observed at higher intensities of mixing or a more open media structure, the concentrations of substrate in the biofilm can increase as well as the substrate flux rate. Additionally, a greater biofilm thickness increases the depth into the biofilm to which the substrate can be used. There is also a concomitant increase in the flux rate if electron acceptors are available in these deeper layers. A higher biomass density (MLVSS of the biofilm, in mg/L) also generates an increase in the COD flux rate.

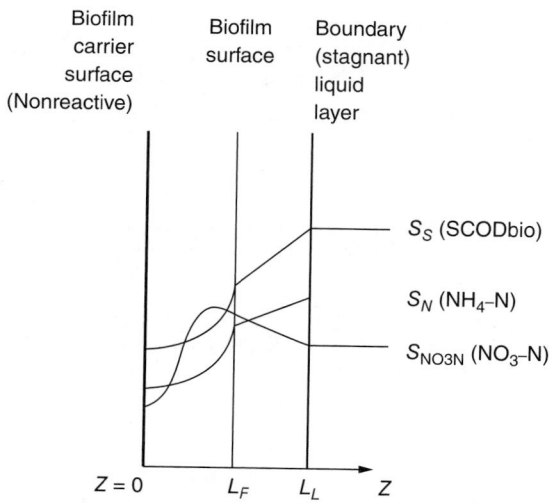

Typical Biofilm in the Middle of Aerobic Zone
(example shows a biofilm that removes COD, nitrifies, and denitrifies)

Figure 13.9   Importance of external substrate concentrations and thickness of stagnant liquid layer on substrate concentrations (and use rates) inside the biofilm.

4.2.2 Removals in Biofilm per Unit of Tank Volume

In addition to the biofilm flux rate, the removal that takes place per unit of tank media volume depends on the biofilm specific surface area (m² of biofilm surface/m³ of tank volume) and the media fill fraction. The removal per unit of tank volume (kg/m³·d) is:

$$\frac{\text{kg pollutant}}{\text{m}^3\text{tank vol./day}} = \frac{\text{kg pollutant}}{\text{m}^2\text{biofilm surf.area/day}} \times \frac{\text{m}^2\text{biofilm surf. area}}{\text{m}^3\text{biofilm carrier}} \times \frac{\text{m}^3\text{biofilm carrier}}{\text{m}^3\text{tank op. vol.}}$$

The biofilm media or carrier fill fraction is the fraction of the activated sludge basin volume occupied by the media or carriers. At 100% fill, the value of media fill fraction equals 1.0. For fixed-bed media, the media fill fraction is the fraction of the tank floor and height covered by media frames. For plastic and sponge-type free-floating media, the fill fraction is the fraction of the empty basin volume occupied by the carriers. The biofilm specific surface area at 100% media fill fraction depends on several factors. These include:

- Type of media used. The specific surface area at a certain fill volume fraction increases from fixed-bed media to sponge moving-bed media to certain types of plastic cylinders (Table 13.2).

- The thickness of the biofilm on the media increases with an increase in organic substrate (soluble biodegradable COD) levels in the bulk liquid. For most media, the surface area decreases when the biofilm thickness increases above a certain optimal level (Sen et al., 2007). In those situations, the thickness can be decreased by more vigorous mixing to induce a higher rate of biofilm shear.

- The extent of bare media surface that is covered by the biofilm. This can change with the external COD concentration (Figure 13.10).

Table 13.2 shows the biofilm specific surface areas and typical fill fractions applied in IFAS systems for secondary treatment. For free-floating media, the upper limit on fill

Type of System	Media	Media Fill Volume Percentage*	Biofilm-Specific Surface Area (m²/m³)	Recommended MLSS (mg/L)	Minimum Aerobic HRT (h) at 12°C
Activated sludge	None	0	0	3000	7
IFAS, fixed bed	Bioweb, Accuweb	70–80	50–100	3000	5
IFAS, sponge MBBR	Linpor, Captor	20–40	100–150	2500	4
IFAS, plastic carrier MBBR	K1 (Kaldnes), Entex, ActiveCell (Hydroxy)	20–65	150–3002	3500	4

*External volume of frame for cord-type fixed-bed media; external volume of cuboids or cylinders with biofilm for moving bed media. Note that fill volume fraction is not the fraction of liquid volume in the activated sludge tank displaced by the media.

TABLE 13.2 Typical Ranges of IFAS Design Parameters

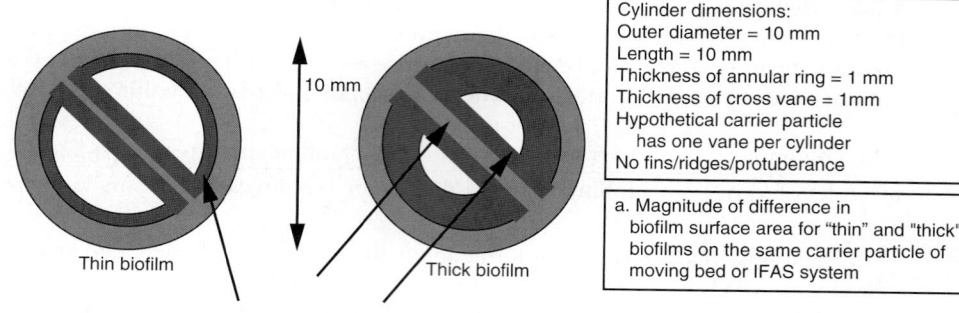

Cylinder dimensions:
Outer diameter = 10 mm
Length = 10 mm
Thickness of annular ring = 1 mm
Thickness of cross vane = 1mm
Hypothetical carrier particle
  has one vane per cylinder
No fins/ridges/protuberance

a. Magnitude of difference in
   biofilm surface area for "thin" and "thick"
   biofilms on the same carrier particle of
   moving bed or IFAS system

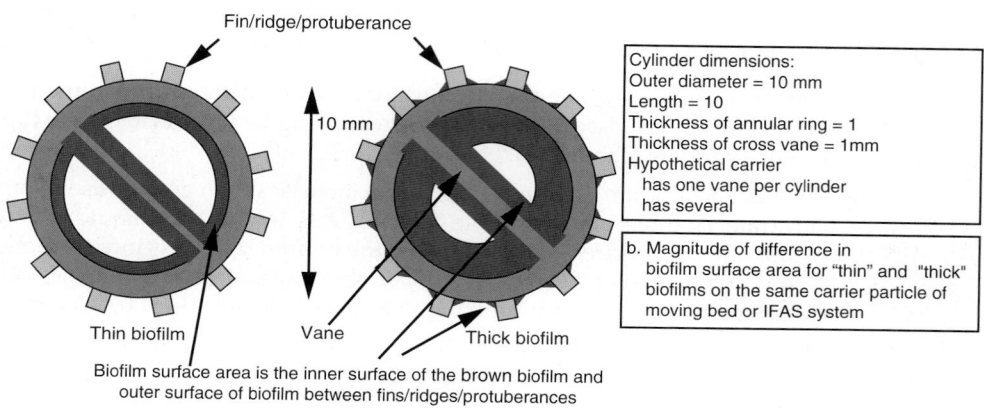

Cylinder dimensions:
Outer diameter = 10 mm
Length = 10
Thickness of annular ring = 1
Thickness of cross vane = 1mm
Hypothetical carrier
  has one vane per cylinder
  has several

b. Magnitude of difference in
   biofilm surface area for "thin" and "thick"
   biofilms on the same carrier particle of
   moving bed or IFAS system

FIGURE 13.10   Effect of biofilm thickness on locations of growth and surface area of biofilm (from Sen et al., 2007).

fractions is determined by the in-basin aeration system or anoxic mixing equipment's ability to create a completely mixed or moving bed. Factors that affect the ability to move the bed include carrier fill fraction, carrier buoyancy, airflow per unit basin volume or floor area, mixing power in anoxic zones, and the basin's hydraulic end-wall velocity, that is, the velocity of the wastewater and all recycle streams toward the end wall of the IFAS basin(s). The end-wall velocity is determined by dividing the total hydraulic throughput under peak conditions by the wetted end-wall area. For plastic free-floating media, a typical acceptable end-wall velocity is 30 to 40 m/h (e.g., m³/h of peak flow per m² wetted end-wall area).

The IFAS designers must also consider biofilm thickness and control of shear. If the biodegradable sCOD within the IFAS basins containing media or carriers is higher than 15 to 25 mg/L, then the biofilm will increase in thickness unless the shear (mixing or air scour) is increased for the type of media. If the shear is not changed, then specific surface area may remain constant over a certain range and then decrease (Sen et al., 2007). The range over which the surface area remains constant will depend on the type of media.

For plastic free-floating media, growth on the outer surface of media, which is otherwise barren, can add to the specific surface area, and the increase in thickness reduces the biofilm specific surface area inside the carrier. For cord media, the increase in the concentration of biodegradable sCOD in the cell may result in development of a tubular biofilm surface enveloping the loops rather than a serrated biofilm surface. This results in a reduction in effective surface area. Biofilm media and carrier design influences the changes in specific surface area with biodegradable sCOD for a given level of mixing.

Design curves can be obtained from the manufacturers that cover the range of biodegradable sCOD expected. For example, in an IFAS system design, the biodegradable sCOD can be decreased by adjusting parameters such as increasing nominal HRT of the first aerobic cell, media fill fraction in the first cell, recycle rates (nitrate recycle, sludge recycle), and the MLSS.

4.3　Parameters Influencing Removals in the Mixed-Liquor Suspended Solids

As with activated sludge systems, removal of COD per unit of tank volume in the IFAS system increases with increases in MLSS, concentrations of substrate (biodegradable sCOD), and dissolved oxygen (or oxidized-N) in the mixed liquor. For the same tank volume, an increase in the suspended-growth SRT increases the concentration of MLVSS. Increasing SRT and subsequent increase in MLVSS can result in a decrease in the biofilm surface area required for removal of organics and for nitrification. The SRT needed for nitrification in an activated sludge process is a function of the maximum specific growth rate (which is related to temperature, the reactor dissolved oxygen concentration, and pH [U.S. EPA, 2009]).

4.4　Interaction between Suspended Biomass and Biofilm

In designing an IFAS system, it is important to understand the interaction between the suspended growth and the biofilm. An increase in removal by the suspended growth, as observed at a higher suspended-growth SRT (or simply SRT), decreases the sCOD concentrations at any location along the aerobic zone. This decreases the COD uptake rates (flux rates) for the biofilm at that location. For the same SRT, an increase in temperature has a similar effect. On the other hand, reducing the SRT or lowering the temperature has the opposite effect, and the rate and the fraction of COD removal in the biofilm will increase.

Additionally, in designing an IFAS system, one can take advantage of the available clarifier capacity to operate the system with a higher MLVSS and SRT and reduce the amount of biofilm required. This may allow one to change the type of media required to a lower biofilm specific surface area (Table 13.2). It also may allow use of a smaller aeration basin volume compared to an activated sludge system or an MBBR.

4.5　Interaction between Heterotrophs and Nitrifiers

Understanding the interaction between heterotrophs and nitrifiers (autotrophs) is a challenge in IFAS systems. As with activated sludge systems, factors that increase soluble biodegradable COD levels in the bulk liquid (from 5 to 20 mg/L) result in a substantial increase in growth rates of heterotrophs preferentially over growth of nitrifiers. This decreases the fraction of nitrifiers in both the suspended growth and the biofilm, and decreases the nitrification rate in each.

In an activated sludge system, a reduction in SRT close to the washout SRT of nitrifiers decreases the nitrifier biomass until nitrification is lost, and addition of a biofilm carrier typically allows nitrifiers to survive. The nitrifier biomass in the biofilm will depend on the soluble biodegradable organic concentration and dissolved-oxygen levels in the bulk liquid outside the biofilm.

When nitrifiers are present in the biofilm of the IFAS system, biofilm sloughing seeds the mixed liquor with nitrifiers. At SRTs approaching washout SRT, this sloughing and seeding can increase significantly the fraction of nitrifiers in the MLSS as compared to an activated sludge system operating at the same SRT. Therefore, the presence of biofilm helps increase nitrification rates per unit of tank volume. If nitrification in the biofilm is not compromised during a facility upset or suspended-growth process washout, then the time required for an IFAS system to recover full nitrification is reduced compared to a system with activated sludge only, due to seeding.

4.6   Design Tools/Procedures

Two types of design procedures for IFAS systems have been developed:

1. Empirical methods based on experience of each manufacturer or process supplier; and

2. Process kinetics models that rely on different levels of integration of activated sludge and biofilm kinetics (discussed in Chapter 4 of this manual).

Each manufacturer customizes its own empirical methods, which may be based partially on activated sludge and biofilm kinetics. Typically, the contribution of the biofilm and its kinetics are simplified by the manufacturer or the process supplier based on observations made at their existing facilities. Models based on empirical methods should be applied with caution. Further, their application should be limited to the type of media and system for which they were developed and tested. For example, if nitrification was observed in a tertiary system with low influent COD, then rates and models should not be applied to a secondary system.

In addition to the kinetic models discussed in Chapter 4 of this manual, a Water Environment Research Foundation (2000) report on hybrid systems presented a method that has been in development since 2000. This method was verified in full scale at the City of Broomfield (Colorado) Wastewater Treatment Facility, an IFAS facility, and is available in the public domain. Researchers and practitioners collaborated to create design tools for IFAS systems. The AQUIFAS model has been verified in full scale against data from that facility (Rutt et al., 2006).

The AQUIFAS model was also used to verify the largest fixed-bed media facility, located in Annapolis, Maryland. As part of the IFAS design for fixed-bed media, the biofilm specific surface area increased gradually to 70 m²/m³ in the three cells of the aerobic zone, which is also the upper limit of certain types of fixed-bed media (Table 13.2). The mean cell residence time (MCRT) was increased to 7 days, which increased the MLSS from 2000 mg/L to an upper limit of the secondary clarifiers, which in this instance was 3600 mg/L. In this application, the IFAS fixed-bed system achieves an effluent ammonium-N limit of 1 mg/L and a total N of 8 mg/L, which were the goals of this design.

In several instances, limitations on the secondary clarifier surface area may not allow an increase in MLSS. Table 13.2 shows how free-floating media can increase the biofilm specific surface area once filling degree is considered, to a range of 100 to more than 500 m^2/m^3.

4.7 Empirical Design Methods

4.7.1 Sludge Age Approaches

Some suppliers of sponge and plastic media systems use the equivalent sludge age approach. The quantity of biomass presented in the media is added to the quantity in the mixed liquor. An equivalent SRT is calculated by dividing the total biomass (in the biofilm and mixed liquor) by the daily wasting rate. In this instance, the process supplier develops specific curves to equate equivalent SRT to facility performance based on knowledge of sloughing. The method is reliable as long as it is applied to the same conditions and performance as the system(s) from which the data were collected. It is akin to doing a scalable pilot study and applying the results to the design.

Some suppliers of plastic, free-floating media systems determine the load that the suspended growth can remove, given site constraints (e.g., clarifier limitations, aeration basin volume available, the desire for BNR with anoxic and/or anaerobic selectors, etc.), and then determine the amount of biofilm required to remove the remaining load. Typically, in this approach, the SRT of the suspended growth will be 3 to 5 days such that the mixed liquor will have good sludge-settling characteristics, but such that nitrification preferentially occurs in the biofilm under winter conditions, and seeding from the biofilm into the mixed liquor allows partial nitrification in the suspended growth during summer conditions. The designer should take seasonal effluent quality requirements into consideration.

4.7.2 Quantity of Media Approach

Some suppliers of fixed-bed media use the quantity (length or web surface area) of media approach. The media is sold based on length of media or surface area, such as for webbed cord media. Suppliers have measured removal rates as well as the mass of biomass attached to the media in the facilities and pilot plants they operated. This information is then applied to new designs. This method should be limited to application where process conditions in the design facility match that under which the data were collected.

4.7.3 Rates Based on Pilot Studies

Another approach is to use an on-site pilot study to represent the proposed full-scale application. This works if pilot study data are representative of the conditions in the full-scale system. There are, however, challenges in replicating fixed-bed media systems in pilot studies that do not accommodate the type of diffusers and depth that would be used in full-scale systems.

4.8 Disadvantages of Integrated Fixed-Film Activated Sludge Systems

4.8.1 Energy Efficiency

Coarse-bubble aeration typically is used to ensure an equal distribution of the buoyant media throughout the mixed-liquor column and to reduce the need for and frequency of access to aeration basins. Coarse-bubble aeration is not as efficient as fine-bubble aeration. In addition, some suppliers prescribe high DO concentrations, further increasing aeration requirements and power demand. A life-cycle evaluation of an IFAS system supplier's aeration power requirements to that of an activated sludge facility providing the same removals using fine-bubble aeration will assist in effective process selection. Life-cycle costs should include capital and operating costs for both systems. Operating costs should include blower power consumption (as discussed in Chapter 12), in-basin maintenance, repair, and replacement costs.

4.8.2 Red Worms
As mentioned above, the fixed-bed media process may require higher aeration rates to maintain sloughing rates causing high DO concentrations. This higher DO concentration and fixed static media may result in the proliferation of red worms in the media. The worms are predators of the biofilm, and their presence means a reduction in the biofilm available to perform treatment. The worms may affect system capacity and performance. In addition, if aeration is lost or access to the tank is required, the red worms may die and become a source of odors, which may be difficult to cleanup and remove.

4.8.3 Storage of Media
The moving and fixed-media products occupy large volumes within the aeration basin. During construction, contractors will need a significant site area for staging. During operation, if access is needed to an aeration tank, then some method of storage of the media or transfer to and from an adjacent tank is required. Also, free-floating media and sponges are typically delivered in plastic sacks, and sunlight will degrade the integrity of the sacks with long-term, outdoor storage. Contractors should cover the sacks with tarps or store those within shipping containers or within warehouse(s) until installation.

4.8.4 Foaming
The designer also should evaluate the risk of foam accumulation in basins and zones designed to retain free-floating media, which may not allow foam to leave the basin or zone. Foam accumulation may get worse with increasing SRT. If the facility has a propensity to accumulate foam, either seasonally or because of changes in the influent characteristics, then the designer may want to limit the suspended-growth SRT and use more biofilm media instead. This is not a concern in fixed-bed or sponge-type systems because they do not use in-basin screens to retain media (Figure 13.8a).

The IFAS systems appear to be prone to *Nocardia*-form foaming, particularly with nitrogen-removal configurations. Every method recommended for foam control must be available for use by operations in these systems including surface chlorine sprays, surface wasting, RAS chlorination, redirection of all scum and surface waste to prevent reseeding, and defoaming polymer.

4.9 Process Kinetics Design Methods (Process Models)

4.9.1 Biofilm Rate Model
The following approach may be used for preliminary sizing of IFAS systems. It is based on rates observed in full-scale and pilot studies and in calibrated process models.

4.9.2 Define Range of Flux Rates

Primary effluent (reactor influent) with a COD/total Kjeldahl nitrogen (TKN) of 7.5:1 to 15:1 can use the following rates at a mixed-liquor temperature of 15°C and an aerobic zone dissolved oxygen of 3 mg/L:

- Aerobic COD uptake: 0.5 to 5 kg/1000 $m^2 \cdot d$; and
- Nitrification: 0.05 to 0.5 kg/1000 $m^2 \cdot d$.

To adjust rates for temperature, an Arrhenius adjustment coefficient of 5% for each 1°C change in temperature can be used. The breadth of these ranges is indicative of the applicability of this system to accommodate a wide range of conditions. The actual value of flux rate for the media location and application is discussed in the following section.

4.9.3 Quantify Removal

This procedure is in keeping with the approach discussed above, in which the load that the MLSS can remove is determined and the remainder of the load must be removed by the biofilm. The recommended fraction of removals in mixed liquor and biofilm to quantify the biofilm surface area required in IFAS systems are based on analysis performed with process kinetics models. These removals at 15°C are given below and may be increased by 3% for every 1°C increase in temperature:

- At 2-day SRT, 50% of COD and 80% of nitrification is on the biofilm; the rest is in the suspended biomass.
- At 4-day SRT, 25% of COD and 50% of nitrification is on the biofilm.
- At 8-day SRT, 20% of nitrification is on the biofilm.

The surface area of biofilm required is the higher of the two needed for COD removal and nitrification. The rates based on BOD typically would be half the rates based on COD.

4.9.4 Select Flux Rates Based on Location along Aerobic Zone

If this approach is taken, it is recommended that the aerobic zone be divided into one-thirds and the following load application is used along the aeration basin length:

- Apply rates of 75%, 50%, and 25% of the maximum rate for COD uptake for the first third, second third, and final third of the aerobic zone, respectively;
- Apply rates of 25%, 50%, and 75% of the maximum rates for ammonium-N oxidation for the first third, second third, and final third of the aerobic zone, respectively, when targeting and effluent ammonium-N of 1 mg/L. Nitrification rates in the middle and final thirds of the aerobic zone should be decreased in proportion to ammonium-N concentration in the mixed liquor when the effluent quality required is less than 1 mg/L. For example, at an effluent ammonium-N of 0.5 mg/L, the rate applied in the final third should be 0.375 of the maximum rate (i.e., $0.75 \times 0.5/1.0$).

4.9.5 Additional Analysis to Finalize a Design

It is recommended that designers use one of the software models discussed in Chapter 4. For IFAS systems, it is also recommended more than one software application should be used to confirm results.

4.9.6   Design Example: Upgrading a Conventional Activated Sludge Facility

Design engineers are often faced with the need to upgrade an existing, non-nitrifying CAS system to increase facility capacity and/or meet new nutrient limitations. Figure 13.11 presents a method of conversion of CAS to IFAS or MBBR. The following examples illustrate this situation.

- Fixed bed IFAS (Annapolis, Maryland).

    The original treatment facility was a CAS process rated for 38 000 m³/d (10 mgd). An initial upgrade to fixed-bed IFAS (using Ringlace media) was operated until 2005, when the IFAS system was removed, as BNR requirements warranted pre/post zones to meet low TN and TP requirements (3 mg/L TN and 0.3 mg/L TP at 50 000 m³/d [13 mgd] flow), and the facility opted for a suspended-growth system.

    The fixed-bed IFAS process operated until 2005 with 244 000 m of Ringlace media with a suspended-growth MCRT of 6.2 to 6.5 days, an HRT of 8.1 to 8.8 hours, and provided effluent ammonia-N concentrations ranging from 0.35 to 0.9 mg/L. The CAS basins retrofitted with the fixed-bed media totaled 9460 m³ (2 497 440 gal) with 27% of the bioreactor volume operated as pre-anoxic

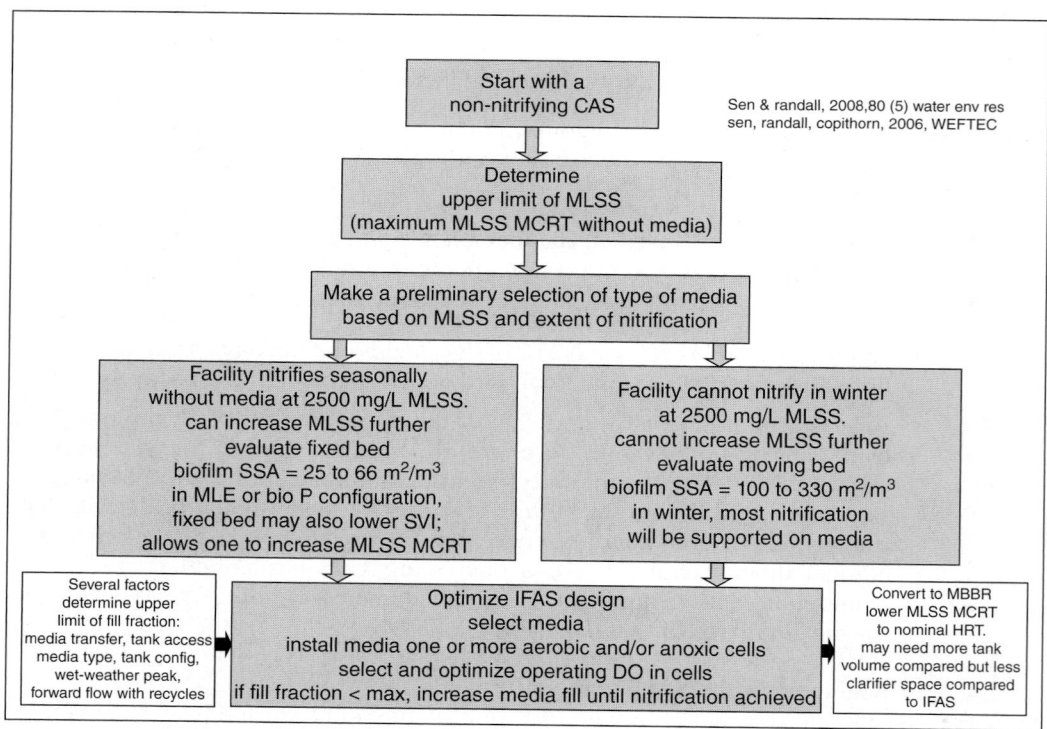

FIGURE 13.11   Method to convert a high-rate CAS system design to and MBBR processes (Sen et al., 2008).

selectors and 73% of the bioreactor volume operated aerobically, with secondary clarification by two 32-m-diameter, 3-m-deep clarifiers.

Improvements in SVI have been observed at several fixed-bed media facilities following installation of media when combined with proper liquid and air flows. Reduction in SVI can be because of (1) denser biofilm sloughing off fixed-bed media and (2) plug-flow kinetics created by the installation of media frames in the basin and improved mixing of liquid through the media frames when a liquid roll pattern is induced. This pattern occurs when diffusers are placed under the frames with media. In addition, improved aeration provides better control of the DO levels along the length of the tank. Observations in basins with media frames show that the placement of frames reduces back-and-forth mixing along the length of the tank, resulting in a multi-CSTR or plug-flow pattern. The frames can act like compartments within the length of the basin.

Results from the Annapolis fixed-bed facility and from facilities in Colony, Texas, and Geisselbullach, Germany, show that proper design of fixed-bed media system lowers the SVI (Copithorn et al., 2006; Lessel, 1994). Interestingly, in Annapolis, the SVI decreased from 130 to less than 100 mL/g with IFAS fixedbed, completely eliminating the need for periodic control of SVI with RAS chlorination. Once the IFAS fixedbed was removed after 10 years of operation, the SVI increased.

- The IFAS with free-floating media (Broomfield, Colorado).

In 2003, the City of Broomfield, Colorado WRF upgraded from a capacity of 20 400 m³/d (5.4 mgd) CAS facility to 30 300 m³/d (8 mgd) using IFAS with plastic free-floating media in the aerobic zones. The facility was designed to meet 10 mg/L BOD and TSS, 1.5 mg/L ammonia-N in summer and 3.0 mg/L ammonia-M in winter, 10 mg/L total inorganic nitrogen (TIN), and 1 mg/L TP. Because the influent stream contained nitrate, the process configuration includes 364 m³ of pre-anoxic basins, 604 m³ of anaerobic basins, 1302 m³ of anoxic basins to which internal nitrate is recycled, and 4550 m³ of aerobic basins containing 30% fill of the plastic media carriers. The facility reportedly meets all design performance requirements with suspended-growth SRT ranging from 3 to 4 days. Additionally, SVIs improved after the IFAS upgrade, with 100 to 200 mL/g prior to the upgrade and 100 to 150 mL/g after the upgrade.

- Moving bed IFAS with sponge carriers (Westerly, Rhode Island).

The Westerly, Rhode Island WRRF was a non-nitrifying CAS process upgraded in 2003 to IFAS using moving-bed Linporfree-floating sponge media with a design capacity of 12 500 m³/d (3.3 mgd) and to meet 30 mg/L BOD and TSS, a seasonal nitrogen limit of 15 mg/L (June-October), and 5.5 mg/L ammonia-N (June-October), and 30.9 mg/L (November-May). The two parallel IFAS trains measure 10.5 m (34 ft) wide by 31 m (102 ft) long, with approximately 33% of the bioreactor volume operated as pre-anoxic MLSS selectors and 67% of the bioreactor volume operated as aerobic IFAS basins. The system is reported to meet TN targets using amixed-liquor internal recycle rate of only 100% of the influent flow, so simultaneous nitrification and denitrification appears to be occurring. The SVI is reportedly better with an operational IFAS system than with CAS alone.

5.0   References

Albertson, O. E.; Eckenfelder, W. W. (1984) Analysis of Process Factors Affecting Plastic Media Trickling Filter Performance. *Proceedings of the Second International Conference on Fixed-Film Biological Processes*, Washington, D.C.

Balmer, P.; Ekfjorden, L.; Lumley, D.; Mattsson, A. (1998) Upgrading for Nitrogen Removal under Severe Site Restrictions. *Water Sci. Technol.*, **37** (9), 185.

Boller, M.; Gujer, W. (1986) Nitrification in Tertiary Trickling Filters Followed by Deep Bed Filters. *Water Res. (G.B.)*, **20**, 1363.

Copithorn, R. R.; Sturdevant, J. H.; Farren, G. D.; Sen, D. (2006) Case Study of an IFAS System–Over 10 Years Experience. *Proceedings of the 79th Annual Water Environment Federation Exposition and Conference* [CD-ROM], Dallas, Texas, October 21–25; Water Environment Federation: Alexandria, Virginia.

Daigger, G. T.; Boltz, J. P. (2011) Trickling Filter and Trickling Filter-Suspended Growth Process Design and Operation: A State-of-the-Art Review. *Water Environ. Res.*, **83**, 388–404.

Daigger, G. T.; Harrison, J. (1987) A Comparison of Trickling Filter Media Performance. *J. Water Pollut. Control Fed.*, **59**, 679.

Daigger, G. T.; Norton, L. E.; Watson, R. S.; Crawford, D.; Sieger, R. B. (1993) Process and Kinetic Analysis of Nitrification in Coupled Trickling Filter/Activated Sludge Processes. *Water Environ. Res.*, **65**, 750.

Daigger, G. T.; Norton, L. E.; Watson, R. S.; Crawford, D.; Sieger, R. B. (1993) Closure to Discussion of: Process and Kinetic Analysis of Nitrification in Coupled Trickling Filter/Activated Sludge Processes. *Water Environ. Res.*, **65**, 750–758.

Flamming, J. J.; Woodman, D. L.; Kulick, F. M. (2011) Application of Structured Sheet Media IFAS for Ammonia Polishing from a Trickling Filter Solids Contact Process. *Proceedings of the 84th Annual Water Environment Federation Technical Exposition and Conference* [CD-ROM]; Los Angeles, California, Oct 15–19; Water Environment Federation: Alexandria, Virginia.

Gharagozian, A.; Swanback, S.; Narayanan, B.; Chan, R. (2011) Solids Yield Surprise, Research on Trickling Filter/Activated Sludge Systems Produces Unexpected Results.

Harrison, J. R.; Daigger, G. T. (1987) A Comparison of Trickling Filter Media. *Water Environ. Technol.*, **23** (6), 46–51.

Harrison, J. R.; Timpany, P. L. (1988) Design Considerations with the Trickling Filter Solids Contract Process. *Proc. Joint Can. Soc. Civ. Eng.; Am. Soc. Civ. Eng. Natl. Conf.; Environ. Eng.*; Vancouver, B. C., Canada.

Harrison, J. R.; Daigger, G. T.; Filbert, J. W. (1984) A Survey of Combined Trickling Filter and Activated Sludge Processes. *J. Water Pollut. Control Fed.*, **56**, 1073

Hu, Z.-R.; Wentzel, M. C.; Ekama, G. C. (2000) External Nitrification in Biological Nutrient Removal Activated Sludge Systems. *Water SA*, **26**, 2, 225.

Kalb, K.; Williamson, R.R.; Frazer, W. M. (1990) Nitrified Sludge: An Innovative Process for Removing Nutrients from Wastewater. *Proceedings of the 63rd Annual Water Pollution Control Conference*, Washington, D.C., Oct 7–11; Water Environment Federation: Alexandria, Virginia.

Krumsick, T. A.; et al. (1984) Trickling Filter Solids Contact Process Demonstration, Salt Lake City, Utah. *Proceedings of the Annual Conference of the Utah Water Pollution Control Association*; Salt Lake City.

Lessel, T. H. (1994) Upgrading and Nitrification by Submerged Bio–Film Reactors–Experiences from a Large Scale Plant. *Water Sci. Technol.*, **29** (10–11), 167–174.

Matasci, R. N.; Kaempfer, C.; Heidman, J. A. (1986) Full–Scale Studies of the Trickling Filter/Solids Contact Process. *J. Water Pollut. Control Fed.*, **58**, 955.

Matasci, R. N.; Clark, D. L.; Heidman, J. A.; Parker, D. S.; Petrik, B.; Richards, D. (1988) Trickling Filter/Solids Contact Performance with Rock Filters at High Organic Loadings. *J. Water Pollut. Control Fed.*, **60**, 68.

Matasci, R. N.; Clark, D. L.; Heidman, J. A.; Parker, D. S.; Petrik, B.; Richards, D. (1989) Author's Response to Discussion of Trickling Filter/Solids Contact Performance with Rock Filters at High Organic Loadings. *J. Water Pollut. Control Fed.*, **61**, 371.

Morgan, S.; Farley, R.; Pearson, R. (1999) Retrofitting an Existing Trickling Filter Plant to BNR Standard–Selfs Point, Tasmania's First. *Water Sci. Technol.*, **39** (6), 143.

Motta, E. J. L.; Jimenez, J. A .; Josse, J. C.; Manrique, A. (2003) The Effect of Air–Induced Velocity Gradient and Dissolved Oxygen on Bioflocculation in the Trickling Filter/Solids Contact Process. *Adv. Environ. Res.*, **7**, 441.

Muller, A. W; Wentzel, M. C.; Ekama, G. A. (2006) Estimation of Nitrification Capacity of Rock Media Trickling Filters in External Nitrification BNR. *Water SA*, **32** (5), 611–618.

Newbry, B. W.; Daigger, G. T.; Taniguchi-Dennis, D. (1988) Unit Process Tradeoffs for Combined Trickling Filter and Activated Sludge Processes. *J. Water Pollut. Control Fed.*, **60**, 1813.

Norris, D. P.; Parker, D. S.; Daniels, M. L.; Owens, E. L. (1982) Production of High Quality Trickling Filter Effluent Without Tertiary Treatment. *J. Water Pollut. Control Fed.*, **54**, 1087.

Parker, D. S.; Matasci, R. N. (1989) The TF/SC Process at Ten Years Old: Past, Present, and Future. *Proceedings of the 62nd Annual Water Pollution Control Federation Conference, San Francisco, California*, Water Pollution Control Federation: Washington, D.C.

Parker, D. S.; Richards, J. T. (1994) Discussion of: Process and Kinetic Analysis of Nitrification in Coupled Trickling Filter/Activated Sludge Systems, G. T. Daigger, et al., **65**, 750 (1993). *Water Environ. Res.*, **66**, 934–935.

Parker, D. S.; Brischke, K. V.; Matasci, R. N. (1993) Upgrading Biological Filter Effluents Using the TF/SC Process. *J. Inst. Water Environ. Manage.*, **7**, 90.

Parker, D. S.; Krugel, S.; McConnell, H. (1994) Critical Process Design Issues in the Selection of the TF/SC Process for a Large Secondary Treatment Plant. *Water Sci. Technol.*, **29** (10–11), 209.

Parker, D. S.; Romano, L. S.; Horneck, H. S. (1998) Making a Trickling Filter/Solids Contact Process Work for Cold Weather Nitrification and Phosphorus Removal. *Water Environ. Res.*, **70**, 181.

Parker, D. S.; Bratby, J. R. (2001) Review of Two Decades of Experience with TF/SC Process. *J. Environ. Eng.*, **127** (5), 380.

Richards, T.; Reinhart, D. (1986) Evaluation of Plastic Media in Trickling Filters. *J. Water Pollut. Control Fed.*, **58**, 774.

Rutt, K.; Seda, J.; Johnson, C. H. (2006) Two Year Case Study of Integrated Fixed Film Activated Sludge (IFAS) at Broomfield, CO, WWTP. *Proceedings of the 79th Annual Water Environment Federation Technical Exposition and Conference* [CD–ROM], Dallas, Oct 21–25; Water Environment Federation: Alexandria, Virginia.

Sen, D.; Randall, C. W.; Copithorn, R. R.; Huhtamaki, M.; Farren, G.; Flournoy, W. (2007) Understanding the Importance of Aerobic Mixing, Biofilm Thickness Control and Modeling on the Success or Failure of IFAS Systems for Biological Nutrient Removal. *Water Pract.*, **1** (5), 1–18.

Sen, D.; Randall, C. W. (2008). Improved Computational Model (Aquifas) for Activated Sludge, IFAS and MBBR Systems, Part I: Semi–Empirical Model Development. *Water Environ. Res.*, **80** (5), 439–453.

Slezak, L. A.; Fries, M. K.; Pickard, L. R.; Palsenbarg, R. A. (1998) Liquid Stream Secondary Treatment Process Design at the Annacis Island Wastewater Treatment Plant of the Greater Vancouver Sewerage and Drainage District. *Water Sci. Technol.*, **3**, 51.

Tekippe, T. R.; Hoffman, R. J.; Matheson R. J.; Pomeroy, B. (2004) A Simple Solution to Big Snail Problems - A Case Study at VSFCD's Ryder Street Wastewater Treatment Plant. *Proceedings of the 77th Annual Water Environment Federation Exposition and Conference* [CD-ROM], New Orleans, Louisiana, Oct 2–6; Water Environment Federation: Alexandria, Virginia.

U.S. EPA (2009) *Nutrient Control Design Manual*, EPA/600/R-09-012, January.

Vestner, R. J.; Gunthert, F. W. (2001) Upgrading of Trickling Filters for Biological Nutrient Removal with an Activated Sludge Stage. *GWF Water Wastewater*, **142** (15), 39.

Water Environment Research Foundation (2000) *Investigation of Hybrid Systems for Enhanced Nutrient Control*; Water Environment Research Foundation: Alexandria, Virginia.

Witzgall, R.; Parker, D.; Waterman, N.; Sen, S.; Hetherington, M. (2013) How a Trickling Filter/Solids Contact Plant Was Designed and Optimized for Handling Greater Flow Variability While Requiring Much Lower Energy Than OCSD's Activated Sludge Plants. *Proceedings of the 86th Annual Water Environment Federation Technical Exposition and Conference* [CD-ROM]; Chicago, Illinois, Oct 5–9; Water Environment Federation: Alexandria, Virginia.

CHAPTER **14**

Physical and Chemical Processes for Advanced Wastewater Treatment

Val S. Frenkel, Ph.D., P.E., D. WRE; and Onder Caliskaner, Ph.D., P.E.

This chapter describes technologies and processes that can be applied to treat secondary effluent from biological treatment to achieve certain quality requirements for water discharge, reuse, recycle, or other advanced treatment purposes.

Engineers often are required to design water resource recovery facilities (WRRFs) for pollutant removal that exceeds the usual definition of secondary treatment. For example, removal of facility nutrients such as nitrogen or phosphorus may be required to prevent eutrophication of the receiving water body or meet tightened effluent requirements. Lower effluent concentrations of biochemical oxygen demand (BOD) or total suspended solids (TSS) may be required to achieve local water quality standards. Sometimes, trace pollutants such as heavy metals or refractory organics must be reduced because of their toxicity to aquatic life or interference with downstream potable water supplies.

There may be treatment requirements for three types of microconstituents: pharmaceutically active compounds (PhACs); endocrine disrupting compounds (EDCs); and personal care products (PCPs).

As effluent toxicity grows in importance, new discharge permits or renewals are likely to include toxicity-based limits. Traditional secondary treatment methods have achieved limited success in removal of toxic substances.

Climate change, global warming, and recent natural disasters triggered close attention to water reuse, which normally requires better water quality than secondary treatment can deliver. The latest trends such as indirect potable reuse (IPR) and direct potable reuse (DPR) are gaining popularity due to the water scarcity in many places. Advanced treatment processes are an integral part of the water reuse, which may include low-pressure membranes (LPMs), high-pressure membranes (HPMs), UV, and/or ozonation. The simplified schematic of IPR reuse is shown in Figure 14.1. The IPR is a similar concept to the DPR utilizing similar treatment schemes. The major difference is that IPR is utilizing natural environmental barrier as shown in Figure 14.1, while the DPR concept is utilizing man-made barrier delivering advance treated effluent to the source of the drinking water supply.

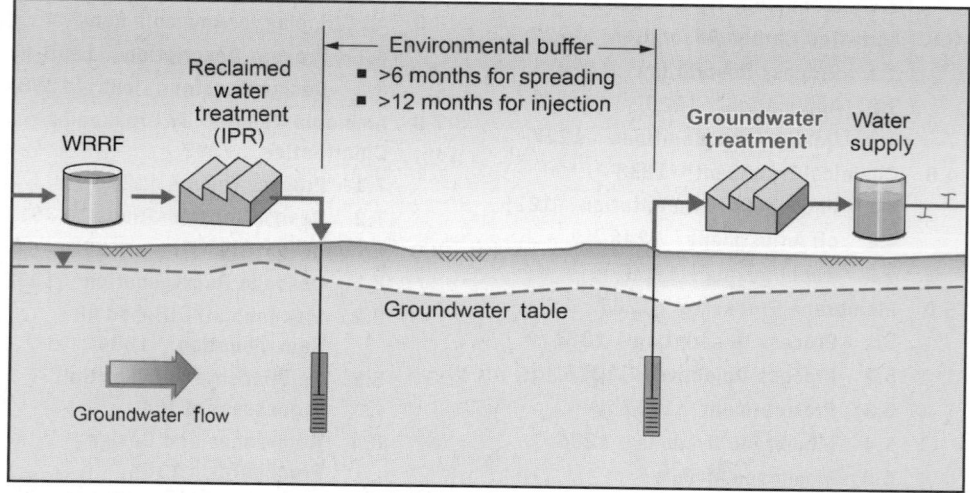

Figure 14.1   Schematic of Indirect Potable Reuse (courtesy of Val S. Frenkel).

These advanced wastewater treatment needs may be addressed as an integral part of a new treatment flowsheet or as an add-on to the existing secondary treatment train. This chapter presents design information for unit processes that provide effluent polishing, nutrient removal, or removal of toxic constituents. The six unit processes discussed are filtration, adsorption, chemical treatment, membrane processes, air stripping, and breakpoint chlorination. Effluent reoxygenation is briefly discussed. Unit processes that may achieve some of these objectives but are not yet accepted for municipal wastewater treatment applications will not be discussed in this chapter.

1.0   Process Selection Considerations

Each unit process is added to a conventional secondary treatment process to achieve certain design objectives and capabilities based on treated effluent limits. Further, there are many variations of systems and equipment. As a minimum, the following factors should be considered:

- Effluent goals;
- Process capabilities;
- Process compatibility with overall treatment flowsheet;
- Operational factors;
- Process control;
- Sidestreams and recycles;
- Solids production;
- Air emissions;
- Energy requirements;
- Space requirements;
- Worker health and safety; and
- Cost.

The relative weight assigned to these considerations in the selection decision process is project specific. Table 14.1 summarizes comments for each of the above criteria relevant to each add-on process. Treatability studies to screen candidate processes typically are necessary; pilot studies to verify performance of proprietary systems are advisable.

2.0   Secondary Effluent Filtration

2.1   Background

In advanced wastewater treatment, filtration is used to remove the residual TSS in secondary effluent. With the numerous filtration systems available, not all systems (many of which are proprietary) can be reviewed in this chapter. Common secondary effluent filtration technologies are depth filtration (involving use of granular, compressible medium, or cloth fiber material); surface filtration (involving use of cloth-woven or stainless steel microscreen material). This section discusses the design considerations of

Treatment Process	Process Capability	Process Control	Operational Factors	Sidestreams and Recycle	Solids	Air Emissions	Energy Requirement	Space Requirement
Filtration	Suspended solids removal to <10 mg/L	Control typically automated based on headloss, filter run time, and turbidity	Headloss, flowrate control, backwash, and effluent turbidity	Backwash water	Sludge from solids contained in backwash	None	Low to moderate	Low
Adsorption	Dissolved organics removal to >99%; some toxic metal removal	Head loss, flow rate, and pollutant breakthrough control	Backwash, column downtime for carbon replacement	Backwash water	None[c]	None	Low to moderate	Low
Chemical treatment	Phosphate removal to <1 mg/L; metals removal to <1 mg/L, acid–base neutralization	pH, ORP;[b] chemical dosage	Chemical storage, chemical feed, mixing	Drainage from chemical sludge dewatering	Hydroxide carbonate, and phosphate precipitates; other chemical sludges	Dust from chemical handling and storage	Moderate to high	Moderate to high
Membrane process	Removal of TSS, TDS,[c] microorganisms, viruses, and some organic compounds	Pretreatment, transmembrane pressure, concentrate flow, membrane flux	Pretreatment requirements, concentrate disposal	Concentrate, cleaning solutions	Solids from backwash on MF/UF[d] processes	Fugitive from chemical storage and use	Low to high emissions	Low to moderate

Air stripping	Volatile organic carbon removal to >99% Ammonia removal to >99%	Minimal air-to-water ratio Minimal pH and air-to-water ratio		Possible column was liquid	Possible carbonate precipitates	Volatile organics, ammonia, or both	Low to moderate	Low
Reoxygenation (postaeration)	Effluent dissolved oxygen increased to nearly saturation concentration	Dissolved oxygen level	Aeration rate	None	None	Minimal, volatile chemicals	Moderate to high	Moderate

aRegeneration of spent carbon can produce air emissions.

bORP = oxidation–reduction potential.

cTDS = total dissolved solids.

dMF/UF = microfiltration/ultrafiltration.

TABLE 14.1 Add-on Processes—General Considerations

these filtration technologies. Membrane filtration is also used for secondary effluent filtration or as pretreatment for further advanced treatment and discussed in more detail in this chapter.

2.1.1 Goals

The main objective of secondary filtration is to reduce TSS and turbidity levels to comply with more stringent effluent requirements (compared to secondary effluent limitations). Typically, filtration is used where the effluent TSS or turbidity limit is equal to or less than 10 mg/L or 2 nephelometric turbidity unit (NTU), or both. Filtration also may be applied following secondary biological treatment to further remove particulate carbonaceous BOD_5 and deposits produced by alum, iron, or lime precipitation of phosphates in secondary effluent. Another subobjective of filtration is to increase the downstream disinfection efficiency by removing particulates, which can shade and shield pathogens. Filtration itself also achieves pathogen removal, but efficiency greatly depends on (1) the type of pathogen (e.g., *Giardia*, *Cryptosporidium*, coliforms); (2) the filtration technology; and (3) operational conditions such as chemical addition or filtration rate (Caliskaner et. al., 2011b; Levine et al., 2004; Vaughn et al., 2004).

2.1.2 Applications

Typical secondary effluent filtration applications include (1) discharge to a receiving water body that requires higher effluent standards (e.g., less than 10 mg/L TSS and/or 2 NTU turbidity); (2) wastewater reuse (e.g., irrigation of residential parks, golf courses, food crops, groundwater recharge, pretreatment for IPR or DPR); and (3) pretreatment for further advanced treatment such as carbon adsorption, membrane filtration, or reverse osmosis (Tooker et al., 2004).

2.1.3 Process Flow Diagram

Filtration has been used to treat municipal wastewater in a variety of processing sequences. Filtration is used to reduce TSS and turbidity following various combinations of secondary biological treatment. Filtration can be followed by another advanced treatment (e.g., downflow carbon adsorption column or clinoptilolite ammonia exchange columns) before disinfection. The only process to return/recycle water from filtration is the filter backwash reject water (BRW), which typically is routed back to the primary clarification stage. Some example secondary effluent filtration systems with different backwash requirements are:

- Filtration facility that includes a BRW storage basin and backwash water supply (BWS) tank;
- Filtration facility that does not require either the BRW storage basin or the BWS tank; and
- Filtration facility that includes a BRW storage basin but not the BWS tank.

2.2 Design Considerations

An effective filtration facility design fulfills three goals:

1. Obtain consistently the desired filtrate quality when treating an influent with a variable TSS concentration exhibiting a wide spectrum of particle sizes and composition.

2. Maintain continuous service under a variety of loading and filtration conditions.

3. Clean successfully the filter medium to restore clean filter medium headloss and solids storage conditions.

Considerations for design of a filtration system are discussed in this section.

2.2.1   Secondary Effluent Characteristics

The degree of TSS removal when filtering secondary effluent without the use of chemical coagulation depends on the degree of bioflocculation achieved during secondary treatment, as shown in Table 14.2. Bioflocculation, which can be defined as aggregation of bacterial flocs, has important implications on the physical characteristics of sludge. It is especially critical to settling and dewatering systems, which impacts the overall economy of the process greatly. For example, the presence of significant amounts of algae impedes filtration of lagoon effluent. Pretreatment with a primary coagulant or dissolved air floatation is considered good practices for lagoon effluent filtration. The particles to be filtered can behave quite differently when using chemical precipitation (e.g., using alum, iron, or lime) of phosphates in secondary treatment compared to filtration of secondary effluent without chemical precipitation of phosphates. It is essential to consider the upstream biological treatment type and characteristics of the secondary effluent when designing filtration facilities. The secondary effluent characteristics that play an important role in filtration design are TSS, particle-size distribution (PSD), and strength and electrostatic charge of flocs.

The secondary effluent TSS typically varies between 4 and 30 mg/L depending on many factors, such as secondary treatment type and the facility's physical and operational conditions. The analysis for TSS is straightforward but can be time consuming. Turbidity is, therefore, often used to estimate TSS concentration. The TSS-to-turbidity ratio for secondary effluent typically varies between 2 and 3 depending on the analytical methods and instruments used and the PSD of secondary effluent.

The filtration removal mechanisms—such as straining, interception, and sedimentation—are affected significantly by particle size. The secondary effluent PSD typically is bimodal. The filtration removal efficiency for a 1-μm-size particle is different

Filter Influent Type	Without Chemical Addition		With Chemical Addition		
	Effluent TSS (mg/L)	Turbidity (NTU)	Effluent TSS (mg/L)	PO$_4$ (mg/L)	Turbidity (NTU)
High-rate trickling filter effluent	10–20	4–10	0–3	0.1	0.1–2
Two-stage trickling filter effluent	6–15	2.5–8	0–3	0.1	0.1–2
Contact stabilization effluent	6–15	2.5–8	0–3	0.1	0.1–2
Conventional activated-sludge effluent	3–10	1–5	0–5	0.1	0.1–2
Extended aeration effluent	1–5	0.5–3	0–5	0.1	0.1–2
Aerated/facultative lagoon effluent	10–50	4–20	0–30*	0.1	N/A*

*Poor removal efficiency can result from filtering lagoon effluent because of the presence of algae.

TABLE 14.2   Typical Average Day Effluent Concentrations from Filtration of Secondary Effluent

than for 50-μm particles. Therefore, the PSD of the secondary effluent has a direct effect on filter removal efficiency.

The secondary effluent floc strength changes with the biological treatment process type and operational mode. For example, secondary effluent filtration is better suited to nitrifying facilities compared to nonnitrifying facilities because of the increase of floc strength with solids retention time. It also has been observed that the filtration of pure oxygen facility effluent is more challenging. This can be attributed to PSD, floc strength, and other secondary effluent physical/chemical characteristics.

2.2.2   Filtered Effluent Quality Characteristics

The primary design objective is to consistently obtain the desired filtrate quality. The filtered effluent quality can be defined in terms of both turbidity and TSS. A typical filtered effluent turbidity requirement is 2 NTU for wastewater reuse applications. The same effluent turbidity requirement also can apply for inland surface-water discharge applications. The filtered effluent TSS requirement typically varies between 5 and 10 mg/L. The PSD of the filtered effluent also influences downstream disinfection efficiency and, therefore, is as important as filtered effluent TSS and turbidity requirements.

2.2.3   Filtration Removal Mechanisms

There are three main removal mechanisms:

1. Physical—straining (mechanical and chance contact), sedimentation, inertial impaction, and interception;

2. Physical and chemical adsorption (bonding and chemical interaction) and physical adsorption (electrostatic and van der Waals forces); and

3. Biochemical—surface biological adsorption.

The relative importance of these removal mechanisms would differ depending on the secondary effluent characteristics, filtration conditions (e.g., interstitial velocity, operational conditions), and filter medium properties.

2.2.4   Filter Medium Properties

There are many filter media available for use with different technologies. This section discusses primary important media properties of each, including:

- Grain effective size, uniformity coefficient, shape, and density (in case of granular medium filters);

- Effective pore or collector size;

- Porosity; and

- Media depth (in case of depth filtration).

Medium selection is probably the most important consideration in filtration design because it affects the entire design and all the operational factors including removal efficiency, hydraulics, footprint requirements, and capital and operational costs.

2.2.5   Backwash Requirements

The filtration cycle is terminated for backwashing when the headloss across the filter medium reaches a preset maximum allowable level (typically referred to as terminal headloss) or the effluent quality deteriorates to an unacceptable level (typically referred to as breakthrough). To maintain the desired effluent quality at all times, filters are designed to backwash when the terminal headloss or breakthrough is reached. The BRW is returned to primary clarifiers, but if necessary can also be returned to head-works or secondary WRRFs depending on site-specific conditions. The ratio of the total BRW to the total filtered water in one day is important because the BRW needs to be treated again. The BRW ratio can be as low as 1% and can be as high as 20%, depending on the filter technology and site-specific filtration conditions. The frequency, period, flowrate, and BRW ratio all need to be considered because of their effects on facility operational costs and capacity. Other backwash-related basin or component require-ments, such as the BRW storage basin (typically known as a mudwell) and BWS tank, also vary based on filtration technologies.

2.2.6 Site Requirements

The existing secondary WRRF layout—including the space available for the filtration facilities—is one of the first design considerations. The space required for the filtration units depends on the design filtration rate, filter configuration, and driving force. The design filtration rate depends on the type of filtration technology and site-specific conditions. If needed, filter pump station or backwash-related facilities (e.g., BRW stor-age basin, BWS tank) should be accounted for in space requirements. Backwash and head requirements also are dependent on the type of technology.

Either gravity or pressure filters may be used for driving force. It should be noted that states have design criteria and regulatory requirements for filtration processes that should be reviewed before starting preliminary design. It is important to consider the footprint requirements in conjunction with the design filtration rate during the feasibil-ity study or preliminary design stage because the design filtration rate varies signifi-cantly (e.g., from 80 to 120 $L/m^2 \cdot min$ [2 to 3 gpm/sq ft] to 1200 to 1600 $L/m^2 \cdot min$ [30 to 40 gpm/sq ft]) depending on the filtration technology.

Multiple filter units are used for redundancy and reliability purposes and to allow continuous treatment during backwash or maintenance. The number of units typically must be minimized to reduce piping and construction costs, but should be sufficient to avoid excessive backwash flows and to ensure flow can be accommodated.

2.2.7 Hydraulic Requirements

Hydraulic criteria that should be included in filtration design include the hydraulic profile of secondary facility, filter medium headloss development, and filter terminal headloss value. The need for a filter feed pump station will be based on available head at the facility and additional head required by the filtration system. The hydraulic pro-file of the secondary facility needs to be updated during early stages of the design pro-cess to include filtration facilities, considering filter headloss development and terminal headloss value.

2.2.8 Filter Controls and Appurtenances

General types of filter operation and control include local manual, remote manual, and fully automatic. All should be provided. Available controls can receive a triggering signal of high filter headloss (or filter effluent turbidity), remove a filter from service, and start backwash (filter to waste period at the end of wash cycle also may be required depending on the specific regulatory requirement or filter technology), and return the filter to duty. Any delay or malfunction in the cycle interrupts the procedure and issues an alarm.

Controls should have the flexibility to allow field programming for optimum backwash sequencing and rate duration (e.g., one-minute increments) of unit operations. Continuity of all mechanical functions that depend on equipment such as blowers, compressors, and pumps should be ensured by permanently installed and operable standby units.

Each filter needs, at a minimum, gauging for headloss and/or pressure, for turbidity, and for flow of influent, backwash water, and air. Backwash water and air should have rate-of-flow controllers. If a rotary surface wash device (for granular medium) is used, then an external indicator should be included in the design to show that the wash arms are actually rotating during backwash. Turbidity meters that monitor continuously will provide valuable information, in order of priority, on effluent, influent, and wash-water quality. The importance of PSD analysis to monitor or improve filtration performance have been well established in Miska et al. (2006), Caliskaner and Tchobanoglous (2005a), and Naddeo and Belgiorno (2007). In-line TSS and PSD sensors can also be used to monitor and increase filter performance in conjunction with turbidimeters. Use of the TSS and PSD sensors should be considered based on project budget and size and operational and maintenance requirements of these sensors. The designer needs to consider that the sensors need to be regularly calibrated and cleaned to avoid biological growth and particle clogging in instruments.

2.2.9 Pretreatment/Use of Filtration (Chemical) Aids

Each secondary effluent filters differently. A pilot-scale study of a process stream can provide a good approximation of the filterability of secondary effluent; however, such studies do not duplicate exactly full-scale stress conditions. Regardless of documented filter performance, good design practice provides provisions to add filter aid chemicals upstream of the filter influent. Proper chemical types and addition facilities differ based on the other design considerations discussed above.

Jar testing is advisable to determine the optimum filter chemical aid dosage for a particular wastewater. Overdosing or use of unsuitable chemical may impair operations as much as or more than adding no filter chemical aids. The overdosing of coagulant may harm downstream-located equipment as well, which needs to be taken into the account. As an example, overdose of the ferric salts may affect UV equipment in case of using it.

Use of proper chemical dosages typically can be expected to produce effluent TSS of 5 mg/L or less (with proper filter medium selection). Filter medium headloss development and backwash frequency increase with addition of chemicals. Therefore, it is important to apply proper chemical dosages.

Typical filter influent TSS values vary between approximately 10 and 25 mg/L depending on the upstream secondary treatment processes and wastewater characteristics. If a daily average filter influent TSS concentration of less than 25 to 30 mg/L cannot be anticipated realistically, then the design engineer should consider enhancements in upstream processes (e.g., secondary treatment processes and/or designated filter pretreatment facilities such as coagulation and flocculation). If the average filter influent quality is anticipated to be good (e.g., TSS less than 10 to 15 mg/L), then rapid-mix or in-line-static-mix-type chemical addition facilities may be sufficient without flocculation facilities. Even though filter influent TSS and turbidity values can be used for preliminary or conceptual design of chemical addition facilities, a detailed analysis of filter influent characteristics (e.g., PSD, floc strength) are recommended for the final design of chemical facilities especially for challenging filtration applications. In addition, during some periods of the day and year, the TSS concentration in the filter influent might exceed significantly the design average concentration. Therefore, the design engineer should assess filter performance under stress loading conditions. Care should be given to prevent medium blinding as a result of chemical addition. Common causes of medium blinding include wrong chemical selection, overdosing, and excessive chemical addition duration.

2.2.10 Design Optimization

Pilot-scale and computer modeling studies typically are conducted to optimize or improve design. Filter removal performance and headloss development are greatly affected by many factors, including PSD, TSS, and floc strength of the secondary effluent, and filter media characteristics (e.g., depth, effective size/pore size porosity, porosity, configuration) and flowrate. There are no generalized design rules to achieve certain effluent criteria because of considerable variation of these parameters between different applications. Pilot-scale studies are, therefore, often recommended to confirm or improve the preliminary design criteria before design of the full-size filtration facility. One main objective of the pilot filtration study is to evaluate performance of the filter with specific reference to the effluent quality requirements (e.g., river discharge, wastewater reuse criteria). Some of the other typical objectives are (1) the quantification of the operating characteristics; (2) evaluation of filter reliability and performance as affected by variations in filter influent quality and flowrate; (3) backwash system requirements; (4) evaluation/confirmation of the design and operational criteria; and (5) determination of chemical aid requirements (e.g., chemical type, dose, application length). Pilot testing of alternative filtration technologies in parallel also can help inform selection of the final technology. Pilot filtration studies typically decrease overall project costs and improve design. The length of the pilot study often is determined by the project budget and schedule. If the pilot study needs to be conducted in a short period, such as 1 to 2 weeks, care should be given to the statistical significance of the pilot results. Testing of varying filtration conditions is not recommended for short-term pilot studies. The design engineer should determine and test for the most typical design condition in a short-term pilot test to obtain statistically valid performance results. It is recommended to conduct longer term pilot tests (i.e., more than 4 to 6 weeks) to obtain statistically valid pilot performance results at different design conditions. The period of the pilot study also is another important consideration because filterability characteristics typically change between seasons.

Computer modeling can help the design engineer correlate the interrelated design considerations with many of the filtration factors discussed above. A variety of equations relating to the filter process variables and removal mechanisms can be found in the literature. Equations characterizing filtration typically account for (1) an initial transport stage, where the particles are brought into contact with the surface of the filter medium, and (2) a physical/chemical stage, where the particles become attached and retained within the body of the filter media. Promising results have been obtained using comprehensive filtration models to accurately predict the filter performance for actual field conditions (Caliskaner and Tchobanoglous, 2005b). The design engineer also can elect to use a simpler model, which can provide some degree of approximation. In the absence of a pilot or modeling study, the design must be based on the engineer's experience with similar filter influent characteristics and upstream processes.

2.2.11   Protective Structures

The filtering unit should be designed for expected temperature conditions. As a good practice, the main control station and all auxiliary piping and valves should be located inside a heated, ventilated environment for ease of operation and maintenance. The increased cost and operational/maintenance requirements associated with the structure, however, need to be considered. These may include a heating, ventilation, and air-conditioning system, odor control, or increased vector control. If the filter is located outside and exposed to cold weather, then a covering superstructure can be considered to ease filter inspection. A minimum clearance of 2.1 m (7 ft) from the top of the filter to the covering structure is advisable. The minimum clearance requirement may be more for different filtration technologies.

2.2.12   Typical Design Deficiencies

The most common design deficiencies in filtration systems are listed below:

- Loss of media (for granular filtration) during backwashing. This can be a serious, ongoing problem caused by improper backwash design and operation.

- Mud-ball formation caused by long filter run time, inadequate backwashing, or both. This can be corrected by adjusting the backwash rates and applying chlorine to influent.

- Inadequate consideration of rapid-flow fluctuations (especially for smaller facilities).

- Insect problems. If insects are anticipated, then periodic addition of chlorine or insecticides in the filter influent can be considered.

- Deficit of chlorination system to chlorinate (occasionally as required) the filter influent flow, backwash water, or both. Chlorination capability is imperative to avoid biological growth and mud-ball formation.

- Inadequate design and piping configuration associated with effluent valve throttling (if used for flow control).

- Selection of wrong filter medium. In proprietary systems, the filter medium can change significantly between manufacturers for similar technology. Care should be given to medium selection when proprietary systems are used for design.

- Inadequate design of required BRW facilities (e.g., BRW storage basin, BWS tank, or backwash pumping/piping not large enough to handle worst-case conditions when multiple filters need to be backwashed simultaneously).

- Inadequate underdrain design (for granular media filtration), which can cause hydraulic and operational/maintenance problems.

- Hydraulic design deficiencies (e.g., not incorporating all of the headloss components) that result in less available head than the required values.

- Inadequate design consideration of pumping facilities that may be required upstream or downstream of filtration facilities.

- Inadequate consideration of overall height requirements specific to the filter technology.

- Uneven distribution of flow. When multiple units are designed and backwash start is based on basin water level, it is important that each filter has its own influent weir or a splitter weir box upstream so that flow can be evenly distributed and each unit can function independently.

- Inadequate consideration of future expansion or upgrade.

- Improper selection of control valves and flow meters (i.e., not accommodating design-flow range).

2.3 Technology Types

Several wastewater filtration technologies now exist to remove TSS remaining in the secondary clarifier effluent. Alternative filtration technologies involve use of various filtering media and configurations. Filtration technologies use granular or synthetic nongranular-type filter media such as compressible media, disk filters, and membrane filters. There are significant design and/or operational differences between different types of filtration technologies. The basic principle of filtration is the same though regardless of the technology used. Particle removal is achieved by passing wastewater through a porous filtering medium. As the suspension flows through the filtering medium, particles are transported to the surface of the collecting medium where they are removed by various mechanisms. Membrane filtration is considered more advanced compared to compressible and disk filtration technologies because of higher effluent quality. The capital and operational costs are also higher, but membrane filtration may be the feasible filtration alternative for certain applications such as pretreatment for desalination. Membrane filtration is discussed in more detail in Section 5.

2.3.1 Granular Media Filtration

Granular media filtration (GMF) has been the sole filtration technology used for wastewater treatment until the 1990s and has numerous different types and configurations. All GMF technologies are designed and operated based on depth filtration principles.

2.3.1.1 Granular Media Filtration Classifications   Granular media filters may be classified according to the direction of flow, type, and number of media composing the bed, driving force, and method of flow control.

2.3.1.1.1 Flow Direction   Downflow systems are the most common filtration processes for municipal applications. Some proprietary systems pass flow upward through the media, and others, called biflow systems, combine downflow and upflow. Schematics of these three variations are shown in Figure 14.2. Alternate configurations to downflow were developed to accommodate higher solids loads and to achieve filtration throughout the bed depth. In single-medium filters, stratification occurs after backwash and the concentration of the fine-grained portion of the medium in the upper portion of the bed prevents penetration and full use of the bed depth. Upflow is one way to achieve fuller use of the bed. Another approach involves use of dual-media or multimedia beds.

2.3.1.1.2 Driving Force   The driving force for filtration may be either gravity or applied pressure through pumping. Gravity filters typically are used in larger WRRFs; pressure filters are often used in industrial waste applications and can be considered for smaller municipal facilities. Pressure filters can accommodate higher hydraulic loading rates (e.g., 410 L/m²·min [10 gpm/sq ft]) and higher terminal headlosses (e.g., 9 m [30 ft]).

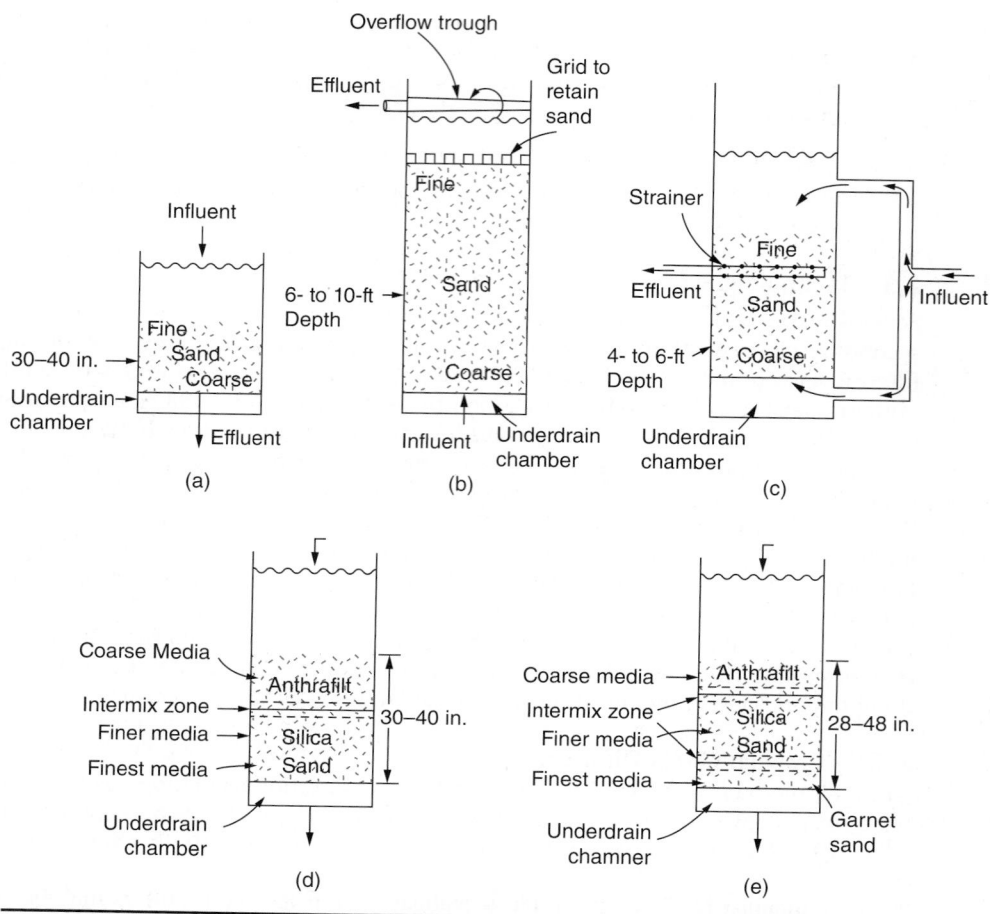

FIGURE 14.2   Filter configurations: (a) single-medium conventional, (b) single-medium upflow, (c) single-medium biflow, (d) dual-media conventional, and (e) mixed-media conventional (ft × 0.3048 = m; in. × 25.4 = mm) (U.S. EPA, 1971c).

Theoretically, this results in longer filter runs and lessened backwash requirements. However, these advantages may be offset by increased energy requirements and operational and mechanical complexity (sometimes resulting in decreased removal performances). Pressure filters must be selected with caution because internal inspection, maintenance, and media replacement are complicated by size and space limitations and concurrent problems of lighting and ventilation. In practice, the use of pressure filters for municipal applications is not common; however, their use is encountered more frequently in industrial wastewater treatment applications.

2.3.1.1.3 Filtration Rate   Filtration rate for GMFs varies significantly depending on the type of the technology. In slow sand filtration, the rate is approximately 1 to 5 L/m²·min (0.025 to 0.125 gpm/sq ft). Filtration rates typically range from 80 to 240 L/m²·min (2 to 6 gpm/sq ft) for rapid sand filtration. Higher filtration rates up to 400 L/m²·min (10 gpm/sq ft) have been reported for GMF, but nominal maximum application rates typically range from 200 to 300 L/m²·min (5 to 7.5 gpm/sq ft) in high-rate applications. Gravity filters involving the use of granular medium have been shown to perform successfully at higher filtration rates (e.g., 300 L/m²·min [7.5 gpm/sq ft]) especially with optimized chemical addition (Holden et al., 2006; Williams et al., 2007).

2.3.1.1.4 Flow Control Methods The main methods used to control the rate of flow through gravity filters may be classified as (1) constant-rate filtration with fixed head, (2) constant-rate filtration with variable head, and (3) variable, declining-rate filtration.

2.3.1.2 Granular Media Selection and Characteristics During media selection, media size, shape, composition, density, hardness, size, and depth relationships and the effects of these characteristics on filter performance and operation should be considered.

2.3.1.2.1 Size The depth of solids penetration to a filter depends on the size of the filter medium. If the medium is too large, then poor filtrate will result. If the medium is too small, then filtrate quality will be better, but removed solids will accumulate near the surface and result in short filter runs.

2.3.1.2.2 Shape Medium shape affects both filtration and backwash. Sharp, angular grains may interlock and require increased backwash pressure. A nonuniform medium increases the tendency for channelization during backwash. Rounded grains break up and fluidize more readily; they also tend to rotate during backwash, scouring adjacent grains and freeing adhered solids. Angular grains with flat, plain surfaces tend to stick together, resist rotation, and may float out of the system when backwashing with air. Filter sand should have a sphericity ratio of approximately 0.9; filter anthracite should have a sphericity ratio of approximately 0.7.

2.3.1.2.3 Composition Anthracite, a largely organic medium, has a surface that preferentially adsorbs other organic substances such as fats and oils. This produces an oily film that may resist removal by conventional cleaning techniques and, in turn, accelerate formation of agglomerates that reduce system effectiveness if they are not flushed from the filter during backwash. Because anthracite is relatively soft, it tends to deteriorate from abrasion, especially during backwash. This deterioration may lead to production of smaller particles that, if not lost during backwash, will diminish solids penetration in the bed. Oily films and deposits can be removed from silica sand using less scouring intensity than for anthracite.

2.3.1.2.4 Density Specific gravities of typical materials used as filtering media are 4.2 for garnet sand, 2.6 for silica sand, and 1.6 for anthracite. Lighter media require greater freeboard to be retained in the filter tanks. However, greater freeboard also increases the tendency for heavy process solids to be retained in the filter.

2.3.1.2.5 Hardness Silica sand has a Moh scale hardness of 7; anthracite ranges from 3.0 to 3.75. High-intensity scouring, needed for effective filter cleaning to remove tenacious solids typically found in wastewater effluent, tends to erode media grains. This attrition, if excessive, shortens filter runs, increases backwash losses, and results in a continuously widening size gradation of media. As a result, the substance with the most hardness is best for wastewater filtration.

2.3.1.2.6 Size and Depth Relationship   The media size and depth are the most important criteria in design of the GMF systems. These parameters are also interrelated significantly impacting the filter performance. Typically, filtrate quality is better for smaller media size and higher media depths. The media headloss development also increases as the depth increases or the media size decreases. Therefore, it is essential to select the correct media depth/size combination to achieve the required removal performances without creating excessive headloss development.

No specific design guides relating media size and depth to TSS removal performance exist. The design approach varies significantly based on project needs (e.g., size, complexity, schedule, and budget). The design can be based on comprehensive filtration mass balance and kinetics models. Or the design can be based on more simplified approaches using typical media depth and size values (e.g., used for similar installations). One common simplified approach is to use media depth-to-size ratio (also referred as media length [L]/media depth [D] ratio) as a design guide. Pilot tests are strongly recommended, if basic design guides are used in conjunction with typical media depth and sizes.

Use of mass conservation and filtration kinetics equations is considered a good design practice especially for larger systems. The mass conservation equation is derived by writing a mass balance equation for the particles, which are removed from the suspension and deposited within the filter medium. One common final form of this well-known equation is:

$$\frac{\partial C}{\partial x} + \frac{1}{v}\frac{\partial \sigma}{\partial t} = 0 \tag{14.1a}$$

where C = concentration of particulate material;

$\quad\quad \sigma$ = mass amount of particulate material accumulated within the medium (usually referred as specific deposit);

$\quad\quad t$ = time;

$\quad\quad x$ = medium depth in the direction of the flow; and

$\quad\quad v$ = filtration velocity.

Several filtration kinetic equations were introduced to describe the filter removal performance as a function of media depth. A filter coefficient is typically used to quantify the impacts of medium properties, wastewater characteristics, and flow conditions. The basic form of filtration kinetics equations is expressed as follows:

$$-\frac{\partial C}{\partial x} = \lambda C \tag{14.1b}$$

where λ is the filter coefficient and changes with filtration time. A governing equation (i.e., for mass balance and kinetics) which can be used to design for media characteristics is given below (Caliskaner and Tchobanoglous, 2005b):

$$\frac{\partial^2 C}{\partial t \partial x} - \frac{1}{C}\frac{\partial C}{\partial x}\frac{\partial C}{\partial t} + \frac{V}{\sigma_{max}}\lambda C\frac{\partial C}{\partial x} - \frac{1}{\lambda}\frac{\partial \lambda}{\partial t}\frac{\partial C}{\partial x} = 0 \tag{14.1c}$$

The filtration governing equations are nonlinear partial differential equations and solved using numerical methods to determine the required media depth and size for specific filtration conditions.

As discussed above, the design can also be based on design guides such as L/D ratio using typical media depth and size values. As a good practice, the minimum depth of the finest medium is at least 150 mm (6 in.). The minimum medium particle diameter will be at least 0.35 mm (0.4 to 0.5 mm is more typical for the minimum size in secondary effluent filtration applications). However, operating systems typically have media depths and sizes within the ranges shown in Table 14.3. Filters for municipal wastewater treatment rarely have a medium coarser than 2.0 mm if effluent TSS residuals of 10 mg/L or less are necessary.

Figure 14.3 illustrates a typical grain size distribution as a function of diameter on U.S. standard sieve sizes. Media specifications typically include the effective size and uniformity coefficient. Uniformity coefficient is defined as the ratio of the sieve size that will pass 60%, by weight, of the sand divided by the size that will pass 10%. The medium shown in the figure has an effective size of 0.5 mm (0.02 in.), with a uniformity coefficient of 1.5. In general, a uniformity coefficient no greater than 1.7 is best for all filters, except those that require some intermixing. Uniform media or media with a uniformity coefficient less than 1.3 are unnecessary, except for deep-bed, single-medium filters, especially if the bed is cleaned with an air scour.

Filtration in GMFs occurs throughout the medium depth, which can vary from 0.3 to 2.1 m (1 to 7 ft) depending on the technology. Filter beds may consist of (1) a single medium in stratified form, (2) a single medium in unstratified form, (3) dual media, or (4) multimedia. Single-medium stratified beds, used in the past, are no longer designed for municipal wastewater applications because of their unfavorable headloss buildup characteristics. Single-medium unstratified beds, now in use, have bed depths typically up to 2.0 m (6.5 ft). The random pore size allows filtration throughout the bed depth, resulting in longer filter runs before headloss buildup requires backwash. The unstratified form for the single medium can be achieved by using a single- or variable-sized medium with a combined air-and-water backwash. The combined air-and-water backwash scours accumulated material from the filter bed without fully fluidizing the medium. This avoids stratification of the nonuniform medium.

Use of two or more layers of media, each with a different specific gravity, prevents stratification and promotes filtration with inherently longer filter runs. Typical combinations of media used in dual-media beds include:

- Anthracite and sand;
- Activated carbon and sand;
- Resin beds and sand;
- Resin beds and anthracite; and
- Garnet or ilmenite (often used in multimedia beds).

Table 14.3 presents typical characteristics of some granular filter media. Table 14.4 presents a design example for a depth filter involving the use of granular medium.

2.3.1.3 Granular Medium Filtration Technology Types   Both semicontinuous and continuous operation granular medium filtration technologies are used for secondary effluent filtration. Conventional and some propriety semicontinuous and continuous operation granular filters are discussed in this section.

Characteristics	Values	
	Range	Typical
Dual media		
Anthracite		
Depth, mm[a]	300–600	450
Effective size, mm	0.8–2.0	1.2
Uniformity coefficient	1.3–1.8	1.6
Sand		
Depth, mm	150–300	300
Effective size, mm	0.4–0.8	0.55
Uniformity coefficient	1.2–1.6	1.5
Filtration rate, m/h[b]	5–24	12
Multimedia		
Anthracite (top layer of quad-media filter)		
Depth, mm	200–400	200
Effective size, mm	1.3–2.0	1.6
Uniformity coefficient	1.5–1.8	1.6
Anthracite (second layer of quad-media filter)		
Depth, mm	100–400	200
Effective size, mm	1.0–1.6	1.2
Uniformity coefficient	1.5–1.8	1.6
Anthracite (top layer of trimedia filter)		
Depth, mm	200–500	400
Effective size, mm	1.0–2.0	1.4
Uniformity coefficient	1.4–1.8	1.6
Sand		
Depth, mm	200–400	250
Effective size, mm	0.4–0.8	0.5
Uniformity coefficient	1.3–1.8	1.6
Garnet or ilmenite		
Depth, mm	50–150	100
Effective size, mm	0.2–0.6	0.3
Uniformity coefficient	1.5–1.8	1.6
Filtration rate, m/h	5–24	12

[a]$mm \times 0.039\ 37 = in.$
[b]$m/h \times 0.409\ 2 = gpm/sq\ ft.$

TABLE 14.3 Media Characteristics for Dual-Media and Multimedia Filters

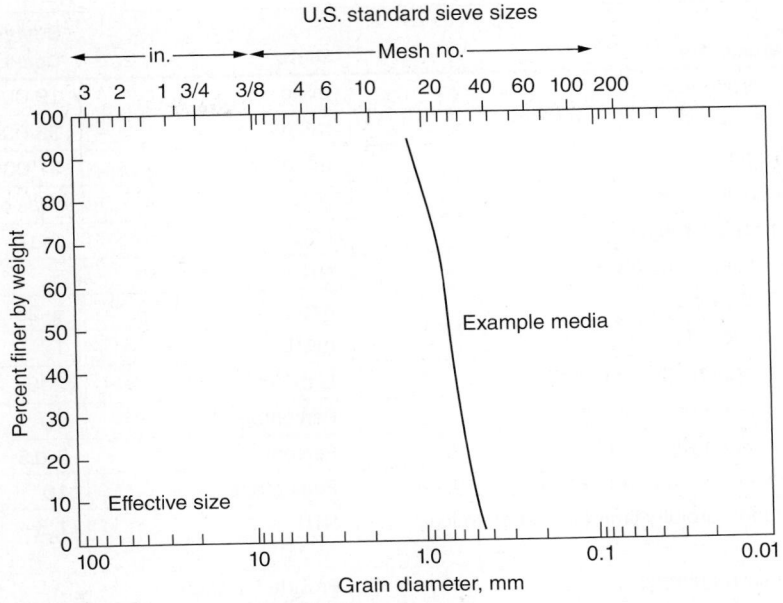

U.S. standard sieve sizes

FIGURE 14.3   Grain size distribution chart.

2.3.1.3.1 Semicontinuous Operation Granular Media Filters   Semicontinuous operation GMFs can be classified by medium characteristics such as number of media, gradation, and mixing.

2.3.1.3.1.1 SINGLE, NONGRADED MEDIUM FILTERS   In slow sand filtration, the filtration rate is approximately 1 to 5 L/m²·min (0.025 to 0.12 gpm/sq ft). Slow sand filters typically consist of a sand layer 150 to 380 mm (6 to 15 in.) in depth placed on an equivalent depth of coarser material that rests on an underdrain system. Slow sand filters are regenerated by remaining idle for a period and then being cleaned manually. They typically can achieve 60% suspended solids removal from a trickling-filter effluent containing 20 to 90 mg/L TSS. Although simple to construct, slow sand filters have high maintenance and space requirements. They also are temperature sensitive and subject to rapid clogging. Because these filters are not considered competitive with other techniques for TSS removal, they are seldom used.

The University of California at Davis pilot-tested several filtration technologies (using granular and nongranular media) at the same activated sludge facility. Figure 14.4 summarizes the influent and effluent turbidity values obtained during these pilot tests.

Effluent turbidity values in Figure 14.4 would vary for different activated sludge facilities because of the effects of secondary effluent characteristics. Figure 14.5 presents probability distributions for filter influent and effluent TSS concentrations obtained from the mono-medium filters at the Dallas Central and Southside Wastewater Treatment Plants (nitrifying activated sludge). The mono-medium filters use 910 mm (36 in.) of 1.2-mm anthracite. Filter influent and effluent TSS concentrations averaged 4.4 and 2.2 mg/L, respectively. The results shown in Figures 14.4 and 14.5 were obtained

Input Parameters	Units	Design Values/Considerations
Annual average flow	m³/d	19 000
Peak-day flow	m³/d	38 000
Peak-hour flow	m³/d	57 000
Basis of design flow*		Peak-day flow
Influent turbidity range	NTU	2–10
Average influent turbidity	NTU	5
Influent TSS range	mg/L	5–25
Average influent TSS	mg/L	12
Maximum design filtration rate	L/m²·min	200
Average water production ratio	Percentage	90
Backwash water reject ratio range	Percentage	5–15
Average backwash water reject ratio	Percentage	10
Filter effluent turbidity limit to start chemical addition	NTU	1.8
Redundancy requirement	Provide filtration capacity with largest filter unit out of service	
Design output		
General		
Filtration type		Depth
Flow direction during filtration cycle		Downflow
Flow direction during backwash cycle		Upflow
Unit sizing		
Minimum total filtration surface area required	m²	130
Number of filter units	#	8
Filtration surface area, width (each filter unit)	m	4.25
Filtration surface area, length (each filter unit)	m	4.25
Filtration surface area for each filter unit	m²	18
Total filtration surface area provided	m²	145
Filtration rate		
Annual average flow	L/m²·min	90.0
Peak-day flow	L/m²·min	180
Annual average flow, one filter out of service	L/m²·min	105
Peak-day flow, one filter out of service	L/m²·min	210
Peak-day flow, one filter out of service and one filter in backwash	L/m²·min	240

TABLE 14.4   Secondary Effluent Filtration Design Example

Input Parameters	Units	Design Values/ Considerations
Medium properties		
Medium type	Mono medium–sand, unstratified	
Medium depth	mm	1200
Effective size	mm	2.00
Uniformity coefficient	Unitless	1.20
Porosity	Percentage	40
Backwash requirements		
Backwash type	Combined air-water backwash	
Design backwash average rate, water	L/m²·min	400
Design backwash rate range, water	L/m²·min	200–600
Auxiliary air scour rate	L/m²·min	2000
Backwash period	min	20
Assumed backwash frequency, average loading	h	18
Assumed backwash frequency, peak loading	h	6
Total daily backwash water required, average loading	m³/d	1500
Total daily backwash water required, peak loading	m³/d	4600
Backwash reject ratio, average conditions	Percentage	8%
Backwash reject ratio, peak conditions	Percentage	12%
Clearwell volume (sufficient volume for two backwashes)	m³	290
Backwash reject water TSS concentration range	mg/L	50–250
Backwash reject water typical TSS concentration	mg/L	150
Backwash reject water storage basin volume (sufficient volume for six filter backwashes)	m³/d	900
TSS/turbidity removal		
Effluent turbidity, average conditions	NTU	<1.6
Effluent turbidity, cannot exceed more than 5% of the time	NTU	4
Effluent turbidity, cannot exceed anytime	NTU	8
Effluent TSS, average conditions	mg/L	<4
Effluent TSS, maximum daily	mg/L	15

TABLE 14.4 Secondary Effluent Filtration Design Example (*Continued*)

Input Parameters	Units	Design Values/ Considerations
Headloss development		
Clean medium headloss at maximum design filtration rate	m of water	1
Terminal headloss	m of water	3
Ancillary units required		
Chemical addition		
Assumed chemical addition requirement (ratio to total operational time)	Percentage	15%
Chemical storage	Days	30
Mixing method		In-line static mixer
Chemical pump type		Mechanical diaphragm
Pump speed control		Variable speed
Chemical type		Polymer
Dose range, assumed	mg/L	0.1–1.5
Typical dose, assumed	mg/L	0.3
Chemical type		Coagulant
Dose range, assumed	mg/L	2.5–25
Typical dose, assumed	mg/L	10.0
Air scour blowers	No (duty, redundant)	1 + 1
Inline turbidity analyzers	No (duty, spare)	2 + 1

*In this design example, it is assumed that the flows greater than peak day are equalized upstream of filtration facilities.

TSS = total suspended solids.

TABLE 14.4 Secondary Effluent Filtration Design Example (*Continued*)

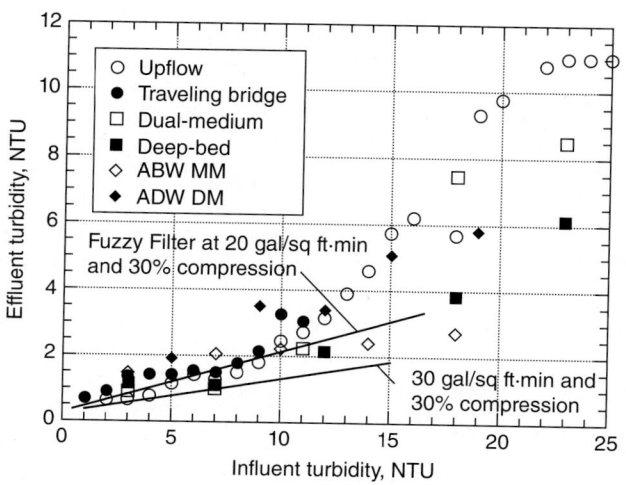

FIGURE 14.4 Comparison of effluent versus influent turbidity for Fuzzy Filter operated at 820 and 1230 L/m²·min with a bed compression of 30% with several filters commonly used for effluent filtration operated at 205 L/m²·min (Caliskaner et al., 1999).

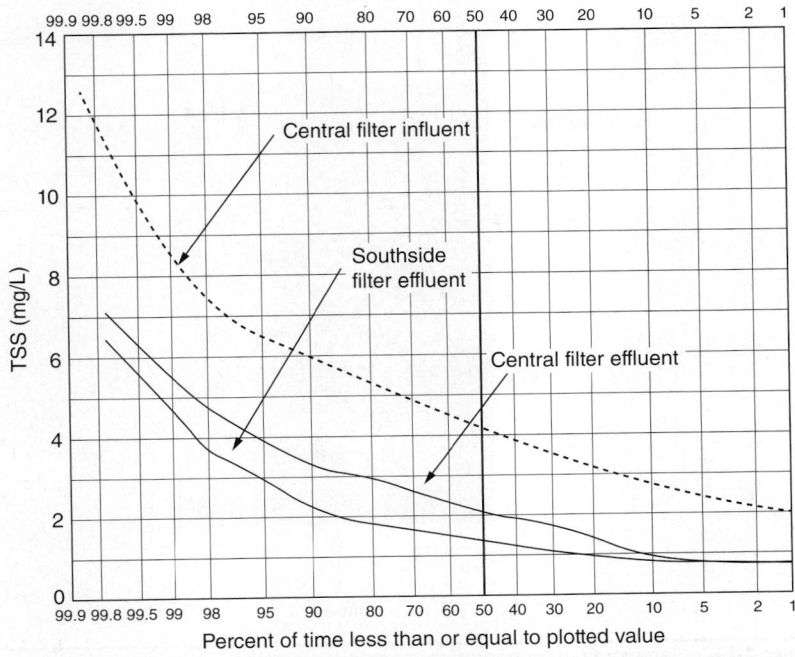

FIGURE 14.5 Probability distribution function for filter influent and effluent total suspended-solids concentrations at Dallas Central and Southside WRRFs.

without chemical addition. The effluent turbidity and TSS values can further be reduced 50% to 70% with optimized chemical addition (i.e., correct chemical type, dose, and addition duration).

Deep-bed, coarse-medium filters typically involve single-medium bed depths of 1.2 to 1.8 m (4 to 6 ft). The proper size for sand medium depends on secondary effluent characteristics and required filtered effluent quality. A diameter of 1.5 mm or larger is typical for secondary effluent filtration. A highly uniform sand (e.g., uniformity coefficient of 1.1) is best for allowing full use of the bed.

Early problems with effective bed cleaning and economical volumes of backwash water seem to have been solved by using an air backwash at 0.015 to 0.025 m³/m²·s (3 to 5 scfm/sq ft), with a water rinse at 240 to 320 L/m²·min (6 to 8 gpm/sq ft). Total rinse water consumption per unit of filter surface per wash varies between 4000 and 8000 L/m² (100 to 200 gal/sq ft) and is typically approximately 6000 L/m² (150 gal/sq ft). Coarser media decrease headloss development and provide a higher solids storage capacity. Figure 14.6 provides a typical schematic of a single, nongraded medium filter.

2.3.1.3.1.2 DUAL AND MULTIMEDIA FILTERS Dual-media and multimedia filter beds are commonly used in WRRFs. These systems consist of two or more media, such as anthracite and sand, in layers (Figure 14.7). The anthracite layer may be distinct or mixed with sand. The specific gravity of the anthracite is typically 1.6; that of silica sand is 2.65. Garnet sand, used with the multimedia system, has a specific gravity of approximately 4.2.

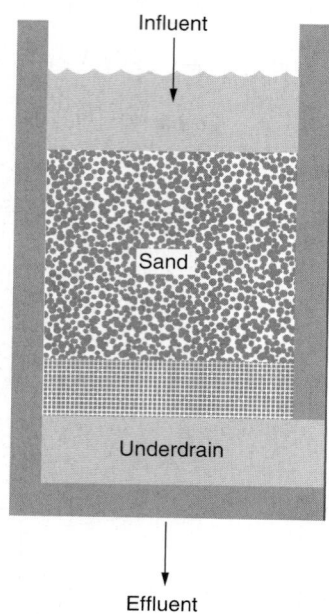

FIGURE 14.6   Semicontinuous operation, nongraded single-medium filter.

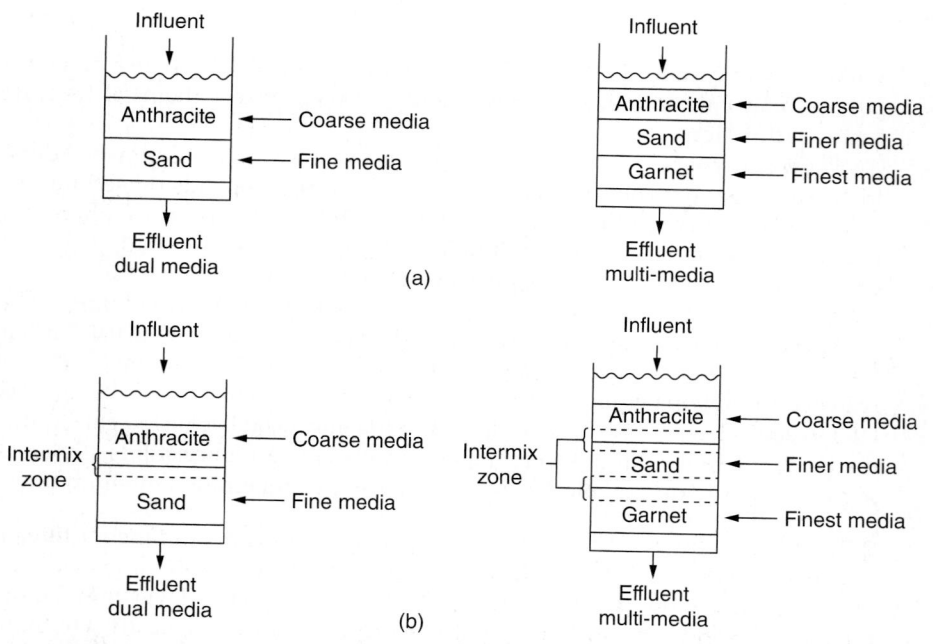

FIGURE 14.7   Dual-media and multimedia filter configurations: (a) nonintermixed and (b) intermixed media.

2.3.1.3.1.2.1 *NONINTERMIXED MEDIA* The minimum uniformity coefficient commercially available for nonintermixed media is 1.3. A lower uniformity coefficient may be achieved by specifying percentages of retention on adjacent sieve sizes. For example, 10% to 15% by weight may be allowed for finer than the specified smaller sieve; and 10% to 15% by weight may be allowed for coarser than the specified larger sieve. Using a medium with a low uniformity coefficient can result in significant savings in backwash water for equivalent degrees of bed expansion when the medium effective size is held constant. This advantage for uniform media exists only if complete bed fluidization and partial expansion is a requirement for bed cleaning. Higher filtration rates (e.g., up to 240 to 320 L/m²·min [6 to 8 gpm/sq ft]) can be applied, but influent TSS must remain below 10 to 15 mg/L. Typically, a maximum filtration rate of 200 L/m²·min (5 gpm/sq ft) is used for handling secondary effluent.

2.3.1.3.1.2.2 *INTERMIXED MEDIA* Some intermixing occurs in all dual-media or multimedia systems unless low uniformity coefficients are strictly maintained. In most dual-media systems, media are intermixed fractionally at their boundary layer. In at least one investigation, intermixing has been shown to offer little tangible benefit in dual-media systems (Baumann and Huang, 1974); however, this conclusion should not be extrapolated to mixed-media systems.

The multimedia system attempts to achieve ideal situation uniform decrease in pore space with increasing filter depth. Typically, filters have a particle size gradation from approximately 2 mm (0.08 in.) in diameter at the top to 0.15 to 0.5 mm (0.006 to 0.02 in.) at the bottom. The size and depth of anthracite, sand, and garnet will change as a function of the loading and strength of the solids to be removed. Proper selection of backwash rates produces an intermixing of media layers so that the pore size of the filter decreases somewhat uniformly from top to bottom (American Society of Civil Engineers [ASCE], 1986).

The theoretical advantages of multimedia filtration often are proclaimed. However, several studies have shown little difference in TSS or turbidity removal between single-medium and multimedia designs in tertiary filtration applications (Brown and Caldwell, 1978; Russell et al., 1980; Tchobanoglous and Eliassen, 1970). Operational parameters such as run times and backwash water consumption, however, do vary between media designs. Allowable hydraulic loads, as with all filters, vary inversely with anticipated TSS load. Primary disadvantages of multimedia systems are their inherent need for high backwash rates for bed expansion and the tendency of finer media to migrate to the surface of the filter. Media are properly classified by a 1- to 2-minute wash at 200 L/m²·min (5 gpm/sq ft) at the end of the backwash period.

Figure 14.8 is a time-series plot of filter effluent TSS concentrations for the dual-media filters used at the Dallas Southside WRRF. The filters use 300 mm (12 in.) of 1-mm (0.04-in.) anthracite over 600 mm (24 in.) of 0.5-mm (0.02-in.) sand for a medium. Median filter effluent TSS concentration is approximately 2 mg/L.

2.3.1.3.1.3 *HEADLOSS DEVELOPMENT* Headloss development in a GMF depends on hydraulics and solids retention capacity. The total headloss at any time is the sum of the hydraulic headloss for a clean bed and additional headloss caused by the deposition of wastewater solids. Both are influenced largely by media size.

Hydraulic headloss encompasses losses associated with piping, valving, metering, bends, constrictions, underdrains, collection systems, and media. Of these, the media bed composes the predominant headloss. For laminar flow conditions that occur for

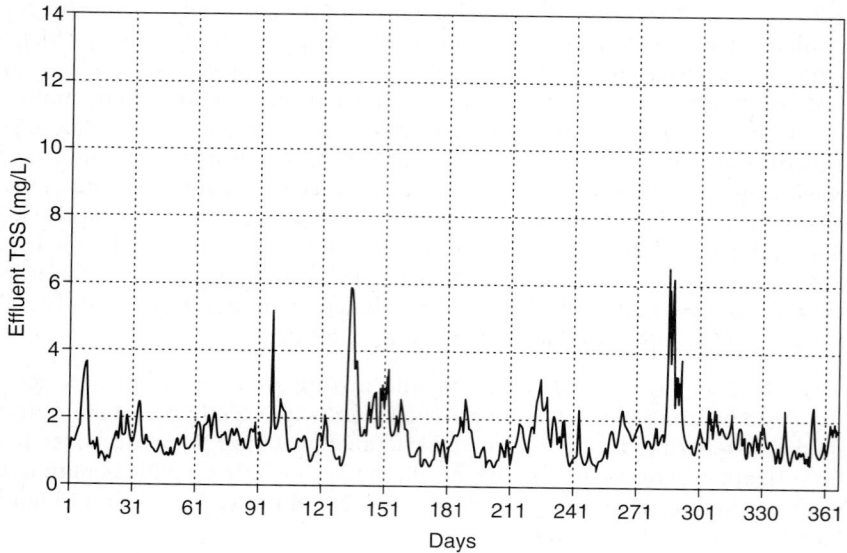

normal operating conditions of media filtration, the Reynolds number based on super-ficial velocity is less than 3 and headloss through the filter media varies directly with flow velocity (American Water Works Association, 1990). For any media filtration sys-tem, a headloss development curve for clean water as a function of filtration rate can be obtained either from pilot-plant study or directly from the manufacturer.

Clean water headloss curves for a given hydraulic load are influenced by media type, size, uniformity, and depth. Figure 14.9 shows that as the filtration rate increases, clean water headloss increases, and the available headloss development for the filtra-tion run decreases. The filtration run length is directly related to the difference between the clean filter headloss and the terminal headloss value. If the filter characterized in the figure were operated with a terminal headloss of 1.5 m (5 ft), then the backwash cycle would start almost instantly when filtration cycle begins at a hydraulic load of slightly more than 400 L/m²·min (10 gpm/sq ft).

At filtration rates of 160 and 320 L/m²·min (4 and 8 gpm/sq ft), corresponding available headloss development values would be approximately 1.4 and 0.5 m (4.6 and 1.6 ft), respectively.

The solids capture efficiency and solids storage capacity of the filter determine its effective run time for a given headloss. A substantial solids breakthrough typically does not occur when filtering secondary effluent if the medium grain size is smaller than 2 mm (0.08 in.). All filters containing a medium of this size or smaller are likely capable of producing water with approximately 5 mg/L TSS, especially if the system includes upstream coagulant addition. The solids storage capacity of a filter will vary with the nature of the solids applied. Headloss development should be associated with accumu-lated solids in addition to other hydraulic considerations. Figure 14.10 shows such data, obtained by compiling operating results (Baumann and Huang, 1974; Tchobanoglous and Eliassen, 1970). The filtering media used in both investigations had uniformity

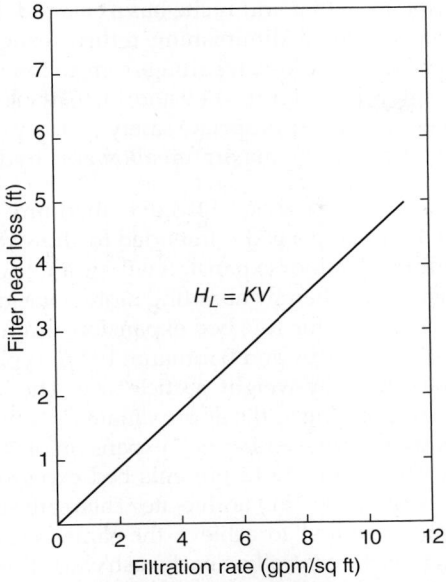

FIGURE 14.9   Typical clean water headloss development curve for granular media filter
(ft × 0.3048 = m; gpm/sq ft × 2.444 = m/h).

coefficients approaching unity. Sand media data with the headloss development termi-
nating at 1.8 m (6 ft) were derived from processing activated sludge effluent at a filtration rate of 240 L/m²·min (6 gpm/sq ft) (Tchobanoglous and Eliassen, 1970). Other data
were developed from processing trickling-filter effluent at a filtration rate of
160 L/m²·min (4 gpm/sq ft) (Baumann and Huang, 1974).

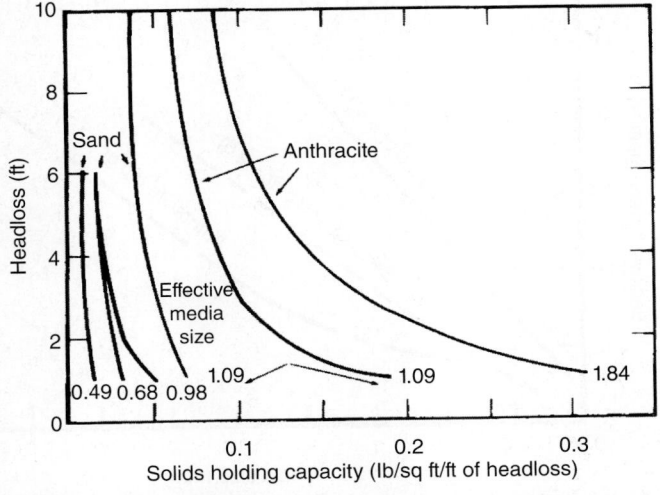

FIGURE 14.10   Solids storage capabilities of filter media (ft × 0.3048 = m; lb/sq ft/ft × 16.02 =
kg/m²·m).

Figure 14.10 data show that anthracite has a clear advantage in terms of solids storage capacity. It also reveals a diminishing return (exaggerated as the medium size increases) as higher filter headloss, resulting from solids accumulation, is achieved.

Information displayed in Figures 14.9 and 14.10, if used with a fixed total allowable headloss requirement and an appropriate safety factor, provides a basis for developing a trial-and-error solution for the maximum allowable hydraulic loading rate of a filter.

2.3.1.3.1.4 BACKWASH AND CLEANING   Effective medium cleaning is not ensured by expanding the bed. Fluidizing a bed is intended to allow trapped particles to be washed away from the medium. A minor expansion will promote cleaning by allowing particles to work against each other, thereby abrading more tenacious solids. Figure 14.11 shows backwash rate requirements for 10% bed expansion as a function of medium size, type, and water temperature (Cleasby and Baumann, 1974). Typically, rates provide up to 10% expansion of the 60% finer-by-weight particle size. The bottom of the wash trough is used as a guide, corresponding to the approximate rise anticipated for the effective size.

Most operators try to achieve filter-bed expansion of 30% to 50%, without the use of supplemental air wash. Figure 14.12 presents bed expansion data for a typical mixed-media filter with 1.6-mm (0.063-in.) anthracite. There are substantial differences in backwash water flowrates required to achieve the same bed expansion at differing water temperatures. Therefore, the controlling backwash water flowrate must provide the appropriate degree of bed expansion at the highest backwash water temperature anticipated.

Some researchers recommend selecting the maximum backwash rate from the 90% finer medium size (Cleasby and Baumann, 1974). Backwash pump sizing should account for the warmest water temperature anticipated. The total backwash requirement will typically be approximately 4000 to 10 000 L/m² (100 to 250 gal/sq ft) and will be largely independent of the backwash rate (Cleasby and Baumann, 1974). This observation is typical for conventional U.S. wash-trough spacing with edges approximately

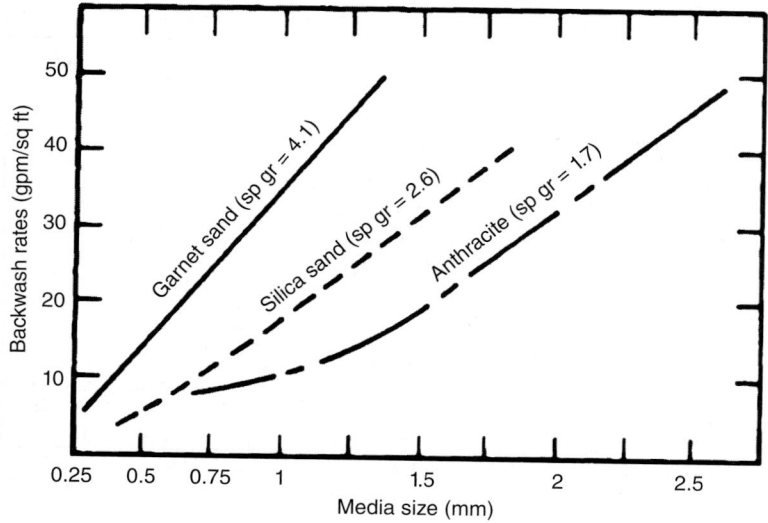

FIGURE 14.11   Minimum backwash rates for a bed expansion of 10% at 25°C (77°F) (gpm/sq ft × 2.444 = m/h).

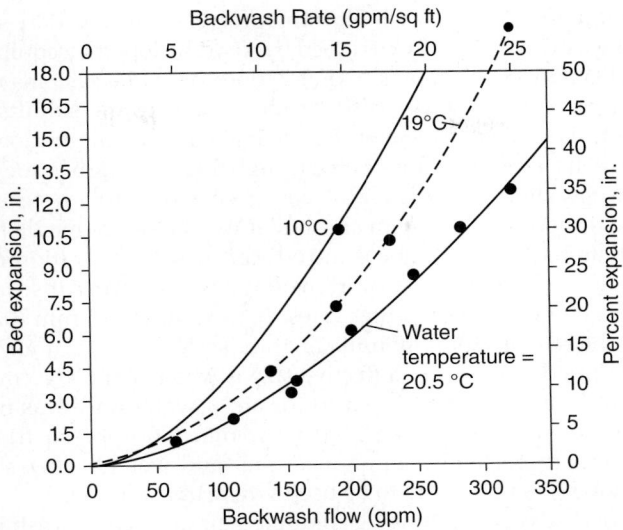

FIGURE 14.12   Bed expansion for sand/anthracite filter at 19 and 29.5°C (gpm × 3.785 = L/min; in. × 25.4 = mm; gpm/sq ft × 2.444 = m/h).

0.9 m (3 ft) above the filter media. This requirement should not be applied to proprietary systems without rigorous review of the proposed system. Selection of relative grain sizes for components of dual-media and multimedia filters typically should allow equal fluidization among bed components. Failure to provide equal fluidization may result in unplanned intermixing or, in the worst extreme, bed turnover.

Relying solely on an upward flow of water to clean granular filters is inadequate because abrasive action among particles is the most effective cleaning mechanism during backwash (Babbitt et al., 1962; Fair and Geyer, 1954). Fluidized media are separated by the flow of water, thus reducing the frequency of grains rubbing together (Amirtharajah, 1979). Also, biological solids in secondary effluent are strongly attached to media because of their "sticky" nature. Consequently, it is not surprising that fluidization alone cannot effectively clean the bed (Amirtharajah, 1979; Cleasby and Lorence, 1978; Cleasby et al., 1975, 1977). Many investigators have reported and cited benefits associated with air scour before and during backwash with both subcritical and fluidizing backwash rates (Huang, 1979). Surface and subsurface washing or agitating equipment also are used to assist cleaning, and are common in surface water treatment.

Air scour of approximately 0.015 to 0.025 m³/m²·s (3 to 5 scfm/sq ft) is advisable as a minimum for conventional systems; deeper beds may require more and shallow beds may require less. Although most of the beneficial action of air scour will occur during the first 2 minutes, sufficient operational flexibility should be provided for 10 or more minutes of scour with some control of the applied rate. Designers also should consider concurrent washing and air scouring. This requires a capability of draining the filter to just above its surface, followed by a brief simultaneous period of air-and-water backwash until the water reaches within 150 to 200 mm (6 to 8 in.) of the wash trough. Then, either the air or the water wash is stopped. The design of the backwash reservoir should provide for some flexibility by affording between a 3-minute and a 15-minute capacity at the maximum backwash rate.

The simultaneous backwash filtration process is sustained by simultaneous air-and-water backwash and specially designed baffles developed by equipment manufacturers. Figure 14.13a and 14.13b shows two variations on baffle systems required at backwash water troughs to prevent loss of filter media during the simultaneous air and water wash. Figure 14.13a shows one of the original baffle configurations developed to separate filter media and air bubbles at the trough. Figure 14.13b is suggested for those facilities requiring a shorter profile for the backwash water trough system. Separation media used in the trough/baffle system are similar to tube settler media found in clarifiers.

To initiate backwash, water is introduced at subfluidization rates and air is introduced at rates sufficient to transport media to the surface. Air scour adds energy to the system, reducing backwash water rates. For example, 1.5-mm (0.059-in.) anthracite is washed at the rate of 480 L/m²·min (12 gpm/sq ft) for water and 70 m/h (4 scfm/sq ft) for air. Sand media of 1.5-mm effective size is washed at 600 L/m²·min (15 gpm/sq ft) for water and 110 m/h (6 scfm/sq ft) for air. If water only was used for these media, then backwash rates would be 1200 L/m²·min (30 gpm/sq ft) and 1800 L/m²·min (45 gpm/sq ft), respectively. Simultaneous air-and-water backwash eliminates rate limitations, allowing large media to be used where appropriate.

A reported advantage of using combined air-and-water wash is that the bed can be cleaned more effectively than with water-or-air scour followed by water wash. Evidence of this improved cleanliness is shown by the results of abrasion tests, which measure the TSS remaining in filter media after backwashing. Results show that 10% of the solids remain in a combined washed filter compared to water wash or air scour followed by water wash filters.

Another benefit is that simultaneous backwash leaves the filter bed unstratified. Conventional water wash or air scour systems leave filter beds stratified. Stratified beds have small particle sizes at the surface of the bed and large particles at the bottom of the bed. This causes rapid accumulation of solids on the surface of the bed, shortening filter run lengths. An unstratified filter bed has uniform media grain sizes throughout the filter bed allowing deep-bed penetration of TSS and increasing filter run lengths significantly. It also increases the net production of filtered water.

A rotary subsurface wash capability appears most necessary when two or more filter media are used. The subsurface washer is located at the expected depth of the expanded interface. Rotary surface wash facilities typically are designed with 280 to 340 kPa (40 to 50 psig) of water pressure and 40 to 120 L/m²·min (1 to 3 gpm/sq ft) of water supply. A strainer with 3-mm (0.1-in.) perforations in the surface wash line may prevent jet clogging.

Filter effluent is typically the source of backwash water. Chlorine contact basins can be sized to provide a convenient reservoir with sufficient capacity for backwash requirements. Clear wells that are used for chlorination contact should not be located below filtering equipment because of possible corrosion. They also should not receive any bypassed, untreated flow from the process sequence preceding the filter. The clear well should have adequate capacity for at least two consecutive backwashes and, in small facilities, the capability to receive supplementary clean water from a source such as a high-volume fire hydrant. Filter backwash should not return directly to the facility if the instantaneous recycle rate can adversely affect the treatment processes. If a BRW storage capability is provided, then the reservoir should be designed for maximum filtration with the longest duration of complete, consecutive filter washes.

Subsequent backwash reject recycle to the facility should extend over a period long enough to minimize hydraulic and loading effects on the facility. Design engineers also

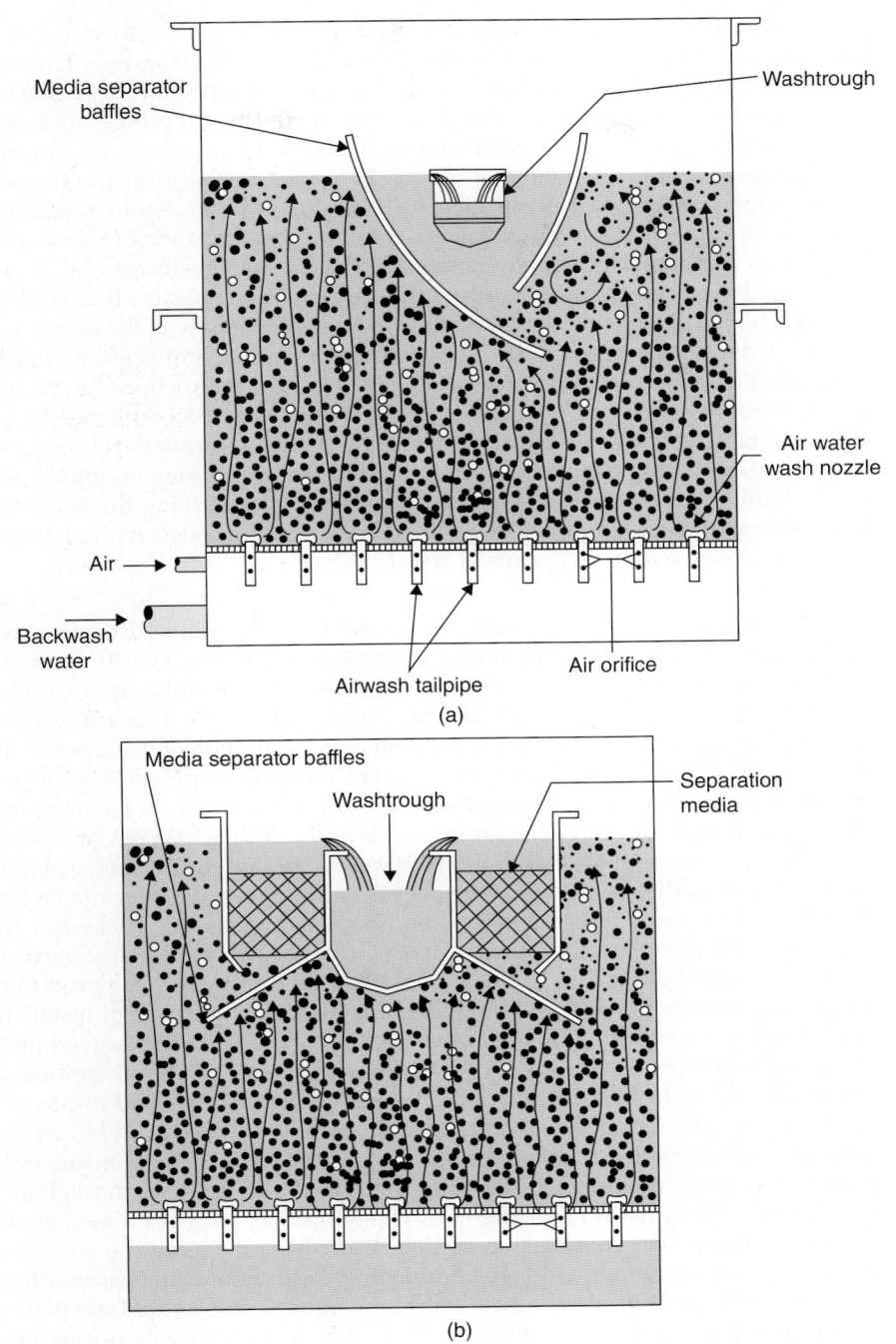

Figure 14.13   (a) Representative baffle configuration for simultaneous air and water backwash filters. (b) Alternate baffle configuration for simultaneous air and water backwash filters with low-profile trough and baffle system.

should consider treating this backwash reject stream separately to avoid the possible recycling of light solids and oil and grease through the facility. If backwash water receives treatment for solids-liquid separation, it may be safely returned directly to the filter influent. Otherwise, it may be returned to the nearest upstream unit process such as the influent of the secondary sedimentation tank or, if lack of flocculation is a concern, to some intermediate point in the biological reactor. Returning the backwash to the front of the facility offers no significant advantage. The BRW storage basin should be stirred mechanically or aerated gently to prevent solids deposition and formation of a scum layer.

Grease that is not effectively removed by biological treatment and its associated sedimentation tanks may remain in suspension as an emulsion that coats the filter media. Chlorination before the filter allows some of the grease in the influent stream or on the media to pass through the filter and into the facility's final effluent. All filtering units should be installed with provisions and operating instructions for the removal of grease accumulated in the media. Grease removal may be accomplished by superdosing with chlorine, hypochlorite, or commercially available industrial cleaning solutions. This entails applying either the concentrated or the appropriately diluted solution to the bed, allowing it to stand, and then vigorously backwashing the bed. Weak acids have been used successfully to free bed encrustation associated with upstream use of metal salts or lime for phosphorus removal.

2.3.1.3.1.5 COLLECTION AND DISTRIBUTION SYSTEMS In conventional downflow filters, the underdrain system collects the filtrate and distributes the backwash water. The latter function controls the design. The traditional system uses several layers of graded gravel under the filter bed and a lateral header positioned on the filter floor. This header, equipped with orifices, provides the preliminary distribution of backwash water. Final distribution occurs as water percolates upward through the gravel. Typical gravel bed designs for a pipe underdrain system can be found in standard references, which also provide accepted rules for pipe lateral underdrain design (AWWA and ASCE, 2012; Eliassen and Bennett, 1967; Fair et al., 1968; Weber, 1972). A disadvantage of gravel-pipe lateral system is the tendency for the gravel to intermix with filter media because of high local velocities during backwash, introduction of air to the backwash system, or excessive backwash flowrates. Gravel-pipe systems also require a large vertical space.

Other underdrain systems have been developed to eliminate the need for a gravel supporting bed (see Figure 14.14). Potential benefits include ease of installation, and smaller vertical footprint (allowing more depth for media) and improved distribution of air and water (for effective backwash). Plastic or steel nozzles of various sizes and shapes are available. Nozzles inserted into precast channels, poured-in-place concrete, or steel plates are designed with built-in orifices to distribute backwash water and air evenly. Nozzles selected for the air-water backwash system should ensure even distribution of air and water and that the finest filter media do not pass through or lodge in the opening. The underdrain plenum below the strainers must be cleaned meticulously before use to prevent construction dirt or debris from clogging the strainers during backwash. At a minimum, a clogged nozzle may result in unequal backwash; at worst, excessive uplift pressures may cause structural failure of the underdrain plate or slab.

Early models of polyvinyl chloride (PVC) nozzles broke easily during installation or medium placement; however, recent designs have improved the mechanical integrity of nozzles. Porous plates made of aluminum oxide or stainless steel also are available. Porous plates offer more perfect backwash distribution but are vulnerable to

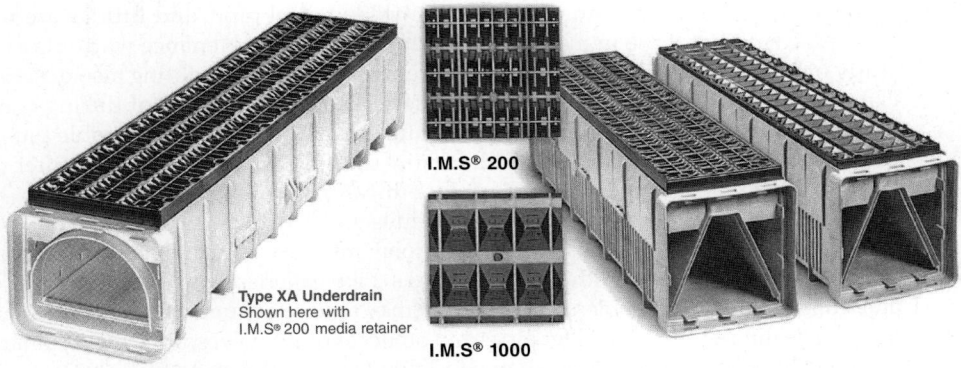

Figure 14.14   Underdrain systems without the need for gravel supporting beds.

clogging (Culp and Culp, 1978). Chlorinated backwash water is used to prevent biological growth in the porous plates.

After moving through the bed, wash water exits the filter through troughs. To prevent nonuniform flow, the maximum lip-to-lip distance between troughs is typically set at 1.8 m (6 ft). Troughs, consisting of rectangular channels, are sized by assuming a convenient width and computing the depth using a momentum analysis of the backwater curve in the trough and adding 50 to 80 mm (2 to 3 in.) for freeboard. The design engineer should cautiously approach combining an auxiliary air scour system with a dual-media or multimedia filter because of the tendency of anthracite to float out of the filter. Mechanical surface and subsurface washing equipment may offer a better alternative. If not, then the designer should consider a sequential mode of media cleaning that incorporates air scour alone in a slightly submerged filter, followed by backwash at the required rate until the filter is clean.

Rubber-seated butterfly valves with an electric motor or air-activated hydraulic operator have almost completely replaced the hydraulic-activated and -operated gate valve that was the standard portable filter valve for many years. The butterfly valve is smaller, lighter, easier to install, is better for throttling service, and can be installed in any position. The seat material should be checked for its susceptibility to corrosion and wear if exposed to cleaning solutions. Electric solenoid actuators and air- or motor-operated valves also are being used more frequently in filtration applications.

Each filter should have the capability to operate when isolated completely from other parallel service units. Typically, for proper operation, a filter includes valves for influent; effluent; wash water supply; wash water drain; surface wash; and, in some facilities, filter-to-waste lines. Filter influent is admitted to the bed so that the medium surface is not disturbed. This entails directing the influent stream to an initial gullet to dissipate the velocity head, using a throttling control on the influent valve, or using a series of distribution troughs and splash plates. The filter-to-waste connection, if used, should have positive air gap protection against back siphoning from the filter drain to the filter bottom. Filter to waste phase is typically exercised at the end of backwash cycles (or after a maintenance/cleaning event) to ensure that backwash reject or cleaning water does not mix with the filtered effluent. The effluent wash water supply and filter-to-waste lines typically are manifolded for a common connection with the filter underdrain system.

Ductile iron and coal tar, enamel-lined welded steel pipe, and fittings are used in filter pipe galleries. Ease in installation, operation, and maintenance warrants a careful study of piping and valve detail. Space, access, isolation, and lighting also deserve special attention. The use of dresser couplings assists in pipe alignment during construction and valve removal for repair. At structure walls, wall sleeves or flexible pipe joints prevent pipeline breaks caused by differential settlement. Color coding of filter piping is advisable for appearance and ease of identification.

Reinforced concrete flumes and box conduits and concrete-encased pipe may be used for wash water drains or other lines on the floor, but they are not suggested for overhead use because of difficulties with cracks and leaks. Pipe galleries should have positive drainage, good ventilation, and plenty of light, and may require dehumidification, air-conditioning, and heating equipment. Filter pipes and other conduits, valves, and gates ordinarily are designed for velocities and flows shown in Table 14.5 (AWWA and ASCE, 2012).

2.3.1.3.1.6 FLOW CONTROL METHODS Design of a filtration system requires an understanding of system operation and requirements and common problems. The main methods used to control the rate of flow through downflow gravity filters may be classified as (1) constant-rate filtration with fixed head, (2) constant-rate filtration with variable head, and (3) variable, declining-rate filtration.

In constant-rate filtration with fixed head, the flow through the filter is maintained at a constant rate. Constant-rate systems are either influent controlled or effluent controlled. Pumps or weirs are used for influent control systems. A modulating valve that can be operated manually or mechanically is used for effluent control systems. In effluent control systems, at the beginning of the run, a large portion of the available head is dissipated at the valve (almost closed position). The effluent valve is opened as the headloss builds up within the filter during the run. This method is not as common in WRRFs because the required control valves are expensive and they have malfunctioned on a number of occasions. Disadvantages of this system include initial and operating costs of the rate control system, which may require frequent maintenance. Alternative methods of flowrate control involving pumps and weirs are more commonly used.

In constant-rate variable head filtration, the flow through the filter is maintained at a constant rate. Pumps or weirs are used for influent control. When the head or effluent turbidity reaches a preset value, the filter is backwashed.

Flow Description	Velocity (ft/s)[a]	Maximum Flow Per Unit of Filter Area (gpm/sq ft)[b]
Influent	1–4	3–8
Effluent	3–6	3–8
Wash water supply	5–10	15–25
Wash water drain	3–8	15–25
Filter to waste	6–12	1–6

[a]ft/s × 0.304 8 = m/s.
[b]gpm/sq ft × 2.444 = m/h.

TABLE 14.5 Design Velocities and Flow Volumes

In variable, declining-rate filtration, flow through the filter decreases as the rate of headloss builds up with time. Declining-rate filtration systems are either influent controlled or effluent controlled. When flow decreases to the minimum design rate, the filter is removed from service and backwashed.

As an alternative, an influent weir box located above the design terminal headloss can be used to achieve constant-rate filtration. This weir box splits flows approximately equally to all operating filter units and smoothly adjusts to flow changes and headloss indications by changing water levels. Variable, declining-rate filtration has advantages similar to those of constant-rate filtration with influent flow splitting. An effluent control weir at an elevation above the surface of the filtering media may prevent accidental bed dewatering and the possibility of a negative head in the filter from air binding caused by gases escaping the solution. The principal difference between this arrangement and the former in the constant-rate method is the location and type of influent arrangement, which maximize working headloss.

Filter influent travels through a common conduit shared with other filters and enters below the wash-trough level. By suitable vertical baffling, constant-rate filtration of influent flow occurs until the water level in the filter rises above the level of the wash trough and baffle; thereafter, the unit provides variable, declining-rate filtration. The water level will remain the same in all operating filters at all times, but filtration rates will vary among filters and with time, depending on the headloss buildup in each. Providing a flow-restricting valve or orifice in the effluent pipe is advisable to prevent excessive filtration rates when the filter is clean. Variable, declining-rate operation sometimes produces better filter effluent quality than constant-rate filtration if the filter effluent quality tends to deteriorate toward the end of a run. Further, a variable declining-rate configuration requires less available headloss because the filtration rate declines toward the end of a filter run.

Typically, declining-rate filters provide the best mode of gravity filter operation unless the design terminal headloss exceeds 3 m (10 ft). In this situation, constant-level control of pressure filters may be a more economical choice. This prospective economy, however, must be weighed against a potential disadvantage because of occurrence of terminal headloss conditions simultaneously in all filters. This could cause a system failure under a declining-rate filtration mode unless backwash is consistently initiated before terminal conditions, an upstream reservoir is provided, or both. Figure 14.15 shows a typical filter-and-clear well arrangement for both influent splitting, constant-rate and variable, declining-rate filtration operations.

2.3.1.3.2 Semicontinuous Operation, Single-Graded Medium, Upflow Filters Upflow filters take advantage of the natural size gradation achieved hydraulically where the media contain different-sized particles of the same specific gravity. One example of an upflow filter medium consists of 100 mm (4 in.) of gravel with a diameter of 30 to 40 mm (1.25 to 1.5 in.); 250 mm (10 in.) of 10- to 16-mm (0.38- to 0.63-in.) gravel; 300 mm (12 in.) of 2- to 3-mm (0.05- to 0.08-in.) sand; and 1.5 m (60 in.) of 1- to 2-mm (0.03- to 0.05-in.) sand. Uniformity coefficient is of minor importance for this system. More important is a slightly buried restraining grid that allows formation of sand arches to provide a tight bed. Approximately 41 distribution nozzles/m² (4/cu ft) supply the influent. Figure 14.16 provides a schematic of an upflow single-medium filter.

The bed is cleaned by draining (approximately 4 minutes), followed by a sequential application of low-pressure air to break up the sand arches (3 minutes), and a flush water backwash (10 minutes). In some installations, air is used in the flush

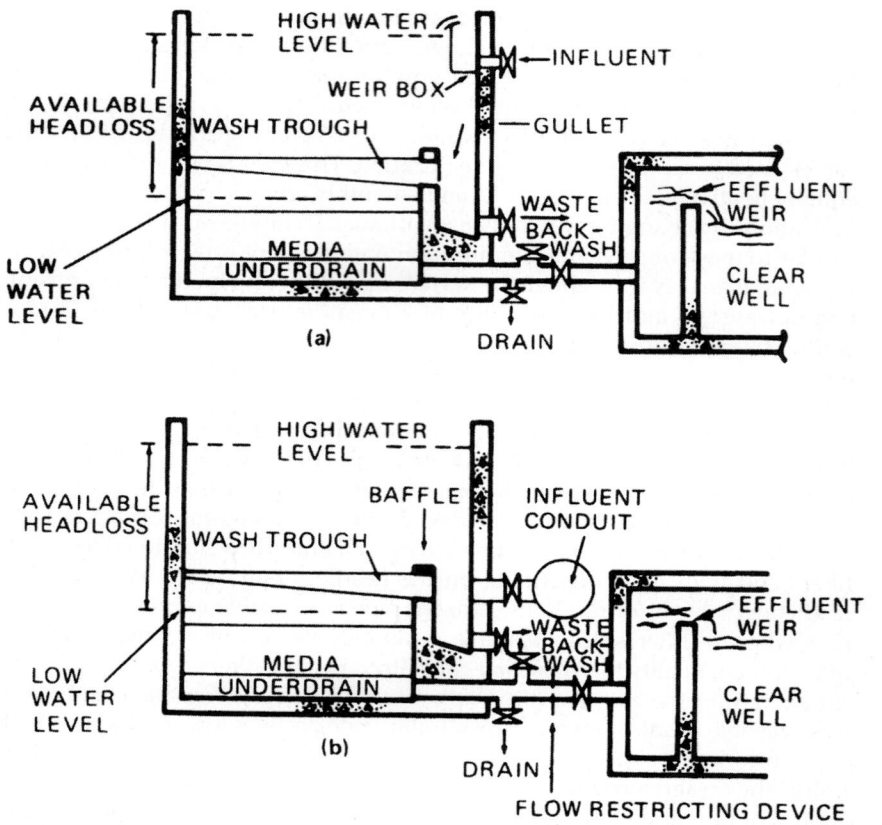

FIGURE 14.15 Typical filter and clear well arrangement elevations for (a) influent splitting, constant-rate filtration and (b) variable, declining-rate filtration.

cycles. The backwash supply is the influent feed stream. Rates are selected to provide 7% to 10% expansion of the upper sand layer. Temperature will dictate what rates are necessary to obtain such expansion. For example, a backwash rate of 560 L/m²·min (14 gpm/sq ft) at 20°C (68°F) corresponds to a rate of 800 L/m²·min (20 gpm/sq ft) at 36°C (97°F). Following backwash, a 5-minute settling classifying period is allowed. Then, a 10-minute prefiltration run establishes effluent clarity before the unit returns to service.

Typical flowrates of 240 to 500 L/m²·min (6 to 12 gpm/sq ft) are reported. Studies at Chicago's Hanover facility indicate that filtration rates of 205 to 220 L/m²·min (5.0 to 5.3 gpm/sq ft) successfully reduced unconditioned secondary effluent TSS from 13 to 28 mg/L to 4 to 10 mg/L, with run times of 45 to 50 hours.

Advantages of the system include natural gradation of media in the direction of flow, minimal headloss caused by accumulated solids, and the use of influent water as a backwash supply. Possible disadvantages include the need for a pump storage feed capability and the potential for solids unloading with diurnal flowrate variations.

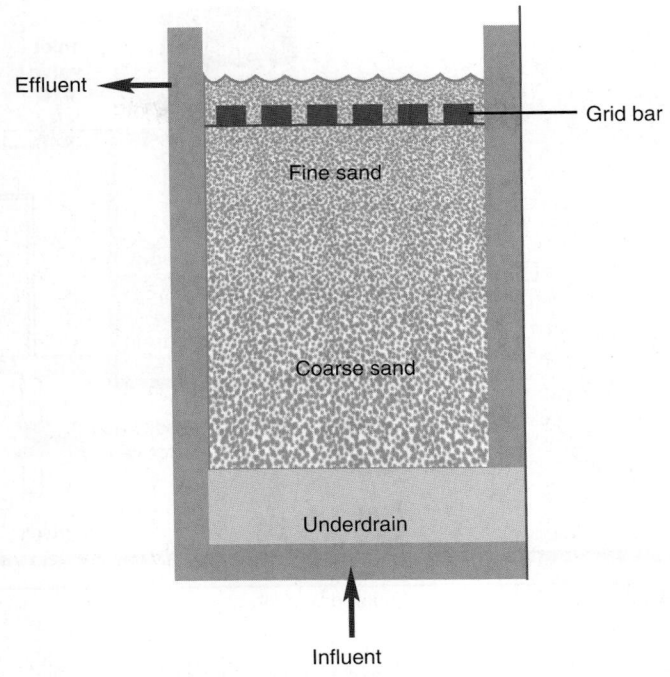

FIGURE **14.16**   Semicontinuous operation, upflow single-medium filter.

2.3.1.3.3   Semicontinuous Operation, Single-Medium, Pulsed Bed Filters   A typical pulsed bed filter has semicontinuous operation with a 300-mm (10-in.) bed depth and media with a 1.5 uniformity coefficient. Media selection depends on effluent standards and filter influent characteristics. This system incorporates an air-and-pulse mix feature that operates if the bed becomes clogged. Once the water level builds up to the air-mix probe, low-pressure air is supplied to the diffuser just above the bed surface. Fine air bubbles leaving the diffuser create a gentle rolling motion above the bed that captures large particles on the surface and holds them in suspension to extend the filtration cycle. As the process continues, solids continue to deposit on the bed; this causes the operating water level to rise to the pulse-mix probe. Once energized, the effluent valve is closed, and the backwash pump turns on to force the trapped air in the underdrain through small orifices in individual subcompartments and, ultimately, through clogged media. After expelled solids are trapped in the gentle-rolling mixture above the bed surface, the backwash pump is shut down. This relieves the bed and operation continues. When the pulse-mix system accumulator ultimately reaches a number of preset pulses—typically 6 to 10—the system is deactivated. The admixture level then rises and activates the true backwash. Figure 14.17 provides a schematic of a pulsed bed filter.

2.3.1.3.4   Continuous Operation, Nongraded Media Filters   Several nongraded-media, continuous-operation filter configurations exist specifically for wastewater treatment operations. Most of these systems have some proprietary features, and several are briefly described in this section. The most common of these filtration processes are the automatic backwashing (ABW) filter and moving bed filter (MBF).

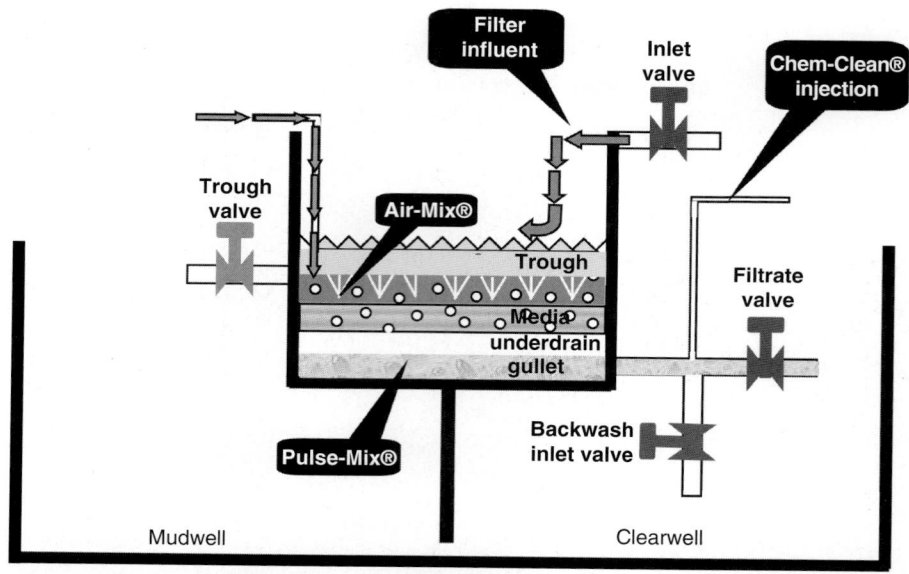

Figure 14.17   Typical schematic of pulsed bed filter.

Moving bed filters provide a continuous supply of filtered water without the interruption of backwash cleaning cycles. In a typical downflow configuration, influent enters the top of the filter and flows down through layers of increasingly finer sand. Filtered water is collected in a central filtrate chamber before exiting. Solids captured in the filter bed are drawn downward with the sand into an air-lift pump. The turbulent, upward flow in the air lift provides scrubbing action that effectively separates sand and solids before discharging to the filter wash box. The wash box is a baffled chamber that allows gravity separation of cleaned sand from concentrated waste solids. From the wash box, regenerated sand returns to the top of the filter bed, and the solids and BRW are piped to a suitable disposal site. Filter media size is tailored to individual application requirements, and effective size ranges from 0.6 to 1.0 mm (0.02 to 0.04 in.).

In a typical upflow moving bed filter (also known as upflow continuous backwash filter), influent entering the bottom of the filter flows upward through a series of riser tubes. Wastewater spreads evenly into the sand bed through the open bottom of an inlet distribution hood and flows upward through the downward-moving sand bed as solids are removed. Clean filtrate exits the sand bed and overflows a weir as it leaves the filter. Simultaneously, the sand bed, with the accumulated solids, is drawn downward into an air-lift pipe positioned in the center of the filter. A small volume of compressed air is introduced to the bottom of the air lift. Sand, dirt, and water surge upward through the pipe at approximately 8000 L/m^2·min (200 gpm/sq ft). This violently turbulent upward flow scours impurities from the sand. On reaching the top of the air lift, dirty slurry spills over into a central reject compartment. Sand returns to the bed through the gravity washer separator that allows fast settling sand to penetrate, but not the dirty liquid. The sand bed is continuously cleaned while a filtrate and reject stream are produced. A schematic of a moving bed filter is given in Figure 14.18.

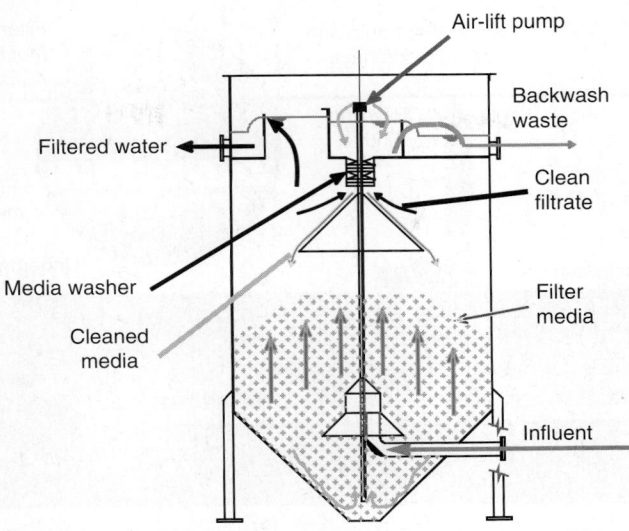

FIGURE 14.18   Typical schematic of a moving bed filter.

The ABW filter—which is often referred to as a traveling bridge filter, continuous backwash filter, or low head filter—typically is constructed in modules that are 4.9 m (16 ft) wide. The length of the unit and the number of units are then adjusted to provide the required filter surface area. The filter uses a relatively shallow depth of media, typically less than 300 mm (12 in.), and the filter is backwashed relatively continuously. The change in backwash procedure eliminates the need for BRW storage basins and reduces capital costs.

Figure 14.19a shows a longitudinal section through an ABW filter. The filter underdrain system is divided into compartments, or cells, and each cell is individually backwashed as the traveling bridge and backwash hood are positioned over the cell to be cleaned. Figure 14.19b is a transverse section through an ABW filter. Water enters through the influent channel, flows into the filter box, and is collected in the filtrate channel. Headloss from the influent to the effluent channel is typically less than 1.5 m (4.9 ft), which is the source of the term "low head filter." Typical configurations for the influent and filtrate channels are indicated.

Original designs for ABW filters did not incorporate a supplemental washing mechanism, such as air wash or surface wash, and filters performed primarily as surface filters. This resulted in considerable sensitivity to TSS loadings; high TSS concentrations in influent to the filters (e.g., more than 40 to 50 mg/L) often resulted in overflow. Modern designs incorporate an air wash mechanism that improves the performance of the ABW filter. Figure 14.20 presents a time-series plot of effluent TSS concentrations for the traveling bridge filters used at the Houston, Texas, WRRF (oxygen-activated sludge process).

Figure 14.21 is a time-series plot of filter influent and effluent TSS concentrations for the ABW filters used at the San Antonio, Texas, WRRF (nitrifying activated sludge). The performance difference between the two facilities emphasizes the importance and effects of upstream biological treatment processes on filtration efficiency.

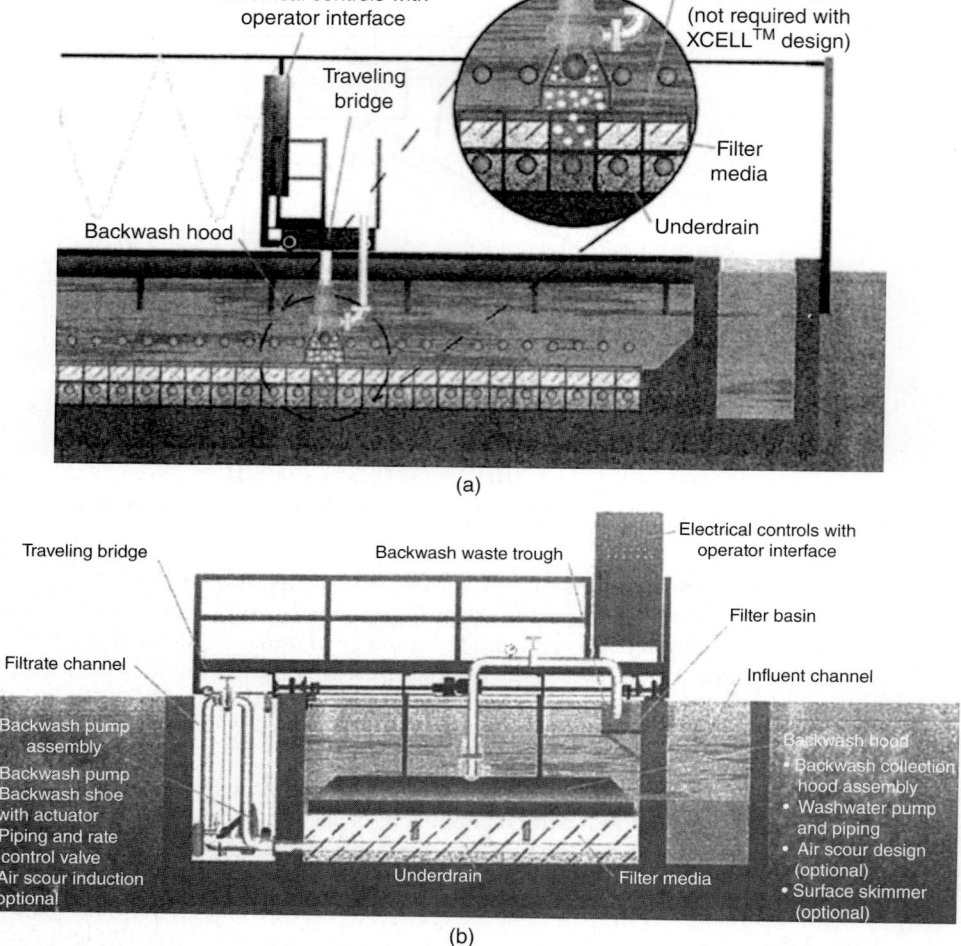

Figure 14.19 (a) Longitudinal section through automatic backwashing filter, (b) transverse section through an automatic backwashing filter.

2.3.2 Nongranular Media Filtration

Several filtration technologies using nongranular media have been developed in the last 20 years to minimize the needs of the backwash operation and the BRW ratio and to provide different operational and design advantages. These filtration technologies can be grouped under two main categories: (1) compressible media filtration and (2) disk filtration.

2.3.2.1 Compressible Media Filtration One of the filtration technologies using nongranular (i.e., synthetic material) is the compressible media filter (CMF), which was first developed in Japan and has been installed in many facilities around the world since its development. The filter bed consists of a large number of compressible balls, which are formed by shrinking synthetic fibers into a quasi-spherical shape

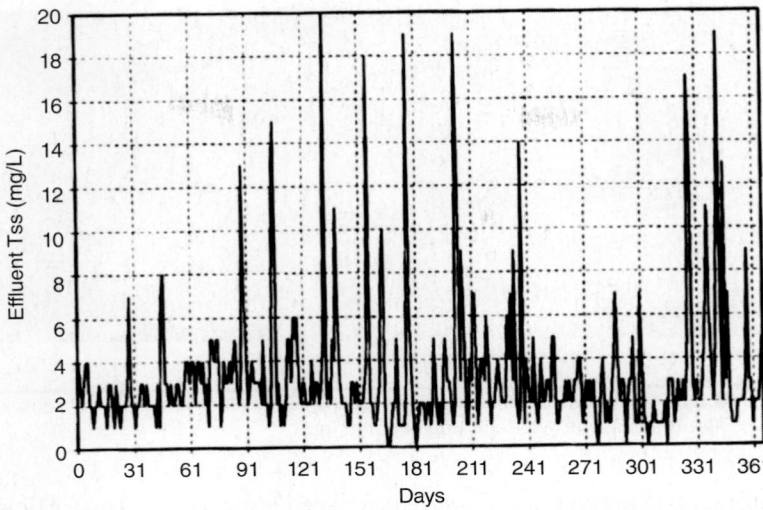

Figure 14.20   City of Houston's 69th Street filter effluent total suspended-solids concentrations.

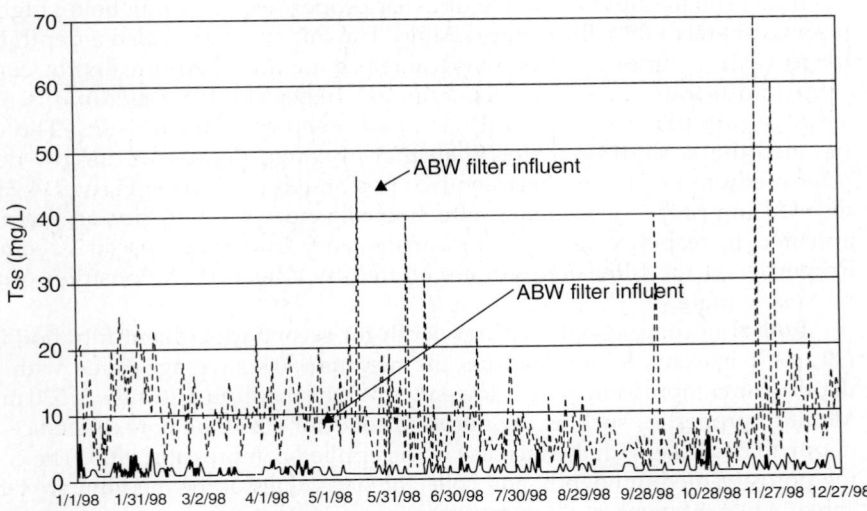

Figure 14.21   San Antonio Dos Rios facility filter influent and effluent total suspended-solids concentrations.

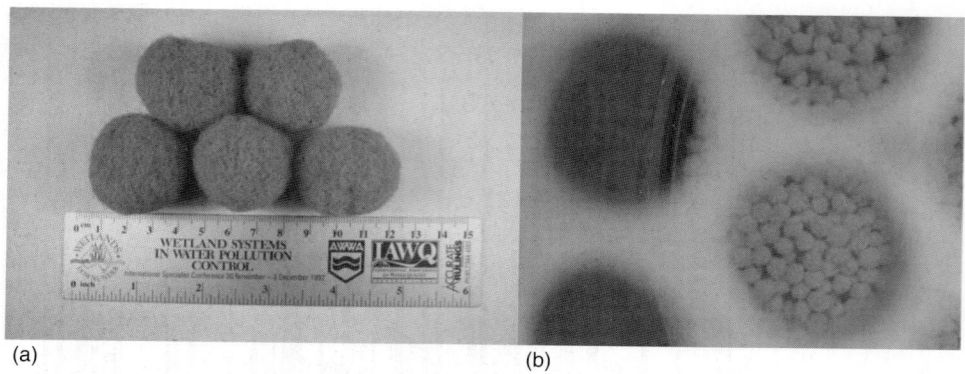

FIGURE 14.22   Compressible filter medium for: (a) Fuzzy Filter (courtesy of Schreiber LLC), (b) FlexFilter (courtesy of WesTech Engineering, Inc.)

resulting in a complex woven structure (see Figure 14.22). The CMF was first introduced in the United States about 20 years ago, and its application has increased in the last 10 to 15 years.

2.3.2.1.1 Mechanical Compression Type   One of the CMF technologies, also known as the Fuzzy Filter™, involves the use of mechanical compression of the filter medium. The medium has some unusual properties, including being highly porous (approximately 90%) and compressible. The Fuzzy Filter is also a depth filter similar to GMFs. Important design parameters include medium depth, compression ratio, and filtration rate (Caliskaner et al., 1999). The filter medium is resistant to temperature, pH, and acid and has good memory characteristics. The density of the medium is slightly greater than that of water. Because of its low density, the filter medium is retained between two perforated plates (see Figure 14.23). The filter medium properties, such as effective collector size, porosity, and depth, can be adjusted in response to changing influent conditions because it is compressible. Properties of the filter medium are altered by adjusting the position of the upper moveable plate.

Typical uncompressed medium depth for secondary effluent filtration is 760 mm (30 in.). For example, to compress the medium 30% (average value with respect to depth), the compression plate is lowered until the medium thickness is 530 mm (21 in.). Medium properties such as the porosity, collector size (i.e., pore/interstice size), and depth all decrease with the increase in the applied compression ratio. The corresponding porosity, medium depth, and collector size values of the medium are tabulated in Table 14.6 at different medium compression ratios.

The CMF is able to operate at filtration rates up to 1600 L/m²·min (40 gpm/sq ft) because of reduced headloss development resulting from high porosity of the medium (Caliskaner et al., 2011a). Typical average design filtration rates are between 800 and 1200 L/m²·min (20 to 30 gpm/sq ft). Footprint requirements are reduced significantly (approximately 75%) compared to the granular filters because of the high filtration rates. Influent and effluent turbidity values and headloss development curves (without chemical addition) are presented in Figure 14.24 (Caliskaner et al., 1999). The medium compression ratio was 40% for the filtration cycle shown in Figure 14.24.

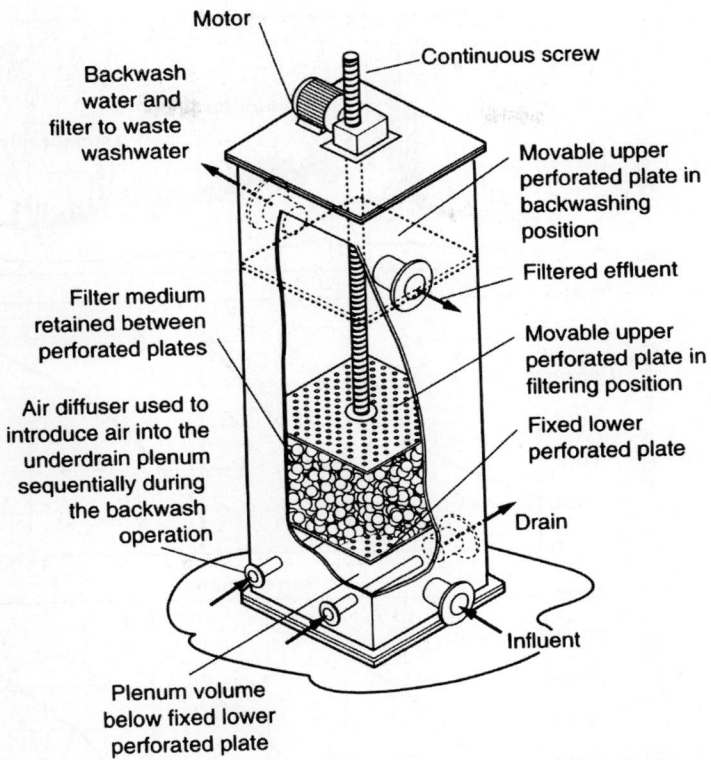

Motor

Continuous screw

Backwash
water and
filter to waste
washwater

Movable upper
perforated plate in
backwashing
position

Filtered effluent

Filter medium
retained between
perforated plates

Movable upper
perforated plate in
filtering position

Fixed lower
perforated plate

Air diffuser used to
introduce air into the
underdrain plenum
sequentially during
the backwash
operation

Drain

Influent

Plenum volume
below fixed lower
perforated plate

Figure 14.23 Schematic view of filter with a compressible filter medium. The properties, porosity, collector size, and depth, of the filter bed can be adjusted by compressing the filter material (Caliskaner et al., 1999).

Removal performance can be increased approximately 50% to 70% with chemical addition. Chemical type, dose, and addition duration should be selected carefully to prevent medium blinding.

During the filtration cycle, secondary effluent is introduced in the bottom of the filter into a plenum. The influent wastewater flows upward through the filter medium, and is discharged from the top of the filter. On a regular operational basis, the filtra-

Bed Compression (%)	Porosity (%)	Depth (mm)	Effective Collector Size (mm)
0	90	760	0.44
15	88	645	0.38
30	86	530	0.23
40	83	460	0.17

Table 14.6 Medium Properties at Different Medium Compression Ratios

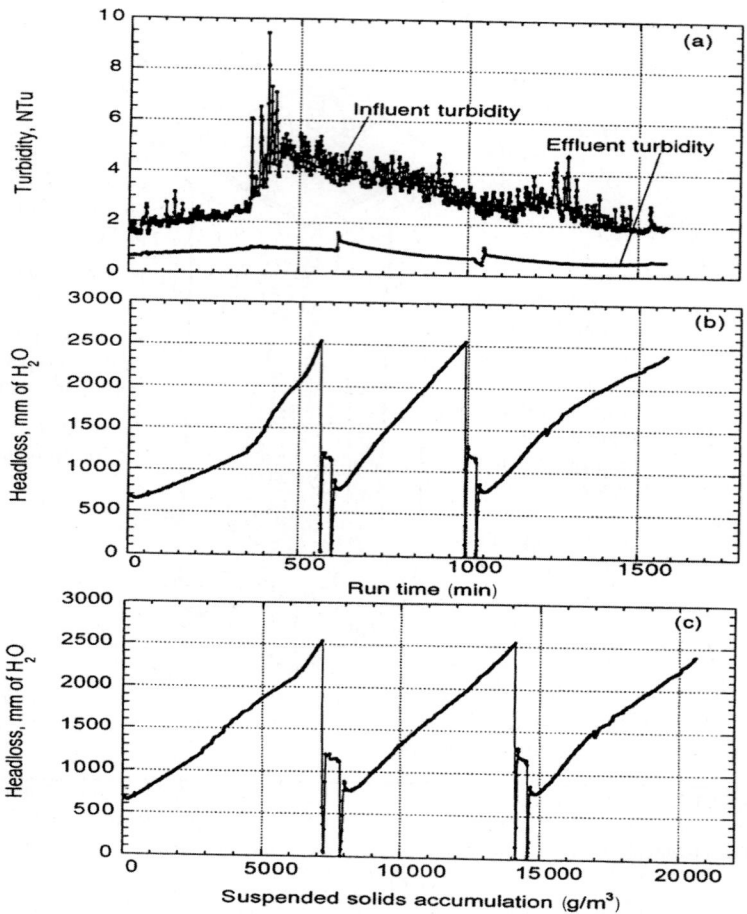

FIGURE 14.24 Typical performance of the compressible medium filter at a filtration rate of 820 L/m²·min: (a) influent and effluent turbidity data versus time, (b) headloss development versus time, and (c) headloss development versus suspended-solids accumulation (Caliskaner, 1999).

tion cycle is interrupted for backwashing when the terminal headloss or breakthrough (as measured in effluent turbidity) is reached. At the inception of the backwash cycle, the flow is diverted to the backwash water line, and the upper plate is moved up mechanically to increase the bed volume for backwashing. Secondary effluent typically is used to backwash the filter, eliminating the need for a clean water tank. While secondary effluent continues to flow up through the filter, air is introduced below the lower perforated plate first from the left side and then from the right side of the filter (through separate air pipes in both sides). The complete backwash cycle takes approximately 20 minutes. To start the next filtration cycle, the upper perforated plate is returned to its desired position to obtain the required level of medium compression. The typical backwash rate is 410 L/m²·min (10 gpm/sq ft), therefore secondary influent flowrate is reduced at the beginning of the backwash cycle (e.g., from 1230 L/ m²·min [30 gpm/sq ft] to 410 L/m²·min [10 gpm/sq ft]). The BRW ratio ranges between 1% and 5 % depending on the specific filtration conditions, with average typical values around 2% to 3%. Underdrain systems are eliminated in the CMF system because a granular medium is not used. The filtration and backwash cycles are illustrated in Figure 14.25.

Clean filter headloss increases linearly with filtration rate. Headloss also increases as porosity and collector size of the medium decrease as compression ratio is increased. The clean filter headloss values as a function of the filtration rate and medium compression ratio is shown in Figure 14.26.

The typical terminal headloss value is between 1.8 and 3.7 m (6 and 12 ft). Figure 14.24 illustrates the development of headloss with time and TSS accumulation for a filtration rate of 820 L/m²·min (20 gpm/sq ft) and a medium bed compression ratio of 40%.

2.3.2.1.2 Hydraulic Compression Type   Another prominent-type CMF technology for tertiary filtration, known as "FlexFilter™," is configured based on downflow filtration. FlexFilter uses an engineered bladder coupled with the hydraulic pressure of the influent water to laterally compress the media. Influent water fills the area behind the bladder causing the bladder to compress the media as illustrated in Figure 14.27 resulting in a conical-shaped bed. Full compression is achieved as the water level reaches the inlet weir elevation. Filtration starts as influent water discharges over the influent weir and passes downward through the compressed medium. The lateral compression provides a porosity gradient with loose media at the top and compressed media at the bottom. As a result, large particles are removed at the top and fine particles are removed at the bottom. As solids are removed within the media, the influent level increases until a backwash cycle is triggered. When the influent pressure is released, the bladder relaxes forming a curved bottom allowing smooth rotation when backwashing. During backwash, a rotational fluidized bed and air scrub is used to clean the media, with air-lift of the spent backwash water into troughs. Secondary influent at a rate of 205 L/m²·min (5 gpm/sq ft) is used for backwash. Backwash airflow rate is 0.05 m³/m²·s (10 scfm/sq ft). Typical backwash cycle ranges between 20 and 25 minutes. The operation of FlexFilter is illustrated in Figure 14.28.

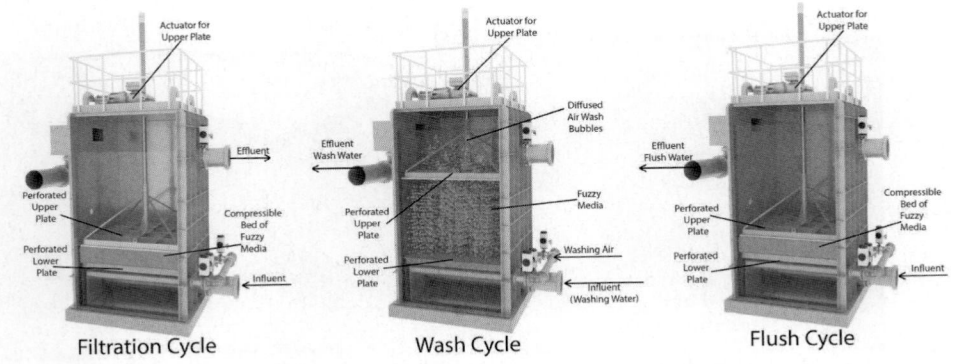

FIGURE 14.25   Filtration and wash cycles of compressible medium filter (courtesy of Schreiber, LLC).

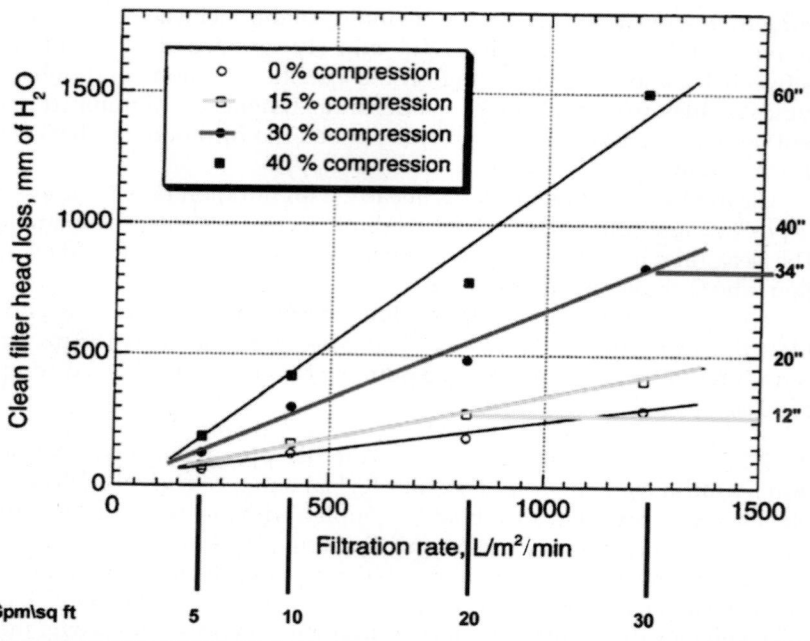

FIGURE 14.26   Initial clean bed headloss across the filter medium versus filtration rate and bed compression (Caliskaner and Tchobanoglous, 2005a).

Main design considerations for FlexFilter include medium depth, hydraulic and solids loading rates, and backwash management. The typical medium depth is 760 mm (30 in.) for secondary effluent filtration applications. Conical-shaped media bed provides a porosity (and pore size) gradient from 0% to 50% compression ratio. Design filtration rates depend on solids and hydraulic loading and headloss development. Filter hydraulic rates up to 800 L/m²·min (20 gpm/sq ft) are used for

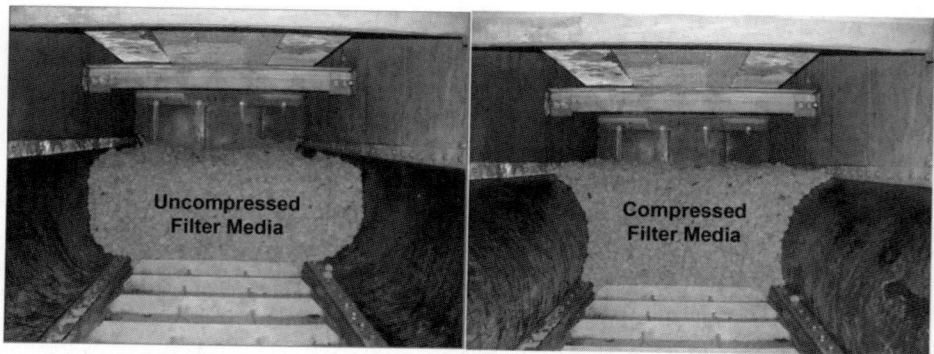

FIGURE 14.27   Compression of the FlexFilter media (courtesy of WesTech Engineering, Inc.).

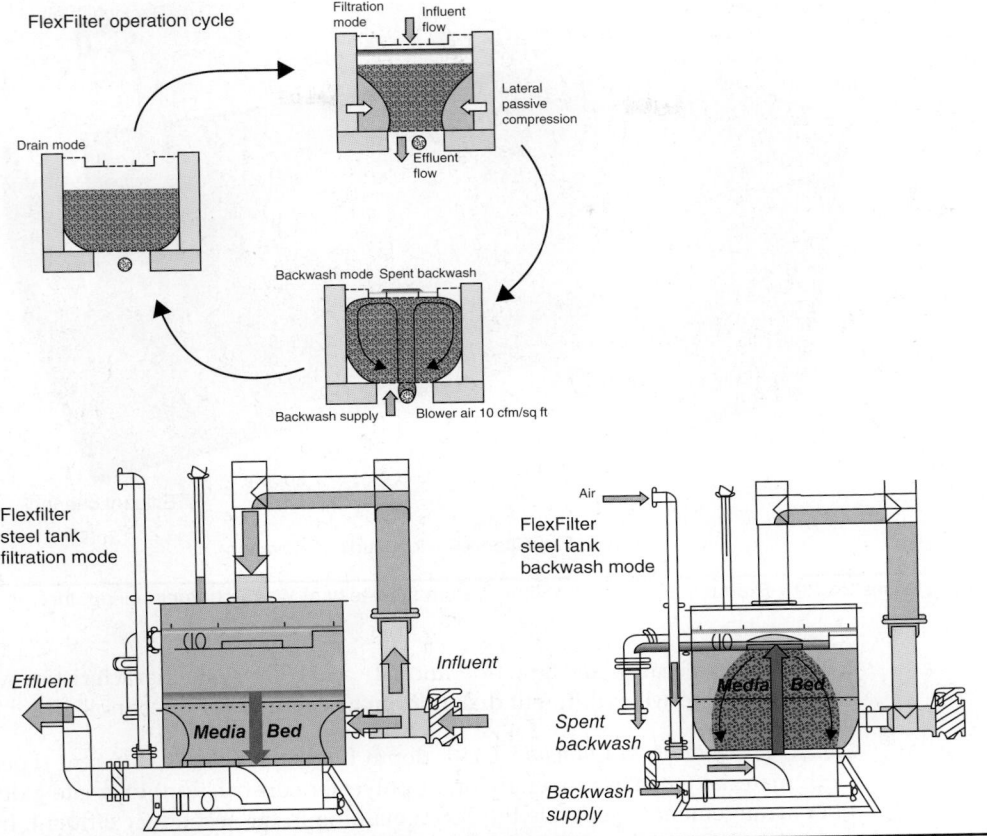

FIGURE 14.28   Operational cycles of FlexFilter (courtesy of WesTech Engineering, Inc.).

secondary effluent filtration. An example layout for a FlexFilter system is given in Figure 14.29.

For both CMF technologies, design ranges for medium depth, the medium compression ratio, and filtration rates depend on specific wastewater characteristics such as filter influent TSS loads/concentrations.

2.3.2.2   Disk Filtration   Like the CMF, disk filtration was introduced for secondary effluent filtration in the last 20 years to reduce the requirements of backwash operation (and BRW ratio) and to simplify filter operation. The two other main advantages of disk filters are the reduced footprint and low operational head requirements. As a result of these advantages, disk filters are frequently used to retrofit into basins formally used for GMF to increase capacity.

Disk filter concept employs either depth or surface filtration mechanisms depending on the technology used. Filtration occurs primarily on the surface of the filter media in case for surface filtration (i.e., as opposed to throughout the media depth). Several proprietary disk filters exist for secondary effluent filtration with different configurations, media, and backwash methods. The design engineer should review the available

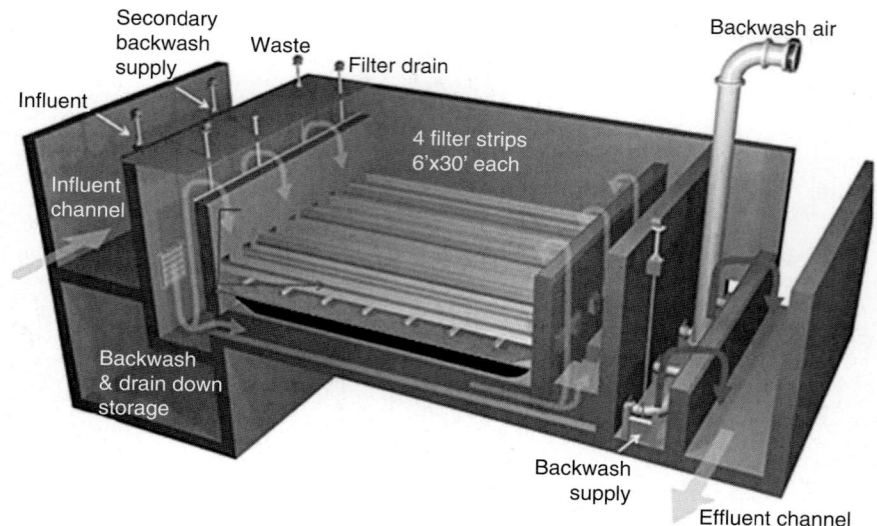

10-mgd FlexFilter cell – longitudinal elevation

FIGURE 14.29   Example layout of FlexFilter system (courtesy of WesTech Engineering, Inc.).

technologies because design, operational, and filtration characteristics vary considerably between different disk filter manufacturers and technologies.

2.3.2.2.1   Cloth Depth Filtration   Cloth depth filters (CDFs) use different types of cloth material (such as woven nylon or polyester construction) to create a dense fiber arrangement to separate the TSS remaining in the secondary effluent. In its operational state, the cloth medium provides an approximate 5-mm (0.2-in.) thickness, which allows depth filtration to occur in addition to surface filtration. The pore size of the media is significantly smaller compared to the GMFs and CMFs resulting in similar medium depth-to-media diameter (L/d) ratios (i.e., similar medium depth-to-medium size ratios observed for GMF technologies). The CDF is one of the most common disk filtration technologies used for secondary effluent filtration.

Different types of cloth media are used depending on the manufacturer. Manufacturers may also offer different cloth media in order to address site-specific conditions (e.g., chemical resistance, different pore size characteristics) (see Figure 14.30). Cloth media technologies can vary in their construction with materials that can be woven, knitted, or produced by needle felting. Materials can be selected for the application and often include nylon, acrylic, or polyester material. The use of woven pile cloth materials has emerged as the most common type of CDF due to improvements in backwash efficiency. The density and thickness of the filter cloth can also be selected according to the characteristics of the influent wastewater and desired effluent quality. The pore size of the cloth media is not absolute because of the random arrangement of the CDF fibers. Nominal pore size ranges between 5 and 10 µm for different type of cloth materials, but significant removals can be realized in smaller particle size ranges.

FIGURE 14.30 Cloth depth filter: (a) media options, and (b) microscopic view of one type of media (courtesy of Aqua-Aerobic Systems, Inc.).

Typical maximum design filtration rates are between 240 to 280 L/m²·min (6 and 7 gpm/sq ft). Although testing has shown that these filters can operate at hydraulic loading rates up to 800 L/m²·min (20 gpm/sq ft) for short periods. The maximum hydraulic loading rate can also be limited by the influent TSS when the solids loading rate exceeds the manufacturer's recommendation (typically 400 to 700 g/m²·h).

Turbidity and TSS removal performance of CDF is similar to GMF and CMF technologies as illustrated in Figure 14.31 (Caliskaner et al., 2011b). The removal performance results presented in Figure 14.31 were obtained (without chemical addition) from a 6-month side-by-side pilot testing of CDF, CMF, and GMF technologies. CDF removal performance can be increased approximately 50% to 70% with chemical addition, but chemical type, dose, and addition duration should be selected carefully to prevent medium blinding. The footprint requirement is 70% to 75% less compared to GMFs (Caliskaner et al., 2011; Haecker and Healy, 2006a and 2006b). The vertical configuration of the filtration disks provides a relatively large filter surface area in a small footprint (see Figure 14.32). An individual filter can contain 1 to 24 disks. The number of disks in a filter unit depends on the design flow and filter loading rates and the size of an individual disc. The diameter of an individual disk ranges from 0.8 m (2.8 ft) to 3.05 m (10 ft) depending on the disk manufacturer, model, and design flow and loading rates. As seen in Figure 14.32, different CDF systems such as the diamond or MegaDisk configurations that use the same type of cloth filter medium are available. The diamond configuration was manufactured to accommodate low-profile/shallow basin designs (Baumann et al., 2006).

During the filtration cycle, the wastewater flow is from the outside to the inside of the disks. Several cloth disks covered by cloth media are mounted vertically to a common hollow tube, which conveys filtered effluent from the filter. Wastewater passes through the cloth media by gravity and enters inside filter disks that are connected to the effluent line by the hollow tube. A total hydraulic head between 0.75 and 1.2 m

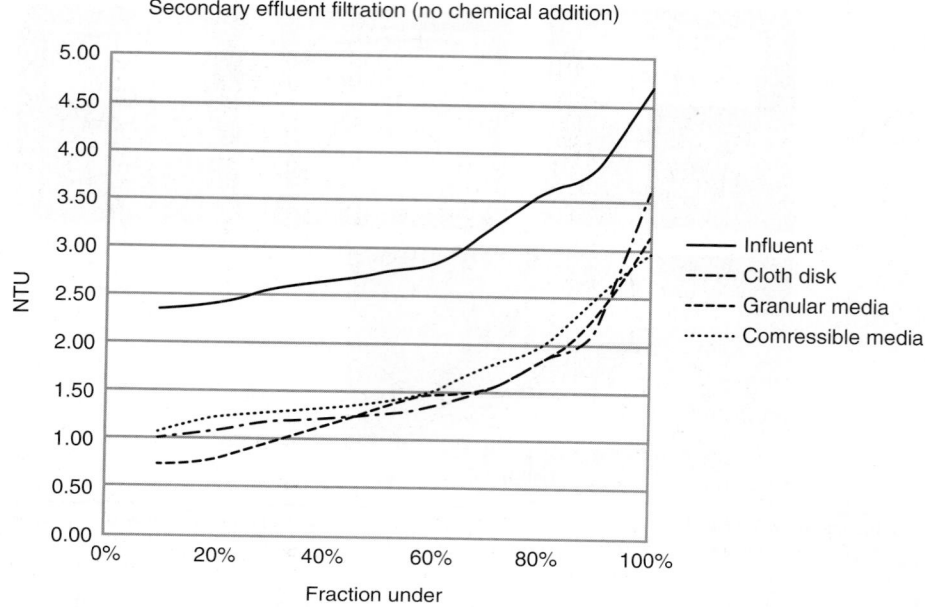

FIGURE 14.31   Probability distribution function for filter influent and filter effluent from parallel testing of GMF, CMF, and CDF technologies at Eugene/Springfield (Oregon) water pollution control facility (Caliskaner et al., 2011b).

(2.5 and 4 ft) is required for the operation of the disk filters. It is common to use weirs on the influent and effluent sides of the filter to maintain a constant hydraulic profile outside of the filter. The filter is at rest during the filtration operation, which allows larger particles to settle to the bottom of the tank. The prior sedimentation of larger size particles decreases the amount of solids to be filtered. The settled solids are pumped periodically to the headworks (or to the solids processing facilities). The design should also accommodate handling of the floating and settling material in the filter tanks outside of the disks.

Backwash cycle starts when the terminal headloss or a certain run time is reached. The disk filters backwash more frequently (e.g., compared to CMFs or GMFs) because of the low head operational characteristics and low terminal headloss design values. Clean medium headloss ranges between 5 and 10 cm (2 and 4 in.). The pressure loss across the membrane increases as more particles are accumulated mostly on the surface and some within the cloth medium depth; a mat forms on the cloth surface. Backwash typically is initiated based on headloss development because the terminal headloss value is set at relatively low levels of approximately 0.3 to 0.5 m (12 to 18 in.). Accumulated particles are removed from the cloth media surface by the liquid suction applied to each side of the disc. A vacuum apparatus is mounted on each side of the filter disk for the liquid suction backwashing. Depending on the disk technology, the cloth disk or vacuum backwash apparatus rotates slowly at approximately 1 rpm during backwash allowing each disk segment to be vacuumed. A typical schematic of a CDF system is illustrated in Figure 14.33.

(a)

(b)   (c)

FIGURE 14.32   Different cloth depth filter configurations: (a) Aquadiamond configuration; (b) Aquadisk configuration; and (c) MegaDisk configuration in package and concrete systems (courtesy of Aqua-Aerobic Systems, Inc.).

Filtration operation is continuous because only a small portion of the media is out of service during backwash. Filtration continues for an individual disk during the backwash cycle. The portion of the disk in contact with the vacuum apparatus (approximately 5%) is cleaned while the remainder of the disk continues to filter. Filtered water is used for backwash; therefore, a separate BWS tank is not required. The need for a BRW storage basin also is eliminated with the disk filter technology because of continuous filtration operation. The BRW ratio typically ranges between 1% and 3%

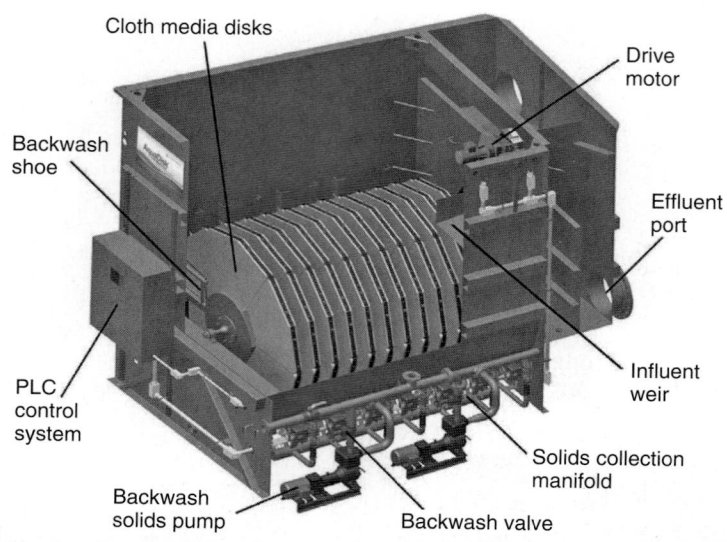

Cloth media disks

Drive motor

Backwash shoe

Effluent port

PLC control system

Influent weir

Backwash solids pump

Solids collection manifold

Backwash valve

FIGURE 14.33 Typical schematic of a cloth depth filter system (courtesy of Aqua-Aerobic Systems, Inc.).

depending on the specific filtration conditions (Bourgeous et al., 2003; Caliskaner et al., 2011b). Within a particular filter, the BRW ratio will be approximately directly proportional to the influent solids concentration.

2.3.2.2.2   Cloth Surface Filtration   Like CDFs, filtration is continuous in cloth surface filtration (CSF) disk filters. The main differences between CDFs and CSFs are medium type, filtration removal mechanisms, flow direction, and submergence. The disks are partially submerged in common CSF technologies used for secondary effluent filtration.

The CSF technologies involve use of a cloth medium made from woven polyester. The medium looks like membrane material and has an absolute pore size between 10 and 20 μm (e.g., depending on the manufacturer) with insignificant depth. The disk panel can be a flat or pleated surface depending on the manufacturer. Pleated panel design was offered to provide more filtration surface area per disk. Different types of CSF technologies are shown in Figures 14.34 through 14.36.

The CSFs can be operated up to 480 L/m²·min (12 gpm/sq ft). Typical maximum design filtration rates are between 120 to 240 L/m²·min (3 and 6 gpm/sq ft). The vertical configuration of the filtration disks provides a relatively large filter surface area in a small footprint (see Figure 14.32). The diameter of an individual CSF disk ranges between 1.7 m (5.6 ft) and 2.6 m (8.4 ft) among different manufacturers and models. One filter tank can include 24 to 32 disks. The footprint requirement is similar to that of CDF systems. Total hydraulic head requirement for the operation is typically between 0.75 and 1.2 m (2.5 and 4 ft).

In contrast to the CDF system, influent wastewater flows by gravity into the filter disks from the center influent collection drum. Filter media mounted on two sides of the disks separate TSS. Filter media retain solids while filtered water flows outside the disks and into the effluent collection tank. Only filtered effluent passes out of the disk

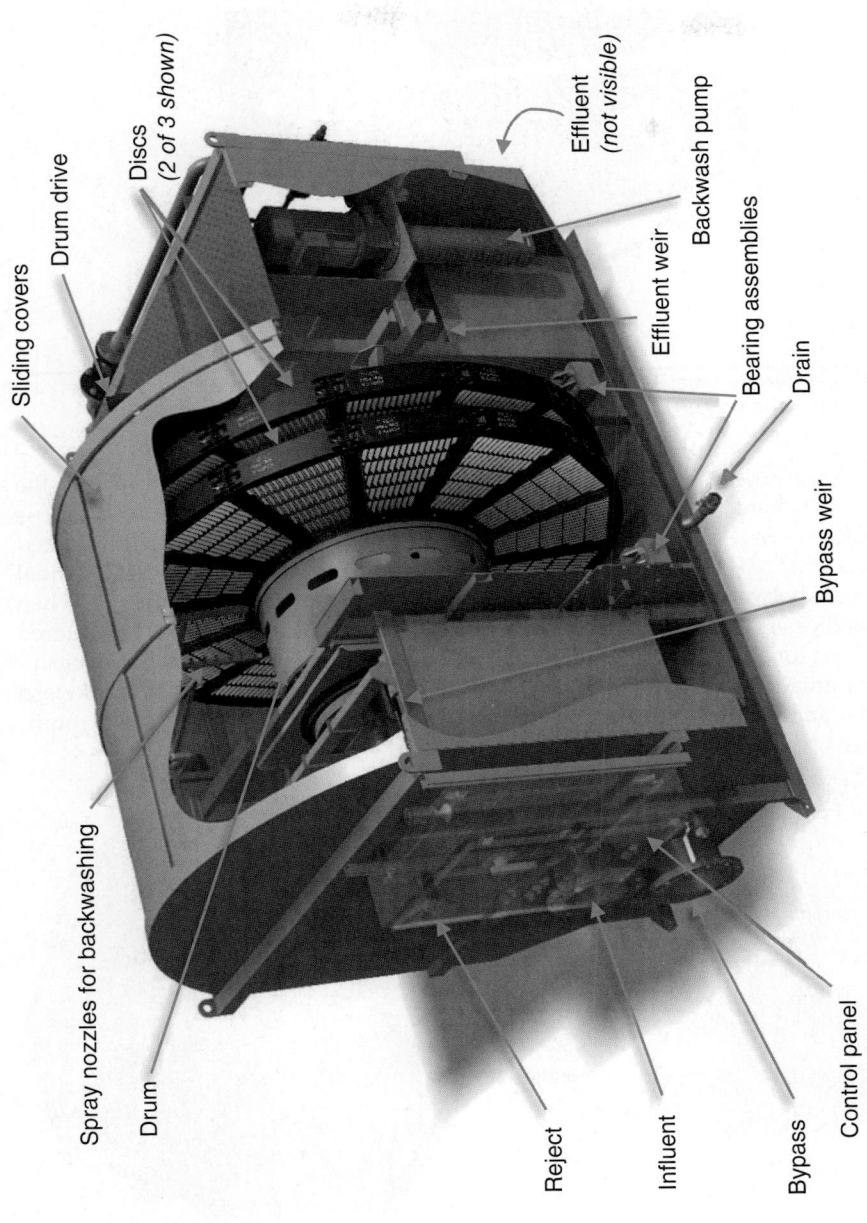

Spray nozzles for backwashing

Sliding covers

Drum drive

Discs
(2 of 3 shown)

Effluent
(not visible)

Backwash pump

Effluent weir

Bearing assemblies

Drain

Bypass weir

Drum

Reject

Influent

Bypass

Control panel

Evoqua Forty-X disk filter components, model E1403TLSS shown, one disc removed for clarity.

FIGURE 14.34   Schematic of Forty-X Filter (courtesy of Evoqua, Inc.).

FIGURE 14.35   Partially submerged disk filters in single filter units. Courtesy of Veolia Water Technologies, Inc. (dba Kruger).

filter with this arrangement. During normal operation, the disks are at rest until the water level in the inlet channels rises to a specific point, such as terminal headloss value. Headloss across the filter medium increases as more particles are accumulated on the disk surface. Clean filter headloss values are similar to the CDF systems. Terminal headloss varies between 0.3 m (1 ft) and 0.6 m (2 ft) depending on the technology. When terminal headloss is reached, the backwash cycle is initiated automatically. The filtered effluent is used for backwash water, eliminating the need for a separate source of cleaning water or an additional BWS tank. The filtered effluent is pumped to the backwash spray header and nozzles, washing solids into the collection trough as the disks rotate.

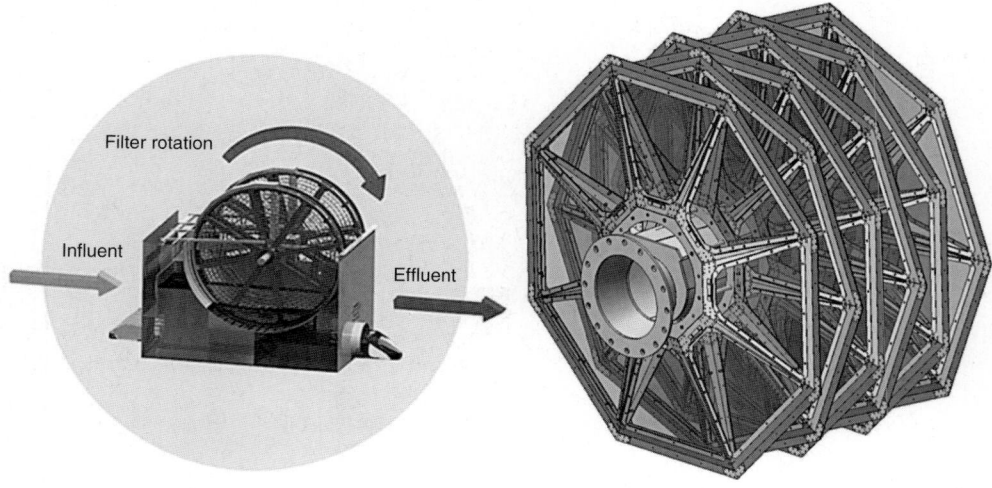

FIGURE 14.36   Schematic of stainless steel disk filters: (a) UltraScreen filter (courtesy of Xylem, Inc.); (b) Quantum filter (courtesy of Nova, LLC).

In normal operation, the disks are approximately 60% to 65% (depending on the disk filter manufacturer) submerged. The need for a BRW storage basin is eliminated because of the continuous filtration operation. The typical BRW ratio is between 1% and 3%.

2.3.2.2.3 Stainless Steel Disk Filtration   Another emerging disk filter uses woven stainless steel as the filtering medium. Filtration occurs mainly via surface filtration mechanisms. For secondary effluent filtration applications, the woven-wire stainless steel mesh pore size ranges between 10 and 20 μm. The steel disk filter was first used for secondary effluent filtration in Europe in the early 2000s, and its use has increased worldwide (and in United States) within the last 5 to 10 years.

The steel medium disk filter is a continuous operation filter like other disk filters. Secondary effluent is introduced in the middle of the disks. Filter disks retain the solids while the filtered water flows outside the disks into the collection well. Disks operate in slow rotation during filtration cycle (also referred as the dynamic tangential filtration). The rotational speed is variable to provide operational adjustments for hydraulic and TSS load changes. Individual disks range in diameter between 1.0 and 1.6 m (3.1 and 5.1 ft) between different manufacturers and models, and an individual unit can contain up to 16 pairs of disks.

As filtration continues, particulates accumulate on the surface of the stainless steel medium, increasing headloss. Typical total operational head requirement is between 0.75 and 1.2 m (2.5 and 4 ft). When the headloss increases to a preset terminal headloss limit, backwash is started. Each disk has a dedicated spray header for backwashing. Typical backwash cycle is between 30 seconds and 1 minute. The BRW is collected in a common stainless steel trough below the washing assembly and exits the filter through a drain line. Effluent water is used for backwash; therefore, a separate BWS tank is not required. The BRW ratio is also very low similar to other disk filtration technologies (e.g., typically between 0.5% and 2%). Figure 14.36 illustrates schematics of two stainless steel disk technologies used for secondary effluent filtration.

The steel media disk filters can be operated up to 640 L/m²·min (16 gpm/sq ft). Typical maximum design filtration rates are between 160 to 280 L/m²·min (4 and 7 gpm/ sq ft). Similar to CDF and CSF, vertical configuration of the disks provides a large filter surface area for a small footprint.

3.0 Activated Carbon Adsorption

The use of activated carbon in wastewater treatment systems is a proven process for removal of organic compounds, while this is not a widely adopted practice due to the high cost of the process and carbon replacement and/or regeneration. Activated carbon can be used to adsorb residual chlorine in the water, although this is not a primary application. Activated carbon can be used in two forms: granular activated carbon (GAC) and powdered activated carbon (PAC).

There are three types of activated carbon available for treatment:

1. Coconut shell—small pore sizes, used for low-molecular-weight organics removal;

2. Coal—moderate pore size, used for moderate-molecular-weight organics removal; and

3. Lignite—large pore size, used for large organics removal.

As a tertiary treatment method, carbon adsorption and regeneration have been used for several years to process domestic wastewater contaminated with industrial waste of organic origin and biologically treated wastewater. Design and operating considerations have been well defined and documented (Culp and Culp, 1978; U.S. EPA, 1970, 1973). Somewhat less frequently, activated carbon also has been used in chemical-physical WRRFs that use chemical coagulation and filtration to remove phosphorus and TSS and carbon adsorption to remove organics. The chemical-physical process appears to be an alternative to biological systems in some instances. To determine full-scale process requirements, however, both laboratory studies and pilot column tests should be incorporated into the design process.

Use of PAC in wastewater treatment was neglected in the past, in part because of the absence of established operational methods. New techniques to recover and regenerate PAC and new applications might, however, result in increased use in wastewater facilities.

3.1   Process Description

Activated carbon removes organic material from water through a combination of adsorption of the less polar molecules, filtration of larger particles, and partial deposition of colloidal material on the exterior surface of the activated carbon (Snoeyink et al., 1969; U.S. EPA, 1970; Weber, 1972). The extent of removal of soluble organics by adsorption depends on diffusion of the particle to the external surface of the carbon and diffusion within the porous adsorbent. For colloidal particles, internal diffusion is relatively unimportant because of particle size. Organic substances are refractory to adsorption, which means that dissolved molecules passing through the column consist of strongly hydrophilic organic molecules such as carbohydrates and other highly oxygenated organic compounds.

Two factors lead to adsorption: (1) forces of attraction at the surface of a particle that cause soluble organic materials to adhere to the particle surface; and (2) limited water solubility of many organic substances. Activated carbon has a large, highly active surface area that results from the activation process; this produces numerous pores within the carbon particle and creates active sites on the surface of pores.

Adsorption occurs in three basic steps: film diffusion, pore diffusion, and adhesion of solute molecules to carbon surfaces. Film diffusion is the penetration of the solute molecule, the adsorbate, through the surface film of the carbon particle. At a molecular level, the adsorbate molecule overcomes the resistance of the carbon particle to mass transfer; pore diffusion involves the migration of solute molecules through carbon pores to an adsorption site; and then adhesion occurs when the solute molecule adheres to the carbon pore surface.

Theory suggests that adsorption is a dynamic rather than static process, or that adsorption/desorption occurs continuously as different organic molecules or solutes approach adsorption sites on particle surfaces. More simplistically, adsorption is a selective process because different organic molecules, or structures, bond to carbon in varying degrees. A loosely bonded organic structure can be displaced at an adsorption site by a molecule with a functional group that more tightly adheres to the carbon. Thus, the displaced solute "desorbs," or is released by the carbon, when more preferred solute species are adsorbed.

Two types of adsorption have been hypothesized. Physical adsorption occurs when solute molecules are held loosely to carbon surfaces by van der Waals forces.

Theory suggests that molecules are mobile and migrate on the carbon surface. Chemisorption, or chemical adsorption, occurs because molecular functional groups of the adsorbate and the carbon interact to form a stable carbon bond. Desorption is more applicable to adsorbates that are physically adsorbed than to those that are chemically adsorbed.

Active carbon typically is considered to consist of rigid clusters of microcrystallites, each of which is made up of a stack of graphitic planes. Each carbon atom within a particular plane is bonded to four adjacent carbon atoms; carbon atoms at the edges of graphitic planes have highly reactive (active) radical sites. At these sites, which consist of a heterogeneous mix of basal planes and microcrystallite edges, adsorption takes place. The adsorbent capacity of carbon is reached when active sites have been filled. As these sites fill, sorption equilibrium is approached, and effluent quality deteriorates to an unacceptable level. Then, the carbon is considered spent and removed for regeneration to a reactivation furnace.

The carbon transport and regeneration system provides for the movement of spent carbon to and from the carbon regeneration furnace, regeneration of the carbon, and the introduction and transport of makeup carbon within the system. Methods to regenerate granular carbon include:

- Passing low-pressure steam through the carbon bed to evaporate and removing adsorbed solvent;
- Extraction of the adsorbate with a solvent;
- Regeneration by thermal means; and
- Exposure of the carbon to oxidizing gases.

Carbon regeneration is accomplished primarily by thermal means (Hassler, 1963, 1974). The two most widely used reactivation methods use rotary kilns and multiple-hearth furnaces. In rotary kilns, carbon moves countercurrent to a mixture of combustion gases and superheated steam. Carbon recovery is reported at more than 90% to 95% adsorptive capacity of the regenerated carbon similar to that of new carbon.

The multiple-hearth furnace is heated to a temperature sufficient to burn off carbon monoxide and hydrogen produced by the regeneration reaction. A shaft with rabble blades moves the carbon continuously to bring fresh granules to the surface and to transport the carbon toward the hearth outlet opening. Thus, carbon can be transferred from hearth to hearth.

Closely controlled heating in a multiple-hearth furnace is the most successful procedure for removal of adsorbed organics from activated carbon. Attrition of activated carbon during the regeneration process is a significant design concern.

If a local furnace is available, then a typical regeneration scheme includes:

- Hydraulic transport of the carbon slurry to the regeneration unit;
- Carbon dewatering and feed to the furnace for volatilization and oxidation of adsorbed impurities;
- Water cooling of the carbon;
- Water washing for fines removal;
- Hydraulic transport of the carbon back to columns for reuse; and
- Scrubbing of furnace off-gases.

In offsite regeneration, spent carbon typically is transferred directly by hydraulic means from the absorbers to a waiting truck or other containment vehicle. Fresh, or virgin, carbon is then used to replace the spent material by hydraulic transfer from a second containment vehicle to the empty adsorber. The spent carbon is transported to a commercial reactivation facility, and the vehicle that supplied the virgin carbon returns to the manufacturing facility for a refill of virgin carbon.

3.2   Application

Two fundamental approaches to the use of GAC in wastewater treatment are (1) as a tertiary process following conventional secondary treatment (Figure 14.37) and (2) as one of several unit processes composing chemical-physical treatment.

3.2.1   Tertiary Treatment

Activated carbon treatment is one of many processes that may be used for advanced wastewater treatment. Processes upstream of activated carbon typically are designed to remove all of soluble, biodegradable organics associated with suspended solids in the secondary effluent. Chemical clarification precedes the carbon adsorption step.

In tertiary treatment, the role of activated carbon is to remove relatively small quantities of refractory organics and inorganic compounds such as nitrogen, sulfides, and heavy metals remaining in an otherwise well-treated wastewater (Pretorius, 1972; Sollo et al., 1976; U.S. EPA, 1973). Effluent inorganic concentrations can be greater than those in influent. This phenomenon can be overcome by the addition of air or, in the case of sulfides, chlorine and by the addition of a supplemental carbon source for denitrification.

3.2.2   Chemical

Activated carbon may be used to remove soluble organics following chemical-physical treatment. Chemical-physical treatment systems typically rely on chemical coagulation, sedimentation, and filtration for removing suspended solids and associated organic materials from primary treated wastewater. Typically, the chemical-physical

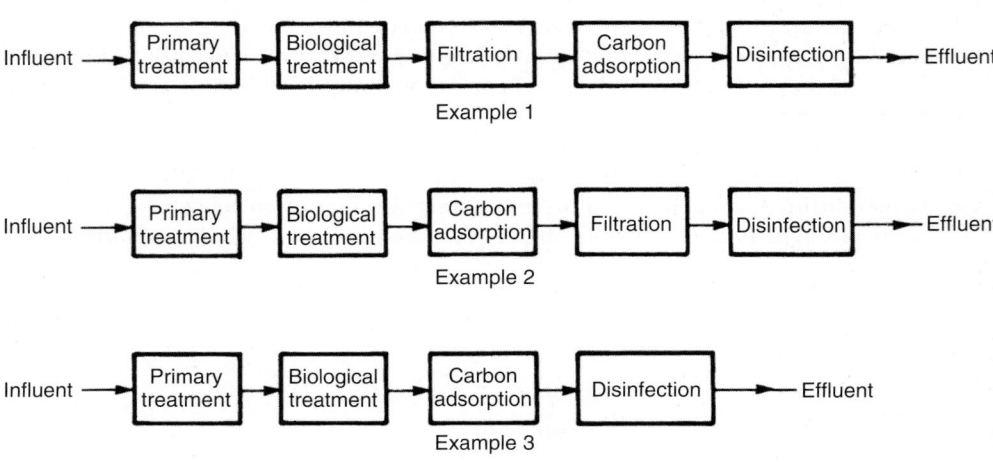

Figure 14.37   Typical flow diagrams for tertiary treatment with carbon adsorption.

system with activated carbon is designed to produce the same effluent quality as that achieved with tertiary treatment. Purely chemical-physical treatment received is not currently in full-scale use for municipal wastewater treatment.

3.3   Design Considerations

3.3.1   Wastewater Quality

The usefulness and the efficiency of carbon adsorption for municipal wastewater treatment depend on the quality and quantity of the delivered wastewater. To be effective, feedwater to the carbon unit should be of uniform quality and consistent flow. Wastewater constituents that can cause potential problems are suspended solids, BOD_5, organics such as methylene blue-active substance or phenol, and dissolved oxygen. Environmental parameters of importance include pH and temperature.

The exact effect of suspended solids on the efficiency and life of carbon is not known; nonetheless, channeling and short-circuiting can occur and reduce bed life and increase carbon losses because of higher localized velocities. The absorptive capacity of carbon can be reduced by (1) restriction of pore openings, or (2) buildup of ash from thermal regeneration of carbon and other materials within the pore structure, which is caused by presence of colloidal materials. Such restrictions or buildups interfere with diffusion processes or reduce effective adsorption sites. To avoid or minimize such impairment, pretreated wastewater of the highest level of clarity is fed to activated carbon columns.

In cases where control of wastewater influent is not manageable, carbon unit performance is affected. With high influent suspended-solids concentrations (20 mg/L and greater in secondary effluent), solids can deposit on carbon granules as a floc, resulting in pressure loss and flow channeling or blockages. Also, if a high level of soluble organic removal in secondary treatment is not maintained, then more frequent carbon regeneration may be required. Similarly, lack of consistency in pH, temperature, or flowrate may adversely affect carbon adsorption. For these reasons, it is good practice to precede activated carbon treatment with GMF.

3.3.2   Carbon Characteristics

The amount of substance that can be removed from wastewater by carbon adsorption depends on conditions that provide optimum adsorption. A useful expression relating the amount of impurity in solution to that adsorbed is the empirically derived Freundlich equation:

$$x/m = kc^{1/n} \tag{14.2}$$

where x = weight of impurity adsorbed;

$\quad\quad m$ = unit weight of adsorbing material (carbon);

$\quad\quad c$ = unadsorbed concentration of impurity left in solution (the equilibrium concentration);

$\quad\quad k$ = constant (the log x/m intercept in the graph of log x/m versus log c); and

$\quad\quad n$ = constant (where $1/n$ is the slope of the curve on the log x/m-versus-log c graph).

To use this equation, the quantity x/m is measured for a number of influent concentrations. The log x/m is plotted versus log c for each of the influent concentrations, and the constants k and n are determined. This produces an adsorption isotherm that allows determination of the degree of removal achieved by the adsorption processes and the adsorptive capacity of the carbon (Liptak, 1974; U.S. EPA, 1973).

However, isotherm data are developed by achieving equilibrium conditions, and field adsorption systems operate in a dynamic environment that is not necessarily in equilibrium. Because differences in adsorptive capacities between equilibrium and dynamic conditions exist, isotherm data typically overestimate the capability of operating systems.

During development of adsorption isotherms, the type of activated carbon available should be considered (Table 14.7) (U.S. EPA, 1973). Activated carbons produced from different base materials and by different activation processes will have varying adsorptive capacities (AWWA, 1974; Mattson and Kennedy, 1971; U.S. EPA,

Properties and Specifications	ICI America Hydrodarco 3000	Calgon Filtrasorb 300 (8 × 30)	Westvaco Nuchar WV—L (8 × 30)	Witco 517 (12 × 30)
Surface area, m²/g	600–650	950–1050	1000	1050
Apparent density, g/cm³	0.43	0.48	0.48	0.48
Density, backwashed and drained, lb/cu ft[a]	22	26	26	30
Real density, g/cm³	2.0	2.1	2.1	2.1
Particle density, g/cm³	1.4–1.5	1.3–1.4	1.4	0.92
Effective size, mm	0.8–0.9	0.8–0.9	0.85–1.05	0.89
Uniformity coefficient	1.7	≤1.9	≤1.8	1.44
Pore volume, cm³/g	0.95	0.85	0.85	0.60
Mean particle diameter, mm	1.6	1.5–1.7	1.5–1.7	1.2
Sieve size (U.S. standard series)				
Larger than No. 8, maximum %	8	8	8	—
Larger than No. 12, maximum %	—[b]	—	—	5
Smaller than No. 30, maximum %	5	5	5	5
Smaller than No. 40, maximum %	—[b]	—	—	—
Iodine No.	650	900	950	1000
Abrasion No., minimum	Not available	70	70	85
Ash, %	Not available	8	7.5	0.5
Moisture as packed, maximum %	Not available	2	2	1

[a]lb/cu ft × 16.02 = kg/m³.
[b]Not applicable for this carbon size.

TABLE 14.7   Properties of Commercially Available Carbons

1971a, 1971b). Some factors influencing adsorption at the carbon and liquid interface are:

- Attraction of carbon for solute;
- Attraction of carbon for solvent;
- Solubilizing power of solvent for solute;
- Association;
- Ionization;
- Effect of solvent on orientation at interface;
- Competition for interface in presence of multiple solutes;
- Coadsorption;
- Molecular size of molecules in the system;
- Pore size distribution in carbon;
- Surface area of carbon; and
- Concentration of constituents.

When an adsorbable solute molecule (adsorbate) contacts an unoccupied adsorption site on the carbon surface, the molecule adheres almost instantly. As the solution passes over a bed of granular carbon, the adsorption rate initially is rapid and subsequently becomes slower. Granular carbons require more time for their adsorptive potential to be exhausted than pulverized or powdered carbons. The time increment for a 0.5- to 1.00-mm (12 × 30 mesh) carbon to use an appreciable portion of the total adsorption capacity can be several hours; a powdered carbon, if mixing is adequate, will achieve its adsorptive potential within 1 hour.

In addition to the adsorptive behavior of activated carbon, other properties should be considered in designing an adsorption system. Physical properties of interest include density, PSD, porosity, surface area, hardness, ignition temperature, and total ash. Other characteristics, such as oil retention, conductivity, and carbon composition, also may deserve consideration, depending on the application.

Two typical size ranges of commercial GAC are available: 8 × 30 mesh and 12 × 30 mesh. Granular carbons also can be purchased in sizes as small as 20 × 50 mesh or as large as 4 × 6 mesh. In most instances, an 8 × 30 mesh carbon is used in upflow packed (fixed) beds and downflow carbon columns to minimize operating headloss. For the same reason, a 12 × 30 mesh carbon typically is used in upflow expanded carbon columns.

3.3.3 Types of Carbon Adsorption Units

Several types of activated carbon contactor systems are used in facility designs. Carbon adsorption columns either of the pressure or of the gravity type can be used as an upflow countercurrent type operated with packed or expanded carbon beds or as an upflow or downflow fixed bed unit having two or three columns in series.

3.3.3.1 Upflow Columns Upflow columns are arranged so that the liquid moves vertically upward. The wastewater inlet is at the bottom of the column and the liquid outlet at the top. As carbon adsorbs organics, the apparent density of carbon particles increases, encouraging migration of heavier or more spent carbon to the bottom of the column. Upflow columns may have more carbon fines in the effluent than downflow columns because the bed tends to expand, not compress, the carbon. Bed expansion creates fines because carbon particles collide, causing particle attrition. These fines then escape through passageways created by the expanded bed.

3.3.3.2   Downflow Columns   Downflow carbon columns typically consist of two or three columns operated in series. Columns are piped and valved to allow operation of contactors in a countercurrent mode. The operational advantage of this design is that two processes—adsorption of organics and filtration of suspended solids—are accomplished in a single step. Claims that capital costs are lower for downflow columns are questionable, as they require extensive interconnecting valves and piping to permit interchanging the relative positions of individual contactors in the adsorption system. Also, because a filter composed solely of carbon is a surface-type filter, increased headloss requiring more frequent backwashing often results. Finally, physical plugging of carbon pores with suspended materials may require premature removal of the carbon for regeneration and increase the probability of ash buildup, thus decreasing the useful life of the carbon.

3.3.3.3   Fixed and Expanded Beds   Process water flows through steel or concrete contactor beds in which carbon granules remain fixed in the downflow mode or in which carbon separates to form an expanded bed in the upflow mode. Fixed beds remove particulates (if present in wastewater influent), thus requiring backwashing to dispose of accumulated particulate material. Typically, fixed beds use downward flow to lessen the chance of accumulating particulate material at the bottom of the bed where it would be difficult to remove by backwashing. Sand and gravel resting on a filter block form the supporting media for downflow contactors.

The upflow column leads to development of the moving, or pulse, bed in which wastewater flows upward through a descending fixed bed of carbon. When the adsorptive capability of the carbon at the bottom of the column is exhausted, it is removed, and an equivalent quantity of regenerated or virgin carbon is added to the top of the column.

Expanded beds provide a degree of organic removal similar to that achieved by fixed beds, but they require less pumping pressure and downtime. Aeration of carbon surfaces also may be provided more readily.

3.3.3.4   Countercurrent Adsorption   When impurities that are difficult to adsorb are present in large concentrations, removal efficiency requirements for a single-stage unit may dictate the use of more carbon than is practical. Countercurrent techniques may reduce the amount of carbon used, however, because carbon in equilibrium with a dilute solution of partially treated wastewater may remove more contaminants from concentrated wastewater.

The operating principle of this process involves using two separate beds of carbon (sometimes more than two are used). A quantity of wastewater is partially treated by a once-used carbon. This twice-used carbon is discarded, and the once-treated wastewater receives a second treatment with sufficient virgin or regenerated carbon to produce the required effluent quality. The process then is repeated for additional wastewater. By exposing the carbon to different concentrations of wastewater constituents, the amount of material held by a unit weight of carbon is increased, thus decreasing the amount of carbon required.

3.3.4   Unit Sizing

The sizing of carbon contactors is based on four factors: contact time, hydraulic loading rate, carbon depth, and number of contactors.

The carbon contact time, typically calculated on the basis of the volume of the column occupied by the activated carbon, typically ranges from 15 to 35 minutes, depending on the application, wastewater constituents, and desired effluent quality. For tertiary treatment applications, carbon contact times of 15 to 20 minutes typically are used where effluent quality limits require a chemical oxygen demand (COD) of 10 to 20 mg/L; 30- to 35-minute contact times are used where the requirement is 5 to 15 mg/L. For chemical and physical treatment facilities, carbon contact times of 20 to 35 minutes typically are used, with a typical contact time of 30 minutes.

A standard column (Figure 14.38) is a vessel having a flat, conical, or dished head and a carbon-retaining screen and supporting grid installed in the bottom. The minimum height-to-diameter ratio of a column is typically 2:1.

Hydraulic loading rates of 10 to 20 m/h (4 to 10 gpm/cu ft) of the cross section of the bed typically are used for upflow carbon columns. For downflow carbon columns,

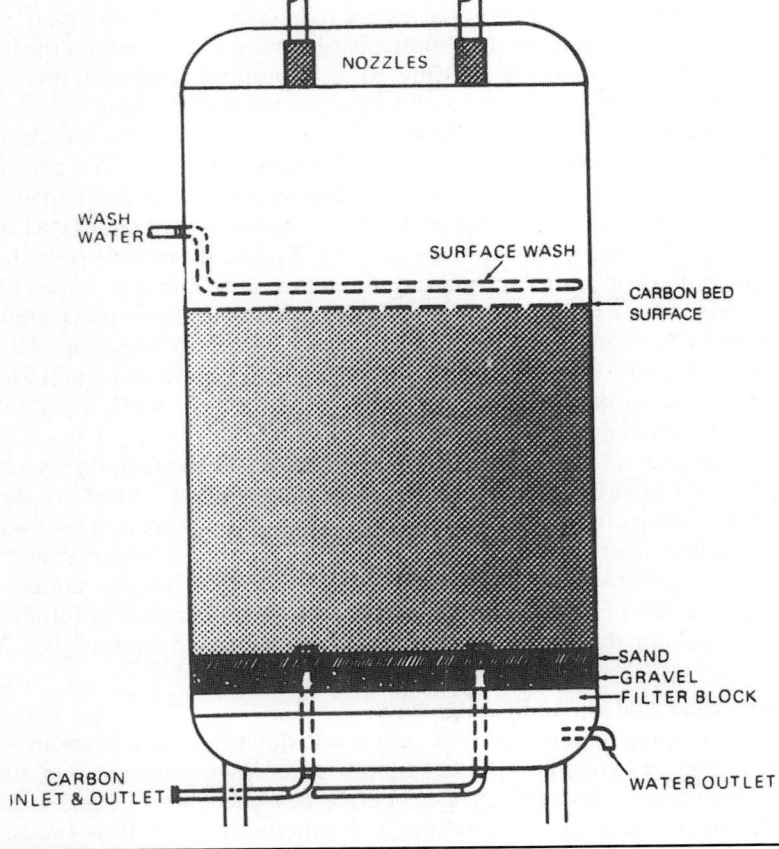

FIGURE 14.38   Typical downflow contactor (U.S. EPA, 1973).

hydraulic loading rates of 7 to 12 m/h (3 to 5 gpm/cu ft) are used. Actual operating pressure seldom is more than 7 kPa (1 psi) for each 0.3 m (1 ft) of bed depth.

Bed depths vary, typically within a range of 3 to 12 m (10 to 40 ft), depending primarily on carbon contact time. A minimum carbon depth of 3 m is advisable. Typical total carbon depths range from 4.5 to 6 m (15 to 20 ft). Freeboard must be added to the carbon depth to allow a bed expansion of 10% to 50% during backwash or expanded bed operation. Carbon particle size and water temperature determine the required quantity of backwash water to attain the desired level of bed expansion.

A minimum of two parallel carbon contactor units is necessary for a facility of any size. Two units in series also are advisable to permit removal of a spent carbon unit while the facility remains in operation. The number of contactors should be sufficient to ensure enough carbon contact time to maintain effluent quality while one column is offline during removal of spent carbon for regeneration or maintenance.

3.3.5 Backwashing

Backwashing consists of exposing the column media to a fluid flow sufficient to remove solids that either have accumulated in the bed or have been created by abrasive action. The rate and the frequency of backwash depend on hydraulic loading, the nature and concentration of suspended solids in the wastewater, carbon particle size, and adsorber type (expanded or fixed bed). Backwash frequency may be specified arbitrarily (each day at a specified time) or determined based on operating criteria (headloss or turbidity). Backwash duration is typically 10 to 15 minutes. The equipment used for backwashing and control is similar to that used for the GMF system.

One of the alternatives for carbon contacting systems is the downflow of wastewater through the carbon bed (Figure 14.39) (Pretorius, 1972). The principal reason for selecting a downflow contactor is dual-purpose use of carbon for adsorption of organics and filtration of suspended materials. A disadvantage associated with downflow systems is that provisions are necessary for backwashing beds periodically to relieve the pressure drop resulting from accumulation of suspended solids. Otherwise, continuous operation for several days without thorough backwashing eventually compacts or fouls beds. Normal quantities of backwash water are less than 5% of the product water for a filter 0.8 m (2.5 ft) deep and 10% to 20% for a filter 4.5 m (15 ft) deep. Typical backwash flowrates for granular carbons of 8 × 12 or 12 × 30 mesh are 29 to 50 m/h (12 to 20 gpm/cu ft) (U.S. EPA, 1973).

Removal of solids trapped in a packed upflow bed may require two steps: (1) relieving the bottom surface plugging by temporarily operating the bed in a downflow mode; and (2) flushing out by expansion suspended solids entrapped in the middle of the bed. When upflow filters are backwashed, additional time and a proportionally larger volume of higher quality water may be required to avoid plugging the bottom of deep beds. Often, backflow of packed upflow carbon contactors, preceded by filtration, merely consists of doubling the normal contactor flowrate for 10 to 15 minutes (U.S. EPA, 1973).

3.3.6 Valve and Piping Requirements

Valve and piping requirements for upflow and downflow contactors are similar. Upflow units are piped to operate either as upflow or as downflow units and allow backwashing. Downflow units are piped to operate downflow and in series. Each column is valved to be backwashed individually. Furthermore, downflow series contactors are

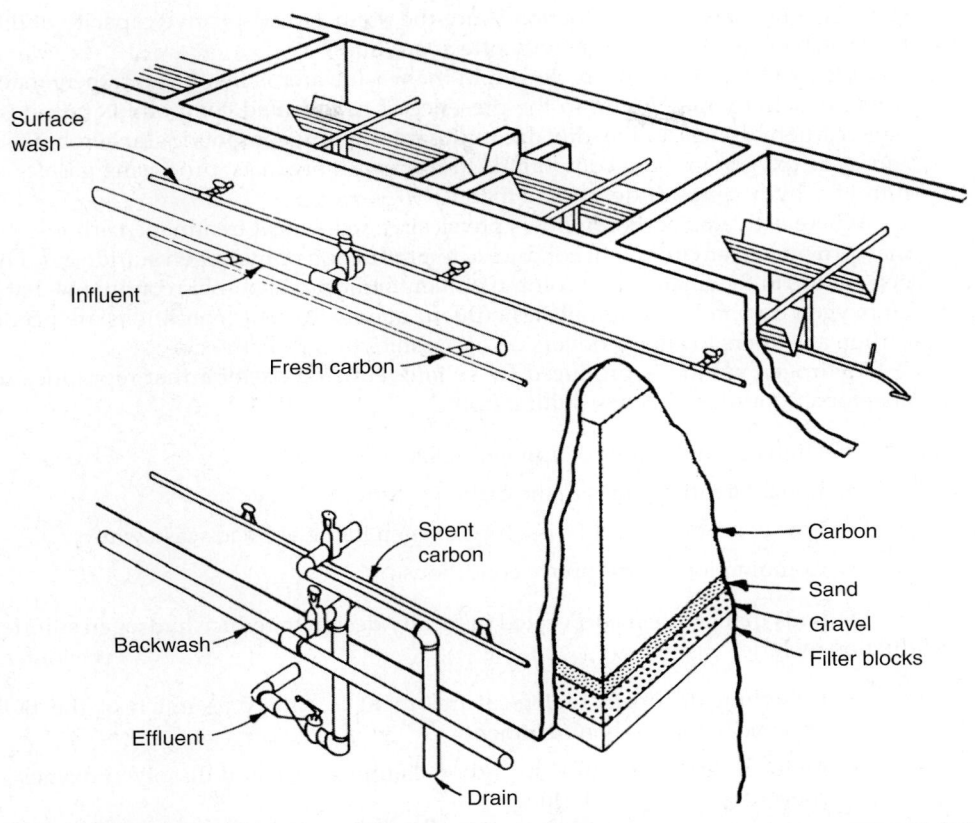

FIGURE 14.39 Typical downflow gravity contactor (U.S. EPA, 1973).

valved and piped so that the respective position of individual contactors may be interchanged, typically referred to as a lead-lag configuration.

3.3.7   Instrumentation
Individual carbon columns should be equipped with flow and headloss-measuring devices. Flow measurements serve to equalize flows through the individual carbon columns and determine the actual carbon contact time.

3.3.8   Control of Biological Activity
A close relationship between adsorption phenomena and some aspects of biological behavior has been observed because both biomass and activated carbon can adsorb and remove certain dissolved substances from wastewater. Therefore, removal of 50% to 100% more organics in full-scale columns than that predicted by laboratory tests is not unusual because of biomass accumulation within the column.

Carbon adsorbents may catalyze biological processes, thereby increasing buildup. Therefore, removal of organics in full-scale columns frequently is greater than that suggested by laboratory evaluation. After the predicted adsorptive capacity of the carbon is exhausted, biological activity often continues.

If the supply of dissolved oxygen in the wastewater is insufficient, then anaerobic bacterial activity may occur. In the presence of oxygenated compounds (nitrates, sulfates, carbohydrates) and readily decomposable organic compounds, anaerobic bacteria cause the oxygen in these compounds to react with organics, producing gases such as nitrogen, hydrogen sulfide, and methane.

Where activated carbon follows physical and chemical treatment, carbon columns may provide an environment conducive to production of hydrogen sulfide gas. Hydrogen sulfide in the final carbon column effluent indicates anaerobic conditions that exert an oxygen demand that diminishes effluent quality. Aerobic conditions are needed in carbon absorbers to allow conversion of organics to carbon dioxide.

Hydrogen sulfide is produced by sulfate-reducing bacteria that reproduce under anaerobic conditions likely resulting from:

- High concentrations of applied BOD_5;
- Long detention times in the carbon column;
- Low concentrations of dissolved oxygen in the applied wastewater; or
- Combinations of the above conditions.

Methods that can be incorporated in facility design to reduce hydrogen sulfide production include:

- Providing upstream biological treatment to satisfy as much of the BOD as possible before carbon treatment;
- Reducing detention time in carbon columns based on dissolved oxygen (DO) concentrations of the effluent;
- Ensuring a higher DO concentration in the influent to carbon columns;
- Backwashing columns frequently;
- Chlorinating carbon column influent (this is not the preferred alternative because chlorine carbon reactions destroy carbon, thereby increasing carbon usage); and
- Introducing an oxygen source such as air or hydrogen peroxide in upflow expanded beds to keep columns aerobic. (In expanded upflow columns, some biological growth is flushed through the bed and, if necessary, may be removed by downstream filters or settling basins. Based on cell mass produced by the introduction of gaseous or liquid oxygen compounds, use of other chemical additives as electron acceptors instead of oxygen in packed columns merits consideration.)

3.3.9   Carbon Transport

Spent carbon must be removed from carbon absorbers. In downflow contactors, removal provisions are straightforward because all of the carbon is removed at the same time and the column is refilled with fresh or regenerated carbon. Care should be taken to prevent entry of the gravel or stone-supporting media used in downflow contactors to the carbon transport system.

In upflow pulsed beds, only 5% to 25% of total carbon column charge is removed at any given time to achieve maximum loading. Spent carbon is removed from the bottom of the vessel and a fresh carbon charge, equal in volume to the spent carbon removed, is added to the top of the vessel. This procedure ensures that fresh carbon is positioned to "polish" the effluent before it leaves the adsorber, and that the carbon containing the most adsorbate of the most spent carbon is removed for regeneration. It is important to obtain a uniform withdrawal of carbon over the entire horizontal surface area of the carbon bed. A cone on the base of the column (Figure 14.40) with water jets can achieve uniform carbon removal (Pretorius, 1972).

Activated carbon typically is transported hydraulically as shown in Figure 14.41. Spent carbon is transported from carbon columns and dewatered in a tank. The dewatered carbon is then conveyed by an inclined screw conveyor to the carbon regeneration furnace. In this screw conveyor, additional dewatering occurs as the carbon moves along the incline from the bottom toward the top, allowing water to flow against the direction of carbon transport. Water from the bottom of the screw conveyor returns to the process, while dewatered carbon discharges to the top of the furnace. Regenerated carbon exits the furnace to a quench tank where the carbon is cooled, wetted, and then transported to carbon de-fining tanks. After carbon fines are removed, the carbon is returned to the carbon columns. Makeup carbon is introduced

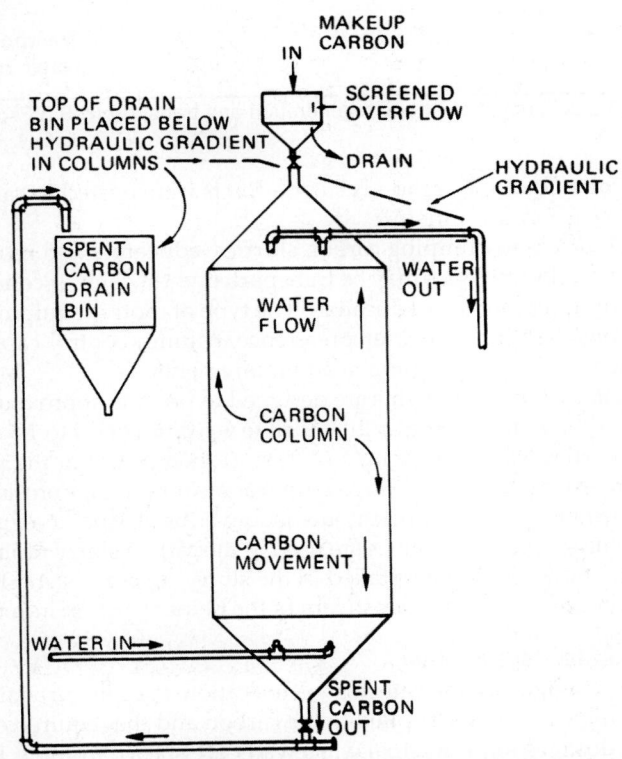

Figure 14.40   Carbon transfer with upflow column in service (U.S. EPA, 1973).

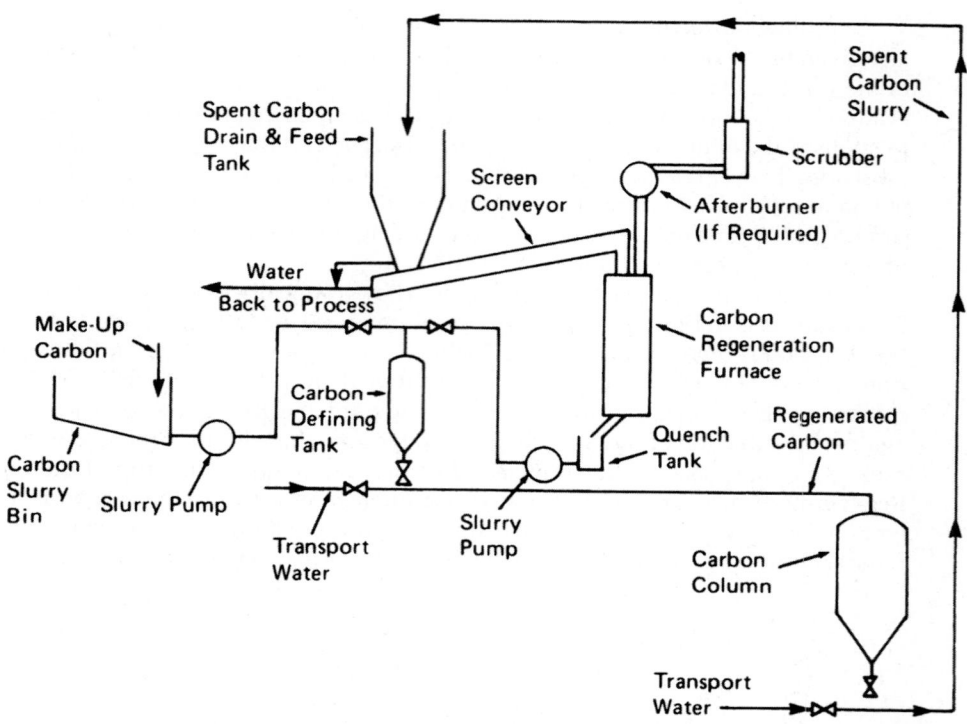

FIGURE 14.41   Schematic of carbon regeneration and transport (U.S. EPA, 1973).

to a slurry bin. Once the carbon is slurried, it is transported to carbon de-fining tanks and then to carbon columns.

As alternatives to pumping carbon slurries, educators and motive water can transport carbon. Carbon slurries may be transported with water or compressed air, centrifugal or diaphragm pumps, or educators. The type of motive equipment selected requires balanced consideration of owner preference, column control capabilities, capital and maintenance costs, and pumping head requirements.

Carbon slurry piping systems are designed to provide approximately 8 L of transport water/kg carbon removed (1 gal/lb). Pipeline velocities of 0.9 to 1.5 m/s (3.5 to 5 ft/s) are best. At velocities less than 0.9 m/s (3 ft/s), carbon settles in the pipeline; at velocities greater than 3 m/s (10 ft/s), excessive carbon abrasion and pipe erosion occur. Long-radius elbows or tees and crosses with cleanouts should be used at points of pipe direction change. Typically, plug- or ball-type valves are used in the carbon slurry piping system. No valves for throttling flow should be installed in the slurry piping system. Flow control is maintained by throttling inlet water upstream of the point of carbon introduction.

3.3.10   Carbon Regeneration

The carbon dosage or use rate for regeneration equipment sizing depends on the strength of the wastewater applied to the carbon and the required effluent quality. Typical carbon dosages for municipal wastewater are shown in Table 14.8.

Prior Treatment	Typical Carbon Dosage Required/mil. gal Column Throughoutput,[a] lb/mil. gal[b]
Coagulated, settled, and filtered activated-sludge effluent	200–400
Filtered secondary effluent	400–600
Coagulated, settled, and filtered raw wastewater (physical-chemical)	600–1800

[a]Loss of carbon during each regeneration cycle is typically 5% to 10%. Makeup carbon is based on carbon dosage and the quality of the regenerated carbon.
[b]lb/mil. gal × 0.119 8 = g/m³.

Table 14.8 Typical Carbon Dosages for Various Column Wastewater Influents

3.3.10.1 Carbon Dewatering Dewatering of spent carbon slurry before thermal regeneration typically is accomplished in drain vessels. Screens in drainage vessels allow transport water to flow from the carbon. Gravity drainage vessels dewater the carbon to 40% to 55% moisture content. Typically, two drain bins are provided to permit continuous carbon feed to the furnace.

Dewatering screws also may be used to dewater activated carbon to approximately 50% moisture. Such systems must include a bin to provide a continuous supply of carbon to the screw and maintain a positive seal on the furnace.

3.3.10.2 Regeneration Furnace Partially dewatered carbon may be fed to the regeneration furnace with a screw conveyor (Figure 14.42) (Bishop et al., 1967; U.S. EPA, 1973) equipped with a variable-speed drive to control precisely the rate of carbon feed. Carbon feed rate is controlled because moisture content and quantities of adsorbed organics vary.

Anticipated carbon dosage governs theoretical furnace capacity. For multiple-hearth furnaces, approximately 5 mm² of hearth area is required for each gram (dry weight) of carbon per day (0.025 cu ft/d/lb). Actual furnace capacity typically includes an allowance for furnace downtime of approximately 40% beyond theoretical capacity.

Based on operating experiences at two full-scale facilities, the furnace should have provisions to add approximately 1 kg of steam per kilogram of carbon regenerated (1 lb/lb). Fuel requirements for the furnace are 7000 kJ/kg (3000 Btu/lb) of carbon when regenerating spent carbon for tertiary and secondary effluent applications. To this value, energy requirements for steam and an afterburner should be added if required.

The furnace is designed to control carbon feed rate, rabble arm speed, and hearth temperature. Off-gases from the furnace must be within acceptable air pollution standards. Air pollution control equipment, designed as an integral part of the furnace, includes a scrubber for removing carbon fines and an afterburner to ensure complete combustion of gases. Regenerated carbon exits the bottom of the furnace to a quench tank. The quench tank cools the regenerated carbon and provides a positive seal on the furnace. Water jets are provided in the quench tank to assist removal of regenerated carbon, which is then conveyed to the de-fining carbon tanks.

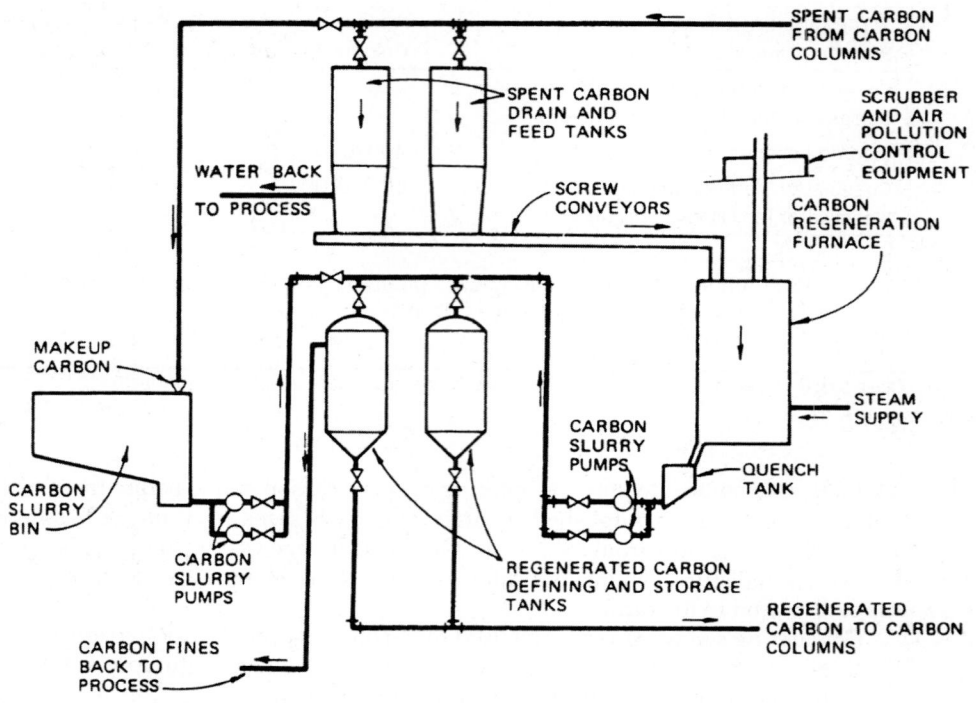

FIGURE 14.42   Schematic of carbon regeneration system (U.S. EPA, 1973).

Carbon losses vary considerably in the carbon transport and regeneration system. System carbon losses are typically 5% to 10% per regeneration cycle, depending on facility size and furnace type. Furnace losses may vary from 3% to 7% during regeneration, depending on furnace type and operating efficiencies.

4.0   Chemical Treatment

Chemicals are used for a variety of municipal treatment applications, including to enhance flocculation/sedimentation, condition solids, add nutrients, neutralize acid-base, precipitate phosphorus, and disinfect or to control odors, algae, or activated sludge bulking. This section describes chemicals use for advanced treatment processes. For chemicals used for conventional processes and biological treatment, refer to Chapter 7.

Chemicals also can be used for precipitation of heavy metals, but are better controlled at the source through an industrial pretreatment program. Therefore, metals precipitation will not be further discussed in this chapter. Similarly, nutrient addition is principally an industrial waste problem that will not be discussed in this chapter. Flocculation, sedimentation, solids conditioning, odor control, and disinfection are discussed in other chapters of this manual. Discussion of chemical treatment in this chapter will, therefore, be limited to phosphorus precipitation and neutralization.

4.1   Phosphorus Precipitation

As an essential element in the metabolism of organic matter, phosphorus is necessary for the proper functioning of biological waste treatment processes. However, when excessive phosphorus is present, it may pollute the receiving body of water by causing excessive growths of rooted or floating aquatic facilities that deplete dissolved oxygen and have adverse aesthetic effects.

Phosphorus may occur in the form of organic phosphorus found in organic matter and cell protoplasm, as complex inorganic phosphates (polyphosphates), such as those used in detergents, and as soluble inorganic orthophosphate (PO_4^{3}). As the final breakdown product in the phosphorus cycle, the orthophosphate form is most readily available for biological use or precipitation by a metal salt.

When considering removal of phosphorus from wastewater, its form and solubility are important. Phosphorus enters a WRRF in all three forms. During the treatment process, most of the organic phosphate and complex phosphates are converted to inorganic orthophosphate. In wastewater that has been treated by biological processes, most of the phosphorus is in the soluble form, although a small amount of insoluble organic phosphorus also exists in the form of cell protoplasm. For example, the average amount of orthophosphate discharged in the effluent of municipal trickling-filter facilities without chemical precipitation has been estimated at approximately 24 mg/L PO_4^{3} (8 mg/L phosphorus) (Griffith et al., 1973), which needs to be controlled in light of recent regulations. Phosphorus removal via chemical (metal) precipitation should include more details concerning metal-to-P ratio (mole), the use of online instrumentation to achieve low effluent limits. In effluent from facilities with chemical precipitation for removal, most of the remaining phosphorus is in the insoluble form (calcium, aluminum, or iron phosphate). Insoluble phosphorus compounds typically do not release phosphorus in other units of the WRRF or in the receiving water body.

In activated sludge, the required minimum phosphorus-to-BOD_5 ratio has been estimated as 1:100, with similar ratios expected for other biological processes. Municipal wastewater has a BOD_5 in the range of 175 to 250 mg/L and a phosphorus content of 3 to 5 mg/L or higher, thus exceeding the nutrient requirements for aerobic biological treatment.

4.1.1   Phosphorus Removal Methods

Phosphorus removal may be part of primary, secondary, or tertiary treatment processes. Physical and chemical treatment techniques include chemical precipitation and flocculation of phosphorus followed by sedimentation, flotation, or filtration. Incidental uptake by biological processes typically accounts for relatively little phosphorus removal unless the systems are designed for phosphorus removal. Chapter 12 discusses biological phosphorus removal methods.

4.1.2   Precipitants

Phosphorus precipitation typically requires addition of chemicals and widely used coagulant aids (flocculant), and a coagulant (Daniels, 1975a, 1975b) can be used. Chemicals typically used for phosphorus precipitation are lime, alum, sodium aluminate, ferric chloride, ferrous sulfate, and polyaluminum chloride. Some examples on chemical reactions when using lime, alum, sodium aluminate and ferric chloride are shown below.

4.1.2.1 Lime

In addition to its reactions with carbonate species, lime reacts with orthophosphate to precipitate hydroxyapatite, according to the following stoichiometric reaction:

$$5Ca^{2+} + 4OH^{-2} + 3HPO_4^- \rightarrow Ca_5OH(PO_4)_3 + 3H_2O$$

The theoretical molar calcium-to-phosphorus ratio is 5:3. The above equation is representative, but not exact, because the composition of the apatite precipitate varies. As a result, the calcium-to-phosphorus mole ratio may vary from 1.3 to 2.0. Hydroxyapatite solubility decreases rapidly with increasing pH and, typically, phosphate removal increases with increasing pH. Almost all orthophosphate converts to the insoluble form if pH is greater than 9.5.

The actual pH required to precipitate a given amount of phosphate and the amount of lime required to raise the pH to the desired level vary with wastewater composition. These parameters should be determined by laboratory jar tests. Figure 14.43 shows a typical curve with phosphorus removal as a function of lime dosage.

The chief variable that affects the lime dose for phosphorus removal is the wastewater alkalinity. Unless a high pH is used, waters with low alkalinity (150 mg/L or less) form a poorly settleable floc because of the small fraction of dense calcium carbonate $(CaCO_3)$ precipitate. Sufficient quantities of calcium carbonate act as a flocculant that enhances settling of the hydroxyapatite. For wastewater with high alkalinity, a pH of 9.5 to 10 can result in excellent phosphorus removal.

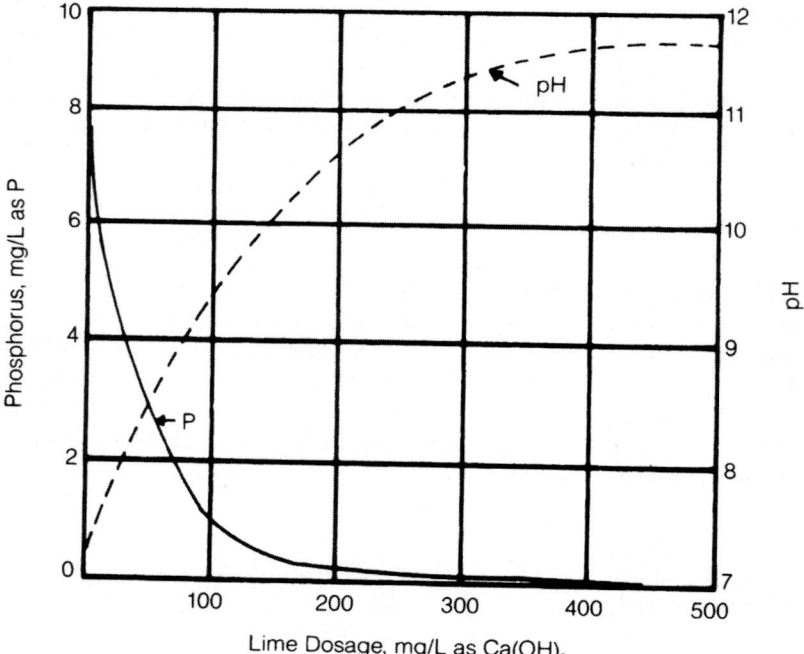

FIGURE 14.43   Typical phosphorus removal curve with an influent phosphorus concentration of 10 mg/L as P assumed.

Magnesium hardness also affects the efficiency of phosphorus removal. At higher pH, magnesium hydroxide precipitates according to the following reaction:

$$Mg^{2+} + Ca(OH)_2 \rightarrow Mg(OH)_2 + Ca^{2+}$$

This reaction begins at a pH of approximately 9.5 and is complete at a pH of 11. The magnesium hydroxide precipitate is gelatinous and removes fine suspended solids as it settles. However, these same gelatinous properties can impair solids dewatering.

Because chemical requirements impose the principal operational cost associated with phosphorus removal, proper chemical dosages are critical to process economy. The lime dosage required to reach a given effluent phosphorus level is independent of influent phosphorus concentration. The degree of phosphorus removal is a function of pH. Lime requirements to reach the required pH correlate more closely with wastewater alkalinity than with any other variable. For other phosphorus precipitants (alum or ferric salts), chemical requirements are proportional to influent phosphorus concentration. In contrast, phosphorus removal is decidedly nonstoichiometric with other chemicals.

4.1.2.1.1 Process-Train Considerations Various lime precipitation schemes exist for phosphorus removal (Figure 14.44). Lime may be added before the primary sedimentation tank in a biological WRRF. Because an excessively high pH interferes with the biological process, lime addition to the primary sedimentation tank ahead of an activated sludge system is limited to a pH of approximately 9.0, with a maximum phosphorus insolubilization of approximately 80%. At this limited pH, 2 to 3 mg/L of remaining

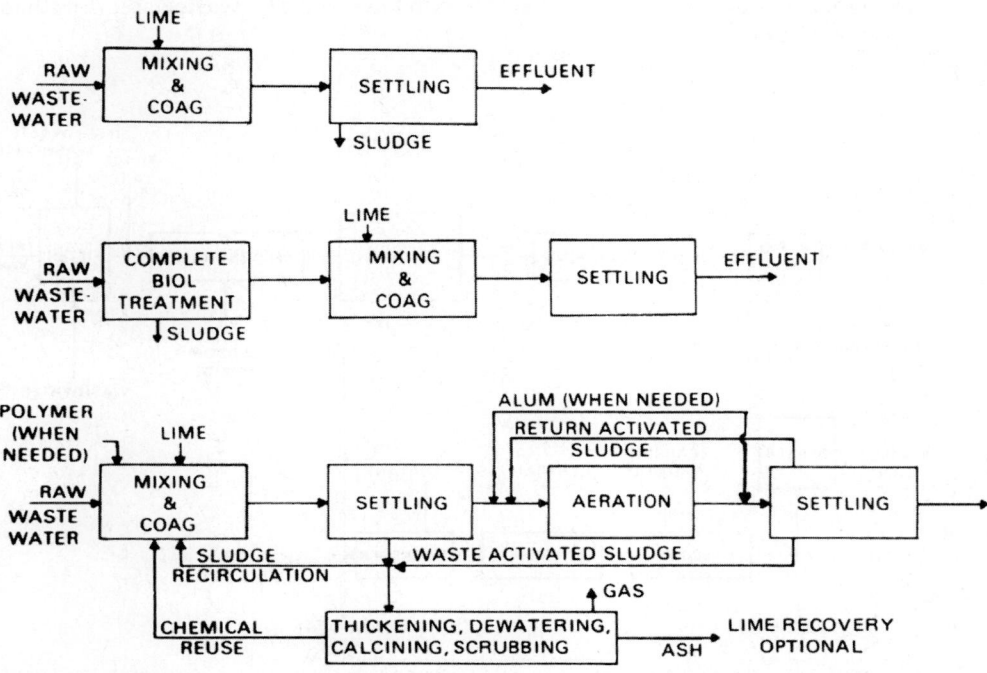

FIGURE 14.44   Various schemes of single-stage lime precipitation for phosphorus removal.

soluble phosphorus is not unusual. Additional phosphorus may be removed, if necessary, by using aluminum or iron addition in aeration tanks or the final sedimentation tank. The use of lime in primary treatment has the added advantage of increasing organics and suspended-solids removal efficiencies in the primary sedimentation tank, thereby decreasing the load on the aeration system.

A second alternative consists of lime treatment following biological treatment. Phosphorus removal from the secondary effluent ensures enough phosphorus in the aeration stage to meet the nutrient demand of the biological floc. In addition, the biological system breaks down many of the complex phosphates to the more readily precipitated orthophosphate form. However, the high pH of returned solids may affect biological treatment.

The tertiary lime treatment system may be either the one-stage or the two-stage type. Typical process flowsheets for one- and two-stage treatment systems are shown in Figures 14.45 and 14.46. The choice of a particular process depends on the degree of phosphorus removal required, the alkalinity of the wastewater, and lime decalcination. Typically, when high percentages of phosphorus removal are required or wastewater alkalinity is low, a high process pH is required. High phosphorus removal efficiencies call for the two-stage process because it allows recovery of calcium carbonate sludge for recalcination. Two-stage removal also allows better control of clarification. With low-alkalinity water, solids settling may be difficult in the first stage, even with a high pH. The second-stage sedimentation tank, with its dense calcium carbonate precipitate, may help eliminate solids carryover.

A variation of the two-stage process involves recirculation of settled solids from the second stage to the mixing tank before first-stage settling. The recirculated solids act as a weight agent that is especially effective with low-alkalinity wastewater (less than 150 mg/L as $CaCO_3$).

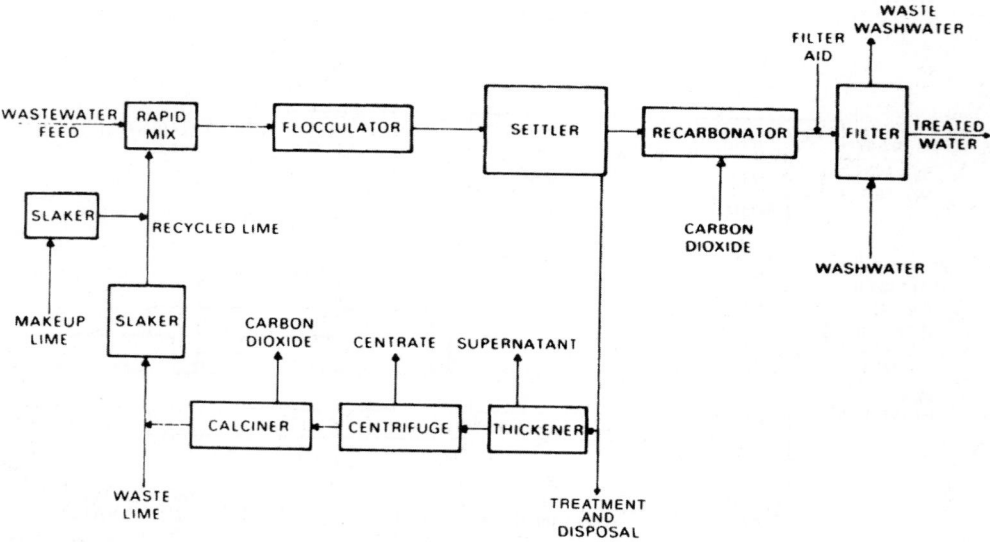

FIGURE 14.45   Single-stage lime treatment system.

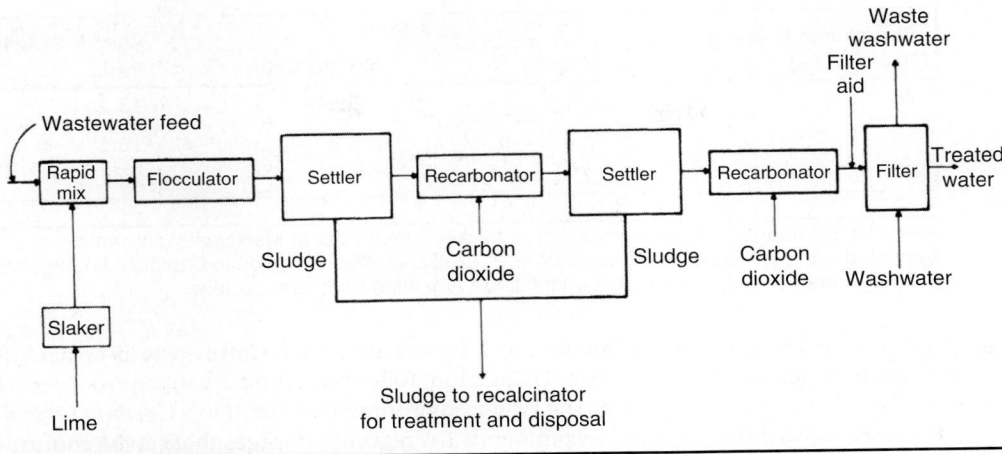

Figure 14.46 Two-stage lime treatment system.

4.1.2.1.2 Dosages and Removals Lime precipitation of phosphorus may require filtration to ensure continuous compliance with effluent requirements. Even with a process pH as high as 11.4, the effluent may contain some residual phosphorus. Although high pH values ensure that virtually all phosphorus is insolubilized, effluent total phosphorus depends on suspended-solids removal efficiency. In cases where the precipitate floc is difficult to settle, granular-medium filtration can ensure an extremely high degree of suspended-solids and phosphorus removal. Typically, a residual phosphorus concentration of 0.1 to 0.2 mg/L may be achieved with granular-medium filtration.

4.1.2.2 Alum Phosphorus may be precipitated from wastewater by chemical reaction with most mineral salt coagulants, including various salts of iron and aluminum. The principal source of aluminum in phosphorus precipitation is aluminum sulfate, typically known as alum. The theoretical stoichiometry of the chemical reaction is:

$$Al_2(SO_4)_3 \cdot (14H_2O) + 2H_2PO_4^- + 4\,HCO_3^- \rightarrow 2AlPO_4 + 4CO_2 + 3SO_4^{2-} + 18H_2O$$

The sulfate ion remains in solution, and pH is depressed. From the above equation, the calculated weight ratio of alum to phosphorus is 0.87Al:1.0P. However, much more alum actually is required because of side reactions involving wastewater alkalinity and organic matter (Table 14.9). The actual weight ratio of alum to phosphorus is up to 3.0Al:1.0P and higher.

The $Al_{0.8}(H_2PO_4)(OH)_{1.4}$ (molecular weight = 142.4 g/mol) is assumed to represent the precipitate formed after aluminum addition, and $Al(OH)_3$ (molecular weight = 5 78 g/mol) is the excess aluminum hydroxide formed.

The solubility of aluminum phosphate is a function of pH. The most efficient chemical use occurs at a process pH near the range of minimum solubility, approximately 5.5 to 6.5 (Table 14.10). Because alum use results in a small pH depression and most existing treatment systems operate at near-neutral wastewater pH, addition of alum almost automatically results in a pH in the range of minimum aluminum phosphate solubility. Sometimes excess alum is required to depress the pH sufficiently to reach the

Phosphorus Reduction Required (%)	Al:P Ratio		Alum:P Weight Ratio
	Mole Ratio	Weight Ratio	
75	1.38:1	1.2:1	13:1
85	1.72:1	1.5:1	16:1
95	2.3:1	2.0:1	22:1

TABLE **14.9**   Ratios of Alum to Phosphorus for Alum Treatment of Municipal Wastewater. Reprinted with permission from Rushton, J. H. (1952) Mixing of Liquids in Chemical Processing. *Ind. Eng. Chem.*, **44** (12), 2931. Copyright 1952 American Chemical Society.

optimal pH range for phosphate removal. Excess alum then simply acts as an acid that might be replaced by another acid. Using alum following a lime treatment process is an example of such an application. The Water Environment Federation's *Operation of Nutrient Removal Facilities*, MOP 37, recommends the required dosages that can be confirmed by jar tests at the project site (WEF, 2013).

4.1.2.2.1 Process-Train Considerations   Alum addition offers flexibility in the location of chemical application points. Alum can be added before the primary settling tank, in the aeration tank, or between aeration and final sedimentation.

Alum addition before the primary settling tank provides additional removal of suspended solids and organic matter in addition to phosphorus precipitation. These removals, however, result from reactions that compete with soluble phosphorus for available alum, thereby increasing the dosage required to obtain a given phosphorus residual level. In addition, some polyphosphates that are more difficult to treat may be present in raw wastewater.

Providing alum addition in the activated sludge aeration tank allows use of the mixing already provided for that system. The best point of addition for alum in an activated sludge facility is likely to be in the aeration basin effluent channel that carries mixed liquor to the final settling basin. Turbulence in this channel adequately mixes the chemical.

Addition of alum after the biological system takes advantage of wastewater stabilization with the attendant hydrolysis of the complex phosphates to the more readily reacted orthophosphate form. Figure 14.47 illustrates the effect of point of addition on added aluminum. An added advantage is that dosage can be based on the phosphorus remaining after biological uptake.

4.1.2.2.2 Dosages and Removal   The Water Environment Federation's *Operation of Nutrient Removal Facilities* (WEF, 2013) describes information on the practical method for dosages and removal, the basics of which are provided below. The starting point

pH	Approximate Solubility of AlPO$_4$ (mg/L)
5	0.03
6	0.01
7	0.3

TABLE **14.10**   Solubility of Aluminum Phosphate versus pH. Reprinted with permission from Rushton, J. H. (1952) Mixing of Liquids in Chemical Processing. *Ind. Eng. Chem.*, **44** (12), 2931. Copyright 1952 American Chemical Society.

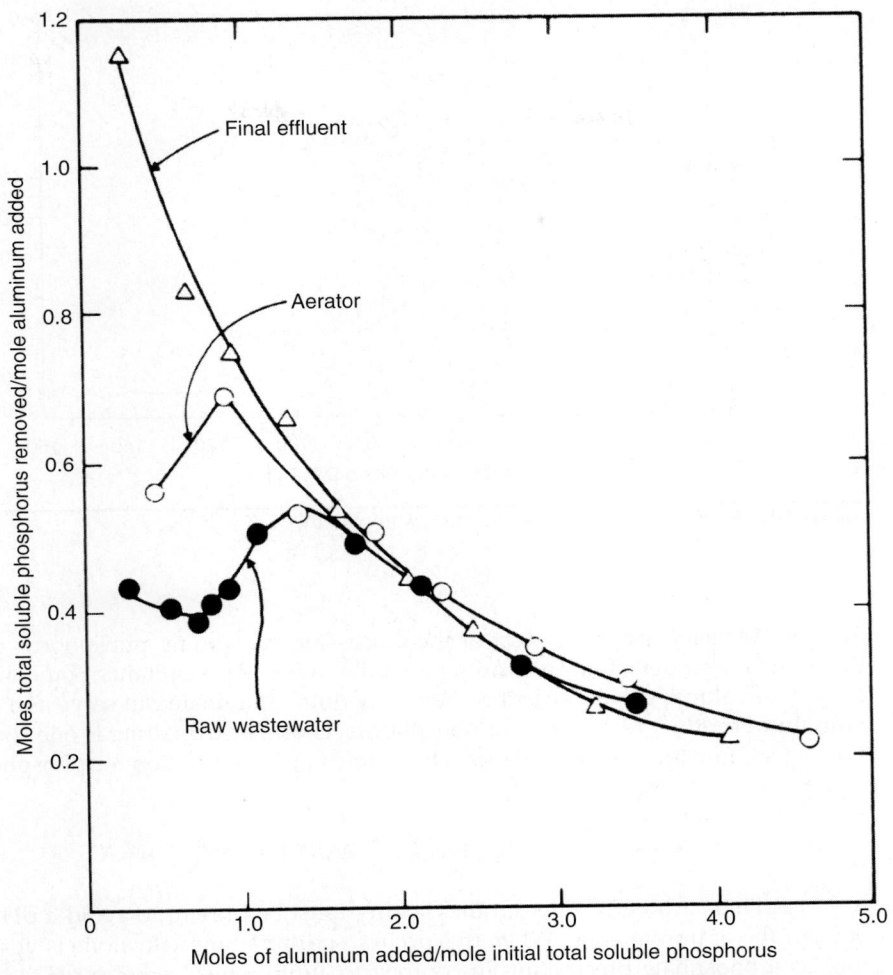

FIGURE 14.47   Influence of point of addition on phosphorus removal with aluminum.

for determining chemical dosages for alum and other mineral precipitants is the stoichiometry of the reactions involved. In the case of lime, the degree of phosphorus removal depends directly on system pH. For aluminum and iron salts, phosphorus removal efficiency varies directly with chemical dosage up to the point where mole requirements (molecular weight in grams of any particular compound) for phosphate precipitation and side reactions have been satisfied. Optimum dosages cannot be calculated readily because of ambiguity of the reactions involved. As a result, laboratory jar tests may be used to determine actual chemical requirements (Figure 14.48). Visible flocculation tends to occur when sufficient coagulant has been added to remove all phosphorus and lower the pH to less than 6.5. More detailed information on the required dosages can be found in the literature on the nutrient removal published by the WEF and others.

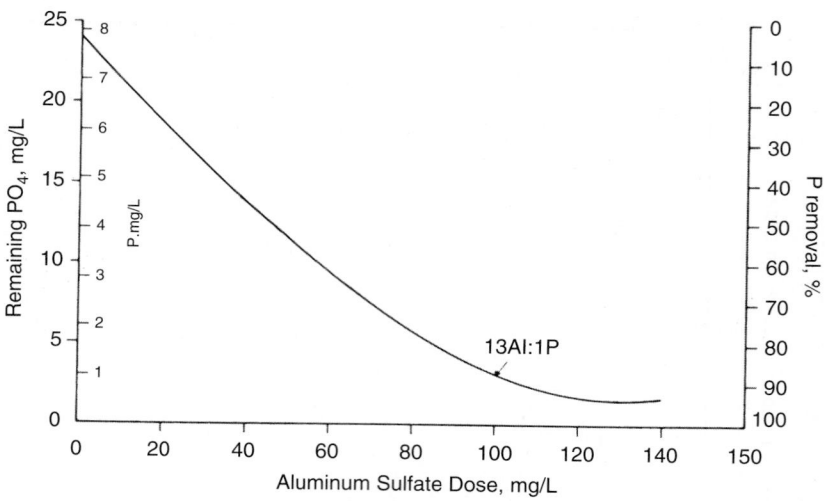

FIGURE 14.48   Typical phosphorus reduction with alum.

4.1.2.3   Sodium Aluminate   Most of the discussion concerning phosphorus precipitation with aluminum sulfate also applies to other mineral precipitants. Some differences for sodium aluminate are described herein. Sodium aluminate can serve as a source of aluminum for the precipitation of phosphorus. Granular trihydrate is one commercial form of sodium aluminate. The theoretical stoichiometric reaction for phosphorus precipitation is:

$$Na_2Al_2O_4 + 2H_2PO_4^- + 4CO_2 \rightarrow 2AlPO_4 + 2NA^+ + 4HCO_3^-$$

Dissolved carbon dioxide or other acidity must be present to avoid a pH increase beyond the optimum zone. When this occurs, sodium aluminate alone is not satisfactory as a phosphate precipitant. In contrast to alum, which reduces pH, addition of sodium aluminate results in a slight increase in pH. Therefore, use of this chemical may be appropriate where wastewater pH is already low, and further depression should be avoided. From the above equation, the calculated sodium aluminate-to-phosphorus weight ratio is approximately 3.6:1. In practice, side reactions will require a higher dosage.

The $Al_{0.8}(H_2PO_4)(OH)_{1.4}$ (molecular weight = 142.4 g/mol) is assumed to represent the precipitate formed after sodium aluminate addition, and $Al(OH)_3$ (molecular weight = 78 g/mol) is the excess aluminum hydroxide formed.

Typically, experience has shown the performance of sodium aluminate to be somewhat inferior to that of alum based on equivalent Al:P mole ratios. Typical curves for a moderately alkaline wastewater comparing phosphorus removal achieved by the two chemicals are shown in Figure 14.49. The effect illustrated, however, does not universally apply to all wastewaters. Therefore, laboratory jar tests are necessary for comparative evaluation of the two aluminum compounds. In either case, the final pH must be 5.5 to 6.5 to achieve optimal phosphate precipitation results.

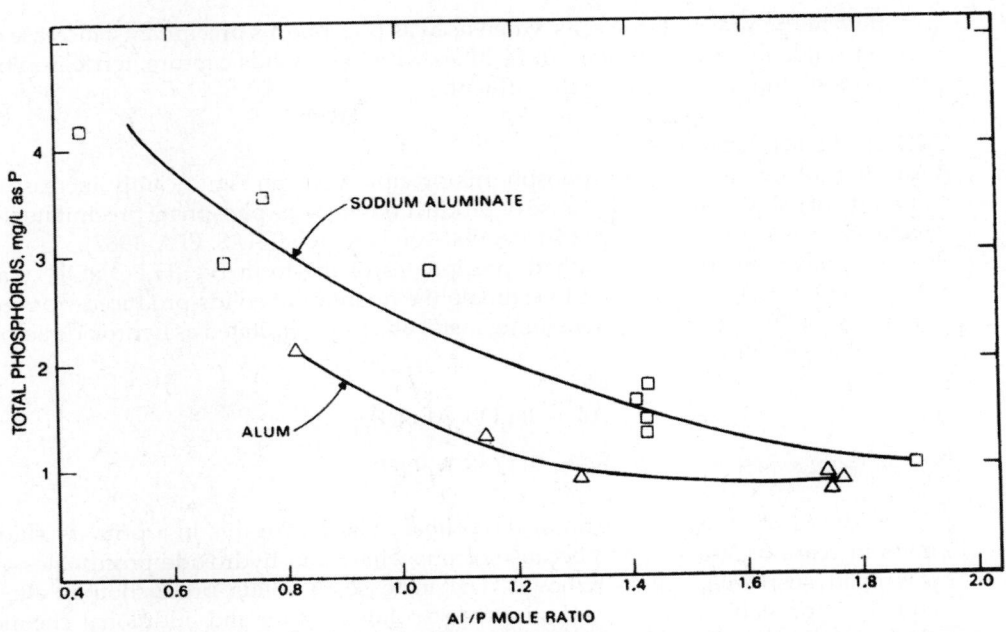

FIGURE 14.49   Comparison of phosphorus removal effectiveness of alum and sodium aluminate.

4.1.2.4 Ferric Chloride   Both ferric and ferrous iron compounds may be used in the chemical precipitation of phosphorus. Although both types of compounds produce equivalent results, ferric chloride is used more often.

The dominant theoretical stoichiometric reaction between ferric chloride and phosphorus is believed to be similar to that of phosphate and alum:

$$FeCl_3 \cdot (6H_2O) + H_2PO_4 + 2HCO_3^- \rightarrow FePO_4 + 3Cl^- + 2CO_2 + 8H_2O$$

This equation indicates that an iron-to-phosphorus mole ratio of 1:1 is required. This corresponds to an iron-to-phosphorus weight ratio of 1.8:1. However, similarly to alum, practically much greater amounts of ferric chloride are required to satisfy side reactions with alkalinity that produce $Fe(OH)_3$. Alkalinity reactions occur because the well-flocculating ferric hydroxide precipitate aids settling of the colloidal ferric phosphate.

Experience has shown that efficient phosphorus removal requires the stoichiometric amount of iron to be supplemented by at least 10 mg/L of iron for hydroxide formation. Typically, iron requirements for municipal wastewater are 15 to 30 mg/L as Fe (45 to 90 mg/L as $FeCl_3$) to reduce phosphorus by 85% to 90%. Dosages vary with influent phosphorus concentration. The optimum pH range for iron precipitation of phosphorus is between 4.5 and 5.0. As with aluminum salts, iron salts may be added during primary, secondary, or tertiary treatment. The actual weight ratio of iron to phosphorus is up to 10.0 to 15.0 Fe:1.0 P and higher.

The $Fe_{1.6}(H_2PO_4)(OH)_{3.8}$ (molecular weight = 251 g/mol) is assumed to represent the precipitate formed after ferric iron addition, and $Fe(OH)_3$ (molecular weight = 106.8 g/mol) is the excess ferric hydroxide formed.

Both aluminum and iron salts, when used as phosphorus precipitants, increase dissolved solids in facility effluent. In facilities with poor solids capture, ferric iron may impart a slight reddish color to the effluent.

4.1.3   Solids Considerations

Addition of mineral salts for phosphorus precipitation can significantly increase the quantity of solids generated because of production of metal-phosphate precipitates and metal hydroxides and improved suspended-solids removal (U.S. EPA, 1987).

As noted previously, phosphate precipitates of the form $Fe_{1.6}(H_2PO_4)(OH)_{3.8}$ and $Al_{0.8}(H_2PO_4)(OH)_{1.4}$ can be used to estimate the quantity of solids produced from precipitation of phosphorus. The remaining metal will be precipitated as hydroxide according to these reactions:

$$Al^{3+} + 3H_2O \times Al(OH)_3 + 3H^+$$

$$Fe^{3+} + 3H_2O \times Fe(OH)_3 + 3H^+$$

Addition of metals upstream of the primary clarifier results in a primary sludge mass increase of 50% to 100% because of phosphate and hydroxide precipitates and improved suspended-solids removal. The increase in solids production is almost equally attributable to improved suspended-solids capture and additional chemical sludge generation (Knight et al., 1973). Overall facility solids mass increase is smaller because of decreased secondary sludge production from improved primary removals. For example, a 60% to 70% increase is typical across the entire facility.

For metal addition to secondary processes, waste mixed-liquor solids mass may increase by 35% to 45%, and the overall facility solids mass increase may be 5% to 25%. Metal addition to either primary or secondary treatment units not only increases solids mass but also solids volume because settled solids concentration in clarifiers may decrease by as much as 20%.

As shown in the previous two reactions, each mole of cation beyond the quantity precipitated with phosphorus should react with three moles of water to produce one mole of metal hydroxide and three moles of hydrogen ions. Therefore, one milligram of alum, $Al_2(SO_4)_3 \cdot 14H_2O$, will react to produce 0.26 mg of insoluble aluminum hydroxide while consuming 0.5 mg/L of alkalinity as calcium carbonate. One milligram of ferric chloride $FeCl_3 \cdot (6H_2O)$ will produce approximately 0.4 mg of ferric hydroxide and consume 0.56 mg of alkalinity as calcium carbonate. Alkalinity reductions are important design considerations for low-alkalinity waters or nitrified effluent. During nitrification, significant alkalinity reductions occur and additional chemical treatment that further reduces alkalinity should be evaluated carefully.

The metals dosing needs to be controlled to minimize potential negative effects in case of overdosing. As an example, the overdose of iron coagulants affects the performance of facilities, which are using UV disinfection due to the high UV lights absorptivity by iron in case of overdose.

4.2   pH Adjustment

One of the most common types of chemical processes used in WRRFs is pH adjustment. Adjustment of pH, frequently used for coagulation and phosphorus precipitation, simply raises or lowers pH to a more acceptable value. For example, wastewater that is excessively acidic or alkaline is objectionable in collection systems, WRRFs, and natural streams. Removal of excess acidity or alkalinity by chemical addition to provide a final pH of approximately 7 is called neutralization. Most effluents must be neutralized to a pH of 6 to 9 before discharge.

There are three critical components of any pH control system: mixing intensity or turnover time in the reactor; response time of the control system; and the ability of the chemical metering system to match process requirements. If any one of these components is not properly designed, then significant problems in system performance may occur.

Other factors that complicate the design of pH control systems include the amount of buffering capacity in wastewater, the change in mass flowrate of the hydrogen ion, and variations in wastewater flowrate or temperature.

Before designing a pH control system, a thorough wastewater characterization study should be performed. If characteristics are not clearly defined, additional sampling and analysis should be considered. However, in almost all cases, titration curves should be developed to provide definitive data with respect to chemical demands and consumption.

Methods used for pH adjustment are selected on the basis of overall cost because material costs and equipment needs vary widely. The volume, kind, and quantity of acid or alkali to be neutralized or partially removed are also variables that influence selection of a chemical agent.

Neutralization capabilities of alkaline reagents used to treat acidic wastes vary. For comparison, each alkali may be assigned a basicity factor, defined as the weight of a specific alkali equivalent in acid-neutralizing power to a unit weight of calcium oxide (CaO). Basicity factors for more common alkaline reagents are shown in Table 14.11. To

Chemical	Formula	Acidity or Alkalinity (Expressed as $CaCO_3$) Required to Neutralize 1 mg/L, mg/L	Neutralization Factor, Assuming 100% Purity of All Compounds
			Basicity
Calcium carbonate	$CaCO_3$	1.0	1.0/0.56 = 1.786
Calcium oxide	CaO	0.560	0.56/0.56 = 1.000
Calcium hydroxide	$Ca(OH)_2$	0.740	0.74/0.56 = 1.321
Magnesium oxide	MgO	0.403	0.403/0.56 = 0.720
Magnesium hydroxide	$Mg(OH)_2$	0.583	0.583/0.56 = 1.041
Dolomitic quicklime	$[(CaO)_{0.6}(MgO)_{0.4}]$	0.497	0.497/0.56 = 0.888
Dolomitic hydrated lime	$[Ca(OH)_2]_{0.6}[Mg(OH)_2]_{0.4}$	0.677	0.677/0.56 = 1.209
Sodium hydroxide	NaOH	0.799	0.799/0.56 = 1.427
Sodium carbonate	Na_2CO_3	1.059	1.059/0.56 = 1.891
			Acidity
Sulfuric acid	H_2SO_4	0.98	0.98/0.56 = 1.75
Hydrochloric acid	HCl	0.72	0.72/0.56 = 1.285
Nitric acid	HNO_3	0.63	0.63/0.56 = 1.125

TABLE 14.11 Neutralization Factors for Common Alkaline and Acid Reagents

properly use this table, the neutralizing capacity and solubility of the reagent must be considered when comparing alkaline reagents. No direct correlation between pH and acidity exists. Adjusting the pH of or neutralizing an acidic process waste stream by adding an alkaline reagent requires development of a titration curve (American Public Health Association [APHA] et al., 2017). From the titration curve, the acidity values as milligrams per liter of calcium carbonate may be determined for any desired pH level. For example, to determine the amount of base required to neutralize an acidic process waste stream to a pH of 7, the acidity at the pH 7 endpoint is used. Concentration factors for various alkaline reagents shown in Table 14.11 may then be used to design neutralization systems for each reagent.

In those processes that demonstrate a significant variation in hydrogen ion mass flowrate, it is often necessary to use cascade neutralization processes of two to four stages. This is a problem most frequently encountered in industrial wastewater treatment; however, if tight pH control is required, cascade control offers significant improvements in process control and reliability.

4.2.1 Neutralization of Acidity
Many acceptable methods may be used to neutralize or adjust overacidity in a raw waste or process waste stream. These methods include:

1. Mixing acid and alkaline waste so that the net effect is a nearly neutral pH.
2. Passing acid wastewater through beds of limestone (if the waste stream does not contain metal salts or sulfuric or hydrofluoric acids that coat the limestone).
3. Mixing acid waste with lime slurries or dolomitic lime slurries.
4. Adding the proper amounts of concentrated caustic soda (NaOH), soda ash (Na_2CO_3), or magnesium hydroxide $Mg(OH)_2$ to acid wastewater.

Mixing acid and alkaline waste typically is not possible in municipal facilities, and the use of limestone beds requires bed replacement, which is a significant drawback. Therefore, only the third and fourth methods are used in municipal wastewater treatment and discussed here.

4.2.1.1 Limes Alkalies
The most common lime alkalies used for acid neutralization and pH adjustment are high calcium lime or quicklime (CaO); hydrated or slaked lime $[Ca(OH)_2]$; and dolomitic limes, mixtures, or compounds of calcium and magnesium oxides or hydroxides.

Calcium and magnesium oxides are considerably less expensive than sodium alkalies and are used more widely. Because these oxides are only moderately soluble in water, they are typically slurried. Because of their limited solubilities, contact with the waste stream must be maintained for a significant period, with agitation, for the required reactions to occur. Many insoluble calcium salts formed by neutralization reactions coat unreacted lime particles, thereby stopping the reaction before the reagent is completely used. The insolubility of the products leads to the formation of a large volume of solids that must be settled or filtered and then either reclaimed or disposed. The formation of this sludge, however, also can act as a coagulant to entrain organic or colloidal material, thus removing pollutants.

4.2.1.2 Sodium Alkalies   Caustic soda and soda ash are the two primary sodium alkalies of interest. Both are highly soluble in water; thus, handling and feeding are convenient and readily adaptable to automatic control. They rapidly react with acidic waste streams.

When combined with most acids, sodium alkalies produce soluble neutral salts and far less sludge than other neutralizing alkaline reagents. This advantage is offset by their higher cost. Caustic soda is a stronger alkali than soda ash and may be used to produce high pH levels in a waste stream. However, either alkali is suitable for the removal of free acidity. Caustic soda is available in anhydrous form or solutions of various concentrations. Soda ash is purchased as a dry granular material. Caustic soda requires precautions in handling to avoid contact burns. Soda ash is a mild alkali, but some precautions also must be taken in its handling and use.

Anhydrous caustic soda is available but typically is not considered practical in water and wastewater treatment applications. Thus, it will not be further discussed.

4.2.1.2.1 Caustic Soda   Designers of a caustic soda system must consider the nature of liquid caustic soda solutions (viscosity, vapor pressure, concentration, solubility, solidification temperature, and heat evolved during mixing) and shipping, unloading, storage, transmission (piping), and feed systems. Liquid caustic soda is produced and shipped in concentrations of 50% and 73%—the percentage denoting the approximate content of sodium hydroxide. Solutions solidify at different temperatures, thus necessitating certain variations in unloading, handling, and storage methods. The most typically used liquid caustic soda contains approximately 50% actual sodium hydroxide; the concentration actually varies between 48% and 52%. The liquid, approximately 50% heavier than water, has a specific gravity that decreases from 1.53 to 1.47 as temperature increases from 10°C to 100°C (50°F to 212°F). Because the actual sodium hydroxide concentration in commercial 50% solutions may be as high as 52%, storage tanks, when required, should be kept higher than 21°C (70°F). At 16°C (60°F), the 50% grade will contain 0.784 kg NaOH/L caustic soda solution (6.54 lb/gal). The transfer piping should be insulated and heat traced to avoid crystallization.

Both 50% and 73% liquid caustic soda can be delivered in bulk quantities by tank cars and tank trucks. Shipments by rail are in well-insulated 30 000- or 38 000-L (8000- or 10 000-gal) tank cars lined with a caustic-resistant material that protects the chemical from contamination in transit. Most cars are equipped with steam coils for heating. Bulk liquid caustic soda shipped by truck is available in 11 000- or 13 200-L (3000- or 3500-gal) capacities. Because trucks typically are used only for deliveries within 300 km (200 mi) of a production point, coils for steaming are not required.

Most tank cars in liquid caustic service may be unloaded either through a bottom discharge nozzle or through a dome by means of an interior siphon pipe extending to a depression in the bottom of the car. The solution can flow through the bottom discharge by gravity to a pump or directly to the storage tank either with or without the aid of air pressure. Air pressure typically assists unloading through the interior siphon pipe.

When unloading caustic liquor, proper precautions and safety measures should be observed. Because a 50% caustic soda solution begins to freeze at approximately 12°C (54°F) and a 73% caustic soda solution begins to freeze at approximately 62°C (144°F), steaming (typically at a maximum pressure of 172 kPa [25 psig]) may be necessary to aid in tank car unloading. For ease of unloading, the temperature of the caustic soda should be higher than 21°C (70°F) in "50% cars" and higher than 77°C (170°F) in "73% cars." Typically, vertical cylindrical steel tanks, which minimize heating, insulation, and field-erection problems, are used for storing liquid caustic soda, typically as a 50% solution. Where storage tanks are located indoors, limited headroom or the presence of other equipment may dictate the use of prefabricated horizontal tanks, typically limited to no more than a 80 000-L (20 000-gal) capacity.

Because of the heat and corrosive nature of caustic soda, transfer lines, fittings, pumps, heat exchangers, and feed lines should be selected carefully.

Further dilution of liquid caustic soda below storage strength may be desirable for feeding by volumetric feeders. A typical feeding system is shown in Figure 14.50.

4.2.1.2.2 Sodium Carbonate (Soda Ash)   Four forms of sodium carbonate that are commercially available are anhydrous, monohydrate, heptahydrate, and decahydrate compounds. Only solid forms are commercially available.

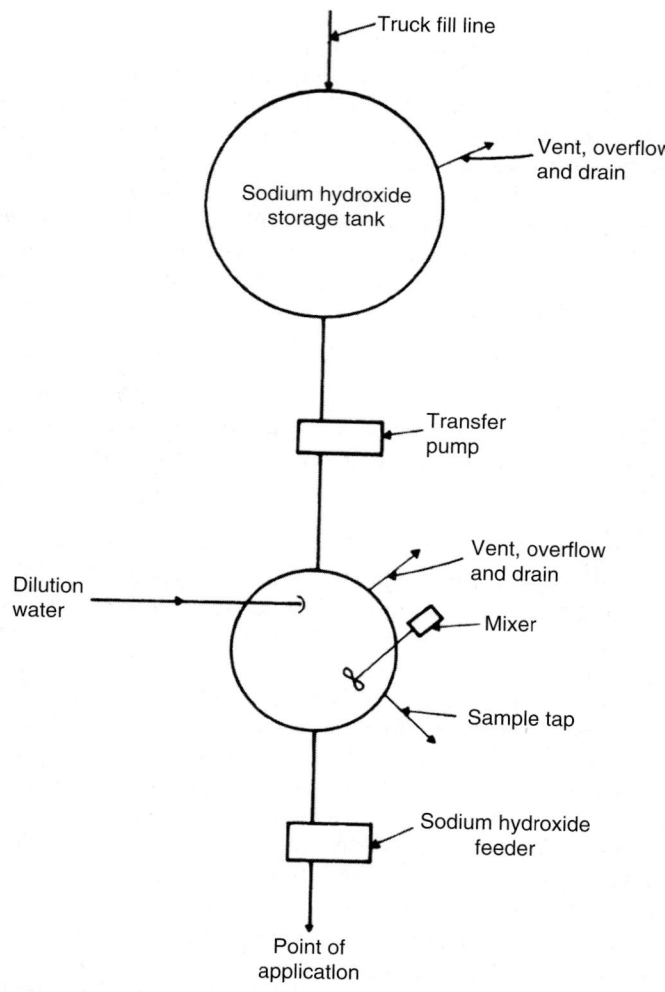

FIGURE 14.50   Typical feed system for caustic soda.

Sodium carbonate monohydrate, $NaCO_3(H_2O)$, contains 85.48% sodium carbonate and 14.52% water of crystallization. It separates as small crystals from saturated aqueous solutions higher than 36°C (96°F). It loses water when heated, and its solubility decreases slightly with increasing temperature. Sodium carbonate heptahydrate, $Na_2CO_3 \cdot (7H_2O)$, contains 45.7% sodium carbonate and 54.3% water of crystallization.

Sodium carbonate decahydrate, $Na_2CO_3(10H_2O)$, typically called washing soda, contains 37.06% sodium carbonate and 62.94% water of crystallization. It can be crystallized from saturated aqueous solutions lower than 32°C (90°F) and higher than 2°C (28°F) or by wetting soda ash with the calculated quantity of water in this temperature range. Crystals readily effloresce in dry air, forming principally monohydrate.

Designers of sodium carbonate systems must consider the nature of the sodium carbonate solutions (viscosity, concentration, solubility, crystallization, heat of formation), shipping, storage, transmission, and feeding.

Dissolving anhydrous sodium carbonate or the monohydrate form in water generates heat. When heptahydrate or decahydrate forms are dissolved, heat is absorbed. The quantity of heat evolved or absorbed depends on the concentration of the solution. The solubility of sodium carbonate in water varies irregularly with temperature. The density of sodium carbonate solution decreases with increasing temperature.

Commercial soda ash is available packaged in 50-kg (100-lb) bags or in bulk by truck, hopper car, or barge. Bulk transport by rail is the typical mode of shipment. Hopper cars range in size from 60 m³ (2000 cu ft), holding 32 000 kg (70 000 lb) of light soda ash or 64 000 kg (140 000 lb) of dense soda ash, to more than 140 m³ (5000 cu ft), holding up to 82 000 kg (180 000 lb) of light or 90 000 kg (200 000 lb) of dense soda ash. Most hopper cars are unloaded by gravity.

Handling bagged soda ash is best accomplished by conveyors or on pallets using forklift trucks. Gravity conveying, pneumatic, or vacuum systems can handle bulk soda ash.

For large quantities of soda ash, bins or bunkers are used. Bins may be filled by an open-bottomed screw conveyor or a system of such conveyors arranged longitudinally across the top. These conveyors load progressively from one end of the bin to the other. The cover or roof of the bin should be tight. Dust control for the entire conveyance system may be provided at the bin.

When shipped, commercial soda ash contains 99.2% sodium carbonate. The accepted commercial standard for soda ash is expressed in terms of the equivalent sodium oxide, Na_2O, content. On a weight percentage basis, 100% sodium carbonate contains 58.5% sodium oxide. Thus, 99.2% soda ash is equivalent to 58.0% sodium oxide.

When soda ash is to be stored as a slurry, pumping it directly through a pipeline from the unloading point to the storage tank is frequently convenient. Slurries containing 35% to 40% suspended solids by weight (50% to 60% total soda ash) can be pumped. The more common concentration is 10% to 20%.

Weak solutions containing 5% to 6% dissolved soda ash may be handled the same as water. Strong solutions, in the 20% to 32% range, require special care (heat loss prevention) to avoid crystallization.

Bagged soda ash should not be stored in a damp or humid place. Excessive air circulation in the storage area should be avoided. If bagged soda ash is to be stored for extended periods under adverse conditions, stacks should be covered by a tight-fitting, impermeable sheet. Large quantities of bulk soda typically are stored in steel bins or bunkers. Bins typically are built longer than their width.

Arching, bridging, or chimneying—sometimes encountered when storing light soda ash—may be avoided by installing electric or pneumatic vibrators mounted on the outside of bin bottoms, just above the outlet. The use of vibrators is inadvisable with dense soda ash.

Basic components of a slurry storage system include a tank, means of slurring the bulk soda ash and transferring it to storage, and means of reclaiming solution from the tank and replenishing it with water. One of the most important requirements for successful operation of a slurry storage system is to maintain an operating temperature higher than 9°C (48°F) for a 10% slurry and 23°C (74°F) for a 20% slurry. Cooler temperatures in the slurry bed may result in the formation of hepta-hydrate and decahydrate forms, which are difficult to rediscover. Water used for operating the system should be preheated. Heating coils may be immersed in the bottom of the slurry tank. If the slurry tank is located outdoors, insulation may also be necessary.

Pipelines carrying strong soda ash solutions (30% to 32% at 32°C [90°F] minimum) should be insulated. Where the use point is distant from the storage tank and the use rate is low or intermittent, pipelines may be constructed as a continuous loop, with most of the solution recirculated to the storage tank.

Soda ash will precipitate insoluble compounds from hard water. If hard water is used for slurry and solution handling, undesirable scaling may result.

For continuous feeding of dry soda ash, volumetric, gravimetric, or loss-in-weight gravimetric mechanical feeders may be used. Solution feed may be pumped.

4.2.2 Neutralization of Alkalinity

In some cases, downward adjustment of the pH of a wastewater stream is necessary. Discharge of effluent with a pH greater than 8.5 typically is undesirable and, in many cases, not permissible. High pH in municipal wastewater typically is caused by industrial waste contributions or by the use of lime for phosphorus removal during the treatment process. The pH may be decreased by adding carbon dioxide (recombination) or acid. Recombination is not discussed here because it is rarely practiced in municipal wastewater treatment.

Sulfuric (H_2SO_4), hydrochloric (HCl), nitric (HNO_3), or phosphoric (H_3PO_4) acids are used in wastewater neutralization. Sulfuric acid is the most widely used, but local economic considerations or availability may favor the use of hydrochloric acid. The use of nitric acid is restricted because of effluent nutrient considerations. Although the following discussion typically is based on sulfuric acid system needs, it applies, with minor modifications, to other acids.

No direct correlation between alkalinity and pH exists. Therefore, to determine acid requirements for design purposes, a laboratory titration curve using a pH meter and acid tartan of standardized normality should be performed using a representative sample of the wastewater to be treated (APHA et al., 2017). Typically, the titration endpoint is at approximately pH 7 or 8, not at pH 5 as in a total-alkalinity determination. To neutralize 1.0 mg/L of alkalinity, the required amounts of 100% acid would be 0.98 mg/L for sulfuric acid, 0.72 mg/L for hydrochloric acid, and 0.63 mg/L for nitric acid. Because acids are commercially available in various strengths, these figures must be adjusted by a dilution factor, depending on acid strength. Concentrations of commercial acids in use are 93.19% for sulfuric acid, 31.45% for hydrochloric acid, and 67.2% for nitric acid.

4.2.2.1 Sulfuric Acid   Sulfuric acid used in wastewater treatment may be the 60° B (strength in Baume), 77.7% concentration, or the 66° B, 93.2% concentration (approximate specific gravity of 1.83). This acid is viscous, with a rating of approximately 0.035 N/m² (0.350 Pa) at 10°C. When diluted with water, the significant heat generated calls for precautions. In particular, the acid must always be added to water and not vice versa.

Iron is not vulnerable to attack from 66° B sulfuric acid. Therefore, this acid may be stored in unlined steel tanks with an air vent that is fitted with a dryer. If the concentration of the acid solution is more than 66° B, then a special tank lining is advisable.

At small facilities, acid reagents typically are supplied and stored in carboys, cans, or drums. At larger facilities, they are delivered by road or rail tankers and transferred to storage tanks by gravity, air compressors, or pumps.

4.2.2.2 Hydrochloric Acid   Hydrochloric acid, 22° B, has an average specific gravity of 1.17 and a content of 33% by weight. Polyvinyl chloride tanks or lined steel tanks may be used to store 22° B hydrochloric acid. Polyvinyl chloride tanks may be reinforced externally with polyester glass as an additional precaution.

4.2.2.3 Acid System Design Considerations   Options for acid addition systems or equipment include proportional feed or constant rate, gravity or pressure feed, and concentrated acid feed or dilute acid feed. Proportional feeders may be controlled according to flow or pH using suitable metering and control equipment. Many variables influence the design of an acid-feed system, including type and quantity of acid to be fed, purchase and installation costs, labor requirements, and method of control.

Constant-rate, acid-feed systems may consist of a simple drip or siphon apparatus for small acid quantities such as 8 to 11 L/d (2 to 3 gpd). A filter, preferably of glass wool, should be included in the line because commercial acid typically contains some suspended matter that may clog a drip-feed system. Educator units constructed of acid-resistant materials such as unplasticized PVC also are used for constant feed of small acid quantities.

Larger capacity systems may regulate acid flow from a storage tank to the point of discharge in the wastewater stream by use of controlled air pressure or by a variety of specialized acid-feed pumps. All materials directly contacting the acid must be acid resistant and noncorrosive.

Commercial acid with a variety of strengths is shipped in 19-L (5-gal) glass carboys, tank trucks, or railroad tank cars. The hazardous and corrosive nature of concentrated acids demands special handling, maintenance, and safety considerations.

4.3   Rapid Mixing

Associated with each type of chemical treatment process is a series of chemicals that can be used. In most cases, the success of the chemical process depends on mixing the rapid dispersal of the chemical reagent throughout the waste stream.

Typically, separate mix tanks are preferred for chemical mixing. If the process lacks a separate tank, then the chemical must be added at a point that offers sufficient agitation and time for mixing. Pump suction and discharge lines have been used for this process.

From the engineer's point of view, mixing may be defined as the unit operation used to blend or mingle coagulating chemicals or other materials with water or wastewater to create a nearly homogeneous single-phase or multiphase system. Rapid mixing is a brief operation that is designed to create a quick response. It often precedes flocculation. The computational fluid dynamics (CFD) model can be used to design and select mixers and to estimate dosages of the chemicals.

4.3.1 Impeller Mixers

The most common type of mixing device for wastewater treatment is the rotating impeller mixer. Three groups of impeller mixers are used: paddles, turbines, or propellers. Of these, only turbine and propeller mixers are used for rapid mixing applications.

4.3.1.1 Turbine Mixers   A turbine impeller mixer submerged in a liquid, shown in Figure 14.51, operates like a centrifugal pump without a casing. Most turbine mixers resemble multibladed paddle mixers with short blades turning at high speeds on a shaft typically located in the center of the mixing chamber. Figure 14.52 presents several variations of turbine mixers. Blades may be straight or curved, pitched, or vertical. The impeller may be open, semienclosed, or shrouded. The diameter of the impeller is typically within a range from 30% to 50% of the diameter of the mixing vessel.

In thin liquids, such as domestic wastewater, turbine mixers generate strong currents that persist throughout the chamber. Near the impeller is a zone of rapid currents, high turbulence, and intense fluid shear. Principal currents are radial and tangential. Tangential components induce vortexing and swirling that, for efficient operation, must be stopped by baffles or a diffuser ring.

4.3.1.2 Propeller Mixers   Propeller mixers, similar to the one shown in Figure 14.53, have high-speed impellers and are used for thick solutions. Small propeller mixers revolve at full motor speed, typically 1750 r/min; larger mixers turn at 400 to 800 r/min. Propellers generate currents that are primarily axial and continue through the liquid in a given direction until deflected by the flow or wall of the mixing chamber. Propeller blades vigorously cut or shear the liquid.

Propeller mixers are smaller in diameter than either paddle or turbine mixers, rarely exceeding 460 mm (18 in.) in diameter, regardless of mixing chamber size. In a deep

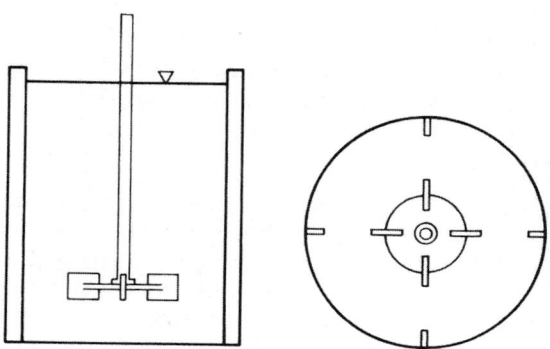

Figure 14.51   Turbine mixer in a baffled tank.

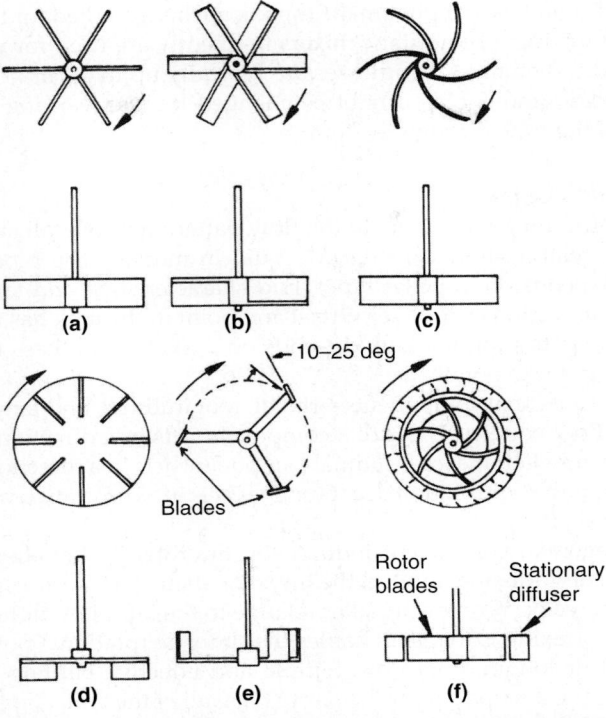

FIGURE 14.52 Turbine impellers: (a) straight blade, (b) 45-degree pitched blade, (c) straight curved blade, (d) vaned disk, (e) radial impeller, and (f) shrouded curved blade with diffuser ring.

chamber, two or more propellers may be mounted on the same shaft; they typically direct the liquid in the same direction.

4.3.2 Other Mixing Devices

Mixing may be accomplished by several other devices, including baffled channels, hydraulic jump mixers, pneumatic mixing by the injection of compressed air, and in-line static mixing devices.

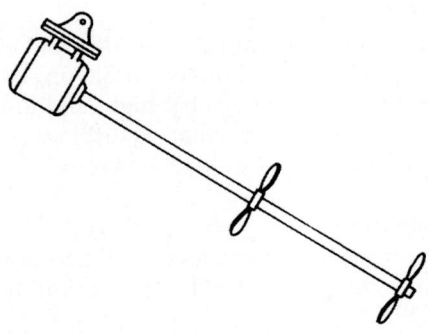

FIGURE 14.53 Propeller mixer.

Baffled channels and pneumatic mixing are better suited for flocculation operations than rapid mixing. In-line static mixers frequently are used for rapid mixing, but they have two disadvantages: headlosses are typically up to 0.9 m (3 ft) and the mean temporal velocity gradient, G, cannot be changed to meet varying requirements but is a function of flowrate through the unit.

4.3.3 Fluid Regimes

The term "fluid regime" refers to the flow pattern and overall summation of the mass flow shear relationships existing in a fluid in motion. The type of flow in a mixing chamber depends on impeller type; fluid characteristics; and tank, baffles, and mixer sizes and proportions. Fluid velocity at any point in the tank has three components; the overall flow pattern in the tank depends on variations in these three velocity components from point to point.

Three orthogonal components—radial, longitudinal, and tangential—conveniently define the flow pattern. The radial component acts in a direction perpendicular to the shaft of the impeller; the longitudinal component acts in a direction parallel to the shaft; and the tangential, or rotational, component acts in a direction tangent to a circular path around the shaft.

The tangential component induced by the rotating impeller promotes rotational movement, or vortexing, around the impeller shaft. This vortexing impedes mixing by reducing the velocity of the impeller relative to the liquid. A preferable method to minimize this vortexing is to install baffles that impede rotational flow without interfering with radial or longitudinal flow. Simple and effective baffling may be achieved by installing vertical strips perpendicular to the wall of the tank. Typically, four baffles suffice to prevent vortex formation, except in large mixing chambers. For turbine mixers, the width of the baffle need not exceed 8.3% of the tank diameter; for propeller mixers, the baffle width should be 5.5% or less of the tank diameter. With side-entering, inclined, or off-center propellers, baffles are unnecessary.

If the combined effects of velocity components of the moving fluid produce laminar flow, then no mixing occurs within the fluid except that caused by diffusion. If flow is turbulent, then fluid particles move in all directions and mixing results primarily from connective displacement. The momentum transfer associated with such displacement generates strong shear stresses within the fluid. Almost all wastewater mixing regimes are in the turbulent flow range.

4.3.4 Design Considerations

The design of a functional, economically feasible rapid mixing system requires consideration of power requirements, laboratory scale-up, batch and continuous systems, hydraulic retention time, vessel geometry, high- and low-speed mixers, propeller and turbine mixers, mixer mounting, top-entering turbines and side-entering propeller mixers, and single-propeller and multipropeller mixers.

4.3.4.1 Power Requirements

The following formula has been developed for determining the power requirements of an impeller mixer to maintain turbulent hydraulic conditions (Reynolds number greater than 10^5) (Rushton, 1952):

$$P = \rho K_T n^3 D^5 / g_c \qquad (14.3)$$

where P = power requirement, N/m·s or Watt (ft/lb$_f$/s);

ρ = mass density of the fluid, 1000 kg/m^3 (62.4 lb/cu ft) for water;

n = impeller revolutions per second, s^{-1};

D = diameter of the impeller, m (ft);

g_c = gravitational acceleration factor, $\dfrac{19.79 \text{ m/kg}}{\text{N} \cdot \text{s}^2} \dfrac{32.17 \text{ ft/lb}_f}{\text{sec}^2 \cdot \text{b}_1}$; and

K_T = constant.

The K_T depends on the impeller shape and size, the number of baffles used to elimi-nate vortexing, and other variables not included in the power equation. Table 14.12 presents K_T values for mixing impellers rotating at the centerline of cylindrical vessels with a flat bottom, four baffles at the vessel wall, baffle widths equaling 10% of the vessel diameter, liquid depth equal to one tank diameter, impeller diameter equal to 30% of the tank diameter, and the impeller positioned one diameter above the tank floor.

Several empirical parameters have been developed to describe performance charac-teristics of the rapid-mixing unit operation. These include the power dissipation func-tion, mean temporal velocity gradient, mixing opportunity parameter, and mixing loading parameter.

The power dissipation function, or power input per unit of mixer volume, provides a rough measure of the mixing effectiveness of the system because more power input creates greater turbulence and greater turbulence leads to better mixing. The mean tem-poral velocity gradient, G (s^{-1}), describes the degree of mixing of the system. As G increases, the degree of mixing increases. In domestic wastewater treatment, values of G typically range from 300 to 1500 s^{-1} for rapid mixing. The power dissipation function in W/m^3 (or ft/lb$_f$/s/cu ft) is expressed as

$$P/V = \mu G \tag{14.4}$$

where V = mixing chamber volume, m^3 (cu ft); and

μ = absolute viscosity of fluid, N/s·m^2 (lb/s/sq ft).

Impeller Type	K_T
Propeller (square pitch, three blades)	0.32
Propeller (pitch of two, three blades)	1.00
Turbine (six flat blades)	6.30
Turbine (six curved blades)	4.80
Turbine (six arrowhead blades)	4.00
Fan turbine (six blades)	1.65
Flat paddle (two blades)	1.70
Shrouded turbine (six curved blades)	1.08
Shrouded turbine (with stator, no baffles)	1.12

TABLE 14.12   Values for K_T Used for Determining Impeller Power Requirements

The mixing opportunity parameter is, in a sense, a ratio of power-induced rate of flow to hydraulic-induced rate of flow. In domestic wastewater treatment, this parameter typically ranges from 9000 to 180 000 (dimensionless) for rapid mixing. The mixing opportunity parameter is expressed as:

$$Gt_d = (PV/\mu)^{0.5}/Q \qquad (14.5)$$

where t_d = hydraulic retention time of mixing basin, s; and
     Q = design flowrate, m³/s.

The mixing loading parameter implies that the hydraulic loading of mixing units is not merely a function of their hydraulic retention time but, more significantly, also a function of power input and viscosity. In domestic wastewater treatment, values of this parameter typically range from 0.03 to 0.0075 s⁻¹. The mixing loading parameter may be expressed (from eq 14.5) as:

$$t_d^{-1} = Q/V = G/(PV/\mu)^{0.5} \qquad (14.6)$$

4.3.4.2 Laboratory Scale-Up

"Scale-up" from a laboratory or bench-scale mixing operation to a full-scale unit poses a significant problem in mixer design. With geometric similarity, one procedure determines, in the laboratory, the hydraulic retention time, power dissipation function, and power function values. These values then are used for the design of a full-scale unit. Typically, the hydraulic retention time and power dissipation function may be determined directly. A graph of power function and Reynolds number also may be prepared on the basis of laboratory testing. The power function corresponding to the Reynolds number that represents best mixing may be determined:

$$\text{Power function } (O) = Pg_c/\rho n^3 D^5 \qquad (14.7)$$

$$\text{Reynolds number } (N_{Re}) = \rho n D^2/\mu' \qquad (14.8)$$

where μ' = absolute viscosity of fluid (lb$_m$/ft·s).

With laboratory data and calculations for O and N_{Re}, the engineer may determine the power requirement, volume, mixer speed, and mixer diameter for a full-scale unit.

4.3.4.3 Batch and Continuous Systems

Batch-mixing systems typically are used in the makeup of chemical solutions. Selection of the proper chamber size depends, to a large extent, on the volume to be mixed and the time allowed. For the makeup of specific chemical solutions requiring rapid mixing, the chemical manufacturer should be consulted to obtain the economical volume and suggested mixing times.

Continuous systems nearly always are required for mixing process waste streams. Where required in the treatment of domestic wastewater, multiple rapid-mixing chambers operated in parallel should be used.

4.3.4.4 Hydraulic Retention Time

Hydraulic retention time in mixing reactors typically varies from 0.5 to 2 minutes. Caution should be used to prevent either undermixing or overmixing. Undermixing results in inadequate dispersal of additives and uneven dosing. Overmixing may rupture wastewater solids already present in the waste stream or cause excessive dispersal of newly formed floc.

4.3.4.5 Vessel Geometry The shape and the size of the tank often are dictated by process considerations. As a rule, circular mixing tanks are more efficient for rapid mixing than square or rectangular tanks. For circular tanks, liquid depths equal to the tank diameter are a good practice.

For tanks less than 4000 L (1000 gal) in capacity, portable or compact turbine mixers are most practical. For larger tanks, heavy-duty, top-entry turbine mixers typically are used.

Squat tanks, that is, tanks with top dimensions greater than liquid depth, have one or more side-entry mixers positioned next to each other. For rectangular tanks, the mixer(s) should enter through one of the narrow walls of the chamber. For all tanks, one or more side-entry mixers are used in lieu of top-entry mixers with extremely long shafts.

4.3.4.6 High- and Low-Speed Mixers Modifying propeller size or pitch can help keep power constant even when speed changes. Typically, the choice between high- or low-speed mixers of equal power depends on the application. High-speed mixers are best for mixing fluids of low viscosity such as domestic wastewater. Low-speed mixers are best for thick and highly viscous fluids or solutions that have a strong tendency to foam.

4.3.4.7 Propeller and Turbine Mixers Propellers are high-speed mixers operated with low horsepower. They are best used as side-entering mixers because they optimize flow more than fluid shear. Propeller speeds range from 400 to 1750 r/min. When the mixer is mounted properly, flow is axial with good top-to-bottom solution turnover. If used as top-entry mixers, propellers should be mounted off-center or at an angle. Baffles are not required for an angle mount, but they are essential for a vertical center mount. Through-bladed propellers typically are used.

Turbines are used primarily in low-speed applications where great energy is required. Turbine speeds range from 56 to 125 r/min. Primary flow is radial and provides good top-to-bottom turnover in a baffled tank. Turbine mixers typically are mounted vertically and in the center of the mixing chamber, 50% to 100% of a diameter above the chamber floor.

4.3.4.8 Mixer Mounting Where top entry is required, the propeller mixer is mounted angled and off-center. Where side entry is necessary, the mixer is mounted horizontally, offset from the centerline of the tank. Flow should run parallel to the long axis of the basin. Power must be sufficient to create a stream velocity strong enough to reach the opposite tank wall with enough momentum to produce fluid flow at that point. For extra large volumes, two or more mixers should be mounted in the same quadrant.

The turbine mixer is always mounted exactly vertically. In an unbaffled tank, the unit is mounted off-center. In a baffled tank, the unit is centered.

4.3.4.9 Top-Entering Turbines and Side-Entering Propeller Mixers For small, open tanks of less than 4000 L (1000 gal), top-entry mixers are best. In this case, either an angle-mounted propeller mixer or a vertically mounted turbine mixer will provide satisfactory service.

Top-entering turbine mixers may be used for tanks of more than 4000 L in capacity. Side-entering mixers typically are used for nonstandard tank geometry and larger tanks.

4.3.4.10   Single-Propeller and Multipropeller Mixers   Propeller size and pitch affect power output. The same power level may be maintained with either single or dual propellers. For equal volumes, vessel shape determines the choice between single or multiple propellers.

Typically, single propellers are used at lower speeds with larger tank sizes and higher flow applications. Dual propellers or multipropellers are used at higher speeds and smaller diameters in vessels with a liquid depth-to-tank top dimension ratio greater than unity.

4.3.5   Chemical Feed Systems

Feeding systems are necessary for the addition of reagents in the form of solid, liquid, or gas to the waste stream at a controlled rate. Design of a chemical feed system must consider the desired state of each chemical to be fed and its particular physical and chemical characteristics, maximum and minimum waste stream flowrates, and the reliability of feeding devices. Table 14.13 provides generalized information about chemical feed systems. Information specific to certain chemical processes is included in the phosphorus precipitation section of this chapter and elsewhere (Daniels, 1975b; Novak and O'Brien, 1975; Priesing, 1962; Stumm and Morgan, 1962; U.S. EPA, 1975b).

Chemicals used in municipal facilities may be received in either liquid or solid form. Coagulants in solid form typically are converted to solution or slurry before entering the waste stream. The dry feeder has numerous forms to handle wide ranges in chemical characteristics, feed rates, and degree of accuracy required. Solution feeding of coagulants depends primarily on liquid volume and viscosity.

Water-soluble coagulant aids, available as dry granular powders or concentrated liquids, are compatible with most dissolved inorganic salts at low concentrations typically found in tap water. Water for preparing flocculation solutions should have a low suspended-solids content to avoid chemical sludge formation. Concentrated solutions of coagulants and flocculants should never be prepared consecutively in the same vessel unless residues are thoroughly removed between uses. Dry coagulant aids should be given enough time to solubilize completely because long, entwined molecules must hydrate fully before they will uncoil completely. Preparation time is decreased when dry particles initially are distributed evenly without large lumps. Then, only minimum agitation is required after dispersion to ensure a uniform solution. Flocculant solutions, unlike primary coagulants, are viscous and exhibit non-Newtonian flow. Therefore, ordinary regulating devices are inadequate, and pumps and lines should not be sized on the basis of Newtonian flow properties.

The capacity of a chemical feed system is an important consideration for storage and feeding. Storage capacity designs must take into account the economy of quantity purchase versus the disadvantages of construction cost and chemical deterioration with time. Potential delays and chemical use rates also merit careful consideration. Design of storage bins or tanks for chemicals must account for the angle of repose of the chemical and its necessary environmental requirements such as temperature and humidity. The size and slope of feeding lines and construction materials are other important considerations.

Chemical feeders must accommodate minimum and maximum feeding rates required. Manually controlled feeders have a common range of 20 to 1, but this range may be increased to approximately 100 to 1 with dual-control systems. Chemical feeder control may be manual, automatically proportioned to flow, dependent on some form of process

| Feeder Type | Use | Equipment Limitations | | Feed Rate Range[b] |
		General	Capacity (m³/h[a])	
Dry feeder				
Volumetric				
Oscillating plate	Any material, granules or powder		0.001–3.1	40–1
Oscillating throat (universal)	Any material, any particle size		0.002–9.0	40–1
Rotating disk	Most materials, including NaF, granules or powder	Use disk unloader for arching	0.001–0.09	20–1
Rotating cylinder (star)	Any material, granules or powder		0.7–180 0.65–27.0	10–1 or 100–1
Screw	Dry, free-flowing material, powder or granular		0.005–1.7	20–1
Ribbon	Dry, free-flowing material, powder, granular, or lumps		0.0002–0.015	10–1
Belt	Dry, free-flowing material up to 40 mm (1.5 in.) in diameter, powder or granular		0.009–270	10–1
Gravimetric				
Continuous-belt and scale	Dry, free-flowing, granular material or floodable material	Use hopper agitator to maintain	0.002–0.18	100–1
Loss in weight	Most materials—powder, granular, or lumps	constant density	0.002–7.2	100–1
Solution feeder				
Nonpositive displacement				
Decanter (lowering pipe)	Most solutions, light slurries		0.009–0.9	100–1
Orifice	Most solutions	No slurries	0.015–0.45	10–1
Rotameter (calibrated valve)	Clear solutions	No slurries	0.0005–0.015 0.0002–0.018	10–1
Loss in weight (tank with control valve)	Most solutions	No slurries	0.0002–0.018	30–1
Positive displacement Rotating dipper	Most solutions or slurries		0.009–2.7	100–1
Proportioning pump Diaphragm	Most solutions (special unit for 5% slurries)[c]		0.0004–0.014	100–1
Piston	Most solutions, light slurries		0.0001–15.3	20–1

[a]Volumetric feed capacities are given because chemical specific gravities must be known to specify mass feed capacity.
[b]Ranges apply to purchased equipment. Overall feed ranges can be extended more.
[c]Use special heads and valves for slurries.

TABLE 14.13 Types of Chemical Feeders

feedback, or a combination of any two of these. More sophisticated control systems are feasible if proper sensors are available. Standby units should be included for each type of feeder used. For proper operational flexibility, points of chemical addition and associated piping should be capable of handling all possible changes in dosing patterns.

Chemicals must always enter the system at areas of active agitation rather than at dead spots. Chemicals should not be added at locations where they may escape to the effluent stream before completion of proper mixing and reaction. A visible flow of reagent, which helps the operator, may be achieved by discharge above the surface of the receiving liquid. However, many designers prefer discharge below the surface to promote faster and more complete mixing. Both objectives can be accomplished by feeding to a funnel that is connected to a line entering below the surface.

4.3.5.1 Dry Feed   A dry feed installation (Figure 14.54) (U.S. EPA, 1975b) consists essentially of a hopper, a feeder, and a dissolver tank. All three units are sized based

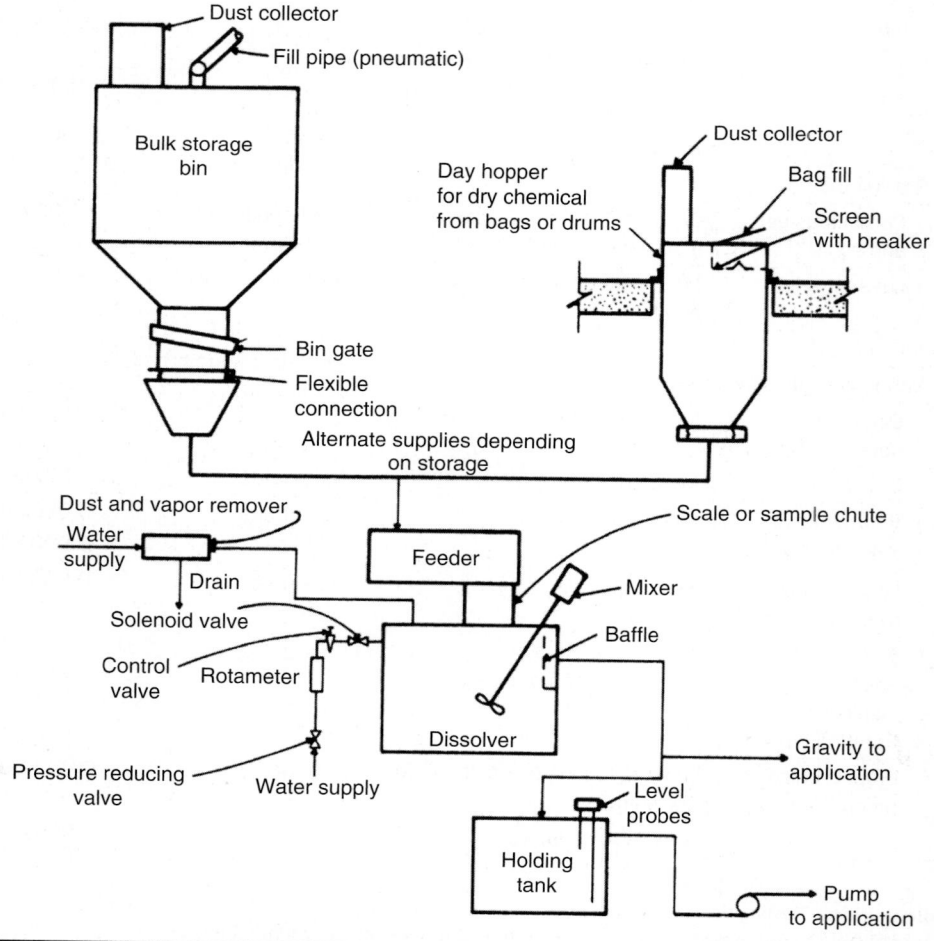

Figure 14.54   Typical feed system for dry chemicals.

on waste volume, treatment rate, and an optimum length of time for chemical feeding and dissolving. The best applications of dry feed systems have high treatment rates, more stable chemicals, and more fluid materials. Less fluid materials can be handled, but feeder accessories are needed. Because a powdered or granular material may arch or bridge in a hopper, it needs vibration for continuous flow. To prevent flooding by some powders that are too free-flowing, a rotor below the hopper exit ensures flow control.

Dry feeders are either of the volumetric or of the gravimetric type. Most types of volumetric feeders are the positive-displacement type, incorporating some form of moving cavity of a specific or variable size. Accurate feeding depends on the chemical having a constant specific weight per unit volume. In operation, the chemical falls by gravity into the cavity and is then enclosed and separated from the hopper's feed. The size of the cavity and its rate of movement govern the material feed rate.

One type of volumetric dry feeder has a continuous belt of specific width moving from beneath the hopper to the dissolving tank. A mechanical gate regulates the depth of material on the belt, and the feed rate is governed by the belt speed, height of the gate opening, or both. The hopper typically is equipped with a vibratory mechanism to reduce arching of powdered chemicals. Granular chemicals such as alum require no vibration. This type of feeder is unsuitable for easily fluidized materials.

Another type of volumetric feeder uses a screw or helix at the bottom of the hopper placed in a tube with an opening slightly larger in diameter. The speed of screw or helix rotation governs the feed rate.

The primary disadvantage of the volumetric feeder is its inability to compensate for changes in material density, which may be avoided by modifying the volumetric design to include a gravimetric or loss-in-weight controller. This modification allows continuous weighing of the material as it is fed. A beam-balance controller measures the actual mass of material. This is considerably more accurate, particularly over a long period, than a spring gravimetric design. Gravimetric feeders are used where a feed accuracy of approximately 1% is required for economy, as in large-scale operations, and where materials are used in small, precise quantities. Many volumetric feeders can be converted to loss-in-weight devices by placing the entire feeder on a platform scale tared to offset the weight of the feeder.

With dry feeders, the dissolving operation is critical. Numerous small particles dissolve more rapidly than larger granules or lumps. Warm water and efficient mixing tend to increase the solution rate. Too much heat, however, may cause excessive vapors that impede flow of the dry chemical. Solution temperature must be taken into account when dissolving certain chemicals.

The capacity of a dissolver is based on detention time, which is directly related to the wettability, or rate of solution, of a chemical. Therefore, the dissolver must be large enough to provide the necessary retention time for both the chemical and the water at the maximum feed rate. At lower feed rates, the strength of the solution leaving the dissolver is less, but the detention time is approximately the same unless the water supply to the dissolver is reduced. When the water supply to any dissolver is reduced to form a constant-strength solution, mechanical mixing within the dissolver is necessary because the mixing jets do not provide sufficient power at low rates of flow.

4.3.5.2 Solution Feed Liquid feed systems are best applied for chemical treatment with lower treatment rates, less stable chemicals, chemicals better fed as liquids to

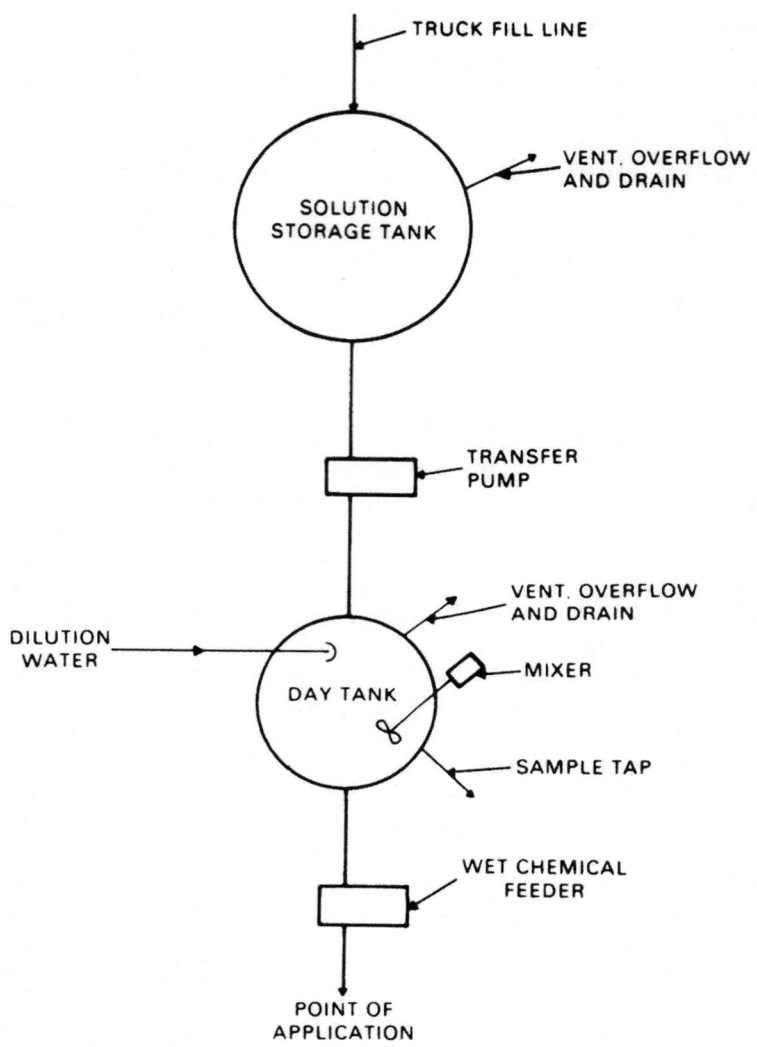

FIGURE 14.55 Typical feed system for wet chemicals.

avoid handling of dusty or more dangerous chemicals, or materials available only as liquids (Figure 14.55) (Huang, 1979).

Liquid feed units include piston, positive-displacement diaphragm, and balanced diaphragm pumps and liquid gravity feeders (rotating dippers). The unit best suited for a particular application depends on feed pressure, chemical corrosiveness, treatment rate, accuracy desired, viscosity and specific gravity of the fluid, other liquid properties, and type of control.

Piston-plunger and positive-displacement pumps are available with low to high capacities at pressures up to 41 000 kPa (6000 psig). Liquids with viscosities up to 10 N/s·m² (100 P) may be handled by the plunger pump, depending on the rate of feed.

Mechanically actuated diaphragm pumps are designed for low discharge pressures less than 830 kPa (120 psig). Flow capacities of 0.076 L/s (72 gph) may be handled, although process fluids are limited to liquids with viscosities under 0.1 N/s·m² (1 P).

Hydraulically actuated, balanced diaphragm pumps can operate against discharge pressures up to 21 000 kPa (3000 psig). They can handle capacities up to 0.4 L/s (400 gph), and liquids with viscosities of 1 N/s·m² (10 P) and higher can be handled at lower flowrates.

Liquid gravity feeders (rotating dippers) can handle acids, alkalies, and slurries over a 100-to-1 range and at flowrates up to 1.90 L/s (1800 gph).

Simple homemade devices can be built in the facility to provide liquid reagent feed. An orifice tank, consisting of a constant head tank and an orifice, is such a device. A Sutro weir can be varied in width, manually or automatically, in accordance with liquid flowrate as indicated by a measuring weir or other control device.

5.0   Membrane Processes

5.1   Process Description

Membrane processes can be pressure or vacuum driven or depend on electrical potential gradients, concentration gradients, or other driving forces. Some facilities that are utilizing the vacuum-driven membranes are applying the gravity-assisted mode of operation when membranes are operated under the gravity (static pressure) until the desired permeability is achieved.

Four main membrane categories, classified by the size of the separated particles or materials, are commercially used at the present time and can be expressed in microns (mm) and/or Daltons molecular weight cutoff (Da MWCO) (Frenkel, 2004):

- Microfiltration separates particles from 0.1 to 10.0 microns (mm) (>100 000 Da);

- Ultrafiltration rejects materials from 0.01 to 0.1 microns (mm) (2000 to 100 000 Da);

- Nanofiltration rejects materials from 0.001 to 0.01 microns (mm) (200 to 1000 Da); and

- Reverse osmosis ranging in molecular size less than 0.001 microns (mm) (<200 Da).

The first two listed, microfiltration and ultrafiltration membranes, are LPMs, and the last two, nanofiltration and reverse osmosis, are HPMs.

These four classes of membranes have been established primarily based on the size range of particles or ions each type will remove from water or wastewater. This characteristic is described in terms of nominal pore size for microfiltration and ultrafiltration membranes and by molecular weight cutoff for nanofiltration and reverse osmosis membranes. Microfiltration and ultrafiltration membranes typically are used as a particle (suspended and colloidal matters) separation processes, whereas nanofiltration and reverse osmosis can be classified as ion rejection processes.

Membrane processes originally were developed in the early 1960s as desalination or demineralization processes. Membranes initially were commercially applied in reverse osmosis, and then in the 1990s were developed for microfiltration and ultrafiltration. Membrane shapes include spiral wound, hollow fiber, flat sheet, and tubular.

There are two primary membrane types based on driven pressure (Frenkel, 2008):

- Pressure driven—low-pressure microfiltration, ultrafiltration, and high-pressure nanofiltration and reverse osmosis; and
- Immersed, vacuum driven—low-pressure microfiltration, ultrafiltration only.

The separate category is the forward osmosis membranes driven by the osmotic power difference between solutes with different salt concentrations.

Figure 14.56 illustrates the relative rejection properties of four types of membranes for a range of contaminants that may be found in secondary wastewater effluent. Although LPMs are designed to remove suspended and colloidal matter from water, HPMs are designed to remove dissolved constituents.

Compared to other membranes, microfiltration membranes have the largest pore size and are designed to remove relatively large suspended particles such as colloids, bacteria, cysts, and others. Ultrafiltration membranes typically have a one-order-of-magnitude smaller pore size than microfiltration membranes and can achieve greater logs of virus removal than the microfiltration membranes. Nanofiltration and reverse osmosis membranes are applied to remove dissolved constituents of different sizes and charges from the water.

5.1.1   Low-Pressure Membranes: Microfiltration and Ultrafiltration

The membrane separation process is a surface process that removes suspended and colloidal matter efficiently and produces high-quality effluent. With proper selection,

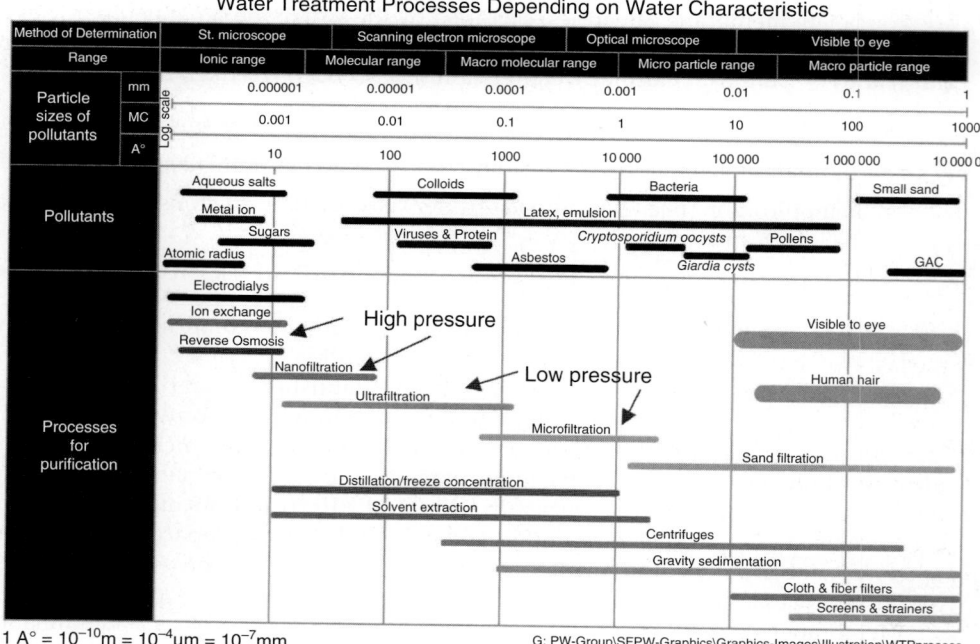

FIGURE 14.56   Size ranges for various treatment processes and membrane separation (courtesy of Val S. Frenkel).

membranes can remove pathogenic microorganisms, including bacteria, viruses, protozoa, and cysts. Additionally, the membrane filtration process can remove certain organic species, provided the molecular size and membrane pore sizes are properly determined. As indicated in Figure 14.56, membranes used in separation processes have pore sizes that are many orders of magnitude larger than those used in reverse osmosis processes.

5.1.2 High-Pressure Membranes: Nanofiltration and Reverse Osmosis

Both HPM types are similar processes and essentially use the same or similar membrane materials. Sometimes nanofiltration is called "loose" reverse osmosis, because these membranes are designed for lower salt rejection than reverse osmosis. Nanofiltration membranes were developed primarily to reduce water hardness caused by sparingly soluble ions or divalent cations such as calcium and magnesium and to remove color caused by organics. For simplicity, only reverse osmosis process is described below.

The natural phenomenon of osmosis occurs when pure water flows from a dilute saline solution through a membrane into a higher concentrated saline solution. Figure 14.57 illustrates the phenomenon of osmosis. A semipermeable membrane is placed between two compartments. With a highly concentrated salt solution in one compartment and low-concentration salt solution in the other compartment, the membrane will allow water to permeate through it. This system will try to reach equilibrium so that water from the low-concentrated solute will flow to the high-concentrated solute until the osmotic pressure in the high-concentrated solute reaches equilibrium with the weight of the water column in the high-concentrated solute. The only possible way to reach equilibrium is for water to pass from the low-concentrated solute compartment

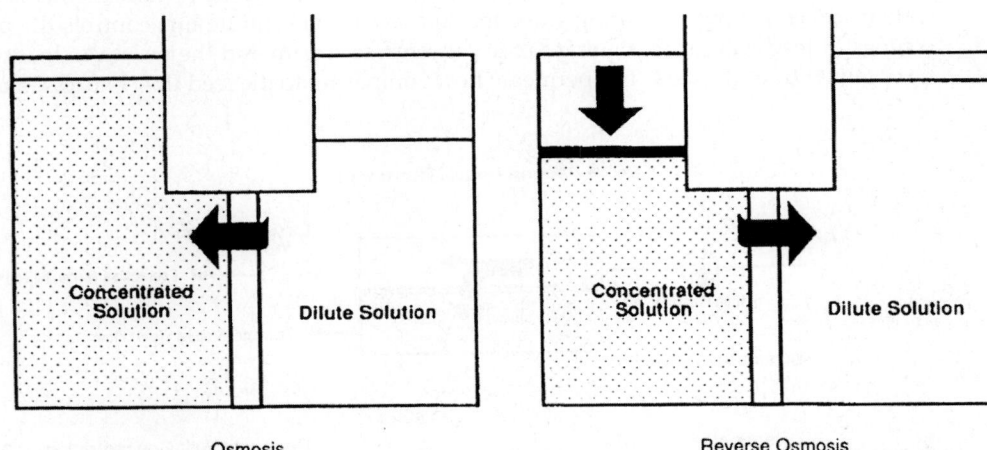

Osmosis

Water diffuses through a semi-permeable membrane toward region of higher concentration to equalize solution strength. Ultimate height difference between columns is "osmotic" pressure.

Reverse Osmosis

Applied pressure in excess of osmotic pressure reverses water flow direction. Hence the term "reverse osmosis."

Figure 14.57 Graphical depiction of osmosis and reverse osmosis.

to the high-concentrated solute compartment to dilute the salt solution. This process is called the natural osmosis.

Figure 14.57 also shows that osmosis can cause a rise in the height of the salt solution. This height will increase until the pressure of the column of water (salt solution) is so high that the weight of the water column stops the water flow. The equilibrium point of this water column height in terms of water pressure against the membrane is called osmotic pressure.

The direction of water flow through the membrane can be reversed by applying pressure to the high salt solution greater than the osmotic pressure. This is the basis of the term reverse osmosis. Note that this reversed flow theoretically produces pure water from the salt solution because osmotic pressure retains ions (cations and anions) in the high-concentrated solute; water discharged out of the solute under pressure that exceeds osmotic pressure. Theoretically, only the water molecules should be displaced from the high-concentrated solution; however, in practice, the process is more complicated. Because of diffusion and other forces, the leakage of ions other than water molecules can be found in the water. Nevertheless, salt rejection by reverse osmosis membranes is high because of the osmosis phenomena. Most commercially available reverse osmosis membranes provide salt rejection up to 99.8% by one single element at standard conditions. Unlike standard conditions, however, real-world operating conditions typically contain more than one element. As a result, most reverse osmosis systems realistically provide overall salt rejection in the range of 95.0% to 98.0 %, which is considered high.

Any membrane process is a separation process. Feedwater is separated into two streams: product water (permeate) and reject water (concentrate, brine). The simplified reverse osmosis process is shown in Figure 14.58. A high-pressure pump is used to feed saline feedwater to the module system. Within the module—consisting of a pressure vessel or housing, membrane elements, or some combination of these—the feedwater is split into a low-saline product, called permeate, and a high-saline brine, called concentrate or reject. A flow-regulating valve located on the concentrate line controls the percentage of feedwater that is going to the concentrate stream and the permeate that will be obtained from the feed. The permeate flow comparing to the feed flow expressed as a

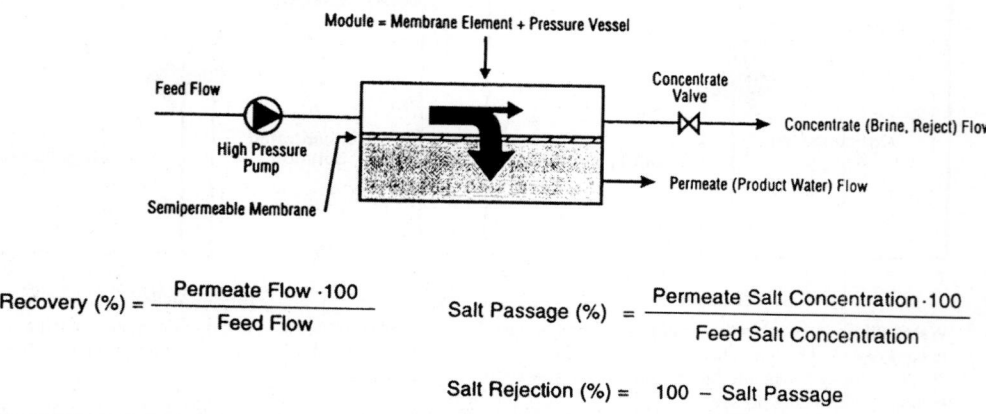

$$Recovery\ (\%) = \frac{Permeate\ Flow \cdot 100}{Feed\ Flow}$$

$$Salt\ Passage\ (\%) = \frac{Permeate\ Salt\ Concentration \cdot 100}{Feed\ Salt\ Concentration}$$

$$Salt\ Rejection\ (\%) = 100 - Salt\ Passage$$

FIGURE 14.58   Operation of a typical membrane filtration or reverse osmosis process.

Process	Materials Removed	Applications	Transmembrane Pressures, kPa (psi)
Microfiltration	Suspended solids and large colloids	Removal of bacteria, flocculated materials, and TSS	69–173 (10–25)
Ultrafiltration	Colloids, proteins, microbiological contaminants, and large organic molecules	Virus removal, removal of colloids and some organic molecules	
Nanofiltration	Organic molecules with weights greater than 200 to 400, some TDS[a] reduction	Removal of color, TOC,[b] hardness, radon, and TDS reduction	345–1550 (50–225)
Reverse osmosis	Dissolved salts, inorganic molecules, organic molecules with molecular weights greater than 100	Desalination, wastewater reuse, food and beverage processing, industrial process water	1379–6895 (200–1000)

[a]TDS = total dissolved solids.
[b]TOC = total organic carbon.

TABLE 14.14   Comparison of Membrane Processes

percent reflects the system recovery. As an example, when a system fed by 22.7 m³/h (100 gpm) produces 17 m³/h (75 gpm) permeate, the system recovery is 17 m³/h:22.7 m³/h×100% = 75%.

Table 14.14 summarizes the four primary membrane processes, including differences, applications, and materials removed by each. Transmembrane pressure (TMP) is the pressure differential between the feed stream and the product stream in the process. Typically, the TMP increases as the pore size of the membrane becomes smaller. The TMP is more applicable to LPMs; it is not characteristic of HPMs because osmotic pressure is of much higher magnitude than TMP.

5.1.3   Electrical Current-Driven Membranes

Electrical current-driven membranes are represented by the electrodialysis and electrodialysis reversal (EDR) technologies that separate ions and molecules down to less than 10 Da MWCO units and are effective in removing TDS, hardness, and other charged ionic contaminants (Frenkel, 2002).

Electrodialysis and EDR are voltage-driven membrane separation processes that separate dissolved ions from a fluid (water, wastewater, or industrial fluid). Electrodialysis and EDR processes use a voltage potential, instead of pressure, to drive the salt ions through a semipermeable membrane. The EDR membranes have a high removal efficiency for multivalent ions such as calcium and magnesium (hardness) and will remove monovalent salts such as sodium and chloride depending on the voltage potential and the selectivity of the EDR membrane.

The electrodialysis and EDR process schematic is shown in Figure 14.59. The electrodialysis/EDR membrane separation surface is a thin, synthetic polymer manufactured into a flat sheet. The flat sheets are assembled into stacks of membranes, spacers,

Electrodialysis

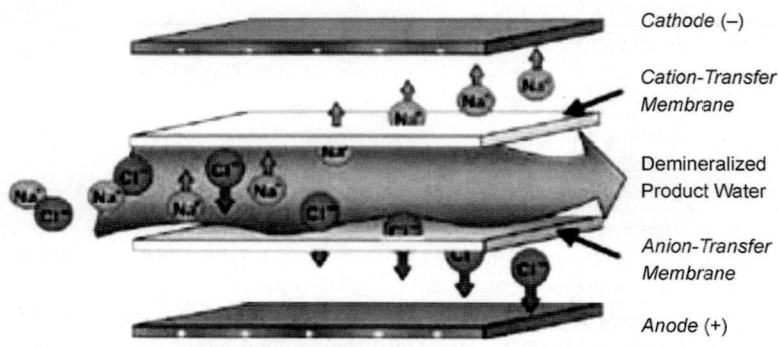

Cathode (−)

Cation-Transfer Membrane

Demineralized Product Water

Anion-Transfer Membrane

Anode (+)

FIGURE 14.59   Schematic diagram of electrodialysis (courtesy of Val S. Frenkel).

and electrodes. As low-pressure source water passes through the electrodialysis/EDR stacks, the voltage potential attracts ions through the membrane and out of the source water. The ions are concentrated into a reject stream and the source water exits the system as a low-TDS product water. Unlike reverse osmosis and nanofiltration separation processes, the product water from an electrodialysis/EDR system does not pass through the membrane.

The membrane sheets and electrodes within a stack are replaceable. The electrodialysis/EDR stacks can be square or cylindrical, and different sizes and heights to accommodate different flows or water quality. Electrodialysis/EDR stacks are grouped in series to achieve the desired amount of demineralization and are supported and manifolded together with common piping, valves, and instrumentation to form an integrated unit or train. The electrodialysis/EDR train has independent flow control. It is cleaned chemically and operated as a complete unit. Multiple electrodialysis/EDR trains are manifolded together into banks to meet overall capacity requirements for a system.

Electrodialysis/EDR membranes require pretreatment ahead of the membranes to protect them from solids and to prevent fouling of the membrane surface. Although EDR membranes are not as sensitive to particulate fouling as reverse osmosis membranes, large solids must be removed ahead of the system. Depending on the source-water quality, microfiltration and/or cartridge filtration is typically used. Acids or antiscalants, or both, also are added to reduce scale formation on the membrane surface. Depending on the water chemistry and system operational parameters, the average brackish water electrodialysis/EDR recovery range is 75% to 85%.

5.1.4   Nanocomposite-Membrane Processes

Current membranes are remarkably engineered material. But they have several significant drawbacks that include a tendency to be energy intensive, nonselective, and prone to fouling. This is one of the reasons why several studies have been initiated on how to improve membrane properties such as permeability, which is reflected in the water throughput, energy demand, and tendency to foul. From one hand the membranes are desired to be hydrophobic to accumulate less fouling materials, from the other hand membranes are desired to have hydrophilic properties to produce more permeate or clean water achieving higher permeability as a result. Because reverse osmosis membranes are the most energy intensive of the membrane technologies, most research is focused on improving reverse osmosis membrane properties. Current reverse osmosis membranes have been prepared from a thin polyamide film and can transform polluted water into clean water after a single pass through the membrane. The polyamide membrane was a huge improvement over the first generation reverse osmosis membranes that were made of cellulose acetate materials that degraded and lost performance relatively quickly. A fatal flaw in the polyamide chemistry is caused by the incompatibility of the nitrogen/carbon bond in the polyamide with chlorine-based chemicals used to control biogrowth in water. The resulting breakdown of the polyamide structure leads to an increase in membrane pore size and a loss of contaminant rejection. As a result, both thin film composite (TFC) and cellulose acetate membranes are used today.

One of the ways in which nanotechnology is being used is in the modification of polyamide reverse osmosis membranes by introducing zeolite molecular-sieve nanoparticles into the polyamide thin film. As a result, once the polyamide nanoparticle film is formed, the nanoparticles increase the ability of the membrane to attract water. These nanocomposite membranes contain a cross-linked matrix of polymers and super-hydrophilic nanoparticles. The addition of nanoparticles enables the reverse osmosis membrane to become more hydrophilic, more negatively charged, and smoother. It is expected that the super-hydrophilic, porous nanoparticles will provide preferential flow paths for water permeation while continuing to reject contaminants. The nanoparticles suck up water like a sponge, while their negative charge repels contaminants more effectively than polymer membranes.

The expected significant advantage of nanocomposite membrane technology is the ability to fine-tune the nanoparticle chemistry and pore structure, the polymer chemistry, and the film structure thickness. The technology should be compatible with existing manufacturing infrastructure and could be used as a direct replacement for conventional reverse osmosis membranes.

The change in the hydrophilicity also may have a significant effect in reducing membrane fouling. Fouling resistance is expected be improved when the water contact angle declines to approximately 40°. Such membranes may resist adhesion of bacteria, colloidal particles, and dissolved organics better that traditional reverse osmosis membranes currently in use. Less fouling of the membranes should lead to fewer membrane cleanings, extending the life of the membranes as a result. Cleanings reduce membrane lifespan rather than filtration modes, and the fewer frequencies of cleanings, the longer the membranes lifespan.

While promising, the nanocomposite membrane is not currently a widely adopted commercial technology in the municipal market. However, it may be commercialized in the near future and will be described in more detail at that time.

5.1.5 Forward Osmosis, Pressure-Retarded Osmosis

In reverse osmosis, high pressure is used to drive water through a semipermeable membrane that holds back unwanted salts. The pressure is needed to oppose or overcome the natural tendency of less salty (clean) water to move across such a membrane via osmosis to dilute the salty water (seawater). Energy also is required to overcome a pressure drop across the membrane.

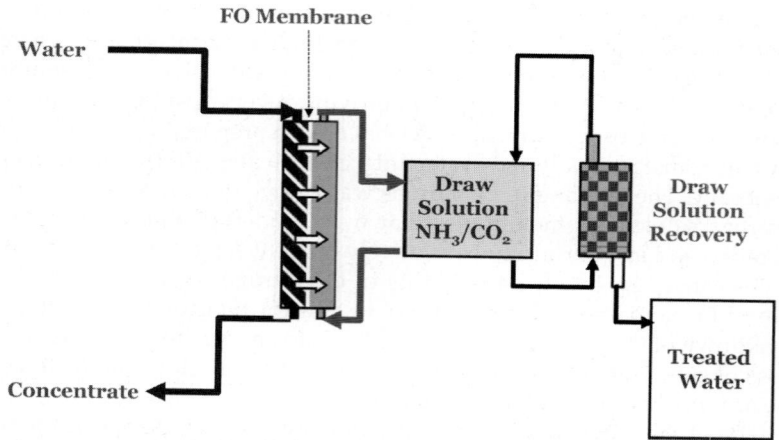

FIGURE **14.60**   Forward Osmosis (FO).

The forward osmosis system takes advantage of this natural tendency, and is shown in Figure 14.60 schematically. Higher osmotic pressure water sits on one side of the membrane, and the lower osmotic pressure water on the opposite side is transformed into a high-concentration solution by adding ammonia (NH_3) and CO_2 or other draw solution. Water naturally flows from the lower osmotic pressure water to what is now the "draw solution," which can have a solute concentration as much as 10 times that of the lower osmotic pressure water. Upon reaching a certain dilution, the draw solution is then heated to about 58°C to evaporate the CO_2 and NH_3 for reuse, leaving behind clean water that has been extracted from the lower osmotic pressure water. Several companies are developing forward osmosis systems to treat wastewater and to clean water leaching from landfills. Forward osmosis found several niche applications and in fracking, and produced water recovery and treatment particularly, while it is not commercially entered the municipal market. One of the benefits of the forward osmosis process reported is the fewer fouling and organic fouling particularly comparing to the other membrane processes. Challenges associated with forward osmosis that recent projects, pilot programs, and researchers report are as follows:

- Scaling from both sides of the membranes (double-side scaling compared to reverse osmosis).

- Low flux compared to reverse osmosis. Low flux requires significantly higher membrane area to achieve similar throughput and can lead to accumulation of microconstituents in the draw solution that remains in the clean water after evaporation, which may require an additional treatment barrier downstream.

Forward osmosis found niche applications in the industry, while it is not adopted in the municipal market at this time.

The pressure-retarded osmosis is shown schematically in Figure 14.61. It has similarity with forward osmosis when water is passing through the membrane from the

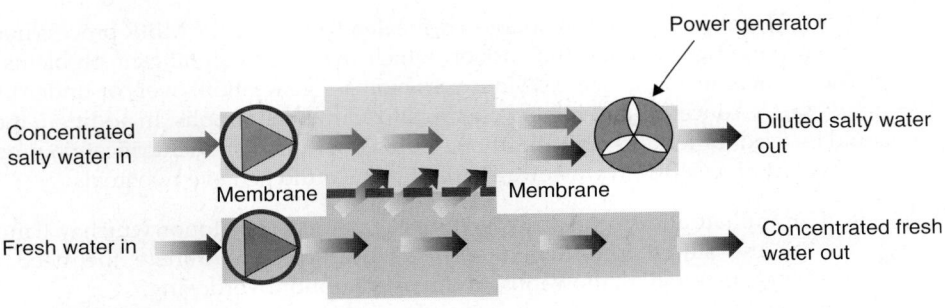

FIGURE 14.61 Pressure Retarded Osmosis.

lower concentrated solute to the higher concentrated solute increasing volume of the diluted higher concentrated solute. When exiting the system, this increased flow of water generates energy using an energy generator by converting kinetic energy to electrical energy. The pressure-retarded osmosis is used to generate energy rather than clean water. There are several projects underway to test pressure-retarded osmosis process viability. While it is a promising method, the economics of the current efforts didn't justify commercialization of pressure-retarded technology at this time. However, it may be commercialized in the near future, at which time it will be described in more detail.

5.1.6 Membrane Bioreactor, Aerobic and Anaerobic

In some places the membrane bioreactor (MBR) is considered as an emerging technology, it became fully established commercial technology adopted around the globe. The MBR is an advanced technology to treat wastewater that uses LPMs, microfiltration, and ultrafiltration in place of the secondary clarifier in the biological activated sludge process. Because of the unique features of microfiltration/ultrafiltration to retain and separate activated sludge from water, the MBR typically operates with a higher mixed-liquor suspended solids (MLSS) than conventional processes, achieving several significant benefits as a result. For the main liquid stream, the aerobic MBR is a well-adopted process. The newer version of MBR, the anaerobic MBR (AnMBR) has been applied, piloted, and studied for high-strength industrial streams and for sludge treatment. AnMBR is not well adopted by the municipal market, while this situation may change as technology develops.

This chapter does not describe the MBR process in depth; Chapter 12, however, provides more details.

5.1.7 Autotrophic Membrane-Biofilm Reactor (MBfR)

The membrane-biofilm reactor (MBfR) is still a relatively new technology in which H_2 gas is delivered as an electron donor to autotrophic bacteria that reduce NO_3^{2-} or other oxidized contaminants and use membranes as a place for bacteria accumulation and biofilm formation. This section introduces MBfR; Chapter 11 provides a more detailed discussion of the processes and application. The H_2-based MBfR delivers H_2 gas to a biofilm that naturally accumulates on the outer surface of bubble-less membranes. Although the MBfR has been tested for reduction of nitrate, perchlorate, and other constituents in drinking water and groundwater, it also may be used for advanced nitrogen removal in wastewater treatment.

By using H_2 gas as the electron donor to drive denitrification, the MBfR process may eliminate an added organic electron donor, which overcomes significant problems. These problems include a large increase in excess biomass generation, over- or underdosing of donor, safety concerns, and relying on specialized methanotrophs. In addition, the MBfR could be used for tertiary denitrification or placed within a pre-denitrification process.

Autotrophic MBfR in wastewater treatment should achieve two goals:

1. Eliminate addition of an exogenous organic electron donor, which will minimize excess sludge production and chemical costs, eliminate the need to use specialized methanotrophs, and remove donor overdosing.
2. Provide a simple system that is easily integrated into existing wastewater treatment systems.

The MBfR can be integrated into existing or new activated sludge designs in two distinct ways:

1. Using the MBfR for tertiary denitrification or to remove NO_3^2 remaining after conventional biological treatment, such as pre-denitrification.
2. Placing MBfR units directly in a pre-denitrification system to enhance its performance without constructing a tertiary-treatment process.

Although promising, MBfR is not a proven commercial technology in wastewater treatment, but likely will be commercialized in the near future and will be described in more detail at that time.

5.2 Process Objectives

The primary applications for the two groups of membranes are:

1. Low-pressure membranes—remove suspended and colloidal matter; and
2. High-pressure membranes—remove dissolved species.

The objectives of membrane processes will vary considerably based on the facility, location, and water quality goals. Membranes can be manufactured to reject many different species, and it is important that treatment objectives be determined early in the project development stage. Typical process objectives could include:

- Removal of TSS and microorganisms by microfiltration or ultrafiltration as a pretreatment process for nanofiltration or reverse osmosis;
- Removal of inorganic cations and anions for water reuse applications or TDS removal by nanofiltration or reverse osmosis;
- Removal of specific species of organic compounds by nanofiltration or reverse osmosis for industrial water quality management issues;
- Production of product water by microfiltration or ultrafiltration that is relatively free of microorganisms for industrial reuse applications; and
- Removal of certain microconstituents by nanofiltration or reverse osmosis.

The following summarizes average representative membrane characteristics for four membrane categories:

Parameter	Reverse Osmosis	Nanofiltration	Ultrafiltration	Microfiltration
Molecular weight cutoff (MWCO), Da	<100	100–1000	1000–100 000	>100 000
Pore size average range, microns (µm)	<0.0001	0.0001 –0.001	0.001–0.01	0.1–0.5
Operational pressure, psi*	150–1200	70–250	35–50	20–40
Operational recovery, average %	60–80	75–85	85–95	85–95
Level of treatment/ removal	TDS	Hardness, TDS partially	Viruses	Suspended solids, bacteria, pathogens
Rejection/removal, %	Up to 99.7% TDS	Up to 50% TDS, > 80% hardness	Up to 99.9999% particles, up to 99.99% pathogens and viruses	Up to 99.9999% particles, up to 99.99% pathogens, and up to 99% viruses
Membrane shape/ type membrane materials	Spiral wound TFC, polysulfone (PSF), cellulose acetate (CA)	Spiral wound TFC, CA	Hollow fiber, flat sheet PVDF, PSF, polytetrafluoro-ethylene (PTFE), polyether sulfone (PES)	Hollow fiber, flat sheet PVDF, PSF, PTFE, PES, polypropylene, nylon

*psi × 6.895 = kPa.

5.3   Pretreatment

In addition to significant differences, LPMs may require minimal pretreatment or no pretreatment at all for several applications compared to HPMs, which do not tolerate suspended and colloidal solids, and their properties are significantly degraded even when short spikes of suspended solids and/or colloids occur.

Pretreatment requirements for various membrane processes vary tremendously. Pretreatment for HPM processes is the most demanding and are discussed below. Typically, LPM is less demanding, but the needs of each application should be reviewed carefully.

To increase the efficiency and life of a membrane process, effective pretreatment of feedwater for nanofiltration/reverse osmosis is required. Selection of proper pretreatment minimizes fouling, scaling, and membrane degradation. This leads to optimization of product quality, salt rejection, product recovery, and operating costs.

The membrane industry has adopted the following classification of four major categories of membrane fouling:

1. Inorganic scale—control with antiscalant, system recovery;
2. Particulate fouling—minimize feedwater turbidity and silt density index (SDI);
3. Organic fouling—minimize feedwater dissolved organics; and

4. Biofouling—control with flux, shock chlorination.

Fouling refers to the entrapment of particulates such as iron floc or silt; scaling refers to the precipitation and deposition within the system of sparingly soluble salts such as calcium sulfate ($CaSO_4$) and barium sulfate ($BaSO_4$).

Pretreatment of feedwater must involve a total system approach for continuous and reliable operation. The proper treatment scheme for feedwater will depend on:

- Feedwater source;
- Feedwater composition;
- System operational protocol; and
- Application.

Once the feedwater source has been determined, a complete and accurate analysis of the feedwater should be made. The importance of this analysis cannot be overemphasized. It is critical in determining the proper pretreatment and nanofiltration/reverse osmosis system design.

The application often determines the type or extent of nanofiltration/reverse osmosis pretreatment.

5.3.1 Scale Control

Scaling of a membrane may occur when sparingly soluble or bivalent ions are concentrated beyond their solubility limit. For example, if a process is operated at 50% recovery, then ion concentration in the concentrate stream will be almost double the concentration in the feed stream. As recovery of a facility increases, so does the risk of scaling. Therefore, care must be taken not to exceed the solubility limits of slightly soluble salts, or precipitation and scaling may occur in the concentrate. System recovery is one of the parameters to control scaling formation. Silica (SiO_2) is also one of the factors that controls system recovery for many waters in groundwater treatment.

The practices discussed below are known to control scaling of the membranes.

5.3.1.1 Acid Addition
Acid addition to control scaling in the nanofiltration/reverse osmosis process is one of the oldest approaches. Many waters are almost saturated with calcium carbonate. The solubility of calcium carbonate depends on the pH because it is less soluble at higher pH values.

Accordingly, by adding H^+ as acid, the equilibrium can be shifted to the low-pH side to keep calcium carbonate dissolved. Because of safety and security issues, scale inhibitor (antiscalant) is a newer, while well adopted by years, approach that is gaining in popularity as a way to control nanofiltration/reverse osmosis scaling.

5.3.1.2 Scale-Inhibitor Addition
Scale inhibitors (antiscalants) can be used to control carbonate scaling, sulfate scaling, and calcium fluoride scaling. Scale inhibitors have a threshold effect, which means that minor amounts adsorb specifically to the surface of microcrystals, thereby preventing further growth and precipitation of crystals. There are many types of scale inhibitors available on the market from several suppliers.

Polymeric organic is one type of scale inhibitor that is in gaining popularity. Precipitation reactions may occur, however, with cationic polyelectrolytes or multivalent cations such as, for example, aluminum or iron. This may result in gum-like products that are difficult to remove from membrane elements.

5.3.1.3 Softening with a Strong-Acid Cation Exchange Resin   Softening with a weak-acid cation exchange resin is one technique that can be used to control scaling by sparingly soluble ions such as calcium and magnesium.

When using the strong-acid cation exchange resin, the scale-forming cations such as Ca^{++}, Ba^{++}, and Sr^{++} are removed and replaced with Na^+. The resin has to be regenerated with sodium chloride at hardness breakthrough. The pH of the feedwater is not changed by this treatment. Therefore, no degassifier downstream is needed to strip out excessive carbon dioxide (CO_2). Only a small amount of carbon dioxide passes from raw water into the product water. The softening process consumes bicarbonate alkalinity.

5.3.1.4 Lime Softening
Lime softening can be used to remove carbonate hardness by adding hydrated lime according to the chemical formula as follows:

$$Ca(HCO_3)_2 + Ca(OH)_2 \rightarrow 2CaCO_3 + 2H_2O$$

$$Mg(HCO_3)_2 + 2Ca(OH)_2 \rightarrow Mg(OH)_2 + 2CaCO_3 + 2H_2O$$

The noncarbonate calcium hardness can be further reduced by adding sodium carbonate (soda ash) as follows:

$$CaCl_2 + Na_2CO_3 \rightarrow 2NaCl + CaCO_3$$

With lime softening, barium, strontium, and organic substances also are reduced significantly. The process requires a reactor with a high concentration of precipitated particles serving as crystallization nuclei. This typically is achieved by solids-contact clarifiers. The effluent from this process needs media filtration or other filtration technology to remove particles followed by pH adjustment prior to the reverse osmosis process. Iron or other coagulants with or without polymeric flocculants (anionic and nonionic) may be used to enhance solid-liquid separation.

5.3.1.5 Preventive Cleaning   In some applications, scaling is controlled by preventive membrane cleaning. This allows the system to run without softening or injection of chemicals. Typically, these systems operate at low recovery or approximately 25%, and membrane elements are replaced after 1 to 2 years of operation. Accordingly, those systems are typically small, single-element facilities for potable water from tap water or seawater. The simplest way to clean is by using a low-pressure forward flush by opening the concentrate valve. Short cleaning intervals are more effective than long intervals, for example, 30 seconds every 30 minutes compared with 1 minute every 60 minutes.

The larger systems can consider the forward flush preventive cleaning by periodic release of concentrate triggering forward osmosis effect for cleaning.

5.3.1.6 Adjusting Operating Variables   When other control methods do not work, facility operating variables have to be adjusted to prevent scaling. The precipitation of dissolved salts can be avoided by keeping the concentration below the solubility limit by reducing the system recovery.

Solubility also depends on temperature and pH. In the case of silica, increasing temperature and pH increases silica solubility. Silica typically is the only reason to consider adjusting operating variables as a scale-control method; all adjustments have economic drawbacks (energy consumption) or other scaling risks (calcium carbonate at high pH).

For small systems, low recovery and a preventive cleaning program might be a convenient way to control scaling.

There are a number of scale inhibitors on the market that allow oversaturated silica in the water while maintaining relatively high recovery of the reverse osmosis facility.

5.3.2 Colloidal Fouling Prevention

Colloidal fouling of membrane elements can seriously impair performance by lowering productivity and sometimes salt rejection. An early sign of colloidal fouling often is either increased pressure differential across the system or reduction in the product flow.

The source of silt or colloids in reverse osmosis feedwater varies and often includes bacteria, clay, colloidal silica, and iron corrosion products. Pretreatment chemicals used in a clarifier such as alum, ferric chloride, or cationic polyelectrolytes also can cause colloidal fouling if not removed either in the clarifier or through proper media or other filtration technology. In addition, cationic polymers may coprecipitate with negatively charged antiscalants and foul the membrane.

The simplest and most widely used technology for determining the colloidal fouling potential of reverse osmosis feedwater is the SDI, sometimes referred to as the fouling index. This is an important measurement to take before designing a reverse osmosis pretreatment system and on a regular basis during operation. There are several methods to measure colloidal fouling, and most have a good correlation to SDI measurements. That is why SDI measurement is still the most highly used technique.

5.3.2.1 Media Filtration

The removal of suspended and colloidal particles by media filtration is based on their deposition on the surface of filter grains. A well-designed and operated filter can typically treat water with a relatively low SDI that is acceptable to feed the reverse osmosis system. For more details on media filtration, refer to the beginning of Chapter 14.

5.3.2.2 In-line Filtration

The efficiency of media filtration to reduce the SDI value can be improved significantly if colloids in the feedwater are coagulated or flocculated before or during filtration. In-line filtration, also named in-line coagulation or in-line coagulation/flocculation, is described in ASTM Standard D 4188. A coagulant is injected to the raw water stream. Effectively mixed and formed microflocs are removed immediately by media filtration. Ferric sulfate and ferric chloride are the most popular and effective, used to destabilize the negative surface charge of colloids and to entrap them into freshly formed ferric hydroxide microflocs. Aluminum coagulants are also used but need to be carefully examined because of possible fouling problems with residual aluminum in case of applying membrane treatment downstream.

Rapid dispersion and mixing of the coagulant is extremely important. An in-line static mixer or injection on the suction side of a booster pump is recommended. The optimum dosage is typically in the range of 5 to 30 mg/L but should be determined case by case depending on the water chemistry to be treated.

5.3.2.3 Coagulation/Flocculation For feedwater containing high concentrations of suspended matter resulting in a high or even nonmeasurable SDI, the classic coagulation/flocculation process is preferred. The hydroxide flocs are allowed to grow and settle in specifically designed reaction chambers. Hydroxide sludge is removed and supernatant water is further treated by media filtration.

For the coagulation/flocculation process, either a solids-contact-type clarifier or a compact coagulation/flocculation reactor may be used.

5.3.2.4 Cross-Flow Microfiltration/Ultrafiltration Cross-flow filtration through a microfiltration or ultrafiltration membrane removes virtually all suspended matter. Hence, an SDI well below 3.0 can be achieved with a well-designed and properly maintained microfiltration or ultrafiltration system. At the same time, a high microfiltration/ultrafiltration system recovery and a high specific permeate flow are required for economic reasons. These objectives typically are achieved by periodic cleanings such as either a forward flush or, preferably, a backflush. If a chlorine-resistant membrane material is used—for example, polyvinylidene fluoride (PVDF), polysulfone (PS), or ceramic membrane—then chlorine can be added to the wash water to prevent biological fouling.

The Orange County Water District in California has conducted research on waste-water reuse issues, including studying the use of microfiltration as a pretreatment process for the reverse osmosis process at Water Factory 21. That research found that the performance and economics of the microfiltration as pretreatment for reverse osmosis are favorable (Leslie et al., 1998). One of the most important observations was that microfiltration reduced reverse osmosis process operating pressures significantly. Based on the positive experience collected over several years, the Orange County Water District commissioned new ultrafiltration-reverse osmosis capacity of 265 ML/d (70 mgd) in 2007, which has been in operation for the last 10 years. Comparable simplified treatment schematic of the Orange County Factory 21 built in the early 1970s and new Orange County ultrafiltration-reverse osmosis facility commissioned in 2007 is presented in Figure 14.62 and more detailed components of the new facility are shown

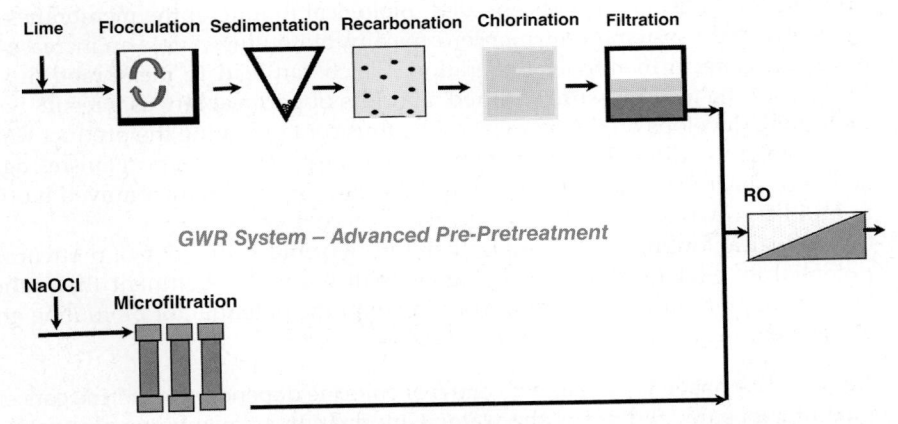

FIGURE 14.62 Water Factory 21–Conventional Pretreatment.

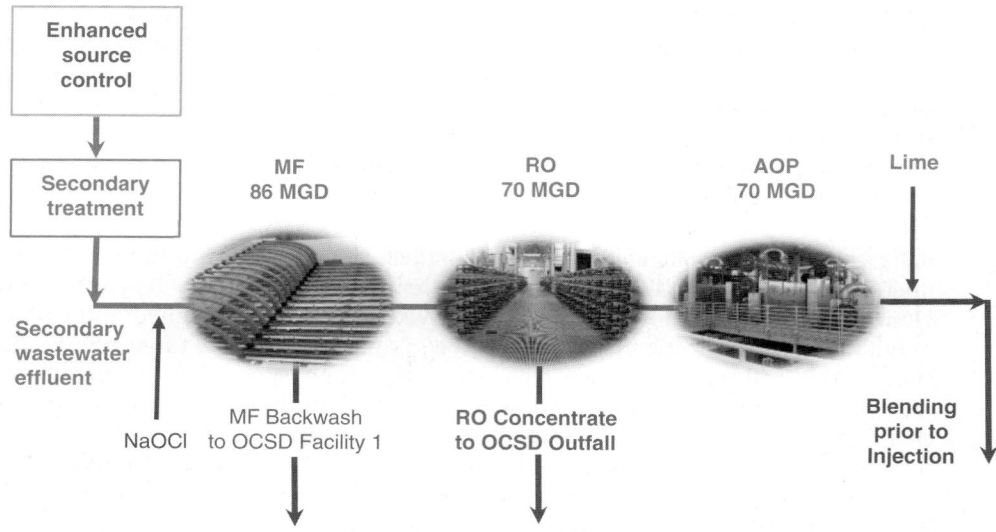

Figure 14.63   OCSD GWR System Advanced Water Treatment Facility, 2007–Present.

in Figure 14.63. The simplified new process using ultrafiltration membranes as pretreatment to reverse osmosis provided more reliable and efficient operation.

5.3.3   Biological Fouling Prevention

Effluents contain microorganisms such as bacteria, algae, fungi, viruses, and many others. Most of the microorganisms can be regarded as organic fine colloidal matter and removed by pretreatment to reverse osmosis as discussed previously. These live particles differ from nonliving particles in their ability to reproduce and form a biofilm under favorable living conditions.

Microorganisms flourish on the membrane surface where the dissolved organic nutrients of the water are concentrated. Biological fouling of the membranes may significantly affect system performance. Typically, biofouling causes an increase in differential pressure in membranes operation, which can lead to membrane flux decline, increase of the feed pressure required, and loss of permeability as a result. Sometimes biofouling develops on the permeate side, thus contaminating the product water.

A biofilm is difficult to remove because it protects its microorganisms against the action of shear forces and disinfection chemicals. Incompletely removed biofilms lead to rapid regrowth.

Biological fouling prevention is, therefore, a primary objective of pretreatment. The potential for biological fouling is higher with wastewater effluent than other water because of high nutrient content. Assessment of the potential for biofouling and possible preventive measures are discussed below.

5.3.3.1   Chlorination   The effectiveness of chlorine depends on chlorine concentration, time of exposure, and pH of the water. Chlorination as membrane pretreatment typically is applied where biological fouling prevention is required. Chlorine is added at the intake, and a reaction time of 15 to 30 minutes should be allowed. A free residual chlorine concentration in the range of 0.5 to 1.0 mg/L should be maintained through the entire pretreatment line. Dechlorination upstream of the nanofiltration or reverse osmosis membranes is required, however, to protect membranes from oxidation. In some instances, the pulsing or periodic chlorination can be applied to maintain proper operation depending on the organics type in the water.

5.3.3.2 Dechlorination The HPMs normally not tolerating free and residual chlorine and the feed to the nanofiltration or reverse osmosis membranes must be dechlorinated to prevent oxidation of the membrane. Residual free chlorine can be reduced to harmless chlorides by activated carbon or chemical reducing agents.

An activated carbon bed is effective in dechlorination of feedwater according to the following reaction:

$$C + 2Cl_2 + 2H_2O \rightarrow 4HCl + CO_2$$

5.3.3.3 Sodium Metabisulfite Sodium metabisulfate (SMBS) typically is used as a biostatic or for quenching free chlorine. Other chemical reducing agents exist such as, sulfur dioxide.

Although dechlorination itself is rapid, good mixing is required to ensure completion; as a result, static mixers are recommended. Ideally, the injection point is located downstream of the pretreatment filters to protect them with chlorine. In this case, the SMBS solution should be filtered through a separate cartridge before being injected to the reverse osmosis feed. Dechlorinated water must not be stored in tanks due to the potential secondary contamination.

The absence of chlorine should be monitored using an oxidation-reduction potential (ORP) electrode downstream of the mixing line. The electrode signal shuts down the high-pressure pump when residual or free chlorine is detected.

5.3.3.4 Chloramination Chloramination has become a popular technique for control of biofouling. Ammonia is injected into the chlorinated water in a concentration greater than the chlorination break point to make sure there is no free chlorine in the water to damage nanofiltration or reverse osmosis membranes. When formed this way, chloramines do not damage nanofiltration or reverse osmosis membranes and help to control nanofiltration and reverse osmosis membrane biofouling.

5.3.3.5 Shock Treatment Shock treatment is the addition of a biocide to the feed stream during normal facility operation for a limited time. Sodium bisulfite is the most typically used biocide for this purpose. In a typical application, sodium bisulfite in the range of 500 to 1000 mg/L is dosed for about 30 minutes.

5.3.3.6 Microfiltration/Ultrafiltration The advantages of microfiltration/ultrafiltration include that it can remove particles, microorganisms, and algae that are sometimes difficult to remove by standard techniques. Microfiltration/ultrafiltration permits the production of feedwater to the reverse osmosis process that has a low SDI value, which reduces operating costs. These membranes, however, should be carefully evaluated for the specific application.

5.3.3.7   Ozone   Although ozone is an even stronger oxidizing agent than chlorine, it decomposes readily. A certain ozone level must be maintained to kill all microorganisms. The resistance of construction materials against ozone has to be considered. Typically, stainless steel is used. De-ozonation must be performed carefully to protect membranes. Ultraviolet irradiation has been used successfully for this purpose. Additional, thermal destruction units frequently are required for ozone off-gas treatment to comply with local ambient air quality requirements.

5.3.3.8   Ultraviolet Irradiation   Ultraviolet irradiation at 254 nm is known to have a germicidal effect. Its application has come into use especially for small-scale facilities. No chemicals are added, and equipment needs little attention other than periodic cleanings or replacement of mercury vapor lamps.

Table 14.15 presents a graphical summary of various pretreatment options.

5.3.3.9   Advanced Oxidation   Advanced oxidation combines injecting peroxide either with UV irradiation or ozonation or with a combination of all together, creating radicals. Advanced oxidation also can use peroxide with both ozonation and UV, the latter being one of the most effective, economic techniques to reduce certain organic concentrations in water.

5.4   Membrane Systems

Although natural, semipermeable membranes with osmotic characteristics are common in living organisms, it was not until the 1950s that a synthetic reverse osmosis membrane with commercial possibilities was developed at the University of California Los Angeles (UCLA).

Two properties are important for a reverse osmosis membrane. Water permeability determines the production rate of desalted water per unit membrane area, typically called flux and measured in gallons per square foot per day or liter per square meter per hour (1 gal/sq ft/d is approximately 1.7 L/m²·h). Salt rejection determines the capability of a membrane to reject dissolved solids. Salt rejection typically is indicated by a ratio of concentrations: salt concentration in feed/salt concentration in brine (reject) (expressed as a percent). Another reverse characteristic to salt rejection is the salt passage, which is indicated by a ratio of concentrations: salt concentration in permeate/ salt concentration in feed (expressed as a percent)

One of the difficulties of the original reverse osmosis design was confinement of large hydraulic pressures associated with the process. Over the years, several novel designs have been developed to cope with this problem.

The first design, known as tubular configuration (see Figure 14.64), incorporates a porous wall tube with the membrane inserted or cast on the inner wall. The brine flow is axial within the tube, while the product-water flow is radial through the membrane and the porous structure. Tubes typically are arranged in bundles, each of which is called a module. All module tubes are connected in series so that the module has only one brine entrance and exit. The product water is collected in a trough or shell. Tubes are approximately 13 mm (0.5 in.) in diameter and are perforated or made from epoxy-bound fiberglass, woven nylon or polyester, or other porous materials of sufficient strength to withstand hydraulic pressure.

The second design is the spiral-wound or spiral-wrap configuration (see Figure 14.65). The process can best be understood if one visualizes a sealed envelope made from a

TABLE 14.15 Summary of Pretreatment Alternatives for Reverse Osmosis Membrane Process

Pretreatment	$CaCO_3$	$CaSO_4$	$BaSO_4$	$SrSO_4$	CaF_2	SiO_2	SDI	Fe	Al	Bacteria	Oxid. Agents	Org. Matter
Acid addition	●[a]							○[b]				
Scale inhibitor	○	●	●	●	●							
Softening with IX[c]	●	●	●	●	●							
Dealkalization with IX	○	○	○	○	○							○
Lime softening	○	○	○	○	○							○
Preventive cleaning	○											
Adjustment of operation parameter		○	○	○	○	○				○		
Media filtration						○	○	○	○			
Oxidation–filtration						○	○	●				
In-line coagulation						●	○	○	○			
Coagulation–flocculation						○	○	○	○			●
Micro-/ultrafiltration						●	●	○	○	○		●
Cartridge filter						○	●	○	○	○		
Chlorination							○	○	○	●		
Dechlorination											●	
Shock treatment										○		
Preventive disinfection										○		
GAC[d] filtration										○	●	●

[a] ● = very effective.
[b] ○ = possible.
[c] IX = ion exchange.
[d] GAC = granular activated carbon.

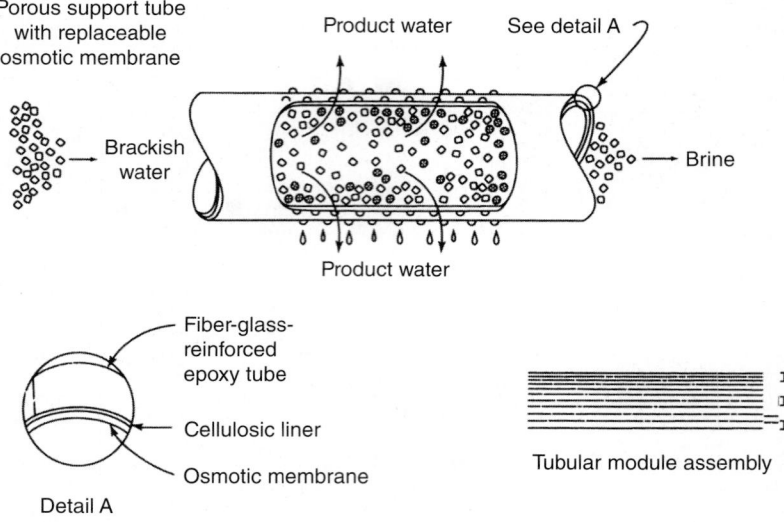

Porous support tube
with replaceable
osmotic membrane

Product water

See detail A

Brackish
water

Brine

Product water

Fiber-glass-
reinforced
epoxy tube

Cellulosic liner

Osmotic membrane

Detail A

Tubular module assembly

FIGURE 14.64   Reverse osmosis tubular model configuration (Cohen, 1971).

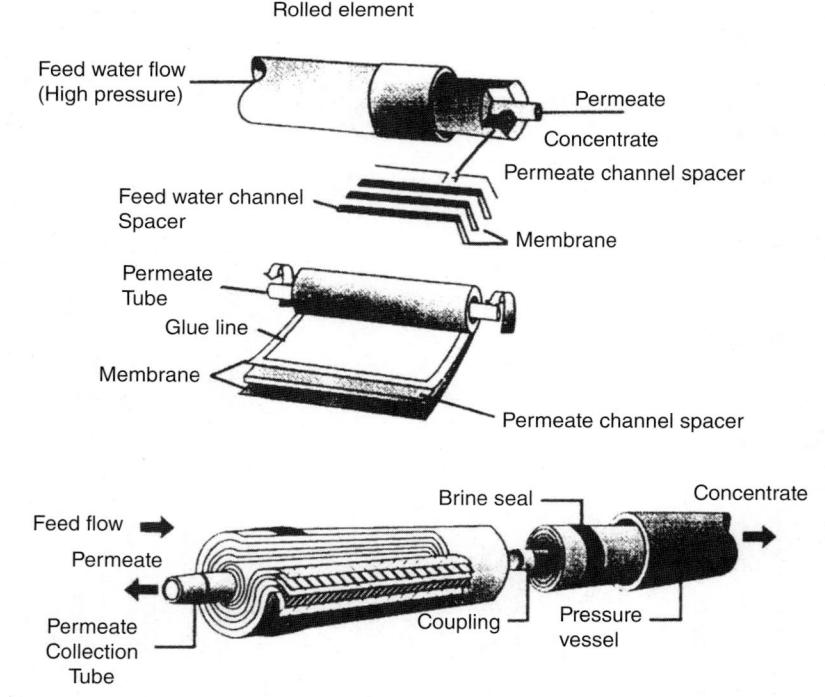

Rolled element

Feed water flow
(High pressure)

Permeate

Concentrate

Permeate channel spacer

Feed water channel
Spacer

Membrane

Permeate
Tube

Glue line

Membrane

Permeate channel spacer

Brine seal

Concentrate

Feed flow

Permeate

Permeate
Collection
Tube

Coupling

Pressure
vessel

FIGURE 14.65   Spiral-wound module configuration.

membrane with a porous backing material inside the envelope. As feedwater pressure is applied to the outside of the envelope, product water traverses the membrane and collects in the porous material within the envelope. The product water is tapped off the envelope along one open edge that is glued to a plastic collection tube. Holes in the side of the tube provide an entrance for product water. One or more envelopes (up to 30) are connected to the same tube. After the brine-side separator screens are placed on each side of the envelope, the whole assembly is wrapped around the collection tube. The rolled-up unit (a cylindrical module) is then placed in a pressure vessel. The pressure vessel is a plastic-lined pipe with feedwater entrance and brine exit at the ends. A product-water exit is provided at the center of one end. Several modules typically are placed end to end in the pressure vessel with O-ring connectors used to fit product collection tubes together. A seal is provided between the outer wrap of each module and the inside of the pipe. Feedwater entering the pipe is directed to the brine-side separator screens in the module and the outer-wrap seal prevents short-circuit brine flow around the module. Brine flows axially through the module by way of separator screens. The product collection tube is well sealed so that contamination by salt water cross leakage is not possible.

The third design is the hollow fine fiber configuration presented in Figure 14.66. As the name implies, a hollow fiber of 25 to 250 μm (micrometer) in diameter (approximately the size of a human hair) is manufactured. The fiber has a wall thickness of 5 to 50 μm (micrometer) and is composed of unsupported membrane material. The use of unsupported membrane is possible because of the small diameter of the fibers, which experiences relatively low stress, even under high pressure. In the original design, the brine flows external to the fibers and the product flows through the fibers. Millions of fibers are looped into a U-shape and the ends are potted into a special epoxy resin, which serves as a tube sheet. The whole mass of fibers is then inserted into a containment vessel. Feedwater, under pressure, enters one end of the vessel and exits at the other end. The flow of water through the shell side of the containment vessel is countercurrent to the flow of product water inside the hollow fibers.

There are several reasons why reject pressure external to the fibers is preferable to internal pressure. First, the fiber wall will withstand greater pressure under compression than under tension. Second, the probability of the fibers clogging is less, and the configuration is easier to clean. In addition, this design functions in a fail-safe mode. If a fiber fails, then it is closed by external pressure, thereby preventing product-water contamination.

One of the outstanding features of the hollow-fiber design is the extremely high membrane area that can be installed in a relatively small space. Packing densities as high as 40 000 m^2/m^3 (10 000 sq ft/cu ft) are possible compared to 300 to 1000 m^2/m^3 (100 to 300 sq ft/cu ft) for the spiral-wound configuration and 130 to 300 m^2/m^3 (40 to 100 sq ft/cu ft) for the tubular design. The hollow fine fibers are manufactured from polyamide, polypropylene, and other organic materials. Their permeability is approximately two orders of magnitude less than typical cellulose acetate membranes.

The tubular design has one distinct advantage over other configurations. It has large, well-defined flow passageways and, therefore, is less affected by feedwater containing a high degree of particulate matter, which means less tendency for clogging and channeling. In addition, slimes and scales are relatively easy to remove by chemical cleaning. The disadvantage of the tubular design is the large number of tubes and fittings required per unit of surface area and relatively high initial cost.

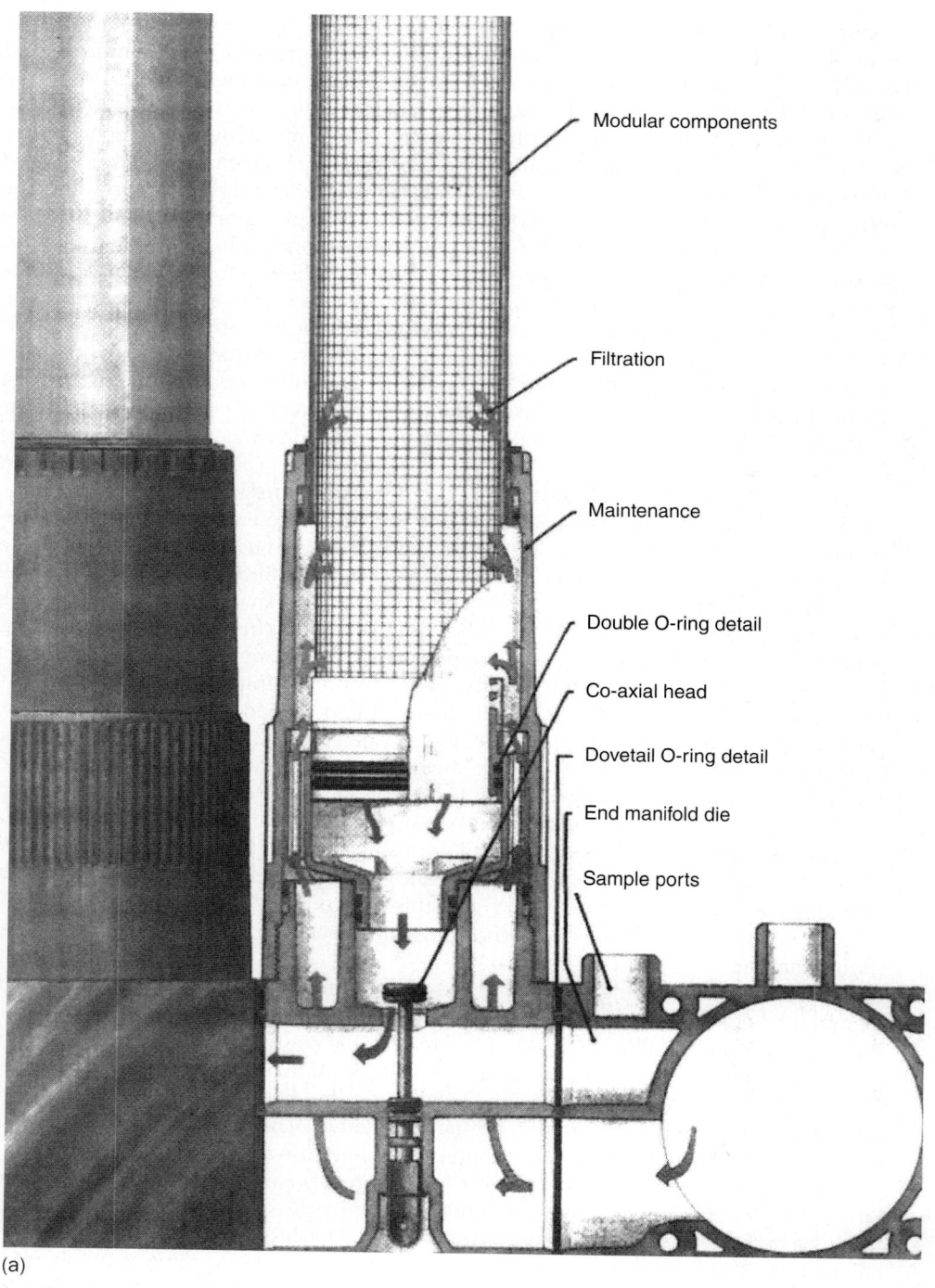

Modular components

Filtration

Maintenance

Double O-ring detail

Co-axial head

Dovetail O-ring detail

End manifold die

Sample ports

(a)

FIGURE 14.66 Hollow fiber module configuration: (a) lower half; (b) upper half.

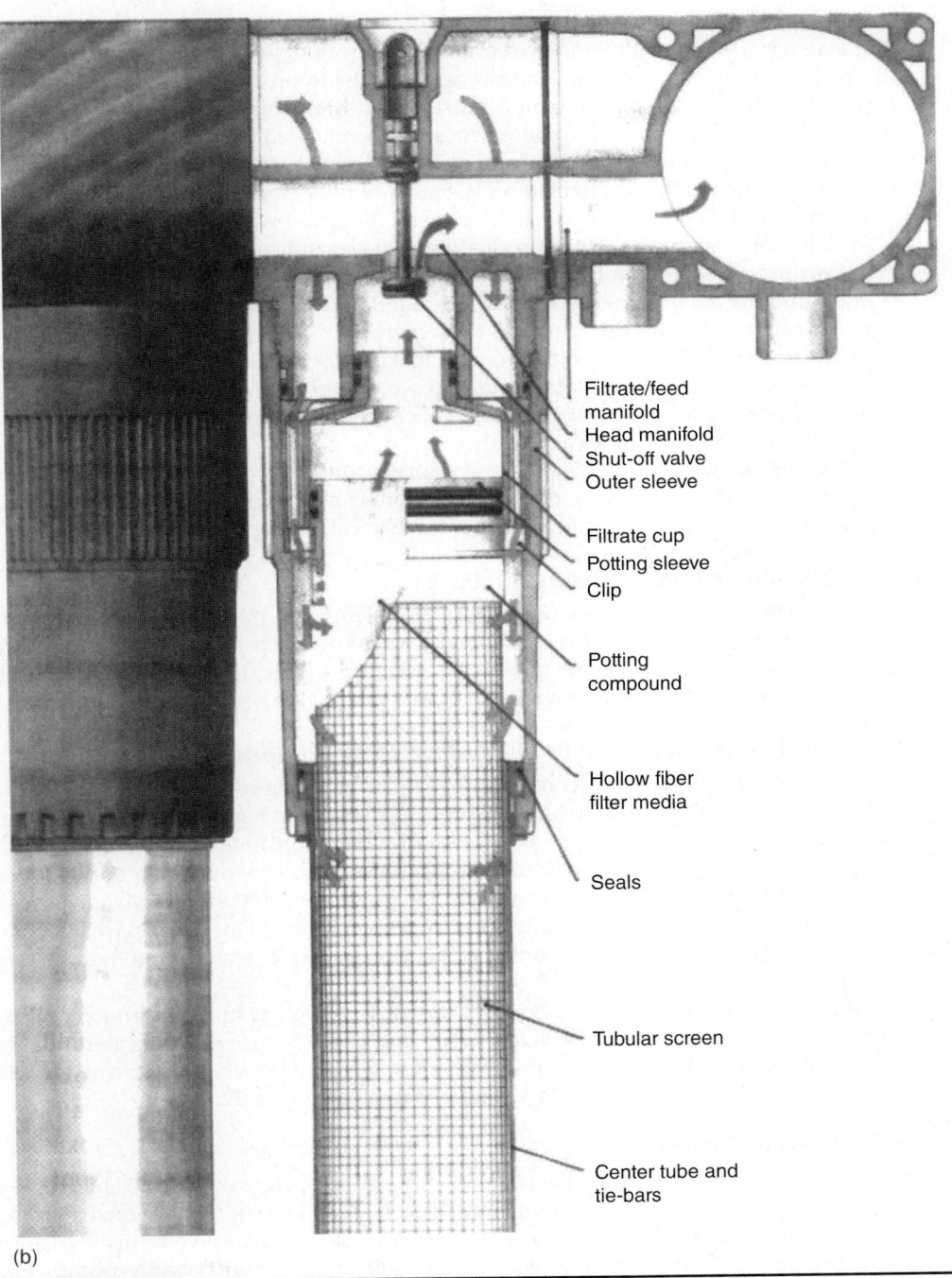

(b)

FIGURE 14.66 *(Continued)*

The spiral-wound and hollow-fine-fiber designs have the advantage of high membrane area packing density and a record of leak-proof operation. It is essential when using these designs for reverse osmosis applications to provide a high degree of pretreatment (filtration). Close attention must be paid to pretreatment, and slimes and other organic and inorganic formations must not be allowed to accumulate. The spiral-wound design permits easy field replacement of membranes. Similarly, the hollow-fine-fiber design lends itself to field replacement of fiber elements.

Currently, the HPM market is dominated by the spiral-wound membranes.

During operation, the water production rate per unit membrane area (flux) of a reverse osmosis facility tends to decline when operational pressure is kept constant. It is important for the designer to keep this in mind and provide for additional capacity where constant water production rates are required.

Flux decline has been traced to two principal causes: (a) membrane compaction and compression related to the sustained high-pressure fluid crushing porous substructures of membranes and (b) membrane fouling, which requires more frequent cleanings (and shorter lifespan of membranes as a result).

Salt rejection decline rates typically are not major concerns with reverse osmosis equipment, especially if cross leakage from brine to product can be controlled. Systems now on the market indicate that mechanical and nonmechanical seals appear to be adequate.

5.5   Membrane Module Configuration

Because there are several types of membrane systems and many types of membrane materials, the designer has several different ways in which to configure membrane modules to accomplish desired treatment objectives. Table 14.16 provides a design example for secondary effluent membrane filtration.

5.6   Membrane Separation by Microfiltration or Ultrafiltration

A typical membrane separation process for use in pretreating secondary effluent before demineralization with a reverse osmosis process is shown in Figure 14.67. In this example, secondary effluent is pumped through a strainer before entering membrane modules. The membranes produce two streams: (1) filtrate, which would be pumped to the reverse osmosis process, and (2) reject, or backwash, stream, which typically is returned to the headworks. Membranes in this example use compressed air for backwash operation, and the removed water is contained in the backwash water surge tank before returning to the headworks.

Membrane bioreactor technology uses the same or similar microfiltration or ultrafiltration membranes as the tertiary treatment process, providing water quality comparable to the microfiltration or ultrafiltration membranes when producing tertiary effluent. The MBR technology is described in detail in Chapter 12.

5.7   Reverse Osmosis

Figure 14.68 shows a simple single-array reverse osmosis process with three membrane modules as an example. The current standard pressure vessels are designed for one, two, and up to eight HPM elements. The feedwater is treated by reverse osmosis after pretreatment and boosted to the required pressure by the high-pressure pump. The modules produce two process streams: (1) permeate, which is the product water, and (2) concentrate or reject, which is a waste stream. Throttling with the concentrate control valve controls the ratio of permeate to reject. The ratio of the permeate flow to the feed flow is called system recovery and is expressed in percent.

Input Parameters	Units	Design Values/ Considerations
Annual average flow	m³/d	19 000
Peak-day flow	m³/d	38 000
Peak-hour flow to the WRRF	m³/d	57 000
Peak-hourly flow to the treatment from the equalization ponds	m³/d	26 500
Influent turbidity range	NTU	2–10
Average influent turbidity	NTU	5
Influent TSS range	mg/L	5–25
Average influent TSS	mg/L	12
Average design membrane flux	L/m²·h	34
Average water production ratio (system recovery)	Percent	90
Range of the reject ratio	Percent	5–15
Average reject ratio	Percent	10
Redundancy requirement	Provide filtration capacity with one membrane unit out of service	
Design output		
General		
Membrane type		Microfiltration
Membrane system configuration		Immersed, OUT–IN
Flow direction during backwash cycle		IN–OUT
Unit sizing		
Minimum total active membrane area required	m²	22 500
Active area of one membrane element	m²	31
Total number of membrane elements	Number	735
Total number of membrane basins	Number	5 (4 + 1)
Number of the membrane elements in one membrane basin	Number	147
Active membrane area in one membrane basin	m²	4500
Filtration rate		
Annual average flow, five basins in service	L/m²·h	34
Annual average flow, four basins in service	L/m²·h	42.5
Peak-hour flow to the WRRF	L/m²·h	N/A
Peak-hour flow to WRRF from equalization ponds, five basins in service	L/m²·h	47.6
Peak-hour flow to WRRF from equalization ponds, four basins in service	L/m²·h	59.5

TABLE 14.16 Design Example for Secondary Effluent Membrane Filtration (*Continued*)

Input Parameters	Units	Design Values/ Considerations
Membrane characteristics		
Membane material	Polyvinylidene fluoride	
Nominal pore size	Microns	0.04
Cleaning protocol		
Back-pulse	1 time every 0.5 hour	
Chemical enhanced cleaning		1 time every 24 hours
Recovery clean		1 time every 3 months
Transmembrane pressure (vacuum)		
Maximum allowed	Bar-vacuum	−0.8
Typical operation	Bar-vacuum	<−0.7
Recovery clean recommended at TMP	Bar-vacuum	−0.3
Ancillary units required		
Chemical addition		
Assumed chemical addition requirement (ratio to total operational time)	Percent	15%
Chemical storage	Days	30
Mixing method		Inline static mixer
Chemical pump type		Mechanical diaphragm
Pump speed control		Variable speed
Chemical type		Polymer
Dose range, assumed	mg/L	0.1–1.5
Typical dose, assumed	mg/L	0.3
Chemical type		Coagulant
Dose range, assumed	mg/L	2.5–5
Typical dose, assumed	mg/L	3.0
Membrane air scour blowers	No (duty, redundant)	4 + 1
Inline turbidity analyzers	No (duty, spare)	4 + 1

Note: In this design example, it is assumed that the flows above peak day are equalized upstream of filtration facilities and membrane are not loaded above 26.5 L/min (7.0 mgd); WRRF = water resource recovery facility; TSS = total suspended solids.

TABLE 14.16   Design Example for Secondary Effluent Membrane Filtration (*Continued*)

Figure 14.69 shows a typical two-array reverse osmosis facility layout, with the modules arranged in two stages. Incoming feedwater is pretreated, filtered, and raised to design pressure with high-pressure feed pumps. Feedwater then flows through by entering four modules (array 1) and then two modules (array 2), which is called a

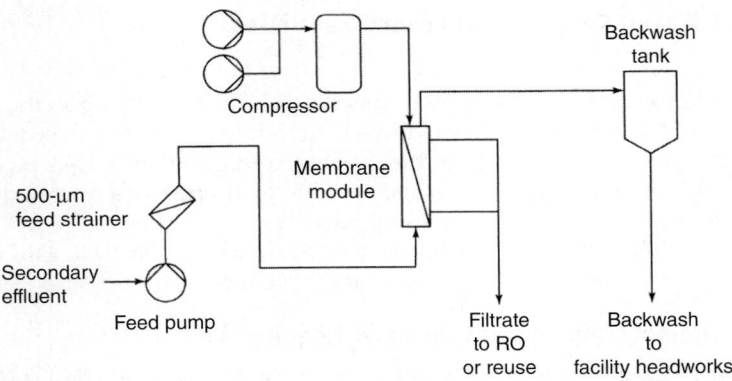

FIGURE 14.67 Simplified process schematic diagram for microfiltration process as pretreatment for reverse osmosis process.

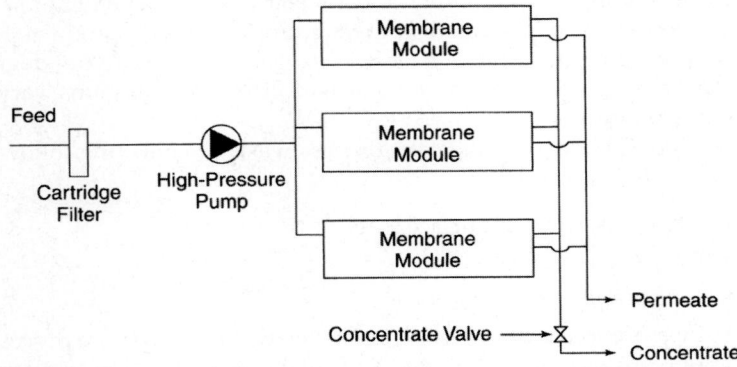

FIGURE 14.68 Simplified schematic diagram of a single-array reverse osmosis process.

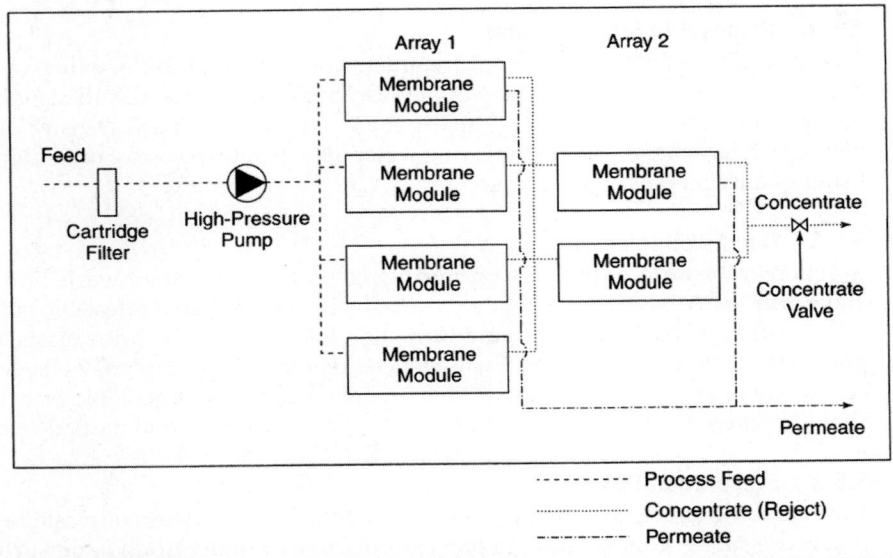

FIGURE 14.69 Simplified schematic diagram of a two-array reverse osmosis process.

tapered flow pattern. The tapered flow keeps brine velocity up as the product water is removed. A concentrate valve reduces the module pressure of the concentrated brine.

A significant advantage of the tapered configuration is that product recovery is increased compared to single-stage operations. If array 1 is operated to produce 75% recovery, and array 2, which is using reject from array 1 as its feed, is operated at 75% recovery, then total process recovery will be approximately 94%. This results in a reject flow of 6% for the two-array configuration versus 25% for the single-array process.

5.8   Reject/Concentrate Disposal Requirements

Disposal of wastewater brine presents significant engineering and economic problems. The old approach of freely discharging brine waste to streams, rivers, oceans, ponds, or underground reservoirs is carefully regulated today.

The HPM waste brines may vary from 2000 to 20 000 mg/L TDS or greater concentrations depending on the salinity of the water treated. Depending on the process, the concentrate may also be heated, contain corrosion-erosion products from the facility, have a pH ranging from 4 to 9, be low in oxygen, or have varying degrees of turbidity. The LMP may produce reject streams with TSS concentrations varying from 50 to 1000 mg/L or greater.

For high TDS demineralization process waste streams, the following methods of disposal are recommended:

- Direct discharge to surface water;
- Deep-well injection;
- Evaporation-pond disposal;
- Evaporation to dryness and crystallization by a separate process;
- Brine lines; and
- Blending with the other streams.

5.8.1   Disposal to Surface Water

Typically, direct discharge without treatment to a stream, lake, or other water body degrades water quality. Permitting of the discharge would require sufficient investigation and analysis of the effect of the discharge on water quality. Density differences between the discharge and the receiving water may dictate the use of three-dimensional water quality models.

5.8.2   Deep-Well Injection

Injection to subsurface strata is a promising method of disposal for waste brine. The oil and gas industry has used the method extensively. Such disposal is feasible only at locations where a suitable underground formation for receiving the brine exists, and each potential site must be evaluated. State agencies regulate brine disposal wells. A properly designed system should be based on sound engineering and geologic principles and should prevent the waste from adversely affecting freshwater and natural resources.

5.8.3   Evaporation Ponds

Evaporation from surface ponds is another method of disposing of waste brine. This practice is subject to regulation to prevent pollution of underground and surface environments. Brine disposal ponds can be costly to permit and construct because of the liner systems required in many locations and the land costs associated with solar evaporation. The use of brine disposal ponds will find application primarily in relatively warm, dry climates with high evaporation rates, level terrain, and low land costs.

5.8.4 Evaporation to Dryness and Crystallization
Evaporation to dryness and crystallization of waste brine to disposable solid salts is probably the most expensive approach to waste brine disposal. It would be used only if legal or site restrictions eliminated other disposal techniques or if a valuable byproduct could be recovered. This method is in use in process industries on products of greater value than are typically recoverable from saline water. The solid salts produced would either be sold or, more likely, require transportation to a disposal area.

5.8.5 Local Brine Management (Brine Lines)
The use of brine lines is a relatively new approach in which agencies combine their high TDS streams for discharge to a common brine line. The combined high TDS water is discharged to a receiving water body with a high TDS concentration such as a sea or ocean.

5.8.6 Blending with Other Streams
Blending with other streams can be implemented when the combined blended water TDS is less concentrated than the receiving stream. An example would be blending of the high TDS stream with discharge water before discharging it to a receiving water body with a higher TDS concentration than the blended water.

5.8.7 Brine Recovery Reverse Osmosis
Brine recovery reverse osmosis (BRRO) is an efficient technology for reducing the concentrate stream. The idea is to treat the primary reverse osmosis stream using BRRO to reduce the volume of the produced concentrate. Most designed BRROs can be operated with recovery in the range of 40% to 60%. For example, the discharged volume of reject from the primary reverse osmosis can be reduced by 40% to 60%, minimizing the effect on brine disposal, while not changing the mass balance of the discharged solids.

6.0 Air Stripping for Ammonia Removal

6.1 Process Description
Under properly controlled conditions, air stripping can remove ammonia nitrogen (NH_4-N) from wastewater because ammonia (NH_3) exists predominantly in the unionized, gaseous form at high pH levels. No ammonia stripping facilities are known to be routinely in use in the United States. Existing installations have been abandoned because of the inappropriate nature of the technology for year-round standards, air emission constraints, and maintenance difficulties. Nevertheless, this chapter briefly summarizes ammonia stripping because this process may be considered in certain regions with moderate climates and in combination with high-lime phosphorus removal. This section deals exclusively with ammonia removal through air stripping towers.

In most wastewaters, ammonia nitrogen is present in the form of ammonium ions. By increasing pH, ions in solution are converted to the molecular form:

$$NH_4^+ \rightarrow NH_3 + H^+$$

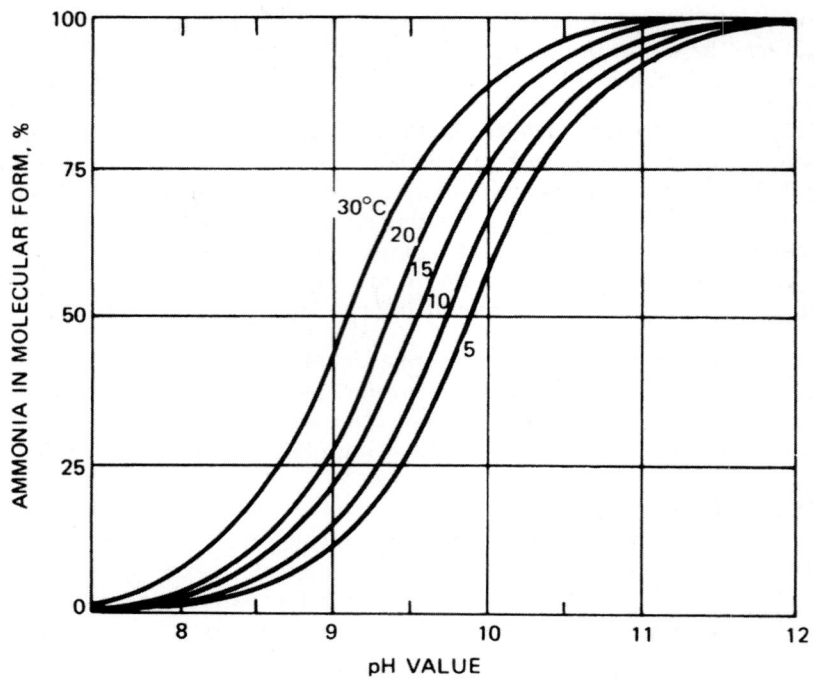

FIGURE 14.70   Relationship between pH value and percentage of ammonia, molecular or free volatile forms, in aqueous solutions.

Figure 14.70 illustrates the effects of pH and temperature on the conversion of ammonium ions to ammonia gas. At a pH of 7, nearly all ammonia nitrogen is in the form of ammonium ions; at a pH of 12, nearly all is in the form of dissolved ammonia gas. To optimize removal of ammonia by air stripping, wastewater pH must be elevated to between 10.8 and 11.5, depending on liquid temperature, to ensure that ammonia is in gaseous form. Following stripping, effluent pH is reduced by recarbonation or acid addition, with the liquid passing to the next unit process.

If ammonia is in the gaseous form, it may be liberated from solution by passing the liquid over a stripping tower. The transfer rate of dissolved ammonia gas to air depends on surface tension at the air and water interface and the difference in ammonia concentration between the water and air.

Surface tension is at a minimum in water drops when the surface film is being formed. At the instant of droplet formation, ammonia release is greatest. Little additional ammonia release occurs after the water drop completely forms. The concentration of ammonia in the air surrounding water droplets is minimized by circulating air through the tower.

The ammonia stripping process is most compatible with a physical and chemical wastewater treatment process such as high-lime coagulation that provides the high pH necessary for the downstream stripping tower. Pure physical and chemical treatment is used rarely in municipal wastewater applications, but ammonia stripping may be used with the two-stage lime process for precipitation of phosphate or the typically used biological PhoStrip process.

Where a biological process precedes the ammonia stripping process, however, nitrification must not occur in the biological process. If nitrification does occur, then ammonia is converted to nitrites and nitrates through oxidation by nitrifying bacteria, thus reducing the amount of molecular ammonia in solution. This nitrogen conversion restricts the total combined nitrogen removal capability of the stripping tower. Thus, with biological nitrification, other nitrogen removal techniques, such as anaerobic denitrification, can remove nitrates if total nitrogen reduction is required.

As illustrated in Figure 14.71, air stripping towers may be either countercurrent (air inlet at base) or cross flow (air inlet along entire depth of fill). A fan mounted on top or at the bottom of the tower forces air through the tower packing. A drift eliminator located near the top of the tower provides headloss so that air is uniformly distributed throughout the tower.

Wastewater leaving the fill collects in a basin and then flows to the next unit process. Filtration may be needed because of precipitation of calcium carbonate sludge. Recarbonation is necessary. Typically, the design provides the capability for recycling tower effluent to increase the removal of ammonia and nitrogen during cool weather.

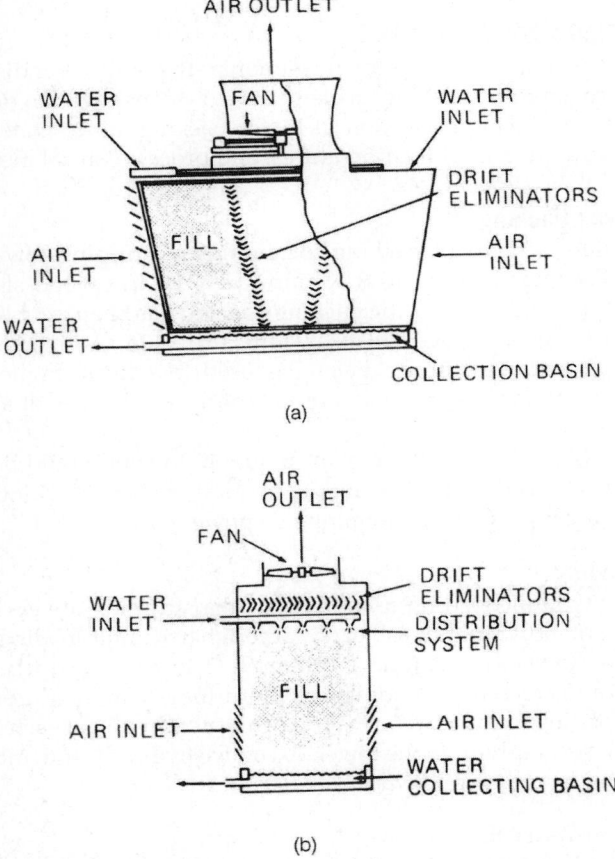

Figure 14.71 Types of Ammonia Stripping Towers: (a) Cross-Flow and (b) Countercurrent.

One severe limitation of the ammonia stripping process must be considered: operating a stripping tower at air temperatures lower than 0°C (32°F) is impossible because of freezing within the tower, which significantly reduces ammonia removal.

Experience with ammonia stripping towers has shown that scaling on the tower media is troublesome. Although it promotes efficient ammonia removal, high pH, and absorption of carbon dioxide from air, leads to buildup of calcium carbonate solids scale on the tower packing. Carbon dioxide absorption reduces pH, thus destabilizing the aqueous system.

The resulting encrustation is categorized as hard or soft, depending on the work required for its removal. Soft scale can be removed readily with a light spray nozzle or acid washing. If the tower fill is cleaned before hard-scale formation, tower packing can be restored to its original condition and operational efficiency. If hard scale forms, removal of the tower packing for cleaning often is necessary. Scale-formation problems may be partially solved by use of countercurrent rather than cross-flow towers and by arranging the tower packing to facilitate removal for cleaning. Performance reliability requirements for this process may necessitate duplicate equipment.

6.2 Design Considerations

Selection of design parameters for an ammonia stripping tower depends on the required level of nitrogen removal. Critical elements to be considered in designing an ammonia stripping tower include selection of tower packing, air-to-water flowrate, hydraulic loading rate, air and liquid temperatures, and process control measures.

6.2.1 Tower Packing

Several different types of packing for ammonia stripping towers are commercially available. Packings include 10×38 mm (0.4×1.5 in.) wood slats, plastic pipe, or a polypropylene grid. No specific packing spacing has been established. Typically, individual splash bars are spaced 38 to 100 mm (1.5 to 4 in.) horizontally and 50 to 100 mm (2 to 4 in.) vertically. Tighter spacing is used to achieve higher levels of ammonia removal and more open spacing is used where lower levels of ammonia removal are acceptable.

Because of the large volume of air required, towers should be designed for a total air headloss of less than 50 to 80 mm (2 to 3 in.) of water. Packing depths of 6 to 7.6 m (20 to 25 ft) are typically used to minimize power costs.

6.2.2 Loading

Allowable hydraulic loading used in ammonia stripping towers depends on the type and spacing of individual splash bars. Although hydraulic loading rates used in ammonia stripping towers range from 2 to 7 m/h (1 to 3 gpm/sq ft), removal efficiency is significantly decreased at loadings of more than 9 m/h (2 gpm/sq ft). A suitable hydraulic loading rate allows formation of a water droplet at each individual splash bar as the liquid passes through the tower. Excessive hydraulic flowrates cause water sheeting, reducing ammonia removal.

6.2.3 Air-to-Water Ratio

Large volumes of air must pass through the tower to achieve high levels of ammonia removal. Required air-to-water ratios range from 2200:1 to 3800:1. The 6 to 7.6 m (20 to

NH$_4$-N Removal (%)	Air Supply (cfm/gpm waste)[a]	Hydraulic Loading[b] (gpm/sq ft)[c]
80	200	3.9
85	210	3.5
90	250	3.0
95	400	2.0
98	800	0.8

[a]cfm/gpm $\times$ 7.48 $\times$ 10^{-6} = m^3/min air per m^3/d wastewater.
[b]Fill depth equals 24 ft (7.2 m); influent pH was raised to 11.5; and wastewater temperature was 20°C.
[c]gpm/sq ft $\times$ 2.444 = m/h.

TABLE 14.17   Effect of Air and Hydraulic Loading on Ammonia Removal

25 ft) of tower packing typically produces a pressure drop of 0.12 to 0.37 kPa (0.5 to 1.5 in. H$_2$O). Table 14.17 illustrates an example of the air requirements necessary to achieve various levels of ammonia removal using 7.3 m (24 ft) of 38 $\times$ 50 mm (1.5 $\times$ 2 in.) wood packing (Slechta and Culp, 1967).

6.2.4 Temperature

Air and liquid temperatures influence design of ammonia stripping towers. Minimum operating air temperature and associated air density should be considered when sizing the fans and air blowers to provide the desired air supply. Liquid temperature also affects the level of ammonia removal. Table 14.18 illustrates the effect of effluent water temperature in a tower with a packing depth of 7.3 m (24 ft), a hydraulic loading of 9 m/h (2 gpm/sq ft), and an air-to-water ratio of 3600:1 (Smith and Chapman, 1967).

7.0 Ammonia Removal By Breakpoint Chlorination

7.1 Process Theory

Oxidation by chlorine can remove ammonia from wastewater. This process, known as breakpoint chlorination, is practical only as an effluent polishing technique, not for

Effluent Water Temperature (°C)	NH$_4$-N Removal (%)*
17	91
12	79
9	71
3	67

*Hydraulic loading was 2 gpm/sq ft (5 m/h); air supply was 480 cfm/gpm (0.003 59 m^3/min air per m^3/d wastewater); fill depth was 24 ft (7.3 m); pH of influent was raised to 11.5; packing was 38 mm $\times$ 50 mm (1.5 in. $\times$ 2 in.) redwood slats.

TABLE 14.18   Effect of Water Temperature on Ammonia Removal

removing high levels of influent nitrogen. When chlorine is added to wastewater containing ammonia and nitrogen, the ammonia initially reacts with hypochlorous acid to form chloramines. Continued addition of chlorine to the water beyond the "breakpoint" forms free chlorine, which converts the chloramines to nitrogen gas. The following series of reactions represent chlorine oxidation of ammonia to nitrogen gas:

$$Cl_2 + H_2O \rightarrow HOCl + HCl$$

$$NH_4^+ + HOCl \rightarrow NH_2Cl + H_2O + H^+$$

$$2NH_2Cl + HOCl \rightarrow N_2 + 3HCl + H_2O$$

The extent to which various chloramines are formed depends on pH, contact time, temperature, and reactant concentrations. The overall reaction may be expressed as:

$$2NH_3 + 3Cl_2 \rightarrow N_2 + 6HCl$$

A typical breakpoint chlorination curve is shown in Figure 14.72. Breakpoint occurs when the ammonia is reduced to zero, the total residual chlorine is minimized, and a free chlorine is detectable. The theoretical amount of chlorine necessary to oxidize 1 mg/L of ammonia and nitrogen to nitrogen gas is 7.6 mg/L chlorine.

The time required for the above reactions to proceed to completion depends on pH, concentration of each reacting substance, and temperature. In practice, reactions require approximately 9 mg/L of chlorine for each 1 mg/L of ammonia and nitrogen.

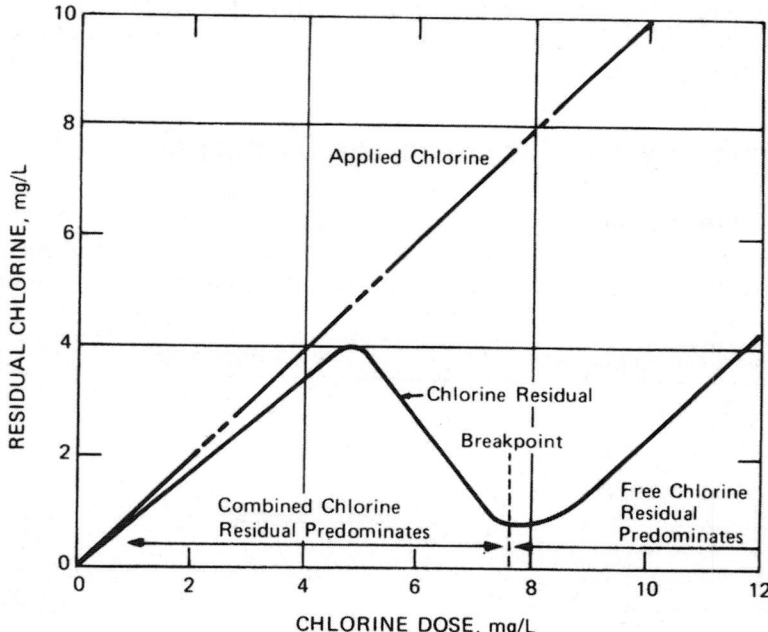

FIGURE 14.72   Typical curve for breakpoint chlorination.

Typical wastewater requires a large amount of chlorine that appreciably depresses the pH. Reactions are slower at lower pH values; therefore, wastewater high in ammonia and nitrogen content requires buffering to a pH of approximately 7 to reach breakpoint with short contact times. Consequently, the process may be impracticable for wastewater high in ammonia and nitrogen if breakpoint chlorination is the sole means of ammonia removal. In addition to altering the required contact time, variables such as pH, temperature, ammonia and nitrogen concentration, and amount of chlorine applied affect the nature of chloramine reaction products. For example, low pH values favor formation of toxic, malodorous nitrogen trichloride (NCl_3). Therefore, design of an effective breakpoint reaction chamber must provide for control of these variables. When the free chlorine residual process has the dual objective of viral inactivation and nitrogen removal, the chlorination chamber design must provide the reaction time required for breakpoint to occur and contact time required for disinfection.

Two undesirable side reactions may occur in breakpoint chlorination. The first is that monochloramine (NH_2Cl) can react with hypochlorous acid (HOCl) to form dichloramine ($NHCl_2$) and ultimately nitrogen trichloride (NCl_3), a noxious gas. Nitrogen trichloride forms at a pH of less than 6.5 according to the following reaction:

$$NHCl_2 + HOCl \rightarrow NCl_3 + H_2O$$

The second is oxidation of ammonia and nitrogen to nitrate ions. Chlorination beyond the breakpoint encourages formation of nitrate ions rather than nitrogen gas (U.S. EPA, 1975a):

$$NH_3 + 4Cl_2 + 3H_2O \rightarrow HNO_3 + 8HCl$$

7.2 Design Considerations

Design of the mechanical components of a breakpoint chlorination system is simple. Both chlorine and sufficient alkali must be added in the same rapid mixer. The problem of pH control is crucial. Approximately 30 mg/L of alkalinity must be present for each 1 mg/L of ammonia and nitrogen to maintain the proper pH range. The first reaction between ammonia and chlorine occurs rapidly, and no special design features are necessary except ensuring complete, uniform mixing of chlorine and wastewater. Good mixing can best be attained with in-line mixers or back-mixed reactors. A minimum contact time of 10 minutes is a good practice.

Sizing of the chlorine-producing and feed device depends on influent ammonia concentration and degree of treatment the wastewater has received. As the quality of influent to breakpoint improves, the required amount of chlorine decreases and approaches the theoretical amount required to oxidize ammonia to nitrogen (7.6 mg/L Cl_2:1 mg/L NH_3-N). Table 14.19 summarizes quantities of chlorine required based on operating experience and suggested design capabilities. These ratios apply to the maximum anticipated influent ammonia concentration. If insufficient chlorine is added to reach the breakpoint, then no nitrogen forms, and the chloramines formed must be destroyed before discharge. Design should include provisions to monitor effluent continuously following chlorine addition for free chlorine residual and to pace the chlorine feed device to maintain a set-point free chlorine residual.

Wastewater Type	Chlorine:NH$_3$-N Ratio of Reach Breakpoint	
	Experience	Suggested Design Capability
Raw	10:1	13:1
Secondary effluent	9:1	12:1
Lime settled and filtered secondary effluent	8:1	10:1

TABLE 14.19   Quantities of Chlorine Required for Three Wastewater Types

The chemical feed assembly used for ammonia removal by breakpoint chlorination should be considered in the preliminary design of the chlorination system, including requirements for pre-, intermediate-, and post-chlorination applications. Depending on use of continuous chlorination at points within the system, use of standby chlorination equipment deserves some consideration for the ammonia removal system. Reliability needs and maximum dosage requirements for various application points also should be examined when sizing equipment.

Other equipment needs for the breakpoint chlorination process stem from the formation of excess acid that some wastewaters do not neutralize and the elimination of active chlorine residual from the effluent by dechlorination. Theoretically, 14.3 mg/L of alkalinity as calcium carbonate is required to neutralize the hydrochloric acid produced by the oxidation of 1 mg/L of ammonia and nitrogen gas. To keep the pH higher than 6.3, at least twice the theoretical alkalinity is needed. Except for wastewater having a high alkalinity and treatment systems using lime coagulation before chlorination, designs typically include means of feeding an alkaline chemical to keep the pH of the wastewater in the proper range. Also, a method for measuring and pacing the alkaline chemical feed pump to keep the pH in the desired range is necessary. Note that if chlorine and alkali are not paced together, the whole process is upset. If the pH lowers momentarily so that nitrogen trichloride forms, raising the pH does not correct this problem; if the pH goes higher than 8, the breakpoint will occur slowly. If NaOCl is used instead of chlorine, the alkalinity requirement is reduced by 75%.

Dechlorination (Chapter 11) may be required as a companion process to breakpoint chlorination. Equipment needs depend on the method of treatment selected.

8.0   Effluent Reoxygenation

Reoxygenation of treated wastewater effluent is often necessary if permit limits require high dissolved oxygen concentrations. Two principal categories of reoxygenation systems are cascade reoxygenation and mechanical or diffused air reoxygenation.

8.1   Cascade Reoxygenation

These systems rely on air and water interfaces at weir overflows, flumes, spillways, and similar hydraulic structures. Precise measurements of oxygen supplied (mass) or changes in dissolved oxygen are difficult to obtain. The process also is difficult to control. Typically, these systems are not energy efficient, but they work well and cost little if hydraulic head is sufficient and the design cannot be modified to conserve the head or use it for other more efficient purposes.

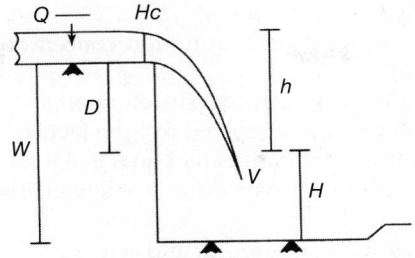

Figure 14.73 Graphical presentation of the terms in cascade reaeration equations (Nakasone, 1987).

In weir systems, reaeration occurs at the weirs crest during water-surface formation and is enhanced by bubble entrainment and splashing in the lower pool. Studies by Nakasone (1987) indicate that reaeration over a single weir can be estimated by the following equations (see Figure 14.73). The similar results were shown by numerous studies that followed.

$$\ln r_{20} = 0.0785(D + 1.5H_c)^{1.31}q^{0.428}H^{0.310} \qquad (14.9)$$

for $(D + 1.5H_c) > 1.2$ m and $q </= 235$ m³/m·h

$$\ln r_{20} = 0.0861(D + 1.5H_c)^{0.861}q^{0.428}H^{0.310} \qquad (14.10)$$

for $(D + 1.5H_c) </= 1.2$ m and $q > 235$ m³/m·h

$$\ln r_{20} = 5.39(D + 1.5H_c)^{1.31}q^{-0.363}H^{0.310} \qquad (14.11)$$

for $(D + 1.5H_c) > 1.2$ m and $q > 235$ m³/m·h

$$\ln r_{20} = 5.92(D + 1.5H_c)^{0.816}q^{-0.363}H^{0.310} \qquad (14.12)$$

where D = drop height, m;

H_c = critical water depth on the weir;

q = discharge per width of weir, m³/m·h;

H = tailwater depth for downstream channels having horizontal beds, m; and

r_{20} = deficit ratio at 20°C.

For deficit ratios at other water temperatures (T), the following equation may be used (Gameson et al., 1958):

$$\ln r_T = \ln r_{20} [1 + 0.0168(T - 20)] \qquad (14.13)$$

Another method used for determining required cascade height is based on the following equations developed by Barrett et al. (1960):

$$H = R - 1/0.361 \, ab \, (1 + 0.046T) \quad \text{(SI units)} \qquad (14.14)$$

$$H = R - 1/0.11 \, ab \, (1 + 0.046T) \quad \text{(U.S. customary units)} \qquad (14.15)$$

where R = deficit ratio = C_s & C_0/C_s & C;
  C_s = dissolved oxygen saturation concentration of wastewater at temperature T, mg/L;
  C_0 = dissolved oxygen concentration of postaeration, mg/L;
  C = required final dissolved oxygen level after postaeration influent, mg/L;
  a = water quality parameter equal to 0.8 for a WRRF effluent;
  b = weir geometry parameter (for a weir, $b = 1.0$; for steps, $b = 1.1$; for step weir, $b = 1.3$);
  T = water temperature, °C; and
  H = height through which water falls, m (ft).

A key element in the use of this method is the proper selection of the critical wastewater temperature that affects the dissolved oxygen saturation concentration, C.

8.2 Mechanical/Diffused Air Reoxygenation

These techniques encompass a wide range of mechanically assisted reoxygenation systems, including mechanical surface aeration (fixed or floating), jet diffusion, pump aerators, agitator-sparged systems, U-tube aerators, and diffused air (coarse- or fine-bubble) systems. These systems are amenable to engineering analysis and can typically be controlled to conserve energy, prevent overaeration, and reduce costs. Chapter 8 contains a discussion of these techniques.

8.3 Relationship to Other Unit Processes

Typically, reoxygenation should be the last step in the treatment process. Reoxygenation upstream of effluent filters, in the absence of compelling considerations to the contrary, is inadvisable because:

- Some filter designs allow negative heads to occur in filter media. High dissolved gases in the unfiltered effluent may escape from the solution in the filter bed, producing bubbles and air binding.
- For nonnitrified effluents, nitrification by attached growths in the filter media can consume dissolved oxygen in the reoxygenated effluent, offsetting, at least partially, the reoxygenation.

8.4 References and Design Procedures

The U.S. EPA (1971d) offers useful descriptions of various reoxygenation systems available. It contains calculation methods for estimating operational variables for cascade reoxygenation systems. Calculation procedures for mechanical/diffused air systems are presented elsewhere (Kormanik, 1969, 1970; Lin, 1979; Pincince et al., 1999).

9.0 References

American Public Health Association; American Water Works Association; Water Environment Federation (2017) *Standard Methods for the Examination of Water and Wastewater*, 23rd ed.; American Public Health Association: Washington, D.C.

American Society of Civil Engineers (1986) Tertiary Filtration of Wastewater. *J. Environ. Eng.*, **112** (6), 1008.

American Water Works Association (1974) *Standards for Granular Activated Carbon*, No. B604-74; American Water Works Association: Denver, Colorado.

American Water Works Association (1990) *Water Quality and Treatment*, 4th ed.; McGraw-Hill: New York.

American Water Works Association; American Society of Civil Engineers (2012) *Water Treatment Plant Design*, 5th ed.; McGraw-Hill: New York.

Amirtharajah, A. (1979) Discussion of "Effectiveness of Backwashing for Wastewater Filters," by J. L. Cleasby and J. C. Lorence. *J. Environ. Eng.*, **105** (EE2), 59.

Babbitt, H. E.; Doland, J. J.; Cleasby, J. L. (1962) *Water Supply Engineering*, 6th ed.; McGraw-Hill: New York.

Barrett, M. J.; Gameson, A. L. H.; Ogden, C. G. (1960) Aeration Studies of Four Weir Systems. *Water Water Eng.*, **64** (9), 407.

Baumann, E. R.; Huang, J. Y. C. (1974) Granular Filters for Tertiary Wastewater Treatment. *Water Pollut. Control*, **46** (8), 1958.

Baumann, P.; Kieffer, C.; Morrall, T.; Bauer, R (2006) Fox Metro Water Reclamation District Solves Filtration Problems with Innovative Aquadiamond® Cloth Media Filter. *Proceedings of the 79th Annual Water Environment Federation Technical Exposition and Conference* [CD-ROM], Dallas, Texas, October 21–25; Water Environment Federation: Alexandria, Virginia.

Bishop, D. F.; Marshall, L. S.; O'Farrell, T. P.; Dean, R. B.; O'Connor, B.; Dobbs, R. A.; Griggs, S. H.; Villiers, R. V. (1967) Studies on Activated Carbon Treatment. *Water Pollut. Control*, **39** (2), 188.

Bourgeous, K.; Riess, J.; Tchobanoglous, G.; Darby, J (2003) Performance Evaluation of a Cloth-Media Disk Filter for Wastewater Reclamation. *Water Environ. Res.*, **75** (6), 532–538.

Brown and Caldwell (1978) *West Point Pilot Plant Study, Volume V, Granular Media Filtration*. Prepared for Municipal Metropolitan Seattle, Washington; Brown and Caldwell: Walnut Creek, California.

Caliskaner, O. (1999) Depth Filtration of Activated Sludge Effluent Using a Synthetic Compressible Medium. Doctoral Dissertation, University of California at Davis.

Caliskaner, O.; Tchobanoglous, G. (2005a) Investigation of the Effects of the Filtration Parameters on the Removal Performance of the Compressible Medium Filter. *Proceedings of the 78th Annual Water Environment Federation Technical Exposition and Conference* [CD-ROM], Washington, D.C., October 29–November 2; Water Environment Federation: Alexandria, Virginia.

Caliskaner, O.; Tchobanoglous, G. (2005b) Modeling Depth Filtration of Activated Sludge Effluent Using a Compressible Medium Filter. *Water Environ. Res.*, **77**, 3080.

Caliskaner, O.; Tchobanoglous, G.; Carolan, A. (1999) High-Rate Filtration of Activated Sludge Effluent with a Synthetic Compressible Media. *Water Environ. Res.*, **71** (6), 1171.

Caliskaner, O.; Tchobanoglous, G.; Carolan, A.; Downey, M. (2011a). Evaluation of the New Compressible Media Filter at Filtration Rates between 30 and 40 gpm/ft² for Unrestricted Water Reuse. *Proceedings WEFTEC 2011*, Los Angeles, CA.

Caliskaner, O., Bennett, G. B.; Celeste, S. (2011b) Comparison of Three Filtration Technologies to Meet Tertiary Level Discharge and Unrestricted Reuse Requirements. *Proceedings WEFTEC 2011*, Los Angeles, California.

Cleasby, J. L.; Baumann, E. R. (1974) *Wastewater Filtration Design Considerations*, U.S. Environmental Protection Agency Technology Transfer Seminar Publication; U.S. Environmental Protection Agency: Washington, D.C.

Cleasby, J. L.; Lorence, J. C. (1978) Effectiveness of Backwashing Wastewater Filters. *J. Environ. Eng.*, **104** (EE4), 749.

Cleasby, J. L., Stangl, E. W.; Rice, G. A. (1975) Developments in the Backwashing of Granular Filters. *J. Environ. Eng.*, **101** (5) (EE5), 713.

Cleasby, J. L.; Arboleda, J.; Burns, D. E.; Prendiville, P. W.; Savage, E. S. (1977) Backwashing of Granular Filters. *J. Am. Water Works Assoc.*, **69** (2), 115.

Cohen, J. M. (1971) Demineralization of Wastewaters. Water Reuse Symp., Dallas, Texas, Sponsored by U.S. EPA.

Culp, R. L.; Culp, G. L. (1978) *Handbook of Advanced Wastewater Treatment*, 3rd ed.; Van Nostrand Reinhold Co.: New York.

Daniels, S. L. (1975a) *Coagulation/Flocculation: State-of-the-Art*; Dow Chemical Co.: Midland, Michigan.

Daniels, S. L. (1975b) Solid-Liquid Separation by Coagulation and Flocculation. In *Particle Science and Solid Fluid Separation*; Chemical Engineering Department, University of Houston: Houston, Texas.

Eliassen, R.; Bennett, G. (1967) *Progress Report, Reclamation of Re-Usable Water from Sewage, Federal Water Pollution Control Association Demonstration Grant WPD 21-05*; U.S. Department of the Interior: Washington, D.C.

Fair, G. M.; Geyer, J. C. (1954) *Water Supply and Waste-water Disposal*; John Wiley and Sons: New York

Fair, G. M.; Geyer, J. C.; Okun, D. A. (1968) *Water and Wastewater Engineering*, vol. 2; Wiley and Sons: New York.

Frenkel, V. (2002) Use Pre-Treatment to Improve Process Water Treatment. *Chem. Eng. Mag.*, New York. http://www.globalspec.com/reference/9433/349867/use-pretreatment-to-improve-process-water-treatment (accessed June 2017).

Frenkel, V. (2004) Membranes vs. Conventional Treatment in Municipal and Industrial Applications. Paper Presented at American Membrane Technology Biennial Conference, San Antonio, Texas; August 5–7.

Frenkel, V. (2008) Membrane Technologies: Past, Present and Future: North American Perspectives. Plenary Keynote Presentation. *International Water Association Conference*, Moscow, Russia; June 2–6.

Gameson, A. L. H.; Vandyke, K. G.; Ogden, C. G. (1958) The Effect of Temperature on Aeration. *Water Water Eng. (G.B.)*, **62** (753), 489.

Griffith, E. J.; Beeton, A.; Spencer, J. M.; Mitchell, D. T., Eds. (1973) *Environmental Phosphorus Handbook*; Wiley and Sons: New York.

Haecker, S.; Healy, J. (2006) Innovative Technology to Implement a Reuse Water Program. *Proceedings of the 79th Annual Water Environment Federation Technical Exposition and Conference* [CD-ROM], Dallas, Texas, October 21–25; Water Environment Federation: Alexandria, Virginia.

Hassler, J. W. (1963) *Activated Carbon*; Chemical Publishing Co.: New York.

Hassler, J. W. (1974) *Purification with Activated Carbon: Industrial, Commercial, Environmental*; Chemical Publishing Co.: New York.

Holden, B.; Nelson, K.; Crook, J.; Cooper, R. C.; Williams, G.; Kouretas, T.; Sheikh, B. (2006) Higher Filter Loading Rates for Greater Water Reuse Capacity. *Proceedings of the 79th Annual Water Environment Federation Technical Exposition and Conference* [CD-ROM], Dallas, Texas, October 21–25; Water Environment Federation: Alexandria, Virginia.

Huang, J. Y. C. (1979) Filter Backwashing by Ejector. *J. Environ. Eng.*, **105** (EE5), 915.

Knight, C. H.; Mondox, R. G.; Hambley, B. (1973) Thickening and Dewatering Sludges Produced in Phosphate Removal. *Proceedings of the Phosphorus Removal Design Seminar*, May 28–29, Toronto. Conference No. 1, Canada–Ontario Agreement on Great Lakes Water Quality. Training and Technology Transfer Division (Water), Environmental Protection Service, Environment Canada: Ottawa, Canada.

Kormanik, R. A. (1969) Simplified Mathematical Procedure for Designing Post Aeration Systems. *Water Pollut. Control*, **41** (11), 1956.

Kormanik, R. A. (1970) Design of Plug-Flow Post-Aeration Basins. *Water Pollut. Control*, **42** (11), 1922.

Leslie, G. L.; Mills, W. R.; Dunivin, W. R.; Wehner, M. P.; Sudak, R. G. (1998) Performance and Economic Evaluation of Membrane Processes for Reuse Applications. *North American Biennial Conference and Exposition: American Desalting Association*, Williamsburg, Virginia; August 2–6, 1998; American Desalting Association: Sacramento, California

Levine, A.; Harwood, V.; Scott, T; Rose, J. (2004) Effectiveness of Secondary Effluent Filtration for Removal of Bacteria, Enteroviruses, and Protozoan Pathogens in Wastewater Reclamation Facilities. *Proceedings of the 77th Annual Water Environment Federation Technical Exposition and Conference* [CD-ROM], New Orleans, Louisiana, October 4–7; Water Environment Federation: Alexandria, Virginia.

Lin, S. H. (1979) Axial Dispersion in Post-Aeration Basins. *Water Pollut. Control*, **51**, 985.

Liptak, B. G. (1974) *Environmental Engineers' Handbook*, Vol. I; Chilton Book Co.: Radnor, Pennsylvania.

Mattson, J. S.; Kennedy, F. W. (1971) Evaluation Criteria for Granular Activated Carbon. *Water Pollut. Control*, **43** (11), 2210.

Miska, V.; Van Der Graaf, J. H. J. J.; De Koning, J. (2006) Improvement of Monitoring of Tertiary Filtration with Particle Counting. *Water Sci. Technol.*, **6** (1), 1–9.

Naddeo, V.; Belgiorno, V. (2007) Tertiary Filtration in Small Wastewater Treatment Plants. *Water Sci. Technol. (Small Water and Wastewater Systems VII)*, **55** (7), 219.

Nakasone, H. (1987) Study of Aeration at Weirs and Cascades. *J. Environ. Eng.*, **113** (1), 64.

Novak, J. T.; O'Brien, J. H. (1975) Polymer Conditioning of Chemical Sludges. *Water Pollut. Control*, **47** (10), 2397.

Pincince, A. B.; Paschka, M.; Ghosh, R.; Dzombak, D. (1999) Effect of Multiple Compartments on Oxygen Transfer in Postaeration Tanks. *Water Environ. Res.*, **71** (6), 1129.

Pretorius, W. A. (1972) The Complete Treatment of Raw Sewage with Special Emphasis on Nitrogen Removal. *Sixth International Water Pollution Research Conference*, June 21; International Association on Water Pollution Research and Control; 685.

Priesing, C. P. (1962) A Theory of Coagulation Useful for Design. *Ind. Eng. Chem.*, **54** (8), 38.

Rushton, J. H. (1952) Mixing of Liquids in Chemical Processing. *Ind. Eng. Chem.*, **44** (12), 2931.

Russell, J. S.; Micclebrooks, E. J.; Reynolds, J. H. (1980) *Wastewater Stabilization Lagoon: Intermittent Sand Filter Systems*, EPA-600/2-80-032; U.S. Environmental Protection Agency: Cincinnati, Ohio.

Slechta, A. F.; Culp, G. L. (1967) Water Reclamation Studies at the South Tahoe Public Utility District. *Water Pollut. Control*, **39** (5), 787.

Smith, C. E.; Chapman, R. L. (1967) Reco of Coagulant, Nitrogen Removal, and Carbon Regeneration in Wastewater Reclamation, Final Report of South Tahoe Public Utilities District to Federal Water Pollution Control Administration, Demonstration Grant WPD-85; South Tahoe Public Utilities District: South Lake Tahoe, California.

Snoeyink, V. L.; Weber, W. J., Jr.; Mark, H. B., Jr. (1969) Sorption of Phenol and Nitrophenol by Active Carbon. *Environ. Sci. Technol.*, **3** (10), 918.

Sollo, F. W., Jr.; Mueller, H. F.; Larson, T. E. (1976) Communication: Denitrification of Wastewater Effluents with Methane. *Water Pollut. Control*, **48** (7), 1840.

Stumm, W.; Morgan, J. J. (1962) Chemical Aspects of Coagulation. *J. Am. Water Works Assoc.*, **54** (8), 971.

Tchobanoglous, G.; Eliassen, R. (1970) Filtration of Treated Sewage Effluent. *J. Sanit. Eng. Div., Proc. Am. Soc. Civ. Eng.*, **96** (SA2), 243.

Tooker, N.; Darby, J.; Tchobanoglous, G.; Mikkelson, K.; Johnson, L. (2004) A Pilot Study Designed to Define the Operating Protocol for a Multiple Barrier Membrane System. *Proceedings of the 77th Annual Water Environment Federation Technical Exposition and Conference* [CD-ROM], New Orleans, Louisiana, October 2–6; Water Environment Federation: Alexandria, Virginia.

U.S. Environmental Protection Agency (1970) *Carbon Column Operation in Waste Water Treatment*, Project No. 17020 DZO, WPCR Series 11/70; U.S. Environmental Protection Agency: Washington, D.C.

U.S. Environmental Protection Agency (1971a) *Effect of Porous Structure on Carbon Activation*, Project No. 17020 DDC, WPCR Series 06/71; U.S. Environmental Protection Agency: Washington, D.C.

U.S. Environmental Protection Agency (1971b) *Improving Granular Carbon Treatment*, Project No. 17020 GDN, WPCR Series 07/71; U.S. Environmental Protection Agency: Washington, D.C.

U.S. Environmental Protection Agency (1971c) *Process Design Manual for Suspended Solids Removal*, Technology Transfer; U.S. Environmental Protection Agency: Washington, D.C.

U.S. Environmental Protection Agency (1971d) *Process Design Manual for Upgrading Existing Wastewater Treatment Plants*, Technology Transfer, Program No. 17090 GNQ, Contract No. 14-12-933; U.S. Environmental Protection Agency: Washington, D.C.

U.S. Environmental Protection Agency (1973) *Process Design Manual for Carbon Adsorption*, EPA-625/1-71-002a, Technology Transfer; U.S. Environmental Protection Agency: Washington, D.C.

U.S. Environmental Protection Agency (1975a) *Nitrogen Control*, Technology Transfer; U.S. Environmental Protection Agency: Washington, D.C.

U.S. Environmental Protection Agency (1975b) *Process Design Manual for Suspended Solids Removal*, EPA-625/1-75-003a, Technology Transfer; U.S. Environmental Protection Agency: Washington, D.C.

U.S. Environmental Protection Agency (1987) *Design Manual Phosphorus Removal*, EPA-625/1-87-001, Technology Transfer; U.S. Environmental Protection Agency: Cincinnati, Ohio.

Vaughn, S. D; Hurley, W.; Madhanagopal, T.; Kunihiro, K.; Slifko, T. (2004) Effect of Hydraulic Loading and Chemical Addition on Filter Performance and Protozoa Removal; a Side-by-Side Comparison. *Proceedings of the 77th Annual Water Environment Federation Technical Exposition and Conference* [CD-ROM], New Orleans, Louisiana, October 2–6; Water Environment Federation: Alexandria, Virginia.

Water Environmental Federation (2013) *Operation of Nutrient Removal Facilities*, WEF Manual of Practice No. 37; Water Environment Federation: Alexandria, Virginia.

Weber, W. J., Jr. (1972) *Physiochemical Processes for Water Quality Control*; Wiley-Interscience: New York.

Williams, G.; Sheikh, B.; Holden, R.; Kouretas, T.; Nelson, K. (2007) The Impact of Increased Loading Rate on Granular Media, Rapid Depth Filtration of Wastewater. *Water Res.*, **41** (19), 4535–4545.

CHAPTER 15
Sidestream Treatment

Dimitri Katehis; Timothy A. Constantine; Anthonie Hogendoorn, MSc.;
Jose Christiano Machado Jr., Ph.D., P.E.; and Samir Mathur, P.E., BCEE

1.0   Introduction

With the ongoing transition from wastewater treatment to resource recovery, water resource recovery facilities (WRRFs) are deploying new process configurations that increasingly result in the production of concentrated sidestreams that can present new challenges for the facility, but also lend themselves opportunities to target and transform or harvest key constituents from within the water/nutrient/energy cycle.

Sidestreams are defined as flows containing nutrients, solids, and organic or inorganic constituents that are generated within a WRRF, typically during biosolids processing. Common sidestreams include overflow from gravity thickeners, filtrate from belt filter presses (BFP) or gravity belt thickeners (GBT), centrate from thickening or dewatering centrifuges, and scrubber water from incinerators. Although filter backwash water and other sidestreams containing solids should not be neglected in mass balances, this chapter focuses on a more critical design consideration: nutrient- and organics-rich sidestreams that are generated during dewatering of digested or otherwise processed biosolids that release nitrogen, phosphorus, and potentially soluble organic carbon. When this centrate or filtrate is returned to the head of the facility, it can result in significant increases in a facility's transient nitrogen, phosphorus, and organic loading, depending on the nature of the solids processing scheme. A typical example is the return of the dewatering sidestream to the head of the facility during single-shift dewatering operations that occur only a few days per week, as is commonly practiced in smaller to medium-sized facilities. Designs must account for these loads and be able to identify the optimal location (mainstream treatment processes or sidestream treatment) and process that needs to be deployed to manage them. Understanding the characteristics of the particular sidestream and what components can be sustainably harvested from it are at the core of the decision-making process of how it should be managed. The decision-making process, design approach, and considerations associated with sidestream treatment systems are the focus of this chapter.

1.1   Sidestream Nutrient and Organics Management

The designer has several options for addressing nutrient- and organics-rich sidestreams:

- Diverting to the mainstream processes,
- Exporting to specialized facilities, or
- Treating and harvesting or removing specific compounds such as nutrients prior to returning the liquid back to the mainstream process.

The strategy selected will depend on the type and averaging periods of the effluent discharge limits and biosolids use options and the utility's environmental stewardship goals; for example, a facility that is only required to remove biochemical oxygen demand (BOD) and total suspended solids (TSS) is less affected by sidestream loads than a facility with stringent total nitrogen and phosphorus limits. Cost and impacts on other elements of facility operation will also play an important role in sidestream management strategies. For example, the impact of the sidestream process on the biosolids characteristics, such as changes in dewaterability, increased nitrogen, or reduced phosphorus content in the biosolids may drive the process selection decision. It is thus important to develop a whole-facility mass balance and evaluate changes in nutrient and organic loadings as well as impacts to other unit processes, through both the liquid and solids processing facilities and address capital, life cycle costs, and sustainability considerations during the decision-making process.

1.1.1   Nutrient and Organics Removal

As of this update, the dominant technologies in the field rely on removal of nitrogen and of soluble and particulate organics from the sidestream, rather than their recovery. Biological oxidation of the organics and nitritation/anammox processes for conversion of ammonia to inert nitrogen gas are used. The application of nitrification and nitritation/denitritation processes has been eclipsed by deammonification due to the elimination of the supplemental carbon requirement and significantly reduced energy requirements. Limited application of nitrititation/denitritation still exists in facilities where bioaugmentation of ammonia oxidizers to the mainstream process is necessary to enhance low-temperature mainstream nitrification or restart nitrification after a process upset. Examples include the City of Richmond, Virginia, WWTP and New York City's AT-3 process used at the Hunts Point, 26th Ward, and Wards Island facilities. Nitrification-based processes are also used in facilities where nitrate can be applied to manage odors in the mainstream headworks and primary treatment processes such as in Phoenix's 91st Avenue facility.

Where phosphorus removal (rather than recovery) is practiced from sidestreams, it is typically via precipitation with metal salts such as ferric or alum. In such instances, the metal salt is typically added upstream of dewatering (and possibly digestion) to maximize the benefits it imparts throughout the facility. Induced precipitation of struvite (without downstream recovery) is being more widely implemented upstream of dewatering, targeting the sludge stream both prior to and downstream of digestion, thereby reducing the orthophosphate content in the produced dewatering liquor (centrate or filtrate) and enhancing the dewaterability of the liquid biosolids via multivalent cation manipulation (Higgins and Novak, 1997).

1.1.2   Nutrient and Organics Recovery

The greater than fourfold increase in nitrogen and phosphate fertilizer prices in 2007 and 2008 resulted in the rapid commercialization of primarily phosphate recovery technologies, with limited efforts also directed toward nitrogen removal. Struvite, a fertilizer composed of approximately 15% ammonia, phosphate, and magnesium has been the primary recovery product although calcium-based phosphate products are also being recovered. Multiple facilities in North America using a variety of struvite recovery flowsheets are in operation, with improvements to the process being commercialized on a routine basis.

The direct recovery of nitrogen has been more problematic, although advances have been made in the application of vacuum distillation, ion exchange, and membrane-based processes. However, as of this update, the relatively poor economics and energetics of ammonia recovery relative to ammonia production via the Haber–Bosch process have resulted in no full-scale nitrogen recovery facilities in operation in North America.

1.2   Designing for Facility-Wide Integration

Deployment of a sidestream treatment process will affect the nutrient balance, the solids balance, and the carbon balance, and thus energy consumption and generation potential for the facility. Additionally, the dewatering characteristic of the liquid biosolids may be affected. A structured approach to defining the impact of alternative sidestream treatment processes is necessary to build a solid business case that will allow a concept to move forward to implementation. Key elements are discussed in the following sub-sections.

1.2.1   Mass Balance

As part of a sidestream system design, the impact of the reduction in nutrient and potentially solids loadings to the main facility should be evaluated. The main facility's oxygen requirements, supplemental carbon requirements, nuisance struvite, energy demand, and sludge production are among the many operating characteristics that will be affected. In facilities that have a marginal alkalinity balance and require supplemental alkalinity addition, provision of nitrogen removal in the sidestream may be adequate to allow for a significant reduction in alkalinity addition or even demobilization of the supplemental alkalinity facilities. Use of a whole facility simulator that incorporates biological and physicochemical reactions should be considered to allow a holistic assessment of alternative sidestream treatment options.

1.2.2   Energy Balance

Sidestream treatment has been identified as a potent tool for facilities that are attempting to materially reduce their energy consumption as well as for increased energy generation potential. By reducing the energy requirements for nitrogen removal and reducing sludge production, energy consumption can be materially reduced. In facilities where phosphorus is removed from the sidestream via induced precipitation (whether as a metal phosphate or as a struvite product) the organics required in the biological phosphorus removal process are reduced, potentially allowing organic carbon to be diverted to the digestion process where it can be used to produce biogas or to other high-value uses. The designer should consider using integrated mass and energy flow models to ascertain the impacts of alternative sidestream treatment configurations on the facility.

1.2.3   Carbon Footprint

Sidestream treatment provides multiple opportunities to reduce the carbon footprint of water resource recovery operations. Removal of nutrient- and organic-rich loadings from the main facility will serve to reduce aeration and biosolids processing energy demands as deammonification-based nitrogen removal processes use approximately 1/3 the energy of conventional mainstream nitrification/denitrification processes and exhibit negligible biomass yields. For facilities that use supplemental carbon sources for denitrification in the main facility, the carbon footprint reduction will be even greater as supplemental carbon use will be reduced, but only if the supplemental carbon source is not biogenic. However, the reduction in WAS production that sidestream treatment of both nitrogen and phosphorus typically achieves, will reduce solids processing energy and chemicals, reducing the carbon footprint of treatment.

1.2.4   International Experiences Gained

Adoption of sidestream treatment technologies in Europe preceded widespread adoption in the United States by more than a decade. This has resulted in a significant body of lessons learned and evolution of designs that can be used in the United States. However, the reader is cautioned to be cognizant of the differences in upstream processes and facility design approach that can affect sidestream characteristics and thus treatment system design, requiring that processes that have been successfully deployed in European facilities be adapted to U.S. norms. For example, this may include the level of pretreatment required (as fine screen is less prevalent in U.S. facilities), the level of automation being provided, and the operational expertise necessary, as specialized labor costs are materially higher in the United States, and even access to a specialized labor pool is problematic for many U.S. utilities.

International design and operations experiences are included and incorporated in this chapter. For example, the Sharon reactor at WWTP Dokhaven in Rotterdam was the first full-scale nitritation/denitritation reactor using the proprietary SHARON® concept, and later on retrofitted into a SHARON-annamox® system. The lessons learned from several other technologies that are emerging in North America as of this writing, such as the deammonification DEMON® systems at WWPT Apeldoorn and Amersfoort (the Netherlands) and deammonification EssDe® system at WWTP Strass (Austria), Velsen and Hertogenbosch (the Netherlands) are incorporated.

Finally, the reader must be aware of the overall trend toward expanding sustainable processes such as deammonification from the sidestream to the mainstream. Where possible, a technically viable path toward integration of the mainstream and sidestream processes must be provided, that may drive the selection of a particular type of process. In general, granular type deammonification technologies appear to have the best chance of integration into the mainstream.

2.0 Sidestream Nutrient Removal and Recovery Design Considerations

2.1 General Design Considerations

Successfully deployed sidestream treatment requires that both the sidestream reactor and the ancillary support systems that convey and pretreat the dewatering reject liquor be carefully designed. The unique characteristics of the dewatering reject stream make minimization of precipitation and fouling critical considerations for design. The decision to move forward with a sidestream treatment scheme will require definition of the costs of managing the dewatering reject liquor from the point of production to the point of return back to the mainstream process.

The following list summarizes key rationale for the implementation of sidestream treatment for N-removal/recovery:

- It will increase the C:N of bioreactor feed, making mainstream nutrient removal more viable/reliable while reducing/eliminating the need for supplemental carbon for denitrification;
- It is cost competitive and more energy efficient in sidestream systems (e.g., anammox-based processes) than in the mainstream; and
- It provides biological seed (e.g., nitrifying organisms and/or anammox) that can be subsequently directed to the mainstream facility to enhance nitrification and nutrient removal and potentially lead to mainstream process intensification through reduction in mainstream SRT requirements or anammox organism seeding.

The following list summarizes key rationale for the implementation of sidestream treatment for P-removal/recovery:

- Increase in carbon to C:P ratio of bioreactor feed, making mainstream biological phosphorus removal (if practiced) more viable/reliable and allowing reduction in metal salt addition requirements (if chemical P-removal is practiced).

- Reduced need for supplemental VFA addition/production to maintain reliable biological phosphorus removal.

- Reduced recycle of phosphorus can have a concomitant reduction in unintentional struvite production in digesters and/or piping to and from dewatering.

- Potential revenue stream through sale of product (e.g., struvite).

- Enhancement of dewaterability of biosolids reducing transportation and beneficial reuse costs.

2.2   Sidestream Characteristics

The sidestream resulting from anaerobic digestion will have high ammonia, alkalinity, phosphorus, and possibly BOD concentrations compared to typical wastewater. Nutrient and organics concentrations in the reject water sidestream are a function of the type of digestion, upstream solids processing, and liquid stream processes used. For example, nutrient concentrations will be elevated under one or more of the following conditions: (1) the liquid stream employs biological phosphorus removal, (2) a waste activated sludge (WAS) pretreatment method is used to enhance digestion, (3) thermal hydrolysis or advanced anaerobic digestion technology is used to enhance volatile solids (VS) destruction (4) concentrated external organic carbon sources, such as slurried food scraps are fed directly to the digester.

To appropriately select and design the sidestream treatment process, a thorough investigation of the sidestream characteristics is required. Many of the sidestream processes are based on biological treatment; hence, alkalinity, pH, and temperature are critical constituents that must be identified. For example, a typical sidestream from anaerobic digestion may have a BOD concentration of 300 mg/L, ammonia concentration of 1000 mg/L, alkalinity concentration of 3500 mg/L, and a phosphorus concentration of 500 mg/L. If a sidestream treatment process is selected that provides full nitrification, approximately 1000 mg/L of nitrate will be formed and 7100 mg/L of alkalinity will be consumed. Roughly half of this alkalinity will be present in the sidestream, and about 3500 mg/L of alkalinity will have to be added to maintain a stable pH. Example sidestream characteristics that could be observed for various liquid and solids treatment scenarios are presented in Table 15.1.

When various sidestreams are comingled, care must be exercised to avoid diluting or cooling down the anaerobically digested biosolids dewatering stream. Reduction of the influent concentration, or even attempting to operate over a wide range of concentrations, can be deleterious to some configurations. For example, the presence of elevated soluble biodegradable organics in the dewatering sidestream is suspected as the primary cause of granular deammonification reactor failures. The provision of a dedicated BOD removal zone in the deammonification reactor (whether granular or moving bed biofilm reactor [MBBR]) is required to prevent this.

Source	BOD$_5$ (mg/L)	TSS (mg/L)	Ammonia (mg NH$_3$-N/L)	Orthophosphorus (mg P/L)
Thickening				
Gravity thickening supernatant	100–1200	200–2500	1–30	1–5 (higher if bio P)
Dissolved air flotation subnatant	50–1200	100–2500	1–30	1–5
Centrifuge centrate	170–3000	500–3000	1–30	1–5
Stabilization				
Aerobic digestion decant	100–2000	100–10 000	1–50	2–200
Anaerobic digestion supernatant	100–2000	100–10 000	300–2000 (higher end for retreated/ advanced processes)	50–200
Incineration Scrubber water	30–80	600–10 000	10–50	1–5
Dewatering				
Belt filter press filtrate	50–500	100–2000	1–1500	2–200
Centrifuge centrate	100–2000	200–20 000	1–1500	2–200
Sludge drying beds underdrain	200–500	20–500	1–1500	2–200

TABLE 15.1   Sludge Processing Sidestreams Characteristics (U.S. EPA, 1987)

In facilities that conduct phosphorus removal (whether enhanced biological or through metal salts addition), micronutrient addition may be required to supplement trace minerals that are precipitated by the phosphate complexes.

Where biosolids processing technologies that operate with atypically concentrated sludges and may include thermal or chemical pretreatment steps, the presence/impact of inhibitors should be investigated for deammonification. Free ammonia levels in excess of 20 mgN/L may require modification of operating pH or reactor configuration to reduce inhibitory effects for biological processing; similarly excessive hydrogen sulfide levels may require that offgasses be treated.

2.3 Nuisance Struvite Precipitation Control

MAP or Struvite (MgNH$_4$PO$_4$-6H$_2$O) is a white crystal that contains equal molar amounts of magnesium, ammonium, and phosphorus. Struvite precipitation occurs when the concentration of magnesium, ammonium, and phosphate ions exceed the solubility product constant (K_{sp}). Nuisance struvite precipitation clogs dewatering and dewatering liquor conveyance and treatment equipment posing significant operational challenges and costs to both BNR and non-biological nutrient removal (BNR) facilities. The precipitation reaction is affected by the pH, temperature, concentration of magnesium, ammonium and phosphate, and ionic strength of the process stream. Struvite solubility decreases as pH increases and a small portion of the orthophosphate is converted to phosphate ion, rather than the hydrogen phosphate and dihydrogen phosphate forms, which are the dominant forms at near neutral pH (6.5–7.5). Thus, the pH/phosphate relationship dominates the struvite precipitation potential in typical digestion and dewatering systems.

Section 4 of this chapter provides an overview of struvite formation fundamentals and design considerations as it applied to using struvite as a vehicle for phosphorus recovery under controlled conditions. In typical systems struvite precipitate is formed in the digestion, dewatering, and reject water conveyance/processing facilities. However, with the exception of severe nuisance struvite formation conditions (such as frequently observed in EBPR facilities), digestion facilities are not materially impacted. Nuisance struvite precipitation is typically observed in dewatering facilities, dewatering reject liquor conveyance systems and in reject liquor treatment processes.

Preventing nuisance struvite precipitation in the conveyance pumping and piping is a key consideration for biosolids processing and sidestream treatment systems. The buildup of struvite in piping and pumps reduces conveyance capacity and can be a severe bottleneck for not just sidestream treatment but dewatering operations as well. The incidents of nuisance struvite formation in North American facilities has increased over the past decade as an increasing number of facilities have converted to enhanced biological phosphorus removal (EBPR) for phosphorus removal. Struvite can also be aggravated in BOD removal and nitrogen removal facilities, as anaerobic conditions in the reactors can occur during periods of heavy organic loadings or when nitrate mass recycled to anoxic zones trails off diurnally, triggering inadvertent EBPR and thus accumulation of phosphate in the facility's biomass and release of that phosphate in the anaerobic digestion process.

A multi-tiered approach to struvite management is required in facilities that will have the combination of constituent concentrations and pH to trigger nuisance struvite precipitation. This should include:

- Mass balance evaluation of NH_3, Mg, PO_4, and alkalinity levels with an integrated whole facility model that incorporates liquid and solids processing. For initial iterations precipitation reactions should be turned off in the model, and then sequentially turned on to allow the designer to assess process sensitivity and gauge where the bulk of the nuisance struvite precipitation will occur.

- Recognizing the sensitivity of struvite precipitation to pH shifts, evaluate the process flow path from the digestion facility (including digester withdrawal and sludge conveyance/storage prior to dewatering) to the dewatering facility and the flow of the dewatering reject stream to the sidestream treatment process or the head of the facility. Maintain the liquid digested biosolids stream under pressure at all times, eliminating points where offgassing of carbon dioxide may occur (increasing pH and inducing precipitation) or where the energy in the process stream is rapidly increased (centrifugal pumping for example) that may cause initial precipitate nucleation. Influent feeding and withdrawal points into unheated sludge storage tanks operated at atmospheric pressure or shallow wet wells are frequent problem areas.

- Identify mitigation mechanisms where the struvite precipitation potential will be exceeded. These may include ferric choride, denucleation/antiscalant chemicals specific to struvite, or carbon dioxide injection. Struvite recovery may be used on the sludge or dewatering liquour stream to remove the phosphate and alter the pH so as to prevent downstream inadvertent struvite precipitation. Restabilization of the treated stream with chemical addition is typically needed after struvite recovery to avoid nuisance struvite precipitation in downstream processes. Each of the above chemicals operates on different mechanisms; the designer must chose the correct tool based on the specific conditions and the downstream impacts of each chemical.

- Ferric chloride reduces struvite precipitation potential by binding phosphate and removing carbonate alkalinity, thereby reducing pH. This dual action mode makes it an effective alternative and it is widely used due to its operational/handling simplicity. It is best applied to facilities with a positive alkalinity balance as the destruction of alkalinity may require addition of supplemental alkalinity for sidestream and/or mainstream nitrogen removal. Approximately 0.9 kg of alkalinity as $CaCO_3$ is destroyed per kg of ferric chloride added, whereas approximately 1 kg of precipitate is produced.
- Denucleation polymers prevent initial precipitate formation. They have been used successfully in European and US facilities; long-term trials are recommended that include biological assays to ensure that they do not negatively impact biological processing and mass transfer.
- Carbon dioxide addition prevents struvite precipitation by reducing the pH of the process stream. Carbon dioxide has been used in the United States for struvite prevention in sludge and reject liquor lines, and for the prevention of struvite in lines that feed struvite recovery equipment as the other options are not viable in those applications.

2.4 Equalization

Flow equalization is often the easiest and most cost-effective way to manage dewatering sidestreams. Although flow equalization does not necessarily break the nutrient cycle, it enables operators to manage the additional nutrient loads by returning sidestreams in a controlled manner. Equalization can be accomplished either operationally with a continuous dewatering schedule (e.g., 24 hours per day, 7 days per week), or by design by incorporating an equalization basin. Several strategies have been utilized including:

- Storage and Return During Low Flow—This strategy is particularly beneficial for facilities that practice dewatering only during daytime shifts and where the high ammonia loading during this period (which often coincides with the high influent diurnal loading) can result in elevated effluent ammonia concentration. In this case, return of stored sidestream flow often occurs during nighttime hours;
- Maintaining Optimal Carbon to Nutrient Levels—Rather than equalizing based on flow, another strategy is to return the sidestream based on influent loading. In particular, systems requiring tight effluent total nitrogen (TN) and total phosphorus (TP) can often vary the sidestream return to maintain an optimal carbon to nutrient ratio in the bioreactor feed.

Use of a dynamic process model that incorporates diurnal flow and concentrations profiles can assist in determining the benefits, limitations, preferred strategy (if any), and expected outcomes of incorporating equalization. Given the volumes required to achieve meaningful equalization and the relative simplicity and cost benefits of side-stream nitrogen and phosphorus removal/recovery ,the range within which an equalization only approach is viable will be very limited, likely applicable to facilities of less than 19 ML/d (5 mgd) capacity.

Equalization is also required for most sidestream biological nitrogen removal processes, as shifts in characteristics and shutdowns can affect process stability, especially with respect to some of the deammonification processes. In centralized/regional dewatering facilities where dewatering sludge characteristics can change over a few hours, equalization of characteristics, rather than flow is required. To maximize the equalization of dewatering reject water characteristics, upstream processing should be evaluated, as digestion/storage upstream of dewatering can be used to equalize characteristics (while also enhancing dewatering operations). Where facilities operate with limited dewatering schedules (say 3 or 4 days per week), spreading out dewatering over the course of the week, or the selection of a sidestream process (particularly with respect to deammonification) that is affected minimally by lack of influent feed, such as moving bed bioreactor processes, should be considered.

2.5   Pretreatment

While sidestream treatment systems have now been shown to be quite robust with appropriate control strategies in place, there are certain circumstances where pretreatment of the sidestream is required to prevent either inhibition or process instability.

Key considerations for biological N-removal configurations include the following:

- Screenings:
  - Processes that use plastic media for nitrogen removal biomass growth such as moving bed bioreactors, or fluidized beds for phosphorus recovery, can be catastrophically impacted by the long-term accumulation of fibrous materials. Granular deammonification reactors will also accumulate screenings over time, particularly where fine mesh screens are used for granular retention. Even where fine screens are used in the main facility flow, hair and other fines that pass through mainstream fine screens will accumulate in sidestream reactors.

Solutions:

- Screening to approximately 2 to 3 mm is recommended, particularly where fine screening to this level is not provided in the upstream liquid or solids processing steps. Steps to minimize the potential for struvite formation on the screens will likely be required. Alternatively, screening of thickened sludge using screen presses prior to digestion/processing can be used to remove screenings and provide dividends throughout the biosolids management program, including production of a higher quality biosolids product for beneficial reuse.
- High TSS:
  - Those processes that operate with a required minimum or design suspended growth solids retention time and MLSS can be negatively affected by high influent TSS, as this increases the sidestream tankage requirements and reduces deammonification biomass stability.

- Sidestream processes that utilize a separate separation device such as a hydrocyclone or fine mesh screen to retain granular biological material (i.e., anammox granules) can be negatively impacted by grit and other heavier TSS as this material may also be inadvertently retained in the system, which will ultimately limit capacity.

- Intermittent high TSS has been found to negatively impact the biological reaction rates in some sidestream treatment facilities.

Solutions:

- Consider diverting centrate away from sidestream treatment during startup and shutdown of dewatering equipment, when high levels of TSS can be present in the centrate. This is typically not an issue with other dewatering technologies.

- Consider implementing a solids removal step, with potential options including gravity sedimentation, dissolved air floatation, or a filtration technology. Ensure compatibility of the selected technology with polymer/solids as plugging has been widely reported.

- Polymer:

  - High residual polymer in the feed can lead to excessive foaming events in the sidestream treatment facility. In some facilities that include anammox granules and gravity sedimentation, this can create conditions where excessive loss of anammox can occur, which may impact treatment performance.

  - High polymer levels may contribute to reduced biological activity, perhaps due to the creation of oxygen/substrate diffusional limitations.

Solutions:

- Provide interlocks to ensure polymer dosing tracks dewatering equipment throughput.

- Consider incorporating a scum/foam removal step in the sidestream pretreatment step, or utilize dissolved air floatation for solids removal.

- Sidestreams arising from anaerobic digestion of thermally hydrolyzed sludge may contain high levels of ammonia or soluble biodegradable organics.

Solutions:

- Consider dilution water or internal recirculation with sidestream process effluent to reduce toxicity to biologically based sidestream treatment processes.

- Include organics oxidation step in sidestream reactor process configuration.

2.6 Supporting Chemical Systems

Whereas sidestream treatment will in general result in chemical and energy savings in the main facility, a variety of chemicals may be required and will depend on the type and degree of treatment used to reduce nutrients in the solids processing sidestream.

The impact of any chemical addition on mainstream processes should be considered. For example, chemical phosphorus removal works primarily on the principle of converting soluble phosphorus (orthophosphate) to an insoluble salt that can be removed using a solids separation process. One of three metal ions is commonly used to form this insoluble material: iron, aluminum, or calcium. The ability to remove soluble phosphate to very low concentrations is a function of the solubility of the metal phosphate formed by the metal ion. The addition of metal ions to decrease phosphorus concentration will increase the solids quantities produced onsite and, depending on the chemical used, may result in a reduction in pH, which will require alkalinity supplementation to minimize adverse effects on downstream biological processes.

For nitrogen reduction, the type of sidestream treatment process used dictates the chemical requirements. For the processes based on achieving oxidation of ammonia (via nitritation or nitrification), alkalinity supplementation may be required to balance the alkalinity consumed even if denitritation/denitrification is used. This is exacerbated by the use of chemicals such as ferric chloride for struvite control. Typically, the organic content in the anaerobic digestion reject water is significantly less than the quantity required to allow for denitritation or denitrification. Consequently, for those processes that require denitrification or denitritation, an external supplemental carbon source is required. The most common carbon supplementation is methanol; however, operation at elevated temperatures allows the use of various sources. With the additional carbon, there is always the potential to dose the chemical in excess of the denitrification requirements. Overdosing carbon can result in an increased aeration requirement to metabolize the carbon and the associated increase in solids production.

Deammonification-based sidestream treatment systems generally do not require supplemental chemicals to operate stably from a pH perspective, although in some cases lower ammonia conversion rates can arise (i.e., the residual ammonia maintains stable and near neutral pH).

Recent experience with full-scale deammonification systems has shown that some systems can suffer from micronutrient deficiencies in the feed stream. Such deficiencies have generally been attributed to insufficient iron and copper, although insufficient quantities of cobalt, zinc, manganese, and molybdenum have also been cited as potential causes of less than optimal performance. Considering the fact that micronutrient deficiencies can impact the growth of nitrifying bacteria (e.g., ammonia-oxidizing bacteria [AOB]), this potential problem should not be solely attributed to anammox-base systems. The addition of small quantities of a micronutrient mixture can has been successful in returning these systems to optimal performance.

2.7   Process Control Instrumentation

Earlier iterations of sidestream treatment technologies, particularly for nitrogen removal using nitritation/denitritation, included the use of multiple online analyzers (DO, pH, ammonia, nitrite, nitrate, alkalinity), which drove proprietary setpoint algorithms. The maintenance of these systems proved problematic for early adopters. As part of the maturation of sidestream treatment technologies and shifting of the focus on deammonification for nitrogen removal, simplified control schemes have been developed that use surrogates based on an enhanced understanding of the underlying fundamentals. Online nutrient analysis is now only necessary for enhancing the information collection to maximize efficiency rather than for basic operation. For example, the use of an online ammonia analyzer may be used in conjunction with a pH/DO based control scheme to modify the pH and or DO operating set point to maximize alkalinity utilization or reduce the potential for nitrite oxidation through nitrite oxidizing biomass (NOB) activity.

The primary control variables used in deammonification-based nitrogen removal are flowrates, liquid levels (for sequencing batch reactors [SBRs]), pH, temperature, and DO. In the case of phosphorus recovery using struvite, the use of an online orthophosphate analyzer may be required, particularly for treatment systems that use low HRT reactors; for high HRT (greater than four hour detention times) reactors daily sampling may be adequate.

2.8 Heat Management

One of the key advantages of sidestream treatment is the elevated liquid temperature at which they operate, a result of the heating provided for anaerobic digestion, which typically operates at 35°C for mesophilic anaerobic digestion. Reject water from thermophilic anaerobic digestion processes, which are operated at 55°C, may be at temperatures of greater than 38°C even after heat recovery, requiring cooling prior to biological treatment. For nitrogen oxidation, the rate of nitrification is maximized at 30 to 35°C; however, operating above 38°C will begin to inhibit the rate of nitrification and complete inhibition can occur at temperatures approaching 40°C. Consequently, it is critical to have the ability to control temperature to allow the sidestream treatment process to operate optimally at 30 to 35°C. With mesophilic anaerobic digestion, natural cooling, which will occur through dewatering, should yield sidestream temperatures in the 30 to 33°C range. For thermophilic treatment processes, however, a cooling step should be considered within the solids handling or sidestream treatment process.

Conversely, to prevent the reject liquor from losing heat, the addition of insulation may be required; for example the use of a "ball blanket" can reduce both heat and gas emissions in a dewatering reject liquor holding tank.

The heat generated by the biological reactions, particularly in processes that use supplemental carbon, may necessitate cooling of the reactor contents. In deammonification systems, cooling is not required as the heat generated by the oxidation and reduction of ammonia and the oxidation of the reject stream organics is typically minimal. This may not be the case if the dewatering reject stream is co-treated with other concentrated streams such as dryer condensate, which is typically rich in soluble organics. Nitritation/denitritation processes will typically always require a cooling heating exchanger as the oxidation of the supplemental carbon source for denitritation will likely result in temperatures above 38°C in the reactor. This may be avoided by adding dilution water or preferably, return activated sludge from the main facility in nitritation/denitritation reactors. The impact of the cooling liquid on detention time and performance must be accounted as detention times and operating concentrations will be reduced.

In deammonification applications where the dewatering centrate is at temperatures less than 20°C, heating the sidestream may be needed to maintain the optimized operating temperature, particularly during startup. For example, the deammonification MBBR facility deployed at Hampton Roads Sanitation District's James River WWTP included heaters for startup (although they were not used). After a stable deammonification, biomass is developed operation at temperatures of as low as 18 to 20°C has been reported in Scandinavian facilities without any negative impacts.

For larger facilities, opportunities to recover low-grade heat immediately downstream of the sidestream reactor may exist, allowing reuse of the recovered heat for building heating purposes in conjunction with a heat pump, with paybacks that are superior to those of a geothermal heat pump system. Retrofit of facilities from nitritation/denitritation or nitrification/denitrification to deammonification may provide additional heat recovery opportunities, as these types of facilities typically have process cooling heat exchangers that would no longer be required.

2.9 Safety

Safety considerations in sidestream treatment facilities are similar to most activated sludge BNR processes. In process configurations that require supplemental carbon, elevated temperatures typically encountered allow the process flexibility to use a range of supplemental carbon sources, including methanol, ethanol, and a range of waste organic products. Fire suppression systems are required where flammable carbon sources such as ethanol, methanol, or flammable organic waste products are intended to be used. Care must be exercised in selecting construction materials for storage tanks, piping, and instrumentation elements exposed to the supplemental carbon source and the influent dewatering liquor.

Although odors from sidestream treatment facilities are typically minimal, air collection and odor control should be provided at the point where aeration of the flow is initiated because volatile organics will be quickly stripped out of the flow. High oxygen transfer efficiency systems that minimize aeration rates and process configurations that expose the influent to anoxic treatment before aeration can help minimize stripping of volatile organics and ammonia, minimizing or eliminating odor control requirements.

2.10 Materials of Construction

Depending on the process configuration being used, special consideration may need to be given to materials of construction. The addition of chemicals such as ferric chloride to prevent struvite precipitation can accelerate corrosion of tank internals such as metals and plastics. The higher operating temperature and high total dissolved solids (TDS) levels of dewatering reject liquours will affect selection of materials for all submerged equipment. Hydrogen sulfide accumulations can damage concrete.

2.11 Incorporating Retrofit Flexibility

Sidestream treatment of anaerobic digestion dewatering flows provides an opportunity to treat a range of organic compounds of concern. Sidestream treatment processes, however, can also introduce new challenges in limiting nitrous oxide emissions from WRRFs. It is generally accepted that nitritation/denitritation processes emit higher levels of nitrous oxide than deammonification processes.

Hydrophobic microcontaminants, such as pharmaceuticals, preferentially partition into the sludge and can be found at elevated levels in the biosolids dewatering stream, relative to the main facility flow. Consideration should be provided for future processes to remove and destroy such compounds from the digestion dewatering flow downstream of the nutrient removal process.

Elevated levels of soluble and colloidal non-biodegradable organics in the dewatering stream, particularly where thermal sludge processing has been used, can pose challenges for facilities that must meet stringent nitrogen discharge permits. Pre- or post-treatment of the solids and or dewatering reject liquor stream can reduce non-biodegradable organic nitrogen levels, minimizing effects on the WRRF.

3.0 Nitrogen Removal Sidestream Process Design

Sidestream treatment processes for nitrogen removal have evolved rapidly over the past ten years in North America. Whereas nitrification/denitrification and nitritation/denitritation systems were dominant through the early 2000s, deammonification (partial nitritation/anammox) systems have gained traction and represent effectively all of the new designs in North America. The reader is directed to the previous edition of this manual for information on sidestream nitrification/denitrification and nitritation/denitritation systems.

Deammonification is the main process used in three basic configurations:

- Granular sludge sequencing batch reactor
- Granular sludge continuous flow reactor
- Moving bed bioreactor (MBBR) with continuous flow

The designer has significant flexibility in deploying new reactors or retrofitting tankage to support these processes. It is assumed that the reader has reviewed the upstream treatment requirements for the dewatering stream and is providing a dewatering reject stream that is compatible with the sidestream treatment process.

3.1 Available Reactor Configurations

One of the primary advantages of deammonification-based sidestream treatment systems is the simplicity of the reactors used to deploy them. This section discusses design considerations for sidestream reactors using both, granular and moving bed configurations. The specifics of the reactor configuration will be proprietary for most of the designs, however non-proprietary options may also be deployed, such as SBR and conventional reactor/clarifier configurations. The focus of proprietary designs is primarily on the granule retention method and the instrumentation/control philosophy used. Because of the rapid and ongoing evolution of these technologies, an intellectual property specialist should be consulted if a non-proprietary design is being considered.

3.1.1 Granular Sludge Reactors

Granular sludge reactors for sidestream treatment are similar to those provided for conventional activated sludge at elevated temperatures and high total dissolved oxygen liquid streams. The same considerations for the design of sequencing batch reactor systems and continuous flow reactor/clarifier combinations are applicable to sidestream designs. However, the designer has opportunities to simplify the design of the sidestream reactor, while incorporating the necessary design elements that will allow for successful operations.

Key components of the granular reactors include:

- Biological reactor
- Granule retention system
- Mixing system
- Fine bubble aeration system
- Decanting/Effluent withdrawal (SBR only)
- Instrumentation and controls

The provision of an internal recycle loop between the reactor effluent and the equalization zone will minimize the potential for nuisance struvite precipitation in both zones, while also providing an opportunity for removal of nitrate produced during deammonification and preventing hydrogen sulfide formation. However, internal recycles should be limited to levels that will not reduce effluent ammonia concentration to less than 50 to 100 mgN/L, as NOB suppression is reported to be affected at lower levels.

The presence of elevated soluble or particulate biodegradable chemical oxygen demand (COD) in the reactor influent has been shown to negatively affect granular reactor operations within a few days. Whereas the pretreatment provided upstream as described in the previous section will serve to minimize this, consideration should be given to providing a zone for biodegradable COD removal, whether in the reactor itself, or if possible in the upstream equalization tank.

A key consideration is the impact of maintaining a consistent level of carbon dioxide within the reactor. Deeper reactors with higher oxygen transfer efficiencies are preferred. Reactors with depths of 20 feet or greater will allow for greater retention of carbon dioxide and a more stable operating pH level, allowing for better utilization of the reject water stream's alkalinity content, thereby also reducing or eliminating the need for a supplemental alkalinity source. Reactor depths of greater than approximately 7.6 m (25 ft) are difficult to justify due to the economics of tankage construction and the design/operational challenges with aeration systems operating at pressures in excess of 90 to 96 kPa (13–14 psi) (excessive process air temperatures, lack of high efficiency blowers, accelerated diffuser aging). The use of relatively low operating DO concentrations, on the order of 0.2 to 0.5 mg/L minimizes carbon dioxide stripping from these reactors, as well as energy use when coupled with fine bubble aeration.

The retention of the biomass granules is a primary goal of the granular reactor design. This can be achieved via conventional sedimentation including lamella clarifiers, however the general design trend is to augment retention through the provision of an external retention mechanism, such as hydrocyclones or ultrafine (200–500 micron) screens. Whereas the first generation of granular sludge reactors exhibited operational instabilities, the inclusion of equipment to enhance granule retention has significantly increased the reliability of granular deammonification technology.

When selecting granule retention equipment, the designer must be cognizant of the upstream liquid and solids treatment train, so as to prevent clogging of granule retention screens and hydrocyclones. Screening to 2 to 3 mm should be provided if upstream processes don't provide screening to this level. Prevention of nuisance struvite formation on the reactor internals should be incorporated in the design.

The reduced energy consumption and higher retention efficiencies of granule retention screens has resulted in increased interest in this approach to granule retention, particularly where hydraulic loadings will be significant. At this time, there are no active facilities in the United States using screens, however granule retention screens are in use in European facilities and design for similar facilities in the United States is ongoing as of this update.

Design loadings on the order of 1 kgN-d/m³ can be sustained in reactors with an external granule retention system. It is anticipated that higher loadings will be achievable as granule retention efficiency is enhanced; the designer should provide the flexibility to readily increase the loadings to granular reactor systems to the level of approximately 1.5 kgN-d/m³. Where only conventional sedimentation is used for granule retention, loadings on the order of 0.5 to 0.8 kgN-d/m³ are advisable.

3.1.2   Moving Bed Biofilm Reactors

MBBRs have gained popularity due to their resilience to shifts in dewatering reject water characteristics and a perception of increased deammonification process robustness. Sedimentation is not provided in MBBRs, the anammox biomass grows on the suspended carrier media.

Key components of MBBRs for deammonification are:

- Influent and effluent media retention
- Carrier media
- Coarse bubble aeration system
- Mixers
- Instrumentation and controls

Screening to prevent blinding of screens and media may be needed for the MBBR. Screen apertures of no larger than 4 mm are required, with considerations regarding preventing struvite formation on the screens as noted in the pretreatment section.

Design loadings for MBBRs are typically higher than those used for granular reactors. Loadings of 1.0 to 1.5 kgN-d/m³ can be readily applied with higher rates possible; resulting in a 30% to 50% reduction in reactor volumetric requirements versus granular reactors. Operational trends with MBBR reactors have shown them to be very resilient with respect to maintaining anammox activity, with the limiting performance parameter being the ammonium oxidizing biomass. Where soluble biodegradable organics excursion may occur (such as in facilities that a history of digesters going sour or dewatering from sources with poorly digested sludge or thermally processed sludge) an organics oxidation zone upstream of the deammonification MBBR is required. Coupled with an internal recycle from the MBBR effluent to the reactor influent to recycle nitrate and nitrite, minimal additional aeration may be required.

Operational controls to minimize NOB activity need to be provided. In reject liquor streams that have marginal alkalinity to ammonia ratios (below ~3.5) provision of supplemental alkalinity can be used to increase the operating pH setpoint to allow AOB a growth advantage while selective pressures (increased hydraulic loading, lower DO, etc.) are applied.

3.2   Aeration System Design

The combination of elevated operating temperatures and TDS levels, coupled with the need to turn off aeration as part of the control logic or in order to prevent formation of excess nitrite requires the deployment of unique design elements in the aeration system design of sidestream treatment systems.

High temperature operations and the potential for precipitate fouling result in rapid aging of fine bubble aeration systems; provision of Polytetrafluoroethylene (PTFE) or silicone fine bubble diffusers designed for high temperature operation is required. Regardless of the material used and care in design and installation, due to the operating conditions, over the life of the diffuser system, some diffusers will fail prematurely and result in process liquid entering the diffuser grid. Failure of positive displacement blowers due to high pressure startups has been experienced to date. Provisions must be made for "soft starts" that will allow process liquid to be expelled from the diffuser grid and prevent damage to the blowers and grid. This may include allowing for a slow increase in the blower speed and use of check/blowoff valves within the diffuser grid to help expel process water.

3.3   Mechanical Mixing/Shearing System Design

In conventional systems, mechanical mixing systems are used to maintain biomass and/or carriers in suspension when aeration systems are turned off. However, in granular deammonification systems, evidence to date shows shear providing a mechanism for selection of ammonia oxidizing over NOB. Thus, whereas one may be tempted to provide mixers sized at approximately 0.1 W/m^3, mixing energies of 0.2 W/m^3 may be required to achieve adequate shear and minimize NOB activity in granular systems. Provisions of multiple mixers that can alter the applied shear in the bioreactor are recommended.

3.4   Effluent Management

Significant nitrogen may still be present in discharge stream, potentially on the order of 100 mgN/L or more. The designer must recognize this recycle loading and identify opportunities to direct sidestream reactor effluent where it can maximize benefits to the main facility. For example, an aerated return activated sludge (RAS) channel can provide for residual ammonia oxidation and potentially oxidized nitrogen species reduction.

Effluent may also provide for specialty biomass that can be beneficial to the process. Particularly in systems where pretreatment is being provided, that includes solids removal from the centrate, the effluent from the sidestream reactor should be directed to the biological process. Redirection of the sidestream reactor effluent to an upstream solids settling process should be considered only if process calculations/modeling show that additional solids loading would be deleterious to the mainstream biological process.

Where hydraulic limitations do not govern process design, the inclusion of a recycle loop from the effluent to the reactor influent should be considered. The recycle loop will bring nitrate (a normal byproduct of the deammonification reaction) to the head of the sidestream reactor, allowing for its removal via conventional denitrification, while also reducing the concentration of ammonia and potentially the pH of the influent, thereby also reducing struvite precipitation potential with the reactor. Where conditions allow, the recycle stream can be directed into the equalization tank or even the screening facility, allowing for an increased margin of safety from struvite precipitation in these facilities.

3.5   Water Resource Recovery Facility Operations Capabilities

Operation of deammonification systems has strong parallels with conventional nitrification/denitrification, but also provides for some significant departures in operating philosophy, including the parameters that need to be monitored and the operators responses. As part of the design of a sidestream treatment system, it is recommended that a structured training program be provided for the facility's process engineering staff and the operators that will have daily responsibility for the facility.

4.0   Phosphorus and Nitrogen Recovery Process Design

4.1   Phosphorus Precipitation and Recovery

Phosphate precipitation and recovery is a sustainable alternative to manage sidestream loads with high concentrations of soluble phosphate. Phosphate recovery methods are typically based on precipitation and crystallization, or adsorption. This sidestream management strategy became popular with the development and implementation of magnesium ammonium phosphate (struvite or MAP) and calcium phosphate (apatite or CaP) precipitation and recovery systems. This section will focus on sidestream management aspects of phosphorus recovery. A more comprehensive review of phosphate recovery technologies can be found in the literature (Sartorius et. al., 2012; Morse et al., 1998; Barnard et. al., 2012).

4.1.1   Adsorption Methods for Phosphate Separation and Recovery

Processes based on adsorption such as the REM NUT® process are available, however they are not common for sidestream management applications. These processes include adsorbent packed columns, mixed reactors with media recovery, or ion exchange systems for phosphate removal. The phosphate recovery process follows downstream the adsorption process through media regeneration and phosphate recovery. These systems are typically more complex than precipitation and crystallization methods and have not been widely adopted by the industry. This section will focus on common precipitation and crystallization methods.

4.1.2   Precipitation and Crystallization Methods for Phosphate Recovery

Precipitation and crystallization methods aim to produce phosphate-based precipitates that can be crystalized and readily utilized as fertilizer products or can be post-processed to produce fertilizer. The most common precipitation products for recovery applications are magnesium ammonium phosphate—MAP (struvite) and calcium phosphate—CaP (e.g., hydroxyapatite). These systems provide multiple benefits including substantial reduction of sidestream phosphate load, scaling control, improved sludge dewaterability, and production of a sustainable fertilizer product. Increased dewaterability of the biosolids has also been achieved depending on the phosphorus recovery process configuration applied. Despite systems based on CaP recovery being in operation in Europe for several years, systems based on struvite recovery have dominated the North American market. Some of the proprietary systems available include: DHV-Crystalactor (CaP), P-RoC (CaP), Ostara Pearl (MAP), Muiltiform Harvest (MAP), PhosNIX (MAP), PRISA (MAP), AirPrex (MAP), however non-proprietary systems also exist, particularly in configurations that induce struvite precipitation prior to dewatering.

4.1.3   Precipitation and Crystallization of MAP and CaP

MAP or struvite ($MgNH_4PO_4 \cdot 6H_2O$) is a white crystal that contains equal molar amounts of magnesium, ammonium, and phosphorus. Struvite precipitation occurs when the concentration of magnesium, ammonium, and phosphate ions exceed the solubility product constant (K_{sp}). This reaction is affected by pH and ionic strength of the sidestream water. Therefore, a conditional solubility product is calculated from the K_{sp} value to take into account these parameters. In general, struvite solubility decreases as pH increases (Figure 15.1). However, as pH increases, the availability of the ionic species that are needed to form struvite changes. For example, for pH values higher than 9, ammonium ion concentrations will decrease as the equilibrium shifts toward ammonia. Optimum pH for effective struvite precipitation is reported to be between 8.0 and 10.7 (Doyle and Parsons, 2002).

Estimating struvite formation potential is complex since it requires an estimate of the concentrations of each ionic species available at a given pH. These estimates are often times performed using chemical equilibria software (Doyle and Parsons, 2002). Struvite precipitation occurs in two steps:

1. Nucleation—combination of ions to form embryonic crystals and,
2. Growth—crystal development until equilibrium is attained.

Operating pH and temperature affect nucleation time. In general, an increase in pH and temperature decrease nucleation time. For pH values lower than 8.0, nucleation is

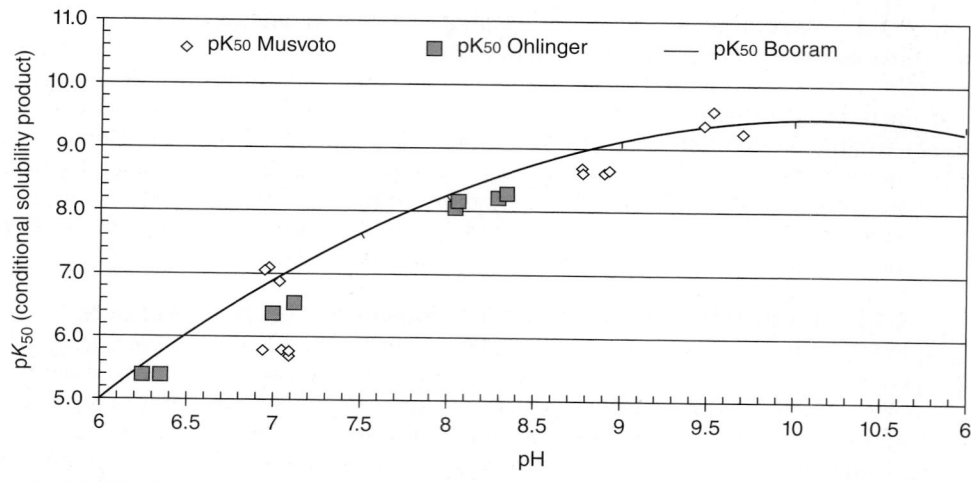

FIGURE 15.1   Typical struvite solubility versus pH.

slow and can take days to occur. For pH values higher than 8.0, nucleation occurs substantially faster taking less than 1.0 hour for pH values higher than 8.5 (Doyle and Parsons, 2002). Mixing energy also affects nucleation time as more turbulent systems promote more release of CO_2 and therefore and increase in pH. Therefore, a higher mixing energy tends to reduce nucleation time (Doyle and Parsons, 2002). However, struvite recovery reactors typically operate at pH levels of 6.8 to 7.2 as struvite nuclei are either already present or are recycled into the reactor.

CaP can crystalize as different compounds depending on the solution. Hydroxyapatite ($Ca_5H(PO_4)O$) is the most thermodynamically stable and is typically present in CaP recovery processes. The precipitation of CaP in wastewater requires a very high oversaturation of the solution and for this reason a seed is typically applied to the system to drive the crystallization process. The recovery of CaP may be attractive to specific applications depending on the target end product market. Contrary to MAP-based products, CaP-based products can be used not only in the fertilizer industry but also in the lower value phosphate market since CaP is essentially phosphate rock (Nieminen, 2010).

4.1.3.1   Struvite Precipitation and Recovery Processes   Struvite recovery systems are typically pre-engineered packaged systems, which may vary slightly from manufacturer to manufacturer. Two approaches are becoming predominant in water resource recovery applications: (1) systems that use sidestream liquors (e.g., solids dewatering recycled water) as a base source of nutrients (e.g., Ostara Pearl, Multiform Harvest, PHOSNIX) and (2) systems that use digested sludge liquors as a base source of nutrients (e.g., AirPrex[R], Bio-Stru[R] Schwing Bioset). In the former systems, struvite recovery is an integral part of the process. In the latter processes, some utilities chose not to recover the struvite separately, but rather allow it to be removed from the facility with the biosolids stream during dewatering.

A typical struvite recovery processes from sidestream liquors is illustrated in Figure 15.2 and includes:

(a) Equalizations and pretreatment system—Some processes may require an equalization and pretreatment step. Dewatering reject streams have a high amount of impurities that can affect the performance of the recovery process or create operational problems. The removal of suspended organic and inorganic material benefits the downstream recovery process. Equalization and sedimentation may be an option for facilities that have intermittent dewatering schedules. Fine screens can be used as noted earlier for fibrous material removal. Although, this is a desirable process component, some manufacturers will not include it as part of the packaged system. Therefore, it is up to the designer and owner's discretion to evaluate the need for equalization and pretreatment.

(b) Fluidized bed reactors—In most processes, the nucleation and crystallization process takes place in a fluidized bed reactor. This reactor is commonly made of 304 stainless steel and has a conical shape, which may vary slightly from manufacturer to manufacturer. These reactors are generally tall with heights between 7.6 and 9 m (25–30 ft), which makes the retrofit of existing buildings challenging and construction of new buildings costly. However, they may be exposed to the environment and constructed without a building enclosure. Struvite is collected in the bottom of the reactor and conveyed for post-processing steps. Chemicals and the influent (e.g., centrate, filtrate) are added in the lower part of the reactor while effluent is collected at the top. Some designs will include a recirculation line to optimize hydraulic conditions of the reactor and promote formation and separation of larger particles.

(c) Chemical addition system—Two chemicals are commonly added to the reactor: (1) A magnesium source such as magnesium chloride or magnesium hydroxide and (2) sodium hydroxide for pH control. Although magnesium hydroxide can be used as both a magnesium source and pH control agent, it may limit the opportunities for separate optimization of Mg addition and operating pH, but may reduce overall chemical costs. Chemical addition systems are, in general, comprised of bulk delivery system, storage tank, and metering system and pump.

(d) Product dewatering and drying—Where recovery is practiced, the struvite product comes out of the reactor it needs to be dewatered and dried for handling and post-processing. The approach to dewatering and drying varies depending on the manufacturer. At a minimum a dewatering or separation stage is added before post-processing. Product drying may be added to the process train. The heat associated with drying represents the most significant energy input into the process; waste heat utilization opportunities will significantly affect the economic viability of drying.

(e) Classifier, storage, and bagging—Some systems may include onsite struvite handling and storage. This would typically include a conveyor system, a product classifier, storage silos, and bagging system. When selecting the process site, attention must be paid to the existing truck routes within the facility.

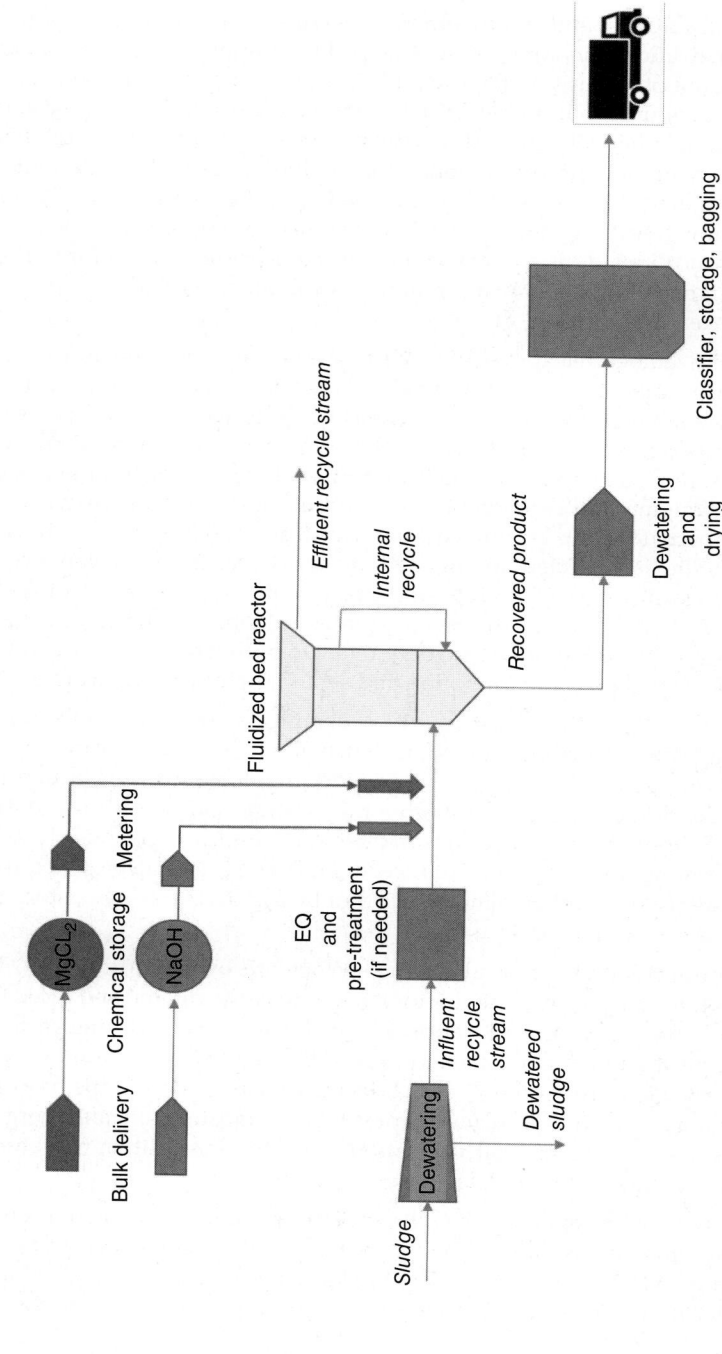

Figure 15.2 Typical struvite recovery processes from sidestream liquors.

A typical struvite recovery processes from digested sludge is illustrated in Figure 15.3 and includes

(a) Struvite precipitation reactor(s)—The precipitation reactor is a tall cylindrical tank with a conical bottom typically of stainless steel construction. In this system, air via a coarse bubble diffuser is used to strip CO_2 and increase the pH of the sludge for optimal precipitation. The use of caustic for pH adjustment is typically not required, reducing the chemical handling requirements. Where only struvite precipitation is required only one reactor will be used with a hydraulic retention time of 4 to 6 hours. Where recovery of the struvite is also targeted, a two-reactor system in series may be used, with an additional 4 to 6 hours of detention time to allow for growth of the struvite crystal, enhancing downstream separation efficiency. In some processes, air is introduced in the first reactor for both pH control and mixing. Others decouple pH control from the recovery reactor and introduce air in the first reactor and supplemental chemicals in the second reactor. Recirculation of struvite crystals can be provided to increase recovered crystal size.

(b) Chemical addition system—Where pH control can be accomplished with CO_2 stripping, only magnesium chloride is added to the recovery reactor as a magnesium source. Caustic or alternatively $Mg(OH)_2$ may be used where an alkalinity source is required. Similarly, chemical addition systems are comprised of bulk delivery system, storage tank, and metering system and pumps.

(c) Struvite product washing system—The recovered product is withdrawn in from the bottom of the reactor and discharged into a classifier and washing system similar to a grit washer and classifier commonly used in grit handling applications.

(d) Product handling system—The washed product is conveyed through a screw conveyor into a dumpster or storage container.

4.1.3.2 CaP Recovery Processes   The general concept of a CaP recovery process is similar to MAP recovery processes. Some of the technologies available include the DHV-Crystalactor and the P-RoC processes. An important difference in these processes is the need of seed to drive the precipitation of CaP. A general CaP recovery process from sidestream liquors includes similar elements described for the MAP processes with the seed addition system.

4.1.4 Benefits of Sidestream Management Through Phosphate Recovery
The benefits of sidestream management through phosphate recovery go beyond the optimization of biological phosphorus removal in the main stream by reducing the phosphate loading to the head of the facility. The following added benefits can be listed; others can be found in the literature (Mayer et. al., 2016):

(a) Scaling control and mitigation—Struvite and vivianite can form and accumulate in pipelines, mechanical equipment, and process tanks. The scaling of mechanical equipment (e.g., pipes, heat exchangers, dewatering equipment, centrate handling equipment, and pipelines) is not only a process performance issue, but also a critical maintenance problem. Scaling removal is a labor intense process and often times costly as it involves equipment/pipeline disassembly, acid washings, and the use of special descaling solutions. The reduction of phosphate levels in the facility through controlled precipitation of MAP or CaP can considerably reduce deposits and scaling issues. Where struvite scaling

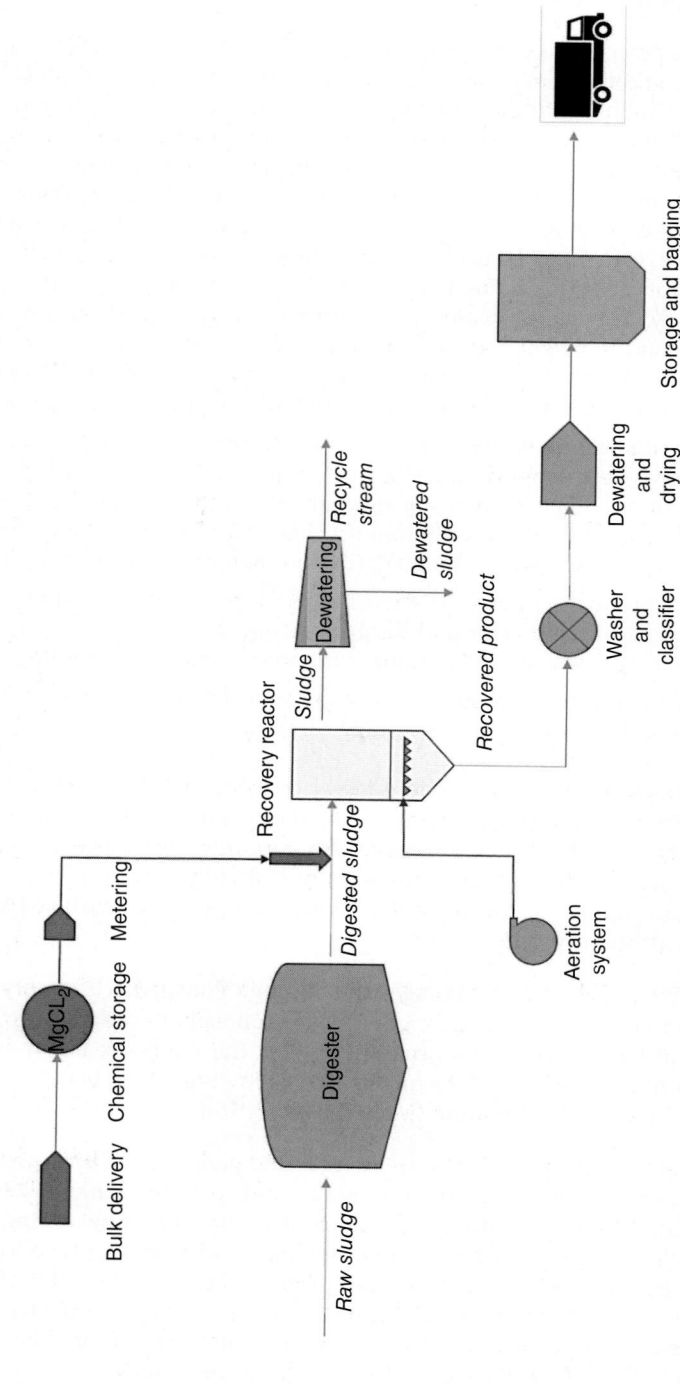

Figure 15.3 Typical struvite recovery processes from digested sludge liquors.

control and dewaterability enhancement is the primary goal, the use of magnesium hydroxide upstream of dewatering in lieu of magnesium chloride is preferred as it has been shown to minimize unintended struvite precipitate formation on reactor internals and piping.

(b) Reduction of solids disposal—Although, the amount of solids recovered is small compared to the total solids disposed, a reduction on solids disposal quantities is noted in processes which include phosphate recovery. The reduction in solids disposal will depend on the treatment process and can vary considerably from facility to facility. No reduction in biosolids is achieved in facilities that do not recover the phosphate; however, the overall solids production of the facility may be reduced if struvite recovery reduces or eliminates the need for ferric chloride addition.

(c) Impacts to solids dewatering processes—The digestion of waste-activated sludge from EBPR systems results on the release of PO_4^{3-}, Mg^{2+}, and K^+, which change the ionic balance of the sludge mixture. High ratios of monovalent to divalent cations (M:D) are reported to reduce dewaterability and decrease cake solids (Higgins, 2014). High concentrations of phosphate result in a decrease of divalent ions (Mg^{2+} and Ca^{2+}) through precipitation and therefore increase the relative concentrations of monovalent cations (K^+) and consequent increase the M/D ratio. The removal of phosphate in a controlled manner before the digestion process can alleviate some of the negative impacts EBPR sludge on dewaterability (Higgins, 2014). Increases of 1% to 5% in cake solids and reduced dewatering polymer consumption has been observed in EBPR facilities in particular, however with the underlying fundamentals still in question, the owner/designer may find pilot testing useful.

(d) Production of sustainable fertilizer product—Phosphate, which is a key element for food production, is a limited resource and readily accessible global phosphate reserves are being depleted. The recovery of phosphate from human excreta is one of the measures proposed in the literature to overcome a potential future global phosphorus scarcity (Cordell and White, 2013; Bird, 2015). The recovery of phosphate as struvite produces a slow release fertilizer product that requires minimum postprocessing. Some manufacturers work with a business model where the fertilizer product is purchased from the resource recovery facility for resale.

4.1.5 Maximizing Benefits Through Phosphate Stripping

The benefits of sidestream management through phosphate recovery can be maximized by the implementation of a phosphate stripping process upstream the digestion process. This approach leads not only to a higher product yield, but also tends to reduce phosphate loadings to the digesters, reducing struvite precipitation potential, and improving sludge dewaterability (Shaw et al., 2014). WASSTRIP® is the most popular process of this kind in North America. This process includes a stripping tank followed by a thickening system upstream of the digester(s). Volatile fatty acids (VFA) are added to the stripping tank to facilitate and maximize phosphate release. VFAs can be produced by fermentation of primary sludge. Fermentation of WAS has also been reported as a feasible alternative for VFA addition (Schauer and Laney, 2015), however longer detention times and thus larger reactor volumes are required. Other proprietary processes include PRISA and MultiWAS. A typical process configuration for phosphate stripping systems is presented in Figure 15.4. Similar benefits can be obtained through

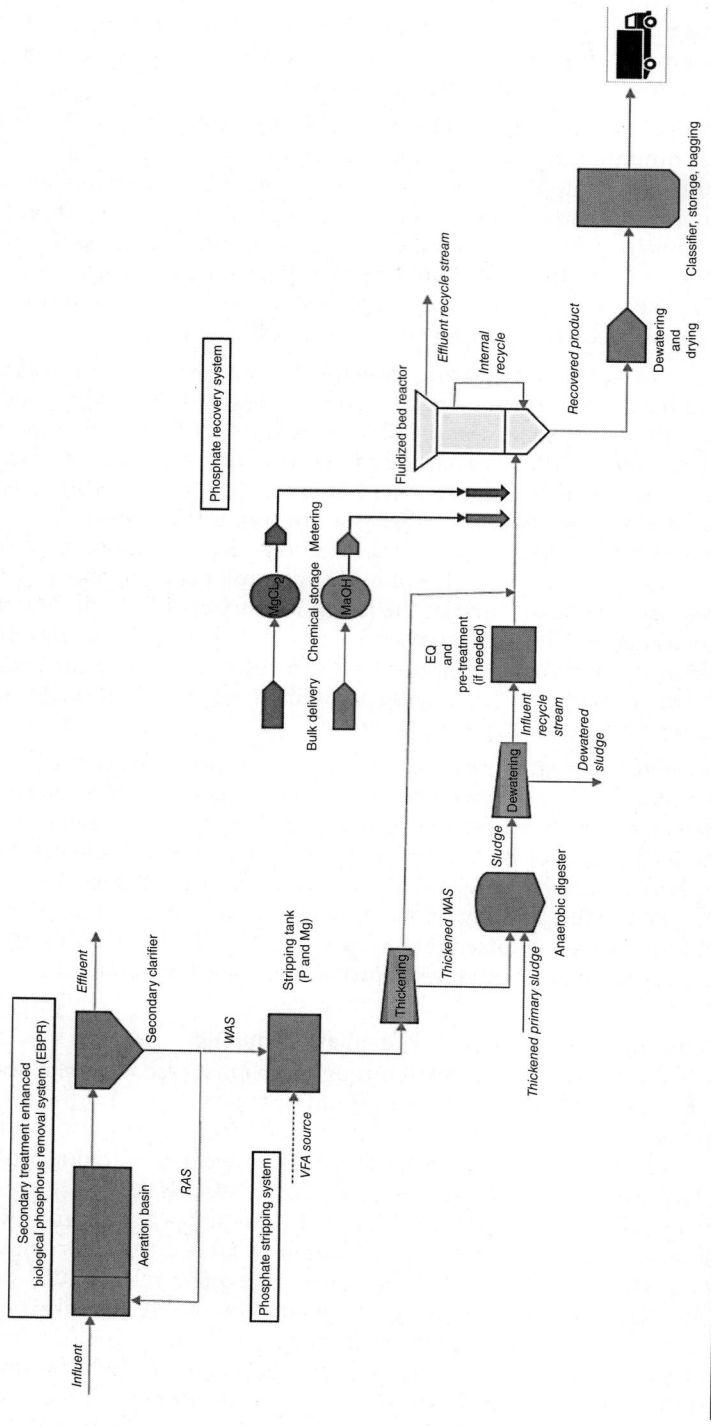

FIGURE 15.4 Typical process configuration for phosphate stripping systems followed by phosphate recovery.

Technology	Key Basis of Design
Recovery from dewatering reject liquor	Phosphate load 65 (145) to 1250 (2750) kg PO_4/d (lb PO_4/d)
Precipitation/Recovery from digested liquid biosolids	Hydraulic retention time 2 to 4 hours (w/o recovery) 5 to 13 hours (w/ recovery)

TABLE 15.2 Typical Basis of Design for Recovery Systems from Sidestream Liquors and Sludge Liquors

the addition of magnesium hydroxide to the digested biosolids upstream of dewatering; however, this will not provide the same level of protection from unintended struvite precipitation in the digesters.

4.1.6 Design Considerations
Design of phosphate recovery systems are highly dependent of supplier and manufacturer input since most systems are proprietary and customized designs are still rare. The role of design consultants may be limited to process integration with existing facilities and design of supporting systems such as structural, electrical, and utilities.

Recovery processes from sidestream liquors are typically designed based on the phosphate load to the fluidized bed reactor. Suppliers offer a range of predesigned reactors to cover multiple applications. Table 15.2 illustrates typical basis of design for recovery systems from sidestream liquors and sludge liquors.

Recovery systems from sludge liquors are designed based on the hydraulic retention time (HRT) within the precipitation/recovery reactor. For this reason, there is a higher degree of flexibility on reactor sizes. In cases where the system is intended to precipitate struvite without subsequent recovery (i.e,. phosphate sequestration), shorter HRTs can be used.

4.2 Ammonia Recovery

4.2.1 Ammonia Stripping and Recovery
Ammonia stripping and recovery has been applied on dewatering streams, using conventional packed towers, steam strippers or vacuum distillation towers coupled with recovery into a dilute sulfuric acid solution resulting in production of an ammonium sulfate solution. However, the economics of the process configurations developed to date limit the applicability of such processes to specialized situations. Examples where consideration of ammonia stripping/recovery may be viable include:

- Where nitrogen levels in biosolids need to be increased to enhance acceptance of biosolids for land application. By mixing the recovered ammonia (typically as ammonium sulfate) into the biosolids, nitrogen content can be increased by a factor of 2 or more.
- If ammonia removals on the order of 50% or lower are adequate from the dewatering reject liquor stream, as the mass transfer rates are greater when residual ammonia is elevated, enhancing economics of treatment.

- Where the dewatering reject stream's temperature is elevated due to the use of thermophilic digestion or other thermal processing.
- If lime addition is used on the digested biosolids as part of the land application program's requirements, increasing the pH of the reject water stream, facilitating increased mass transfer rates.
- Where chlorine for disinfection is produced on-site resulting in an alkaline wastestream that can be used for pH adjustment.

Pretreatment is a key to successful implementation of stripping/recovery processes. Key process would include:

- Fine screening—To prevent accumulation of fibers onto the media the dewatering reject stream, fine screens of 2 mm or less opening are required. The considerations discussed previously under pretreatment also apply in this configuration.
- pH Adjustment/CO_2 removal—Depending on the temperature and removal targets increase in the pH of the high ammonia reject stream may be required, to increase the fraction of ammonia in the molecular form within the reject water stream. Prior to alkali addition, reduction of pH with acid addition and stripping of CO_2 will reduce acidity and precipitation potential within the packed tower. The economics of acid addition will be site specific, however CO_2 stripping should be considered unless downstream processing obviates its use. Increasing pH will result in significant precipitate formation that will need to be removed prior to introduction into the stripping tower.
- Temperature adjustment—If temperature is to be adjusted, this step should occur prior to solids removal as additional precipitate may form, depending on the target operating pH.
- Solids removal/clarification—Removal of the reject water stream solids and precipitate is required prior to introduction of the stream into the air stripper. Lamella-type clarifiers may be used, but with caution as high solids levels in the dewatering stream have been shown to result in plugging and failure.
- Hot air stripper—Packed countercurrent towers using antifouling media are required. Acid cleaning of the packed tower needs to be provided to combat precipitation on the tower packing. Tower sizing is provided using conventional stripping tower design equation sets. High-efficiency mist eliminators are required to minimize water carryover to the downstream acid absorber.
- Acid absorber—The air stream from the stripping tower is brought into an acid absorber; in which the ammonium in the air is converted to an ammonium sulfate solution. Heating of the acid absorber may be required to achieve the target ammonium sulfate solution concentration.

5.0 Design Example

A 500 000 pollution equivalent WWTP is considering sidestream treatment since the current influent load (excl. dewatering liquor) is close to the WWTP maximum. The facility is comprised of a primary settling facility and activated sludge BNR process. Sludge processing includes gravity belt co-thickening of primary and secondary sludge, anaerobic digestion, and centrifuge dewatering.

A design example is provided for the following processes:

- Equalization of dewatering liquor;
- Phosphorus precipitation in a struvite reactor; and
- Nitrogen removal via deammonification of the dewatering liquor.

The sludge flow from the anaerobic digester is 150 000 m³/yr and has the following characteristics:

- NH4-N (soluble): 900 mg N/L;
- Temperature: 35°C;
- Alkalinity: 3300 mg/L as $CaCO_3$, (Alkalinity/NH_3-N > 3.5; i.e., suitable for deammonification without alkalinity addition);
- Ortho-P: 320 mg P/L.

The dewatering centrifuges are operated 8 hours per day on weekdays. The dewatering flow is 135 000 m³/yr.

5.1 Equalization Tank Design

The yearly average flow is 15.5 m³/hr, which is chosen as the design flow for the subsequent treatment steps. The longest period when the centrifuges are not in operation is during the weekends, that is, during 64 hours. The equalization capacity required is calculated as

$$\text{Equalization capacity} [m^3] = 15.5 \left[\frac{m^3}{hr} \right] \cdot 64 \, [hr] = 992 \, [m^3]$$

Consideration should be given to providing odor control for equalization tank and on-demand mechanical mixing for resuspension of settled solids. Coarse-bubble aeration is not used for mixing because of high struvite precipitation potential on tank internals.

5.2 Phosphorus Precipitation in a Struvite Reactor

5.2.1 Reactor Size Selection

A vendor supplied reactor from the types available as indicated in Table 15.3 is selected. The ortho-P load in the reject flow is calculated at

$$\text{ortho-P load} \left[\frac{kg}{d} \right] = 15.5 \left[\frac{m^3}{hr} \right] \cdot 24 \left[\frac{hr}{d} \right] \cdot 320 \left[\text{gram} \frac{\text{ortho-P}}{m^3} \right] = 112 \left[\frac{kg}{d} \right]$$

Reactor	Ortho-P Capacity (kg/d)
R-1	65
R-2	250
R-3	1250

TABLE 15.3 Available Induced Struvite Precipitation Reactor Types

Deammonification System	Design Loading Rate (kg NH$_4$-N/m³ Reactor Volume/d)	Required Tank Volume (m³)
Granular sludge reactor	0.4–0.6	560–840
Granular with cyclone on waste sludge line	0.6–1.0	340–560
MBBR	1.0–1.5	170–255

TABLE 15.4 Types of Nitritation/Anammox Systems

A R-2 reactor is chosen. Note that no redundancy is provided in this design. When the struvite precipitation reactor is offline, metal salt addition to the reject liquor (alum or ferric) will be practiced to reduce loadings to the mainstream process.

5.2.2 Chemical Dosing

The MgCl$_2$-demand is 3.1 kg MgCl$_2$/kg P$_{recovered}$. The assumed recovery of phosphorus is 90%. This results in the following chemical demand:

$$\text{MgCl}_2\text{-demand [kg/d]} = 112 \text{ [kg ortho-P/d]} \times 0.9 \text{ [Recovery ratio]}$$

$$\times 3.1 \text{ [kg MgCl}_2/ \text{ kg P}_{recovered}] = 313 \text{ [kg MgCl}_2]$$

5.3 Deammonification Reactor

5.3.1 Tank Dimensions

The dimensions of the deammonification system depend on the design loading rate of the chosen technology and the actual ammonium load of the flow to be treated. The ammonium load is calculated as:

$$\text{NH}_4-\text{N}-\text{load} \left[\frac{\text{kg}}{\text{d}}\right] = 15,5 \left[\frac{\text{m}^3}{\text{hr}}\right] \cdot 24 \left[\frac{\text{hr}}{\text{d}}\right] \cdot 900 \left[\text{gram} \frac{\text{NH}_4-\text{N}}{\text{m}^3}\right] = 335 \left[\frac{\text{kg}}{\text{d}}\right]$$

Three general types of deammonification systems are considered in this example (Table 15.4). The nitrogen loading rates and the required tank volume are also shown.

5.3.2 Oxygen Demand

Assume that 85% of the influent ammonium is removed in the nitritation/anammox system, the oxygen demand is calculated as follows:

3.43 kg O$_2$ is required for the oxidation of 1 kg NH$_4$-N to NO$_2^-$. For design purposes it can be assumed that 1.7 kg of O$_2$ will be utilized per kg of NH$_4$-N in the deammonification reaction. The oxygen demand is calculated as:

$$\text{O}_2\text{-demand [kg O}_2/\text{d]} = 335 \text{ [kg NH}_4\text{-N /d]} \times 0.85 \text{ [Removal ratio]}$$

$$\times 1.7 \text{ [kg O}_2/\text{kg NH}_4\text{-N oxidized]} = 484 \text{ [kg O}_2/\text{d]}$$

6.0 References

Barnard, J.; Phillips H.; Steichen, M. (2012) State-of-the-art Recovery of Phosphorus from Wastewater. *Proceedings of the 85th Annual Water Environment Federation Technical Exposition and Conference* [CD-ROM]; New Orleans, Louisiana, Sep 29–Oct 3; Water Environment Federation, Alexandria, Virginia.

Bird, A. R. (2015) Evaluation of the Feasibility of Struvite Precipitation from Domestic Wastewater as an Alternative Phosphorus Fertilizer Resource. Master's Projects. Paper 141.

Cordell, D.; White, S. (2013) Sustainable Phosphorus Measures: Strategies and Technologies for Achieving Phosphorus Security. *Agronomy*, **3** (1), 86–116.

Doyle, J.; Parsons, S. (2002) Struvite Formation, Control, and Recovery. *Water Res.*, **36**, 3925–3940.

Higgins, M.J.; Novak, J.T. (1997) The Effect of Cations on the Settling and Dewatering of Activated Sludges: Laboratory Results. *Water Environ. Res.*, **69** (2), 215–224.

Higgins, M.J. (2014) Does Bio-P Impact Dewatering after Anaerobic Digestion? Yes, and Not in a Good Way! *WEF Biosolids and Residuals Conference*, Austin TX. Water Environment Federation, Alexandria, VA.

Mayer, B. K.; Baker, L. A.; Boyer, T. H.; Drechsel, P.; Gifford, M.; Hanjra, M.A.; Parameswaran, P.; Stoltzfus, J.; Westerhoff, P.; Rittmann, B. (2016) Total Value of Phosphorus Recovery. *Environmental Sci. & Technol.*, **50**, 6606–6620.

Morse, G. K.; Brett, S. W., Guy, J. A., Lester, J. N. (1998) Review: Phosphorus Removal and Recovery Technologies. *The Science of Total Environment*, **212**, 69–81.

Nieminen, J. (2010) Phosphorus Recovery and Recycling from Municipal Wastewater Sludge. M.Sc. Thesis. Aalto University, School of Science and Technology, Department of Civil and Environmental Engineering.

Sartorius C.; von Horn J.; Tettenborn F. (2012) Phosphorus Recovery from Wastewater–Expert Survey on Present Use and Future Potential. *Water Environ. Res.*, **84** (4), 313–322.

Schauer, P.; Laney, B. (2015) Continued Evolution of the WASSTRIP Process, *Proceedings of the 88th Annual Water Environment Federation Technical Exposition and Conference* [CD-ROM]; Chicago, Illinois, Sep 26–30; Water Environment Federation, Alexandria, Virginia.

Shaw, A.; Koch, D.; Wirtel, S.; Britton, A. (2014) The Multiple Benefits of Phosphorus Recovery. *Proceedings of the 87th Annual Water Environment Federation Technical Exposition and Conference* [CD-ROM]; New Orleans, Louisiana, Sep 27–Oct 1; Water Environment Federation: Alexandria, Virginia.

U.S. Environmental Protection Agency (1987) Design Manual: Dewatering Municipal Wastewater Sludges. EPA/625/1-87/014: U.S. Environmental Protection Agency: Washington, D.C.

7.0 Suggested Readings

Lackner et al. (2014) Full-scale Partial Nitritation/Anammox Experiences—An Application Survey, *Water Research*, **55**, 292–303.

STOWA (2018) Sharon-Anammox-systemen, Amersfoort, The Netherlands.

Wett and Hell, Betriebserfahrungen mit dem DEMON®-Verfahren zur Deammonifikation von Prozesswasser, KA Abwasse Abfall, 2008.

Wett et al. (2007) Key Parameters for Control of DEMON Deammonification Process, *Water Practice*, **1** (5), 1–11.

CHAPTER 16

Natural Systems

A. Robert Rubin, Ph.D., and Michael Hines, P.E.

1.0 History

Natural systems for wastewater treatment include a variety of terrestrial and hydric systems. These are represented as subsurface soil absorption systems, ponds, land treatment, floating aquatic plants, and constructed wetlands. Natural systems may be developed with discharge to surface water or as a system that assimilates liquid and nutrients through the plant–soil system and results in no surface water discharge. If liquid is discharged to waters of the state, a National Pollutant Discharge Elimination System (NPDES) permit is required. State and local government may address the land-based system through individual permits or coverage under a general permit. Where sufficient land of suitable character is available, these natural systems can be the most cost-effective option for achieving stringent water quality limits. They typically are better suited for small communities and rural areas because of the need for and availability of suitable land area. Natural systems range in size from typical, small 1-m³/d (264-gpd) individual home soil absorption systems to large 190 000-m³/d (50-mgd) rapid infiltration land treatment system such as in Orlando, Florida.

The common element in the use of natural systems for wastewater treatment is the significant contribution made by the "natural" environmental components that provide the desired treatment. Typically, these responses by the vegetation; soil; microorganisms (terrestrial and aquatic); and, to a limited extent, higher animal life proceed at their "natural" rates.

Table 16.1 presents the five types of natural systems and the number of existing systems in each category. Waterless systems, such as composting or incinerating toilets, are described elsewhere (Leverenz et al., 2002).

2.0 Subsurface Soil Absorption Systems

These systems typically are limited to wastewater flows of approximately 190 m³/d (0.05 mgd) or less. The following two sections describe these systems; other publications describe system design [Crites and Tchobanoglous, 1998; U.S. Environmental Protection Agency (U.S. EPA, 2002; U.S. EPA Management Guidelines, 2003; NOWRA Guidelines, 2006)]. Subsurface soil absorption systems have typically been used to treat domestic wastewater from homes. Increasingly these systems are utilized to accommodate the distributed or decentralized wastewater needs in periurban communities. Unlike many NPDES systems, there are no federal rules for these systems and state

Type of System	Number of Systems
Onsite soil absorption	26 000 000
Land treatment	1287
Ponds	8614
Floating aquatic plants	20
Constructed wetlands	497

TABLE 16.1   Number of Natural Wastewater Treatment Systems in the United States (U.S. Environmental Protection Agency, 2002, 2004; Wallace and Knight, 2006)

rules vary leading to a variety of design options. The design challenge in the development of subsurface systems include infiltrating liquid into the soil, ensuring the soil permeability is appropriate to move liquid through the soil, and assuring the transmissivity of the site boundary allows liquid to move off site. Historically subsurface systems (septic systems) have received effluent from septic tanks. Today increasing levels of effluent pretreatment are utilized to address site and soil limitations and the soil is utilized as a receiver to further treat and assimilate liquid, often through pressurized liquid dispersal systems designed to optimize assimilation.

The most common natural system for wastewater management is the subsurface soil absorption system on which approximately 20% of U.S. households rely for wastewater treatment and disposal (U.S. EPA, 2002). Review of permits issued by local agencies suggests an increase in pressure distribution networks and utilization of advanced pretreatment to overcome site- and soil-based limitations. Many communities rely on these soil absorption systems; they are the infrastructure for these populations and infrastructure must be managed.

Although most soil absorption systems serve individual homes, in recent years, small communities have begun to adopt the technology. Several small towns use community septic tanks followed by large soil absorption systems. For example, the community of Taylorsville, California (population 400), uses a community septic tank and a series of leach lines for its wastewater treatment and disposal system. Many larger communities still depend on individual onsite systems for wastewater management.

Subsurface soil absorption systems are typically permitted through local health agencies. State agencies typically utilize a site and soil assessment process to develop the design criteria for subsurface soil absorption systems. Liquid loadings to the soil are expressed as gallons per square foot of trench bottom per day. Typical loading rates range from 1 to 1.5 gallons per square foot per day in sandy soils to as low as 0.1 gallons per square foot per day in clay rich soils. Consult state criteria for actual rates in each state.

2.1 Pretreatment

Historically septic tanks typically were used to provide primary treatment to remove solids, oil, and grease that could lead to accelerated plugging of the soil absorption system. Depending upon the type of wastewater dispersal of the system and any special discharge requirements, additional treatment consisting of anything from a simple outlet screen to a packaged advanced treatment system can be part of pretreatment before dispersal to the soil absorption system.

2.2 Typical Absorption Systems

The typical absorption system is a series of gravel-filled trenches preceded by a septic tank. Effluent from the septic tank flows by gravity into the trenches, which are often referred to as leach lines or drain lines. The basis for the system design is the ability of applied wastewater to infiltrate and then percolate through the soil profile. Conventional percolation tests measured both horizontal and vertical percolation rates and may have overestimated actual vertical percolation rates. Today, permitting agencies rely increasingly on comprehensive site and soil assessments to determine liquid loading rates to the soil. In some jurisdictions, utilization of an advanced wastewater pretreatment device allows designers to increase liquid load to the soil because solids removal is increased through the pretreatment.

Alternative System	Applicability Under Otherwise Restrictive Conditions
Mounds	High groundwater; shallow impervious layer; low permeability soil; shallow soil profile
At grade	Shallow fractured bedrock; high groundwater
Deep trench, bed, or seepage pit	Shallow (but rippable) impermeable layer in the soil profile, with more permeable soil below (must not have high groundwater)
Sand-lined beds and fill systems	Shallow fractured bedrock; high permeability soil
Evapotranspiration beds	Net positive annual evaporation; high groundwater; low permeability soil; shallow fractured bedrock; percolation disallowed by regulatory agency

TABLE 16.2   Alternative Soil Absorption Systems

2.3 Alternative Systems

Conventional leach lines cannot function adequately under adverse site conditions. Alternative systems have been developed, however, that can overcome adverse site conditions (see Table 16.2).

Mound systems work well if appropriately sited, designed, and constructed. Problems relating to poor design, construction, and inadequate hydraulic considerations of percolate flow have been discovered with large-scale (more than 25 equivalent single homes) mound systems [Water Environment Federation (WEF, 2001)].

The "at grade" alternative is similar to a mound system except that the bottom of the aggregate (gravel envelope for the distribution piping) is located on the tilled soil surface. An aggregate depth of 0.3 m (1 ft) topped by a synthetic fabric and 0.3 m (1 ft) of soil is often used. Pressure distribution with dosing from one to four times per day is typical for all types of soil absorption systems (WEF, 2010).

Sand-lined beds and fill systems can be used above fractured bedrock or highly permeable soils to improve pollutant removal. A typical trench or bed is lined with 0.6 m (2 ft) of sand around and below the distribution systems.

Evapotranspiration absorption systems can be used where annual evaporation rates exceed precipitation rates. Evapotranspiration beds typically are filled with 0.6 to 0.75 m (2 to 2.5 ft) of fine sand and covered with topsoil and vegetation. Wastewater is drawn from the sand bottom (underlain by a liner for evapotranspiration systems) to the soil and vegetative surface by capillary forces, where it then evaporates.

2.4 Drip Application

Low loading rates of wastewater at shallow depths (0.1–0.3 m [0.33–1 ft]) can be achieved using subsurface drip irrigation for wastewater dispersal in a soil absorption system. The low loading rates and shallow application depths provide increased soil residence time for treatment and greater opportunity for water and nutrient uptake by plants compared with most other soil absorption systems. Drip lines typically are installed using vibratory or chisel plows spaced at 0.3 to 0.6 m (1–2 ft) [Electric Power Research Institute (EPRI) and Tennessee Valley Authority (TVA, 2004)]. Drip lines are dosed 2 to 12 times

FIGURE 16.1   Installation of subsurface drip lines.

per day depending upon design. Pretreatment filtration and the use of drip tubing specifically designed for subsurface wastewater dispersal are important for minimizing plugging in the drip system. A drip application system is shown in Figure 16.1.

2.5   Soil-Absorption System Design Example

This section outlines how to determine the dimensions of a gravel-filled trench style gravity soil absorption system. Many assumptions are based on state plumbing codes or local ordinances:

Given:

Q = design flow = 1.9 m³/d (500 gpd) (estimated or from code based on building served);

Soil = sandy loam;

R = design application rate = 0.025 m/d (0.6 gpd/sq ft) (based on soil type and/or percolation tests);

A = percolation area = Q/R = (1.9 m³/d)/(0.025 m/d) = 76 m² (820 sq ft);

W = gravel-filled trench width = 0.91 m (3 ft) (from code and construction ease);

S = side-wall infiltrative depth = 0.3 m (1 ft) (from code and trench configuration);

L = total trench length needed = $A/(W + 2S)$ = 76/[0.91(2)(0.3)] = 50 m (164 ft);

Drains = assume one drain line per trench;

D_{max} = maximum drain line length = 30 m (100 ft) (from code and/or hydraulics);

N = number of drain lines = L/D_{max} = 50 m/30 m = 1.67 (round up to 2);

I = length of individual drain lines = L/N = 50 m/2 = 25 m (82 ft);

X = drain line spacing (center-to-center) = 2.4 m (8 ft) (minimum is from code); and

Gross leach field area = $(N)(I)(X)$ = 2(25 m)(2.4 m) = 120 m² (1290 sq ft).

Gravity soil absorption systems typically use 0.1 m (4 in) diameter distribution pipe with 0.01-m (0.5-in) diameter holes every 0.1 m (4 in). Gravel-filled trench style pressurized distribution system dimensions would be determined in the same manner as the gravity system, but using smaller pipes selected based on friction loss calculations with 3.1-mm (0.125-in.) openings. Pressurized distribution systems enable greater layout flexibility and much more uniform distribution.

3.0 Pond Systems

The second most prevalent natural system is the wastewater treatment pond. Wastewater treatment ponds can be classified based on their depth and the biological reactions that occur in the pond. Using this classification, the four main types of ponds are

- Aerobic,
- Facultative,
- Aerated, and
- Anaerobic.

Aerobic ponds are relatively shallow with typical depths ranging from 0.3 to 0.6 m (1–2 ft). Oxygen is provided by algae during photosynthesis and wind-aided surface aeration. These ponds are often mixed by recirculation to maintain dissolved oxygen throughout the entire depth. Aerobic ponds typically are limited to warm, sunny climates and are used mostly in the southern United States and similar climates (Crites et al., 2006; U.S. EPA, 1983).

Facultative ponds, the most prevalent pond type, also are referred to as oxidation ponds. These ponds typically are 1.5 to 2.5 m (5–8 ft) deep, with detention times ranging from 25 to more than 180 days. Depths are kept at 1.5 m or more to avoid the growth of emergent plants. Surface layers of the ponds are aerobic with an anaerobic layer near the bottom. Photosynthetic algae and surface aeration through wind action supply oxygen. Facultative ponds are usually designed in series with a minimum of three cells to reduce short circuiting. Facultative ponds produce algae that remain in the effluent, sometimes causing effluent suspended solids to exceed discharge requirements.

Aerated ponds can be either partially or completely mixed. Oxygen is supplied by mechanical floating aerators or diffused aeration. Aerated ponds typically are 3 to 6 m (10–20 ft) deep, with detention times ranging from 5 to 30 days. Aerated ponds accept higher biochemical oxygen demand (BOD) loadings than facultative ponds, are less susceptible to odors, and typically require less land. Aerated ponds can be followed by a facultative pond or a settling pond (one-day detention or less) to reduce suspended solids before discharge.

Anaerobic ponds are heavily loaded with organics and do not have an aerobic zone. They are 2.5 to 6 m (8–20 ft) deep and have detention times of 20 to 50 days. Biological activity is typically low when compared with that of a mixed anaerobic digester. Anaerobic ponds have been used as pretreatment to facultative or aerobic ponds for strong industrial wastewater and for rural communities with a significant organic load from industries such as food processing. They are not used widely for municipal wastewater in the United States, thus are not further discussed in this chapter.

Another method of classifying ponds is based on the duration and frequency of their effluent discharges:

- Total containment ponds;
- Controlled discharge ponds;
- Hydrograph-controlled release ponds; and
- Continuous discharge ponds.

The use of total containment pond or evaporation ponds is generally limited to climates in which the evaporation exceeds the precipitation on an annual basis. The controlled discharge concept is to discharge only once or twice per year when stream conditions are satisfactory. The hydrograph-controlled release (HCR) pond, a variation of the controlled discharge concept, is designed with a discharge rate correlated to the stream flow rate. As with the controlled discharge ponds, the HCR pond only discharges when stream flow is higher than some acceptable minimum value. Most controlled discharge and HCR ponds are facultative. In the continuous discharge pond, the influent displaces an equal amount of effluent (less evaporation and seepage).

The following section presents performance and design criteria for facultative and aerated ponds, the two types of ponds most commonly used in wastewater treatment. It also includes information on controlled discharge ponds.

3.1 Facultative Ponds

Facultative ponds typically receive no more pretreatment than screening; therefore, they store heavy solids and grit in the first, or primary, ponds. Typical practice is to operate three or more ponds in series, although flexibility to discharge raw wastewater to different ponds and recycle treated effluent back to the primary pond deserves consideration.

3.1.1 Treatment Performance

Properly designed and operated facultative ponds typically can convert essentially all of the incoming BOD. However, some portion ultimately becomes new algal cells that can persist into the effluent receiving stream. Simply stated, incoming brown BOD is converted to energy, carbon dioxide, and green BOD. Typical effluent BOD values range from 30 to 40 mg/L. Sedimentation removes influent suspended solids. Algae, however, contribute to the 40 to 100 mg/L of effluent suspended solids that typically are found during periods of maximum algal growth.

Facultative ponds can remove significant amounts of nitrogen and pathogens because of long detention times (U.S. EPA, 2009). Nitrogen removal often ranges from 40% to 95% (Crites et al., 2006). Phosphorus removal is low, typically less than 40%. Sedimentation, predation, natural die-off, and adsorption remove bacteria and viruses. Table 16.3 presents data on removal of fecal coliforms for both facultative and partial-mix aerated ponds.

A U.S. EPA pilot study (Hannah et.al., 1986) found that removal of toxic organic compound in facultative and aerated ponds varied depending on the volatility of the compound. A facultative pond removed between 77% and 96% (average 86%) of volatile organics, whereas an aerated pond removed between 61% and 80% (average 68%).

Location	Number of Cells	Detention Time (days)	Fecal Coliform, No./100 mL	
			Influent	Effluent
Facultative ponds				
Corinne, Utah	7	180	1.0×10^6	7.0×10^0
Eudora, Kansas	3	47	2.4×10^6	2.0×10^2
Kilmichael, Mississippi	3	79	12.8×10^6	2.3×10^4
Peterborough, New Hampshire	3	37	4.3×10^6	3.6×10^5
Partial-mix aerated ponds				
Edgerton, Wisconsin	3	30	10^6	3.0×10^1
Gulfport, Mississippi	2	26	10^6	1.0×10^5
Pawnee, Illinois	3	60	10^6	3.3×10^1
Windber, Pennsylvania	3	30	10^6	3.0×10^2

TABLE 16.3   Removal of Fecal Coliform in Pond Systems (Reed et al., 1995; U.S. EPA, 1983)

Corresponding removals of semivolatile compounds by the two systems were found to be from 25% to 80% and from 22% to 77%, respectively. Only the activated-sludge process removed more of the organic compounds (Hannah et al., 1986).

3.1.2   Design Procedures
At least five methods have been used to design facultative ponds (WEF, 2010). For each method, *Natural Systems for Wastewater Treatment* (WEF, 2010) presents the calculated detention time, surface area, depth, and BOD loading rate for 5000 m³/d (1.32 mgd) flow, a BOD of 200 mg/L, required effluent BOD of 30 mg/l, k_{20} of 0.15 d⁻¹ and a water temperature of 10°C (50°F). Selection of rate constant and ensuring no short circuiting are the two most critical aspects of lagoon design. Using actual operating data, evaluation of the various design methods failed to show any of the methods to be superior.

The areal loading rate method is the most commonly used design method as it typically produces the most conservative design and can be adapted to specific standards. Table 16.4 presents the recommended BOD loading rates based on average winter air temperatures.

To calculate the area needed for the facultative pond, the designer must divide the BOD load (kilograms per day) by the appropriate loading rate from Table 16.4 for appropriate average winter air temperature or from specific state standards. The first

Average Winter Air Temparature (°C)	Depth (m)	BOD Loading Rate (kg/ha·d)
<0	1.5–2.1	11–22
0–15	1.2–1.8	22–45
>15	1.1	45–90

TABLE 16.4   Facultative Pond Biochemical Oxygen Demand Loading Rates (U.S. EPA, 1975)

cell in a series of cells should not be loaded at more than 100 kg/ha·d (90 lb/ac-d) [for warm climates, average winter air temperatures higher than 15°C (50°F)] and 40 kg/ha·d (36 lb/ac-d) [for cold climates, average air temperatures lower than 0°C (32°F)]. When ice cover forms for extended periods of time, facultative ponds perform as cold anaerobic ponds that remove particulate BOD by settling with little biological activity.

Properly designed inlets and outlets can help avoid hydraulic short-circuiting, either by outfitting them with manifolds or diffusers, spacing them as far apart as possible, or by including multiple sets. In-pond baffles and multiple ponds (typically three or four in a series) will reduce short circuiting. Recirculation from the last pond to the first helps distribution and also reduces short circuiting.

3.1.3 Controlled Discharge Ponds

These are typically facultative ponds with detention times of 120 days or more. Ponds with seasonal discharge have been operated in the North-central United States with the following design criteria (Crites et al., 2006):

- Overall organic loading of 22 to 28 kg/ha·d (20–25 lb/ac-d);
- Water depth of 2 m (6.5 ft) or less in the first cell and 2.5 m (8 ft) or less in subsequent two to three cells; and
- Minimum detention time of 6 months above a minimum depth of 0.6 m (2 ft) (in Alberta, Canada, and other cold climate areas, the required detention time is typically 1 year, and fall discharge is preferred to meet secondary effluent criteria).

3.1.4 Hydrograph-Controlled Release Ponds

A variation of the controlled discharge pond concept has been developed to optimize the dilution of pond effluent in receiving water. Effluent from facultative or aerated ponds is stored in an HCR pond and is discharged based on stream flow. Discharge rate varies—above some minimum flow—with the actual stream flow. The HCR is for storage only and is not designed with any treatment credit. HCR ponds must be able to retain incoming effluent during periods of low stream flow. Assimilative capacity of receiving stream must be established from historical data or estimated (Zirschky, 1987).

3.2 Partial-Mix Aerated Ponds

Data is lacking on actual reaction rates in operating systems. Therefore, partial-mix ponds are designed using complete-mix kinetics. The basic equation for design of partial-mix aerated ponds is given below (WEF, 2010).

$$C_n/C_o = 1/(1+kt/n)^{-n} \tag{16.1}$$

where C_n = effluent BOD_5 in cell n, mg/L;

$\quad\quad C_o$ = influent BOD_5, mg/L;

$\quad\quad k$ = first-order reaction rate constant (0.14–0.3), d^{-1};

$\quad\quad t$ = detention time, d; and

$\quad\quad n$ = number of cells in series.

Equation 16.1 is based on equal-size ponds at the same temperature. Detention time is total for pond system.

In the partial-mix aerated pond system, no attempt is made to completely mix the pond. Aeration demand is calculated to provide an adequate oxygen supply, typically by providing 1 to 1.5 kg of oxygen/kg of BOD loading or more. A portion of the suspended solids and, therefore, some of the particulate BOD settle to the bottom and degrade anaerobically. Because this bottom reaction resembles that of a facultative pond, the partial-mix pond is often referred to as a facultative aerated pond.

The reaction rate constant k depends on the pond temperature (see Chapter 12 for temperature coefficients). At 20°C (68°F), the typical k value used for domestic wastewater is 0.276 d^{-1}; at 1°C (34°F), the k value should be reduced to 0.14 d^{-1} (Crites et al., 2006; U.S. EPA, 1983).

For most municipal systems, detention times range from 5 to 30 days, and water depths range from 3 to 6 m (10–20 ft). BOD loading rates of 100 to 400 kg/ha·d (90–356 lb/d-ac) are typical. A settling pond with a half- to one-day detention often follows the last partially aerated pond in series. The number of cells ranges from three to five or more. The advantages provided by a series of partially mixed aerated ponds diminish after the third or fourth cell. Treatment performance is typically good, with effluent BOD values ranging from 20 to 40 mg/L and effluent suspended solids values ranging from 20 to 60 mg/L. Using several iterations of equations given in WEF (2010), one can estimate the water surface and dimensions of partially-mixed ponds.

The area calculated should be increased to account for the side slope of the pond berm. In partially mixed aerated ponds, the aeration requirements are based on the oxygen demand of the wastewater, not on the mixing requirements for ponds, which are approximately 3 W/m³ (15 hp/mil. gal). Oxygen requirements and aeration system requirements are given elsewhere (Crites and Tchobanoglous, 1998).

Most existing treatment ponds are not lined but have soil conditions that minimize percolation. The trend in designing wastewater ponds is to include liners to minimize percolation. Some states now are requiring a permanent barrier along with groundwater monitoring wells down gradient of the pond. Typical materials used for liners are native clay, bentonite-amended soil, geosynthetic clay liners, and geomembranes. Many states reference the Recommended Standards for Sewage Works (10 States Standards) requirement that the permeability of the liner shall not exceed the value derived from the following equation (Great Lakes, 2014):

$$k = FL \qquad (16.2)$$
$$k = 3.0 \times 10^{-9} \, (L)$$

where k = permeability, cm/s;

      $F = 3.0 \times 10^{-9}$, s^{-1}; and

      L = thickness of the seal, cm.

For water balance calculations, note that permeability and seepage rates per unit area differ. Seepage rate can be calculated using Darcy's law:

$$Q = \frac{kAh}{L} \qquad (16.3)$$

where k = permeability, cm/s;

L = thickness of the seal, cm;

Q = flow through the liner, cm^3/s;

A = liner area, cm^2; and

h = hydraulic head over the liner, cm.

Installation of a pond liner is shown in Figure 16.2. To prevent erosion and desiccation of clay or bentonite liners, interior slopes of the pond should have soil cover and riprap. Suggested minimums are 0.5 m (1.6 ft) above and below the pond water level or twice the impinging wave height calculated for twice the maximum wind velocity anticipated. With surface aerators, the basin may need to be protected directly beneath aerators from the vortex or other scouring action. Concrete pads, anti-erosion plates on the aerators, or 150 mm (6 in.) of crushed rock provide adequate protection. Riprap for erosion control on the slopes of a facultative pond is shown in Figure 16.3.

Embankment tops with a minimum width of 2.5 m (8 ft) permit access of maintenance vehicles. Embankment outer slopes typically are designed to be no steeper than 3:1 to allow grass growth and tractor mowing. A minimum freeboard of 0.6 m (2 ft) represents typical practice.

3.3 High-Performance Aerated Ponds

The concept of high-performance aerated pond system (HPAPS)—also known as dual-power multicellular (DPMC) aerated ponds—is a series of partial-mixed, aerated ponds. The first cell has a depth of 3 m (10 ft) and is aerated with a surface aerator at a rate of 6 W/m^3 (30 hp/mil. gal). The subsequent three cells serve as sludge storage and

FIGURE 16.2 Pond liner installation.

FIGURE 16.3   Oxidation pond with riprap for erosion control.

conditioning cells and are aerated at a rate of 1 W/m^3 (5 hp/mil. gal). Detention time in the first cell is 1.5 to 2 days, and the overall detention time of all four cells is 4.5 to 5 days (Rich, 1980).

Complete mixing in aerated ponds requires between 15 and 30 W/m^3 (75 and 150 hp/mil. gal), typically 20 W/m^3 (100 hp/mil. gal) of aeration. The recommended 6 W/m^3 is more than the 2 W/m^3 (10 hp/mil. gal) of most partial-mix aerated ponds; however, it is less than the mixing required to keep all solids in suspension. The combination of two levels of aeration meets the oxygen requirement for biological conversion while minimizing algae production by the reduced hydraulic detention times and turbulence of mixing. A number of DPMC systems are in operation in South Carolina (Rich, 1996; U.S. EPA, 2009), Tennessee, Oregon, and Louisiana. An existing 1 mgd aerated pond/subsurface flow constructed wetland system at Mandeville, Louisiana, was converted to a 4 mgd DPMC/surface flow wetland system in 2001 (Reed et al., 2003).

3.4   Advanced Integrated Pond Systems

The advanced integrated pond system concept combines multiple ponds with recycle (Oswald, 1991). The system consists of a deep, primary facultative pond followed by a shallow aerobic pond. The primary pond has fermentation pits for anaerobic treatment of settled solids. The fermentation pits must be unaerated and unmixed and will then serve as upflow anaerobic digesters. An example of an advanced integrated pond system is provided to illustrate the concept.

The city of St. Helena, California, located in the Napa Valley north of San Francisco, installed an integrated pond system in 1966 to treat 1900 m^3/d (0.5 mgd) of municipal wastewater. Three shallow, floating aerators have been added to the primary pond to supplement the recycled, algae-laden pond-two water that provides an aerobic cap on the primary pond. A third pond is used in series for settling and the treated effluent is

Design Factor[a]	Value
Design flow rate, mgd	0.5
Average flow rate (1994), mgd	0.4
Primary lagoon	
Aeration, hp	5
BOD loading, lb/d/ac	345
Depth, ft	10
Area, ac	3
Detention time, days	19
High-rate aerobic lagoon	
Depth, ft	3
Area, ac	5.1
Detention time, days	10
Settling lagoon	
Depth, ft	9
Area, ac	2.5
Detention time, days	15
BOD_5 influent (1994), mg/L	250–300
BOD_5 effluent (1994), mg/L	15–40
TSS[b] influent (1994), mg/L	200–250
TSS effluent (1994), mg/L	20–40

[a]ac × 0.404 7 = ha; ft × 0.304 8 = m; hp × 745.7 = W; lb/d/ac × 0.112 1 = g/m²·d; mgd × 3 785 = m³/d.
[b]TSS = total suspended solids.

TABLE 16.5 Design Factors and Performance for St. Helena, California, Advanced Integrated Lagoon System (from Crites, R.W., Tchobanoglous, G., *Small and Decentralized Wastewater Management Systems*. Copyright © 1998, The McGraw-Hill Companies, New York, N.Y., with permission)

chlorinated before irrigation. The fourth and fifth ponds are maturation or storage ponds. Design factors and performance of the St. Helena system are summarized in Table 16.5. The BOD loading rate based on design flow rate and a typical influent BOD of 300 mg/L is 387 kg/ha·d (345 lb/d/ac).

After 27 years of operation, there has been little accumulation of solids in the primary pond. No solids have been deliberately removed from the pond and their accumulated depth in 1993 was approximately 0.3 m (1 ft) on average, with maximum accumulations near the wastewater inlets of approximately 1.2 m (4 ft).

3.5 Ponds Design Example

This section presents two multi-stage, facultative pond design examples. See Figures 16.4 and 16.5 for design drawing.

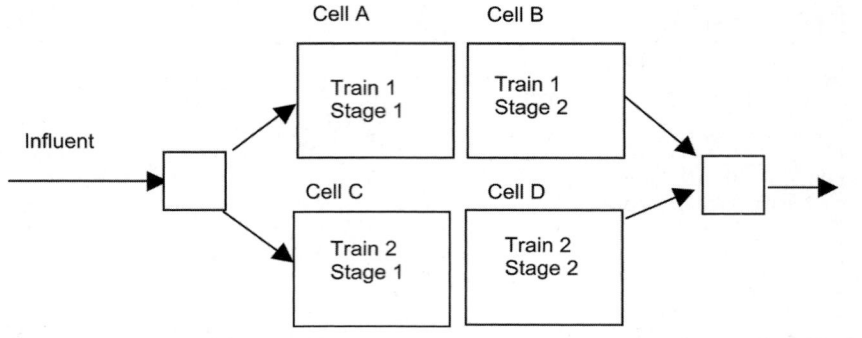

FIGURE 16.4   Option 1: Two-train operation in parallel.

Design conditions:

- 880 m³/d (232 000 gpd) annual average flow (AAF);
- 2320 m³/d (612 000 gpd) peak hourly flow (PHF); and
- Winter water temperature is approximately 0°C (32°F).

Pollution loading:

- 180 kg/d (395 lb/d) BOD_5 (205 mg/L at AAF);
- 293 kg/d (646 lb/d) total suspended solids (TSS) (334 mg/L at AAF);
- 23 kg/d (50 lb/d) ammonia (25 mg/L at AAF); and
- 35 kg/d (77 lb/d total nitrogen) (40 mg/L at AAF).

Discharge criteria:

- 25 mg/L BOD (monthly average) and 45 mg/L (daily peak); and
- 30 mg/L TSS (monthly average) and 45 mg/L (daily peak).

Design calculations:

- Facilities designed for two trains, two-stage operations and for one train, four-stage operations.

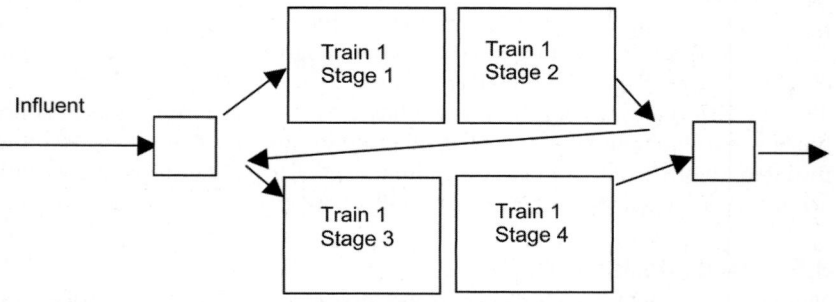

FIGURE 16.5   Option 2: Single-train operation in series.

Solve for total efficiency to achieve BOD removal:

> BOD in (influent BOD) 5180 kg/d;
>
> BOD eff (effluent BOD) 5 22 kg/d [from (880 m³/d × 25 mg/L)/1000];
>
> BOD in $(1 - \text{Eff})^2$ < BOD eff;
>
> [180 kg/d (1 2 Eff)²] , 22 kg/d; and
>
> Eff ~ 66 %.

where Eff = desired efficiency of BOD removal in the stage, represented by $(C_o - C_n)/C_o$.

   T = temperature

From eq 16.1, k = coefficient of BOD reduction (0.06/d at T = 0°C and 0.20/d at T = 20°C).

Combined with eq 16.1 and solved for detention time (DT).

Solve for volume:

> DT = 66%/2.3 (0.06/d) (100 − 66); and
>
> 14.1 days at T = 0°C.

Therefore, the lagoon size for either stage is approximately 6200 m³.

> DT = 66%/2.3 (0.2/d) (100 − 66); and
>
> 4.2 days at T = 20°C.

Therefore, the lagoon size for either stage is approximately 1750 m³.

To solve for total efficiency to achieve BOD removal:

> BOD in = 180 kg/d, (consider potential for short circuiting and review PHF and load);
>
> BOD eff = 22 kg/d;
>
> [180 kg/d $(1 - \text{Eff})^4$] > 22 kg/d; and
>
> Eff ~ 42% (assumes all cells same size).

Solve for volume:

> DT = 42%/2.3 (0.06/d) (100 − 42);
>
> 5.3 days at T = 0°C for all cells;
>
> DT = 42%/2.3 (0.2/d) (100 − 42); and
>
> 1.6 days at T = 20°C for all cells.

Sizing can be reviewed to provide for both series and parallel operation. Table 16.6 provides volume requirements of the two flow options.

This system was placed in operation in 1993. It originally included mechanical surface aeration and separate mixing design to standards included within this text (1.5 kg O_2/kg BOD applied for aeration and 3 W/m³ for mixing). In this case, each cell was supplied with four, 3.73-kW (5-hp) surface aerators and one, 7.46-kW (10-hp) floating mixer. The lagoon operated successfully as permitted except for periodic TSS excursions associated with spring and fall "turnover" (single day values approaching 50 mg/L) and normal maintenance issues.

Operation	Cell A	Cell B	Cell C	Cell D
Parallel, cold	6200 m³	6200 m³	6200 m³	6200 m³
Parallel, average temperature	1750 m³	1750 m³	1750 m³	1750 m³
Series, cold	4700 m³	4700 m³	4700 m³	4700 m³
Series, average temperature	1400 m³	1400 m³	1400 m³	1400 m³

Note: size cells at 4700 m³ each to provide operational flexibility under typical conditions while allowing for series operation in cold winter months.

TABLE 16.6   Volume Requirements Yielded by Two-Flow Options

4.0   Land Treatment Systems

Land treatment is the controlled application of wastewater to land at rates compatible with the natural physical, chemical, and biological processes that occur on and in the soil. The three types of land treatment systems are:

- Slow rate,
- Overland flow, and
- Rapid infiltration.

In slow-rate and rapid-infiltration systems, wastewater is treated as it enters the soil profile and percolates through the soil. In overland flow, treatment occurs in a thin film on the vegetated slopes established on slowly permeable soil. Rapid infiltration systems are frequently utilized to facilitate aquifer recharge and, more recently, recovery. The features of the three types of land treatment are presented in Table 16.7.

Feature	Slow Rate	Overland Flow	Rapid Infiltration
Treatment goals	Secondary or advanced wastewater treatment, zero discharge	Secondary, nitrogen removal	Secondary, advanced wastewater treatment, groundwater recharge, zero discharge
Vegetation	Yes, various crops	Yes, water-tolerant grasses	Only for soil stabilization
Climate restrictions	Storage needed for cold weather and heavy precipitation	Storage needed for cold weather	No storage needed when properly designed and operated
Hydraulic loading, m/a	0.5–6	3–20	6–100
Area needed, ha*	23–280	7–46	1.4–23

*For design flow of 3785 m³/d (1 mgd).

TABLE 16.7   Features of Land Treatment Systems

Of the three types of land treatment, slow-rate systems typically achieve the highest level of pollutant removal. Surface runoff typically is contained onsite, although rainfall-induced runoff is typically allowed to exit the site. With typical loading rates of 1 to 2 m/yr (3–7 ft/yr) much of the applied wastewater can be lost to evapotranspiration, particularly in arid climates. The technology is similar to that which is used for crop irrigation, varying from drip and sprinkler operations to surface flooding. Recent design standards utilized in several states impose strict nutrient management planning requirements on slow-rate systems. These are developed to assure assimilation of nutrients in the plant–soil system and to reduce potential for nitrate transport to groundwater (USDA NRCS 590 program).

Overland-flow systems, similar to other fixed-film biological treatment systems, remove significant amounts of BOD, suspended solids, and nitrogen. Removals of phosphorus, trace elements, and pathogens, however, are not as effective. Hydraulic loading rates range from 3 to 20 m/yr (10–70 ft/yr). Overland flow is best suited to slowly permeable soils that can be graded to mild slopes (2%–8%) and planted with water-tolerant grasses. Overland flow produces an effluent of better than secondary quality, depending on the application rate. This technology emerged in the United States in the 1970s and is now relatively well developed, although it is less frequently used outside the Southeast or Southwest. Overland-flow systems may require NPDES permits depending on the fate of the liquid applied.

Rapid infiltration (also known as soil aquifer treatment), which is considered to be an established treatment technology, consists of shallow spreading basins in permeable soils to which wastewater is intermittently applied. Each basin is dosed for 1 to 7 days and then rested for 6 to 20 days. Treatment is accomplished by physical, chemical, and biological means as wastewater infiltrates through the surface and percolates through the soil. Rapid-infiltration systems discharge to groundwater or can be underdrained. Recently, high levels of effluent treatment have been utilized and the liquid entering the groundwater system is recovered for beneficial use.

The following section describes the preapplication treatment, site requirements, design criteria, and expected performance of each of the three land treatment systems.

4.1   Preapplication Treatment

Historically, land treatment systems have provided a method of treating and dispersing treated water to the environment. As the performance of these systems was documented and became better understood, the need for preapplication treatment was reexamined. Table 16.8 presents U.S. EPA's recommended preapplication treatment guidance for land treatment systems.

Slow-rate systems typically require the preapplication treatment appropriate to end-use. Dedicated land treatment sited frequently utilizes lagoon effluent as the irrigation resource. When site access is not controlled such as on a turf field, park or golf course, or when private farmer contracts are used and liquid is applied to edible crops, high levels of treatment are necessary. Increasingly, slow-rate irrigation is utilized to benefit crop production. Crops destined for human consumption or utilized as athletic fields are irrigated with reclaimed water and, where human contact is probable, tertiary treatment and advanced disinfection are utilized. Many states impose effluent BOD limits of 5 mg/l or less and coliform levels of 14/100 ml or less where human contact is probable.

Land Treatment System	Guidance
Slow-rate systems	Primary treatment is acceptable for isolated locations with restricted public access and when limited to crops not used for direct human consumption
	Biological treatment using ponds or processes plus control of fecal coliforms to less than 1000 MPN/100 mL is acceptable for controlled agricultural irrigation except for human food crops to be eaten raw
	Biological treatment by ponds or in-plant processes with additional BOD or suspended solids control as needed for aesthetics plus disinfection to log mean fecal coliform count of 200 MPN/100 mL is acceptable for application in public access areas such as parks and golf courses
Overland-flow systems	Screening or comminution* is acceptable for isolated sites with no public access
	Screening or comminution* plus aeration to control odors during storage or application is acceptable for urban locations with no public access
Rapid infiltration	Primary treatment is acceptable for isolated locations with restricted public access
	Biological treatment by ponds or in-plant processes is acceptable for urban locations with controlled public access

*Comminution is mentioned in this original U.S. EPA guidance (1981). Current considerations of comminution are presented in Chapter 7.

TABLE 16.8   U. S. EPA Guidelines for Minimum Preapplication Treatment for Land Treatment Systems (U.S. EPA, 1981)

For overland flow, the use of a partially aerated pond with a one-day detention time for pretreatment has been successful. The pond removes larger solids and adds dissolved oxygen to the wastewater. Short-term detention minimizes algal growth that otherwise would not be removed efficiently.

Rapid-infiltration systems can operate year-round using primary or Imhoff tank effluent if the underlying groundwater is of limited resource value. When aquifer storage and recovery is proposed, high levels of treatment and nutrient removal are necessary.

4.2   Site Requirements

Table 16.9 presents requirements for suitable sites for land treatment systems. Sites suitable as receivers for treated wastewater share some common characteristics. Site information can be assessed from review of published information such as USGS topographic information, USDA Soil Survey information, crop production potential from USDA, and published geologic information.

Characteristics	Slow Rate	Overland Flow	Rapid Infiltration
Soil depth, m	>0.6	>0.3	>1.5
Soil permeability class range	Slow to moderately rapid	Very slow to moderately slow	Rapid
Soil permeability, mm/h	1.5–500	<5.0	>50
Depth to groundwater, m	0.6–1	Not critical[a]	1 during flood cycle[b]; 1.5–3 during drying cycle
Slope, %	<20 on cultivated land; 40 on noncultivated land	0–15; finished slopes 2 to 8[c]	<10; excessive slopes require much earthwork

[a]Effect on groundwater should be considered for more permeable solids.
[b]Underdrains can be used to maintain this level at sites with high groundwater table.
[c]Slopes as low as 1% and as high as 10% may be considered.

TABLE 16.9   Site Requirements for Land Treatment Processes

Detailed site and soil investigations are necessary to select the most appropriate land treatment process and the best available site (U.S. EPA, 2006). For slow-rate systems, investigations concentrate on topography, soil characteristics, groundwater depth, and surface drainage features. Site hydrologic boundary conditions must be defined. If a detailed USDA Natural Resources Conservation Service (NRCS) soil survey is available, the soil investigation typically serves as a suitability assessment, not a detailed investigation. Regardless of the availability of a USDA survey, a detailed site and soil assessment is necessary to optimize liquid and nutrient loadings to the site. Evaluation may include examination of profiles with backhoe pits and analysis of the soil profile by a soil scientist or experienced land treatment specialist, measurement of soil permeability and evaluation of crop production potential. These measures are necessary since the fate of the liquid applied is ultimately to shallow groundwater and that resource must be protected with no violation of groundwater standards.

The same site features listed for slow-rate systems are important for overland-flow systems. The soil depth and slope are more important than permeability for overland flow because site grading to uniform slopes is necessary, and deep percolation is not desired. Liquid applied to the site is intended for discharge to surface water.

For rapid-infiltration systems, the most important site features are soil depth, permeability, and depth to groundwater. Site investigations should be performed to determine the soil depth, depth to groundwater, and, most importantly, infiltration rate of the limiting soil layer in the soil profile. The field investigation may need to extend beyond the proposed application site to ensure that the percolate will flow away from the application point and will not emerge as surface seepage at undesirable locations.

4.3   Slow-Rate Systems

Slow-rate systems have been demonstrated as effective wastewater treatment systems. Slow-rate systems have been utilized in watershed protection programs as efficient nutrient removal technologies. In the Chowan River Basin in North Carolina, land-based wastewater systems were employed throughout the basin to eliminate the direct discharge of nutrients to the river. This section describes the performance, design objectives, crop selection, loading rates, and land-area requirements.

4.3.1 Treatment Performance

Table 16.10 summarizes treatment performance of slow-rate systems for BOD, nitrogen, and phosphorus. Filtration, soil absorption, and bacterial oxidation remove BOD. Slow-rate systems effectively remove BOD at loading rates of 500 kg/ha·d (450 lb/ac/d) and more (Jewell, 1978). For the systems in Table 16.10, BOD loading rates range from 3 to 11 kg/ha·d (2.7–10 lb/ac/d), which is significantly less than the rate of 500 kg/ha·d (450 lb/ac/d). Effective BOD removal (more than 90%) can be expected for slow-rate systems loaded up to 500 kg/ha·d (Crites et al., 2006). Careful management should be practiced with organic loadings of more than 300 kg/ha·d (270 lb/ac/d) to avoid odor production. In practice, municipal slow-rate systems will rarely be loaded beyond 10 kg/ha·d (33 lb/ac/d).

In slow-rate systems, a combination of nitrogen mineralization, plant uptake, denitrification, and soil storage removes nitrogen and reduce potential for migration to shallow groundwater. Soil adsorption, chemical precipitation, and limited plant uptake remove phosphorus. Adsorption, chemical precipitation, ion exchange, limited plant uptake, and complexation remove metals. Soil filtration, adsorption, desiccation, radiation, predation, and exposure to other adverse environmental conditions remove pathogens.

Photodecomposition, volatilization, sorption, and biological degradation remove trace organics. At Muskegon County, Michigan, the slow-rate system receives stable organics from many industrial sources and effectively removes them from wastewater (U.S. EPA, 2006). Of the 59 organic pollutants identified in municipal wastewater, the percolate contained only 10 organic compounds, all at low levels (1–10 mg/L) (U.S. EPA, 2006). Based on these results, slow-rate systems appear to be effective at removing trace organics.

Location	BOD (mg/L)		Total Nitrogen (mg/L)		Total Phosphorus (mg/L)	
	Applied	Percolate	Applied	Percolate	Applied	Percolate
Dickinson, North Dakota	42	<1	11.8	3.9	6.9	0.05
Hanover, New Hampshire						
Primary effluent	101	1.4	28.0	9.5	7.1	0.03
Secondary effluent	36	1.2	26.9	7.3	7.1	0.03
Muskegon, Michigan	34	1.3	8.2	2.5	3.8	1.10
Roswell, New Mexico	43	<1	66.2	10.7	8.0	0.39
Yarmouth, Massachusetts	85	<2	30.8	1.8	12.0	0.04

TABLE 16.10 Removal of Biological Oxygen Demand, Nitrogen, and Phosphorus in Slow-Rate Systems (Reed et al., 1995; U.S. EPA, 1981)

4.3.2 Design Objectives
Slow-rate systems, classified according to the design objective, are either Type 1 (slow-rate infiltration) or Type 2 (crop irrigation). The objective of Type 1 systems is wastewater treatment. Design of Type 1 systems is based on the limiting design factor, which typically is either the soil permeability or the allowable loading rate for a particular wastewater constituent, such as nitrogen. Type 1 systems typically are found in humid areas of the United States and are managed by municipal wastewater agencies, private utilities, or industries.

The objective of Type 2 systems is water and nutrient reuse. Crop production is a primary objective in Type 2 systems and wastewater treatment is a secondary objective. Design of Type 2 systems is based on applying sufficient water to meet crop irrigation requirements for water and nutrients. Type 2 systems typically are found in the arid areas of the United States and are managed by municipal wastewater agencies through farmer contracts or leases or through private utilities associated with planned communities where treated wastewater is utilized to irrigate community amenities, golf courses, and common areas.

4.3.2.1 Crop Selection Crop selection and management are important steps in the design of slow-rate systems because the crop can affect the level of preapplication treatment, type of distribution system, and hydraulic loading rate. For Type 1 systems, compatible crops have high nitrogen uptake capacity and evapotranspiration rates and are tolerant to moisture and wastewater constituents. Type 1 system crops include perennial forage grasses, turf grasses, some tree species, and some field crops. Table 16.11 provides annual crop nitrogen uptake rates for various crops. Recent changes associated with NRCS 590 Nutrient Management Planning recommend development of comprehensive nutrient management plans specific to sites, soils, climate, and realistic crop yield.

For Type 2 systems, a broader variety of crops can be considered, including those presented in Table 16.11. Table 16.12 presents nutrient requirements for various crops. Double cropping increases revenue potential. In warm climates, short-season summer crops (such as corn or sorghum) can be combined with winter grains (such as barley, oats, or wheat). Again, the NRCS nutrient management criteria must be utilized to optimize design and prevent migration of soluble nutrients to groundwater. Mineralization of nitrogen is a particularly contentious issue. Organic nitrogen (which is typically not available to crops) applied to a site does mineralize to soluble forms (nitrate) available to plants. Application of excessive levels of organic nitrogen must be controlled carefully since mineralization may contribute nitrogen to shallow groundwater in subsequent years.

Comprehensive nutrient management planning (CNMP) requires system designers and regulatory agency personnel to tailor nutrient loads to specific crops and demonstrable crop yield data. Using the new NRCS nutrient management criteria, specific crop data is obtainable for specific crops raised on a specific soil series in a county.

4.3.2.2 Distribution System The choice of distribution system depends on the crop, topography, and soil. Sprinkler systems are commonly used in wastewater applications because they are adaptable to different soil and topographic conditions. Variations in sprinkler systems include fixed, impact-type sprinklers, continuous-move (center pivots and linear) systems, and move-stop systems (wheel line and

Crop	Nitrogen Uptake (kg/ha·a)[a]
Forage crops	
Alfalfa[b]	225–675
Bromegrass	130–224
California grass	2000
Coastal Bermuda grass	400–675
Kentucky bluegrass	200–270
Orchard grass	250–350
Quack grass	235–280
Reed canary grass	335–450
Ryegrass	200–280
Sweet clover[b]	175–300
Tall fescue	150–325
Timothy	150
Vetch	390
Field crops	
Barley	125–160
Corn	175–250
Cotton	75–180
Grain sorghum	135–250
Oats	115
Sugar beets	255
Wheat	160–175
Tree crops	
Mixed hardwoods (eastern)	220
Mixed hardwoods (southern)	340
Hybrid poplar (lake states)	155
Hybrid poplar (west)	300–400
Southern pine with understory	320

[a]Range indicates yield variation.
[b]Legumes will take a minimal amount of nitrogen from the atmosphere when under nitrogen fertilization.

TABLE 16.11　Nitrogen Uptake of Selected Crops (U.S. EPA, 1981)

traveling gun) (Crites et al., 1988). Design guidance can be found elsewhere [Irrigation Association, 1983; State Water Resources Control Board (SWRCB), 1984 and product literature from irrigation equipment manufacturers]. Surface irrigation is a low-cost, moderately labor-intensive technique, which was typically associated with level 1 land treatment (Hansen et al., 1979). Additional labor may be necessary during crop

Crop	N required (lb/ac)	P required (lb/ac)	N as pounds/yield	P as pounds/yield
Forage				
Alfalfa	225–675	75–300	50 lb/ton	10 lb/ton
Bromegrass	130–225	50–80	40 lb/ton	8 lb/ton
Ky. Bluegrass	200–270	60–90	40 lb/ton	8 lb/ton
Orchard grass	250–350	50	40 lb/ton	8 lb/ton
Reed canary grass	325–450	60	40 lb/ton	8 lb/ton
Rye grass	150–200	50	20 lb/ton	5 lb/ton
Clover	150–250	50	30 lb/ton	5 lb/ton
Fescue	150–300	50	40 lb/ton	8 lb/ton
Timothy	150	60	40 lb/ton	8 lb/ton
Bermuda grass	300–450	60	50 lb/ton	10 lb/ton
Field Crops				
Barley	125–160	60	2 lb/bu	0.5 lb/bu
Corn Grain	150–250	50	1.25 lb/bu	0.4 lb/bu
Corn stover	200–300	75	50 lb/ton	10 lb/ton
Cotton	75–125	40	0.75 lb/lb	0.25 lb/lb
Sorghum	125–200	40	2 lb/bu	0.5 lb/bu
Oats	110–150	30	0.75 lb/bu	0.25 lb/bu
Wheat	150–180	50	0.5 lb/bu	0.2 lb/bu
Forest Crops				
Southern hardwood	250			
Hybrid poplar	300			
Pine	150			
Northern hardwood	200			

TABLE 16.12   Nutrient Requirements for Selected Crops as unit/ac or Yield Based unit/ac/yield

harvest and management operations. The advent of precision gear drive sprinklers for surface application and drip irrigation have expanded the role of land-based irrigation into high-value landscapes such as crop irrigation, golf course irrigation, and corporate campus beautification projects. These high-value operations require advanced treatment, multi-barrier disinfection and are used with in Type 2 slow-rate systems (Reed et al., 1998).

4.3.2.3 Hydraulic Loading Rate   For Type 1 systems, the hydraulic loading rate (Crites and Tchobanoglous, 1998) can be calculated from the following water balance equation:

$$L_w = \text{ET} - P + Wp \qquad (16.4)$$

where L_w = wastewater hydraulic loading rate based on soil permeability, m/yr;

ET = design evapotranspiration rate, m/yr;

P = design precipitation rate, m/yr; and

Wp = design percolation rate, m/yr.

The design evapotranspiration rate is the estimated average rate for the selected crop. The design precipitation rate is typically the total for the wettest year in a 10-year period. The design percolation rate should be measured in the field using a cylinder infiltrometer, sprinkler infiltrometer, field portable constant head permeameter, or basin flooding technique (U.S. EPA, 2006; Amoozegar and Wilson, 1999).

For Type 2 systems, the hydraulic loading rate equation for a specific crop use is

$$L_w = \left(\frac{ET - P}{1 + LR} \right) \left(\frac{100}{E_u} \right) \tag{16.5}$$

where L_w = annual wastewater loading, m/yr;

ET = crop evapotranspiration rate, m/yr;

P = precipitation, m/yr;

LR = leaching requirement, 15% to 25%; and

E_u = irrigation efficiency, 65% to 85%.

The specific crop and its sensitivity to wastewater total dissolved solids determines the leaching requirement, which may range from 10% to 40% but is typically between 15% and 25%. Table 16.13 presents a list of crops and the electrical conductivity (EC$_w$) of the applied wastewater for different yields (SWRCB, 1984).

Irrigation efficiency for sprinklers ranges from 70% to 80%; for surface irrigation it ranges from 65% to 85%. The total percolation is a combination of leaching fraction and irrigation inefficiency fraction $(1 - E_u/100)$.

4.3.2.4 Nitrogen Loading Rate If the slow-rate system percolate enters a potable groundwater aquifer, the percolate nitrogen quality is often limited to 10 mg/L or less (as nitrate nitrogen). This is the nitrate nitrogen standard in drinking water and, in many areas, groundwater is drinking water. The nitrogen balance is

$$L_n = U + fL_n + C_p P_w F \tag{16.6}$$

where L_n = nitrogen loading rate (Crites and Tchobanoglous, 1998), kg/ha/yr;

U = crop uptake of nitrogen from Table 16.11, kg/ha·yr;

f = fraction of applied nitrogen lost to denitrification, volatilization, and soil storage;

C_p = percolate nitrate nitrogen concentration, mg/L; and

P_w = percolate flow, m/yr;

F = conversion factor, 10 kg·m²/g·ha

The value of f depends on the BOD to nitrogen ratio in wastewater and the air temperature during application season. High-strength wastewater (BOD:N) has the

Crop	EC$_w$ Values, mmhos/cm (Electrical Conductivity of Applied Wastewater), for a Reduction in Crop Yield		
	0%	25%	100%
Forage crops			
Alfalfa	1.3	3.6	10.3
Bermuda grass	4.6	7.2	15.0
Clover	1.0	2.4	6.7
Corn (forage)	1.2	3.5	10.3
Orchard grass	1.0	3.7	11.7
Perrennial ryegrass	3.7	5.9	12.7
Tall fescue	2.6	5.7	15.3
Vetch	2.0	3.5	8.0
Tall wheat grass	5.0	8.9	21.0
Field crops			
Barley	5.3*	8.7	18.8
Corn	1.1	2.5	6.7
Cotton	5.1	8.7	18.0
Potatoes	1.1	2.5	6.7
Soybeans	3.3	4.1	6.7
Sugar beets	4.7	7.3	16.0
Wheat	4.0*	6.3	13.3

*Barley and wheat are less tolerant during germination and seeding stage when EC$_w$ should not exceed 2.7 mmhos/cm.

TABLE 16.13 Electrical Conductivity Values Resulting in Reductions in Crop Yield (Ayers and Westcot, 1984)

highest f value, as shown in Table 16.14; lower f values apply to cold climates. Misrepresentation of this value may lead to subsequent nitrate contamination of shallow groundwater.

By combining the water balance and nitrogen balance equations, the hydraulic loading rate based on nitrogen limits can be calculated as follows:

$$L_{wn} = \frac{C_p(P - ET) + 0.1U}{(1 - f)C_n - C_p}$$

(16.7)

where L_{wn} = hydraulic loading rate based on nitrogen limits, m/yr; and
C_n = wastewater nitrogen concentration, mg/L.

The design limiting loading rate is the lower of the two calculated values, L_w or L_{wn}, for Type 1 systems.

Wastewater Type	f value
High strength	0.5–0.8
Primary effluent	0.25–0.5
Secondary effluent	0.15–0.25
Advanced treatment effluent	0.10–0.15

TABLE 16.14   Ranges of f Values for Municipal Wastewaters

4.3.2.5 Land Requirements   The land area needed for a slow-rate site includes the field application area plus space for roads, buffer zones, and any required storage (Crites and Tchobanoglous, 1998). The field area can be expressed as:

$$A = \frac{Q}{L_w F} \tag{16.8}$$

where A = field area, ha;

Q = wastewater flow, mil. gal/yr;

L_w = wastewater design hydraulic loading rate, in/yr; and

F = conversion factor, 0.027 mil. gal/ac·in.

4.3.2.6 Storage Requirements   Most slow-rate systems require wastewater storage during cold or wet weather periods and during routine crop management operations. In addition, the application rate will vary during the year while the wastewater supply remains relatively constant. A storage pond can store excess wastewater whenever the allowable application rate is lower than average. Storage needed based on climatic data can be estimated from maps or by using computer programs from the National Oceanic and Atmospheric Administration (U.S. EPA, 1976; 2006). A detailed water balance is necessary for final design to determine storage volume. Design details of storage facilities may be found elsewhere (Crites and Tchobanoglous, 1998; U.S. EPA, 2006).

4.3.3   Slow-Rate Land Treatment Design Example
An equation can be used to determine the land area requirement for a slow rate land treatment site. Based on an average annual flow of 378 m³/d and limiting hydraulic loading rate of 2.5 m/yr, with 100 days of storage and a net loss of water from evaporation and seepage from the storage pond of 3000 m³.

The solution is:

$$A = \frac{365\,Q + V_s}{10\,000\,L_l} \tag{16.9}$$

4.4   Overland-Flow Systems
Overland-flow systems (OLF) can be designed to achieve secondary treatment, advanced treatment, or nitrogen removal. Phosphorus removal requires either pre- or post-application treatment. Typically OLF facilities are designed as systems that discharge to surface water and an NPDES permit may be required.

4.4.1 Treatment Performance

Table 16.15 presents the removal of BOD, TSS, and nitrogen in overland flow. As Table 16.15 shows, treated runoff concentrations of BOD and suspended solids differ little among raw, primary, and secondary effluent applications. Most overland-flow systems do not effectively remove algae from pond effluent (Witherow and Bledsoe, 1983). For raw and primary effluent applications at Ada, Oklahoma, nitrogen removal was better than for secondary effluent application. This is primarily because of the lower BOD to nitrogen ratio in the secondary effluent (1:1) compared to the raw wastewater (6.4:1) and the primary effluent (3.7;1). The primary mechanism for nitrogen removal in overland flow is nitrification/denitrification, which requires a BOD:nitrogen ratio of 3:1 to be effective (Crites et al., 1998).

Land treatment requires thorough soil–water contact to provide effective phosphorus, metals, and pathogen removal. As a result of the limited soil contact, overland flow removes approximately 40% to 60% of applied phosphorus, 60% to 90% of trace metals, and 99% of bacteria and viruses (U.S. EPA, 2006). Trace organics are adequately removed in overland flow by the same mechanisms as in slow-rate systems.

4.4.2 Design Factors

Overland-flow design factors include application rate, slope length, slope grade, and application period. The application rate, expressed in cubic meters per meter per hour, applied to the top of the slope or terrace, ranges from 0.03 to 0.37 m^3/m·h (0.04–0.5 gpm/ft). The length of the slope or terrace is typically 30 to 60 m (100–200 ft). Slope grades are between 1% and 12%, with a preferred range of 2% to 8%. Application periods are typically 6 to 12 hours/day and five to seven days/week. Table 16.16 presents design factors for overland-flow systems.

Location	BOD* (mg/L)		SS* (mg/L)		Total Nitrogen (mg/L)	
	Applied	Effluent	Applied	Effluent	Applied	Effluent
Ada, Oklahoma (raw)	150	8	160	9	23.6	2.1
Ada, Oklahoma (primary)	70	8	56	7	19	5
Ada, Oklahoma (Secondary)	18	6	12	5	16	8.5
Easley, South Carolina (raw)	200	23	186	8	30.5	7.7
Easley, South Carolina (pond)	28	15	60	40	6.7	2.1
Hanover, New Hampshire (primary)	72	9	74	10	45	9.4
Melbourne, Australia (primary)	507	12	233	19	55.6	39.7

*SS = suspended solids; BOD = biological oxygen demand.

TABLE 16.15 Treatment Performance of Overland-Flow Systems

Location	Wastewater Applied	Application Rate (m³/h·m)	Slope Length (m)	Slope Grade (%)	Application Period (h/d)	Loading Rate* (mm/d)
Ada, Oklahoma	Raw wastewater	0.075	36	4	8–12	11.6
	Primary effluent	0.065	36	4	12	25
	Secondary effluent	0.12–0.2	36	4	12	42
Easley, South Carolina	Raw wastewater	0.22	55	6	6	24
	Lagoon effluent	0.23	46	6	7	36
Hanover, New Hampshire	Primary effluent	0.075	30.5	5	5	12.5
	Secondary effluent	0.075	30.5	5	5	12.5
Utica, Mississippi	Lagoon effluent	0.032–0.13	46	4	6–18	12.7–50.8
Melbourne, Australia	Primary effluent	0.24	250	0.3	24	23

*Loading rate is the total daily flow divided by total field

TABLE 16.16   Design Factors for Overland-Flow Systems (U.S. EPA, 1981)

4.4.3   Design Procedures

The following equation presents the relationship between BOD removal and application rate that has been developed and validated by existing demonstration projects (U.S. EPA, 2006).

$$\frac{C_z - R}{C_o} = A \exp\left(-KZ/q^n\right) \tag{16.10}$$

where A = constant;
C_z = effluent BOD concentration at point Z, mg/L;
R = residual BOD at end of slope, mg/L;
C_o = applied BOD concentration, mg/L;
Z = slope length, m;
q = application rate, m³/m·h; and
K, n = empirical constants.

Figure 16.6 graphs the equation that has been validated for screened raw wastewater and primary effluent but not for high-strength industrial wastewater (U.S. EPA, 1984). The graph can be used by finding the required BOD fraction remaining and looking right to the longest slope length within the validated range and noting the application rate indicated by the family of lines. As good practice, the application rate can be reduced by dividing by a safety factor of 1.5 before calculating the field area.

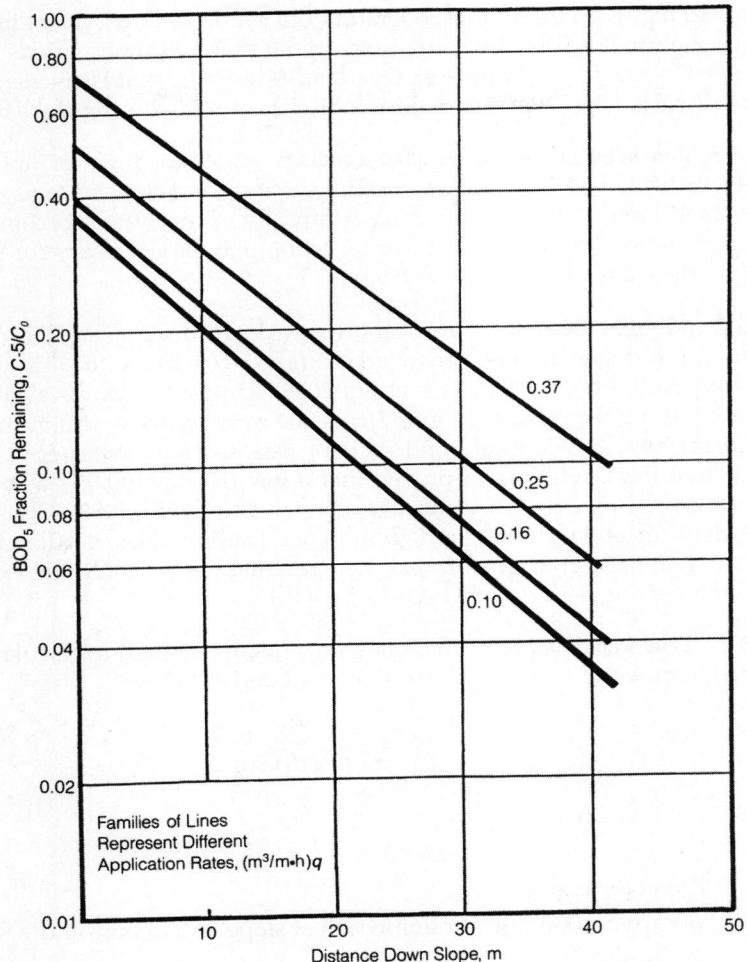

FIGURE 16.6 Biochemical oxygen demand fraction remaining versus distance downslope for overland-flow treatment of primary effluent.

4.4.3.1 Suspended Solids Loadings Except for algae, wastewater solids typically will not be limiting in overland-flow system designs. Suspended solids effectively are removed on overland-flow slopes because of low velocity and shallow depth of flow. For high-strength and high-solids content wastewater, sprinkler applications will distribute suspended solids uniformly over the upper 65% of the slope.

Removal of algae by overland flow varies based on application rate and type and concentration of algae. Removal rates range from 45% to 83% (Witherow and Bledsoe, 1983). Buoyant or motile algae resist removal by sedimentation or filtration (WEF, 2001). Where facultative ponds are used as pretreatment, the loading rate on the overland-flow slopes should not exceed 0.09 $m^3/m \cdot h$ (0.12 gpm/ft). If algae concentrations cause TSS values to exceed 100 mg/L, then the overland-flow system will not reduce the TSS below 30 mg/L. In these cases, operating the overland-flow system in a nondischarge mode might be possible by using repeated short application periods (15–30 minutes) followed by one to two hours rest. Overland-flow systems at Heavener, Oklahoma, and Sumrall, Mississippi, operate in the nondischarge mode during algae blooms.

4.4.3.2 Biochemical Oxygen Demand Loadings BOD loadings of up to 100 kg/ha·d (90 lb/ac/d) have been used successfully in overland-flow systems. When the BOD exceeds 800 mg/L, the oxygen-transfer capacity of the system becomes limiting, and preapplication treatment or effluent recycling may be necessary for successful treatment (Crites and Tchobanoglous, 1998).

4.4.3.3 Nitrogen Removal Nitrification and denitrification are the primary mechanisms for nitrogen removal in overland-flow systems, with 60% to 90% removal reported (U.S. EPA, 2006; Crites et al., 2006). Up to 90% removal of ammonia was reported at 0.10 m³/h·m (0.13 gpm/ft) at the overland-flow site in the city of Davis, California, where oxidation pond effluent was applied (Crites et al., 2006). Further research at the Davis system proved that a low (less than 0.5;1) wetting period:total cycle is critical to ammonia removal (Johnston and Smith, 1988). Garland, Texas, conducted nitrification studies and found that loading rates needed to be less than 0.43 m³/h·m (0.56 gpm/ft) and that the operating period could not exceed 10 hours/day for a wetting period/total cycle ratio of 0.42.

4.4.3.4 Land Requirements The following equation is used to calculate the field area needed (Crites and Tchobanoglous, 1998) for overland flow:

$$A = \frac{QZ}{qP\,(10\,000)} \tag{16.11}$$

where A = field area, ha;

Q = wastewater flow rate, m³/d;

Z = slope length, m;

q = application rate per unit width of slope, m³/m·min; and

P = period of application, h/d

10 000 m² = 1 ha

If wastewater storage is required, then the field area can be expressed as:

$$A = \frac{(365\,Q + V_s)\,Z}{qP\,(10\,000)\,D} \tag{16.12}$$

where V_s = net loss or gain from evaporation, seepage, or precipitation on the storage pond, m³/yr; and

D = operating time, d/yr.

4.4.3.5 Vegetation Selection Water-tolerant grasses are used in overland-flow systems to provide a support medium for microorganisms, minimize erosion, and remove nitrogen. The crop is cut periodically and either removed as hay or green chop or left on the slope. Sod-forming, water-tolerant grasses such as reed canary grass typically are selected. Some older varieties of reed canary grass exhibit alkaloid concentrations that may be harmful to animals and designers must select appropriate varieties. Other cool-season grasses include tall fescue, perennial ryegrass, and redtop. Warm-season grasses include common and coastal Bermuda grass and bahia grass.

4.4.3.6 Distribution System   Municipal wastewater can be surface applied to overland-flow systems using gated pipe; however, industrial wastewater should be sprinkler applied. Sprinkler systems for municipal wastewater should be located 30% of the distance down the slope. Typical distances from the edge of the sprinkler wetted diameter to the runoff collection ditch range from 15 to 20 m (50–65 ft). Top-of-the-slope distribution methods, in addition to gated pipe, include low-pressure sprays, bubbling orifices, and perforated pipe (Crites and Tchobanoglous, 1998).

4.4.4 Overland-Flow Land Treatment Design Example
A calculation can be used to determine the land area required for an overland-flow land treatment site. The application rate, slope length, wetting period, and area needed for treatment can also be calculated as:

$$A = QZ/qP(10\ 000)$$

4.5 Rapid-Infiltration Systems or Rapid Infiltration Beds
Rapid-infiltration systems or Rapid Infiltration Beds (RIBS) require deep, rapidly permeable soils to receive wastewater for final treatment or dispersal. This section describes expected treatment performance, design procedures, hydraulic loading rates, organic loading rates, and land requirements.

4.5.1 Treatment Performance
Rapid-infiltration systems effectively remove BOD and suspended solids through filtration, adsorption, and bacterial decomposition. BOD loadings and removals for rapid infiltration are presented in Table 16.17. Suspended solids are typically removed to low levels, approaching 1 mg/L.

Nitrogen removal of rapid-infiltration systems varies from 40% to 90% as a result of biological denitrification. Important design criteria are the BOD to nitrogen ratio, hydraulic loading rate, and ratio of flooding to drying periods. The design objective is

Location	Applied Wastewater BOD		Percolate (mg/L)	Removal (%)
	kg/ha·d	mg/L		
Brookings, South Dakota	13	23	1.3	94
Fort Devens, Massachusetts	87	112	12	89
Hollister, California	177	220	8	96
Lake George, New York	53	38	1.2	97
Phoenix, Arizona	45	15–30	0–1	93–100

TABLE 16.17 Biological Oxygen Demand (BOD) Loadings and Removal in Rapid-Infiltration Systems

Location	Total Nitrogen Applied		Percolate Nitrogen (mg/L)	Applied BOD*: Nitrogen Ratio	Removal (%)
	kg/ha·a	mg/L			
Brookings, South Dakota	1330	10.9	6.2	2:1	43
Calumet, Michigan	4170	24.4	7.1	3.4:1	71
Fort Devens, Massachusetts	15 250	50.0	10–20	2.4:1	60–80
Hollister, California	6110	40.2	2.8	5.5:1	93
Lake George, New York	6960	12.0	7.5	2:1	38
Phoenix, Arizona	16 710	27.4	9.6	1:1	65

*BOD = biological oxygen demand.

TABLE 16.18 Total Nitrogen Removal at Rapid Infiltration Systems (from Crites, R.W. [1985] Nitrogen Removal in Rapid Infiltration Systems, *J. Environ. Eng.*, 111, 865, with permission of the American Society of Civil Engineers, Reston, Virginia)

to manage these factors to obtain nitrification/denitrification and allowing escape of nitrogen as a gas. The BOD to nitrogen ratio should be greater than 3:1 for effective denitrification. The loading rate, if kept in the range of 15 to 30 m/yr (50–100 ft/yr), should provide adequate detention time within the soil profile for effective nitrogen removal (Crites, 1985). The soil profile should be 3 m (10 ft) or deeper to ensure adequate detention time at a 30 m/yr (100 ft/yr) loading. Wetting and drying is also critical for nitrogen removal (Crites, 1985; U.S. EPA, 2006). Table 16.18 summarizes nitrogen removal in rapid infiltration. Figure 16.7 shows a typical, rapid-infiltration basin.

FIGURE 16.7   Rapid-infiltration basin.

Location	Years of Operation	Applied Concentration (mg/L)	Distance to Sample Point (m)	Percolate Concentration (mg/L)	Removal (%)
Brookings, South Dakota	5	3.0	0.8	0.45	85
Calumet, Michigan	88	3.5	1700	0.03	99
Dan Region, Israel	7	2.1	150	0.03	99
Fort Devens, Massachusetts	31	9.0	45	0.10	99
Lake George, New York	38	2.1	600	0.014	99
Phoenix, Arizona	15	5.5	30	0.37	93
Vineland, New Jersey	50	4.8	530	0.27	94

TABLE 16.19   Phosphorus Removal in Rapid-Infiltration Systems (from Crites, R.W. [1985] Nitrogen Removal in Rapid Infiltration Systems. *J. Environ. Eng.*, 111, 865, with permission of the American Society of Civil Engineers, Reston, Virginia)

Phosphorus removal is accomplished by absorption and chemical precipitation. Detention time—critical for chemical precipitation—is a function of the percolation rate through the soil and the aquifer and the flow distance to the point of monitoring. Table 16.19 summarizes phosphorus removal in rapid-infiltration systems. Although phosphorus removal declines with time, the removal rate might remain high for many years. For example, at Calumet, Michigan, the phosphorus removal rate was 99% after 88 years (Crites, 1985).

Rapid-infiltration systems are also effective in removing metals, pathogens, and trace organics (U.S. EPA, 2006). In Phoenix, Arizona, 90% to 99% of the applied viruses were removed within 0.1 m (0.3 ft) of travel through soil; 99.99% was removed after travel through 9 m (30 ft) of soil (Gilbert et al., 1976).

4.5.2   Design Objectives
Design objectives for rapid-infiltration systems include:

- Treatment and avoidance of direct discharge to surface water by discharging to groundwater.
- Treatment and groundwater recharge.
- Treatment and recharge of streams by interception of groundwater.
- Treatment and recovery of treated water by wells or underdrains for reuse.
- Treatment and temporary storage of water in the aquifer.

4.5.3   Design Procedures
An outline of the basic procedure for design is as follows:

1. Determine the field-measured infiltration rate.
2. Predict the hydraulic pathway of treated water.
3. Determine overall treatment requirements.
4. Select the appropriate level of preapplication treatment.
5. Calculate the annual hydraulic loading rate.
6. Calculate the needed field area.

7. Check the potential for groundwater mounding.

8. Select the final hydraulic loading cycle.

9. Calculate the application rate.

10. Determine the number of individual basins needed.

11. Locate monitoring wells.

If nitrogen removal is required, add the following seven steps:

1. Calculate the mass of ammonium nitrogen that can be adsorbed on the cation exchange sites in the soil (Lance, 1984).

2. Calculate the loading period that can be used without exceeding the mass loading from Step 1 based on ammonium concentration and daily application rate.

3. Compare ammonium and organic nitrogen loading rate to the maximum nitrification rate of 70 kg/ha·d (60 lb/ac/d) to check feasibility of complete nitrification (U.S. EPA, 2006).

4. Select the loading cycle based on Step 2 and the guidance in Table 16.20.

5. Check the BOD: nitrogen ratio in the applied wastewater.

6. Consider limiting the infiltration rate (typically to approximately 30–45 m/yr [100 to 150 ft/yr]).

7. Eliminate the nitrogen issue through development of a pretreatment system capable of high-level nitrogen removal.

Moderate infiltration rates are conducive to higher nitrogen removal rates. If the requirement for nitrogen removal is stringent (nitrate values of 5 mg/L or less or more than 80% removal), then pilot studies are needed for optimization or pretreatment is required to lower level sufficiently.

Objective	Applied Wastewater	Season	Application Period (d)	Drying Period (d)
Maximize infiltration rate	Primary effluent	Summer	1–2	6–7
		Winter	1	7–12
	Secondary effluent	Summer	1–3	4–5
		Winter	1–3	5–10
Maximize nitrification	Primary effluent	Summer	1–2	6–7
		Winter	1	7–12
	Secondary effluent	Summer	1–3	4–5
		Winter	1–2	7–10
Maximize nitrogen removal	Primary effluent	Summer	1–2	10–14
		Winter	1–2	12–16
	Secondary effluent	Summer	7–9	10–15
		Winter	9–12	12–16

TABLE 16.20   Loading Cycles for Rapid Infiltration (U.S. EPA, 1981)

4.5.3.1 Hydraulic Loading Rate The design hydraulic loading rate is based on the soil infiltration rate, subsurface flow rate, or loading of BOD or nitrogen. Each of these loading rates must be calculated and the lowest value must be selected for design.

The procedure for calculating hydraulic loading based on infiltration rate includes converting the hourly infiltration rate into an annual rate (multiply by 8760 h/yr) and multiplying the result by a factor to account for the wetting-and-drying cycle, variability of the soils, and type of infiltration rate field test. The field-measured infiltration rate used in design is the steady-state rate measured during one hour or more at the end of a test. The equation for the annual design loading rate is

$$L_w = al \tag{16.13}$$

where L_w = annual design loading rate, m/yr;
  a = design factor ranging from 0.02 to 0.15; and
  l = measured steady-state infiltration rate, m/yr.

The design factor should be 0.02 to 0.04 for small-scale tests (cylinder infiltrometers or air entry permeameters). For larger-scale, basin-flooding tests, the design factor can be increased to 0.07 to 0.15, depending on soil variability, number of test results, and degree of conservatism used. The design factor must not exceed the fraction of the loading cycle during which the basins are flooded. For example, if the application period is one day and the drying period is nine days (total cycle of 10 days), then the design factor must be less than 0.10.

4.5.3.2 Organic Loading Rate For municipal rapid-infiltration systems, the BOD loading rate typically will range from 10 to 200 kg/ha·d (9180 lb/ac-d) (see Table 16.17). The suggested maximum rate is 670 kg/ha·d (598 lb/ac/d) (Crites and Tchobanoglous, 1998). The BOD loadings in Table 16.17 are typical of those for successfully performing systems.

4.5.3.3 Land Requirements Equation 16.14 can be used to calculate the basin bottom area. Basin berms, roads, buffer area, or expansion must be taken into consideration.

$$A = \frac{365\,Q}{10000\,L_w} \tag{16.14}$$

where A = net field area, ha;
  Q = wastewater design average flow, m³/d; and
  L_w = limiting loading rate, m/yr.

4.5.4 Rapid Infiltration Design Example
The example below calculates the field area for a rapid-infiltration system treating 378 m³/d of secondary effluent. The minimum infiltration rate of the soil profile is 0.10 m/h using the basin infiltration test. The hydraulic loading rate is calculated using 7% design factor for a:

$$L_w = aI = 0.07\,(0.10\text{ m/h})(24\text{ h/d}) = 0.168\text{ m/d}$$

The calculation below determines the land area needed:

$$A = Q/L_w = 378\text{ m}^3/\text{d}/(0.168\text{ m/d}) = 2250\text{ m}^2$$

5.0   Floating Aquatic Plant Systems (Constructed Wetlands)

Floating aquatic plants have been used for wastewater treatment in several processes, including upgrading facultative pond effluent. In this case, the plants can achieve degrees of advanced wastewater treatment depending on loading and management. Water hyacinths and duckweed are the most studied and used floating plants. Water hyacinth can be invasive if allowed to exit a system and enter into adjacent surface waters.

The concept of using floating aquatic plants in wastewater treatment arose, in part, from an attempt to control suspended solids concentrations in aerobic and facultative pond discharges. The floating plants shield the water from sunlight and reduce the growth of algae. Floating-plant systems can also reduce BOD, nitrogen, metals, and trace organics (Crites and Tchobanoglous, 1998).

5.1   Water Hyacinths

Water hyacinth systems are used at full-scale systems in Lanai, Hawaii; Headlands, Alabama; and San Benito, Texas. Cold weather restricts the growth of water hyacinths, limiting their suitability to warm climates. In some instances, hyacinths have exited the treatment wetland and entered surface water. Duckweed (*Lemmna* sp.) is more cold tolerant and can survive, at least seasonally, in most U.S. locations. Duckweed is a common aquatic plant in most of the country. Plants may need to be obtained from certified vendors if they are not available locally.

The primary characteristics of water hyacinths that make them an attractive biological support medium for bacteria are their extensive root systems and rapid growth rate. The primary characteristic that limits their widespread use is their temperature sensitivity (that is, they are rapidly killed by frost conditions).

Water hyacinths have been used in Florida for removal of nitrogen (U.S. EPA, 1988). Nitrogen removal can be accomplished by optimizing nitrification/denitrification and by crop harvesting. Phosphorus removal by water hyacinths is not typically practiced.

5.2   Duckweed Systems

Duckweed (family *Lemmaceae*) systems have been studied alone and together with water hyacinths in polyculture systems. The primary advantage of duckweed is its lower sensitivity to cold climates; its primary disadvantages are its shallow root systems and sensitivity to movement by winds. Table 16.21 summarizes several projects that have provided valuable performance data for water hyacinth and duckweed systems.

5.3   Design Criteria for Water Hyacinth and Duckweed Systems

Design criteria for water hyacinth systems are presented in Table 16.22 and design criteria for duckweed systems are presented in Table 16.23. The pilot work at San Diego, California, with water hyacinths has led to the development of recycle, step feed, and wraparound configurations of the long rectangular basins (Crites et al., 1988). The San Diego work has also demonstrated odor-free and mosquito-free conditions using a supplemental aeration system to overcome anaerobic conditions stemming from high sulfate concentrations in the screened wastewater (Tchobanoglous et al., 1989). The operation at Koele on Lanai, Hawaii is shown in Figure 16.8.

Project	Flow (m³/d)	Plant Type	Hydraulic Loading Rate (m³/ha·d)	BOD₅* Influent/ Effluent (mg/L)	SS* Influent/ Effluent (mg/L)
San Diego, California	378	Water hyacinth	590	130/10	107/10
NSTL* Mississippi	8	Duckweed and pennywort	504	35.5/3	47.7/11.5
Austin, Texas	1700	Water hyacinth	140	42/12	40/9
N. Biloxi, Mississippi (Cedar Lake)	49	Duckweed	700	30/15	155/12
Disney World, Florida	30	Water hyacinth	300	200/26	50/14

*SS = suspended solids; BOD = biological oxygen demand; NSTL = National Space Technology Laboratories.

TABLE 16.21 Summary of Wastewater Treatment Performance of Aquatic Plant Systems

Factor	Secondary Treatment (Nonaerated)	Advanced Secondary Treatment (Aerated)	Nutrient Removal (Nonaerated)
Design criteria			
Influent wastewater source	Screened, settled, or lagoon effluent	Screened or settled	Secondary
Influent BOD₅*, mg/L	130–180	130–180	30
BOD₅ loading rate, kg/ha·d	40–80	150–300	10–40
Water depth, m	0.5–0.8	0.9–1.2	0.6–0.9
Detention time, days	10–36	4–8	6–18
Hydraulic loading, m³/ha·d	200–800	550–1000	<800
Harvest schedule	Annual	Twice monthly to continuous	Twice monthly to continuous
Expected effluent quality, mg/L			
BOD₅	<30	<15	<10
Suspended solids	<30	<15	<10
Total nitrogen	<15	<15	<5
Total phosphorus	<6	<6	<1–2

*BOD = biological oxygen demand.

TABLE 16.22 Design Criteria for Water Hyacinth Systems (WEF, 2001)

Item	Value
Design criteria	
Wastewater input	Facultative lagoon effluent
BOD_5* loading, kg/ha·d	22–28
Hydraulic loading, m³/ha·d	22–28
Water depth, m	1.5–2.0
Hydraulic detention time, days	20–25
Water temperature, °C	>7
Harvest schedule	Monthly for secondary treatment, weekly for nutrient removal
Expected effluent quality secondary, mg/L	
BOD_5	<30
Suspended solids	<30
Total nitrogen	<15
Total phosphorus	<6
Nitrient removal, mg/L	
BOD_5	<10
Suspended solids	<10
Total nitrogen	<5
Total phosphorus	<1–2

*BOD = biological oxygen demand.

TABLE 16.23 Design Criteria and Effluent Quality for Effluent Polishing With Duckweed Treatment Systems (WEF, 2001)

FIGURE 16.8 Water hyacinth treatment at Koele in Lanai, Hawaii.

The area needed for water hyacinth or duckweed systems can be estimated based on criteria for design of facultative lagoons. Select the detention times and depths from Table 16.22 or 16.23.

Both water hyacinths and duckweed need to be harvested. Harvested duckweed is typically land applied or composted while the water hyacinth at San Diego are chopped and composted. Attempts to feed harvested duckweed as an animal feed (replacement for hay) have encountered regulatory concern because of time requirements or lay-by associated with harvest and feeding. Typical time requirement imposed between harvest and feeding associated with vegetation raised on land treatment sites are 30 days.

6.0   Constructed Wetlands

Constructed wetlands are recognized as an effective and economically attractive technology for treatment of agricultural, municipal, and industrial wastewater. These systems are designed to treat wastewater using emergent plants such as cattails, reeds, and rushes. Applications for constructed wetlands include treatment of stormwater, acid mine waste, landfill leachate, agricultural runoff, and food-processing wastewater (Crites et al., 2006; Kadlec and Wallace, 2008). Constructed wetland systems vary in design from surface flow systems to a variety of subsurface flow options. Wetland systems have been demonstrated to lower organic strength, nutrient levels, regulated metal content, microbial constituent, and pharmaceutical and personal care product (PPCP) levels in a variety of source wastewaters.

A variant on the constructed wetland system involves reed bed systems for handling sludges. In this system, a reed-based system is typically developed in a sandy substrate and the bed is flooded with a thin layer of sludge. The reeds grow through the sludge and the liquid drains through the sand. When filled to a desired level, the reed/sludge mix is composted on-site or the material is removed and composted off-site into a class "a" sludge compost.

6.1   Types of Constructed Wetlands

There are three primary categories of constructed wetlands: free water surface (FWS); subsurface flow (SSF); and vertical flow (U.S. EPA, 1999; 2000). For FWS wetlands, the flow path of the applied wastewater is above the soil surface. For SSF wetlands, the flow runs lateral through the root zone and the medium, which ranges from sand to coarse gravel to rocks. For vertical-flow wetlands, the application is either by spray or surface flooding, and the flow path is down through the medium and out through the underdrains.

6.2   Free Water Surface Wetlands

Free water surface systems are more widely used than SSF systems and are found throughout the United States, including in Gustine and Arcata in California; Cannon Beach, Oregon; Benton, Kentucky; Ouray, Colorado; and Minot, North Dakota. The technology is still developing. Pilot-plant operations are conducted before large, full-scale designs are implemented, when treatment objectives have not been met, or when wastewater characteristics are unique. Figure 16.9 illustrates a typical FWS wetland system.

FIGURE 16.9   Free water surface constructed wetland at Cle Elum, Washington.

6.3   Subsurface Flow Wetlands

Subsurface flow systems consist of beds or channels filled with gravel, sand, or other permeable medium planted with emergent vegetation. Wastewater is treated as it flows horizontally through the medium/plant filter. Alternate nomenclature for SSF wetlands include rock/reed filters and vegetated submerged beds. Modifications to the SSF process have been employed in indoor applications to generate reclaimed water at Emory University, The San Francisco Public Utility Commission (SFPUC), and the Portland, Oregon airport terminal.

Several research pilot-scale and small full-scale facilities are evaluating the performance of subsurface flow wetlands. One of the more complete research tests using gravel-filled trenches was conducted in Santee, California. Full-scale systems are located in Benton and Houghton, Louisiana; Mesquite, Nevada; and Hardin, Kentucky. Subsurface flow wetlands are less developed primarily because they have only recently been demonstrated in the United States (Reed et al., 1998). A technology assessment was completed for U.S. EPA in 1993 (U.S. EPA, 1993b).

6.4   Vertical-Flow Wetlands

Vertical-flow wetlands are a variant of intermittent packed-bed filter technology. Influent that has already received primary settling treatment (at a minimum) is introduced into a recirculation tank, where it is mixed with water that already has been treated in the wetland. A dosing pump intermittently sends water from the recirculation tank to a distribution piping network located on the surface of the wetland. The applied water percolates vertically through the wetland's pea-gravel media. Wetland vegetation growing in the media improves the aesthetics of the packed-bed filter while supplying a small amount of oxygen to bacteria in the pea-gravel media via the plant roots. Water is collected at the bottom of the wetland cell and returned to the recirculation tank in which a valve controls the effluent flow.

Vertical-flow wetlands are able to achieve BOD and TSS concentrations of less than 10 mg/L and to remove approximately 50% of total nitrogen. Vertical-flow wetlands are much more effective at converting ammonia into nitrate (nitrification) than free water surface and subsurface flow wetland systems because of the intermittent dosing of water and continuously drained media. Vertical-flow wetlands placed in series or combined in a stepwise manner with horizontal flow cells can achieve full nitrification of even high concentrations of ammonia (Crites et al., 2006).

6.5 Design and Performance of Constructed Wetlands

Issues that must be addressed in the design of constructed wetland systems include climate, plant selection, hydraulic loading and hydraulic retention time, and the method of operation. The various operational methods have been described, above. Ideally plant selection involves utilization of native species of wetland tolerant vegetation. System designers must exercise caution to preclude introduction of invasive plant materials into an area. This requires designers, operators, and regulators possess some basic knowledge of appropriate plant communities in the various geographic/climactic settings where the wetland system will be utilized.

Constructed wetland systems can be designed to utilize natural aeration as supplied through convection or can be designed to utilize induced or mechanical aeration as supplied by blowers. The natural aeration can be developed as simple air transfer between a surface and the atmosphere or a fill and draw system as developed through tidal or "fill and draw" wetlands. Those systems utilizing mechanical aeration rely on standard transfer calculations to match aeration needs with the characteristics of the liquid requiring treatment.

Recently a modification of the constructed wetland process has been utilized in indoor settings to treat and renovate domestic wastewater to reclaimed water standards. These decentralized or distributed water reclamation facilities are utilized in San Francisco, California; Portland, Oregon; and at the Water Hub at Emory University in Atlanta, Georgia.

The San Francisco Public Utility Commission (SFPUC) headquarters building in downtown San Francisco is a LEED Platinum certified structure. A series of constructed wetlands are located on the site in both outdoor settings and within the building lobby as part of the SFPUC decentralized wastewater treatment and reuse program. This advanced treatment process generates approximately 5000 gallons of reclaimed water per day, and the reclaimed water is recycled through the building for toilet and urinal flushing. This volume represents approximately 60% of the water required in the building. Through the treatment processes, BOD, TSS, and nitrogen levels are reduced significantly. Total nitrogen in the effluent is reported as reduced from 150 mg/l to less than 5 mg/l. Information on this program is available on the SFPUC website at SFWater.org.

Table 16.24 summarizes performance of selected FWS constructed wetlands systems. Free water surface systems remove BOD using bacteria attached to the plants and vegetative litter. Suspended solids are removed by entrapment in vegetation and sedimentation. In SSF systems, filtration is a primary mechanism for suspended solids removal. Table 16.25 summarizes removal of BOD and TSS in SSF wetlands.

Location	BOD* (mg/L)		TSS* (mg/L)	
	Influent	Effluent	Influent	Effluent
Arcata, California	26	12	30	14
Benton, Kentucky	25.6	9.7	57.4	10.7
Cannon Beach, Oregon	26.8	5.4	45.2	8.0
Ft Deposit, Alabama	32.8	6.9	91.2	12.6
Gustine, California	75	19	102	31
Iselin, Pennsylvania	140	17	380	53
Listowel, Ontario	56.3	9.6	111	8
Ouray, Colorado	63	11	86	14
West Jackson Co., Mississippi	25.9	7.4	40.4	14.1
Sacramento Co., California	23.9	6.5	8.9	12.2

*BOD = biological oxygen demand; TSS = total suspended solids.

TABLE 16.24   Typical BOD and TSS Removals Observed in FWS Constructed Wetlands (from Crites, R.W., and Tchobanoglous, G., *Small and Decentralized Wastewater Management Systems.* Copyright © 1998, The McGraw-Hill Companies, New York, N.Y., with permission)

Nitrogen removal can be effective in both types of constructed wetlands depending on preapplication treatment, detention times, and loading rates. In cold weather, the ability to nitrify decreases when water temperatures fall below 5°C (Crites et al., 2006). When plants go into senescence, nutrients are released into the water column. Phosphorus removal in most FWS wetland systems is not effective because of limited contact between wastewater and soil. For SSF wetlands, the potential for phosphorus removal is greater than for FWS systems, depending on the medium and detention time.

Location	Pretreatment	Concentration (mg/L)		Removal (%)	Nominal Detention Time (d)
		Influent	Effluent		
Benton[a], Kentucky	Oxidation lagoon	23	8	65	5
Mesquite[b], Nevada	Oxidation lagoon	78	25	68	3.3
Santee[c], California	Primary	118	1.7	88	6
Sydney[d], Australia	Secondary	33	4.6	86	7

[a]Full-scale operation from March 1988 to November 1988 operated at 80 mm/d (Watson et al., 1989).
[b]Full-scale operation, January 1994 to January 1995.
[c]Pilot-scale operation, 1984, operated at 50 mm/d (Gersberg et al., 1985).
[d]Pilot-scale operation at Richmond. New South Wales, near Sydney, Australia, operated at 40 mm/d from December 1985 to February 1986 (Bavor et al., 1987).

TABLE 16.25   Total Biological Oxygen Demand Removal Observed in a Subsurface Flow Wetlands

6.6   Land Requirements
Equation 16.17 can be used to calculate the area for FWS wetlands:

$$A_{fw} = \frac{q(\ln C_o - \ln C_e)}{k_t d\, n\,(10\ 000)} \qquad (16.15)$$

where A_{fw} = surface area of FWS wetland, ha;
q = wastewater flow, m³/d;
C_o = influent BOD concentration, mg/L;
C_e = effluent BOD concentration, mg/L;
n = porosity, fraction; and
k_t = first-order rate coefficient, d⁻¹.

The wastewater flow is the average of the influent and effluent flow. The porosity, as a decimal, can range from 0.70 for heavily vegetated wetlands to 0.9 for lightly vegetated wetlands. The 20°C k factor is 0.678/d; Q factor is 1.06.

Equation 16.17 can be used to calculate the area for an SSF wetland:

$$A_{sf} = \frac{q(\ln C_o - \ln C_e)}{k_t d\, n\,(10\ 000)} \qquad (16.16)$$

where A_{sf} = SSF wetland area; ha;
q = flow, m³/d;
C_o = influent BOD concentration, mg/L;
C_e = effluent BOD concentration, mg/L;
k_t = first-order rate coefficient, d⁻¹;
d = depth of media, m;
n = drainable voids, fraction; and
10 000 = conversion from square meters to hectares.

For a bed of medium-to-coarse gravel, a typical value of k_{20} would be 1.1 and a typical value of n would be 0.38 (Crites et al., 2006). Typical bed depths for SSF wetlands range from 0.5 to 0.75 m (20–30 in.). Design details may be found elsewhere (Crites and Tchobanoglous, 1998; Kadlec and Wallace, 2008; Crites et al., 2006).

Reed et al. (1998) has compared the area required for treatment by an FWS and an SSF and found that the FWS wetland is 1.52 times larger than the SSF wetland for the same temperature and the same performance and flow.

Free water surface wetlands can provide significant wildlife habitat (U.S. EPA, 1993a). Alternating shallow (less than 0.6 m) and deep (greater than 1 m) water areas can provide supplemental oxygen for aerobic treatment, provide open water for waterfowl, and reduce the need for planting and harvesting of emergent plants.

Distribution to an FWS wetland should be designed as a manifold or equivalent method. Outlets can be by manifolds, adjustable standpipes or adjustable weirs to allow variation in the water depth.

Mosquito control in FWS wetlands is necessary; whereas, with SSF wetlands, water is not exposed to adult mosquitoes. Control methods for mosquitoes include chemical and biological controls, mixing, water-level management, and encouragement of predators (Williams et al., 1996).

6.7   Wetlands Design Example

Below is an example of a wetlands design based on several assumptions:

- Temperature = 15°C;
- Flow, Q = 1136 m³/d (414 640 m³/yr);
- Septic tank or some other preliminary settling tank precedes the wetland; and
- BOD influent = 175 mg/L.

Using basic kinetic models:

$$A = \frac{q\,(\ln C_o - \ln C_e)}{k_t d\,n\,(10\,000)} \qquad (16.17)$$

$$K_T = K_{20}\,(\theta)^{(T-20)} \qquad (16.18)$$

where C_o = wetland influent concentration, mg/L; and

C_e = wetland effluent concentration, mg/L (assume C_e = 25 mg/L for this example).

K_T = reaction rate constant at temperature T;

U = Temperature coefficient (assume U = 1.06);

K_{20} = reaction rate constant at 20°C, 0.678;

A = treatment area of wetland, m²;

q = annual influent wastewater flow rate, m³/yr;

d = 0.6 m; and

n = 0.8

To correct the reaction rate constant for temperature, use the conversion below:

$$K_T = K_{20}\,(u)^{(T-20)}$$

$$K_T = 0.678\,(1.06)^{(15-20)}$$

$$K_T = 0.507$$

Based on the equation and assumptions above, the area needed for treatment is A = 0.91 ha.

7.0   Summary and Conclusion

Natural systems for managing wastewater and stormwater continue to evolve as an integral component of the water infrastructure. These systems are designed and managed to mimic the physical, chemical, and biological processes ongoing in the plant-soil system and in wetland or hyporheic environments. Natural systems have been developed to address the wide variety of pollutants introduced through municipal, industrial, residential, and agricultural activities. Properly designed, operated, and managed natural systems can be cost-effective options for managing wastewater flows ranging from hundreds of gallons per day to flows exceeding many millions of gallons per day. The land treatment system serving Muskegon, Michigan, accommodates wastewater flows of up to 40 mgd on a land area of over 10 000 acres. Land-based and natural systems do require sufficient land to accommodate specified liquid and nutrient levels present in wastewater. Design guidelines and regulations available through USEPA and state agencies help assure these systems continue to serve the infrastructure niche where they are used.

8.0 References

Amoozegar, A.; Wilson, G. V. (1999) Methods for Measuring Hydraulic Conductivity and Drainable Porosity. In *Agricultural Drainage*; Skaggs R. W., Schilfgaarde, J. van, Eds.; Monograph No. 38; American Society of Agronomy, Crop Soil Science of America, and Soil Science Society of America: Madison, WI; 1149–1205.

Bavor, H. J.; Roser, D. J.; McKersie, S. A. (1987) Nutrient Removal Using Shallow Lagoon-Solid Matrix Macrophyte Systems. In *Aquatic Plants for Water Treatment and Resource Recovery*; Reddy, K. R., Smith, W. H., Eds.; Magnolia Publishing: Orlando, Florida; 228.

Crites, R. W. (1985) Nitrogen Removal in Rapid Infiltration Systems. *J. Environ. Eng.*, **111** (6), 865.

Crites, R.W.; Tchobanoglous, G. (1998) *Small and Decentralized Wastewater Management Systems*; McGraw-Hill, Inc.: New York.

Crites, R. W.; Kruzic, A. P.; Tchobanoglous, G. (1988) Aquatic Treatment Systems for Wastewater Management. *Proceedings of the Joint Canadian Society of Civil Engineers and the American Society of Civil Engineers National Conference on Environmental Engineering*; Vancouver, British Columbia, Canada, Jul 13–15; University of British Columbia: Vancouver, British Canada.

Crites, R. W.; Middlebrooks, E. J.; Reed, S. C. (2006) *Natural Wastewater Treatment Systems*; CRC Press: Boca Raton, Florida.

Electric Power Research Institute; Tennessee Valley Authority (2004) Wastewater Subsurface Drip Distribution: Peer Reviewed Guidelines for Design, Operation, and Maintenance; Electric Power Research Institute: Palo Alto, California; Tennessee Valley Authority, Chattanooga, Tennessee.

Gersberg, R. M.; Elkins, B. V.; Lyons, R.; Goldman, C. R. (1985) Role of Aquatic Plants in Wastewater Treatment by Artificial Wetlands. *Water Res.* (G.B.), **20**, 363.

Gilbert, R. G.; Gerba, C. P.; Rice, R. C.; Bouwer, H.; Wallis, C.; Melnick, J. L. (1976) Virus and Bacteria Removal from Wastewater by Land Treatment. *Appl. Environ. Microbiol.*, **32**, 333.

Great Lakes Upper Mississippi River Board of State Sanitary Engineering Health Education Services (2014) *Recommended Standards for Sewage Works*; Great Lakes Upper Mississippi River Board of State Sanitary Engineering Health Education Services: Albany, New York.

Hannah, S. A.; Austern, B. M.; Eralp, A. E.; Wise, R. H. (1986) Comparative Removal of Toxic Pollutants by Six Wastewater Treatment Processes. *J. Water Pollut. Control Fed.*, **58** (1), 27.

Hansen, V. E.; Israelsen, O. W.; Stringham, G. E. (1979) *Irrigation Principles and Practices*, 4th ed.; Wiley & Sons: New York.

Irrigation Association (1983) *Irrigation*, 5th ed.; Pair, C.H., Ed.; Irrigation Association: Silver Spring, Maryland.

Jewell, W. J. (1978) *Limitations of Land Treatment of Wastes in the Vegetable Processing Industries*; Cornell University Press: Ithaca, New York.

Johnston, J.; Smith, R. (1988) Operating Schedule Effects on Nitrogen Removal in Overland Flow Treatment Systems. *Proceedings of 61st Annual Conference of the Water Pollution Control Federation*; Dallas, Texas, Oct 3–6; Water Pollution Control Federation: Alexandria, Virginia.

Kadlec, R. H.; Wallace, S. D. (2008) *Treatment Wetlands*, 2nd ed.; CRC Press: Boca Raton, Florida.

Lance, J. C. (1984) Land Disposal of Sewage Effluents and Residue. In *Groundwater Pollution Microbiology*; Britton, G.; Gerba, C. P., Eds.; John Wiley and Sons: New York.

Leverenz, H.; Tchobanoglous, G.; Darby, J. (2002) *Review of Onsite Technologies for the Onsite Treatment of Wastewater in California*, Report to the California State Water Resources Control Board, Center for Environmental and Water Resources Engineering, 2002-1; University of California Davis: Davis, California.

National Onsite Wastewater Reycling Association (2006) *Recommended Guidance for the Design of Wastewater Drip Dispersal Systems*, March, NOWRA, Alexandria, Virginia.

Oswald, W. J. (1991) Introduction to Advanced Integrated Wastewater Ponding Systems. *Water Sci. Technol.*, **24** (5), 1.

Reed, S. C.; Crites, R. W.; Middlebrooks, E. J. (1998) *Natural Systems for Waste Management and Treatment*, 2nd Ed.; McGraw-Hill: New York.

Reed, S. C.; Hines, M.; Ogden, M. (2003) Improving Ammonia Removal in Municipal Constructed Wetlands, *Proceedings of the 76th Annual Water Environment Federation Technical Exposition and Conference* [CD-ROM]; Los Angeles, California, Oct 11–15; Water Environment Federation: Alexandria, Virginia.

Rich, L. G. (1980) *Low Maintenance Mechanically Simple Wastewater Treatment Systems*; McGraw-Hill: New York.

Rich, L. G. (1996) Low-Tech Systems for High Levels of BOD and Ammonia Removal. *Public Works*, **127** (4), 41.

SF Water Guide to Non-Potable Water Treatment and Reuse, SFWater.org

State Water Resources Control Board (1984) *Irrigation with Reclaimed Municipal Wastewater A Guidance Manual*, Report 84-1; Pettygrove, G. S., Asano, T., Eds.; State Water Resources Control Board: Sacramento, California.

Tchobanoglous, G.; Maitski, F.; Thompson, K.; Chadwick, T. H. (1989) Evolution and Performance of City of San Diego Pilot-Scale Aquatic Wastewater Treatment System Using Hyacinths. *J. Water Pollut. Control Fed.*, **61** (11/12), 1625.

U.S. Department of Agriculture Natural Resources Conservation Service (2013) Conservation Practice Standard Nutrient Management, NRCS 590; U.S. Department of Agriculture: Washington, D.C.

U.S. Environmental Protection Agency (1975) *Wastewater Treatment Lagoons*; EPA-430/9-74-001; MCD-14; U.S. Environmental Protection Agency: Washington, D.C.

U.S. Environmental Protection Agency (1976) Use of Climatic Data in Estimating Storage Days for Soil Treatment Systems, EPA-600/2-76-250; U.S. Environmental Protection Agency, Office of Research and Development: Cincinnati, Ohio.

U.S. Environmental Protection Agency (1983) *Design Manual on Municipal Wastewater Stabilization Lagoons*, EPA-625/1-83-015; Center for Environmental Research Information, U.S. Environmental Protection Agency: Cincinnati, Ohio.

U.S. Environmental Protection Agency (1984) *Process Design Manual for Land Treatment of Municipal Wastewater: Supplement on Rapid Infiltration and Overland Flow*, EPA-625/1-81-019a; Center for Environmental Research Information, U.S. Environmental Protection Agency: Cincinnati, Ohio.

U.S. Environmental Protection Agency (1988) *Constructed Wetlands and Aquatic Plant Systems for Municipal Wastewater Treatment*, EPA-625/1-88-022; Center for Environmental Research Information, U.S. Environmental Protection Agency: Cincinnati, Ohio.

U.S. Environmental Protection Agency (1993a) *Constructed Wetlands for Wastewater Treatment and Wildlife Habitat*, EPA-832R/93-005; U.S. Environmental Protection Agency: Washington, D.C.

U.S. Environmental Protection Agency (1993b) *Subsurface Flow Constructed Wetlands for Wastewater Treatment: A Technology Assessment*, EPA-832/R-93-008; U.S. Environmental Protection Agency, Washington, D.C.

U.S. Environmental Protection Agency (1999) *Free Water Surface Wetlands for Wastewater Treatment: A Technology Assessment*; Office of Water Management, U.S. Environmental Protection Agency: Washington, D.C.

U.S. Environmental Protection Agency (2000) *Constructed Wetlands Treatment of Municipal Wastewater*, EPA/625/R-99/010; Office of Research and Development, U.S. Environmental Protection Agency: Cincinnati, Ohio.

U.S. Environmental Protection Agency (2002) *Design Manual Onsite Wastewater Treatment and Disposal Systems*, EPA-625/R-00/008; Center for Environmental Research Information, U.S. Environmental Protection Agency: Cincinnati, Ohio.

U.S. Environmental Protection Agency (2003) *Voluntary National Guidelines for Management of Onsite and Clustered (Decentralized) Wastewater Treatment Systems*, EPA-832-B-03-001, March 2003: Washington, D.C.

U.S. Environmental Protection Agency (2004) *Needs Survey*, 2004. Office of Water Management; U.S. Environmental Protection Agency; Washington, D.C.

U.S. Environmental Protection Agency (2006) *Process Design Manual Land Treatment of Municipal Wastewater Effluents*, EPA-625/R-06/016; Center for Environmental Research Information, U.S. Environmental Protection Agency: Cincinnati, Ohio.

U.S. Environmental Protection Agency (2009) *Design Manual: Municipal Wastewater Stabilization Ponds*, Office of Research and Development; U.S. Environmental Protection Agency: Cincinnati, Ohio.

Wallace, S. L., and Knight, R. L. (2006) *Small Scale Constructed Wetland Treatment Systems: Feasibility, Design Criteria and O&M Requirements*, WERF Report 01-CTS-5.

Water Environment Federation (2010) *Natural Systems for Wastewater Treatment*, Manual of Practice No. FD-16; Water Environment Federation: Alexandria, Virginia.

Water Pollution Control Federation (1989) Technology and Design Deficiencies at Publicly Owned Treatment Works. *Water Environ. Technol.*, **1** (4), 515.

Watson, J. T.; Reed, S. C.; Kadlec, R. H.; Knight, R. L.; Whitehouse, A. E. (1989) Performance Expectations and Loading Rates for Constructed Wetlands. In *Constructed Wetlands for Wastewater Treatment*; Hammer, D. A., Ed.; Lewis Publishers: Chelsea, Michigan; 319.

Williams, C. R.; Jones, R. D.; Wright, S. A. (1996) Mosquito Control in a Constructed Wetland. *Proceedings of the 69th Annual Water Environment Federation Technical Exposition and Conference* [CD-ROM]; Dallas, Texas, Oct 5–9; Water Environment Federation: Alexandria, Virginia.

Witherow, J. L.; Bledsoe, B. E. (1983) Algae Removal by the Overland Flow Process. *J. Water Pollut. Control Fed.*, **55** (10), 1256.

Zirschky, J. (1987) State of the Art Hydrograph Controlled Release (HCR) Lagoons, *J. Water Pollution Control Federation*, **59** (7), 695–698.

9.0   Suggested Readings

Bavor, H. J.; Roser, D . J.; McKersie, S. A. (1987) Nutrient Removal Using Shallow Lagoon-Solid Matrix Macrophyte Systems. In *Aquatic Plants for Water Treatment and Resource Recovery.* Reddy, K. R.; Smith, W. H., Eds.; Magnolia Publishing, Inc.: Orlando, Florida.

Mitsch, W. J. (1994) *Global Wetlands: Old World and New.* Elsevier: Amsterdam, Holland.

Mitsch, W. J.; Gosselink, J. G. (2000) *Wetlands,* 3rd ed.; John Wiley & Sons: New York.

Reed, S. C. (1991) Constructed Wetlands for Wastewater Treatment. *BioCycle,* **32,** 44.

Reed, S. C.; Crites, R. W. (1984) *Handbook of Land Treatment Systems for Industrial and Municipal Wastes;* Noyes Publications: Park Ridge, New Jersey.

San Francisco Public Utilities Commission Design Guide for Water Reuse (2015) SFWater .org

Tchobanoglous, G.; Crites, R.; Gearheart, R.; Reed, S. C. (2003) A Review of Treatment Kinetics for Constructed Wetlands. In *The Use of Aquatic Macrophytes for Wastewater Treatment in Constructed Wetlands;* Dias, V.; Vymazal, J., Eds.; Instituto da Conservacao da Natureza and Instituto Nacional da Agua: Lisbon, Portugal.

Water Environment Research Foundation (2006) *Small-Scale Constructed Wetland Treatment Systems,* Water Environment Research Foundation: Alexandria, Virginia.

CHAPTER 17

Disinfection

Gary Hunter, P.E., BCEE, ENV SP; Naomi Eva Anderson;
Keith Bourgeous, Ph.D., P.E.; Leonard W. Casson, Ph.D., P.E., BCEE;
Ludwig Dinkloh; Kristin Frederickson; Samuel S. Jeyanayagam, Ph.D., P.E., BCEE;
Morgan Knighton; Garrison W. Myer; Andrew Salveson, P.E.; and Jay L. Swift, P.E.

1.0   Introduction

This chapter is intended to serve as guidance for engineers, scientists, and wastewater resource recovery facility (WRRF) operators in the comparison, selection, and design, of various commonly used wastewater disinfection processes. Disinfection is the most critical component of wastewater treatment for the protection of public health. Improperly disinfected water and wastewater have been responsible for major disease outbreaks in both the developing and developed worlds. Readers are encouraged to consult other literature published by WEF like *Disinfection Guidance* and *Ultraviolet Disinfection for Wastewater* and other literature need to be consulted to obtain additional information on the importance of the disinfection practice on public health.

This chapter is organized into a section that discusses design considerations for disinfection systems in that then specific information for the following commonly used disinfection alternatives:

- Chlorine gas
- Bulk hypochlorite
- UV
- Ozone

In addition, design requirements for the dechlorination process are also described within this chapter.

A key concept in disinfection design is the concept of the indicator organism or target organism—an organism whose concentration serves as a conservative indicator of the presence of other pathogens, and the reduction of such organisms indicates a reduction of a broader group of pathogens and thus the success of disinfection or relative safety of the water. The target pathogens, regulatory standards applied, and treatment technologies used have evolved in recent years, as more is understood about the risks associated with the various pathogens. In the United States, the Federal Water Pollution Control Act Amendments of 1972 provided a unified basis for determining secondary treatment standards, including disinfection standards mandated through the establishment of fecal coliform criteria. Increasingly, however, other indicator organisms have been considered to be more conservative indicators of disinfection and thus are increasingly used as a basis of regulatory limits and disinfection design, including *Escherichia coli* (*E. coli*), Enterococci, total coliform, viruses, and protozoa. Tables 17.1 through 17.8 provide background information on density and removal rates of various coliform, virus, and protozoa in wastewater.

Permitting agencies may limit the maximum allowable concentration of bacterial and viral indicator organisms based on daily, monthly, or seasonal requirements. Currently, regulators are using many methods to monitor and control the disinfection processes at municipal WRRFs and maintain adherence to design standards. Historically, standards have been directed toward the control of chlorine-based disinfection systems. In addition, compliance with specific water-quality discharge or receiving water limitations commonly is required in WRRF discharge permits. Such limitations vary, but often consist of fecal coliform concentrations of 200 or 400 most probable number (MPN) fecal coliforms (FC)/100 mL, 240 MPN total coliforms (TC)/100 mL, or 2.2 MPN TC/100 mL. Also, limitations on effluent chlorine residuals and toxicity are common in many areas of the United States.

With the heightened awareness of sustainability, there is a growing interest in improving wastewater-disinfection practices, while continuing to develop effective and affordable low-impact disinfection alternatives. Applying sustainability principles to the disinfection system requires the careful consideration of technologies that not only meet regulatory requirements, but also achieve the economic, social, and environmental goals of the project. Details of sustainable design can be found at http://sustainableinfrastructure.org/. Chapter 2 of this manual provides additional information on sustainable design.

Organism	Minimum (MPN/100 mL)	Maximum (MPN/100 mL)
Total coliform	1 000 000	—
Fecal coliform	340 000	49 000 000
Fecal streptococci	64 000	4 500 000
Virus	0.5	10 000
Cryptosporidiun oocysts	85	1370
Giardia cysts	80	320

TABLE 17.1 Typical Wastewater Influent Concentration Ranges for Pathogenic and Indicator Organisms (Casson et al., 1990; Rose, 1988; U.S. EPA, 1979)

Waterborne Diseases	Transmission Route as Reported in US		
	Drinking Water	Recreational Water	Shellfish
Bacterial diseases			
Bacillary dysentery (*Shigella* spp.)	Yes	Yes	No
Cholera (*Vibrio cholerae*)	No	No	Yes
Diarrhea (enteropathogenic *Escherichia coli*)	Yes	No	No
Leptospirosis (*Leptospira* spp.)	Yes	Yes	No
Salmonellosis (*Salmonella* spp.)	Yes	Yes	Yes
Typhoid fever (*Salmonella typhosa*)	Yes	Yes	No
Tularemia (*Francisella tularensis*)	Yes	No	No
Yersinosis (*Yersinia pseudotuberculosis*)	Yes	No	No
Unknown etiology			
Diarrhea, acute undifferentiated	Yes	Yes	Yes
Gastroenteritis, acute, benign, self-limiting*	No	Yes	No
Viral diseases			
Gastroenteritis (Norwalk-type agents)	Yes	No	No
Hepatitis A (hepatitis virus)	Yes	Yes	Yes
Parasitic disease			
Amoebic dysentery (*Entamoeba histolytica*)	Yes	No	No
Ascarariosis (*Ascaris lumbricoides*)	No	No	No
Balantidial dysentery (*Balantidium coli*)	No	No	No
Giardiasis (*Giardia lamblia*)	Yes	No	No
Cryptosporidiosis (*Enteric coccidian*)	Yes	Yes	No

*Not a reportable disease, but transmission via recreational water has been demonstrated by an epidemio logical investigation.

TABLE 17.2   Waterborne Diseases and Transmission Routes (Pipes, 1982. Sorvillo et al., 1992)

Organism	Minimum (MPN/100 mL)	Maximum (MPN/100 mL)
Total coliform	45 000	2 020 000
Fecal coliform	11 000	1580 000
Fecal streptococci*	2000	146 000
Viruses	0.05	1000
Salmonella sp.	12	570

*Assuming removal efficiencies for fecal streptococci similar to fecal coliform removal efficiencies.

TABLE 17.3   Secondary Effluent Ranges for Pathogenic and Indicator Organisms Before Disinfection (U.S. EPA, 1986a)

Wastewater	Million Organisms/100 mL			
	Total Coliform	Fecal Coliform	Fecal Streptococci	Ratio FC: FS
"A"	17.2	17.20	4.00	4.3
"B"	33.0	10.90	2.47	4.4
"C"	1.94	0.34	0.064	5.3
"D"	6.30	1.72	0.20	8.6

TABLE 17.4   Bacterial Densities in Domestic Wastewater

Wastewater	Total Coliforms (no./100 mL)	Fecal Coliforms (no./100 mL)
Raw	10^7–10^8	10^6–10^7
Primary effluent	10^7–10^8	10^6–10^7
Secondary	10^5–10^6	10^4–10^5
Filtered secondary	10^4–10^5	10^3–10^5
Nitrified	10^4–10^5	10^3–10^5
Filtered nitrified	10^4–10^5	10^3–10^5

TABLE 17.5   Typical Levels of Coliform Bacteria in Domestic Wastewater After Various Wastewater Treatment Steps (Hubley et al., 1985)

Microorganism	Primary Treatment Removal (%)	Secondary Treatment Removal (%)
Total coliforms	<10	90–99
Fecal coliforms	35	90–99
Shigella spp.	15	91–99
Salmonella spp.	15	96–99
E. coli	15	90–99
Viruses	<10	76–99
Entamoeba histolytica	10–50	10

TABLE 17.6   Microbial Reductions by Conventional Treatment Processes (U.S. EPA, 1986a)

Pathogen	Soil		Plants	
	Absolute Maximum[a]	Common Maximum	Absolute Maximum	Common Maximum
Bacteria	1 year	2 months	6 months	1 month
Viruses	1 year	3 months	2 months	1 month
Protozoan cysts[b]	10 days	2 days	5 days	2 days
Helminth ova	7 years	2 years	5 months	1 month

[a]Greater survival time is possible under unusual conditions such as consistently low temperatures or highly sheltered conditions (for example, helminth ova below the soil in fallow fields).
[b]Few, if any, data are available on the survival times of *Giardia* cysts and *Cryptosporidium* oocysts.

TABLE 17.7   Survival Times of Pathogens in Soil and on Plant Surfaces (U.S. EPA, 1992a)

Disease or agent	1980		1981	
	Outbreaks	Cases	Outbreaks	Cases
Dermatitis[a]	5	78	7	642
Shigella	4	335	NR[b]	NR
Acute gastrointestinal illness	2	83	NR	NR
Adenovirus	1	15	NR	NR
Legionella	NR	NR	1	34
Total	12	511[a]	8	676

[a]Four outbreaks of dermatitis were attributed to *Pseudomonas aeruginosa* and the agent for one outbreak was undetermined.
[b]NR = none reported.

TABLE 17.8   Disease Outbreaks Related to Recreational Water During 1980 and 1981 (WPCF, 1984)

2.0   Reactor Design Considerations

2.1   Reactor Dynamics

Wastewater disinfection efficiency may be influenced by a number of variables, from the following three general categories:

- Wastewater and other surrounding conditions (chemical/physical),
- Disinfectant properties (kinetics), and
- Hydraulic characteristics of the reactor vessel.

As with most municipal wastewater treatment processes, the first category is difficult or impossible to control, because it is often a function of the influent and other natural characteristics at the WRRF. Treatment of the wastewater and surrounding conditions should be addressed by the processes upstream of disinfection, not by the disinfection process. Designers also need to examine possible impacts from upstream industrial and commercial users. The second category primarily depends on the type of disinfectant selected by the designer. Disinfection agents will vary in their flow rated (low hour, average, peak day, peak hour, and diurnal pattern), water temperatures, reaction rates, decay reactions, and mechanisms of biological inactivation. This section focuses on the hydraulic characteristics of a reactor vessel and shows how these characteristics can influence disinfection efficiency and, therefore, should be considered at the design level.

2.2   Typical Wastewater Disinfection Reactors

Serpentine or plug-flow reactors used for chlorine contact chambers and certain ozone contactors are designed to approach plug-flow conditions, but may contain zones of back-mixing (particularly at the entrance zone), short-circuiting, and dead zones. This type of reactor has been found to be most successful in terms of approaching a plug-flow condition when length-to-width ratios exceed 10:1. A length-to-width ratio of 40:1 or greater is preferred (Metcalf & Eddy, Inc./AECOM, 2013) for chlorine contact basins.

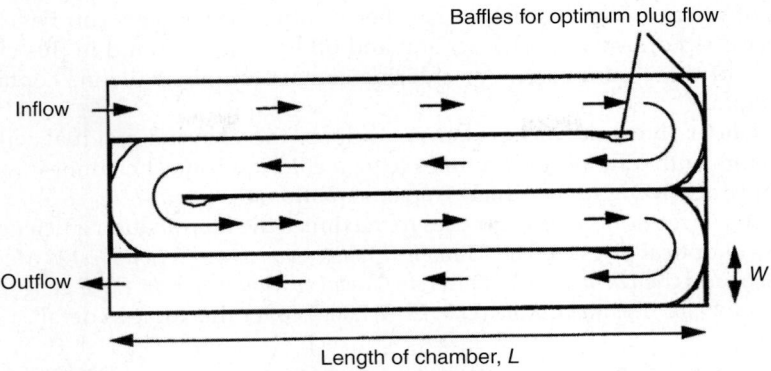

FIGURE 17.1 Serpentine flow reactor (length of flow in this chamber is four times the length of the chamber) (Black & Veatch Corporation, 2010).

Modifications for existing units with lower length-to-width ratios have been suggested to improve the hydraulic pattern (Hart, 1979; Louie and Fohrman, 1968). Figure 17.1 shows a typical chlorine contact chamber.

Ultraviolet-disinfection units, with their high length-to-width ratios, are designed to closely follow a plug-flow pattern. Inlet and outlet conditions for these reactors are important because of relative short detention times in reactor units (Black & Veatch Corporation, 2010). UV dose monitoring relies upon reactor-specific validations following NWRI (2012), U.S. EPA UVDGM (2003), or other methods. Note that U.S. EPA UVDGM document is specific to drinking water and not wastewater.

2.3   Initial Mixing

Mixing conditions before the inlet of serpentine contactors (especially for chemical-disinfection processes) are extremely beneficial, because uniform distribution of the chemical will be ensured (Calmer, 1993). A uniform chemical distribution becomes very critical in wet weather applications where the contact time is limited. For UV, maximizing radial mixing (mixing perpendicular to flow) is a desirable feature of these disinfection units. This configuration is unique to UV reactors, because radiation dose is proportional to the distance from the radiation source.

2.4   Combining Tracer Analysis and Disinfection Kinetics

A useful analytical procedure for predicting the efficiency of a wastewater-disinfection reactor using a chemical alternative is to assume that the reactor behaves according to a simplified model (e.g., the ideal plug-flow or complete-mix, segregation, dispersion, or tanks-in-series model). Tracer studies can be used to confirm the effective detention time. Combining equations that describe disinfection kinetics and contact time, one can predict the effluent quality through the disinfection process.

2.5   Factors that Influence Disinfection Efficiencies

The means by which hydraulic characteristics influence the subsequent disinfection efficiency emphasize the significance of hydraulics when evaluating and designing a disinfection reactor. Clearly, a number of other parameters can have an effect on a reactor's performance. This section and other sections found in this chapter identify some of these parameters for chlorine contact chambers, ozone contactors, and UV reactors.

Under dynamic wastewater conditions, the chlorine residual will decay. Thus, reactors with long residence times (e.g., a chlorine contact chamber) will not perform simply according to the Chick-Watson equation.

In UV reactors, a designer tries to maximize the illumination efficiency, based on the laws of optical physics. The *Municipal Wastewater Disinfection Design Manual* (U.S. EPA, 1986a) and the *Ultraviolet Disinfection Guidance Manual for the Final Long-Term 2 Enhanced Surface Water Treatment Rule* (U.S. EPA, 2006) cover this topic in detail.

2.6 Reactor Design for Chemical Disinfectants

Specialized cross-references regarding the design of disinfection reactors include *White's Handbook of Chlorination and Alternative Disinfectants* (Black & Veatch Corporation, 2010); *Recommended Standards for Wastewater Facilities* (Great Lakes–Upper Mississippi River Board of State and Provincial Public Health and Environmental Managers, 2014), *Design Manual for Municipal Wastewater Disinfection,* (U.S. EPA, 1996); and *Wastewater Engineering: Treatment and Resource Recovery* (Metcalf & Eddy, Inc./AECOM, 2013).

2.6.1 Designing to Achieve a Specific Contact Time "t"

One of the keys to effective disinfection with chemical disinfection systems is to apply the proper dose of chemical for the appropriate contact time. For chemical disinfection systems, the concepts of dose, residual, and demand are complicated by the fact that chlorine reacts with ammonia, resulting in the formation of other disinfectants, collectively called combined chlorine (e.g., monochloramine and other chloramine species). Such reactions are known to depend on the ratio of the chlorine concentration to ammonia. The dose–residual relationship is known to be non-linear and is described by the breakpoint curve (Figures 17.2 and 17.3). The kinetics of chlorine disinfection is also complicated by the fact that various species of chlorine resulting from the reactions with ammonia have varying inactivation potentials for microorganisms.

One of the first steps is to define hydraulic uniformity goals and approaches in the design of a chemical disinfection system are to use computational fluid dynamics (CFD) and tracer testing methods. For UV systems, when coupled with irradiance modeling, CFD can track trajectories of individual microorganisms and calculate their resultant UV dose (fluence). Using CFD in this manner can minimize the overdesigning of UV reactors and, thereby, lower the construction and operating costs of UV systems (Nisipeanu and Sami, 2004).

2.6.2 Designing to Facilitate Reactor Maintenance

The major maintenance activity associated with disinfection reactors is keeping the reactor relatively free of microbial films/slimes and accumulations of settled particles. The principles used in the design of large-scale chlorine-contact reactors apply to much smaller ozone reactors. The following chlorine-contact-reactor-design features help the operators maintain a reactor free of films, slimes, and settled particles:

- The materials and surfaces used in reactors must be smooth, to facilitate easy cleaning, such as with a high-pressure-water jet and a squeegee-type wiper.

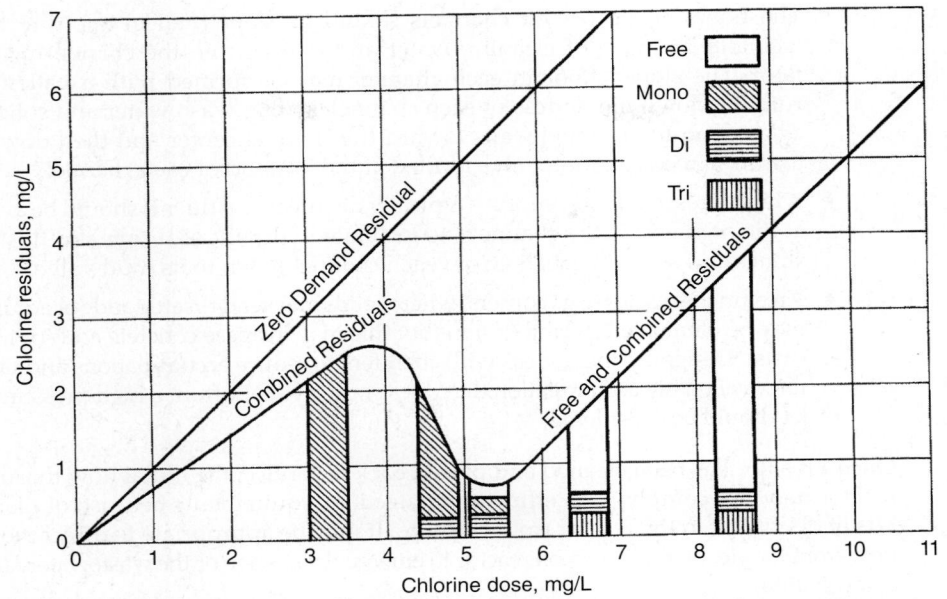

FIGURE 17.2   Chlorine residual with ammonia-nitrogen (Black & Veatch Corporation, 2010).

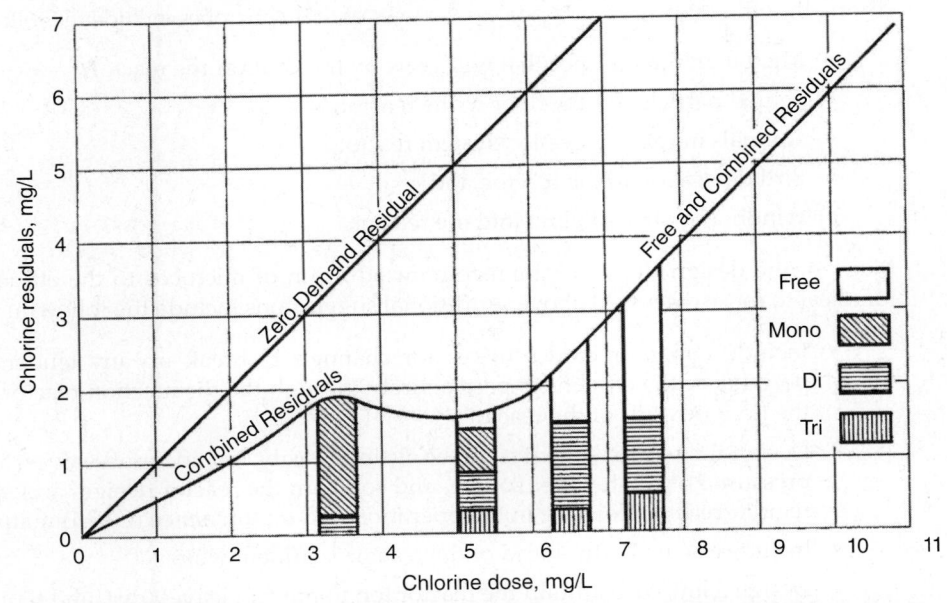

FIGURE 17.3   Chlorine residual with ammonia-nitrogen and organic nitrogen (Black & Veatch Corporation, 2010).

- The floors of the reactor channels should be sloped, all in one direction, to facilitate drainage of cleaning water and debris from the channel walls and floor. The sloped floor in each channel may be formed with a valley gutter running down the middle of each channel, so that wash water and solids flow away from the channel walls toward the channel center and then downslope toward a recessed mud valve located at the low end of each channel.

- A high-pressure water supply (typically disinfected effluent) should be available at the high end of the channels, to facilitate hydraulic scouring and flushing of slimes and settled particles from each channel down to its mud valve.

- Areas in the reactor (e.g., corners) where solids may tend to settle and/or accumulate may be eliminated during design (but this may increase concrete and/or forming costs). Designing these areas with chamfers to reduce accumulations and facilitate removal of any accumulation that does occur by scour from a high-pressure water jet should be considered.

Once a reactor has been cleaned and placed back in service, the initial flow through the reactor may not comply with effluent disinfection requirements because of disturbed bits of debris left in the reactor from cleaning. It may be appropriate to direct the initial flow from the cleaned disinfection reactor to emergency storage or the wastewater facility's drain system.

2.6.3   Designing to Minimize Reintroduction of Microbes to the Effluent

A major cause of exceedances of effluent disinfection limitations, particularly for stringent limits, such as are applied for water reuse, is reintroduction of microbes to the effluent during the effluent-disinfection process. Sources of these microbes include the following:

- Microbial films/slimes that will grow on the walls of the reactor,
- Settled particles on the floor of the reactor,
- Animals in the disinfection system reactor,
- Bird defecation in the reactor, and
- Windborne debris falling into the reactor.

Many of the design steps to minimize reintroduction of microbes to the effluent are identical to those discussed above. Additional suggestions include the following:

- Include bends or baffles in reactor channels to break up any tendency for portions of the effluent to pass entirely through the disinfection reactor along the floor or walls of the reactor.

- Design the reactor so that it can be cleaned easily. Include mud valves, a high-pressure hot wash water source, and so on, in the reactor design, and specify regular reactor cleaning in the operation and maintenance (O&M) manual.

- Include a scum baffle ahead of the reactor's effluent weir.

- Design a low curb around the reactor top to prevent leaves, dirt, and tools, and so on from falling into the reactor.

- Keep the reactor away from landscaping that may shed debris into the reactor or provide significant habitat for birds.

- Discourage birds from frequenting the disinfection reactor area by use of wires over the tops of the handrails to stop perching (and possibly a few wires stretched over the top of the reactor if waterfowl are landing on the reactor's water surface) or install equipment indoors.

As noted previously, covers over disinfection-reactor channels are not always recommended, because they hinder the reactor's cleaning process.

2.6.4 Designing to Control Disinfection-Byproduct Formation

Design considerations in chemical disinfection system are to reduce disinfection-byproduct (DBP) formation to the extent feasible include the following:

- Avoid adding more chemical disinfectant than necessary to achieve the desired level of effluent disinfection (i.e., Chemical residual × contact time = C_t). This practice is more a matter of proper programming of the chemical disinfectant dosing system than the reactor design itself, but the reactor design and its resulting contact-time curve have a bearing on that programming.
- Once the reactors are designed, the operation of the reactors needs to be specified (or programmed, if automated) to avoid unnecessarily long chemical disinfectant contact times. Having multiple, parallel, reactors provides the operator with some control over contact time, t, as a function of effluent flow. This control prevents *it* from becoming large at low flows (e.g., startup), which may cause excessive DBP formation in chemical disinfection systems.

2.7 Surrounding Conditions

The importance of initial mixing in chlorination and ozonation is noted later in this chapter. Initial mixing may be accomplished at the inlet zone using both static and mechanical mixers.

In open-channel reactors, the designer may need to consider wind effects that may cause surface currents, short-circuiting, and the disruption of an acceptable plug-flow condition. When monochloramine is used as the disinfection agent, turbulence within the chamber should be minimized, because it may cause back-mixing and reduce the concentration of volatile monochloramine (Black & Veatch Corporation, 2010).

3.0 Chlorination

Chlorine is the most widely used chemical for the disinfection of wastewater. It can be applied either as gaseous chlorine (elemental chlorine, Cl_2), a hypochlorite compound, or as chlorine dioxide.

3.1 Chemistry of Chlorine as a Disinfectant

When chlorine is added to water or wastewater, hydrolysis occurs, and a mixture of hypochlorous acid and hydrochloric acid is formed. The reaction is pH- and temperature-dependent, completed within milliseconds, and reversible. The hypochlorous acid formed is a weak acid and dissociates or ionizes to form an equilibrium solution of hypochlorous acid and hypochlorite ion (OCl^-). The equilibrium approaches 100% dissociation ($H^+ + OCl^-$) when the pH exceeds 8.5 and approaches 100% hypochlorous acid when the pH is less than 6.0.

For water and wastewater, the hypochlorous acid and hypochlorite ions are the oxidizing agents that provide disinfection. Hypochlorite solutions of sodium hypochlorite and calcium hypochlorite also dissociate in water to form hypochlorite ions yielding, in essence, the same chemistry as chlorine gas. The addition of chlorine gas will decrease the pH, because it forms both hypochlorous acid and hydrochloric acid, while sodium and calcium hypochlorite solutions have a minimal effect on pH. Gaseous chlorine also reduces alkalinity by as much as 2.8 parts chlorine/part calcium carbonate ($CaCO_3$) for the same reason. Further discussion of chlorine chemistry can be found in Black & Veatch Corporation (2010).

The chlorine present as both hypochlorous acid and hypochlorite ions is defined as *free available chlorine*, or simply *free chlorine*.

3.1.1   Inorganic Reactions

In addition to oxidizing and disinfecting pathogens, chlorine compounds will react with and oxidize many inorganic compounds present in wastewater, including hydrogen sulfide (H_2S), nitrite (NO_2^-), ferrous iron (Fe^{+2}), and manganous manganese (Mn^{+2}). These reducing agents exert a chlorine demand and react rapidly with the free chlorine.

When chlorine is added to water containing hydrogen sulfide or sulfite compounds, these compounds are oxidized to elemental sulfur or sulfate, depending on the conditions of the reaction. Nitrite is oxidized to nitrate (NO_3^-), ferrous iron to ferric (Fe^{+3}), and manganous manganese to manganic (Mn^{+4}) ions. The chlorine product in the reaction is the chloride ion.

3.1.1.1   Chloramines
Chloramines are products of the reactions between chlorine and ammonia-nitrogen found in wastewater. Ammonia-nitrogen may be present in appreciable amounts, typically 10 to 40 mg/L, in wastewater as either dissolved ammonia gas (NH_3) or the ammonium ion (NH_4^+).

Three chloramines are formed—monochloramine (NH_2Cl), dichloramine ($NHCl_2$), and trichloramine or nitrogen trichloride (NCl_3)—in a stepwise process, beginning with the reaction of hypochlorous acid and ammonia in dilute aqueous solutions. Chlorine and ammonia can also react to produce nitrogen gas, although this reaction should not be considered a representation of the reaction mechanism.

How far these stepwise reactions proceed and how much of each compound is formed depends on the pH, temperature, time of contact, and ammonium ion and hypochlorous acid concentrations. In general, low pH levels and high chlorine-to-ammonia ratios favor dichloramine ($NHCl_2$) formation. At a pH of greater than approximately 8.5, monochloramine (NH_2Cl) exists almost exclusively. At a pH between 8.5 and 5.5, monochloramine and dichloramine exist simultaneously; between a pH of 5.5 and 4.5, dichloramine exists almost exclusively. At a pH of less than 4.4, nitrogen trichloride will be produced (Jafvert and Valentine, 1992).

Chloramines have disinfecting properties and are part of the combined available chlorine (CAC) measurement, also called *total chlorine*. Chloramines are a weaker oxidant compared with hypochlorite and are more specific. Their application for disinfection generally is limited to providing drinking-water-distribution-system disinfectant residuals, because they are not effective as a primary disinfectant.

3.1.1.2   The Breakpoint Phenomenon
Breakpoint chlorination is the process of using chlorine's oxidative capacity to oxidize ammonia. Although breakpoint chlorination is a practice used in water treatment to obtain a free chlorine residual in the presence of ammonia-nitrogen, it is not used in conventional wastewater treatment because of the

large quantities of chlorine required (approximately a 10:1 mass ratio of chlorine to ammonia-nitrogen). An exception to this occurs when a WRRF knowingly or unknowingly nitrifies most of its ammonia-nitrogen, leaving little or no ammonia present to form chloramines. In such a case, the addition of chlorine can produce the breakpoint reaction. In breakpoint chlorination, ammonia is progressively oxidized, until a point is reached, beyond which, combined chlorine or chloramines and ammonia react with chlorine to produce nitrogen gas, and excess chlorine is present as free chlorine. In this case, ammonia is no longer present as ammonia gas or ammonium ion, and only a minimum is present in combined chlorine forms.

Breakpoint reactions proceed at a rate that is highly pH-sensitive. At pH 7 to 8, and when the mass ratio of chlorine to ammonia-nitrogen is 5:1 or less, all free chlorine is converted to monochloramine. At a lower pH, dichloramine is formed. When the applied chlorine-to-ammonia-nitrogen ratio exceeds 5:1, the total chlorine residual decreases, and the breakpoint is approached, with nitrogen (N_2) as a product. When the breakpoint is reached (a chlorine-to-ammonia mass ratio from 8:1 to 10:1), ammonia-nitrogen is oxidized, and free chlorine begins to appear in small amounts in the resultant residual. Figures 17.2 and 17.3 show typical breakpoint curves for selected effluents.

3.1.1.3   Other Chlorine/Nitrogen Reactions   Organic chloramines are formed when chlorine reacts with amino acids, proteinaceous material, and other organic nitrogen forms. Organic chloramines are not effective disinfectants, and the presence of organic nitrogen can exert considerable chlorine demand, in addition to interfering with differentiation between free and combined forms. Chlorine existing in chemical combination with ammonia or organic nitrogen chloramines is termed as *combined available chlorine* (CAC).

3.1.2   Organic Reactions

Chlorinated organic compounds, or chloro-organics, are formed by the action of chlorine on organic carbon in wastewater, just as organic chloramines are products of the chlorination of the organic nitrogen fraction. Organic nitrogen compounds contain nitrogen atoms, in addition to carbon atoms, and include urea, amino acids, and proteins. Most of the reactions between chlorine and organic compounds produce compounds that no longer have oxidizing potential and do not contribute to the CAC.

Trihalomethanes are some of the DBPs of complex chemical reactions of chlorine with a group of organic acids known as *humic acids*. The trihalomethanes (THMs) are single carbon molecules containing three halogen atoms present in varying combinations. There is concern regarding the formation of THMs and other DBPs because of their effects on the environment and human health. Chloroform, a well-known THM, is a documented animal carcinogen, and all haloforms are believed to act in a similar manner. U.S. EPA has issued regulations for the drinking water industry (Safe Drinking Water Act) targeted at minimizing public exposure to this class of compounds as well as state and national water quality standards for the discharge of these types of compounds. WRRF are required to meet individual THM for discharges to receiving streams so the concentrations may be relatively low.

3.2   Chlorination and Dechlorination Chemicals

3.2.1   Elemental Chlorine

At standard conditions, chlorine (Cl_2) is a greenish-yellow gas. When cooled and compressed to $-34.5°C$ ($-30.1°F$) and 100 kPa (1 atm), respectively, it condenses to a clear, amber-colored liquid. Commercial chlorine is classified as a nonflammable, toxic, compressed gas. It typically is shipped in steel containers that are designed, constructed, and handled in accordance with strict government regulations.

3.2.1.1 Physical Properties In the gaseous state, chlorine is 2.5 times as heavy as air. In liquid form, chlorine is approximately 1.5 times as heavy as water. Liquid chlorine vaporizes rapidly. One volume of liquid yields approximately 450 volumes of gas. Thus, 1.0 kg (2.2 lb) of liquid vaporizes to approximately 0.31 m^3 (11 cu ft) of gas.

Chlorine is only slightly soluble in water, with a maximum solubility at 100 kPa (1 atm) of approximately 10 000 mg/L (1%) at 9.6°C (49.3°F) and a solubility of 6500 mg/L at 25°C. The solubility of chlorine, like all gases, decreases with increasing temperature (Chlorine Institute, 1986). Practical solubility is approximately 50% of theoretical.

3.2.1.2 Chemical Properties Chlorine is highly reactive and, under specific conditions, chlorine can rapidly react with and oxidize many compounds and elements. Because of its affinity for hydrogen, chlorine removes hydrogen from some compounds, as in the reaction with hydrogen sulfide to form elemental sulfur (S) or the sulfate ion (SO_4^{-2}), depending on the chlorine-to-sulfur ratio and reaction conditions. Chloramines are formed when ammonia or other nitrogen-containing compounds react with chlorine.

3.2.2 Hypochlorites

Hypochlorites are salts of hypochlorous acid. Sodium hypochlorite (NaOCl) is the only liquid hypochlorite form in current use. There are several grades available. Calcium hypochlorite [Ca(OCl)$_2$] is the predominant dry form.

3.2.2.1 Physical Properties Sodium hypochlorite, often referred to as *liquid bleach*, is commercially available only in liquid form, typically in concentrations between 5% and 15% available chlorine.

Calcium hypochlorite, sometimes referred to as *powder bleach*, is a dry material typically consisting of 65% available chlorine. Often called *high-test hypochlorite* (HTH), 1 kg (2.2 lb) is equivalent to 0.65 kg (1.43 lb) of elemental chlorine.

3.2.2.2 Chemical Properties Hypochlorites are strong oxidants. All sodium hypochlorite solutions are unstable. Heat, light, storage time, and impurities, such as iron, accelerate product degradation. Hypochlorites are destructive to wood, corrosive to most common metals, and will adversely affect the skin, eyes, and other body tissues with which they come in contact.

The dry form (calcium hypochlorite) is unstable under normal atmospheric conditions. Reactions may occur spontaneously with numerous chemicals, including turpentine, oils, water, and paper. Therefore, calcium hypochlorite should be stored in dry locations and used only with equipment that is free of organics. Serious fire and explosion hazards exist when using this material.

Most common metals are not affected at normal temperatures by dry chlorine in the gas or liquid state (dry chlorine typically contains less than 150 mg/L of water). However, chlorine reacts with titanium and ignites carbon steel at temperatures greater than 232°C (450°F).

3.2.2.3 Toxicity Chlorine gas is a respiratory irritant and is classified as a toxic gas. A concentration in air greater than approximately 1.0 ppm by volume can be detected by most people because of its characteristic odor. Chlorine causes varying degrees of irritation of the skin, mucous membranes, and respiratory system.

Liquid chlorine will cause skin and eye burns on contact. As noted earlier, liquid chlorine vaporizes rapidly when unconfined and produces the same effects as the gas.

Complete recovery can occur following mild, short-term exposures to chlorine. The current OSHA permissible-exposure level is 0.5 ppm, the short-term-exposure level is 1.0 ppm, and the immediately-dangerous-to-life-or-health level is 30 ppm. Higher concentrations can be fatal (U.S. EPA, 1993).

The corrosivity of chlorine–water solutions can create handling problems. Most chlorine solutions are corrosive to common metals, with the exception of gold, silver, platinum, and certain specialized alloys. Hard rubber, unplasticized polyvinyl chloride (PVC), lined metal pipe, and certain other plastics are also resistant to the corrosivity of chlorine-water solutions.

3.2.3 Sulfur Dioxide

Sulfur dioxide is commonly used for dechlorination. Sulfur dioxide is classified as a nonflammable, corrosive, liquefied gas and is shipped commercially in steel containers designed, constructed, and handled in accordance with strict government regulations.

3.2.3.1 Physical Properties In the gaseous state, sulfur dioxide is colorless, with a suffocating, pungent odor, and is approximately 2.25 times as heavy as air. Liquid sulfur dioxide is approximately 1.5 times as heavy as water. Commercially, sulfur dioxide is supplied as a pressurized, colorless, liquefied gas. The solubility of sulfur dioxide gas in water (approximately 20 times greater than that of chlorine) is approximately 18.6% at 0°C (32°F). In solution, sulfur dioxide hydrolyzes to form a weak solution of sulfurous acid (H_2SO_3).

At a pH greater than 8.5, 95% of the sulfur dioxide gas dissolved in water exists as the sulfite ion (SO_3^{-2}). The solubility of sulfur dioxide in water decreases at elevated temperatures.

3.2.3.2 Chemical Properties Dry sulfur dioxide (liquid or gas) is not corrosive to steel and most other common metals. However, galvanized metals should not be used to handle sulfur dioxide, and, in the presence of sufficient moisture, sulfur dioxide is corrosive to most common metals. Because sulfur dioxide does not burn or support combustion, there is no danger of fire or explosion.

3.2.3.3 Toxicity Sulfur dioxide is an extremely irritating gas. The gas may cause varying degrees of irritation to the mucous membranes of the eyes, nose, throat, and lungs because of sulfurous acid formation. Contact with the liquid results in freezing of the skin, because the liquid absorbs its latent heat of vaporization from the skin. Worker exposures, on an 8-hour, time-weighted average, are currently limited by OSHA to 5 ppm by volume in air, or approximately 13 mg/m³. Concentrations of 500 ppm are acutely irritating to the upper respiratory system and cause a sense of suffocation after several inhalations (Compressed Gas Association, 1988).

3.2.4 Sulfite Salts

Sodium sulfite (Na_2SO_3), sodium bisulfite ($NaHSO_3$), and sodium metabisulfite ($Na_2S_2O_5$) are also used in dechlorination. On dissolution in water, these salts produce the same active ion, sulfite (SO_3^-). All three compounds typically are more expensive than sulfur dioxide per kilogram (pound) of active reducing agent (sulfite formed);

however, of the three, sodium metabisulfite is less costly and more stable than sodium sulfite and sodium bisulfite.

3.2.5 Chlorine Dioxide

Chlorine dioxide rarely is used as a disinfectant for secondary or tertiary wastewater effluent in the United States, as a result of traditional concerns associated with its unstable nature and the handling of hazardous chemicals required for its generation. However, advancements and introduction of chlorine-dioxide-generating systems that produce the disinfectant on-site, by combining sodium chlorite and hydrochloric acid, have helped to reduce these concerns.

Some testing indicates that chlorine dioxide is only slightly superior to chlorine as a bactericide, but is a much superior viricide (Black & Veatch Corporation, 2010). Chlorine dioxide was also shown to be more effective than chlorine against protozoa, including the cysts of *Giardia* and the oocysts of *Cryptosporidium* in wastewater streams. When Poliovirus I and a native coliphage were subjected to chlorine and chlorine dioxide disinfectants, a 2-mg/L dose of chlorine dioxide produced a much lower survival rate than a 10-mg/L dose of chlorine (Roberts, 1980). Chlorine dioxide has also been shown to be effective in killing other infectious bacteria, such as *Staphylococcus aureus* and *Salmonella*.

Apart from being a more potent viricide, disinfection of wastewater by chlorine dioxide offers additional advantages over chlorination, as follows:

- It exhibits relatively constant biocide power in a pH range between 6 and 9;
- It dissolves readily in water;
- It may enhance coagulation; and
- It reduces formation of organohalogenated compounds, including THMs.

The disadvantages of using chlorine dioxide include the following:

- It can form potentially toxic DBPs, such as chlorite and chlorate;
- The production cost is high;
- Chlorine dioxide needs to be produced on-site, because it is highly explosive gas;
- Chlorine dioxide decomposes in sunlight; and
- The chemicals used in the generation of chlorine dioxide are hazardous.

3.2.5.1 Properties   Chlorine dioxide (ClO_2) is a neutral compound of chlorine in the +IV oxidation state. It is a relatively small, volatile, and highly energetic molecule, and exists as a free radical, even while in dilute aqueous solutions. At high concentrations, it reacts violently with reducing agents. However, it is stable in dilute solution in a closed container in the absence of light (AWWA, 1990). Chlorine dioxide functions as a highly selective oxidant, as a result of its unique, one-electron transfer mechanism, where it is reduced to chlorite (ClO_2^-) (Hoehn et al., 1996). The pK_a for the chlorite ion, chlorous acid equilibrium, is extremely low, at pH 1.8. This indicates that the chlorite ion will exist as the dominant species in drinking water.

One of the most important physical properties of chlorine dioxide is its high solubility in water, particularly in chilled water (U.S. EPA, 1999a). In contrast to the hydrolysis of chlorine gas in water, chlorine dioxide in water does not hydrolyze to any appreciable extent, but remains in solution as a dissolved gas (Aieta and Berg, 1986). Above 11 to 12°C, the free radical is found in gaseous form. This characteristic may affect chlorine dioxide's effectiveness when batching solutions and plumbing appropriate injection points. Another concern is difficulty in performing chemical analysis on treated solutions.

3.2.5.2 Generation Chlorine dioxide cannot be compressed or stored commercially as a gas, because it is explosive under pressure. Therefore, it is never shipped. Chlorine dioxide is considered explosive at concentrations higher than 10% by volume in air, and its ignition temperature is approximately 130°C (266°F) at partial pressures (National Safety Council Data Sheet 525–ClO_2, 1967). Strong aqueous solutions of chlorine dioxide will release gaseous chlorine dioxide into a closed atmosphere above the solution at levels that may exceed critical concentrations. Some newer generators produce a continuous supply of dilute gaseous chlorine dioxide in the range 100 to 300 mm Hg (absolute) rather than in an aqueous solution.

Chlorine dioxide is produced by activating sodium chlorite with an oxidizing agent or an acid source. Sodium chlorite is converted to chlorine dioxide through a chlorine dioxide generator and applied as a dilute solution. Chlorine dioxide solutions should be applied to the processing system at a point, and in a manner, which permits adequate mixing and uniform distribution. The feed point should be well below the water level, to prevent volatilization of the chlorine dioxide. Precautions should be taken to avoid coincident feeding of chlorine dioxide with lime or powdered activated carbon.

Most commercial generators use sodium chlorite ($NaClO_2$) as the common precursor feedstock chemical to generate chlorine dioxide. However, production of chlorine dioxide from sodium chlorate ($NaClO_3$) recently has been introduced as a generation method wherein sodium chlorate is reduced by a mixture of concentrated hydrogen peroxide (H_2O_2) and concentrated sulfuric acid (H_2SO_4). Chlorate-based systems traditionally have been used in pulp and paper applications, but recently have been tested full-scale at two United States municipal WRRFs (U.S. EPA, 1999a). However, a recent Water Environment Research Foundation (Alexandria, VA) (WERF) survey found that no major municipal WRRFs were using chlorine dioxide as a primary disinfectant (Leong et al., 2008).

For wastewater applications, residual chlorine dioxide concentrations up to 5 mg/L generally may be adequate. Residual chlorine dioxide concentrations must be determined by substantiated methods that are specific for chlorine dioxide. Two suitable methods—the DPD (N, N-diethyl-p-phenylenediamine) glycine method and the amperometric method—are published in *Standards Methods for the Examination of Water and Wastewater* (APHA et al., 2012).

3.3 On-Site Generation of Sodium Hypochlorite

On-site generation systems are increasingly popular for wastewater disinfection and include both unseparated and separated electrolyzer systems.

3.3.1 Unseparated Electrolyzer System Types and Principles of Operation

Unseparated electrolyzer systems can be classified into two basic types—brine electrolysis and seawater electrolysis. The basis for classification is that the feedstock is derived from either crystallized salt for brine systems or seawater for seawater electrolysis systems. Although the product of each system is the same sodium hypochlorite disinfectant, differences in the electrolysis method exist, as a result of the variations in the calcarious hardness and other properties of the feed material. Because crystallized salt is dissolved and used for electrolysis in brine systems, control of the calcarious components may be achieved using water softening or by selecting the desired quality of the crystallized salt. Seawater does not allow for easy methods of calcarious-component control. Thus, an entirely different approach to electrolysis is used in seawater systems.

To electrolyze a brine solution for sodium hypochlorite production, brine electrolysis cells are designed for very low brine feed flowrates and narrow electrode gaps, and produce sodium hypochlorite concentrations approaching 1%. The seawater-system approach is to use very high seawater flowrates and wide electrode gaps, and produce sodium hypochlorite concentrations of less than 0.3%, to reduce the rate of deposit formation on the cathodes. Brine systems have an average current efficiency of 65%, while seawater systems have an average current efficiency greater than 80%. This difference in current efficiency has an effect on power consumption and, for the brine system, on salt consumption.

Brine systems can be used for any application requiring chorine or chloramines as a part of the disinfection regimen. These systems almost always are installed inland and are designed to provide substantial quantities of stored sodium hypochlorite. Systems generally are configured with the following components: water softeners, salt dissolvers, electrolyzer cell or cells, direct-current-power rectifier, storage tanks, hydrogen dilution blowers, dosing pumps with dosing control, cell-cleaning system, and central control panel.

Electrolytic cells may be designed having monopolar or bipolar electrode configurations and may include plate electrode designs and/or tubular electrode designs, or tubular cells with plate electrodes. Electrolytic cell modules consist of a group of cells connected together hydraulically and electrically, in series, to form a complete cell circuit. Additionally, corrosion must be prevented, because humidity and salt air will create electrical problems in tiny crevices that cannot be seen or foreseen until they occur.

3.3.2 Separated Electrolyzer Systems (Membrane Systems) and Principles of Operation

The overall process of sodium hypochlorite generation in a separated cell is a two-part process; the first part involves simultaneous production of caustic soda solution and chlorine gas, followed by recombining these two chemicals to produce the final product. In this process, a semipermeable membrane acts as a physical barrier between the anode and cathode sections of the process, as shown in Figure 17.4. This design allows sodium ions to pass through pores into the cathode from the anode. Hydroxide ions, formed by electrolysis of water, combine with sodium ions in the cathode to produce caustic soda solution.

The brine solution is fed through the inlet of the anode compartment. With applied current, a part of the feed salt is electrolyzed into chloride and sodium ions. The chloride ions plate out as chlorine molecules on the anode and exit with the remaining brine solution. Sodium ions pass through the membrane into the cathode compartment under the influence of electrical potential. In the cathode compartment, a weak solution of caustic soda is fed through the inlet line. A portion of the water in the entering caustic soda solution is electrolyzed to hydrogen and hydroxide ions. Because of the negative charge associated with hydroxide ions, they try to pass through the membrane to the anode.

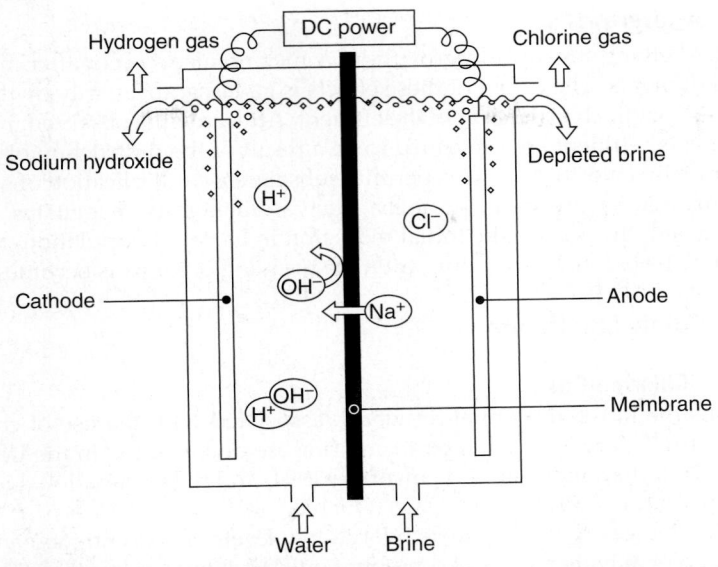

FIGURE 17.4 Typical separated electrolytic cell schematic.

However, the membrane is ion selective and only permits the positive sodium ions to pass through. It deflects any hydroxide ions back into the cathode compartment.

Theoretically, 1 Faraday of electrical energy should produce one equivalent of chlorine gas at the anode, one equivalent of hydrogen gas at the cathode, one equivalent of hydroxide ions in the cathode compartment, and one equivalent of sodium ions in the anode compartment. With the transport of the sodium ions from anode to cathode, the cell should produce one equivalent of sodium hydroxide in the cathode compartment. *Current efficiency* is defined as the fraction of the total produced hydroxide ions that passes through the membrane into the anolyte.

The purity of the feed salt affects the quality of the brine, which has a great influence on the performance and life of the membrane. Therefore, membrane-type electrolyzers require high-purity salt as the raw material. Power consumption by an electrolyzer assembly, for a given rate of production of caustic soda and chlorine, is a function of two electrical operating variables—operating current, and the voltage drop across the electrolyzer required to make the current flow through the electrolyzer.

3.4   Chlorine and Disinfection Byproduct Toxicity

Chlorine is an extremely reactive element, rapidly undergoing chemical reactions with inorganic and organic substrates. When the organic substrate is part of a living organism, the reaction can have a toxic effect on the organism. This toxicity may affect the organism's ability to reproduce or metabolize, cause genetic dysfunctions (mutations), or ultimately kill the organism. The active agents are hypochlorous acid, hypochlorite ion, monochloramine, and dichloramine. Hypochlorous acid reacts with organic nitrogen compounds and may alter the chemical structure of the organism's organic materials and change the genetic information.

Undesirable effects of chlorination on receiving water biota have been documented.

3.5   Aftergrowths

In the receiving stream, aftergrowths of some organisms occur after discharge of chlorinated effluents. The extent of these effects is governed primarily by the amount of biologically oxidizable material in the effluent. Aftergrowths observed in waters receiving chlorinated effluent are presumed to be a result of the destruction of large numbers of protozoa by chlorination. This permits subsequent multiplication of surviving bacteria unhampered by predatory protozoa, such as ciliates and flagellates. When effluent is chlorinated, the greater the initial reduction in bacterial population, the longer the lag time will be before the multiplication of surviving organisms becomes apparent.

3.6   Safety and Health

3.6.1   Chlorine Gas

Perhaps the most substantial drawback associated with the use of chlorine gas is the safety risk. Safety issues related to chlorine are documented in the Uniform Fire Code (UFC), (International Fire Code Institute, 1994) and the Occupational Safety and Health Administration (Washington, D.C.) (OSHA).

Chlorine gas has a detectable odor at low levels of concentration and has a greenish yellow color at higher levels of concentration. At volumes of less than 0.1 ppm in air, chlorine gas is undetectable, except by instruments. The maximum contaminant level established for chlorine gas by OSHA is a 1-ppm, time-weighted average over 8 hours. The harmful effects of chlorine gas exposure begin to become evident at approximately 5 ppm and higher. Between 5 and 10 ppm, however, these effects (choking, coughing, watery eyes, mild skin irritation, and lung irritation) are often temporary. At higher concentrations, the effects become more long lasting and can result in serious health consequences or death.

High-capacity storage and transportation facilities should be located in isolated areas and equipped with scrubbers. The designer should consult the latest edition of *The Chlorine Manual* (Chlorine Institute, 1986); Compressed Gas Association (Chantilly, VA) data; the appropriate U.S. Department of Transportation (Washington, D.C.) and OSHA regulations; and any applicable fire codes, including local codes and the Uniform Fire Code, Standard Fire Prevention Code, and National Fire Code. U.S. Department of Transportation regulations govern the use of tank cars, trucks, and 900-kg (1-ton) containers. Additional regulations pursuant to the Clean Air Act require risk-management plans for some facilities depending upon the amount of chlorine gas stored at the WRRF.

3.6.2   Hypochlorites

Eye protection and access to an emergency eyewash and showers are recommended for operators handling sodium hypochlorite. As with any form of hypochlorite, the undiluted chemical can cause severe burns on the skin and clothing. It is recommended that operators working with any of the hypochlorites wear protective clothing.

Operators who use calcium hypochlorite should wear eye protectors and dust masks when transporting the powder or mixing it with water. All areas exposed to hypochlorite should be washed thoroughly. Rubber gloves are recommended to provide hand protection.

3.6.3   Shipment and Handling

Operators of WRRFs in which hazardous chemicals are used must be thoroughly familiar with U.S. Department of Transportation regulations regarding the transportation of these chemicals. Calcium hypochlorite is classified as a corrosive and rapid oxidant. Sodium hypochlorite is a corrosive agent. Chlorine and sulfur dioxide are nonflammable, corrosive, toxic, liquefied gases under pressure. Various sulfite solutions are classified as corrosive.

3.6.3.1 Cylinders Precautions for chlorine or sulfur dioxide containers should be observed during design as outlined by

- The National Fire Code, Standard Fire Prevention Code, and Chlorine Institute guidance; and
- Applicable local, state, and national codes.

3.6.3.2 Containers Chlorine and sulfur dioxide are supplied in steel pressure vessels of 45 and 70 kg (100 and 150 lb) and in 900-kg (1-ton) containers, tank trucks and tank cars, and barges. For large WRRFs, a rail siding or stationary bulk storage tank may be provided.

The 70-kg (150-lb) cylinders are moved, stored, and used in the upright position. The cylinder valve is equipped with a fusible plug in its body. This plug is designed to melt between 70 and 73.9°C (158 and 165°F), to prevent hydrostatic rupture of the cylinder.

The 900-kg (1-ton) containers are moved, stored, and used in a horizontal position. The container valve is similar to the upright standard cylinder valve, except that it has no fusible plug. Three fusible plugs are located on each end of the 900-kg container. Each valve on a 900-kg container is connected to a tube inside the container. The valves on these containers must be aligned in the vertical position, to enable gas withdrawal from the upper valve and liquid withdrawal from the lower valve.

3.6.3.3 Facility Design Recommended Standards for Wastewater Facilities (Great Lakes–Upper Mississippi River Board of State and Provincial Public Health and Environmental Managers, 2014) The Chlorine Institute provide good references for chlorine facility design.

Many jurisdictions require the installation of gas-scrubbing equipment for chlorine and sulfur dioxide storage rooms. Regulations are changing rapidly and are subject to wide variations in local interpretation. Planning and design of such facilities should be coordinated carefully with local authorities.

Appropriate OSHA-approved warning signs should be posted at the entrance and any other exposed side. The appropriate container repair kit and self-contained breathing apparatus should be located at a convenient external location.

Gas storage and use areas should be dedicated rooms. In these facilities, nothing should be stored, and no work should be performed that is not related directly to handling chlorine or sulfur dioxide.

It is recommended that chlorine or sulfur dioxide equipment be located in isolated rooms separate from cylinder storage. Remote vacuum regulators should be used to convert gas to a vacuum, preferably at the source. This practice will maintain any gas piping in the equipment room under vacuum, which improves safety and will permit the use of plastic piping, such as schedule 80 PVC, from the regulator to the feeding equipment and ejector. To facilitate operation, a gas-tight observation window should be provided in the common wall between the storage/vacuum regulator rooms.

3.6.3.4 900-kg (1-Ton) Containers Facilities handling 900-kg (1-ton) containers must meet the same requirements of adequate space, safety devices, light, and ventilation as facilities handling cylinders. A properly designed 900-kg-container-handling facility will also have an overhead monorail hoist and motorized trolley of at least 1800 kg (2 tons) capacity. Slow-speed cranes typically are used to prevent jerky movements of containers.

Containers that are not presently in use may be stored on either trunnions or storage cradles. Any container scale and container in use must incorporate trunnions with rollers as part of its design. Trunnions are necessary for both stored and in-use containers, to allow the container to be rotated in the event of a leak, so that the leaking area is at the top of the container, resulting in the release of gas rather than liquid. Trunnions are also necessary to permit rotating the container, so that the valves on the 900-kg container are vertical. Hold-down chains for each container, both stored and in use (including those on scales), are recommended to prevent container movement, especially in earthquake-prone areas.

3.6.3.5 Vaporizer Facilities
Vaporizers, sometimes referred to as *evaporators,* for chlorine and sulfur dioxide, typically are electrically heated water baths that contain pressurized vaporization chambers. Design requirements for vaporizers are found in *Recommended Standards for Wastewater Facilities* (Great Lakes–Upper Mississippi River Board of State and Provincial Public Health and Environmental Managers, 2014), and The Chlorine Institute.

3.7   Analytical Determination of Chlorine Residuals
A critical aspect in controlling the chlorination process is the accurate measurement of chlorine residuals. The reader is referred to the current edition of *Standard Methods* (APHA et al., 2012) for full details on the tests. Many of these analytical methods have been successfully adapted to on-line process control.

3.8   Free Versus Combined Chlorine Residual
Toxicity is related to the individual sensitivities of target organisms. A residual toxicity is also related to the degree of chemical reactivity of the compound. Free chlorine residual, hypochlorous acid, and hypochlorite ion are more reactive compounds than the combined residuals, monochloramine and dichloramine. Figure 17.5 relates residual concentration and contact time to 99% destruction of *E. coli,* a typical coliform. Although wastewater would require significantly higher chlorine doses, the relationship between relative efficiencies of various residuals remains.

Free chlorine residuals, hypochlorous acid, and hypochlorite ion exist only momentarily when added to wastewater containing ammonia-nitrogen. Because ammonia is a significant constituent of most effluents, and chlorine is able to form chloramines, it follows that combined residuals, which typically are predominately monochloramine, will be responsible for most of the germicidal activity of chlorine present in chlorinated effluent.

Free chlorination does not necessarily lead to improved disinfection. Presumably, the reactive free residual is dissipated in organic reactions and no longer available for disinfection, even though it titrates as a free residual. Combined residuals do not undergo such side reactions and remain available for disinfection. Free residuals are more reactive than combined residuals, although they may be consumed in organic reactions that do not contribute to disinfection. Both free chlorine and combined residual can produce undesirable byproducts that may be harmful to humans or to biota of receiving streams; combined residuals can also kill biota in the receiving stream.

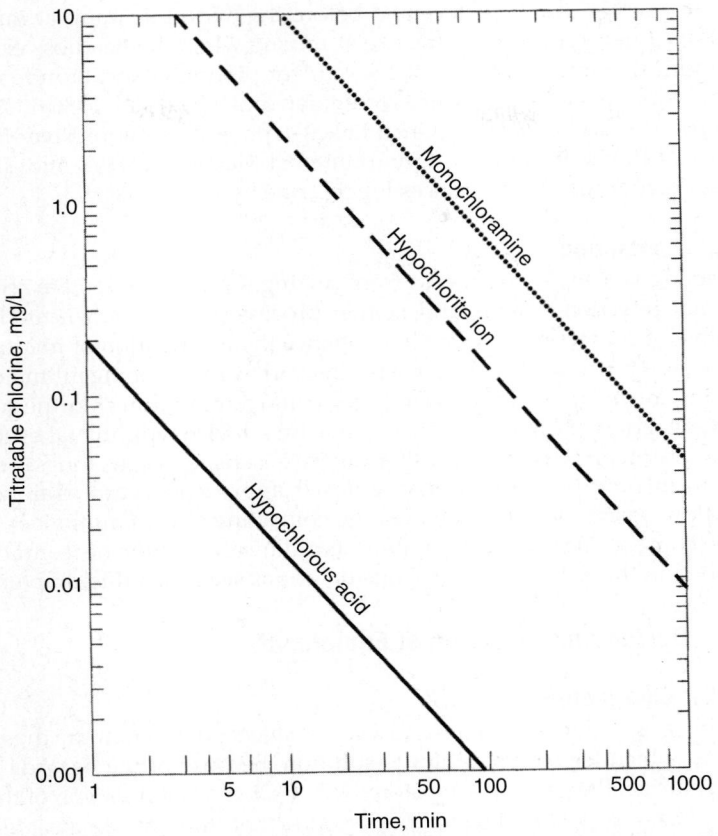

FIGURE 17.5 *Escherichia coli* kill times versus residual concentration (Clarke et al., 1964).

3.9   Process Design Requirements

3.9.1   Mixing

3.9.1.1   Closed Conduits   Proper mixing may be achieved in a closed conduit with turbulent flow, by placing a properly designed chlorine diffuser in the center of the cross-section of the conduit's flow field. Design considerations include construction materials, perforation design, and flow velocity. This type of mixing is considered to be less effective, however, than mechanical mixers or proprietary submerged propeller/turbine direct-gas mixing systems.

3.9.1.2   Hydraulic Devices   A simple hydraulic jump may be a satisfactory mixing device. The chlorine diffuser, perforated and positioned perpendicular to the water flow, can be located in the quiet zone upstream of the turbulent zone created by the hydraulic break or directly in the turbulent zone. The disadvantage of locating the diffuser in the turbulent zone is that the zone shifts position when the flow changes. When using a hydraulic jump, the submergence of the diffuser should not be less than 230 mm

(9 in.) below the water surface and before the hydraulic jump at minimum flow. The hydraulic jump typically is effective at mixing when the headloss exceeds 0.6 m (2 ft). Minimum Reynolds numbers of 1.9×10^4 for pipe flow and Froude numbers between 4.5 and 9 for open channels are recommended (U.S. EPA, 1986a). *Recommended Standards for Wastewater Facilities* (Great Lakes–Upper Mississippi River Board of State and Provincial Public Health and Environmental Managers, 2014) and The Chlorine Institute also provide good references for chlorine hydraulic design.

3.9.2 Contacting

Contacting is a separate process from mixing. Both processes are required in an optimized disinfection system, and neither process can be a substitute for the other. The objective of contacting is to further enhance the inactivation of microorganisms by the disinfection process. This objective is achieved by maintaining intimate contact between microorganisms in the wastewater stream and a minimum chlorine concentration for a specified period of time. A chlorine-contacting device typically takes the form of a pipeline or a serpentine chamber; either device is satisfactory, as long as short-circuiting is minimized, plug-flow conditions are closely approached, corners are rounded to reduce dead-flow areas, and the velocity of the contacting stream minimizes solids deposition in the contact chamber or pipeline. Guidance on chlorine contact tank design is provided in the Reactor Design Considerations section in this chapter.

3.10 Design and Selection of Equipment

3.10.1 Chlorinators

Chlorine gas feeders are referred to as *chlorinators*, to differentiate them from hypochlorinators, which feed a hypochlorite solution. Because of the hazards involved in handling chlorine, the gas tanks are stored in a well-protected section of the WRRFs, which is out of easy access by the facility personnel not directly responsible for O&M of this area. *Recommended Standards for Wastewater Facilities* (Great Lakes–Upper Mississippi River Board of State and Provincial Public Health and Environmental Managers, 2014) and The Chlorine Institute also provide good references for design practices.

3.10.2 Chemical-Feed for Hypochlorite Solutions

The risk involved in the transport, storage, and handling of gaseous chlorine has forced many WRRFs to switch to the application of hypochlorite solutions from gaseous chlorine. Typically, the application of sodium or calcium hypochlorite solution to the treated effluent is done by a set of chemical-feed pumps that are referred to as *hypochlorinators*. The basic components are a storage reservoir or mixing tank for the hypochlorite solution; a metering pump, which consists of a positive displacement pumping mechanism, motor or solenoid, and feed-rate-adjustment device; and an injection device. Depending on the size of the system, a plastic or fiberglass vessel is used to hold a low-strength hypochlorite solution. Hypochlorite solutions are corrosive to metals commonly used in the construction of storage tanks. Feeding of calcium hypochlorite will require a mixing device, typically a motorized propeller or agitator located in the tank. Also in the tank is a foot-valve and suction strainer connected to the suction inlet of the hypochlorinator. Mixing takes place in a separate tank. *Recommended Standards for Wastewater Facilities* (Great Lakes–Upper Mississippi River Board of State and Provincial Public Health and Environmental Managers, 2014) and The Chlorine Institute also provide good references for hypochlorite system design.

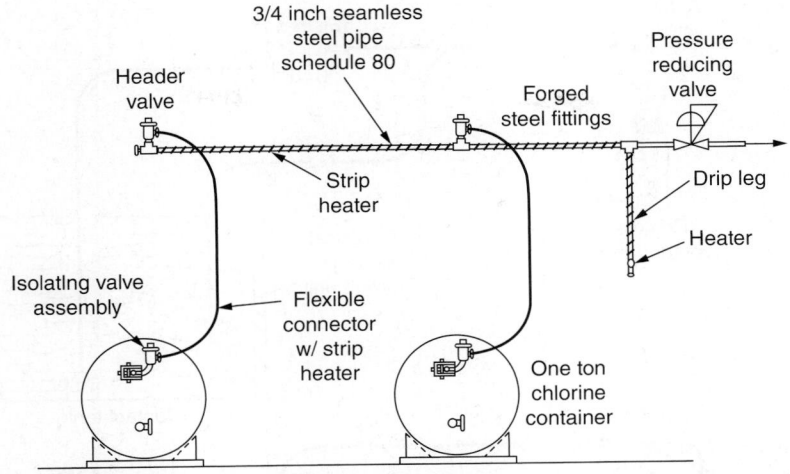

FIGURE 17.6 Typical gas manifold with heaters (in. × 2.54 = mm; ton × 907.2 = kg).

Selection of proper materials of construction for all the wetted parts is important to avoid corrosion and the potential of degradation of sodium hypochlorite solution. Most metals catalyze the degradation of sodium hypochlorite. Therefore, metering pump internals that come in contact with sodium hypochlorite solution are made up of synthetic materials. Peristaltic pumps use hoses made of flexible, polymeric material.

3.10.3 Manifolds and Vacuum-Regulator Location

If the vacuum regulator in a chlorination system is not mounted directly on the cylinder, a manifold or pressure piping is required. Gas manifolds consist of a flexible copper connector, typically with an isolation valve(s), and a rigid pipe section of carbon steel with a drip leg (Figure 17.6). If pressure manifolds are in use, reliquefaction must be prevented. Tracing of pressure lines, the use of pressure-reducing valves, insulation of the pressure line, sloping of the line back toward the source, and the use of drip legs at points of direction change and low points in the piping are recommended.

Often, for pressure piping that extends for long distances [more than 6 m (20 ft)], a pressure-reducing valve, preceded by a gas strainer, may be helpful. This requirement is unnecessary if the pressure/vacuum regulator is located as close as possible to the container. The determination of line size should be calculated so that pressure in the line at the regulator meets minimal requirements for operation of the regulator. In vacuum systems, the drop should be such that the vacuum immediately downstream of the feed-rate controller is greater than 371 mm (14.6 in.) for sonic feed systems and 150 mm (6 in.) for non-sonic systems.

Facility designs should include the use of 100% standby and automatic switching from online to standby equipment. The use of vacuum-operated, automatic switchover devices that change from an empty to a full supply of chlorine is recommended (Figure 17.7). These devices are available at operating rates up to 80 kg/h (4000 lb/d) and are often built into vacuum regulators. When requirements exceed these values, pressure-type switchover systems are used.

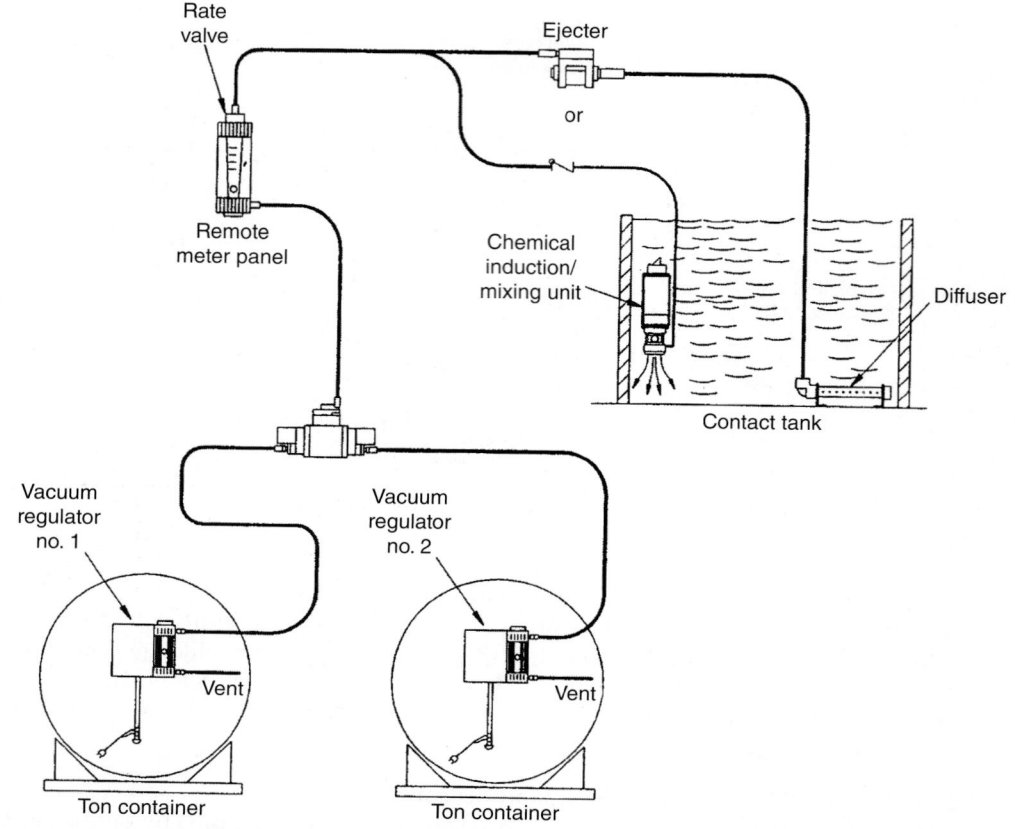

FIGURE 17.7 Automatic vacuum switchover system.

3.10.4 Vaporizers

Vaporizers for chlorine and sulfur dioxide (used for dechlorination) are similar, and both are addressed in this section. *Recommended Standards for Wastewater Facilities* (Great Lakes–Upper Mississippi River Board of State and Provincial Public Health and Environmental Managers, 2014) and The Chlorine Institute also provide good references for vaporizers design.

The difference between operation of vaporizers for sulfur dioxide and chlorine is the vapor pressure of each gas. The vapor pressure of sulfur dioxide is approximately 50% that of chlorine. Chlorine can be produced at the rate of 8 kg/h (400 lb/d) by natural vaporization within a container. Therefore, systems requiring feed rates up to 30 to 40 kg/h (1500–2000 lb/d) capacity may have several containers manifolded and would not require chlorine vaporizers. However, a sulfur dioxide container can vaporize approximately 4.5 to 5 kg/h (225–250 lb/d) of sulfur dioxide gas. Thus, sulfur dioxide vaporizers commonly are used on systems with capacities as low as 4.5 kg/h. Alternatively, two or more 900-kg containers can be manifolded together to achieve the desired gas-flowrate without resorting to use of a liquid chlorine or sulfur dioxide vaporizer.

3.11 Feed-Control Strategies

There are several ways to control the feed rate of chlorine gas or hypochlorite solutions—manual control, automatic-flow proportioning or open-loop control, automatic residual or closed-loop control, or automatic compound-loop control, which combines flow and residual signals to vary the gas-feed rate. Flow proportioning is sometimes referred to as *feed-forward control*, and residual control is sometimes referred to as *feedback control*.

Flow pacing is based on the concept that varying the chlorine feed rate in proportion to flowrate will provide adequate quantities of chlorine at any flow. A typical arrangement found in some WRRFs involves the use of a flow control devices to measure the secondary clarifier effluent flow and send the flow signal to an automatic controller, which effectively controls the pacing of the sodium hypochlorite or chlorine-solution-dosing pump. However, in wastewater, this is not always correct. The chlorine demand will vary with flow and can vary independently of flow, depending on the constituents in it.

Residual control involves varying the chlorine feed rate based on the deviation of concentration from a setpoint on a controller. For systems in which the flowrate is nearly constant on a daily basis or is strictly seasonal, this type of control system works well. For systems in which the flowrate varies often and demand is variable also, residual control may not be as effective as flow pacing, because residual control systems do not react well to large variations in flowrate over short periods of time. These methods can be applied to control based on free chlorine residual, total chlorine residual, or ORP. The ORP has been used as the control variable for disinfection with oxidants. Proponents of this control approach maintain that, because different disinfectants have different disinfecting powers, relating the microorganism inactivation with the residual concentration can be unreliable. Oxidation-reduction potential has been able to serve as a combined surrogate parameter for disinfection strength.

Compound-loop control provides the ability to use both flow and residual input to control gas feed. Flow is the primary drive, while residual is used to trim the gas feed. When the setpoint in a compound-loop control system is further controlled automatically, the configuration is referred to as *cascade control*. Cascade control requires the use of another analyzer downstream of the compound-loop control analyzer. Output from the cascade-control analyzer is used to regulate the compound-loop controller.

The choice of control strategy for a particular installation is based on regulatory requirements, existing facilities, wastewater-treatment-system design, economics, cost effectiveness, and required system maintenance. The more complicated the selected control system, the more likely it is that service requirements will be more exacting and, therefore, will require more training and an increase in the skill level of operating personnel.

3.11.1 Manual Control

Manual control of chlorine feed is the simplest strategy and, because of its simplicity, may often be the most effective method. A manual-control system requires less maintenance and operator expertise than any form of automatic control. The basic chlorinator feeds chlorine at a predetermined constant feed rate, which is changed by the operator as required. In the case of hypochlorination, the chemical-dosing rate is controlled at the desired level by manually adjusting the pump stroke length and/or the motor speed. Manual control has a low capital cost, but it is prone to either overfeeding or underfeeding of chlorine and, therefore, excessive dechlorination, insufficient disinfection, or overchlorination. Manual-control systems are used where flowrate and demand are fairly constant. An example of where this may be appropriate is the discharge from a pond treatment system.

3.11.2 Semiautomatic Control

The same equipment used in manual-control systems can often be used to partially automate the operation of a system.

One such system is on–off control. The chlorinator can be turned on and off automatically in response to a signal, such as a wet-well level or pump activation. The feeder can be turned on and off by controlling the following: the booster pump, a solenoid valve in the water supply line to the injector or ejector, or a solenoid valve located in the gas vacuum line between the rate-control valve and the injector.

Another option for semiautomatic control is a method known as *band control.* In this technique, two chlorine gas-flow metering tubes are used in the gas vacuum line in conjunction with two vacuum line solenoid valves and a chlorine residual analyzer. The analyzer, or a recorder receiving a signal from the analyzer, is equipped with two setpoint alarm contacts. The contacts are preset at points of maximum and minimum levels of residual. These contacts activate the vacuum line solenoid valves as follows:

- When the residual is lower than the low setpoint, both valves are open;
- When the residual is higher than the high setpoint, both valves are closed; and
- When the residual is between the two setpoints, one valve is closed.

The above control actions can also be achieved in the case of hypochlorinators, by turning the dosing pumps on or off in response to the signal from the controller.

3.11.3 Flow-Proportional Control

In most WRRFs, flow is variable, and it is not possible or practical to construct equalization basins. Therefore, control of chlorination feed is often set in proportion to flow, which enables the ratio to be varied by adjusting the dosage. In this strategy, a flow signal is transmitted by a primary flow element to the chlorinator, where an automatic valve opens or closes, depending on the signal level (typically 4 to 20 mA direct current) from the flow meter.

Flow-proportional chlorinators are sized by establishing a design dosage for the chlorine feed rate in grams per hour (pounds per day). This design sets the maximum wastewater flow and maximum chlorine flow dosage at a 1:1 ratio. This means that, at a 10% signal from the wastewater flow meter, the chlorinator flow meter will read 10% of full scale; at 90% of the wastewater flow rate, the chlorine flow meter will read 90%; and so on. Flow-proportional controllers have a dosage-control adjustment that allows the operator to vary the design dosage ratio from 10:1 turndown to 1:4 turn-up.

Flow measurement for hypochlorinator control commonly is performed by a Parshall flume, and the signal is recorded continuously and transmitted by a transmitter (Flow Indicator Transmitter, FIT) to a programmable logic controller (PLC). The controller sends a controlled signal to the speed controller of the dosing pump, to deliver the desired amount of hypochlorite solution in proportion to the effluent flow. The mathematical relationship between the output (pumping rate of hypochlorite solution) in response to an input (effluent flow) signal is programmed into the PLC, such that the controller maintains the dosing rate in proportion to the flow. The setpoint for the controller is the proportion of flow between the input and the output, which it tries to maintain.

3.11.4   Residual Control

One of the criteria for successful design of residual-control systems is to minimize lag time. Lag time consists of the following four primary components:

- Time required for the flow to pass from the injection point to the sampling point,
- Time required for the flow to pass from the sample point to the analyzer (including the speed of response of the analyzer),
- Time required for the chlorine gas or chlorine solution to reach the diffuser, and
- Response speed of the control valve in the chlorinator.

The first component of lag time depends on the flowrate and distance between the injection point and sample point. To limit lag time, this distance should be minimized. For optimal control, the sample point should be located at a distance that corresponds to an approximately 90-second travel time at maximum flowrate from the injection point.

The distance from the sample point to the analyzer and speed of response of the analyzer can be minimized also, by locating the analyzer as close as possible to the contact chamber and sample point. The use of a dedicated analyzer produces a consistent, rapid reading, because recalibration is unnecessary, and sample-line cleaning is not required. The sample line should include the capability of periodic cleaning with high concentrations of chlorine, to remove any buildup of algae and slime. Also, the sample line should be sized to maintain a sample velocity of approximately 3 m/s (10 ft/s) to minimize lag time and provide for plug flow.

When the injector and control valve are located at the diffuser site, any change in chlorine feed called for by the control signal is sensed rapidly, because the distance to the point of addition is minimized. The speed of response of the control valve in the chlorinator is relatively insignificant compared with the other three factors.

This discussion does not suggest that the chlorination reaction takes 90 seconds or that contact for a minimum of 30 minutes is unnecessary. Rather, the implication is that, after 90 seconds, chlorination reactions are sufficiently complete to be measured for control purposes, and, if the control system is functioning properly, the chlorine residual after 30 minutes of contact may be obtained by manual sampling and analysis.

If manual sampling is insufficient to provide an adequate safeguard against improper disinfection, an additional dedicated residual analyzer can be installed at a sample point at least 30 minutes detention downstream of the injection point. The use of dedicated analyzers allows continuous measurement and control, and the additional cost is negligible compared with the benefit derived.

With a typical feedback-control loop when using a hypochlorinator for the residual chlorine concentration, the analyzer probe and signal transmitter (Analysis Indicator Transmitter, AIT) transmits a signal to a PLC. The controller will send a signal to pace the dosing pump, to adjust the dosing rate, so that the desired residual chlorine concentration, which is the set point of the controller, is maintained.

3.11.5   Compound-Loop Control

Because chlorine demand is not exactly proportional to flowrate and because it is often impossible to regulate flowrate to the extent that a manual- or residual-control system becomes practical, many WRRFs are equipped with a chlorination system that combines the advantage of gross regulation of chlorine feed using flow proportioning and that of residual control (Figure 17.8).

A compound-loop system consists of two interlocking control loops—the flow loop and the residual loop. These loops can be interlocked in one of several ways. One way is to have the signal from a residual controller modulate the dosage adjustment of the chlorinator. Alternatively, signals from the analyzer and flow meter can be sent to a multiplier, which then sends a composite signal corresponding to combined measured feedback to the chlorinator. The chlorinator then injects the appropriate amount of chlorine based on this mass signal.

An advantage of the compound-loop controller is that it can introduce damping to the system, enabling the system to avoid uncontrollable oscillation that results from excess lag time. The multiplier type of compound-loop controller has the advantage of being able to react quickly to rapid changes in residual (and, therefore, demand). In WRRFs, a compound-loop system not only provides more accurate control, but, equally important, it can save costs, by minimizing the overfeeding of chlorine by operators and reducing the amount of chlorine and chlorine byproducts in the effluent. Figure 17.8

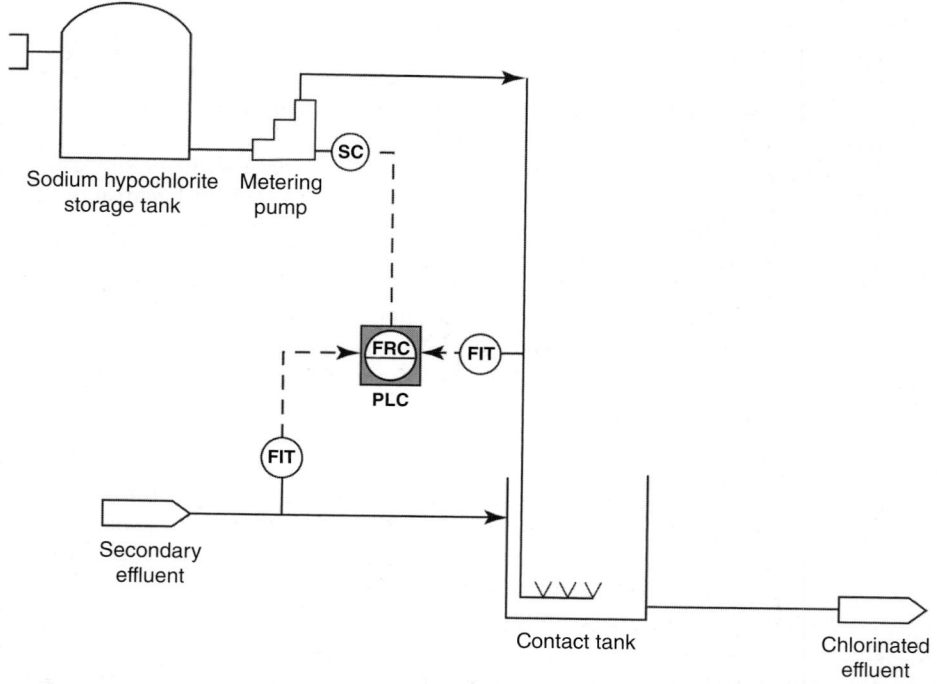

FIGURE 17.8   Schematic process and instrumentation diagram for compound control of sodium hypochlorite feed.

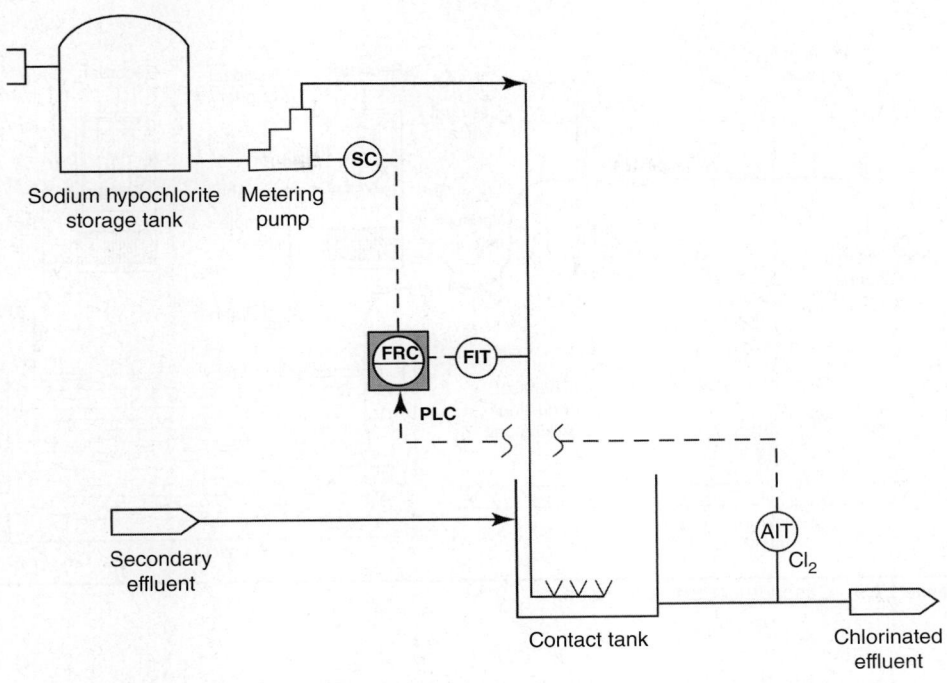

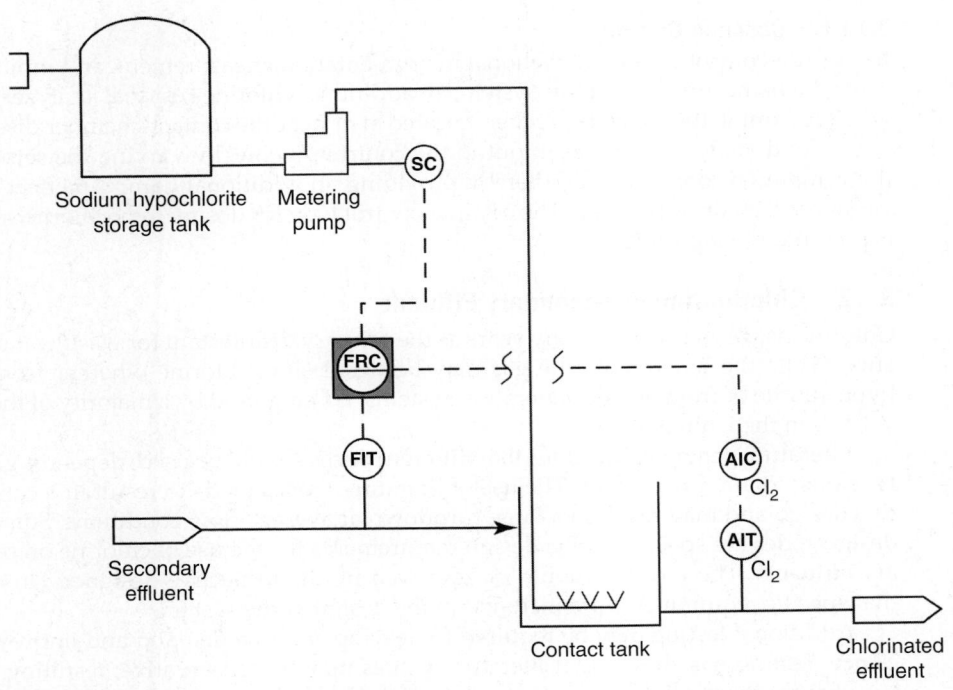

FIGURE 17.8 (*Continued*)

1421

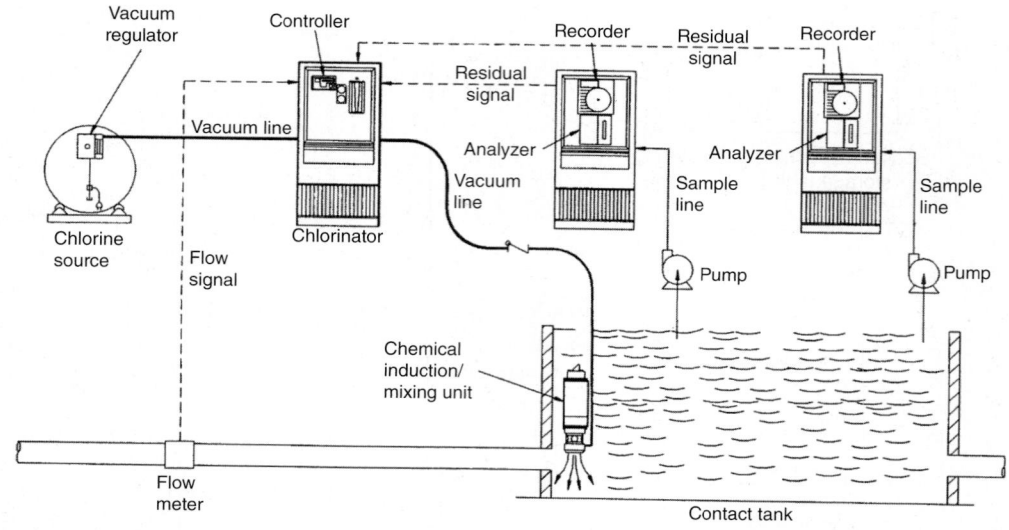

FIGURE 17.9   Cascade control schematic.

represents the compound-control loop for hypochlorite-solution dosing, based on the effluent flow and residual chlorine concentrations.

3.11.6   Cascade Control

In cascade-control systems, additional instrumentation, measurement, and input to the control scheme are used. In this system, an additional chlorine-residual analyzer is provided to sample the contact chamber. Located at or near the contact-chamber discharge, this second analyzer provides input to the control scheme, by varying the setpoint of the compound-loop analyzer, thereby providing an additional damping effect on the variation of chlorine residual. The cascade-control loop for dosing hypochlorite solution is presented in Figure 17.9.

3.12   Chlorination of Secondary Effluent

Chlorine has been used for many years as the primary disinfectant for wastewater effluents. While the use of chlorine gas may be decreasing, chlorine whether from bulk hypochlorite or from on-site generation systems, is being used by a majority of the large WRRFs in the United States.

Literature generally indicates that different doses should be used, depending on the type of secondary treatment. The use of literature values tends to result in a conservative design and may result in a large turndown at average flow conditions. Some state design codes are specific on the design requirements for the use of chlorine on secondary effluents. These codes should be reviewed by the respective designers, to ensure that specific requirements are included in the design of the system.

Additional testing may be required to develop chlorine demand and improve efficiency. Testing has shown that literature values may be conservative, resulting in the use of more chemical than may be needed. These bench-scale tests can assist operations staff in optimizing the chemical-addition system.

3.13 Chlorination of Reclaimed Water

Title 22 of the California Code of Regulations (Title 22) governs recycled water treatment in California and provides a model for recycled water regulation nationwide. Title 22 requires that recycled water disinfection practices result in a minimum of a 5-log (99.999%) reduction in viruses, with poliovirus being the standard. Title 22 requirements are based on a chlorination CT (combination of chlorine residual and modal contact time) of 450 mg·min/L providing 4 logs of poliovirus kill without media filtration and 5 logs of poliovirus kill with media filtration. Thus, Title 22 requires the following:

- Approved media filtration of clarified secondary effluent followed by chlorination (free or combined chlorine) at a CT of 450 mg·min/L; or

- An alternative filtration/disinfection process that results in 5-log inactivation of MS2 (a non-pathogenic indicator virus used in seeding studies) or poliovirus.

3.14 Factors Influencing Chlorination Efficacy

In general, the more efficiently a WRRF is operated, the easier it will be to disinfect the effluent. Any failure to provide adequate treatment will increase the level of pathogens and the chlorine requirements. High solids content increases the chlorine requirement, as does the soluble organic load. For example, care should be exercised in returning digester supernatant liquor to the primary tank. Such supernatant will increase the chlorine demand should it reach the chlorine contact chamber.

Increases in the proportion of industrial waste in the influent will often increase the amount of chlorine required for adequate disinfection. Cyanide in plating waste is a particularly troublesome constituent. In general, when fluctuating percentages of industrial waste are contained in the influent, difficulty may be anticipated in maintaining a chlorine feed that will ensure adherence to a specified bacterial standard.

Approximate chlorine requirements for disinfecting normal domestic wastewater are listed in Tables 17.9 and 17.10. The roles of mixing, contacting, and control strategies in maximizing the effectiveness of chlorination are important factors to consider.

Treatment	Removal Expected (%)	Virus Concentration in Effluent (no./L)*
Primary sedimentation (without chemicals)	0	7000
Secondary treatment		
Trickling filters	50	3500
Activated sludge	90	700
Physical/chemical treatment		
Precipitation of phosphate and suspended solids	90	
Activated carbon adsorption	10	630

*Assumes a virus concentration of 7000/L in raw wastewater.

TABLE 17.9 Expected Virus Concentration in Effluent

Treatment	Chlorine Design Dose (mg/L)	
	Great Lakes (2014)	Black & Veatch Corporation (2010)
Prechlorination (mechanically cleaned tanks)	20–25	—
Raw wastewater, depending on age	—	8–15
		15–30
Primary effluent	20	8–15
Trickling filter effluent	15	3–10
Activated-sludge effluent	8	2–8
Sand filter effluent	6	1–5

TABLE **17.10**   Chlorine Design Requirements to Disinfect Normal Domestic Wastewater as Listed in Various References

Because the flow of effluent in stabilization ponds or pond treatment systems is placid and often channeled, these treatment processes can produce an effluent that creates serious problems in the disinfection process. A few issues regarding challenges of chlorinating effluent from pond treatment systems are discussed in a later section. In areas where ice cover is present, effluent may have an excessively high chlorine demand because of stagnant conditions.

Gong (2002) conducted experiments on effluents from four different WRRFs. Gong observed that WRRF effluents tended to become less stable after disinfection processes than undisinfected samples, in terms of regrowth potential, when fecal coliforms were used as the indicator. In other words, the regrowth potential of fecal coiforms is much higher for disinfected WRRF effluents than undisinfected ones. However, fecal coliform concentrations in disinfected samples, even after regrowth, generally were lower than that for undisinfected samples. Gong also observed that the addition of substrate did not increase regrowth potential significantly. The study covered both chlorination/dechlorination and UV disinfection processes.

Under certain process conditions for nitrification, the second step of conversion of nitrite to nitrate can become much slower than the first, leading to an accumulation of nitrite ions, leading to "nitrite lock". This can happen under various biochemical conditions affecting the activity and growth of the nitrite-oxidizing bacteria. Oxidation of nitrite to nitrate is a thermodynamically favored reaction, and, in the absence of an enzymatic pathway, nitrite ions seek an alternate route to convert to nitrate. Chlorine added for disinfection to a treated wastewater containing large amounts of nitrite ions will oxidize nitrite and become partially consumed by the nitrite oxidation reaction. Stoichiometrically, one nitrite ion consumes one hypochlorite ion (produced by one molecule of chlorine) to complete the oxidation reaction. Approximately 14 g NO_2^--N consumes 71 g chlorine. Therefore, when several milligrams per liter of nitrite-nitrogen are found in the effluent, large dosages of chlorine are required before any residual is obtained.

Pond treatment systems can present complex chlorination issues, as a result of the ever-changing environment present within the treatment process. These issues typically include the following:

• Changes in pH. Pond treatment system pH will vary seasonally, daily, and even hourly, as a result of biochemical oxygen demand (BOD) removal, algae growth, and nitrification. Issues with changes in pH most commonly are the result of algae. Algae in large numbers will result in diurnal variations in the distribution of inorganic carbon species that provide pH buffering. Daily maximum pH values of 9.5 or greater are common in pond treatment systems during algae blooms. This affects disinfection, because a high pH results in the hypochlorite ion being the predominant form present and is less effective as a disinfecting agent.

- Incomplete nitrification/denitrification. The rate of oxidation of nitrite to nitrate decreases as alkalinity is depleted or as temperature decreases. As nitrite accumulates and places a high demand on chlorine, maintaining chlorine residual can become difficult.

- Pond-treatment-system environments can produce areas of low or no oxygen at different depths and times of the day and can lead to nitrite accumulation. In these areas, even at temperatures above 17°C, any nitrates will be reduced to nitrites and subsequently to nitrogen (N_2). The second step is the slowest, especially if the carbon source is limiting growth and can lead to rapid nitrite accumulation.

- High effluent TSS resulting from algae growth. Pond treatment systems have an abundance of nitrogen and phosphorus; when this is combined with long hydraulic retention time (HRTs), an optimal environment for algae growth occurs. Typically, alkalinity (inorganic carbon) is the only nutrient likely to be limiting for algae growth. High effluent TSS, as a result of algae blooms, may interfere with disinfection.

Rapid mixing at the point of chlorine injection is very important and can improve disinfection effectiveness. Mixing is also critical for the formation of chloramines when ammonia is being added. The chlorine contact tank should be tested for obvious signs of short-circuiting, and tracer testing may be done to determine if adequate contact time is provided. The U.S. EPA *Municipal Wastewater Disinfection Design Manual* (U.S. EPA, 1986a) provides some insight to analyzing residence time and distribution curves.

Algae levels in pond treatment systems affect the effectiveness of the chlorine-disinfection process. Investigations to diurnal variations in the algae concentrations (and/or BOD or COD or TSS concentrations) at various depths within the pond treatment system may prove beneficial. Cleaning of sludge accumulation in the pond treatment system bottom should help to reduce nutrients that are being released as a source of food for the algae. Periodically monitoring the pond treatment system for nutrients beneficial to algae growth would be useful to determine cycles associated with sludge accumulation and seasonal changes within the pond treatment system. Other control mechanisms, such as mixing, may also warrant investigation as long-term solutions.

Depending on nitrite levels, alkalinity addition may be used seasonally, to prevent nitrite lock. Effluent ammonia, nitrite, and nitrate should be tested on a regular basis to determine the cycle of nitrification/denitrification.

4.0   Dechlorination

Much attention has been focused on the toxic effects of chlorinated effluent. Both free chlorine and chloramine residuals are toxic to fish and other aquatic organisms, even at concentrations less than 0.02 mg/L. Although fish are repelled by low levels of chlorine and frequently escape harm, other aquatic organisms in the food chain may be killed by chlorine discharges. Dechlorination—the removal of remaining chlorine—is required in most states. Stream standards have been established, in most parts of the country, which limit TRC. However, carcinogenic chlorination byproducts will not be reduced by dechlorination.

4.1 Dechlorination Reactions and Kinetics

Free and combined chlorine residuals can be effectively reduced by sulfur dioxide and sulfite salts. The sulfite ion reacts rapidly with free and combined chlorine. The sulfite ion is the active agent when sulfur dioxide or sulfite salts are dissolved in water. Their dechlorination reactions are identical. Sulfite reacts instantaneously with free and combined chlorine, yield small amounts of acidity, which is neutralized by the alkalinity of the wastewater alkalinity as calcium carbonate is consumed per milligram chlorine reduced). From the above equations, the amount of sulfur dioxide required per part chlorine is 0.9, but typical actual practice calls for the use of a 1:1 ratio.

It should be noted that previous research (Helz and Nweke, 1995) has shown that complete dechlorination may not be achieved or may be delayed in the presence of monochloramine.

Granular and powdered carbon may be used to dechlorinate free, and some combined, chlorine residuals. Carbon requirements for dechlorination typically are determined by on-site pilot testing. Parameters of significance include mean particle diameter of the carbon (pressure drop within a contactor) and influent quality (pH, organics, and colloids). For typical municipal effluent, costs are high. Doses in the range 30 to 40 mg/L have been reported.

4.2 Sulfur Dioxide

Sulfur dioxide is more soluble than chlorine in water. Sodium sulfite and bisulfite, typically provided in solution form, are reducing agents like sulfur dioxide. Detailed information on safe handling procedures for all chemicals is available from various chemical suppliers. Training of all personnel in the handling and use of sulfur dioxide is available from the Compressed Gas Association (1988) and individual manufacturers.

4.3 Shipment and Handling Safety

Sulfur dioxide gas is shipped in containers that are similar to those used for chlorine, and similar safety precautions need to be followed for handling of the containers as stated above for chlorine gas. Similarly, the handling and storage facility for these containers should be designed with the same requirements as stated above for the chlorine containers.

The sulfur dioxide cylinders should be located in separate rooms and stored in a well-ventilated, temperature-controlled area, so that their temperature is maintained between 18 and 37°C. Gas-leak detectors are necessary in the storage area and the sulfonator area. An emergency eyewash shower and self-contained breathing apparatus should also be provided. All personnel should receive emergency-response training. Facilities with more than 454 kg (1000 lbs) of sulfur dioxide stored on-site must abide by the Process Management Safety Standard in OSHA regulations (OSHA, 1996).

4.4   Design and Selection of Equipment

Sulfur dioxide gas feeders are referred to as *sulfonators*. The four basic components of the system are discussed in *Recommended Standards for Wastewater Facilities* (Great Lakes–Upper Mississippi River Board of State and Provincial Public Health and Environmental Managers, 2014), and The Chlorine Institute.

4.5   Dechlorination Control

In general, many of the same principles identified for chlorination systems also apply to dechlorination systems. The continuous measurement of sulfite residuals can be accomplished directly or indirectly. The commonly accepted practice is to use a chlorine residual analyzer and shift the zero point, by adding a known amount of oxidant (chlorine). This enables a residual of chlorine or sulfur dioxide to be determined and used in the control scheme. Alternately, the Renton system, which is a variation of the zero-shifted analyzer, has been used successfully in some areas (Finger et al., 1985). Finally, proprietary analyzers are available that use an iodine bias, making it possible to measure very low residuals continuously.

Two types of control systems for sulfonation are often used. In WRRFs that are not required to completely dechlorinate their effluent, it is possible to use a feedback control system (Figure 17.10), whereby the analyzer measures the chlorine residual a short

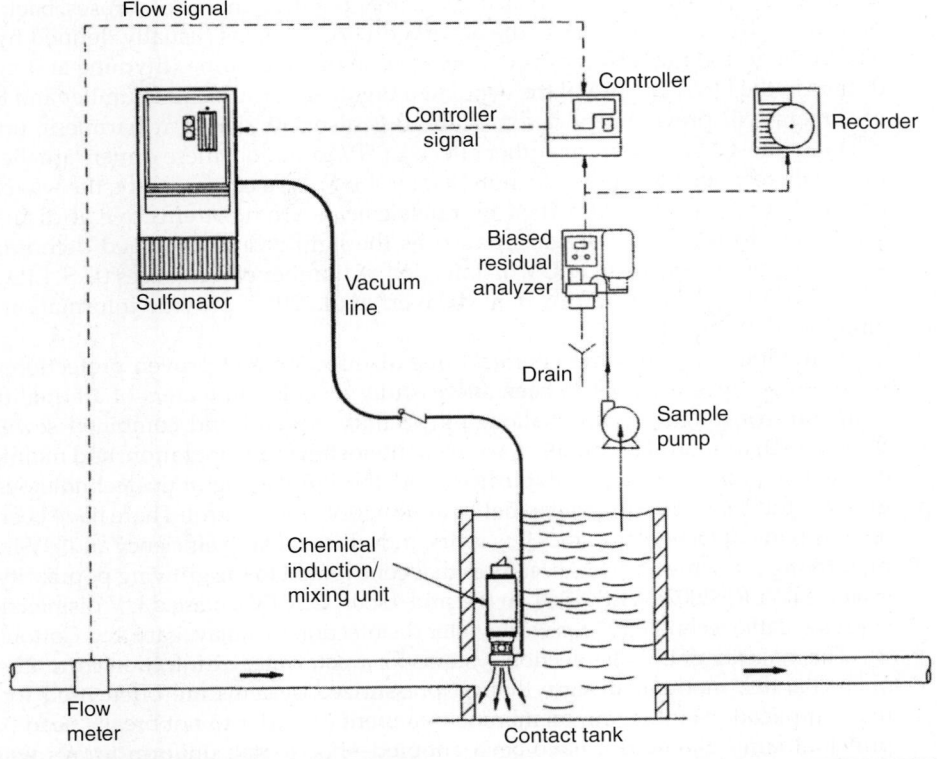

FIGURE 17.10   Dechlorination feedback schematic.

time after the injection and mixing of the sulfur dioxide. Lag time between the injection point and sample point is minimal, because the dechlorination reaction is almost instantaneous. The setpoint signal is used as an inverse controller; as the chlorine residual increases, the feed rate of sulfur dioxide also increases. Figure 17.10 shows an example of a dechlorination system schematic.

In those facilities that must completely dechlorinate their effluent and do not have a biased or direct-reading analyzer, the feedback control system may not be practical. Facilities of this type can use a feed-forward system, with a multiplier to send a mass-flow signal to the sulfonator, based on the signal from the analyzer located at the end of the contact chamber. A modification of the standard feed-forward design, in which a gas-flow transmitter is installed in the vacuum line of the sulfonator, can also be used. The transmitter measures the flow of sulfur dioxide through the feeder and transmits this signal to a ratio controller. The controller compares the multiplied signal, in kilograms per hour (pounds per day) of chlorine in the water, with the measured feed rate of sulfur dioxide, and provides a control signal output that is proportional to the ratio required between the two values for proper dechlorination.

5.0   UV Disinfection

5.1   General UV Description

Ultraviolet disinfection acts primarily by damaging the genome of viruses, bacteria and parasites. The primary effect of the absorbed UVC photons (usually defined by wavelength range 100 nm–280 nm) is the production of pyrimidine (thymine and cytosine) dimers in the DNA or RNA of the irradiated organism. A sufficient number and location of dimers will prevent replication of the organism, rendering it harmless, unable to infect or infest. While there are other effects of UV radiation, these dimers are the factors primarily responsible for disinfection (Jagger, 1967). As a consequence, the wavelengths of UV that are strongly absorbed by nucleic acids are most effective at disinfection. The extent to which the organism absorbs the light and is damaged determines the sensitivity of the organism to UV disinfection. A number of references (U.S. EPA, 1986a; WEF, 2015; IUVA News, 2016; Haji Malayeri et al., 2017) provide information on the kinetics of UV disinfection.

Ultraviolet systems offer chemical-free disinfection and proven protection against waterborne illnesses. UV has been successfully installed in waters of all qualities and temperatures ranging from challenging primary effluent and combined sewer overflows (CSO) to secondary and tertiary effluent. Installation, operation, and maintenance of the UV systems is well-established, and the latest generation technologies bring greater efficiency, less lamps, and better automation and control. There have been many advances in UV technology since it was first introduced. Lamp efficiency and UV-intensity monitoring are some of the key features that contributed to the growing popularity of UV systems (WERF 2007; WEF 2015). In the mid-1980s, U.S. EPA named UV disinfection as a "best available technology" for wastewater disinfection (Jeyanayagam and Cotton, 2002).

The majority of UV disinfection systems for wastewater disinfection currently use an open-channel, modular design, though pressurized systems are often used for water reuse applications following membrane treatment (in order to not break "head"). Three principal lamp geometries have been adopted—horizontal, uniform arrays with flow directed parallel to lamp axes; vertical, staggered arrays with flow directed perpendicular to lamp axes; and inclined, staggered arrays with flow directed at an angle to lamp axes.

Literature varies on when ultraviolet disinfection was implemented for use in wastewater treatment in North America. Some authors indicate that the rapid growth of this type of disinfection technologies has occurred since the later 1970s to early 1980s (Henri et al., 1910).

A fundamental variable that strongly impacts the amount of UV equipment that needs to be installed in any application is the UV transmittance (UVT – units %/cm). This measure is an indication of how much light the water lets through—the higher the number, the further the UV light is able to penetrate. Waters with low UVT absorb more UV light and therefore require higher energy for a desired inactivation: hence, the capital and operating and maintenance costs are generally higher for lower UVT installations. For high UVT applications, the UV reactor efficiency can be improved by increasing the lamp spacing, which would result in lower headloss across the UV system. Such system design changes can only be made by or in consultation with system suppliers who fully understand the operational range (UVT, headloss, disinfection target dose/fluence) of their system designs. Arbitrary changing of lamp spacing is not recommended.

UV system installations can sometimes be optimized by improving upstream water quality (such as UVT, TSS) through improved biological treatment performance [e.g., higher solids retention times (SRTs), nutrient removal, improved settling for more TSS reduction], secondary process optimization, and/or tertiary filtration (media and membrane), all leading to increased UVT of the water (WERF, 2007). The UVT of the wastewater to be disinfected is one of the most important parameters in determining the cost of UV disinfection, with reduced UVT having an exponential (not linear) impact on equipment and power to maintain a target dose. Therefore, frequent (ideally online) monitoring of UVT is critical to optimize UV system operation, and maintaining higher effluent quality is important to maintain efficient UV operation. However, the assumption of a conservative UVT can be made to allow systems to run very safely even without constant on-line UVT monitoring, where concerns around energy optimization are not paramount, such as in small systems.

5.2 UV Lamps

Practical UV disinfection requires an efficient and low-cost source of UVC photons. The mercury-vapor lamp is typically used for disinfection, since these lamps are very efficient and the physics and manufacturing were well developed in the 1960's for fluorescent lamps which rely on the same basic process, the ionization of mercury. (Fluorescent lamps use a fluorescent coating which is absent in disinfection lamps.) There are three main types of mercury lamp used in disinfection applications: low-pressure, amalgam, and medium-pressure lamps. These lamps are used in both open channel and pressurized (closed vessel) applications. The lamps are enclosed within UV-transparent quartz sleeves to protect the lamps from exposure to water and to control lamp temperature.

Mercury vapor lamps require power supplies that are suited to the unique characteristics of the mercury vapor gas discharge. These power supplies are sometimes referred to as ballasts, and older designs rely primarily on large inductors and capacitors. Modern electronic power supply designs incorporate sophisticated and efficient solid-state circuit designs able to precisely control lamp power.

5.2.1   Low-Pressure Lamps

The low-pressure, low-intensity, mercury-arc lamp principle is used in germicidal and standard fluorescent-lighting lamps. Both produce UV radiation by means of an electric discharge through a mixture of mercury vapor and inert gases at subatmospheric pressure (0.007 mm Hg [torr]). The primary emission from the ionized mercury vapor occurs at 253.7nm, which is very close to the maximal UVC absorption of DNA at 260 nm.

In a low-pressure mercury vapor lamp, the mercury vapor is at equilibrium with a small liquid-mercury reservoir in the lamp. Lamp efficiency is a function of the vapor pressure of the mercury, which is in turn a function of the "cold-spot" temperature. For optimum lamp efficiency, the cold spot should be maintained near 40°C. Lamp efficiency will decrease at lower or higher temperatures. The relatively small temperature difference between the lamp and the surrounding water (acting as the heat sink) mean that the lamp output can be affected by changes in water temperature. Wall temperature is a function of the quartz sleeve diameter (i.e., the thickness of the air gap between the quartz sleeve wall and the lamp wall), water temperature, and power driving the lamp. UV system manufacturers carefully balance these factors during system design.

Two standard lamp lengths are typically used in conventional disinfection systems—0.9 m (0.7-m arc length) [36 in. (30-in. arc length)] and 1.6 m (1.5-m arc length) [64 in. (58-in. arc length)]. This lamp configuration is used in pressurized closed-vessel systems and in open-channel systems. While the low-pressure lamp is efficient at producing effective germicidal radiation, its output intensity is relatively low.

5.2.2   Amalgam Lamps

Amalgam lamps decrease and control the vapor pressure of the mercury by combining it with other metals to create a solid amalgam. As a result, the mercury vapor can be maintained near its optimum value over a broader range of higher temperatures. This allows these lamps to be operated at higher power, resulting in increased UV output for a given lamp size, while maintaining the nearly monochromatic output spectrum of the low-pressure lamp. Accordingly, amalgam lamps (also known as "low-pressure, high-output") have become increasingly popular. The amalgam lamps generally operate at higher electrical current, higher temperature, and higher total internal pressure than low-pressure lamps. The actual operating pressure is as much as 40% higher than that of its low-pressure counterpart. Operating temperatures for high-intensity lamps are in the 180 to 200°C range. The high-intensity lamp is driven by currents as high as 5 A, which is 10 to 15 times higher than those of conventional low-pressure, low-intensity lamps. Amalgam lamps are used in enclosed pressurized reactors, and in open-channel reactors in vertical, horizontal, or inclined orientations. Figures 17.11 through 17.13 provide exampled pictures of low-pressure high output UV alignments.

5.2.3   Medium-Pressure Lamps

Medium-pressure (MP) lamps do not use amalgams, but operate at higher power density and temperature, so that the mercury is completely vaporized and many additional energy levels and emission transitions are populated. As a result, these lamps have a complex polychromatic broad emission of wavelengths from vacuum UV through to the infrared region. The UV emission power density is higher than for other mercury lamp types, but the efficiency of conversion from electrical power to germicidal output (output capable of inactivating microbes) is lower than that of low-pressure or amalgam lamps.

Figure 17.11 Tubular alignment low-pressure, high-output system.

Figure 17.12 Horizontal alignment low-pressure, high-output system.

Figure 17.13 Inline alignment low-pressure, high-output system.

The medium-pressure lamp operates in the 10^2- to 10^4-mm Hg (torr) range, which is at or near atmospheric pressure. Lamp operating temperatures range from 600 to 800°C, which is much higher than the standard operating temperature range of 40 to 60°C for low-pressure, low-intensity lamps. Since the temperature difference between the MP lamp temperature and the water temperature is so great, changes in the water temperature do not significantly affect the lamp temperature or output.

The arc length of an MP lamp can range from less than 1 inch to more than 48 inches, with proportional changes in total lamp power. Medium-pressure lamps have a rated life of 4000 hours or more, although experience has shown an expected life exceeding 8000 hours. Actual lamp life depends on lamp operating power. A higher operating power results in higher lamp temperatures and typically lower lamp life. Medium-pressure lamps are generally more expensive than low-pressure or amalgam lamps, but this is offset by their greater power, which requires fewer lamps for a given output.

Medium-pressure systems typically include automatic in-place cleaning systems (wipers). One system incorporates mechanical and chemical cleaning in a single unit. It operates while the system is in operation, without affecting disinfection performance. This is accomplished by a wiper mechanism that circulates cleaning solution under pressure within the wiper as it moves along the lamp length. Medium-pressure lamps are available in either closed vessel or open channel systems. Figures 17.14 and 17.15 provide example medium-pressure UV systems.

5.2.4 Alternative Light Emission Technologies

A number of alternative light emission technologies such as pulsed power and light emitting diodes (LED) exist in the market, though have yet to be installed in large municipal applications. These technologies need to be evaluated (and validated) on a case-by-case basis for application in the wastewater industry. In general, these technologies have lower electrical-to-germicidal conversion efficiency and higher cost than mercury-based lamps with the same germicidal output.

5.3 UV-System-Reactor Design

5.3.1 System Hydraulics

The hydraulic behavior of a UV-disinfection system is a critical element to ensure adequate UV-disinfection performance. The hydraulic behavior influences the retention

FIGURE 17.14   Open-channel, medium-pressure UV system.

FIGURE 17.15 Closed-vessel, medium-pressure UV system.

time of the various fluid elements passing through the complex UV-intensity profile within the system and, thus, directly affects UV dose. Inadequate hydraulic conditions (especially those caused by incorrect installation of the UV system in the field, e.g., lack of attention to the manufacturer's installation specifications, etc.) are the most common cause of UV-disinfection-system failures.

Important hydraulic influences include longitudinal and radial mixing and turbulence, the inlet structure, outlet structure, and headloss through the system. Each will be addressed in the sections that follow.

5.3.1.1 Longitudinal Dispersion, Axial Dispersion, and Turbulence One inherent characteristic of UV-disinfection systems, due to the absorption of light by water, is a non-uniform internal UV-intensity profile. Longitudinal and axial dispersion and turbulence (together referred to as *mixing*, herein) affects the retention time of the various fluid elements in the associated light field. Only laboratory-scale UV-disinfection apparatus (e.g., collimated beam) approximate the application of a single and consistent UV dose among all fluid elements that can be ensured to be free of extraneous hydraulic influences. Municipal-scale systems are affected by longitudinal and axial mixing. A resulting distribution of UV doses exists, and any attempt at assigning a single UV dose for a specific hydraulic condition is only for regulatory, operational, or design convenience.

Manufacturers use many tools, including bioassays, optical models and CFD models, in their efforts to optimize the longitudinal and axial mixing conditions associated with their systems before their installation to maximize disinfection performance. In fact, pilot testing with scale-up limitations before full-scale installation is required for producing recycled water suitable for unrestricted reuse, per guidelines published by the NWRI (NWRI, 2012) and is recommended for any UV disinfection project (*Ultraviolet Disinfection for Wastewater*, WEF, 2015). An UV-disinfection system should not be designed/specified, regardless of the intended disinfection objective, if that system has not undergone some type of performance-validation process, unless performance testing is part of the installation process. This validation process can be formal (e.g., per the NWRI guidelines) or informal (e.g., reference to other similar operating facilities or pilot-testing on-site). However, the goal of performance testing is largely an attempt to ensure proper accounting of the longitudinal and axial mixing that will be present in full-scale design, in order to ensure proper system dose distributions.

It may be intuitive to assume that a reduction in flowrate through a given UV-disinfection system will result in an increase in the applied UV dose, as a result of an increased residence time within the UV-disinfection system. However, there have been numerous examples whereby UV-disinfection performance efficiency deteriorates with a reduction in flowrate through the system, because the reduction in flowrate also decrease radial mixing, leaving many fluid elements limited to flow through low-UV-intensity microenvironments. Using manufacturer-supplied design curves, which were developed through third-party empirical testing of UV reactor performance (also known as bioassay validation), is the best means, to date, of ensuring adequate accounting of longitudinal and radial mixing (and hence adequate doses and reactor headloss) in the UV-disinfection-system design.

When designing a UV-disinfection facility, it is important to ensure that the wide range of hydraulic conditions expected during operation is evaluated during design. Design for peak flows alone is insufficient, as, sometimes, low flows can significantly affect disinfection performance, as described above. For example, in addition to evaluating peak-hour as well as low hour flows at facility build-out, the velocities and corresponding doses should be evaluated that result during the diurnal low-flow period, which might occur during facility startup, when the facility is far from hydraulic design capacity. Also, intermediate flows should be evaluated. The number of channels and the operating schedule for those channels should be selected to ensure operation at a velocity that has been tested in some manner and is known to provide adequate UV-dose delivery. It is possible to recirculate flow from the effluent of the UV-disinfection system to the head end of the UV-disinfection system during extreme low-flow events, to ensure maintenance of acceptable velocity conditions, if necessary.

Finally, the designer should account for hydraulic effects from the upstream treatment process on the UV-disinfection system. Equalization basins/ponds can be advantageous, because they minimize hydraulic fluctuations, leading to more stable disinfection performance. Conversely, some filters can pulse to maintain filtration effectiveness. If these very high pulsed flows are outside of the operating range of the UV systems, then they may impair UV-disinfection performance. If present, pulses resulting from backwashing or filter-bed maintenance may need to be dampened before their introduction to the UV-disinfection system.

5.3.1.2   Inlet Structure   Inlet-flow conditioning can be achieved via the use of stilling plates and submerged dams. These structures impose a controlled energy loss on the system influent and are effective in achieving an even distribution of momentum throughout all channels. By positioning inlet structures far enough upstream of the zone exposed to UV, flow irregularities induced by the inlet structure are given ample time for dissipation, thereby allowing a uniform velocity profile to be incident on the first bank of UV lamps. A perforated stilling plate (or similarly engineered flow modifier) can be installed, if sufficient head is available. The flow modifiers must be deliberately engineered and installed properly as shown in Figure 17.16. Improper stilling plates may not work, and can even make things worse, depending on the orifice distribution and local Reynolds number. When designed properly, flow modifiers can distribute the flow and equalize the velocities across the cross-section of the channel. Flow modifiers are often recommended to be placed at least 1.5 m (5 ft) in front of the first bank of lamps in open channel systems. In closed systems, flow modifiers can be an integral part of the UV system itself. The industry has demonstrated the value of CFD to accurately

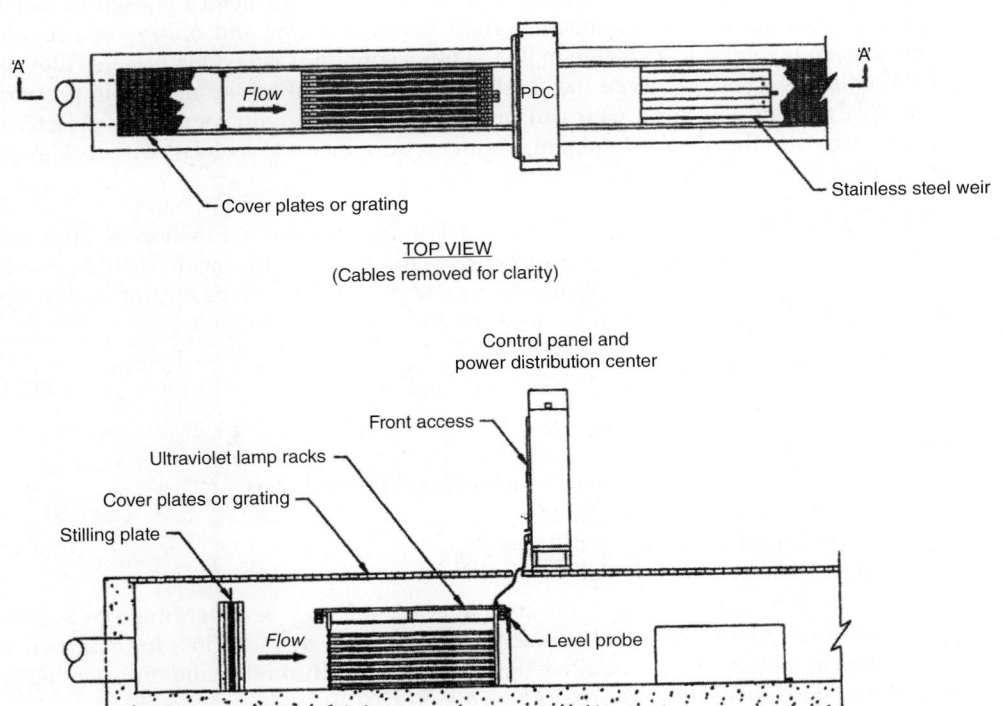

FIGURE 17.16   Schematic illustration of UV disinfection system with stilling plate for flow conditioning and elongated weir for level control.

model flow split and hydraulic approach conditions, allowing for greater confidence in system performance in advance of operation.

5.3.1.3 Outlet Structure A similar logic applies to outlet structures; flow patterns leaving the irradiated zone should be uniform if the irradiated zone has a relatively uniform cross-sectional dose (fluence rate) distribution. Outlet structures must also allow liquid-level control over the range of expected flow conditions. Several alternatives have been used to achieve these performance objectives, including elongated weirs and flap gates (Figure 17.17) or automatically modulated overflow weir gates. Flap gates are typically used on larger systems where an elongated weir cannot be used. Elongated weirs have the advantage of no mechanical components and are also potentially advantageous in systems with low overnight flows, because they are less likely to allow channel draining than flap gate systems.

As discussed above, the placement of inlet and outlet structures relative to lamp arrays is critical to achieving uniform flow leading to uniform dose distribution and high photon utilization efficiency, for systems designed with the assumption of uniform influent velocity profile. Measurements of velocity profiles in full-scale systems (Blatchley et al., 1995) suggest that a minimum of approximately 2 m (6 ft) should be allowed between inlet/outlet structures and the closest lamp array. Lamp arrays placed within these inlet/outlet zones may be used suboptimally because of induced abnormalities in the flow structure.

It is critical that, in multi-channel systems, a water-tight device is used to isolate channels when not in use. Channels that are brought online and offline on a regular basis (e.g., in response to a change in flows) often contain wastewater, because they are not drained when offline. Even the smallest amount of leakage around an improperly sealed valve/gate or over a weir can prevent compliance with regulatory objectives, particularly with the most stringent regulatory requirements associated with reclaimed-water-type systems (e.g., 2.2 TC/100 mL).

5.3.1.4 Headloss Energy (head) losses in UV systems are a function of approach velocity, and consistent with physics increases as the square of the incident fluid velocity. As in many fluid-mechanics applications, a general equation can be written to describe the functional relationship between headloss and velocity, as follows:

$$\frac{\Delta H}{L} = aV + b\rho V^2 \tag{17.1}$$

where ΔH = headloss (cm);
   L = channel distance over which ΔH is expressed (cm);
   V = approach velocity (cm/s);
   ρ = liquid density (g/cm^3); and
   a, b = empirical constants.

For a given system geometry, the constants a and b may be determined by experiment and are available from the manufacturer. Typically, the head loss for a particular UV system is based upon modeling that is calibrated through validation testing in accordance with specific standards such as NWRI (2012).

Headloss can have a significant effect, particularly in open-channel UV systems. Headloss is manifested as a drop in the water-free surface through the system. In open channels, while the headloss may be inconsequential compared with other losses in a

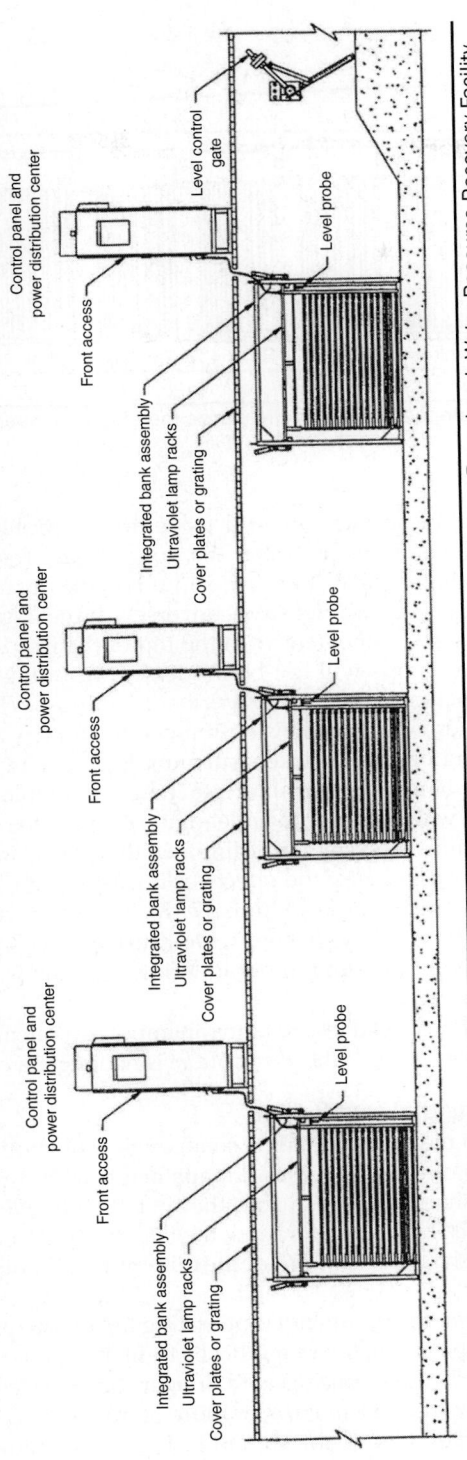

Figure 17.17 Schematic illustration of flap gate and submerged dam used as an outlet structure at Bonnybrook Water Resource Recovery Facility, Calgary, Alberta, Canada.

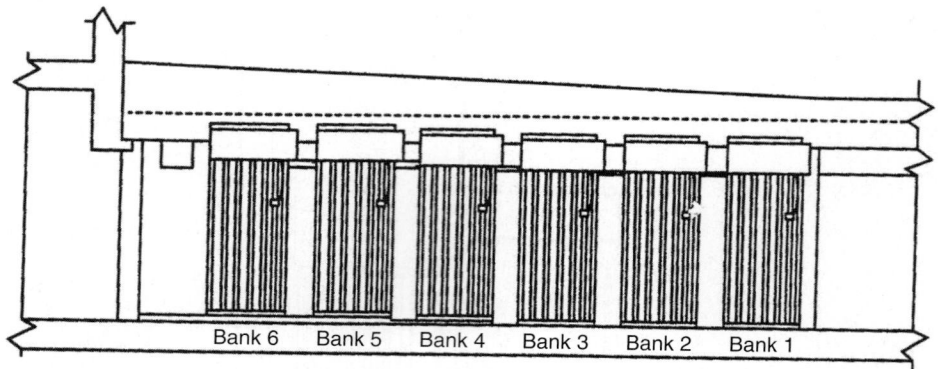

FIGURE 17.18 Use of a stepped channel to minimize the effects of headloss in an open-channel UV disinfection system.

WRRF, the drop in the free surface can induce operational problems in the disinfection process. If the liquid level is set such that the downstream free surface is coincident with the top of the irradiated zone, then some liquid on the upstream end of the system will pass through a region of low intensity. Conversely, if the liquid level is set such that the upstream free surface is coincident with the top of the irradiated zone, then a portion of the downstream lamps will not be immersed, resulting in loss of energy efficiency. With diurnal fluctuations in flow experienced at most WRRFs, this allows some lamps to experience alternate conditions of immersion and dryness, which can lead to increased rate of fouling of quartz jackets surrounding the lamps. Empirical observations with horizontal lamp systems indicate that acceptable performance can be achieved by setting the water level at the inlet of the channel to one-half of the distance of adjacent lamps within the system and limiting total head losses, such that lamps remain submerged. In some cases, the effects of head loss can be minimized by construction of a stepped channel (see Figure 17.18). Designers should use caution in adopting this practice for facilities where wide diurnal flow variations are expected because of the possibility of flooding under low-flow conditions, when head losses are relatively small.

For pressurized systems, head loss remains an important calculation and engineering issue, but the exposure of lamps to the atmosphere is no longer a concern.

5.3.2 Factors Affecting UV-Lamp Output

Ultraviolet output from mercury-arc lamps changes as a function of time. As the lamps are operated, the lamp output decreases. Lamps begin with a relatively high output power, followed by a sharp decline during the first 1000 to 2000 hours of operation. After approximately 2000 hours of operation, the lamp output gradually declines, until it reaches a point called the *end of lamp life*, or the minimum UV output that was designed for (see Figure 17.19).

The UV manufacturer's recommended operating life of low-pressure and amalgam mercury-arc lamps is typically in the range 10 000 to 14 000 hours; however, lamps have been operated effectively for considerably longer times. Medium-pressure, high-intensity lamps typically have 5000 hours or more of guaranteed lamp life. In practice, the lamp replacement interval decision should be based on a comparison of the lamps'

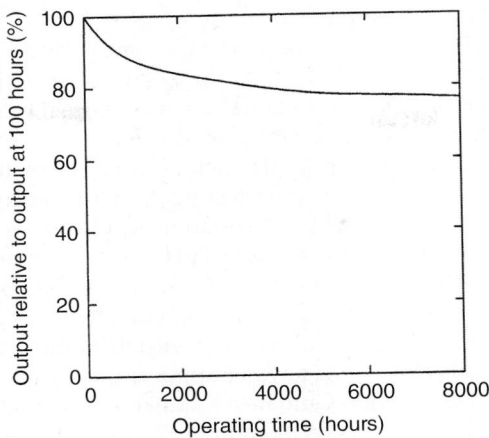

FIGURE 17.19 Typical UV lamp output as a function of time.

replacement costs and the added cost of operating the aged lamps with their lower efficiency. System output can be kept relatively uniform by implementing a schedule of staged lamp replacement. If performed in a logical and orderly manner, a system with staged lamp replacement can provide relatively consistent UV output.

For low-pressure, low-intensity lamps, manufacturers can provide lamps with internal coatings that reduce the rate of mercury absorption into the lamp wall. This advance has been found to significantly reduce the rate of decline in lamp output and increase the operating life. Some lamps do not include these coatings, which can result in more rapid decay of lamp output.

With some lamp drivers, the lamp life can be affected adversely by restarting the lamps more than 3 or 4 times per day. Badly controlled lamp startup can result in sputtering and premature failure of lamp filaments.

As mentioned in the section on UV lamps, the efficiency and output of lamps can be affected by water temperature, particularly for low-pressure mercury vapor lamps. Water temperatures outside the system manufacturer's design range can adversely affect lamp output.

Organic and inorganic fouling and scaling of the quartz sleeve can block and significantly reduce the UV energy that passes through the sleeve. The age and quality of the quartz sleeve may also affect the passage of UV energy through the sleeve, because aged or lesser quality sleeves can absorb UV light.

Electronic ballasts provide variable power output to modulate lamp output. By modulating output, lamps can be operated at less-than-maximum powers and at lower UV intensities, saving energy when possible. Lamp impedance can vary with manufacturing tolerance, with age and with operating temperature. If the power supply is designed to maintain constant current, the lamp power and output will vary with these factors and may fluctuate. Some electronic power supplies can measure and control lamp power, compensating for some of this effect. Electronic ballasts can optimize the lamp output over a broad range of operating conditions. Electronic ballasts have given designers the ability to better control the factors that affect lamp output, by designing for the expected range of liquid temperatures, lamp manufacturing tolerance, and lamp age.

5.3.3   UV-Reactor Validation

Techniques and standards for conducting UV system bioassays have been the subject of much recent research and debate. Literature (WEF, 2015) provides a good discussion of the issues surrounding various validation approaches, the use of MS2 bacteriophage and other challenge organisms, and provides guidance on UV transmittance, velocity profiling, design doses, lamp-aging factors, and sleeve-fouling factors collected or determined as part of the validation testing. Other protocols exist, including the Uniform Protocol for Wastewater UV Validation Applications from the International UV Association (IUVA), (IUVA News, July 2011). Several standards exist for testing UV systems for drinking water, but caution must be used in applying these to wastewater applications due to the higher UV transmittance in drinking water, and the general absence of particles in the treated water. Hence drinking water testing techniques may not be directly applicable in wastewater applications. Drinking water protocols include the U.S. EPA UV Disinfection Guidance Manual (UVDGM 2006), NWRI-WRF and ETV guidelines, the German DVGW standard (2006) and Austrian ÖNORM which use *Bacillus subtilis* (spores) as the test organism (Swift et al., 2002) (in distinction to U.S. approaches which use MS2 bacteriophages and other organisms such as T1, T7, Qβ, φX174, *aspergillus brasiliensis*, *B. pumilis*, *B. subtilis*, and others).

The joint effort between NWRI and WRF was an expansion and revision of guidelines once commonly referred to as *UV Guidelines for Disinfecting Reuse Waters*. The NWRI UV Guidelines (NWRI, 2012) have become, in some form, an unofficial "standard" in several states with active wastewater reuse initiatives, such as Hawaii, Florida, Arizona, Washington, and Texas, in addition to California. The NWRI 2012 Guidelines are not an American national standard *per se*, and only California strictly follows this revision; some states such as Hawaii operate systems with control equations generated according to the previous version of the guidelines (NWRI 2003 UV Guidelines).

U.S. EPA's ETV program was created to accelerate the development and commercialization of better environmental technologies through third-party verification and performance reporting using objective and quality-assured data. U.S. EPA's protocol was written to closely mimic the NWRI-WRF guidance protocol for verifying UV systems. The intent was to provide a common ground for manufacturers to verify their respective claims of conformance with the NWRI-WRF guidance under the umbrella of a credible verification organization and program. The NWRI-WRF 2012 UV Guidelines and EPA ETV protocols allow manufacturers to demonstrate that their equipment meets emerging industry standards for UV disinfection for effluent reuse. An important part of the protocols are bioassays.

Because the goal of UV disinfection in reuse applications typically is to inactivate 99.999% or more of the target pathogen(s), failure to provide an adequate UV dose to as little as 0.001% of the wastewater flow can result in effluent regulatory exceedances. Doses in UV reactors are often estimated by mathematical models (increasingly supplemented by CFD and optical modeling), but a bioassay or UV reactor validation is considered to be the most accurate means to establish the dose and is often used to develop data for calibrating and verifying UV reactor models.

A bioassay implicitly estimates the delivered UV dose in a reactor through the use of a technically defensible test procedure in which an appropriate indicator organism—typically a bacterial virus or bacteriophage (see list above)—is subjected to varying UV doses in the laboratory using a collimated-beam apparatus. The apparatus, combined with an accurate calibrated UV intensity meter and closely timed exposure allows for precise delivery of UV doses to a water sample. The end result is a defined biological response (e.g., log survival ratio) at each UV dose. The UV dose-response regression equation is then used to quantify the challenge organism-specific reduction equivalent dose (RED) using the log reduction of the organism (sensitivity to UV) as determined from reactor influent and effluent samples taken under a given test condition (flow rate, UVT, power setting, etc.).

According to the NWRI–WRF 2012 UV Guidelines, bioassays are performed using MS2 in wastewater (or simulated wastewater). Influent and effluent samples are taken under steady-state conditions and at locations proven to be representative (well-mixed) The dose delivered by the UV system under a given test condition (MS2 reduction equivalent dose) must be quantified using an NWRI 2012 UV Guidelines standard dose-response regression equation, which is a significant departure in protocol relative to other guidance, including the NWRI 2003 UV Guidelines, which used water quality and MS2 stock-specific UV dose-responses generated at least once per test day.

MS2 is used because it provides several advantages over other organisms, including the following:

- It is easy to cultivate in large quantities and count using standard techniques;
- It is relatively resistant to UV light, so a dose–response relationship can be developed that encompasses dose levels required for most target pathogens in disinfection applications; It has a consistent response to UV light; and
- It is not pathogenic to humans and is harmless in the aquatic environment.

Other UV challenge organisms with similar inactivation kinetics as common target pathogens are often considered for use in bioassays (see list above). and Q-beta. Multi-organism bioassays have been adopted to take advantage of UV sensitivity "bracketing", where performance equations generated with multiple organisms allow for sizing with pathogen-specific UV sensitivities (UVDGM 2006).

5.3.4 Design Criteria for Reuse–NWRI-WRF Guidelines

The NWRI-WRF UV Guidelines are based on UV system performance testing at various flowrates, yielding a relationship of UV dose delivered by the UV reactor, and hydraulic loading rate [typically expressed in U.S. gallons per minute per lamp (gpm/lamp)]. The hydraulic loading rate that corresponds to the desired dose is used to design the UV system. Other NWRI-WRF design factors include wastewater quality and variability, aging and fouling characteristics of lamps and sleeves, and the application statistical analyses to the bioassay data through a "confidence ratio" factor applied to the system performance equation.

UV system designs must also acknowledge the variability of effluent UVT (e.g., using the 5th or 10th percentile UVT) as well as equipment robustness (e.g., sensor uncertainty). Some have criticized this compounding of attenuation factors as too conservative, resulting in oversized UV systems that waste electrical power and municipal funds.

The NWRI–WRF UV Guidelines specified MS2 design doses and transmittance values include 100 mJ/cm^2 and 55% UVT for effluent treated with media filtration, 80 mJ/cm^2 and 65% UVT for membrane filtration, and 50 mJ/cm^2 and 90% UVT for reverse osmosis treated waters upstream of the UV.

Reclamation facilities with water-quality characteristics that are outside the required limits would need to test UV-reactor performance under site-specific conditions to validate that the UV system disinfects effluent sufficiently. If the facility plans to use lower design UVTs (at 254 nm) than those shown in the table, then 6 months of effluent analysis is recommended to justify the higher values (based on the 10th percentile of data collected).

The NWRI guidelines apply to the design, validation, and operation of UV systems and include the following requirements:

- At least two reactors (defined as an independent combination of single or multiple banks in series) must be operated simultaneously in any single reactor train (a combination of reactors in series).

- Flowrate, UV intensity, UVT, and turbidity must be monitored continuously. Doing so will also allow continuous monitoring of calculated operating dose (another guideline requirement).

- UV-intensity monitors must be calibrated at least monthly. The UVT and turbidity monitors must be calibrated in accordance with manufacturer recommendations, and laboratory measurements of UVT must be used weekly to verify the accuracy of online transmittance monitoring equipment.

- Effluent must be sampled for coliform bacteria and other microorganisms when water-reuse characteristics are most demanding on the treatment and disinfection facilities.

- Operators must operate the UV system at the same velocity range and flow per lamp as used for performance validation and with total headloss less than or equal to that measured during equipment validation testing.

- Water-reclamation-facility operators also must meet their state's specific water-reuse requirements. The NWRI UV Guidelines specify a UV effluent quality of 2.2 total coliforms (MPN, 7-day median) per 100 ml.

5.3.5 Validation Testing for Reuse–NWRI-WRF Guidelines

Validation testing procedures under the NWRI guidelines are outlined in the literature (NWRI, 2012; WEF, 2015). The procedures outlined in the literature need to be followed strictly in California to ensure regulatory approval.

The following are some issues that may require resolution on a site-specific basis:

- Unfortunately, the systems designed under prior versions of the NWRI Guidelines (e.g., the 2003 UV Guidelines (NWRI, 2003)) are likely to require retrofit to comply with the current Guidelines. The statistical analyses of data in the two documents are different, and this change alone can cause systems that were compliant in 2003 to be non-compliant in the 2012 version. Nonetheless, the 2012 Guidelines specifically state that the industry continues to evolve, and designers should account for potential retrofits when planning UV-disinfection facilities.

- The 2012 UV Guidelines does not assign a scale-up limit from validation to full-scale systems. Instead, any differences between validation and installation are assumed to be vetted during the site-specific spot-check bioassay. Emerick and Borroum (2005) report that larger facilities appear to perform better than pilot facilities, provided the theoretical velocity between the two facilities remains constant. The larger facilities likely perform better because larger facilities make use of larger lamp arrays that exhibit higher internal UV intensities (or less surface volume around the outer surface of the array of lamps compared with the entire volume irradiated). They report concern that full-scale installations that are smaller than those pilot-tested will not provide an equivalent dose as was observed during the pilot testing.

5.4 UV Dose (Fluence)

Dose, also known as *fluence*, is defined as the product of fluence rate and exposure time provided to a pathogen, where the fluence rate is the UV flux, from all directions, through an infinitesimally small sphere with a given cross-sectional area, divided by the cross-sectional area of that sphere. Fluence rate is sometimes called UV intensity. Because intensities and exposure times vary for every fluid parcel or particle in a UV reactor, a distribution of doses exists, and measurement of dose is not straightforward.

5.4.1 UV Dose Estimation Methods

In general, four methods have been used to estimate the UV dose in flow-through reactors; the first two are computational models, and the last two are empirical. Models need to be validated, as they are merely mathematical representations of the actual UV-dose conditions (WERF, 2007). The most accurate dose estimation models are currently based upon detailed UV system validation testing using bioassay methods. For further discussion of dose, the reader is directed to the literature (WEF, 2015).

5.4.2 Factors Affecting UV Dose

A summary of the factors are known to affect the dose provided by UV reactors (WERF, 2007):

- Lamp type—Emission spectrum for various lamp types is different and need to be examined as part of the determination of the UV dose.
- Lamp UV flux (power) output—Various types of UV lamps provide different UV flux output per unit lamp surface area. The higher the UV flux output, the larger the intensity at the lamp surface. In addition, the UV flux of the lamp will increase with lamp length for a given UV flux per unit lamp surface area.
- Power setting of the lamp—The UV intensity is lower if the lamp is operated at less-than-maximum power (modulated by the power supplies.
- Lamp age—After the initial burn-in period, the lamp output and resultant UV intensity typically decreases gradually toward the end of the lamp's useful life.
- Quality and age of the quartz sleeve—The extent of absorption by the quartz sleeve depends on its quality and may increase with sleeve age.
- Fouling on the quartz sleeve—Organic or inorganic fouling can reduce the UV radiation transmission into the wastewater.
- UVT of the wastewater—As UV light passes through wastewater, its intensity is reduced exponentially with distance from the source, because water and the substances in wastewater absorb some of the UV light.
- Type and size of suspended solids—The concentration, type and the size distribution of suspended solids will affect the UV intensity distribution, because suspended solids can absorb and scatter UV light.

- Particles can shield microorganisms from UV radiation—The UV intensity received by a free-flowing microorganism is expected to be greater than that received by a microorganism embedded in a particle at the same location.
- Distance from the center of the lamp—The UV intensity decreases with increasing distance from the lamp, because the UV flux is distributed over a larger area as the distance from the lamp increases.
- Hydraulics—Dead space reduces the effective reactor volume and shortens the HRT. Also, without good radial mixing, a microorganism may pass through the UV reactor between lamps and be exposed to a smaller UV dose.

The reader is directed to additional information on factors impacting UV dose in the literature (WEF, 2015).

5.4.3   UV-Inactivation Kinetics

UV-inactivation of microorganisms is described by U.S. EPA (1986a, 2006) and within other literature (WEF, 2015). A very comprehensive, curated compilation of the sensitivity of many microorganisms from many classes can be found in the literature (Haji Malayeri et al., 2017.) To date, all microorganisms that have been studied appear to respond to UV treatment and do so over a narrower dose range (from most sensitive to most resistant organism) compared to other disinfectants.

Organisms respond to UV irradiation by first-order kinetics, with sensitivities that range from ~<2 mJ/cm² per log of reduction for UV-sensitive organisms such as *E. coli* or *Cryptosporidium parvum* (depending on the type) to >40 mJ/cm² per log of reduction for certain types of UV-resistant organisms such as adenoviruses; however, most pathogens tested so far are shown to be quite sensitive to UV disinfection, and the adenoviruses are generally considered outliers. Designers of UV systems and regulators need to consider the target design pathogen of concern and its kinetics when determining the design dose of a system.

5.4.4   Photoreactivation and Dark Repair

Microorganisms have effective systems for repairing damage caused by hostile environmental conditions, such as exposure to disinfectants. Repair and recovery from inadequate damage is known to occur following all disinfection operations. Designers and operators of disinfection processes should understand these repair processes and their potential consequences When the inflicted damage is small due to the application of a very low UV dose relative to the target design dose, the photochemical damage to an organism's nucleic acids caused by UV irradiation can be repaired. These repair mechanisms can allow UV-treated microorganisms to regain viability (ability to reproduce) following the disinfection process. Two principal repair mechanisms have significance relative to UV disinfection—photorepair, which occurs in the presence of visible light, and dark repair, which can occur even in the absence of light.

Photorepair can occur in some organisms when only minor damage has occurred, and as such is an indication of underdosing. At higher doses, the same organisms will not be able to photorepair. The design dose must be carefully selected to ensure that adequate damage to the target pathogen's nucleic acids is provided. Historic evidence suggests that at least 4 logs inactivation or higher are required to minimize the impact of photorepair on final counts of pathogens that can photorepair (Lindenauer and

Darby, 1994). In the case of the protozoan parasites and *Legionella*, and other UV-sensitive bacteria, this could mean a UV dose of at least 10 mJ/cm^2, but when attempting to provide redundancy for chlorination by targeting the majority of the photorepairable pathogens, this would mean a UV dose of closer to 25 mJ/cm^2. *Cryptosporidium* has been shown not to photorepair, and laboratory lines of *Giardia* have been shown (Linden et al., 2002) to have no manifestations of repair at UV doses of 16 and 40 mJ/cm^2; that is, at UV doses used respectively to polish municipal potable water by POU/POE UV technologies and disinfect municipal potable water respectively. The historical target dose for drinking water of 40 mJ/cm^2 results from the European target for the pathogen rotovirus, and both the original assessment of the target design dose and subsequent epidemiological assessments through years of use in municipal applications suggests that photorepair is adequately controlled at this dose.

Much of the design work for secondary WRRFs has been at dose levels of less than 40 mWs/cm^2, with consequential significant increases in residual organisms being measured after disinfection (via the static light/dark bottle technique) (U.S. EPA, 1986a).

Lehrer and Cabelli (1993) pointed out that etiologic agents of the most common waterborne diseases are Norwalk-like viruses. Viruses lack significant metabolic activity and cannot self-repair, but genetic damage to viruses may be repaired by the host and the virus itself may code for repair in the host. Many of these viruses are thought not to undergo repair of UV-induced damage. Whitby and Palmateer (1993) suggest that the reactivation phenomenon is not observed *in situ*. Using labeled *E. coli* bacteria, they demonstrated a lack of reactivation in UV-irradiated wastewater effluent after release to a receiving stream. These same bacteria were shown to undergo photoreactivation when exposed to a sufficient dose of photoreactivating radiation under controlled conditions. The depth of the receiving water and the receiving water's absorbance of photoreactivating light likely account for a failure to see repair phenomenon. Additionally, particle settling or adhesion to receiving water surfaces such as banks and facilities, as well as consumption by aquatic organisms could diminish any impacts of photoreactivation of microbes within particles or particle disintegration.

A consensus does not exist within the engineering, scientific, or regulatory communities regarding the inclusion of repair in UV-disinfection-system design, although the work of Lindenauer and Darby (1994) suggests that targeting 4 logs of inactivation will eliminate concerns of photorepair because the damage inflicted will be too severe for organisms to withstand. Although many operating WRRFs have been designed and are operating successfully with and without consideration of repair, photorepair can only occur with very low doses. Photorepair is a concern in systems that are underdesigned with respect to particulates management and/or UV dose design (resulting from inadequate consideration of water quality and/or inadequate characterization of hydraulic behavior of the UV reactor or its installation within the treatment train with consequential underdosing of part of the flow). Variations from normal of equipment in less than optimal condition (even with peripheral equipment like flow meters, UVT meters), or with significant transient changes in water quality, could lead to operational doses that are less than the validated doses under the same nominal conditions. The risk of inadequately protecting public health due to these factors becomes higher as target design doses are lowered. When systems are properly designed with adequate dose, issues of photorepair are not a concern.

5.5 Role of Computational Fluid Dynamics in UV Design

Computational fluid dynamics modeling combined with optical modelling of the light field within reactors and UV inactivation kinetics is increasingly an important tool for UV-reactor design by UV equipment vendors, because it can overcome the limitations associated with physical prototyping and overdesign or underdesign. Though not intended to replace physical prototyping, such numerical modelling enables the designer to test alternative scenarios before building a prototype. Computational modeling can be used to force failure modes and identify design limitations, including short-circuiting and dead zones, which may have otherwise been impossible to test, as a result of physical or financial constraints. Models developed by vendors have been shown to replicate the results obtained through validation testing. CFD can also be used by the engineering community to optimize flow split to multiple UV trains, to improve approach and exit hydraulic conditions to and from the UV reactor, and to assess the efficacy of UV reactors in series including dose additivity between UV banks.

5.6 Fouling and Sleeve-Cleaning Systems

The ability to deliver UV photons from the source (UV lamp) to the target pathogen is critical to the performance of UV disinfection systems. The accumulation of insoluble materials on the surfaces of the sleeves that house UV lamps can limit photon delivery. Sleeve fouling matter can contain organic and/or inorganic constituents. Organic fouling is largely attributable to floatable materials that accumulate on lamp sleeves near the free surface in open-channel systems. Additionally, extracellular polymeric substances are thought to play a role (Swift et al., 2000). Control of organic fouling can be augmented by removal of these wastewater constituents in upstream processes.

The inorganic components of a fouling material will accumulate over the entire wetted surface of a sleeve. Chemically, these materials are similar to inorganic scale, which can form in plumbing or on heated surfaces (e.g., heating elements). Empirical observations of sleeve fouling have suggested that water containing high hardness and/or high iron concentrations is likely to promote fouling. However, Swift et al. (2001) noted that some effluents with higher concentrations of hardness and iron may have substantially lower fouling rates than effluents with substantially lower concentrations. Elemental analysis (with an electron microprobe) of the scales on the quartz sleeves of several UV systems indicated that the scale was predominantly inorganic. Iron, phosphorus, and aluminum were the major components, with minor amounts of calcium and silicon.

Control of lamp fouling is achieved by a variety of techniques. There are several sleeve-cleaning options that are used currently.

- Manual cleaning strategies, which require periodic removal of the sleeves for soaking in a chemical bath or manual wiping with a chemical cleaner;
- Automated online strategies, which use mechanical cleaning devices that wipe frequently and require periodic manual chemical cleaning; and
- Automated chemical/mechanical cleaning systems.

As part of the design process the cleaning process needs to be validated. Sleeve fouling tests should be performed and witnessed to substantiate fouling factors used in full-scale system sizing. The test water used should have a higher fouling potential that the desired usage. For example, a fouling test conducted in tertiary treated water should not be used to size a system for secondary treated water.

Automated mechanical cleaning technology typically consists of metal brushes, rubber or synthetic rubber wipers, or Teflon rings, which mechanically remove the foulant from the sleeve. Mechanical cleaning alone is not effective for all effluents and, in some cases, may require periodic offline chemical cleaning. Automated chemical/mechanical cleaning systems using phosphoric-acid-based cleaning solutions have proven to be very effective (Salveson et al., 2004).

Automated chemical and/or mechanical cleaning can help remove most fouling materials. However, some deposits may stay permanently on the quartz sleeves after cleaning. Some research suggests that the wipers used in long-term automated cleaning systems may damage sleeve surfaces and scratch the sleeves or create holes, which consequently trap foulants, which attach more tightly to the surfaces of the scratches. In specific case studies, chemical cleaning in addition to mechanical cleaning was more effective in removing foulants and avoiding the accumulation of permanent deposits than mechanical cleaning alone (WERF, 2007; WEF, 2015).

5.7   Design Considerations

5.7.1   Liquid Level Control
The liquid level control device and its impact on the hydraulics of the UV System is discussed in Section 5.3.1. Maintaining precise liquid level control is critical for the effectiveness of the UV system.

5.7.2   Transmittance Monitoring
Aside from hydraulics, UVT is likely the most critical water-quality parameter in determining disinfection effectiveness. If a UV-disinfection system was designed to perform under a water transmittance of 65%, and an actual transmittance of 55% was observed, the system may underperform by 33% or more. Industrial discharges and wet-weather events often exhibit significant effects on UVT.

It is recommended that site-specific determination be made of UVT, rather than relying on default values offered by the NWRI–WRF guidelines (2012). Filtered secondary effluent often exhibits a UVT that is much higher than the default 55% recommended by NWRI (NWRI and WRF, 2012), which could lead to significant overdesign at considerable expense. The best way to determine a design UVT is through monitoring over a minimum period of 6 months at a frequency of three samples per day, including wet-weather periods.

5.7.3   Intensity Monitoring
Because of the non-uniform UV-intensity profile within a UV-disinfection system, the monitoring of UV intensity is used primarily to report relative differences in performance. UV intensity sensors are useful in estimating lamp aging or fouling effects over time. An intensity sensor reading is recorded under new lamp and unfouled conditions, with subsequent recording of sensor readings over time. Sensor readings can be compared pre- and post-cleaning, to estimate cleaning effectiveness. They can be compared over time on clean systems, to discern the need to replace lamps.

5.7.4   System Controls
System control is typically in the form of a PLC coupled with an operator interface display screen supplied by the UV manufacturer and should be a function of the system

type and the size of the WRRF. Control interfaces should be simple; the objective is to ensure that system loading can be maintained and disinfection can be accomplished while conserving the operating life of the lamps. This becomes increasingly important in larger systems. In smaller systems, it may be best to have the full unit in operation at all times, excluding redundant units incorporated to the design. Manual control and flexibility should be available as the system increases in size, enabling the operator to bring portions of the system (e.g., channels and banks) into and out of operation as needed, to adjust for changes in flow or water quality. Automation of this activity is increasingly beneficial as the system becomes larger and incorporates multiple channels.

System control varies from minimal to fully automatic. Fully automatic systems enable system control from a remote location, such as a central operations center. System controls typically provide, at minimum, system power, system hours, and lamp status indicators. Fully automatic designs can integrate flow and wastewater conditions and pace the UV system, by dimming lamps, shutting down banks, or taking channels out of service.

5.7.5 Facility Requirements

Planning of the UV facility should include adequate space allocation, based on the required disinfection system as well as the associated operational facilities. One of the key design considerations is good access to the channels (or pipes) for ease of maintenance and especially lamp removal. There should also be adequate space for cleaning and chemical infrastructure, availability, and location of the power supply, and provision of emergency power. Lifting devices provided by some UV vendors should be provided to facilitate removal of lamps, modules, and banks. Adequate space for electrical equipment, including control panels, needs to be considered, assuming maximum allowable separation distance between the UV reactors and electrical controls. Proper design should include easy access to the lamp modules for cleaning and other maintenance tasks. Ultraviolet systems should be installed in an area that is large enough for maintenance activities and for handling the modules when taken out of the channels, or for lamp replacement in closed vessel systems that are space constrained in buildings.

5.7.6 System Redundancy

In most designs, the UV facility will consist of multiple (minimum two), parallel channels or pipe trains of the same capacity. Each channel would be equipped with multiple banks of the same capacity. A decision regarding the required number of UV channels (trains) should include adequate redundancy, considering that full disinfection capacity is provided with the largest UV reactor out of operation.

5.7.7 Bypass Channels and Designing for Flood Conditions

A provision for UV-system bypass should be incorporated to the design to prevent the UV channels from potential flooding.

5.7.8 Channel Drains

Reactors, channels, and related tankage should be equipped with drains, to allow for complete and rapid dewatering. Drainage should be directed back to the headworks of the WRRF. A clean-water system should be permanently available for rinsing and cleaning needs. Consideration should be given to providing a bypass around the UV system, particularly in WRRFs that have seasonal disinfection requirements.

5.7.9   Screens

Screening should be considered upstream of the UV units, to remove any debris from the wastewater if the final clarifier launders are not covered. Algae, in particular, have caused problems resulting from sloughing from upstream clarifiers and channels. Leaves and plastic debris have also been observed in UV channels. These materials tend to catch on lamps and cause difficulties. Cleaning can present a maintenance problem. Screens can range from simple mesh inserts, which are removed and maintained manually, to self-cleaning, mechanical moving screens.

5.7.10   Module Lifting

For open channel systems, devices, such as cranes, are recommended to lift the UV banks out of the channel. At small- to medium-sized facilities, jib cranes typically are used. For large facilities, an overhead crane can be advantageous. Provisions for cranes require careful planning of space and access. Sufficient vertical clearance for crane operation should be provided. Several of the latest UV systems now include an inherent mechanical bank or module removal system, minimizing the need for an overhead crane.

5.7.11   Spare Parts

Operators should maintain adequate spare parts, including ballasts, lamps, and quartz sleeves. In general, the spare parts stock will be based on the manufacturer's recommendations, operators' preference, and the size of the facility.

5.7.12   Power Supply and Harmonic Distortion

UV equipment requires a significant amount of electrical power to illuminate the lamps; hence, accurate assessment of the available capacity of the electrical-power-distribution system is important. To ensure continuous operation, the UV disinfection system should be connected to emergency power (a generator or uninterruptible power supply) of adequate size (WERF, 2007). A simple backup-power supply (generator) may be sufficient, if power-quality variations are infrequent. Because power-quality fluctuations may result in jeopardized disinfection, an evaluation of power supply should be undertaken, to identify the best approach for power conditioning (U.S. EPA, 2006). Ultraviolet systems also require a reliably continuous power supply. In designing the backup-power supply, consideration should be given to the frequency and duration of power interruptions. Manufacturers of UV disinfection systems design their power supplies to ride through small power "glitches" and an assessment of technology ride-through ability may be warranted for some locations. If there is an existing chlorination system at the facility, it can be considered for backup disinfection, as long as the permit will allow the use of chlorine in an emergency.

5.8   Effects of Water Quality and Wastewater Design

Data should be collected to obtain a thorough characterization of effluent quantity and quality. For existing facilities, direct sampling and testing should be conducted and should address seasonal and diurnal variations. If the facility is new, an effort should be made to develop design effluent characteristics from similar WRRFs and collection systems. The most significant data that can be collected includes UVT and collimated beam data. Sufficient data of this kind ensures confidence in the UV dose required to meet a specified pathogen disinfection limit in any specific water.

5.8.1   Effects of Upstream Processes

Upstream treatment processes have a significant effect on the disinfection of secondary effluent. Because of "shading," particle-associated coliforms (PACs) are often difficult to inactivate to levels required for reuse (e.g., 2.2 total coliforms/100 mL). Additionally, certain upstream processes, such as fixed-film biological treatment, yield recalcitrant effluent with a high concentration of PAC in large particles. Loge et al. (1999) reported that PACs were found to decline exponentially with increasing mean cell residence times (MCRTs). The factors influencing the formation of PAC included the concentration of particles, concentration of dispersed (non-particle-associated) coliform bacteria, and the MCRT. The concentration of dispersed coliform bacteria was found to decline with increasing MCRTs. The rate of decline was greater than the typical half-life attributed to endogenous decay, suggesting that other factors (e.g., predation by protozoa) influence the concentration of dispersed coliform bacteria and subsequently the formation of PAC (Loge et al., 1999).

5.8.2   Effects of Industrial and Water Resource Recovery Facilities' Chemicals

A number of industries have been implicated as discharging wastewater with high dissolved UV absorbance (low filtered UVT), as a result of the presence of organic compounds not readily degraded (Swift et al., 2007), including sunblock, coffee, pharmaceutical, and chemical manufacturers; centralized WRRFs; and printed-circuit-board manufacturers. Several WRRFs have been documented as having potentially violated effluent disinfection standards, as a result of the presence of refractory organic compounds passing through the WRRF and lowering effluent UVT. Compounds in the wastewater responsible for lowering effluent UVT can be considered to be "pollutants of concern" for source control programs and thus appropriate for regulatory control through the use of local limits established through an evaluation of maximum allowable headworks loading using methods analogous to those used for other pollutants (Swift et al., 2007).

5.9   Safety and Health

Operators of the UV system should be familiar with its O&M manual and with any safety requirements. Operators should follow equipment manufacturer's recommended safety precautions and procedures, OSHA regulations, and state guidance and regulations for UV reactor operations. In addition to the standards and procedures established for WRRF UV-disinfection system operations, the following safety issues pertain specifically to the design of UV systems (U.S. EPA, 2006):

- Exposure to UV light
- Electrical hazards

5.9.1   Exposure to UV Light

To minimize the danger of exposure, warning signs regarding UV radiation should be posted. For open-channel systems, the UV lamps are often covered with "light locks" and the channels have solid grating or checkered plates, to further protect workers from any UV light. Lamps should not be operated in air, to prevent overexposure of skin and eyes to UV radiation. This means that UV systems should be equipped with safety interlocks that will automatically shut down lamp modules, if they are taken out of the reactor or the water level falls below the top of the lamps in the reactor (WERF, 2007). For in-line systems, if viewing ports are provided, they should be fitted with UV-filtering windows. Workers near open channel UV disinfection systems should wear proper PPE including UV-blocking safety glasses at all times.

5.9.2 Electrical Hazards

To prevent electrical hazard, all safety and operational precautions required by the National Electric Code (National Fire Protection Association and American National Standards Institute, 2008), OSHA, local electric codes, and the UV manufacturer should be followed and include the following precautions (U.S. EPA, 2006):

- Proper grounding,
- Lockout/Tagout procedures,
- Use of proper electrical insulators, and
- Installation of safety cutoff switches.

According to the U.S. EPA *Ultraviolet Disinfection Guidance Manual for the Final Long-Term 2 Enhanced Surface Water Treatment Rule* (2006), proper grounding and insulation of electrical components are critical for protecting operators from electrical shock and protecting the equipment. To minimize electrical hazards, ground-fault-interruption (GFI) circuitry should be provided with each module. For a GFI to function properly, the transformer in the UV reactor ballast must not be isolated from the ground.

6.0 Ozone Disinfection

Although ozone and chlorine were both developed as disinfection methods at roughly the same time in the late nineteenth and early twentieth centuries (von Sonntag and von Gunten, 2012), chlorine became the dominant wastewater disinfection technology due to its substantially lower capital cost. Moreover, many early ozone disinfection systems were abandoned because of maintenance and operational problems (Leong et al., 2008). However, concerns about the residual oxidizing effects of chlorine and the production of chlorinated DBPs have caused WRRF owners to look again at alternative treatment technologies. Improvements in ozone-production technologies, combined with the capacity of ozone to oxidize micropollutants and color- and odor-causing compounds, has brought ozone back into consideration as a disinfection method, especially in applications involving water reuse.

Leong et al. (2008) reported that, as of 2007, there were seven major WRRFs in the United States using ozone with a median design flow of 37 900 m^3/d (10 mgd) and a range of 11 400 to 129 000 m^3/d (3–34 mgd). As of 2016, ozone disinfection systems are in use or in construction at approximately 20 U.S. WRRFs, many of which cite color removal or the destruction of trace organics as deliberate supplementary goals (International Ozone Association, 2016).

6.1 General Description of Ozone Disinfection

Ozone acts as an effective disinfection agent through the following four mechanisms:

- It causes cell lysis by oxidation of cell walls;
- It breaks down purines and pyrimidines, the building blocks of nucleic acids;

- It breaks down carbon-nitrogen bonds, leading to depolymerization of organic molecules; and

- It produces hydroxyl radicals in water, which are powerful oxidants.

The effectiveness of ozone for disinfection depends on several factors, including target pathogen, the pathogen's exposure to ozone, particulate shielding, and temperature. In municipal wastewater effluent, ozone exposure, defined as the integral of ozone concentration over time, is influenced by the transferred ozone dose and the concentration and type of dissolved organic matter (DOM) present.

The disinfection process can be summarized as follows. Ozone, an unstable gas, must be generated on-site and then transferred into the treated water. Once transferred, a large portion of the initial ozone concentration rapidly decays within the first few seconds ("instantaneous demand"); the remaining concentration ("residual") decays more slowly, with a typical half-life of less than 30 seconds in municipal wastewater effluent (Metcalf & Eddy, Inc./AECOM, 2013). However, since ozone decays so quickly, and much of the actual CT is accounted for before measurement is practical, treatment trains should be designed using site-specific bench and pilot studies.

6.2   Analytical Methods for Ozone Measurement

Measurement of ozone concentration is required to determine the initial demand and subsequent decay rate of ozone in a given wastewater matrix. This information is useful both for preliminary process design and for monitoring the performance of full-scale systems. Many ozone system manufacturers offer testing services to establish required ozone doses during the design process.

Gottschalk et al. (2010) provide an excellent description of the available measurement techniques, listing the advantages and disadvantages, detection lmits, and potential interferences of each test method. A brief summary is provided in Table 17.11.

6.3   Ozone Kinetics

For a given pathogen, inactivation performance is partially a function of CT, which is in turn determined by ozone kinetics.

Ozone is unstable in aqueous solution, and will proceed along one of the two reaction pathways: the direct pathway or the indirect pathway. Relatively few substrates react directly with ozone, so the direct pathway is the slower and more selective of the two. However, in some cases ozone will form hydroxide radicals (•OH) or other radicals in its reaction products. These radicals react quickly and non-selectively with

Method	Advantage	Disadvantage
Iodometric	Inexpensive	Not selective; time consuming
UV or visible-light absorption	Easy	Interference from aromatics
Indigo trisulfate	Inexpensive, quick, relatively selective	Needs calibration
DPD	Inexpensive	Not selective, difficult to perform
Chemiluminescence	Low detection limit	Difficult to perform
Membrane ozone electrode	Easy, selective	Expensive

TABLE 17.11   Summary of Methods of Ozone Analysis

neighboring molecules, sometimes forming new radicals and perpetuating the reaction cycle. This indirect pathway greatly increases the ozone reaction rate and range of potential substrates. The following section is a summary of the mechanisms behind aqueous ozone kinetics. For a more detailed discussion, the reader is referred to von Sonntag and von Gunten (2012).

6.3.1   Factors Influencing Ozone Kinetics

The major factors influencing ozone reaction kinetics are DOM, carbonaceous alkalinity, ozone concentration, pH, and temperature.

1. DOM: By far the most significant factor affecting ozone kinetics is DOM, which competes with pathogens for ozone exposure through both the direct and indirect pathways. There is increasing evidence that the ozone demand in municipal wastewater effluent scales linearly with dissolved organic carbon (DOC), which can be taken as a reasonable proxy for DOM (von Sonntag and von Gunten, 2012).

2. Alkalinity: Inorganic carbon is a major inhibitor to the indirect pathway because the radical forms of carbonate and bicarbonate are not reactive with most organic compounds. Hence, carbonate alkalinity generally increases the stability of dissolved ozone (von Sonntag and von Gunten, 2012).

3. Ozone concentration: An increase in the concentration of ozone in the bulk liquid will cause an increase in the reaction rate constants for direct (Bellamy et al., 1991; Prados et al., 1995) and indirect (Prados et al., 1995) reactions. This increase will result in an overall increase in the ozonation reaction rate constant.

4. pH: Since hydroxide ions are the primary initiators of the indirect pathway, increasing pH promotes the indirect pathway. The direct pathway dominates for acidic conditions (pH < 4), while the indirect pathway dominates above pH 10 (Staehelin and Hoigne, 1983). However, since pH is typically within a narrow range in wastewater effluent, it is not typically a factor of concern.

5. Temperature: Based on the Arrhenius expression (eq 17.4), increasing the temperature will increase the ozone reaction rate, but will decrease the solubility of ozone in the liquid phase, which decreases gas-to-liquid transfer efficiency.

6.3.2   The Role of the Hydroxyl Radical in Disinfection

While $\cdot$OH is a much stronger oxidant than ozone, and is the primary mechanism for the destruction of most micropollutants, its effectiveness as a disinfectant is debated. Several studies have shown that increasing the ratio of $\cdot OH/O_3$ in solution correlates with lower inactivation rates against *E. coli* and other heterotrophic bacteria (Wolfe, 1990; Hunt and Mariñas, 1999; von Gunten, 2003). Other studies have shown improved inactivation rates against *C. perfringens* (a spore-forming bacterium) and *B. subtilis* spores (Cho et al, 2003; Lanao et al, 2008). It may be that bacterial spores are particularly resistant to ozone and require attacks via the indirect pathway.

6.4   Non-Kinetic Factors in Ozone Disinfection Efficiency

6.4.1   Particulate Shielding

Suspended solids can inhibit the performance of ozone disinfection by shielding pathogens (Gottschaulk et al., 2010). Xu et al (2002) observed a 1-log improvement in inactivation for the same CT values after wastewater effluent had been filtered at 10 μm.

6.4.2   Temperature

Generally, decreasing temperatures will require a higher CT to achieve the same inacti-vation rates (Gottschaulk et al., 2010), although *B. subtilis* spores appear to exhibit a local maximum inactivation rate at 10°C, with decreasing efficiency at temperatures above and below (von Gunten, 2003).

6.5   Ozonated Effluent Concerns

Environmental issues associated with ozone include:

- Formation of disinfection byproducts, including bromate (formed from bromide), and aldehydes.
- Increase in the BOD of the effluent, due to breakdown of recalcitrant compounds associated with natural organic matter (NOM) and the formation of biodegradable smaller molecules. For this reason, ozonation is often followed by a tertiary biological treatment step (such as biological aerated filtration), which may, in turn, be followed by additional downstream "post-ozonation."

Ozonation will often result in high concentrations (10–20 mg/L) of effluent dis-solved oxygen, typically obviating the need for post-aeration filtration.

6.6   Safety and Health

The severity of injury from exposure to ozone in the gas phase depends on concentration and time of exposure. Ozone is a mild irritant to humans exposed for 8 hours at 0.1 ppm, but can be dangerous and even fatal at 30 minutes of exposure to 50 ppm (Gottschalk et al., 2010). While ozone can be detected by smell at 0.02 ppm, the perception of continuous exposure fades over time, so any detection should be addressed promptly.

To minimize ozone contact, ozone systems are contained in sealed piping and ves-sels. Areas that vent to atmosphere should be equipped with ozone-destruction units. Reactor off gas vents should also be equipped with ozone destruction units when the off gas contains ozone in concentrations above those required for human health and safety. The ozone-destruction units should be checked periodically for performance, to ensure that off gas ozone levels are not problematic.

Ozone facilities should also be equipped with integrated ventilation systems and ambi-ent-air-ozone monitors, to ensure the ozone levels in the working environment are kept below exposure limits. If ozone levels are detected above 0.1 ppm$_v$, the ventilation system can be called on to provide room ventilation. If the ozone level continues to increase, the monitor can produce an alarm. The alarm can be integrated to the facility supervisory con-trol and data acquisition system and can be used to shut down the ozone system.

Potential locations for leaks include piping, fittings, and equipment. Small leaks can be noticed easily by smell near the leak. The leak can be pinpointed further by using a portable ozone detector, applying a soap solution to the leak area, or by soaking a white rag in 2% potassium iodide solution (the rag will turn brown in the presence of ozone) (Rakness et al., 2005).

Ozone systems that use oxygen also present safety considerations. Oxygen is an oxidant and, therefore, supports the combustion of flammable materials. Although on-site generation is a viable option for many facilities, most existing ozone systems use liquid oxygen (LOX). Safety information for oxygen is available from compressed-gas suppliers. Ozone systems that use LOX also have cryogenic hazards associated with the compressed liquid.

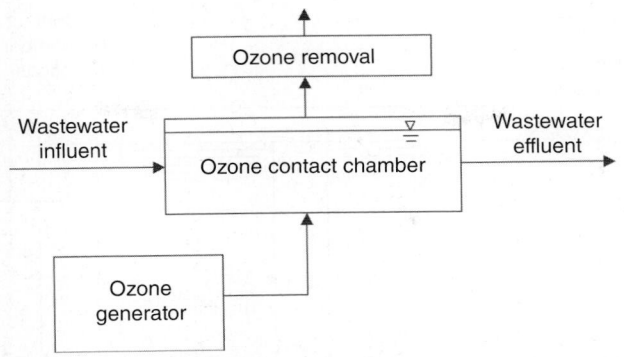

FIGURE 17.20   Components of ozone-disinfection system.

6.7   Process Design

The three most important components of the ozonation process are the ozone generator, ozone contactor, and ozone-exhaust-gas destruction as shown in Figure 17.20.

6.7.1   Ozone Generators

Ozone is produced in an ozone generator by using oxygen or air as a feed gas. The feed gas is passed through electrodes carrying a high voltage (6000–20 000 V), which produces a corona. The feed gas is converted to ozone, and the efficiency depends on the type of feed gas—air (1%–4%) or oxygen (1%–10%). The use of pure oxygen yields ozone between 4% and 12% by weight. Approximately 80% to 95% of the energy will be converted to heat and must be removed at the ground electrode, typically through cooling water. The operational variables are the applied power, efficiency of the generator, flow of feed gas, and temperature.

The feed-gas treatment should be designed to remove moisture from the feed gas. The recommended moisture concentration in the generator is <1 ppm by weight (Rakness et al., 2005).

6.7.2   Ozone Contactors

For the ozone to do its work of disinfection and oxidation, it must be brought into the water and dispersed as finely as possible. This is accomplished generally through fine-bubble diffusers located in baffle chambers or in a turbine-type contactor. Sidestream systems with venturi injectors are also becoming increasingly popular.

A typical ozone contactor has several compartments, in series, with bubble diffusers at the bottom. In the first compartment, the water flows downward against the rising bubbles, and, in the second compartment, the water flows upward as shown in Figure 17.21. The chambers are covered to prevent the escape of ozone and to increase the partial pressure of the ozone in the contactor. Additional chambers follow to ensure necessary contact time between the ozone and the water. Each of the chambers has sampling ports, so that the ozone concentration in each chamber can be determined. This capability is needed to calculate the product of the concentration and detention time, to obtain the required CT value.

Figure 17.22 shows a turbine-diffuser contactor, which mixes the ozone with the water. Contact chambers to establish contact time must follow.

Increasingly, sidestream injection is used for ozonation, which involves splitting off a portion of the main flow into a sidestream. Ozone is injected into this sidestream and then

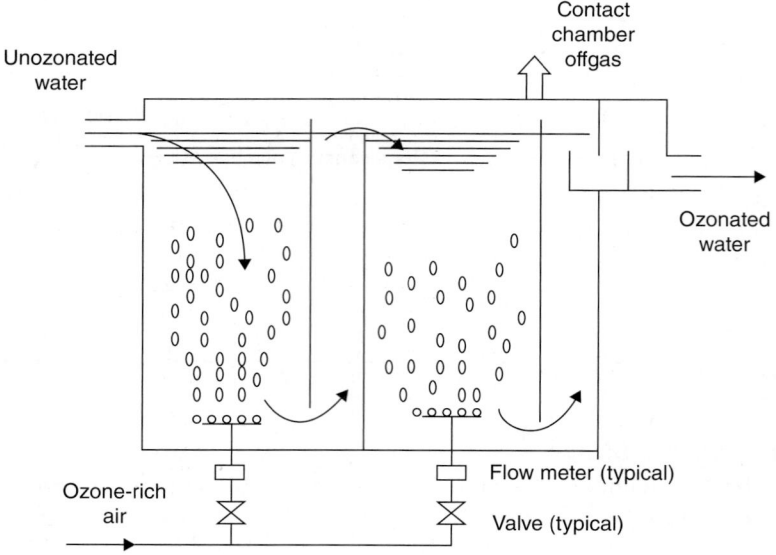

Figure 17.21   Baffled chamber (Deininger et al., 2008).

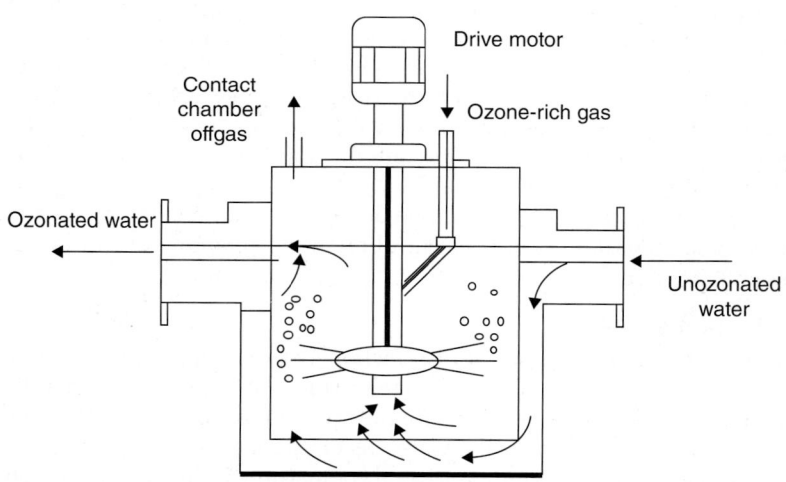

Figure 17.22   Turbine diffuser (Deininger et al., 2008).

the sidestream is mixed back into the main flow. The sidestream injection system is typically composed at a minimum of a booster pump, a venturi and nozzles for introducing the sidestream back into the main flow. A sidestream injection system may also include degassing equipment and in-line mixers.

6.7.3   Ozone Demand and Absorption

Ozone demand and absorption for wastewater applications are determined using laboratory analysis or online sensors. Online ozone analyzers are now accurate to parts per billion levels (EPA LT2ESWTR Toolbox Guidance 11.10.1).

6.7.4 Ozone Exhaust Destruction

The off-gas from ozone contactors generally exceeds the OSHA limit of 0.1 ppm by volume; therefore, the remaining ozone has to be recycled or destroyed. The off-gas is first passed through a demister, which traps small water droplets on stainless-steel mesh. The gas then is heated and passed through a destruction unit, which contains a catalyst to speed up the process. The power requirement is between 1 and 3 kW per 100 scfm (3 m³/min) of gas flow.

7.0 Peracetic Acid

Peracetic acid (CH_3COOOH) (PAA), also known as *peroxyacetic acid, ethaneperoxoic acid,* or *actyl hydroxide,* is a strong oxidant. Laboratory research is underway to determine the disinfection and cost-effectiveness of this strong oxidant when compared with other disinfection-unit processes. Peracetic acid is generated by creating a mixture of acetic acid and hydrogen peroxide. Because this mixture is somewhat unstable, PAA must be generated on-site and cannot be transported once it is generated.

PAA is a strong disinfectant with a wide spectrum of antimicrobial activity. Because of its bacterial disinfection effectiveness, the use of peracetic acid as a disinfectant for wastewater effluents has been drawing increasing interest in recent years. Researchers debate whether PAA's disinfection action occurs as a result of active oxygen release or the hydroxyl radical. Regardless, it is an effective disinfectant that is not mutagenic or carcinogenic; decomposes to harmless acetic acid, oxygen, and water; and thus does not yield harmful DBPs. In addition, no subsequent processes (e.g., dechlorination) are required.

The advantages of PAA for wastewater disinfection are the potentially low capital costs, broad spectrum of germicidal effectiveness, absence of persistent toxic or mutagenic residuals or byproducts, no quenching requirement (e.g., no dechlorination), little dependence on pH, and short contact time. Major disadvantages associated with PAA disinfection are the residual acetic acid and thus the potential microbial regrowth (acetic acid is already present in the mixture and is also formed after peracetic acid decomposition). Another disadvantage of the use of peracetic acid is its high cost, which is partly as a result of limited production capacity worldwide.

Currently research projects are being conducted to validate the design parameters for PAA systems.

8.0 Other Disinfection Methods

The majority of disinfection systems in the United States use chlorination or UV disinfection. However, other technologies exist, often either as methods common in other parts of the world or as emerging techniques. These emerging technologies exist for consideration by designers for incorporation to WRRFs. Some of these technologies are taken directly from drinking-water applications; some are new developments; and others are combinations of conventional physical- or chemical-unit processes. These emerging technologies are discussed in the following sections.

8.1 Bromine

Bromine is a chemical with a similar nature as chlorine. Oxidizing forms of bromine in water, which have been used for disinfection purposes, include bromide chloride (BrCl); 1-bromo-3-chloro-5,5-dimethylhydantoin (BCDMH); and sodium bromide (NaBr) activated with chlorinated water (Boner et al., 2002).

Both bromine and chlorine are in the halogen family of chemicals and react similarly in water and wastewater. However, chlorine and bromine may react differently in the presence of ammonia- and nitrogen-based compounds at elevated pH values. Although chlorine and bromine form chloramines and bromamines in the presence of nitrogen-based compounds, bromamines (mono and dibromamine) have been shown to be effective disinfectants when compared with chloramines and free bromine (Mills, 1973; Wyss and Stockton, 1947).

Also, the dissociation of hypobromous acid into hypobromite ion is less affected by pH than the similar reaction with chlorine. Thus, bromine remains active as an effective disinfectant over a larger pH range than chlorine. It is also affected less by high temperature and nitrogen wastes than chlorine (Boner et al., 2002).

BCDMH is a disinfectant that is used rarely in the wastewater treatment industry. It has been used for the disinfection of cooling water systems. However, reports about the disinfection effectiveness of BCDMH are somewhat confusing and contradictory (Boner et al., 2002).

Studies by Boner et al. (2002) have investigated the effectiveness of using bromine and BCDMH for CSO disinfection. These studies indicate that oxidizing forms of bromine may be applicable for CSO disinfection where wide variability in flows occurs routinely.

8.2   Ferrate

Ferrate Treatment Technologies, LLC (Orlando, FL) has patented ferrate, which is a powerful oxidant and disinfectant and is being investigated for use in water and wastewater treatment systems. In addition to disinfection properties, ferrate also may be an effective coagulant.

Laboratory studies are underway to determine the effectiveness of ferrate as a disinfectant in WRRF. The results of these studies will inform design engineers on the merits of Ferrate allow designers to din WRRF applications.

8.3   Electron-Beam Irradiation

One of the emerging technologies in the 1990s was electron beam irradiation. This process used a stream of high-energy electrons that were directed into a thin film of water. The electrons would break apart the water molecules and produce a large number of highly reactive chemical species. This unit process was once thought to be a competitor to chemical-disinfection processes. However, it is no longer being considered as a viable wastewater-disinfection process.

8.4   Solar Disinfection

Solar disinfection is most often used by developing countries that lack the financial and technical resources needed to implement conventional disinfection systems. Because solar disinfection is a naturally occurring disinfection method, its effectiveness and reliability are influenced heavily by environmental factors, including weather and water transmittance.

Solar disinfection is classified as an advanced oxidation process (AOP). Like all AOPs, solar disinfection generates hydroxyl radicals and other reactive oxygen species (ROS), which inactivate microorganisms by damaging cell constituents. In solar disinfection, solar irradiation produces the ROS, which can lead to the inactivation of both viral and bacterial pathogens in wastewater (Kohn and Nelson, 2007; Kohn et al., 2007).

The ability of solar disinfection to inactivate pathogens varies between species. *E. coli* is widely accepted as a measure of disinfection and is inactivated readily by solar radiation. However, it has been shown that enterococci are inactivated less rapidly, and some bacterium and cysts are altogether resistant to solar-UV irradiation (Blanco-Galvez et al., 2007). On the other hand, it has been demonstrated that the MS-2 coliphage, a surrogate for human enteric viruses, can be inactivated by applying solar irradiation to surface waters (Kohn and Nelson, 2007).

8.5 Pond Treatment System Die-Off

In wastewater-pond-treatment systems, natural processes, including starvation, predation, toxicity, sunlight-induced stress, and sedimentation, can lead to fecal coliform and pathogen die-off (Richard, 2001). The most common limit for final effluents that are being discharged to surface water is 200 or 400 FC/100 mL. When detention time and temperature variations are correctly factored into the design, pond treatment systems can achieve this level of disinfection.

By maintaining a water temperature above 15°C and a detention time of at least 80 days, facultative pond treatment systems can produce effluent with less than 200 FC/100 mL. However, when mean temperatures fall below 10°C, much longer detention times are required to achieve this same level of treatment. Because die-off is heavily influenced by changes in temperature, pond treatment systems should be designed using wintertime water temperatures.

Unless a pond treatment system effluent is to be disinfected chemically, pond treatment systems typically are sized around the HRT needed to provide the desired level of fecal coliform reduction. Pond-treatment-system die-off can be enhanced by using multiple cells in series and maintaining aerobic conditions (Richard, 2001).

8.6 Pasteurization

Pasteurization uses the exhaust heat from an engine or turbine driven generating unit as a heat source to inactivate pathogens. The heat is passed through plate and frame heat exchangers to transfer the heat from the air to the bulk liquid. Pasteurization recently was approved for Title 22 disinfection in California, joining chlorine and UV as approved technologies. The process can also result in the co-generation of electricity.

Developed and named after Louis Pasteur in 1865, pasteurization is the process of heating liquids to destroy bacteria, protozoa, molds, and yeast. The effectiveness of the process is heavily dependent on liquid temperature and exposure time. Pasteurization has been used, for a long time, by the food industry for inactivating microorganisms in products like milk, eggs, and wine. Other applications include the treatment of wastewater sludge and disinfection of wastewater.

The State of California has approved the use of specific pasteurization technologies for disinfecting reclaimed water under the Title 22 regulation. The Title 22 approved technology uses "flash pasteurization" (short exposure time and higher temperature) to disinfect treated wastewater. The wastewater is heated for a time period and temperature specific to the microorganisms, <10 seconds and ~163°F to meet "tertiary recycled water" standards for Title 22 in California. The gas turbine or engine is powered by a natural gas supply. The external natural gas supply can be supplemented with methane gas generated on-site by anaerobic digesters. In appropriate applications, the turbine can also be used to co-generate electricity, providing electricity for the facility that is separated from the power grid. Not all heat is transferred from the air to the bulk liquid, and some remaining heat can be used for other applications. Typically, waste heat can be used to increase the efficiency of anaerobic digesters by raising the temperature in the anaerobic digesters (Leong et al., 2008).

Currently, there is little to no information on the environmental effects of using pasteurization disinfection. However, the formation of DBPs is not anticipated, and, because no chemicals are used in the process, the effects are expected to be few. In some areas where receiving waters are temperature-sensitive or where the discharge makes up a majority of the receiving stream, the heated effluent may be a concern to receiving waters with a temperature total maximum daily load, strict dissolved oxygen limits, or a thermal-discharge plan (Leong et al., 2008). Preliminary economic analyses indicate that the Title 22 approved pasteurization disinfection process may be competitive with chlorination and UV-disinfection technologies in parts of the United States with high power costs and/or at locations with substantial waste heat from engine or turbine use (Ryan Pasteurization & Power, 2006).

8.7   Tertiary Filtration and Membrane Treatment

Disinfection is often differentiated from liquid–solid treatment processes, including clarification, tertiary filtration, and membrane treatment. However, these processes can be effective at removing pathogens, often resulting in the ability to decrease the size of downstream disinfection processes. Asano et al. (2007) reported typical log removals of various pathogens as <0.1 to 1.7 for primary sedimentation, <0.1 to 2 for secondary treatment, 0 to 4 for depth filtration, 0 to 6 for microfiltration, and 4 to 7 for reverse osmosis (enteric viruses showed the poorest removals among the bacteria, protozoa, and viruses studied). These processes and their pathogen removal characteristics are discussed in more detail in other chapters.

8.8   Advanced Oxidation

In water reclamation systems, advanced chemical-oxidation processes are used increasingly to remove trace organic pollutants. However, they have not yet found widespread use in wastewater treatment. Often, these processes use high doses of UV light and chemical oxidants, such as ozone or peroxide. Examples include the following:

- Peroxone (ozone + hydrogen peroxide),
- Ozone + UV,
- Peroxone + UV,
- Hydrogen peroxide + UV,
- Hydrogen peroxide + metal catalyst + UV,
- Fenton's reagent, and
- Sunlight + metal catalyst.

Use of these technologies often results in a high level of disinfection, in addition to chemical oxidation, because the doses required for chemical oxidation are often higher than for disinfection. Advanced oxidation is discussed in more detail in Chapter 16.

8.9   Combined Processes

The combination of two or more disinfection processes in the treatment train has some advantages. As previously discussed, the combination of UV and various oxidants can result in advanced oxidation processes to remove trace organic pollutants. Additionally, multiple processes in series can provide an additional margin of safety through additional log removals; if a pathogen is not inactivated as it passes through one process, the next process may inactivate it. Finally, combined processes may provide complementary pathogen inactivation, such that one process that provides effective inactivation in certain classes of pathogens, but not others, is complemented by another process that is known to be effective for the other pathogens.

Rose et al. (2004) evaluated the relative effect of loading conditions, process design, and operating parameters on the removal/inactivation of various pathogens at six WRRFs and water-reclamation facilities. Operation of biological treatment with higher levels of mixed-liquor suspended solids and longer MCRTs and under nitrification conditions tended to result in increased removal of pathogens.

9.0 References

Aieta, E. M.; Berg, J. D. (1986) A Review of Chlorine Dioxide in Drinking-Water Treatment. *J. Am. Water Works Assoc.*, **78** (6), 62–72.

American Public Health Association; American Water Works Association; Water Environment Federation (2012) *Standards Methods for the Examination of Water and Wastewater*, 22nd ed.; American Public Health Association: Washington, D.C.

American Water Works Association (1973) *Water Chlorination Principles and Practices*, 2nd ed., M20; American Water Works Association: Denver, Colorado.

American Water Works Association (1990) *Water Quality and Treatment*, 4th ed.; American Water Works Association: Denver, Colorado.

Asano, T.; Burton, F. L.; Leverenz, H. L.; Tsuchihashi, R.; Tchobanoglous, G. (2007) *Water Reuse: Issues, Technologies, and Applications*; Metcalf & Eddy: New York.

Beltran, F. J.; Encinar J. M., et al. (1992) Kinetic-Study of the Ozonation of Some Industrial Wastewaters. *Ozone-Science & Engineering* **14** (4): 303–327.

Bellamy, W.; Awad, J.; Wei, J.; Gramith, J. (1991) In-Line Ozone Dissolution Demonstration-Scale Evaluation. *Ozone Sci. Eng.*, **13** (5), 559–591.

Bingham, A. K.; Jarroll, E. L.; Meyer, E. A.; Radulescu, S. (1979) Induction of *Giardia* Excystation and the Effect of Temperature on Cyst Viability as Compared by Eosin-Exclusion and In Vitro Excystation. In *Waterborne Transmission of Giardiasis*, EPA-600/9-79-001; Jakubowski, W.; Hoff, J. C. Eds.; U.S. Environmental Protection Agency: Cincinnati, Ohio.

Black & Veatch Corporation (2010) *White's Handbook of Chlorination and Alternaive Disinfectants*, 5th ed.; John Wiley & Sons: Hoboken, New Jersey.

Blanco-Galvez, J.; Fernández-Ibáñez, P; Malato-Rodríguez, S. (2007) Solar Photocatalytic Detoxification and Disinfection of Water: Recent Overview. *J. Solar Energy Eng.*, **129**, 4–14.

Blatchley, E. R. III; Bastian, K. C.; Duggirala, R.; Hunt, B. A.; Alleman, J. E.; Wood, W. L.; Moore, M.; Anderson, B. L.; Gasvoda, M.; Schuerch, P. (1993) Large-Scale Pilot Investigation of Ultraviolet Disinfection. *Proceedings of the Water Environment Federation Specialty Conference: Planning, Design, and Operation of Effluent Disinfection Systems*, Whippany, New Jersey, May 23–25; Water Environment Federation: Alexandria, Virginia, 417.

Blatchley, E. R. III; Gong, W. L.; Rose, J. B.; Huffman, D. E.; Otaki, M.; Lisle, J. T. (2005) *Effects of Wastewater on Human Health*, Water Environment Research Foundation Report 99-HHE-1; Water Environment Research Foundation: Alexandria, Virginia.

Blatchley, E. R. III; Hunt, B. A. (1994) Bioassay for Full Scale UV Disinfection Systems. *Water Sci. Technol.*, **30** (4), 115–123.

Blatchley, E. R. III; Wood, W. L.; Schuerch, P. (1995) UV Pilot Testing: Intensity Distributions and Hydrodynamics. *J. Environ. Eng.*, **121**, 258–262.

Boner, M.; Kim, J. Y.; Muller, R. J. (2002) Is Bromine Disinfection a Viable Wet Weather Solution? *Proceedings of the Water Environment Federation Disinfection Specialty Conference,* St. Petersburg, Florida, Feb 17–20; Water Environment Federation: Alexandria, Virginia.

Buffle, M. O.; Schumacher, J.; Salhi, E.; Jekel, M.; Von Gunten, U. (2006) Measurement of the initial phase of ozone decomposition in water and wastewater by means of a continuous quench-flow system: application to disinfection and pharmaceutical oxidation. *Water Research,* **40,** 1884–1894.

Buffle, M. O.; Schumacher, J.; Meylan, S.; Jekel, M.; von Gunten, U. (2006). Ozonation and advanced oxidation of wastewater: Effect of O3 dose, pH, DOM and HO•-scavengers on ozone decomposition and HO• generation. *Ozone: Science and Engineering,* **28,** 247–259.

Cabelli, V. J. (1977) Indicators of Recreational Water Quality. In *Bacterial Indicators/Health Hazards Associated with Water,* Technical Publication 635; Hoadley, A. W.; Dutka, B. J. Eds.; ASTM International: West Conshohocken, Pennsylvania.

Cabelli, V. J.; Dufour, A. P.; Levin, M. A.; Haberman, P. W. (1976) The Impact of Pollution on Marine Bathing Beaches: An Epidemiological Study. In *Middle Atlantic Continental Shelf and the New York Bight, Limnology and Oceanography, Special Symp.,* Vol. 2, Gross, G. (Ed.); American Society of *Limnology and Oceanography: Waco, Texas, 424.*

Cabelli, V. J.; Dufour, A. P.; Levin, M. A.; McCabe, L. J.; Haberman, P. W. (1975a) Relationship of Microbial Indicators to Health Effects at Marine Bathing Beaches. Paper presented at *Annual Meeting of the American Public Health Association,* Chicago, Illinois: American Public Health Association: Washington, D.C.

Cabelli, V. J.; Levin, M. A.; Dufour, A. P.; McCabe, L. J. (1975b) The Development of Criteria for Recreation Water. In *Discharge of Sewage from Sea Outfalls*; Gameson, A. L. H. Ed.; Pergamon Press: New York.

California DHS (2006 California Wastewater Reclamation Criteria, Title 22, Division 4, Chapter 3, of the California Code of Regulations002E.

Calmer, J. C. (1993) Chlorine Mixing Energy Requirements for Disinfection of Municipal Effluents. *Proceedings of the Water Environment Federation Specialty Conference: Planning, Design, and Operation of Effluent Disinfection Systems,* Whippany, New Jersey, May 23–25; Water Environment Federation: Alexandria, Virginia.

Calmer, J. C.; Adams, R. M. (1977) *Design Guide Chlorination-Dechlorination Contact Facilities*; Kennedy/Jenks Engineers: San Francisco, California.

Calmer, J. C., et al. (1994) Dynamics of Coliform Regrowth in a Dechlorinated Secondary Effluent. *Proceedings of the 67th Annual Water Environment Federation Technical Exposition and Conference* [CD-ROM], Chicago, Illinois, Oct 15–19; Water Environment Federation: Alexandria, Virginia.

Camper, A. K.; McFeters, G. A. (1979) Chlorine Injury and the Enumeration of Waterborne Coliform Bacteria. *Appl. Environ. Microbiol.,* **37,** 633–641.

Casson, L. W.; Sorber, C. A.; Palmer, R. H.; Enrico, A.; Gupta, P. (1992) HIV Survivability in Wastewater. *Water Environ. Res.,* **64,** 213–215.

Casson, L. W.; Sorber, C. A.; Sykora, J. L.; Gavaghan, P. D.; Shapiro, M. A.; Jakubowski, W. (1990) Giardia in Wastewater—Effect of Treatment. *J. Water Pollut. Control Fed.,* **62,** 670–675.

Center for Disease Control (CDC) (2008) CDC Surveillance for Waterborne Disease and Outbreaks Associated with Recreational Water Use and Other Aquatic Facility-Associated Health Events—United States, 2005–2006, CDC website, CDC, Atlanta, Georgia.

Chlorine Institute (1986) *The Chlorine Manual,* 5th ed.; Chlorine Institute: Washington, D.C.

Cho, M.; Chung. H.; Yoon, J. (2003) Disinfection of Water Containing Natural Organic Matter using Ozone-Initiated Radical Reactions, *Appl. Environ. Microbiol.,* **69** (4), 2284–2291.

Chramosta, N.; Delaat, J., et al. (1993) Rate Constants for Reaction of Hydroxyl Radicals with S-Triazines. *Environ. Technol.,* **14** (3), 215–226.

Clarke, N. A.; Berg, J.; Kabler, P. W., Chang, S. L. (1964) Human Enteric Viruses in Water: Source, Survival and Removability. *Advances in Water Pollution Research,* Vol 2, Pergamon Press: London, U.K., 523.

Compressed Gas Association (1988) *Sulfur Dioxide,* 4th ed., Pamphlet G-3; Compressed Gas Association: Chantilly, Virginia.

CRC (1990) *Handbook of Chemistry and Physics,* 71st ed.; Lide D. L. Ed.; CRC Press: Boca Raton, Florida.

Crockett, C. S. (2007) The Role of Wastewater Treatment in Protecting Water Supplies Against Emerging Pathogens. *Water Environ. Res.,* **79,** 221–232.

Curds, C. R. (1992) *Protozoa in the Water Industry;* Cambridge University Press: Cambridge, United Kingdom.

Darby, J.; Emerick, R.; Loge, F.; Tchobanoglous, G. (1999) *The Effect of Upstream Treatment Processes on UV Disinfection Performance;* Water Environment Research Foundation: Alexandria, Virginia.

Darby, J. L.; Snider, K. E.; Tchobanoglous, G. (1993) Ultraviolet Disinfection for Wastewater Reclamation and Reuse Subject to Restrictive Standards. *Water Environ. Res.,* **65,** 169–180.

DuPont, H. L.; Chappell, C. L.; Sterling, C. R.; Okhuysen, P. C.; Rose, J. B.; Jakubowski, W. (1995) The Infectivity of *Cryptosporidium parvum* in Healthy Volunteers. *N. Engl. J. Med.,* **332,** 855–859.

DVGW (2006) *UV Devices for the Disinfection for Drinking Water Supply;* German Association for Gas and Water: Bonn, Germany.

Emerick, R. W.; Borroum, Y. (2005) Bioassay Comparison Of Similar Pilot- And Full-Scale UV Disinfection Systems, *Proceedings of the 78th Annual Water Environment Federation Technical Exposition and Conference* [CD-ROM], Washington, D.C., Oct 31–Nov. 2; Water Environment Federation: Alexandria, Virginia.

Emerick, R. W.; Darby, J. L. (1993) Ultraviolet Light Disinfection of Secondary Effluents: Predicting Performance Based on Water Quality Parameters. *Proceedings of the Water Environment Federation Specialty Conference: Planning, Design, and Operation of Effluent Disinfection Systems,* Whippany, New Jersey, May 23–25; Water Environment Federation: Alexandria, Virginia, 187.

Finger, R. E.; Harrington, D.; Paxton, L. E. (1985) Development of an On-Line Zero Chlorine Residual Measurement and Control System. *J. Water Pollut. Control Fed.,* **57,** 1068–1073.

Gates, F. L. (1929) A Study of the Bactericidal Action of Ultraviolet Light II. The Effects of Various Environmental Factors and Conditions. *J. Gen. Physiol.,* **13,** 249–260.

Gaudy, A. F.; Gaudy, E. T. (1980) *Microbiology for Environmental Scientists and Engineers;* McGraw-Hill: New York.

Gehr, R.; Pinto, D.; Santamaria, M.; Brenner, B. G. (2000) Fouling of UV Lamps with Varying Influent Water Quality. *Proceedings of the 2000 Water Environment Federation Disinfection Specialty Conference,* New Orleans, Louisiana, March 15–18; Water Environment Federation: Alexandria, Virginia.

Gong, G. (2002) Ph.D. Dissertation, Department of Civil and Environmental Engineering, Purdue University: West Lafayette, Indiana.

Gottschalk, C.; Libra, J. A.; Saupe, A. (2010) *Ozonation of Water and Waste Water: A Practical Guide to Understanding Ozone and Its Applications*; Wiley-VCH: Weinheim, Germany.

Great Lakes–Upper Mississippi River Board of State and Provincial Public Health and Environmental Managers (2014) *Recommended Standards for Wastewater Facilities*; Health Education Services: Albany, New York.

Haas, C. N.; Englebrecht, R. S. (1980) Physiological Alterations of Vegetative Microorganisms Resulting from Aqueous Chlorination. *J. Water Pollut. Control Fed.*, **52**, 1976–1989.

Haji Malayeri, A.; Mohseni, M.; Cairns, B.; Bolton, J. R. (2017) Fluence (UV Dose) Required to Achieve Incremental Log Inactivation of Bacteria, Protozoa, Viruses and Algae, *IUVA News*, **18** (3), 4–6 + Supp. Tables.

Hart, F. (1979) Improved Hydraulic Performance of Chlorine Contact Chambers. *J. Water Pollut. Control Fed.*, **51**, 2868–2875.

Helz, G. R.; Nweke, A. C. (1995) Incompleteness of Wastewater Dechlorination. *Environ. Sci. Technolol.*, **29** (4), 1018 -1022.

Henri, H.; de Recklinghausen et al. (1910) *Compt. Rend. Acad. Sci.*, **151**, 677–680.

Hoehn, R. C.; Rosenblatt, A. A.; Gates, D. J. (1996) Considerations for Chlorine dioxide Treatment in Drinking Water. Proceedings American Water Works Association Conference on Water Quality Technology, Boston, Massachusetts; American Water Works Association: Denver, Colorado.

Hunt, N. K.; Marinas, B. J. (1999) Kinetics of *Escherichia Coli* Inactivation with Ozone: Chemical and Inactivation Kinetics. *Water Res.*, **33**, 2633–2641.

HydroQual, Inc. (1994) *Disinfection Effectiveness of Combined Sewer Overflows*, Draft report; U.S. Environmental Protection Agency: Washington, D.C.

International Fire Code Institute (1994) *Uniform Fire Code*; International Fire Code Institute: Austin, Texas.

Jafvert, C. T.; Valentine, R. L. (1992) Reaction Scheme for the Chlorination of Ammoniacal Water. *Environ. Sci. Technol.*, **26**, 577–586.

Jagger, J. (1967) *Introduction to Research in Ultra-Violet Photobiology*; Prentice-Hall, Inc.: Englewood Cliffs, New Jersey.

Jeffcoat, S., CH2M Hill, Inc. (2005) Personal communication with Christine Cotton, Malcolm Pirnie, Inc., regarding fouling in Clayton County. Tucson, Arizona.

Jeyanayagam, S.; Cotton, C. (2002) Practical Considerations in the Use of UV Light for Drinking Water Disinfection. *Proceedings of the CASE/ASCE Joint Conference: An International Perspective on Environmental Engineering*.

Kim, C. K.; Min, K. H. (1979) Inactivation of Bacteriophage f2 with Chlorine. *Misaengmul Hakhoe Chi*, **16** (2), 62.

Kohn, T.; Grandbots, M.; McNeill, K; Nelson, K. (2007) Associations with Natural Organic Matter Enhances the Sunlight-Mediated Inactivation of MS2 Coliphage by Singlet Oxygen. *Environ. Sci. Technol.*, **41** (13), 4626–4632.

Kohn, T.; Nelson, K. (2007) Sunlight-Mediated Inactivation of MS2 Coliphage via Exogenous Singlet Oxygen Produced by Sensitizers in Natural Waters. *Environ. Sci. Technol.*, **41** (1), 192–197.

Lanao, M.; Ormad, M. P.; Ibarz, C.; Miguel, N.; Ovelleiro, J. L. (2008) Bactericidal Effectiveness of O_3, O_3/H_2O_2 and O_3/TiO_2 on *Clostridium perfrigens*. *Ozone: Sci. Eng.*, **30** (6), 431–438.

Lee, Y.; Gerrity, D.; Lee, M.; Bogeat, A. E.; Salhi, E.; Gamage, S.; Von Gunten, U. (2013) Prediction of Micropollutant Elimination During Ozonation of Municipal Wastewater Effluents: Use of Kinetic and Water Specific Information. *Environ. Sci. Technol.*, **47** (11), 5872–5881.

Leong, L. Y. C.; Kuo, J.; Tang, C. (2008) *Disinfection of Wastewater Effluent—Comparison of Alternative Technologies*; Water Environment Research Foundation: Alexandria, Virginia.

Lin, L.; Johnston, C. T.; Blatchley E. R. III (1999) Inorganic Fouling at Quartz: Water Interfaces in Ultraviolet Photoreactors—I. Chemical Characterization. *Water Res.,* **33** (15), 3321–3329.

Linden, K. G. (2000) UV Dose Verification Using Chemical Actinometry and Biodosimetry Methods. *Proc. UV 2000: A Technical Symposium, 2000.*

Linden, K.G.; Shin Gwy-am; Faubert G.; Cairns, W; and Sobsey M. D. (2002) UV Disinfection of *Giardia lamblia* Cysts in Water. *Environ. Sci. Technol.,* **36**, 2519–2522.

Lindenauer, K. G.; Darby, J. L. (1994) Ultraviolet Disinfection of Wastewater: Effect of Dose on Subsequent Photoreactivation. *Water Res.,* **28**, 805–817.

Loge, F.; Emerick, R.; Thompson, D.; Nelson, D.; Darby, J. (1999) Factors Influencing UV Disinfection Performance --Part 1: Light Penetration Into Wastewater Particles. *Water Environ. Res.,* **71**, 377–381.

Louie, D.; Fohrman, M. (1968) Hydraulic Model Studies of Chlorine Mixing and Contact Chambers. *J. Water Pollut. Control Fed.,* **40**, 174–184.

Marske, D. M.; Boyle, V. D. (1973) Chlorine Contact Chamber Design—A Field Evaluation. *Water Sewage Works,* **120**, 70–77.

McDougal, J. S. et al. (1985) Immunoassay for the Detection and Quantification of Infectious Human Retrovirus, Lymphadepathy-Associated Virus (LAV). *J. Immunol. Methodol.,* **76,** 171.

Metcalf & Eddy, Inc./AECOM (2013) *Wastewater Engineering: Treatment and Resource Recovery,* 5th ed.; McGraw-Hill: New York.

Meulemans, C. C. E. (1987) The Basic Principles of UV-Disinfection of Water. *Ozone Sci. Eng.,* **9,** 299–313.

Mills, J. (1973) *The Disinfection of Sewage by Chlorobromination.* American Chemical Society, Division of Water, Air and Waste Chemistry: Dallas, Texas.

Moore, A. C.; Herwaldt, B. L.; Craun, G. F.; Calderon, R. L.; Highsmith, A. K.; Juranek, D. D. (1994) Waterborne Disease in the United States, 1991 and 1992. *J. Am. Water Works Assoc.,* **86,** 87–98.

Nagy, R. (1964) Application and Measurement of Ultraviolet Radiation. *Am. Ind. Hyg. Assoc. J.,* **25,** 274–281.

Najm, I.; Trussell, R. R. (2000) NDMA Formation in Water and Wastewater. *Proc. 2000 WQTC,* Salt Lake City, Utah.

National Fire Protection Association; American National Standards Institute (2008) *National Electrical Code, an American National Standard,* NFPA No. 70-2008 ANSI C1-2008; National Fire Protection Association: Quincy, Massachusetts.

National Water Research Institute (1993) *UV Disinfection Guidelines for Wastewater Reclamation in California and UV Disinfection Research Needs Identification;* National Water Research Institute: Fountain Valley, California.

National Water Research Institute; American Water Works Association Research Foundation (2012) *Ultraviolet Disinfection Guidelines for Drinking Water and Reuse;* U.S. Environmental Protection Agency: Washington, D.C.

Nisipeanu, E.; Sami, M. (2004) Lighting the Way to Better Disinfection. https://www.edie.net/library/Lighting-the-way-to-UV-disinfection/707 (accessed July 2017).

O'Brien, R. T.; Newman, J. (1979) Structural and Compositional Changes Associated with Chlorine Inactivation of Polioviruses. *Appl. Environ. Microbiol.,* **38,** 1034–1039.

Occupational Safety and Health Administration (1996), Process Safety Management of Highly Hazardous Chemicals, Regulation 1910.119, 57 FR 23060, June 1, 1992; 61 FR 9227, March 7, 1996.

Olivieri, V. P. et al. (1980) Reaction of Chlorine and Chloramines with Nucleic Acids under Disinfection Conditions. In *Water Chlorination: Environmental Impact and Health Effects,* Vol. 3; Jolley, R. L.; Brungs, W. A.; Cumming, R. B.; Jacobs, V. A. Eds.; Ann Arbor Science: Ann Arbor, Michigan.

Prados, M.; Paillard, H.; Roche, P. (1995) Hydroxyl Radical Oxidation Processes for the Removal of Triazine from Natural Water. *Ozone Sci. Eng.,* **17** (2), 183–194.

Rakness, K. L.; Najm, I.; Elovitz, M.; Rexing, D.; Via, S. (2005) Cryptosporidium Log-Inactivation with Ozone Using Effluent CT10, Geometric Mean CT10, Extended Integrated CT10 and Extended CSTR Calculations. *Ozone Sci. Eng.,* **27** (5), 335–350.

Richard, M. (2001) Wastewater Pond Treatment System Operations Troubleshooting and Upgrade Workshop, July 18, Marysville, Washington; Pacific Northwest Pollution Control Association Northwest Washington Operators Section, Everett, Washington.

Roberts, P. V.; Aieta, E. M.; Berg, J. D.; Cooper, R. C. (1980) Comparison of Chlorine Dioxide And Chlorine In Wastewater Disinfection. *J. Water Pollut. Control Fed.,* **52**, 810–824.

Robertson, L. J.; Campbell, A. T.; Smith, H. V. (1992) Survival of *Cryptosporidium parvum* Oocysts Under Various Environmental Pressures. *Appl. Environ. Microbiol.,* **58**, 3494–3500.

Rose, J. B.; Farrah, S. R.; Harwood, V. J.; Levine, A. D.; Lukasik, J.; Menendez, P.; Scott, T. M. (2004) *Reductions of Pathogens, Indicator Bacteria, and Alternative Indicators by Wastewater Treatment and Reclamation Processes,* Water Environment Research Foundation Report 00-PUM-2T; Water Environment Research Foundation: Alexandria, Virginia.

Ryan Pasteurization & Power (2006) RP&P Wastewater Pasteurization System Validation Report Submitted to the California Department of Health and Hospitals. Ryan Pasteurization & Power: Geyserville, California.

Salveson, A.; Oliver, M.; Bourgeous, K.; and Mahar, E. (2004) Has something gone foul with your UV? The impact of sleeve fouling on delivered UV dose. *Proceedings,Water Environment Federation 77th Annual Technical Exhibition and Conference,* [CD-ROM] New Orleans, LA., Water Environment Federation, Alexandria, Virginia.

Spire, B.; Dormont, D.; Barré-Sinoussi, F.; Montagnier, L.; Chermann, J. C. (1985) Inactivation of LAV by Heat, Gamma Rays and Ultraviolet Light. *Lancet,* **1,** 188–189.

Spire, B.; Montagnier, L.; Barre-Sinoussi, F.; Chermann, J. C. (1984) Inactivation of LAV by Chemical Disinfectants. *Lancet,* **2,** 899–901.

Staehelin, J.; Hoigne, J. (1985) Decomposition of Ozone in Water in the Presence of Organic Solutes Acting as Promoters and Inhibitors of Radical Chain Reactions. *Environ. Sci. Technol.,* **19** (12), 1206–1213.

Swift, J. L.; Emerick, R.; Scheible, K.; Soroushian, F.; Putnam, L. B.; Sakaji, R. (2002) Treat, Disinfect, Reuse: New Guidelines for Water Reuse are Intended to Ensure that Ultraviolet (UV) Disinfection Systems are Effective. *Water Environ. Technol.,* Nov.

Swift, J. L.; Wilson, J. P.; Hunter, G. (2007) Implementing Local Limits for the Control of WWTP Effluent Ultraviolet Transmittance. *Proceedings of the 80th Annual Water Environment Federation Technical Exposition and Conference* [CD-ROM], Anaheim, California, Oct 13–17; Water Environment Federation: Alexandria, Virginia.

Swift, J. L.; Wilson, J. P.; Johnson, M.; Jacobsen, B. (2001) The Impact of UV-Absorbing Wastewater from a Printed Circuit Board Manufacturing Facility on the Performance of a Municipal UV Disinfection System. *Proceedings of the 74th Annual Water Environment Federation Technical Exposition and Conference* [CD-ROM], Atlanta, Georgia, Oct 13–17; Water Environment Federation: Alexandria, Virginia.

Swift, J. L.; Wilson, J. P.; Welch, D.; Johnson, M.; Conley, P.; Bowman, B. (2000) An Assessment of Operation and Maintenance Costs for Ultraviolet Disinfection Systems. *Proceedings of the 73rd Annual Water Environment Federation Technical Exposition and Conference* [CD-ROM], Anaheim, California, Oct 14–18; Water Environment Federation: Alexandria, Virginia.

Tchobanoglous, G.; Stensel, H.D.; Burton, F. L. (Metcalf and Eddy) (2003) *Wastewater Engineering: Treatment and Reuse,* 4th ed.; McGraw-Hill: New York.

U.S. Environmental Protection Agency (1979) *Health Effects Criteria for Fresh Recreational Waters,* EPA-600/1-84-004; U.S. Environmental Protection Agency: Cincinnati, Ohio.

U.S. Environmental Protection Agency (1984) *Ambient Water Quality Criteria for Bacteria,* EPA-44D/5-84-002; U.S. Environmental Protection Agency: Cincinnati, Ohio.

U.S. Environmental Protection Agency (1986a) *Municipal Wastewater Disinfection Design Manual,* EPA-625/1-86-021; U.S. Environmental Protection Agency: Cincinnati, Ohio.

U.S. Environmental Protection Agency (1986b) *Quality Criteria for Water,* EPA-440/5-86-001; U.S. Environmental Protection Agency: Washington, D.C.

U.S. Environmental Protection Agency (1992a) *Control of Pathogens and Vector Attraction in Sewage Sludge,* EPA-625/R-92-013; U.S. Environmental Protection Agency: Washington, D.C.

U.S. Environmental Protection Agency (1992b) Draft Report, March 1992. U.S. Environmental Protection Agency: Washington, D.C.

U.S. Environmental Protection Agency (1992c) *Ultraviolet Disinfection Technology Assessment,* EPA-832/R-92-004; U.S. Environmental Protection Agency: Washington, D.C.

U.S. Environmental Protection Agency (1993) *Code of Federal Regulations*, Title 29.

U.S. Environmental Protection Agency (1996) *Design Manual for Municipal Wastewater Disinfection;* U.S. Environmental Protection Agency: Washington, D.C.

U.S. Environmental Protection Agency (1999a) Chapter 4, *Alternative Disinfectants and Oxidants Guidance Manual;* U.S. Environmental Protection Agency: Washington, D.C.

U.S. Environmental Protection Agency (1999b) *Inactivation of Cryptosporidium parvum oocysts in Drinking Water, Calgon Carbon Corporation's Sentinel™ Ultraviolet Reactor,* EPA-600/R-98-160; U.S. Environmental Protection Agency: Washington, D.C.

U.S. Environmental Protection Agency (2002) *Generic Verification Protocol for Secondary Effluent and Water Reuse Disinfection Applications.* U.S. Environmental Protection Agency: Washington, D.C.

U.S. Environmental Protection Agency (2006) *Ultraviolet Disinfection Guidance Manual for the Final Long-Term 2 Enhanced Surface Water Treatment Rule,* EPA-815/R-06-007; U.S. Environmental Protection Agency: Washington, D.C.

von Gunten, U. (2003) Ozonation of Drinking Water: Part 1 Ozonation Kinetics and Product Formation. *Water Res.,* **37** (7), 1443–1467.

Von Sonntag, C.; Von Gunten, U. (2012) *Chemistry of Ozone in Water and Wastewater Treatment: From Basic Principles to Applications;* IWA Publishing: London, U.K.

Water Environment Federation (1996) *Wastewater Disinfection,* Manual of Practice FD-10; Water Environment Federation: Alexandria, Virginia.

Water Environment Federation (2009) *An Introduction to Process Modeling for Designers, Manual of Practice No. 31*; Water Environment Federation: Alexandria, Virginia.

Water Environment Federation (2015) *Ultraviolet Disinfection for Wastewater*; Water Environment Federation: Alexandria, Virginia.

Water Environment Research Foundation (2007) *Disinfection of Wastewater Effluent— Comparison of Alternative Technologies*, Report No. 04-HHE-4; Water Environment Research Foundation: Alexandria, Virginia.

Water Pollution Control Federation (1984) *Wastewater Disinfection, A State-of-the-Art Report*; Water Pollution Control Federation: Alexandria, Virginia.

Whitby, G. E.; Palmateer, G. (1993) The Effects of UV Transmission, Suspended Solids, Wastewater Mixtures and Photoreactivation in Wastewater Treated with UV Light. *Proceedings of the Water Environment Federation Specialty Conference: Planning, Design, and Operation of Effluent Disinfection Systems,* Whippany, New Jersey, May 23–25; Water Environment Federation: Alexandria, Virginia, 24.

Wolfe, R. L. (1990) Ultraviolet Disinfection of Potable Water. *Environ. Sci. Technol.*, **24**, 768–773.

Wyss, O.; Stockton, J. R. (1947) The Germicidal Action of Bromine. *Arch. Biochem.*, **12**, 267–271.

Xu, P.; Janex, M.; Savoye, P.; Cockx, A.; Lazarova, V. (2002) Wastewater Disinfection by Ozone: Main Parameters for Process Design. *Water Res.*, **25**, 761–773.

Yip, R. W.; Konasewich, D. E. (1972) Ultraviolet Sterilization of Water—Its Potential and Limitations. *Water Pollut. Control*, **14**, 14–18.

10.0 Suggested Readings

Aieta, E. M.; Berg, J. D.; Robert, P. V.; Cooper, R. C. (1980) Comparison of Chlorine Dioxide and Chlorine in Wastewater Disinfection. *J. Water Pollut. Control Fed.*, **52** (4), 810–822.

Asbury, C.; Coler, R. (1980) Toxicity of Dissolved Ozone to Fish Eggs and Larvae. *J. Water Pollut. Control Fed.*, **52,** 1990–1996.

Ashley, R. M.; Souter, N.; Butler, D.; Davies, J.; Dunkerley, J.; Hendry, S. (1999) Assessment of the Sustainability of Alternatives for the Disposal of Domestic Sanitary Waste. *Water Science Technol.*, **39** (5), 251–258.

Blatchley, E. R. III; Duggirala, R.; Chiu, K. P.; Noesen, M.; Jaques, R.; Schuerch, P. (1994) Macro-Scale Hydraulic Behavior in Open Channel UV Systems. *Proceedings of the 67th Annual Water Environment Federation Technical Exposition and Conference* [CD-ROM], Chicago, Illinois, Oct 15–19; Water Environment Federation: Alexandria, Virginia.

Blatchley E. R. III; Schmude, B. M.; Cole, K. A.; Hamilton, D. (2000) Analysis of Process Performance in Polychromatic UV Disinfection Systems. *Proceedings of the 73rd Annual Water Environment Federation Technical Exposition and Conference* [CD-ROM], Anaheim, California, Oct 14–18; Water Environment Federation: Alexandria, Virginia.

Boliden Intertrade (1979) *Sulfur Dioxide Technical Handbook*; Boliden Intertrade: Atlanta, Georgia.

Cabelli, V. J. (1980) *Health Effects Quality Criteria for Marine Recreational Waters*, EPA-600/1-80-031. U.S. Environmental Protection Agency: Cincinnati, Ohio.

Chick, H. (1908) An Investigation of the Laws of Disinfection. *J. Hyg.*, **8**, 92–158.

Collins, H. F.; Selleck, R. E. (1972) *Process Kinetics of Wastewater Chlorination*, SERL Rep. 72-5. University of California: Berkeley, California.

Comptroller General of the United States (1977) *Unnecessary and Harmful Levels of Domestic Sewage Chlorination Should Be Stopped*, CED-77-108, Report to Congress; U.S. General Accounting Office: Washington, D.C.

Craun, G. F. (1988) Surface Water Supplies and Health. *J. Am. Water Works Assoc.*, **80** (2), 40–52.

Emerick, R.; Loge, F.; Ginn, T.; Darby, J. (2000) Modeling the Inactivation of Particle-Associated Coliform. *Water Environ. Res.*, **72**, 432–438.

Emerick, R.; Loge, F.; Thompson, D.; Darby, J. (1999) Factors Influencing UV Disinfection Performance—Part 2: Association of Coliform Bacteria with Wastewater Particles. *Water Environ. Res.*, **71**, 1178–1187.

Emerick, R.; Salveson, A.; Tchobanoglous, G.; Sakaji, R.; Swift, J. L. (2003) Is It Good Enough for Reuse? New Guidelines for Water Reuse Spell Out How to Test and Design Ultraviolet Disinfection Systems. *Water Environ. Technol.*, **15**, 40–45.

Fogler, H. S. (1993) *Elements of Chemical Reaction Engineering*; Prentice Hall: Englewood Cliffs, New Jersey.

Francey, D. S.; Hart, T. L.; Virostek, C. M. (1996) Effects of Receiving Water Quality and Wastewater Treatment on Injury, Survival and Regrowth of Fecal Indicator Bacteria Implications, *Water Resources Investigation Report 96-4199*; U.S. Geological Survey: Reston, Virginia.

Giuliano, L. (1997) Nitrite Lock Phenomenon. Power Point Presentation, City of Las Vegas Water Pollution Facility: Las Vegas, Nevada.

Gottschalk, C.; Libra, J. A.; Saupe, A. (2010) *Ozonation of Water and Waste Water: A practical guide to understanding ozone and its applications*; Wiley-VCH: Weinheim, Germany.

Green, D. E.; Stumpf, P. K. (1946) The Mode of Action of Chlorine. *J. Am. Water Works Assoc.*, **38**, 1301–1305.

Haas, C. N.; Hornberger, J. C.; Anmangandla, U.; Heath, M.; Jacangelo, J. G. (1994) A Volumetric Method for Assessing *Giardia* Inactivation. *J. Am. Water Works Assoc.*, **86**, 115–120.

Haas, C. N.; Rose, J. B.; Gerba, C. P. (1999) *Quantitative Microbial Risk Assessment*; John Wiley and Sons, Inc.: New York.

Hunt, B. A. (1992) *Ultraviolet Dosimetry Using Microbial Indicators and Theoretical Modelling*. M.S. Thesis, School of Civil Engineering, Purdue University: West Lafayette, Indiana.

HydroQual, Inc. (1992a) A Review of UV Disinfection Process Design Consideration, Draft; U.S. Environmental Protection Agency: Washington, D.C.

Jacob, S. M.; Dranoff, J. S. (1970) Light Intensity Profiles in a Perfectly Mixed Photoreactor. *Am. Inst. Chem. Eng. J.*, **16**, 359–363.

Kreft, P.; Scheible, O. K.; Venosa, A. (1986) Hydraulic Studies and Cleaning Evaluations of Ultraviolet Disinfection Units. *J. Water Pollut. Control Fed.*, **58**, 1129–1137.

Lev, O.; Regli, S. (1992) Evaluation of Ozone Disinfection Systems: Characteristic Time T. *J. Environ. Eng.*, **118**, 268–285.

Linden, K. G.; Oliver, J. D.; Sobsey, M. D., Shin, G. (2004) *Fate and Persistence of Pathogens Subjected to Ultraviolet Light and Chlorine Disinfection*, Water Environment Research Foundation Report 99-HHE-1; Water Environment Research Foundation: Alexandria, Virginia.

Loge, F.; Emerick, R.; Thompson, D.; Nelson, D.; Darby, J. (1999) Factors Influencing UV Disinfection Performance—Part 1: Light Penetration Into Wastewater Particles. *Water Environ. Res.*, **71**, 377–381.

National Fire Protection Association (2008) National Fire Codes. National Fire Protection Association: Quincy, Massachusetts.

Nieuwstad, T.; Havelaar, A. H.; Van Olphen, M. (1991) Hydraulic and Microbiological Characterization of Reactors for Ultraviolet Disinfection of Secondary Wastewater Effluents. *Am. Soc. Civ. Eng. J. Water Resour. Plann. Manage. Div.,* **25,** 775–783.

Oliver, B. G.; Cosgrove, E. G. (1975) The Disinfection of Sewage Treatment Plant Effluents Using Ultraviolet Light. *Can. J. Chem. Eng.,* **53,** 170–174.

Oliver, M. (2002) UV Cleaning System Performance Validation. *Proceedings of the 2003 Water Environment Federation Disinfection Conference,* Saint Petersburg, Florida; Water Environment Federation: Alexandria, Virginia.

Ongerth, J. E.; Stibbs, H. H. (1987) Identification of *Cryptosporidium* Oocysts in River Water. *Appl. Environ. Microbiol.,* **53,** 672–676.

Petri, B.; Sealey, L.; Mohamed, O. (2007) Validated CFD Models of UV Disinfection; *Proceedings of the Water Environment Federation Disinfection Specialty Conference;* Pittsburgh, Pennsylvania, Feb 4–7; Water Environment Federation: Alexandria, Virginia.

Rokjer, D.; Valade, M.; Keesler, D.; Borsykowsky, M. (2003) Computer Modeling of UV Reactors for Validation Purposes. *American Water Works Association Annual Conference and Exposition,* Anaheim, California, June 15–19; American Water Works Association: Denver, Colorado.

Rose, J. B.; Gerba, C. P.; Jakubowski, W. (1991) Survey of Potable Water Supplies for *Cryptosporidium* and *Giardia. Environ. Sci. Technol.,* **26,** 1393–1400.

Sabin, A. B. (1957) Properties of Attenuated Poliovirus and Their Behavior in Human Beings. In *Cellular Biology, Nucleic Acids and Viruses,* Vol. 5; New York Academy of Sciences: New York.

Santoro, D.; Bartrand, T.; Greene, D.; Farouk, B; Haas, C.; Notarnicola, M.; Liberti, L. (2005) Use of CFD for Wastewater Disinfection Process Analysis: *E. coli* Inactivation with Peroxyacetic Acid (PAA). *Int. J. Chem. Reactor Eng.,* **3** (A46), 1–12.

Schulz, C. R.; Knatz, C. L.; Yelpo, J. (2004) Optimizing the Design of a Medium Pressure UV Reactor Using Computational Fluid Dynamics and Irradiance Modeling. *Proceedings of the American Water Works Association Water Quality Technology Conference,* San Antonio, Texas, Nov 14–18; American Water Works Association: Denver, Colorado.

Severin (1980) Disinfection of Municipal Effluents with Ultraviolet Light. *J. Water Pollut. Control Fed.,* **52,** 2007–2018.

Sobsey, M.; Olsen, B. (1983) Microbial Agents of Waterborne Disease. In *Assessment of Microbiology and Turbidity Standards for Drinking Water,* EPA-570/4-83-001; U.S. Environmental Protection Agency: Washington, D.C.

Southern Building Congress International (1991) Standard Fire Prevention Code. Southern Building Congress International: Birmingham, Alabama.

Teefy, S.; Singer, P. (1990) Performance and Analysis of Tracer Tests to Determine Compliance of a Disinfection Scheme with the SWTR. *J. Am. Water Works Assoc.,* **82,** 88–98.

U.S. Environmental Protection Agency (1976) *Disinfection of Wastewater,* EPA-430/9-75-012; U.S. Environmental Protection Agency: Washington, D.C.

U.S. Environmental Protection Agency (2003) *Ultraviolet Disinfection Guidance Manual,* EPA-815-D-03-007, U.S. Environmental Protection Agency: Washington, D.C.

Wyss, O.; Stockton, J. R. (1947) The Germicidal Action of Bromine. *Arch. Biochem.,* **12,** 267–271.

Xin, Z. (2004) *Disinfection Development: The Rise of UV in China;* IWA Publishing: London, United Kingdom.

CHAPTER **18**

Introduction to Solids Management

K. Richard Tsang, Ph.D., P.E., BCEE; Amber Batson, P.E.;
Matt Seib, Ph.D.; and Eric Spargimino, P.E., LEED AP

1.0   Introduction

Solids management is an important aspect of wastewater treatment design because of the interrelationships between the liquid and solids processes. Chapters 18 to 25 cover the solids generated during sedimentation and/or biological and chemical treatment of raw wastewater. [For information on minor residuals streams (e.g., scum, grit, and screenings) and their removal from wastewater, see Chapter 9 on preliminary treatment.]

It is impossible to completely separate liquid- and solids-handling processes, and engineers must consider their relationships when designing water resource recovery facilities (WRRFs). The liquid treatment processes chosen will affect both the amount of solids generated and their characteristics, which in turn will affect the choice of thickening, conditioning, and dewatering processes. Furthermore, the recycle streams from solids treatment processes can affect liquid processes. For example, dewatering anaerobically digested biosolids produces a sidestream with high ammonia and phosphorus concentrations, which will increase these loadings to the facility's liquid treatment processes. This chapter includes a mass balance example that illustrates these relationships.

Chapters 18 to 25 describe accepted methods and procedures for planning, designing, and constructing solids-handling processes and equipment. It discusses current U.S. regulations, methods for determining solids quantities, and descriptions of typical characteristics associated with the various residuals generated in WRRFs. Finally, although there is a thorough discussion of degritting and screening as it relates to raw wastewater in Chapter 9, there is a brief discussion in chapters from this point forward of how these processes relate to solids management.

Chapter 19 discusses methods for transporting and storing residuals and biosolids. Chapter 20 discusses solids conditioning, including the types of chemicals involved, factors that affect conditioning, chemical-feed systems, and dose optimization. Chapters 21 and 22 describe thickening and dewatering processes; they include information on process design conditions and criteria, key process variables, and ancillary equipment. Chapter 23 covers biological- and chemical-stabilization processes (e.g., aerobic and anaerobic digestion, composting, and alkaline stabilization). Chapter 24 discusses thermal processes (e.g., incineration) for stabilizing, drying, or destroying solids. Chapter 25 describes land application and other biosolids use and disposal practices.

2.0   Definitions

The Water Environment Federation has adopted the following terminology for wastewater residuals.

Sludge is any residual produced during primary, secondary, or advanced wastewater treatment that has not undergone any process to reduce pathogens or vector attraction. Another common term for this is raw sludge. The term sludge should be used with a specific process descriptor (e.g., primary sludge, waste activated sludge, or secondary sludge).

Biosolids is any sludge that has been stabilized to meet the criteria in the U.S. Environmental Protection Agency's (U.S. EPA's) 40 CFR 503 regulations and, therefore, can be beneficially used. Stabilization processes include anaerobic digestion, aerobic digestion, alkaline stabilization, and composting. Additionally, heat drying produces biosolids that can also be used beneficially.

Solids and residuals are terms used when it is uncertain whether the material meets Part 503 criteria (e.g., during thickening, because stabilization may occur either before or after this process). In this manual, the terms solids and residuals will be used as general references and for general descriptors (e.g., solids handling).

Land application is the process of adding bulk or bagged biosolids to soil at agronomic rates—the amount needed to provide enough nutrients (e.g., nitrogen, phosphorus, and potash) for optimal plant growth while minimizing the likelihood that they pass below the root zone and leach to groundwater. Land application can involve agricultural land (e.g., fields used to produce food, feed, and fiber crops); pasture and rangeland; nonagricultural land (e.g., forests); public-contact sites (e.g., parks and golf courses); disturbed lands (e.g., mine spoils, construction sites, and gravel pits); and home lawns and gardens.

3.0   Regulations

When designing any solids or biosolids project, engineers should take into account the prevailing local, state, and federal regulations. Most, if not all U.S. states, have adopted the federal regulations for managing wastewater residuals (40 CFR 503); some have promulgated regulations that impose even stricter requirements. Designers should begin by evaluating the regulatory consequences of any proposed action, because the choice of treatment process(es) is often governed as much by regulatory constraints as by process performance and cost.

3.1   40 CFR 503 (Part 503)

3.1.1   Background

The U.S. Environmental Protection Agency's biosolids regulations (40 CFR 503) address the use and disposal of solids generated during the treatment of domestic wastewater and septage. They are organized into five subparts: general provisions, land application, surface disposal, pathogen and vector-attraction reduction, and incineration (Fed. Reg., Feb. 19, 1993).

Part 503 biosolids standards are typically incorporated into a WRRF's National Pollutant Discharge Elimination System (NPDES) permit. Such permits are issued by U.S. EPA or by states with agency-approved biosolids programs. As of 2016, eight states are delegated authority to administer the NPDES biosolids program—Arizona, Michigan, Ohio, Oklahoma, South Dakota, Texas, Utah, and Wisconsin. Any facility that treats domestic wastewater (e.g., facilities that generate, treat, or provide disposal for solids, including nondischarging and "sludge-only" facilities) must have a permit. That said, the Part 503 rule was written to be self-implementing, which means that WRRFs are expected to follow it even before their permits are issued. It can be enforced either by U.S. EPA or via citizen lawsuits.

The agency continues to review Part 503—especially the land-application provisions to ensure that current regulations protect public health and the environment.

If the WRRF's solids are disposed in municipal solid waste landfills or used as landfill cover material, they must comply with the requirements of 40 CFR 258 (municipal solid waste landfill regulations) rather than with Part 503.

3.1.2   General Requirements

Biosolids generators are responsible for complying with Part 503. The regulation establishes two sets of criteria for heavy metals—Pollutant Concentrations and Pollutant Ceiling Concentrations—and two sets of criteria for pathogen densities—Class A and Class B. It also allows for two approaches to reduce vector attraction: treating the solids or using physical barriers.

Biosolids that meet the higher-quality criteria have fewer restrictions. The minimum requirements for a biosolids to qualify for land application are Pollutant Ceiling Concentrations, Class B requirements, and vector-attraction reduction requirements. Biosolids that meets Pollutant Concentration limits, Class A requirements, and vector-attraction reduction requirements can be land-applied without the additional precautions (site restrictions) required for the Class B biosolids. However, all land-appliers must meet the minimum monitoring, recordkeeping, and reporting requirements (no matter which type of biosolids is used).

3.1.3   Pollutant Limits

Before biosolids can be land-applied, its level of heavy metals cannot exceed either Pollutant Concentration limits or Pollutant Ceiling Concentrations and Cumulative Pollutant Loading Rates (see Table 18.1). Bulk biosolids that will be applied to lawns and home gardens must meet Pollutant Concentration limits. Biosolids sold or given away in bags or other containers must meet either Pollutant Concentration limits or Pollutant Ceiling Concentrations. Users should be directed to apply this material at rates based on Annual Pollutant Loading Rates.

3.1.4   Pathogen Limits

Part 503 labels biosolids either Class A or Class B based on their pathogen levels. Both types have been treated to reduce pathogens and minimize their ability to attract vectors

Pollutant	Ceiling Concentration Limits[a] (mg/kg)	Cumulative Pollutant Loading Rates (kg/ha)	"High-quality" Pollutant Concentration Limits[b] (mg/kg)	Annual Pollutant Loading Rates (kg/ha·a)
Arsenic	75	41	41	2.0
Cadmium	85	39	39	1.9
Copper	4300	1500	1500	75
Lead	840	300	300	15
Mercury	57	17	17	0.85
Molybdenum	75	—	—	—
Nickel	420	420	420	21
Selenium	100	100	100	5.0
Zinc	7500	2800	2800	140

[a]Absolute values.
[b]Monthly averages.

TABLE 18.1   Pollutant Limits for Land-Applied Biosolids (All Limits are on a Dry Weight Basis)

(e.g., rats). However, Class B biosolids still contain detectible levels of pathogens, while Class A biosolids are essentially pathogen-free. [If its metals levels are also low, then a Class A material is labeled an "exceptional quality" (EQ) biosolids.]

Both Class A and Class B biosolids can be land-applied per Part 503, but land-applying a Class B material involves buffer requirements, public-access limits, and crop-harvesting restrictions (see Table 18.2). These rules are intended to protect public health and enable microorganisms in the soil to degrade remaining pathogens.

Any biosolids being applied to lawns and home gardens—or sold or given away in bags or other containers—must meet Class A criteria. For information on selling or distributing biosolids (e.g., composted or heat-dried products), see Chapter 25. For information on product characterization and marketing approaches, also see Chapter 25.

3.1.4.1 Class A Requirements   To be considered Class A, biosolids must meet specific fecal coliform and salmonella limits at the time of use or disposal. Also, the requirements of one of the following alternatives must be met:

Alternative 1. Thermal treatment (one of four time/temperature regimes as seen in Table 18.3),

Alternative 2. High pH, high temperature treatment,

Alternative 3. Prior testing for enteric virus/viable helminth ova to demonstrate process performance then follow similar operating criteria,

Class	Allowable Uses	Site Restrictions[b]
A and A EQ	• Home lawns and gardens • Public contact sites • Urban landscaping • Agriculture • Forestry • Soil and site rehabilitation • Landfill disposal • Surface land disposal	Unrestricted
B	• Agriculture • Forestry • Soil and site rehabilitation • Landfill disposal • Surface land disposal	• Food crops: no harvesting after application for 14 to 38 months • Feed crops: no harvesting for 30 days after application • Public access: restricted access for 30 to one year • Turf: no harvest for one year after application

[a]Note state regulations, local ordinances, and agricultural practices may limit or restrict the use of Class A Class A EQ, and/or Class B further than Part 503.

[b]Site restrictions listed herein are based upon pathogen and vector attraction reduction requirements alone. There are numerous use restrictions based upon nutrient loading and agricultural practice issues that apply to both Class A and Class B biosolids.

TABLE 18.2   Allowable Uses of Class A and Class B Biosolids per Part 503[a]

Total solids (%)	Temperature, t (°C)	Time (D)	Equation	Notes
≥7	≥50	≥20 min	$D = \dfrac{131\ 700\ 000}{10^{0.1400t}}$	No heating of small particles by warmed gases or immiscible liquid
≥7	≥50	≥15 sec	$D = \dfrac{131\ 700\ 000}{10^{0.1400t}}$	Small particles heated by warmed gases or immiscible liquid
<7	≥50	≥15 sec to <30 min	$D = \dfrac{131\ 700\ 000}{10^{0.1400t}}$	
<7	≥50	≥30 min	$D = \dfrac{50\ 070\ 000}{10^{0.1400t}}$	

TABLE 18.3   Time and Temperature Guidelines for Producing Class A Biosolids

Alternative 4. Ongoing Testing for additional pathogens at the time the biosolids are used/disposed,

Alternative 5. Treatment by a process to further reduce pathogens (PFRP) (see Table 18.4), and

Alternative 6. Treatment by a process equivalent to one of the PFRPs as determined by permitting authority.

3.1.4.2 Class B Requirements

Biosolids must meet at least Class B pathogen requirements before being used or disposed through one of three alternatives: (1) the monitoring of indicator organisms, (2) the use of one of the Processes to Significantly Reduce Pathogens (PSRPs), or (3) the use of a process deemed equivalent to the PSRPs as determined by the permitting authority. PSRPs have been included in Table 18.4. Biosolids that do not meet Class B criteria cannot be land-applied, but they may be placed in a surface-disposal unit that is covered daily per Part 503.

If Class B biosolids or domestic septage are land-applied, site restrictions must also be met (see Section 3.1.6).

3.1.4.3 Discussion on Typical Pathogen Treatment Processes

Municipal WRRFs typically use one of the following four processes to produce Class A or Class B biosolids: heat drying, digestion, composting, and alkaline stabilization as discussed in the following subsections.

3.1.4.3.1 Heat Drying

Heat drying and pelletizing processes typically produce Class A biosolids. Dryers typically have temperatures higher than 70°C, and retain biosolids for at least 30 minutes, thereby meeting the requirements for Class A pathogen reduction. If recycling shrinks retention times to less than 30 minutes, the PFRP time and temperature criteria for heat dryers may apply. Drying processes must also meet the requirements of the thermal equation in Part 503.

Processes to Significantly Reduce Pathogens (PSRPs)	
Aerobic Digestion	The process is conducted by agitating sludge with air or oxygen to maintain aerobic conditions at residence times ranging from 60 days at 15°C to 40 days at 20°C.
Air Drying	Liquid sludge is allowed to drain and/or dry on underdrained sand beds, or on paved or unpaved basins. A minimum of 3 months is needed, for 2 months of which temperatures average on a daily basis above 0°C.
Anaerobic Digestion	The process is conducted in the absence of air at residence times ranging from 60 days at 20°C to 15 days at 35°C to 55°C.
Composting	Using the within-vessel, static aerated pile, or windrow composting methods, the solid waste is maintained at minimum operating conditions of 40°C for 5 days. For 4 hours during this period the temperature exceeds 55°C.
Lime Stabilization	Sufficient lime is needed to produce a pH of 12 after 2 hours of contact.
Processes to Further Reduce Pathogens (PFRPs)	
Composting	Using the within-vessel or static aerated pile composting method, the sewage sludge is maintained at operating conditions of 55°C or greater for 3 days. Using the windrow composting method, the sewage sludge attains a temperature of 55°C or greater for at least 15 days during the composting period. Also, during the high temperature period, there will be a minimum of five turnings of the windrow.
Heat Drying	Dewatered sludge cake is dried by direct or indirect contact with hot gasses, and moisture content is reduced to 10% or lower. Sludge particles reach temperatures in excess of 80°C, or the wet bulb temperature of the gas stream in contact with the sludge at the point where it leaves the dryer is an excess of 80°C.
Heat Treatment	Liquid sludge is heated to temperatures of 180°C for 30 minutes.
Thermophilic Aerobic Digestion	Liquid sludge is agitated with air or oxygen to maintain aerobic conditions at residence times of 10 days at 55°C to 60°C.
Beta Ray Irradiation	Sludge is irradiated with beta rays from an accelerator at dosages of at least 1.0 megarad at room temperature (ca. 20°C).
Gamma Ray Irradiation	Sludge is irradiated with gamma rays from certain isotopes, such as ^{60}Cobalt and ^{137}Cesium, at dosages of at least 1.0 megarad at room temperature (ca. 20°C).
Pasteurization	Sludge is maintained for at least 30 minutes at a minimum temperature of 70°C.

Source: 40 CFR 503, Appendix B.

TABLE 18.4 Pathogen Reduction Processes

Heat dryers that produce a marketable biosolids can easily meet vector-attraction reduction requirements. Basically, if the material does not contain unstabilized primary sludge, then it must be at least 75% solids. If it does, then it must be at least 90% solids.

3.1.4.3.2 Digestion   Aerobic and anaerobic digestion systems typically can produce Class B biosolids if operated as designed. Modified anaerobic digestion systems (e.g., thermophilic anaerobic digestion) may produce Class A biosolids; such systems are considered PFRPs.

3.1.4.3.3 Composting In-vessel composting or static aerated-pile systems can meet Class A pathogen-reduction requirements if the temperature is maintained at 55°C or higher for 3 days. Windrow composting systems can meet Class A requirements if the temperature is maintained at 55°C or higher for at least 15 days and the windrow is turned at least five times in that period. Other composting systems may produce Class A biosolids if they meet the time and temperature or pathogen-testing requirements.

To meet vector-attraction reduction requirements, composting systems must heat biosolids to more than 40°C for 14 days; the average temperature during that period must be higher than 45°C.

Also, after treatment, composted biosolids must be monitored periodically for pathogen regrowth. If the pathogen level increases, the material may not meet land application requirements and may require other means of disposal.

3.1.4.3.4 Alkaline Stabilization One patented alkaline-stabilization process meets Class A pathogen-reduction requirements by elevating pH above 12 for 72 hours while elevating temperature above 52°C for 12 hours or longer, followed by air drying to produce a material with more than 50% solids. Other alkaline-stabilization approaches meet Class A standards by pasteurizing biosolids via the time and temperature criteria. Still others meet Class A standards via PFRP equivalency requirements.

3.1.5   Vector-Attraction Reduction Requirements

Vectors (e.g., flies, rodents, and birds) are attracted to volatile solids. Materials with lower volatile solids concentrations are less likely to attract vectors, which spread infectious disease agents. There are 10 options for reducing vector attraction. All biosolids must meet at least one of them before they can be beneficially used. The options are as follows:

- Option 1: Meet 38% volatile solids reduction.
- Option 2: Demonstrate vector-attraction reduction (VAR) with additional anaerobic digestion in a bench-scale unit.
- Option 3: Demonstrate VAR with additional aerobic digestion in a bench-scale unit.
- Option 4: Meet a specific oxygen uptake rate for aerobically digested biosolids.
- Option 5: Use aerobic processes at >40°C for 14 days or longer.
- Option 6: Alkali addition under specified conditions.
- Option 7: Dry biosolids with no unstabilized solids to at least 75% solids.
- Option 8: Dry biosolids with unstabilized solids to at least 90% solids.
- Option 9: Inject biosolids beneath the soil surface.
- Option 10: Incorporate biosolids into the soil within 6 hours of application.

Part 503's pathogen and vector-attraction reduction requirements are complex. For more information, see U.S. EPA's related guidance documents [especially A Plain English Guide to EPA Part 503 Biosolids Rule (1994)], the regulation itself, and the preamble that accompanied the rule when it was originally published in the Federal Register (1993).

3.1.6   Management Practices

When using bulk biosolids that have not met Pollutant Concentration Limits, Class A pathogen requirements, and vector-attraction reduction (Section 3.1.5), Part 503 restricts application

- to flooded, frozen, or snow-covered ground, where the material can enter wetlands or other U.S. waters (unless authorized to do so by the permitting authority);
- at rates above agronomic rates (except in reclamation projects when authorized to do so by the permitting authority);
- where they could adversely affect a threatened or endangered species; or
- within 10 m of U.S. waters (unless authorized to do so by the permitting authority).

Also, biosolids sold or given away in a container must come with a label or an information sheet that provides the name and address of the person who prepared the analysis information on proper use, including the annual application rate (which ensures that annual pollutant loading rates are within regulatory limits).

When using Class B biosolids, land-appliers must ensure that

- public access to the site is restricted for 30 days after application if the land has little public exposure (e.g., agricultural lands, reclamation sites, and forests) and for 1 year after application if the land has great public exposure (e.g., public parks, golf courses, cemeteries, and ball fields);
- animals are not grazed on the site for 30 days after application;
- no food, feed, or fiber crops are harvested from the site for 30 days after application;
- no food crops whose harvested parts are aboveground but touch the soil (e.g., melons, cucumbers, and squash) are harvested for 14 months after application;
- no food crops whose harvested parts are underground (e.g., potatoes, carrots, and radishes) are harvested for 20 months after application if biosolids remain on the surface for at least 4 months before being incorporated into soil, or for 38 months after application if biosolids are incorporated into soil in less than 4 months;
- turf is not harvested for 1 year after application if the turf will be put on land with high public exposure (e.g., a lawn) unless the permitting authority specifies otherwise.

3.1.7   Monitoring Requirements

The minimum frequency of pollutant, pathogen, and vector-attraction reduction monitoring depends on the amount of solids used or disposed annually. Permitters may impose more frequent monitoring requirements, but after 2 years of monitoring, they may reduce the monitoring frequencies for pollutants (and sometimes for pathogens). However, monitoring frequencies may not drop below once per year.

3.1.8   Recordkeeping Requirements

Recordkeeping requirements depend on which pathogen-reduction option, vector-attraction reduction method, and pollutant limits are met, as well as the ultimate use of the biosolids or product derived from biosolids. In general, the biosolids or product preparer is responsible for certifications and records related to pollutant concentrations, pathogen reduction option, and vector-attraction reduction method. Meanwhile, the biosolids or product applier is responsible for certifications and records concerning field operations, application rates, management practices, and site restrictions. Unless otherwise noted, records should be kept for 5 years.

3.1.9   Reporting Requirements

Once a year, all Class I solids management facilities and publicly owned treatment works (POTWs) with a design flowrate of at least 4000 m³/d (1 mgd) or a service population of at least 10 000 people should submit their annual biosolids report to the permitting authority. As of 2016, U.S. EPA is implementing electronic reporting requirements for all NPDES permitted biosolids programs.

3.1.10   Incineration

Part 503's requirements for solids incinerators address feed solids, the furnace itself, furnace operations, and exhaust gases. The rule does not apply to the incineration of hazardous solids (as defined in 40 CFR 261) or solids containing more than 50 ppm of polychlorinated biphenyls. It also does not apply to incinerators that cofire solids with other wastes, although an incinerator can burn a mix of solids and municipal solid waste (up to 30% as an "auxiliary fuel") and still be regulated under Part 503.

Furthermore, this rule does not apply to the ash produced by a solids incinerator. Design engineers should be aware that ash disposal can be a significant problem. Some states regulate the ash as a hazardous waste (although federal regulations do not).

Recently, U.S. EPA promulgated new regulations for emissions from existing and proposed Sewage Sludge Incinerators (SSIs) under the Clean Air Act (40 CFR 60). These new SSI Maximum Achievable Control Technology (MACT) standards include limits for all regulated pollutants, visible emission limits for ash handling, requirements for annual inspections of emissions control devices, annual testing/monitoring/recordkeeping/reporting requirements, and a schedule for compliance. The additional costs for meeting the SSI MACT standards have rendered incineration to no longer be a cost effective solids management option for many utilities.

3.1.11   Prohibited Disposal Method (Ocean Disposal)

The Part 503 regulations do not address disposing solids in the ocean. Ocean disposal once was acceptable in the United States and practiced widely by communities on the Atlantic coast. However, in 1988, the U.S. Congress passed the Ocean Dumping Ban Act, which made this practice unlawful after 1991. Although some still argue about the scientific basis of this decision and the environmental effects of ocean disposal, this method has ceased to be an option in the United States and so is not discussed in this manual. (Ocean disposal may be an option in other countries if studies indicate that the environmental effects are either negligible or beneficial.)

3.2   Additional Regulations

Many states and local communities have promulgated biosolids regulations and/or local codes that are as strict or stricter than Part 503. These additional regulations can restrict land application of Class B biosolids; impose additional permitting, management, and recordkeeping requirements for generators, haulers, and land appliers;

establish phosphorus limits restricting land application; require fertilizer registration for Class A (EQ) biosolids; or even impose additional requirements for landfilling biosolids. That is why design engineers must review state and local regulations before making decisions about the solids-management train.

For details on state biosolids regulations, inquire with the state environmental management agency or check the National Biosolids Partnership's website (www .biosolids.org).

4.0 Environmental Management Systems

The National Biosolids Partnership (NBP)—an alliance between the Water Environment Federation, the National Association of Clean Water Agencies, and U.S. EPA—has developed an environmental management system (EMS) for facilities that produce biosolids intended for beneficial use, similar to the ISO 14001 standard. The program is designed to help organizations establish good biosolids management practices and become certified for following them consistently. The NBP Code of Good Practice requires that certified organizations agree to

- commit to compliance with all applicable federal, state, and local requirements for biosolids production at the wastewater treatment facility, and management, transportation, storage, and use or disposal of biosolids away from the facility;
- provide biosolids that meet the applicable standards for their intended use or disposal;
- develop an EMS for biosolids that includes a method for independent third parties to verify that ongoing biosolids operations are effective;
- better monitor biosolids production and management practices;
- maintain good housekeeping practices for biosolids production, processing, transport, and storage, as well as during final use or disposal operations;
- develop response plans for unanticipated events (e.g., inclement weather, spills, and equipment malfunctions);
- enhance the environment by committing to sustainable, environmentally acceptable biosolids management practices and operations via an EMS;
- prepare and implement a preventive-maintenance plan for equipment used to manage solids and biosolids;
- seek continual improvement in all aspects of biosolids management; and
- provide effective communication methods with gatekeepers, stakeholders, and interested citizens about the key elements of each EMS, including information about system performance.

There is a tiered program that allows organizations to choose their level of participation with varying requirements by tier for the organization's efforts to commit to the program, develop and implement the program, and have the program audited by internal and external teams. Full program information can be found at www .biosolids.org.

5.0   Solids Quantities

The amount of solids generated during wastewater treatment is an important design parameter because it affects the sizing of solids treatment processes and all related equipment. Solids generation rates also affect the size of liquid treatment processes. For example, the size of a secondary treatment process to maintain a desired solids retention time (SRT) depends on how much solids will be produced.

5.1   Estimating Solids Quantities

While engineers generally recognize the importance of solids production in WRRF design, many still do not understand the wide variation of quality and quantity of solids produced at WRRFs and how difficult it is to estimate solids quantities accurately. The best source of information for estimating solids production is facility-specific data that reflects the nature of the wastewater being treated and the treatment processes being used. If such data do not exist, default approaches or sophisticated mathematical models can be used; however, designers should understand that these estimates may differ significantly from actual results and, therefore, apply conservative safety factors to them.

In general, domestic wastewater typically produces about 0.23 kg/m³ (1 dry ton/ mil gal) of solids. The WRRFs using processes that destroy solids (e.g., digestion or heat treatment) will generate less, and those using chemical addition will produce more. That said, 0.25 kg/m³ is a convenient benchmark for cursory comparisons.

A good approach to estimating solids production is to provide a mass balance for the entire WRRF that relates solids production to design parameters for each treatment process (see Figure 18.1). The mass balance should show key constituents [e.g., flow, total suspended solids (TSS), and biochemical oxygen demand (BOD)] and the process assumptions used in the calculations. It should also include solids generated during nitrogen- and phosphorus-removal processes.

Recycle streams can be included in one of two ways. In the first approach, engineers assume that a fixed percentage of solids or BOD is recycled from downstream processes to the head of the facility. They then iterate the solids balance until the recycled quantities assumed at the head of the facility equal the sum of recycled quantities computed for each process.

The second approach is to estimate the WRRF's net solids production based on historical data, anticipated influent strength, or experience at similar facilities. Engineers then use this information to determine the amount of solids leaving the WRRF, and typically apply it to the output end of the dewatering process. They then back-calculate solids loading to a specific process via the mass balance.

In either approach, engineers must typically estimate the quantities of primary, secondary, and chemical solids separately. They must also take into account expected fluctuations in wastewater characteristics that result from changes in industrial contribution, stormwater flows, seasonal weather conditions, and an expanded collection area. Engineers need to understand peak solids production and diurnal variations to size solids-handling processes properly.

5.2   Primary Solids Production

Most WRRFs use primary sedimentation tanks to remove settleable solids from wastewater. Primary sedimentation is a relatively efficient method for reducing BOD and TSS loading to secondary treatment processes. The amount of solids removed via primary

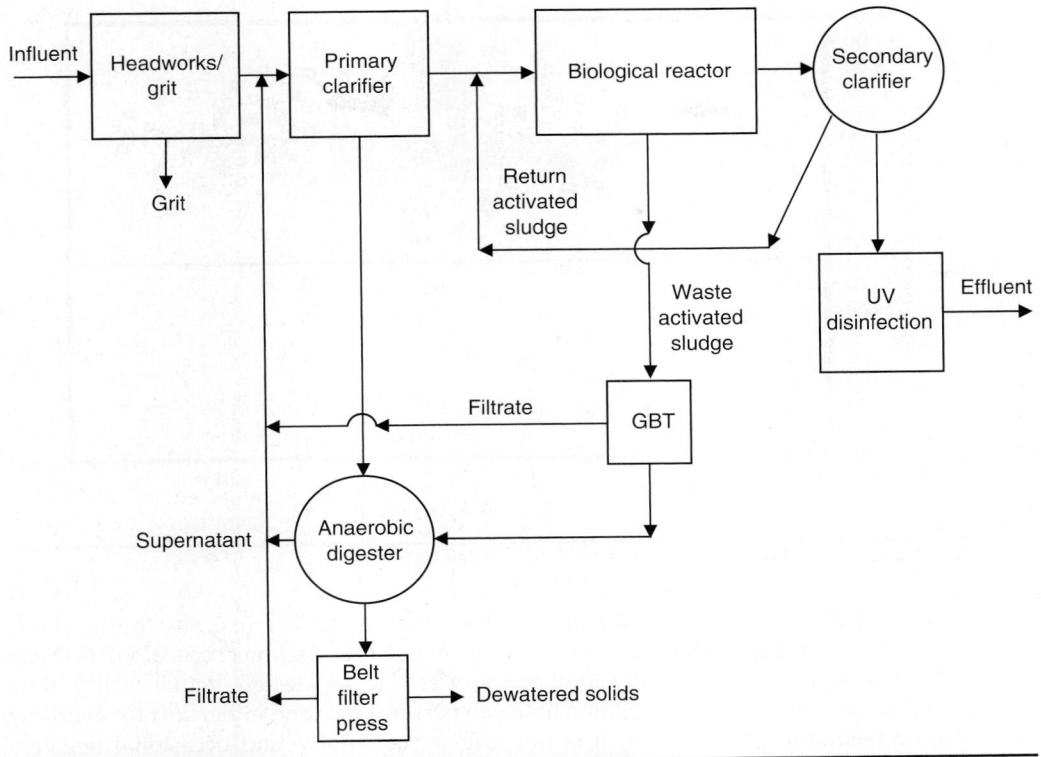

FIGURE 18.1 Process flow schematic for mass balance example (GBT = gravity belt thickening).

sedimentation is typically related to either the surface overflow rate or hydraulic retention time (HRT). The relationship between primary solids production and HRT is as follows (Koch et al., 1990):

$$\text{Primary production} = \text{Facility flow} \times \text{Influent TSS} \times \text{Removal rate} \qquad (18.1)$$

$$\text{Removal rate} = T/(a + bT) \qquad (18.2)$$

where Removal rate = removal (%);

   T = detention time (minutes);

   a = constant (0.406 minutes); and

   b = constant (0.0152).

This expression was developed by fitting a curve to data from 18 large WRRFs. While a suitable model for many WRRFs, its predicted value can vary greatly (see Figure 18.2). In fact, the equation provides a reasonable approximation when data from many facilities are used, but the correlation coefficient is often rather poor when plotted for only one facility.

Figure 18.3 presents some typical curves relating primary solids production to the surface overflow rate (Great Lakes, 2014). This is the most common method for

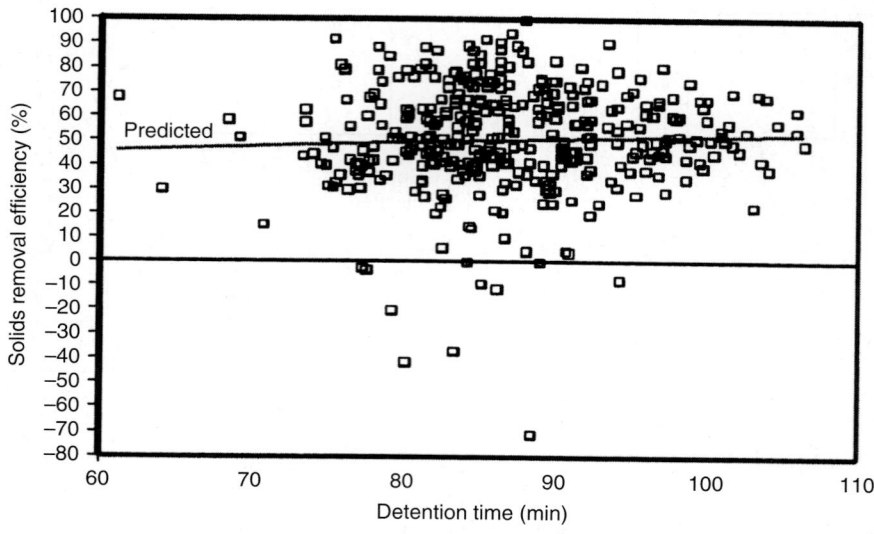

Figure 18.2   Primary tank performance, Cedar Creek daily data.

estimating primary solids production. However, other factors (e.g., hydraulic short-circuiting, poor flow distribution, density currents, and other mechanical factors) can affect performance significantly. Without proper clarifier design, actual facility data may indicate only a weak correlation between performance and either surface overflow rate or detention time. In fact, it is not unusual to briefly find occasional negative removal efficiencies for highly loaded primary sedimentation tanks. (For primary sedimentation design guidance, see Chapter 10.)

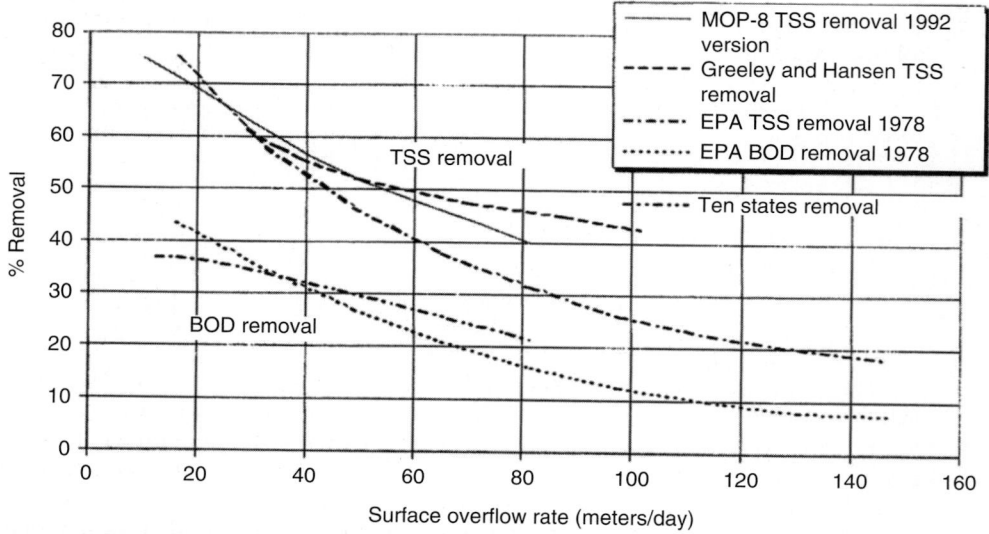

Figure 18.3   Primary treatment performance—TSS and BOD removal.

The degree of BOD or chemical oxygen demand (COD) removal across a primary sedimentation tank affects organic loading to the secondary treatment process and, hence, secondary solids production. Typically, BOD removal is about 50% of TSS removal, although wastewater characteristics can alter this ratio in either direction (Koch et al., 1990).

The retention time and surface overflow rate approaches are essentially equivalent for sedimentation tanks with similar depths. For tanks less than 4 or 5 m (12–15 ft) deep, the retention time approach may be better. Facilities that add chemicals to enhance primary treatment performance or remove phosphorus will produce more solids.

In chemically enhanced primary treatment (CEPT), chemicals (e.g., ferric chloride) are used to remove more suspended solids and BOD from wastewater. Adding about 20 mg/L of ferric chloride and 0.2 mg/L of polymers to the headworks before primary sedimentation has been shown to increase primary solids production by about 45% (Chaudhary et al., 1989). About 30% of this increase was the result of better suspended solids removal; 65% stems from chemical precipitation and removal of colloidal material (Chaudhary et al., 1989). (For a more detailed discussion of this subject, see Chapter 10.)

5.3   Secondary Solids Production

Secondary solids are produced by biological treatment processes [e.g., activated sludge, biological nutrient removal, trickling filters, rotating biological contactors (RBCs), and other attached-growth systems] that convert soluble wastes or substrates (measured as BOD or COD) into microorganisms or biomass. Secondary solids also include some of the particulate that remains after primary sedimentation and becomes incorporated into the biomass. The quantity of secondary solids produced is a function of many factors [e.g., the efficiency of the primary treatment process, the ratio of TSS to 5-day BOD (BOD_5), the amount of soluble BOD or COD in the wastewater, and the design parameters of the secondary treatment process].

In an activated-sludge process, the length of time that secondary solids remain in the process (i.e., SRT) significantly affects the amount of secondary solids produced because the longer solids are retained, the more endogenous decay (self-destruction of biomass) occurs. Temperature also affects secondary solids production. At higher temperatures, solids production should be decreased by a higher growth rate and more endogenous respiration.

The kinetic relationship between secondary solids production and SRT can theoretically be expressed as follows:

$$Y_{obs} = \frac{Y}{1 + k_d(\theta_c)} \tag{18.3}$$

where Y_{obs} = observed yield (g biomass/g substrate);

   Y = yield (g biomass/g substrate);

   k_d = endogenous decay rate (g biomass/g biomass-d^{-1}); and

   θ_c = SRT or MCRT (d).

The substrate concentration is represented by either BOD_5 or COD; it is typically expressed as grams BOD_5 or COD per liter consumed by the process, although some designers prefer to express it as the concentration applied rather than removed. The biomass concentration is expressed in either TSS or volatile suspended solids (VSS).

Coefficient	Basis	Range	Typical
Y	g VSS/g BOD$_5$	0.4–0.8	0.6
Y	g VSS/g COD	0.25–0.4	0.4
k_d	d^{-1}	0.04–0.075	0.06

TABLE 18.5 Typical Values For Solids Yield Coefficients

Typically, the ratio of VSS to TSS is in the range of 0.7 to 0.8. The yield can also include effluent biomass and biomass wasted from the secondary treatment process. Some design engineers use the phrases total yield (including effluent solids) and net yield (excluding effluent solids) to distinguish between types of yields. Unfortunately, these phrases have been used interchangeably in the literature, along with observed yield and apparent yield, which can make plant comparisons confusing. The ranges listed in Table 18.5 have been reported for the yield and endogenous decay coefficients and are expressed as net production (i.e., 1 g TSS versus 1 g substrate removed). Engineers can also obtain values for yield and endogenous decay coefficients by plotting solids production versus SRT (see Figure 18.4).

Figure 18.5 includes the Monod curve, which is named for the scientist who pioneered the application of Michaelis-Menten enzyme kinetics to microbial growth (Monod, 1949). This curve fits the above equation to the data using linear-regression techniques. The Monod curve's values for yield and endogenous decay rate coefficients are 0.731 g VSS/g BOD$_5$ removed and 0.055 d^{-1}, respectively. These values are within the range of reported typical values.

The data in Figure 18.5 emphasize the data variations typical of operating plants. In many plants, it is difficult to see a clear relationship between SRT and solids production. In fact, some researchers have reported that solids production does not appear to be affected by SRT (Wilson et al., 1984; Zabinski et al., 1984). A relationship between yield and SRT typically becomes apparent, however, when multiple plants are plotted.

In theory, the coefficients' values should also vary with temperature. For a given SRT, the growth rate and amount of endogenous respiration should increase with

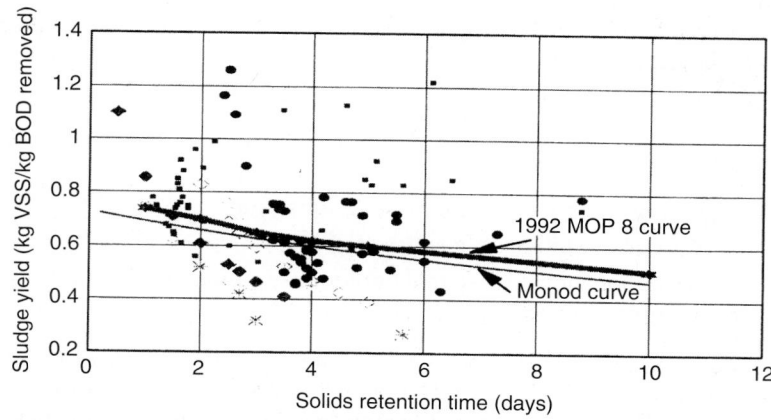

FIGURE 18.4 Solids yield versus solids retention time (with primary treatment).

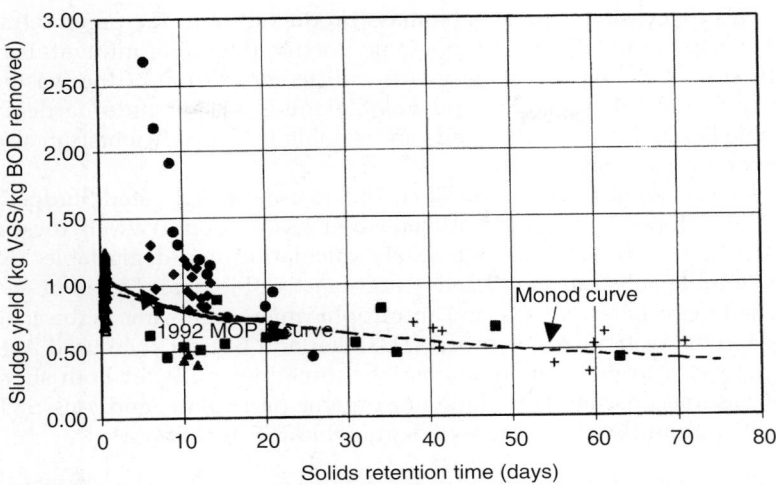

FIGURE 18.5   Solids yield versus solids retention time (without primary treatment).

temperature, thereby lowering solids production. The 1992 edition of this manual showed a series of curves relating solids production to SRT at three temperatures. The curve for 20°C in that edition was also plotted in Figure 18.5; it closely follows the curve fitted to the operating data. Although wastewater temperature varies over the year, the effect on solids production may be masked because many plants adjust SRT seasonally. Many plants see little difference in solids production throughout the year, and some plants have reported higher solids production during the warmer summer months (Koch et al., 1990).

One method for deriving yield coefficients is to plot the solids production per kilogram of aerated volatile solids versus the food-to-microorganism (F:M) ratio (U.S. EPA, 1979). A plot of these two expressions will show the value of yield as the slope and the value of endogenous decay rate as the intercept. A plot of typical data presented by U.S. EPA (1979) in Process Design Manual Sludge Treatment and Disposal yields values within the range shown in Table 18.5. This approach is often used to obtain kinetic parameters from pilot-plant data.

Another approach for estimating secondary solids production is to separate production into three terms (Koch et al., 1990):

$$\text{Net production} = \text{Inerts} + a\,\text{VSS} + b\,\text{SBOD} \qquad (18.4)$$

where a = volatile solids coefficient;
   b = soluble BOD coefficient;
   VSS = volatile suspended solids (mg/L); and
   SBOD = soluble BOD (mg/L).

The volatile solids coefficient varies between 0.6 and 0.8, and the soluble BOD coefficient varies between 0.3 and 0.5. Both coefficients decrease with increasing SRT and are dimensionless.

This expression separates biomass production into the organic fraction (volatile solids and soluble BOD) and inorganic fraction (inerts) of influent. Each term on the right side of the equation represents a different portion of the wastewater influent strength (nonvolatile solids and volatile solids that include particulate BOD_5 and soluble BOD_5). The volatile solids and soluble BOD coefficients can also be related to temperature and SRT.

A more sophisticated form of eq 18.4 is used in Activated Sludge Models Nos. 1 and 2 developed by the International Water Association (IWA). In these models, solids production is modeled by separately calculating non-degradable TSS, degradable TSS, soluble substrates, and active and inactive biomass. Different growth rates are applied to the heterotrophic and autotrophic microorganisms in the anaerobic, anoxic, and aerobic sections of the process. Daigger and Butzz (1992) used Activated Sludge Model No. 1 to develop equations for estimating solids for both short SRTs, where they assumed partial degradation or organic particulate, and long SRTs, where they assumed complete destruction of degradable TSS as follows:

For SRTs longer than 3 days,

$$Y_{obs} = TSS/BOD_5 + Y_H SBOD_5/BOD_5 \tag{18.5}$$

If SRT is shorter than 3 days,

$$Y_{obs} = \frac{TSS[1 - f_v + f_v f_{nv} + (f_v - f_v f_{nv})(1 - f_D)]}{BOD_5} + \frac{Y_H(SBOD_5) + (BOD_5 - SBOD_5)f_D}{BOD_5} \tag{18.6}$$

where Y_{obs} = g TSS generated/g BOD_5 removed;
TSS = influent suspended solids concentration (mg/L);
BOD_5 = influent BOD_5 concentration removed (mg/L);
$SBOD_5$ = influent soluble BOD_5 concentration (mg/L);
f_v = volatile fraction of TSS (%);
f_{nv} = non-biodegradable VSS fraction (%);
f_D = fraction biodegradable VSS that is degraded at a given SRT (%); and
Y_H = heterotrophic cell yield (g suspended solids generated/g soluble BOD_5 removed).

Typical values for the parameters in eqs 18.5 and 18.6 can be found elsewhere (Daigger & Butzz, 1992). It should also be noted that newer models for solids production that have seen widespread use, such as BioWin and GPS-X, are based on these models. However, these newer models utilize different input nomenclature and variations in calculations to estimate solids production. Further discussion of modeling is included in Chapter 4.

For a TSS:BOD_5 ratio of 0.6, which indicates a facility with primary treatment, eqs 18.5 and 18.6 yield values that typically follow the curves in Figure 18.4 for an SRT longer than 2 days. For an SRT shorter than 2 days, eqs 18.5 and 18.6 yield values higher than the Figure 18.4 curves, although close to the data points in Figure 18.4. Using two models to predict solids production at different SRTs may provide a better fit to the data in Figure 18.4. However, it is necessary to assume values for many more parameters to apply to eqs 18.5 and 18.6.

Design engineers should also consider the effect of COD:BOD$_5$ ratio on solids yield. Facilities with high COD:BOD$_5$ ratios tend to yield higher quantities of solids (U.S. EPA, 1987). Another factor that can affect secondary solids production is the need for phosphorus removal. If the secondary treatment process includes chemical or biological phosphorus removal, yields typically will be higher than those in Figure 18.4. Methods for estimating biological phosphorus yield can be found in other references (U.S. EPA, 1987) and are discussed in the "Combined Solids Production" section below.

All of these approaches can be used to estimate secondary solids production from suspended-growth activated-sludge systems. A similar approach can be applied to attached-growth secondary treatment systems (e.g., RBCs and trickling filters). In general, solids production for attached-growth systems can be expressed as follows (U.S. EPA, 1979):

$$\text{Net production} = Y - k_d$$
$$= Y \, \text{BOD}_r + k_d \, \text{AB} \tag{18.7}$$

where Y = observed yield (g biomass/g substrate);
k_d = endogenous decay rate (g biomass/g biomass·d);
BOD_r = BOD removed (kg/d); and
AB = media area with attached biomass (m^2).

The yield coefficient has units similar to the BOD-removed coefficient for activated-sludge systems (i.e., grams VSS per gram BOD$_5$ removed). The amount of attached biomass typically is directly related to the amount of surface area available to support attached growth. In addition, plots of solids production per unit of surface area versus BOD$_5$ removal per unit of surface area can be used to derive facility-specific kinetic constants from facility data. The values of the yield coefficient for attached-growth systems are similar to those for suspended-growth systems; however, the values of the attached-biomass coefficient for attached growth tend to be higher to reflect the longer effective SRT in attached systems. The attached-biomass coefficient for an attached-growth system typically ranges from 0.03 to 0.40 d^{-1} (U.S. EPA, 1979).

In hybrid systems that use both suspended and attached growth, solids production varies depending on how much biomass is attached and how much is suspended. Some researchers have reported that the character and amount of solids produced is dominated by the attached growth (Newberg et al., 1988).

5.4 Combined Solids Production

The WRRFs without primary treatment processes generate combined solids. This mix of primary and secondary solids will be significantly greater in quantity than the secondary solids produced at facilities with both treatment processes (see Figure 18.5) (Koch et al., 1997; Schultz et al., 1982). Design engineers can estimate how much combined sludge will be produced by adjusting the yield coefficients used to estimate secondary solids production to account for the additional solids. The values of the yield and endogenous decay rate coefficients for the Monod curve (see eq 18.3), as shown in Figure 18.5, are 0.975 g VSS/g BOD_5 removed and 0.017 7 d^{-1}, respectively.

The high SRTs represent extended aeration facilities and oxidation ditch facilities. Although Figure 18.5 shows that high SRTs tend to lower solids production, several facilities have solids-production levels significantly above this curve, so engineers should be cautious when using this information to design. A comparison of Figures 18.4 and 18.5 shows that the absence of primary treatment increases the solids yield coefficients. In IWA's model-derived equations, the absence of primary sedimentation is accounted for by the TSS/BOD_5 term in the equation, which is higher for facilities without primary sedimentation.

Biological nitrogen removal typically will increase solids production as a result of the nitrification and denitrification processes (see Table 18.6) (U.S. EPA, 1987). However, the extra solids produced are often offset by the additional endogenous respiration that occurs at higher SRTs. If a second substrate (e.g., methanol) is added for denitrification, even more solids will be produced. Engineers can estimate this extra mass based on the amount of substrate added and expected biomass yield using the methods previously described in Section 5.3.

Biological phosphorus removal may generate more solids as a result of the inorganic salts that accumulate with phosphorus in the biomass (U.S. EPA, 1987). To estimate how much more solids will be produced, engineers can multiply the mass of additional phosphorus removed by 4.5, which is based on a molecular weight of 140 for phosphorus crystals in biomass (U.S. EPA, 1987).

5.5   Chemical Solids Production

Design engineers can estimate the quantity of chemical solids produced based on anticipated chemical reactions. Solids production typically increases in direct proportion to the amount of chemical added; however, competing reactions must be considered. For example, adding ferric chloride will generate more solids than estimated by considering the reaction of ferric chloride to form ferric hydroxide, which preferentially reacts with phosphate, yielding more precipitates than the hydroxide reaction alone. As a general rule, engineers can assume 1 g more solids per 1 g of ferric chloride added. Similarly, adding lime can significantly increase solids production.

Both chemicals are also used in the CEPT process. Adding inorganic chemicals for CEPT or phosphorus removal increases the mass and characteristics of primary solids (see Chapter 14, Section 4.1 for further discussion).

5.6   Mass Balance Example

Mass balance calculations yield the data that engineers need to design solids thickening, dewatering, and stabilization processes. The following is a simplified example of a mass balance for an activated sludge facility with headworks degritting, primary

Process	Bases	Typical Values
Nitrification	g VSS/g NH_4-N removed	0.17
Dentrification	g VSS/g NO_3-N removed	0.8

TABLE 18.6   Nitrification and Denitrification Yield Factors

clarification, gravity belt thickening, anaerobic digestion, belt filter press dewatering, and ultraviolet disinfection (see Figure 18.1 and Table 18.7). Mixed-liquor suspended solids are wasted from the biological reactors.

Mass balance is an iterative process, and this example shows two iterations. The first establishes the recycle flow and concentration. If the second iteration's results are not within ±5% of those of the first iteration, engineers should do a third iteration. It is easy to set up a spreadsheet that incorporates the various formulae needed for numerous iterations. (NOTE: mg/L is identical to g/m³)

5.6.1 Step 1: Determine the Mass of BOD and TSS in Influent

a. Mass (kg/d) = Concentration (g/m³) × Q (m³/d)/1000 g/kg

  BOD = 300 g/m³ × 90 850 m³/d/1000 g/kg = 27 255 kg/d

  TSS = 335 g/m³ × 90 850 m³/d/1000 g/kg = 30 435 kg/d

  TSS after grit = 286 g/m³ × 90 850 m³/d/1000 g/kg = 25 983 kg/d

b. Mass (lb/d) = Concentration (mg/L) × Q (mgd) × 8.34

  BOD = 300 mg/L × 24 mgd × 8.34 = 60 048 lb/d

  TSS = 335 mg/L × 24 mgd × 8.34 = 67 054 lb/d

  TSS after grit = 286 mg/L × 24 mgd × 8.34 = 57 246 lb/d

	In SI Units	In U.S. Customary Units
Q [m³/d (mgd)]	90 850	24.0
Q_p	227 125	60.0
Influent		
BOD [g/m³(mg/L)]	300	300
TSS [g/m³(mg/L)]	335	335
TSS after grit [g/m³(mg/L)]	286	286
Solids characteristics	%	%
Primary	4.8	4.8
Thickened WAS	5.5	5.5
TSS digested	5.3	5.3
Specific gravity	1	1
Biodegradable fraction of WAS	65	65
Effluent characteristics		
BOD [g/m³(mg/L)]	10	10
TSS [g/m³(mg/L)]	14	14
UBOD	1.42 g/g	1.42 g/g

*BOD = biochemical oxygen demand; Q = influent flow; Q_p = peak flow; TSS = total suspended solids; UBOD = ultimate BOD; and WAS = waste activated sludge.

TABLE 18.7 Solids Characteristics for Mass Balance Example*

5.6.2   Step 2: Estimate Soluble BOD in Effluent

a. Assume TSS biodegradable portion = Effluent TSS × 65%

b. Biodegradable portion of Effluent TSS = 14 g/m³ × 65% = 9.1 g/m³

c. UBOD of Effluent TSS = Biodegradable portion × 1.42 = 9.1 g/m³ × 1.42 = 12.9 g/m³

d. BOD of Effluent TSS = UBOD × 0.68 (assuming $k = 0.23$ d⁻¹) = 8.8 g/m³

e. Soluble BOD in Effluent = Effluent BOD – BOD of Effluent TSS = 10 g/m³ – 8.8 g/m³ = 1.2 g/m³

5.6.3   Step 3: Conduct First Iteration

5.6.3.1   Step 3.1: Primary Settling

a. Assume 33% removal of BOD and 70% removal of TSS

b. Calculate mass of BOD and TSS removed by primary settling

BOD removed = 27 255 kg/d × 33% = 8994 kg/d (19 816 lb/d)

TSS removed = 25 983 kg/d × 70% = 18 188 kg/d (40 072 lb/d)

c. Calculate mass of BOD and TSS in primary effluent

BOD in primary effluent = 27 255 kg/d × 67% = 18 261 kg/d (40 232 lb/d)

TSS in primary effluent = 25 983 kg/d × 30% = 7795 kg/d (17 174 lb/d)

d. Calculate concentration of BOD in primary effluent

BOD in primary effluent = 18 261 kg/d × 1000 g/kg/90 850 m³/d = 201 g/m³

(assume primary effluent flow is much greater than primary solids flow)

e. Calculate volatile fraction of primary solids

   i.   Assume volatile fraction of influent TSS = 67%

   ii. Assume volatile fraction of grit = 10%

   iii. Assume volatile fraction of influent TSS going to biological reactor = 85%

   iv. Influent VSS = 30 435 kg/d × 67% = 20 391 kg/d (44 926 lb/d)

   v.   VSS removed in grit = (30 435 – 25 983) kg/d × 10% = 445 kg/d (981 lb/d)

   vi. VSS to biological reactor = 7795 kg/d × 85% = 6626 kg/d (14 598 lb/d)

   vii. VSS in primary solids = (20 391 – 445 – 6626) kg/d = 13 320 kg/d (29 347 lb/d)

   viii. Volatile fraction of primary solids = 13 320 kg/d/18 188 kg/d = 73%

5.6.3.2   Step 3.2: Secondary Process

a. Set operating parameters

MLSS = 3500 g/m³

Volatile fraction = 80%

$Y_{obs} = 0.3125$ g/g

MLVSS $= 2800$ g/m^3

b. Calculate mass quantities of BOD and TSS in effluent

Effluent BOD $= 10$ g/m$^3 \times 90\ 850$ m^3/d/1000 g/kg $= 909$ kg/d (2002 lb/d)

Effluent TSS $= 14$ g/m$^3 \times 90\ 850$ m^3/d/1000 g/kg $= 1272$ kg/d (2802 lb/d)

c. Estimate the amount of VSS produced in the bioreactor (assume primary solids flow is small relative to facility flow)

VSS produced $= [Y_{obs} \times Q$ (m^3/d) $\times (S_o - S)]/1000$ g/kg

$= [0.3125 \times 90\ 850$ m^3/d $\times (201 - 1.2)$ g/m$^3]/1000$ g/kg $= 5672$ kg/d (12 497 lb/d)

$S_o =$ concentration of BOD in primary effluent

$S =$ concentration of soluble BOD in the final effluent

d. Estimate mass of TSS wasting from the bioreactor

$= 5672$ kg/d/80% $= 7090$ kg/d (15 621 lb/d)

e. Estimate mass of FSS wasting from the bioreactor

$= 7090$ kg/d $- 5672$ kg/d $= 1418$ kg/d (3124 lb/d)

f. Estimate mass of waste activated sludge (WAS)

$= 7090$ kg/d $- 1272$ kg/d $= 5818$ kg/d (12 804 lb/d)

g. Estimate flow of WAS from the bioreactor

$= 5818$ kg/d/3500 g/m$^3 \times 1000$ g/kg $= 1662$ m^3/d (0.439 mgd)

5.6.3.3   Step 3.3: Gravity Belt Thickening

a. Set operating parameters

Thickened solids $= 5.5\%$

Solids recovery $= 92\%$

Specific gravity $= 1000$ kg/m^3

b. Calculate flowrate of thickened solids

$=$ Mass of WAS $\times$ Solids recovery/(Specific gravity $\times$ Thickened solids)

$= (5818$ kg/d $\times 92\%)/(1000$ kg/m$^3 \times 5.5\%) = 97$ m^3/d (25 710 gal/d)

c. Calculate flows from GBT

i. Filtrate recycle flowrate $=$ WAS flowrate $-$ Thickened solids flowrate

$= 1662$ m^3/d $- 97$ m^3/d $= 1565$ m^3/d (0.413 mgd)

ii. Mass of TSS to digester $=$ Mass of WAS $\times$ Solids recovery

$= 5818$ kg/d $\times 92\% = 5353$ kg/d (11 793 lb/d)

iii. Mass of TSS to headworks $=$ Mass of WAS $-$ Mass to digester

$= 5818$ kg/d $- 5353$ kg/d $= 465$ kg/d (1025 lb/d)

iv. Concentration of TSS to headworks

$= 465$ kg/d $\times 1000$ g/kg/1565 m^3/d $= 297$ g/m^3

    v. Concentration of BOD in TSS to headworks

        $= 297 \text{ g/m}^3 \times 65\% \text{ degradable} \times 1.42 \times 0.68 = 187 \text{ g/m}^3$

        (Same steps as 5.6.2.a-c)

    vi. Mass of BOD to headworks

        $= 187 \text{ g/m}^3 \times 1565 \text{ m}^3/\text{d}/1000 \text{ g/kg} = 292 \text{ kg/d} (644 \text{ lb/d})$

5.6.3.4 Step 3.4: Anaerobic Digestion

a. Set operating parameters

VS destruction = 47%

Biogas production = 0.9 m³/kg VS destroyed (15 ft³/lb)

Biogas density = 1.204 kg/m³ (86% of air density)

BOD in digester supernatant = 1000 g/m³

TSS in digester supernatant = 5000 g/m³

TSS concentration in digested solids = 5.3%

b. Calculate total solids flow fed to the digester from primary settling and GBT

TS mass = 18 188 kg/d + 5353 kg/d = 23 541 kg/d (51 865 lb/d)

Flowrate = 18 188 kg/d/(4.8% × 1000 kg/m³) + 5353 kg/d/(5.5% × 1000 kg/m³)

= 476 m³/d (125 810 gal/d)

c. Calculate VS fed to the digester

= 18 188 kg/d × 73% + 5353 kg/d × 80% = 17 560 kg/d (38 687 lb/d)

d. Calculate % VS of digester feed

= 17 560 kg/d/23 541 kg/d = 75%

e. Calculate FS mass of digester feed

= 23 541 kg/d – 17 560 kg/d = 5981 kg/d (13 178 lb/d)

f. Calculate VS destroyed

= 17 560 kg/d × 47% = 8253 kg/d (18 183 lb/d)

g. Calculate TS of digested solids

= 17 560 kg/d – 8253 kg/d + 5981 kg/d = 15 288 kg/d (33 682 lb/d)

h. Calculate digester biogas production

= 8253 kg/d × 0.9 m³/kg VS destroyed = 7428 m³/day (272 744 ft³)

  i. Perform mass balance around digester

  ii. Mass flow to digester (Primary Sludge + WAS)

      = 18 188 kg/d/4.8% + 5353 kg/d/5.5% = 476 243 kg/d (1 049 254 lb/d)

  iii. Biogas mass flow leaving digester

      7428 m³/day ×1.204 kg/m³ = 8943 kg/d (19 716 lb/d)

  iv. Solid and liquid mass leaving digester

      = 476 244 kg/d – 8943 kg/d = 467 300 kg/d (1 029 538 lb/d)

5.6.3.5 Step 3.5: Flowrate Distribution of Supernatant and Digested Solids

a. Determine mass of supernatant solids (S)

Mass output = (S/supernatant concentration) + (Total digested sludge mass − S)/ solids in sludge

(5000 mg/L TSS digester supernatant = 0.5%)

467 301 kg/d = S/ 0.5% + (15 288 kg/d − S)/5.3%

2337 kg/d = S + 1442 kg/d − 0.0943S

S = 987 kg/d (2175 lb/d)

b. Determine mass of digested solids

= 15 288 kg/d − 987 kg/d = 14 301 kg/d (31 507 lb/d)

c. Determine supernatant flow

= 987 kg/d/(0.5% × 1000 kg/m³) = 197 m³/d (52 167 gal/d)

d. Determine digested solids flow

= 14 301 kg/d/(5.3% × 1000 kg/m³) = 270 m³/d (71 279 gal/d)

e. Determine BOD of supernatant

= 197 m³/d × 1000 g/m³/1000 g/kg = 197 kg/d (435 lb/d)

5.6.3.6 Step 3.6: Solids Dewatering

a. Set operating parameters

Solids cake = 22%

Specific gravity = 1.06

Solids capture = 96%

Filtrate BOD concentration = 2000 g/m³

b. Determine solids cake characteristics

Solids = 14 301 kg/d × 96% = 13 729 kg/d (30 247 lb/d)

Volume = 13 729 kg/d/(1.06 × 22% × 1000 kg/m³) = 59 m³/d (15 552 gal/day)

c. Determine filtrate characteristics

Flow = 270 m³/d − 59 m³/d = 211 m³/d (55 727 gal/d)

BOD = 211 m³/d × 2000 g/m³/1000 g/kg = 422 kg/d (943 lb/d)

TSS = 14 301 kg/d × (100% − 96%) = 572 kg/d (1260 lb/d)

5.6.3.7 Step 3.7: Summary of Recycle Flows

a. Liquid flow = 1565 m³/d + 197 m³/d + 211 m³/d = 1973 m³/d (0.521 mgd)

b. BOD mass = 292 kg/d + 197 kg/d + 422 kg/d = 912 kg/d (2022 lb/d)

c. TSS mass = 465 kg/d + 987 kg/d + 572 kg/d = 2025 kg/d (4461 lb/d)

5.6.4 Step 4: Conduct Second Iteration

5.6.4.1 Step 4.1: New Influent Concentration and Mass of BOD and TSS to Primary Sedimentation

a. Calculate TSS mass entering primary sedimentation (Influent after grit + Recycle)

= 25 983 kg/d + 2025 kg/d = 28 008 kg/d (61 707 lb/d)

b. Calculate BOD mass entering primary sedimentation (Influent after grit + Recycle)

= 27 255 kg/d + 912 kg/d = 28 167 kg/d (62 070 lb/d)

5.6.4.2 Step 4.2: Primary Setting

a. Calculate TSS removed and sent to bioreactor (assuming 70% removal)

TSS removed = 28 008 kg/d × 70% = 19 606 kg/d (43 195 lb/d)

TSS to bioreactor = 28 008 kg/d × 30% = 8402 kg/d (18 512 lb/d)

b. Calculate BOD removed and sent to bioreactor (assuming 33% removal)

BOD removed = 28 167 × 33% = 9295 kg/d (20 483 lb/d)

BOD to bioreactor = 28 167 × 67% = 18 872 kg/d (41 587 lb/d)

5.6.4.3 Step 4.3: Secondary Process

a. Set operating parameters

SRT = 10 d

F:M = 0.35 d^{-1}

Y = 0.5 g/g

k_d = 0.06 d^{-1}

SRT = 10 d

b. Using the target F:M ratio and original MLVSS concentration, calculate bioreactor volume (V)

Bioreactor volume = Mass BOD to bioreactor/(MLVSS × F:M)

V = 18 872 kg/d/(2800 g/m³/1000 g/kg × 0.35) = 19 257 m³ (5.09 MG)

c. Calculate new flow rate (Q), (Influent + Recycle)

Q = 90 850 m³ + 1973 m³/d = 92 823 m³/d (24.5 mgd)

d. Calculate new bioreactor influent BOD concentration

= 18 872 kg/d × 1000 g/kg/92 823 m³ = 203 g/m³

e. Calculate new concentration of MLVSS

MLVSS = [(SRT × Q)/V] × [Y × (S_o − S)]/[1 + (k_d × SRT)]

= [(10 d × 92 823 m³/d)/19 257 m³] × [0.5 × (203 − 1.2) g/m³]/[1 + (0.06 × 10 d)]

= 3044 g/m³

f. Calculate new concentration of MLSS (assuming 80% VSS)

   MLSS = MLVSS/80% = 3044 g/m³/80% = 3805 g/m³

g. Calculate new cell growth (TSS of cells = $[Q \times Y_{obs} \times (S_o - S)]/1000$ g/kg)

   = 92 823 m³/d × 0.3125 × (203 − 1.2) g/m³/1000 = 5862 kg/d (12 916 lb/d)

h. Calculate mass of WAS to thickening (Cell mass TSS − Effluent TSS)

   = 5862 kg/d − 1272 kg/d = 4590 kg/d (10 113 lb/d)

i. Calculate flowrate of WAS (Flowrate = WAS mass × 1000 g/kg/MLSS)

   = 4590 kg/d × 1000 g/kg/3805 g/m³ = 1206 m³ (318 668 gal/d)

5.6.4.4   Step 4.4: Gravity Belt Thickening

a. Determine flowrate of thickened sludge

   = 4590 kg/d × 92%/(1000 g/kg × 5.5%) = 77 m³/d (20 284 gal/d)

b. Calculate flows from GBT

   i. Filtrate recycle flow rate = WAS flowrate − Thickened solids flowrate
       = 1206 m³ − 77 m³/d = 1130 m³/d (298 384 gal/d)
   ii. Mass of TSS to digester = Mass of WAS × Solids recovery
       = 4590 kg/d × 92% = 4223 kg/d (9304 lb/d)
   iii. Mass of TSS to headworks = Mass of WAS − Mass to digester
       = 4590 kg/d − 4223 kg/d = 367 kg/d (809 lb/d)
   iv. Concentration of TSS to headworks
       = 367 kg/d × 1000 g/kg/1130 m³/d = 325 g/m³
   v. Concentration of BOD in TSS to headworks
       = 325 g/m³ × 65% degradable × 1.42 × 0.68 = 204 g/m³
   vi. Mass of BOD in TSS to headworks
       = 204 g/m³ × 1130 m³/d/1000 g/kg = 230 kg/d (508 lb/d)

5.6.4.5   Step 4.5: Anaerobic Digestion

a. Calculate total solids flow fed to the digester from primary settling and GBT

   TS mass = 18 188 kg/d + 4223 kg/d = 22 411 kg/d (49 376 lb/d)

   Flowrate = 18 188 kg/d/(4.8% × 1000 kg/m³) + 4223 kg/d/(5.5% × 1000 kg/m³)

   = 481 m³/d (126 929 gal/d)

b. Calculate VS fed to the digester

   = 18 188 kg/d × 73% + 4215 kg/d × 80% = 16 656 kg/d (36 696 lb/d)

c. Calculate % VS of digester feed

   = 16 656 kg/d/22 411 kg/d = 74%

d. Calculate FS mass of digester feed

   = 22 411 kg/d − 16 656 kg/d = 5755 kg/d (12 680 lb/d)

e. Calculate VS destroyed

   = 16 656 kg/d × 47% = 7828 kg/d (17 247 lb/d)

f. Calculate TS of digested solids

= 16 656 kg/d − 7828 kg/d + 5755 kg/d = 14 583 kg/d (32 129 lb/d)

g. Calculate digester biogas production

= 7828 kg/d × 0.9 m³/kg VSS destroyed = 7045 m³/day (258 706 ft³)

h. Perform mass balance around digester

   i. Mass flow to digester (Primary Sludge + WAS)
= 18 188 kg/d/4.8% + 4223 kg/d/5.5% = 455 703 kg/d (1 004 001 lb/d)

   ii. Biogas mass flow leaving digester
7045 m³/day × 1.204 kg/m³ = 8483 kg/d (18 701 lb/d)

   iii. Solid and liquid mass leaving digester
= 455 703 kg/d − 8483 kg/d = 447 220 kg/d (985 300 lb/d)

5.6.4.6 Step 4.6: Flowrate Distribution of Supernatant and Digested Solids

a. Determine mass of supernatant solids (S)

447 220 kg/d = S/ 0.5% + (14 583 kg/d − S)/5.3%

2235 kg/d = S + 1376 kg/d − 0.0943S

S = 950 kg/d (2093 lb/d)

b. Determine mass of digested solids

= 14 583 kg/d − 950 kg/d = 13 633 kg/d (30 036 lb/d)

c. Determine supernatant flow

= 950 kg/d/(0.5% × 1000 kg/m³) = 190 m³/d (50 189 gal/d)

d. Determine digested solids flow

= 13 633 kg/d/(5.3% × 1000 kg/m³) = 257 m³/d (67 952 gal/d)

e. Determine BOD of supernatant

= 190 m³/d × 1000 g/m³/1000 g/kg = 190 kg/d (419 lb/d)

5.6.4.7 Step 4.7: Solids Dewatering

a. Determine solids cake characteristics

Solids = 13 633 kg/d × 96% = 13 088 kg/d (28 835 lb/d)

Volume = 13 088 kg/d/(1.06 × 22% × 1000 kg/m³) = 56 m³/d (14 826 gal/day)

b. Determine filtrate characteristics

Flow = 257 m³/d − 56 m³/d = 201 m³/d (53 126 gal/d)

BOD = 201 m³/d × 2000 g/m³/1000 g/kg = 402 kg/d (899 lb/d)

TSS = 13 633 kg/d × (100% − 96%) = 545 kg/d (1201 lb/d)

5.6.4.8 Step 4.8: Summary of Recycle Flows (GBT + Digester + Belt Filter Press)

a. Liquid flow = 1130 m³/d + 190 m³/d + 201 m³/d = 1521 m³/d (0.402 mgd)

b. BOD mass = 230 kg/d + 190 kg/d + 402 kg/d = 823 kg/d (1826 lb/d)

c. TSS mass = 367 kg/d + 950 kg/d + 545 kg/d = 1863 kg/d (4103 lb/d)

5.6.5 Step 5: Create Summary of Recycle Flows and Loadings

	First Iteration		Second Iteration		Percent Difference (%)
	SI Units	U.S. Units	SI Units	U.S. Units	
Recycle flow	1973	521 328	1521	401 700	−23
Recycle TSS	2025	4461	1863	4103	−8
Recycle BOD	912	2022	823	1826	−10

6.0 Solids Characteristics

When designing solids-handling facilities (e.g., conveyance, conditioning, thickening, dewatering, or drying systems), engineers must know the characteristics and volumes of the solids involved. There are several types of solids (e.g., primary, secondary, mixed primary, and chemical), and their characteristics depend on many factors (e.g., liquid stream influent characteristics, the use of chemical precipitants and coagulants; process control; peak loads and weather conditions; and the treatment process chosen).

There are also numerous references available that can help designers obtain detailed information on solids sources, characteristics, and quantities. This manual, however, focuses on WRRF design, so it only addresses topics that significantly impact the design process.

6.1 Primary Solids

Most WRRFs use primary sedimentation to remove settleable solids, which thicken via gravity. Called primary solids, this material consists of a combination of organic solids, grit, and inorganic solids. Primary solids typically are conveyed downstream for further processing (e.g., thickening, digestion, dewatering, etc.) prior to beneficial re-use and/or disposal.

The composition of primary solids varies widely—from day to day, hour to hour, within a facility, and between facilities. Table 18.8 notes the typical composition (adapted from ASCE, 1998; U.S. EPA, 1979).

Parameter	Concentration (Dry-Weight Basic)
Total Solids, %	1.0–8.0
Total volatile solids, % of TS	60–85
Grease, % of TS	5.0–8.0
Phosphorus, % of TS	0.8–2.8
Protein, % of TS	20–30
Cellulose, % of TS	8–15
Nitrogen, % of TS	1.5–4.0
pH	pH 5.0–8.0

TABLE **18.8** Primary Solids Characteristics

The total solids concentration depends on the characteristics of the solids, and the rate at which solids are removed from the primary sedimentation tank(s). If they are removed rapidly, a lower solids concentration can be expected. Some facilities remove solids more slowly or add chemicals to achieve any combination of type I, II, III, and IV settling. Furthermore, primary solids with more grit and heavier material will settle and thicken more quickly than other primary solids.

Facilities that receive both sanitary and stormwater or have a high contribution of infiltration and inflow will produce primary solids that vary greatly in both volume and volatile solids concentration. Those with inadequate grit removal may produce more primary solids, but they only contain 60% VSS because of all the inorganics and grit. Also, facilities that add inorganic chemicals to primary sedimentation tanks will produce more solids and may have lower VS concentrations.

The heavy metal content of primary solids depends on the types of industries that discharge to the facility. Also, chemical addition can also play a large role; it is higher for facilities that add inorganic chemicals (e.g., ferric chloride, alum, or lime) to primary tanks.

6.2 Secondary Solids

Secondary solids are those generated when soluble wastes and other particles in primary effluent are converted to biomass via aerobic biological-treatment processes (e.g., activated sludge, trickling filters, and RBCs). Typically, biological sludges are more difficult to thicken or dewater than primary solids and most chemical solids.

Table 18.9 indicates the typical composition of secondary solids (adapted from ASCE, 1998, and U.S. EPA, 1979). The WRRFs with high F:M ratios tend to produce secondary solids with higher nitrogen levels than conventional activated-sludge facilities do because less endogenous respiration occurs. Facilities with biological phosphorus-removal processes produce solids with higher phosphorus levels. Facilities that treat significant amounts of industrial wastewater can produce solids with higher heavy metal concentrations.

6.3 Tertiary Solids

As NPDES permits in the United States become more stringent the need for tertiary forms of treatment has become much more popular. Tertiary treatment systems vary dramatically based on the intended treatment. In some cases, systems can be implemented to treat for metals or pharmaceuticals, but for purposes of this section, tertiary

Parameter	Concentration (Dry-Weight Basis)
Total Solids, %	0.4–2.5
Total volatile solids, % of TS	60–85
Grease, % of TS	5–12
Phosphorus, % of TS	1.5–3.0
Protein, % of TS	32–41
Nitrogen, % of TS	2.4–7.0
pH	pH 6.5–8.0

TABLE **18.9** Secondary Solids Characteristics

Parameter	Value
Total Solids, %	0.1–1.0
Total volatile solids, % of TS	10–85
Grease, %	0–5
pH	6.5–8.0

Tᴀʙʟᴇ **18.10**   Tertiary Sludge Characteristics

treatment is referring to nutrient treatment (nitrogen and/or phosphorus). These systems are typically biological, chemical, or a combination of the two. As such, the sludge derived from these processes varies greatly. Table 18.10 indicates typical composition of tertiary sludge found in some of the widely used tertiary systems for nitrogen and phosphorus removal.

When treating for phosphorus, the primary mechanism for removing phosphorus from the liquid stream is to precipitate it in the sludge. So these tertiary systems typically produce much higher quantities of sludge then nitrogen treatment system. When treating for nitrogen the primary mechanism for removing nitrogen is by converting it to N_2 gas, so these systems typically have lower quantities of sludge (see Chapter 14 for further discussion).

6.4   Combined Solids

The characteristics of combined solids are dependent on the proportions of each type and their compositions. Table 18.11 indicates typical characteristics of combined, unthickened sludges.

6.5   Chemical Solids

Chemical solids are the result of adding metal salts, polymer, or lime to wastewater to improve suspended solids removal or precipitate phosphorus. Typically, these chemicals improve thickening and dewatering performance. However, modern polymers have become more effective and thus more popular than metal salts for improving thickening and dewater performance. The characteristics of chemical solids are affected by wastewater chemistry, pH, mixing, reaction time, and opportunities for flocculation. That said, they typically contain more heavy metals than other solids because of the heavy metals in coagulant and those that co-precipitate with iron and aluminum. See Chapter 20 for further discussion.

Parameter	Value
Total Solids, %	3–10
Total volatile solids, % of TS	50–80
Grease, %	5–10
pH	6.0–8.0

Tᴀʙʟᴇ **18.11**   Combined sludge

6.6   Thickened Solids

Solids concentrations and constituents vary greatly depending on the treatment processes used upstream, strength of the wastewater, and other characteristics of the wastewater influent. Typically, storing, conveying, treating, and/or disposing of solids at low concentration is not economical, because the media is primarily water. So many facilities thicken their solids from the concentrations listed in the tables mentioned earlier, to 4%–10% total solids, depending on the thickening technology and polymers used. Thickening systems are described in more detail in Chapter 21. These systems generally take advantage of gravity to concentrate solids and employ types III and IV settling to remove the free water. Some of these systems are passive, like a gravity thickener, and some are mechanical like a gravity belt thickener or rotary drum thickener. In many cases polymer is used to enhance settling and capture rate.

6.7   Digested Solids

Some most common forms of stabilization are anaerobic digester, aerobic digestion, and lime stabilization. New methods of stabilization are growing in popularity, and are discussed further in Chapter 23. Digestion is a biological process that stabilizes organic matter, reducing the total solids inside the digesters by volatilizing a portion of those solids into the gaseous state. Digested sludge is also often less odorous and has few vector attractants than raw sludge.

During anaerobic digestion, in the absence of oxygen, biodegradable organic matter is converted to water and biogas containing methane (CH_4) and carbon dioxide (CO_2). Aerobic digestion is a biosolids stabilization process in which biodegradable organic matter is converted to water and gas containing carbon dioxide (CO_2) and nitrogen gas (N_2) in the presences of oxygen. Table 18.12 indicates typical characteristics of traditional high-rate mesophilic digested sludge, this does not include advanced digestion and pre-digestion processes like thermal hydrolysis and/or recuperative thickening.

Table 18.13 shows typical characteristics of traditional aerobic digestion.

6.8   Dewatered Solids

Dewatering solids is generally intended to remove as much water as possible to prepare solids for the most economical means of disposal. In some instances, downstream processes require some degree of water to remain in the solids (incineration, drying, soil amending, etc.), however in most cases the more water removed results in the least water transported and thus more economical disposal. Dewatering is typically done by mechanically enhancing type III and IV settling. For this reason, dewatering processes typically require some level of thickening upstream. Otherwise, most dewatering systems'

Parameter	Value
Total Solids, %	2.0–5.0
Volatile Solids Reduced, %	55–65
Total volatile solids, % of TS	30–58
Grease, %	0–5
pH	6.8–7.2

TABLE 18.12   Mesophilic Anaerobically Digested Sludge

Parameter	Value
Total Solids, %	2.0–5.0
Volatile Solids Reduced, %	38–50
Total volatile solids, % of TS	38–67
Grease, %	0–5
pH	6.0–7.2

TABLE **18.13**   Aerobically Digested Sludge

throughput becomes hydraulically limited by the excessive flow and lack of solids. For example, most belt filter presses have a thickening section where solids are spread across a porous belt and water is allowed to drain from the solids by gravity. Next there is a dewatering section where solids are squeezed between two porous belts that pass through a set of rollers placing pressure on the belts, thus squeezing the free water from the solids. Similarly, a centrifuge uses centrifugal force to squeeze water from the solids. While the slope of the beach and dimensions of the scroll and bowl are designed in such a way to provide optimal dewatering. Polymer is typically added to aid dewatering and capture. Similar to other solids handling processes, dewatering is highly dependent on the upstream treatment systems and characteristics of the solids. High ratios of primary to secondary sludge can typically achieve higher percent solids, while solids from phosphorus remove processes are generally more difficult to dewater. One could expect to see 18%–28% total solids from traditional processes, but in some instances facility have been able to achieve 30%–40% total solids.

6.9   Thermally Dried Solids

Thermal drying is the process in which heat is applied to solids to promote the evaporation of water, increase the total solids content, and further reducing pathogens in the solids. Methods of thermal drying include convection (direct drying), conduction (indirect drying), radiation, or a combination therein.

Convection, or direct dryers, include technologies like rotary drum dryers, flash dryers, or fluidized bed dryers where the heating medium is in direct contact with the solids. Conduction dryers, or indirect dryers, include technologies that utilize a barrier between the heating media (steam, oil, or other) and the sludge. These systems typically employ paddles, hollow flights, or disks mounted on rotating shaft(s) that move solids through the unit while transferring heat to the solids.

7.0   Pretreatment Options

Solids are removed from primary and sedimentation tanks after being pretreated to facilitate pumping and subsequent handling. The most common pretreatment processes include degritting, grinding, and screening. Grinding and screening are becoming more popular as facilities strive to increase the quality of their biosolids.

7.1   Degritting

Grit typically can be removed more easily and efficiently from raw wastewater rather than from the primary sludge stream. So, current design practice favors installing grit-removal and -processing facilities at the headworks, where raw wastewater first enters a WRRF. This practice reduces wear on influent pumping systems (if grit removal is upstream of the wet well) and primary solids pumping, piping, thickening, and digestion systems. Efficiently removing grit from raw wastewater also reduces grit accumulation in thickening and digestion tanks.

However, some WRRF operators have found it more convenient to capture grit and primary solids together, rather than separately. These facilities typically use hydrocyclones (or other induced tangential-flow devices) to remove grit from primary solids. To remove 100- to 150-mesh grit particles effectively via hydrocyclones, the solids concentration should be less than 1% (preferably closer to 0.5%). Some of the grit-removal devices discussed in Chapter 9 can also be used to remove grit from primary sludge.

7.2   Grinding

Grinding processes cut or shear large solids into smaller particles to prevent operating problems in downstream processes. To accomplish this, several types of systems are used. One style is by means of a chopper or grinder pump, where the impeller of a pump reduces the particle size passing through it, while pumping it. Another style similarly reduces particle sizes as it passes through the unit, but it does not add any head to the system. These systems are referred to as grinders, macerators, and/or comminutors. These systems can be inline on a pressurized piping system, or open to the atmosphere on an in-channel system. Styles of each vary depending on the application and manufacturer, but in all instances the purpose is to protect downstream equipment and reduce particle sizes.

Today, more WRRFs are using inline grinders to try to reduce equipment maintenance downtime. They can shear solids into 6- to 13-mm (0.25- to 0.5-in.) particles, depending on design requirements, and can handle either dilute or thickened solids. However, operators are finding that rags have a tendency to pass through some styles of grinder and accumulate ("re-rag") in downstream processes.

A solids macerator–grinder works like a meat grinder (see Figure 18.6). Its multiple-blade cutter rotates rapidly over a perforated grid plate through which solids are forced. The size of the holes and the speed of the blade determine how small the particles will become. Holes range from 11 mm (0.44 in.) in diameter to slots 26 to 38 mm (0.6 to 1.5 in.) wide.

Some styles of grinder can produce nominal pressure increases, however most have head losses associated with them that should be accounted for during design. They

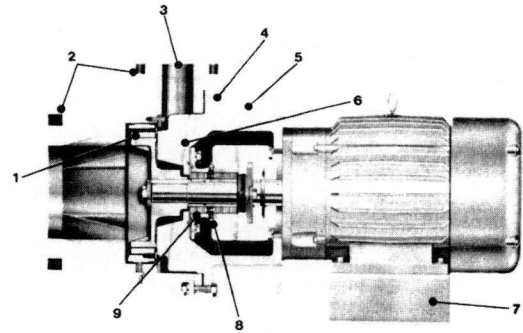

FIGURE 18.6   Macerator-grinder: (1) carbide impeller tips, (2) discharge and suction flanges, (3) discharge port, (4) canopy construction, (5) lifting holes, (6) deflection surface, (7) mounting pedestal (horizontal or vertical), (8) seal-flushing connection, and (9) seal.

typically are installed on the suction side of solids pumps to prevent pump clogging. If the grinder must be located on the discharge side of the pump, the discharge pressure must be low (consult the manufacturer for installation guidelines). A high-pressure pump discharge will shorten the lives of grinder seals and shafts.

Although grinders will handle large organic particles readily, rocks and metal objects can cause damage and/or excess wear. So, they must be protected from tools or rocks dropped into the open tanks in upstream processes. One grinder design uses a sump, formed by a standard cross in the solids line ahead of the grinder. The bottom of the sump has a basket that periodically can be lifted out through the top section of the cross. Heavy objects flow along the bottom of the pipe and drop into the basket, thereby protecting the grinder.

More recently, vendors have developed slow-speed hydraulic or electric grinders that sense blockages. The macerator/inline solids grinder has two sets of counter-rotating, intermeshing cutters that trap and shear solids; producing a consistent particle size (see Figure 18.7). The cutters are stacked on two steel or stainless-steel drive shafts with intermediate spacers between successive cutters. The spacers are made of the same material as the cutters which are commonly hardened steel or tungsten. The drive shafts counter-rotate at different speeds, producing a self-cleaning action in the cutters.

Figure 18.7   Sludge grinder.

7.3 Screening

Screening the raw wastewater, or influent flow, has the benefit of removing particles and inert material from the influent that may damage downstream equipment and/or cannot be broken down in the liquid and solids handling processes. The smaller the screen size openings, the less material passes through (discussed in more detail in Chapter 9).

Dedicated inline sludge screens are also available for situations that require a higher level of screening in the sludge then headworks or influent screens allow. These screens are pressurized vessels that pump sludge through a perforated plate or wire mesh to remove inert material. This creates two separate outlets, one of presumably inert screenings that can be disposed of in a landfill, and a separate stream of clean sludge for downstream processing. The clean sludge reduces the downtime and maintenance required to remove rags and other inert materials that typically accumulate in downstream processes like digestion and mixing. It also allows for more advanced processes that may have otherwise clogged without screening upstream. Inline screens are available with capacities up to about 32 L/s (68 cu ft/min). Screen openings are typically 5 mm to 10 mm. Screenings can contain up to 30% to 40% dry solids; solids throughput decreases as solids concentration increases (See Figure 18.8). So, for example, if a WRRF was screening primary sludge with a dry solids concentration of about 5%, the compacted screenings would be produced at a rate of about 0.3 to 1.0 kg of dry solids per cubic meter of solids flow. Maximum inlet pressure is about 100 kPa (14.5 lb/in.^{2}). Although the pressure drop across the screen depends on the feed rate, solids concentration, and screen openings, it typically ranges from about 15 to 50 kPa (2–7 lb/in.^{2}). If the pressure drops too much, causing solids to blind the inlet side of the screen, design engineers can add an inline booster pump downstream of the screen.

7.4 Lysis

Cell lysis processes paired with digestion have become popular at European WRRFs. The successes have attracted the attention of municipalities within the United States, and as a result many facilities are choosing to install lysis systems.

Lysis as it relates to wastewater treatment solids is the process by which the bacteria cells are stressed to a point where the cell walls rupture. The lysed cells have proven to be more digestible and increase the volatile solids reduction in a digester, and many of these lysis processes also inherently pasteurize the solids. The lysed solids have also been found to pump, thicken, and dewater more effectively than traditional solids.

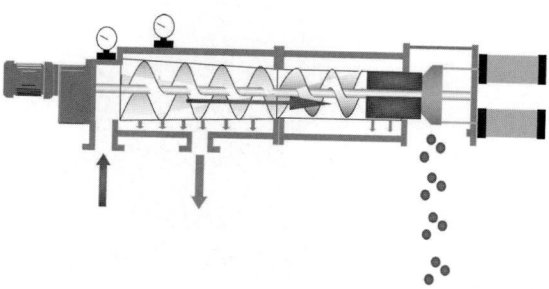

FIGURE 18.8   Inline sludge screen.

Several companies have developed their own methods and systems to lyse waste-water solids. Cell lysis treatment systems consist of mechanical, thermal, chemical processes, or some combination thereof to efficiently disintegrate the bacteria cell walls. Each offers its own advantages and disadvantages that need to be considered for a specific installation.

8.0   References

Chaudhary, R.; et al. (1989) Evaluation of Chemical Addition in the Primary Plant at Los Angeles 11 Hyperion Treatment Plant. *Proceedings of the 62nd Annual Water Pollution Control Federation Technical Exposition and Conference*; San Francisco, California, Oct 15–19; Water Pollution Control Federation: Washington, D.C.

Criteria for Municipal Solid Waste Landfills, 40 C.F.R. pt 258 (1991).

Daigger, G. T.; Butzz, J. A. (1992) Upgrading Wastewater Treatment Plants, 2nd ed.; *Water Quality Management Library*, Vol. 2; CRC Press: Boca Raton, Florida.

Great Lakes–Upper Mississippi River Board of State and Provincial Public Health and Environmental Managers (2014) *Recommended Standards for Wastewater Facilities*; Health Research Inc.; Health Education Services Division: Albany, New York.

Identification and Listing of Hazardous Waste, 40 C.F.R. pt 261 (1980).

Koch, C.; et al. (1990) Spreadsheets for Estimating Sludge Production. *Water Environ. Technol.*, **2** (11), 65.

Koch, C.; et al. (1997) A Critical Evaluation of Procedures for Estimating Biosolids Production. *Proceedings of the Joint Water Environment Federation and American Water Works Association Specialty Conference; Residuals and Biosolids Management: Approaching 2000*; Philadelphia, Pennsylvania; Water Environment Federation: Alexandria, Virginia; American Water Works Association: Denver, Colorado.

Monod, J. (1949) The Growth of Bacterial Cultures. *Annu. Rev. Microbiol.*, **3**, 371.

Newberg, J. W.; et al. (1988) Unit Process Trade-Offs for Combined Trickling Filters and Activated Sludge Processes. *J. Water Pollut. Control Fed.*, **60**, 1863.

Schultz, J.; et al. (1982) Realistic Sludge Production for Activated Sludge Plants without Primary Clarifiers. *J. Water Pollut. Control Fed.*, **54**, 1355.

Standards for New Source Performance for New Stationary Sources, 40 C.F.R. pt. 60 (1971).

Standards for the Use and Disposal of Sewage Sludge, 58 Fed Reg. 9248 (February 19, 1993) (codified as 40 C.F.R. pt. 503).

U.S. Environmental Protection Agency (1979) *Process Design Manual: Sludge Treatment and Disposal*; EPA-625/1-79-011; U.S. Environmental Protection Agency: Washington, D.C.

U.S. Environmental Protection Agency (1987) *Design Manual: Dewatering Municipal Wastewater Sludges*; EPA-625/1-87-014; U.S. Environmental Protection Agency: Washington, D.C.

U.S. Environmental Protection Agency (1994) *A Plain English Guide to the EPA Part 503 Biosolids Rule*; EPA/832-R-93-003; U.S. Environmental Protection Agency: Washington, D.C.

Wilson, T.; et al. (1984) Operating Experiences at Low Solids Retention Time. *Water Sci. Technol.* (G.B.), **16**, 661.

Zabinski, A.; et al. (1984) Low SRT: An Operator's Tool for Better Operation and Cost Savings. *Proceedings of the 57th Annual Water Pollution Control Federation Technical Exposition and Conference*; New Orleans, Louisiana, Sep 30–Oct 5; Water Pollution Control Federation: Washington, D.C.

CHAPTER **19**

Storage and Transport

Bruce DiFrancisco, P.E.

1.0   Introduction

This chapter discusses methods for storing and transporting sludge and biosolids through water resource recovery facilities (WRRFs) and to the disposal location. Headworks solids—screenings and dewatered grit—are covered in Chapters 9 and 10. Chapters 20, 21, 22, and 23 discuss methods for thickening, dewatering, and treating biosolids.

This chapter includes storage and transport of mechanically thickened or dewatered biosolids; it does not discuss thermally conditioned biosolids or ash generated during incineration. Chapter 24 discusses methods for thermal drying of dewatered solids and biosolids. Chapter 25 discusses use and disposal of residuals and biosolids, including ash.

The sludge and biosolids transport methods discussed here are currently prevalent in the United States. When designing solids transportation systems, engineers should consider service life and favor equipment that is flexible enough to remain useful despite changing technology, regulations, economics, and solids characteristics. They should also investigate full-scale working systems whenever possible to determine actual operating conditions and costs, and then make allowances for uncertainties that are specific to the project.

It is noted that raw and thickened residuals are referred to in MOP 8 as "sludges" and that stabilized solids are referred to as "biosolids." The term "residuals" is used in this chapter when the text refers to information applicable to both nonstabilized sludge and stabilized biosolids.

1.1   Flow Characteristics

The flow characteristics (rheology) of sludge and biosolids (residuals) cannot be defined simply. They vary widely from process to process and from facility to facility. (Wagner, 1990; Levine, 1986; 1987; Borrowman, 1985; Carthew et al., 1983; Mulbarger et al., 1981; and U.S. EPA, 1979). As a result, the characteristics and resulting headloss are difficult to predict.

Solids content is an important rheological parameter. Generally, the higher a fluid's solids concentration, the higher its shear stress, density, and viscosity. Viscosity increases exponentially as solids concentration increases (Brar et al., 2005, and references therein).

A residual's rheological characteristics are strongly affected by the kind of treatment the material has undergone (Guibaud et al., 2004; Brar et al., 2005). For example, a raw, fresh, nonhydrolyzed solids stream at 3% solids has a higher apparent viscosity than a 4% material that has been thermally alkaline hydrolyzed (Brar et al., 2005). Similarly, digested biosolids can be pumped more easily than raw or undigested sludge with the same moisture content (see Figure 19.1) (U.S. EPA, 1979). To derive approximate headlosses for residuals, design engineers should calculate the headloss for water and then multiply the result by the factor in Figure 19.1 corresponding to the residual's solids concentration. This will provide a rough estimate when velocities are between 0.76 and 2.4 m/s (2.5 and 8 ft/s) and no thixotropic behavior or serious obstructions, such as grease, are anticipated.

For the case where an existing water resource reclamation facility wants or needs more precise control of the transport process but is not changing the biosolids treatment method, field testing can be performed to determine the headloss in the existing system. Once this is known, the controls system can be designed around this value.

For design purposes, residuals can be divided into several distinct categories based on solids concentration as presented in Table 19.1. For more detailed information on

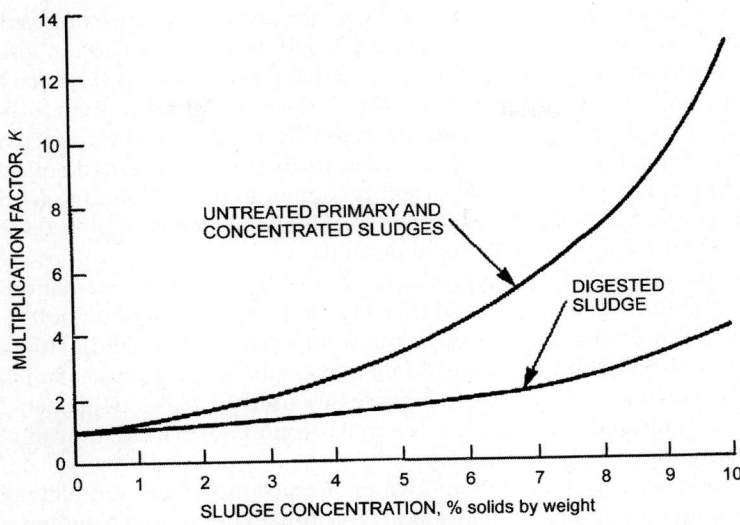

FIGURE 19.1 Approximate multiplication factors (based on solids concentration) to be applied to headlosses calculated for water in laminar flow (taken from U.S. EPA, 1979).

Solids Type[a]	Temperature (°C)	Thickened Solids, % TS	Thixotropic Solids		Granular-Compactable	
			Low-Medium Viscosity, % TS	Medium-high Viscosity, % TS	Wet Solids, % TS	Dry Solids
RPS	10–20	5–10	20–26	24–40	35–65	65+
RWAS	10–25	3–7	10–16	14–15	22–65	65+
R (PS + WAS)[b]	10–25	4–8	13–20	18–31	28–65	65+
DPS	20–30	5–10	20–28	24–40	35–65	65+
SWAS	20–30	3–7	10–16	14–25	22–65	65+
D (PS + WAS)[b]	20–30	4–8	13–20	18–31	28–65	65+
Alum [Al(OH)^3][2]	5–25	3–7	8–15	13–30	25–60	65
Iron [Fe(OH$_3$)][2]	5–25	3–7	8–15	15–35	30–60	65
Lime (CaCO$_3$)	5–25	15–30	8–15	25–50	70–80	80+

[a]R = raw, D = digested, PS = Primary sludge, and WAS = waste activated sludge.
[b]50 : 50 mixture of PS and WAS.

TABLE 19.1 Classification of Water and Wastewater Thixotropic Residuals by Type and Solids Content

solids concentrations and residuals behavior, see Conveyance of Residuals from Water and Wastewater Treatment (ASCE, 2000). Dilute residuals contain less than 5% solids. Such residuals typically include waste activated sludge (WAS), which typically contain less than 2% solids, and primary sludge, which contains less than 5% solids.

Thickened residuals, which are typically produced via a mechanical thickening process, have higher solids concentrations than gravity settled solids. Such residuals range from WAS with a 3% total solids content to primary sludge with up to 10% total solids content. Thickened residuals typically have a much higher viscosity and cannot be handled reliably by centrifugal pumps.

Dewatering processes typically used at water resource reclamation facilities tend to make residuals thixotropic, and their rheology is dependent on both time and applied stress. While engineers can design pumping systems that handle thixotropic materials, the limited availability of related data makes site-specific studies important.

Further dewatering, such as processes used to make pelletized fertilizer, makes residuals granular. Such residuals cannot be pumped; instead, they must be transported via conveyors or similar devices.

Fluids may be Newtonian (water, for example) or non-Newtonian. Residuals containing more than 3% total solids typically do not follow Newtonian behavior.

The Herschel–Bulkley model is an equation that models the rheological behavior of both Newtonian and non-Newtonian fluids:

$$S = S_y + K \, (dv/dy)^n \tag{19.1}$$

where S = shear stress (in Pa, N/m^2, or kg/m-s^2),

$\quad S_y$ = yield stress (in Pa),

$\quad dv/dy$ = shear rate (velocity gradient in s^{-1}), and

K and n = fluid constants.

When $S_y = 0$ and $n = 1$, the model describes a Newtonian fluid and eq 19.1 becomes:

$$S = \mu \, (dv/dy) \tag{19.2}$$

where μ = absolute viscosity (in Pa-s).

When $S_y \neq 0$ and $n = 1$, the model describes a Bingham plastic fluid:

$$S = S_y + Rc(dv/dy) \tag{19.3}$$

where Rc = coefficient of rigidity (Pa-s).

When $S_y = 0$ and $n \neq 0$, the model describes a pseudoplastic, Ostwald de Vaele, or shear-thinning fluid, and the equation becomes:

$$S = K \, (dv/dy)^n \tag{19.4}$$

where K = fluid consistency index, and

$\quad n$ = flow behavior index.

Researchers are almost equally divided on whether the Bingham plastic model or the pseudoplastic model describes the rheological behavior of thickened and dewatered residuals more appropriately.

When designing solids transport systems with kinetic pumps (e.g., centrifugal), engineers need to be accurate rather than conservative. Such systems are most efficient when the system curve matches the pump curve. However, engineers should be conservative when designing transport systems for thicker residuals, either liquid or dewatered, that rely on positive-displacement pumps.

1.2 Flow and Solids Monitoring

The flow rate of residuals with total solids concentrations of 7% to 8% can be measured using electromagnetic meters, , but it is recommended that for concentrations above 5% that the manufacturer provide a history of successful operation. Optical sensors can be used to measure mass flow rates for residuals containing up to 4% total solids.

Except for the equipment listed above, flow rates for thickened or dewatered residuals transported in a pipe cannot be reliably measured, but efforts continue to develop this technology. If designers are presented with flow measuring equipment, a history of successful use should be reviewed and verified prior to specifying such equipment. It is recommended that engineers estimate solids quantities by performing a mass balance around a particular process. In a dewatering process, for example, they can calculate the quantity of biosolids cake produced based on influent loading, removal efficiency and solids concentrations. Suspended solids meters can register air bubbles as solids (Radney, 2008), so to avoid interference, designers should include degassing tanks to release entrapped air.

Design engineers can also indirectly measure the flow rate of thickened residuals based on positive-displacement pump operations. For example, if designers know the rotational speed and volumetric discharge per rotation of a progressing cavity pump, they can calculate the fluid flow rate, anticipating that the pump cavity fill rate is 100%. Manufacturers of hydraulically driven reciprocating piston pumps offer an internal flow measuring system that is accurate within 5%.

1.3 Energy Usage and Sustainability Review

The storage and transport of thickened residuals dewatered cake and dried biosolids is an energy intensive process, potentially involving such equipment as screw pumps, piston pumps, diaphragm pumps, conveyors, augers and others described in this manual. Design engineers are encouraged to perform a life cycle power consumption analysis, including a cost review, as part of any storage and transport system design. This analysis will give the owner and operator the information they need to make an informed decision as to how power consumption will impact their biosolids processing facility over the life of the proposed improvement.

In addition, many facilities are adding sustainability practices into their operating guidelines. LEED and Envision evaluations are becoming popular in the industry and energy consumption is a key component of these evaluations. Please refer to *Sustainability and Energy Management for Water Resource Recovery Facilities* (WEF et al., 2018) for detailed information on this emerging field.

2.0 Sludge and Biosolids Storage

2.1 Storage Requirements

The storage needs for liquid residuals depend on where and why they are being stored. Storage can be used to equalize flows and provide operational flexibility. For example, if dewatering equipment is operated only periodically, solids need to be stored between operating periods. The volume of storage required is process-specific. If biosolids are stored between stabilization and land application, the storage volume required will depend on agricultural schedules and climatic issues.

If sludge or biosolids will be land-applied or surface disposed, the reclamation facility must have enough storage available to allow for periods when the material cannot be used. The Great Lakes—Upper Mississippi River Board, in its 2014 Recommended Standards document (GLUMRB, 2014) offers the following items for design consideration:

- Inclement weather effects on access to the application land
- Temperatures including frozen ground and stored biosolids cake conditions
- Haul road restrictions including spring thawing conditions
- Area seasonal rainfall patterns
- Cropping practices on available land
- Potential for increased sludge volumes from industrial sources during the design life of the facility
- Available area for expanding sludge storage
- Appropriate pathogen reduction and vector attraction reduction requirements

A 120- to 180-day storage capacity is recommended, but this can be reduced if the facility has an available alternate disposal option.

Design engineers need to take these situations into account when designing the storage systems. For example, Figure 19.2 indicates the approximate number of days per year when climatic conditions do not allow effluent application (U.S. EPA, 1995). Anticipating that effluent and biosolids applications would be affected by the same climatic conditions, designers can use the information in this figure as a basis for estimating storage requirements and to assist with determining alternate management requirements. Likewise, Table 19.2 shows the months in which biosolids can be applied in the north central United States (data that can be extrapolated to other areas). Designers can also obtain climatic data for US sites from the National Oceanic and Atmospheric Administration's National Climatic Data Center in Asheville, North Carolina (http://lwf.ncdc.noaa.gov/oa/ncdc.html).

Chapter 23 identifies storage requirements specific to certain stabilization processes. For example, many autothermal thermophilic aerobic digestion systems have a pre-digestion holding tank, so the digester(s) can be batch-fed. In addition, they often have a post-digestion tank where the biosolids can cool and further stabilize. Sizing requirements for such tanks are process-specific. This information is often included in the design guidelines for stabilization processes.

Regulations may also affect process storage systems. Some regulations require storage volumes ranging from 3 to 60 days of biosolids storage (FDEP, 2008). Such storage may simply be excess capacity in the digesters or other process tanks, or it may be separate storage tanks.

2.2 Storage Tanks

2.2.1 Typical Design Criteria
Although more costly per unit volume than earthen basins, storage tanks can be a good choice when residuals volumes are small, land costs are high, or other restrictions make

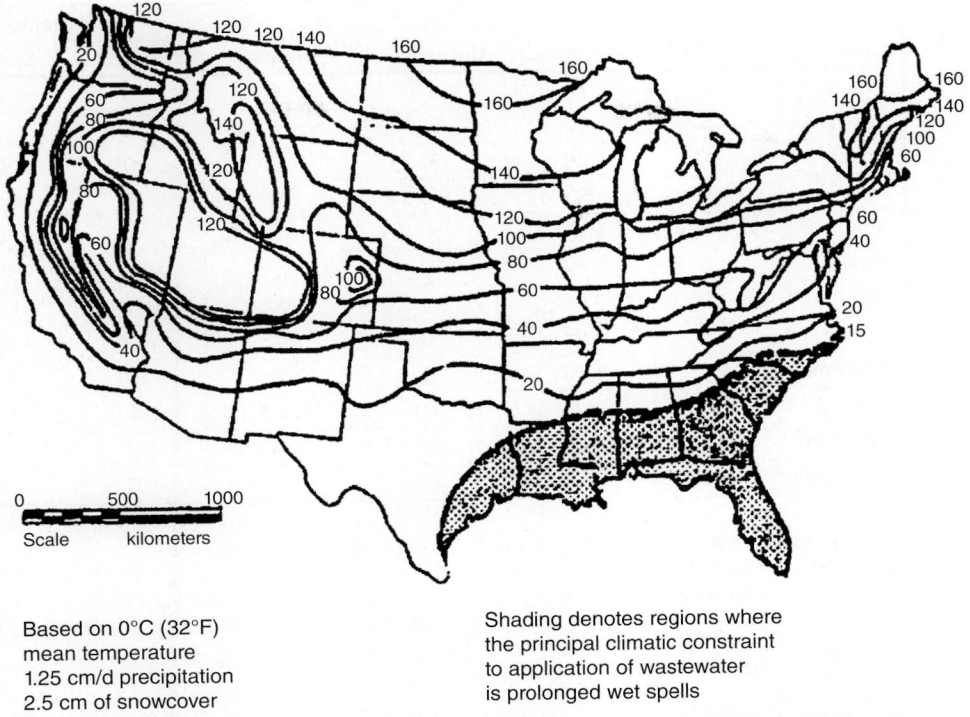

Based on 0°C (32°F)
mean temperature
1.25 cm/d precipitation
2.5 cm of snowcover

Shading denotes regions where
the principal climatic constraint
to application of wastewater
is prolonged wet spells

FIGURE 19.2   Storage days required as estimated from the use of the EPA-1 computer program for wastewater-to-land programs. Estimated storage based only on climatic factors (U.S. EPA, 1979).

earthen basins infeasible. These tanks typically are cylindrical with either a flat or sloped bottom (see Figure 19.3) (U.S. EPA, 1979). The U.S. Army Corps of Engineers recommends a 4:1 floor slope and a minimum depth of 4.5 m (15 ft) (U.S. Army Corps, 1984). Cylindrical tanks are preferred because they do not have corners, which may become "dead spots." The tanks can be constructed of either concrete or steel. Steel tanks are susceptible to corrosion, however, which design engineers should consider when designing a sludge storage system.

Storage tanks should be mixed to ensure that the discharged solids are homogeneous (Spinosa and Vesilind, 2001). Mixer manufacturers recommend mixing energies ranging from 10 to 12 kW/ML (40–50 hp/mg) (Lottman, 2008). The key is to keep solids suspended without inducing excessive air into the sludge.

The tank(s) may also require aeration, especially if the solids are unstabilized or aerobically digested (U.S. EPA, 1979). If so, the tank's oxygen requirements should be similar to those for aerobic digesters. Information regarding the sizing aerobic-digestion mixing equipment is provided in Chapter 23. If the material was stabilized before storage, however, the tank's oxygen requirements to maintain aerobic conditions will be significantly less. Maintaining a minimum dissolved oxygen level of about 0.5 mg/L should prevent anaerobic activity as long as the basin has adequate mixing. Otherwise, nuisance odors may be generated and more odor control may be required.

Month	Corn	Soybeans	Cottons[c]	Forages[d]	Small grains[b]	
					Winter	Spring
January	S[e]	S	S/I	S	C	S
February	S	S	S/I	S	C	S
March	S/I	S/I	S/I	S	C	S/I
April	S/I	S/I	P, S/I	C	C	P, S/I
May	P, S/I	P, S/I	C	C	C	C
June	C	P, S/I	C	H, S	C	C
July	C	C	C	H, S	H, S/I	H, S/I
August	C	C	C	H, S	S/I	S/I
September	C	H, S/I	C	S	S/I	S/I
October	H, S/I	S/I	S/I	H, S	P, S/I	S/I
November	S/I	S/I	S/I	S	C	S/I
December	S	S	S/I	S	C	S

[a]Application may not be allowed due to frozen, flooded, and snow-covered soils.
[b]Wheat, barley, oats, or rye.
[c]Cotton, only grown south of southern Missouri.
[d]Established legumes (alfalfa, clover, trefoil, etc.), grass (orchard grass, timothy, brome, reed canary grass, etc.), or legume-grass mixture.
[e]S = surface application
Note: S/I = surface/incorporation application
    C = growing crop present; application would damage crop
    P = crop planted; land not available until after harvest
    H = after crop harvested, land is available again: for forages (e.g., legumes and grass), availability is limited and application must be light so regrowth is not suffocated.

TABLE 19.2   General Guide to Months Available for Applying Biosolids to Various Crops in the North Central United States (U.S. EPA, 1979)[a]

2.2.2   Spill Prevention

Where residuals pumps, piping, and valves are above ground, spills can occur. Failed gaskets, poor installation, failed valves, and high pressure relief valve activation are just some of the possible spill sources. When the piping penetrates the tank below the maximum water level in the tank, this possibility increases because there will always be water in the pipe under head pressure. Designers should, at a minimum, consider the following when designing above-ground residuals systems:

- Design the above ground portion of the system as a containment area and install drains to collect any spilled fluids.
- Design the inflow and outflow pipes to enter and exit the tank at a level higher than maximum tank level. Include appropriate air-release and freeze protection, if needed.
- Add emergency shutoff and automatic flow cutoff controls in the design, designed to minimize any spillage.

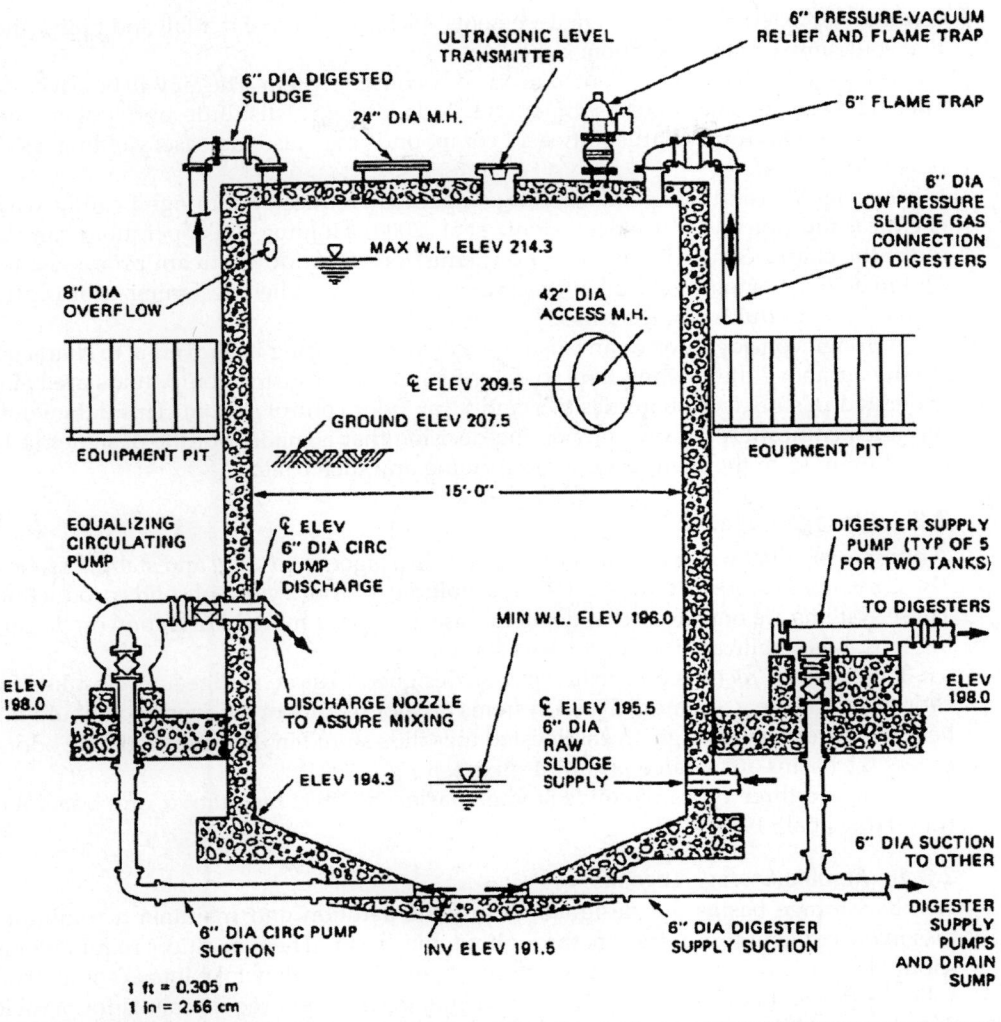

6" PRESSURE-VACUUM
RELIEF AND FLAME TRAP

ULTRASONIC LEVEL
TRANSMITTER

6" FLAME TRAP

6" DIA DIGESTED
SLUDGE

24" DIA M.H.

6" DIA
LOW PRESSURE
SLUDGE GAS
CONNECTION
TO DIGESTERS

MAX W.L. ELEV 214.3

8" DIA
OVERFLOW

42" DIA
ACCESS M.H.

₵ ELEV 209.5

GROUND ELEV 207.5

EQUIPMENT PIT

EQUIPMENT PIT

15'-0"

EQUALIZING
CIRCULATING
PUMP

₵ ELEV
6" DIA CIRC
PUMP
DISCHARGE

DIGESTER SUPPLY
PUMP (TYP OF 5
FOR TWO TANKS)

TO DIGESTERS

MIN W.L. ELEV 196.0

ELEV
198.0

ELEV
198.0

DISCHARGE NOZZLE
TO ASSURE MIXING

₵ ELEV 195.5
6" DIA
RAW
SLUDGE
SUPPLY

ELEV 194.3

6" DIA SUCTION
TO OTHER

6" DIA CIRC PUMP
SUCTION

INV ELEV 191.5

6" DIA DIGESTER
SUPPLY SUCTION

DIGESTER
SUPPLY
PUMPS
AND DRAIN
SUMP

1 ft = 0.305 m
1 in = 2.56 cm

FIGURE 19.3   A 98-m³ (26 000-gal) solids equalization tank (taken from U.S. EPA, 1979).

If storage facilities are near surface waters or other sensitive areas, a containment wall or berm may be advisable. The berm should be designed to retain or retard the movement of spilled residuals. A structural wall may not be necessary; an earthen berm may be sufficient. The containment berm should contain a spill long enough for it to be cleaned up but also include some means of removing excess water from rainfall or other sources.

2.2.3 Odor Control

Odors may be an issue, depending on the type of residuals and the storage time. Well-stabilized biosolids may be stored for several days without odor control. However, residuals characteristics may change over time, so it is preferable to cover storage tanks to contain and minimize offsite odors. Covers provide the additional benefit of improved control over external environmental elements, such as excessive rainfall and high winds that could impact tank operation.

The air in the tank can be vented to an odor-control system. The air will be adversely affected by such compounds as dimethyl sulfide, dimethyl disulfide, and longer-chain mercaptans. Odorous volatile fatty acid compounds may also increase, yielding a sour odor (WEF, 2004).

Storing WAS and raw primary sludge separately reduced odors significantly, while chemical addition had little effect (Hentz et al., 2000). Holding-tank operations can also affect the character and intensity of odor emissions from downstream processes. The design engineer should take all of this into consideration when designing odor-control systems for residuals storage tanks.

A design consideration at an existing facility that is being improved is to characterize the air above the existing tank to determine if odor causing compounds are being generated in sufficient quantities to mandate an odor control system. This data would provide justification for any engineering decision that is made, but the designer must have buy-in from the client prior to performing any such testing.

2.3   Storage Lagoons

A number of small water reclamation facilities use lagoons to store and stabilize solids. These treatment lagoons typically have the volume to store 2 years of solids production. These systems are not covered in detail in this section; for more information on designing biosolids stabilization lagoons, see Chapter 16.

This sections focuses on earthen basins designed to store sludge and biosolids for shorter periods (e.g., winter). These systems are not treatment lagoons and should not be used to treat raw sludge. Even digested biosolids stored in them can generate odors, unless the basins are aerated and well-mixed.

There are three types of solids storage basins: aerobic, facultative, and anaerobic (Lue-Hing et al., 1998).

2.3.1   Aerobic Storage Lagoons

Aerobic storage basins are designed to provide aeration and maintain a minimum dissolved oxygen concentration throughout the basin. Their aeration requirements should be similar to those for aerobic digesters, except that they take into account prior solids stabilization. (For information on calculating aeration requirements for aerobic digesters, see Chapter 23.) Because of their size and earthen construction, a piped aeration system is typically not used; subsurface or surface free-standing mechanical aerators are most often used.

2.3.2   Facultative Storage Lagoons

Facultative basins are unmixed and typically consist of three layers: a 0.3 to 1.0 m (1–3.3 ft) deep aerobic surface layer, a deeper anaerobic zone, and a sludge storage zone at the bottom. Both the aerobic and anaerobic zones are biologically active; anaerobic stabilization substantially reduces solids volume. The aerobic zone receives oxygen via surface transfer from the atmosphere, algal photosynthesis, and (if provided) surface-mix aerators which are operated periodically to break up scum on the pond surface and optimize oxygen transfer. Most of the satisfactory installations use brush-type surface mixers, according to U.S. Environmental Protection Agency (U.S. EPA, 1979). The oxygenation rate is low, however, so the U.S. EPA recommends that these basins only be used for anaerobically digested solids.

These basins are typically designed based on a volatile solids loading rate of 0.097 kg/m²/d (0.023 lb/sf/d) (Lue-Hing et al., 1998). They are typically 5 m (16 ft) deep to provide enough space for aerobic and anaerobic layers. Because facultative solids lagoons' "sour," odor can be a major issue, when designing such storage basins, engineers should consider prevailing wind patterns, and minimize odor potential via proper loading and surface mixing.

2.3.3 Anaerobic Basins
Anaerobic storage basins are similar to earthen basins but differ in that oxygen transfer from the surface is not considered in design. Therefore, earthen basins can be deeper than facultative ponds. Also, because maintaining an aerobic zone is not a key parameter, solids loading to earthen basins can be higher than with facultative ponds. Anaerobic ponds have essentially the same advantages and disadvantages as facultative ponds (Lue-Hing et al., 1998).

3.0   Liquid Sludge and Biosolids Transport
There are two basic methods for transporting liquid residuals at water reclamation facilities: trucking and pumping. Residuals are typically trucked offsite and pumped onsite.

3.1   Trucking
Although not typically part of the design process, trucking is often used to transport liquid biosolids, especially to land-application sites. Liquid sludge and biosolids may also be trucked to another site for further treatment. Trucking liquids can be larger challenge than transporting dewatered or dried biosolids, and there is some basic information designers should know.

Liquid sludge and biosolids are typically is transported via tanker trailer trucks. Such trucks typically have nominal capacities ranging from 22 680 to 34 020 L (6000–9000 gal). However, depending on weight restrictions, a tanker trailer may not be filled completely. In the United States, the maximum overall weight of a tractor-trailer is limited to 36 288 kg (80 000 lb). So, if a tractor weighs between 5443 and 6804 kg (12 000 and 15 000 lb), and an empty trailer weighs between 4990 and 7711 kg (11 000 and 17 000 lb), the contents cannot weigh more than approximately 21 773 to 25 855 kg (48 000–57 000 lb). This is equivalent to 21 000 to 25 500 L (5700 to 6800 gal) of dilute residuals.

Also, a 5850-kg (12 900-lb), 26 460-L (7000-gal) trailer typically is 13.1 m (40 ft) long and 2.4 m (8 ft) wide. Some are equipped with baffles to control movement and resulting surges of the liquid.

3.2   Pumping
Pumping systems are an intrinsic part of solids management at water reclamation facilities. They typically transport solids from

- primary and secondary clarifiers to thickening, conditioning, or digestion systems;
- thickening and digestion systems to dewatering operations;
- biological processes to further treatment units; and
- fegritting facilities to temporary storage areas.

Principle	Common Types	Typical Applications
Kinetic (rotodynamic) pumps	Nonclog mixed-flow pump[a] Recessed-impeller pump (vortex pump, torque flow pump)	Grit slurry,[b] incinerator ash slurry[c] Unthickened primary sludge[b,c]
	Screw centrifugal pump	Return activated sludge[c,d]
	Grinder pump	Waste activated sludges from attached-growth biological processes[c] Circulation of anaeronic digester[b] Drainage, filtrate, and centrate Dredges on sludge lagoons

Positive-displacement pumps	Plunger pumps
	  Progressing cavity pump
	  Air-operated diaphragm pump
	  Rotary lobe pump
	  Pneumatic ejector
	  Peristaltic pump
	  Reciprocating piston
		Waste activated sludge
		Thickened sludges (all types)
		Unthickened primary sludge
		Feed to dewatering machines
		Unthickened secondary sludges
		Dewatered cakes[e]

Other	Air lift pump Archimedes screw pump	Return activated sludge

[a]Limited solids capability; useful in larger sizes for return activated sludges. In most other applications, recessed-impeller pumps are more common.
[b]Abrasion is moderate to severe. Abrasion-resisting alloy cast iron is usually specified.
[c]May contain precipitates from aluminum or iron salts added for phosphorus removal.
[d]Particular need for reliable flow meters, for process control, in this application.
[e]Reciprocating piston pumps and progressing cavity pumps only.

TABLE 19.3 Sludge Pump Applications by Principle

While specifying only one type of pump for all of a facility's solids-transport systems might seem advantageous, the wide range of conditions involved typically exceeds the capabilities of any given pump. Fortunately, many types of pumps are available (see Table 19.3).

3.2.1 Design Approach

When designing pumping systems, design engineers should begin by asking: What sort of residuals will be pumped? Kinetic pumps—especially recessed-impeller pumps—can handle some types of residuals, but other types may require positive-displacement pumps.

Kinetic pumps have lower capital costs (especially in large sizes), lower maintenance costs, and smaller footprints. They are also available in submersible form (although conventional dry-well pumps are preferred for most applications).

Positive-displacement pumps have better process control because the pumping rate is less affected by fluid viscosity. They function better over the entire head range from zero to maximum without damaging the pump or motor, or changing drive speed. They work better under high pressure and at low flows. They are also less sensitive to non-ideal suction conditions (e.g., entrained air and gas) and less likely to disrupt fragile floc particles in flocculated solids.

The traditional approach to designing residuals transport systems is to minimize the pumping distance and apply a conservative multiplier to headlosses calculated for equivalent flows of water. However, this approach can be inaccurate, especially at higher solids concentrations with non-Newtonian characteristics. Such inaccuracies may not matter for short pumping distances, but they can be problematic for longer distances or critical applications.

The need to pump residuals long distances has increased in the last 20 years, so researchers have been developing methods to predict site-specific friction losses in pumping systems more accurately (Mulbarger et al., 1981; Carthew et al., 1983; Wagner, 1990; Honey and Pretorius 2000; Murakami et al., 2001). Study results have shown that, once rheological properties have been determined, the Bingham plastic model (Carthew et al., 1983; Mulbarger et al., 1981) or the pseudoplastic model (Honey and Pretorius, 2000; Murakami et al., 2001) for non-Newtonian fluids may describe how reclaimed water residuals flow. They can also predict the critical velocity at which laminar flow changes to turbulent flow. [In the turbulent flow range, dilute residuals obey conventional flow relationships for Newtonian fluids (Mulbarger et al., 1981).]

When designing solids pumping systems for smaller facilities, engineers should be careful to ensure that velocities will be adequate without undersizing piping, which increases the risk of line blockage.

3.2.1.1 Dilute Sludge Clarifiers often produce a relatively dilute settled solids (maximum concentrations of 1.2% to 1.5% are typical for activated sludge). At velocities greater than 0.3 to 0.6 m/s (1–2 ft/s), such solids are in the turbulent flow regime and have a headloss essentially equal to that of water (Mulbarger, 1997). At lower velocities, the flow becomes laminar, and headlosses increase sharply. So, engineers should design dilute solids pumping systems to maintain a minimum velocity of 0.6 to 0.75 m/s (2–2.5 ft/s) whenever possible to ensure turbulent flow.

3.2.1.2 Thickened Residuals The concentration at which sludge and liquid biosolids can be defined as "thickened" depends on the type of solids and the preceding treatment processes (see Table 19.1)(ASCE, 2000).

When designing pumping systems for thickened solids, design engineers can use the Darcy–Weisbach and Manning equations for water to determine headloss—regardless of the solids' flow regime (laminar, transition, or turbulent)—and apply a solids correction factor to the final calculation. The correction factor for residuals with up to 10% solids may be found in Figures 19.4 and 19.5 [taken from Sanks et al. (1998) and Metcalf and Eddy (2003), respectively]. As a simplified alternative, designers can use Figures 19.6 and 19.7, which indicate the headloss multiplier for worst-case design conditions in 150- and 200-mm (6- and 8-in.) forcemains, respectively.

If design engineers use the curves in Figures 19.6 and 19.7, they should choose pumps and motors that will operate satisfactorily over the entire headloss range from "water" to "worst-case." Head changes affect centrifugal pumps much more than positive-displacement pumps, so if centrifugal pumps (e.g., recessed-impeller) are used, engineers should also check the motors to avoid overloading if operating head drops significantly below design head. Motors may be overloaded if a pump becomes "runaway" (operates beyond the right terminus of its characteristic curve). Also, residuals occasionally can exceed the worst-case headloss curve. So in some instances, oversized motors and variable-frequency drives should be specified to provide the operational flexibility needed.

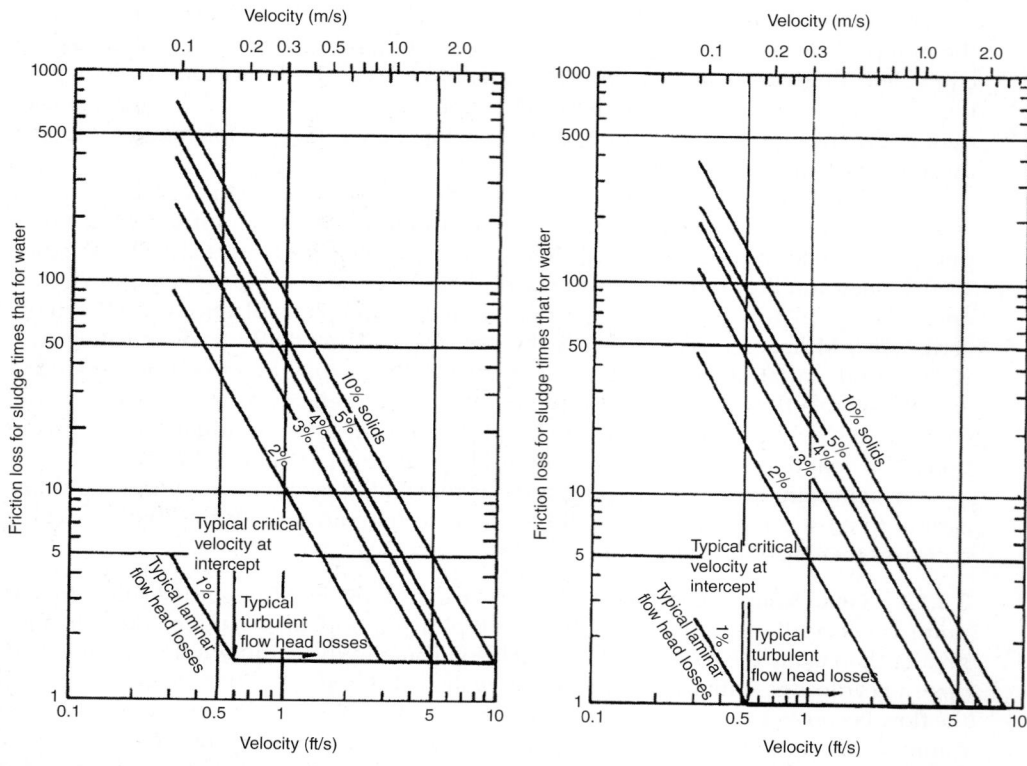

FIGURE 19.4   Multiplication factor for residuals headloss: (a) routine design and (b) worst-case design (Sanks et al., 1998).

In addition to headloss, design engineers should consider the nature of the process that will receive the solids. Many positive-displacement pumps suitable for thickened solids produce a pulsating flow, which may not be acceptable if the downstream process depends on steady flow or flow-proportioned chemical addition to operate properly.

Researchers have derived equations for long-distance pumping of WAS, thickened solids, and digested biosolids in the 2% to 5% total solids range (Murakami et al., 2001). Such materials behave like pseudoplastic fluids. Based on the assumption that fluid viscosity depends solely on percent solids, residual density is 1000 kg/m³ (1690 lb/cy) and temperature is 15°C, researchers proposed the following equations (eqs 19.5, 19.6, and 19.7):

Laminar flow

$$H_f(m) = 1.90 \times 10^{-3} k^{0.88} \frac{L\,V^{0.20}}{D^{1.20}} \qquad (19.5)$$

where $k = 0.059\,C^{2.74}$ for digested biosolids,

$k = 0.052\,C^{2.91}$ for thickened solids, and

$k = 0.050\,C^{3.06}$ for waste activated sludge.

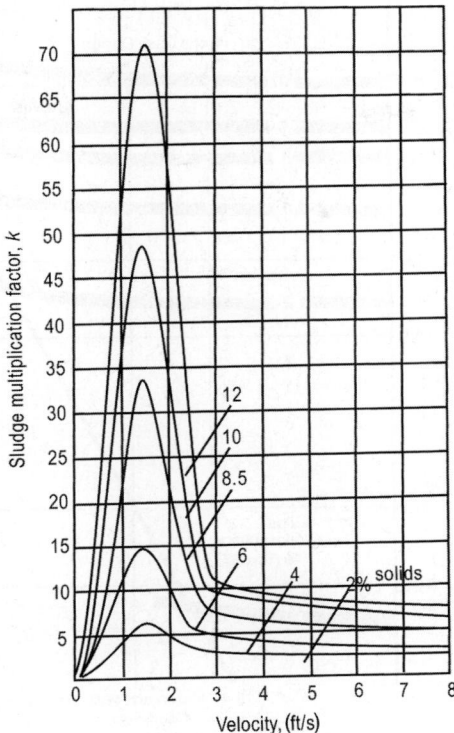

FIGURE 19.5 Multiplication factor for residuals headloss (from Metcalf & Eddy, *Wastewater Engineering: Treatment and Reuse*, 4th ed. Copyright © 2003, The McGraw-Hill Companies, New York, N.Y., with permission).

Turbulent flow

$$1.93 \times (1 - C/100) \tag{19.6}$$

$$H_f\,(\text{m}) = 9.06 \frac{(1)}{C_H} \frac{L\,V^{1.82}}{D^{1.18}} \tag{19.7}$$

where $CH = 110$ for mortar-lining, cast iron pipe, and

$CH = 95$ for carbon steel pipe.

Solids concentration is approximately 5% or less. The designer must also account for pipe age in this equation and make an engineering judgment to determine an appropriate c value for a specific project.

Critical velocity c

$$V_C\,(\text{m/s}) = 1.20 \frac{C_H}{100} k^{0.52} \tag{19.7.5}$$

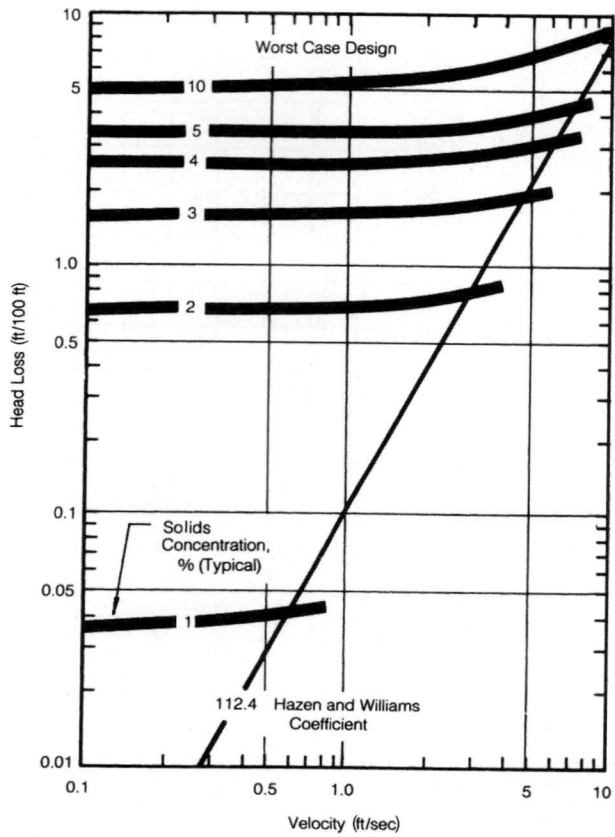

FIGURE 19.6 Predicted frictional headlosses for worst-case design of a 150-mm-diameter (6-in.-diameter) solids forcemain (in. × 25.4 = mm; ft × 0.3048 = m) (Mulbarger et al., 1981).

where H = Headloss (m),

V = Velocity (m/s),

C = percent solids concentration,

L = pipe length (m), and

D = pipe diameter (m).

Another design approach is based on the assumption that the flow of thickened sludge and biosolids follows the Bingham plastic model. To use this model, design engineers need to know a solids' yield stress (S_y) and coefficient of rigidity (R_c), which may be determined experimentally. If solids-specific data are not available, then designers can use Figures 19.8 and 19.9 (ASCE, 2000) to estimate these values. [Similar graphs have been created by Battistoni (1997); Guibaud et al. (2004); Laera et al. (2007); and Mori et al. (2007).]

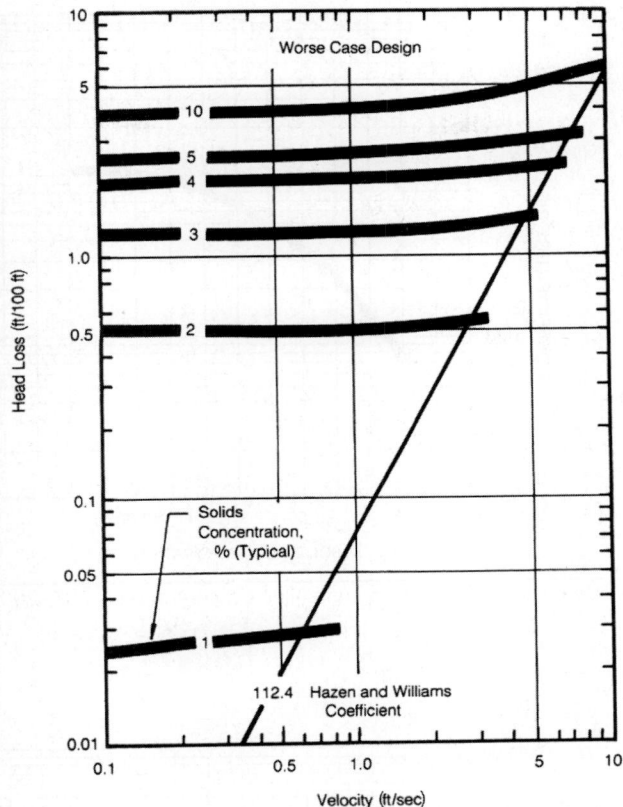

FIGURE 19.7   Predicted frictional headlosses for worst-case design of a 203-mm-diameter (8-in.-diameter) solids forcemain (in. × 25.4 = mm; ft × 0.3048 = m) (Mulbarger et al., 1981).

Once the rigidity coefficient and yield stress are known, designers can use the following two equations to calculate the upper and lower critical velocities:

$$V_{uc} = 1500\, R_c/\rho D/1500/\rho D\ (R_c^2 + S_y\, \rho D^2/4500)^{1/2} \tag{19.8}$$

$$V_{lc} = 1000\, R_c/\rho D/1000/\rho D\ (R_c^2 + S_y\, \rho D^2/3000)^{1/2} \tag{19.9}$$

where V_{uc} = upper critical velocity (m/s),
   V_{lc} = lower critical velocity (m/s),
   ρ = fluid density (kg/m^3),
   D = pipe diameter (m),
   S_y = yield stress, and
   R_c = rigidity coefficient (N-s/m^2).

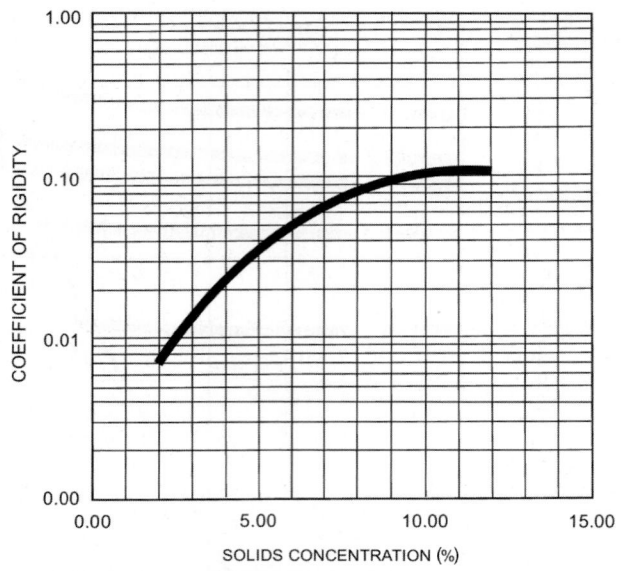

FIGURE 19.8 Coefficient of rigidity versus solids concentration (ASCE, 2000).

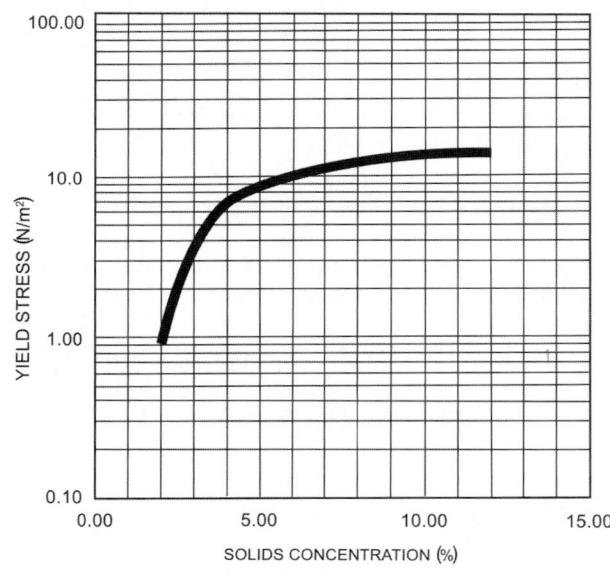

FIGURE 19.9 Yield stress versus solids concentration (ASCE, 2000).

Alternatively, designers can calculate the Reynolds number as follows:

$$\text{Re} = \rho V D / R_c \qquad (19.10)$$

If Re < 2000, the flow is laminar. If Re > 3000, the flow is turbulent.

3.2.1.2.1   Laminar Flow   At velocities less than the lower critical velocity, or when Re < 2000, the material's flow will be in the laminar range and designers can calculate headloss using the Buckingham equation:

$$H/L = 32 \, (S_y/6 \, \rho g D / + R_c V / \rho g D^2) \qquad (19.11)$$

where H/L = headloss per unit length (m/m),

  S_y = yield stress (Pa or N/m^2),
  ρ = fluid density (kg/m^3),
  g = gravitational acceleration (m/s^2),
  D = pipe diameter (m),
  R_c = rigidity coefficient (Pa-s or N-s/m^2), and
  V = velocity (m/s).

Honey and Pretorius (2000) experimentally determined the rheological parameters of settled waste activated sludge. They measured solids concentration, particle density, and torque, and then derived shear stress and shear rate from the torque data. They then determined the fluid consistency coefficient (K) and the pseudoplastic model's flow-behavior index (n). (They suggest that the pseudoplastic model more accurately indicates the behavior of 5% settled activated sludge in laminar flow.) From there, they compared the following generalized Reynolds number with a critical Reynolds number for pseudoplastic fluids to determine the flow regime.

$$\text{Re(g)} = \rho V D / (8V/D)^{n-1} K [(3n + 1)/4n]^n \qquad (19.12)$$

$$\text{Re(critical)} = 6464 \, n / (1+3n)^2 \, [(1/(2 + n)]^{(2+n)/(1+n)} \qquad (19.13)$$

They then used the Darcy–Weisbach equation:

$$H_f = 4 f L \, V^2 / 2gD \qquad (19.14)$$

where f = the dimensionless Fanning friction factor, which is 16/Re(g) for pseudoplastic fluids in laminar flow.

In their study, Honey and Pretorius assumed that residuals behaved as a thixotropic fluid. The fluid exerted a maximum headloss when the pump was turned on, and dropped to a lower, constant headloss after a certain time or travel distance in the pipeline. Then the thixotropic behavior disappeared.

3.2.1.2.2   Transition and Turbulent Flow   At velocities greater than the upper critical velocity, or when Re > 3000, solids flow will be turbulent.

When designing pumping systems for turbulent conditions, engineers can solve the Hazen–Williams equation for turbulent water and apply the solids correction factor to the result (Sanks et al., 1998; Metcalf & Eddy 2003). When using this equation, design engineers should assume that C equals 140 under normal conditions and 112.4 under worst-case design conditions.

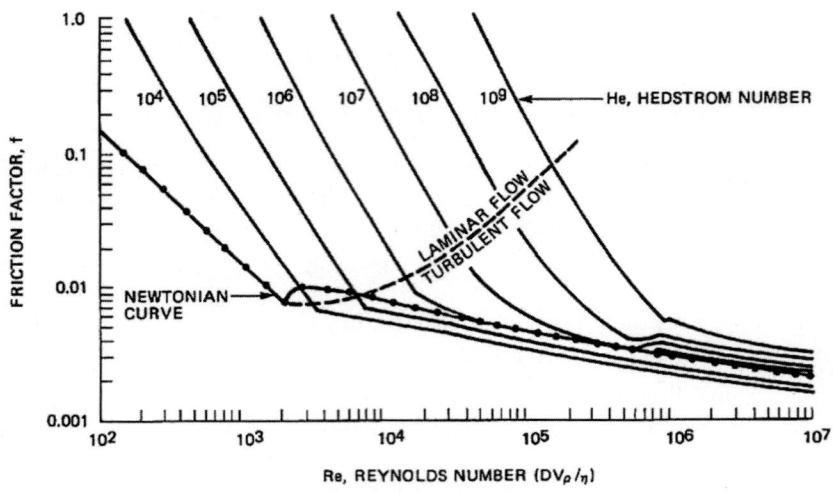

FIGURE 19.10   Friction factor (*f*) for solids, assuming Bingham plastic behavior (U.S. EPA, 1979).

Designers can also use Reynolds and Hedstrom numbers to calculate headlosses. To find the Reynolds number, see eq 19.10. The Hedstrom number is calculated as follows:

$$He = D^2 \rho\, S_y/R_c^2 \tag{19.15}$$

After calculating the Hedstrom and Reynolds numbers, the friction factor (also called the Fanning friction factor) is then calculated using Figure 19.10. Design engineers then should use the Darcy–Weisbach equation to calculate headloss:

$$\Delta P = 2 f \rho L V^2/D \tag{19.16}$$

where ΔP = pressure headloss (Pa).

Design engineers should make sure they use the correct friction factor for residuals, because the friction factor for water taken from a Moody diagram is often quoted as four times that of residuals (Figure 19.10).

Chilton and Stainsby (1998) used both analytical methods and numerical techniques to determine headlosses of residuals at four different total solids concentrations flowing through a 150-mm (6-in.) pipe. They used the rheological parameters noted in a 1980 paper by Ackers and Allen. The materials were characterized only by their density, not by type or solids concentration.

Recently, Bechtel (2003; 2005) used computational fluid dynamic (CFD) methods to analyze pipeline flow and then compared his results with

- an analytical solution and an early work of Mulbarger to determine pipeline headlosses in the laminar-flow range, and

- the same analytical solution, the early work of Mulbarger, a graphical approach from Metcalf and Eddy (1981), the equations proposed by Chilton and Stainsby (1998), and Steffe's 1996 work to determine pipeline headlosses in the turbulent-flow range.

The analytical solution involved calculating the Reynolds and Hedstrom numbers and then determining the Fanning friction factor as follows:

$$f^4 + f^3 (-16/\text{Re} - 8He/3\text{Re}^2) + 16He^4/3\text{Re}^8 = 0 \qquad (19.17)$$

Bechtel found that that the Mulbarger curves overpredicted headlosses for solids at laminar flows. When solids were at turbulent flows, all of the models—except Mulbarger curves—predicted similar results. One conclusion to be drawn from this information is that designers must exercise caution when using the Mulbarger data and curves.

3.3   Example 19.1

Calculate the friction-related headloss associated with pumping 5% thickened WAS 150 m at a laminar flow rate. The pump's design flow rate is 400 L/min. The inside diameter of a mortar-lined ductile iron pipe is 155 mm (0.155 m) (D). So, the fluid velocity (V) is

$$V = 24/60/60/\pi \times 0.155^2 \times 4$$

$$= 0.353 \text{ m/s}$$

3.3.1   Option 1: Using the Darcy–Weisbach Equation

Design engineers could use the Darcy–Weisbach equation and the Moody Diagram for water and then multiply the result by the appropriate solids multiplication factor.

The Darcy–Weisbach equation is

$$H_f = fL\, V^2/D\, 2g$$

where f = the friction factor for water as derived from a Moody Diagram.

Unlike the = 140 used in the Hazen–Williams equation for ductile iron, the Moody Diagram for ductile iron pipe lacks an explicit ε coefficient. This coefficient may range from 0.13 to 0.33 mm.

In this case, the median of the range is selected, so:

$\varepsilon = 0.23$ mm

$D = 155$ mm

$\varepsilon/D = 0.001484$, relative roughness

The Reynolds number for water is:

$$\text{Re} = VD/v$$

where $v = 1.14 \times 10^{-6}$ m²/s, kinematic viscosity of water at 15°C

so, Re = 47 996 and from the Moody Diagram

$f = 0.026$

Substituting all the values from above, we get:

$$H_f = 0.026 \times 150 \text{ m} \times (0.353 \text{ m/s})^2/(0.155 \text{ m} \times 2 \times 9.81 \text{ m/s}^2)$$

$$H_f = 0.1598 \text{ m (for water flows)}$$

From Figure 19.4b (Sanks et al., 1998; Figure 19-4), for worst-case design, a sludge multiplication factor equal to 35 is derived, so the final headloss for sludges containing 5% solids is:

$$H_f = 0.1598 \times 35 = 5.59 \text{ m of water column}$$

3.3.2   Option 2: Using the Buckingham Equation

The Reynolds number (Re) for sludge is:

$$\text{Re} = \rho \, V D / R_c$$

where ρ = 1020 kg/m³, solids density

R_c = 0.035 kg/m/s, coefficient of rigidity determined using Figure 19.8

Re = 1595, which is less than 2000, so the flow regime is laminar and eq 19.11 may be used. Substituting all the values from above gives

H/L = 32 (9.5 Pa/6 × 1020 × 9.81 × 0.155 + 0.035 × 0.353/1020 × 9.81 × 0.155²)

H/L = 0.034 m of head per meter of pipe

because the pipe is 150-m long, total headloss is

$$H_f = 5.10 \text{ m of water column}$$

3.3.3   Option 3: Using Figure 19.6 (155 mm. pipe, worst case)

V = 0.353 m/s, so Figure 19.6 indicates that headloss is 3.5 m/100 m. Because the total length is 150 m, the total headloss is

$$H_f = 3.5 \times 1.5 = 5.25 \text{ m of water column}$$

In addition to friction-related headloss, design engineers should calculate static head and "minor" headlosses from valves and fittings. The sum of all these headlosses is the total dynamic head that the pump must provide. Design engineers also need to ensure that the available net positive suction head is sufficiently more than is needed, should consider changes in thickened solids characteristics, and evaluate multiple duty points, if needed.

3.4   Kinetic Pumps

Kinetic (dynamic) pumps continuously add energy to the pumped fluid to make the velocity in the pump higher than the velocity at the discharge point and, therefore, increase pressure. Several types of these pumps and their common applications are discussed in the following subsections.

3.4.1   Solids-Handling Centrifugal Pumps

A wide variety of centrifugal pumps are available. Except for special designs (e.g., recessed impeller), however, these pumps should only be used with relatively dilute (less than 1% solids), trash-free residuals. They typically are used to transport return activated sludge (RAS) because of the pump's high volumetric flow rate and excellent efficiency. The minimal debris in RAS typically does not clog the pumps.

Centrifugal pumps are not recommended for primary solids, primary scum, or thickened residuals for two reasons. First, there is no means to ensure that the pump's suction will positively draw thickened material to the pump impeller. Second, the system head curve depends on solids concentration that is often inconsistent, leading to significant variations in liquid flow rate and pump power requirements. GLUMRB recommends that, if these pumps are used for these applications, positive displacement pumps be provided in parallel to be used when the solids concentration increases past the capabilities of the centrifugal pumps (GLUMRB, 2014)

3.4.2 Recessed-Impeller Pumps

The recessed-impeller pump (also called a torque-flow, vortex, or shear-lift pump) has a standard concentric casing with an axial suction opening and a tangential discharge opening. The impeller, which is recessed into the pump casing, can be open or semi-open with either straight radial blades or ones tapered to the shaft.

In solids pumping applications, design engineers typically choose pumps with fully recessed, open impellers. When it rotates, the impeller creates a spiraling vortex field in the fluid within the casing. This vortex moves residuals through the pump, allowing large solids to pass easily. Most of the solids do not pass through the impeller vanes, thereby minimizing abrasion.

Recessed-impeller pumps work well on untreated sludge containing no more than 2.5% solids or on digested biosolids with total solids as high as 4%. Although they can pump thicker residuals, varying friction losses cause erratic flow rates and heavy radial thrusts on the pump shaft. Positive-displacement pumps perform better in such applications. If design engineers use recessed-impeller pumps to transport thickened solids or biosolids, they should provide flow meters and variable-speed drives to maintain a relatively constant flow. They should also specify the heaviest possible shafts and bearings. In addition, the pumps should be horizontally mounted to simplify maintenance, and include adequate clean-outs and flushing connections.

Although contact between solids and impeller vanes is minimal, design engineers should consider specifying abrasion-resistant, cast-iron (ASTM A532) volutes and impellers, especially if the grit and abrasives content is high or unknown. However, such impellers cannot be trimmed, so if using them, designers must size the pump(s) accurately.

Recessed-impeller pumps are available in both vertical (close-coupled or extended-shaft) and horizontal configurations that are suitable for either wet or dry wells. Wet-well pumps are available with hydraulic drives or submersible electric motors. They typically are available in sizes from 50 to 200 mm (2–8 in.), with capacities from 180 to 1800 L/min (50–500 gpm) at up to 64 m (210 ft) total dynamic head.

The primary drawback of recessed-impeller pumps (compared to other nonclog centrifugal units) is their significantly lower efficiency. A recessed-impeller pump's efficiency typically is between 5% and 20% lower than that of other nonclog centrifugal pumps.

3.4.3 Screw/Combination Centrifugal Pumps

Screw/Combination centrifugal pumps combine a screw-type impeller with a normal centrifugal impeller. They typically have a relatively high efficiency and relatively low net positive suction head (NPSH) requirements. In addition, the corkscrew action of screw impellers may provide more positive feed to the suction, so the pump handles thicker sludge better. These pumps are commonly used for primary sludge pumping and are capable of handling higher total solids concentrations up to 8%.

3.4.4   Disc Pumps

Disc pumps operate on the principles of boundary layer and viscous drag. Their impellers are basically parallel discs installed at a certain distance apart. Fluid flows through the gap between the rotating discs, which transfer energy to the fluid and generate velocity and pressure gradients that force the fluid to flow through the pump. Pump wear is minimized because the fluid moves parallel to the discs and does not touch other pump parts.

Disc pumps traditionally are specified for residuals with total solids concentrations up 6% including slurries, viscous materials, and solids with high entrained-air content (e.g., from DAF units). They can run dry and handle abrasive materials, which makes them excellent candidates for pumping grit.

3.4.5   Grinder/Chopper Pumps

Special combination centrifugal pump grinders are also available (see Figure 19.11). These pumps combine a hardened steel cutting bar with a relatively typical centrifugal

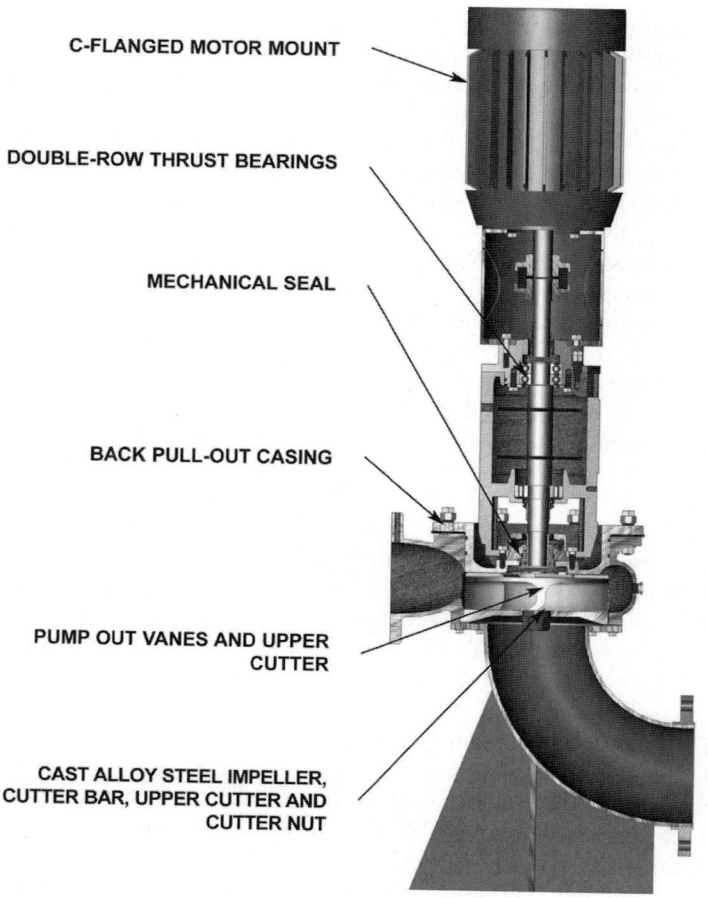

C-FLANGED MOTOR MOUNT

DOUBLE-ROW THRUST BEARINGS

MECHANICAL SEAL

BACK PULL-OUT CASING

PUMP OUT VANES AND UPPER CUTTER

CAST ALLOY STEEL IMPELLER, CUTTER BAR, UPPER CUTTER AND CUTTER NUT

FIGURE 19.11   Chopper–grinder pump (Courtesy of Vaughan Company, Inc.).

vortex-type pump. They can be used as digester recirculation pumps and are gaining popularity for use in pumping thickened residuals to dewatering processes. These pumps are beneficial in existing water reclamation facilities that do not have proper screening facilities to remove rags and stringy substances. These substances find their way into the thickening process and form rag balls that can clog other pump types. These pumps will act as a grinder would to cut the rag balls and allow the residuals to pass to the dewatering process in a form that is typically not detrimental to the process. However, operating experience indicates that such pumps require as much maintenance as grinders. Depending on the site-specific severity of the ragging, grinders and grinder/chopper pumps can be used in series.

3.5 Positive-Displacement Pumps
There are several types of positive-displacement pumps that can be used to transport sludges.

3.5.1 Plunger Pumps
Plunger pumps use pistons driven by either an exposed eccentric crank shaft or a walking beam to pump residuals. They are available in simplex, duplex, triplex, and quadplex configurations. Plunger pumps have an output of 150 to 225 L/min (40–60 gpm) per plunger and can develop up to 70 m (230 ft) of discharge head. These pumps typically are designed for an efficiency of 40% to 50%, which leaves a power reserve to overcome changes in pumping head (Sanks et al., 1998).

Plunger pumps have several advantages:

- They can transport residuals containing up to 15% total solids if the equipment is designed for load conditions;
- They are available in cost-effective options up to 30 L/s (500 gpm) and 60 m (200 ft) of discharge head;
- Units with large port openings can operate at low pumping rates;
- They provide positive delivery unless some object prevents the ball check valves from seating;
- They provide constant-but-adjustable capacity in spite of large variations in pumping head;
- They can operate for short periods of time under "no-flow" conditions (e.g., a plugged suction line) without damage (designer should confirm with specified manufacturers concerning guaranteed operating time under this condition);
- The pulsating action of low-velocity simplex and duplex pumps sometimes helps concentrate residuals in feed hoppers and re-suspend solids in pipelines; and
- They have relatively low operations and maintenance (O&M) costs.

Changing the stroke length changes the pump output. However, the pumps typically operate most efficiently at or near full stroke, so designers typically provide a variable-pitch V-belt drive or a variable-speed drive to control pumping capacity.

Plunger pumps have paired ball or flap check valves on the suction and discharge sides. A connecting rod joins the throw of the crankshaft to the piston. The piston is housed in an oil-filled crankcase (for lubrication) and sealed in a stuffing box gland and packing, which is kept moist by an annular pool of water directly above it. Unless the pool receives a constant supply of water, the packing will dry and fail rapidly, which can cause sludge to spray throughout the immediate area. As such, a seal water monitoring and control system with interlocking alarm and shutoff capability is recommended.

Plunger pumps can operate with up to 3 m (10 ft) of suction lift, but this can reduce the solids concentration they can handle. Using a pump whose suction pressure is higher than its discharge pressure is impractical because flow would be forced past the check valves. Using special intake and discharge air chambers reduces noise and vibration and dampens pulsations of intermittent flow.

If designers use pulsation-dampening air chambers, they should be glass-lined to avoid destruction via hydrogen sulfide corrosion. If the pump operates while the discharge pipeline is obstructed, the pump, motor, or pipeline can be damaged; a simple shear pin arrangement can prevent this problem.

The number of pistons directly influences the variation in downstream flow rates. If an application requires a relatively constant flow rate, design engineers should consider using triplex or quadplex pumps.

3.5.2 Progressing Cavity Pumps

Compared to plunger pumps, progressing cavity pumps provide a more consistent flow rate, even with changes in discharge head. Design engineers should guard against pump operation in no-flow conditions because it can quickly damage the stator. Monitoring of the pump influent line, with shutoff control if a no-flow condition occurs in the line (high-pressure switch, for example), is one method to protect the pump.

A progressing cavity pump uses a worm-shaped metal rotor that turns eccentrically inside a pliable elastomeric stator. The stator's axial pitch is about 50% that of the rotor. The rotor seals against the stator, forming a sealing line or lines that move down the pump as the rotor turns. Cavities progress axially between these lines, moving residuals from the suction end of the pump to the discharge end. As the stator wears, some "slippage" flow occurs at the sealing lines; this slippage causes further wear. To minimize slippage, design engineers should use enough cavities (multistage construction) to limit the pressure difference across the sealing lines.

The elastomeric stator is relatively soft and subject to abrasion, so progressing cavity pumps should be used in facilities with good grit-removal facilities. They should not be used to transport grit. Also, design engineers should minimize the rotor's rotational speed. In some applications (particularly with variable-speed drives), designers should consider selecting a pump with greater than design flow capacity to ensure that the pump still can meet design flow requirements after the stator has begun to wear.

One advantage of a progressing cavity pump is that the stator acts as a check valve, preventing backflow under most conditions. An actual check valve or anti-reverse ratchet is only required if the pump's static backpressure is more than 50 m (165 ft) or stator wear is expected due to significant grit concentrations. However, design engineers should always include isolation valves on both suction and discharge sides so the pump can be removed from service for routine maintenance.

Most progressing cavity pumps are tested with water. When used to transport residuals , the pumps may need more motor horsepower. Design engineers should consult with pump manufacturers on each application to ensure that adequately sized motors are specified.

Solids capacity depends on pump size. Pumps sized for at least 3 L/s (50 gpm) at suitably low rotating speeds typically pass solids of about 20 mm (0.8 in.), so grinders are unnecessary. Smaller pumps, however, typically need grinders. If grinders are not included, then design engineers should specify protective covers on any required universal joints.

To minimize pump maintenance costs,

- make sure prior processes remove grit effectively;
- limit rotating speeds to approximately 250 rpm;
- make suction lines as short as possible and use open-throat, hopper-type suction ports (Jones, 1993);
- limit the pressure per stage to about 170 kPa (25 psi) (if higher pressures are needed, most manufacturers offer multistage pumps);
- carefully specify rotor material, stator material, and design of universal joints (where applicable);
- provide room to dismantle the pump efficiently;
- consider including reversing starters, which allow the pump to reverse flow direction and possibly clear minor blockages in suction piping; and
- reverse the pump's flow direction in high suction lift applications.

Progressing cavity pumps can continue to operate as discharge pressure increases above the rated point, as in the case of a blocked discharge line. As such, the pump discharge must have pressure safety switches to prevent blocked discharge lines from rupturing. Also, design engineers should use flow indicator switches or proprietary devices to prevent the pumps from running dry. Designers should also consider using pressure-relief assemblies or rupture disks to protect downstream piping.

3.5.3 Diaphragm Pumps

Diaphragm pumps typically transport solids from primary sedimentation tanks and gravity thickeners. These pumps are a relatively simple means of pumping thickened solids and can handle grit with minimum wear. Manufacturers claim that the pump's pulsating action increases solids concentrations when transporting gravity-thickened solids, but designers are cautioned about trying to use this as a design parameter in thickening systems. Also, pulsating flow may not be acceptable for some downstream treatment processes, so designers must consider this when considering diaphragm pumps.

An air-operated diaphragm pump typically consists of a single-chambered, spring-return diaphragm, an air-pressure regulator, a solenoid valve, a gauge, a muffler, and suction and discharge isolation valves (see Figure 19.12). Compressed air flexes a membrane that is pushed or pulled to contract or enlarge an enclosed cavity. Unless the water resource recovery facility already uses compressed air, providing this service can significantly increase pumping costs. Also, the air exhausted from the pump valves creates significant noise, which can be detrimental to operations staff.

Hydraulically or electric-motor driven diaphragm pumps are also available, and should be considered. Key considerations would include the availability of compressed air, power consumption of the various systems, and operator familiarity with air-operated and hydraulically operated equipment.

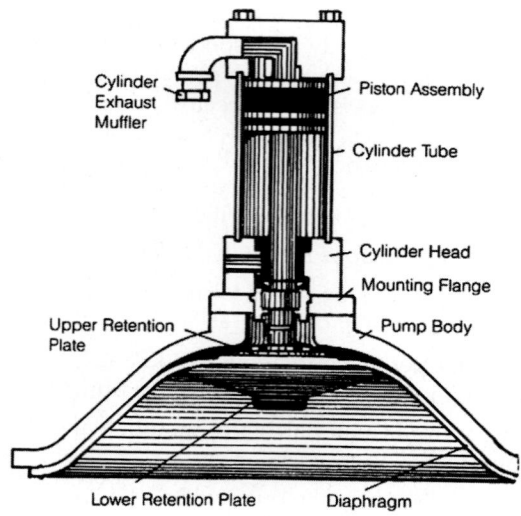

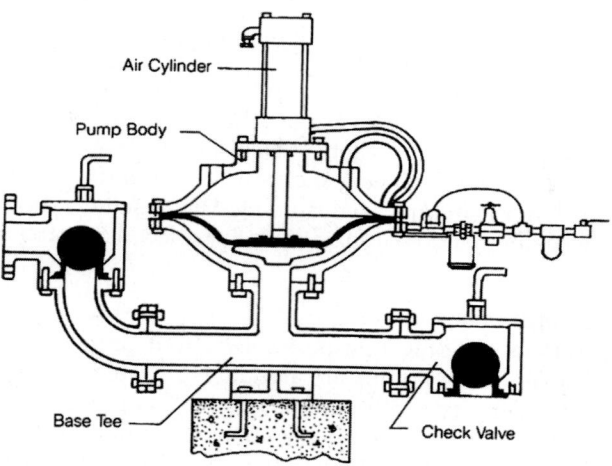

FIGURE 19.12   Air-operated diaphragm pump.

3.5.4   Rotary Lobe Pumps

A rotary lobe pump uses multilobed, intermeshed rotating impellers to transfer residuals containing up to 10% total solids. Like progressing cavity pumps, rotary lobe pumps offer a relatively smooth flow and do not require check valves in many applications with low-to-moderate discharge static heads. However, both suction and discharge isolation valves are needed so the pump can be removed from service for maintenance.

Pumping efficiency depends on maintaining relatively close tolerances between rotating lobes, so all large or abrasive material should be removed from residuals before they enter the pumps. Rotary lobe pumps are more suitable for applications with efficient grit removal and should not be used to transport grit. Engineers should also design the pumps to rotate at the lowest possible speed to minimize abrasion. Both rotary lobe and progressing cavity pumps react similarly to abrasion, so engineers can apply many of a progressing cavity's design considerations to a rotary lobe pump. Designers should also select appropriate lobe material for the residuals to be transported; otherwise, the lobes may fail prematurely. An advantage of these pumps over progressing cavity pumps is their ability to handle short periods of no flow without significant damage (designer should confirm this attribute with specified pump manufacturers to confirm that this operating feature is warrantied).

3.5.5 Pneumatic Ejectors

Rather than rotating elements and electric motors, a pneumatic ejector has a receiving container, inlet and outlet check valve and isolation valve, air supply, and liquid level detector. When liquid reaches a preset level, air is forced into the container and the stored volume is ejected. Then, the air supply cuts off and liquid flows through the inlet into the receiver.

Pneumatic ejectors can be used to convey residuals and scum. They have also been used in some facilities to transport grit and screenings. They are available in capacities from 110 to 570 L/min (30–150 gpm) and heads up to 30 m (100 ft).

3.5.6 Peristaltic Hose Pumps

Although more widely applied in the industrial sector, peristaltic hose pumps have been used to transport municipal wastewater solids. These self-priming pumps are available in capacities of 36 to 1250 L/min (9–330 gpm) and heads up to 152 m (500 ft). They can be used to meter flow because their output is directly proportional to speed at either high or low discharge pressures. Peristaltic hose pumps are suitable for suction lift applications [up to 44.7 kPa (15 ft of water)] and can pump abrasive fluids. These devices are relatively simple, requiring only common tools and basic mechanical skills for assembly, servicing, and repair.

A peristaltic hose pump has no seals, valves, or bearings; it moves sludge by alternately compressing and relaxing a specially designed resilient hose. The hose is compressed between the inner wall of the pump housing and the compression shoes on the rotor. A liquid lubricant may be used to minimize sliding friction. The residuals only touch the hose's thick inner wall, which cushions entrained abrasives during compression; abrasives are released after compression. Replacement hoses can be expensive, however, so to maximize hose life, the pump's maximum rotational speed should be limited to 25 rpm.

The primary disadvantage of this pump is its pulsed flow (because the rotor typically only has two compression shoes). Depending on the rotational speed required to obtain the design pumping rate, the pulsing flow may not be suitable for downstream processes. This can be offset, however, by using pulsation dampeners on the discharge.

3.5.7 Reciprocating Piston Pumps

Reciprocating piston pumps are useful and cost-effective when dewatered biosolids must be transported to cake storage or loading facilities. They typically are not used ahead of the dewatering process. However, because these pumps can achieve discharge pressures up to 1.5×10^4 kPa (2200 psi), they are the primary choice for pumping thickened solidslong distances. Given these pumps' high potential discharge pressures, however, engineers must design downstream piping systems properly.

3.6   Other Pumps

Other types of pumps used to transport residuals are discussed in the following subsections.

3.6.1   Air-Lift Pumps

An air-lift pump has an open riser pipe, the lower end of which is submerged in the liquid to be pumped. When an air-supply tube introduces compressed air at the bottom of the pipe, air bubbles form and mix with the residuals in the pipe. As the density of the air–residuals mixture decreases, denser material outside the pipe pushes the mixture up and out of the riser pipe.

Air-lift pumps are often used to transport WAS, RAS and similar solids in smaller facilities, where high efficiency and a precisely controlled flow rate are not required. Air-lift pumps typically are used in high-volume, low-head applications, those with lifts less than 1.5 m (5 ft). Their capacity can be varied by optimizing the air-supply rate. Increasing the air supply beyond its optimum level, however, only decreases the volume of liquid discharged. The main advantages of air-lift pumps are the absence of moving parts and their simple construction and use.

The primary disadvantages of air-lift pumps are their inefficiency (approximately 30%), which leads to higher energy costs to operate, and their head limitations (often limited to 5 ft due to compressed air limitations) (Metcalf and Eddy, 1981).

The air-supply arrangement governs the solids-handling capability. Air-lift pumps with an external air supply and circumferential diffuser can pass solid particles as large as the riser pipe's internal diameter without clogging. Those with air supplied by a separately inserted pipe lack this nonclog feature.

3.6.2   Archimedes Screw Pumps

Archimedes screw pumps occasionally are used to transport RAS (see Figure 19.13). This pump has an open design for lifts up to 9 m (30 ft) and an enclosed design for lifts up to 12 m (40 ft) or more. It automatically adjusts its discharge rates in proportion to the depth of liquid in the inlet chamber until the water gets to the "fill point," and then becomes constant. In other words, the pump has an inherent variable-flow capacity and does not need motor-speed controllers.

An Archimedes screw pump has a fairly constant efficiency (70%–75%) within 30% to 100% of its rated design capacity. The screw spirals' peripheral tip speeds are typically less than 229 m/min (750 ft/min); those for centrifugal or recessed-impeller pumps are 1070 to 1220 m/min (3500–4000 ft/min). Also, the screw pumps are not pressurized. These characteristics are advantageous in RAS applications because they make the screw pump less likely to shear the activated sludge floc.

The pump's principal disadvantage is its space requirements. If exposed to the sun and left idle for extended periods or unequipped with cooling water sprays, the pump can warp due to thermal expansion. Off-line units may also freeze in cold weather. Another potential disadvantage is that RAS often aerates in these systems. In some RAS applications, Archimedean screw pumps were no longer used because the RAS high dissolved oxygen content was interfering with biological nutrient removal. Designers

Figure 19.13   Archimedes screw pump (Courtesy of Evoqua).

should account for this aeration potential and its impact on downstream systems when considering the use of these pumps for RAS/WAS pumping.

3.7   Long-Distance Pipelines

Many facilities successfully pump solids long distances. The work of Carthew et al. (1983) was driven by the design of a 29-km (17.7 mi) pipeline. Honey and Pretorius (2000) solved their example using a 2-km (1.2 mi) pipeline, and Murakami et al. (2001) quote a distance of 1 km (0.6 mi) in their manuscript. However, engineers must develop special design criteria to minimize potential operating problems. They should carefully determine sludge characteristics (e.g., viscosity, solids percentage, and type), study the effects of flow velocity on fluid viscosity and pipe-friction losses (Mulbarger et al., 1981), and minimize friction losses wherever possible. Minimizing bends and fittings in piping design and using PVC or glass-lined pipe are two friction-reducing possibilities.

When designing long-distance systems, actual field data are critical. Several studies provide detailed information on the analysis and design of long-distance solids transport systems (Carthew et al., 1983; Mulbarger et al., 1981; Setterwall, 1972; Spaar, 1972; U.S. EPA, 1979). Figure 19.14 shows the test systems used in the field by Carthew et al. (1983) and Murakami et al. (2001).

Long-distance pumping typically creates high-pressure losses, so design engineers should choose pumps that can generate the high pressures needed. In the United Kingdom, for example, a vertical, positive-displacement, hydraulically driven ram pump transfers primary and waste activated sludge over a 2.2-km (1.3 mi) long pipeline, working against pressures of up to 2600 Pa (377 psi) (Ram Pumps, 1999).

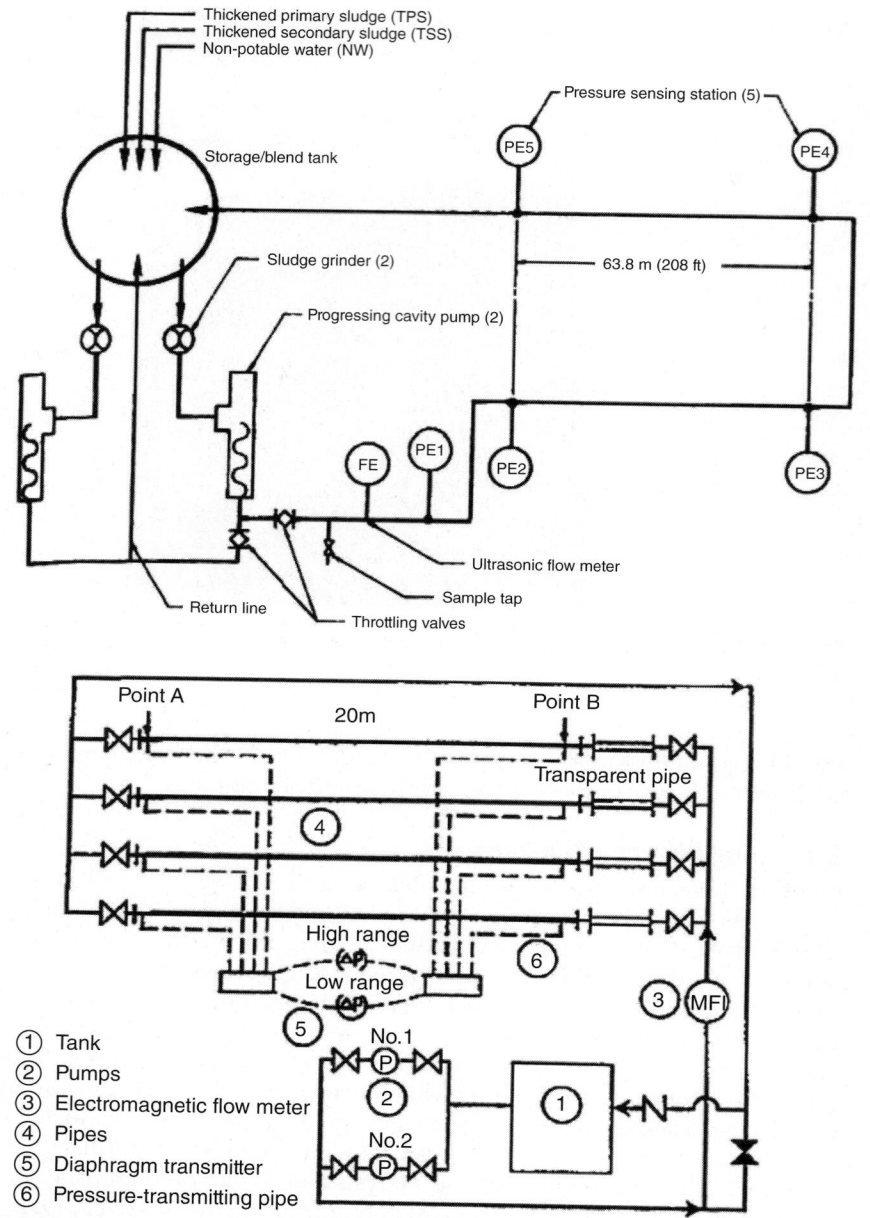

FIGURE 19.14 Experimental setups used by (top) Carthew et al. (1983) and (bottom) Murakami et al. (2001).

3.8 Common Design Deficiencies in Pumps and Piping

Several design errors in pumping and piping systems are particularly noteworthy:

- Incorrectly calculating friction head and not providing enough allowance for variations that occur during operation. Calculating friction using non-Newtonian formulae, calculating for multiple friction conditions (multiple flow rates, in other words) and obtaining historical records to predict headloss are three methods to improve accuracy.

- Not providing adequate flushing and cleaning lines. Many residuals form grease deposits or scale in pipe, and flushing water connections and cleanout ports become more critical as the material become thicker.

- Not providing enough suction to handle thickened solids. Thixotropy and plasticity can greatly affect friction, so a good design includes a straight, short suction pipe to a pump set low enough to allow for substantial positive suction pressure.

- Operating progressing cavity pumps at excessive speed or pressure per stage will increase maintenance costs. Designer is encouraged to take the effort to accurately predict the flow and pressure requirements of the system.

- Burying or encasing pipe elbows. Grit slurries and residual streams with poor upstream degritting processes can wear out elbows. If this is necessary, specify extra thickness of pipe material to increase the life of the fitting.

- Using one pump to withdraw solids from two or more tanks simultaneously; this is not recommended. Ideally, each tank should have a dedicated pump, with interconnections that allow another pump to be used when the dedicated pump is out of service. Otherwise, the system should be valved so one pump can draw from multiple tanks sequentially.

- Creating a pipeline route with high spots, which trap air or gas. Designers should avoid high spots wherever possible because air-relief valves have often been shown to be an unreliable long-term mechanism for this application. When the valves are not properly maintained, as is often the case, valves do not operate properly when needed.

- Using the wrong valves. In residuals pumping applications, design engineers should use "full port" plug valves. Pinch valves may also be applicable, but the designer should carefully weigh their advantages against their disadvantages.

- Lack of rupture disks or other pressure-relief devices between isolation valves. These devices should be installed on all residuals pumping systems.

- Other errors are cited in publications by Sanks et al. (1998) and U.S. EPA (1982).

The following are general design guidelines for any sludge pumping system:

- The minimum desirable size for residuals piping is 150-mm (6-in.), although some designers prefer 200-mm (8-in.) piping.

- For smaller facilities, the designer should consider intermittent pumping to ensure that velocities are maintained.

- In any pipe size, using a smooth lining minimizes the formation of struvite crystals in pumping anaerobically digested biosolids. Typically, ductile iron piping can be lined with cement, glass, or polyethylene. While other materials, such as polyvinyl chloride (PVC), may not require lining, operating pressures should be carefully considered.
- Determine the solids concentration to be pumped and calculate pumping requirements based on whether the fluid is Newtonian or non-Newtonian.
- Consider power consumption as part of the analysis of pumping alternatives.
- Consider existing facility equipment and operator experience with equipment types in the analysis. A well-designed pump may not perform in the field if the operations staff is not familiar with equipment operation and has difficulty learning how to operate it properly.

When designing a transport system for thickened solids, engineers should also consider both the process it is coming from and the one that will receive it. As an example, dissolved air flotation units produce sludge that contains a lot of entrained air, which can be problematic for many pumps. Another example is if a mechanical thickening process will be discharging solids directly into an open-throat progressing cavity pump, then design engineers must choose a pumping rate that exceeds the thickening unit's maximum discharge rate.

3.9 Standby Capacity

Design engineers should consider the facility size, system's function, arrangement of units, anticipated service period, and time required for repair to determine the level of standby capacity to provide. For example, standby capacity for RAS pumping is important because a service interruption could quickly impair effluent quality. Primary and secondary solids pumping are also critical functions, so designers typically either provide dual units or use units that can perform dual duty. For example, primary sludge pumps also typically serve as standbys for scum pumps.

The designer should always propose standby capacity in all residuals pumping systems. If an owner has to make the decision to remove standby capacity from the design, the pumps used should be heavy duty, have readily available spare parts, and be easy to repair quickly (preferably in place). Design engineers should ensure that the pump comes with adequate spare parts.

4.0 Dewatered Biosolids Cake Storage

4.1 Storage Requirements

Dewatered biosolids cake (biosolids) typically is stored somewhere before receiving more treatment (e.g., heat drying) or being transported offsite for use or disposal. The NFPA Report on Comments A2007—NFPA 820 notes that there appears to be only minimal hazard potential when storing dewatered biosolids. Most flammable liquids appear to be removed during dewatering, and methane-generating microorganisms do not thrive in dry aerobic environments, so special safety precautions are not required. However, the dewatered cake's viscous, sticky nature can complicate storage designs.

For short-term disposal operations (further offsite treatment or direct to landfill hauling, for example) biosolids will only be held for a few days or weeks. In this case, storage alternatives include large roll-off containers, 18-wheel dump trailers, concrete bunkers with push walls, or bins with augers.

If the biosolids will be land-applied or surface disposed, however, long-term storage may be required. Weather-related application impacts, regulatory limits on land-applied nutrient loading, and landfill restrictions are three instances that may dictate long-term storage need. In these cases, biosolids are often stockpiled on concrete slabs or other impervious pads. When designing long-term storage facilities, engineers need to consider facility buffering requirements (minimum distances between the facility and the nearest house or industry), whether or not odor control is required, and accessibility of machines to distribute the biosolids during storage and remove biosolids when removal is an available option. Designers also need to determine whether the storage facility should be open or covered, as re-wetting of dewatered biosolids is often undesirable for land application and can cause biological reactivation of pathogens.

For more information on calculating storage requirements for land application, see Section 2.1. For further guidance, see U.S. EPA's Guide to Field Storage of Biosolids (U.S. EPA, 2000).

4.2   Odor-Control Issues

Odor control can be an issue with dewatered biosolids—especially when larger quantities are stored or the solids storage area is relatively close to neighboring businesses or residences. The odors that dewatered, anaerobically digested solids produce are primarily organo-sulfur compounds.

Storing the cake for 20 to 30 days at 25°C significantly cuts odor generation (Novak et al., 2004; WERF, 2008). Adding alum after digestion also reduced storage-related odors (Novak et al., 2004). However, the best way to minimize dewatered biosolids odors is to optimize the solids treatment processes before dewatering, according to the Water Environment Research Foundation 2008 study.

The decision to provide odor control at any given biosolids storage facility can be made after the facility is placed into service and actual operating conditions are determined. A viable design approach, if approved by a facility owner, is to delay design of odor control systems until after operation starts and the need becomes apparent. If this design path is chosen, the designer should design the system to allow for the ability to "add on" an odor control system without the need to reconstruct a newly-designed facility.

5.0   Dewatered Biosolids Cake Transport

Modern dewatering operations can produce biosolids containing 15% to 40% total solids or more, depending on the conditioning chemicals and dewatering equipment used. The consistency of such biosolids ranges from pudding to damp cardboard, so they will not exit the dewatering equipment by flowing via gravity into a pipe or channel. Instead, they must be transported via mechanical devices, such as positive displacement pumps or belt or screw conveyors. Where there are receiving containers directly below dewatering equipment, the biosolids can be dropped by gravity into the containers.

Before choosing a cake transportation method, design engineers should analyze various options based on solids-management requirements (i.e., end use of the biosolids product), site or building constraints, reliability, O&M, and life-cycle costs.

5.1   Pumping/Conveyors

Both progressing cavity pumps and hydraulically driven reciprocating piston pumps can handle dewatered biosolids. Compared to belt or screw conveyors, these pumps better control odors (because the biosolids travel in an enclosed pipe), eliminate spills, and have fewer maintenance requirements. The pumps also have much smaller foot-prints and, therefore, are suitable in buildings with space constraints. They can even reduce noise levels in some cases. However, pumps often need more electricity than conveyors to move a given volume of biosolids.

Choosing which pump to use depends on the application. Progressing cavity pumps provide a steady flow, while hydraulically driven reciprocating piston pumps pulsate (List et al., 1998). Progressing cavity pumps typically are preferred in applications where the biosolids are thinner and transport distances are short. Hydraulically driven reciprocating piston pumps are more expensive, but may handle greater pressures and thicker cake. (Biosolids process discharge piping can be a high-pressure environment.)

5.2   Hydraulics

The hydraulic characteristics of dewatered biosolids (with more than 15% solids) have not been extensively studied or widely reported. Likewise, headloss-calculation methods for pumping such solids are limited. However, researchers have shown that dewatered biosolids may exhibit both plastic and pseudoplastic (thixotropic) behavior (List et al., 1998; Barbachem and Pyne, 1995; Bassett et al., 1991). For a Bingham plastic, a minimum shearing stress is required to initiate flow. For thixotropic materials, the apparent viscosity and headloss gradient (dH/dL) decrease as the rate of shear increases or as the fluid travels a certain distance inside a pipe until time-independent behavior is reached (Honey and Pretorius, 2000). These two behaviors complicate hydraulic design.

Nonetheless, dewatered biosolids can be pumped—even though experts only rec-ommend it for relatively short distances. There are many successful pumping installa-tions in North America and Europe, and many others are being designed or constructed.

5.3   Flow and Headloss Characteristics

In most dewatered biosolids applications, headlosses are high-up to 6900 kPa (1000 psi). It depends on the length, diameter, and configuration of the discharge piping. Headloss also depends on biosolids type and solids concentration, as well as the conditioning and dewatering methods that produced the biosolids. Current experience indicates that headlosses typically are sensitive to velocity and piping constrictions, particularly if the dewatered biosolids concentration exceeds 30%.

Typical piping headlosses in biosolids pumping applications can be as high as 79.1 kPa/m (3.5 psi/ft); design engineers often use these values as a general guideline during preliminary design. Ideally, the final design would keep headlosses below 45.2 kPa/m (2 psi/ft).

5.4   Design Approach

Engineers should avoid constrictions (e.g., smaller-than-line-size valves, valves that do not have full-port openings, and short radius bends) and minimize any type of fitting when designing discharge piping; the focus should be on maximizing straight run of pipe to minimize head loss. The piping should be large enough that theoretical biosolids flow velocities never exceed 0.15 m/s (0.5 ft/s)—although maximum velocities of 0.08 m/s (0.26 ft/s) are preferred, especially if dewatered biosolids concentrations exceed 30%. The pipe should be designed to allow flushing and pigging.

Unlike liquid solids, where pumping design equations are available for solids concentrations up to 12%, design approaches for pumping dewatered biosolids are more site-specific. Field testing is highly recommended, especially in relatively long-distance applications. For existing facilities that are being upgraded, using the existing system to determine flow characteristics can be a valuable design tool if the project upgrades are not impacting the biosolids characteristics. A pipe pumping biosolids containing more than 30% total solids should not be more than 150-m (500 ft) long, and line lubrication using high-pressure fluid delivery systems is highly recommended.

Four peer-reviewed case studies with actual field data were published in the 1990s. These studies explored pipeline headlosses for the following types of cake:

- Anaerobically digested, centrifugally dewatered biosolids containing 20% total solids (Bassett et al., 1991);

- Anaerobically digested, centrifugally dewatered biosolids containing 22% total solids (Barbachem and Pyne, 1995);

- Unspecified dewatered biosolids containing 28% total solids with and without polymer injection for line lubrication (List et al., 1998); and

- Undigested, plate and frame pressed biosolids containing 34% total solids (Barbachem and Pyne, 1995).

Comparing actual field data, researchers developed a simple headloss equation for biosolids with 20% total solids concentration based on the pseudoplastic model, warning that it might only apply to the specific case (Bassett et al., 1991). Equation 19.18 may be valid for all pipe sizes from 100 to 300 mm (4–12 in.) with velocities ranging from 0.015 to 0.43 m/s (0.05–1.4 ft/s):

$$\Delta P = 15.68/D^{1.28} + 0.245\, Q/D^2 \qquad (19.18)$$

where ΔP = pressure drop (psi/ft),

D = inside diameter of the pipe (in.), and

Q = biosolids flow (gpm).

Table 19.4 illustrates the effect of pipe size on pipeline headloss (based on eq 19.18). It shows that headloss increases almost 50% for each decrease in nominal pipe size. It also shows that pumping thicker biosolids (from 20%–28% total solids) causes a minimal increase in headloss (approximately 10%) when a large enough pipe (e.g., 300-mm or 12 inches) is used. (NOTE: The previous statement is based on a comparison of the results from two investigations.)

Barbachem and Pyne (1995) used the Re, He, and f number methodology described in eqs 19.10, 19.15, 19.17, and 19.14 to model their actual field data. Figure 19.15 combines headloss data reported from Bassett et al. (1991) and Barbachem and Pyne (1995) for flow in a 150-mm (6-in.) pipe. Although the data come from two investigations and so differences in data may be due to different experimental procedures, certain trends and results may be derived. Table 19.5, which one data point from each curve in

Pipe Size	Percent solids Concentration	Velocity (m/s)	H_f (m of water/10 m)	H_f (psi/ft)	Change in H_f from One Size Smaller
100 mm (4 in.)	20 (Bassett, 1991)	0.08	64.0	2.77	57%
150 mm (6 in.)	20 (Bassett, 1991)	0.08	41.0	1.77	46%
200 mm (8 in.)	20 (Bassett, 1991)	0.08	28.0	1.21	55%
300 mm (12 in.)	20 (Bassett, 1991)	0.08	18.0	0.78	Baseline
300 mm (12 in.)	28 (List et al., 1998)	0.10	19.5	0.86	10%

TABLE 19.4   Headloss Data for Pumping Cake at Near-Maximum Recommended Fluid Velocities

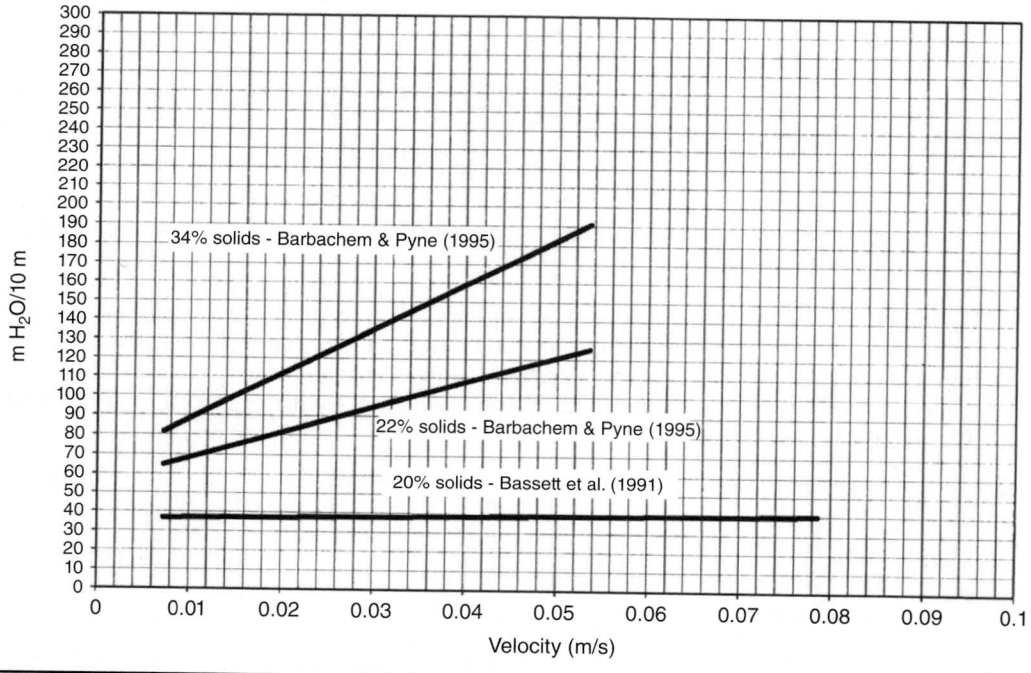

FIGURE 19.15   Headloss in a 150-mm-diameter (6-in.-diameter) pipe for three types of residuals at various velocities below the recommended maximum [created based on Bassett et al. (1991) and Barbachem and Pyne (1995)].

Figure 19.15, leads to two significant conclusions. First (as expected), friction headlosses through a 150-mm (6-in.) pipe increase dramatically as solids concentrations increase. Second, the 150-mm (6-in.) pipe is too small and inappropriate for pumping dewatered biosolids containing more than 20% total solids; excessive headlosses are created.

   In addition to actual field data and considering the high compressibility of biosolids, List et al. (1998) offered a specific method for determining headlosses created in biosolids

Type of Residuals	Velocity (m/s)	H_f (m of water/10 m)	H_f (psi/ ft)	Change in H_f
20% anaerobically digested, centrifuged	0.08	41	1.77	Baseline
22% anaerobically digested, centrifuged	0.05	120	5.19	293%
34% undigested, plate and frame pressed	0.05	180	7.79	50.0%

TABLE 19.5 Headloss Data for Pumping Cake Through a 150-mm (6-in.) Pipe at Near Maximum Recommended Velocity

pumping applications. Assuming steady, nonaccelerating pumping applications and a Bingham plastic behavior, the method is as follows:

1. Collect field data to determine a pumping pressure for a given flow.
2. Construct a graph depicting $(\Delta P/L)(D/4)$ on the y-axis and $8V/D$ on the x-axis [where ΔP = pressure loss (Pa), L = pipe length (m), D = pipe diameter (m), and V = fluid velocity (m/s)].
3. The intercept ($V = 0$) determines the critical shear stress {stress at the pipe wall [τ_w (Pa or N/m²)]}.
4. The slope of the graph is the fluid's dynamic viscosity [m (Pa-s or kg/m-s)]
5. Calculate a dimensionless factor Z as follows:

$$Z = 8\,\mu Q_m/\rho_a A D \tau_w \tag{19.19}$$

where Q_m = mass flow rate (kg/s),
ρ_a = cake density at atmospheric pressure (kg/m³), and
A = pipe cross-sectional area (m²).

Engineers then solve the following equation by trial and error until the left side equals the right side:

$$\rho^* - 1 + Z \ln[(Z+1)/(Z+\rho^*)] = 4\,\tau_w L/KD \tag{19.20}$$

where ρ^* = is the density ratio equal to ρ/ρ_a, and
K = bulk modulus of biosolids (N/m²) determined from a graph of bulk density versus consolidating pressure and assumed constant.

They then calculate the pressure drop:

$$\Delta p = K(\rho^* - 1) \tag{19.21}$$

Assuming a pseudoplastic fluid behavior under the same conditions involved solving a differential equation, but the results were less than 4% different than those derived from the Bingham plastic method above.

As reported in List's article, the results obtained for pumping biosolids with a total solids concentration of 28% through a 305-mm-diameter (12 in.-diameter), 150-m (500 ft) long pipe using eq 19.21, using a differential equation assuming Bingham plastic fluid (not shown) and one for a pseudoplastic fluid (not shown) were all very similar.

If the compressibility effect is disregarded, the following two equations may be solved for a Bingham plastic and a pseudoplastic material, respectively:

$$\Delta P/L = 4\tau_w/D - 32\,\mu V/D^2 \tag{19.22}$$

$$\Delta P/L = -4\tau_w/D - 4k/D(8/D)^m \tag{19.23}$$

where k and m are empirical parameters describing the material's properties.

The actual application involved accelerating flow generated by a reciprocating piston pump. The actual field tests of pumping 28% total solids biosolids through a 250-mm (10-in.) pipe showed that maximum pressure near the pump was 3100 kPa (450 psi), which was not very far from the calculated values for steady flow.

Table 19.6 is a compilation of data from a progressing cavity pump manufacturer on transporting biosolids with different solids percentages (Bourke, 1997). Headlosses ranged from 6 to 68 KPa/m (0.25–3.0 psi per foot) of straight 150-mm (6-in.) pipe (Bourke, 1999). The table shows that centrifugally dewatered biosolids is pumped more easily than rotary-drum dewatered cake. Centrifugally dewatered biosolids is also pumped more easily than belt-pressed dewatered biosolids. That said, the headlosses reported in Table 19.6 seem somewhat underrated; designers probably should use more conservative values or require performance warranties from the pumping-system supplier.

A progressing cavity pump may put significant shear stress on biosolids, resulting in a thixotropic, shear-thinning behavior that decreases the fluid's apparent viscosity, when compared to biosolids transported by a hydraulically driven reciprocating piston pump. Progressing cavity pumps can handle biosolids with low solids contents, while hydraulically driven reciprocating piston pumps handle thicker biosolids more reliably. However, a progressing cavity pump may achieve near 100% cavity fill, so it pumps more efficiently than a hydraulically driven reciprocating piston pump. It is noted that progressing cavity pump manufacturers typically quote headlosses in the neighborhood of 22.6 kPa/m (1 psi/ft), while hydraulically driven reciprocating piston pump manufacturers typically use 45.2 kPa/m (2 psi/ft) in their designs.

Method	Percent solids	100 mm (4 in.)	150 mm (6 in.)	200 mm (8 in.)
Rotary drum	20–30	Slightly more than double of 150 mm	7.62–11.5 (0.33–0.50)	About half of 150 mm
Centrifuge	20–30	Slightly more than double of 150 mm	50% less than rotary drum	About half of 150 mm
Rotary drum and then heat treated	45	Slightly more than double of 150 mm	23.0–35.0 (1.00–1.50)	About half of 150 mm
Filter press	65	Slightly more than double of 150 mm	35.0–46.0 (1.50–2.00)	About half of 150 mm

TABLE 19.6   Headloss Data [in m of Water/10 m (Psi/Ft)] as Reported in Bourke (1997) for Fluid Velocities Less Than 0.06 M/S

Type of Residuals	Percent Solids	Density (kg/m³)	Reference
Diluted dewatered solids	3.51	1015	Spinosa and Lotito, 2003
Settled activated sludge	5.00	1015	Honey and Pretorius, 2000
Diluted dewatered solids	5.15	1020	Spinosa and Lotito, 2003
Diluted dewatered solids	6.81	1026	Spinosa and Lotito, 2003
Diluted dewatered solids	8.40	1031	Spinosa and Lotito, 2003
Manure 1	9.10	1037	El-Mashad et al., 2005
Diluted dewatered solids	9.49	1034	Spinosa and Lotito, 2003
Diluted dewatered residuals	10.5	1038	Spinosa and Lotito, 2003
Manure 2	10.7	1044	El-Mashad et al., 2005
Dewatered solids	28.3	1062	List et al., 1998

TABLE 19.7   Residuals Density at Various Solids Concentrations

Density is an important parameter in all biosolids calculations. Biosolids density at a certain temperature may be correlated with solids concentration; this is often helpful in design. Table 19.7 lists data correlating density with solids concentration at unspecified temperatures (presumably near room temperature). The designer should consider the temperature variations and ranges in the design of their facility and determine if there is an impact to pumping operation.

5.4.1 Example 19.2: Pumping Biosolids with 28% Total Solids

Determine the friction headloss when pumping Biosolids with 28% total solids concentration solids through a 300-mm-diameter, 150-m-long pipeline. This example was adapted from List et al. (1998). Other experimentally determined input parameters include:

- Bulk modulus of biosolids $(K) = 2550$ kN/m²,
- Biosolids density at atmospheric pressure $(\rho_a) = 1060$ kg/m³,
- Biosolids mass flow rate $(Q_m) = 8.012$ kg/s,
- Wall shear stress $(\tau_w) = 1468$ N/m²,
- Biosolids's dynamic viscosity $(\mu) = 12$ kg/m·s, and
- Fluid velocity $(V) = 0.1036$ m/s.

5.4.1.1 Solution Option 1   Following the methodology described previously in Section 5.4, engineers should use eq 19.19 to calculate the following dimensionless factor:

$$Z = 0.222$$

They then solve eq 19.20 by trial and error:

$$\rho^* = 2.15$$

Engineers then use eq 19.21 to calculate the pressure drop:

$$\Delta P = 2933 \text{ kPa (299 m of water column)}.$$

5.4.1.2   Solution Option 2   Anticipating that the biosolids are behaving like a Bingham plastic fluid and disregarding the compressibility effect, engineers could use eq 19.22 to get the following result:

$$\Delta P = 2826 \text{ kPa (288 m of water column)}$$

5.4.1.3   Solution Option 3   If the biosolid's rheological parameters are difficult to determine, an estimated conservative headloss value may be sufficient for design. This practice is valid because biosolids are transported by positive-displacement pumps that can deliver the same flow rate over a wide range of pressures; it is even more valid when pumping biosolids short distances through sufficiently large pipelines.

Typical headlosses in biosolids pumping systems can be as high as 79.1 kPa/m (3.5 psi/ft). Their headlosses are larger (in the high end of the range) when pipe diameter is small (e.g., 155 mm or 6 in.) and fluid velocity is near the maximum recommended (0.08 m/s or 0.25 ft/s). In this example, the pipe diameter is sufficiently large (300 mm or 12 in.) and fluid velocity is not much higher than the recommended maximum, so system headlosses should be in the low end of the range. On the other hand, the biosolids content is 28%, and engineers should be careful not to select a design whose values would be too small. A headloss value of 45.2 kPa/m (2.00 psi/ft) may be safely assumed.

Also, engineers could take Table 19.4's value for pumping biosolids with 20% solids through the 300-mm (12-in.) line and double it (to approximate the value for a biosolids with 28% total solids concentration). The estimated headloss for design then becomes 35.2 kPa/m (1.56 psi/ft).

Another approach is to extend the 34% and 22% curves of Figure 19.15 to a velocity of 0.1 m/s (0.33 ft/s) and record the headloss for each case at that particular velocity (300 m/10 m and 190 m/10 m, respectively). Pumping biosolids with 28% total solids would fall somewhere between the two values. If the median value is selected, the headloss is 245 m/10 m of 6-in.-diameter (0.4 m) pipe. The headlosses calculated from Options 1 and 2 are 0.86 and 0.83 psi/ft (19.4 to 18.7 kPa), respectively. These are well below the estimated values in Option 3.

5.5   Line Lubrication for Long-Distance Pumping

A method for reducing pipe-friction losses (mainly for long-distance pumping applications), called boundary layer injection, involves injecting a liquid lubricant into the discharge pipe via an annular ring, which distributes the liquid equally around the pipe's perimeter to create a "boundary layer." The lubricant could be water-, polymer-, or oil-based. The designer should research the impact of adding this fluid on the end use of the dewatered biosolids as part of the design process. Field tests indicate that such lubrication can cut discharge pressure up to 80% (List et al., 1998). For a detailed description of this process, see Conveyance of Wastewater Treatment Plant Residuals (ASCE, 2000).

Consider the experience of a water reclamation facility in Georgia, which was pumping biosolids with 25% total solids through a glass-lined pipe. The original pressure drop was 67.8 kPa/m (3 psi/ft). After adding a small amount of water (0.5% of total solids flow) via boundary layer injection, the pressure drop lowered to 22.5 kPa/m, according to a study by a manufacturer of hydraulically driven reciprocating piston pumps (Crow and Cortopassi, 1994).

There are two types of boundary layer injection rings: the original configuration, which has individual injection points, and a newer design, in which lubricating media is injected through an annular groove around the pipe perimeter. Typically, the annular groove design is preferred because it requires less lubricant and has minimal effect on the biosolids' percent solids content (Wanstrom, 2008).

5.6   Controls

Controls typically are used to match the pumping and biosolids production rates. For example, a progressing cavity pump's speed may be controlled by a variable-frequency drive, while a piston pump's speed is controlled by fluid flow in the hydraulic power unit. Meanwhile, ultrasonic or radar level sensors monitor the level in the biosolids collection hopper and send high and low-level signals to the pump controls, which automatically adjust pump speed to maintain a preset hopper level.

Note that sensor placement in a dewatered biosolids storage tank is very important. Dewatered biosolids typically form pyramidal shapes when deposited by gravity into storage bins, so obtaining an accurate level reading is difficult to achieve. Storage hoppers often have leveling devices in the hoppers, but historical operation indicates that these devices do not always provide adequate leveling because dewatered biosolids have varying angles of repose based on solids concentration. Knowing the angle of repose of the dewatered biosolids is a key to sensor placement. Field testing of the in-place system should be performed, and storage tank design should allow for moving the sensor based on the field test results.

Most facility operators report stable biosolids production from belt presses or centrifuges; spot checks are used to prevent bridging or clogging. Little operator time is required to adjust the system to match biosolids production rates.

Automatic pump controls involving capacitance probes, pressure switches, and no-flow sensors typically have proved unreliable. If the control fails to shut off the pump when the feed hopper holds little biosolids, the pump can run dry. If it fails to start the pump, biosolids can spill over the hopper or push back and pack centrifuges or other dewatering units. Either condition can result in expensive repairs and production loss.

5.7   Progressing Cavity Pumps

Progressing cavity pumps have low capital costs and can consistently transport thinner dewatered biosolids over short distances. To ensure effective operations, pump manufacturers and WRRF personnel recommend that design engineers

- use progressing cavity pumps to transport biosolids containing about 20% total solids or less. About 20% to 25% biosolids applications can be researched by the design engineer, but it is recommended that the design engineer get references from facilities that are successfully pumping this more dense biosolids and speak directly to the facility operators to gain as much data as possible.

- use a large-diameter pipe to reduce friction loss to more suitable levels.

- limit the pumping distance to 50 m (164 ft) or less. Further distances (up to 90 m or 295 ft) can be researched by the design engineer, but it is recommended that the design engineer get references from facilities that are successfully pumping solids through this length of pipe and speak directly to the facility operators to gain as much data as possible.

- minimize suction piping length or eliminate this pipe.
- restrict pump rotational speeds to 200 rpm or less (Bourke, 1997).
- limit the pressure per stage to 52 kPa (7.5 psi).
- after determining final design criteria, ask pump manufacturers for specific recommendations on pump materials, number of stages, and models.
- use either adjustable-frequency or adjustable-hydraulic motor drives.

The expected service life of the pump's stator and rotor assembly is facility-specific, and drops markedly as the biosolids' solids concentration increases. Other factors that affect equipment wear include pump speed, grit content, running time, and operating pressure. Parts are expensive and labor-intensive to replace (compared to servicing kinetic pumps).

When progressing cavity pumps transport biosolids containing more than 20% total solids, the biosolids can bridge over the screw auger in the feed hopper. It can also clog the throat section (interface of auger feed screw and rotor and stator assembly of pump). If uncorrected, the problems can cause the pump to run dry quickly, ruining the stator and rotor. To avoid this, designers can put temperature sensors in the stator to monitor for the high temperatures that indicate the pump is running dry.

Depending on the pumping system configuration, bridging could pack a centrifuge bowl or overflow a supply hopper. Some manufacturers use paddle-type bridge breakers to combat this problem (with moderate success). However, they may require intensive maintenance. A bridge breaker can be powered by the pump motor via gears or chains, or by a dedicated motor. Using a dedicated motor lets operators adjust paddle speed independent of the pump. A newer design uses a ribbon auger attached to a plate fixed either to the pump drive shaft or to a separate variable speed drive; it allows the pump to transport biosolids with higher solids concentrations (Dillon, 2007). Another design uses a twin screw feeder also powered by a dedicated variable-speed drive (Doty, 2005). Both designs address biosolids bridging problems and maximize the pump cavity's fill rate.

5.8   Hydraulically Driven Reciprocating Piston Pumps

In the United States and Canada, hydraulically driven reciprocating piston pumps are the standard for transporting high solids and dewatered biosolids long distances. Municipal water resource recovery facilities have used them for this purpose for more nearly 30 years, and many of the units are still operating. Several manufacturers sell them in the United States.

Developed from concrete pumping technology, a hydraulically driven reciprocating piston pump consists of a twin screw auger feeder, a pumping assembly, and a hydraulic power unit (see Figure 19.16). It can handle both screenings and biosolids containing 5% to more than 40% dry solids that were dewatered via belt presses, centrifuges, plate and frame presses, screw presses, or rotary presses.

The pump's principal advantage is that it can move high solids materials. Hydraulically driven reciprocating piston pumps have higher capital costs than other pumps, but can move very thick dewatered biosolids that other types of pumps cannot.

The pump has two product-delivery cylinders with pistons powered by two isolated hydraulic-drive cylinders (rams). The delivery cylinders are synchronized so while one

FIGURE 19.16   A hydraulically driven reciprocating piston pump moving a centrifugally dewatered cake (containing 31% solids) at the Norman Cole Jr. facility in Lorton, Virginia (Courtesy of Schwing Bioset, Inc.).

is being filled with dewatered biosolids, the other is delivering biosolids to the discharge line. This reduces the pulsing effect of one cylinder and piston, maintaining essentially uninterrupted flow.

Nearly all the hydraulically driven reciprocating piston pumps used at U.S. sites have twin screw auger feeders. These feeders develop 35 to 205 kPa (5–30 psi) of pressure in the pump assembly's charging unit. This pressure helps push biosolids into the emptying cylinder while the piston returns to its starting position. The feeders typically are driven by a hydraulic motor but can also be used with electric motor drives.

The pumping assembly's charging unit typically has either a poppet valve or a transfer tube. Each delivery cylinder has a suction intake and discharge exhaust poppet valve. The valves or tubes are hydraulically driven and synchronized to the piston strokes; they permit one cylinder to fill while the other discharges. Poppet valves provide a positive shutoff to prevent backflow, cost less, and need less maintenance than transfer tubes. The poppet valve can also be equipped with an internal flow monitor that will measure the volume of pumped biosolids, but reported accuracy is only about 65% (Wanstrom, 2008).

Hydraulically driven reciprocating piston pumps can run dry for indefinite intervals at slightly faster wear rates but without catastrophic damage. So if bridging or clogging occurs, facility operators have time to react before severe problems develop. A water-filled isolation box (e.g., water box) between the hydraulic and delivery halves of each cylinder allows this capability. The water cools the connecting rods, flows into the delivery cylinder, and lubricates it as the delivery piston moves forward on its discharge stroke.

5.8.1   Operating Experience and Design Considerations

Based on nearly 30 years of operating experience with hydraulically driven reciprocating piston pumps at U.S. and Canadian water resource recovery facilities, design engineers should

- minimize line pressure in the piping system when pumping material containing more than 25% total solids (either through minimizing fittings, lining the pipe, using boundary layer fluid injection, or a combination of any or all of these methods),

- use a boundary layer injection system in long-distance or high-solids applications,
- maximize volumetric efficiency (see Section 6.8.3), and
- minimize stroke rate (see Section 6.8.2).

Hydraulically driven reciprocating piston pumps are available with capacities up to 1500 L/min (400 gpm) and discharge-pressure capabilities up to 13 800 kPa (2000 psi). There are also a wide range of suction feed hopper-to-pump configurations. In addition, hydraulic power units are available in broad output ranges, depending on the discharge-line pressure required.

When designing such pump systems, engineers should discuss suitable equipment sizing, features, and options with manufacturers and staff at installations with similar requirements. They should ensure that the piping design and components (valves, etc.) are suitable for high-pressure service. When sizing the pump(s), they should

- determine the biosolids production rate;
- estimate the pump's volumetric efficiency given the biosolids characteristics;
- reduce stroke speed in accordance with the type of pump service (e.g., intermittent or continuous) expected;
- examine the pump curves and check the stroke speed turndown ratio; and
- choose a pump based on the volumetric-efficiency and service factors needed.

5.8.2   Biosolids Production Rate

Biosolids production rate depends on installation-specific requirements, including the following:

- The volume of solids the facility generates for thickening and dewatering;
- Number of hydraulically driven reciprocating piston pumps and dewatering machines installed to dewater the thickened sludge;
- Operating schedule of both dewatering equipment and pumps (e.g., 24-hour service or single shift);
- Constraints downstream of piping (e.g., furnace capacity, storage capacity, or trucking schedules);
- Variable capacity requirements imposed on the pumps by dewatering processes; and
- Standby considerations during maintenance or emergencies.

5.8.3   Volumetric Efficiency

Volumetric efficiency is the ratio of solids volume pumped per piston stroke to the total volume displaced per piston stroke. If a hydraulically driven reciprocating piston pump were pumping water or residuals containing 1% to 4% total solids, its volumetric efficiency would be essentially 100% because wastewater is nearly incompressible. This behavior is typical of a true Newtonian fluid.

Dewatered biosolids, however, neither physically resembles nor behaves like a true Newtonian fluid. It typically contains air, other entrained or dissolved gases, and concentrated organic material, all of which are compressible. So, when the piston begins applying pressure, the biosolids tend to compress. Until squeezed against the downstream resistance, the biosolids do not move forward with the pumping cylinder. When the biosolids finally move forward, the piston already has displaced a certain volume of the cylinder. This displaced volume is part of the "lost" volumetric efficiency; the rest is due to the inability to completely fill the cylinder as the piston returns to its starting position. Even with the slight pressure provided by a twin screw auger or conical plow feeder [35 to 205 kPa (5–30 psi)] and the partial vacuum in the cylinder, dewatered biosolids resist moving into the cylinder bore. Such resistance typically increases as biosolids dryness increases, further lowering volumetric efficiency.

Using a pressure sensor in the transition between the twin screw auger and poppet housing can ensure cylinder-filling efficiency. This sensor will monitor pressure in the transition and, via a programmable logic controller, automatically increase or decrease auger speed to maintain a preset pressure. So, regardless of pump speed or fluctuations in solids concentration, optimum pressure is maintained on the suction poppet to promote the highest filling efficiency possible for a given material (Wanstrom, 2008).

Meanwhile, designers must account for volumetric-efficiency loss when sizing a hydraulically driven reciprocating piston pump. There is no theoretical model for predicting volumetric efficiency, but it typically ranges from 60% to 90%. Once design engineers know or can estimate biosolids characteristics, they should ask manufacturers to recommend an appropriate volumetric efficiency. As a rough, conservative estimate, a volumetric efficiency of 70% can be used for pumping dewatered biosolids containing 20% to 30% solids.

5.8.4 Service Factor

As with most recovered water treatment equipment in continuous service, operating hydraulically driven reciprocating piston pumps at lower speeds makes them more reliable and extends equipment life. So, design engineers should limit the pump-stroking speed (strokes per minute) to 50% of the maximum recommended for intermittent operation or 75% of the maximum recommended for continuous service, whichever is less.

5.8.5 Hydraulic Power Unit Sizing

Power unit sizing is principally a function of the pump's discharge pressure and the hydraulic fluid flow needed to achieve the desired solids pumping rate. If pumping tests cannot be made and data are not available from other sources, design engineers should ask manufacturers for specific recommendations and rate the power unit conservatively.

5.9 Conveyors

Conveyors typically move wet or dry solids (e.g., primary grit, screenings, and dewatered biosolids) that are not easily pumped. Municipal-recovered water facilities typically use either belt or screw conveyors.

5.9.1 Belt Conveyors

Belt conveyors move material on top of a moving, flexible belt (see Figure 19.17). Such belts typically are supported by rollers spaced 0.9 to 1.5 m (3–5 ft) apart on the carrying side and about 3 m (10 ft) apart on the return side of the conveyor. The rollers on the carrying side are called load-side rollers; those on the return side are called idlers. To increase capacity, load-side rollers are often angled so the belt will form a concave carrying surface.

FIGURE 19.17   Belt conveyor (troughed with idlers) transporting screw press dewatered, anaerobically digested solids at the City of Tallahassee's T. P. Smith facility in Florida.

The belt is driven by one or more drive drums or pulleys connected to a motor via a belt or chain drive. In simple conveyor systems, the drive pulley is located at the discharge (head) end of the belt and the tail pulley is at the loading end.

The belt must maintain a minimum tension to reduce sag between carrying idlers, provide contact force, and prevent slippage at the drive pulley. This tension can be maintained by several take-up devices, including a weighted pulley (called a gravity take-up), a spring-loaded pulley, or a screw adjustment for pulley position. The least costly option is a screw take-up on the tail pulley; it is typically used in conveyors less than 90 m (300 ft) long.

5.9.1.1 Belt Conveyor Applications   Conventional belt conveyors move biosolids via a continuous loop of reinforced rubber belt. They typically transport relatively dry material (15% or more total solids). For this method to be economical, biosolids must be dry enough not to flow freely or seek a constant level (like a liquid does). Biosolids with a high angle of repose (i.e., the slope of the solids pile when measured from the horizontal) are suitable for transport via belt conveyors. Digested and belt-pressed primary and secondary biosolids can have an angle of repose of 40° or more. Because belt movement vibrates the material, design engineers must consider the biosolids' flow tendencies when deciding whether a belt conveyor is suitable. The angle that a pile of material retains while moving is called the surcharge angle. Belt-pressed biosolids may have a surcharge angle of more than 30%. The designer should determine a material's characteristics before deciding which conveyor to use.

The transport distance and elevation change also influence the choice of conveyor. Belt conveyors have been used to move mining ores and construction spoil solids more than 14 km (8 mi), but in typical recovered water facilities, the distance could be less than 200 m (660 ft). If the distance is less than 6 m (20 ft), other conveyors may be more suitable.

Conventional belt conveyors are limited by both the rate of elevation change and horizontal direction changes that require multiple belts. The conveyor's maximum incline depends on the material involved and belt speed required. Faster speeds allow for higher angles so long as the speed exceeds the rate at which material flows or rolls back down the incline. However, faster speeds also increase O&M costs because they increase friction and shorten belt life. When moving dewatered biosolids, a belt conveyor's maximum incline angle is limited to about 15° to 20° above horizontal. The maximum incline is much less for biosolids that are watery or tend to flow easily.

Elevation gains can also be limited by curvature radius as a horizontal belt becomes an inclined one. The radius must be long enough so the belt will not lift from the idlers under any operating condition. Depending on the specific design, this radius could be 15 to 75 m (50–250 ft) or more. So, engineers need to consider an existing facility's physical dimensions when deciding what type of conveyor to use.

Belt conveyors have a low cost per linear meter (foot) of transport distance, but they may require significant space and be maintenance intensive. They can also be a source of odors. In addition, if the conveyor will be installed outside, weather conditions can affect operations.

5.9.1.2 Belt Conveyor Design and Operation Considerations When considering a belt conveyor for a new or existing facility, design engineers should begin by establishing the following criteria:

- Biosolids characteristics (e.g., angle of repose and surcharge, degree of matting or stickiness, average density, and range of variation of these characteristics);
- Biosolids volumes and transport rates (i.e., daily or weekly variations in solids rate and hours of operation) so they can determine conveyor capacity;
- Belt construction material (acid, oil, and abrasion resistance); and
- Conveyor layout and power requirements so they can determine belt width and speed (conveyor activity); loading arrangement (chutes or conveyor skirtboards); curve radius, incline angles, and total elevation gain (multiple conveyors may be warranted if these factors are limiting); idler type, spacing, and pulley and take-up arrangements (which influence belt friction and power requirements); and motor horsepower and belt tension.

The Conveyor Equipment Manufacturers Association (CEMA) publishes a handbook that includes procedures for establishing such criteria and sizing belts (CEMA, 1979). First, however, engineers must know the characteristics of the material to be conveyed. Some recovered water facility biosolids are sticky, for example, so the belt should be cleaned to prevent spillage on return runs and the consequent loss of drive-pulley friction. Other facilities may have site-specific problems, such as water breakout, odors, or spillage.

The U.S. Environmental Protection Agency offers the following guidelines for problems unique to WRRF biosolids (U.S. EPA, 1979):

- Belt transfer points should have both minimum drop heights and skirtboards with wipers to minimize splashing and spillage.

- Belt cleaning is potentially troublesome. Counterweighted rubber scrapers below the head pulley have been ineffective and require intensive maintenance. Scrapers with multiple "fingers" and adjustable tensions are suggested. Another option is a water spray followed by a rubber scraper (if the water can be collected and disposed easily).

- Design engineers should avoid accessories (e.g., snubber or counterweight pulleys) that touch the dirty side of the belt. Snubbers are pulleys positioned to increase the angle of contact between the belt and the drive pulley, thereby increasing friction and reducing drive slippage. Instead of snubbers or gravity take-ups, designers should use manual screw take-ups and, if necessary, multiple shorter belts.

- When designing conveyors, design engineers should include housekeeping facilities (e.g., frequent hose stations); oversized floor or paving drains with exaggerated grades below the conveyor; and nonskid tread plates rather than grates).

Because a water resource recovery facility is a humid or wet environment and biosolids typically are corrosive and abrasive, engineers need to design belt conveyors carefully. Conveyor framing should be made from corrosion-resistant materials (e.g., 6061-T6 aluminum alloy). Idlers can be made of neoprene or PVC, and cable-supported neoprene rollers can be used for troughed sections. Roller bearings should be sealed with external grease fittings. Drive chains and motors require removable splash guards to protect them from spillage. Belt materials should include abrasion- and oil-resistant covers.

To prolong belt life, designers should set the belt's actual running tension conservatively below its rated tension and check the loaded conveyor's initial tension to avoid overstressing the belt. A vulcanized belt splice provides longer life than mechanical joints. Slower belt speeds also typically lengthen belt life, so about 30 m/min (100 ft/min)—approximately 50% of CEMA's maximum speed guideline—is suggested.

When designing conveyor sections that are outside buildings, engineers should provide for weather and wind protection. At a minimum, they should provide a half-diameter rain cover; however, a three-quarter cover with open access on the downwind side prevents wind-induced spillage and belt-training problems while allowing access to parts for maintenance. If odors must be controlled, designers can completely enclose the conveyor and provide ventilation. The enclosure must include hinged or easily removable partial cover plates to allow access for maintenance, cleaning, and periodic observation of the conveyor.

5.9.1.3 Special Belt Conveyors Manufacturers have developed belt conveyors that overcome some of the limitations previously mentioned. For example, conveyors with cleats (See Figure 19.18), buckets, or sidewalls attached to the belt can move material up steeper inclines. One patented conveyor allows both horizontal and vertical curves. Another uses two flat, converging belts to completely enclose the material and to permit steady inclined or vertical lifts. Design engineers considering one of these conveyors should discuss the application with various manufacturers and check similar installations to compare operating costs and avoid potential design problems.

FIGURE 19.18   Cleated belt conveyor.

5.9.2   Screw Conveyors

Screw conveyors push material via a helical blade (flight) mounted in a U-shaped trough or enclosed in a tube. The flights may be attached to a center shaft, or the screw may be shaftless. In shafted conveyors, a drive mechanism turns the center shaft, which is supported by the end bearings and intermediate hanger bearings needed to reduce shaft deflection. Both shafts and flights can be tapered.

Flights are manufactured in a wide variety of designs and can have full or partial cross-sections (or both). Two flights that are cut, folded, or otherwise shaped can mix or fold the material during transport. The pitch—the horizontal distance between flight blades—can vary along the shaft length.

5.9.2.1   Screw Conveyor Applications
Screw conveyors (augers) can be used to move biosolids horizontally, vertically, or along an incline. When properly designed and used, they are an economical and reliable transport method. Before selecting a screw conveyor, design engineers must evaluate the material to be moved. Its water content and flowability are particularly important for inclined and vertical conveyors.

Standard screw conveyors work best when moving material horizontally over a relatively short distance. Although some operating screw conveyors are more than 150 m (500 ft) long, most of the ones in WRRFs are 9 to 12 m (30–40 ft) long. The conveyors are available in sections that are about 3 to 4 m (10–12 ft) long, depending on shaft and flight size. Longer sections are either custom made or formed by joining standard lengths. They typically need intermediate hanger bearings to reduce shaft deflection.

Inclined screw conveyors are less efficient than horizontal ones and have different design criteria. For every degree of elevation beyond 10°, a screw conveyor's capacity declines about 2% and its speed must increase significantly to compensate. Inclined and vertical conveyor speeds typically are 200 rpm or more, while a horizontal conveyor's velocity is 20 to 40 rpm in an abrasive application. Inclined conveyors also use different flight designs than horizontal ones.

Vertical screw conveyors are designed for uniform flight loading to avoid packing or binding the material. Like inclined conveyors, vertical units have faster shaft speeds and the screw's centrifugal action helps provide lift. These systems typically include special horizontal feeders. Engineers should consult with manufacturers when designing lifts taller than 6 m (20 ft). Although at least one manufacturer allows vertical lifts up to 21 m (70 ft) high, another recommends a practical limit of 7.5 m (25 ft).

Screw conveyors typically move grit or biosolids horizontally. Inclined conveyors are sometimes used for dewatered biosolids. Screws are also used as truck-loading hopper dischargers to spread a load across the entire truck trailer. Manual or automatic knife gates on the conveyor bottom function as multiple discharge points. Screws can control biosolids feeding from hoppers to either belt conveyors or the suction side of biosolids pumps.

5.9.2.2 Screw Conveyor Design and Operation Considerations

When designing screw conveyors, engineers first must define material properties, volume, and variability. Conveyor capacity is a direct function of screw speed, flight size or diameter (assuming shaft size remains fixed), and amount of trough loading. Conveyor flight pitch and any folding, cutting, or other special flight designs also affect capacity. Some manufacturer catalogs include tables and charts that help design engineers determine preliminary conveyor size based on a range of these variables. Designers should consult with manufacturers when developing a design for a specific application.

Some design criteria specific to water resource recovery facilities deserve consideration. For example, designers should avoid screw conveyors if the biosolids contain sticks, large objects, or rope-type materials. Enclosed screw conveyors can reduce or eliminate spillage and housekeeping problems but are slightly more susceptible to jamming and are difficult to access for maintenance. For sticky solids like dewatered biosolids, a designer should avoid intermediate support or hanger bearings because they could cause plugging as material packs against them. Larger shafts, heavier shaft-wall thicknesses, or both allow greater screw lengths between support bearings; typically, enlarging the shaft is more effective than increasing the shaft wall thickness. If intermediate bearings are unavoidable, the flight design near the hanger can be modified to minimize the packing problem.

Other design criteria include construction materials and drive configuration. For example, a conveyor that will transport abrasive and corrosive materials (e.g., dewatered biosolids and grit) should have a flight facing made of hardened materials. Steel flights with hot-dipped galvanized troughs have been used successfully for dewatered cake. Where exposed to biosolids, outlet knife gates should have stainless steel parts because any free water released from the solids will collect on the parts.

Ideally, the conveyor drive will be mounted at the unloading end so the shaft is in tension during operation and will not buckle during a jam. The motor can be connected directly or via a belt or drive chain. If the conveyor's daily capacity varies significantly, designers can use variable-speed drives to match capacity to transport requirements. End bearings should be heavy-duty roller bearings located outside the conveyor. Design engineers also typically specify shaft seals with a compression-type packing gland to prevent abrasive material or corrosive liquids from migrating to the outside of the conveyor or to the shaft bearing.

If the material packs, sticks, and has a high angle of repose (e.g., dewatered biosolids), the feed portion of the conveyor deserves special consideration. If the screw is fed from a hopper above, the hopper's sides must be steep enough and its opening large enough to prevent material from bridging across it. Exact figures depend on the cake, but as a general guideline, the hopper wall should be no more than 30° to 35° from the vertical and the bottom opening area be approximately 1 m² (11 sq ft), with the smallest dimension about 0.6 m (2 ft) long. To help remove material evenly across the hopper bottom, the flight diameter should gradually shrink as it travels across the screw feed area or the flight pitch should increase.

When used properly, screw conveyors have fewer O&M requirements than a belt conveyor. Because they can be completely enclosed, screw conveyors have substantially fewer housekeeping and odor-control requirements. Any intermediate hangers should have a hinged or other easily removed cover for bearing inspection or replacement. If odor control is a significant problem, the screw conveyor can be connected to a vent system at a point just past the discharge end.

Access to the screw conveyor for inspection, maintenance, or replacement can be from above or below, depending on installation requirements. Typically, the screw conveyor cover is attached in bolted sections that can be removed as required. For more frequent access to certain areas, such as intermediate hangers, the cover can be hinged on one side. A variety of easily removable cover arrangements are available. Access from below can be provided by removing trough sections or by installing special hinged trough sections available from most manufacturers.

Designers should provide hose stations and oversized drains so staff can clean up the conveyor when parts must be replaced. Inclined or vertical conveyors should have a low-point drain so any backflow liquid can be drained manually.

5.10 Standby Capacity

Design engineers should consider several factors when determining the need for standby transport capacity. These factors include the function involved, facility size, anticipated service period, repair time, and arrangement of units. At larger water resource recovery facilities, dewatering operations are often critical and cannot be out of service for long periods, so biosolids pumping or conveying systems should be designed with standby or quick-replacement capability.

The designer should always propose standby capacity in biosolids pumping systems. If an owner has to make the decision to remove standby capacity from the design, the pumps used should be heavy duty, have readily available spare parts, and be easy to repair quickly (preferably in place). Design engineers should specify what spare parts are required and provide quantities of spare parts to be provided.

6.0 Dried Biosolids Storage

Dried biosolids described in this chapter are defined as post-dewatering process dried solids. Thermally dried solids are discussed in Chapter 24. Dried ash solids are discussed in Chapter 25.

6.1 Design Considerations

Dried biosolids typically are stored either onsite or at a land-application site before disposal or beneficial use. They may be stored in stockpiles or silos.

Because dried biosolids contain a significant amount of combustible organic material that can be released as dust, temperature control is important. If silos are used, engineers should design them to promote cooling and maximize heat dissipation. Tall, narrow silos are more effective at heat dissipation than wide ones. Narrow silos also make fires easier to control. However, if the silo is too narrow, it will make relief venting problematic. If multiple silos are used, there should be procedures to ensure that they are emptied cyclically to avoid exceeding safe residence times. Also, designers need to consider the stored product's thermal stability in case a prolonged facility shutdown or silo blockage occurs.

6.2 Safety Issues

It is critical that dried biosolids be stored safely. Dried biosolids have the ability to self-ignite and have done so on several occasions. It has been reported that one of these ignition events resulted in the discharge of biosolids to a surface waterbody. Explosions have also occurred in dried biosolids storage silos.

If the storage area's temperature rises above a critical point, dried biosolids can begin to self-heat. Design engineers can calculate the critical temperature using isothermal basket tests and considering the effects of storage volume and residence time. In England, water recovery professionals performed a series of tests on 1 m³ (35 cu ft) of biosolids and found that the self-heating temperature was typically above 60°C (HSE, 2005). When they extrapolated the data to 27 m³ (950 cu ft) of biosolids, however, the critical temperature range dropped to between 50°C and 60°C. Results depend on the actual product and contaminants (e.g., oil), which can lower the critical temperature. Therefore, Health and Safety Executive recommends that biosolids should be cooled to no more than 40°C before being sent to a silo. (If the silo is particularly large, the critical temperature may be even lower.) Controlling biosolids temperature not only prevents silo fires but also avoids self-heating further along in the biosolids use or disposal process. Biosolids that are hotter than the critical temperature must not be removed from the water resource recovery facility until they are so cool that there is no risk of a fire either in transit or at the use or disposal site. There are cooling devices manufactured specifically for biosolids, and design engineers may want to consider using them in larger installations.

Dust generation is another issue, because biosolids dust can be an explosion hazard. Biosolids may generate little dust if the material is within specifications and the transport method minimizes attrition. However, if the storage silos could contain significant levels of dust, design engineers should try to minimize any potential for explosion. The Occupational Safety and Health Administration (OSHA) requires that silos be designed to either contain the maximum explosion pressure or include passive explosion-relief vents (1995 Hazard Information Bulletin). If explosion-relief vents are used, it is important that they discharge to a safe area—preferably to an outdoor location away from normal working areas.

If biosolids are not stabilized before drying, water condensation may lead to bacterial decomposition inside the silo. Bacterial activity produces heat, which could cause a silo fire. Also, wet pellets may become sticky and difficult to handle. To minimize condensation, designers should ventilate the silo via small volumes of dehumidified air or larger volumes of atmospheric air.

Because fires can be a concern in storage silos, design engineers should include systems to identify and contain them. Multipoint temperature probes can monitor stored material, but they provide only localized measurements and so may miss a hot area. A biosolids fire will produce carbon monoxide and consume atmospheric oxygen, so engineers should include carbon monoxide monitoring in their silo designs. At first, the burning material will only produce small quantities of carbon monoxide, which may be further diluted by aspirated air, so engineers must set the detector to identify low carbon monoxide levels. The slow initial exothermic reaction will be followed by a rapid exothermic reaction producing large quantities of carbon monoxide, so engineers should determine the alarm set point for carbon monoxide detection (this typically is on the order of 100 mg/L).

Spraying water into a burning silo may only produce a surface cake that bars further water penetration. Designers should also consider using an inert gas to contain a fire. The Occupational Safety and Health Administration specifically requires that all dried, hot biosolids be transported and stored in a nitrogen inert atmosphere that contains less than 5% oxygen. While this will help prevent fires, injecting an inert gas into a burning silo will not necessarily extinguish one. Such injections may have limited effect; it will prevent further propagation but thermal currents may divert the gas away from the hottest parts of the stored material. The temperature will drop right after the cold gas enters the silo, but this does not mean the fire has been extinguished. So, design engineers should include provisions for monitoring temperatures over time to determine whether a fire has been brought under control. Such "worst case" scenarios need to be considered in both design and operations.

7.0   Dried Solids Transport

7.1   Belt Conveyors

Belt conveyors are one of the most widely used and efficient means of transporting bulk materials. A chute deposits material onto the top of the belt at one end, and the belt transports it to the other end, where the material is discharged into another chute.

Belt conveyors range from 0.35 to 1.5 m (1.15–5 ft) wide and can accommodate a broad range of capacities, speeds, and distances. The smallest belt conveyors can handle up to 76 kg/s (30 ton/hr) and operate at speeds up to 100 m/min (325 ft/min). The largest ones can handle up to 1134 kg/s (4500 ton/hr) and operate at speeds up to 200 m/min (650 ft/min). They work well transporting material 6 to 200 m (20–660 ft); if the material needs to move less than 6 m (20 ft), other conveying devices are preferred.

7.1.1   Belt Conveyor Applications

Belt conveyors are appropriate for transporting dried biosolids because they are gentle, efficient, and durable. They also prevent material degradation. When installing a belt conveyor, the belt should be angled to form a trough to prevent material from rolling off the belt. Belts used to transport dried biosolids typically are no more than 1 m (3.3 ft) wide.

While belt conveyors are preferred when transporting dried biosolids horizontally, other conveyors should be considered when moving them vertically. Typically, a belt conveyor should not be inclined more than 20° when transporting dried biosolids—especially dried biosolids pellets, which tend to be free flowing. If dried materials must be moved at steeper angles, consider using sidewalls, cleats, and/or cover belts.

Although belt conveyors may have higher capital costs, they can outlast most other conveyors. In addition, they typically need less energy to move material than augering or dragging it.

7.1.2   Belt Conveyor Design and Operation Considerations

Several important material characteristics and design criteria should be considered when designing belt conveyors. Engineers can find procedures for establishing design and other criteria and belt sizing in the Conveyor Equipment Manufacturers Association's Belt Conveyors for Bulk Materials (1979). But first, they should determine the following:

- The material's bulk density, lump size, temperature, and moisture content;
- The conveyor's peak capacity (metric tons per hour);
- Conveyor layout, length, and elevation gain; and
- Loading and unloading requirements.

This information will enable design engineers to calculate the conveyor's size (width) and belt speed. They can also determine the energy required to move material on the conveyor, taking into account the transport distance, any elevation gain, and any friction loads induced by required belt accessories and cleaners.

There are several types of conveyor belts:

- Solid-woven cotton (layers of woven threads that may be used with or without treatment);
- Solid-woven PVC (a single-ply made from nylon and polyester and impregnated or coated with PVC);
- Stitched canvas [separate plies of fabric (usually cotton) stitched together and treated];
- Multiple ply (made from three or more plies of fabric bonded together by elastomeric material);
- Rubber;
- Reduced ply (one or two plies of nylon or polyester); and
- Metal (wire or steel band).

Multiple-ply belts convey dried biosolids effectively. Three to five plies typically are sufficient for narrow belts, while wide belts require anywhere from 6 to 16 plies. The belts may be hot-vulcanized or mechanically spliced in the field. Engineers should ask equipment manufacturers about standard design features.

Carrying idlers typically consist of three sealed cylinders made of painted steel. The center cylinder (roller) is horizontal, while the two side rollers are inclined to force the belt into a trough shape. Return idlers typically consist of six to eight equally spaced neoprene or PVC disks mounted on painted steel rods beneath the conveyor's deck pans. Idlers are spaced to minimize belt sag but are closer together at loading points to minimize the related impact on the belt. Typically, they are about 0.9 to 1.5 m (3–5 ft) apart on the carrying side of the conveyor and about 3.1 m (10 ft) apart on the return run.

A take-up device is an adjustable pulley or roller arrangement that compensates for belt-length changes caused by wear, stretching, and varying loads. Adjustments may be manual or automatic via counterweights, springs, pneumatics, or hydraulics. Tail pulleys are often used; these devices are on the end of the conveyor opposite the drive pulley.

Designers should avoid counterweight pulley take-ups because they require roller contact on the carrying side of the belt, thereby increasing housekeeping requirements.

There are many drive arrangement options. In most belt conveyor systems, the drive is at the discharge (head) end of the belt to limit the tension required and enhance belt service life. The most common drive is a gear motor, a system in which one or more drive pulleys or drums is connected to the motor via a chain, belt, or direct drive.

When transporting dried biosolids, belt speed typically should be no more than 40 m/min (125 ft/min) to keep material stationary, minimize belt wear, and increase belt life. Moving dried materials more rapidly can generate excessive fugitive dust and associated housekeeping demands.

Belt conveyors have options that allow for design flexibility. If the material must be weighed, for example, a scale can be mounted on the conveyor so operators can monitor material flow or control chemical addition. A single conveyor can have multiple loading chutes. Plow stations may be installed to divert material off the belt at intermediate points. Also, the conveyors may be designed with reversible motors.

Odors are a significant disadvantage of using a belt conveyor to transport dried biosolids. Enclosing the conveyor can keep dust and odors in and weather out. Also, proper and regular housekeeping is essential in reducing the buildup of dust and other materials.

A clean discharge is vital to the O&M of a belt conveyor. If the belt does not discharge cleanly, then when the carrying side of the belt contacts the return idlers, it may deposit material on them, causing excessive wear and extensive cleanup requirements. So, design engineers should provide a belt-cleaning system. Several cleaning systems are commonly available, including urethane cleaning blades (also available in tungsten carbide, ceramic, stainless steel, or rubber); brush cleaners; scrapers; and spray-wash systems. If using scrapers, designers should avoid the counterweighted rubber type mounted below the head pulley; they are ineffective. Scrapers with multiple "fingers" and adjustable tensioners are preferred.

All conveyors must include emergency stop switches and pull cables. They should also have speed switches, especially when system control is highly automated. Misalignment switches may also be desirable to indicate potential problems promptly.

Design engineers should make provisions for oiling and greasing all pulley and sheave axles or shafts. Facilities using belt conveyors should also include hose stations and drains to permit frequent washdowns of the area.

7.2 Screw Conveyors

Screw conveyors are one of the most economical options for moving dried biosolids. Ranging from 150 to 600 mm (6–24 in.) in diameter, these conveyors use a shaft-mounted spiral helix or a self-supporting helix to move material in a covered trough. They typically are installed horizontally or at inclines less than 45°. Inlet and discharge openings may be located where needed; the system typically is supported at the ends, loading points, and intermediate points by either feet or saddles.

Shafted screw conveyors typically are available in 3- to 4-m (10–13 ft) long sections that are bolted together. Internal intermediate hangers provide support, maintain alignment, and serve as bearing surfaces.

Shaftless screw spirals typically are furnished in one piece (either fabricated or welded) that can be up to 50-m (165-ft) long, depending on the ability to ship to the site. They rely on polyethylene liners or steel wear bars for intermediate support.

Conveyor length is limited by the center shaft's torsional capacity, as well as the couplings (shafted screws) or spiral (shaftless screws). As a result, screw conveyors typically are no more than 14-m long, unless exceptional design considerations are addressed.

7.2.1   Screw Conveyor Applications

Screw conveyors can be used to feed, distribute, collect, or mix dried materials. They also can heat or cool the material while transferring it.

If material degradation is a concern (e.g., dried biosolids slated for beneficial reuse), shaftless screw conveyors are better than shafted screw conveyors because their slow turning spiral and continuous support along the trough generates less dust.

7.2.2   Screw Conveyor Design

When designing screw conveyors, engineers should begin by identifying capacity requirements, origination and terminus points, and conclude with determining components and final layouts. Following are brief summaries of each design step.

7.2.2.1   Known Factors   Design engineers should identify the material's characteristics, such as type, incidence, and lump size, as well as the conveyor's capacity requirements (cubic meters (cubic feet) per hour), transport distance, and elevation change. They should also note whether any other operation, such as mixing or cooling, is required during transfer and choose an appropriate conveyor (e.g., one with a ribbon flight or jacketed trough).

7.2.2.2   Materials Classification   Conveyed material should be classified according to CEMA standards (see Tables 19.8 and 19.9). The dewatered grit, dried biosolids, and other dried granular solids typically encountered at wastewater facilities would be classified as C1/235 [C1/2 = 13 mm (0.5 in.) and under granules; 35 average flowability; 55 mildly abrasive]. Designers also should identify other miscellaneous properties that may affect conveyor design, such as whether the sludge is especially corrosive.

7.2.2.3   Determine Conveyor Diameter and Speed   Conveyor manufacturers publish charts and tables listing the capacities of various screw conveyors based on trough loadings and conveyor speeds [revolutions per minute (rpm)]. Shafted screw conveyors transporting dried biosolids are typically lightly loaded (trough loads less than 30% total solids), while shaftless ones can handle loads up to 70% to 80% total solids. Also, the area of a shafted screw conveyor's trough is equal to the area of the screw spiral minus the shaft area. The area of a shaftless screw conveyor's trough is equal to that of the screw spiral. In other words, a shaftless screw conveyor has more capacity at slower rotational speeds than an identically sized one with a shaft (see Table 19.10).

If the application involves inclination, design engineers should consult with the manufacturer.

Typically, a conveyor's screw diameter is equal to the pitch of its spiral (helix), so a 300-mm (12 in.) diameter screw would have a 300-mm (12 in.) pitch. In theory, one rotation of the screw will transport the material one pitch (minus some inefficiency), meaning a conveyor operating at 30 rpm would move material 30 pitch lengths in 1 minute.

7.2.2.4   Compare Conveyor Diameter to Lump Size   Conveyor diameters typically are at least four to six times the diameter of 75% of the lumps encountered at the WRRF.

Major Class	Material Characteristics Included	
	Density	**Bulk Density, Loose**
Size	Very fine	No. 200 sieve (0.074 mm) and under No. 100 sieve (0.15 mm) and under No. 40 sieve (0.41 mm) and under
	Fine	No. 6 sieve (3.35 mm) and under
	Granular	No. 6 sieve to 13 mm 13 to 80 mm 80 to 180 mm
	Lumpy	0 to 410 mm Over 410 mm to be specified X = actual maximum size
	Irregular	Stringy, fibrous, cylindrical, slabs, etc.
Flowability	Very free flowing Free flowing Average flowability Sluggish	
Abrasiveness	Mildly abrasive Moderate abrasive Extremely abrasive	
Miscellaneous properties or hazards	Builds up and hardens Generates static electricity Decomposes-deteriorates in storage Becomes plastic or tends to soften Very dusty Aerates and becomes fluid Explosiveness Stickiness-Adhesion Contaminable, affecting use Degradable, affecting use Gives off harmful or toxic gas or fumes Highly corrosive Mildly corrosive Hygroscopic Interlocks, mats or agglomerates Oil present Packs under pressure Very light and fluffy—may be windswept Elevated temperature	

TABLE 19.8 Material Classification Code Chart

Material	Material Characteristics		
	Weight (kg/m³)	Material Code	Trough Loading (%)
Dewatered cake (compactable)	800–960	E–450V	15
Grit	1600–2 180	B6–47	15
Solids, dried	640–800	E–47TW	15
Solids, dry ground	720–880	B–46S	30

Material: dried sludge solids

$C_{1/2}$	3	5	G
Size			Other possible characteristics
	Flowability	Abarsiveness	

TABLE 19.9 Material Classification Codes for Typical Wastewater Residuals

Type	Size (mm)	Speed (r/min)	Trough area (cm²)	Loading (%)	Capacity (m³/h)
Shafted	300 mm	30	650	30%	35
Shaftless	300 mm	30	700	70%	88

TABLE 19.10 Capacity Differences Between Shafted and Shaftless Screw Conveyors

Design engineers should confirm that the conveyor diameter can move the maximum lump size safely.

7.2.2.5 Determine Conveyor Horsepower Drive horsepower is calculated as a function of capacity (cubic meters (cubic feet) per hour), density [kilograms per cubic meter (pounds per cubic foot)], transport length [meters (feet)], diameter [meters (feet)], flighting design, elevation change, and friction losses.

7.2.2.6 Select Components for Torsional and Horsepower Requirements Once design engineers determine the horsepower needed, they should select components (e.g., pipe shafts, drive shafts, and bearings) that resist or transmit loads induced by the conveyors.

The torsional limits of conveyor shafts and flighting may require designers to divide long transport distances among two or more short conveyors.

7.2.3   Other Considerations

Screw conveyors are used to transport dried biosolids because they can be sealed to completely contain odors and dust. Carbon steel components typically are suitable, but galvanized steel also may be considered. Although the dried biosolids eventually will wear away most of the galvanizing or paint in the interior, the areas not continuously in contact with the conveyed material should be protected against corrosion.

The conveyors should not be operated with exposed shafting or flighting; all covers and lids should be kept closed. The covers may provide significant structural integrity.

Screw conveyors should include speed switches at the tail or nondrive ends to verify auger movement. Running a conveyor into nonoperating equipment can cause severe damage.

These conveyors require relatively little maintenance or housekeeping, compared to other biosolids transport options. Routine maintenance typically consists of checking and adjusting the drive unit, and greasing hanger bearings (shafted screw conveyors) or replacing polyethylene liners (shaftless screw conveyors).

7.3   Drag Conveyors

Drag (en masse) conveyors have a wide range of uses in numerous industries. They have a long history of conveying such materials as biosolids, coal, grit, logs, rock salt, sawdust, and wood chips. These conveyors are highly adaptable and can be customized to transport most, if not all, recovered water facility solids.

Drag conveyors have a slow moving chain-and-flight assembly that typically pushes material along a steel pan or trough, which may be rectangular or U-shaped. The troughs typically are constructed from structural "C" channels or bent plate. The chain typically is made of cast forged or fabricated steel. The flights, which typically are flat plates, "C" channels, or tubes, are shaped to match the trough and are bolted directly to chain links. Wider chains [up to 600 mm (24 in.) wide] can act as both chain and flight.

7.3.1   Drag Conveyor Applications

Most drag conveyors are designed to transport material horizontally. Special designs, however, can move material vertically or in "Z" lifts. A "Z" lift conveyor system is designed to transport material horizontally, then vertically, and then horizontally again.

Drag conveyors are used in many recovered water facility processes, such as primary clarification, grit removal, and solids treatment. They may also be used as live bottom feeders in hoppers.

Drag conveyors are exceptionally strong and robust. Their lengths typically are only limited by the chain's strength and the weight of the material being transported. In practical terms, drag conveyors typically are not designed to move material more than about 40 m (130 ft) horizontally or 10 to 15 m (33–50 ft) vertically.

7.3.2   Drag Conveyor Design

Drag conveyor design primarily depends on the chain and the pulling loads encountered when pushing material the length of the conveyor. Following are design issues to consider for a drag conveyor's major components.

7.3.2.1 Chain Type   Chain designs and material characteristics vary greatly. Rollerless chains typically are suitable for most WRRF applications. However, "Z" lift applications sometimes require roller chains to reduce overall chain load and kilowatt requirements. Also, drag chains—whose links can be up to 600 mm (24 in.) wide and can serve as both chain and flight—are suitable for horizontal and slightly inclined installations.

7.3.2.2 Chain Material   Malleable cast-iron chains are suitable for most drag conveyors used in water resource recovery facilities. However, steel chains with hardened components are more wear resistant than most cast chains.

7.3.2.3 Chain Pitch   Chain pitch is the distance between two successive rollers on a conveying chain. This distance typically depends on the desired size or spacing of cross rods or flights. Short conveyors can have 100-mm (4-in.) pitches, while long or heavily loaded conveyors often have 150- to 300-mm (6- to 12-in.) pitches.

7.3.2.4 Sprocket Size   Head and tail sprockets should be designed with as many teeth as practical because the number of teeth greatly influences chain and sprocket wear, as well as how smoothly the conveyor operates. As a general rule for optimum results, sprockets for pitch chains up to 150 mm (6 in.) should have between 12 and 21 teeth; those for larger pitch chains should have between 6 and 14 teeth (Link-Belt Industrial Chain Division, 1983).

7.3.2.5 Drive   The drive typically is installed at the head (discharge) end of the conveyor so only the chain's carrying run is under maximum tension.

7.3.2.6 Take-Ups   Take-up devices are used to maintain proper chain tension. Screw take-ups typically are acceptable for most drag conveyors used at water resource recovery facilities. Spring take-ups are useful if shock loads are anticipated. Gravity and catenary take-ups are also available.

7.3.2.7 Head and Tail Sections   Head and tail sections are custom made from steel plate to match the conveyor and installation requirements. They should be designed to support the loads induced by chain tension, which may be substantial. The tail section typically includes the take-up mechanism unless a catenary take-up method is used, in which case the take-up is in the head section. The end without the take-up should have shaft bearing mounts with slotted holes and jack screws or other mechanisms so operators can align conveyor sprockets accurately.

7.3.2.8 Troughs   Most drag conveyor troughs are fabricated of steel plate and angle iron sections. They can also be constructed of concrete. All troughs should be equipped with replaceable wear bars for the chain or flights to rest on and slide along.

7.3.3 Other Considerations

Drag conveyors typically cost more to install than other conveyors used for dried biosolids, but they allow for more flexibility in layouts and configurations then belt or screw conveyors and can handle a much larger spectrum of materials. They also have more capacity per cross-sectional area and can handle higher-impact loads, but are less energy efficient because of the high frictional forces of the biosolids, flights, and chain against the trough.

For safety reasons, drag conveyors only should be operated when all covers, enclosures, and other safety appurtenances are in place. Inspection ports and access doors should have a metal screen or welded wire fabric that prohibits operators from inserting arms, legs, or other appendages into moving conveyors.

Most drag conveyors are designed with top covers that are bolted every 100 to 200 mm (4–8 in.) down the length of the conveyor. While this may complicate maintenance efforts, O&M personnel should be aware that such covers significantly increase the conveyor's structural strength. Tightening only one or two bolts per section when replacing covers may result in conveyor buckling and catastrophic failure.

Dried biosolids are so abrasive that it may be best to avoid chain lubrication. Using conventional lubrication on the drag chain actually may accelerate wear by adhering abrasive particles to the chain, where they then act as a lapping or grinding compound.

For drag conveyors, the most important maintenance consideration is routinely checking and adjusting chain tension. Improper chain tension, whether too much or too little, significantly shortens the lives of the chain, sprockets, and bearings. Other maintenance checks include bearing lubrication, wear bar thicknesses, chain lubrication (if applicable), bolt torques, and alignment of head and tail shafts.

7.4   Bucket Elevators

A bucket elevator is a simple, dependable device for vertically transporting dry materials. It consists of a series of buckets mounted on a belt or chain within a housing. The buckets are filled with material at the base of the unit and discharge it at the top. They are available in a wide range of capacities [10–350 kg/s (4–140 ton/hr)] and are totally enclosed to prevent dust and odors from escaping.

There are three types of bucket elevators: centrifugal-discharge, positive-discharge, and continuous. Centrifugal-discharge elevators are the most common and are best suited for handling fine, free-flowing materials that can be dug from the elevator boot at the base of the unit. These units have the fewest buckets, which are mounted on either a chain or belt. The buckets are easily loaded and travel rapidly enough [up to 90 m/min (300 ft/min)] to discharge material via centrifugal force as they pass around the head pulley or sprocket.

Positive-discharge elevators are designed for sticky materials that tend to pack. They are similar to centrifugal-discharge units, except that their buckets are mounted on two strands of chain, are large and closely spaced, and are snubbed back under the head sprocket to invert them for positive discharge. (As the snub sprockets engage the chain, the slight impact helps free materials from the buckets.) The buckets also move slower [35 m/min (120 ft/min)] than those in centrifugal-discharge units.

Continuous elevators are recommended for sluggish, aerated, and friable materials—applications in which product degradation is a concern. They have closely spaced buckets mounted on either belts or chains that travel at 38 m/min (125 ft/min). The buckets are often direct-loaded and are designed so the fronts and extended sides form a chute as they pass around the head pulley or sprocket. Gravity allows the material to flow gently out of the buckets and down the chute (formed by the preceding bucket) into the discharge spout.

7.4.1   Bucket Elevator Applications

At water resource recovery facilities, bucket elevators are used to transport dried materials vertically within solids-processing units and to load dried biosolids products into trucks or railcars.

For most dried materials, centrifugal-discharge elevators are generally acceptable and the most cost-effective option. However, if the dried biosolids are intended for beneficial reuse, then dust content and degradation are concerns, so designers should consider continuous bucket elevators.

7.4.2 Bucket Elevator Design and Operation Considerations

Before selecting a bucket elevator, designers should determine the material's characteristics (abrasiveness, flowability, etc.), its maximum lump sizes, density of the bulk material, capacity needed, and transport height.

Manufacturers sell standard-sized units, and tables are available to help designers size bucket elevators that will convey materials vertically up to 30 m (100 ft). The tables provide elevator dimensions, bucket sizes, and energy requirements.

Bucket elevators vary in casing thickness, bucket type and thickness, belt or chain quality, and drive equipment.

The casing is constructed of either heavy-gauge steel sections or steel plate and angle iron that are continuously welded for the full length of the unit. The steel is either mild or galvanized. Galvanizing may be preferred over painted casings because it provides corrosion protection for the casing interior. (Most bucket elevator casings are too small for the interior to be painted properly.) Casings can be up to 7 mm (0.25 in.) thick. A heavier casing may be recommended if the elevator will be outside and exposed to harsh weather conditions. Casings can also be made dust tight. A split, removable hood is recommended for ease of service and maintenance at the top end of the unit.

Buckets are available in a variety of styles, and designers should ask manufacturers for recommendations. Centrifugal elevators typically have malleable iron buckets, which are appropriate for heavy-duty abrasive applications (e.g., dried biosolids). Ductile iron or steel buckets can also be used if desired. Continuous elevators typically have steel buckets. Polymer and nylon buckets are also available; they resist corrosion and promote discharge of sticky materials. In addition, buckets can be perforated to handle dusty materials. The perforations allow air to be released from the buckets during loading and improve material discharge by eliminating blowing.

Buckets can be mounted on single or dual strands of chains. Class "C" combination chains, which have alternating cast-iron black links, are sufficient for normal applications, according to CEMA. Class "S" chains are stronger and wear less quickly; they are recommended for great heights [up to 45 m (150 ft)] or when transporting abrasive materials. Rubber-covered belts are acceptable for most applications involving belt-mounted buckets. Such belts can be made of impregnated canvas or fabric.

Bucket elevators typically are driven from the head shaft and have take-up bearings in the boot. A shaft-mounted gear reducer with a V-belt drive is recommended for economy and versatility. Another option is a gear motor connected to the elevator head shaft via a chain drive; it is supported on a bracket mounted to the elevator casing. Backstops prevent backward rotation when the elevator stops under load; they may be added to either the head shaft or countershaft. The tail shaft should include a zero-speed switch to indicate motor or conveyor problems, or an overloaded elevator. Drive guards are also required for safety reasons.

Bucket elevators typically use less energy than other types of vertical conveyors. As a general guideline, design engineers can estimate an elevator's electricity requirements as follows:

$$\text{Power needed (kW)} = \text{Capacity (tonne/h)} \times \text{Conveyance height (m)}/103$$

If the bucket elevator is more than 10 m (33 ft) tall, engineers should include guy wires or structural steel members to provide lateral bracing. Also, designers should include a service platform so operators can inspect and maintain the head terminal and drive more easily. The platform should be accessed by a ladder with a safety cage. Likewise, design engineers should provide a clean-out door in the boot—especially in continuous elevators—so operators can periodically remove any material that has accumulated in the base. In centrifugal units, the casing corners may fill with material that should be removed regularly.

In applications involving dust, designers should ventilate bucket elevators. The ventilation system should be designed in accordance with guidelines established by the American Conference of Governmental Industrial Hygienists in Industrial Ventilation (1982). These guidelines require an exhaust point at the top of the elevator and a second one at the bottom if the elevator is more than 10-m (33 ft) high. They also recommend a flow of 30 $m^3/m^2/min$ (100 $ft^3/ft^2/min$) of casing, with a minimum duct velocity of 18 m/s (60 ft/s).

Bucket elevators should be designed with explosion relief or suppression mechanisms because organic dusts—including the fine material generated during biosolids drying—could explode under certain conditions. Explosion relief directs such forces through expendable panels (rupture plates) in the elevator casing and then into the room or outside the facility. Engineers should be extremely careful when designing explosion vents because the force of these explosions could hurt or kill operators who are next to the equipment when an explosion occurs.

7.5 Pneumatic Conveyors

Pneumatic conveyors use air to move material through a pipeline. There are two types of these conveyors: dilute-phase and dense-phase.

Dilute-phase conveyors have low material-to-air ratios [less than 5:1 (5 kg material/ kg air)]; they use a large volume of air to move a small amount of material. Rotary airlocks feed material into the pipeline, and positive-displacement blowers or fans supply low-pressure air [less than 100 kPa (14.5 psi)]. The air velocity typically ranges from 20 to 40 m/s (65–130 ft/s)—high enough to suspend the material in the air stream. The systems then use either positive or negative pressure to push or pull the material through the pipeline.

Dense-phase conveyors have high material-to-air ratios (up to 100 kg material/kg air). A pressure tank and a high-pressure air compressor can provide 350 to 700 kPa (50–100 psig) of air, which typically moves at velocities less than 2.5 m/s (8 ft/s). In these systems, material enters the pressure tank or transporter via gravity and settles in the pipeline. Once a certain volume has settled, the transporter inlet valve closes, the vessel is pressurized using compressed air, and the material flows out of the vessel into the pipeline. As the pressure increases, the material forms a plug that the air pushes to its destination.

There are two types of dense-phase systems: conventional batch and full-line. In a conventional batch system, air is introduced at the pressure vessel with enough force to transport a batch of material in the pressure vessel to the final receiving bin. The pipe is completely purged before the next cycle begins.

A full-line system introduces air in both the pressure vessel and via low-pressure air-booster fittings spaced along the length of the pipeline. The air is introduced at the lowest possible velocity. Once the pressure vessel is emptied, any material left in the pipeline remains there until the next batch is conveyed. The booster fittings serve to move that material when the next batch begins.

7.5.1   Pneumatic Conveyor Applications

Pneumatic conveyors have been used to transport dried biosolids and grit in continuous processes, load railcars or trucks, and help collect and remove fugitive dust.

Dilute-phase conveyors may be used to transport lime, sawdust, and other chemicals typically used at water resource recovery facilities. They are ideal for transporting nonabrasive, powdered materials over short distances. Although successfully used to convey a wide variety of bulk solids, dilute-phase conveyors may not be suitable for dried biosolids meant for beneficial reuse because they can be abrasive and degrade when exposed to high-velocity air. Also, dilute-phase conveyors have low capital costs, but their energy costs can be quite high because of the large air requirements.

Both conventional-batch and full-line, dense-phase conveyors are appropriate for transporting dried biosolids. Conventional batch conveyors are preferred for short distances [less than 35 m (115 ft)]. Full-line conveyors are preferred for longer distances and easily degraded material, such as dried biosolids pellets intended for beneficial reuse.

7.5.2   Pneumatic Conveyor Design and Operation Considerations

Pneumatic conveyor design depends on such parameters as material bulk density, capacity or flow rate, and equivalent pipeline length.

When designing dilute-phase pneumatic conveyors, engineers should be aware that much information is available in manufacturer's brochures, data sheets, and monographs. They can use these resources to determine the optimum pipeline diameter and air volume based on recommended solids ratios and system pressure drop [less than 70 kPa (10 psi)]. Then they can determine the blower or fan size needed based on the calculated conveying air volume.

Pressure systems typically are used when flow rates exceed 9000 kg/h (20 000 lb/h). Vacuum systems are used when flow rates are less than 7000 kg/h (15 500 lb/h) and the equivalent length is less than 300 m (1000 ft).

The rotary airlock is sized based on desired flow rate. Carbon steel construction is acceptable for most materials encountered in a WRRF.

The system will need a dust-collection device to clean the conveying air before it is exhausted. This device, which may be a cyclone separator or fabric filter, should be sized to handle the air flows determined in accordance with manufacturer's recommendations.

When designing dense-phase pneumatic conveyors, engineers will need more input from manufacturers because there are no readily available monographs and such designs are often considered more an art than a science. Given a set of design parameters, the system manufacturer should be able to provide the optimum pipeline diameter for the conveying air volume and pressure required. The manufacturer should also recommend a standard transporter size (typically corresponding to 5–15 cycles per hour) and spacing requirements for air-booster fittings [typically every 1.5–6 m (5–20 ft)].

Transporter size depends on conveying distance; transport-cycle frequency lessens as conveying distances exceed 150 m (500 ft). Also, the transporter should include a 60° hopper at the bottom to encourage dried solids to flow out of the vessel.

If continuous conveying is required (e.g., in solids treatment processes), design engineers should provide a dedicated air compressor, an air receiver, and a compressed air dryer.

The pressure vessel typically is made of carbon steel. The pipeline should be constructed of either Schedule 40 carbon steel or galvanized steel.

When designing dense-phase conveyors, engineers should place vertical runs as early as possible and avoid using back-to-back bends. Flat-back elbows may also be considered when transporting heavy-duty, abrasive material. Designers should also be aware that full-line dense-phase conveyors have less pipeline wear because they function at the lowest velocity.

Because dried biosolids are abrasive, all pipeline bends in dilute-phase or dense-phase systems should be long-radius, sweeping bends. Also, status lights and pressure indicators should be included to help facility staff monitor operations.

The pneumatic conveyor's blower, fan, or air compressor typically only requires routine maintenance. However, the exhaust air is odiferous and may require further treatment before discharge. The air will also contain some dust that must be removed before discharge. So, design engineers should equip the pressure vessel with a vent line connected to a dust-collection device and provide for odor control.

Because of the energy requirements, pneumatic conveyors are one of the least efficient methods to transport dried biosolids and other dry granular materials. However, a properly designed and operated pneumatic conveyor will have fewer O&M requirements than mechanical conveyors. Because the conveyor is totally enclosed, housekeeping and odor control are simpler, but problems are more difficult to identify. The conveyor's operating costs are high, but it has a small footprint, easily retrofits into existing facilities, and can handle long distances and multiple discharge points.

8.0 References

American Conference of Governmental Industrial Hygienists (1982) *Industrial Ventilation*, 17th Ed.; American Conference of Governmental Industrial Hygienists: Lansing, Michigan.

American Society of Civil Engineers (2000) *Conveyance of Wastewater Treatment Plant Residuals*; American Society of Civil Engineers: Reston, Virginia.

Barbachem, M. J.; Pyne, J. C. (1995) Pipeline Hydraulics of Dewatered Non-Newtonian Cakes. *Proceedings of the 68th Annual Water Environment Federation Technical Exposition and Conference*, Miami Beach, Florida, Oct 21–25; Water Environment Federation: Alexandria, Virginia; pp. 41–49.

Bassett, D. J.; Howell, R. D.; Haug, R. T. (1991) Hydraulic Properties Evaluation for Sludge Cake Pumping. *Proceedings of the 64th Annual Water Environment Federation Technical Exposition and Conference*; Toronto, Ontario, Oct 7–10; Water Environment Federation: Alexandria, Virginia.

Battistoni, P. (1997) Pretreatment, Measurement Execution Procedure and Waste Characteristics in the Rheology of Sewage Sludges and the Digested Organic Fraction of Municipal Solid Wastes. *Wat. Sci. Tech.*, **36**, 33.

Bechtel, T. B. (2003) Laminar Pipeline Flow of Wastewater Sludge: Computational Fluid Dynamics Approach. *J. Hydr. Eng.*, **129** (2), 153.

Bechtel, T. B. (2005) A Computational Technique for Turbulent Flow of Wastewater Sludge. *Water Environ. Res.*, **77**, 417.

Borrowman, D. (1985) *Wastewater Sludge Characteristics and Pumping Application Guide*; WEMCO Pump Co.: Sacramento, California.

Bourke, J. D. (1992) *Pumping Abrasive Slurries with Progressing Cavity Pumps*; Moyno Industrial Products: Springfield, Ohio.

Bourke, J. D. (1997) *Handling High Solids Content Non-Newtonian Fluids*; Moyno Industrial Products: Springfield, Ohio.

Brar, S. K.; Verma, M.; Tyagi, R. D.; Valéro, J. R.; Surampalli, R. Y. (2005) Sludge Based *Bacillus huringiensis* Biopesticides: Viscosity Impacts. *Water Res.*, **39**, 3001.

Carthew, G. A.; Goehring, C. A.; Van Teylingen, J. E. (1983) Development of Dynamic Head Loss Criteria of Raw Sludge Pumping. *J. Water Pollut. Control Fed.*, **55**, 472.

Chilton, R. A.; Stainsby, R. (1998) Pressure Loss Equations for Laminar and Turbulent Non-Newtonian Pipe Flow. *J. Hydr. Eng.*, **124** (5), 522.

Conveyor Equipment Manufacturers Association (1979) *Belt Conveyors for Bulk Materials*, 2nd ed.; CBI Publishing Co.: Boston, Massachusetts.

Crow, H.; Cortopassi, R. (1994) *Schwing KSP 17V(K) Pump Demonstration from July, 1994 through August 3, 1994*; Schwing America Inc., Environ. Div.: Danbury, Connecticut.

Dillon, M. L. (2007) Comparing PD Pump Designs for Transferring Dewatered Sludge Cake. *Pumps & Systems*, September, 84.

Doty, D. (2005) New Progressing Cavity Pump Developments in Sludge Transfer. *World Pumps*, October, 24.

El-Mashad, H. M.; van Loon, W. K. P.; Zeeman, G.; Bot, G. P. A. (2005) Rheological Properties of Dairy Cattle Manure. *Bioresource Technol.*, **96**, 531.

Florida Department of Environmental Protection (2008), *Chapter 62-640: Biosolids*, Draft Version; Florida Department of Environmental Protection: Tallahassee, Florida.

Great Lakes—Upper Mississippi River Board of State and Provincial Public Health and Environmental Managers (2014) Chapter 80, Sections 87-13 and 89-12, Recommended Standards for Wastewater Facilities; Health Research Inc, Albany, NY.

Guibaud, G.; Dollet, P.; Tixier, N.; Dagot, C.; Baudu, M. (2004) Characterisation of the Evolution of Activated Sludges Using Rheological Measurements. *Process Biochem.*, **39**, 1803.

Health and Safety Executive (2005) *Control of Health and Safety Risks at Sewage Sludge Drying Plants*; HSE 847/9; Health and Safety Executive: London, United Kingdom.

Hentz, L.; Cassel, A.; Conley, S. (2000) The Effects of Liquid Sludge Storage on Biosolids Odor Emissions. *Proceedings of 14th Annual Residuals and Biosolids Management Conference*; Boston, Massachusetts, Feb 27–29; Water Environment Federation: Alexandria, Virginia.

Honey, H. C.; Pretorius, W. A. (2000) Laminar Flow Pipe Hydraulics of Pseudoplastic-Thixotropic Sewage Sludges. *Water SA*, **26**, 19.

Jones, H. (1993) *Solving Sludge Handling Problems with Progressive Cavity Pumps*. Robbins & Myers Inc., Fluid Handling Group: Springfield, Ohio.

Laera, G.; Giordano, C.; Pollice, A.; Saturno, D.; Mininni, G. (2007) Membrane Bioreactor Sludge Rheology at Different Solid Retention Times. *Water Res.*, **41**, 4197.

Levine, L. (1986) *Coming to Grips with Rheology*; Viscous Products.

Levine, L. (1987) *An Introduction to the Measurements of Viscosity*; Viscous Products.

Link-Belt Industrial Chain Division (1983) *Link-Belt Chains and Sprockets for Drives, Conveyors and Elevators*; Link-Belt Industrial Chain Division: Homer City, Pa.

List, E. J.; Hannoun, I. A.; Chiang, W.-L. (1998) Simulation of Sludge Pumping. *Water Environ. Res.*, **70**, 197.

Lue-Hing, C.; Zenz, D.; Tata, P.; Kuchenrither, R.; Malina, J.; Sawyer, B. (1998) *Municipal Sewage Sludge Management: A Reference Text on Processing, Utilization and Disposal*; Technomic Publishing Co. Inc.: Lancaster, Pennsylvania.

Lottman, S. (2008) Personal communication; Siemens: Berlin, Germany.

Metcalf and Eddy, Inc. (1981) *Wastewater Engineering: Collection and Pumping of Wastewater;* Tchobanoglous, G.; McGraw-Hill, Inc.: New York, New York

Metcalf and Eddy, Inc. (2003) *Wastewater Engineering: Treatment and Reuse;* Tchobanoglous, G.; Burton, F. L.; Stensel, H. D., Eds; 4th ed.; McGraw-Hill Inc.: New York, New York.

Mori, M.; Isaac, J.; Seyssiecq, I.; Roche, N. (2008) Effect of Measuring Geometries and of Exocellular Polymeric Substances on the Rheological Behaviour of Sewage Sludge. *Chem. Eng. Res. Des.,* **86,** 554.

Mulbarger, M. C.; Copas, S. R.; Kordic, J. R.; Cash, F. M. (1981) Pipeline Friction Losses for Wastewater Sludges. *J. Water Pollut. Control Fed.,* **53,** 1303.

Mulbarger, M. C. (1997) Selected Notions about Sludges in Motion, and Movers. Paper presented at Central States Water Environment Association Education Seminar, Madison, Wisconsin.

Murakami, H.; Katayama, H.; Matsuura, H. (2001) Pipe Friction Head Loss in Transportation of High-Concentration Sludge for Centralized Solids Treatment. *Water Environ. Res.,* **73,** 558.

National Fire Protection Association (1995) *Report on Comments A2007—NFPA 820 Standard for Fire Protection in Wastewater Treatment and Collection Facilities;* National Fire Protection Association: Quincy, Massachusetts.

Novak, J.; Adams, G.; Chen, Y.-C.; Erdal, Z.; Forbes, R. H, Jr.; Glindemann, D.; Hargreaves, J. R.; Hentz, L.; Higgins, M. J.; Murthy, S. N.; Witherspoon, J.; Card, T. (2004) Odor Generation Patterns from Anaerobically Digested Biosolids. *Proceedings of the Joint WEF/A&WMA Odors and Air Emissions Conference;* Bellevue, Washington, Apr 18–24; Water Environment Federation: Alexandria, Virginia.

Pilehvari, A. A.; Serth, R. W. (2005) Generalized Hydraulic Calculation Method Using Rational Polynomial Model. *J. Energy. Res. Technol.,* **127,** 15.

Radney J. (2008) Personal communication; Cerlic USA: Atlanta, Georgia.

Ram Pumps Take the Drudgery out of Sludge Transfer (1999). *World Pumps,* Feb, 18–19.

Sanks, R. L.; Tchobanoglous, G.; Bosserman, B. E., II; Jones, G. M., Eds. (1998) *Pumping Station Design;* 2nd ed.; Butterworth-Heinemann: Boston, Massachusetts.

Setterwall, F. (1972) Discussion/Communication on Pumping Sludge Long Distances. *J. Water Pollut. Control Fed.,* **44** (1), 648.

Spaar, A. (1972) Pumping Sludge Long Distances. *J. Water Pollut. Control Fed.,* **43** (1), 702.

Spinosa, L.; Lotito V. (2003) A Simple Method for Evaluating Sludge Yield Stress. *Adv. Environ. Res.,* **7,** 655.

Spinosa, L.; Vesilind, P. A. (2001, reprinted 2007) *Sludge into Biosolids—Processing, Disposal, Utilization;* IWA Publishing: London, United Kingdom.

U.S. Army Corps of Engineers (1984) *Engineering and Design—Domestic Wastewater Treatment Mobilization Construction;* EM-1110-3-172; U.S. Army Corps of Engineers: Washington, D.C.

U.S. Environmental Protection Agency (1979) *Process Design Manual, Sludge Treatment and Disposal;* EPA-625/-79-011; U.S. Environmental Protection Agency, Munic. Environ. Res. Lab.: Cincinnati, Ohio.

U.S. Environmental Protection Agency (1982) *Handbook: Identification and Correction of Typical Design Deficiencies at Municipal Wastewater Treatment Facilities;* EPA-625/6-82-007; U.S. Environmental Protection Agency, Munic. Environ. Res. Lab.: Cincinnati, Ohio.

U.S. Environmental Protection Agency (1983) *Process Design Manual—Land Application of Municipal Sludge;* EPA-625/1-83-016; U.S. Environmental Protection Agency: Washington, D.C.

U.S. Environmental Protection Agency (1995) *Process Design Manual—Land Application of Sewage Sludge and Domestic Septage*; EPA-625/R-95/001; U.S. Environmental Protection Agency: Washington, D.C.

U.S. Environmental Protection Agency (2000) *Guide to Field Storage of Biosolids and Other Organic By-Products Used in Agriculture and for Soil Resource Management*; EPA/832-B-00-007; U.S. Environmental Protection Agency: Washington, D.C.

Wagner, R. L. (1990) *Sludge Digester Heating*; Alfa-Laval Thermal Co.: Ventura, California.

Wanstrom, C. (2008) Personal communication. Schwing Bioset: Somerset, Wisconsin.

Water Environment Federation (2004) *Control of Odors and Emissions from Wastewater Treatment Plants*; WEF Manual of Practice No. 25; McGraw-Hill: New York.

Water Environment Federation; American Society of Civil Engineers; Environmental and Water Resources Institute (2018) *Sustainability and Energy Management for Water Resource Recovery Facilities*, WEF Manual of Practice No. 38; Water Environment Federation: Alexandria, Virginia.

Water Environment Research Foundation (2008) *Identifying and Controlling Odor in Municipal Wastewater Environment Phase 3: Biosolids Processing Modifications for Cake Odor Reduction*; Water Environment Research Foundation: Alexandria, Virginia.

CHAPTER 20
Chemical Conditioning

Peter Loomis, P.E.; Samir Mathur, P.E., BCEE; Eric Spargimino, P.E.,
LEED AP; Anthony Tartaglione, P.E., BCEE; and Tanush Wadhawan, Ph.D.

1.0 Introduction

Conditioning does not reduce the water content of solids; it alters the physical properties of solids to facilitate the release of water during thickening and dewatering. Mechanical thickening and dewatering techniques are rarely economical for a utility without chemical conditioning upstream.

For purposes of this chapter, chemical conditioning is a treatment used to improve the efficiency of downstream processes (e.g., thickening or dewatering). Chemical conditioning processes use inorganic chemicals, organic polymers, or both to improve solids' thickening and dewatering characteristics. Physical conditioning techniques (e.g., thermal conditioning) use heat to condition and stabilize solids. (For more information on thermal conditioning, see Chapters 23 and 24.)

This chapter discusses chemical conditioning, and includes theory and design considerations. Some thickening and most dewatering of wastewater residuals particularly those containing solids from biological treatment processes (e.g., fixed-film and suspended growth activated solids treatment systems) typically are not practical without some type of conditioning.

Conditioning can significantly alter the characteristics of the incoming solids, and effectiveness of thickening or dewatering processes. Typically, chemical conditioning paired with the appropriate thickening and dewatering process can increase the total solids content from approximately 0.4% to between 18% and 40%.

In the United States, most water resource recovery facilities (WRRFs) use polymers for thickening and dewatering optimization, and no longer use inorganic chemicals like lime or metal salts (e.g., ferric chloride, Alum etc.). These chemicals are now more frequently used for other types of treatment like stabilization or phosphorus treatment, respectively.

The use of organic polyelectrolytes (polymers) in municipal WRRFs was introduced during the 1960s and was rapidly adopted for both thickening and dewatering processes. The primary advantage of polymers is that they do not significantly increase solids production. Alternatively, every kilogram of inorganic chemicals added during conditioning creates an additional kilogram or more of solids that must be managed, in addition to the solids added by enhancing the capture efficiency.

All polymers are intended to increase the capture rate, reducing the total solids in the filtrate/centrate/supernatant sent back to the liquid process. Additionally, polymers have less of an impact on the water chemistry then lime or metal salts. Metal salts and inorganic chemicals such as lime have a significant impact on the overall alkalinity, pH, and hardness of the liquid fraction of the solids.

The conditioning method must be compatible with the proposed thickening or dewatering method. For example, centrifuges use centrifugal force to compact the solids, whereas belt filter presses permit the water to pass through the void spaces utilizing gravity and pressure; therefore, a single type of conditioning agent cannot be expected to be useful for all applications (WEF, 2014). Furthermore, varying types of organic and synthetic polymers are developed for use with solids of varying characteristics and water chemistry. Cross-linked polymers and varying charges and molecular weights of cationic and anionic polymers exist to suit the needs of the respective solid and thickening or dewatering process.

For final disposal of solids after dewatering, Inorganic chemicals such as lime can be added to stabilize solids and increase the final product total solids to 20% to 30% (dry solids). Fly ash or other bulking agents can also be added increase the final product total solids to 50% to 100% (dry solids) and create a product more suitable for certain end uses.

2.0 Factors Affecting Conditioning

The type and dosage of conditioning agent needed depends on the residuals' characteristics, solids handling and processing before and after conditioning, and the mixing process after agent addition.

2.1 Residuals Characteristics

Several residuals' characteristics that affect conditioning requirements (adapted from U.S. EPA, 1979d) include

- The source of residuals,
- Solids concentration,
- Alkalinity and pH,
- Biocolloids and biopolymer production,
- Particle size and distribution,
- Degree of hydration,
- Particle surface charge,
- Volatile suspended solids (VSS) content, and
- Phosphate content.

Also reference To (2015).

2.1.1 Source of Residuals

To some extent, the conditioning method depends on the type of solids that must be treated. For example, cationic polymers typically work best on primary, secondary, and digested solids, while anionic polymers may be effective with inorganic solids.

An examination of published data for a variety of thickening and dewatering devices suggests that primary solids require lower doses than secondary solids do, and that fixed-film secondary solids require lower doses than suspended growth secondary solids do (U.S. EPA, 1979d). Depending on the thickening or dewatering method used, aerobically and anaerobically digested solids typically require conditioning doses comparable to those for secondary solids. Similarly, combined solids (primary and secondary solids) have properties that are closer to those of secondary solids, although they are affected by the respective composition of each type. More importantly, characteristics of solids vary from facility to facility and can also vary seasonally, so the conditioner dose depends on the specific conditioning agent used and the goal (thickening or dewatering solids). Constituents like sodium or phosphate can also inhibit the dewaterability of sludge, such that adding more polymer or different types of polymers has negligible benefit. Also, some sludges cannot be dewatered beyond a certain point due to their cellular structure. For example, typical waste activated sludge by itself limited to 20% total solids due to the bacteria's cellular structure's inherent ability to bind water at 4 or 5 to 1.

Chemical solids are solids that have been mixed with an inorganic conditioning agent (e.g., the addition of aluminum, iron salts, or lime). These types of solids can be more variable than traditional primary and secondary solids, and so are hard to categorize with respect to dose, and their conditioning requirements are often qualitatively different from those for primary and secondary solids. For example, adding lime to mixed-liquor suspended solids before secondary clarification may improve suspended solids removal; however, the resulting solids may require an anionic polymer, not the cationic one typically used for primary or secondary solids, because the positively charged calcium has neutralized some of the negative surface charge. While charge neutralization is a fundamental part of the process, an equally significant piece is the interconnection of solids particles via the polymer chain.

2.1.2   Solids Concentration

In many applications, conditioning neutralizes the colloidal surface charge by adsorbing oppositely charged organic polymers or inorganic complexes. The residuals' solids concentration will affect the dosage and dispersal of the conditioning agent. Therefore, for a given particle size distribution, increasing the suspended solids concentration increases the required coagulant dose for effective surface coverage (on a volumetric basis). For this reason, polymer dosing and performance is typically discussed on a dry weight basis, for example, X lbs of active polymer per Y dry pounds or dry tons of solids.

The suspended solids concentration also affects two additional aspects of conditioning agents. First, the process is less susceptible to overdosing at higher solids concentrations. Second, the solids and the conditioning agent are more difficult to mix at higher solids concentrations.

2.1.3   Alkalinity and pH

When inorganic conditioning agents are used, alkalinity and pH are the most important chemical parameters affecting conditioning. Coagulation occurs when coagulant interacts with the surface of solids' colloids, and nature of the charged surface and the coagulant charge are both pH-dependent. When inorganic conditioners are used, the solids' pH determines which chemical species are present.

Iron and aluminum salts behave like acids when added to water (i.e., these conditioners reduce pH), so the pH of the conditioning process will depend on the solids' alkalinity and the dosage of iron or aluminum salts. This pH, in turn, determines the predominant coagulant species and the nature of the charged colloidal surface. The high alkalinity typically associated with anaerobically digested solids is one reason for the higher coagulant doses required. Low-molecular-weight coagulants tend to be more effective over a broader pH range than inorganic conditioning agents.

2.1.4   Biocolloids and Biopolymers

Although more commonly measured by researchers than by design engineers, these fundamental parameters provide some insight into the conditioning process. The biopolymers in activated solids flocs seem to affect the physico-chemical properties of flocs (e.g., floc density, floc particle size, specific surface area, charge density, bound water content, and hydrophobicity).

Other studies have shown that cations can affect bioflocculation and change the settling and dewatering properties of activated sludge flocs (Eriksson and Alm, 1991; Bruus et al., 1992; Higgins and Novak, 1997a, b). Divalent cations bridge across negatively charged biopolymers to form a dense, compact floc structure. Monovalent cations tend to prevent proper flocculation by forming a much weaker structure. As a result, divalent cations promote bioflocculation and produce subsequent improvements in settling and dewatering properties. Monovalent cations tend to degrade settling and dewatering properties. It seems that settling and dewatering properties are further improved when the two divalent cations are added to the feed rather than superficially added to the settling tank (Higgins and Novak, 1997a).

A series of laboratory-scale studies were conducted using waste activated sludge (WAS) to gain insight into the floc-destruction mechanisms that account for changes in solids conditioning and dewatering properties after anaerobic or aerobic digestion. The data indicated that biopolymer was released from solids under both anaerobic and aerobic conditions, but much more was released under anaerobic conditions. In particular, four to five times more protein was released into solution under anaerobic conditions than under aerobic conditions (Novak et al., 2003). Both the dewatering rate (as characterized by the specific resistance to filtration) and the polymer dose depend directly on the amount of biopolymer (protein 1 polysaccharide) in solution.

2.1.5 Particle Size and Distribution

Particle size distributions affect the total particle surface area and the porosity of cakes formed from these particles. These properties affect required coagulant doses and dewaterability. Several researchers (Karr and Keinath, 1978; Novak et al., 1988; Sorensen and Sorensen, 1997) studied the effect of particle size on dewaterability and concluded that particle size was one of the most important parameters in determining dewaterability. Smaller particles (colloidal and supracolloidal) can blind filters and solids cakes (Novak et al., 1988; Sorensen and Sorensen, 1997) and deter the release of water from the solids cake. Also, another study has suggested that an increase in floc density improves dewatering properties via a decrease in bound water associated with the flocs (Kolda, 1995). These studies concluded that dewaterability improvements often associated with other factors (e.g., pH, mixing, biological degradation, and conditioning) all could be explained by the effects of these factors on particle size distributions.

2.1.6 Degree of Hydration

Excessive bound water has been suggested as the cause of dewatering difficulties. The percent of bound water associated with the floc also indicates the maximum dryness that can be achieved in the solids cake by mechanical means (Robinson, 1989). Additionally, Vesilind (1979) reviewed the work of several investigators on water distribution in activated sludge. This water was described as free water, floc water, capillary water, and bound (particle) water. These categories are defined based on the amount of centrifugal acceleration required to release a given portion of water. Vesilind suggested that the water distribution in a given solids could determine the applicability of a specific thickening or dewatering operation.

2.1.7 Particle Surface Charge

Solids particles (e.g., subcolloidal and macromolecular constituents) typically have a negative surface charge, so they tend to repulse each other. The resulting spaces between these constituents are occupied by cations and water. If the charge can be eliminated, thickening or dewatering improves. This is why chemical conditioners are positively charged, or become positively charged when added to water.

In most cases, polymer conditioning is optimal when the charge is neutralized, so measuring charge can be useful in laboratory comparisons of polymers and doses. A streaming current detector (zeta meter) can be used to measure this charge. It also can be used to monitor or control polymer dose real-time in dewatering processes (Dentel et al., 1995).

Shear during mixing or dewatering tends to open up new negative surfaces in the biocolloids, thereby undoing the charge effects of cationic polymers. So, increases in mixing shear increase the required polymer dose (Dentel, 2001).

2.1.8 Wastewater Cations

Numerous studies have suggested that cations interact with the negatively charged biopolymers in activated solids to change the structure of the floc (Higgins, 1995; Bruus et al., 1992; Eriksson and Alm, 1991; Novak and Haugan, 1979; Tezuka, 1969). One study indicated that monovalent cations tend to deteriorate settling and dewatering characteristics, while divalent cations tend to improve them (Higgins, 1995). The effects of charge density on activated sludge properties could decrease the polymer dose needed to condition secondary solids.

Researchers have studied influent concentrations of cations (e.g., aluminum, ammonium, calcium, iron, magnesium, potassium, and sodium) extensively. They postulated that cations play a critical role in bioflocculation. Cations have been found to influence the thickening and dewatering characteristics of biological solids. For example, high concentrations of sodium typically resulted in poor dewatering; however, if the floc contained enough aluminum and iron concentrations, it typically offset the deleterious effects of sodium (Park et al., 2006). The data associated with aluminum further revealed that WAS with low aluminum levels contained high concentrations of soluble and colloidal biopolymer (proteins and polysaccharides), resulting in a high effluent COD concentrations, a need for larger doses of conditioning chemical, and poor solids dewatering properties (Park et al., 2006). Studies have shown that iron may contribute to floc strength, and it seems that the reduction and solubilization of iron during anaerobic digestion may be a reason why digested solids dewater poorly. Also, the presence of proteins in solution contributes to poor dewatering and larger conditioning chemical doses (Novak et al., 2001).

2.1.9   Rheology

Many studies have focused on the rheological properties of wastewater solids in an attempt to correlate solids properties with chemical conditioning requirements (Ormeci et al., 2004). For example, yield strength and viscosity have been used to optimize chemical conditioning. Another study demonstrated that mixing considerably affected the rheological characteristics of conditioned solids (Abu-Orf and Dentel, 1999). Another showed that solids conditioning could be improved by monitoring centrate or filtrate viscosity (Bache and Dentel, 2000). In another study, both laboratory- and full-scale testing showed that the network strength of the sludge could be used to optimize chemical conditioning and achieve drier solids (Abu-Orf and Ormeci, 2005). In general, the type of conditioner (e.g., polymers or fly ash) and wastewater solids (e.g., chemical, WAS, or biosolids) involved will determine which rheological parameters should be used.

2.2   Handling and Processing Conditions Before Conditioning

The efficiency of any conditioning process depends to a large degree on the solids' chemical and physical characteristics (e.g., origin, solids concentration, inorganic content, chemistry, storage time, and mixing) before conditioning. The solids' physical characteristics are a function of the physical stresses they were exposed to before conditioning. For example, any process that damages the flocculant nature of solids particles typically either increases chemical conditioning demand or reduces performance in the final treatment stage. The extent of mixing and shear stress before and after conditioning can significantly affect conditioning efficiency and, ultimately, solids treatment performance.

2.2.1   Storage

There are two types of storage for liquid residuals: long-term and short-term. Long-term storage may occur in stabilization processes with long detention times (e.g., aerobic and anaerobic digestion; see Chapter 23) or in specially designed tanks (see Chapter 19). Short-term storage may occur in wastewater treatment process (e.g., increasing solids inventory) or in smaller, specially designed tanks. Storage helps smooth out fluctuations in solids production, make the solids feed rate more uniform. It also provides a place to keep solids during equipment downtime. However, long-term storage has been reported to negatively affect the dewaterability of solids.

Unstabilized solids that have been stored for long periods typically require more conditioning chemicals than fresh solids because the degree of hydration and percentage of fine particles increased. Also, storing activated sludge increased sludge's specific resistance to filtration and, subsequently, conditioning requirements (Karr and Keinath, 1978). Storing aerobically or anaerobically stabilized solids for extended periods typically lowers temperature significantly and can change pH and alkalinity. Temperature drops typically increase conditioning requirements. However, if the temperature decrease is small, the negative effect on conditioning may be more than offset by an increase in solids concentration.

2.2.2 Pumping

Pumping subjects solids to shear forces; the level of shear depends on the type of pump and the flow rate. Solids particles are fragile, and pumping typically causes some of them to fragment. Researchers have shown that the major demand for chemical conditioning is associated with the fraction of particles in the colloidal and supracolloidal range (Karr and Keinath, 1978; Roberts and Olsson, 1975), so any process that reduces particle size will increase conditioning chemical requirements.

Conditioned solids should not be pumped, because pumping introduces shear forces that tend to break down flocs. If required, however, then the pump should be designed to minimize shear.

2.2.3 Mixing

During conditioning, the solids and added chemicals must be mixed enough to ensure that the chemical is evenly dispersed throughout the solids. However, the mixer must not break the floc once it has formed. Design engineers should optimize the mixing time with these two goals in mind. Mixing requirements depend on the thickening or dewatering method used. In-line mixers typically are used with most modern thickening and dewatering units. A separate mixing and flocculation tank is provided with some older thickening and dewatering devices.

For many municipal solids, intense mixing (a mean velocity gradient in the range of 1 200 to 1500 s^{-1}) should be followed by much gentler agitation (a mean velocity gradient less than 200 s^{-1}), so fine particles can flocculate into particle aggregates that settle or can be readily filtered. Anaerobically digested solids need mean velocity gradients up to 12 000 s^{-1}. Once the solids and conditioning chemical have been mixed thoroughly, a hydraulic retention time (HRT) of 15 to 45 seconds in the pipeline or flocculation tank will complete the flocculation process before solids enter the thickening or dewatering system.

2.2.4 Solids Concentration

The residuals' solids concentration can significantly affect conditioning system performance and cost. In most cases, as the influent solids concentration increases, the conditioning cost decreases to a certain level. However, it becomes increasingly difficult to evenly mix coagulant in residuals containing 4% solids (or more) before further dewatering.

2.2.5 Stabilized and Unstabilized Solids

Polymer requirements also are affected by the type of solids to be conditioned. In fact, this may have the most effect on the quantity of chemical needed. Solids that are difficult to dewater require the largest doses of chemicals, typically yield a wetter cake, and result in poorer-quality sidestreams (filtrate, centrate, etc.). The following types of solids are listed in increasing order (approximately) of conditioning chemical requirements (Metcalf and Eddy, Inc./AECOM, 2013):

- Untreated (raw) primary solids,
- Untreated mixed primary and trickling filter solids,
- Untreated mixed primary solids and WAS,
- Anaerobically digested primary solids,
- Anaerobically digested mixed primary solids and WAS,
- Untreated WAS, and
- Aerobically digested primary solids.

Digestion changes solids' chemical and physical characteristics, increasing alkalinity while reducing mass. However, stabilized solids typically are more difficult to dewater than unstabilized solids. Anaerobically digested solids contain considerably more colloidal and supracolloidal solids than primary solids or activated sludge does (Karr and Keinath, 1978). Aerobic digestion detains solids for 30 days or more, greatly reducing the dewatering characteristics of the resulting biosolids. Digested solids typically have higher specific resistance values (Karr and Keinath, 1978), which, in turn, mean higher chemical conditioner doses to achieve a specified dewatered solids concentration.

Solids with inorganic contents in the range of 15% to 35% (e.g., biological solids) typically have cationic charge-neutralization requirements. Digestion also produces solids with cationic charge-neutralization requirements, although the inorganic solids content may increase to between 30% and 50%. Occasionally, lime-stabilized or chemically treated solids contain higher levels of inorganic solids, which respond better to an anionic or non-ionic polymer. As a general rule, residuals containing less than 50% inorganic solids have a cationic charge demand, while those containing more than 50% inorganic solids have an anionic or non-ionic charge demand (regardless of whether they are accompanied by a pH shift).

Whenever possible, raw, undigested, or unprocessed solids should remain separate from biological or chemically treated solids until just before dewatering. This is especially true if biological solids are generated at a facility that practices biological nutrient removal. Septic conditions cause bacteria to release their bound phosphorus to the filtrate, which then is recycled back to the influent. If such biological solids will be dewatered with primary solids, they should not be mixed until just before they enter the thickening and dewatering device.

2.3 Purpose of Conditioning: Thickening and Dewatering

The fundamental objective of conditioning is to cause fine solids to aggregate via coagulation with inorganic or organic coagulants, flocculation with organic polymer, or both (IWPC, 1981). It should improve the efficiency of thickening, dewatering, and other subsequent treatment processes. Also, conditioning is a significant item in a solids-management O&M budget, so it is desirable to select the most cost-effective method that produces acceptable liquid and solid output streams.

To be effective, the conditioning method must be compatible with the proposed methods of solids thickening, dewatering, and ultimate use or disposal. For example, belt filter presses, gravity belt thickeners, and rotary drum thickeners perform better when the solids are a uniform floc size that increases the voids between particles, thus allowing free water to filter more rapidly through the porous belt or drum. Polymer conditioning is the easiest way to produce such a floc. Other dewatering methods (e.g., pressure filtration and sand bed filtration) performed well when the solids were conditioned via the addition of chemical solids (organic and inorganic) or bulking materials.

Design engineers should take all of the subsequent solids treatment processes into account. For example, if the dewatered solids will be sent to a composting system or thermal dryer, then the conditioning and dewatering systems must produce a cake with maximum solids content. However, in concept, the conditioning and dewatering systems should not reduce the fraction of volatile solids; use an exotic, expensive polymer; or add inorganic chemicals that dramatically increase the volume of material to be dried or composted.

3.0   Ultimate Disposal or Use of Biosolids

Chapter 40 of the Code of Federal Regulations Part 503 addresses the use and disposal of sewage sludge (solids) generated during the treatment of domestic wastewater. Wastewater solids disposed in municipal landfills or used as landfill cover material must comply with the requirements of 40 CFR 257 and 258 as well as state and local requirements. For example, it is becoming increasingly common for local or state authorities to require that residuals contain 35% to 40% solids before they can be codisposed with municipal solid waste. Landfilled solids may have to meet certain levels of biological stability, soil engineering properties, or both. A sufficiently high dose of lime can both stabilize and condition solids, while a dose of lime, fly ash, or other bulking materials can improve a cake's mechanical properties. The solids' mechanical strength can be measured by a slump test (similar to that used for concrete).

Due to continuing public pressure regarding land application of biosolids use, many farmers will only accept Class A EQ (Exceptional Quality; see Chapter 25) solids, which in turn affects overall conditioning and treatment options. Also, some farmers only accept solids that were treated with specific conditioners. In addition, biosolids characteristics and site conditions (e.g., groundwater and soils) may limit the use of certain conditioners and treatment methods. For example, certain crops are better cultivated in acidic soils, and land-applying lime-treated biosolids to such fields would not help the overall agricultural operation. This is discussed in further detail in Chapter 25.

Disposal alternatives must also factor in the cost of each alternative, which can be significantly impacted by the hauling distance from the generating facility. In addition to hauling distance, the total mass of solids is can impact costs so it may be advantageous to utilize different dewatering methods and dewater to a higher solids content since less water will be hauled.

4.0   Types of Chemical Conditioning

Solids can be conditioned via a number of methods (e.g., chemical, heat, pressure, electrolysis, freeze–thaw, etc.). Historically, the most popular has been chemical conditioning (e.g., polymers, inorganic chemicals, or both). Other conditioning methods have been used and have grown in popularity for a variety of purposes (e.g. thermal hydrolysis paired with digestion), discussed further in Chapter 23.

In the mid-20th century, wastewater treatment professionals used various inorganic chemicals and natural organics to condition solids. The most common inorganic chemicals used were lime and metal salts. When synthetic organic polymers were introduced in the late 1960s, they were quickly adopted for solids conditioning because they did not significantly increase the amount of solids to be thickened and dewatered, and were able to achieve better capture with less side effects (e.g., fluctuations in alkalinity and pH).

4.1   Inorganic Chemicals

Inorganic chemical conditioning is principally associated with recessed-chamber plate-and-frame filter presses, although they have also been used for belt filter presses. Lime and liquid ferric chloride are the two most widely used inorganic conditioning agents. These conditioning agents are readily available and can condition a wide range of solids. In addition, the resulting biosolids are suitable for land-application or composting. Compared to polymers, larger doses of inorganic chemicals are required to condition solids, and this affects the volume of solids to be managed. For example, adding iron salts and lime can increase the solids mass (and volume) by as much as 20% to 40% (WEF, 2014).

Less commonly used inorganic coagulants include liquid ferrous sulfate, anhydrous ferric chloride, aluminum sulfate, and aluminum chloride. Other inorganic materials (e.g., fly ash, power plant ash, cement kiln dust, pulverized coal, diatomaceous earth, bentonite clay, and sawdust) have been used to improve dewatering, increase cake solids, and in some cases, reduce the required dosage of other conditioning agents.

In addition to increasing the volume of solids to be managed, inorganic chemical conditioners reduce the solids' heat value. However, cake combustibility depends on the ratio of water to dry volatile solids, not the level of chemical precipitates, in the cake.

4.1.1   Lime and its Characteristics

Lime is often used to control pH and improve settling in wastewater treatment processes, as well as condition and stabilize solids. Lime is commercially available in two main dry forms:

- Pebble quicklime (CaO) and
- Powdered hydrated lime [$Ca(OH)_2$]

In either form, lime is caustic, tends to produce dust, and tends to precipitate when slurried, forming a calcium carbonate scale on conveyance equipment.

As a conditioner, lime is typically used to raise the pH, which was lowered by ferric chloride addition. It also forms calcium carbonate and calcium hydroxide precipitates, which improve dewatering by acting like a bulking agent, increasing porosity while resisting compression. Some dissolved calcium hydroxide also is available at high pH levels.

Quicklime is typically 85% to 95% pure; it is typically called calcined lime because it is manufactured by burning crushed limestone (calcium carbonate) in high-temperature kilns to drive off carbon dioxide, leaving calcium oxide (quicklime). It is typically purchased in pebble form to minimize dust problems during handling. However, it rarely is applied in dry form, except to stabilize solids. Instead, it is typically mixed with water and converted to the more reactive hydrated form (calcium hydroxide) before application. This hydration reaction (typically called slaking) emits heat as part of the reaction:

$$CaO + H_2O \rightarrow Ca(OH)_2 + Heat \qquad (20.1)$$

The quicklime pebbles rupture during slaking, splitting into microparticles of hydrated lime, which have a large total surface area and are highly reactive. A high-grade quicklime produces a quick-slaking, highly reactive, calcium hydroxide slurry, while a low-grade quicklime produces a slow-slaking, less reactive slurry (see Table 20.1). Low-grade quicklime requires critical water control to maximize slaking efficiency and minimize calcium hydroxide particle size.

Quicklime must be stored under controlled conditions, because prolonged contact with carbon dioxide in moist air causes quicklime to air slake, cake, and become less reactive. Likewise, air or excessively hard water (alkalinity more than 180 mg/L as calcium carbonate) in a hydrate slurry encourages carbonate scaling, which can eventually lead to plugging problems in conveyance pumps and piping.

From a quality control standpoint, quicklime should be highly reactive, quick-slaking, and able to disintegrate without producing objectionable amounts of dissolved or unslaked products. Medium-slaking limes are not preferred. Low-slaking and run-of-kiln quicklimes are unacceptable.

Hydrated lime is a powdered form of calcium hydroxide; its composition and characteristics depend on the quality of its parent quicklime (see Table 20.2). Hydrated lime typically costs 30% more than quicklime with the same calcium oxide content because of its higher production and transportation costs. However, at small

CaO Content (%)	Degree of Burn	Slaking Ability
High	Soft	Very quick
	Normal	Medium
	Over to hard	Medium to slow
Medium	Soft	Quick to medium
	Normal	Medium
	Over to hard	Slow
Low	Soft	Quick to medium
	Normal	Medium
	Over to hard	Slow to very slow

TABLE 20.1 Relative Slaking Ability of Quicklimes (NLA, 1982)

Material	Available Forms	Containers and Requirements	Appearances and Properties	Weight	Commercial Strength	Solubility in Water
Quicklime	Pebble, 6–19 mm	80–100 lb moisture-proof bags, barrels, and container cars. Store dry, maximum 60 days in tight container and 3 months in moisture-proof bag.	White (light grey to tan). Lumps to powder. Unstable caustic irritant slakes to hydroxide slurry evolving heat. Saturated volume pH approximately 12.5.	3.4–4.7 kg/m^3 sp gr: 3.2–3.4	70%–96% CaO	Reacts to form $Ca(OH)_2$; 1 lb of quicklime will form 1.16 to 1.32 lb of $Ca(OH)_2$ with 2–12% grit, depending on the purity.
Hydrated lime	Powder, <200 mesh	50-lb bags, 100-lb barrels, and container cars. Store dry, maximum 1 year	White. Powder free of lumps. Caustic dust irritant absorbs H_2O and CO_2 to form $Ca(HCO_3)_2$. Saturated volume pH approximately 12.4.	1.6–2.5 kg/m^3 sp gr: 2.3–2.4	82%–98% $Ca(OH)_2$ 62%–74% CaO	10 lb/1000 gal at 70°F 5.6 lb/1000 gal at 175°F

*gal × 3.785 = L and lb × 0.4536 = kg.

TABLE 20.2 Characteristics of Quicklime and Hydrated Lime (Wang et al., 2007)*

facilities where daily requirements for lime are intermittent or minimal, hydrated lime often is preferred because it does not require slaking. The storage and mixing operations are relatively simple (e.g., typically a dedicated storage area and minimal manual labor). Hydrated lime is more stable than quicklime, so storage precautions are satisfied more easily. However, because of its dusting characteristics, handling is more difficult.

4.1.2 Ferric Salts

Both ferric chloride and ferric sulfate react with the bicarbonate alkalinity in solids to form ferric hydroxide precipitates. The precipitate can lead to both charge neutralization and floc aggregation. The chemical reaction may be written as follows:

$$Fe + 3H_2O \rightarrow Fe(OH)_3 + 3H \tag{20.2}$$

$$2FeCl_3 + 3Ca(HCO_3)_2 \rightarrow 2Fe(OH)_3 + 3CaCl_2 + 6CO_2 \tag{20.3}$$

Ferric-chloride coagulation is pH-sensitive; it works best above pH 6. Below pH 6, floc formation is weak and dewaterability is sometimes poor. So, lime is used to adjust the pH to optimize ferric chloride use and solids dewatering.

The acid formed during the reaction causes the pH to drop to 6.0. Adding lime raises the pH as high as pH 8.5, thus allowing the ferric chloride reaction to be more efficient in forming hydroxides. Lime also reacts with bicarbonate to form calcium carbonate, a granular structure that provides the porosity needed to increase the water-removal rate during pressure filtration. This chemical reaction is as follows:

$$Ca(OH)_2 + Ca(HCO_3)_2 \rightarrow 2CaCO_3^2 + 2H_2O \tag{20.4}$$

Depending on the type of solids involved, the ferric chloride dosage ranges from 2% to 10% (dry solids basis), and lime dosages range from 5% to 40% (dry solids basis). Activated sludge requires high ferric chloride dosages, anaerobically digested solids require mid-range dosages, and fresh raw primary solids require low dosages (see Table 20.3).

Typically used to flocculate solids, ferric chloride is sold as an orange-brown liquid—containing between 30% and 40% (by weight) ferric chloride. At 30°C (86°F) and a specific gravity of 1.39, a 30% ferric chloride solution typically contains 1.46 kg (3.24 lb) of ferric chloride. Liquid ferric chloride is corrosive, so it must be handled and stored properly. In colder climates, for example, the shipping strength is reduced to prevent a crystalline hydrate from forming on cold rail cars.

Liquid ferric sulfate typically is sold as a reddish-brown liquid—water containing 50% to 60% of ferric sulfate. It is a cationic coagulant and flocculant typically used with another conditioning agent (e.g., lime or polymers). It has been reported that using ferric sulfate before solids thickening or dewatering will reduce the amount of polymer needed and improve the filtrate or centrate quality. However, its use as a solids conditioner is limited; it is primarily used in water and wastewater treatment to remove turbidity, color, suspended solids, and phosphorus.

Ferrous sulfate ($FeSO_4 \cdot H_2O$, also called copperas) is similar to ferric chloride in terms of handling, storage, and stoichiometry. Ferrous sulfate is available in granular form in bags, barrels, and bulk. The product has a bulk density of about 1000 to 1100 kg/m³ (62–66 lb/cu ft). Dry ferrous sulfate will begin to cake when stored at

Application	Sludge Type	Ferric Chloride (g/kg)[a]	Lime (g/kg)[a]
Vacuum filter	Raw primary	20–40	70–90
	Raw WAS	60–90	0–140
	Raw (primary + TF[b])	20–40	80–110
	Raw (primary + WAS)	22–60	80–140
	Raw (primary + WAS + septic)	25–40	110–140
	Raw (primary + WAS + lime)	15–25	None
	Anaerobically digested primary	30–45	90–120
	Anaerobically digested (primary + TF)	40–60	110–160
	Anaerobically digested (primary + WAS)	30–60	140–190
Recessed-chamber filter press	Raw primary	40–60	100–130
	Raw WAS	60–90	180–230
	Anaerobically digested (primary + WAS)	40–90	100–270
	WAS + TF	40–60	270–360
	Anaerobically digested WAS	70	360
	Raw primary + TF + WAS	75	180

[a]All values shown are mass of either $FeCl_3$ or CaO per unit mass of dry solids pumped to the dewatering unit.
[b]Trickling filter.

TABLE 20.3   Typical Dosages of Ferric Chloride and Lime for Dewatering Wastewater Solids (U.S. EPA, 1979)

temperatures above 20°C (68°F) and will further oxidize and hydrate in moist and humid conditions. Ferrous sulfate should be stored in a dry area, and care should be taken to control dust, which can stain and also irritate skin, eyes, and the respiratory tract. Ferrous sulfate forms an acidic solution, so manufacturer precautions should be followed when storing, feeding, and transporting the material. Ferrous sulfate in granular (dry) form may be fed using gravimetric or volumetric feeding equipment; it also may be fed as a solution.

The effectiveness of ferric coagulation depends on pH and alkalinity. A lower pH favors the formation of positively charged hydroxoiron (III) complexes; a higher pH favors the solid species $Fe(OH)_3(s)$ (see Figure 20.1). Because hydroxoiron (III) complexes are effective coagulants, a lower pH should produce better results (see Figure 20.1). In a study conducted by Tenney et al. (1970), ferric iron was most effective between pH 5 and 8, which is near the pH of maximum precipitation (shown in Figure 20.2).

Alkalinity is important in ferric solids conditioning because it controls solids' pH during conditioning. The ferric ion functions as an acid, lowering pH, while alkalinity maintains the existing pH. For a given solid, the pH decreases as ferric doses increase.

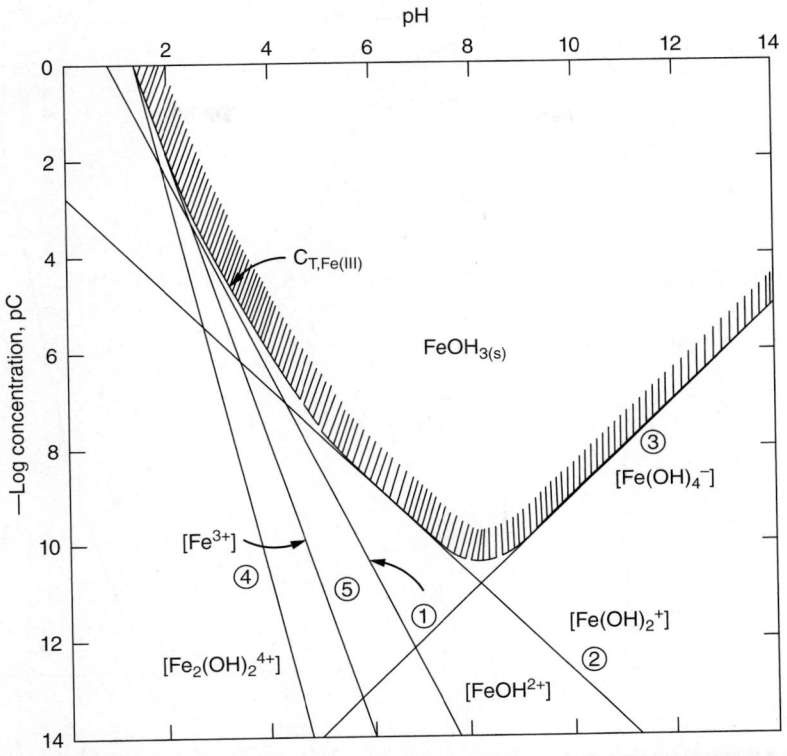

FIGURE 20.1   Equilibrium concentrations of hydroxoiron (III) complexes in a solution in contact with freshly precipitated $Fe(OH)_{3(s)}$ at 25°C (Snoeyink and Jenkins, 1980; reprinted with permission from Wiley & Sons, Inc.).

4.1.3   Ferric Salts with Lime

Precipitation of $Fe(OH)_3$ can neutralize charge and lead to effective aggregation and filtration in the pH range 6 to 8. Practically, however, the majority of wastewater solids cannot be adequately conditioned unless lime is added after the ferric salt (Christensen and Stulc, 1979). The iron neutralizes and precipitates organic constituents, but the lime creates a much more rigid framework of calcium carbonate, which provides a rigid shell around the organic material (Denneux-Mustin et al., 2001). Full-scale filtration involves much more pressure than that typically applied in laboratory tests (e.g., Figure 20.2), which is why lime is required at full scale. The key ingredients are a pH of 11 to 12, a high calcium ion concentration (10^{22} M), and the presence of solid ferric species (Christensen and Stulc, 1979). Conditioning with ferric salts and lime is not practiced in centrifugation because the solids cannot resist the imposed shear stresses but corrode and abrade metallic surfaces. Ferric should be added before lime at a separate addition point because adding ferric and lime to thickened solids together in the same tank (or in close proximity) adversely affects ferric conditioning (Christensen and Stulc, 1979; Webb, 1974). When conditioning with both ferric and lime, it typically takes two to four times more quicklime than ferric chloride to reach a pH between 11 and 12.

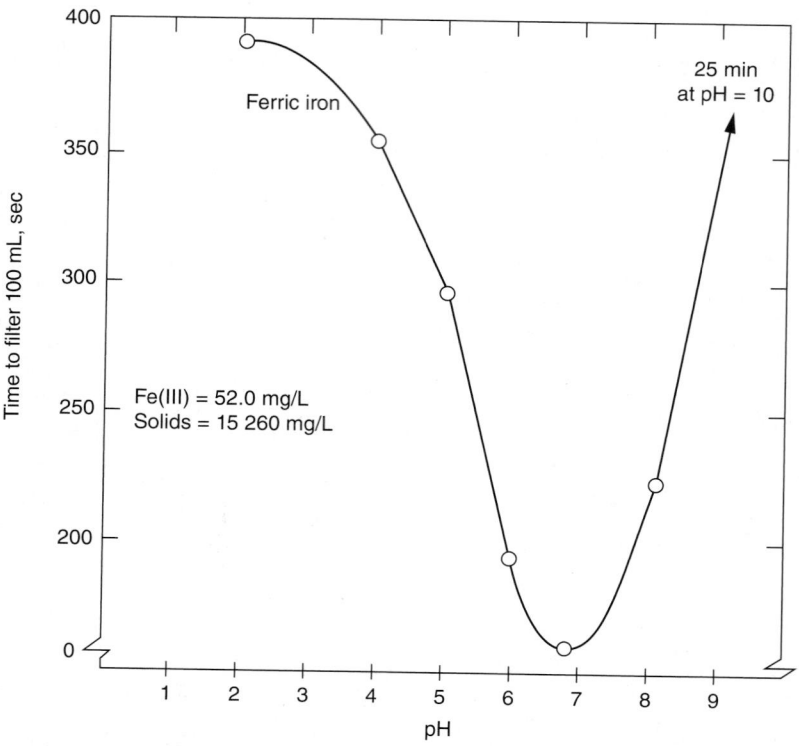

FIGURE 20.2   Effectiveness of ferric iron as a function of pH (Tenney et al., 1970).

The choice of ferric salt for conditioning is more significant when the ferric salt is followed by lime (see Table 20.4). Ferric sulfate followed by lime deteriorates more rapidly and produces a poorer result than ferric chloride followed by lime. The deterioration of solids dewaterability seems to be associated with the formation of insoluble calcium sulfate.

4.1.4 Aluminum Salts

Aluminum salts typically are not used for solids conditioning in the United States, although they have been used at some facilities with limited degrees of success. Coagulants such as polymerized aluminum chloride (PACl) and aluminum chlorydrate (ACH) that are widely used in the water treatment industry are also being used in the wastewater treatment industry for phosphorus removal and, to a limited degree, solids conditioning. While aluminum salts are not widely used as a solids conditioner in the United States, aluminum chlorohydrates have been a popular conditioner in Great Britain for some time.

The primary differences between aluminum and ferric chemistry are the relative solubility of aluminum above pH 7 and the relative insolubility of ferric above pH 7. The practical significance is that ferric hydroxide is relatively insoluble at the highest pH values used in ferric and lime conditioning (i.e., pH 12–12.5), while aluminum hydroxide is quite soluble above pH 10. Aluminum salts, therefore, are unlikely to be effective with the same lime doses often used with iron salts.

Sludge[a] Total Solids (%)	Iron Conditioner	Iron Dose (% Fe)	CST[b] after Iron (Seconds)	Lime Dose (% CaO)	Specific Resistance after Line (Tm/kg[c])
5.5	$FeSO_4 \cdot 7H_2O$	1.72	208	15	1.40
5.5	$FeCl_2 \cdot 4H_2O$	1.72	157	15	0.79
5.5	$Fe_2(SO_4)_3 \cdot 6H_2O$	1.72	41	15	0.50
5.5	$FeCl_3 \cdot 6H_2O$	1.72	26	15	0.26
5.5	$FeSO_4 \cdot 7H_2O$	3.44	180	30	0.60
5.5	$FeCl_2 \cdot 4H_2O$	3.44	139	30	0.29
5.5	$Fe_2Cl_4 \cdot 6H_2O$	3.44	27	30	0.23
5.5	$FeCl_3 \cdot 6H_2O$	3.44	19	30	0.12
7.0	$FeSO_4 \cdot 7H_2O$	3.44	480	20	1.1
7.0	$FeCl_2 \cdot 4H_2O$	3.44	—	20	0.56
7.0	$Fe_2(SO_4)_3 \cdot 6H_2O$	3.44	117	20	0.53
7.0	$FeCl_3 \cdot 6H_2O$	3.44	58	20	0.18

[a]Mixture of raw primary sludge and WAS.
[b]Capillary suction time.
[c]Terameters/kg (Tm/kg) = 10^{12} m/kg.

TABLE 20.4 Comparison of Iron Conditioners Used with and without Lime (Lewis and Gutschick, 1988; Reprinted with permission)

4.1.5 Process Design Considerations for Thickening and Dewatering

The following subsections describe the use of inorganic coagulants for thickening and dewatering. However, because the inorganic conditioners discussed in this chapter increase the total solids to be managed by about 20% to 40%, their use in thickening and dewatering applications is limited. Therefore, only one thickening and one dewatering application that may use inorganic chemicals are discussed.

4.1.5.1 Conditioning for Gravity Thickening Gravity thickening characteristics depend on the concentration and flocculant nature of the solids being thickened. In many cases, conditioning agents are not used; it depends on the type of solids being thickened. While polymers are the first choice if chemical conditioning is required, alum and ferric salts—with or without lime—could also be used (see Table 20.5).

The primary mechanism when using these inorganic chemicals is coagulation and flocculation. Efficient flocculation increases solids loading rates, improves solids capture, improves supernatant clarity, and may increase underflow concentrations from conventional gravity solids thickeners. When designing any thickener, engineers should determine, whenever possible, appropriate coagulants and their dosage rates by using bench-scale tests to evaluate the effectiveness of the conditioning agents during thickening operations.

4.1.5.2 Conditioning for Belt Filter Press Dewatering Inorganic chemicals typically are not used to condition solids before dewatering in a belt filter press, nor is it recommended

Solids	Nature of Solids/Dosage of Chemical			
	Raw		Anaerobically Digested	
	FeCl$_3$ (mg/L)	CaO (mg/L)	FeCl$_3$ (mg/L)	CaO (mg/L)
Primary	1–2	6–8	1.5–3.5	6–10
Primary + trickling filter	2–3	6–8	1.5–3.5	6–10
Primary + WAS*	1.5–2.5	7–9	1.5–4	6–12
WAS	4–6	No data	No data	No data

*WAS = waste activated sludge.

TABLE 20.5   Typical Chemical Dosages for Gravity Sludge Thickening (WEF, 1996)

because of chemical deposits that can "blind" the belt, as well as excessive wear on the rollers and belt (reducing the equipment's overall life expectancy). So, the amount of information available on using inorganic conditioners with belt filter presses is limited. Alum is sometimes used, and other inorganic chemicals (e.g., lime) may be important for chemical stabilization before land application.

As with other dewatering processes, the optimal dose depends on feed solids concentration and type, mixing intensity, and mixing time. The limited information available indicates that the chemical dosage required varies directly with the ratio of secondary to primary solids. At a secondary-to-primary ratio of 1:1, an approximate dosage of 5% for ferric chloride and 15% for lime can be expected. Doubling the secondary-to-primary ratio to 2:1 could double the required lime and ferric chloride dosages.

While using inorganic coagulants may be advantageous when routinely dewatering a widely varying solid, there will be a significant increase in solids to be disposed of and, therefore, increased hauling, handling, and use or disposal costs to consider. Also, design engineers should consider ventilating the belt filter press rooms because of the strong ammonia odor that may result from lime addition.

4.2   Organic Polymers

The organic chemicals used to condition solids are primarily long-chain, water-soluble, synthetic organic polymers. Polyacrylamide, the most widely used polymer, is formed by the polymerization of a monomer acrylamide. Polyacrylamide is non-ionic. To carry a negative or positive electrical charge in aqueous solution, the polyacrylamide must be combined with anionic or cationic monomers. Because most solids carry a negative charge, cationic polyacrylamide copolymers typically are the polymers most used to condition biological solids. Polymers are further categorized by the following characteristics: molecular weight (varies from 0.5–18 million), charge density (varies from 0%–100%), active solids levels (varies from 2–100), and form (e.g., dry, liquid or solution, emulsion, or gel).

High-molecular-weight, long-chain polymers are highly viscous in liquid form, extremely fragile, and difficult to mix into aqueous solution. Unmixed polymers in a diluted solution look like fish eyes. As the polymer's molecular weight increases, so does the difficulty in mixing and diluting it.

Unlike the inorganic chemicals discussed earlier, polymers have become attractive because they do not appreciably add to the volume of solids to be used or disposed of. Nor do they lower the fuel value of thickened or dewatered solids. Also, polymers are safer and easier to handle, and result in easier maintenance than inorganic chemicals, which require frequent cleaning of equipment, typically via acid baths. However, polymers are not completely stable, can plasticize at high temperatures, and are slippery when spilled on floors.

4.2.1 Properties of Organic Polymers

Polymers are classified by the polymer compound's charge (e.g., anionic, non-ionic, or cationic), molecular weight, and form (when received). A combination of the polymer molecule's charge and molecular weight is useful in product identification.

4.2.1.1 Polymer Charge To some extent, the chemical reactions for polymers and inorganic chemicals are similar (e.g., they neutralize surface charges and bridge particles). Neutralizing a particle's negative electrical charge via the polymer's positive charge reduces the electrostatic repulsion between particles and, therefore, encourages aggregation. In polymer bridging, a long-chain polymer molecule attaches itself, via adsorption, to two or more particles at once. Flocs formed by particle bridging tend to resist shear more than flocs formed by charge neutralization.

Charge is developed by ionizable organic constituents distributed throughout the polymer molecule. Measuring the charge of a specific polymer under field conditions is nearly impossible, so its relative charge (sometimes called the application charge) can be used to measure its charge capability.

For anionic and non-ionic polymers, the application charge does not change significantly because the usual levels of dissolved materials present do not overcome the ionic equilibrium among anionic-charged particles in the solids. For cationic polymers, charge neutralization brought about by the influence of water alkalinity counter-ions depletes the cationic-charged species of the polymer. This effect typically causes deteriorating charge levels over time. Some polymers seem to be more charge-stable than others; however, polymer-charge stability typically is a manufacturer trade secret.

Most polymer manufacturers use the phrase relative charge to describe the measured titratable charge level of their products under specified test conditions. So, comparative charge levels among different manufacturers may be practically meaningless, and users should be wary of claims relating to charge in applications without onsite testing under controlled conditions. Charge is not the sole governing criterion that determines a polymer's effectiveness in a given application.

4.2.1.2 Polymer Molecular Weight Conditioners typically can be categorized as low, intermediate, and high molecular weight. A polymer's molecular weight is a rough indication of the length of polymer chain that holds the charged sites apart. It also affects other product attributes (e.g., solubility, viscosity, and charge density in aqueous solution).

Although there are exceptions, lower-molecular-weight products tend to be more soluble, less viscous, and have higher charge density in water. Low-molecular-weight polymers are often called primary coagulants, a term typically reserved for products ranging from 2.0×10^4 to 1.0×10^5 (Kemmer and McCallion, 1979). These water-soluble products typically are marketed in concentrations of 30 to 50%. They have low viscosities (close to the viscosity of water) and can be easily diluted and mixed with water at the application point. These polymers are useful for clarification applications where there are many small dispersed particles to be destabilized and settled. They are typically in oily waste and biological waste treatment applications where low concentrations of solids are being treated. They also are sometimes used as the first part of a two-polymer program in which high-charge density is required to break the suspension.

Intermediate-molecular-weight products are available as solutions and in dry and liquid-emulsion forms. It is difficult to generalize about the entire class of intermediate-molecular-weight products; however, most require wetting (e.g., mixing activation to disperse the polymer) and aging to develop full-product activity in application. Solutions of intermediate-molecular-weight products are typically more viscous than lower-molecular-weight products. In fact, product handling of feeding characteristics typically limits commercial solutions of these products to 1% (dry solids basis) or less. Consequently, supplemental dilution water is typically needed to improve polymer disbursement in the solids being conditioned.

Intermediate-molecular-weight products are common in thickening and dewatering systems treating wastewater solids, especially those with high concentrations of secondary solids. Virtually all charge variations are available in the intermediate molecular weight range.

High-molecular-weight polymers can be cationic, anionic, and non-ionic, and are available as liquid viscous solutions, emulsions, or dry powder. Their molecular weights vary from 2×10^6 to more than 12×10^6. Solubility and viscosity considerations typically dictate the solution concentrations available. Product solutions are made up at 0.25% to 1.0% solids concentration and allowed to age for several hours before further dilution at the application point.

4.2.2   Polymer Cross-Linkage

A relatively recent development in polymer formulation is the use of controlled degrees of cross-linkage. Such polymers can be highly branched, rather than linear, and are called structured polymers. Larger doses of structured polymers may be needed to reach an optimum performance, but the resulting floc is stronger (Dentel, 2001). High-shear dewatering applications (e.g., centrifuges and some recessed-plate filter presses) can benefit from such flocculants. Some suppliers use such terms as XL, FS, FL, and FLX to indicate the cross-linked forms (Dentel et al., 2000a).

4.2.3   Polymer Forms and Structure

Polymers are available in two physical forms: dry and liquid. Dry polymers can be delivered in a microbead or gel powder form, while liquid polymers can be delivered as a solution or emulsion. All dry and liquid polymers can be prepared with three charge types—cationic, anionic, and non-ionic—and can be purchased in a wide array of molecular weights, charge densities, and active solids levels. The form, charge, and activity level of the polymer can greatly affect their reactivity with solids.

A polymer's "activity" relates to the percent of the molecular weight that is available to react with and flocculate solids particles; it can greatly vary with the form of the polymer. The polymer dosing criteria are stated in grams of active polymer per kilograms of dry solids. This method allows polymer types with different activity levels to be compared on an equivalent basis. For example, a polymer with an activity of 9% will require 10 times more grams of bulk polymer than a similar polymer with an activity of 90%.

4.2.3.1   Dry Polymers   Dry polymers can have an active solids level as high as 94% to 100%. The shelf life of dry polymers is typically 2 years. Storage areas that are susceptible to wet and humid conditions should be avoided, because dry polymers will tend to cake and deteriorate.

Most dry polymers are difficult to dissolve. To make up a working solution, an eductor is used as a pre-wetting device to disperse polymers in water. The solution is slowly mixed in a mixing tank until the dry polymer particles are dissolved, and then aged in accordance with manufacturer recommendations. Aging time typically ranges from 30 minutes to 2 hours. Aging allows polymer particles to "unfold" into long chains. However, once the dry polymer is diluted and converted into a solution, it is only stable for about 24 hours.

The quality of the water used to dissolve dry polymer particles is important. Hard water (greater than 120 mg/L as calcium carbonate) or water containing more than 0.5 mg/L of free chlorine can cause the solution to deteriorate within a few hours.

4.2.3.2 Emulsion Polymers Emulsions are dispersions of polymer particles in a hydrocarbon oil or light mineral oil. Surface active agents typically are applied to prevent the polymer–oil phase from separating from the water phase. Provisions must be made to mix the bulk storage tank regularly to prevent the oil and water from separating. With emulsions, it is possible to achieve a high molecular weight and maintain an active solids level of 30% to 50% without producing a solution that has a high viscosity. The approximate viscosity of emulsion polymer in its as-delivered state ranges from 300 to 5000 cP. The shelf life of emulsion polymers is typically 6 months to 1 year. The initial breaking of the emulsion and aging are critical for optimum performance, which can be accomplished with a static mixer, high-speed mixer, or wet dispersal unit.

Emulsion polymers can have higher molecular weights and higher charges than dry polymers without the operating problems. The primary disadvantages of emulsion polymers are the potential for oil and water separation and the higher cost per volume of active material.

One concern about these types of polymers is the adverse environmental impacts of the surfactants used in them. Such surfactants include alkylphenoethoxylates, which decompose to nonylphenol, a known endocrine disruptor. Recent developments in emulsion polymer manufacturing have been to abandon the use of mineral oils and surfactants for a new class of water-soluble emulsions. The process essentially involves dissolving the polymers in an aqueous salt of ammonium sulfate. A low-molecular-weight dispersant polymer is added to prevent aggregation of polymer chains.

Additionally, some of the typically used copolymers are susceptible to chemical hydrolysis at high pHs. If dewatered solids will be stabilized using an alkaline chemical (e.g., kiln dust, lime, etc.), odor problems could occur from the generation of trimethylamine, which has a "fishy" odor (Chang et al., 2005).

4.2.3.3 Mannich Polymers A Mannich polymer typically contains 3% to 8% active polymer; it is produced by using a formaldehyde catalyst to promote the chemical reaction to create the organic compound. Because vapors from formaldehyde pose a safety hazard and can be carcinogenic, Mannich polymers should be stored carefully and only used in well-ventilated areas. Mannich polymers are viscous (from 50 000 to more than 150 000 centipoise), difficult to pump, and have a relatively short shelf life. However, they can be effective and economical for large WRRFs, depending on the shipment cost.

4.2.4 Polymer Dosage
Various polymers can enhance the performance of thickening or dewatering processes. The dose needed depends on the specific process used and the solids or biosolids to be thickened (see Table 20.6).

Application	Sludge Type	Polymer Dosage (g/kg)
Gravity thickening	Raw primary	2–4
	Raw (primary + WAS + TF*)	0.8
	Raw WAS	4.3–5.6
Dissolved air flotation	WAS (oxygen)	5.4
	WAS	0–14
	P + TF	0–3
	P + WAS	0–14
Solid-bowl centrifuge	Raw WAS	0–3.6
	Anaerobically digested WAS	2–7.2
Rotary drum	WAS	6.8
Gravity belt	Digested secondary	5

*Trickling filter.

TABLE 20.6   Typical Dosages of Polymer for Thickening Wastewater Solids (U.S. EPA, 1979)

Most dewatering processes (except recessed-chamber filter presses) require polymer addition (see Table 20.7). (Recessed-chamber filter presses typically use ferric chloride and lime as solids conditioners, either alone or with fly ash or polymers, and typically produce a slightly thinner cake with polymer conditioning than with ferric chloride and lime conditioning.) Centrifuges and belt filter presses cannot achieve optimum dewatering performance without polymer addition. Both applications require polymers with high positive charge and high molecular weight to produce a strong and durable floc.

4.2.5 Application of Polymers

Because of the wide range of polymers now available, the performance of almost any conditioning or dewatering process can be enhanced by their use. Depending on the application, polymers may improve unit throughput, solids capture, filtrate quality, thickened or dewatered solids, or a combination of these parameters.

As with inorganic chemical conditioning, proper organic-chemical conditioning centers on four basic requirements:

- Correct dosage of polymer,
- Proper wetting, mixing, and aging of raw polymer,
- Correct solids and polymer mixing procedures, and
- Continuous observation of results and response to those observations.

Adhering to these requirements is more critical to conditioning performance when using polymers than when using inorganic chemicals. The polymers perform under a narrower range of operating conditions than inorganic agents do, so they are more sensitive to dosage and mixing. Although this sensitivity requires more operator attention, it can promote efficiency because the dewatering process will not work if conditioning is not closely controlled.

Application	Sludge Type	Polymer Dosage (g/kg)
Belt filter press	Raw (primary + WAS)	2–5
	WAS	4–9
	Anaerobically digested (primary + WAS)	6–10
	Anaerobically digested primary	4–7
	Raw primary	2–3
	Raw (primary + TF*)	3–6
Solid-bowl centrifuge	Raw primary	0.5–2.3
	Anaerobically digested primary	2.7–5
	Raw WAS	5–10
	Anaerobically digested WAS	1.4–2.7
	Raw (primary + WAS)	2–7
	Anaerobically digested (primary + WAS + TF)	5.4–6.8
Vacuum filter	Raw primary	1–5
	Raw WAS	6.8–14
	Raw (primary + WAS)	5–8.6
	Anaerobically digested primary	6–13
	Anaerobically digested (primary + WAS)	1.4–7.7
Recessed chamber filter press	Raw (primary + WAS)	2–2.7

*Trickling filter.

TABLE 20.7 Typical Dosages of Polymer for Dewatering Solids (U.S. EPA, 1979; WPCF, 1983)

4.2.5.1 Dosage The correct chemical dosage is critical to proper operation. Chemical conditioning tests [e.g., Buchner funnel or capillary suction time (CST)] should be conducted frequently to determine conditioning requirements. Maintaining the correct dosage requires knowledge and control of the solids stream and chemical feed(s). Continuous metering equipment should be used and the solids content monitored to determine the mass flow. Chemical mass feed rates also should be continuously monitored and controlled to maintain desired dosages. Flow-measuring devices should be used to determine the chemical volume being fed. The consistency of chemical concentration also should be monitored and maintained throughout the process. Although feed-solution concentrations can be measured using total solids or viscosity measurements, an accurate flow-metering system on the feed solution makeup system is the preferred option.

Polymers should be used at specific solution strengths based on the manufacturer's recommendation and the results of any laboratory tests performed. Dilute solutions may be required because of the polymer's chemical activity or to allow good contact of relatively small chemical quantities with large solids volumes. In some cases, poor quality dilution water (e.g., secondary effluent) affects polymer solution activity. Although this is not a common concern, high-quality water typically should be used to prepare feed solutions.

4.2.5.2 Mixing Procedure   Proper mixing is critical to conditioning. It has two primary components: intensity and duration. Getting a viscous material thoroughly dispersed in the solids is of utmost importance when using polymers. Supplemental dilution water is used to reduce the polymer's viscosity, and high-intensity, short-duration mixing is needed to disperse the polymer. This high-intensity mixing is often accomplished with a polymer-injection ring and an adjustable check-valve device that delivers a high shearing action. Some facilities have also had success injecting polymer immediately upstream of the sludge feed pumps to achieve high energy mixing; however, this can result in other complications if not done properly. The flocculation phase requires 15 to 45 seconds of gentle agitation to allow the chemical reaction between the polymer and solids to occur.

Mixing duration typically is on the order of 15 to 60 seconds, and can be accomplished a number of ways (e.g., in the pipeline to the equipment or in a separate flocculation tank). For example, multiple addition points located at HRTs of 15, 30, and 45 seconds from the inlet to the thickening or dewatering equipment. At each addition point in the solids feed pipe, a flexible coupling and a polymer pipe drawoff point should be provided to allow the insertion of a polymer-injection ring and an in-line, high-intensity mixing unit (see Figure 20.3). This arrangement provides the most flexibility in allowing operators to fine-tune the process feed rate.

Another method of providing HRT is to provide a flocculating tank directly upstream of the dewatering unit. This tank provides 15 to 30 seconds of flocculating time based on the design loading rate. A disadvantage of the flocculant tank is that it can create dead zones, which could result in improper conditioning.

4.2.5.3 Process Monitoring and Control   To ensure that thickening or dewatering performance is optimal, both processes should be monitored and frequently tested. The thickened or dewatered cake's total solids and released water should be analyzed at least once per shift to monitor solids loading and polymer performance. Although one sample per shift is sufficient, taking composite samples each shift would allow an operator to detect any operational changes and potential equipment problems.

Figure 20.3   Polymer injection ring and inline high-intensity mixing unit (courtesy of the City of Orlando, Florida).

Polymer monitoring includes performance and quality control checks. Polymer use should be recorded during each shift via drawdown readings in bulk or solution-storage tanks, timers on polymer-transfer pumps, or flow meters on polymer feed lines. To confirm solution strengths, total solids tests should be conducted regularly. From these data, product performance or process variations may be detected.

There has been much technological development on a variety of automated process sensors, controllers, and related software for managing thickening and dewatering operations to optimize performance and polymer use. These systems typically include a solids probe in the drain line from the dewatering unit's filtrate line and a controller on the solids feed pump and polymer feed equipment. Typically, when the solids probe detects a sudden increase in filtrate solids concentration, the controller first tries to increase the polymer flow rate. If this does not clear the filtrate after a specific time period, the controller sends a signal to decrease the solids feed rate. These alternate steps are repeated until the filtrate clears. Other control systems monitor the viscosity or particle charge in filtrate. The particle charge measurement informs the control system whether to increase or decrease the polymer dose (Dentel et al., 2000a). The primary benefit of such automated systems is that they tend to smooth out variations in unit operations and decrease polymer use by 20% to 50%. Systems that control dewatering based on filtrate or centrate properties typically will not provide optimal cake solids (Abu-Orf and Dentel, 1999). So, if cake dryness significantly affects handling costs, a control system should not be based solely on polymer savings.

Several publications provide detailed information on automating thickening and dewatering operations that use polymers as conditioning agents (Gillette and Scott, 2001; Pramanik, et al., 2002; WERF, 1995, 2001). For example, liquid stream current monitors have been studied and show promise to continuously optimize a facility's chemical conditioning requirements. One study concluded that streaming current detectors (SCD) using different conditioning agents to dewater undigested solids and biosolids are suitable for monitoring and optimizing chemical conditioning requirements (Abu-Orf and Dentel, 1997).

4.3 Process Design Considerations for Thickening and Dewatering

Below are summaries of design considerations for several types of thickening and dewatering processes. Typical polymer dosages for thickening and dewatering applications are provided in Tables 20.8 and 20.9, respectively.

4.3.1 Conditioning for Gravity Thickening

Conventional gravity thickening typically does not require the use of organic polymers. That said, using these chemicals increases solids and hydraulic loading rates by a factor of two to four and improves solids capture. However, they have minimal effect on the resultant underflow solids concentration. Also, using polymer increases the overall cost of gravity thickening, so it only should be used to prevent operating problems caused by solids carryover.

Bench tests and other laboratory or field investigations should be performed to test the relative effectiveness of flocculating aids (either alone or in combinations). Also, care should be taken during feeding and mixing to prevent the overfeeding or poor mixing that causes "islands" to form.

Using about 2 to 4.5 g of active polymer/kg of dry solids (4–9 lb/ton) can produce a solids-loading rate of about 22 to 34 kg/m² d (4.5–7.0 lb/d/sq ft) when thickening

Facility Name	Type of Sludge	Influent Feed Solids (%)	Method of Thickening	Polymer Dosage (lb/ton)	Thickened Solids (%)	Method of Stabilization
Water Reclamation Facility, Bend, Oregon	Secondary (100%)	0.4–0.6	Gravity belt thickening	8–10	4–5	Anaerobic digestion
Irwin Creek WRRF Charlotte, North Carolina	Primary (60%) Secondary (40%)	3.2 0.5–1.0	Conventional gravity thickening	0	4–5	Anaerobic digestion
City of Greeley WPCF Greeley, Colorado	Secondary (100%)	0.6–0.8	Centrifuge	1.7–2.5	4.5–6.5	Anaerobic digestion
JEA Buckman Street WRRF, Jacksonville, Florida	Primary (60%) Secondary (40%)	1–3	Gravity belt thickening	9–12	3–5	Anaerobic digestion
Greenfield Road WRP Mesa, Gilbert and Queen Creek, Arizona	Primary and secondary	1–1.25	Centrifuge	0.5	5	Anaerobic digestion
Iron Bridge Water Reclamation Facility, Orlando, Florida	Secondary (100%)	0.4–0.6	Gravity belt thickening	7–9	2–3	Lime stabilization
Northwest Water Reclamation Facility, Orange County Utilities, Orlando, Florida	Secondary (100%)	0.7–0.8	Conventional gravity thickening	0	1–2	Contract lime stabilization
Water Conserv I, Orlando, Florida	Secondary (100%)	0.7–1.2	Gravity belt thickening	10–12	4–5	Transported to larger facility for further stabilization

Facility	Solids type	Dosage	Thickening process			Stabilization
Water Conserv II, Orlando, Florida	Secondary (100%)	0.4–0.8	Gravity belt thickening	10–12	2–4	Anaerobic digestion
Orange County Utilities, South Water Reclamation Facility, Orlando, Florida	Secondary (100%)	0.4–1.25	Gravity belt thickening	2.4–6.5	3.1–6.4	Anaerobic digestion
91st Avenue Wastewater Treatment Plant, Phoenix, Arizona	Secondary (100%)	1.0–1.5	Centrifuge	2.0–3.0	5.7–6.2	Anaerobic digestion
McAlpine Creek WRRF, Pineville, North Carolina	Primary (50%) Secondary (50%)	1.0	Conventional gravity thickening	0	3–5	Anaerobic digestion
			Centrifuge	3–5	0.9–1	
Roger Road Wastewater Reclamation Facility	Secondary	0.2–0.4	Gravity belt thickening	9–10	5–6	Anaerobic digestion
Pima County, Tucson, Arizona	Primary	0.4–0.6	Conventional gravity thickening	0	4–5	Anaerobic digestion
Columbia Boulevard WRRF, Portland, Oregon	Secondary (100%)	0.5–1.0	Gravity belt thickening	5–9	4–6	Anaerobic digestion

WRRF = Water resource recovery facility.

TABLE 20.8 Polymer Dosages Associated With Various Solids Thickening Processes

Facility Name/Location	Type of Sludge	Type of Stabilization	Influent Feed Solids (%)	Method of Dewatering	Polymer Dosage (g/kg)	Dewatered Cake Solids (%)
Water Reclamation Facility Bend, Oregon	Primary (55%) Secondary (45%)	Anaerobic digestion	1.8–2.1	Belt filter press	3.75–6	12–15
Irwin Creek WRRF Charlotte, North Carolina	Primary (60%) Secondary (40%)	Anaerobic digestion	1.4	Belt press	3.5–4	18.75
City of Greeley WPCF Greeley, Colorado	Primary (60%) Secondary (40%)	Anaerobic digestion	1.5–2.0	Centrifuge	5–8	19–22
Buckman Street WRRF, Jacksonville, Florida	Primary (60%) Secondary (40%)	Anaerobic digestion	2–4	Centrifuge	3.75–6.25	19–22
Greenfield Road WRP Mesa, Gilbert and Queen Creek, Arizona	Primary and Secondary	Anaerobic digestion	2.75–3.0	Centrifuge	5.75	22–23
Water Conserv II Orlando, Florida	Secondary	Anaerobic digestion	2.0–2.5	Belt filter presses	3.5–4	12
Eastern Water Reclamation Facility Orange County Utilities Orlando, Florida	Secondary (100%)	Contract lime stabilization	<1	Belt filter presses (three belt)	3–3.5	16–17
Iron Bridge Water Reclamation Facility Orlando, Florida	Secondary	Lime stabilization	2.0–3.0	Belt filter presses	3.5–4	17
Iron Bridge Water Reclamation Facility Orlando, Florida	Secondary	Lime stabilization	0.4–0.6	Belt filter presses (three belt)	3.5–4	17
Northwest Water Reclamation Facility Orange County Utilities Orlando, Florida	Secondary (100%)	Contract lime stabilization	1–2	Belt filter presses	3.75–6.25	14–16

Facility						
South Water Reclamation Facility Orange County Utilities Orlando, Florida	Secondary	Anaerobic digestion	2.3–3.7	Belt filter presses	0.5–1.1	9.3–19.7
McAlpine Creek WRRF Pineville, North Carolina	ORC reported a 1:1 ratio of primary to secondary solids	Anaerobic digestion	2.4	Centrifuge	3.75	20
Columbia Boulevard WRRF Portland, Oregon	Freshly digested (65–75%) thickened WAS (20%) Primary (80%) and previously digested, lagoon stabilized thickened WAS (25–35%)	Anaerobic digestion	1.5–2.0	Belt filter presses	3.75–5	19–22
Thomas P Smith Water Reclamation Facility Tallahassee, Florida	Secondary (100%)	Anaerobic digestion	2.81–4.39	Screw presses	3.75–6.75	13–20

Notes:

WAS = waste activated sludge;

WPCF = water pollution control facility;

WRP = water reclamation plant;

WRRF = water resource recovery facility.

TABLE 20.9 Polymer Dosages Associated With Various Solids Dewatering Processes

primary solids. Adding about 4.5 to 6.0 g/kg (9.5–12.5 lb/ton) of polymer can increase the thickener's loading rate to 12 to 16 kg/m² d (2.4–3.2 lb/d/sq ft) when treating WAS (Ettlich et al., 1978; U.S. EPA, 1978c, 1979d).

4.3.2   Conditioning for Dissolved Air Flotation Thickening

Chemical conditioning is unnecessary for dissolved air flotation (DAF) thickening if low hydraulic and solids-loading rates are used. However, if high loading rates are required, or compaction is poor and the sludge volume index is high, chemical conditioning improves solids capture and can increase the float solids concentration. Although the increase in float solids is typically small (in the range of 0.5%), polymers may be required for WAS if a 4% float solids concentration is to be achieved. Float solids can routinely be 6% or higher when co-thickening mixtures of primary solids and WAS. Unless problems exist, solids capture without polymers is typically about 95%. With polymers, solids capture can increase to 97 or 98%, thereby improving subnatant quality and lessening the effect of recycle solids on facility performance. Also, with polymer addition, it is possible to as much as double the solids loading rate. [A typical rate is 10 kg/m²/h (2 lb/sq ft/hr).]

Typically, a cationic polymer with a moderate charge and high molecular weight is used; however, lower-charge cationic polymers are starting to show better performance. Typical dosages are from between 2 and 5 g/kg (4 and 10 lb/ton) up to 7.5 g/kg (15 lb/ton) of dry solids.

A common problem when conditioning solids for DAF thickening is improper mixing of conditioner and solids. To mitigate this problem, a more dilute polymer solution (0.25–0.5%) should be used, or the flocculant should be mixed with pressurized recycle before contacting the solids (Ettlich et al., 1978; U.S. EPA, 1978c).

4.3.3   Conditioning for Centrifugal Thickening

A solid-bowl conveyor centrifuge has been used to thicken a wide variety of solids. Centrifugal thickening typically does not require polymer addition when treating biological and aerobically stabilized solids. Well-digested solids, however, have little natural flocculating tendency, and require polymer additions to achieve acceptable solids recovery levels. So, engineers should make provisions for polymer addition in the initial design, even if chemical conditioning is not planned. The design should be flexible enough to allow conditioning chemical to be added at one of several points in the influent piping.

Dry or liquid high-molecular-weight cationic polymers are effective thickeners. When dry polymers are used, a 0.05% to 0.1% feed solution is used, while liquid polymers can range in concentration up to 0.5% on an active basis. It is important that a solids capture of at least 95% is obtained to prevent recycling filamentous bacteria and fines to the wastewater treatment process. Waste activated sludge produces a weak floc that tends to shear inside the centrifuge; a dose of up to 4 g/kg (8 lb/ton) of polymer can be used to formulate a tougher floc. Aerobically and anaerobically digested solids have little natural floc and, therefore, require about 4 to 8 g/kg (8–16 lb/ton) of polymer.

4.3.4   Conditioning for Gravity Belt Thickening

Gravity belt thickening works well with many types of solids. Difficult-to-thicken solids only require minor modifications of polymer dosages and solids loading rates to keep the effluent solids concentration and percent solids capture high. Gravity belt thickeners have been used to treat solids containing as little as 0.4% solids or as much as 10% solids with polymer addition. Polymer dosages range from between 1.5 and 3 g/kg (3 and 6 lb/ton) (dry weight basis) for raw primary solids up to between 4 and 6 g/kg (8 and 12 lb/ton) for anaerobically stabilized solids. In all cases, solids capture remained above 95%.

4.3.5   Conditioning for Rotary Drum Thickening

Rotary drum thickeners work much like gravity belt thickeners: in both systems, a moving, porous media retains conditioned solids while allowing free water to drain through. A polymer is injected into the feed line and mixed with incoming solids before entering the flocculation tank.

Drum speed, mixer speed, and spray water cycling is adjustable to ensure maximum performance with minimal polymer and water use. Polymer requirements are about 10% to 20% greater than those associated with gravity belt thickeners. Rotary drum thickeners are suited for high-fiber solids, as well as raw and digested solids with a significant fraction of primary solids. Their success with municipal WAS is variable and depends on solids characteristics. Residuals typically can be thickened to 5% to 7% total solids (in some cases, more than 10% total solids) with up to 99% capture of feed solids at polymer dosages ranging from 4 to 6 g/kg (8–12 lb/ton) (dry weight basis).

4.3.6   Conditioning for Centrifugal Dewatering

Polymers have been used with solid-bowl conveyor centrifuges to increase machine throughput without lowering cake dryness, to improve solids recovery, or both. Typically, a moderate-to-high charge, high-molecular-weight cationic polymer is used. Pilot studies are needed to determine the correct conditioning agent and dosage. Designs should include facilities for feeding both dry and liquid polymers. Polymer use typically increases solids capture; however, too much polymer can lead to a wetter cake because more fines are captured. Therefore, the relationship between recycled solids and cake dryness determines the dosage of polymer to be used.

4.3.7   Conditioning for Belt Filter Press Dewatering

The performance of belt filter presses depends on proper conditioning, and organic polymers traditionally are used. A properly conditioned product has a 95% to 98% solids recovery rate. However, the quantity of polymer required for proper conditioning varies widely; it depends on solids type, solids concentration, and ash content (typically, less polymer is needed when the ash content is high). For example, primary solids require a polymer dose of 3.5 to 5 g/kg (7–10 lb/ton), anaerobically digested solids require a dose of 7 g/kg (14 lb/ton on a dry weight basis), and autothermal thermophilic aerobically digested solids require a dose between 18 to 23 g/kg (37–47 lb/ton).

Insufficient conditioning causes inadequate dewatering in the initial sections of the press, which, in turn, can cause solids to extrude from the press section, overflow in the drainage section, or blind the belt. Over-conditioning can cause belt blinding and over-flocculation, which causes solids to drain too fast and mound on the belt, resulting in poor dewatering. The goal is to remove as much water as possible in the gravity section of the press. Over-flocculation may be mitigated by using deflection plates to even out the mounds before pressing begins, or by selecting a belt filter press with an extended gravity table.

Because of the shearing action between belts, Novak and Haugan (1980) have suggested using turbulent mixing when adding polymers for conditioning before dewatering. The best dosage and overall system performance depend on solids concentration, mixing intensity, and mixing time.

4.3.8   Conditioning for Screw Press Dewatering

The screw press is a simple, slow-moving mechanical device that gradually compresses conditioned, thickened solids as they move through the unit. Dewatering is continuous; it begins with gravity drainage at the inlet end of the screw and then dewatered the result of increasing pressure at the end of the unit. Proper screw design is critical, because different solids require different polymer dosages, screw speeds, and configurations to maintain a desired dewatered cake concentration and solids capture rate.

Proper solids conditioning is essential to produce a consistent dewatered cake. Slower operations will produce a dryer cake but also will reduce solids throughput. Therefore, it is important that a relationship between polymer dosage, solids throughput, and cake dryness be established. Depending on the influent solids characteristics, the polymer dosage may range from 8 to 12 g/kg (16–24 lb/ton) to produce a cake containing between 12% and 25% dry solids and a solids capture rate of 90% to 95%.

4.3.9   Conditioning for Rotary Press Dewatering

The polymer dosage for a rotary press depends on the type of solids to be dewatered. Work performed at the Daniels and Plum Island facilities in Charleston, South Carolina, indicated that 4.5 to 6 g/kg (9–12 lb/ton) of polymer was needed for a mixture of raw primary and secondary solids, resulting in an average dewatered cake concentration of 25%. However, in St. Petersburg, Florida, polymer dosages ranging from 15 to 18 g/kg (31–37 lb/ton) were required to dewater aerobically digested secondary solids to an average of 15% solids. In both cases, solids capture was more than 95%.

4.3.10   Conditioning for Drying Beds

Conditioning solids before sending them to drying beds is not widely practiced. In fact, the "Ten State Standards" (Great Lakes, 2004) and other design guidelines do not consider using conditioning chemicals in this application, even though they may significantly reduce drying time and, therefore, the bed area required. Relatively small polymer doses (as little as 50 mg/L) can considerably improve the drainage capabilities of properly digested solids by flocculating smaller particles. Flocculation speeds up the drainage period of the dewatering cycle and maintains a porous cake that is more readily susceptible to evaporation.

Studies have indicated that conditioning significantly increases the loading rate for digested primary solids and WAS; the unconditioned solids-loading rate was 73 kg/m^2/yr (15 lb/sq ft/yr), while the conditioned solids-loading rate was 270 kg/m^2·yr (55 lb/sq ft/yr). A well-conditioned product will dry in about one-third the time (approximately 10 to 15 days) required for unconditioned solids. Such performance improvements typically result from a dosage of about 15 to 23 g/kg (31–46 lb/ton) of a cationic polymer with a moderately high or high charge and a high molecular weight.

The conditioning system must be designed to avoid rupturing the conditioned floc during transport to the drying beds. Rupturing typically occurs during pumping; it can be overcome by locating the flocculation tank close to the drying beds, and allowing conditioned solids to flow by gravity from the flocculating chamber. Excessive holding increases the percentage of fines, which will impair flocculation and dewatering.

Both wedge-wire and vacuum-assisted drying beds use polymers to coagulate fines and promote rapid cake formation. The polymer is injected into the solids in the inlet line or in a flocculation tank next to the bed. Typical doses are between 1.5 and 3 g/kg (3 and 6 lb/ton).

5.0   Chemical Storage and Feed Equipment

Critical to any thickening or dewatering design is the decision of what chemicals are to be used, how they are shipped and stored, and what type of feed equipment should be used. For details on designing chemical conditioner-handling facilities, see Chapter 7, which includes a discussion on sizing the various unit operations and processes, as well as the necessary appurtenances. Because many of the chemicals are corrosive and available in various forms (e.g., liquid, dry, and gel), design engineers need to pay special attention to the design of chemical storage, feeding, piping, and control systems. For example, dry conditioners are typically converted to solution or slurry form before being introduced to solids. Liquid chemicals typically are delivered in a concentrated form and must be diluted before being mixed with solids. Other issues that must be considered when designing these systems are local building codes and the need to maintain operations during natural disasters (e.g., earthquakes, floods, hurricanes, and tornados).

As noted in Chapter 7, the sizing of storage facilities begins with an investigation of the chemicals to be used and their dosage requirements. Many conditions must be evaluated to determine the appropriate range of feed rates, which determine the feed-equipment capacities for each chemical. However, most facilities are limited by subsequent thickening or dewatering equipment capabilities, number of shifts (operating times), and desired final product.

5.1   Inorganic Chemicals

Because ferric chloride and lime have different chemical characteristics, they require different storage, pumping, piping, and handling procedures. The most important consideration when designing facilities for both chemicals is providing enough flexibility to accommodate variations in solids characteristics.

5.1.1   Ferric Chloride

Ferric chloride is corrosive and can be delivered in either liquid or dry form. Liquid ferric chloride is dark brown and has a shipping weight of 1.3 to 1.5 kg/L (11.2 to 12.4 lb/gal) for a 35% to 45% solution. Dry ferric chloride shipments should be stored in a dry room. Once opened, the chemical immediately should be used or mixed with water and stored in solution.

Storage tanks for ferric chloride typically are made of fiberglass, rubber- or plastic-lined steel, polypropylene, or spiral-wound extruded high-density polyethylene. Storage tanks must be insulated and, if holding a 45% solution, heated when the ambient temperature is expected to fall below 16°C (60°F).

Liquid ferric chloride feed equipment includes transfer pumps, day tanks, and metering pumps (see Figure 20.4). Rubber- or plastic-lined, self-priming centrifugal transfer pumps are used to convey bulk solution from storage tanks to day tanks. Double-diaphragm metering pumps or peristaltic pumps are used to control the chemical feed rate at the application points. Chemical feed rates typically are paced according

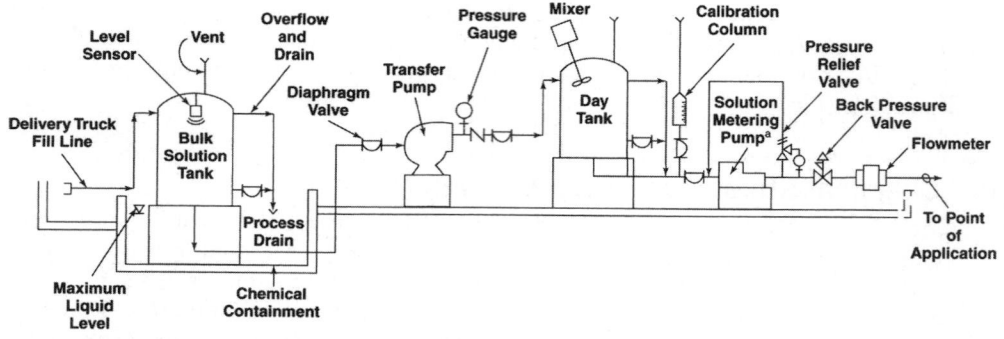

FIGURE 20.4   Simplified polymer-solution feed system (provide pressure relief on the discharge side of all positive-displacement polymer pumps).

to solids feed rates. Dilution water should not be added because of the potential for hydrolysis. Aboveground piping and valves typically are made of polyvinyl chloride (PVC), and rubber- or plastic-lined steel is used for buried applications.

The choice of feed bulk solution (30%–45%) or diluted solution (20%) typically depends on total ferric chloride use and the expected ambient temperature. If this temperature is below the bulk solution's freezing temperature, feed facilities should be insulated and heat traced. Diluting a bulk solution may lower its freezing temperature below the lowest expected ambient temperature (thereby avoiding insulation and heat tracing), but it increases the size of day tanks, piping, valves, and feed pumps.

5.1.2   Lime

Large facilities use pebble quicklime, while small ones use hydrated lime. For lime-application rates in excess of 1800 to 2700 kg/d (2–3 ton/d), bulk quicklime is typically more economical than hydrated lime. Bagged lime requires a waterproof, well-ventilated storage building; bulk lime requires watertight and airtight storage bins. Bagged lime should be stored on pallets in a dry place for no longer than 60 days. Bulk lime can be pneumatically transferred in bins or conveyed to the bins via conventional bucket elevators or screw conveyors.

5.1.2.1   Lime Silos   Quicklime bins typically have a 55° to 60° slope to the bin outlet; hydrated lime bins have a 60° to 66° slope. Tall, slender structures with a height-to-diameter ratio (H:D) ratio of 4:2.5 are preferred. The design volume should be based on the average bulk density of the chemical, with an allowance for 50% to 100% extra capacity beyond that required to accommodate a typical delivery. Quicklime and hydrated limes are abrasive, but not corrosive, so steel or concrete bins can be used. It is imperative that the storage bins be airtight and watertight to prevent the effect of air slaking.

Hydrated lime bins should be equipped with bin agitation and a non-flooding rotary feeder at the bin outlet.

Other required appurtenances include air-relief valves, access hatches, and a dust-collector mechanism.

5.1.2.2   Lime Feed System   A typical lime storage and feed system is illustrated in Figure 20.5. Bulk quicklime typically is fed to a slaking device, where oxides are converted to hydroxides, producing a paste or slurry that is further diluted before being

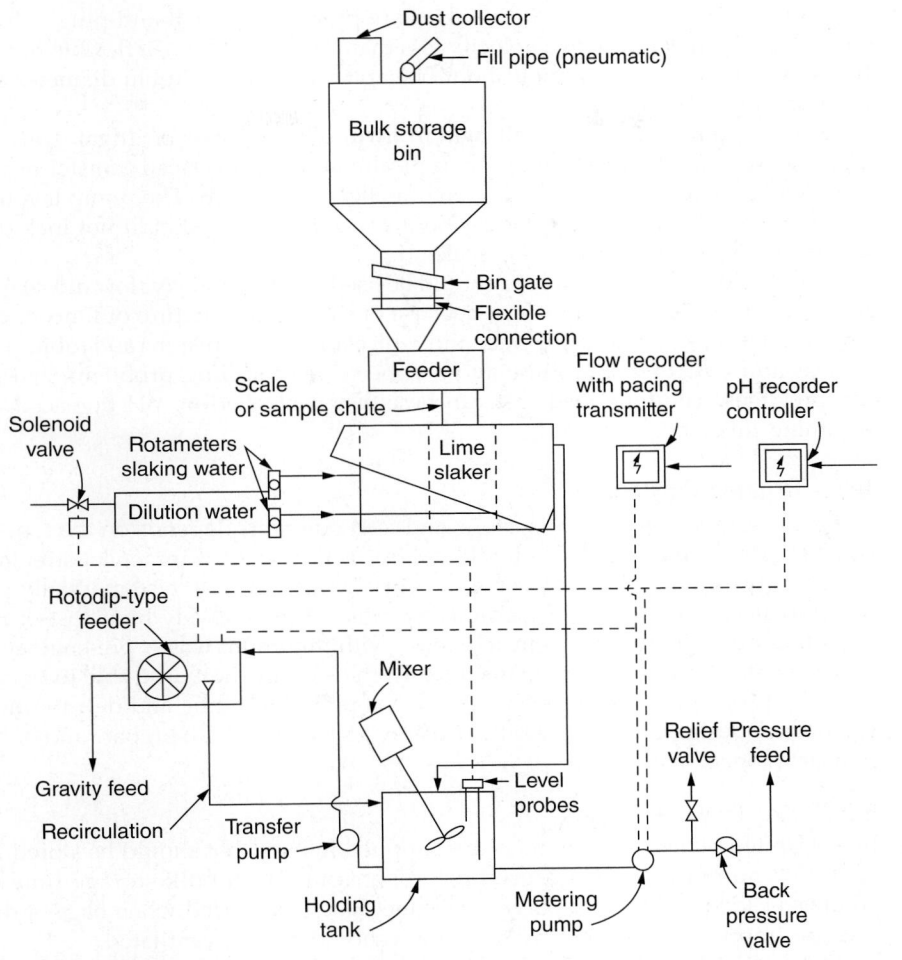

FIGURE 20.5   Typical lime feed system (vapor remover not shown).

piped or pumped to the application points. There are several manufacturers of suitable dry feeders; the choice depends on facility capacity and the degree of accuracy desired. For example, if a gravimetric-type feeder and slaker combination is indicated, larger facilities can use pebble lime because gravimetric feeders are the most accurate (0.5%–1% of set rate). This translates into cost reductions in large operations. Large, medium, and small facilities also will find volumetric feeders satisfactory, with their accuracy range of 1% to 5%.

A reasonably pure lime slurry is not corrosive and is relatively easy to keep in suspension, provided that it has been stabilized once all chemical reactions between the water and quicklime were completed. The suggested method for transferring slurry is via gravity and open trough, as long as the slurry is stabilized. If piping and transfer pumping cannot be avoided, the feed loop should be designed with a minimum velocity of 0.9 to 1.5 m/s (3–5 ft/s). Pinch valves are preferable to ball-and-plug valves. For a short transfer distance with a velocity less than 0.9 m/s (3 ft/s), a flexible fire hose can be used. In general, feed piping should be at least 50 mm (2 in.) in diameter and have minimal turns and bends.

Slurry pumps typically fall within two categories: centrifugal and positive-displacement. Centrifugal pumps are typically used for low-head transfer or recirculation. Replaceable liners and semi-open impellers are desired. The pump layout should provide for easy dismantling for cleanout and repairs, and should not include water-flushed seals because they tend to scale.

Positive-displacement pumps should be used when the slurry flow must be metered or positively controlled. However, because of the abrasive nature of lime slurry, these pumps are subject to excessive wear and replacement (e.g., pistons and tubing). Turbine pumps and eductors should be avoided because of scaling problems that occur in the pipelines. The lime feed rate can be either controlled by pH or paced with the incoming solids flow.

5.2   Organic Polymers

The feed system needed to mix, store, and feed polymers depends on the type of polymer to be delivered (e.g., dry or liquid). Many facilities feed commercial-strength liquid polymer direct from shipping containers or storage tanks, or else manually prepared dry polymer solutions from batch mixing tanks. The relatively high cost of chemical conditioning requires maximum activation with minimum waste; pre-engineered feed systems can accomplish both goals. Ideally, the system should be able to handle both dry/emulsion and solution polymers. Also, if both thickening and dewatering will be performed, design engineers should consider a system that can prepare and deliver two products concurrently.

5.2.1   Dry Polymer Feeders

In the United States, dry polymers are supplied in bags that should be stored in a dry, cool, low-humidity area and used in proper rotation. A bulk storage time of 15 to 30 days is adequate for dry polymers. Some dust is produced when bags of dry polymer are emptied, so polymer-makeup areas should be well ventilated.

Batch-mixing and solution feed equipment consists of a dry storage hopper, dispenser and conveyor (pneumatic or hydraulic), dust collector, mix tank and agitator, aging tank, flow control valves and polymer metering feed pumps. A typical batch mixing for dry polymer is presented in Figure 20.6. Figure 20.7 presents the dry polymer mixing system at the City of Knoxville WRRF. The system can be semiautomatic or fully automatic. The dry polymer can be dispensed either by hand or via a volumetric dry feeder (e.g., screw or vibrator) to a wetting jet (eductor). The polymer then is sent to a mixing (aging) tank that produces a working solution (stock solution) in 30 minutes to 2 hours. Metering pumps dispense the polymer to the solids stream. In most cases, the solution is further diluted with secondary dilution water and mixed in a static mixer to produce polymer concentrations as low as 0.01%.

Polymer feeders should be flexible enough to accommodate any type and grade of polymer. The aging tank's mixer should be variable-speed, with a maximum speed exceeding 500 rpm. The metering pump should be positive-displacement with a variable-speed controller. In general, diaphragm pumps are used for applications of about 380 L/h (100 gal/h) and less. Progressing-cavity or gear pumps are used in applications

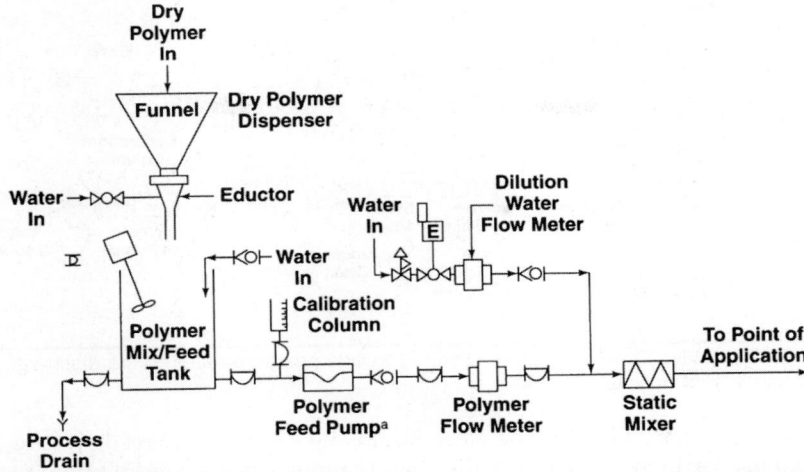

FIGURE 20.6 Typical dry polymer batch system (provide pressure relief on the discharge side of all positive-displacement polymer pumps).

greater than 380 L/h. The speed controller can be adjusted manually or set to automatically change in response to solids flow variations. The dilution water should have a flow meter and a control valve for adjustment.

Tanks, piping, and valves should be constructed of PVC or fiberglass. Any metal parts that contact polymer solution should be constructed of stainless steel. Floors, platforms, and steps should be provided with anti-slip patterns to prevent hazardous working conditions.

5.2.2 Liquid Polymer Feeders

Liquid polymers should be stored in a heated building or in heat-traced tanks. If it is stored in a building, harmful fumes and unpleasant odors can occur, so the building should be well ventilated.

FIGURE 20.7 Dry polymer feed system (City of Knoxville Water Resource Recovery Facility; courtesy of VeloDyne–Velocity Dynamics, Inc.).

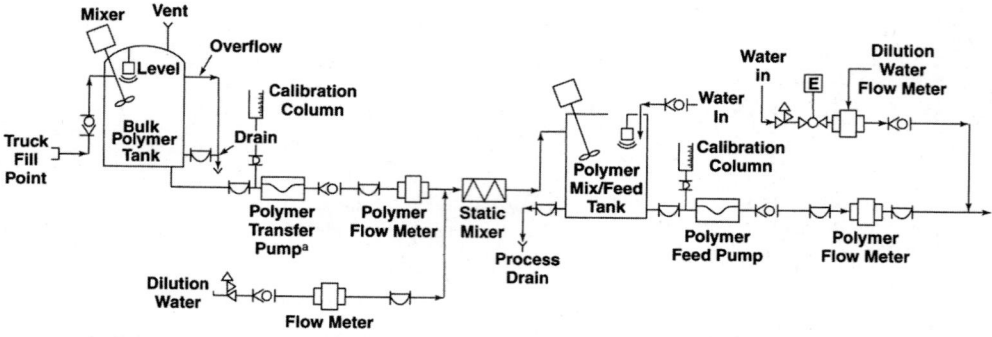

FIGURE 20.8   Typical liquid polymer batch system (provide pressure relief on the discharge side of all positive-displacement polymer pumps).

The primary difference between liquid and dry polymer-feed systems is the equipment used to blend polymers with water to prepare a working solution (see Figure 20.8) when using dry polymer. Dry polymer storage, polymer transfer, the use of a preparation and polymer maturing tank are additionally used with dry polymer. Typically, powder polymer is transferred by a vacuum conveyor or blower to the point where it is mixed with water. Solution preparation typically is a hand-batching operation in which the mixing and aging tank is manually filled with water and polymer, although in large systems this batching operation can be automatic. Variable-speed metering pumps may control the dose of liquid polymer to the aging tank.

Compact polymer-blending units can automatically mix and dilute polymers and deliver the resulting solution to the application point (see Figures 20.9 and 20.10). These pre-engineered equipment packages include a flow metering pump, valves (e.g., check, pressure-relief, and back-pressure), a dilution-water flow-control valve, an integral mixing chamber, and instrumentation and controls. They use a high-shear mixing energy zone rather than a conventional aging tank. However, there is some question as to whether this zone fully activates the polymer. Some facility report more polymer efficiency when an aging tank is provided.

5.2.3   Emulsion Polymers

Emulsion polymers consist of a high-molecular-weight polymer concentrated in a hydrocarbon solvent (oil) dispersed in water. This form allows a manufacturer to provide a high-solids organic polymer in liquid form without high-solution viscosity or limited solubility. Anionic, non-ionic, and cationic polymers are available in this form.

The storage and handling facilities for emulsion polymers are similar to those for liquid polymers. Except for the solution-preparation area, the feed system is also similar. The critical issues are aging and the initial breaking of the emulsion. Emulsion polymers must be activated—dispersed in water—before they are used. Activation is a two-step process. The first step, called inversion, involves a brief period of strong mixing to disperse the oil (continuous phase) in water (dissolving phase). The second step is a quiescent aging period, which allows the flocculant to become fully active. Anionic latex polymers require 3 to 15 minutes of aging to be completely active. Non-ionic latex polymers typically require up to 20 to 30 minutes (even longer in colder water). Some cationic latex polymers only need a few minutes to be fully active, while others need as much as 30 minutes.

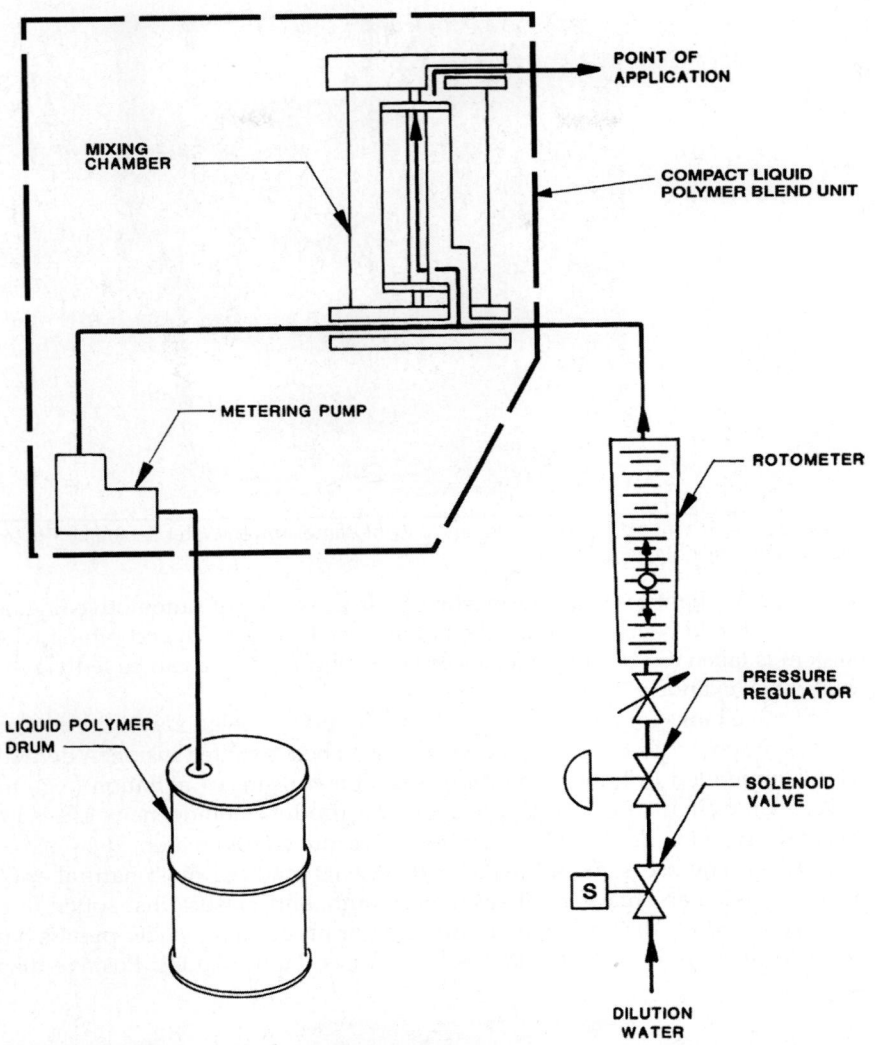

FIGURE 20.9   Compact blending system for liquid polymers.

It is possible to invert latex emulsion flocculants in a batch makeup system. A measured amount of neat polymer [about 20 kg (40 lb)] is dissolved in makeup water [about 1800 L (480 gal)] in the vortex of an agitated tank. Inversion by this method takes 30 to 60 minutes to complete, so a separate aging tank is recommended. Typical makeup concentrations for anionic, non-ionic, and cationic polymers are 0.5%, 1.0%, and 0.5% to 2.0% (as neat product), respectively.

Neat emulsion-polymer piping can "cake up" with dried polymer when not in use. To minimize this problem, piping should be at least 30 mm (1.25 in.) in diameter, sloped away from the polymer feed system, and include appropriately placed diaphragm or ball valves to isolate sections, as well as appropriately placed unions and blanked-off tees.

FIGURE 20.10 Compact polymer feed unit (City of Winter Haven Water Resource Recovery Facility No. 3; courtesy of PolyBlend).

In addition, a light-to-moderate machine oil [e.g., Society of Automotive Engineers (SAE) 10W-30] should be used to flush the polymer makeup system and piping whenever the system is taken out of service for more than a week. The oil can be fed via the system's calibration cylinder.

Storage tanks for emulsion flocculants should be designed with vents and breather tubes outdoors to keep fumes and vapors from being vented inside. A dehydration cell is recommended in humid environments. Some means of agitation (e.g., mechanical mixers or recirculation pumps) to maintain product homogeneity is also advisable, because emulsion flocculants tend to separate into oil and water.

Design engineers should avoid components made of most natural and synthetic rubber elastomers, brass, mild steel, aluminum, and plastics that soften in petroleum solvents. Positive-displacement, rotary gear, or progressing-cavity pumps typically are used to feed emulsion flocculant solutions (see Figure 20.11). Positive-displacement

FIGURE 20.11 Polymer feed pumps (courtesy of the City of Orlando, Florida, Conserv II).

pumps should have low-level alarm and shutoff controls to avoid running dry and damaging feed equipment.

5.3   Safety

Most of the chemicals used as conditioning agents can cause eye burns, skin irrigation, and possibly serious burns. Appropriate safety equipment [e.g., personal protective equipment (safety glasses, filter mask, rubber gloves, boots, aprons, etc.); safety showers; water hoses; and eyewash stations] should be clearly marked and easily accessible in the unloading, storage, and feeding locations. Other safety provisions include a dust-collection system at dry-chemical handling points (e.g., a dry pickup vacuum around feeders and slakers).

Dry chemical bags should be stored in clean, dry places to avoid picking up moisture. (The intense heat generated if quicklime accidentally contacts water could ignite flammable materials nearby.)

A vital slaker safety measure is a thermostatic valve to prevent overheating and possible explosion. This danger can occur if the controlled water supply fails while the lime feed continues, thereby allowing lime to overheat and produce excessive steam. A safety valve delivers a supply of cold water as soon as the maximum safe temperature is exceeded.

Design engineers should avoid using one conveyor or bin to handle both quicklime and other coagulants containing water of crystallization (e.g., copperas, alum, and ferric sulfate). Quicklime could withdraw the crystallization water and generate enough heat to cause a fire. When lime mixes with alum in an enclosed bin, the intense heat (greater than 590°C) generated during the reaction may release enough hydrogen to cause an explosion. Any facilities that must be alternately used should be cleaned thoroughly between applications.

6.0   Dose Optimization for Organic Conditioners

Selecting the right dosage of a chemical conditioner is critical to optimum performance. Dosage affects not only cake dryness but also the solids capture rate and solids disposal costs. Dosage is determined based on pilot-plant tests, bench tests, and on-line tests. The dosage should be re-evaluated periodically because solids characteristics can change.

6.1   Cost-Effectiveness of Chemical Conditioner and Dosage

Economic factors often are a consideration when selecting a chemical conditioner and dosage. Vendors typically are willing to conduct the testing and using their expertise to set up the tests (e.g., chemicals tested, dosage ranges, and injection locations) could significantly reduce facility personnel's workload. Once testing conditions are established, however, the vendor's involvement should end; all actual performance testing should be done by facility personnel.

When analyzing the cost-effectiveness of a polymer-enhanced dewatering technology, for example, investigators should begin by establishing minimum performance standards (e.g., a specified cake solids, feed rate, and solids capture rate) for the dewatering unit involved. Polymers that cannot meet these standards should be eliminated from further consideration.

Then investigators should calculate a recycle-reduction credit for polymers whose solids capture rates exceed the minimum standard, because it can cost as much to reprocess recycled solids as it does to process influent solids the first time through the liquid treatment process. This reduction credit is the product of the recycled solids volume multiplied by the reprocessing cost.

Naturally, investigators need to estimate the costs associated with re-processing recycled solids, as well as the anticipated biosolids use or disposal method (hauling, landfilling, incineration, land-application, etc.). Such costs typically depend on the percentage of solids in residuals, and investigators can develop a cost curve illustrating this relationship (i.e., solids management cost per kilogram of dry solids as a function of solids percentage). A good record of the O&M and energy costs for solids management is critical for this step.

Investigators then should conduct onsite prequalification tests, using identical operating conditions and solids feed characteristics for all polymers. First, they should adjust the operating conditions, polymer application rate, and dilution water feed rate to obtain the best performance for each polymer. Because it is difficult to maintain constant solids feed conditions from day-to-day, each polymer should be tested against a "standard" polymer. If the performance of the standard polymer changes during the test, a ratio can be developed to correct the performance of the polymer being tested. Second, investigators should analyze how various doses of each polymer affect cake solids, throughput, and filtrate quality. These tests should range from smallest dose that has any effect to those that clearly overdose the solids (i.e., produce a complete dosage curve). Third, investigators should determine the minimum polymer dosage that produces acceptable conditions (e.g., the driest cake with the best filtrate quality).

Investigators then should analyze test results to determine the lowest net operating condition for each polymer. Any dosage that results in an acceptable solids recovery rate and cake dryness should be used for the cost-effectiveness analysis.

Next, investigators should give each vendor the performance data for their specific products to obtain unit prices for the polymers meeting the minimum standards. The WRRF's polymer cost is the product of polymer dosage multiplied by polymer unit price. Also, any special equipment needed to apply a particular polymer should be added to the polymer cost.

The net cost of the optimum dosage is calculated as follows:

$$\text{Net cost} = CP * DC * RC \tag{20.5}$$

where CP = cost of polymer,
DC = disposal costs, and
RC = reduction credit.

When the annual net cost and polymer dosage are tabulated for all of the tested polymers, the one with the lowest annual net cost is the most cost-effective polymer type and dosage.

This procedure can be modified to fit any thickening or dewatering process or any condition.

6.2   Tests for Selecting Conditioning Agents and Dosages

Conditioning agents are critical to the optimum performance of any thickening and dewatering processes. The choice of conditioning agent and dosage affects solids capture, product dryness, and use or disposal costs. Bench-, pilot-, or full-scale conditioning tests typically are used to determine the best method for conditioning solids.

Also, the dosage should be re-evaluated periodically because changes in other wastewater treatment processes may influence conditioning requirements.

Numerous laboratory tests are available to determine the effectiveness of conditioning agents in thickening and dewatering processes. Test objectives include

- evaluating various conditioning and dewatering chemicals to determine which provides the best dewaterability;
- developing design criteria for pilot- or full-scale dewatering processes;
- comparing and evaluating different conditioning techniques; and
- using different conditioning techniques to control the dewatering process.

For the results to be useful, a representative solids sample must be tested. The sample must be fresh (i.e., tested within 24 hours of collection) because storage can affect solids properties and result in erroneous conditioning data. If the sample must be stored or shipped before testing, an acceptable preservative should be used. The conditioning agents also must be fresh (i.e., storing a diluted polymer sample too long can decrease its activity).

For a detailed explanation of each test and the procedures used to select the most cost-effective conditioning agents, see Operation of Water Resource Recovery Facilities (WEF, 2016).

7.0   Design Example

A belt filter press system used to dewater anaerobically digested solids operates under the following conditions:

- Two belt filter presses with an effective belt width of 2 m (one unit as a standby);
- Operation is 5 d/week, 7 h/d;
- Peak weekly solids production is 110 m³/d (0.001 27 m³/s);
- Total solids concentration of the belt filter press feed is 35 000 mg/L;
- Specific gravity of the solids feed is 1.03; and
- Polymer solution is 0.2% and is added before the belt filter press at a rate of 25 L/min.

Calculate the polymer dosage requirements.

7.1   Step 1: Calculate the Peak Weekly Solids to Be Dewatered

$$\text{Wet solids} = 110 \text{ m}^3/\text{d} \times 7 \text{ d/week} \times 1000 \text{ kg/m}^3 \times 1.03$$
$$= 793\ 100 \text{ kg/week}$$
$$\text{Dry solids} = 793\ 100 \text{ kg/week} \times 0.035$$
$$= 27\ 760 \text{ kg/week}$$
$$= (27\ 760 \text{ kg/week})/(5 \text{ d/week})$$
$$= 5560 \text{ kg/d}$$
$$= (5560 \text{ kg/d})/(7 \text{ h/d})$$
$$= 795 \text{ kg/h}$$

7.2   Step 2: Determine Whether Solids Loading and Hydraulic Loading Rates Are Within Operating Parameters

Solids loading = (795 kg/h)/2 m

= 398 kg/h·m (determined to be within acceptable range of 230 to 455 kg/h m)

Hydraulic loading = 110 m³/d × (1 d/1440 min) × 1000 L/m³

= 76 L/min

= 76 L/min × (7 d/5 d) × (24 h/7 h)/2 m

= 183 L/min·m (determined to be within acceptable range of 150 to 190 L/min·m)

7.3   Step 3: Calculate the Polymer Dosage

Dosage = 25 L/min × 60 min/h

= 1500 L/h

= (1500 L/h × 0.002 × 1 kg/L)/(2 m × 398 kg/h·m) × (1 tonne/1000 kg)

= 3.77 kg/tonne

8.0   References

Abu-Orf, M. M.; Dentel, S. K. (1997) Polymer Dose Assessment Using the Streaming Current Detector. *J. Water Environ. Res.*, **69** (6), 1075–1084.

Abu-Orf, M. M.; Dentel, S. K. (1999) Rheology as Tool for Polymer Dose Assessment and Control. *J. Environ. Eng.*, **125** (12), 1133–1141.

Abu-Orf, M. M.; Ormeci, B. (2005) Measuring Sludge Network Strength Using Rheology and Relation to Dewaterability, Filtration, and Thickening—31 Laboratory and Full-Scale Experiments. *J. Environ. Eng.*, **131** (8), 1139–1146.

Aichinger, P.; Wadhawan, T.; Kuprian, M.; Higgins, M.; Ebner, C.; Fimmi, C.; Murthy, S.; Bernhard, W. (2015). Synergistic Co-Digestion of Solid-Organic-Waste and Municipal-Sewage-Sludge: 1 Plus 1 Equals More Than 2 in Terms of Biogas Production and Solids Reduction. *Water Res.*, **87**, 416–423.

Bache, D. H.; Dentel, S. K. (2000) Viscous Behaviour of Sludge Centrate in Response to Chemical Conditioning. *Water Res.*, **34** (1), 354–358.

Bruus, J. H.; Nielsen, P. H.; Keiding K. (1992) On the Stability of Activated Sludge Flocs with Implication to Dewatering. *Water Res.*, **26**, 37 1597–1604.

Cassel, A. F.; Johnson, B. P. (1978) Evaluation of Filter Presses to Produce High-Solids Solids Cake. *J. New Eng. Water Pollut. Control Assoc.*, **12**, 137.

Chang, J. S.; Abu-Orf, M. M.; Dentel, S. K. (2005) Alkylamine Odors from Degradation of Flocculant Polymers in Sludges. *Water Res.*, **39**, 3369–3375.

Christensen, G. L.; Stulc, D. A. (1979) Chemical Reactions Affecting Filterability in Iron–Lime Sludge Conditioning. *J. Water Pollut. Control Fed.*, **51**, 2499.

Dentel, S. K.; Abu-Orf, M. M.; Griskowitz, N. J. (1995) *Polymer Characterization and Control in Biosolids Management*; Publication 11 D43007; Water Environment Research Foundation: Alexandria, Virginia.

Dentel, S. K.; Gucciardi, B. M.; Griskowitz, N. J.; Chang, L.; Raudenbush, D. L.; Arican, B. (2000a) Chemistry, Function, and Fate of 14 Acrylamide-Based Polymers. In *Chemical Water and Wastewater Treatment VI*; Hahn, H. H.; Odegaard, H.; Hoffmann, E., Eds.; Springer Verlag: Berlin, Germany. pp. 35–44.

Dentel, S. K.; Abu-Orf, M. M.; Walker, C. A. (2000b) Optimization of Slurry Flocculation and Dewatering Based on Electrokinetic and Rheological Phenomena. *Chem. Eng. J.*, **80** (1–3), 65–72.

Dentel, S. K. (2001) Conditioning. In *Sludge into Biosolids*; Spinosa, L.; P.A. Vesilind, P. A., Eds; IWA Publishing: London.

Eriksson, L.; Alm, B. (1991) Study of Bioflocculation Mechanisms by Observing Effects of a Complexing Agent on Activated Sludge Properties. *Water Sci. Technol.*, **24**, 21–28.

Ettlich, W. F.; Hinrichs, D. J.; Lineck, T. S. (1978) Operations Manual: Sludge Handling and Conditioning; EPA-68/01-4424; U.S. Environmental 27 Protection Agency: Washington, D.C.

Gillette, R. A.; Scott, J. D. (2001) Dewatering System Automation: Dream or Reality? *Water Environ. Technol.*, **13** (5), 44–50.

Higgins M. J. (1995) The Roles and Interactions of Metal Salts, Proteins, and Polysaccharides in the Settling and Dewatering of Activated Sludge. Ph.D. dissertation, Virginia Polytechnic Institute and State University, Blacksburg, Virginia.

Higgins M. J.; Novak J. T. (1997a) The Effect of Cations on the Settling and Dewatering of Activated Sludge: Laboratory Results. *J. Water Environ. Res.*, **69**, 215–224.

Higgins M. J.; Novak J. T. (1997b) Dewatering and Settling of Activated Sludges: The Case for Using Cation Analysis. *J. Water Environ. Res.*, **69**, 225–232.

IWPC (1981) *Sewage Sludge II: Conditioning, Dewatering and Thermal Drying*; Manual of British Practice in Water Pollution Control; IWPC: Maidstone, Kent, Great Britian.

Karr, P. R.; Keinath, T. M. (1978) Influence of Particle Size on Sludge Dewaterability. *J. Water Pollut. Control Fed.*, **50**, 1911.

Kemmer, F. N.; McCallion, J. (1979) *The NALCO Water Handbook*; McGraw–Hill: New York.

Kemp, J. S. (1997) Just the Facts on Dewatering Systems: A Review of the Features of Three Mechanical Dewatering Technologies. *Water Environ. Technol.*, **9** (12), 47–55.

Kolda, B. C. (1995) Impact of Polymer Type, Dosage, and Mixing Regime and Sludge Type on Sludge Floc Properties. Master's thesis, Virginia Polytechnic Institute and State University, Blacksburg, Virginia.

Lewis, C. J.; Gutschick, K. A. (1988) *Lime in Municipal Sludge Processing*; National Lime Association: Washington, D.C.

Metcalf and Eddy, Inc./AECOM (2013) *Wastewater Engineering: Treatment and Resource Recovery*, 5th ed.; McGraw-Hill: New York.

Mysels, K. J. (1951) *Introduction to Colloid Chemistry*; Interscience Publishers: New York.

National Lime Association (1982) *Lime Handling, Application, and Storage in Treatment Processes*, 4th ed.; Bulletin 213; National Lime Association: Arlington, Virginia.

Novak, J. T.; Haugan, B. E. (1979) Chemical Conditioning of Activated Sludge. *J. Environ. Eng.*, **105**, EE5, 993.

Novak, J. T.; Haugan, B. E. (1980) Mechanisms and Methods for Polymer Conditioning of Activated Sludge. *J. Water Pollut. Control Fed.*, **52**, 2571.

Novak J. T.; Goodman G. L.; Pariroo, A.; Huang, J. C. (1988) The Blinding of Sludges during Filtration. *J. Water Pollut. Control Fed.*, **60**, 206–214.

Novak, J. T.; Miller, C. D.; Murthy, S. N. (2001) Floc Structure and the Role of Cations. *Water Sci. Technol.*, **44** (10), 209–213.

Novak, J. T.; Sadler, M. E.; Murthy, S. N. (2003) Mechanisms of Floc Destruction During Anaerobic and Aerobic Digestion and the Effect on Conditioning and Dewatering of Biosolids. *Water Res.*, **37**, 3236.

Ormeci, B.; Cho, K.; Abu-Orf, M. M. (2004) Development of a Laboratory Protocol to Measure Network Strength of Sludges Using Torque Rheometry. *J. Residuals Sci. Technol.*, **1** (1), 35–44.

Park, C.; Muller, C. D.; Abu-Orf, M. M.; Novak, J. T. (2006) The Effect of Wastewater Cations on Activated Sludge Characteristics: Effects of Aluminum and Iron in Floc. *Water Environ. Res.*, **78**, 31–40.

Pitman, A.R.; Deacon, S.L.; Alexander, W.V. (1991) The thickening and treatment of sewage sludges to minimize phosphorus release. *Water Res.*, **25** (10), 1285–1294.

Pramanik, A.; LaMontagne, P.; Brady, P. (2002) Automation Improvements: Installing an Integrated Control System Can Improve Sludge Dewatering Performance and Cut Costs. *Water Environ. Technol.*, **14** (10), 46–50.

Roberts, K.; Olsson, O. (1975) The Influence of Colloidal Particles on the Dewatering of Activated Sludge with Polyelectrolyte. *Environ. Sci. Technol.*, **9**, 945.

Robinson, J. K. (1989) The Role of Bound Water Content in Designing Sludge Dewatering Characteristics. Master's thesis, Virginia Polytechnic Institute and State University, Blacksburg, Virginia.

Saveyn, H.; Curvers, D.; Thas, O.; der Meeren, P.V. (2008). Optimization of Sewage Sludge Conditioning and Pressure Dewatering by Statistical Modelling. *Water Res.*, **42** (4-5), 1061–1074.

Snoeyink, V. L.; Jenkins, D. (1980) *Water Chemistry*; Wiley and Sons: New York.

Sorensen, B. L.; Sorensen, P. B. (1997) Applying Cake Filtration Theory to Membrane Filtration Data. *Water Res.*, **31** (3), 665–670.

Tenney, M. W.; Echelberger, W. F., Jr.; Coffey, J. J.; McAloon, T. J. (1970) Chemical Conditioning of Biological Sludges for Vacuum Filtration. *J. Water Pollut. Control Fed.*, **42**, R1.

To, V.H.P. (2015) Improved Conditioning for Biosolids Dewatering in Wastewater Treatment Plants. University of Technology, Sydney.

Tezuka, Y. (1969) Cation-Dependent Flocculation in *Flavobacterium* Species Predominant in Activated Sludge. *Appl. Microbiol.*, **17**, 222.

Trung, Le.; Al-Omari, A.; Wadhawan, T.; Salil, K.; Higgins, M.; Novak, J.; Murthy, J. (2015) The Use of Free Chlorine and PolyDADMAC Coagulants to Improve Filtrate Quality and Reduce Dose in Thermally Hydrolyzed Anaerobically Digested Biosolids. *Proceedings of the 88th Annual Water Environment Federation Technical Exposition and Conference* [CD-ROM]; Chicago, Illinois, Sep 24–30; Water Environment Federation: Alexandria, Virginia.

U.S. Environmental Protection Agency (1978a) *Innovative and Alternative Technology Assessment Manual*; EPA-430/9-78-009; U.S. Environmental Protection Agency, Office of Water Program Operations: Washington, D.C.

U.S. Environmental Protection Agency (1978b) *Operations Manual for Sludge Handling and Conditioning*; EPA-430/9-78-002; U.S. Environmental Protection Agency: Washington, D.C.

U.S. Environmental Protection Agency (1978c) *Sludge Treatment and Disposal, Sludge Treatment, Vol. 1*; EPA-625/4-78-012; U.S. Environmental Protection Agency: Cincinnati, Ohio.

U.S. Environmental Protection Agency (1979a) *Chemical Aids Manual for Wastewater Treatment Facilities*; EPA-430/9-79-018; U.S. Environmental Protection Agency: Washington, D.C.

U.S. Environmental Protection Agency (1979b) *Chemical Primary Sludge Thickening and Dewatering*; EPA-600/20-79-055; U.S. Environmental Protection Agency, Municipal Environmental Research Laboratory, Office 36 of Research and Development: Cincinnati, Ohio.

U.S. Environmental Protection Agency (1979c) *Evaluation of Dewatering Devices for Producing High-Solids Sludge Cake*; EPA-600/2-79-123; U.S. Environmental Protection Agency, Water Resources Management Administration, Municipal Environmental Research Laboratory: Cincinnati, Ohio.

U.S. Environmental Protection Agency (1979d) *Process Design Manual for Sludge Treatment and Disposal*; EPA-625/1-79-011; U.S. Environmental Protection Agency, Municipal Environmental Research Laboratory, Office of Research and Development: Cincinnati, Ohio.

U.S. Environmental Protection Agency (1979e) *Review of Techniques for Treatment and Disposal of Phosphorus-Laden Chemical Sludges*; EPA-600/2-6 79-083; U.S. Environmental Protection Agency, Municipal Environmental Research Laboratory, Office of Research and Development: Cincinnati, Ohio.

U.S. Environmental Protection Agency (2000) *Biosolids Technology Fact Sheet Recessed-Plate Filter Press*; EPA-832/F-00-058; U.S. Environmental Protection Agency, Office of Water: Washington, D.C., Sep.

Vesilind, P. A. (1979) *Treatment and Disposal of Wastewater Sludges*; Ann Arbor Science Publishers: Ann Arbor, Michigan.

Wang, L. K.; Pereira, N. C.; Hung, Y. T. (2007) *Handbook of Environmental Engineering Biosolids Treatment Processes*, 6th ed.; Humana Press: Totowa, New Jersey.

Water Environment Federation (2014) *Introduction to Water Resource Recovery Facility Design*, 2nd ed.; Water Environment Federation: Alexandria, Virginia.

Water Environment Federation (2016) *Operation of Water Resource Recovery Facilities*, 7th ed.; Manual of Practice No. 11; Water Environment Federation: Alexandria, Virginia.

Webb, L. J. (1974) A Study of Conditioning Sewage Sludges with Lime. *J. Water Pollut. Control Fed.*, **73**, 192.

Yan, Ze.; Ormeci, B.; Zhang, J. (2016) Effect of Sludge Conditioning Temperature on the Thickening and Dewatering Performance of Polymers. *J. Residuals Sci. Technol.* **13** (3), 215–224.

Steven Swanback; Hany Gerges, Ph.D., P.E., P. Eng; Rashi Gupta, P.E.;
Anthony Tartaglione, P.E., BCEE; and Adam Evans, P.E.

1.0   Introduction

Water resource recovery facilities (WRRFs) typically use solids thickening processes to concentrate primary solids, secondary solids, or a combination of primary and secondary solids (typically waste activated solids). The reason for solids thickening is generally to increase the solids concentration of waste solids and reduce the hydraulic loading (total volume) to subsequent solids treatment processes such as anaerobic or aerobic digestion.

There are a number of widely accepted solids thickening technologies including: gravity thickeners, solids-flotation (DAFTs), centrifuges, gravity-belt thickeners, and rotary drum thickeners. These processes differ significantly in process configuration; size/footprint, degree of thickening provided; and chemical, energy, and labor requirements. Newer technologies such as disk thickeners, volute thickeners, membrane thickening systems, and recuperative thickening process are also gaining more attention as their use increases in the United States and abroad.

This chapter primarily describes the solids thickening technologies and presents process design criteria, operational considerations related to design, process control, performance assessment and optimization, a design example and a general comparison of these technologies. It also includes a discussion, where pertinent, of the liquid side streams from thickening processes that are often recycled back to the WRRF upstream (usually a process upstream of the primary clarifiers) and their impact on the water resource recovery process. There is also a brief discussion of co-thickening (primary and secondary solids) and its advantages and disadvantages where appropriate.

For additional information, see other references [e.g., *Process Design Manual for Sludge Treatment and Disposal* (U.S. EPA, 2013) and *Solids Process Design and Management* (WEF et al., 2012)].

2.0   Gravity Thickener

2.1   Operating Principle

Gravity thickeners function much like settling tanks: solids settle via gravity and compact on the bottom, while water or supernatant flows up over weirs. They also provide some solids equalization and storage, which may be beneficial to downstream processes.

Gravity thickeners work best on primary and lime-conditioned solids, but are also effective on primary solids combined with trickling filter secondary solids, primary and activated sludge, anaerobically digested solids, and to a lesser degree, waste activated sludge (WAS) and chemically enhanced primary solids. Primary and lime solids typically settle quickly and achieve a high underflow concentration without chemical conditioning. Biological solids—particularly WAS—typically have lower capture rates and underflow solids concentrations. Chemically enhanced primary solids settle fast but don't compact very well (fluffier than conventional primary solids), and occupy larger volumes leading to lower underflow concentrations and/or higher overflow concentrations.

2.2   Physical and Mechanical Features

The most common gravity thickener configuration is a circular tank with a side water depth designed at 3 to 4 m (10–13 ft). Such tanks typically range from to 21 to 24 m (70–80 ft) in diameter. Larger-diameter tanks increase solids detention time, which can cause anoxic and anaerobic activity, leading to problems with gasification and solids flotation. The tank floor typically has a slope between 2:12 and 3:12, steeper than that of a standard settling tank. The steep slope allows for minimized solids detention and maximized solids depth over the withdrawal pipe in the center of the floor for efficient removal of settled solids. This configuration also reduces raking transport problems.

Combination clarifier–gravity thickener units are typically circular tanks with a deeper center section that functions as a gravity thickener. Combined units are seldom rectangular because of difficulties associated with sludge removal.

A basic gravity thickener typically has the following main components: center cage and column, center feed well, rake arm with squeegees and scraper blades, pickets, drive unit, weir plate and overflow launder. Each gravity thickener has unique physical and mechanical features that can affect throughput, capture efficiency, polymer dose, and solids concentration (see Figure 21.1).

2.2.1   Center Cage and Column

Circular gravity thickeners are configured with a center cage and column, which comprise the structural components of the gravity thickener, supporting the walkway, feed well, rake arms with squeegees and scraper blades, pickets, and the drive unit.

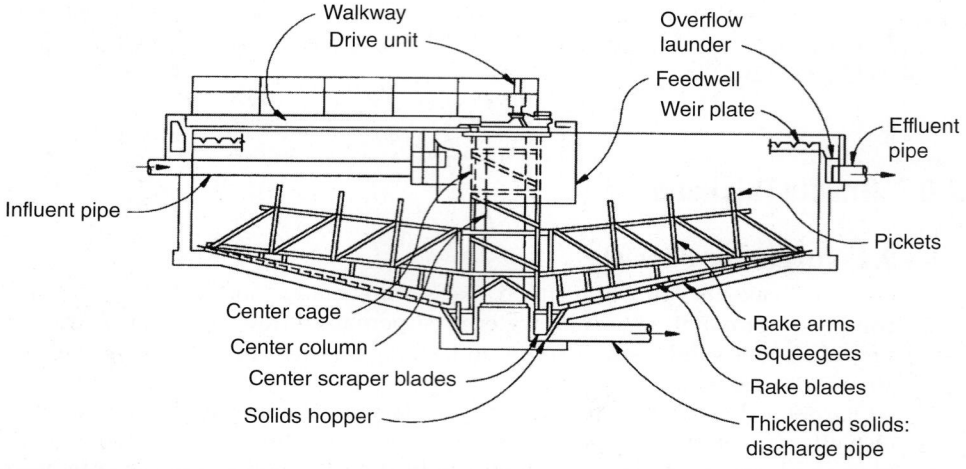

Figure 21.1   Example of a gravity thickener.

2.2.2   Feed Well
The center feed well is supported off of the center cage and column. It is circular, can be up to 40% of the diameter of the gravity thickener and is intended to act as a stilling well for the influent feed solids.

2.2.3   Rake Arm with Squeegees and Scraper Blades
Rake arms extend across the radius of the gravity thickener, and are typically provided as a pair on opposite sides of the feed well. Scraper blades with squeegees are attached to the bottom of the rake arms that rotate on the bottom of the gravity thickener in order to move thickened solids to the center of the gravity thickener for removal.

2.2.4   Pickets
Some gravity thickeners have pickets attached to the top of the rake arm and are intended to help release water from the settling solids and improve the thickening process.

2.2.5   Drive Unit
Each gravity thickener has a drive unit to rotate the center cage, column, and rake arms. The drive unit includes a motor and a gear reducer that is attached to the drive gear for the center cage and column. Because the solids are typically denser or heavier than in a primary settling process, torque control is critical.

2.2.6   Weir Plate and Overflow Launder
As the solids settle and thicken, the overflow or supernatant flows over a perimeter weir plate at the periphery of the gravity thickener. The weir plate is typically a V-notch weir. The supernatant then flows into the overflow launder for conveyance and additional treatment. Some gravity thickeners are also equipped with scum baffles if scum treatment is provided.

2.3   Process Design Conditions and Criteria

When designing gravity thickeners, important design considerations include:

- Solids' source and characteristics,
- Nature and extent of flocculation (including flocculation induced by chemical additives such as polymer or lime),
- Concentration of suspended solids in the supernatant overflow and effect of recycling lines on facility performance,
- Solids loading,
- Solids retention time in thickening zone or blanket,
- Blanket depth,
- Hydraulic retention time and surface loading rate,
- Solids withdrawal rate,
- Tank shape (including bottom slope),
- Physical arrangement of feed well and inlet pipe, and
- Arrangement of withdrawal pipe and local velocities around the piping.

2.3.1   Loading Rate

The critical design parameter for gravity thickening is the loading rate in terms of weight of total solids per unit area per unit time. Design loadings are determined via one of the methods given in Section 2.8. A thickener's capacity (allowable solids loading rate) typically is expressed in kilograms per square meter per day or pounds per square foot per day. For specific feed solids, the capacity is primarily a function of removal rate and desired underflow solids concentration. To increase the underflow concentration, the solids removal and loading rates both must be reduced. For any given feed solids, engineers can establish an operating range, with capacity expressed as a function of underflow solids concentration.

2.3.2   Overflow Rate

The second most important parameter when designing gravity thickeners is the thickener overflow rate. The maximum overflow rate for primary solids is typically 15.5 to 31.0 m^3/m^2 d (380–760 gal/d sq ft); the maximum for secondary solids is typically 4 to 8 m^3/m^2 d (100–200 gal/d/sq ft). If the hydraulic loading is too high, solids carryover can be excessive. If hydraulic loading is too low, detention times lengthen and anaerobic (septic) conditions (floating solids and odors) can occur. When thickening primary solids, design engineers often select a feed pumping rate that will maintain a desired overflow rate. They also add a dilution-water supply (e.g., facility effluent) also called elutriation to maintain aerobic conditions and may add chlorine, potassium permanganate, or hydrogen peroxide (typically via the dilution-water supply) to control odor and septicity.

See Table 21.1 for typical operating results for gravity thickeners at various overflow rates.

2.3.3   Inlet

The thickener inlet should be designed to minimize turbulence in the feed well. Most circular gravity thickeners at non-industrial WRRFs use bottom-feed inlets to a center

Location	Type of Solids[b]	Influent Solids Concentration (Percent Solids)	Hydraulic Loading Rate (L/m²·h)	Mass Loading Rate (kg/m²·h)	Thickened Solids Concentration (Percent Solids)	Overflow Suspended Solids (mg/L)
Port Huron, Michigan	P + WAS	0.6	330	1.67	4.7	2500
Sheboygan, Wisconsin	P + TF	0.3	760	2.35	8.6	400
	P + (TF + Al)	0.5	780	3.58	7.8	2000
Grand Rapids, Michigan	WAS	1.2	180	2.06	5.6	140
Lakewood, Ohio	P + (WAS + Al)	0.3	1050	2.94	5.6	1400

[a]Values shown are average values only. For example, at Port Huron, Michigan, the hydraulic loading varies between 300 and 400 L/m²·h, the thickened solids in the underflow varies between 4.0% and 6.0% solids, and the suspended solids in the overflow ranges from 100 to 10 000 mg/L.

[b]Al = alum solids; P = primary solids; TF = trickling filter solids; and WAS = waste activated sludge.

TABLE 21.1   Reported Operating Results for Gravity Thickeners at Various Overflow Rates (U.S. EPA, 1979)[a]

feed well; the feed flows vertically and then laterally with low turbulence. Most industrial and some non-industrial WRRFs use other configurations such as an overhead feed. Tangential entries or opposing tangential feed entries via a T-connection are preferred to a system that directs feed solids straight down. A horizontal solids feed entry just under the liquid surface that is directed toward the center of the feed well typically will produce satisfactory results. Air entrainment should be avoided in the feed entry to reduce froth formation on the gravity thickener surface. In general provisions for achieving good mixing without entraining air in the center well feed well should be provided.

2.3.4   Pickets
Gravity thickening mechanisms often include pickets to help release water from solids. Pickets are typically constructed of 0.6- to 2-m high (2- to 6-ft high) angle irons or pipes spaced 150 to 460 mm (6–18 in.) apart, and are designed based upon the type of solids being handled. The pickets are attached to the rake arms to provide the necessary agitation in the lower part of the tank. If the rake only consists of one pipe arm (or similar construction), pickets can improve thickening performance.

For maximum benefit, pickets should be operated in dense solids zones. They should not be used in gravity thickeners treating WAS from pulp and paper facilities or fibrous wastes from other systems. Fibrous material tends to collect on pickets, eventually causing the entire mass to rotate in the gravity thickener. Nor should pickets be used when thickening thermally conditioned solids because they will increase torque unnecessarily.

Note that reports of the effectiveness of pickets have been varied, and studies have produced contradictory results. Ettelt and Kennedy (1966), Voshel (1966), Sparr and Grippi (1969), and Dick and Ewing (1967) all indicate that using devices like pickets to stir the solids blanket improves thickening performance. Additionally, Dick and Ewing (1967) found that pickets seemed to help destroy the solids' macrostructure in static areas of the gravity thickener, thereby permitting subsidence and consolidation to continue. On the other hand, Vesilind (1968) and Jordan and Scherer (1970) reported that mixing was not beneficial; and in fact, pickets actually could hinder thickening. Likewise, if the thickener mechanism provided enough agitation on its own, pickets may have been redundant. Therefore, application of pickets on gravity thickening mechanisms should be determined on a case-by-case basis. Bench and pilot testing and more extensive research on similar applications are recommended.

2.3.5 Rake Arms and Drive Units

Lifting devices on rake arms have been typically unnecessary when treating non-industrial solids, but in certain instances—particularly when handling lime or heat-treated solids—hinged lift mechanisms are used so the rake arms lift when the torque exceeds a preset limit. The machine continues to operate with the rakes lifted [up to 0.3–1 m (1–3 ft) above the bottom] until torque drops. However, most of a gravity thickener's severe loads (e.g., those caused by islands of highly viscous sludge blankets) actually prevent the self-lifting raking arm from functioning properly, making them unreliable in this application. Cables and other lifting mechanisms have also been used to prevent high torque conditions.

Because high solids loadings lead to high torque on the rake arm, the drive units for gravity thickeners are often larger than those for primary settling processes (primary clarifiers). In some cases, it may be more desirable to simply provide an oversized drive unit to handle the torque created by intermittent, extraordinary solids loads.

Operating continuously at torques greater than the drive unit's rated capacity greatly shortens the operating life of gears and bearings. So, a gravity thickener's normal operating torque should not exceed 10% of the rated (maximum) torque value.

Design engineers typically calculate the torque for a typical gravity thickener as follows (Boyle, 1978):

$$T = Kr^2 \tag{21.1}$$

where T = torque (kg·m) (lb/ft),
  K = constant (kg/m) (lb/ft), and
  d = radius of the thickener (m) (ft).

K is a function of the material being thickened and is application-specific (see Table 21.2).

2.3.6 Skimmers and Scrapers

Gravity thickeners typically utilize skimmers, scrapers, and baffling to remove scum and other floating material. However, some installations don't have provisions to remove scum but let scum escape with the overflow. Skimmer and scraper speeds depend on the gravity thickener's diameter. Peripheral velocities typically are kept between 4.6 and 6 m/min (15 and 20 ft/min), which is substantially greater than the velocities in clarifiers. Scum drainage pipe should be sized to allow proper flow of scum with minimum size of 7.5 cm (3 inches).

In seismic zones, baffles, skimmers, and scrapers are vulnerable to additional forces—particularly forces associated with liquid sloshing. To overcome these forces,

Type of Sludge	Diameter (m)	K (kg/m)	Pickets	Overflow Rate (m³/m²*d)
Primary sludge—no to little grit	3–24	45	No	15–31
Primary sludge—with grit	3–24	60	No	15–31
Primary sludge with lime	3–30	60–90	No	40
WAS	3–15	30	Yes	4–8
Trickling filter sludge	3–15	30	Yes	15–25
Thermal conditioned sludge	3–18	120	No	10–16
Primary sludge and WAS	3–18	30–45	Yes	8–20
Primary sludge and trickling filter sludge	3–18	30–45	Yes	15–25

Note: 1 lb/ft = 1.49 kg/m

TABLE 21.2   Torque Constant K Values for Different Types of Sludges

engineers should design the rake arms, center wells, and the gearing and motor in the drive unit to prevent equipment failure during seismic events.

2.3.7   Underflow Piping
Underflow piping is a critical design element for gravity thickeners. Headlosses get higher as solids concentrations increase, so the thickened solids discharge pipe should be designed as short as possible with a diameter large enough [minimum diameter of 15 cm (6 in.)] to avoid clogging, keep the headloss low, while maintaining a minimum velocity of 0.6 m/s (2 ft/s). For operation and maintenance O&M purposes, the underflow pump should be located close to the gravity thickener and below the gravity thickener's water surface level to provide a flooded suction condition.

In addition to being as short as possible, the thickened solids discharge pipe between the sludge hopper and pump suction should have adequate access points for cleanout. Design engineers should include access for cleaning the thickened solids discharge pipe from the solids hopper to the underflow pumps. If excessive plugging is anticipated, especially with lime solids, parallel thickened solids discharge pipes may be necessary so normal operations can continue while the plugged pipe is being cleaned.

2.3.8   Rectangular Gravity Thickener Considerations
The most common problems with rectangular gravity thickeners are rat-holes and mechanical equipment failure. (A rat-hole is a conical hole in the solids that is as deep as the thickened solids layer.) For these reasons, circular gravity thickener designs are preferable. However, rat holing may be prevented by covering the hopper with a solids canopy, a concept that has been widely used in settling tanks. Deep solids hoppers can also be used to prevent rat holing but they increase the potential for anaerobic conditions.

When sizing rectangular units, design engineers should use many of the same principles and criteria as for circular units. However, two additional factors should be considered: mechanism strength and the potential for increased solids inventory and load at the sludge hopper. The design should include a mechanism to move solids laterally or transversely to the sludge hopper(s). This equipment should be strong enough to handle the added load that results from solids buildup and concentration near the sludge hopper(s).

2.4   Operational Considerations Related to Design

2.4.1   Feed Solids Source and Characteristics

The source and characteristics of feed solids greatly influence gravity thickener design and applicability. Depending on temperature, primary solids can be retained in the gravity thickener for 2 to 4 days before upset conditions develop. However, a solids retention time (SRT) of 1 to 2 days is best.

Waste activated sludge (WAS) settles slowly and resists compaction, significantly reducing mass loading rates. Waste activated sludge also tend to stratify because the continued biological activity produces gas, which creates a flotation effect.

The following precautionary measures apply when considering using a gravity thickener to thicken waste activated sludge:

- In climates where wastewater temperatures exceed 20°C (68°F), gravity thickening should be avoided unless the SRT of the activated sludge in the upstream aeration basins exceed 20 days;

- Solids retention time should be less than 18 hours to reduce the undesirable effects of continued biological activity;

- Gravity thickener diameter should be 10.7 to 13.7 m (35–45 ft) or less; and

- Solids should be wasted directly from the aeration basin to the thickener.

2.4.2   Polymer

A polymer can be added to gravity thickeners to improve solids capture. Synthetic polyelectrolytes work better than inorganic coagulants (e.g., alum and ferric chloride) in this application because they do not yield metal hydroxides that add to the solids volume. While adding chemicals improves the overflow concentrations, it may not lead to higher under flow concentrations but rather increase the solids inventory in the gravity thickener. Jar testing or pilot testing of polymers is recommended prior to design or implementation to determine removal efficiency and cost effectiveness.

2.4.3   Underflow Withdrawal

Gravity thickener operations are most effective if underflow is withdrawn continuously. If intermittent withdrawal is necessary, a time-controlled system will allow operators to achieve efficient thickening performance. Pumping should be frequent and brief rather than for longer periods only once or twice per shift. Frequent pumping minimizes the solids blanket variation required to maintain a suitable average underflow concentration.

The effect of compacting solids depth can be significant. A certain minimum depth [typically about 1–2 m (3–6 ft)] is required to achieve a desired thickened solids underflow. A deeper solids blanket can increase the underflow concentration and, to a lesser extent, the capacity of a given thickening area.

2.5 Ancillary Equipment

The ancillary equipment associated with gravity thickeners includes; feed and under-flow pumps, chemical conditioning equipment, odor control and ventilation, and process control monitoring equipment such as blanket-depth indicators, flow meters, and solids-density monitors.

2.5.1 Feed System

The preferable feed system design utilizes a positive displacement-type pump such as a progressing cavity pump, diaphragm pump, or disc pump for primary solids and some type of centrifugal pump (usually with an open impeller design to prevent clogging) for waste activated sludge solids.

It is a good design practice to use flow meters with either type of pump, particularly when other measurement controls (e.g., density sensors) will be used to maintain a relatively constant solids load to the gravity thickener.

2.5.2 Chemical Conditioning

Chemical conditioning can be utilized to increase solids capture and can consist of either a polymer or inorganic coagulants. Their characteristics vary widely, so design engineers should consult manufacturers about their properties, as well as chemical preparation and delivery techniques. For detailed information on chemical mixing and feeding systems, see Chapters 7 and 20.

The chemical conditioning system should be designed to allow operators to easily test new chemicals, chemical solution characteristics, and feed points on a regular basis. Because chemical conditioning represents a significant operating cost, optimization is necessary to minimize overall costs. For example, aging is required for dry polymers and sometimes recommended for emulsion polymers to maximize effectiveness and reduce operating costs. If emulsion polymer is used, the costs of an aging system and downstream pumps should be balanced against the expected savings from maximum polymer activation. In some instances, an aging system may not be economical compared to expected savings.

Chemicals other than polymers are delivered in trucks on regular intervals and stored in bulk tanks. Polymer storage can be provided with bulk tanks or polymer totes, depending on system size and acceptable frequency of truck deliveries. If exposed to moisture or water, polymer becomes slippery and difficult to deal with. Hence, systems should be designed to minimize such exposure and equipment areas should be provided with non-skid surfaces.

2.5.3 Underflow Pumps

Gravity thickened solids is typically between 3% and 6% solids (by weight), depending on the application. This cake can be viscous and thixotropic. It is directly discharged from the solids hopper to a collection well or directly to a pump suction for subsequent pumping. It is common to use some type of positive displacement pump such as a progressing cavity pump, diaphragm pump, or disc pump to convey thickened solids.

2.5.4 Odor Control and Ventilation

By their design gravity thickeners have a large exposed surface that will generate odors. The most common odors are related to reduced sulfur compounds but ammonia compounds can also occur.

If odor treatment is provided, it will be necessary to enclose the gravity thickener, ventilate the air space in the enclosure, and treat the exhausted air. The odorous air exhausted from the enclosure should be treated before atmospheric discharge if the area adjacent or near the WRRF is expected to be impacted by odors. Odor-treatment options (see Chapter 6) include packed-tower scrubbers and mist scrubbers (chemical scrubbers), activated-carbon beds, and biofilters. In chemical scrubbers, hypochlorite and sodium hydroxide are typically used to scrub sulfur-related odors and sulfuric acid solutions can be used to scrub ammonia-related odors. Another option is adding an odor-control compound (e.g., hydrogen peroxide) to the feed solids, but the compound first should be tested to determine whether it interferes with polymer efficiency and thickening.

If the gravity thickener is enclosed, close attention must be paid to materials of construction of all components to provide long life, limit maintenance requirements, and prevent excessive corrosion of the components. Typical cover materials include aluminum and fiber-glass-reinforced plastic (FRP). The cover consists of multiple panels that could be easily removed for quick inspection and/or maintenance. For rectangular gravity thickeners, sliding dome FRP covers have been used that provide fast and easy access to facility staff.

2.6   Process Control

2.6.1   Overview
Given the simplicity of a gravity thickener system there is little opportunity for significant process control. The components of the system that can be adjusted to affect process performance are the rake arm, the chemical conditioning system, and the sludge feed and withdrawal pump system. The rake arms are typically of constant speed but could be designed with variable speed control to improve the ability to match the solids withdrawal rate to the pumping capacity of the underflow pumps. Process control of the chemical conditioning system is key to solids separation and this involves selecting the best performing polymer (or other chemical conditioning aid) and the optimum flocculation time. The sludge feed should be at a rate that does not disturb the existing solids blanket or prevent process upset. The underflow should also be withdrawn at a constant rate to prevent process upset. Controls should be provided to allow the ability to change the sludge feed rate and the solids underflow rate.

2.6.2   Process Control Monitoring
Maximizing cake solids, maintaining reasonable capture efficiency, and minimizing polymer requirements are difficult tasks to accomplish without proper instrumentation and controls. Equipment such as solids flow meters designed for a high-solids application are necessary to adequately gauge the hydraulic feed rate and the underflow rate. Solids blanket-depth indicators are used to monitor the effectiveness of thickening, measure solids inventory, and control the number and speed of the underflow pumps. Suspended solids monitors on the overflow may be used to monitor solids capture in the gravity thickener.

2.7   Performance Assessment and Optimization

2.7.1   Performance Parameters
The process performance of a gravity thickener is primarily assessed through thickened sludge concentration, polymer consumption, and overflow quality. Other parameters

of importance are operational consistency, equipment longevity, ease of operation, and ease of maintenance.

The expected process performance for a specific feed sludge and a gravity thickener are difficult to estimate through pilot-testing as discussed above. Expected operational performance assessment and optimization will largely come through operational testing or learning from others to determine how gravity thickeners have performed elsewhere on similar feed solids.

2.7.2   Optimization Measures

System optimization focuses on maintaining consistent thickened sludge concentration, minimizing polymer consumption, and improving overflow quality. The control system features discussed previously such as chemical conditioning, use of pickets, rotational arm speed, and torque control should be tested to optimize performance. Specifically, chemical conditioning systems should be tested on a regular basis to identify those that provide the best value (cost versus performance) throughout the year. Many facilities' sludge characteristics change in winter versus summer and different chemicals or chemical doses may perform differently during each of those seasons.

2.8   Evaluation and Scale-Up Procedures

Experience has shown that thickening characteristics vary considerably—not only among various types of solids, but also among samples at a single facility taken at different locations. These variations can be caused by a wide range of factors (e.g., physical properties of solids particles, type, and volume of industrial wastes treated, processes used and their operating conditions, and solids-handling practices before thickening). So, engineers should design a thickening process based on criteria developed via a specific test program.

The two main parameters in gravity-thickener tank design are depth and area. Engineers can calculate depth based on solids volume and storage requirements; it is not controlled by the type of sludge being thickened. Tank area, on the other hand, depends greatly on solids type and is typically determined via one of four methods: existing data, batch-settling tests, bench-scale testing, or pilot-scale testing. (For information on depth requirements, clarification function, and other design considerations, see Section 2.3.)

2.8.1   Determining Area Based on Existing Data

Engineers can use empirical data from similar applications to determine the area of a gravity thickener. However, two facilities using the same upstream processes can produce solids with very different characteristics, so using empirical data may not always provide the desired results.

Table 21.3 presents typical surface-area design criteria for various types of solids. The mass loading rate is the quantity of solids allowable per unit area of gravity thickener per unit time (kg/m² h) (lb/sq ft/hr) to achieve the indicated underflow solids concentration. This table can be used to determine the gravity thickener area by dividing the actual solids loading rate by the mass loading rate associated with the type of solids and desired underflow concentration. That said, design engineers should carefully evaluate site-specific conditions, particularly with respect to the quantity of wastes treated.

Type of Solids	Influent Solids Concentration (% solids)	Expected Under Concentration Flow (% solids)[a]	Mass Loading Rate (kg/m²·hr)[b]
Separate solids			
Primary (PRI)	2–7	5–10	4–6
Trickling filter (TF)	1–4	3–6	1.5–2
Rotating biological contractor (RBC)	1–3.5	2–5	1.5–2
Waste activated sludge (WAS)			
WAS-air	0.5–1.5	2–3	0.5–1.5
WAS-oxygen	0.5–1.5	2–3	0.5–1.5
WAS-extended aeration	0.2–1.0	2–3	1.0–1.5
Anaerobically digested sludge from primary	8	12	5
Thermally conditioned solids			
PRI only	3–6	12–15	8–10.5
PRI+WAS	3–6	8–15	6–9
WAS only	0.5–1.5	6–10	5–6
Tertiary solids			
High lime	3–4.5	12–15	5–12.5
Low lime	3–4.5	10–12	2–6.5
Alum	—	—	—
Iron	0.5–1.5	3–4	0.5–2
Mixed solids			
PRI+WAS	0.5–1.5	4–6	1–3
	2.5–4.0	4–7	1.5–3.5
PRI+TF	2–6	5–9	2.5–4
PRI+RBC	2–6	5–9	2–3.5
PRI+iron	2	4	1.5
PRI+low lime	5	7	4
PRI+high lime	7.5	12	5
PRI+(WAS+iron)	1.5	3	1.5
PRI+(WAS+alum)	0.2–0.4	4.5–6.5	2.5–3.5
(PRI+iron)+TF	0.4–0.6	6.5–8.5	3–4
(PRI+alum)+WAS	1.8	3.6	1.5
WAS+TF	0.5–2.5	2–4	0.5–1.5
Anaerobically digested PRI+WAS	4	8	3
Anaerobically digested PRI+(WAS+iron)	4	6	3

[a]Data on supernatant characteristics is covered later in this chapter.

[b]This term is typically given in kg/m²/day. Because wasting to the thickener is not always continuous, it is more realistically to use kg/m²/hr.

TABLE 21.3 Typical Surface Area Design Criteria for Gravity Thickeners (U.S. EPA, 1979)

2.8.2 Determining Area Based on Batch Settling Tests

Another method for determining the area of a gravity thickener is the solids flux theory. It requires that design engineers determine the relationship between settling flux and solids concentration. This relationship is based on batch-settling test results and the premise that a suspension's settling rate is solely a function of solids concentration. Because this premise is not true for wastewater solids with high solids concentrations, the method is not completely valid, but it may give satisfactory results if batch-settling conditions resemble those in a full-scale continuous gravity thickener.

To develop the relationship between settling flux and solids concentration, batch-settling tests at various solids concentrations should be conducted. For each concentration, plot the depth of the solids–liquid interface and the time required for it to develop. Once enough data have been collected, engineers can plot a subsidence curve (see Figure 21.2).

Engineers may use the graphical method of Yoshioka et al. (1957) to determine the area needed to accomplish a desired degree of thickening (see Figure 21.3), but plotting an operating line as a tangent to the settling flux curve. The intercept on this line's abscissa is the underflow solids concentration, and the intercept on the ordinate is the limiting solids flux (G_t)—the maximum solids flux that can be transported to the bottom of the gravity thickener. The required thickener area is then calculated as follows:

$$A = \frac{c_o Q_o}{G_t} \tag{21.2}$$

where A = thickener area (m²),
$\quad Q_0$ = influent flow (m³/d),

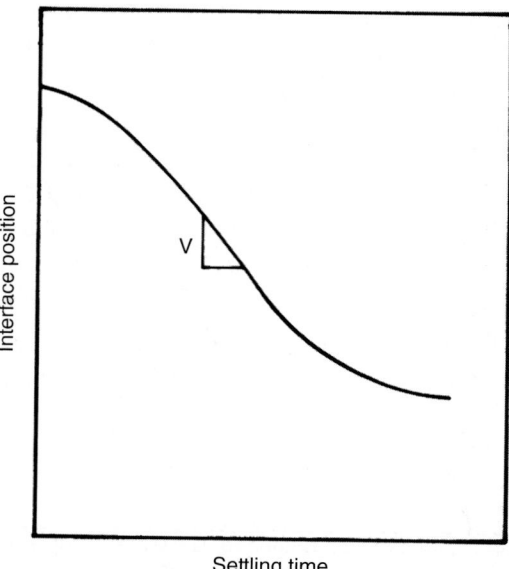

FIGURE 21.2 Example of a subsidence curve for a liquid–solids interface.

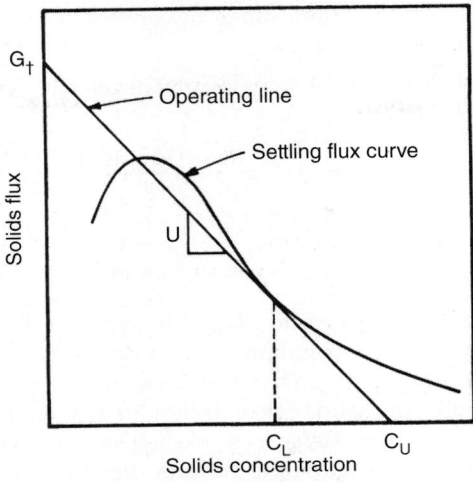

Figure 21.3 An example of the graphical method of Yoshioka et al. (1957).

c_0 = influent solids concentration (kg/m³), and
G_t = limiting solids flux (kg/m² d).

In this procedure, gravity thickener operation is assumed to be strictly one-dimensional (i.e., solids are distributed uniformly and horizontally at the feed level, and thickened-underflow removal produces equal downward velocities throughout the tank). However, full-scale gravity thickeners typically cannot meet these conditions, because of the relatively small feed-well and central withdrawal of thickened solids. No data are available on the effects of non-uniform solids distribution and removal, although these factors should be considered when sizing a gravity thickener. Design methods based on a single-batch settling test are available in the literature (Talmage and Fitch, 1955; Wilhelm and Naide, 1979; Purchas, 1977).

2.8.3 Determining Area Based on Bench-Scale Testing

William and Naide (1979) developed a useful method when using bench-scale studies to help design a gravity thickener. It has three basic steps:

- Compute the settling velocity based on settling curves taken at several feed solids concentrations (at least three).

- Obtain the constants a and b using the following equation:

$$V = aC^{2b} \tag{21.3}$$

where V = settling velocity (m/d),
$\quad C$ = solids concentration (kg/m³), and
$\quad a,b$ = constants.

The constant is a measure of the relative ease of settling; it is a function of particle size and shape, liquid and solid densities, liquid viscosity, and attractive or repulsive forces between particles. The exponent b is calculated from the slope of the line. It is typically constant over a certain range of concentrations, but gradually increases as particle-to-particle contact increases.

- For each straight line on a log–log plot of velocity versus concentration, calculate the unit area as follows:

$$UA = \frac{[(b-1)/b]^{b-1}(C_u)^{b-1}}{ab}$$

(21.4)

where UA = unit area (m²/kg d) and
C_u = the underflow's solids concentration (kg/m³).

Also, bench tests have been developed to evaluate the significance of flocculating agents during thickening. Coagulants (e.g., alum, ferric salts, or organic polyelectrolytes) can enhance flocculating characteristics and reduce the required settling area. Polymers may double the solids concentration in a given unit area.

Once suitable flocculants have been selected, engineers can conduct additional testing in a 1- to 2-L cylinder to determine the underflow concentration that can be achieved. In this test, they should add a relatively dilute concentration of polymer (less than 1000 mg/L) to solids. Then they should insert picket rakes in the cylinder and continue thickening for a standard time (1–24 hours, depending on the flocculant's effectiveness). The ultimate density achieved will be a fair but conservative measure of what to expect in a full-scale unit. For more accuracy during scale-up, engineers should use a test cylinder with a depth closer to that of the full-scale unit.

2.8.4   Determining Area Based on Pilot-Scale Testing

If a WRRF can provide enough solids for pilot-testing, reliable design data for a gravity thickener can be obtained by operating a continuous pilot unit. When sizing the pilot unit, engineers should consider the availability of test solids and the means for withdrawing thickened solids at the low flowrates required. If at all practicable, the unit should be at least 2 m (6 ft) in diameter, have a side water depth of at least 2 m (6 ft), and a bottom slope ratio of 70 mm to 3010 mm (2.75:12 in.) (vertical distance to tank radius). It should also be equipped with a feed well and a mechanism for directing solids to the withdrawal point on the tank bottom.

Engineers should conduct pilot-scale tests at several solids loading rates to determine the effect on required solids-withdrawal rates and resulting underflow solids concentrations. During each test run, ensure that the gravity thickener is operated under steady state, fully loaded conditions. A gravity thickener is operating at steady state when the solids feed and withdrawal rates are equal and do not change the unit's solids inventory. It is fully loaded when it has a solids blanket but does not lose solids in the overflow. Attaining such conditions is difficult and time consuming. One approach is to start with a slightly overloaded gravity thickener, gradually increase the solids-withdrawal rate until the overflow is solids-free, and then maintain these conditions to stabilize the blanket level. Ideally, the blanket level should be constant under all solids loading conditions. (Engineers can conduct a separate study at a convenient solids loading rate to identify any effects blanket depth may have on thickener performance.)

Once steady state, fully loaded conditions have been maintained for a certain period of time as determined by site conditions (typically 0.5–2 hours of hydraulic retention time), gravity thickener performance should be monitored by measuring the following parameters at convenient intervals:

- Solids feed rate (as determined from feed flowrate and solids concentration),
- Volumetric underflow rate and solids concentration,
- Volumetric overflow rate and suspended solids concentration, and
- Concentration profile of thickener at the end of the run.

2.9   Design Example

Design engineers need to size a circular gravity thickener for a WRRF with a primary sludge solids loading rate of 22 680 kg/d. Using Table 21.3, designers should select the higher solids loading rate to allow for operation with one unit out of service. Design engineers typically use a side water depth of 3 to 4 m.

Using Table 21.3, however, recent designs have favored deeper side water depths. Design engineers typically select a loading rate of 6 kg/m²·h.

$$\text{Solids loading of primary sludge (dry weight)} = 22\,680 \text{ kg/d}$$

$$\text{Number of operating units} = 1 \text{ unit}$$

$$\text{Design loading rate (6 kg/m}^2\cdot\text{h)} = 147 \text{ kg/m}^2\text{/d}$$

Equation used:

$$\text{Surface area} = \frac{\text{Solid Loading}}{\text{Design Loading Rate}} \qquad (21.5)$$

$$\text{Radius} = \sqrt{\text{Surface area} * \frac{1}{\pi}} \qquad (21.6)$$

$$\text{Surface area} = \frac{22\,680 \text{ kg/d}}{147 \text{ kg/m}^2\cdot\text{d}} = 155 \text{ m}^2 \text{ (use eq 21.5)}$$

$$\text{Radius} = \sqrt{155 \text{ m}^2 \times 1/\pi} = 7 \text{ m (use eq 21.6)}$$

$$\text{Diameter} = 14 \text{ m}$$

Assumption:

(1) Thickening facility operates continuously.

(2) Typically, two tanks operate simultaneously, however, this calculation allows for operations with one out-of-service.

Note: Gravity thickener diameter is based on solids loading per unit from manufacturer.

3.0   Dissolved Air Flotation Thickener

Dissolved air flotation (DAF) can be used to either clarify liquids or concentrate solids. The quality of the liquid effluent (subnatant) is the primary performance factor in clarification applications (e.g., refinery, meat-packing, meat-rendering, and other oily wastewaters). The concentration of floating solids is the main performance criteria in concentration applications (e.g., waste solids of biological, mining, and metallurgical processes). This section focuses on concentrating solids or sludge thickening. Particular advantages of dissolved air flotation thickener (DAFT) include the following:

- Ability to successfully co-thicken primary and secondary sludges
- Amenability of thickening scum from both the primary and secondary scum collection systems
- Allowing scum and sludge to be transported to the thickening process with maximum hydraulic flow
- Allowing the separation and capture of grit from a continuous bottom sludge removal system when co-thickening primary and secondary sludges
- Continuously producing a homogeneously mixed thickened sludge product that is of ideal quality for feeding digesters
- Allowing all solids processing recycle loads to be concentrated into one stream
- Achieving a significant soluble BOD reduction in the DAFT subnatant stream

3.1 Operating Principle

In contrast with gravity thickeners, the solids are floated in the DAFT process by using air bubbles to alter their specific gravity. A DAFT, which essentially consists of a DAFT tank (or flotation unit) and a pressurization system (or saturation system), uses a solids/ liquid separation process to achieve a thickened product. The purpose of the process is to provide a source of air for the flotation process by pressurizing a stream of liquid, saturating the liquid with air, and depressing the liquid at a location where the bubbles that form upon release of pressure will come in contact with the solids entering the DAFT. One can observe the depressurization effect by removing the cap from a bottle of soda water. The bubbles that form in the liquid when the cap is removed represent the excess gas that can no longer remain in solution at atmospheric pressure. The DAFT tank serves to separate the solid phase from the liquid phase. The pressurization system dissolves air into the liquid stream, typically recycled subnatant, under pressure. As the pressure saturated recycled subnatant is introduced into the DAFT tank, its pressure is reduced causing the air to precipitate out of solution in the form of very small bubbles, which are blended with the DAFT feed. The precipitated bubbles are blended with the DAFT feed and become attached to the feed solids forming bubble-particle agglomerates with a density lower than water.

The buoyant bubble-particle agglomerates rise to the liquid surface and accumulate as a float while the heavier particles settle out as bottom sludge. The difference in density between the float and the liquid causes the top of the float to rise above the liquid surface. The float and bottom sludge are removed from the DAFT tank by a surface skimmer and a bottom collector, respectively. Drainage of interstitial water from the float above the liquid surface increases the solids concentration. This process is termed as thickening. Introducing primary sludge during co-thickening promotes more porous float structure, thereby enhancing thickening. The advantages and disadvantages of co-thickening are discussed later in this section.

A DAFT is typically used to thicken WAS, aerobically digested solids, and contact-stabilized, modified activated, or extended aeration solids without primary settling. It is typically not used for primary solids because gravity thickening is more economical. However, it can effectively co-thicken primary sludge with WAS or trickling filter solids. The main components of a DAFT are the pressurization system and DAFT tank (see Figure 21.4).

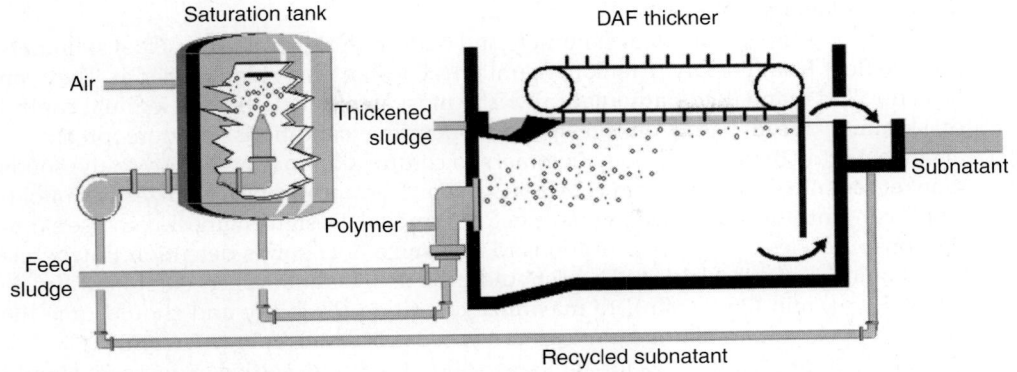

Figure 21.4 Schematic of a dissolved air flotation thickener.

The pressurization system has a recycle pressurization pump, an air compressor, an air saturation tank, and a pressure-release valve. The DAFT tank has a surface skimmer for float blanket depth control and float removal from the DAFT tank to a float collection box. The settled heavier solids are removed by a bottom collector. Most DAFT tanks are baffled and equipped with an overflow weir. Clarified effluent passes under an end baffle (rectangular units) or peripheral baffle (circular units) and then flows over the weir to an effluent launder. The weir controls the liquid level within the DAFT tank with respect to the float collection box and helps regulate the capacity and performance of the DAFT.

3.2   Physical and Mechanical Features

3.2.1   DAFT Tanks
The number and configuration of DAFTs tanks to be installed at a WRRF depends on the facility size, method of operation, the quantity of solids to be thickened under average and peak loading conditions, and the requisite operating flexibility.

DAFT tanks can be rectangular or circular. Both rectangular and circular units have been used in WRRFs ranging from 38 to more than 380 000 ML/d (1 to more than 100 mgd).

The typically surface area for rectangular DAFTs varies from 9 to 167 m² (100–1800 sq ft). Length-to-width ratios typically are between 3:1 and 4:1. Their float skimmers can be closely spaced and designed to skim the entire surface. The bottom sludge collector typically has a separate drive, so it can be operated independently of the skimmer. The liquid surface can be adjusted more easily because of the straight-end weir configuration.

The typical surface area for circular DAFTs varies from 29 to 130 m² (300–1400 sq ft). They are often used when land availability is not a constraint.

DAFT tanks can be constructed of concrete or steel. Typically, larger tanks are made of concrete, while rectangular tanks up to 41.8 m² (450 sq ft) [2.4–3 m (8–10 ft) wide] and circular tanks up to 9 m² (100 sq ft) are made of steel. The size of steel DAFT tanks is limited by structural and shipping considerations; they are typically completely assembled and only require a concrete foundation pad, piping, and wiring hookups. Steel tank systems have higher equipment costs but avoid field-installation costs (e.g., structural, labor, and equipment components). Concrete tanks are typically more economical for a large installation requiring multiple or large tanks (U.S. EPA, 1979).

3.2.2   Skimmers and Rakes

DAFTs are equipped with float skimmers and bottom sludge collectors. Float skimmers remove float from the DAFT tank to maintain a constant float-blanket depth. They can be controlled manually or automatically. The most common method is manual control of skimmer speed based on site-specific operating conditions. A more preferable arrangement is the use of automatic timers to control skimmer operation so the solids blanket remains 300 to 500 mm (12–18 in.) deep. This approach maximizes both float-solids concentration and float drainage before removal. Design engineers can use skimmer on–off cycles of variable durations to maximize float-solids detention time while maintaining a stable blanket. They should also use variable-speed skimmers [up to about 7.6 m/min (25 ft/min)] to maximize operating flexibility and should time the skimmer cycle so the skimmer's maximum speed is 300 mm/min (1 ft/min).

The bottom sludge collector removes solids that have settled. Engineers should design the bottom sludge collectors and float skimmers as separate systems. Bottom sludge collectors that are operated at excess speeds for the application may adversely affect DAFT performance. Providing a separate drive system for the bottom rake collector allows operators to operate them only as required.

3.2.3   Overflow Weir

DAFTs are baffled and equipped with an overflow weir. The weir controls the liquid level in the DAFT tank with respect to the float collection box, thereby regulating the DAFT's capacity and performance. To maximize capacity and performance under widely fluctuating conditions, the overflow weir should be adjustable and or the float skimmer should be on a variable speed drive.

3.2.4   Pressurization System

Pressurization systems dissolve gas (typically air) into the liquid used during DAF. The theoretical principles of pressurization systems are well known and have been discussed by several researchers (e.g., Vesilind, 1974b; Bratby and Marais, 1975a, 1975b, and 1976; Speece et al., 1975).

Historically, three methods have been used to provide gas bubbles for a DAFT system: total, partial, and recycle pressurization flow schemes (see Figures 21.5, 21.6, and 21.7). The total-pressurization flow scheme pressurizes the entire waste stream entering the DAFT; it is only practical for small flowrates, oily liquids, or other situations where turbulence in the pressurization systems will not degrade solids enough to impair DAFT performance. This approach should not be used when the influent contains flocculated solids because the turbulence in the tank and pressure-relief valve would destroy flocs. This approach should also not be used when the influent contains abrasive or large solids, which can wear eductors and clog pumps; rather, recycle pressurization should be used instead.

Partial pressurization systems pressurize a fraction of influent; how much depends on the air-to-solids ratio needed for optimal performance. This flow scheme is typically only practical for small rates of non-flocculated oily wastewaters. Its limitations are the same as those for total pressurization.

Most DAFT units thickening WAS use recycle pressurization systems, in which some of the subnatant is pressurized. Influent solids do not pass through the pressurization system but are mixed with the pressurized recycle stream before entering the DAFT unit.

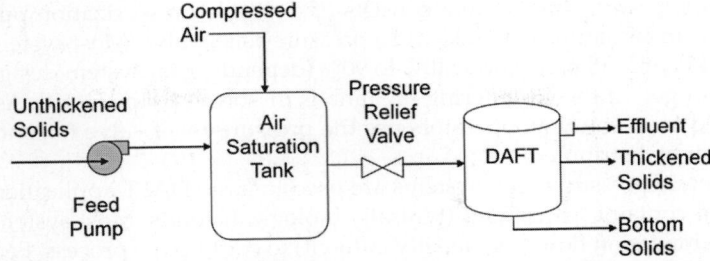

FIGURE 21.5   A dissolved air flotation thickener using a total-pressurization-of-solids flow scheme to produce gas bubbles.

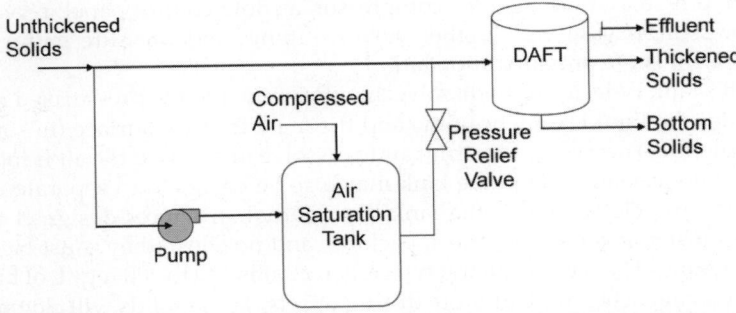

FIGURE 21.6   A dissolved air flotation thickener using a partial-pressurization-of-solids flow scheme to produce gas bubbles.

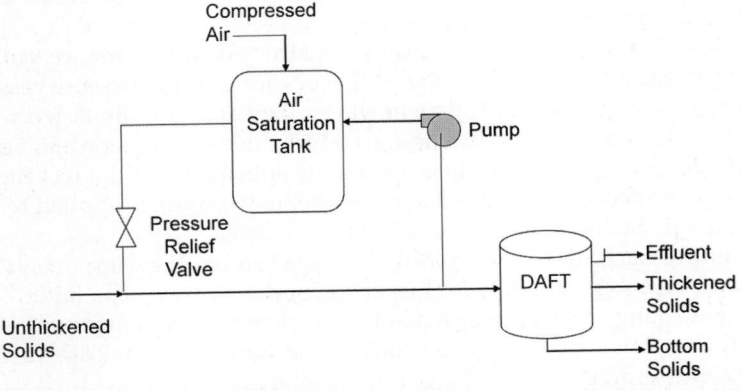

FIGURE 21.7   A dissolved air flotation thickener using a recycle-pressurization-of-solids flow scheme to produce gas bubbles.

The pressurization system consists of a recycle pressurization pump, an air compressor, an air-saturation tank, and a pressure-relief valve. Most systems operate at 380 to 520 kPa (55–75 psi). About 40% to 90% (depending on system design) of the oxygen and nitrogen in the air entering the tank is dissolved in the liquid. As dissolved air is released from solution, operators use the pressure-relief valve to control pressure loss and evenly distribute flow.

Recycle pressurization systems are used in large DAFT applications and when the influent contains flocculated (typically biological) solids. Most systems include auxiliary recirculation flow (e.g., facility effluent) to start up the process. Because the system is complex and consists of numerous valves and fittings, staff training programs are essential to ensure proper operation.

3.2.5 Pressurization Tanks

Each pressurization system includes one or more of the following: a pressurization pump, pressurization tank, air compressor, airflow control panel, recycle-flow indicator, pressure-release valve, other valves, piping, and pressure gauges. The primary component is the pressurization tank.

This tank is designed to dissolve air efficiently into the pressurized recycle liquid. It provides the liquid residence time and the mass-transfer surface (in some cases, internal structures) necessary to permit air to dissolve in liquid. If the air is injected upstream of the pressurization tank, the tank may also be designed to separate undissolved air from the recycle stream. If the tank has internal structures designed to create liquid mass-transfer surface (e.g., trays, packing, and nozzles), they must be designed to be nonclogging. The recycle stream typically contains 100 to 200 mg/L of biological solids; it can contain 3000 mg/L or more during upsets. These solids will clog most traditional mass-transfer packing surfaces.

The pressurization tank should be next to the DAFT and pressurization pump(s) to minimize piping requirements and headloss via interconnecting piping. Any pressure loss downstream of the pressurization tank tends to release dissolved air from solution. Released air can enter the DAFT as entrained air bubbles and create disruptive turbulence in the inlet section.

Pressurization tanks are typically constructed in accordance with the American Society of Mechanical Engineers (ASME) code for unfired pressure vessels with a working pressure of 700 kPa (100 psi); however, they are typically tested hydrostatically to 1300 kPa (150 psi). The ASME's design code includes a corrosion allowance whose magnitude depends on the specific constituents anticipated in the recycle stream. If more corrosion protection is required, a layer of epoxy coating is applied to the tank's internal surfaces. Stainless steel vessels can also be used.

The pressurization tank typically has steel legs or other support systems, a drainage opening, an access manhole for inspection and maintenance, a liquid-level sight glass, a pressure gauge protected by a diaphragm element, and a pressure-relief safety valve. It may also have one or more air-inlet connections, an air-release valve, and a liquid-level control valve.

3.3 Process Design Conditions and Criteria

The capacity of DAFT units is an order of magnitude greater than that of gravity thickeners, so their space requirements are typically low. At large WRRFs with an influent 5-day biochemical oxygen demand (BOD$_5$) of 150 to 200 mg/L, a DAFT process with polymer addition needs 0.37 to 0.0005 m^2 d /m^3 (15–20 sq ft/mgd); without polymer addition, the process needs 0.7 to 0.001 m^2 d /m^3 (30–40 sq ft/mgd). At small WRRFs with the same BOD$_5$, a DAFT process with polymer addition needs 0.5 to 0.0007 m^2 d/m^3 (20–30 sq ft/mgd); without polymer addition, the process needs 1.0 to 0.0015 m^2 d/m^3 (40–60 sq ft/mgd). Because they do not need much space, DAFT thickeners can be located inside buildings. This is especially desirable in locales where odor control is required, or cold or wet climates could adversely affect a DAFT unit's mechanical performance.

Before feed solids enter a DAFT tank, they typically are mixed with a recycled flow. The recycled flow is pressurized up to 520 kPa (75 psi) and added at a rate that depends on feed-solids concentration and the air-to-solids ratio. The intent of the pressurization system is to deliver air to the system under pressure. When the pressure is released into the DAFT tank fine bubbles attach to the incoming solids. This recycle flow that is pressurized is typically DAFT subnatant. Recycled flow first is pumped to an air-saturation tank, where compressed air dissolves into the flow. When returned to the DAFT tank (whose surface is at atmospheric pressure), the pressure release creates the air bubbles used for flotation. These bubbles typically range from 10 to 100 micrometers in diameter, which is approximately the diameter of human hair, pollen, plant spores, and fog.

The air combines with solids particles and floats, forming a blanket on the DAFT tank surface. Meanwhile, clarified effluent flows under the tank baffle and over the effluent weir. A properly designed and operated DAFT typically captures between 94% and 99% of suspended solids.

Other DAFT pressurization systems do not use recycled flow; instead, they pump all or part of feed solids through an air-saturation tank and then into the DAFT tank. Such systems are inadvisable for water resource recovery applications because they subject solids to high-shear conditions and the solids can clog various pressurization-system components.

Polymers can enhance DAFT performance by significantly increasing applicable solids-loading rates and solids capture; and they can also increase the concentration of floating-solids concentrations to some degree. If used, a polymer is typically introduced at the point where feed solids and recycle flow are mixed. For the best results, design engineers should introduce polymer to the recycle flow just as the bubbles are being formed (before it is mixed with feed solids). Good mixing, enough to ensure chemical dispersion while minimizing shearing forces, will provide the best solids–air bubble aggregates.

Table 21.4 presents operating data from selected DAFT thickener installations. Numerous factors affect DAFT process performance, including

- Type and characteristics of feed solids,
- Mixed liquor sludge volume index,
- Hydraulic loading rate,
- Solids loading rate,
- Feed-solids concentration,
- Air-to-solids ratio,
- Float-blanket depth,

Location	Activated-Sludge Type	Feed Solids Concentration (mg/L)	Solids Loading Rate (kg/m²·h)	Float Concentration (%)	Polymer Dosage, g Active Polymer/kg of Solids	Solids Capture (%)
Green Bay, Wisconsin	Contact stabilization	4000	1.5	3–4	None	80–85
San Francisco, California Southeast Plant	High-purity oxygen	6000	3.4	3.7	1.6	98.5
Salem, Oregon	High-purity oxygen	14 800– 20 300	19.5	5	48–59	95+
Milwaukee, Wisconsin South Shore Plant	Conventional	5000	4.9	3.2	1.5–2.5	90–95
Tri-Cities, Oregon	Conventional	11 300	—	3.9	1.5–2.5	98
Arlington, Virginia	Conventional	10 000	8.5	2.6	1.5–2	95+
Kenosha, Wisconsin	Conventional	8600	5.4	4.3	None	99+

TABLE 21.4   Typical Operating Data for Dissolved Air Flotation Thickeners

- Chemical conditioning,
- Float-solids concentration,
- Solids capture,
- Solubilization efficiency, and
- Co-thickening primary and secondary solids.

These factors often act synergistically to produce a net positive or negative effect on DAFT performance. Isolating each factor's effect is often difficult, but Bratby and Marais (1975a) have proposed a model to predict DAFT performance as a function of various conditions.

3.3.1   Type of Solids

DAFTs can thicken a variety of solids, including conventional WAS, solids from extended aeration and aerobic digestion, pure-oxygen activated sludge, and solids from dual biological processes (trickling filter plus activated-sludge). The performance characteristics of each type of solids are difficult to document because site-specific conditions [e.g., type of process, SRT, and sludge volume index (SVI) in the aeration basin] affect DAFT performance more than flotation-equipment adjustments (e.g., air-to-solids ratio). Gulas et al. (1978) and Wood and Dick (1975) discuss the effects of some facility operating parameters on DAFT performance in considerable detail.

3.3.2 Mixed-Liquor Sludge Volume Index

One of the solids characteristics affecting DAFT performance is activated sludge mixed-liquor SVI. The floating-solids concentration typically decreases as SVI increases. To produce a 4% floating solids concentration with nominal polymer doses, SVI should be less than 200 mL/g. If SVI is low, solids are compacting well, and a broad band of floating solids exists, then other factors are likely influencing DAFT performance. At higher values, SVI has a deleterious effect on floating solids. Large doses of polymer are typically required when thickening WAS from systems with excessively high SVI.

3.3.3 Hydraulic Loading Rate

The hydraulic loading rate is the sum of the feed and recycle flowrates divided by the net available flotation area. Engineers typically design DAFTs for hydraulic loading rates of 30 to 120 m^3/m^2 d (0.5–2 gpm/sq ft), with a suggested maximum daily hydraulic loading of 120 m^3/m^2 d (2 gpm/sq ft) if no conditioning chemicals are used. If the hourly hydraulic loading rate exceeds 5 m^3/m^2 h (2 gpm/sq ft), the added turbulence may prevent a stable float blanket from forming and reduce the attainable floating-solids concentration. Also, fewer solids may be captured because increased turbulence forces the flow regime to convert from plug flow to mixed flow. A polymer flotation aid is typically required to maintain satisfactory performance when hourly hydraulic loading rates are greater than 5 m^3/m^2 h (2 gpm/ sq ft).

3.3.4 Solids Loading Rate

The solids loading rate for a DAFT denoted in terms of solids weight per hour per effective flotation area (see Table 21.5). Without chemical conditioning, the loading rates for DAFT processes thickening WAS range from about 2 to 5 kg/m^2 h (0.4–1 lb/hr sq ft); this produces a thickened underflow of 3% to 5% total solids (Ashman, 1976; Burfitt, 1975; Jones, 1968; Mulbarger and Huffman, 1970; Reay and Ratcliff, 1975; U.S. EPA, 1974; Walzer, 1978). With polymer, the solid loading rate can typically be increased from 50% to 100%, producing a thickened underflow that contains up to 0.5% to 1% more solids.

Type of Solids	Solids Loading Rate (kg/m²·h)	
	No Chemical Addition	Optimal with Chemical Addition
Primary only	4–6	up to 12

WAS
  Air
  Oxygen
	
	2
	3–4
		
		up to 10
		up to 11

Trickling filter	3–4	up to 10
Primary + WAS (air)	3–6	up to 10
Primary + trickling filter	4–6	up to 12

TABLE 21.5 Percent Suspended Solids Captured When Using Dissolved Air Flotation to Thicken WAS (U.S. EPA, 1974; Komline, 1976)

Operating difficulties may arise when the solids loading rate exceeds approximately 10 kg/m² h (2.0 lb/hr sq ft). These difficulties typically are caused by coincidental operation at excessive hydraulic loading rates and by float-removal difficulties. Even when the hydraulic loading rate can be kept below 120 m³/m² d (2 gpm/sq ft), operating at solids loading rates more than 10 kg/m² h (2.0 lb/hr sq ft) can cause float-removal difficulties. The extra floating material created at high solids loading rates necessitates continuous, often rapid, skimming.

Faster skimming, however, can disturb the float blanket and lead to a subnatant with unacceptable solids levels. In these circumstances, a polymer aid can increase the solids' rise rate and float-blanket consolidation rate, thereby alleviating some of the operating difficulties. Although stressed conditions (e.g., mechanical breakdown, excessive solids wastage, or adverse solids characteristics) may make it necessary to operate in this manner periodically, the flotation system should not be designed on this basis.

3.3.5   Feed-Solids Concentration

Feed-solids concentration affects DAFT processes in two ways. As in sedimentation processes, feed-solids concentration directly affects the floating solids' characteristics in terms of initial and, to a lesser extent, hindered rise rate. Within the normal range of feed-solids concentration between 0.5% and 1% (5000–10 000 mg/L), more dilute feed solids results in more rapid initial and hindered rise rates. However, this phenomenon only has a minor effect on DAFT sizing and performance because the solids blanket's hindered rise rate and compression rate govern design and performance for most thickening applications.

Feed-solids concentration also indirectly affects DAFT performance via resulting changes in operating conditions. For example, if the feed flowrate, recycle flow, pressure, and skimmer operations remain constant, then increasing feed-solids concentration decreases the air-to-solids ratio. Changes in feed-solids concentration also change float-blanket inventory and depth. Float skimmer speed may need adjustments when the operating strategy involves maintaining a specific float blanket depth or range of depths.

3.3.6   Air-to-Solids Ratio

The air-to-solids ratio—the ratio by weight of air available for flotation to the floatable solids in the feed stream—is the most important factor affecting DAFT performance. Reported ratios range from 0.01:1 to 0.4:1 (U.S. EPA, 1979) at most municipal WRRFs; adequate flotation occurs at ratios of 0.015:1 to 0.03:1. Design engineers size pressurization systems based on many variables (e.g., design solids loading, pressurization-system efficiency, system pressure, liquid temperature, and dissolved solids concentration). Pressurization-system efficiencies vary among manufacturers and system configurations; they can range from as low as 50% up to more than 90%. The U.S. EPA (1979) provides detailed information on designing, specifying, and testing pressurization systems.

Because the solids blanket in a DAFT thickener contains a considerable amount of entrained air, design engineers should use positive-displacement or centrifugal pumps that do not air bind, and consider suction conditions. Initially, the density of skimmed solids is about 700 kg/m³ (6 lb/gal). After they are held for a few hours, the air escapes and solids return to normal densities.

Up to a point, solids blankets increase as air-to-solids ratios increase; then, further increases in air-to-solids ratios result in little or no increase in floating solids (Gehr and Henry, 1978; Gulas et al., 1978; Maddock, 1976; Mulbarger and Huffman, 1970;

Turner, 1975). The solids blanket is typically maximized when air-to-solids ratios are in the range of 0.1:1.

There are several explanations for this wide range. First, the optimum air-to-solids ratio is related to the type of feed solids and its characteristics. For example, activated sludge with low SVIs require lower air-to-solids ratios than those with high SVIs.

Second, evaluating the effects of air-to-solids ratios is difficult because other DAFT operating conditions (e.g., blanket depth) can vary as the air-to-solids ratio changes. So, the effect of a change in air-to-solids ratio is often masked by other changes.

Finally, differences among the DAFT systems researched (e.g., the pressurization system's air-dissolving efficiency, gas-bubble size distribution, and feed-recycle mixing methods) undoubtedly are responsible for some of the differences in optimum air-to-solids ratios.

Although the optimum air-to-solids ratio is probably related to solids type and characteristics, lower air-to-solids ratios seem to be required to maximize the performance of systems that operate at a high air-dissolving efficiency, produce optimum air-bubble size distribution, and correctly contact the feed solids and minute air bubbles at the proper time.

3.3.7 Float-Blanket Depth

The floating solids produced during DAFT operation must be removed from the tank. This solids-removal system typically consists of a variable-speed float skimmer and a beach arrangement. The volume of floating solids that must be removed during each skimmer pass depends on the solids loading rate, the chemical dose rate, and the consistency of floating solids.

A blanket of waste biological solids consists of two sections: one above the nominal water level and one below it. When evaluating a DAFT system, Bratby and Marais (1975b) found that its ratio of float depth above the surface to float depth below the surface was 0.2:1 when the air-to-solids ratio was 0.02:1. They stated that the optimal ratio of above-surface and below-surface solids will differ according to the type of feed solids involved.

The concentration of solids on the surface of the solids blanket is always greater than the average concentration of solids within the blanket. Bratby and Marais (1975b) also suggested that DAFT thickening occurs as water drains from the section above water to that below it. Maddock (1976) found that the solids concentration at the blanket surface was nearly twice that at the blanket–subnatant interface.

Blanket skimmers are designed and operated to maximize float drainage time by incrementally removing only the top (driest) portion of the blanket and preventing the blanket from expanding to the point where solids exit the system in the subnatant. The optimal float depth varies from installation to installation. A float depth of 300 to 600 mm (1–2 ft) is almost always sufficient to maximize floating-solids content.

3.3.8 Chemical Conditioning

Chemical conditioning can enhance DAFT performance. Conditioning agents can improve clarification or increase the floating solids concentration. Design engineers should determine the amount of conditioning agent required, the point of addition (in the feed stream or recycle stream), and the intermixing method for each installation. Bench- or pilot-scale tests are the most effective methods of determining the optimal chemical-conditioning scheme for a particular installation.

Typical polymer doses range from 2 to 5 g dry polymer/kg dry feed solids (4–10 lb/ton) as active content. Adding polymer typically affects solids capture more than floating-solids content. For example, adding dry polymer at a dose of 2 to 5 g/kg dry feed solids typically increases floating solids content by no more than 1%.

If design engineers use the lower ranges of hydraulic and solids loadings, properly designed and operated DAFTs typically do not need polymer. Maintaining proper design and operating conditions results in stable operations and satisfactory solids capture and floating-solids concentration. Routine additions of polymer should only be considered for designs with extreme loading conditions or when solids are expected to have poor compaction characteristics such as high SVI.

Without polymer addition, a properly sized DAFT unit will typically recover more than 90% of solids. High loadings or adverse solids conditions can cut solids recovery to 75% to 90%. Polymer-aided recovery can exceed 95%.

Under normal operations, solids recycled from the DAFT unit will not damage the treatment system but rather increase WAS. However, if solids or hydraulic loading are already excessive, recycled solids pose an additional burden on the system. Under these conditions, polymers should be used to maximize solids capture from the DAFT unit.

3.3.9 Floating Solids Concentration

As with any thickening process, flotation performance strongly depends on the type and characteristics of the solids being thickened. Although municipal WRRFs typically use DAFT to thicken WAS, they also have used it to thicken raw primary solids, trickling filter humus, and various combinations of these.

The floating solids concentration that a DAFT treating WAS can obtain is influenced by various factors, the most important of which are innate solids characteristics (i.e., SVI), solids loading rate, air-to-solids ratio, and polymer application. Test results demonstrate that the floating solids concentration typically decreases as solids loading rates increase (see Figure 21.8). They also indicate (with few exceptions) that polymers must be used to achieve the higher loading rates. Although high loadings of 15 to 29 kg/m² h (3–6 lb/hr sq ft) can be achieved, these results are neither typical of the average facility nor a relevant basis for new designs. In some cases, a lot of expensive chemicals are necessary to achieve a loading level in excess of 10 kg/m² h (2 lb/hr sq ft).

The curve in Figure 21.8 does not indicate the effect of polymers on floating solids concentration. Polymers can improve poor float concentration up to 1% (2%–3% TSS), but their effect lessens as the concentration of untreated floating solids increases.

DAFTs are typically designed for floating-solids concentrations of 3.5% to 4.0% total solids—a reasonable goal based on the data presented (see Table 21.4) and other published information (U.S. EPA, 1974; Wanielista and Eckenfelder, 1978). However, DAFT performance, like other solids processing equipment performance, is influenced by factors beyond the design engineers' control. Therefore, designers should anticipate variations in float-solids concentration when sizing downstream unit operations.

3.3.10 Solids Capture

Overall solids capture measures how efficiently a DAFT unit recovers solids at a fixed set of operating conditions. The solids-capture calculation is based on a material balance about the DAFT unit (Mulbarger and Huffman, 1970). The flows of interest include feed solids, subnatant, and floating solids. Overall solids capture is defined as:

$$R = \frac{TS_p(TS_F - TSS_S)}{TS_F(TS_P - TSS_S)} * 100 \tag{21.7}$$

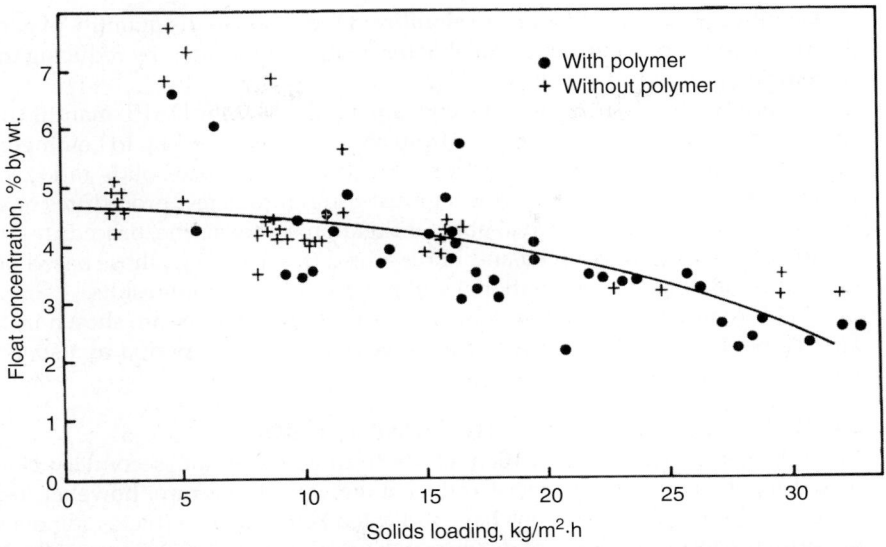

Figure 21.8   Floating-solids concentration versus solids loading rates (Noland and Dickerson, 1978).

where R = percentage recovery;

   TS_p = thickened sludge total solids concentration (% by weight);

   TS_F = feed sludge total solids concentration (% by weight); and

   TSS_S = suspended solids concentration in subnatant (% by weight).

Published results of suspended solids capture at numerous DAFT facilities indicate that they can capture at least 95% of solids without using polymer (see Table 21.5). With polymer, they typically capture at least 97% to 98%. When recycling subnatant, most thickening operations need to capture at least 95% of solids to minimize adverse effects on other treatment processes.

3.3.11   Solubilization Efficiency

The most cumbersome procedure associated with DAFT equipment is determining the solubilization efficiency of the dissolving tank in the pressurization system. Solubilization efficiency is the ratio of the amount of air (oxygen and nitrogen) actually dissolved in the tank to the amount that theoretically could be dissolved in the tank under existing conditions. Data collection and calculation procedures have been developed and published (APHA et al., 2012; Leininger and Wall, 1974; U.S. EPA and ASCE, 1979).

The pressure/saturation curve for liquids follows Henry's Law and is easily computed from information available in standard references. The most important consideration is that saturation concentration varies inversely with liquid temperature. Therefore, it is important to establish the maximum operating temperature for the pressurization system (or saturation system). In addition, it is important to note that the saturation constant in water for nitrogen is substantially less than that for oxygen. Unless the pressurization tank gas space is vented, the gas above the liquid will become rich in nitrogen. The result will be a significant reduction in the quantity of gas absorbed in a unit volume of water, and available for floating solids thereby reducing the dissolving efficiency.

A variety of pressurization systems are available from DAFT manufacturers, and their air-dissolving efficiency ranges from 65% to 85%, according to Leininger and Wall (1974). Dissolving efficiency is important because the air-to-solids ratio is critical to DAFT performance. Design engineers must use rigorous test procedures to accurately determine the amount of air available for flotation. For example, procedures that do not distinguish between dissolved and undissolved (free) air (e.g., those based on conventional air mass-balance calculations) will not provide accurate results.

Solids loadings to DAFT systems for various facility sizes are shown in Table 21.6. DAFT operations and performance averages over a 2-year period at a specific WRRF are shown in Table 21.7.

3.3.12 Co-thickening Primary and Secondary Solids

In co-thickening processes, settled solids from primary and secondary clarifiers are mixed together then thickened. Co-thickening used to be rare; however, recent pilot testing and full-scale operations have indicated benefits of co-thickening over thickening primary and secondary solids separately (Butler et al., 1997). Some of these include:

- Ability to increase DAFT solids loading rate (could double the solids loading rate per surface area compared to separate thickening);
- Ability to reduce soluble BOD and chemical oxygen demand (COD) as much as 80% and 60%, respectively;
- Lower present-worth and operating costs; and
- Significantly reduced secondary BOD loading while reducing grit because of thickened solids recycling.

A float total solids content of 6% approximates a typical minimum for DAFTs handling a 50:50 mixture of secondary and primary sludge compared to 4% for DAFTs handling solids without primary solids. Under optimum conditions, minimum solids contents ranging from 6% to 8% can be expected from thickening such mixtures (Bratby et al., 2004). Co-thickening can be used at small or large WRRFs but must include

| Plant Size (m³/d) | Solids Loading—Winter Operation | | | |
	kg/wk	kg/Operating Day (5 day/wk)	kg/Operating Hour	Area (m²)
4000	2130	430	108	12.5
20 000	10 640	2130	532	62.0
40 000	21 280	4300	716	84.0
200 000	106 400	21 280	1480	173.0

*Summer operation at 0.65 to 0.8 of winter conditions; DAF loading at 8.6 kg/m²·h.

TABLE 21.6 Design Solids Loading Rate for Dissolved Air Flotation (DAF) Systems*

| Location | Loading | | Polymer | % Suspended Solids |
	kg/m²·h	m³/m²·h		
Kenosha, Wis.	53	—	No	99+
Chicago, Ill.	28.3	—	No	99+
Amarillo, Texas	14.2	—	No	92+
East Fitchburg, Mass.	42.5–85.0	—	No	99.5
Xenia, Ohio	42.5	—	No	99.5
Eugene, Ore.	35.4	—	No	90+
Bernardsville, N.J.	62.3	2.9	No	94.5
Morristown, N.J.	48.1	1.1	No	97.0
Bay Park, N.Y.	36.8	0.7	No	94.0
East Wenatchee, Wash.	42.5–0.8	2.9	No	97.3
Incline Vill., Nev.	31.2	2.5	No	96.4
Fairfax, Va.	17	3.2	No	93.0
Plano, Texas	36.8	5.8	No	95.0
San Pablo, Calif.	17	2.9	No	99.0
Richmond, Calif.	42.5	3.6	No	98.9
Tenneco, Texas	5.7	2.2	No	98.8
Adolf Coors, Golden, Colo.	104.8	—	Yes	99.4
Springdale, Ark.	70.8	—	Yes	99+
Biddeford, Maine	90.6	—	Yes	99+
East Fitchburg, Mass.	85	—	Yes	98.3
Athol. Mass.	90.6	—	Yes	99.4
Somerset, Mass.	99.1	—	Yes	99+
Dartmouth, Mass.	90.6	—	Yes	98
The Dalles, Ore.	68	—	Yes	99.3
Denver, Colo.	104.8	—	Yes	99
Amarillo, Texas	65.1	—	Yes	99.2
Warren, Mich.	59.5	—	Yes	99+
Atlanta, Ga.	135.9	—	Yes	99.7
Chicago, Ill.	70.8	—	Yes	99+
Abington, Pa.	82	2.9	Yes	96.2
Hatboro, Pa.	83.5	1.8	Yes	96.0
Omaha, Neb.	87.8	1.8	Yes	99.4
Bellview, Ill.	107.6	1.1	Yes	98.7
Indianapolis, Ind.	59.5	3.6	Yes	95.0

TABLE 21.7 Performance Averages When Using a Dissolved Air Flotation Thickener to Cothicken Solids (1994–1995) (Butler et al., 1997)

Location	Loading		Polymer	% Suspended Solids
	kg/m²·h	m³/m²·h		
Frankenmuth, Mich.	184.1	3.2	Yes	99.1
Oakmonth, Pa.	85	2.5	Yes	98.7
Columbus, Ohio	93.5	2.5	Yes	99.5
Levittown, Pa.	82.1	2.5	Yes	99.4
Bay Park, N.Y.	138.8	2.9	Yes	99.6
Nashville, Tenn.	144.4	1.4	Yes	99.6
East Wenatchee, Wash.	45.3	3.2	Yes	98.6
Plano, Texas	53.8	5	Yes	98.8
Richmond, Calif.	42.5	3.6	Yes	98.0
San Pablo, Calif.	31.2	2.9	Yes	98.6

TABLE 21.7   Performance Averages When Using a Dissolved Air Flotation Thickener to Cothicken Solids (1994–1995) (Butler et al., 1997) (*Continued*)

polymer to improve clarification and thickening (see Table 21.7). Design engineers should ensure that primary solids and WAS are mixed thoroughly because variable concentrations of mixed solids can cause operating problems and result in poor thickening performance. Mixing is important to maintain consistent polymer doses in the thickening process.

3.4   Ancillary Equipment

In addition to the thickening tank, a complete DAFT system includes a number of appurtenances (e.g. pressurization-system; chemical conditioning; and various control elements).

3.4.1   Pressurization System

3.4.1.1   Pipes, Valves, and Instruments   Typical recycle-flow DAFT systems have numerous valves and fittings (e.g., interconnecting pipe and pipe fittings, liquid and gas flow-control valves, gas and liquid flowrate indicators, and a level-control valve). All must be properly designed to ensure proper DAFT operations.

Pressurization system components must be spaced as closely as possible to reduce costs and minimize pressure loss and air release in the pipes. Liquid recirculation piping is typically sized to produce a liquid velocity of 0.9 to 1.5 m/s (3–5 ft/s) and manufactured with Schedule 40 or 80 carbon steel or coal-tar, epoxy-coated, carbon steel materials.

Designers should utilize traditional piping practice, including the installation of eccentric reducers and expanders on the suction and discharge sides of the recirculation pump. Isolation valves (e.g., ball, plug or gate valves) with a maximum open passage and minimum pressure drop in the full-open position, should be installed on the influent and effluent side of the flotation vessel (feed, float discharge, and subnatant); recirculation pump; and pressurization tank.

Air supply piping should include oil and moisture traps, a pressure-regulating valve, a rotameter with appropriate temperature and pressure gauges as well as isolation valves and bypass lines, and a check valve next to the air-injection port in the pressurization tank or recirculation piping. Isolation valves should be either ball or gate valves. The pressurization tank's air supply piping should include a solenoid valve that is wired to shut off process air when the pressurization pump is off.

The air supply line should also include a pressure regulating valve, which is typically set to discharge air at 70 kPa (10 psi) above the air absorption tank's pressure to ensure a constant airflow despite small fluctuations in tank pressure. The airflow rotameter should be direct reading in standard volumetric units and equipped with a stainless steel float and a safety shield. The design should include a needle valve downstream of the rotameter to control the airflow rate. All valves in the air supply system and interconnecting air piping should be made of stainless steel.

If the pressurization system is designed to accommodate a variety of flowrates, a recycle liquid flow indicator, and control valve can be useful. The flow indicators should be able to handle solids laden streams. Venturi and vortex shedding indicators work well in this application. The flow indicator and control valve should be installed in the pump discharge piping upstream of the pressurization tank. Ball, eccentric-plug, and diaphragm valves are effective flow-control valves that can also serve as isolation valves for pump discharge.

All pressurization systems use a pressure relief valve, which is typically located next to the DAFT tank in the pressurization tank's discharge line. The valve reduces recycle-liquid pressure to atmospheric conditions; the air dissolved under pressure is precipitated at the valve in the form of microscopic air bubbles. These air bubbles contact the solids to be floated. Sometimes the pressure relief valve can be used to control recycle liquid flows. Design engineers should consult pressurization system manufacturers in each instance.

Operators can use a float-controlled air bleed off valve to maintain the liquid level in the pressurization tank. It typically bleeds off a small amount of excess air. If the water level rises, the float closes the bleed port so the air will force the liquid level back down, after which the air bleed resumes. If an alarm circuit is used to indicate a high water level, a float switch can be wired to an air-bleed solenoid valve that bleeds off excess air.

The DAFT tank feed line and subnatant recirculation piping should include provisions for feed and subnatant sampling. Polymer addition taps should be installed in both feed and subnatant lines and should be far enough upstream of the discharge point to allow for thorough mixing. The proper location of these taps is site-specific and should be tailored to the design. Drain plugs should be installed in all low points in feed and subnatant piping. Tee joints should be used for cleanouts rather than elbows so that operators can remove any debris that becomes lodged in process piping.

3.4.1.2 Pumps and Compressors Design engineers can use positive-displacement, diaphragm, piston, or progressing cavity pumps to feed solids to the DAFT, although centrifugal pumps have been preferred. The pumps should possess variable capacity and an operating range wide enough to accommodate expected variations in solids-production and thickening requirements, as well as variations in feed solids characteristics. They should also be equipped with a flow totalizer or monitor so operators can maintain records of the amount of solids thickened and control DAFT operations. Each DAFT should have its own dedicated pump.

A key element of any DAFT system is the pressurization pump, which feeds enough liquid into the pressurization tank to ensure that the flotation tank will receive the desired amount of dissolved air. Open-impeller, centrifugal pumps typically are used for this purpose. Single- and two-stage pressurization pumps have also been used. Most currently operating DAFT thickeners use single-stage pressurization systems. Two-stage pumps reportedly provide more air-dissolving efficiency than single-stage ones. If using two-stage pumps, compressed air is delivered to the suction end of the second stage.

For system flexibility, design engineers should use pressurization pumps with a relatively steep head-capacity curve. This allows operators to adjust pump flow by throttling the pump isolation valve between the pressurization pump and air-dissolving tank. Throttling provides controlled discharge from pumps with steep head-capacity curves. This is not true for centrifugal pumps with a flat curve; in this case, using a throttling valve could induce pump surging.

Pressurization pumps typically use single-speed motors. The use of two-speed motors or adjustable sheaves for variable head and flow capability depends on several factors (e.g., number of DAFT tanks, operating method, quantity of solids to be thickened under both average and peak conditions, and degree of flexibility desired). Although initial costs are higher, variable-speed pressurization pumps can lower power costs and enhance flexibility.

A variety of air compressors (e.g., reciprocating piston, rotary vane, and screw) can be used to provide air for the DAFT process. Some WRRFs use central compressors to meet DAFT air requirements as well as other needs in the facility.

Most flotation systems have their own air compressors. Reciprocating piston-type units are the most common and are typically sized to deliver at least twice the maximum air theoretically required for saturation so the compressor can operate in an unloaded condition about 50% of the time.

In addition to the compressor, a pressure reservoir, air filter, oil trap, pressure regulator, and airflow meter are required.

3.4.2 Chemical Conditioning
Dissolved air flotation processes often have polymer systems, which include mixing and storage tanks and chemical feed pumps. The systems can be purchased as a unit from polymer suppliers or designed by engineers. Either way, they should use variable-capacity, positive-displacement pumps so operators can accurately control the amount of polymer used. Each flotation thickener should have its own chemical pump (see Chapter 20).

3.4.3 Thickened Sludge Conveyance
DAFT float and underflow can be between 3% and 8% solids (by weight), depending on the application. This cake is highly viscous and often thixotropic. When skimmed as float or scraped as underflow it is typically

- directly discharged to a collection well for subsequent pumping, or
- directly discharged to an open-throat progressing cavity pump.

It is common to use some type of positive displacement pump such as a progressive cavity pump, diaphragm pump or disc pump to convey thickened solids.

3.4.4 Odor Control and Ventilation

By design DAFTs have a large liquid/solids exposed surface, which will generate odors. The most common odors are related to reduced sulfur compounds but ammonia compounds can also occur.

If odor treatment is provided it will be necessary to enclose the DAFT, ventilate the air space in the enclosure, and treat the exhausted air. Required ventilation rates may be as high as 20 air changes per hour to prevent corrosion and maintain evacuation of all odors. The odorous air exhausted from the enclosure should be treated before atmospheric discharge if the area adjacent or near the WRRF is expected to be impacted by odors. Odor-treatment options (see Chapter 6) include packed-tower scrubbers and mist scrubbers (chemical scrubbers), activated-carbon beds, and biofilters. In chemical scrubbers, hypochlorite and sodium hydroxide are typically used to scrub sulfur-related odors and sulfuric acid solutions can be used to scrub ammonia-related odors. Another option is adding an odor-control compound (e.g., hydrogen peroxide) to the feed solids, but the compound first should be tested to determine whether it interferes with polymer efficiency and solids thickening.

If the DAFT is enclosed, close attention must be paid to materials of construction of all components to assure long life and prevent excessive corrosion of the components.

3.5 Process Control

3.5.1 Overview

A DAFT system typically includes sludge feed pumps, the chemical conditioning system, the DAFTs, the pressurization system, and the thickened solids conveyance system. The criteria that can be monitored for process control are the hydraulic feed rate, the solids loading rate, and the air-to-solids ratio. These criteria should be checked against the original design criteria if there are process control issues. The DAFT systems that can be adjusted for process control include the chemical conditioning system, the pressurization system, and the skimmers and rake arms.

If using chemical conditioning the chemical effectiveness should be monitored to determine if thickened sludge and solids capture goals are being met. The control valve and the air compressors are critical to control of the pressurization system and ensure there is an air cushion inside the pressurization tank. The speed of the skimmer arms can be changed and the height of squeegees above the ramp adjusted depending on factors such polymer dose, hydraulic feed rate, solids loading rate, and air-to-solids ratio to improve solids thickening performance.

3.5.2 Process Control Monitoring

Maximizing cake solids, maintaining reasonable capture efficiency, and minimizing polymer requirements are difficult tasks to accomplish without some instrumentation and controls. Equipment such as flow monitors are necessary to understand the hydraulic feed rate and the solids discharge rate. Solids blanket-depth indicators are used to monitor the effectiveness of solids thickening, measure solids inventory, and control the number and speed of the thickened sludge conveyance pumps. Suspended solids monitors on the overflow may be used to monitor solids capture in the DAFT. Pressure gauges are necessary to monitor air pressure and control dissolution in the pressurization tank.

3.6 Performance Assessment and Optimization

3.6.1 Performance Parameters

The process performance of a DAFT is primarily assessed through thickened sludge concentration, polymer consumption, and overflow quality. Other parameters of importance are operational consistency, equipment longevity, ease of operation, and ease of maintenance.

The expected process performance for a specific feed sludge and a DAFT are difficult to estimate through pilot-testing as discussed in Section 3.7. Expected operational performance assessment and optimization will largely come through operational testing or learning from others to determine how DAFTs have performed elsewhere on similar feed solids.

3.6.2 Optimization Measures

System optimization focuses on maintaining consistent thickened sludge concentration, minimizing polymer consumption, and improving overflow quality. The DAFT systems or components that can be adjusted and should be periodically tested to optimize performance include: the chemical conditioning system, the skimmers and rakes, the pressurization systems, and the thickened sludge conveyance pumps. The chemical conditioning systems should be tested on a regular basis to identify the polymer (or other chemical aid) that provides the best value (cost versus performance) throughout the year. Many facilities' sludge characteristics change in winter versus summer and different chemicals or chemical doses may perform differently during each of those seasons. The elevation of the skimmer arms should be periodically compared to the elevation of the beach to optimize float removal and prevent subnatant carryover. The pressurization system air pressure and the recycle stream flow rate can affect the amount of dissolved air. This should be checked/tested periodically to ensure optimal performance (i.e., float solids thickness and solids capture). The thickened sludge conveyance pumps such as a progressing cavity pump naturally wear over time affecting the ability to maintain a constant, required withdrawal rate.

3.7 Evaluation and Scale-Up Procedures

DAFTs have been widely used to thicken WAS since the mid-1960s. Engineers can typically size DAFT equipment based on design experience at comparably sized facilities. However, bench- or pilot-scale performance investigations can provide valuable information, such as

- Thickened solids concentration, solids recovery rates, and chemical needs;
- DAFT designs that can satisfy performance requirements;
- Acceptable loading rates with polymer addition; and
- The causes of poor or suboptimum DAFT performance.

Before conducting any bench- or pilot-scale tests, however, engineers should collect a representative sample of the solids to be thickened. Then they should determine its suspended solids content, volatile solids content, and SVI for WAS.

3.7.1 Bench-Scale Evaluations

Bench-scale tests provide insight into the thickening characteristics of specific solids. Manufacturers have designed and built bench-scale units that are available for such evaluations. They also have scale-up criteria for their own equipment that enable engineers to predict full-scale operations with reasonable accuracy.

A typical bench-scale unit consists of a pressurization chamber, a flotation chamber, a pressure-release valve, and ancillary equipment. The test is typically conducted as follows:

- Introduce a sample of the fluid (typically clarified liquid) to the pressurization chamber (a full-scale unit typically uses subnatant).

- Adjust an air-bleed valve to allow compressed air to bubble through the liquid. After a suitable pressurization period (typically 10 minutes), close the air-bleed valve.

- Place a measured sample of the material to be floated (e.g., WAS) in the flotation chamber.

- Open a pressure-relief valve, allowing pressurized fluid to enter the flotation chamber and be distributed about the space. Close the pressure-relief valve when the total volume in this chamber reaches a predetermined level. The material is permitted to float in the chamber for a suitable period (typically 10 minutes).

- Collect samples of subnatant and floating material.

To identify the optimum value, engineers should perform enough tests to determine system performance at several air-to-solids ratios. Air-to-solids ratios can be varied by changing the solids concentration or volume of sample to be floated. Further tests may be required to assess the efficacy of chemical conditioning and the effects of feed-solids concentration.

Bench-scale tests are especially useful for predicting float-solids content and solids capture, as well as for evaluating the effects of chemical flotation aids on float-solids and solids capture. However, they are seldom used to establish design loading rates because of uncertainties in scale-up.

One option for using bench-scale test data to size full-scale DAFT units involves applying batch or limiting flux methods to the interface height-versus-time data obtained during flotation (Wood, 1970). Engineers must develop separate flux curves for each air-to-solids level of interest. As with gravity thickening, scale-up uncertainties have limited application of this procedure. Also, engineers have accumulated experience in designing DAFT systems to thicken WAS and developed other means to thicken the solids.

3.7.2 Pilot Flotation Units

The flotation performance at a given installation depends on the interaction of many factors. In most situations, pilot-scale flotation units are the best way to identify this performance. Results obtained from pilot and field equipment are analogous when the devices are geometrically, kinematically, and dynamically similar. However, complete similarity is seldom achieved because of innate physical differences between pilot- and full-scale equipment, so the goal is to be as similar as practical.

Two different-sized systems are geometrically similar if they are proportional in all corresponding dimensions (e.g., the length, width, and depth of the flotation unit). They are dynamically similar when the ratios of all corresponding forces are equal. They are kinematically similar if velocities at corresponding points have the same ratio. Kinematic similarity is approached when geometrically similar pilot- and full-scale equipment have identical hydraulic loading rates, and when the pilot-scale unit's pressure-relief valve is a properly scaled-down version of the valve on the full-scale unit.

Ideally, both units should treat the same feed material, create the same size gas bubbles, and operate at the same pressure. Also, the loading rate and air-to-solids ratio used during pilot-scale tests must be applicable to full-size equipment.

When scaled-up, pilot-testing data can only reveal the full-scale unit's probable performance because pilot units are not completely similar to full-scale units. Equipment manufacturers have scale-up information specific to their own equipment.

3.8   Design Example

A municipal WRRF is planning on using DAFTs to co-thicken PS and WAS. The facility has primary clarification and complete-mix aeration basins.

Given:

1 kg = 1 L of water (8.34 lb = 1 gal water)

Primary sludge, at 0.5% solids

   Average = 2000 kg/h (4409 lb/hr)

   Peak = 3450 kg/h (7606 lb/hr)

WAS, at 0.5% solids

   Average = 1400 kg/h (3088 lb/hr)

   Peak = 2400 kg/h (5292 lb/hr)

Operation is continuous; 24 hours per day, 7 days per week

Wastewater temperature

   Minimum: 22°C (72°F)

   Maximum: 35°C (95°F)

Determine:

The number of DAFTs, size DAFTs and pressurization system components

Assumed:

1. Initial guess of the combined solids loading rate of 6 kg/m² h (1.23 lb/hr sq ft) for sizing of the DAFT tanks.

2. Maximum solids loading rate 10 kg/m² h (2 lb/hr sq ft)

3. DAFTs to operate without polymer with all units in service

4. Use 448 kPa (65 psig) pressurization system

5. Use air to solids ratio of 0.03

6. Circular cast-in-place concrete DAFT tanks will be constructed

Calculations:

Maximum rate:

   3450 kg/h + 2400 kg/h = 5850 kg/h (12 898 lb/hr)

Average rate:

   2000 kg/h + 1400 kg/h = 3400 kg/h (7497 lb/hr)

Gross thickener area required:

5850 kg/h / 6 kg/m² = 975 m² (10 495 sq ft)

2 DAFT tanks at 25-m diameter (82-ft diameter)

3 DAFT tanks at 20-m diameter (65.6-ft diameter)

4 DAFT tanks at 17.5-m diameter (57.4-ft diameter)

Use 3 DAFT tanks at 20-m diameter (65.6-ft diameter)

Maximum loading rate = 6.21 kg/m² h (1.27 lb/hr sq ft)

With one out of service

Maximum loading rate = 9.31 kg/m² h (1.91 lb/hr sq ft)

Size all system components on the basis of one unit out of service.

Maximum rate to one unit:

5850 kg/h / 2 units in service = 2925 kg/h (6449 lb/hr)

Average rate to one unit:

3400 kg/h / 2 units in service = 1700 kg/h (3749 lb/hr)

Hydraulic load to each unit with one out of service:

[(5850 kg/h) / (2 units in service*0.5% / 100)] / [(60 min/s) * (kg/L)] = 9750 L/min (2578 gpm)

Hydraulic loading rate to each unit with one out of service:

[9750 L/min * 1.44 m³/d / (L/min)] / (3.142 * 20 m * 20 m / 4) = 44.7 m³/m² d (0.76 gpm/sq ft)

Pressurization system:

Design for two (2) pressurization systems per DAFT tank

Two (2) in service for maximum load

One (1) in service for average load

By inspection the average load rate governs. Therefore, size each of the two (2) pressurization systems per unit (per DAFT tank) for the average load to each DAFT tank with one of the three (3) DAFT tanks out of service

Required air at an air to solids ratio of 0.03

(3400 kg/h / 2 units in service) * (0.03) = 51 kg/h (113 lb/hr)

Select 448 kPa (65 psig) system pressure

Assume 85% air saturation efficiency in the pressurization system

Assume 95% release pressure efficiency at the DAFT tank

51 kg/h / (0.85*0.95) = 63 kg/h (139 lb/hr)

Saturation constant (C_s) based upon Henry's Law at 35°C (95°F) system pressures

C_s 448 kPa (5.42 atm) = 105 mg/L

C_s 101 kPa (1 atm) = 19.5 mg/L

Delta C_s = 85 mg/L

Pressurized flow required to saturate 63 kg/h (139 lb/hr) at 35°C (95°F)

(63 kg/h * 1000 g/kg) / (0.085 g/L * 60 min/h) = 12 353 L/min (3263 gpm)

Total hydraulic load at maximum flow with one unit out of service

9750 L/min + 2 * 12 353 L/min = 34 456 L/min (9102 gpm)

Total overflow rate

[34 456 L/min * 1.44 m³/d / (L/min)] / (3.142 * 20 m * 20 m / 4) = 158 m³/m² d (2.69 gpm/sq ft)

Since this is above the suggested maximum daily hydraulic loading of 120 m³/m² d (2 gpm/sq ft) if no conditioning chemicals are used per Section 3.3.3, above. It is recommended to either (1) increase the DAFT tank diameter, or (2) provide a polymer flotation aid to maintain satisfactory performance. For this design example it is recommended to increase the DAFT tank diameter. By increasing the DAFT tank diameter from 20 m (65.7 ft) to 23 m (75.5 ft) reduces the total overflow rate from 158 m³/m² d (2.69 gpm/sq ft) to 119 m³/m² d (2.0 gpm/sq ft). The maximum loading rate with all units in service and one out of service is reduced from 6.21 kg/m² h (1.27 lb/hr sq ft) to 4.69 kg/m² h (0.96 lb/hr sq ft) and 9.31 kg/m² h (1.91 lb/hr sq ft) to 7.04 kg/m² h (1.44 lb/hr sq ft), respectively.

Air flow; density of dry air at 20°C (68°F) and 101.325 kPa (14.7 psi) is 1.2041 kg/m³

63 kg/h / (1.2041 kg/m³ * 60 min/h) = 0.87 m³/min (31 SCFM)

4.0   Centrifuge

Centrifugal thickening is analogous to gravity thickening except that centrifuges can apply a force 500 to 3000 times that of gravity. The centrifugal force causes suspended solids particles to migrate through the liquid toward or away from the centrifuge's rotation axis, depending on the difference between the liquid's and solids' densities. The increased settling velocity and short particle-settling distance accounts for a centrifuge's comparatively high capacity.

Centrifuges have been used to thicken waste solids since the early 1920s, with solid-bowl conveyor centrifuges being the most widely used in this application. Variables affecting centrifuge thickening are grouped into three basic categories: performance, process, and design. Performance is measured by the thickened solids concentration, polymer and power consumption, and the suspended solids recovery in the centrate. The recovery is calculated from the thickened dry solids as a percentage of feed dry solids. Using the commonly measured solids concentrations, recovery is calculated as follows in eq 21.8:

$$\text{Percent Recovery} = (T/F) \times [(F - C)/(T - C)] \times 100 \qquad (21.8)$$

where T = Thickened sludge total solids (mg/L),
F = Feed total solids (mg/L),
C = Filtrate TSS (mg/L)

Process variables that affect thickening include feed flowrate, the centrifuge's rotational speed, differential speed of the conveyor relative to the bowl, pond depth, chemical use, and the physicochemical properties of the liquid and suspended solids (e.g., solids concentration, variability in feed solids concentration, particle size and shape, particle density, temperature, and liquid viscosity). These variables are the tools that WRRF operators have to optimize centrifuge performance.

4.1 Operating Principle

A centrifuge's main components are the bowl and the scroll. The bowl is mounted horizontally and turns rapidly to create the centrifugal force. The scroll is mounted inside the bowl and conveys solids from one end of the bowl to the other. Other important features of a centrifuge are identified in Figure 21.9.

The bowl consists of a cylindrical section and a conical section (see Figure 21.10). Both are typically cast stainless steel but can also be made of rolled stainless steel plate. The sections are assembled in the factory, machined, and balanced at a high speed. Figure 21.10 shows the primary components that comprise the centrifuge's rotating assembly.

The scroll consists of a stainless steel screw conveyor mounted on a hollow shaft (see Figure 21.11). It can either be an open design mounted to the shaft via spokes or a closed design mounted directly to the shaft. The entire scroll is mounted inside the bowl and can turn independently.

To thicken solids, the bowl and scroll typically operate at more than 1500 rpm; the scroll rotates just a few rpm faster (or slower) than the bowl to create a differential speed. Feed solids are injected into the scroll's hollow shaft and discharged into the spinning bowl. A conditioning agent, typically polymer, is added either at the centrifuge feed nozzle or further upstream within the feed solids piping depending on solids characteristics and need for mixing. Injection at the feed nozzle is typically sufficient but should be evaluated and determined on a case-by-case basis. The bowl's centrifugal force causes solids to migrate toward the bowl wall. The scroll's screw conveyor moves the solids up the conical section of the bowl and discharges them. Meanwhile, the liquid is discharged at the opposite end of the bowl via openings in the end plate (see Figure 21.12).

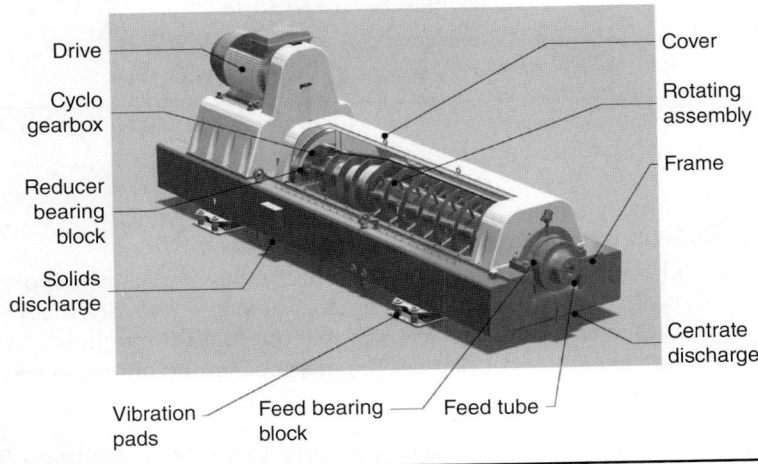

FIGURE 21.9   Main components of centrifuge (courtesy of Andritz Separation).

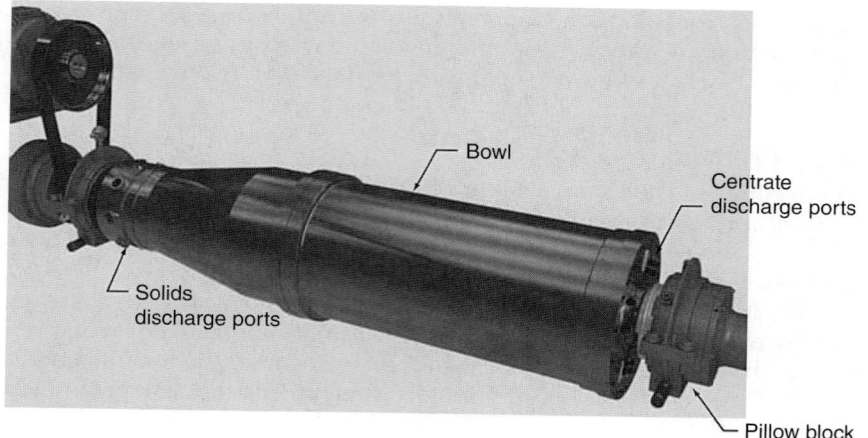

FIGURE 21.10   Centrifuge rotating assembly, showing cylindrical and conical sections of bowl, discharge ports, and pillow block bearings (courtesy of Andritz Separation).

FIGURE 21.11   An example of a centrifuge scroll (courtesy of GEA Westfalia Separator, Inc.).

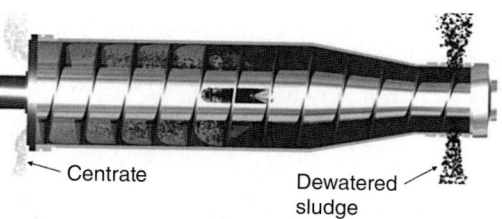

FIGURE 21.12   An example of a centrifuge thickening solids (courtesy of GEA Westfalia Separator, Inc.).

4.2   Physical and Mechanical Features

A basic solid-bowl conveyor centrifuge has the following main components: base, case, bowl, conveyor, feed pipe, main bearings, gear unit, and back drive. Each centrifuge has unique physical and mechanical features that can affect throughput, capture efficiency, polymer dose, cake solids concentration, power consumption, and system longevity.

4.2.1   Base

The base provides a solid foundation on which to mount and support the centrifuge's main components. Vibration isolators between the base and the machine foundation reduce the transmission of vibrations from the unit to the structural foundation. Vibration is discussed further below.

4.2.2 Case

The case completely encloses the rotating assembly, acting as a safety guard for personnel and a noise dampener. [A solid-bowl centrifuge's noise typically ranges from 80 to 90 dbA at 0.9 m (3 ft).] The case also contains and directs cake solids and centrate as they are discharged from the rotating assembly.

4.2.3 Bowl

A solid-bowl centrifuge's bowl typically resembles a cylinder or cone. Proportions vary, depending on manufacturer. Bowl diameters range from 0.23 to 1.38 m (9–54 in.), and the bowl length-to-diameter ratio ranges from 2.5:1 to 4:0.1. Thickening centrifuge capacity typically ranges from 40 to 3000 L/min (10–800 gpm).

Bowls used for wastewater treatment applications are typically made of carbon steel, duplex, or 300 series stainless steel with strips or grooves on the inside that retain a protective layer of solids. Sometimes the bowl has a stainless steel or ceramic liner.

4.2.4 Bowl Geometry

The bowl is one of the centrifuge's most critical features. Bowl geometry significantly affects throughput, capture efficiency, and cake solids concentration.

The bowl consists of two major sections: the cylinder and the cone (see Figure 21.13). The critical dimensions that manufacturers use to describe a particular centrifuge are bowl diameter (d_{cyl}), cylindrical bowl length (L_{cyl}), overall bowl length ($l_{cyl} + l_{co}$), discharge diameter (d_{isdd}) and beach angle (β).

These dimensions affect centrifuge performance as follows:

- Together, bowl diameter and bowl speed dictate the centrifugal force at the bowl wall. At a given bowl speed, centrifugal force at the bowl wall increases as bowl diameter increases.

- The discharge diameter dictates the pond depth of solids in the centrifuge. This is associated with the maximum volume of solids that the centrifuge can hold.

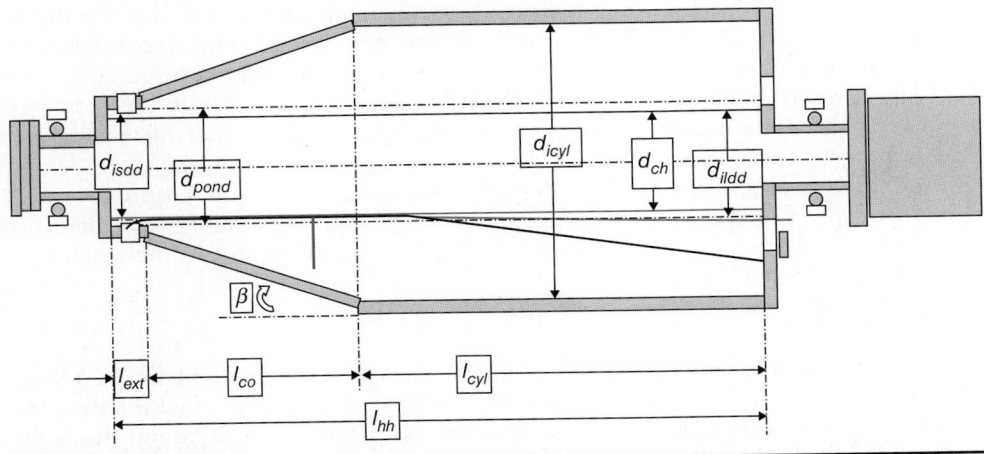

FIGURE 21.13 Important dimensions within centrifuges (courtesy of Centrisys Centrifuge Systems).

At a given bowl diameter, pond depth and maximum solids volume decrease as discharge diameter increases.

- Thickened solids are conveyed up the beach (conical section of the centrifuge) and then discharged. Manufacturers have found that a 15- to 20-degree beach angle is optimum for thickening centrifuges.

4.2.4.1 Bowl Volume   Table 21.8 presents the data used to calculate the total volume of a centrifuge bowl. However, the centrifuge's usable volume—total volume minus the air space associated with the discharge diameter (the space that the scroll occupies)—is the actual volume of solids that a centrifuge can hold (see Figure 21.14).

4.2.4.2 Cylinder Volume   Thickening centrifuges use the conical section of the bowl to convey solids to the discharge point. So, the only part of the bowl separating solids from liquids is the cylindrical portion. The capacity of a centrifuge can be determined by G-volume. G-volume is a function of the solids acted on by G-force while inside the cylindrical section of the centrifuge and is proportional to the speed. G-volume can be calculated per eq 21.9 with parameters defined per Figure 21.13.

$$\text{G-volume} = kN^2 d_{cyl} l_{cyl} (d_{cyl}^2 - d_{isdd}^2) \tag{21.9}$$

where $k = 4.83 \times 10^{-8}$
N = operating bowl speed
d_{cyl} = bowl diameter, mm (in.)
l_{cyl} = bowl cylindrical length, mm (in.)
d_{isdd} = solids discharge diameter, mm (in.)

G-volume can be used to compare capacity of different manufacturer's proposed units for a specific operating condition. If this is done, it is important to assess G-volume at the units' operating speed, which may not be the same as the units' nameplate speed.

4.2.5   Scroll/Conveyor

A helix or screw-conveyor (scroll) conveys solids along the bowl and up the beach, where they are discharged. The scroll assembly consists of a central core or hub, a feed compartment, and feed ports lined with abrasion-resistant ceramic or tungsten carbide. The helical flights leading surfaces and blade tips are coated with abrasion-resistant materials. Modern solid-bowl centrifuges use replaceable tiles that are made of ceramic or tungsten carbide. Some manufacturers use flame-sprayed tungsten carbide coatings rather than tiles. The entire assembly fits concentrically into the centrifuge bowl. Conveyor speed is controlled by the gear unit and back-drive assembly. Flocculent aids are added to the feed compartment, into a separate injection port in the machine, or in upstream sludge feed piping.

The scroll may be open or closed.

4.2.5.1 Open Scroll   An open scroll consists of a steel ribbon flight attached to a scroll shaft by spokes. Manufacturers that use this type of scroll claim that it reduces turbulence in the bowl because the centrate does not have to travel around the flight and agitate solids along the bowl wall. This could reduce polymer use and improve capture efficiency, according to the manufacturers.

Manu-facturer	Bowl Diameter (mm)	Bowl Circum-Ference Length (m)	Bowl Length (m)	Cone Length (mm)	Cone Angle (Deg)	Cylinder Length (m)	Cylinder Volume (m³)	Discharge Diameter (mm)	Usable Bowl Volume (m³)	Unusable Bowl Volume (m³)	Bowl Speed (s/rev)	Normal Operating Speed (rpm)	Angular Velocity (m/s)	Centripetal Acceleration (m/s²)	G-Force at Bowl Wall (m³)	G-Force at Bowl Wall (gal)
A	740	2.32	3.05	450	20	2.6	1.1	410	0.8	0.3	0.024	2500	97	25 500	2600	600 000
B	690	2.15	2.92	470	15	2.5	0.9	430	0.5	0.4	0.023	2600	93	25 000	2600	380 000
C	740	2.32	3.1	490	20	2.6	1.1	380	0.8	0.3	0.026	2300	89	22 000	2200	500 000
D	760	2.39	3.07	380	20	2.8	1.3	480	0.8	0.5	0.027	2200	89	21 000	2100	500 000

TABLE 21.8 Characteristics of Various Centrifuge Bowls

FIGURE **21.14**   The usable volume in a centrifuge bowl.

4.2.5.2   Closed Scroll   A closed scroll consists of a flight directly attached to the scroll shaft. Manufacturers that use this type of scroll claim that it allows solids inventory to be built up higher than the open scroll does. (The open scroll permits more cake compression, potentially producing higher solids.) One manufacturer has used closed scrolls successfully in several dewatering applications.

4.2.6   Scroll Configuration
The scroll can be configured to lead or lag (see Figure 21.15). A leading scroll runs slightly faster than the bowl, while a lagging one runs slightly slower.

4.2.7   Scroll Drive/Back-drive
The gear unit and back-drive assembly allow the bowl and the conveyor to maintain different speeds. The drive typically consists of a planetary or cyclo gear and a mechanical, hydraulic, or electrical back-drive. The gear unit is typically lubricated by either an oil-bath or grease-lubrication system.

4.2.8   Differential Speed Adjustment
The scroll drive (back drive) system turns the scroll relative to the bowl creating a differential speed that can range from 1 to 15 rpm. At higher differential speeds, thickened solids are removed from the centrifuge more rapidly and have lower solids concentrations. At lower differential speeds, thickened solids are removed from the centrifuge more slowly and have higher solids concentrations.

Thickening centrifuges operate at low differential speeds to keep solids in the bowl as long as possible. Differential speed is a fine adjustment that operators use to achieve a desired thickened solids concentration while providing the required throughput and capture efficiency, and minimizing polymer dose. They make fine adjustments via a separate scroll drive operating system that converts a high motor speed to a slow scroll

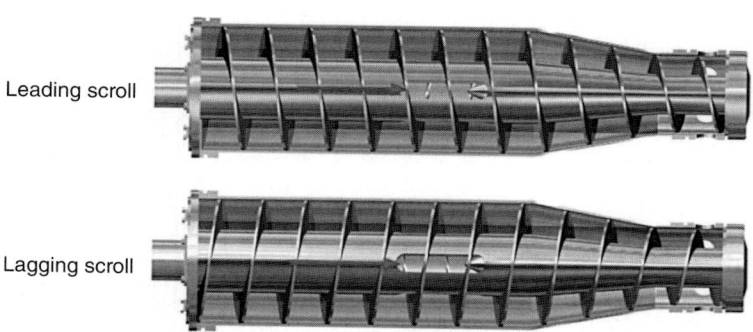

Leading scroll

Lagging scroll

FIGURE **21.15**   Possible scroll configurations (courtesy of GEA Westfalia Separator, Inc.).

speed. The finer the scroll speed adjustment, the more control operators have to optimize the centrifuge.

4.2.9 Motor Type and Size

Most centrifuge manufacturers provide two motors and two VFDs with their systems: a scroll drive motor and VFD, and a main drive motor and VFD. The AC motors receive current directly from the VFDs. The scroll drive motor speeds up or slows down the scroll; it does not seem to affect the total connected horsepower of either VFD. However, when the scroll drive motor is used to speed up the scroll, the size requirement is typically larger than when it is used to slow down the scroll. Conversely, when the scroll drive motor is used to speed up the scroll, the main drive motor size requirement is typically smaller than when the scroll drive motor is used to slow down the scroll.

4.2.10 Feed Pipe

The feed pipe is removable, and design engineers should determine the length of the pipe to ensure there is sufficient space to remove the feed pipe if necessary. Several polymer-feed locations can be used and should be provided to allow for polymer optimization: into the feed pipe, directly into the centrifuge, or into the feed piping upstream of the centrifuge. Engineers should make sure the feed pipe has a flexible inlet connection (rather than a valve or fitting) to protect against damage due to vibration.

4.2.11 Bearings

Depending on machine size and speed, three types of main bearings—ball, spherical, and cylindrical—support the entire rotating assembly. The bearings are lubricated by grease, a static oil bath, an oil/air mist system, or an external circulating-oil system and typically have an L10 life of 100 000 hours. External oil lubrication systems consist of an oil pump, oil reservoir, heat exchanger, cooling water (typically potable), and various oil filters, valves and instruments. Appropriate lubrication is critical to centrifuge operation and equipment life. Lubrication minimizes friction and associated temperature increases at the bearings. Improper lubrication can quickly damage the main bearings, resulting in costly repairs, potential rebuilds, and system rebalancing. Bearings can also be damaged by excessive vibration.

4.2.12 Abrasion Protection

Centrifuges have a number of areas susceptible to abrasion (e.g., the bowl's interior wall, the conveyor blades, the feed compartment, the feed ports, and the solids discharge area). Such areas are typically protected by various hard-facing materials (e.g., sintered tungsten carbide or ceramic). Abrasion protection can include tiles on conveyor tips, flame-sprayed tungsten carbide coatings, and replaceable inserts. Modern techniques have increased conveyor lives to between 10 000 and 20 000 hours. Tiles are tack-welded to the conveyor and require replacement when worn. Experienced maintenance personnel can replace tiles onsite if equipped to do so. Otherwise, the rotating assembly is shipped to the manufacturer's service center for replacement. Flame-sprayed coatings, when worn, should be reapplied at the manufacturer's facilities.

4.2.13 Vibration

The high-speed revolution of a centrifuge creates significant vibration that must be addressed. To dampen vibrations transmitted to the foundation or piping, the centrifuge base should be supported on vibration isolators and directly connected piping and conduit should have flexible connectors. Designers can also consider an isolated foundation for the centrifuges that is independent of the surrounding structure. Centrifuge vibration is a significant parameter that must be considered in the structural design of the thickening facility. Excessive vibration can damage the centrifuge itself, the surrounding structure, and equipment. While vibration isolators reduce the vibration imparted on the surrounding structure, they do not eliminate vibration completely. Equipment resonance, especially during centrifuge startup and shutdown, can result in substantial vibration and movement. Oil-filled isolators may reduce vibration more than isolators that rely only on elastomeric or spring dampening characteristics. It is critical to properly address the vibration created by a centrifuge in the design of the support systems and connected equipment. Failure to do so can lead to major damage and a dangerous operating area.

4.3 Process Design Conditions and Criteria

4.3.1 Overview

Centrifuge manufacturers offer designs with substantially different features. Table 21.9 lists the major design and operating variables that influence the operation of a horizontal solid-bowl centrifuge. These variables are discussed at length in the literature (U.S. EPA, 1979; WEF et al., 2012). A desirable characteristic of the centrifuge is that its performance—as measured by thickened solids, polymer consumption, power demand, and solids capture—can be adjusted to desired values by modifying control variables [e.g., feed flowrate, bowl and conveyor differential speed, conditioning-chemical (polymer) dose, and pond depth].

Table 21.10 indicates how a horizontal solid-bowl centrifuge's capabilities relate to basic rotating assembly size and operating speed. (Specific design recommendations are omitted because anticipated performance ranges vary widely due to design differences and solids characteristics.)

Sometimes polymer addition can increase a centrifuge's hydraulic loading while maintaining its solids-capture and thickening characteristics. Polymer use typically can improve solids-capture efficiencies to between 90% and more than 95%.

Basic Machine Design Parameters	Adjustable Machine and Operational Features	Solids Characteristics
Flow geometry	Bowl speed	Particle and floc size
Countercurrent	Bowl and conveyor differential speed	Particle density
Cocurrent		Consistency
Internal baffling	Pool depth and volume	Viscosity
Bowl/Conveyor geometry	Feed rate	Temperature
Diameter	Hydraulic loading	SVI
Length	Solids loading	Volatile solids
Conical Angle	Flocculant use	Solids retention time
Pitch and lead		Septicity
Maximum pool depth		Floc deterioration
Solids and flocculant feed points		
Maximum operating speed		

TABLE 21.9 Factors Affecting Centrifugal Thickening

Table 21.10 Operating Results Reported For Horizontal Solid-Bowl Centrifuges

Location	Activated-Sludge Type	Feed Solids Concentration (mg/L)	SVI (% VSS)	Feed Flow Rate (L/min)	Thickened Solids Concentration (%)	Solids Capture (%)	Polymer Use, g Active Polymer/Dry kg of Solids	Machine Size (Bowl Diameter × Length) (mm)	Bowl Speed (r/min)	Centrifuge Configuration
Atlantic City, New Jersey	Conventional	3000	100 (60)	1230	10	95	2.5	740 × 2340	2600	Countercurrent
Los Angeles, California, Hyperion	Conventional	4 8000–6000	110–190	2300–3000	3.7–5.7	88–91	None	1100 × 4190	1600	Cocurrent
					3.6–6.0	77–96	0.2–2.2	1100 × 4190	1600	Cocurrent
	Conventional	4800–6000	110–190	2300–3000	1.9–7.9	47–89	None	1100 × 3600	1995	Countercurrent
					1.7–8.2	57–97	0.4–1.4	1100 × 3600	1995	Countercurrent
Oakland, California, East Bay MUD	High-purity oxygen	5000	250–400	4200	7	66	6	1000 × 3600	1995	Countercurrent
Naples, Florida	Conventional	10 000–15 000	70–80	380	6	90–92	None	740 × 3050	2000	Countercurrent
Milwaukee, Wisconsin, Jones Island	Conventional	6000–8000	80–150 (75)	1100–1900	3–5.5	92–93	—	—	1000	Cocurrent
Littleton, Colorado	Conventional	6000–8000	100–300	570–1100	6–9	88–95	3–3.5	740 × 2340	2300	Countercurrent
Lakeview, Ontario (Canada) (PM 75 000)	Conventional	7560	80–120	840	4.7	77	None	740 × 2340	2300	Countercurrent
Lakeview, Ontario (Canada) (XM-706)	Conventional	7120	80–120	1350	6.1	65	None	740 × 3050	2600	Countercurrent

Following are significant design considerations for centrifuge thickening applications:

- Provide effective wastewater degritting and screening upstream of the centrifuge. If wastewater screening or grinding is inadequate, feed solids should be sent through grinders before entering the centrifuge to avoid plugging problems.

- Use a feed source with a relatively uniform consistency (a mixed storage or blend tank is often appropriate) and feed it to the centrifuge via an adjustable-rate pump with positive flowrate control and consistent flow.

- Consider handling thickened solids via one of the following methods: direct discharge to a collection well followed by transport using a positive-displacement pump, direct discharge to an open-throat progressing cavity pump, or discharge to a screw conveyor.

- Consider recycling centrate to either primary or secondary treatment processes and providing the ability to vent and/or suppress foam in the centrate piping. Direct connecting centrate piping to a drain system may limit the formation of foam and the emission of odors.

- Consider structural aspects (e.g., static and dynamic loadings from the centrifuge, resonance, vibration isolation, and provision of an overhead hoist for equipment maintenance). Provide snubbers (to isolate vibrations), piping flexibility, and flexible connections to auxiliary equipment. In seismic regions or sites with poor soils, consider foundation design that mitigates these issues.

- Provide water to flush the centrifuge during equipment shutdowns, reduce foaming in the centrate chute, and cool lubrication oil in forced oil bearing lubrication systems.

- Consider whether a heated water supply will be needed to periodically flush grease buildup.

- Ensure that the centrifuge is vented properly and consider whether odor controls are needed. Improper venting can allow solids to accumulate around the bowl and cause wear damage.

- When thickening anaerobically digested solids, consider the potential for struvite (ammonium magnesium phosphate) to form.

- Pay attention to polymer feed-system design. Refer to chemical conditioning for additional information on polymer systems.

4.3.2   Process Design Criteria

Engineers typically define process design criteria by establishing which solids characteristics are significant in a given application and determining how they affect process performance. Unfortunately, specific design criteria are not possible for centrifugal thickeners because of all the variations in both solids characteristics and centrifuge designs.

Centrifuges have independent hydraulic loading and solids loading capacities. For typical WAS thickening applications, hydraulic loading capacity governs over solids loading. Hydraulic loading to the centrifuge controls the liquid-phase residence time, which typically ranges from 30 to 60 seconds in thickening applications. For situations where thicker feed solids are present, solids loads must be considered. Solids concentration determines the specific solids load applied to the centrifuge (kilograms or pounds of dry solids per day) and the thickened solids output volume (cubic meters or gallons per day). These measurements help centrifuge designers determine the machine's parameters. They also help operators adjust machine and process variables (e.g., conveyor differential and feed rates) to balance load demands.

The density of the feed material's solids and liquid fractions is important, and operators have limited control over this characteristic. Activated sludge and mixed liquor typically have similar low floc densities. Operators frequently need to add chemical conditioners to increase the effective density of the aggregate floc, thereby increasing sedimentation or centrifugal settling rates.

The size and distribution of particles in the feed solids significantly affect the centrifuge's thickening performance, but these characteristics are difficult to measure accurately. For the most part, operators simply measure solids concentration rather than particle size or density. One concern—particularly with WAS—is that these naturally well-flocculated materials consist of small particles loosely bound together in one aggregate floc. This naturally occurring aggregate often breaks up easily, particularly under the high shearing forces in a centrifuge. So, polymers may be necessary to make the floc aggregate more cohesive.

4.3.3   Impacts of Different Feed Sludge Types
Solid-bowl conveyor centrifuges are versatile; they can be used to thicken a variety of waste streams. Most municipal WRRFs use them to thicken WAS; however, they can also be used to thicken primary solids, and to co-thicken primary solids and WAS. The type of solids to be thickened impacts some of the equipment design and performance parameters because of solids characteristics.

Primary solids settle rapidly and require lower bowl speed than WAS to thicken. It also requires more wear protection and screening/grinding due to the grit and debris typically present in primary sludge. Primary solids feed can be highly variable due to diurnal flows and primary sludge pump cycling. Substantial variation in feed solids can result in wide shifts in thickened sludge concentration and sub-optimal polymer and power consumption. The variability in concentration can be accommodated through upstream sludge blending tanks or automated pond depth adjustment systems. If provided, solids blending tanks can be used to mix primary solids with WAS for co-thickening. These tanks should be mixed to maintain homogeneity of centrifuge feed. Automated pond depth adjustment can be provided by certain manufacturers to maintain thickened solids concentration at setpoint values.

Some of the differences in thickening primary solids, WAS, or co-thickening both are indicated in Table 21.11.

4.4   Operational Considerations Related to Design
When determining the required size and number of units required for an application, the designer must consider operating conditions and redundancy requirements. Units that operate continuously (24 hours/day; 7 days/week) will be smaller than equipment required to operate only during staffed hours (e.g., 8 hours/day; 7 days/week). Whereas larger centrifuges have high capacities, they tend to cost substantially more than smaller units and they represent a significant loss in capacity upon failure. In addition, spare parts for very large centrifuges may have long lead times. For these reasons, some situations may be better served with more small units or provision of critical spare parts during initial equipment purchase.

	Primary Sludge	WAS	Co-Thickening Primary Sludge and WAS	Comments
Feed concentration	0.2 to 4% TSS	0.4–1.5% TSS	0.2–5% TSS	Thin feed should be measured in TSS due to effect of dissolved solids; thick feed (>1.5% TS) can be measured in TS
Target thickened sludge solids concentration	5 to 6% TS, but can easily increase to 12% TS with excessive polymer	5 to 6% TS	5 to 6% TS, but can easily increase to 12% TS with excessive polymer	Cake will overthicken if bowl speed and polymer dose are too high
Polymer dose	0–2.2 kg active/tonne (0-5 active lb/dry ton)	0–2.7 kg active/tonne (0–6 active lb/dry ton)	0–4.5 kg active/tonne (0–10 active lb/dry ton)	For co-thickening, the polymer use may increase due to ineffectiveness with either the primary sludge or the WAS
Target centrate quality	Less than 500 mg/L TSS	Less than 500 mg/L TSS	Less than 500 mg/L TSS	
Abrasion protection	Robust abrasion projection (scroll tiles) required due to grit	100% WAS should have very little grit; tiles are beneficial but other types of abrasion protection may suffice	Robust abrasion projection (scroll tiles) required due to grit	
Torque	0–5, but may increase if slugs of sand/rags enter bowl	Typically consistent and near 0	0–5 with variation due to blend ratios	Gearboxes can be smaller than for dewatering, but must still have enough torque to clear a bowl if a plug should occur
Differential speed variation	10–20 rpm	1–15 rpm	1–15 rpm	For systems with automated pond depth adjustment, differential speed is not typically a control variable

TABLE 21.11 Differences in Thickening Primary Solids, WAS, Or Co-Thickening With Centrifuges

Many thickening systems include a thickening centrifuge located on an upper level, discharging to the thickened solids system below. Proper access to the equipment below the centrifuge should be provided. When laying out such a system, the designer should consider all of the piping, ductwork, and electrical conduit that are connected to the centrifuge and ensure that sufficient workspace, headroom, and lighting are provided around and below the centrifuge. Accessible sample ports for solids feed, centrate, and thickened solids are important for system operations. Rapid solids analyzers (e.g., microwave moisture/solids analyzers) are useful tools for quick sample analysis.

Centrifuges produce a high-pitched noise that typically requires ear protection within the centrifuge room. Provision of noise attenuation panels along the walls and roof of the centrifuge room can reduce noise levels. Rooms with concrete floors and walls to separate the centrifuges from other system equipment can also reduce noise. Operating floors with metal grating allow noise to travel to other areas within the building.

If facility operations staff members are unfamiliar with centrifuge operations, robust training before system handover to the facility is a necessity. This training should cover centrifuge components and operation, ancillary system components, preventative maintenance, rebuilds, electrical and mechanical maintenance, optimization, and troubleshooting. Staff presence during performance testing allows operators to understand proper startup, shutdown, and clean-in-place procedures and sequences. Hands-on training, in addition to classroom training, is most beneficial. Maintenance service agreements with centrifuge manufacturers can alleviate some of the burden on staff and provide time for staff to become familiar with maintenance tasks.

4.5   Ancillary Equipment

4.5.1   Feed System
Design engineers typically prefer to use progressive cavity pumps for small-to-medium centrifuges because the positive displacement and steady pumping rate allow for both effective metering and close control of the solids-feeding rate. Progressive-cavity drives should be variable speed with at least a five-fold range [e.g., 80–400 L/min (20–100 gpm)] so that operators can meet solids-loading goals regardless of variations in feed solids concentrations. Rotary lobe pumps have also been used in some facilities for centrifuge feed.

If centrifugal pumps are selected, design engineers should be aware that changes in solids consistency will affect the pumping rate. Therefore, it is important to choose appropriate flow meters and controllers to maintain centrifuge loadings.

It is a good design practice to use flow meters with either type of pump, particularly when other measurement controls (e.g., density sensors) will be used to maintain a relatively constant solids load to the centrifuge.

4.5.2   Chemical Conditioning
Polymers characteristics vary widely, so design engineers should consult manufacturers about their properties, as well as preparation and hauling techniques. For detailed information on polymer mixing and feeding systems, see Chapter 20.

The polymer system should be designed to allow operators to easily test new polymers, polymer solution characteristics, and feed points on a regular basis. Because polymer represents a significant operating cost for thickening centrifuges, optimization is necessary to minimize overall costs. Aging is required for dry polymers and sometimes recommended for emulsion polymers to maximize effectiveness and reduce operating costs. If emulsion polymer is used, the costs of an aging system and downstream pumps should be balanced against the expected savings from maximum polymer activation. In some instances, an aging system may not be economical compared to expected savings.

Polymer storage can be provided with bulk tanks or polymer totes, depending on system size and acceptable frequency of truck deliveries. If exposed to moisture or water, polymer becomes slippery and difficult to deal with. Hence, systems should be designed to minimize such exposure and equipment areas should be provided with non-skid surfaces.

4.5.3   Thickened Solids Conveyance

Centrifuges discharge a thickened cake containing between 3% and 15% solids (by weight), depending on the application. This cake is highly viscous and often thixotropic. When discharged from the centrifuge's directional chute, the cake can be

- directly discharged to a collection well for subsequent pumping,
- directly discharged to an open-throat progressing cavity pump, or
- discharged to a horizontal screw conveyor, which carries it to a sump or open-throat pump.

The last alternative could work well when the centrifuge is used for both thickening and dewatering. A reversing screw conveyor can direct thickened solids in one direction while operating in thickening mode and dewatered cake in the opposite direction when operating in dewatering mode. However, conveyance of wet material may result in spillage and housekeeping issues. Consult conveyance system manufacturers to assess suitability of equipment for specific thickened solids characteristics.

4.5.4   Odor Control and Ventilation

Odor control, especially if thickening or co-thickening primary sludge, may be required. Refer to Chapter 6 for additional information on odor control. The designer should connect the centrate and conveyance system to the odor control system. Foul air withdrawal rates must be coordinated with centrifuge venting requirements. Ductwork should be aligned to avoid entrainment of centrate foam in the odor control duct.

The centrifuge room must be ventilated as required for electrical classification standards per NFPA 820. The electrical and controls rooms housing the centrifuge power and control panels should be air conditioned.

4.6   Process Control

4.6.1   Overview

A thickening system typically includes sludge feed pumps, the polymer system, the thickeners, and the thickened solids conveyance system. The thickening centrifuge is usually provided with a programmable logic controller (PLC) that can control and/or communicate with the system components during automated startup, shutdown, and cleaning sequences. At various points of these cycles, the centrifuge PLC controls starting, stopping, and speed of the bowl and scroll. The PLC typically also controls the unit's clean-in-place cycle to flush the bowl out during normal shutdown and most alarm conditions that require the centrifuge to stop.

Maximizing cake solids, maintaining reasonable capture efficiency, and minimizing polymer requirements are difficult tasks to accomplish without instrumentation and controls. During operation, the centrifuge is usually operated by its PLC to maintain a specific torque or differential speed setpoints. This helps maintain consistent cake solids but does not directly address capture efficiency and polymer dose. Capture efficiency is maintained by manually changing the polymer dose or torque setpoint until the centrate is clear. This system works well, but requires significant operator attention. The current state-of-the-art control systems offer a combination of sensors and software to control differential speed, bowl speed, pond depth, and polymer dose.

4.6.2   Feed-Forward System
A feed-forward system uses a sensor to track the feed solids concentration and a flow monitor to track the flowrate. This information is used to determine the solids loading rate. Then, the system automatically adjusts the polymer dose to match an operator-specified dose per unit mass of solids processed. It also automatically adjusts the torque (differential speed) based on historical information to produce the desired cake solids concentration.

4.6.3   Feed-Backward System
The feed-backward system uses a sensor to measure the centrate's solids concentration and uses this information to adjust the polymer dose until the desired centrate solids concentration is achieved. It also automatically adjusts the torque to produce a consistent cake solids concentration. If the desired centrate quality cannot be achieved regardless of polymer dose, the system automatically reduces the torque until the centrate clears. Measuring solids in centrate can be difficult due to air entrainment in the process. Deaerators may be required for sensors to work accurately and consistently.

4.6.4   Variable-Speed Bowl and Scroll
Most centrifuge manufacturers offer variable-frequency drives on both the main and scroll drives. The drives can be changed while the centrifuges are operating (i.e., both differential and bowl speed can be adjusted). This allows operators to reduce bowl speed and differential speed simultaneously until the desired solids concentration is obtained. Also, operating at lower bowl speeds can reduce power consumption and may reduce both wear and polymer requirements.

4.6.5   Pond-Depth Adjustments
The characteristics of primary sludge or co-mingled primary solids and WAS tend to change throughout the day. Solids concentration, for example, could change from 1.0% to as high as 4.5%. Such significant changes in solids concentration may require changes in centrifuge pond depth to maintain the desired solids-capture efficiency and thickened-solids concentrations at a minimum polymer dose.

Manually changing pond depth is a trial-and-error process that involves shutting the unit down, unbolting and adjusting the centrate weirs, starting the centrifuge back up, collecting and analyzing thickened-solids and centrate samples, and repeating as necessary until the desired results are obtained. Some manufacturers now offer automated pond depth adjustment systems that change pond depth while the centrifuge is operating to maintain a set point thickened sludge concentration.

These systems include an automated disk located near the centrate weirs that can move axially to increase or reduce the gap through which the centrate discharges. Gap adjustments subsequently change pond depth within the centrifuge. The automated pond depth adjustment system includes the throttling disk and a sensor to measure solids concentration of the thickened sludge. To maintain a thickened sludge concentration, the disk is throttled to change pond depth and the bowl speed is modulated. When used to minimize bowl speed while maintaining required concentration of thickened

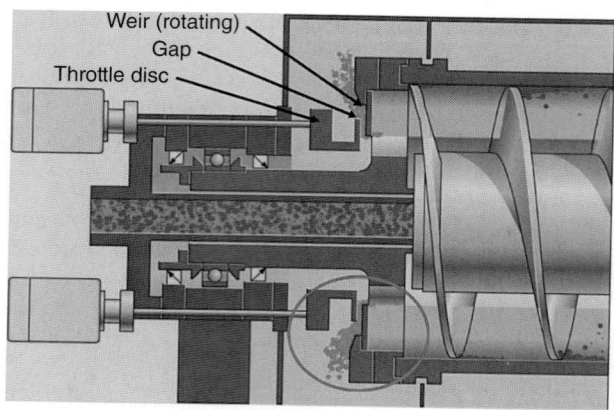

Weir (rotating)
Gap
Throttle disc

FIGURE 21.16   Automated pond depth adjustment system components (courtesy of GEA Westfalia Separator Inc.).

sludge, power consumption can be reduced and the equipment sees less wear. One automated pond depth system is shown in Figure 21.16.

4.6.6   Monitoring and Alarms

A solid-bowl centrifuge's control circuitry and programming is typically designed to protect the centrifuge from malfunctions (e.g., torque overload, loss of bearing lubrication, excessive vibration, high bearing temperature, and motor overload). Modern centrifuges are programmed with monitoring and interlocks to prevent equipment damage and to pause or stop the equipment upon specific alarm conditions. Design engineers should provide interlocks that shut down ancillary systems and initiate a water-flush sequence during centrifuge shutdowns. Because centrifuge startup and shutdown can take up to 30 minutes, all alarm conditions do not typically result in complete shutdown. Rather, the centrifuge can go into a "pause" mode where the bowl continues to spin for a specific time during which the alarm condition may be cleared. Upon clearing of the alarm, the centrifuge can quickly return to solids processing.

4.7   Performance Assessment and Optimization

4.7.1   Performance Parameters

The process performance of a thickening centrifuge is primarily assessed through thickened sludge concentration, polymer consumption, power consumption, and percent capture. Other parameters of importance are operational consistency, equipment longevity, ease of operation, and ease of maintenance.

The expected process performance for a specific feed sludge and centrifuge can be estimated through pilot-testing in the field. Most manufacturers maintain trailer-mounted centrifuge systems that include ancillary pumps, polymer, and thickened sludge conveyance equipment. In addition to projecting potential performance, such tests also provide O&M staff an opportunity to become familiar with the system and ask questions.

Expected operational performance is harder to project. Requiring the provision of references and checking with those references offers an opportunity to determine how the equipment has performed elsewhere.

4.7.2 Optimization Measures

System optimization focuses on maintaining consistent thickened sludge concentration, minimizing polymer consumption, minimizing power consumption, and maximizing percent capture. Many of the control system features discussed previously seek to automatically optimize centrifuge performance. Additional optimization beyond automated control includes selection of the polymer that provides the best value and determining the optimal polymer solution concentration and injection point. Polymers should be tested on a regular basis to identify those that provide the best value (cost versus performance) throughout the year. Many facilities' sludge characteristics change in winter versus summer and different polymers often work best during each of those seasons.

4.8 Evaluation and Scale-Up Procedures

There are relatively few centrifuge manufacturers and they typically guard their evaluation and scale-up technologies, so basic, widely accepted design criteria for centrifugal thickeners do not exist. Instead, design engineers typically rely on experience, laboratory testing, and pilot tests to estimate centrifuge performance. Feed solids are sometimes not available for experimentation, so designers must make judgments based on past experiences with similar solids under identical conditions. If solids are readily available, however, designers should make every effort to base their designs on pilot-test results.

Design engineers typically use bench-scale tests to determine whether centrifugation is feasible, choose a chemical conditioner, and select an appropriate dose of that chemical. They use pilot tests to generate centrifugal design data because equipment manufacturers are reluctant to guarantee performance without such data.

Pilot tests should be conducted on a full-scale centrifuge whose design and proportions are similar to the commercial unit being considered. Many manufacturers have test units available on a rental or trial basis; several of them have portable units that can move readily from site to site. Operating a pilot unit through a broad range of machine and process variables allows design engineers to assess the effects of normal variations in feed-solids flow and quality. In particular, they should evaluate how the following parameters affect centrifuge performance:

- Mass-based and hydraulic feed rate, including polymer solution flow;
- Thickened solids (cake) discharge rate;
- Polymer dose rate;
- Clarification area;
- Pool depth and volume;
- Solids retention time;
- Conveyor differential speeds;
- Centrifugal force (bowl speeds);
- Percentage solids recovery; and
- Cake concentration.

Testing results typically provide enough accurate information for designers to select and size the most economical full-scale equipment. That said, the final selection of a solid-bowl, scroll-type centrifuge is typically a compromise between two operations that are intrinsic to successful centrifuge operations: solids separation (which is a function of clarification area) and solids consolidation and removal (via the screw conveyor). If these cannot be balanced, then centrate clarity and/or solids concentration will deteriorate.

4.8.1 Theoretical Capacity Factors

Researchers have developed certain theoretical equations for use in scaling up pilot data to the full-scale commercial unit. [Complete derivations of these scale-up factors can be found in literature by Perry and Chilton (1963), Vesilind (1974a, 1974b), and Purchas (1977).]

The two most important criteria for successful centrifuge operations are solids separation (hydraulic or clarification capacity) and solids removal (cake-conveying capacity). A solid-bowl centrifuge's hydraulic capacity (Σ) is determined as follows:

$$\Sigma = 2\Pi\, l\, \frac{\omega^2}{100g}(0.75r_2^2 + 0.25r_1^2) \tag{21.10}$$

where Σ = theoretical hydraulic capacity (cm²);
l = centrifuge bowl's effective clarifying length (cm);
ω = centrifuge bowl's angular velocity (rad/s);
g = acceleration from gravity (m/s²);
r_1 = radius from centrifuge centerline to the liquid surface in the centrifuge bowl (cm); and
r_2 = radius from centrifuge centerline to the inside wall of the centrifuge bowl (cm).

Below is a simpler way to calculate Σ (applicable only to solid-bowl centrifuges) (Vesilind, 1974b):

$$\Sigma = \frac{V\omega^2}{g\ln(r_2/r_1)} \tag{21.11}$$

where V = centrifuge pool volume (cm³).

When scaling up solids-handling (cake-conveying) capacity, the assumption is that if two geometrically similar (but different-sized) machines have the same ratio of solids discharge rate to theoretical solids-handling capacity, then their performance will be similar for a given solids feed (Vesilind, 1974a). The scale-up relationship is as follows:

$$\frac{Q_{SP}}{\beta_p} = \frac{Q_{SF}}{\beta_f} \tag{21.12}$$

where Q_{SP} = the pilot-scale unit's solids discharge rate (m³);
Q_{SF} = the full-scale unit's solids discharge rate (m³);
β_p = the pilot-scale unit's solids-handling capacity factor (m³/h); and
β_f = the full-scale unit's solids-handling capacity factor (m³/h).

The best approach is to develop full-scale requirements based on both capacity and solids-loading considerations. The limiting criterion would govern machine selection.

There are limitations with the theoretical relationships just cited. Neither Σ nor β takes into consideration interactions between the clarification and cake-storage zones. Certain wastewater solids (e.g., WAS) are thixotropic and may have difficulty moving up the conical section of the bowl before discharge. Also, the full-scale unit may not be able to achieve the theoretical solids depth (cake pile) because of the theoretical nature of cake solids.

Design engineers should consult with centrifuge manufacturers to properly identify limiting design factors and develop full-scale requirements based on both hydraulic and solids loading considerations. Some limitations may be overcome by altering the centrifuge design (e.g., bowl or conveyor speed, conveyor pitch, number of conveyor leads, or pool depth) to provide a more conducive environment for solids separation and removal.

4.9 Design Example

Suppose design engineers have calculated that a maximum of 10 883 kg/d (23 900 lb/d) of WAS (dry weight) must be thickened. The following operating criteria applies:

- Thickening facility operates 7.5 h/d
- WAS solids concentrations is 0.5% solids
- New facility with no pilot testing
- Design for three operating units and one standby unit

Maximum daily WAS production = 10 883 kg/d

No. of operating units = 3 units

Hours operating per day = 7.5 hours

WAS minimum solids concentration = 0.5%

WAS specific gravity = 1

Equations used:

$$\text{Maximum net hourly load} = \frac{\text{Maximum Daily WAS Load}}{\text{No. of Operating hours per day}} \qquad (21.13)$$

$$\text{Maximum net hourly load per unit} = \frac{\text{Maximum Net Hourly Load}}{\text{No. of Units}}$$

$$\text{Unusable Cylinder Volume (m}^3) = \pi \times \left(\frac{\text{Discharge Diameter (m)}}{2} \right)^2 \qquad (21.14)$$
$$\times \text{Cylinder Length (m)}$$

$$\begin{array}{l} \text{Volume per operating} \\ \text{minute per unit} \end{array} = \frac{\text{Net Hourly Load per unit}}{\begin{array}{c} \text{WAS Solids Conc.} * \text{Specific Gravity (WAS)} * \\ \text{No. of Units} \end{array}} \qquad (21.15)$$

$$\text{Bowl Circumference (m)} = [\text{bowl diameter (m)}] \times \pi \qquad (21.16)$$

$$\text{Bowl Speed}\left(\frac{\text{sec}}{\text{rev}}\right) = \frac{\text{No. of seconds/min}}{\text{Bowl Speed}\left(\dfrac{\text{rev}}{\text{min}}\right)} \qquad (21.17)$$

$$\text{Angular Velocity}\left(\frac{\text{m}}{\text{sec}}\right) = \frac{\text{Bowl Circumference (m)}}{\text{Bowl Speed}\left(\dfrac{\text{sec}}{\text{rev}}\right)} \qquad (21.18)$$

$$\text{Centripital Acceleration}\left(\frac{\text{m}}{\text{sec}^2}\right) = \frac{\left(\text{Angular Velocity}\left(\dfrac{\text{m}}{\text{sec}}\right)\right)^2}{\text{Bowl Radius (m)}} \qquad (21.19)$$

$$G-\text{Force at Bowl Wall} = \frac{\text{Centrifugal Acceleration}\left(\dfrac{\text{m}}{\text{sec}^2}\right)}{\text{Acceleration due to Gravity}\left(\dfrac{\text{m}}{\text{sec}^2}\right)} \qquad (21.20)$$

$$\text{Cylinder Length (m)} = \text{Bowl Length (m)} - \text{Cone Length (m)} \qquad (21.21)$$

$$\text{Cylinder Volume (m}^3) = \pi \times \left(\frac{\text{Bowl Diameter (m)}}{2}\right)^2 \times \text{Cylinder Length (m)} \qquad (21.22)$$

$$\text{Unusable Cylinder Volume (m}^3) = \pi \times \left(\frac{\text{Discharge Diameter (m)}}{2}\right)^2$$
$$\times \text{Cylinder Length (m)} \qquad (21.23)$$

$$\text{Usable Volume of Bowl Cylinder (m}^3) = \text{Cylinder Volume (m}^3)$$
$$- \text{Unusable Cylinder Volume (m}^3) \qquad (21.24)$$

$$\text{Usable Volume of Bowl Cylinder (gal)} = \frac{\text{Usable Volume of Bowl}}{\text{Cylinder (m}^3)} * \frac{\text{No. of gal}}{\text{m}^3} \qquad (21.25)$$

$$G-\text{volume} = G-\text{force at Bowl Wall} \times \text{Usable Volume of Bowl Cylinder (gal)} \qquad (21.26)$$

Maximum net hourly load = 1451 kg/h (3200 lb/h); use eq 21.13

Maximum hourly load per unit = 483 kg/h/unit (1066 lb/h/unit); use eq 21.14

Volume per operating minute = 97 m³/h/unit (427 gal/min/unit); use eq 21.15

Contact several manufacturers, give them information about the application, along with the desired mass and flow rate criteria for each unit and request specific references for the size unit recommended. For further information, refer to Table 21.8.

5.0   Gravity Belt Thickener

Introduced in 1980, a gravity belt thickener has a gravity drainage zone much like the upper gravity drainage zone of a belt filter press that allows water to drain through a moving, fabric-mesh belt while coagulating and flocculating solids (see Figure 21.17). It originally was designed to be a dewatering pretreatment method, but subsequent improvements have made it suitable for solids thickening. Gravity belt thickeners are currently used to treat aerobically or anaerobically digested solids, alum and lime solids, primary solids, WAS, and blended solids that initially contain between 0.4% and 8% solids. However, in most applications the feed solids concentrations are less than 1.5%.

Gravity belt thickeners are often selected because of their relatively small footprint, low power use, and moderate capital costs. Experience has shown that gravity belt thickeners work well with many types of wastewater solids and are less affected by facility operating problems than other thickening processes. They typically handle even difficult-to-thicken solids with minor modifications to polymer dose, hydraulic loading rates, and solids loading rates.

Gravity belt thickeners are less commonly used for thickening primary sludge. Two reasons for this are odor control and the possibility of belt blinding due to amount of grease, oils, and other debris in primary sludge. If design engineers intend to use a gravity belt thickener as part of a co-thickening process, they should ensure that primary solids and WAS are mixed properly before the material is loaded onto the belts. This will help to control belt blinding caused by grease and debris in the primary solids.

5.1   Operating Principle

A gravity belt thickener consists of a flocculation/retention tank with a slow-speed mixer, a discharge chute, a gravity belt, chicanes or plows, a ramp, a spray-wash system, and a variable- or constant-speed belt drive (see Figure 21.18). A conditioning agent (typically polymer) is injected into the feed sludge and the mixture is sent to a flocculation/retention tank. The tank is mixed slowly to allow sludge to flocculate, and the conditioned sludge flows over the discharge chute for an even distribution across the width of the gravity belt. The gravity belt is driven by a variable- or constant-speed belt drive and as the sludge moves along the belt the plows spread the sludge and facilitate the separation of free water from the solids. As the free water releases from the sludge it passes through the porous gravity belt to a filtrate drain. Near the end of the

Figure 21.17   An example of a gravity belt thickener (courtesy of Alfa-Laval, Inc.).

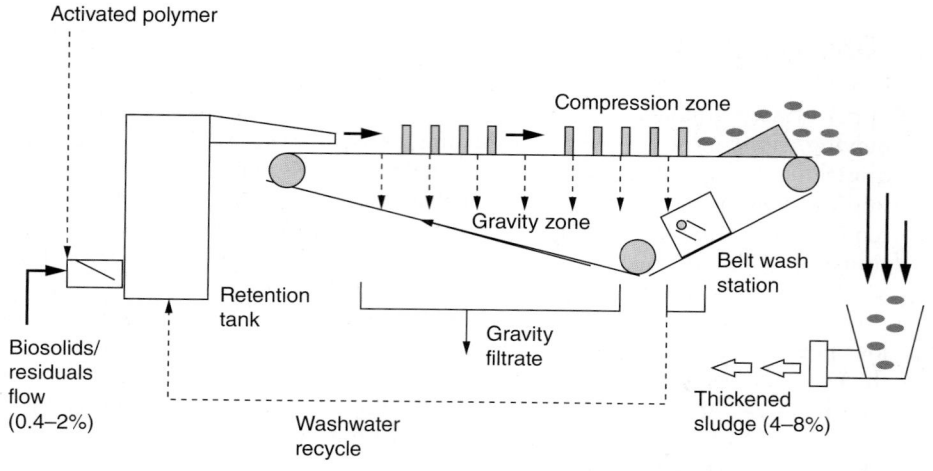

Activated polymer

Compression zone

Gravity zone

Belt wash
station

Retention
tank

Gravity
filtrate

Biosolids/
residuals
flow
(0.4–2%)

Washwater
recycle

Thickened
sludge (4–8%)

FIGURE 21.18   Gravity belt thickener schematic (courtesy of Alfa-Laval, Inc).

gravity belt discharge there is typically a ramp or plate that is used to slow the solids discharge to allow build-up of solids and further pressing free water from the thickened solids.

The most commonly sized gravity belt thickener has a belt width of 2 to 3 m. A gravity belt thickener typically captures more than 95% of solids when 1.5 to 5 g/kg (3 to 10 lb/ton) dry weight of active polymer dose is used to concentrate the material and avoid excessive solids losses. When treating municipal WAS and biosolids, gravity belt thickeners can thicken the feed solids up to 6%–8% solids.

5.2   Physical and Mechanical Features

5.2.1   Frame
A gravity belt thickener has a frame that supports and holds its components, except for the polymer-injection ring, polymer–solids mixer, flocculation/retention tank, and tracking and tensioning power unit. The frame is typically galvanized or otherwise coated with a durable corrosion-resistant surface; it can also be constructed of stainless steel if required. Some manufacturers offer gravity belt thickeners constructed of stainless steel plate that encloses the typically open areas in order to provide for better odor control.

5.2.2   Flocculation/Retention Tank and Feed Distribution
Unless the feed line offers proper flocculation conditions, a flocculation/retention tank is required. Polymer-conditioned solids flow into the bottom of this tank so enough time is provided to flocculate before overflowing onto the belt. As the solids leave the feed tank, they should be distributed across the entire working width of the belt for optimal operation and thickening. Some manufacturers use a feed chute with wedges for this purpose. The tank should be equipped with a drain valve so it can be emptied when feed is discontinued to the gravity belt thickener.

5.2.3 Gravity Drainage Area

The feed chute distributes solids onto the continuously moving, horizontal filter belt, which retains solids but allows free water pass through. The belt is made of a porous, woven mesh and is seamed to form a continuous loop between the feed chute and discharge point. A series of adjustable plows along the belt ensure that solids are distributed evenly across the mesh, turn solids over to promote water separation, and create solids-free areas that enable free water to drain through the belt. Adjustable retention plates with seals prevent solids from spilling off the sides of the belt. A drainage grid underneath the belt supports it and enables filtrate to drain into collection trays below.

5.2.4 Discharge Area

The back end of the unit often has a ramp or wedge, depending on the manufacturer. If the unit has a ramp, its leading edge contacts the top of the belt allowing solids to roll as they move onto the ramp, further squeezing out free water. The angle of the ramp also acts as a dam, increasing product depth and retention time on the belt, further thickening the solids before discharge. The ramp angle is usually adjustable to optimize thickening and can be rotated out of the way if it hinders thickening operations.

If the unit has a wedge, its leading edge has a wide clearance from the belt that decreases as the belt progresses toward the end of the table. Solids pass under the wedge and are squeezed as the clearance between the wedge and belt decreases. This removes a little more water just before discharge.

Thickened solids may be discharged to a wet well, open-top pump, or another conveyor. Once the belt clears the ramp or wedge, it moves past a scraper blade, which separates remaining solids from the belt. These solids are discharged to the same location as the rest of the thickened material.

5.2.5 Belt Washing

After being scraped, the belt passes through a wash station that removes embedded particles from that portion of the belt before it begins another thickening cycle. The wash station consists of a washwater supply pipe, spray nozzles, and a housing to contain spray. The station typically is constructed of stainless steel.

5.2.6 Filtrate and Washwater

Filtrate is the free water that passes through the belt. It collects in trays below the belt and then discharges to drain pipes, which return the filtrate to the liquids treatment process.

Used washwater (from the wash station) also discharges to the sump. Another option is to collect washwater and recycle it to the thickener feed tank. This reduces the load on downstream treatment processes and improves solids capture because the solids in washwater are combined with the incoming floc.

5.2.7 Gravity Belt Thickener Drive, Tracking, and Tensioning

Gravity belt thickeners require multiple rollers to drive, steer, adjust the tension, and guide the belt. The drive roller pulls the belt through the machine. The roller-drive motor typically has either mechanically or electrically adjustable speed controls to vary belt speed. The steering roller maintains proper alignment of the belt in response to sensing devices on the gravity belt thickener. The tensioning roller pulls on the belt to create the belt tension required for traction between the belt and belt-drive roller. The guide roller directs the belt through the wash station.

A gravity belt thickener requires a power unit that provides pressure to the belt-tracking and -tensioning systems. Some manufacturers use recirculating hydraulic fluid for this purpose, while others use compressed air. Unlike the belt-drive motor, which is connected to the gravity belt thickener, the power units may be remotely located.

5.2.8 Solids Polymer Injection and Mixer

Good polymer distribution optimizes polymer use and improves thickening, so immediately after being added, polymer should be well mixed with the feed solids. Manufacturers of gravity belt thickeners often provide polymer-injection and -mixing devices. A device with multiple injection points around the solids feed pipe probably will provide better distribution. One option is a polymer manifold; it connects to the end of the polymer-feed pipe and has feed tubes that connect to an injection ring on the solids feed pipe. Design engineers should avoid injection points that extend into the solids path, because they may cause fibrous or stringy material to entangle and plug the pipe. Some manufacturers provide manifolds with clear, flexible feed tubes that make installation easier and let operators see whether polymer is flowing to each point on the injection ring. If one of the tubes is plugged, operators sometimes can clear it by using tubing clamps on one or more open lines, which increases the pressure and scouring velocity in the plugged line. To keep the tubing transparent, it occasionally must be removed and cleaned or replaced.

The polymer mixer should be immediately downstream of the injection point. Design engineers should select adjustable mixers that allow mixing optimization both during initial operations and whenever solids feed rates or characteristics change. Adjustable orifice mixers can include an adjustable counterweight lever that changes the orifice size so that operators can adjust the mixer according to solids variations to optimize mixing.

5.3 Process Design Considerations and Criteria

The primary design components of a gravity belt thickener are

- feed solids pumps and feed flow control,
- polymer system and feed control,
- gravity belt thickener,
- belt washwater supply,
- thickened-solids pumps, and
- odor control.

Before designing these components, engineers need to determine potential operating modes. For example, the process may need to thicken 7 days' worth of solids in a shorter timeframe. Common operating modes include continuous thickening 7 days per week, continuous thickening 5 days per week, one-shift thickening 7 days per week, and one-shift thickening 5 days per week.

All components must be sized to handle both minimum and maximum potential solids-feed rates. Other important design considerations and criteria are noted in the following subsections. (For design information related to ancillary equipment and controls, see Section 5.5.)

Belt Size (Effective Dewatering Width) (m)	Hydraulic Loading Range* (L/min)
1.0	400–950
1.5	570–1420
2.0	760–1900
3.0	1100–2800

*Assumes 0.5 to 1.0% feed solids for municipal sludges. Variations in sludge density, belt porosity, polymerreaction rate, and belt speed will act to increase or decrease the rates of flow for any given size belt.

TABLE 21.12   Typical Hydraulic Loading Ranges for Gravity Belt Thickeners (MacConnell et al., 1989)

5.3.1   Unit Sizing

Engineers can design gravity belt thickeners based on pilot-test results and manufacturers' flow and solids loading capacity criteria (see Tables 21.12 and 21.13). They can rely on manufacturer criteria for most municipal WAS, anaerobically digested solids, aerobically digested solids, drinking water solids, and pulp and paper (recycled paper) solids because of the abundance of historical operating data available. However, some testing is recommended to verify that a given solids can be thickened at typical polymer doses.

In lieu of pilot-test data, engineers could use a conservative design value of 800 L/m min (200 gpm/m) for the hydraulic loading rate. As feed flowrates increase, more operator attention is needed to maintain stable operations.

Operating experience demonstrates that some gravity belt thickeners can treat WAS with 0.6% to 1.5% solids at a hydraulic loading rate of up to 1 500 L/m min (400 gpm/m) and a solids loading rate of up to 500 kg/m h (1100 lb/hr m). They also can treat digested solids with 2% to 4% solids at a hydraulic loading rate of up to 1100 L/m min (300 gpm/m) and a solids loading rate of up to 770 kg/m h (1700 lb/hr m). In both cases, they can produce a thickened material containing 4% to 7% solids. Solids capture typically ranges from 90% to 98%.

Optimizing operations and increasing the polymer dose sometimes can produce a thickened material containing 10% solids; however, it can be difficult to pump and treat

	Primary Sludge	WAS	Co-thickening Primary Sludge and WAS
Typical feed solids concentration	1%–4% TS	0.5%–1% TS	1.5%–3% TS
Typical thickened sludge solids concentration	7%–10% TS	4%–8% TS	5%–9% TS
Typical polymer dose	2.2–4.4 kg/tonne (5–10 active lbs/dry ton)	2.2–4.4 kg/tonne (5–10 active lbs/dry ton)	2.2–4.4 kg/tonne (5–10 active lbs/dry ton)
Typical percent capture	95%–99%	95%–99%	95%–99%

TABLE 21.13   Performance Differences in Thickening Primary Sludge, WAS, or Co-Thickening with Gravity Belt Thickeners

in downstream processes, so gravity belt thickeners typically are designed for a maximum of 5 to 7% thickened solids. Also, engineers must design thickened-solids and filtrate-conveyance systems to handle a range of possible conditions.

Pilot testing is recommended for atypical solids or facilities receiving atypical industrial contributions. Such tests allow engineers to determine flow and solids-loading capabilities, solids capture, and required polymer dose. Typically, the only difference between pilot- and full-scale models is belt width, so testing results are determined on a per-meter belt-width basis to allow for a directly proportional scale up.

As part of pilot testing, design engineers should select a suitable polymer. Polymers are typically screened via bench-scale tests and selected based on pilot-test performance.

5.3.2   Other Design Considerations

When designing gravity belt thickeners, engineers also need to consider mixing, flocculation, belt speed, plow, and discharge design. This information relates to items often supplied with the thickener and so affects the choice of manufacturer and equipment model.

Other details to consider include mechanical durability, corrosion resistance, availability and cost of replacement parts, and service assistance provided by the manufacturer. To determine mechanical durability and corrosion resistance, engineers should evaluate a manufacturer's engineering drawings and specifications, as well as interview O&M staff at other installations. To evaluate service-assistance and parts issues, they should survey O&M staff at other installations.

5.3.3   Mixing Design

The process design should provide for adequate mixing of polymer and solids followed by enough flocculation time before solids are discharged onto the belt. Poorly flocculated particles will blind the belt, hinder thickening, and result in poor solids capture. The polymer must be injected immediately upstream of the mixing device. Mixing must be intense enough to provide good contact without breaking the floc apart or shearing the polymer. Adjustable mixers (e.g., an adjustable orifice mixer) are recommended so mixing conditions can be optimized. The mixer should be installed next to a feed tank so operators can visually evaluate the floc immediately after adjustments. The feed tank typically provides enough time for flocculation before solids enter the gravity belt thickener.

Once the mixer is properly adjusted, it typically will not require adjustments to react to small changes in solids characteristics or feed rates. Larger changes, however, will require mixer adjustments to minimize polymer use and improve thickening and solids capture.

5.3.4   Flocculation Design

After mixing, the polymer-conditioned solids need time to agglomerate into larger floc. Operating experience has shown that the conditioned solids need at least 30 seconds to flocculate before entering the gravity belt thickener. So, design engineers should provide for 30 seconds of contact time under peak-flow conditions.

Flocculation time can occur in the piping or in a feed tank, and turbulence should be avoided. One manufacturer suggests that the piping downstream of the mixing point include no more than three 90-degree elbows or similar fittings. Although the significance of such fittings is not well documented, design engineers may need to adhere to such manufacturer recommendations to enforce performance guarantees. It typically is difficult to detain flow in the feed piping for 30 seconds without using more than three elbows, so manufacturers offer a feed tank for this purpose. Design engineers should size this feed tank to provide about 30 seconds of detention time at the peak feed rate.

Whenever the gravity belt thickener will be shut down for more than a few hours, the feed tank should be drained to avoid excessive odors. If the drain valve is inaccessible, it should be automated.

5.3.5 Belt Speed Design

Operators need to be able to adjust belt speed so they can control thickening and maximize solids capture. Gravity belt thickeners typically perform well within a certain range of belt speeds, depending on manufacturer's specification. Below this range, the solids feed rate must be limited to avoid flooding the belt. Above this range (without other performance improvements), solids capture degrades because more residuals are being washed off the belt in the wash station. With proper polymer conditioning, however, a faster-moving belt can accommodate higher mass loadings that can offset the solids loss due to higher speed. Design engineers should perform a mass balance to confirm this. They should also survey O&M staff at existing installations about the performance records of the manufacturers being considered.

If operators can adjust the belt speed, they can maintain operations at the slowest speed that will accommodate the feed rate without the possibility of flooding. Belt speed can be changed mechanically or electrically; manufacturers often provide a mechanical adjustment mechanism as standard equipment and offer a variable-frequency drive as an option. If solids characteristics and feed rates are expected to be fairly constant, the mechanical adjustment mechanism may be sufficient. If frequent belt-speed adjustments are anticipated, adjustable-frequency drives can be used. Also, the thickener's control panel should include a potentiometer so that operators can adjust belt speed and monitor the result more easily.

5.3.6 Plow Design

The clearance between the plows and the belt must be large enough for the belt seam, which typically protrudes above the rest of the belt, and yet small enough for the plows to clear drainage pathways right down to the belt. It should be adjustable to allow for proper installation, plow wear over its lifetime, and manufacturing variations among replacement belts. If the plows press too hard on the belt, excessive wear will reduce belt life.

5.3.7 Discharge Design

Some gravity belt thickeners have an adjustable ramp over which solids must flow before discharging from the belt. The ramp can act as a quasi-dam, causing the solids to thicken further as they roll up and over it. A steeper ramp typically results in thicker solids. Some installations, however, may require the exit ramp to have little or no angle to avoid flooding the belt.

Rather than a ramp, some manufacturers use a wedge under which solids must squeeze before discharging from the belt. Side-by-side pilot tests have indicated that, under identical operating conditions, the ramp design can produce slightly thicker solids. Put another way, the ramp design would need less polymer to produce a thickened material with the same solids content as that produced by the wedge design. Design engineers would need to conduct side-by-side pilot tests of the two systems to quantify differences in polymer use and solids thickening for a given solids stream and compare them to differences in capital costs.

5.3.8 Impacts of Different Feed Sludge Types

Gravity belt thickeners can be used to thicken a variety of sludge types. Most municipal WRRFs use them to thicken WAS. However, the units can also be used to thicken primary sludge, and to co-thicken primary sludge and WAS. The type of sludge to be thickened impacts some of the equipment design and performance parameters because of sludge characteristics.

Primary sludge thickens readily and requires different types of polymer, lower polymer doses, and lower polymer solution concentrations than WAS. Grease often found in primary sludge can blind gravity belt thickener belts, necessitating higher pressure water sprays or hot water sprays to clean the belts. The added effort to clean the belts might reduce belt life and increase operating costs. Primary sludge usually has higher solids content than biological sludge so it may require better mixing or longer mixing time with the polymer to optimize results. However, because there is less bound water in primary sludge solids the performance should be better. Odor control is also a bigger issue when thickening primary sludge on a gravity belt thickener. The presence of primary sludge in a co-thickening application has similar impacts to thickening primary sludge.

Because the solids content impacts unit capacity this must be discussed with the manufacturers during system sizing.

Some of the performance differences in thickening primary sludge, WAS, or co-thickening both are indicated in Table 21.13.

5.4 Operational Considerations Related to Design

Gravity belt thickeners provide the flexibility of operating at variable loading rates (gpm/meter or # TSS/meter) but at some point a higher loading rate will reduce solids capture and solids thickening and/or increase polymer consumption. When determining the required size and number of units required for an application, the designer must consider operating conditions, redundancy requirements, and capacity of downstream processes to receive a variable solids stream. Units that operate continuously (24 hours/day; 7 days/week) will be smaller (1-m vs., 2-m or 3-m gravity belt thickeners) than equipment required to operate only during staffed hours (e.g., 8 hours/day; 7 days/week). Fewer or larger units can be selected provided there is the understanding that loss of a redundant unit may decrease performance. Whereas larger gravity belt thickeners (3 m) have higher capacities and therefore do represent a significant loss in capacity upon failure, given the similarity in design configuration spare parts for very larger gravity belt thickeners do not have any longer lead times. For these reasons, the designer needs to consider if the added reliability of a larger number of smaller units is more important than the potential cost savings of fewer larger units.

Critical to a good layout is proper access to the gravity belt thickener both from the top and bottom. One common layout is to elevate the gravity belt thickener above a drainage sump to provide access to bottom bearings, wash boxes, etc. The height above the drain should be sufficient for access without increasing building height or making it difficult for operator access. When laying out such a system, the designer should consider all of the piping, ductwork, and electrical conduit that are connected to the gravity belt thickener and ensure that sufficient workspace, headroom, and lighting are provided near and below the gravity belt thickener. The width of the grating around the GBT should be sufficient for an operator to perform all necessary O&M tasks.

Locations should be provided to allow easy access for sampling sludge feed, filtrate, and thickened sludge streams. Rapid solids analyzers (e.g., microwave moisture/solids analyzers) are useful tools for quick sample analysis.

Typically, the gravity belt thickener is open, and not enclosed for odor control. This creates an opportunity of significant aerosols and/or spray in the area of the gravity belt thickener belt, wash box, etc. It is important to provide a significant number of hose bibs at convenient locations for operator wash downs. If the plan is to locate a control panel or control equipment near the gravity belt thickener pay close attention to materials of construction such as stainless steel, FRP, or other material to avoid excessive corrosion.

If performance testing is required staff should be present to understand proper startup, shutdown, and clean-in-place procedures and sequences. Hands-on training, in addition to classroom training, is of significant value.

5.5 Ancillary Equipment

A gravity belt thickener's ancillary components primarily include the following:

- Feed pumps and feed flow control,
- Polymer system and feed control,
- Belt washwater supply,
- Thickened-solids pumps, and
- Odor control.

All components must be sized to handle both maximum and minimum potential feed rates. Ideally, all controls should be near the thickener in a place where operators can see the top of the gravity belt while making process adjustments. See Figure 21.19 for a typical process and instrumentation diagram (P&ID)/flow schematic showing a gravity belt thickener and major ancillary components. Also, design engineers should include an emergency stop cord along the unit so operators can stop the thickener, solids feed pump, and polymer feed pump for safety reasons.

5.5.1 Feed Systems

A gravity belt thickener can operate over a large range of feed rates, but each feed rate requires its own polymer dose and belt speed. Feed-rate changes may also require adjustments in discharge-ramp angle, polymer dilution water, and the position of the solids–polymer mixer. So, design engineers should provide a flow meter (e.g., an electromagnetic flow meter) and an adjustable speed pump or a flow-control valve so operators can maintain a constant solids flowrate.

Screw-induced centrifugal pumps work well when enough suction head is available and the discharge head is not too high. These pumps require less maintenance than positive-displacement pumps because of the low contact between the slurry and screw impeller.

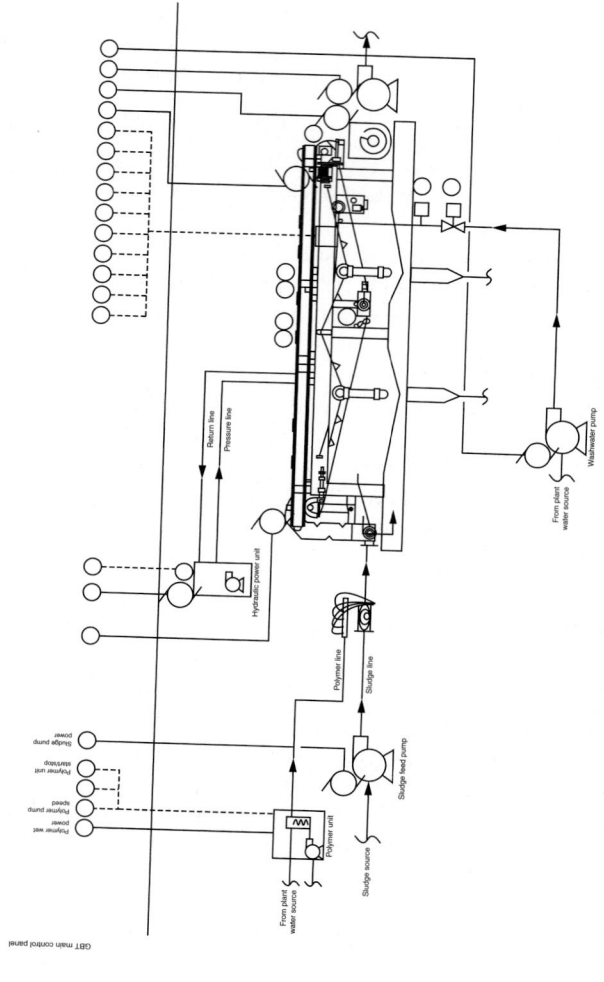

Figure 21.19 Gravity belt thickener process flow diagram (BDP Industries, Inc.).

1696

Recessed-impeller centrifugal pumps are used when the slurry is abrasive but enough suction head is available and discharge head is not too high. Positive-displacement pumps (e.g., progressing cavity pumps) and certain rotary lobe pumps are good choices when pumping slurries with higher friction losses. They provide good suction draw and can pump against higher heads. Design engineers can set up a control loop that uses a flow meter to track solids feed and adjusts pump speed to maintain a setpoint feed rate. This arrangement requires that each gravity belt thickener has a dedicated pump.

Another feed-control strategy involves flow-control valves and centrifugal pumps. It can be used to split flow among multiple thickeners. Design engineers can set up a control loop in which feed-pump speed is adjusted to maintain a setpoint pressure in a header, which serves as a manifold with branches to each thickener. Each branch requires a flow-control valve and flow meter. Alternatively, design engineers can set up a control loop in which the flow-control valve is adjusted to maintain a setpoint feed rate (as monitored by the flow meter). This alternative does not require each thickener to have a dedicated pump, so design engineers can reduce the number of pumps involved. They also may put a flow-control valve downstream of a dedicated constant-speed pump, cutting costs by eliminating the need for an adjustable-speed drive. However, this option requires a pump that can operate at the intended thickener-feed rates and restrict the range of acceptable feed rates. Also, the variable-orifice mixer may have wide swings in pressure drops (depending on the type of solids involved), making stable thickener operations problematic.

5.5.2 Chemical Conditioning

Adding polymer to the feed solids is essential to successful thickening when using a gravity belt thickener. It promotes flocculation of solids and release of free water. Without polymer, the belt would blind (because of fine solids filling belt pores) and flood (because of poor water release). Design engineers should test various polymers to determine which is the most effective. Cationic (positively charged) polymers typically are chosen because wastewater solids often are negatively charged. However, if the solids contain significant amounts of aluminum or ferric salts (which impart a positive charge), then an anionic (negatively charged) polymer may be better.

Design engineers also need to determine the appropriate polymer dose—the minimum and maximum amounts that can be added to get good results. Typically, solids capture and concentration increase as polymer doses rise above the minimum effective level. They eventually level off until the polymer dose exceeds the maximum effective level, when solids capture and concentration can be reduced. Excess polymer can blind belt pores and create a floc more susceptible to breakup.

If gravity belt thickeners will be used to thicken both WAS and digested solids, then design engineers typically select one polymer that is suitable for both feed solids because this is more cost-effective than adding two chemical-feed systems. They typically can find a polymer that is effective for both solids, although it may not be the optimal choice for either material.

Adding higher doses of a less expensive polymer can be cheaper than using lower doses of a polymer with a high charge strength or molecular weight. So, design engineers should specify performance requirements based on dollars of polymer per thousand kilograms of solids rather than on grams of polymer per kilogram of solids. During equipment performance tests, manufacturers should choose the polymer so they have full control of and responsibility for test results. Afterward, however, design engineers can invite chemical companies to test other polymers. It is important to consider the effect of increasing polymer dose on downstream solids handling processes and balance this with potential cost savings of a high-dose, inexpensive polymer.

Polymer is made in batches; it often is made at higher concentrations (up to 1%) to reduce the size and number of batches required. The appropriate concentration depends on use (to avoid excessive storage and keep up with demand). The batch should be diluted downstream of the polymer feed tank before being added to solids, and this dilution rate depends on polymer feed rate and concentration. The polymer's concentration can affect both solids thickening and polymer efficiency, so design engineers should make both the polymer-batching and -dilution systems flexible so operators can adapt them as needed.

5.5.3   Belt Washwater Supply

Before each portion of the belt begins another thickening cycle, it should be washed to remove embedded solids and excess polymer. Belt wash stations are typically installed on the belt's return loop and use about 80 L/min (20 gpm) of water per meter of belt. Washwater flow and pressure recommendations depend on the manufacturer. Pressure recommendations typically range from 517 to 586 kPa (75–85 psi), although some manufacturers recommend pressures of 760 to 830 kPa (110–120 psi). This pressure is created by nozzle losses, minor losses caused by fittings, pipe-friction losses, and any elevation differences between the nozzles and the water source. The actual pressure at the wash station typically depends on nozzle losses (i.e., nozzles create the backpressure). So, design engineers need an accurate curve of nozzle losses versus flow (typically available from the nozzle manufacturer) to design the washwater-supply system properly.

Washwater use depends on thickener size (typically defined by belt width), number of thickener units in operation, and time of operation. If water use is insignificant, a facility's potable water supply can be used. Manufacturers typically provide the control valve, and a booster pump can compensate for inadequate facility water pressure. If water use is significant, facility effluent should be considered. Design engineers will have to add pumps and automatic strainers to remove larger suspended solids if the facility effluent's suspended solids concentration is less than 50 mg/L. Alternatively, some manufacturers offer the option of using the gravity belt thickener's filtrate (which also requires pumps and automatic strainers). The controls for washwater pumps should be interfaced with those for the gravity belt thickener and water-supply valve. Design engineers should also select automatic strainers for uninterrupted operation during cleaning. The strainers are typically either cleaned continuously or based on a timer and/or pressure loss.

5.5.4   Thickened Sludge Conveyance

Headlosses are high when thickened solids flow through a pipe. The amount of headloss depends on pipe size and type, flow velocity, type of solids, and solids concentration. Engineers have used various curves, models, and "general rule" multipliers to estimate headlosses. The estimating procedures developed for certain solids (e.g., paper-stock solids) are fairly reliable, but those for municipal solids are less standardized. Headloss can become more difficult to estimate as piping distance and solids concentration increase. So, unless available data indicate otherwise, design engineers should use conservative models and worst-case assumptions.

Because of the high headlosses, design engineers should use positive-displacement pumps (e.g., progressing cavity pumps, certain rotary lobe pumps, and air-operated diaphragm pumps) to transport gravity belt-thickened solids. These pumps provide good suction draw and can pump against higher heads. Engineers should also design suction pipes to be straight and as short as possible. If thickened solids must be pumped long distances, then special provisions (e.g., multiple-stage progressing cavity pumps) probably will be necessary.

5.5.5   Odor Control and Ventilation

Whether odor treatment is provided or not, high ventilation rates are necessary to provide an acceptable working environment. A minimum of 12 to 15 fresh air changes per hour are common if the gravity belt thickeners are housed in a building and even higher ventilation rates may be necessary near the gravity zone of the gravity belt thickener. The exhaust rate should be slightly faster than the fresh air supply rate to maintain a slight vacuum in the room, which will prevent odor problems in adjacent areas.

The odorous air exhausted from the building should be treated before atmospheric discharge if the area adjacent or near the WRRF is expected to be impacted by odors. Odor-treatment options (see Chapter 6) include packed-tower scrubbers and mist scrubbers (chemical scrubbers), activated-carbon beds, and biofilters. In chemical scrubbers, hypochlorite and sodium hydroxide typically are used to scrub sulfur-related odors and sulfuric acid solutions can be used to scrub ammonia-related odors. Another option is adding an odor-control compound (e.g., hydrogen peroxide) to the feed solids, but the compound first should be tested to determine whether it interferes with polymer efficiency and solids thickening.

As an alternative to odor control some manufacturers can provide an enclosed gravity belt thickener (see Figure 21.20). The enclosed gravity belt thickener will reduce the amount of ventilation required and potentially reduce the capacity of any odor control equipment. However, if an enclosed gravity belt thickener is selected close attention must be paid to materials of construction of all components (frame, enclosure, belt, etc.) to assure long life and prevent excessive corrosion of the components.

5.6   Process Control

5.6.1   Overview

A thickening system typically includes sludge feed pumps, the polymer system, the thickeners, and the thickened sludge conveyance system. The gravity belt thickener is usually provided with a PLC that can control and/or communicate with the system

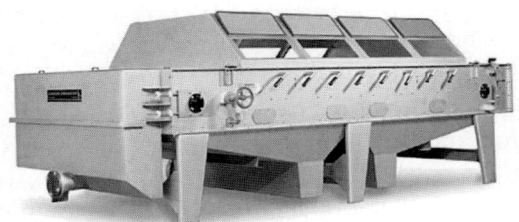

FIGURE 21.20   An example of an enclosed gravity belt thickener (courtesy of Alfa-Laval, Inc.).

components during automated startup, shutdown, and cleaning sequences. At various points of these cycles, the gravity belt thickener PLC controls starting, stopping, and speed of the gravity belt thickener belt. The PLC typically also controls the normal shutdown including run time without sludge feed, belt wash, and most alarm conditions that require the gravity belt thickener to stop.

Maximizing cake solids, maintaining reasonable capture efficiency, and minimizing polymer requirements are difficult tasks to accomplish without instrumentation and controls. During operation, the gravity belt thickener is usually operated by its PLC to maintain a specific belt speed. This helps maintain consistent cake solids but does not directly address capture efficiency and polymer dose. Capture efficiency is maintained by manually changing the polymer dose until the filtrate is relatively clear. But unlike other solids thickening processes the operator observation of process startup and operation are a key factor in process control. Over time, an operator can learn through observation of sludge separation to be key to optimizing polymer dose, sludge feed rate, and belt speed.

5.7   Performance Assessment and Optimization

5.7.1   Performance Parameters

The process performance of a gravity belt thickener is primarily assessed through thickened sludge concentration, polymer consumption, and percent capture. Other parameters of importance are operational consistency, equipment longevity, ease of operation, and ease of maintenance.

The expected process performance for a specific feed sludge and gravity belt thickener can be estimated through pilot-testing in the field. Most manufacturers maintain trailer-mounted gravity belt thickener systems that include ancillary pumps, polymer, and thickened sludge conveyance equipment. In addition to projecting potential performance, such tests also provide O&M staff an opportunity to gain first-hand operational experience and ask questions. However, given the known range of capabilities of gravity belt thickeners on most sludge, pilot testing is not typically done.

5.7.2   Optimization Measures

System optimization focuses on maintaining consistent thickened sludge concentration, minimizing polymer consumption and maximizing percent capture. Many of the control system features discussed previously seek to automatically optimize gravity belt thickener performance. Additional optimization beyond automated control includes selection of the polymer that provides the best value and determining the optimal polymer solution concentration and injection point. Polymers should be tested on a regular basis to identify those that provide the best value (cost versus performance) throughout the year. Many facilities' sludge characteristics change in winter versus summer and different polymers often work best during each of those seasons.

5.8   Evaluation and Scale-Up Procedures

In the past, engineers have tested gravity belt thickener performance to verify design parameters before installation. Over the years, they tested this process on many types of solids (e.g., WAS, anaerobically digested solids, primary solids, trickling filter solids, aerobically digested solids, and pure-oxygen activated sludge) using a trailer-mounted, 1-meter gravity belt thickener that came with all necessary ancillary equipment. Results showed a good correlation between pilot- and full-scale operations.

The process' flexibility and overall good performance have made pilot testing unnecessary for most wastewater solids applications. Design engineers now primarily use pilot tests to compare the cost-effectiveness of various manufacturers' machines, to evaluate an unusual thickening application, or to demonstrate the machine's performance to gain acceptance. Laboratory testing typically consists of sending a solids sample to the manufacturer for jar and free-drainage tests to determine polymer type and dose.

As with other wastewater thickening processes, performance is solids-specific. However, performance and design criteria can be predicted based on similar full-scale installations (see Table 21.14).

5.9 Design Example

Suppose a WRRF plans to install a gravity belt thickener to treat a mix of primary solids and WAS. Design engineers have calculated that the facility produces 1150 L/min of combined solids on a continuous basis. The following operating criteria apply:

- thickening facility operates 8 hr/d, 5 days/week,
- WAS solids concentrations = 0.5% to 1.0% solids,
- the gravity belt thickeners have an effective thickening width of 2 m, and
- the design should allow for one standby unit.

Using the criteria in Table 21.12 to select a hydraulic loading rate based on one unit out of service and one unit undergoing maintenance, choose a rate of 800 L/m/min.

Hydraulic loading rate = 1145 L/min

Effective gravity belt thickening width = 2 m

Design loading rate = 800 L/m min

Type of Biosolids	Initial Concentration (%)	Solids Loading (kg/m·h)	Polymer Dosage (g/kg)	Final Concentration (%)
Primary	2–5	900–1400	1.5–3	8–12
Secondary	0.4–1.5	300–540	3–5	4–6
(50% P)(50% S)	1–2.5	700–1100	2–4	6–8
Anaerobic	2–5	600–790	3–5	5–7
(50% P)/(50% S)				
Anaerobic (100% S)	1.5–3.5	500–700	4–6	5–7
Aerobic (100% S)	1–2.5	500–700	3–5	5–6

TABLE 21.14   Typical Performance of Gravity Belt Thickeners (Reprinted with permission from Alfa Laval)

Equations used:

$$\text{Equivalent hydraulic loading per week (L/min)} = \text{Hydraulic loading rate} * \left(\frac{\text{Total No. of min/week}}{\text{No. of min/week of operation}}\right) \quad (21.27)$$

$$\text{Loading capacity per GBT (L/min)} = \text{Design loading rate} * \text{Belt width (m)} \quad (21.28)$$

$$\text{No. of GBTs to meet hydraulic requirement} = \frac{\text{Equivalent hydraulic loading rate (L/min)}}{\text{Loading capacity per GBT (L/min)}} \quad (21.29)$$

Equivalent hydraulic loading rate per min (L/min)

$$= 1145 \text{ L/min} * \frac{(24 \text{ hours/day} * 7 \text{ days/week} * 60 \text{ min/h})}{(8 \text{ hours/day} * 5 \text{ days/week} * 60 \text{ min/h})} = 4800 \text{ L/min}$$

$$\text{Loading capacity per GBT} = 800 \text{ L/m} \cdot \text{min} * 2 \text{ m} = 1600 \text{ L/min}$$

$$\text{No. of GBTs to meet hydraulic limit} = \frac{4800 \text{ L/min}}{1600 \text{ L/min/GBT}}$$

$$= 3 \text{ GBTs (2 m); Use eq 21.29}$$

Three gravity belt thickeners must be in service during operational hours. Therefore, the system will need four gravity belt thickeners if one will be on standby.

6.0   Rotary Drum Thickener

6.1   Operating Principle

A rotary drum thickener (also called a rotary screen thickener) consists of a flocculation tank with slow-speed mixer, an internally fed rotary drum, an integral internal screw or conveying flights, spray-wash system, filtrate collection tank, and a variable- or constant-speed drum drive (see Figure 21.21). A conditioning agent (typically polymer) is injected into the feed sludge and the mixture is sent to a flocculation tank. The tank is mixed slowly to allow sludge to flocculate and the conditioned sludge flows into the drum. As the drum rotates, the sludge moves up the side of the drum and falls to the bottom. This movement releases free water, which flows through the drum covering material into the filtrate tank. Both gravity belt and rotary drum thickeners allow free water to drain through a moving, porous media while retaining flocculated solids. In rotary drum thickeners, the rotating drum can consist of a filter cloth stretched over a frame or can be constructed as a metal wedge wire, wire mesh, or perforated plate screen itself. The internal screw/flights transport thickened solids out of the drum. An enclosure or chute at the end of the drum directs the thickened sludge into a wet well or pump.

The most established rotary drum thickener manufacturers to date produce units with a nominal maximum capacity of 90 m³/hr (400 gpm) (for a 1% solids WAS feed). However, as the technology becomes more popular, other manufacturers have started producing units with larger capacities up to 180 m³/hr (800 gpm). These larger capacity units often include two drums configured in parallel within one "unit." Advantages of

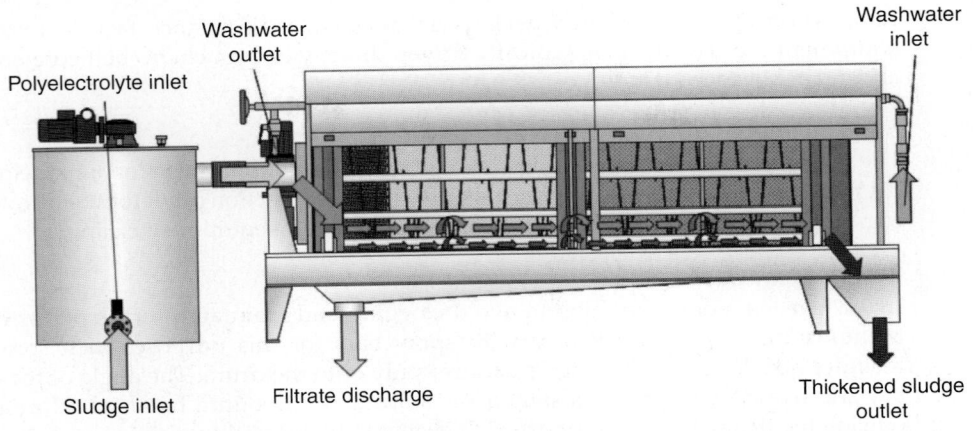

Polyelectrolyte inlet

Washwater outlet

Washwater inlet

Sludge inlet

Filtrate discharge

Thickened sludge outlet

FIGURE 21.21 Schematic of a rotary drum thickener (courtesy of Seimens Water Technologies).

rotary drum thickeners include relatively low power use and ease of enclosure, which improves housekeeping and odor control.

6.2 Physical and Mechanical Features

The primary mechanical components of a rotary drum thickener are the flocculation tank, drum, filtrate collection tank, spray wash system, enclosure, and discharge chute.

6.2.1 Flocculation Tank

Sludge that has been injected with polymer upstream is fed into the flocculation tank. A slow speed mixer circulates the tank contents, allowing the sludge to flocculate and then discharge into the rotating drum. The flocculation tank includes a hinged cover that can be opened to view floc structure.

Polymer injection upstream of the flocculation tank often consists of a polymer injection ring, manifold, and venturi-style mixing valve. Polymer is injected into the sludge line via the injection ring, and this mixture then flows through the mixing valve. The opening of the valve and subsequent mixing energy is adjusted with a lever arm and counterweight. Some manufacturers use static mixers for sludge/polymer mixing rather than this valve assembly.

6.2.2 Drum

The drum consists of a support structure fitted with either filter fabric (polyester) or a metal screen (stainless steel). Different manufacturers produce metal screens with woven-wire, wedge-wire, or perforated plate. The pore openings may change in size along the drum, but this is specific to certain manufacturers. Within the drum, radial flights move the solids toward the end of the drum while released water flows through the screen into the filtrate collection tank. Some manufacturers include additional features such as roll bars, split augers, and varying mesh size along the drum to improve thickening performance. The drum is supported on lubricated trunnion wheels at specific locations or for some manufacturers, a full-length drive shaft.

Lubrication of the trunnion wheels is an essential maintenance task to maximize equipment life. The drum is typically driven directly or by a chain/belt connected to an adjustable speed drive.

6.2.3 Filtrate Collection Tank
The water released from the sludge is collected in the filtrate collection tank below the drum. A pipe stub off the collection tank provides a connection point for the filtrate piping. Sloping the tank floor to the stub assists staff during equipment cleaning.

6.2.4 Spray Wash System
Drum screens require washing to avoid clogging and maintain process performance. Manufacturers fit the thickeners with spray bars for this purpose. Each spray bar includes nozzles that direct high pressure spray onto the drum. The angle of the spray bar and associated nozzles is sometimes adjustable through a hand wheel mounted outside the drum. Some rotary drum thickeners utilize continuous spray while others require only intermittent washing. Some units require higher pressure water than can be provided with the facility's utility water system. Booster pumps can be provided by the manufacturers for these situations. Treated effluent is often used as a source for spray water.

6.2.5 Enclosure
Rotary drum thickeners are typically enclosed with stainless steel plate. Sections of the plate are removable to allow maintenance and inspection. The enclosures prevent escape of odors, filtrate, solids, and mist. An odor control nozzle on the enclosure allows connection to the facility's odor control system, if necessary.

6.2.6 Discharge Chute
The thickened sludge drops from the end of the drum into a discharge chute. The chute conveys thickened sludge to a wet well or pump. This chute can be specified as a small hopper connected to an open-throat progressive cavity pump. The hopper volume should be sufficient to allow pump control and prevent run dry situations. Level sensors used in the hopper should be carefully selected and situated to mitigate impacts of splashing and inaccurate readings while allowing the entire length of the hopper to be measured.

6.3 Process Design Conditions and Criteria
The unit size required to thicken a given amount of solids depends on machine capacity, solids characteristics, and polymer dose. Rotary drum thickeners have hydraulic loading and solids loading capacities, but the solids loading criteria typically limits hydraulic throughput. Engineers define process design criteria by establishing which solids characteristics are significant in a given application and determining how they affect process performance. Feed sludge type and characteristics impact system design and design criteria.

6.3.1 Impacts of Different Feed Sludge Types
Rotary drum thickeners can be used to thicken a variety of sludge types. Most municipal WRRFs use them to thicken WAS. However, the units can also be used to thicken primary sludge, and to co-thicken primary sludge and WAS. The type of sludge to be

thickened impacts some of the equipment design and performance parameters because of sludge characteristics.

Primary sludge thickens readily and requires different types of polymer, lower polymer doses, and lower polymer solution concentrations than WAS. Grease often found in primary sludge can blind drum screens, necessitating hot water spray water systems or the ability to periodically wash the screens with hot water. The presence of primary sludge in a co-thickening application has similar impacts. Primary sludge usually has higher solids content than biological sludge so it requires longer residence time in the drum, which is provided through lower drum speeds. The solids content also impacts unit capacity so this must be discussed with the manufacturers during system sizing.

Some of the performance differences in thickening primary sludge, WAS or co-thickening both are indicated in Table 21.15.

6.4 Operational Considerations Related to Design

Rotary drum thickeners are relatively simple machines, but require consideration of several items during design. When determining the required size and number of units required for an application, the designer must consider operating conditions and redundancy requirements. Units that operate continuously (24 hours/day; 7 days/week) will be smaller than equipment required to operate only during staffed hours (e.g., 8 hours/day; 7 days/week).

The units can be tall, so platforms near the flocculation tank and at the hopper allow staff to safely observe sludge quality. Frequent maintenance tasks include lubrication of trunnion wheels/bearings and periodic power washing of the drum. Sufficient access and utility connections should be provided to allow for easy completion of these tasks.

The extent to which sludge moves within a rotary drum thickener necessarily impacts floc structure and requires cross-linked polymers. Operating experience has indicated that rotary drum thickeners require more polymer than gravity belt thickeners, so this must be considered in O&M cost analyses.

Many thickening systems include a thickener located on an upper level, discharging to the thickened sludge system below. Proper access to the equipment below the thickener should be provided. When laying such a system out, the designer should consider all of the piping, ductwork, and electrical conduit that are connected

	Primary Sludge	WAS	Co-thickening Primary Sludge and WAS
Typical feed solids concentration	1%–4% TS	0.5%–1% TS	1.5%–3% TS
Typical thickened sludge solids concentration	7%–12% TS	4%–7% TS	5%–9% TS
Typical polymer dose	2.2–6.7 kg/tonne (5–15 active lbs/dry ton)	2.2–6.7 kg/tonne 5–15 active lbs/dry ton)	2.2–6.7 kg/tonne (5–15 active lbs/dry ton)
Typical percent capture	93%–98%	93%–98%	93%–98%

TABLE 21.15 Performance Differences in Thickening Primary Sludge, WAS, or Co-Thickening with RDTs

to the thickener and ensure that sufficient workspace, headroom, and lighting are provided near and below the unit. Accessible sample ports for sludge feed, filtrate, and thickened sludge are important for system operations. Rapid solids analyzers (e.g., microwave moisture/solids analyzers) are useful tools for quick sample analysis.

Odor control may be required, especially if the thickeners are used for primary sludge or for co-thickening. Materials of construction should be suitable for exposure to moisture, hydrogen sulfide, and chlorides if treated effluent is used for spray water.

6.5 Ancillary Equipment

Besides the rotary drum thickener itself, the primary components to be designed in a rotary drum thickening system are

- solids feed pumps and feed flow control,
- polymer system and feed control,
- spray wash system,
- thickened-solids pumps, and
- odor control.

6.5.1 Feed System

The feed pumps, flow meters, and control valves are similar to those described for gravity belt thickeners and thickening centrifuges. Positive displacement pumps (e.g., rotary lobe) and centrifugal pumps (e.g., nonclog or chopper) have both been used for thickener feed. Whereas WAS can be pumped easily by centrifugal pumps, thicker primary sludge may necessitate progressing cavity or rotary lobe pumps. If primary sludge is thin and expected to contain significant amounts of grit, hardened recessed impeller pumps often used for grit pumping can also be used to pump primary sludge.

Before deciding on thickening system capacity, design engineers should consider the WRRF's solids production rate and determine whether it would be more cost-effective for the thickener to run continuously or intermittently. Many operators prefer continuous wasting so this operational need must also be considered. Design engineers must size all components, including feed pumps, to handle both maximum and minimum potential solids feed rates.

6.5.2 Chemical Conditioning

The polymer systems and associated considerations are similar to those described for gravity belt thickeners and thickening centrifuges. For detailed information on polymer mixing and feeding systems, see Chapter 20.

The polymer feed system must be sized to handle both maximum and minimum feed rates, and it must be flexible enough to allow operators to vary the dose to meet flocculation requirements. Both under- and overdosing can degrade thickener performance. It is good practice to provide several locations for polymer injection upstream of the flocculation tank. While the injection assembly (polymer injection ring, manifold, and mixing valve or static mixer) can be installed in one location, other locations can be provided with spool pieces that match the length of the polymer injection assembly. If locations need to be switched for optimization, the assembly and spool are simply swapped.

6.5.3 Spray Wash System

The spray water supply can usually be WRRF effluent, but strainers may be required to avoid clogging of spray nozzles. Manufacturers require significantly different volumes [5–23 m^3/hr (20–100 gpm)] and pressures [2.8–6.7 bar (40–100 psi)] for their spray water systems. If necessary for pressure requirements, centrifugal booster pumps can be included in system supply. The spray water piping is connected to the spray bar with an automated valve that opens and closes as required.

6.5.4 Thickened Sludge Conveyance

The thickened sludge conveyance systems are similar to those described for gravity belt thickeners and thickening centrifuges. If thickened sludge is discharged to a wet well and then pumped, careful consideration is required for suction piping design. This type of sludge can be viscous and exhibit thixotropic characteristics. As such, flooding pump suctions can be difficult without sufficient suction pipe slopes and head.

6.5.5 Odor Control and Ventilation

Odor control is greatly simplified by the enclosures that rotary drum manufacturers provide. Odors can be drawn off these enclosures for treatment in appropriate odor-control units. Refer to Chapter 6 for details on odor control systems.

6.6 Process Control

Rotary drum thickeners offer the flexibility of varying processing performance with solids and polymer feed rate control and speed adjustment for the drum and flocculation tank mixer. Most rotary drum thickeners are not provided with PLCs. Control panels are wired to allow communication and monitoring through the facility's control system, which sends signals to the thickening system components. Thickener control panels do include start/stop/emergency stop push buttons and speed knobs that provide local control of the drum and mixer speeds.

6.6.1 Sludge Feed Rate

Sludge feed pump speed modulates to maintain setpoint feed rates to the rotary drum thickener. The feed rate to each thickener should be maintained between a minimum and maximum value as indicated by the manufacturer. Feed rates outside of this range can impair performance.

6.6.2 Polymer Feed Rate

Polymer feed pump speed modulates to maintain setpoint polymer dose to the feed sludge. Programming is normally required to combine sludge flow, sludge concentration, and polymer characteristics into a required dose and associated pump speed. Sludge concentration can be measured with inline sensors or input by operators based on solids testing. Polymer characteristics such as percent active and specific weight should not change often and are typically entered manually into the facility control system.

Type of Solids	Feed (% TS)	Water Removed (%)	Thickened Solids (%)	Solids Recovery (%)
Primary	3.0–6.0	40–75	7–9	93–98
WAS	0.5–1.0	70–90	4–9	93–99
Primary and WAS	2.0–4.0	50	5–9	93–98
Aerobically digested	0.8–2.0	70–80	4–6	90–98
Anaerobically digested	2.5–5.0	50	5–9	90–98
Paper fibers	4.0–8.0	50–60	9–15	87–99

TABLE 21.16   Typical Performance Ranges for Rotary Drum Thickeners

6.6.3   Drum Speed

Operators can change drum speed in response to changes in feed solids concentrations and flowrate to maintain the desired thickened-solids concentration. When treating WAS, rotary drum thickeners capture between 90% and 99% solids and produce a thickened material containing between 4% to 9% total solids. Typical performance ranges for other types of solids are shown in Table 21.16.

6.6.4   Flocculation Tank Mixer Speed

Operators can change flocculation tank mixer speed in response to changes in feed solids concentrations and flowrate to maintain the desired thickened-solids concentration and floc characteristics.

6.7   Performance Parameters and Optimization

6.7.1   Performance Parameters

The process performance of a rotary drum thickener is primarily assessed through thickened sludge concentration, polymer consumption, power consumption, and percent capture. Other parameters of importance are operational consistency, equipment longevity, ease of operation, and ease of maintenance.

The expected process performance for a specific feed sludge and rotary drum thickener can be estimated through pilot-testing in the field. Most manufacturers maintain trailer-mounted units that include ancillary pumps, polymer, and thickened sludge conveyance equipment. In addition to projecting potential performance, such tests also provide O&M staff an opportunity to become familiar with the equipment and ask questions.

Expected operational performance is harder to project. Requiring the provision of references and checking with those references offers an opportunity to determine how the equipment has performed elsewhere.

6.7.2   Optimization Measures

System optimization focuses on maintaining consistent thickened sludge concentration, minimizing polymer consumption, minimizing power consumption, and maximizing percent capture. Tracking impacts of changing feed flow can illustrate operational "sweet spots." Additional optimization includes selection of the polymer that provides the best

Drum Size (m)	Influent Flow Rate (L/min)	Influent Solids Concentration (%)	Effluent Solids Concentration (%)
0.6	Average = 160 Range = na	Average = 1 Range = 1–2	Average = 6 Range = 5.5–6
1.5	Average = 900 Range = 490–1100	Average = 0.9 Range = 0.001–1.5	Average = 5.5 Range = 4–9

TABLE 21.17 Performance Data for Rotary Drum Thickeners

value and determining the optimal polymer solution concentration and injection point. Polymers should be tested on a regular basis to identify those that provide the best value (cost versus performance) throughout the year. Many facilities' sludge characteristics change in winter versus summer and different polymers often work best during each of those seasons.

6.8 Evaluation and Scale-Up Procedures

Rotary drum thickeners work well on many types of wastewater solids. As with other thickening processes, performance is solids-specific. Projected performance can be estimated by manufacturers at their factories through testing of feed sludge. Performance can also be determined through pilot testing at the WRRF or by using data from similar full-scale installations (see Table 21.17). Design engineers often conduct pilot tests to compare units from various manufacturers or to compare rotary drum thickeners to other thickening devices (e.g., gravity belt thickeners). They also perform pilot tests to select an appropriate polymer and dose. Because of floc sensitivity and potential for shearing, adding relatively large amounts of polymer can be required in rotary drum thickening.

6.9 Design Example

Suppose a WRRF is installing a rotary drum thickener to treat its primary solids and WAS. Design engineers calculated that the WRRF produces 2200 L/min (580 gpm) of combined solids on a continuous basis. The following operating criteria apply:

- Thickening facility operates 8 hours per day, 5 days per week, (2400 min/week)
- WAS solids concentrations is 0.5% to 1% solids,
- Rotary drum thickeners have an effective drum size of 1.5 m (60 in.), and
- The design should allow for one unit undergoing maintenance and one unit in standby.

Using the criteria in Table 21.17 to select a higher hydraulic loading rate based on one unit out of service and one undergoing maintenance, design engineers select a design loading rate of 1100 L/min/unit (290 gpm/unit).

Facility's hydraulic loading rate = 2200 L/min (580 gpm)

Effective drum size = 1.5 m (60 inches)

Design loading rate per unit = 1100 L/min/unit (290 gpm/unit)

Equations used:

$$\text{Facility's hydraulic loading per week} = \text{Hydraulic Loading} * \frac{\text{No. of min}}{\text{week}} \qquad (21.30)$$

$$\text{Weekly loading capacity} = \frac{\begin{array}{c}\text{Design Loading Rate} * \text{Screen width} * \\ \text{No. of operating min}\end{array}}{\text{week}} \qquad (21.31)$$

$$\text{Drums needed to meet hydraulic requirement} = \frac{\text{Hydraulic Loading per Week}}{\text{Weekly Loading Capacity}} \qquad (21.32)$$

Thickening system loading rate per min = Facility's hydraulic loading per week/Number operating minutes per week

Drums needed to meet hydraulic requirement = Thickening system loading rate per min / Design loading rate per unit; Use eq 21.32

Facility's hydraulic loading per week = 22 176 000 L/week (5 846 400 gal/week); Use eq 21.30

Thickening system loading rate per min = 9240 L/min (2436 gpm); Use eq 21.31

Drums needed to meet hydraulic requirement = 9 screen drums requirement; Use eq 21.32

To meet the design loading rate, nine drums must be operational at all times. So, the final system must have eleven drums in case one is on standby and one is undergoing maintenance.

7.0   Emerging Technologies

Several technologies other than those described above have been used in the United States or abroad for a number of years. The number of installations for these technologies in the United States is limited so they are described here as "emerging." As more facilities use them, performance and design criteria for these systems will become better defined. They are described briefly here to provide a fuller picture of thickening options.

7.1   Disk Thickeners

7.1.1   Operating Principle

The disc thickener shown in Figure 21.22 is an inclined, slowly rotating filter disc construction. The filter bottom consists of a perforated carrier disc covered with a micro filter that has a constant mesh size. The filter disc is installed into a closed stainless steel tank and divides the tank into a thickening zone (working area) and a filtrate collection zone.

As with gravity belt thickeners and rotary drum thickeners, polymer is injected into the sludge upstream of the thickener. The pre-flocculated sludge flows from the reaction

Reaction tank   Thickener

FIGURE 21.22   Major disk thickener components (courtesy of Huber).

tank onto the filter disc surface. The flocculated sludge settles on the filter surface while the filtrate flows to the collection zone and leaves the tank through the outlet. The residual solids on the micro filter thicken statically and are transported by disc rotation from the inlet section to a discharge opening. A simple schematic of the disk thickener is included in Figure 21.23.

The rotational speed and inclination of the disc thickener is adjustable to produce the required thickened sludge quality. Chicanes are installed in the thickening zone to increase the thickening efficiency. A scraping device is fitted on the disc surface to promote discharge of the thickened material. The scraper allows for continuous removal of solids from the filter disc. Cleaning of the filter disc is continuous by means of a spray nozzle bar, which is installed under the filter disc. The units are enclosed and can be connected to odor control systems if required.

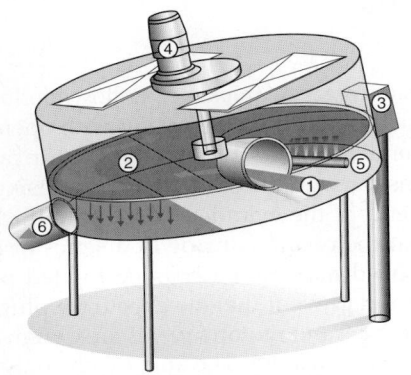

① Sludge inlet
② Sludge liquor
③ Sludge discharge
④ Drive
⑤ Spray bar
⑥ Filtrate outlet

FIGURE 21.23   Disk thickener schematic (courtesy of Huber).

7.1.2   Design Criteria

Disc thickeners can treat WAS with 0.5% to 1.5% solids at a hydraulic loading rate of up to 680 L/min (180 gpm) and a solids loading rate of up to 400 kg/hr (880 lb/hr). Disc thickeners may also be used to thicken primary sludge. For a typical primary sludge with 2% to 3% solids, a disc thickener can process a hydraulic loading rate of up to 355 L/min (94 gpm) and a solids loading rate of up to 465 kg/hr (1025 lb/hr).

The primary components for a disc thickener system are

- solids feed pumps
- inlet solids flow meters
- polymer feed systems
- polymer injection device
- flocculation tank
- disc thickeners
- thickened solids pumps
- washwater supply

The solids feed pumps, flow meters, polymer feed systems, polymer injection device, flocculation tank, and thickened solids pumps are similar in design to what has been described for gravity belt thickeners. The disc thickener washwater requirement is a continuous 29 to 35 L/min (7.7–9.2 gpm) at 300 kPa (43.5 psi).

7.1.3   Performance Characteristics

For thickening of both WAS and primary sludge, a thickened product material containing 5% to 7% solids is typical and may be optimized by the angle of disc inclination. Solids capture typically exceeds 95% in municipal applications.

7.1.4   Ancillary Systems

The ancillary systems for the disk thickener are similar to those required for gravity belt thickeners and rotary drum thickeners.

7.2   Volute/Screw Thickeners

7.2.1   Operating Principle

The volute thickener is a mechanical screw thickener. Following flocculation, the sludge enters the thickener that consists of a cylindrical casing with openings, and a screw or auger that rotates. As the material is moved through the cylindrical casing by the rotating auger, the free water drains through the openings in the casing until the thickened product is discharged at the end of the cylinder. The volute thickener differs from a regular screw thickener in that the casing is made of a series of rings with fine gaps between them rather than a fixed material such as perforated plate or wedge wire. These rings continually move and prevent the fine gaps from plugging with the solids being thickened. Because these gaps and moving plates are designed to eliminate plugging, the unit does not require backwashing. The auger is the only mechanical moving part, rotating at a speed less than 25 RPM. The unit uses relatively little power. In addition, the thickening mechanism can be completely enclosed to allow connection to

FIGURE **21.24**   Volute thickener (courtesy of PWTech).

odor control. The volute thickener is shown in Figure 21.24 and the cutaway in Figure 21.25 identifies equipment components.

7.2.2   Design Criteria
Key design criteria are the hydraulic flow to the thickener and the desired thickened sludge concentration. Because there is a fixed volume the only factor that impacts the throughput is the screw speed. Higher speeds will convey more material through the unit but allow less time for water to drain. As well as screw speed, thickened sludge concentration is also a factor of the nature of the sludge and how well the flocculated sludge releases water.

7.2.3   Performance Characteristics
Most commonly used for thickening WAS, the volute thickener will achieve output solids concentrations between 4% and 8% when starting with 0.3% to 0.7% solids. The presence of primary sludge can significantly increase those numbers. Thickening only primary sludge can yield concentrations up to 16% solids for the thickened sludge.

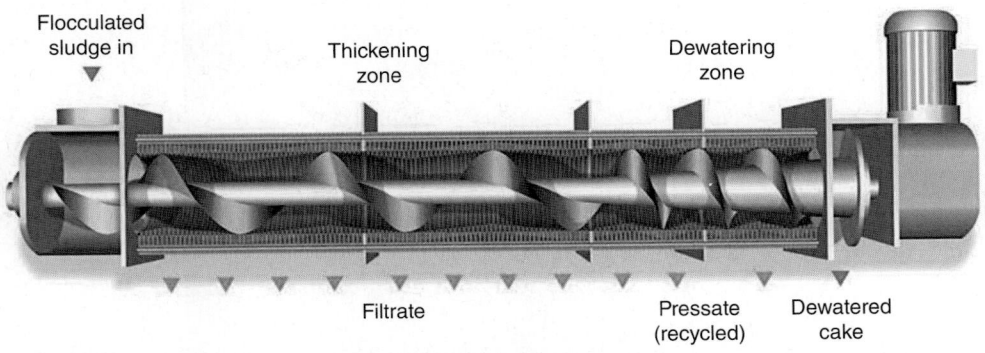

FIGURE **21.25**   Volute thickener schematic (courtesy of PWTech).

7.2.4   Ancillary Systems
The ancillary systems for the volute thickener are similar to those required for gravity belt thickeners and rotary drum thickeners.

7.3   Membrane Thickeners

7.3.1   Operating Principle
Membrane thickeners thicken WAS, often upstream of aerobic digestion systems. While membrane thickening does not produce as concentrated a product as other thickening processes, the thickened sludge concentration is suitable for feed into aerobic digesters. Polymer is not required for the process, but air is necessary for mixing and membrane scour. Either the flat sheet or hollow fiber membranes used for membrane bioreactors can also be used for membrane thickening. One process description for a membrane thickening system follows below. Various manufacturers may have different operating modes and should be consulted for specific requirements.

The WAS is initially held in a holding tank (Figure 21.26). When there is sufficient volume of WAS to thicken, the pumps or valves on the thickener membrane tanks are used to allow WAS to flow into the membrane thickener tank (Figure 21.27). The process pump starts and pulls the permeate through the membranes while leaving the thickened WAS in the thickener membrane tank. Once the target concentration is

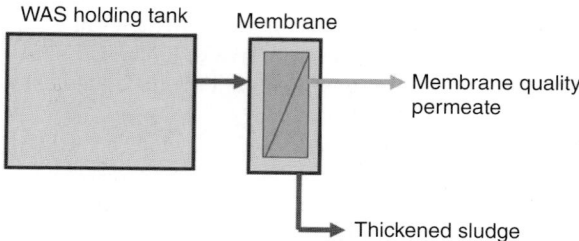

FIGURE 21.26   Schematic of Membrane Thickening (courtesy of GE).

FIGURE 21.27   Representation of a membrane thickening tank (courtesy of Ovivo/Kubota).

reached the system then drains the thickened sludge to either a holding tank or further dewatering process.

It is important to note that a solids thickener membrane process utilizes aeration to keep the tank mixed and to minimize solids accumulation within the membrane fibers. Anytime there is mixed liquor within the system the aeration system is called upon to provide air for scouring.

7.3.2 Design Criteria

Membrane flux rates depend on the solids concentration in the membrane tank. Flat sheet membranes are limited to flux rate of 5 to 10 gallons/day/sq ft for membrane thickening applications. Hollow fiber membrane start thickening at a flux rate of 10 gallons/day/sq ft and gradually decrease flux rate to approximately 2 gallons/day/sq ft as the sludge is thickened to 4% solids.

7.3.3 Performance Characteristics

Membrane thickening can be used for WAS from conventional activated sludge facilities and membrane bioreactors. Typical solids concentrations for WAS from a conventional activated sludge facility range from 0.4% to 1%, whereas WAS from an MBR can have concentrations exceeding 1%. For both sludge types, membrane thickening can produce thickened sludge with 4% solids.

7.3.4 Ancillary Systems

The ancillary systems for the membrane thickener are the feed tanks and valving, permeate pumps, aeration/scour blowers, chemical cleaning, and tank drain provisions. As with all membrane applications, upstream fine screens are required to protect the material and reduce maintenance. Screening for an MBR facility is typically included upstream of the secondary process.

7.4 Recuperative Thickening

7.4.1 Operating Principle

Interest has grown in increasing the capacity of existing digestion tanks to avoid incurring capital costs and to fully leverage infrastructure. Recuperative thickening separates SRT from hydraulic retention time (HRT), allowing for high SRTs within smaller tanks. By maintaining high SRT within the tank, digesters with recuperative thickening can be loaded at higher organic loading rates than conventional digesters. The high SRT and organic loading rates increase digestion capacity of existing tanks or reduce the new tankage required for a specific solids processing need.

Commercial systems are available that use recuperative thickening as part of advanced digestion processes. However, recuperative thickening has been practiced in the United States and abroad for a number of years. The thickening technologies used include rotary drum thickeners, screw press thickeners, and membrane thickeners. In each case, digestate is withdrawn from a digester and fed to the thickener. The thickened material is returned to the digester while the removed water is returned to the liquid treatment process. For all technologies other than membranes, polymer conditioning is required and all of the ancillary systems required for conventional thickening also apply for recuperative thickening. If the digester contents are concentrated sufficiently, special high solids mixers may be required to keep the tanks homogenous.

7.4.2 Design Criteria and Performance Characteristics

One commercially available recuperative thickening process, which includes high solids mixers as well as the thickening equipment, is designed for a maximum organic loading rate of 0.3 lb volatile solids/cf/day and 6% solids concentration within the digester. The thickener is operated intermittently to produce 12% solids that are mixed with the digestate to maintain 6% solids within the tank.

7.4.3 Ancillary Systems

The ancillary systems for most recuperative thickening systems are similar to those required for gravity belt thickeners and rotary drum thickeners. Systems relying on membrane thickening require blowers to compress digester gas that is used to scour the membranes.

8.0 Comparison of Thickening Methods

A model comparing the cost-effectiveness of various thickening processes rarely applies to all situations because many factors that govern the final decision may be site-specific and more qualitative than quantitative. Such factors include sensitivity to upset, the benefits of achieving the highest possible solids concentration, the quality of operation required, installation size, compatibility with existing thickeners, the effect of downstream processing methods, and various personal preferences based on experience. Design engineers should consider all of the thickening alternatives in this chapter (see Table 21.18).

Method	Advantages	Disadvantages
Gravity	Simple Low operating cost Low operator attention required Ideal for dense rapidly settling sludges such as primary and lime Provides a degree of storage as well as thickening Conditioning chemicals not typically required Minimal power consumption	Odor potential Erratic for WAS Thickened solids concentration limited for WAS High space requirements for WAS Floating solids
Dissolved air flotation	Effective for WAS Will work without conditioning chemicals at reduced loadings Relatively simple equipment components	Relatively high power consumption Thickening solids concentration limited Odor potential Space requirements compared to other mechanical methods Moderate operator attention requirements Building corrosion potential, if enclosed Requires polymer for high solids capture or increased loading

TABLE 21.18   The Advantages and Disadvantages of Various Thickening Technologies

Method	Advantages	Disadvantages
Centrifuge	Space requirements Control capability for process performance Effective for WAS Contained process minimizes housekeeping and odor considerations Will work without conditioning chemicals High thickened concentrations available	Relatively high capital cost and power consumption Sophisticated maintenance requirements Best suited for continuous operation Moderate operator attention requirements Vibration and noise
Gravity belt thickener	Space requirements Control capability for process performance Relatively low capital cost Relatively lower power consumption High solids capture with minimum polymer High thickened concentrations available	Housekeeping Polymer dependent Moderate operator attention requirements Odor potential Building corrosion potential, if enclosed
Rotary drum thickeners	Space requirements Low capital cost Relatively low power consumption High solids capture Can be easily enclosed	Polymer dependent Sensitivity to polymer type thickener Moderate operator attention requirements Odor potential if not enclosed Maintenance/regular lubrication for trunnion wheels/bearings critical to equipment life

TABLE 21.18   The Advantages and Disadvantages of Various Thickening Technologies (*Continued*)

9.0   References

American Public Health Association; American Water Works Association; Water Environment Federation (2012) *Standard Methods for the Examination of Water and Wastewater*, 22nd ed.; American Public Health Association: Washington, D.C.

Ashman, P. S. (1976) Operational Experiences of Activated Sludge Thickening by Dissolved Air Flotation at the Aycliffe Sewage Treatment Works. Paper presented at the Conference on Flotation in Water and Waste Treatment; Felixstowe, Suffolk, Great Britain.

Boyle, W. H. (1978) Ensuring Clarity and Accuracy in Torque Determinations. *Water Sew. Works*, **125** (3), 76.

Bratby, J.; Marais, G. V. R. (1975a) Dissolved Air (Pressure) Flotation, An Evaluation of the Interrelationships Between Process Variables and Their Optimization for Design. *Water SA*, **1**, 57.

Bratby, J.; Marais, G. V. R. (1975b) Saturation Performance in Dissolved Air (Pressure) Flotation. *Water Res.* (G.B.), **9**, 929.

Bratby, J.; Marais, G. V. R. (1976) A Guide for the Design of Dissolved Air (Pressure) Flotation Systems for Activated Sludge Systems. *Water SA*, **2**, 87.

Bratby, J.; Jones, G.: Uhte, W. (2004) State-of-Practice of DAFT Technology–Is There Still a Place For It? Paper presented at the Water Environment Federation's 77th Annual Technical Exposition and Conference, New Orleans, Louisiana, October 2–6; Water Environment Federation: Alexandria, Virginia.

Burfitt, M. L. (1975) The Performance of Full-Scale Sludge Flotation Plant. *Water Pollut. Control* (G.B.), **74**, 474.

Butler, R. C.; Finger, R. E; Pitts, J. F.; Strutynski, B. (1997) Advantages of Cothickening Primary and Secondary Sludges in Dissolved Air Flotation Thickeners. *Water Environ. Res.*, **3**, 69.

Dick, R. I.; Ewing, B. B. (1967) Evaluation of Activated Sludge Thickening Theories. *J. Sanit. Eng.*, **93** (EE4), 9.

Ettelt, G. A.; Kennedy, T. J. (1966) Research and Operational Experience in Sludge Dewatering at Chicago. *J. Water Pollut. Control Fed.*, **38**, 248.

Gehr, R.; Henry, J. G. (1978) Measuring and Predicting Flotation Performance. *J. Water Pollut. Control Fed.*, **50**, 203.

Gulas, V.; et al., (1978) Factors Affecting the Design of Dissolved Air Flotation Systems. *J. Water Pollut. Control Fed.*, **50**, 1835.

Jones, W. H. (1968) Sizing and Application of Dissolved Air Flotation Thickeners. *Water Sew. Works*, **115**, R-177.

Jordan, V. J., Jr.; Scherer, C. H. (1970) Gravity Thickening Techniques at a Water Reclamation Plant [part I]. *J. Water Pollut. Control Fed.*, **42**, 180.

Komline, T. R. (1976) Sludge Thickening by Dissolved Air Flotation in the USA. Paper presented at the Conference on Flotation in Water and Waste Treatment; Felixstowe, Suffolk, Great Britain.

Leininger, K. V.; Wall, D. J. (1974) Available Air Measurements Applied to Flotation Thickener Evaluations. *Highlights/Deeds Data*, **11**, D1.

MacConnell, G. S.; et al. (1989) Full Scale Testing of Centrifuges in Comparison with DAF Units for WAS Thickening. Paper presented at the 62nd Annual Water Pollution Control Federation Technical Exposition and Conference; San Francisco, California, Oct 15–19; Water Pollution Control Federation: Alexandria, Virginia.

Maddock, J. E. L. (1976) Research Experience in the Thickening of Activated Sludge by Dissolved Air Flotation. Paper presented at the Conference on Flotation in Water and Waste Treatment; Felixstowe, Suffolk, Great Britain.

Mulbarger, M. C.; Huffman, D. D. (1970) Mixed Liquor Solids Separation by Flotation. *J. Sanit. Eng.*, **96** (SA4), 861.

Noland, R. F.; Dickerson, R. B. (1978) *Thickening of Sludge*, Vol. 1; EPA-625/4-78-012; U.S. EPA Technology Transfer Seminar on Sludge Treatment and Disposal; U.S. Environmental Protection Agency: Washington, D.C.

Perry, R. H.; Chilton, C. H. (1963) *Chemical Engineer's Handbook*, 4th ed.; McGraw-Hill: New York, New York.

Purchas, D. B. (1977) *Solid/Liquid Separation Equipment Scale-up*; Uplands Press Ltd.: Croydon, United Kingdom.

Reay, D.; Ratcliff, G. A. (1975) Experimental Testing on the Hydrodynamic Collision Model of Fine Particle Flotation. *Can. J. Chem. Eng.*, **53**, 481.

Sparr, A. E.; Grippi, V. (1969) Gravity Thickeners for Activated Sludge. *J. Water Pollut. Control Fed.*, **41**, 1886.

Speece, R. C.; et al. (1975) Application of a Lower Energy Pressurized Gas Transfer System to Dissolved Air Flotation and Oxygen Transfer. *Proceedings of the 30th Purdue Industrial Waste Conference*; West Lafayette, Indiana; 465.

Talmage, W. P.; Fitch, E. B. (1955) Determining Thickener Unit Areas. *Ind. Eng. Chem. Fundam.*, **47**, 38.

Torpey, W. N. (1954) Concentration of Combined Primary and Activated Sludges in Separate Thickening Tanks. *Proc. Am. Soc. Civ. Eng.*, **80**, 443.

Turner, M. T. (1975) The Use of Dissolved Air Flotation for the Thickening of Waste Activated Sludge. *Effluent Water Treat. J.* (G.B.), **15** (5), 243.

U.S. Environmental Protection Agency (1974) *Process Design Manual for Sludge Treatment and Disposal*; EPA-625/1-74-006; U.S. Environmental Protection Agency, Office of Technology Transfer: Cincinnati, Ohio.

U.S. Environmental Protection Agency (1985) *Handbook of Estimating Sludge Management Costs*; EPA-625/6-85-010; U.S. Environmental Protection Agency, Water Engineering Research Laboratory: Lancaster, Pennsylvania.

U.S. Environmental Protection Agency (2013) *Process Design Manual for Sludge Treatment and Disposal*; EPA-625/1-79-011; U.S. Environmental Protection Agency, Municipal Environmental Research Laboratory, Office of Research and Development: Cincinnati, Ohio.

U.S. Environmental Protection Agency; American Society of Civil Engineers (1979) *Proceedings of the Workshop Towards Developing an Oxygen Transfer Standard*; EPA-600/9-78-021; U.S. Environmental Protection Agency: Washington, D.C.

Vesilind, P. A. (1968) The Influence of Stirring in the Thickening of Biological Sludge. Ph.D. thesis, University of North Carolina, Chapel Hill.

Vesilind, P. A. (1974a) Scale-Up of Solid Bowl Centrifuge Performance. *J. Environ. Eng.*, **100**, 479.

Vesilind, P. A. (1974b) *Treatment and Disposal of Wastewater Sludges*; Ann Arbor Science Publishers Inc.: Ann Arbor, Michigan.

Voshel, D. (1966) Sludge Handling at Grand Rapids, Michigan, Wastewater Treatment Plant. *J. Water Pollut. Control Fed.*, **38**, 1506.

Walzer, J. G. (1978) Design Criteria for Dissolved Air Flotation. *Pollut. Eng.*, **10**, 46.

Wanielista, M. P.; Eckenfelder, W. W. (1978) *Advances in Water and Wastewater Treatment, Biological Nutrient Removal.* Ann Arbor Science Publishers Inc.: Ann Arbor, Michigan.

Water Environment Federation; Water Environment Research Foundation; U.S. Environmental Protection Agency (2012) *Solids Process Design and Management*; Water Environment Federation: Alexandria, Virginia.

Wilhelm, J. H.; Naide, Y. (1979) Sizing and Operating Continuous Thickeners. Paper presented at the American Institute of Mechanical Engineers Meeting; New Orleans, Louisiana.

Wood, R. F. (1970) The Effect of Sludge Characteristics upon the Flotation of Bulked Activated Sludge. Ph.D. thesis, University of Illinois, Urbana–Champaign.

Wood, R. F.; Dick, R. I. (1975) Factors Influencing Batch Flotation Tests. *J. Water Pollut. Control Fed.*, **45**, 304.

Yoshioka, N.; et al. (1957) Continuous Thickening of Homogeneous Flocculated Slurries. *Chem. Eng.* (Jpn.), **21**, 66.

10.0 Suggested Readings

Albertson, O. E.; Vaughn, D. R. (1971) Handling of Solid Wastes. *Chem. Eng. Prog.*, **67** (9), 49.

Ashbrook-Simon-Hartley (1992) *Aquabelt Operations & Maintenance Manual*; Ashbrook-Simon-Hartley: Houston, Texas.

Coe, H. S.; Clevenger, G. H. (1916) Methods for Determining the Capacities of Slime Settling Tanks. *Trans. Am. Inst. Min. Eng.*, **55**, 356.

Dick, R. I. (1970) Thickening. In *Advances in Water Quality Improvement, Physical and Chemical Processes*; Gloyna, E. F., Eckenfelder, W. W., Jr., Eds.; University of Texas Press: Austin.

Dick, R. I. (1972a) Gravity Thickening of Waste Sludges. *Proc. Filtr. Soc., Filtr. Sep.*, **9**, 177.

Dick, R. I. (1972b) Thickening. In *Water Quality Engineering: New Concepts and Developments*; Thackson E. L., Eckenfelder, W. W., Jr., Eds.; Jenkins Publishing Co.: New York.

Eckenfelder, W. W., Jr. (1970) *Water Quality Engineering for Practicing Engineers*; Barnes & Noble: New York.

Fitch, B. (1966) A Mechanism of Sedimentation. *Ind. Eng. Chem., Fundam.*, **5**, 129.

Fitch, B. (1974) *Unresolved Problems in Thickener Design and Theory*; Dorr-Oliver Inc.: Stamford, Connecticut.

Fletcher, N. H. (1959) Size Effect in Heterogeneous Nucleation. *J. Chem. Phys.*, **29**, 572.

Flint, L. R.; Howarth, W. J. (1971) The Collision Efficiency of Small Particles with Special Air Bubbles. *Chem. Eng. Sci. (G.B.)*, **26**, 1155.

George, D. B.; Keinath, T. M. (1978) Dynamics of Continuous Thickening. *J. Water Pollut. Control Fed.*, **50**, 2561.

Hassett, N. J. (1958) Design and Operation of Continuous Thickener [parts I, II, and III]. *Ind. Chem.*, **34/116/169**, 489.

Javaheri, A. R. (1971) Continuous Thickening of Non-ideal Suspensions. Ph.D. thesis, University of Illinois, Urbana–Champaign.

Kos, P. (1977) Gravity Thickening of Water Treatment Plant Sludges. *J. Am. Water Works Assoc.*, **69**, 272.

Kynch, G. J. (1952) A Theory of Sedimentation. *Trans. Faraday Soc. (G.B.)*, **48**, 166.

Shin, B. S.; Dick, R. I. (1975) Effect of Permeability and Compressibility of Flocculent Suspensions on Thickening. *Prog. Water Technol.*, **7**, 137.

Tarrer, A. R.; et al. (1974) A Model for Continuous Thickening. *Ind. Eng. Chem., Process Des. Dev.*, **13**, 341.

Vaughn, D. R.; Reitwiesner, G. A. (1972) Disk-Nozzle Centrifuges for Sludge Thickening. *J. Water Pollut. Control Fed.*, **44** (9), 1789.

Vesilind, P. A. (1968) Design of Thickeners from Batch Tests. *Water Sew. Works*, **115**, 9.

Angela Hintz, P.E.; Jeovanni Ayala-Lugo, P.E.;
Harold E. Schmidt, Jr., P.E., BCEE; Kristen Waksman;
and Jason J. Williams, P.E.

1.0   Introduction

1.1   Objectives

Sludge dewatering systems convey mixed and/or thickened liquid sludge, stabilized biosolids, and/or scum streams through a process that reduces the amount of excess water in the liquid/solids stream, thereby reducing the volume of sludge and/or biosolids for disposal. Dewatering produces a material suitable for further processing, beneficial use, or disposal. Additional processing may be required after dewatering to create a marketable product (such as drying and making fertilizer pellets) and/or prepare for other ultimate disposal methods (landfilling, land application, incineration, etc.). Dewatering differs from thickening in that thickening processes generate a product that behaves like a liquid (2% to 10% solids), while dewatering processes produce a denser material (i.e., cake) that behaves like a semisolid or solid material (12% to 50% solids).

The dewatering process consists of several major components: liquid sludge/ biosolids conveyance, dewatering, and cake conveyance. Conveyance of both liquid sludge/biosolids and the resulting dewatered cake are discussed further in Chapter 19.

Dewatering systems typically require a large capital investment and a substantial share of a facility's annual budget for operation and maintenance (O&M). To design cost-effective dewatering facilities, engineers need to evaluate a wide array of dewatering options, solids characteristics, and site-specific variables (e.g., other treatment processes and sidestreams, disposal costs, energy costs, operator labor costs, and availability).

1.2   Key Process Performance Indicators

Wastewater sludge dewatering processes generate two products: a solids cake that concentrates the solids in a smaller volume and a liquid stream that consists of water removed from the feed sludge stream and minimal residual solids (see Figure 22.1). The liquid stream removed from the sludge, which can be called many things based on the

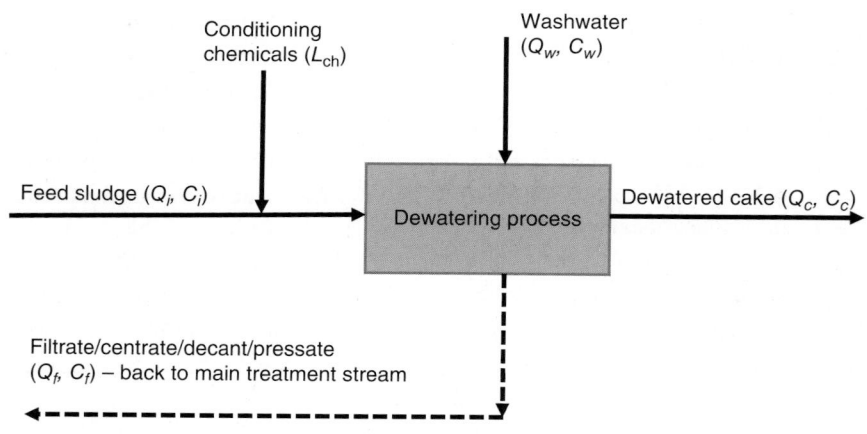

FIGURE 22.1   Dewatering solids and water balance.

type of dewatering process used (e.g., supernatant, decant, underdrainage, filtrate, centrate, and evaporated moisture), is typically recycled back to some point within the main wastewater treatment process and is treated alongside the incoming facility flow or routed to a separate sidestream process prior to reintroduction into the main wastewater process stream (sidestream processes are discussed in Chapter 15). The performance of the dewatering process is measured using indicators common to all dewatering processes: cake solids content, solids capture, and cake bulk density.

1.2.1   Cake Solids Content

Cake solids content is a measure of solids and moisture content of the dewatered cake; it is the weight of total dry solids in the cake divided by the total weight of the cake (solids plus remaining water in the cake matrix), expressed as a percentage. Because the amount of dissolved solids in the water trapped in the dewatered cake is minimal when compared to the amount of suspended solids in the dewatered cake, a cake's total solids and total suspended solids (TSS) are often considered equal. Depending on the characteristics of the particular sludge being dewatered, the type of dewatering used, and the type and amount of pretreatment, the solids content of dewatered sludge cake typically ranges from around 12% to as high as 50%.

1.2.2   Solids Capture

Solids capture rate represents the percentage of suspended solids in the sludge feed to the dewatering process that remain in the dewatered cake. If a chemical is added to condition the solids prior to dewatering, then the weight of the added chemical should be included in the weight of the feed solids when calculating the solids capture rate. The solids capture rate is calculated performing a solids–and-water balance around the dewatering process (see Figure 22.1). The solids capture rate for a dewatering process can be expressed by the following equations, derived from the mass balance around the dewatering process:

$$E = M_c/M_f = C_c \times (C_f - C_r)/[C_c \times (C_c - C_r)]$$ (22.1)

where E = solids capture, %;

M_c = mass of solids within the cake, kg (lb);

M_s = mass of solids within the feed sludge, kg (lb);

C_c = cake solids concentration, mg/L;

C_f = feed solids concentration, mg/L; and

C_r = filtrate/centrate/decant solids concentration, mg/L.

The capture rates of various dewatering processes typically range from 90% to 99%. Once the target cake solids content and solids capture rate have been determined, design engineers can use eq 22.1 to determine the solids concentration in the liquid recycle stream returned to the main treatment process.

1.2.3   Cake Bulk Density

Cake bulk density is the weight of the cake divided by its volume. The bulk density takes into account air voids that are present as the cake solids concentration increases and the product takes on soil-like or granular properties. The bulk density will change as the cake is handled and conveyed. For example, a cake's bulk density as it leaves a dewatering device will be different than the bulk density when the cake is being trucked off-site, as some cake compaction and water release may occur as the cake is conveyed or loaded onto trucks. Bulk density is useful when determining what volume of dewatered solids will be generated in a plate-and-frame press, sizing cake conveyance systems, and determining the number of trucks needed to transport dewatered cake off-site.

1.3   Other Considerations Affecting Dewatering

There are many factors that will influence how easily a particular sludge stream may be dewatered, including the chemical and physical characteristics of the feed sludge, type of dewatering system used, and extent of chemical conditioning and pretreatment. Sludge characteristics are discussed also in Chapter 18, while information on sludge conditioning is included in Chapter 20.

1.3.1   Sludge Type and Characteristics

There are several sludge parameters that impact the dewaterability of sludge and the ability of conditioning chemicals to improve the dewatering process performance:

- *Type of sludge*: In general, waste activated sludge (WAS) is more difficult to dewater than primary sludge, as much of the water in WAS is attached to bacterial cells or tied up chemically in the cell structures. The larger the WAS-to-primary-sludge ratio is, generally the harder the sludge is to dewater.

- *Amount and type of organic matter present*: Other organic components of sludge, such as organic acids and exocellular biopolymers, also impact dewatering. These components typically are removed from wastewater via charge neutralization or adsorption to solid surfaces. Polymers are generally ineffective at removing these organics; in fact, polymers combine with the organics to create a material with poorer dewatering characteristics. Organic molecules are present in wastewater undergoing biological activity. For example, primary solids typically contain lower levels of organic biopolymers during colder seasons, when biological activity is low. Increased biological growth during warmer seasons may generate enough biopolymer to greatly alter process performance.

- *Presence of industrial organics*: Industrial organics also may alter solids properties. They typically have the same effect as naturally occurring organics, but may be present at higher concentrations and may not coagulate as easily, particularly if they adsorb poorly. However, industrial discharges that are more fibrous in nature may dewater better than discharges not containing fibrous material.

- *Stabilization method*: Well-digested WAS are often difficult to dewater, while well-digested primary solids dewater more easily. Much of this variability has to do with the solids constituents involved and how water bonds to them.

- *pH*: The pH of the solids can have a significant effect on dewatering. Studies by Chen et al. (2001) have shown that activated sludge dewaters better at lower pH values. Ferric chloride lowers the pH to a more desirable level, while polymers raise the pH above 9. Only lime conditioning optimally performs at high pH because mixing dense, porous calcium carbonate with solids provides a matrix that promotes rapid water removal.

- *Feed sludge particle size*: The particle size of the feed solids influences solids dewaterability (Shao et al., 2009). As the average particle size decreases (which may result from excessive mixing and shear), the capillary suction time increases, resulting in greater difficulty in dewatering.

- *Sludge temperature*: Temperature influences biological activity and can alter solids properties. Lower temperatures slow chemical reactions, affecting the performance of both metal ion conditioners and polymers. Typically, reactions will be less complete and chemical doses may need to be increased when temperatures drop, although some of these reactions may be offset by a decrease in biological activity.

- *Feed solids concentration*: A higher solids concentration in the feed solids has been shown to directly increase cake solids that can be achieved in most mechanical dewatering processes. A 1985 survey of more than 100 municipal belt-press installations showed a strong correlation between cake solids, feed solids, and the percentage of WAS in the feed (Koch et al., 1988). Similar results have been reported for centrifuges and plate-and-frame filter presses (Koch et al., 1989). However, there are a few exceptions: (1) when the gel point of the feed material has been reached and (2) when large amounts of liquid are required to provide good interaction between the solids and conditioning chemicals (e.g., polymers).

When determining the solids loading to the dewatering device, the effect of recycled solids should be taken into account. For example, if the centrate or filtrate is recycled to the head of the facility, the solids in the filtrate will be removed during the wastewater treatment process, increasing the solids generated beyond those solids received in the facility's influent, increasing the total solids loading to the dewatering device. At steady-state conditions, the mass of solids leaving the dewatering device should be equal to the solids generated during wastewater treatment, assuming that none of the recycled solids were destroyed, but rather added to the solids being managed.

1.3.2 Pretreatment

While many solids streams require chemical conditioning before being dewatered, some additional pretreatment systems have been found to be useful to protect equipment, improve cake solids, or enhance the quality of the dewatered cake product:

- *Grinding/maceration*: Grinders can be used upstream of the dewatering process to reduce the size of the solids entering the dewatering equipment and prevent entry of elongated or jagged pieces of material that could tear the filter belts or cloths or clog dewatering equipment feed nozzles. Grinders should be added to the suction side of the dewatering feed pumps.

- *Thermal treatment*: Some stabilization processes combine thermal conditions with anaerobic digestion to improve the dewatering characteristics of solids (see Chapter 23).

- *Addition of bulking materials*: There is limited experience with adding bulking materials (e.g., wood chips or ash) to increase the fiber content of the feed solids and improve the dewaterability and bulking properties. However, the addition of bulking materials has been largely limited to solids dewatered using filtration devices (e.g., belt presses and recessed-plate filter presses).

- *Other pretreatment processes*: These processes use elevated temperatures or pressures, ultrasonic waves, high-shear devices (e.g., ball mills), or chemicals (acids or bases) to break down WAS, reduce chemical conditioning requirements, and produce a high-solids cake. Often these processes are utilized prior to digestion, to increase the amount of volatile solids destroyed in the digestion process, resulting in lower solids loadings to the dewatering process.

1.3.3   Chemical Conditioning

All dewatering methods benefit from some form of chemical conditioning. Proper chemical conditioning improves cake solids and solids capture rates. The chemicals used for conditioning can be either inorganic or organic; the most commonly used are inorganic salts and organic polymers.

Inorganic chemicals typically used are metal salts (e.g., ferric chloride, aluminum chloride, or ferrous sulfate). Their activity is pH dependent, so pH may need to be adjusted for optimal use of the chemical. Organic chemicals typically used in dewatering are high-molecular-weight (HMW), water-soluble organic polymers. Past evaluations indicate that organic polymers are cost-effective and have better maintenance, performance, and safety records than inorganic chemicals.

Optimal locations for adding polymer depend on the conditioning chemical, the specific solids involved, and the dewatering device used. Typically, the optimum location for chemical addition (i.e., at the equipment, in the feed sludge pipe, at the suction of the feed pump, etc.) is determined through trial and error. To determine the suitability of individual polymers on dewatering, bench-scale tests can be used. Other bench tests (e.g., piston filter presses and batch centrifuges) can provide indications of likely cake solids content. Likewise, polymer testing can be done during pilot-testing of dewatering devices to determine the most appropriate chemical and dose.

For plate-and-frame filter presses, ferric chloride and lime (with or without ash) are used to improve solids particle size distribution, while ash may reduce the conditioned-solids compressibility. While the use of ferric chloride and lime increases the amount of inert material in the dewatered solids, it improves compressibility and enhances cake release. More information on chemical conditioning can be found in Chapter 20.

1.3.4   Effect of Recycled Streams on Treatment Processes

Water removed from solids streams during dewatering typically is returned to the main treatment plant process, typically to the head of the facility or to the head of the main biological treatment process. The liquid waste stream from dewatering processes contains constituents that add to the facility's influent loading (e.g., biochemical oxygen demand [BOD], ammonia, and phosphorus). The extent of the impact of dewatering recycle streams will depend on solids capture efficiency and the nature of the wastewater and solids treatment processes used upstream of the dewatering process. For example, if the facility uses biological phosphorus removal and anaerobic digestion, then soluble phosphate may be released. This phosphate can increase the influent phosphate loading to the facility or cause magnesium-based scale (e.g., struvite) to precipitate in the pipeline or pumping system that conveys recycled filtrate or centrate. The recycled centrate or filtrate can also be a source of additional nitrogen loading to the plant as a result of the ammonia released during anaerobic digestion. (For a more detailed discussion of the effect and treatment of recycle streams, see Chapter 15.)

1.4   Overview of Dewatering Technologies

The physical and chemical characteristics of solids affect the choice of dewatering process to be used at specific facilities. For example, belt filter presses typically rely on in-line polymer feeding (as opposed to using a sludge conditioning tank), which may not work well with solids with high pH or salt levels. Using metal salts instead of polymers may increase equipment and storage requirements, thereby losing some of the advantages of belt filter presses. Also, while a centrifuge may handle one type of solids better than a belt press, the reverse may be true when treating another type of solids.

Centrifuges also are affected by solids variations, but they can be more readily adjusted to achieve different combinations of cake solids and solids capture. Polymer addition typically improves the solids capture rate.

Fixed-volume filter presses using ferric chloride and lime conditioners are less sensitive to physical and chemical variations in solids, but often require large (expensive) quantities of chemicals.

Dewatering performance can be optimized by adjusting process variables and testing different types and concentrations of conditioning chemicals. Full-scale pilot-testing is desirable so systems can be optimized and then compared.

1.5   Design Considerations

1.5.1   Pilot Testing

The selection and design of a dewatering process can be based on performance observed at similar installations, on bench-scale testing, or on full-scale pilot-testing. Because solids vary from facility to facility, the most accurate prediction of how a dewatering process will perform for a particular sludge stream is best determined via full-scale pilot-testing.

When evaluating the performance of a dewatering device, the quantity and quality of the filtrate and backwash and their effects on the wastewater treatment system should be considered. However, while a pilot test generally provides good data for the selection of dewatering equipment and conditioning chemicals, they are operated with consistent feed conditions and significant operator attention. Designs based on pilot-test data should account for variable feed conditions and should not require constant operator attention to produce acceptable performance.

The most effective pilot-testing programs involve more than one machine or type of dewatering device running concurrently. Side-by-side operations alleviate concerns of variable operating conditions or solids characteristics between tests. Most dewatering equipment manufacturers have mobile trailer-mounted pilot units that can be brought to a facility for testing. Most of these units are smaller production machines and provide performance comparable to larger models. Data collected as part of the pilot-testing include hydraulic- and solids-loading rates, polymer type and use, percent solids in the dewatered cake, and capture efficiency. However, care must be exercised when scaling up the data from production units used during pilot-testing to ones recommended for the full-scale installation. Design engineers should work with the equipment manufacturer to develop full-scale criteria.

Pilot-testing also allows an opportunity to identify the appropriate dewatering conditioning chemical type and dose. Different types of chemicals can be tested and the relationship between dose, hydraulic and solids loading, solids capture efficiency, and cake solids can be ascertained.

In many cases, side-by-side pilot-testing of different types of dewatering equipment enables the project team to select a process that minimizes capital and O&M costs while meeting process objectives. However, this is not possible if the wastewater treatment process is being changed and/or a representative solids sample is not available for testing.

As an alternative to a full-scale pilot test, many dewatering equipment manufacturers have in-house testing equipment that can be used to predict performance. A sample of the sludge to be dewatered is sent to the manufacturer, who typically provides design criteria based on the results on their in-house testing. Care needs to be taken to minimize both high temperatures and excessive vibration of the solids during transport as both have the ability to significantly change the characteristics of the solids. Tests should be performed as soon as possible to ensure that the solids do not change properties as they age.

1.5.2   Design Example
The following design example outlines the key steps in selecting and designing dewatering equipment (see Figure 22.2). The example takes the solids generation rate and applies peaking factors and the proposed number of operating hours of the dewatering

Given Information:	Average Sludge Generation Rate	16 kg/h dry solids
	Assumed Peaking Factor	1.5
	Design Sludge Generation Rate	24 kg/h dry solids = average generation rate × peaking factor
	Operating Hours per Week	40
	Design Dewatering Rate	101 kg/h dry solids = design generation rate × 168 hours per wk/operating hours per wk
Feed Sludge:	Solids Concentration	2%
	Wet Sludge Feed Rate	5040 kg/h = design dewatering rate/solids concentration
	Water in Feed	4939 kg/h = wet sludge feed rate × (1-solids concentration)
	Density of Water	1 kg/L (we can use this because the feed sludge is mostly water)
	Volumetric Feed Rate	5040 L/h = Wet sludge feed rate × density of water

Performance: (from pilot testing)	Dewatering Technology Performance	% Capture	% Cake Solids	Added Chemical	Cake Bulk Density (kg/m³)
	Belt Press or Centrifuge	92%	20%	1%	900
	Recessed Plate Filter Press	97%	45%	35%	1200

| Other Assumptions: | All recycle from dewatering devise returned to and removed in primary clarifiers and returned to dewatering |
| | Solids leaving dewatering are equal to solids generated at steady state |

FIGURE 22.2   Dewatering equipment design example.

Calculations:		Belt Filter Press/ Centrifuge	Recessed Plate Filter Press	
	Dry Solids Entering Dewatering Process	102	136	kg/h = Design dewatering rate + (1 + added chemical weight)
	Wet Cake Leaving Dewatering Process	509	302	kg/h = Dry solids entering process/% cake solids
	Water Weight in Cake	407	166	kg/h = Wet cake leaving process × (1 − % cake solids)
	Cake Volume	1.0	0.1	m^3/h
	Dry Solids Entering Device Based on Capture	110.7	140.3	kg/h = Dry solids/Capture rate
	Wet Solids Entering Device	5533	7014	kg/h = Dry solids entering/Feed solids %
	Water Entering Device	5422	6874	kg/h = Wet solids entering device − Dry solids entering device based on capture
	Recycle Stream Dry Solids	8.9	4.2	kg/h = Dry Solids entering device based on capture − Dry entering dewatering
	Recycle Water	5015	6708	kg/h = Water entering device = water weight in cake
	Recycle Percent Solids	0.2%	0.1%	Recycle stream/Recycle water
Sizing:	No. of belt filter presses and centrifuges based on hydraulic and solids loading rates for each unit, based on manufacturer-specific information			
	No. of Plate and Frame Presses sized based upon size of plates, no. of plates, pressed cake volume, and cycle time			
	Additional units added for redundance as required			

FIGURE 22.2 Dewatering equipment design example. (*Continued*)

process to determine design solids- and hydraulic-loading rates. The chemical dose, percent solids capture, and cake solids typically are obtained from pilot-testing or from the performance of similar units treating similar solids. The example establishes the projected solids and water balances across the units. Once calculated, the hydraulic- and solids-loading rates are used to select the appropriate size and number of dewatering devices. Manufacturers should be contacted to determine acceptable hydraulic and solids loadings for a specific piece of equipment. Some flexibility should be provided to allow for the use of different types of conditioning chemicals.

1.5.2.1 Input Parameters
Input parameters assumed for the design example are as follows:

- Solids generation rate of 16 dry kg/h (35.2 lbs/hr);
- Feed solids concentration of 2%;

- Peaking factor of 1.5; and
- Dewatering operations are assumed to run 40 h/week.

1.5.2.2 Assumptions General assumptions for the design example include:

- Actual solids loading to the equipment is increased above the solids generation rate to account for chemical addition and the device's solids capture efficiency. Solids in the recycle steam returned to the main wastewater treatment process will be removed from the wastewater and dewatered again and contribute to the overall solids loading of the dewatering process. If a digestion or other sludge minimization process precedes dewatering, some recycle solids may be destroyed and this assumption may be overly conservative. A more thorough mass balance around the entire wastewater treatment process can be used to more accurately estimate the solids loading to the dewatering process.
- The assumptions for the dewatering technologies considered are as follows:
  - Dewatering via belt filter presses or centrifuges:
    - Solids capture efficiency = 92%,
    - Cake solids = 20%,
    - Chemical conditioning using polymer,
    - Solids added from chemical conditioning = 1%, and
    - Cake bulk density = 900 kg/m^3;
  - Dewatering via recessed-plate filter presses:
    - Solids capture efficiency = 97%,
    - Cake solids = 45%,
    - Chemical conditioning using lime and ferric chloride,
    - Solids added from chemical conditioning = 35%, and
    - Cake bulk density = 1200 kg/m^3.

1.5.2.3 Calculations The loading rate to the dewatering equipment is calculated by applying the peaking factor and hours of operation of the dewatering process to the solids generation rate. The feed solids concentration is used to calculate the hydraulic- and solids-loading rates to the dewatering device.

Next, the calculated solids-loading rate is increased to account for extra solids added through chemical addition. Finally, the loading rate is divided by the solids capture efficiency to obtain the effective loading rate to the dewatering equipment. This assumption sets the solids leaving the dewatering device equal to the amount of solids generated in the wastewater treatment process.

Using the assumed loading rates, solids capture efficiency, cake solids, and bulk density, the dry and wet weight and the volume of the cake produced by the dewatering device can be calculated. The volume and the solids concentration of the centrate or filtrate are computed by performing a mass balance around the dewatering equipment. Even though the plate-and-frame press will produce more cake on a dry-solids basis, the volume of the cake is considerably less than that produced by the belt press or centrifuge because of the dryer cake achieved (see Figure 22.2).

1.5.2.4 Output The number of dewatering units is selected based on the processing rates of each specific dewatering technology and the amount of redundancy required.

If a standby unit is required, smaller-capacity units may be preferable. Having a larger number of units with lower throughput per unit allows operators to take some units out of service during periods where lower amounts of solids are generated. On the other hand, some facilities may choose to use fewer, larger units and respond to changes in solids production by changing the hours of operation of the dewatering process.

Most dewatering units are rated based on hydraulic-loading rates, although they may become solids loading limited at higher feed solids concentrations. Centrifuge manufacturers typically offer units having different hydraulic processing rates. Belt press manufacturers offer units with different belt widths to accommodate different hydraulic-loading rates. Manufacturers of plate-and-frame presses offer different plate sizes and can vary the number of plates to accommodate different hydraulic-loading rates. The throughput of rotary presses can be adjusted by varying the number of channels within the rotary press. Ancillary facilities (e.g., feed pumps, chemical conditioning, washwater, odor control, conveyors, and cake storage) are then sized and coordinated with the number of dewatering units selected.

1.5.3   Other Considerations

1.5.3.1   Layout, Disposal, and Operation Schedule   Often, the selection of a dewatering process is dependent upon the space available at the treatment facility or within the dewatering building, especially for those projects involved with replacement of an existing dewatering system. For example, drying beds are not a feasible option for facilities that are land locked. Likewise, belt filter presses generally would not fit in a dewatering building that may have held screw presses or centrifuges previously, without significant expansion of the building. Therefore, consideration of how much space is available for dewatering should be taken into account during design.

The method of disposal will often dictate the feasible alternatives available for dewatering. For example, if the dewatered cake is to be trucked to an off-site landfill that is a distance away, a drier cake produced by recessed-plate filter presses may be preferable to a wetter cake produced by rotary presses. Designers should consider the end use when evaluating dewatering technologies.

Hours of operation have an impact on the type and size of dewatering units used in a particular application. Facilities that operate dewatering continuously can use smaller units than those that operate intermittently. However, at some facilities, the dewatering process may only run 40 hours a week to correspond to the day shift, and the facility is unable to expand the hours of operation as they do not have staff available during other shifts. In that case, larger dewatering units are necessary as the equipment needs to handle solids produced continuously over a shorter time span. The current and future intended hours of operation for dewatering activities must be known in order to properly size equipment.

1.5.3.2   Structural Requirements   In all cases, the structural requirements needed for implementing a specific dewatering process should be considered. Dewatering equipment with higher potential for vibration, such as centrifuges, may require significant structural changes to the dewatering area to adequately support the weight and vibration loads expected from the centrifuge. Access openings to the dewatering building may be required to accommodate equipment installation and ongoing maintenance of the equipment. Regardless of the dewatering technology chosen, a structural evaluation should be completed to provide adequate support and access for the equipment.

1.5.3.3 Utility Requirements   Each type of dewatering process will require a different amount of supporting utilities, such as clean water supply and electrical power. Centrifuges typically have the highest energy demand of any dewatering device, and an electrical evaluation should be completed to determine the adequacy of the existing facility electrical supply system to accommodate the increased demand of the centrifuges.

Dewatering equipment may also require a washwater supply with certain pressures. The design professional should consider if the flows and pressures necessary to operate the dewatering equipment are present and, if not, how to achieve the flows and pressures necessary. Washwater booster pumps may be required as part of the dewatering design in some cases.

1.5.3.4 Odor Control   During the dewatering process, off-gases and odors may also be released. The likelihood and extent of odor production depends on how solids are processed, how long they are held and stored before dewatering, how they are stored, the dewatering technology used, and the type of cake conveyance following the dewatering process. If solids are stored in unaerated tanks, they begin to anaerobically digest and release odorous compounds. In some dewatering devices (e.g., centrifuges, screw presses, and rotary presses), the odors are contained in the equipment; therefore, it is only necessary to control odors from the dewatered cake conveyance and any off-gases from the equipment. In other devices (e.g., belt presses and recessed-plate filter presses), it may be necessary to install a hood over the equipment to provide negative pressure to ventilate the dewatering process and collect the air for odor control and/or isolate the entire dewatering area for odor control. Ventilation should be designed according to NFPA 820, *Standard for Fire Protection in Wastewater Treatment and Collection Facilities*, and Chapter 6, Odor Control (NFPA, 2016).

2.0 Centrifuges

This section discusses solid bowl centrifuges, and not basket centrifuges or disk nozzle decanter centrifuges, which are typically not used to dewater municipal wastewater solids.

2.1 Description of Technology

Centrifuges are used for both thickening (discussed in Chapter 21) and dewatering solids, although the internal design dimensions of thickening centrifuges may vary from dewatering centrifuges. Table 22.1 presents the advantages and disadvantages of using dewatering centrifuges.

Advantages	Disadvantages
- Clean appearance, good odor containment, fast startup and shutdown capabilities	- Scroll wear potentially a high maintenance problem
- Produces a relatively dry sludge cake	- Requires grit removal and possibly a sludge grinder in the feed stream
- Low capital cost–to–capacity ratio	- Skilled maintenance personnel required
- High installed capacity to building area ratio	- Moderately high suspended solids content in centrate

TABLE 22.1   Advantages and Disadvantages of Centrifuge Dewatering [Metcalf & Eddy, Inc. (2003) *Wastewater Engineering Treatment and Reuse*, 4th ed.; McGraw–Hill: New York]

Like thickening centrifuges, dewatering centrifuges comprise of the following major components:

- *Bowl*: The dewatering centrifuge bowl is a horizontal, hollow, solid-walled cylinder, tapered at one end that is rotated by an electromechanical main motor at approximately 2000 to 3000 rotations per minute (rpm).

- *Scroll*: The centrifuge scroll (also called a "conveyor") rotates in the opposite direction of the bowl by the centrifuge's back-drive system and directs solids through the bowl to the solids discharge ports.

- *Main drive*: The main drive, including an electromechanical motor typically driven by a variable frequency drive (VFD), rotates the bowl.

- *Back drive*: The back drive rotates the scroll. Centrifuge back drives typically include an electromechanical motor driven by a VFD, or a hydraulically driven system consisting of a constant-speed electromechanical motor that drives a low-speed, high-pressure oil hydrostatic pump located within an oil reservoir, high-pressure oil filter, hydrostatic slip drive unit, liquid cooling system, instrumentation and monitoring, electronic regulation controls, and appurtenances.

- *Control panels*: A centrifuge is typically furnished with a control panel that includes motor starters, VFDs, programmable logic controller (PLC), operator interface terminal (OIT), hardware modules for control and status signals to the PLC, and appurtenances. Control panels can be located adjacent to centrifuges on the operating floor or remote from the centrifuges. Control panels control and monitor centrifuge functions, including speed, torque, cleaning, alarms, failures, and other functions, and certain aspects of appurtenant equipment, possibly including the flowrate set point of dewatering sludge feed and/or polymer systems, operation of the cake discharge equipment, and other ancillary equipment. Control systems are customized based on the owner's preferences.

- *Instrumentation*: Various instruments are used to monitor, indicate alarms, and automatically shut down the centrifuge system, and typically include, motor temperature, bowl and scroll bearing temperature and vibration, hydraulic oil reservoir level, temperature, and pressure (for hydraulic back-drive systems), and others.

Centrifuges are typically operated in one of two modes:

- *Manual mode* is not intended to be an operational model for a centrifuge, but rather for maintenance purposes only.

- *Automatic mode* is used to start, stop, and operate the sludge dewatering system (i.e., sludge feed pumps, polymer system equipment, centrifuge systems, cake handling equipment, and/or other ancillary systems), as well as start and stop a centrifuge bowl clean-in-place (CIP) cycle.

In automatic mode, scroll torque is typically used to control cake dryness.

2.2   Process Design Conditions and Criteria

Dewatering centrifuges are sized based on sludge hydraulic or solids-loading rates to the centrifuge, and are typically designed and specified to meet process goals, such as cake dryness (often related to disposal costs), solids capture/centrate quality, polymer usage, and power usage. To ensure process goals are met by the furnished equipment, guarantees or performance-based damages can be specified or purchased. Consideration should be given to the facility's solids generation/influent, sludge/solids storage availability, preferred operating schedule (e.g., full-time versus part-time dewatering), footprint/space availability, solids disposal method (e.g., incineration, landfill, etc.), required equipment redundancy, and other facility-specific considerations when designing and specifying centrifuge equipment. An existing, operating piece of dewatering equipment or pilot-testing is a useful benchmark for setting process goals and expectations for a dewatering centrifuge system.

Due to various centrifuge sizes and options, specifications for centrifuges should be clear, concise, complete, and correct to assist manufacturers in bidding and furnishing equipment that will meet the facility's needs.

2.2.1 Bowl Speed

Centrifuge bowls rotate to create a centrifugal force, which drive separation of solids from liquid. The acceleration term is referred to as "g-force." The acceleration of the earth's gravity (g) is 9.81 meters per second squared (m/s^2) (32.2 feet per second squared [ft/sq s]), typically referred to as "$1g$," and centrifuge manufacturers use this as a unit of acceleration. Most dewatering centrifuges operate between 1500- and 3000g. In practical terms, if a centrifuge operating at 3000g has an object with a mass of 1.0 kilogram (kg) (2.2 pounds-mass [lbm]) within the bowl, the centrifuge would impart a force of 29 400 newtons (N) or 3000 kilograms-force (kgf) (6600 pounds-force [lbf]) on the object. The relationship between g-force and bowl speed is calculated as:

$$g-\text{factor} = \frac{a_e}{g} = \frac{r\omega^2}{g} = \frac{r}{g}\left(\frac{2\pi}{60}N\right)^2 \tag{22.2}$$

where a_e = centrifugal acceleration = $r\omega^2$;

g = acceleration of earth's gravity;

r = distance of an object from the axis of rotation (i.e., the inner bowl radius);

ω = rotational velocity (radians per unit of time) = $\left(\frac{2\pi}{60}N\right)^2$; and

N = rpms.

The "best" operating g-force is application specific; greater is not necessarily better. In general, centrifuges designed to operate at higher g-forces require better materials of construction and manufacturing methods, and result in higher capital costs. Vibration and noise levels of an operating centrifuge can be used to evaluate the quality of the centrifuge's mechanical design, and consideration should be given to specifying a noise and vibration requirement at the predicted operating g-force of the centrifuge.

2.2.2 Pond Depth

As common with thickening centrifuges, greater pond depths provide more force to compress solids. Centrifuge features (e.g., solids discharge port diameter, weir opening sizes and locations, and adjustable weir plates) differ from manufacturer to manufacturer, and are specifically designed by manufacturers to influence pond depth, affecting dewatering efficiency, power consumption, differential pressure, torque, and other factors. Centrifuges operate under two different methods: (1) positive pond (i.e., liquid level is "below" the solids discharge radius and (2) negative pond (i.e., liquid level is "above" the solids discharge radius). Centrifuges have negative ponds when starting up, resulting in sludge feed (start-up liquid/semisolid "slop") discharging from the solids discharge port until a "seal" is established (typically within 5 to 15 minutes). Handling of start-up "slop" discharged from the solids discharge port should taken into account during design of a centrifuge system.

2.2.3 Cake Solids Content and Capture Efficiency
Cake solids content should be determined by the solids disposal method being used (e.g., incineration, landfilling, etc.). Solids recovery or capture efficiency, as determined by eq 22.1, should target approximately 95% to minimize recycle. Dewatering centrifuges can achieve up to and greater than 99% capture, but can result in excessive polymer and energy usage costs with little process benefit. Table 22.2 presents typical cake solids and solids capture efficiencies for various types of dewatering feed sludge.

2.3 Design Considerations

2.3.1 Mechanical Features and Materials of Construction
Design of a centrifuge system should consider the following:

> *Process connections*: All electrical and process connections to a centrifuge should include flexible couplings to minimize the transfer of centrifuge vibration loads to electrical and process connections.

Type of Sludge	Cake Solids (%)	Solids Capture	
		Without Chemicals	With Chemicals
Untreated:			
Primary	25–35	75–90	95+
Primary and trickling filter	20–25	60–80	95+
Primary and air activated	12–20	55–65	92+
Waste sludge:			
Trickling filter	10–20	60–80	92+
Air activated	5–15	60–80	92+
Oxygen activated	10–20	60–80	92+
Anaerobically digested:			
Primary	25–35	65–80	92+
Primary and trickling filter	18–25	60–75	90+
Primary and air activated	15–20	50–65	90+
Aerobically digested:			
Waste activated	8–10	60–75	90+

TABLE 22.2 Performance Characteristics of Centrifuges [Metcalf & Eddy, Inc. (2003) *Wastewater Engineering Treatment and Reuse*, 4th ed.; McGraw–Hill: New York]

Start-up slop: All centrifuges start-up and shutdown with a "negative" pond, resulting in discharge of liquid/semisolid "slop" out the solids discharge chute. Two methods are typically used for handling start-up/shutdown slop:

1. Diverter gate system: With a diverter gate system, cake discharges vertically into a chute that includes a knife gate to isolate the chute from equipment below and a side-mounted pipe discharge above the gate. During start-up, the diverter gate remains closed and the slop is diverted to the side-mounted drain that is typically connected to the centrate piping and recycled. As the centrifuge begins to separate solids from liquid, the torque increases. When a torque set point is reached, the diverter gate will automatically open and the cake will be directed to the facilities below.

2. Inclined screw conveyor: With an inclined screw conveyor, cake enters at the lower end of the inclined conveyor. Upon start-up and shutdown, the conveyor auger runs in reverse and liquid/semisolid slop discharges to a drain at the bottom of the conveyor that is typically connected to the centrate piping and recycled. Draining of slop should be aided by flush water. As the centrifuge begins to separate solids from liquid, the torque increases. When a torque set point is reached, the conveyor auger switches to run forward and the flushing water is shut off.

Sample taps: Sample taps should be provided for sampling of sludge, centrate, polymer, and dewatered sludge cake. Sampling locations should be easily accessible for operators and adjacent to a floor drain for easy disposal or cleanup of sampling activities. Centrate quality is important to gauge dewatering efficiency; thus it is common for a continuously flowing sample stream to view.

Cake conveyance: Dewatered solids can be conveyed by four methods (refer to Chapter 19 for additional information regarding dewatered cake conveyance systems):

1. Gravity: Dewatered solids "fall" out of the centrifuge discharge chute. Ductwork and/or piping can be used to contain and direct dewatered solids to a container or piece of equipment located below the centrifuge. Bends in ductwork/piping should be avoided to prevent "bridging" of dewatered solids, causing solids to build up into the centrifuge.

2. Cake pumps: Positive displacement piston-style or progressing cavity pumps can be used to convey dewatered solids. Centrifuges can discharge directly to a cake pump inlet hopper or into a separate "bin," where it can be fed into a cake pump with other conveyance equipment (e.g., twin auger screw feeder, screw conveyor, etc.).

3. Belt conveyors: Centrifuges that discharge directly to a conveyor belt should have adequate vertical distance between the conveyor and the solids discharge chute outlet to prevent "bridging." In addition, consideration should be given to "splash shields" on both sides of the belt conveyor adjacent to the area of the centrifuge solids discharge to prevent solids from spilling onto the operations floor and/or surrounding areas. Centrifuges are typically interlocked with the downstream conveyor so that the centrifuge will not start or run if the downstream conveyor is not running.

4. Screw conveyors: Screw conveyors are most common conveyance devices used with centrifuges. Screw conveyors are less expensive than cake pumps, are enclosed systems (less messy and odorous), and can be used in an inclined layout to handle centrifuge start-up/shutdown "slop." Shaftless augers are typically used for cake conveyance applications. Metering/quantifying sludge cake conveyance quantities is difficult with a screw conveyor, however. Like belt conveyors, centrifuges are typically interlocked with the downstream conveyor so that the centrifuge will not start or run if the downstream conveyor is not running.

Design of dewatered solids conveyance systems can be challenging. Elevating centrifuges, whether on the upper floor of a building or on a platform, is recommended so that solids can be transported horizontally or downward, avoiding having to convey solids upward.

Ventilation/Foaming: Restricting the flow of air by sealing the centrate or cake discharge chutes may result in foaming of centrate and solids being carried into the casing, possibly causing bowl corrosion. Both chutes should be properly vented and positive ventilation (i.e., providing a vent pipe to the atmosphere or a compressor intake pipe connection) should be provided on centrate discharge chutes to prevent foaming within the centrate discharge chute and centrifuge, and carry odors away from the operating floor.

Bearings: Centrifuges include main bearings that support the bowl (high-speed), conveyor bearings (low-speed), and a thrust bearing that carries the axial load of the centrifuge. Centrifuge bearings have an L_{10} basic rating life (or minimum expected life), which is a bearing life associated with a 90% reliability when operating under conventional conditions (i.e., after a stated amount of time, 90% of a group of identical bearings will not yet have developed metal fatigue). However, improper lubrication, contamination, or misalignment—all common causes of bearing failure—will affect bearing life. Typical bearing types used in centrifuges are spherical roller, cylindrical roller, and ball bearings. Main bearings are typically lubricated by forced-oil, grease, or oil-mist lubrication systems. Conveyor bearings are typically grease lubricated.

Torque reducer: All centrifuges have a torque reducer; the alternatives include a planetary or cyclo gearbox for VFD back-drive systems, or a slip drive system for hydraulically back-drive systems. Torque reducers should last between 20 000 and 40 000 hours without mechanical problems, and can last up to 100 000 hours.

Materials of construction: All wetted parts of a centrifuge should be fabricated of stainless steel, to avoid corrosion and maintain tolerances (see Figure 22.3).

Areas of the scroll and other parts of the centrifuge in contact with solids (e.g., sludge feed zone nozzles, conveyor flights, solids discharge ports, etc.) are protected with abrasion-resistant materials. Urethane- and plasma-applied hard surfacings work well in feed zones. The edges of the conveyor flights typically have field-replaceable sintered tungsten carbide tiles that can be replaced without the need for rebalancing the scroll. Tungsten carbide protection is typical at the solids discharge ports. The casing wear liner should be long enough to cover the flexible boots on the solids chute. Urethane and rubber material is typically used for casing wear liners, which should be long enough to cover the flexible connections on the solids discharge chute, but hard-faced steel or stainless steel is sometimes used.

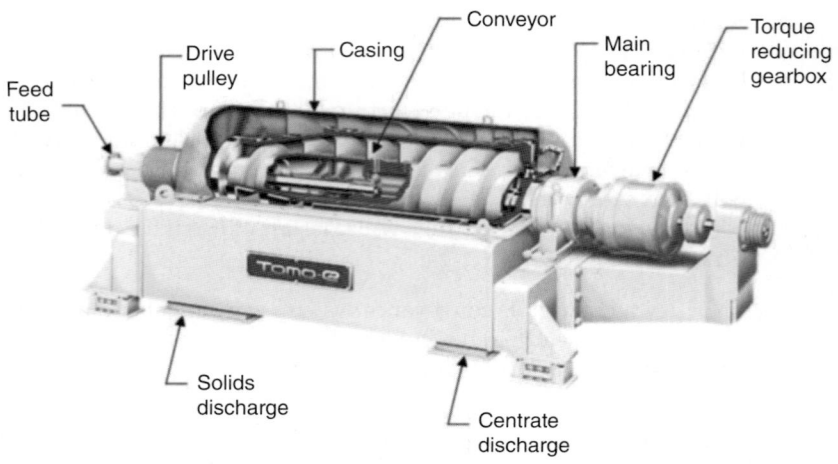

FIGURE 22.3   Cutaway view of a centrifuge.

Centrifuges are designed to meet the same vibration standard (6- to 7-millimeters per second [mm/s]) (0.24 to 0.28 inches per second [in./s] velocity at their maximum rated speed). Imbalances in the rotor create eccentric loads, which are transferred to the main bearings and to the centrifuge frame. Excessive vibration leads to shortened bearing life. Ideally, the weight of the rotor would be equal to the weight of the frame (i.e., rotor–weight-to-frame-weight ratio is equal to 1). A cost-efficient way for centrifuge manufacturers to meet the vibration standard is to load "dead" weight (e.g., additional steel or iron that does not provide structural support) into the stationary frame, resulting in dampening the vibration, but not reducing bearing loads. Designers should evaluate or request information from centrifuge manufacturers regarding rotor and frame weights, and/or specify a maximum centrifuge rotor-weight-to-frame-weight ratio.

Designers will sometimes require vibration (and noise) certifications from centrifuge manufacturers for centrifuges running dry at the factory. If included in centrifuge specifications, the speed or g-force at which the vibration (or noise) measurements are taken should also be specified.

2.3.2   Instrumentation and Controls

Proper instrumentation is necessary to provide operators with sufficient data to operate the dewatering system efficiently and make timely operation decisions. Magnetic flow meters on sludge feed and positive-displacement rotary abutment flow meters on the polymer solution pipelines are recommended. As pumps age and wear, flow proportionality to pump speed can diverge from a linear relationship and pump curves become less accurate to interpolate flow. Flow meters can be used to provide more accurate data on dewatering pumping systems.

Pressure gauges should be considered on sludge and polymer feed piping, washwater, and cooling water (if necessary for oil-lubricated main bearings or hydraulic back drives), along with high-level pressure switches to prevent damage to equipment and piping from excessive pressures.

2.3.2.1 Scroll/Bowl Differential Speed-Control Drives It is recommended that controls have "one-button starts" in automatic mode, the ability to "pause" centrifuge operation (i.e., stop flowrates to the centrifuge, but maintain centrifuge bowl and scroll rotation), the ability to do a "hot" restart after a centrifuge pause to resume dewatering operations, the ability to initiate a CIP cycle during operation, and a "coasting" feature, so that, in the event of power failure, power is maintained to pertinent systems to allow the centrifuge to ramp down (or "coast") to a stop. Uninterrupted power supply (UPS) devices should be considered to maintain operations during "brown-outs" or provide sufficient power for a proper shutdown during "blackouts." Emergency stop features are typically provided for a centrifuge, but should only be used as a last resort.

Typical automatic controls operate based on differential speed and torque. Differential speed or torque set points are operator adjustable, and the centrifuge will automatically adjust bowl/scroll speeds to maintain the differential speed or torque set points.

All centrifuge manufacturers include standard automation packages for their centrifuges, have various options that can be purchased, and can customize controls per specifications and/or owner preference. Sophisticated systems can use sensors to measure centrate solids and torque as a measure of cake dryness, performing much of the automatic adjustment without operator input or control. For example, operators can set the sludge feed rate and torque, and the control system uses the centrate solids measurement to adjust the polymer rate to maintain a centrate solids set point. Feed solids concentration and feed rate can also be measured to automatically adjust polymer dosing rates. Adding automation features increases the capital cost of the system, and ultimately should be used to help operators, not replace them. Operators are required to troubleshoot system issues, maintain the equipment and instruments, and optimize the system.

2.3.2.2 Safety Features All centrifuges should be protected by interlocks for excessive torque, motor amperage, vibration, and motor and bearing temperature, resulting in an alarm and/or shutdown of the centrifuge system. Flow switches are also recommended to interlock operation of the centrifuge system with essential process flows (e.g., cooling water) to prevent damage to equipment.

2.3.3 Area/Building Requirements

It is recommended to elevate a centrifuge, either via a platform or on higher floors in a building, to better control solids conveyance. Including a temperature-controlled control room to house centrifuge starter and/or control systems (e.g., motor control centers, motor starters, VFDs, SCADA terminals, etc.) is recommended to provide a proper environment for operators to efficiently control, optimize, and record centrifuge operations. Hoisting and ingress/egress systems also need consideration. The centrifuge bowl and scroll need to be periodically maintained, and extracting them from the centrifuge requires a hoisting system. A permanent hoisting system, such as a bridge crane or monorail/hoist/trolley, is recommended to assist maintenance crews in disassembling centrifuges. Facilities for moving centrifuge equipment into and out of the building is also required and can include access hatches, wall transom panels, overhung monorails through a double door, etc. Hoisting systems should be adequately designed to accommodate equipment weights with hook height ranges so that removed equipment can "clear" permanently installed equipment and lowered to proper heights for loading onto a flatbed truck. Trucks should have proper access to centrifuge equipment ingress/egress systems.

2.3.3.1 Structural Requirements Centrifuge installations should consider structural requirements. Larger centrifuges can weigh up to 18 000 kilograms (kg) (19.8 tons) and may require substantial design effort. Facilities in more stringent earthquake zones have more complexity because the rotating mass tends to remain in place, while the building under the centrifuge moves. During design, centrifuge manufacturers should be asked for information regarding relevant dimensions, static and dynamic loadings, and process requirements. Thick concrete floors to absorb vibration or structural systems that isolate the centrifuge from the building structural system are common. Platforms that include grating should be designed to minimize transfer of centrifuge vibration to the platform grating to prevent "rattling."

A centrifuge imbalance—magnified by the g-force at which the unit is operating—can create large dynamic loads on the bearings, which are transferred to the centrifuge frame and possibly the building structure. Dynamic loadings are in the x-y direction, not in the z (axial) direction. The vibration frequency is the rotational speed of the centrifuge. Centrifuge manufacturers typically provide vibration isolators for mounting the centrifuges on, which help to dampen vibrations caused by the centrifuge. When centrifuges are starting up or coasting to a stop, they will typically "travel" through the building's natural harmonic frequencies, which can result in vibration of building systems (e.g., floor slabs or walls). Structural components should be installed to reduce vibration transferences to the building during centrifuge start-up and shut down to a minimum.

2.3.3.2 Energy Requirements Energy consumption can be calculated and tested during shop and field tests, and can be included in specifications and/or performance requirements. Power is proportional to the square of speed, so bowl speed is a critical factor. Power also is proportional to the radius of discharge and feed rate. The efficiency of the drives and gear reducers also factors into power use. Energy use requirements for the centrifuge system should account for and specify the rated speed and/or g-force, and loading for determination of power use.

2.4 Ancillary Systems

2.4.1 Feed System

Sludge feed conditions are variable, and design of sludge pumping and polymer systems, and other systems, should be robust enough to account for extreme variables. Causes of variability that can pose dewatering challenges include the following:

- Ratio of primary to secondary sludge can alter dewaterability efficiency as it decreases.

- Insufficient mixing in sludge storage tanks can result in inconsistent sludge characteristics fed to the centrifuge system, resulting in inconsistent dewatering results.

- Systems where digested sludge is pumped directly to centrifuges rather than to a mixing/storage tank.

- Inadequately or oversized sludge storage tanks, undersized dewatering equipment, or insufficient dewatering schedules can cause sludge to become septic, resulting in decreased dewaterability efficiency and poor dewatering results.

- Excessive sludge mixing intensity in storage tanks may reduce the solids particle size, resulting in increased polymer usage, and reduced dewaterability efficiency.
- Excessive polymer mixing intensity can cause long-chain polymer molecules to shear resulting in poor dewatering results.

Low-shear positive displacement pumps such as progressing cavity or rotary lobe pumps are recommended for sludge and polymer feed pumps.

2.4.2 Cake Discharge

Sludge cake conveyance equipment should be selected based on owner preference, facility requirements, cost, and other factors. Start-up and shutdown "slop" should be accounted for in design of the sludge cake conveyance system, as indicated in previous sections.

2.4.3 Chemical Conditioning

All dewatering centrifuges require polymer. Chapter 20 of this manual of practice includes information on dewatering chemical conditioning, including different polymer types used in dewatering applications. The type of polymer used is largely based on capital and long-term costs. Various polymer types and polymer solutions should be pilot-testing prior to the design of polymer and dewatering systems. Chapter 7, Chemical Systems, also provides useful information for designing polymer storage and feed systems.

Day/aging tanks are recommended for dewatering applications that have high variability in the sludge. "Batches" of polymer can be made and fed from the day/aging tank to accommodate current sludge conditions. A day/aging tank may not necessarily be required for direct polymer dosing applications (i.e., feeding polymer directly from the makeup units into the sludge feed stream). Pilot-testing can be used to determine the effectiveness of aging polymer in dewatering.

Multiple polymer addition points should be provided, including the following: (1) at the centrifuge manufacturer-provided polymer inlet port for polymer discharge inside the centrifuge or immediately upstream of the centrifuge (depending on the centrifuge manufacturer); (2) into the sludge feed piping approximately 8 meters (m) (25 ft) upstream of the centrifuge; and (3) into the sludge feed piping approximately 16 m (50 ft) upstream of the centrifuge. The optimum polymer discharge location depends on the solids characteristics, polymer type and dosing concentration, and dewatering objectives. One way to easily switch and optimize polymer dosing locations is to install a manifold in a convenient location. To change the addition point, operators can manipulate valves from the same location. Ball valves are recommended for polymer service so that throttling of flowrates at different dosing locations is also possible. Periodically, operators need to optimize polymer application points to account for sludge variability.

Inorganic chemicals used for purposes other than dewatering may affect dewatering. For example, the use of ferric chloride reduces polymer demand, while the use of alum and lime increase polymer demand. Ferric chloride is acidic and high in chlorides. When diluted by sludge, ferric chloride acidity and chloride content may not be an issue, but an interlock should be considered to ensure that ferric chloride is added only when the feed solids are entering the centrifuge. Peroxides and permanganate are typically used in such small quantities that they typically have no discernable effect on dewatering. Most inorganic chemicals cannot justify their cost based on a corresponding reduction in polymer demand, but may have other benefits.

2.4.4 Washwater

Each time a centrifuge is shut down, the bowl must be flushed with water for approximately 5 to 20 minutes. Flush water is typically fed to the centrifuge through the sludge feed line; flushing water piping connects to the feed sludge line near the centrifuge and should include a check valve to prevent cross-contamination. Approximately 3.15 liters per second (L/s) (50 gallons per minute [gpm]) at a minimum of 276 kilopascals (kPa) (40 pounds per square inch, gauge [psig]) of flushing water is required during a CIP cycle. Shutting a centrifuge down without a CIP cycle may result in a centrifuge becoming plugged and will require service to empty solids out of the centrifuge before its next restart. Improper cleaning of a centrifuge is the main cause of excessive vibration.

In rare cases where the facility water has high chloride levels, letting the flush water evaporate on "wetted" parts of the centrifuge may concentrate chlorides and result in corrosion. In these instances, a short final flush with potable water is advised before shutting down.

3.0 Belt Presses

3.1 Description of Technology

Belt presses dewater solids in three stages: chemical conditioning, gravity drainage, and compaction in a pressure and shear zone. Dewatering operations begin when polymer-flocculated solids enter the gravity drainage zone. The conditioned solids are evenly applied to the gravity feed belt via a distribution system, which typically is provided. The continuous, porous belt provides a large surface area through which gravity drains free water. Filtrate from the gravity zone is collected and piped to a drainage system. Belt filter press machines evolved from paper-making applications to dewatering municipal wastewater solids. The belt filter press was introduced to North America in the 1970s as a lower-energy alternative to centrifuges and vacuum-filter equipment. Belt presses are used throughout the United States and are available from more than a dozen manufacturers.

Compared to other mechanical dewatering devices, belt presses still have the lowest energy consumption per volume of solids dewatered. However, energy consumption is not the only factor to consider when selecting, sizing, and designing a dewatering system. Table 22.3 summarizes the advantages and disadvantages of belt filter presses.

Advantages	Disadvantages
- Low energy requirements	- High odor potential
- Relatively low capital and operating costs	- Requires sludge grinder in feed stream
- Less complex mechanically and easier to maintain	- Very sensitive to incoming sludge feed characteristics
- High–pressure machines are capable of producing very dry cake	- Automatic operation generally not advised
- Minimal effort required for system shutdown	

TABLE 22.3 Advantages and Disadvantages of Belt Press Dewatering [Metcalf & Eddy, Inc. (2003) *Wastewater Engineering Treatment and Reuse*, 4th ed.; McGraw–Hill: New York]

3.2   Process Design Conditions and Criteria

The main design elements for belt filter presses include performance requirements:

- Discharge cake, % dry solids;
- polymer dose, kg/Mg sludge (lb/ton sludge); and
- solids capture rate, %.

and loading criteria:

- Hydraulic-loading rate, L/s·m of belt width (gpm/m of belt width); and
- Solids-loading rate, kg/h·m of belt width (lb/hr/m of belt width).

Other features to evaluate during the design of belt filter press systems are: wash-water for belt washing, filtrate collection/drainage, dewatered-cake conveyance, equipment access/layout, and odor control requirements. The end use or disposal options for the dewatered solids (and the solids characteristics needed for that option) also must be considered.

Solids characteristics, origin, and degree of stabilization all significantly affect belt press loading and obtainable dewatering performance. Dilute solids require more gravity drainage, more polymer, and a longer dewatering time than more concentrated feed material. The type of process used for stabilization has a direct bearing on the maximum solids content achievable. Although manufacturers' estimates differ regarding the amount of dewatering that can be achieved, they typically agree that anaerobically digested residuals are easier to dewater than aerobically digested residuals. Typically, digested solids with lower volatile solids contents produce thicker dewatered cakes. In addition, as the primary-to-secondary solids ratio increases, dewatering becomes easier and the cake solids concentration increases.

The best method for evaluating belt filter press performance on a specific material is to dewater the residuals using a pilot unit. Several belt press manufacturers have mobile trailer-mounted pilot units that can be rented for testing. Most pilot units are small production machines that can perform comparably to larger models.

As an alternative to a pilot test, bench testing can be used to predict belt press performance. A sample of the material to be dewatered can be sent to the manufacturer, who will provide design criteria and expected performance based on their tests. These tests have been shown to provide adequate projections of belt filter press performance.

New press designs have been developed (with more rollers or a separate gravity-drainage or gravity-thickening deck, vertical compression zones to reduce footprint) to produce higher dewatered-cake solids concentrations.

3.2.1   Loading Rates

The throughput capacity of a belt press is the primary design criterion when sizing a belt press system. Throughput capacity typically is considered to be either hydraulically limited (<2.5 % dry solids) or solids limited (≥2.5% dry solids), depending on feed solids concentration. Belt presses have a maximum hydraulic or solids-loading capability for a given unit of width that can be attained only when solids are conditioned correctly. However, these typical loading rates vary significantly depending on the source of the residuals to be dewatered and must be taken into account when designing the belt press system. Some manufacturers use mixed units (gpm/meter, lb/day/meter) when discussing the loading capacity of their machine. Because of this, all three sets of units are included in the following sections.

3.2.1.1 Hydraulic-Loading Rates   Typical design hydraulic-loading rates for a belt press range from 3 to 12.5 L/s·m of belt width (15 to 61 gpm/ft of belt width or 50 to 200 gpm/m of belt width). Recommended loading limits vary per machine manufacturer and should be verified when performing sizing calculations and comparing units.

3.2.1.2 Solids-Loading Rates   Typical solids-loading rates range from 135 to 1136 kg/m·h (92 to 760 lb/ft/hr dry-solids basis). Recommended loading limits vary per type of sludge and machine manufacturer and should be verified when performing sizing calculations and comparing units.

3.2.1.3 Loading Rates per Sludge Type   As previously mentioned in this section, the source of sludge also has an effect on the loading and performance of dewatering equipment, including belt filter presses. The type of sludge (primary, secondary, blend) and/or treatment upstream of dewatering (undigested, anaerobically digested, thermal hydrolysis digestion, etc.) can significantly affect the capacity and performance of the dewatering system. Table 22.4 includes some examples of loading rates (both hydraulic and solids) and expected performance based on the type of sludge and solids treatment upstream of the belt filter presses. These recommended loading limits vary per machine manufacturer and should be verified when performing sizing calculations and comparing units.

3.2.2   Solids Capture Rates

Capture efficiency is affected by the hydraulic-loading rate, the solids-loading rate, the nature of the solids being dewatered, the mesh size of the belt, and chemical conditioning. It is important to test polymers to determine the best type of polymer, as well as the relationship between polymer doses, solids capture efficiency, and cake solids. Capture efficiency will decline if the solids are not properly conditioned. The solids capture rate for belt filter presses (total solids recovery, including washwater solids) is between 92% and 98%. However, most manufacturers guarantee solids capture rates exceeding 95% when the system is operated as designed. Table 22.4 includes the expected capture rates for a number of municipal sludges.

3.2.3   Dewatering Performance

Belt press performance data indicate significant variations in the dewaterability based on the different types of residuals or biosolids being processed. Although the press typically can produce a dewatered cake containing anywhere from 12% to 35% solids when treating a typical combination of primary and secondary solids, these values vary significantly based on the type of sludge (primary, waste activated sludge) and treatment (aerobically digested, autothermal thermophilic aerobic digestion, etc.). Table 22.4 includes some examples of expected performance for belt filter presses based on the type of sludge and treatment. These values are particular to each facility and highly dependent on volatile suspended solids (VSS) content of the residuals. Therefore, the residuals characteristics should be carefully evaluated during the design stage to establish design and performance criteria.

	Primary	WAS	Undigested Blend	Aerobically Digested	Anaerobically Digested	Anaerobically Digested BNR	Anaerobic Digested THP	ATAD
Primary/secondary blend	100/0	0/100	50/50	0/100	50/50	70/30	50/50	50/50
Feed solids (% TS)	2.0–5.0	2.0–6.0	0.5–2.0	1–2.5	2.0–4.0	1.5–3.0	3.0–6.0	2.0–3.0
Hydraulic loading rate								
(gpm-m)	60–120	60–150	60–100	60–120	30–100	30–80	30–60	40–80
(L/s-m)	3.79–7.57	3.79–9.46	3.79–6.31	3.79–7.57	1.89–6.31	1.89–5.05	1.89–3.79	2.52–5.05
(gpm-ft)	18.3–36.6	18.3–45.7	18.3–30.5	18.3–36.6	9.1–30.5	9.1–24.4	9.1–18.3	12.2–24.4
Solids loading rate								
(lb/m/hr)	1000–2500	300–800	1000–2000	400–800	600–1100	400–800	800–1800	400–800
(kg/m/h)	453.6–1134.0	136.1–362.9	453.6–907.2	181.4–362.9	272.2–499.0	181.4–362.9	362.9–816.5	181.4–362.9
(lb/ft/hr)	305.0–762.5	91.5–244.0	305.0–610.0	122–244	183.0–335.5	122.0–244.0	224.0–549.0	122.0–244.0
Polymer dosage (active lb/ton sludge)	10–20	20–30	15–25	20–30	15–25	25–40	15–35	20–50*
Expected cake solids (% ds)	10–20	10–20	10–20	10–20	10–20	10–20	10–20	10–20
Expected capture rate (%)	95–98	92–96	94–97	94–96	95–97	94–96	96–98	95–97

*May require 200 to 600 lb/ton of ferric chloride to reduce polymer demand.

TABLE 22.4 Performance Characteristics of Belt Filter Presses (Andritz Separation Inc.)

Other variables that affect the maximum solids-loading rate include the amount and type of fiber in the solids, shear strength of the solids, type of chemical conditioning, belt type, and maximum pressure applied to the solids.

3.3 Design Considerations

As previously mentioned, belt presses dewater solids in three stages: chemical conditioning, gravity drainage to about 4.5% to 5.0% dry solids concentration, and compaction in a pressure and shear zone. The following sections provide additional details on the component of each of these stages.

3.3.1 Mechanical Features and Materials of Construction

Each belt press manufacturer produces machines with slightly different mechanical features and operating characteristics. Presses are available in widths ranging from about 0.5 to 3.5 m. Most municipal presses use 1- to 2-m belt widths. Care should be taken when specifying presses exceeding the 2.5-m width as deflection in the rollers could reduce machine performance (lower % dry solids). The main components of a belt filter press system include feed equipment and piping, frame, belts, belt-tracking and tensioning systems, belt wash system, rollers and bearings, cake-discharge blades, chutes, cake conveyance, drive system, belt-speed control, and chemical conditioning and flocculation Some of these systems (for a two-belt press setup) can be seen in Figure 22.4.

3.3.1.1 Gravity Drainage Zone The gravity drainage zone is a flat or slightly inclined belt that is unique to each manufacturer. The effectiveness of this zone is a function of solids type, concentration, chemical conditioning, belt fabric, and detention time.

Typically, multiple rows of chicanes on the upper surface of the filter belt are used to move the sludge solids and remove excess liquids. The chicanes continuously turn the sludge over, freeing up entrapped liquid, and creating an open area to allow the entrapped water to exit to the dewatering belt.

Two-belt machines (for sludge concentrations ≥1.5% dry solids) have a continuous flow path for the solids (i.e., the belt where gravity drainage occurs carries the material

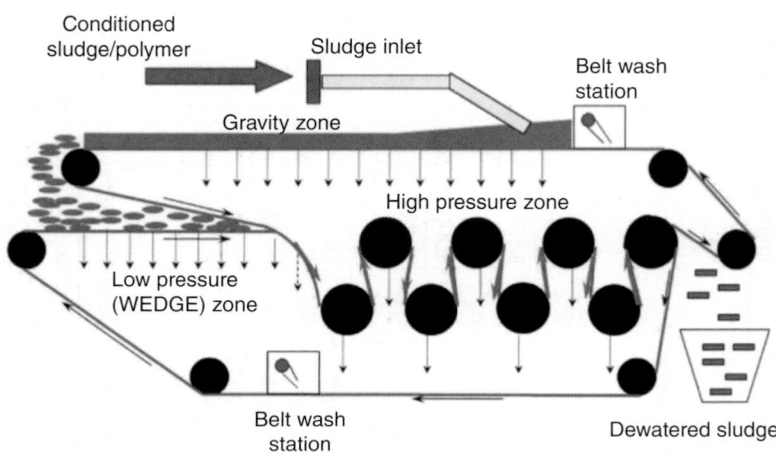

FIGURE 22.4 Two-belt unit diagram (courtesy of Alfa-Laval, Inc.).

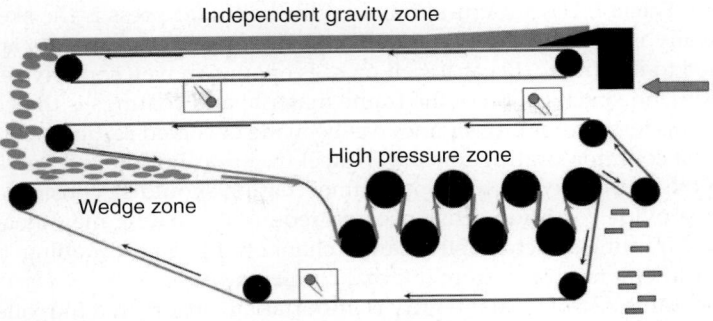

Independent gravity zone

High pressure zone

Wedge zone

FIGURE **22.5**   Three-belt unit diagram (courtesy of Alfa-Laval, Inc.).

directly into the pressure zone). Three-belt machines (typically for sludge concentrations <1.5% dry solids, see Figure 22.5) have a single-belt gravity drainage zone (similar to a gravity belt thickener) mounted above, and discharge to a two-belt pressure zone. Some systems use a rotary screen for gravity drainage rather than a single-belt gravity zone, but these are not frequently used due to additional maintenance requirements. The separate gravity zone provides a larger gravity zone that likely benefits more dilute solids. The three-belt machine also allows for different hydraulic-loading rates and belt speeds for the gravity-drain and dewatering sections of the press. Equipment manufacturers select the length of the gravity drainage zone based on the inlet solids concentration and the relative drainage rate required to obtain a 4% to 5% dry solids concentration when entering the pressure zone. Gravity-zone lengths typically range from 2 to 4.1 m (6.6 to 13 ft).

3.3.1.2 Pressure Zones   After gravity thickening, solids move into the pressure zones. Typically, a low-pressure zone is followed by a high-pressure zone.

The low-pressure zone is the area where the two belts first come together with the gravity-drained solids between them. This is a "wedge zone" where the solids are sandwiched between the two belts. The low-pressure zone provides enough dewatering to form a cake that can withstand the additional pressure and shear in the high-pressure zone without extruding out the edges of the belts. A common operational mistake is applying too much pressure before enough water has been removed from the flocs.

In the high-pressure zone, forces are exerted on the solids by the movement of the upper and lower belts relative to each other as they move over and under rollers with decreasing diameters. Some machines have an extended roller section that provides higher pressures and sustains those pressures for a longer time.

Pressures continue to increase as the solids pass through the wedge zone and enter the high-pressure (drum-pressure) stage of the belt filter press. The belt tension squeezes the cake as the belts proceed around several drums or rollers (of varying diameters) to maximize shearing action. As solids moves through the press, increasingly smaller-diameter rollers progressively increase the pressure. Average pressures applied typically are 35 to 105 kPa (5 to 15 psi), although they can range up to 210 kPa (30 psi) depending on the size and arrangement of the rollers. Arbitrarily increasing belt tension to increase cake solids may reduce belt life and solids capture, and embed more solids in the belt.

3.3.1.3   Frame   The structural frame of the belt filter press is the skeleton of the unit; it typically is constructed of steel. All belt filter press components are supported and attached to this frame. Because belt presses operate in wet, corrosive environments, the selection and specification of the frame material and coatings is the key to the installation's long-term durability. Frames can be made of coated carbon steel or stainless steel. The most common coating for carbon steel frames is hot-dipped galvanizing. Depending on the site, epoxy or baked-on enamel coating should be considered. Stainless steel frames provide corrosion resistance and require no coating maintenance at additional cost. The framing structural steel can be channels, I-beams, or tubing. (However, tubing can be difficult to protect from internal corrosion.)

The frame's structural integrity is important to ensure that the rollers are supported and function properly. The frame should be designed (specified) to accommodate operating and static loads with a factor of safety not less than 5, so the machine can operate without deflection, deformation, or vibration. Seismic design of conduit and piping connections and anchorage of the frame are important considerations.

Access to the belt press building, room, or area needs to be considered when specifying the frame. The specification should include whether the machine can be installed in one piece. If the installation needs to be in pieces because of limited access, the size and/or weight of the largest piece needs to be defined. Dismantling the frame and rebuilding it in the field can affect the frame's critical protective coating. Lifting lugs should be specified to facilitate placement or removal of the units. An overhead crane, hoist, or portable lifting device that is sized to handle the largest equipment component should be included in the building design.

The structural design of the frame should include platforms or walkways so an operator can observe the gravity portion of the belt press and perform routine maintenance. Structural members of the walkways must be clear of the rollers and bearings.

In addition, the layout of the belt press needs to provide enough clear space between units to remove individual rollers.

3.3.1.4   Rollers   Rollers support the porous cloth belts and provide tension, shear, and compression throughout the pressure stages of the belt press. Rollers can be made of a variety of materials, including stainless steel. Corrosion and structural considerations are important. The most common coating systems include rubber for the drive rollers and thermoplastic nylon for the others. Roller deflection at the rated belt tension of at least 8.75 kN/m (50 lb/in.) should be limited to 1 mm (0.05 in.) at roller midspan. Belt tension should be based on at least 5.4 kPa/cm of belt width (200 lb/in. of belt width), and drive tension should be calculated based on a belt speed of at least 4.6 m/min (15 ft/min). Some manufacturers use perforated stainless steel rollers in the initial pressure stages to enhance drainage.

3.3.1.5   Belts   Most belt presses have two operating belts, but there are three-belt units available that provide a separate gravity thickening section before the high-pressure dewatering section. Belts are made of woven synthetic fibers, typically monofilament polyester. Nylon belts are available, but typically are used for specific applications (e.g., high-pH solids or abrasive slurries). Both seamed and seamless belts are available. Seamed belts have either stainless steel clipper-type seams or zipper-type seams; they tend to wear quickly at the seam because of a high degree of discontinuity and stress concentration at that point. The raised metal seam also causes wear on the rollers and the doctor blade (i.e., a belt scraper). Zipper-type seams have a lower profile and provide less discontinuity than clipper seams, and have a longer life. Seamless belts are continuously woven, endless belts that have a longer service life than any other belt type. However, seamless belts are more costly and difficult to change out. Several manufacturers market belt presses that accommodate seamless belts. Available in various materials and weave combinations, belts should be evaluated relative to the expected solids characteristics, solids capture required, and durability.

3.3.1.6 Bearings   Bearings are an important part of the belt press. Many manufacturers mount the bearings directly on the structural mainframe so they are accessible for maintenance and service on the exterior of the units. These bearings typically are pillow-block construction and should be rated for at least an L10 life of 300 000 hours based on forces and loads (e.g., belt tension, roller mass, and drive torque loads). Bearings should be double- or triple-sealed to prevent contamination and wear resulting from press washdown and solids penetration. Bearings should be self-aligning. A split-housing type of bearing is necessary if ready access is unavailable outside of the mainframe. A centralized lubrication system is an option offered by some manufacturers.

3.3.2 Controls and Drives

A control panel (typically custom designed for the site) is necessary to control the belt presses and their ancillary systems. The panel should provide for automatic, semiautomatic, and manual starting or stopping of the system's components. Sequencing relays or programmable controllers can be provided, as well as electrical and safety interlocks. Critical alarms should be annunciated at the panel and at a central location, and a system-wide emergency power shutdown should be provided. The controls should be in a dry area within sight of the belt press but away or protected from the potentially corrosive atmosphere and spray from equipment washdown. Control panels should meet National Electrical Manufacturers' Association (NEMA) 4X standards to protect components from the moist, corrosive environment.

The controls for each part of the dewatering system should be interconnected to ensure that system operations are coordinated. Solids feed, polymer feed, and belt press and conveyor startup and shutdown must be properly sequenced for either automatic or manual operation. At a minimum, polymer feed should keep pace with the solids feed rate. The dewatering equipment (and sludge/polymer feed pumps) should automatically shut down for a belt-drive failure, belt misalignment, insufficient belt tension, loss of pneumatic or hydraulic system pressure, low belt-washwater pressure, emergency stop (trip wire and/or pushbutton), and stoppage of the cake-conveyance system.

3.3.2.1 Belt Speed   Compressive and shear forces are exerted as the solids pass between the belts and wind through the belt press. Belt speed directly relates to solids retention time (SRT) in various sections of the press, the dryness of cake solids, and throughput. Belt speed should be adjustable at the belt press control panel.

3.3.2.2 Belt Tracking   The belt-tracking system maintains proper belt alignment by keeping the belts centered on the rollers. It uses sensing arms connected to a limit switch to sense movement in the belt position. A continuously adjustable roller senses the shift and automatically adjusts the belt to compensate. This roller is connected to a pneumatic, hydraulic, or electrically operated response system. An automatic, continuous modulating control must be an integral part of the system. This system should include the capability of shutting down the belt filter press and associated systems if the belt veers too far off-center to prevent damage to the belt and other components.

3.3.2.3   Belt Tensioning   Belt-tension adjustments can be one of the operators' process-control variables. During operations, belt tension is maintained and controlled pneumatically, mechanically, or hydraulically. Increasing belt tension will increase dewatering pressure, increasing cake solids percent, increasing the chance for extrusion of sludge through the belt, and decreasing solids capture in the filtrate. Several manufacturers offer separate control systems for the upper and lower belts so the tension of each can be adjusted independently. An automatic adjustment system, similar to the one for the tracking system, is necessary. A pressure gage (or similar device) is recommended to indicate belt tension. The belt-tensioning system should be able to accommodate at least a 3% increase in belt length. The system should adjust to maintain the desired belt tension as the belt stretches under normal use and wear. (Note that belt life decreases as belt tension increases.).

3.3.2.4   TSS/Polymer Dosing Control   One of the major disadvantages of belt filter presses versus other dewatering systems is the inability of the machine to detect changes in the feed solids and make automatic changes to maintain operation. For example, changes in sludge concentration would cause a typical belt filter press system to under/overdose polymer. Not only is this inefficient, it could cause the belt filter press operation to fail to dewater the sludge to the expected quality. Sludge and polymer feed pumps are normally "tied" together to match changes in pump speed, but this does not account for changes in sludge quality. Improved TSS sensors now allow monitoring of the quality of the incoming sludge and adjustment of the polymer feed to maintain the established dose.

Adding TSS sensors on the incoming sludge allows monitoring of sludge quality and, when combined with the flow meter, calculates real-time loading to the machine and modifies polymer feed to maintain the preestablished dose. In addition, adding TSS sensors to the filtrate drainage allows facility personnel to monitor and calculate solids capture rate real time providing early warning of operations upset. Instrument manufacturers provide these TSS sensors/polymer dosage controls as complete control systems that interact with the belt filter press, sludge, and polymer feed systems to optimize operations. However, these sensors can also be purchased individually and be integrated into the dewatering system.

3.3.3   Area/Building Requirements

3.3.3.1   Structural Requirements   Structural loads for a belt filter press are limited to dead load requirements for the unit when "wet" (full of sludge) as vibration and rotational loads are negligible for this type of equipment. This is a significant reduction in cost when compared with other type of dewatering equipment (i.e., centrifuges).

Sufficient space should be provided between adjacent units to allow for removal of the rollers. Overhead cranes typically are provided to facilitate roller removal, although portable cranes and hoists also can be used.

Because of the continuous backwash and the potential for occasional solids spillage from the belts, belt presses typically are enclosed with containment walls or grating to capture the water running off the press. This also permits the units to be hosed off when they are taken out of service. This typically involves installing the

FIGURE 22.6   Uncovered belt filter press (courtesy of Alfa-Laval, Inc.).

belt press several feet above the floor. Access to the belt press for lubrication of bearings, and inspection of bearings, belts, and rollers, typically is provided by placing elevated metal platforms and walkways around the presses. These walkways need to allow room for removing belts and rollers and to not interfere with the moving parts of the press.

Typically, a belt press dewatering room will be at least three or four times larger than the footprint of the presses to accommodate all of these requirements.

3.3.3.2   Press Enclosures   Because of the open nature of a belt press (see Figure 22.6), there is a significant potential for odors and sprays. Workers in the belt press areas can be exposed to aerosols from the belt-wash spray nozzles, as well as pathogens and hazardous gases (i.e., hydrogen sulfide). One alternative for containing odors is installing a ventilation hood above the belt press, as well as enclosures that surround the machine. Ventilation hoods reduce the amount of foul air to be treated, compared with presses in an open room. However, they can restrict lifting-equipment access. Some manufacturers offer enclosed belt presses (see Figure 22.7). While more expensive, enclosing the units contains odors,

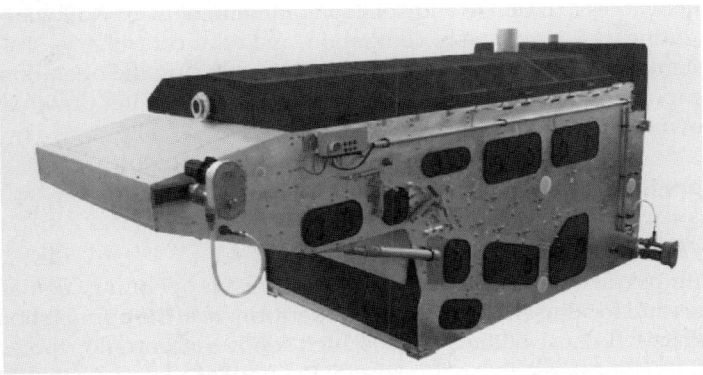

FIGURE 22.7   Enclosed belt filter press (courtesy of Alfa-Laval, Inc.).

reduces odor-handling volumes, and better contains sprays and spills. However, the enclosed system is more susceptible to moisture and chemical corrosion, which must be considered when establishing ventilation rates and selecting materials and coatings. Enclosures also limit visual and physical access to the machine and to the solids being processed.

In large installations, another option is to house the belt presses in a separate room. This helps reduce ventilation requirements and improve the overall building environment. If the belt filter presses are not enclosed, closed-circuit television cameras can be used to allow the operators to monitor the condition of the sludge in the gravity deck, without exposing them to the aerosols and odors associated with sludge dewatering.

3.3.3.3 Drainage Requirements As previously mentioned, belt presses typically are installed above a containment area or grating to capture the water running off the press and allow the units to be hosed off when they are taken out of service. Drainage pans, shield, and piping should be designed to confine spray and splashed liquids and should discharge to a sump or floor-drainage system directly below the unit. Drainage connections should be self-venting to prevent overflow. If possible, the drainage piping should be hard piped to the floor drainage system to minimize turbulence, and thereby reduce odors, although this is rarely seen. Drainage capacity must be sufficient to allow filtrate, washwater, and washdown of the unit. It is important that the drainage systems account for the maximum drainage flow possible, not expected operational flow. This means maximum sludge pump capacity plus maximum washwater booster pump capacity.

This drainage flow typically is routed back to the headworks or primary clarifiers for re-treatment.

3.3.3.4 Energy Requirements A belt filter press' energy requirements are relatively low compared to some other types of dewatering equipment. Typically, a 2.5-m-wide belt press will require around 7 kW (10 hp) per machine. The power required for the polymer conditioning system, washwater pressure boosting, conveyors, and any hoist to remove the rollers is installation specific and must be added to the total energy requirements. Building ventilation and odor-control energy requirements also need to be taken into account when determining the total energy needed for the belt press installation.

3.3.3.5 Safety Personnel safety must be fully considered and incorporated into the design. The design must provide for and facilitate maintenance, and provide safety stops and trip wires around the belt press and any cake conveyors, convenient and safe equipment access, drainage and spill containment, nonslip walkways and floors, sufficient lighting, noise reduction, ventilation, and odor control.

System interlocks should be provided to safely stop the operation if any of the system components (sludge and polymer feed pumps, belt filter press, cake conveyance) is shut down individually.

3.4 Ancillary Systems

3.4.1 Feed System
Feed pumps run continuously while the belt press is operating. To match solids production rates and to adjust or optimize press performance, the pumps should have variable-speed drives. Because of the residuals' high solids concentration, potential variability in feed solids characteristics, and the desire to pump at a known or selected rate, positive-displacement pumps are recommended. Centrifugal pumps are not recommended

because of their potential to damage floc formation and the difficulty maintaining a constant feed rate when using a variable orifice mixer. As a good practice, one pump per press should be provided for uniform loading to each press. For multiple-press installations, interconnecting piping and valves are needed for redundancy and reliability. Feed controls typically are incorporated into the main belt press panel.

As with other solids-handling systems, smooth-lined pipe (e.g., glass-lined ductile iron or steel, PVC ceramic epoxy) can be used for the dewatering system's piping. Pressures, velocities, and plugging all require consideration. Velocities should be maintained at 1 m/s (3 ft/s) or higher to prevent solids deposition and clogging problems. Cleanouts and flushing connections are needed at bends and tees.

Piping systems should include multiple locations for polymer injection so operators can vary the detention time between polymer addition and dewatering, as needed, for best results. Ideally, polymer-injection locations should be spaced at 15-second intervals along the piping system.

3.4.2 Conditioning System

The upstream feed-piping system should include several taps or spool pieces (i.e., injectors and/or mixing equipment). The contact time between conditioning chemical and sludge affects dewatering performance. If the feed piping is too short to provide adequate mixing and flocculation, design engineers should consider adding a flocculation tank ahead of the belt press.

Variable-output, positive-displacement pumps are recommended for chemical metering. Pump output can be either manually or automatically adjusted via speed controllers or stroke-length positioners. For automated systems, the chemical-pump control would be integrated with the belt press control panel.

While polymer is the most common conditioner used with belt filter presses, other chemicals (e.g., ferric chloride) have been used. Lime also has been added for stabilization before dewatering, which affects press performance. These alternate designs often require nonstandard press components (e.g., special material for the belts) and should be reviewed with press manufacturers.

3.4.3 Cake Discharge

The cake produced on the belt filter press falls freely into a steep angle chute to direct it to a conveyor or directly to a disposal bin/truck. The cake from the belt press tends to be laminar in nature, but easily breaks when exposed to the energy from a conveyance system. The interface between the steep angle chute and the discharge point should be located to prevent accumulation of cake in the chute. Although no damage would occur to the belt filter press if cake accumulates in the steep angle chute, the excess cake could rub into the belt, requiring more cleaning pressure/flow.

As with the drainage system, the cake conveyance system associated with a belt filter press must be sized to handle the maximum capacity of the unit, not the expected operating capacity. In addition, conveyance systems that service multiple units must be sized to handle the maximum combined capacity.

3.4.4 Belt Cleaning System

3.4.4.1 Discharge (Doctor) Blade Often called a doctor blade, the discharge (scraper) blade is typically a knife edge constructed of ultrahigh-molecular-weight plastic. It typically is located at the outlet end of the high-pressure section to scrape or peel dewatered solids from the belt into the cake disposal or conveyance system. Worn or poorly adjusted blades reduce belt life and deteriorate the belt seam. A blade-tension system can adjust the pressure exerted by this blade against the belt, as well as the angle at which the blade touches the belt. The blade-tension system's components should be made of corrosion-resistant material (e.g., polycarbonate) and inspected frequently.

Doctor blades are considered a wear item and should be removable for easy replacement.

3.4.4.2   Belt-Wash System   After cake is discharged, the part of the belt that was in contact with solids should be washed before returning to the pressing zones. This belt-washing system consists of piping, nozzles, drip pans, and spray-containment shields. A belt-wash station typically is provided for each belt. The belt-wash pipe and nozzle, housed in either a stainless steel or a fiberglass enclosure, provides a high-pressure water spray to cleanse the belt of any dried or residual solids, grease, polymer, or other material that blinds the mesh. Self-cleaning nozzles are suggested; however, most manufacturers provide a manual cleaning feature that includes a handwheel-operated brush internally mounted in the nozzle header pipe. Spray piping and nozzles should be adequately braced and pressure rated to withstand the pressure transients caused by sudden valve closures.

3.4.5   Washwater

A reasonably clean washwater supply is needed to ensure adequate belt cleaning, especially when dewatering secondary WAS and scum, which tends to rapidly clog the belt. This water supply, which amounts to 25% to 50% of the flowrate to the machine, typically is pressurized up to 700 kPa (100 psi). Booster pumps are needed if the source water pressure is below 85 psi (these pressures may vary slightly with each manufacturer). Belt washwater can be potable water, secondary effluent, or even recycled filtrate water, although a clean supply is preferable.

A good rule of thumb to determine the amount of washwater required is 1.3 L/s (20 gpm) per meter of belt per belt. Therefore, a 3-meter, 3-belt unit would require about 11.4 L/s (180 gpm), while a 3-meter, 2-belt unit would require 7.6 L/s (120 gpm).

Some manufacturers now provide washwater recycling systems. The system generally consists of a washwater recycle tank, pressure boosting pump, strainer, and miscellaneous controls. Belt filter presses equipped with these systems are able to isolate the washwater from the filtrate. The washing stations in the belt filter presses are modified to collect the spray water on each spray box and redirect it to the recycle tank instead of being sent with the filtrate to the head of the plant for re-treatment. The strainers remove the larger solids particles in the recycled water. Any solids that pass the strainer accumulate in the recycle tank and need to be flushed out periodically. This flushing water constitutes the washwater demand, significantly reducing the volume of washwater needed and filtrate returned to the wastewater facility.

4.0   Recessed-Plate Filter Presses

4.1   Description of Technology

Pressure filter presses (commonly called recessed-plate filter presses or plate-and-frame filter presses) have been in widespread use in the United States since the 1970s, evolving from a labor-intensive batch process to a partially automated operation in the presses of today. The reduction in overall labor requirements has made the filter press a more practical solids dewatering option for many facilities.

Pressure filtration uses a positive pressure differential to separate suspended solids from the liquid slurry. Recessed-chamber filter presses are operated as a batch process, whereas solids are pumped to the filter press and subjected to pressures ranging from 700 to 2100 kPa (100 to 300 psi) to force the liquid through a filter medium. This high-pressure action leaves a highly concentrated solids cake trapped between the filter cloths covering a series of recessed plates. The filtrate drains into internal conduits and collects at the end of the press for discharge. The plates are separated at the end of the fill cycle and the cake is discharged via gravity into a conveyor, collection hopper, or truck.

At operating pressures greater than or equal to 1600 kPa (225 psi), the recessed-plate filter press' pressures are expressed in atmospheres (bars). A machine rated at 1600 kPa (225 psi) would be a 15-bar unit. Similarly, a pressure filter press rated at 2100 kPa (300 psi) is a 20-bar filter press.

Filter press dewatering is both a constant-rate and a constant-pressure process because of the boundary conditions set by the type of equipment used and the complex, unpredictable interrelationships of the process variables. The beginning of the cycle uses a constant filtration rate up to the maximum pumping head available and then switches over to constant-pressure filtration until the rate diminishes to a predetermined low level.

Advantages and disadvantages of recessed-plate filter presses are shown in Table 22.5.

Most of the pressure filter press installations in the United States are semi-mechanized and use a fixed-volume chamber. The mechanized and automated systems typically are being replaced with more reliable manual systems, although other fully mechanized and automated filter presses continue to be developed and marketed.

In terms of both capital and O&M costs, the pressure filter press system typically remains more expensive than other dewatering alternatives; however, when disposal requirements dictate drier cakes, pressure filter presses often have been proven cost-effective. Moreover, many landfills have adopted more stringent criteria for the moisture content of solids cakes; often, dewatered cake is required to contain more than 35% solids before it can be landfilled. Other dewatering devices cannot reliably and routinely meet this requirement. Pressure filter presses also are cost-effective when the dewatered cake must be incinerated. Often, the drier filter press cake (which increases

Advantages	Disadvantages
- High cake solids achievable	- Batch operation
- High capture rates	- High equipment cost
- More cost effective than other technologies when drier solids are required	- Higher operating costs
- Can adapt to wide range of solids characteristics	- Special support structure requirements
	- Large footprint
	- High labor requirements

TABLE 22.5 Advantages and Disadvantages of Recessed–Plate Filter Press Dewatering

the ratio of volatile matter to water content) enables autogenous combustion in incinerators, thus reducing the need for other fossil fuels (e.g., natural gas or fuel oil).

The typical filtration cycle is characterized by temporal variations in flowrate, pressure, and solids loading. Design capacities and controls further define these relationships by limiting the maximum pressure under which the system will operate, and the maximum and minimum flowrates based on the design limitations of the feed system selected. Figure 22.8 illustrates the typical relationships between feed rate and pressure during a filter press cycle, occurring at various times within the cycle, represented by the x-axis, θ_c, in the figure. The portion of the cycle in which wastewater solids are actually applied to the filter is called the form cycle (P_c) and consists of the initial fill and increasing cake formation stages. During the form cycle, resistance to filtration remains relatively low and constant, until enough solids collect on the media to fill the pressure chambers. The form cycle is characterized by high, constant feed rates and relatively low pressure.

As solids accumulate and resistance to filtration builds, the flowrate declines and pressure increases up to the design maximum pressure. Solids accumulate at a relatively high but steadily declining rate until the cake experiences a significant change in porosity, which severely restricts the amount of flow discharged as filtrate. Thereafter, because of the increased resistance to flow through the cake, pressure will continue to increase (but more slowly) while the flowrate will continue to decrease (also more slowly). The system's pressure will continue to increase until the set point pressure is reached; thereafter, system pressure remains relatively constant, while the flowrate continues to decline. Meanwhile, water will still trickle through the cake and solids will continue accumulating in proportion to the flowrate (if the particle concentration remains constant).

Pressure filtration is affected by several factors (e.g., particle size, specific gravity, and particle concentration). If solids particles were all the same size, the resulting cake would be loosely packed and relatively unstable (like a stack of marbles), particularly if the cycle incurred a large pressure drop. If the particles were relatively flat (platelike),

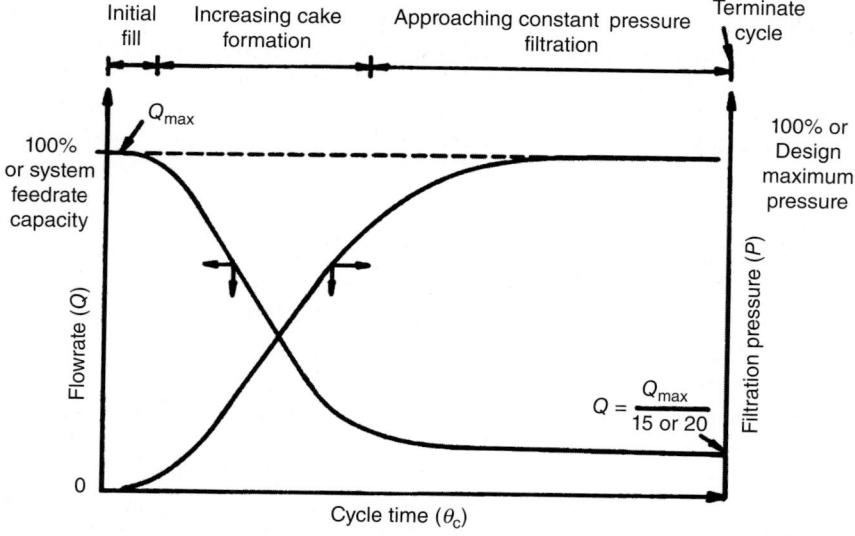

FIGURE 22.8 Filter-press cycle relationships.

the resulting cake would resemble a relatively impervious envelope with a highly fluid, moist center. However, wastewater solids consist of a wide variety of particle sizes and shapes that, under certain conditions, help keep an open matrix of particles, thus promoting free filtration. Ideally, the voids between larger particles can be filled with smaller particles, and adequate flow channels will exist between individual particles and throughout the entire physical structure to promote and maintain free filtration. For most wastewater solids, the solids must be conditioned before entering the filter press. Creating an open matrix of biological solids is difficult; biological solids are gelatinous, leaving relatively small void spaces for filtrate even when they are well conditioned. If the feed solids consist of a large fraction of biological solids, lime may be needed to ensure that the solids matrix will have adequate flow channels.

If low concentrations of solids particles with a wide range of specific gravities are pumped into the filter, they may settle in the lower chambers of the press, resulting in poor cake formation and unbalanced cake pressures. As solids concentration increases, the viscous drag created among particles inhibits coarser solids from settling out, unless there is a significant difference in specific gravity. This effect becomes less important when the feed consists of fine particles.

Particle concentration in the solids has a significant effect on the filtration cycle time. A feed with a higher solids concentration will increase cake yield and decrease cycle time.

4.1.1 Type of Press

Two types of filter presses typically are used to dewater wastewater solids. The most common is the fixed-volume, recessed-chamber filter press. The other is the variable-volume, recessed-chamber filter press (also called the diaphragm filter press).

4.1.1.1 Fixed Volume   The fixed-volume, recessed-chamber filter press basically consists of a number of plates rigidly held in a frame to ensure alignment. These plates are pressed together either hydraulically or electromechanically between a fixed end and a moving end (see Figure 22.9). The plates contain drainage surfaces, drainage ports for

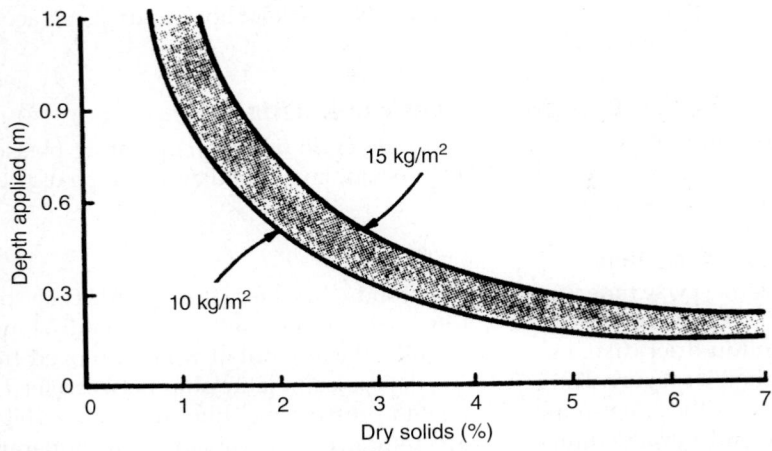

FIGURE 22.9   Bed depth required to obtain optimum loading at various solids concentrations.

discharging filtrate, and a large centralized port for solids feed. A filter cloth covers the drainage surface of each plate and provides a filter medium. A closing device presses and holds the plates closely together while feed solids are pumped into the press through the inlet port at pressures of 700 to 2100 kPa (100 to 300 psi). The filter cloth on each plate captures suspended solids and permits the filtrate to drain through the plate drainage channels. A backing cloth (also referred to as underdrainage), made of a rigid cloth or PVC, is used to keep the filter cloth separate from the drainage channels and ports during the high-pressure cycle. The backing cloth is the size of the recessed portion of the plate and is held in place by pins. Solids collect in the chambers until a practical low feed rate limit is reached (typically 5% to 7% of the initial flowrate) and the filter cycle is terminated. Then, the filter press feed pump is stopped, the individual plates are shifted, and the sludge cakes are discharged.

4.1.1.2 Variable Volume The variable-volume press incorporates a flexible membrane across the face of the recessed plate. As with fixed-volume presses, the initial stage involves feeding sludge to the press. However, once the plate chamber is filled and the filter-cake formation has started, the membrane is pressurized (between 600 and 1000 kPa [85 and 150 psi]) with compressed air or water, compressing the filter cake within the plate chamber. Typically, the squeeze pressure is kept relatively low and water is used as the pressurizing media for safety reasons. Diaphragm units are operated at filtration pressures of 700 kPa (100 psi). The physical compression of the filter cake increases the dewatering rate and shortens the cycle time. The results are higher production rates and more flexibility in achieving desired cake solids concentrations.

The variable-volume press significantly differs from the fixed-volume press in that the volumetric capacity is typically less, cakes are much thinner, and the press is more highly automated.

A variation of the variable-volume pressure filter press involves pumping hot water through the plate, warming the solids to 60°C (140°F) or higher, while pulling a vacuum on the chamber (to 135 kPa [4 in. mercury] absolute) where the solids are being dewatered. Thermodynamics show that water evaporates at lower temperatures under a vacuum. This variation takes advantage of this principle to obtain a drier cake. The filter press is operated identically to a conventional variable-volume press, including the compression. The length of time the solids are kept hot and under a vacuum determines how dry the material will be.

4.2 Process Design Conditions and Criteria
The principal design elements include cycle time (θ_c), operating pressure, number of plates, feed method, type of feed system, layout and access, type of press, mechanical features, and safety.

4.2.1 Cycle Time
The total cycle time is governed by solids characteristics, desired cake solids concentration, and the relationship between feed rate and pressure. Generally, longer cycle times result in drier cake, but after a while, the amount of water removed from the cake is negligible. Figure 22.8 illustrates the typical relationship between feed rate (which is equal to the water-removal rate) and pressure in a filter press cycle. Filtrate discharge rate and filtrate volume for each filter cycle are valuable control parameters in filter press operations, so a flow-measurement system with a recorder to plot filtrate flow

versus time and a totalizer to sum filtrate volume should be included with each filter press. The shape of the filtrate flow curve indicates the feed solids' dewatering characteristics and allow operators to note changes in the filtrate curve over multiple filter runs that may suggest chemical dose adjustments or filter media blinding. The filtrate flow curve also can be used to indicate when to terminate the filter cycle. For specific solids at constant conditioning-chemical dosages, a desired cake solids concentration can be calculated as follows:

$$S_T = S_I + \frac{V_T}{V_F}(S_F - S_I) \qquad (22.3)$$

where S_T = cake solids concentration (% dry solids);

S_I = initial feed solids concentration (% dry solids);

S_F = reference cake solids concentration (% dry solids);

V_T = cumulative filtrate volume at S_T (L); and

V_F = cumulative filtrate volume at S_F (L).

Using a family of cumulative filtrate curves, operators terminate a cycle when its flowrate matches the proportion of the initial feed rate corresponding to the preselected cake solids concentration. Alternatively, for a given feed solids concentration, filter press chamber or cake volume, and desired cake solids concentration, operators can use eq 22.3 to estimate the required filtrate volume (solids feed volume) and terminate the filter cycle when the filtrate totalizer reaches the calculated value.

Because of the wide range in filtrate flowrates, it is important to have a flow-measuring device that has a reasonable degree of accuracy over the full range—particularly at low flows. Parshall flumes and V-notch weirs have good characteristics for measuring a wide range of flows and are well suited for filter press use. Bubbler tubes, ultrasonic sensors, and capacitance probes have been used to accurately read the levels produced in the Parshall flume or V-notch weir for conversion to the respective flowrate. Depending on the filtrate characteristics and the filtrate piping configuration, foam can be generated and occasionally can cause difficulty in obtaining accurate level readings. Generally, foam is caused when the filtrate from the press goes through severe turbulence. Bubbler tubes are not affected by foam and produce reliable results. Ultrasonic sensors are affected by foam, however, producing inaccurate level readings.

4.2.2 Operating Pressure

Pressure is the driving force for filtration in these systems. Filter press systems typically are designed to operate at 700, 1600, or 2100 kPa (100, 225, or 300 psi). Researchers cite successful applications at all operating pressures; however, increases in filtration pressure can simultaneously increase cake resistance if the solids are compressible. Excessively high pressures can inhibit the process by tightly packing the solids, thereby reducing cake porosity and increasing resistance. Higher cake resistance reduces filtration flowrates. Experience at several installations has demonstrated that close attention to filter media selection and proper solids conditioning will often overcome these problems.

Difficult-to-dewater industrial solids sometimes respond well to increased dewatering pressure. Increasing pressure on highly compressible solids often can decrease the rate of compression.

4.2.3   Number of Plates

The number of plates in a filter press affects the overall efficiency of recessed-plate filter presses. When well-conditioned solids are filtered in a press with a large number of plates, solids distribution throughout the filter chambers may be poor. In this situation, chambers nearest the entry points begin filling and start filtering, while chambers toward the center or end of the press have not yet received solids. As a result, unequal pressures develop throughout the length of the press during the filter cycle, producing cakes of randomly poor quality that often do not meet design criteria. The unbalanced forces created by poorly formed cakes can warp and eventually break plates made of weaker materials.

4.2.4   Type of Feed System

The feed system must deliver conditioned, flocculated feed solids to the filter presses under various flow and pressure requirements. There are two ways to feed a pressure filter press, and the system should be capable of both. In the first (more typical) method, the feed system will complete the initial fill cycle by achieving an initial system pressure of 70 to 140 kPa (10 to 20 psi) within 5 to 15 minutes to minimize uneven cake formation. This can be done using separate fast-fill pumps, or running two feed pumps to one press during the initial fill cycle.

As the cake forms, the resistance to filtration increases, requiring higher pressures to feed the press. During this period, the feed system should provide a relatively constant high-solids feed rate at continually increasing pressure until the maximum design pressure is reached. Then, the solids feed rate decreases to maintain a constant system pressure.

The second method, although slower, achieves the same result. A lower flowrate is used to fill the press (typically less than half of the feed pump's capability). When pressure starts building to about half of the operating pressure, the feed pumps are ramped up to full flow and then are controlled by pressure (similar to the first method). This method has been used with coarser filter cloths to prevent blinding at the high initial flows used in the first method.

4.2.5   Cake Solids and Capture Efficiency

Recessed-plate filter presses typically can achieve capture efficiencies from 95% to more than 99%, depending on the nature of the solids. These capture efficiencies do not include the water used to wash the plates at the end of a filtration recycle, which also contributes solids to the recycle stream from the dewatering operation. Table 22.6 summarizes typical performance data of recessed-plate filter presses for different types of solids.

4.3   Design Considerations

4.3.1   Mechanical Features and Materials of Construction

Both fixed-volume and variable-volume filter presses are reliable when appropriate operations and maintenance activities are performed. The main operational difficulty encountered in pressure filter installations is inconsistent separation of the cake from the filter media. This problem may indicate the need to wash the filter media or increase chemical doses.

The main mechanical components of filter press equipment include the structural frame, filter press plates, diaphragms, filter cloths, and plate shifters. Various options are available for each component. In many cases, individual manufacturers provide

Type of Sludge	Cycle Time[c] (min)	Conditioning Chemicals		Feed Solids Concentration (%)	Cake Solids (%)
		Lime[d] (%)	Ferric Chloride (%)		
Raw primary	120	10	5	5 to 10	45
Primary/WAS (50:50)	150	10	5	3 to 6	40 to 45
Primary + Tricking filter sludge	120	20	6	5 to 6	38
Primary + Ferric chloride[a]	90	10	—	4	40
Primary + WAS[b]	180	10	5	8	45
WAS	150	15	7.5	5	45
Digested primary	120	30	6	8	40
Digested primary + WAS	120	10	5	6 to 8	45
Digested primary + WAS[b]	180	10	5	6 to 8	40
Digested primary + 50% WAS	120	10	5	6 to 10	45
Digested primary + 50% WAS	150	15	7.5	1 to 5	45

[a]Ferric chloride added at primary settling tank as a coagulant.
[b]Ferric chloride used as coagulant aid in the secondary treatment process.
[c]Cycle time is time from initiation of feed to feed pump discharge; time excludes cake discharge time.
[d]As CaO.

TABLE 22.6 Performance Characteristics of Recessed-Plate Filter Presses (Adapted from Table 15.3 in WEF, 2012)

only one option for a particular component. Design engineers should carefully evaluate all requirements of a contract that may either specify or exclude a specific option. Figure 22.10 illustrates the main parts of the recessed-plate filter press.

4.3.1.1 Structural Frame The filter press structural frame has a fixed end, a moving end, and either plate-support bars on each side (sidebar type) or one or more beams on the top of the unit (overhead type). The fixed end anchors one end of the filter press, as well as the plate-support bars. The moving end anchors the opposite end of the filter press and houses the closing mechanism. The plate supports span the fixed and moving ends to carry the filter press plates and the shifting mechanism that moves individual plates for cake discharge. Larger filter presses use intermediate supports midway between the fixed end and the moving end to provide more rigidity and strength for the plate-support bars.

The sidebar support bar supports each side of the plate at a point above the plate center. The overhead support bar suspends each plate at the top center of the plate. The sidebar option allows removal of individual plates from the structural frame by lifting directly out of the structural frame. The overhead option allows easier access to, and observation of, individual plates mounted on the structural frame. Because the overhead option supports each plate from one point rather than two, it simplifies plate closing and shifting operations.

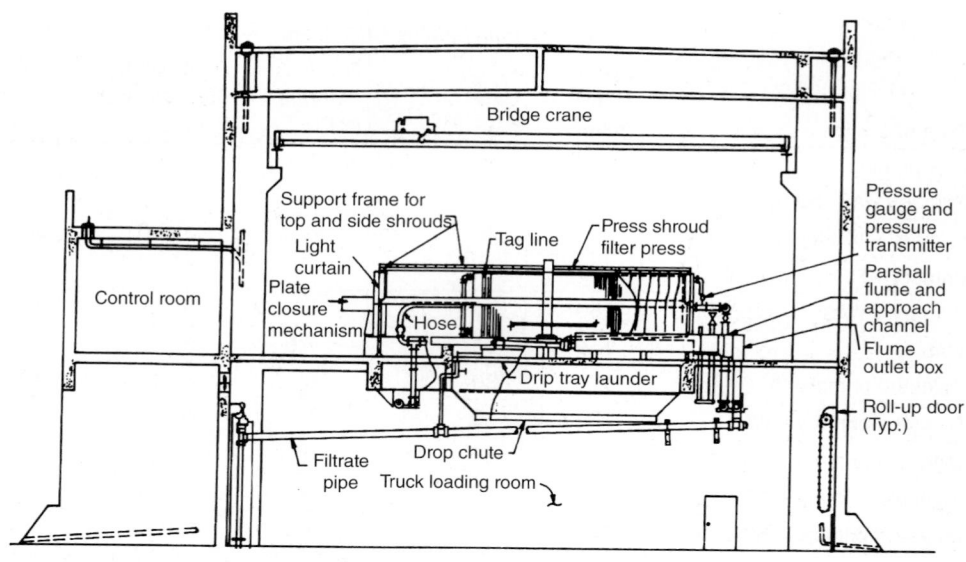

FIGURE 22.10   Sectional view of a filter press building.

4.3.1.2 Filter Press Plates

Filter press plates are available in several types of constructions, dimensions, and materials. Recessed plates are used almost exclusively in solids dewatering applications. Recessed plates are fabricated with a constant recess depth and area formed on both sides of the plate and the filter cake forms in the volume contained within the recessed area of two adjacent plates. Additional interior supports (called stay-bosses) have the same overall thickness as the plate perimeter to prevent deflection of the plate in the recessed area. The number and size of stay-bosses, which have a truncated cone shape, are a function of the dimensions and structural material of the plates. The face of the plates is machined to close tolerances, and the stay-bosses are similarly machined. Uneven surfaces will cause filter-cloth wear, plate-stack shifting during press closure, and frequent blowouts.

Filter press plates can be round, square, or rectangular, and range in length and width from 0.5 to 2.6 m (1.6 to 8.5 ft). The plates typically are constructed with a top-center or center feed port with filtrate ports located at the corners of the recessed area. The surface of the plates at the perimeter and stay-bosses is flat, to provide a seal when the plates are closed. The surface of the recessed area of ductile iron plates is typically constructed with rows of drainage channels, while the recessed area of plastic plates typically is constructed with cylindrical pipes. The channels or pipes provide support for the filter cloths, and the gaps between them provide paths for filtrate to drain to the filtrate ports.

Filter press plates can be made of epoxy-coated steel, rubber-covered steel, cast-iron, ductile iron, and polypropylene. Epoxy-coated steel plates have a low initial cost with good strength and moderate weight, but are susceptible to corrosion if the coating is not maintained. Rubber-covered steel plates also have a higher initial cost than epoxy-coated plates, but also provide good strength and moderate mass. The rubber covering, which is molded to the plate, offers excellent chemical and corrosion resistance provided that its integrity is maintained. Pinholes and delaminations in the rubber covering can result in corrosion problems. Cast-iron (not readily available) and ductile iron plates offer superior strength with reasonable chemical and corrosion resistance. However, they have the highest initial cost and weigh considerably more than steel plates. Polypropylene plates have the lowest initial cost of all types of plates, offer excellent chemical and corrosion resistance, and are lightweight for easier handling. Although the inherent strength of polypropylene is less than steel or ductile iron, the plates are thicker and have more or larger stay-bosses to compensate.

Design engineers should consider both plate mass and strength during the design phase. Because the mass and the strength of each plate material are interrelated, design engineers must consider the trade-off between more mass–less strength or less mass–greater strength.

The mass of plates affects not only the ease of handling for inspecting, cleaning, and changing filter cloths but also the overall cost and mass of the filter press and associated structural requirements.

Plate strength is critical. Filter presses operating between 700 to 2100 kPa (100 to 300 psi) can impart tremendous forces on plates if feed solids are distributed unequally, causing voids on one side of the plate. Unequal distribution can cause plate deflection and deformation, blowouts, and filter-cloth wear. This effect is magnified as plate size increases. Plates constructed of lower-strength material tend to be thicker and have greater stay-boss area, reducing the volume for filter-cake formation and requiring a longer plate stack to dewater a given volume of sludge solids. The longer the stack, the larger the structural frame and the greater the equipment footprint.

4.3.1.3 Diaphragms A variable-volume press uses a flexible membrane to apply additional pressure to the dewatered cake. Toward the end of the filtration cycle, the membrane is inflated with air or water to apply the extra pressure and subsequently deflated before the plates are separated. The membranes are typically made of rubber or a synthetic plastic and are subject to wear.

The variable-volume recessed plate is a newer development in filter press plate construction. With variable-volume recessed plates, the diaphragms can either be incorporated with the plates (less expensive initially) or removable (more expensive, but a diaphragm failure will not cause the entire plate to be replaced). In addition, the diaphragm may be installed on one side of the plate or on both sides. If replacing fixed-volume plates with variable-volume plates, it may be desirable to use plates with the diaphragms on one side only. This reduces the overall stack length compared to those with diaphragms on both sides of the plates.

4.3.1.4 Filter Cloths Filter media require routine cleaning via high-pressure water spray, closed-circuit acid wash, or both. Although many factors can affect the overall performance of filter cloths, the initial selection of filter-cloth media influences subsequent filter-cloth performance. Important factors to consider when selecting filter-cloth media are durability, cake release, blinding, and chemical resistance. The durability of filter cloths and the cake release from the cloths is affected by the media material and construction.

Filter cloths are available in a variety of materials, weaves, and air permeabilities. The most widely used material for solids dewatering is polypropylene with nylon; saran is also used. Filter cloths are fabricated with monofilaments into a plain or twill weave, with the fabric strands having different monofilament diameters in the warp and waft to achieve particular filter-cloth characteristics. The air permeability of filter cloths is a measure of the openness of the weave as determined by airflow through a unit area of media at a given pressure drop (e.g., 1800 m^3/m^2·h at 0.1 kPa [100 cu ft/min/sq ft at 0.5 in. water]). Although the permeability changes during use as solids impregnate the filter cloth, swelling the material and distorting the weave, it does serve as a useful parameter for the initial selection of filter media. A low-permeability rating (less than 900 m^3/m^2·h [50 cu ft/min/sq ft]) will yield high solids captures, but has an increased tendency toward media blinding, poorer cake release, and cleaning difficulties. Medium permeability ratings (900 to 5500 m^3/m^2·h [50 to 300 cu ft/min/sq ft]) yield good solids captures without excessive cloth blinding and provide good cake release. Higher-permeability ratings (greater than 5500 m^3/m^2·h (300 cu ft/min/sq ft]) provide advantages when treating difficult-to-dewater solids, where cloth blinding and cake release are critical.

After the initial stage of operation in a filter press cycle, solids buildup begins and the filter cake itself starts to serve as the filtration media. As a result, although a low-permeability cloth will result in relatively higher solids capture in the initial stage of a filter press cycle, once cake formation has started, the efficiency of solids capture is generally independent of the filter media.

Filter cloths are sometimes reinforced at the stay-bosses and plate perimeter to improve wear resistance. Such reinforcement can consist of a double layer of filter-cloth material, impregnation of the media with a coating, or insertion of a different material. Care must be exercised to ensure that the thickness of the material at the stay-bosses and the perimeter remain the same if the filter cloth is altered. Otherwise, the closing mechanism on the press can impart unequal forces on the plates, causing blowouts or plate deformation.

Proper attachment of filter cloths to the plates is critical to filter press performance. The cloths must drape across the face of the plate without creases and must remain in place through multiple filter press cycles. This is particularly important if the cloth has been reinforced at the stay-bosses and perimeter, and the cloths must be exactly aligned. A sewn tube, which connects two filter cloths, is inserted through the feed port of each plate. This tube is often impregnated with a waterproof coating to prevent pre-filtration from occurring in the feed port; rubber feed tubes are also available. The filter cloths are secured with grommets around the perimeter and fastened with ties to the filter cloth on the opposite face of the plate. It is desirable to use a sewn loop at the top edge of the filter cloth through which a rod is inserted to support the filter cloth. This method provides uniform support across the top of the filter cloth rather than point supports at the grommet openings, which may promote creases in the cloth.

The choice of filter media is the most important equipment variable affecting cake quality and release, filtrate quality, and filter yield. Tightly woven cloths improve initial filtrate quality, but may extend filter cycles and result in difficulties in cake release. Open-weave media typically facilitate cake release during cake discharge but extend the initial cake-formation time and reduce initial filtrate quality. Moreover, although multifilament-fabric media typically improve filtrate quality by entrapping solids, they are susceptible to blinding and may result in poor cake release during discharge. Mono-filament-type fabric construction facilitates cake releases during discharge and typically is easier to clean and to maintain. Larger plates (greater than 2 m × 2 m) require special consideration to prevent stretching of the cloth. In addition, the filter chambers can be precoated with porous materials (e.g., fly ash) to serve as a filter body for fine solids particles and to promote release of the finished cake from the filter cloth.

4.3.1.5 Plate Shifters At the end of the press cycle, a plate-shifting mechanism (referred to as plate shifters) moves each filter press plate one by one to release the filter cake enclosed in the recessed areas between the plates. The shifting mechanism is housed in the plate-support bars and operates via either a continuous chain or a reciprocating bar. Pawls attached to the chain or bar automatically engage the plate at the end of the plate stack and slide it along the plate-support bar 0.6 to 1.0 m (2 to 3 ft). As each successive plate is separated from the end of the plate stack, the filter cake in the corresponding chamber is freed and drops from the filter plate. Typically, reciprocating-bar shifters are used on sidebar filter presses, and the endless-chain shifter is used on overhead-beam filter presses.

4.3.2 Instrumentation and Controls

Typically, the recessed-plate filter press comes with a control system for the pressure hydraulic opening and closure system, the plate shifter, and an interlock for the filter press feed pumps. Controls for the target feed rate and discharge pressure are typically included in the controls for the sludge feed pumps to allow variation in feed pressures and flows to accommodate varying feed sludge influent quality. Otherwise, automatic controls are provided to automatically increase and maintain the filter press pressure up to a set point maximum pressure and then shut down the system, so that the plates can be opened and the sludge cake discharged.

4.3.3 Area/Building Requirements

The design of filter press dewatering facilities should consider the size and weight of the filter press equipment and associated support systems. The size of the filter press room is dictated not only by the size of the filter press but also by the clear space around the filter press necessary to facilitate cake release, remove plates, and perform appropriate maintenance operations. Generally, at least 1 to 2 m (4 to 6 ft) of clearance is required at the ends of the filter presses, and clearances of 2 to 2.5 m (6 to 8 ft) between filter presses are desirable. Height clearances must be sufficient for removal of the plates via a bridge crane. Filter press installations with a sidebar design may use the bridge crane to remove each plate to make cloth removal and replacement easier; this critical maintenance procedure typically is an annual event.

Design engineers should consider how the filter presses will be installed in and removed from the building at the end of their service lives (which can be more than 20 years). In addition, provisions for installing future filter presses typically are included in building designs. The building should have openings large enough to allow major filter press components (e.g., the fixed end, moving end, and plate support bars) to be passed through. An overhead bridge crane rated to lift the heaviest individual filter press component for maintenance should also be installed. The bridge crane is also used to lift filter press plates for removal, replacement, or inspections.

An elevated platform on one or both sides of the filter press may be necessary to facilitate cake release and equipment inspection. If the press is mounted at floor level, structural supports to mount the press over discharge openings would be required. Sufficient floor area near the filter presses should be provided for storage of spare filter plates, filter cloths, and other spare parts. Clearance for truck-loading facilities (where applicable) should be provided for a wide range of possible vehicles, with a minimum vertical clearance of 4 m (12 ft) recommended.

All building areas should be prevented from freezing, localized heaters should be provided where work activities are concentrated, and control rooms should be designed to meet office-environment conditions. If rubber-coated, steel filter plates are used, the filter press and plate-storage areas must be kept above 4°C (40°F) to avoid damage to the rubber-covered plates via thermal contraction.

Cake-conveying systems consistently pose housekeeping problems for filter press installations as cake has the tendency to be scattered inadvertently on some systems. Design engineers minimize conveyance problems by reducing the number of cake-transfer points, reducing cake drop distance, adding flexible discharge chutes at each transfer point, and installing skirt boards to contain cake on the conveyor.

Cake-breaking cables or bars typically are provided beneath the filter presses to break up the filter cakes as they drop from the plate chambers. The cables or bars typically are spaced 300 to 600 mm (12 to 24 in.) apart and aligned parallel to the length of the filter press.

Blowout curtains are a highly desirable housekeeping feature for filter press operations. Filter press blowouts occur when the plates do not seal properly. During blowouts, a high-pressure stream of solids can be emitted in any direction. Blowout curtains are positioned over the top and sides of the filter press and can be mounted on a frame supported from the filter press. The top curtain or canopy is fixed in place, and the side curtains are designed to slide to the side to provide ready access to the filter press plates at the end of a filter press cycle. Larger filter presses typically have considerably fewer plate blowouts than smaller filter presses.

4.3.3.1 Structural Requirements The structural load imposed on buildings by recessed-plate filter presses is substantial. Some filter presses use the building structure to provide the support to close the plate stack and maintain sufficient pressure during high-pressure operations. Other systems are designed to withstand all forces internally so only loads in the vertical direction are imparted to the building structure. It is desirable that horizontal loads be self-contained within the structural frame. Improper alignment of filter presses can warp the structural frame and twist anchor bolts. The manufacturer should verify that the filter press is properly installed and aligned before it is placed in service.

4.3.3.2 Drainage Requirements Under the filter press and cake conveyors, drip troughs wider than the filter and conveyors can collect any spillage. Such troughs should be U- or V-shaped to facilitate washdown and drainage while conveying drainage from the filter press to the building drain system. Liquid is released at the end of each filter press cycle from the filtrate ports, filter-cloth cleaning operations, plate leakage during filtration, and during general equipment washdown. Drip trays are hinged single-leaf or double-leaf type that are sloped to one or both sides for drainage to a launder trough parallel to the length of the filter press. If filter presses discharge filter cake directly to outdoor loading areas, the drip trays also serve as a barrier to the outdoors.

4.3.3.3 Energy Requirements Significant amounts of energy are required to pressurize the units, typically on the order of 0.04 to 0.07 kWh per kilogram of dry solids processed.

4.3.3.4 Safety With filter presses, the paramount safety consideration is avoiding inadvertent plate shifter or moving-head operation while an operator is physically between the plates assisting cake discharge. Most filter press installations use an electric light curtain on both sides of the unit. A light curtain consists of several vertically stacked photoelectric (or infrared) cells to guard each side of the filter press that are automatically activated when the plate-shifting mechanisms are engaged. If an operator interrupts the light beam from the photocells during filter press opening or closing or during a plate-shifting cycle, controls will temporarily stop the mechanism until the light beam is restored to protect any foreign object—including parts of a worker's body—from being caught between the plates. In addition, a tag line along the operating side of the filter press enables operators to stop the plate shifter manually and then resume operation at their discretion.

4.4   Ancillary Systems

The difficulty of designing and controlling a pressure filter press system is the number of ancillary systems that must be coordinated and operated for successful system performance. In fact, the ancillary equipment sometimes requires more space and operational effort than the press itself.

4.4.1   Feed System

Filter press feed systems must deliver conditioned, flocculated feed solids to filter presses under varying flow and pressure requirements. Feed system components include precoat, rapid fill, pressurization, and cake removal.

4.4.1.1   Precoat System   Solids with high biological content or difficult-to-dewater industrial sludges tend to stick to the filtration media. The precoat system aids cake release from the media and protects the media from premature blinding. Precoat material can be fly ash, incinerator ash, diatomaceous earth, cement-kiln dust, buffing dust, coal, or coke fines. A thin layer of this material is deposited over the filtration surface before each filtration cycle begins. There are two types of precoat systems: dry-material and wet-material feeding.

Dry-material precoat feeding systems are used at larger installations, particularly those that operate continuously. A precoat pump draws clear water from a filtrate-storage tank (or other reasonably clean source), circulates it through the filter press, and returns it to the tank. Once the filter press is full of water and all air has been evacuated, a predetermined amount of precoat material is transferred from a storage hopper to a closed precoat tank. The water stream is diverted through the precoat tank and, aided by a baffle arrangement inside the tank, forces the slurry of precoat material out of the tank and deposits it on the filter medium. To achieve uniform precoating, clean water must be circulated through the filter at high rates. The entire precoat cycle lasts between 3 and 5 minutes before each filtration cycle. Precoat material requirements range from 0.2 to 0.5 kg/m^2 (5 to 10 $lb/100$ sq ft) of filter area; 0.4 kg/m^2 (7.5 $lb/100$ sq ft) is typical.

Wet-material feed systems are used at smaller filter press installations. This system consists of a precoat preparation tank into which water is metered and the proper amount of dry precoat material is added, while an agitator keeps the precoat material in suspension. The precoat-material pump circulates the material from the bottom outlet and discharges it back into the same tank. At the beginning of the precoat cycle, water from a filtrate-storage tank is pumped through the filter press and returned to the tank. After the filter is completely filled and all air is expelled, the precoat-material pump injects the precoat slurry into the piping on the suction side of the precoat pump, and the precoat material is uniformly distributed throughout the filter.

4.4.1.2 Rapid Fill Two rapid-fill methods have been developed for filter press systems. The first method uses one pump or a combination of pumps with variable-speed drives to achieve the required flow and pressure characteristics. The second method uses a combination of pumps and pressure tanks to achieve the required flow and pressure characteristics.

The first method typically uses one variable-speed feed pump for each filter press. Automated controls vary the pump speed (maximum flow until system pressure is reached and decreasing flow to maintain system pressure). For large filter presses where initial flow requirements are high and available turndown of the variable-speed pump is too limited to operate from the minimum to the maximum flow, a second pump (either constant- or variable-speed) operates in parallel with the first pump to achieve the initial high-flow requirements. The second pump is controlled to shut down when flow requirements drop within the first pump's capacity range.

The second method typically uses one feed pump and one pressure tank for each filter press. At the start of the filter press cycle, the pressure tank is filled with feed solids and pressurized with air. To initiate the cycle, an automatic valve is opened, releasing the feed solids from the pressure tank into the filter press. This method achieves a rapid initial fill because the working volume of the pressure tank is designed to exceed the solids feed volume required for the initial fill. When the solids level in the pressure tank drops, the solids feed pump starts and runs until the working volume in the pressure tank is replenished. Air controls associated with the pressure tank operate to add or release air to or from the pressure tank to maintain the desired system pressure. The automatic outlet valve from the pressure tank closes at the end of the filter cycle to terminate the solids feed to the filter press.

Design engineers should note that the first method requires less building space (floor area and room height) than the second. However, the second method provides more rapid, positive initial fill. If pressure tanks are used, design engineers should take precautions to properly handle the air released from the pressure tank because of possible odor and safety problems.

4.4.1.3 Pressurization Plate pressurization is either hydraulic or electromechanical, closing and maintaining the necessary force to hold the plates closed during a filter press cycle. The hydraulic system consists of one or more hydraulic rams and a hydraulic power pack. The electromechanical system consists of a single or twin screw and an electric gear motor. Either system can be equipped with automatic controls to maintain a constant closing force throughout the filter press cycle. This feature is desirable because the closing force required may vary as the solids feed pressure increases, filter cloths and plates compress, and materials of construction expand or contract as the temperature changes.

Smaller filter presses typically only need one piston, mounted against a rigid support at the opposite end of the unit. This is a push-to-close unit, which imposes a large load on the building structure. One piston should not be used on a large plate stack, because the stack can shift if the plate faces are uneven.

Larger presses typically use a pull-to-close design, in which two or four rods extend from the front to the tail stand, and hydraulic cylinders "pull" the tail stand toward the front. In this design, only vertical loads are imposed on the supporting structure.

4.4.2 Chemical Conditioning Requirements

Solids conditioning involves adding lime and ferric chloride, polymer, or polymer combined with either inorganic compound to the solids before filtration. Most existing installations use lime and ferric chloride. Therefore, the solids conditioning system typically has lime slakers, lime-transfer pumps, lime slurry equipment, ferric chloride equipment, and a conditioning tank. Lime and ferric chloride are added to the conditioning tank on a batch basis. For installations that use only polymer, polymer is added in-line. However, polymer must be fed to match the flowrate of the solids feed pumps, so instrumentation and control are essential for polymer condition systems.

Several installations use only polymer for conditioning because their experience has shown that a slight decrease in performance is offset by lower chemical costs, reduced ammonia odors, and smaller volumes of dewatered cake. However, cake release from the cloth during the discharge cycle may be a problem with polymer-only systems. Several installations in Europe use ferric chloride to enhance cake release; however, combining ferric chloride with polymer may result in severe corrosion of piping and the press. When lime and ferric chloride are used together, lime neutralizes the corrosive ferric chloride. Installations that use polymer and ferric chloride should line all metallic surfaces with rubber.

Solids transfer pumps convey feed solids to a conditioning tank, where lime and ferric chloride are added. Ideally, the pumps withdraw solids from a holding or storage tank equipped with a mixing system. Such tanks allow operators to maintain an inventory of solids rather than scheduling solids wasting, allowing solids dewatering to be independent of the biological treatment process. The mixing system for the solids holding tank ensures a uniform solids feed concentration, ensures consistent chemical dosing requirements, and minimizes the possibility of chemical overdosing or underdosing.

The solids conditioning tank requires a mechanical mixer to thoroughly and gently mix solids with lime and ferric chloride to develop flocculated solids. Minimum retention time in the conditioning tank should be 5 to 10 minutes. A longer retention time of 20 to 30 minutes may be desirable to ensure that the lime completely reacts with the solids, minimizing lime scale in the solids feed piping and filter media. Longer retention times (greater than 30 minutes) tend to promote floc deterioration and breakdown as the solids floc ages. Conditioning is further complicated for installations with multiple filter presses and a single conditioning tank, as the retention time is affected by the number of filter presses in operation and the solids feed rates to each filter press at various stages of the filter cycle.

The conditioning system must be designed to match the filter presses' feed requirements. Two methods have been frequently used. The first method uses variable-speed solids-transfer pumps that maintain a nearly constant level in the conditioning tank by matching the tank's inflow and outflow rates. To maintain a constant chemical dose, the lime slurry and ferric chloride feed pumps also must be variable speed to proportionately match the rate of the variable-speed solids-transfer pumps. This method maintains a constant retention time in the conditioning tank, but requires a wider range in pump capacity, variable-speed drives and controls, resulting in additional operational complexity. The second method uses constant-speed solids-transfer pumps and varying levels in the conditioning tank. Solids are intermittently pumped to the conditioning tank at a constant rate, and lime and ferric feed pumps operate at constant, but manually adjustable, rates in parallel to the solids-transfer pumps. By varying retention time in the conditioning tank, this method increases conditioning-tank capacity, but requires less-complex controls.

The solids-transfer pump capacity and conditioning tank volume must be carefully sized to ensure that the filter press feed requirements are met. The lime slurry and ferric chloride feed pumps also must be carefully sized to meet the ranges of solids feed flow, solids feed concentration, and chemical feed concentration and provide the appropriate chemical doses.

4.4.3   Cake Removal

Filter press cake is a thixotropic material that can change from relatively firm, discrete pieces to a gelatinous, homogenous mass if the material is allowed to settle and compact over time. Consequently, the characteristics of stored cake rarely are the same as those of fresh cake. Storage bins that slope to a small opening are susceptible to bridging across the outlet opening, preventing cake release. Storage bins with vertical sidewalls and helical screws in the bin bottom were rendered useless at one installation because the distance between the outside edges of the screws was sufficient to allow the cake to mass and bridge over the screws. The screws "tunneled" through the mass and could not remove any cake. Therefore, cake-storage bins should be designed with steep sidewalls (vertical-to-horizontal slope greater than 5:1) and true "live bottoms" operating over the full width and length of the bin (e.g., chain-and-flight mechanisms or gauged helical screws with a minimum clearance of 625 mm [1 in.] between the outside edges of the screw flights). The live-bottom mechanisms should be provided with variable-speed drive capability to control loading to the next solids management process, or to trucks for conveyance off-site.

If the cake will be incinerated, then two approaches are common. The first approach is providing storage capacity beneath each filter press and metering the cake onto the incinerator feed system. A second approach provides intermediate storage between the filter presses and the incinerator.

A "core blow" is an optional feature for filter press applications. The "core" is the annular space through the press formed by the feed ports of the plates. At the end of a filter press cycle, the core is filled with liquid residuals that have not been dewatered. When the plates shift to drop the filter cakes, the residuals in the core also are discharged. These residuals tend to run down the face of the filter cloths, creating blind spots that may result in nonuniform cake formation. A core blow uses compressed air to force the residuals out of the core and back to the solids conditioning tank. Design engineers should consider the duration, flowrate, and air pressure required to provide such core blows, and carefully weigh the cost of the equipment, piping, and building space required for this option.

4.4.4   Cleaning Systems

4.4.4.1   Washwater Requirements   Filter media periodically are washed with water or acid when the filter press is shut down. As cakes degrade during normal operation, operators can determine when washing is required. The washing cycle depends on solids characteristics, the conditioning system, and filter-cloth weave. Some facilities wash with water after 20 cycles and wash with acid after 100 cycles. If solids are conditioned with polymer (and not with lime), acid washing is not required.

Filter media washing is essential to remove residual cake left over after normal cake discharge, liquid residuals from the feed-port core, solids and grease impregnated in the media, and scale and solids buildup on the filter-plate drainage surface. These materials must be removed to prevent the filter blinding and maintain atmospheric pressure between the filter medium and filtrate discharge. Backpressure reduces the effect of the applied pressure on the filtration rate.

Design engineers typically provide facilities for both water spray wash and acid wash. The water spray wash removes accumulated solids on the filter media, while the acid wash is used periodically to remove impregnated solids and scale buildup not readily removed by the water spray wash.

The least expensive and most typically used water spray-wash method is a portable spray-wash unit, which consists of a hydraulic reservoir, high-pressure wash pump, and a portable lance to direct the spray water. Operators direct the high-pressure spray (up to 13 800 kPa [2000 psi]) wherever buildup is observed. However, this method is labor intensive and tedious when cleaning larger filter presses.

Automatic spray-wash systems use a control system to automatically shift plates while an overhead spray-wash mechanism washes the filter media surface. High-pressure water-booster pumps provide satisfactory surface wash pressures. Although more expensive and complicated than the portable spray-wash system, the automatic system provides more thorough, efficient, and frequent media washing with less labor.

4.4.4.2 Acid Cleaning The acid-wash method cleans filter media in situ. A dilute solution of hydrochloric acid is pumped into the empty filter press with the plate pack in the closed position. The acid is either circulated through the plate chambers or detained in the plate chambers to clean the filter cloths. The acid-wash system typically includes a bulk-acid-storage tank, acid-transfer pump, dilution appurtenances, dilute-acid-storage tank, acid-wash pump, and associated valves and piping. Acid is furnished in carboy containers, tank trucks, or tank car shipments as 32% hydrochloric acid (muriatic acid) solution. A cleaning solution strength of about 5% is recommended, although specific experience may warrant slightly higher concentrations (up to a maximum of 10%).

5.0 Drying Beds and Lagoons

5.1 Description of Technology

Drying beds and lagoons use a combination of drainage, evaporation, and time to dewater solids. Both require considerable area as compared to other dewatering methods.

If well-designed and properly operated, drying beds are typically less sensitive to influent solids concentration and can produce a drier product than most mechanical devices (see Table 22.7). Particularly suited to small facilities in the southwestern United States, they can be used successfully in water resource recovery facilities of all sizes and in widely varying climates.

Compared to mechanical dewatering technologies, drying beds are more labor intensive. However, at smaller facilities, the higher capital and operating costs of mechanical dewatering systems have caused designers to take a second look at drying beds when adequate land is available and environmental conditions are acceptable.

The use of sand drying beds may be affected by concerns of groundwater contamination. Regulations in certain states prohibit unlined drying beds in many areas. The additional costs of bed lining and groundwater quality monitoring may make mechanical dewatering more cost-effective for all but the smallest facilities. Drying beds also contribute to higher wet-weather flows to the water resource recovery facility because they consist of relatively large areas that drain to the facility. Drying beds may, however, be a useful backup to mechanical dewatering methods.

Advantages	Disadvantages
- Low capital cost if elaborate lining and leachate is not needed and land is available	- Lack of rational design approach for sound economic analysis
	- Large land requirement
- Low operator attention and skill level	- Stabilized sludge requirement
- Low electric power consumption	- Climate significantly impacts design and performance
- Low sensitivity to sludge variability	- Labor intensive sludge removal
- Low chemical consumption	- Fuel and equipment costs for bed cleaning systems
- High dry cake solids concentrations	- Real or perceived odor and visual nuisances

TABLE 22.7   Advantages and Disadvantages of Using Drying Beds (adapted from Wang et al., 2007)

5.2   Sand Drying Beds

Sand drying beds are the oldest, most widely used drying bed dewatering method. The beds typically consist of a layer of sand underlain by a gravel layer and perforated drain piping. The beds are contained by concrete walls around the perimeter. To prevent the free liquid in the sludge from percolating into the groundwater, sand drying beds can be lined. Drying beds also may be enclosed to prevent rainwater from rewetting dewatered cake and to control odors.

Solids on sand beds are dewatered via drainage and evaporation. Initially, free water drains into the sand layer and is collected and removed via the underdrains. This step, which lasts a few days, continues until the sand becomes clogged with fine particles or all free water has drained. Once a supernatant layer forms, decanting can remove surface water. Decanting is important for removing rainwater, which can slow the drying process if allowed to accumulate on the surface. Decanting is also useful for removing free water released by chemical treatment. Solids drying in beds also can be enhanced by using auger-mixing vehicles in paved beds. Water remaining after initial drainage and decanting is removed via evaporation.

5.2.1   Process Design Considerations and Criteria

The performance of a sand drying bed for dewatering depends on:

- Feed solids concentration;
- Depths of sludge applied;
- Loss of water via the underdrain system;
- Extent of conditioning and digestion provided;
- Evaporation rate (which is affected by many environmental factors);
- Method of dewatered cake removal; and
- Solids disposal method used.

These site-specific considerations determine the optimum solids loading, area requirements, and other design criteria for a given bed.

5.2.1.1 Area Requirements Per capita area criteria initially used to size sand drying beds are shown in Table 22.8a. These criteria are based largely on empirical studies of primary solids conducted in the early 1900s by Imhoff and Fair (1940), who recommended a range from 0.1 to 0.3 m²/capita (1.0 to 3.0 sq ft/capita), depending on the type and solids concentration of the residuals applied to the bed. Other factors such as applied solids depth and number of yearly applications were also considered in this work. States within each of the U.S. Environmental Protection Agency (U.S. EPA) regions have also published drying bed sizing criteria on a per capita basis, as summarized in Table 22.8b.

As a result of today's more stringent effluent standards, these criteria may or may not suffice for sizing drying beds. The greater quantity of solids produced from achieving lower effluent suspended solids, chemical reactions associated with the use of polymer, additional solids contributed to the system through garbage grinders, and the use of advanced treatment processes may require larger drying areas.

5.2.1.2 Solids-Loading Criteria Sizing sand drying beds based on solids loading is a better approach than sizing based on per capita area. Typical solids-loading rates for sand drying beds vary (see Table 22.9). Furthermore, because the total quantity of solids produced daily by the overall treatment process can generally be predicted accurately, the risk of error is minimal. The best criteria to take into consideration are climatic conditions (e.g., temperature, wind velocity, humidity, and precipitation). Other design criteria can be found in commonly used design guidance manuals—such as "Ten States Standards" (GLUMRB, 2014); *Technical Report 16, Guides for the Design of Wastewater Treatment Works* (New England Interstate Water Pollution Control Commission, 2011)—or in state-specific manuals, such as the Pennsylvania Department of Environmental Protection's *Domestic Wastewater Facilities Manual* (1997).

Several models have attempted to mathematically describe the complex relationships involved in sand drying beds. Although early models were empirical, they have been used extensively to size drying beds. The advantage of using mathematical models is that they take local weather conditions (e.g., the amount of rainfall received) into consideration.

Using a rational engineering design approach, Rolan (1980) developed a series of equations to determine the design criteria and optimal operation for sand drying beds. Rolan found that the optimum application depth for a given percentage of dry solids is a function of the desired dry cake thickness, dry solids content, evaporation rate, and number of applications per year. The cost of removing solids (labor, equipment, and sand replenishment) primarily depends on the number of applications per year rather than the volume of solids.

Walski (1976) developed a similar mathematical model to account for major solids drying mechanisms, but indicated that the area required is relatively independent of the depth of the solids applied over the range of bed operation. Because neither Rolan nor Walski account for the removal of chemically bound water, these models are not valid where such water must be removed. Nevertheless, their studies show that many design standards do not adequately address the environmental and mechanical factors involved in operating sand drying beds and may result in inadequate designs.

When applying models, design engineers must recognize that cleaning a bed depends on variables other than just cake dryness. For example, operators at a large plant in Albuquerque, New Mexico, adopted a policy that no beds would be poured

Initial Sludge Source	Uncovered Beds		Covered Bed Area (m²/cap %)
	Area (m²/cap %)	Solids Loading (kg/m² · yr %)	
Primary			
Imhoff and Fair (1940)	0.09	134	—
Rolan (1980)	0.09–0.14	—	0.07–0.09
Walski (1976)			
N45°N latitude	0.12	—	0.09
Between 40° and 45°	0.1	—	—
S40°N latitude	0.07	—	0.05
Primary plus chemicals			
Imhoff and Fair (1940)	0.2	110	—
Rolan (1980)	0.18–0.21	—	0.09–0.12
Walski (1976)			
N45°N latitude	0.23	—	0.173
Between 40° and 45°	0.18	—	0.139
S40°N latitude	0.14	—	0.104
Primary plus low–rate trickling filter sludge			
Quon and Johnson (1966)	0.15	110	—
Imhoff and Fair (1940)	0.15	110	—
Rolan (1980)	0.12–0.16	—	0.09–0.12
Walski (1976)			
N45°N latitude	0.173	—	0.145
Between 40° and 45°	0.139	—	0.116
S40°N latitude	0.104	—	0.086
Primary plus waste activated sludge			
Quon and Johnson (1966)	0.28	73	—
Imhoff and Fair (1940)	0.28	73	—
Rolan (1980)	0.16–0.23	—	0.12–0.14
Walski (1976)			
N45°N latitude	0.202	—	0.156
Between 40° and 45°	0.162	—	0.125
S40°N latitude	0.122	—	0.094
Randall and Koch (1969)	0.32–0.51	35–39	—

TABLE 22.8a   Summary of Recognized, Published Sand Bed Criteria for Anaerobically Digested, Unconditioned Solids

EPA Region

Type of Sludge	Region 1		Region 2		Region 4[b]	Region 6		Region 7[b]	Region 8		Region 9[b]		Region 10[c]	
	Uncovered	Covered	Uncovered	Covered	Uncovered	Uncovered	Covered	Uncovered	Uncovered	Covered	Uncovered	Covered	Uncovered	Covered
Anaerobically digested primary only	1.5	1	1.5	0.75	0.5–1	1	—	1	—	—	—	—	1.5	1
Primary + low rate trickling filter	1.75	1.25	1.5	0.75	0.75–1.2	0.5–1	0.25	1.5	1	1	1	1	1.5–2	1–1.25
Primary + sand filter	—	—	—	—	1	1	—	—	—	—	0.5	—	—	
Primary + high rate trickling filter	—	—	—	—	1	1	—	—	1.25	1.25	1	1	2	1.25
Primary + WAS	2.5	1.5	2	1	1.5–2.5	1–1.5	1	—	1.35	1.35	1	1	1.5–2.5	1–1.5
Primary + chemical sludge	—	—	2	1	1–1.33	1	—	—	1.5	1.3	—	—	3	2

[a]Taken from individual state design criteria for states not using Ten States Standards.

[b]States in Regions 4, 7, and 9 do not have published requirements for covered sand beds.

[c]For the state of Idaho, values shown are for rainfall of 76 to 114 cm (30 to 45 in.); for rainfall of 25 to 76 cm (10 to 30 in.), reduce values by 50%.

Table 22.8b Summary of Recognized, Published State Bed Sizing Criteria by U.S. EPA Regions (cu ft/cap) (adapted from Wang et al., 2007)[a]

Sludge Solids Concentration (g/L)	Water Loss for Applied Sludge Depth (%)		
	100 mm	150 mm	200 mm
7.55	—	—	85.7
14.6	77.6	—	79.3
17.3	86.4	92.2	79.4
18.6	85.4	85.5	80.3
19.5	85.5	—	74.5
20.45	80.0	85.0	78.0
21.3	—	—	73.6
24.1	78.3	—	71.0
25.2	74.0	—	72.0
25.4	72.8	77.8	71.0
28.75	77.7	80.–	73.4
28.8	86.3	85.0	70.5
29.2	78.7	82.1	70.5
29.7	77.5	70.7	70.3
34.0	—	—	76.3
38.0	69.0	71.8	67.2
Average	79.2	80.0	73.4

TABLE 22.9   Total Water Drained from Aerobically Digested Sludge When Dewatered in a Sand Drying Bed

during the hottest month of the year because experience had shown that odors were greatest then, even though bed productivity was highest during the same month. Figure 22.11 illustrates the effect of evaporation rate on bed loading at various percentages of dry solids applied.

Despite the fact that design parameters such as those included in Table 22.8a and 22.8b were developed long ago, they continue to be actively used in the design of sludge drying beds.

5.2.1.2.1 Drying Time   The total drying time required depends on the desired final moisture content, the solids removal method, and final use of the dewatered biosolids. The time required to achieve a liftable cake depends more on the initial solids content and percentage of total water drained than on the initial drainage rate. This is particularly significant from a dewatering standpoint as the time required for evaporation is considerably longer than for drainage. Therefore, the total time that solids must remain on the bed is controlled by the amount of water to be removed by evaporation, which, in turn, is determined by the amount removed by drainage and decanting. The percentage of drainable water strongly depends on initial solids concentration (see Table 22.8). Quon and Johnson (1966) demonstrated that the percentage of drainable water in aerobically digested activated solids often considerably exceeds that reported for anaerobically digested solids.

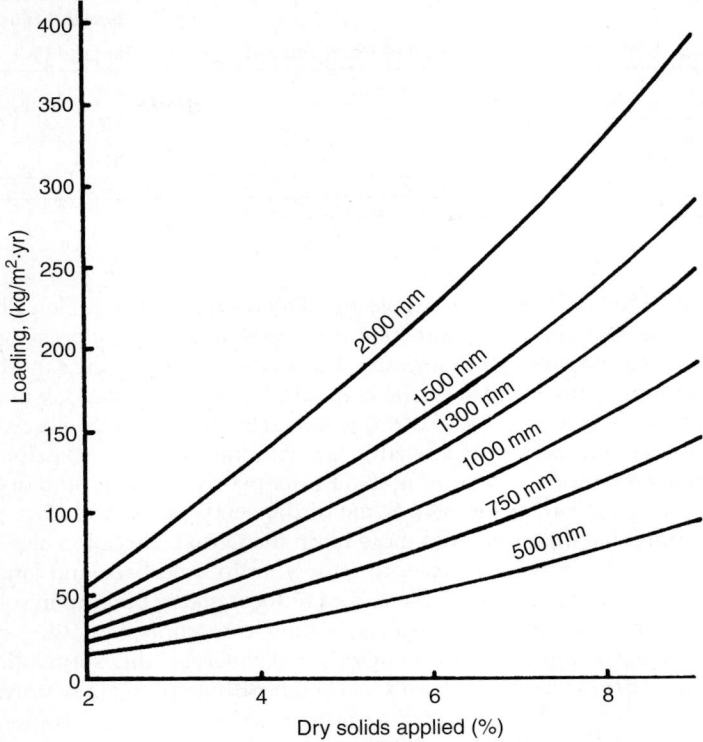

FIGURE 22.11 Effect of evaporation rate on bed loading.

5.2.1.2.2 Application Depth Quon and Johnson (1966) reported that the depth of applied solids affects the drainage rate and that the depth should not exceed 200 mm (8 in.). Haseltine (1951) reported an optimum depth of 230 mm (9 in.), depending on drying time and solids removal method; his suggested application depths ranged from 200 to 400 mm (8 to 16 in.). The applied depth should result in an optimum loading of 10 to 15 kg/m^2 (2 to 3 lb/sq ft). Even today, most design guidance documents recommend a maximum depth of applied sludge of 200 to 300 mm (8 to 12 in.).

Randall and Koch (1969) found that for a given solids concentration and depth, the solids drainage rate was constant after 8 hours. In addition, a typical applied solids depth of 200 mm had been reduced to a total depth of less than 25 mm (1 in.) when it was ready to be removed from the bed. The thickness of the dried cake primarily is a function of the solids concentration and the depth applied (see Figure 22.11).

The solids concentration of removed cake ranges from 44.5% to 95.5% dry solids, with higher cake moisture contents typically corresponding to higher initial solids concentrations. A cake with 40% to 45% may be achieved in approximately 2 to 6 weeks in good weather with well digested sludges (Wang et al., 2007).

According to Coackley and Allos (1962), drying occurs at a constant rate until a critical moisture content is reached; then, it proceeds at a declining rate. In general, the lower the initial concentration of solids and the required final moisture content are, the longer the drying time will be (Wang et al., 2007).

Removal Method	Raw Sludge Removed (%)	Digested Sludge Removed	
		Poor Drainage (%)	200 mm
Drainage	48–52	28	72
Decantation	4–9	22	2
Evaporation	43–44	50	27

TABLE 22.10   Effect of Digestion on Sand Bed Dewatering

5.2.1.2.3 Effect of Digestion on Dewatering   Dried, digested solids on drying beds contain small cracks that allow for greater surface exposure to the drying air, greater drainage of water, and easier passage of rainwater directly to the underlying sand-bed drains as compared to typical raw solids (see Table 22.10). According to Randall and Koch (1969), the dewatering properties of aerobically digested activated solids are closely related to oxygen-use characteristics. Solids obtained from aerobic digesters where the dissolved oxygen concentration remained less than 1 mg/L dewater poorly. Drainage and drying properties are improved by extending the SRT. Some of the solids studied reach a point, however, at which additional digestion does more harm than good. Digestion also increases friability of air-dried cake, making it easier to remove from sand beds and land-apply (mix with soil). It also minimizes odor problems and reduces grease buildup in soil.

Another advantage of digestion is the destruction of pathogens. The U.S. EPA guidelines for disinfecting solids suggest that anaerobic digestion followed by dewatering on sand drying beds may destroy enough pathogens to allow unrestricted use of the dried cake, assuming that the end use is not restricted by concentrations of heavy metals or other regulated parameters. Neither anaerobic nor aerobic digestion alone destroys pathogens as effectively as either would in combination with dewatering on sand drying beds. Reimers et al. (1981) noted that the inactivation of viable parasite eggs in raw solids increases as moisture content decreases.

5.2.1.2.4 Climatic Effects   Regional climatic conditions greatly affect dewatering performance in drying beds. In general, drying time is shorter in regions with abundant sunshine, low rainfall, and low humidity. Natural freezing in northern climates has been reported to improve dewaterability, but can also deactivate a bed for the winter. The prevalence and velocity of wind affects evaporation rates. Therefore, climatic conditions may warrant some modifications of design criteria. For example, storage of liquid residuals should be included if drying beds may be unavailable for extended periods because of climatic conditions.

5.2.2   Design Considerations

5.2.2.1   Structural Requirements   Each drying bed typically is designed to hold, in one or more sections, the full volume of solids removed from a digester or aerobic reactor at one drawing. Structural elements of the bed include the sidewalls, underdrains, gravel and sand layers, partitions, decanters, solids distribution channel, runway and ramps, and bed enclosures, if used. Figure 22.12 shows a schematic of a typical sludge drying bed.

5.2.2.1.1 Sidewalls   Construction above the sand surface should include an embankment (vertical wall) with above-sand freeboard of 0.5 to 0.9 m (20 to 36 in.). Walls can be

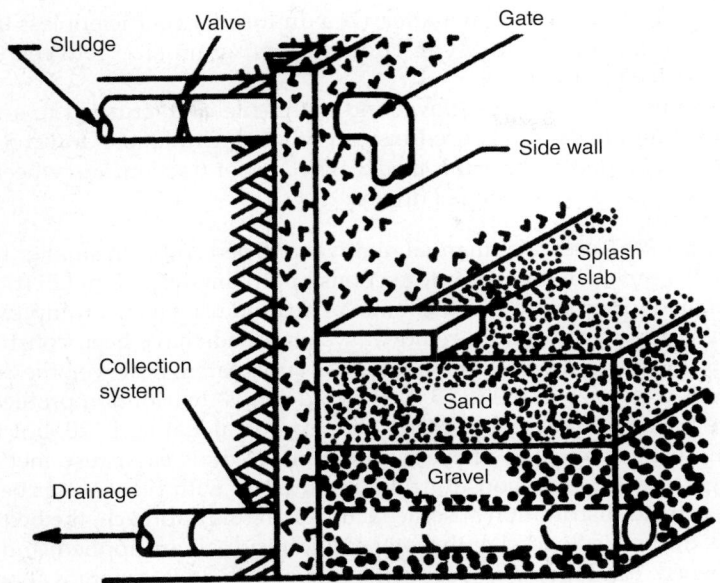

FIGURE 22.12　Schematic of a typical sand drying bed.

constructed of earth sodded with grass, wooden planks (preferably treated to prevent rotting), concrete planks, or reinforced concrete or concrete blocks that are set on an edge around the extremities of the sand surface and extended to the underdrain gravel to help prevent weed and grass encroachment.

5.2.2.1.2　Underdrains　Underdrains are constructed of perforated plastic pipe or vitrified clay tile and sloped toward a main collection pipe, known as the outlet drain. The outlet drain should be no less than 100 mm (4 in.) in diameter and have a minimum slope of 1%. Spacing should range from 2.5 to 6 m (8 to 20 ft) and take into account the type of solids-removal vehicles used to avoid damage to the underdrain.

　　Lateral pipes feeding into the outlet drain should be spaced from 2.5 to 3 m (8 to 10 ft) apart, with the shorter distance preferred. If infiltration would endanger groundwater, then the earth floor should be sealed with an impervious membrane system approved by local regulators. The area around the drain tiles should be backfilled with coarse gravel; disturbing or breaking the tiles should be avoided. Heavy equipment should be excluded from the bed after the underdrains are laid, unless the beds are designed to accommodate such heavy loads.

5.2.2.1.3　Gravel Layers　Gravel layers are graded to an overall depth of 200 to 460 mm (8 to 18 in.), with the relatively coarser materials at the bottom. The gravel particles range from 3 to 25 mm (0.1 to 1.0 in.) in diameter.

5.2.2.1.4　Sand Layer　Sand depth varies from 200 to 460 mm (9 to 18 in.). However, a minimum depth of 300 mm (12 in.) is suggested to achieve suitable sludge cake, while reducing the frequency of sand replacement caused by cleaning-related losses. A good-quality sand has (1) particles that are clean, hard, durable, and free from clay, loam, dust, or other foreign matter; (2) a uniformity coefficient less than 4.0, but preferably less than 3.5; and (3) an effective sand grain size between 0.3 and 0.75 mm (0.01 and 0.03 in.).

In some instances, pea gravel and anthracite coal, crushed to an effective size of about 0.4 mm (0.02 in.), is used instead of sand. Gradations toward water filter sand should be avoided because this media affords poor traction, and wheeled cake-removal vehicles might become bogged down.

5.2.2.1.5 Partitions For manual removal of dried solids in smaller facilities, the drying bed is typically divided into sections approximately 7.5 m (25 ft) wide. The width is designed to accommodate the removal method used (e.g., multiples of loader bucket width and span of vacuum-removal system). Beds have been constructed as long as 30 to 60 m (100 to 200 ft). If polymer use is anticipated, however, the bed length should not exceed 15 to 25 m (50 to 75 ft) to avoid solids-distribution problems. The angle of repose for many polymer-treated solids can be as flat as 1:120, but the angle can be much greater; therefore, unevenly distributed solids can cause inefficient use of the drying bed area. Provisions for flooding the bed with plant water before introducing solids can aid distribution at some facilities. In this approach, the bed drain valves are closed, the bed is flooded with water, liquid residuals are applied, and the drain valves are opened. Preflooding increases the rate of initial water removal because it adds to the hydraulic head by creating a vacuum under the bed when the drain valves are opened. This vacuum holds until air begins to leak through the bed into the underdrain system.

The partitions may be earth embankments (where land is plentiful) or walls constructed of concrete block, reinforced concrete, or planks and supporting grooved posts. The posts can be made of wood, although the preferable material is reinforced concrete planks fitted into grooves in reinforced concrete posts. If used, partition planks should extend about 80 to 100 mm (3 to 4 in.) below the top of the sand surface, and the posts should extend 0.6 to 0.9 m (2 to 3 ft) below the bottom of the gravel. The placement of partitions and other structural elements should be designed to accommodate mechanical removal equipment, if used. For example, including at least one solid, vertical wall in each bed against which a wheeled front-end loader can push will speed bed cleaning.

5.2.2.1.6 Decanters Supernatant can be removed continuously and/or intermittently from the drying bed using decanters provided on the perimeter of the bed (see Figure 22.13). Decanters are particularly useful for relatively dilute secondary and polymer-treated solids and for removing rainwater. Properly performed, decanting can reduce drying time significantly.

5.2.2.1.7 Solids Distribution Channel Liquid residuals can be applied to the sand-bed sections via a closed conduit or a pressurized pipeline with valved outlets at each sand-bed section, or via an open channel with side openings controlled by sluice gates or hand slide gates. The open channel is easier to clean after each use. With either type, a concrete splash slab 130-mm (5-in.) thick and 0.9-m (3-ft) square is necessary to receive falling solids and prevent erosion of the sand surface. If a pressurized pipeline with valved outlets is used, a 90° elbow should direct the solids trajectory against the splash slab at all pumping rates. Piping and valves should be protected from freezing because draining completely after each bed is filled is impractical.

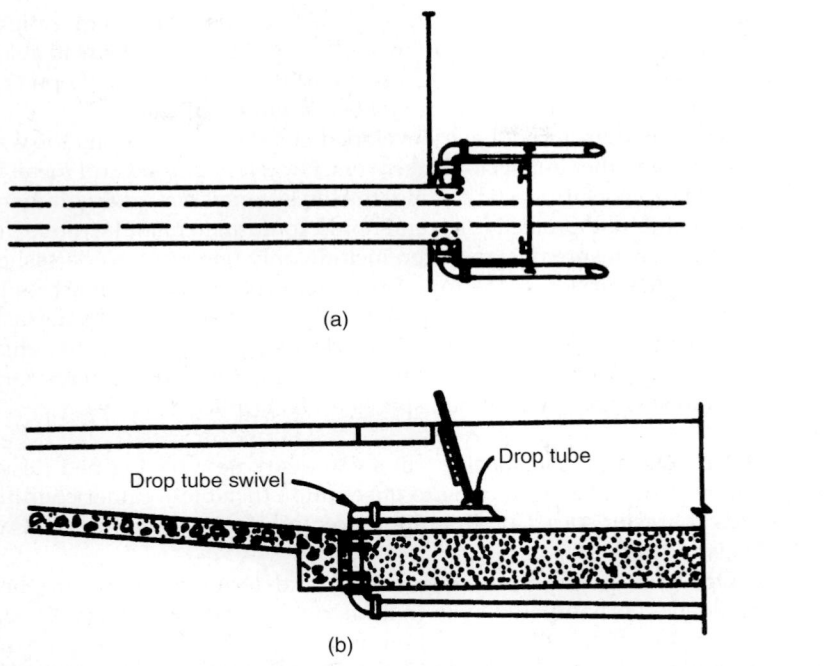

Figure 22.13 Typical decant piping: (a) plan and (b) elevation.

Preferably, the distribution channel runs between two series of 7.5-m (25-ft)-wide sand-bed sections. This bed width typically suffices for manual removal of cake; however, another width may better accommodate mechanical removal equipment.

5.2.2.1.8 Runway and Ramps To remove cake via truck, narrow concrete runways are needed along the central axis of each bed section. Concrete runway slabs minimize compaction of the sand filter surface; they keep truck wheels from touching the sand layer. In addition to reducing sand compaction and protecting the underdrain system from damage, multiple runway strips reduce the loss of sand and provide a good gauge for sand replacement.

If ramps are included in the entrance design and mechanical removal devices are used, design engineers should consider full-width ramps to avoid problems involving corner access and maneuvering of equipment on the sand.

5.2.2.1.9 Enclosures Enclosures can be provided for drying beds to prevent rainwater from re-wetting the dewatered cake, while also providing greater temperature control and containment of odors. Continuously sloped shed roofs are best in northern latitudes to prevent snow buildup.

Drying beds can be fully covered on all sides or only on the top with glass or polyester glass-fiber roofs. The roof-only configuration covers the top of the drying bed but leaves the sides open to the atmosphere, protecting the drying cake from precipitation but providing little temperature control. Completely enclosed drying beds, on the other hand, permit more cake withdrawals per year in most climates because of better temperature control. As a result, enclosed beds typically require less area than open beds. Favorable weather conditions, however, allow open beds to evaporate cake moisture faster than enclosed ones. Consequently, a combination of open and enclosed beds can achieve the most effective use of bed drying facilities.

Most manufacturers have developed standard dimensions for width, length, truss spacing, and other details for bed covers. For interior wood and metal work, paints (e.g., a coal-tar-epoxy bitumastic coating) must resist moisture and hydrogen sulfide. Application of paints and protective covers should conform to manufacturers' recommendations.

Older enclosure designs often include only one row of side sashes that opens out, diverting the air across the top of the enclosures rather than across the surface of the solids. Newer enclosure designs typically have two rows of side sashes, with the top row opening out and the bottom row opening in. Mechanical ventilation of enclosed beds is suggested in humid climates. Ventilation requirements should ensure that enclosures will not be confined spaces under applicable codes.

5.2.2.2 Chemical Conditioning   In some cases, new drying bed designs may need to include chemical conditioning to offset unpredictable weather conditions and variable solids characteristics. Conditioning also can help improve the solids drying capacities of existing beds.

Optimum polymer dosages should be determined with care, because polymer's effectiveness is hampered by both underdosing and overdosing. The net and gross bed loadings for chemically treated and untreated beds should be compared in laboratory tests and under actual field conditions. Blinding of the sand can result if chemical use is excessive.

If polymer addition is included in the design, at least three points of addition are required for optimum effectiveness: one near the suction side of the feed pump, one at the feed pump discharge, and one near the discharge point of each bed. Where possible, provisions also should be made for recirculating polymer-treated solids to allow dose optimization before discharging the initial solids to the bed. This will prevent blinding of the sand-bed surface by poorly treated solids.

5.3   Other Types of Drying Beds

5.3.1   Polymer-Assisted Filter Bed

A polymer-assisted bed consists of a sand layer over specialized drainage panels. Physically, the filter beds look similar to conventional beds, but have a much quicker turnover time. The key to polymer-assisted filter beds is an upstream polymer activation system designed to properly mix polymer with feed solids and then properly distribute the mixture across the surface of the filter bed.

Sizing a polymer-assisted filter bed system is similar to sizing conventional solids drying beds, but should be confirmed with the manufacturer. Solids are applied to the system much as they would be on a conventional drying bed, but are first conditioned with polymer. The underdrain system provides a siphoning effect that provides vacuum assistance to the drying process. The plant water system (see Figure 22.14) is used to apply plant water to the underdrain system, which is key to the siphoning effect. Together, polymer conditioning and the siphoning effect allow solids to dewater more rapidly than in a conventional bed. A specially designed articulated vehicle is used to remove dewatered solids from the polymer-assisted filter bed.

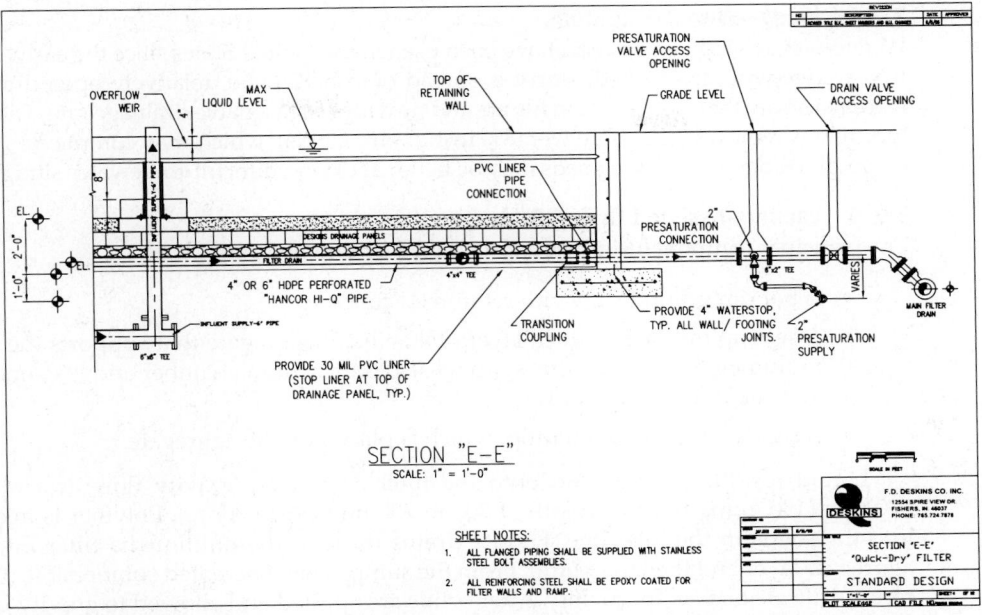

FIGURE 22.14 Schematic of a typical polymer-assisted filter bed system (courtesy of Deskins).

The polymer-assisted filter bed also can be used as part of an U.S. EPA-approved Class A biosolids process, in which the beds are used to dewater biosolids to 40% solids. The dewatered solids are then removed, placed in windrows, and turned each day using the air drying (back-blending) method for a period of 15 to 21 days. However, the final product must be tested in a laboratory to meet the pathogen reduction requirements because this is not approved as a process to further reduce pathogens.

5.3.2 Mechanically Assisted Solar Drying System

Solar drying systems consist of a series of drying beds covered by a translucent, climatically controlled chamber. Sensors monitor the atmosphere in the chamber and control air louvers and ventilation fans to optimize drying conditions. The monitors also control a mobile, electric "mole" that plows the solids cake during the drying cycle up to 10 times per day. Such systems can produce dried cake containing 50% to 90% solids. The advantages of this system are protection from rain, elevated temperatures because of the "greenhouse effect" in the chamber, potentially drier cake solids, and shorter drying time.

5.3.3 Paved Drying Beds

Paved drying beds are constructed with concrete, asphalt, or soil cement liners to help front-end loaders more easily remove cake and mix solids to speed up drying. A series of tests conducted by Randall and Koch (1969) indicated that drying beds with sand bottoms perform better than beds with impervious bottoms, although they may require more area than traditional sand drying beds for a given amount of sludge.

5.3.4 Wedge-Wire Drying Beds

Wedge-wire drying bed systems have been used in the United States since the early 1970s. In a wedge-wire drying bed, slurry is spread onto horizontal, relatively open drainage media in a way that yields a clean filtrate and provides a reasonable drainage rate. The cake typically is removed relatively wet (8% to 12% dry solids), which may complicate use or disposal. Wedge-wire drying beds may be better at drying difficult-to-dewater sludge.

5.3.5 Vacuum-Assisted Drying Beds

The principal components of vacuum-assisted drying beds are:

- A bottom ground slab with reinforced concrete;
- A several-millimeter-thick layer of stabilized aggregate that supports the rigid multimedia filter top (this space is also the vacuum chamber and is connected to a vacuum pump); and
- A rigid multimedia filter top, which is placed on the aggregate.

Liquid residuals are spread onto the filter surface by gravity flow at a rate of 570 min (150 gpm) and to a depth of 300 to 750 mm (12 to 30 in.). Polymer is injected into the solids in the inlet line. Filtrate drains through the multimedia filter into the aggregate layer and then to a sump. From the sump, a level-actuated submersible pump returns filtrate back to the facility. After solids are applied and allowed to gravity-drain for about 1 hour, the vacuum system is started and maintains a vacuum of 34 to 84 kPa (10 to 25 in. Hg) in the sump and under the media plates.

Under favorable weather conditions, this system can dewater a dilute, aerobically digested solids feed to 14% solids concentration in 24 hours. The dewatered solids can be lifted from the bed by mechanical equipment. It will further dewater to an about 18% solids concentration in 48 hours.

5.4 Reed Beds

5.4.1 Description of Technology

Reed beds to treat stabilized solids from secondary water resource recovery facilities have been used in a hand full of U.S. locations as well as in many developing countries. This method was developed in the 1960s by the Max Planck Society of Germany and was recognized by the U.S. EPA as an alternative and innovative system (Riggle, 1991).

The system combines the action of conventional drying beds with the effects of aquatic plants on water-bearing substrates. While conventional drying beds drain more than 50% of the water content from solids, the resulting cake must be hauled away for further treatment or disposal at designated sites. When the drying beds are planted with reeds of the genus *Phragmites communis*, further desiccation results from the demand for water by these plants. To satisfy this demand, the plants continually extend their root system into the solids deposits. This extended root system establishes a rich population of microflora that feed on the organic content of the solids. This microflora is also partly kept aerobic by the action of the plants. Degradation by the microflora is so effective that eventually up to 97% of solids are converted into carbon dioxide and water, with a corresponding volume reduction. The beneficial result is that these planted drying beds can reportedly be operated for up to 10 years before the accumulated residues have to be removed. This represents a considerable monetary savings.

5.4.2 Design Considerations

The reed-bed treatment system consists of a set of rectangular, parallel basins with concrete sidewalls. The bottom of each bed is lined and provided with two underdrains. In addition, a 230-mm (9-in.) layer of 19-mm (0.75-in.) washed river gravel is topped with a 102-mm (4-in.) layer of filter sand. The reeds are planted in the gravel, with 11 plants per square meter (1 plant per square foot) of filter area. A freeboard of 1.0 to 1.5 m (3.5 to 5 ft) is often provided, depending on storage design requirements. Basins are cyclically loaded. In each cycle, the first basin is loaded over a 24-hour period and then allowed to absorb the loading over a 1-week resting period before the cycle is repeated (Banks and Davis, 1983).

Figure 22.15 shows a typical reed drying bed system. Hydraulic design loadings for residuals containing 3% to 4% solids are 0.0042 $m^3/m^2 \cdot h$ (2.5 gal/d/sq ft) or 35 $m^3/m^2 \cdot a$ (86 gal/yr/sq ft). At this loading rate, about 1.0 m (3.5 ft) of product with

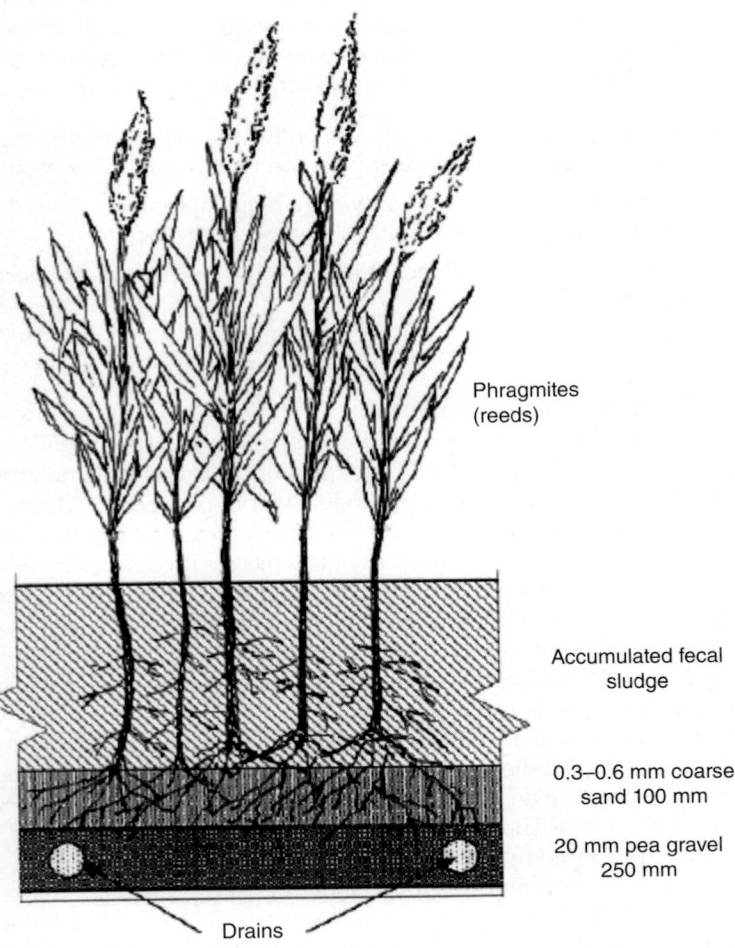

Phragmites (reeds)

Accumulated fecal sludge

0.3–0.6 mm coarse sand 100 mm

20 mm pea gravel 250 mm

Drains

FIGURE 22.15 Schematic of a typical reed drying bed system (courtesy of Crolla et al., 2007).

70% moisture will accumulate over 10 years. When solids are removed for disposal, the top layer of sand also is removed and replaced. Generally, the root system remaining in the gravel bed will allow the reed plants to regenerate without the need for replanting (Banks and Davis, 1983).

These systems are most applicable in climates where winter temperatures ensure at least one prolonged frost. The planted reeds are harvested in the fall once they have become dormant.

The use of reed-bed systems is a reasonable application for water resource recovery facilities with less than 5.7- to 7.6-ML/d (1.5- to 2.0-mgd) capacity. The primary benefits of this system are the reduced O&M requirements and the frequency of disposal handling is once every 10 years.

5.5 Lagoons

Lagoons, which are natural or artificial earth basins, can be used for both solids drying and storage. However, the use of lagoons typically starts as a temporary expedient to handle solids volumes in excess of the original plant design. Climatic conditions have a decided effect on the functioning of a lagoon, with warmer, arid climates producing the best results.

Lagoon operations typically involve the following processes:

- Pumping liquid solids to the lagoon for several months or more. The pumped solids are stabilized before application to minimize odor problems.

- Continuously or intermittently decanting supernatant from the lagoon surface and returning it to the water resource recovery facility.

- Removing the dewatered material with some type of mechanical removal equipment.

- Repeating the cycle.

5.5.1 Environmental Considerations

The location of natural or artificial dewatering lagoons must be carefully considered before selecting lagoons as a treatment method. The proposed site should be sufficiently removed from dwellings and other areas where odors would produce problems. Because they go through a series of wet and dry conditions, large lagoons can produce nuisance odors.

The use of deep lagoons to dry raw solids has resulted in severe odor problems, but the use of shallow lagoons for drying typically has not produced odors more intense than those experienced with conventional sand drying beds. However, odors produced from lagoons used to store solids can be more of a problem because wet treated municipal solids retain a higher moisture content far longer than solids treated on conventional sand drying beds.

Odor problems with lagoons have been described by Zablatzky and Peterson (1968). Both the intensity and the type of odors produced from wastewater lagoons have varied greatly, depending on the condition of the solids and the depth of the lagoon. Odors can vary from a gas- and tar-like odor produced by a well-digested material to the putrid odors produced by decomposing raw solids.

The possibility of polluting groundwater or nearby surface waters should be investigated thoroughly. If the subsurface soil is permeable, the potential for groundwater contamination exists. Clay and/or membrane liners and underdrain systems can minimize this potential.

Baxter and Martin (1982) found that the application depth also can affect groundwater contamination. Their results indicate that a 250-mm (10-in.) application of liquid residuals to earthen drying lagoons can lead to significant quantities of polluted water moving toward the groundwater. However, applications of 0.6 to 1.0 m (2 to 3 ft) of liquid solids lead to a rapid sealing effect by producing an impermeable layer of solids that prevents contaminants from moving toward groundwater.

Finally, all lagoon areas should be fenced to keep animals and other trespassers out and to prevent vandalism and potential liability problems.

5.5.2 Storage Lagoons

Storage lagoons can be 1.5 m (5 ft) deep or more and are primarily designed for storage rather than drying. To allow for cleaning, the dikes should be constructed to be about 3 m (10 ft) wide across the top to accommodate trucks and other mechanical equipment used for solids removal or for maintenance. The side slopes of the dikes should be a maximum of 3:1 (horizontal-to-vertical) to provide a slope surface that can be mowed by mechanical equipment. Residuals in storage lagoons typically do not dry or condense enough to permit removal by anything but a dragline, so the maximum width of lagoons must be less than twice the length of a dragline boom or other equipment to be used.

The design volume of lagoons depends on intended use and length of storage. Generally, solids lagoons are designed to provide for the emptying of one or more digesters and can be sized in reference to digester volume and frequency of emptying. Local climate and the proposed disposal of supernatant liquor and rainwater drawoffs influence the length of time between the initial placing of solids and their eventual concentration to a degree that permits mechanical removal.

Properly designed drawoff piping should be provided for the removal of supernatant liquid and rainwater from storage lagoons. The drawoff piping should be arranged to discharge these liquids to the water resource recovery facility's headworks. Removing these liquids prolongs the life of the lagoon, helps prevent insect breeding, and hastens dewatering.

Nuisances resulting from storage lagoons depend on the type and condition of the solids placed in the lagoon; they can be reduced by chemicals to control or mask odors and prevent insect breeding. If poorly digested wastewater solids are added to a storage lagoon, it may not be possible to control odors. If this situation is suspected during design, one solution to the problem is to adequately isolate the lagoon site. Another solution entails adding clean water on top of the lagoon to provide an aerobic layer exposed to the atmosphere.

5.5.3 Drying Lagoons

Drying lagoons are used for dewatering when sufficient land is available (Vesilind, 1979). Lagoons are similar to drying beds; however, solids are placed in depths three to four times greater than in a drying bed. The solids are allowed to dewater and dry to a predetermined solids concentration before removal; this process may require 1 to 3 years.

Dewatering in lagoons occurs via evaporation and transpiration. Studies by Jeffrey (1959, 1960) indicate that evaporation is the most important dewatering factor.

Drying lagoons should be shallower than storage lagoons, with about 0.6 to 1.2 m (2 to 4 ft) of dike provided above the bottom of the lagoon. Drying lagoons typically do not include an underdrain system because most of the drying is accomplished by decanting supernatant liquor and by evaporation. However, groundwater pollution is a potential problem. The depth of solids in drying lagoons should not exceed 400 mm (15 in.) after excess supernatant liquor has been removed.

Dikes should be a shape and size that permits maintenance and mowing. The hydraulic loading against the dikes and the possibility of dike leakage are not great because the depth of liquid seldom exceeds about 0.5 m (1.5 ft). Dikes should be constructed of compacted material to provide stability on the slopes, and the bottom of the lagoons should be level or have a small slope away from the liquid-residuals inlet opening. The dikes also should be large enough to allow trucks and front-end loaders to enter the lagoons for cleaning and to permit easy mowing with mechanical equipment.

The outlet into the lagoon, supernatant drawoff lines, and other piping should be 0.3 m (1 ft) below the original bottom of the lagoon. Supernatant liquor and rainwater drawoff points should be provided, and the drawn-off liquid should be returned to the water resource recovery facility for further processing. In addition, surrounding areas should be graded to divert surface water around and away from the lagoons.

Wet solids typically will not dry enough to be removed with a fork except in an arid climate or when an extremely long, warm dry spell has occurred. The concentrated cake typically can be removed with a front-end loader.

The potential odor problem from lagoons used to dewater well-digested solids is about the same as it is for sand drying beds. If supernatant liquor and rainwater are removed promptly from the solids surface so the cake is exposed to oxygen in the air and can rapidly dry, there should be minimal odors.

The actual depth and area requirements for drying lagoons depend on several factors (e.g., precipitation, evaporation, type of solids, volume, and solids concentration). Solids-loading criteria specify 35 to 38 kg/m^3·a (2.2 to 2.4 lb/yr/cu ft) of capacity (Zacharias and Pietila, 1977). The area provided for drying lagoons varies from 0.1 m^2/ capita (1 sq ft/capita) for primary digested solids in an arid climate to as high as 0.3 to 0.4 m^2/capita (3 to 4 sq ft/capita) for activated sludge plants in areas where the annual rainfall is about 900 mm (36 in.).

6.0   Rotary Presses

6.1   Description of Technology

Rotary presses can achieve cake solids and solid captures similar to belt presses and centrifuges (Crosswell et al., 2004). Rotary press and rotary fan press dewatering technologies rely on gravity, friction, and pressure differential to dewater solids.

Figure 22.16 illustrates the principal components of a rotary press. Solids are dosed with polymer and fed into a channel bound by screens on each side. The channel curves with the circumference of the unit, making a 180° turn from inlet to outlet. Free water passes through the screens, which move in continuous, slow, concentric motions. The motion of the screens creates a "gripping" effect toward the end of the channel, where cake accumulates against the outlet gate, and the motion of the screens squeezes out more water. The cake is continuously released through the pressure-controlled outlet.

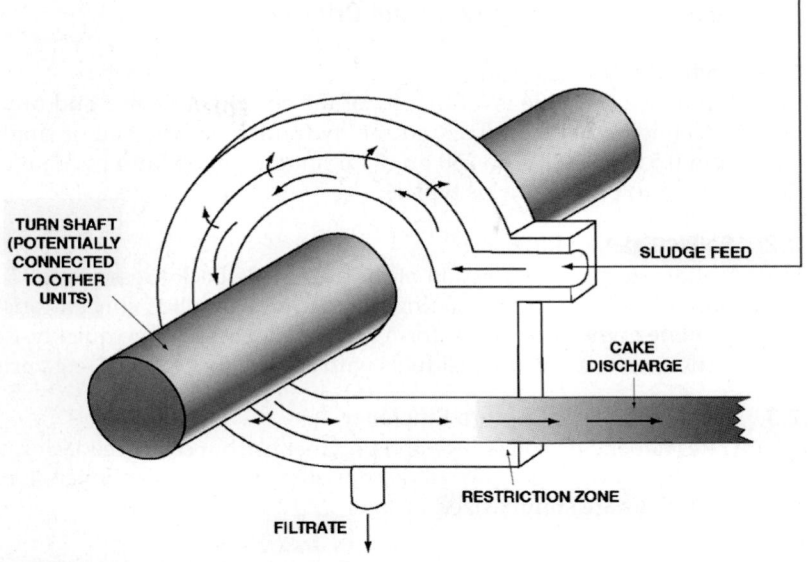

FIGURE **22.16**   Schematic of a rotary press system.

The flow path through the units differs between the rotary press and the rotary fan press. Rotary presses are fed from the top and discharge on the bottom, while rotary fan presses are fed on the bottom and discharge from the top.

Table 22.11 summarizes the advantages and disadvantages of rotary-type presses.

Advantages	Disadvantages
- Uses less energy than centrifuges or belt filter presses	- May be more dependent on polymer performance than centrifuges or belt filter presses
- Small footprint	- Low throughput compared to other mechanical dewatering processes
- Containment of odors	- Screen clogging potential
- Low shear	- Need for heavy rated overhead crane to lift and maintain channels
- Minimal moving parts	- High capital cost
- Minimal building requirement	
- Minimal start–up and shutdown time	
- Uses less wash water than belt filter presses	
- Low vibration	
- Low noise	
- Modular design	

TABLE **22.11**   Advantages and Disadvantages of Rotary Presses

6.2   Process Design Conditions and Criteria

6.2.1   Hydraulic-Loading Rate
The hydraulic-loading rate is a function of the equipment's size and number of channels. The technology is modular, and the hydraulic-loading rate of single-drive units ranges from 0.5 to 15 L/s (7 to 250 gpm), although a maximum hydraulic-loading rate of 3 L/s (50 gpm) per channel is typical.

6.2.2   Solids-Loading Rate
Because solids capture is a function of the adjustable back pressure, the solids-loading rate varies with the hydraulic-loading rate. At higher solids concentrations, residuals will accumulate in the outlet zone, form cake, and extrude more quickly. Rotary presses provide better performance on residuals with higher fiber content (e.g., primary solids).

6.2.3   Cake Solids and Capture Efficiency
Capture rates depend on solids type and polymer use but can exceed 95% solids capture. Performance depends largely on solids consistency. The performance characteristics of select installations are summarized in Table 22.12.

Equipment	Facility Size ML/d (mgd)	Solids Type	Incoming % TS	Discharge % TS
Rotary fan press	4.2 (1.1)	Extended aertion	1.6–2.5	19–22
	12 (3.3)	Thermophilic anaerobic digestion	3.5	25
	59 (15.5)	Conventional anaerobic digestion	2	17
Rotary press	1.9 (0.5)	Conventional anaerobic digestion	1.0–1.5	25
	75 (19.8)	Thickened PS/WAS	3–6	19–25
	9.5 (2.5)	Septage/PS/WAS	6–8	26–28
	61 (15)	PS/WAS	0.8–1	12–14
	5.3 (1.4)	Thickened PS/WAS	3	28
	53 (14)	Conventional aerobic, thickened digestion	4.4	35
	7.6 (2)	BNR anaerobic digestion	2.7	23
	26 (6.8)	Thickened anaerobic lagoons	2.6	24
	31 (8.1)	BNR thickened	2.6	20
	23 (6)	BNR anaerobic digestion	2.3	19
	7.6 (2)	BNR thickened	4.5	25
	67 (23)	BNR thickened	3.7	26
	23 (6)	WAS aerobic digestion	1.9	16.3
	3.8 (1)	BNR WAS aerobic digestion	1.5	12
	5.7 (1.5)	BNR WAS aerobic digestion	1.4	16.9
	4.5 (1.2)	BNR WAS thickened	3.3	15.5

TABLE 22.12   Performance Characteristics of Rotary Presses (WEF, 2012)

6.3   Design Considerations

6.3.1   Mechanical Features and Materials of Construction

The major elements of a rotary press are the polymer feed and mixing system, parallel filtering screens, a circular channel between the screens, the rotation shaft, and a pressure-controlled outlet. The key differences between the rotary press and rotary fan press are the screens, drive mechanism, and pressure differential. In the rotary press, the screens consist of two layers of perforated stainless steel, with each layer having different sieve sizes. The rotary fan press' screens consist of fabricated wedge wire with small openings and linear gaps. The rotary press drive configuration allows up to six rotary press channels to be operated on a single drive. Each channel has bearings, and the combined unit has an outboard bearing cantilevered on one end. The rotary fan press drive configuration uses a maximum of two rotary press channels on a single drive with isolated bearings in a sealed gearbox.

The entry zones of rotary presses and rotary fan presses function much like the gravity phase of belt press dewatering. Free water "falls" through the filtering screen pores and is collected in a filtrate channel. Pressure builds gradually as the solids travel toward the machine outlet. Because the outlet controls the pressure at which cake can be released, cake solids accumulate against the outlet and are further dewatered via friction from the continuous motion of the screens. In the rotary press, the friction generated between the screens and the cake plug translates into mechanical pressure that deflects the cake away from the center and forces it sideways against the restricted outlet. In the rotary fan press, frictional force also is imparted in the outlet zone to dewater solids, but the mechanical pressure is not generated to the same magnitude. In both designs, water is released via friction and is collected in the filtrate channel along with water released by gravity in the entry zone.

A key feature of both rotary press and rotary fan press dewatering technology is their slow rotational speed. Typical installations use speeds of 1 to 3 revolutions per minute (rpm). This provides low vibration, low shear, low noise, and low energy use.

6.3.2   Area/Building Requirements

6.3.2.1   Structural Requirements   Rotary and rotary fan presses can be mounted on concrete floors, a concrete pad, or a metal skid. The structural housing for the rotating equipment is built into the unit. A hopper, conveyor, or additional cake conveyance must be supplied at the cake discharge outlet. Building requirements are minimal because rotary and rotary fan presses are enclosed and have small footprints. The dewatering area for the presses themselves can be as small as 9.3 m² (100 sq ft) for a small system, depending on the model size and number of channels for a given application.

6.3.2.2   Energy Requirements   Rotary press and rotary fan press systems have a connected horsepower of about 3.7 to 15 kW (5 to 20 hp).

6.3.3   Safety

In general, because the rotary press and the rotary fan press technology is enclosed, the operator is protected from moving parts. However, as with any piece of equipment, electrical power should be disconnected before attempting to service the unit.

6.4   Ancillary Systems

6.4.1   Operational Controls
Rotary presses and rotary fan presses are typically provided with control systems by the manufacturer. Optimum feed rates and discharge pressures are determined at startup and programmed into the controls.

6.4.2   Chemical Conditioning Requirements
Chemical conditioning is required in rotary press or rotary fan press dewatering. Polymer feed systems can be supplied by the press manufacturer or can be procured independently. In both cases, the feed systems typically include a polymer storage tank and metering pump, which feeds the polymer into the mixing or flocculation tank, where it is blended with the solids. Dry or emulsion polymers can be used.

6.4.3   Washwater Requirements
Rotary presses and rotary fan presses include a self-cleaning system that must run for 5 minutes per day at the end of use to flush all lines and equipment. Typically, the normal in-plant water source has sufficient pressure, but in some cases, high-pressure booster pumps may be required.

7.0   Screw Presses

7.1   Description of Technology
Although screw presses have been in existence since the 1960s, the technology originally was used solely in agricultural (e.g., dairy manure) industrial applications (e.g., pulp and paper mills and food processing plants). In the municipal market, the application is relatively new in the United States, with most facilities installed after the year 2000. The increasing number of installations is still relatively small, compared to other solids dewatering technologies.

There are presently two major types of screw presses used in municipal dewatering applications: horizontal and inclined. Inclined screw presses are at angles 10° to 20° from the horizontal. Other areas of difference pertain to solids inlet configuration, screen basket design (perforated plate or wedge wire), basket cleaning from the inside and outside (brushes and rotating wash system), and filtrate water collection. One manufacturer also provides an option in which lime and heat are added to the screw press, which then both dewaters solids and reduces pathogens to produce biosolids that potentially meet the Class A standards in 40 CFR 503. Advantages and disadvantages of screw press systems are summarized in Table 22.13.

The major elements of a screw press dewatering system are a solids feed pump, polymer makeup and feed system, polymer injection and mixing device (injection ring and mixing valve), flocculation vessel with mixer (if applicable), solids inlet headbox or pipe, screw drive mechanism, shafted screw enclosed within a screen, rectangular or circular cross-section enclosure compartment, and outlet for dewatered cake (see Figure 22.17). Some horizontal screw press systems (e.g., the combined dewatering and pasteurization process) include a thickening unit before the screw press, which may be desirable for reducing the hydraulic load to the screw press, given certain feed solids characteristics in conventional applications.

Advantages	Disadvantages
- Low rotational speed results in low maintenance and noise	- Lower cake concentrations, especially when there are no primary clarifiers at the facility
- Low energy consumption	- Large footprint
- Odors are contained	- Requires washwater
- Low operator attention required	- Lower solids captures than other dewatering processes
- Lower washwater demand and pressure than belt filter presses	- Lower capacity per unit requiring more units than other technologies for larger facilities

TABLE 22.13 Advantages and Disadvantages of Screw Presses

A screw press is a simple, slow-moving device that achieves continuous dewatering (see Figures 22.18 and 22.19). Polymer is combined with solids in flocculation vessels upstream of the screw press to enhance the solids' dewatering characteristics. Screw presses dewater solids first by gravity drainage at the inlet section of the screw and then by squeezing free water out of the solids as they are conveyed to the discharge end of the screw under gradually increasing pressure and friction. The increased pressure to compress the solids is generated by progressively reducing the available cross-sectional area for the solids. The released water is allowed to escape through perforated plate or wedge-wire screens surrounding the screw while the solids are retained inside the press. The liquid forced out through the screens is collected and conveyed from the press, and the dewatered solids are dropped through the screw's discharge outlet at the end of the press. Screw speed and configuration, as well as screen size and orientation, can be tailored for each dewatering application.

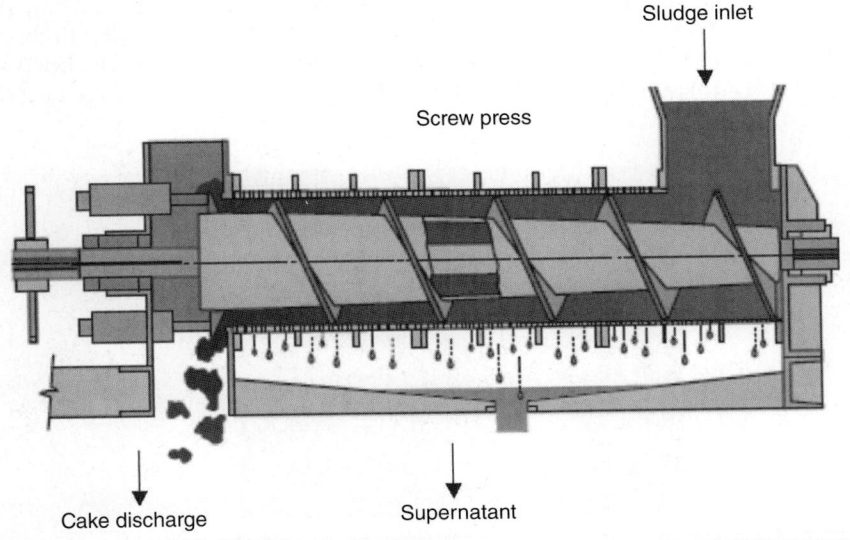

FIGURE 22.17 Cutaway view of a horizontal screw press (courtesy of FKC Co., Ltd., Port Angeles, Washington).

FIGURE 22.18   Cutaway view of an inclined screw press (courtesy of Huber Technology, Inc.).

Solids are combined with polymer and pumped into the flocculation vessel. After flocculation, solids are transferred to the screw press. In the horizontal and some inclined screw press configurations, solids are fed by gravity from the flocculation tank into the screw press headbox. If a thickening unit is used, solids flow from the flocculation tank to the thickener and then to the screw press headbox. Solids then flow from the headbox into the inlet of the screw press.

In most inclined configurations, solids are pumped through the polymer injection and mixing device into the flocculation tank and then enter the inlet pipe to the screw press. The polymer injection and mixing device is designed to intensely mix solids and polymer; the flocculation tank is designed to provide the reaction time needed to create the appropriate floc size. The inclination also facilitates consistent cake solids at startup, because discharge is above the inlet elevation.

A slow-rotating screw conveys flocculated solids through a wedge-wire drum. Solids compaction increases gradually by reducing the pitch of the flights, increasing the center shaft diameter, and reducing the flight diameter. The inclined screw press includes a pneumatically adjusted dewatering cone at the discharge end of the screw

FIGURE 22.19   Indoor installation of an inclined screw press (courtesy of Huber Technology, Inc.).

press. It can be adjusted to provide an opening between 0.95 and 1.9 cm (3/8 and ¾ in.) and used to regulate the solids pressure, which in turn provides a balance between increasing cake dryness and an associated decrease in pressate clarity (solids capture). The pressure of the solids in the inclined screw press is typically between 34 and 152 kPa (5 and 22 psi), with a max. of 276 kPa (40 psi), depending on the type of unit and solids characteristics. In the horizontal screw press, the screen is cleaned from the inside by the rotating screw flight, which has a nominal clearance of 0.5 mm (1/50 in.). In the inclined screw press, the wedge-wire screen is cleaned from the inside by brushes or wipers fitted on the edge of the screw flights to prevent solids from attaching to the inner surface of the screen. The screen is cleaned from the outside via a spray wash system.

A screw press also is used in a process that combines dewatering and pasteurization. In this patented process, lime is added to solids to raise the pH to 12. The lime-treated solids then are flocculated with polymer and fed to the screw press. Steam is then used to heat the screw press to achieve the pathogen-reduction requirements for Class A biosolids. The resulting biosolids typically contain 30% to 50% solids. The process has more odor-control requirements because of the increased temperature and pH. The owner must decide early in the project if a Class A system is, or ever will be, a necessary part of the design because this system's design characteristics are different from those for standard screw presses and it is expensive to upgrade a dewatering screw press to the dewatering-pasteurization screw press. The first dewatering-pasteurization screw press installations started up in 2003, and there is presently little experience with this system in the United States.

Based on a survey of existing installations, solids concentrations in screw press cake vary widely depending on polymer use, solids characteristics, and dewatering application (see Table 22.14). The higher solids concentrations typically are achieved in applications with primary solids or with the combined dewatering-pasteurization process. Typical solids concentrations are 15% to 28% for secondary WAS, and 13% to 40% for a blend of primary and secondary solids. Digested solids typically result in higher concentrations than undigested solids, but require a larger polymer dosage. The dewatering-pasteurization screw press process produces cake containing up to 50% solids because of the added lime. Cake dewatering performance for solids with poor dewatering properties (e.g., in treatment facilities without primary clarifiers) can be significantly lower compared to other dewatering processes (Kabouris et al., 2005).

In addition to extrapolating data from other similar facilities, pilot-testing provides useful information for screw press designs. Pilot-testing can be used to establish trends and relationships among various performance measures (e.g., hydraulic-loading rate, solids-loading rate, polymer dose, solids capture, and cake solids concentration). However, because of significant variations in screw design and operating parameters, scaling up pilot information to a full-scale design requires careful evaluation and, potentially, an appropriate full-scale performance-guarantee contract (when pilot-scale units are used). If available, full-scale screw presses should be used in pilot-testing to better predict full-scale performance.

Screw presses run continuously at low speeds and do not require close operator supervision; therefore, they are easy to maintain and have low power consumption. The manual cleaning schedule ranges from once per week to once every 30 days.

Screw Press Type	Facility Size, ML/d (mgd)	Solids Type	Incoming % TS	Discharge % TS	Typical Feed Rate (L/m)	Polymer Dose (g/dry kg)
Horizontal	3 (0.8)	Oxidation ditch secondary sludge	4–6	20–45	57–114	11.5
Horizontal	7.6 (2)	Oxidation ditch secondary sludge	2–3	15–20	106	Not given
Horizontal	1.1 (0.3)	Secondary sludge	1.5	30	14–16	Not given
Horizontal	Not given	Secondary sludge	1.2	16	13	Not given
Horizontal	Not given	Secondary sludge	1.2	16	13	Not given
Inclined	3.8 (1)	Activated sludge secondary sludge	0.07	not given	38	Not given
Inclined	17 (4.5)	Oxidation ditch secondary sludge	0.5–1	12–17	up to 280	7.5–10
Inclined	7.7 (2)	Oxidation ditch secondary sludge	3.5–4	18–20	up to 150	7.5
Inclined	17.8 (4.7)	RBC, combined primary/ secondary sludge	2/5	18–30	up to 150	Not given
Inclined	17 (4.5)	RBC, combined primary/ secondary sludge	1.1–1.2	30–40	300–340	Not given
Horizontal	0.9 (0.2)	Aerobically digested secondary TF Sludge	1.2	18	26–38	Not given
Horizontal	16 (4.2)	Primary/secondary digested sludge	3–4	>18	190–378	9–11
Horizontal	31 (8.2)	Aerobically digested secondary sludge	1	15–19	95–397	8.2
Inclined	9.9 (2.6)	Aerobically digested oxidation ditch secondary sludge	0.9	12–14	170	Not given
Inclined	1.0 (0.3)	Aerobically digested SBR secondary sludge	Not given	22–25	30–34	Not given
Horizontal	6.1 (1.6)	Aerobically digested secondary sludge	0.75–1.0	21–28	22–64	Not given
Horizontal	83 (22)	Anaerobically digested primary/ secondary sludge	3.2	17–21	114	Not given
Inclined	23 (6.1)	Anaerobically digested oxidation ditch primary/secondary sludge	3–3.5	13–16	90–227	Not given
Horizontal	3.7 (1.0)	Anaerobically digested activated sludge primary/secondary sludge	1	20–24	26–30	Not given
Horizontal	Not given	Anaerobically digested primary/ secondary sludge	2–4	25	156–312	Not given
Horizontal	Not given	Anaerobically digested primary/ secondary sludge	2	22	163	Not given
Inclined	7.6 (2)	Aerobically digested RBC primary/secondary sludge	2	19–24	106–170	Not given

TABLE 22.14   Performance Characteristics of Screw Presses

Screw Press Type	Facility Size, ML/d (mgd)	Solids Type	Incoming % TS	Discharge % TS	Typical Feed Rate (L/m)	Polymer Dose (g/dry kg)
Horizontal	0.9 (0.2)	Aerobically digested secondary TF sludge	1.2	18	26–38	Not given
Horizontal	1.1 (0.3)	Waste activated sludge	1.5	30	14–16	Not given
Inclined	1.7 (0.46)	Aerobically digested sludge	1.0–2.0	16–25	228	12.5
Inclined	2.8 (0.75)	Aerobically digested sludge	1.5	20	95	7.5–10
Horizontal	3 (0.8)	Waste activated sludge	4–6	20–45	57–114	11.5
Inclined	3.3 (0.88)	Aerobically digested sludge	1.5–3.0	16	137–186	12.5
Horizontal	3.7 (1.0)	Anaerobically digested sludge	1	20–24	26–30	Not given
Inclined	3.8 (1)	Primary/secondary sludge	2.3–2.4	19–20	27–31	11–Oct
Inclined	5.3 (1.4)	Aerobically digested sludge	1.25	21–25	141	8.5
Inclined	5.5 (1.45)	Anaerobically digested sludge	0.5–4.0	18–25	76–152	5–12.5
Horizontal	6.1 (1.6)	Aerobically digested secondary sludge	0.75–1.0	21–28	22–64	Not given
Inclined	6.2 (1.65)	Primary/secondary sludge	1.98–2.4	21.5	31–38	10.5
Inclined	7 (1.85)	Waste activated sludge	0.5–1.5	15–20	190–243	Not given
Horizontal	7.6 (2)	Waste activated sludge	2–3	15–20	106	Not given
Inclined	7.7 (2)	Waste activated sludge	3.5–4	18–20	up to 150	7.5
Horizontal	8.3 (2.2)	Waste activated sludge	1.6	21.5	204	16.5
Inclined	11.4 (3)	Waste activated sludge	0.25	17	605	15
Horizontal	16 (4.2)	Anaerobically digested sludge	3–4	>18	190–378	9–11
Inclined	17 (4.5)	Waste activated sludge	0.5–1	12–17	up to 280	7.5–10
Inclined	17 (4.5)	RBC, combined primary/ secondary sludge	1.1–1.2	30–40	300–340	Not given
Inclined	17.8 (4.7)	RBC, combined primary/ secondary sludge	2–5	18–30	up to 150	Not given
Inclined	18.9 (5)	Aerobically digested sludge	1.5–2.0	18–22	219–254	15
Inclined	23 (6.1)	Anaerobically digested sludge	3–3.5	13–16	90–227	Not given
Inclined	29.5 (7.8)	Aerobically digested sludge	0.75–4.0	15	144	Not given
Inclined	30.3 (8)	Waste activated sludge	2.0–3.0	15	341	Not given
Horizontal	31 (8.2)	Aerobically digested secondary sludge	1	15–19	95–397	8.2
Inclined	34.1 (9)	Primary/Secondary Sludge	0.8	18	474	10
Inclined	34.1 (9)	Anaerobically digested sludge	2.2–2.7	22–25	225–250	17
Inclined	37.9 (10)	Aerobically digested sludge	1.5	17	95	7.5
Horizontal	83 (22)	Anaerobically digested sludge	3.2	17–21	114	Not given
Inclined	90.9 (24)	Aerobically digested sludge	1.5–2.5	18	284–360	13
Horizontal	92.6 (24.5)	Anaerobically digested sludge	3.1–4.0	15.9–19.0	132	17.9–28.5
Inclined	106 (28)	Anaerobically digested sludge	1.0–4.0	25	379	19

TABLE 22.14 Performance Characteristics of Screw Presses (*Continued*)

7.2   Process Design Conditions and Criteria

Screw press designs are determined by solids characteristics, proper chemical conditioning, and hydraulic and solids loads, which affect the detention time (speed of the screw). It is essential to select the proper conditioning agent (polymer); flocculation should agglomerate solids into large, strong flocs that release as much water as possible. The released water drains via gravity in the thickening section of the screw press. Screw press performance improves when less water must be removed in the dewatering zone.

7.2.1   Hydraulic Loading

In addition to solids loading, solids type, and desired discharge dryness, the hydraulic-loading rate is a factor that should be considered in screw press sizing. Higher hydraulic-loading rates typically require larger-diameter presses or a coarser screen. An important factor affecting the hydraulic-loading rate is solids conditioning, because the screw press can only attain maximum capacity with the optimal polymer type and application. Optimal operation occurs when flocculated solids have sufficient time to fully gravity drain before being conveyed beyond the inlet end of the screw press. Typical hydraulic-loading rates for a horizontal screw press range from 3.8 to 2 081 L/min (1 to 550 gpm), depending on the screw press model. Typical hydraulic-loading rates for an inclined screw press are between 18.9 and 454 L/min (5 and 120 gpm), depending on the model and solids concentration.

7.2.2   Solids Loading

The solids-loading rate for screw press dewatering varies, depending on solids characteristics, screw size, and rotational speed. The solids-loading rate capacity for a horizontal screw press ranges from 0.91 to 703 kg/h (2 to 1550 lb/hr), depending on screw press model. The typical solids-loading rate for an inclined screw press is 22.7 to 454 kg/h (50 to 1000 lb/hr), depending on screw press model. The inclined screw press design is mainly controlled by solids loading, but hydraulic loading is an important factor if the residuals have a low solids content.

Screw presses can operate under a wide range of load rates just by changing the screw speed. The design involves balancing an increased solids or hydraulic-loading rate with increasing screw speed and higher cake solids concentrations (which improve at slower screw speeds). Screw press operations are determined by solids flow (hydraulic or solids loading) and auger speed (retention time of conditioned solids in dewatering system). Cake solids will improve as auger speed decreases, provided that solids flow is constant. The speed can be reduced until incoming flow exceeds the amount of water that drains by gravity in the lower part of the screw press. Then, solids will start backing up.

If screw speed is constant, then the screw press can be operated with a variable solids-feed flowrate as follows. Because cake solids will improve as solids flow increases, the flowrate can be increased until the incoming flow exceeds the amount of water that drains by gravity at the front of the screw press and solids start to back up. The flowrate then is decreased a little to keep the system in a steady state.

The screw press will operate at its maximum capacity as long as the water drainage is not limited by high solids load and the volume of the auger is filled properly with solids to ensure maximum pressure in the dewatering zone. The loading highly depends on proper conditioning, so it is impossible to predict maximum loading-rate values. Maximum loading rate is plant specific and must be determined on-site (e.g., during pilot tests or startup).

7.2.3 Rotation Speed

Typical rotation speeds range from 0.1 to 2.0 rpm for horizontal screw presses and from 0.5 to 2.0 rpm for inclined screw presses. In general, an increase in screw rotation speed increases production capacity but decreases cake solids concentration. In a full-scale application, increasing rotational speed from 1 to 1.25 rpm reduced the cake concentration from 23% to 20% (Atherton et al., 2005).

7.3 Design Considerations

7.3.1 Mechanical Features and Materials of Construction

Significant variations exist in screw design among manufacturers, and detailed screw design information is proprietary. Primary features include the screw, screen, flocculation system, cleaning, and drive. Stainless steel (either 304 or 316L) typically is used for all wetted screw surfaces.

7.3.1.1 Screw Screw press design parameters include the inlet-end screw-flight outside diameter; inlet-end shaft/shell diameter; inlet-end flight pitch or flight-to-flight dimension; screw length; discharge-end screw-flight outside diameter; discharge-end shaft/shell diameter; discharge-end flight pitch or flight-to-flight dimension; and combinations of single-helix and double-helix designs. Common designs include constant flight, outside diameter with tapered screw shaft/shell diameter, and varied flight, outside diameter with constant screw shaft/shell diameter.

The inclined screw design (flight thickness, pitch, shaft diameter) is determined by the machine size and screen configuration. The diameter of the screw is 28 to 80 cm (11 to 31.5 in.), and the pitch of the flights typically is 15 to 25 cm (5.9 to 9.8 in.), although up to 40 cm (15.7 in.) is possible. Inclination varies from 10° to 20°. The inclination of the screw press allows the cake conveyor to fit under the discharge so the screw press does not need another support/pedestal. The inclination also facilitates a consistent cake during startup, because the discharge is above the inlet elevation.

7.3.1.2 Screen Screw-press screen configurations also vary significantly among manufacturers. Screens can be either perforated stainless steel sheet or wedge wire. The holes used in perforated plate screens range from 1.0 to 3.0 mm in diameter, depending on solids type and press size. The open area of perforated plate screens ranges from 2% to 48%, depending on solids type, inlet consistency, and location on the screw press. The drums that support the screens can be either two-piece or one-piece drums. The advantage of two-piece (split) drums is ease of assembly and disassembly, because the drums (screens) can be removed from the press without removing the screw. Two-piece (split) drums allow an easy changeout of the inner perforated plate screens to fine-tune press performance, while removing one-piece drums requires disassembling the press.

The inclined screw press is only available with wedge-wire screens made of stainless steel (304 or 316 Ti [optional]). The bar spacing varies from 0.05 to 0.5 mm, depending on screen configuration, solids quality, and pilot-testing results. There are up to three wedge-wire sections in a screw press. The open area of the screen sections is 3% to 30%. The support structure for the basket does not affect the open area. Additionally, the drums that support the screens can be either two-piece or one-piece drums, and provide the same advantages as noted above. The main goal is to ensure a capture rate of at least 95% regardless of what type of solids are dewatered.

There is the possibility of nonstandard press designs for solids with high fiber content (e.g., primary solids). The fiber content allows for bigger openings, which may allow higher loading rates without compromising capture efficiency. Nonstandard screw-press designs are always based on pilot tests.

7.3.2 Area/Building Requirements

7.3.2.1 Structural Requirements Structural elements for screw presses include a concrete base (pad) or elevated support system for horizontal screw-press installations, a screw-press support system for inclined screw presses, flocculation vessel, screw press, and support for the rotary screen thickener, if present. Depending on the screw-press installation configuration, a stair and landing system or movable ladder may be required to gain access to all parts of the screw press.

Because the fully enclosed design contains odors and minimizes operating noise, screw presses can be installed outdoors in mild climates. If installed indoors, there must be adequate room around the perimeter of the press to accommodate normal maintenance and manual washdown. Adequate overhead space also must be considered for maintenance, as well as lifting equipment. Lifting equipment can be either fixed (e.g., a crane rail centered above the press) or mobile equipment. If mobile equipment will be used, then equipment access needs to be considered.

Because screw presses are fully enclosed, ventilation requirements are minimal (refer to NFPA 820, *Standard for Fire Protection in Wastewater Treatment and Collection Facilities*, and Chapter 6, Odor Control). If odors are a concern, screw-press covers can be fitted with ventilation connections. Such connections should have flexible ductwork that can be removed by operators without tools (e.g., a flex hose slipped over a pipe stub, held in place by a hose clamp with a thumbscrew). Typical ventilation airflow rate requirements range from 340 to 680 m³/h (200 to 400 cu ft/min) for direct connection to the screw press. The screw-press housing, screw-press discharge, and/or cake conveyor system also can be attached to a ventilation system for odor control.

Following are other design and layout considerations for screw presses:

- Provide curb around the screw-press area to protect other areas from spills and washdown water.

- Provide access to all parts of screw press via platforms or movable stepladders.

- Provide a hoist or crane for the installation, removal, and/or repair of screw-press components.

- Provide a plant water connection for the automatic washdown system.

- Provide connections to the plant's odor-control system if odor ducts off of the screw press are desired. The dewatered solids conveying system also should be connected to the odor-control system.

- Provide adequate space between screw-press units for maintenance and manual cleaning operations.

- Storage tanks from which solids are being fed to the screw press should have enough mixing to ensure a near-constant feed solids concentration during dewatering.

- Primary and secondary solids should be blended in an upstream storage tank; mixing solids in the pipe feeding the dewatering system is not recommended.

Unlike many other dewatering systems, screw presses typically are designed for continuous operation; because of the automation of the screw press, operations can range from 4 to 24 hours per day, 5 to 7 days per week depending upon facility operations, staffing, volume of sludge treated, and capacity of the screw press units. In some cases, screw-press systems are designed without a redundant (backup) unit due to the infrequent maintenance required; however, the facility must have facilities to store solids for several days or alternative means to handle the sludge to be dewatered, if the screw press is off-line. Multiple screw presses may be necessary for larger facilities to provide redundant capacity and increase system reliability. Multiple screw presses also allow for a smaller number of screw presses to be operated during extended periods of lower solids production.

7.3.2.2 Energy Requirements Screw presses have relatively low power requirements. Screw-press motor horsepower ranges from 0.67 to 10 kW (0.5 to 7.5 hp) for horizontal screw presses and from 0.67 to 2.7 kW (0.5 to 2 hp) for inclined screw presses, depending on screw-press size. Screw-press installations also require a solids feed pump and polymer system, whose power requirements depend on system size. The flocculation tank's mixer motor is typically 2 kW (1.5 hp) or less.

7.4 Ancillary Systems

Ancillary equipment needed for screw-press dewatering includes a solids feed pump, polymer makeup and feed system, flocculation vessel, and dewatering control system. The screw-press manufacturer may provide these components as part of a package system.

7.4.1 Controls

The dewatering control system is provided by the manufacturer; it includes an operator interface to monitor and adjust screw press operations. The main control panel operates the entire dewatering system (i.e., the solids feed pump, polymer system, flocculation vessel, rotary screen thickener [if applicable], and screw press) as a complete system. The control system is equipped with operating and warning lights for various system monitoring points, audio alarms, emergency shutoff devices, and a display panel. Screw-press dewatering systems typically are operated in an automatic mode.

7.4.2 Feed System

Screw-press performance depends on a consistent inflow of solids. Therefore, positive-displacement or progressing-cavity pumps are highly recommended because their performance is not affected by solids consistency or changing water levels in the storage tanks. The progressing-cavity pumps must be designed properly to minimize the wear and tear caused by grit and other abrasive material.

The polymers used in screw-press dewatering are not specially designed for this application; they are widely available from various suppliers. The choice of polymer needs to be determined via jar testing. There are no specific requirements for the polymer makeup system. It needs to be sized properly for the maximum solids load. System controls should allow automatic operation and include a feature to pace polymer feed rate with solids flow. Using an aging tank will maximize polymer efficiency and, therefore, minimize polymer consumption.

The system always needs some operator attention, especially if the residuals' solids concentration is inconsistent. The control system can only adjust the polymer dosing rate when solids flow (hydraulic load) changes; any change in solids loading due to solids concentration must be addressed manually by operators. So, it is essential to operating reliability that the solids concentration is constant (i.e., proper mixing occurs in the storage tank). Solids content could be monitored continuously, but such sensors often require a lot of maintenance and operator attention. Therefore, it may be easier to readjust the dewatering system's settings by changing the "solids content of raw solids" parameter at the operator interface.

For the horizontal screw press, solids pump speed typically is adjusted automatically based on headbox-level analog input. The goal is to maintain a constant headbox level.

7.4.3   Chemical Conditioning

Polymer addition promotes particle flocculation and increases the dewatering and solids-capture rates. Jar testing and pilot-testing can be used to estimate the type and quantity of polymer necessary for each application, because it may vary significantly depending on solids characteristics. Polymer consumption is affected by multiple parameters (e.g., grit content of the solids, the presence or absence of primary clarifiers, the type of biological treatment, and the type and duration of solids digestion). Polymer doses for screw-press systems can range from 3 to 17.5 g of active polymer per kilogram of dry solids (g/kg) (6 to 35 lb of active polymer per dry ton of solids), with a typical range of 6 to 10 g/kg (12 to 20 lb/dry ton). For industrial wastewaters, the polymer dosages can range from 15 to 40 g/kg (30 to 80 lb/dry ton), which is dependent upon the wastewater characteristics to be dewatered. The dewatering-pasteurization process also requires lime addition. Lime dosage typically ranges between 100 and 400 g/kg of dry solids (200 and 800 lb/dry ton of solids). In general, an increase in polymer dose increases cake solids concentration, although other factors (e.g., screw speed) can affect performance as well.

The polymer type and dosage also largely determines the solids capture rate. Solids capture also is affected by the efficiency of polymer injection and mixing, the screen design (size of opening), and the pressure inside the dewatering zone.

Flocculation tanks typically are used to mix solids and polymer to condition solids before dewatering. The horizontal screw-press flocculation tank is a vertical cylindrical tank with the solids-polymer feed at the bottom and an overflow at the top. The tank typically is sized for a retention time of 3 to 10 minutes, depending on solids type and inlet consistency. Flocculation tanks may be provided with variable-speed agitators to allow operators to optimize mixing energy. Undermixing results in undispersed polymer, while overmixing results in broken floc. If the flocculation tank is undersized, then a non-clog in-line static mixer can be used to blend solids and polymer before the mixture enters the flocculation tank.

In the inclined screw-press design, polymer is injected into the solids flow immediately before they enter the mixing device. The mixing valve is equipped with a manually adjustable weight to adjust the mixing energy in response to solids characteristics and to minimize polymer consumption. The flocculation tank typically is sized for a retention time of 0.5 to 1.0 minute to control floc formation.

7.4.4   Cleaning System

Screw-press systems have automatic cleaning systems that involve facility water and spray nozzles. During automated wash cycles, washwater from solenoid valves sprays onto the screw-press screen to remove built-up solids. The washwater system minimum design pressure for a horizontal screw press is 276 kPa (40 psi).

The inclined screw press has two cleaning processes. First, the screen is cleaned continuously from the inside via brushes or wipers mounted on the edge of the rotating screw flight. The brushes are made of nylon with stainless steel mounting hardware. The wipers are made of polyurethane with stainless steel mounting hardware. This mainly cleans the screen to allow water to drain by gravity (especially in the lower part of the screen) and minimize resistance to water filtration. Clean screens require less dewatering pressure, which improves the solid capture rate. The second cleaning process is an automatic spray wash system, which cleans the screen from the outside. It comprises of a rotating spray-bar washing system and spray nozzles fed by solenoid valves. The spray-bar system is made of stainless steel piping (304, 316 on request) with flat fan nozzles (made of polyvinylidene difluoride [PVDF]), and the washwater system design pressure is 414 to 517 kPa (60 to 75 psi).

Pressate quantity is a factor of inlet flow, inlet consistency, and washwater flow. Screw presses do not require continuous washwater, and most have automatic, intermittent showers. Typically, a horizontal press' automated wash cycle occurs for 1 minute each hour. The resulting washwater flow volume is 2% to 5% of solids feed rate (on average). Depending on the size of the screw press, the instantaneous flowrate can range from 76 to 454 L/min (20 to 120 gpm). The presses also require manual washdown at a frequency ranging from once per week to once per month. Typical washwater volume required for manual washdown is 0.2% to 0.5% of solids feed rate (on average).

The inclined screw press' automatic spray wash system is controlled by a timer and runs intermittently, typically for one revolution (about 60 seconds) every 10 to 15 minutes, but there are installations where the wash cycle only operates once every 30 minutes. Typical washwater demand for an inclined press is 49 to 163 L (13 to 43 gal) per wash cycle per screw press, depending on the screw press size. The actual flowrate is 79 to 132 L/min (21 to 35 gpm). Total washwater volume is 4% to 9% of the processed solids volume (on average), with a maximum of 15%. Water demand depends on solids flocculation performance.

Manual spraydown of the flocculation vessel, screw, and screen is done as needed, depending on installation conditions and dewatering schedule. Manual washdown takes about 1 to 2 hours and is recommended from about once per week to once per month, depending on screw-press runtimes.

There are limited data available documenting the full-scale solids capture for screw presses under various field conditions. Solids capture typically ranges from 85% to 97%. Captures of up to 99% have been reported in certain applications, but a design capture rate of 95% is more common. Pressate typically is clear when a screw press dewaters WAS. It is slightly cloudy when primary or digested solids are processed. The particles are typically small, and the solids content varies depending on screw-press status: low solids content during dewatering, and higher solids content during the wash cycle.

8.0 References

Atherton P. A.; Steen, R.; Stetson, G.; McGovern, T.; Smith, D. (2005) Innovative Biosolids Dewatering System Proved a Successful Part of the Upgrade to the Old Town, Maine, Water Pollution Control Facility. *Proceedings of the 78th Annual Water Environment Federation Technical Exhibition and Conference* [CD-ROM], Washington, D.C., October 29–November 2; Water Environment Federation: Alexandria, Virginia.

Banks, L.; Davis, S. (1983) Desiccation and Treatment of Sewage Sludge and Chemical Slimes with the Aid of Higher Plants. *Proceedings of the 15th National Conference on Municipal and Industrial Sludge Utilization and Disposal*, Atlantic City, New Jersey; Hazardous Materials Control Research Institute: Silver Springs, Maryland.

Baxter, J. C.; Martin, W. J. (1982) Air Drying Liquid Anaerobically Digested Sludge in Earthen Drying Basins. *J. Water Pollut. Control Fed.*, **54**, 16.

Chen, Y.; Yang, H.; Gu, G. (2001) Effect of Acid and Surfactant Treatment on Activated Sludge Dewatering and Settling. *Water Res.*, **35** (11), 2615–2620.

Coackley, P.; Allos, R. (1962) The Drying Characteristics of Some Sewage Sludges. *J. Proc. Inst. Sew. Purif. (G.B.)*, **6**, 557.

Crolla, A.; Goulet, R.; Kinsley, C.; Ho, T., (2007) Septage Treatment Pilot Project, Drying Bed and Reed Bed Filters. *Proceedings of the Ontario Onsite Wastewater Association Conference*, March 26–28; Ontario Onsite Wastewater Association: Cobourg, Ontario, Canada.

Crosswell, S.; Young, T.; Benner, K. (2004) Performance Testing of Rotary Press Dewatering Unit under Varying Sludge Feed Conditions. *Proceedings of the 77th Annual Water Environment Federation Technical Exhibition and Conference* [CD-ROM], New Orleans, La., October 2–6; Water Environment Federation: Alexandria, Virginia.

Great Lakes—Upper Mississippi River Board of State and Provincial Public Health and Environmental Managers (2014) *Recommended Standards for Wastewater Facilities "Ten States Standards"*; Health Research, Inc.: Albany, New York.

Haseltine, T. R. (1951) Measurement of Sludge Drying Bed Performance. *Sew. Ind. Wastes*, **23**, 1065.

Imhoff, K.; Fair, G. M. (1940) *Sewage Treatment*; Wiley & Sons: New York.

Jeffrey, E. A. (1959) Laboratory Study of Dewatering Rates for Digested Sludge in Lagoons. *Proceedings of the 14th Purdue Industrial Waste Conference*, West Lafayette, Indiana; Purdue University: West Lafayette, Indiana.

Jeffrey, E. A. (1960) Dewatering Rates for Digested Sludge in Lagoons. *J. Water Pollut. Control Fed.*, **32**, 1153.

Kabouris J. C.; Gillette, R. A; Jones, T. T.; Bates, B. R. (2005) Evaluation of Belt Filter Presses, Centrifuges, and Screw Presses for Dewatering Digested Activated Sludge at St. Petersburg's Water Reclamation Facilities. *Proceedings of the 78th Annual Water Environment Federation Technical Exhibition and Conference* [CD-ROM], Washington, D.C., October 29–November 2; Water Environment Federation: Alexandria, Virginia.

Koch, C. M.; Chao, A; Semon, J. (1988) Belt Filter Press Dewatering of Wastewater Sludge. *ASCE J. Environ. Eng.*, **114** (5), 991–1005.

Koch, C. M.; McKinney, D. E.; Fagerstrom, A. A.; Palmer, E. W. (1989) Comparison of Centrifuge Performance on Oxygen Activated Sludge. *Proceedings of the Environmental Engineering Division, American Society of Civil Engineers in Cooperation with the University of Texas, at Austin, Civil Engineering Department*, Austin, Texas, July 10–12; American Society of Civil Engineers: New York.

National Fire Protection Association (2016) *NFPA 820: Standard for Fire Protection in Wastewater Treatment and Collection Facilities*.

New England Interstate Water Pollution Control Commission (2011) *Technical Report 16 (TR-16, 2011), Guides for the Design of Wastewater Treatment Works*; New England Interstate Water Pollution Control Commission: Lowell, Massachusetts.

Pandey, M. K.; Jenssen, P. D. (2015) Reed Beds for Sludge Dewatering and Stabilization. *J. Environ. Prot.*, **6**, 341–50.

Pennsylvania Department of Environmental Protection (1997) *Domestic Wastewater Facilities Manual*; PADEP: Harrisburg, Pennsylvania.

Quon, J. E.; Johnson, G. E. (1966) Drainage Characteristics of Digested Sludge. *J. Sanit. Eng. Div., Proc. Am. Soc. Civ. Eng.*, **92**, 4762.

Randall, C. W.; Koch, C. T. (1969) Dewatering Characteristics of Aerobically Digested Sludge. *J. Water Pollut. Control Fed.*, **41**, R215.

Reimers, R. S.; et al. (1981) *Parasites in Southern Sludges and Disinfection by Standard Sludge Treatment*, Project Summary; U.S. Environmental Protection Agency: Washington, D.C.

Riggle, D. (1991) Reed Bed System for Sludge. *BioCycle*, **32** (12), 64–66.

Rolan, A. T. (1980) Determination of Design Loading for Sand Drying Beds. *J. N.C. Sect., Am. Water Works Assoc., N.C. Water Pollut. Control Assoc.*, L5, 25.

Shao, L.; He, P.; Yu, G.; He, P. (2009) Effect of Proteins, Polysaccharides, and Particle Sizes on Sludge Dewaterability. *J. Environ. Sci.*, **21**, 83-88.

Vesilind, P. A. (1979) *Treatment and Disposal of Wastewater Sludges*, Revised ed.; Ann Arbor Science Publishers: Ann Arbor, Michigan.

Walski, T. M. (1976) Mathematical Model Simplifies Design of Sludge Drying Beds. *Water Sew. Works*, **123**, 64.

Wang, L. K.; Li, Y.; Shammas, N .K., Sakellaropoulos, G. P. (2007). Chapter 13, Drying Beds. In *Handbook of Environmental Engineering*, Volume 6: Biosolids Treatment Processes, The Humana Press: Totowa, New Jersey.

Water Environment Federation (2012) *Wastewater Treatment Plant Design Handbook*; Water Environment Federation: Alexandria, Virginia.

Zablatzky, H. R.; Peterson, S. A. (1968) Anaerobic Digestion Failures. *J. Water Pollut. Control Fed.*, **40**, 581.

Zacharias, D. R.; Pietila, K. A. (1977) Full-Scale Study of Sludge Process and Land Disposal Utilizing Centrifugation for Dewatering. *Proceedings of the 50th Annual Meeting of the Central States Water Pollution Control Association*, Milwaukee, Wisconsin; Central States Water Pollution Control Association: Milwaukee, Wisconsin.

CHAPTER **23**

Stabilization

Matthew J. Williams, P.E.; Paul Bizier, P.E., F.ASCE, D.WRE;
Jim Groman; and Rachelle Tippetts

1.0 Introduction

With growing focus on renewable energy production and biosolids utilization, stabilization processes have gained more attention in recent years. Stabilization treats the solids generated in the liquid treatment train, converting them to a stable (i.e., not readily putrescible) product for beneficial use or disposal. Pathogens and odors are reduced, making the resulting biosolids appealing for beneficial use. This chapter discusses the four most common stabilization processes used in the United States today: anaerobic digestion, aerobic digestion, composting, and alkaline stabilization. (Thermal drying also is considered a stabilization process, but it is covered in Chapter 24.)

When designing a stabilization process, engineers should start by identifying the reasons for stabilization and all the stabilization options that could be easily integrated into the existing wastewater treatment scheme. Before selecting a process, the end use must be considered in light of local demand and regulations. Once a process is chosen, engineers should review all aspects (e.g., sidestream management, effectiveness in producing the desired biosolids quality, safety, ease of operation, and ancillary equipment needs) to ensure that the system is well designed. Design engineers should use a systematic approach that addresses both the economic and noneconomic ramifications of any proposed processing. This includes evaluating all parts of the system to ensure that they are flexible and reliable enough to consistently produce biosolids with the required characteristics. Furthermore, any potential effects on the liquid treatment processes (e.g., nutrient recycle loads) that may come as a result of the selected system must be carefully evaluated and mitigated where necessary.

The quality of the biosolids produced and the overall process operation can be improved if debris is removed from the solids prior to processing. Solids screens may be used to remove plastics and other debris from the solids. Grinders are also acceptable to help prevent problems with pumps and clogging, but these may leave small pieces of debris in the finished product, thus diminishing its quality.

Not all water resource recovery facilities (WRRFs) stabilize their solids. Those who do, however, typically stabilize them for one or more of the following reasons:

- Aesthetic reasons (e.g., product appearance and odor);
- Mass reduction;
- Volume reduction;
- Biogas (renewable energy) production;
- Reduction of pathogens (typically to comply with 40 CFR Part 503);
- Vector-attraction/volatile solids reduction (VSR) (typically to comply with 40 CFR Part 503); and
- Product usefulness and marketability.

Facility planners should consider biosolids management when designing or upgrading WRRFs. They should start by deciding how the solids will be used or disposed since this will determine whether and what type of stabilization is needed. For example, if solids will be landfilled with routine cover, federal and many state agencies may not require that the material be stabilized. Stabilization also may not be desirable if the solids will be thermally oxidized, because doing so will reduce the energy content. On the other hand, when treated biosolids are used in agriculture or silviculture, or distributed commercially, the material is stabilized to reduce pathogens, odor, and vector attraction in accordance with local and federal requirements.

1.1 Comparison of Processes

The implementation of 40 CFR Part 503 and the public's growing concern for the environment have increased research into new technologies for beneficially using biosolids. They also have prompted many engineers and municipalities to investigate or design more effective stabilization systems. Tables 23.1 and 23.2 summarize many of the advantages and disadvantages of the principal stabilization processes used today. (Drying is not included in these tables because it is evaluated in Chapter 24.)

Anaerobic digestion is a very common stabilization process, typically utilized only at facilities that generate primary sludge. It produces relatively stable biosolids as well as biogas to fuel:

- Boilers to heat digesters and buildings;
- Cogeneration systems (e.g., generators or microturbines with heat recovery); and
- Thermal hydrolysis and/or biosolids dryers.

Compared to other stabilization options, these systems may be expensive to build; require a significant tankage (see Figure 23.1) and equipment (particularly if the digester gas is beneficially used); produce a strong ammonia and phosphorus sidestream; and need extra heat to maintain the desired temperature; and the process biology is slow growing and somewhat sensitive. Such disadvantages to this widely used process are largely resolved via proper design, careful operations, and pretreatment programs.

Aerobic digestion is typically used at smaller WRRFs (smaller than about 19 000 m³/d [5 mgd]) and those that only produce biological solids or waste activated sludge (WAS). Compared to anaerobic digestion, aerobic digestion is a power-intensive process (because of the power needed for oxygen transfer), but it typically is less expensive to construct and simpler to operate than anaerobic digestion.

Researchers have developed many methods for increasing VSR, and pathogen destruction in aerobic digestion processes. Some use higher temperatures to destroy pathogens and reduce volatile solids (VS), while others use phasing or mechanical disruption to improve performance. Sometimes aerobic digestion is not a separate process; many extended aeration facilities (e.g., oxidation ditches) have a long-enough solids retention time (SRT) to provide at least partial digestion via endogenous respiration. However, Part 503 does not permit the VSR achieved in aeration tanks to be included as part of the 38% VSR required for biological solids stabilization.

Autothermal thermophilic aerobic digestion (ATAD) is an advanced, self-heating aerobic digestion process that operates at 50°C to 65°C (131°F to 149°F). Early ATAD experience in the United States established the process and allowed it to evolve with improvement of subsequent systems.

Composting often is used to convert solids into a soil amendment or conditioner. The feedstock can be either raw solids or biosolids, but should contain at least 40% solids. A bulking agent frequently is added to increase solids content, provide carbon for the process, improve the material's structural properties, and promote adequate air circulation. Composting typically is a labor-intensive process (e.g., adding bulking agent, turning the material, and recovering the bulking agent). It also can emit odors, especially if the

Process	Advantages	Disadvantages
Anaerobic digestion	Good volatile suspended solids destruction (40% to 60%)	Requires skilled operators
		May experience foaming
	Net operational cost can be low if gas (methane) is used	Methane formers are slow growing; hence, "acid digester" sometimes occurs
	Broad applicability	Recovers slowly from upset
	Biosolids suitable for agricultural use	Supernatant strong in ammonia and phosphorus
	Reduces total sludge mass	
	Low net energy requirements	Cleaning is difficult (scum and grit)
		Can generate nuisance odors resulting from anaerobic nature of process
		High initial cost
		Potential for struvite (mineral deposit)
		Safety issues concerned with flammable gas
Advanced anaerobic digestion (many process options)	Excellent volatile solid destruction	Requires skilled operators
	Can produce Class A biosolids using time and temperature based batch operations	Can be maintenance intensive (see Anaerobic digestion for other disadvantages)
	Can increase gas production	
	Can reduce solid retention time	
Aerobic digestion	Low initial cost, particularly for small plants	High energy cost
	Supernatant less objectionable than anaerobic	Generally lower volatile suspended solids destruction than anaerobic
	Simple operational control	
	Broad applicability	Reduced pH and alkalinity
	If properly designed, does not generate nuisance odors	Potential for pathogen spread through aerosol drift
	Reduces total sludge mass	Biosolids typically are difficult to dewater by mechanical means
		Cold temperatures adversely affect performance
		May experience foaming

TABLE 23.1   Comparison of Stabilization Processes

Process	Advantages	Disadvantages
Autothermal thermophilic aerobic digestion	Reduced hydraulic retention compared with conventional aerobic digestion	High energy costs
		Potential of foaming
	Volume reduction	Requires skilled operators
	Excess heat can be used for building heat	Potential for odors
	Pasteurization of the sludge pathogen reduction	Requires 18% to 30% dewatered solids
Composting	High-quality, potentially saleable product suitable for agricultural use	Requires bulking agent
		Requires either forced air (power) or turning (labor)
	Can be combined with other processes	
	Low initial cost (static pile and window)	Potential for pathogen spread through dust
		High operational cost: can be power, labor, or chemical intensive, or all three
		May require significant land area
		Potential odors
Lime stabilization	Low capital cost	Biosolids not always appropriate for land application
	Easy operation	
	Good as interim or emergency stabilization method	Chemical intensive
		Overall cost very site specific related to product management costs
		Volume of biosolids to be managed is increased
		pH drop after treatment can lead to odors and biological growth in product odorous operations
Advanced alkaline stabilization	Produces a high-quality Class A product	Operator intensive
	Can be started quickly	Chemical intensive
	Excellent pathogen reduction	Potential for odors
		Volume of biosolids to be managed is increased
		May require significant land area

TABLE 23.1 Comparison of Stabilization Processes (*Continued*)

| | Degree of Attenuation | |
Process	Pathogens	Putrefaction and Odor Potential[a]
Anaerobic digestion	Fair	Good
Advanced anaerobic digestion	Excellent[b]	Good
Aerobic digestion	Fair	Good
Autothermal thermophilic aerobic digestion	Excellent	Good
Lime stabilization	Good	Good
Advanced lime stabilization	Excellent	Good
Composting	Excellent	Good

[a]In addition to the stabilization process, putrefaction and odor potential also depends on post-processing and storage practices.
[b]For Class A time–temperature processes.

TABLE 23.2   Attenuation Effect of Well-Conducted Treatment Processes on Stabilizing Wastewater Solids

site is poorly designed or operated. In addition, the process may increase the mass of biosolids to be used or disposed, and could spread pathogens via dust from the material.

Lime or alkaline stabilization frequently is used to meet the 40 CFR Part 503 requirements for Class B biosolids. In some cases, this process can produce a soil amendment or conditioner that meets Class A requirements. Alkaline stabilization typically is simpler to operate than digestion or composting. However, the resulting biosolids can become

FIGURE 23.1   Anaerobic digesters at Loudon Water in Loudon, VA (photo courtesy of Crom Corporation).

unstable if the pH drops after treatment and organisms regrow. Also, the lime or alkaline agent often is costly and can significantly increase the mass and cost of solids to be used or disposed. Some facilities have experienced odors and undesirable working conditions.

A number of advanced alkaline-stabilization technologies now used in the wastewater treatment field include chemical additives in addition to, or instead of, lime.

Alkaline-stabilization processes produce a rich, soil-like product containing a few pathogens. The biosolids also have a higher pH, which is desirable at farms with acidic soils. However, alkaline stabilization increases the mass of biosolids to be managed, as well as generating strong ammonia and amine odors that may need to be treated. One of the more important parameters for alkaline stabilization is mixing efficiency, which depends on the raw materials used in the process (rheology of dewatered cake and gradation of lime). Improper mixing results in variable biosolids characteristics and odors during storage and land application.

2.0   Anaerobic Digestion

2.1   Process Development

Anaerobic digestion is the most common and established technology used to stabilize wastewater solids and has been in use for nearly a century. The major objectives of the technology have historically been as follows:

- Stabilize primary and secondary/tertiary solids;
- Reduce pathogens;
- Reduce the mass of material (hauling cost savings); and
- Produce usable biogas (typically for boilers).

As sustainable practices have developed in the United States, the role of anaerobic digesters at WRRFs has evolved to include the following objectives:

- Generate a biosolids product with fertilizer value;
- Recover resources by co-digesting solids with other organic wastes; and
- Develop power and energy via biogas use in cogeneration facilities.

As the importance of anaerobic digestion has increased in the wastewater industry, a number of new process alternatives, designs, and fundamental understandings have evolved.

Anaerobic digestion is a relatively complex process biochemically, but mechanically it is quite straightforward. It requires both proper design and careful operation. A better understanding of the fundamental aspects of design and process control is required to ensure reliable and efficient operation, especially for more advanced, high-rate systems.

Drawbacks of anaerobic digestion include the following:

- Handling potentially explosive and corrosive gases.
- A more complex system than aerobic digestion.
- Recycle streams high in nitrogen and phosphorus.

- Decreased dewaterability (where EBPR liquid processes are used).
- Completely closed tanks make process monitoring more challenging.

2.2   Process Fundamentals

2.2.1   Microbiology and Biochemistry

Anaerobic digestion is driven by a series of syntrophic relationships that convert complex organic matter via a series of intermediate compounds to a variety of low-molecular-weight reduced compounds. The primary products of anaerobic digestion are methane (CH_4), carbon dioxide (CO_2), hydrogen (H_2), hydrogen sulfide (H_2S), ammonia (NH_3), phosphate (PO_4^{-3}), and residual organic matter and biomass. Digestion can be viewed as a series of steps in which the waste products of one organism are the substrate for another. Solids destruction is the result of a balanced coupling of a variety of metabolisms.

Figure 23.2 is a simplified flow diagram of the major metabolic processes in anaerobic digestion for converting organic matter to methane and carbon dioxide. Each metabolic pathway represents myriad microorganisms, many of which have yet to be speciated.

The microorganisms responsible for digestion are bacteria and archaea. Each group provides a unique and indispensable biotransformation. Hydrolysis, acidogenesis, and methanogenesis are the three major metabolic steps in anaerobic digestion. Each step involves several biochemical reactions to convert complex organics to intermediates, such as short-chained organic acids, and final products, such as methane and carbon dioxide.

2.2.2   Process Rates and Kinetics

Process rates are impacted by several external factors, including temperature, substrate, interspecies competition, and the presence of toxicants. While the environmental conditions affect the observed process rates, the fundamental limits of a process are regulated by kinetics.

Two kinetic models dominate the fundamental description of process performance: Michaelis-Menten and Monod. Michaelis-Menten kinetics describe the kinetics of enzymatic reactions, which are primarily responsible for the first step in anaerobic digestion: hydrolysis. Monod kinetics describe the process reactions mediated by specific microorganisms (e.g., acetogenesis and methanogenesis).

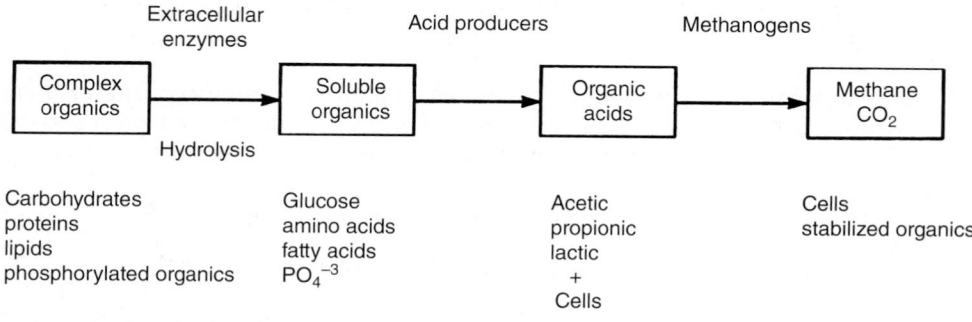

FIGURE 23.2   The major metabolic processes and products of anaerobic digestion.

Microbial Population or Metabolic Group	μ max (h^{-1})	K_s (mg substrate COD/L)	Source
Amino acid and sugar fermenting bacteria	0.25	20–25	Grady et al. (1999)
Long chain fatty acid oxidation	0.01	800	Grady et al. (1999)
Propionate oxidation	0.0065	250	Gujer and Zehnder (1983)
Propionate oxidation	0.0033	800	Bryers (1984)
Methanosarcina sp.	0.014	300	Grady et al. (1999)
Methanosaete sp.	0.003	30–40	Grady et al. (1999)
H$_2$ oxidizing methanogens	0.06	0.6	Grady et al. (1999)
Acetoclastic methanogenesis	0.0173	166	McCarty and Smith (1986)

TABLE **23.3** Summary of Common Microbial Populations Associated with Anaerobic Digestion and Reported Kinetic Parameters (Muller, 2006)

Table 23.3 provides a small sample of kinetic rates for different microbial populations associated with anaerobic digestion. As with enzymes, the different organisms have significantly different maximum growth rates and half-saturation coefficients. The minimum retention time is set by the slowest-growing organisms in a system.

When designing a system, a sufficient level of conservatism must be built into the design. Depending on substrate characteristics, complex organic matter will hydrolyze at different rates, and the microbial consortia that form in the digester will be a function of initial substrate characteristics. So the observed loading rates and retention time required for one facility will not necessarily translate to another because of differences in substrate characteristics.

2.2.2.1 Hydrolysis In the first stage (hydrolysis), the proteins, cellulose, lipids, and other complex organics are cleaved into lower-molecular-weight components that can pass through the cell wall for conversion to energy and additional biomass. Hydrolysis is thought to be the rate-limiting step of anaerobic digestion (Pavlostathis and Gossett, 1988, 2004).

Several factors can affect hydrolysis, including the organisms present, growth condition, temperature, particle surface area (Sanders et al., 2000), and solids composition.

2.2.2.2 Acidogenesis In the second stage (acid formation), the products of the first stage are converted to complex soluble organic compounds (e.g., long-chained fatty acids), which in turn are broken down into short-chained organic acids (e.g., acetic, propionic, butyric, and valeric acids). The concentration and relative proportions of these acids can indicate the overall condition of a digester.

2.2.2.3 Methanogenesis In municipal solids digesters, methanogenesis occurs via two primary metabolic pathways: acetoclastic methanogenesis and hydrogenotrophic methanogenesis. Acetoclastic methanogenesis—methane formed via acetate reduction—is the primary route of methane formation in mesophilic anaerobic digestion (McHugh et al., 2006).

Although hydrogenotrophic methanogenesis—methane formed via hydrogen reactions—may produce less of the methane in a digester, it plays a critical role in preventing feedback inhibition. McCarty and Smith (1986) reported that when the partial pressure of hydrogen exceeds 5 mPa, fatty acid hydrolysis becomes thermodynamically unfavorable. Volatile acids then accumulate, reducing pH and souring the digester. Hydrogenotrophic methanogenesis ensures that the system remains in balance.

Another methanogenic population is methylotrophic methanogens, which convert simple methylated compounds into methane and a reduced product. These organisms typically consume methyl mercaptan, trimethyl amine, dimethyl disulfide, etc. While not significant in overall digester performance, they play a critical role in controlling organic methylated odorants.

2.2.3 Microbial Ecology

An anaerobic digester can be described as an ecosystem whose environmental conditions are dictated by design engineers, operators, and the composition of the feedstock.

Feed solids composition, SRT, hydraulic retention time (HRT), temperature, and mixing regime all exert selective pressures on the microbial populations in the digester. Such pressures lead to the proliferation of some species and the recession or absence of others. Selective pressure can be to such a degree that overall digestion capacity can be affected by the relative population of species, such as has been reported for *Methanosarcina* spp. and *Methanosaeta* spp. (Conklin et al., 2006).

The relationship among microorganisms in an anaerobic digester can best be described as a syntrophic relationship. The metabolic activity of one population supports another, though not for the mutual benefit of either. Often the waste products of one group of organisms serve as the substrate of another. These relationships result in some distinct control points in the digestion process.

The production of hydrogen from fatty acid metabolism is thermodynamically not a highly favorable reaction. When hydrogen accumulates in the system at partial pressures above 5 mPa, there is a feedback inhibition of the acid oxidation process (McCarty and Smith, 1986). For the process to continue, as it does under stable digestion conditions, the hydrogenotrophic methanogenic population must be well established and respiring. What is evident from this one example is that a stress or toxin that affects one population may be enough to retard or upset the entire digestion process.

Feed solids characteristics will affect which populations are dominant and can set up conditions where there is competition between a desired population and one that is less desirable. For example, acetate, the substrate from which about 75% of the methane in biogas is generated, is also the preferred substrate of sulfate-reducing bacteria. If the feed solids have high sulfate concentrations, conditions may exist in which the methanogenic population is in direct competition with sulfate-reducing bacteria. When this happens, the biogas methane content or overall biogas production may decrease.

Anaerobic digestion can be adversely affected by loading changes, both quantity and quality. Given the complex nature of the different microbial interactions and the potential for process upset via stress on the weakest population, sufficient care must be taken when changing loading conditions, both quantity and quality.

2.2.4 Feedstock Characteristics

Key objectives of anaerobic digestion are to stabilize raw solids and reduce the mass of the material. Raw, primary, and secondary solids are primarily composed of the following compounds: proteins, polysaccharides, nucleic acids, fatty acids, and lipids. The relative concentrations of these compounds are a direct function of influent wastewater characteristics and the liquid treatment train used.

Contrary to the prevailing perception, fats, oil, and grease (FOG)/high-strength waste (HSW) co-digestion can make building new anaerobic digesters and cogeneration facilities economically feasible without needing primary clarification and primary sludge (Kabouris et al., 2013).

2.2.5 Hydraulic and Solids Residence Time

Anaerobic digesters are sized to provide enough residence time in well-mixed reactors to meet regulatory requirements, allow significant VSR to occur, and prevent slower growing microbial populations from washing out. Sizing criteria are defined as follows:

- SRT (days) is equal to the mass of solids in the digester (kilograms) divided by the solids removed (kilograms per day).
- HRT (days) is equal to the working volume (liters) divided by the amount of solids removed (liters per day).

Typically, HRT is calculated based on either digester inflow or outflow rate. However, in older, existing systems where supernatant is removed from the digester, SRT is calculated based on the solids volume removed. SRT and HRT are equal in digestion systems without recycle or supernatant withdrawal.

The SRT (or HRT) and the extent of hydrolysis, acid formation, and methane formation during anaerobic digestion are directly related: an increase in SRT increases the extent of each reaction; a decrease in SRT decreases the extent of each reaction (see Figure 23.3). Each reaction has a minimum SRT; if the SRT is shorter, bacteria cannot grow rapidly enough to remain in the digester, the reaction mediated by the bacteria will cease, and the digestion process will fail. Excessively long SRTs would prevent washout, but the

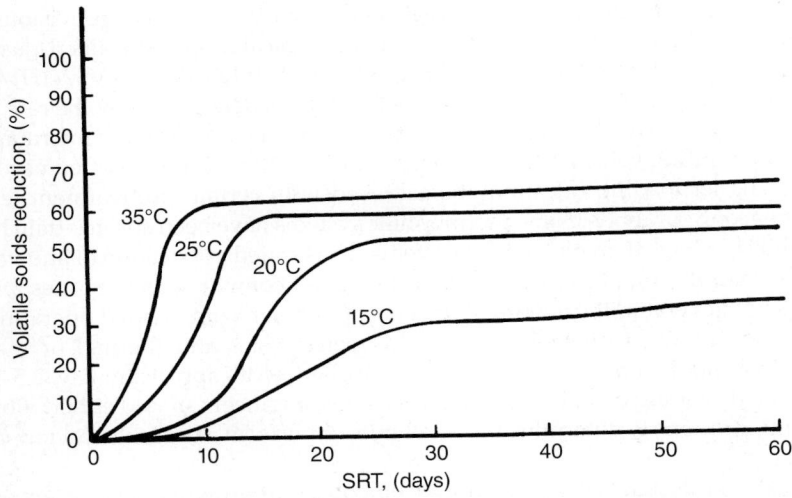

Figure 23.3 The effect of temperature and SRT on COD removal and methane production in anaerobic digesters.

Parameter	Value	Units
Volatile solids destruction	45–55	%
pH	6.8–7.2	
Alkalinity	2500–5000	mg CaCO₃/L
Methane content	60–65	percent by volume
Carbon dioxide content	35–40	percent by volume
Volatile acids	50–300	mg VA/L
Vol. acid:alkalinity ratio	<0.3	mg CaCO₃/mg VA
Ammonia	800–2000	mg N/L

TABLE 23.4   Typical Operating Parameters for Mesophilic Anaerobic Digestion of Wastewater Solids

extra equipment and infrastructure costs typically are not justified by the marginal increase in process performance. Reference Table 23.4 for typical SRT values.

Mesophilic anaerobic digestion of solids has been characterized through years of experience in operation and design, although design engineers always should consider the fundamental microbiology when designing a digester to optimize it toward maximum efficiency. Integrating process fundamentals into a design becomes increasingly important when the technology has a shorter operational history (e.g., thermophilic and phased systems). Applying inappropriate process parameters can result in improperly sized systems, which perform poorly and result in odors or other consequences.

2.2.6 Organic Loading Rate and Frequency

The phrase *volatile solids loading rate* refers to the mass of volatile solids (VS) added to the digester each day divided by the digester's working volume (kg VS/m³·d [lb VS/cu ft/d]). Loading criteria typically are based on sustained loading conditions (typically peak month or peak week solids production), with provisions for avoiding excessive loading during shorter periods. A typical design sustained-peak loading rate for mesophilic digesters is 1.9 to 2.5 kg VS/m³·d (0.12 to 0.16 lb VS/cu ft/d). The upper limit of the volatile solids loading rate typically is determined by the rate at which toxic materials—particularly ammonia—accumulate or slow-growing microorganisms wash out. A limiting value of 3.2 kg VS/m³·d (0.20 lb VS/cu ft/d) is often used.

Higher loading rates may be achieved with certain pretreatment technologies or other configurations. Some thermophilic systems have been able to attain higher volatile solids loadings than mesophilic systems. The limited application of and variable performance of thermophilic systems (and few good comparisons with mesophilic systems) has not generated the empirical data needed for a recommended operating range. With pre-dewatering to increase feed solids percentages and thermal or thermochemical pretreatment, some digesters may be loaded with approximately 2.5 kg VS/m³·d. When designing these highly loaded systems, engineers should either pilot-test digester configurations to determine loading limits or perform a thorough review of comparable systems.

While design engineers should be conservative when setting an upper loading limit for a design, excessively low volatile solids loading rates (e.g., due to poor thickening) may result in designs that are expensive to construct (large tank volume) and operate (high heating demand of low percent solids feed). Thickening of solids before digestion is an efficient method for increasing a digester's HRT and SRT and decreasing the energy used to heat the process. Advances in thickening technology (e.g. thickening centrifuges) may allow digesters to be fed at 6% to 8% solids with a higher volatile solids loading rate. When considering thickening improvements, pumps and mixing systems should be evaluated to ensure they will handle the increased viscosity and care should be taken to avoid overloading the digester (excessive volatile solids loading rate).

The loading frequency can affect operations as well as design. Microorganisms typically prefer to be maintained at a constant metabolic state (steady state), which is achieved via consistent loading. Constant loading also results in a more constant gas-production rate and may reduce gas storage requirements when gas is being utilized. However, it does involve some significant additional design considerations.

To maintain constant loading to the system, the feed and wastage rates need to be balanced, which may require continuous thickening and dewatering and/or sufficient storage upstream and downstream of the digesters to equalize flows.

Continuous feed or frequent incremental feeding is the ideal operation, but may be challenging because of the cost and operational constraints. Slug or pulse feeding (semi-continuous feeding) is often used; it involves feeding and wasting solids at set intervals, typically at peak staffing times. This type of operation has been implemented at many utilities but has some process risks (e.g., overloading, upsets, rapid volume expansion [RVE], and foaming). However, some research has shown that pulse feeding may increase the diversity and resilience of microbial communities within an anaerobic digester (De Vrieze et al., 2013).

2.2.7 Process Stability

Stable anaerobic digester operations can provide significant benefits (e.g., consistent solids destruction, pathogen reduction, and biogas generation). Stability is achieved through consistent loading, effective temperature control, and proper mixing.

A well-functioning, stable anaerobic process will exhibit specific digester and biosolids characteristics (see Table 23.5). Unstable digesters are more likely to foam, go acidic, and have microbial populations that are more susceptible to toxins.

Foaming is a common problem for many digesters. It can be caused by several factors (e.g., unstable digester operations, high concentrations of filamentous organisms in raw solids, and surfactants and other agents). Foaming associated with filaments and chemical additives can be remedied via source control. Foaming associated with digester stability is a result of microorganisms responding to environmental stress.

Chapman and Krugel (2011) suggest that RVE is an alternative explanation to upsets traditionally considered to be foaming. RVE occurs when gas bubbles become entrained in the solid-liquid matrix, often observed as a result of a change in apparent viscosity that occurs when mixing is suddenly stopped (or direction is reversed). This can occur over the course of several hours, and the digester fluid volume may increase by 10% to 15%. Willis et al. (2016) takes this a step further in suggesting that gas consistently present in the fluid effectively reduces the specific gravity of the digester contents, possibly as low as 0.6 to 0.7, causing problems with level sensing, digester cover operation, overflow systems. A simple solution to this problem is to ensure that solids are removed from the top of the digester as well as the bottom and/or the mixing system is able to pump down from near the surface (e.g., mechanical draft tube).

Substance	Moderately Inhibitory Concentration (mg/L)	Strongly Inhibitory Concentration (mg/L)
Na⁺	3500–5500	8000
K⁺	2500–4500	12 000
Ca⁺⁺	2500–4500	8000
Mg⁻⁺	1000–1500	3000
Ammonia-nitrogen	1500–3000	3000
Sulfide	200	200
Copper (Cu)	—	0.5 (soluble)
		50–70 (total)
Chromium VI (Cr)	—	3.0 (soluble)
		200–250 (total)
Chromium III	—	180–420 (total)
Nickel (Ni)	—	2.0 (soluble)
		30.0 (total)
Zinc (Zn)	—	1.0 (soluble)

TABLE 23.5   Concentrations of Selected Inorganic Compounds that Inhibit Anaerobic Processes (Parkin and Owen, 1986)

A good design can help promote stable digestion operations. For example, to ensure a constant temperature, boilers and heat exchangers should be sized to meet both the heat demands of raw solids and the shell losses that will occur under the coldest influent and ambient conditions expected.

Blending tanks improve process stability by homogenizing raw solids and metering them more constantly to the digester. Minimizing fluctuations in solids strength and loading rate help the microorganisms in the digester maintain a constant metabolic state, which minimizes their stress. Though some research indicates that minor variations in feed may help promote more robust microbial populations, wide, frequent fluctuations in loading can stress the biomass—especially if loadings, substrates, and nutrients are insufficient.

Mixing improves the contact between the biomass and raw solids. Dispersing solids in the digester ensures that its entire volume is used and all of the biomass is engaged in stabilization.

2.2.8 Temperature

An anaerobic digester's operating temperature significantly affects its observed performance and stability. Temperature affects growth rates (Lawrence and McCarty, 1969; van Lier et al., 1996; Salsali and Parker, 2007); substrate half-saturation constants (Lawrence and McCarty, 1969; van Lier et al., 1996); and microbial diversity (Chen et al., 2005; Wilson et al., 2008a). Lawrence and McCarty (1969) observed that growth rates increased as temperature increased in the mesophilic operation range. Salsali and Parker (2007) evaluated anaerobic digestion performance at 35°C, 42°C, and 49°C; they observed VSR increased as temperature increased. They did not attempt to derive growth rates from their experiments. Van Lier et al. (1996) evaluated volatile fatty acid (VFA) degradation by methanogens and suggested that acetate conversion in mesophilic and thermophilic digestion was described by an Arrhenius relationship, suggesting an increase in growth rates as temperature increased. Both Lawrence and McCarty (1969) and van Lier et al. (1996) suggested an increase in the substrate (VFA) half-saturation constant for acetoclastic methanogenesis with an increase in temperature (i.e., rising temperatures increased residual acetic and propionic acids).

Selecting a fixed operating temperature affects not only digester design but also day-to-day operations. From a design standpoint, the ability to achieve and maintain that temperature is critical to process stability and optimization, so appropriate safety factors should be employed.

The design operating temperature establishes the minimum SRT (or HRT) required to destroy a given amount of volatile solids (see Figure 23.3). Currently, most anaerobic digesters are designed to operate in the mesophilic temperature range (about 35°C [95°F]). Some systems have been designed to operate in the thermophilic temperature range (about 55°C [131°F]). Many new digesters are being designed so they can operate at both thermophilic and mesophilic temperatures, allowing future process flexibility.

Regardless of which temperature is selected, keeping it constant is of utmost importance. The microorganisms involved (particularly methanogenic populations) are sensitive to temperature changes; fluctuations in temperature can stress the organisms, thereby destabilizing the process. Temperature changes greater than 1°C/d can impact the digestion process. A good design avoids temperature changes greater than 0.5°C/d.

This is a critical consideration when determining feed schedules. However, for start-up of a thermophilic digester, a rapid increase in the temperature from mesophilic to thermophilic conditions has been reported to be an effective means of establishing a stable anaerobic population (Griffin et al., 2000; De la Rubia et al., 2013).

Temperature stability affects not only microbial stability but also the classification of the resulting biosolids. Under 40 CFR 503 Alternative 1 (time and temperature), solids must be maintained at a specific temperature above 50°C for a set period of time to achieve Class A status. In this instance, time is separate from HRT because every particle must be treated, requiring a batch held in an isolated tank, unless process equivalency has been granted. The effect of temperature on pathogen inactivation has made it one of the core mechanisms for achieving Class A biosolids, so it is important to design the system to maintain temperature (within close tolerances) under varying loads.

2.2.9 Volatile Fatty Acids, Concentration and Composition

When designing anaerobic digestion systems, engineers need to understand how volatile acid concentrations affect system design. Volatile fatty acids are the primary intermediates between complex organic matter and methanogenesis. The gross concentration of volatile acids can indicate how complete digestion is, while the composition of the acids can indicate or cause a process upset or disturbance. Some acids have been reported to be variably inhibitory to anaerobic consortia (Franke-Whittle et al., 2014). Volatile fatty acids also can lead to on-site odors because of fugitive emissions from digesters or dewatering processes.

In many cases, the concentration of VFAs in a digester is also a function of operating conditions. For example, the residual VFA concentration increases when the operating temperature is high (van Lier et al., 1996). Higher ammonia concentrations also can result in higher residual VFA concentrations (Nielsen and Angelidaki, 2008). The temperature and ammonia effects represent normal operating conditions as long as pH is not depressed.

As volatile acid concentrations increase, alkalinity is consumed. Once the buffer capacity is consumed, pH will decrease, leading to process upset and failure. Monitoring acid production and concentration can provide evidence of impending upset or recent disturbance, so operators can take remedial actions.

Volatile acid concentrations of 50 to 300 mg/L are considered normal for an anaerobic digester operating at mesophilic temperatures. This is not necessarily true in other anaerobic digestion systems, however; thermophilic systems (e.g., temperature-phased anaerobic digestion [TPAD]) and phased systems (e.g., acid-gas phasing) will have vastly different volatile acid concentrations in their reactors.

2.2.10   Alkalinity and pH

Anaerobic microorganisms—particularly methane formers—are sensitive to pH. Optimum methane production typically occurs when the pH is maintained between 6.8 and 7.2. Acid forms continuously during digestion and tends to lower pH. However, methane formation also produces alkalinity—primarily carbon dioxide and ammonia, which buffer changes in pH by combining with hydrogen ions.

A reduction in pH (by various causes) promotes more acid formation and inhibits methane formation. As acid production continues, methane and alkalinity formation are further inhibited, possibly leading to process failure. Mixing, heating, and feed-system designs are important in minimizing the potential for such upsets. Design engineers also should include provisions for adding chemicals (e.g., lime, sodium bicarbonate, or sodium carbonate) to neutralize excess acid in an upset digester.

2.2.11   Toxicity in Digesters

If concentrations of certain materials (e.g., ammonia, heavy metals, light metal cations, and sulfide) increase sufficiently, they can create unstable conditions in an anaerobic digester (see Tables 23.6 and 23.7). A shock load of such materials in facility influent or a sudden change in digester operation (e.g., overfeeding solids or adding excessive chemicals) can create toxic conditions in the digester.

Typically, excess concentrations of such toxicants inhibit methane formation, which typically leads to volatile acid accumulation, pH depression, and digester upset. Depending on the concentration and type of toxicant, the effect can be acute (e.g., instant process failure) or chronic (e.g., depressed performance). Chemicals can be added to control the concentrations of dissolved forms of some toxicants (e.g., using iron salts to control sulfide). Sometimes, toxicity is a result of more than one factor (e.g., ammonia toxicity increases at higher pH) (Finger and Butler, 1996).

Design engineers typically can only address toxicity by mitigating a known impact. Identifying and monitoring process toxicity (e.g., sampling and analytical techniques and practices) is typically an operational issue and beyond the scope of this text. However, a sound monitoring and control program, and an understanding of toxic agents, can greatly improve the design of mitigation systems.

2.2.12   Volatile Solids and Chemical Oxygen Demand

Volatile solids and chemical oxygen demand (COD) are common measures of the substrate entering a digester. Volatile solids are the ignitable (550°C) fraction of total solids. They typically are thought of as the organic fraction. Volatile solids measurements typically are used as part of determining overall process performance and regulatory compliance, and for mass-based calculations. Figure 23.4 shows the effect of temperature on

Compound	Concentration Resulting in 50% Activity (mM)
1-Chloropropene	0.1
Nitrobenzene	0.1
Acrolein	0.2
1-Chloropropane	1.9
Formaldehyde	2.4
Lauric acid	2.6
Ethyl benzene	3.2
Acrylonitrile	4
3-Chlorol-1,2-propandiol	6
Crotonaldehyde	6.5
2-Chloropropionic acid	8
Vinyl acetate	8
Acetaldehyde	10
Ethyl acetate	11
Acrylic acid	12
Catechol	24
Phenol	26
Aniline	26
Resorcinol	29
Propanol	90

TABLE 23.6 Concentrations of Select Organic Chemicals that Reduce Anaerobic Digester Activity by 50% (Parkin and Owen, 1986)

COD removal and methane production. One must be careful when using data from a volatile solids test. There are artifacts to the test, because a significant amount of inorganic salts—especially ammonium-based salts—can volatilize in analytical tests, skewing the volatile solids concentration higher and the VSR lower (Beall et al., 1998; Wilson et al., 2008b).

Material	Specified Gas Production Per Unit Mass Destroyed	
	m³/kg	Methane Content (%)
Fats	1.2–1.6	62–72
Scum	0.9–1.0	70–75
Grease	1.1	68
Crude fibers	0.8	45–50
Protein	0.7	73

TABLE 23.7 Gas-Production Rates from Various Organic Substrates (Buswell and Neave, 1939)

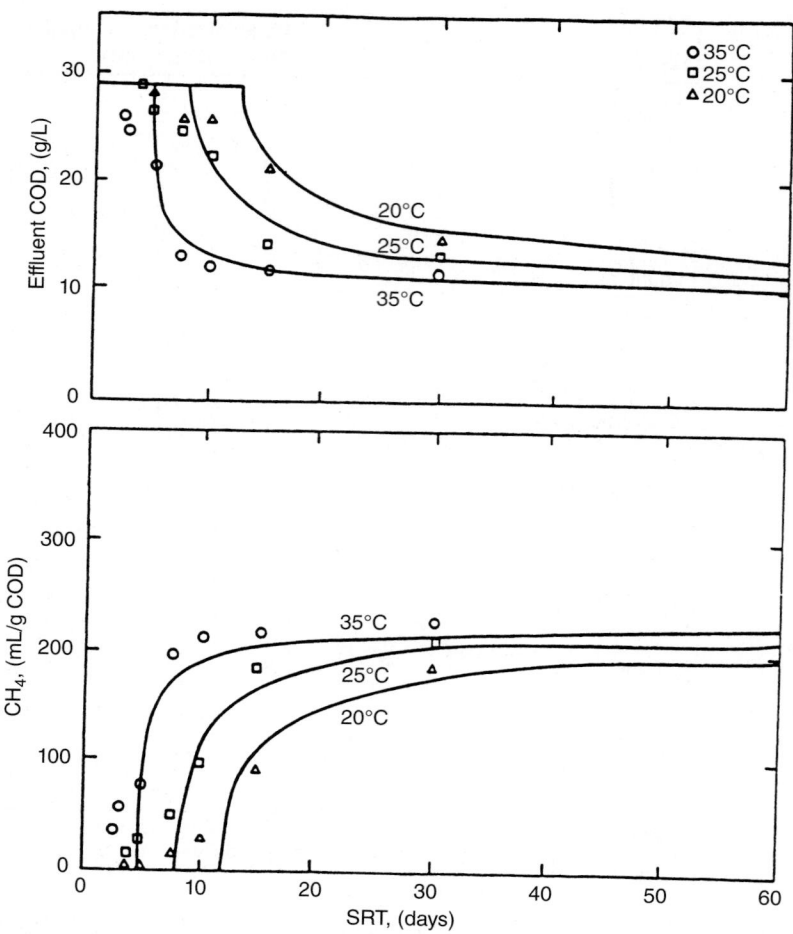

Figure 23.4 The effect of SRT and temperature on the rate and extent of VSR during anaerobic digestion (O'Rourke, 1968).

COD is a measure of the chemically oxidizable material in solids. As with volatile solids, this measure has limitations (e.g., the accounting of non-substrate components). COD typically is used in kinetic models (e.g., ADM 1) because all of the parameters are based on COD coefficients (WEF, 2013; Batstone et al., 2002).

Most facilities provide volatile solids data rather than COD data in their regulatory reports because of the relative ease of conducting the volatile solids test.

2.2.13 Biogas Production and Characterization

The quality and quantity of digester gas (biogas) produced also can be used to evaluate digester performance. Biogas production is directly related biochemically to the amount of volatile solids destroyed; it often is expressed as volume of gas per unit mass of volatile solids destroyed. The gas-production rate is different for each organic

Constituent	Values for Various Plants, % by Volume[a]							
Methane	42.5	61.0	62.0	67.0	70.0	73.7	75.0	73–75
Carbon dioxide	47.7	32.8	38.0	30.0	30.0	17.7	22.0	21–24
Hydrogen	1.7	3.3	[b]	—	—	2.1	0.2	1–2
Nitrogen	8.1	2.9	[b]	3.0	—	6.5	2.7	1–2
Hydrogen sulfide	—	—	0.15	—	0.01	0.06	0.1	1–1.5
Heat value, Btu/cu ft[c]	459	667	660	624	728	791	716	739–750
Specific gravity (air = 1)	1.04	0.87	0.92	0.86	0.85	0.74	0.78	0.70–0.80

[a]Except as noted.
[b]Trace.
[c]Btu/cu ft × 37.26 = kJ/m^3.

TABLE 23.8 A Survey of the Characteristics of Biogas from Anaerobic Digesters (Tortorici and Stahl, 1997)

substance in the digester (see Table 23.8). The gas-production rate of fats ranges from about 1.2 to 1.6 m^3/kg (20 to 25 cu ft/lb) of volatile solids destroyed; the gas-production rate of proteins and carbohydrates is 0.7 m^3/kg (12 cu ft/lb) of volatile solids destroyed. The gas-production rate of a typical anaerobic digester treating a combination of primary solids and WAS should be about 0.8 to 1 m^3/kg (13 to 18 cu ft/lb) of volatile solids destroyed. The amount of gas produced is a function of temperature, solids retention time (SRT), and volatile solids loading. Specific gas production should be measured until an average value can be obtained and used for monitoring.

The two main constituents of digester gas are methane and carbon dioxide; it also contains trace amounts of nitrogen, hydrogen, and hydrogen sulfide. Performance data from healthy digesters suggest that methane concentrations should be 60% to 70% (by volume) and carbon dioxide concentrations should be 30% to 35% (by volume). Tortorici and Stahl (1977) have published data on typical digester-gas characteristics (see Table 23.9). An increase in carbon dioxide levels (percent) often indicates an upset digester. Excessive concentrations of hydrogen sulfide can indicate unbalanced digestion, industrial waste sources, or saltwater infiltration. Hydrogen sulfide may be responsible for odor problems and excessive corrosion in the digester and adjacent piping. Heavy metals can precipitate as metallic sulfide, thereby minimizing hydrogen sulfide concentrations in biogas.

2.2.14 Pathogens

Pathogen and pathogen-indicator reductions are major disinfection criteria and often a component of stabilization processes. The rate and the extent of pathogen or pathogen-indicator reduction (inactivation) are process specific. The degree of pathogen or pathogen-indicator reduction required depends on the biosolids quality desired (Class B or Class A, assuming beneficial use is desired). Design engineers should consult 40 CFR Part 503 (U.S. EPA, 1999) and other applicable regulations for pathogen or pathogen-indicator reduction requirements for anaerobic processes. (For more information on pathogen-reduction regulations, see Chapter 18.)

Parameter	Low Rate	High Rate
Solids retention time, days	30–60	15–20
Volatile suspended solids loading, lb/cu ft/d (kg/m³·d)	0.04–0.1	0.12–0.16
	(0.64–1.6)	1.9–1.5
Volume criteria, cu ft/cap (m³/cap)		
Primary sludge	2–3 (0.06–0.08)	1.3–2 (0.03–0.06)
Primary sludge + trickling filter sludge	4–5 (0.11–0.14)	2.6–3.3 (0.07–0.09)
Primary sludge + waste-activated sludge	4–6 (0.11–0.17)	2.6–4 (0.07–0.11)
Combined primary + waste biological sludge feed concentration, % solids-dry basis	2–4	4–6
Anticipated digester underflow concentration, % solids-dry basis	4–6	4–6

TABLE 23.9   Typical Design Parameters for Low- and High-Rate Digesters (Burd, 1968)

Design engineers should carefully consider which digestion system to use. While multiple methodologies meet the desired degree of pathogen reduction, each comes with costs and degrees of process complexity.

2.3 Process Options

Process options for anaerobic digestion of wastewater solids have advanced significantly since the early 1990s. This section provides some historical context and discusses process options that are being considered and implemented more frequently in the 21st century (e.g., staged and phased systems, and mesophilic and thermophilic processes).

2.3.1 Low-Rate Digestion

Before the 1950s or 1960s, solids were anaerobically digested in "low-rate" systems characterized by intermittent feeding, low organic loading rates, little to no external heating or mixing, and detention times of 30 to 60 days. The tanks were large because grit and scum accumulated on the bottom and top, respectively, thereby decreasing the effective volume. Optimum digestion conditions were not maintained. Relatively few low-rate digestion systems are in service today because technology advancements have made them uneconomical and unattractive.

2.3.2 High-Rate Digestion

Wastewater treatment professionals kept tinkering with the low-rate system, making improvements that eventually led to the "high-rate" anaerobic digestion system that was more common in the 1960s (see Figure 23.5). This system is still widely used for mesophilic digestion.

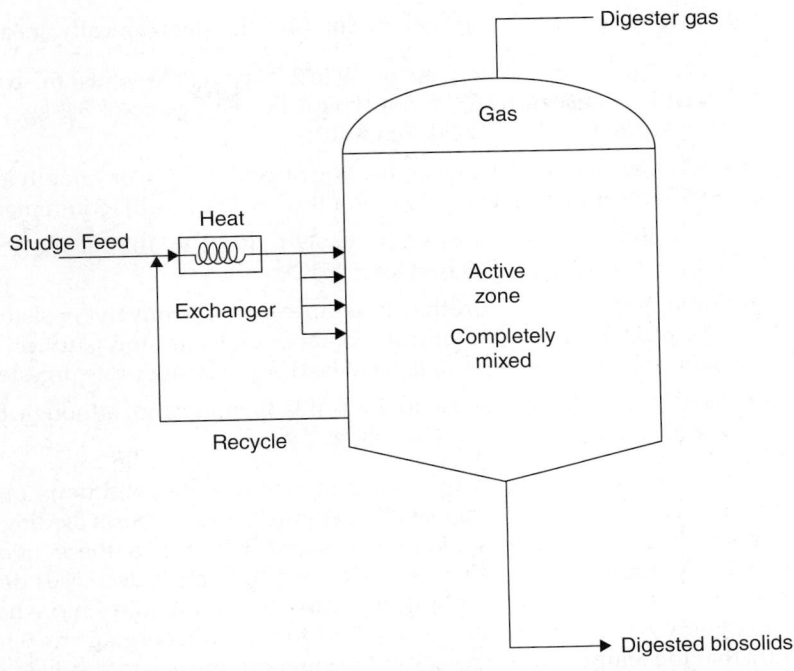

FIGURE 23.5 Simplified flow schematic of high-rate anaerobic digestion.

2.3.2.1 Process Development High-rate anaerobic digestion was developed after research demonstrated the benefits of controlling environmental conditions in the digester. High-rate digestion is characterized by supplemental heating and mixing, relatively uniform feed rates, and thickening of solids (solids feedstock typically should contain 4% to 5% solids, although some recent improvements allow for thicker or thinner solids). These factors result in relatively uniform conditions throughout the reactor, leading to lower overall tank volume requirement and increased process stability.

Several heating methods (e.g., steam injection, internal heat exchangers, and external heat exchangers) have been used for anaerobic digesters. External heat exchangers are the most popular because of their flexibility and easily maintained heating surfaces. Internal coils are not recommended as they can foul and the digester must be emptied to clean them. Internal or external draft tube heat exchanger jackets can provide reliable service as long as they are constructed of stainless steel. Steam injection dilutes the contents of the digester without heat exchangers but may be prone to ragging.

2.3.2.2 Design Criteria—Mesophilic Typically, high-rate systems achieve increased gas production, solids destruction, and overall process stability when compared to the increasingly rare low-rate systems. In the 1970s, the U.S. Environmental Protection Agency (EPA) extensively evaluated single-stage, high-rate anaerobic digesters operated at mesophilic temperatures with SRTs exceeding 15 days; regulators found that the process achieves significant pathogen reduction and solids stability. The agency defined it as a process to significantly reduce pathogens (PSRP) in its 1979 rule (40 CFR 257), and it essentially became a baseline for wastewater solids stabilization.

The basic design criteria for such mesophilic digesters typically are as follows:

- A volatile solids loading rate of 1.9 to 2.5 kg volatile solids/m³·d (0.12 to 0.16 lb volatile solids/cu ft/d), and a typical limiting value of 3.2 kg volatile solids/ m³·d (0.20 lb volatile solids/cu ft/d);

- SRT of at least 15 days when feeding at peak 15-day or -month loads (a 15-day SRT is the minimum allowed under the PSRP [Class B] requirement in Part 503);

- Mesophilic temperatures (35°C to 39°C [95°f to 102°F]; the PSRP [Class B] requirement in Part 503 is at least 35°C);

- Enough mixing to ensure that the temperature is relatively consistent throughout the reactor (and to minimize bottom deposits and surface scum/debris, although this is only partially achieved in many high-rate digesters); and

- Feed solids containing 4% to 5% solids (historically), although more facilities are now aiming for 5% to 7% solids.

Frequent solids feeding helps maintain steady-state conditions in the digester. Methanogens are sensitive to changes in substrate levels; uniform feeding and multiple feed-point locations in the tank reduce shock loading to these microorganisms. Excessive hydraulic loading should be avoided because it decreases detention time, dilutes the alkalinity needed for buffering capacity, and requires more heat to achieve process goals. Good mixing also is required to mix microorganisms with fresh feed, ensure that the temperature is consistent throughout the reactor, and aid in preventing grit and scum/foam accumulation.

Improvements in mixing, heating, and solids loading enhanced anaerobic digestion performance. Mixing and heating provide better contact between substrates and microorganisms, increasing stabilization while reducing short-circuiting (making pathogen kill more consistent and increasing biosolids stability).

The relative success of high-rate mesophilic digestion has made this process the most common means of solids stabilization in the world. It also is the standard for evaluating future process variations.

2.3.2.3 Design Criteria—Thermophilic

Although most anaerobic digesters are operated at mesophilic temperatures (i.e., 35°C [95°F]), they also can be operated at thermophilic temperatures (typically between 50°C and 57°C [122°F and 135°F]). Thermophilic digesters have somewhat different design and performance criteria than those for mesophilic digestion. For example, volatile solids loading rates can be higher and SRTs can be lower (Schafer et al., 2002). Thermophilic digesters also reduce more volatile solids than identical-size mesophilic digesters, as suggested by the Arrhenius relationship. Because the temperature is higher, however, more energy is needed to provide heat. To reduce heating costs, energy recovery heat exchangers may be employed.

Thermophilic digestion has a number of advantages over mesophilic digestion. For example, solids from thermophilic digesters generally have better dewatering characteristics, so heating costs may be offset by reduced dewatering costs. Thermophilic digestion systems also destroy more volatile solids (Schafer et al., 2002) and typically produce biosolids containing fewer pathogens. However, Part 503 classifies both mesophilic and thermophilic anaerobic digestion (non-batch, non-phased systems) as PSRPs (Class B processes), so continuous-flow thermophilic digesters do not get regulatory credit for their pathogen-reduction performance. Part 503 also specifies that any process to further reduce pathogens (PFRP) (Class A process) must precede or be concurrent with the vector-attraction reduction (VAR) process (e.g., mesophilic or thermophilic anaerobic digestion) to allay concerns about post-pasteurization regrowth (Clements, 1982; Keller, 1980). So, to meet Class A requirements, a thermophilic digester may need to be designed to be partly or wholly operated in batch mode, where every particle meets the time and temperature relationship established by the U.S. EPA.

According to Part 503, a thermophilic digestion process typically meets Class A requirements by maintaining its temperature at or above 50°C (more typically, 55°C) for a specific period of time in a batch operation. The amount of time is calculated using the formula (U.S. EPA, 1999) below:

$$D = \frac{50\,070\,000}{10^{0.14T}} \tag{23.1}$$

where D = time (days); and

   T = temperature (°C).

This equation can be applied to sludge containing less than 7% solids. Based on this equation, one point of compliance would be 55°C for a minimum of 24 hours. Another point of compliance would be 50°C for a minimum of 120 hours. As the thermophilic digestion temperature increases, the batching time (and thus tank volume) required to destroy pathogens or pathogen indicators is considerably reduced.

If a utility chooses to disinfect sludge containing more than 7% solids, the time and temperature requirements are determined using the following equation:

$$D = \frac{131\,700\,000}{10^{0.14T}} \tag{23.2}$$

where D = time (days) and

   T = temperature (°C).

So, for a sludge containing more than 7% solids, the minimum batching time is 63.1 hours at 55°C. In practice, however, thermophilic digestion reactors are unlikely to be operated at such high solids concentrations because of the higher viscosity involved. Digesters need much more energy to pump and mix high-viscosity solids.

Early full-scale tests at Los Angeles indicated that thermophilic digesters were difficult to operate (Garber, 1982). However, thermophilic digester operations can be reliable when temperatures are constant and good mixing and feeding systems are used (Krugel et al., 1998). Inadequate mixing and temperature control systems may have contributed to challenges with early thermophilic digesters.

In summary, thermophilic digestion destroys more volatile solids and pathogens, and produces more biogas, but can be more expensive to implement and operate than mesophilic digestion. For further comparisons between thermophilic and mesophilic, see Gebreeyessus and Jenicek (2016). Design engineers may wish to perform pilot tests with the actual feedstock before deciding to use thermophilic digestion.

2.3.3   Primary-Secondary Digestion

Now, mostly obsolete, in the primary-secondary digestion system, the primary tank is a typical mixed and heated anaerobic digester and the secondary tank is a solid-liquid separator (supernatant withdrawn from the top). The secondary reactor traditionally did not have mixing or heating but this configuration is not common anymore. In fact, in modern systems or facilities that have been upgraded, the second tank may serve several other functions including providing storage capacity and insurance against process short-circuiting, standby digester capacity. Primary-secondary digestion worked well on primary clarifier solids (settled solids), because the second tank typically provided good separation. The now-common practice of digesting WAS or primary/WAS blends has made this process impractical.

2.3.4 Recuperative Thickening
Recuperative thickening is a process that increases SRT relative to HRT in an anaerobic digester. It also returns anaerobic microorganisms to the digester to potentially increase biological activity.

Thickening options include centrifuges, screw presses, gravity belt thickeners (GBTs), and dissolved gas flotation thickeners (both air flotation [DAF] and anoxic gas flotation have been used). GBTs are used at facilities in Pennsylvania and Wisconsin. DAF recuperative thickening was tested at the Spokane, Washington, WRRF, and the bacteria survived the oxygenation effect (Reynolds et al., 2001).

Recuperative thickening has been used for several reasons (e.g., temporarily increasing digester SRT while some of the facility digestion capacity is off-line for maintenance or construction). It also can be used to delay construction of more digestion tank capacity or provide increased solids storage capacity in existing tankage (Kabouris et al., 2015).

The disadvantages of recuperative thickening include operational complexity, polymer use, ventilation requirements (for odor and explosive gas), maintenance, and costs for an additional process.

2.3.5 Staged Digestion
The concept of staged (phased) digestion has been used in various ways over the years (e.g., to stage metabolisms, operating temperatures, and redox conditions), but it is increasingly being recognized for its pathogen-control benefits. Primary-secondary digestion, for example, is basically a two-stage mesophilic process that produces well-stabilized solids because of the relatively long retention times and reduction of solids short-circuiting. Other staged mesophilic digestion options also are being recognized and used.

2.3.5.1 Two-Stage Mesophilic Digestion
Two-stage mesophilic digestion is an extension of single-stage high-rate anaerobic digestion that uses two complete-mix digesters in series. The first stage must be designed to provide reliable mesophilic digestion (e.g., sufficient SRT and reasonable volatile solids loading). Because much of the process considerations are met in the first reactor, the second stage can operate with a relatively shorter SRT. Both stages are heated and mixed. Placing the tanks in series causes the reaction kinetics to behave more like a plug-flow rather than a complete-mixed process, thus reducing short-circuiting and improving process efficiency.

Schafer and Farrell (2000) and Chapman and Muller (2010) reported that, compared to single-stage digestion, two-stage mesophilic digestion can:

- Improve product stability (because more volatile solids are destroyed) and
- Reduce short-circuiting of raw solids and pathogens.

Zahller et al. (2005) and McCarthy et al. (2004) reported that two-stage digestion provided better VSR and biogas composition (i.e., more methane content) than a single-stage digester with an equivalent SRT. However, the study also noted that the two-stage system seemed to have less capacity to absorb large variations in loadings than the single-stage system, as measured by additional acetate use capacity.

2.3.5.2 Multiple-Stage Thermophilic Digestion Compared to mesophilic processes, thermophilic digestion offers more gas production, solids reduction, and pathogen destruction. Likewise, two-stage thermophilic digesters are more effective than single-staged systems. The Annacis Island Wastewater Treatment Facility in Vancouver, British Columbia, is an example of multiple-stage thermophilic digestion. Krugel et al. (1998) predicted, based on the equations describing time and temperature relationships for batch systems, that Annacis Island's process would achieve pathogen and pathogen-indicator reductions equivalent to a Class A batch process, and the agency's monitoring results confirm the predictions. The system also has been reported to show low organic sulfur release from centrifugally dewatered biosolids, as well as no *Escherichia coli* regrowth following dewatering—something not observed in other thermophilic or mesophilic anaerobic processes (Chen et al., 2008).

2.3.6 Temperature-Phased Anaerobic Digestion

Temperature-phased anaerobic digestion (TPAD) uses both thermophilic and mesophilic digestion to improve digestion performance. Such systems are not nearly as common as conventional mesophilic systems. The Western Lake Superior Sanitary District in Duluth, Minnesota, has a TPAD system (see Figures 23.6 and 23.7). Constructed in 2001, this system has four tanks: one that operates as a thermophilic digester, followed by three tanks operating in parallel as mesophilic digesters.

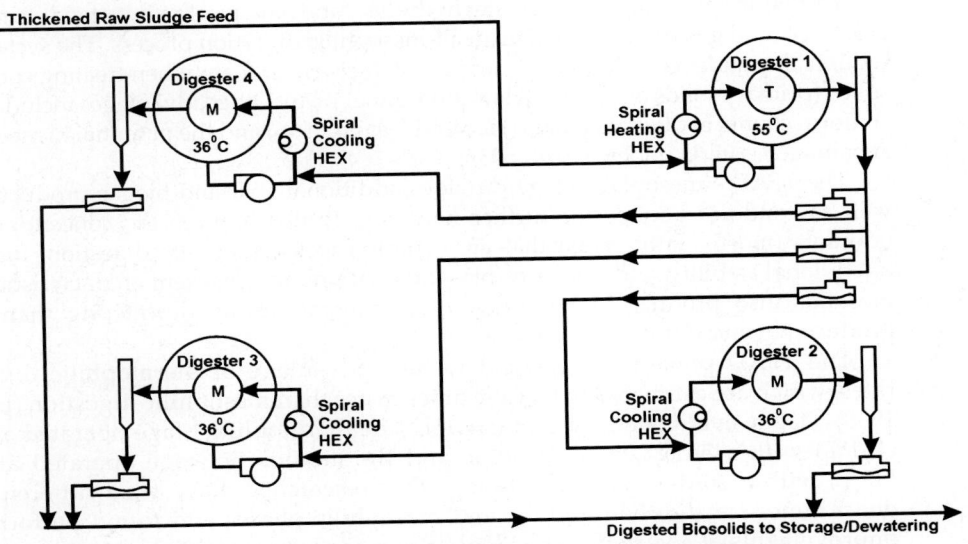

Figure 23.6 Process flow schematic of the TPAD installation at Western Lake Superior Sanitation District in Duluth, Minnesota (Krugel et al., 2006).

FIGURE 23.7 Photograph of the TPAD installation at Western Lake Superior Sanitation District in Duluth, Minnesota (courtesy of Brown and Caldwell).

2.3.6.1 Process Development Researchers in Germany identified the potential advantages of TPAD. Anaerobic digesters in Cologne, Germany, have been operated in a temperature-phased mode since August 1993 (Dichtl, 1997). In the United States, Han and Dague (1996) conducted laboratory studies documenting the advantages of TPAD, and a patent for the TPAD process was issued to Iowa State University in the 1990s based on Dague's work.

The thermophilic digester's greater hydrolysis and biological activity tends to provide more VSR and gas production than an all-mesophilic digestion process. The system also reduces the tendency of high-rate mesophilic digesters to foam when treating combined solids (primary solids and WAS) (Han and Dague, 1996). Other advantages include lower coliform counts in digested solids (Han and Dague, 1996) and the potential to meet Class A pathogen criteria under 40 CFR 503.

The TPAD's mesophilic stage provides additional VSR and biogas production, as well as conditions solids for further handling. It also reduces the concentration of odorants (mostly fatty acids) that are common to thermophilic digestion, increases operational stability, and produces biosolids with more consistent characteristics. The biosolids also produce higher cake solids content during dewatering than those produced by mesophilic digesters.

The TPAD process is designed to take advantage of thermophilic digestion rates, which are estimated to be four times faster than mesophilic digestion (Dague, 1968). Dague evaluated a system in which the thermophilic stage operated at 55°C (131°F) with a 5-day detention time and the mesophilic stage operated at 35°C (95°F) with a 10-day detention time. Other researchers have tried different residence times for the thermophilic and mesophilic phases and found performance improvements at a variety of residence times for each stage. Few full-scale WRRFs have operated the thermophilic phase at a 5-day SRT or less, but research shows that this can be successful.

2.3.6.2 Design Criteria Design criteria for TPAD vary because performance success has been demonstrated under various situations. However, based on most of the research and full-scale experience to date, design criteria for a typical TPAD system are:

- Thermophilic temperatures of 50°C to 57°C;
- Thermophilic residence times of 4 to 10 days (existing TPAD systems may have longer SRTs because a large tank was available for the thermophilic stage, or early year loads were less than design loads);
- Mesophilic temperatures of 35°C to 40°C; and
- Mesophilic residence times of 6 to 12 days (again, existing TPAD systems may have longer SRTs because of the tankage used or loads that are less than design loads).

When designing a TPAD system, engineers should choose design criteria based on project objectives, solids feed characteristics and variability, and existing facilities (because most TPAD systems are modifications of existing digestion systems). If there are wide variations in feedstock quantity and characteristics, design engineers may want to use a longer residence time in the first-stage thermophilic reactor—perhaps a 10-day SRT or longer. If an existing tank can be used for thermophilic digestion, but its SRT is only 4 or 5 days, the TPAD system may work well as long as the mesophilic system's SRT is long enough to adequately handle some variable performance from the first-stage thermophilic reactor. Total SRTs of about 15 days (minimum) are considered good design practice for peak 15-day or peak month loads. If the facility needs to ensure that its biosolids meet Class B standards, then a 15-day total SRT (minimum) typically is required. Another Part 503 requirement is that digester temperatures in the mesophilic stage must remain above 35°C (if mesophilic SRT is needed to meet Class B requirements).

2.3.6.3 Performance The performance of TPAD systems often is measured based on VSR or biogas production. Schafer et al. (2002) reported significant improvement in VSR at several facilities using the TPAD process, compared to the performance of a mesophilic system with a similar SRT. (If the mesophilic and TPAD systems have different SRTs, then direct performance comparisons are more difficult to quantify without additional information.)

Schafer et al. (2002) also reviewed pilot- or demonstration-scale studies and found that TPAD outperformed mesophilic digestion systems when fed the same feedstock and using similar total SRTs. Improvements in VSR are often cited as follows:

- A high-rate mesophilic digester with a 20-day SRT achieved 50% VSR, while
- A TPAD system with a 20-day total SRT and identical or similar feedstock achieved 57% VSR.

2.3.6.4 Heating, Cooling, and Other Design Considerations TPAD can be heated (and cooled) in several ways, with some precautions:

- For energy efficiency, the heat from thermophilic solids can be recycled to heat cold feedstock and partially cool the thermophilic solids. Various arrangements have been used for this heat recycling concept (e.g., solids-solids heat exchangers and solids-water/solids heat exchangers). This approach requires supplemental heat (e.g., heat exchangers or steam addition) to ensure that feedstock reaches thermophilic conditions.

- Solids can be heated to thermophilic temperatures via heat exchangers and/or steam addition without heat recycling. In this case, thermophilic solids typically must be cooled before entering the mesophilic stage, and this heat can be transferred to facility effluent, blown into the atmosphere, or used to heat water for building heating or other purposes.

- In colder seasons, and if the mesophilic stage's SRT is long enough, a purposeful cooling system may not be needed because the mesophilic stage can cool itself via thermal losses to the atmosphere and ground. Design engineers should calculate whether this is possible and if the mesophilic stage may be too hot for reliable performance under certain conditions.

Engineers should ensure that the thermophilic stage is operated at a consistent temperature because its microorganisms are more sensitive to temperature changes, particularly increases. For example, if the design temperature is 55°C, the control system should ensure that this temperature is maintained (close tolerances). This may be difficult due to the poor heat transfer efficiency of sludge, reduced temperature differential between hot water and sludge, and challenges with sludge-to-sludge heat exchangers (for heat recovery). Also, a mixing system is required to ensure that all tank contents are close to the measured thermophilic temperature. The temperature tolerances for the mesophilic stage are not as exacting, but must be evaluated carefully for reliable performance.

Other design considerations include protection against odor release from thermophilic digesters. For example, floating covers typically are avoided because of odor release from the annular space. Also, the gas-handling system must be designed to handle the much larger production rate and moisture content of thermophilic biogas. In addition, design engineers may need to evaluate whether an existing mesophilic digester is adequate for thermophilic service. This includes a structural evaluation of the tank, its mechanical systems, piping, and coating/lining systems.

2.3.7   Lagoon Digestion

It used to be common to digest solids via open lagoons, but anaerobic lagoons typically caused odor problems and were phased out of use. Today, various wastewater treatment agencies use facultative solids lagoons (FSLs), which have an aerobic cap layer to help oxidize and control the odorous decomposition products that rise from the anaerobic activity below. The Sacramento (California) Regional Wastewater Treatment Plant has a 50-ha (125-ac) FSL system of lagoons filled with liquid 4.5 m (15 ft) deep. Considerable research was conducted when this system was developed in the 1970s (Schafer and Wolfenden, 1982).

2.3.7.1   System Performance   Large-scale lagooning is conducted at ambient temperatures, which typically encompass the psychrophilic temperature range. A key feature of such digestion is that more digestion occurs in warmer seasons, and less occurs in colder seasons.

The approach used at Sacramento, Chicago, and other facilities is to achieve maximum VSR and solids stabilization via long-term digestion (1 to 5 years). Sacramento achieves almost 60% VSR in its mesophilic digesters, and has documented another 40%

to 45% VSR in its FSL (Schafer and Wolfenden, 1982). Such long-term stabilization results in biosolids with relatively little product odor.

Design and operating criteria vary widely for FSLs, but most systems currently are fed either mesophilically digested biosolids or aerobic solids from extended aeration facilities. Facility effluent is often used for the cap water layer.

At Chicago and other facilities, dredged solids are air dried in warmer-weather months, producing biosolids that contain at least 60% solids. The biosolids are used for land application, land reclamation, landfill cover material, and other beneficial purposes. FSLs have been reported to produce Class A biosolids when batch storage is used to prevent short-circuiting (WERF, 2004).

2.3.7.2 Covered Lagoons for Methane Emission Control Since the 1990s, concerns about odors and methane emissions from open waste lagoons (mostly animal waste lagoons) have increased. In response, the animal-waste treatment industry has begun covering some lagoons to collect the biogas generated during anaerobic digestion and use it to produce power. Such systems are becoming more common in North America, Australia, and Asia.

For example, the Western facility in Melbourne, Australia, has used an extensive floating cover system on its wastewater ponds for almost a decade; the high-density polyethylene (HDPE) system covers 7.8 ha (19 ac) of anaerobic digestion ponds (DeGarie et al., 2000). The collected biogas is used to generate more than 2 MW of electricity, which powers other portions of the facility. This system reduces direct methane emissions to the atmosphere (cutting greenhouse gas emissions) and generates renewable power (offsetting carbon emissions from fossil fuel-based generators elsewhere). Memphis, Tennessee, also has covered some of its FSLs since the 1990s to control odor and collect biogas for energy use.

2.4 Pretreatment

2.4.1 Pre-Pasteurization

Pre-pasteurization is a pathogen reduction process involving a pasteurization step before digestion. This was developed primarily in Europe to allow biosolids to be applied directly to sensitive croplands or pastures.

2.4.1.1 Process Development In an early paper on pasteurization, Clements (1982) reported that Switzerland had issued regulations in 1971 requiring biosolids to be treated to reduce pathogens before they could be applied to grazing land. The initial concept was to pasteurize the biosolids after digestion. About 70 post-pasteurization facilities were constructed in the following 6 years; they typically processed anaerobically digested biosolids (i.e., biosolids were digested, pasteurized, and then either used or stored). When veterinary scientists investigated the material, they reported that the stored products often contained extremely high densities of pathogens, even though they had contained few enteric bacteria and no *Salmonella* immediately after pasteurization. Further investigation showed that regrowth was due to surviving organisms or to contamination. Investigators concluded that, because the pasteurized material had no remaining vegetative bacteria, any bacteria present later could grow explosively in the absence of competitors.

Keller (1980) presents data on regrowth of bacteria in biosolids from a post-pasteurization process. The data show that, after pasteurization, biosolids contained between 20 and 75 colony-forming units (CFU) of Enterobacteriaceae per gram of solids.

When it was transported from the treatment facility, however, the material contained between 207 000 and 35 million CFU/g. Shortly afterward, WRRFs modified the process to pasteurize solids before digestion (called pre-pasteurization), and the regrowth problem disappeared. The practice has successfully sustained itself over several decades in many countries (mostly in Europe). A few U.S. facilities now use the process.

2.4.1.2 Design Criteria   The main reason to use a pasteurization process is to disinfect solids; it has not been reported to enhance VSR significantly. The pre-pasteurization process meets Part 503's Class A pathogen standards by typically maintaining its temperature above 65°C (more typically 70°C) for a specific period of time in a batch operation. The amount of time is calculated using eq 23.1, which applies to sludge containing less than 7% solids.

Calculations indicate that solids must be maintained at 65°C for approximately 1 hour. As pasteurization temperature increases, the required time (and, therefore, tank volume) shrinks. Most systems are designed to achieve at least 70°C for 0.5 hour, even though operating at slightly lower temperatures may lower operations and maintenance (O&M) costs.

If a utility chooses to pasteurize sludge containing more than 7% solids, the time and temperature requirements are determined using eq 23.2.

Part 503 specifies that the pasteurization process must precede the VAR process (e.g., mesophilic or thermophilic anaerobic digestion) and should be operated in a batch mode to allay regrowth concerns. Hence, post-pasteurization is not allowed under this rule.

2.4.1.3 Pre-Pasteurization Vessel   The U.S. EPA has indicated that to produce a Class A biosolids that meets the requirements in Alternative 1 under Part 503, every particle of solids should be exposed to a minimum temperature for a minimum time. So design engineers should avoid using completely mixed systems or systems with potential for back-mixing or short-circuiting as pre-pasteurization tanks. Most vendor-supplied systems are batch tanks; only one vendor supplies a plug-flow tank for pre-pasteurization. Design engineers should consult U.S. EPA staff or other pertinent regulators before using a non-batch system.

Batch pre-pasteurization systems are operated in a fill/hold/draw mode, with several batch vessels used to perform each cycle if continuous operation is desired. The vessels should be well mixed to ensure that the monitored temperature reflects the entire contents (i.e., every solids particle meets the time and temperature requirements). If the downstream anaerobic digestion process uses an intermittent feed cycle and upstream storage is adequate, the system needs fewer than three batch vessels for the required fill, hold, and draw cycles.

2.4.1.4 Ancillary Equipment for Pre-Pasteurization   Design engineers should consider three important ancillary features when installing a pre-pasteurization process:

- Solids heating and cooling;
- Solids screening; and
- Temperature monitoring and control.

Because the temperature of pre-pasteurization systems typically is maintained at 70°C, solids must be heated and then cooled. Heat exchangers are the most common method for heating and cooling solids. If desired, design engineers could include a heat-recovery step to use heat from cooling biosolids to preheat raw solids. The recovery step will require a substantial amount of heat-exchange capacity. If heat exchangers are used, the solids may need to be screened before pre-pasteurization—even if fine screens are used in the facility headworks. Screening also helps produce a more aesthetically pleasing product.

Good temperature monitoring and control are required to maintain Class A compliance. It is critical to have a well-automated system to both ensure pasteurization and prevent downstream contamination, which would take months to remedy. It should prohibit unpasteurized solids from passing through, and either waste or recirculate material that did not meet the time and temperature requirements. If necessary, standby equipment should be included to maintain time and temperature, because compromising these parameters could contaminate the downstream anaerobic digestion process.

2.4.1.5 Performance The pre-pasteurization process can meet the *Salmonella* criteria in Part 503. Ward et al. (1999) showed that pre-pasteurized solids resisted regrowth even after they were seeded with *Salmonella*; instead, the organisms died off. Chen et al. (2008) also observed that pre-pasteurization effectively destroyed *Salmonella*, even though fecal coliforms regrew (suggesting that it may be necessary to measure the actual pathogen rather than the indicator to ensure Part 503 compliance).

In summary, pre-pasteurization is an effective method for destroying pathogens in solids and is commonly used throughout the world. Several U.S. facilities use this process, including a 204 400-m³/d (54-mgd) in Alexandria, Virginia.

2.4.2 Thermal Hydrolysis

Thermal hydrolysis is a predigestion conditioning process where solids are exposed to elevated temperatures and pressures. The process improves the digestibility of biological solids (e.g., WAS), in particular, while reducing the size of digestion tankage and improving dewatering performance. This is more commonly achieved through a batch process but some continuous processes are available as well.

2.4.2.1 Process Development Thermal hydrolysis was first developed in the United States (Haug et al., 1978; 1983), but successful implementation occurred in Europe. The first full-scale system was implemented at the Hias Wastewater Treatment Plant in Norway in 1996. The largest facility in operation and the first in the United States is at the DC Water Blue Plains facility (Figure 23.8). More than 20 large and small systems are currently in operation, mostly in northern Europe. There are several additional WRRFs in the United States that are in various stages of thermal hydrolysis process (THP) design.

Another aspect of the full-scale THP is the rapid depressurization step. It occurs after the reaction step and is reported to help burst cells, further promoting hydrolysis and disinfection.

When used before anaerobic digestion, thermal hydrolysis achieves one or more of the following:

- Enhances digestion rates and gas production;
- Reduces the size of the anaerobic digestion system (increases allowable loading rates);
- Disinfects solids;
- Reduces the viscosity of the solids feed to the digester;
- Improves dewaterability of the biosolids; and
- Prepares solids for thermal processing downstream of anaerobic digestion.

FIGURE 23.8   The thermal hydrolysis process and anaerobic digesters at DC Water's Blue Plains WRRF (courtesy of Cambi, Norway).

2.4.2.2 Design Criteria—Thermal Hydrolysis Vessels   Thermal hydrolysis vessels are made of Type 316 stainless steel and are built to withstand both pressure and vacuum. The vessel configuration varies between manufacturers. Pressure vessels require annual inspection that may require a reactor to be shut down for up to a week so this must be factored into the design through the use of redundant trains, storage, or alternative destinations for the solids.

Operated at temperatures between 150°C and 170°C for 30 minutes and a pressure of about 827 kPa (120 psi), thermal hydrolysis solubilizes and hydrolyzes solids (Stuckey and McCarty, 1984; Li and Noike, 1992), and disintegrates biological cells (e.g., bacteria and viruses). According to Li and Noike (1992), maximum solubilization occurs at 170°C and the optimal SRT is between 30 and 60 minutes. In practice, a 30-minute SRT optimizes reactor size and delivers a solubilized and hydrolyzed product.

Thermal hydrolysis consists of a preheating step, a heating and batch-reaction step, and a rapid depressurization step for further solubilization and rupturing of microbial cells (see Figure 23.9). The preheating step is used to conserve spent heat from the reaction and depressurization steps, as well as produce a favorable energy balance. The system's feedstock is dewatered cake containing 14% to 18% solids. Dewatering considerably improves the heat balance and reduces the volume of the downstream anaerobic digestion process by about 50%.

If a Class A product is desired, design engineers and operators should ensure that every particle of solids in the reactor meets the time and temperature requirements and that any dilution water utilized after the THP process is disinfected. Solids screening (5-mm opening) is also required to protect the system components and ensure that no debris remains in the biosolids product.

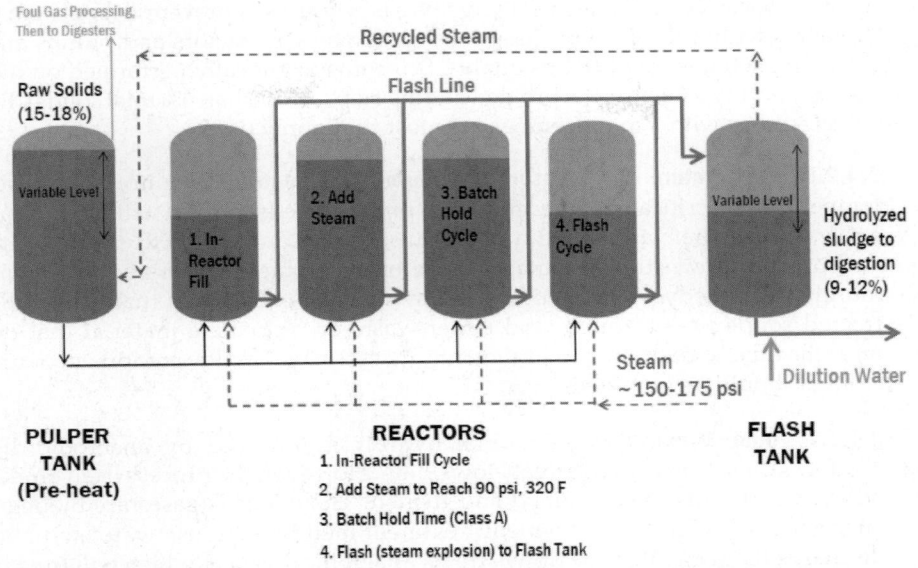

Foul Gas Processing,
Then to Digesters

Recycled Steam

Flash Line

Raw Solids
(15-18%)

Variable Level

1. In-Reactor Fill

2. Add Steam

3. Batch Hold Cycle

4. Flash Cycle

Variable Level

Hydrolyzed sludge to digestion
(9-12%)

Steam
~150-175 psi

Dilution Water

**PULPER
TANK
(Pre-heat)**

REACTORS
1. In-Reactor Fill Cycle
2. Add Steam to Reach 90 psi, 320 F
3. Batch Hold Time (Class A)
4. Flash (steam explosion) to Flash Tank

**FLASH
TANK**

FIGURE 23.9   Schematic of the thermal hydrolysis process (courtesy of Brown & Caldwell).

Because the THP feed is dewatered and contains more than 7% solids, time and temperature requirements are determined using eq 23.2. Under Regime A, solids must reach 150°C and stay that hot for 20 minutes, so the system's 30-minute SRT exceeds the U.S. EPA's time and temperature requirements. This retention time is used more to optimize hydrolysis and solubilization, and to ensure that the required temperature has diffused to the interior of all solids particles in the solids mass.

2.4.2.3 Ancillary Equipment for Thermal Hydrolysis   The following four ancillary features should be considered when designing a thermal hydrolysis system: cooling and heat recovery, screening, process control, and odor management.

2.4.2.4 Solids Cooling and Heat Recovery   After depressurization, thermally hydrolyzed solids are about 100°C and must be cooled before entering the anaerobic digestion process. This is typically achieved by first blending the THS with recirculating digested solids (typically approximately 1:3, respectively) and then cooling to the appropriate temperature in the digester. According to Qui (2016), the blending step is important because it:

- Dilutes THS to reduce the risk that it congeals as it cools and blocks the pipe (unblended THS should not be cooled below 70°C);
- Reduces the total solids % and improves pumping characteristics; and
- Creates turbulent flow, which improves heat transfer in the heat exchanger.

Cooling is typically achieved using facility effluent. Heat recovery can be performed if desired, but is not typically as it increases heat exchanger size and may complicate maintenance and operations.

2.4.2.5 Solids Screening   Solids screening is required to prevent debris from causing problems within the THP process, causing damage to reactors and pumps and negatively impacting final biosolids quality. Screening is typically performed on thickened primary and WAS before the pre-dewatering step (at less than 6% total solids). Enclosed solids screens with 5-mm openings are most commonly used.

2.4.2.6 Temperature and Pressure Monitoring and Control   Monitoring pressure and temperature is critical. Furthermore, the process needs to be installed with pressure- and vacuum-relief valves. The systems must be well automated to ensure disinfection and prevent downstream contamination of anaerobic digesters, which would take months to remedy. A well-automated system should ensure that only fully heat-treated solids pass through, and either waste or recirculate material that does not meet the time-temperature requirements. If necessary, standby equipment or upstream storage should be provided.

2.4.2.7 Odor Management   Thermal hydrolysis followed by anaerobic digestion produces biosolids with relatively low odors. However, the process itself emits strong odors, which must be contained and treated. The odorous gases are biodegradable and water soluble, so a convenient treatment method is to use water scrubbers and discharge the water into the downstream anaerobic digester, which will treat the process odors. The valving design for thermal hydrolysis vessels is critical to minimize vented odors.

2.4.2.8 Sidestream Treatment   The return liquor from thermal hydrolysis contains colloidal material that will contribute organic nitrogen, phosphorus, COD, and color to the mainstream process. For example, it will increase the facility effluent's organic nitrogen content by 0.75 to 1.5 mg/L. If the WRRF has low limits for any of these constituents, then they should be removed before the liquor enters the mainstream process. Treating these constituents with chemical conditioners (e.g., iron or aluminum) in the dewatering step has been proposed by Wilson et al. (2008b) but other biological sidestream treatment processes (e.g., anammox and struvite formation) are becoming more common.

2.4.2.9 Process Mode Variations   Two full-scale versions of thermal hydrolysis are currently in use. One process mode consists of a preheat tank, a reactor tank, and a flash tank (i.e., three tanks in series). The other process mode consists of one tank in which preheating, reaction, and depressurization occur. This mode could be designed with multiple parallel vessels.

2.4.2.10 Anaerobic Digestion Performance   The solubilized and hydrolyzed solids are easier to digest and, therefore, can increase digestion rates or reduce digester SRT. For example, Li and Noike (1992) report that the digester reached a stable maximum methanogen population and degraded most of the substrate in a 5-day SRT, suggesting that this was the minimum digester SRT to prevent washout or process instability. However, their work did not evaluate thermal hydrolysis performance under the high ammonia concentrations observed in highly loaded systems today, so full-scale processes are operated at a minimum 15-day SRT under average conditions. Design engineers can decrease the required anaerobic digester SRT by up to 10% to 25% (compared to conventional high-rate digestion) and expect similar process performance. For example, Wilson et al. (2008b) operated parallel conventional high-rate anaerobic digesters with and without thermal hydrolysis, using solids from DC Water Blue Plains facility. Results showed that both digesters had similar volatile solids and COD destruction, but the digester with thermal hydrolysis achieved these results at a 25% shorter digestion SRT (i.e., in 15 days rather than 20).

The solubilized solids from thermal hydrolysis are much less viscous (Kopp and Ewert, 2006). Hydrolyzed sludge contains 10% solids or more (compared to a typical digester feed sludge containing 5% solids). Through pre-dewatering and dilutions, THP feed is 16% solids, where it is heated with steam. The steam both increases the temperature and dilutes the solids. The hydrolyzed cake typically contains between 9% and 12% total solids, which can be adjusted with facility-dilution water, if needed, to maintain fairly constant solids loading to the anaerobic digester. The maximum solids content depends on the concentration of ammonia-nitrogen produced during digestion; ammonia-nitrogen concentration is typically kept at less than 2500 mg/L to prevent process inhibition.

The THP, digested, and dewatered biosolids can achieve significantly higher total solids (6% to 9% more) (Kopp and Ewert, 2006; Wilson et al., 2008b) than that produced via conventional digestion. So thermal hydrolysis is attractive when the resulting bio-solids will be hauled long distances for land application (to reduce hauling costs) or will be thermally processed (e.g., heat drying) since influent cake dryness and process evaporative capacity are major considerations when determining size and energy demand of a thermal process. For example, Kopp and Ewert (2006) report that cake solids increased from 25% to 34% when thermal hydrolysis pretreatment was used. Wilson et al. (2008b) had similar results.

2.4.3 Aerobic Pretreatment
Aerobic pretreatment of solids has been practiced in North America and Europe for more than a quarter century. It involves adding air or oxygen to solids at thermophilic temperatures as an initial "conditioning" step before anaerobic digestion.

2.4.3.1 Process Development   Aerobic pretreatment developed differently in Europe and North America; the aerobic thermophilic pretreatment (ATP) process is mainly practiced in Europe, while dual digestion is mainly practiced in North America. Both processes are intended to enhance VSR and pathogen reduction. The main difference between the two processes is the method of heating solids. In ATP, the heat used to attain thermophilic temperatures is waste heat from cogeneration of digester gas (not an autothermal process). ATP's SRT mainly depends on both process and Part 503 requirements. It typically is 24 hours or fewer, depending on the digestion temperature and the results of time-temperature requirements in eq 23.1 to attain Alternative 1 in Part 503 (U.S. EPA, 1999b). The minimum temperature is typically 55°C and the maximum is 65°C (when biological conditioning of raw solids is encouraged to enhance VSR).

In dual digestion, the aerobic step is an autothermal step in which heat generated during microbial aerobic metabolism is used to increase the process' temperature to thermophilic conditions. The temperature range for this process is also between 55°C and 65°C. Dual digestion's SRT depends on two factors: the U.S. EPA's time and temperature equation, and the time needed to autothermally raise the temperature of raw solids to the required set point. In most cases, the time needed to meet the temperature set point is greater than that demanded by the U.S. EPA's time-temperature equation. Using oxygen rather than air reduces SRT requirements and improves the heat balance.

Also, heat recovered from thermophilic solids can be used to help raise the input solids' temperature. All of the North American installations have been at WRRFs that already used high-purity oxygen in their activated solids process. Dual digestion's SRT is about 1 to 2 days, depending on the raw solids' volatile solids content. Feed containing more volatile solids significantly helps the heat balance to achieve autothermal temperatures. Several dual-digestion facilities were commissioned in the 1980s, and three U.S. facilities are operating the process successfully today. The 143 800-m³/d (38-mgd) Central Treatment Plant in Tacoma, Washington, has used this process for more than a decade, producing and marketing a Class A soil amendment product (called Tagro) from the resulting biosolids (Eschborn and Thompson, 2007).

2.4.3.2 Design Criteria Aerobic pretreatment has two primary goals: pathogen reduction and enhanced VSR. Both dual digestion and ATP are designed to operate in the thermophilic temperature range, so the tanks should be well insulated to maintain a favorable heat balance. In addition, solids screening may be required if heat exchangers are used for heat recovery. Under Part 503, processes must meet pathogen reduction requirements to achieve the Class A biosolids status. Aerobic pretreatment meets these requirements by typically maintaining a temperature between 55°C and 65°C for a specific period of time in a batch (plug-flow) operation. The amount of time is calculated using eq 23.1 (the equation for solids content less than 7%). Two possible options are 60°C for a minimum of 4.8 hours or 55°C for 24 hours. Hotter temperatures typically reduce time and, therefore, tank volume. In dual digestion, however, the minimum SRT depends more on the time required to achieve the desired autothermal temperature than on the U.S. EPA's time-temperature equation.

If a utility chooses to disinfect feedstock containing more than 7% solids, the time and temperature requirements are determined using eq 23.2.

ATP typically is heated using waste heat from a cogeneration process. The heat balance largely depends on the insulation of the preheat tank and the decision to use heat-recovery heat exchangers.

Dual digestion is mainly heated autothermally (the mixers introduce some heat). Three important parameters in dual-digestion designs are the decay rate, the biological heat of reaction (BHR), and oxygen demand. Gemmell et al. (1999) estimated an average decay rate of 0.087 ± 0.010 d⁻¹ at an average temperature of 37°C. Gemmell et al. (1999) determined a BHR of 16.6 MJ/kg volatile solids destroyed for the dual-digestion process at Barrie, Ontario, in initial trials with a 26% VSR. Grady et al. (2011) suggest a BHR value of 18.8 MJ/kg volatile solids destroyed (for a design involving autothermal thermophilic aerobic digesters). Messenger et al. (1993) determined a BHR of 18.6 MJ/kg volatile solids destroyed. Haas (1984) found that BHR ranged from 17.4 to 23.3 MJ/kg volatile solids destroyed at 20% and 10% VSR, respectively, during trials conducted at Hagerstown, Maryland. The oxygen demand depends on the type of solids and the ratio of primary and biological solids. Values in the range of 1.7 kg O_2/kg volatile solids destroyed have been reported by Pitt and Ekama (1996) and Gemmell et al. (1999). This value can easily be determined experimentally and should be tested, because it can vary from facility to facility.

2.4.3.3 Aerobic Vessel Design To produce a Class A biosolids, the treatment process must meet the time and temperature requirements in Alternative 1 under Part 503, which specify that every particle of solids should be exposed to a minimum temperature for a minimum period of time (U.S. EPA, 1999b). So complete-mixed systems or systems that could back-mix or short-circuit should be avoided. Design engineers should consult EPA staff or other pertinent regulators before using a non-batch plug-flow system. The batch systems should be designed to operate in a fill/hold/draw mode and be well mixed to ensure that every particle is maintained at the required temperature (for the required time) and that the monitored temperature reflects the entire contents of the batch. Typically, if continuous operation is desired, three batch vessels are needed (one for each cycle). However, if the downstream anaerobic digester can handle an intermittent feed cycle and upstream storage is adequate, then fewer batch vessels can be supplied.

2.4.3.4 Ancillary Equipment for Aerobic Pretreatment Design engineers typically consider three important ancillary features when installing an aerobic pretreatment process:

- Solids heating and recovery (ATP) or oxygen system (dual digestion);
- Solids screening; and
- Temperature monitoring and control.

The aerobic vessel for ATP and dual digestion typically is maintained between 55°C and 65°C. In the ATP process, solids must be heated and then cooled. The most common method for heating and cooling solids is heat exchangers. In dual digestion, solids are heated autothermally so an external heat source is unnecessary. Both processes must cool treated solids before digestion—unless the anaerobic digester also is operated at thermophilic temperatures. The cooling method could include a heat-recovery step in which the heat transferred from cooling solids is used to preheat raw solids. This step depends on owner, engineer, and vendor preference, because it will require a substantial amount of heat-exchange capacity. If heat exchangers are used, the solids may need to be screened first. Screening also helps produce an aesthetically pleasing biosolids. Good temperature monitoring and control are required to maintain Class A compliance. If necessary, standby equipment should be provided to maintain time and temperature, because compromising these parameters could contaminate the anaerobic digester with inadequately disinfected solids. The solids should be thickened to at least 5% total solids for successful operations; therefore, downstream solids pumps and pipes should be designed to handle thicker solids adequately.

Also, dual digestion will need an oxygen supply (hence, dual digestion typically is installed at facilities that already use high-purity oxygen in their activated sludge processes).

2.4.3.5 Performance Preconditioning of solids with air is intended to increase the overall VSR. Researchers (Pagilla et al., 1996; Cheunbarn and Pagilla, 1999, 2000) have extensively evaluated ATP performance and compared it to mesophilic digestion performance. They confirmed the European full-scale observations of Baier and Zwiefelhofer (1991) that ATP enhances VSR and gas production. Cheunbarn and Pagilla (1999) showed that VSR increased as the ATP's SRT (0.6 to 1.5 days) and temperature (55°C to 65°C) increased. In pilot-testing work at Sacramento, Pagilla et al. (1996) compared ATP to conventional mesophilic digestion and determined that ATP enhanced VSR from 53% to 59% for combined solids (primary solids and WAS). They also showed that ATP could meet the U.S. EPA's Part 503 requirements for fecal coliform, *Salmonella*, enteric virus, and helminth ova. In addition, they determined that ATP effectively controlled

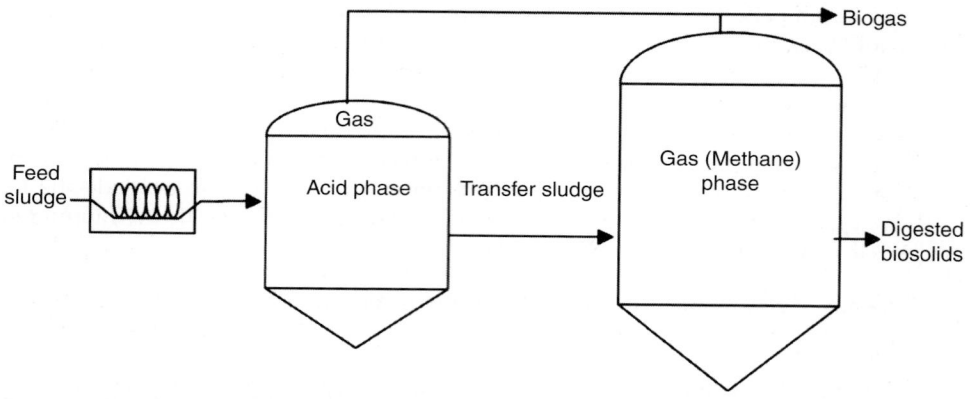

Figure 23.10 Schematic of the two-phase anaerobic digestion process.

and destroyed *Nocardia* filaments. Finally, they observed that ATP-treated, centrifuged biosolids contained between 32% and 36% total solids, compared to mesophilically digested, centrifuged biosolids, which only contained 30% total solids.

Gemmell et al. (1999) suggested that dual digestion could achieve stable performance when the first high-rate aerobic reactor had an HRT of 1 to 2 days and the second high-rate anaerobic reactor had an HRT of 9 to 12 days. Overall, this retention time was 6 to 10 days shorter than the 20-day SRT typically required for conventional high-rate digestion. Also, Gemmell et al. (2000) suggested that the full-scale dual-digestion process at the Barrie Wastewater Treatment Plant achieved 60% VSR. Operators at a 143 800-m^3/d (38-mgd) WRRF in Tacoma, Washington, have been producing and marketing a biosolids-based soil amendment for more than a decade; they attribute much of its high quality to the dual-digestion process (Eschborn and Thompson, 2007).

2.4.4 Acid-Phase Hydrolysis

Acid-phase hydrolysis (sometimes called two-phase or acid-methane digestion) separates two major anaerobic reactions—acid formation (acidogenesis) and methane generation (methanogenesis)—to benefit the overall stabilization process (see Figure 23.10). The most practical way to separating phases is via kinetic control, by regulating the detention time and loading rate for each reactor. Increasing loadings to the first-stage digester and reducing SRT (HRT) favors acidogenic organisms because the low pH and retention time are unfavorable to acetoclastic methane formers. In the second stage, a larger digester (or multiple digesters) increases SRT, so methanogens proliferate. High influent concentrations of short-chain fatty acids also promote the growth of methanogens.

2.4.4.1 Process Development Early work on the process was largely completed by Professor Sam Ghosh (Ghosh et al., 1975, 1987; Lee et al., 1989). Raw solids initially are fed to a reactor with 1- to 2-day SRT, called an acid-phase digester. In this reactor, low-pH environment (typically 5.5 to 6.2) is established, suspended organic matter is hydrolyzed, and then low-molecular-weight fatty acids are formed. Methane generation is limited in this phase. This first phase has been tested at both mesophilic and thermophilic temperatures, although few full-scale systems have operated the acid phase at thermophilic temperatures. Some research has suggested that there is no true phase separation (Shimada et. al., 2011)

Acid-phase sludge then is fed to a second vessel with a 10- to 15-day SRT (called a methane-phase digester). This phase also can be operated at mesophilic or thermophilic temperatures. Conditions in this phase are similar to those found in conventional high-rate digesters, which are operated to maintain an optimum environment for methanogenic microorganisms.

2.4.4.2 Design Criteria Laboratory- and small-scale work led to the development of larger-scale systems for wastewater solids (Ghosh et al., 1991, 1995). Experience indicated that anaerobic digester performance could be improved by optimizing the acid-forming and methane-generation phases separately. Compared to the single-phase systems, two-phase anaerobic digestion systems have higher rates of VSR and biogas production, produce biogas that contains more methane, inactivate more pathogens, minimize foam, and are overall more resilient and stable.

The key design criteria are loading rates and retention times. The recommended process design criteria for the acid-phase digester typically are as follows:

- Volatile solids loading rate of 25 to 40 kg volatile solids/m³·d (1.5 to 2.5 lb volatile solids/cu ft/d);
- Feedstock that contains 5% to 6% solids;
- SRT of 1 to 2 days (at mesophilic temperature);
- Total volatile fatty acid (VFA) concentrations of 7000 to 12 000 mg/L; and
- pH range of 5.5 to 6.2.

The methane reactor in two-phase digestion can be loaded at higher rates than conventional high-rate digestion systems because of the hydrolysis that occurred in the acid-phase reactor. Volatile solids loading rates for the methane reactor often are similar to those for conventional high-rate mesophilic digestion. Residence times of about 10 days have been tested and promoted by process proponents; however, most full-scale systems have longer residence times (often 15 days or more). The total SRT for the entire two-phase digestion process is rarely less than 15 days, which is required if the resulting biosolids must meet Class B criteria under Part 503.

2.4.4.3 Performance Two-phase digestion performance often has been measured via VSR or biogas production. Schafer et al. (2002) reported that this process significantly improved VSR at the DuPage County, Illinois, facility, but other agencies had seen less improvement. Barnes et al. (2007) reported that Denver, Colorado's two-phase digestion facility had not shown any significant increase in VSR compared to its prior high-rate mesophilic system, but digester foaming was no longer a major problem. The DuPage County facility also had a major reduction in digester foaming (Ghosh et al., 1995).

2.4.4.4 Process Variation—Three-Phase Digestion Three-phase digestion is a variation of two-phase digestion that uses both thermophilic digestion and a third reactor, which may have variable temperatures. This process has been used at the DuPage County, Illinois, facility and at the Inland Empire Utilities Agency in Chino, California. The primary objectives of this approach are to ameliorate the higher VFA levels that can occur in thermophilic digestion and provide another phase of digestion to reduce short-circuiting and allow for more pathogen control. Inland Empire's three-phase digestion process has been reported to produce biosolids that meet Class A requirements under Part 503 (Drury et al., 2002).

2.4.4.5 Process Variation—Enzymic Hydrolysis and Digestion Enzymatic hydrolysis expands the acid phase of the system into as many as six tanks in series at 42°C. The goal is to shift reactor kinetics away from complete mix to plug flow, which can provide more treatment. According to proponents, enzymatic hydrolysis proponents increased biogas and solids destruction, and enhanced enzymatic hydrolysis greatly improved pathogen destruction.

2.4.5 Other Pretreatment Technologies

Solids disintegration technologies are designed to increase the rate and extent of anaerobic solids digestion by applying external energy to render solids more bioavailable. These processes typically are applied to WAS because it is considered the most difficult to digest.

The means of energy application is technology specific, but the reported effects are consistent:

- More biogas production;
- Increased VSR; and
- Reduced mass of solids for disposal.

Table 23.10 lists various disintegration technologies that currently are commercially available.

2.4.5.1 Ultrasonic Technologies—Process Development Full-scale implementation of ultrasonic technologies to enhance anaerobic digestion began in Europe in the mid- to late 1990s. The technology was developed as a means of increasing biogas production while reducing the mass of biosolids for disposal; it essentially increases the digester's gasification rate.

Ultrasonics generate transient acoustic cavitation, which improves anaerobic digestion of WAS, in particular. Acoustic cavitation occurs when ultrasonic waves compress and rarefy the liquid. During rarefaction, enough energy may be applied to exceed intermolecular forces, forming a void (cavitation bubble). The bubble's subsequent collapse generates significant amounts of heat (4000°C), pressure (about 1000 atm) (Christi, 2003), and shear forces from the liquid jets formed (Mason and Lorimer, 1988). Particles near or within a collapsing bubble can be exposed to one or more of these forces, breaking them down to a size of 40 000 Da (Portenlänger and Heusinger, 1997).

Process	Example Manufacturer	Disintegration Method
Ultrasonics	Sonix™, Enpure	Acoustic cavitation
High-pressure homogenization	Microsludge™, Paradigm Environmental	Chemical pretreatment with hydrodynamic cavitation and shear
Pulsed electric field	OpenCel	Electroporation

TABLE 23.10 Examples of Solids Disintegration Technologies

Pressure, temperature, and shear forces are the primary forces responsible for disintegrating WAS. Chemical transformations also are theoretically possible. The environment in the cavitation bubble could lead to the formation of free radicals, which can interact with various elements in the surrounding fluid. Depending on the ultrasonic probe's frequency, either mechanical or sonochemical forces can dominate. At lower frequencies (around 20 kHz), mechanical forces typically dominate; free radical formation becomes more common at higher frequencies.

2.5 Digestion Processing

This section discusses some key processes that can improve digestion performance and biosolids characteristics.

2.5.1 Prethickening

Thickening solids before digestion is a means of preserving or expanding digester capacity. There are a variety of technologies for prethickening solids (see Chapter 21), and their design and operation should be coordinated with the anaerobic digesters. A thicker cake will preserve volumetric capacity and reduce heating requirements. Because thicker feed typically results in longer HRT, organic overloading of a digester is unlikely. Excessively thick solids can increase the sludge viscosity, thus increasing the energy required for digester mixing and sludge pumping. One notable exception is the thermal hydrolysis process.

2.5.2 Debris Removal

Debris typically is removed via screening and grit removal at the facility headworks; however, subsequent screening of primary solids, scum, and other digestion feedstocks may be necessary. If debris enters the digester, several of the following process issues could arise:

- Loss of digester volume (because of accumulated debris in the tank);
- Excessive wear on pumps;
- Clogging and additional cleaning of heat exchangers; and
- Ragging and binding of mixing and pumping equipment.

Debris also affect biosolids quality. Large quantities of debris (e.g., plastic materials) can degrade the aesthetic qualities of biosolids, making it undesirable or even unfit for beneficial use.

A variety of technologies (e.g., rotary drum screens and strain presses) can remove debris from solids. Screening (straining) has been done on raw solids, raw scum, and unthickened, thickened, and digested solids in slurry form. When evaluating technologies, design engineers should consider the intended application, the material to be processed, and other site-specific constraints (e.g., whether the sewer is combined or separated, extent of debris removal from wastewater, whether grinders have already been used, whether stringy material could bind pumps or other equipment, desired use for the biosolids, biosolids aesthetics, and regulatory requirements).

2.5.3 Debris Size Reduction (Reduction in "Identifiables")

Reducing the size of debris protects process equipment from blockages and binding; it also makes debris less identifiable in biosolids. Debris can be reduced by grinding solids before pumping or dewatering them. In-line solids grinders often will be placed in front of recirculation, feed, or wasting pumps. Some states (e.g., Washington) required debris to be made unidentifiable before biosolids could be land applied or otherwise beneficially used.

2.5.4   Batch and Plug-Flow Systems

Most existing digestion systems can be classified PSRPs, which meet Class B requirements for land application (assuming vector-attraction requirements are met). The systems can be upgraded to PFRPs, which produce Class A biosolids.

One upgrade option typically involves higher temperatures (e.g., thermophilic) and batch or plug-flow operations. To meet the time and temperature requirements of Part 503, every particle must be treated at a temperature higher than 50°C for a prescribed period of time. Typically, at least three tanks are required to meet this requirement: one in fill mode, one in hold mode, and one in draw mode.

Another option is using a plug-flow reactor followed by a complete-mix reactor—both operating at thermophilic temperatures. Developed by the Columbus (Georgia) Water Works, this combination is thought to achieve similar pathogen reductions as the batch system but has fewer control points and less process complexity (Willis et al., 2003). The U.S. EPA recently determined that this process is a conditional, site-specific, PFRP-equivalent process.

2.6   Post-Digestion Processing

2.6.1   Process Development

After digestion, solids are considered "stabilized" because the available substrate has largely been depleted and microbial activity has largely been reduced, so the product is far less likely to emit odors, attract vectors, or regrow pathogens. However, because of its biological origin, any perturbation of biosolids characteristics can "destabilize" the material. So the methods used to store and handle biosolids can be extremely important.

Murthy et al. (2003), Chen et al. (2005), Chen et al. (2006), and Higgins et al. (2006a) evaluated the headspace of bottle-stored anaerobically digested biosolids and found that destabilizing biosolids could increase odor production by increasing the available substrate and decreasing methanogenic activity. Research has shown that a key group of odorants are the volatile organic sulfur compounds (VOSCs), which are mainly methanethiol (or methyl mercaptan), dimethyl sulfide, and dimethyl disulfide. When present in air samples, VOSCs correlate well with odor panel measurements from biosolids (Adams et al., 2004). Higgins et al. (2006b, 2008b) also showed that fecal coliform regrowth was possible from post-digestion solids processing. These authors suggested that biosolids shearing caused both fecal coliform regrowth and odorant production.

Post-digestion processes should minimize conditions that would destabilize biosolids or the population dynamics within the material.

2.6.2   Storage of Biosolids

Biosolids in liquid and cake forms can be stored under different conditions (see Chapter 22). One aspect of storage that affects microbial populations is freeze-thaw, in which stored biosolids alternately freeze and thaw during cold-weather seasons. Freezing biosolids can disrupt cells, thus releasing substrate and inhibiting methanogens. Thawing them could increase biological activity, resulting in odors. Eschborn et al. (2006) showed that the internal temperature of a field storage pile dropped during winter, and the outer layer of the pile froze. As temperatures increased the following spring, odorant production also increased. When designing biosolids storage, engineers should take these factors into account, especially in regions where long-term winter storage is anticipated. Higgins et al. (2003) simulated freeze-thaw conditions in laboratory headspace experiments and showed that odorant production could be substantial when frozen cake samples were thawed (see Figure 23.11). Freezing biosolids led to a delay in methanogen recovery when the material thawed (Figure 23.12).

Managing biosolids storage before land application can help reduce nuisance odors. The goal of storage is to allow solids to restabilize once odorant production begins, thus allowing VOSC-associated odors to dissipate. For example, once odorant production begins, it typically peaks about a week or two later—although this depends on temperature (Higgins et al., 2003). Cooler temperatures increase the time needed for odors to peak and VOSC concentrations to reduce (Figure 23.13).

When designing storage systems for biosolids, engineers should consider the following:

- If frozen biosolids are stored for several days once thawing begins, odors can dissipate before the material is beneficially used;
- Fresh biosolids emit more odors during land spreading than biosolids that had been in long-term field storage (to get past the peak concentrations of odorants), so proper storage is important to managing odors (Eschborn et al., 2006);

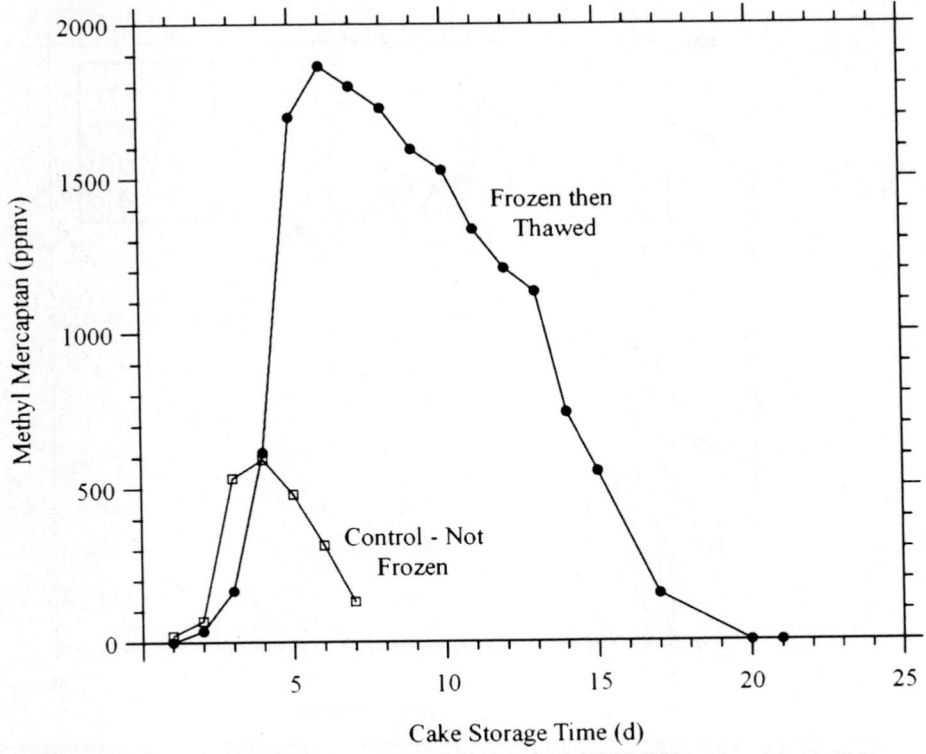

FIGURE 23.11 The effect of freeze-thaw conditions on methyl mercaptan emissions from biosolids (Higgins et al., 2003).

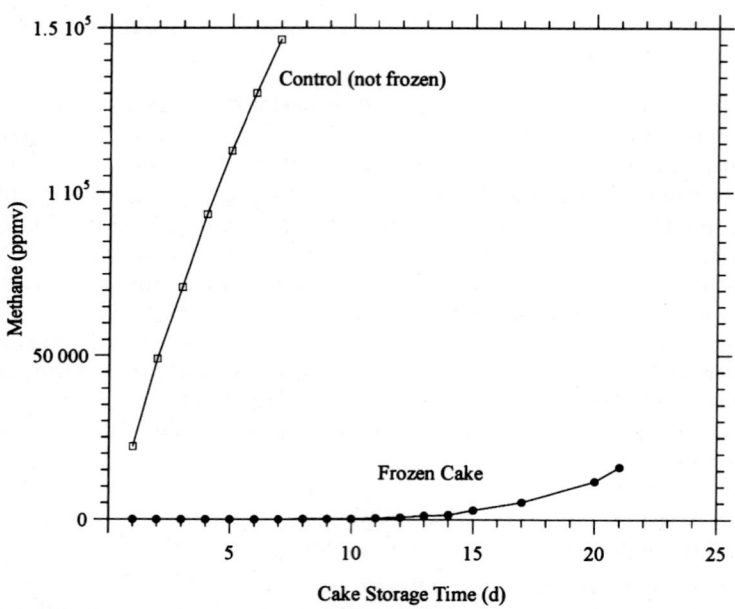

Figure 23.12 The effect of freeze-thaw conditions on methanogen recovery from dewatered biosolids (Higgins et al., 2003).

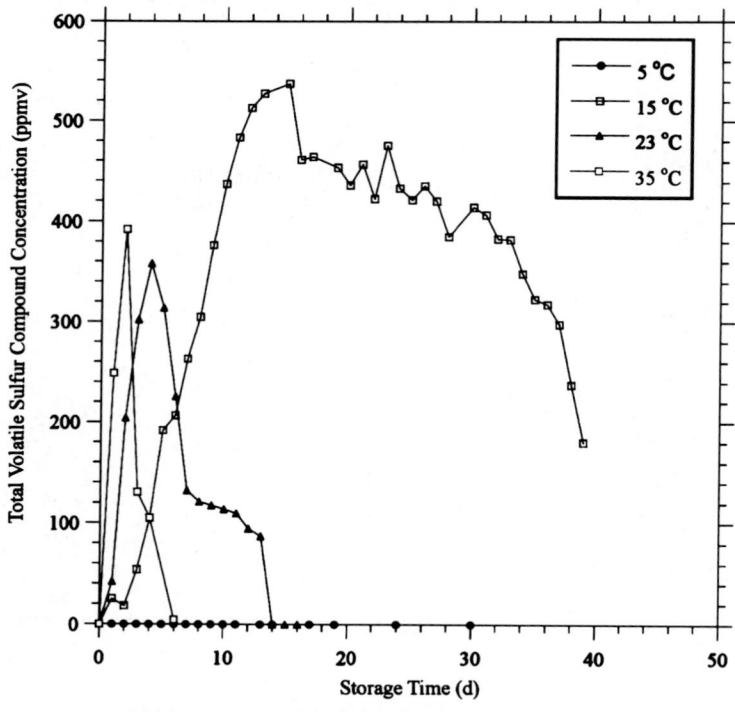

Figure 23.13 The effect of incubation temperature on emissions of total volatile sulfur compounds from dewatered biosolids (Higgins et al., 2003).

- Mixing old and new biosolids mitigates odorant production by bioaugmenting fresh biosolids with active methanogens that can degrade VOSCs (Chen et al., 2005; Williams et al., 2008); and

- Storage conditions should minimize freezing and perturbations that can destabilize population dynamics in biosolids (EPA, 2000).

2.6.3   Cake Conveyance Effects

The effect of high-shear conveyance has not been studied in great detail. Murthy et al. (2002) report that high-shear conveyance methods increase the biosolids' odorant production profile. Headspace experiments involving anaerobically digested biosolids confirm this research (see Figure 23.14). So, when designing solids conveyance systems, engineers should consider the following:

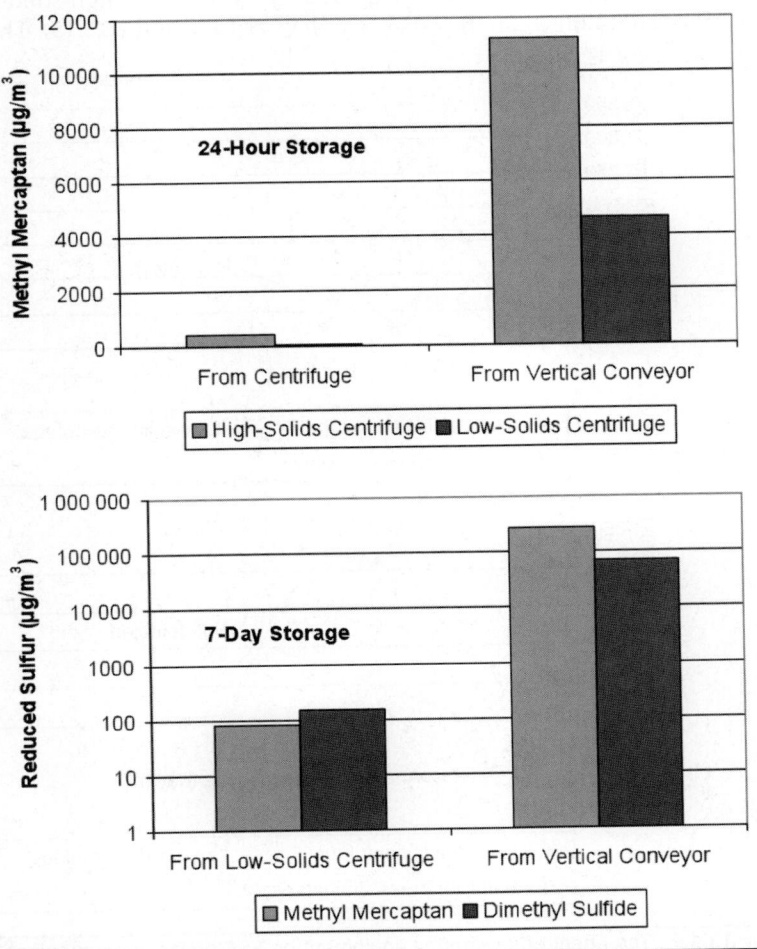

FIGURE 23.14   The effect of a vertical screw conveyor on odorant production in dewatered biosolids (Murthy et al., 2002a).

- Keep transport distances as short as possible;
- Use low-shear conveyance methods, if possible;
- Minimize the use of vertical high-shear conveyance; and
- Maintain a top-down design philosophy, if possible (i.e., install dewatering equipment at the top of a building and storage silos at the bottom to minimize conveyance distance and shear).

2.6.4 Dewatering Effects

Murthy et al. (2002) and Higgins et al. (2002) have suggested that VOSC production is influenced by a combination of factors. They conducted side-by-side tests on three dewatering systems: a high-solids centrifuge, a low-solids centrifuge, and a belt filter press simulator. The centrifuges were adjusted to produce lower cake solids (similar to the belt press device), so researchers could compare cakes from the same facility with similar solids content. Results showed that cakes from the high-solids centrifuge produced more odorants than the other two devices (see Figure 23.15). The authors found

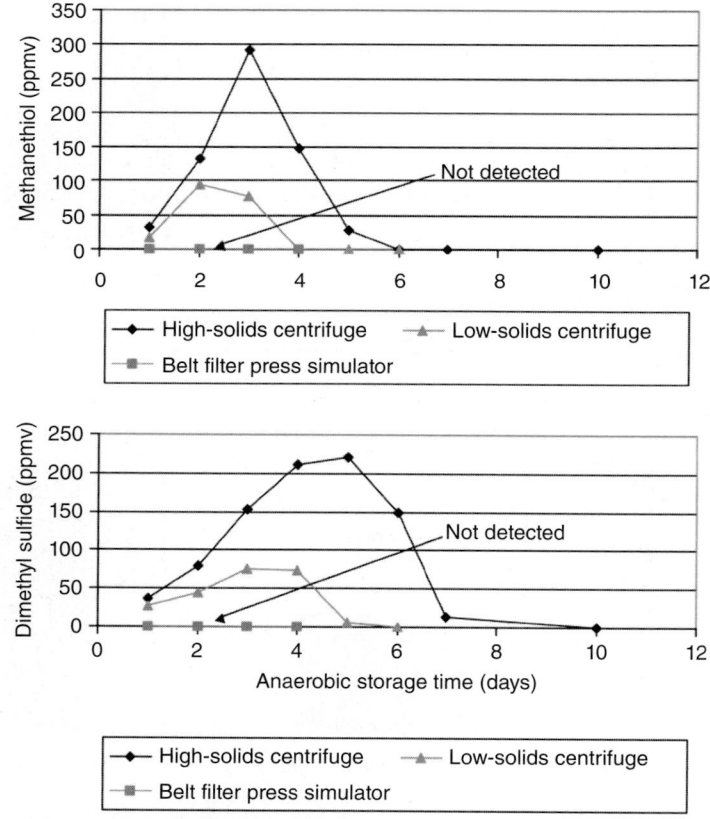

Figure 23.15   The effect of dewatering equipment on VSC emissions from biosolids (centrifuge cake contained 26% solids, belt filter press cake contained 25% solids, and detection limit is 1 ppmv) (Murthy et al., 2003).

that solids sheared during centrifugation released both labile protein and inhibited methanogenesis, thus increasing odorant production.

Higgins et al. (2006b) also showed that centrifuging biosolids promoted the regrowth of fecal coliforms. In some cases, this regrowth exceeded fecal coliform concentrations in raw solids. Further work by Higgins et al. (2008) showed that *Salmonella* regrew in Class B biosolids during storage. The biosolids had been mesophilically digested and centrifuged. However, *Salmonella* did not regrow in stored Class A and Class B biosolids that had been thermophilically digested and centrifuged.

Design engineers should consider pilot-testing dewatering equipment, and monitoring for possible odorant production and regrowth before selecting a unit.

2.6.5   Digestion Process Effects

While low-shear dewatering equipment produces fewer odors regardless of the anaerobic digestion process used, this is not the case for high-shear dewatering. More complete digestion (e.g., enhanced or thermophilic digestion) can affect overall odorant production and in some cases reduce it, especially if high-shear post-processing is used. Figures 23.16 and 23.17 show the total VOSC production profile for several full-scale digestion processes. These figures suggest that enhanced digestion processes can produce fewer odors than a conventional mesophilic digestion process.

If high-shear post-digestion processes are being proposed or already exist at a facility, design engineers should consider using enhanced digestion processes to mitigate overall odorant production if land application is proposed. Processes that have been shown to reduce odorant production from high-shear solids processing include thermophilic digestion, temperature-phased anaerobic digestion (TPAD), and thermal hydrolysis (THP).

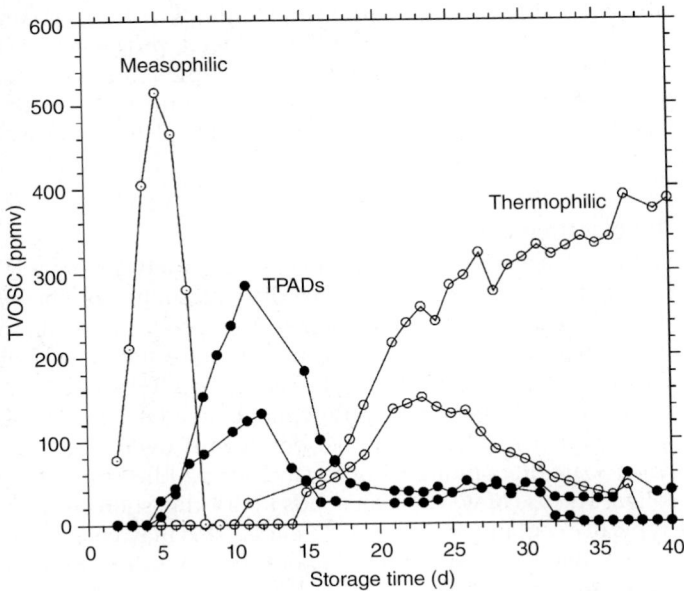

Figure 23.16   The effect of two enhanced digestion processes on odorant production (TVOSC = total volatile organic sulfur compound) (courtesy of Dr. Matthew Higgins, Bucknell University).

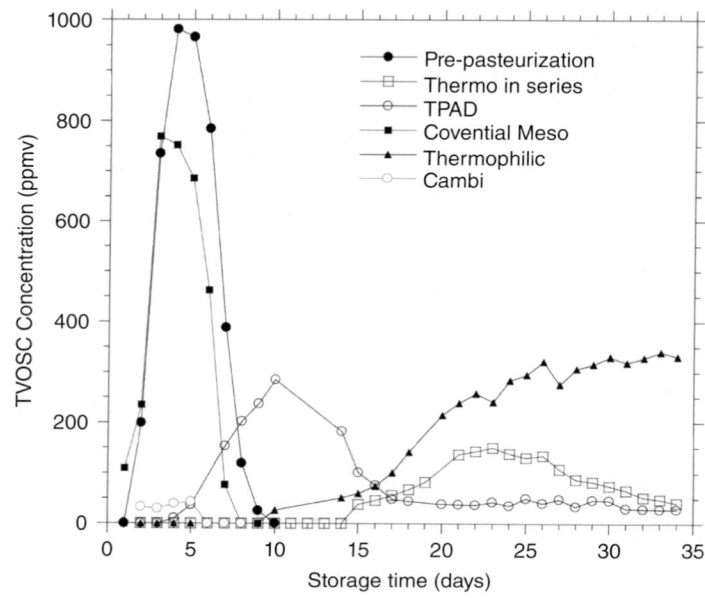

Figure 23.17   The effect of five enhanced digestion processes on odorant production (TVOSC = total volatile organic sulfur compound) (courtesy of Dr. Matthew Higgins, Bucknell University).

2.7   Co-Digestion Processing

Historically, anaerobic digestion was used to minimize solids volume, stabilize solids, and reduce pathogens. As process options evolved, process efficiency improved, and the focus on biogas production and energy use grew, interest in anaerobic digestion expanded. Some wastewater utilities would like to further expand biogas production by adding other feedstocks (fats, oil, and grease [FOG], food wastes, and organic liquid wastes) to anaerobic digesters.

2.7.1   Dry Digestion and Wet Digestion

Co-digestion of wastes can encompass a wide variety of organic feedstocks. Some organic feedstocks are more amenable to "dry digestion," in which feedstocks contain 15% or even 20% solids. Such systems were developed primarily for organic solid waste or bulk-waste materials, and are seen primarily in technologies coming from Europe. Most of these systems originally were applied to the solid waste industry. The digestion vessels are often developed for plug-flow movement (e.g., feedstocks are added to the top of a silo-shaped reactor and move downward over a 20-day digestion period). Mixing systems are often limited, but can include paddles and other systems that work in thick solids. The use of wastewater solids in dry-digestion systems is limited because the cake typically contains less than 10% solids. This may change over time, so readers may wish to gather more information about dry digestion from the Internet or other literature sources.

In this manual of practice, the discussion of anaerobic digestion is devoted to liquid-slurry ("wet digestion") in which feedstocks contain less than 10% to 15% solids (they frequently contain 5% or 6% solids). In wet-digestion systems, feedstocks are pumpable materials and the mixing systems are compatible with slurries that are typically less than 5% solids. Wet-digestion systems typically are operated as complete-mix reactors that minimize deposits on the bottom of the tank and minimize floating layers of solids on the liquid surface. Wastewater utilities in North America, Europe, and elsewhere are adding not only solids to their wet-digestion systems but also a variety of compatible organic feedstocks.

2.7.2 FOG and Grease Wastes

Fats, oil, and grease are generated in a variety of locations in an urban environment. To protect the collection system, many municipalities strictly limit FOG discharges, so the material is accumulated at point sources (e.g., waste drums, grease traps, and grease interceptors). While municipalities do not want this material in the collection system, its use in anaerobic digesters can provide major benefits. FOG is an energy-rich substrate that is readily degradable in anaerobic digestion systems.

Digesters with FOG loads as high as 30% of feedstock or volatile solids (VS) loading still can maintain stable operations (Schafer et al., 2008). Kester et al. (2008) has reported that some facilities digesting both solids and FOG have both enhanced biogas production and enhanced VSR.

The benefits of FOG addition are well documented, but there are challenges as well. At ambient temperatures, FOG tends to be highly viscous and adheres to metallic and concrete surfaces, making it difficult to introduce to a digester via a conventional hauled-waste receiving station. Also, FOG tends to stratify during transport and storage. Receiving stations used for FOG have a wide variety of tankage. Tanks of FOG used to feed digesters need to be mixed (to obtain consistent feedstock characteristics) and may need to be heated via steam injection, hot water circulation, or heat exchanger. Also, screening and grinding often are required to remove debris (e.g., stones, rags, and metallic objects). Because of its association with the food service industry, FOG can contain other extraneous food debris that must be removed to protect pumps and downstream equipment.

To minimize stratification and eliminate a possible grease layer on the liquid surface in the digester, FOG must be properly mixed with the mass in the tank. Utilities often introduce FOG into a recirculating slurry of digesting mass, so it becomes mixed and diluted quickly. Also, the digester's mixing system should be carefully evaluated to ensure that a surface layer of scum or grease does not develop and cause operational problems.

2.7.3 Liquid and High-Strength Wastes

Liquid and slurry wastes are typically easiest for a digestion system to accept because its infrastructure is designed to pump liquids. Some liquid wastes may require screening or grit removal, depending on its composition. These wastes often come from food processing industries (e.g., fruit and vegetable processing), the beverage industry, pharmaceutical industry, and others. While liquid wastes are relatively easy to accept, they can have significant negative effect on digester capacity. Most digesters do not use solids-liquid separation for SRT control, so the added volume of liquid waste takes up a digester's hydraulic capacity. So high-strength wastes should be used before lower-strength wastes because they produce the most biogas per unit hydraulic capacity consumed. When characterizing liquid wastes, design engineers should ensure that their components will not negatively affect digestion (e.g., promote struvite formation or produce a compound that will affect the liquid treatment train). Also, increasing levels of total dissolved solids or other dissolved constituents can be introduced via digestion and typically will be returned to the liquid treatment train via the post-digestion dewatering system. Such dissolved material may be inert and proceed directly to the outfall or can affect the liquid treatment process.

Liquid or slurry wastes being considered for co-digestion should be carefully evaluated for their compatibility with the digestion process, digestion-feeding system, mechanical systems used in digestion, and biogas management-system capacity. High-protein wastes, for example, could greatly affect foam production in digestion vessels. Sudden or large foam production is often debilitating for digester operations; it has caused digesters to overflow material onto the ground and caused accidents that resulted in major structural damage to tanks and tank covers.

These wastes often need to be carefully metered into the digestion system to prevent overfeeding microbes and causing digester upsets and foaming. Rapid feed of highly digestible waste also can lead to large spikes in gas production that the gas piping cannot accommodate. Instead, it is directly released to the atmosphere via gas-relief valves. In extreme cases, gas production could outpace the gas system's ability to discharge gas (especially if foam is blocking gas-relief valves), leading to gas pressure buildup in the vessel and potential catastrophic tank or cover failure.

2.7.4   Food Waste Materials

As with FOG, food scraps or post-consumer food waste materials are a source of renewable energy when anaerobically digested and converted to methane. One of the challenges of food scraps (the organic fraction of municipal solid waste) is removing contaminants to protect the digestion process. Also, the waste itself typically requires preprocessing to make it pumpable and amenable to efficient digestion without compromising biosolids aesthetics. Preprocessing can include screening; manual separation of debris and extraneous materials; removal of metals, aluminum, glass, grit, and plastics; and pulping—depending on the specific situation. Technologies that preprocess food scraps or food-waste materials are being developed primarily in Europe. Gray et al. (2008) describe one of these systems, which was tested at the East Bay Municipal Utility District in Oakland, California.

2.8   Design Considerations

2.8.1   Design Data and Parameters

Before designing the anaerobic digestion process, engineers should consider modeling it using available models (Batstone et al., 2002; Jones et al., 2008a). Jones et al. (2008b) recommend using a full-facility model to simulate influent wastewater characteristics and quantify fractions of solids produced to predict anaerobic digester performance. The Water Environment Federation (2013) provides more details on anaerobic digester modeling.

Historically, anaerobic digesters have been designed based on SRT, organic loading rate (volatile suspended solids [VSS] per volume), and volume per capita (see Table 23.11). In the absence of operating data (or estimates of projected facility flows) and calibrated system modeling, volume figures per capita can be used to estimate influent volumes

Substrate	35°C	30°C	25°C	20°C
Acetic acid	3.1	4.2	4.2	—
Propionic acid	3.2	—	2.8	—
Butyric acid	2.7	—	—	—
Long-chain fatty acid	4.0	—	5.8	7.2
Hydrogen	0.95[a]	—	—	—
Wastewater sludge	4.2[b]	—	7.5[b]	10

[a]For 37°C.
[b]Computed value.

TABLE 23.11   Minimum Values of Solids Retention Time (θ_c) for Anaerobic Digestion of Various Substrates (Reprinted with permission from Lawrence (1971). Copyright © 1971 American Chemical Society)

at WRRFs. Low-rate digesters typically have organic loading rates of about 0.5 to 1.5 kg VSS/m³·d (0.04 to 0.1 lb VSS/d/cu ft). High-rate digesters with mixing and heating typically have organic loading rates of 1.9 to 2.5 kg VSS/m³·d (0.12 to 0.16 lb/d/cu ft).

Typical SRTs are about 30 to 60 days for low-rate digestion and 15 to 20 days for high-rate digestion at mesophilic temperatures. The SRT is the ratio of the total mass of solids in the system to the quantity of solids withdrawn per day. In two-stage digestion, the typical SRT above refers to that of the first reactor in the system. For anaerobic digesters with no internal recycle, the SRT equals the HRT. If settled solids are recycled, the SRT would be larger than the HRT. Recycling is characteristic of the anaerobic contact or two-phase digestion process or recuperative thickening.

A minimum SRT is essential in anaerobic digestion; it ensures that the necessary microorganisms are being produced at the same rate as they are wasted daily. It also is different for various constituent groups. For example, lipid-metabolizing bacteria grow most slowly and, therefore, need a longer SRT, while cellulose-metabolizing bacteria require a shorter SRT (see Figure 23.18).

If the SRT is too short, then the microbial population of methanogens will wash out and the system will fail. Lawrence (1971) has published the minimum SRT needed to reduce several specific substances; a function of temperature, these SRTs range from less than 1 day for hydrogen to 4.2 days for wastewater solids (see Table 23.12). Hotter temperatures reduce the SRTs needed for maximum performance because they increase specific gas production (see Figure 23.19).

However, digester SRT is not only a function of system microbiology; the selected use for biosolids also must be considered, especially for Class B biosolids. Both thermophilic and mesophilic anaerobic digestion are considered PSRPs. One method for achieving this status, along with maintaining temperature, mixing, and anaerobic conditions, is maintaining a minimum SRT of 15 days. Furthermore, for overall process stability, ease of control, to account for grit and scum accumulation, imperfect mixing, variability in solids-production rates, and biosolids stability, most digesters operate at 15 days.

Later sections will provide equations for process parameter estimation. Given the variability in reactor configurations and solids composition, generic and lumped kinetic

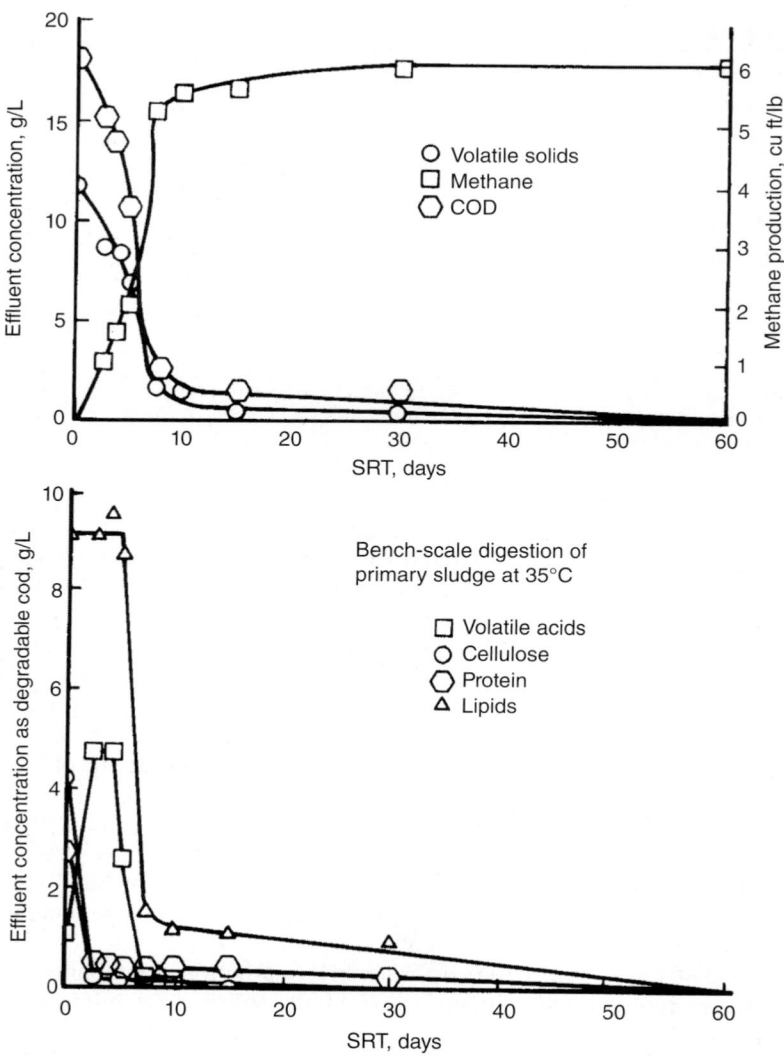

Figure 23.18 The effect of SRT on the relative breakdown of degradable waste components and methane production (cu ft/lb × 0.062 43 = m³/kg).

parameters may not sufficiently describe system performance. Pilot or full-scale testing in conjunction with using newer more complex process models, with sufficient background data, should be considered when the accuracy of the projected process performance is critical.

Pilot-testing, before design, should also be considered when implementing a new process configuration or the raw solids contain unconventional substrates such as industrial inputs. The effect of industrial wastes on anaerobic digestion is always questionable until the specific waste is characterized and tested.

SRT, Days	Percent of Facilities in Each Range	
	Primary Sludge Only	**Primary Plus Secondary**
0–5	0	9
6–10	0	15
11–15	0	9
16–20	11	12
21–25	45	25
26–30	11	3
31–35	11	15
36–40	0	6
41–45	0	0
46–50	22	0
Over 50	0	6
Total number of plants	(12 plants)	(132 plants)

TABLE 23.12 Solids Residence Times Reported for anaerobic digestion operations in the United Substrates (Reprinted with permission from ASCE, 1983)

2.8.2 Process Design

2.8.2.1 Sizing Criteria When sizing digesters, the key parameters are SRT and VS loading rate. For digestion systems without recycle, there is no difference between SRT and HRT. The VS loading rate is also used and is important in situations, like co-digestion, where the VS content in the feed may vary. The selection of the design SRT and VS loading rate should consider several factors (e.g., process microbiology, stability, biosolids regulatory requirements, biosolids stability, and industrial inputs). Generally, at higher TS concentrations, VS loading rate is limiting and, at lower TS concentrations, SRT is limiting.

Presently, the design minimum SRT typically is selected based on experience, general rules, and regulatory requirements. Design engineers should note that a range of solids-production conditions must be considered when developing appropriate SRT design criteria.

Researchers are developing more quantifiable approaches to understanding digestion and its limits. The selection of a design SRT directly affects system kinetics and, consequently, process performance. Both complex models (e.g., ADM-1) and simplified models (e.g., that presented by Parkin and Owen [1986]) have been suggested as a means of estimating or predicting the limiting (minimum) digester SRT, although the data required for its application are limited. The Parkin and Owen approach is based on applying a safety factor to a limiting SRT to establish the design SRT. If the limiting SRT is based on a desired digestion efficiency and the digester approaches complete-mix conditions, the limiting SRT can be estimated as follows:

$$\text{SRT}_{\min} = \frac{Y k S_{\text{eff}}}{K_c + S_{\text{eff}}} - b^{-1} \tag{23.3}$$

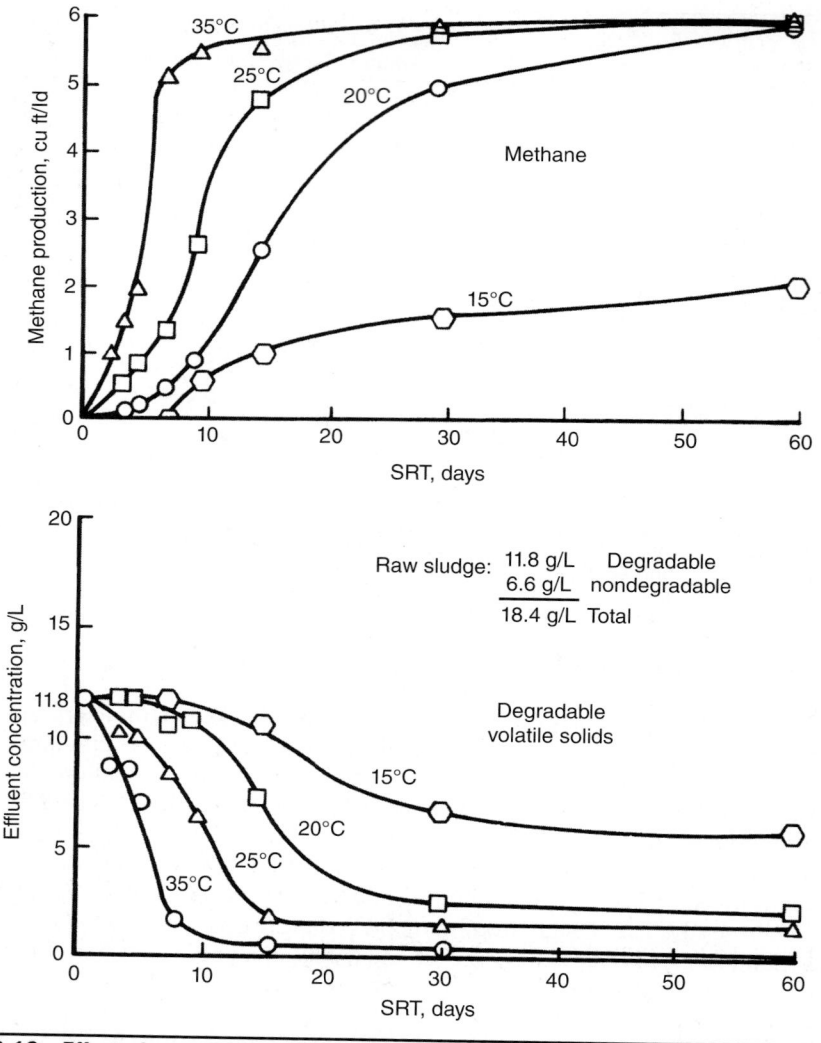

Figure 23.19 Effect of temperature and SRT on methane production and VSR (O'Rourke, 1968).

where SRT_{min} = limiting SRT for required digester performance;

Y = yield of anaerobic organisms resulting from growth (g VSS/g COD destroyed);

k = maximum specific substrate use rate (g COD/g VSS·d);

S_{eff} = concentration of biodegradable substrate in digested solids (therefore, in digester) (g COD/L); which is equivalent to $S_o(12e)$, where S_o is the concentration of biodegradable substrate in the feed solids, g COD/L, and e is the digestion efficiency at removing S_o;

K_c = half saturation concentration of biodegradable substrate in feed solids (g COD/L); and

b = endogenous decay coefficient (d^{-1}).

Values for constants in eq 23.3 have been proposed for typical municipal primary solids within a temperature range of 25°C to 35°C (77°F to 95°F). The following proposed values (Parkin and Owen, 1986) are based on laboratory experiments (full-scale data are not currently available):

- $k = 6.67$ g COD/g VSS·d (1.035^{T-35});
- $K_c = 1.8$ g COD/L (1.112^{35-T});
- $b = 0.03$ d^{-1} (1.035^{T-35});
- $Y = 0.04$ g VSS/g COD removed; and
- T = temperature (°C).

Using the calculated value of the limiting SRT for required digester performance, the safety factor (SF) for anaerobic digestion then can be calculated as:

$$SF = \frac{SRT_{actual}}{SRT_{min}} \qquad (23.4)$$

For example, suppose engineers were designing a new anaerobic digestion system with the help of data from similar existing systems (see Table 23.4) (ASCE, 1983). The data indicate a median average SRT of about 20 days. Assuming the new system would have a digestion efficiency of 90%, a design temperature of 35°C (95°F), and feed solids with a biodegradable COD concentration of 19.6 g/L, they used eq 23.3 to calculate that the minimum SRT would be 9.2 days (Parkin and Owen, 1986). (Biodegradable COD is assumed to represent the degradable fraction of VS in feed solids and liquor.) Then, using eq 23.4 and the median SRT of the existing systems, engineers calculate that the new system's safety factor is 2.2. This safety factor obtained should be used to estimate if short-term or dynamic increase in hydraulic loadings could be accommodated within the process.

Pilot-testing may be required to determine a system's actual limits because using generic and/or lumped kinetic parameters can over- or underestimate SRT requirements, depending on actual operating conditions and the composition of the substrate. For example, if feedstocks contain significant amounts of materials (e.g., lipids) that are more difficult to degrade than typical municipal primary solids, then the constants given for eq 23.3 will not apply. In such cases, or if consistently high VSR is critical, engineers may need to use higher design SRT values than those listed in Table 23.13. If solids readily degrade (e.g., typical primary solids without biological solids), then engineers may be able to use slightly lower design SRT values than those listed in Table 23.13.

Other limits (e.g., regulatory limits to meet PSRP requirements) need to be considered as well. While a shorter design SRT may reduce tank and ancillary equipment size and cost, the ease of biosolids use or disposal also must be considered. Any initial cost savings may be lost if the biosolids are poor quality, do not meet regulatory standards, or incur significant O&M costs to ensure compliance.

Digestion Type	Digestion Time (Days)	Volatile Solids Destruction (%)
High-rate (mesophilic range)	30	55
	20	50
	15	45
Low-rate	40	50
	30	45
	20	40

TABLE 23.13   Estimated Volatile Solids Destruction

Furthermore, to minimize the likelihood of a digester upset, design engineers should select the design SRT based on a critical operating period (e.g., a high-solids-loading period when grit and scum has accumulated, or when a digester is out of service). The choice of critical operating period will depend on facility size, anticipated solids production, and other site-specific factors.

2.8.2.2 Loading Rates and Frequency   A digester's loading rate and frequency significantly affects digester performance. Constant loading will produce the most stable operations because the microorganisms will reach and maintain steady-state conditions. Relative loading will affect the process' overall stability because both over- and underloading can impair process performance. The design loading rate should be coupled with the process selected (suggested values can be found in the sections discussing each process).

Furthermore, design engineers also should consider the potential effects of a specific feeding regime. Continuous feeding may necessitate storage tanks or 24-hour thickening and dewatering operations. Slug loading may lead to foaming.

In addition, design engineers should ensure that upstream processes are adequate to meet design loading. Thickening performance could prevent a design loading rate or SRT from being met consistently. Design engineers should take this into account when sizing tanks because it could lead to process limitations that could reduce the expected life of the system.

2.8.2.3 Solids Blending   Adding a solids blending tank before the reactor can improve digestion stability. The blend tank homogenizes solids before digestion, particularly when multiple solids streams are being loaded to a digester. Primary solids and secondary solids degrade differently, so without homogenization and flow pacing, the digester can experience a wide fluctuation of organic loads throughout the day, even at constant pumping speeds. (The effective organic loading rate is a function of degradable VS rather than the total VS load.) Blending tanks reduce diurnal loading variability by providing a "wide spot" in the line. The tank absorbs high solids flows, allowing the digester to maintain more consistent operations, rather than peaking with the solids wasting protocols.

Blending tanks typically are mixed by mechanical mixers or pumps. The degree of mixing selected must be balanced by equipment costs and power use. When sizing blending tanks, design engineers should take care not to oversize them because that can promote acid-reactor conditions if the detention time is long enough. Sometimes blending tanks are heated to partially heat solids before they enter the digester. Heating also can exacerbate acid formation in excessively large blending tanks.

2.8.2.4 Solids Destruction and Gas Production Design engineers can estimate the expected VSR using previous data (40% to 70%) or equations relating VSR to detention time (Liu and Liptak, 1997). For a standard-rate system,

$$V_d = 30 + \frac{t}{2}$$ (23.5)

where V_d = VSR (%), and
t = time of digestion (days).

For a high-rate digestion system,

$$V_d = 13.7 \ln (u_d^m) + 18.94$$ (23.6)

where V_d = VSR (%), and
u_d^m = design SRT.

Additional estimates can be made using Table 23.13. The concentration of fixed solids entering the digester will remain constant. Table 23.14 compares the alternative methods for estimating VSR for a single source; it shows that VSR estimates depend on

Description	Value	Units	Source or Notation
Process parameters for estimated solids destruction			
Raw solids biodegradable COD	19.3	mg-COD/L	Example in section 2.7.2.1. Parkin and Owen (1986)
COD removal efficiency	90	%	Example in section 2.7.2.1. Parkin and Owen (1986)
Digester temperature	35	°C	
Design solids retention time (SRT)	20	Days	
Process type	HR		HR = high rate
Volatile solids destruction based on Equations 23.8 and 23.15			
SRT_{min}	9.2	Days	Equation 23.8, using kinetic parameters in section 2.7.2.1
V_d	49	%	Equation 23.15
Volatile solids destruction based on Table 23.13			
V_d	60	%	
Volatile solids destruction based on Figure 23.38			
V_d	42	%	Solids age × temperature = 700

TABLE 23.14 Comparison of Methods for Estimating Digester Volatile Solids Destruction

the method selected, even for the same data set. Design engineers should understand the underlying assumption(s) behind each method before choosing one. If operational data are available, engineers should use them rather than these methods.

A close estimation of the solids (kg/d) that would enter the second-stage digester of a two-stage system is given by the following equation:

$$\text{Solids} = TS = (A \times \text{Total solids} \times V_d) \tag{23.7}$$

where TS = total solids entering digester (kg/d [lb/d]),

   A = volatile solids (%), and

   V_d = volatile solids destroyed in primary digester (%).

Equation 23.7 can be used to estimate the solids load to a second-stage digester and the degree of thickening (%). However, the secondary digester's volume often is equal to that of the primary digester to allow units to be taken out of service.

Design engineers can estimate the specific gas production at WRRFs by using the relationship of approximately 0.8 to 1.1 m³/kg (13 to 18 cu ft/lb) of VSR. Gas production increases as the percentage of FOG in the feedstock increases (as long as adequate SRT and mixing are provided) because FOG is the slowest to metabolize. The total gas volume produced is as follows:

$$G_v = (G_{sgp})\, V_s \tag{23.8}$$

where G_v = volume of total gas produced (m³ [cu ft]);

   V_s = VSR (kg [lb]); and

   G_{sgp} = specific gas production, taken as 0.8 to 1.1 m³/kg VSR (13 to 18 cu ft/lb VSR).

The total amount of methane produced can be estimated from the amount of organic material removed each day:

$$G_m = M_{sgp}\, [\Delta OR - 1.42(\Delta X)] \tag{23.9}$$

where G_m = volume of methane produced (m³/d [cu ft/d]),

   M_{sgp} = specific methane production per mass of organic material (COD) removed (m³/kg COD [cu ft/lb COD]),

   ΔOR = organics (COD) removed daily (kg COD/d [lb COD/d]), and

   ΔX = biomass produced (kg VSS/d [lb VSS/d).

Because digester gas is about two-thirds methane, the total digester gas produced is equal to the following:

$$G_T = \frac{G_m}{0.67} \tag{23.10}$$

where G_T = total gas produced (m³/d [cu ft/d]).

Expected methane concentrations can range from 45% to 75%; typical methane concentrations range from 60% to 75% (by volume). Typical carbon dioxide concentrations range from 25% to 40% (by volume). Biogas typically includes hydrogen sulfide, but excessively high concentrations should be investigated (e.g., by determining any sources of industrial wastes or saltwater infiltration). The expected heat value of digester gas depends on the biogas' composition.

2.8.3   Tank Configuration and Shape

The tank configuration (shape) significantly affects the operating characteristics of anaerobic digestion, as well as the cost of construction and O&M. Typically, digesters are available in three basic shapes: short cylinder ("pancake"), tall cylinder ("silo"), and egg shaped.

2.8.3.1   Egg-Shaped Digesters

Egg-shaped digesters (Figure 23.20) have been in service for more than 50 years in Europe, and since the early 1990s in the United States (Volpe et al., 2004). This shape typically is considered the optimal shape for a digester, providing excellent mixing characteristics, very few dead zones, and good grit suspension.

This digester's shape improves mixing. The tapered base, with centrally located mixing, is designed to promote the resuspension and removal of grit and other heavy materials (see Figure 23.21). This minimizes the amount of material retained in the digester, increasing the active fraction and reducing the out-of-service time for cleaning. In a review of egg-shaped digesters, Volpe et al. (2004) reported that some have been in service for 20 years without needing to be cleaned.

The digester's small gas dome provides operational advantages and disadvantages. First, the degree of liquid agitation in the dome due to mixing and the foam-suppression system is typically high, preventing a scum layer from forming; instead, it remains entrained in the liquid phase and can be withdrawn. Depending on tank configuration, however, withdrawal from the small gas dome can be problematic. This design may be susceptible to significant liquid-level variation during rapid volume expansion events but this can be overcome through properly sized overflows.

The German style of solids withdrawal—bottom withdrawal—can be problematic for systems with solids with a high propensity to foam. In such cases, foam (or a low-density sludge) can accumulate at the top of the digester or in the gas dome

Figure 23.20   Egg-shaped digesters at the back river WRRF in Baltimore, Maryland (courtesy of the Baltimore Sun).

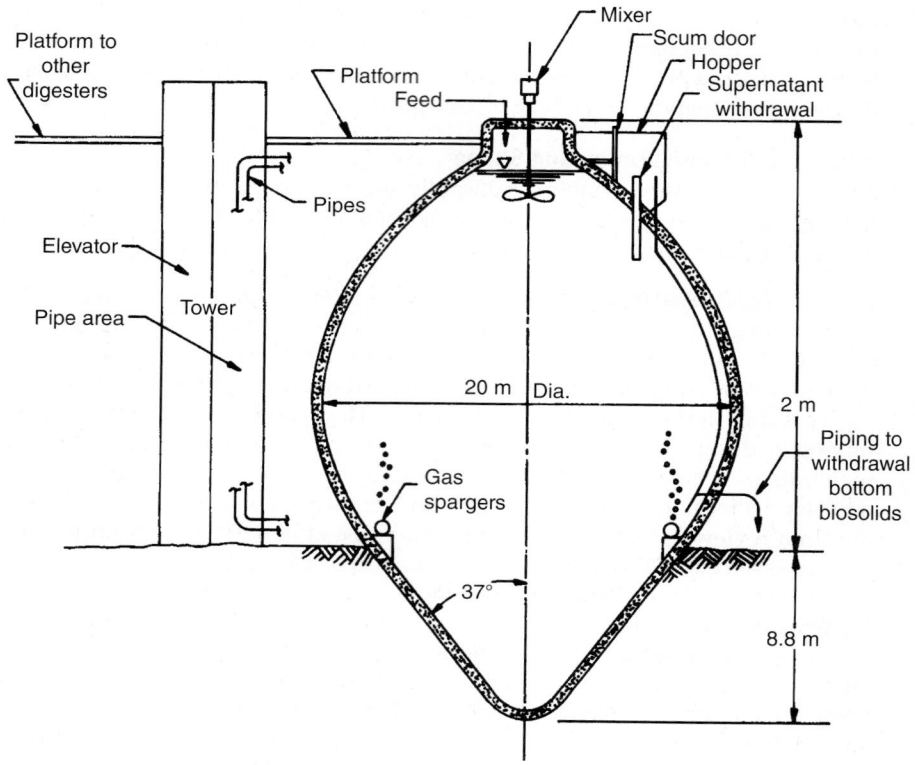

FIGURE 23.21   Schematic of a typical egg-shaped digester.

despite foam suppression. If the foaming event is great enough, the only means of exit is via the gas-handling system. So both surface and bottom withdrawals should be provided.

Egg-shaped digesters typically are made of either steel or concrete. Both are vulnerable to corrosion by the digester contents, so a corrosion-resistant coating typically is applied, especially in the headspace and several feet below the normal operating level. Egg-shaped digesters made of steel could use stainless steel in areas prone to corrosion, although this increases the cost. Steel eggs also need a corrosion-resistant coating on the outside of the digester. The insulation system often serves as part of the exterior coating, preventing corrosion by minimizing moisture contact.

Their relatively tall profile makes these digesters visible from a distance but allows for a larger volume in a relatively small footprint. Some facilities including the Back River in Baltimore, Maryland, and Newtown Creek in New York City, New York, have applied architectural features to make their eggs landmarks. Deer Island in Boston, Massachusetts, and the Socol facility in Napa, Califorina, have half eggs used as holding tanks with gasholder covers installed on top.

The shape and construction constraints of these unique tanks make their construction a specialty. A limited number of firms have the equipment and experience to build them, so egg-shaped digesters typically are more expensive to construct than a similar-sized cylindrical or silo digester. Unique construction techniques such as monolithic concrete airforms that are used to construct large free-span concrete domes have the potential to decrease costs of egg-shaped digester construction (Brinkman and Voss, 1997).

Volatile solids loading in these digesters ranges from 0.64 to 2.4 kg volatile solids/m$^3 \cdot$d (0.040 to 0.150 lb volatile solids/cu ft/d), with a maximum loading of 1.6 to 2.9 kg volatile solids/m$^3 \cdot$d (0.106 to 0.175 lb volatile solids/cu ft/d). SRT ranges from 15 to 45 days, and VSR ranges from 42% to 65%, depending on the intended use for the resulting biosolids. Gas production ranges from 0.68 to 1.08 m^3/kg volatile solids destroyed (11 to 17.5 cu ft/lb volatile solids destroyed). Though traditionally suited to mechanical mixers, gas, mechanical, and pump mixing systems can be used, and the mixing energy varies from 3.16 to 13.19 W/m^3 (0.12 to 0.5 hp/1 000 cu ft). Heating systems primarily are heat exchangers, although a few use steam injection.

2.8.3.2 Silo Digesters (Tall Cylinders)   Silo digesters are a newer version of cylindrical (pancake) digesters, with a greater height-to-diameter ratio. This achieves some of the mixing advantages of egg-shaped digesters without the cost (see Figure 23.22). For this reason, the DC Water Blue Plains facility, despite early plans for new egg-shaped digesters, ended up with silo digesters in 2014.

Silo digesters can resemble egg-shaped digesters when equipped with a submerged fixed cover (see Section 2.8.4.5). The small liquid-gas interface of the submerged cover is similar to the gas dome of the egg-shaped digester, and provides many of the same advantages. Other cover options include gasholder (membrane or steel), fixed and floating, allowing design engineers more flexibility to customize the tank for a specific use or even multiple uses (something egg-shaped digesters are not well suited for).

2.8.3.3 Cylindrical Digesters   Cylindrical (pancake) digesters are by far the most common tank design for anaerobic digesters in the United States (see Figure 23.23). Their relatively low height-to-diameter ratio makes them easier and less expensive to construct. However, this tank does not have the process benefits realized by egg-shaped and silo digesters.

Pancake digesters are more prone to dead zones and poor mixing regimes, resulting in lower VSR and more grit deposition. So cleanings are needed more often, increasing O&M costs.

However, pancake digesters can be fitted with a variety of cover configurations, which can allow for significant process flexibility and multiple roles (e.g., gas storage and variable liquid level). Egg-shaped digesters are only suited for service as anaerobic reactors.

These tanks typically are made of reinforced concrete, with sidewall depths ranging from 6 to 14 m (20 to 45 ft) and diameters ranging from 8 to 40 m (25 to 125 ft) (U.S. EPA, 1979). Conical bottoms are preferred for cleaning purposes, with slopes varying between 1:3 and 1:6. Slopes greater than 1:3, although desirable for grit removal, are difficult to construct and difficult to stand on while cleaning. Conical bottoms also minimize grit accumulation, instead providing for relatively continuous grit removal. The floors can have either one central withdrawal pipe or be divided into wedges, each with its own withdrawal pipe (waffle-bottom digester). The latter are more expensive to construct but may reduce cleaning costs and frequency.

Where necessary, cylindrical digesters have been insulated using brick veneer and an air space, earth fill, polystyrene plastic, fiber glass, or insulation board.

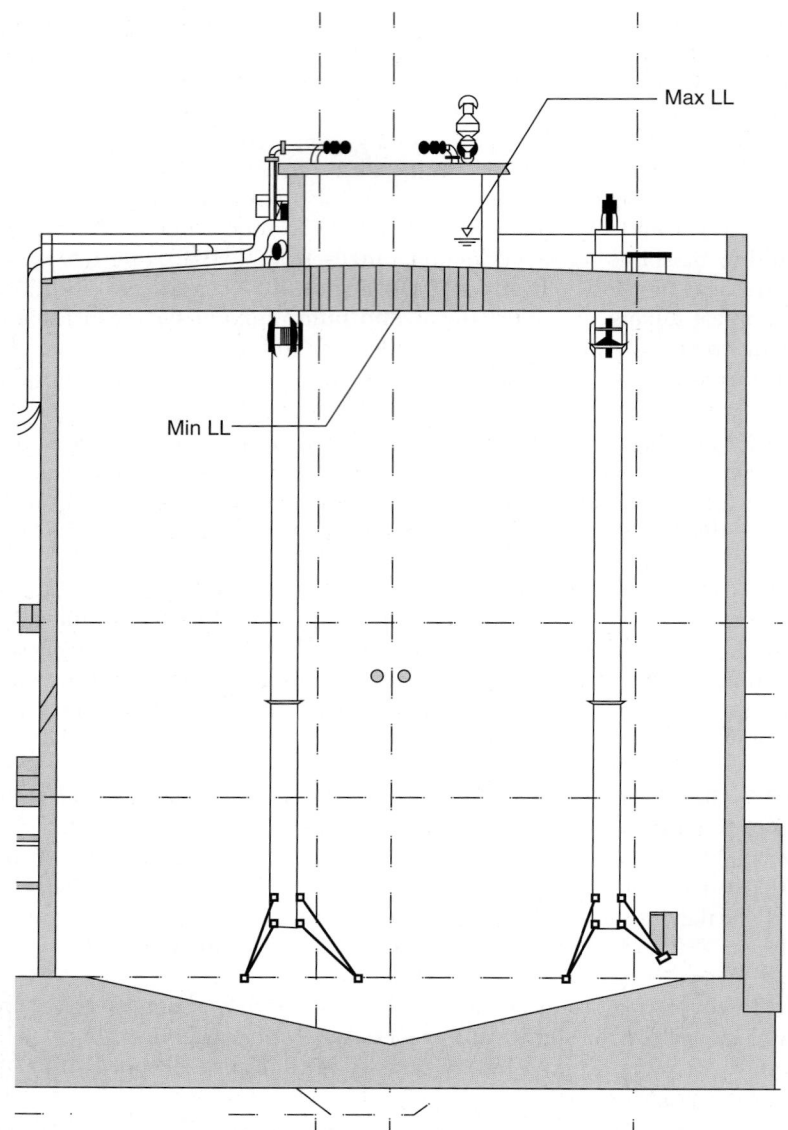

Max LL

Min LL

FIGURE 23.22   Cross section of a typical silo digester with internal draft tube mixers (courtesy of Brown and Caldwell and King County Brightwater WWTP).

2.8.4   Digester Cover Types

2.8.4.1   Fixed Digester Covers   Fixed digester covers can be constructed of either concrete or steel (see Figure 23.24). In some cases, stainless steel has been used to eliminate the need for coatings and to speed installation. The fixed cover minimizes fugitive

FIGURE 23.23   The anaerobic digesters at King County's South Wastewater Treatment Plant in Renton, Washington (courtesy of Brown and Caldwell and King County, Washington).

odors, which can be released from the annular space between any floating cover and the digester wall. Since it rests on the tank wall and not on the liquid, it is not vulnerable to tipping; however, provisions must be made to prevent overpressurization. To prevent recurring damage from foaming incidents, several facilities installed emergency pressure-relief manhole covers.

FIGURE 23.24   Photograph of an anaerobic digester with a fixed cover at East Bay Municipal Utility District, Oakland, California (courtesy of WesTech).

These covers do not require the ballasting that floating or steel gasholder covers do; however, they must be anchored to the top of the digester to resist the uplift caused by the operating gas pressure.

2.8.4.2 Floating Digester Covers   A floating cover is traditionally used to allow for significant liquid-level variation since it floats on the digester liquid. Floating covers require ballasting to ensure that the cover floats at the proper level in the liquid. Weights may also need to be added to balance uneven loads on the cover because piping and equipment on the cover could make it float unevenly on the liquid surface. Design engineers should ensure that the operating liquid level is above the corbels to maintain the perimeter gas seal. Floating covers are inherently immune to hydraulic overloading because during an overpressure event, if the gas safety devices are blocked, the cover will effectively "burp" gas, typically at the lightest (or highest) radial location on the cover.

All connections on floating covers between the cover and the digester wall must be flexible enough to allow full movement of the cover without stressing any of the equipment (e.g., gas pipes, or stairways/walkways on and off the cover). Also, guides are needed to ensure that the cover moves up and down without getting bound and tipped (see Figure 23.25). Original designs involving rollers and spring shoes have been largely replaced by slide guides: a series of cover-mounted plastic bars that slide vertically in a wall-mounted channel.

Floating covers typically are made of steel. So they typically are coated or covered with some type of insulation (e.g., spray-on polyurethane foam with a protective coating or a modular insulation system). Without insulation, shell losses would be significant—particularly in cold environments—increasing heating requirements.

2.8.4.3 Steel Gasholder Covers   Like floating covers, gasholder covers are not fixed in place. The cover has a skirt that extends into the liquid, allowing the cover to float on a bubble of gas stored for off-peak use (Figure 23.26). As with floating covers, hydraulic overpressurization of the tank is difficult because solids will escape from the annular space.

FIGURE 23.25   Photograph of a slide guide used for floating and gas holder covers (courtesy of WesTech).

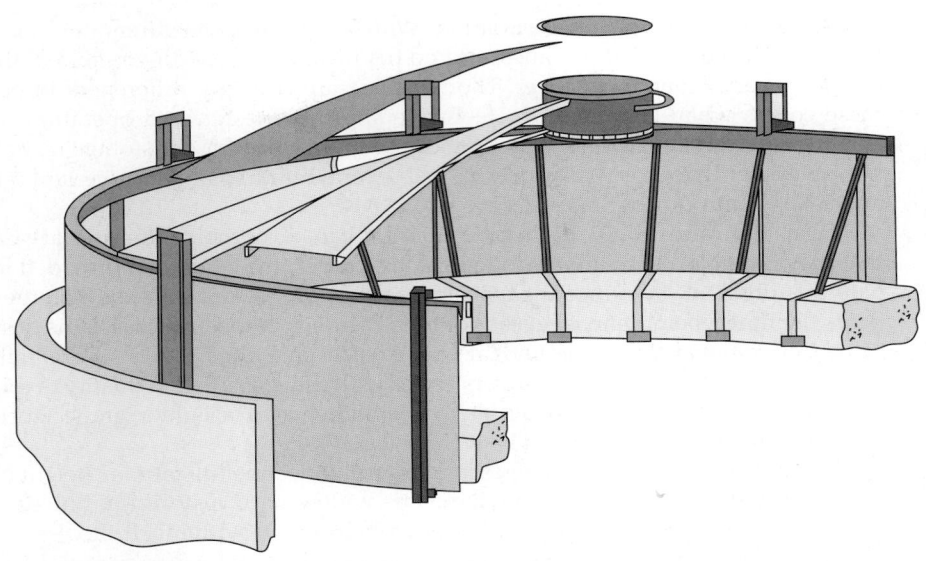

Figure 23.26 Schematic of a typical gas holder cover (courtesy of OTI).

While gasholder covers provide extra gas storage, they have some distinct disadvantages. They have no surface withdrawal above the corbels, so foam and scum become entrapped and accumulate in the digester. The cover is effectively a large gas-liquid interface, which can lead to corrosion. Furthermore, mixing alternatives are limited because the cover moves and must be ballasted to ensure that it is balanced. Also, because the cover is riding on a gas bubble, the total volume of the reactor is not used for solids digestion (i.e., overall maximum hydraulic capacity is reduced).

2.8.4.4 Membrane Gasholder Covers Membrane digester covers (Figure 23.27) provide maximum gas-storage capacity, ensure absolute gas containment and provide for the

Figure 23.27 Photograph of a membrane gas holder cover at Janesville, Wisconsin (courtesy of WesTech).

highest degree of liquid level variation. With several manufacturers now in the U.S. market, reduced installation times, and no field welding or coating required, these can also be a very economical cover. Though there are certainly differences in operating these covers when compared to steel covers, proper design and operation can help ensure a successful outcome. With an increase in the need for gas storage for cogeneration and other uses, it is not surprising that several hundred are in successful operation at WRRFs in the United States.

Gas is typically drawn off the side, and this can be a problem for primary digesters that may be subject to foaming. Top gas draw-off designs have been provided to eliminate this problem, but since the air and gas membranes are clamped together at the peak, the inner membrane level cannot be measured. In the best cases, this design has caused operational challenges and, at worst, has been at the heart of several failures.

The life expectancy of these covers varies with the strength and quality of the fabric. Traditionally, cable-supported designs have allowed for a lighter grade fabric to be used, which may decrease fabric life.

Originally, there were challenges in measuring the inner membrane height (indicating gas volume). Ultrasonic technology was widely used historically but since 2012, several manufacturers have been using laser instruments with much success.

2.8.4.5 Submerged Fixed Covers A concrete submerged fixed cover converts the top of a cylindrical or silo digester into a system resembling the top of an egg-shaped digester (see Figure 23.28). The sloping top directs gas, foam, and scum to the small gas dome. Because the cover is fixed, surface overflow can occur in the gas dome, which serves as a key solids-wastage point. Typically, solids can be circulated from the digester to the dome for foam suppression, and gas is withdrawn from the very top of the gas dome.

As with standard fixed covers, there is the danger of hydraulic overpressurization by wasting at a slower rate than the feed rate to the digester. To help alleviate this

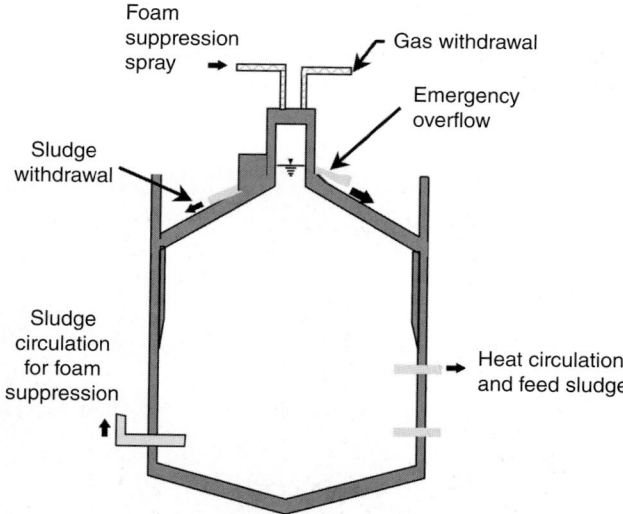

Figure 23.28 Schematic of a digester with a submerged fixed cover (courtesy of Brown and Caldwell, Seattle, WA).

concern (especially because the small gas dome has limited volume), an emergency overflow typically is added to the gas dome. A dual U-trap in the line keeps gas from exiting via the emergency overflow (see Figure 23.29).

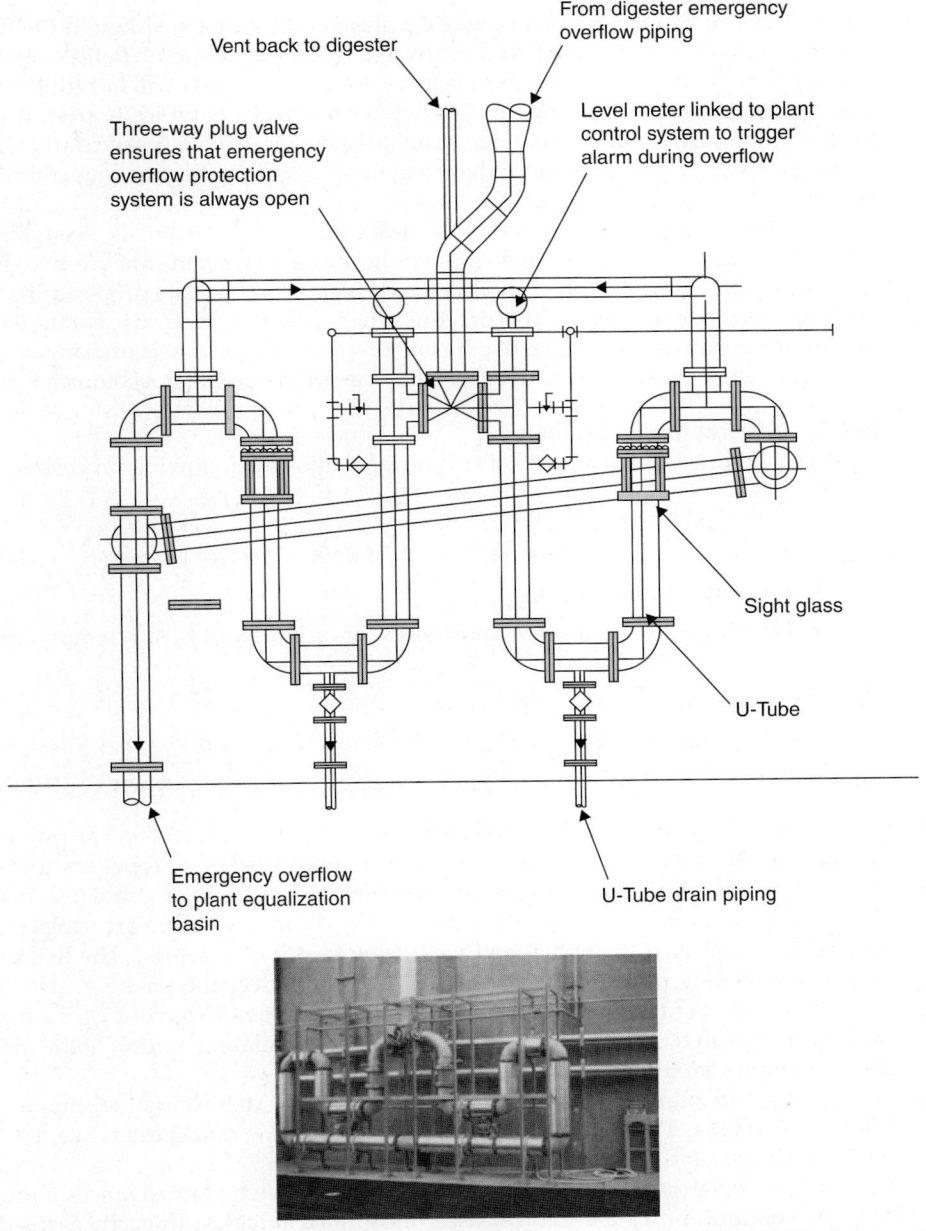

FIGURE 23.29 Emergency overflow apparatus for submerged fixed covers (courtesy of Brown and Caldwell).

While submerged fixed covers have some obvious benefits, they also have some limitations. The small gas dome provides limited gas storage, and any change in liquid volume is quickly realized at the overflow points.

2.8.5   Digester Feeding Systems

Raw solids can be introduced to digesters at several locations, although the key criterion is to avoid short-circuiting feed to any exit (withdrawal) point. Solids can be added at the top, at the bottom, or in a recycle loop; however, if solids will be withdrawn from the top and bottom, feed through the digester sidewalls is typically best. If possible, feed location also should maximize solids dispersion. This is particularly easy with pumped systems if there is more than one return location. Sequencing control valves will allow for a good distribution of solids.

Feeding to a recycle loop ahead of the heat exchangers may initially seem like a good way to increase the "delta-T" and improve heat exchanger performance but should be done with care as raw sludge may affect the rheology and pressure drop causing the flow and heat transfer rates to vary. This practice, which may increase grease and fouling of the heat exchangers, is not recommended by manufacturers of spiral heat exchangers. Tube-in-tube heat exchangers may be less sensitive to these problems but not completely immune.

2.8.6   Digester Mixing Systems

Auxiliary mixing of digester contents is beneficial for the following processes:

- Reducing thermal stratification;
- Dispersing substrate for better contact with the active biomass;
- Reducing scum buildup;
- Diluting any inhibitory substances or adverse pH and temperature feed characteristics;
- Increasing the reactor's effective volume;
- Allowing biogas to separate and rise more easily; and
- Keeping in suspension inorganic material that has a tendency to settle.

Three types of mixing methods typically have been used: mechanical, pumped, and gas recirculation. Mechanical or internal mixers use impellers, propellers, and turbine wheels. While some technologies have proven very reliable, problems may arise because of the internal surfaces of the mixers. Shafts and impellers are subject to vibration (due to collected materials) and wear (due to grit and debris). The linear motion mixer is relatively new technology used in digesters over the past 5 years. Typically mounted in the center of the tank cover, this mixer does not spin but uses a cam-drive to operate a shaft to move a disk up and down. The oscillating motion of the disk in the fluid promotes mixing in the tank.

Mixing via pumped recirculation involves using an external pump to recycle digester contents. The efficiency of this system depends on digester size, net energy input, viscosity, and turnover rate.

Gas-recirculation systems may use tubes, sequentially operated lances, diffusers on the tank bottom, or a tube that releases unconfined bubbles. In each case, the gas is produced, compressed, and circulated through the tank to promote mixing. There are basically two types of gas-recirculation systems: unconfined or confined. Unconfined systems include top-mounted lances and diffusers on the tank bottom. Confined systems discharge gas through draft tubes. Each system has advantages and disadvantages, and the degree of mixing typically depends on the energy input.

2.8.6.1 Mixing Requirements Most manufacturers of digester mixing equipment can suggest the appropriate type, size, and power level, which depend on the digester volume and geometry. These suggestions typically are based on in-house studies and the successful experiences of similar installations. Anaerobic digesters can be mixed via gas, mechanical, or pumped mixing systems (various mixing systems have different advantages and disadvantages). Selection of a mixing system is based on costs; maintenance requirements; process configuration; and the screenings, grit, and scum content of the feed. Suggested parameters for sizing digester mixing systems include unit power, velocity gradient, unit gas flow, and digester volume turnover time. These four parameters are related and can be used to equate manufacturers' recommendations but no single technology should be expected to meet all of these criteria.

Some newer technologies, including the linear motion mixer, fall outside these conventional recommendations and, therefore, use less energy than other mixers. The digestion process can be sustained with less mixing energy; however, it is unclear if other problems may arise (e.g., grit/scum accumulation). Long-term results are not yet available so it is too soon to draw many conclusions about this approach.

Unit power is defined as delivered motor watts per cubic meter (horsepower per 1000 cu ft) of digester volume. Actual energy applied, viscosity, and digester configuration are not accounted for. Several values have been suggested for unit power selection, ranging from 5.2 to 40 W/m³ (0.2 to 1.5 hp/1000 cu ft) of reactor volume. Using laboratory data, Speece (1972) predicted a level of 40 W/m³ to be sufficient for a complete-mix reactor.

The velocity gradient parameter as a measure of mixing intensity was presented by Camp and Stein (1943). It is expressed as the following equation:

$$G = (W/m)^{1/2} \tag{23.11}$$

where G = root-mean-square velocity gradient (s^{-1});

W = power dissipated per unit volume (W/m³ or N/m²·s [lb·s/sq ft]); and

m = dynamic viscosity (N·s/m² or Pa·s [lb·s/sq ft]) (for water, 7.2×10^{-4} N·s/m² or Pa·s at 35°C [$1.5/10^{-5}$ lb·s/sq ft at 95°F]).

and

$$W = E/V \tag{23.12}$$

where E = power dissipated (Watts [ft·lb/s]), and

V = tank volume (m³ [cu ft]).

The power for gas injection can be determined from the following equation:

$$E = P_1(Q)\,(\ln P_2/P_1)\ (\text{SI units}) \tag{23.13}$$

or

$$E = 2.40P_1(Q)\,(\ln P_2/P_1)\ \text{(U.S. customary units)}$$

where Q = gas flow (m³/s [cu ft/min]);

$\quad P_1$ = absolute pressure at surface of liquid (Pa [psi]);

$\quad P_2$ = absolute pressure at depth of gas injection (Pa [psi]); and

$\quad 2.40$ = lumped conversion factor for U.S. customary units.

These equations can be used to determine the necessary power and gas flow of compressors and motors for a gas-injection system. Viscosity is a function of temperature, total solids concentration, and volatile solids concentration. As temperature increases, viscosity decreases; as solids concentration increases, viscosity increases. In addition, as volatile solids increase to more than 3.0%, viscosity increases. Appropriate values of the root-mean-square velocity gradient are 50 to 80 s⁻¹. The lower values can be used for a system using one gas port, or where grease, oil, and scum are suspected problems.

By rearranging the preceding equations, the unit gas flow relationship to the root-mean-square velocity gradient can be solved by the following equation:

$$\frac{Q}{V} = \frac{G^2}{P_1\left(\ln \dfrac{P_2}{P_1}\right)} \tag{23.14}$$

Suggested values of gas flow/tank volume (Q/V) for a free-lift system range from 76 to 83 mL/m³·s (4.5 to 5.0 cu ft/min/1000 cu ft). For a draft tube system, the suggested values range from 80 to 120 mL/m³·s (5 to 7 cu ft/min/1000 cu ft).

Turnover time is defined as digester volume divided by the flowrate through the draft tube. This concept typically is used only with draft tube gas and mechanically pumped recirculation systems, where such a flowrate actually can be determined. Typical digester turnover times range from 20 to 30 minutes.

2.8.6.2 System Performance A specific definition of "adequate digester mixing" has not yet been formulated. Various methods (e.g., solids concentration profiles, temperature profiles, and tracer studies) have been used to evaluate mixing system performance.

Solids concentration profiles are used to determine the effectiveness of digester mixing. To use this method, samples are collected at specified depth intervals in the tank (typically 1.0 to 1.5 m [3 to 5 ft]) and analyzed for total solids concentration. Mixing is considered adequate if the solids concentration does not deviate from the average concentration in the digester by more than a specified amount (often 5% to 10%) over the entire digester depth. Allowances are sometimes made for greater deviations in the scum and bottom solids layers. A drawback of the solids concentration profile method, particularly for systems digesting secondary or combined primary and secondary solids, is that these solids often do not stratify significantly, even without mixing, so inefficient mixing cannot be shown by solids concentration profiles alone.

Temperature profiles also have been used to assess mixing effectiveness. The temperature profile method is similar to the solids profile method. Temperature readings are taken at specified depth intervals in the digester. Mixing is considered adequate if the temperature at any point does not deviate from the average by more than a specified amount (often 0.5°C to 1.0°C [1.0°F to 2.0°F]). A drawback of this method is that the digester may have enough heat dispersion without effective mixing to maintain a relatively uniform temperature profile, particularly in digesters with a long SRT.

The most reliable method currently available for evaluating mixing effectiveness is the tracer test method. In this method, a carefully measured amount of a conservative tracer material (e.g., lithium) is injected as a slug to the digester. (Continuous feed methods also can be used but are typically impractical because of the large amounts of tracer required and the long time required to perform the test.) Samples of digested solids are collected and analyzed for tracer content. For an "ideal" (i.e., completely mixed) digester, the tracer concentration in digested solids leaving the digester at any time is calculated as follows:

$$C = C_0 e^{-\frac{t}{\text{HRT}}} \tag{23.15}$$

where C = tracer concentration at time t (mg/L);

$\quad C_0$ = theoretical initial tracer concentration at time $t = 0$ (total mass of tracer injected/total digester volume) (mg/L);

$\quad t$ = elapsed time since injection of tracer (hours); and

HRT = digester hydraulic retention time (hours).

Substituting and taking natural logs, this equation becomes the following:

$$\ln C = \ln C_0 - \frac{v}{V_0} \tag{23.16}$$

where v = total volume of solids fed in time t ($F \times t$, where F = average solids feed rate [m³/h]) (m³); and

$\quad V_0$ = total digester volume (m³).

Plotting ln C (y axis) versus v/V_0 (x axis) gives the "tracer washout curve." The slope of this line gives an estimate of the effective digester volume, as follows:

$$V_e = \frac{1}{\text{slope}} \tag{23.17}$$

where V_e = estimated effective digester volume (m³).

The percentage active volume then is calculated by the following equation:

$$V_{\text{act}} = \frac{V_e}{V_0} \times 100 \tag{23.18}$$

where V_{act} = estimated percent active volume.

This method of estimating mixing effectiveness is the most accurate of the methods discussed (Chapman, 1989). However, because it requires careful monitoring of digester feed and withdrawal rates and a large number of tracer concentration analyses in digested solids, this method is considerably more expensive than any of the other methods discussed.

2.8.7   Digester Heating Systems
To be effective, anaerobic digesters need a consistent, reliable heating system.

2.8.7.1   Digester Heating Needs   Anaerobic digesters must be heated to provide suitable environmental conditions for optimal biological activity. Mesophilic digestion need to operate between about 35°C and 39°C (95°F and 102°F), and thermophilic digestion needs to remain between 50°C and 56°C (122°F and 133°F). The amount of heat needed varies seasonally, mainly in relation to the raw solids temperature, and in relation to heat losses from the reactor to the environment.

2.8.7.2   Solids Heating   The lion's share of a digester's total heating load is the energy needed to heat raw solids to the temperature needed for anaerobic digestion. This energy is calculated as follows:

$$q = m \times C_p \times T \tag{23.19}$$

where q = heat load (J/h or MJ/h [Btu/hr or million Btu/hr]);

$\quad$ m = mass flowrate of the cake's sludge's liquid, treated as water (kg/h [lb/hr]);

$\quad$ C_p = solids' specific heat or heat capacity (J/kg·°C or MJ/kg·°C [Btu/lb/°F]);

$\quad$ T = temperature difference between the cold, raw solids and desired heated solids temperature (°C [°F]).

To accurately compute the solids' heating needs, design engineers need to know the actual solids temperature, which typically is rarely recorded. However, the WRRF's influent and effluent temperatures typically are known and are representative of the raw solids temperature. (A WRRF does not appreciably change the average temperature of wastewater because of the large mass of water involved.)

2.8.7.3   Digester Heat Losses   In all but the very hottest weather, digesters lose heat to the environment via their roofs, walls, sides, and bottom. However, because digester temperatures typically are near ambient, radiant heat loss is small; virtually all heat is lost via convection. The general formula for heat loss (heat transfer) from these areas is as follows:

$$q = U \times A \times \Delta T \tag{23.20}$$

where q = heat load (J/h or MJ/h [Btu/hr or million Btu/hr]),

$\quad$ U = overall heat-transfer coefficient (J/h·m²·°C [Btu/hr/sq ft/°F]),

$\quad$ A = surface area (m² [sq ft]), and

$\quad$ ΔT = temperature difference between the digester and the environment (°C [°F]).

The coefficient U is also the inverse of resistance to heat transfer. For multiple layers, $1/U$ can be expressed as a series of resistances (Bird et al., 1960):

$$\frac{1}{U} = \frac{1}{h_0} + \frac{x_1}{k_1} + \frac{x_2}{k_2} \cdots + \frac{1}{h_3} \tag{23.21}$$

where h_0 and h_3 = film coefficients (J/h·m²·°C) or (Btu/hr/sq ft/°F),

$\qquad x_2$ = thickness of material (consistent units), and

$\qquad k_2$ = thermal conductivity of material (J/h·m·°C [Btu/hr/ft/°F]).

Coefficient values and the application of these equations to heat loss from digesters can be found in American Society of Heating, Refrigerating, and Air Conditioning Engineers (2013), Avallone and Baumeister (1996), and Green and Perry (2007).

2.8.7.4   Heat Sources   The following types of heat sources are available:

- Fired boilers. Boilers (e.g., steam boilers and hot-water boilers) typically are used at WRRFs to produce heat from fuel. Fired boilers burn a fuel such as digester gas, natural gas, or propane to supply heat.

- Cogeneration. Cogeneration is the production of both usable heat and electric power from one fuel. These systems, also called combined heat and power (CHP) can deliver heat for solids digestion.

- Water-source heat pumps. A heat pump is a mechanical device that extracts heat from one source, and elevates its temperature to make it usable for other applications. Water-source heat pumps can heat water to between 68°C and 76°C (155°F and 170°F).

- Solar radiation. Solar energy is an increasingly popular form of clean, carbon-free, renewable energy that can provide a portion of solids-heating needs. Because it is not available at night or during inclement weather, solar power is typically is not adequate as a sole source of heat. However, it is a carbon-free, renewable energy source that can provide a portion of solids-heating needs.

2.8.8   Heat Exchangers

2.8.8.1   Heat Exchanger Types   The following types of heat exchangers have been used in anaerobic digestion systems:

- Concentric tube. One of the oldest, most widely used heat exchangers is the concentric-tube (often called the tube-in-tube or concentric-pipe) heat exchanger (see Figure 23.30).

- Spiral plate. Spiral (spiral-plate) heat exchangers also are common (see Figure 23.31). Water temperatures are typically kept below 68°C (154°F) to prevent caking.

- Multiple tubes in a box. A variation of the concentric-tube heat exchanger consists of multiple tubes in a box. The small-diameter tubes have a common inlet and outlet.

FIGURE 23.30 Tube-in-tube heat exchanger at the Littleton-Englewood Wastewater Treatment Plant in Colorado (courtesy of Brown and Caldwell).

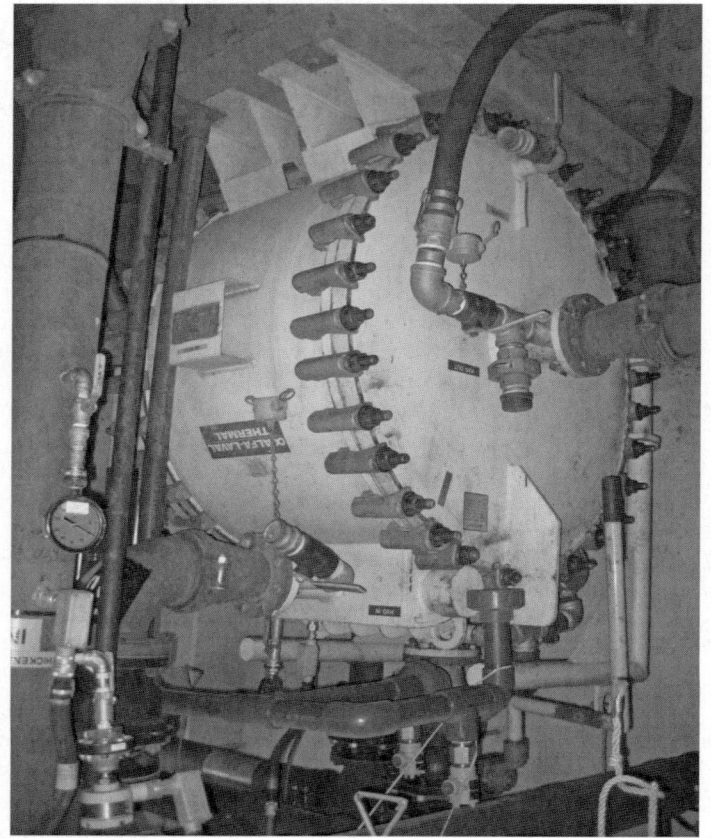

FIGURE 23.31 A spiral heat exchanger (courtesy of Brown and Caldwell and City of Tacoma, Washington).

- Interior submerged coils. Older digesters may have internal heating coils attached to the walls. The coils circulate hot water. They are subject to fouling, which can decrease the heat transfer. Maintaining the coils requires operators to remove the digester from service and empty it.

- Jacketed external solids mixers. A heat-jacketed draft-tube mixer also can be used to internally heat solids. However, internal heating systems are seldom used because of the difficulty associated with providing maintenance inside the reactor.

2.8.8.2 Heat Exchanger Characteristics Heat-transfer coefficients for external heat exchangers range from 0.9 to 1.6 kJ/m²·°C. Transfer coefficients for internal heating coils range from 85 to 450 kJ/m²·°C (15 to 80 Btu/hr/sq ft/°F) depending on the biomass' solids content.

2.8.9 Steam Heating
The following two types of steam heating systems have been used in anaerobic digestion systems:

- Submerged pipes. The submerged steam pipe (also called a steam lance) is a vertical, open-ended, small-diameter pipe that discharges at least 3 m (10 ft) below the liquid surface of the digester. Some WRRFs (e.g., the Hyperion Wastewater Treatment Plant [City of Los Angeles], the JWPCP [of the Los Angeles County Sanitation Districts], and the Rancho Las Virgenes digester complex in Calabasas, California) heat via submerged steam.

- Steam injection. Steam injectors are precision devices that blast a small jet of steam into a stream of solids. The device is outside the digester, and the steam flow is precisely controlled to produce a specific discharge temperature. The amount of steam is adjusted by throttling the plug in the steam injector throat. A few facilities (e.g., the Back River Wastewater Treatment Plant in Baltimore, Maryland, the Crystal Lake [Illinois] Wastewater Treatment Plant, and the Spokane [Washington] Wastewater Treatment Plant) use external steam jet injectors to warm their solids.

2.8.10 Heat Recovery
The following three types of heat-recovery systems have been used in anaerobic digestion systems:

- Cogeneration equipment. These systems typically withdraw heat from hot exhaust gases or cooling equipment. Heat-recovery percentages for specific cogeneration equipment are discussed in later sections of this chapter.

- Solids heat recovery. At press time, recent increases in fuel costs had increased the value of heat, so recovering heat from digested solids and using it to warm raw solids is becoming more practical and common.

- Gas compression and equipment cooling. At least one WRRF is capturing the relatively low-temperature heat (40°C to 50°C) from other large processes and using it, along with some water-source heat pumps, to warm solids.

2.8.11   Additional Equipment Options

2.8.11.1   Debris Buildup and Foam Control   Given the heterogeneous nature of raw solids, the accumulation of debris and foam are part of normal digester operations. However, the rate and extent of accumulation can be controlled via proper digester design, best management practices, and wastewater treatment controls.

Debris can affect digesters in two ways. Floating debris can accumulate on the surface of digesters at the gas-liquid interface, forming a thick blanket that can affect mixing and reduce digester capacity. Heavy debris (e.g., grit) can accumulate on the bottom of the digester, consuming process capacity and possibly causing short-circuiting (if the draw-off points are at the bottom of the tank).

Debris can be reduced via prescreening or grinding solids, or via better screening at the headworks. Most debris and grit enter the solids-handling system via primary solids and scum-removal systems (floating materials). If debris removal can only occur in the digester, several solutions are available (e.g., better mixing to keep material suspended, spray bars to entrain material back into solution, surface withdrawal for floating materials, or grinding and screening). In most cases, a combination of these solutions would be used to mitigate the effects of debris.

Digester foaming can be caused by several factors (e.g., surfactants in influent or hauled wastes, filaments from the secondary treatment system, or digester perturbation). Surfactants can be removed via source identification testing. Filaments can be controlled in the activated sludge system via selectors or chlorine dosing. Foaming typically is associated with erratic feeding after digester perturbation, which can be mitigated via changes in operating practices. Foaming is more difficult to control when associated with a transient event (e.g., a toxic compound), which is difficult or impossible to predict. (In the case of toxic compounds, efforts should be made to identify chronic sources.)

Several design strategies can be used to limit foaming, many of which are limited to the type of cover associated with the digester. A fairly common approach is to add spray nozzles, which entrain foam back into the bulk solution. If the cover or tank design allows it, surface withdrawal is an effective means of removing foam (similar to a selector in secondary systems). Typically, surface withdrawal systems are augmented with spray bars to direct flow. Another alternative is to change the mixing system. Gas systems may exacerbate foaming because the material interacts with the gas bubbles carrying it out of solution. Other mixing systems that do not sufficiently mix the liquid surface may cause areas of process instability and foaming.

Foaming can greatly increase O&M costs for a utility. As foam can escape digesters through the annular space on floating covers or enter the gas system requiring it to be taken off-line for cleaning.

Both foam and debris reduce digester capacity either by directly displacing volume or by causing operating levels to be reduced to contain them. Anaerobic digestion systems should be designed to minimize foam accumulation and have safety provisions (e.g., manhole pressure-relief valve) to protect digesters and covers from damage resulting from foaming events.

2.8.11.2   Scaling (Struvite)   Struvite (magnesium ammonium phosphate, $MgNH_4PO_4 \cdot 6H_2O$) is a white crystalline solid or scale that typically forms in anaerobic digesters, digested solids and centrate piping, solids lagoons and lagoon piping, and dewatering

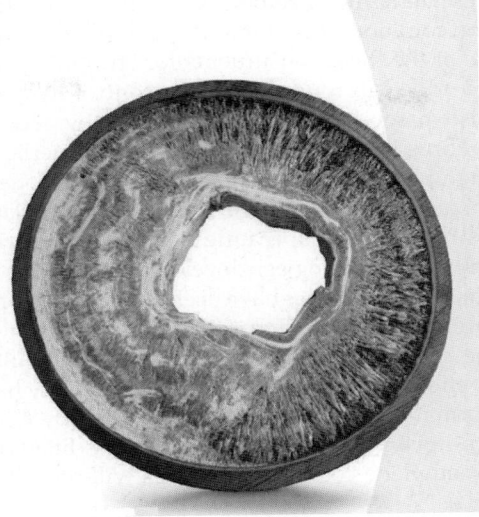

FIGURE 23.32   Struvite precipitate in piping (courtesy of CNP).

equipment. It forms a hard, tenacious precipitate that adheres to pipes and equipment, reducing pipe flow capacity and overloading motors serving brush aerators (see Figure 23.32).

Struvite typically forms when the concentrations of magnesium, ammonium, and phosphate exceed the solubility limit of struvite. Its formation also is influenced by various conditions in the digester (e.g., pH, temperature, and other chemicals that can compete primarily for phosphate). Anaerobic digestion can promote struvite formation via the release of phosphate and ammonia during stabilization. The amount of phosphorus and ammonium released depends on the digestion process and the wastewater treatment process(es) that generated the raw solids. For example, the phosphorus taken up during the enhanced biological phosphorus removal (EBPR) process is much more likely to be released in an anaerobic digester than in a conventional activated solids system. The amount of magnesium in solids also will affect struvite formation. Because struvite formation (as the chemical formula suggests) is based on equimolar stoichiometric concentrations of magnesium, ammonium, and phosphate, the limiting concentration of each component typically determines the amount of struvite that eventually forms. For example, removing or reducing the concentration of any of these constituents can reduce struvite scaling (buildup).

Systems should be designed to give operators access to pipes and equipment so they can remove struvite. In particular, if a facility being upgraded has a history of struvite, design engineers should ensure that the changes provide access and minimize deposition points.

Struvite control can be both a design and a process consideration. However, most options for addressing struvite scaling (e.g., smooth-lined piping and more maintenance) are preventive.

Innovative research is needed to better control and reduce struvite scaling in digestion systems. One commonly used method is adding an iron compound to precipitate phosphorus (one of the three constituents for struvite formation). Staff at a California digestion system added ferrous chloride to control hydrogen sulfide generation and found that it also precipitated phosphate, thereby reducing struvite precipitation. However, phosphorus is an essential nutrient for bacteria, so design engineers should ensure that enough phosphorus remains to meet nutritional needs.

Other control methods that may mitigate struvite precipitation include dilution (to reduce ion concentrations) and operating at a lower pH. Neither is a desirable operating procedure. Emerging technologies, developed to reduce scaling of boiler pipes by applying electrical signals to pipes have been suggested to prevent struvite precipitation; however, there are very few reference installations.

Once struvite deposits have formed, they are difficult to remove. Acid washing removes struvite effectively but can be costly and a safety hazard (Barker, 1996). During early stages of formation, struvite can be controlled by frequent cleaning (pigging) of pipelines. Smooth-lined pipes made of PVC or glass-lined materials, and polyethylene- or polytetrafluoroethylene-coated plug valves will resist struvite accumulation better than other materials.

Several facilities have found vivianite ($Fe_3(PO_4)_2 \cdot 8(H_2O)$, also called hydrated iron phosphate) in their systems, especially when the digesters contain high levels of phosphate and also contain iron. Vivianite loses solubility when temperatures rise; it forms quickly when solids containing iron and phosphate are heated. Vivianite is blue, green, or gray-black; turns opaque or dark when exposed to light; and is soluble in hydrochloric acid or nitric acid (HNO_3).

2.8.11.3 Piping and Cleaning Maintenance

Piping configurations should be designed to promote maximum flexibility for feeding, recirculating, and discharging solids. Piping should be arranged to provide several points for solids feed and solids withdrawal. Older designs typically also had multiple ports near the liquid surface for supernatant withdrawal. Because solids pumping is characterized by low velocities and possible solids accumulation in pipelines, design engineers should make provisions for cleaning out and backflushing lines (using treated effluent when available). They also should consider the choice of valves and where they should be placed for most utility (e.g., easy access and manual operations). Design engineers also should include provisions to allow all tanks and pumps to be isolated for maintenance and safety purposes.

Piping should be arranged to accommodate the following operating modes for two-stage digestion: transferring biomass via gravity from the first stage to the second, pumping the biomass from one digester to another, recirculating solids via suction and discharge ports, and providing redundancy/backup.

2.8.11.4 Corrosion

An anaerobic digester is a highly corrosive environment, especially because of the release of hydrogen sulfide. Tank interiors, equipment, piping, and any other elements that may contact biogas should be designed to be corrosion resistant, and all seals and gaskets should be compatible with the material they contact. A system's life expectancy will be significantly reduced if its equipment and materials lack corrosion protection. For a detailed discussion of corrosion, see Chapter 8.

2.8.11.5 Pumping Pumping is the primary means of conveying solids to and from digesters. For a detailed discussion of solids pumping, see Chapter 19.

2.8.11.6 Sampling and Process Monitoring Anaerobic digestion is sensitive to changes in operating conditions. If uncontrolled, such changes can result in digester upsets and failure. Proper monitoring techniques promote successful operation and ensure process stability and methane production. All process streams should be available for sampling and analysis. Feed, digested solids, digester gas, and the heating fluid (hot water) should be analyzed for various constituents and physical conditions. Sampling ports should be incorporated into the design to ensure that operators have adequate access for sampling. Feed typically is analyzed for the following: total solids, volatile solids, pH, alkalinity, and temperature. Digester content and effluent solids should be analyzed for the same parameters and for volatile acids. Digester gas should be analyzed for volume and percentage of methane, carbon dioxide, and hydrogen sulfide. Heating fluid should be analyzed for total dissolved solids, pH, and levels of other hydronic system chemical additives.

The flowrates of all streams should be monitored by accurate meters. Additional monitoring requirements (e.g., those for toxics) should be determined on a case-by-case basis.

2.8.11.7 Alkalinity and pH Control The methanogens in anaerobic digesters are affected by small pH changes, while the acid producers can function satisfactorily in a wide range of pH values. The effective pH range for methane producers is about 6.5 to 7.5, with an optimum range of 6.8 to 7.2. Maintaining this optimum range is important to ensure effective gas production and eliminate digester upsets. Digestion stability depends on the buffering capacity of the digester's contents (i.e., the digester contents' ability to resist pH changes). Alkalinity is important in anaerobic digestion; higher alkalinity values indicate more capacity for resisting pH changes. It is measured as bicarbonate alkalinity and ranges from 1500 to 5000 mg/L as calcium carbonate in anaerobic digesters. The volatile acids produced by the acid producers tend to depress pH. Under stable conditions, volatile acid concentrations range from 50 to 100 mg/L. By maintaining a constant ratio of volatile acids to alkalinity that is less than 0.3, the system's buffering capacity can be maintained.

The bicarbonate alkalinity concentration can be calculated from the total alkalinity (which also includes the alkalinity of volatile acids [e.g., acetate] and ammonium) as follows:

$$\text{Bicarbonate alkalinity (mg/L as } CaCO_3) = \text{total alkalinity (mg/L as } CaCO_3)$$
$$- [0.71 \times \text{volatile acids (mg/L as acetic acid)}] \tag{23.22}$$

where 0.71 is a conversion factor to mg/L as $CaCO_3$.

Barber and Dale (1978) developed the following equation to predict how much bicarbonate alkalinity is needed to raise the total alkalinity:

$$D_d = D_{max}\left(1 - \frac{1}{\theta}\right) \tag{23.23}$$

where D_d = amount added daily to reach set level (mg/L as $CaCO_3$),

D_{max} = required increase (mg/L as $CaCO_3$), and

$1/\theta$ = reciprocal of average detention time or SRT (d^{-1}).

Sodium bicarbonate, lime, sodium carbonate, and ammonium hydroxide all have been used successfully to increase the alkalinity of digester contents, especially during startup or upset conditions. However, most well-designed and well-operated digester facilities do not require alkalinity addition as long as the wastewater has sufficient buffering capacity. Design engineers should evaluate the wastewater's alkalinity before determining whether an alkalinity feed system should be constructed.

2.9 Digester-Gas Handling

This section covers a wide range of issues with respect to digester-gas (biogas) characteristics, processing equipment, handling equipment, and beneficial use. Design of systems for safe transport and maintainability are also covered.

The digester-gas collection and distribution system (Figure 23.33) must be maintained under positive pressure to avoid the possibility of explosion by allowing air into the system. A mixture of 5% to 15% gas to air is considered to be in the explosive range and is dangerous.

The gas intake line off the digester should be located a minimum of 1.2 meters above the maximum liquid level in the tank. A greater distance may be necessary to minimize the amount of solids and foam entering the piping system.

2.9.1 Characteristics and Contaminants

Because of the energy inherent in the methane, biogas is valuable as a renewable energy source and natural gas replacement. Biogas also has many contaminants and constituents that can cause problems if careful consideration is not given during the design stage. Hydrogen sulfide (H_2S), siloxanes, carbon dioxide (CO_2), and water may need to be treated or removed, depending on the ultimate use of the gas.

Typical composition of biogas is given in Table 23.15. Note that the values are typical for municipal mesophilic digesters treating primarily domestic sewage. The addition of FOGs, industrial waste streams, and food waste as well as the use of

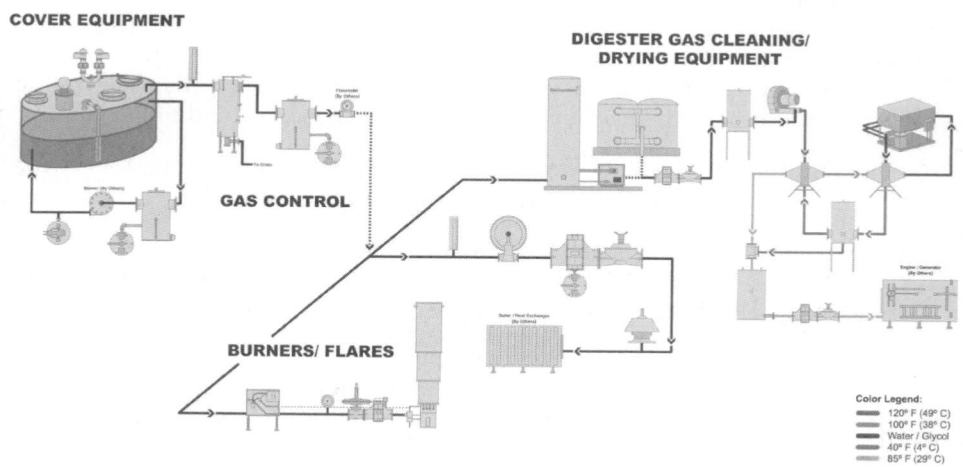

FIGURE 23.33 Diagram of a gas control system (courtesy of Varec Biogas).

Sludge line valves
Sludge line (permanent)
Sludge line (temporary)
Digester access
Explosion-proof vent fan
Explosive level meter
Safe ladder
Self-contained breathing apparatus
Safety harness
Nonskid boots
Explosion-proof lights
Water source
Washdown hose
Nozzle with shutoff
Wash water pump
Fixed sludge pump
Portable sludge pump
Turret nozzle
Tripod or hoist
Tank truck
Crane

TABLE 23.15   Digester Cleaning and Safety Equipment

thermophilic digestion, thermal hydrolysis, and chemicals for struvite control can greatly affect these values.

2.9.2   Gas Collection, Transport, and Safety

2.9.2.1   Commonly Referenced Standards   The following North American standards are recognized as either best design practice or, in some jurisdictions, code. Though not comprehensive these documents should be referenced and adhered to when designing digester-gas systems.

1. ANSI/CSA B149.6-15 Code for Digester Gas, Landfill Gas, and Biogas Generation and Utilization;

2. NFPA 820 Standard for Fire Protection in Wastewater Treatment and Collection Facilities; and

3. GLUMRB (Ten States Standards) Recommended Standards for Waste Water Facilities.

2.9.2.2 Materials of Construction Because of the corrosive nature of digester gas, primarily due to H_2S, CO_2, and water, careful consideration should be given to the materials of construction. Although lined or coated carbon steel options, most designers use 316/316L stainless steel piping for above-ground applications. Below ground, some designers choose to use plastic or fiber-reinforced plastic as a cost-effective alternative. Care should be taken in choosing these materials, and consideration should be given to the practices outlined in NFPA 820 regarding flammability, and those listed in ANSI/CSA B149.6-15 regarding installation.

Equipment used in-line (flame arresters, valves, sediment traps, drip traps) should be either low-copper aluminum (Type 356) or 316 SS. In general, cast products should be aluminum, which has been proven in digester service for over 70 years and is generally less expensive than stainless steel. Stainless steel should be used in fabricated products as it is much more readily available.

Care should be taken in the use of steel in digester-gas service. When rust (Fe_2O_3) forms on the steel, it can react with the H_2S to form Fe_2S_3. When the pipe is taken apart for maintenance, the reaction of Fe_2S_3 in air is exothermic, which can cause high temperatures and create a dangerous situation.

2.9.2.3 Flame Arresters and Flame Traps Flame arresters (Figure 23.34) are devices that prevent the propagation of flame in gas-piping systems. They work by utilizing two different principles. The first is what is called the mean experimental safe gap (MESG). The MESG is an orifice of a size such that a flame of a particular gas will not pass through. It is used for determining the hazardous classification of a gas, which for biogas or methane is Group D.

The second principle is by the use of a large material area that dissipates the heat such that the protected side of the flame arrester does not rise above the auto-ignition temperature of the gas.

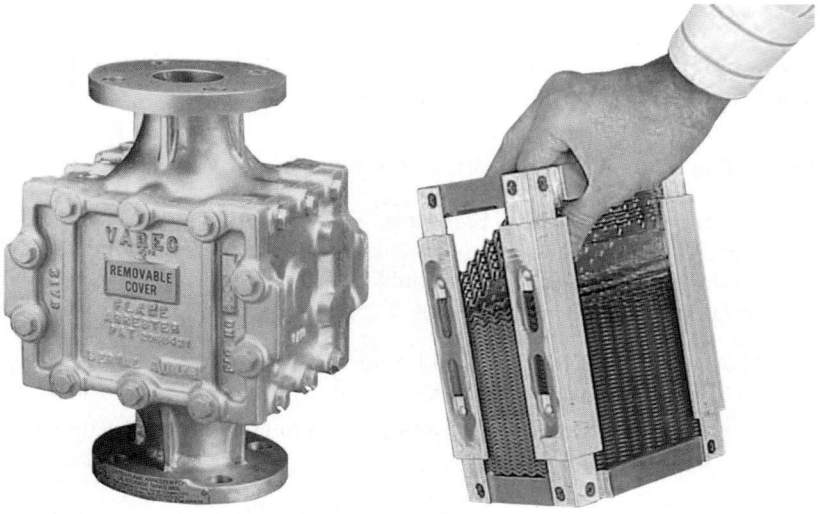

FIGURE 23.34 Flame arrester and removable bank assembly (courtesy of Varec Biogas).

Because the dissipation of heat is only effective for a limited time, in some applications a thermal shutoff valve is added to a flame arrester to shut off the gas flow in case of a continuous flame. This will stop the flame. A thermal shutoff valve is a spring-loaded valve whose pallet is held open by a fusible element. If a flame is present, the fusible element will melt, closing the valve.

A flame arrester combined with a thermal shutoff valve is called a flame trap assembly. Flame arresters should be used where there is the possibility of air entering the system, such as under the pressure- and vacuum-relief valves. Flame traps should be used wherever there is a combustion source such as flares and boilers.

Because of the small openings required to maintain the MESG and digester gas is wet and dirty, flame arresters are subject to fouling and need to be maintained on a regular basis. Depending on the application, this can be as often as every 6 months. Therefore, flame arrester bank assemblies should be easily removable and, in the case of in-line arresters, removal should not change the stress on the surrounding piping. In horizontal applications, flame arresters should not only have flow paths for gas but also for water, so that it can be drained through drip traps without becoming trapped in the element, which will cause premature fouling.

2.9.2.4 Climate In colder climates, especially those subject to freezing, gas piping and appurtenances need to be protected from the weather. Because biogas is saturated with water, exposure to freezing temperatures will cause freezing, which will prevent flow and could cause catastrophic failures of the piping and equipment. Insulation and, where indicated, heat tracing need to be evaluated. Devices such as regulators and sediment traps should be installed indoors. Equipment requiring maintenance should be fitted with removable insulation covers.

2.9.2.5 Digester Cover Equipment

2.9.2.5.1 Pressure- and Vacuum-Relief Valves and Flame Arresters with Three-Way Valve Pressure- and vacuum-relief valves protect the digester from structural damage related to over- or under-pressure (See Figure 23.35). These conditions can be caused by a number of occurrences including rapid pumping of sludge into or out of the digester, failure of the digester heating system, blocked gas draw-off lines, and rapid rise or foaming events. Flame arresters are included as part of the assembly in order to prevent a flame igniting the vapors in the tank during a vacuum condition. ANSI/CSA B149.6-15 requires that two sets of valves and arresters be mounted on a three-way valve to insure that maintenance can be performed while still leaving the tank protected. Depending on operating conditions, maintenance may need to be formed as frequently as every 6 months. Many three-way valves have low C_v values, so it is critical that pressure drop of the three-way valve be considered when sizing the valves for service. Pressure drops greater than 3% of the set point have been known to cause valve chatter and excessive pressure drops can cause reduced capacity of the valve, leading to failure of the digester.

ANSI/CSA B149 prohibits the use of liquid-based relief devices. Conventional pressure-relief devices are weight loaded with a Teflon seal. These devices are not bubble tight by design, as the low sealing force needed prohibits a tight seal. Although these devices, when new, can have very low leakage rates at 90% or above, during operation they can experience leakage as low as 80% of the set point. It is therefore recommended that these valves be set at least 20% above the normal operating pressure of the system as long as they can achieve full relief capacity at or below the design pressure of the digester.

Figure 23.35   Digester cover gas safety equipment (courtesy of Varec Biogas).

If the digester is expected to experience foaming based on past experience or other digester facilities in the vicinity, it is recommended to install a 24-in. pressure-relief manhole cover. These devices act as a backup to the normal pressure-relief valve, and relieve pressure or foam if the primary pressure-relief device gets clogged.

2.9.2.5.2 Sampling Hatches and Manways   Sampling wells should be mounted with 200-mm quick-opening, gas-tight, non-sparking sampling hatch covers. To facilitate entry into the digester during cleaning operations, one 915-mm and one 1.05-m manway covers should be installed for ventilation of the digester and to allow easy access of personnel equipped with breather packs to enter. These devices should be quick opening and easily operated to prevent the need for cutting bolts that have become seized, which could be a potential flame source.

2.9.2.6   Sizing and Installation   Most digester systems operate at a pressure less than 3.5 kPa. Because of the low operating pressures, and moisture and particulates in the gas, special consideration should be given to line losses, preventing condensate buildup, and ease of maintenance of valves and flame arresters.

2.9.2.6.1 Pressure Drop and Moisture Considerations   Gas pipe slopes of 20 mm/m are recommended, with a minimum allowable slope of 10 mm/m for drainage of condensate. In order to minimize pressure drop and condensate carryover, lines should be sized so that gas velocity is no more than 3.5 m/s. Velocities should be calculated based on maximum gas production from the digester or digesters. The Ten States Standards prohibits the use of smaller than 4-in. piping for digester-gas service.

2.9.2.6.1.1 Condensate and Sediment Traps   Condensate and sediment traps operate by lowering the gas velocity to drop out moisture and particulates via gravity. In addition, they are fitted with curved inlets, which allows centrifugal force to move heavier particles to the outside. These devices should be placed at the digester and the end of long pipe runs to remove free liquid. Note that condensate and sediment traps are not replacements for filters or moisture-removal systems required for a higher level of gas treatment. They are simple devices used to protect downstream equipment.

2.9.2.6.1.2 DRIP TRAPS Drip traps are devices that remove condensate from the line without allowing gas to escape. They can be manual, electrically actuated, or continuous flow type (condensate accumulator). Both Ten States Standards and ANSI/CSA B149.6-15 prohibit the use of float-operated drip traps. In no case should valves be used, except in the case where there are two valves between the gas system and the drain, and positive mechanical or electrical interlocks are provided to prevent both valves from being opened at the same time.

One of the most common flaws in gas-collection-system designs is an insufficient number of drip traps to remove condensate from piping. Design engineers should install drip traps at all low spots in the gas system and on each sediment trap. Low-pressure drip traps typically are made of low-copper aluminum castings. High-pressure devices should be made of steel or stainless steel. Corrosion and freezing can be minimized by judicious use of drip traps.

2.9.3   Digester-Gas Storage

Some WRRFs use one or more forms of digester-gas-storage systems (e.g., low-pressure gasholder or higher-pressure compressed-gas storage) to help them use their gas more effectively.

2.9.3.1   Low-Pressure Digester-Gas Storage   In an operating digester-gas system, biogas constantly is being evolved and used. This is a dynamic system in an essentially constant-volume arrangement of piping and vessels. Gas is produced in the digesters at a variable rate that depends heavily on how recently each digester was fed with raw solids. Meanwhile, the devices using digester gas (e.g., engines and boilers) may have variable or relatively constant gas-consumption rates.

Low-pressure gas-storage systems include flexible-membrane dome covers, dry-seal cylindrical steel gasholder tanks, and gasholder digester covers.

2.9.3.2   Flexible Membrane Covers   One newer fabric gas-storage option is the flexible-membrane gasholder cover. Some of these covers have been used successfully for 20 years. Flexible-membrane covers provide short-term gas storage to equalize gas pressure, an important consideration when cogeneration systems are involved. Flexible-membrane digester-gas-storage systems are available in sizes up to 34 m (110 ft) and for gas pressures up to 4 kPa (16 in. H_2O). They often are less expensive than traditional steel digester covers, and they contain gas and allow for liquid-level variation better than floating digester covers do since they are sealed to the top of the wall and not by the liquid.

2.9.3.3   Flexible-Membrane Cover Comparison   A summary of the advantages and disadvantages of flexible-membrane covers is noted here. Table 23.16 provides a summary of membrane cover design considerations.

Item or Parameter	Digester Gas		Natural Gas
	Range	Common Value	
Methane (%) (dry basis)	50–73	60	80–98
Carbon dioxide (%) (dry basis)	30–48	39	0–2
Nitrogen (%) (dry basis)	0.2–2.5	0.5	0.2–10
Hydrogen (%) (dry basis)	0–0.5	0.2	~0
Hydrogen sulfide (ppm_v) (dry basis)	200–3500	500	<16
Ethane (%) (dry basis)	0	0	0.3–5
Propane (%) (dry basis)	0	0	0.6–5
Butane (%) (dry basis)	0	0	0.5–3
Specific gravity (based on air = 1.0)	0.8–1.0	0.91	0.58
Ignition velocity, maximum (ft/s)	0.75–0.90	0.82	1.28
Wobbe number			
Higher heating value (HHV) (Btu/cu ft)	600–650	620	1030–1050
Lower heating value (LHV) (Btu/cu ft)	520–580	560	930–950

*All percentages listed above are percentages by volume; ppm_v = parts per million, by volume; the higher heating value includes the heat of the water of vaporization; the fuel's lower heating value does not include the heat of the water of vaporization; ft/s × 0.3048 = m/s; and Btu/cu ft × 37.26 = kJ/m³.

TABLE 23.16   Typical Digester Gas Compared to Typical Natural Gas*

2.9.3.4 Dry-Seal-Type Cylindrical Steel Gasholder Vessels The dry-seal (piston) gasholder is a vertical steel tank-within-a-tank gas-pressurization device designed to use its ample weight to keep digester-gas pressure virtually constant while gas production or use varies. It is a non-powered technique for supplementing the limited gas-storage volume between the liquid surface and the digester cover. The weighted, movable piston helps maintain a constant digester-gas pressure in the low-pressure gas piping.

2.9.3.5 Dry-Seal-Type Gasholder Dry-seal (piston) welded-steel gasholders have been used at many U.S. locations for more than 150 years. Table 23.17 lists some advantages and disadvantages of replacing secondary digester covers with similar welded-steel covers and then adding dry-seal digester gasholder vessels.

2.9.3.6 Gasholder Digester Covers Digester covers were common years ago, but are less common today because of concerns about gas escaping from the annular space between the outer edge of the cover and the inner digester wall. Another concern is the effect of such gas leakage on air quality. Most digester covers in California are fixed covers or membrane gasholders.

Advantages	Disadvantages
• Could function both as a digester cover and as a gas holder. • Might be available quicker.	• Has no proven method of reporting status of percent storage used. • No clear method for operating the secondary digesters with balanced storage covers in parallel. • Is much more mechanically complex. • Has a long-term durability concern. • Some safety concerns. May not be completely impervious to methane gas leakage. • Requires operation and maintenance on the pressuring air blowers, accessories, and controls.

TABLE 23.17   Advantages and Disadvantages of Flexible Membrane Covers

2.9.3.7 High-Pressure Compressed Digester-Gas Storage   Some WRRFs effectively use a higher percentage of their digester gas via a system of medium- or high-pressure gas compressors and gas storage spheres or horizontal storage tanks (pressure vessels).

When biogas is compressed via high pressure, smaller storage vessels are needed but more electricity and a more expensive compressor are required. Several facilities use high-pressure digester-gas-storage systems; they frequently operate at pressures of about 7 to 22 kPa (50 to 150 psi).

A few WRRFs have medium-pressure digester-gas-storage systems, which operate at a pressure of about 3 kPa (20 psi). Medium-pressure systems sometimes are used only to compress the excess gas produced at night. This gas is used in cogeneration engines the following day during on-peak hours, when electric rates are higher. Utilities vary in their willingness to participate in these types of programs that facilitate renewable energy generation.

2.9.4 Gas Processing, Utilization, and Combustion Equipment
One of the primary advantages of anaerobic digestion is the fact that it produces renewable energy in the form of digester gas. For example, it can be used in boilers to supplant natural gas usage, and in combined heat and power (CHP) sets to generate electricity for either on-site use or export, or it can be upgraded to natural gas for injection into the grid or for vehicle fueling. Each of these uses requires a different level of treatment to remove contaminants, which may otherwise damage or reduce the life of the gas utilization equipment.

Because there are times when the gas conditioning and utilization equipment is undergoing maintenance, or the digester is producing more gas than can be utilized, there is often a need to dispose of the gas. The safest, most efficient and cost-effective way of doing this is through flaring.

2.9.4.1 Flaring   Flares are normally divided into two types: open or candlestick, and enclosed. The open type is considerably less expensive but also less efficient, but the choice between the two normally has less to do with economics than environmental regulations and facility locations. Because open flares have combustion occur in the atmosphere, the combustion process is considered uncontrolled, and therefore, emissions cannot be guaranteed and will generally not meet EPA permitting requirements in stricter regulatory areas like California, Washington, or Massachusetts. In addition, they produce a flame that, depending on the amount of gas flared and the proximity to the property line, can cause public relations issues with neighbors.

2.9.4.1.1   Flare Pilots   Flare pilots with flame verification should be used to insure that all gas going to the flare is combusted. In addition, high-temperature, pre-mixed pilots can assist in the conversion of hydrogen sulfide, thereby eliminating odors.

40 CFR 60.118 requires that the flame be verified. The most reliable way to do this is by monitoring the pilot flame, as varying flow rates inhibit the ability to verify the main flame. ANSI/CSA B149.6-15 requires that a flare be operated at all times when flaring, and this standard and Ten States Standards require that the pilot gas be propane or natural gas. Although pilot systems have been developed that can utilize digester gas, and some cost savings can be achieved, the inherent variability as well as the wet and dirty nature make these systems potentially unreliable, regardless of the design.

2.9.4.1.2   Open Flares   Open flares typically consist of a pipe, a windshield, and a pilot nozzle. Items provided by the flare manufacturer should include the flare stack, pilot valve and regulator system including pre-mixed pilot controls, and a control panel to start and stop the pilot system and control valves. Typical inputs are for the start/stop signal, which can either be a pressure switch or, in the case of systems with a constant-pressure gasholder, a signal from the facility distributed control system (DCS) based on the gasholder position. Typical outputs are power on, pilot on, and pilot failure. These types of flares can be used in most cases, except as described above.

2.9.4.1.3   Enclosed Flares   Enclosed flares are substantially more expensive than open flares, but may be required based on regulatory or facility requirements described above. Enclosed flares are normally divided into three types: natural draft temperature controlled, forced draft temperature controlled, and natural draft venturi pre-mixed. Each of these have their advantages and disadvantages, but care should be taken to choose the right technology based on whether the flare is going to operate continuously or intermittently, and also the flow range that the flare is going to be required to operate over. This is especially true if the flow at initial construction is significantly lower than the future flow the flare may be designed for.

2.9.4.1.3.1   NATURAL DRAFT TEMPERATURE-CONTROLLED ENCLOSED FLARES   These are the oldest style of enclosed flares and were initially developed for the landfill market. Although the composition of landfill gas and digester gas is similar, landfill gas often contains volatile organic compounds (VOCs) and other toxic compounds that require a longer retention time to destroy.

These types of flares consist of a refractory-lined stack with a thermocouple at the stack exit and motorized dampers at the bottom. They work on a time-and-temperature basis to destroy methane and provide low NO_x and CO emissions. As the temperature at the stack exit increases above the set point, the dampers open further to allow in more air to cool the combustion process.

These flares are primarily designed for continuous operation and a narrow flow range. In intermittent operation, the refractory tends to wick moisture from the atmosphere. When these flares are started after a long period of inactivity, the moisture in the refractory turns to steam, which can cause pieces of the refractory to come off. If this condition is not monitored, and proper repairs are not made, which in most cases includes refractory replacement, the stack can suffer irreparable damage. In addition, because these flares need to go through a purge cycle of 15 to 20 minutes each time the flare is called upon, their use may not match with facility requirements and may lead to venting due to flare unavailability.

The flow range the flare can operate under, commonly referred to as "turn-down" is limited by the heat loss in the stack and the ability to maintain stack exit temperature. This number is generally approximately 5:1 (maximum flow:minimum flow), although this can be extended by placing thermocouples at multiple elevations on the stack, and controlling from lower thermocouples during low-flow conditions. Care must be taken as these lower thermocouples become a maintenance item. In addition, at lower flows, the limited number of burner nozzles typically provided in this style of flare can have low velocities, which will not provide sufficient mixing energy to achieve good combustion. This can lead to long, lazy flames, which may impinge on the refractory and cause damage.

2.9.4.1.3.2 FORCED DRAFT TEMPERATURE-CONTROLLED ENCLOSED FLARES This type of flare is similar to the natural draft temperature enclosed flares in that they include a refractory-lined stack and a thermocouple at the exit to control temperature at the exit. The main difference is that instead of using dampers to control airflow, these units use an air blower controlled by a variable-frequency drive (VFD) to provide combustion and cooling air. The advantage of this is more accurate control of the air, which can provide higher destruction of the methane and lower emissions of NO_x and CO. In addition, because of the control of the air, the flames are normally shorter and have a high temperature, which allows the flare stacks to be shorter. However, as with any piece of rotating equipment, this requires a higher level of maintenance, especially if the flare is going to be operated intermittently. Also the parasitic power load is much higher than the other two styles of flares, which may be a concern when calculating the economy of a gas utilization system.

As with all temperature-controlled or refractory-lined flares, these units have a limited turn-down ratio (typically 7:1) and can experience the same issues with intermittent operation as described above.

2.9.4.1.3.3 NATURAL DRAFT VENTURI PRE-MIXED ENCLOSED FLARES These types of flares (see Figure 23.36) use multiple venturi nozzles to pre-mix the air and gas, giving a short, high-temperature, robust flame. Instead of a refractory-lined stack, these flares have stack sections of gradually increasing diameter, which permit a boundary layer of cooling air that protects the stack.

Because these flares do not rely on temperature control and the smaller-diameter venturi orifices allow for higher velocities at lower flows, these types of flares are not subject to the turn-down limitations of other types of flares. Since there is no refractory, there is no possibility of refractory failure. In addition, because there are multiple openings in the stack, there is no need for a purge cycle, and these flares can be ready for gas flow within 1 to 2 minutes after being called on.

2.9.4.2 Moisture Removal When produced, digester gas is saturated with water. However, an increasing number of digester-gas treatment technologies and use equipment require dry gas. Digester gas can be dried via several techniques (e.g., refrigerant dryers, desiccant dryers, coalescent filers, and glycol systems). All moisture-removal

FIGURE 23.36   Natural draft venturi premixed enclosed flares (courtesy of Varec Biogas).

equipment should be made of corrosion-resistant materials and preceded by sediment and condensate trips. Consideration must be given for the safe drainage of the large amount of condensate generated by moisture-removal systems.

Refrigerated dryers are the most common and often the most successful technique for drying digester gas. They use gas heat exchangers and mechanical chillers to cool the gas and condense water, so it can be easily removed via physical separation. The gas is then reheated utilizing the heat of compression to give the gas a low dew point and a low relative humidity. The ability to control the exit temperature of the gas makes them ideal for systems where downstream adsorbers are used to remove siloxanes or other compounds.

Desiccant dyers are sometimes used in packaged compressed-air dryers or natural gas applications. Because digester gas typically is saturated and often contains other contaminants, desiccant gas dryers must be specifically designated for these conditions to be effective.

Coalescent filters can be used to remove water from digester gas, but not molecular water vapor. So these filters are used with other water-removal equipment to prevent carryover of the water and to help with drainage.

2.9.4.3 Gas-Pressure Boosters   Anaerobic digesters typically produce gas at a pressure of 1 to 2.5 kPa (4 to 10 in.) H_2O, which may be insufficient for the gas-use equipment. However, many boilers, flares, and some gas-use equipment only require an inlet gas pressure of 0.3 to 2 kPa (2 to 14 psi). A gas booster often can make up the difference. Centrifugal gas-booster blowers can increase gas pressure by approximately 0.6 to 1.5 kPa

(4 to 10 psi) (depending on size), and this increase in gas pressure is typically adequate for many applications.

Centrifugal gas-booster blowers are available both as open machines and as hermetically sealed gas blowers. They are available in stainless steel and cast iron. When using centrifugal gas blowers in indoor applications, design engineers should make sure to provide gas-tight seals.

2.9.4.4 High-Pressure Gas Compressors Rotary screw compressors, sliding vane compressors, and reciprocating piston gas compressors all have been used to successfully compress digester gas up to 50 kPa (350 psi). All have limitations, and all must be designed for continuous duty with a typically contaminant-laden wet gas. Light-duty, intermittent-use air-compression machinery should not be used in digester-gas applications. Important design issues with high-pressure digester-gas compressors include:

- Robust structural foundations for the compressors;
- Gas-flow pulsation attenuation;
- Lubricating oil carryover and removal from the gas;
- Moisture condensation in the gas during shutdown; and
- Compressor cooling.

2.9.4.5 Gas Metering and Gas-Pressure Monitoring Facility personnel need accurate, reliable measurement of gas flow. Gas production is a measurement of digester performance. A reliable metering system enables the facility to optimize both the digester and the gas-use system. It promptly alerts operators to gas-system leaks and process fluctuations. It also allows operators to store excess gas properly and helps them plan process schedules. In addition, flow data are needed to calculate digester efficiency and the fuel savings obtained by using digester gas.

Digester gas, which is moist, dirty, and corrosive, is produced at fluctuating rates. The piping and appurtenances are designed to convey gas at low velocities and pressures. These characteristics can cause numerous maintenance problems for metering devices that engineers need to address when designing the gas-collection system and selecting gas meters.

Thermal mass dispersion meters are the most common type currently in use and are favored for the fact that they do not create pressure drop and have no moving parts, minimizing maintenance. Other technologies that have made recent advancements are sonic flow meters and venturi-style meters.

Other flow meters that have monitored gas successfully include positive displacement bellows, shunt flow, turbine, differential-pressure venturi tube, orifice plate, and flow tube. Although these can be potentially more accurate, the high amounts of maintenance required have generally precluded their use.

The meters should measure each digester's gas production, total gas production (after recirculation), gas sent to each engine and/or boiler, and gas wasted to the flares. They should resist corrosion and be easily serviced. Meters should be placed downstream of condensate and sediment traps to minimize the amount of free liquids and particulates they see.

When selecting the device, design engineers also should consider both startup and design flow conditions. Startup (low-flow) conditions may be below the operating range of a meter sized for design flows.

Gas-pressure gauges for local indication are available in both dial and manometer designs; the manometer typically is used on low-pressure lines due to its lack of moving parts, which make it well suited to wet, dirty gas.

Facilities may use pressure transmitters with remote indication in conjunction with local readouts as part of the facility's overall DCS. Gauges indicate the pressure available in the system; they help O&M personnel locate line blockages.

2.9.4.6   Isolation Valves   Several types of isolation valves (e.g., butterfly valves, plug valves, and knife gate valves) have been used successfully on digester-gas piping. For information on valve location requirements, see NFPA 320 and NFPA 54.

2.9.4.7   Gas Analysis and Conditioning   The sampling and analytical methods required for digester gas depend on the type and concentration of compound(s) of interest, which generally depends on the gas utilization. Many relevant methods and techniques are available, but only should be used under the correct conditions.

Digester gas often is sampled via Tedlar bags or by collecting a gas sample in a methanol impinger.

There are several methods for analyzing digester gas (e.g., gas chromatography and mass spectrometry) (see Table 23.18). If the gas will be tested for siloxanes, or similar organic compounds typically found in small concentrations, then more analytical work may be required.

Complex sample collection and analysis (typically due to reactivity and low detection-limit requirements) require the use of certified and accredited labs staffed by analysts with the appropriate skills.

Gas conditioning is one of the most rapidly developing fields in biogas handling. Entire books could be written on different technologies currently and commonly in use as of the publication date. It is recommended that designers use outside resources, including WERF research and WEF conference papers when evaluating gas conditioning systems for the latest technologies.

2.9.4.8   Hydrogen Sulfide Removal   Unless controlled, the concentration of hydrogen sulfide in digester gas can range from 150 to 3000 ppm or more, depending on both the influent's composition and the digester feedstock's characteristics. Major sources of sulfur compounds in the influent are the potable water supply and industrial discharges.

Advantages	Disadvantages
• Is mechanically simple; easy to operate and to understand. • The dry-seal vessel height provides a complete and accurate indication of digester gas-storage status. • Emergency repairs, if required, could be performed by local staff.	• When the costs of digester covers and dry seal gas holders are added together, it almost certainly will total more than a gas-membrane cover solution alone. • Will probably not be available as quickly as a gas-membrane cover could be available.

TABLE 23.18   Advantages and Disadvantages of Dry-Seal Gas Holders

Sulfates occur naturally in water when urine and protein decompose; they also result from alum treatment in the water supply system. Industries can discharge various sulfur materials to the collection system. In addition, trucked wastes, which are fed directly to anaerobic digesters, often contain sulfur material.

The hydrogen sulfide in digester gas is formed when anaerobic bacteria reduce sulfates and other sulfur material. It may need to be removed from digester gas to reduce corrosion in boilers and engine parts. It also may need to be removed to satisfy local air emissions standards. Hydrogen sulfide is a toxic air pollutant that can create both odor and safety issues, even in minute concentrations. Biogas with a high hydrogen sulfide content can contribute to air pollution. Flaring eliminates the odor problem but produces sulfur dioxide, which is a significant cause of acid rain.

There are a number of methods for removing hydrogen sulfide from digester gas, and the method chosen usually depends on the sulfur loading. When choosing a technology, the designer should keep in mind both capital and operating costs, as well as space requirements.

2.9.4.8.1 Dry Scrubbers (Absorbers and Adsorbers) Dry scrubbers usually consist of a tank made either of stainless steel or fiber-glass-reinforced plastic (FRP), which contains a medium that removes the hydrogen sulfide.

A common dry scrubber (called the iron sponge) uses iron oxide hydrated wood chips (see Figure 23.37). Hydrogen sulfide reacts with iron oxide to form elemental iron, elemental sulfur, and water. The iron sponge is periodically regenerated by removing the sulfur and oxidizing the iron to form iron oxide. Such regeneration could be hazardous because spontaneous combustion is possible if the iron is oxidized too rapidly. There are systems where this is performed in situ, and the vessel is filled with water to prevent high heat during the oxidation process. Iron sponge can also be fitted with continuous regeneration systems, which add a small amount of air to control the reoxidation process. These systems should be provided with safety systems to prevent a runaway reaction, although at the level air is added this is rare.

Figure 23.37 Iron sponge purifier (courtesy of Varec Biogas).

Other manufactured iron compounds have been developed that do not have the exothermic oxidation reaction iron sponge has, and give off various inert substances as a by-product.

Activated carbon can also be used as an adsorber for hydrogen sulfide and is referred in some areas due to its ready availability. However, operating costs tend to be much higher than with iron-based systems.

Dry scrubbers are typically best suited to relatively small sulfur loadings.

2.9.4.8.2 Wet Scrubbers   A conventional wet scrubber (see Figure 23.38) uses a liquid that is maintained at a high pH (via caustic) to enhance hydrogen sulfide absorption. It also may contain an oxidant (e.g., sodium hypochlorite or potassium permanganate) to reduce adsorbent disposal problems and increase its useful life. Wet scrubbers use nozzles or diffuser plates, which periodically require cleaning. The gas leaving the wet scrubber is saturated with moisture, which must be condensed and removed downstream. Because the headloss through wet scrubbers typically is too high for the low-pressure digester-gas system, the gas must be compressed before being scrubbed.

One of the major costs of these systems is the caustic chemical costs. Recently systems have been developed where the caustic is recovered biologically from the drain water, which will lower operating costs.

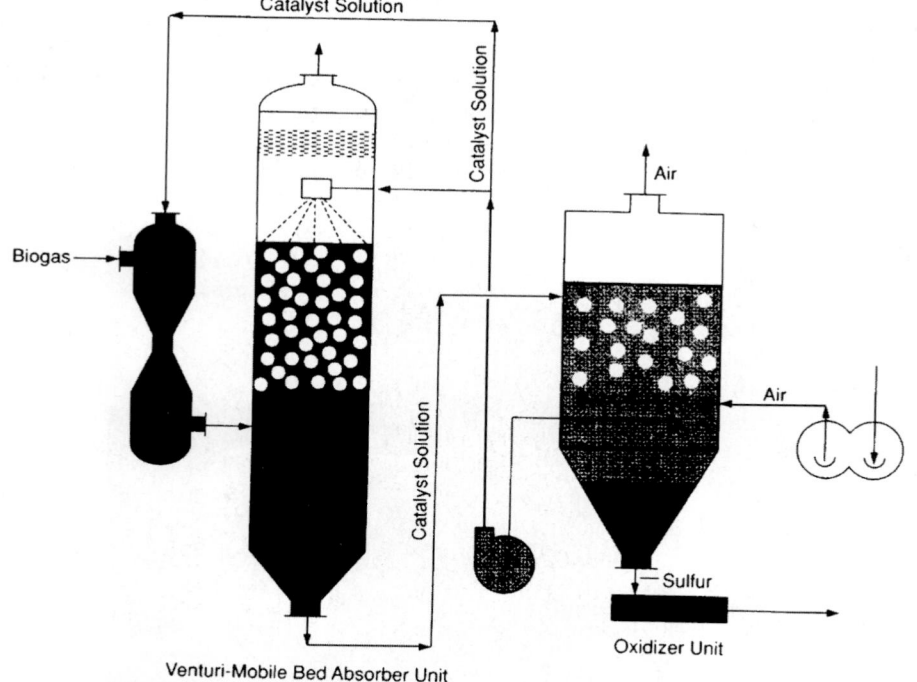

FIGURE 23.38   Schematic of a typical catalytic scrubber used to remove hydrogen sulfide from biogas.

2.9.4.8.3 Biological Treatment Over the past 15 years, advances have been made to make biological H$_2$S removal more reliable. In these systems, an FRP tower is filled with packed media (see Figure 23.39), which is then seeded with bacteria. These bacteria, when operated in a slightly aerobic environment, work to scavenge the sulfur from the gas.

Because these units require oxygen to operate, blowers are required to add air into the system. Sufficient safeguards are added to prevent the mixture from becoming flammable. In addition to air, the bacteria will also require nutrients to operate. These can often be provided by facility effluent, although fertilizer may also be required.

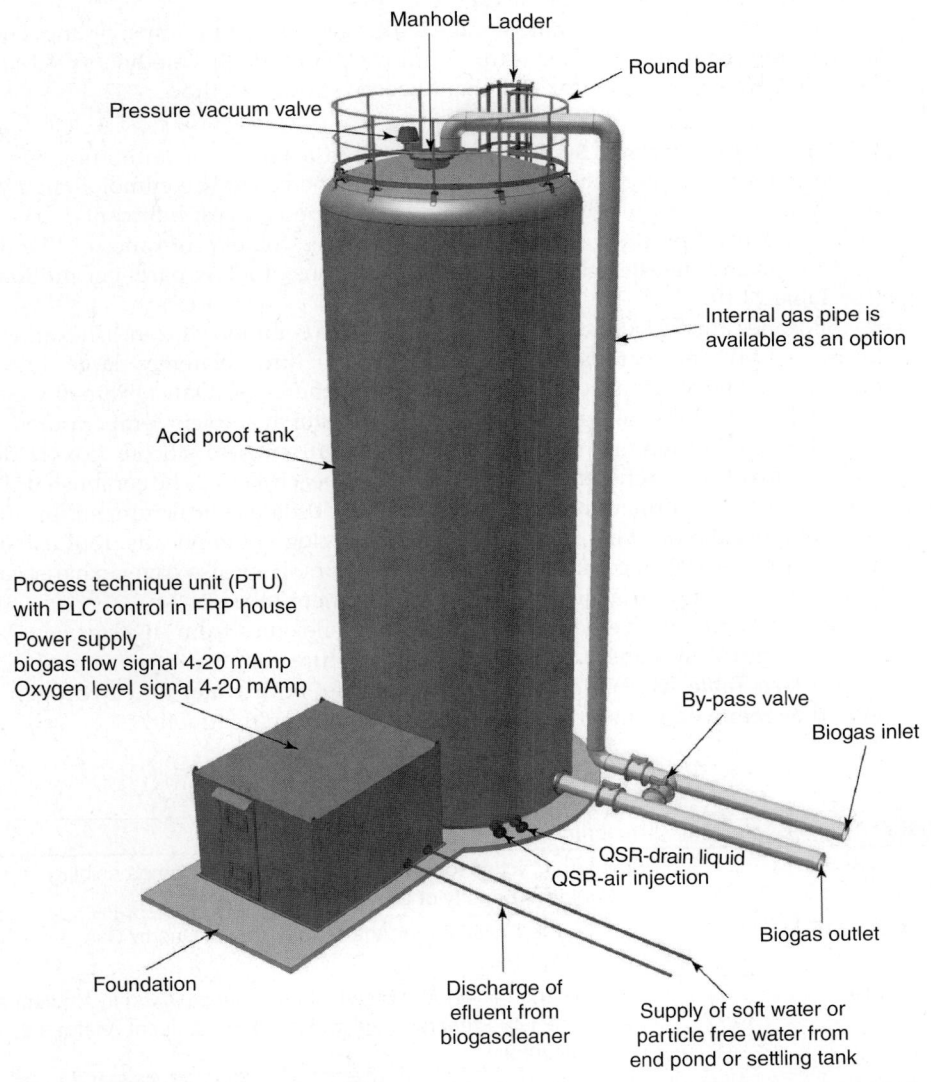

FIGURE 23.39 Biological H$_2$S removal (courtesy of Varec Biogas).

Because sulfur and bacteria will build up on the media, the units require periodic cleaning. This is normally done by filling the vessel with water and blowing compressed air into the tank to agitate the media. Because these tanks are typically 12 to 15 m tall, the manufacturer needs to consider the weight loading on the media supports as well as having air nozzles of sufficient size and quantity to insure satisfactory agitation.

Instead of gas scrubbing, some facilities have added iron salts directly to the digester or facility influent. Iron reacts with sulfide to form insoluble iron sulfide. However, iron salts should not be added to heated solids lines because this results in a buildup of vivianite (ferrous phosphate) scale. Iron salts also can reduce digester alkalinity, so design engineers must make provision to monitor and control the solution strength and dosing rate to avoid lowering the digester's pH.

This method requires a bulk-storage tank, chemical feed pumps, piping, and monitoring equipment. Its chief cost is for chemicals. Although this method's O&M costs are low, it requires more operator skill than the iron sponge method.

2.9.4.9 Siloxane-Removal Systems Siloxanes are a family of anthropogenic organic compounds containing silicon that are becoming increasingly common in many household products (e.g., deodorants, cosmetics, shampoos, dyes, lubricants, dry-cleaning fluids, and waterproofing compounds). As a result, volatile siloxanes can be found in landfill gas and digester gas, often at concentrations of a few parts per million or less (see Table 23.19).

Siloxanes are difficult to detect and control. Two common siloxanes, hexamethyldisiloxane (MM) and octamethyldisiloxane (MCM), are relatively large linear-chain molecules while others are cyclical (similar to benzene rings). Only hexamethylcyclotrisiloxane (D3) and MM are significantly soluble in water at ambient temperatures.

When combusted, siloxanes form tough, often abrasive silicon dioxide deposits. (Silicon dioxide is the chemical name for ordinary beach sand.) The combusted siloxanes also promote the formation of other chemical deposits (e.g., calcium, sulfur, iron, and zinc compounds) on them. These deposits often clog engine heads, foul exhaust and intake valves, and coat combustors and fuel injectors. They also cover exhaust catalysts, boiler surfaces, and exhaust heat-recovery equipment tubes.

Several WRRFs have had success removing siloxanes from digester gas. The best siloxane-control systems typically include moisture removal upstream of activated carbon (see Table 23.20). To maximize media life, both water and hydrogen sulfide should be removed from biogas before activated carbon treatment.

Method Number	Description
ASTM D-3588	Standard Practice for Calculating Heat Value, Compressibility Factor, and Relative Density of Gaseous Fuels
ASTM D-1945	Standard Test Method for Analysis of Natural Gas by Gas Chromatography
EPA TO-14	Determination of Volatile Organic Compounds (VOCs) in Ambient Air Using Specially Prepared Canisters With Subsequent Analysis by Gas Chromatography

TABLE 23.19 Methods Typically Used to Analyze Biogas

Siloxane Species	Formula	Common Abbreviation	Molecular Weight	Vapor Pressure (mm Hg at 77°F)*	Boiling Point (°F)	Water Solubility (mg/L at 25°C)
Hexamethyldisiloxane	$C_6H_{18}Si_2O$	MM	162	31	224	0.93
Octamethyltrisiloxane	$C_8H_{24}Si_3O_2$	MCM	236	3.9	N/A	0.035
Hexamethylcyclotrisiloxane	$C_{12}H_{18}O_3Si_3$	D3	222	10	275	1.56
Octamethylcyclotetrasiloxane	$C_8H_{24}O_4Si_4$	D4	297	1.3	348	0.056
Decamethylcyclopentasiloxane	$C_{10}H_{30}O_5Si_5$	D5	371	0.02	412	0.017

*0.5556 (°F – 32) = °C.

TABLE 23.20 Typical Volatile Organic Siloxanes Found in Digester Gas

2.9.4.10 Carbon Dioxide Removal As global concern about climate change increases, some WRRFs have begun to monitor carbon dioxide. Dry digester gas typically contains about 60% carbon dioxide by volume, or about 55% to 60% carbon dioxide on a mass basis. Techniques for removing carbon dioxide from digester gas include pressure swing adsorption, temperature swing adsorption, cryogenic refrigeration, and amine treatment.

2.9.4.10.1 Pressure Swing Adsorption In pressure swing adsorption, gas constituents adsorb to the surface of a media at one pressure (typically high), and are released at another pressure (typically much lower).

2.9.4.10.2 Temperature Swing Adsorption Temperature swing adsorption might be the most common technique for removing carbon dioxide. In this process, carbon dioxide adsorbs to a media at a low temperature {typically at or near ambient temperatures (10°C to 32°C [50°F to 90°F]). Once the adsorption media is saturated, carbon dioxide is expelled from it by heating the media to typically 150°C to 200°C (300°F to 400°F).

2.9.4.10.3 Cryogenic Refrigeration Cryogenic refrigeration takes advantage of the fact that carbon dioxide freezes at a warmer temperature (279°C [2110°F]) than methane does (2182°C [2297°F]). Cryogenic systems can work well but require a significant amount of mechanical energy to refrigerate the gas.

2.9.4.11 Amines Amines (e.g., monothanolamine [MEA] and diethanolamine [DEA]) are a class of substances derived from ammonia. They frequently are used to remove both hydrogen sulfide and carbon dioxide from raw or sour natural gas.

2.9.5 Gas Use—Boilers

Boilers extract usable heat energy from a fuel, typically via combustion. Historically, this is the most common technique for capturing a digester gas' energy. Even WRRFs that use engine generators or gas turbines need boilers as standby or supplemental heating equipment. Boiler sizes range from about 100 000 to more than 1 billion kJ/h (Btu/hr).

Emissions controls are becoming increasingly important features for boilers throughout the United States. Engineers should address air quality regulations as part of all boiler designs. If designed appropriately, specially modified boilers can meet nearly all air quality regulations.

2.9.5.1   Fire-Tube Boiler   Packaged fire-tube boilers are the most common type of boilers used in WRRFs. They are available in sizes from about 2 to 30 million kJ/h (2 to 30 million Btu/hr).

2.9.5.2   Fire-Box Boilers   Fire-box boilers are a special type of fire-tube boiler with an oversized combustion chamber. This chamber may help properly combust the relatively low caloric (Btu) content of digester gas. Fire-box boilers range in size from about 2 to 10 million kJ/h (2 to 10 million Btu/hr).

2.9.5.3   Water-Tube Boilers   Water-tube boilers are similar to fire-box boilers, except that the combustion chamber is a horizontal insulated gas-tight compartment containing multiple water-filled steel tubes that are heated by the combustion gases. They contain less water internally, so that water-tube boilers often can warm up quicker and start up faster than fire-tube boilers. Those with flexible tubes are particularly resistant to thermal shock and less vulnerable to siloxane-caused silicon deposits on the tubes. Water-tube boilers are available in sizes ranging from less than 8 to more than 20 million kJ/h (8 to 20 mil Btu/hr).

2.9.5.4   Cast-Iron Boilers   Cast-iron sectional boilers are sometimes used in retrofit applications because they can fit through small doorways. These boilers are smaller, typically available in sizes from 300 000 to 10 million kJ/h (300 000 to 10 million Btu/hr).

2.9.6   Gas Use—Combined Heat and Power (Cogeneration)

Facilities larger than approximately 20 to 40 ML/d (5 to10 mgd) are possible candidates for digester-gas cogeneration (combined heat and power systems). Typical cogeneration systems are based on internal-combustion engines, microturbines, or gas turbines.

2.9.6.1   Reciprocating Internal-Combustion Gas Engines   Most WRRFs that have digester-gas cogeneration systems use reciprocating internal-combustion engines. Major engine manufacturers have recently developed many advanced internal-combustion engines to improve fuel economy, reduce maintenance, and lower exhaust emissions.

2.9.6.1.1   Reciprocating engines   Reciprocating engines are the most widely used technology in digester-gas cogeneration applications.

2.9.6.1.2   Advanced reciprocating engine systems   With the goals of significantly better fuel economy and lower exhaust emissions, several manufacturers developed models that they used to create technically modern, progressive spark-ignition, lean-burn engines. These engines are called advanced reciprocating engine systems (ARES). With their higher fuel efficiency, these technologies could enable WRRFs to produce substantially more electric power to offset energy costs using existing digester-gas production.

ARES have been in service since 2005 or 2006; they have an electrical output of about 1000 to 3000 kW. The more fuel-efficient engines produce more power using less digester gas than the older engines now in service at some WRRFs. They also operate at a gas pressure less than 0.6 kPa (4 psi) and so often can be used with many existing digester-gas systems.

2.9.6.1.3   Dual-Fuel Engine Generator   Dual-fuel (gas-diesel) engines are compression-ignition, not spark-ignition, engines. To ignite, they simultaneously burn gas and a small amount of diesel fuel pilot oil. These engines must use some diesel fuel as pilot fuel, but their controls also allow automatic switchover to 100% diesel fuel operation without changing load if the gaseous fuel supply is interrupted. This capability is a beneficial feature for standby units because they can start and operate even during power failures.

Dual-fuel engines typically use 1% to 5% diesel fuel oil, but many can, if necessary, operate on 1% to 100% diesel fuel. Such fuel flexibility is an excellent advantage, especially if the gaseous fuel supply is disrupted. This option includes storage and handling equipment for diesel fuel, along with 11-kPa (75-psi) gas compressors to supply gaseous fuel to these engines.

2.9.6.2 Combustion Gas Turbine Generators   Combustion gas turbines are available in sizes ranging from 250 to 250 000 kW. They typically are used at WRRFs with influent flows of 300 ML/d (80 mgd) or more. They also are widely used in new large commercial electric-power plants. Gas turbines are an attractive option for generating electricity because they have several important characteristics (see Table 23.21).

A few U.S. WRRFs have been successful in using gas turbines fueled by low-energy digester gas. Most probably will require some form of exhaust emissions control. There are three types of emission-control systems available for turbines: wet technologies; catalytic converters; and dry, low-nitrogen oxides (NO_x) combustors. Wet technologies (e.g., water or steam injection directly into the turbine's combustion zone) can substantially reduce exhaust emissions, but they require a continuous flow of ultraclean water. This is both expensive and time-consuming. Catalytic converters, which often follow water or steam injection, are expensive and simply not appropriate for digester-gas fuel

General Type	Advantages	Disadvantages
Water scrubber	Continuous process Can be used upstream of others technologies	Only removes water soluble siloxanes
Refrigeration (to 40°F*)	Continuous process Can be used upstream of others technologies	Only removes a fraction of the siloxanes
Refrigeration (to below 0°F)	Continuous process Also removes water	All units to date have had significant freezing problems Consumes more electricity
Activated carbon adsorption	Proven technology that works very effectively Also removes H_2S and other trace organics	Batch type process Media must be replaced or regenerated
Silica gel systems	An alternative to carbon Can be used upstream of others technologies	Very limited operation experience

*0.5556 (°F − 32) = °C.

Table 23.21   Advantages and Disadvantages of Various Siloxane-Removal Systems

without extensive, reliable fuel treatment. Dry, low-NO$_x$ combustors are the newest and most attractive technology, and may be the only one appropriate for digester-gas operations. Not all gas turbine manufacturers offer this technology, and many have dry-NO$_x$ units that are, at best, experimental. Most of the newer, more advanced gas turbines are available with low-NO$_x$ combustors.

Gas turbines require slightly less maintenance than reciprocating engines, but service is highly specialized and expensive. So it is important that the WRRF have local service support.

The fuel should be free of condensation and particles larger than 5 mm. The inlet pressure should be 17 to 36 kPa (120 to 250 psi). A high-pressure gas compressor and a moisture separator or filter probably would be needed to meet these requirements.

Gas turbines require a fuel gas-booster compressor to supply the required 29 kPa (200 psi) to the combustion chamber. Some turbine models are available in a "dry low emissions" version. Other gas turbines can meet NO$_x$ emission standards via a selective catalytic reduction system.

Catalysts are not suitable for use with digester gas unless the gas has been thoroughly and reliably cleaned of impurities. It has been tried unsuccessfully at the Los Angeles County Sanitation Districts' Carson plant, and at the Sacramento Regional Wastewater Treatment Plant's cogeneration facility. Various contaminants in digester gas quickly poison the noble metals in the catalyst.

2.9.6.3 Microturbines A highly publicized new technology, microturbines are small, high-speed gas turbines ranging from 30 to 250 kW (see Table 23.22). Many were originally developed from large engine turbochargers and use new technologies (e.g., extended-surface recuperators, air bearings, and ultra-fast operating speeds). Recently, interest has grown in using microturbines for distributed generation and cogeneration.

All gas turbines—including microturbines—generate less power when installed at high elevations and when ambient temperatures exceed 15°C (59°F). If installed at a site with an elevation of 1295 m (4250 ft), for example, the gas turbine's performance would be about 20% to 25% less than that of one installed at sea level, depending on the inlet combustion air temperature.

Advantages	Disadvantages
Suitably equipped gas turbines can burn several fuels, including digester gas.	Gas turbines are less-efficient electricity generators than engines.
Gas turbines are available from several experienced manufacturers in sizes from about 250 to 250 000 kW.	Gas turbines lose power and fuel efficiency in ambient air temperatures above 60°F*.
Gas turbines have few moving parts and typically require less maintenance than internal-combustion engines.	Gas turbines lose power and fuel efficiency at high elevations.
High-pressure steam can be produced from the hot turbine exhaust gasses sufficient in large gas turbines to provide another 50% electric generation capacity.	Gas turbines require high-pressure, clean fuel.

*0.5556(°F − 32) = °C.

TABLE 23.22 Advantages and Disadvantages of Gas Turbines

Because of their comparatively small size and output, microturbines have been attractive to WRRFs with smaller flows (average flows as low as 15 ML/d [4 mgd]) than typically suitable for digester-gas-fueled cogeneration systems. For example, a 57-ML/d (15-mgd) WRRF in southern California installed a 250-kW microturbine to process its digester gas. Additionally, between 2000 and about 2004, several California WRRFs installed microturbines without sufficient digester-gas treatment, and the facility staffs had difficulty operating and maintaining them. Several of the microturbines have been shut down.

2.9.6.4   Steam Turbines and Steam Boilers   A few U.S. WRRFs are large enough (more than 400 ML/d [100 mgd]) to produce and burn digester gas in large steam boilers and then generate high-pressure steam and electricity via a steam-driven rotating steam turbine generator.

For smaller WRRFs, this steam boiler-turbine technology is an inefficient power generator. Superheated, very high-pressure steam, typically above 130 kPa (900 psia) is required for efficient steam turbine generator performance, and steam turbines less than about 10 MW in size are not physically large enough to be built with the relatively close matching tolerances and thus higher mechanical efficiencies of much larger steam turbines. Also, the high-pressure steam boiler must be continuously staffed (around the clock) by a licensed steam-boiler operator.

2.9.7   Gas Cleanup and Sale

The cost of natural gas in the United States has risen over time and various renewable energy credits are available. This has made it more economical to clean up and more fully utilize digester gas.

Table 23.23 characterizes pipeline-quality digester gas. Once carbon dioxide, hydrogen sulfide, and water are removed, digester gas also can be used for direct pipeline injection. The applicability of scrubbing and selling digester gas to the local natural gas utility depends entirely on the gas utility's interest and willingness to offer a competitive price for the methane derived from digester gas.

Alternatively, this gas may be upgraded and stored as compressed natural gas (CNG) for vehicle fuel. Due to the high costs of vehicle fuels, this option is one of the

Advantages	Disadvantages
Are available is smaller sizes, down to 30 kW for small-capacity plants.	Are inefficient electricity producers. Their electrical-generation efficiency typically is only 24% to 30%.
Produce fewer exhaust emissions than some other types of digester gas cogeneration equipment.	Require significant gas cleanup, including moisture and siloxane removal.
Are available as modular, fully packaged equipment.	Even with appropriate heat-recovery equipment, they supply relatively little heat for the quantity of digester gas consumed.
Are quiet and can be used outdoors.	The digester gas must be compressed to between 75 and 100 psi*.

*psi × 6.895 = kPa.

TABLE 23.23   Advantages and Disadvantages of Microturbines

most financially viable options for biogas end use (Knight et al., 2016). Gas upgrading may also be used in conjunction with cogeneration (Kemp, 2014).

2.9.8 Solids Drying
Reference Chapter 24.

2.9.9 Emerging Technologies—Fuel Cells
A fuel cell is an electrochemical device that combines hydrogen with oxygen to continuously produce electricity. The hydrogen is extracted from the fuel delivered to the unit, while the oxygen is simply obtained from the air. The popularity of fuel cells is due to their high power-generation efficiency, vibration-free operation, clean exhaust emissions, and technical novelty. Fuel cells are quiet; their accessories generate what little noise they produce. Stationary fuel cells are available as fully modular units in sizes of 200 kW and larger. They are readily installed outdoors.

Fuel cells are used today at some municipal and industrial treatment facilities. The technology continues to develop but their use is not yet as widespread as was anticipated. The current economic viability of fuel cells, however, depends largely on funding assistance, often via available grants.

2.9.9.1 Types of Fuel Cells
Four types of fuel cells are in development or commercially available: phosphoric acid fuel cells, carbonate fuel cells, solid oxide fuel cells, and proton exchange membrane (PEM) fuel cells.

2.9.9.1.1 Phosphoric Acid Fuel Cells Phosphoric acid fuel cells were the first commercial fuel cells used at WRRFs. The phosphoric acid system is the most mature technology. At least 10 municipal WRRFs have installed 200-kW digester-gas phosphoric acid fuel cells, and some have more than 7 years of operating experience with this technology. So phosphoric acid fuel cells should be considered a proven technology, not a developing or experimental one.

Several of the initial digester-gas phosphoric acid fuel-cell installations at WRRFs are best characterized as developmental or experimental applications. These early units might not accurately characterize current fuel-cell offerings.

2.9.9.1.2 Carbonate Fuel Cells Many newer fuel-cell installations use the carbonate fuel cell (sometimes called the molten carbonate fuel cell or direct carbonate fuel cell) (see Table 23.24). Portions of the molten carbonate fuel cell (e.g., the reformer and the inverter) are similar to those in phosphoric acid fuel cells. One important difference is the lithium and potassium carbonate electrolyte solution, which allows electrons to transfer within the unit.

Like phosphoric acid fuel cells, a carbonate fuel cell is a mature technology with a proven track record. Design engineers should consider both technologies when evaluating digester-gas applications (see Table 23.25).

2.9.9.2 Fuel-Cell Components
A fuel cell consists of several main process modules: the gas-cleanup unit, reformer, cell stack, and inverter.

2.9.9.2.1 Gas-cleanup unit This module purifies digester gas or natural gas, removing all potential contaminants. Fuel cell stacks are exceptionally sensitive to certain impurities, so only exceptionally pure, clean, and pressurized methane gas leaves this module for the reformer.

Item or Parameter	Digester Gas	
	Range	Common Value
Methane (%) (dry basis)	50–70	60
Carbon dioxide (%) (dry basis)	30–45	39
Nitrogen (%) (dry basis)	0.2–2.5	0.5
Hydrogen (%) (dry basis)	0–0.5	0.2
Water vapor (%)	5.9–15.3	6
Hydrogen sulfide (ppm_v) (dry basis)	200–3500	500
Siloxanes, total (ppb_v)	100–4000	800
Ammonia (ppb_v)	100–2000	1000
Carbon disulfide (ppb)	200–900	500
Specific gravity (based on air = 1.0)	0.8–1.0	0.91
Higher heating value (HHV) (Btu/cu ft)	600–650	620
Lower heating value (LHV) (Btu/cu ft)	520–580	560

*All percentages listed above are percentages by volume
ppm_v = parts per million, by volume.
ppb_v = parts per billion, by volume.
As produced, mesophilic digester gas at 37°C (98°F) is saturated with water and contains 5.9% to 6% water vapor. Thermophilic digester gas at 55°C (131°F) contains about 15% water vapor. A fuel's higher heating value includes the heat of the water of vaporization.

TABLE 23.24 Characteristics of Typical Digester Gas*

2.9.9.2.2 Reformer This device combusts a tiny amount of fuel to produce steam. The reformer mixes this pressurized, high-temperature steam with pure methane from the gas-cleanup module to produce the hydrogen gas essential to fuel-cell operations.

2.9.9.2.3 Cell stack The cell stack uses hydrogen gas to produce electricity. Hydrogen gas and oxygen ions in the carbonate, or similar electrolyte, react to produce the constant flow of electrons needed to produce electricity.

2.9.9.2.4 Inverter The inverter consists of electrical devices that convert the direct current (DC) electric power created by the fuel cell into alternating current (AC) and transforms this AC power into the required system voltage.

2.9.9.3 Solid Oxide and Proton Exchange Membrane Fuel Cells Solid oxide fuel cells and PEM fuel cells have not yet been widely tested for long-term digester-gas use.

2.9.10 Emerging Technologies—Stirling Engines
A Stirling engine, invented in 1816, uses an external combustion process to convert heat into mechanical power. The manufacturer(s) claims that the engines require only limited fuel treatment and can operate on a low-energy fuel source (e.g., digester gas with less than 40% methane). They also have fewer emissions than reciprocating engines. Within the last decade, several 55-kW Stirling engines were installed, but the manufacturer is no longer in business.

There is a new generation of smaller-scale (7- to 10-kW) Stirling engines that provide a modular, scalable solution, especially for smaller facilities. While reliability shows promise due to the external combustion process and large-scale manufacturing, time will tell whether this technology is a cost-effective method of generating heat and electrical power from biogas.

2.9.11   Digester-Gas Use Technology and Heat Recovery

Anaerobic digesters require a constant, reliable supply of heat to ensure optimal biological activity in the reactors. So the first requirement of any digester-gas use technology is to reliably satisfy the WRRF's heating needs. This means providing:

- An adequate quantity of digester heat;
- A reliable heating source or heat-recovery system; and
- A consistent heating supply when used with the facility's heating water loop (typically a heating loop at 60°C to 80°C [140°F to 180°F]).

The type of heat and relative amount of recoverable heat is shown in Table 23.25.

2.9.11.1   Internal-Combustion Engine Heat Recovery
Large stationary internal-combustion engines sometimes are used to drive big air blowers, electric generators, and large pumps at WRRFs. Recovering heat from gas-fueled reciprocating engines is an established practice used at many WRRFs. Traditionally, engine-jacket water at 80°C to 110°C (180°F to 230°F) and heat from hot-engine exhaust gases (340°C to 540°C [650°F to 1000°F]) are common sources of recovered engine heat. Lower-temperature engine lubricating oil heat and turbocharger aftercooler heat at only about 50°C to 60°C (120°F to 140°F) typically are wasted to an air-cooled radiator or a water-cooled waste-heat exchanger.

The newer lean-burn reciprocating engines (e.g., ARES) incorporate two-stage aftercoolers or intercoolers. In these advanced gas-fueled engines, heat from the first turbocharger aftercooler is combined with the engine-jacket cooling-water system for better use in heat-recovery applications. This arrangement improves engine turbocharger performance and makes more of the engine's total heat available at a higher, more economically usable temperature.

Process or Component	Chemical Reaction
Within the internal steam reformer, the pure methane is converted to hydrogen gas via a reaction with steam.	$CH_4 + 2H_2O_{(steam)} \rightarrow 4H_2 + CO_2$
After the reformer, hydrogen combines with carbonate at the anode. The reaction produces water and carbon dioxide.	$H_2 + CO_3^= \rightarrow H_2O + CO_2 + 2e-$
The carbonate is the electrolyte media allowing electrons to flow from anode to cathode.	No chemical reaction
At the cathode, oxygen from the air completes the carbonate balance, and the flow of electrons is finished.	$2CO_2 + O_2 + 2e- \rightarrow 2CO_3^=$
At the electric power inverter, direct current is converted to alternating current at 480 V.	No chemical reaction

TABLE 23.25   How the Carbonate Fuel Cell Works

2.9.11.2   Fuel-Cell Heat Recovery   The chemical reactions in the current generation of fuel cells are exothermic, and they generate enough heat to vaporize the water chemically produced during the reactions. Excess fuel-cell heat is often captured and used productively.

Based on vendor performance data, a 1400-kW (1.4-MW) fuel-cell assembly produces about 8300 kg/h (18 300 lb/hr) of 370°C (700°F) exhaust consisting of steam and clean hot gases. When passed through a heat-recovery heat exchanger, this exhaust can produce about 2.1 million kJ/h (2.2 million Btu/hr) of fuel-cell heat while cooling to about 120°C (250°F). Although slightly more fuel-cell exhaust heat could be captured, it would only be about 50°C to 60°C (120°F to 140°F), which would be too cool for the WRRF's heating water loop and its digester heat exchangers. The 2.1 million kJ/h (2.2 million Btu/hr) of fuel-cell heat is enough to meet summer heating needs, but the heating boiler must be used for the rest of the year. The boiler is digester gas fueled, so for most of the year, a portion of the available digester gas must be diverted to operate the boiler, which supplements the fuel cell's heat-recovery process.

2.9.12   Air Emissions; Limits and Control Options, Greenhouse Gases

2.9.12.1   Criteria Pollutants   The traditional air pollutants of concern (criteria pollutants) from gas-combustion equipment are NO_x, carbon monoxide, SO_x, non-methane hydrocarbons (NMHC), and PM10:

- Nitrogen oxides (e.g., nitrogen dioxide, NO_2, and nitric oxide, NO) traditionally have been the most significant criteria pollutants. They are formed via combustion from nitrogen in the air. This group typically does not include nitrous oxide (N_2O).
- Carbon monoxide is formed via the partially complete combustion of methane (CH_4). Its emissions are controlled by combustion modifications.
- Sulfur oxides (e.g., sulfur dioxide) typically form when the hydrogen sulfide in digester gas combusts. It is controlled by eliminating hydrogen sulfide from the gas.
- Non-methane hydrocarbons typically are not found in significant quantities in digester gas.
- Particulate matter can include both PM10 and PM2.5. PM10 are particles larger than 10 mm, while PM2.5 are particles larger than 2.5 mm.

2.9.12.2   Greenhouse Gases   One project consideration that has become a significant public concern is the reduction of greenhouse gases (climate change emissions). There are three greenhouse gases of concern in digester-gas use evaluations: carbon dioxide (CO_2), methane (CH_4), and nitrous oxide (N_2O).

2.9.12.2.1   Carbon dioxide   Probably the best known greenhouse gas, carbon dioxide is a relatively heavy gas. Digester gas can contain as much as 40% carbon dioxide by volume or about 60% carbon dioxide on a per-weight basis. For most of the digester-gas-use processes under consideration, the carbon dioxide initially in digester gas passes through unreacted and unchanged.

Carbon dioxide is formed from the complete combustion of any fuel that contains carbon (e.g., methane). Any boiler, flare, incinerator, or power-generation technology that combusts methane will produce a corresponding predictable amount of carbon dioxide. The basic reaction for this chemical reaction is as follows:

$$CH_4 + 2O_2 \rightarrow CO_2 + 2H_2O \tag{23.24}$$

In other words, when 1 mole of methane is combusted completely, it will form exactly 1 mole of carbon dioxide. This conversion is essentially the same in a boiler, engine, gas turbine, or flare. Because methane has a molecular weight of 16 and carbon dioxide has a molecular weight of 44, each completely combusted kilogram of methane will produce 44/16 = 2.75 kg of carbon dioxide.

Biogenic carbon dioxide is carbon dioxide produced by life processes. It is not included in greenhouse gas inventories.

2.9.12.2.2   Methane   Methane is the principal component of both digester gas and natural gas. It is a light gas with a specific gravity of less than 1.0. Methane is an exceptionally important greenhouse gas; its global warming potential is 21 to 23 times that of carbon dioxide. From a greenhouse gas perspective, completely combusting all methane without atmospheric release is vital.

In fuel cells, methane gas first reacts with steam to produce hydrogen gas, as follows:

$$CH_4 + H_2O(steam) \rightarrow 3H_2 + CO \tag{23.25}$$

Then, the carbon monoxide (CO) combines with atmospheric oxygen (O_2) to produce carbon dioxide:

$$CO + \tfrac{1}{2}O_2 \rightarrow CO_2 \tag{23.26}$$

When totally combusted, 1 mole of methane produces 1 mole of carbon dioxide.

Low-NO_x boilers, lean-burn engines, and combustion gas turbines operate with an abundant amount of excess air (up to 70% or more) in their carefully controlled combustion chambers to ensure virtually complete oxidation of all methane in their fuel.

However, traditional waste-gas burners are not precisely controlled combustion devices, and a substantial portion of the methane in digester gas passes unburned through the flare. The amount flared is difficult to measure because the conditions in one part of the flame differ greatly from another, according to the wind direction. Getting an accurate and true sample is next to impossible.

Unburned methane is an exceptionally powerful greenhouse gas, so alternatives that flare some of the digester gas via a conventional waste-gas burner contribute much more greenhouse gas than alternatives that fully combust all the digester gas to generate electricity or that consume all the gas via other means.

Methane also is released to the atmosphere whenever digester gas is discharged from pressure-relief valves. Again, such discharges are difficult to measure; but fortunately, they are unusual.

2.9.12.2.3   Nitrous Oxide Emissions   Nitrous oxide indirectly serves as a greenhouse gas because it produces tropospheric ozone when its molecules break down. With a global warming potential of 310 (based on carbon dioxide = 1), nitrous oxide is particularly a concern, even in small quantities. It can be formed during methane combustion as an intermediate combustion by-product. Nitrous oxide emissions are a function of many complex combustion dynamics and combustion equipment. For example, higher combustion-zone temperatures destroy nitrous oxide.

Emissions factors for nitrous oxide are varied. Most gas equipment has little or no information about nitrous oxide emissions, partially because it typically is produced in tiny amounts during combustion. One common guideline (AP-42) lists emissions factors for nitrous oxide from natural gas combustion as 0.01 to 0.034 g/m^3 (0.64 to 2.2 lb/million cu ft). A nitrous oxide emissions factor of 0.1 kg nitrous oxide per tetrajoule (TJ) also is used for natural gas. Few actual, reliable nitrous oxide emissions factor data are available for digester gas, but they typically should be similar to natural gas emissions. Many combustion authorities do not consider nitrous oxide to be a component of traditional nitrogen oxides emissions.

2.9.12.3 Greenhouse Gases and Power-Generation Efficiency   Another metric is based on the amount of electricity produced per pound of carbon dioxide released. As expected, more energy-efficient power-generation technologies (e.g., fuel cells and ARES) are the leaders in this area among fuel-burning power generators. Also, the net electrical output—after subtracting auxiliary electrical loads and the digester gas used to fire a supplemental heating boiler—is much more important than the gross power-generation efficiency.

2.9.12.4 Digester-Gas-Use Greenhouse Gas Concerns   When selecting a digester-gas-use application, design engineers should address the following greenhouse gas emission concerns:

- The digester-gas-use application should be selected to avoid venting or indirectly releasing any gas because of methane's high greenhouse gas potential.
- Digester-gas-use applications should flare as little of the gas as possible due to the incomplete combustion characteristics of virtually all traditional waste-gas burners.
- The digester-gas-use application should be designed to produce—directly or indirectly—as little nitrous oxide as possible.
- In cogeneration applications, the gas-use technology should produce as much net usable electricity as possible to reduce the electric utility's carbon dioxide emissions.

2.10 Physical Facilities

The choice of anaerobic digestion tanks and equipment, and sometimes the configuration itself, is often affected by the physical space available, height restrictions, as well as cost and preference.

2.10.1 Tanks and Materials

The tanks, tank configuration, and system geometry involved depend on the situation. Pancake digesters, common in the United States, are relatively short and require the most land for a given volume. If land is limited, then silo or egg-shaped digesters may be more appropriate. However, such digesters are often expensive as well as more complex to design and build, and height restrictions could be an issue.

Digester tanks typically are made of steel-reinforced concrete that is either cast in place or post-tensioned. They often are designed to provide 40- to 50-year service lives, or even longer. Some tanks are made of bolted or welded steel. Most egg-shaped digesters in the United States have been made of steel but typically reinforced concrete in Europe.

Tank shape can influence digestion performance. For example, the egg-shaped has superior mixing characteristics, but silos may provide similar mixing performance at a reduced cost. The tank bottom, top slope, and tank configuration can be critical for good mixing and treatment performance. Tall tanks with a small free liquid surface will see a substantial liquid level increase during rapid volume expansion (RVE) events, so emergency overflows need to be sized adequately. Tanks with shallow floor slopes and limited means of grit removal can be expected to have more frequent need for shutdown and cleaning.

Design engineers should select the tank type, shape, bottom and top configuration, and construction materials and methods based on each system's criteria and needs. Such criteria include costs, available area, future expansion needs, desired life expectancy, specific digestion process and temperature regime, specific foundation and structural needs (i.e., seismic requirements), contractor and specialty firm availability, and schedule constraints.

2.10.2 Pumps and Piping

Pumps and piping systems should have enough clearance for staff to maintain the equipment, move equipment in and out, and allow for easy cleaning. Piping systems should have cleanouts at periodic intervals, with drains and associated flushing systems nearby. Hot-water flushing is particularly effective for adhesive materials such as grease.

2.10.3 Mixing Equipment

The physical considerations for mixers are process dependent. Pump-based systems should have enough space for pump maintenance and piping cleanout. When designing draft-tube mixers, engineers must ensure that there is enough space between and around digesters to allow a crane to remove and replace equipment. When designing gas-mixing systems, engineers need to consider pipe materials and routes and be aware of confined and classified space requirements.

2.10.4 Heating and Heat-Transfer Equipment

The primary issue for all heat exchangers is cleaning and maintenance. The frequency depends on the type of heat exchanger used, the solids being conveyed, and the specific operating conditions. An effective design will include convenient wash stations, drains, and ample space to clean the heat exchanger. This is particularly critical for tube-in-tube heat exchangers, which often require clearance at one end to remove the pipes.

2.10.5 Cleaning and Safety

The decision to clean a digester tank is based on several factors: the degree to which grit and scum accumulation has reduced the digester's effective volume, the condition of internal heating and mixing equipment, and the availability of alternate solids-handling equipment. The tanks, mixers, and heaters should be designed for easy access during cleanout operations. At a minimum, there should be access manholes on the top and sides of the digester. The manholes should be at least 0.9 m (36 in.) in diameter, or large enough to enable an operator to use grit- and scum-removal equipment.

Heating and mixing equipment must be maintained throughout the life of the digester, so ideally, most of the critical equipment should be outside the tanks. However, interior equipment that can be removed during digester operations is often satisfactory. Digesters can be cleaned by in-house staff or by a contractor that specializes in such services. They typically are cleaned every 5 years, but this is only a general guideline—the timing should be based on the facility's specific situation that may depend on grit/scum accumulation rates, coating and equipment inspection intervals, or overall digester performance.

Safety is of primary importance during digester cleaning. All systems must be designed so that they can be made safe for O&M personnel to perform their work, often in permit-required confined spaces. Design engineers also need to specify the appropriate personal safety equipment needed when entering the tank for inspection and cleanout. Several pieces of equipment are available to perform safe cleaning operations (see Table 23.26); the items needed depend on the size of the operation.

Other safety equipment should be included to prevent falls, infection, and injuries during system operations and to comply with any safety regulations.

The gas-collection and -piping system design must include vacuum- and pressure-relief valves, flame arrestors, and automatic thermal-shutoff valves where appropriate. Biogas safety also includes protecting against suffocation, asphyxiation, and explosions, so proper enclosures and ventilation must be provided.

For more information on safety features, see *National Electric Code* (NFPA, 2017), *Standard for Fire Protection in Wastewater Treatment and Collection Facilities* (NFPA, 2016), *Recommended Standards for Wastewater Facilities* (Great Lakes), and *Safety, Health, and Security in Wastewater Systems* (WEF, 2012).

Item	Phosphoric Aid	Molten Carbonate	Remarks
Representative fuel cell manufacturer	United Technology Corporation (UTC)	Fuel Cell Energy	UTC purchased ONSI fuel cell business
Modular sizes, electrical output	200 kW	300-kW and 1200- kW units	FCE recently increased their unit capacity
Electrical efficiency, percent	36–40	45–47	New equipment performance, typical
Operating electrolyte temperature (°F)	375	1200	Operating temperature affects start-up time
Recoverable heat output at 180°F to 200°F (Btu/hr)	2.0 million (for units totaling 1 MW)	1.49 million (for 1.2 MW unit)	Per manufacturer's performance claims
Expected performance degradation (%)	About 2% yearly capacity decrease	2% to 3% yearly capacity decrease	Capacity drops off as the cell stack fouls
Water consumption, Continuous (gpm)	1.6 gpm for 1-MW unit	2 gpm for 1.2-MW unit	Water is used to make steam in the reformer
Required digester-gas fuel pressure (psi)	20–30	About 25	Gas compressors are required
Supplemental natural gas to the digester gas	Not required	Recommends 10% natural gas	Strongly recommended by fuel cell supplier

*0.5556 (°F − 32) = °C; Btu/hr × 0.2931 = W; gpm × 5.451 = m³/d; psi × 6.895 = kPa.

TABLE 23.26 Comparison of Two Types of Digester-Gas Fuel Cells*

3.0 Aerobic Digestion

Stabilization during aerobic digestion occurs from the destruction of degradable organic components and the reduction of pathogens by aerobic, biological mechanisms. Aerobic digestion is a suspended-growth biological treatment process based on biological theories similar to those of the extended aeration modification of the activated sludge process. The objectives of aerobic digestion, which can be compared to those of anaerobic digestion, include producing a stable biosolids product, reducing mass and volume, reducing pathogens, and conditioning solids for further processing.

Advantages of the aerobic process compared to anaerobic digestion are the production of an inoffensive, biologically stable product; lower capital costs; simpler operational control; safer operation with no potential for gas explosion and less potential for odor problems; and discharge of a supernatant with a 5-day BOD (BOD_5) concentration typically less than that found in the anaerobic process. There is also some evidence that aerobic digestion is helpful in reducing some antibiotics (Burch et al., 2013). In addition, it is less prone to upsets and less susceptible to toxicity.

The primary disadvantage typically attributed to aerobic digestion is the higher power cost associated with oxygen transfer. Recent developments in aerobic digestion (e.g., highly efficient oxygen-transfer equipment and research into operation at elevated temperatures) may reduce this concern. Other disadvantages cited include the process' reduced efficiency during cold weather; its inability to produce a useful by-product, such as methane gas; and the mixed results achieved during mechanical dewatering of aerobically digested solids.

Conventional, or mesophilic, aerobic digestion has been used for more than 60 years. The autothermal thermophilic aerobic digestion (ATAD) process modification has been used since the late 1970s in Europe, and is in its second generation of development in the United States. This section will also discuss methods of optimizing mesophilic aerobic digestion, based on recent research.

3.1 Process Applications

Aerobic digestion has been used successfully in facilities with capacities up to 1.89×10^5 m³/d (50 mgd), but is most common in facilities with design capacities of less than 19 000 m³/d (5 mgd). It has been used successfully in extended aeration and nutrient-removal activated sludge facilities, both with and without primary settling, and in many package treatment facilities. In larger facilities, mixed primary and biological solids are most often handled, and their oxygen requirements are greater than those of waste biological solids alone. Because of the high energy cost for aeration in such facilities, and the potential for resource recovery, it may be more economical to anaerobically digest primary solids separately while aerobically digesting biological solids.

In cases where other disposal methods are not readily available, grease and skimmings can be treated in aerobic digesters. These streams should have a low inorganic content and pass through grinders or screens before they are added to the aerobic digestion system. Even with thorough grinding, recombination of stringy material, resulting in clogging or fouling of aeration equipment, is a potential problem. Grease and skimmings are likely to accumulate as digester scum unless special provisions are made to keep this material in suspension (e.g., relatively intense surface mixing).

3.2 Process Theory

Aerobic digestion is based on the biological principle of endogenous respiration. Endogenous respiration occurs when the supply of available substrate (food) is depleted and microorganisms begin to consume their own protoplasm to obtain energy for cell-maintenance reactions.

Aerobic digestion actually involves two steps: direct oxidation of biodegradable matter and subsequent oxidation of microbial cellular material by organisms. These processes are illustrated by the following formulas (U.S. EPA, 1979):

$$\text{Organic matter} + NH_4 + O_2 \xrightarrow{\text{bacteria}} \text{Cellular material} + CO_2 + H_2O \quad (23.27)$$

$$\text{Cellular material} + O_2 \xrightarrow{\text{bacteria}} \text{Digested biosolids} + CO_2 + H_2O + NO_3 \quad (23.28)$$

Equation 23.27 describes the oxidation of organic matter to cellular material, which then is oxidized to digested biosolids. The process represented by eq 23.28 is typical of endogenous respiration and is the predominant reaction in aerobic digesters.

As shown in eq 23.28, during digestion, cell tissue is oxidized aerobically. Because aerobic oxidation is exothermic, heat is released during the process. Although digestion should theoretically go to completion given an infinite SRT, in actuality only 75% to 80% of cell tissue is oxidized. The remaining 20% to 25% is composed of refractory solids and organic compounds that are not biodegradable via aerobic digestion (Park et al., 2006). The material that remains after digestion is complete exists at such a low energy state that it is essentially biologically stable. So it is suitable for a variety of disposal options.

To optimize aerobic digestion and maintain the process in the endogenous respiration phase, aerobic digestion typically is used to stabilize WAS, where the organic matter has already been oxidized to biomass (eq 23.27). Because primary solids contain little cellular material, most of the organic and particulate material in primary solids is an external food source for the active biomass in biological solids. So longer retention times are required to accommodate the metabolism and cellular growth that must occur before endogenous respiration conditions are achieved.

Using the formula $C_5H_7NO_2$ as representative of a microorganism's cell mass, the stoichiometry of aerobic digestion can be represented by either of the following equations:

$$C_5H_7NO_2 + 5O_2 \rightarrow 5CO_2 + 2H_2O + NH_3 + \text{Energy} \quad (23.29)$$

$$C_5H_7NO_2 + 7O_2 \rightarrow 5CO_2 + 3H_2O + NO_3 + H^+ + \text{Energy} \quad (23.30)$$

Equation 23.29 represents a system designed to inhibit nitrification (because it is oxygen limited); nitrogen appears in the form of ammonia. The stoichiometry of a system in which nitrification occurs is represented by eq 23.30, where nitrogen appears in the form of nitrates.

As indicated by eq 23.30, nitrification during aerobic digestion increases the concentration of hydrogen ions and subsequently decreases pH if the solids have insufficient buffering capacity. As in the activated sludge process, about 7 kg of alkalinity is destroyed per kilogram of ammonia oxidized (7 lb/lb). The pH may drop as low as 5.5 during long aeration times, but aerobic digestion does not seem to be adversely affected.

If excessive pH depression is a problem (as a result of alkalinity consumption by nitrification), it may be possible to control this problem through alkaline addition (e.g., lime) or periodic denitrification. Theoretically, about 50% of the alkalinity consumed by nitrification can be recovered by denitrification in an aerobic-anoxic cycle. This is discussed further in Section 3.6.3.

Equations 23.29 and 23.30 indicate that, theoretically, 1.5 kg of oxygen is required per kilogram of active cell mass (1.5 lb O_2/lb) in the non-nitrifying system, while 2 kg of oxygen per kilogram of active cell mass (2 lb O_2/lb) is required when nitrification occurs. Actual oxygen requirements for aerobic digestion depend on such factors as operating temperature, inclusion of primary solids, and the SRT in the activated sludge system.

3.3   Process Fundamentals

3.3.1   General

This discussion on the design of aerobic digestion systems is directed toward conventional aerobic systems (i.e., systems operating at temperatures between 20°C and 30°C that use air as the oxygen source for biological activity). Thermophilic aerobic digestion systems are discussed later in this chapter. Cryophilic (low-temperature) aerobic digestion systems have also been used in some situations, but are covered in other references (Koers and Mavinic, 1977; WEF et al., 2012).

Factors that govern the design of aerobic digestion systems include feedstock characteristics, desired reduction in volatile solids, process operating temperature, oxygen-transfer and mixing requirements, tank volume/detention time, and method of system operation. However, since the U.S. EPA's 40 CFR 503 regulations (typically called Part 503 or the 503 regulations) went into effect, the overriding factor in the design of these systems has been meeting the requirements for vector attraction and pathogen reduction.

3.3.2   Feedstock Characteristics

As noted previously, aerobic digestion typically is used to stabilize biological solids (e.g., WAS). When the process is used to stabilize mixtures of primary and biological solids, the oxygenation requirements are increased by the additional demand of oxidizing the organic matter in primary solids to biomatter. In addition, longer detention times are required for these mixtures, as compared to a pure WAS feedstock. Because the mechanism of aerobic digestion is similar to that of the activated sludge process, the same concerns regarding variations in influent characteristics and levels of biologically toxic materials apply, although a dampening effect will occur as a result of upstream treatment processes.

The heavy metals accumulated in activated solids via precipitation and adsorption (which can occur when pH is greater than 7) can resolubilize under low-pH conditions in the digester, resulting in toxicity.

The influent concentration is important in the design and operation of an aerobic digestion process. Sludge digestion at concentrations of 1% to 2% allows the use of conventional equipment for mixing and aeration, but requires a greater volume to achieve treatment objectives. Thickening solids to higher concentrations will reduce oxygen-transfer efficiency, but will result in smaller digester volume requirements, and potentially more volatile solids reduction (VSR).

3.3.3 Reduction in Volatile Solids

The primary purpose of aerobic digestion is to produce biosolids that are stabilized and amenable to various disposal options. Here, stabilized implies that the biological organisms, particularly pathogens, have been reduced to a level at which the use or disposal of the biosolids will not result in a significant adverse environmental impact.

Depending on feedstock characteristics, aerobic digestion can reduce volatile solids by 35% to 50%. Part 503 regulations require that a 38% VSR across the digester be met to attain the VAR requirements. In facilities with shorter sludge ages, and more volatile solids in the digester feed, this requirement can be achieved. On the other hand, if significant biodegradable solids reduction has already occurred in a secondary treatment system with a long SRT, it can be difficult to achieve this 38% reduction. Prior to the implementation of the 503 regulations, credit was sometimes given by regulators for the VSR that occurred in extended aeration facilities. Therefore, the overall VSR took into account the reduction through both the activated sludge and solids-handling processes.

With the promulgation of Part 503, however, this option is no longer available. If the aerobic digester does not achieve 38% reduction, the regulations provide other alternatives to demonstrate compliance with the VAR criteria. Although some of these alternatives are specific to other stabilization methods or pertain to land application practices, there are methods that can demonstrate compliance in the WRRF. Compliance can be demonstrated by aerobically digesting a portion of the previously digested material that has a solids concentration of 2% or less in the laboratory in a bench-scale unit for 30 more days at 20°C. If the sample's volatile solids concentration is reduced by less than 15%, VAR is achieved. Another alternative from the Part 503 is to meet the specific oxygen uptake rate (SOUR) criteria of less than 1.5 mg of oxygen per gram per hour total solids at a temperature of 20°C for digested solids instead of the 38% VSR criterion.

Researchers have suggested that VSR may not be a valid indication of stabilization (Hartman et al., 1979; Matsch and Drnevich, 1977). Other parameters (e.g., the residual rate of oxygen demand, pathogen levels, odor-producing potential, SOUR, or oxidation/reduction potential) may be more indicative of stabilized, aerobically digested biosolids. Therefore, these alternative tests may be valuable to demonstrate stabilization of solids from long sludge-age secondary treatment processes.

3.3.4 Temperature

The aerobic digester's operating temperature is a critical parameter. Conventional aerobic digestion systems are operated in the mesophilic zone of bacterial action (between 10°C and 40°C). Operating temperatures are typically closely related to ambient temperatures because most aerobic digestion systems use open tanks.

A frequently cited disadvantage of the aerobic process is the variation in process efficiency that results from changes in operating temperature. Because aerobic digestion is a biological process, the effects of temperature can be estimated by the following equation:

$$(K_d)_T = (K_d)_{20°C}\, q^{T-20} \tag{23.31}$$

where K_d = reaction rate constant (time);

q = temperature coefficient; and

T = temperature (°C).

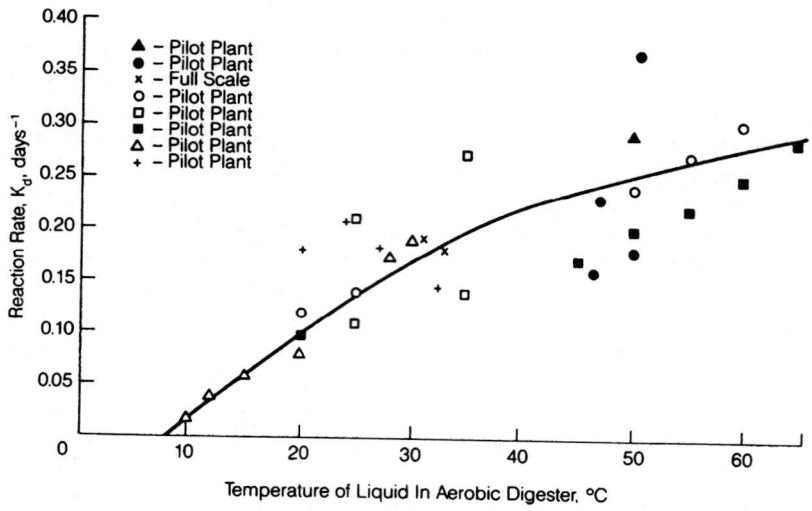

Figure 23.40 Experimentally determined reaction rate (K_g) versus aerobic digester liquid temperature. The value of K_d depends on solids characteristics and digester operating conditions (e.g., pH, TSS, and oxygen level) (U.S. EPA, 1978).

The reaction rate constant indicates the destruction rate of volatile solids during digestion. Temperature coefficients ranging from 1.02 to 1.10 have been reported. A study by Krishnamoorthy and Loehr (1989) on WAS and mixed solids found a value of 1.092. As can be seen from eq 23.31, the reaction rate constant increases when the system's temperature increases, implying an increase in the digestion rate. This is shown graphically in Figure 23.40. However, above a critical temperature, the process will be inhibited. One study showed a maximum rate at 30°C, with a rate reduction at higher temperatures (Hartman et al., 1979). This is in contrast with the data in Figure 23.40 and indicates the importance of obtaining rate data applicable to the system being designed.

3.3.5   Oxygen-Transfer and Mixing Requirements

The biological reaction that occurs during aerobic digestion requires oxygen for the respiration of cellular material in activated sludge and, in the case of mixtures with primary solids, the oxygen needed to convert organic matter to cellular material. In addition, proper system operations require adequate mixing of the contents to ensure contact of oxygen, cellular material, and organic matter (food source). Because introducing oxygen to maintain the biological process typically mixes the contents in the process, these parameters are interrelated.

In aerobic digestion systems that strictly treat biological solids with a total solids concentration in the range of 1% to 2%, the need for adequate mixing typically will govern the capacity of the oxygenation equipment. Systems treating primary and biological mixtures require more oxygen for the biological oxidation process, and in most cases, this requirement will govern aeration equipment sizing. For designs with a higher total solids concentration (3% to 6%), process air requirements usually control aeration equipment sizing and design.

Actual mixing requirements typically range from 10 to 100 W/m³ (0.5 to 4.0 hp/1000 cu ft) of digester volume; however, this value will vary depending on tank geometry and type of mixing device. If using mechanical mixing, design engineers should consult an experienced equipment manufacturer to determine actual mixing requirements.

As noted above, aerobic digestion theoretically requires 1.5 to 2.0 parts of oxygen per part of applied organic cell mass, depending on whether nitrification is inhibited or allowed to proceed. Design experience has shown that 2.0 parts of oxygen per part of organic cell mass destroyed is a standard minimum value for biological stabilization. Including primary solids in the digestion process requires another 1.6 to 1.9 parts of oxygen per part of volatile solids destroyed to convert the organic matter to cell tissue and satisfy the endogenous demand of the resulting cell mass. Some U.S. state regulators (e.g., Wisconsin) stipulate in their design standards for WRRFs that the aeration system should account for another 0.91 kg (2 lb) of oxygen per pound of BOD_5 applied by primary solids in the aerobic digestion design.

Oxygen requirements for aerobic digestion systems typically represent airflow rates of 0.25 to 0.33 L/m³·s (15 to 20 cu ft/min/1000 cu ft) for WAS. Airflow rates increase to a range of 0.40 to 0.50 L/m³·s (25 to 30 cu ft/min/1000 cu ft) for a mixture of primary solids and WAS (Benefield and Randall, 1980). These airflow rates are valid for solids concentrations between 1% and 2% and will increase with increased solids content. Dissolved oxygen levels in the aerobic digester typically are maintained at about 2 mg/L; however, this level may be reduced if the oxygen uptake rate is less than 20 mg/L·h.

After the requirements for adequate mixing and oxygen transfer have been separately computed, the larger of the two requirements will govern overall system design. If the mixing requirement exceeds the oxygen-transfer requirement, design engineers should consider providing supplemental mechanical mixing rather than overdesigning the oxygen-transfer system. The increased capital cost of supplemental mechanical mixers must be balanced against the power and maintenance costs of more aeration to determine the optimum configuration.

3.3.6 Detention Time and Tank Volume Requirements

Prior to the implementation of the Part 503 regulations, the required volume of an aerobic digester was typically governed by the detention time needed to achieve a desired reduction in volatile solids. In other cases, as in the Ten States Standards (Great Lakes Upper Mississippi River Board of State Sanitary Engineering Health Education Services Inc., 2014), the volume was based on a per capita loading.

The Part 503 regulations have two requirements, either of which may govern the required detention time, and thus the aerobic digester volume. The regulations require compliance with both VAR criteria and pathogen reduction requirements. Typically, the pathogen reduction requirements will govern. To comply with EPA requirements for aerobic digestion as a PSRP requires a residence time of 60 days at 15°C and 40 days at 20°C.

Full-scale aerobic digestion studies have shown that a total aeration time (including time in the extended aeration process) of 35 to 50 days was required to consistently meet Part 503's VAR requirement for a SOUR of less than 1.5 mg oxygen/g·h volatile solids at WRRFs located at higher altitudes with colder climates (Maxwell et al., 1992). The total required aeration time depended significantly on operating temperature and WAS biodegradability. Although VSR will continue as detention time increases, the oxidation rate significantly decreases, and continuing digestion past the typical detention time is not economical. It has been reported that retention time in mesophilic aerobic digesters has a significant impact on dewaterability (Zhou et al., 2001).

The reduction in biodegradable solids during digestion typically is described by a first-order biochemical reaction at constant volume conditions, similar to the following:

$$dM/dt = K_d M \qquad (23.32)$$

where dM/dt = rate of change of biodegradable volatile solids per unit of time (mass/time);

K_d = reaction rate constant (time^{-1}); and

M = concentration of biodegradable volatile solids remaining at time t (mass/volume).

The time factor in eq 23.32 represents the SRT in the aerobic digester. Research (Krishnamoorthy and Loehr, 1989) has found that first-order kinetics adequately represent digestion of WAS only and WAS/primary sludge mixtures, but must be modified for digestion of primary solids.

Such factors as the method of digester operation, solids concentration, operating temperature, and the SRT of the activated sludge system may affect the rate constant and make the time factor equal to or greater than the system's theoretical HRT. Figure 23.40 depicts a graph of the change in reaction rate constant versus increasing operating temperature. Other studies have shown that the reaction rate constant declines as the digester's suspended solids level increases (D'Antonio, 1983) (Figure 23.41). Given these variables, several methods of calculating SRT have been derived and remain in use.

The product of temperature and SRT appears to correlate with the percentage of VSR that can be achieved during digestion. Figure 23.42 (U.S. EPA 1979) shows the effect of the temperature-SRT product on VSR. The selection of a desired percentage of

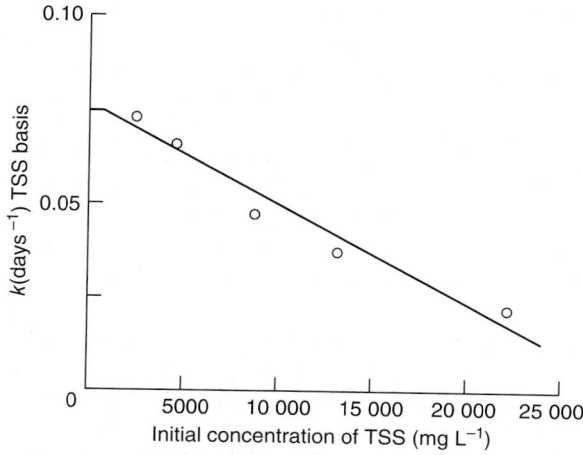

FIGURE 23.41   Impact of solids concentration on reaction rate (d'Antonio, 1983).

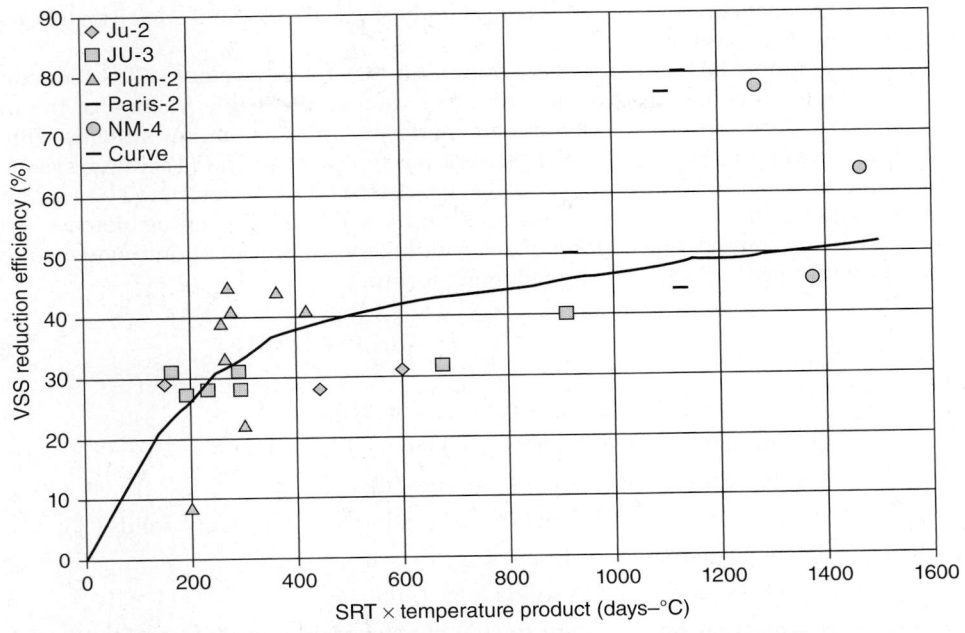

FIGURE 23.42 Volatile solids reduction as a function of digester liquid temperature and solids retention time—update of U.S. EPA curve (Daigger et al., 1999).

reduction in volatile solids, coupled with an assumed operating temperature for the system, can be used to estimate the required digester SRT.

Benefield and Randall (1980) developed equations for determining required digester detention times. These equations resulted from an analysis of the kinetics associated with the digestion process, and the understanding that a portion of the volatile solids in the process are refractory and a portion of the nonvolatile solids is solubilized from microbial cells contained within the solids. The basic equation is as follows:

$$SRT = (X_i - X_e)/(K_d)(D)(X_{oad})(X_i) \qquad (23.33)$$

where SRT = digester detention time (days);

$\quad X_i$ = TSS concentration in influent (mg/L);

$\quad X_e$ = TSS concentration in effluent (mg/L);

$\quad K_d$ = reaction rate constant for the biodegradable portion of the active biomass (d^{-1});

$\quad D$ = biodegradable active biomass in the influent that appears in the effluent (%); and

$\quad X_{oad}$ = percentage of active biomass that is biodegradable in the influent.

The relationships shown in eq 23.33 are further defined and stated in the Activated Sludge Model 1 (ASM1) terminology by Grady et al. (2011).

In instances in which there is a mixture of primary solids and WAS, the equation should be modified with the inclusion of factors describing the primary solids component.

Refer to Benefield and Randall (1980) or Grady et al. (2011) for more detail on this less common aerobic digestion design.

Equation 23.33 supports the assumption that for equivalent solids reduction and constant solids loading, SRT must be increased as the active fraction of the influent biomass decreases. Actual operating experience indicates that for systems with a low fraction of active biomass in the feed, which is typical of extended aeration systems, this trend does not hold. For these systems, the detention times computed by eq 23.33 may be reduced in proportion to the decrease in the active fraction of the biomass. After the SRT is determined, one method of determining the volume of a continuously operating aerobic digester is to apply the following formula:

$$V = Q_i(X_i + YS_i)/X\,(K_dP_v + 1/\text{SRT}) \qquad (23.34)$$

where V = volume of the aerobic digester (L [cu ft]);

Q_i = influent average flowrate (L/d [cu ft/d]);

X_i = TSS concentration in influent (mg/L);

Y = portion of the influent BOD consisting of raw primary solids (%);

S_i = BOD_5 in digester influent (mg/L);

X = digester suspended solids (mg/L);

K_d = reaction rate constant for the biodegradable portion of the active biomass (d^{-1});

P_v = volatile fraction of digester suspended solids (%); and

SRT = solids retention time (days).

The term YS_i in eq 23.34 can be disregarded if no primary solids are included in the load to the aerobic digester. Equation 23.34 should not be used to compute digester volumes in systems where significant nitrification will occur.

3.3.7 Supernatant Quality of Recycled Sidestreams

As noted previously, one of the benefits of aerobic digestion is that the sidestream from the process typically has less effect on facility loadings. On the other hand, depending on how the digesters are operated, there can be significant release of both nitrogen and phosphorus in the supernatant. Careful monitoring of solids-liquid separation in continuous and batch-feed digesters and proper operation both increases aerobic digester performance and reduces the recycle loadings to the facility influent. Table 23.27 lists "acceptable" supernatant values from aerobic digestion processes (Metcalf and Eddy, Inc./AECOM, 2013).

Although the Metcalf and Eddy, Inc./AECOM data are widely used, other data developed in additional studies provide some additional guidelines in managing sidestream loadings. Jenkins and Mavinic (1989) found the following results when comparing aerobic and aerobic/anoxic digestion.

- Aerobic digestion produces high levels of nitrates (>200 mg/L) in the supernatant, with some digesters with longer detention times also producing nitrites and ammonia.

Parameter	Combined Heat and Power Technology				
	Microturbine	**Fuel Cells**	**Combustion Gas Turbines**	**Advanced Internal-Combustion Engines**	**Steam Boilers + Steam Turbines**
Forms of heat energy available from this technology	Hot water	Hot water	Low-pressure steam	Hot water	Low-pressure steam
	Hot exhaust gases	Hot exhaust gases	High-pressure steam and/or hot exhaust gases	Some low-pressure steam	High-pressure steam and/or hot exhaust gases
Portion of the biogas fuel energy available as heat energy		20–25%	40–50%	35–45%	45–55%

TABLE **23.27**   Heat Recovery from Cogeneration Systems

- Aerobic/anoxic digestion produced low levels of nitrates (<2 mg/L) and also produced little ammonia.
- Aerobic digestion resulted in the release of significant amounts of both ortho- and total phosphorus (>100 mg/L), while aerobic/anoxic digestion produced concentrations of 42 and 46 mg/L, respectively.
- COD in the aerobic/anoxic system supernatant was less than one-half (41 mg/L) that of the aerobic systems (118 mg/L).

This study indicates that the key steps to improving supernatant quality are:

- Operation in aerobic-anoxic mode to promote nitrogen removal. As shown in Figure 23.43, Al-Ghusain (Al-Ghusain et al., 2004) found that approximately 8 hours of anoxic time reduced the total nitrogen significantly.
- Maintaining necessary quantities of carbon source for denitrification.
- Maintaining a neutral pH of 7.0 to enhance nitrification and denitrification.

For more information, refer to Figures 23.43 and 23.44 on the effect of both anoxic-cycle length and temperature on total nitrogen levels in filtrate.

Table 23.28 shows data from an installation in Stockbridge, Georgia, that uses a gravity thickener in loop with an aerobic digester to allow nitrification to occur in the digester and denitrification and recovery to occur in the thickener (Stege and Bailey, 2003). Thickener blanket data indicate good settling. Average data for TSS, ammonium, and phosphorus indicate both nitrification and denitrification, and phosphorus removal with TSS removal. The solids blanket during the collection of these data points varied between 2.4 and 4.1 m (8.0 and 13.5 ft).

3.3.8   Summary of Design Parameters

The implementation of 40 CFR 503 has substantially changed the typical design parameters used for the standard aerobic digestion process. Although shorter detention times

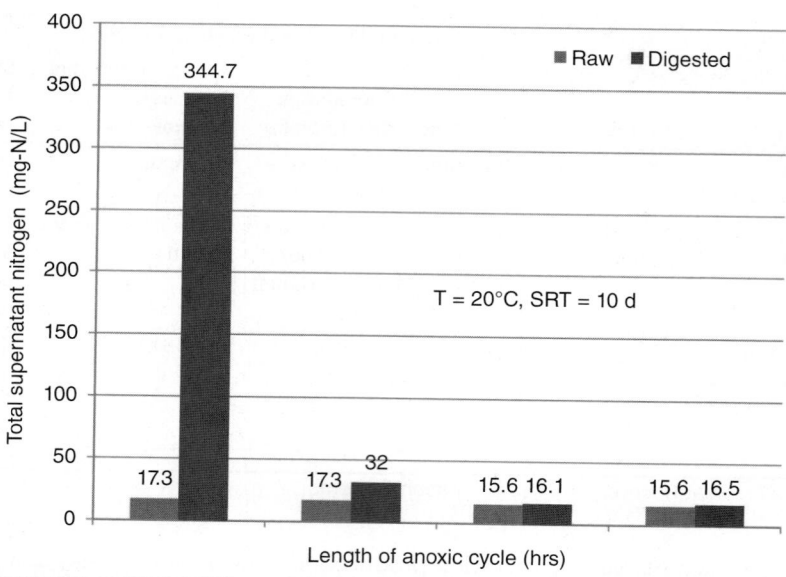

FIGURE 23.43   Effect of anoxic-cycle duration on total nitrogen concentration in filtrate (Al-Ghusain et al., 2004).

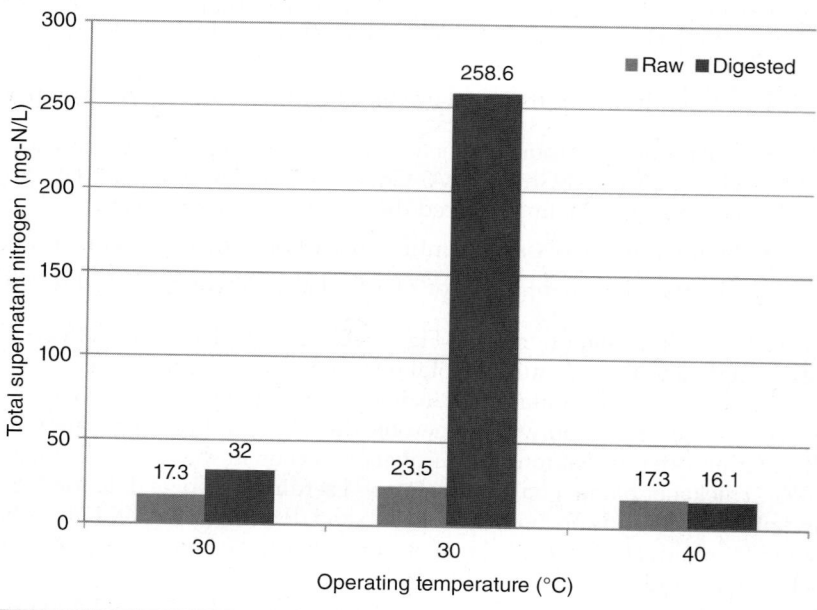

FIGURE 23.44   Effect of digestion temperature on total nitrogen concentration in filtrate (Al-Ghusain et al., 2004).

Monitoring Parameter*	Frequency	Operating Range		
		Minimum	Nominal	Maximum
Temperature (°C)	Daily	15	20	37
pH	Daily	6.0	7.0	7.6
Dissolved oxygen (mg/L)	Daily	0.1	0.4 to 0.8	2.0
Alkalinity (mg/L as CaCO₃)	Weekly	100	>500	—
Ammonia-nitrogen (mg/L)	Weekly	—	<20	40
Nitrate (mg/L)	Weekly	—	<20	—
Nitrite (mg/L)	As required	—	<10	—
SOUR (mg oxygen/h/g of total solids)	As required	—	<1.5	—
Phosphorus (mg/L)	As required		<5	

*$CaCO_3$ = calcium carbonate and SOUR = specific oxygen uptake rate.

TABLE 23.28   Monitoring Parameters for Aerobic Digestion Performance (Adapted from WEF, 2007; Primary source is Stege and Bailey, 2003)

may meet VSR requirements, the U.S. EPA requirements for pathogen reduction and the specific time-temperature guidelines for aerobic digestion as a PSRP typically result in a design residence time of 40 days at a temperature of 20°C and 60 days at 15°C. For temperatures between these two extremes, the relationship that the EPA requires (U.S. EPA, 2003) is:

$$\frac{\text{Time @ } T \ (°C)}{40 \ \text{days}} = 1.08(20 - T(°C)) \qquad (23.35)$$

If design engineers use aerobic digestion for stabilization and do not meet the specified time-temperature requirements, then the facility operator must monitor fecal coliform and demonstrate compliance with the PSRP requirements through testing. In both cases, design engineers must demonstrate that the VAR criteria have been achieved by a 38% reduction in volatile solids, an oxygen uptake rate of less than 1.5 mg of oxygen per gram of total solids per hour, or a VSR of less than 15% on further testing after 30 days of additional aerobic digestion. Alternately, the VAR criteria may be met by managing land application practices.

An important deviation from this rule is obtained in the design of two-stage or batch operation. The EPA's guidance manual on control of pathogens (U.S. EPA, 2003) recognizes the higher efficiency of two-stage or batch systems and allows a 30% reduction in the time required to obtain the pathogen reduction criteria specified by the U.S. EPA regulations. Acceptance of this EPA guidance varies with state regulatory agencies, but utilizing this credit reduces the time required to achieve PSRP standards from 40 days to 28 days at 20°C (68°F), and from 60 days at 15°C (59°F) to 42 days at 15°C (59°F).

The Class B biosolids criteria that a conventional mesophilic aerobic digestion system typically is designed to meet can be summarized as follows. A system with one digester, or with multiple digesters operating in parallel, would be designed to meet the following criteria.

1. Meet one of the following pathogen-reduction requirements:
   - 60-day SRT at 15°C or 40-day SRT at 20°C, or
   - Fecal coliform density of less than 2 million most probable number (MPN)/g total dry solids.

2. Meet one of the following VAR requirements:
   - At least 38% VSR during biosolids treatment, or
   - A SOUR of less than 1.5 mg/g·h of total solids at 20°C (68°F), or
   - Less than 15% additional VSR after 30 days of further batch digestion at 20°C (68°F).

A system with multiple digesters in series or true batch configuration could be designed to meet the following criteria (U.S. EPA, 2003), taking advantage of the credit offered in the EPA document. However, the acceptability of this credit should be verified with local regulators.

1. Meet one of the following pathogen-reduction requirements:
   - Fecal coliform density of less than 2 million MPN/g total dry solids, or
   - 42-day SRT at 15°C or 28-day SRT at 20°C.

2. Meet one of the following VAR requirements:
   - At least 38% VSR during biosolids treatment, or
   - A SOUR of less than 1.5 mg/g·h of total solids at 20°C (68°F), or
   - Less than 15% additional VSR after 30 days of further batch digestion at 20°C (68°F).

3.4 Conventional (Mesophilic) Aerobic Digestion

Conventional aerobic digestion is a simple treatment process that may be used to treat WAS, mixtures of WAS or trickling filter solids and primary solids, waste solids from extended aeration facilities, or waste solids from membrane bioreactors (MBRs). Aerobic digestion is most efficient at treating solids that are primarily biological mass from wastewater treatment. Aerobically digested biosolids are less likely to generate odors and have fewer bacteriological hazards than unstabilized solids.

3.4.1 Process Design

The design of conventional aerobic digestion facilities is based on the principles described in Section 3.3.

3.4.1.1 VSR and Solids Reduction   Aerobic digestion destroys VSS. Primary solids and WAS from a system with a short SRT will contain relatively high fractions of biodegradable material, as opposed to WAS from a system with a long SRT, which will contain a low fraction of biodegradable material and a high fraction of biomass debris (Grady et al., 2011).

In a study performed to determine the minimum SRT required to meet Class B requirements, by meeting both pathogens and VSR operating at minimum temperatures as concluded by Lu Kwang Ju (Daigger et al., 1999), two systems were evaluated. One system included two basins in series and the other three basins in series. Table 23.29 shows VSRs at different temperatures and SRTs (Daigger et al., 1999).

Parameters	Range	Typical
Number of reactors	2–3	2
Prethickened solids	4–6%	4%
Reactors in series		Yes
Total HRT* in reactors	4–30 days	6–8 days
Temperature—stage 1	35–60°C	40°C
Temperature—stage 2	50–70°C	55°C

*HRT = hydraulic retention time.

TABLE 23.29   Recommended Design Parameters for ATAD Digester Systems (Stensel and Coleman, 2000)

The solids used in this study had a very low digestible fraction. It was concluded that, even though all systems met the pathogen-destruction requirement, an SRT of 29 days was required to exceed the required 38% VSR. This confirms that VSR depends on the source of the solids, and, if the solids have a low fraction of digestible organic content, it is difficult to meet the U.S. EPA's minimum requirements.

3.4.1.2   Solids Retention Time-Temperature Product   A significant factor in the effective operation of aerobic digesters, SRT is the total mass of biological solids in the reactor divided by the average mass of solids removed from the process each day. Typically, an increase in SRT increases VSR.

Based on the discussion above, with respect to temperature, degradable solids, non-degradable solids, and the effect on VSR, the SRT-temperature product (days · °C) curve can be used to design digester systems, taking into consideration not only the total days · °C that the digester system will have to meet but also the quality of the source. The original U.S. EPA SRT-temperature curve (Figure 23.45), developed in the late 1970s (U.S. EPA, 1978, 1979), was updated by incorporating data from two extensive pilot studies conducted by Lu Kwang Ju over 3 years, as well as data from three full-scale installations (Daigger et al., 1999). Figure 23.45 shows a process design based on 600 days · °C, assuming the feed has relatively high degradable solids content (Daigger et al., 1999). Figure 23.46 shows another operation if the feed contains low degradable solids (Daigger et al., 1999). Both systems can coexist, from a design standpoint, if prethickening is incorporated into the design to allow an increase or decrease of SRT, as necessary.

3.4.1.3   Specific Oxygen Uptake Rate   The microorganisms' rate of oxygen use depends on the biological oxidation rate. The oxygen uptake rate is used to determine the level of biological activity and resulting solids destruction in the digester. The U.S. EPA selected an SOUR of 1.5 mg/g · h of oxygen of total solids at 20°C (68°F) to indicate that aerobically digested solids have been adequately reduced in vector attraction. Because of the issues in meeting a 38% VSR with solids from long sludge-age secondary treatment processes, the SOUR test is becoming more common.

The SOUR is a quick test and is independent of the initial value in the system or reduction of SOUR in the upstream process. During active aerobic digestion for staged

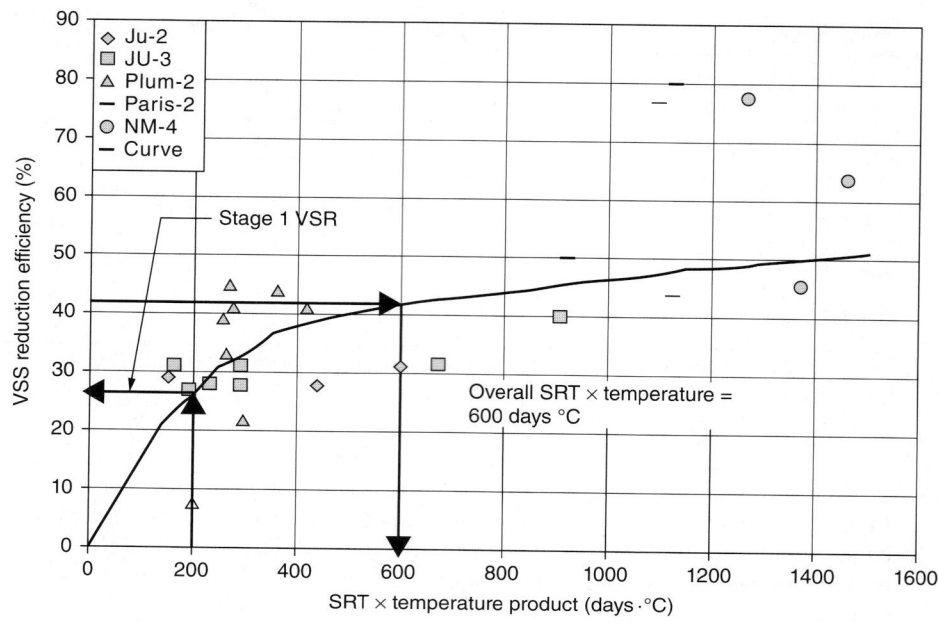

FIGURE 23.45 Selection of SRT × temperature (days·°C) product for feed with high degradable solids content (Daigger et al., 1999).

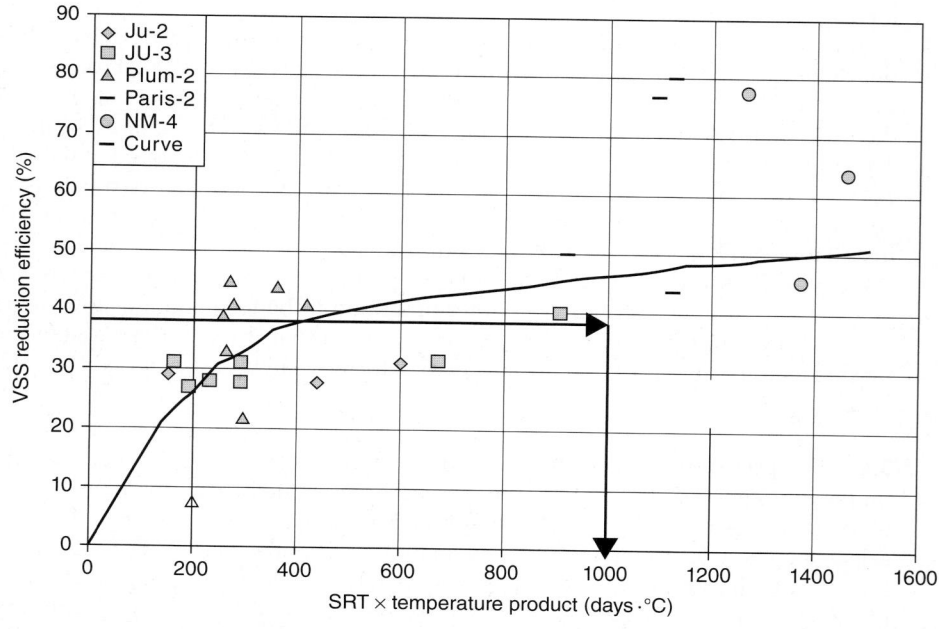

FIGURE 23.46 Selection of SRT × temperature (days·°C) product for feed with low degradable solids content (Daigger et al., 1999).

operation, the typical oxygen uptake rate is between 3 and 10 mg/g·h of total solids in the first-stage digester, compared with a range of 10 to 20 mg/g·h in the active phase of the activated sludge process. If primary solids are added to the first digester, its uptake rate may range from 10 to 30 mg/g·h of total solids. The oxygen uptake rate for aerobically digested biosolids ranges from 0.1 to 1.0 mg/g·h of total solids, well below the required 1.5 mg/g·h by U.S. EPA standards.

One potential issue is that the SOUR test is only accepted by the EPA with a specific range of testing parameters of solids concentrations and temperatures. Extrapolation of the test to concentrations and temperatures outside these ranges can be problematic.

3.4.1.4 Pathogen Reduction   Like solids reduction, little pathogen reduction can be expected at temperatures less than 10°C (50°F). On the other hand, significant reduction may be achieved at temperatures higher than 20°C (68°F), up to temperatures where inhibition begins to occur. Although U.S. EPA standards allow for operations at 15°C (59°F), the detention time must be increased to account for the slower rate of pathogen reduction. The exothermic reactions of digestion and the heat derived from compressing air in blowers will help maintain digester temperature, but this may be insufficient in colder climates. Therefore, in these areas, it may be desirable to either thicken the sludge (allowing longer detention times in the same volume) or to insulate the digesters to help retain the generated heat.

3.4.2 Equipment Design
Design of conventional aerobic digesters is similar in many ways to the design of activated sludge systems. Much of the equipment will be similar, and may even be shared.

3.4.2.1 Aeration and Mixing Equipment   Several devices (e.g., diffused air, mechanical surface aeration, mechanical submerged turbines, jet aeration, and combined systems) have been used successfully to provide oxygenation and mixing in aerobic digesters. Each has advantages and disadvantages as outlined below. For more information on the relative merits and design features of aeration systems from the perspective of their more common use for oxygenating activated sludge systems, see Chapter 12.

The design of diffused-air systems for aerobic digesters is similar to the design of those used in standard activated sludge systems. Diffusers typically are located near the tank bottom. They can be placed along one side of the tank to produce a spiral or cross-roll pattern, or they may be installed as a floor-mounted grid system. Both fine-bubble and coarse-bubble diffusers have been used in aerobic digesters. Airflow rates of 0.33 to 0.67 L/m³·s (20 to 40 cu ft/min/1000 cu ft) typically are required to ensure that mixing is adequate. The airflow rates needed to meet oxygen-transfer requirements depend on digester loading, diffuser type, system layout, and overall oxygen-transfer efficiency.

Diffused-air systems provide the following advantages: oxygen transfer is controlled by varying the air-supply rate; the introduction of compressed air to the digester typically adds heat to the system, which minimizes temperature loss during cold weather; and overall heat loss from the system is minimized because of the relatively small degree of surface turbulence.

Advantages of diffused-air systems may be outweighed by clogging problems that can occur in aerobic digesters, especially in those whose operation includes periodic settling and supernatant removal. While the air is turned off, solids can enter the air piping and adhere to the inner walls of piping or diffusers. Nonclogging and porous media devices are more resistant to this type of plugging than large-bubble, orifice diffusers.

However, surface fouling of porous diffusers can occur. If a diffused-air system is to be used, it is imperative that provisions be included for easy cleaning of the diffusers and air drop pipes.

For higher solids concentrations, there is a significant decrease in oxygen-transfer efficiency with conventional diffusers (Krampe and Krauth, 2003; WEF et al., 2012). In addition, the clogging issues with conventional diffusers may be exacerbated. An alternative to the floor-covering diffuser systems is high-shear, nonclogging aeration equipment designed specifically for high solids concentrations (4% to 8% solids). This aeration system combines draft tubes or shear tubes with an adjustable, above-water orifice for airflow control and nonclogging point-source diffusers. The shear tube and draft tubes provide mixing and shearing to transfer oxygen and achieve volatile reduction with high solids (see Figures 23.47 and 23.48). The limitation of this system is that it works better when the liquid is more than 6.1 m (20 ft) deep (Daigger et al., 1997).

Mechanical surface aerators typically are floating, pontoon-mounted devices of either low- or high-speed design. Low-speed aerators are more often used in aerobic digesters. Compared to diffused-air systems, mechanical surface-aeration systems typically are

FIGURE 23.47   A typical draft tube system (in this case used to treat a mixture of primary and secondary waste at Paris, Illinois). The picture was taken after conversion from anaerobic to a prethickened, two-stage-in-series, aerobic digestion (Daigger et al., 1997).

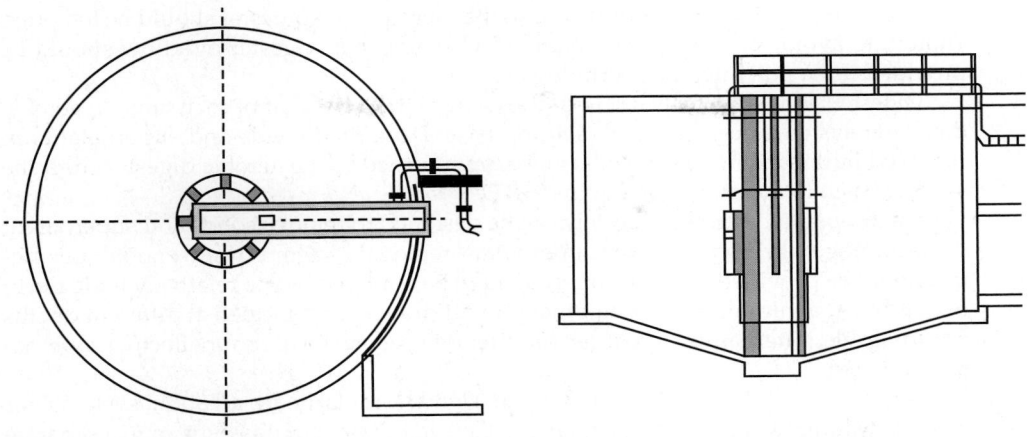

FIGURE 23.48   Plan and section views of a digester basin at Paris, Illinois. Basin is 14 m diameter × 9.4 m deep (45 ft diameter × 31 ft deep) (Daigger et al., 1997).

simpler and easier to maintain, and less prone to fouling. Disadvantages typically attributed to surface aeration include the lack of control of the oxygenation rate, performance deterioration if excessive foam is present, more potential for foaming because of high surface turbulence, increased heat loss from the system, and the potential for ice accumulation during winter in cold climates as a result of the device's splashing.

Mechanical submerged turbine aerators (and other combined mechanical mixing and diffused-air systems) provide several advantages and eliminate some disadvantages of the diffused-air and surface-aerator devices. Oxygenation rates can be controlled by varying the airflow rate to the submerged impeller. Because impellers are submerged, they are not as sensitive to foaming conditions as surface aerators and avoid the ice and heat dissipation problems associated with surface aerators. Additionally, the submerged unit can be operated as a mixer only, thereby promoting denitrification. Although the oxygen-transfer efficiency is similar to that of fine-bubble diffusers, the penalty resulting from two motors (turbine and blower) reduces the aeration efficiency (SAE) (Stenstrom and Rosso, 2008).

Jet aeration devices provide many of the advantages of submerged turbines. These devices typically are more easily installed and have a similar oxygen-transfer efficiency to submerged turbines. Problems with device plugging have occurred in the past when liquid flow paths were not large enough to pass the stringy solids typically found in aerobic digesters. The use of jets may also promote floc shear and subsequent dewatering difficulties.

3.4.2.2 Piping Arrangements   Specific piping requirements for aerobic digesters include provisions for feeding solids, withdrawing digested solids, decanting supernatant, and supplying air for aeration, when applicable. All piping should be designed large enough to provide for passage of solids, and should consider the need for future maintenance or repairs to the piping and the digester. See Chapter 19 for a discussion of design of sludge pumping systems and associated piping.

Only one solids-feed inlet per basin is necessary if the digester is designed with adequate mixing. If feasible, it may be preferable to have the feed occur where the operator can physically monitor flow into the digester. The digester should be fed often enough to avoid localized shock loading. An emergency digester overflow should be provided if the potential for overfilling exists.

Digested solids typically are withdrawn from the low point of each tank. In aerobic digestion systems designed with settling basins, digested solids and supernatant are removed in the settling basin. Solids also are returned to the aerobic digester from the settling basin to maintain the required SRT.

Batch-operated aerobic digesters can be designed to remove solids and supernatant via pumping or gravity. If a fixed supernatant-removal system is used, enough flexibility should be provided to allow supernatant to be removed over a relatively wide range of depths. At a minimum, two supernatant withdrawal lines located at different depths are advisable. Alternatively, floating decanter devices can be used for effective supernatant removal.

Air piping, if required, should be designed similarly to aeration systems for activated sludge systems (see Chapter 12). Consideration should be given to a separate air supply to the aerobic digester, especially if the liquid level varies because of supernatant decanting. When the liquid level in the digester is lower than that in the aeration tank, the digester will "rob" the aeration tank of air unless the air supply is separate or there is pressure compensation. Also, if centrifugal blowers are used for aeration, the impact of varying water levels on the blowers should be considered.

Provide strategically located hose connections to allow flushing out solids lines with facility effluent. Water (effluent) sprays, although providing some level of foam control, are seldom used because of the quantities of fluid that would be added to the system. Drains or sumps for dewatering and cleaning basins should be provided.

3.4.2.3 Instrumentation and Controls

Aerobic digestion typically is controlled manually. The operating variables that currently lend themselves to automatic control are dissolved oxygen, ORP, and tank level.

The dissolved oxygen signal can be used to control the aeration system so it maintains an optimum dissolved oxygen level (typically between 1 and 2 mg/L). This can conserve energy. Low- (and occasionally high-) dissolved oxygen conditions can trigger an alarm to allow operators to take corrective action. However, except for inadvertent digester overloads, dissolved oxygen changes in aerobic digesters typically are minimal, and maintaining dissolved oxygen monitors may be time-consuming.

An alternative that may be more beneficial, especially for either thermophilic systems or systems operated in aerobic/anoxic modes, may be the use of ORP probes. Research (Peddie et al., 1990) has found that the ORP profile in a cycled digester is characteristic and reproducible. There are distinct changes in the profile as the system moves from aerobic to anoxic respiration, and the system can be used to monitor and control systems even with very low levels of dissolved oxygen. Therefore, use of ORP probes, combined with aerobic-anoxic cycling, offers one potential means of improving the energy efficiency of the aerobic digestion process.

A tank-level signal, useful in preventing overfilling, can be used for on-off control of digester feed pumps. Intermittent feeding of primary solids to digesters via timer-controlled pumps has been used. This controlled-feeding technique also has been used for waste biological solids feeding. Automatically controlled feeding is successful when care is taken to establish the proper time as facility-operating conditions change. With manually controlled feed systems, a tank-level signal can be used to warn of a high-level condition.

Withdrawal of solids from aerobic digesters typically is manually controlled and done intermittently. Manual withdrawal allows rational reaction to variable solids production, solids concentration, digestion rates, and capacity of subsequent processing.

3.4.2.4 Considerations of Equipment Selection Flexibility and maintainability are key criteria when selecting aerobic digester equipment. The major equipment items of concern are piping and aeration and mixing equipment.

Piping systems require valves that resist clogging (e.g., eccentric plug valves). All feed and withdrawal piping (e.g., feed solids, digested solids, and supernatant lines) should have provisions (e.g., visible discharge points, cleanable sight glasses, or flow meters) for confirming that liquid or solids are flowing through the line during operations. Flow meters may not be effective with some positive-displacement pumps because pulsations can give the impression of positive flow when no net solids movement is occurring.

The aeration and mixing system(s) should be designed to facilitate maintenance. Swing-arm (knee-joint) or lift-out diffuser assemblies simplify the maintenance and cleaning of diffusers. Access to remove surface aerators or mixers also should be provided. Consideration also should be given to multiple tanks, so one tank can be completely drained for maintenance without interrupting the process.

3.4.2.5 Design for Safety Although subject to the safety hazards typically associated with mechanical and electrical equipment, aerobic digestion does not involve the explosive and toxic gases generated in anaerobic digestion. Safety considerations for aerobic digesters are similar to those for activated sludge basins. For example, placing life preservers with safety lines at intervals around the digesters can prevent drowning accidents. Adequate lighting around the tanks should be provided to allow for safe nighttime O&M. Nonslip, corrosion-resistant grating should be used for access walkways.

3.4.2.6 Design for Operability All systems, including aerobic digestion, should be designed with operability in mind. Design engineers should consider the following operability issues:

- Aeration system selection for ease of maintenance (periodic diffuser cleaning);
- Location, number, and type of monitoring instruments to enhance control capability;
- Location, number, and type of supernatant-withdrawal devices;
- Aboveground or belowground installation of digestion reactor (ease of temperature control versus accessibility);
- Ability to mix and aerate the system independently; and
- Access to the tank and other equipment for maintenance.

3.4.3 Process Performance and Operation

Conventional aerobic digestion typically produces Class B biosolids. Class B biosolids are biosolids in which the pathogen levels are unlikely to pose a threat to public health and the environment under specific-use conditions (U.S. EPA, 2003). Class B biosolids cannot be sold or given away in bags or other containers or applied on lawns or home gardens. They typically are land applied or landfilled.

The important factors when controlling aerobic digestion operations are similar to those for other aerobic biological processes (see Table 23.30) (Stege and Bailey, 2003).

Volatile Solids Reduction Data from Both Studies in All Basins						
°C	8–10°C	12°C	21°C	21°C	23°C	31°C
Two basins in series (no. of days)	19.25	13.75	13.75	13.75	19.25	19.25
Volatile solids removal	27%	31%	31%	31%	28%	31%
Three basins in series (no. of days)	29.25				29.25	29.25
Volatile solids removal	28%				32%	40%

TABLE 23.30   Volatile Solids Reduction at Minimum Operating Temperatures and Minimum Solids Retention Time (Daigger et al., 1999)

Operators should monitor the primary process indicators (e.g., temperature, pH, dissolved oxygen, odor, and settling characteristics, if applicable) daily. Monitoring helps control process performance and serves as a basis for future improvements. The secondary indicators (e.g., ammonia, nitrate, nitrite, phosphorus, alkalinity, SRT, and SOUR) are useful in monitoring long-term performance and for troubleshooting problems associated with the primary indicators. While monitoring and controlling these parameters is important, the degree of control that can be exercised on each parameter varies. Analysis frequency should be increased during startup and whenever large changes are made to operating conditions, such as solids flowrate and large increase or decrease in feedstock's solids concentration.

3.5   Autothermal Thermophilic Aerobic Digestion

As noted in Section 3.2, aerobic digestion is an exothermic reaction, generating heat. The ATAD process uses this heat to help achieve operating temperatures of 40°C to 80°C (see Figure 23.49). It relies on having efficient oxygen-transfer systems (to provide oxygen without heat stripping), insulated vessels (to minimize heat loss), and higher concentrations of solids to provide enough heat generation to maintain the thermophilic temperatures. If sufficient insulation and adequate solids concentrations are provided, the process can be controlled at thermophilic temperatures to achieve greater than 38% VSR and meet Part 503's Class A pathogen requirements (U.S. EPA, 1990).

ATAD has been studied since the 1960s. Much of the developmental work was done by Popel (1971a, 1971b), who, along with his coworkers, studied animal manure and wastewater residuals in Germany. They developed an aspirating aeration device that was key to the process success. Research in the United States was done by Matsch and Drnevich (1977) using pure oxygen and by Jewell and Kabrick (1980) using air with submersible aeration devices. The initial full-scale installations in the United States occurred during the 1990s. Some of the early systems experienced odor and other issues. Despite this, ATAD digestion gained popularity because it can produce Class A biosolids. In 2003, there were 35 ATAD systems operating in North America. More than 40 facilities are operating in Europe (Stensel and Coleman, 2000). This experience has led to changes in the ATAD process, with second-generation systems designed to address these issues through changes in operating parameters, addition of appropriate odor control systems, and provisions of mesophilic stabilization (Smith et al., 2012).

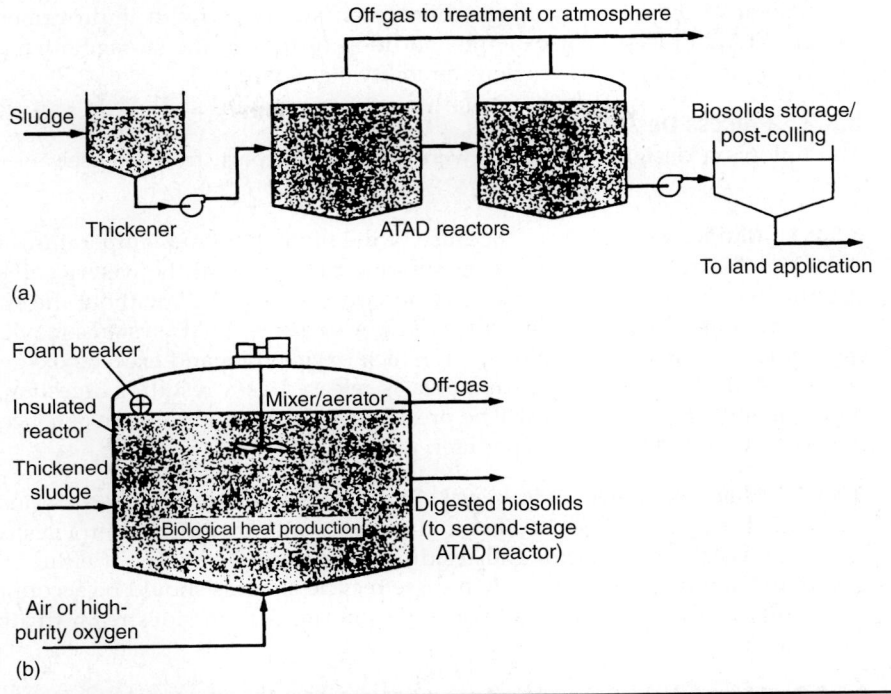

FIGURE 23.49 Schematic (a) and individual reactor configuration (b) for a typical autothermal thermophilic aerobic digestion system (from Metcalf & Eddy, *Wastewater Engineering: Treatment and Reuse*, 4th ed. Copyright © 2003, The McGraw-Hill Companies, New York, NY, with permission).

3.5.1 Advantages and Disadvantages

The major advantages of ATAD are as follows:

- Shorter retention times of approximately 5 to 6 days to achieve 30% to 50% VSR, resulting in a (smaller volume required to achieve a given suspended solids reduction).

- Lower oxygen requirements than mesophilic aerobic digestion, as nitrification does not occur.

- When the reactors are well mixed and maintained at 55°C and above for 10 days, the process meets the EPA requirements for processes to further reduce pathogens, yielding Class A biosolids.

The major disadvantages of ATAD are as follows:

- Potential for poor dewatering characteristics (Daigger et al., 1998).

- Potential for objectionable odors.

- Lack of nitrification and/or denitrification (Daigger et al., 1998).

- A foam layer must be managed to ensure effective oxygen transfer (Metcalf and Eddy, Inc./AECOM, 2013).

Several of these disadvantages have been addressed with improvements in the design ATAD process. For example, addition of mesophilic storage after the ATAD system significantly improves dewaterability (Scisson, 2009).

3.5.2   Process Design

The following design parameters were adapted, in part, from Stensel and Coleman (2000).

3.5.2.1   Nitrification Inhibition
Because of the high operating temperatures involved, ATAD inhibits nitrification and so the system's pH is typically between 8 and 9. Aerobic destruction of volatile solids occurs as described by eq 23.27, without the subsequent nitrification reactions described in eq 23.28. Also, some ATAD systems may be operating under microaerobic conditions, in which oxygen demand exceeds oxygen supply (Stensel and Coleman, 2000). Ammonia is released as a result of digestion, and the ammonia-nitrogen produced will be present in both gas and solution at concentrations of several hundred milligrams per liter.

3.5.2.2   Effect of Liquid Sidestreams That Contain Ammonia-Nitrogen
Most of the ammonia-nitrogen will be recycled to the wastewater treatment train via sidestreams from the ATAD odor-control and residuals dewatering systems. If effluent nitrogen and phosphorus limits are low, then these recycle streams should be accounted for in evaluating facility performance. For more information on sidestream treatment, see Chapter 15.

3.5.2.3   Foam
Autothermal thermophilic aerobic digestion generates a substantial amount of foam as cellular proteins, lipids, and FOG are broken down and released into solution. The foam contains high concentrations of biologically active solids, which provide insulation. It is important to manage foam effectively (via foam cutters or spray systems) to ensure effective oxygen transfer and enhanced biological activity. A freeboard of 0.5 to 1 m (1.65 to 3.3 ft) is recommended (Stensel and Coleman, 2000).

3.5.2.4   Equipment Design
Table 23.31 shows recommended design parameters for ATAD systems (Stensel and Coleman, 2000).

Parameter	Acceptable Range	Acceptable Values
pH	5.9–7.7	7.0
5-day BOD (mg/L)	9–1700	500
Filtered 5-day BOD (mg/L)	4–173	50
Suspended solids (mg/L)	46–2000	1000
Kjeldahl nitrogen (mg/L)	10–400	170
Nitrate-nitrogen (mg/L)	0–30	10
Total phosphorus (mg/L)	19–241	100
Soluble phosphorus (mg/L)	2.5–64	25

TABLE 23.31   Acceptable Characteristics for Supernatant from Aerobic Digestion Systems (from Metcalf & Eddy, *Wastewater Engineering: Treatment and Reuse*, 4th ed. Copyright © 2003, The McGraw-Hill Companies, New York, N.Y. with permission)

3.5.2.5 Prethickening As noted previously, an important consideration in maintenance of ATAD conditions is maintaining a sufficiently high feed concentration to allow the exothermic reactions to provide necessary heat. Typically, the ATAD influent should contain more than 4% solids, with a minimum of 2.5% biodegradable volatile solids. Therefore, thickening facilities may be required prior to the ATAD process.

3.5.2.6 Basin Configuration The system should include two or more enclosed, insulated reactors in series. The reactors are typically either cylindrical or rectangular, depending upon the system design and the type of equipment selected. Cylindrical structures can be either steel or concrete, while rectangular structures are more commonly concrete. Both reactors need mixing, aeration, and foam-control equipment.

Both continuous and batch processing are acceptable. To comply with the EPA definition of ATAD as a process to further reduce pathogens (PFRP), a batch process should be used. In this case, pumps should be designed to withdraw and feed the daily allotment of solids in 1 hour or less. A specific volume of solids is removed on a daily basis from the second-stage reactor (which is operating in a range of 55°C to 65°C). After the solids are removed, biomass from the first-stage reactor is transferred into the second-stage reactor. The second-stage reactor is then isolated for the remaining 23 hours each day, at a minimum temperature of 55°C. After the biomass is transferred from the first-stage reactor to the second stage, raw feed is then introduced to the first stage to make up the volume removed. This feeding approach isolates the reactors from each other and reduces the potential for contamination of the product.

3.5.2.7 Post-Process Storage and Dewatering Post-process cooling is necessary to consolidate solids and enhance dewaterability. Typically, solids exiting the second-stage digester will be at a temperature in excess of 60°C. Allowing 14 to 20 days of SRT in post-digestion cooling allows cooling the solids to less than 35°C, improving dewatering characteristics. Some of the second-generation systems also use this as a nitrification/denitrification reactor. Even with storage, it may be necessary to provide heat exchangers to cool the biosolids.

3.5.3 Process Performance and Operation

3.5.3.1 Volatile Solids Reduction The VSR achieved by the process depends on the feedstock(s), SRT, operating temperature, and reactor loading. Table 23.32 (Spinosa and

Parameter	Actual Data from a Gravity Thickener–Aerobic Digester in-Loop System	Compared with Acceptable Typical Values from Table 23.31
pH	6.5–7.1	7.0
Suspended solids (mg/L)	10–50	1000
Total Kjeldahl nitrogen (mg/L)	2.5–4	170
Nitrate-nitrogen (mg/L)	—	30
Total phosphorus (mg/L)	0.3	100
Thickener blanket	8–13.5	10

TABLE 23.32 Data from a Gravity Thickener–Aerobic Digester in-Loop Process at the Stockbridge, Georgia, Wastewater Treatment Plant (Courtesy of Stantec Consulting)

Vesilind, 2001) shows reported VSRs in ATAD systems. The Bowling Green, Ohio, facility, which is a second-generation facility, reported a VSR of 75% (Scisson, 2006).

3.5.3.2 Pathogen Reduction   German regulations require ATAD systems to produce biosolids containing no more than 1000 enterobacteria/mL. The German government considers ATAD to be a process capable of producing "pasteurized (hygienic) solids"— a status similar to the PFRP designation in Part 503. The U.S. EPA has approved one set of operating conditions for the ATAD system as a PFRP. For systems not meeting these requirements, PFRP equivalence can be demonstrated through time and temperature requirements or by testing.

The Haltwhistle, U.K., facility reported a more than 4-log pathogen reduction via its ATAD system (Murray et al., 1990). Canadian facilities using ATAD had less than 100 MPN/wet gram of fecal coliform and fecal streptococci in 7 of 12 samples of their biosolids (Kelly, 1991), whereas *Salmonella* was not detected in any of the samples. Tests by Jewell and Kabrick (1980) in Binghamton, New York, showed that an ATAD system with a 24-hour SRT at 45°C reduced *Salmonella* and virus concentrations below detection limits.

3.5.3.3 Odor Control   Odor control has been a key concern for ATAD systems, with varying success in addressing the issue. Older facilities, especially first-generation systems, tend to experience issues more often.

The Banff facility uses a water scrubber on ATAD exhaust. Its dewatered biosolids exhibited no odors and seemed well stabilized. The Haltwhistle facility had no odor complaints. The Salmon Arm facility sends exhaust gases to a trickling filter; no odor problems have been reported. The Ladysmith and Gibson's facilities discharge ATAD exhaust to biological filters. The Bowling Green, Ohio, facility reports no odors after a year of operation (Scisson, 2006).

Glenbard, British Columbia, has a pure-oxygen ATAD system that emits "rotten broccoli" odors during operations. When analysts tested the offgas, they found dimethyl sulfide, which is an indicator of anaerobic conditions. At the Salmon Arm and Whistler facilities, testing showed that the offgas contained hydrogen sulfide, methyl disulfide, dimethyl sulfide, ammonia, and unidentified organic compounds (Kelly et al., 1993). Reports on facilities in Colorado and Pennsylvania cite odor issues and the need for odor control (Bowker and Trueblood, 2002; Hepner et al., 2002). Meanwhile, several facilities in North America and Europe emitted odors and needed to implement odor controls (Layden et al., 2007).

Typically, odors can be minimized if the ATAD system maintains proper operating temperatures and is adequately mixed and aerated. Further odor-control measures are now recommended practice (Kelly et al., 2003; Kelly 2006) and include devices such as water scrubbers, biofilters, or thermal oxidizers.

3.5.3.4 Dewaterability   Autothermal thermophilic aerobic digestion produces biosolids with small flocs and, therefore, a large surface area that requires more polymer during dewatering operations (Kelly et al., 2003). In fact, conditioning chemical costs could offset the benefits of ATAD (Agarwal et al., 2005) if the goal is a dewatered biosolids containing 20% to 30% solids, because it can cost 5 to 10 times more to chemically condition ATAD solids than undigested solids (Murthy et al., 2000a), and about 2 to 3 times more to chemically condition ATAD solids than anaerobically digested solids (high-rate mesophilic) (Spinosa and Vesilind, 2001).

The system's high temperature contributes to dewatering challenges because it promotes cell lysis and the release of proteins to liquid. These proteins, along with extracellular polymeric substances, alter the biosolids' conditioning polymer requirements. However, if operating temperatures exceed 70°C, the dewatering properties of biosolids actually improve because the production of extracellular substances decreases (Zhou et al., 2002).

Investigators have tried several methods for improving the dewaterability of ATAD solids:

- Sequential polymer dosing using iron and anionic polymer, or cationic and anionic polymers (Murthy et al., 2000a; Agarwal et al., 2005);
- Post-ATAD mesophilic aeration of biosolids (Murthy et al., 2000b; Scisson, 2009); and
- Electrical arc treatment (Abu-Orf et al., 2001).

Mesophilic holding, based on the work in Bowling Green (Scisson, 2009), appears to offer significant benefits in dewatering.

3.6   Design Techniques to Optimize Aerobic Digestion

3.6.1   Thickened Aerobic Digestion

3.6.1.1   Advantages of Thickening   Typically, waste activated solids from secondary clarifiers will have solids concentrations in a range from 7500 to 20 000 mg/L. If WAS is being digested without primary sludge, it may be beneficial to prethicken the solids. Thickening is required for ATAD systems, and can be beneficial for conventional mesophilic digestion. The main advantages of this technique include:

- Increased SRT and VSR for a given volume; and
- The higher solids concentrations can yield auto-heating.

The oxidation of biodegradable organic matter elevates digester temperatures via its heat of combustion (about 3.6 kcal/g [6500 Btu/lb] of VSS destroyed) and accelerates digestion and pathogen destruction rates (Grady et al., 2011). If the heat generated by this process can be maintained in the digester, it can be used to control the reactor's temperature. This could be beneficial in cold climates.

3.6.1.2   Disadvantages of Thickening   Thickened aerobic digestion may require an additional unit process, which may increase in labor and O&M. Solids can only be prethickened to a maximum solids concentration wherein oxygen can be successfully transferred in the solids using available aeration equipment, and to the maximum extent that solids rheology does not significantly affect mixing characteristics in the aerobic digester. It is also important to avoid excessive temperatures, especially in summer months.

3.6.1.3   Categories of Thickening   Thickened aerobic digestion is divided into five major categories (as described below) based on the thickening treatment processes used to increase the feed cake's solids concentration and the position of thickening in the digestion process.

3.6.1.3.1 Batch Operation or Decanting of Aerobic Digester Batch operation involves the practice of manually decanting digested solids. Originally, aerobic digestion was operated as a draw-and-fill process, a concept still used at many facilities. Solids are pumped directly from the clarifiers or sequencing batch reactors (SBRs) to the aerobic digester. The time required to fill the digester depends on the tank volume available and the volume of solids. When a diffused-air aeration system is used, the solids being digested are aerated continually during the filling operation. When the solids are removed from the digester, aeration is discontinued, and the biosolids are allowed to settle. The clarified supernatant is then decanted and returned to the treatment process. The removed biosolids can contain up to 2.5% solids, depending on the sludge and the time allowed for settling.

Advantages of this process are that no additional tanks are required, and it is possible to utilize existing tanks to both digest and thicken. Disadvantages of this process include the following:

- Basins may be sized based on low solids concentration and high water content (i.e., large volumes are required).
- Larger basins raise the capital cost.
- Varying liquid levels may impact aeration efficiency.
- No control of alkalinity, temperature, ammonia, nitrates, and phosphorus.
- Difficult to meet stringent limits on supernatant quality.

3.6.1.3.2 Continuous-Feed Operation with Post-Sedimentation This mode of thickening treatment process consists of a continuous-feed operation using sedimentation (e.g., a gravity thickener) after digestion. This is typically a continuous aerobic digestion process that closely resembles the activated sludge process. Solids are pumped directly from the clarifiers, SBR, or MBR into the aerobic digester. The digester operates at a fixed level, with the overflow going to a solid-liquid separator. The design and operation of these sludge thickening devices is discussed further in Chapter 21. Thickened and stabilized solids are removed for further processing. Continuous operation typically produces biosolids with lower solids concentrations. This process has some advantages over the semi-batch operation, because the aerobic digestion basin is operated at a fixed level and the aeration-transfer efficiency is optimized.

For continuous-feed digesters, the process can be improved by adjusting the rate of settled return solids to obtain the best balance between return solids concentration and supernatant quality.

Disadvantages of this process include the following:

- Digester basins are sized based on low solids concentration and high water content (i.e., large volumes are required).
- Larger basins increase the capital cost.
- Higher O&M costs associated with aerating and mixing the larger tank volumes.
- No control of alkalinity, temperature, ammonia, nitrates, and phosphorus.
- Difficult to meet stringent limits on supernatant effluent.
- If nitrification and denitrification are not controlled between the digester and thickener, it can lead to anaerobic conditions in the thickeners and undesirable odors.

3.6.1.3.3 Gravity Thickener in Loop with Aerobic Digestion This process typically consists of two main phases (in-loop and isolation) and four main basins (two digesters, one pre-mix basin, and a gravity thickener). For feeds from SBRs and MBRs, more basins can be incorporated into the design to optimize flexibility; however, the four basins are still the main components of the process. During the in-loop phase, a premix basin, a digester, and a thickener operate in a loop, which reduces volatile solids, reduces ammonia, and increases solids concentration. The in-loop thickener has two main functions: thickening and denitrification. The in-loop digester, or volatizer, acts as a stabilization step and reduces most volatile solids. The in-loop digester is fed in batches 8, 16, or 24 times per day for a period that typically lasts 10 to 20 days. The digester then enters the isolation phase. During the isolation phase, no solids are introduced to the digester, which completes the additional pathogen reduction needed to meet regulatory requirements. The process is considered a "modified batch process" because of the multiple feedings to the loop digester. This process can produce biosolids containing 2.5% to 3% solids.

Advantages of this process include the following:

- Process provides the benefits of aerobic-anoxic operation (see Section 3.6.3).
- Process provides the benefits of staged operation (i.e., less detention time required).
- Denitrification in the thickener provides good control of alkalinity.
- Low concentrations of ammonia, nitrates, phosphorus, and total suspended solids (TSS) in supernatant.
- Process provides better SOUR and pathogen reduction compared to the processes in Sections 3.6.1.3.1 and 3.6.1.3.2 as a result of true isolation.
- Moderate capital cost.
- Low O&M cost.

3.6.1.3.4 Membranes for In-Loop Thickening with Aerobic Digestion Membrane technology has developed rapidly in the United States over the past 20 years. Its use in digestion is more recent; the oldest effective installations date back to 1998. The process incorporates a wastewater membrane suitable for high solids (e.g., flat plate, or hollow fiber).

Membrane thickening can be used in any process listed in Sections 3.6.1.3.1 through 3.6.1.3.3. Applications range from 3% to 5% solids concentrations; however, it is not recommended that design solids exceed 3.5% for single-stage systems. Membranes can operate in continuous or batch mode, in isolation and in series. The air required to scour the membranes can also provide oxygen for digestion, allowing membranes fitted into existing basins to provide both thickening and digestion at the same time. Designs include two-, three-, four-, or five-basin configurations, operating in batch or in series (see Figures 23.50 and 23.51) (Daigger et al., 2001).

Advantages of this process include the following:

- The physical barrier of the membrane provides the best control of supernatant quality.
- The process requires small footprint, which is ideal for high-rate digestion.
- If operated in staged mode, the process provides the benefits of staged operation.
- The process can provide the benefits of aerobic-anoxic operation, depending on the operational mode.

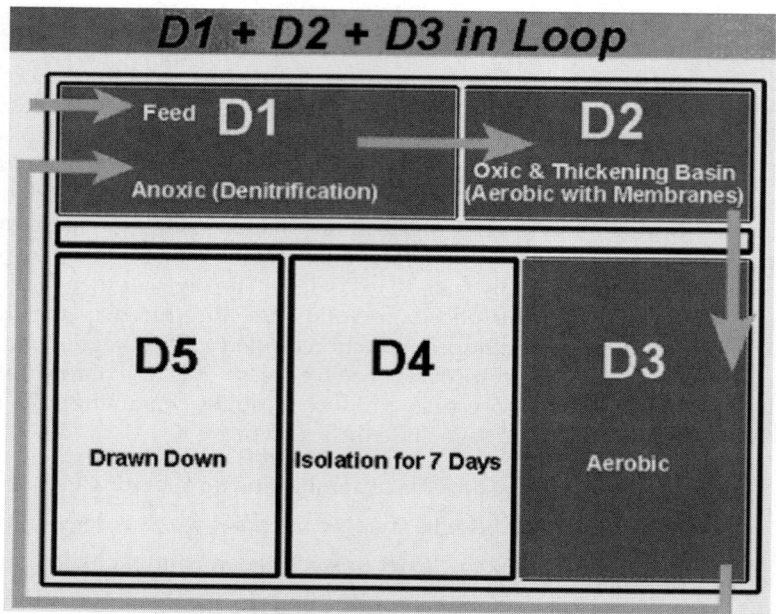

FIGURE 23.50   A five-stage batch operation setup using membranes for in-loop thickening as part of an aerobic digestion system (Daigger et al., 2001).

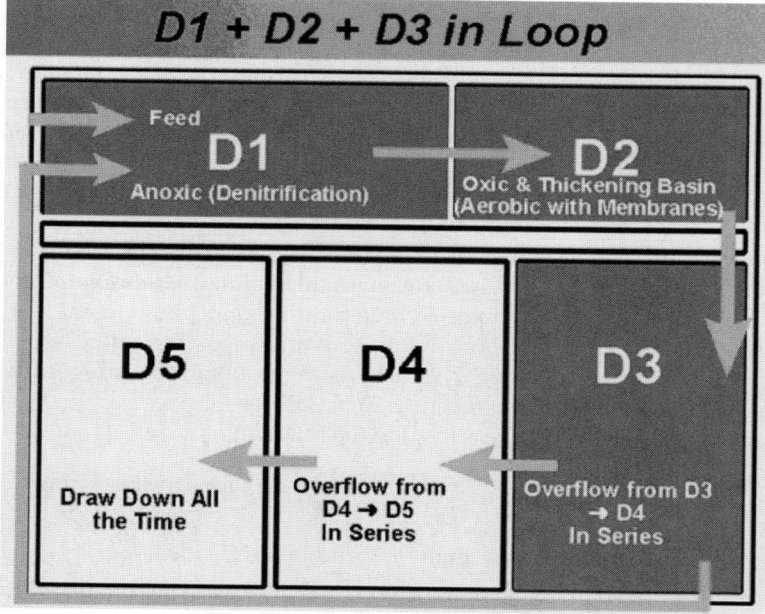

FIGURE 23.51   A five-stage, in-series operation setup using membranes for in-loop thickening as part of an aerobic digestion process (Daigger et al., 2001).

As noted in Section 3.3.7, phosphorus release occurs during aerobic digestion. Membranes offer several advantages in reducing phosphorus recycles in liquid streams. Membranes typically produce a filtrate with a low solids content, reducing the potential for particulate phosphorus discharges. If a biophosphorus permit applies on the liquid sidestream, the permeate can be treated with alum or ferric chloride to fix phosphorus so it can be removed with the solids. The permeate from a membrane digester is collected in the aerobic phase, minimizing the impact of release under anoxic or anaerobic conditions. Another factor that affects phosphorus release is pH. Because membrane systems include an anoxic zone to balance alkalinity, pH balancing is integral to the process.

3.6.1.3.5 Using Any Mechanical Thickener before Aerobic Digestion In this process, a mechanical prethickening device (e.g., a gravity belt thickener, DAF mechanism, centrifuge, or drum thickener) is used before aerobic digestion. This process uses. The designer can choose the ideal mechanical device (see Chapter 21) and desired operating solids concentration (e.g., 4%, 5%, or 6%) to minimize the aerobic digestion basins. This process gives flexibility to meet performance requirements in summer and winter by modifying the mechanical device's operating schedule as desired. For this process, two digesters in series are recommended, as a minimum (series operation will be addressed later in this chapter). Because of the flexibility and reliability, and the capital and O&M cost savings, this process is preferred in WRRFs designed for average daily flows greater than 7600 m³/d (2.0 mgd).

Advantages of this process include the following:

- Provides operational flexibility;
- Provides ability to optimize equipment selection and solids concentration;
- Minimizes footprint, which is ideal for high-rate digestion;
- Provides temperature control when flexibility is included in the design (cold weather not an issue with these systems, but provisions may needed to prevent thermophilic conditions in summer);
- Can provide the benefits of staged operation, depending on system design; and
- Can provide the benefits of aerobic-anoxic operation, depending on system design.

One concern with prethickening is the reduction in aeration efficiency, which occurs with higher solids concentrations. The alpha values and transfer efficiency are lower in digesters operating at 4% to 6% than in those operating at 2% to 3% (WEF et al., 2012). However, because of the reduced basin volume required for these systems, the airflow required for both process and mixing is comparable. Mixing requirements typically are higher than process air requirements for systems with lower solids concentration, so the overall operating horsepower for higher solids is less.

3.6.2 Basin Configuration—Staged or Batch Operation (Multiple Basins)

Traditionally, aerobic digesters have been designed with one basin, or with multiple basins operated in parallel. Multiple tanks in series or in isolation operated in a batch operation have proven to improve both pathogen destruction and compliance with VSR requirements.

According to the U.S. EPA, solids can be aerobically digested using a variety of process configurations, including continuous-flow or batch systems, in one or multiple stages (U.S. EPA, 2003). From a process basis, single-stage completely mixed reactors with continuous feed and withdrawal are the least effective option for bacterial and viral destruction, mainly because of the potential for short-circuiting.

Operation in series is simpler, as it reduces the complexity of controlling a batch system. Using two or more completely mixed digesters in series reduces the potential for short-circuiting, and improves the process kinetics.

Farrah et al. (1986) have shown that the decline in densities of enteric bacteria and viruses follow first-order kinetics. Assuming that first-order kinetics are correct, it can be shown that for the same total volume, an additional 1-log reduction of organisms is achieved in a two-stage reactor compared to a one-stage reactor. Direct experimental verification of this prediction has not been done, but Lee et al. (1989) have qualitatively verified the effect.

Although calculations indicate that a two-stage system would theoretically allow a 50% reduction in volume and detention time, not all factors involved in the decay of microorganism densities are known. Therefore, to allow a factor of safety, for staged operation (using two stages with about equal volume), it is recommended that the required time be reduced to 70% of that needed for single-stage aerobic digestion in a continuously mixed reactor. The same reduction is recommended for true batch operation or for more than two stages in series. This is consistent with the credit allowed by the EPA (U.S. EPA, 2003) for the improved efficiency of both batch and multi-stage operation. Thus, the time required would be reduced from 40 to 28 days at 20°C (68°F) and from 60 to 42 days at 15°C (59°F). These reduced times are also more than sufficient to achieve adequate VAR. (For more information on this topic, see Section 3.4.1.2.)

The benefits of a two-stage reactor system (in series or in isolation) include:

- Improvement in pathogen destruction;
- Improvement in VSR;
- Smaller tank volumes required; and
- Lower airflows required because of smaller tank volumes.

3.6.3 Aerobic-Anoxic Operation

As described in Section 3.2, the oxidation of biomass produces carbon dioxide, water, and ammonia (see eq 23.29). The sludge age in aerobic digesters is typically long enough to allow for nitrification. As the system nitrifies, alkalinity is consumed. Depending on the feed alkalinity, the pH may drop until it begins to inhibit nitrification. In this case, partial nitrification occurs, and a portion of the nitrogen remains as ammonia. This is primarily a concern with poorly buffered wastewater.

If the oxygen in nitrate can be used to stabilize biomass, using anoxic reactions, then both nitrification and denitrification can occur in the reactor. Oxidizing biomass with nitrates both releases ammonia and produces nitrogen gas plus bicarbonate (a form of alkalinity).

Equation 23.36 is a balanced stoichiometric equation of combined nitrification and denitrification. It illustrates how oxidized biomass is converted through the process to carbon dioxide, nitrogen gas, and water.

$$C_5H_7NO_2 + 5.7O_2 \rightarrow 5CO_2 + 3.5H_2O + 0.5N_2 \tag{23.36}$$

In Kuwait, investigators studied aerobic digestion in a controlled environment at 20°C and a 10-day SRT and optimized the duration of the anoxic stage at 8 to 16 h/d (Al-Ghusain et al., 2004). This is shown in Figure 23.43. Peddie et al. (1990) have also studied optimizing the aerobic-anoxic cycle and have found that ORP can be used to automate the process.

Aerobic-anoxic operation also has the benefit of reducing operating costs. With some of the oxygen required for digestion coming from the nitrates, there is an oxygen credit that reduces the additional oxygen that must be provided via aeration. This is similar to the denitrification credit in secondary treatment processes. Finally, as discussed in Section 3.3.7, aerobic-anoxic cycling can have benefits in reducing sidestream loadings.

3.7 Nutrient Removal in Aerobic Digestion

3.7.1 Nitrogen Removal

The conventional aerobic digestion process typically operates in a nitrifying mode. Supernatant from these digesters will typically have high concentrations of nitrate; however, at longer detention times, the supernatant can also contain significant concentrations of nitrite and ammonia (Jenkins and Mavinic, 1989). When the conventional digesters are operated in the aerobic-anoxic mode, nitrification and denitrification occurs with a reduction in total nitrogen. Typically, nitrate levels will be reduced to less than 5 mg/L (Jenkins and Mavinic, 1989). Because nitrification is inhibited in the ATAD process, the supernatant from the ATAD process will have low nitrates, but will have very high ammonia concentrations.

For many areas, the annual design loading rates for land application of biosolids are limited by the nitrogen loading rate. Therefore, reducing the total nitrogen in the biosolids is beneficial in maximizing the use of the land application sites. For this case, the conventional process operated in the aerobic-anoxic mode provides the greatest benefit in terms of land application.

3.7.2 Phosphorus Reduction in Biosolids and Biophosphorus

From an overall mass-balance prospective, phosphorus entering the digester ends up either in waste solids or in effluent. If it cannot move forward with the solids, then phosphorus will be recycled back to the head of the facility. Phosphorus will be released into the liquid phase in both anaerobic and aerobic digestion; however, the release is lower in aerobic processes than anaerobic processes.

The amount of phosphorus released in aerobic digestion is dependent on the digestion time and temperature, as well as the operational mode of the digester (Jenkins and Mavinic, 1989). Operating the system in an aerobic-anoxic mode or under low-dissolved-oxygen conditions (which provides simultaneous nitrification and denitrification) reduces the phosphorus release (Daigger et al., 2001). Based on research (Jenkins and Mavinic, 1989), the release in a continuously aerated digester is 2 to 3 times the release in intermittently aerated systems.

As shown in Figure 23.52, when solids from an EBPR facility (Ozark, Kansas) were digested under fully aerobic conditions, the phosphorus release was in the range of 120 to 150 mg/L (Daigger et al., 2001). If they were digested under cyclic operations, phosphorus releases ranged from 70 to 90 mg/L after 500 hours of operation.

Another factor that affects phosphorus release is pH. Several studies show that a pH less than 6.0 should be avoided, because it encourages inorganic metal phosphates

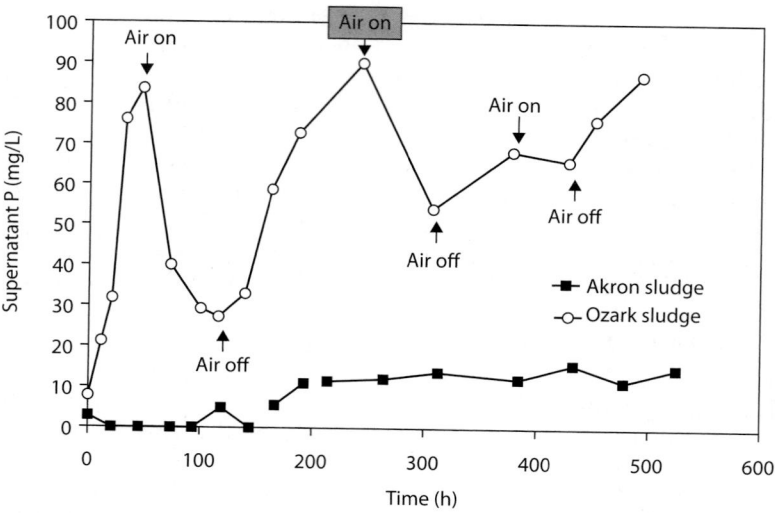

FIGURE 23.52   Polyphosphorus release and uptake of phosphorus during solids digestion (Daigger et al., 2001).

to dissolve (Jenkins and Mavinic, 1989; Daigger et al., 2001). Alternatively, feeding lime to the digester to raise the pH significantly lowers the phosphorus release.

Dealing with phosphorus in the digesters depends both on the overall nutrient requirements for the facility and on any phosphorus restrictions on land application. Depending upon the controlling parameters, there are options that can be used to manage this phosphorus release.

When controlling phosphorus in the recycle stream is the primary concern, there are several potential options for reducing phosphorus loads in the sidestreams. First, solids can be prethickened using a mechanical thickener. Figure 23.53 shows a typical prethickened application for liquid disposal with no phosphorus limit

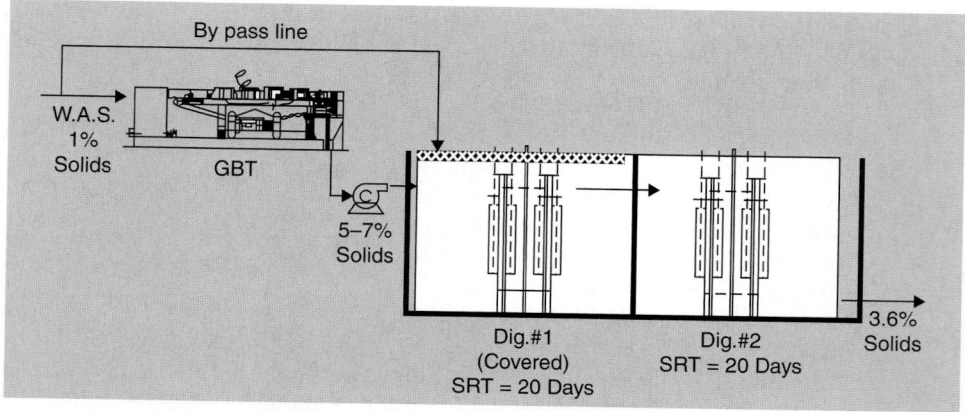

FIGURE 23.53   Option I: prethickened liquid disposal with no phosphorus limit restriction on land application. (GBT = gravity belt thickener) (Daigger et al., 2000).

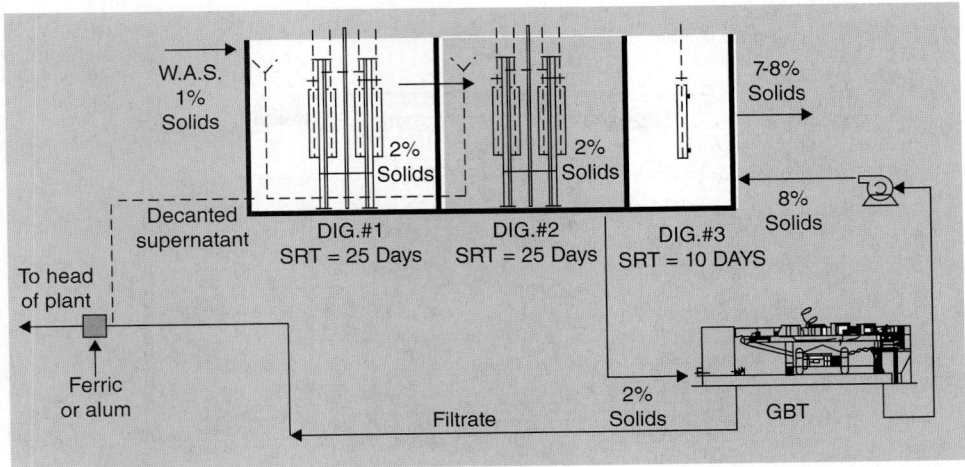

FIGURE 23.54 Option II: dewatering post-thickening, with no phosphorus limit restriction on land application. (GBT = gravity belt thickener) (Daigger et al., 2000).

restriction on land application (Daigger et al., 2000). The key, in this case, is maintaining the solids being fed to the prethickening device in an aerobic state. To prevent anaerobic conditions, solids should be wasted directly from the liquid sidestream to the prethickening device; or if storage is required before prethickening, the detention time should be minimized. If the solids can be prethickened while still in an aerobic state, and no further decanting or dewatering occurs, the phosphorus will be wasted as part of the biosolids.

When thickening or dewatering occurs after digestion (Figure 23.54), there is a greater potential for return of phosphorus to the influent of the facility. In this case, operating an aerobic/anoxic system and maintaining a neutral pH will help reduce phosphorus release. In a multi-stage system, the first stage can be operated with limited aeration to control nitrification and promote the formation of struvite (Daigger et al., 2000), which can be separated and recovered (see Chapter 15). Remaining phosphorus in the supernatant or filtrate from the post-digestion dewatering can be chemically bound, passing through the secondary process and allowing the phosphorus to be removed as part of the final biosolids.

The final alternative occurs when there is a limit on phosphorus loading for land application sites (see Figure 23.55). As in the previous alternative, limiting nitrification in the first state will promote the formation of struvite. The supernatant or filtrate from post-digestion dewatering is again chemically treated, but is sent to an inclined plate separator (Daigger et al., 2000). This allows the bound phosphorus to be removed for separate disposal, while the final biosolids have a lower phosphorus concentration.

3.8 Process Variations

Investigators have tested several variations on standard mesophilic aerobic digestion. Two of the more notable variations are high-purity oxygen aeration and dual digestion. These and other variations are discussed further in *Solids Process Design and Management* (WEF et al., 2012).

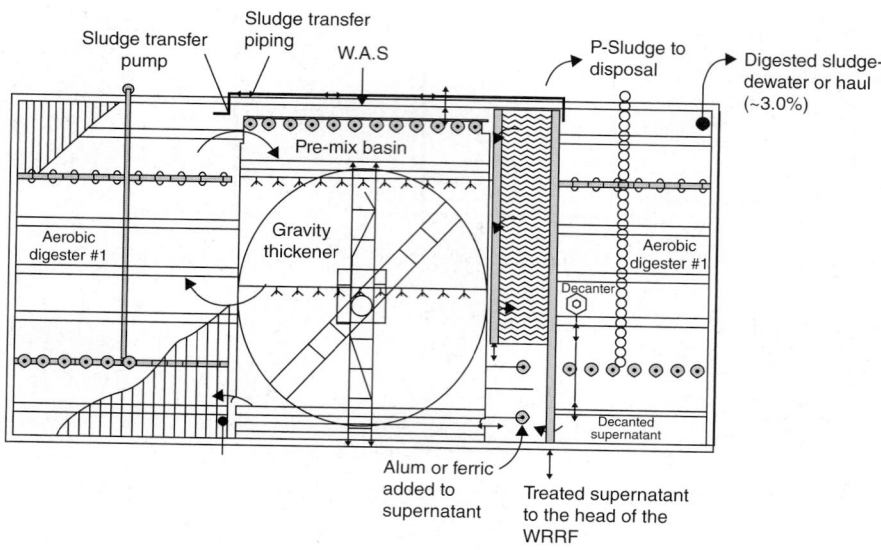

FIGURE 23.55 Option III: chemical precipitation of phosphorus, when phosphorus limits land application (Daigger et al., 2000).

3.8.1   High-Purity-Oxygen Aeration

This aerobic digestion process uses high-purity oxygen rather than air. Recycle flows and the resultant biosolids are similar to those obtained via conventional aerobic digestion. Typical influent solids concentrations may vary from 2% to 4%. High-purity-oxygen aerobic digestion works well in cold-weather climates because the use of pure oxygen, rather than air, reduces the effect of changes in ambient air temperatures. Because colder temperatures do not impact the digester, higher digester temperatures and an increased rate of biological activity are maintained.

High-purity-oxygen aerobic digestion can be conducted in either open or closed tanks. Because the digestion process is exothermic in nature, the use of closed tanks will result in a higher operating temperature and a significant increase in the VSR rate. Design of these systems will be similar to activated sludge systems using pure oxygen (see Chapter 12). The operating costs associated with separating oxygen will in most cases be more than conventional aeration. This makes high-purity-oxygen aerobic digestion most feasible when used with a high-purity-oxygen activated sludge system. Because the system will nitrify, but will not denitrify, neutralization of the pH may be required (Metcalf and Eddy, Inc./AECOM, 2013).

3.8.2   Combined Stabilization Processes

3.8.2.1   Combined Aerobic and Anaerobic Digestion
The fusion of two stabilization processes in ATAD and conventional mesophilic anaerobic digestion are covered in the dual-digestion subsection of the anaerobic digestion section. Recent research has focused on using aerobic digestion as a posttreatment for conventional mesophilic anaerobic digestion and has been demonstrated by Kumar et al. (2006a, 2006b), Parravicini et al.

(2008), and Novak et al. (2010). The advantages of these systems reportedly include improved VSR, nitrogen removal from return streams, and improved dewaterability.

3.8.2.2 Aerobic Digestion and Drying Aerobic digestion has been used as a conditioning step for solids that will undergo drying to stabilize them to a level that will minimize the risks for odor production during the drying process. This also reduces issues with regrowth when the dried solids get wet.

3.9 Aerobic Digester Design Examples

3.9.1 Standard Design: Single Tank
Design a mesophilic aerobic digester processing solids from a non-primary solids secondary biological treatment process. The following conditions are applicable to the design:

Secondary solids concentration	0.8%
Total solids	1144 kg/d (2522 lb/d)
Volatile solids	894 kg/d (1971 lb/d) or 78.15%
Decanted solids concentration	1.5%
Minimum liquid temperature (winter)	15°C (59°F)
Maximum liquid temperature (summer)	30°C (86°F)

The biosolids must meet Class B conditions in a single-tank configuration.

3.9.1.1 Determine the Digester Volume

3.9.1.1.1 Determine the SRT Required to Meet Class B Requirements Using the worst-case condition (the winter temperature of 15°C), the SRT required to meet Class B requirements at this temperature is 60 days.

3.9.1.1.2 Determine the Decanted Solids Volume

In SI units: $\qquad$ 1144 kg/d/1000 kg/m^3/1.75% = 65.37 m^3/d

In U.S. customary units: $\qquad$ 2522 lb/d/8.34 lb/gal/1.75% = 17 268.91 gpd

3.9.1.1.3 Determine the Digester Volume

Daily decanted solids volume × Required SRT = Digester volume

65.37 m^3/d × 60 days = 3922.2 m^3

or 17 268.91 gpd × 60 days = 1.04 million gal

3.9.1.2 Determine the Oxygen Requirements

3.9.1.2.1 Mixing Air Requirements From Section 3.7.1, we use the average value of 0.5 L/m^3·s (30 cu ft/min/1000 cu ft):

In SI units: $\qquad$ 3922.2 m^3 × 0.5 L/m^3·s = 1961.1 L/s

In U.S. customary units: 1.04 million gal/(7.48 gal/cu ft)/1000 × (30 cu ft/min/ 1000 cu ft) = 4155.3 cu ft/min

3.9.1.2.2 Process Air Requirements   From Figure 23.45, determine the amount of VSR expected:

Winter: (Solids temperature × SRT required per regulation at temperature) 15°C × 60 days = 900°C · days, which yields a 45% VSR

Summer: (Solids temperature × SRT to meet Class B in winter) 30°C × 60 days = 1800°C · days, which yields a 55% VSR

Calculate the VSR:

Winter: 894 kg/d × 45% = 402.3 kg/d (SI units)

1971 lb/d × 45% = 887 lb/d (U.S. customary units)

Summer: 894 kg/d × 55% = 491.7 kg/d (SI units)

1971 lb/d × 55% = 1084 lb/d (U.S. customary units)

Calculate the oxygen demand at 2 kg O_2/kg VSR (2 lb O_2/lb VSR):

Winter: 402.3 kg/d × 2 kg O_2/kg VSR = 804.6 kg O_2/d

887 lb/d × 2 lb O_2/lb VSR = 1774 lb O_2/d

Summer: 491.7 kg/d × 2 kg O_2/kg VSR = 983.4 kg O_2/d

1084 lb/d × 2 lb O_2/lb VSR = 2168 lb O_2/d

Calculate the process air requirement, assuming 0.56 actual oxygen requirement (AOR)/specific oxygen requirement (SOR) and 14% oxygen-transfer efficiency (OTE) for coarse-bubble diffusers:

$$\text{Winter}: \frac{(1774 \text{ lb } O_2 \text{ / day})(0.56 \text{ AOR/SOR})(1440 \text{ min/day})(14\% \text{ OTE})}{(0.2315 \text{ lb } O_2 \text{ /lb air})(0.075 \text{ lb air/cu ft})}$$

= 905 cu ft/min (427 L/s)

$$\text{Summer}: \frac{(2168 \text{ lb } O_2 \text{ /day})(0.56 \text{ AOR/SOR})(1440 \text{ min/day})(14\% \text{ OTE})}{(0.2315 \text{ lb } O_2 \text{ /lb air})(0.075 \text{ lb air/cu ft})}$$

= 1106 cu ft/min (522 L/s)

The air requirements that govern the design are based on mixing, so the blower is sized based on 1961.1 L/s (4155.3 cu ft/min).

3.9.1.3 Determine the Blower Power   Assuming an approximation of 12.657 L/s · kW (20 cu ft/min/blower hp), we get:

1961.1 L/s/12.657 L/s · kW = 154.93 kW

or

4155 cu ft/min/20 cu ft/min/hp = 207.76 hp

3.9.2   Optimizing the Single-Tank Conventional Design by Thickening

As discussed in Sections 3.3.7 and 3.4.1.2, aerobic digester design can be optimized by setting a batch or a staged design comprising at least two tanks. The design example for a two-tank design is as follows:

Secondary solids concentration	0.8%
Total solids	1144 kg/d (2522 lb/d)
Volatile solids	894 kg/d (1971 lb/d) or 78.15%
Thickened solids concentration	3.0%
Minimum liquid temperature (winter)	15°C (59°F)
Maximum liquid temperature (summer)	30°C (86°F)

The biosolids must meet Class B requirements in a two-tank-in-series configuration.

3.9.2.1   Determine the Digester Volume

3.9.2.1.1   Determine SRT Required for Class B Solids Regulations   Using the worst-case condition, which is the winter temperature at 15°C, determine the SRT required to meet Class B (as expressed in Section 3.7). Based on the U.S. EPA guidance manual (U.S. EPA, 2003), at this temperature and applying the 30% credit for staged operation, we can use 42 days.

3.9.2.1.2   Determine the Thickened Solids Volume

In SI units: 1144 kg/d/1000 kg/m³/3.0% = 38.13 m³/d

In U.S. customary units: 2522 lb/d/8.34 lb/gal/3.0% = 10 073 gpd

3.9.2.1.3   Determine the Digester Volume

38.13 m³/d × 42 days = 1601.46 m³

or 10 073 gpd × 42 days = 423 066 gal

Each digester volume is 1601.46 m³ (423 066 gal)/2 = 800.73 m³ (211 533 gal).

3.9.2.2   Determine the Oxygen Requirements

3.9.2.2.1   Mixing Air Requirements Per Digester, Assuming Equal VSR in Both Digesters   From Section 3.7.1, we use the average value of 0.5 L/m³·s (30 cu ft/min/1000 cu ft):

In SI units: 800.73 m³ × 0.5 L/m³·s = 400.27 L/s

In U.S. customary units: 211 533 gal/(7.48 gal/cu ft)/1000 × (30 cu ft/min/1000
cu ft) = 848 cu ft/min

The total air requirement for both digesters is 800.73 L/s (1696 cu ft/min).

3.9.2.2.2   Process Air Requirements   From Figure 23.45, determine the amount of VSR expected:

Winter: (Solids temperature × SRT required per regulation at temperature) 15°C × 42 days = 630°C·days, which yields a 42% VSR

Summer: (Solids temperature × SRT to meet Class B in winter) 30°C × 42 days = 1260°C·days, which yields a 49% VSR

Calculate the VSR (assuming equal VSR in both digesters):

Winter: 894 kg/d × 42% = 375.5 kg/d (SI units)

1971 lb/d × 42% = 827.8 lb/d (U.S. customary units)

Summer: 894 kg/d × 49% = 438 kg/d (SI units)

1971 lb/d × 49% = 965.8 lb/d (US customary units)

Calculate the oxygen demand at 2 kg O_2/kg VSR (2 lb O_2/lb VSR):

Winter: 375.5 kg/d × 2 kg O_2/kg VSR = 751 kg O_2/d

827.8 lb/d × 2 lb O_2/lb VSR = 1656 lb O_2/d

Summer: 438 kg/d × 2 kg O_2/kg VSR = 876 kg O_2/d

965.8 lb/d × 2 lb O_2/lb VSR = 1931.6 lb O_2/d

Calculate the process air requirement. Assume 0.56 AOR/SOR and 14% OTE for coarse-bubble diffusers:

$$\text{Winter}: \frac{(1656 \text{ lb } O_2 / \text{day})(0.56 \text{ AOR/SOR})(1440 \text{ min/day})(14\% \text{ OTE})}{(0.2315 \text{ lb } O_2 / \text{lb air})(0.075 \text{ lb air/cu ft})}$$

$$= 844.6 \text{ cu ft/min (398 L/s)}$$

$$\text{Summer}: \frac{(1932 \text{ lb } O_2 / \text{day})(0.56 \text{ AOR/SOR})(1440 \text{ min/day})(14\% \text{ OTE})}{(0.2315 \text{ lb } O_2 / \text{lb air})(0.075 \text{ lb air/cu ft})}$$

$$= 985.4 \text{ cu ft/min (465 L/s)}$$

The air requirements that govern the design are based on mixing again, but with much lower air demand than the single-tank design with thinner solids.

3.9.2.3 Determine the Blower Power
Assuming an approximation of 20 cu ft/min/blower hp (12.657 L/s·kW) we get:

For both digesters   1697 cu ft/min/20 cu ft/min/hp = 84.85 hp

For both digesters   800.73 L/s/12.657 L/s/kW = 63.26 kW

The optimization results via thickening are clearly evident in the smaller digester volume, air requirements, and blower power. The only caution on using this technique is the limited selection of diffusers that can handle 3% solids in continuous service.

Also, this savings could be increased by reducing air requirements in the second stage, because the VSR in the first digester is significant.

4.0 Composting

Composting is a biological process in which organic matter is decomposed under controlled, aerobic conditions to produce humus. Operators can accelerate the process by using the proper blend of materials and controlling the temperature, moisture content, and oxygen supply. The resulting compost is stable and can be safely used in many landscaping, horticulture, or agriculture applications.

Composting can be used to treat both unstabilized solids and partially stabilized biosolids. Odors would generally be reduced in the latter case due to a lower oxygen demand that allows aerobic conditions to be more easily maintained. In both solids and biosolids composting, operators control several essential process variables to optimize the material's decomposition/stabilization rate:

- Solids content;
- Carbon-to-nitrogen ratio (C:N);
- Aerobic conditions; and
- Temperature.

Via process-control methods, operators typically can cause the composting mass to achieve thermophilic temperatures, which destroy pathogens. Well-stabilized compost can be stored indefinitely and has minimal odor, even if rewetted. It is suitable for a variety of uses (e.g., landscaping, topsoil blending, potting, and growth media) and can be distributed to the public for gardening. It also can be used in agriculture to control erosion, improve the soil's physical properties, and revegetate disturbed lands. Local markets may be developed in urban and nonagricultural areas, as well as in agriculture and mine revegetation.

4.1   Process Variables

Although a wide variety of composting technologies are available, they all are designed to control the essential variables mentioned above.

4.1.1   Solids Content

The initial solids content depends on how much amendment or bulking agent is mixed with dewatered cake. For good process performance, the dewatered cake should contain between 14% and 30% solids. It then is blended with drier materials (e.g., wood chips, sawdust, shredded yard waste, and ground pallets) to achieve a solids content of about 38% to 45%.

The target solids content depends on the composting technology used. Solids content is controlled throughout the process via aeration, material agitation, or both.

4.1.2   Carbon-to-Nitrogen Ratio

The amount of carbon and nitrogen used by microorganisms depends on the composition of the microbial biomass. Ideally, the ratio of available carbon to nitrogen is between 25:1 and 35:1. If the ratio is less than 25:1, excess nitrogen will be released as ammonia, reducing the compost's nutrient value and emitting odor. If the ratio exceeds 35:1, organic material will break down more slowly, remaining active well into the curing stage (Poincelot, 1975). Wastewater residuals typically have a carbon-to-nitrogen ratio between 5:1 and 20:1. Adding an amendment or bulking agent increases the carbon content, improving both the energy balance and the mixture's carbon-to-nitrogen ratio.

Calculating the carbon-to-nitrogen ratio is complicated, because some of the carbon becomes available more slowly than the nitrogen (Kayhanian and Tchobanoglous, 1992).

If wood chips are the bulking agent, for example, only a thin surface layer of the wood provides available carbon. The carbon in sawdust, on the other hand, is more readily available to degradation.

4.1.3 Maintaining Aerobic Conditions

Microbial oxygen demand during composting can reduce the available oxygen in air to as low as 3% to 5% in as little as 15 minutes. Aerobic conditions are maintained via forced or convective aeration, material agitation, or both, depending on the composting technology used.

4.1.4 Maintaining Proper Temperatures

At first, the challenge is to heat the material up to the thermophilic range as quickly as possible. Then, the challenge is removing excess heat to maintain the process in the thermophilic range. It is also difficult to achieve uniform temperatures throughout the pile. Covers may help in this regard. Near the end of composting, the goal is to dry the material without removing too much heat. All of these are achieved using aeration, agitation, or both.

4.1.5 Microbiology

Three major categories of microorganisms involved in composting are bacteria, actinomycetes, and fungi. Bacteria are responsible for decomposing a major portion of organic matter. At mesophilic temperatures (lower than 40°C [104°F]), bacteria metabolize carbohydrates, sugars, and proteins. At thermophilic temperatures (higher than 40°C), they decompose proteins, lipids, and the hemicellulose fractions. Bacteria also are responsible for much of the heat produced.

Actinomycetes are microorganisms common to soil environments. They metabolize a wide variety of organic compounds (e.g., sugars, starches, lignin, proteins, organic acids, and polypeptides). Their role in composting is unclear. Waksman and Cordon (1939) indicated that this group attacks hemicellulose but not cellulose. Stutzenberger (1971) isolated a thermophilic actinomycete that may be important in cellulose degradation.

Fungi are present at both mesophilic and thermophilic temperatures. Chang (1967) indicated that mesophilic fungi metabolize cellulose and other complex carbon sources. Their activity is similar to that of actinomycetes; both typically are found in the exterior portions of compost piles. Golueke (1977) suggested that this phenomenon is related to the organisms' aerobic nature, because most fungi and actinomycetes are obligate aerobes.

Microbial activity during composting occurs in three basic stages: mesophilic, when temperatures in the pile range from ambient to 40°C (104°F); thermophilic, when temperatures range from 40°C to 70°C (104°F to 158°F); and a cooling period associated with a reduction in microbial activity and the completion of composting. The optimum temperature in the thermophilic range seems to be between 55°C and 60°C (131°F and 140°F) where the maximum rate of VSR occurs.

Biological solids, newly harvested wood wastes and yard wastes provide a diverse population of microflora that can respond to changes in temperature and substrate. Under most circumstances, an inoculum of pure cultures does not significantly enhance composting. Sawdust decomposition, however, can be accelerated by inoculating a cellulose-decomposing fungus and adding nutrients.

4.1.6 Energy Balance

Heat is generated when organic carbon converts to carbon dioxide and water vapor. The fuel is provided by rapidly degraded volatile solids. Heat primarily is removed by the evaporative cooling promoted by aeration and agitation. Some heat also is lost at the pile surface. The process temperature will not rise if heat is lost faster than it is generated.

Haug (1980) provides a detailed discussion of the energy balance, concluding with the following relationship:

$$W = \frac{\text{Weight of water evaporated}}{\text{Weight loss of volatile solids}} \tag{23.37}$$

If W is below 8 to 10, enough energy should be available for heating and evaporation. If W exceeds 10, the mix will remain cool and wet. This generalization is based on heat of vaporization and does not consider the effect of ambient conditions on evaporation and surface cooling.

4.2 Process Objectives

The primary objective of composting is to produce a nutrient-rich soil amendment that complies with federal, state, and local requirements for beneficial use of biosolids. The compost must meet both environmental and public health requirements, and be attractive for use. This primary objective is met via the following process objectives: pathogen reduction, maturation, and drying.

4.2.1 Pathogen Reduction

There are five types of pathogens in wastewater residuals: bacteria, viruses, protozoa cysts, helminthic (parasitic worm) ova, and fungi. The first four groups often are called primary pathogens because they can invade typically healthy persons and cause diseases. Fungi are called secondary pathogens because they typically only infect persons with weakened respiratory or immune systems.

Heat is one of the most effective methods for destroying pathogens. Table 23.33 summarizes time-and-temperature relationships for inactivating pathogens in actual composting operations. Note that temperatures measured in a composting pile or vessel may not be uniform because of variations in heat loss, solids-mixture characteristics, and airflow.

Microorganisms	Exposure Time for Destruction at Various Temperatures Hours		
	45–55°C	60°C	65°C
Salmonella Newport		25	
Salmonella	168	116	
Poliovirus type 1		1.0	
Candida albicans		72	
Ascaris lumbricoides		4.0	1.0
Mycobacterium tuberculosis			336

TABLE 23.33 Temperature Exposure Required for Pathogen Destruction in Compost (Knoll, 1964; Morgan and MacDonald, 1969; Shell and Boyd, 1969; Wiley and Westerberg, 1969)

Composting in the thermophilic range should eliminate practically all viral, bacterial, and parasitic pathogens (WEF, 2016). However, some fungi (e.g., *Aspergillus fumigatus*) are thermotolerant and, therefore, survive.

Data on windrow composting in Los Angeles showed that bacterial concentrations were markedly reduced within 15 days (Iacoboni et al., 1980). At 20 days, no *Salmonella* was detected. Fecal and total coliforms survived windrow composting in cool, humid climates, but *Salmonella* was eliminated after 14 days. Studies using an F_2 bacteriophage virus (as an indicator of virus destruction) showed it could survive for as long as 45 days in digested solids and more than 55 days in undigested solids.

Static-pile composting data show that total coliforms, fecal coliforms, and *Salmonella* were not detected after 10 days of composting when temperatures exceeded 55°C (131°F) for several days. Further studies using an F_2 bacteriophage virus revealed that static-pile composting destroyed the indicator in 14 days.

Salmonella can regrow in finished compost. However, parasite ova and virus cannot. Regrowth can be reduced by not using the same equipment to handle both raw feed and finished compost or by cleaning the equipment before handling finished compost.

Many microorganisms can function as secondary pathogens, although composting conditions favor the growth of some more than others. Millner et al. (1977) report that the fungus *A. fumigatus* Fres has been isolated at relatively high concentrations from finished compost and from compost-pile zones at less than 60°C (140°F). Other secondary fungi occasionally isolated from compost are *M. pusillus* and *M. miebei*. Common to composting operations, these fungi typically are found in backyards, decayed leaves, grass, commonly available organic soil amendments, and ventilation ducts.

During certain composting operations, more *A. fumigatus* spores are released to the atmosphere. In windrow and reactor studies at the Los Angeles County Sanitation District in California, LeBrun (1979) found that compost feedstock contained 1000 to 10 000 colony-forming units (CFU)/g; after composting, biosolids contained 10 CFU/g. Exposure to airborne spores can be minimized by controlling dust. So compost should not be allowed to become too dry, and workers should be provided with dust masks when working in dusty areas.

4.2.2 Maturation

Maturation refers to the conversion of a solids-amendment mixture's rapidly biodegradable components into substances similar to soil humus, which decomposes slowly. Insufficiently mature compost will reheat and generate odors when stored and rewetted. It also may inhibit seed germination (by generating organic acids) and facility growth (by removing nitrogen as it decomposes in soil). Stability refers to the reduction in microbial-degradation rate of the mixture's biodegradable components. Stabilization is achieved by maintaining optimal conditions for a sufficient period of time. Cellulose materials (e.g., wood and yard wastes) take longer to decompose than wastewater residuals, so screening out the bulking agent may improve stability.

There are a number of testing methods and standards for measuring compost stability or maturity, but none is universally accepted (Jimenez and Garcia, 1989). The standards associated with each test are still tentative, and much work needs to be done to correlate test results with odor generation and facility growth. A complete assessment of maturity may require multiple tests.

Volatile solids (as a percentage of total solids) is not a good measure of stability because it fails to account for the biodegradation rate and materials added or removed from the compost during processing (e.g., bulking agents).

Respiration tests, which measure carbon dioxide production or oxygen demand, better represent stability but are sensitive to test conditions. Carbon dioxide production typically is measured directly on the mixture in an incubator. Incubators are useful compost simulators that can effectively measure carbon dioxide productions in both highly unstable samples (from early in the process) and highly stable samples such as finished compost. Oxygen uptake rates can be measured on the mixture, or in an aqueous extract via the specific oxygen uptake rate (SOUR) test. Mature compost should have a carbon-to-nitrogen ratio that is less than 20:1. Available carbon in compost can deplete the nitrogen in soil that microorganisms typically use.

Seed-germination and root-elongation tests measure phytotoxicity caused by organic acids in compost. They are performed by germinating seeds (e.g., cress) in a filtered extract of compost and comparing them with a control using distilled water.

4.2.3   Drying

To dry compost, operators provide enough aeration or agitation to facilitate the removal of water vapor. This increases the solids content from about 40% to 55% or more. Drying is critical in processes that include screening, because screens do not perform well if the compost contains less than 50% to 55% solids.

4.3   Description of Composting Methods

Although composting is a naturally occurring biological process, the degree of control imposed on a system can range from periodically turning a pile or windrow to the more involved enclosed or in-vessel system with mechanical agitation and forced aeration.

In an attempt to respond to local and regional needs, a number of composting methods have evolved (Mussari et al., 2013). These methods offer the following benefits: accelerating a naturally occurring biological process; providing for process control over variables such as moisture, carbon, nitrogen, and oxygen; containing odors and particulates; reducing land area requirements; reliably producing consistent product quality; and integrating aesthetically pleasing facilities into local and regional sites.

4.3.1   Aerated Static-Pile Composting

Aerated static-pile composting is also called the Beltsville method because it was developed in Beltsville, Maryland, in the 1970s by the U.S. Department of Agriculture. As the name suggests, it involves aerating piled feedstock (see Figure 23.56). This flexible method is popular in the United States.

In this method, the solids-amendment mixture is constructed into a 2- to 4-m-deep (6- to 12-ft-deep) pile over an aeration floor (plenum) and then covered with a 150- to 300-mm-deep (6- to 12-in.-deep) insulating blanket of wood chips or unscreened finished compost to ensure that all of the mixture meet the temperature standards for pathogen and VAR. Small operations may construct individual piles, while large ones may divide a continuous pile into sections representing each day's contribution. The mixture typically remains in the pile for 21 to 28 days while the plenum forces air through the material to provide an aerobic composting environment. Then the piles are broken down, and the material is either moved directly to a curing area or screened and then moved to the curing area. Compost must contain at least 50% to 55% solids before screening. In some facilities, an intensive drying step (with a higher aeration rate than active composting) precedes screening. Compost typically is cured for at least 30 days to further stabilize the material. Some facilities screen the compost after curing (rather than before curing).

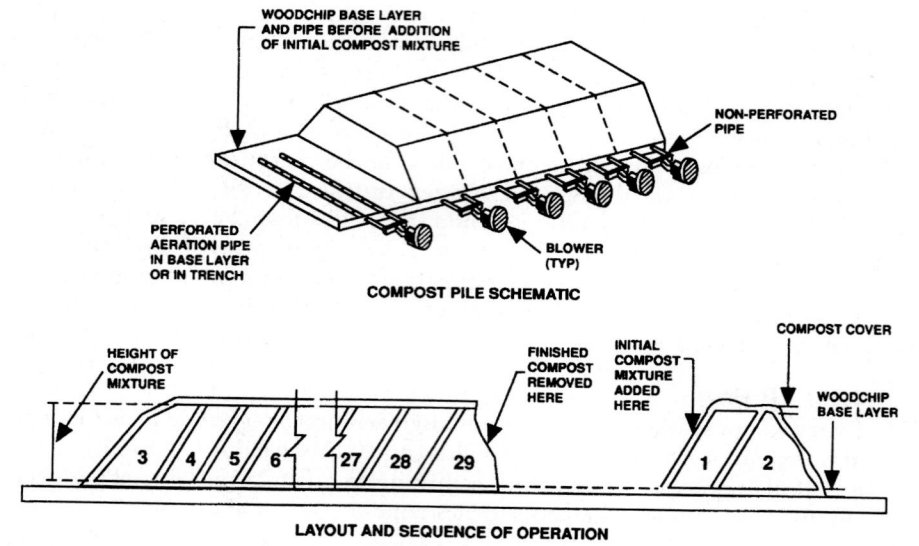

FIGURE 23.56   Schematic of an aerated static-pile composting system.

Aerated static-pile composting originally was developed for outdoor sites, but many systems are either partially or fully enclosed to control odors or facilitate operations during unfavorable environmental conditions (e.g., temperature or rainfall extremes).

4.3.2   Windrow Composting

In windrow composting, the solids-amendment mixture is formed into long parallel windrows whose cross sections are either trapezoidal or triangular (see Figure 23.57). The material then is turned periodically by a front-end loader or a dedicated windrow-turning machine to release moisture, expose more particles to the air, and loosen (fluff) the material to facilitate air movement through the windrow.

In the aerated windrow method, windrows are constructed over air channels to protect aeration piping from the turning equipment. Air can either be forced up through the windrow or be pulled down through the windrow into the channel. The windrows are turned periodically to expose more particles to air. Aeration and turning optimize the composting rate and release of moisture.

Windrow composting occurs at open outdoor sites or covered sites. This system needs more space than other composting technologies because of pile geometry and the room needed to maneuver a windrow-turning machine.

4.3.3   In-Vessel Composting

In-vessel systems typically combine aeration with some type of automated material movement in a reactor. A wide variety of such systems has been developed over the years, but only a few have been installed in more than one or two sites.

The SRT ranges from about 10 to 21 days, depending on system-supplier recommendations, regulatory requirements, and costs. It also should be based on desired

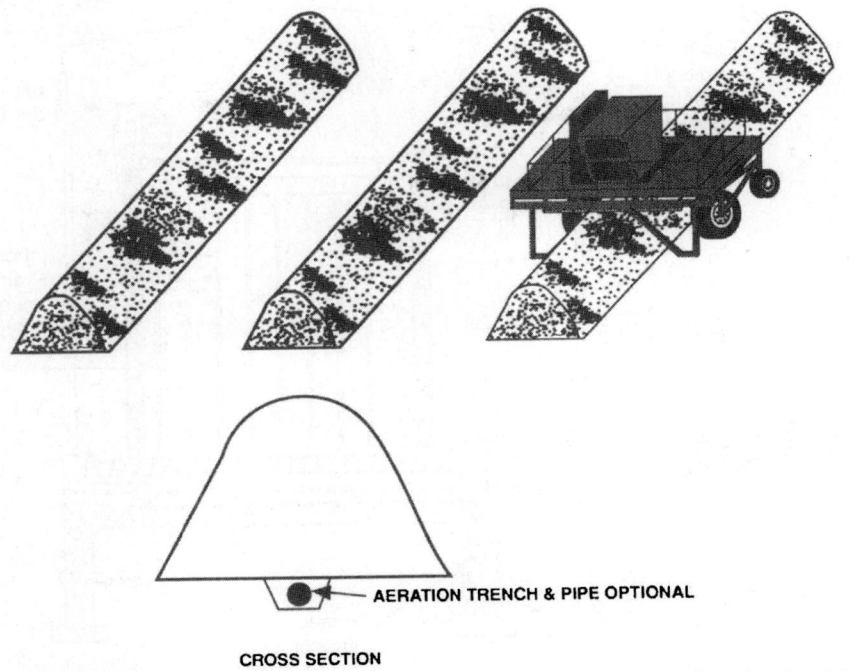

AERATION TRENCH & PIPE OPTIONAL

CROSS SECTION

FIGURE 23.57   Schematic of a windrow composting system.

product characteristics—especially stability—and take into account the overall solids residence time in the entire composting operation (all process phases). Once discharged from the reactor, the composted biosolids typically must be further stabilized for 30 to 60 days to achieve the desired product stability.

There are basically three types of in-vessel composting systems: vertical plug-flow reactors, horizontal plug-flow reactors, and agitated bay systems. Vertical plug-flow reactors are made of steel, concrete, and/or reinforced fiber-glass panels (see Figure 23.58). A mix of dewatered cake, amendment, and recycled solids is loaded in the top of the reactor, where it is aerated but not agitated (mixed). It moves as a plug to the bottom of the reactor, where it is removed via a traveling auger.

Horizontal plug-flow reactors are similar to vertical ones, except that the solids-amendment mixture is moved laterally through the reactor by a hydraulic ram (see Figure 23.59).

Agitated-bay reactors are open-topped bays with blowers and piping systems that supply air from the bottom (see Figure 23.60). Unlike plug-flow reactors, they also have mechanical devices that periodically agitate the mixture during its stay in the reactor. These systems are designed to function much like aerated windrows. A variety of methods are used to transfer compost from the reactors.

The most commonly used in-vessel system is the horizontal agitated-bed reactor. These reactors are rectangular, aerated from the bottom with independently programmable aeration zones, and enclosed in a building. A loader places the solids-amendment mixture into the front end. The agitation device is completely automatic, operates only

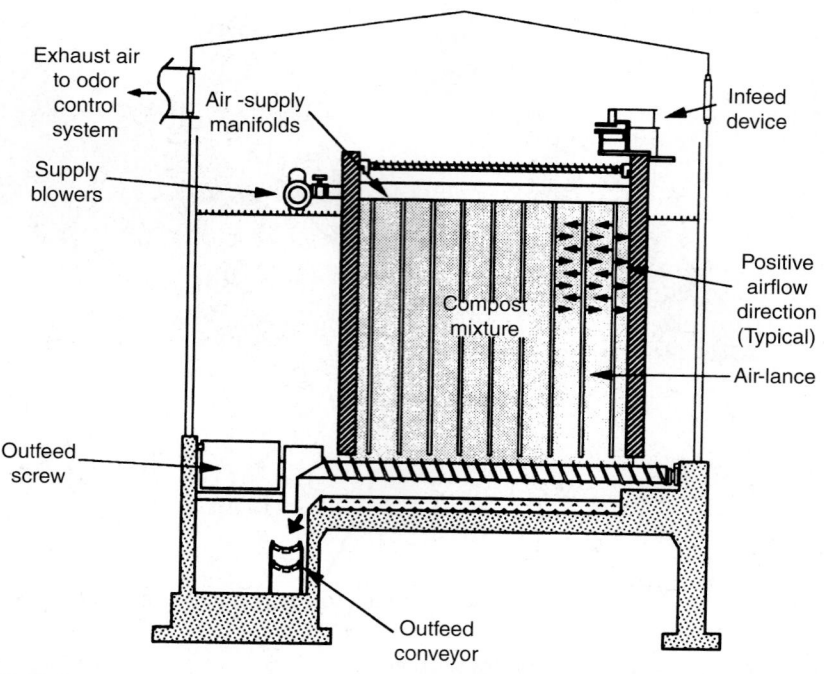

FIGURE 23.58   Cross-section of a vertical plug-flow reactor (rectangular design, made of steel).

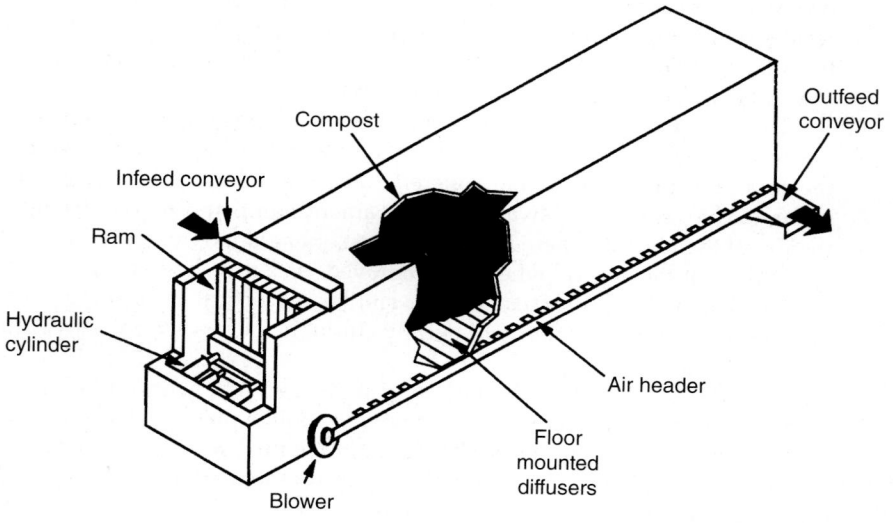

Note: Not to scale

FIGURE 23.59   Schematic of a horizontal plug-flow reactor.

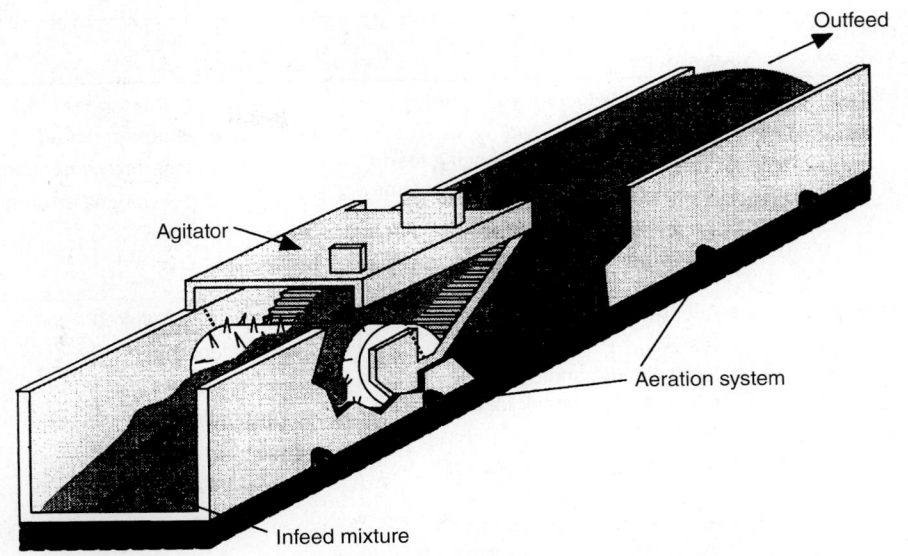

FIGURE 23.60   Schematic of a horizontal agitated-bed reactor.

in agitation mode, and typically makes one pass through the reactor each day. The composting material is dug out and redeposited about 4 m (11 ft) behind the machine until it has moved through the entire length of the reactor.

4.3.4   Comparison of Composting Methods
Of the three technologies discussed above, aerated static-pile composting is the most commonly used (see Table 23.34) (NEBRA, 2007).

Table 23.35 lists the advantages and disadvantages of five composting technologies based on physical facilities, processing aspects, and O&M. None is appropriate for

Facility	Type	Capacity (Dry Ton/d)*
Inland Empire Regional Composting Facility, California	Enclosed ASP	100
Davenport, Iowa	Enclosed ASP	25
Columbus, Ohio	Outdoor ASP	28
Rockland County, New York	Agitated bay	25
Hamilton, Ohio	Horizontal plug flow	15
Schenectady, New York	Vertical plug flow	50
Hawk Ridge, Unity, Maine	Tunnel (ASP)	15

*ton/d × 0.9072 = Mg.

TABLE 23.34   Representative List of Composting Facilities

Composting Technology	Advantages	Disadvantages
Aerated static pile	• Adaptability to various bulking agents • Flexibility to handle changing feed conditions and peak loads (volume not fixed) • Relatively simple mechanical equipment	• Relatively labor intensive • Relatively large area required • Operators exposed to composting piles • Potentially dusty working environment
Windrow	• Adaptability to various bulking agents. • Flexibility to handle changing feed conditions and peak loads (volume not fixed) • Relatively simple mechanical equipment • Requires no fixed mechanical equipment	• Very large area required • Relatively labor intensive • Operators exposed to composting piles • Dusty working conditions
Vertical plug flow	• Completely enclosed reactors in some systems improve ability to control odors • Relatively small area required • Operators not exposed to composting material	• Single outfeed device per reactor (large reactors), potential bottleneck • Potential inability to maintain uniform aerobic conditions throughout reactor • Relatively maintenance intensive. • Limited flexibility to handle changing conditions • Materials-handling system may limit choice of bulking agents
Horizontal plug flow (tunnel)	• Completely enclosed reactors improve ability to control odors • Relatively smaller area required (composting mix compacted) • Operators not exposed to composting material	• Fixed-volume reactors (no flexibility) • Limited ability to handle changing conditions. • Relatively maintenance intensive • Materials-handling system may limit choice of bulking agents
Agitated bin	• Mixing enhances aeration and uniformity of compost mixtures • Ability to mix compost (advantage in handling some bulking agents • Adaptability to various bulking agents	• Fixed-volume reactors (no flexibility) • Relatively large area required • Potentially dusty working environment • Operators exposed to composting piles • Relatively maintenance intensive

TABLE 23.35   Key Advantages and Disadvantages of Composting Systems

every situation. The choice depends on many factors (e.g., climate, siting considerations, operational concerns, and sensitivity to odors). Design engineers should consider the following factors when selecting a composting technology:

- Physical facilities (availability of space, material-handling system complexity, aeration equipment, and degree of enclosure);
- Process considerations (e.g., uniform aeration, aeration type, availability of different bulking agents, adaptability to changes in volume of feed solids, and odor emissions/odor control); and
- O&M issues (e.g., labor requirements, energy requirements, operator exposure, dust generation, and degree of maintenance).

4.4   Process Considerations for Designers

This section provides ranges of design parameters for each stage of the composting process and identifies the design criteria essential to successful operation. Consideration is made for each type of composting technology. For additional information, see Williams (2014).

4.4.1   Bulking Agents and Amendments

All composting technologies require mixing sufficient quantities of bulking agent with dewatered solids to adjust the initial solids content and provide porosity. Bulking agents also provide supplemental carbon to adjust the carbon-to-nitrogen ratio and energy balance. Unfortunately, these bulking agents also increase the quantity of solids that must be handled. Table 23.36 lists some typically used bulking agents and their characteristics.

Although yard debris can be used as a bulking agent, grass clippings and substantially green yard waste are unsuitable because of their high water and nitrogen content, and lack of porosity. If grass clippings and substantially green yard waste are composted, they also will require supplemental bulking agent.

4.4.2   Characteristics of the Solids-Amendment Mixture

The ratio of bulking agent to biosolids depends on the available agent's characteristics and the desired solids content. For example, if dewatered cake contains 18% to 24% solids and the agent (a blend of woody yard debris) contains 55% to 65% solids, then the bulking agent-to-biosolids ratio must be 3:1 or 4:1 (by volume) to produce a mixture containing 40% solids. To produce a mixture containing 45% solids, the ratio should be 5:1 or 6:1. To produce a mixture containing 38% solids, the ratio may be as low as 2.5:1. (Below 2.5:1, the mixture probably will not be porous enough to promote decomposition.)

The initial solids content needed depends on the composting technology used (specifically, the amount of agitation and aeration involved):

- Aerated static-pile systems need a mixture containing 40% to 45% solids. Wetter mixtures will lose heat energy to evaporation, thereby slowing the process. Drier mixtures may not provide enough moisture to complete the biological process.
- Turned windrow systems need a mixture containing about 45% solids. In wet climates, however, the mixture should be slightly drier to compensate. Wetter mixtures will not be porous enough to allow for convective airflow.

Bulking Agent	Characteristics
Wood chips (1 to 2 in.*)	• Must typically be purchased • High recovery rate in screening (60% to 80%) • Good source of supplemental carbon
Chipped yard or land-clearing debris	• May be available as waste material • Low recovery rate in screening (40% to 60%) because of higher percentage of fines • Green waste fraction adds nitrogen; more may be needed for C:N ratio • Good source of supplemental carbon
Ground waste lumber	• May be available as waste material • May be poor source of supplemental carbon if old and extremely dry because more volatile forms of carbon will be missing
Leaves	• Insufficient porosity to be used alone • Rapidly available source of supplemental carbon • Available as waste material • Not recovered by screening and adds to compost volume
Sawdust	• Insufficient porosity to be used alone • Rapidly available source of supplemental carbon • Must be purchased and typically is expensive • Not recovered by screening and adds to compost volume
Shredded paper	• Insufficient porosity to be used alone • Rapidly available source of supplemental carbon • Available as waste material • Not recovered by screening and adds to compost volume

*in. × 25.4 = mm.

TABLE 23.36   Types and Characteristics of Bulking Agents

- Automated loading tunnel and vertical plug-flow systems need a mixture containing 40% to 45% solids. Agitated bay systems, however, need a mixture containing 38% to 40% solids because the frequent agitation and forced aeration will dry the material much faster than other systems. Experience has shown that an agitated bay system can lose as much as 2% moisture during one agitation period.

It is critical that the mixture has uniform porosity and that all particles of cake be in close contact with the bulking agent. Dewatered cake with 18% to 25% solids should be mixed with a bulking agent, so each wood chip or other bulking agent particle is coated with a thin layer of solids. Dewatered cake with 30% to 35% solids will break into clumps that must be uniformly small and mixed with the bulking agent. (Large clumps and balls will become anaerobic, leading to excessive odors.) If mixing is not uniform, zones with a disproportionate amount of bulking agent will divert the flow of air, allowing other zones to become anaerobic.

4.4.3 Materials Balance Calculations

Materials balance calculations track the weight and volume of each material through each stage of the composting process. Table 23.37 shows a typical materials balance for 1 dry ton of biosolids (20% solids) in an aerated static-pile process (see Figure 23.56).

In this process, solids were mixed with yard waste, stacked over a layer of yard waste to provide air distribution, and covered with a layer of unscreened compost. The entire pile (except for the volume reserved for the cover layer) was screened after composting, and the oversized particles were recycled as a bulking agent. Screening typically recovers between 50% and 80% of the bulking agent (by volume), so it must be supplemented with makeup bulking agent. The recovery rate depends on the compost's moisture content (stickiness), the bulking agent's particle size, and the screen's loading rate. Because some of the bulking agent is recycled, it is important to account for all of this material and balance recycled and new bulking agent so all recycled agent is used.

The required input assumptions are the density of each material, the VSR of each input, and the screen's recovery efficiency.

4.4.4 Temperature Control and Aeration

In the United States, each state is responsible for regulating biosolids use within its borders. However, the federal government has issued minimum guidelines that all states must meet: 40 CFR 503 regulations. These regulations require that solids treatment processes meet certain requirements to produce biosolids that will not endanger the environment or public health. The specific requirements for composting depend on the technology used.

In addition to meeting regulatory requirements, composting systems also need to control temperatures to optimize decomposition. The optimum temperature range for VSR is about 55°C to 60°C (131°F to 140°F). Part 503 regulations require pathogen kill temperatures of 55°C for aerated static pile and in-vessel systems, for 14 days for windrow systems with 5 turnings during the 14-day period. Fourteen days with an average temperature of 45°C with a minimum of 40°C are required for VAR. In addition to maintaining certain regulatory dictated temperatures, it is also desirable to prevent material temperatures from climbing too high. Pile temperatures in excess of 70°C inhibit the biological decomposition process. Also, if high temperatures persist for periods longer than several weeks, the potential of spontaneous combustion can occur in very dry material (>75% solids).

In turned windrow operations, the temperature and the oxygen content are controlled by the porosity of the windrows and the frequency of turning. Initial porosity is controlled by thorough blending of the feedstock and having the proper bulking-agent-to-biosolids mix ratio. Once the windrows are in place, both temperature and oxygen content are controlled by turning of the windrow. Turning incorporates oxygen, and releases heat and moisture. Although turning releases heat, the pile temperature will spike upward shortly after turning. This is the result of the redistribution of feedstock and the infusion of oxygen. These spikes are typically short lived (a few hours).

Material	Volume (cu yd)	Total Weight (Ton)	Dry Weight (Ton)	Volatile Solids (Ton)	Bulk Density (lb/cu yd)	Solids Content	Volatile Solids
Biosolids	6.3	5.0	1.0	0.5	1600	20.0%	53.0%
Yard waste (processed)	10.9	3.3	1.8	1.3	600	55.0%	70.0%
Wood waste	0.0	0.0	0.0	0.0	500	60.0%	95.0%
Screened recyled bulking agent	9.9	3.5	1.9	1.8	695	55.0%	93.0%
Unscreened recycle	0.0	0.0	0.0	0.0	780	55.0%	88.6%
Mixture	25.7	11.7	4.7	3.6	911	40.1%	75.7%
Base (recycled bulking agent)	1.4	0.5	0.3	0.3	695	55.0%	93.0%
Cover (unscreened)	2.9	1.1	0.6	0.5	780	55.0%	88.6%
Composting losses		9.4	0.4				
Cover (unscreened)	2.9	1.1	0.6	0.5	780	55.0%	88.6%
Screen feed	21.5	8.4	4.6	3.5	780	55.0%	74.8%
Recycled bulking agent	11.4	4.0	2.2	2.0	695	55.0%	93.0%
Curing	9.9	4.4	2.4	1.4	900	55.0%	58.6%
Curing losses		0.2	0.1	0.1			
Compost to storage	9.5	4.3	2.3	1.3	900	55.0%	56.9%

Assumptions:
Recovery by screening
- Yard waste   50% by volume
- Wood waste   70% by volume
- Recycled bulking agent   50% by volume
- Pile base   95% by volume

Processing losses
- Losses during composting   10% of volatile solids
- Losses during curing   5% of volatile solids

*cu yd × 0.7646 = m³; lb/cu yd × 0.5933 = kg/m³; ton × 0.9072 = Mg.

TABLE 23.37   Materials Balance for 1 Dry Ton of Biosolids in Aerated Static-Pile Composting*

In aerated static-pile and in-vessel composting systems, forced aeration is used to supply oxygen and maintain aerobic conditions within the material, control temperatures, and remove moisture. In the first 1 to 2 days of composting, increasing airflow typically kick-starts the process and causes pile temperatures to rise quickly. However, throughout the rest of the process as the rate of airflow is increased in a forced aeration system, the pile temperature decreases and the rate of water vapor removal increases. As with a turned windrow system, agitation releases heat and water vapor.

Higgins et al. (1982) reported that an aeration rate of 34 m³/Mg·h (1100 cu ft/hr/dry ton) provided adequate drying and high-enough temperatures for pathogen destruction. Early in the composting process, higher aeration rates may be needed to prevent excessive pile temperatures.

To maintain temperatures less than 60°C during peak activity, aeration rates may need to approach 300 m³/Mg·h (10 000 cu ft/hr/dry ton). Such aeration capacity may be impractical in large systems. Practical aeration capacities are in the range of 90 to 160 m³/Mg·h (3000 to 5000 cu ft/hr/dry ton) of wastewater solids. Aeration in this range will control temperatures throughout most of the composting period and provide adequate moisture-removal capacity. Higher aeration rates are possible, but require more energy, larger and more closely spaced piping or trenches, and larger odor-collection and -treatment systems (if provided). If highly reactive bulking materials are used (e.g., ground-up leaves), the mass of bulking agent and dewatered cake may enter into the sizing of aeration capacity.

4.4.5   Detention Time

The time required to stabilize organic material typically is divided between an active composting stage and a curing stage (see Figure 23.61). When the U.S. Department of Agriculture (USDA) developed the aerated static-pile process, researchers found that 21 days of aerated composting followed by 30 days of unaerated curing would adequately stabilize a raw feed with wood chips as the bulking agent. However, to create fully stabilized compost suitable for any use, another 20 days or more of detention time is recommended. This detention time criterion has been codified in a number of state regulations and incorporated into some design standards. Most horizontal agitated-bed systems are designed for 21 days of aerated composting followed by curing. However, other in-vessel systems use shorter active composting times (often 14 days)

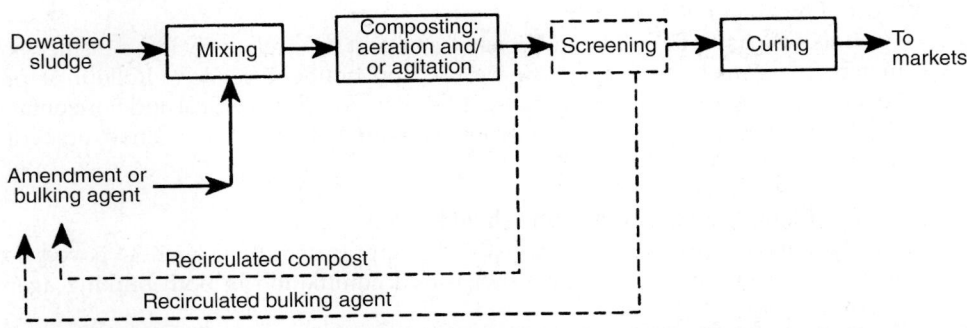

FIGURE 23.61   Generalized composting flow chart (dashed lines indicate optional steps; screening may follow curing; drying step may precede screening).

to minimize the system's capital costs. Additional detention times outside the vessel (in the form of windrow or static pile systems) are typically added to these systems.

Detention time is affected by the bulking agent or amendment, carbon-to-nitrogen ratio, and pH. An amendment that is not screened out may continue to decompose, prolonging the curing period. An excessively high carbon-to-nitrogen ratio may have the same effect. The composting process has no fixed end point because the organic materials continue to decompose after the compost is considered stable. One test for stability is based on a respiration rate measured as a rate of carbon dioxide evolution. A respiration rate of 3 mg CO_2/g organic carbon per day typically indicates that compost will be free of fecal odor and phytotoxic effects.

Another test measures oxygen consumption. Jimenez and Garcia (1989) report that a compost taking up 0.96 mg O_2/g of organic carbon per day is considered stable. This is equivalent to 1.4 mg CO_2/g of carbon per day.

4.5 General Design Considerations

The essential elements of composting facility designs involve handling large volumes of material and air. The relative importance of each depends on the composting technology used. For windrow operations, air handling is negligible or nonexistent; for enclosed operations, proper air handling is vital.

4.5.1 Site Layout

As with any facility, layout is dictated by the available site; however, there are a few items to keep in mind. Because of the large volumes of material handled, all composting operations involve the use of heavy equipment (e.g., front-end loaders and trucks). Concrete or high-durability asphalt-paved pads serve best for bulking agent storage, mixing, pile construction, screening, curing, and finished compost storage areas. Runoff from any areas exposed to raw feedstock must be collected and treated. Typically, facilities are designed with covered areas for bulking agent storage and composting aerated static-pile systems. Covered facilities can be operated under adverse weather conditions and generate minimal runoff.

4.5.2 Material-Handling Systems

Material typically is moved around a composting facility by either a front-end loader or a conveyor. In-vessel technologies have special equipment for moving material in the vessel for portions of processing, but they still rely on the loaders and conveyors for most of the material movement.

Bulking agents, biosolids, and finished compost have relatively low densities, so light-material, large-volume buckets can be used on front-end loaders. Rollout or pushout buckets are also advantageous because they allow for more vertical and horizontal reach.

The two most commonly used conveyors are belt and screw conveyors. For more information, refer to Chapter 19.

4.5.3 Bulking Agent Storage and Handling

Ideally, storage is provided for a 15- to 30-day supply of bulking agent. A paved, covered storage area minimizes excessive moisture accumulation in both bulking agent and finished compost.

Enclosing the unloading and conveying facilities minimizes the spread of dust and particulate, and protects the equipment from the adverse effects of wet and cold weather. Because the dust could explode, the design of any enclosure should adhere to applicable explosion hazard standards.

4.5.4 Mixing

As previously indicated, dewatered solids must be well mixed with a bulking agent to ensure uniformity and good airflow characteristics during composting. Therefore, a mechanical mixing system typically is included in the design. A good mixture consists of bulking agent particles uniformly coated with solids containing no balls of dewatered solids that are more than 126 mm (5 in.) in diameter. Immediately mixing dewatered solids with a bulking agent minimizes storage-facility size and the potential for odor generation. A solids-bulking agent mixture can be stacked and conveyed more easily than dewatered cake alone.

Several types of mechanical mixing systems are available:

- Front-end loaders portion feedstock in discrete piles and "toss" the material several times until it is blended (much like how a salad is tossed). Mixing is time consuming, and not particularly effective. Loaders are best suited for small facilities and as backup for another mixing system.

- Batch mixers are stationary, truck-mounted, or trailer-mounted hoppers equipped with internal paddles or augers that mix the material. The blended batch is discharged via a short, side-mounted conveyor with a slide gate. Batch mixers also have internal scales and a weight display to help operators portion the material. They typical are loaded by front-end loaders but also can be loaded via a conveyor from a live-bottom hopper. Batch mixers are well suited for small and medium facilities.

- Continuous mixers (e.g., pug mills and plow mixers) are the most automated, complex mixing systems. In these systems, feedstocks first are loaded into separate live-bottom hoppers, which have variable-speed augers to meter the correct portions of each feedstock. (Feedstocks are weighed by elements in the hoppers' discharge conveyors.) The material is conveyed to the mixers for blending, and afterward, conveyors discharge the blended material into or near the composting piles or vessels. Continuous mixers are found only at medium and large facilities, where their capital costs are offset by labor cost savings.

- Windrow turners are mobile machines designed to mix materials that have been layered on a concrete pad. The machines vary in size and complexity. Small machines towed by a tractor can stack material 0.6 to 0.9 m (2 to 3 ft) high. Large self-propelled machines can form piles about 2.4 m (8 ft) high. This technology works well in windrow composting operations.

- A horizontal agitated-bed reactor also provides mixing. However, the material should be premixed before being loaded into such reactors to optimize the reactors' SRT.

Mixing and storage areas are odorous and, if enclosed, typically need at least six air changes per hour for effective odor control and personnel safety. Design engineers should consider treating exhaust airstreams from these process areas before discharging to the atmosphere.

4.5.5 Leachate

All composting processes produce leachate that must be treated. Some common sources of leachate sources in composting operations include:

- Aeration pipes and ducts;
- Building ventilation ductwork;
- Composting piles;
- Washdown water for all mobile and stationary equipment;
- Biosolids and recycled bulking agent storage areas; and
- Site drainage from areas exposed to unfinished compost, recycled bulking agent, and biosolids.

Aeration fans and ductwork—especially in negative-mode aeration—must be equipped with drains and cleanouts in all low areas. Even in positive aeration, ventilation fans and ducts will collect condensation, so they must have adequate drains and cleanout access. Drains for compost piles are often part of the aeration floor and must be equipped with traps to prevent air from short-circuiting the process.

All stationary and mobile equipment must be washed down periodically to keep it in good working order. For mobile equipment, a designated washdown area is often part of the facility. For stationary equipment, drains must be provided. All conveyor and other equipment pits also should have drains both for washdown and condensation, which occurs in enclosed facilities.

All of the water from the above sources, as well as any water that contacts unfinished compost or raw materials, must be collected and treated. Water is a by-product of decomposition, so leachate can contain soluble organics, nutrients, and other material that cannot be released to the environment. Leachate should be discharged to a sanitary sewer, recycled to the WRRF's headworks, or treated onsite.

4.5.6 Aeration and Exhaust Systems

Compost typically can be aerated either by forcing air up through the material (positive aeration) or by pulling it down through the material (negative aeration). Positive aeration typically requires less energy than negative aeration to move the same volume of air. In positive aeration, the air is cooler, is drier, and, therefore, has less volume.

Negative aeration is better in enclosed and worker-occupied operations because it captures most of the material's odors and moisture, preventing them from entering the air above the pile (where greater airflows are required to capture and treat such emissions). However, condensation accumulates in the ductwork and blowers, so ample drainage must be provided.

Most in-vessel systems use positive aeration for system-specific reasons. For example, in-tunnel systems have little headspace above the piles and are not occupied by workers during active composting, so there is no advantage to negative aeration. Negative aeration is popular in aerated static-pile operations because it directly captures odors and moisture. Many aerated static-pile operations are configured to allow for both negative and positive aeration. During decomposition, negative aeration captures odors and moisture. Afterward, positive aeration can provide more air to accelerate drying before the compost is screened.

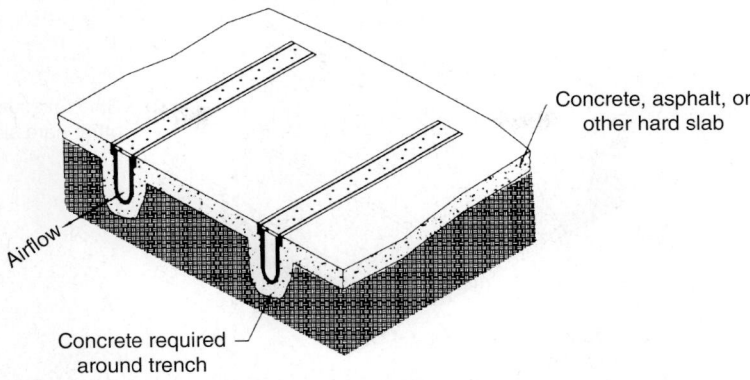

FIGURE 23.62   Composting floor with aeration trenches.

Figures 23.62 to 23.65 show several air-floor configurations for aerated static-pile systems. For example, agitated bay systems use a perforated pipe embedded in a gravel plenum. No matter which configuration is used, it is vital that it be designed to deliver air evenly the entire length of the pile. Three methods are used to accomplish this:

- Provide progressively more air outlets along the pipe or trench so friction headloss is offset by reduced velocity loss through the outlets;

- Change the cross section of the pipe or trench to provide a constant air velocity along the entire length; or

- Use a combination of the two.

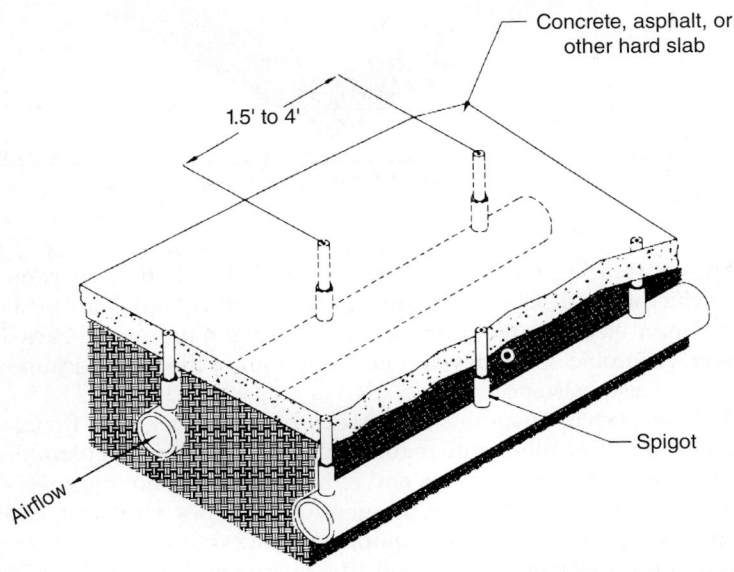

FIGURE 23.63   Composting floor with embedded pipe and spigot aeration system.

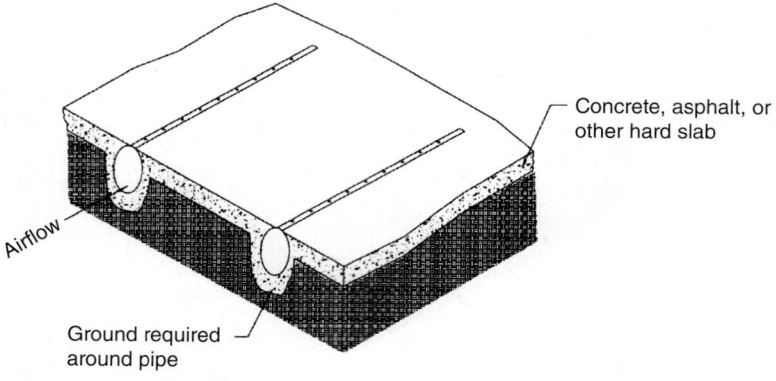

FIGURE 23.64   Composting floor with embedded pipe aeration system.

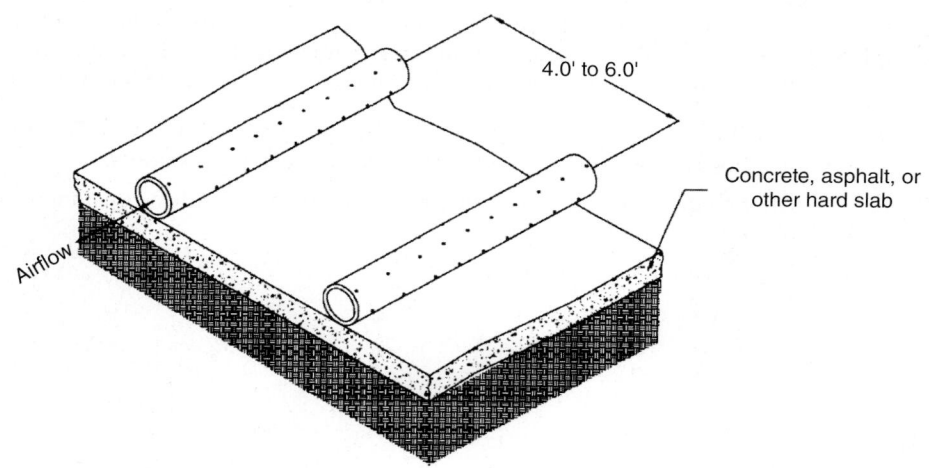

FIGURE 23.65   Composting floor with aeration pipes on slab.

Pipes or trenches typically are spaced 1 to 2 m (3 to 6 ft) apart on a layer of wood chips, which help distribute airflow. The spacing depends on the size of the pipe or trench; larger elements require more space. If the pipes or trenches are too far apart, however, anaerobic zones develop in the bottom of the piles because air always seeks the path of least resistance.

In-vessel systems may use continuous plenums or gravel floors with permanent piping. All of them will require regular cleaning. Many gravel plenums develop a hard pan on the surface that will block and redirect airflow if not routinely removed.

In aerated static-pile systems, the air outlets in pipes or trenches can become blocked with material, particularly when equipment moves over the outlets to add and remove material. The outlets must be cleaned after every one or two uses, using either water or compressed air.

In negative aeration, the air initially is hot and virtually saturated, but cools slightly as it moves through the duct. So, condensation forms and must be removed via frequent drains and cleanout. Drain traps also will be needed to prevent airflow from short-circuiting. Because of the heat and moisture, negative aeration systems need corrosion-resistant ductwork. Fiberglass, PVC, polyethylene, and stainless steel have all been used.

When operating composting systems with forced aeration, O&M personnel need to be able to:

- Monitor and record pile temperatures; and
- Control aeration quantities based on oxygen demand, temperature, and moisture-removal requirements.

The simplest control system involves measuring and recording pile temperatures manually and controlling aeration blowers via a manually adjusted cycle timer. The most complex system includes a temperature feedback control where outputs from temperature probes in the piles are connected to a computer, which adjusts the aeration rates based on temperature readings using a preset control strategy.

All control systems used there are certain rules that should be observed in controlling the aeration:

- During active composting (when the material is heating up), the blowers typically should not be off for more than 15 minutes in a cycle. Murray and Thompson (1986) reported significant oxygen depletion after 12 to 15 minutes without aeration (see Figure 23.66).
- The temperature of the composting material should be measured directly, whenever possible. Although this seems obvious, some systems measure temperatures via sensors in the ductwork or in the walls contacting the piles (to avoid damage from agitation). Such sensors consistently provide measurements that are lower than the actual pile temperature, causing the material to be underaerated.

Because the material's aeration demand constantly changes during composting, blowers must be able to provide a varying amount of airflow to the material. Design engineers can either provide single- or two-speed blowers, which will operate intermittently, or provide variable-frequency drives so the blowers can run continuously but the airflow is varied based on process needs. Another option is equipping single- or two-speed blowers with motorized dampers to regulate the amount of airflow supplied to each pile based on process needs.

4.5.7 Ventilation
The National Fire Protection Association issues ventilation rules related to fire prevention; however, in enclosed composting facilities, ventilation rates typically must be larger to control odors, control moisture (reduce fog and condensation), and ensure that workers are safe. In cold climates, heavy building insulation is needed to prevent condensation in winter and avoid worker heat stress in summer. Ventilation is also an important element in reducing corrosion resulting from a humid environment.

Because the material often is moved around a facility by a front-end loader, design engineers must consider what happens when doors are left open for extended periods.

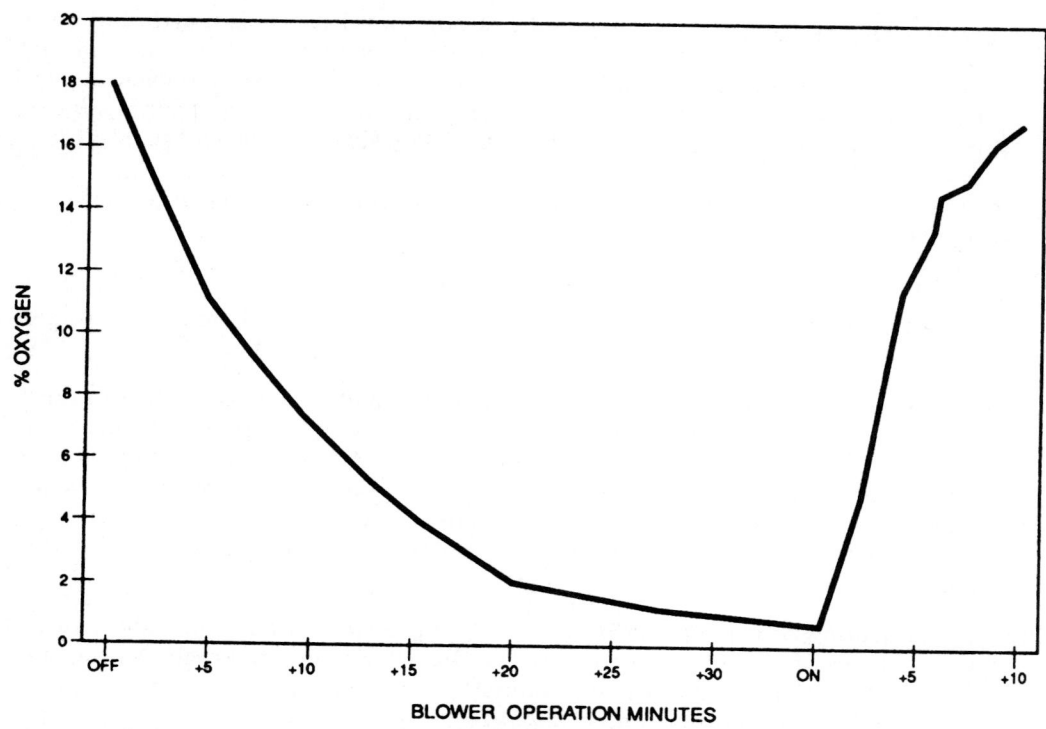

FIGURE 23.66   Oxygen depletion and regeneration in an active compost pile (Murray and Thompson, 1986, with permission from *BioCycle*).

How will that affect ventilation rates and ductwork design? They also must locate the air-collection points in the building to avoid dead air spaces, where ammonia and other compounds can accumulate.

4.5.8   Screening

Except for leaves and sawdust, bulking agents can be screened out of the finished compost and reused. This reduces bulking agent costs by 50% to 80%. Screening also produces more uniform, aesthetically pleasing compost, thereby improving its marketability. Vibrating deck screens and rotating screens typically are used. All screens must have a self-cleaning feature (e.g., rotating brushes in rotating trommel screens, or a layer of balls between the decks of a vibrating deck screen).

Vibrating deck screens and rotary trommel screens can separate material into multiple sizes, which can be useful if some markets (e.g., turf top dressing) demand a product with fine particles.

4.5.9   Product Curing and Storage

Composting basically has two phases: a rapid decomposition period (14 to 21 days) followed by a longer, slower one with significantly lower oxygen and moisture-removal demands. This second phase (called curing) is typically about 30 days long and is needed to produce a stable, usable product. Sometimes, curing consists of merely stockpiling the material, but this can prolong curing time and increase the danger of fires. Low-rate aeration better controls curing time and product stability. Also, the curing material should be covered to control its moisture content and thereby prevent the material from compacting and going anaerobic. This is especially important if the material was screened first.

4.5.10   Odor Control

Odor control may be the composting industry's greatest challenge. Most conflicts over and suspensions of composting operations have been caused by odors or concerns about potential odors. Composting is inherently odorous as a result of the production and removal of volatile products of decomposition. Current design practices include more emphasis on enclosing operations, capturing and treating exhaust air, and improving process control to reduce odors at the source.

4.5.10.1   Odor Sources in Composting   Every stage in the composting process is a potential source of odors (see Tables 23.38 and 23.39). Odor sources can be divided into the following three categories:

- Active sources are those that exist when material is being actively handled (e.g., during mixing, screening, and dewatering). Odors from these sources occur during working hours.
- Continuous sources are those that originate in the aeration and storage areas. These may be point sources (e.g., blower exhaust) or area sources (e.g., pile and windrow surface emissions). Odors from these sources may occur 24 hours a day.
- Housekeeping sources are those related to material spills, unclean equipment, and condensate on ground surfaces. Such odors can persist after daily activity has stopped, so they are continuous sources.

4.5.10.2   Odor Measurement   The concentrations of individual compounds can be measured via standard analytical methods. For example, a simple apparatus consisting of a manual pump and a colorimetric adsorption tube can be used in the field. (Tubes are available for a number of the compounds listed in Table 23.39.) For more accurate and complete results, samples should be collected (in bags, stainless steel vacuum canisters, or tubes filled with adsorbent) and analyzed via gas chromatography in a laboratory.

However, the odor of composting typically is a mixture of compounds that cannot be quantified as a sum of individual constituents. Such odors can only be directly measured by the human nose (sensory analysis). Odor samples can be captured in Tedlar bags for sensory analysis at another location. Several methods have been developed to quantify odor concentrations using a panel of human subjects; these are described in detail in Chapter 6.

4.5.10.3   Containment and Treatment   The level of odor containment and control is dictated by the proximity of neighbors and local regulations. Design engineers must take care to provide for adequate capture of emissions under all operating conditions. For example, failing to account for material-movement operations that require open doors lead to fugitive emissions.

Odor Source	Category*
Dewatered sludge transport and storage	
Open trucks en route	A
Open trucks parked on site	A
Dumping operations	C
Untreated ventilation from storage facilities	C
Open conveyors	A
Spillage from trucks	H
Spillage around storage facilities	H
Residue on empty trucks	H
Puddles from truck washing	H
Tire trucking of spillage sludge	H
Mixing	
Surface emissions from mixing by front-end loader or batch mixer	A
Untreated ventilation from pug mills	A
Mix left on paved surface after day's activities	H
Residue on equipment	H
Pile building	
Surface emission from materials-handling activities	A
Spillage of mix left on paved surface after day's activities	H
Residue on equipment	H
Sludge balls from poor mixing	H
Surface emissions from pile before placement of blanket layer	A
Composting	
Surface emission from active piles	C
Leachate puddles at base of piles	H
Aeration	
Blowers exhaust	C
Leakage of condensate from aeration piping	H
Leakage of exhaust from piping and blower housing	H

*A = active source; C = continuous sources; and H = housekeeping sources.

TABLE 23.38   Typical Odor Sources in Composting Operations

Class Compounds	Odor Threshold, ppm	Likely Source at Treatment Plant	Pathway of Formation/Release
Inorganic sulfur			
Hydrogen sulfide	0.000 47	Septic wastewater or sludge	Anaerobic reduction of sulfate to sulfide or anaerobic breakdown of amino acids
Organic sulfur			
Mercaptans			
Ethyl mercaptan	0.000 19	Sludge or wastewater subjected to anaerobic conditions	Anaerobic and aerobic breakdown of amino acids
tert-Butyl mercaptan	0.000 08		
Allyl mercaptan	0.000 05		
Organic sulfides			
Dimethyl sulfide	0.001	Composting	Aerobic oxidation of mercaptans
Dimethyl disulfide	0.002		
Inorganic nitrogen			
Ammonia	0.037	Composting; processing of anaerobically digested biosolids	Anerobic decomposition of organic nitrogen volatilization at high pH, temperature
Organic nitrogen			
Methylamine	0.021	Solids processing	Anaerobic decomposition of acids
Ethylamine	0.83		
Dimethylamine	0.047		
Fatty acids			
Acetic acid	0.001–1.0	Sludge subjected to anaerobic conditions	Anaerobic decomposition
Propionic acid	0.005–0.05		
Gutyric acid	0.000 01–0.01		
Aromatics			
Acetone	0.05–400	Preliminary and primary wastewater treatment processes, solid processing, and composting	Present in wastewater contribution from industry; breakdown of lignins
Methylethyl ketone	1–12	Composting, wood-based bulking agents	
Terpenes	Varies		Present in wood products, such as wood chips, sawdust

TABLE 23.39 Odor Compounds and Sources (Verschueren, 1983; WEF, 1995)

Once contained and captured, odors can be treated or exhausted. Treatment typically is required. A wide variety of treatment technologies are available (see Chapter 6). Organic media biofilters have been used extensively at composting facilities for several reasons:

- They have proven effective in treating compost odors.
- They are inexpensive and easy to operate.
- Composting facilities typically are large enough to have space for the biofilter.
- The materials used for biofilter media are readily available at composting facilities.
- The equipment used to replace biofilter media (e.g., front-end loaders) is available at any composting facility.

For more details on odor removal, see Chapter 6.

4.5.11 Design Example

When designing any in-vessel system, engineers need details from vendors; in fact, a vendor often is selected before the detailed design proceeds. When designing an aerated static-pile system, on the other hand, the details are not vendor dependent. Below is an example of an aerated static-pile system design. The following design criteria apply to this example:

- 20 dry ton/d of cake containing 20% solids.
- Operations occur 7 days per week.
- The bulking agents are yard waste supplemented by ground wood waste.
- All storage, mixing, active composting, and screening operations are fully enclosed.
- Enough covered storage space for 30 days' worth of new bulking agent.
- Enough storage space for 1 day's worth of feedstock biosolids.
- Enough covered storage space for 7 days' worth of recycled bulking agent.
- 21-day minimum SRT in active composting area.
- 28-day minimum SRT in aerated curing area.
- Enough outdoor storage for 90 days' worth of finished product.

In this exercise, the various areas of the aerated static-pile system will be sized. Each area's size depends on the types of vehicles expected and the site topography.

First, design engineers must develop a materials balance for the facility (see Table 23.40). The total amount of bulking agent that needs to be recycled is:

Total recycled bulking agent = Screened recycled input + Base (recycled bulking agent)

$$= 204 \text{ cu yd} + 28 \text{ cu yd} \tag{23.38}$$

$$= 232 \text{ cu yd } (177 \text{ m}^3)$$

Comparing the materials balance with Table 23.37 and multiplying those values by 20, design engineers determine that adding drier ground-wood waste reduced the

Material	Volume (cu yd)	Total Weight (Ton)	Dry Weight (Ton)	Volatile Solids (Ton)	Bulk Density (lb/cu yd)	Solids content	Volatile solids
Biosolids	125.0	100.0	20.0	10.6	1600	20.0%	53.0%
Yard waste (processed)	180.0	54.0	29.7	20.8	600	55.0%	70.0%
Wood waste	26.7	6.7	4.0	3.8	500	60.0%	95.0%
Screened recycled bulking agent	204.1	70.9	39.0	36.3	695	55.0%	93.0%
Unscreened recycle	0.0	0.0	0.0	0.0	780	55.0%	88.6%
Mixture	508.9	231.6	92.7	71.5	910	40.0%	77.1%
Base (recycled bulking agent)	28.3	9.8	5.4	5.0	695	55.0%	93.0%
Cover (unscreened)	56.5	22.1	12.1	10.7	780	55.0%	88.6%
Composting losses		182.8	7.1	7.1			
Cover (unscreened)	56.5	22.1	12.1	10.7	780	55.0%	88.6%
Screen feed	424.0	165.4	91.0	69.3	780	55.0%	76.2%
Recycled bulking agent	232.2	80.7	44.4	41.3	695	55.0%	93.0%
Curing	188.2	84.7	46.6	28.1	900	55.0%	60.3%
Curing losses		3.8	2.1	2.1			
Compost to storage	179.8	80.9	44.5	26.0	900	55.0%	58.4%

Assumptions:
Recovery by screening
- Yard waste   50% by volume
- Wood waste   70% by volume
- Recycled bulking agent   50% by volume
- Pile base   95% by volume

Processing losses
- Losses during composting   10% of volatile solids
- Losses during curing   5% of volatile solids

*cu yd × 0.7646 = m³; lb/cu yd × 0.5933 = kg/m³; ton × 0.9072 = Mg.

TABLE 23.40   Materials Balance for Design Example*

overall amount of active-composting feedstock from 393 to 389 m³ (514 to 509 cu yd). It also reduced the amount of product produced from 145 to 138 m³ (190 to 180 cu yd). The effect on product production is larger than that on feedstock volume and, therefore, facility size.

The following areas are constructed with concrete walls on three sides: active composting, curing, and storage areas for biosolids and all bulking agents.

For most front-end loaders, the maximum height will be 3 to 3.6 m (10 to 12 ft). In this example, the maximum height (H) is 3 m (10 ft). The following equation represents the total volume for a given area; it is manipulated to determine the desired value. For example, a narrow site may limit the allowable length (L). As a general rule, the width (W) should be at least 4.6 m (15 ft) for each day's worth of material. This is wide enough for a front-end loader to dig out the material while leaving the piles around it intact.

$$\text{Pile volume/d} \times \text{Number of days} \times 27\ \text{cu ft/cu yd} = H \times (L - H/2) \times W \qquad (23.39)$$

For the bulking agent storage area (for 30 days' worth of ground yard waste) with an assumed length of 30 m (100 ft) and height of 3 m (10 ft),

$$W = \frac{204 \times 27 \times 30}{10\left(100 - \dfrac{10}{21}\right)} = 157\ \text{ft} \qquad (23.40)$$

For a biosolids storage area with an assumed width of 9 m (30 ft) and height of 0.9 m (3 ft),

$$\frac{L}{D} = \frac{127 \times 27 \times 1 + \dfrac{3}{2}}{3 \times 30} = 38\ \text{ft (11 m)} \qquad (23.41)$$

Biosolids typically are dense and gelatinous, so they do not stack well. Biosolids containing 18% to 24% solids will only stack about 0.9 or 1.2 m (3 or 4 ft) high. Wetter biosolids will not stack more than 0.3 m (1 ft) high.

When sizing the active compost area, design engineers should keep in mind that most composting facilities will put 1 day's worth of material in each bay. (Small facilities may put 2 or 3 days' worth of material in a bay.) The number of bays depends on the desired SRT (21 days is the usual minimum). Two extra bays should be provided to allow for one bay to be torn down and another to be constructed without reducing SRT. Aerated static-pile facilities typically have active composting areas that are constructed with multiple bays on either side of a center aisle. A bay on one side typically serves as a mixing surge area (depending on the mixing method selected). There is no physical obstruction between bays.

Below is the length of the active compost hall, based on an assumed width of 6 m (20 ft) and a mixture depth of 2.4 m (8 ft). Although the overall pile depth will be 3 m (10 ft), design engineers need to allow for 0.3 m (1 ft) of plenum layer and 0.3 m (1 ft) of cover layer. The minimum bay width should be 4.6 m (15 ft) so front-end loaders have enough space to build and tear down 1 day's worth of material.

$$\frac{L}{\text{bay}} = \frac{509 \times 27 \times 1 + 8/2}{8/20} = 86\ \text{ft (26 m)} \qquad (23.42)$$

When calculating the composting hall's overall width, design engineers need to include allowances for the piles, the center aisle, and the aeration blowers (which typically are housed behind the piles). The minimum allowance for the blower gallery depends on the size of the blowers and ductwork, and the access to the area. If the only access to the blower gallery is from the ends of the compost building, the gallery must be wide enough to move blowers without dismantling them. If access doors can be put closer to the blowers, the hall can be narrower. In this example, a 4.6-m-wide (15-ft-wide) gallery is assumed.

The center aisle must be at least 9 m (30 ft) wide so front-end loaders have enough maneuvering space to construct and tear down piles. If the material from the active compost pile will be loaded directly onto trucks, which will deliver it to another location for curing, then the center aisle should be at least 13.7 to 15.2 m (45 to 50 ft) wide.

$$\text{Hall width} = 2 \times (86 + 15) + 30 + 4 \times 1$$
$$(\text{allowance for concrete walls}) = 236 \text{ ft } (72 \text{ m}) \tag{23.43}$$

$$\text{Compost hall length} = 20 \times 12 + 2$$
$$(\text{allowance for concrete walls}) = 242 \text{ ft } (74 \text{ m}) \tag{23.44}$$

In most aerated static-pile facilities, each bay (1 day's worth of material) is aerated separately (typically one blower per bay). This configuration provides the most flexibility and least interruption in operations if a blower goes out of service. At small facilities, however, one continuously running blower can serve several bays. In this example, one blower will be used for each bay.

$$\text{Required aeration} = \frac{5000 \, \frac{\text{cu ft}}{\text{hr} \cdot \text{dry ton}} \times 20 \text{ dry ton biosolids}}{60 \text{ min/hr}}$$
$$= 1667 \text{ cfm } (787 \text{ L/s}) \tag{23.45}$$

4.6   Health and Safety Considerations

Potential dangers associated with composting systems include poorly ventilated areas, areas where exhaust gas is discharged, conveyors, and heavy equipment traffic. The primary concerns include:

- Fog generation in cold weather: Dense fog in a building with heavy equipment is an obvious hazard; it also may prevent others from seeing an injured worker.
- Worker heat stress: Composting generates significant amounts of heat. During warm weather, enclosed composting facilities can easily exceed 100°F for prolonged periods.
- Unsafe chemical concentrations: If not properly ventilated, pockets of dead air can develop unhealthy concentrations of compounds (e.g., ammonia).
- Dust—Near screening operations and in high-traffic areas: Dust levels can exceed Occupational Safety and Health Administration (OSHA) limits if not properly contained and captured. Design engineers should provide screens with hoods connected to dust collectors. High-traffic areas should be cleaned regularly to prevent dust buildup.

Material-handling equipment (e.g., conveyors and screens) has exposed moving parts and poses a worker hazard. The main safety concerns are at the points of material transfer and locations of exposed belts. To minimize the possibility of material spilling or accumulating at transfer points, design engineers should provide emergency pull-cords along the full length of conveyors, as well as interlocks to shut down all material-handling operations in the event of an emergency.

Wood chips and compost piles may contain high concentrations of the airborne fungus *A. fumigatus*, which naturally occurs in grass and leaves. Although typically not harmful, *A. fumigatus* may cause aspergillosis in individuals with extreme susceptibility. Personnel with respiratory problems, that exhibit adverse physical reactions, or who have histories of suppressed immune response should not work in a composting or wastewater treatment facility.

5.0   Alkaline Stabilization

Adding alkaline chemicals to solids is a reliable stabilization method that WRRFs have practiced since the 1890s. The chemicals traditionally used are quicklime and hydrated lime.

In recent years, a number of advanced alkaline-stabilization technologies have emerged. These technologies, which use new chemical additives, special equipment, or special processing steps, all claim advantages over traditional lime stabilization (e.g., enhanced pathogen control and a more publicly acceptable product). They also produce a biosolids that sometimes is called artificial soil because it has been successfully used as a soil substitute.

Lime is the most widely used and one of the least expensive alkaline materials available in the wastewater industry. It has been used to reduce odors in privies, increase pH in stressed digesters, remove phosphorus in advanced wastewater treatment processes, treat septage, and condition solids before and after mechanical dewatering. It is also the principal stabilizing chemical at WRRFs with capacities ranging from 379 m^3/d to approximately 1.13 million m^3/d (0.1 to 300 mgd) (U.S. EPA, 1979). Larger facilities that have used the process include those in Pittsburgh, Pennsylvania; Memphis, Tennessee; and Toledo, Ohio; as well as DC Water Blue Plains Advanced Wastewater Treatment Plant in Washington, D.C. According to the U.S. EPA's *1988 Needs Survey of Municipal Wastewater Treatment Facilities*, more than 250 facilities use lime stabilization (U.S. EPA, 1989). According to a 2007 Northeast Biosolids Management Association survey, 900 of the 4800 facilities surveyed—18% of facilities surveyed and 12% of the total volume of biosolids produced—used some form of alkaline stabilization (NEBRA, 2007). These results emphasize that alkaline stabilization primarily is used by smaller treatment facilities.

Alkaline-stabilized biosolids can be beneficially used in many ways, depending on the particular quality requirements and associated standards. Traditional lime stabilization is classified in the U.S. EPA's Standards for the Use or Disposal of Sewage Solids as a Class B process (PSRP) (U.S. EPA, 1999b). Many of the advanced alkaline-stabilization technologies meet the U.S. EPA's definition of a Class A process (PFRP).

Many of the beneficial use and disposal options for alkaline stabilized biosolids are further discussed in Oerke (1999).

5.1 Stabilization Objectives

The purposes of alkaline stabilization may include:

- To substantially reduce the number and prevent the regrowth of pathogenic and odor-producing organisms, thereby preventing biosolids-related health hazards;
- To create a stable product that can be stored; and
- To reduce the short-term leaching of metals from biosolids not incorporated with natural soil.

Several studies have demonstrated that both liquid and dry lime stabilization achieve significant pathogen reduction, provided that a sufficiently high pH or temperature is maintained for an adequate period of time (Bitton et al., 1980; Christensen, 1982). Table 23.41 lists bacteria levels measured during full-scale studies at the Lebanon, Ohio, WRRF; it shows that liquid lime stabilization at pH 12.5 and a 25% dose (dry-weight) reduced total coliform, fecal coliform, and fecal streptococci concentrations by more than 99.9%. Also, the numbers of *Salmonella* and *Pseudomonas aeruginosa* were reduced below the level of detection. In addition, Table 23.41 shows that pathogen concentrations in liquid lime-stabilized biosolids ranged from 10 to 1000 times less than those in anaerobically digested biosolids from the same WRRF.

Christensen (1987) researched the pathogen-reduction performance of dry lime stabilization using dry quicklime doses of 13% and 40% (as calcium hydroxide;

Type of Solids	Bacterial Density, Number/100 mL				
	Total Coliforms[a]	Fecal Coliforms[a]	Fecal Streptococci	*Salmonella*[b]	*Ps. aeruginosa*
Raw sludge					
Primary	2.9×10^9	8.2×10^8	3.9×10^7	62	195
Waste activated	8.3×10^8	2.7×10^7	2.7×10^7	6	5.5×10^3
Anaerobically digested biosolids					
Mixed primary and waste activated	2.8×10^7	1.5×10^5	2.7×10^5	6	42
Lime stabilized biosolids[c]					
Primary	1.2×10^5	5.9×10^3	1.6×10^4	<3	<3
Waste activated	5.2×10^5	1.6×10^4	6.8×10^3	<3	13
Anaerobically digested	18	18	8.6×10^3	<3	<3

[a]Millipore filter technique used for waste activated sludge. Most probable number technique used for other sludges.
[b]Detention limit = 3.
[c]To pH equal to or greater than 12.0.

TABLE 23.41 Bacteria Reduction Via Liquid Lime Stabilization at Lebanon, Ohio (U.S. EPA, 1979)

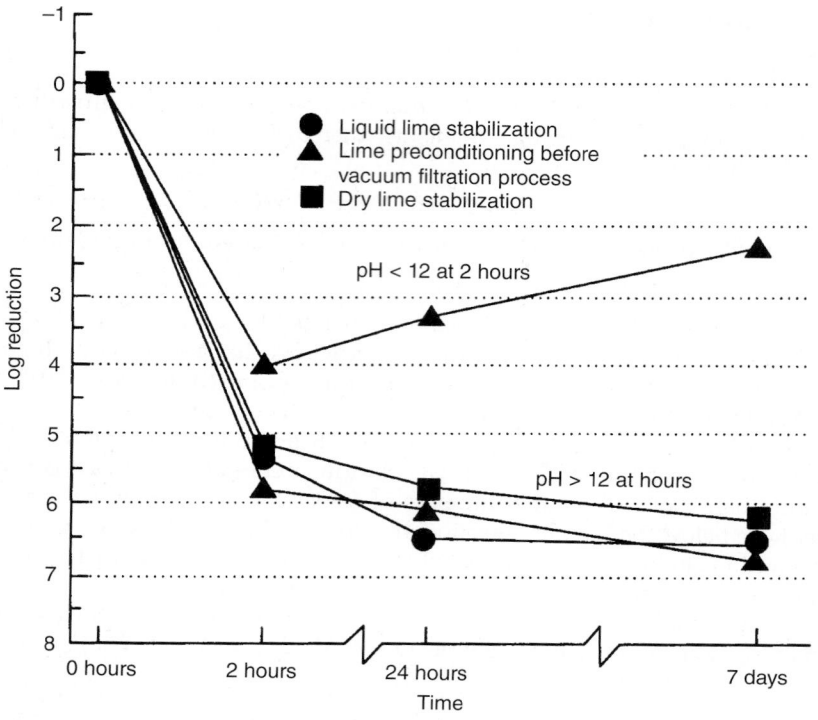

FIGURE 23.67 Average fecal coliform inactivation via two liquid lime stabilization processes and one dry lime stabilization process (Westphal and Christensen, 1983).

dry-weight basis). His results indicated that dry lime stabilization can reduce fecal coliform and streptococcus pathogens by at least two orders of magnitude. This was as good as, and in some cases better than, the results of standard liquid lime stabilization and liquid lime conditioning followed by vacuum filtration (see Figures 23.67 and 23.68). No growth of either fecal coliforms or fecal streptococci occurred by the seventh day (Westphal and Christensen, 1983). Westphal and Christensen (1983) also reported that alkaline-stabilization processes used to reduce the densities of fecal coliform and fecal streptococcus performed as well as or better than mesophilic aerobic digestion, anaerobic digestion, and mesophilic composting (see Table 23.42). Additional discussions of lime treatment and the control of bacterial, viral, and parasitic pathogens are reviewed in reports by Christensen (1987) and Reimers et al. (1981).

Another study of dry lime stabilization showed a 4- to 6-log reduction of fecal streptococcus at about pH 12 (Otoski, 1981). Such a treatment scheme can yield Class A biosolids by using a 2:1 lime dose at 20% solids (dry-weight basis) to raise the solids temperature to more than 70°C for 30 minutes and meet pathogen-reduction requirements via the heat of lime hydration.

Treating dewatered cake with cement-plant kiln dust (alone or with a small amount of quicklime) reduces pathogenic microbe populations below the U.S. EPA's Class A standard (Burnham et al., 1992). Both laboratory- and large-scale field tests have shown

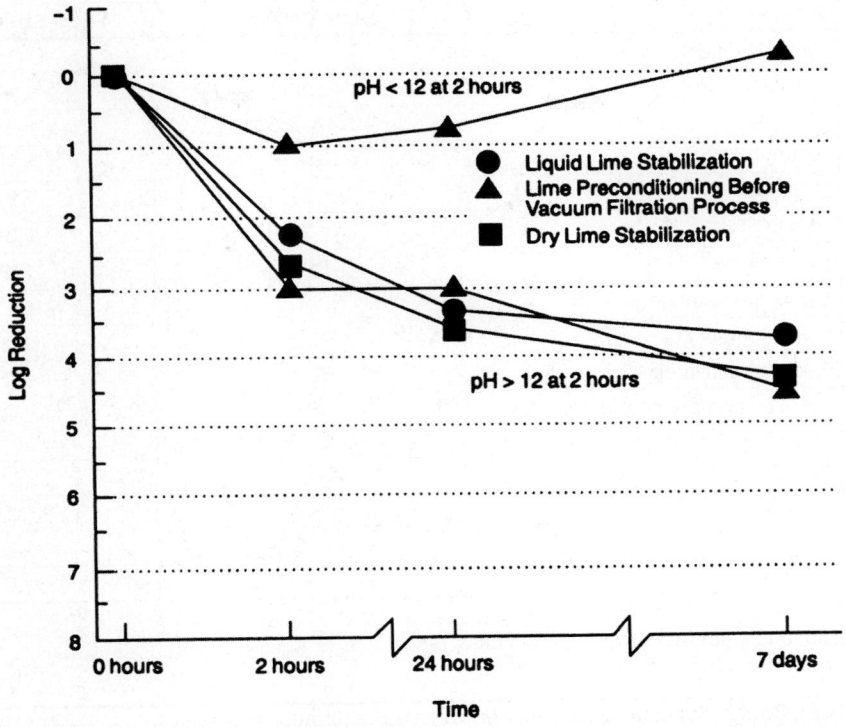

FIGURE 23.68 Average fecal streptococci inactivation via two liquid lime stabilization processes and one dry lime stabilization process (Westphal and Christensen, 1983).

that indigenous and seeded populations of *Salmonella*, poliovirus, and *Ascaris* ova can be eliminated within 24 hours if the treated biosolids are contained at pH 12 and 52°C for 12 hours.

Although there is little information quantifying virus reduction during lime stabilization, lime has been identified as an effective viricide. Qualitative analysis has indicated substantial survival of higher organisms (e.g., hookworms and amoebic cysts) after 24 hours at high pH (Farrell et al., 1974). It is unknown whether prolonged contact eventually destroys these organisms. Class A alkaline-stabilization processes that maintain 70°C for 30 minutes have been shown to kill *Ascaris* ova. Studies have shown that a high pH has little effect on parasites (e.g., toxocara, mites, and nematodes) (U.S. EPA, 1975). Comparisons of parasite types in lime-stabilized and anaerobically digested solids showed similar parasite types and densities in both solids.

Alkaline stabilization is a simple process. An alkaline chemical is added to feed solids to raise its pH, and adequate contact time is provided. At pH 12 or higher, with sufficient contact time and thorough lime–feed cake mixing, pathogens and microorganisms are either inactivated or destroyed. The chemical and physical characteristics of the resulting biosolids also are altered. The chemistry of the process is not well understood, although it is believed that some complex molecules are split by reactions (e.g., hydrolysis and saponification) (Christensen, 1982). It is also now understood that

Process	Fecal Coliform	Fecal Streptococci
Anaerobic digestion (35°C)		
Mean	1.84	1.48
Range	1.44–2.33	1.1–1.94
Anaerobic digestion		
20°C[a]	1	1
30°C[a]	2	1.64
Composting	≥4	2.9
Liquid lime stabilization		
Raw primary	5.1	2.4
Waste activated	3.2	3.2
Mixed primary and trickling filter humus, 4% solids	2.6	1.8
Storage[b]		
10°C	—	1
20°C	—	1.5
30°C	—	2.0

[a]Laboratory study, 35-day detention time.
[b]Laboratory study, 30-day detention time.

TABLE 23.42   Bacteria Reduction Via Various Stabilization Processes

high pH releases gaseous ammonia from biosolids. Gaseous ammonia has been shown to be an effective disinfectant.

To meet Class B stabilization requirements, the pH of the feed cake–chemical mixture must be elevated to more than pH 12.0 for 2 hours and then maintained above pH 11.5 for another 22 hours to meet VAR requirements. To meet Class A stabilization requirements, the elevated pH is combined with elevated temperatures (70°C for 30 minutes or other U.S. EPA-approved time and temperature combinations listed in U.S. EPA, 1999b). As long as the pH remains above 10 to 10.5, microbial activity and the associated odorous gases are greatly reduced or eliminated (U.S. EPA, 1979). However, other odorous gases (e.g., ammonia and trimethyl amine) may be produced under high-pH and -temperature conditions.

5.1.1   Process Application

Although both small and large WRRFs have used lime stabilization, this process is more common at small facilities. It typically is more cost-effective than other chemical stabilization options. Relatively large facilities have typically used lime stabilization as an interim process when their primary stabilization process (e.g., anaerobic or aerobic digestion) was temporarily out of service. Lime stabilization also has been used to supplement the primary stabilization process during peak solids-production periods.

Lime-stabilized biosolids may be land applied, benefiting large agricultural areas with acidic soils. However, because of the inert solids and reactions involved, lime-stabilized biosolids have lower concentrations of available nutrients (e.g., nitrogen and phosphorus) than a comparable mixture of biologically stabilized primary and WAS. (For more information on biosolids use and disposal considerations, see Chapter 25.)

5.1.2 Process Fundamentals

5.1.2.1 pH Elevation Effective lime stabilization depends on raising the pH high enough and maintaining it at that level long enough to halt or substantially retard the microbial reactions that otherwise could lead to odor production and vector attraction. The process also can inactivate viruses, bacteria, and other microorganisms.

Lime stabilization involves a variety of chemical reactions that alter the chemical composition of solids. The following equations (simplified for illustrative purposes) show the types of reactions that may occur:

Reactions with inorganic constituents:

Water: $$CaO + H_2O = Ca(OH)_2 \qquad (23.46)$$

Calcium: $$Ca^{2+} + 2HCO_3^- + CaO \rightarrow 2CaCO_3 + H_2O \qquad (23.47)$$

Phosphorus: $$2PO_4^{-3} + 6H^+ + 3CaO \rightarrow Ca_3(PO_4)_2 + 3H_2O \qquad (23.48)$$

Carbon dioxide: $$CO_2 + CaO \rightarrow CCO_3 \qquad (23.49)$$

Reactions with organic constituents:

Acids: $$RCOOH + CaO \rightarrow RCOOCaOH \qquad (23.50)$$

Fats: $$\text{"Fat"} + CaO \text{ fatty acids} \qquad (23.51)$$

Lime initially raises the pH of solids. Then, reactions occur (e.g., those in the equations above) that will lower the pH unless excess lime was added. The amount of excess lime needed depends on the length of time that a high pH must be maintained (e.g., during extended storage).

Biological activity produces compounds (e.g., carbon dioxide and organic acids) that react with lime. If biological activity is not sufficiently inhibited during alkaline stabilization, these compounds will reduce the pH, which could result in incomplete stabilization.

5.1.2.2 Heat Generation If quicklime (or any compound with high quicklime concentrations) is added to solids, it initially reacts with the water in solids to form hydrated lime. This exothermic reaction releases about 15 300 cal/g·mol (2.75×10^4 Btu/lb/mol) (U.S. EPA, 1982). The reaction between quicklime and carbon dioxide is also exothermic, releasing about 4.33×10^4 cal/g·mol (7.8×10^4 Btu/lb/mol).

Both reactions can raise the temperature substantially, particularly in solids cake with a low moisture content. For example, adding 45 g (0.1 lb) of quicklime per gram of solids to a cake containing 15% total solids can result in a temperature increase of more than 10°C (18°F), as the following formula demonstrates:

$$(0.1 \text{ lb CaO})(1 \text{ lb mol}/56 \text{ lb})(27\,500 \text{ Btu/lb mol}) = 49 \text{ Btu } (52 \text{ kJ}) \qquad (23.52)$$

$$(49 \text{ Btu})(1/0.85 \text{ lb H}_2\text{O})(1°F/\text{lb H}_2\text{O/Btu}) = 58°F \ (14°C) \qquad (23.53)$$

In practice, temperature increases will be smaller, although they can be substantial. Sometimes they can be sufficient to contribute to pathogen destruction during lime stabilization.

5.1.3 Process Description
Several alkaline-stabilization technologies are available. Each system has advantages and disadvantages, so design engineers should evaluate them and select the appropriate process on a case-by-case basis.

5.1.3.1 Liquid Lime (Pre-Lime) Stabilization
In liquid lime (pre-lime) stabilization, a lime slurry is added to feed solids to meet Class B stabilization requirements (see Figure 23.69). The lime typically is added to thickened solids at WRRFs that land-apply liquid biosolids (e.g., subsurface injection on agricultural land). This practice typically has been limited to smaller WRRFs or those with nearby land-application or use sites. That said, a Washington Suburban Sanitary Commission facility in Piscataway, Maryland, has used pre-lime stabilization followed by belt filter-press dewatering to create a biosolids suitable for hauling longer distances. Because the biosolids were pre-limed, Piscataway operators claim that they have low odor characteristics. However, equipment scaling remains a concern at this facility.

Another liquid lime stabilization method involves conditioning solids or septage with lime before dewatering. The lime typically is combined with other conditioners (e.g., aluminum or iron salts) to improve solids dewatering. This method primarily has been used with vacuum filters and recessed-plate filter presses; in such cases, the lime dose needed to condition solids typically exceeds that required to stabilize them.

5.1.3.2 Dry Lime (Post-lime) Stabilization
In dry lime (post-lime) stabilization, dry quicklime or hydrated lime is added to dewatered cake. This process has been practiced

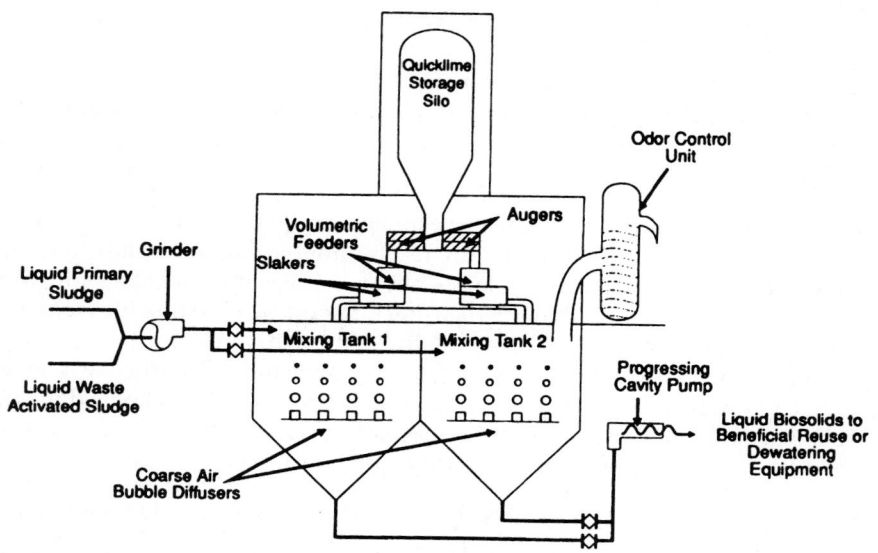

Figure 23.69 Typical liquid lime stabilization system (U.S. EPA, 1979).

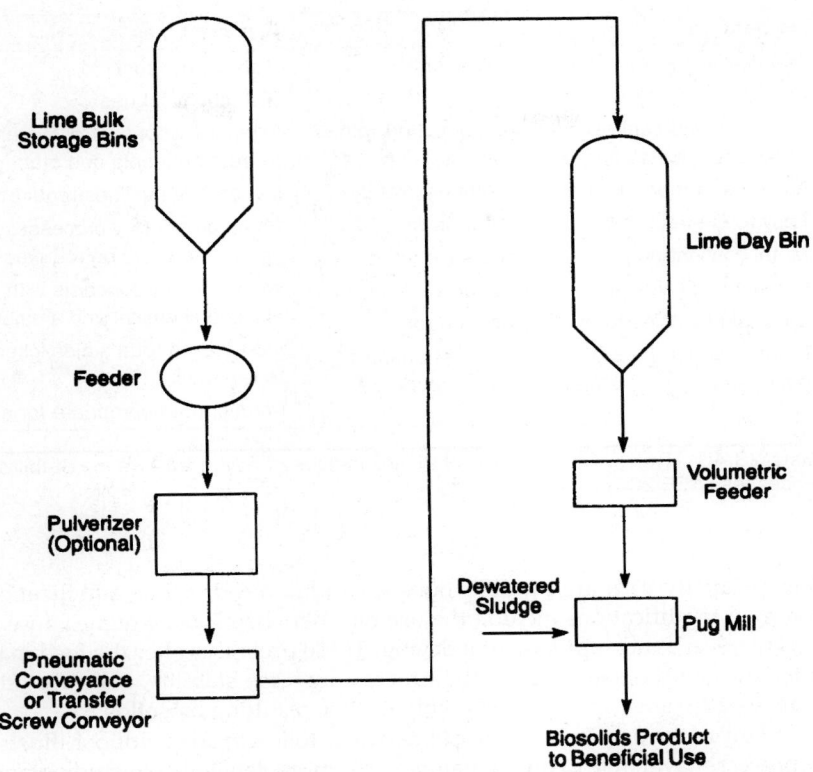

FIGURE 23.70   Process schematic of a typical dry lime stabilization system (Oerke and Rogowski, 1990).

at WRRFs since the 1960s (Stone et al., 1992). The lime typically is mixed with the cake via a pug mill, plow blender, paddle mixer, ribbon blender, screw conveyor, or similar device. Figure 23.70 is a process schematic for a typical dry-lime stabilization system with a pneumatic lime-conveyance system.

Quicklime, hydrated lime, or other dry alkaline materials can be used in this process, although the use of hydrated lime typically is limited to smaller installations. Quicklime is less expensive and easier to handle than hydrated lime, and the heat of hydrolysis released when quicklime is added to dewatered cake can enhance pathogen destruction.

If enough dry alkaline material is added to feed solids, the resulting biosolids can meet either Class B or Class A requirements.

5.1.3.3   Advanced Alkaline-Stabilization Technologies   Typical advantages and disadvantages of advanced alkaline stabilization are shown in Table 23.43.

In the last 30 years, alkaline-stabilization methods have been developed that use materials other than lime; these methods are being used by a number of municipalities. Most of those that rely on additives (e.g., cement kiln dust, lime kiln dust, Portland

Advantages	Disadvantages
Meets Class A stabilization requirements	High annual cost
Multiple product markets	High chemical use
Typically lower capital cost when compared with other Class A stabilization processes	Extensive odor-control systems required to treat ammonia and other offgases
Proven with more than 40 installations in U.S.	Dewatering facilities required
Easy to operate, start up, and shut down	Some proprietary processes; annual patent fee could be required
Metal concentrations in biosolids are diluted	
Product has value as a liming agent	Worker safety concerns with dust from alkaline chemical and ammonia offgas
Enclosed facilities for better odor control	
Properly stabilized product is easy to transport handle and can be stored in smaller storage	Increase on total solids/chemical mass to facilities
	Product not appropriate for alkaline soils

TABLE 23.43   Typical Advantages and Disadvantages of Advanced Alkaline Stabilization Processes (WEF, 2007)

cement, or fly ash) are modifications of traditional dry lime stabilization. The most common modifications include the use of other chemicals, a higher dose (depends on the chemical), and supplemental drying. These processes alter the feed material's characteristics and, depending on the process, increase stability, decrease odor potential, reduce pathogens, and otherwise enhance the resulting biosolids.

Many of the processes are proprietary. The following descriptions illustrate the scope of processes available to municipalities. [For more detailed case-study planning, design, and operational considerations on advanced alkaline-stabilization processes, see Technology Evaluation Report: Alkaline Stabilization of Sewage Sludge (Engineering–Science, Inc. and Black and Veatch, 1991).]

Pasteurization processes use the exothermic reaction of quicklime with water to raise process temperatures above 70°C. They then maintain this temperature for more than 30 minutes, as required by federal regulations for add-on pasteurization to meet Class A criteria. This pasteurization reaction must occur under carefully controlled and monitored mixing and temperature conditions to ensure that all solids particles are uniformly treated and pathogens are inactivated by the heat generated during the reaction.

The process produces a soil-like material that is nonviscous and, therefore, not subject to liquefaction under mechanical stress. Varying the process additives and mixing ratios results in a range of biosolids-derived materials suitable for use as daily, intermediate, and final landfill cover or in land reclamation (Sloan, 1992). Figure 23.71 is a process schematic for a typical pasteurization process. In a variation of this process, pasteurization occurs in a heated and insulated vessel reactor, where temperatures are maintained at 70°C or higher for at least 30 minutes.

A chemical stabilization/fixation process typically involves adding pozzolanic materials to dewatered cake (see Figure 23.72). Such materials cause cementitious reactions and produce, after drying, a soil-like material containing about 35% to 50% solids. To date, this soil-like product has been used only as landfill cover material. In many cases, the treated material is further dried at the landfill for 2 to 3 days in small

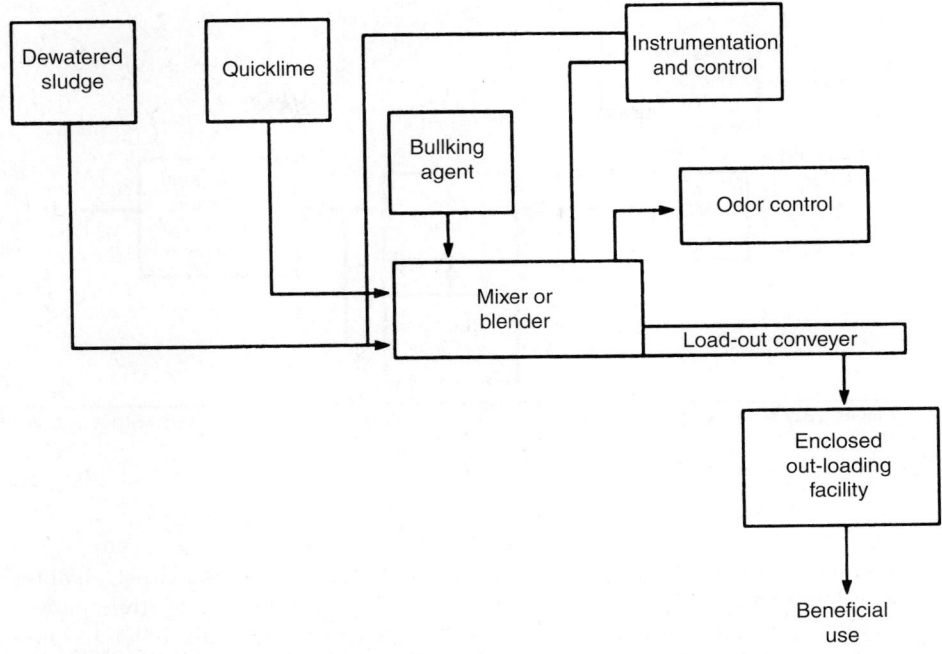

FIGURE 23.71   Process schematic of a typical pasteurization system.

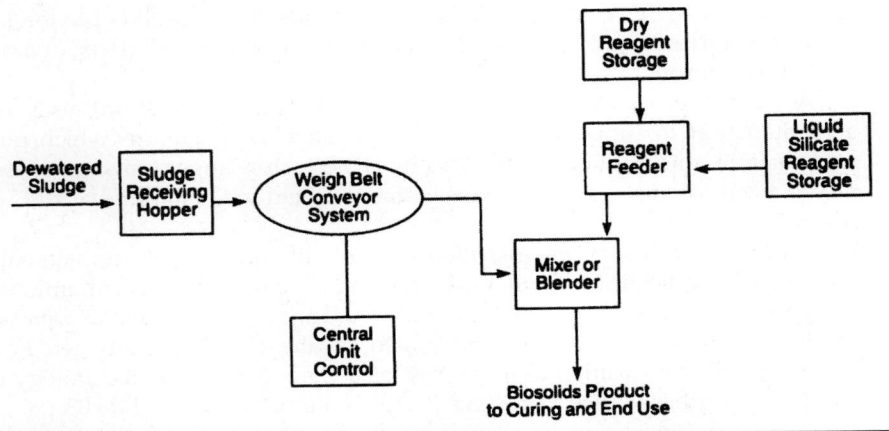

FIGURE 23.72   Process schematic of a typical chemical stabilization system.

windrows. Class A or PFRP equivalency has not yet been proven (Oerke and Rogowski, 1990; Reimers et al., 1981).

One proprietary process (the N-Viro process) combines advanced alkaline stabilization with accelerated drying (AASAD) (see Figure 23.73). The U.S. EPA has approved two versions of this technology as systems that produce Class A biosolids. Both versions

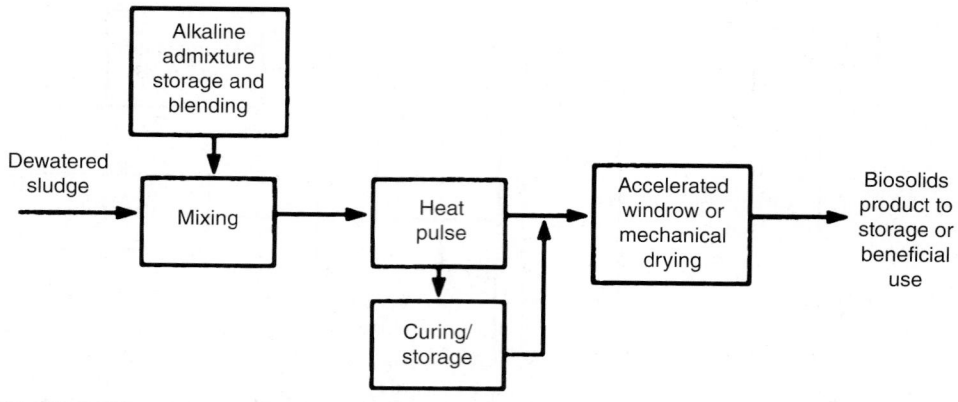

FIGURE 23.73 Process schematic of a typical alkaline stabilization system with a subsequent drying process.

involve adding quicklime, cement-plant kiln dust, lime-plant kiln dust, alkaline fly ash, or other alkaline admixtures and further processing the solids to stress pathogens via pH, temperature, ammonia, salts, and dryness (Burnham et al., 1992). In one version, chemical addition is followed by raising the material's temperature to between 52°C and 62°C for at least 12 hours so the heat generated by the chemical reaction can further reduce pathogens. The second version uses chemical addition to raise the solids' pH above 12 and then mechanically dries the material in windrows or a rotary-drum dryer to produce biosolids containing 50% to 60% solids. The biosolids predominantly are used as an agricultural liming agent, a soil conditioner, a landfill cover, or a component of blended topsoil.

A second proprietary process (RDP envessel pasteurization) uses electrically generated heat to supplement the heat generated by quicklime, which purportedly reduces lime consumption. An electrically heated screw auger transfers the solids-lime mix to an enclosed reactor, where the material is held for 30 minutes at 70°C to achieve pasteurization.

A third proprietary process (Bioset) uses sulfamic acid to supplement the heat produced by quicklime. (Both lime and water, and lime and sulfamic acid, react exothermically.) The process occurs in a pressurized vessel to achieve pasteurization conditions. Bioset has been recommended by the U.S. EPA Pathogen Equivalency Committee for certification of its process as a Class A biosolids technology under the "process to further reduce pathogens (PFRP)" alternative in 40 CFR 503.

5.1.4 Process Variations

Several alternative approaches or modifications to the basic alkaline-stabilization process have been developed. Some evolved from other treatment processes. For example, lime-treated primary solids have been combined with raw secondary solids to remove phosphorus (Paulsrud and Eikum, 1975). Existing digesters (or other available tanks) have been used to thicken alkaline-stabilized biosolids before dewatering and disposal (Farrell et al., 1974).

Another alternative uses two mixing vessels: the pH is raised above 12 in one, and the other provides adequate contact time and excess lime addition to keep pH within the desired range (Counts and Shuckrow, 1975).

Waukegan, Illinois, mixes fly ash and dewatered cake at ratios between 2.0:1 and 2.5:1 to produce Class B alkaline-stabilized biosolids. Personnel used this structurally stable material to "build" a biosolids-only monofill, rather than buying and importing fill material (Byers and Jensen, 1990).

5.2   Advantages and Disadvantages

Both liquid and dry lime stabilization processes are reliable, compact, relatively inexpensive to install, and easier to operate than many other stabilization processes. Many wastewater utilities that use lime stabilization have indicated that the process greatly reduces odors if the mixing is thorough (Kampelmacher and van Noorle Jansen, 1972; Westphal and Christensen, 1983). However, odor experiences with lime stabilization have been mixed and are typically the result of variations in operating procedures. This process' pathogen reduction has been reported to be as effective as or better than digestion processes (U.S. EPA, 1979).

Nevertheless, there are disadvantages. Compared to digestion, alkaline stabilization does not reduce solids mass. In fact, it increases mass because of the added lime and resulting chemical formations; the amount to be handled is essentially proportional to the chemical dose. The increase in mass may increase transportation costs for bisolids use or disposal, but such costs may be offset by capital and O&M savings (from using alkaline stabilization rather than another process). Also, the weight typically increases more than the volume, which actually may shrink because of lime slaking. Slaking raises the temperature of solids, causing water to evaporate.

Stabilized solids are a source of nitrogen, phosphorus, and beneficial organic matter that can be land-applied on farms. However, alkaline-stabilized biosolids typically contain less soluble nitrogen and phosphorus (on a dry-weight basis) than aerobically or anaerobically digested biosolids. The biosolids also may partially or fully replace liming agents on acid soils because it elevates soil pH and, therefore, restricts facility uptake of metals. However, metal ions only are immobilized as long as the biosolids' pH remains high. Also, alkaline-stabilized biosolids may not be appropriate in areas where the soils are naturally alkaline (e.g., many parts of the western United States).

Another disadvantage is that the system has difficulty consistently providing thorough mixing. Also, alkaline stabilization produces ammonia and possibly other odorous gases that should be treated before being exhausted.

5.3   Applicability

Alkaline stabilization has been used in numerous biosolids-management programs (Oerke, 1999). Below are some typical situations in which alkaline stabilization has been used:

- Traditional dry lime stabilization is a cost-effective technology for land-applied or landfilled biosolids. However, because biosolids are not destroyed, it is more cost-effective when hauling distances are short.
- Traditional liquid lime stabilization is appropriate at small WRRFs, where the small volume of biosolids produced can be readily land-applied. It is also

practical at small facilities that store biosolids for later transportation to larger facilities for further treatment or disposal.

- Because chemicals are the main O&M expense in this process and because the process has great flexibility, alkaline stabilization may be a cost-effective option for facilities that only operate seasonally or whose solids production are variable.

- Advanced alkaline stabilization may allow municipalities to operate a biosolids distribution and marketing program at a lower capital cost than other technologies (e.g., in-vessel composting or heat drying).

- Because a well-maintained alkaline-stabilization system can be quickly started (or stopped), it can be used to supplement existing solids treatment capacity or substitute for incineration and drying facilities during fuel shortages. It also can treat the total solids production when existing facilities are out of service for cleaning or repair.

- Alkaline-stabilization systems have comparatively low capital costs, so they may be cost-effective for facilities with short service lives.

- Alkaline stabilization typically is used to treat septage, reducing odors before the material is land-applied or discharged to WRRFs. (The U.S. EPA's Standards for the Use or Disposal of Sewage Solids (1993) require that septage be treated with lime and maintained at pH 12 for 30 minutes before land application.)

- Alkaline stabilization may be added to processes (e.g., overloaded digesters) that have inadequate pathogen reduction. However, strong ammonia odors typically are generated when anaerobically digested solids are treated with alkaline materials.

5.4   Design Considerations

Because product quality and process design are interdependent, the importance of defining both process and product goals cannot be overemphasized. Engineers should evaluate a number of design criteria before implementing an alkaline-stabilization process (see Table 23.44). Although they vary from site to site, typical design criteria include:

- Sources and characteristics of feed cake (e.g., quantity, type, quality, and solids content);
- Contact time, pH, and temperature;
- Alkaline chemical types and doses;
- Solids concentration of the feed cake–chemical mixture;
- Energy requirements;
- Storage requirements; and
- Pilot-scale test results.

The desired product is also an important design criterion. For more information on biosolids use considerations, see Section 7.8.

Item	Description or Equipment	Parameter	Units	Range of Value Minimum	Range of Value Maximum	Selected Design Value
Materials	Sludge	Solids	Percent	20	30	25
		Density	lb/cu ft[a]	45	55	50
	Alkaline bulking chemical	Solids	Percent	90	98	95
		Density	lb/cu ft[a]	50	65	65
	Lime	Solids	Percent	90	96	95
		Density	lb/cu ft[a]	55	60	60
	Stabilized product	Solids	Percent	55	65	60
		Density	lb/cu ft[a]	65	75	75
Curing	Technology	Windrow				
	Detention time	Average	Days	3	7	6
		Peak	Days	3	7	4
	Temperatures		°C	—	—	52 to 12 hr
	Pile dimensions	Bottom width	ft[b]	6	14	10
		Mix height	ft[b]	2	3	3
		Top width	ft[b]	4	8	6
		Area/unit length	sq ft/ft[c]	10	33	24
		Pile spacing	ft[b]	—	—	5
	Pile turning		lb/d[d]	—	—	1 (typical)
Odor control	Building air	Number of stages	Number	1	3	1
		Air changes	Number/hr	6	15	12
	Product storage	Number of stages	Number	1	3	1
Storage	Sludge	Days of storage	Days	0	1	1
	Chemicals	Days of storage	Days	5	30	5
	Product	Days of storage	Days	80	180	60

[a] lb/cu ft × 16.02 = kg/m³.
[b] ft × 0.3048 = m.
[c] sq ft/ft × 0.3048 = m²/m.
[d] lb/d × 0.4536 = kg/d.

TABLE 23.44 Typical Advanced Alkaline Stabilization Design Criteria (Fergen, 1991)

5.4.1   Feed Characteristics

The amount, sources, and composition of the feed cake determine the overall size of the alkaline-stabilization system. Variable thickening or dewatering performance is an important consideration because poor performance significantly increases the size of the stabilization system. The dewatered cake's solids concentration affects both chemical dose and system size. Equipment capacities must be able to accommodate the volume of feed cake to be processed. The system will need larger equipment and more alkaline chemical to process a "wet" cake (10% to 15% solids) than a drier one (20% to 25% solids).

The feed cake's nutrient content affects the biosolids characteristics. The agronomic benefit of an alkaline-stabilized biosolids depends on the amount of nutrients it contains and the need for a liming agent at the application site. Alkaline stabilization may be advantageous for untreated solids with relatively high metal concentrations because alkaline additives dilute metals (on a dry-weight basis) and immobilize some trace metals.

The type of solids also should be considered. For example, anaerobically digested biosolids contain five to eight times more ammonia-nitrogen than other solids. All of this ammonia-nitrogen would volatilize at the elevated pH required for alkaline stabilization, increasing the potential for odors. Anaerobically digested biosolids treated in alkaline-stabilization systems also may release odors related to other nitrogen compounds (e.g., amines). Alkaline-unstable polymers also can contribute to the formation of odorous methyl amines. As with all solids-processing systems, odor-control facilities typically are required at alkaline-stabilization systems near residences or sensitive commercial areas.

5.4.2   Contact Time, pH, and Temperature

Contact time and pH are directly related because the pH must be maintained at the required level for enough time to destroy pathogens. The treatment chemical must have enough residual alkalinity to maintain a high pH in the biosolids until they are used or disposed. The high pH prevents odorants and pathogenic organisms from growing or reactivating.

A drop in pH (pH decay) occurs when biosolids absorb atmospheric carbon dioxide or acid rain (which forms a weak acid when dissolved in water), which gradually consumes the residual alkalinity. The pH gradually decreases, eventually dropping below 11.0. Bacterial action then resumes, and the renewed production of organic acids causes the pH to continue decaying (similar to the reactions in anaerobic digestion).

The pH typically drops during stabilization, so it should be raised to and maintained at more than pH 12. Biosolids do not have to be inside a contact vessel as long as the pH can be monitored to ensure that it remains at the desired value for the desired time.

5.4.3   Alkaline Chemical Types and Doses

The types and doses of alkaline chemicals are important design criteria. The quality of the chemicals (e.g., lime, cement-plant kiln dust, Portland cement, and lime-plant kiln dust) should be consistent. Different types or sources of additives produce different biosolids textures and granularities. Lime is available from numerous sources, ranging from a high-calcium lime in oyster or clam shells to a relatively low-calcium dolomitic lime. Major considerations when selecting a chemical include economics, availability, desired mixing, and desired product characteristics.

Some alkaline reagents (e.g., cement kiln dust, lime kiln dust, and fly ash) are considered industrial by-products, and design engineers must ensure that this material does not introduce contaminants or additional pollutants that jeopardize biosolids quality. Cement kiln dust from hazardous waste kilns, for example, should be avoided. Also, the characteristics of a by-product can vary from one location to the next, so consistent vendor quality-control procedures are essential. The material from one kiln or furnace will remain fairly consistent, provided that operating conditions do not drastically change.

Facility personnel should develop a quality assurance/quality-control program that includes frequent sampling and analysis to ensure that biosolids quality is consistent. Because the quality of alkaline additives may directly affect biosolids quality, adequate monitoring and proper management are important. More importantly, pilot- or bench-scale testing should be performed to determine how variations in alkaline additives will affect product quality and how the process and chemical doses should be adjusted to compensate for such variations.

The two predominant types of lime are quicklime (calcium oxide) and calcium hydroxide. Slaked lime in a liquid slurry (carbide lime) is also available. Carbide lime is a by-product of manufacturing welding-grade acetylene from calcium carbide. Its application principles are the same as those for calcium hydroxide or quicklime in slurry form, so carbide lime is not specifically discussed here.

Design engineers should select the type of lime based on economics and material-handling characteristics (e.g., alkaline-material particle size). Calcium hydroxide costs about 30% more to produce and transport than quicklime, but it requires less equipment on-site because it already has been hydrated (slaked). Calcium hydroxide typically is economical for use at small facilities, but if more than 9000 to 13 000 m³/d (3 to 4 ton/d) is needed, quicklime should be considered.

Quicklime typically requires slaking equipment on-site. Dry lime stabilization (i.e., adding quicklime directly to dewatered cake) does not require that the chemical be slaked first, but additional handling precautions must be addressed because of the exothermic reaction of quicklime and water. Dry lime stabilization also eliminates lime sidestreams and the related abrasion and scaling of piping and mechanical equipment.

The required doses of specific chemicals will depend on the type of feed solids (e.g., primary, WAS, trickling filter, or septage), its quality and chemical composition (including organic content), its solids concentration, the desired final product characteristics, and the type and quality of the alkaline material.

Table 23.45 shows the range of liquid lime doses required to maintain pH 12 for 30 minutes (U.S. EPA, 1979). Numerous researchers have confirmed these doses (Ramirez and Malina, 1980).

The chemical dose is affect by the feed cake's chemical composition, which depends on the type of solids and the treatment process used (e.g., chemical coagulation). Another factor that affects chemical dose is solids concentration (see Figure 23.74) (U.S. EPA, 1975). Table 23.46 shows a wide range of lime doses (from 10% to 60% on a dry-weight basis). As the solids concentration increases, the required dose typically increases. The required dose per unit mass of solids tends to be somewhat higher for dilute feeds (less than 2.0% solids) because more lime is required to raise the pH of water. However, liquid lime requirements are more closely related to the feed cake's total mass than to its volume when its solids concentration ranges from 0.5% to 4.5% (U.S. EPA, 1979). Thickening solids to reduce the volume may have little or no effect on lime requirements because the mass is not significantly changed.

Type of Sludge	Average Solids Concentration, %	Average Lime Dosage, lb Calcium Hydroxide/lb[b] Dry Solids	Average pH	
			Initial	Final
Primary sludge[c]	4.3	0.12	6.7	12.7
Waste activated sludge	1.3	0.30	7.1	12.6
Anaerobically digested combined	5.5	0.19	7.2	12.4

[a]Dose required to maintain pH 12 for 30 minutes.
[b]lb/lb × 1000 = g/kg.
[c]Includes waste activated sludge.

TABLE 23.45   Lime Dose Required For Liquid Lime Stabilization at Lebanon, Ohio[a] (U.S. EPA, 1979)

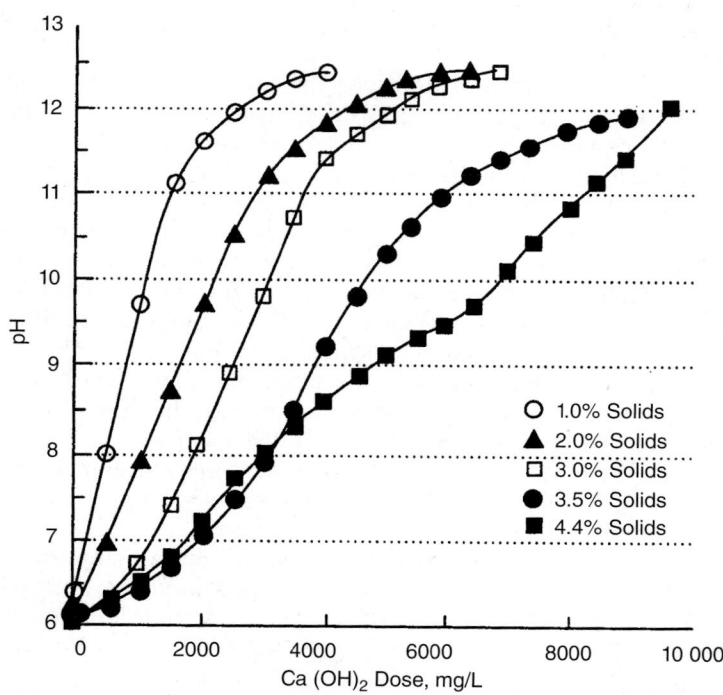

FIGURE 23.74   Dose of liquid lime required to raise the pH in a stabilization system feedstock (primary solids and trickling filter humus) with various solids concentrations (U.S. EPA, 1975).

Type of Raw Sludge	Lime Dose, lb Calcium Hydroxide/lb Suspended Solids[a]
Primary sludge	0.10–0.15
Activated sludge	0.30–0.50
Septage	0.10–0.30
Alum sludge[b]	0.40–0.60
Alum sludge[b] plus primary sludge[c]	0.25–0.40
Iron sludge[b]	0.35–0.60

[a]lb/lb × 1000 = g/kg.
[b]Precipitation of primary treated effluent.
[c]Dry-weight basis.

TABLE 23.46   Liquid Lime Stabilization Doses Required to Keep pH Above 11.0 for at Least 14 Days (Farrell et al., 1974)

Minimum lime doses of 25% to 40% (on a dry-weight basis as calcium hydroxide) typically are required for liquid lime Class B stabilization before vacuum filtration. The curves in Figures 23.75 to 23.77 show the characteristic pH drop that occurs when not-enough liquid lime is added. When the dose is too low, the pH of the feed cake–lime mixture initially may reach 12 but then rapidly decay.

Minimum doses of 13% to 40% (on a dry-weight basis as calcium hydroxide) typically are required for effective dry lime stabilization (see Figure 23.78). Figure 23.79 shows the theoretical dry lime stabilization dose for both Class B and Class A stabilization. The lower line shows maximum pH requirements, and the upper line shows Class A temperature requirements. Figure 23.79 is based on a quicklime dose requirement of 25% (dry-weight basis). Design engineers should note that while the quicklime requirement for Class B stabilization theoretically increases as the solids concentration increases, the quicklime requirement for Class A stabilization decreases as the solids concentration increases, because lime is used to heat the cake to achieve Class A biosolids (a lower solids concentration will mean that more mass of water needs to be heated using quicklime to the required temperature), whereas lime is used to raise pH for Class B (Lue-Hing et al., 1992).

The following assumptions were used for the Class A temperature requirements in Figure 23.79:

- The feed cake's temperature was 20°C (68°F).
- All of the quicklime reacted with water in the feed cake to produce heat (1140 kJ/kg [490 Btu/lb] of quicklime).
- Quicklime is 100% calcium oxide (this value typically is 90%).
- The feed solids' specific heat is 0.25.
- There was no heat loss from the feed to the air or the equipment.

Such conditions rarely exist in practice, so the amount of quicklime actually needed to meet Class A requirements can be up to 50% more than that indicated in Figure 23.79.

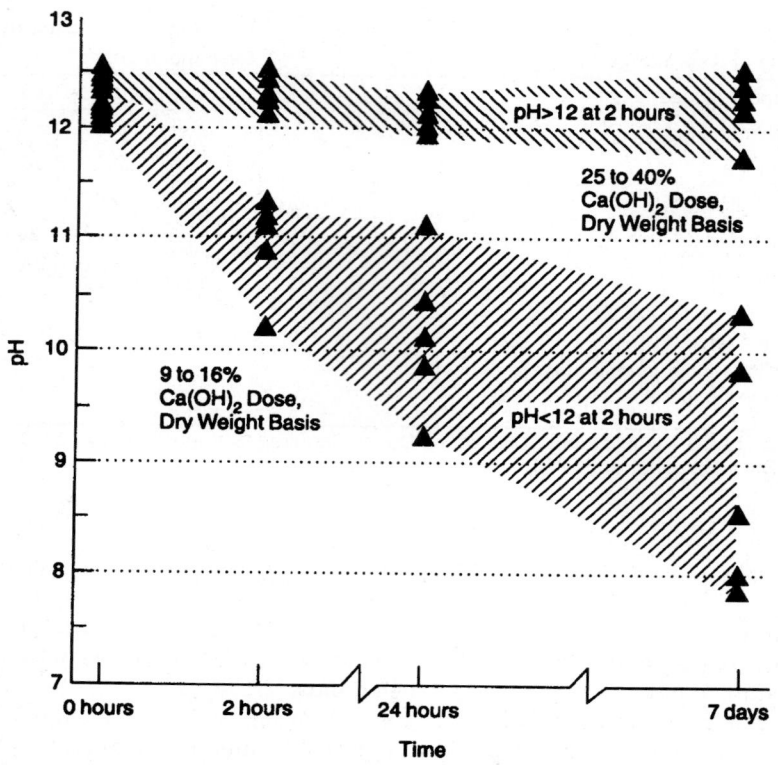

Figure 23.75   Example of pH decay following liquid lime stabilization before vacuum filtration (Westphal and Christensen, 1983).

To produce a drier, more easily crumbled biosolids, design engineers should increase quicklime dose by as much as twice the value shown in the table.

Chemical doses for advanced alkaline-stabilization technologies depend on the process, chemical, and biosolids requirements. Material balances should be used to size alkaline-stabilization facilities and determine initial and final solids characteristics. Table 23.47 shows a typical material balance for an advanced alkaline-stabilization facility, assuming a 65% chemical dose (wet-weight basis). Design engineers should note that a lime dose expressed on a wet-weight basis is four times greater than a dose expressed on a dry-weight basis for a dewatered cake with a 25% solids concentration. For example, a chemical dose of 65% (wet-weight basis) is equal to about 245% on a dry-weight basis.

Design engineers can use the data in Tables 23.44 to 23.47 for preliminary design of liquid and dry lime stabilization facilities; however, the required dose should be determined on a case-by-case basis because of the many factors involved (Farrell et al., 1974). To prevent pH decay and the associated regrowth of organisms, the lime dose may have to be higher than necessary for stabilization (Ramirez and Malina, 1980). The exact dose for any particular feed cake can be estimated via laboratory testing.

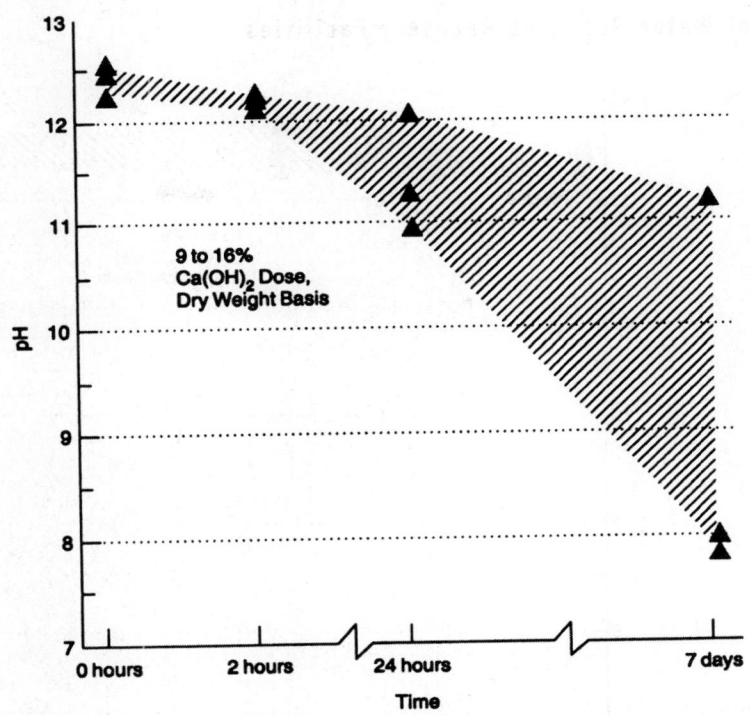

Figure 23.76 Example of pH decay following liquid lime stabilization (Westphal and Christensen, 1983).

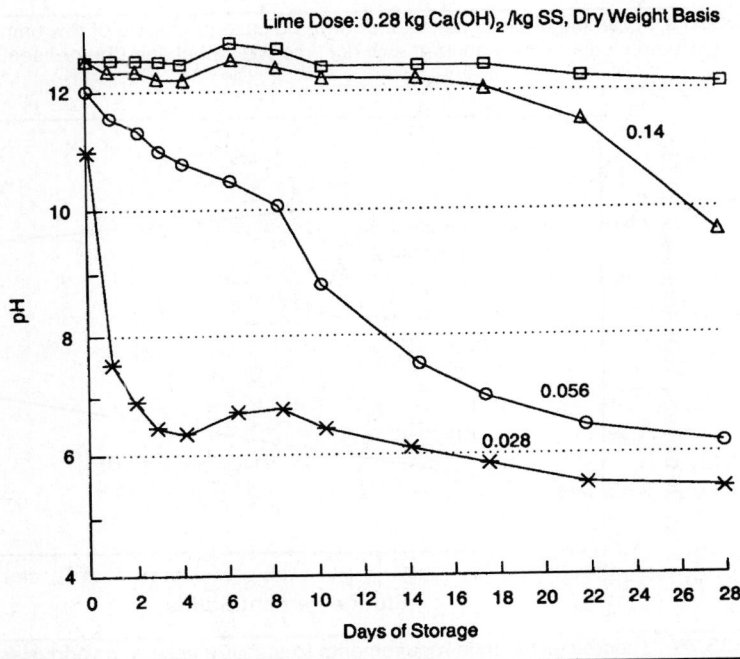

Figure 23.77 Change in pH during storage of raw primary solids that had been stabilized using various liquid lime doses (Farrell et al., 1974).

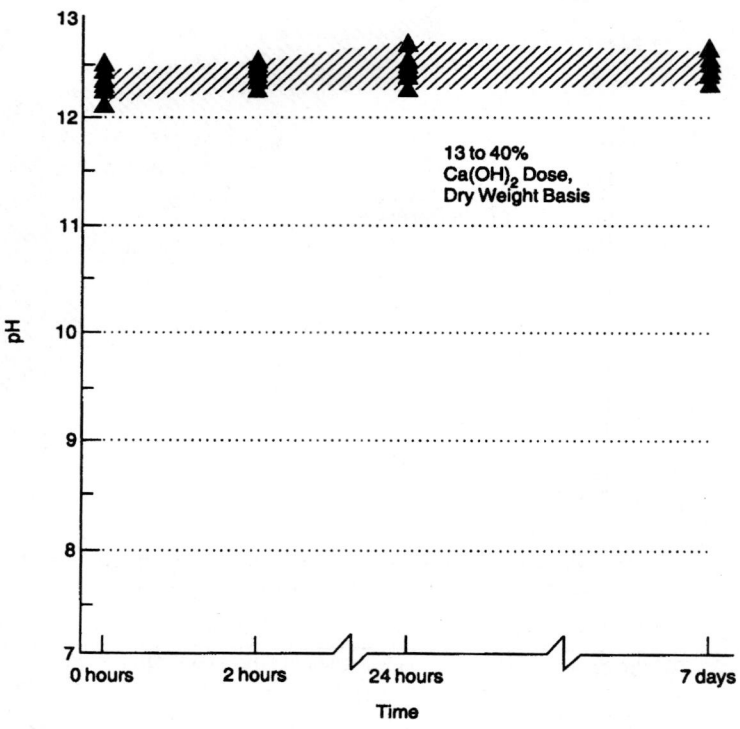

FIGURE 23.78 Example of pH decay after dewatered cake (a mixture of raw primary solids and waste activated sludge) was stabilized with dry lime (Westphal and Christensen, 1983).

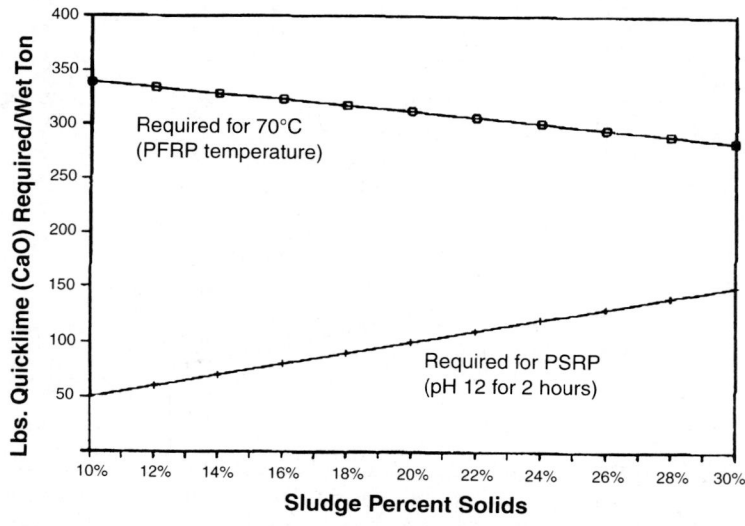

FIGURE 23.79 Theoretical dry lime requirements to stabilize cake with various solids concentrations so it meets Class B or Class A standards (PFRP = process to further reduce pathogens and PSRP = process to significantly reduce pathogens) (Lue-Hing et al., 1992).

		Solids Balance				
Process	Item	Solids Content (%)	Volume (cu ft)[b]	Total Weight (Ton)	Dry Weight (Ton)	Bulk Density (lb/cu ft)[c]
Mixing	Sludge cake	25.0	6400	160.0	40.0	50.0
	Chemicals	95.0	3186	103.5	98.4	65.0
	Initial mix	52.5	9586	263.5	138.4	55.0
Windrow	Initial mix	52.5	9586	263.5	138.4	55.0
	Evaporation loss	—	3437	32.9	—	—
	Product	60.0	6149	230.6	138.4	75.0

Chemical dose:
  Wet-weight basis 65%
  Dry-weight basis 245%
[a]For peak conditions multiply all the values by peaking factor (except density and percent solids).
[b]cu ft × 28.32 = L.
[c]lb/cu ft × 16.02 = kg/m^3

TABLE 23.47 Typical Materials Balance for Advanced Stabilization Facilities (Fergen, 1991)[a]

5.4.4 Solids Concentration of Feed–Chemical Mixture
The solids concentration of the feed cake–chemical mixture is an important design consideration for material-handling purposes. Regulations may require a minimum solids concentration (e.g., for landfilling or extended storage). The final product solids concentration (dryness) and granularity also affect the type of biosolids trucks and application/disposal equipment needed.

The solids concentration of the initial feed cake–chemical mixture also affects any supplemental drying step in advanced alkaline-stabilization processes. The alkaline additive causes chemical reactions to occur that increase the mix's apparent solids content. This increase in solids is caused by the addition of solids (treatment chemical), chemical binding, and evaporation of water from the feed cake. The alkaline material—particularly quicklime—produces a fast reaction that increases temperature in a matter of minutes. Thorough mixing of feed cake and alkaline material is important to achieve the target solids content and pathogen destruction, and to reduce residual odors (e.g., ammonia) in biosolids.

A high chemical dose can produce the desired solids concentration, thereby reducing or eliminating the need for supplemental drying, but this practice may be prohibitively expensive. Adding other bulking materials (e.g., fly ash, wood ash, sawdust, sand, and soil) can increase biosolids dryness and improve handling characteristics without increasing the chemical dose. Mechanical mixing in a windrow operation enhances drying, blends the material, and releases trapped ammonia and other volatile gases created during dewatering, resulting in a more homogeneous product. The final design should reflect the best balance between the chemical dose and the amount of subsequent drying required.

5.4.5 Energy Requirements
In liquid lime stabilization processes, energy principally is needed to mix solids with the lime slurry. In dry alkaline stabilization, mixing energy requirements are minimal; they depend on solids throughput, chemical dose, and mixer type.

Energy also may be needed for transport vehicles (e.g., feed cake, chemicals, and biosolids), and air ventilation and scrubbing equipment (for ammonia and odor control).

5.4.6 Storage Requirements

The system's storage facilities should be tailored to the facility's actual needs. Both intermediate and final storages should be provided.

5.4.6.1 Intermediate Storage Some advanced alkaline-stabilization processes (e.g., N-Viro) require intermediate storage for the heating step to achieve Class A stabilization requirements. The objective of this step is to contain the heat produced during the exothermic reaction, so less chemical is needed. Intermediate storage units can include insulated steel, live-bottom hoppers; concrete bunkers; or an uninsulated stockpile in an open concrete pad.

5.4.6.2 Product Storage Biosolids storage is another important design consideration. Facilities need adequate storage capacity if its biosolids markets are seasonal or have not been established. The amount of storage needed depends on both the type of biosolids and the distribution and marketing methods involved. At least 30 to 90 days' worth of storage should be provided if biosolids curing is required, it also is needed to accommodate road and weather conditions, as well as fluctuations in the biosolids-marketing and -distribution schedule. Facility personnel try to develop markets for biosolids as an agricultural fertilizer, liming agent, or soil amendment. Until such markets have been established, however, the biosolids must be stockpiled or discarded. Also, the demand for agricultural products is seasonal, so the facility must have provisions for stockpiling during low-demand periods.

On the other hand, if the material will be used as landfill cover, storage requirements probably will be minimal (e.g., weekend storage may be needed if the landfill only operates 5 days per week).

The solids concentration and long-term stability of biosolids are important considerations when designing storage facilities. Biosolids storage facilities should be sized to meet each facility's actual needs, including storage during scheduled and unscheduled equipment maintenance, if maintenance includes downtime. The storage facility should be designed to prevent deterioration of product quality during inclement weather. In many climates, covered storage may be desirable. If uncovered storage is used, provisions should be made for leachate and runoff collection to avoid ponding and, in some instances, treatment. Runoff from stockpiles of alkaline-stabilized biosolids can result in stagnation and septic odors (Engineering–Science, Inc. and Black and Veatch, 1991).

5.4.7 Pilot-Scale Testing

Because the quality and the consistency of feed cake are site specific, engineers must perform qualitative and quantitative analyses to determine the appropriate chemical doses and process design parameters. Pilot-scale testing should be used to determine optimum chemical doses and mixer performance. It also allows a municipality to evaluate various operating procedures and end-use products.

Engineers should conduct bench- and pilot-scale tests before implementing an alkaline-stabilization process. Four primary areas to be evaluated include:

- Process requirements (e.g., alkaline material types and doses);
- Equipment (e.g., energy requirements);
- Biosolids quality (e.g., desired solids concentration and granularity); and
- Odor generation and control.

Process concerns include the chemical types and doses; solids concentration in feed cake and biosolids; and other process steps (e.g., supplemental heating and drying), as required. Engineers must determine the chemical dose that will meet pH, solids content, heat rise, and biosolids requirements. It can be estimated in bench-scale tests using carefully measured volumes of feed cake mixed with various doses of chemicals. All pilot-scale testing should include generating mass-balance calculations to ensure consistency between chemical doses and solids concentration in biosolids. Where possible, full-scale pilot equipment should be used to assess actual chemical doses and mixing performance.

It is extremely important that testing conditions be controlled to simulate field conditions to the greatest extent possible. During winter, for example, the system may need a different dose or a modified formula of the chemical to achieve the desired biosolids. In large-scale windrow drying operations, carbon dioxide mixed into the product can lower the pH, so the alkaline-material dose may have to be increased to compensate. Samples should be cured in the same type of closed or open containers that will be used in the full-scale system.

Engineers should test the initial solids concentration of the feed-chemical mixture for compatibility with the proposed drying technique. It also may be useful to investigate various chemical doses in different drying/curing configurations. The chemical dose can significantly affect the drying rate and corresponding drying-area requirements.

Engineers also can use pilot-scale tests to evaluate equipment requirements. The goal of such testing is to determine the equipment, energy, and chemical needed to produce biosolids compatible with the next processing step or desired use. For example, an inappropriate paddle configuration or operating speed on a pug mill at a dry lime stabilization system resulted in an undesirable material. Proper mixing is necessary not only to achieve the desired biosolids characteristics but also to ensure that the alkaline additive has been thoroughly blended. Excessive mixing energy can result in a non-granular mass that is difficult to handle.

Other process parameters that should be considered during pilot- or bench-scale testing include odor emissions; concentrations of nutrients, metals, and organic chemicals; and compatibility of the alkaline material dose with the dewatering polymer. Some polymers may deteriorate in high-alkaline conditions, exhibiting strong trimethyl amine ("dead fish") odors (Jacobs and Silver, 1990). Engineers should test various doses of alkaline material with different polymers to determine their effects on biosolids odors and physical characteristics (e.g., compaction and granularity).

The final item evaluated in pilot-scale tests is the product. Pilot- and bench-scale testings provide excellent opportunities to investigate biosolids quality and marketability before beginning full-scale production. It is helpful to invite prospective users to observe pilot-scale tests or implement small-scale demonstration programs to encourage interest in the product. Physical characteristics (e.g., solids content, pH decay, leachability, permeability, or unconfined compressive strength) should be evaluated if the product will be landfilled or to stabilize slopes. Biosolids quality also should be tested to provide the data and documentation required for regulatory approval.

5.5   Description of Physical Facilities

5.5.1   Solids-Handling and Feed Equipment

Cake-handling equipment chiefly consists of belt and screw conveyors and pumps. Belt conveyors typically are used to move solids horizontally or at gentle slopes. Belt conveyor problems typically include minor spills, slips, and frequent bearing maintenance. Screw conveyors also are used to transfer dewatered cake to the alkaline-stabilization mixer or storage hopper. Screw conveyors and high-pressure cake pumps can physically "condition" dewatered cake, making it difficult (sometimes impossible) to homogeneously mix with a dry alkaline chemical. Some screw conveyors tend to roll the cake–chemical mixture into "balls." Pumps can compact dewatered cake into a long tube that must be broken up during mixing. The rolled balls and compacted cake, which may be desirable or undesirable depending on the final objective, can be especially critical for the resulting biosolids (Oerke and Stone, 1991).

Although alkaline-stabilization processes are relatively simple, a regular inspection and maintenance program is essential. The conveyance system and other moving parts must be closely monitored for wear. If only one conveyor feeds the alkaline-stabilization process, it must be routinely inspected, maintained, and calibrated because conveyance-system downtime can delay or halt stabilization. If multiple process trains are used, bypasses and crossovers should be provided to avoid excessive downtime. Also, engineers should design the alkaline-stabilization system to be as close as possible to both the dewatering equipment and the storage system.

Design engineers should seriously consider using redundant process and storage trains to allow for routine maintenance and calibration, as well as operational flexibility, without downtime. Another option (although less desirable) is using temporary portable units, which can be placed in operation in a matter of hours or days, if necessary. Storage hoppers or bunkers may be placed between the dewatering and alkaline-stabilization systems to dampen variations in dewatering system output and to allow each process to operate independently.

5.5.2   Alkaline Material Storage and Feeding

Alkaline stabilization requires special chemical-storage and -feeding equipment. Traditionally, an alkaline chemical storage system should be able to meet at least 7 days' worth of demand (although a 2- to 3-week supply is preferred). Calcium hydroxide can be stored up to 1 year. Quicklime deteriorates more rapidly; it should not be stored longer than 3 to 6 months.

Because some advanced alkaline-stabilization processes have high chemical demand, traditional design criteria can result in an excessively large storage capacity; however, design engineers can consider a smaller capacity (2 to 3 days of chemical use) so long as chemical-delivery arrangements are reliable. The costs associated with daily chemical delivery should be compared to those of extra storage capacity.

Quicklime can be stored in lump or pebble form and ground on-site to reduce the potential for reaction with moisture during storage, especially if the alkaline material will be stored for up to 6 months.

Alkaline material is stored in steel silos with hoppers that have a side slope of at least 60°. Bulk-storage silos and day chemical bins, if used, should be equipped with dust collectors and live-bottom bins, hopper agitation, or air pads to facilitate unloading and reduce clogging or bridging.

There are potential problems with any chemical, however. During storage, lime can react with carbon dioxide in the air to form a calcium carbonate coating on lime particles, making them less reactive. Quicklime and other alkaline materials readily react with moisture from the air (slake), leading to caking that can interfere with feeding and slaking. Therefore, lime should be stored in dry facilities and protected against moisture to prevent accidental slaking. Also, because slaking generates heat, quicklime should not be stored near combustible materials.

Dry alkaline materials can be conveyed mechanically via a screw conveyor if the distance from the bulk-storage silo to the chemical-addition point is short. Dry alkaline materials also can be pneumatically conveyed under either pressure or vacuum. Each type has its benefits. Vacuum systems have fewer dust problems because any leaks are into the system, not out of it. Pressure systems can move more material. Pneumatically conveyed air should be predried to reduce hydration and other moisture-related problems. Pneumatic conveyance systems may have problems maintaining homogeneous chemical bulk densities, however, if a variety of alkaline materials is used (Rubin, 1991).

A wide variety of chemical feed equipment (e.g., volumetric screw feeders, rotary airlock feeders, and gravimetric feeders) is available. A volumetric feeder delivers a constant volume of alkaline material, regardless of its density. A gravimetric feeder delivers a constant mass of alkaline material and provides more accurate control. However, it costs about twice as much as the volumetric type. Design engineers should evaluate feeders to determine which is appropriate for a given application (Rubin, 1991). The feed equipment should be isolated from the storage silo via a slide gate or similar device so the metering equipment can be removed easily if it becomes jammed.

Most chemical feed systems have dust problems. Poorly fitting slide gates and leaking feeders are obvious sources of dust. Also, the vertical drop between the feeding equipment and the process mixers should be reduced or enclosed to reduce dust problems.

Moisture can be generated during mixing that may rise into the chemical feed and storage equipment. For example, lime backups in pipes primarily are caused by moisture generated during the mixing process in the pug mill. Powdery lime is hygroscopic and tends to pack in the corners of the storage hopper. Venting the mixer away from chemical feed and storage equipment can reduce such problems.

5.5.3 Liquid Lime Chemical Handling and Mixing Requirements
Lime typically is fed to liquid solids in slurry ("milk of lime") form. Dry lime cannot be added to liquid solids effectively because caking will occur.

After being mixed into a slurry, both calcium hydroxide and slaked quicklime are chemically the same, and the same feeding processes can be used for both. Lime slurry can be prepared via either the batch or the continuous method. The batch method consists of dumping bagged lime into a mixing tank. The contents of slurry tanks are agitated by compressed air, water jets, or mechanical mixers. To ensure initial wetting and dispersion, a mechanical mixer needs about 200 kW/m³ to handle a calcium hydroxide slurry at a concentration of 120 kg/m³ (Beals, 1976).

The slurry then is metered into the mixing tank. This may be the most troublesome step in the process. The slurry can react with bicarbonate alkalinity in the makeup water and with atmospheric carbon dioxide to form calcium carbonate scale that can plug lines. The magnitude of this problem increases as transfer distances increase and more bicarbonate or carbon dioxide contacts the slurry. So slurry tanks should be as close to the mixing basin as possible, and design engineers should avoid using cascading weirs or other equipment that causes turbulence.

The basic difference between using quicklime and calcium hydroxide to stabilize dewatered cake is that quicklime requires slaking equipment. Slaking can be done on either a batch or a continuous basis. The batch method is more appropriate for small-scale facilities; however, the use of quicklime typically is less advantageous for such facilities. Slaking consists of mixing quicklime and water to create either a lime paste (water-to-lime ratio of 2:1) or a slurry (water-to-lime ratio of 4:1). The paste should be held for approximately 5 minutes to allow complete hydration in the slaking chamber; the slurry should be held for 30 minutes. The hydration reaction is exothermic (i.e., releases heat). Proper slaking requires heat, but localized boiling and spattering could make conditions hazardous. After slaking, the paste enters a chamber where grit is removed and the paste is diluted to the desired concentration.

The appropriate automation equipment for continuous slaking largely depends on the proportion of lime to water, which in turn depends on the type of lime and mixing equipment used.

A stabilization tank is recommended downstream of the slaker to ensure that all chemical reactions between calcium hydroxide and dissolved solids in the water have been completed. This reduces scaling in downstream portions of the system. Slakers should discharge lime slurry directly to the stabilization tank, if possible, and it should be detained in the tank for at least 15 minutes. Adequate mixing is required to keep particles in suspension and prevent short-circuiting.

If baffles are required to prevent vortex formation, they should be designed to prevent solids from building up in the corners (depending on tank geometry).

A cleaning system should be provided that uses dilute hydrochloric acid to remove calcium carbonate scale from pumps and piping. So the pumps' and pipes' materials of construction must be compatible with both acidic and caustic environments.

To facilitate scale removal, design engineers should use flexible piping or open troughs to convey lime slurry whenever possible. Lime slurry may be abrasive, particularly if low-grade pebble lime is used, so equipment and materials should be selected accordingly.

The mixing tank's primary purpose is to provide adequate mixing and contact time for the dewatered solids and lime slurry. The recommended contact time is about 30 minutes after the pH reaches pH 12.5. Mixing time is site specific, so engineers should conduct bench- or pilot-scale tests whenever possible.

The tank can be constructed of mild steel. Its size depends on whether mixing will be done on a batch or a continuous basis.

Batch mixing tanks typically are used at smaller facilities. Such tanks should be sized to treat a day's worth of solids in one batch because many small facilities only have one staffed shift. With adequate capacity, these tanks also can thicken the solids via gravity after stabilization. If a tank is used for both stabilization and thickening, then special equipment must be used to withdraw the thickened biosolids.

In continuous mixing systems, the pH and the volume are held constant, and automated lime-feeding equipment is required. The primary advantage of continuous mixing facilities is that a smaller tank may be used than is required in batch mixing. Because pH is important, tank contents should be closely monitored and maintained at a pH above 12 for at least 2 hours after mixing.

Both systems must provide enough mixing to keep solids in suspension and distribute lime efficiently. The two most common mixing systems are diffused-air and mechanical. Although both have been successful, diffused air is more widely used.

Diffused-air mixers have at least two important advantages over mechanical mixers. The first is more aeration, which in batch operations, helps keep dewatered solids fresh before the lime is added. The second is less potential for debris to foul the equipment (however, "nonclog" mechanical mixers are available).

Diffused-air systems also have several disadvantages. One is that ammonia stripping creates odors and reduces the biosolids' fertilizer value. Ammonia release also can be hazardous, so adequate ventilation must be provided. Another disadvantage is that the mixture absorbs carbon dioxide from air, so more lime is needed (because some of it reacts with the carbon dioxide). Finally, because gases (e.g., ammonia) are stripped, the facilities must be enclosed and the offgas may require treatment.

The design criteria for mixing facilities are similar to those for aerobic digestion systems. If design engineers select a diffused-air mixer, coarse-bubble diffusers should be used. Diffusers typically are mounted along one wall of the tank to induce a spiral-roll mixing pattern. Airflow rates of 0.3 to 0.5 L/m^3·s have successfully been used for mixing (Beals, 1976). Airflow requirements may be higher if mixing thickened feeds.

The design criteria for mechanical mixers are based on bulk fluid velocity and impeller Reynolds number. Table 23.48 lists the various sizes of mechanical mixers required for various volumes. The data are based on both maintaining bulk fluid velocity (i.e., turbine agitator pumping capacity divided by cross-sectional area of mixing tank) at more than 0.13 m/s and an impeller Reynolds number at more than 1000. The mixer sizes listed are adequate for mixing feeds with concentrations of up to 10% dry solids and viscosities up to 1 Pa/s (1000 cP).

When feed solids are conditioned in mixing tanks before thickening or dewatering, engineers must carefully consider the mixing design to prevent floc shearing. Typically, lower mechanical mixer speeds and larger turbine diameters are required. Mechanical mixers also should have variable-speed drives to allow for process control.

The American Water Works Association (2013) and the National Lime Association (1988) have published several documents on selecting lime and lime-handling equipment, as well as on designing lime-application systems. These should be consulted for more design information.

5.5.4 Dewatered Cake-Chemical Mixing for Dry Alkaline Stabilization

The most critical component of dry alkaline stabilization is mixing (blending) dewatered cake and alkaline material. The goal is to provide intimate contact between cake and chemical, so the pH of the entire mixture is adjusted. Inadequate mixing has led to incomplete stabilization, odors, and dust problems at several dry alkaline stabilization facilities in the United States (Oerke and Stone, 1991).

Tank Size		Tank Diameter		Motor Size		Shaft Speed (rpm)	Turbine Diameter	
m³	gal	m	ft	kW	hp		m	ft
19	5000	2.9	9.5	6	7.5	125	0.8	2.7
				4	5	84	1.0	3.2
				2	3	56	1.1	3.6
57	15 500	4.2	13.7	15	20	100	1.1	3.7
				11	15	68	1.3	4.4
				7	10	45	1.6	5.3
				6	7.5	37	1.7	5.6
114	30 000	5.2	17.2	30	40	84	1.5	4.8
				22	30	68	1.6	5.1
				19	25	56	1.7	5.5
				15	20	37	2.1	6.8
284	75 000	7.1	23.4	75	100	100	1.6	5.2
				56	75	68	1.9	6.2
				45	6	56	2.0	6.6
				37	50	45	2.2	7.3
380	100 000	7.8	25.7	93	125	84	1.8	6.0
				75	100	68	2.0	6.5
				56	75	45	2.4	7.8

Bulk fluid velocity >0.13 m/s (26 ft/min).
Impeller Reynolds > 1000.
Mix tank configuration:
  Liquid depth equals tank diameter.
  Baffles with a width of 1/12 the tank diameter placed at 90-deg spacing.

TABLE 23.48 Mechanical Mixer Specification for Liquid Lime Stabilization (Counts and Shuckrow, 1975)

Both batch and continuous mixing systems are available. A mechanical mixer (e.g., a pug mill or plow blender) typically is used (see Figures 23.80 and 23.81). Diffused air is not used for mixing lime with cake. Mixers typically are selected based on experience and trial-and-error testing. Many mixer manufacturers have mobile pilot-scale units available, and engineers should use such equipment whenever possible to evaluate and select the most effective mixer.

Thorough mixing is an art; many variables affect the mixing process and, therefore, the resulting biosolids characteristics. Dewatered cake and chemicals are added together at the "head" of the mixer, and the proportions are important. The mixing characteristics of a dewatered cake depend on the solids concentration, polymer used to condition solids before dewatering, stabilization chemical and dose, temperature, mixing intensity, SRT, and mixer's surface area per volume of exposed cake. When selecting a mixer, design engineers also should consider minimum and maximum cake production, hours of operation, and other operating conditions.

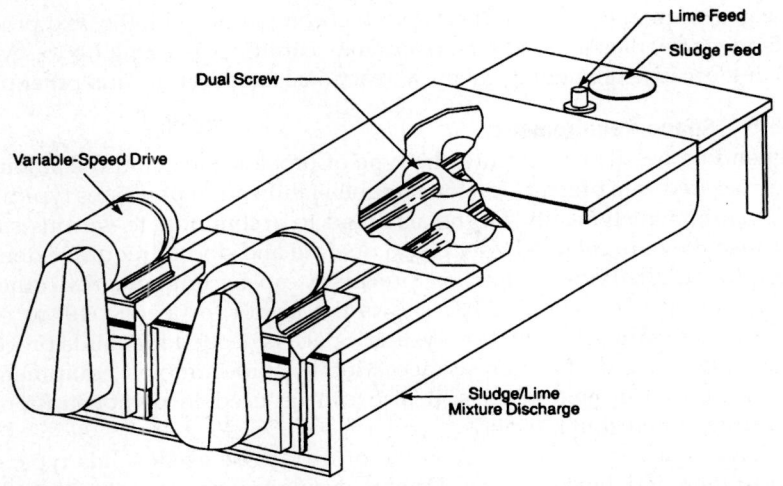

FIGURE 23.80 Typical dual-screw pug mill.

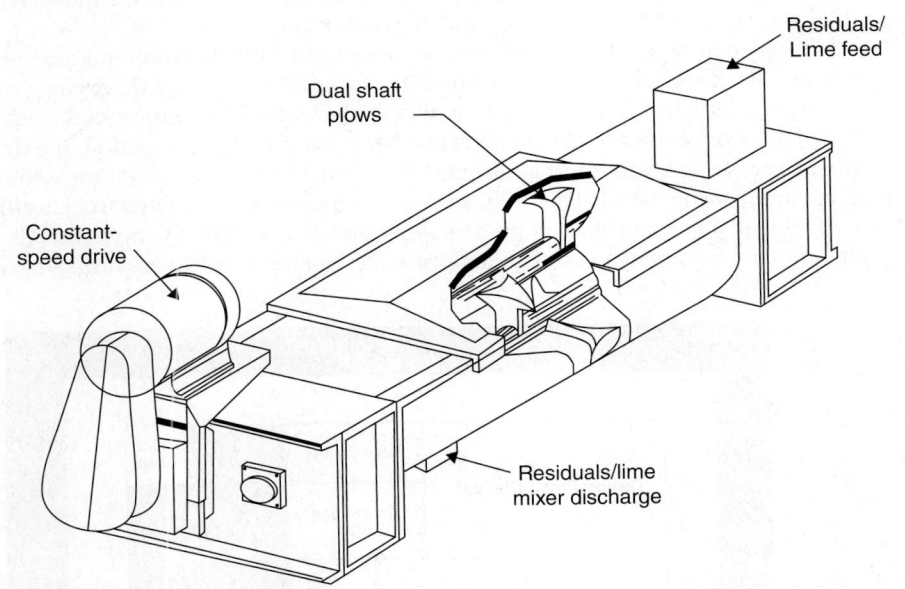

FIGURE 23.81 Typical plow blender.

To adapt to variations in mixing conditions, mixers can be equipped with variable-speed drives, adjustable paddle configurations, weir plates, and other options that adjust mixing intensity and retention time (Christy, 1992).

The physical characteristics of the resulting biosolids depend on the mixing parameters. Its physical consistency can range from sticky and plastic to granular and dusty.

The goal of mixing is to produce a product compatible with the next processing step or intended use. Biosolids characteristics may continue changing up to several days after mixing because of ongoing chemical reactions, temperature, and other parameters.

5.5.5   Space Requirements

Depending on site constraints, the type of process used, and the amount of solids to be processed, site preparation for alkaline-stabilization processes typically is minimal. The equipment typically can be arranged to accommodate various site constraints. Because they are relatively simple to operate and do not require extensive, complex equipment, alkaline-stabilization processes can be implemented quickly in a relatively small space. Figure 23.82 shows the layout for a 3.4×10^5 m³/d (100-ton/d) advanced alkaline-stabilization system. Space is needed for solids processing, drying (if necessary), and biosolids storage. Mobile, skid-mounted equipment can be used for backup or in emergencies; it also can be used in demonstration programs to encourage interest in biosolids.

Land requirements depend on the process to be used; solids type, characteristics and volume; and the specific site. Drying/curing area needs are typically 25 to 34 m²/Mg (300 to 400 sq ft/wet ton) of processed cake, but the area needed also depends on the overall amount of material to be dried or cured, and the drying method used. The size of the drying/curing building can be significantly reduced by increasing the alkaline chemical dose or using a mechanical dryer.

Design engineers should consider storing product offsite if not-enough area is available on-site. Landfills typically can provide space to accommodate drying/curing, but the drying and storage areas must be relocated as landfilling progresses. Also, outdoor drying/curing sites at landfills can cause odor complaints. In addition, the drying area must be easily accessible and large enough for trucks to unload biosolids without excessive maneuvering. Land also is required to accommodate additional truck traffic on-site.

If the alkaline-stabilization process is located at a WRRF, the access roads probably already exist. Sufficient access should be provided for regular delivery of alkaline materials.

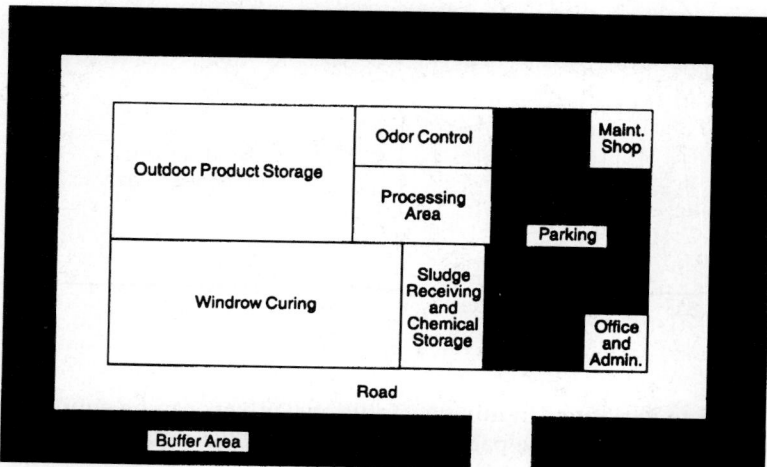

FIGURE 23.82   Layout for a 91-Mg/d (100-ton/d) advanced alkaline stabilization facility (Fergen, 1991).

This truck access should not interfere with the traffic associated with the process or with biosolids distribution.

5.5.6 Economic Considerations

Economics is another important factor when selecting a solids-management option. Design engineers should evaluate the costs of an alkaline-stabilization process based on total life-cycle costs via present-worth, equivalent annual cost, or similar approaches. The cost of hauling and land-applying biosolids can be significant and must be included in the cost analysis (Jacobs et al., 1992). In addition, the costs of a privatized option (if the preferred procurement method) should be compared to those of publicly owned and operated options.

Annual O&M costs include labor, chemical costs, fuel, utilities, maintenance costs, and transportation (e.g., chemicals to the facility and biosolids to use or disposal sites). Other annual costs may include public education, public relations, biosolids marketing, soil testing, agronomic testing, and analyses. An owner should exercise caution when examining annual costs at other facilities because they include a number of site-specific factors (e.g., power, labor, and distance of the chemical supplier from the facility). Moreover, minimum biosolids-production amounts specified in the contract also affect total annual costs. When alkaline-stabilization technologies are operated under private contracts, the negotiated contract must accurately reflect actual biosolids production.

Many site-specific factors (e.g., physical layout, solids type and characteristics, biosolids use, local regulations, local biosolids market, and local climate) influence costs and make economic evaluations and comparisons difficult. Also, at some facilities, existing equipment has been retrofitted for use in alkaline stabilization.

Design engineers should consider the flexibility (adaptability) of alkaline stabilization and the use of existing facilities when evaluating solids-management options. Although not always possible, municipalities can save money if existing equipment is used in the process train.

5.6 Other Design Considerations

This section highlights and summarizes some of the O&M issues pertaining to alkaline stabilization. In general, most alkaline-stabilization technologies are relatively simple and not equipment intensive; in addition, staff requirements are low compared to those for other stabilization processes. Operating considerations that must be addressed include startup issues, labor requirements, health and safety considerations, feed and product quality monitoring, maintenance, odors, dust, drying, procurement options, and process performance (Oerke, 1991).

5.6.1 Startup Issues

Startup issues associated with alkaline stabilization are installation specific. The greatest concerns include equipment performance; process verification; physical, chemical, and biological product quality (to verify regulatory compliance and ensure product acceptance); and, if privatized, contractor performance.

While alkaline-stabilization processes are not as equipment intensive as other stabilization processes, some equipment problems and operating difficulties may occur during startup. To make the most of the dry chemical dose and produce the desired product, operators may need to vary the mixer-paddle speed and retention times. All process equipment should be tested at rated capacity during the startup period.

The project team also should test a significant amount of representative feed and verify that dose measurement is accurate and mixing is homogeneous. If several types of alkaline materials are to be used, each should be tested with the system storage and feeding equipment to verify acceptable operation.

Regulators should be consulted during the design process to verify the parameters to be monitored for process approval. Monitoring results should be submitted to them as soon as possible to initiate the approval process. Permit delays are not uncommon, and appropriate measures should be taken to avoid them if at all possible. Frequent, continued communication with key regulators can facilitate the approval process. Also, the federal permitting authority (U.S. EPA region or delegated state agency) must be appropriately notified before startup.

Startup operations provide an opportunity to vary process parameters and evaluate the effects of these changes on product quality. Although the effect of various chemical doses and drying times should have been evaluated during pilot- or bench-scale testing, pilot-scale test conditions do not always adequately simulate full-scale operations.

An advantage of alkaline stabilization is the ability to start up operations quickly. A mobile, outdoor processing unit can be fully operational in about 10 days or less, depending on the amount of material to be processed.

5.6.2   Health and Safety Considerations

Dust generated from alkaline materials probably is the most significant health concern. Alkaline materials are caustic and cause skin burns and irritation and discomfort to moist surfaces (e.g., eyes, lips, and sweating arms); therefore, readily accessible eye-washes and showers should be provided at various locations throughout a WRRF. Operators working in dusty environments or servicing alkaline storage and feeding equipment should be supplied with proper work clothing and safety equipment (e.g., gloves, proper respirators, and eye protection).

Ammonia is another safety concern, especially if anaerobically digested solids are processed, because it is likely that considerable ammonia gas will be released (about 6 to 10 times more likely than from raw solids). Ammonia emissions can be controlled via proper ventilation of mixers, storage hoppers, and loading areas. Strong releases of ammonia may be experienced during mechanical aeration or mixing, and during drying. So mixing equipment should be enclosed and vented to odor-control facilities if at all possible. In some areas, it may be necessary to provide operators with respirators, depending on the amount of ammonia released to meet OSHA requirements.

Special safety measures may be required for drying areas. The layer of fine, operations-related dust that tends to settle in the drying area can be slippery on concrete or asphalt surfaces. During wet weather, a layer of mud may form outdoors on the drying pad. Mud is slippery and may pose a hazard to pedestrians and vehicle traffic. Special precautions should be taken to improve safety via good housekeeping practices.

5.6.3   Process Monitoring and Control

Feed cake and alkaline materials must be monitored frequently, so operators can adjust the process, as needed, to achieve adequate stabilization and a consistent product. The effects of incomplete stabilization are not readily apparent and may not be seen at a WRRF; therefore, proper process control is important. Operators must be aware that acceptable dewatering characteristics and the absence of odors alone are not good indicators of adequate stabilization.

Monitored characteristics include the total solids concentration, pH, and temperature of both feed cake and biosolids. For Class A (PFRP) products, fecal streptococci also must be monitored at the frequency specified in 40 CFR 503.16 (U.S. EPA, 1993). In addition, metals must be monitored if the product will be used for agricultural purposes. If the product will be landfilled, toxic characteristics leaching procedure (TCLP) tests must be performed. Quality-control data may be required for regulatory approval; the method and the frequency depend on regulatory requirements. In some cases, odor characterization and emissions monitoring also may be required.

Operators can adjust the chemical dose in response to manual measurements of temperature and pH or visual inspections of the feed-chemical mixture. However, some automatic process control can be incorporated if desired. For example, thermocouples can be used to measure the heat pulse in an enclosed vessel. The chemical feed rate can be controlled by pacing it with the incoming-feed flowrate or dewatered cake via a weigh belt or similar means.

Using programmable logic controllers to monitor the chemical feed system helps produce a consistent product. A typical system may include an electronic chemical meter linked to feed-cake belt weigh scales; then, chemical feed is automatically controlled based on the weight of feed cake. Special care should be taken to keep the weigh scales frequently calibrated and correctly operating to ensure that the appropriate dose of alkaline materials is added. Solids weighing and volumetric systems should be calibrated every month. An automated system also decreases the number of personnel needed to operate the process.

Sensors—particularly pH electrodes (both laboratory and automatic process-control units)—must be properly cleaned, calibrated, and maintained. Special pH electrodes are necessary for routine measurements of more than pH 10. The pH must be monitored carefully to ensure that it is kept high enough for long enough to meet regulatory requirements. Portable pH pen probes are acceptable for process monitoring. A qualified laboratory should perform microbiological examinations for indicator organisms (e.g., fecal coliforms and fecal streptococci) regularly.

5.6.4 Odor Generation and Control

Odors and odor control are important issues when evaluating alkaline stabilization as a solids management option. Inadequate control and treatment of odors can be detrimental to a solids management program. Local conditions (e.g, weather, other sources of odors, and the characteristics of the odor-causing compounds) will influence the selection and design of an odor-treatment system. There are many site-specific factors that should be considered when developing a publicly acceptable odor-control program. A successful odor-control effort includes the following elements:

- Initial site selection;
- Proper process performance;
- Reduced biosolids storage time and volume;
- Identification of odor sources and odor-causing compounds;
- Meteorological modeling at different heights;
- Distance to nearest receptors; and
- Appropriate odor-control technology and equipment.

Ammonia is the odor most typically encountered at alkaline-stabilization facilities. Adding alkaline materials raises the pH, which causes the dissolved ammonia in dewatered cake to volatilize. Although the odors tend to dissipate quickly, the ammonia levels in mixing and drying areas can be high if the gas is not collected and treated. Also, if adequate ventilation is not provided, operators may need to wear respirators. So appropriate odor-control equipment should be provided to ventilate and scrub the air to remove ammonia, thereby reducing odor problems and increasing public acceptance. As pH and temperatures rise, the intensity of ammonia emissions in the processing area may mask other, more prevalent odors that do not readily dissipate (e.g., trimethyl amines). So an odor survey should be performed to identify the sources of odors and characterize the odorants.

In addition to an odor survey, an assessment of meteorological conditions and atmospheric dispersion should be performed. Atmospheric data should be collected on wind speed and direction, temperature, and inversion conditions. This information typically is available from local weather stations and can be used to determine the effect of odor on residents near the alkaline-stabilization facility and the degree of odor control needed to meet community odor standards.

An effective odor-control program involves operational monitoring and may include bench-scale testing (to determine ammonia emissions at various chemical doses) and gas chromatography/mass spectrometry testing. After odors have been characterized, they must be collected and treated. Pilot-scale testing helps check the effectiveness of a proposed treatment option and its chemistry. At many alkaline-stabilization facilities, odor control primarily consists of diluting odors via open-air drying. If drying operations are enclosed, odors can be diluted and dispersed via rooftop ventilation. However, if large quantities of materials are processed in a densely populated area, a combination of dispersion and chemical scrubbing should be seriously investigated. It is important that the odor-control program be responsive to odor complaints. Depending on meteorological conditions and the sources and types of odors, operational or process modifications may be necessary to resolve the problem.

Initially, wastewater treatment professionals thought that alkaline- and advanced alkaline-stabilization facilities did not need odor-control systems. However, numerous odor concerns and complaints have made it clear that odor-control systems should be strongly considered and may be necessary for alkaline-stabilization systems near populated areas. Such systems may consist of enhanced ventilating systems and simple one-stage chemical scrubbers designed to remove ammonia only in the feed–chemical mixing area. They also can be state-of-the-art, three-stage, packed tower–mist scrubber–packed tower systems with air dispersion stacks, designed to treat high volumes of foul air containing particulate, ammonia, amines, dimethyl disulfide, mercaptans, and hydrogen sulfide generated from all areas of the solids treatment train. These sophisticated odor-control systems use sulfuric acid, sodium hypochlorite, and sodium hydroxide to neutralize and oxidize odorants (see Chapter 6 for more information).

5.6.5   Dust

Dust is inherent in alkaline-stabilization systems. Alkaline material-handling systems can create significant dust problems, particularly if fine-textured materials (e.g., hydrated lime, cement-plant kiln dust, or lime-plant kiln dust) are used. The alkaline material-handling system should be designed with provisions for reducing dust production. Excessive alkaline dust also affects odor-control scrubber performance (e.g., acid chemical requirements).

5.6.6   Sidestream Effects
Alkaline-stabilization processes typically have little effect on WRRF operations. Minimal sidestreams result from site drainage, product stockpile leachate, and runoff if the storage area is not covered. However, a potential sidestream facility load is ammonia recycle if ventilation and odor-control acid scrubbers are installed.

5.6.7   Drying
Supplemental drying, if required in the process, also requires special consideration. Drying may make the product easier to handle, and the type of drying system will affect biosolids characteristics. For example, if the material is set out on a pad for solar drying without mechanical turning, it may dry in large clumps that would be incompatible with land application via granular fertilizer spreaders.

The duration of drying or curing depends on environmental conditions, chemical dose, windrow configuration, and initial and final solids concentrations. It also depends on the time required to achieve the process goal (i.e., for the heat of reaction to occur and for the pH to rise enough to destroy pathogens). Both drying and curing modify physical properties to attain the desired solids concentration and biosolids characteristics. The duration of drying depends on windrow size and weather conditions (if the drying facility is not enclosed).

5.6.8   Process Performance
Properly designed and operated alkaline-stabilization systems reduce odors, odor-production potential, and pathogen levels.

5.6.8.1   Odor Reduction   With proper mixing, alkaline-stabilization systems substantially reduce odor. One source of odors in solids-processing facilities, hydrogen sulfide, essentially is eliminated after the alkaline chemical is added and the pH rises to 9 or higher, because hydrogen sulfide is converted to nonvolatile ionized forms (see Figure 23.83). When air mixing systems are used, ammonia odors initially increase as a result of ammonia stripping. Once these odors have been emitted and dispersed or treated, odors can be reduced by a factor of 10 (Westphal and Christensen, 1983).

Other odorous gases emitted at high pH and temperature (e.g., trimethyl amine) must be considered and dispersed or treated.

5.6.8.2   Settling and Dewatering Characteristics   Lime stabilization improves solids settling and dewatering characteristics. Lime alone has been used in the past as a conditioner before dewatering (although lime conditioning and lime stabilization are different processes). Precipitates associated with excess lime addition (primarily $Ca(CO_3)$ and unreacted $Ca(OH)_2$) act as bulking agents, increasing porosity while resisting compression.

Limited reports of lime-stabilized thickening and dewatering processes show mixed results. One study showed improved thickening (U.S. EPA, 1975). Two studies showed slightly better to slightly poorer dewatering on sand drying beds, compared to solids that were not lime stabilized (Novak et al., 1977; U.S. EPA, 1975).

Design engineers should use caution when designing mechanical dewatering systems for lime-stabilized solids. If the design does not include proper preventive measures, scaling problems (e.g., deposition of $CaCO_3$ and other precipitates) can occur, resulting in higher O&M costs.

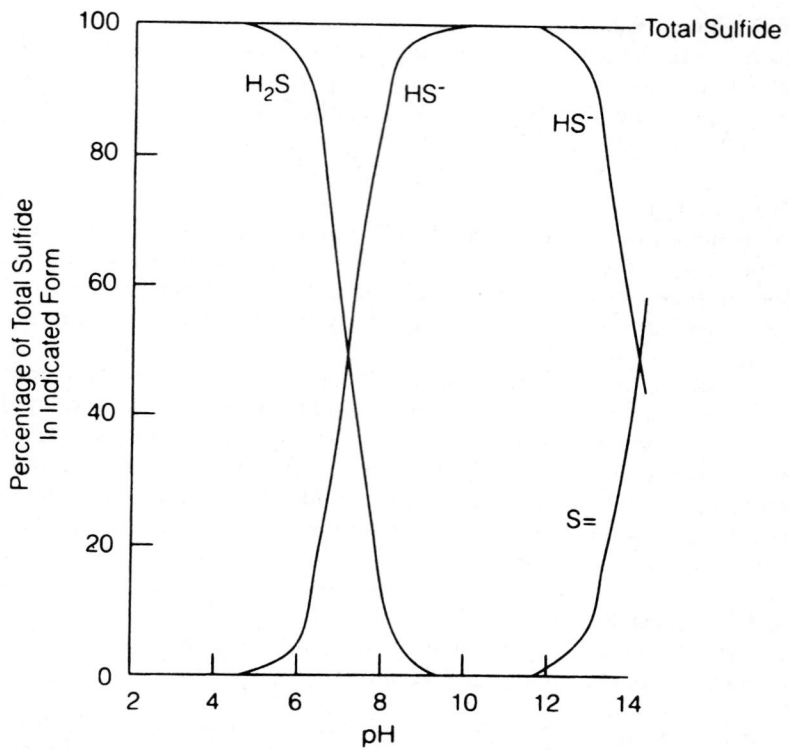

FIGURE 23.83   Effect of pH on speciation of hydrogen sulfide.

5.6.9   Procurement Options

Private firms offer many advanced alkaline-stabilization technologies involving propri-etary processes or specialized equipment. Such technologies also involve royalty fees, quality-control fees, or sole-source equipment. Additionally, some firms may offer turn-key design–build facility procurement options or require privatization of various types of solids-processing services.

5.7   Process Considerations for Designers

5.7.1   Dosage Criteria

5.7.1.1   Class B Stabilization   Class B stabilization is achieved by adding enough lime (or its equivalent if using alkaline by-products) to raise the pH to 12 for 2 hours and then hold it at 11.5 or higher for another 22 hours. The pH must be measured at a tem-perature of 25°C or corrected to 25°C.

Figure 23.79 shows the theoretical lime dose rates needed to achieve the design pH criteria. However, design engineers always should conduct bench-scale tests with the lime type and grade to be used during full-scale operations. Dose rates depend on the cake's solids content; more lime is needed when solids content is low (13% to 18%) or high (>25%). Limed biosolids should be tested to ensure that they meet both the pH criteria for Class B pathogen reduction and the Class B coliform limit.

5.7.1.2   Class A Stabilization   Lime stabilization meets Class A pathogen requirements by using the exothermic reaction of CaO and water in the biosolids to generate heat. Alkaline-stabilization processes can meet Class A requirements under Alternative 1 (time and temperature) or Alternative 5 (pasteurization); both are based on the assumption that every particle of biosolids will be exposed to 70°C for 30 minutes. This requirement can be met by treating batches of solids with lime in a closed container. Alkaline-stabilization processes that operate in continuous mode may need the specific approval of the EPA's Pathogen Equivalency Committee to be accepted as a Class A process.

Several proprietary technologies have been approved by the committee (e.g., N-Viro's AASAD process) or achieve pasteurization via a combination of lime and other sources of heat (e.g., RDP, Bioset). RDP envessel pasteurization uses an electrically heated screw to provide more heat. The Bioset process uses sulfamic acid to generate extra heat via an exothermic reaction in a pressurized reactor. The alkaline doses for these processes are given in Table 23.49 (U.S. EPA, 2007).

5.7.1.3   Class B Odor Control   Raising pH into the high alkaline range not only stabilizes solids but also provides short-term odor control. However, the lime doses for Class B only raise pH above 12 temporarily. To control odors days or several weeks, the dose should be above the minimum for Class B stabilization. Although bench-scale testing is the best way to determine the optimum lime dose for odor control, a good general rule is to double the dose required for pathogen reduction. Odors also can be controlled effectively by adequate mixing to ensure that there are no pockets of biosolids not in contact with lime.

5.7.2   Lime Type and Gradation

The suitable treatment agents are all lime-based materials. Lime is an alkaline earth material that produces a pH of 12.4 at 25°C when mixed with water. It is found in two forms: calcium oxide (CaO) and calcium hydroxide [$Ca(OH)_2$]. Calcium oxide (also called quicklime or hot lime) is the result of heating limestone [calcium carbonate ($CaCO_3$)] enough to drive off carbon dioxide (CO_2). When mixed with water, CaO forms a fine white powder [$Ca(OH)_2$, also called hydrated lime] and gives off considerable heat (called heat of hydration).

B	Generic	2–5	None	15–30
A	Generic	10–20	None	23–35
A	N-Viro	12–20	FA, CKD, LKD	50–65
A	RDP	15	Electrical heat	23–35
A	Bioset	15	Sulfamic acid	20–30

Final solids based on cake solids of 25–25%. Dose rate in % wet wt.

TABLE 23.49   Mass Balance for Various Alkaline-Stabilization Alternatives

Many industrial processes have by-products that contain usable amounts of lime (e.g., industrial scrubber sludge, fly ash [from incinerators that burn coals containing limestone], cement-plant kiln dust, lime-plant kiln dust, and dry industrial flue-gas scrubbing by-products). If used to treat solids, however, these alkaline agents must be carefully evaluated and monitored because their concentrations of free (active) lime content and contaminants vary.

Commercial quicklime grades can vary from several inches in diameter to material passing a #100 sieve. The National Lime Association (1988) lists the following five grades:

- Lump lime (50.8 to 203.2 mm [2 to 8 in.] in diameter);
- Pebble lime (the most common form, ranging from 6.35 to 50.8 mm [0.25 to 2 in.] in diameter);
- Granular lime (100% passes through a #8 sieve, and 100% is retained on a #100 sieve);
- Ground lime (100% passes through a #8 sieve, and 40% to 60% passes through a #100 sieve); and
- Pulverized lime (100% passes through a #20 sieve, and 85% to 95% passes through a #100 sieve).

The following quicklime definitions will help in relieving the confusion of so many terms:

- Unslaked quicklime fines (calcium oxide fines) are quicklime particles that typically are less than 9.5 mm (3/8 in.) in diameter and have not been mixed with water.
- Pulverized calcium oxide is quicklime that has been mechanically ground into particles that typically are less than 60 mesh.
- Granular calcium oxide fines are quicklime that have been ground into particles that are larger than pulverized calcium oxide (i.e., there are no dust-sized particles).
- Unslaked CaO fines are small quicklime particles that have not been mixed with water.
- Unhydrated calcium oxide is any quicklime that has not been hydrated (slaked).

Lime's reactivity with water is measured by the slaking rate (as defined in AWWA specification B202-93, Sec. 5.4). Small-pore limes need 20 to 30 minutes to fully react with water, forming $Ca(OH)_2$ with a slow heat rise. A moderately reactive lime needs 10 to 20 minutes to react with water, forming $Ca(OH)_2$ and raising the temperature to 40°C in 3 to 6 minutes. A highly reactive lime fully reacts with water within 10 minutes and raises the temperature to 40°C within 3 minutes. Design engineers can use the slaking rate to evaluate the suitability of various industrial by-products.

Solids should be treated with a moderately or highly reactive lime to ensure that the CaO fully converts to $Ca(OH)_2$. For the reaction to generate a high pH that migrates throughout the solids, there must be a continuous film of water throughout the material. Otherwise, the lime may not be fully hydrated or the hydroxide ions may not migrate throughout the solids. This can and does result in improper pH measurements, improper doses, and, therefore, unstabilized solids. If calcium oxide must be pulverized, it should be pulverized at the point of application to prevent air slaking and ensure the desired reactivity.

5.7.3   Mixing Requirements

In a survey of 19 WRRFs in Pennsylvania, the Pennsylvania Department of Environmental Protection examined process variables (e.g., biological treatment and lime dose) and their effects on odor, as determined by an odor panel (U.S. EPA, 2007).

Results showed a wide range in lime dose and in solids content before treatment. Centrifuged solids tended to be more odorous than belt-pressed solids, but there was no clear relationship between other process variables and odor.

The agency selected two of the surveyed WRRFs to study the effect of lime dose and mixing time on pH decay and on odor. Researchers used two parameters to indicate mixing efficiency: total Ca (as measured by EDTA titration) and pH (as measured by a flat-surface pH electrode) (U.S. EPA, 2007). Higher, relatively stable Ca concentrations throughout solids in the mixing vessel indicated that solids and lime were well mixed. The flat-surface pH electrode measured actual pH in the solids-lime mixture more accurately than the traditional slurry method. (In the slurry method, water is added to the solids-lime mixture before pH measurement; this dissolves any unreacted lime, producing a falsely high pH reading.)

In the first study, researchers added CaO to cake at 4.5% and 11.7% (wet weight) and mixed them for 15 and 45 seconds. Results showed that 15 seconds were inadequate; there was much higher variability in Ca and pH at 15 seconds of mixing. Results also showed that a CaO dose of 4.5% would raise pH above 12, but only at the longer mixing time.

The slurry method indicated that pH dropped below 12 after 15 days at the lower CaO dose and shorter mixing time. The flat-surface electrode, however, showed that the lower CaO dose and shorter mixing time never achieved pH 12.

Increasing CaO dose and mixing time decreased odor generation. They also reduced the generation of NH_3 and amines, an indicator of biological decomposition. Biological decomposition can result in increased odors.

Odor increased in all limed solids up to 15 days, but decreased thereafter for solids with the higher CaO dose and the longer mixing time. The study also showed that NH_3 and amines greatly increased after 15 days in the solids with the lower CaO dose and shorter mixing time.

In the second study, researchers examined the facility-scale effect of optimizing CaO and solids mixing on pH decay and odor generation. To optimize mixing, researchers added CaO to the solids upstream of the mixer to increase contact time. They then compared samples of limed solids from the existing operation with those from the optimized operation.

Results showed that optimizing mixing reduced variations in Ca levels in solids, prevented pH from decaying, and decreased odor, NH_3, and amine generation for up to 20 days.

5.7.3.1   Measuring Mixing Efficiency

5.7.3.1.1 Identifying Issues   The DC Water (formerly DCWASA) Blue Plains Advanced Wastewater Treatment Plant has used lime stabilization to achieve Class B pathogen standards for many years. While fecal coliform results always met the regulatory limit (2 million CFU/g), they were inconsistent. Odors also were inconsistent, according to empirical evidence gathered in the field. Field inspectors said the odors resembled those of rotten eggs or rotting cabbage (methyl mercaptan, dimethyl sulfide, dimethyl disulfide); rancid meat (volatile fatty acids); and fecal matter (indole, skatole). Some odorants were confirmed via a gas chromatograph (Kim et al., 2003). All of these odors are the products of anaerobic microbial activity, indicating less than optimum microbial inactivation. Personnel suspected these inconsistencies in odor and fecal coliform destruction were related, and that poor mixing might be responsible. So they implemented changes based on a series of studies, and DC Water solids now consistently contain less than 1000 CFU fecal coliforms and emit few odors.

Most facilities using lime stabilization rely on the results of pH tests at 2 and 24 hours to indicate if the material is in compliance with U.S. EPA Class B standards. However, the U.S. EPA assumes that complete, efficient mixing occurs and that a pH >12 after 2 hours and a pH >11.5 after 24 hours indicate a stabilized, low-odor material. The standard pH test (the slurry method) involves adding water to and stirring the sample before measurement, so although the test is a good indicator of whether the sample contains enough lime, it does not indicate whether the sample was well mixed before testing. So pH results may be consistent while the final product has wide swings in quality (fecal coliform levels and odors).

Facilities experiencing odor complaints should determine whether the solids have consistent concentrations of fecal coliforms and odorants. If results indicate that fecal coliform levels are inconsistent or considerably above 1000 CFU, or that odors (measured either by a nose [qualitatively] or by a reduced-sulfur meter or tubes [quantitatively]) are inconsistent or intolerably offensive, then the lime was not thoroughly incorporated into the solids. A set of simple, inexpensive tests can help identify solutions.

Efficient, adequate mixing is affected by at least five factors: lime gradation, cake dryness, residence time in the mixer, mixer type, and conveyance method before mixing. Once operators have a tool to measure mixing efficiency, they can adjust one or more of these factors to achieve the desired product quality.

5.7.3.1.2   Establishing a Benchmark for Good Mixing   If investigators suspect poor mixing, they should start by establishing parameters consistent with sufficient mixing that they can use when comparing results. A simple means of determining mixing efficiency is a calcium test, which requires a 1-g sample. In well-mixed solids, each 1-g sample would contain solids and calcium in the required ratio (i.e., 15% lime on a dry-weight basis). In poorly mixed solids, one sample might contain no calcium, another might contain a high percentage of calcium, and others would bear results in between. A large sample set (e.g., 12 to 15 samples) with a high standard deviation would indicate poor mixing, while one with a low standard deviation would indicate well-mixed biosolids. Staff can conduct a bench-scale test in which they mix with solids with lime (as delivered) and determine parameters for well-mixed material. The results then are compared to facility results to grade the performance of full-scale operations.

A bench-scale setup can use a simple bread mixer. Start with unlimed dewatered material, and add lime at the prescribed dose (e.g., 15% on a dry-weight basis). Operate the mixer, stopping and sampling after 10, 20, 30, 40, 60, and 90 seconds. Each time

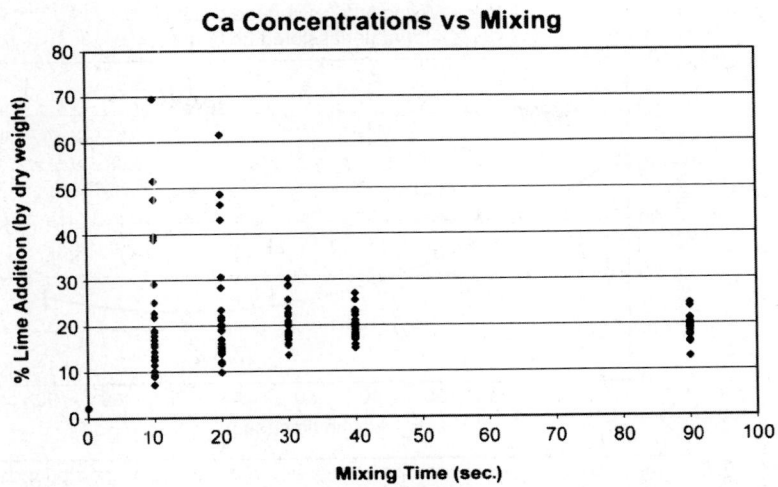

FIGURE 23.84 Results of DCWASA bench-scale mixing test for calcium content.

the mixer is stopped, take fifteen 1-g samples for calcium analysis. (Fewer samples may be adequate, but calcium tests are inexpensive and more data will provide clearer results). Mixing probably is inadequate at 10 seconds and probably sufficient at 90 seconds. The sample set with the smallest standard deviation is the facility-specific benchmark for a well-mixed product. It is important to conduct this bench test on cake collected just before it enters the mixers because the dewatering and conveyance methods will affect mixing results. The data in Figure 23.84 were generated during the bench-scale testing phase of DC Water's research (North et al., 2008a); they show that standard deviation decreased as mixing time increased.

5.7.3.1.3 Measuring Performance of Full-Scale Facility Operations The next step is to take 15 samples from the full-scale operation, analyze them, and calculate their standard deviation. If this standard deviation is higher than the minimum achieved in the lab, then the mixing system can be improved. If the full-scale and minimum bench-scale standard deviations are identical, then better mixing and product quality are unlikely. DC Water found that when odors were high, the standard deviations of its full-scale sample sets were close to that for the 15- to 20-second samples in bench-scale testing, indicating that the full-scale mixer was far from providing optimum mixing during these periods (North et al., 2008b).

5.7.3.1.4 Mixing Energy and Odor Suppression At the Blue Plains facility, the minimum standard deviation of the sample sets was about 2.6 (which occurred at about 40 seconds of bench-scale mixing). Results are facility specific, but this number gave DC Water operators a tool to measure lime-solids mixing and mixer performance, as well as improve product quality. Figure 23.85 shows the relationship between mixing energy (time, in this case) and reduced sulfur compounds (odors) for the samples in Figure 23.84. Not surprisingly, odors are minimized when good mixing occurs.

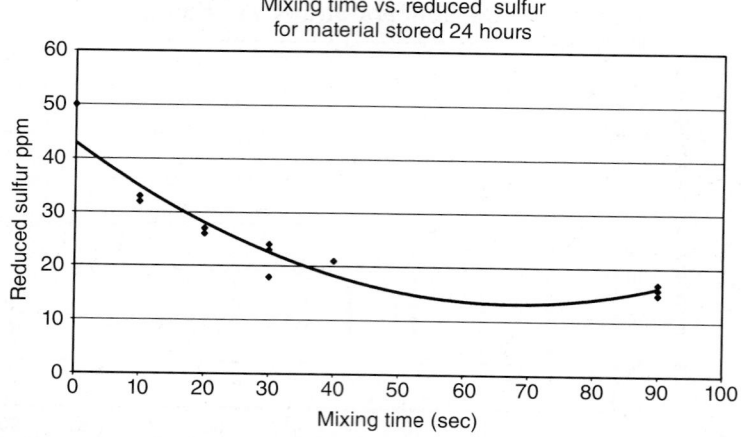

FIGURE 23.85   Relationship between mixing energy and reduced sulfur compounds in DCWASA bench-scale mixing test.

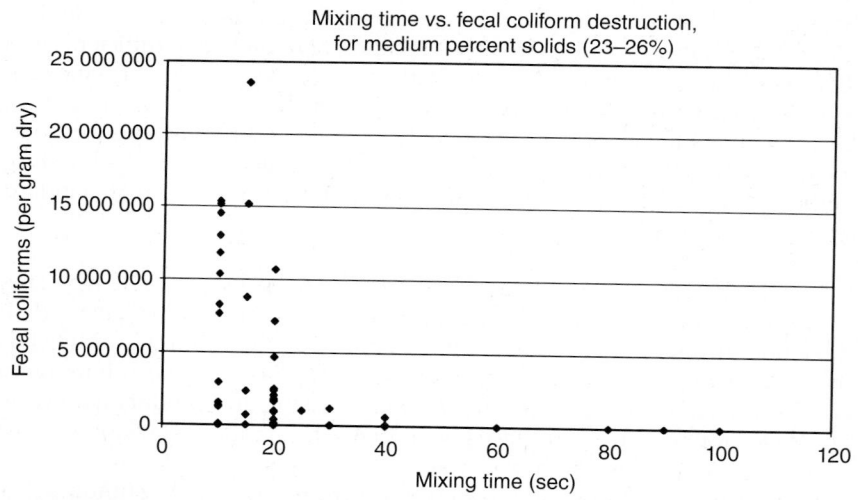

FIGURE 23.86   Effects of mixing on fecal coliform concentrations in DCWASA bench-scale mixing test.

5.7.3.1.5   Mixing Energy and Fecal Coliform Destruction   Figure 23.86 shows the relationship between mixing energy (time) and fecal coliform results for the samples in Figure 23.84. Again, fecal coliforms are minimized when good mixing occurs. Surprisingly, minimizing fecal coliforms in this Class B stabilization process yielded results (CFU, 1000) consistent with Class A biosolids.

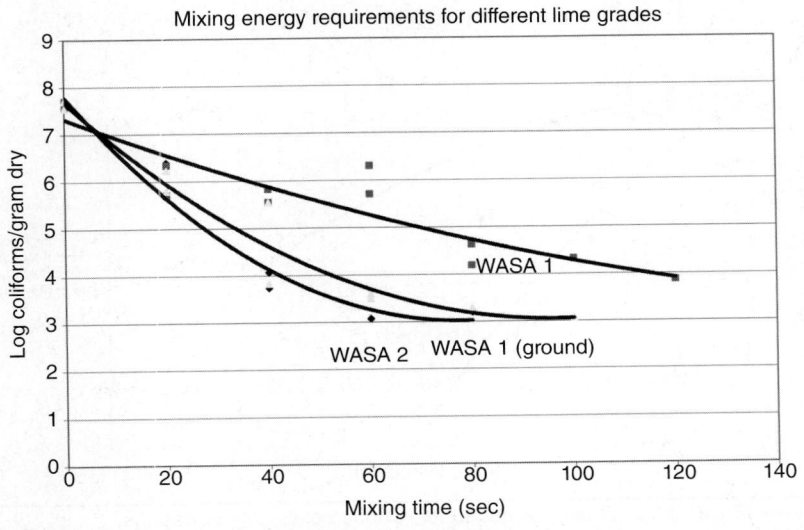

FIGURE 23.87   Mixing energy required for various grades of lime.

5.7.3.2   Optimization of Mixing—Examining Five Factors Affecting Mixing

5.7.3.2.1   Factor 1: Lime Gradation   To ensure proper mixing, personnel periodically must test lime deliveries via a sieve analysis and compare results to the required lime specifications. If the delivered lime is too coarse, mixing energy may be inadequate. This simple test can help ensure adequate mixing, low odors, and proper stabilization.

Coarser lime requires more mixing energy for adequate incorporation. Figure 23.87 shows results from the DC Water bench tests for mixing solids with different grades of lime. The lime used in both DC Water dewatering trains was subjected to sieve analysis, and results showed that the lime used in the WASA 1 train (operated by DC Water employees) was coarser than that used in the WASA 2 train (lime supplied by and equipment operated by contractor). Both limes were mixed with raw solids from one source, and a third sample was mixed with the coarser lime ground to match the sieve analysis of the WASA 2 lime. The results of the coliform analysis show that the samples with finer lime stabilize much sooner than the others. Also, the sample with Lime 1 ground to the fineness of Lime 2 stabilized at a similar rate, showing that the difference between Lime 1 and Lime 2 was primarily due to gradation, not other characteristics.

5.7.3.2.2   Factor 2: Cake Dryness   A dewatered cake's solids content can dramatically affect mixing energy requirements. Drier solids require more energy for adequate lime mixing. Small differences in percent solids (3% to 4 %) can double the required mixing energy. Figure 23.88 shows the mixing energy required for proper stabilization of low, medium, and high cake solids. Cakes with higher solids concentrations require much more mixing energy to minimize fecal coliform concentrations. This is an important consideration when assessing mixing problems, because dewatering facilities some-

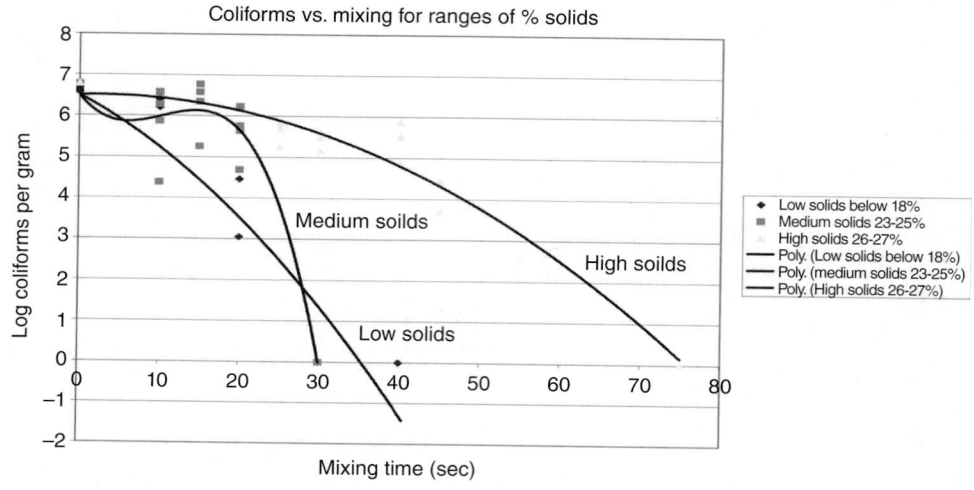

FIGURE 23.88 Mixing energy required for various solids concentrations.

times can produce inconsistent cake solids. Inconsistencies in odors and fecal destruction might be attributable to inadequate mixing when the cake had high solids concentrations.

Engineers should design lime-mixing facilities to handle the maximum solids content expected in dewatered cake. If hauling costs are not paramount, a WRRF might consider scaling back the dryness of the cake solids slightly to help ensure adequate mixing and stabilization. If hauling costs are a major portion of the budget, better mixing equipment (or another fix mentioned in this section) might be required.

5.7.3.2.3   Factors 3 and 4: Mixer Residence Time and Mixer Type   A system's mixing efficiency can be affected by the type of mixer used and the equipment configuration. A facility that adequately mixes lime with a specific piece of equipment for years may run into problems if dewatering-system upgrades (e.g., high solids centrifuges or changes in polymers) produce cake containing more solids. To ensure that enough mixing energy can be provided, operators need to examine whether existing mixers should be modified or replaced.

Often, existing mixers can be modified to increase agitation or residence time. For example, plow blenders have removable weirs that are designed to keep the material in place longer. Other blenders come with openings, so chopper blades can be easily installed to enhance mixing. If a unit is not achieving optimum mixing, operators should install all optional equipment designed to enhance mixing and residence time. If the unit still cannot achieve optimum mixing after considering Factors 1, 2, and 3, operators must consider replacing it. Before considering larger mixers, however, staff should examine the conveyance system.

5.7.3.2.4   Factor 5: Conveyance Method before Mixing   Figure 23.89 shows a plan of part of the solid conveyance system at the DC Water Blue Plains Advanced Wastewater

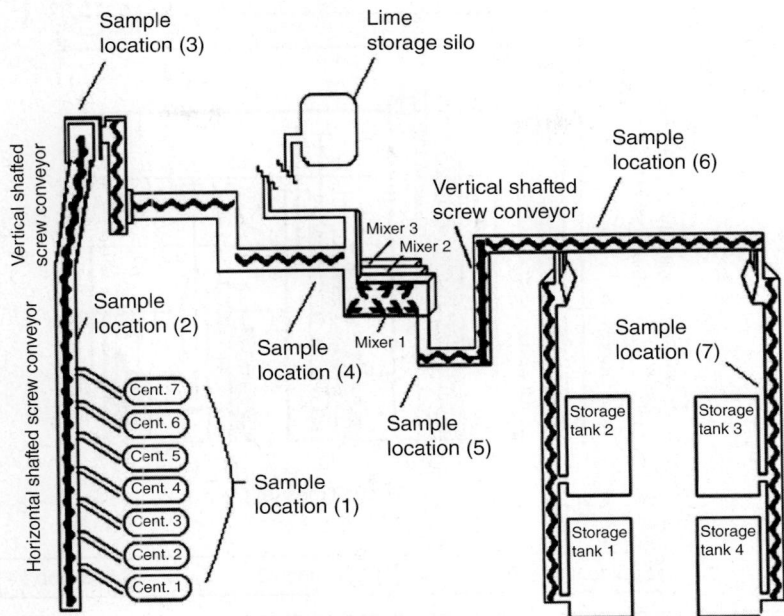

FIGURE 23.89   Plan view of conveyors and sample locations in the DCWASA lime stabilization system.

Treatment Plant. Sample Location 1 is at the discharge of the high-solids centrifuge, Location 2 is on the horizontal screw conveyor, Location 3 is after a vertical screw conveyor, and Location 4 is just before the lime mixer. Location 5 is at the discharge end of the lime mixer, Location 6 is on the horizontal screw conveyors moving material to the storage bunkers, and Location 7 is at the point of discharge to the bunkers. Research results show that using screw conveyors before mixing adds mixing energy (thereby changing the material's rheology), making it more difficult to stabilize. Using screw conveyors after mixing, however, adds more mixing energy and thereby reduces odors and improves product quality.

Unlimed material conveyed a longer distance requires more energy for proper stabilization. Staff grabbed unlimed material from four locations before mixing and subjected it to bench-scale mixing tests. Results show materials that have been conveyed longer (Location 4) consistently have more residual coliforms than those conveyed for a shorter distance (Location 2) (see Figure 23.90). This shows that conveyance changes the material's rheology and affects its ability to mix properly. Visual observations showed that the material changed from a crumbly consistency to a toothpaste consistency between Locations 1 and 4.

When considering equipment changes (if no other intervention has helped), operators should compare the costs of replacing screw conveyors (with belt conveyors, which do not affect a material's rheology) to the cost of upgrading mixers. Existing mixers might be adequate for material that is not screw conveyed over a long distance.

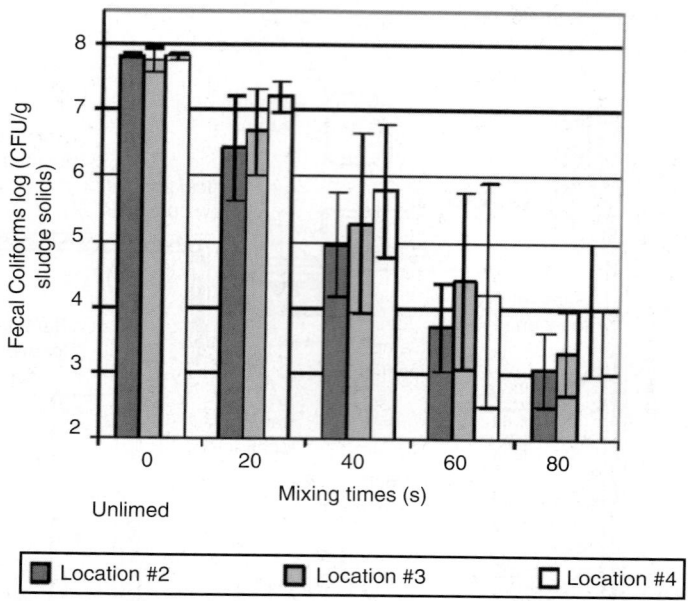

FIGURE 23.90   Effect of conveyance distance on solids stabilization.

After mixing, screw conveyors can add mixing energy and improve product quality. Figure 23.91 shows that, the farther biosolids were conveyed, the less reduced sulfur it generated. These data also show the importance of sampling at the end of conveyor runs, rather than at the discharge end of the lime mixer. Fecal coliform samples from

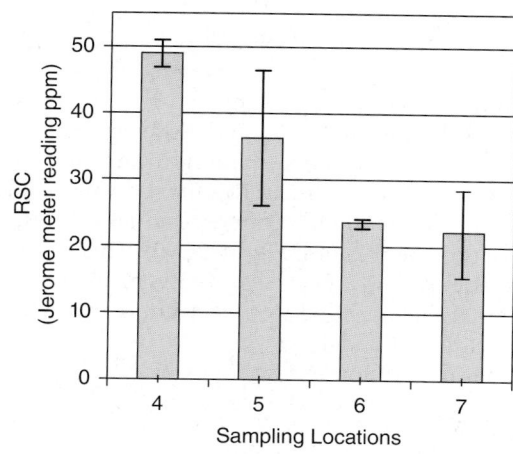

FIGURE 23.91   Effect of conveyance distance on odors.

Locations 5 and 7 will bear strikingly different results, again showing that screw conveyors further stabilize biosolids after lime mixing.

5.7.4 Class B Lime Stabilization Design Example

To design a system to meet EPA Class B standards (2 000 000 CFU/g fecal coliforms), a facility must adequately mix into raw solids an appropriate amount of lime on a consistent basis to produce a useable, low-odor biosolids product. This requires a design for storage and conveyance of an appropriate amount of lime for the biosolids produced, and a mixing system with adequate mixing energy to match the material (which can vary considerably from facility to facility, depending on dryness, conveyance, and lime gradation). Lime dosing and lime mixing have been addressed in this manual. The following example uses the concepts outlined in earlier sections of this manual.

5.7.4.1 Design Example—Part I Design a solids stabilization system that meets Class B pathogen requirements using quicklime and a lime mixer. The wastewater treatment facility's peak production is 20 dry tonne/d of solids. Solids are dewatered by a belt filter press that can produce a cake containing 18% solids. The facility initially will produce biosolids 24 hours a day, 7 days a week, loading directly into trucks that each hold 21 wet tonne. So it must be staffed round the clock. Suppose that eventually the facility expands its settling capacity and wants to reduce solids treatment system operations (and staffing) to 8 hours a day, 5 days per week. Assume the chosen mixer has a rated capacity of 100 kg/min, and provide about 15% to 25% extra mixing capacity (beyond the manufacturer-suggested rated capacity) in mixer and standby mixer equipment. Assume a lime dose of 15% on a dry-weight basis.

5.7.4.1.1 Design a Lime Mixing System for 24/7 Operations   Using the average daily production of dewatered cake (7 days per week), design engineers first should calculate how much lime and how many mixers are needed:

$$20 \text{ tonne/d solids} \times 15\% \text{ lime} \times 1000 \text{ lb/tonne} = 3000 \text{ kg/d lime} \qquad (23.54)$$

$$20 \text{ dry tonne/d/18\% dry cake solids} = 111 \text{ wet tonne/d (77 kg/min)} \qquad (23.55)$$

so two mixers are required to achieve the desired redundancy.

Design engineers should consider for sizing mixers with extra capacity to take into account changes in the solids' rheology characteristics during the project's design life. In this example, one mixer provides enough capacity for peak conditions, but the standby mixer is available for O&M considerations.

5.7.4.1.2 Design a Lime Mixing System for 8 hr/d, 5 d/wk Operations   After facility expansion and staff reduction, how much cake is produced during the 8-hour, 5-day-per-week operational shift? How much lime is required? How many mixers are required?

$$111 \text{ wet tonne/d} \times 7/5 \times 156 \text{ wet tonne/8-hour shift} \qquad (23.56)$$

$$156 \text{ wet tonne} \times 18\% \text{ dry cake solids} \times 15\% \text{ lime} \times 1000 \text{ lb/tonne}$$
$$= 4200 \text{ kg lime per shift (9 kg/min)} \tag{23.57}$$

$$156 \text{ wet tonne}/8 \text{ hr} \cdot 60 \text{ min} \times 1000 \text{ lb/ton} = 324 \text{ kg/min to mixer(s)} \tag{23.58}$$

$$324 \text{ kg/min} + 9 \text{ kg/min} = 333 \text{ kg/min solids and lime to mixers,}$$
$$\text{so four mixers are required at 83 kg/min per mixer,} \tag{23.59}$$
$$\text{plus one standby mixer for O\&M considerations.}$$

5.7.4.2 Design Example—Part II Several years into the life of the project, facility personnel decide to convert from belt filter presses to centrifuges, which produce a cake containing 24% solids. Before this conversion, facility operators test the mixers' adequacy using the calcium test method. Results showed that the bench-scale mixer provided a minimum standard deviation of 2.6 after 40 seconds of mixing, beyond which the standard deviation did not further improve. The full-scale mixer also achieved a standard deviation of 2.6 for solids dewatering with a belt filter press, indicating mixing was adequate.

Knowing that as solids content increases, mixing energy requirements also increase (a 3% increase in solids nearly doubles [100% more] the required mixing energy), staff tested the mixers again after the centrifuges were installed. The new results had a standard deviation of 8.4, indicating less-than-optimal mixing.

Personnel needed to determine the answers to the following questions:

- What is the flow to each of the four existing mixers?
- What is the effect of the higher solids on the material's rheology?
- Does a decrease in solids flowrate (because of the centrifuge installation) compensate for an increase in mixing energy requirement?

While the extra capacity originally designed into the mixing system was sufficient to account for changes in rheology in belt-pressed solids, it may not be enough for the drier cake produced by a centrifuge.

Calculations showed that the cake flowrate from a centrifuge is lower than that from a belt filter press:

$$156 \text{ wet tonne} \times 18\% \text{ belt press dry cake solids}/24\% \text{ centrifuge dry cake solids}/8 \text{ h}/60 \text{ min}$$
$$\times 1000 \text{ kg/tonne} = 244 \text{ kg/min solids} + 9 \text{ kg/min} \tag{23.60}$$
$$= 253 \text{ kg/min to four mixers, or 63 kg/min per mixer}$$

The flow to the mixers decreases by 25% (from 83 to 63 kg/min) with about 40% excess capacity (100 kg/min rated capacity), but mixing energy requirements also change:

$$24\% \text{ dry cake solids} - 18\% \text{ dry cake solids} = 6\% \text{ increase}$$
$$\text{(or a 200\% increase in required mixing energy)} \tag{23.61}$$

Knowing that inadequate mixing can increase odors and fecal coliforms, what can staff do to ensure that the existing mixers provide enough mixing energy to maintain the product quality of past production years?

If the mixers are left untouched, biosolids quality probably will decline (i.e., produce more fecal coliforms and odors) once the centrifuges are installed. Calcium test results confirm this assumption; they show that the standard deviation is not minimized after dewatering improvements.

Facility personnel can install weirs in the mixer to increase detention time. If this does not provide adequate mixing, they also can consider installing another mixer. Finally, personnel may need to consider detuning the new centrifuge to produce cake with lower solids content.

This example shows that changes in upstream rheology should be considered when sizing lime stabilization equipment. In many cases, significant changes in dewatering equipment may require pilot-tests and changes in mixer design to compensate.

5.8   Product End-Use Considerations

Wastewater solids contain organic matter and plant nutrients, making them a valuable crop fertilizer and soil conditioner. However, adding alkaline material dilutes some plant nutrients and volatilizes the ammonia-nitrogen content. Also, the alkaline material adsorbs a substantial portion of mineralized organic nitrogen, further reducing the amount of nitrogen available to plants (Logan, 1990). The net result may be a relatively low-grade fertilizer, but a good lime substitute and organic soil amendment.

That said, alkaline materials can be custom blended with solids and other feedstocks (e.g., sand, topsoil, yard waste, and leaves) to produce a specific, marketable product. Some municipalities (e.g., Warren, Ohio) have done this successfully, creating a more publicly acceptable product with a lower pH. Such products may be called artificial soils.

Applications for alkaline-stabilized biosolids include:

- Agriculture (e.g., organic fertilizer, agricultural-lime substitute, or soil amendment);
- Horticulture (e.g., nurseries and sod farms);
- Residential lawns and gardens (e.g., manufactured organic topsoil blends);
- Bulk fill (e.g., slope stabilization and dike construction);
- Nonagricultural land application, land reclamation, or dedicated land disposal; and
- Landfill (e.g., disposal or daily, intermediate, final, and vegetative cover).

Each has particular quality requirements and standards. For example, the application rate of alkaline-stabilized biosolids for agronomic purposes may be limited based on calcium carbonate equivalence or alkalinity content, rather than its low plant-nutrient content. If the material will be used at landfills, most regulators require extensive testing and documentation first.

An advantage of an alkaline-stabilized biosolids compared to other biosolids (e.g., compost) is that it can partially or fully satisfy the liming requirements of many soils.

Also, it may contain small amounts of plant nutrients. For example, most cement-plant kiln dust contains significant amounts of potassium and smaller amounts of trace nutrients. Some alkaline-stabilization processes use mineral by-products (e.g, cement-plant kiln dust, lime kiln dust [lime-plant kiln dust], and alkaline fly ash). Nutrient content is biosolids specific and should be carefully monitored. Biosolids also may contain regulated trace elements that should be carefully monitored to avoid exceeding regulatory limits.

For more information on biosolids uses, see Chapter 25.

6.0   References

Abu-Orf, M. M.; Griffin, P. P.; Dentel, S. K. (2001) Chemical and Physical Pretreatment of ATAD Biosolids for Dewatering. *Water Sci. Technol.*, **44** (10), 309–314.

Adams, G. M.; et al. (2004) Identifying and Controlling the Municipal Wastewater Odor Environment Phase 2: Impacts of In-Plant Operational Parameters on Biosolids Odor Quality; Report No. 00HHE5T; Water Environment Research Foundation: Alexandria, Virginia.

Agarwal, S.; Abu-Orf, M.; Novak, J. T. (2005) Sequential Polymer Dosing for Effective Dewatering of ATAD Sludges. *Water Res.*, **39**, 1301–1310.

Al-Ghusain, I.; Hamoda, M. F.; El-Ghany, M. A. (2004) Performance Characteristics of Aerobic/Anoxic Sludge Digestion at Elevated Temperatures. *Environ. Technol.*, **25**, 501–511.

American Society of Civil Engineers (1983) *A Survey of Anaerobic Digester Operations*; American Society of Civil Engineers: New York.

American Society of Heating, Refrigerating, and Air Conditioning Engineers (2013) *ASHRAE Handbook Fundamentals*, Inch-Pound Edition; American Society of Heating, Refrigeration, and Air-Conditioning Engineers: Atlanta, Georgia.

American Water Works Association (2013) *Quicklime and Hydrated Lime*; AWWA B202-13: American Water Works Association: Denver, Colorado.

Avallone, E. A.; Baumeister, T. (1996) *Marks' Standard Handbook for Mechanical Engineers*, 10th ed.; McGraw-Hill: New York.

Baier, U.; Zwiefelhofer, H. P. (1991) Effects of Aerobic Thermophilic Pretreatment. *Water Environ. Technol.* **3**, 56–61.

Barber, N. R.; Dale, C. W. (1978) Increasing Sludge Digester Efficiency. *Chem. Eng.*, **85** (16), 147–149.

Barker, J. C. (1996) Crystalline (Salt) *Formation in Wastewater Recycling Systems*; Publication EBAE 082-81; North Carolina Cooperative Extension Service: Raleigh, North Carolina.

Barnes, C.; Walker, S.; Anderson, W.; Papke, S. (2007) Implementation of a Two-Phase Anaerobic Digestion System. *Proceedings of the 21st Annual Water Environment Federation Residuals and Biosolids Conference*; Denver, Colorado, April 15–18; Water Environment Federation: Alexandria, Virginia.

Batstone, D. J.; Keller, J.; Angelidaki, I.; Kalyuzhnyi, S. V.; Pavlostathis, S. G.; Rozzi, A.; Sanders, W. T. M.; Siegrist, H.; Vavilin, V. A. (2002) The IWA Anaerobic Digestion Model No. 1 (ADM1). *Water Sci. Technol.*, **45** (10), 65–73.

Beall, S. S.; Jenkins, D.; Vidanage, S. A. (1998) A Systematic Analytical Artifact that Significantly Influences Anaerobic Digestion Efficiency Measurement. *Water Environ. Res.*, **70**, 1019.

Beals, J. L. (1976) Mechanics of Handling Lime Slurries. *Proceedings of the Int. Water Conference*, Pittsburgh, Pennsylvania, October 26–28; Engineers' Society of Western Pennsylvania.

Beecher, N.; Kuter, G.; Petroff, B. (2009) Another Reason Not to Landfill: Composting Can Help Reduce Greenhouse Gas Emissions; *Water Environ. Technol.*, **21**, 4.

Benefield, L. D.; Randall, C. W. (1980) *Biological Process Design for Wastewater Treatment*; Prentice-Hall Inc.: Englewood Cliffs, New Jersey.

Bird, E. B.; Stewart, W. E.; Lightfoot. E. N. (1960) *Transport Phenomena*; Wiley & Sons: New York.

Bitton, G.; Damron, B. L.; Edds, G. T.; Davidson, J. M. (1980) *Sludge Health Risks of Land Application*; Ann Arbor Science Publishers: Ann Arbor, Michigan.

Bowker, R. P. G.; Trueblood, R. (2002) Control of ATAD Odors at the Eagle River Water and Sanitation District. *Proceedings of the 2002 Water Environment Federation Odors and Toxic Air Emissions Conference*; Albuquerque, New Mexico, April 29–30; Water Environment Federation: Alexandria, Virginia.

Brinkman, D. G.; Voss, D. (1997) *Egg-Shaped Digesters: Are They All They're Cracked Up to Be?; Operational Survey*; Black and Veatch: Gaithersburg, Maryland.

Bryers, J. D. (1984) Structured Modeling of the Anaerobic Digestion of Biomass Particulates. *Biotechnol. Bioeng.*, **27**, 638–649.

Burch, T. R.; Sadowsky, M. J.; LaPara, T. M. (2013) Aerobic Digestion Reduces the Quantity of Antibiotic Resistance Genes in Residual Municipal Wastewater Solids. *Front. Microbiol.*, 12 February 2013.

Burd, R. S. (1968) *A Study of Sludge Handling and Disposal*; Publication No. WP-20-4; U.S. Department of the Interior, Federal Water Pollution Control Administration: Washington, D.C.

Burnham, J. C.; Hatfield, N.; Bennett, G. F.; Logan, T. J. (1992) *Use of Kiln Dust with Quicklime for Effective Municipal Sludge Treatment with Pasteurization and Stabilization with the N-Viro Soil Process*; Stand. Tech. Publication 1135; American Society for Testing and Materials: Philadelphia, Pennsylvania.

Buswell, A. M.; Neave, S. L. (1939) Laboratory Studies of Sludge Digestion. *Ill. State Water Surv. Bull.*, 30. Institute for Natural Resource Sustainability, University of Illinois, Champagne, IL.

Byers, H. W.; Jensen, B. (1990) Stabilizing Sludge with Fly Ash-Sludge. Paper Presented at Dep. Eng. Professional Development; University of Wisconsin: Madison, Wisconsin.

Camp, T. R.; Stein, P. C. (1943) Velocity Gradients and Internal Work in Fluid Motion. *J. Boston Soc. Civ. Eng.*, **30**, 219-237.

Chang, Y. (1967) The Fungi of Wheat Straw Compost: Part II—Biochemical and Physiological Studies. *Trans. Br. Mycol. Soc.*, **50**, 667.

Chapman, D. T. (1989) Mixing in Anaerobic Digesters: State of the Art in Encyclopedia of Environmental Control Technology; Vol. 3, *Wastewater Treatment Technology*; Cheremisinoff, P. N., Ed.; Gulf Publishing Co.: Houston, Texas.

Chapman, T.; Krugel, S. (2011) Rapid Volume Expansion—An Investigation into Digester Overflows and Safety. *Proceedings of the Water Environment Federation Annual Residuals and Biosolids Management Conference*, Sacramento, California.

Chapman, T. and Muller, C. (2010) Impact of Series Digestion on Process Stability and Performance. *Proceedings of the 24th Annual Water Environment Federation Residuals and Biosolids Conference*; Savannah, Georgia, May 23-26; Water Environment Federation. Alexandria, Virginia.

Chen, Y. C., Higgins, M. J.; Murthy, S. N.; Beightol, S. M. (2008) The Link between Odors and Regrowth of Fecal Coliforms after Dewatering. *Proceedings of the 22nd Annual Water Environment Federation Residuals and Biosolids Conference*; Philadelphia, Pennsylvania, March 30–April 2; Water Environment Federation: Alexandria, Virginia.

Chen, Y.; Higgins, M. J.; Murthy, S. N.; Maas, N. A.; Covert, K. J.; Toffey, W. E. (2006) Production of Odorous Indole, Skatole, p-Cresol, Toluene, Styrene, and Ethylbenzene in Biosolids. *J. Residuals Sci. Technol.*, **3** (4), 193–202.

Chen, Y.; Higgins, M. J.; Maas, N. A.; Murthy, S. N.; Toffey, W. E.; Foster, D. J. (2005) Roles of Methanogens on Volatile Organic Sulfur Compound Production in Anaerobically Digested Wastewater Biosolids. *Water Sci. Technol.*, **52**, 67–72.

Cheunbarn, T.; Pagilla, K. R. (1999). Temperature and SRT Effects on Aerobic Thermophilic Sludge Treatment. *J. Environ. Eng.*, **125** (7), 626–629.

Cheunbarn, T.; Pagilla, K. R. (2000). Aerobic Thermophilic and Anaerobic Mesophilic Treatment of Sludge. *J. Environ. Eng.*, **126** (9), 790–795.

Christensen, G. L. (1982) Dealing with the Never-Ending Sludge Output. *Water Eng. Manag.*, **129**, 25.

Christensen, G. L. (1987) Lime Stabilization of Wastewater Sludge. In *Lime for Environmental Uses*, Gutschick, K. A., Ed.; American Society for Testing and Materials: Philadelphia, Pennsylvania.

Christi, Y. (2003) Sonobioreactors: Using Ultrasound for Enhanced Microbial Productivity. *Trends Biotechnol.*, **21** (2), 89–93.

Christy, R. W. (1992) Process and Mechanical Design Considerations for Sludge/Lime Mixing. *Proceedings of the 6th Annual Water Environment Federation Residuals Management Conference: Future Directions in Municipal Sludge (Biosolids) Management: Where We Are and Where We're Going*; Portland, Oregon, July 26–30; Water Environment Federation: Alexandria, Virginia.

Clements, R. P. L. (1982) Sludge Hygienization by Means of Pasteurization Prior to Digestion. *Disinfection of Sewage Sludge: Technical, Economic and Microbiological Aspects: Proceedings of a Workshop in Zurich*; Zurich, Switzerland, May 11–13; Bruce, A. M.; Havelaar, A. H.; Hermite, P. L., Eds.; D. Riedel Pub. Co.: Dordrecht, Switzerland; 37–52.

Conklin, A.; Stensel, H. D.; Ferguson, J. (2006) Growth Kinetics and Competition Between Methanosarcina and Methanosaeta in Mesophilic Anaerobic Digestion. *Water Environ. Res.*, **78**, 486–496.

Counts, C. A.; Shuckrow, A. J. (1975) *Lime Stabilized Sludge: Its Stability and Effect on Agricultural Land*; EPA-670/2-75-012; Battelle Memorial Institute: Richland, Washington.

D'Antonio, G. (1983) Aerobic Digestion of Thickened Activated Sludge - Reaction Rate Constant Determination and Process Performance. *Water. Res.*, 17 (11), 1525–1531.

Dague, R. R. (1968) Application of Digestion Theory to Digester Control. *J. Water Pollution Control Fed.*, **40**, 2021.

Daigger, G. T.; Bailey, E. (2000) Improving Aerobic Digestion by Prethickening, Staged Operation, and Aerobic–Anoxic Operation: Four Full-Scale Demonstrations. *Water Environ. Res.*, **72**, 260–270.

Daigger, G. T.; Ju, L. K.; Stensel, D.; Bailey, E.; Porteous, J. (2001) Can 3% SS Digestion Meet New Challenges? *Proceeding of the Aerobic Digestion Workshop, Volume V; Featured Presentation Sponsored by Enviroquip, Inc.* (Austin, Texas) at the Water Environment Federation 74th Annual Exposition and Conference, Atlanta, Georgia, Oct. 14–19.

Daigger, G. T.; Novak, J.; Malina, J.; Stover, E.; Scisson, J.; Bailey, E. (1998) Panel of Experts. *Proceedings of the Aerobic Digestion Workshop, Volume II*; Sponsored by Enviroquip, Inc.; Orlando, Florida, Oct. 3.

Daigger, G. T.; Scisson, J.; Stover, E.; Malina, J.; Bailey, E.; Farrell, J. (1999) Fine Tuning the Controlled Aerobic Digestion Process. *Proceedings of the Aerobic Digestion Workshop, Volume III*; Sponsored by Enviroquip, Inc.; New Orleans, Louisiana, Oct. 10.

Daigger, G. T.; Stensel, D.; Ju, L. K.; Bailey, E.; Porteous, J. (2000) Experience and Expertise Put to the Test. *Proceedings of the Aerobic Digestion Workshop, Volume IV*; Sponsored by Enviroquip, Inc.; Anaheim, California, Oct. 15.

Daigger, G. T.; Yates, R.; Scisson, J.; Grotheer, T.; Hervol, H.; Bailey, E. (1997) The Challenge of Meeting Class B While Digesting Thicker Sludges. *Proceedings of the Aerobic Digestion Workshop, Volume I*; Sponsored by Enviroquip, Inc.; Chicago, Illinois, Oct. 18.

DeGarie, C. J.; Crapper, T.; Howe, B. M.; Burke, B. F.; McCarthy, P. J. (2000) Floating Geomembrane Covers for Odour Control and Biogas Collection and Utilization in Municipal Lagoons. *Water Sci. Technol.*, **42**, 291–298.

De la Rubia, M.A.; Riau, V.; Raponso, F.; Borja, R. (2013) Thermophilic Anaerobic Digestion of Sewage Sludge: Focus on the Influence of the Start-up. A Review. *Crit. Rev. Biotechnol.*, **33** (4), 2013.

De Vrieze, J.; Verstraete, W.; Boon, N. (2013) Repeated Pulse Feeding Induces Functional Stability in Anaerobic Digestion. *Microb. Biotechnol.*, **6** (4), 414–424.

Dichtl, N. (1997) Thermophilic and Mesophilic (Two Stages) Anaerobic Digestion: Innovative Technologies for Sludge Utilization and Disposal. *J. Chartered Inst. Water Environ. Manag.*, **11**, 98–104.

Drury, D. D.; Lee, S.; Baker, C. (2002) Comparing Pathogen Reduction in Three Different Anaerobic Thermophilic Processes. *Proceedings of the 75th Annual Water Environment Federation Technical Exposition and Conference* [CD-ROM]; Chicago, Illinois, Sep. 28–Oct. 2; Water Environment Federation: Alexandria, Virginia.

Engineering–Science, Inc.; Black and Veatch (1991) Technology Evaluation Report: Alkaline Stabilization of Sewage Sludge; Report prepared for U.S. EPA; Contract No. 68-C8-0022; Work Assignment No. 01-08; U.S. Environmental Protection Agency: Washington, D.C.

Eschborn, R.; Higgins, M. J.; Johnston, T.; Toffey W.; Chen, Y. C. (2006) Philadelphia's Experience Using Static, Non-Aerated Curing to Produce Low Odor Biosolids. *Proceedings of the 20th Annual Water Environment Federation Residuals and Biosolids Management Conference*, Cincinnati, Ohio, March 12–15; Water Environment Federation: Alexandria, Virginia.

Eschborn, R.; Thompson, D. (2007) The Tagro Story—How the City of Tacoma, Washington, Went Beyond Public Acceptance to Achieve the Biosolids Program Words We'd Like to Hear: Sold Out. *Proceedings of the 21st Annual Water Environment Federation/American Water Works Association Joint Residuals and Biosolids Management Conference*; Denver, Colorado, April 15–18; Water Environment Federation: Alexandria, Virginia.

Farrah, S. R.; Bitton, G.; Zan, S. G. (1986) *Inactivation of Enteric Pathogens during Aerobic Digestion of Wastewater Sludge*; EPA-600/2-86-047; U.S. Environmental Protection Agency, Water Engineering Research Laboratory: Cincinnati, Ohio.

Farrell, J. B.; Smith, J.; Hathaway, S.; Dean, R. (1974) Lime Stabilization of Primary Sludge. *J. Water Pollution Control Fed.*, **46**, 113.

Fergen, R. E. (1991) Stabilization and Disinfection of Dewatered Municipal Wastewater Sludge with Alkaline Addition. *Proceedings of the 5th Annual American Water Works Association/Water Pollution Control Federation Joint Residuals Management Conference*; Durham, North Carolina, Aug. 11–14; Water Pollution Control Federation: Alexandria, Virginia.

Finger, R. E.; Butler, R. C. (1996) The Effect of Sampling Procedures on Digester pH Measurement. *69th Annual WEFTEC*. Dallas, Texas.

Franke-Whittle, I.; Waltera, A.; Ebnerb, C.; Insama H. (2014) Investigation into the Effect of High Concentrations of Volatile Fatty Acids in Anaerobic Digestion on Methanogenic Communities. *Waste Manag.*, 34 (11), Nov., 2080–2089.

Garber, W. F. (1982) Operating Experience with Thermophilic Anaerobic Digestion. *J. Water Pollution Control Fed.*, **54**, 1170.

Gebreeyessus, G. D.; Jenicek, P. (2016) Thermophilic versus Mesophilic Anaerobic Digestion of Sewage Sludge: A Comparative Review. *Bioengineering*, June 18, 2016 (www.mdpi.com/2306-5354/3/2/15/pdf).

Gemmell, R.; Deshevy, R.; Elliott, M.; Crawford, G.; Murthy, S. (1999) Full Scale Demonstration of Dual Digestion: Thermodynamic and Kinetic Analysis. *Proceedings of the 72nd Annual Water Environment Federation Technical Exposition and Conference* [CD-ROM]; New Orleans, Louisiana, Oct. 10–13; Water Environment Federation: Alexandria, Virginia.

Gemmell, R.; Deshevy, R.; Elliott, M.; Crawford, G.; Murthy, S. (2000) Design Considerations and Operating Experience for a Full Scale Dual Digestion System with Separate Sludge Thickening. *Proceedings of the 73rd Annual Water Environment Federation Technical Exposition and Conference* [CD-ROM]; Anaheim, California, Oct. 14–18; Water Environment Federation: Alexandria, Virginia.

Ghosh, S.; Conrad, J. R.; Klass, D. L. (1975) Anaerobic Acidogenesis of Wastewater Sludge. *J. Water Pollution Control Fed.*, **47**, 30.

Ghosh, S.; Henry, M. P.; Sajjad, A.(1987) *Stabilization of Sewage Sludge by Two-Phase Anaerobic Digestion*; EPA-7600/2-87-040; U.S. Environmental Protection Agency, Water Engineering Research Laboratory: Cincinnati, Ohio.

Ghosh, S.; et al. (1991) Pilot- and Full-Scale Studies on Two-Phase Anaerobic Digestion for Improved Sludge Stabilization and Foam Control. *Proceedings of the 64th Annual Water Pollution Control Federation Technical Exposition and Conference*; Toronto, Ontario, Canada, Oct. 7–10; Water Environment Federation: Alexandria, Virginia.

Ghosh, S.; Buoy, K.; Dressel, L.; Miller, T.; Wilcox, G.; Loos, D. (1995) Pilot- and Full-Scale Studies on Two-Phase Anaerobic Digestion of Municipal Sludge. *Water Environ. Res.*, **67**, 206.

Golueke, C. G. (1977) *Biological Reclamation of Solid Waste*; Rodale Press: Emmaus, Pennsylvania.

Grady, C. P. L., Jr.; Daigger, G. T.; Love, N. G.; Filipe, C. D. M. (2011) *Biological Wastewater Treatment*, 3rd ed.; CRC Press: Boca Raton.

Gray, D. M.; Suto, P. J.; Chien, M. H. (2008) Producing Green Energy from Post-Consumer Solid Food Wastes at a Wastewater Treatment Plant Using an Innovative New Process. *Proceedings of the Water Environment Federation Sustainability Conference*; Washington, D.C., June 22–25. Water Environment Federation: Alexandria, Virginia.

Great Lakes Upper Mississippi River Board of State Sanitary Engineering Health Education Services Inc. (2014) *Recommended Standards for Wastewater Facilities*; Great Lakes Upper Mississippi River Board of State Sanitary Engineering Health Education Services Inc.: Albany, New York.

Green, D.W.; Perry, R. H. (2007) *Perry's Chemical Engineers' Handbook*, 8th ed.; McGraw-Hill: New York.

Griffin, M. E.; McMahon, K. D.; Mackie, R. I.; Raskin, L. (2000) Methanogenic Population Dynamics during Start-Up of Anaerobic Digesters Treating Municipal Solid Waste and Biosolids. *Biotechnol. Bioeng.*, **57**, 342–355.

Gujer, W.; Zehnder, A. J. B. (1983) Conversion Process in Anaerobic Digestion. *Water Sci. Technol.*, 15, 127–167.

Haas O. (1984) *Demonstration of Thermophilic Aerobic-Anaerobic Digestion at Hagerstown, MD*; Grant S-805823-01-0; Final Report; EPA-600/S2-84-142; U.S. Environmental Protection Agency, Municipal Environmental Research Laboratory: Cincinnati, Ohio.

Han, Y.; Dague, R. (1996) Heat Control: Temperature-Phased Anaerobic Digestion Reduces Foaming and Produces Class A Biosolids Without the Odors. *Oper. Forum*, 13, 19–23.

Hartman, R. B.; Smith, D. G.; Bennett, E. R.; Linstedt, K. D. (1979) Sludge Stabilization through Aerobic Digestion. *J. Water Pollution Control Fed.*, 51, 2353.

Haug, R. T.; LeBrun, T. I.; Tortoriei, L. D. (1983) Thermal Pretreatment of Sludges—A Field Demonstration. *J. Water Pollution Control Fed.*, 55, 23–34.

Haug, R. T. (1980) *Compost Engineering*; Ann Arbor Science Publishers: Ann Arbor, Michigan.

Hepner, S.; Striebig, B.; Regan, R.; Giani, R. (2002) Odor Generation and Control from the Autothermal Thermophilic Aerobic Digestion (ATAD) Process. *Proceedings of the 2002 Water Environment Federation Odors and Toxic Air Emissions Conference*; Albuquerque, New Mexico, April 29–30; Water Environment Federation: Alexandria, Virginia.

Higgins, M.; Murthy, S.; Toffey, W.; Striebig, B.; Hepner, S.; Yarosz, D.; Yamani, S. (2002) Factors Affecting Odor Production in Philadelphia Water Department Biosolids. *Proceedings of the 2002 Water Environment Federation Odors and Toxic Air Emissions Conference*; Albuquerque, New Mexico, April 29–30; Water Environment Federation: Alexandria, Virginia.

Higgins, M.; Yarosz, D.; Chen, Y.; Murthy, S. (2003) Mechanisms for Volatile Sulfur Compound and Odor Production in Digested Biosolids. *Proceedings of the 17th Annual Water Environment Federation/American Water Works Association Joint Biosolids and Residuals Conference*; Baltimore, Maryland, Feb. 19–22; Water Environment Federation: Alexandria, Virginia.

Higgins, M. J.; Yarosz, D. P.; Chen, D. P.; Murthy, S. N.; Maas, N.; Cooney, J.; Glindemann, D.; Novak, J. T. (2006a) Cycling of Volatile Organic Sulfur Compounds in Anaerobically Digested Biosolids and Its Implications for Odors. *Water Environ. Res.*, 78, 243–252.

Higgins, M. J.; Chen, Y. C.; Murthy, S. N.; Hendrickson, D. (2006b) *Examination of Reactivation of Fecal Coliforms in Anaerobically Digested Biosolids*; Report No. 03-CTS-13T; Water Environment Research Foundation: Alexandria, Virginia.

Higgins, M. J.; Chen, Y. C.; Murthy, S. N.; Hendrickson, D. (2008b) *Evaluation of Bacterial Pathogen and Indicator Densities After Dewatering of Anaerobically Digested Biosolids: Phase II and III*; Report No. 04-CTS-3T; Water Environment Research Foundation: Alexandria, Virginia.

Higgins, M. J.; Murthy, S. N.; Aynur, S; Beightol, S. (2012) WERF ROSI Project – Do We Need a Revised Time-Temperature Requirement to Achieve Class A Biosolids. *Proceedings of the Water Environment Federation, Residuals and Biosolids 2012*, pp. 726–733(8).

Iacoboni, M. D.; Leburn, T. J.; Lingston, J. (1980) Deep Windrow Composting of Dewatered Sewage Sludge. *Proceedings of the National Conference of the Municipal and Industrial Sludge Composting Hazardous Materials Control Research Institute*; Silver Spring, Maryland; pp. 88–108.

Jacobs, A.; et al. (1992) Odor Emissions and Control at the World's Largest Chemical Fixation Facility. *Proceedings of the 6th Annual Water Environment Federation Residuals Management Conference: Future Directions in Municipal Sludge (Biosolids) Management: Where We are and Where We're Going*, Portland, Oregon; Water Environment Federation: Alexandria, Virginia.

Jacobs, A.; Silver, M. (1990) Sludge Management at the Middlesex County Utilities Authority. *Water Sci. Technol.*, **22**, 93.

Jenkins, C. J.; Mavinic, D. S. (1989) Anoxic-Aerobic Digestion of Waste Activated ludge: Part II-Supernatant Characteristics, ORP Monitoring Results and Overall Rating System. *Environ. Technol. Letters*, 10 (4), 371–384,

Jewell, W. J.; Kabrick, R. M. (1980). Autoheated Aerobic Thermophilic Digestion with Air Aeration. *J. Water Pollution Control Fed.*, **52**, 512.

Jimenez, E. I.; Garcia, V. P. (1989) Evaluation of City Refuse Compost Maturity: A Review. *Biol. Wastes*, **27**, 115.

Jones, R.; Parker, W.; Khan, Z.; Murthy, S.; Rupke, M. (2008a) Characterization of Sludges for Predicting Anaerobic Digester Performance. *Water Sci. Technol.*, **57**, 721–726.

Jones, R.; Parker, W.; Zhu, H.; Houweling, D.; Murthy, S. (2008b) Predicting the Degradability of Waste Activated Sludge. *Proceedings of the 81st Annual Water Environment Federation Technical Exhibition and Conference* [CD-ROM]; Chicago, Illinois, Oct. 18–22; Water Environment Federation: Alexandria, Virginia.

Kabouris, J.; et al. (2013) Positive Business Case for Codigestion and Resource Recovery Facilities in Australia without Pre-Existing Anaerobic Digesters. *Proceedings of the 86th Annual Water Environment Federation Technical Conference and Exhibition*, Chicago, Illinois, Oct. 5–9.

Kabouris, J.; et al. (2015) Anaerobic Digestion with Recuperative Thickening Minimizes Biosolids Quantities and Odors in Sydney, Australia. *Proceedings of the Water Environment Federation Annual Residuals and Biosolids Conference*; Water Environment Federation: Alexandria, Virginia.

Kampelmacher, E. H.; van Noorle Jansen, L. M. (1972) Reduction of Bacteria in Sludge Treatment. *J. Water Pollution Control Fed.*, **44**, 309.

Kayhanian, M.; Tchobanoglous, G. (1992) Computation and Importance of Carbon to Nitrogen (C/N) Ratios for Various Organic Fractions of Municipal Solid Waste. *BioCycle*, **33**, 58–60.

Keller, U. (1980) Klarschlammpasteurisierung in der Abwasserreinigungsanlage Altenrhein. Wasser, Energie, Luft. 72 Jahrgang. Heft 1/2 (side-by-side article in French).

Kelly, H. G. (2006) Emerging processes in biosolids treatment, 2005, *Jour. Environ. Eng. Sci.*, 5. pp. 176-186.

Kelly, H. G.; Mavinic, D. S.; Trueblood, B.; Zhou, J.; Hystad, B.; Frese, H.; Cheshuk, J. (2003) Autothermal Thermophilic Aerobic Digestion Research Application and Operational Experience. *Proceedings of the 76th Annual Water Environment Federation Technical Exhibition and Conference*, Workshop W104—Thermophilic Digestion: Hot Update!, Los Angeles, California, Oct. 11–15; Water Environment Federation: Alexandria, Virginia.

Kelly, H. G.; Melcer, H.; Mavinic, D. S. (1993) Autothermal Thermophilic Aerobic Digestion of Municipal Sludge: A One-Year Full-Scale Demonstration Project. *Water Environ. Res.*, **65**, 849.

Kelly, H. G. (1991) Autothermal Thermophilic Aerobic Digestion: A Two Year Appraisal of Canadian Facilities. *Proceedings of the American Society of Civil Engineers Environmental Engineering Specialty Conference*, Reno, Nevada, July 10–12; American Society of Civil Engineers: New York, New York.

Kemp, J. (2014) Renewable CNG with Combined Heat and Power Provides Flexible End Use for Biogas. *Proceedings of the Water Environment Federation Residuals and Biosolids Management Specialty Conference*, Austin, Texas, May 18–21; Water Environment Federation: Alexandria, Virginia.

Kester, G.; Schafer, P.; Gillette, B. (2008) Using Treatment Plant Digesters to Process Fats, Oils and Grease. *BioCycle*, **49**, 47.

Kim, H.; Murthy, S.; Peot, C.; Ramirez, M.; Strawn, M.; Park, C.; McConnell, L. (2003) Examination of Mechanisms for Odor Compound Generation during Lime Stabilization. *Water Environ. Res.*, **75**, 121.

Knight, G.; Lackey, K.; Polo, C.; Carr, S.; Kemp, J; Brower, A.; Lynch, T. J. (2016) Exploring The Best and Highest Use of Biogas From Wastewater Utilities. *Proceedings of the Water Environment Federation Residuals and Biosolids Management Specialty Conference*, Milwaukee, Wisconsin, April 3-6; Water Environment Federation: Alexandria, Virginia.

Knoll, K. H. (1964) *Information Bulletin No. 13-20*; Int. Research Group Refuse Disposal, U.S. Public Health Service, Rockville, Maryland.

Koers, D. A., Mavinic, D. S., (1977). Aerobic digestion of waste activated sludge at low temperature. *J. Water Pollution Control Federation*, **49** (3): 460–468.

Kopp, J.; Ewert, W. (2006) New Processes for the Improvement of Sludge Digestion and Sludge Dewatering. *Proceedings of the 11th European Biosolids and Organic Resources Conference*, Wakefield, United Kingdom, Nov. 13–15; Chartered Institution of Water and Environmental Management: London, U.K.

Krampe, J.; Krauth, K. (2003) Oxygen Transfer into Activated Sludge with High MLSS Concentration. *Water Sci. Technol.*, **47** (11), 297–303.

Krishnamoorthy, R.; Loehr, R.C. (1989) Aerobic Sludge Stabilization – Factors Affecting Kinetics. *J. Env. Eng.*, **115**, 283–301.

Krugel, S.; Nemeth, L.; Peddie, C. (1998) Extending Thermophilic Anaerobic Digestion for Producing Class-A Biosolids at the Greater Vancouver Regional Districts Annacis Island Wastewater Treatment Plant. *Water Sci. Technol.*, **38** (8), 409–416.

Krugel, S.; Parella, A.; Ellquist, K.; Hamel, K. (2006) Five Years of Successful Operation–A Report on North America's First New Temperature Phased Anaerobic Digestion System at the Western Lake Superior Sanitary District (WLSSD). *Proceedings of the 79th Annual Water Environment Federation Technical Exposition and Conference* [CD-ROM], Dallas, Texas, Oct. 21–25; Water Environment Federation: Alexandria, Virginia.

Kumar, N.; Novak, J. T.; Murthy, S. N. (2006a) Sequential Anaerobic-Aerobic Digestion for Enhanced VSR and Nitrogen Removal. *Proceedings of the 20th Annual Water Environment Federation Residuals and Biosolids Management Conference,* Cincinnati, Ohio, March 12–14; Water Environment Federation: Alexandria, Virginia.

Kumar, N.; Novak, J. T.; Murthy, S. N. (2006b) Effect of Secondary Aerobic Digestion on Properties of Anaerobic Digested Biosolids. *Proceedings of the 79th Annual Water Environment Federation Technical Exhibition and Conference* [CD-ROM], Dallas Texas, Oct. 21–25; Water Environment Federation: Alexandria, Virginia.

Lawrence, A. W. (1971) Application of Process Kinetics to Design of Anaerobic Processes. In *Anaerobic Biological Treatment Processes,* Pohland, F. G., Ed.; Advances in Chemistry Series; American Chemical Society: Washington, D.C., 105.

Lawrence, A.W.; McCarty, P. L. (1969) Kinetics of Methane Fermentation in Anaerobic Treatment. *J. Water Pollution Control Fed.,* **41,** 1–17.

Layden, N. M.; Kelly, H. G.; Mavinic, D. S.; Moles, R.; Bartlett, J. (2007) Autothermal Thermophilic Aerobic Digestion (ATAD) Part II: Review of Research and Full-Scale Operating Experiences. *J. Environ. Eng. Sci.,* **6** (6), 679–690.

LeBrun, T. (1979) Memorandum to the LA/OMA Project on Status of Aspergillus Monitoring. Los Angeles County Sanitation Districts, California.

Lee, K. M.; Brunner, C. A.; Farrell, J. B.; Ealp, A. E. (1989) Destruction of Enteric Bacteria and Viruses during Two-Phase Digestion. *J. Water Pollution Control Fed.,* 61, 1421–1429.

Li Y. Y.; Noike T. (1992) Upgrading of Anaerobic Digestion of Waste Activated Sludge by Thermal Pre-Treatment. *Water Sci. Technol.,* **26** (3–4), 857–866.

Liu, D. H. F.; Liptak, B. G. (Eds.) (1997) *Environmental Engineers' Handbook,* 2nd ed.; CRC Press: Boca Raton.

Lue-Hing, C.; Zenz, D. R.; Kuchenrither, R., Eds. (1992) *Municipal Sewage Sludge Management: Processing, Utilization and Disposal*; Technomic Publishing Co. Inc.: Lancaster, Pennsylvania.

Mason, T. J.; Lorimer, J. P. (1988) *Sonochemistry: Theory, Applications and Uses of Ultrasound in Chemistry*; Ellis Horwood: Chichester, U.K.

Matsch, L. C.; Drnevich, R. F. (1977) Autothermal Aerobic Digestion. *J. Water Pollution Control Fed.,* 49, 296.

Maxwell, M. J.; et al. (1992) Impact of New Sludge Regulations on Aerobic Digester Sizing and Cost-Effectiveness. *Proceedings of the 65th Annual Water Environment Federation Exposition and Conference*; New Orleans, Louisiana, Sep. 20–24; Water Environment Federation: Alexandria, Virginia.

McCarthy, W. C.; Nelson, C. J.; Vandenburgh S.J.; Butler, R.C. (2004) Improved Digester Performance without the Acroynms: Full-scale Evaluations of Series Mesophilic Digestion at King County's South Treatment Plant. *Proceedings of the Water Environment Federation, Residuals and Biosolids Management 2004,* pp. 42–69 (28).

McCarty, P. L.; Smith, D. P. (1986) Anaerobic Wastewater Treatment: Fourth of a Six-Part Series on Wastewater Treatment Processes. *Environ. Sci. Technol.,* **20** (12), 1200–1206.

McHugh, S.; Carton, M.; Mahony, T.; O'Flaherty, V. (2006) Methanogenic Population Structure in a Variety of Anaerobic Bioreactors. *FEMS Microbiol. Lett.*, **219**, 2297–2304.

Messenger, J. R.; de Villiers, H. A.; Ekama, G. A. (1993) Evaluation of the Dual Digestion System, Part 2: Operation and Performance of the Pure Oxygen Aerobic Reactor. *Water SA*, 19 (3), 193–200.

Metcalf and Eddy, Inc./AECOM (2013) *Wastewater Engineering: Treatment and Resource Recovery*, 5th ed.; McGraw-Hill: New York.

Millner, P. D.; Marsh, P. B.; Snowden, R. B.; Parr, J. F. (1977) Occurrence of *Aspergillus fumigatus* during Composting of Sewage Sludge. *Appl. Environ. Microbiol.*, **34**, 6.

Morgan, M. T.; MacDonald, F. W. (1969) Tests Show MB Tuberculosis Doesn't Survive Composting. *J. Environ. Health*, **32**, 101.

Muller, C. D.; Novak, J. T. (2007) The Influence of Anaerobic Digestion on Centrifugally Dewatered Biosolids. *Proceedings of the 21st Annual Water Environment Federation Residuals and Biosolids Conference*; Denver, Colorado, April 15–18. Water Environment Federation: Alexandria, Virginia.

Murray, C. M.; Thompson, J. L. (1986) Strategies for Aerated Pile Systems. *BioCycle*, **27** (6), 22–26.

Murray, K. C.; Tong, A.; Bruce, A. M. (1990) Thermophilic Aerobic Digestion: A Reliable and Effective Process for Sludge Treatment at Small Works. *Water Sci. Technol.*, **22**, 225.

Murthy, S. N.; Novak, J. T.; Holbrook, R. D. (2000a) Optimizing Dewatering of Biosolids from Autothermal Thermophilic Aerobic Digesters (ATAD) Using Inorganic Conditioners. *Water Environ. Res.*, **72**, 714–721.

Murthy, S. N.; Novak, J. T.; Holbrook, R. D.; Surovik, F. (2000b) Mesophilic Aeration of Autothermal Thermophilic Aerobically Digested Biosolids to Improve Plant Operations. *Water Environ. Res.*, **72**, 476–483.

Murthy, S. N.; et al. (2002) Impact of High Shear Solids Processing on Odor Production from Anaerobically Digested Biosolids. *Proceedings of the 75th Annual Water Environment Federation Technical Exposition and Conference* [CD-ROM]; Chicago, Illinois, Sep. 28–Oct. 2; Water Environment Federation: Alexandria, Virginia.

Murthy, S. N.; Higgins, M. J.; Chen, Y. C.; Toffey, W.; Golembeski J. (2003) Influence of Solids Characteristics and Dewatering Process on Volatile Sulfur Compound Production from Anaerobically Digested Biosolids. *Proceedings of the 17th Annual Water Environment Federation/American Water Works Association Joint Residuals and Biosolids Conference*; Baltimore, Maryland, February 19–22; Water Environment Federation: Alexandria, Virginia.

Mussari, F.; Smith, J., Midlane, D.; Finch, R.; Norris, M. (2013) Accelerated Composting Methods and Equipment. *Proceedings of the Water Environment Federation Annual Residuals and Biosolids Conference*; Water Environment Federation: Alexandria, Virginia.

National Fire Protection Association (2016) *Standard for Fire Protection in Wastewater Treatment and Collection Facilities*; NFPA 820; National Fire Protection Association: Quincy, Massachusetts.

National Fire Protection Association (2017) *National Electric Code*, NFPA 70; National Fire Protection Association: Quincy, Massachusetts.

National Lime Association (1988) *Lime: Handling, Application, and Storage in Treatment Processes*; Bulletin 213; National Lime Association: Arlington, Virginia.

Nielsen, H. B.; Angelidaki, I. (2008) Strategies for optimizing recovery of the biogas process following ammonia inhibition. *Bioresour. Technol.*, **99**, 7995–8001.

North, J. M.; Becker, J. G.; Seagren, E. A.; Ramirez, M.; Peot, C. (2008a) Methods for Quantifying Lime Incorporation into Dewatered Sludge. I: Bench-Scale Evaluation. *J. Environ. Eng.*, **134** (9), 750–761.

North, J. M.; Becker, J. G.; Seagren, E. A.; Ramirez, M.; Peot, C.; Murthy, S. N. (2008b), Methods for Quantifying Lime Incorporation into Dewatered Sludge. II: Field-Scale Application. *J. Environ. Eng.*, **134** (9), 762–770.

Novak, J. T.; Becker, H.; Zurow, A. (1977) Factors Influencing Activated Sludge Properties. *J. Environ. Eng.*, **103**, 815.

Novak, J. T.; Banjade, S.; Murthy, S. N. (2010) Combined anaerobic and aerobic digestion for increased solids reduction and nitrogen removal. *Water Res.*, **45** (2011), 618–624.

O'Rourke, J. T. (1968) Kinetics of Anaerobic Treatment at Reduced Temperatures. Ph.D. Thesis, Stanford University, Palo Alto, California.

Oerke, D. W. (1999) Alkaline Stabilization of Biosolids Can Save Money, Space. *Water World*, March, 14–16.

Oerke, D. W; Rogowski, S. M. (1990) Economic Comparison of Chemical and Biological Sludge Stabilization Processes. *Proceedings of the 4th Annual Water Pollution Control Federation Specialty Conference on the Status of Municipal Sludge Management*; New Orleans, Louisiana; Water Pollution Control Federation: Washington, D.C.

Oerke, D. W.; Stone, L. A. (1991) Detailed Case Study Evaluation of Alkaline Stabilization Processes. *Proceedings of the 5th Annual American Water Works Association/Water Pollution Control Federation Joint Residuals Management Conference*; Durham, North Carolina, Aug. 11–14; Water Pollution Control Federation: Washington, D.C.

Otoski, R. M. (1981) *Lime Stabilization and Ultimate Disposal of Municipal Wastewater Sludge*; EPA-600/S2-81-076; U.S. Environmental Protection Agency, Municipal Environmental Research Laboratory, Center for Environmental Research: Cincinnati, Ohio.

Pagilla, K. R.; Craney, K. C.; Kido, W. H. (1996) Aerobic Thermophilic Pretreatment of Mixed Sludge for Pathogen Reduction and Nocardia Control. *Water Environ. Res.*, **68**, 1093–1098.

Park, C.; Abu-Orf, M. M.; Novak, J. T. (2006) The Digestibility of Waste Activated Sludges. *Water Environ. Res.*, **78**, 59.

Parkin, G. F.; Owen, W. F. (1986) Fundamentals of Anaerobic Digestion of Wastewater Sludge. *J. Environ. Eng.*, **112**, 5.

Parravicini, V.; Svardal, K.; Hornek, R.; Kroiss, H. (2008) Aeration of Anaerobically Digested Sewage Sludge for COD and Nitrogen Removal: Optimization at Large-Scale *Water Sci. Technol.*, **57**, 257.

Paulsrud, B.; Eikum, A. S. (1975) Lime Stabilization of Sewage Sludge. *Water Res.* (G.B.), **9**, 297.

Pavlostathis, S. G.; Gossett, J. M. (1988) Preliminary Conversion Mechanisms in Anaerobic Digestion of Biological Sludges. *J. Environ. Eng.*, **114**, 575–592.

Pavlostathis, S. G.; Gossett, J. M. (2004) A kinetic Model for Anaerobic Digestion of Biological Sludge. *Biotech. Bioeng.*, **28**, 1519–1530.

Peddie, C. C., Mavinic, D. S., Jenkins, C. J. (1990) Use of ORP for Monitoring and Control of Aerobic Sludge Digestion. *J. Environ. Eng.*, **116** (3), 461–471.

Pitt, A. J.; Ekama, G. A. (1996) Dual Digestion of Sewage Solids Using Air and Pure Oxygen. *Proceedings of the 69th Annual Water Environment Federation Technical Exposition and Conference* [CD-ROM]; Dallas, Texas, Oct. 5–9; Water Environment Federation: Alexandria, Virginia.

Poincelot, R. P. (1975) *The Biochemistry and Methodology of Composting*, Bulletin 754. Connecticut Agricultural Experiment Station: New Haven, Connecticut.

Popel, F. (1971a) Die Theoretischen und Praktischen Grunlagen der Flussig-Kompostierung Hockkonzentrierten Substrate. Ausgearbeitet fur die Bad-ische Anilin-und Sodafabrik Ludwigshafen, Stuttgart, November.

Popel, F. (1971b) Energieerzeugung Beim Biologischen Abbau Organischer Stoffe. *Gewaesser Abwaesser* (Ger.), **112**.

Portenländer, G.; Heusinger, H. (1997) The Influence of Frequency on the Mechanical and Radical Effects for the Ultrasonic Degradation of Dextranes. *Ultrasound Sonochem.*, **4**, 127–130.

Qui, Y. (2016) So Now You Need to Cool the Sludge? *Proceedings of the Water Environment Federation Biosolids Specialty Conference*, Milwaukee, Wisconsin. April 3-6; Water Environment Federation: Alexandria, Virginia.

Ramirez, A.; Malina, J. (1980) Chemicals Disinfect Sludge. *Water Sew. Works*, **127** (4), 52.

Reimers, R. S.; Little, M. D.; Englande, A. J.; Leftwich, D. B.; Bowman, D. D.; Wilkinson, R. F. (1981) *Parasites in Southern Sludge and Disinfection by Standard Sludge Treatment*; EPA-600/2-81-166; U.S. Environmental Protection Agency: Washington, D.C.

Reynolds, D. T.; Cannon, M.; Pelton, T. (2001) Preliminary Investigation of Recuperative Thickening for Anaerobic Digestion. *Proceedings of the 74th Annual Water Environment Federation Technical Exhibition and Conference*; Atlanta, Georgia, Oct. 13–17; Water Environment Federation: Alexandria, Virginia.

Rubin, A. R. (1991) Agricultural Limitations and Criteria for Lime Stabilized Sludge (PSRP or PFRP). *Proceedings of the 5th Annual American Water Works Association/Water Pollution Control Federation Joint Residuals Management Conference*; Durham, North Carolina, Aug. 11–14; Water Pollution Control Federation: Washington, D.C.

Salsali, H. R.; Parker, W. J. (2007) An Evaluation of 3 Stage Anaerobic Digestion of Municipal Wastewater Treatment Plant Sludges. *Water Practice*, **1**, 1–12.

Sanders, W. T. M.; Geerink, M.; Zeeman, G.; Lettinga, G. (2000) Anaerobic hydrolysis kinetics of particulate substrates. *Water Sci. Technol.*, **41**, 17–24.

Schafer, P., Wolfenden, A. (1982) *Odor Control Features Make Lagoons an Acceptable Sludge Process Sixth Mid-America Conference on Environmental Engineering Design*; American Society of Civil Engineers, Kansas City, Missouri, June.

Schafer, P. L.; Farrell, J. (2000) Performance Comparisons for Staged and High-Temperature Anaerobic Digestion Systems. *Proceedings of the Annual Water Environment Federation Technical Exposition and Conference* [CD-ROM]; Anaheim, California, Oct. 17; Water Environment Federation: Alexandria, Virginia.

Schafer, P. L.; Farrell, J.; Newman, G.; Vandenburgh, S. (2002) Advanced Anaerobic Digestion Performance Comparisons. *Proceedings of the 75th Annual Water Environment Federation Technical Exposition and Conference* [CD-ROM]; Chicago, Illinois, Sep. 28–Oct. 2; Water Environment Federation: Alexandria, Virginia.

Schafer, P. L.; Trueblood, D.; Fonda, K.; Lekven, C. (2008) Grease Processing for Renewable Energy, Profit, Sustainability, and Environmental Enhancement. *Proceedings of the Water Environment Federation Biosolids Specialty Conference*, Philadelphia, Pennsylvania. April. Water Environment Federation: Alexandria, Virginia.

Schwinning, H. G.; Cantwell, A. (1999) Thermophilic Aerobic Digestion and Hygienisation. Paper Presented at Sludge Workshop, Wakefield.

Scisson, J. P. (2006) ATAD, the Far Country: Improvements to the 3rd, 2nd Generation ATAD Yield Better than Expected Returns. *Proceedings of the Water Environment Federation Residuals and Biosolids Management Specialty Conference*, Cincinnati, Ohio, Mar 12–15; Water Environment Federation: Alexandria, Virginia.

Scisson, J. P. (2009) "As Good as the Hype: An Overview of the Second Generation ATAD Performance." *Proceedings of the Residuals and Biosolids Management Conference 2009*, Portland, Oregon, May 3-6, 2009. Water Environment Federation, Alexandria, Virginia.

Shell, G. L.; Boyd, J. L. (1969) *Composting Dewatered Sewage Sludge*; SW-12c; U.S. Department of Health, Education, and Welfare: Washington, D.C.

Shimada T.; et al. (2011) Syntrophic Acetate Oxidation in Two-Phase (Acid–Methane) Anaerobic Digesters. *Water Sci. Technol.*, 64.9 (2011), 1812-1820.

Sloan, D. (1992) Design and Process Considerations for Advanced Alkaline Stabilization with Subsequent Accelerated Drying Facilities. Paper Presented at the 5th Annual International Conference on Alkaline Pasteurization Stabilization, Somerset, New Jersey.

Smith, J. E.; Bizier, P.; Sobrodos-Bernardos, L. (2012) Global Development of the ATAD Process and its Significant Achievements in Energy Recovery and Utilization. *Proceedings of the Residuals and Biosolids Conference 2012*, Raleigh, North Carolina, April 25-28, 2012. Water Environment Federation, Alexandria, Virginia.

Speece, R. E. (1972) Anaerobic Treatment. In *Process Design in Water Quality Engineering*, E. L. Thackston and W. W. Eckenfelder (Eds.); Jenkins Publishing Company: New York.

Spinosa, L.; Vesilind, P. A.; Eds. (2001) *Sludge into Biosolids. Processing, Disposal, and Utilization*; IWA Publishing: London, U.K.

Stege, K.; Bailey, E. (2003) *Aerobic digestion Operation of Stockbridge Wastewater Treatment Plant, GA*; Georgia Water Pollution Control Association, Marrietta, Georgia.

Stensel, H. D.; Coleman, T. E. (2000) Assessment of Innovative Technologies for Wastewater Treatment: Autothermal Aerobic Digestion (ATAD), Preliminary Report, Project 96-CTS-1; U.S. Environmental Protection Agency: Washington, D.C.

Stenstrom, M. K.; Rosso, D. (2008) Aeration and Mixing, Chapter 9. In *Biological Wastewater Treatment*, Henze, M., van Loosdrecht, M. C. M., Ekama, G. A., Eds.; Biological Wastewater Treatment: Principles, Modeling and Design. International Water Association: London, U.K.

Stone, L. A.; et al. (1992) The Historical Development of Alkaline Stabilization. Paper Presented at the Water Environment Federation Specialty Conference on Future Directions in Municipal Sludge (Biosolids) Management: Where We Are and Where We're Going, Portland, Oregon. Water Environment Federation: Alexandria, Virginia.

Stuckey, D. C.; McCarty, P. L. (1984) The Effect of Thermal Pretreatment on the Anaerobic Biodegradability and Toxicity of Waste Activated Sludge, *Water Res.*, **18** (11), 1343–1353.

Stutzenberger, F. J. (1971) Cellulase Production by Thermomonospora curvata Isolated from Municipal Solid Waste Compost. *Appl. Microbiol.*, **22**, 2, 147.

Tortorici, L.; Stahl, J. F. (1977) Waste Activated Sludge Research. *Proceedings of the Sludge Management, Disposal, and Utilization Conference*, Information Transfer, Inc., Rockville, Maryland.

U.S. Environmental Protection Agency (1975) *Lime Stabilized Sludge: Its Stability and Effect on Agricultural Land*, EPA-670/2-75-012; National Environmental Research Center; U.S. Environmental Protection Agency: Washington, D.C.

U.S. Environmental Protection Agency (1978) *Sludge Treatment and Disposal—Volume I*, EPA-625/4-78-012, Technology Transfer; U.S. Environmental Protection Agency: Washington, D.C.

U.S. Environmental Protection Agency (1979) *Process Design Manual for Sludge Treatment and Disposal*, EPA-625/1-79-011; U.S. Environmental Protection Agency: Cincinnati, Ohio.

U.S. Environmental Protection Agency (1982) *Guide to the Disposal of Chemically Stabilized and Solidified Waste*, SW-872; Office of Solid Waste Emergency Response, U.S. Environmental Protection Agency: Washington, D.C.

U.S. Environmental Protection Agency (1989) *1988 Needs Survey of Municipal Wastewater Treatment Facilities*; EPA-430/09-89-001; U.S. Environmental Protection Agency: Cincinnati, Ohio.

U.S. Environmental Protection Agency (1990) *Autothermal Thermophilic Aerobic Digestion of Municipal Wastewater Sludge*, EPA-625/10-90-007; U.S. Environmental Protection Agency: Cincinnati, Ohio.

U.S. Environmental Protection Agency (1993) Standards for the Use or Disposal of Sewage Sludge. *Fed. Regist.*, **58**, 32.

U.S. Environmental Protection Agency (1999) Standards for the Use or Disposal of Sewage Sludge. *Code of Federal Regulations*, Part 503, Title 40.

U.S. Environmental Protection Agency (2003) *Control of Pathogens and Vector Attraction in Sewage Sludge*, EPA-625/R-92013; U.S. Environmental Protection Agency: Cincinnati, Ohio.

U.S. Environmental Protection Agency (2007) *Alkaline Treatment of Municipal Wastewater Treatment Plant Sludge Technical Guide*; National Risk Management Research Laboratory; Office of Research and Development; U.S. Environmental Protection Agency: Cincinnati, Ohio.

Van Lier M. J. B.; Martin M. J. L. S.; Lettinga, M. G. (1996) Effect of Temperature on the Anaerobic Thermophilic Conversion of Volatile Fatty Acids by Dispersed and Granular Sludge. *Water Res.*, **30**, 199–207.

Verschueren, K. (1983) *Handbook of Environmental Data on Organic Chemicals*, 2nd ed.; Van Nostrand Reinhold Co.: New York.

Volpe, G.; Keaney, J.; Schlegel, P.; Tyler, C.; Carr, J.; Nagel, J. (2004) Large Egg-Shaped Digesters: Issues and Improvements. *Proceedings of the 77th Annual Water Environment Federation Technical Exhibition and Conference* [CD-ROM]; New Orleans, Louisiana, Oct. 2–6; Water Environment Federation: Alexandria, Virginia.

Waksman, S. A.; Cordon, T. C. (1939) Thermophilic Decomposition of Plant Residues in Composts by Pure and Mixed Cultures of Microorganisms. *Soil Sci.*, **47**, 217.

Ward, A.; Stensel, H. D.; Ferguson, J.; Ma, G.; Hummel, S. (1999) Preventing Growth of Pathogens in Pasteurized Digested Sludge. *Water Environ. Res.*, 71, 176.

Water Environment Federation (2012) *Safety, Health, and Security in Wastewater Systems,* 6th ed.; Manual of Practice No. 1; Water Environment Federation: Alexandria, Virginia.

Water Environment Federation (2013) *Wastewater Treatment Process Modeling,* 2nd ed., Manual of Practice No. 31; Water Environment Federation: Alexandria, Virginia.

Water Environment Federation (2016) *Operation of Municipal Wastewater Treatment Plants,* 7th ed.; Manual of Practice No. 11; Water Environment Federation: Alexandria, Virginia.

Water Environment Federation; Water Environment Research Foundation; U.S. Environmental Protection Agency (2012) *Solids Process Design and Management;* Water Environment Federation: Alexandria, Virginia.

Water Environment Research Foundation (2004) *Producing Class A Biosolids with Low-Cost, Low-Technology Treatment;* Report No.99-REM-2; Water Environment Research Foundation: Alexandria, Virginia.

Westphal, A.; Christensen, G. L. (1983) Lime Stabilization: Effectiveness of Two Process Modifications. *J. Water Pollution Control Fed.,* **55,** 1381.

Wiley, J. S.; Westerberg, S. C. (1969) Survival of Human Pathogens in Composted Sewage. *Appl. Microbiol.,* 18, 944.

Williams, T. (2014). Biosolids Composting Technology: Where it has come from and where it is going. *Proceedings of the Water Environment Federation Residuals and Biosolids Management Specialty Conference;* Austin, Texas. May 18–21; Water Environment Federation: Alexandria, Virginia.

Williams, T. O.; Forbes, Jr., R. H.; Wagoner, D. L.; Hahn, J. T. (2008) Control of Biosolids Cake Odors Using the New Biosolids Odor Reduction Selector Process. *Proceedings of the 22nd Annual Water Environment Federation Residuals and Biosolids Management Conference,* Philadelphia, Pennsylvania, March 30–April 2; Water Environment Federation: Alexandria, Virginia.

Willis, J.; Carrio, L.; Chapman, T., Keaney, J., Newman, G.; McNeal, T.; Muller, C.; Pianelli, P.; Salerno, L.; Schafer, P. (2016) Are Your Digesters Burping, Frothing, or Otherwise Not Behaving? – Let's talk about Solutions and NOT REPEAT the "Digester Foam" Paper you've already Heard Five Times. *Proceedings of the 30th Annual Water Environment Federation Residuals and Biosolids Management Conference,* Philadelphia, Pennsylvania, April 2016; Water Environment Federation: Alexandria, Virginia.

Willis, J.; Aiken, M.; Arnett, C.; Hull, T.; Matthews, J.; Schafer, P.; Sobsey, M.; Turner, B. (2003) Cost to Convert to Class A: Columbus Biosolids Flow-Through Thermophilic Treatment (CBTF3) as a Cost Effective Option. *Proceedings of the Water Environment Federation Technical Exhibition and Conference* [CD-ROM], Los Angeles, California, Oct. 11–15; Water Environment Federation: Alexandria, Virginia.

Wilson, C. A.; Fang, Y.; Novak, J. T.; Murthy, S. N. (2008a) The Effect of Temperature on the Performance and Stability of Thermophilic Anaerobic Digestion. *Water Sci. Technol.,* **57,** 297–304.

Wilson, C. A.; Murthy, S. N.; Novak, J. T. (2008b) Laboratory Digestibility Study of Wastewater Sludge Treated by Thermal Hydrolysis. *Proceedings of the 22nd Annual Water Environment Federation Residuals and Biosolids Conference: Traditions, Trends & Technologies,* Philadelphia, Pennsylvania, March 30–April 2; Water Environment Federation: Alexandria, Virginia.

Zahller, J. D.; Bucher, R. H.; Ferguson J. F.; Stensel, H. D. (2005) Performance and Stability of Two-Stage Anaerobic Digestion. *Proceedings of the 78th Annual Water Environment Federation Technical Exhibition and Conference* [CD-ROM]; Washington, D.C., Oct. 29–Nov. 2; Water Environment Federation: Alexandria, Virginia.

Zhou, J., Kelly, H. G., Mavinic, D.S., Ramey, W.D. (2001) Digestion effects on dewaterability of thermophilic and mesophilic Aerobically digested biosolids. *Proceedings of the 74th Annual Water Environment Federation Technical Exhibition and Conference;* Oct. 14–19, Water Environment Federation: Atlanta, Georgia.

Zhou, J.; Mavinic, D. S.; Kelly, H. G.; Ramey, W. D. (2002) Effects of Temperature and Extracellular Proteins on Dewaterability of Thermophilically Digested Biosolids. *J. Environ. Eng. Sci.*, **1**, 409–415.

CHAPTER **24**

Thermal Processing

Lee A. Lundberg, P.E.; Eric Auerbach; Stan Chilson;
Emma Cooney; Greg Homoki, P.E.; Andrew Jones;
F. Michael Lewis; Ashley Pifer, Ph.D., P.E.; Marcel Pomerleau;
Ray Porter; Kimberly Schlauch; and Steve Waters, P.E., P. Eng

2051

1.0   Introduction

The importance of municipal wastewater sludge processing and disposal has grown since the establishment of secondary treatment standards by the Clean Water Act of 1972 and subsequent construction of water resource recovery facilities that have generated increasingly larger volumes of sludge. At the same time, increasingly strict regulations governing wastewater solids disposal combined with the decreasing availability of disposal sites have limited disposal options and increased disposal costs.

Consequently, interest continues in thermal processing methods as a means of producing a marketable product or reducing sludge or biosolids volumes. Economic, environmental, and sociopolitical analyses provide the best basis for deciding whether to use thermal processing methods. Such an analysis considers all available reuse and disposal options and their costs and revenues. Costs include process costs, energy costs, and the cost of further handling and management (hauling, tipping fees, and so on). Revenues include those generated by production of a marketable product (fertilizer, electricity, fuel, and so on). The selection of a thermal processing system should recognize new, rapid advancements in dewatering technology. Some of these advancements may eliminate the future need for, or economic feasibility of, some thermal processing alternatives. Further, the selection process must weigh the likely effect of present and future regulations on the thermal processing system, associated air pollution control equipment, current and future energy prices, and the reuse or disposal option. The process should also assess whether the options are flexible enough to accommodate likely regulation changes. Such forethought may prevent the installation of a system that becomes obsolete or uneconomical, or requires major modifications to comply with future regulations.

A sound analysis of thermal processing options requires realistic expectations for advanced technology. Some existing sludge-processing systems were planned with high expectations for advanced technology that have often not been realized. As a result, the best reuse or disposal option available for these facilities might have been overlooked.

The approach for operating advanced technology systems for thermal processing differs from that for operating typical water resource recovery facility systems. Successful operation of some thermal processing systems requires highly trained and skilled operators or process engineers to be available 24 hours per day, especially during initial startup of the facility. Some municipalities may not offer a salary structure sufficient for such staff. A good analysis will account for these specialized staffing needs. A comprehensive cost/benefit analysis will help with selection of the best sludge-processing strategy for the water resource recovery facility.

The design and procurement of a thermal processing system often depends on proprietary vendor information, which may not be readily accessible. As a result, a designer should avoid specifying equipment or processes in a manner that may unnecessarily restrict some vendors from bidding on the project. In some cases, however, restrictive specifications reflect good engineering judgment. An engineer must set basic requirements such as solids and liquid loading capacities to the unit, types of support equipment, and minimum standards for materials.

The following four sections describe the prevalent thermal processing methods for municipal wastewater sludge, including conditioning and pretreatment; drying; oxidation; and gasification and pyrolysis. It should be noted that since the last edition of this manual, low-pressure wet-air oxidation (Zimpro process) as a thermal conditioning technology for municipal wastewater sludge has generally been phased out, although there are some promising emerging and proven thermal conditioning technologies that have been applied in this market, notably thermal hydrolysis processes, which are often used in conjunction with anaerobic digestion and are covered in Chapter 23— Stabilization. Gasification and pyrolysis remain as emerging technologies for wastewater solids management. The fifth section discusses the regulations that govern particulate and gaseous emissions and emission-control technology.

2.0 Thermal Conditioning and Pretreatment

2.1 Thermal Conditioning

Thermal conditioning is the simultaneous application of heat and pressure to solids to enhance dewaterability without adding conditioning chemicals. During thermal conditioning, heat lyses the cell walls of microorganisms in biological solids, releasing bound water from the particles. This process further hydrolyzes and solubilizes hydrated particles in biological solids and, to a limited degree, organic compounds in primary solids. Conventional mechanical dewatering devices can then readily separate released water from particles as long as the solids have enough fibrous solids for cake structure.

The two basic modifications of thermal conditioning that were historically used in wastewater treatment were heat treatment and low-pressure oxidation. At one time, about 31 U.S. water resource recovery facilities were using heat treatment systems. About 78 U.S. facilities used low-pressure oxidation systems. Few currently operate today, and vendors are not actively marketing these systems in the United States.

Presently, thermal hydrolysis as a thermal conditioning process prior to anaerobic digestion is experiencing renewed interest, with a number of installations in Europe and one of the largest thermal hydrolysis systems having recently been installed at DC Water's Blue Plains facility. Thermal hydrolysis is covered in Chapter 23.

2.2 Thermal Pretreatment

There are two main thermal pretreatment processes in use today, including thermal dewatering and thermal hydrolysis. Thermal hydrolysis is a process used to enhance the digestibility of sludge in a digestion system and is covered in Chapter 23—Stabilization. Thermal dewatering is used as a pretreatment process upstream of a thermal oxidization system to reduce or eliminate auxiliary fuel usage. Essentially thermal dewatering is a partial thermal drying process, sometimes called "scalping" that involves removing sufficient water content from the thermal oxidation system feed material to make the combustion process self-supporting or "autogenous," to the extent practical. This is done using waste heat recovered from the thermal oxidation system as the energy source for the dewatering process. Thermal dewatering is a useful way to both minimize auxiliary fuel usage and take advantage of more of the recoverable energy available in the thermal oxidation system exhaust gases provided that the thermal oxidation system can handle the higher feed solids concentration. The utility of thermal dewatering is greater for fluidized-bed systems than it is for multiple-hearth systems, as the latter are not as amenable to operation with higher feed solids concentrations.

Figure 24.1 provides a schematic of a thermal dewatering system in conjunction with a fluidized-bed thermal oxidation system. Fluidized-bed systems normally use a

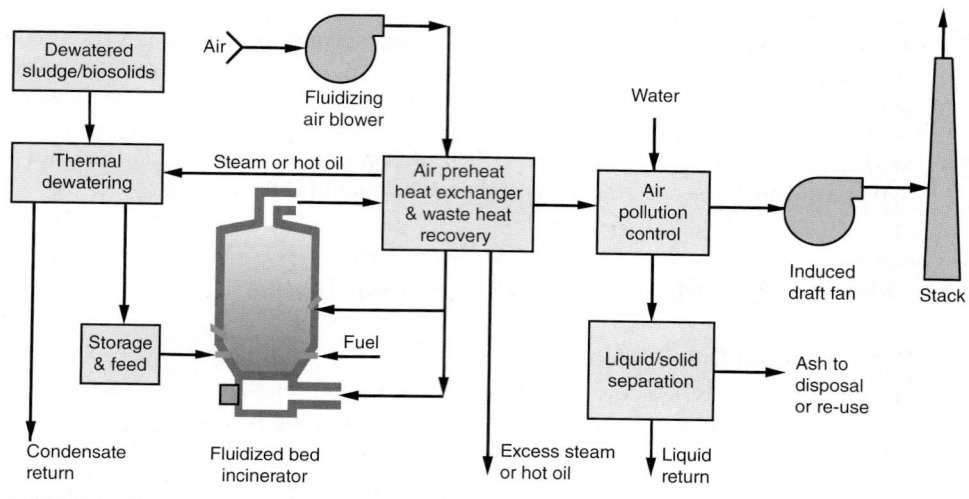

Figure 24.1   Thermal dewatering system schematic.

portion of the recoverable energy in the system exhaust gases to preheat combustion air, which also serves to reduce auxiliary fuel usage. Depending on the solids content from the mechanical dewatering system and the thermodynamic characteristics of the feed material, air preheating alone may not be sufficient to enable the designer to achieve autogenous operation. Generally speaking, air preheating only requires 40% to 60% of the recoverable energy in the system exhaust gases, so there is still a significant amount of energy remaining downstream of the air preheater to support thermal dewatering. Technically, this energy can be recovered to produce either a separate heated air stream, steam, or a heated thermal fluid stream. The designer should evaluate these options in conjunction with the requirements of the thermal dewatering equipment, along with other planned uses for the supplemental recovered energy, such as to meet building or other process heating loads. Generally speaking, indirect drying systems, such as disk or paddle dryers, are most appropriate for thermal dewatering applications, as the solids content of their product can be more closely controlled. Indirect drying systems can utilize energy in the form of steam or heated thermal fluids. More detail on the various dryers is provided in the next section.

3.0   Thermal Drying

3.1   Overview of Technology

Thermal drying involves heating biosolids to evaporate water and thereby reduce the moisture content further than conventional mechanical-dewatering methods can achieve. Thermally dried biosolids cost less to transport because of the mass and volume reduction that occur. Thermal drying meets the requirements to be a Class A process to further reduce pathogen (PFRP). There are criteria associated with this PFRP, so all drying technologies do not automatically meet the Class A requirements. While most do, the system must be designed to meet the PFRP requirements. A Class A product is more marketable than dewatered cake. However, thermal drying processes are more complex and expensive to operate than conventional dewatering methods. There also are safety concerns addressed later in this section.

Thermal drying has been practiced since the 1920s when the Milwaukee (WI) Metropolitan Sewerage District began producing and selling Milorganite®, its thermally dried biosolids fertilizer. For decades, only a few U.S. municipalities used this technology, but its popularity has grown since the early 1990s because of technology improvements, increasingly stringent regulations, and social factors affecting land application and landfilling. As of 2016, there were more than 105 thermal drying facilities operating or under construction in the United States and more than 375 worldwide. Their throughputs range from less than 1 to more than 100 dry metric ton/d (110 dry ton/d).

Thermally dried biosolids can be used as a fertilizer, soil conditioner, or biofuel. The use depends on the source of the solids, the pretreatment and drying systems used, and local conditions.

Some equipment manufacturers call thermally dried biosolids "pellets," while others call them "granules." In this chapter, the term *granules* will be used.

3.2   Process Fundamentals

The drying rate depends on both the solids flow and evaporation rates. During drying, a temperature gradient develops from the heated surface inward, causing moisture to migrate from inside wet solids to the surface via diffusion, capillary flow, and internal pressures generated as the solids dry and shrink. As heat transfers from the machine to the solids, the solids' temperature rises and water evaporates from the surface. Other conditions that affect the process include temperature, humidity, rate and direction of gas flow, exposed surface area, physical form of solids, agitation, detention time, and the support method used. Design engineers need to understand these conditions and their effects when investigating solids' drying characteristics, choosing the correct dryer, and determining the optimal operating conditions.

3.2.1   Stages of Thermal Drying

There are three stages of thermal drying: warm-up, constant-rate, and falling-rate (see Figure 24.2).

3.2.1.1   Warm-Up Stage   During the warm-up stage, both the solids temperature and the drying rate increase until they reach steady-state conditions. This stage typically is short and results in little drying.

3.2.1.2   Constant-Rate Stage   During the constant-rate stage, the surface of the solids remains saturated as interior moisture moves outward to replace the moisture that has evaporated. The drying rate depends on how much heat is transferred to the solids' surfaces (the heat- or mass-transfer coefficient). It also depends on how much area is exposed to the drying medium, as well as temperature and humidity differences between the drying medium and the solids' surfaces. This stage typically is the longest and when most drying occurs.

3.2.1.3   Falling-Rate Stage   During the falling-rate stage, moisture evaporates faster from the solids' surfaces than it can be replaced by the moisture within the solids.

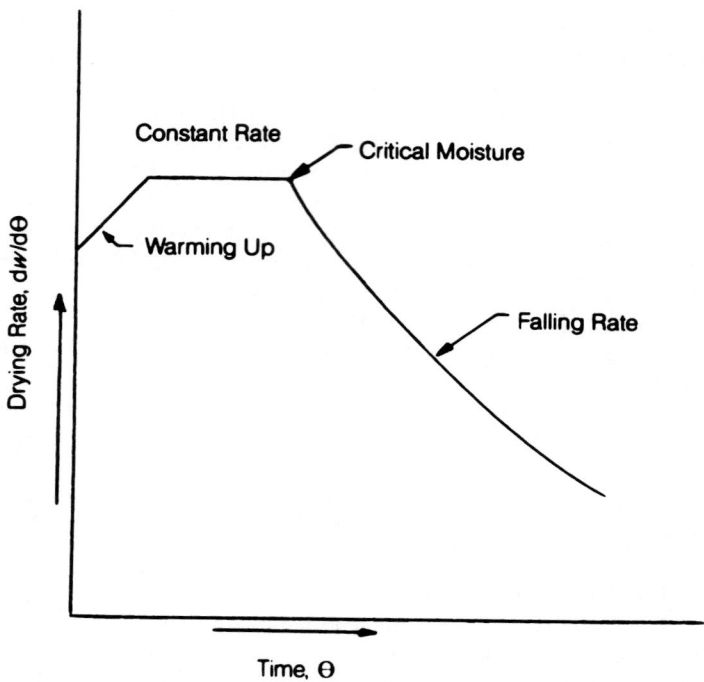

FIGURE 24.2   Three stages of drying.

Because the exposed surface is no longer saturated, solids do not transfer latent heat as rapidly as they receive sensible heat from the drying medium and the temperature of the solids increases. This stage is when the drying rate decreases.

The drying rate at the transition point between the constant- and falling-rate stages is called critical moisture.

3.2.2 Heat-Transfer Methods

Dryers are classified based on the method (convection or conduction) predominantly used to transfer heat to wet solids. Although most systems actually use multiple heat-transfer methods, one method typically is more dominant than the others.

3.2.2.1   Convection   In convection (direct-drying) systems, the heat-transfer medium (e.g., hot gas) contacts the wet solids directly. Convection heat transfer is expressed mathematically as follows (U.S. EPA, 1979):

$$q_{conv} = h_c A(t_g - t_s)$$

(24.1)

where q_{conv} = convective heat transfer (kJ/h [Btu/hr]);

h_c = convective heat-transfer coefficient (kJ/h/m²/°C [Btu/hr/sq ft/°F]);

A = area of wetted surface exposed to gas (m² [sq ft]);

t_g = gas temperature (°C [°F]); and

t_s = temperature at sludge-gas interface (°C [°F]).

Manufacturers have developed convective heat-transfer coefficients for their proprietary systems that they use to size thermal drying equipment.

3.2.2.2 Conduction Conduction (indirect-drying) systems transfer heat from the heat-transfer medium (e.g., hot oil or steam) to the solids via a material (e.g., a steel plate) that separates the solids from the medium. The mathematical expression for conduction heat transfer is as follows (U.S. EPA, 1979):

$$q_{cond} = h_{cond} A(t_m - t_s) \qquad (24.2)$$

where q_{cond} = conductive heat transfer (kJ/h [Btu/hr]);
$\quad h_{cond}$ = conductive heat-transfer coefficient (kJ/h·m²/°C [Btu/hr/sq ft/°F]);
$\quad A$ = area of heat-transfer surface (m² [sq ft]);
$\quad t_m$ = temperature of heating medium (°C [°F]); and
$\quad t_s$ = temperature of solids at drying surface (°C [°F]).

The conductive heat-transfer coefficient is a composite term that includes the effects of both the solids' and the medium's heat-transfer surface films. Manufacturers have developed system-specific coefficients that they use to size their proprietary systems.

3.3 Process Description

As with most technologies, thermal dryers have continually evolved in the more than 75 years since they first were used to treat solids. Some systems have adapted to the industry's changing needs, while others have come and gone. Some systems ceased being viable technologies because:

- Adverse commercial issues affected the manufacturers;
- The technology was overly complex;
- The technology could not produce a product that was easy to handle; or
- Full-scale installations performed poorly.

Below are basic descriptions of both established and emerging heat-drying processes.

3.3.1 Established Systems

3.3.1.1 Convection/Direct-Drying Systems Convection dryers that have been used successfully to dry municipal solids include the flash dryer, rotary drum dryer, fluidized-bed dryer, and belt dryer.

3.3.1.1.1 Flash Dryer At one point, flash dryers were being used in as many as 50 U.S. facilities. Currently, however, only Houston, Texas, still operates eight flash dryers at its water resource recovery facilities, along with two rotary drum dryers that were installed more recently. The reasons for the decline of this technology include safety concerns, high energy and O&M costs, and the primary equipment supplier's limited interest in the solids-treatment sector. Therefore, flash drying will not be further reviewed.

3.3.1.1.2 Rotary Drum Dryer Rotary drum dryers have been used to treat solids since Milwaukee first installed them in the 1920s. In 2016, rotary drum dryers were in use or under construction at more than 30 U.S. and 75 European and Asian facilities.

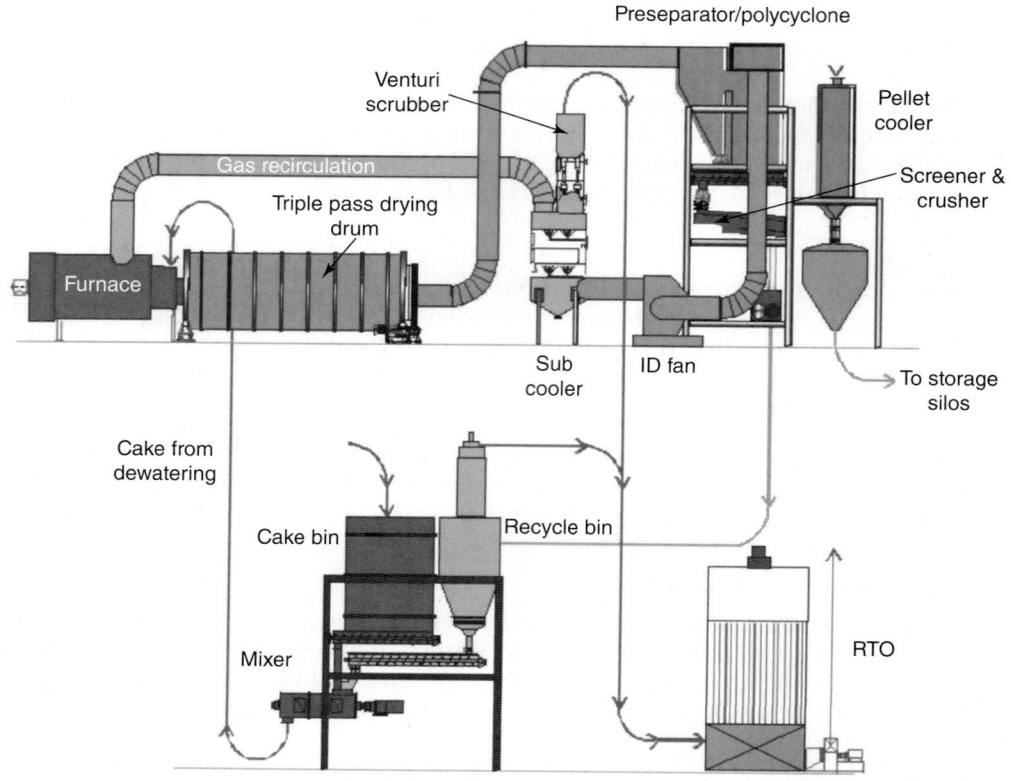

FIGURE 24.3   Schematic of rotary drum dryer (courtesy of Andritz Separation, Inc.).

Although some system components are manufacturer specific, all rotary drum dryers have long, cylindrical steel drums that rotate on bearings (see Figure 24.3). Some have one drum through which solids make a single pass, while others have concentric drums to allow solids to make multiple passes through the drum (e.g., triple-pass systems). Hot gases dry solids and help move them through the system.

Some of the granules are recycled to the feed solids (dewatered solids), where a mixer blends them to reduce the feed material's moisture content to typically less than 35% to avoid the "plastic" or "sticky" material phase and has better handling characteristics. Actual influent moisture content requirements depend on the dryer supplier and solids characteristics. Feed solids and hot furnace gases (400°C to 650°C [750°F to 1200°F]) continuously enter the upper end of the drum dryer and are conveyed (typically concurrently) to the discharge end (see Figure 24.4). As they travel through the drum, axial flights along the slowly rotating interior wall pick up and cascade solids through the dryer. This thin sheet of falling solids directly contacts the hot gases, which rapidly dry it.

In most systems, about 80% of furnace gases are recycled to enhance the system's thermal efficiency and help maintain an inert atmosphere. (See Section 3.5.4 on why inertization is important.) Recycling gas also reduces the volume of exhaust gases to

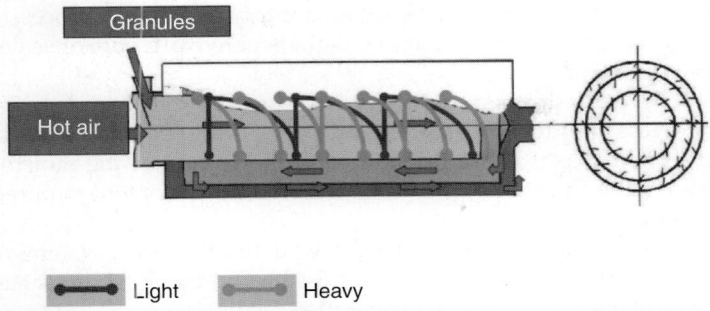

Light   Heavy

FIGURE 24.4 Schematic of solids movement through rotary drum dryer (courtesy Andritz Separation, Inc.).

be treated. Exhaust gases exit the dryer at 66°C to 105°C (150°F to 220°F) and travel to emissions-control equipment for particulate removal and odor control (Niessen, 1988).

Rotary drum dryers have been successfully used to dry both digested and undigested solids. In Europe, where granules typically become a biofuel, dryers often process undigested solids. In the United States, where the granules primarily are used as a fertilizer or soil amendment, the dryers typically process digested solids. Digestion can break down much of the fiber in raw solids that can cause operating problems for dryers. It also reduces the putrescible compounds that can cause odor problems in granules.

3.3.1.1.3 Fluidized-Bed Dryer Fluidized-bed technology first was applied to drying biosolids in Europe in the early 1990s; now, more than 40 facilities use this technology worldwide. This technology was first used in the United States in the late 1990s, and at least two U.S. systems were operating in 2016.

The main component of a fluidized-bed dryer is a stationary, vertical chamber segregated into three zones. The first zone is the wind box (gas plenum), which distributes hot fluidizing gases (typically air at about 85°C [185°F]) evenly through a bed of granules. The gases are forced or induced through the system by a blower arrangement in a closed-loop system.

The second zone is a heat exchanger. A heat-transfer medium (typically steam or hot oil) circulates through the heat exchanger tubes, while fluidizing gas keeps the granules in suspension, helps transfer heat from the tubes to the granules, and removes moisture from the system.

The third zone is a hood, where fluidizing gases exit the unit and most solids are separated from the gas flow. Exhaust gas then is sent to emissions-control equipment (e.g., cyclone separator, scrubber, and condenser), which removes particulates and moisture.

The method for introducing dewatered solids to the dryer is manufacturer specific. Some systems pump solids into the dryer, where a rotating device cuts dewatered cake into small, irregular pieces that fall into the fluidized bed. The fluidizing gas intimately contacts and highly intermixes the solids, resulting in high mass and heat transfer from gas to solids. Agitation in the fluidized bed removes rough edges, resulting in granules with a relatively uniform shape. Granules exit the dryer by overflowing from the chamber via an adjustable weir with a rotary airlock. All processing units are typically enclosed and maintained under slight negative pressure or sealed to capture dust and potential odors.

As with rotary drum systems, fluidized-bed dryers will reduce the solids moisture content to less than 10%. These systems are not designed to operate as scalping (partial drying) systems, as it is not possible to properly control the moisture content of the product to yield a partially dried cake with the solids content required from a scalping application (28% to 35%).

Because of the low temperatures involved, fluidized-bed systems are well suited for applications where energy recovery is possible. European systems have been operated by waste streams from garbage incineration facilities.

3.3.1.1.4 Belt Dryers   In belt dryers, solids are placed on a porous conveyor belt, and large volumes of heated gas are blown across or drawn through them. The belt speed and configuration control the solids retention time. The belts are made of either steel mesh or synthetic material (similar to the filter media used in a belt filter press). The unit may have one or two belts, and its configuration depends on the manufacturer. The granules typically are less uniform in size than those from more complex systems because of the limited materials handling. Screens and conditioners can be used to improve uniformity.

The gas-management system is also supplier specific. To remove moisture from the process, a fan typically removes a portion of the gas from the process air loop and sends it to a condenser. One supplier returns most of the scrubbed gas to the dryer, but diverts a portion to mix with combustion air. Other suppliers recommend that the scrubbed gas be treated by a biofilter or chemical scrubber to control odors before discharge. Recycling most of the process air improves the system's thermal efficiency and reduces the exhaust needing treatment.

The odor-control method is supplier specific. Biofilters traditionally have been used in Europe to control odors from these systems, but chemical scrubbers also can be used. One supplier uses the furnace to combust exhaust gases.

Traditional belt dryers use a relatively low-temperature gas (about 120°C to 177°C [250°F to 350°F]), so low-grade waste heat can be used as an energy source. Some very low temperature designs have evolved that can operate at temperatures as low as 65°C to 80°C (150°F to 175°F). These very low-temperature dryers require increased effective belt area to achieve the same capacity as the higher temperature designs, but can utilize more energy from low-grade waste-heat sources. This allows for a wide range of energy-recovery options. European facilities have used waste heat from gas-fired engines, municipal solid waste incinerators, and solids incinerators. One U.S. system that was installed and started up in 2009 combusts its dried product granules to supply heat for the dryer. A number of other facilities in the United States and Canada were considering coupling belt dryers with other external waste-heat sources, such as from an external cogeneration system or from heat recovered from cement kilns.

As of 2016, there were at least 14 municipal belt dryer installations in the United States, with at least 4 more expected to be operational in 2017. There were more than 100 municipal belt dryer installations worldwide.

3.3.1.2 Conduction/Indirect-Drying Systems   Over the last decade, conduction dryers have continued to become more popular in the United States as more small and medium-size utilities have begun drying solids. The simple materials-handling and emissions-control systems associated with conduction dryers appeal to smaller utilities. Most facilities use the dryers to reduce the solids' moisture content to less than 10%, but some use these systems as scalping dryers to reduce the solids' moisture content to about 50% to 70%, depending on the requirements of downstream equipment or the ultimate market use.

Conduction dryers that have been used successfully to dry solids include the paddle dryer, hollow-flight dryer, rotary chamber dryer, tray dryer, and pressure filter/vacuum dryer.

3.3.1.2.1 Paddle and Hollow-Flight Dryers There are three types of hollow-flight dryers: paddle, disk, and rotary chamber. Paddle dryers have been used in industrial applications for many decades, but were first applied to drying biosolids in Japan in the late 1980s. Paddle and disk dryers have been used to treat biosolids in the United States since the mid-1990s. In 2016, more than 65 U.S. facilities were building or operating paddle, disk, or rotary chamber dryers.

Both paddle and disk dryers consist of a stationary horizontal vessel (trough) and a series of agitators (paddles, disk, or flights) mounted on a rotating shaft (rotor). In some systems, the trough has a jacketed shell through which a heat-transfer medium (typically hot oil or steam) circulates. Rotary chamber dryers are similar, except that the trough's outer shell rotates and is heated by gas-fired burners. All of these systems have a hollow rotor and agitators, through which the heat-transfer medium circulates. The surfaces of the trough, agitators, and rotors conduct heat to the biosolids.

Dewatered biosolids typically are fed directly to the dryer by a positive-displacement pump or screw conveyor. (In these systems, feed solids typically are not mixed with recycled granules before treatment.) Granules exiting the dryer are irregularly shaped and vary in size because the agitation associated with paddles or disks can increase the concentration of fines in the product. Screens or conditioners can be used to produce a more uniformly sized product.

Rotor and agitator assemblies are fabricated of carbon steel, stainless steel, or a combination of both. They also can be made of various special alloys if highly corrosive elements are present. In some systems (primarily the paddle type), the agitators are arranged on the shaft to intermesh, which enhances the relative contact between heated surfaces and solids, thereby increasing the heat-transfer coefficient. They also are self-cleaning. In other systems, stationary agitator ploughs or breaker bars are placed between rotating agitators to improve mixing and prevent cake buildup on the agitator surface.

The temperature of the heat-transfer medium ranges from about 150°C to 205°C (300°F to 400°F). After transferring its available energy to solids, the heat-transfer medium typically is recirculated through the heating system. A high-pressure condensate return system can be used for steam to reduce energy consumption.

The water vapor concentration in the vessel creates an inert environment, but enough gas (air) is drawn through the dryer to remove evaporated water so it does not condense in the system. Instead, it is exhausted from the dryer and conveyed to a condenser. Noncondensable gases then can be routed to the odor-control system. Some facilities use chemical scrubbers for odor control, while others convey the exhaust to the activated sludge process' aeration system. Thermal oxidation also can be used. The volume of exhaust gas involved is less than that produced by convection systems.

Paddle, disk, and rotary chamber dryers can be either continuous or batch fed. In continuous-feed systems, solids are conveyed through the dryer via the paddles' agitation of the bed, which causes volume displacement. Some systems use a weir at the discharge end of the dryer to help submerge the heat-transfer surface in the solids. In batch-fed systems, a bulk charge is loaded into the dryer; then this load is processed to completion and unloaded from the dryer. Additional material cannot be introduced into a batch-fed dryer until the cycle is complete.

3.3.1.2.2 Tray Dryers   Tray dryers are similar to disk dryers, except that they are configured vertically and mix recycled granules with dewatered solids before feeding them to the dryer. The blended solids are fed through a top inlet in the vertical multistage dryer. The dryer has a central shaft with attached rotating arms that move solids from one heated stationary tray to another in a rotating zigzag motion until they exit at the bottom as granules. The rotating arms have adjustable scrapers that move and tumble solids in 10- to 30-mm layers over the heated trays.

After spiraling through the trays, solids exiting the dryer typically contain less than 10% moisture. These granules are screened and then either sent to a recycle bin for further processing or cooled to 30°C (90°F) and pneumatically conveyed to a storage silo.

Tray dryers have had limited growth in North America over the last decade. The Baltimore, Maryland, Back River Wastewater Treatment Plant has used a system of three tray dryers to treat solids since 1995. In addition, there is a system of two tray dryers at the Ashbridges Bay Wastewater Treatment Plant in Toronto, Ontario and a system of four tray dryers at the Stickney Water Reclamation Plant in Chicago that have been operational since 2009 and a small single tray dryer system at the water resource recovery facility in Kingston, New York. By 2016, there were at least 8 tray dryer installations in the United States and Canada, and 15 in Europe.

3.3.1.2.3 Pressure Filter/Vacuum Dryer   A pressure filter/vacuum dryer is a variation on a standard pressure filter that combines dewatering and drying in one process. First, the process uses the standard pressure cycle to mechanically dewater solids, which typically have been conditioned with lime and ferric chloride before entering the unit. After the pressure cycle, hot water (80°C [180°F]) is circulated through the system to heat the chambers while a vacuum is drawn across the system to lower the boiling point of the water still in the solids. This is a batch process, which allows operators to control the level of dryness achieved, making the units suitable for partial or complete drying applications.

To date, the pressure filter/vacuum dryer has not been widely used to treat solids; only two installations existed in 2009. Mountain City, Tennessee, has been using a pressure filter/vacuum dryer to treat solids since 2000. This utility uses the unit to produce solids containing 50% moisture, although it could dry solids further.

3.3.1.3 Solar Dryers   Solar dryers represent a technology that has evolved from a somewhat-emerging one to a more established technology over the past decade. These systems started out in Europe, have been used there for more than 20 years, and have a more-than-8-year history of operation in the United States. Solids have been dewatered naturally (e.g., sand drying beds) for decades, but such systems were susceptible to weather and typically produced solids containing about 70% moisture (for ease of disposal). In the 1990s, German researchers developed systems based on radiant (solar) energy and convective (air) drying theory that could produce solids containing no more than 10% moisture. The first full-scale facility began operating in Europe in the early 1990s; by 2008, more than 100 installations were in use there, and 10 facilities were being built or operated in the United States. The U.S. systems often have been designed to produce solids containing less than 25% moisture so they will comply with 40 CFR 503's vector-attraction reduction criteria.

Solar dryers typically consist of a concrete pad with low walls that is surmounted by a greenhouse-type structure. Unthickened, thickened, or dewatered solids are trucked or pumped onto the pad. A program monitors climatic variables (e.g., humidity and temperature) both inside and outside the structure, and adjusts fans and louvers to provide enough ventilation for drying. An electrically powered, mobile mixer tills the solids to expose more surface for evaporation. Some tiller designs include the ability to convey granules to either end of the structure to facilitate product removal or automatic transport to the next process, to storage or to hauling vehicles. Solids may be spread in a relatively thin layer or arranged into windrows.

In the United States, this technology typically has been used by smaller treatment facilities. Some of the European installations, however, serve relatively large treatment facilities. For example, one system with a 2-ha (5-ac) footprint handles solids from a city with 650 000 inhabitants. Another is designed to dewater solids from a 4970-m^3/h (31.5-mgd) facility.

Most of these dryers obtain their thermal energy solely from solar radiation, but about 20 use auxiliary heat (e.g., burned digester gas or waste heat from a power cogeneration facility) to increase drying performance. An Austrian facility burns its solar-dried solids on-site and uses this heat to speed up the drying process.

3.3.2 Emerging Drying Systems

Emerging processes are those that have not been in full-scale, commercial use inside or outside the United States for at least 5 years. Chemical drying is an emerging process that produces a value-added fertilizer product using biosolids through a combination of chemical and thermal processes.

3.3.2.1 Chemical Drying Chemical drying is a process that incorporates wastewater biosolids in the production of an organic/inorganic ammonium sulfate fertilizer. Chemical drying produces a value-added biosolids product that is high in nutrient value. In chemical drying, biosolids are added to other traditional fertilizer components and, because of chemical exothermic reactions, a significant percentage of the contained water is removed before later physical thermal drying. In this process, the biosolids are only a small fraction (5% to 30%) of the dry weight of the resulting fertilizer product.

Chemical drying occurs when dewatered biosolids, sulfuric acid, and ammonia are mixed in a pressure vessel. The exothermic reaction drives off water, and an organic ammonium sulfate fertilizer is produced. The reaction occurs at 135°C to 150°C (275°F to 302°F) at pressures of 207 to 480 kPa (30 to 70 psi). The high-temperature/pressure process meets the standards of the EPA for Class A pathogen reduction.

Chemical drying has been proven to be a feasible treatment process for biosolids. The complexity of the technology and the need to market the product as a high-end fertilizer will likely require the system suppliers to also operate facilities and market the product.

3.4   Process Design Guidelines

3.4.1   Sizing Parameters

3.4.1.1   Evaporative Capacity   Dryers are evaporation devices, so a design engineer's first step is to calculate how much water needs to be evaporated in a given time. The key parameters when calculating the desired evaporative capacity are the solids concentration of both the feed solids and product granules, as well as operating time. Evaporative capacity typically is expressed in terms of kilograms (pounds) of water to be evaporated per hour.

Then, designers should consult with manufacturers, who have established their systems' evaporative capacity based on heat-transfer calculations, lab tests, and demonstrated full-scale performance using various solids. A drying system may have multiple evaporative capacities based on the type of solids (e.g., undigested, digested, or combined primary and secondary), which affect heat-transfer coefficients and materials-handling characteristics.

Design engineers should keep in mind that the performance of the mechanical dewatering process substantially affects the size of a thermal dryer and its energy requirements. It typically is more cost-effective to remove water mechanically than thermally (to the extent possible).

3.4.1.2   Hours of Operation   Another key parameter in sizing thermal dryers is the acceptable hours of operation. Dryers have a fixed evaporative capacity, so the longer they operate each day, the more solids they can process. A dryer that operates 24 h/d only needs one-third of the evaporative capacity to process the same amount of solids as a dryer that operates 8 h/d. Because thermal dryers are expensive, many are sized to run almost continuously (e.g., 24 h/d for 5 days each week [about 100 to 105 hr/wk]) to fit with typical work schedules while leaving some down time for preventive maintenance. Also, running nearly continuously reduces equipment wear. Frequent heating and cooling cycles increase metal fatigue and, therefore, the rate of component failure. Longer run times also save on energy and resource costs associated with heating up and cooling down the dryer.

3.4.1.3   Solids Residence Time   Solids residence time (SRT) depends on the type of dryer and the manufacturer's design. Manufacturers determine SRT based on system capacity and the heat transfer rate. System temperature does not affect SRT in thermal dryers; a rotary drum dryer and belt dryer may have similar SRTs even though their operating temperatures are significantly different. Some manufacturers also can use the dryer's SRT to demonstrate compliance with 40 CFR 503's pathogen-reduction criteria.

3.4.1.4   Operating Temperatures   The heat-transfer equations (presented earlier in this chapter) show that the heat-transfer rate depends on the temperature of the drying medium (e.g., gases, oil, or steam). When two similar systems are operating at different temperatures, the one with hotter temperatures typically will dry solids faster. Although operating temperature may affect the size of the unit, the heat-transfer rate is affected by several factors (e.g., area of exposed, wetted surface, and the unit's heat-transfer coefficient), so temperature will not directly indicate the size.

3.4.1.5   Storage   Storage is an important but often overlooked element of a thermal drying system. It typically is needed both before and after the dryer. The system needs enough storage before the dryer to attenuate solids variations, to accommodate intermittent operating schedules, and to allow for system shutdowns. It also needs enough storage after the dryer to handle changes in hauling schedules and seasonal fluctuations in market demand.

Granules often are stored in silos that are sealed from the outside environment. Because of the cost of silos, design engineers should size them carefully. Alternative use or disposal methods (e.g., landfilling) may be more cost-effective than significant storage capacity. Small and medium-sized facilities can use other storage methods (e.g., bulk bags or covered pads) as long as dust control and other safety issues are addressed. For more information on safety and solids storage, see Section 3.5.4.

3.4.2 Selection

3.4.2.1 Product Quality and Use Design engineers should carefully consider the planned use for granules when selecting a thermal dryer. Some uses (e.g., high-end fertilizer) will require a higher quality product than others (e.g., soil amendment, bulk land application, or biofuel). Although thermal dryers typically produce similar granules in terms of solids/moisture content, physical characteristics of the product granules (size distribution, granule shape, etc.) vary significantly based on the dryer technology and associated processing. Accordingly, design engineers should note the variations among suppliers, which may affect product use. For example, one belt dryer supplier uses back-mixing, while another extrudes material directly onto the belt, resulting in granules with different characteristics.

It is recommended that utilities considering thermal drying perform preliminary market/use analyses before selecting a system to determine which outlets would be most viable. (For more information on biosolids marketing and use issues, see Chapter 25.)

3.4.2.2 Processing Train Unit Capacity When selecting a thermal drying system, engineers should consider the desired design capacity and the number of processing trains needed to supply it. Thermal dryers should be sized to provide an optimal balance among capital costs, redundancy, and space constraints. Design engineers also should carefully consider turndown capacity when sizing a system that is expected to accommodate significant increases in throughput capacity over time. Thermal dryers typically have limited turndown capability, so multiple trains may be needed to meet future processing needs or else the system may need shorter operating cycles initially to optimize throughput. (Throughput optimization also improves the system's thermal efficiency.)

Typically, it is more cost-effective to install one processing train rather than several to meet capacity requirements because this reduces space and ancillary support-system requirements and takes advantage of equipment-related economies of scale. However, one train may not meet the utility's needs for redundancy. To address this issue, numerous U.S. systems have one processing train (to reduce capital costs) and another dewatering and disposal option (e.g., landfilling) when the thermal dryer is out of service.

Convection systems (e.g., rotary drum and fluidized bed) typically have larger single-train capacities than conduction systems (e.g., paddle, disk, and rotary chamber). Rotary drum and fluidized-bed systems typically are sized for capacities in the range from 2000 to 11 000 kg H_2O/h (4400 to 24 250 lb H_2O/hr). Tray dryers also are considered large-capacity systems. Conduction systems, on the other hand, typically are sized for capacities in the range from 500 to 5440 kg H_2O/h (1100 to 12 000 lb H_2O/hr). The exception is belt dryers. One supplier has developed belt dryers with capacities comparable to those of rotary drum systems.

3.4.2.3 Labor Requirements   Labor requirements depend on the type of thermal dryer and its design. Some systems require more supervision and O&M staff skill sets than others. For example, rotary drum systems require continuous supervision and well-trained operators. Conversely, many convective and conductive drying systems operate with minimal operator attention, requiring largely walk-through operator oversight and remote monitoring. Several fluidized-bed and belt drying systems in Europe, on the other hand, operate in automatic mode with minimal supervision during off-shifts. A fluidized-bed system at Houthalen, Belgium, operates 24 hours a day, 7 days a week in two 12-hour shifts. During one of these shifts, the dryer operates unattended in automatic mode.

Some municipal water resource recovery facilities hire contractors to handle solids drying and marketing operations. These private firms have the expertise for system O&M and product marketing.

3.4.3 Utility Requirements

3.4.3.1 Electrical   Thermal dryers may need substantial electrical power for the fans and solids-handling components. Requirements depend on system type and manufacturer. More complex systems will need more power because of their additional processing equipment.

3.4.3.2 Thermal   Thermal energy traditionally was supplied via the combustion of fossil fuels (e.g., natural gas and fuel oil); however, alternative fuels (e.g., digester biogas, landfill gas, and wood waste gasification) have become more common, helping reduce or eliminate fuel costs. A number of thermal dryers have been designed to use waste heat recovered from other processes, such as power generation or incineration as the primary heat source for drying, with auxiliary fuel only used to make up for any energy deficit from the waste-heat recovery system.

The net thermal energy requirement quoted for drying equipment alone typically ranges from about 2900 to 3250 kJ/kg (1250 to 1400 Btu/lb) H_2O evaporated. Allowing for the efficiencies of fuel-burning equipment and other system losses, gross thermal energy requirements will be higher in the field and could increase to as much as 3500 to 3900 kJ/kg (1500 to 1700 Btu/lb) H_2O evaporated. System thermal energy needs will depend on the type of thermal dryer and the manufacturer, as well as the particulars of the overall system layout and design. Numerous factors affect energy efficiency, including many proprietary features that system owners and designers cannot control. Manufacturers can provide guaranteed energy-consumption factors based on tested operating conditions and feed solids characteristics.

One factor that owners can control to improve thermal efficiency is the mode of operation. Continuously operated systems with weekly startup and shutdown cycles will be more energy efficient than those with daily startup and shutdown cycles. Also, operating close to design capacity will enhance overall thermal efficiency.

3.4.3.3 Water   Most thermal dryers require substantial supplies of water to cool granules and remove moisture from process gas (via a condenser), particularly those designs that operate at high temperatures and yield a warm or hot product. Product cooling is not always required, depending on the dryer system design, though some form of off-gas cooling or condensing is normally required for all designs except for solar dryers. Water resource recovery facility effluent can be used for these purposes. Potable water may be required for some safety systems and emissions-control systems (e.g., demisters), but such demands typically are not significant.

3.4.3.4 Sidestreams Design engineers should consider the effects of all thermal dryer sidestreams on the water resource recovery facility. For example, condensers produce odorous liquid sidestreams that contain both organic oils and ammonia. Wet scrubbers and other ancillary equipment also generate sidestreams.

3.4.3.5 Emissions and Odor Control It is good design practice to enclose dryer equipment, as well as handling and storage areas, and vent these spaces to air pollution-control equipment to control fugitive odors. Some dryers are operated at a slight vacuum, with all exhaust gas/vapors directed to a scrubber/condenser. These subsystems are incorporated into the dryer system design to minimize fugitive odors. Odors typically are controlled via chemical scrubbing and thermal oxidation. Cyclone separators, wet scrubbers, baghouses, or a combination of these technologies can remove particulate from exhausted air.

All system components that come in contact with the dryer exhaust gases will be exposed to a variety of chemical compounds that may accumulate on the exposed surfaces and adversely affect system performance. Siloxane deposits have been observed in several installations on dry exposed surfaces in heat exchanger and emission-control equipment that can lead to periodic system downtime and potentially expensive maintenance activities. As siloxane content in biosolids increase, more of these problems are likely to be reported. Consideration should be given during system design to providing suitable means for maintenance access to facilitate inspection and cleaning of potentially exposed surfaces.

3.5 Design Practice

3.5.1 Pre-Processing Equipment

Mechanical dewatering systems significantly affect thermal dryer requirements. It is more cost-effective to remove water mechanically (to the extent possible) than thermally. Slight improvements in dewatering-system performance can significantly reduce the overall thermal dryer system capacity requirements and energy-consumption needs. Figure 24.5 is a graph of the amount of water evaporation required as a function of feed solids concentration to the dryer. This figure illustrates the dramatic reduction in dryer evaporative load that can be achieved by merely improving dewatering system performance by as little as 1% to 2% solids. In some dryer designs, particularly those that use recycled products to serve as a nucleus for granule growth, feeding dewatered solids that are too dry may cause problems in forming the "skim coat" for granule growth. Consequently, for certain dryer designs, there is a sweet spot in dewatering to achieve optimal dryer system performance. This varies based on feed characteristics, but is often in the range of 22% to 25% solids, but could be higher or lower. To minimize energy consumption, one should try to operate with as high feed cake solids as is possible, consistent with maintaining acceptable product characteristics.

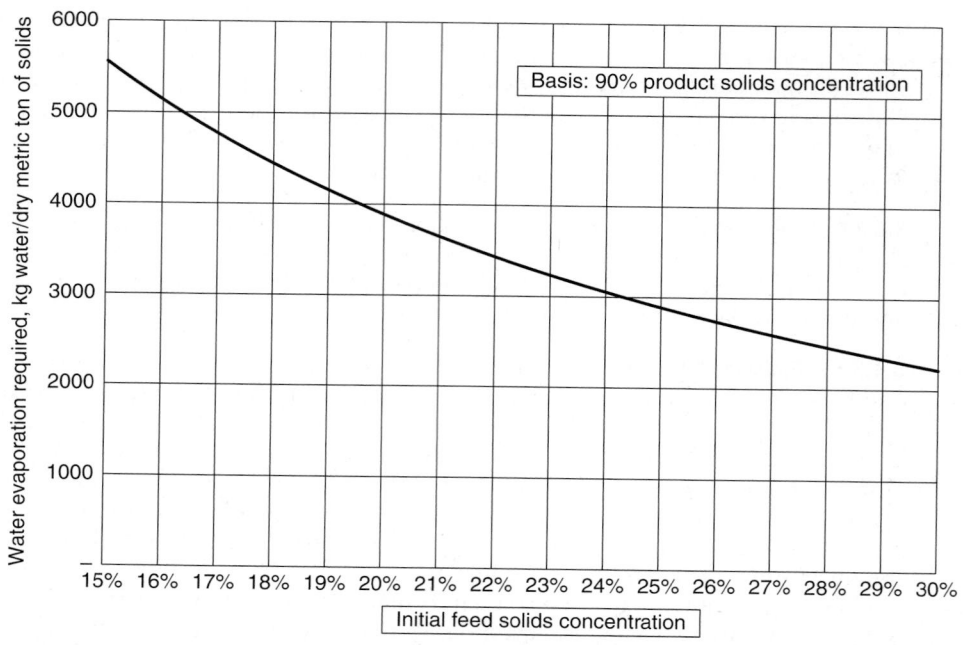

FIGURE 24.5   Dryer evaporative load versus feed solids concentration.

Consistent dewatering performance is also important. Large variations in feed solids concentrations can be problematic for some systems (e.g., drum dryers). The variability can affect the ratio of recycled granules to wet cake, thereby degrading the performance of drum dryers and subsequent processes. Batch-run systems (e.g., fluidized-bed systems and some conduction systems) can better process feed cake with varying solids concentrations. Some dryer systems can be equipped with sensors to monitor the drying process that can be used to adjust the dryer operating parameters to respond to the effects of changes in feed cake solids content.

3.5.2   Post-Processing Equipment

Post-drying systems (e.g., screening, cooling, and dust control) depend on dryer type and manufacturer. These systems can fulfill processing requirements, improve granule-handling characteristics, and make the overall system safer. For example, screening makes the granules more uniform, and its reject can be recycled for blending with feed cake. Granules must be cooled before storage to reduce the risk of auto-oxidation (see Section 3.5.4). Dust-control systems (e.g., mineral-oil spray and mixing systems) agglomerate dust particles to improve granule quality and enhance system safety.

Depending on the type of product granules from the dryer, it may be appropriate to install pelletization or briquetting equipment to produce a final product in a form that meets the specifications and requirements of the ultimate market user. This adds cost and complexity to the system, but may be worthwhile economically if it opens up a greater market for the product and/or increases the product value appropriately.

3.5.3 Materials of Construction

The choice of construction materials for a thermal dryer and its ancillary equipment depends on the characteristics of the solids being treated. Mild steel typically is appropriate for components that only touch granules, but stainless steel or other corrosion-resistant materials may be required for components that contact wet cake. Solids are abrasive, so design engineers should consider hard surfaces in areas especially prone to wear (e.g., agitators in indirect dryers) to avoid frequent equipment replacement. Also, all equipment should be airtight or operate under a slight vacuum and be well insulated to minimize heat and efficiency losses as well as to contain odors.

3.5.4 Safety

Dried biosolids are combustible and, therefore, a fire hazard. Under certain conditions (even when wet), the material can auto-oxidize—generate enough heat via either biological or chemical processes to combust. A fire requires three components: a fuel source, oxygen, and an ignition source. Because the fuel source (granules) obviously cannot be eliminated, fire-prevention measures for thermal dryers should focus on minimizing oxygen levels (inerting) and eliminating ignition sources.

Thermal dryers also have explosion hazards (e.g., combustible dust generated in solids-handling equipment, and CO and other combustible gases created during a smoldering fire). An explosion also requires three components: combustible dust or gases, oxygen, and an ignition source. Therefore, explosion-prevention measures should focus on keeping combustible dust and gases below explosive concentrations, minimizing oxygen levels (inerting), and eliminating ignition sources.

Because design engineers cannot eliminate every condition that could lead to fires or explosions, they should provide mitigation measures (e.g., explosion venting) designed to limit potential damage to people and equipment, as well as isolation systems to prevent fires and explosions from spreading among drying-system components. Most dryer manufacturers incorporate these safety features into their dryer designs as standard practice. During design, discussions should be held with each manufacturer covering any safety concerns, project-specific issues, and how the dryer design can be configured to minimize these risks.

3.5.4.1 Dryer Equipment To prevent fires and explosions in thermal drying systems, engineers should make sure the final design includes the following safety equipment, as appropriate to the dryer design:

- Gas-inerting (typically nitrogen-based) systems in granule storage silos, solids-handling equipment, and the drying area that either provide continuous inerting or activate in response to incipient fire conditions;

- Water sprays or other suppression systems designed to quench fires;

- Some means of temperature monitoring to alert when temperatures rise and means of control and response measures to keep temperatures from rising to unsafe levels;

- A system for cooling granules before storage to prevent auto-oxidation;

- Ventilation systems for solids-handling areas to remove condensation and keep dust concentrations below explosive limits;

- Means for monitoring oxygen content within or at the exit of the dryer to alert when oxygen concentration rises and means of control and response measures to keep oxygen concentration from rising to unsafe levels;

- Explosion venting in granule-storage areas and higher-risk solids-handling equipment and transfer points, where air (oxygen) may leak into the system, to relieve the pressure from an explosion and minimize equipment damage; and

- Isolating devices (e.g., rotary valves) that are provided between drying-system components—particularly areas with large amounts of granules (e.g., storage bins) or high dust concentrations (e.g., baghouses)—to prevent fires or explosions from spreading.

3.5.4.2 Dryer Operation   Because design engineers cannot eliminate all hazardous conditions associated with drying biosolids, thermal dryers must be operated to minimize such conditions. For example, controlling the solids' moisture content throughout the system is critical to eliminating auto-oxidation and excessive-dust hazards. In a solids-recycling system, granules that are too wet can provide the water needed for auto-oxidation to occur, as well as allow solids to build up inside equipment. However, granules that are too dry can form excessive amounts of dust. Maintaining steady-state operations and optimal moisture levels throughout the system can reduce such hazards.

3.5.4.3 Industry Standards   To ensure that proven safety measures are implemented, design engineers should consult the latest editions of relevant industry standards. Those addressing the fire and explosion risks associated with drying biosolids are issued by the National Fire Prevention Association (NFPA) and include:

- *Standard for the Prevention of Fire and Dust Explosions from the Manufacturing, Processing and Handling of Combustible Particulate Solids* (NFPA 654);

- *Standard on Explosion Prevention Systems* (NFPA 69);

- *Standard on Explosion Protection by Deflagration Venting* (NFPA 68); and

- *Standard for Fire Protection in Wastewater Treatment and Collection Facilities* (NFPA 820).

4.0 Thermal Oxidation

4.1 Process Fundamentals

Thermal oxidation systems totally or partially convert organic solids into oxidized products (primarily carbon dioxide and water) or else partially oxidize and volatilize organic solids via starved air combustion into products with a residual caloric value. Thermal oxidation refers to high-temperature oxidation in the presence of excess air and is generally referred to as incineration. Starved air combustion is thermal oxidation that restricts airflow (oxygen) to produce three potentially energy-rich products: gas, oil or tar, and a char. Pyrolysis and gasification are examples of processes using starved air combustion.

Although wastewater solids typically are organic, they only will sustain combustion without auxiliary fuel if enough water has been removed. Hence, dewatering should always precede thermal oxidation. Dewatered cake with 20% to 35% total solids can be burned with or without auxiliary fuels, depending on the volatile solids-to-moisture ratio, heating value of volatile solids, amount of excess air, and the temperature of the combustion air. Dewatered cake with a water content of 70% to 74% (total solids of 26% to 30%) and volatile solids content of 75% to 80% have supported autogenous combustion conditions in both fluidized-bed incinerators (FBIs) and multiple-hearth furnace (MHF)-style thermal oxidizers.

Combustion (burning) is the rapid combination of oxygen with fuel, resulting in the release of heat. The typically combustible elements of wastewater solids—carbon, hydrogen, and sulfur—chemically combine in organic solids as grease, carbohydrates, and protein. The combustible portion of wastewater solids typically has a heating value about equal to that of lignite coal. Adding air provides oxygen to support the combustion. The major products of complete combustion are carbon dioxide, water vapor, sulfur dioxide, and inert ash. Good combustion requires proper proportioning and thorough mixing of fuel and air and the initial and sustained ignition of the mixture. To achieve complete combustion, three fundamental principles must be followed: temperature, time, and turbulence (see Section 4.3).

The gaseous products of thermal oxidation include dry combustion gases, excess air, and water vapor from the moisture in the solids and the oxidation of hydrogen. The heat content of dry combustion gases (which is predominantly CO_2 and N_2 with trace amounts of SO_2) is determined by multiplying the individual weight of each gas by its specific heat content at the exit temperature. The weight of moisture in the exit gases (resulting from the combustion of hydrogen) typically is combined with the weight of the moisture in the wet feed to calculate the total heat content of the total water vapor exiting in combustion gases. If supplemental fuel is used, the volumes and heat contents of stack gases resulting from fuel combustion and solids combustion must be calculated separately.

The main objective of thermal oxidation is to reduce the quantity of solids requiring disposal. This process reduces the volume and weight of feed solids by about 80% to 95%, oxidizes or reduces toxics, can operate without consuming fossil fuels (autogenous combustion), and can yield exhaust heat that can be recovered for use in beneficial heating applications (process heat or building heat) or used to produce electrical power or mechanical energy. While the capital costs for thermal oxidation may be higher than those for other solids use and disposal options, the overall economics have shown to be favorable due to reduction of final solids product, bioenergy potentials, and the flexibility, in some cases, for economic revenue from contract burning. Incorporation of energy recovery and utilization systems can greatly mitigate O&M costs and often make thermal oxidation more sustainable and competitive with other options.

4.1.1   Solids Calorific Values

The composition and quantity of fuels to be burned are fundamental inputs for the design of any combustion system. The composition determines the fuel's calorific or heating value, and the fuel's quantity determines the required size of the unit.

A solid's heating value is the amount of heat that can be released per unit mass of the solid. Its gross heating value—the prime indicator of combustion potential—depends on how much carbon, hydrogen, and sulfur it contains. Carbon burned to carbon dioxide has a gross heating value of 3.4×10^4 kJ/kg (1.46×10^4 Btu/lb). Hydrogen has a gross heating value of 1.44×10^5 kJ/kg ($6.2\ 3 \times 10^4$ Btu/lb). Sulfur has a gross heating value of 1.0×10^4 kJ/kg (4500 Btu/lb). So, any changes in the solid's carbon, hydrogen, or sulfur content will raise or lower its heating value.

Each fuel has a characteristic ("effective") heating value and a gross heating value. Effective heat (available heat) is the amount of heat released in a combustion chamber minus both the dry flue-gas loss and the moisture loss. It depends on the solids' characteristics and the combustion chamber's design and operating characteristics; it cannot be determined via a bomb calorimeter alone. Incinerators that need large amounts of excess air offer less available heat because of the energy needed to heat the excess air (see Figure 24.6) (North American Manufacturing Co., 1965; Los Angeles County Sanitation Districts, 1988).

The typical heating values of wastewater solids obtained via bomb calorimetry are shown in Table 24.1 (U.S. EPA, 1979). Generally, the higher heating value of the combustible fraction of dry primary and WAS solids will range from 2.2×10^4 to 2.6×10^4 kJ/kg (9500 to 11 200 Btu/lb). The literature contains several methods for calculating the heating values of organic materials using proximate and ultimate analyses, but none of these, including the Dulong formula, accurately approximate the heating value of wastewater solids. Laboratory determination of the higher heating value of the wastewater solids is recommended when considering a thermal oxidation system.

4.1.2   Oxygen Requirements

The oxygen needed for combustion typically comes from air. Because air contains a large amount of nitrogen, thermal oxidizers need much more air than would be necessary if

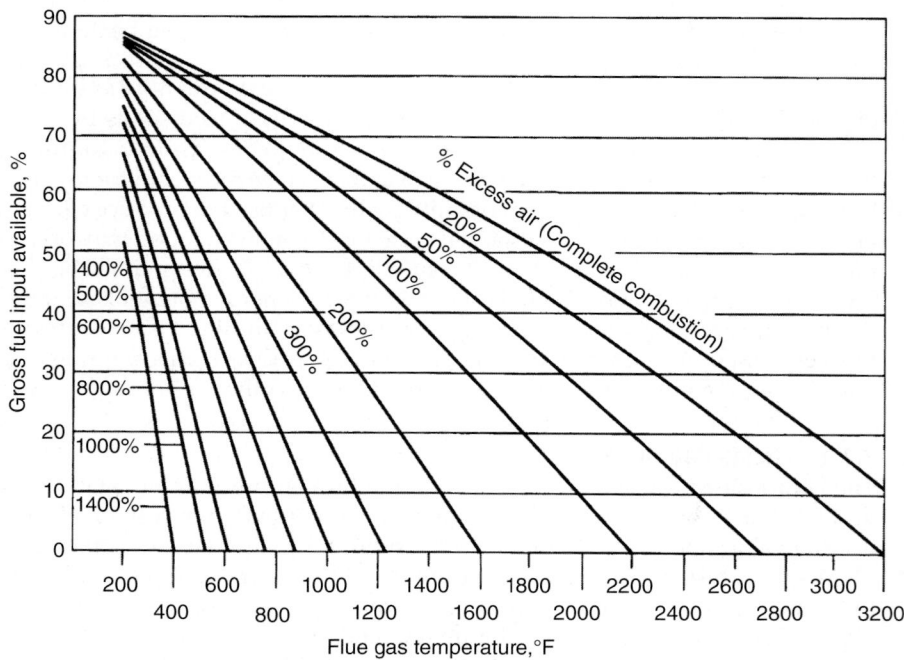

Figure 24.6   Generalized available heat chart.

Material	Combustible Solids (%)	High Heating Value (kJ/kg [Btu/lb] of Dry Solids)
Grease and scum	88	38 816 (16 700)
Raw wastewater solids	74	17 678 (7600)
Fine screenings	75	15 700 (6750)
Ground garbage	75	13 095 (5630)
Biosolids	60	12 793 (5500)
Grit	30	9304 (4000)

TABLE 24.1   Representative Heating Values of Some Solids (courtesy of Suez)

Substance	kg/kg of Substance	
	Air	Oxygen
Carbon	11.53	2.66
Carbon monoxide	2.47	0.57
Hydrogen	34.34	7.94
Sulfur	4.29	1.00
Hydrogen sulfide	6.10	1.41
Methane	17.27	3.99
Ethane	16.12	3.73
Ammonia	6.10	1.41

TABLE 24.2   Theoretical Air and Oxygen Requirements for Complete Combustion (courtesy of Suez)

pure oxygen were used (see Table 24.2). For every kilogram of oxygen required, about 4.6 kg of air must be supplied.

4.1.2.1 Stoichiometric Air   Insufficient quantities of oxygen lead to partial combustion and the resulting soot, carbon monoxide, and gaseous hydrocarbon emissions. Heat and material balances indicate the stoichiometric amount of oxygen needed to completely combust solids, and design engineers typically should add 25% to 100% more air to ensure that the system has enough oxygen for complete combustion. However, they should recognize that increasing excess air amounts decreases thermal efficiency of the process, thereby increasing fuel needs from heating unneeded oxygen and inert nitrogen. Most MHFs are designed to operate with 75% to 100% excess air, though some innovative multiple-hearth designs have been able to operate as low as 40% excess air. FBIs typically operate with about 40% excess air (Dangtran et al., 2000). Turndown of a FBI is limited by the air velocity required to fluidize the media bed, so FBIs always must be operated at or above 65% of design solids loadings.

4.1.2.2 Oxygen Content The design oxygen content of exhaust gases typically ranges from 6% to 8% on a dry-weight basis (WEF, 2009). To control combustion, the exhaust gas' oxygen content must be measured and combustion air flows should be varied to maintain an appropriate set-point value. Since the process conditions in the combustion zone include the water vapor, it is important to assure that wet oxygen content is sufficiently high to maintain proper system performance and ensure safe operations. Normal operating wet-oxygen content may range from 3.5% to 5.5% in the furnace exhaust, depending on furnace type (fluidized bed or multiple hearth). For safety purposes, the system should include interlocks to shut down solids and fuel feed if wet-oxygen content gets too low (~2.0 to 2.5%) for even brief periods of time. Two types of analyzers are used to measure this oxygen content: extraction and in situ. Both types measure wet-oxygen concentration. Measurement of dry-oxygen concentration can be done using extraction-type analyzers if sufficient gas cooling and/or drying is provided, but is often calculated based on psychrometric and temperature (at the point of analysis) data. The extraction analyzer withdraws a sample of exhaust gas and then cools it, filters it, and analyzes it with a zirconium oxide element. The in situ analyzer uses a zirconium oxide element surrounded by a ceramic filter to directly measure the oxygen content of the exhaust gas. Regulators require a correction to flue gas emission for oxygen content, dry, often 7% or 12% oxygen, to discourage dilution as a means of concentration-based emission limits compliance.

4.1.3 Heat and Temperature

When fuel, oxygen, and an ignition source combine, the resulting combustion releases energy that is called heat when it transfers from the burning material to cooler solids and gases. This transfer is driven by the temperature differential—the difference in temperature between two objects.

A combustion system releases two types of heat: sensible and latent. Sensible heat is heat that, when gained or lost by a body, is reflected by a change in that body's temperature. This heat is computed by multiplying the material's heat capacity (specific heat) by its temperature above some reference point (typically 0 or 15.5°C [32°F or 60°F]) as follows:

$$Q_s = (M_{cp})(W)(T - t) \qquad (24.3)$$

where Q_s = sensible heat (kJ [Btu]);

M_{cp} = mean specific heat (kJ/kg·°C [Btu/lb/°F]);

W = weight of material (kg [lb]);

T = temperature above a reference temperature (°C [°F]); and

t = reference temperature (°C [°F]).

Examples of sensible heat include the heat content of ash and the heat required to raise the temperature of flue gases.

Latent heat is the heat required to evaporate moisture from solids. The latent heat of vaporization is a function of temperature; in combustion calculations, it is assumed to occur at a temperature of either 0 or 15.5°C (32°F or 60°F).

A substance's theoretical flame temperature is:

$$T = [Q_s/(W \times M_{cp})] + t \qquad (24.4)$$

This equation assumes all heat is recovered and used to elevate the temperature of combustion products; none is lost (e.g., radiation) to the combustion environment. In practice, however, theoretical flame temperatures are impossible to reach because several factors limit the temperature to a level somewhat below the calculated value. For example, when wastewater solids are burned, the theoretical flame temperature will be reduced by the following major heat losses to the combustion environment:

- Moisture in the stack gases (latent heat);
- Unburned carbon;
- Excess air;
- Radiation;
- The ash's heat content; and
- Heats caused by reaction or formation.

Combustion actually seeks a final equilibrium temperature at which the heat input from the fuel equals the heat losses. When self-supporting, this equilibrium temperature is called the autogenous burning temperature.

Wastewater solids dried to about 40% solids will ignite at less than 540°C (1000°F). Operating temperatures greater than 760°C (1400°F) are required to combust all organic material completely. Operating MHFs at temperatures below 980°C (1800°F) prevents the damage of furnace insulation and structural components, and the slagging of ash (U.S. EPA, 1979, 1983a).

Exhaust temperature leaving the stack is an important indicator of the efficiency of a combustion reaction and one of the main operating variables monitored and controlled to conserve fuel use. The MHFs with a recycle air flow can achieve exhaust temperatures in the range of 540°C to 650°C (1000°F to 1200°F) while exhaust from FBIs is typically higher, in the range of 760°C to 870°C (1400°F to 1600°F). Exhaust temperatures are also often purposefully raised in afterburner chambers to ensure complete destruction of visible emission products and comply with air emissions regulations.

4.2   Process Description

4.2.1   Established Technologies

Three thermal oxidation processes are discussed in this section: FBIs, MHFs, and multiple-hearth enhancement processes.

4.2.1.1   Fluidized-bed Incinerator   The principle of using a high-temperature gas to fluidize solid materials was first commercially developed by the oil-refining industry in a technology called fluidized-bed catalytic crackers. While FBIs only vaguely resemble catalytic crackers, they function similarly. In a FBI, combustion air is passed through an enclosed space (fluidized-bed zone) in a way that sets all the particles in that zone in a homogeneous, boiling motion (i.e., fluidizes them) (see Figure 24.7). In this state, particles are separated from each other by an envelope of fluidizing gas (combustion air), thereby presenting more surface for a gas-to-solid reaction (e.g., air to carbon). The maximized surface area is what gives most FBIs their high thermal efficiency.

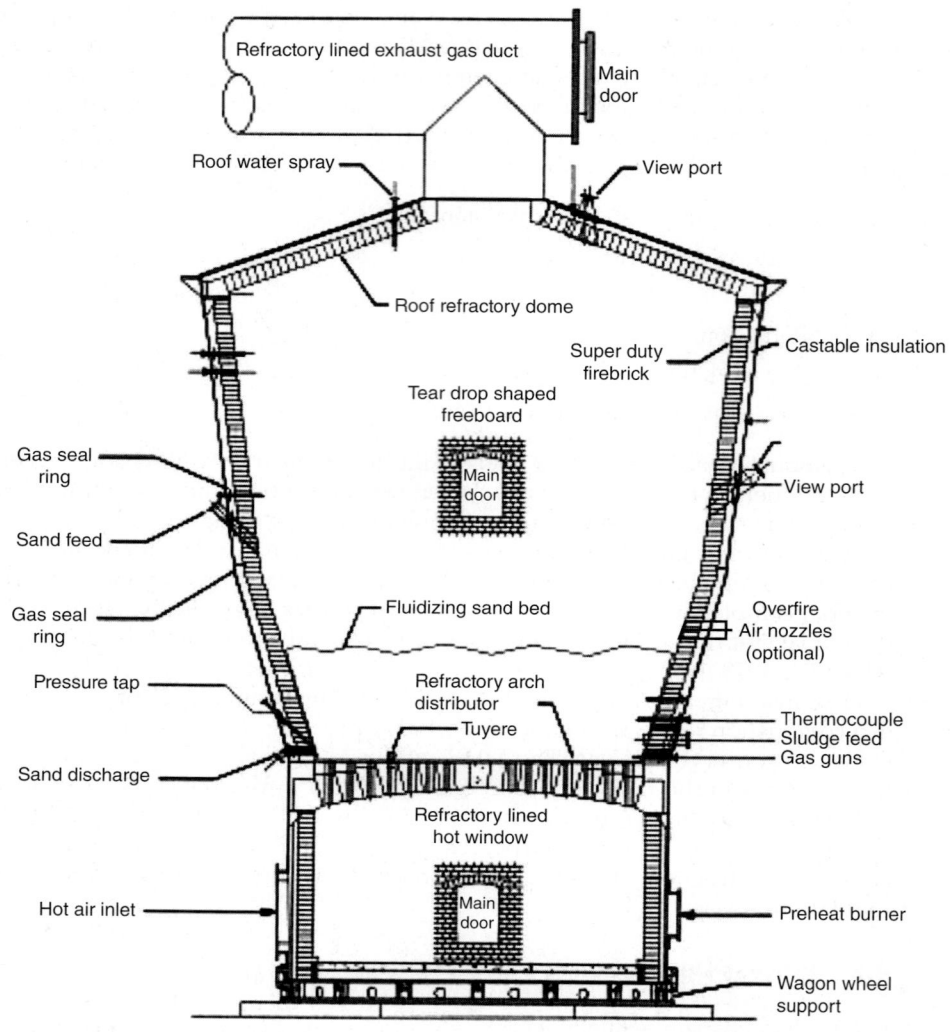

FIGURE 24.7 Typical section of a fluidized bed furnace (courtesy of Suez).

Treatment capacity, therefore, is a function of total reactor bed volume (although typically expressed as either fluidized-bed surface area or freeboard area) and the maximum gas velocity permitted because of particle entrainment.

At combustion equilibrium, the fluidized bed resembles a boiling liquid and obeys most hydraulic laws. A specially designed orifice plate or series of nozzles (tuyeres) are used to disperse fluidizing gas (air) throughout the fluidized-bed zone to ensure complete mixing. The orifice or nozzles (tuyeres) are designed to have sufficient pressure drop to provide even distribution throughout the cross section of the bed. When the bed is properly fluidized, temperature variations between any two spots in the fluidized bed typically will not exceed 6°C to 8°C (10°F to 15°F).

Dewatered cake is fed directly into the fluidized bed, which is preheated with aux-iliary fuel to about 650°C to 760°C (1200°F to 1 400°F) (WEF, 2009). In the bed, solids moisture evaporates, and the volatile fraction burns at temperatures of 760°C to 815°C (1400°F to 1500°F) (U.S. EPA, 1985). The fluidized-bed medium is kept at the combus-tion temperature via oxidation of organic material in the feed or combustion of auxil-iary fuel. Rather than flaming, the medium glows. Combustion is rapid, and the fluidized bed itself will contain negligible amounts of unburned organic matter. (Complete oxidation of organic matter is the first step in controlling air pollution.)

There typically is a temperature difference between the fluidized bed and the freeboard (disengaging zone) because a portion of volatile organic compounds (VOCs) will be combusted in the area above the bed, causing a temperature increase of 80°C to 170°C (150°F to 300°F). In other words, while the bed temperature typically ranges from 730°C to 815°C (1350°F to 1500°F), the freeboard temperature will be significantly hotter (840°C to 900°C [1550°F to 1650°F]). One means that has been developed to address the phenomenon of delayed combustion in fluidized-bed systems is to provide over-fire air nozzles to introduce a portion of the combustion air to the freeboard zone instead of the bed zone (Lundberg, 2004). These over-fire air nozzles can improve system perfor-mance, reduce auxiliary fuel usage, and improve gas-phase mixing in the freeboard zone, thus enhancing combustion efficiency.

Combustion gases and ash discharge from the top of the furnace. Entrained ash typically is separated from the combustion gases by a venturi scrubber or another particulate-removal device.

Fluidized-bed incinerators have thermally oxidized various wastewater solids successfully. They have several advantages:

- The smooth, liquid-like flow of particles allows continuous, automatically controlled operations with ease of handling.
- They are well suited to large-scale operations.
- Heat- and mass-transfer rates between gas and particles are high.
- Heat exchangers inside FBIs require relatively small surface areas because the heat-transfer rate between the incinerator and an object immersed in the bed zone is high.
- The incinerator can be shut down and started up daily without a preheating period because the bed cools slowly (3 to 8°C/h [6 to 15°F/hr]).

There are two types of FBIs: hot wind-box and cold (warm) wind-box. The hot wind-box systems typically incinerate wastewater solids with typically low heat values that require intensive air preheating to minimize auxiliary fuel consumption. The cold wind-box system typically is used to burn heat-treated solids, dried solids, primary solids, scum, grease, and other materials (e.g., wood chips or sawdust) that can burn autoge-nously without heat recovery or with moderate heat recovery.

4.2.1.1.1 *Hot Wind Box* The hot wind-box fluidized bed is designed for maximum tem-peratures of about 650°C to 980°C (1200°F to 1800°F) for most applications. It is used when the wind-box air temperature is greater than about 400°C (750°F). A cross section of a typical hot wind-box fluidized bed is shown in Figure 24.7. The unit is a vertical steel shell made of carbon steel. The inside lining is made of refractory and insulating brick. The refractory lining is necessary because of the inside temperature of approximately

980°C (1800°F). Combustion air is usually not preheated to more than about 675°C (1250°F), but these systems often use an under-bed burner to preheat the bed or support combustion, so the refractory lining must be designed for the higher temperatures that occur when the burner is in operation. The fluidized bed is composed of four sections: the wind box, the bed support and air distributor, the bed, and the freeboard.

The lower section of the furnace is the hot wind box, a refractory-lined plenum in which hot combustion air is received. It serves as a distribution chamber for fluidizing air and a combustion chamber for the preheat burner. The wall of the wind box has openings for fluidizing air supply, preheat burner, observation port, and instrument ports.

The roof of the wind box serves as a bed support and air distributor. Typically called the dome, it can be made of refractory or metal alloys, depending on design requirements. Its refractory arch (or metal bottom) is constructed to be self-supporting. The refractory arch has a special shape because of the refractory elements used. The dome (or metal bottom) supports the weight of the unfluidized-bed material and distributes fluidizing air via a number of air nozzles (typically called tuyeres). The tuyeres' special shape and material prevent sand drainage, provide uniform air distribution, and withstand furnace operating temperatures. Both dome and wind box are designed for about 980°C (1800°F); the combustion air typically is preheated to about 675°C (1250°F) or less.

The bed (combustion zone) contains the fluidized mass of sand. Air from the distributor causes the sand to fluidize. The height of the fluidized sand layer depends on the amount of sand in the bed. An expanded bed of about 1.5 m (5 ft) typically is used to thermally oxidize wastewater solids. The side walls slope outwardly from the bottom of the bed to ensure that water vapor expands and to keep gas velocity within acceptance limits. The walls are also equipped with nozzles for solids and auxiliary fuel injections, and ports for various instruments.

The space above the bed is called the freeboard (disengagement zone). It acts as both a gas-retention and particle-separation chamber. To ensure complete combustion of any volatile hydrocarbons escaping from the bed, the freeboard must be large enough to provide about 6.5 seconds of gas-residence time at a minimum. It is typically at least 4.6 m (15 ft) high. The freeboard could be shaped as a cylinder or a teardrop. The cylinder typically is designed based on a gas velocity of 0.76 m/s (2.5 ft/s). The teardrop is designed with the narrowest part at the bottom to maximize residence time and further reduce gas velocity. The gas velocity at the top of the teardrop is 0.64 m/s (2.1 ft/s). An exhaust-gas duct is installed in the center of the roof dome to minimize gas bypassing, minimize dead zones, and maximize residence time in the freeboard. To minimize sand losses, gas velocity is designed to decrease in the freeboard so particles drop out of suspension.

4.2.1.1.2   Cold (Warm) Wind Box   A cold (warm) wind-box thermal oxidizer is used for solids that can be incinerated without heat recovery (or with moderate heat recovery). The air temperature in the wind box typically is limited to less than 400°C (750°F). The designs for hot and cold (warm) wind-box systems are similar, except that in cold wind-box systems,

- The wind box is not refractory lined;
- The bed support and air distributor can be a metal alloy plate, which typically is refractory lined to sustain the bed's high temperature; and
- The preheat burner is installed in the freeboard and angled downward to heat the top of the fluidized sand bed during startup.

4.2.1.2   Multiple-Hearth Furnace   A MHF consists of a vertical, refractory-lined steel cylinder containing a series of stacked horizontal refractory shelves (hearths) (see Figure 24.8) (U.S. EPA, 1979, 1983b). A hollow cast-iron rotating shaft runs through the center of the hearths. Rabble arms are attached to this shaft above each hearth, and cooling air may flow through them. Metal blades (rabble teeth) on each arm may be angled in the direction of rotation (forward rabbling) or may be reversed (back rabbling). Back rabbling increases the cake's detention time and improves drying. Combustion air enters at the bottom of the hearth and circulates upward through drop holes in the hearths, countercurrent to solids flow. It exhausts through the top hearth to the waste-heat exchanger (if present) and then to the air pollution-control equipment.

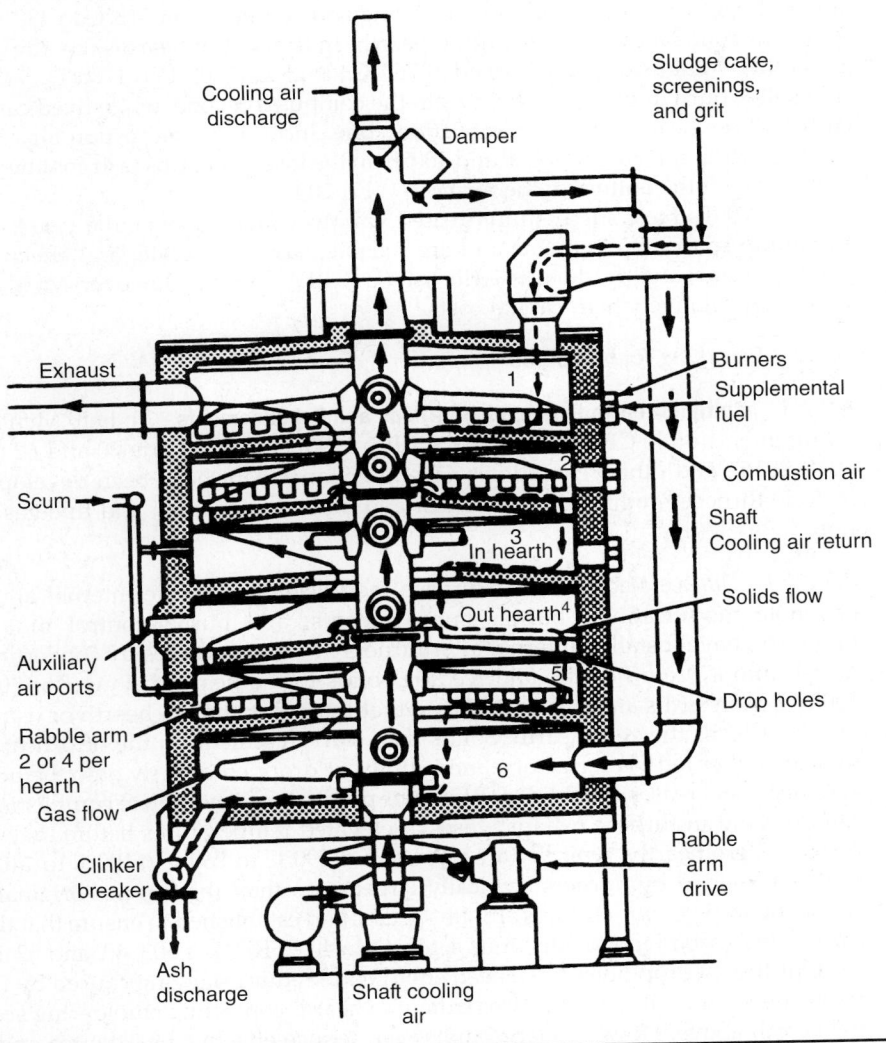

FIGURE 24.8 Typical cross section of a multiple hearth furnace.

Solids flow down through the furnace in a countercurrent pattern to the combustion gas and airflow, motivated by the rabble teeth, or plows, mounted on the rabble arms that are attached to the rotating central shaft. Solids move outward and inward on successive hearths, dropping through either peripheral holes in a hearth ("out-hearth") or an annular opening at the center ("in-hearth"). The rabble arms and teeth sweep solids forward and radially several times on each hearth, resting after each plowing action. In addition to moving solids, the rabble teeth cut, furrow, turn, and open them, thereby exposing new surfaces for drying. So, the effective drying surface is estimated to be 130% of the hearth's actual area (Niessen, 1988; U.S. EPA, 1979).

A MHF has three zones: drying, combustion, and char burning and cooling. Most of the water in solids evaporates in the upper hearths (drying zone). The heat-transfer mechanisms in this zone are about 15% convection and 85% radiation (Lewis and Lundberg, 1988). The temperature in the drying zone ranges from 315°C to 480°C (600°F to 900°F) as solids move from hearth to hearth. In the central hearths (combustion zone), the solids' combustibles are burned at 760°C to 925°C (1400°F to 1700°F). Volatile gases and solids burn in the upper hearths in the combustion zone, while fixed carbon burns in the lower hearths in the combustion zone. Incoming combustion air cools ash to between 95°C and 205°C (200°F and 400°F) in the lowest hearths (ash cooling zone). Ash exits through the bottom of the furnace (WEF, 2009).

The MHFs have successfully treated various solids for over 100 years. They have minimum space requirements and are reliable, easy to operate, and effective; reduce solids volume; and produce a sterile ash (U.S. EPA, 1983a). However, MHF-based systems have relatively high capital costs.

4.2.2   Emerging Technologies

4.2.2.1   Multiple-Hearth Furnace Innovations and Enhancements   To help MHFs meet the requirements of 40 CFR Part 503 (U.S. EPA, 1993), 40 CFR, Parts 60 and 62 (U.S. EPA, 2011, 2016), and other regulations, several enhancement have been developed. These include furnace modifications, a reheat-and-oxidation process, and flue-gas recirculation.

4.2.2.1.1 Furnace Modifications   Furnace modifications include internal afterburners, drophole modifications, burner improvements, and burner-control modifications. Engineers have installed internal afterburners in many furnaces by converting the top hearth into a "zero hearth" and feeding solids to the hearth below (Dangtran et al., 2000). Zero hearths are constructed by either removing the top hearth or using the top two hearths as the zero hearth. Solids then start treatment on the next hearth below; they are either fed via refractory-lined chutes or dropped directly onto the hearth. The zero hearth increases the residence time for products of incomplete combustion so they can burn out in furnace exhaust gases at elevated temperatures before they leave the furnace. Zero hearths typically are operated at 590°C to 760°C (1100°F to 1400°F); this heat is provided by burners (typically larger ones than the furnace originally used). Actual operating temperatures are site specific and established to ensure that the hydrocarbon and carbon monoxide limits set forth in 40 CFR, Parts 503, 60, and 62 are met.

Out-hearth drop holes have been enlarged to reduce slagging caused by the "blow torch" effect of high-velocity gases from one hearth contacting smoldering solids from the hearth above. However, larger holes can reduce effective hearth area and has met with mixed success.

Manufacturers have improved burner design to reduce burner tile and refractory slagging. One of these burners, which can operate on dual fuel, has been used in many furnaces (Nuss et al., 2008). Also, various improvements to burner controls have been implemented. One modification, called the "on-the-fly air:fuel ratio control," keeps airflow to the burner at the maximum rate and adjusts the fuel rate to match heat requirements. This maintains maximum turbulence for good solids combustion (O'Kelley et al., 2006).

Another furnace enhancement is installing variable-frequency drives on induced draft fans to replace dampers and improve furnace-draft control and electrical efficiency. Yet another improvement is replacing venturi/impingement scrubber systems with post-quench/impingement multiple venturi scrubbers to improve particulate removal efficiency and reduce power consumption.

4.2.2.1.2 Reheat and Oxidation Process   The reheat and oxidation process is a practical, economical means of producing high-quality exhaust gas while maintaining the conditions needed to combust wet solids (RHOX, 1989). In this process, a regenerative thermal oxidizer (RTO) or recuperative heat exchanger with multiple separate vessels and a fired afterburner is installed downstream of the scrubber to reduce emissions of total hydrocarbons (THCs), carbon monoxide, dioxin, and furan (see Figure 24.9). The fired afterburner is normally equipped with a low-NO_x burner to avoid the introduction of excessive "burner NO_x" into the exhaust gas stream. The RTO uses its high heat-transfer efficiency to raise the scrubber's exhaust gas temperature from about 38°C (100°F) to about 540°C to 650°C (1000°F to 1200°F).

The RTO is more fuel efficient than a conventional external or internal afterburner. In a conventional afterburner, the mass load through the afterburner includes the water evaporated in the MHF. So, if the furnace exhausts gas that is about 480°C (900°F) and the afterburner temperature is about 730°C (1350°F), the temperature of the entire mass

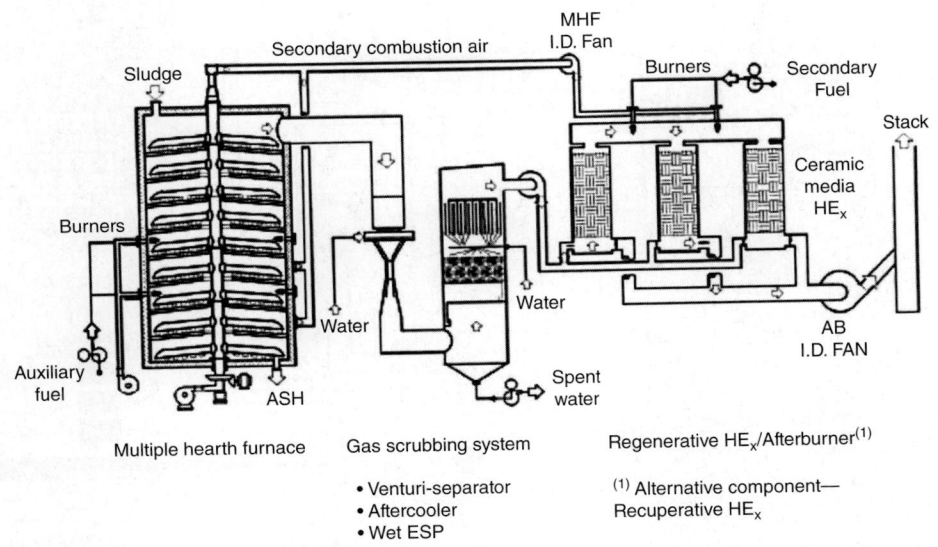

FIGURE 24.9   Reheat and oxidation process flow sheet for a multiple-hearth furnace with regenerative heat exchanger (RHOX, 1989).

of exhaust gas has to be raised about 250°C (450°F). The RTO, on the other hand, only has to raise the temperature of the saturated gas, which contains much less water, from about 540°C to 650°C (1000°F to 1200°F)—a difference of about 110°C (200°F). Also, the mass loading of scrubber exhaust gas is about 70% of the mass loading of furnace exhaust gas.

The exhaust gas must have low particulate concentrations to avoid fouling the ceramic surfaces of the heat exchangers. So existing scrubbers may have to be augmented by a wet electrostatic precipitator or replaced by a scrubber.

4.2.2.1.3 Flue-Gas Recirculation   In flue-gas recirculation, ducting and fans recirculate flue gas from the top hearth to one near the bottom (see Figure 24.10). Typically, two sets of ducts and fans are provided. The temperature and flowrate of recirculated gas is measured by a flow meter in the duct. The flowrate is controlled by dampers or a variable-speed fan. Cooling air is provided to the flue-gas recirculation system to prevent overheating. Flue-gas recirculation offers several benefits (Sapienza et al., 2007):

- More stable furnace operations;
- Lower NO_X emissions (elimination of yellow plume);
- Less slag formation in the MHF;

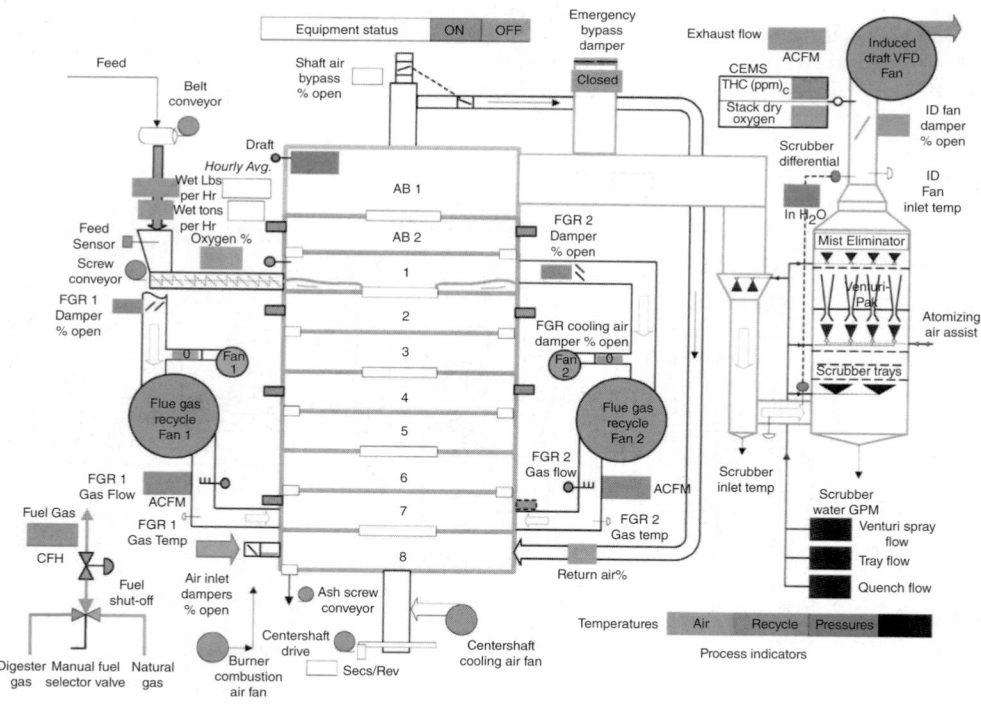

FIGURE 24.10   Schematic of a multiple-hearth furnace with flue gas recirculation (Porter and Mansfield, 2002).

- Higher throughput capacity (because of less downtime due to slag removal);
- Reduced THC emissions (because of more stable furnace operations); and
- Complete ash burnout (because recirculated gas raises lower hearth temperature).

4.3 Process Design Guidelines

Because thermal oxidizers should produce minimal air-pollution emissions, FBIs typically are preferred to MHFs. For the past few decades, wastewater utilities have installed far more FBIs than MHFs, and even have replaced multiple-hearth units with fluidized-bed ones. So this section focuses on FBIs.

4.3.1 Furnace Sizing Parameters

Combustion (burning) is the rapid union of oxygen with carbon, hydrogen, and sulfur. Complete combustion primarily depends on temperature, time, and turbulence.

4.3.1.1 Temperature Feed solids should be introduced to the fluidized bed when the bed's operating temperature exceeds the material's ignition temperature. Ignition temperatures vary. Some compounds (e.g., chlorinated compounds) require temperatures hotter than 980°C (1800°F) to oxidize. Wastewater solids typically can be completely combusted when the bed temperature ranges from 650°C to 760°C (1200°F to 1400°F) and the freeboard temperature is between about 815°C and 870°C (1500°F and 1600°F).

The bed's huge thermal inventory ensures that feed solids quickly reach the bed temperature, at which point organics volatilize and most VOCs oxidize, releasing heat and maintaining the thermal inventory. The small amount of fixed carbon in solids typically is combusted in the bed. The bed temperature is stable because the large heat inventory and the ease of automatic control ensure that the necessary combustion temperature is maintained.

4.3.1.2 Time Combustible materials must have sufficient time to react. Fluidized-bed thermal oxidizers are designed to allow enough time for feed solids and any auxiliary fuels to react with oxygen in fluidizing air. The bed is designed to completely disintegrate feed solids in a fraction of second and combust some of its volatiles to keep temperatures above 650°C (1200°F).

Residual volatiles finally burn out in the elevated temperature zone above the bed (the freeboard). With temperatures between 840°C and 900°C (1550°F and 1650°F), the freeboard is designed to completely combust any volatiles that escape from the bed. Typically, gas residence times are 2 to 3 seconds in the bed and 6 to 7 seconds in the freeboard. Although high combustion efficiencies can be achieved at lower freeboard residence times, design engineers should provide enough disengagement height (5 to 6 m [15 to 18 ft]) in the freeboard to reduce sand carryover in exhaust gases, and doing this lengthens freeboard residence times.

4.3.1.3 Turbulence To optimize combustion (operate efficiently with low excess air), combustible materials and fluidizing air must be mixed so enough of the combustible material's surface area will contact oxygen and react. Turbulence better distributes feed solids in the bed and exposes every particle of the fluidized bed to fluidizing air. The highly agitated hot sand quickly fragments feed material into small particles, which in turn are quickly heated to volatilization temperature without significantly lowering bed temperature (because of the sand bed's large heat inventory).

Without turbulence, feed material will be poorly distributed and more volatilized organics will reach the freeboard before being oxidized in the bed. This phenomenon can lead to excessive over-bed burning and, subsequently, higher emissions of hydrocarbon volatiles and other products of incomplete combustion.

A certain amount of freeboard burning is inescapable; however, freeboard-to-bed differentials should not exceed 140°C to 170°C (250°F to 300°F). The goal should be less than 110°C (200°F) unless temperatures higher than 840°C (1 550°F) are required to completely combust VOCs.

4.3.1.4 Space Velocity   When designing fluidized-bed systems, the selection of bed material is critical because particle size directly affects fluidization quality, as illustrated in Figure 24.11 (Geldart, 1973). Thermal oxidizers treating wastewater solids typically use a sand-like material with a median size of 550 μm (30 mesh). At bed operating conditions, such particles have a minimum fluidization gas velocity (U_{mf}) of 0.33 m/s (1 ft/s). The optimal bed fluidizing gas velocity is between $2.5U_{mf}$ and $3U_{mf}$, so the particles' gas spatial velocity (fluidization gas superficial velocity) would be 0.75 to 1 m/s (2.5 to 3 ft/s). When sizing the bed, design engineers should use gas velocities corrected to bed temperature and pressure. Bed velocities greater than 1.0 m/s (3.0 ft/s) are not recommended for solids combustion because fluidization is less stable, and bed losses, caused by entrainment, are higher. If higher velocities are used, coarser bed material is required.

Freeboard gas velocity typically ranges from 0.76 to 0.64 m/s (2.5 to 2.1 ft/s). To minimize sand loss, freeboard gas velocity must be kept as low as possible.

4.3.1.5 Evaporation/Heat Release Limitations   Fluidized-bed furnaces typically burn up to 552 kJ/svm² (1.75×10^5 Btu/hr/sq ft) of freeboard area. High-velocity FBIs can process

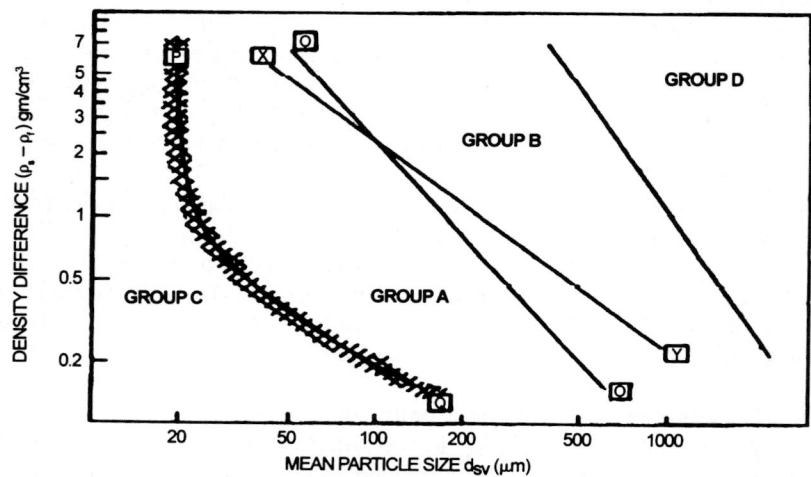

FIGURE 24.11   Geldart classification of powders.

up to 1260 kJ/s·m² (4.0×10^5 Btu/hr/sq ft) in some installations. Coal-burning boilers, on the other hand, only consume 126 kJ/s·m² (4.0×10^4 Btu/hr/sq ft) of grate area.

The feed material's water content significantly affects thermal oxidizer operations. Water does not release heat in a combustion process. Quite the opposite is true: water in the feed material requires a large amount of heat to be vaporized and heated to the oxidizer's operating temperature. This heat must be supplied by combustible materials in the feed solids or by auxiliary fuel.

Changes in water content will affect the thermal oxidation system's operating temperature and capacity. When feed solids are wetter than the design condition, more auxiliary fuel will be required to maintain bed temperature, reducing system capacity. To maintain design capacity and optimize operations, the percentage of total solids in feed cake should be monitored twice per day.

4.3.1.6 Sufficient Air A fluidized-bed thermal oxidizer receives oxygen in the form of fluidizing (combustion) air, which must be supplied at amounts slightly greater than that theoretically required for complete combustion. Operators typically monitor excess air levels measured as the percent of free oxygen in the air released to the atmosphere. Depending on feed solids and combustion temperature, the oxygen content in exhaust gases should be at least 4% by volume on a dry basis (about 2% on a wet basis) just before the gas enters the scrubber.

For efficient operations, the oxygen content in exhaust gases should not exceed 5% (on a wet-gas basis). Anything more than 5% will quench (cool) the bed and reduce overall efficiency.

4.3.1.7 Summary If thermal oxidizer operations violate one or more of the six principles of combustion, then combustion will be poor or incomplete. The following criteria indicate that a fluidized-bed system is being operated inefficiently and, therefore, ineffectively:

- Excess oxygen below prescribed minimum;
- Black and smoky exhaust gases;
- High residual carbon content in bed material;
- High carbon content (black color) in scrubber effluent; and
- Excessive freeboard-to-bed differential temperatures.

4.3.2 Selection and Sizing

Selecting and designing a thermal oxidation system to treat solids is a complex, technical, and highly specialized task. A computer can best handle the iterative heat-transfer calculations. Design procedures include empirical methods that rely on extensive pilot-plant kinetic data. Such information typically is regarded as proprietary and may be patented by the manufacturers who typically perform the mass and energy balance and design calculations after receiving input data, conditions, and functional specifications from the operating agency (see Table 24.3).

A thermal oxidation system should be able to burn dewatered cake, concentrated scum, and other feed materials to an inert ash without emitting smoke, flash, or objectionable odors. Through a variety of regulations (40 CFR Parts 503, 60, and 62), the U.S.

Dewatered solids characteristics		
Description of the feed, including its nature, material handling, pumpability and hazardous constituents POHCs if applicable.		
Furnace design capacity (kg DS/h [lb DS/hr])		
Moisture content (%)		
Volatile solids (%)		
Heating value (kcal/kg [Btu/lb] volatile)		
Bulk density (kg/m³ [lb/cu ft])		
Ultimate analysis (volatile basis): concentrations of the following volatiles: C, H, O, N, S, Cl		
Trace metals: concentration of regulated metals in dry solid in ppm weight		
Solids ash chemical compositions, including soluble and insoluble parts of the following elements: Al, Na, K, P, Mg, Ca, Fe, Si, S		
Auxiliary fuel characteristics		
Description of the fuel, including its nature, viscosity (if liquid), material handling, pumpability		
Heating value (kcal/kg [Btu/lb] of the fuel)		
Ultimate analysis (volatile basis): concentrations of the following volatiles: C, H, O, N, S, Cl		
Moisture content (%)		
If steam is desired, the following are required:		
Steam temperature (°C [°F])		
Steam pressure (barg [psig])		
Air pollution-control system		
Maximum pollutant emissions		
Maximum or minimum allowable flue gas temperature		
Process water analysis (e.g., flowrate, temperature, alkalinity, and TSS)		
Environmental conditions (e.g., seismic zone, elevation, design air temperature and moisture, and design windload [if outdoor])		

TABLE 24.3 Typical Parameters for Selecting and Designing a Thermal Destruction System for Municipal Solids (courtesy of Suez)

Environmental Protection Agency and/or local agencies may require that thermal oxidation systems:

- Combust dewatered cake, concentrated scum, and other feed materials satisfactorily within an exhaust gas temperature range of 427°C to 870°C (800°F to 1600°F);

- Burn dewatered cake, concentrated scum, and other feed materials to a sterile ash that contains no more than 2% to 5% combustibles;

- Produce flue gas that contains between 6% and 11% oxygen;

- Limit the temperature of discharged exhaust gases to the atmosphere to a range from 65°C to 120°C (120°F to 250°F);

- Prevent the discharge of exhaust gases whose opacity exceeds 20% to the atmosphere; and

- Do not emit more than 0.65 g/kg of feed solids (1.3 lb/dry ton of feed solids) of particulate to the atmosphere or from 9.6 to 80 mg/dscm (specific value dependent on furnace design type and vintage), whichever is less.

4.3.2.1 Furnace Selection When selecting a furnace, design engineers should consider plant size, solids composition, air emission regulations, facility maintenance, sustainability, and labor requirements. Thermal oxidization typically is feasible for facilities with flows more than 3.79×10^4 m^3/d (10 mgd); that said, site-specific factors (e.g., facility location, land available for solids reuse or disposal, and air emission regulations) may make the other option more attractive. Fluidized-bed furnaces typically are more economical if high-temperature combustion of exhaust gases is required, because MHFs require afterburners to achieve high temperatures. A fluidized-bed furnace has lower capital and O&M costs (amortized to 20 years) than a MHF equipped with RTO (Dangtran et al., 2000).

Since 1988, 53 fluidized-bed systems and one multiple-hearth system have been installed at North American water resource recovery facilities. Of the fluidized-bed installations, 18 replaced existing MHFs.

4.3.2.2 Equipment Sizing Design engineers must size thermal oxidation systems based on realistic design loadings. Using overly conservative peaking factors for solids loadings and flows results in oversized systems that fail to operate efficiently during the initial years, when solids quantities and contents are lower than design values. Oversized systems must be operated intermittently, which increases auxiliary fuel costs (due to repeated shutdowns and startups, or maintaining a constant furnace temperature in standby mode). To avoid such problems, design engineers should consider furnace systems that can be modified incrementally to increase capacity, or multiple units that can enter service incrementally as solids production increases.

4.3.2.3 Effect of Dewatered Solids Characteristics on Selection and Sizing

4.3.2.3.1 Moisture Content Wastewater solids with high moisture contents have complicated solids processing at older facilities, particularly those constructed before 1980 (U.S. EPA, 1985). High moisture contents reduce a furnace's equivalent dry solids throughput capacity and require more auxiliary fuel to evaporate water before and during combustion. Water resource recovery facilities that dewater solids via efficient belt presses, recessed plate filter presses, or high-solids centrifuges produce drier cakes. When designing thermal oxidation systems, engineers should consider replacing existing, inefficient dewatering systems with more effective ones. Alternatively, consideration could be given to recovering thermal energy from the furnace exhaust gases and using it in a thermal dewatering system to "scalp" sufficient moisture from the dewatered cake to make the feed to the furnace self-supporting (autogenous) to eliminate or significantly reduce auxiliary fuel input requirements of the thermal oxidation system.

4.3.2.3.2 Solids Composition Feed cake properties that affect thermal oxidizers include solids content, the percentage of combustibles (volatile solids plus fixed carbon), the heating value of combustibles, and the presence of chemicals that react endothermically (e.g., lime). Because water resource recovery facility screenings tend to clog feed mechanisms, they are ground or shredded before thermal oxidation. Wastewater solids with high percentages of volatile solids (e.g., grease and scum) have high heating values (HHVs) that can cause localized heat release and slagging, resulting in clinkering and air emissions associated with incomplete combustion. A fluidized-bed furnace can accommodate grease, scum, and other high-caloric materials better than a MHF because it provides better contact with combustion air.

Grit and chemical precipitates (e.g., lime and ferric chloride) contain a large percentage of inert materials with low heating values and high ash production, so they should not be thermally oxidized unless necessary for odor control. Also, because the heating value of combustibles affects the amount of auxiliary fuel required for complete combustion, engineers should examine ranges of expected values to ensure that the thermal oxidation system will meet present and future needs.

4.3.2.3.3 Inorganic Sludge Conditioning   Conditioning solids with inorganic chemicals can increase O&M costs, fuel consumption, and corrosion. Inert conditioning chemicals produce more ash and promote the formation of metal salts, resulting in slagging and clinkering; this can increase O&M costs. Adding inert conditioning chemicals with low heating values or an endothermic lime reaction probably will increase fuel consumption. Polymers may cause uneven burning in a MHF (unlike a fluidized-bed furnace, because of its turbulence and huge heat reservoir), leading to the formation of combustion-resistant solids balls. Ferric chloride can produce a chlorine-laden exhaust gas that is extremely corrosive to steel surfaces at high temperatures. To reduce these adverse effects, design engineers should optimize the use of conditioning chemicals, improve the dewatering process, use polymers rather than lime or metal salts, and use ferric sulfate rather than ferric chloride.

Design engineers also should evaluate the solids' chemical composition. Sodium and potassium chlorides have low melting points, so large quantities of them can lead to vitrification of bed media (i.e., the bed media can become sticky, agglomerates [clinkers] can form, bed materials segregate, and eventually the bed de-fluidizes). Iron, phosphorus, and chlorides can lead to the deposition of iron oxides. Iron oxide scaling can obstruct the exhaust gas duct, potentially resulting in excessive backpressure, operating difficulties, and system shutdown. These problems can be eliminated by chemical addition (Dangtran et al., 1999).

Kaolin clay (a mixture of hydrous aluminum silicates) can neutralize sodium and potassium. It typically is available as a fine powder and is a convenient source of both silicate oxide and aluminum oxide, which react with sodium and potassium chlorides to form crystalline sodium and potassium aluminum silicates. These silicates have a melting point above 1093°C (2000°F).

Lime can convert iron phosphate into iron oxides in the sand bed at bed temperature. This conversion prevents iron from forming gaseous iron chloride, which can precipitate and form scales in the freeboard and exhaust gas duct.

To calculate the appropriate dose(s) of such chemical additives, design engineers must perform a complete ash analysis, including both soluble and total concentrations of the components (Dangtran et al., 1999).

4.3.2.4 Fuel   Optimization   One of the most important criteria in designing a thermal oxidation system is minimizing fuel consumption. Engineers calculate auxiliary fuel needs based on heat and mass balances; the result depends on the feed cake's water content, volatile solids content, the heating value of the combustible solids, the amount

Basis: Excess air 40%, feed rate 450 kg DS/h, 75% VS, 5550 kcal/kg VS, 850°C
freeboard, auxiliary fuel HHV 10 500 kcal/kg

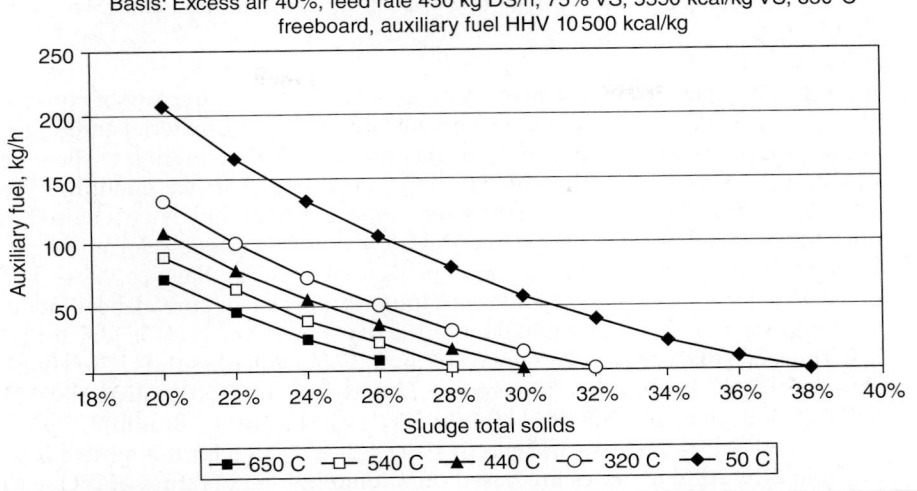

FIGURE 24.12   Theoretical fuel requiment versus cake solids content at various combustion air temperatures (courtesy of Suez).

of excess air used for combustion, combustion air supply temperature, and the exhaust temperature exiting the thermal oxidizer. The feed cake's solids content depends on the dewatering equipment used and the quantity of polymer used as a dewatering aid. Volatile solids content depends largely on whether the cake is digested or undigested. The combustion air supply temperature depends on the amount and efficiency of heat exchange from the exhaust to the incoming combustion air. Typically, a heat exchanger can recover up to about 50% of the flue gas' available enthalpy to preheat combustion air to about 675°C (1250°F).

A theoretical curve of supplementary fuel consumption is presented in Figure 24.12. The calculations were based on a combustion gas temperature of 843°C (1550°F) and a throughput capacity of 454 kg/h (1000 lb/hr) of dry solids. The feed cake contains 75% volatile solids and a HHV of 23 260 kJ/kg (10 000 Btu/lb) of volatile solids (values typical for wastewater solids).

The auxiliary fuel requirement drops as the solids content and combustion air temperature increase. When the wind-box temperature is 648°C (1200°F), feed cake containing 27% solids burns autogenously (i.e., is thermally self-supporting).

Meanwhile, to lower polymer consumption and minimize NO_x emissions, which increase as dry solids increase (Dangtran and Holst, 2001), a thermal oxidizer typically is designed based on autogenous combustion at maximum air temperature and minimum dry solids content.

4.3.2.5   Air Emission Objectives   If inadequately designed or operated, thermal-processing facilities for wastewater solids can contribute significantly to air pollution. Two categories of air contaminants associated with thermal processing are odors and combustion emissions. Engineers should design thermal oxidation systems to generate no odors and to meet the requirements of the air permit. Odors are controlled by keeping the furnace's exhaust temperatures above the odor-destruction threshold temperature. This will be site specific. Combustion emissions are controlled by maintaining good combustion efficiency and by installing appropriate air pollution-control equipment (see Section 6.0).

4.3.2.6 Mass and Energy Balance   A mass balance applies the law of conservation of mass to the analysis of physical systems. By accounting for material entering and leaving a system, design engineers can identify mass flows that might have been unknown or difficult to measure otherwise. Similarly, an energy balance quantifies the energy used or produced by a system. However, while a system may have a closed mass balance, its energy balance may be open. Also, while it is possible for a system to have more than one mass balance, it can have only one energy balance. Mass and energy balances typically are performed via computer programs or spreadsheet models.

Suppose a FBI is designed to handle 105 dry metric ton/d (115.5 dry ton/d) of solids. The feed cake contains 28% total solids and 68% volatile solids. The volatile solids consist of 55.83% carbon, 7.52% oxygen, 8.13% hydrogen, 27.10% nitrogen, 1.42% sulfur, and 0.1% chlorine. The volatiles' HHV is 5560 kcal/kg (10 007 Btu/lb).

A summary of the mass and energy balances calculated for this system is presented in Table 24.4. All enthalpies are based on a reference temperature of 0°C (32°F). Feed cake provides about 74% of the inlet heat required to sustain combustion; the remaining 26% is provided by preheated air. With a hot wind-box temperature of 650°C (1200°F) and a feed material containing 28% total solids, no auxiliary fuel is required during normal operations. Most of the exhaust-gas heat is in the water vapor (63%). The rest is in the dry combustion gas (34%), ash (1%), and heat loss (2%).

Mass and energy balances for MHF systems are significantly more complex to perform, since this type of thermal oxidizer is a multistage process. As a result, mass and energy balances must be conducted on a hearth-by-hearth basis and are most often provided by the furnace manufacturer. Hearth-by-hearth analysis is necessary to assure that temperatures in the combustion zone (middle hearths) do not exceed 870°C (1600°F) and can also be used to determine the appropriate points (hearths) in the system to introduce combustion air and fuel to optimize performance (Lewis and Lundberg, 1988). Limiting the temperature of the "hottest hearth" is necessary both for emissions control as well as to prevent the formation of slag or "clinkers" in the furnace.

4.3.2.7 Operating Schedule and Redundancy   Thermal equipment typically should not be heated and cooled frequently because cycling temperatures increase the material's fatigue and rate of failure. Longer run times also save on energy and resource costs associated with heating up the system. However, fluidized-bed furnaces have minimal heat loss during shutdown (around 5°C/h [10°F/hr]) because of their thick refractory insulation and mass of granular sand. The hot granular bed material can act as a thermal reservoir, retaining the heat during shutdown, so the system can be operated intermittently. Existing fluidized-bed furnace installations have varied operating schedules, ranging from 24 h/d in some facilities to 16 h/d or 5 d/wk. This allows downtime for preventive maintenance without exposing too much of the refractory layer to temperature cycling.

Redundancy is not common at fluidized-bed furnace systems, except for very large installations, where an entire redundant process train may be provided. Normally, backup or redundant components (fans, blowers, pumps) are not installed unless they are critical to enhance system safety or to protect major system equipment under abnormal operating conditions, such as a power outage or a major fan or blower failure.

Mass and Energy Input	kg/h (lb/hr)	kcal/h (Btu/hr)	Heat (%)
Cake solids	4375 (9645)	16 560 068 (65 710 349)	73
Cake water	11 250 (24 802)	236 250 (937 440)	1
Aux fuel	0 (0)	0 (0)	0
Quench water	33 (74)	701 (2781)	0
Purge air (dry)	1085 (2392)	12 931 (51 309)	0
Fluidizing air (dry)	32 630 (71 931)	5 405 955 (21 450 830)	24
Air moisture	635 (1400)	574 696 (2 280 394)	2
Sand	51 (112)	236 (935)	0
Total in	**50 059 (110 360)**	**22 790 836 (90 434 038)**	**100%**
Mass and Energy Output	**kg/h (lb/hr)**	**kcal/h (Btu/hr)**	**Heat (%)**
Dry flue gases	34 516 (76 095)	7 642 654 (30 326 049)	34
Water vapor	14 095 (31 073)	14 398 243 (57 132 228)	63
Ash	1400 (3086)	261 800 (1 038 822)	1
Sand	51 (112)	9537 (37 843)	0
Heat losses		478 603 (1 899 095)	2
Total Out	**50 062 (110 367)**	**22 790 836 (90 434 038)**	**100%**

TABLE 24.4 Example of a Mass and Energy Balance for a Fluidized Bed Furnace (courtesy of Suez)

4.3.2.8 Ash Handling Thermal oxidation facilities generate two types of ash—wet and dry—so there are two types of ash-handling systems, which are briefly described below. For more information on ash-handling systems, see *Wastewater Solids Incineration Systems* (WEF, 2009).

4.3.2.8.1 Wet Ash-Handling System Wet ash-handling systems are hydraulic and remove ash as a slurry. They typically are used in fluidized-bed systems with wet scrubbers because most of the ash removed from the process train is wet. In fluidized-bed thermal oxidizers, ash slurry is drained from the bottom of the wet scrubber and conveyed (via pump or gravity) to a lagoon or to mechanical thickening and dewatering equipment.

4.3.2.8.2 Dry Ash-Handling System   Dry ash-handling systems may be mechanical or pneumatic. Pneumatic systems may be either pressure or vacuum and either dilute or dense phase. They are suitable for fluidized-bed thermal oxidizers with waste-heat boilers, fabric filters, multiple cyclone separators, or electrostatic precipitators because most of the captured fly ash is dry. Bottom ash systems may need a grinder to ensure that ash can be transported effectively and protect downstream conveyance equipment. Dry ash typically is kept in storage bins until it can be hauled offsite. While being discharged from bins into disposal trucks, the ash typically is conditioned (wetted with water) to reduce fugitive dust during loading. Then, it is transported to the ultimate use or disposal site.

4.3.3   Electricity Requirements
Suppose a FBI treats 100 dry metric ton/d (110 dry ton/d) under positive pressure, with primary heat recovery for air preheat and a venturi scrubber. It would have an operating horsepower of 684 kW (917 hp) and, therefore, consume 164 kW/dry metric ton of electricity (see Table 24.5). The fluidizing air blower would use the most power.

Equipment	Time Operation (%)	Motor Horsepower	Average Operating Horsepower
Air			
Fluidizing blower	100	700	626
Preheat burner air blower	0	20	0
Oil injection blower	100	20	17
Gas injection blower	100	20	17
Sand air compressor	20	15	2.5
Instrument air compressor	100	40	34.5
Oil			
Preheat burner pump	0	2	0
Bed injection pump	10	2	0.1
Water			
Roof spray pump	20	20	3.4
Scrubber booster pump	10	20	1.7
Solids			
Hopper sliding frame	100	20	17
Hopper extraction conveyor	100	20	17
Feed pump	100	200	177
Pipe lube pump	100	5	3.7
Total operating horsepower			917 hp (684 kW)

TABLE 24.5   Horsepower List for a 100-Dry Metric Ton/d Fluidized Bed Furnace with Push System, Primary Heat Exchanger, and Wet Scrubber (courtesy of Suez)

Suppose instead that this system included superheated steam production via a water-tube waste-heat boiler and associated boiler feedwater, condensate, and water treatment pumps (see Table 24.6). The boiler also had a bottom screw to remove ash from the economizer and boiler, as well as induced draft fan to keep negative pressure in the boiler and avoid ash leakage. Then, it would use 50% more electricity

Equipment	Time Operation (%)	Motor Horsepower	Average Operating Horsepower
Air			
Fluidizing blower	100	600	536
Induced draft fan	100	500	447
Oil injection blower	100	20	17
Gas injection blower	100	20	17
Sand air compressor	20	15	2.5
Instrument air compressor	100	40	34.5
Oil			
Preheat burner pump	0	2	0
Bed injection pump	10	2	0.1
Water			
Roof spray pump	20	20	3.4
Scrubber booster pump	10	20	1.7
Solids			
Hopper sliding frame	100	20	17
Hopper extraction conveyor	100	20	17
Feed pump	100	200	177
Pipe lube pump	100	5	3.7
Steam			
Boiler feed water pump	100	75	66
Boiler sootblowers	40	0.17	0.2
Chemical feed pump	100	1.5	0.8
Chemical feed mixer	20	1.5	0.2
Condensate pump	100	15	12
Dearator recirculation pump	100	7.5	6
Water treat feed pump	100	7.5	6
Water treat concentrate pump	100	.75	0.3
Ash screw conveyor	100	1	0.4
Total operating horsepower			1365 hp (1018 kW)

TABLE 24.6 Horsepower List for a 100-Dry Metric Ton/d Fluidized Bed Furnace with Push System, Primary Heat Exchanger, and Wet Scrubber (courtesy of Suez)

(244 kW/dry metric ton). However, if some of the generated steam were used in a turbine to drive the fluidizing air blower, its electricity consumption would drop to 141 kW/dry metric ton.

4.4 Design Practice

4.4.1 Feed Equipment and Systems

Continuous, even transport and distribution of feed cake to the furnace is important. Past practice was to use screw extrusion feeders for dry cake and progressing cavity pumps for wet cake. Now, hydraulic piston or progressing cavity pumps typically are used to convey cake from the dewatering equipment to the furnace. A piston pump is preferred because of its flexibility (insensitivity to feed solids quality).

Two types of solids-feeding systems can be found in the literature: over-bed and in-bed. In over-bed feeding, solids either drop via gravity or are air sprayed onto the bed from the freeboard sidewall or furnace roof. In in-bed feeding, solids are conveyed at high pressure directly into the sand bed about 1.2 m (4 ft) below the bed surface.

Over-bed feeding is simple but prone to bypassing. Solids can end up unburned in the exhaust, leading to excessive fuel use in the bed, explosions in the freeboard, or explosions in the heat-recovery and air pollution-control equipment. This method typically is used in other applications (e.g., fluidized-bed boilers burning coal or another solid-waste fuel); a cyclone at the exhaust end of the furnace recycles unburned carbon back to the bed.

In-bed feeding typically is used for thermal oxidation of solids because the combustion process is slower and in two stages (evaporation and combustion). The feeding location (1.2 m [4 ft] under the bed surface or 30 cm [1 ft] above the distributor) ensures that the maximum possible SRT is obtained before solids particles reach the bed surface, where carryover may occur. Solids should release their maximum energy to the sand bed, to counteract the quenching effect due to water evaporation.

The number of feed points needed to evenly distribute solids throughout the bed depends on the furnace's diameter. Feed ports typically are arranged in pairs (e.g., two or four).

4.4.2 Process Train Equipment

Fluidized-bed thermal oxidation systems could be located indoors or outdoors. Puerto Nuevo, Puerto Rico, and Naugatuck, Connecticut, have outdoor systems that consist of a hot wind-box fluidized bed, a heat exchanger to preheat combustion air to about 675°C (1250°F), a quench section followed by a wet scrubbing system, a wet electrostatic precipitator, and a stack (see Figure 24.13). The Puerto Nuevo system processes 2315 kg/h (5325 lb/h) of dry solids. In addition, the fluidized-bed system in Naugatuck, Connecticut has a thermal oil-based heat recovery system (located just downstream of the air preheater) that provides thermal energy that can be used to thermally dewater the furnace feed and/or to preheat the feed to the dewatering centrifuges.

The storage and feed system typically consists of a live-bottom bin and piston pumps. The thermal oxidation process could be either a hot or a cold wind box. The heat-recovery system could consist of heat exchangers (to preheat combustion air and suppress plumes), a waste-heat boiler (to generate steam), or both. The air pollution-control system could be either a dry-ash or a wet-ash system.

FIGURE 24.13   Puerto Nuevo, Puerto Rico's fluidized bed incinerator system (courtesy of Suez).

In the past, common practice in the United States was to provide a fully pressurized combustion and gas-cleaning system, from the wind box to the stack. European practice typically involves a "push and pull" concept, in which the reactor's freeboard is under negative pressure and the zero-pressure point is in the top portion of the bed. More recently, the push and pull approach has become more common in the United States as additional energy recovery and pollution control systems have been added to the configuration. There are advantages to each approach.

The fully pressurized system, in which a fluidizing blower provides total pressurization and gas flow, is a simpler and less costly design. It probably is best used in situations where the scrubbing systems have high headloss (e.g., a venturi scrubber and tray cooler producing pressure differentials of up to 10 kPa [40 in.] of water column). A disadvantage of this system is that the feed systems, whether in the freeboard or in the bed zone, must be completely sealed to prevent hot gases and ash from discharging into the combustion working area.

The push-and-pull system initially was used for top-feed fluidized-bed reactors. Controlled air leakage can help cool the feeder without disrupting fluidization if a fluidizing air blower provides the air. The induced draft fan must be paced with the fluidizing air blower and the amount of air leakage into the reactor freeboard to maintain a negative pressure of about 0.5 kPa (2 in.) of water column. Design engineers should keep in mind that fluidized-bed reactors used to combust solids are prone to bed surging, which causes pulsations in freeboard and bed pressure. These, in turn, can cause backflow unless the feeder is adequately sealed.

A typical process flow diagram of a fluidized-bed thermal oxidation system is shown in Figure 24.14. Solids are dewatered by belt filter presses and then transported by piston pumps to the furnace, where they enter through two or four feed ports. This thermal oxidation system uses a wet-ash system with a hot wind-box furnace and heat recovery via heat exchangers. The hot wind-box furnace has a refractory arch that supports the sand bed and evenly distributes air. Makeup sand may be fed to the bed pneumatically during operation, if required. The reactor has an expanded freeboard

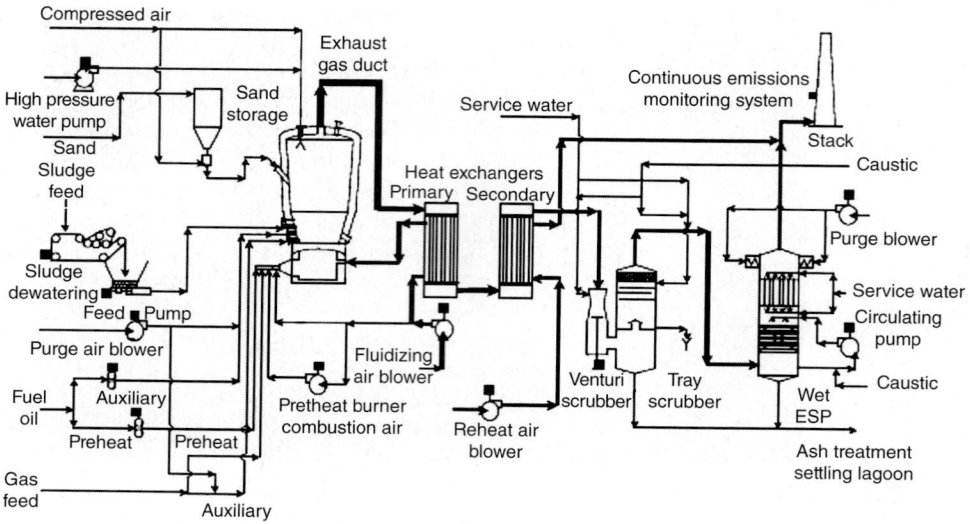

FIGURE 24.14   Typical process flow diagram for a wet ash system (courtesy of Suez).

so the larger particles can decelerate to minimize sand carryover and maximize carbon burnout. The freeboard operates at a design temperature of about 843°C (1550°F). No. 2 fuel oil or natural gas is used as auxiliary fuel during startup and operation as needed. To minimize its use, fluidizing air is preheated to about 675°C (1250°F) in an external tube-and-shell heat exchanger heated by the reactor's exhaust flue gas.

The air pollution-control system includes a venturi scrubber followed by a tray tower. A wet electrostatic precipitator could be used to eliminate submicrometer particulate matter. The venturi scrubber, which has a high pressure drop, removes ash and fine sand particles from flue gas, creating an ash slurry that is sent (via pump or gravity) to an outdoor ash-settling lagoon for dewatering. Dry ash (about 50% total solids) is removed from the drying lagoon about once per month, depending on the size of the lagoon. Meanwhile, hot air (260°C [500°F]) can be added to the stack gas to suppress plumes. This air is preheated in a secondary heat exchanger using exhaust flue gas from the primary heat exchanger.

In the wet-ash system, acid gases (e.g., SO_2 and HCl) are removed by water in the venturi scrubber and cooling tray. These gases are soluble in water, which means that up to 95% of the acids can be removed by effluent alone. To meet stricter limits, a caustic solution can be added to the cooling tray to further reduce acid gas concentrations. Mercury and dioxins can be removed from flue gas via an activated carbon adsorption system installed before the stack.

More than 90% of North American installations currently use the wet-ash system because of its simplicity and the availability of effluent and space at the water resource recovery facility. However, other types of heat recovery (e.g., waste-heat boilers and dry air pollution-control systems) could be used. In the dry-ash system (Figure 24.15), the flue-gas temperature has to be in the range of 150°C to 205°C (300°F to 400°F) before entering a wet scrubber, fabric filter, or dry electrostatic precipitator. A waste-heat boiler or an economizer can be installed between the fluidized bed or heat exchanger and the

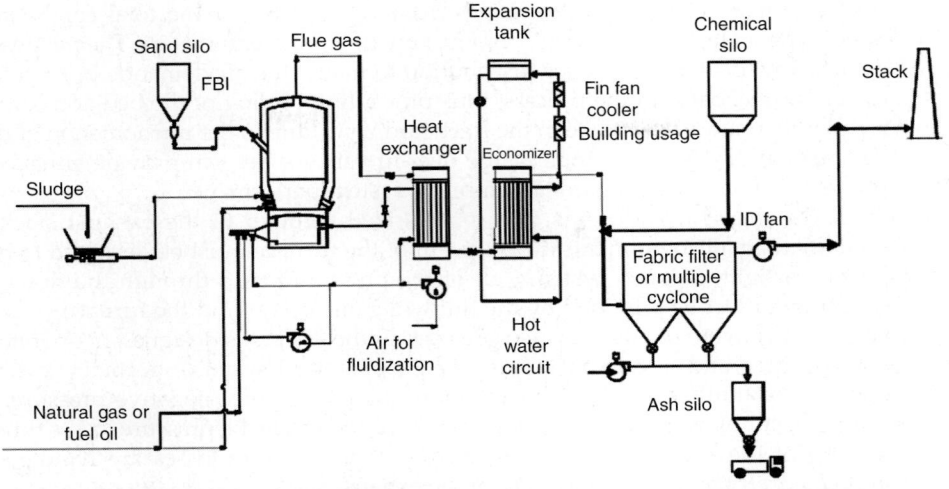

FIGURE 24.15 Typical process flow diagram for a dry ash system (courtesy of Suez).

air pollution-control system to generate steam, hot water, or heated thermal fluid (oil) that can be used for further processing operations or building/structure heating. To remove acid gases, mercury, and dioxins, chemical sorbent can be injected into the duct or into a reactor chamber installed between the heat-recovery and air pollution-control systems.

4.4.3   Fans and Blower Equipment
A thermal oxidation system needs three types of air: fluidizing (combustion) air, purge air, and atomizing air.

4.4.3.1   Fluidizing Air
To calculate the amount of fluidizing (combustion) air needed, design engineers typically need to know the maximum amount of solids and fuel heat anticipated and add an average excess air level of 30% to 50%. They also typically add a 10% to 15% peaking (safety) factor. In some systems, over-fire or quench air is also provided and is introduced into the freeboard zone of the furnace through a series of over-bed air nozzles. Over-fire air is either drawn from the fluidizing air upstream of the bed zone (before or after preheating) or introduced through a separate low-pressure centrifugal fan with an ambient (cold) air intake.

Fluidizing air typically is supplied by a multistage centrifugal blower with impellers made of stainless steel or aluminum. The blower is designed with several stages to provide the required outlet pressure. Thermal oxidation capacity is regulated by the amount of fluidizing air entering the wind box and over-fire air nozzles, if provided. The total fluidizing air blower flowrate is controlled by a damper on the blower inlet side and measured by a flow meter. When over-fire air is provided, it should be measured by a second flow meter, so that the ratio of over-fire air to total air can be monitored and the actual amount of fluidizing air entering the wind box can be determined. It is normal practice to maintain wind-box airflow at or above a preset minimum flowrate, so that adequate bed fluidization is maintained at all times when solids are being fed to the furnace. Over-fire air typically ranges from 10% to 30% of the total combustion air, depending on feed characteristics and furnace operating parameters. The purpose of the over-fire air is to adjust the air distribution to match the air demands in each location (bed and freeboard) in the process, thus preventing cooling of the bed zone and a significant temperature increase in the freeboard zone, due to the phenomenon of delayed combustion. If designed properly, the over-fire air nozzles can provide enhanced gas-phase mixing in the freeboard and improve system performance.

If the thermal oxidizer is fully pressurized through to the exhaust stack (push system), then all components downstream of the furnace must be designed to be completely airtight. In push systems, all of the pressure losses through the downstream equipment must be provided by the fluidizing air blower and the furnace vessel must be designed to handle the elevated pressure. If the system is designed to operate under both positive and negative pressure (i.e., a push-pull system), normal practice is to operate the fluidized-bed freeboard under zero or slightly negative pressure, so the fluidizing air blower only needs to be designed to handle the pressure losses through to the top of the bed zone. The remaining system pressure losses are handled by an induced draft fan located upstream of the exhaust stack.

4.4.3.2   Purge Air   Purge air is used to keep all thermal-oxidizer ports, ductwork expansion joints, and ductwork pressure taps cool and free of sand and ash. It can be either high-pressure or low-pressure air. For example, a compressed air system can supply high-pressure air to keep roof spray nozzles and pressure ports cool and free of deposits. A fluidizing air blower can provide air for the annular sleeve of all ports, including site ports; bed oil guns; sand inlet and outlet; solids inlet; and roof sprays.

4.4.3.3   Atomizing Air   If the auxiliary fuel is fuel oil, then about 0.48 to 0.55 bar (7 to 8 psi) of atomizing air is required. Atomizing air can be provided by an injection-purge air blower, which typically is a lobe-type blower.

4.4.3.4   Induced Draft Fan   If the system is designed to operate under both positive and negative pressure (i.e., a push-pull system), then an induced draft fan is needed to ensure that negative pressure is maintained from the fluidized-bed freeboard zone through all subsequent equipment and subsystems. Major components downstream of the FBI may include air preheaters, waste-heat recovery subsystems (e.g., boilers, heat exchangers), dry-ash removal subsystems, air pollution-control subsystems (e.g., fabric filters, multiple cyclones, electrostatic precipitators, venturis, tray towers, carbon beds), and/or exhaust gas reheating equipment. Induced draft fans typically are heavy-duty industrial centrifugal fans made of stainless steel or higher grades of alloy (e.g., Hastelloy C). The fan typically is controlled by a variable-speed drive and/or variable-inlet vane damper to maintain a set-point operating pressure in the freeboard zone of the furnace.

4.4.4   Other Auxiliary Equipment

4.4.4.1   Auxiliary Fuel System   Auxiliary fuel is used in a preheat burner during startup operations. It also is directly injected (via fuel guns) into the fluidized sand bed or through operating fired burner systems during normal operations. A wide variety of fuels (e.g., coal, sawdust, and digester gas) can be used, as long as the fuel can be fed to the system in a reliable, controlled manner. The most commonly used auxiliary fuels are natural gas and No. 2 fuel oil.

4.4.4.2   Preheat Burner   When starting up a cold thermal oxidizer, operators need a preheating method (e.g., a burner that heats the fluidized bed to a temperature at which fuel injected into the fluidized bed will ignite). In a hot wind-box system, preheating is done by a standard industrial oil (or gas) burner placed in the sidewall of the wind box. Fluidizing air from the heat exchanger (preheater) is thus heated by mixing with the hot gases from the burner, and the resulting hot air then fluidizes the sand bed. Heat from the air is then intimately transferred to the fluidizing sand.

In a cold (warm) wind-box system, the preheat burner is in the freeboard. It transfers heat from the gas to the bed less efficiently, so more time and fuel are needed to get the unit to operating temperatures. The preheat burner gets its air supply from the outlet side of the fluidizing air blower. The air pressure is about 14 kPa (2 psi) above the pressure inside the wind box.

Preheat burners should come with the burner-management package required by process and local insurance regulations. Operators manually adjust the fuel's flowrate from the control panel. Flow adjustments are based on required temperature and firing rates.

4.4.4.3   Bed Fuel Injection   During startup and other modes of operation when there are not enough solids for autogenous combustion, an auxiliary fuel (e.g., No. 2 fuel oil, natural gas, or digester gas) must be added to the system. It can be added via the wind-box preheat burner or via guns directed into the fluidized sand bed.

Fuel oil can be injected into a working fluidized bed via parallel, positive-displacement rotary-gear pumps driven by a common variable-speed drive. An automatic temperature-control loop senses bed temperature and adjusts the pumping speed accordingly. Oil is injected by oil guns installed at the periphery of the bed about 30 cm (1 ft) above the distributor. To prevent fouling, the oil is mixed with purging air at the supply end of the gun, and blown into the bed in the form of a coarse mist. Purge air is provided by a blower. Fluidizing air can be used to cool and continuously purge the gun sleeves. Each oil-injection nozzle can be shut down and withdrawn for cleaning and inspection during system operations or shutdown periods.

Natural gas is directly injected into the bed via gas guns. To maximize in-bed combustion (and avoid over-bed burning), the gas must be distributed homogenously throughout the sand. The guns must be located in the bottom against the distributor and deliver gas throughout the sand bed in different lengths. For safety reasons, gas should only be fired into beds with temperatures higher than 730°C (1350°F), so some water resource recovery facilities use the wind-box gas burner to control the bed temperature instead.

It is important to maintain airflow in oil or gas guns when they are inserted into the bed, so the air-supply line should be equipped with a flow indicator.

4.4.4.4   Water System   To protect the heat exchanger from excessive temperatures, the fluidized bed has water quench spray nozzles that are installed through the roof of the reactor above the freeboard. The high-pressure nozzles (about 2070 kPa [300 psi]) create a fine water mist that will evaporate and quench exhaust gas as fast as possible. The nozzles are used in sequence, depending on the temperature of the gas at heat-exchanger inlet. Evaporation occurs close to the freeboard exhaust-gas duct to limit the cooling effect to flue gas leaving the thermal oxidizer.

The spray nozzles have very small orifices and swirl grooves. To maintain clear internal passages and cool them when not in use, the nozzles are purged with air from the compressed-instrument air system at a pressure of about 415 kPa (60 psi). Both air and water supplies have check valves to prevent one medium from backflowing into the pipes of the other.

The water supply system also includes a water pump, pressure regulator, filter, and relief valve.

4.4.4.5 Sand System   The fluidized bed typically is filled with a sand-like material to a static depth of 0.9 m (3 ft). During operations, fluidizing air will expand the bed material to about 1.5 m (5 ft). The sand will become abraded over time, and makeup sand will be required. Makeup sand typically is pneumatically fed to the furnace by dense-phase conveyors during normal operations. (The reactor typically is equipped with a pneumatic sand makeup system and a bed drain system.)

Because the furnace is sized based on gas flowrate, the hydrodynamics of the fluidized layer depends on the size and density of the media. The media should have a bulk density of about 1601 kg/m³ (100 lb/cu ft) and a particle size distribution similar to that shown in Table 24.7.

It must be angular, dry, and free of sodium and potassium. It also must not grind into fines at 871°C (1600°F) or fuse at 982°C (1800°F).

Two types of fine bed media can be used: silica or olivine sands. The choice of media depends on feed solids quality. Silica sand is less expensive but more abrasive than olivine sand. Olivine sand is not desirable when:

- Feed solids contain grit, which can build up in the bed until it is too deep and excess material must be removed; or

- Feed solids contain high alkali metals, which can accumulate on the bed and lead to low melting eutectics.

4.4.5   Materials of Construction

The choice of construction materials for fluidized-bed furnace and ancillary equipment depends on the composition of both the solids and flue gas. Although mild steel typically is appropriate, stainless steel or special alloy steel are highly recommended for equipment that directly contacts the flue gas. The furnace's shell is carbon steel

Particle Size (μm [U.S. Mesh])	Distribution (%)
2380–841 (8–20)	0 to 20
841–500 (20–30)	10 to 30
500–350 (30–40)	20 to 25
350–295 (40–50)	20 to 25
295–210 (50–70)	0 to 5

TABLE 24.7   Typical Sand Particle Size Analysis (courtesy of Suez)

ASTM-A36. Design engineers should carefully consider the thickness of this steel, which can range from 9.5 to 12.7 mm (0.375 to 0.5 in.), depending on the unit's size.

Refractory materials are selected in accordance with three basic criteria:

- The material's interior face must be strong and wear resistant enough to withstand temperature and abrasion in the three zones of the vessel.
- The material must be backed by an insulating material that can sustain solids temperatures within accepted norms.
- The refractory layers must be designed and installed to comply with gas-tight parameters.

The third criterion is particularly critical in the fluidized-bed zone of a sand bed reactor, where the pressure is highest. If this zone is not constructed to be gas-tight, gas leaks can occur and form pockets behind the refractory layer. The pockets fill with combustible mixtures that can ignite and cause "hot spots" on the shell. Both the reactor and any accompanying nozzles should be fitted with gas baffle rings. Also, the entire vessel's refractory lining should be designed with enough expansion provisions to accommodate the thermal stresses developed by the shell or refractory layer.

An air preheater typically has a severe operating environment: hot, corrosive, and erosive. The material used for the heat-exchanger tubes depends on the concentration of hydrochloric acid in the flue gas. Type 300 austenitic stainless steels, Alloy 20, and Alloy 625 have been used in past projects. Experience indicates that stainless steels have considerably more problems when chloride levels in the flue gas reach about 100 ppmv; such problems become progressively worse as chloride levels further increase. Intermediate alloys (e.g., Alloy 20, 800H, and Type 825) work when chloride levels are above 100 ppmv. Alloy 625 is required if chloride levels exceed 1000 ppmv.

Hot tube sheets follow the same pattern and should be compatible with welding the tubes to the sheets. The lower tube sheet typically is made of carbon steel. It is sufficiently cool and protected from flue-gas exposure by refractory and insulation layers, so failures there are rare.

Expansion joints now typically are furnished with Alloy 625 bellows, whose high nickel content effectively resists corrosion cracking from chloride stress. Although the alloy is costly, the thin bellows only use a small amount, and the extra cost is easily justified when compared to the cost of a typical heat exchanger and the longer life it has when this alloy is used.

4.4.6 Process Control

The control system is designed with alarms and interlocks to ensure safe operation. Interlocking typically is based on the following fundamental philosophies. No combustion operation can be started until the various safety checks (e.g., airflow rates as per design condition, water flowrates to venturi and tray tower scrubbers) are cleared. The control system typically consists of programmable logic controller and personal computer based controls with screen monitors and process graphics used as interface. All process information recorded by the instrument and control equipment is displayed on the operator's graphics computer screen for facility monitoring.

For the safety of the operation, the facility is fitted with temperature elements (thermocouples) that give control signals to various combustion-control loops associated with thermal oxidizer operation. The thermocouples also determine the bed temperature span, which is an indication of fluidization quality. The thermal oxidizer also is fitted with pressure taps. Differential pressures in the bed indicate bed height, and are also used to monitor the quality of fluidization. A wide span of bed-pressure differential typically indicates a well-fluidized bed. Thermocouples and pressure taps also are used in the heat recovery and air pollution-control systems. Water flow and airflow are measured by a mass flow meter.

Fluidized-bed exhaust is supplied with an oxygen sampling and monitoring system to help operators monitor combustion and function as a source of interlocks and alarm.

The operation of the fluidized bed and its performance depend on the feed rates of the three major flows (air, solids, and supplemental fuel) to the thermal oxidizer. While airflow can be constant, the solids composition—especially solids content—and, therefore, cake feed rate vary with time. Variations in solids feed rate is the major reason operators must observe and control the process continuously. The control is simple and based on only two parameters—temperature and excess air (or oxygen)—that are continuously monitored.

4.4.7   Safety

When designing safety measures for a thermal oxidation system, the first priority is to protect facility personnel, contractors, and visitors. The second priority is to protect the thermal oxidation system itself (i.e., its equipment and structures). Also, while the following safety issues are particular to thermal oxidation, this system cannot be considered separately from overall facility safety. Adequate thermal oxidation safety procedures depend on a strong overall facility safety program, which creates a culture that emphasizes safety in all decisions and procedures.

4.4.7.1   Regulations, Codes and Standards   The design and operation of thermal oxidation systems are governed by numerous federal, state, and local regulations, codes, and standards. Some are guidelines, while others are enforceable. In all cases, engineers should consider the underlying safety principles and incorporate them into the design.

4.4.7.1.1   Occupational Safety and Health Standards   The Occupational Safety and Health Act established general national Occupational Safety and Health Standards (29 CFR, 1910), which apply directly to privately owned water resource recovery facilities. It is common practice for publicly owned treatment facilities to meet these standards also. Some of the items that apply to thermal oxidation systems include:

- Walking-working surfaces;
- Means of egress;
- Occupational health and environmental control (ventilation and noise exposure);
- Personal protective equipment;
- Permit for entry to confined spaces;
- Control of hazardous energy (lockout/tagout);
- Fire protection;
- Machinery and machine guarding;
- Electrical; and
- Fall protection.

4.4.7.1.2 Building, Fire, and Mechanical Codes Local codes typically include safety requirements that apply to thermal oxidation systems. Code requirements are location specific, but the model codes listed below illustrate the types of codes that would apply to thermal oxidation systems:

- International Building Code, which addresses egress requirements for thermal oxidizer rooms;

- International Fire Code, which addresses thermal oxidizer requirements related to fires and means of egress for thermal oxidizer rooms; and

- International Fuel Gas Code, which addresses commercial-industrial thermal oxidizers constructed and installed in accordance with NFPA 82, gas-piping installation requirements (e.g., sizing, materials, support, shutoff valves, and flow controls).

These model codes were issued by the International Code Council.

4.4.7.1.3 National Fire Protection Association The NFPA maintains codes and standards that can apply to various aspects of fire and explosion safety for thermal oxidation equipment:

- Flammable and Combustible Liquids Code (NFPA 30), which applies to the storage, handling, and use of liquid fuels, including fuel oil used as auxiliary fuel for thermal oxidizers;

- Standard for the Installation of Oil-Burning Equipment (NFPA 31), which applies to the installation of stationary oil-burning equipment and appliances;

- National Fuel Gas Code (NFPA 54), which addresses safety issues (e.g., piping materials, operating pressures, over- and under-pressure protection, need for shutoff valves, and combustion air) relating to the design, sizing, and installation of gaseous fuel systems for thermal oxidizers that use natural gas, propane, or other similar fuel;

- Standard on Thermal Oxidizers and Waste and Linen Handling Systems and Equipment (NFPA 82), which applies to the installation and use of thermal oxidizers. (Although seemingly directed at thermal oxidizers burning solid waste, the explanatory material notes that there are many types of thermal oxidizers burning a wide range of wastes, so the standard is not intended to address all design details for each thermal oxidation technology. However, it includes many requirements that seem applicable to solids thermal oxidizers [e.g., requirements for auxiliary fuel, air for combustion and ventilation, thermal oxidizer design, placement, and clearances].)

- Standard for Ovens and Furnaces (NFPA 86), which applies to ovens, dryers, or furnaces that process industrial materials at about atmospheric pressure. (Although it contains no direct reference to combustion of wastewater solids, this standard typically is applied to municipal thermal oxidizers and used to specify the safety equipment and practices [e.g., purging requirements] associated with using fuel-firing burners.)

- Fire Protection in Wastewater Treatment and Collection Facilities (NFPA 820), which establishes minimum requirements for preventing and protecting against fire and explosions in water resource recovery facilities. This code addresses hazard classifications for specific processes (e.g., thermal oxidizers) and addresses the buildings in which they are housed as follows:

- Does not require ventilation in the thermal oxidizer area because ventilation typically is provided for other purposes (e.g., heat removal);
- Requires an unclassified area for electrical equipment;
- Requires limited-combustion, low-flame-spread, or noncombustible building materials; and
- Requires a fire-suppression system (e.g., sprinklers).

4.4.7.1.4 Insurance and Other Industry Standards   Insurance standards often are more stringent than local codes or NFPA regulations and should be considered when selecting and designing safety provisions for thermal oxidation systems, especially when particular hazards are present. When designing the fuel valve train and burner safety systems, for example, engineers should identify the relevant insurance standards, including those issued by Industrial Risk Insurers (IRR) or Factory Mutual (FM), and industry standards, including those issued by Underwriters Laboratories (UL). They also should identify the water resource recovery facility's insurance carrier and its requirements to ensure that the design will meet them.

4.4.7.2 Thermal Oxidizer Safety Considerations   Thermal oxidation involves high temperatures, fuel supply and combustion, and solids combustion. They should be designed and operated with adequate safety features and procedures to address the related hazards.

4.4.7.2.1 Hot Equipment Surfaces, Personnel Protection   Thermal oxidizers and some of their exhaust breeching are refractory lined (insulated) but must operate at surface temperatures above 60°C (140°F). The reactor walls are often designed to keep shell temperatures above 100°C (212°F) to prevent condensation from exhaust leaks to the inside wall of the metal. To protect facility personnel against burns from such hot surfaces:

- They should have protective gear (e.g., gloves, clothing, and eye shields) while operating or servicing a hot thermal oxidizer;
- Ducts and equipment surfaces should be insulated, where possible, to keep surface temperatures at 60°F (140°F) or lower while operating in an ambient temperature of 32°C (90°F) or higher; and
- Barriers (e.g., expanded metal shields or barrier fences with locked gates) should be used to protect personnel from contact with hot ducts and equipment, and prevent unauthorized access.

4.4.7.2.2 Fuel Safety Provisions   Thermal oxidizers use an auxiliary fuel (e.g., oil or natural gas) to heat the unit during startup and to supplement operations when needed. Fuel supply and combustion systems should include the following safety features:

- Supplemental fuel-supply piping whose size, materials, configuration, support, shutoff valves, and pressure and flow controls are in accordance with applicable standards; and

- Supplemental fuel-safety systems whose components (e.g., gas conditioner, burner, pilot, ignition, flame monitor, combustion air-pressure monitor, fuel-pressure control and monitor, emergency-fuel shutoff, venting, and purging equipment) are in accordance with applicable standards.

4.4.7.2.3 Fire and Explosion Protection Fluidized-bed thermal oxidizers have a good safety record in preventing fire and explosions. When designing thermal oxidizers, engineers should make sure that:

- Solids-handling systems are configured to reduce the risk of spills and accumulation of solids that could dry and produce combustible dust;
- The preheat burner is purged before startup;
- Instrumentation (e.g., temperature, pressure, and feed monitors) is adequate;
- All equipment is sized and controlled to ensure proper combustion conditions;
- Flue-gas ductwork and equipment are designed to prevent exhaust gas from leaking into the building;
- Dewatered cake combusts under slightly positive pressure in a reactor vessel that meets the applicable structural and welding standards for the pressures to be encountered;
- Natural gas, fuel oil, or other fuels cannot accumulate in the reactor—particularly when cooled—because appropriate interlocks, equipment (e.g., block and bleed valves on fuel-supply lines and removable oil lances), and operating practices are in place;
- The reactor's refractory layer is installed to prevent pockets from developing between it and the shell (such pockets could allow combustible gases to accumulate and cause minor explosions, or permit condensate to collect on the interior of the shell and promote corrosion).

If the thermal oxidizer is a MHF, design engineers also should make sure that:

- The system has an emergency bypass damper and ductwork to vent combustion gases directly to atmosphere, if the power fails; and
- The thermal oxidizer is purged before the burners are started.

4.4.7.2.4 Hazard and Operability Reviews Some North American and European industries use hazard and operability (HAZOP) reviews to identify major operability problems and significant hazards to health, safety, and the environment in a facility design. Once potential problems have been identified, they must be resolved, mitigated, or eliminated via design changes.

HAZOP reviews also are beginning to be used for North American solids-processing systems. This is a systematic, structured method for identifying potential safety problems and taking appropriate mitigation measures to reduce the risks. HAZOP reviews can be performed during all stages of design, from conceptual design through completion of final contract documents. They also may be used on existing facilities. Ideally, though, the review should be timed so design changes can be made before construction begins.

HAZOP reviews typically proceed as follows:

- The HAZOP review should be facilitated by an individual qualified in HAZOP reviews. The HAZOP review team should include, as a minimum, process design engineers, discipline design engineers, operators, construction engineers, safety engineers, and an owner's representative.

- The team determines what will be reviewed (e.g., the entire system, a part of it, or specific items of equipment) and gather the drawings (e.g., process and instrumentation diagrams) needed to identify nodes for study and chart the review's progress.

- The team selects parts of the facility (nodes) for review, and reviewers evaluate them using parameters (e.g., flow, temperature, pressure, and level) and deviation guidewords (e.g., more, less, obstructed, reverse, higher, or lower) to describe the causes and effects of identified deviations.

- Reviewers discuss the consequences of deviations; identify safeguards (anything that could prevent or alleviate the consequences); request more information, if necessary; and recommend appropriate modifications (e.g., design changes).

- The review is complete when all nodes have been examined.

After completing the review, engineers evaluate the identified safeguards to determine which should be incorporated into the design. They must follow up on reviewers' recommendations to ensure that concerns are mitigated. If design changes are recommended, a follow-up HAZOP review should be considered.

The primary benefit of HAZOP reviews is that they raise more awareness among designers, owners, equipment suppliers, and operators of the potential risks associated with process operations. They force all parties to have a thorough discussion of all facility functions, helping them understand facility operations and related safety issues before construction is completed.

4.4.8 Heat Recovery and Use Opportunities

Hot flue gas is the chief source of recoverable energy in thermal oxidation systems. This gas contains most of the heat energy added to the system by feed cake, auxiliary fuel, and combustion air.

Heat-transfer technology is either direct or indirect. In direct heat-transfer processes, the heat source comes in direct contact with the material being heated. In indirect heat-transfer processes, a physical barrier separates the heat source from the material being heated. A heat exchanger that uses a thermal oxidizer's flue gases to preheat combustion air is an indirect heat-transfer process. Most municipal thermal oxidation systems use indirect heat transfer.

The following heat-recovery methods typically are used with thermal oxidation systems:

- A recuperative air preheater is a shell-and-tube heat exchanger in which hot flue gases indirectly preheat fluidizing air. This helps maintain autogenous operation.

- A secondary heat exchanger is a shell-and-tube heat exchanger used to cool flue gas before it enters the air pollution-control equipment. The recovered heat typically is used to raise the temperature of cleaned exhaust gases before they are discharged to the stack to suppress stack plumes.

- A waste-heat recovery boiler (WHRB) is a fire- or water-tube boiler used to cool flue gases; it can be installed after the furnace or a recuperative air preheater. The heat it recovers is transferred to a heat-transfer medium (e.g., water or steam) for external use. Water or steam can be used to heat processes or buildings, pre-dry solids, drive steam turbines, or generate electrical power using steam turbine/generator sets.

- A waste-heat recovery system can also be designed to heat thermal fluids, using various heat exchanger designs, often similar in design to the heat exchangers used as boiler economizers. Thermal fluid can be used to heat processes, to heat buildings, or alternatively to provide heat for use in an organic Rankine cycle (ORC) system to generate power.

- Exhaust gas reheaters for mercury control systems regulate the temperature of the gas to be cleaned, allowing the system to operate at optimal conditions.

If WHRBs are used to produce steam, design engineers should select a unit that can operate at saturation pressures or superheated temperatures. This process has been installed as part of FBI systems (Quast, 2006) and retrofitted to MHF systems (DiGangi et al., 2008). For more information on heat recovery, see *Wastewater Solids Incineration Systems* (WEF, 2009).

4.4.8.1 Combined Heat and Power from Exhaust Heat Recovery   The heat energy contained in exhaust gases from thermal oxidation can be used to generate electricity and heat in a combined heat and power (CHP) application. Under this type of system, hot exhaust gases are passed through a WHRB and used to generate steam at pressures of 30 to 50 barg (450 to 750 psig). This high pressure or "main steam" is fed to a steam turbine generator where expansion within the turbine blades provides the momentum to create mechanical energy to turn the generator (or directly drive mechanical equipment). A portion of the steam flow can be extracted from a middle turbine stage ("extraction steam") at the desired steam heating system pressure, typically about 5 to 8 barg (70 to 115 psig), and be used for facility heating applications. The remaining steam flow at the low-pressure end of the turbine ("exhaust steam") is diverted to a surface condenser where cooling water flow condenses the steam under partial vacuum conditions.

Energy recovery CHP systems can be installed for both multiple-hearth and fluidized-bed thermal oxidation systems. The main process variable driving the viability of energy recovery CHP is the temperature of the exhaust gases leaving the furnace or combustion air preheater, if one is provided. For efficient steam generation, WHRBs optimally require exhaust gas temperatures as high as 760°C to 815°C (1400°F to 1500°F), but significant heat recovery and steam generation potential is still available at exhaust gas temperatures in the range of 480°C to 600°C (900°F to 1100°F). These lower exhaust gas temperatures are typical for hot wind-box fluidized-bed systems, which may use 40% to 60% of the recoverable energy in the exhaust gases to preheat combustion air. Cold wind-box FBI systems are ideal, however, as exhaust temperatures typically fall in the higher temperature range at the inlet to the WHRB.

The MHF systems can also be compatible with energy recovery CHP, although exhaust gas temperatures are typically somewhere between those of cold and hot wind-box fluidized-bed systems, unless afterburners are used to raise furnace exhaust temperatures for emissions control.

Some design variables and considerations for energy recovery CHP systems are listed below:

- Main steam pressure and temperature—required pressures are in the 30 to 50 barg (450 to 750 psig) range and steam must be superheated at least somewhat to prevent excessive formation of liquid droplets within the turbine. The selection of design steam conditions must be coordinated with energy available from exhaust and the requirements of the steam turbine design.

- Boiler feedwater systems—for significantly higher steam pressures, contaminants in boiler feedwater can be much more problematic in boiler tubes. Robust makeup water treatment systems including reverse osmosis and careful monitoring of condensate return water quality are typically required. Multistage boiler feedwater pumps delivering feedwater at high pressures are also required.

- Extraction steam pressure and flow—extraction steam pressure must match the existing steam heating system pressure at the facility. Since lowering extraction steam pressure enhances turbine electrical efficiency, facilities may look to lower steam heating system pressures that are unnecessarily high. Also, extraction steam demands tend to be seasonal with larger demands in the winter. To an extent, steam turbines can be designed to accommodate flow swings by isolating some of the nozzles within the turbine during low-steam-flow conditions (heating season) and opening them during the warmer months, but this requires balancing system efficiency and output with the varying flow envelope. Facilities may look to use auxiliary steam generation in high winter demand periods to avoid large increases in extraction steam and thus decreases in electric production.

- Exhaust steam pressure and cooling water—exhaust steam pressure is an important variable that significantly affects electrical efficiency of the turbine. Lower exhaust pressures lead to greater electrical production. Exhaust pressures are below atmospheric pressure and are typically on the range of 75 to 300 mm Hg absolute (3 to12 in. of mercury absolute). The temperature and flow of the cooling water passing through the surface condenser shell and tube heat exchanger are the variables controlling exhaust pressure. Large availability of facility effluent water tends to give facilities a high degree of control over exhaust pressure. However, quality of facility effluent water needs to be evaluated to avoid excessive fouling of heat exchanger tubes.

5.0 Gasification and Pyrolysis

5.1 Introduction

Gasification and pyrolysis processes have many similarities to incineration processes, but there are key differences that are important to note. Like incineration, gasification and pyrolysis of wastewater solids offer significant volume reduction, but the products

Parameter	Incineration	Gasification	Pyrolysis
Oxygen provided	Surplus oxygen, in excess of the stoichiometric requirement	Limited oxygen, slightly less than the stoichiometric requirement	No oxygen provided
Sulfur emissions	Oxidized sulfur compounds (SO_x)	Reduced sulfur compounds (H_2S, etc.)	Reduced sulfur compounds (H_2S, etc.)
Nitrogen emissions	Oxidized nitrogen compounds (NO_x)	Reduced nitrogen compounds (ammonia, N_2)	Reduced nitrogen compounds (ammonia, N_2)
Particulate emissions (PM)	Significant potential PM emissions without proper controls	Significant potential PM and tar emissions without proper control	Some potential PM and tar emissions without proper control
Feed preparation required	Mechanical or thermal dewatering	Dry feed (~80% to 85% + solids) and potentially some size reduction	Dry feed (~80% to 85%+ solids) and potentially some size reduction

TABLE 24.8 Comparison of Incineration, Gasification, and Pyrolysis Processes

of each of these processes can be very different from one another. Gasification, pyrolysis, and incineration processes are compared and contrasted in Table 24.8.

Gasification literally means to convert a solid or liquid substance into a gas by oxidization-reduction reactions that thermochemically convert biomass energy. Biomass gasification is a two-step or combined-step process. Pyrolysis plus gasification principles are employed to convert the energy potential from solid to gas.

In the first reaction, pyrolysis, thermochemical changes occur at elevated temperatures, in an atmosphere devoid of oxygen, to release the lighter volatile molecules termed "synthesis gas," or "syngas." The pyrolysis process predominantly involves volatilization and thermal cracking/decomposition, producing mainly C, CO, CO_2, CH_4, C_2H_4, C_2H_6, longer-chain hydrocarbons, and tars. Pyrolysis is the precursor to combustion: in direct combustion the volatile gases released by the pyrolysis reactions come into contact with excess stoichiometric oxygen, to support a flame and release heat.

The pyrolysis-produced char and combustible gases are then acted upon by gasification reactions. Conventional air/oxygen-blown gasification thermochemical reactions occur at temperatures of 600°C to 900°C (1100°F to 1650°F) and with about 30% of the stoichiometric oxygen required for complete combustion.

The gasification reactions are generally understood to follow chemical reactions 1 through 7 followed by combustion reactions, which follow chemical reactions 8 through 12 below.

Gasification:

(1) $C + H_2O \rightarrow CO + H_2$

(2) $C + 2H_2O \rightarrow CO_2 + 2H_2$

(3) $C + CO_2 \rightarrow 2CO$

(4) $C + 2H_2 \rightarrow CH_4$

(5) $CO + H_2O \rightarrow H_2 + CO_2$

(6) $CO + 3H_2 \rightarrow CH_4 + H_2O$

(7) $C + H_2O \rightarrow 1/2CH_4 + 1/2CO_2$

Combustion:

(8) $C + O_2 \rightarrow CO_2$

(9) $C + 1/2O_2 \rightarrow CO$

(10) $2CO + O_2 \rightarrow 2CO_2$

(11) $H_2 + 1/2O_2 \rightarrow H_2O$

(12) $CH_4 + 2O_2 \rightarrow CO_2 + 2H_2O$

In the first series of reactions, pyrolysis-produced by-products are gasified through reaction with air/oxygen, steam, and hydrogen. Some char is combusted to release the heat needed for the continuing and endothermic (heat absorbing) pyrolysis reactions.

Pyrolysis and gasification technologies are not new. Ancient Egyptians practiced wood distillation by collecting the tars and pyro-ligneous acid for use in their embalming processes. The process of making charcoal from wood dates back to 2000 BCE. In World War II, small downdraft woodchip fired gasifiers produced fuel gas for motor vehicles.

Incineration is still widely utilized in water resource recovery to reduce the amount of solids that must be disposed of by landfilling or other measures. However, increasing regulatory scrutiny of wastewater solids incinerator stack emissions under the sewage sludge incineration rule (SSI rule) has motivated many facilities that utilize incineration to seek out alternative measures for solids reduction. A significant development for sludge gasification occurred in 2013 in a letter from the U.S. EPA written in response to an inquiry from a company operating a biosolids gasifier in Sanford, Florida. In the letter the U.S. EPA indicated that the gasifier did not fall under the guidelines imposed by the SSI rule. From a regulatory standpoint, this opened the door to achieving a reduction of sludge solids comparable to that of an incinerator while eliminating many of the difficulties with emissions compliance. As a practical matter, though, it is anticipated that the U.S. EPA would likely impose similar emission limits on gasifiers to those in the SSI rule.

5.2 Gasification

Gasification is the name given to the set of thermodynamic processes in which carbonaceous feedstock materials are converted from a solid state into a low-heating-value synthesis gas, or "syngas." The solids that are not converted into synthesis gas leave the reactor as "ash" or "slag" and occasionally some tars or oils.

A gasification process applies heat, pressure, and sometimes steam to convert feedstock solids into syngas, which is composed primarily of carbon monoxide and hydrogen (see Figure 24.16) (California Integrated Waste Management Board, 2001). Variations in operating temperatures, pressures, and the physical configuration of the gasifier affect the composition of the products produced during gasification.

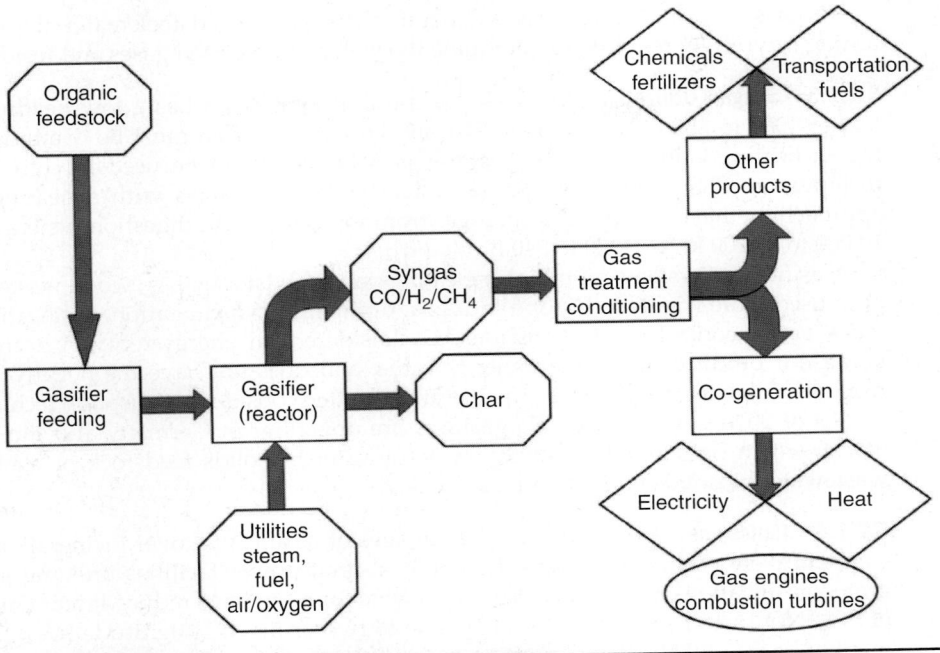

Figure 24.16 Typical process schematic for gasification.

5.2.1 Process Fundamentals

5.2.1.1 Dewatering and Drying One of the primary challenges with utilizing municipal sludge as a feedstock material is the high water content of the sludge. The drier a feedstock material is, the higher the potential net energy output of the system will be. Water contained within feedstock materials must be evaporated prior to feeding to the gasifier, thereby consuming a portion of the energy output of the system and reducing the net heat energy output of the gasifier. This places a high degree of importance on the performance of mechanical dewatering equipment (centrifuges, belt filter presses, etc.). Depending on the type of sludge and the configuration of the dewatering equipment, mechanical dewatering can produce a sludge cake that is between 20% and 35% dry solids. A gasifier requires a feedstock with a dry solids content of at least 75% dry solids. This can be achieved by blending mechanically dewatered sludge with drier feedstock materials like wood waste, applying thermal drying to the sludge, or a combination of the two.

5.2.1.2 Gasification There are many different chemical reactions and processes in a gasifier that have important implications for performance. These vary slightly depending on the gasifier. Operating temperatures may be in the range of 815°C to 1815°C (1500°F to 3300°F) and pressures may be up to 200 kPa (400 psi), for coal/tar oil gasification. Biosolids gasifiers are typically operated at about 600°C to 850°C (1100°F to 1550°F) and operating pressures less than or equal to that of a FBI.

Process dynamics and products depend on the type of feedstock material used. Pilot testing is typically required to determine the yields of produced gases and residues.

5.2.1.3 Syngas Utilization   Also, the resulting syngas often has a low heating value (about 4000 to 8000 kJ/m³ [105 to 210 Btu/cu ft]) and often must be combined with higher-heating-value fuels (e.g., natural gas) before it can be used in conventional turbines for CHP generation. Some systems produce syngas with a heating value approaching that of landfill gas or biogas from some anaerobic digestion systems (about 10 000 to 20 000 kJ/m³ [270 to 540 Btu/cu ft]).

Gasification systems with higher-heating-value feedstocks (e.g., wood wastes) have been used more widely in Europe and Asia. Compared to incineration, gasification provides better control of air emissions, is considered an energy-recovery technology because it produces items with energy value, and does not have the negative public perception that incineration has. (Incineration is often considered a disposal technology.)

As of 2016, several full-scale gasifiers are operating in Germany and the United States, which operate either solely on wastewater biosolids feedstock or a blend of wastewater biosolids with other feedstocks.

5.2.1.4 Emissions   One of the key advantages of gasification over incineration is the comparatively cleaner emissions. In fact, coal-fired power facilities utilizing an integrated gasification combined cycle (IGCC) were found to have reduced their emissions of SO_x, NO_x, and particulate matter by one to two orders of magnitude (U.S. Department of Energy, 2000) after converting to gasification. The reason for this has to do with the fate of the nitrogen and sulfur present in the feedstock material. In an incinerator these materials are oxidized to NO_x and SO_x, but in a gasifier the reducing environment leads to the formation of ammonia (NH_3) and reduced sulfur compounds like hydrogen sulfide (H_2S) and organic sulfides. The majority of the ammonia can be scrubbed out of the gas using an acid scrubbing process. The sulfur compounds can be scrubbed out either by applying a caustic scrubbing process to the syngas or by adding a small amount of limestone to the gasifier. The calcium in the limestone reacts at high temperature to form calcium sulfide and the sulfur can then be removed from the gasifier along with the ash.

5.2.2 Process Description

Several different thermodynamic processes are at work simultaneously within a gasifier, but the bulk of these reactions can be described by the following general categories of reactions.

5.2.2.1 Dehydration   As the feedstock material enters the gasifier, the material is heated rapidly and water within the feedstock is evaporated as steam. Heat energy produced by other reactions within the gasifier is consumed to achieve this evaporation.

5.2.2.2 Pyrolysis   Dehydrated feedstock material is heated further in the absence of oxygen. This causes organic molecules, including tars and oils, to volatilize. The feedstock solids that remain after pyrolysis are referred to as "char."

5.2.2.3 Combustion   In areas of the gasification reactor where oxygen is available in excess of the stoichiometric requirement for full oxidation of the feedstock materials, combustion will occur. These processes are highly exothermic and supply the heat energy that drives other types of reactions within the gasifier (e.g., dehydration, reduction).

5.2.2.4 Reduction The carbonaceous materials that were not volatilized in the pyrolysis reactions (char) can be further converted to produce additional syngas when it is reacted at high temperature with water vapor or other gases present in the gasifier. These reactions increase the efficiency of the gasifier by facilitating further conversion of feedstock carbon into syngas carbon.

Arguably the primary challenge of operating a gasifier is avoiding the deposition of tar in the piping that conveys the syngas produced. This is described in detail later in this section. Another challenge is the control of temperature within the gasifier. Increasing the operating temperature can benefit the kinetics of the reactions in the gasifier, but if the temperature inside of the gasifier exceeds the melting point of the ash (also called the ash fusion point), the resulting liquid ash will agglomerate and form solid masses within the gasifier. These solid "clinkers," which can also form inside of incinerators, can lead to significant operation and maintenance difficulties with the equipment and result in extended system shutdown periods.

5.2.3 Gasification Configurations

The configuration of a gasifier determines how the feedstock materials and their derived solid and gaseous materials will interact within the gasifier. This has important implications for how the gasifier must be operated and the products produced by the gasifier. For instance, the combustion reactions occur in the locations of the gasifier that have the most oxygen available. The physical location of this combustion relative to the location where the syngas leaves the gasifier has important implications in the thermal efficiency of the gasifier and the composition of the resulting syngas. Each variation in the physical configuration of a gasifier also has a significant effect on the amount of tars produced in the syngas (Milne et al., 1998). A few commonly used gasifier configurations are discussed below.

5.2.3.1 Updraft Gasifiers A gasifier in which feedstock materials are fed into the top of the gasifier while air is introduced at the bottom of the gasifier and travels upward through the bed of feedstock material is referred to as an updraft gasifier. This type of gasifier offers a very high thermal efficiency because the heated syngas produced by the gasifier passes through the entire bed of feedstock material before exiting the gasifier. The primary downside of this configuration is that the syngas produced has a relatively high tar content.

5.2.3.2 Downdraft Gasifiers In a downdraft gasifier, the feedstock material is introduced into the top of the gasifier, but the air supplied to the gasifier is introduced at a point in the middle of the gasifier. The syngas produced is drawn down from this position and exits at the bottom of the gasifier. The primary advantage of this gasifier is that the syngas produced has a low tar content. The primary disadvantage of this gasifier configuration is the difficulty in controlling the temperature in all areas of the gasifier. This means that there is a comparatively higher potential to form slag or clinkers in a downdraft gasifier.

5.2.3.3 Fluidized-Bed Gasifiers The configuration of a fluidized-bed gasifier is similar to a FBI. The feedstock materials are introduced into the fluidized bed and are immediately mixed into the fluidized bed of materials in the gasifier. Air is introduced at the bottom of the gasifier along with fluidizing gases. These fluidizing gases can be steam, inert gases like nitrogen gas, or exhaust from combusted syngas. The air and fluidizing gases maintain mixing within the gasifier. The primary advantage of a fluidized gasifier is the ease of temperature control within the gasifier. The primary disadvantage is the relatively high tar content of the syngas produced by the gasifier.

5.2.4   Supplemental Feedstock Materials

5.2.4.1 Wood Waste   If available, wood waste is a dry feedstock material that can be combined with wastewater sludge to reduce the overall moisture content of the mixed feedstock. Pre-processing of the wood is necessary to control the size of the wood chips and to eliminate any nails or other metallic materials present in the raw feedstock.

5.2.4.2 Wastewater Screenings   Wastewater screenings can potentially be used as a high-heat-value feedstock material in some gasifier configurations. However, the heterogeneous nature of wastewater screenings can lead to "hot spots" in the gasifier and thereby promote the formation of clinkers. Additionally, pre-processing may be necessary to ensure that the screenings are properly macerated and free from rocks or other objects that might otherwise interfere with the operation of the gasifier.

5.2.5   Synthesis Gas Composition

5.2.5.1 Energy Content   The heating value of the syngas generated by a gasifier (energy per volume of gas) depends largely on how the supply of oxygen to the gasifier is configured. Oxygen fed to the gasifier can be introduced either as air or as pure oxygen. Pure oxygen is of course more expensive to supply to the gasifier than air, but a gasifier that is provided oxygen will produce a syngas with a higher heating value because the nitrogen gas present in air will not be mixed into the produced syngas and dilute the heating value of the gas. The energy content of syngas can vary widely, based on a variety of system and process parameters, ranging from as low as 4000 kJ/m^3 (105 Btu/cu ft) to as high as 20 000 kJ/m^3 (540 Btu/cu ft), still much lower than the energy content of natural gas at 38 000 kJ/m^3 (1028 Btu/cu ft) (U.S. Department of Energy, 2010).

5.2.5.2 Gas Composition   The main components of syngas are carbon monoxide, hydrogen gas, methane, and carbon dioxide. There are wide variations in the relative fractions of these gases, which depend on the configuration and operation of the gasifier and the composition of feedstock material. Other minor gas constituents include sulfides, ammonia, particulate matter, and tar. The majority of these components can be removed by scrubbing with water, acid, and caustic prior to using the syngas for beneficial use.

5.2.5.3 Steam Addition   Research has shown that the addition of steam into the gasifier has the effect of shifting the production of tar materials away from refractory tars to tar materials that can be more easily removed by other methods such as catalytic cracking (Milne et al., 1998).

5.2.6   Tar Formation and Control Measures
The volatile tars produced in gasifiers consist of a variety of different high-molecular-weight compounds. Each of the compounds within the tar has a condensation point,

but when the temperature of a gas stream containing volatile tars falls below the condensation point, the tars will condense and fall out of the gas stream. This can lead to significant maintenance difficulties as the tar can lead to blockages. The amount of tar formed during gasification depends primarily on the configuration of the gasifier.

There are numerous measures available for controlling tars. The mechanisms by which tar materials are eliminated is referred to as "cracking," which is the process of breaking high-molecular-weight tars into smaller molecules. This process of breaking tars into smaller molecules can potentially produce organic molecules with lower condensation points and thereby reduce their potential to deposit within the gasifier piping. In addition to introducing steam into the gasifier to enable cracking, there are several metal catalysts that can be used. Along with facilitating cracking, tar deposition can be mitigated by insulating and heating the piping conveying the syngas in order to prevent tar condensation.

5.2.7   Gasifier Installations
There are relatively few gasifier systems in operation that process municipal sludge, either alone or in combination with other feedstocks. One of the earliest gasifier installations involved the three fluidized-bed gasifier systems installed at the Hyperion Energy Recovery System in the City of Los Angeles. These systems operated for approximately 10 years (1986 to 1996), processing a mixture of dried biosolids and sludge oil, a by-product of the drying process used at the facility. Each process train had a design capacity of 120 dry metric ton/d (132.5 dry ton/d).

Other fluidized-bed gasifiers operating on municipal sludge alone include relatively small installations in Balingen, Germany and Mannheim, Germany. Downdraft gasifiers that process municipal sludge along with wood waste and/or waste tires have been installed in Covington, Tennessee and Lebanon, Tennessee. The gasifier in Lebanon, Tennessee has a design capacity of 58 dry metric ton/d (64 dry ton/d).

5.3   Pyrolysis
Pyrolysis, literally meaning to "break with fire," is the process of breaking down the organic components of a solid organic feedstock into VOCs in the absence of oxygen at high temperatures. The portions of the feedstock that are not volatilized comprise of oils and char. Pyrolysis provides pathogen inactivation and substantial volume reduction. The resulting pyrogasor syngas can also be used for heat and power generation or for generating biofuels.

5.3.1   Process Fundamentals
Preparing wastewater sludge as a viable feedstock for pyrolysis involves dewatering and sludge drying, similar to the measures needed for gasification. The composition of the resulting pyrolysis gas and solid char produced depends largely on the composition of the sludge and the operating temperature of the pyrolysis reactor (pyrolyzer). Pyrolysis processes include various operating conditions (time, temperature) and can include torrefaction, slow pyrolysis, fast pyrolysis, and carbonization over the temperature range of 300°C to 600°C (570°F to 1100°F) or higher. Each of these variations will produce a characteristically different proportional mix of solids (char), liquids (oil), and gases (pyrogas or syngas). Lower-temperature processes, like torrefaction, tend to yield higher proportions of char. Higher-temperature processes, like fast pyrolysis, tend to yield higher proportions of oils. Gasification processes normally operate at even higher temperatures and tend to yield the higher proportions of syngas.

5.3.2 Process Description

The process is mostly commonly designed in a continuous flow reactor configuration. In one commercially available system, the dried wastewater biosolids are introduced at one end of a heated cylindrical reactor. The material is then heated as a screw auger gradually advances the material along the length of the reactor. The pyrolysis gas is collected as it is released and is pulled from the reactor for further processing. The retention time of the biosolids is controlled by the speed of the auger. The process schematic of a pyrolysis system is very similar to that of a gasification system (see Figure 24.16), except that no air or oxygen is introduced to the pyrolysis zone.

When the biosolids reach the terminal end of the reactor, the raw feedstock has been transformed into biochar. The pyrogas produced in the reactor can also be used for beneficial use. The energy contained within the pyrogas can be extracted using a thermal oxidizer. Alternatively, the VOCs in the pyrogas can be used to synthesize biodiesel by applying a Fischer-Tropsch process. This process is common in the oil and gas industry and combines low-molecular-weight organic molecules with hydrogen gas to produce liquid hydrocarbons.

5.3.3 Pyrogas Composition

Pyrogas produced by a pyrolyzer is significantly different from the syngas produced by a gasifier. It primarily comprises of hydrogen gas, methane, ethane, other alkanes, and higher-molecular-weight compounds referred to generally as "tar."

5.3.4 Biochar

5.3.4.1 Physical Composition The physical composition of biochar depends largely on the feedstock material. Wood chips and pelletized wastewater biosolids tend to produce granular biochars while general wastewater sludges tend to produce relatively fine particles of char. The physical characteristics of char, such as surface area and porosity, depend on the characteristics of the raw material being processed and the type of equipment used for processing.

5.3.4.2 Soil Amendment Biochar from biosolids is an economically desirable by-product. Biochars are beneficially used in agricultural applications to improve soil moisture retention. Other non-soil amendment uses of biochar include the following:

- Filtration—The high sorption capacity of some biochars makes them a good option for filtering toxics from the air (e.g., smokestacks), water, and soil.
- Biochars can adsorb and bind heavy metals, making them less environmentally available.
- Biochar with absorbed toxics in its matrix can be disposed of in a safe location, or chemically processed to flush toxins, essentially recharging biochar for reuse.

6.0 Emissions Control

6.1 Odors

The most noticeable form of air pollution, odors are the reason for about half of all citizen complaints to local air pollution-control agencies. Thermal processing odors are particularly offensive, so design engineers must make every effort to minimize them.

For specific information on odor control, see Chapter 6 and various manuals on the subject (Water Environment Federation, 2004).

6.2   Combustion Emissions

The quantity and quality of combustion emissions from thermal processes depend on the thermal processing method used and the composition of the solids and auxiliary fuel. Completely combusted solids produce carbon dioxide, sulfur dioxide, water vapor, hydrogen chloride, and hydrogen fluoride. The exhaust gases will contain nitrogen, excess oxygen, and an inert ash that contains heavy metals, some of which (e.g., mercury) are volatile and present in both solid and gaseous forms. Exhaust gases could also contain ammonia or other compounds evaporated from the biosolids or formed during processing.

In real life, however, combustion is never complete, so thermal processes release both nitrogen oxides and products of incomplete combustion. Such products include carbon monoxide, VOCs, and polycyclic organic matter.

6.2.1   Carbon Monoxide

Carbon monoxide is a product of incomplete combustion that occurs when carbon in the feed cake partially oxidizes with oxygen in the combustion air. It is the result of one or more of the following combustion-system deficiencies: inadequate temperatures, inadequate residence time for combustion gases, or inadequate mixing or turbulence (needed to bring combustion gases in dynamic contact with oxygen in the air supply).

In general, multiple-hearth and fluidized-bed furnaces have markedly different combustion environments. In a MHF, the feed cake dries on the upper hearths before being combusted on the middle hearths. While this arrangement is efficient (using the heat from combusting dried cake to dry incoming wet feed cake), it releases partially oxidized combustion gases and products of incomplete combustion from the upper hearths of the furnace when dried feed cake begins burning. The slow, stratified flow of combustion gases (i.e., lack of turbulence) in this part of the furnace leads to emission of high levels of CO and products of incomplete combustion.

Carbon monoxide emissions from a MHF without an afterburner can range from 900 to 2700 mg/Nm^{3}dv$_{11}$ (1000 to 3000 ppmdv$_7$); the typical CO mass emission rate is 15.5 g/kg (31 lb/dry ton) of solids incinerated. High CO and VOC emissions are one reason why the use of MHFs has declined steadily in recent decades. A high-temperature afterburner can control such emissions. Multiple-hearth furnaces that have been retrofitted with top-hearth ("zero-hearth") afterburners have significantly lower CO and VOC emissions. However, afterburners also increase fuel requirements. Post-scrubber afterburners have been shown to reduce CO and VOC emission with a fraction of the fuel demand required by zero hearth afterburners (Chilson, 1994).

A fluidized-bed furnace is a completely mixed, highly turbulent system in which drying and combustion take place concurrently and in a matter of seconds. Turbulence provides complete, intimate contact of feed cake, volatilized gases, and oxygen in the fluidizing air. As hot combustion gases rise from the bed, they enter the freeboard, which provides a long residence time that allows CO and other volatilized organics to fully burn out. Carbon monoxide emissions from a fluidized-bed furnace are invariably less than 45 mg/Nm^{3}dv$_{11}$ (50 ppmdv$_7$) and often less than 9 mg/Nm^{3}dv$_{11}$ (10 ppmdv$_7$). Mass emission rates are typically less than 0.5 g/kg (1.0 lb/dry ton) of solids incinerated.

State regulators typically require a new facility to meet a CO emission limit of 90 mg/Nm^{3}dv$_{11}$ (100 ppmdv$_7$). A fluidized-bed furnace can easily meet this limit. A MHF would require an afterburner operating at a minimum of 816°C (1500°F) to meet this standard.

6.2.2   Volatile Organic Compounds

The VOCs are products of incomplete combustion that occur when organic matter in feed cake vaporizes and volatilized compounds partially oxidize. Incomplete combustion happens when the temperature, residence time, oxygen content, and/or mixing in the thermal oxidizer is insufficient.

Chemically, a wide variety of compounds (e.g., straight and branched-chain aliphatic hydrocarbons [methane, ethane, acetylene, etc.]; oxygenated hydrocarbons [acids, aldehydes, ketones, etc.]; chlorinated hydrocarbons [perchloroethylene, trichloroethane, etc.]; and saturated and unsaturated ring compounds [benzene, toluene, phenols, etc.]) are volatile and organic. Such compounds are regulated under 40 CFR Part 503 (U.S. EPA, 1993) and 40 CFR, Parts 60 and 62 (U.S. EPA, 2011, 2016), which require that thermal oxidizers limit concentrations of total non-methane hydrocarbons to less than 100 ppm as propane on a dry volume basis corrected to 7% oxygen (i.e., 100 ppmdv$_7$ [140 mg/Nm^{3}dv$_{11}$]).

Because multiple-hearth and fluidized-bed furnaces have different combustion conditions, the VOC emissions from each are quite different. The upper drying hearths of a MHF typically will be hot enough to volatilize organic compounds but not fully oxidize them. Also, VOC emissions from a MHF depend on the cake's feed rate and combustion characteristics (percent solids, percent volatile solids, and heating value), as well as the furnace's operating conditions (hearth temperatures, excess air, and burner firing rates on different hearth levels). Some MHFs can meet the 140-mg/Nm^{3}dv$_{11}$ (100-ppmdv$_7$) standard without an afterburner. Maintaining a top hearth temperature of at least 595°C (1100°F) typically is important for achieving the standard (Waltz, 1990; Baturay, 1990).

In a fluidized-bed furnace, temperatures and turbulence are high, and total hydrocarbon (THC) emissions are low, typically less than 14 mg/Nm^{3}dv$_{11}$ (10 ppmdv$_7$) as propane.

6.2.3   Polycyclic Organic Matter

Polycyclic organic matter (e.g., polychlorinated biphenols [PCBs], polychlorinated dibenzo-p-dioxin [PCDD], and polychlorinated dibenzo furan [PCDF]) is a subset of VOCs that is of particular concern to regulators because of its potentially high health-risk effects. The U.S. Environmental Protection Agency does not have specific emission limits for PCBs; however, total PCDD and PCDF emissions are regulated under the SSI maximum achievable control technology (MACT) rule (U.S. EPA, 2011). In addition, some state regulatory agencies have included emission criteria for a few of these compounds in permits for new thermal oxidizers. The U.S. Environmental Protection Agency's Compilation of Emission Factors (AP-42) indicates that both multiple-hearth and fluidized-bed furnaces emit low levels of these pollutants, in the range from 0.5×10^{-7} to 0.5×10^{-9} g/kg (U.S. EPA, 1998). To minimize emissions of polycyclic organic matter, system design should maximize the thermal oxidizer's combustion efficiency. The MHFs may need high-temperature afterburners if more control of polycyclic organic matter is necessary. This would likely not be necessary for a fluidized-bed furnace.

6.2.4 Nitrogen Oxides

Nitrogen oxides (e.g., nitrogen dioxide, nitrous oxide, and nitric oxide) are formed by the oxidation of nitrogen during combustion. The two types of nitrogen oxides are thermal nitrogen and fuel nitrogen. Thermal nitrogen is formed by atmospheric nitrogen in the combustion air. Fuel nitrogen is formed by the chemical oxidation of fuel- or solids-bound nitrogen. Emissions of NO_x attributable to furnace burner firing can be significant and should not be overlooked in the design of a thermal oxidation system. Low-NO_x burners should be used where feasible, particularly for multiple-hearth systems, where burner operation may be continuous in the furnace and an afterburner, if provided.

6.3 Emission Regulations

Federal, state, and local governments all promulgate air pollution regulations, and they require emission sources to have permits to operate. For information on permitting thermal processes, see Chapter 6. The rest of this section discusses regulations applicable to thermal processes.

The Clean Air Act (CAA) of 1970 gave the U.S. EPA the responsibility and authority to establish a nationwide program to abate air pollution and enhance air quality, and the 1990 amendments strengthened that authority. State governments have primary responsibility for implementing the program under federal supervision, and each state has developed its own plan for achieving and maintaining the National Ambient Air Quality Standards (NAAQS). State governments often delegate the responsibility and authority for implementing the plan to local pollution-control agencies.

The U.S. Environmental Protection Agency promulgated Standards for the Use or Disposal of Sewage Sludge (40 CFR, Part 503) in 1993. This rule, which falls under the Clean Water Act (CWA), includes emission standards and management practices for solids incinerators.

There are two types of air pollution regulations: limiting and administrative. Limiting regulations establish specific limits for specific air pollutants. Such regulations include the New Source Performance Standards, National Emission Standards for Hazardous Air Pollutants, local prohibitory rules, and source-specific standards. Administrative regulations establish requirements for gathering data (i.e., source testing and ambient monitoring), controlling the growth of emission sources, and using pollution-control technologies. Such regulations include prevention of significant deterioration (PSD) regulations for attainment areas (areas that meet NAAQS) and new-source review regulations for nonattainment areas (areas that do not meet NAAQS).

6.3.1 New Source Performance Standards

Under federal standards for water resource recovery facilities (40 CFR 60, Subpart O), solids incinerators with capacities larger than 1000 kg/d (2205 lb/d) on a dry basis are limited to particulate discharges of 0.65 g/kg (1.30 lb/ton) of dry solids input, and to an opacity of 20%. State standards already in effect may limit particulate emissions to less than 0.2 g/kg (0.4 lb/ton) of dry solids input.

6.3.2 National Emission Standards for Hazardous Air Pollutants

The U.S. Environmental Protection Agency established National Emission Standards for Hazardous Air Pollutants (NESHAPs) to address listed hazardous pollutants (e.g., carcinogens, mutagens, and toxicants) emitted from specific new and existing sources.

For example, solids incinerators must emit less than 10 g (0.022 lb) of beryllium over a 24-hour period, according to 40 CFR, Part 61, Subpart C. Similarly, solids incinerators must emit less than 3.2 kg (7.1 lb) of mercury over a 24-hour period, according to 40 CFR Part 61, Subpart E. More recent standards (U.S. EPA, 2011, 2016) have established concentration-based limits for mercury emissions that are significantly more stringent than the NESHAPs standards. The newer concentration-based mercury limits are different for fluidized-bed and multiple-hearth systems, as well as for new or existing systems.

6.3.3   Prevention of Significant Deterioration

These regulations are designed to prevent the degradation of existing air quality beyond an allowable increment. For example, new stationary sources of a regulated pollutant that are located in a pollutant attainment area and emit more than 620 kg/d (250 ton/yr) of that pollutant are subject to PSD regulations (40 CFR, Part 61, Subpart A, Section 52.21). Modifications of major sources also are subject to PSD if they emit more than the specified minimum amount of any pollutant.

6.3.4   New Source Review

The CAA Amendments of 1977 required all states to establish a permit program for major new, modified, or reconstructed stationary sources in nonattainment areas. The goal is to prevent potential new sources of pollutants from increasing the net air pollution or delaying the area's ability to comply with NAAQS.

To obtain a permit in a nonattainment area, a major source must use the best available control technology that will produce the lowest achievable emission rate, demonstrate a net air quality improvement by offsetting new emissions with emission reductions from other combustion sources in the area (emission-reduction credits), and certify that all other similarly regulated sources are in compliance with all applicable emission regulations.

Sales of emission-reduction credits have developed into a full-fledged commercial "trade" in the past 5 years.

6.3.4.1   Clean Air Act Amendments of 1990   The CAA Amendments of 1990 expand the scope of the 1970 Clean Air Act and 1977 amendments. For example, they divide nonattainment areas into categories based on the severity of the nonattainment and set category-specific timetables for attainment.

The 1990 amendments also directed the U.S. EPA to change its approach to regulating air toxics. Rather than analyzing data for a chemical emitted from a specific source and setting standards to protect human health, the amendments require all major sources—facilities that emit at least 25 kg/d (10 ton/yr) of one of the 189 chemicals listed in the 1990 amendments or whose emissions of all of these chemicals total 60 kg/d (25 ton/yr) or more—to install MACT. The MACT is the control technology used by the cleanest 12% of facilities in the same source category. Once these controls are in place, the U.S. EPA plans to review scientific data to assess the health risk of each chemical and issue new standards if assessment results indicate that MACT did not provide an ample margin of safety for the people most exposed to high-risk emissions.

The 1990 amendments also expand the U.S. EPA's authority to enforce the provisions of the CAA and its amendments.

6.3.5 Standards for the Use and Disposal of Sewage Sludge

Federal standards (40 CFR 503, Subpart E) limit emissions of beryllium, mercury, lead, arsenic, cadmium, nickel, and chromium from solids incinerators and prescribe an operational standard of 100 ppm for THC or CO. The rule also prescribes management practices, including continuous monitoring of THC or CO, oxygen, moisture, solids feed rate, and furnace temperature. In addition, it prescribes the frequency of monitoring, recordkeeping, and reporting.

6.3.6 Local Prohibitory Rules

Local air-quality agencies also can promulgate various regulations to limit specific air pollutants, thereby protecting local air quality and public health. For example, the South Coast Air Quality Management District of California imposes a "nuisance rule" prohibiting emissions from causing injury or being a detriment, nuisance, or annoyance to the public.

6.3.7 MACT Standards for Sewage Sludge Incinerators

Recent changes to the federal regulations governing emissions from thermal oxidation systems, designated as sewage sludge incinerators (SSI) in the regulations, have affected both fluidized-bed (FB is used in the regulations) and multiple-hearth (MH is used in the regulations) systems to a significant degree. In June 2010, the EPA issued a proposed rulemaking on the definition of "solid waste." One of the key aspects of this rulemaking was the inclusion of sludge under the solid waste definition. The effect of this change was that sludge incinerators, which had been regulated under Section 112 of the CAA, would be regulated under the more stringent CAA Section 129 MACT standards. In October 2010, the EPA issued proposed MACT standards for sewage sludge incinerators (SSI), which included very strict limits on a variety of emissions.

During the comment period, more than 80 agencies and individuals submitted detailed comments on the proposed regulations, along with thousands of comments on the related boiler MACT rules. As a result, the EPA requested an extension until July 15, 2011, from the court-ordered deadline of January 14, 2011, for issuance of final SSI regulations, to allow sufficient time to address the substantial number of comments and supporting data received on the proposed regulations. This request was denied and the EPA was ordered to issue final regulations within one month. On March 21, 2011, the EPA published the final rules relating to the definition of solid waste and the MACT standards for SSI units.

The SSI MACT standards are divided into two subcategories, one for MH and one for FB systems. In addition, there are separate regulations for existing units and for new units. These regulations retained many of the original requirements, but did make modifications to limits for certain pollutants and eliminated beyond the floor level of control for mercury emissions that were included in the original rule. Some of the adjustments to emission limits were apparently made to make them more consistent with the existing EPA 503 regulations governing sludge incinerators. Lastly, while the EPA indicated that it will be voluntarily reconsidering many of the other rules being finalized related to the boiler MACT, it decided not to reconsider the final SSI MACT rule. This decision is still being adjudicated. Table 24.9 presents a summary of the proposed limits for existing and new MH and FB incinerators.

The implications of the proposed regulations on existing and future SSIs are still being evaluated and no consensus has been reached regarding the costs or feasibility of making the modifications that will undoubtedly be required to modify some existing

Pollutant	Units	Existing MH Incinerators	New MH Incinerators	Existing FB Incinerators	New FB Incinerators
Cd	mg/dscm	0.095	0.0024	0.0016	0.0011
CO	ppmvd	3800	52	64	27
HCl	ppmvd	1.2	1.2	0.51	0.24
Hg	mg/dscm	0.28	1.15	0.037	0.0010
NO_x	ppmvd	220	210	150	30
Pb	mg/dscm	0.30	0.0035	0.0074	0.00062
PM	mg/dscm	80	60	18	9.6
SO_2	ppmvd	26	26	15	5.3
PCDD/PCDF TEQ*	ng/dscm	0.32	0.0022	0.1	0.0044
PCDD/PCDF TMB*	mg/dscm	5.0	0.045	1.2	0.013

*Must meet one of the two limits for PCDD/PCDF.

TABLE 24.9   Summary of MACT Floor Limits for Existing and New SSI Units (Corrected to 7% Oxygen, Dry Basis) (40 CFR, Part 60 *Standards of Performance for New Stationary Sources and Emission Guidelines for Existing Sources: Sewage Sludge Incineration Units; Final Rule*, Federal Register [Vol. 76, No. 54, 15372-15454], March 21, 2011.)

MH and FB incinerators. Many SSIs will likely have to add some additional air pollution-control equipment to meet one or more of these limits, notably mercury and possibly NO_x and/or SO_2. Systems with very old-style wet scrubbers may also need to upgrade these systems. It would appear that the effect of these regulations may be that some SSI owners may elect to shut down their systems in favor of another biosolids management approach, rather than to implement costly capital improvements required to meet these regulations. Owners considering the installation of new thermal oxidation systems will have to either design to meet these regulations or await final resolution of the legal challenges in anticipation of future potential changes.

6.3.8   Source-Specific Standards

Air emissions can be divided into criteria pollutants and noncriteria pollutants. Criteria pollutants are air contaminants for which the U.S. EPA has established NAAQS (e.g., nitrogen, carbon monoxide, sulfur oxides, reactive organic gases, particulate, and lead). Noncriteria pollutants are air contaminants for which the U.S. EPA has not yet established NAAQS (e.g., hydrogen chloride, hydrogen sulfide, vinyl chloride, trace metals, polynuclear aromatic hydrocarbons, dioxins, and polychlorinated biphenyls).

Air contaminant concentrations can be determined via emission factors or source testing. Emission factors for criteria pollutants from fuel combustion are well established and readily available from incinerator manufacturers and other references. However, emission factors for noncriteria pollutants or solids combustion are not readily available; source testing often is required.

6.3.9   Nonregulated Emissions (Greenhouse Gases)

Greenhouse gases are compounds that contribute to global warming; they currently are unregulated. The four principal greenhouse gases of concern are carbon dioxide (CO_2),

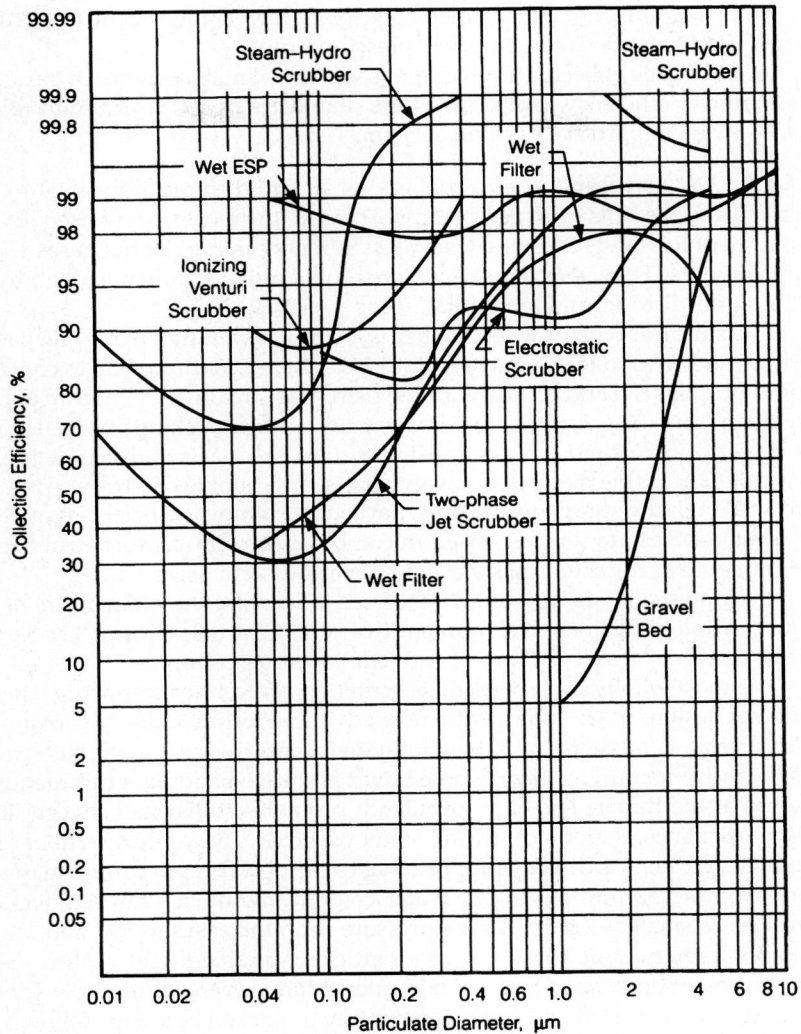

FIGURE 24.18   Efficiency curves for novel air pollution control equipment.

Impingement separators direct gas against collecting bodies, where particles lose momentum and drop out of the gas. Typically designed with pressure drops between 2 and 40 mm (0.1 and 1.5 in.) of water column, they remove particles larger than 10 mm.

Cyclone separators are the most widely used mechanical collectors. Gas enters the device tangentially at the top of a cylindrical shell and then spirals downward toward the narrowest part of a conical section. Particles exit the bottom via an airlock, while the gas is directed back up through the center of the vortex and discharged from the top. Cyclones have top diameters ranging from 0.7 to 3 m (2 to 10 ft) and can remove 85% of particles as small as 10 mm at pressure drops between 10 and 80 mm (0.5 and 3 in.)

of water column. Smaller-diameter cyclones have higher removal efficiencies than larger-diameter ones with the same pressure drop.

A multi-tube collector (multiclone) uses many small cyclones in parallel to improve the removal efficiency. Units with tube diameters of 100 to 300 mm (4 to 12 in.) can remove 95% of particles as small as 5 μm.

6.4.1.2 Wet Scrubbers   Wet scrubbers use a liquid (typically water) to separate particulate and aerosols from gas streams. They are the most widely used emission-control equipment for solids incinerators. Wet scrubbers can handle hot gases with high moisture contents. They also remove water-soluble air contaminants (e.g., hydrogen chloride, sulfur dioxide, and ammonia).

The simplest type of wet scrubber is a spray tower. In this device, gas enters at the bottom and rises to the top, where liquid is sprayed continuously to contact the gas. As liquid droplets fall and collide with the rising gas stream, they capture particles, removing them from the gas. Spray towers typically contain packing material (packed towers) or trays to increase the mass-transfer surface area. Many different tray designs are available, including sieve plates, valve trays, and bubble cap trays. Although pressure drop through the spray tower is low, particulate-removal efficiencies are less than 50%, so these devices are typically used in combination with venturi scrubbers for particulate removal or for chemical scrubbing of various acid gases.

Cyclone scrubbers use centrifugal force to increase the momentum of particle-and-liquid-droplet collisions. They can remove 90% of particles as small as 5 μm at pressure drops between 50 and 150 mm (2 and 6 in.) of water column.

The most widely used particulate scrubbers are venturi scrubbers. There are several different designs in use today, with more advanced technologies now required to comply with stringent emission limits. In a traditional venturi scrubber, the gas stream is accelerated across a venturi (orifice), where liquid is sprayed and then turbulently mixed with the gas at the throat. The high turbulence promotes collisions between liquid droplets and particulates, thereby capturing small particles. The venturi scrubber's particulate-removal efficiency is directly proportional to the power input (pressure drop). At a pressure drop of 250 mm (10 in.) of water column, the device can remove about 90% of particles as small as 1 mm. As the pressure drop increases to 500 mm (20 in.) of water column, it removes up to 98% of such particles. Variable-throat venturi scrubbers allow efficient operations at a range of flowrates. More advanced multiple-venturi scrubber designs have a quench and impingement tray or packed bed stage followed by multiple venturi tubes arranged in parallel. In these advanced designs, the quench stage and tray or packed bed stages remove larger particulate, and the multiple-venturis remove submicron particulates. Multiple-venturi scrubbers have particulate-removal efficiencies that are similar to those of traditional venturi scrubbers, but at lower pressure drops.

6.4.1.3 Fabric Filters   Fabric filters (baghouses) collect solid particles by passing dust-laden gas through a filter medium or fabric that most particles cannot penetrate. As the dust layer accumulates, the pressure drop increases, so accumulated dust must be removed periodically. Operators can clean bags via mechanical shaking, pulse jet, or reverse airflow.

Fabric filters can remove more than 99% of particles down to submicron sizes with pressure drops typically between 50 and 100 mm (2 and 4 in.) of water column. These devices typically are suitable for solids incinerators when the gas temperature can be reduced reliably to about 149°C to 177°C (300°F to 350°F) before gas enters the baghouse. A heat-recovery boiler can be used for this purpose.

6.4.1.4 Electrostatic Precipitators In a dry electrostatic precipitator, exhaust gases pass through a large chamber, where a negative charge is imparted to particulate. It then is attracted to positively charged collector plates that parallel the gas flow through the chamber. Collected particles are removed periodically via vibration, rapping, or rinsing.

Some precipitators use pipes as collectors rather than plates. These units remove liquid droplets and particulate fumes, as well as solid particles.

Wet electrostatic precipitators are similar to dry ones except that they contain a washing mechanism to counteract the buildup of volatile or particulate matter on the plates. They also typically perform better than dry ones when treating certain types of pollutants, removing 99% or more of particulate with negligible pressure drops. In many incineration systems, wet electrostatic precipitators have been installed downstream of wet scrubbers to help the system comply with 40 CFR, Part 503 (U.S. EPA, 1993). Wet electrostatic precipitators have not been required after multiple-venturi scrubbers to meet the old 40 CFR Part 503 regulations, but may be required for the more stringent 40 CFR Part 60, subpart LLLL emission requirements for new source FBIs and MHFs (U.S. EPA, 2011).

6.4.2 Nitrogen Oxides

Two types of nitrogen-control technologies are available, including combustion modification and flue gas treatment.

6.4.2.1 Combustion Modification Systems can be designed to reduce the formation of nitrogen oxides through operation at reduced combustion temperature and lower amounts of oxygen. There are five principal methods for modifying combustion to limit nitrogen oxide formation: low excess air, staged air combustion, staged fuel combustion, low-nitrogen burners, and flue-gas recirculation. The simplest option is operating burners with low excess air. Doing this reduces nitrogen oxides by about 20%. The degree of control is constrained by increases in CO emissions.

Staged air combustion can reduce the formation of nitrogen oxides by up to 70%. In this option, combustion occurs in two stages: In Stage 1, fuel is burned with insufficient air, and in Stage 2, more air is mixed with the fuel to complete combustion. Rotary kilns and FBIs often are designed with staged air combustion.

In staged fuel combustion, the first stage of combustion occurs with insufficient fuel, and the cooling effect of excess air suppresses the formation of nitrogen oxides. The rest of the fuel is injected and burned in the second stage. Here, inerts from the first stage depress peak temperatures and reduce oxygen concentration, thereby reducing the formation of nitrogen oxides. Reserving about 20% of the fuel for the second stage can reduce nitrogen oxide formation by 50% to 70%.

Low-nitrogen burners limit the exposure of fuel to oxygen in the immediate flame zone. As a result, fuel and air mix gradually, reducing peak flame temperatures. This option can reduce nitrogen oxide formation by 30% to 50%.

Recirculating some of the flue gas as inlet combustion air lowers both the peak flame temperature and the formation of thermal nitrogen. Recirculating up to 15% of flue gas can reduce nitrogen oxide formation by 50%.

6.4.2.2 Flue-Gas Treatment   Nitrogen oxide emissions can be reduced by injecting the appropriate reducing agents to the post-combustion region (flue), where they reduce flue-gas nitrogen to molecular nitrogen. Three types of flue-gas treatment are the thermal denitrification process, urea injection process, and selective catalytic reduction process. In the thermal denitrification process, ammonia is injected into an incinerator heated to between 870°C and 1095°C (1600°F and 2000°F) to reduce nitrogen oxides. This process can reduce nitrogen oxide emissions by 35% to 70%, depending on flue-gas temperature, residence time, degree of mixing, and the ammonia-to-nitrogen mole ratio.

The urea injection process is similar to thermal denitrification, except that it uses urea as the reducing agent. Urea breaks down to form CO and ammonia, which reacts with nitrogen oxides to form nitrogen and water. This process also can reduce nitrogen oxide emissions by 35% to 70%, depending on flue-gas temperature, residence time, degree of mixing, and the ammonia-to-nitrogen mole ratio.

In the selective catalytic reduction process, a catalyst bed operates at lower flue-gas temperatures, allowing ammonia to react with nitrogen. It can treat flue gases with temperatures of 315°C to 425°C (600°F to 800°F) and reduce nitrogen oxide levels by 90% or more. However, the process is expensive and the catalyst can be adversely affected by trace metals in flue gas.

6.4.3 Acid Gases, Including Sulfur Oxides

Acid gases include sulfur oxides, hydrogen chloride, hydrogen fluoride, and hydrogen bromide. The emission concentration of such gases is a direct function of the sulfur, chloride, fluoride, and bromide levels in the fuel and solids to be combusted. Acid gas emissions can be reduced via fluidized-bed combustion, wet scrubbers, and dry scrubbers. In fluidized-bed combustion, fuel can be combusted in a bed of granular limestone (dolomite), though this is not common for sludge incinerators. Acid gases react with limestone to form solid calcium compounds, which are removed with the ash. Fluidized-bed combustion also has relatively low nitrogen oxide emission levels because combustion temperatures are moderate (815°C to 925°C [1500°F to 1700°F]).

Wet scrubbers remove both particulate and acid gases from flue gas. They reduce acid gases by diffusing them into liquid droplets. Removal efficiency may be improved by adding an alkaline reagent.

In dry scrubbers, a dilute slurry of lime or another reagent is sprayed into the flue-gas stream. The water fraction of the slurry instantly evaporates, while the atomized reagent reacts with acid gases to form sulfate, sulfite, and chloride precipitates. These precipitates then are removed by particulate-control devices.

6.4.4 Mercury Control

6.4.4.1 Mercury in Flue Gas   Mercury is volatilized into the flue gas as elemental mercury (Hg^0) during the combustion of sludge. As the gas cools in the heat recovery and wet scrubber, the mercury can oxidize to form ionic mercury (Hg^{2+}) compounds (e.g., $HgCl_2$). The mercury can also be adsorbed onto particulate matter emitted with the flue gas (Hg_p). Wet scrubbers remove much of the water-soluble ionic mercury and the particulate-bound mercury species; however, the elemental mercury remains in the gas phase due to its relative insolubility in water.

Note that if scrubber underflow is returned to the water resource recovery facility, however, the mercury it contains will enter water resource recovery facility flows and the water-soluble mercury species could affect effluent concentrations.

6.4.4.2 Mercury Control Technologies Gaseous mercury emissions can be effectively reduced through adsorption onto activated carbon. Removal of the elemental mercury can be effected by chemisorption onto specially impregnated activated carbon, with sulfur-impregnated carbon being the most commonly used. The high active surface area and microporous structure of the carbon also provides adsorption of the ionic mercury species in the flue gas. Depending upon the removal efficiencies required to meet the emission limits, the proven technology options include fixed bed adsorbers, granular activated carbon (GAC), a sorbent polymer composite (SPC) material or flue-gas injection of powdered activated carbon upstream of a particulate capture device (e.g., bag filter).

The highest mercury removal efficiencies (up to >99.9%) can be achieved by fixed bed GAC adsorber systems. Fixed bed GAC adsorbers are either single or multi-compartment vessels located downstream of air-pollution-control equipment. The adsorber typically contains sulfur-impregnated activated carbon that afterward, the flue gas passes through a conditioning chamber where the humidity and temperature are controlled to remove the mercury species from the flue gas via both chemisorption and adsorption of the elemental and ionic mercury respectively. The very high mercury removal efficiencies require prior conditioning of the gas to control flue-gas water content, humidity, and temperature. Dioxins/furans can also be substantially removed (>99%) in a GAC process designed with suitable gas conditioning.

The design of the flue gas conditioning system and GAC adsorber depends upon the target removal efficiencies and final emissions limits to be achieved. A typical schematic of the fixed bed GAC adsorber system for mercury control is presented in Figure 24.19. The flue gas exiting the scrubber/wet electrostatic precipitator is treated through a droplet separator to remove free water and associated dust, and then reheated to up to 45°F

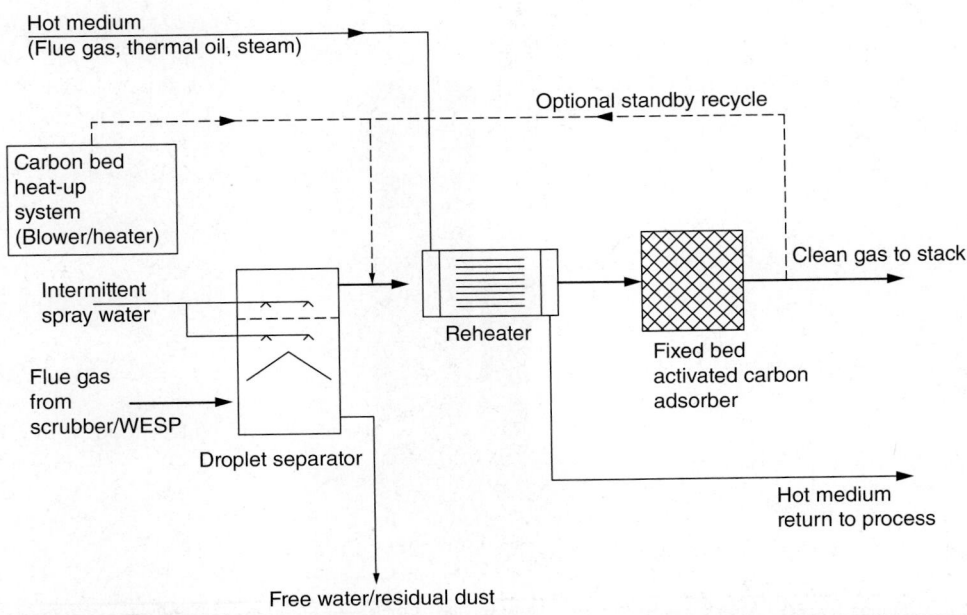

FIGURE 24.19 Schematic of mercury-removal system.

above its dew-point temperature, resulting in a feed gas to the adsorber that is free of moisture and has a low relative humidity (less than 40%). The GAC adsorber-removal efficiency is maintained through proper conditioning of the flue gas. Apart from the quality and size of the activated carbon, and its impregnation with sulfur, the main criterion for removal efficiency is the contact time of mercury with the activated carbon. The carbon bed depth and gas velocity are therefore predominant for removal efficiency, as well as for pressure drop and lifetime of the activated carbon.

A relatively new mercury-control technology being effectively applied, when moderate mercury removal is required to achieve the emission limits, is a sorbent polymer composite (SPC) technology. The scrubbed saturated flue gas is presented to the SPC system, which comprises a series of discrete modules with open-channel design providing lower pressure drops compared with a fixed bed GAC adsorber. The SPC is capable of adsorbing both elemental and any remaining ionic mercury species in the flue gas. The process also affects the removal of residual acid gases, through conversion of SO_2 to aqueous solution of H_2SO_4 (sulfuric acid), which is drained from the system as condensate. Regular washing of the SPC stages is required to wash away the potentially corrosive acid. A typical schematic of the SPC system for mercury control is presented in Figure 24.20.

In flue-gas injection, powdered activated carbon is injected into a reaction chamber, where it mixes with flue gas and mercury is transferred to the carbon. This process is used as a pretreatment step for baghouses or dry electrostatic precipitators, which then remove mercury-laden carbon from the gas. Baghouses require less carbon than electrostatic precipitators because the carbon trapped on the bag's dust layer continues to capture mercury. Table 24.12 provides a summary of mercury control technologies appropriate for sludge incinerator applications.

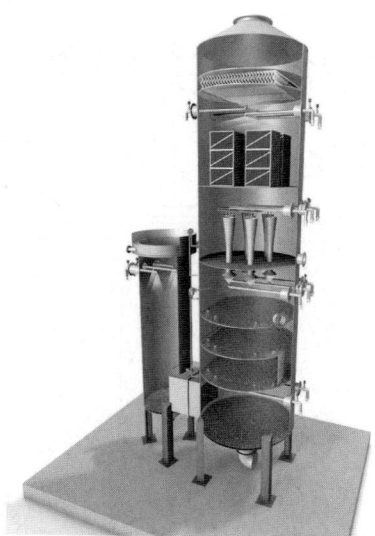

FIGURE 24.20   Multiple venturi set scrubber with SPC technology for mercury removal (courtesy of EnviroCare International).

Technology	Mercury Removal Efficiency	Additional Pollutant Control
Fixed bed GAC adsorber	Up to >99.9%	Dioxins and furans up to 99.9%, other heavy metals, hydrocarbons
Adsorbent (PAC) dosing	Up to 95%	Dioxins and furans, other heavy metals
Sorbent polymer composite (SPC)	Up to 70%	SO_2

TABLE 24.12 Comparison of Mercury Control Technologies

6.4.5 Stack-Gas Reheat

While an occasional steam plume may not hurt the environment, it can hurt public perception of solids combustion. Eliminating such plumes is important for widespread public acceptance of solids combustion.

In an efficient gas-cooling system, exhaust gases typically are not more than 11°C (20°F) above the inlet water temperature (about 20°C to 40°C [70°F to 110°F]) and, therefore, may produce a steam plume under some atmospheric conditions. Reheating stack gas, however, can completely eliminate the steam plume and better disperse gases to the atmosphere.

Water resource recovery facilities typically have short exhaust stacks and other building and topographical conditions that limit the dispersal of gases to the atmosphere. Relatively dry, warm gases rise more rapidly and disperse better.

There are three methods for raising the temperature of exhaust gases from between 20°C and 40°C (70°F and 100°F) up to 90°C to 150°C (200°F to 300°F): a fuel-fired afterburner, indirect reheating with preheated air using furnace off-gases (see Figure 24.21), and direct reheating via furnace off-gases (see Figure 24.22). Water resource recovery facility personnel are reluctant to use fuel-fired afterburners because of the fuel costs as they use natural gas to avoid secondary contamination of stack gases.

Both furnace off-gas methods are effective and require exhaust gases at temperatures of 430°C to 540°C (800°F to 1000°F). In the indirect method, primary exhaust gases are passed through a secondary heat exchanger, where they transfer heat to an ambient airflow, which then is discharged to the stack via a low-pressure blower. The direct heating method is more thermally efficient by transferring heat directly to the clean cool gases within a heat exchanger. In addition, the direct method does not involve this blower.

7.0 Design Example

These examples have been selected to illustrate how design engineers select and size both thermal dryers and incinerators for a water resource recovery facility that produces digested and dewatered solids. The incineration example also illustrates how to select and size an incinerator for dewatered raw solids.

In these examples, the water resource recovery facility produces 50 dry metric ton/d (55 dry ton/d) of digested and dewatered solids at 25% solids. Peak month production is 65 dry metric ton/d (72 dry ton/d) of solids. The equivalent dewatered raw solids production is 75 dry metric ton/d (83 dry ton/d) at 28% solids. Solids characteristics and other relevant parameters are provided in each example.

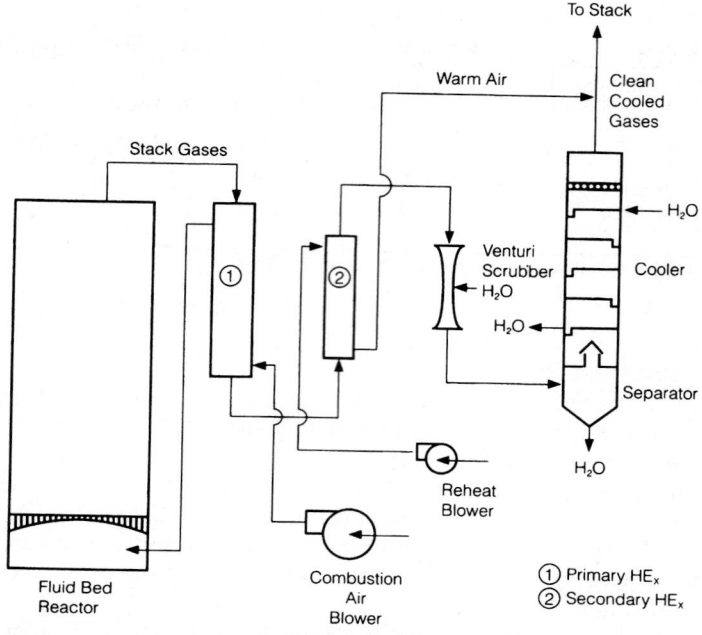

FIGURE 24.21 Indirect reheat of stack gases.

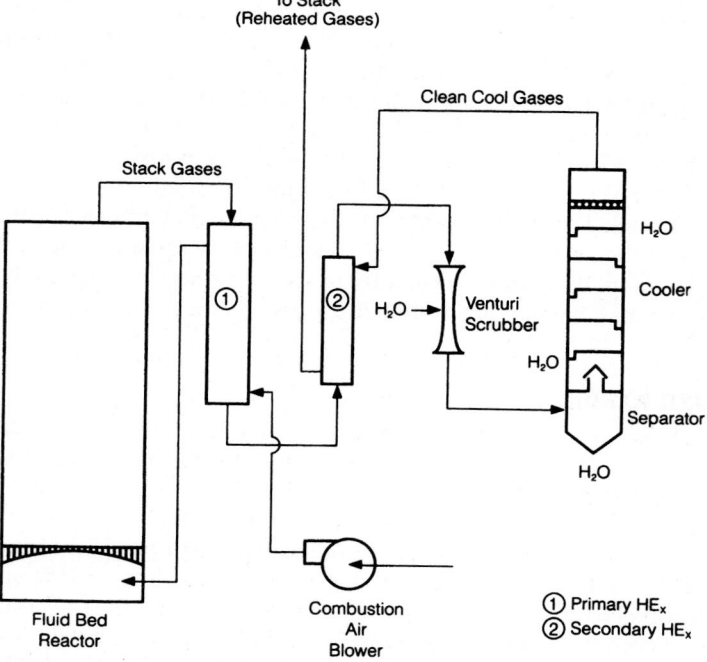

FIGURE 24.22 Direct reheat of stack gases.

7.1 Thermal Drying

Note that many parameters vary by system type and manufacturer. Confirm specific parameters with manufacturer during design.

Input parameters are based on approximate values for a rotary drum system using an RTO for odor control.

7.1.1 Given Information

Anaerobically digested solids

Annual average production	50 metric ton/d	55 dry ton/d
Maximum month production	65 metric ton/d	72 dry ton/d
Cake solids concentration (total solids)	25%	

7.1.2 Procedure

1. Establish hours of operation and redundancy requirements.

2. Establish throughput requirements and product dryness, based on desired processing capacity, operating schedule, and end-product requirements.

3. Calculate required evaporation rate to meet design requirements and contact manufacturers to confirm appropriate equipment sizing and thermal energy efficiency values.

4. Determine capacity for one drying train and confirm number of dryer trains, based on information provided by equipment manufacturers.

5. Determine annual thermal and electrical energy requirements, based on information provided by equipment manufacturers.

6. Establish storage requirements (daily or weekly) and calculate storage capacity requirements.

Each manufacturer will have unique thermal and electrical energy requirements; building footprint and structural requirements; utility loads and demands (e.g., water, drainage); odor-control requirements; and potentially different operator staffing needs and maintenance costs. Accordingly, it is recommended that a life-cycle cost analysis be performed to determine which equipment is the best value for the individual facility.

7.1.3 Assumptions

Establish hours of operation. Drying systems are best suited for continuous operation for extended periods to improve thermal efficiency and reduce equipment wear due to frequent starts and stops. They typically operate continuously throughout the week and allow downtime on weekends.

Sizing considerations:

- Redundancy often is not provided because of high capital cost, but unit capacity can affect this decision. In addition, the availability of an economical alternative system for processing or disposal when the equipment is out of service can affect both the sizing and redundancy design considerations.

- In many cases, the system can be sized conservatively to allow adequate time for preventive maintenance. Design engineers should consider whether in-facility storage can be expanded or if landfilling or another disposal method is available as a backup option.

- In this example, size is based on processing maximum monthly production in a 5-day work week, 24 h/d.
- Assume two units will be provided to meet total drying capacity requirements. This will provide 50% redundancy.

Drying trains installed	2
Actual hours of operation	103 h/wk
Annual operation	52 wk/yr
Dried product's solids	92% concentration (total solids)
Dried product's bulk density	729 kg/m³ (45 lb/cu ft)

(Varies by product type. Contact manufacturer for correct value for system considered.)

Heat energy consumed	3.72 MJ/kg H_2O 1600 BTU/lb H_2O evaporated

(Varies by system. Contact manufacturer for correct value for system considered.)

Electrical energy consumed	85 kW/dryer capacity in metric ton H_2O/h

(Varies by system. Contact manufacturer for correct value for system considered.)

Required days of storage	10 days volume at annual average conditions

7.1.4 Calculations

Determine total evaporation load during the week at maximum monthly design conditions.

$$
\begin{aligned}
\text{Max month evaporation load (metric ton } H_2O/\text{wk)} &= \left(\frac{\text{Max month production rate (metric ton/d)}}{\text{Cake solids concentration}} - \frac{\text{Max month production rate (metric ton/d)}}{\text{Dried product solids concentration}} \right) \times 7 \text{ d/wk} \\
&= 1323 \text{ metric ton } H_2O/\text{wk}
\end{aligned}
$$

$$
\begin{aligned}
\text{Annual average evaporation load (metric ton } H_2O/\text{wk)} &= \left(\frac{\text{Annual average production rate (metric ton/d)}}{\text{Cake solids concentration}} - \frac{\text{Annual average production rate (metric ton/d)}}{\text{Dried product solids concentration}} \right) \times 7 \text{ d/wk} \\
&= 1018 \text{ metric ton } H_2O/\text{wk}
\end{aligned}
$$

Determine hourly evaporation rate at maximum monthly conditions to establish total drying evaporation capacity required.

$$\begin{array}{l}\text{Evaporation rate for} \\ \text{max month load} \\ \text{(kg H}_2\text{O/h)}\end{array} = \dfrac{\begin{array}{c}\text{Weekly evaporation load} \\ \text{(metric ton H}_2\text{O/wk)}\end{array}}{\text{Hours of operation / week}} \times \dfrac{1000 \text{ kg H}_2\text{O}}{\text{Metric ton H}_2\text{O}}$$

$$= \boxed{\begin{array}{l}12\ 845 \text{ kg H}_2\text{O/h} \\ \textit{This is required total drying} \\ \textit{evaporation capacity.}\end{array}} \quad \text{or } 28\ 310 \text{ lb H}_2\text{O/hr}$$

Determine the required evaporative capacity for one drying train.

$$\begin{array}{l}\text{One dryer's evaporative} \\ \text{capacity (kg H}_2\text{O/h)}\end{array} = \dfrac{\begin{array}{c}\text{Design evaporation rate} \\ \text{(kg H}_2\text{O/h)}\end{array}}{\text{Number of drying trains}}$$

$$= \boxed{\begin{array}{l}6422 \text{ kg H}_2\text{O/h} \\ \textit{This is required evaporative} \\ \textit{capacity for one} \\ \textit{drying train.}\end{array}} \quad \text{or } 14\ 155 \text{ lb H}_2\text{O/hr}$$

Determine hours of operation given this capacity and processing annual average throughput.

$$\begin{array}{l}\text{Hours of operation} \\ \text{at annual average} \\ \text{capacity (hr/wk)}\end{array} = \dfrac{\begin{array}{c}\text{Annual average} \\ \text{evaporation load} \\ \text{(metric ton H}_2\text{O/wk)}\end{array}}{\begin{array}{c}\text{Drying capacity} \\ \text{(kg H}_2\text{O/h)}\end{array}} \times \dfrac{1000 \text{ kg H}_2\text{O}}{\text{Metric ton H}_2\text{O}}$$

$$= 80 \text{ hr/wk}$$

Therefore, the units would only need to operate 3 to 4 days per week at average conditions. This may require that the facility be able to store biosolids in liquid form within the facility and/or may require biosolids cake storage upstream to allow for optimum dryer operations.

Note, if initial biosolids quantity is significantly less than design quantity, the designer may want to consider staging installation. Dryers are not thermally efficient if operated at throughput rates significantly lower than design capacity. The designer should consult with the manufacturer to determine optimum turndown capacity.

Determine annual heat energy requirements using annual average biosolids production.

$$\begin{array}{c}\text{Annual heat}\\\text{energy}\\\text{required}\\\text{(MJ/a)}\end{array} = \dfrac{\begin{array}{c}\text{Annual average}\\\text{evaporation load}\\\text{(metric ton } H_2O/wk)\end{array}}{\begin{array}{c}1000 \text{ kg } H_2O/\text{metric}\\\text{ton } H_2O\end{array}} \times \begin{array}{c}\text{Heat energy}\\\text{consumed to}\\\text{evaporate water}\\\text{(MJ/kg } H_2O)\end{array} \times \begin{array}{c}\text{Weeks}\\\text{per Year}\end{array}$$

$$= 196\ 885\ 384 \text{ MJ/a} \qquad \begin{array}{c}186\ 621\\\text{or MMBTU/yr}\end{array}$$

Determine approximate annual electrical energy demand.
Note: This parameter varies significantly per system type and manufacturer. Designer should consult the equipment manufacturer.

$$\begin{array}{c}\text{Annual}\\\text{electrical}\\\text{energy}\\\text{required}\\\text{(kWh/a)}\end{array} = \dfrac{\begin{array}{c}\text{Total drying}\\\text{evaporation}\\\text{capacity}\\\text{(kg } H_2O/h)\end{array}}{\begin{array}{c}1000 \text{ kg } H_2O/\\\text{metric ton } H_2O\end{array}} \times \begin{array}{c}\text{System}\\\text{electrical}\\\text{energy}\\\text{requirement}\\\text{(kW/metric ton}\\H_2O/h)\end{array} \times \begin{array}{c}\text{Hours/Year}\\\text{of operation}\end{array}$$

$$= 5\ 847\ 789 \text{ kWh/a}$$

7.2   Incineration

Fluidized-bed thermal oxidation can be used to dispose of both digested and undigested solids (raw solids or typical mixture of primary and secondary). Typically, digestion can reduce solids production by up to 33%, so it can reduce the size of the incinerator.

However, incinerating digested solids also has disadvantages. Compared to raw solids, digested solids have a lower volatiles content and, therefore, a lower heat content. Furthermore, dewatering digested solids is more difficult than dewatering raw solids.

In the following example, two sizes of FBIs are shown. The first, which is incinerating raw sludge, has a capacity of 75 dry metric ton/d (82.5 ton/d) for raw solids. The second, which is incinerating digested solids, has a capacity of 50 metric dry ton/d (55 ton/d). Raw and digested solids characteristics, including proximate analysis, ultimate analysis, and heating values, are presented in Tables 24.13 through 24.15.

7.2.1   Design Data

A decanter centrifuge is used to dewater solids. The raw and digested solids cake contains 28% and 25% solids, respectively.

Raw solids have a higher combustible content and heat value than digested solids. We use 74% and 61% as combustible contents for raw and digested solids, respectively.

Component	Raw Solids	Digested Solids
Dry metric ton/d	75	50
Wet metric ton/d	268	200
Total solids	28%	25%

TABLE 24.13 Solids Feed Rates

Component	Raw Solids	Digested Solids
Ash (wt%, dry basis)	24.0	39.0
Volatiles (wt%, dry basis)	70.5	60.0
Fixed carbon (wt%, dry basis)	3.5	1.0
Total (wt%, dry basis)	100.0	100
HHV (kcal/kg volatile solids)	26 050	23 260

TABLE 24.14 Proximate Analyses and Heating Values

Volatile Composition (Ash-Free Solids)	Raw Solids	Digested Solids
Carbon (wt%, dry, basis)	55.8	53.3
Hydrogen (wt%, dry basis)	8.1	7.7
Oxygen (wt%, dry basis)	27.5	29.7
Nitrogen (wt%, dry basis)	7.4	7.1
Sulfur (wt%, dry basis)	1.2	2.2
Total (wt%, dry basis)	**100.0**	**100.0**

TABLE 24.15 Ultimate Analysis of Primary and Digested Solids

The combustible heating values are 26 050 kJ/kg (11 200 Btu/lb) for raw solids, and 23 260 kJ/kg (10 000 Btu/lb) for digested solids.

Three alternative cases are developed as follows:

- Case No. 1: Incineration of raw solids (or mixture of primary and secondary);
- Case No. 2: Incineration of digested solids, where biogas is used as auxiliary fuel; and
- Case No. 3: Incineration of digested solids, where dryer is used before incineration.

In each case, maximum heat is removed from the incinerator exhaust gas to preheat combustion air and generate steam. Hot air up to 665°C is generated in a tube and shell gas to air heat exchanger and is used as combustion and fluidizing air to minimize auxiliary fossil fuel consumption.

In the first two cases (Nos. 1 and 2), superheated steam at 3500 kPa (507 psi) and 370°C (700°F) is generated in a water-tube waste-heat boiler. The steam will be used in a steam turbine generator to produce electricity (see Figure 24.23a). In these two cases, a push-pull system with a force-draft fluidizing air blower and an induced draft fan is used to maintain negative pressure in the boiler and to minimize risk of ash leakage.

In Case No. 3 where sludge is pre-dried, saturated steam at 850 kPa (125 psi) is generated in a fired-tube boiler and used in a dryer to evaporate enough water so the incinerator can operate autogenously (see Figure 24.23b). Because of the incinerator's size

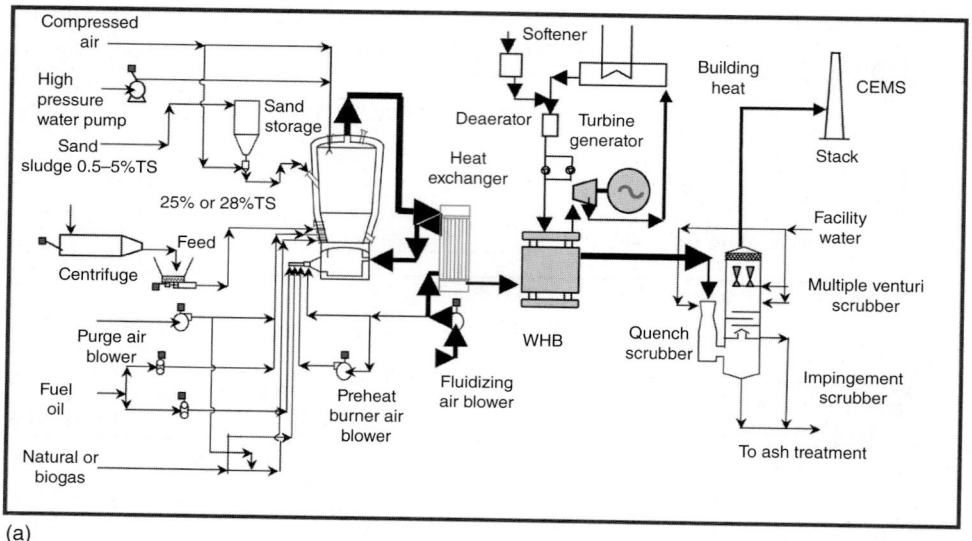

(a)

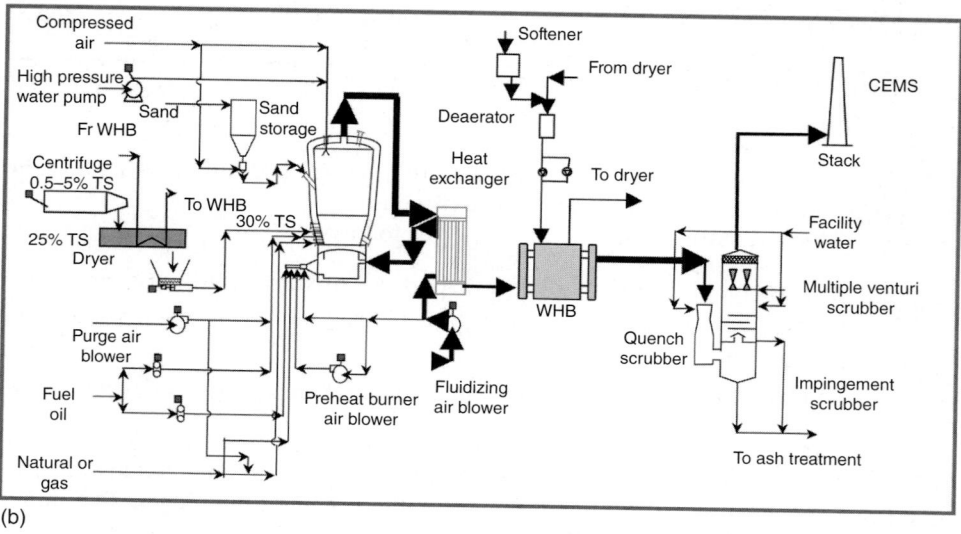

(b)

FIGURE 24.23   (a) Push-pull system with water tube boiler for high-pressure superheated steam production and (b) push system with fired-tube boiler for low-pressure saturated steam production.

and the amount of steam used for pre-drying, no electricity is produced. With low-pressure steam, a fire-tube boiler can be used instead of a water-tube boiler. So a push system can be conveniently used in this case without risk of ash leakage.

In all cases, a venturi scrubber is used to remove particulate, heavy metals, and acid gases.

7.2.2 Procedure

- Establish hourly throughput requirements and feed parameters.
- Perform mass and heat balances. (These are complex and difficult to perform manually. They can be performed by incinerator vendors or via commercially available models. A description of how to perform mass and heat balances is available in the literature [WEF, 2009]).
- Use results of mass and heat balances to select and size equipment. Designers should consult incinerator vendors, who are experienced in selecting and sizing of equipment.

Results of calculations follow (see Tables 24.16 through 24.19).

7.2.2.1 Heat and Mass Balances

For raw solids at 28% total solids, the process is autogenous with a wind-box air temperature at 326°C. The incinerator's thermal capacity is 70.7×10^6 kJ/h.

At 25% total solids and combustion of digested solids with a maximum wind-box air temperature of 650°C, the process still requires 7.9×10^6 kJ/h of biogas. As the

Inputs	kg/h	kJ/h
Waste solid	3125	60 292 323
Waste water	8036	706 388
Aux fuel	0	0
Quench water	12	1059
Purge air	1063	53 058
Fluidizing air	25 074	8 115 775
Air moisture	492	1 519 723
Sand	39	753
Total in	**37 841**	**70 689 078**
Outputs	**kg/h**	**kJ/h**
Dry gases	26 764	24 809 974
Water vapor	10 226	43 728 111
Ash	813	636 012
Sand	39	30 528
Heat losses	0	1 484 456
Total out	**37 841**	**70 689 078**

TABLE 24.16   Case 1: Incineration of Raw Solids (Cake Containing 28% Total Solids)

Inputs	kg/h	kJ/h
Waste solid	2083	29 594 225
Waste water	6250	549 413
Biogas	147	7 892 498
Quench water	0	0
Purge air	1063	53 058
Fluidizing air	15 410	10 591 321
Air moisture	310	1 159 472
Sand	25	486
Total in	**25 290**	**49 840 475**
Outputs	**kg/h**	**kJ/h**
Dry gases	16 846	15 609 632
Water vapor	7607	32 528 623
Ash	813	636 012
Sand	25	19 570
Heat losses	0	1 046 638
Total out	**25 290**	**49 840 475**

TABLE 24.17   Case 2: Incineration of Biosolids Using Biogas (Cake Containing 25% Total Solids)

Inputs	kg/h	kJ/h
Waste solids	2083	29 594 225
Waste water	4861	1 220 918
Biogas	0	0
Quench water	105	26 409
Purge air	1063	53 058
Fluidizing air	12 452	8 557 750
Air moisture	255	946 819
Sand	19	1051
Total in	**20 838**	**40 475 049**
Outputs	**kg/h**	**kJ/h**
Dry gases	13 905	12 882 834
Water vapor	6102	26091 376
Ash	813	636 012
Sand	19	14 873
Heat losses	0	849 955
Total out	**20 838**	**40 475 049**

TABLE 24.18   Case 3: Incineration of Biosolids Using Dryer (Cake Containing 30% Total Solids)

Parameter	Case 1 Raw Solids	Case 2 Digested Using Biogas	Case 3 Digested Using Dryer
Dry feed rate (metric ton/d)	75	50	50
Volatile solids	74%	61%	61%
Total solids	28%	25%	30%
Incinerator			
OD freeboard (m)	9.26	7.82	6.67
Diameter distributor (m)	5.26	4.09	3.51
Operating weight (tonne)	296	215	173
Wind-box temperature (°C)	316	649	649
Combustion airflow rate (kg/h)	25 546	15 701	12 686
Flue gas flowrate (kg/h)	36 990	24 453	20 007
Thermal capacity ($\times 10^6$ kJ/h)	71	50	42
HE capacity ($\times 10^6$ kJ/h)	7	10	8
WHB capacity ($\times 10^6$ kJ/h)	24	11	9
Steam generation (kg/h)	9216[a]	4140[a]	4034[b]
Steam use (kg/h)	9216[a]	4140[a]	1935[b]
Net steam production (kg/h)	0[a]	0[a]	2099[b]
Venturi scrubber gas (m³/h)			
Dry gas rate (kg/h)	26 764	17 556	13 905
Water vapor (kg/h)	13 906	7607	6102
Water vapor condensation (kg/h)	11 833	6464	5304
Auxiliary fuel ($\times 10^6$ kJ/h)	–	8	–
Venturi scrubber facility effluent water rate (m³/h)	228	171	171
Electricity consumption[c] (kW·h/h)	968	692	759
Estimated electricity production (kW·h/h)	1551	697	–
Net production (kW·h/h)	583	5	(759)

[a]Superheated steam (3500 kPag and 370°C).
[b]Saturated steam (850 kPag).
[c]Not including dewatering equipment.

TABLE 24.19 Other Results and Conclusions

facility capacity is reduced to 50 dry metric ton/d, the facility thermal capacity is also reduced to 49.8×10^6 kJ/h (47.2 MM Btu/hr) from 70.7×10^6 kJ/h (67 MM Btu/hr) for raw solids.

In this case, a disk dryer is used to thermally dewater centrifuge cake. The dryer is designed to raise the solids content of digested solids from 25% to 30% total solids. With less water in the solids than in Case No. 2, the incinerator's thermal capacity is reduced from 49.8×10^6 kJ/h (47.2 MM Btu/hr) to 40.5×10^6 kJ/h (38.5 MM Btu/hr). Among the three cases, this case results in the smallest incinerator.

7.2.2.2 Other Results and Conclusions   The calculations have shown that with raw solids, the fluidized bed is larger with more exhaust-gas flowrates. Therefore, more heat can be recovered from exhaust gases in Case No. 1 (31×10^6 kJ/h as opposed to 21 and 17×10^6 kJ/h for Case Nos. 2 and 3, respectively).

Furthermore, because raw solids have a higher heat value, the fraction of heat needed to be recovered in the air preheater is also minimal (23% in Case No. 1, versus over 47% for the two other cases). Therefore, more steam can be generated.

With more steam generated in Case No. 1, more electricity can be produced (1.55 MW-h/h for Case No. 1, versus Case No. 2 with 0.7 MW-h/h). Electricity generation in Case Nos. 1 and 2 is large enough to cover all electricity needed in the incinerator. A net electricity production of 37% (0.6 MW-h/h) is found in Case No. 1 with raw solids.

Case No. 3, with 4034 kg/h of 850 kPa saturated steam generated, 1935 kg/h is used in the dryer, and the remaining 2099 kg/h can be used as building heat. This represents a net steam production of about 52%.

8.0   References

Baturay, A. (1990) *Case Studies of Total Hydrocarbon Emissions (THC) From Multiple Hearth Sewage Sludge Incinerators and THC Reduction Strategies*; Prepared for the Association of Metropolitan Sewerage Agencies Incinerator Workgroup Meetings; New Orleans, Louisiana; Association of Metropolitan Sewerage Agencies: Washington, D.C.

California Integrated Waste Management Board (2001) *Conversion Technologies for Municipal Residuals*; a Background Primer Prepared for the Conversion Technologies for Municipal Residuals Forum; Sacramento, California, May 3–4; California Integrated Waste Management Board: Sacramento, California.

Chilson, S. J. (1994) Hatfield Targets 20% of EPA 503 Regulations at 85% Fuel Savings. *Proceedings of the 67th Annual Water Environment Federation Exposition and Conference*; Chicago, Illinois, October; Water Environment Federation: Alexandria, Virginia.

Dangtran, K.; Jeffers, S.; Mullen, J. F.; Cohen, A. J. (1999) Control Problem Waste Feeds in Fluid Beds. *Chem. Eng. Prog.*, **95**, 59–63.

Dangtran, K.; Mullen, J. F.; Mayrose, D. T. (2000) A Comparison of Fluid Bed and Multiple Hearth Biosolids Incineration. *Proceedings of the 14th Annual Water Environment Federation Residuals and Biosolids Management Conference*; Boston, Massachusetts, February; Water Environment Federation: Alexandria, Virginia.

Dangtran, K.; Holst, T. (2001) Minimization of Major Air Pollutants from Sewage Sludge Fluid Bed Thermal Oxidizers. *Proceedings of the 74th Annual Water Environment Federation Exposition and Conference*; Atlanta, Georgia, October 13–17; Water Environment Federation: Alexandria, Virginia.

DiGangi, D.; Melchiori, E.; Habetz, D. (2008) The Beneficial Reuse of Sludge Incinerator Exhaust Gases to Produce Renewable Energy. *Proceedings of the 81st Annual Water Environment Federation Technical Exhibition and Conference* [CD-ROM]; Chicago, Illinois, Oct 18–22; Water Environment Federation: Alexandria, Virginia.

Geldart, D. (1973) Types of Gas Fluidization. *Powder Technol.*, 7, 285–92.

Intergovernmental Panel on Climate Change (1995) *Climate Change 1995: The Science of Climate Change, Published for the Intergovernmental Panel on Climate Change*; Cambridge University Press: New York.

Intergovernmental Panel on Climate Change (2001) *Climate Change 2001: The Scientific Basis*, Published for the Intergovernmental Panel on Climate Change; Cambridge University Press: New York.

Intergovernmental Panel on Climate Change (2007) *Climate Change 2007: The Physical Scientific Basis*, Published for the Intergovernmental Panel on Climate Change; Cambridge University Press: New York.

Intergovernmental Panel on Climate Change (2013) *Climate Change 2013: The Physical Science Basis*, Published for the Intergovernmental Panel on Climate Change; Cambridge University Press: New York.

Lewis, F. M.; Lundberg, L. A. (1988) Modifying Existing Multiple-Hearth Incinerators to Reduce Emissions. *Proceedings of the National Conference on Municipal Sewage Treatment Plant Sludge Management*; Palm Beach, Florida, June; Hazardous Materials Research Institute.

Los Angeles County Sanitation Districts (1988) *1988 Survey of Thermal Conditioning and Wet Air Oxidation Facilities*; Los Angeles County Sanitation Districts: Whittier, California.

Lundberg, L. A. (2004) The Future of Fluidized Bed Incineration. *Proceedings of the 18th Annual Water Environment Federation Residuals and Biosolids Management Conference*; Salt Lake City, Utah, February 22–25; Water Environment Federation: Alexandria, Virginia.

Milne, T. A.; Evans, R. J.; Abatzoglu, N. (1998) *Biomass Gasifier "Tars": Their Nature, Formation, and Conversion*, NREL/TP-570-25357; National Renewable Energy Laboratory: Golden, Colorado.

NFPA 30 (2015) *Flammable and Combustible Liquids Code*; National Fire Protection Association: Quincy, Massachusetts.

NFPA 31 (2016) *Standard for the Installation of Oil-Burning Equipment*; National Fire Protection Association: Quincy, Massachusetts.

NFPA 54 (2015) *National Fuel Gas Code*; National Fire Protection Association: Quincy, Massachusetts.

NFPA 68 (2013) *Standard on Explosion Protection by Deflagration Venting*; National Fire Protection Association: Quincy, Massachusetts.

NFPA 69 (2014) *Standard on Explosion Prevention Systems*; National Fire Protection Association: Quincy, Massachusetts.

NFPA 82 (2014) *Standard on Incinerators and Waste and Linen Handling Systems and Equipment*; National Fire Protection Association: Quincy, Massachusetts.

NFPA 86 (2015) *Standard for Ovens and Furnaces*; National Fire Protection Association: Quincy, Massachusetts.

NFPA 654 (2017) *Standard for the Prevention of Fire and Dust Explosions from the Manufacturing, Processing and Handling of Combustible Particulate Solids*; National Fire Protection Association: Quincy, Massachusetts.

NFPA 820 (2016) *Standard for Fire Protection in Wastewater Treatment and Collection Facilities*; National Fire Protection Association: Quincy, Massachusetts.

Niessen, W. R. (1988) Thermal Processing of Wastewater Treatment Plant Sludges. *Proceedings of the National Conference on Municipal Sewage Treatment Plant Sludge Management*; Palm Beach, Florida, June; Hazardous Materials Research Institute.

North American Manufacturing Co. (1965) *North American Combustion Handbook*; North American Manufacturing: Cleveland, Ohio.

Nuss, S.; Persinger, D.; Brunner, T.; Netzel, J. (2008) Performance of Instrumentation and Control Upgrades to Multiple Hearth Furnace in Anchorage, Alaska. *Proceedings of the 81st Annual Water Environment Federation Technical Exhibition and Conference* [CD-ROM]; Chicago, Illinois, Oct. 18–22; Water Environment Federation: Alexandria, Virginia.

O'Kelley, S.; Williamson, R.; Lewis, F. M. (2006) Who Says you Can't Teach Old Dogs New Tricks? *Proceedings of the 20th Annual Water Environment Federation Residuals and Biosolids Management Conference*; Cincinnati, Ohio, March; Water Environment Federation: Alexandria, Virginia.

Porter, J.; Lill, W.; Mansfield, W. (2002) Reviewing Multiple Hearth Furnaces: The Atlanta Experience. *Proceedings of the 16th Annual Water Environment Federation Biosolids Conference* [CD-ROM]; Austin, Texas, Feb. 27–29; Water Environment Federation: Alexandria, Virginia.

Quast, D. (2006) Energy Efficiency Improvements for the Metropolitan Wastewater Treatment Plant Solids Processing Facility. Paper Presented at the Conference on the Environment, a Joint Conference by the Minnesota Section of the Central States Water Environment Association and the Air and Waste Management Association; Bloomington, Minnesota, November.

RHOX International, Inc. (1989) *RHOX Process*; Technical Bulletin MHF-1; RHOX International Inc.: Salt Lake City, Utah.

Sapienza, F.; Walsh, T.; Sangrey, K.; Barry, L. (2007) Upgrade of UBWPAD's Multiple-Hearth Furnace Sludge Incinerators. *Proceedings of the Annual Water Environment Federation/ American Water Works Association Joint Residuals and Biosolids Management Conference*; Denver, Colorado, April; Water Environment Federation: Alexandria, Virginia.

U.S. Department of Energy (2000) *A Comparison of Gasification and Incineration of Hazardous Wastes*, March 30; U.S. Department of Energy: Washington, D.C.

U.S. Department of Energy (2010) *Clean Heat and Power Using Biomass Gasification for Industrial and Agricultural Projects*, February; U.S. Department of Energy: Washington, D.C.

U.S. Environmental Protection Agency (1979) *Process Design Manual for Sludge Treatment and Disposal*, EPA-625/1-79-011; U.S. Environmental Protection Agency: Washington, D.C.

U.S. Environmental Protection Agency (1983a) *Assessment of Sludge Processing Problems at Selected POTWs*; U.S. Environmental Protection Agency, Office of Water Management: Cincinnati, Ohio.

U.S. Environmental Protection Agency (1983b) *Municipal Wastewater Sludge Combustion Technology*, EPA-625/4-85/015; U.S. Environmental Protection Agency: Washington, D.C.

U.S. Environmental Protection Agency (1985) *Multiple-Hearth and Fluid Bed Sludge Incinerators: Design and Operational Considerations*, EPA-430/9-86-002; U.S. Environmental Protection Agency: Washington, D.C.

U.S. Environmental Protection Agency (1993) Standards for the Use or Disposal of Sewage Sludge. *Code of Federal Regulations*, 40 CFR, Part 503, Subpart E.

U.S. Environmental Protection Agency (1998) *Compilation of Air Pollutant Emission Factors, Volume I: Stationary Point and Area Sources*, AP-42; Section 2.2; U.S. Environmental Protection Agency, Office of Air Quality Planning and Standards: Research Triangle Park, North Carolina.

U.S. Environmental Protection Agency (2011) *Standards of Performance for New Stationary Sources and Emission Guidelines for Existing Sources: Sewage Sludge Incineration Units*; Final Rule. *Code of Federal Regulations*, 40 CFR, Part 60.

U.S. Environmental Protection Agency (2016) Federal Plan Requirements for Sewage Sludge Incineration Units Constructed on or Before Oct. 14, 2010; Final Rule. *Code of Federal Regulations*, 40 CFR, Part 62.

Waltz, E. W. (1990) *Technical Discussion of Proposed EPA Hydrocarbon Regulation for Sludge Incinerators—Charts and Graphs*; Prepared for the Association of Metropolitan Sewerage Agencies Incinerator Workgroup Meetings; New Orleans, Louisiana; Association of Metropolitan Sewerage Agencies: Washington, D.C.

Water Environment Federation (2004) *Control of Odors and Emissions from Wastewater Treatment Plants*, Manual of Practice No. 25; Water Environment Federation: Alexandria, Virginia.

Water Environment Federation (2009) *Wastewater Solids Incineration Systems*; Manual of Practice No. 30; Water Environment Federation: Alexandria, Virginia.

9.0   Suggested Readings

Baturay, A.; Bruno, J. M. (1990) *Reduction of Metal Emissions from Sewage Sludge Incinerators with Wet Electrostatic Precipitators*. Paper presented at 83rd Annual Meeting of the Air Waste Management Association; Pittsburgh, Pennsylvania, June.

Borghesi, J.; Burrowes, P.; Voth, H.; Flood, R. (2002) A State-of-the-Art Fluid Bed Incineration Process to Meet the Solids Processing Needs of the Twin Cities. *Proceedings of the 16th Annual Water Environment Federation Residuals and Biosolids Management Conference*; Austin, Texas; February; Water Environment Federation: Alexandria, Virginia.

Burrowes, P.; Brady, P. (2005) Development in Emissions Technology for Incinerators. *Proceedings of the 78th Annual Water Environment Federation Technical Exhibition and Conference* [CD-ROM]; Washington, D.C., Oct. 29–Nov. 2; Water Environment Federation: Alexandria, Virginia.

Burrowes, P.; Borghesi, J.; Quast, D. (2007) The Twin Cities Sludge-to-Energy Plant Reduces Greenhouse Gas Emissions. *Proceedings of the 80th Annual Water Environment Federation Technical Exhibition and Conference* [CD-ROM]; San Diego, California, Oct. 13–17; Water Environment Federation: Alexandria, Virginia.

Haug, R. T.; et al. (1983) Thermal Pretreatment of Sludges: A Field Study. *J. Water Pollut. Control Fed.*, **55**, 23.

Los Angeles County Sanitation Districts (1977) *Assessment of Existing and Past Sludge Product Marketing Experiences*; Technical Report 9; Prepared for Los Angeles/Orange County Metropolitan Area Project; Los Angeles County Sanitation Districts: Whittier, California.

Los Angeles County Sanitation Districts (1978) *Carver Greenfield Process Evaluation: A Process for Sludge Drying*; Prepared for Los Angeles/Orange County Metropolitan Area Project; Los Angeles County Sanitation Districts: Whittier, California.

Marshall, D. W.; Gillespie, W. J. (1974) Comparative Study of Thermal Techniques for Secondary Sludge Conditioning. *Proceedings of the 29th Purdue Industrial Waste Conference*; West Lafayette, Indiana; Purdue University: West Lafayette, Indiana.

Metcalf and Eddy, Inc./AECOM (2014) *Wastewater Engineering: Treatment and Resource Recovery*, 5th ed.; Tchobanoglous, G.; Stensel, H. D.; Tsuchihashi, R.; Burton, F. L., Eds.; McGraw-Hill: New York.

Metro Wastewater Reclamation District (2006) *Biosolids Management Program/Facility Study PAR 880*; Metro Wastewater Reclamation District: Denver, Colorado.

Morton, E. L. (2006) A Sustainable Use for Dried Biosolids. *Proceedings of the 79th Annual Water Environment Federation Technical Exhibition and Conference* [CD-ROM]; Dallas, Texas, Oct. 21–25; Water Environment Federation: Alexandria, Virginia.

Orange County Sanitation District (2003) *Long-Term Biosolids Master Plan*; Job No. J-40-7; Orange County Sanitation District: Huntington Beach, California.

U.S. Environmental Protection Agency (1978) *Effects of Thermal Treatment of Sludge on Municipal Wastewater Treatment Cost*, EPA-600/2-78-073; U.S. Environmental Protection Agency, Office of Water Management: Cincinnati, Ohio.

U.S. Environmental Protection Agency (1987a) *Dewatering Municipal Sludges Design Manual*, EPA-625/1-87-014; U.S. Environmental Protection Agency, Office of Water Management: Cincinnati, Ohio.

U.S. Environmental Protection Agency (1987b) *Aqueous-Phase Oxidation of Sludge Using the Vertical Reaction Vessel System*, EPA-600/2-87-022; U.S. Environmental Protection Agency, Office of Water Management: Cincinnati, Ohio.

CHAPTER 25

Use and Disposal of Residuals and Biosolids

Kari Fitzmorris Brisolara, ScD, MSPH, QEP; Lisa Boudeman;
Lauren McDaniel, MPH; Helena Ochoa; and A. Robert Rubin, Ph.D.

1.0 Introduction

The beneficial use and disposal options available for wastewater solids rely heavily on the type of solids involved. Publicly Owned Treatment Works (POTWs, the term used in the regulations) manage wastewater solids in one of three major forms: sludge, biosolids, or ash. The major management options available for each of the solids products include:

- Sludge. Sludge is the term used for raw, unstabilized, primary, and secondary solids. In many states, dewatered, unstabilized sludge has two major end-use options: incineration and landfilling. Other conversion technologies are under development for unstabilized solids, such as gasification and pyrolysis, but have not yet been commercially proven for sludge. All other end-use options, such as land application, require that solids must first meet the U.S. Environmental Protection Agency's (U.S. EPA's) requirements in 40 CFR Part 503, *Standards for the Use or Disposal of Sewage Sludge* (also called Part 503).

- Biosolids. Biosolids are any solids that have been stabilized to meet the criteria in the Part 503 regulations and, therefore, can be beneficially used or landfilled depending upon state requirements. (For more information on Part 503, see Chapter 18.) There are wide varieties of stabilization processes, which produce different types of biosolids (e.g., liquid or dewatered biosolids, compost, heat-dried biosolids, and alkaline-stabilized biosolids). Most of these products can be land applied; only the highest quality biosolids are suitable for commercial marketing and distribution.

- Ash. Ash is a product of incineration. Ash historically was landfilled, but in recent years, there has been more emphasis on finding beneficial uses for this material (e.g., as landfill cover, a soil amendment, an ingredient in concrete, a fine aggregate in asphalt, a flowable fill material, and an additive in brick manufacturing).

However, no matter which beneficial use or disposal option is selected, the project team should review *National Manual of Good Practice for Biosolids* (NBP, 2005) for guidance on developing and implementing biosolids management practices that emphasize environmental stewardship and strong community relations. Published by the National Biosolids Partnership (NBP), this manual is the foundation of NBP's environmental management system (EMS) program (http://www.biosolids.org).

2.0 Land Application

Land application is the practice of adding biosolids to land for beneficial purposes (e.g., to promote crop growth, to promote forest growth, to improve soil quality, and to reclaim former mining sites and other disturbed land). In these applications, both plants and soil benefit from the nutrients and organic matter in biosolids.

The popularity of land application has grown dramatically over the last 30 years. Rising disposal costs and encouragement from the U.S. EPA have substantially increased the number of facilities undertaking land-application programs. According to a 2004 survey conducted by the North East Biosolids and Residuals Association (NEBRA), 55% of the biosolids generated nationwide were "applied to soils for agronomic, silvicultural, and/or land restoration purposes, or were likely stored for those purposes," and 74% of land-applied biosolids were used for agricultural purposes (NEBRA, 2007).

The NEBRA survey also noted the term beneficial use historically referred to "biosolids that are applied to soils to take advantage of the nutrients and organic matter they contain." In the future, it noted, this definition may be too narrow as biosolids are used in ways that provide other benefits (e.g., energy).

2.1   Regulatory Considerations

Federal and state regulations establish controls for the land application of biosolids. Some activities (e.g., the selection and management of land-application sites) also can be governed on a local level. The U.S. EPA has delegated responsibility for compliance with federal regulations to a limited number of states.

2.1.1   Federal Requirements

At the federal level, biosolids are regulated under Section 405(d) of the Clean Water Act. Specific criteria are set forth in 40 CFR Part 503 (U.S. EPA, 1993a), which is the baseline for beneficial use across the country. States and local jurisdictions may add more stringent or site-specific requirements. (For more information on Part 503, see Chapter 18.)

2.1.2   State Requirements

Some states have enacted regulations more stringent than Part 503. Wastewater treatment professionals should consult state regulatory requirements when assessing the feasibility of land application or planning a land-application program.

State requirements vary. However, an increasing number of states require nutrient management planning and some states (e.g., Virginia) may require conservation planning (Evanylo, 1999). In addition, a number of states require permits for land-application sites.

2.1.2.1   Nutrient Management Planning   Land-applied biosolids supply nutrients (e.g., nitrogen, phosphorus, and micronutrients) to plants. However, too much of any nutrients can lead to water quality issues. Excess nitrogen is associated with groundwater concerns, for example, and excess phosphorus is associated primarily with surface water concerns. (For more information on nutrients and biosolids, see *Comparing the Characteristics, Risks and Benefits of Soil Amendments and Fertilizers Used in Agriculture* [Moss et al., 2002] and *National Manual of Good Practice for Biosolids* [NBP, 2005].)

Part 503 requires that biosolids be applied at the agronomic rate for nitrogen (i.e., the rate needed to meet a given crop's nitrogen requirements), and many state regulations share this requirement. However, increasingly more states are requiring that land appliers manage phosphorus as well. Phosphorus-based management can quadruple the area needed for land-application programs. Some of the states that require phosphorus-based management rely on the Nutrient Management Standard (Code 590) issued by the U.S. Department of Agriculture's (USDA's) Natural Resources Conservation Service (NRCS) (USDA-NRCS, 1999). Where used, this national standard (called the P index) typically is modified by states to account for local conditions.

2.1.2.2   Site Permits   Site-permitting programs typically are administered by state agencies, although local entities can be involved in some states. The permits contain the issuing agency's specific requirements for land application, which are designed to

Feature	Setback Distance	
	ft	**m**
Public road	0–50	0–15
House	20–500	6–152
Well	100–500	30–152
Surface water	25–300	8–91
Property line	None listed–100	None listed–30
Intermittent stream	10–200	3–61

TABLE 25.1 Representative Setback Requirements for Land Application

ensure that the biosolids are being beneficially used, not disposed. Permit requirements also are typically designed to protect public health, surface water, and groundwater, as well as address the aesthetic concerns (e.g., odor) of the land-application sites' neighbors.

For example, most state programs impose separation distances (setbacks) between the areas receiving biosolids and adjacent site features (e.g., developments, dwellings, surface water, wells, roads, and rights of way). Setbacks are based primarily on the potential for surface runoff, leaching, and aesthetic concerns and will vary depending on the environmental resource being protected (see Table 25.1). Many states will reduce the setback requirements if the biosolids are immediately incorporated into the soil, applied via subsurface injection, or if the neighboring property owner(s) waive the setbacks (Forste, 1996).

2.2   Project Planning

A successful land-application program is based on the careful consideration of multiple factors (e.g., the biosolids to be applied, the suitability of proposed land-application sites, and the estimated application rates [which determine the amount of land required]).

2.2.1   Biosolids Characteristics and Suitability

When biosolids are used in agriculture, regulations and crop needs typically determine which biosolids characteristics (e.g., pathogen content, vector attraction, and nutrient content [typically nitrogen, phosphorus, and/or calcium]) will help to drive the development of a land-application program (see Table 25.2). Site conditions and economics (e.g., distance to land-application sites, application equipment typically in use) also may determine whether the biosolids should be applied in liquid or cake form.

2.2.1.1   Physical Characteristics   The textures and particle sizes of biosolids are as varied as the processes used to create them. Heat-dried biosolids are typically granular, and can range in size from a fine dust to irregular or spherical particles, the size of a small ball bearing, to larger "noodle-like" pieces up to a few inches long. Composts have a texture similar to that of peat, but depending on how they were made, they may contain some wood fibers or wood chips. Alkaline-stabilized materials may be clay-like or soil-like, depending on moisture content and other factors. Aerobically and anaerobically digested biosolids may be liquid or dewatered into a clay-like material.

Property	Class A Alkaline Stabilized	Compost	Liquid	Cake	Heat-Dried
Total solids (%)	30–65	58	4	22	95
Volatile solids (% of total solids)	12–20	60	61	60	67
Bulk density (kg/m³ [lb/cu ft])	1040–1200 (65–75)	720 (45)	960 (60)	880 (55)	530–740 (33–46)
Organic content (%)	12–20	50–75	60–75	60–75	78
pH	—	—	—	—	6.4–8.0

TABLE 25.2 Typical Biosolids Physical Characteristics (Logan, 2008; Lue-Hing et al., 1998)

2.2.1.2 Pathogens and Vectors Pathogen content of wastewater treatment solids depends on the level of treatment they have received (see Table 25.3). Raw primary solids contain the most pathogens due to lack of treatment. In contrast, biosolids that meet Part 503's Class A standards have very low pathogen concentrations, which can, in some cases, even be below detection limits. Concentrations of specific pathogens (e.g., helminths) depend on the sources in a sewershed, as well as the climate.

Before they can be land applied, biosolids must meet Class A or B pathogen-reduction standards (as well as relevant vector-attraction reduction [VAR] and pollutant concentrations). Land-application sites that receive Class B biosolids have additional management restrictions (e.g., waiting periods before specific crops can be grown) (U.S. EPA, 1994a).

Part 503 requires that Class B biosolids contain less than 2 million colony-forming units (CFU) (most probable number [MPN]) of fecal coliforms per gram of dry biosolids. Class A biosolids must contain less than 1000 MPN/g of fecal coliforms or less than 3

Pathogen	Class A Alkaline Stabilized	Class A Composted	Class A Heat-Dried	40 CFR Part 503 Class A Standards
No. of plants in sample	5	4	5	NA
Fecal coliform (MPN/g mean)	3	76	6	<1000
Fecal coliform (MPN/g median)	1	506	8	NA
Salmonella sp. (MPN/4 g mean)	2	2	0	<3
Salmonella sp. (MPN/4 g median)	2	2	0	NA
No. of plants in sample	55	26	41	NA
Fecal coliform (MPN/g mean)	60	104 600	6521	2 million
Fecal coliform (MPN/g median)	9600	472 600	16 071	NA
Salmonella sp. (MPN/4 g mean)	2070	NA	NA	NA
Salmonella sp. (MPN/4 g median)	4000	NA	NA	NA

TABLE 25.3 Typical Pathogen and Pathogen-Indicator Concentrations in Biosolids (Lue-Hing et al., 1998)

MPN/4 g of *Salmonella* bacteria and meet one of six alternatives. One alternative includes testing for enteric viruses and helminth ova (U.S. EPA, 2003). Larger water resource recovery facilities (WRRFs) may have the personnel, equipment, and accreditation to handle bacterial testing on-site, whereas smaller WRRFs may send samples for bacterial testing to commercial laboratories with appropriate accreditation. Virus and helminth ova testing is nearly always performed by commercial laboratories with appropriate accreditation.

Land-applied biosolids also must meet VAR requirements via thermal drying, volatile solids reduction (VSR), oxygen uptake rate, time, and temperature, or pH increases (U.S. EPA, 2003). Eight VAR options are based on process parameters or testing; two are based on land management (e.g., soil incorporation or injection). In order to qualify for exceptional quality (see description in Section 2.2.1.3), the generator must meet one of the first 8 VAR options.

For details on pathogen and VAR requirements, methodologies, monitoring, and testing, see *Environmental Regulations and Technology: Control of Pathogens and Vector Attraction in Sewage Sludge* (also called the "White House document") (U.S. EPA, 2003).

2.2.1.3   Metals   Biosolids contain a wide range of chemical elements, many of which are trace elements, also called heavy metals. This is a misnomer because the density of some trace elements (e.g., boron and arsenic) is too low to be considered a heavy metal. Likewise, not all trace elements behave as metals in the environment. The ones with metallic chemical behavior are those that are cations in the environment [e.g., copper (Cu^{2+}), cadmium (Cd^{2+}), and lead (Pb^{2+})]. Other trace elements are anions in the environment (e.g., arsenic (AsO_4^{3-}), molybdenum (MoO_4^{2-}), and selenium (SeO_4^{2-}). This distinction is important because cationic trace elements are more bioavailable when soil pH is low, while anionic trace elements are more bioavailable when soil pH is high.

When developing 40 CFR Part 503, the U.S. EPA conducted a risk assessment on a number of trace elements in biosolids and developed a hazard index, which considered various pathways of exposure (e.g., direct ingestion of biosolids, plant uptake, and uptake by soil organisms) from trace elements in land-applied biosolids (U.S. EPA, 1995a). The agency eliminated two trace elements (fluorine and iron) based on the hazard index. Then, regulators developed acceptable concentrations for the remaining 10 elements based on a pathway analysis. The ceiling concentration is the maximum safe concentration for biosolids land applied at agronomic rates. A lower concentration also was established for what was later called exceptional quality (EQ) biosolids (this term refers to biosolids that meet Class A pathogen-reduction requirements, meet VAR requirements [503.33(a)(1) through (8)] at the same time or after pathogen reduction is met, and have low concentrations of regulated pollutants [503.13, Table 3]). The intent of the concentrations for EQ biosolids is that these levels would be safe no matter how much biosolids were applied to land. It is important to note that while biosolids may meet the metals levels associated with EQ biosolids, they cannot be considered EQ biosolids without meeting the Class A pathogen requirements.

Part 503 originally regulated 10 trace elements: arsenic, cadmium, chromium, copper, lead, mercury, molybdenum, nickel, selenium, and zinc. Lawsuits filed against the U.S. EPA resulted in the removal of chromium as a regulated metal and the removal of the EQ limit for molybdenum.

When Part 503 was promulgated, there were a number of municipal WRRFs whose biosolids failed to meet the EQ limits for one or more elements (see Table 25.4). Most,

Metal	Class A Alkaline Stabilized	Compost	Liquid	Cake	Heat-Dried	40 CFR Part 503 Table 3 Limits
No. of treatment plants in sample	7	10	48	117	5	NA
Arsenic	5.79	5.39	7.78	13.86	5.58	41
Cadmium	3.07	4.64	4.90	7.03	8.94	39
Chromium	50.3	72.7	62.0	119.4	152.6	NA
Copper	176	317	448	559	472	1500
Lead	62.0	80.4	74.8	128.6	93.3	300
Mercury	0.73	1.94	2.60	2.01	1.60	17
Molybdenum	8.20	13.8	11.0	15.3	17.9	75
Nickel	38.8	26.5	33.5	69.3	35.0	420
Selenium	2.43	3.73	5.86	6.06	8.32	100
Zinc	878	878	807	886	906	2800

TABLE 25.4 Typical Metals Concentrations in Biosolids (mg/dry kg) (Lue-Hing et al., 1998)

however, met the ceiling concentrations, so their biosolids could be land applied but there was a limit on the cumulative loading to a given site (U.S. EPA, 1995a). By the late 1990s, industrial pretreatment was so successful that most WRRFs could meet EQ limits. Today, trace elements in biosolids are no longer considered an important limitation in land-application programs.

2.2.1.4 Nutrients Biosolids contain various concentrations of macro and micro nutrients. As shown in Table 25.5, the nitrogen content in biosolids ranges from 1.0% (dry weight) for Class A alkaline-stabilized biosolids to up to 6.0% (dry weight) for heat-dried biosolids. Primary and waste activated solids mostly contain organic nitrogen (in the form of protein) while up to a third of total nitrogen in aerobically and anaerobically digested biosolids is ammonia. Biosolids contain very little, if any, nitrate.

The total phosphorus content in biosolids typically ranges from 0.9% to 3.1% (dry weight). This includes both organic phosphorus and various forms of inorganic phosphorus. Municipal WRRFs that use iron and aluminum salts for phosphorus

Nutrient Content (as % of Dry Weight)	Class A Alkaline Stabilized	Compost	Liquid	Cake	Heat-Dried
Nitrogen (N)	1.0	2.8	5.3	4.1	6.0
Total phosphorus (P)	0.4	1.7	2.2	2.2	3.1
Potassium (K)	0.3	0.3	0.3	0.2	0.3

TABLE 25.5 Primary Nutrient Concentrations in Biosolids (adapted from Moss et al., 2002; reprinted with permission from the Water Environment Research Foundation)

removal will have solids that contain inorganic phosphorus in the form of iron and aluminum phosphates. The WRRFs that use lime for dewatering or disinfection will have solids that contain inorganic phosphorus in the form of calcium phosphates. Iron, aluminum, and calcium phosphates are relatively insoluble, which will limit the availability of P in soils receiving repeated applications of biosolids.

Biosolids also contain small amounts of potassium—typically 0.1% to 0.3% (dry weight)—and are not a significant source of potassium for crops. Most of the potassium in biosolids is in a water-soluble form.

In addition, biosolids contain various concentrations of other macro- and micro-nutrients (e.g., calcium, magnesium, sulfur, iron, manganese, as well as the regu-lated elements [copper, molybdenum, selenium, zinc]). Some plant species also require nickel for growth. The plant availability of these nutrients depends on the major chemical phases in the biosolids. Elements like copper and zinc are known to be complexed with the biosolids organic matter, and most sulfur in biosolids is in the form of protein. On the other hand, iron and manganese exist as relatively insol-uble oxides.

2.2.1.5   Other Constituents   Biosolids contain myriad organic and inorganic chemical compounds. Some are discharged into the collection system as household chemicals (e.g., surfactants and conditioning agents used in detergents) and very low levels of medicines (e.g., birth-control chemicals) and concentrate in biosolids. Biosolids also contain natural materials (e.g., silica and aluminosilicate clays) that enter the collection system as sediments. For quantitative estimates, the U.S. EPA has conducted three national sewage sludge surveys with the most recent in 2009. These survey results pres-ent national estimates of the concentrations of more than 500 pollutants in sewage sludge, including metals; dioxins and dioxin-like compounds; inorganic ions; certain organics (e.g., polycyclic aromatic hydrocarbons, semivolatiles); polybrominated diphenyl ethers (flame retardants); and pharmaceuticals, steroids, and hormones in sewage sludge managed by land application (U.S. EPA, 2009).

In the case of alkaline-stabilized biosolids, lime and lime-containing chemicals (e.g., alkaline coal ash, cement kiln dust, and lime kiln dust) are added to biosolids for disin-fection. The lime reacts with water in biosolids to form calcium hydroxide, which has a pH of 12 to 12.5 and adds soil liming value to the biosolids. Alkaline-stabilized biosol-ids often are used as a substitute for agricultural limestone.

2.2.2   Site Suitability

After establishing that the biosolids are suitable for land application, program staff must determine if suitable land is available.

2.2.2.1   Objectives   The objective of a site evaluation is to select application sites that not only meet the technical requirements of land application but also balance prevailing economic and social constraints. When determining how much land is needed for a land-application program, it is important to note that adequate acreage must be avail-able during all but the most inclement weather conditions. Site availability is deter-mined by local farming practices, and there are periods of the year when certain types of cropland are unavailable for the application of biosolids. Additionally, a number of states have seasonal restrictions, such as prohibitions on applications on frozen or snow-covered ground. Therefore, to ensure land availability, it is advisable to maintain two to three times the acreage actually needed in any given year to accommodate all of the biosolids produced. Typically, storage capacity at the WRRF or the beneficial use site will also be necessary at some time during the year.

2.2.2.2 Resources Soil surveys are a useful source of information for initially evaluating a site's suitability for land application. The surveys provide delineated, broad-scale soil maps on a photographic background, a soil description by series and mapping unit, data on the drainage and agronomic properties of soils, and interpretive tables. Soil surveys are prepared by the USDA's NRCS in cooperation with agricultural experiment stations and local government units. They are available from local NRCS offices or online (http://websoilsurvey.nrcs.usda.gov). The soil surveys are used extensively to identify potential sites that meet the regulatory and agronomic requirements for land application.

2.2.2.3 Site Evaluation Criteria While selection criteria for land-application sites vary from state to state, the *National Manual of Good Practice for Biosolids* discusses a variety of factors that should be considered in any land-application program (NBP, 2005). These selection criteria are summarized below.

Soil surveys contain essential information for determining whether a field has appropriate soil characteristics for biosolids application (e.g., soil texture, erodibility, soil drainage characteristics, and slopes). Favorable soil characteristics are discussed in Appendix C of the *National Manual of Good Practice for Biosolids* (NBP, 2005). U.S. Geological Survey (USGS) quadrangle maps are also helpful during preliminary planning and screening to estimate slope, topography, depressions or wet areas, rock outcrops, drainage patterns, and water table elevations. The following topographical and soil characteristics are unfavorable for biosolids application:

- Steep areas with sharp relief;
- Undesirable soil conditions (sandy soils, shallow, highly erodible, or poorly drained);
- Environmentally sensitive areas (e.g., intermittent streams, ponds);
- Rocky, nonarable land;
- Wetlands and marshes; and
- Areas bordered by surface water bodies without appropriate setback areas.

Verification of candidate sites also is recommended, particularly for small parcels of land that are not adequately represented on surveys or maps. Soils should be evaluated by a person skilled in soil science or a certified professional soil scientist depending upon local and/or state requirements.

In addition to topography and soil characteristics, any preliminary site evaluation should assess water quality-based requirements at each farm (e.g., buffer zones, conservation planning, and nutrient management). Part 503 requires a buffer of at least 10 m (33 ft) to surface waters, and states often have additional buffer requirements to surface water bodies, water supply wells, property lines, outcroppings, or public roads. Because the buffer requirements limit the actual land available for application, factoring in these limitations may make some sites unsuitable for land application.

Some states require that farms implement conservation plans to control soil erosion and nutrient management plans to limit excess nutrient transportation to water resources. Certain conservation measures (e.g., no-till or residue management) and nutrient management issues (e.g., fields with a long history of land-applied manure or biosolids) could limit a site's suitability for land application.

Site criteria important to the economic viability of a land-application program include the ease of access to the farm (e.g., road restrictions or traffic limitations) and the hauling distance or travel time (and therefore cost).

2.2.2.4 Public Perception   Assessing the public acceptance of and attitudes toward land application is an important part of the site evaluation process. The attitudes of stakeholders (e.g., neighbors) can be critical to the success of a land-application program, and discussions with farmers or landowners beforehand often can provide key insights in this regard.

Farmer and landowner attitudes are critical to program success as well. According to the *National Manual of Good Practice for Biosolids* (NBP, 2005), the landowner or farmer "must be confident about the benefits of biosolids and be willing to accept and comply with all of the...regulatory requirements, as well as meet biosolids program needs." To maximize program success, POTWs should provide outreach and education to famers and to all stakeholders during the evaluation process.

2.3   Design and Implementation

2.3.1   Application Rates

When determining application rates for biosolids, wastewater treatment professionals must consider metals content, nutrient content (nitrogen and occasionally phosphorus), and, for liming applications, calcium content as well. The application rate selected will be the lowest of those calculated for each appropriate parameter.

2.3.1.1 Metals   Part 503 regulates the loading rates of the elements identified in Section 2.2.1.3. Biosolids whose metal and heavy metal concentrations exceed the EQ limit but are lower than ceiling concentrations can be land applied up to the cumulative loading limit (U.S. EPA, 1993a) defined in Part 503. There also is a maximum annual loading rate. Arsenic, for example, has an EQ limit of 41 mg/kg, a ceiling concentration of 75 mg/kg, an annual loading limit of 2.0 kg/ha·yr, and a cumulative loading limit of 41 kg/ha. Several states have lower loading limits than Part 503, and some regulate other elements (e.g., thallium in New York).

2.3.1.2 Nutrients   Part 503 restricts the land-application rate on agricultural land based on the crop's nitrogen requirements and the available nitrogen in biosolids. A given crop's nitrogen requirement is location specific; the data are readily obtained from state agricultural extension publications. The available nitrogen in biosolids is the sum of its ammonia-nitrogen content (which is assumed to be 100% plant available) and a percentage of its organic nitrogen content. Organic nitrogen becomes plant available by converting to ammonia and then nitrate via mineralization by soil bacteria. A typical mineralization rate for aerobically digested biosolids is 30% in the first year of application. Mineralization rates for anaerobically digested and composted biosolids are typically 20% and 10%, respectively (Sommers et al., 1981). Recent work performed by Gilmour et al. (2000), however, recommends a more site-specific approach that may result in the use of higher mineralization rates for available nitrogen. If biosolids are not incorporated or injected (e.g., at no-till sites), a percentage of the ammonia-nitrogen (typically 50%) is assumed to be lost via volatilization. For specific details on calculating available nitrogen in biosolids, see state biosolids regulations and recommendations.

The phosphorus in biosolids has variable plant availability (see Section 2.2.1.4), which is assessed by evaluating the effect of biosolids application rates on soil test phosphorus. Unlike nitrogen, phosphorus accumulates in soil, so repeated biosolids applications will increase soil test phosphorus concentrations. When biosolids are land applied at nitrogen rates, its phosphorus content typically exceeds the phosphorus requirement of the crops, leading to phosphorus accumulation in some soils. Soil test phosphorus is strongly correlated with phosphorus in surface runoff, and states are beginning to regulate phosphorus applications to agricultural soils to protect water quality.

The NRCS has developed a soil model for phosphorus applications called the phosphorus index, and some states have adopted various forms of it to regulate fertilizer, manure, and biosolids application rates. As used in some states, the phosphorus index includes a factor that reflects the bioavailability of the material being applied, and some reflect the solubility of source-specific P in terms of a P source coefficient (PSC). For example, biosolids with high levels of iron, aluminum, and calcium will have lower phosphorus bioavailabilities than aerobically or anaerobically digested biosolids (although this difference is not always reflected in phosphorus index calculations).

Because phosphorus-based management requirements are state specific, planners should contact state regulators to define potential requirements.

2.3.1.3 Calcium Alkaline-stabilized biosolids contain various concentrations of calcium in the form of calcium hydroxide and calcium carbonate. When lime is added to biosolids, it picks up water from the biosolids and is rapidly converted to calcium hydroxide. Over a longer period (weeks and months), calcium hydroxide is converted to calcium carbonate by interacting with carbon dioxide in the atmosphere. Class B alkaline-stabilized biosolids contain between 2% and 5% lime (dry weight) or its equivalent, while Class A alkaline-stabilized biosolids contain between 12% and 30% lime (dry weight) or its equivalent, depending on which Class A alkaline-stabilization process is used.

The high lime content in alkaline-stabilized biosolids makes these materials excellent substitutes for agricultural limestone (calcium carbonate or dolomite), which is applied to soils that are naturally acidic or have become so as a result of chemical fertilizer and organic matter. In some states, the land-application rates are based on the results of tests that determine the soil's lime requirement and the liming value of alkaline-stabilized biosolids.

Soil pH is not an effective measurement of soil acidity, and separate tests must be conducted to measure soil acidity. The lime-requirement soil test is state specific; state agricultural extension services provide recommendations on test methods and the lime requirements of various crops. The results of the lime-requirement soil test typically include soil acidity and a recommended application rate (in metric ton/ha or ton/ac) of agricultural limestone required to neutralize soil acidity to a target pH value.

The test to determine the liming value of biosolids is based on acid-neutralizing capacity. Results are expressed as calcium carbonate equivalency (CCE), which represents the percentage of pure calcium carbonate, on either a wet-weight or a dry-weight basis. Standard tests to determine CCE are the ASTM International Standard C-25-06, *Standard Test Methods for Chemical Analysis of Limestone, Quicklime, and Hydrated Lime*

(ASTM International, 2006), and one derived from American Water Works Association Standard B202-07 (AWWA, 2008). The alkaline-stabilized biosolids' CCE is used to adjust the lime requirement. For example, alkaline-stabilized biosolids with a CCE of 25% would have to be applied at a rate of 8 metric ton/ha (3.6 ton/ac) to satisfy a lime requirement of 2 metric ton/ha (0.90 ton/ac) of agricultural limestone.

2.3.1.4   Design Examples

2.3.1.4.1   Nutrient and Metals-Based Land Application   A WRRF plans to surface apply biosolids to hay crops. The facility produces 3 dry metric ton/d of liquid, aerobically digested biosolids. The biosolids will not be incorporated. The fields have not received soil amendments or fertilizers in the past. The aerobically digested biosolids have the following characteristics:

- Total nitrogen content = 3%.
- Organic-nitrogen content = 1%.
- Ammonia-nitrogen content = 2%.
- Nitrate content = 0%.
- Total phosphorus content = 1%.
- The limiting metal is copper at 400 mg/dry kg solids.

1. Determine the land-application rates and required acreages on (A) a nitrogen-limiting basis and (B) a phosphorus-limiting basis. Assume that the annual crop requirement is 235 kg nitrogen/ha·yr (210 lb nitrogen/ac/yr) and 73 kg phosphorus/ha·yr (65 lb phosphorus/ac/yr).
2. Determine the metal-limiting land-application rate.
3. Which is more limiting: crop nutrient requirements or metals limits?

2.3.1.4.1.1   SOLUTION (1A)   First, the plant-available nitrogen (N_{PA}) in biosolids can be calculated as follows:

$$N_{PA} = 1000 [(NH_3)K + NO_3 + N_O f] \qquad (25.1)$$

where N_{PA} = plant-available nitrogen (kg/metric ton dry solids);
NH_3 = percent ammonia-nitrogen in biosolids (as a decimal);
K = volatilization factor for ammonia;
NO_3 = percent nitrate in biosolids (as a decimal);
N_O = percent organic nitrogen in biosolids (as a decimal); and
f = mineralization factor (conversion of organic nitrogen to ammonium-nitrogen).

K depends on the application method used (see Table 25.6).

Because the biosolids will not be incorporated, K is assumed to be 0.5 (50% loss of ammonia-nitrogen). The mineralization factor f can be determined via Table 25.7. Because this is the first year of application and aerobically digested biosolids are being applied, f can be assumed to be 0.3. Therefore,

$$N_{PA} = 13 \text{ kg nitrogen/metric ton dry solids (26 lb nitrogen/ton dry solids)}$$

If Solids Are	K Factor Is
Liquid and surface applied	0.5
Liquid and injected into the soil	1.0
Dewatered and applied in any manner	1.0

TABLE 25.6 Volatilization Factor (K) for Ammonia (U.S. EPA, 1994b)

Time After Sludge Application (Year)	Percent of Organic Nitrogen Mineralized from Stabilized Primary Solids and WAS	Percent of Organic Nitrogen Mineralized from Aerobically Digested Biosolids	Percent of Organic Nitrogen Mineralized from Anaerobically Digested Biosolids	Percent of Organic Nitrogen Mineralized from Composted Biosolids
0–1	40	30	20	10
1–2	20	15	10	5
2–3	10	8	5	3
3–4	5	4	3	3
4–5	3	3	3	3

TABLE 25.7 Mineralization Factor (f) for Nitrogen in Various Types of Biosolids (U.S. EPA, 1994b)

Next, the nitrogen-limiting biosolids loading (R_N) can be calculated as follows:

$$R_N = \frac{U_N}{N_{PA} + N_{PM}} \tag{25.2}$$

where U_N = annual crop requirement for nitrogen,
N_{PA} = plant-available nitrogen from this year's sludge application, and
N_{PM} = plant-available nitrogen from mineralization of all previous applications.

As mentioned above, U_N is 235 kg N/ha·yr (210 lb N/ac/yr). In addition, N_{PM} is zero because this is the first year of application. Therefore,

R_N = 18 metric ton dry solids/ha·yr (8.0 ton dry solids/ac/yr)

The total land required on a nitrogen-limiting basis is then

(3 metric ton solids/d) × (365 days)/(18 metric ton solids/ha·yr)
= 61 ha (150 ac)

2.3.1.4.1.2 SOLUTION (1B) First, the total phosphorus percentage must be converted to the percentage of plant-available phosphorus (P_2O_5). This is done as follows:

P_2O_5 = total phosphorus (%) × 2.29
= 22.9 kg P_2O_5/dry metric ton solids (45.8 lb P_2O_5/dry ton solids)

The phosphorus-limiting biosolids loading (R_p) can be calculated using a formula analogous to that for nitrogen. In this case, the phosphorus crop requirement (U_p) is 73 kg phosphorus/ha · yr (65 lb phosphorus/ac/yr), all P_2O_5 is plant available (P_{PA}), and the plant-available phosphorus from previous applications is zero because this is the first year of biosolids application. Thus,

$$R_p = 3.2 \text{ metric ton dry solids/ha · yr (1.4 ton dry solids/ac/yr)}$$

The total land required on a phosphorus-limiting basis is then

$$(3 \text{ metric ton/d}) \times (365 \text{ days})/(3.2 \text{ metric ton/ha · yr})$$

$$= 344 \text{ ha (851 ac)}$$

2.3.1.4.1.3   SOLUTION (2)   Because copper is the limiting metal, the lifetime biosolids application rate is calculated based on the maximum lifetime cumulative loading for copper (1500 kg/ha [1340 lb/ac]) listed in 40 CFR Part 503 Table 2. The biosolids contain 400 mg copper/kg solids, which is equivalent to 0.4 kg copper/metric ton dry solids (0.8 lb copper/ton dry solids). Therefore, the metal-limiting solids loading rate is 3750 metric ton dry solids/ha (1675 ton dry solids/ac).

2.3.1.4.1.4   SOLUTION (3)   A comparison of the nitrogen-, phosphorus-, and metal-limiting solids-loading rates calculated above shows that crop nutrient requirements are the limiting factors when determining appropriate biosolids loading rates:

- Nitrogen-limiting rate = 18 metric ton dry solids/ha·a (8.0 ton dry solids/ac/yr);
- Phosphorus-limiting rate = 3.2 metric ton dry solids/ha·a (1.4 ton dry solids/ac/yr); and
- Metal-limiting rate = 3750 metric ton dry solids/ha (1675 ton dry solids/ac).

2.3.1.4.2 Alkaline Stabilized Biosolids Land Application   An agricultural operation finds that its soil has become more acidic over the years, and it wants to apply Class A alkaline-stabilized biosolids to raise the soil's pH. The biosolids' average solids content is 35%. A laboratory tested the soil's acidity and gave a liming recommendation of 2 metric ton/ha of agricultural limestone (0.9 ton/ac). The biosolids' CCE is 60% on a dry-weight basis. Determine the appropriate land-application rate for the alkaline-stabilized biosolids.

The laboratory provided a recommended limestone application rate assuming that the agricultural limestone has 100% purity as calcium carbonate. Because biosolids do not have 100% purity, the application rate must be adjusted based on the biosolids' CCE. First, the CCE on a dry-weight basis must be converted into a wet-weight basis:

$$(0.60 \text{ kg as } CaCO_3/\text{kg dry biosolids}) \times (0.35 \text{ kg dry biosolids/kg wet biosolids})$$

$$= 0.21 \text{ kg as } CaCO_3 \text{ per kg of wet biosolids}$$

$$= 21\% \text{ on a wet weight basis}$$

Then, the limestone application rate can be used to determine the biosolids application rate:

$$= (2 \text{ metric ton limestone/ha})/(0.21 \text{ metric ton limestone/metric ton wet biosolids})$$

$$= 9.5 \text{ metric ton/ha } (4.3 \text{ ton/ac})$$

It is important to keep in mind that regulations require the biosolids application rate to be based on nitrogen (and phosphorus, if the state has phosphorus limitations), which should be calculated according to crop requirements to ensure that loading limits are not exceeded. If the nutrient-based loading rate is less than the CCE rate, then program staff should use the nutrient-based rate and add supplemental limestone, or else complete the liming over a second year.

2.3.2   Field Storage

Many land-application programs may be subject to seasonal, weather, or other factors that temporarily limit the ability to land-apply material and, therefore, need short-term storage. In most states, biosolids cake, alkaline-stabilized biosolids, compost, or heat-dried biosolids can be stored for up to 6 months to 1 year at a beneficial use site (depending on local regulations), in either stockpiles or constructed storage facilities unless otherwise restricted by state or local regulations. For details on properly siting, managing, and operating stockpiles, see the *Guide to Field Storage of Biosolids* (U.S. EPA, 2000). For information on designing constructed storage facilities, see the guide and Table 25.8 along with any pertinent state and (local) regulations.

2.3.3   Odor Management

Odors are frequently cited as the basis for opposing land-application programs. One approach to minimizing the potential for offsite odors is maximizing the buffers between land-application fields and the public. As many agencies have found, however, development can encroach on previously remote sites, reducing buffers and increasing the potential for odor complaints. Therefore, odor-management plans cannot be based on setbacks and buffers alone. The *National Manual of Good Practice for Biosolids* (NBP, 2005) recommends a comprehensive odor-management approach that addresses liquid and solids treatment at the WRRF, transportation, site storage, field operations, and community relations. Addressing potential odors (and their effects) at each "critical control point" is essential to manage odors at land-application sites effectively. In all cases, generating a stable, low-odor product is essential to minimize the potential for odor-related challenges.

2.3.4   Land-Application Equipment and Methods

Biosolids typically are land applied as a cake or as liquid (exceptions include some Class A alkaline-stabilized biosolids, heat-dried biosolids, and composts). Manure spreaders are used to land-apply biosolids cake and, depending on the crop, disks may be used to incorporate the biosolids into the soil. Liquid biosolids can be applied to the soil surface via a tanker spreader or spray irrigation system; they also can be injected into the soil. For a comparison of application methods, along with method details and equipment, see the *National Manual of Good Practice for Biosolids* (NBP, 2005). The manual also discusses calibration needs, which are critical to ensure that the proper application rate is not exceeded.

Issue	Liquid/ Thickened 1–12% Solids Lagoons	Dewatered/Dry Biosolids Facilities 12–30% Solids/50% Solids (Dry) Pads/Basins	Enclosed Buildings	Liquid/Thickened 1–12% Solids Tanks
Design	Below ground excavation. Impermeable liner of concrete, geotextile, or compacted earth.	Above ground, impermeable liner of concrete, asphalt, or compacted earth	Roofed, open-sided, or enclosed. Flooring: concrete, asphalt, or compacted earth	Above or below ground, concrete, metals, or prefab. If enclosed— ventilation needed
Capacity	Expected biosolids volume + expected precipitation + freeboard	Expected biosolids volume, unless precipitation is retained; then, biosolids volume + expected precipitation + freeboard	Expected biosolids volume	Enclosed: expected biosolids volume. If open-top— expected biosolids volume + expected precipitation + freeboard
Accumulated water management	Pump out and spray irrigate or land apply the liquid, haul to WRRF, or mix with biosolids	Sumps/pumps if facility is a basin for collection of water for spray irrigation, land apply or haul to a WRRF	Roof and gutter system, enclosure, or up-slope diversions	Decant and spray irrigate, land apply or haul to WRRF or mix with biosolids in tank
Runoff management	Diversions to keep runoff out of lagoon	Diversions to keep runoff, out of site, curbs and/or sumps to collect water for removal, or downslope filter strips or treatment ponds	Enclosure or up-slope diversions	Prevent gravity outflows from pipes and fittings. Diversions for open, below ground tanks
Biosolids consistency	Liquid or dewatered— removal with pumps, cranes, or loaders	If no side walls, material must stack without flowing	Material must stack well enough to remain inside	Liquid or dewatered biosolids. If enclosed, material must be liquid enough to pump
Safety	Drowning hazard—post warnings, fence, locked gates, and rescue equipment on-site	Drowning hazard— post warnings, fences, locking gates, and rescue equipment on-site	Post "No Trespassing," signs, remote location, lock doors, gates, and fences	Posted warning, locking access points, e.g., use hatches, controlled access ladders, and confined space entry procedures to access

TABLE 25.8   Key Design Concepts for Constructed Biosolids Storage Facilities (U.S. EPA, 2000)

2.3.5 Biosolids Transportation

When selecting transportation routes between a WRRF and a land-application site, design engineers must consider odor and other nuisance factors (e.g., traffic and noise). Clean, covered trucks will help address odor concerns, as will minimizing waiting in line at the site. With respect to traffic, the ability of selected roads to accommodate the weight, width, and turning radii of hauling vehicles must be considered in addition to nuisance concerns. In addition, site operators may provide a wash station and/or require drivers to carry equipment for cleaning tire treads and trailers. The *National Manual of Good Practice for Biosolids* (NBP, 2005) should be consulted for additional discussions of biosolids transportation issues.

2.3.6 Public Acceptance

While farmers typically accept that land-applied biosolids function as a fertilizer, liming agent, or soil conditioner, concerns about public health and environmental safety remain. With this in mind, the NBP created an environmental management system (EMS) for beneficially used biosolids. The system is based on the premise that education and outreach, coupled with exemplary biosolids programs, are critical to gaining public acceptance. The partnership also has developed a set of planning and management documents that help wastewater utilities develop and improve their overall programs—especially their outreach programs.

The planning tools and guidance provided by the NBP include comprehensive guidance on developing an environmentally sound biosolids management program. These documents, which include the *National Manual of Good Practice for Biosolids*, can be found on the NBP's website (NBP, 2009). The WRRFs also can gain public acceptance of their beneficial use programs by active participation in state and regional WEF-sponsored biosolids programs. These organizations promote sound biosolids management and actively communicate with the public.

3.0 Land Reclamation and Other Nonagricultural Uses

In addition to crop-based agriculture, biosolids can be used for land reclamation, silviculture, and other nonagricultural purposes. When developing programs for such uses, many of the recommendations noted above for agricultural programs are applicable. However, the application rate and frequency will be site specific. For example, biosolids used to reclaim land may be applied only once (or infrequently) at a rate of 50 to 150 dry metric ton/ha (NBP, 2005). Biosolids also may be applied infrequently to silviculture sites, but the application rate will be between those for agricultural and reclamation sites (NBP, 2005).

Nonagricultural applications may be subject to both Part 503 and other state and federal requirements. For more information on the regulatory and other issues that should be considered for these programs, see *National Manual of Good Practice for Biosolids* (NBP, 2005). Additionally, the use of biosolids for reclamation and remediation of disturbed soils is detailed extensively by the University of Washington, Center for Urban Horticulture (2002). There is significant documentation related to the success of biosolids in reclaiming mine sites and establishing sustainable vegetation. According to the U.S. EPA (https://www.epa.gov/biosolids/frequent-questions-about-biosolids), "Not only does the organic matter, inorganic matrix and nutrients present in the biosolids reduce the bioavailability of toxic substances often found in highly disturbed mine soils, but also regenerate the soil layer. This regeneration is very important for reclaiming abandoned mine sites with little or no topsoil. The biosolids application rate for mine reclamation is generally higher than the agronomic rate which cannot be exceeded for use of agricultural soils."

4.0   Landfilling

Landfilling is an option for disposing of residuals from wastewater treatment. This section presents information on planning, designing, constructing, monitoring, and closing such landfills. Important aspects include:

- Landfill siting and capacity needs;
- Liner and leachate-collection system design;
- Surface water control;
- Landfilling methods;
- Daily, intermediate, and final covers;
- Monitoring requirements;
- Gas migration control, collection, and reuse;
- Covers and cap systems; and
- Landfill closure and reuse.

4.1   Landfill Types and Regulatory Considerations

There are two types of landfill sites that typically accept solids. One is a monofill—a landfill that only accepts stabilized or unstabilized municipal wastewater solids. The other is a co-disposal landfill, which accepts both residuals and municipal solid waste (MSW).

4.1.1   Monofills

The U.S. EPA regulates monofill design and operation under Subpart C of 40 CFR Part 503, which governs surface disposal of wastewater solids. Part 503 standards include general requirements; pollutant limits; management practices; pathogen-reduction and VAR alternatives; and monitoring, recordkeeping, and reporting requirements. (For more information on 40 CFR Part 503, see Chapter 18.)

Subpart C addresses two types of monofill operations. Unstabilized solids must adhere to VAR Alternative 11 and must be covered with soil or another material at the end of each operating day. Stabilized solids, on the other hand, must meet one of the other VAR alternatives and do not have to be covered each day. Monofills for stabilized solids operate more like land-application sites with exceptionally high application rates. Further guidance on monofills for raw solids is covered in this section (Section 4), while monofills for stabilized solids (also called dedicated land disposal sites) are addressed in Section 5.

4.1.2   Co-Disposal Landfills

The U.S. EPA regulates co-disposal landfill design and operation under 40 CFR 258, which governs the disposal of MSW (e.g., household wastes) (U.S. EPA, 1991). Because wastewater solids typically are a small percentage of the waste at such sites, Part 258 regulations are not referenced in detail in this section.

The disposal rate partially depends on the residuals' solids content. At co-disposal landfills, solids typically are spread in the active area (the part of the landfill accepting waste) and mixed with incoming solid waste to ensure that the material has acceptable handling characteristics. For example, a cake containing 20% solids might be mixed with solid waste 4:1 (i.e., 4 metric ton of solid waste to 1 metric ton of solids [wet ton basis]). Residuals with lower solids contents would need to be mixed with more solid waste. The mixing process (also called bulking operation) typically depends on the type and quantities of wastes delivered to the landfill.

Solids delivered to a co-disposal site must not contain any free liquids, as defined via the paint filter liquids test (U.S. EPA, 1995b). Dewatered cakes containing 20% solids typically pass this test.

4.2 Planning

The first step in planning a new landfill is to identify a site that meets the related design, regulatory, and cost requirements. Project teams typically evaluate and compare several prospective sites before selecting one. The siting process often involves gathering input from both the public and regulators. It may take several years to identify, design, permit, and construct a new landfill.

4.2.1 Siting

A monofill must meet the siting criteria in Part 503 (U.S. EPA, 1994a). Such criteria include:

- The monofill cannot be likely to affect a threatened or endangered species adversely.
- It cannot restrict the flow of a base flood (i.e., a 100-year flood event).
- It must not be in a geologically unstable area.
- It must be at least 60 m from a fault area that experienced displacement in Holocene time.
- If regulators permit the monofill to be located in a seismic impact zone, then it must be designed to resist seismic forces (a seismic impact zone is an area where the ground-level rock has at least a 10% probability of accelerating horizontally more than 0.10 g once in 250 years).
- The monofill cannot be located in a wetland (unless a special permit is obtained).

Some states may have additional siting criteria (e.g., setbacks from property lines, public or private drinking water wells, surface drinking water supplies, and buildings or residences). Design engineers should use these criteria to screen potential sites and select one that meets all applicable criteria.

Design engineers then need to determine the landfill footprint (where actual disposal activities occur) within the site boundaries. This footprint is determined based on proximity to sensitive receptors (e.g., wetlands, residences, water bodies, property boundaries, and roads). Dikes constructed to both structurally support fill material and contain surface runoff surround this footprint. Dikes are made of soil compacted to a specified strength. The footprint also should include special working areas for use in inclement weather or other contingency operations.

Design engineers also should establish appropriate buffer distances between land-fill footprint and sensitive receptors. Buffers are site specific, based on regulations or general guidelines for mitigating adverse environmental effects. The landfill design also must accommodate support facilities, which may include:

- Access roads;
- Administrative offices and employee facilities;
- Equipment storage and maintenance areas;
- Stockpiling areas;
- Utilities;
- Fencing;
- Lighting;
- Truck-washing facilities;
- Leachate storage and pumping stations;
- Monitoring wells; and
- Stormwater detention basins.

4.2.2 Landfill Capacity Needs

When selecting a new site, the landfill footprint and site geometry must provide enough capacity for its entire operational life, which should be at least 20 years. A landfill's operational life is affected by many variables (e.g., the solids production rate, the volume consumed by liners and capping systems, and the volume consumed by bulking soils and cover materials). The volume lost to liner and cover systems is easy to calculate and is based on the thickness of these layers over the landfill area. The volume consumed by cover and bulking material depends on the characteristics (solids content, bulk density, etc.) of these materials. Solids must be mixed with a certain amount of bulking material (e.g., soil) to improve their strength and handling characteristics. Daily and intermediate cover material may consume 20% of the overall landfill volume. Design engineers need to conduct actual capacity assessments for each site. Assessment results will depend on the type of materials to be landfilled, the proposed method of landfill operation, cover requirements, and other factors.

4.3 Landfill Design

4.3.1 Regulatory Requirements for Liners

Part 503 does not require that all monofills (or dedicated land disposal [DLD] sites) have liner systems. The need for a liner is based on the solids' pollutant concentrations; the U.S. EPA allows for disposal without a liner if the solids' arsenic, chromium, and nickel concentrations are below the pollutant limits. The limits are based on the distance between the monofill footprint and the property line (see Table 25.9). If a liner is not used, a sampling and analysis program must be established in accordance with Part 503.

If the solids exceed the pollutant concentration limits, the project team may be able to obtain a site-specific permit from regulators that allows the monofill to be unlined. To do this, the team must demonstrate that site conditions vary significantly from the criteria the U.S. EPA used to derive pollutant concentrations. Otherwise, the U.S. EPA requires that the monofill be lined.

Location in the Part 503 Rule	Distance from the Boundary of Active Biosolids Unit to Surface Disposal Site Property Line (m)	Pollutant Concentration*		
		Arsenic (mg/kg)	Chromium (mg/kg)	Nickel (mg/kg)
Table 2 of section 503.23	0 to less than 25	30	200	210
	25 to less than 50	34	220	240
	50 to less than 75	39	260	270
	75 to less than 100	46	300	320
	100 to less than 125	53	360	390
	125 to less than 150	62	450	420
Table 1 of section 503.23	Equal to or greater than 150	73	600	420

*Dry-weight basis (basically, 100% solids content).

TABLE 25.9 Pollutant Concentration Limits for Unlined Monofills (U.S. EPA, 1994a)

Increasingly, professionals are considering it best engineering practice to design all monofills with linings to prevent the escape of contaminants, regardless of their concentrations. Many states' regulations for monofills have more stringent requirements than Part 503, including the need for a liner system.

4.3.2 Liner Design

Liner systems are used to contain waste in landfills and prevent the migration of leachate constituents out of the landfill (U.S. EPA, 1993b). Three types of materials typically are used: low-permeability-soil (clay) liners, geosynthetic clay liners, and geomembrane liners. A combination of these materials may be used, depending on regulatory or site-specific conditions. Geosynthetic clay liners, which are substitutes for low-permeability soil liners, are manufactured of bentonite clay supported by geotextiles held together by needling, stitching, or chemical adhesives. Geomembranes typically are placed on top of a clay layer (or other low-permeability layer) to form a composite lining.

Part 503 requires that a monofill liner have a maximum hydraulic conductivity (water vapor transmission) of 1×10^{-7} cm/s. The permeability of a geosynthetic clay liner is about 1×10^{-10} cm/s. Geomembrane liners have an average hydraulic conductivity of about 1×10^{-14} cm/s.

For detailed guidance on designing landfill lining systems, see the U.S. EPA's *Solid Waste Disposal Facility Criteria Technical Manual* (U.S. EPA, 1993b) and *Waste Containment Systems, Waste Stabilization and Landfills: Design and Evaluation* (Sharma and Lewis, 1994).

4.3.3 Leachate-Collection System

As water percolates through a monofill, it dissolves ("leaches") various solids constituents and becomes polluted. One of the most important considerations in the design, operation, and long-term care of a landfill is leachate collection and management. A leachate-collection system should minimize the leachate's hydraulic head on the primary liner during landfill operations; it should be able to maintain a leachate head of less than 0.3 m (1 ft). The collection system also should remove leachate from the landfill through the post-closure monitoring period.

The leachate-collection system consists of a drainage layer (e.g., sand or a geonet), leachate-collection pipes, a sump or a series of sumps, and pumps to transport leachate to an on-site treatment system or to a sanitary sewer. Cleanouts should be provided in the collection-system piping. The design and layout of the collection system must be compatible with the phased development of the monofill. Each phase's liner and collection system should be installed concurrently. In addition, later segments of the collection system should easily connect with previously installed piping.

The leachate-control system should include facilities to monitor leachate leaks at the base of the landfill and to withdraw leachate and, therefore, prevent buildup that would promote leachate migration from the landfill.

For detailed guidance on designing leachate-collection and -removal systems, see the U.S. EPA's *Solid Waste Disposal Facility Criteria Technical Manual* (U.S. EPA, 1993b) and *Waste Containment Systems, Waste Stabilization and Landfills: Design and Evaluation* (Sharma and Lewis, 1994).

4.3.3.1   Leachate Quantity   The amount of leachate generated in both the active and closed areas of a landfill depends on the amount of infiltration from the landfill surface. Various mathematical models are available to predict the amount of leachate produced during landfill operations. One commonly used model is the U.S. EPA's *Hydrologic Evaluation of Landfill Performance (HELP) Model*, which estimates the amounts of surface runoff, subsurface drainage, and leachate that may be expected in both the active and closed areas (U.S. EPA, 1994c). One advantage of this model is that it allows users to differentiate layers within the landfill (e.g., topsoil, sand drainage, barrier soil, and waste layer).

Many default values are available for various liner and leachate-collection components. Required model input parameters (e.g., porosity, field capacity, wilting point, and saturated hydraulic conductivity) can be determined by testing solids samples from the WRRF. For example, Hundal et al. (2005) found that biosolids from the Chicago area had a hydraulic conductivity (saturated potassium) of 67 to 118 mm/h (17 to 30 in./hr); 54% to 74% porosity; and 18% to 25% plant-available water.

4.3.3.2   Leachate Quality   Biological decomposition is the primary solids-degradation mechanism that causes contaminant leaching (see Table 25.10). Biological decomposition begins aerobically, but as oxygen is depleted, aerobic microorganisms give way to anaerobes. Because anaerobic decomposition is slow, organic contaminants may take several decades to degrade. Leachate production may continue for years, but the leachate's strength gradually decreases. Collected leachate must be treated on-site or transported to the WRRF.

Parameter	Concentration
Chemical oxygen demand, mg/L	2258
Total organic carbon, mg/L	737
pH	6.2
Volatile acids	1213
Volatile solids	5555

TABLE 25.10   Average Leachate Values for Solids-Only Test Cells (U.S. EPA, 1995b)

4.3.4 Surface Water Control

Adequate drainage is essential to good monofill operations. If precipitation-related surface water is not controlled, it ponds on the liner, increases leachate generation, and causes operating problems at the working face and cover-material storage areas. To control surface water, design engineers should consider vegetation, compaction, drainage structures, holding ponds, the type and thickness of cover material, and modifications to surface slope and slope length.

Surface slope and slope length determine the expected degree of erosion. The top surface grade should be between 2% and 5% to promote runoff, inhibit ponding, and minimize soil erosion by keeping flow velocities relatively low. Side slopes should be a maximum of three horizontal to one vertical (3:1) and require more care in seeding and runoff protection.

To the maximum extent possible, run-on should be diverted from the landfill footprint. Runoff water from active landfill areas (which may have contacted solids or solids-contaminated soils) must be drained to the leachate system and treated as leachate, or stored separately, tested, and treated to ensure that it meets discharge standards. Meanwhile, the U.S. EPA requires that surface water runoff be collected and disposed of in accordance with National Pollutant Discharge Elimination System (NPDES) requirements. Both runoff and run-on collection systems must be designed to handle a 25-year, 24-hour rain event to ensure that contaminants are not released to the environment.

4.4 Groundwater Monitoring Requirements

Both Part 503 and Part 258 require that a landfill not contaminate an aquifer. Under Part 503, a monofill has contaminated an aquifer if it causes the aquifer's nitrate concentration to exceed the maximum contaminant level (MCL), which is 10 mg/L. To avoid this situation, most states require that the groundwater at monofill sites be monitored. In fact, state or local regulators typically require routine sampling for several parameters.

When designing a groundwater monitoring network, engineers should start by evaluating the site's geological and hydrogeological conditions (e.g., groundwater elevations and flowrate, and soil and bedrock conditions). Groundwater quality should be monitored at up-gradient locations to provide background data. Down-gradient wells are used to determine whether and how the landfill has affected groundwater quality. The number and type of wells is site specific, and may include overburden wells at various depths and bedrock wells.

For more guidance on the selection and placement of groundwater monitoring wells, monitoring procedures, data analysis, and recordkeeping, see the U.S. EPA's *Solid Waste Disposal Facility Criteria Technical Manual* (U.S. EPA, 1993b) and *Waste Containment Systems, Waste Stabilization and Landfills: Design and Evaluation* (Sharma and Lewis, 1994).

4.5 Landfill Gas Management

Landfill gas primarily consists of methane and carbon dioxide. Methane is a combustible gas that is explosive at atmospheric concentrations of 5% to 15%. Under stable anaerobic conditions, landfill gas can contain between 45% and 55% of methane. The remaining 45% to 55% primarily consists of carbon dioxide, along with small amounts of hydrogen, oxygen, nitrogen, and traces of other gases.

Because of the potential explosion hazard, the U.S. EPA has established monitoring requirements for methane gas at landfills. Methane's lower explosive limit (LEL) is 5%

(by volume). (An LEL is the concentration above which a mixture of air and combustible gas is explosive.) Federal regulations governing landfills mandate that the methane concentration shall not exceed 25% of the LEL in on-site structures and shall not exceed the LEL (in subsurface soils) at the property line. Methane must be monitored throughout the landfill's operating life and for 3 years after it closes.

A landfill design also should include an active or passive gas-collection system. For guidance on designing both systems, see the U.S. EPA's *Solid Waste Disposal Facility Criteria Technical Manual* (U.S. EPA, 1993b) and *Waste Containment Systems, Waste Stabilization and Landfills: Design and Evaluation* (Sharma and Lewis, 1994).

The collected gas (called landfill gas or biogas) typically is either flared or used as fuel, where practical. Increasingly, landfill gas is seen as a renewable resource, and efforts to use it are increasing. For information on potential uses for biogas, see Chapter 23. (While the chapter discusses uses for digester gas, the options also apply to landfill gas.)

4.6   Landfill Closure

4.6.1   Closure Plan

Part 503 requires that a surface disposal site (monofill or DLD) owner or operator submit a closure plan up to 180 days before closing the facility. This plan should include a description of closure and post-closure activities. It should document how the leachate-collection system will be maintained for at least 3 years after closure and outline the methane monitoring program, which also must occur for 3 years after closure.

If ownership of the disposal site changes after closure, the new owner must be notified in writing that a surface disposal site was operated on the property.

4.6.2   Cap System

A landfill cap is a multilayered cover for the piled wastes that is added just before a landfill is closed. It is designed to prevent rainwater from infiltrating the waste, thereby minimizing leachate production. The layers of this cap typically include a:

- Subgrade layer, which is used to contour the landfill and provide a base for the other layers;
- Gas-control layer, which transports gas from under the barrier layer to a venting system;
- Hydraulic barrier layer, which limits the water infiltrating the landfilled waste;
- Drainage layer, which collects and transports the water that percolates into the final cover from the surface;
- Biotic layer, which protects the hydraulic barrier layer from biointrusions by animals or plants;
- Filter layer, which prevents particles of a finer material (i.e., the surface layer) from migrating into a coarser material (i.e., the drainage layer); and
- Surface layer, which may be either a soil that can support vegetation or an armored protection layer. Biosolid material may be incorporated into this layer to support vegetation.

These layers perform important complementary functions (see Figure 25.1). For example, the vegetative layer helps prevent erosion by promoting the growth of plants, whose root structure anchors the soil.

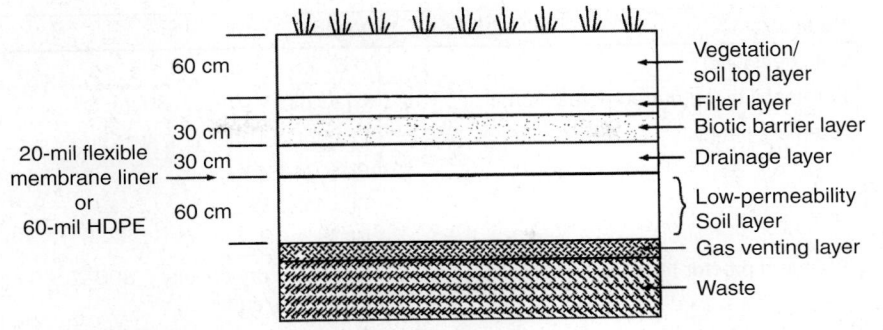

FIGURE 25.1   Example of various landfill cap components.

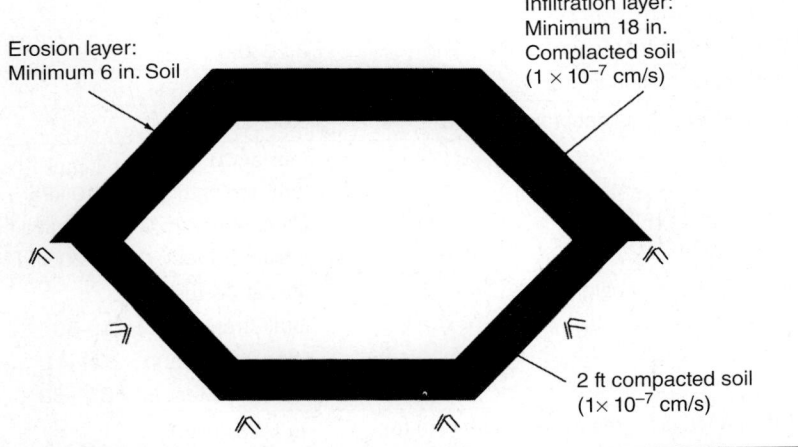

FIGURE 25.2   Minimum requirements for a landfill cap.

The first step in selecting the layers of a landfill cap is to ensure that the proposed cap complies with state and federal regulations (see Figure 25.2). Although the layers and their functions are relatively standard in current engineering practice, each layer's thickness and performance standards may depend on state or local regulations.

In addition, when designing a landfill cap, it is important to consider the strength of the underlying residuals. The geotechnical properties of solids depend on several factors (e.g., the degree and method of dewatering, the polymers added, and age) (see Table 25.11). Such properties should be tested as part of landfill design.

For more guidance on designing a landfill cap, consult the U.S. EPA's *Solid Waste Disposal Facility Criteria Technical Manual* (U.S. EPA, 1993b) and *Waste Containment Systems, Waste Stabilization and Landfills: Design and Evaluation* (Sharma and Lewis, 1994), provides guidance on selection and design of final cover layers.

4.6.3   Post-Closure Considerations

Aftercare costs should be budgeted for as subtitle D of Resource Conservation and Recovery Act in the United States or the European Landfill Directive requires financial provisions for a minimum aftercare period of 30 years once an MSW landfill is closed.

Parameter	Value
Bulk density	0.6–0.8 g/cm³
Permeability (saturated potassium)	67–118 mm/h (17–30 in./hr)
Atterberg limits	LL = 71–119%
	PL = 54–85%
	PI = 17–53%
	Class = OH
Standard proctor test	Maximum dry density = 800–1090 kg/m³ (50–68 lb/cu ft)
	Opt. moisture content (OMC) 37–64%
Modified proctor test	Maximum dry density = 830–1 150 kg/m³ (52–72 lb/cu ft)
	OMC = 31–64%
Compression index (Cc)	0.26–0.5
Recompression index (Cr)	0.03–0.1
Secondary consolidation index (C α)	0.02
Cohesion (C)	Triaxial CU test:
	Total strength: C = 0–40 kPa
	Effec. strength: C = 0–30 kPa
	Triaxial UU test: C = 0–20 kPa
Internal friction angle (α)	Triaxial CU test:
	Total strength: α = 21°–30°
	Effec. strength: α = 32°–41°
	Triaxial UU test: α = 32°–38°
Unconfined compressive strength (Qu) test	Qu = 23–126 kPa
	Strain at failure = 4.9–5.2%

TABLE 25.11   Geotechnical Properties of Dewatered Solids in Chicago Area

The goal of aftercare completion is to reach a state at which the landfill no longer poses a threat as defined by the Health Hazard Evaluation, yet there are not specific evaluation requirements or approaches defined by regulatory agencies as of how to define a completion of landfill aftercare. Laner et al. (2012) suggested three approaches when evaluating and managing landfills after closure has been suggested: (i) target values to evaluate aftercare, (ii) impact/risk assessment to evaluate aftercare, and (iii) performance-based system for aftercare. They concluded that target values were best used as a screening method and that all approaches needed to be made on a site-specific basis. Morris and Barlaz (2011) presented the evaluation of post-closure care (EPCC) methodology that combined target values, impact, and risk assessments, which provided operative assessment protocols to decide on appropriate levels of aftercare.

Heyer and Stegmann (2004) have shown that controlled infiltration and in situ aeration of landfills have become promising technologies to reduce leachate concentrations and landfill gas production, accelerate biodegradation, and complete major settlements within several years of a landfill closure.

When designing landfills, engineers should consider potential uses for the site once landfill operations cease and select cover material, grading, monitoring, and stormwater management options that are compatible with such uses (when possible). A former landfill can become either a passive- or an active-use area. Most become recreational and open-space areas (e.g., athletic fields, game courts, golf courses, playgrounds, and picnic areas).

5.0   Dedicated Land Disposal

Dedicated land disposal is the process of applying municipal solids to land for disposal purposes where nutrient loading criteria typically exceed traditional agronomic loading criteria. The equipment and the process involved are similar to those used in land application. In effect, the site involved is a landfill that only accepts large quantities and heavy loadings of solids, and it is regulated as landfill.

5.1   Regulatory Considerations

Part 503 regulations apply to all surface disposal practices, including dedicated land disposal (and monofills, as discussed previously). The regulation is divided into several subparts, including general requirements, pollutant limits, management practices, and pathogen and vector-attraction controls.

5.1.1   General Requirements

The general requirements that apply to surface disposal of municipal solids include:

- Compliance with all applicable Part 503 requirements;
- Closure by 1994 of active units within 60 m of a fault with displacement in Holocene time, in an unstable area, or in a wetland, unless authorized by the permitting authority; and
- The need for closure and post-closure plans at least 180 days before closing any active units.

In addition, site owners must provide written notification to the subsequent owner that municipal solids were placed on the land.

5.1.2   Pollutant Limits

Surface disposal sites that use liners and leachate-collection systems do not have pollutant-concentration limits because the pollutants that seep from solids will be collected in the leachate and treated, as necessary, to avoid a pollution problem. For the site liner to qualify, it must have a hydraulic conductivity of 1×10^{27} cm/s.

Surface disposal sites without liner and leachate-collection systems have maximum allowable concentration limits established by rule for arsenic, chromium, and nickel (see Table 25.9) because these pollutants are the most likely to leach to groundwater, causing it to exceed the MCL. Different limits for these pollutants can be developed if a site-specific assessment (specified by the permitting authority) demonstrates that the site has different parameters than the ones U.S. EPA used to establish maximum allowable concentration limits.

5.1.3 Nitrate Contamination

Part 503 specifies that surface disposal operations cannot cause groundwater to exceed the MCL for nitrate (or exceed the existing concentration if it already is above the MCL). Groundwater-monitoring results or a statement from a qualified groundwater scientist can be used to demonstrate compliance with this requirement.

5.1.4 Other Management Practices

Other management practices that DLD facilities must follow include:

- Active disposal sites shall not be located within 60 m of a Holocene-period (within last 10 000 years) fault or in a wetland, unless authorized by the permitting authority; when located in a seismic impact zone, an active disposal site shall be designed to withstand the maximum-recorded horizontal ground-level acceleration.

- Surface runoff from a 25-year, 24-hour storm event shall be controlled in accordance with an NPDES permit.

- Active disposal sites shall not restrict base-flood flows, adversely affect threatened or endangered species, or be located in a structurally unstable area.

- Active surface disposal sites with a liner and leachate-collection systems shall operate and maintain them, and dispose of the collected leachate in accordance with applicable requirements for as long as the site is active and for 3 years post-closure.

- No crops shall be grown nor animals grazed on surface disposal sites unless the permitting authority specifically authorizes such activities based on site-specific management practices.

- Public access to the site is restricted during operations and for 3 years post-closure.

5.1.5 Pathogen and Vector-Attraction Reduction Requirements

Solids that will be placed in a surface disposal site must meet one of the Class A or Class B pathogen-control alternatives, unless they will be covered with soil or another material every day. They also must meet one of the first 11 VAR options listed in Part 503.

5.2 Planning

The planning requirements for a new dedicated land disposal site are extensive. Significant hydrogeologic and soils investigations are required. Regulatory requirements and stormwater-management logistics also will affect site selection. Application rates will determine the size of the disposal site.

5.2.1 Groundwater Protection

Water that percolates through the soil of a surface disposal site (because of applied biosolids and precipitation) contains dissolved salts and products of decomposition. The constituents of concern typically are nitrate, sulfate, and total dissolved salts (e.g., chloride); metals remain in the soil unless its pH is low. Design engineers need to evaluate the fate and transport of percolated constituents carefully. In addition, site-specific soil and hydrogeologic evaluations are mandatory.

Many sites require a relatively impermeable layer above the groundwater. This layer, which should be at least a few meters thick, retains and collects percolated liquid. As the liquid turns anaerobic, much of the nitrate is denitrified and lost as nitrogen gas. If the percolation is minimal and shallow, most of the water may migrate back up to the surface and evaporate during the following year. Because the rule requires that surface-disposal operations not cause nitrate concentration in groundwater to exceed the MCL (or the existing concentration if it already exceeds the MCL), groundwater monitoring wells may need to be installed to verify compliance with this requirement.

Design engineers should evaluate lining systems and percolation controls based on biosolids and soil characteristics, site geology, climate, and application rates. Liner and leachate-collection system design requirements are addressed in Section 4.

5.2.2 Site Selection

Design engineers also should consider the following when locating a dedicated land disposal site:

- Site runoff typically is considered wastewater, which needs to be returned to the WRRF.

- The soil's pH must be 5.5 or higher to ensure that metals remain in the soil.

- A site susceptible to flooding or whose runoff could affect adjacent wetlands or surface waters is not a good candidate for dedicated land disposal.

- Flat terrain is preferred because soil slope may affect site operations. A biosolids slurry requires limited slope, while dewatered cake can be applied on slopes that are more variable. Also, because runoff is an issue, dedicated land disposal typically is not considered on anything resembling hilly terrain.

- Buffer zones of about 150 to 300 m (500 to 1000 ft) have been suggested to minimize odor and dust complaints from the neighbors.

5.2.3 Application Rates

All land-application programs are "aerobic systems" at the surface incorporation layer. In this case, however, the goal is not crop or vegetation production but rather maximizing the aerobic stabilization of biosolids by soil bacteria. The application rates typically are much higher than agronomic rates and primarily limited by the evaporation rate of biosolids-related moisture from the applied material and soil.

Application rates typically range from 30 to 250 dry metric ton/ha·a (22 to 110 dry ton/ac/yr). Assuming that 25% of biosolids are lost to long-term products of decomposition, then these rates would increase soil depth by about 2.5 to 13 mm/yr (0.1 to 0.5 in/yr). If the site operates for 50 years, then total loading would be 1500 to 12 500 dry metric ton/ha. Bacteria in the soils can readily handle these application rates as long as enough water evaporates and percolates to allow the aerobic microbe population to thrive. (Aerobic biological activity is about one to two orders of magnitude faster than anaerobic biological activity.)

The application rate typically depends on the evaporation rate [evapotranspiration (ET) rate should exceed precipitation], which affects how soon and how often equipment can be moved onto the site. If a slurry (6% solids) is applied at the above loading rates, it adds 8 to 38 cm (0.25 to 1.25 ft) of water to the site each year. This much water can be evaporated from DLDs in many (but not all) regions of the United States. Evaporation rates depend on temperature, humidity, wind, and rainfall. Because evaporation is a critical component of the operation, areas with limited evaporation seasons may find dedicated land disposal difficult to implement or may need to restrict application rates.

5.3   Design and Implementation

Engineers need a significant amount of site data to design and implement a dedicated land disposal site, including biosolids, hydrologic, topographic, and meteorological and other information.

5.3.1   General Considerations

Solids must be stabilized before they are applied to these sites. Subsurface injection is the preferred application method because it minimizes odor, minimizes vector problems, and ensures consistent application rates. Other general considerations include:

- The biosolids, soil, and groundwater must be monitored.
- Traffic controls (e.g., hours of use, trucks per hour, and proper timing) may be required.
- Dust from the site and from access and internal roads needs to be controlled.
- Biosolids O&M plans must be developed.

5.3.2   Biosolids Storage

Dedicated land disposal typically is a seasonal operation, so biosolids must be stored during the off seasons. Storing biosolids for several months in an environmentally acceptable manner with limited odor is often difficult depending upon factors such as solids content and storage types.

5.3.3   Groundwater Protection

Extensive soil and geologic data need to be collected to determine how the site will react to heavy biosolids applications. This data-collection effort often involves soil borings, groundwater monitoring wells, surface and near-surface soil collection, and laboratory analysis. The primary goals are to:

- Ascertain whether the site has a natural liner (impermeable soil or underlying geologic strata protective of shallow aquifer);
- Determine if a supplemental lining is needed and how to implement it;
- Create an appropriate leachate-collection and -handling system, if necessary; and
- Discover likely limitations to loading rates.

5.3.4   Biosolids Application Rates

Design engineers need to calculate the application rate so the total site area requirements can be determined (see Table 25.12). Application rates are a function of the material applied; its water content; the seasonal evaporation rate; and the methods used to apply biosolids and subsequently aerate or mix them into the soil.

Facility	Size mt/a*	Liquid or Cake	Crop Grown
Evanston, Wyoming	295	Cake	Grasses
Rapid City, South Dakota	1222	Liquid	—
Colorado Springs, Colorado	8939	Liquid	None

*1996 data.

TABLE 25.12 Typical Design Practices for Several Dedicated Land Disposal Facilities

5.3.5 Application Methods

Biosolids can be applied in slurry or dewatered form. Dewatered cake allows for higher application rates (because of its lower moisture content) but can be difficult to spread over the entire site in a consistent layer. Slurries need high evaporation rates but can be pumped and are easier to spread. They also can be injected just below the surface by mobile equipment specifically designed for this purpose.

6.0 Ash Use and Disposal

Ash is the product of solids incineration, essentially consisting of the noncombustible portions of the feed material. There are numerous methods and equipment available for handling ash and various uses for ash. Handling methods and the final destination or use of material is often site specific. Ash-handling equipment—especially conveyance—can be the most troublesome subsystem associated with incinerators.

Multiple-hearth furnaces and fluid-bed incinerators use either wet or dry systems for ash conveyance. Ash is abrasive and sometimes nonuniform, making it difficult to convey as a bulk material, and conveyance systems must be designed accordingly. The design of such systems is not covered in this chapter, but is detailed in the WEF Manual of Practice titled *Solids Incineration Systems* (WEF, 2009). That document also covers another critical aspect of ash handling: storing wet and dry ash.

6.1 Regulatory Considerations

Regulations vary from state to state. Some states do not regulate ash; others treat it as a waste. Design engineers should check local, state, and federal regulations against biosolids characteristics before determining whether to use or dispose incinerator ash. Each facility should conduct its own research on local, state, and federal regulations concerning ash disposal and reuse.

Some landfills will require a toxicity characteristic leaching potential (TCLP) test or a pH test before accepting ash.

6.2 Use and Disposal Options

Biosolids incinerator ash is similar in physical and chemical characteristics to coal ash and can be used in applications where coal ash is permitted. Incinerator ash may be acidic, neutral, or alkaline, depending on the solids characteristics. In addition, incinerator ash will contain more phosphorus than coal ash. While ash has historically been landfilled, interest in beneficial use of this material is increasing (Dominak et al., 2005). Beneficial use options for ash include:

- Landfill—Ash can be used as a landfill cover or blended with soil and used as a cover.

- Fill material—ash can be used as fill material for excavations. One utility, for example, has a contractor using the material to fill old solids lagoons. The material also can be used as a flowable fill.

- Soil amendment—in some areas (e.g., those with high-clay soils), incinerator ash may be used as a soil additive that produces a soil that handles more easily, allows better drainage and airflow, and includes some valuable minerals.

- Concrete fly ash—ash has been used as a fly ash substitute in concrete mixes.

- Asphalt additive—the ash has been used as a mineral filler and fine aggregate in asphalt mixes. New Jersey's Department of Solid and Hazardous Waste (NJDEP) has permitted the use of incinerator ash for this purpose.

Additional valued-added products derived from ash are discussed later in this chapter.

Use options for incinerator ash tend to be site specific. Utilities should pursue all avenues available for using ash. State departments of transportation should be contacted to determine requirements for using incinerator ash as a fly ash substitute in concrete or as a mineral filler or fine aggregate substitute in asphalt mixes. If the incinerator ash is approved for use in mix designs by the state department of transportation, then a considerable market can be opened for incinerator ash.

7.0   Distribution and Marketing

In some of North America, the environment for land-based management of Class B biosolids has become more challenging. As farming areas are developed into residential and commercial areas, biosolids producers must transport their product farther and farther away to reach the rural environments necessary to support agricultural application. Not only are the transportation-related energy costs becoming increasingly burdensome, but also the perceived imposition of urban and suburban "wastes" on rural communities can become a public relations issue for biosolids managers.

To increase reuse options, many WRRFs are evaluating techniques to produce a Class A biosolids product, which can be managed differently than Class B biosolids. Class A biosolids are essentially free of all pathogens, so the product can be distributed to the public. EQ biosolids can be distributed as commercial products with even fewer limitations imposed by Part 503.

When contemplating a Class A process, it is important to determine how the product will be used. The costs associated with upgrading to a Class A process typically are not warranted if the material has no outlet. Recommended options should lead to simple and safe management alternatives that stress reclamation of resources. In addition, once a market for Class A material is identified, developing that market can require 3 to 5 years of effort before the revenues might offset costs. In assessing value-added products, there has to be a product life-cycle analysis. This analysis requires obtaining the true worth of the value-added products and takes into account processing cost and risk assessment or potential liability. Before implementing a biosolids or manure management program, a market survey and the required engineering, economic, and outreach plans must be developed to utilize the value-added products in the market place. Again, the definition of a "value-added product" states that it must be sold at a value that meets the needs of the utility.

7.1 Value-Added Products
The following biosolids products typically are marketed and distributed to various agricultural, horticultural, commercial landscape, and residential markets. See Chapter 15 for information on phosphorus as a potential sidestream product.

7.1.1 Compost
The public already considers compost to be an organic-based product that is a useful soil conditioner. Thus, biosolids compost producers have a less difficult time introducing their product to the marketplace. However, because many other organic byproducts also are composted (e.g., animal wastes, yard or green waste, and food wastes) under different regulatory requirements, biosolids compost may have more competition than other biosolids-derived products.

If biosolids generators produce a good-quality compost and commit the resources needed to develop and sustain the market, there typically is a strong demand for the product. Biosolids compost is successfully marketed throughout North America by generators ranging in size from the Washington (D.C.) Suburban Sanitary Commission and Los Angeles County to Davenport, Iowa, and the Hampton Roads (Virginia) Sanitation District. Each of these producers developed and exploited local markets for bulk and bagged compost through very active long-term marketing programs.

7.1.2 Heat-Dried Product
Heat dryers produce a biosolids product that can meet all U.S. EPA and most state requirements for unlimited distribution as a soil conditioner or fertilizer. Most, though not all, drying systems are designed to meet the time and temperature requirements for Class A pathogen reduction. The dryers typically also meet VAR standards simply because they routinely dry biosolids to 90% solids or greater. In addition, heat drying does not concentrate pollutants in the biosolids; the dry-weight concentrations in both feedstock (dewatered cake) and product (heat-dried biosolids) will be nearly identical. Therefore, if the dewatered cake met the pollutant concentrations in 40 CFR Part 503's Table 3, the heat dryers routinely produce EQ biosolids, which may be suitable for marketing as a soil conditioner or fertilizer.

The primary benefit associated with heat drying has historically been volume reduction. The potential for producing fertilizer with a serviceable nutrient content is another aspect that makes heat drying attractive. The WRRF managers look at examples like the Milwaukee Metropolitan Sewerage District (MMSD), which has produced and sold a heat-dried biosolids since about 1926, and at Houston, Texas, which has produced and distributed heat-dried biosolids since the 1950s. Both organizations developed successful markets for their products, and as a result, sales revenues play an important role in their budgets. However, such successes must be put into context; Milwaukee and Houston invested many years in developing and servicing their markets. New heat-drying facilities are unlikely to emulate these marketing programs fully until they have been operating for a number of years. Likewise, as the number of heat-drying facilities increase, there is likely to be a downward pressure on the prices of these products. Further, dryers require a lot of power, and increasing energy costs may negatively affect the economic feasibility of this option. Different drying technologies generate products of differing quality; revenues will depend on customers' desired quality.

7.1.3 Advanced Alkaline Stabilized Products

Class A alkaline-stabilized products can vary substantially, depending on the process used to generate them. The generic lime process, as well as several proprietary processes (which use relatively low doses of lime and supplemental heat to meet Class A time and temperature requirements), creates products that typically have final solids contents in the range of 30% to 40%, depending on initial cake solids. One proprietary process also dries the biosolids, producing a product with a solids content of 60%. Another process, which uses significant quantities of cement kiln dust and other waste alkaline products, typically produces a material containing 55% to 65% solids.

Alkaline-stabilized products typically are land applied as substitutes for agricultural limestone (see Section 2.3.1.3). They are also excellent aids for reclaiming acidic soils and spoils. Alkaline-stabilized products with higher solids contents can be blended with native soils, composts, spent foundry sand, or dredge spoil to make marketable topsoil. These tend to be regional options for those areas with low soil pH due to geology or agriculture practices, that is, sugarcane farms in south Louisiana.

7.1.4 Ash-Derived Products

In addition to the ash use options discussed earlier, the following value-added products can be generated from ash. See Chapter 20 for more details on ash.

- Brick—ash has been used in brick manufacturing by various utilities quite successfully. The Japanese have used incinerator ash to make water-permeable bricks. Brick manufacturers typically require large quantities of ash at a time. Such quantities could be obtained by emptying a lagoon and incinerating the solids.

- Worm castings (vermiculture)—one of the more innovative uses of incinerator ash is in vermiculture process. The ash is blended with food waste, and then worms are added. The worms are separated from the mixture once enough time has passed, and the remaining material is used as a soil amendment.

- Lightweight glass aggregate—a proprietary technology manufactures lightweight glass aggregates by burning a mixture of dewatered biosolids and fly ash.

7.1.5 Other Products

Any number of processes can create products that meet Class A and VAR standards. Some of these (e.g., vermicomposting and air drying) rely on Class A Alternatives 3 and 4 (see 40 CFR Part 503.32), which involve sampling and analyzing biosolids for helminth ova and enteric viruses. The specific processes may not have been approved by the U.S. EPA as Class A alternatives, but the regulations state that a product is safe for distribution if monitoring demonstrates the process meets reduction requirements in addition to low pathogen densities. That said, the U.S. EPA and a number of states have expressed concern about the use of Alternatives 3 and 4 to meet Class A requirements. Generators considering such approaches should be aware of this and monitor the rule-making activities of the U.S. EPA and their state agency.

Other processes may meet the regulatory criteria for distribution and marketing as EQ biosolids. Technically, any biosolids that meet EQ criteria are suitable for marketing, but generators considering such processes should realize that merely achieving EQ status does not guarantee a market for the product. For example, some advanced digestion techniques may produce EQ biosolids, but if they are simply mechanically dewatered before use, then their appearance, handling properties, and odor may not be remarkably different from a Class B dewatered cake. There is little potential for marketing this material without further processing. A good general rule is that products must have high solids content—typically more than 50%—to be marketable. One example of such a product is a soil blend, where dewatered Class A biosolids are blended with materials like sand and sawdust to create a custom soil blend. The City of Tacoma's TAGRO program is one example of a successful biosolids soil blending program.

Finally, there are emerging technologies (e.g., those aimed at manufacturing a biosolids-derived fuel or an aggregate component of building materials or cement) that may produce EQ biosolids or a marketable product. For more information on these processes, see *Emerging Technologies for Biosolids Management* (U.S. EPA, 2006).

7.2   Regulatory Considerations

Marketed biosolids must meet 40 CFR Part 503 criteria, as well as state regulations.

7.2.1   Federal Regulations

All biosolids-derived products that will be distributed as a soil amendment or fertilizer are regulated under 40 CFR Part 503. Biosolids that meet Class A pathogen-reduction and VAR standards but do not meet Table 3 limits can still be marketed; however, their applications are subject to cumulative pollutant loadings. Such products probably have limited "marketability," although they remain suitable for broad-acre general land application.

7.2.2   State Regulations

State regulatory programs that affect the distribution and marketing of a biosolids-derived product include solid waste and water quality regulations, as well as some different regulatory controls.

Many states had solids or biosolids regulations in place before 40 CFR Part 503 was promulgated. The Part 503 provisions that facilitate the distribution and marketing of EQ biosolids may not have existed in those state regulations, so there was a lag time between 1993, when Part 503 was promulgated, and the year states adopted similar language (as agencies went through the processes necessary to modify their regulations). However, by 2008, most states had adopted the Part 503 provisions that facilitate distribution and marketing. Some states' provisions are somewhat different or more restrictive than Part 503, so any entity considering producing EQ biosolids should carefully evaluate state regulations in the areas where the product will be distributed to determine how those regulations may affect distribution costs.

Most states also have rules controlling how products with fertilizer, soil conditioning, or soil amendment (e.g., raise soil pH) value are distributed. These regulations may not be designed to control how these products are placed in the environment, but rather to ensure that financial transactions are based on the real value for dollars charged. Therefore, state regulations controlling the sales and distribution of such products typically require certain guarantees of analysis (e.g., product contains at least 4% total nitrogen), efficacy of performance (e.g., product has a CCE of 55%), or other attributes (e.g., compost salt index is less than 2). These guarantees typically are embodied in a product label that must be approved by a state agency (e.g., the state department of agriculture) before the product can be sold. Many states also require producers or distributors of such products to be licensed or registered and typically require periodic payment of a tonnage tax—a fee collected on each ton of distributed product that typically pays for product inspections and analyses.

There may be fines and penalties for being out of compliance with the state fertilizer/soil amendment rules. For example, distributing a product that is dramatically different from the claims made on the product label can result in a fine or a "stop sales" order. Therefore, entities planning to distribute and market biosolids-derived products should develop a sound database of their product's claimed attributes before labeling it for distribution. The label guarantees should be based on statistically reliable determinations of the attributes.

7.3   Marketability Criteria

7.3.1   Product Quality
The markets for each biosolids-derived product typically have a set of benchmarks that a product must meet to achieve market success.

7.3.1.1   Compost   The product quality benchmarks for compost have been established by the U.S. Composting Council (USCC) and are included in their *Field Guide to Compost Use* (USCC, 2001). These parameters include (in addition to regulated pollutants) pH, soluble salts (salinity), nutrient content (N-P-K), water-holding capacity, bulk density, moisture content, organic matter content, particle size, growth screening, and stability. The USCC also has a program to test, label, and disclose information on compost called the "seal of testing assurance" (STA).

7.3.1.1.1   pH   Most compost has a pH between 6 and 8. Adding compost can affect the pH of growing media; pH is adjusted by adding such materials as lime (alkaline) and sulfur (acidic).

7.3.1.1.2   Soluble Salts   The soluble salts concentration is the concentration of total soluble ions in a solution. Most plant species have a salinity tolerance rating, and maximum tolerable quantities are known. Excess soluble salts can cause phytotoxicity. Another measure of this parameter is the salt index, which is determined in the lab and compared to a standard (ammonium sulfate = 100). Products with a salt index well below 100 typically are not problematic. Ferric chloride conditioning may be one source of excessive salinity levels.

7.3.1.1.3   Nutrient Content   Nitrogen, phosphorus, and potassium are the three macronutrients required by plants in the greatest amounts, and therefore, the fertilizer components of greatest interest. These three nutrients are measured and expressed on a dry-weight basis as a percentage of the total dry mass; nitrogen typically is broken down in terms of inorganic and organic or slow-release forms, and both phosphorus and potassium are expressed as phosphoric acid (P_2O_5) and potash (K_2O). Laboratory determinations of extractable P_2O_5 and K_2O are necessary to determine plant-available fertilizer.

7.3.1.1.4 Water-Holding Capacity Water-holding capacity is a measure of a given volume of compost's ability to hold water under 101.325 kPa (1 atm) of pressure, measured as a percent of dry weight. It indicates the potential benefit of reducing the required irrigation frequency, as well as gross water requirements. The water-holding capacity should be known to allow users to monitor or estimate the compost's effect on their crop-watering regime and growing media.

7.3.1.1.5 Bulk Density Bulk density is the weight per unit volume of compost. It is used to convert compost application rates from metric tonnage to cubic meters. In a field application, cubic meters per hectare then would be extrapolated to express an application rate represented as a depth (e.g., 30-mm application rate). Bulk density also is used to determine the volume of compost that may be transported on a given occasion, taking into account that most vehicles have a specific maximum gross weight that may not be legally surpassed.

7.3.1.1.6 Moisture Content Moisture content is the measure of the amount of water in a compost product, expressed as a percent of total solids. The moisture content of compost affects its bulk density, and, therefore, transportation costs. Moisture content also affects product handling. Dry compost can be dusty and irritating to work with, while wet compost can be heavy and clumpy, making it difficult to apply and more expensive to transport. The feedstock should be approximately 60% moisture before composting begins as the heat generated tends to aid in evaporation and therefore reduces moisture content. The process is inhibited at moisture content less than 35% (WEF et al., 2012).

7.3.1.1.7 Organic Matter Content Organic matter content is a measure of organic carbon-based materials in compost. It typically is expressed as a percentage of dry weight.

7.3.1.1.8 Particle Size Particle size is a measure (in percent) of how much of the product (sample) will pass through a certain-sized screen opening. The particle size often dictates the product's use. For example, coarse compost is best suited for land reclamation and erosion control, while finer composts are used for topsoil blending and top dressing of turf. For most applications, merely specifying the product's maximum particle size (or the screen size through which it passes) is sufficient. However, some applications (e.g., potting/nursery media component) require a particular particle size distribution. Particle size distribution measures the amount of compost within a specific particle size range. It is determined by pouring a sample of compost through a series of sieves (screens), each having smaller openings than the one before. Results are expressed as the percent (by weight) of sample retained by each sieve size.

7.3.1.1.9 Growth Screening The growth screening test is an indicator of the presence of phytotoxic substances (e.g., volatile fatty acids, alcohol, soluble salts, heavy metals, or ammonia). These substances may cause delayed seed germination, seed or seedling damage or death, or plant damage or death. Growth-screening tests include germination, root elongation, and pot tests; they are not intended to identify which growth inhibitor caused the poor growth response. In addition, a product that passes initial growth-screening tests may fail later if improperly stored. Specific growth inhibitors (e.g., volatile fatty acids and alcohol) may form in compost stored under anaerobic conditions.

7.3.1.1.10 Stability   Stability is a measure of the level of biological activity in compost under a given set of conditions. Unstable compost consumes nitrogen and oxygen in significant quantities to support biologic activity and generates heat, carbon dioxide, and water vapor. Stable compost consumes almost no nitrogen and oxygen and generates almost no carbon dioxide or heat. Unstable compost demands nitrogen; it can cause nitrogen deficiency and be detrimental to plant growth, even killing plants in some cases. If stored, and left unaerated, unstable compost can become anaerobic and emit nuisance odors.

7.3.1.2   Heat-Dried Biosolids   Heat-dried biosolids typically should meet most of the physical, chemical, and microbiological regulatory criteria; however, Class A standards must be met to allow distribution of the product without restriction. Some of the important criteria include particle size, durability (hardness), dust, odor, bulk density, nutrients, salt index, and heating value, and these vary by dryer type and manufacturer.

7.3.1.2.1 Particle Size   As with compost, a measure of the size of the particles of dried biosolids is useful. Screening may be included in drying processes; this allows the producer to specify the size of the particles that will be managed merely by selecting a specific screen size or sizes. Preferred particle size may vary from market to market. For example, users seeking a standard-grade fertilizer might prefer a product with a particle size ranging from 2 to 4 mm in diameter, while a golf course superintendent seeking to use dried biosolids on tees and greens might prefer a smaller product (typically between 0.3 and 0.8 mm).

   Particle size is an important consideration for a number of reasons. If the dried biosolids will be blended with other fertilizer ingredients, then it will be less likely to segregate after blending if it is about the same size as those other ingredients. Particle size also may affect the even distribution of a product. To an extent, particle size even affects the rate at which dried biosolids release nutrients (i.e., a smaller particle should release nutrients more quickly than a larger one).

   Fertilizer users also may be interested in variations on the particle size determination. The size guide number (SGN) is a measure of the average size of granules in a fertilizer. To calculate SGN, a product is screened through a nested series of screens with different-sized openings, screeners determine which screen opening size (in millimeters) retains 50% (by weight) of the material, and they then multiply that opening size by 100. In other words, a product with an SGN of 200 would have an average particle size of 2 mm.

   Another physical measurement that may be important to fertilizer blenders is the uniformity index. In general, the uniformity index is the ratio of the size of the small particles (fines) to the large particles (coarse). A lower uniformity index indicates a broad particle size distribution; a higher one indicates a narrow distribution (i.e., a uniformity index of 100 would mean that all particles are the same size). Users typically prefer a higher uniformity index.

7.3.1.2.2 Durability (Hardness)   Because of the methods of transport of heat-dried biosolids (conveying, dumping, and moving with front-end loaders), the particles' durability (resistance to degradation) is important. Less durable particles will easily fracture, creating finer particles and dust. Three common types of durability or hardness are associated with commercial fertilizers: crushing strength (minimum force required to crush individual particles), abrasion resistance (resistance to granule fracturing as a result of granule-to-granule or granule-to-equipment contact), and impact resistance (resistance of the granule to breakage upon impact to a hard surface) (IFDC/UNIDO, 1998).

There is no universally accepted test method for hardness. The Fertilizer Institute offers one test procedure. It involves placing selected fertilizer particles onto a ratcheting hardness tester, which exerts pressure on the particle and records the "weight" at which the particle crumbles as its "crushing strength." According to this method, urea pills have a crushing strength between 1.5 and 3.5 kg/granule, and ordinary super-phosphate has a crushing strength of 4.5 to 8 kg/granule). Anything with a crushing strength of greater than 2.5 kg/granule is considered highly desirable (IFDC/UNIDO, 1998).

7.3.1.2.3 Dust   A heat-dried biosolids' dust content may be the most important physical parameter when assessing whether the material is well suited for a particular use; a dusty product tends to be poorly received by users. Many users and intermediaries demand that heat-dried biosolids contain virtually no dust.

One aspect of wastewater treatment that affects this characteristic is the degree to which grit, fine fibers, and hairs are kept out of the biosolids. These items tend to diminish the durability of heat-dried biosolids so that they are more susceptible to fracturing during handling and transfer, resulting in many fine particles. If the feedstock contains a substantial fraction of raw primary solids or the headworks equipment is not as efficient as possible, then the product typically will contain extraneous materials and could easily fracture. Generally, undigested solids typically produce a dustier heat-dried biosolids than biosolids that are digested prior to heat drying.

Moreover, some drying technologies produce biosolids more prone to dust than others. Biosolids particles that are more angular tend to break during handling, creating fine particles and dust. Particles that are more spherical typically are less likely to do so.

That said, most heat-drying technologies produce some dust and so many facilities install screening systems to remove fine particles or add dust-suppression oils to the product.

7.3.1.2.4 Odor   Most heat-dried biosolids emit a musty, earthy odor with overtones of ammonia and—depending on the system—sometimes a burnt smell. Anecdotally, heat-dried biosolids created from well-digested feedstock tend to be perceived as less odorous than those produced from raw solids. In some applications, odor does not limit the heat-dried biosolids' suitability for beneficial use. Occasionally, comments about product odor are heard after application and rewetting (e.g., after rainfall or irrigation). As with other biosolids, if heat-dried biosolids are surface applied at a high rate without subsequent incorporation, one can anticipate that some odor will emanate. The intensity of the odor will be a function of the application rate, but in most cases, the odor should dissipate over time. It is prudent, if such applications are planned, to select sites where fewer sensitive receptors might note the odors.

7.3.1.2.5 Bulk Density   Bulk density is a measure of the mass of a product per unit volume (e.g. pounds per cubic yard). This measurement is important in managing heat-dried biosolids for a number of reasons.

The bulk density of dried biosolids typically ranges from 560 to 720 kg/m^3 (35 to 45 lb/cu ft); however, lower bulk densities have been noted (14 to 29 lb/cu ft). It is important to note the product's actual density cannot be determined until after the drying system is in operation; therefore, a reasonable estimation must be calculated when designing product-handling and storage systems. Product-management issues may arise if storage-system designers based the work on a high bulk density and the actual product has a much lower density. In addition, a product with low bulk density may fill up a transport unit before the maximum allowable weight is met, thereby increasing delivery costs.

If heat-dried biosolids will be bagged, either as a standalone fertilizer or as a component of blended dry fertilizers, its bulk density becomes important in bag sizing. Low bulk density can be especially problematic if the material is replacing a denser filler in a blended fertilizer; the manufacturer may need to purchase new bags to accommodate the lighter material.

7.3.1.2.6 Nutrients The nutrient content of heat-dried biosolids typically is its chief value to prospective users. Users are interested in three types of nutrients: primary nutrients (nitrogen, phosphorous, and potassium); secondary nutrients (calcium, magnesium, and sulfur); and micronutrients (boron, chlorine, copper, iron, manganese, molybdenum, and zinc).

In most states, the state department of agriculture (or a similar agency) requires that all fertilizers (e.g., heat-dried biosolids) must be registered and labeled. Most states require that percentages of primary plant nutrients be guaranteed (typically presented on a fertilizer bag as % N–% P_2O_5–% K_2O [e.g., 10-10-10]); they also will accept guarantees of secondary or micronutrients. These guarantees become regulatory standards, and if a product fails to provide the promised amount of nutrients, the generators can be subjected to fines, "stop sales" orders, or both.

Heat drying typically preserves the nutrient content of the feedstock solids, except for nitrogen. Typically, heat drying removes all but a small portion of the soluble ammonia-nitrogen from dewatered biosolids. It does not remarkably change the dry-weight concentrations of organic nitrogen or most of the other inorganic elements.

Utilities typically measure biosolids' phosphorous content on an elemental basis. However, the fertilizer industry assesses a product's phosphorus value based on available phosphoric acid, which is determined by citric acid attraction.

In addition, the fertilizer industry typically does not measure potassium in its elemental form, but rather as K_2O. A biosolids' total potassium content (dry weight basis) can be converted to K_2O by multiplying by 1.2; however, some states are now requiring soluble K_2O to be measured directly in the lab. The fertilizer industry measures potassium via water extraction. Typically, potassium remains soluble through the wastewater treatment processes and does not accumulate in biosolids to any great extent. Most heat-dried biosolids producers do not guarantee K_2O content.

Secondary plant nutrients are typically important on a regional basis (e.g., calcium and magnesium may be important in areas with acidic soils) or for specific crops or cropping rotations. If guaranteed as a component of heat-dried biosolids, these elements typically are reported as the percent concentration of the element (dry-weight basis). Most states have minimum concentration requirements for such guarantees.

Compared to the major elements, plants require small amounts of the micronutrients. Micronutrients are unique among the essential elements because their deficiency frequently is associated with a combination of crop species and soil characteristics. Copper, molybdenum, and zinc also are regulated pollutants under Part 503. As with the secondary nutrients, these elements typically are reported as the percent concentration of the element (dry-weight basis), and most states have minimum concentration requirements for such guarantees. It is important to note there is a difference between regulatory and market requirements for reporting the nutrient content in addition to any other components of the product.

The fertilizer components of heat-dried biosolids typically are either incorporated in the organic matter or precipitated as somewhat-insoluble compounds in the solid matrix. Thus, most of the biosolids' fertilizer value is as a "slow-release" form of the nutrients. What this really means, in terms of marketing heat-dried biosolids, is that most of the fertilizer components are not released until naturally occurring populations of soil microbes break down the organic matter. This typically occurs when soil temperatures are above about 52°F and soil moisture is adequate.

7.3.1.2.7 Salt Index Soluble salts are the concentration of dissolved ions in an extract of the water held within and by the soil matrix. In some parts of the country, soils may accumulate excessive concentrations of soluble salts, especially during summer, because of increased transpiration, insufficient leaching of salts from the soil solution, and over-fertilization. Extremely high salt concentrations may inhibit water uptake and lead to chlorosis, foliage scorching, premature leaf drop, dieback, and inhibition of growth (frequently called burning). The salt index is a measure of a fertilizer's ability to cause these effects in plants. A fertilizer with a high salt index is more likely to lead to salt accumulation and desiccation injury than a fertilizer with a low salt index. High salt index fertilizers tend to cause water to move out of, not into, root cells.

Specifically, the salt index compares a fertilizer's potential to prevent water uptake by roots with that of an equivalent weight of the standard, sodium nitrate. Sodium nitrate, which has a salt index value of 100, was chosen as the standard because it is 100% water soluble and was a commonly used nitrogen fertilizer when the salt index was first proposed in 1943. Fertilizers with salt index values greater than 100 are more likely to prevent roots from absorbing water than sodium nitrate does. Fertilizers with salt index values greater than 20 should be worked into the soil before planting, or else applied to the soil surface and then sufficiently irrigated to reduce the potential of salt injury ("fertilizer burn").

Heat-dried biosolids are expected to have relatively low salt indices, but testing for this parameter and having the data available may facilitate conversations with prospective users.

7.3.1.2.8 Heating Value An emerging market for heat-dried biosolids is as a biomass fuel in industrial applications. In Europe and other parts of the world, biosolids are combusted as a fuel in industrial processes (e.g., cement manufacturing). Biomass fuels are characterized by "proximate and ultimate analyses." The "proximate" analysis includes moisture content, volatile content (when heated to 950°C), the remaining free carbon at that point, the ash (mineral) in the sample, and the high heating value (HHV) based on complete combustion of the sample to carbon dioxide and water. The "ultimate" analysis gives the fuel's composition in weight percentage of carbon, hydrogen, and oxygen (the major components), as well as sulfur and nitrogen (if any).

Typical heating values for dried biosolids range from 15 000 to 19 000 kJ/kg (6500 to 8000 Btu/lb) on a dry-weight basis (Moss et al., 2013). Digested biosolids have about half the HHV of primary solids. In comparison, according to the U.S. Energy Information Administration the average HHV of coal produced in the United States in 2015 was approximately 23 000 kJ/kg (11 000 Btu/lb), while woody biomass might have an HHV approximately 19 000 kJ/kg (8000 Btu/lb).

7.3.2   Product Consistency

Paying customers expect to get the same product and performance with each bag, batch, or load they buy. Therefore, biosolids generators should endeavor to produce the most reliable, consistent product possible.

Because the vagaries of seasons, flow, and process control naturally will result in some inconsistencies in product quality, generators should develop as much representative data as reasonably practical so they can show customers that each product characteristic remains within an expected range. This may involve more frequent sampling and analyses than required under Part 503, especially in the early years of a marketing program, so a database can be developed and statistically reviewed.

7.4   Identifying and Developing Markets

It is rare for generators of a new Class A biosolids product to find a fully developed market waiting for it.

Before investing in a new Class A facility, agencies should carefully assess local and regional markets for the products to be produced, use feedback from that assessment to finalize the design, and then endeavor to maintain communications with those markets once the process is selected and construction commences. If the marketplace is aware that a new product is coming and has some understanding of it, customers will be more receptive when the agency's new product is unveiled.

Doing this may shorten the period required to develop full market opportunities for the new product; however, biosolids generators still should expect at least 3 years to elapse before product revenues begin to offset the costs of marketing meaningfully. Likewise, organizations contemplating a new biosolids process should not presume that revenues from product sales would do more than slightly reduce O&M costs; rather, they should assume that the biosolids management program would remain a cost in the budget.

That being said, there are markets for various biosolids-derived products, and such marketing can support an agency's overall mission of producing clean water while limiting the total costs of residuals management and thereby stabilizing overall management costs in the long term.

7.4.1   Typical Markets

Markets are product specific. Compost typically is sold as a soil conditioner or amendment to add organic matter to the soil or improve the soil's physical properties. Composting tends to reduce the biosolids' total nitrogen content, thus reducing its usefulness as a fertilizer; however, with fertilizer prices at historical highs in 2008, even a minor amount of nitrogen or phosphorus content might improve a compost's value.

Table 25.13 lists typical uses for compost, and the markets associated with each use. Each market has particular needs and preferences. For example, customers that use compost as a seed-germination medium are particularly sensitive to phytotoxicity and salinity.

Uses	Markets
Potting and horticulture mixes	Greenhouses, nurseries, and retail distribution
Soil replacement	Field nurseries and sod farms
Blending to produce topsoil	Topsoil blenders and landscape material suppliers
Turf establishment	Landscape and site contractors, public sector,* and retail distribution
Top-dressing of turf	Gold courses, institutions, public sector, and retail distribution
Amendment of sandy or clay soils	Agriculture, also soil replacement, topsoil, turf establishment, and land reclamation markets
Land reclamation	Landfill operators, mine and gravel pit operators, landscape and site contractors
Landfill cover	Landfill operators
Private gardens	General public

*Public sector may include municipal, county, state, and federal agencies (such as parks departments, highway and public works departments, and airports).

TABLE 25.13   Typical Uses and Markets for Compost

Those that use it as a top dressing are sensitive to particle size. In landfill cover, land reclamation, and erosion-control applications, coarser compost may be preferred because it will be more erosion resistant. Meanwhile, texture and residual odor are key concerns in retail or homeowner applications.

Compost typically is sold in bulk to landscapers, soil blenders, and site-specific end users who may want to spread the product over a larger site (e.g., parks and ball fields). Some agencies also provide bulk compost on consignment to home and garden stores, which in turn allow customers to purchase in small bulk lots. Over time, as the market becomes more familiar with the product, some agencies will install bagging systems or contract with local baggers and begin to sell the end product in 18-kg (40-lb) bags or similarly sized containers. Bag sales typically are directed at the homeowner or smaller users, but it would not be inconceivable that other municipal agencies, golf courses, and the like might buy compost in bags.

Unlike compost, heat-dried biosolids retain most of their initial nitrogen content and, therefore, offer somewhat more fertilizer value. A typical dried biosolids with 5% total nitrogen, in mid-2008, might have the equivalent nitrogen fertilizer value of about $90/metric ton (not taking into account anything other than the total nitrogen concentration). Some portion of that will be "slow release," which represents both a market enhancement and detriment. To customers seeking release of nutrients over time, this is a valuable feature, but if a prospective customer is looking to provide all the nitrogen needs for a crop to be grown, the slow-release property may reduce his or her perceived value of the product.

Heat-dried biosolids can be sold as a component of dried, blended complete-analysis fertilizers (i.e., a fertilizer that offers nitrogen, phosphorus, and potassium). If the fertilizer is a low analysis (e.g., 10-10-10), the blender cannot construct that product merely with chemical fertilizer resources. Thus, the blender typically will add some amount of inert filler to complete the blend; if biosolids can be substituted for this product, the blender can both avoid the cost of filler and potentially reduce their purchase of other fertilizers necessary to complete the blend.

Heat-dried biosolids also can be sold as a standalone product both in bulk and in bags. Most "local" users (e.g., golf courses, parks departments, and turf managers) can often use product in bulk bags (about 450 kg [1000 lb] each). Smaller bags (18 to 20 kg [40 to 50 lb]) can be distributed to homeowners and local users.

An emerging bulk user of dried biosolids is the cement industry. Cement kilns may burn a number of fuels—including "waste" products—so the industry is open to using dried biosolids as another fuel. Each cement company and kiln location probably will have a different perception of this practice and/or local limits on what can be burned. Nonetheless, if heat-dried biosolids can be demonstrated to contribute useful heat, then kiln operators (and other biomass-to-energy processes) may be interested in the product.

7.4.2 Regional and Seasonality Issues

The "peak" seasons for using biosolids-derived products vary throughout the United States. As a general rule, the two primary peaks are spring and fall. These peaks may be extended in areas with longer growing seasons, but agencies typically should plan on slow sales during some portion of the year. In the southwestern United States, for example, sales of biosolids-derived products may be robust from February to April, curtailed during the hot summer months, and peak again from September through December. Conversely, in the northeastern United States, spring sales may peak from April through June, and fall sales may rise in late August and remain strong through October.

A market assessment can help define the selling seasons. Agencies should use such an assessment when developing a product storage (inventory) strategy.

Although most biosolids-derived products can be used in most of the United States, agencies should not anticipate that sales of alkaline-based products would be robust in areas where soil pH levels are well above neutral. The need for liming agents in these areas is minimal at best, and the value of these products may be limited unless they also possess other marketable attributes.

7.4.3 Distribution Approaches

Generators typically have three distribution approaches: They can have staff dedicated to marketing a biosolids-derived product, they can contract a third party to handle the complete distribution/marketing, or a combination of both. The third option is most common in mid-size utilities where a consultant is brought in to offer the marketing expertise, but the staff handles the day-to-day operations. If generators decide to market the product themselves, then the staff assigned to this work should have experience with or expertise in the markets to be developed and served. Generators may find that they need to hire staff specifically for this purpose.

If distribution and marketing are contracted, the agency has a number of approaches to consider, ranging from using a broker to hiring a contractor to remove and manage the product on a daily basis. A broker typically is someone who adds the biosolids-derived product to a suite of products they market (similar to a manufacturer's rep). Brokers work hard to sell a product and receive a commission when they do so, but typically, there is no guarantee that a specific quantity of product will be distributed.

From the contractor perspective, some agencies enter into contracts with service providers that remove and manage products daily (similar to the companies that manage Class B biosolids). These companies are responsible for finding locations to distribute the material. Generators typically see little revenue from such contracts; in fact, they may have to pay the company for the service. The primary benefits of this approach are reduced staff costs and storage requirements.

An intermediate approach is a contract with a third party that will develop and service markets for the product. In this case, the contractor may provide financial support until the market begins to realize the value of the product. Such contracts typically are at least 5 years long, so contractors have an opportunity to realize the fruits of their early labors.

7.4.4   Distribution Methods

Various customers desire product delivery in various ways. The delivery aspect of the product/market development can represent a significant portion of overhead costs. Although a market assessment will indicate how the market prefers delivery, most agencies begin with a bulk distribution program until market demand is established. The bulk distribution can take place in a variety of ways, but it is important to balance the cost-benefit. Other options at this stage include pickup vs. delivery, freight vs. truck, and in-house vs. contractor. Another aspect to consider is the product stability/shelf life. Once an agency has a reliable, consistent set of outlets for the bulk product, they can begin investigating the value of bagging. Over time, sales of bagged products may become a useful income stream as the value of the product on a per pound basis is significantly higher, but that determination is better made when an agency has experience in the marketplace.

If an agency decides that sales of a bagged product are viable, there are essentially two ways to implement the program:

- Invest in the equipment and labor necessary to bag the product; or
- Contract with a third party to bag, pallet, and wrap the product.

In either approach, the agency will have to develop a bag label and purchase bags. The bag label must meet fertilizer label requirements, which are generally set by each state. All fertilizer labels typically have six basic elements: brand, grade, guaranteed analysis, directions for use, name and address of registrant, and net weight.

8.0   References

American Water Works Association (AWWA) (2008) *Quicklime and Hydrated Lime*; B202-07; American Water Works Association: Denver, Colorado.

ASTM International (2006) *Standard Test Methods for Chemical Analysis of Limestone, Quicklime, and Hydrated Lime*; C25-06; ASTM International: Philadelphia, Pennsylvania.

Dominak, R. P. (2005) *Long Term Residuals Management Plan for the Northeast Ohio Regional Sewer District*; Northeast Ohio Regional Sewer District: Cleveland, Ohio.

Evanylo, G. K. (1999) *Agricultural Land Application of Biosolids in Virginia*; Publication No. 452-303; Virginia Cooperative Extension: Blacksburg, Virginia.

Forste, J. (1996) Land Application. In *Biosolids Treatment and Management*; Girovich, M., Ed.; Marcel Dekker, Inc.: Monticello, New York.

Gilmour, J.; Cogger, C.; Jacobs, L.; Wilson, S.; Evanylo, G.; Sullivan, D. (2000) Estimating Plant-Available Nitrogen in Biosolids. Water Environment Research Foundation. Project 97-REM-3.

Heyer, K.-U.; Stegmann, R. (2004) Landfill Systems, Sanitary Landfilling of Solid Wastes, and Long-Term Problems with Leachate. in *Environmental Biotechnology: Concepts and Applications*; Jördening, H.-J., Winter, J., Eds.; Wiley-VCH Verlag GmbH & Co.: Weinheim, Germany.

Hundal, L.; Cox, A.; Granato, T. (2005) Promoting Beneficial Use of Biosolids of Chicago: User Needs and Concerns. Paper Presented at WEF Innovative Uses of Biosolids and Animal and Industrial Residuals Conference; Chicago, Illinois, June 29–July 1, 2005.

International Fertilizer Development Center, United Nations Industrial Development Organization. (1998) *Fertilizer Manual*; Kluwer Academic Publishers: Dordrecht, The Netherlands. ISBN 0-7923-5032-4.

Laner, D.; Crest, M.; Scharff, H.; Morris, J.; Barlaz, M. (2012) A Review of Approaches for the Long-Term Management of Municipal Solid Waste Landfills. *Waste Manage.*, **32** (3), 498–512.

Logan, T. (2008) Personal communication. July.

Lue-Hing, C.; Tata, P.; Granato, T. A.; Pietz, R.; Johnson, R.; Sustich, R. (1998) *Sewage Sludge Survey*; Association of Metropolitan Sewerage Agencies: Washington, D.C.

Morris, J. W. F.; Barlaz, M. A. (2011) A Performance-Based System for the Long-Term Management of Municipal Waste Landfills. *Waste Manage.*, **31**, 649–662.

Moss, L.; Epstein, E.; Logan, T. (2002) *Comparing the Characteristics, Risks and Benefits of Soil Amendments and Fertilizers Used in Agriculture*, Report 99-PUM-1; Water Environment Research Foundation: Alexandria, Virginia.

Moss, L.; et al. (2013) Enabling the Future: Advancing Resource Recovery from Biosolids. Technical Report, Residuals and Biosolids Committee, Water Environment Federation, Water Environment Research Foundation, National Biosolids Partnership. http://www.wef.org/uploadedFiles/Biosolids/PDFs/ENABLING%20THE%20FUTURE.pdf (accessed September 2016).

National Biosolids Partnership Home Page. http://www.biosolids.org (accessed August 2016).

National Biosolids Partnership (2005) *National Manual of Good Practice for Biosolids*; National Biosolids Partnership: Alexandria, Virginia.

North East Biosolids and Residuals Association (NEBRA) (2007) *A National Biosolids Regulation, Quantity, End Use and Disposal Survey*; North East Biosolids and Residuals Association: Tamworth, New Hampshire.

Sharma S. D.; Lewis H. P. (1994) *Waste Containment Systems, Waste Stabilization and Landfills: Design and Evaluation*; Wiley & Sons: New York.

Sommers, L. E.; Parker, C. F.; and Meyers, G. J. (1981) Volatilization, Plant Uptake and Mineralization of Nitrogen in Soils Treated with Sewage Sludge; IWRRC Technical Reports. Paper 133. http://docs.lib.purdue.edu/watertech/133 (accessed August 2016).

U.S. Composting Council (2001) *Field Guide to Compost Use*; U.S. Composting Council: Rokonkoma, New York.

U.S. Department of Agriculture Natural Resources Conservation Service (1999) *Nutrient Management Code 590*; Conservation Practice Standard Publication; U.S. Department of Agriculture, Natural Resources Conservation Service: Madison, Wisconsin.

U.S. Department of Agriculture, Natural Resources Conservation Service (USDA-NRCS), Web Soil Survey Home Page. http://websoilsurvey.nrcs.usda.gov (accessed August 2016).

U.S. Environmental Protection Agency (1991) Solid Waste Disposal Facility Criteria, Final Rule, Part II. *Code of Federal Regulations*, Parts 257 and 258, Title 40.

U.S. Environmental Protection Agency (1993a) Standards for the Use or Disposal of Sewage Sludge. *Code of Federal Regulations*, Parts 405(d) and 503, Title 40.

U.S. Environmental Protection Agency (1993b) *Solid Waste Disposal Facility Criteria Technical Manual*, EPA 530R93017; U.S. Environmental Protection Agency: Washington, D.C.

U.S. Environmental Protection Agency (1994a) *Plain English Guide to the EPA Part 503 Rule*, EPA 832R93003; U.S. Environmental Protection Agency: Washington, D.C.

U.S. Environmental Protection Agency (1994b) *Guidance for Writing Permits for the Use or Disposal of Sewage Sludge*; U.S. Environmental Protection Agency: Washington, D.C.

U.S. Environmental Protection Agency (1994c) *The Hydrologic Evaluation of Landfill Performance Model (HELP)*; *Users Guide Version 3*, EPA600R94168a; U.S. Environmental Protection Agency: Washington, D.C.

U.S Environmental Protection Agency (1995a) *A Guide to the Biosolids Risk Assessment for the EPA Part 503 Rule*, EPA832B93005; U.S. Environmental Protection Agency: Washington, D.C.

U.S. Environmental Protection Agency (1995b) *Process Design Manual: Surface Disposal of Sewage Sludge and Domestic Septage*, EPA 625R95002; U.S. Environmental Protection Agency: Washington, D.C.

U.S. Environmental Protection Agency (2000) *Guide to Field Storage of Biosolids*, EPA832B00007; U.S. Environmental Protection Agency: Washington, D.C.

U.S. Environmental Protection Agency (2003) *Environmental Regulations and Technology: Control of Pathogens and Vector Attraction in Sewage Sludge*, EPA 625R92013; U.S. Environmental Protection Agency: Washington, D.C.

U.S. Environmental Protection Agency (2006) *Emerging Technologies for Biosolids Management*, EPA 832R06005; U.S. Environmental Protection Agency: Washington, D.C.

U.S. Environmental Protection Agency (2009) *Targeted National Sewage Sludge Survey Sampling and Analysis Technical Report*, EPA-822-R-08-016; U.S. Environmental Protection Agency: Washington, D.C.

University of Washington, Center for Urban Horticulture (2002) Using Biosolids for Reclamation and Remediation of Disturbed Soils. Report Prepared for Plant Conservation Alliance, Bureau of Land Management, U.S. Department of Interior, and U.S. Environmental Protection Agency. https://www.nps.gov/plants/restore/pubs/biosolids/biosolids.pdf (Accessed September 2016).

Water Environment Federation (2009) *Wastewater Solids Incineration Systems*, Manual of Practice No. 30; Water Environment Federation: Alexandria, Virginia.

Water Environment Federation; Water Environment Research Foundation; U.S. Environmental Protection Agency (2012) *Solids Process Design and Management*; Water Environment Federation: Alexandria, Virginia.

Index

G

Pumping (*Cont.*):
  transport, 1519–1530
    design approach, 1520–1529
    dilute sludge, 1521
    thickened residuals, 1521–1529
  variable-speed, 201
  viscosity, 200–201
  wet-well types and sizing, 202, 204
Pumping/conveyors, dewatered biosolids cake
  transport, 1544
Pumping station, trickling filters, 773–774, 807–808
Pump-monitoring guidelines, 207
Pure oxygen, 860–862
Pyrogas, composition, 2116
Pyrolysis, 2115–2116
  biochar, 2116
  fundamentals, 2115
  gasification, 2112
  overview, 2108–2110, 2115
  process description, 2116
  pyrogas composition, 2116

Q

Quality Assurance Project Plan (QAPP), 255
Quantifiable uncertainty, 145

R

Rakes, dissolved air flotation thickener, 1646
Rapid-infiltration systems, land treatment, 1373–1377
  design example, 1377
  design objectives, 1375
  design procedures, 1375–1377
  hydraulic loading rate, 1377
  land requirements, 1377
  organic loading rate, 1377
  treatment performance, 1373–1375
Rapid mix, 620–622
Rate-funded demands, 19
Rating frameworks, 14
RCRA. *See* Resource Conservation and Recovery Act
  (RCRA)
Reactor, suspended-growth biological treatment
  system, 850–854
  complete mix, 850–851
  plug flow, 851
  sequencing batch reactors, 852–854
Reactor design considerations, disinfection,
  1396–1401
  achieving specific contact time, 1398
  chemical disinfectants, 1398
  dynamics, 1396
  facilitating reactor maintenance, 1398, 1400
  factors influencing disinfection efficiencies, 1397–1398
  minimizing reintroduction of microbes to effluent,
    1400–1401
  mixing, 1397
  reducing DBP formation, 1401

Reactor design considerations, disinfection (*Cont.*):
  surrounding conditions, 1401
  tracer analysis and kinetics, 1397
  typical reactors, 1396–1397
Receiving station design, septage acceptance and
  pretreatment, 577–580
  screening and grit removal, 578–579
  storage and equalization, 579–580
Recessed-impeller pumps, 1531
Reciprocating piston pumps, 1537–1538
Recirculation, FBBR, 750
Reclaimed water
  CEC, 29
  quality criteria, 26
  storage and distribution, 30
Reclaimed water, chlorination, 1423
Recommended Standards for Wastewater Facilities,
  5, 18, 37
Recordkeeping requirements, 1479–1480
Rectangular clarifiers, 597
Rectangular clarifiers, primary sludge collection and
  removal, 638–640
Recuperative thermal oxidizers, 346
Recuperative thickening, 1715–1716
Recycled streams on treatment processes, 1726
Reduction, gasification, 2113
Reed beds, 1784–1786
Regenerative thermal oxidizers, 344–345
Regional issues, marketing, 2192
Regional protocols, 130
Region of Waterloo (ROW), 147. *See also* City of Galt
  Wastewater Treatment Plant
Regulations, 21
  air quality, 31
  ash use and disposal, 2179
  dedicated land disposal, 2175–2176
    general requirements, 2175
    nitrate contamination, 2176
    pollutant limits, 2175
  emission control, thermal processing, 2119–2124
    Clean Air Act Amendments of 1977, 2120
    Clean Air Act Amendments of 1990, 2120
    local prohibitory rules, 2121
    MACT standards for SSI, 2121–2122
    NESHAPs, 2119–2120
    new source performance standards, 2119
    nonregulated emissions, 2122–2124
    prevention of significant deterioration, 2120
    source-specific standards, 2122
    standards for the use and disposal of sewage
      sludge, 2121
  hazardous substances, 32
    CERCLA, 33
    EPCRA, 33
    RCRA, 32
    TSCA, 34
  land application, 2151–2152
  landfill/landfilling, 2166–2167